COMPREHENSIVE
MEDICINAL CHEMISTRY

IN 6 VOLUMES

COMPREHENSIVE MEDICINAL CHEMISTRY

The Rational Design, Mechanistic Study & Therapeutic Application of Chemical Compounds

Chairman of the Editorial Board
CORWIN HANSCH
Pomona College, Claremont, CA, USA

Joint Executive Editors
PETER G. SAMMES
Brunel University of West London, Uxbridge, UK

JOHN B. TAYLOR
Rhône-Poulenc Ltd, Dagenham, UK

Volume 2
ENZYMES & OTHER MOLECULAR TARGETS

Volume Editor
PETER G. SAMMES
Brunel University of West London, Uxbridge, UK

PERGAMON PRESS
Member of Maxwell Macmillan Pergamon Publishing Corporation
OXFORD • NEW YORK • BEIJING • FRANKFURT
SÃO PAULO • SYDNEY • TOKYO • TORONTO

U.K.	Pergamon Press plc, Headington Hill Hall, Oxford OX3 0BW, England
U.S.A.	Pergamon Press, Inc., Maxwell House, Fairview Park, Elmsford, New York 10523, U.S.A.
PEOPLE'S REPUBLIC OF CHINA	Pergamon Press, Room 4037, Qianmen Hotel, Beijing, People's Republic of China
FEDERAL REPUBLIC OF GERMANY	Pergamon Press GmbH, Hammerweg 6, D-6242 Kronberg, Federal Republic of Germany
BRAZIL	Pergamon Editora Ltda, Rua Eça de Queiros, 346, CEP 04011, Paraiso, São Paulo, Brazil
AUSTRALIA	Pergamon Press Australia Pty Ltd., P.O. Box 544, Potts Point, N.S.W. 2011, Australia
JAPAN	Pergamon Press, 5th Floor, Matsuoka Central Building, 1-7-1 Nishishinjuku, Shinjuku-ku, Tokyo 160, Japan
CANADA	Pergamon Press Canada Ltd., Suite No. 241, 253 College Street, Toronto, Ontario, Canada M5T 1R5

Copyright © 1990 Pergamon Press plc

First edition 1990

Library of Congress Cataloging in Publication Data
Comprehensive medicinal chemistry: the rational design, mechanistic study & therapeutic application of chemical compounds/ chairman of the editorial board, Corwin Hansch; joint executive editors, Peter G. Sammes, John B. Taylor. — 1st ed.
p. cm.
Includes index.
1. Pharmaceutical chemistry. I. Hansch, Corwin. II. Sammes, P. G. (Peter George) III. Taylor, J. B. (John Bodenham), 1939–
[DNLM: 1. Chemistry, Pharmaceutical. QV 744 C737]
RS402.C65
615'.19—dc20
DNLM/DLC
for Library of Congress   89–16329

British Library Cataloguing in Publication Data
Hansch, Corwin
Comprehensive medicinal chemistry
1. Pharmaceutics
I. Title
615'.19

ISBN 0–08–037058–6 (Vol. 2)
ISBN 0–08–032530–0 (set)

Printed in Great Britain by
BPCC Hazell Books Ltd, Aylesbury, Bucks, England

Contents

Preface   vii

Contributors to Volume 2   ix

Contents of All Volumes   xiii

Enzymes

5.1   Enzyme Structure   1
      L. SAWYER, *University of Edinburgh, UK*

5.2   Nomenclature and Classification of Enzymes   33
      I. B. R. BOWMAN, *University of Edinburgh, UK*

5.3   Enzyme Catalysis   45
      M. I. PAGE, *Huddersfield Polytechnic, UK*

5.4   Enzyme Inhibition   61
      M. I. PAGE, *Huddersfield Polytechnic, UK*

5.5   Resistance and Tolerance to Antimicrobial Drugs   89
      L. B. QUESNEL, *University of Manchester, UK*

Agents Acting on Oxygenases, Electron Transport Systems and Pyridoxal Dependent Systems

6.1   Oxygenases   123
      L. I. KRUSE, *Smith Kline & French Research Ltd, Welwyn, UK*

6.2   The Arachidonic Acid Cascade   147
      G. A. HIGGS, E. A. HIGGS and S. MONCADA, FRS,
      Wellcome Research Laboratories, Beckenham, UK

6.3   Agents Acting on Passive Ion Transport   175
      P. M. MAY, *Murdoch University, Perth, Australia*

6.4   Agents Acting on Active Ion Transport   193
      J. L. SUSCHITSKY and E. WELLS, *Fisons Pharmaceutical Division,*
      Loughborough, UK

6.5   Pyriodoxal Dependent Systems   213
      D. GANI, *University of Southampton, UK*

Agents Acting on Metabolic Processes

7.1   Sulfonamides and Sulfones   255
      P. G. SAMMES, *Brunel University of West London, Uxbridge, UK*

7.2   Reductases   271
      J. J. McCORMACK, *University of Vermont, Burlington, VT, USA and*
      National Institutes of Health, Bethesda, MD, USA

7.3   Purine and Pyrimidine Targets   299
      J. B. HOBBS, *The City University, London, UK*

7.4   Therapeutic Consequences of the Inhibition of Sterol Metabolism   333
      J. L. ADAMS and B. W. METCALF, *Smith Kline & French Laboratories,*
      Philadelphia, PA, USA

Agents Acting on Hydrolases and Peptidases

8.1   Hydrolases   365
        A. D. ELBEIN, *University of Texas Health Science Center,*
        San Antonio, TX, USA

8.2   Peptidase Inhibitors   391
        D. H. RICH, *University of Wisconsin, Madison, WI, USA*

8.3   Enzyme Cascades: Purine Metabolism and Immunosuppression   443
        R. B. GILBERTSEN and J. C. SIRCAR, *Warner-Lambert Company,*
        Ann Arbor, MI, USA

8.4   Enzyme Cascades: Coagulation, Fibrinolysis and Hemostasis   481
        M. D. TAYLOR, *Warner-Lambert Company, Ann Arbor, MI, USA*

8.5   Selective Inhibitors of Phosphodiesterases   501
        R. E. WEISHAAR and J. A. BRISTOL, *Warner-Lambert Company,*
        Ann Arbor, MI, USA

8.6   Agents Acting Against Phospholipases A_2 515
        H. VAN DEN BOSCH, *State University of Utrecht, The Netherlands*

8.7   Protein Kinases   531
        K. J. MURRAY and B. H. WARRINGTON,
        Smith Kline & French Research Ltd, Welwyn, UK

Agents Acting on Cell Walls

9.1   Cell Wall Structure and Function   553
        J. B. WARD, *Glaxo Group Research Ltd, Greenford, UK*

9.2   β-Lactam Antibiotics: Penicillins and Cephalosporins   609
        C. E. NEWALL and P. D. HALLAM, *Glaxo Group Research Ltd,*
        Greenford, UK

9.3   Other β-Lactam Antibiotics   655
        A. G. BROWN, M. J. PEARSON and R. SOUTHGATE,
        Beecham Pharmaceuticals, Betchworth, UK

Agents Acting on Nucleic Acids

10.1   DNA Intercalating Agents   703
        L. P. G. WAKELIN, *Peter MacCallum Cancer Institute, Melbourne,*
        Victoria, Australia and M. J. WARING, *University of Cambridge, UK*

10.2   DNA Binding and Nicking Agents   725
        D. I. EDWARDS, *The Polytechnic of North East London, UK*

10.3   Agents Interfering with DNA Enzymes   753
        R. P. HERTZBERG, *Smith Kline & French Laboratories,*
        King of Prussia, PA, USA

10.4   Inhibitors of the Transcribing Enzymes: Rifamycins and Related Agents   793
        P. SENSI, *University of Milan, Italy* and
        G. C. LANCINI, *Lepetit Research Center, Gerenzano, Italy*

10.5   Agents which Interact with Ribosomal RNA and Interfere with its Functions   813
        M. CANNON, *King's College London, UK*

Subject Index   839

Preface

Medicinal chemistry is a subject which has seen enormous growth in the past decade. Traditionally accepted as a branch of organic chemistry, and the near exclusive province of the organic chemist, the subject has reached an enormous level of complexity today. The science now employs the most sophisticated developments in technology and instrumentation, including powerful molecular graphics systems with 'drug design' software, all aspects of high resolution spectroscopy, and the use of robots. Moreover, the medicinal chemist (very much a new breed of organic chemist) works in very close collaboration and mutual understanding with a number of other specialists, notably the molecular biologist, the genetic engineer, and the biopharmacist, as well as traditional partners in biology.

Current books on medicinal chemistry inevitably reflect traditional attitudes and approaches to the field and cover unevenly, if at all, much of modern thinking in the field. In addition, such works are largely based on a classical organic structure and therapeutic grouping of biologically active molecules. The aim of *Comprehensive Medicinal Chemistry* is to present the subject, the modern role of which is the understanding of structure–activity relationships and drug design from the mechanistic viewpoint, as a field in its own right, integrating with its central chemistry all the necessary ancillary disciplines.

To ensure that a broad coverage is obtained at an authoritative level, more than 250 authors and editors from 15 countries have been enlisted. The contributions have been organized into five major themes. Thus Volume 1 covers general principles, Volume 2 deals with enzymes and other molecular targets, Volume 3 describes membranes and receptors, Volume 4 covers quantitative drug design, and Volume 5 discusses biopharmaceutics. As well as a cumulative subject index, Volume 6 contains a unique drug compendium containing information on over 5500 compounds currently on the market. All six volumes are being published simultaneously, to provide a work that covers all major topics of interest.

Because of the mechanistic approach adopted, Volumes 1–5 do not discuss those drugs whose modes of action are unknown, although they will be included in the compendium in Volume 6. The mechanisms of action of such agents remain a future challenge for the medicinal chemist.

We should like to acknowledge the way in which the staff at the publisher, particularly Dr Colin Drayton (who initially proposed the project), Dr Helen McPherson and their editorial team, have supported the editors and authors in their endeavour to produce a work of reference that is both complete and up-to-date.

Comprehensive Medicinal Chemistry is a milestone in the literature of the subject in terms of coverage, clarity and a sustained high level of presentation. We are confident it will appeal to academic and industrial researchers in chemistry, biology, medicine and pharmacy, as well as teachers of the subject at all levels.

CORWIN HANSCH
Claremont, USA

PETER G. SAMMES
Uxbridge, UK

JOHN B. TAYLOR
Dagenham, UK

Contributors to Volume 2

Dr J. L. Adams
F32, Smith Kline & French Laboratories, 1500 Spring Garden Street, PO Box 7929, Philadelphia, PA 19101, USA

Dr I. B. R. Bowman
Department of Biochemistry, University of Edinburgh Medical School, Hugh Robson Building, George Square, Edinburgh EH8 9XD, UK

Dr J. A. Bristol
Parke-Davis Pharmaceutical Research Division, Warner-Lambert Company, 2800 Plymouth Road, Ann Arbor, MI 48105, USA

Dr A. G. Brown
Biosciences Research Centre, Beecham Pharmaceuticals, Great Burgh, Yew Tree Bottom Road, Epsom, Surrey KT18 5XQ, UK

Dr M. Cannon
Department of Biochemistry, King's College, The Strand, London WC2R 2LS, UK

Dr D. I. Edwards
Department of Paramedical Science, Chemotherapy Research Unit, The Polytechnic of North East London, Romford Road, London E15 4LZ, UK

Professor A. D. Elbein
Department of Biochemistry, University of Texas Health Science Center, 7703 Floyd Curl Drive, San Antonio, TX 78284, USA

Dr D. Gani
Department of Chemistry, University of Southampton, Southampton SO9 5NH, UK

Dr R. B. Gilbertsen
Parke-Davis Pharmaceutical Research Division, Warner-Lambert Company, 2800 Plymouth Road, Ann Arbor, MI 48105, USA

Dr P. D. Hallam
Perkin-Elmer Ltd, Maxwell Road, Beaconsfield, Bucks HP7 0HA, UK

Dr R. P. Hertzberg
Smith Kline & French Laboratories L410, PO Box 1539, King of Prussia, PA 19406, USA

Dr E. A. Higgs
The Wellcome Research Laboratories, Langley Court, Beckenham, Kent BR3 3BS, UK

Dr G. A. Higgs
The Wellcome Research Laboratories, Langley Court, Beckenham, Kent BR3 3BS, UK

Dr J. B. Hobbs
Department of Microbiology, 300-6174 University Boulevard, University of British Columbia, Vancouver, BC V6T 1WS, Canada

Dr L. I. Kruse
Sterling Drug Inc., 9 Great Valley Parkway, Malvern, PA 19355, USA

Professor G. C. Lancini
Lepetit Research Center, Merrell Dow Research Institute, Via R. Lepetit 34, I-21040 Gerenzano (Varese), Italy

Dr P. M. May
School of Mathematics and Physical Sciences, Murdoch University, Perth 6150, Australia

Professor J. J. McCormack
Department of Pharmacology, College of Medicine, Given Building, University of Vermont, Burlington, VT 05405, USA

Dr B. W. Metcalf
F32, Smith Kline & French Laboratories, 1500 Spring Garden Street, PO Box 7929, Philadelphia, PA 19101, USA

Dr S. Moncada, FRS
The Wellcome Research Laboratories, Langley Court, Beckenham, Kent BR3 3BS, UK

Dr K. J. Murray
Smith Kline & French Research Ltd, The Frythe, Welwyn, Herts AL6 9AR, UK

Dr C. E. Newall
Microbiological Chemistry Department, Glaxo Group Research Ltd, Greenford Road, Greenford, Middlesex UB6 0HE, UK

Professor M. I. Page
Department of Chemical and Physical Sciences, The Polytechnic, Queensgate, Huddersfield HD1 3DH, UK

Dr M. J. Pearson
Biosciences Research Centre, Beecham Pharmaceuticals, Great Burgh, Yew Tree Bottom Road, Epsom, Surrey KT18 5XQ, UK

Dr L. B. Quesnel
Department of Cell and Structural Biology, University of Manchester, Stopford Building, Oxford Road, Manchester M13 9PT, UK

Professor D. H. Rich
School of Pharmacy, Center for Health Sciences, University of Wisconsin, 425 North Charter Street, Madison, WI 53706, USA

Professor P. G. Sammes
Department of Chemistry, Brunel University of West London, Uxbridge, Middlesex UB8 3PA, UK

Dr L. Sawyer
Department of Biochemistry, University of Edinburgh Medical School, Hugh Robson Building, George Square, Edinburgh EH8 9XD, UK

Professor P. Sensi
Instituto di Chimica Farmaceutica, Università di Milano, Viale Abruzzi 42, I-20131 Milano, Italy

Dr J. C. Sircar
Parke-Davis Pharmaceutical Research Division, Warner-Lambert Company, 2800 Plymouth Road, Ann Arbor, MI 48105, USA

Dr R. Southgate
Biosciences Research Centre, Beecham Pharmaceuticals, Great Burgh, Yew Tree Bottom Road, Epsom, Surrey KT18 5XQ, UK

Dr J. L. Suschitzky
R & D Laboratories, Fisons Pharmaceutical Division, Bakewell Road, Loughborough, Leics LE11 0QT, UK

Dr M. D. Taylor
Parke-Davis Pharmaceutical Research Division, Warner-Lambert Company, 2800 Plymouth Road, Ann Arbor, MI 48105, USA

Professor H. van den Bosch
Biochemisch Laboratorium, Rijksuniversiteit te Utrecht, Padualaan 8, Postbus 80 054, NE-3508 TB Utrecht, The Netherlands

Professor L. P. G. Wakelin
St Luke's Institute for Cancer Research, Highfield Road, Rathgar, Dublin 6, Ireland

Dr J. B. Ward
Glaxo Group Research Ltd, Greenford Road, Greenford, Middlesex, UB6 0HE, UK

Dr M. J. Waring
Department of Pharmacology, University of Cambridge, Tennis Court Road, Cambridge CB2 1QJ, UK

Dr B. H. Warrington
Smith Kline & French Research Ltd, The Frythe, Welwyn, Herts AL6 9AR, UK

Dr R. E. Weishaar
Parke-Davis Pharmaceutical Research Division, Warner-Lambert Company, 2800 Plymouth Road, Ann Arbor, MI 48105, USA

Dr E. Wells
R & D Laboratories, Fisons Pharmaceutical Division, Bakewell Road, Loughborough, Leics LE11 0QT, UK

Contents of All Volumes

Volume 1 General Principles

Historical Perspective

1.1   Medicinal Chemistry: A Personal View
1.2   Chronology of Drug Introductions
1.3   Evolution of the Pharmaceutical Industry
1.4   Development of Medicinal Chemistry in China
1.5   Contribution of Ayurvedic Medicine to Medicinal Chemistry

Targets of Biologically Active Molecules

2.1   Physiology of the Human Body
2.2   The Architecture of the Cell
2.3   Macromolecular Targets for Drug Action
2.4   The Concept of Bioselectivity
2.5   The Immune System
2.6   Selectivity

Bioactive Materials

3.1   Classification of Drugs
3.2   Lead Structure Discovery and Development
3.3   Computer-aided Selection for Large-scale Screening
3.4   Isolation of Bioactive Materials and Assay Methods
3.5   Biomaterials from Mammalian Sources
3.6   Development and Scale-up of Processes for the Manufacture of New Pharmaceuticals
3.7   Genetic Engineering: The Gene
3.8   Genetic Engineering: Applications to Biological Research
3.9   Genetic Engineering: Commercial Applications

Socio-economic Factors of Drug Development

4.1   Industrial Factors and Government Controls
4.2   Organization and Funding of Medical Research in the UK
4.3   Organization and Funding of Medical Research in the USA
4.4   Health Care in the UK
4.5   Health Care in the USA
4.6   Good Pharmaceutical Manufacturing Practice
4.7   Toxicological Evaluation of New Drugs
4.8   Clinical Pharmacology and Clinical Trials
4.9   Post-marketing Surveillance
4.10   Animal Experimentation
4.11   Orphan Drugs
4.12   Patents
4.13   Trade Marks
4.14   Sources of Information

Subject Index

Volume 2 Enzymes and Other Molecular Targets

Enzymes

5.1   Enzyme Structure
5.2   Nomenclature and Classification of Enzymes
5.3   Enzyme Catalysis
5.4   Enzyme Inhibition
5.5   Resistance and Tolerance to Antimicrobial Drugs

Agents Acting on Oxygenases, Electron Transport Systems and Pyridoxal Dependent Systems

6.1   Oxygenases
6.2   The Arachidonic Acid Cascade
6.3   Agents Acting on Passive Ion Transport
6.4   Agents Acting on Active Ion Transport
6.5   Pyridoxal Dependent Systems

Agents Acting on Metabolic Processes

7.1   Sulfonamides and Sulfones
7.2   Reductases
7.3   Purine and Pyrimidine Targets
7.4   Therapeutic Consequences of the Inhibition of Sterol Metabolism

Agents Acting on Hydrolases and Peptidases

8.1   Hydrolases
8.2   Peptidase Inhibitors
8.3   Enzyme Cascades: Purine Metabolism and Immunosuppression
8.4   Enzyme Cascades: Coagulation, Fibrinolysis and Hemostasis
8.5   Selective Inhibitors of Phosphodiesterases
8.6   Agents Acting Against Phospholipases A_2
8.7   Protein Kinases

Agents Acting on Cell Walls

9.1   Cell Wall Structure and Function
9.2   β-Lactam Antibiotics: Penicillins and Cephalosporins
9.3   Other β-Lactam Antibiotics

Agents Acting on Nucleic Acids

10.1 DNA Intercalating Agents
10.2 DNA Binding and Nicking Agents
10.3 Agents Interfering with DNA Enzymes
10.4 Inhibitors of the Transcribing Enzymes: Rifamycins and Related Agents
10.5 Agents which Interact with Ribosomal RNA and Interfere with its Functions

Subject Index

Volume 3 Membranes and Receptors

Membranes, Membrane Receptors and Second Messenger Pathways

11.1   Structure and Function of Cell Membranes
11.2   Quantitative Analysis of Ligand–Receptor Interactions
11.3   Isolation, Purification and Molecular Biology of Cell Membrane Receptors
11.4   Transmembrane Signalling and Second Messenger Analogues

Neurotransmitter and Autocoid Receptors

12.1   α-Adrenergic Receptors
12.2   β-Adrenergic Receptors
12.3   Dopamine Receptors
12.4   Peripheral Dopamine Receptors
12.5   Histamine Receptors
12.6   Cholinergic Receptors
12.7   Amino Acid Receptors
12.8   Ligand Interactions at the Benzodiazepine Receptor
12.9   Serotonin (5-HT) Receptors
12.10 Purinergic Receptors
12.11 Prostanoids and their Receptors
12.12 Platelet Activating Factor Receptors
12.13 Leukotriene Receptors

Peptidergic Receptors

13.1   Design of Drugs Acting at Peptidergic Receptors
13.2   Opioid Receptors
13.3   Hypothalamic and Adenohypophyseal Hormone Receptors
13.4   Neurohypophyseal Hormone Receptors
13.5   Glucagon and Insulin Receptors
13.6   Gastrointestinal Regulatory Peptide Receptors
13.7   Angiotensin and Bradykinin Receptors
13.8   Atrial Natriuretic Factor Receptors
13.9   Tachykinin Receptors
13.10 Calcitonin and Parathyroid Hormone Receptors

Drugs Acting on Ion Channels and Membranes

14.1   Drugs Acting on Ion Channels and Membranes

Lymphokines and Cytokines

15.1   Lymphokines and Cytokines of the Immune System

Intracellular Receptors

16.1   Molecular Mechanism of the Action of 1,25-Dihydroxyvitamin D_3
16.2   Thyroid Hormone Receptors
16.3   Steroid Hormone Receptors

Subject Index

Volume 4 Quantitative Drug Design

Introduction to Drug Design and Molecular Modelling

17.1 History and Objectives of Quantitative Drug Design
17.2 Computers and the Medicinal Chemist
17.3 Chemical Structures and Computers
17.4 Use and Limitations of Models and Modelling in Medicinal Chemistry

Quantitative Description of Physicochemical Properties of Drug Molecules

18.1 Quantum Mechanics and the Modeling of Drug Properties
18.2 Molecular Mechanics and the Modeling of Drug Structures
18.3 Dynamic Simulation and its Applications in Drug Research
18.4 Three-dimensional Structure of Drugs
18.5 Electronic Effects in Drugs
18.6 Hydrophobic Properties of Drugs
18.7 Methods of Calculating Partition Coefficients
18.8 Intermolecular Forces and Molecular Binding

Quantitative Description of Biological Activity and Drug Transport

19.1 Quantitative Description of Biological Activity
19.2 Molecular Structure and Drug Transport

Molecular Graphics and Drug Design

20.1 Introduction to Computer Graphics and Its Use for Displaying Molecular Structures
20.2 Use of Molecular Graphics for Structural Analysis of Small Molecules
20.3 Application of Molecular Graphics to the Analysis of Macromolecular Structures

Quantitative Structure–Activity Relationships

21.1 The Extrathermodynamic Approach to Drug Design
21.2 The Design of Test Series and the Significance of QSAR Relationships
21.3 The Free–Wilson Method and its Relationship to the Extrathermodynamic Approach

Pattern Recognition and Other Statistical Methods for Drug Design

22.1 Substructural Analysis and Compound Selection
22.2 Linear Discriminant Analysis and Cluster Significance Analysis
22.3 Pattern Recognition Techniques in Drug Design
22.4 The Distance Geometry Approach to Modeling Receptor Sites

Subject Index

Volume 5 Biopharmaceutics

Principles of Pharmacokinetics and Metabolism

23.1 Absorption Processes
23.2 Pharmacokinetic Aspects of Drug Administration and Metabolism
23.3 Distribution and Clearance Concepts
23.4 Sites of Drug Metabolism, Prodrugs and Bioactivation
23.5 Metabolic Pathways

23.6 Drug Interactions

23.7 Stereoselectivity in Pharmacokinetics and Drug Metabolism

23.8 Enzyme Induction and Inhibition

23.9 Species Differences in Drug Metabolism

23.10 Developmental Drug Metabolism

23.11 Pharmacogenetics

23.12 Chronokinetics

23.13 Population Pharmacokinetics

23.14 Toxicokinetics

23.15 Pharmacodynamics

Analytical Methodology

24.1 Use of Isotopes in Quantitative and Qualitative Analysis

24.2 Chemical Analysis

24.3 Biological Analysis

24.4 Methods in Drug Metabolism

24.5 Isolation and Identification of Metabolites

24.6 Systems Analysis in Pharmacokinetics

Chemistry and Pharmacy in Drug Development

25.1 Physicochemical Principles

25.2 Formulation

25.3 Routes of Administration and Dosage Regimes

25.4 Delivery System Technology

25.5 Drug Targeting

Subject Index

Volume 6

Cumulative Subject Index

Drug Compendium

5.1

Enzyme Structure

LINDSAY SAWYER

University of Edinburgh, UK

5.1.1	INTRODUCTION	2
5.1.2	ENZYMES AS POLYMERS OF AMINO ACIDS	2
5.1.2.1	*Post-translational Modification*	*3*
5.1.2.2	*Classification*	*5*
5.1.3	ENZYME ISOLATION	5
5.1.3.1	*Criteria for Purity*	*6*
5.1.3.2	*Precipitation Techniques*	*6*
5.1.3.3	*Ion Exchange Chromatography*	*7*
5.1.3.4	*Gel Filtration*	*7*
5.1.3.5	*Affinity Chromatography*	*7*
5.1.3.6	*Crystallization*	*8*
5.1.4	PHYSICAL PROPERTIES OF ENZYMES	9
5.1.4.1	*General Structural Characteristics*	*9*
5.1.4.2	*Protein Structure Classification*	*12*
5.1.4.3	*Hierarchy of Protein Architecture*	*12*
5.1.4.4	*The Folding Problem*	*15*
5.1.5	PRIMARY STRUCTURE DETERMINATION	16
5.1.5.1	*Chemical and Enzymic Cleavage*	*16*
5.1.5.2	*Sequential Degradation of Peptides*	*16*
5.1.5.3	*Mass Spectrometry*	*18*
5.1.5.4	*DNA Sequencing*	*18*
5.1.6	SECONDARY STRUCTURE	20
5.1.6.1	*Circular Dichroism*	*20*
5.1.6.2	*Infrared and Raman Spectroscopy*	*22*
5.1.6.3	*Prediction Methods*	*22*
5.1.7	TERTIARY STRUCTURE	23
5.1.7.1	*X-ray Crystallography*	*23*
5.1.7.2	*NMR Spectroscopy*	*25*
5.1.7.3	*Protein Dynamics*	*26*
5.1.8	QUATERNARY STRUCTURE	27
5.1.8.1	*Crosslinking*	*27*
5.1.9	CHEMICAL MODIFICATION	28
5.1.9.1	*Aromatic Residues*	*28*
5.1.9.2	*Hydroxyl Groups*	*29*
5.1.9.3	*Thiol Groups*	*29*
5.1.9.4	*Acidic Groups*	*30*
5.1.9.5	*Basic Groups*	*31*
5.1.10	CONCLUSION	31
5.1.11	REFERENCES	31

5.1.1 INTRODUCTION

Living cells carry out complex series of chemical reactions, mostly under the mild conditions of neutral pH, 37 °C and 1 atm. In order to do this, they contain extremely efficient catalysts, enzymes, which not only enhance reaction rates by factors of up to 10^{20} but also are capable of varying this enhancement in response to molecular signals, thereby providing the cell with a means of modifying its behaviour. Unlike the Ni catalyst used, for example, in hydrogenating vegetable oil, enzymes are proteins which have relative molecular masses ranging from around 10^4 to more than 10^6. They are characterized by the reactions they catalyze and the thousands of known examples include such varied functions as the hydrolysis of fat, through reduction of nitrate, to the isomerization of glucose to fructose.

Enzymes, albeit not in a purified form, have been used by man since Neolithic times to ferment honey or sugar in bread-making and brewing. Tanning, too, used the hydrolytic enzymes of the bacteria found in faeces. Thus fermentation and putrefaction had been known to produce fundamental changes under moderate conditions for a long time before the 19th century, when the breakdown of starch to sugar was first shown to be facilitated by a precipitated extract of malt. In 1878, Kuhne used the term enzyme to describe the agents in gastric juice responsible for digestion. The first enzyme to be purified significantly was urease, which was crystallized by Sumner in 1926, and the first enzyme crystal structure to be determined was that of lysozyme in 1965. Since then, some 200 three-dimensional structure determinations have been performed and a substantial database is available[1] from which a number of general observations can be made. Nowadays, enzymes are used for a great variety of purposes, including manufacturing detergents and antibiotics, for producing food and, in a purified form, for clinical diagnoses.[2] They are also recognized as crucial targets for a number of drugs that inhibit their action, thereby modifying the metabolic processes mediated by these enzymes.

The specificity of an enzyme for its substrate Fischer likened to a lock-and-key, although perhaps hand-in-glove provides a better analogy since it implies flexibility in both enzyme and substrate. Pauling predicted that the extraordinary catalytic power of enzymes could be explained in general terms if the structure of their active site was such as to be complementary to, and hence stabilize, the activated complex. This chapter will describe the general features of the molecular architecture of proteins or, specifically, enzymes to provide a basis for understanding their catalytic efficiency and often exquisite specificity. Further, an overview will be given of the chemical and physical methods which are particularly suited to studying enzyme structure. Since in a single chapter comprehensive coverage cannot be contemplated, it is on the advantages and limitations of the methods that greater emphasis will be placed, with the reader being left to refer elsewhere for the technical details.

5.1.2 ENZYMES AS POLYMERS OF AMINO ACIDS

Enzymes are proteins and proteins are polymers formed by condensing the carboxyl group of one amino acid with the α-amino group of the next, and so on, producing the polypeptide backbone, the main chain, of the protein. The side chains, which are extensions on the C_α atom of all except the simplest amino acid, glycine, can be of a variety of kinds and it is the linear sequence of these side chains which gives each protein its individual characteristics. Twenty amino acids form the basis of all proteins and, since all save glycine have a chiral C_α atom, they possess a particular stereochemical configuration: they are all L-amino acids. Figure 1 shows a simple tripeptide and Table 1 summarizes[3,4] the properties of the 20 amino acids to show the variation of size, shape and charge which are responsible for the particular three-dimensional fold and function of a given polypeptide sequence. These properties can be classified as shown in the Venn diagram of Figure 2,[5] from which it can be seen that there is a subtle distribution covering a wide range from the hydrophobic to the hydrophilic.

The sequence of amino acids in a protein arises from the information coded in the DNA of the gene present in the cell where the polypeptide is synthesized. The DNA is transcribed to mRNA which may be translated directly or, after further processing, into the protein sequence. This latter process, which is mediated by a molecular assembly, the ribosome, occurs with remarkable fidelity from one generation to the next and is helped to some extent by the fact that the genetic code has built-in redundancy. DNA and mRNA are polymers made of a precise sequence of 4 types of nucleotide. Each amino acid is specified by 3 nucleotides and so there are 4^3, *i.e.* 64, possible triplets coding for the 20 amino acids described above.

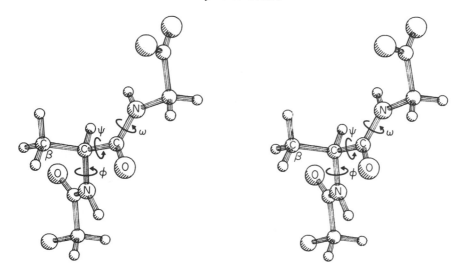

Figure 1 Stereo diagram of the tripeptide Gly-Ala-Gly showing the L-amino acid, alanine, and the conventional names for the main-chain atoms and angles. The atomic radii reflect atomic number

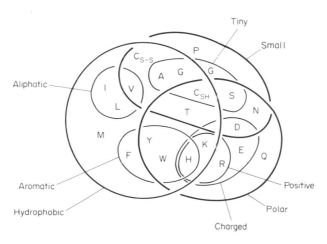

Figure 2 Classification of amino acid side chains according to Taylor.[5] The single letter code is defined in Table 1, where the sizes above can be related to the molecular volume. Notice the domination of the hydrophobic residues (redrawn from ref. 5 with permission of Academic Press)

5.1.2.1 Post-translational Modification

When polypeptide synthesis is complete, the chain of amino acids may be further processed, often as part of a secretory process, and it is this post-translational modification which gives rise to the 'extra' amino acids sometimes listed as appearing in proteins. Examples are 3-hydroxyproline, which is found in collagen, diiodotyrosine, found in thyroglobulin and, possibly the best known, the oxidation of two cysteine residues to produce the disulfide-bridged cystine.

Other forms of post-translational modification can be identified,[6,7] occurring to modify the charge, solubility or viscosity, or to target the molecule for a destination remote from the site of biosynthesis. For the purposes of metabolic control, some modifications are reversible. Of the irreversible changes, four distinct groups can be identified. First, specific hydrolysis is a frequent mechanism whereby the synthesized polypeptide is cleaved in order to activate the protein. Such a mechanism results in the production of active insulin and enkephalin from their respective precursors and an enzyme example is the conversion of prothrombin to thrombin by the specific action of Factor Xa.

Second, a number of group transfer reactions occur, once again catalyzed by specific enzymes. For example, ADP-ribosylation of elongation factor 2 by diphtheria toxin results in the cessation of protein synthesis in the affected cell. Methylation of both acidic and basic amino acid side chains can

Table 1 Some Properties[3,4] of the 20 Amino Acid Residues[a]

Amino acid	Three letter code	One letter code	Side-chain structure	Volume (Å³)	Accessible surface area (Å²)	Hydrophobicity	pK_a	Occurrence in proteins (%)
Alanine	Ala	A	—Me	88.6	115	1.8	—	9.0
Arginine	Arg	R	—CH₂CH₂CH₂NHC(=NH)NH₂	173.4	225	−4.5	12	4.7
Asparagine	Asn	N	—CH₂CONH₂	117.7	160	−3.5	—	4.4
Aspartic acid	Asp	D	—CH₂CO₂H	111.1	150	−3.5	4.5	5.5
Cysteine	Cys	C	—CH₂SH	108.5	135	2.5	9.3	2.8
Glutamine	Gln	Q	—CH₂CH₂CONH₂	143.9	180	−3.5	—	3.9
Glutamic acid	Glu	E	—CH₂CH₂CO₂H	138.4	190	−3.5	4.6	6.2
Glycine	Gly	G	—H	60.1	75	−0.4	—	7.5
Histidine	His	H	—CH₂— (imidazole)	153.2	195	−3.2	6.2	2.1
Isoleucine	Ile	I	—CH(Me)CH₂Me	166.7	175	4.5	—	4.6
Leucine	Leu	L	—CH₂CHMe₂	166.7	170	3.8	—	7.5
Lysine	Lys	K	—CH₂CH₂CH₂CH₂NH₂	168.6	200	−3.9	10.4	7.0
Methionine	Met	M	—CH₂CH₂SMe	162.9	185	1.9	—	1.7
Phenylalanine	Phe	F	—CH₂Ph	189.9	210	2.8	—	3.5
Proline	Pro	P	(pyrrolidine)	122.7	145	−1.6	—	4.6
Serine	Ser	S	—CH₂OH	89.0	115	−0.8	—	7.1
Threonine	Thr	T	—CH(OH)Me	116.1	140	−0.7	—	6.0
Tryptophan	Trp	W	—CH₂— (indole)	227.8	255	−0.9	—	1.1
Tyrosine	Tyr	Y	—CH₂—(phenol)	193.6	230	−1.3	9.7	3.5
Valine	Val	V	—CHMe₂	140.0	155	4.2	—	6.9

[a] The pK_a values for the free amino and carboxyl groups are 6.8–7.9 and 3.5–4.3 respectively.

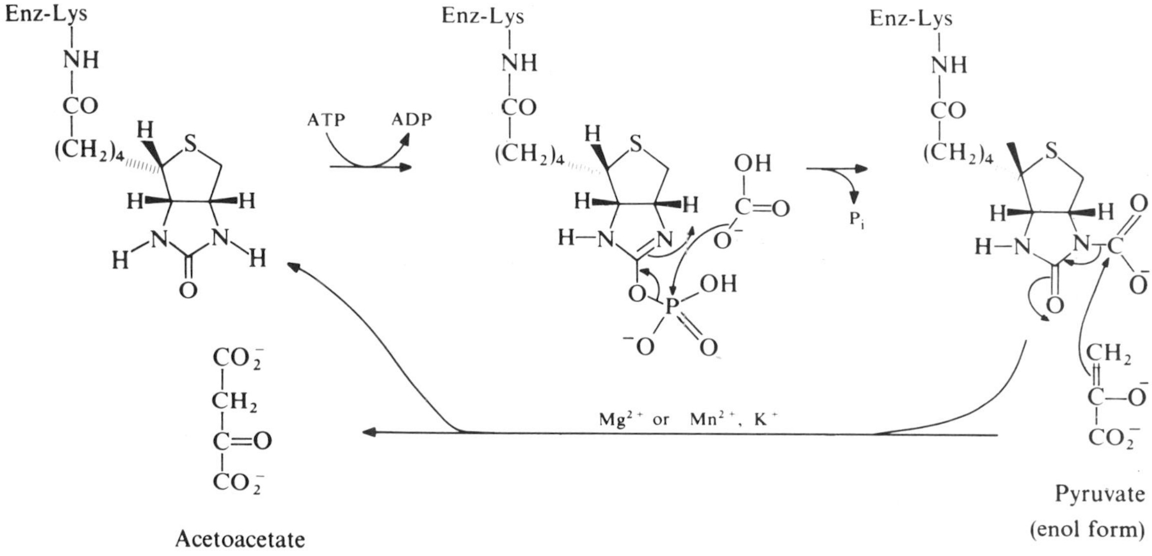

Scheme 1

occur and the defective methylation of some sperm proteins leads to infertility. Also, *N*-acetylation of the amino terminus of a polypeptide chain is moderately common.

Third, various prosthetic groups, or non-protein moieties tightly associated with the enzyme, may be attached to enzymes. These play a key role in the catalysis either acting as carriers or as additional reagents. Often, but not always, prosthetic groups are covalently attached. Thus biotin is found covalently linked to Lys side chains in enzymes whose function is to perform β-carboxylations, *e.g.* pyruvate carboxylase (Scheme 1) with the CO_2 being carried on the biotin. Another example is the covalent attachment of the heme group to cytochrome bc_1 (cytochrome *c* reductase) *via* two Cys residues.

A final type of covalent modification involves the attachment of carbohydrate, lipid or nucleic acid to the polypeptide chain. Such modifications may be effectively irreversible as in the glyco-proteins, where oligosaccharides can be linked to Ser, Thr or Asn residues. Reversible modifications can serve as metabolic switches as, for example, in the phosphorylation–dephosphorylation of glycogen synthase.

5.1.2.2 Classification

The foregoing section on protein modification implies a means of classifying proteins on the basis of their composition. Simple proteins contain only amino acids and conjugated ones contain amino acids plus one or more other type of molecule. Thus, phosphoproteins, metalloproteins and lipoproteins contain phospho, metal and lipid groups, respectively, in addition to the amino acids. Another useful general classification is based upon the shape of the protein: globular or fibrous. It is likely that all enzymes are globular proteins. A final means of classifying enzymes derives from the reaction that they catalyze. For example, an enzyme which catalyzes the transfer of a phosphate from ATP to a substrate can be referred to as a transferase, although other names may be in more general use. In the case given, kinase is more generally used for such a transfer reaction. A fuller discussion of the classification of enzymes is given in the Chapter 5.2.

5.1.3 ENZYME ISOLATION

Whilst it is often possible to determine kinetic constants associated with an enzyme-catalyzed reaction without much purification, structural studies generally require that the enzyme be pure. The methods employed for a particular enzyme purification are often a matter of personal preference and will also depend on the source and abundance of the starting material. Much has been written on the purification of particular enzymes as well as on the techniques to be outlined below.[8,9]

A general procedure which might be appropriate for an intracellular enzyme is first to break open the cells by mechanical, chemical or even enzymic means. It is often necessary to add digestive enzyme inhibitors like EDTA and phenylmethanesulfonyl fluoride and/or protective media such as glycerol. Since the initial extraction usually results in a large volume, the first stage is to precipitate the protein by adding salt or alcohol, the latter necessarily being carried out at low temperature. This then allows dissolution of the precipitate in a smaller volume, simplifying subsequent stages. Adsorption on to a suitable chromatographic medium is an alternative means of achieving this initial concentration. Chromatographic separations on the basis of charge, size or specificity then follow, each stage being monitored by assaying the enzyme activity. In the case of a membrane-bound protein, suitable detergents must be found which will solubilize the material without its denaturing.

Recently, considerable success in enzyme purification has been achieved using reverse phase chromatography where the column packing is a hydrophobic material such as alkylated silica gel. Elution is generally carried out with a gradient of decreasing ionic strength and it is advantageous to use HPLC.[10]

5.1.3.1 Criteria for Purity

Unlike an organic synthesis where success is often judged by obtaining crystals with a constant melting point, crystallization is no criterion for enzyme purity, although well-formed large crystals probably only result from a pure enzyme solution (see Section 5.1.3.6). It is necessary, then, to have other criteria, perhaps the most obvious of which is the specific activity, *i.e.* the number of μmol of substrate converted per min under given conditions of pH, temperature, additives *etc.*, per mg protein. Since the enzyme is characterized by the reaction catalyzed, no purification can sensibly be performed before an enzyme assay procedure has been established.[11] Once a reliable assay is available, and this may involve ingenious coupling to other reactions offering better scope for spectrophotometry, purification is continued until the specific activity is constant.

The other most commonly used means of establishing purity is sodium dodecylsulfate–polyacrylamide gel electrophoresis (SDS–PAGE).[12] The detergent has the property of 'wrapping up' the protein in a micelle so that the rate of migration through a gel is proportional to the log of its relative molecular mass. The presence of minor contaminants, possibly as a result of degradation, can then be seen when the gel is overloaded with the 'pure' protein, and developed with a sensitive stain such as Coomassie Blue. Dissociation of disulfide bridges with a reducing agent such as β-mercaptoethanol, or running the gel without SDS (separation by charge rather than size), will perhaps overcome the problem of the contaminant having the same molecular weight.

Other possible means of determining purity are to observe the shape of the elution profile from a chromatography column (see below) and to determine the constancy of the specific activity across the peak. Isoelectric focusing, whereby the separation of the proteins in a pH gradient depends upon their isoelectric points (pI),[13] is a sensitive but relatively expensive method complementary to SDS–PAGE. Determination of the N- or C-terminal residues can provide a sensitive, if tricky, method for establishing the homogeneity of a preparation but may be misleading if the enzyme has been released from a membrane in a relatively non-specific manner. Similarly, the stoichiometry of any cofactor may be used. Ultracentrifugation and solubility measurements are nowadays regarded as old fashioned for determining purity.

5.1.3.2 Precipitation Techniques

The solubility of an enzyme depends on the protein concentration, the pH, the concentration and type of salt present, the dielectric constant of the medium and the temperature.[14] Often too, the presence of the substrate will affect the solubility. Thus, any of these can be altered to precipitate a protein. There is a solubility minimum at the pI of the protein and, quite often, a protein is less soluble at 4 °C than at room temperature. Most precipitations, however, are carried out by increasing the concentration of a highly soluble salt such as $(NH_4)_2SO_4$ or NaCl in steps of perhaps 20% saturation and separating the centrifuged pellet from the supernatant. Enzyme assay then locates which fraction needs to be worked-up further. The decrease in protein solubility upon the addition of salt, salting out, arises because of the removal of the water necessary to hydrate the protein by the greater charge density of the ions.

A converse argument can be applied to the addition of ethanol or acetone, where association of the charged protein molecules results from their not being effectively hydrated. Because the addition of organic solvent provides a milieu in which the inside and outside of the molecule are less well distinguished, the likelihood of denaturation or unfolding of the protein is increased. It is possible to minimize this by keeping the temperature of the solution always close to its freezing point. A very successful separation scheme based almost entirely on alcohol precipitation is used for large-scale blood plasma fractionation.[15] However, salting out is the more commonly used technique since it is less likely to lead to irreversible denaturation. Several enzymes, *e.g.* rabbit muscle aldolase, can be prepared in a remarkably pure form by salting out alone.[16]

5.1.3.3 Ion Exchange Chromatography

This technique is well known and, since proteins are complex charged molecules, it is readily applicable to enzyme purification. The most commonly used matrices have diethylaminoethyl or carboxymethyl groups as the active species with cellulose, dextran, agarose or polyacrylamide for the support.[9] Quaternary aminoethyl and sulfopropyl exchangers provide more strongly charged groups which may be useful in some applications. Great care must be taken to ensure that the pH and ionic strength of the added protein solution are such as to allow good binding so that subsequent elution, either by pH or salt gradient, releases the molecule cleanly. It is possible to use the ion exchanger either as a batch adsorber by stirring the resin into the enzyme solution, or by the usual column methods. It should be remembered that the process of ion exchange differs from conventional partition chromatography in that it is much more of an 'all-or-nothing' phenomenon: the enzyme binds to the exchanger right at the top of the column and will remain there (and can be seen when purifying a coloured protein) until released by some competing ion. This means that a short column is all that is usually required. For many years, amino acid analyses of hydrolyzed protein mixtures were performed by ion exchange chromatography but, more recently, HPLC techniques based on ion exchange or reverse phase have become routine for amino acids as well as for protein separations.[10,17] Depending on the protein and the particular conditions, capacities up to 0.5 g per g exchanger are fairly typical.

5.1.3.4 Gel Filtration

When a mixture of enzymes of different sizes is added to a column containing a gel, and eluted with dilute buffer, those proteins which are significantly larger than the pore size of the gel will be able to pass through without being able to enter the pores and hence be retarded. Smaller proteins will enter some of the larger pores but be excluded from the smaller ones, whilst small molecules will have the largest volume available to them and hence their permeation through the column will be slowest. Thus large molecules will elute first with the smallest ones eluting last. Although early work used starch gels, it was found that the pore sizes were unreliable and modern media are made by crosslinking dextran, agarose or polyacrylamide to varying degrees: the greater the crosslinking, the smaller the pore size.[9] In contrast to ion exchange, where the columns tend to be shorter and broader, gel filtration uses long thin columns and, for optimum separation, pumping the buffer upwards produces sharper peaks because the diffusion resulting from gravity opposes that of the buffer flow. The technique is applicable both for preparation and for determination of molecular weight since the log of the molecular weight is directly proportional to the eluted volume within the range of partial retardation. Normally a series of known proteins is run to calibrate the column before use.

With gel filtration also, HPLC techniques are now common,[8,18] providing a rapid (minutes rather than hours) separation of complex mixtures in quantities large enough to raise antibodies, to do a partial N-terminal sequence and even to crystallize. It is perhaps wrong to use the term 'gel' in this context since the nature of a gel will change considerably under high pressure. Thus, the matrices used are considerably more rigid and are generally made from porous silica or hydroxylated ethers, the former being unstable above pH 8 and more affected by the net charge on the protein whilst the latter give poorer resolution.

5.1.3.5 Affinity Chromatography

In Section 5.1.3.3, specific interactions between charged groups on an enzyme surface and an immobilized charge were seen to separate an enzyme mixture. However, the technique need not be specific and several steps under a variety of conditions are usually needed to remove contaminating species by ion exchange alone. Using an enzyme's specificity, however, it is possible to achieve dramatic purifications. The enzyme substrate, or an inhibitor, is bound to an inert support such as agarose or polyacrylamide, whereupon addition of a protein mixture will result in only the specific enzyme binding while all other molecules pass through. Removal of the enzyme is then achieved by adding free substrate, changing the pH or possibly increasing the ionic strength.[8,19] Enzymes catalyzing single substrate reactions generally require an inhibitor to be used if the column is to be reused. For enzymes which involve a substrate and a cofactor like ATP or NADH then either substrate or cofactor can be used, although if the latter is used it is likely to be less specific.

Coupling the substrate to the matrix will usually involve insertion of a suitable spacer to facilitate reaction between substrate and enzyme. The basis of the coupling is often CNBr, carboxyl or, better, activated ester and epoxy (Schemes 2–4) as well as amino and thio groups. Since the coupling methods are general, it is possible to attach proteins to the matrix, thus affording protein–protein interactions the most specific of which involve antibodies raised to the enzyme in question. In principle, a column filled with an antibody affinity matrix should purify the protein in one step from as complex a mixture as is likely to be found! Another variation on this theme is to use an immobilized lectin to select for glycoprotein.

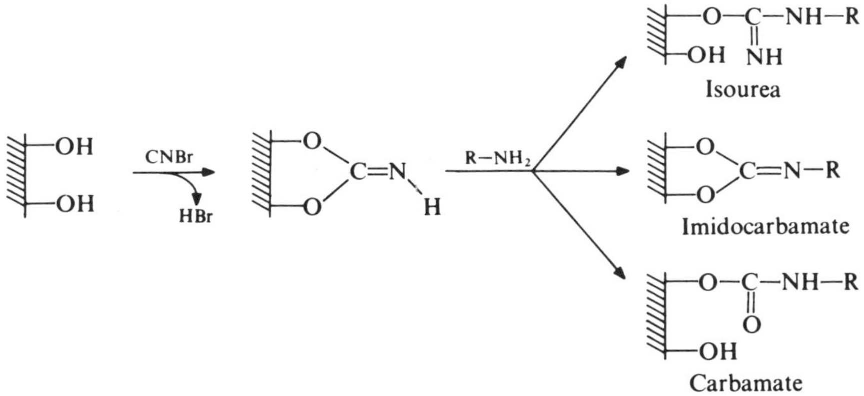

Scheme 2

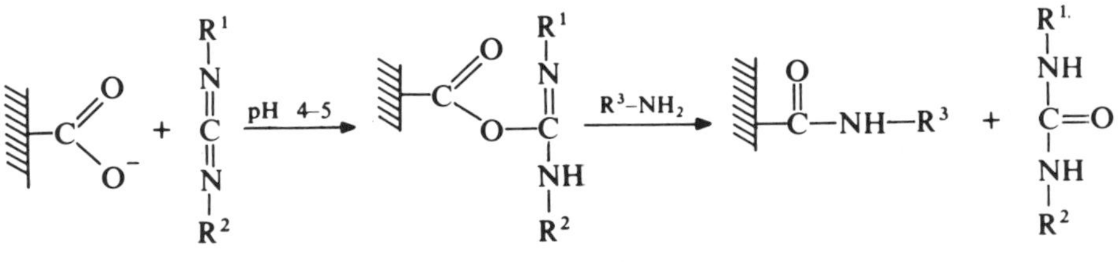

Scheme 3

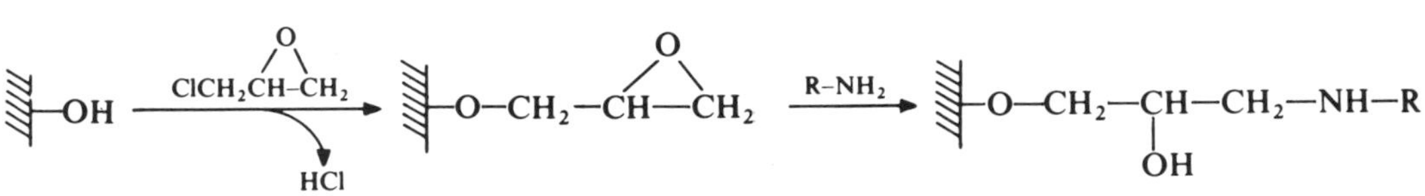

Scheme 4

5.1.3.6 Crystallization

Whilst crystallization is not generally used as a final purification step as it is in organic chemistry, it does form an important part of an enzyme structure project. This is because the principal method available for determining protein tertiary structure is X-ray crystallography and for this technique, large (at least $0.2 \times 0.2 \times 0.2$ mm) single crystals are essential.[20] Crystals grow from a supersaturated solution of the enzyme in a two-stage process: nucleation followed by growth. Supersaturation is achieved by increasing the salt concentration, by adding alcohols such as methylpentanediol, poly(ethylene glycol) or ethanol, or possibly by changing the temperature. Several ingenious crystallization experiments have been devised and, since the initial conditions are only found by trial and error, robotic techniques are being introduced.[21] Variables such as precipitant, pH, temperature, protein concentration, and additives can now be systematically varied and further experiments planned on the basis of the results. Two recent conferences on the topic of macromolecular crystallization have tried to introduce a formalism into what many consider is still an art.[22,23]

5.1.4 PHYSICAL PROPERTIES OF ENZYMES

The physical properties of enzymes arise from the nature of their constituent amino acid side chains together with the forces which hold them together. Apart from the covalent bonds of the main chain and disulfide bridges, the compact structure of the globular protein is maintained by a multitude of hydrogen bonds, ion pairs between charged groups like Asp and Lys, and the weaker charge–charge interactions of dipoles, induced dipoles and transient dipoles, often referred to as van der Waals forces. Mention must also be made of the hydrophobic effect. From the outset, crystal structures of proteins have shown that the charged and polar side chains tend to be on the outer surface of the protein whilst the hydrophobic ones (Figure 2) are more usually found inside. This micellar appearance of the protein can best be explained in entropic terms. The transfer of an alkyl side chain into water can only be achieved by breaking some water–water interactions. Energy is required to do this but is not recovered by favourable side-chain–water interactions and, if the hydrophobic–water interface area can be minimized, then less energy is required. Thus, the hydrophobic residues will tend to cluster in the interior, thereby allowing more extensive water–water and water–polar/charged residue interaction. Stabilization of a native protein structure over its unfolded form is marginal, a fact reflected in the ease with which many proteins can be denatured by mechanical action.

In 1970, when only a few crystal structures were available, two general observations were made.[24] Charged groups found away from the surface were associated with polar residues and had functional importance. Every hydrogen bond donor not available for interaction with water had a conveniently placed acceptor, possibly an internal water molecule. The same was not observed with acceptors. Now that more than 200 structures are available,[1] some further generalizations are possible, but no foolproof set of rules is available. Thus, the internal packing of a protein is found to be similar to that observed in organic crystals, reinforcing the idea that the hydrophobic core is precisely formed. The conformations of the various groups tend to be those of lowest energy and the interactions between different parts of the polypeptide chain tend to be restricted to relatively few types which are governed by the main-chain conformation. Careful analysis of a number of highly refined structures reveals that virtually all of the potential hydrogen bonding groups are satisfied.[25] Further, many main-chain proton donors form H-bonds to main-chain acceptors, reflected by the occurrence of several characteristic main-chain conformations.

The physical properties of proteins are therefore to be seen in the context of molecules stabilized by a complex series of weak interactions which involve the solvent as well as the protein. The distribution of charged amino acids gives the protein a characteristic isoionic point, pI, in the range 4–10, corresponding to the range of pK_a values of the side chains (Table 1). Protein molecules are generally denatured by extreme conditions of pH, temperature, dielectric constant or the presence of heavy metals. Of course, exceptions to all of these exist: acid proteases like renin are stable below pH 4, whilst the thermophilic bacteria which inhabit volcanic springs survive in temperatures of 80 °C. Unlike manufactured polymers, proteins are produced as a unique sequence and so are a single species with a precise molecular mass. They are, however, liable to change conformation under the influence of pH or substrate, a feature which has important functional implications. Further, treatment with urea or guanidine hydrochloride causes the protein to unfold, often reversibly.[26]

5.1.4.1 General Structural Characteristics

Before any definitive structural information was available for proteins, prediction of the likely polypeptide conformations was made by Pauling.[27] Two bonds on each residue allow free rotation, ϕ and ψ in Figure 1. Because the peptide bond has quite a high degree of double-bond character derived from delocalization of the lone pair of electrons on the N atom, rotation about the C—N bond is restricted. Thus, the angle ω is usually near 180° making the main chain *trans*, although *cis* peptides do occur occasionally. Two particular conformations are preferred: the α-helix and the

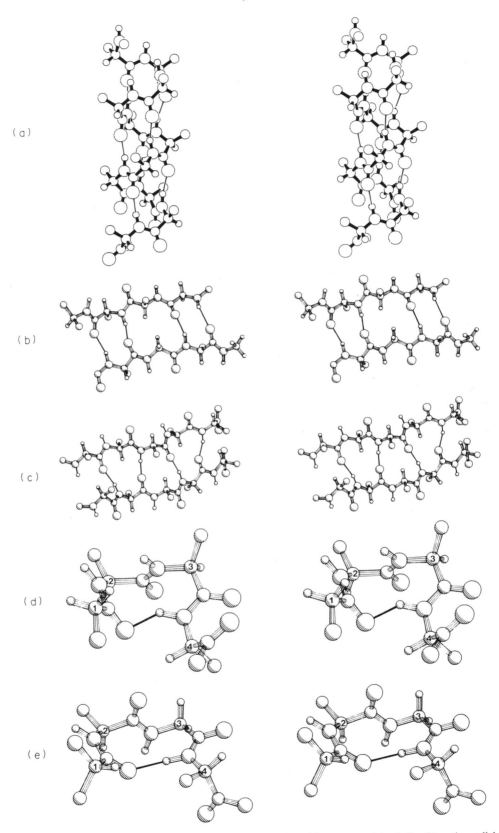

Figure 3 Stereo diagram of the main types of secondary structure found in enzymes: (a) α-helix, (b) anti-parallel β-sheet, (c) parallel β-sheet, (d) type I reverse turn, (e) type II reverse turn. The numbers in the C_α atoms in (d) and (e) indicate the four amino acids (numbering towards the C-terminus) involved in forming the reverse turns. Hydrogen bonding is depicted by the narrower bonds and the atomic radii reflect atomic number

Table 2 Main-chain Conformational Angles for the More Common Secondary Structures[a]

Conformation	$\phi\,(°)$	$\psi\,(°)$	$\omega\,(°)$	R^b	P^c (nm)
α-Helix (right-handed)	−57	−47	180	3.6	0.15
β-Sheet (antiparallel)	−139	+135	−178	2.0	0.34
β-Sheet (parallel)	−119	+113	180	2.0	0.32
3_{10}-Helix	−49	−26	180	3.0	0.20

Reverse Turns[d]	$\phi_2\,(°)$	$\psi_2\,(°)$	$\phi_3\,(°)$	$\psi_3\,(°)$
Type I	−60	−30	−90	0
Type II	−60	120	80	0

[a] The data are from Creighton.[3] [b] R is the number of residues per turn. [c] P is the pitch, or the rise per residue along the helix or strand axis. [d] The subscripts 2, 3 are as shown in Figure 3.

β-pleated sheet, the latter having both parallel and antiparallel forms. These are found frequently together with the reverse or β turn. They are illustrated in Figure 3 with the ideal and observed angles being shown in Table 2, which also includes data on other regular structures which can occur. Notice that type II reverse turns show the effect of the unfavourable side-chain–main-chain interactions. In position 3, only Gly is acceptable since the β-carbon collides with the carbonyl oxygen of the preceding residue.

It can now be seen why hydrogen bonding between main-chain carbonyl and amide groups is important. In the helical conformations the hydrogen bonding is parallel to the helix axis, whilst in the sheet conformations it is perpendicular to the chain direction. A plot of the energy as a function of the ϕ–ψ angles for the tripeptide Gly-Ala-Gly is given in Figure 4, with the positions of the various conformations of lowest energy shown. Figure 4 also contains the ϕ–ψ angles for elastase, showing that the observed conformations do indeed correspond to the lowest energies. The optimum β-sheet conformation is such that a twist develops between adjacent strands which averages about −15°. Side-chain interactions appear in part responsible for this and the outcome is that packing between sheet and helix is facilitated.[28] Finally, the disposition of side-chain type often shows a distinct bias with hydrophobic residues on one side, hydrophilic on the other (Figure 5). Membrane spanning regions are also largely hydrophobic and these features may often be located in the amino acid sequence by hydrophobicity plot (Table 1).[4]

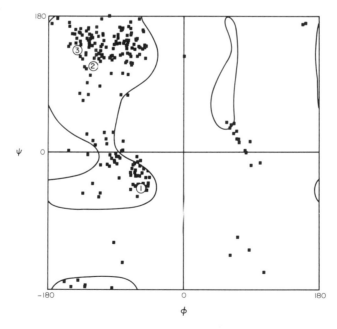

Figure 4 Plot of the ϕ–ψ angles for the enzyme elastase together with the contour of allowed contacts for Ala. The labelled circles show the ideal positions of α-helical (1), anti-parallel sheet (2) and parallel sheet (3) sheet conformations

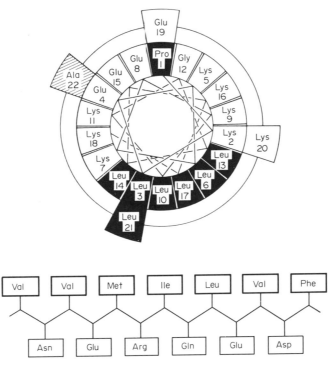

Figure 5 Schematic diagram showing a projection down the axis of an α-helix with 3.6 residues per turn (upper) and perpendicular to the axis of a strand of extended polypeptide with 2 residues per turn, showing bias of side chain character (redrawn from *Trends Biochem. Sci.*, 1987, **27**, 305 with permission of Elsevier)

5.1.4.2 Protein Structure Classification

From the foregoing it is seen that there are only four ways of constructing a protein from the various elements described. This has indeed been observed for the globular proteins whose structures are known.[29] The classification scheme is very general but can be subdivided,[30] although this may give rise to misleading similarities: proteins with similar polypeptide chain folds but with little in common from a functional or evolutionary point of view. Thus, the general classification below has the advantage of simplicity but suffers from putting together, in the same group, proteins of wildly different function, not to mention their actual structure.

α-Proteins are those containing almost no other regular structure except an α-helix. An obvious example of such a protein is myoglobin; for an enzyme example, structures like the second domain of tyrosyl-tRNA synthetase appear to comply with the definition. β-Proteins have mostly a β-sheet and an enzyme example is superoxide dismutase. The next group comprises α/β-proteins where there are alternating segments of helix and strands of sheet. An example of this type is triosephosphate isomerase, where a barrel is formed of eight β-strands flanked by eight helices. Finally, the α + β proteins have both types of structure but in distinct parts of the molecule. Lysozyme, thermolysin and ribonuclease provide striking examples. Figure 6 shows one member of each type, together with a short-hand notation of the topology.

5.1.4.3 Hierarchy of Protein Architecture

It is convenient to have a means of referring to the various recognizable levels of protein structure and the basis of the classification was introduced by Linderstrøm-Lang. The term primary structure refers to the linear sequence of amino acid residues. The covalent structure is sometimes referred to as the primary structure but this would normally include disulfides and any non-amino acid additions. Now that the primary sequence is often inferred from the DNA sequence, such information about post-translational modification is not usually available.

Secondary structure refers to the conformation of the main chain of the polypeptide. Thus, the helix, sheet and reverse turn conformations are its easily identifiable elements. In the original scheme, tertiary structure came next but it has been found convenient to introduce two additional levels. The first is super-secondary structure, sometimes called a folding unit. This refers to the commonly found arrangements of several adjacent secondary structural elements which make up a motif like those shown in Figure 7. Another motif, the Ω-loop, has recently been described.[31] Second are domains, whose identification is somewhat subjective, which appear as separately folded sub-structures forming a distinct part of the enzyme. Connections between other domains in the

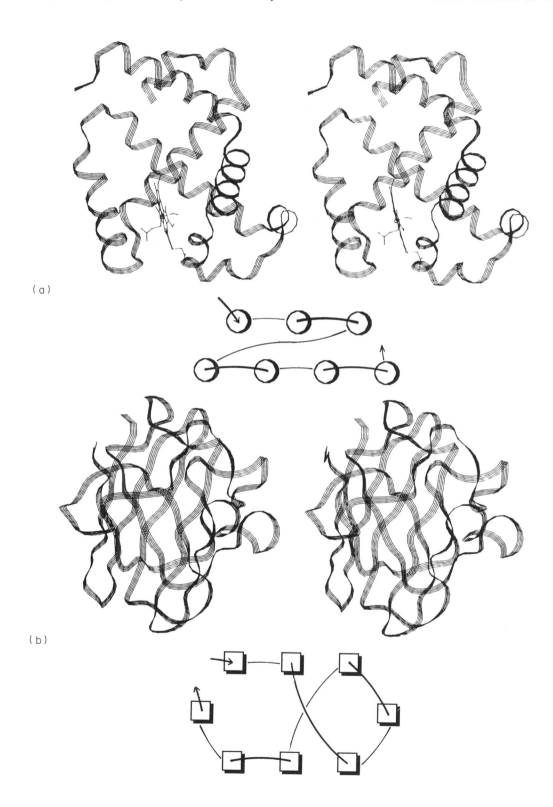

(a)

(b)

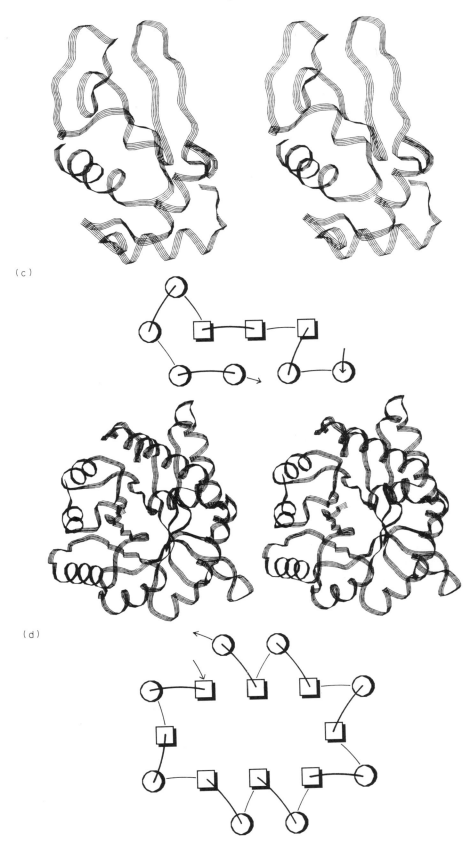

Figure 6 Cartoon diagrams of the backbones of the four protein structure types: (a) all-α (myoglobin); (b) all-β (superoxide dismutase); (c) α + β (lysozyme); (d) α/β (triosephosphate isomerase). Beneath each is an idealized topology diagram of the polypeptide fold, where squares and circles represent strands of sheet and helices respectively

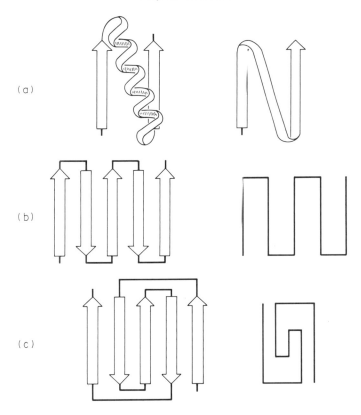

Figure 7 Diagrams of the more common structural motifs or super-secondary structures: (a) β–α–β; (b) β-meander: (c) Greek key. Only the C_α atoms are shown (redrawn from ref. 3 with permission of Freeman)

molecule are few and this has led to the idea that domains can fold independently of each other. Not all of the DNA in a mammalian gene is translated. It is possible, therefore, that domains correspond to the translated parts of the gene called exons[32] since this could account for similarities between some parts of enzymes, whilst other parts appear unrelated.

Tertiary structure is the three-dimensional arrangement of all of the main-chain and side-chain atoms. In other words it is the structure of the polypeptide. Clearly, if one knows the tertiary structure, domains, super-secondary and secondary structures are also known although the primary sequence may still be uncertain. In monomeric proteins this level of organization is the final one. However, many proteins are made up of several distinct polypeptides which may be identical, similar or quite distinct and they are assembled using the same forces as the individual monomers, including covalent linkage through disulfides, as in the immunoglobulins. The aggregation of units into complexes represents the quaternary structure. Some enzyme complexes involve several distinct activities and result in the formation of multi-subunit proteins, *e.g.* pyruvate dehydrogenase which has 60 subunits of three distinct types.

5.1.4.4 The Folding Problem

This chapter has dealt with the forces involved in stabilizing enzyme structure and mention was made in Section 5.1.2 of protein synthesis by translation of the genetic message. However, the protein must attain its final folded form after biosynthesis by a spontaneous process apparently requiring no template or catalyst, although enzymes exist which enhance the formation of the correct disulfide bridges.[33] Thus the primary sequence must contain all of the information necessary for the correct folding of the protein *in vivo*, the driving force being the minimization of the energy of the protein *and* solvent.[3,26] It follows, then, that an understanding of the interactions between the various amino acids should permit the tertiary structure to be predicted from the amino acid sequence. Such attempts were initially limited to predicting the secondary structures[34,35] based on local interactions. Careful analysis of the packing of secondary structural elements has allowed some further progress[28,36] and it is now being found that the effects of side chains more distant in the sequence give rise to particular folding units. These sequence templates can now be used to search a new primary structure and obtain an idea of its fold.[37]

5.1.5 PRIMARY STRUCTURE DETERMINATION

Recently, high-sensitivity automatic amino acid sequencers and the complementary techniques for DNA sequence analysis have led to a revolution in primary structure determination. It is still necessary to sequence polypeptides to establish the DNA reading frame, the start of transcription, or, at an earlier stage, to provide the means of isolating the DNA from a gene library by hybridization with a synthetic oligonucleotide whose sequence is predicted from that of the protein.[38] Also, DNA sequencing gives no data about postsynthetic modification. The basic principle of sequencing a polypeptide is to remove and identify each amino acid in turn from one end. In practice, it is necessary to generate smaller pieces by specific cleavage of the protein because, even with the modern micro methods, continuous runs in excess of 40 amino acid residues are not routinely feasible. At each stage, careful account must be kept of the composition of the fragments in relation to that of the whole polypeptide. Thus an overall strategy might be: isolate the pure polypeptide, then determine the amino acid composition by total hydrolysis with 6 M HCl at 105 °C for various times to allow correction to be made for residues like Thr which are degraded by the conditions. Next, cleave the chain into a few smaller peptides in a specific manner by either chemical or enzymic means and separate the resulting fragments. Further subdivision may be necessary for the larger fragments. Repetitive removal and identification of the terminal residues is then performed. The whole process is repeated with a different series of fragments to provide overlaps and in this way the sequence of the whole chain is established.

5.1.5.1 Chemical and Enzymic Cleavage

Acid hydrolysis to break peptide bonds was mentioned above, and partial acid hydrolysis was used to establish the sequence of insulin. More specific cleavage at infrequently occurring residues is preferred and the use of CNBr to cleave the chain specifically at Met residues (Scheme 5) is one popular method. Hydroxylamine is useful for Asn–Gly cleavage (Scheme 6) while iodosobenzoic acid (1) and 3-bromo-3-methyl-2-(nitrophenylthio)indolenine (BNSP-skatole, 2) may be used for specific chemical cleavage at Trp and Tyr.

(1)   (2)

Enzymes themselves are used to cleave proteins selectively, thus facilitating amino acid sequencing. Trypsin, chymotrypsin, and the V8 protease from *Staphylococcus aureus* are three enzymes capable of limited hydrolysis of a polypeptide on the C-terminal side of Arg and Lys, of Leu, Tyr, Phe and Trp (mainly), and of Glu (pH 4) or Glu and Asp (pH 7.8), respectively. Modification of the Lys residues allows a more limited hydrolysis by trypsin, and several other enzymes are available with different cleavage sites, not to mention the exopeptidases which remove residues from the termini of a peptide. It should be noted that none of these methods works perfectly!

5.1.5.2 Sequential Degradation of Peptides

The key reaction in sequence determination is the stepwise removal of residues from one or other end of the peptide. Mention has been made of the enzymes which cleave residues from the ends: aminopeptidases are not particularly stable and are hard to obtain pure but carboxypeptidases are routinely used, although their abilities to release every amino acid with equal efficiency vary. It is also necessary to perform a time course for such a determination as cleavage of the first residue generates another susceptible terminus.

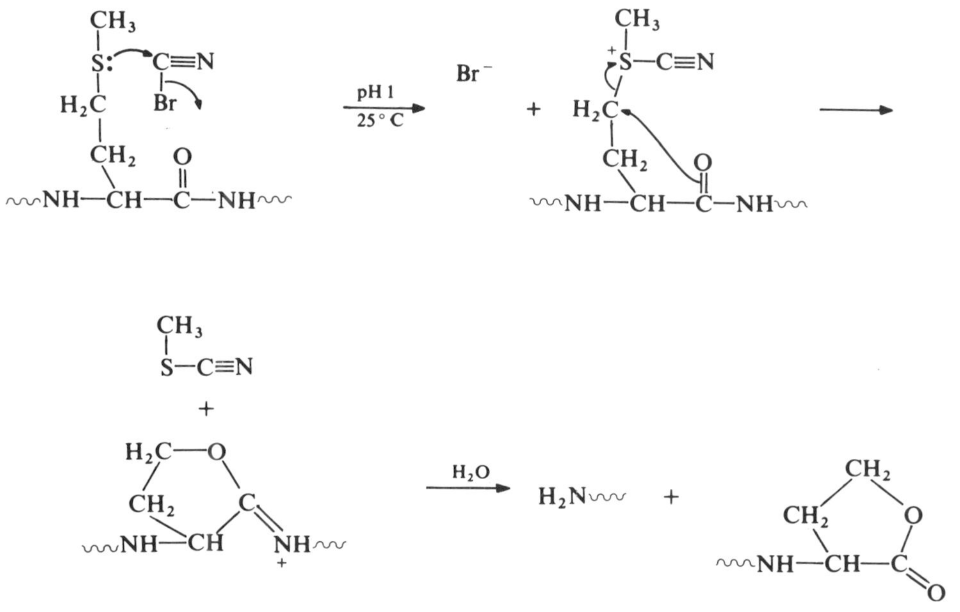

Scheme 5

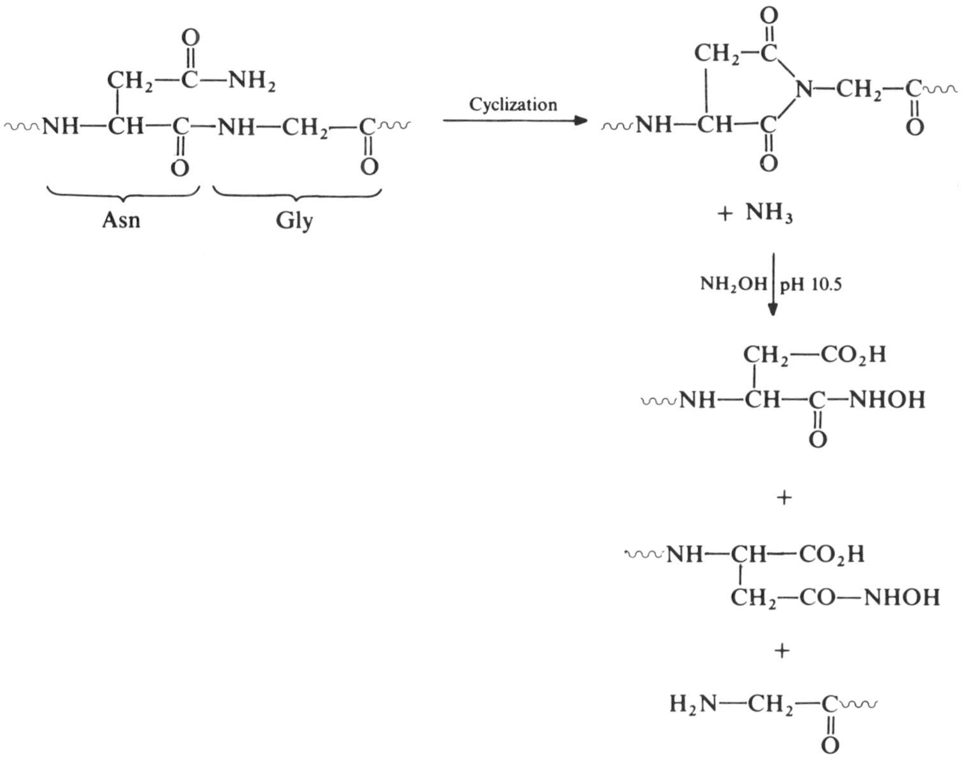

$$\text{Asn} \qquad \text{Gly}$$

$$+ \text{NH}_3$$

$$\text{NH}_2\text{OH} \Big| \text{pH } 10.5$$

Scheme 6

It is, therefore, upon chemical methods that sequence analysis depends and, in particular, upon the Edman degradation. Consecutive reactions must be performed with repetitive yields of nearly 100% in order to be useful over more than a few steps. Reaction of the free N-terminus of a peptide with phenyl isothiocyanate under alkaline conditions produces a thiourea which can be washed free from any unreacted species before being subjected to acid treatment; this cleaves off the modified amino acid, converting it to a thiohydantoin (**3**), easily identifiable by GLC, TLC or HPLC. The new, free amino group is then available for the next round (Scheme 7).

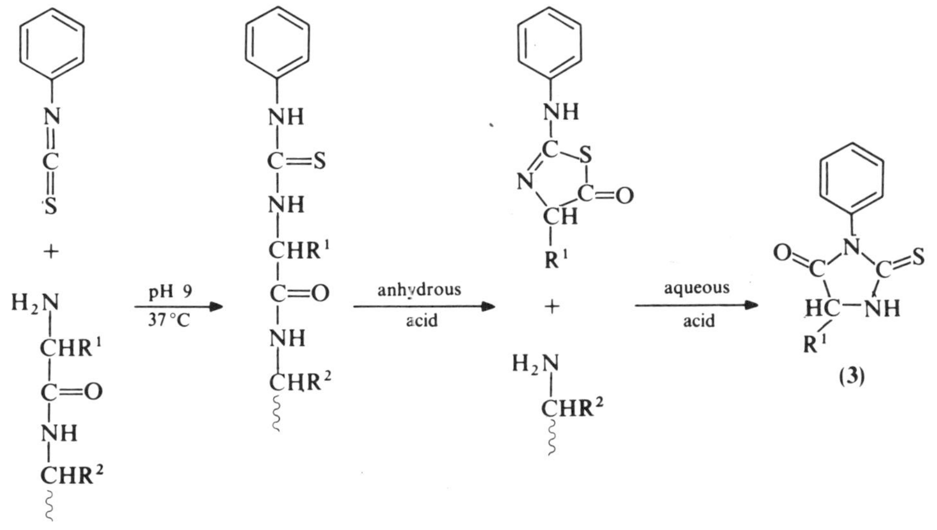

$$\underset{\text{37°C}}{\overset{\text{pH 9}}{\longrightarrow}} \qquad \underset{\text{acid}}{\overset{\text{anhydrous}}{\longrightarrow}} \qquad \underset{\text{acid}}{\overset{\text{aqueous}}{\longrightarrow}}$$

(3)

Scheme 7

This degradation forms the basis of all modern peptide sequencing and is routinely applicable for 10–15 cycles when done manually. Nowadays, however, the process is performed automatically, which makes it even more efficient because optimization of the reaction conditions and quantitative, on-line, HPLC analysis of the thiohydantoins is then possible, allowing continuous runs of at least 40 residues, often 50–60. Further, solid-phase and liquid-phase techniques have been developed which make it possible to sequence pmol quantities.[39,40] Thus, a band eluted from a gel electrophoretogram can provide sufficient material for such analysis.

5.1.5.3 Mass Spectrometry

Mass spectrometry provides a means of sequencing peptides of up to 1500 Da.[41] Conveniently, fragmentation often occurs at the peptide bond and so the principal fragments allow the sequence of the whole peptide to be deduced. The advantages that carboxyl and amide are distinguishable, and that blocked N-termini and pyroglutamate are readily detectable, are offset by the cost of instrumentation and the non-volatile nature of the peptides. Fast atom bombardment mass spectrometry, where the ions are generated from a liquid matrix such as glycerol by bombardment with xenon, has enhanced the applicability. However, derivatization is still an important aspect of the technique. Acetylation will block free amino groups, and the methylation of main-chain amide protons, together with those on amide, carboxyl and hydroxyl side chains, enhances the volatility of the peptide. GLC before the spectrometer, or differing rates of volatilization, allow mixtures of peptides to be sequenced.

5.1.5.4 DNA Sequencing

Since the sequence of an enzyme is dictated by that of the DNA in the gene which encodes it, one approach to protein sequencing is to find the sequence of this DNA. Equally, if the mRNA for the protein can be identified, its sequence, too, corresponds to that of the protein. Space precludes any description of how the gene is identified or how mRNA is converted back to cDNA with reverse transcriptase except to point out that hybridization is the formation of a length of double-stranded DNA by joining complementary single strands; thus a synthetic oligonucleotide deduced from a short amino acid sequence can be hybridized to the appropriate single strand, permitting its identification and purification. There are two possible ways of proceeding: selective cleavage of the DNA, adopting an approach similar to that for amino acid sequence determination, or synthesis with chain-terminating nucleotides.

The Maxam–Gilbert method of selective cleavage[42] applies to segments of single or double stranded DNA which have been ^{32}P-labelled at (usually) the 5′ end. After the strands have been separated, the purines are modified by methylation, thus generating an unstable glycosidic bond at the methylated purine. Conditions exist which lead to the modification of the chain at G (Scheme 8) or A + G. Hydrazine reacts with both pyrimidines, or in the presence of NaCl, only with cytosine (Scheme 9). In each case, piperidine will cleave the modified chain (Scheme 10). Thus, if the reagents are added in limited quantities, families of fragments all ending in A + G, G, C or C + T will be produced. These families can then be separated by gel electophoresis run in adjacent tracks on the same gel, so that the sequence can be read off directly.

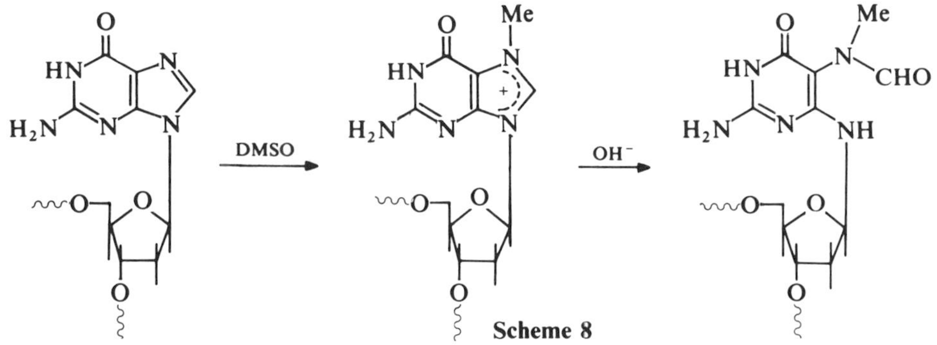

Scheme 8

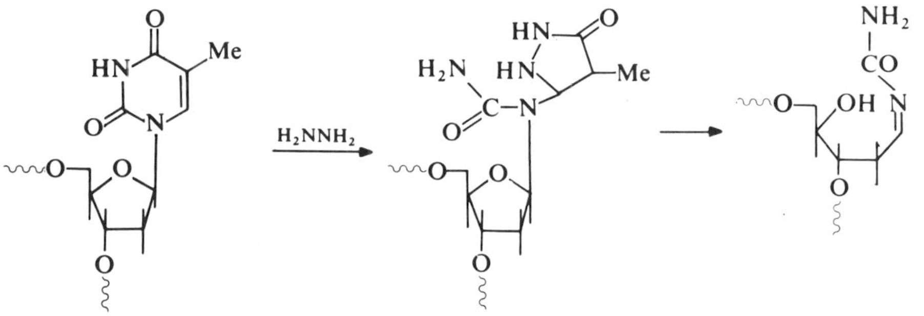

Scheme 9

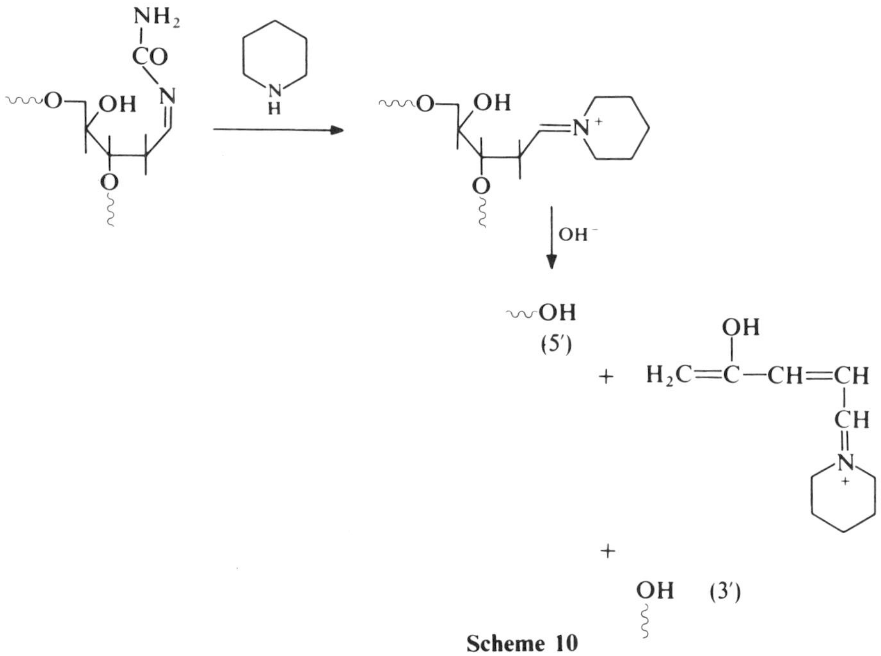

Scheme 10

The Sanger or dideoxy method[43] relies on the synthesis of a strand complementary to that whose sequence is required, adding to the mixture a small amount of dideoxy nucleotide which does not possess the necessary 3'-OH to permit further synthesis (Scheme 11).

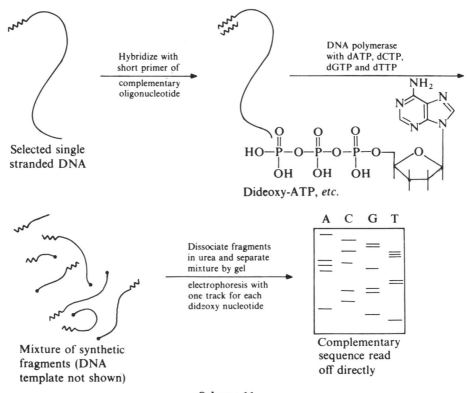

Scheme 11

Thus a family of strands of different length will be generated which can, as in the previous method, be separated by electrophoresis. It is essential to have a single-stranded DNA molecule and to add a short primer with which to initiate the polymerase reaction. Detection of the larger fragments is greatly simplified by adding a radioactive nucleotide in order that the longer fragments will contain the same relative amount of label as the shorter segments. This contrasts with the Maxam–Gilbert method where only a single labelled group is present in each strand of whatever length. To offset this disadvantage in the latter method, which leads to the larger fragments being less easy to identify, there are usually the two strands which allow a complementary check.

Typically, sequences of up to 250 bases can be sequenced in a single experiment with little difficulty. Longer sequences can be subdivided in much the same way as for proteins, using restriction enzymes to cut the DNA at a variety of specific base sequences. To ensure adequate overlapping of the fragments, cleavage with several different restriction enzymes is required. It is possible, too, to use mechanical cleavage of the DNA to generate the necessary small pieces. The time to sequence a 250 base length is about 24 h and this does not include the time for its isolation. The automatic sequencing of a peptide of 40 amino acids takes around 48 h, once again not including the time for isolation.

5.1.6 SECONDARY STRUCTURE

As defined above, the secondary structure refers to the conformation of the main chain and is generally thought of in its more easily identifiable forms: α-helix, β-sheet and reverse turn. Whilst X-ray crystallography *inter alia* reveals the full extent of the secondary structure, valuable information about the main-chain conformation can be obtained by a variety of techniques now to be described. This information is generally in the form of the fraction of each conformational type present but as such serves as a useful guide in the predictions discussed in Section 5.1.6.3.

5.1.6.1 Circular Dichroism

The rotation of plane-polarized light by chiral centres arises from a difference in the absorption of the left- and right-circularly polarized components.[44] The local environment of the various chiral

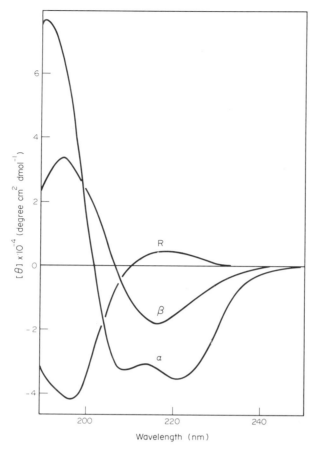

Figure 8 CD spectra of poly-L-lysine at 25 °C under conditions in which it adopts one of three distinct conformations. The unordered form, R, is observed at neutral pH; α, the helical form, occurs at pH 10.8, and β, the sheet form, occurs at pH 11.1 after 15 min annealing at 52 °C (redrawn from *Biochemistry*, 1969, **8**, 4108 with permission of the American Chemical Society)

centres in a protein is affected by the nearby peptide bonds and so there is a marked difference in the absorption of the two components in the region of the peptide bond absorption around 220 nm. This can be seen clearly in the CD spectrum of polylysine under conditions in which it adopts one of the helical, sheet or random conformations (Figure 8). Further, β-turn data have been provided by model compounds so that an estimate of turn content is also possible. Assuming that the optical effects of the various conformations are additive, three possible approaches exist for the extraction of secondary structure content from an observed CD spectrum. The use of reference spectra derived from model polypeptides is essentially the same method analytically as that based on the known conformations of proteins. The mean residue ellipticity $[\theta]$, at a given wavelength λ [a function of $(E_L - E_R)$], equals the sum of the various fractions f_i of secondary structure type i times the appropriate reference value $X(\lambda)$ subject to the condition that the sum of the $f_i = 1$:

$$[\theta]_\lambda = \Sigma f_i X_i(\lambda)$$

The third approach is to use the CD spectra for the reference proteins directly, rather than deriving an ideal spectrum for each conformation. Comparison of the various methods with the X-ray structures reveals that the helix and sheet values are fairly accurate whilst those for turn, and hence random, are much more prone to error.

The complementary technique of optical rotatory dispersion (ORD) has also been used to derive the same information although in a less direct manner,[45] and is not now much used. Both techniques can be employed in studying changes in conformation induced when a substrate binds to an enzyme. Further, the aromatic absorption region at 280 nm provides another means of monitoring structural change.

5.1.6.2 Infrared and Raman Spectroscopy

The application of infrared spectroscopy to determinations of protein secondary structure relies upon the fact that the hydrogen bonding of the main chain affects the vibrational frequencies of the peptide bond and so reflects the conformation.[46,47] Aqueous solutions are complicated in that many bands are obscured by the water. In particular, the amide I band (Figure 9) around 1650 cm^{-1} is particularly sensitive to main chain conformation but corresponds to a strong water absorption. The amide II band at 1550 cm^{-1} is insensitive to the conformation. Whilst difference spectra can be measured, the awkwardness of so doing has meant that most solution work has been carried out in D_2O where the amide I′ band is free of solvent interference and not much affected by ^2D exchange. The amide III band at about 1300 cm^{-1} is both conformation and ^2D dependent but allows distinction between β-sheet and reverse turns to be made. In a similar manner to that for CD measurements, accounting for the observed band in terms of the fractions of the contributions of standard structures allows an estimate of the helix and sheet content to be obtained. Like CD estimates, those from IR give but a semi-quantitative estimate. Recent improvements arising from FT-IR and resolution enhancement by deconvolution have led to better accuracy.

The Raman technique has also been applied successfully to secondary structure estimations.[48] The advantage of Raman over conventional IR spectroscopy in aqueous solution derives from the low intensity of the O—H vibration. In addition, estimates of disulfide conformation can be obtained from the region between 500–550 cm^{-1}. The availability of UV lasers has meant that pre-resonance Raman enhancement of the amide bands can now be observed by excitation at 257 nm.

5.1.6.3 Prediction Methods

As stated in Section 5.1.4.4, it should be possible to predict the tertiary structure of an enzyme from its amino acid sequence. Thus, as a first step, considerable effort has been put into providing algorithms which will predict the elements of secondary structure from a given primary one.[35] It is likely that the techniques so far employed have reached their limit and that further progress will require inclusion of other information such as the templates associated with particular tertiary structural features.[36] Comparisons of the various methods with X-ray structures known, but not revealed until after the prediction, indicated that some 60% of the residues were assigned the correct secondary structure. However, assignment of a particular residue in a crystal structure to a secondary structure type is often rather subjective, although an objective method has now been published.[49] Generally, it is the ends of secondary structural elements which are inaccurate, reflecting the way in which most methods search for nucleating sequences and then move outwards towards both N- and C-termini until feature-breaking sequences occur.

The Chou–Fasman method[50] is one of the most popular methods and is relatively straight-forward to apply. Consideration of known protein X-ray structures allows the frequency of

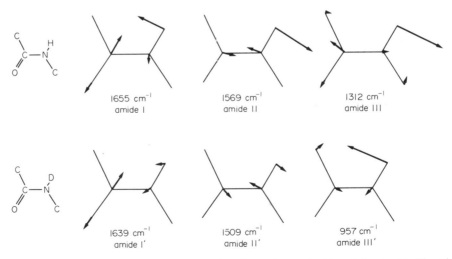

Figure 9 Atomic displacement vectors of the amide modes referred to in the text, for *N*-methylacetamide. The primed values refer to the part deuterated molecule (redrawn from 'Spectroscopy of Biological Systems', 1986, vol. 13, p. 114, with permission of Wiley)

occurrence of each residue in each type of secondary structure to be determined. These figures can be normalized and updated as necessary. The product of four adjacent residues can then be used to search for a nucleation site of a given conformation. Residues are then added at each end until the feature's existence becomes improbable. The process is repeated throughout the sequence for each secondary structure type: helix, sheet, reverse turn. Segments of sequence may predict equally as sheet and helix and here the CD or IR spectroscopic data may help in discriminating. Other criteria, particularly homologous sequences, can also be used.

An information theory approach was adopted by Garnier *et al.* to produce an algorithm which is considered good and which provides not only for some optimization when the overall secondary structure content is known, but also a unique assignment.[51] A different approach has been proposed by Lim in which the hydrophobic nature of the residues is considered. The technique has been refined[52] and is perhaps now one of the best.

More recent developments involve the use of templates and consideration of short segments of the same or similar sequences derived from the structure database.[53] Much careful analysis of supersecondary structure, for example the packing of helices, of sheets and the packing of helices with sheets, has been done.[28] This has led to some attempts at tertiary structure prediction.[54]

5.1.7 TERTIARY STRUCTURE

While tertiary structure prediction is still incapable of producing reliable results, several experimental techniques, particularly X-ray crystallography, have been, and are, at the forefront of providing the structural information necessary to form the basis of current molecular enzymology and are indispensable for modern drug design and protein engineering. However, recent advances in cryomagnet design have increased the sensitivity of NMR spectrometers so that three-dimensional data can now be derived for small proteins.[55] While no other technique provides the detail possible with X-ray (or neutron) diffraction, electron microscope techniques have produced low-resolution structures of membrane-bound proteins[56] and higher resolution is likely to be achieved in the near future.

5.1.7.1 X-ray Crystallography

Since the structure of myoglobin was obtained in 1959, the growth in protein structure determinations has been exponential and, while the Brookhaven Data Bank has perhaps 200 coordinate sets,[1] there are probably at least that number of protein structures solved but as yet not freely available. This section will give an overview of X-ray crystallography as applied to proteins, pointing out the advantages and disadvantages. Fuller accounts are legion.[57-59]

X-rays are scattered by electrons and consequently what is determined in an X-ray diffraction experiment is the electron density distribution of the molecule. The interference of the waves scattered in a particular direction by the electrons of a single atom leads to a spherically symmetrical intensity distribution called the atomic scattering factor. With several atoms in a molecule, there is further interference between the scattered radiation in any given direction, giving a characteristic distribution which is not now spherically symmetrical. In order to observe the scattered intensity, it is convenient to have a crystal because the molecules therein are all in a precise geometrical relationship to one another. The effect of the crystal on the scattering from the molecule is to limit its observation to discrete points related to the dimensions of the crystal lattice, but since at these points the scattered waves interfere constructively, an observable intensity results. It is possible to recombine the amplitudes, the square roots of the intensities, of the scattered waves to produce the electron distribution of the molecule responsible for the scattering if and only if the relative phases of each of the contributing amplitudes is known. This is what the objective lens of a light microscope does to the scattered light from the object. No lens capable of recombining X-rays exists and so the electron distribution must be calculated. Unfortunately, the phase information is lost when the intensities are recorded and must be determined indirectly. This phase problem is at the heart of all X-ray structure determination. Of course, if the atomic positions are known, then the correct phases can be calculated!

Reference has been made to the problems associated with protein crystallization and it is likely that the production of good quality crystals is the rate limiting step in protein crystallography. However, once good crystals have been produced, the following stages constitute a typical structure determination. First, diffraction data are recorded, usually photographically, to assess the quality of the crystals and to determine the space group and unit cell dimensions, the unit cell being the convenient building block of a crystal. After characterizing the crystal, intensity data are collected from the native protein and from at least one, more usually three or four crystals which have been soaked in a variety of heavy-atom containing solutions. Complexes like $KAu(CN)_2$, K_2PtCl_4, $MeHgNO_3$ and UO_2NO_3 are commonly used, although more than 100 have been reported as being useful. Each data set may require many crystals because radiation damage can lead to rapid deterioration of the diffraction pattern. Successful determination of the positions of the added heavy atoms, by Patterson techniques, allows an estimate of the phase for each diffracted wave to be determined, as shown in Figure 10. An electron density map based on the observed amplitudes for the native enzyme, F_P, and the derived phases, α_P, is then calculated and either plotted on clear plastic sheets or displayed on an interactive graphics device. The polypeptide chain is then traced through the map and standard amino acid side chains fitted. The full set of atomic coordinates so produced is then refined by a least squares procedure. The reliability index, R-factor, at the end of a good structure determination should be less than 0.20 for data to better than 0.2 nm (2 Å) resolution. Values above 0.25 indicate a structure still containing significant error, perhaps resulting from incomplete refinement or limited or poor X-ray data. However, the general features are likely to be correct. Thereafter, soaking native crystals in substrate, substrate analogue or inhibitor molecules, collecting diffraction data and calculating difference electron density maps with coefficients $(F_{P+S} - F_P)$ and phases α_P will reveal how a substrate is bound. The following paragraphs expand a few of the above points in the light of recent developments.

Data at the start of an investigation are generally collected on film since it is easier to establish the space group when substantial parts of the diffraction pattern are recorded. However, rotation and Laue photography require that films be digitized and processed further in order to extract the intensity data. With diffractometer techniques, using either the conventional single counter or area detector, numerical data are obtained much more rapidly. Susceptibility to radiation damage means that rapid data collection is desirable, thus favouring the area detector and it is now possible to collect a complete 3-D data set for a protein in minutes rather than days if use is made of synchrotron radiation.[60]

Some crystals do not diffract well for reasons that are not understood. The effect of this is to limit the radiation lifetime of the crystal to such an extent that many crystals must be used to collect the necessary data. Since some of the correction terms used are crystal dependent: the more crystals needed, the greater is the error in the data set. Other crystals may only diffract to a limited resolution and this reduces confidence in the final model, usually reflected in a higher R-factor. Refined enzyme structures determined at resolutions of around 2.8 Å will have a fairly accurate main chain over most of the structure, although parts may be indistinct through disorder or high flexibility. Many side chains should be distinguishable though not necessarily identifiable. At around 3.5 Å resolution only rarely will the main chain be totally correct although helix and sheet may well be obvious; at 5–6 Å, just a general shape for the molecule is the most that should be expected.

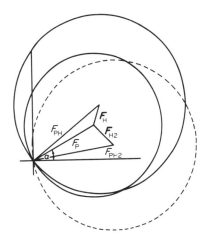

Figure 10 A vector diagram showing how the phase, α, for a diffracted X-ray beam can be determined by isomorphous replacement. The amplitudes of native protein, F_P, and heavy atom derivative, F_{PH}, allow circles to be drawn at each end of the vector $\boldsymbol{F}_H$ whose amplitude and phase are known when the heavy atom positions are known. This leads to an ambiguity which is resolved by a second heavy atom derivative

Effectively every independent crystal structure so far determined has relied upon the isomorphous replacement method for phase determination. The method is enhanced by using the anomalous scattering of the heavy atoms, especially with synchrotron radiation which can be tuned to optimize the effect for each heavy metal. A different approach, that of molecular replacement,[61] has been used in cases where the structure of a related molecule is sought. Positioning the known molecule correctly in the unknown cell can be achieved in a straightforward manner and the phases derived therefrom will be sufficiently accurate to permit interpretation of the unknown structure.

High-speed computers are essential in the latter stages of a protein structure determination since the least-squares refinement process is a monster calculation even for a small protein. The advent of parallel processors or array processors has meant that calculations requiring days of CPU time can now be accomplished in hours. Coupled to these advances in the power of 'number-crunchers' is the progress in interactive molecular graphics (see Part 20 in Volume 4) and both have had a significant effect on protein crystallography.

Protein crystallography provides direct three-dimensional data at the molecular level but the technique has disadvantages. First, the protein must be crystallized, which usually involves far from physiological conditions. This may mean that the crystalline enzyme is in an inactive conformation. Second, data collection usually takes days and this produces a time average structure which, at least initially, appears rigid and inflexible. Third, distortions in the structure may arise from the packing of the enzyme in the crystal and these may involve the active site. Finally, it may not be possible to diffuse into the existing crystals substrates, activators or other biologically relevant molecules because of either crystal instability or active site occlusion. The time taken for a crystal structure analysis varies from several weeks when everything goes well and the structure is related to a known one, to several years when crystals are unstable, heavy atom derivatives hard to obtain and the diffraction pattern is weak.

These disadvantages can all be overcome to some extent and it is possible to counter all of the above objections with established examples. However, perhaps the one objection about which some enzymologists and crystallographers continue to argue is that the crystal does not reflect the structure *in vivo*.

5.1.7.2 NMR Spectroscopy

NMR spectroscopy is in many ways complementary to X-ray crystallography. The structural information derived does not require crystals of the protein and considerable scope for large variation in the solution conditions (pH, ionic strength, substrate, temperature, *etc.*) exists. Both binding studies and molecular dynamics are readily amenable to NMR experiments. The application of the technique to protein structural studies is recent and relies crucially on being able to assign the ^1H resonances to at least some of the specific amino acids in the sequence. This, together with the short range distance information provided by the nuclear Overhauser effect (NOE), forms the basis of determining tertiary structures. Excellent references are available on this fast-moving topic.[55,62] The next paragraphs give an overview of the applications of ^1H NMR to both secondary and tertiary structure determination, together with how dynamic information can be obtained.

The polarization of the ^1H nuclear spins in a strong magnetic field leads to a magnetization in the sample which can be perturbed by the application of a suitable radio frequency pulse. The perturbation takes the form of a transverse (relative to that of the strong field) magnetization which precesses about the strong field direction, thus inducing a current in the detection coil. The current decays as a function of time, giving a free induction decay spectrum. If this is Fourier transformed, a frequency spectrum is produced. The peak positions in the frequency spectrum are generally expressed as a chemical shift in p.p.m. relative to tetramethylsilane. The chemical shift is related to the chemical nature of the particular nucleus, *e.g.* aromatic protons are found around 7 p.p.m. The spin–spin coupling constants, which are independent of field strength and measured in Hz, reflect the effects of neighbouring nuclei transmitted through adjacent covalent bonds. They are orientation dependent such that if the C—H bonds on adjacent carbons are perpendicular, the splitting is at a minimum. Coupling constants can be hard to determine accurately in macromolecules, however. The intensity of the peaks observed in an NMR spectrum in general reflects the number of nuclei contributing to them. Two decay rates are important: the spin–lattice (or longitudinal) and the spin–spin (or transverse) relaxation times, T_1 and T_2 respectively. The first reflects the overall tumbling of the molecule in solution and may be affected also by its flexibility. The second, which can never be greater than the first, measures the decay rate of the components of the transverse magnetization and is correlated with intramolecular motion. As the molecular size increases the transverse relaxation time falls, which puts an effective limit on the application of the technique. This can be seen from the ideal expression for the linewidth (half width at half height), $\Delta v_{1/2} = (2\pi T_2)^{-1}$, which shows that a decrease in T_2 broadens the line to such an extent that it disappears. The linewidth is also affected by chemical exchange arising from the existence of environments like those created when a side chain has two possible orientations, or when a carboxyl group can exist in both protonated and deprotonated forms. The rates of these exchange processes affect the NMR spectrum so that as the temperature is raised, lines arising from a distinct arrangement will broaden and move until the differences between the two environments exchange too rapidly for their discrimination and a single, average peak results. Finally, the NOE is a through-space phenomenon whereby an intensity change occurs in a neighbouring proton when one proton is specifically irradiated by a secondary source. This effect falls off as r^6 in contrast to the effect of an aromatic group or paramagnetic ion on chemical shift which falls off as r^3, where r is the separation of the interacting species.

The major problem is the assignment of resonances to the correct residues, without which no structural data can be deduced. It is also essential that the enzyme is in its native conformation when these assignments are made. What with the generally broad linewidths and considerable overlap of resonances in even a small protein, the assignment of only a few resonances is a major undertaking. Direct experiment is essential and this involves isotope exchange, chemical modification or use of mutants of the enzyme in question. However, two-dimensional techniques have revolutionized the assignment problem and allowed 3-D data to be extracted. In general, a sequence of pulses is applied to the sample: an equilibration period ending with a pulse, an evolution time t_1, a second pulse and, lastly, a recording time t_2. Data can then be collected as a function of both t_1 and t_2 which, when doubly Fourier transformed, will yield a 2-D spectrum of frequencies ω_1 and ω_2. On the diagonal of this frequency map is found the conventional spectrum whilst the off-diagonal peaks represent 'associations' between pairs of protons. Provided at least one assignment has been determined, usually by conventional means, it is then possible to work along the chain adding more assignments. The particular sequence of pulses and evolution time gives rise to several forms of spectra. The most common reveal the through-bond spin–spin couplings and the through-space NOE and are referred to as 2-D correlated spectroscopy (COSY) and NOESY, respectively. The latter, which gives non-bonded information, is particularly valuable for tertiary structure determination.

Because protein secondary structure has regular, short ^1H–^1H distances, Wüthrich was able to show by careful analysis of some well-defined crystal structures that suitable criteria for distinguishing between helix, sheet and turn conformations could be deduced from the peptide proton interactions.[55] Unlike the other solution methods, NOESY, spin–spin coupling constants and amide proton exchange rates can be attributed to particular parts of a sequence as well as giving the overall content of secondary structure. However, reverse turns are not well discerned by the technique. Tertiary structures can be deduced from the secondary structure by using the distance constraints determined by NMR. Alternatively, sufficient pairwise short contacts should provide a set of geometrical data capable of yielding the 3-D fold directly. Interactive graphics programs can also be used to expedite this process.

An important feature of NMR spectroscopy is its ability to supply dynamic information about molecular motion. The rate of change of peak intensity as the ^1H is replaced by ^2H is related to its hydrogen bonding and accessibility. The labile surface ^1H's exchange too rapidly to be observed unless H-bonded; the internal ones exchange slowly. Consequently, amide ^1H's, exchanging at rates up to about 10^4 min^{-1}, can be observed and related to the mobility of the enzyme.[63] Buried aromatic rings provide a heterogeneous environment for the ring protons and an analysis of the temperature dependence of the line shapes will therefore reveal the ease with which the ring can flip. Clearly, too, dynamic information must be contained in the spin-relaxation times.

5.1.7.3 Protein Dynamics

The fact that enzyme structures are dynamic is vividly illustrated by the hinge-bending enzymes, like hexokinase,[64] where a substrate mixture added to the crystal of the native form causes it to explode because the conformational change induced on binding the substrate cannot be accommodated by the lattice. Dynamic data for enzymes can be obtained from at least four sources, one of which, NMR, was discussed in the preceding section.

A second is crystallography which, though it produces a time-average model of the structure, does contain dynamic information in the temperature or B-factors of each atom.[65] The B-factors are related to the mean square atomic displacement from the mean position so that a high value of, say,

30 Å^2 means that the atom is mobile. It is usual to take the average values for main chain, side chain or residue as a measure of the chain flexibility and, in general, the higher values are to be found on the surface of the protein. It is necessary, however, to distinguish between static disorder and thermal motion, the former arising from the existence of several, discrete conformations whose average electron density is too low to be observed above the noise. Crystal structure determination at very low temperature will reveal the mobile but not the disordered regions.

Third, the Raman spectrum at low frequencies contains information on the gross vibrational modes of the enzyme.[47] It should be possible to relate these modes to specific molecular motions but at present this is not often successful.

Lastly, molecular dynamic simulations seek to determine the various molecular motions by calculation.[66] It is possible to approximate the potential energy of the protein as a sum of the various bonded and non-bonded interactions. These interactions are expressed as deviations from the ideal geometry for bonds and angles and as Lennard-Jones type potentials for the non-bonded ones. Various force and dielectric constants, van der Waals radii and partial charges are also required but, once determined, usually from small molecule crystal structures or empirically from highly refined proteins, the energy can be minimized with respect to the atomic coordinates. Such a minimum energy conformation allows the vibrational frequencies and normal modes to be estimated and these can then be related to those observed in the Raman spectrum. Further, starting from this energy-minimized conformation a numerical integration of Newton's equations of motion over small time steps of about 10^{-15} s will reveal the trajectories of the atoms within the structure. The computation of such a simulation requires the most powerful computers available but it can allow the dynamic properties of the enzyme to be examined, including the binding of substrates and inhibitors.

Other techniques which can be applied to study the dynamic nature of enzymes, but which space prevents discussion here, are neutron diffraction studies of amide ^{1}H–^{2}H exchange[67] and fluorescence depolarization.[68]

5.1.8 QUATERNARY STRUCTURE

The association of two or more polypeptide chains leads to the quaternary structure of the protein. The bonds which hold the chains together are those which also hold the individual subunits in shape: mostly non-covalent, weak interactions but occasionally disulfide bridges as in the immunoglobulins. Whilst many proteins can associate in a sequential manner, thus building repetitive rod-like structures of the sort found in the cytoskeleton, enzymes have fairly compact arrangements in which the intersubunit interactions are maximized.[69,70] This imposes a fairly strict symmetry on the molecule. Thus, a tetramer of identical subunits will usually have tetrahedral symmetry with each subunit able to make contact with the other three. However, proteins may comprise subunits which are only similar or which are quite distinct.

Diffraction techniques will reveal the nature of the quaternary structure in detail and, because the molecular symmetry is often reflected in the particular space group in which the enzyme crystallizes, useful size and association data can often be obtained before the structure is determined. Often an oligomer will crystallize with one polypeptide chain in the asymmetric unit, although this is not always the case. NMR, too, can be used to observe the contact regions. Electron microscopy may be used to obtain a direct image of the subunit structure, either in microcrystals or in solution, although distortion during sample preparation by negative staining may be misleading.[71]

Gel filtration under normal and denaturing conditions is a convenient method for observing quaternary structure. The molecular masses from both experiments allow an estimate of the stoichiometry and, if reducing conditions are also used, whether or not disulfides hold the subunits together. Gel electrophoresis can also be used but notice that SDS–PAGE may not give a true result unless the molecular weight is known from some other method, because the SDS dissociates the quaternary structure unless it is stabilized covalently.

5.1.8.1 Crosslinking

For molecules which are made from several different types of subunit or which are membrane bound such that their extraction also results in their disruption, the technique of chemical crosslinking may be useful.[72] Reagents like glutaraldehyde (**4**) and dimethyl suberimidate (**5**) will react with lysine residues and, if a suitable pair are located in adjacent subunits, will link them covalently. Thus, a mild crosslinking of the intact protein will generate a population of crosslinked species which can be separated and characterized by the methods discussed above. An ingenious selection of both homo- and hetero-bifunctional reagents like maleimidobenzoyl-*N*-hydroxysuccin-imide ester (6), which links thiol to amine groups, is available. Photoactivation of one function, often a nitrene from an azide, is often used to minimize unwanted side reactions.

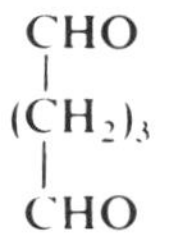

(4) 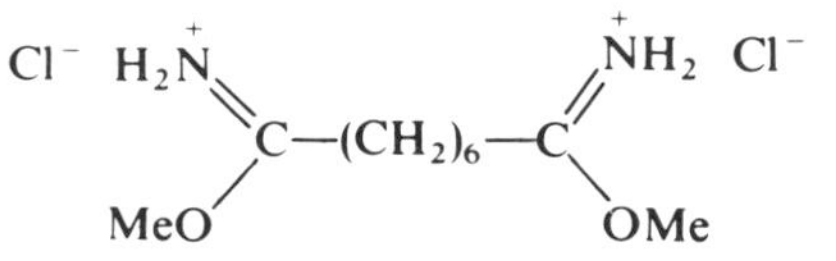   (5) 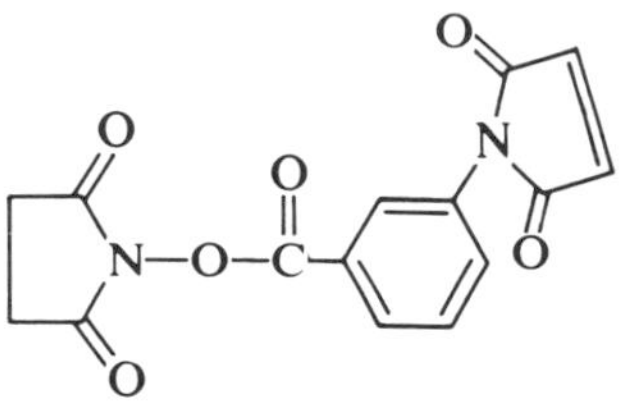   (6)

5.1.9 CHEMICAL MODIFICATION

Much information about an enzyme structure, particularly its active site, can be derived from the experiments which modify chemically its various side chains. In particular, the protection of some residues by enzyme substrates (or subunit–subunit interaction) can be determined by carrying out the reaction both in the presence and absence of the substrate (or denaturant). Similarly, the extent to which a particular residue type is buried or exposed to the solvent can be determined by modifying the side chain covalently, and then sequencing the protein. Obvious targets for reaction are —NH$_2$ and —SH groups, as shown in Sections 5.1.3.5 and 5.1.8.1, although an enzyme active site may impart special reactivity to other groups: Ser-195 of the serine proteases will react readily with sulfonyl fluorides. Reagents designed to react specifically with the active sites of enzymes are usually called affinity labels and will often be substrate analogues or inhibitors. These are more specific labels for enzyme active sites and by making the labels photoactivated, a further degree of control is provided. A large literature exists on chemical modification[73,74] so that only a selection of the more common reagents are described here.

5.1.9.1 Aromatic Residues

Of the aromatic residues, modification is not generally practicable to Phe but the others, provided that they are accessible, will react in a number of ways. Many of the reagents may also react with either thiol or amino groups, making protection a necessary preliminary. For Trp, formylation, sulfenylation and oxidation are feasible modifications (Scheme 12). The reactions of Tyr include nitration and iodination (Scheme 13), whilst His undergoes alkylation with α-haloacid of the ring nitrogens and dye-sensitized photooxidation with, for example, rose bengal. Diethyl pyrocarbonate is another reagent useful for attacking His residues.

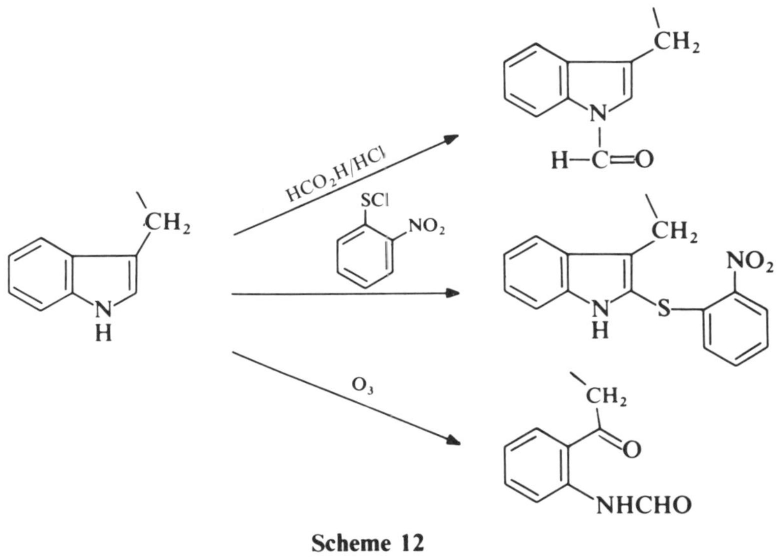

Scheme 12

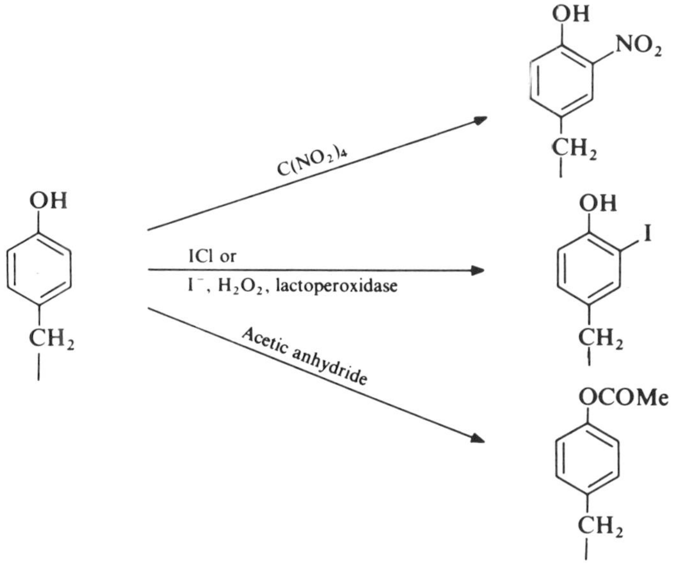

Scheme 13

5.1.9.2 Hydroxyl Groups

The hydroxyl groups of Ser and Thr are not particularly reactive except, as pointed out above, in special cases. Esterification with acetic anhydride, although quite feasible, will also acetylate amino and thiol residues.

5.1.9.3 Thiol Groups

Many reagents react specifically with the thiol of cysteine. For example, heavy metals or heavy metal complexes can be used (Scheme 14). Carboxymethylation with iodoacetate, reaction with *N*-ethylmaleimide and ethylenimine (which renders the product susceptible to trypsin) are convenient

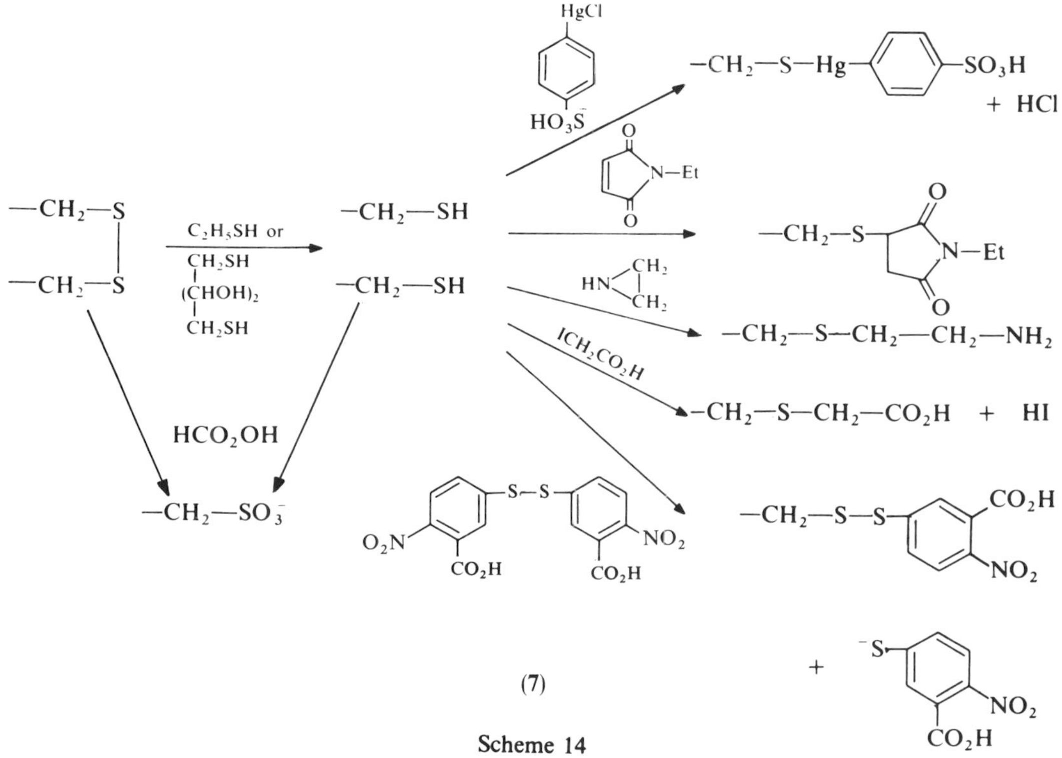

Scheme 14

means of modifying free —SH groups, whilst performic acid will also oxidize disulfide bridges to cysteic acid. Disulfide bridges can be readily reduced with mercaptoethanol or dithiothreitol, whereupon further reaction is possible. The chromophoric Ellman's reagent (**7**), which undergoes interchange, provides a ready means of titrating free —SH.

Methionine residues can be oxidized with performic acid to the corresponding sulfone and sulfoxide, and mention has been made (Section 5.1.5.1) of the reaction with cyanogen bromide.

5.1.9.4 Acidic Groups

The two most common reactions of carboxyl groups are that with diazomethane and those involving coupling with carbodiimides (Schemes 3, 15). A means of distinguishing between protonated and ionized forms of the carboxyl group is also available.

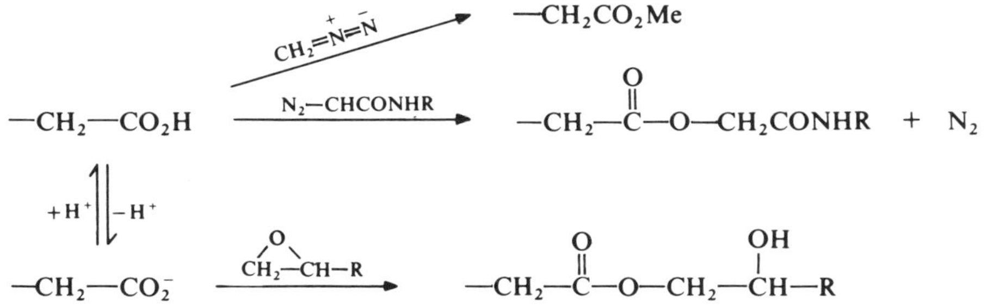

Scheme 15

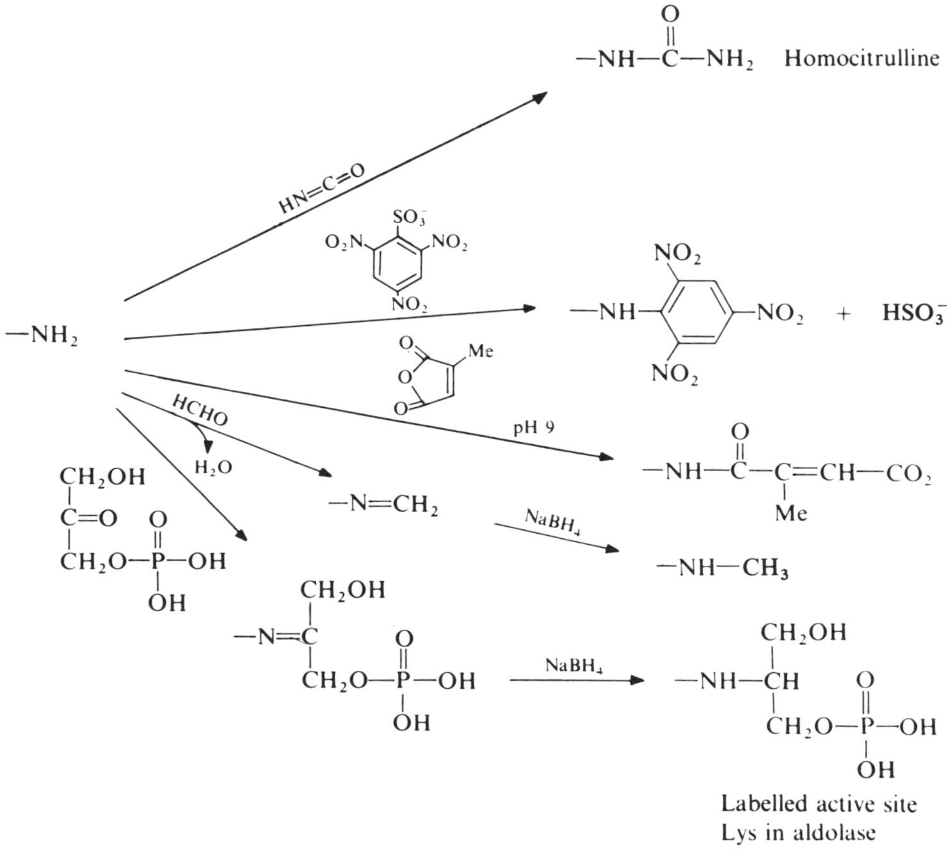

Scheme 16

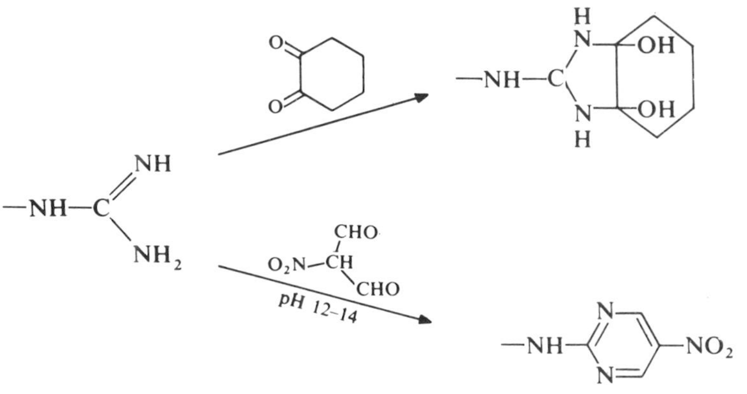

Scheme 17

5.1.9.5 Basic Groups

The highly nucleophilic nature of the —NH$_2$ group has meant that many different reagents have been used to modify Lys (Scheme 16). These include inhibition of chain cleavage by trypsin, introduction of a thiol group or labelling of an active site residue. In the latter case, if the reaction involves a Schiff base it may be reduced by NaBH$_4$. It is also possible to change the charge from positive to negative. With Arg, the reactions are less numerous but can be used to limit trypsin hydrolysis or tag functionally important residues (Scheme 17).

5.1.10 CONCLUSION

The foregoing discussion has attempted to cover a very large amount of ground in order to give an overview of enzyme architecture[75] and the main methods which are available for its study. Whilst X-ray crystallography provides many of the current ideas about structure, essential data are derived from many other sources. The modern ideas of drug design rely heavily upon the notion that a precise description of the enzyme, or receptor, provides the springboard from which new and more effective drug molecules can be devised. However, there is still much to be accomplished before it will be possible to relate enzyme structure to function in anything like the detail necessary to predict, quantitatively, the interaction between substrate and active site. It is therefore essential to continue to amass structural and functional data by both established and new techniques, such as kinetic crystallography,[76] in order to gain deeper insight into the relationships which are beginning to emerge.

5.1.11 REFERENCES

1. F. C. Bernstein, T. F. Koetzle, G. J. B. Williams, E. F. Meyer, M. D. Brice, J. R. Rodgers, O. Kennard, T. Shimanouchi and M. Tasumi, *J. Mol. Biol.*, 1977, **112**, 535.
2. P. S. J. Cheetham, in 'Handbook of Enzyme Technology', 2nd edn., ed. A. Wiseman, Horwood, Chichester, 1985, p. 274.
3. T. E. Creighton, 'Proteins: Structures and Molecular Properties', 1984, Freeman, New York, p. 7.
4. J. R. Kyte and R. F. Doolittle, *J. Mol. Biol.*, 1982, **157**, 105.
5. W. Taylor, *J. Mol. Biol.*, 1986, **188**, 233.
6. F. Wold and K. Moldave (eds.), *Methods Enzymol.*, 1984, **106**.
7. F. Wold and K. Moldave (eds.), *Methods Enzymol.*, 1984, **107**.
8. W. B. Jakoby (ed.), *Methods Enzymol.*, 1984, **104**.
9. E. L. V. Harris and S. Angal, 'Protein Purification: a Practical Approach', IRL Press, Oxford, 1988.
10. M. T. W. Hearn, *Adv. Chromatogr.*, 1982, **20**, 1.
11. A. Cornish-Bowden, 'Fundamentals of Enzyme Kinetics', Butterworths, London, 1979, p. 39.
12. U. K. Laemmli, *Nature (London)*, 1970, **227**, 680.
13. B. J. Radola, *Methods Enzymol.*, 1984, **104**, 256.
14. M. Dixon and E. C. Webb, *Adv. Protein Chem.*, 1961, **16**, 197.
15. E. J. Cohn, L. E. Strong, W. L. Hughes, D. J. Mulford, Jr., J. N. Ashworth, M. Melin and H. L. Taylor, *J. Am. Chem. Soc.*, 1946, **68**, 459.
16. R. Czok and Th. Bücher, *Adv. Protein Chem.*, 1960, **15**, 315.
17. R. F. Pfeifer and D. W. Hill, *Adv. Chromatogr.*, 1983, **22**, 37.
18. G. Seipke, H. Müllner and U. Grau, *Angew. Chem., Int. Ed. Engl.*, 1986, **25**, 535.

19. P. G. D. Dean, W. S. Johnson and F. A. Middle, 'Affinity Chromatography', IRL Press, Oxford, 1985.
20. A. McPherson, 'Preparation and Analysis of Protein Crystals', Wiley, New York, 1982.
21. M. J. Cox and P. C. Weber, *J. Appl. Crystallogr.*, 1987, **20**, 366.
22. Special supplement to *J. Crystal Growth*, 1987.
23. R. Giege (ed.), 'Crystal Growth of Biological Macromolecules', Special Issue of *J. Crystal Growth*, 1988.
24. D. M. Blow and T. A. Steitz, *Annu. Rev. Biochem.*, 1970, **39**, 63.
25. E. N. Baker and R. E. Hubbard, *Prog. Biophys. Mol. Biol.*, 1984, **44**, 97.
26. R. Jaenicke, *Angew. Chem., Int. Ed. Engl.*, 1984, **23**. 395.
27. L. Pauling and R. B. Corey, *Proc. Natl. Acad. Sci. USA*, 1951, **37**, 235, 251.
28. C. Chothia, *Annu. Rev. Biochem.*, 1984, **53**, 537.
29. M. Levitt and C. Chothia, *Nature (London)*, 1976, **261**, 552.
30. J. S. Richardson, *Adv. Protein Chem.*, 1981, **34**, 167.
31. J. F. Leszczynski and G. D. Rose, *Science (Washington, D.C)*, 1986, **234**, 849.
32. C. C. F. Blake, K. Harlos and S. K. Holland, *Cold Spring Harbor Symp. Quant. Biol.*, 1987, **52**, 925.
33. D. A. Hillson, N. Lambert and R. B. Freedman, *Methods Enzymol.*, 1984, **107**, 281.
34. W. R. Taylor, in 'Nucleic Acid and Protein Sequence Analysis', ed. M. J. Bishop and C. J. Rawlings, IRL Press, Oxford, 1987, p. 141.
35. G. E. Schulz, *Annu. Rev. Biophys. Biophys. Chem.*, 1988, **17**, 1.
36. R. K. Wierenga, P. Terpstra and W. G. J. Hol, *J. Mol. Biol.*, 1986, **187**, 101.
37. J. W. Ponder and F. M. Richards, *J. Mol. Biol.*, 1987, **193**, 775.
38. D. M. Glover, 'Gene Cloning', Chapman and Hall, London, 1984.
39. B. Wittmann-Liebold, J. Salnikow and V. A. Erdmann (eds.), 'Advanced Methods in Protein Microsequence Analysis', Springer-Verlag, Berlin, 1986.
40. K. K. Han, D. Belaiche, O. Moreau and G. Briand, *Int. J. Biochem.*, 1985, **17**, 429.
41. K. Biemann, *Biochem. Soc. Trans.*, 1989, **17**, 237.
42. A. M. Maxam and W. Gilbert, *Methods Enzymol.*, 1980, **65**, 499.
43. F. Sanger and A. R. Coulson, *J. Mol. Biol.*, 1975, **94**, 441.
44. W. C. Johnson, Jr., *Annu. Rev. Biophys. Biophys. Chem.*, 1988, **17**, 145.
45. A. J. Adler, N. J. Greenfield and G. D. Fasman, *Methods Enzymol.*, 1975, **27**, 675.
46. S. Krimm and J. Bandekar, *Adv. Protein Chem.*, 1986, **38**, 181.
47. A. T. Tu, in 'Spectroscopy of Biological Systems', ed. R. J. H. Clark and R. E. Hester, Wiley, New York, 1986, vol. 13, p. 47.
48. B. M. Bussian and C. Sander, *Biochemistry*, 1989, **28**, 4271.
49. W. Kabsch and C. Sander, *Biopolymers*, 1983, **22**, 2577.
50. P. Y. Chou and G. D. Fasman, *Adv. Enzymol.*, 1978, **47**, 45.
51. J. Garnier, D. J. Osguthorpe and B. Robson, *J. Mol. Biol.*, 1978, **120**, 97.
52. O. B. Ptitsyn and A. V. Finkelstein, *Biopolymers*, 1983. **22**, 15.
53. T. A. Jones and S. Thirup, *EMBO J.*, 1986, **5**, 819.
54. B. Robson and J. L. Garnier, 'Introduction to Proteins and Protein Engineering', Elsevier, Amsterdam, 1986.
55. K. Wüthrich, 'NMR of Proteins and Nucleic Acids', Wiley, New York, 1986.
56. L. A. Amos, R. Henderson and P. N. T. Unwin, *Prog. Biophys. Mol. Biol.*, 1982, **39**, 183.
57. J. P. Glusker and K. N. Trueblood, 'Crystal Structure Analysis: a Primer', 2nd edn., Oxford University Press, Oxford, 1985.
58. T. L. Blundell and L. N. Johnson, 'Protein Crystallography', Academic Press, New York, 1976.
59. H. W. Wycoff, C. H. W. Hirs and S. N. Timasheff (eds.), *Methods Enzymol.*, 1985, **114** and **115**.
60. J. R. Helliwell, *Rep. Prog. Phys.*, 1984, **47**, 1403.
61. M. G. Rossmann (ed.), 'Molecular Replacement Method', Gordon & Breach, New York, 1972.
62. A. Bax, *Annu.. Rev. Biochem.*, 1989, **58**, 223.
63. K. Wüthrich and G. Wagner, *Trends Biochem. Sci.*, 1984, **9**, 152.
64. W. S. Bennett and T. A. Steitz, *J. Mol. Biol.*, 1980, **140**, 211.
65. D. Ringe and G. A. Petsko, *Prog. Biophys. Mol. Biol.*, 1985, **45**, 197.
66. I. A. McCammon and S. C. Harvey, 'Dynamics of Proteins and Nucleic Acids', Cambridge University Press, Cambridge, 1987.
67. A. A. Kossiakoff, *Annu. Rev. Biochem.*, 1985, **54**, 1195.
68. J. M. Beecham and L. Brand, *Annu. Rev. Biochem.*, 1985, **54**, 43.
69. I. M. Klotz, N. R. Langerman and D. W. Darnall, *Annu. Rev. Biochem.*, 1970, **39**, 25.
70. R. Jaenicke and R. Rudolf, *Methods Enzymol.*, 1986, **131**, 218.
71. E. Spiess, H.-P. Zimmermann and H. Lunsdorf, in 'Electron Microscopy in Molecular Biology', ed. J. Sommerville and U. Scheer, IRL Press, Oxford, 1987, p. 147.
72. P. M. Conn, *Methods Enzymol.*, 1983, **103**, 49.
73. J. O. Thomas, in 'Companion to Biochemistry', ed. A. T. Bull, J. R. Lagnado, J. O. Thomas and K. F. Tipton, Longman, London, 1974, p. 87.
74. A. N. Glazer, R. J. DeLange and D. S. Sigman, in 'Laboratory Techniques in Biochemistry and Molecular Biology', ed. T. S. Work and E. Work, North-Holland, Amsterdam, 1975, vol. 4, p. 10.
75. G. E. Schulz and R. H. Schirmer, 'Principles of Protein Structure', Springer-Verlag, New York, 1979.
76. J. Hadju, P. A. Machin, J. W. Campbell, T. J. Greenhough, I. J. Clifton, S. Zurek, S. Gover, L. N. Johnson and M. Elder, *Nature (London)*, 1987, **329**, 178.

5.2

Nomenclature and Classification of Enzymes

IAIN B. R. Bowman

University of Edinburgh Medical School, UK

5.2.1   SUMMARY   33

5.2.2   INTRODUCTION   34
   5.2.2.1   The Requirement for Enzyme Classification   34
   5.2.2.2   The Formalization of Nomenclature   34

5.2.3   THE ENZYME COMMISSION SYSTEM OF CLASSIFICATION   34
   5.2.3.1   Key to the Numbering and Classification   34
   5.2.3.2   Examples of Systematic Names and Recommended Trivial Names   35
      5.2.3.2.1   Class 1, oxidoreductases   35
      5.2.3.2.2   Class 2, transferases   35
      5.2.3.2.3   Class 3, hydrolases   35
      5.2.3.2.4   Class 4, lyases   36
      5.2.3.2.5   Class 5, isomerases   36
      5.2.3.2.6   Class 6, ligases   36

5.2.4   ISOENZYMES   37

5.2.5   ISOFUNCTIONAL ENZYMES   37

5.2.6   MULTIENZYME SYSTEMS   38

5.2.7   APPENDIX 1: KEY TO NUMBERING AND CLASSIFICATION OF ENZYMES   38

5.2.8   REFERENCES   43

5.2.1   SUMMARY

The need for a formal classification and nomenclature of enzymes is described by referring to the inconsistencies and ambiguities which developed from the inventiveness of biochemists seeking appropriate terminology. Most enzymes were named by placing the suffix '-ase' to the name of the substrate or to the type of reaction. Enzymes are now grouped into six classes according to their type of reaction with subclasses defining the chemical groups or bonds being metabolized. A unique numbering system evolved from this system. Isoenzymes can be fitted into this scheme but where different enzymes catalyze the same reaction (isofunctional enzymes) then the source must be given because the classification does not differentiate them. With multienzyme systems classification and numbering break down but formal nomenclature is applicable to the individual catalytic components.

5.2.2 INTRODUCTION

5.2.2.1 The Requirement for Enzyme Classification

During the development of biochemistry many metabolic pathways were discovered, which resulted in the description of a large number of enzymes. These were named according to their function and led to a commonly acceptable system of trivial nomenclature coined by individual workers. Enzymes catalyzing degradation reactions were named by the apposition of the suffix '-ase' to the name of the *substrate* acted upon by the enzyme. For example, glucose-6-phosphatase catalyzes the hydrolysis of glucose 6-phosphate to glucose and inorganic phosphate, or arginase, which cleaves arginine to urea and ornithine. Other enzymes were named according to their *function*, that is to say the *reaction* they catalyzed, for example alcohol dehydrogenase, which brings about the dehydrogenation of ethyl alcohol to form acetaldehyde, or glucose oxidase, which oxidizes β-D-glucose to gluconolactone. Nondegradative enzymes were named by the addition of '-ase' to the word describing the reaction catalyzed.

These reactions were catalyzed by: kinases (phosphate group transferring), isomerases and synthetases; the enzymes were usually prefixed by the name of the substrate such as phosphofructo-kinase (phosphorylating fructose 6-phosphate to fructose 1,6-diphosphate), glucose-6-phosphate isomerase (converting glucose 6-phosphate to fructose 6-phosphate). Mutases were confused with isomerases though the terms are now usefully separated. Synthetases were named after the *product* of reaction rather than the substrate, as in glutamine synthetase or the various amino acid:tRNA synthetases. The term synthase, which was used synonymously, is now reserved for class 4 lyases to emphasize the synthetic nature of the reaction. Given such biochemical licence, quite colloquial terms emerged: nickase, which hydrolyzes a phosphodiester bond in one strand of double helical DNA, or helicase, which unwinds double helical DNA and could be confused with 'helicase', a commercial name given to a snail extract containing a powerful set of digestive enzymes from *Helix pomatia*. A newcomer to the field could be confused by the terms old yellow enzyme (now NADPH:acceptor oxidoreductase), papain (a protease, not classified), repairase and replicase (function in repair and replication of DNA), and rhodanese (thiosulfate sulfurtransferase). In these cases the rules of trivial nomenclature were not applied. From this it can be seen that an unsystematic terminology developed, which worked but did result in confusion where one enzyme had more than one name or, even worse, where different enzymes had the same name.

5.2.2.2 The Formalization of Nomenclature

In order to resolve the confusion caused by uncontrolled naming of enzymes, the International Union of Biochemistry (IUB) in consultation with the International Union of Pure and Applied Chemistry set up in 1956 an Enzyme Commission (EC) to develop a set of well-defined rules for the formal classification of enzymes. The Commission's recommendations were accepted by the IUB in 1961 and are revised at frequent intervals (see Enzyme Nomenclature 1984,[6] which is the fifth revision). The number of enzymes listed has risen steadily from 712 in 1961 to 2477 in 1984. The EC proposed a systematic name for each enzyme where the reaction catalyzed is well characterized and can be expressed by a formal equation. Furthermore the EC assigned a unique four-number code closely related to the system of classification. It turned out that enzymes and their reaction types fell into six main classes:

1. oxidation–reduction reactions catalyzed by *oxidoreductases*
2. group transfer reactions catalyzed by *transferases*
3. hydrolytic reactions catalyzed by *hydrolases*
4. elimination reactions forming double bonds or conversely adding groups to double bonds catalyzed by *lyases*
5. isomerization reactions catalyzed by *isomerases*
6. bond-forming reactions joining two molecules, the energy for which is supplied by the simultaneous binding of the nucleoside triphosphates to nucleoside phosphates. These reactions are catalyzed by *ligases* (also known as *synthetases*).

5.2.3 THE ENZYME COMMISSION SYSTEM OF CLASSIFICATION

5.2.3.1 Key to the Numbering and Classification

For the sake of completeness in this volume, the key is given in full. From this key an enzyme can be located in the complete enzyme list,[6] which gives the systematic name and which is the basis for classification, the recommended (trivial) name and the reaction catalyzed. The key is to be found at the end of the chapter (Appendix 1, reproduced from ref. 6 by permission of the International Union of Biochemistry and Academic Press).

5.2.3.2 Examples of Systematic Names and Recommended Trivial Names

The benefits of classification in bringing order out of confusion are well known in biology. However, the systematic names of enzymes are often clumsy, though descriptive, and the corresponding code numbers are nondescriptive and difficult to remember. The EC recognized these weaknesses and so proposed a single recommended trivial name for each enzyme, which would be unambiguous. Most journals require that an enzyme referred to in a paper should first of all be fully identified by its systematic name and EC number; thereafter its trivial name can be used.

These points are illustrated by referring to EC 5.4.2.2, which has the systematic name of α-D-glucose-1,6-phosphomutase and the trivial name phosphoglucomutase

$$\text{glucose 1-phosphate} \rightleftharpoons \text{glucose 6-phosphate}$$

The benefit of the trivial name is that it is easy to say and to remember but it is misleading as it implies an *intra*molecular transfer of phosphate between carbons one and six of glucose. The benefit of the formal name is that it describes the reaction as an *inter*molecular transfer of phosphate from glucose 1,6-diphosphate as phosphate donor, placing the enzyme correctly in the isomerase class 5. The EC number characterizes the class (first digit) the subclass (second digit) and subsubclass (third digit) so EC 5.4.2.2 is a transferase (class 5), involves phosphate transfer (subclass 4) and the reaction occurs with the regeneration of a donor molecule (subsubclass 2). The last digit of EC 5.4.2.2 completes the identification of the enzyme phosphoglucomutase.

One enzyme from each class will now be taken to point up the strengths and weaknesses of the classification.

5.2.3.2.1 *Class 1, oxidoreductases*

Lactate dehydrogenase has the systematic name L-lactate:NAD^+ oxidoreductase, EC 1.1.1.27, which implies that L-lactate acts as the electron donor and NAD^+ as the electron acceptor as in

$$\text{L-lactate} + NAD^+ \rightleftharpoons \text{pyruvate} + NADH + H^+$$

However, the classification does not differentiate among the different isoenzymes of lactate dehydrogenase (see later).

5.2.3.2.2 *Class 2, transferases*

Aspartate aminotransferase has the systematic name L-aspartate:2-oxoglutarate aminotransferase, EC 2.6.1.1, which shows aspartate to be the amino group donor and 2-oxoglutarate the amino group acceptor

$$\text{L-aspartate} + \text{2-oxoglutarate} \rightleftharpoons \text{oxaloacetate} + \text{L-glutamate}$$

The trivial name gives no indication of the acceptor molecule.

5.2.3.2.3 *Class 3, hydrolases*

Triacylglycerol lipase has the systematic name triacylglycerol acylhydrolase, EC 3.1.1.3, and adequately describes the reaction

$$\text{triacylglycerol} + H_2O \rightleftharpoons \text{diacylglycerol} + \text{a fatty acid anion}$$

This is a pancreatic enzyme hydrolyzing dietary triacylglycerols at the lipid–water interface of micelles formed by the emulsifying action of bile salts and should not be confused with lipoprotein lipase, *i.e.* triacylglycerol protein acylhydrolase, EC 3.1.1.34, which hydrolyzes triacylglycerols in chylomicrons and very low density lipoproteins (VLDL) during the transport of these lipid particles in the plasma. The EC system does not classify the hormone sensitive lipase of adipose tissue, which catalyzes the same reaction but which is regulated by hormone-controlled phosphorylation–dephosphorylation processes. Thus, to define an enzyme specifically, it is often necessary to state the tissue or the species from which it is derived.

5.2.3.2.4 Class 4, lyases

Fructose-bisphosphate aldolase cleaves a carbon–carbon bond as do most lyases (see key). It has the systematic name D-fructose-1,6-bisphosphate D-glyceraldehyde-3-phosphate lyase, EC 4.1.2.13, and catalyzes the reaction

$$\text{D-fructose 1,6-diphosphate} \rightarrow \text{D-glyceraldehyde 3-phosphate} + \text{dihydroxyacetone phosphate}$$

In the same class are the hydrolyases such as fumarate hydratase, correctly named L-malate hydrolyase, EC 4.2.1.2, which brings about the making or breaking of a double bond.

$$\text{L-malate} \rightleftharpoons \text{fumarate} + H_2O$$

Within this class many of the reactions are freely reversible as in the above-mentioned examples. Hence they are bond-making enzymes and may be referred to as synthases. They should not be confused with *ligases* (class 6), which are bond makers and will be discussed later.

5.2.3.2.5 Class 5, isomerases

Glucose-6-phosphate isomerase has the systematic name D-glucose-6-phosphate ketol isomerase, EC 5.3.1.9, catalyzing the ketose aldose isomerizations

$$\text{D-glucose 6-phosphate} \rightleftharpoons \text{D-fructose 6-phosphate}$$

This class includes the racemases and epimerases. The mutases are specific to this class because they carry out *intra*molecular transfers such as methylmalonyl-CoA mutase, systematic name methylmalonyl-CoA CoA-carbonylmutase, EC 5.4.99.2. The term *mutase* is reserved for this subsubclass of isomerases and also for subsubclass 2, as in the case of phosphoglucomutase, EC 5.4.2.2, which involves an *inter*molecular transfer.

5.2.3.2.6 Class 6, ligases

Alanyl-tRNAligase has the systematic name L-alanyl:tRNA^{Ala} ligase (AMP-forming), EC 6.1.1.7, indicating that L-alanine is bonded (ligated) to its specific tRNA^{Ala} with the concomitant hydrolysis of ATP to AMP and pyrophosphate in a two-stage reaction

$$\text{L-alanine} + \text{ATP} \rightleftharpoons \text{L-alanyl—AMP} + \text{pyrophosphate}$$
$$\text{L-alanyl—AMP} + \text{tRNA}^{\text{Ala}} \rightleftharpoons \text{L-alanyl—tRNA}^{\text{Ala}} + \text{AMP}$$

These reactions are virtually unidirectional in favour of synthesis because the pyrophosphate is subsequently hydrolyzed. Other pyrophosphate-splitting reactions, such as those catalyzed by acid thiol ligases, amide synthetases, peptide synthetases and DNA and RNA synthetases, are similarly unidirectional or 'committed' reactions in favour of synthesis, as a result of hydrolyzing two high energy phosphate bonds in which the free energy change is about $-60\ \text{kJ mol}^{-1}$.

The *carboxylases* fix carbon dioxide into carboxyl groups with the conversion of ATP into ADP and orthophosphate as in the case of pyruvate carboxylase, systematically called pyruvate:carbon dioxide ligase (ADP-forming), EC 6.4.1.1

$$\text{pyruvate} + CO_2 + \text{ATP} + H_2 \rightleftharpoons \text{oxaloacetate} + \text{ADP} + \text{orthophosphate}$$

The difference here from the other synthetases is that ATP is hydrolyzed to ADP and orthophosphate with a corresponding decrease in the free energy change in the overall reaction. Virtually all carboxylases have biotin as their prosthetic groups.

5.2.4 ISOENZYMES

Isoenzymes catalyze the same chemical reaction in the same organism or even in a single cell. They may show close similarity in function and in homology of structure, nevertheless they are distinguishable in other ways. For example, lactate dehydrogenase has five different isoenzymic forms, depending on the subunit composition of this tetrameric protein. The subunits are termed H for heart muscle and M for skeletal muscle. Each tissue expresses the genes for H and M subunits at different rates, allowing different combinations for five different isoenzymes: H_4, H_3M, H_2M_2, HM_3, M_4.

These can be shown to be different in four ways: they are separable by electrophoresis, they display different kinetic parameters, they differ in temperature stability and they differ in the effect of inhibitors.

Whilst the classification system does not differentiate isoenzymes, the Enzyme Commission recommends that isoenzymes be numbered 1, 2, 3 *etc.* in order of their migration to the anode. Different tissues have different proportions of isoenzymes. When these are detected in blood plasma by electrophoretic separation followed by activity staining of the gel, then this is useful in diagnosing which tissue is damaged. Lactate dehydrogenase isoenzyme patterns have been used in the diagnosis of heart disease.

Hexokinase, EC 2.7.1.1, has four isoenzyme forms referred to as I, II, III and IV, each made up of four subunits. Hexokinase is quite distinct from glucokinase, EC 2.7.1.2; each catalyzes the same reaction.

$$\text{D-glucose} + \text{ATP} \rightleftharpoons \text{glucose 6-phosphate} + \text{ADP}$$

Glucokinase and the hexokinase group differ markedly in K_m values and specificity for substrates, with glucokinase being absolutely specific for glucose and hexokinase having a broad specificity for several hexoses and a low K_m for glucose. Glucokinase and hexokinase are not related genetically or in structure; it is therefore more correct to consider them as isofunctional enzymes.

5.2.5 ISOFUNCTIONAL ENZYMES

The two isofunctional enzymes referred to above (glucokinase and hexokinase) are sufficiently different and well characterized to be separately classified. This is not always the case. The system of nomenclature takes account only of the reaction catalyzed but not of the source of the enzyme. It is important in defining an enzyme completely to state the source because the properties may vary from tissue to tissue or may differ even in different compartments of a cell. Such is the case of muscle and liver pyruvate kinase, EC 2.7.1.40, where the muscle (M type) shows no allosteric properties and is antigenically related to one form in the liver, which has three other isoenzyme types (L_1, L_2, L_3) that are, in contrast, allosterically controlled.[1] The L and M types of pyruvate kinase are so functionally different that they must be products of distinct genes where the M-type gene does not encode the allosteric domain. Within the same genotype different enzymes are synthesized with the same general function but the proteins have quite different amino acid sequences and catalytic mechanisms.

In medicinal chemistry the concept of isofunctional enzymes becomes important when comparing them in different species because they are products of different genotypes. Comparison of isofunctional enzymes in the host organism and in the invader or parasitic organism leads to an understanding of the mode of action of drugs and their selection of target proteins. The ideal target for drug attack is an enzyme present in the pathogen and absent from the host. This is rare but one example may suffice. *Trypanosoma brucei* and related trypanosomes have a unique terminal oxidase, L-glycerol-3-phosphate oxidase (unclassified), which is cytochrome independent and is absent in the mammalian host. The trypanocidal drug Suramin has been shown to inhibit this enzyme selectively.[4]

There are numerous examples in which the isofunctional enzyme of the infecting organism is more sensitive than the host enzyme to drug inhibition. Dihydrofolate reductase (DHFR) or 5,6,7,8-tetrahydrofolate:$NADP^+$ oxidoreductase, EC 1.5.1.3, is a good example. The antibacterial properties of trimethoprim and the antimalarial effects of pyrimethamine involve the inhibition of DHFR. This emphasizes the important metabolic role of the enzyme and the diversity of its structure in different organisms (for a review see ref. 2). The isofunctional enzyme of the malarial parasite has a molecular weight of 200 000 *versus* that of 20 000 for the mammalian enzyme and there are different cofactor requirements. The sensitivity of DHFR to pyrimethamine may not be simply due to size difference but to more subtle differences in structure because 2,4-diaminopyrimidines (*e.g.* trimethoprim) discriminate between bacterial and mammalian enzymes which are of very similar size (for a review see ref. 7). Other examples of selective toxicity are to be seen in the following chapters.

5.2.6 MULTIENZYME SYSTEMS

Multienzymes fall into two groups: one consists of several noncovalently linked polypeptide chains, each with a distinct catalytic site; the other group consists of a single protein with more than one catalytic site, each of which has distinct reaction and substrate specificities. The EC recommends that multienzyme complexes should be referred to as *systems*, for example the pyruvate dehydrogenase system has all the catalytic functions for the oxidative decarboxylation of pyruvate to CO_2 and acetyl-CoA. Each component enzyme will have an EC number, a recommended and a systematic name as follows: pyruvate dehydrogenase (lipoamide), formally pyruvate:lipoamide oxidoreductase (decarboxylating and acceptor acetylating), EC 1.2.4.1, followed by dihydrolipoamide acetyltransferase, systematically acetyl-CoA:dihydrolipoamide *S*-acetyltransferase, EC 2.3.1.12, to mention only two of the many catalytic functions of pyruvate dehydrogenase. The advantages of combining unwieldy nomenclature into one term, namely *system*, are obvious.

The other group of multifunctional enzymes is illustrated by homoserine dehydrogenase and aspartate kinase activities, which are found in a *single* protein, aspartokinase I. These are numbered EC 1.1.1.3 and EC 2.7.2.4. In *E. coli* this bifunctional enzyme forms β-aspartyl phosphate, which serves for the biosynthesis of threonine, isoleucine, methionine (*via* homoserine) and lysine. Both activities are inhibited by threonine in a feedback mechanism (see ref. 3).

Finally bifunctional enzyme systems have been synthesized by DNA recombination of two genes and their expression as a chimaeric protein. Bulow and Mosbach[5] have fused β-galactosidase monomer with a galactokinase monomer, which then aggregated to give an active β-galactosidase tetramer carrying on each polypeptide an active galactokinase unit.

A single multifunctional unit will have its component enzymes in different positions in the enzyme list but the complete system will not appear in the classification.

5.2.7 APPENDIX 1: KEY TO NUMBERING AND CLASSIFICATION OF ENZYMES[6]

1. OXIDOREDUCTASES

1.1   Acting on the CH—OH group of donors
  1.1.1   With NAD^+ as acceptor
  1.1.2   With a cytochrome as acceptor
  1.1.3   With oxygen as acceptor
  1.1.5   With a quinone or related compound as acceptor
  1.1.99   With other acceptors

1.2   Acting on the aldehyde or oxo group of donors
  1.2.1   With NAD^+ or $NADP^+$ as acceptor
  1.2.2   With a cytochrome as acceptor
  1.2.3   With oxygen as acceptor
  1.2.4   With a disulfide compound as acceptor
  1.2.7   With an iron–sulfur protein as acceptor
  1.2.99   With other acceptors

1.3   Acting on the CH—CH group of donors
  1.3.1   With NAD^+ or $NADP^+$ as acceptor
  1.3.2   With a cytochrome as acceptor
  1.3.3   With oxygen as acceptor
  1.3.5   With a quinone or related compound as acceptor
  1.3.7   With an iron–sulfur protein as acceptor
  1.3.99   With other acceptors

1.4   Acting on the CH—NH$_2$ group of donors
  1.4.1   With NAD^+ or $NADP^+$ as acceptor
  1.4.2   With a cytochrome as acceptor

1.4.3   With oxygen as acceptor
1.4.4   With a disulfide compound as acceptor
1.4.7   With an iron–sulfur protein as acceptor
1.4.99   With other acceptors

1.5   Acting on the CH–NH group of donors
1.5.1   With NAD$^+$ or NADP$^+$ as acceptor
1.5.3   With oxygen as acceptor
1.5.99   With other acceptors

1.6   Acting on NADH or NADPH
1.6.1   With NAD$^+$ or NADP$^+$ as acceptor
1.6.2   With a cytochrome as acceptor
1.6.4   With a disulfide compound as acceptor
1.6.5   With a quinone or related compound as acceptor
1.6.6   With a nitrogenous group as acceptor
1.6.8   With a flavin as acceptor
1.6.99   With other acceptors

1.7   Acting on other nitrogenous compounds as donors
1.7.2   With a cytochrome as acceptor
1.7.3   With oxygen as acceptor
1.7.7   With an iron–sulfur protein as acceptor
1.7.99   With other acceptors

1.8   Acting on a sulfur group of donors
1.8.1   With NAD$^+$ or NADP$^+$ as acceptor
1.8.2   With a cytochrome as acceptor
1.8.3   With oxygen as acceptor
1.8.4   With a disulfide compound as acceptor
1.8.5   With a quinone or related compound as acceptor
1.8.7   With an iron–sulfur protein as acceptor
1.8.99   With other acceptors

1.9   Acting on a heme group of donors
1.9.3   With oxygen as acceptor
1.9.6   With a nitrogenous group as acceptor
1.9.99   With other acceptors

1.10   Acting on diphenols and related substances as donors
1.10.1   With NAD$^+$ or NADP$^+$ as acceptor
1.10.2   With a cytochrome as acceptor
1.10.3   With oxygen as acceptor
1.10.99 With other acceptors

1.11   Acting on hydrogen peroxide as acceptor

1.12   Acting on hydrogen as donor

1.12.1   With NAD$^+$ or NADP$^+$ as acceptor
1.12.2   With a cytochrome as acceptor

1.13   Acting on single donors with incorporation of molecular oxygen (oxygenases)
1.13.11 With incorporation of two atoms of oxygen
1.13.12 With incorporation of one atom of oxygen (internal monooxygenases or internal mixed
         function oxidases)
1.13.99 Miscellaneous (requires further characterization)

1.14   Acting on paired donors with incorporation of molecular oxygen
1.14.11 With 2-oxoglutarate as one donor and incorporation of one atom each of oxygen into
         both donors
1.14.12 With NADH or NADPH as one donor and incorporation of two atoms of oxygen into
         one donor
1.14.13 With NADH or NADPH as one donor and incorporation of one atom of oxygen

1.14.14 With reduced flavin or flavoprotein as one donor and incorporation of one atom of oxygen

1.14.15 With a reduced iron–sulfur protein as one donor and incorporation of one atom of oxygen

1.14.16 With reduced pteridine as one donor and incorporation of one atom of oxygen

1.14.17 With ascorbate as one donor and incorporation of one atom of oxygen

1.14.18 With another compound as one donor and incorporation of one atom of oxygen

1.14.99 Miscellaneous (requires further characterization)

1.15 Acting on superoxide radicals as acceptor

1.16 Oxidizing metal ions
  1.16.1 With NAD^+ or $NADP^+$ as acceptor
  1.16.3 With oxygen as acceptor

1.17 Acting on $-CH_2$ groups
  1.17.1   With NAD^+ or $NADP^+$ as acceptor
  1.17.3   With oxygen as acceptor
  1.17.4   With a disulfide compound as acceptor
  1.17.99 With other acceptors

1.18 Acting on reduced ferredoxin as donor
  1.18.1   With NAD^+ or $NADP^+$ as acceptor
  1.18.6   With dinitrogen as acceptor
  1.18.99 With H^+ as acceptor

1.19 Acting on reduced flavodoxin as donor
  1.19.6 With dinitrogen as acceptor

1.97 Other oxidoreductases

2. TRANSFERASES

2.1 Transferring one-carbon groups
  2.1.1 Methyltransferases
  2.1.2 Hydroxymethyl-, formyl- and related transferases
  2.1.3 Carboxyl- and carbamoyl-transferases
  2.1.4 Amidinotransferases

2.2 Transferring aldehyde or ketonic residues

2.3 Acyltransferases
  2.3.1 Acyltransferases
  2.3.2 Aminoacyltransferases

2.4 Glycosyltransferases
  2.4.1   Hexosyltransferases
  2.4.2   Pentosyltransferases
  2.4.99 Transferring other glycosyl groups

2.5 Transferring alkyl or aryl groups, other than methyl groups

2.6 Transferring nitrogenous groups
  2.6.1   Aminotransferases
  2.6.3   Oximinotransferases
  2.6.99 Transferring other nitrogenous groups

2.7 Transferring phosphorus-containing groups
  2.7.1 Phosphotransferases with an alcohol group as acceptor
  2.7.2 Phosphotransferases with a carboxyl group as acceptor
  2.7.3 Phosphotransferases with a nitrogenous group as acceptor
  2.7.4 Phosphotransferases with a phosphate group as acceptor
  2.7.6 Diphosphotransferases
  2.7.7 Nucleotidyltransferases
  2.7.8 Transferases for other substituted phosphate groups
  2.7.9 Phosphotransferases with paired acceptors

2.8 Transferring sulfur-containing groups
2.8.1 Sulfurtransferases
2.8.2 Sulfotransferases
2.8.3 CoA-transferases

3. HYDROLASES

3.1 Acting on ester bonds
3.1.1 Carboxylic ester hydrolases
3.1.2 Thiol ester hydrolases
3.1.3 Phosphoric monoester hydrolases
3.1.4 Phosphoric diester hydrolases
3.1.5 Trisphosphoric monoester hydrolases
3.1.6 Sulfuric ester hydrolases
3.1.7 Diphosphoric monoester hydrolases
3.1.11 Exodeoxyribonucleases producing 5′-phosphomonoesters
3.1.13 Exoribonucleases producing 5′-phosphomonoesters
3.1.14 Exoribonucleases producing other than 5′-phosphomonoesters
3.1.15 Exonucleases active with either ribo- or deoxyribo-nucleic acids and producing 5′-phosphomonoesters
3.1.16 Exonucleases active with either ribo- or deoxyribo-nucleic acids and producing other than 5′-phosphomonoesters
3.1.21 Endodeoxyribonucleases producing 5′-phosphomonoesters
3.1.22 Endodeoxyribonucleases producing other than 5′-phosphomonoesters
3.1.25 Site-specific endodeoxyribonucleases: specific for altered bases
3.1.26 Endoribonucleases producing 5′-phosphomonoesters
3.1.27 Endoribonucleases producing other than 5′-phosphomonoesters
3.1.30 Endonucleases active with either ribo- or deoxyribo-nucleic acids and producing 5′-phosphomonoesters
3.1.31 Endonucleases active with either ribo- or deoxyribo-nucleic acids and producing other than 5′-phosphomonoesters

3.2 Glycosidases
3.2.1 Hydrolyzing *O*-glycosyl compounds
3.2.2 Hydrolyzing *N*-glycosyl compounds
3.2.3 Hydrolyzing *S*-glycosyl compounds

3.3 Acting on ether bonds
3.3.1 Thioether hydrolases
3.3.2 Ether hydrolases

3.4 Acting on peptide bonds (peptide hydrolases)
3.4.11 α-Aminoacylpeptide hydrolases
3.4.13 Dipeptide hydrolases
3.4.14 Dipeptidylpeptide hydrolases
3.4.15 Peptidyldipeptide hydrolases
3.4.16 Serine carboxypeptidases
3.4.17 Metallocarboxypeptidases
3.4.18 Cysteine carboxypeptidases
3.4.19 ω-Peptidases
3.4.21 Serine proteinases
3.4.22 Cysteine proteinases
3.4.23 Aspartic proteinases
3.4.24 Metalloproteinases
3.4.99 Proteinases of unknown catalytic mechanism

3.5 Acting on carbon–nitrogen bonds, other than peptide bonds
3.5.1 In linear amides
3.5.2 In cyclic amides
3.5.3 In linear amidines
3.5.4 In cyclic amidines
3.5.5 In nitriles
3.5.99 In other compounds

3.6   *Acting on acid anhydrides*
   3.6.1   In phosphoryl-containing anhydrides
   3.6.2   In sulfonyl-containing anhydrides

3.7   *Acting on carbon–carbon bonds*
   3.7.1   In ketonic substances

3.8   *Acting on halide bonds*
   3.8.1   In C halide compounds
   3.8.2   In P halide compounds

3.9   *Acting on phosphorus–nitrogen bonds*

3.10   *Acting on sulfur–nitrogen bonds*

3.11   *Acting on carbon–phosphorus bonds*

4.   LYASES

4.1   *Carbon–carbon lyases*
   4.1.1   Carboxy-lyases
   4.1.2   Aldehyde-lyases
   4.1.3   Oxo-acid-lyases
   4.1.99   Other carbon–carbon lyases

4.2   *Carbon–oxygen lyases*
   4.2.1   Hydro-lyases
   4.2.2   Acting on polysaccharides
   4.2.99   Other carbon–oxygen lyases

4.3   *Carbon–nitrogen lyases*
   4.3.1   Ammonia-lyases
   4.3.2   Amidine-lyases

4.4   *Carbon–sulfur lyases*

4.5   *Carbon–halide lyases*

4.6   *Phosphorus–oxygen lyases*

4.99   *Other lyases*

5.   ISOMERASES

5.1   *Racemases and epimerases*
   5.1.1   Acting on amino acids and derivatives
   5.1.2   Acting on hydroxy acids and derivatives
   5.1.3   Acting on carbohydrates and derivatives
   5.1.99   Acting on other compounds

5.2   Cis–trans *isomerases*

5.3   *Intramolecular oxidoreductases*

   5.3.1   Interconverting aldoses and ketoses
   5.3.2   Interconverting keto and enol groups
   5.3.3   Transposing $C{=}C$ bonds
   5.3.4   Transposing S—S bonds
   5.3.99   Other intramolecular oxidoreductases

5.4   *Intramolecular transferases*
   5.4.1   Transferring acyl groups
   5.4.2   Phosphotransferases
   5.4.3   Transferring amino groups
   5.4.99   Transferring other groups

5.5   *Intramolecular lyases*

5.99   *Other isomerases*

6. LIGASES (SYNTHETASES)

6.1 Forming carbon–oxygen bonds
   6.1.1   Ligases forming aminoacyl-tRNA and related compounds

6.2 Forming carbon–sulfur bonds
   6.2.1   Acid–thiol ligases

6.3 Forming carbon–nitrogen bonds
   6.3.1   Acid–ammonia (or amine) ligases (amide synthetases)
   6.3.2   Acid–amino acid ligases (peptide synthetase)
   6.3.3   Cyclo-ligases
   6.3.4   Other carbon–nitrogen ligases
   6.3.5   Carbon–nitrogen ligases with glutamine as amido N donor

6.4 Forming carbon–carbon bonds

6.5 Forming phosphate ester bonds

5.2.8 REFERENCES

1. T. Tanaka, Y. Harano, F. Sue and H. Morimura, *J. Biochem. (Tokyo)*, 1967, **62**, 71.
2. G. H. Hitchings and S. L. Smith, *Adv. Enzyme Regul.*, 1980, **18**, 349.
3. H. E. Umbarger, *Annu. Rev. Biochem.*, 1978, **47**, 533.
4. A. H. Fairlamb and I. B. R. Bowman, *Exp. Parasitol.*, 1977, **43**, 353.
5. L. Bulow and K. Mosbach, *Ann. N.Y. Acad. Sci.*, 1987, **501**, 44.
6. International Union of Biochemistry, 'Enzyme Nomenclature', Academic Press, New York, 1984.
7. C. R. Beddell, in 'X-ray Crystallography and Drug Action', ed. A. S. Horn and C. J. De Ranter, 1984, Oxford University Press, Oxford, p. 169.

5.3

Enzyme Catalysis

MICHAEL I. PAGE

Huddersfield Polytechnic, UK

5.3.1	ENZYME STRUCTURE	45
5.3.2	KINETICS AND SPECIFICITY	48
5.3.3	ENERGETICS	51
5.3.4	BINDING ENERGIES	55
5.3.5	MOLECULAR FIT	56
5.3.6	REFERENCES	59

5.3.1 ENZYME STRUCTURE

The problems associated with understanding enzyme catalysis cannot be separated from the questions one asks about protein folding and structure (see Chapter 5.1). The forces responsible for the unique three-dimensional structure of enzymes are the same as those responsible for binding the substrate and for bond making and breaking, leading to chemical reaction.[1]

Enzymes are proteins, being unbranched polymers of S-α-amino acids joined together by peptide bonds between the α-carboxyl of one residue and the α-amino group of the next. The peptide bond is invariably planar and adopts an *anti* configuration ($>99.9\%$).

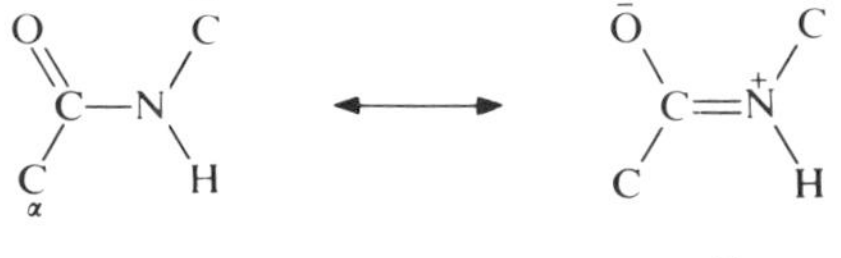

(1a)   (1b)

Resonance of the amide (**1a** and **1b**) gives a polar entity with a considerable dipole moment; it is also responsible for the high energy barrier to rotation about the C—N bond. Proline is an exceptional amino acid residue because the difference in energy between the *syn* and *anti* forms is so small that about 20–30% of proline peptide bonds may exist in the *syn* configuration.[2]

Resonance in the peptide is also responsible for its high solution energy of $80\,\mathrm{kJ\,mol^{-1}}$.[3] The peptide NH is a good hydrogen bond proton donor and a weak acid of $\mathrm{p}K_a \approx 15$ (similar to that of water and alcohols). The peptide carbonyl oxygen is a good hydrogen bond proton acceptor and a weak base, the conjugate acid of which has a $\mathrm{p}K_a \approx -1$ (again similar to that of water). Because the nitrogen lone pair is involved with resonance, it is *not* available as a proton acceptor, which is reflected in the NH^+ conjugate acids of amides having $\mathrm{p}K_a$ values of less than -8. The hydrogen bond donor/acceptor properties of peptides are very important in determining the structure of enzymes and in contributing to the stabilization of bound species (**2**). Some physical properties of the 20 amino acid residues found in enzymes are summarized in Table 1.

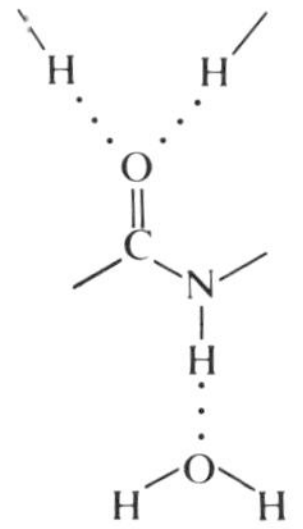

Enzymes

(2)

Table 1 Physical Properties of Amino Acid Residues Found in Enzymes

	Occurrence in proteins[6] (%)	Volume[4] (Å^3)	Partial specific volume[4] (mLg^{-1})	Accessible surface area[5] (Å^2)	Relative hydrophilicity[48] (kcal mol^{-1})[a]	Relative hydrophobicity[48] (kcal mol^{-1})[a]
Ala	9.0	88.6	0.748	115	0.45	0.5
Arg	4.7	173.4	0.666	225	22.31	(−11.2)
Asn	4.4	117.7	0.619	160	12.07	−0.2
Asp	5.5	111.1	0.579	150	13.31	(−7.4)
Cys	2.8	108.5	0.631	135	3.63	(−2.8)
Gln	3.9	143.9	0.674	180	11.77	−0.3
Glu	6.2	138.4	0.643	190	12.58	(−9.9)
Gly	7.5	60.1	0.632	75	0	0
His	2.1	153.2	0.670	195	12.62	0.5
Ile	4.6	166.7	0.884	175	0.24	(2.5)
Leu	7.5	166.7	0.884	170	0.11	1.8
Lys	7.0	168.6	0.789	200	11.91	(−4.2)
Met	1.7	162.9	0.745	185	3.87	1.3
Phe	3.5	189.9	0.774	210	3.15	2.5
Pro	4.6	122.7	0.758	145		(−3.3)
Ser	7.1	89.0	0.613	115	7.45	−0.3
Thr	6.0	116.1	0.689	140	7.27	0.4
Trp	1.1	227.8	0.734	255	8.28	3.4
Tyr	3.5	193.6	0.712	230	8.50	2.3
Val	6.9	140.0	0.847	155	0.40	1.5

[a] 1 cal = 4.18 J. The values in parentheses are calculated from surface areas (see ref. 48).

Most enzymes have very similar amino acid compositions (Table 1) and substantial deviation from this average composition is rare for globular water soluble proteins. Proteins that are membrane bound tend to have higher levels of hydrophobic and lower levels of ionized amino acid residues. Apart from this simple observation there appear to be, at present, no simple general rules for enzyme structure.

The stability of proteins varies tremendously. Although human eye lens proteins are stable for at least 70 years, since lens cells do not die, there is usually a rapid turnover of proteins because they are being constantly hydrolyzed to their constituent amino acids. This degradation acts as a control process for errors made during biosynthesis and for loss of activity due to aging.[7] Enzymes catalyzing metabolic reactions tend to have very short half-lives, for example that for rat liver ornithine decarboxylase is only 11 min. The concentration of this enzyme, which controls the rate of polyamine synthesis, varies 1000-fold and presumably its rapid turnover is the method of regulating enzyme activity.[8]

Enzyme turnover is obviously related to the inherent stability of the protein and the rates of thermal unfolding and rates of dissociation of stabilizing ligands. Although 'chemical' factors such as hydrolysis and deamidation of glutamine and asparagine residues are also important, it appears that the folded state of enzymes is important for minimizing the rate of degradation.[9]

Although an enzyme folds up because its folded state, under physiological conditions, is more stable than the unfolded state, the differences in free energy between the two states is of the order of only 10–15 kJ mol^{-1}.[1] However, it is conceivable that the native folded state is not the lowest free energy state attainable but is the kinetically controlled conformation. The thermodynamic argument is that the polypeptide chain contains all the information required for folding. It has been suggested that a kernel of secondary structure is initially formed and subsequently the side chains interact in a unique series of steps to ultimately form the globular state.[10] There appears to be a unique pathway for folding, involving several intermediates.[11] Some enzymes, such as lysozyme, are simply one polypeptide chain and presumably fold into their unique catalytically active conformation. Other enzymes, such as chymotrypsin, are synthesized as inactive precursors and removal of part of the chain generates the active enzyme but in a thermodynamically unstable state. Similar situations exist for multichain proteins where, after removal of part of the polypeptide chain, the component chains no longer possess all the information required to fold and associate into an active enzyme. Extracellular enzymes which contain disulfide bonds are initially synthesized inside the cell where the environment is not conducive to oxidation of the thiol groups. Consequently, such enzymes are often not folded inside the cell but only after release outside the cell.

X-Ray crystallography of enzymes shows that the active sites of many enzymes are often situated at the hinge regions covering two structural domains.[12] This region is relatively more flexible than the enzyme molecule as a whole and may therefore be more susceptible to denaturants. The binding of substrates leads to the relative movements within the active site of some enzymes[13] and, in fact, the loss of activity during partial unfolding of enzymes caused by denaturants has been attributed to conformational changes in the more flexible active site.[14]

The average size of domains in proteins is about 20 000 Da[15] but it has been suggested that there are distinct binding sites, called modules, within these domains, corresponding to about 5000 Da.[16] Binding sites are required to recognize substrates, coenzymes, allosteric effectors and many cellular components such as lipid bilayers, proteins, carbohydrates and nucleic acids. It has long been realized that catalytic functionality of an enzyme is not related to its size. For example, kinases vary in molecular weight from 18 to 83 kDa and although the chemical step catalyzed is similar for each enzyme, they have different numbers and types of binding sites. At least 20–40 amino acid residues (2300–4600 Da) are required to form a stable folded structure[17] and the molecular weight per binding site appears to be approximately constant at around 5000 Da.[16]

Studies using site-directed mutagenesis are allowing the elucidation of the contribution of individual amino acids to overall binding and protein stability.[18]

The folded structures of most enzymes are generally very compact and roughly spherical in overall shape but with an irregular surface. Typically they have diameters of 40–50 Å, surface areas of 10^4 Å^2 and volumes of 5×10^4 Å^3. Enzymes vary in their subunit molecular mass from about 10 kDa to 250 kDa but most are smaller than 60 kDa with an average of 25–30 kDa per subunit. In volume terms the ratio of enzyme to substrate is roughly 50–100.

Despite the compact nature of the folded enzyme, the conformation is the result of a multitude of weak interactions between atoms within the protein and between the protein and the surrounding medium. The force field is thus relatively weak and at ambient temperatures thermal energy is sufficient to allow considerable motion of most atoms within the enzyme. There is almost certainly a degree of coupling of motions within the protein, which gives rise to 'breathing motions'. The flexibility of globular proteins is suggested by amide proton exchange, and ^{13}C NMR relaxation studies and the observation of aromatic ring 'flips' by ^{1}H NMR.[19] The exploitation of the temperature factors in the single-crystal X-ray data outlines the spatial confinement for high frequency atomic motions. The importance of these fluctuations in atomic positions is relevant to the understanding of enzyme catalysis and the design of enzyme inhibitors. The force field generated by the enzyme is not an inflexible one, although dipole moments generated by the relatively rigid protein are highly orientated, compared with a solvent; the mobility ensures that the dipoles are oscillating. There may be unfavourable dipole–dipole interactions within the enzyme which are relieved upon substrate binding and occur in the native enzyme because there are compensations from favourable interactions upon protein folding.

Bulk water has special dielectric properties due to the large dipole moment of about 1.85 D of the water molecule and the rotational mobility of these molecules in the liquid phase. The importance of the dipole moment for the dielectric constant of a liquid is shown by the values of 80 and 109 for water and formamide compared with that of 2 for cyclohexane. The necessity of rotational freedom is shown by the values of only 4 for solid acetamide and ice II. The estimation of electrostatic interactions between charges and dipoles in enzymes is therefore difficult because of the different dynamic properties of water and protein.

As the atoms within an enzyme occupy relatively fixed positions, the dielectric screening effect is small, estimates of the dielectric constant being as low as 4. The first hydration shell of the protein has unknown dielectric properties,[20] although bulk solvent molecules can freely rotate to take up preferred orientations. X-Ray and neutron diffraction studies show that some water molecules are tightly bound to certain sites in enzymes but the translational and rotational mobility of these water molecules is uncertain. The concept of a 'dielectric constant' therefore cannot really be applied to enzymes. Since it costs energy to create a reaction field around charges in liquid water by orientating the water molecules, the 'preorganized' immobile protein dipoles could present a large intraglobular electric field which compensates for the unfavourable energy changes of taking a charged substrate out of water and binding it to an enzyme.

The fact that the acidity/basicity of many groups within enzymes are similar to that in water implies that there must be significant stabilization of the charged groups by bound water molecules and permanent and induced dipoles of the enzyme.

In protein molecules the interior is densely packed with seriously restricted motions.[21] It is not easy to estimate the space available within molecules. The atomic hard sphere model assumes that the volume occupied by a molecule, that is the volume impenetrable to other molecules, is the van der Waals volume. The radii determined by van der Waals constant distances between molecules required to reproduce packing densities of liquids and solids results in spheres which overlap one another.[22] The hard sphere model has been applied to the packing of atoms in protein molecules and the surface area of solvent–protein interfaces.[21] The shape and area of the exposed surface vary with the dimensions of the molecule used to probe its surface, reaching a limiting value as the size of the probe increases. Consequently Richards[21] has distinguished between (i) the van der Waals envelope, the outer surface as determined by the hard sphere model; (ii) the accessible surface, the continuous sheet defined by the locus of the centre of the probe molecule as it rides over the van der Waals surface; (iii) the contact surface, those parts of the van der Waals surface that are actually in contact with the surface of the probe; and (iv) the reentrant surface, as defined by the interior facing part of the probe when it is simultaneously in contact with more than one atom. The latter two surfaces combine to form a continuous sheet to give a molecular surface described as a solvent accessible surface.[23]

The van der Waals volume is necessarily less than the volume per molecule as determined by the experimentally determined molar volume, the difference between the two giving rise to the concept of a free volume. The fraction of space occupied by atoms in enzymes is very high, about 0.75, and consequently there is a high surface area of contact between atoms of the enzyme and between atoms of the enzyme and its substrate or inhibitor.

5.3.2 KINETICS AND SPECIFICITY

Experimentally, the initial rate (v) of an enzyme-catalyzed reaction is found to show saturation kinetics with respect to the concentration of the substrate (S). At low concentrations of substrate the initial rate increases with increasing concentration of S but, at high or saturating concentrations of S, it becomes independent of the concentration of the substrate. This observation is interpreted in terms of the rapid and reversible formation of a noncovalent complex (ES), from the substrate (S) and enzyme (E), which then decomposes into products (P) (equation 1). This scheme led to the familiar Michaelis–Menten equation (2), where K_m is the Michaelis constant, which is the concentration of substrate at which the initial rate is *half* the maximal rate of saturation V_{max}. The constant $K_m = (k_{-1} + k_2)/k_1$ if k_2 is comparable with k_{-1}, but $K_m = k_{-1}/k_1$, the true dissociation constant of ES, if $k_{-1} \gg k_2$. The first-order rate constant for the decomposition of ES is commonly called k_{cat} (the turnover number), k_2 in equation (1). At low concentrations of substrate, where $[S] \ll K_m$, v is given by equation (3), with k_{cat}/K_m being an apparent second-order rate constant. At high concentrations of S, where $[S] \gg K_m$, v is given by equation (4) and becomes independent of S.

$$\text{E} + \text{S} \underset{k_{-1}}{\overset{k_1}{\rightleftharpoons}} \text{ES} \overset{k_2}{\rightarrow} \text{EP} \rightleftharpoons \text{E} + \text{P} \tag{1}$$

$$v = \frac{k_2[\text{S}]}{(k_{-1} + k_2)/k_1 + [\text{S}]} = \frac{k_{cat}[\text{E}]_0[\text{S}]}{K_m + [\text{S}]} = \frac{V_{max}[\text{S}]}{K_m + [\text{S}]} \tag{2}$$

$$v = \frac{k_{cat}}{k_m}[\text{E}]_0[\text{S}] = \frac{k_1 k_2}{k_{-1} + k_2}[\text{E}]_0[\text{S}] \tag{3}$$

$$v = k_{cat}[\text{E}]_0 = k_2[\text{E}]_0 = V_{max} \tag{4}$$

Although the Michaelis–Menten equation (2) is valid for most enzyme-catalyzed reactions, the scheme outlined in equation (1) is not always followed. The measured K_m and k_{cat} values are not always equal to the dissociation constant K_s for the enzyme–substrate complex and the rate constant for decomposition of ES, respectively. The apparent dissociation constant K_m can be less than K_s, *i.e.* apparent tighter binding of substrate, if additional intermediates, covalently or noncovalently bound, are formed during the reaction pathway and the rate-limiting step is the reaction of one of these intermediates. Similarly, $K_m > K_s$ if the rate of dissociation of ES to E and S is comparable or slower than the forward rate of reaction of ES (Briggs–Haldane kinetics). The measured value of k_{cat} may also be a function of various and several microscopic rate constants. If the chemical steps occur sufficiently quickly then the binding of the substrate or the desorption of the product may become rate limiting.

In vitro 'nonspecific' substrates are sometimes described as 'poor' because they show a low value of k_{cat} or a high value of K_m. However, *in vivo* discrimination results from a *competition* of substrates for the active site of the enzyme. If two substrates, S and S', compete for the same enzyme, wrong conclusions could be reached about the relative rates of product formation if k_{cat} values or the K_m values of the individual substrates are compared instead of their relative values of k_{cat}/K_m.

If the enzyme catalyzes the reaction of both S and S' (equation 5), the relative rates of product formation may be derived from the usual steady state procedure, avoiding any assumptions about the relative values of k_2 and k_{-1}, and is given by equation (6) whether the enzyme is working below or above saturation for both substrates. Furthermore equation (6) is applicable even if the individual K_m values or substrate concentrations are such that the enzyme is working below saturation for one substrate but above saturation for the other. Specificity between competing substrates is therefore given by the relative values of k_{cat}/K_m and not by the individual values of k_{cat} and K_m. The relative rates of product formation depend upon the differences in the free energies of activation of the various substrates. It is incorrect to say that an enzyme catalyzes the reaction of A more efficiently than that of B because it binds A more tightly than B. It is also incorrect to say that the difference is attributable to a higher value of k_{cat} (k_2). If for a series of possible substrates some of the microscopic rate constants are similar, *e.g.* $k_1 \approx k_1'$ and $k_2 \approx k_2'$, then discrimination could be discussed in terms of the relative values of k_{-1} and k_{-1}'. However, in general all the microscopic rate constants in equation (6) will contribute to the relative rates of product formation.

$$P' \rightarrow ES' \underset{k_1'}{\overset{k_{-1}'}{\rightleftharpoons}} E + S + S' \underset{k_{-1}}{\overset{k_1}{\rightleftharpoons}} ES \overset{k_2}{\rightarrow} P \tag{5}$$

$$\frac{v}{v'} = \frac{k_1 k_2 (k_{-1}' + k_{-2}')[S]}{k_1' k_2' (k_{-1} + k_2)[S']} = \frac{(k_{cat}/K_m)[S]}{(k_{cat}'/K_m')[S']} \tag{6}$$

Because enzymes increase the rate of reactions *and* show discrimination between possible substrates, there has been a temptation to treat these two phenomena separately. Classically the rate enhancement is attributed to the chemical mechanism used by the enzyme to bring about transformation of the substrate. The fidelity of enzyme and substrate is accounted for by binding—as in the analogy of the 'lock and key'. It is now apparent that these simple ideas need to be reappraised. Chemical catalysis alone cannot explain the efficiency of enzymes.[1] The forces of interaction between the *nonreacting* parts of the substrate and enzyme may also contribute to a lowering of the activation energy of the reaction.

The enzyme succinyl-CoA-acetoacetate transferase catalyzes reaction (7) ($R^1CO_2^-$ = acetoacetate and $R^2CO_2^-$ = succinate) and proceeds by the initial formation of an enzyme–CoA intermediate in which the coenzyme A is bound to the enzyme as a thiol ester of the γ-carboxyl group of glutamate (equation 8).[24] In turn this intermediate is generated by nucleophilic attack of the glutamate carboxylate on succinyl-CoA to give an anhydride intermediate (3). The second-order rate constant for this reaction is 3×10^{13} times greater than the analogous reaction of acetate with succinyl-CoA (4). It seems unlikely that the chemical reactivity of acetate and the enzyme's glutamate will be vastly different. Similar chemical reactions are therefore being compared and yet the *nonreacting* part of the enzyme lowers the activation energy by $78 \, kJ \, mol^{-1}$ ($RT \ln 3 \times 10^{13}$).

$$R^1CO_2^- + R^2COS\text{-}CoA \rightarrow R^1COS\text{-}CoA + R^2CO_2^- \tag{7}$$

$$R^2COS\text{-}CoA + Enz\text{—}CO_2^- \rightarrow R^2CO_2^- + Enz\text{—}COS\text{-}CoA \tag{8}$$

The same example may be used to illustrate specificity. The enzyme also forms an anhydride with the 'nonspecific' substrate, methyl succinylmercaptopropionate (5). However, the enzyme reacts with succinyl-CoA (3) 3×10^{12} times faster than with (5). The chemical reactivities of the two substrates are similar towards, for example, alkaline hydrolysis. The small substrate, methyl succinylmercaptopropionate, should be able to fit into the active site. Therefore the *nonreacting* part of succinyl-CoA of molecular weight *ca.* 770 lowers the activation energy by $72 \, kJ \, mol^{-1}$ ($RT \ln 3 \times 10^{12}$).[24]

$$
\begin{array}{ccc}
\text{(3)} & \text{(4)} & \text{(5)}
\end{array}
$$

For structures (3), (4), (5):

(3): $R^2-\overset{\overset{O}{\|}}{C}-SCoA$ with $Enz-\overset{\overset{O^-}{}}{\underset{O}{C}}$

(4): $R^2-\overset{\overset{O}{\|}}{C}-SCoA$ with $Me-\overset{O^-}{\underset{O}{C}}$

(5): $R^2-\overset{\overset{O}{\|}}{C}-SCH_2CH_2CO_2Me$ with $Enz-\overset{O^-}{\underset{O}{C}}$

Enzymes may obviously discriminate against substrates which are larger than the specific substrate as the substrate is too large to fit into the active site. However, it is not obvious how to explain why some small nonspecific substrates that can presumably bind to the active site react very slowly. For example, is there enough binding energy available between the extra methylene of the substrate and the enzyme to account for the discrimination between isoleucine (6) as a specific substrate and valine (7) as an alternative? Specificity can be reflected in poor binding (high K_s) and/or slow catalytic steps (low k_{cat}) but it was shown earlier that specificity between competing substrates is controlled by their relative values of k_{cat}/K_m.

(6): $\underset{Me}{\overset{Et}{\diagdown}}CH-CH\underset{CO_2^-}{\overset{\overset{+}{N}H_3}{\diagup}}$

(7): $\underset{Me}{\overset{Me}{\diagdown}}CH-CH\underset{CO_2^-}{\overset{\overset{+}{N}H_3}{\diagup}}$

$$
\begin{array}{cc}
\text{(6)} & \text{(7)}
\end{array}
$$

It is often assumed that discrimination between substrates can occur if undesirable substrates are preferentially bound to a site on the enzyme where catalysis cannot occur (equation 9). The rate of reaction is given by $k_2[ES]$, which in terms of microscopic rate constants is given by equation (10), where K_m° is the Michaelis constant for the reaction in the absence of an additional binding site and K_s', is the dissociation constant for the nonproductive complex ES'. The *observed* binding constant K_m is lower than this because the additional binding site leads to apparently tighter binding. At saturation only a fraction of the substrate is productively bound and so $k_2[ES]$ is lowered. The *in vitro* conclusion could therefore be that the binding energy between the undesirable smaller substrate and a site on the enzyme away from the active site is used to prevent catalysis; the larger specific substrate can only bind at the active site and is sterically prevented from binding to the nonproductive site. Although this is an appealing idea, it cannot explain the discrimination between competing substrates *in vivo*. The free energy difference between the unbound substrate and enzyme, $E+S$, and the transition state, $ES^{\ddagger}$, is unaffected by alternative modes of binding, *e.g.* to ES' (equation 9), *i.e.* the free energy of activation, represented by k_{cat}/K_m, is not changed by nonproductive binding.

$$
ES' \underset{K_m^{\circ}}{\overset{K_s'}{\rightleftharpoons}} E + S \underset{K_m^{\circ}}{\rightleftharpoons} ES \overset{k_2}{\rightarrow} ES^{\ddagger} \tag{9}
$$

$$
v = \frac{k_2[E][S]}{K_m^{\circ}(1 + [S]/K_s') + [S]} \tag{10}
$$

There are also side *reactions* which may prevent the formation of an undesired product, although this may be an energetically wasteful form of specificity. There may be an alternative *reactive* site on the enzyme to which the smaller 'unwanted' substrate, intermediate or product *can bind and react* but which is less efficient towards the desired substrate, intermediate or product. This has been suggested to occur with the aminoacyl-tRNA synthetases.[25]

In the induced fit model of discrimination, the active site of the free enzyme E is in the 'wrong' conformation for catalytic activity. The binding of the desired substrate induces a conformational change in the enzyme making it catalytically active (E'; equation 11).

$$
\begin{array}{ccc}
E & \overset{K}{\rightleftharpoons} & E' \\
\updownarrow & & \updownarrow \\
E & \rightleftharpoons & E'S \overset{k_{cat}}{\longrightarrow} E'S^{\ddagger}
\end{array}
\tag{11}
$$

Poor substrates do not have enough favourable binding energy to compensate for the unfavourable conformational change in the enzyme. 'Good' substrates provide enough binding energy to 'pay for' the conversion to the unfavourable, but active, conformation of the enzyme, E'. For this conformational change to occur the free energy of binding 'good' substrates must be greater than the free energy of distortion of the enzyme from its native inactive state to the catalytically active conformation. Induced fit can explain specificity between very good and very poor substrates but it is less successful when applied to a series of substrates that all have sufficient binding energy to compensate for the unfavourable conformational change. An enzyme using the induced fit mechanism of specificity is therefore less efficient than the active enzyme by a factor of K—the ratio of inactive to active molecules of enzyme, $(k_{cat}/K_m)_{obs} = K k_{cat}/K_m$. If the substrates which must be discriminated against have sufficient binding energy to compensate for the unfavourable conformational change of the enzyme, then induced fit cannot explain specificity between them. Compared with the situation where the enzyme is initially in the active conformation, the induced fit mechanism reduces the value of k_{cat}/K_m for *all* substrates by the *same* fraction and therefore does not affect their relative rates.

Induced fit can therefore only be used to explain the rates of reactions of very poor substrates, *e.g.* the rate of phosphoryl transfer to water, compared with that to glucose, catalyzed by hexokinase. Although water can almost certainly bind to the active site, it must have insufficient binding energy to induce the necessary conformational change in the enzyme.[26]

5.3.3 ENERGETICS

The binding energy between substrate and enzyme may be used in a variety of molecular mechanisms to lower the activation energy of the reaction such as charge neutralization, desolvation, geometrical and entropic effects. The electron density distribution in a molecule determines the nuclear configuration and charge stabilization and geometrical effects are not always easily separable, but as molecules may have similar shapes and yet different charge distributions, there are important differences (see Chapters 18.5 and 20.3).

It is now generally considered that maximum binding energy, *i.e.* stabilization, occurs between the substrate and enzyme in the transition state of the reaction. There is an exception to this generalization for the case in which an enzyme equally stabilizes the ground state and transition state, but catalysis can only occur in this situation if the enzyme is working below saturation. There is a limit to this type of catalysis because if the enzyme binds the transition state and ground state very tightly, although the free energy of activation will be reduced, the concentrations of enzyme and substrate required to maintain nonsaturation conditions will be decreased. These concentrations could be so low that catalysis may not be observed. If a nonreacting substituent of a specific substrate contributes a large amount of binding energy, it is essential that this is not expressed in the ground state or intermediate states in order to avoid saturation conditions and the low concentrations of E and S required to observe nonsaturation.

Maximum catalytic efficiency may be achieved by the enzyme stabilizing all transition states, but not intermediate states, in the pathway between reactants and products. The interconversion of S and P may proceed in the thermodynamically favourable direction with a rate that is limited by the diffusion together of the enzyme and S (Figure 1). In the reverse direction the minimum free energy of activation corresponds to the free energy of activation for diffusion plus $\Delta G°$, the free energy difference between S and P (Figure 1). It is obvious that an efficient enzyme must stabilize the transition state(s) of a reaction, but it is equally important that the enzyme does not excessively stabilize any intermediate. Stable enzyme intermediate states will bring about saturation conditions at a low concentration of substrate and valuable enzyme will be then tied up in an energy well.[1, 18, 26]

For a given amount of binding energy between the substrate and enzyme the most effective catalysis will be obtained if this energy is used to stabilize the transition state, which maximizes the value of k_{cat}/K_m. For a given free energy of activation, $\Delta G^{\ddagger}$, for the process $E + S \rightarrow ES^{\ddagger}$, and for a given substrate concentration, the maximum rate is obtained if the substrate is bound *weakly, i.e.*

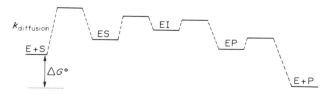

Figure 1 Standard state free energy changes for an enzyme-catalyzed reaction proceeding *via* the formation of intermediates. The enzyme should stabilize the transition state but not excessively stabilize any of the intermediates

shows a high K_m. A low value of K_m, *i.e.* strong binding of the substrate or intermediate state, mediates against catalysis. In agreement with these ideas, the physiological concentrations of most substrates are below their K_m values.[27]

The argument that the enzyme–substrate complex should not be too stable compared with the free enzyme and substrate is sometimes interpreted to mean that the substrate should be *destabilized* in the ES complex. If the energy of the transition state is fixed but the energy of the ES complex is raised, by an unspecified mechanism, the result will be that k_{cat} is increased. However, such destabilization of the enzyme–substrate complex has no effect upon k_{cat}/K_m. This mechanism would therefore make the enzyme more effective above saturation but not below saturation. Specificity, the discrimination between competing substrates, depends upon the value of k_{cat}/K_m and would thus not be affected by having the enzyme–substrate complex destabilized. The introduction of strain or any other mechanism of destabilizing the substrate in the ES complex cannot therefore directly affect specificity. However, a high value of K_m, resulting from weak binding or the 'destabilization' of the ES complex, favours nonsaturation conditions, which can enhance the value of k_{cat}/K_m. Weak binding of the substrate therefore favours efficient catalysis and can contribute to specificity.

The term differential binding has been applied to cases where the enzyme binds substrate and product unequally.[28] Consequently the free energy difference between reactants and products in solution is different from that of the enzyme-bound species. For example, the equilibrium constant for the conversion of tyrosine and ATP to tyrosine adenylate and pyrophosphate changes from 3×10^{-7} to 2.3 when the reactants and products are bound to the enzyme tyrosyl-tRNA synthetase.[29] Differential binding of reactant and products also leads to differential catalysis expressed by k_{cat} for forward and reverse directions although, of course, there is no effect on k_{cat}/K_m.

A major question in understanding the discrimination between substrates is how does the system allow the binding energy between the 'nonreacting' parts of the enzyme and substrate to be expressed in the transition state and not the ground state? If a 'nonspecific' and a 'specific' substrate have identical reaction centres, the intrinsic binding energy between the reacting groups undergoing electron density changes and the enzyme should be similar. Differential binding of the transition states and ground states around the reaction centre is easy to visualize. Recognition of the geometrical and electronic differences between ground states and transition states depends on suitably placed amino acids in the enzyme.[30] This idea has received experimental support by site-directed mutagenesis. For example, the active site arginine in staphylococcal nuclease does not affect the binding of substrate but increases k_{cat} by 10^5 due to transition state stabilization.[31] How can the binding energy of the nonreacting part of the specific substrate stabilize the transition state but not overstabilize the enzyme–substrate or enzyme–intermediate complex? This is very relevant to building in recognition sites in enzyme inhibitors since it is often the binding energy between the nonreacting part of the substrate that accounts for catalysis. The binding energy could be used to 'destabilize' the substrate or to compensate for thermodynamically unfavourable processes necessary for reaction to occur. The importance of this interaction is illustrated by the observation that the *observed* binding constant (usually K_m values) of a specific substrate is often apparently 'weaker' or no 'tighter' than that for a nonspecific substrate.

A word of caution is required to prevent a literal interpretation of the preceding description of enzyme action. That 'the enzyme stabilizes the transition state and destabilizes the ground state' should not encourage us to believe that enzymes are there in a fixed state doing 'something' to the substrate. Stabilization is a mutual process—one could equally well describe the substrate in the transition state stabilizing the enzyme. The important change is the *difference* in energy between the ground *state*—enzyme and substrate in their own environment—and the transition *state* of the enzyme–substrate complex. For example, the statement that 'a positive charge on the enzyme stabilizes the negative charge on the substrate' could easily be inverted. Futhermore, a positive charge on an enzyme would probably be 'neutralized' in some way in the ground state of the isolated enzyme, for example by anionic charges on a neighbouring part of the protein chain or buffers, and by solvation provided by both the rest of the enzyme and water. The rate of reaction would be controlled by the *difference* in free energy provided by these stabilizations in the ground state and the transition state.

Although enzymes are often considered to be anthropomorphic and have found out how 'to get something for nothing', this is a doubtful achievement even for evolutionarily perfect enzymes. Enzymes cannot solvate and distort substrates with no energetic loss to themselves. The molecular recognition between enzymes and ligands depends on the *difference* between interactions of the isolated molecules and their environment, and the binding energy interactions between the enzyme and ligand in the complex. Catalysis by enzymes depends on the *difference* between interactions in the transition state and those in the initial state of the isolated molecules.[1, 26, 32, 33] Very rarely will an interaction take place in one state that is completely absent in the other.

The interaction between the substrate S and its environment X (which is usually the solvent) has a complementary interaction in the enzyme E and its environment Y (which may also be the solvent or another site within the enzyme). The changes in interactions which occur when a complex is formed between E and S include those generated in the pair X and Y (equation 12).

$$S{-}{-}X + E{-}{-}Y \rightleftharpoons E{-}{-}S + X{-}{-}Y \tag{12}$$

The latter are not necessarily the result of a direct exchange if, for example, the interactions in the initial state of E or S are intramolecular. The net change in the overall energetics depends on the *difference* in the free energy resulting from this interchange of the 'bonding' partners. The free energy change reflects the *difference* in enthalpy and entropy of these interactions. For example, S may have a proton donor site which forms a hydrogen bond with the solvent water acting as an acceptor, X, and also with a site in the enzyme. Usually, in the initial state, this same site in the enzyme will be hydrogen bonded to water and $X \cdots Y$ of equation (12) would correspond to the formation of a new hydrogen bond between water molecules which accompanies enzyme–substrate complex formation.[34]

During the chemical reaction catalyzed by the enzyme, there may be changes in the electron density distribution of the substrate and the enzyme such that the polarity or charge is very different between the initial and transition states. Nonetheless, the overall efficiency of catalysis will depend on the *difference* in the energetics of the interactions in the two states.

At the reaction centre electron density and geometrical changes in the substrate, on going to the transition state, may be stabilized by complementary charges and shape of the active site in the enzyme. These electron density changes could be accompanied by a conformational change so that a large nonreacting group not bound to the enzyme in the ground state becomes bound in the transition state. In general, however, it is less easy to see how the binding energy of the nonreactive part of the substrate is prevented from being fully expressed in the ES complex unless it is used to compensate for unfavourable processes. If the environment of the enzyme is unfavourable to the ground state structure of the substrate, the binding energy from the nonreacting part of the substrate could be used to force the substrate into this 'unwelcoming hole'. Examples are: (i) a 'rigid' enzyme that has an active site complementary in shape to the transition state but not the ground state; (ii) an active site which is nonpolar and conducive to stabilizing a neutral transition state but destabilizing a charged substrate; (iii) desolvation or solvation changes of groups on the substrate or enzyme; and (iv) an active site where electrostatic charges on substrate and enzyme are similar.

If the binding energy is used to compensate for the induction of 'strain' in the substrate, it is *essential* that the 'strain' is *relieved* in the transition state in order to increase the rate of reaction. This would be the case if the changes in geometry of the substrate were in the direction which accompanies the reaction mechanism. The observed binding energy would be what is 'left over' after the strain has been 'paid for' and may thus appear to be weak. It is also essential that the nonreacting group only exhibits its binding energy when the *reactive centre* of the substrate is bound to the active site. Alternative binding modes would otherwise result, leading to nonproductive binding which does not affect k_{cat}/K_m and specificity, but does decrease K_m, leading to saturation conditions at lower concentrations of substrate. The binding energy of the substituent of the specific substrate may be used to prevent nonproductive binding.

Amino acid side chains, not necessarily those near the active site, may have their exposure to water changed during catalysis. For example, a polar group may become exposed or protected from water on going from the ground state to transition state, which would make a favourable and unfavourable contribution to the free energy of activation respectively.[1]

An example of enzymes 'getting something for nothing' has been the suggestion that the energetics of reactions within the enzyme are similar to those in the gas phase.[35] Although the rate of ionic reactions is faster when the species are desolvated, the process of desolvation is energetically very expensive and must be paid for.

Probably the most important way that the binding energy is 'used' is to compensate for the unfavourable entropy change that accompanies formation of the ES and ES‡ complex.[36, 37] The entropy loss that is required to reach the transition state may already have been partially or completely lost in the ES complex. The binding energy of a nonspecific substrate may be insufficient to restrict the necessary degrees of freedom of the substrate when bound to the enzyme. Consequently, for this substrate more entropy has to be lost to reach the transition state, which reduces

k_{cat}, compared with a specific substrate. For the latter, the binding energy of the additional substituent may compensate for the required loss of entropy which may occur in ES or ES‡. Thus, although the free energy change accompanying formation of the ES complex may appear to be similar for specific and nonspecific substrates, as measured by the observed K_m, the specific substrate may be physically more 'tightly' bound.

At 25 °C for a standard state of 1 M, the complete restriction of medium-sized substrates requires a decrease in entropy and an increase in energy of about 150 kJ mol^{-1}, an unfavourable factor of 10^8. This entropy change is that typically required to form a covalent bond in which the atoms are confined to a relatively small volume because of the loss of translational and rotational freedom.[36, 38] If the enzyme-catalyzed reaction requires the formation of a covalent bond then the binding energy of the nonreacting part of the substrate may compensate for this necessary but unfavourable entropy change. A nonspecific substrate may have insufficient binding energy to compensate for the required entropy loss, resulting in a reduced value of k_{cat}/K_m. If the chemical mechanism of catalysis requires the involvement of other functional groups such as general acids or bases, metal ions or a change in solvation then a further entropic advantage may be apparent. However, the contribution is smaller than that from covalent catalysis because these secondary effects require less restriction of degrees of freedom because the 'flexibility' of hydrogen bonds and metal ion coordination is greater than that for covalent bonds.[1]

The use of binding energy to compensate for unfavourable solvation changes necessary for chemical reaction is a possible mechanism of catalysis because of the large solvation energies of groups and ions in water.[1] Lone pairs which may act as general bases or nucleophiles will usually be 'solvated' by hydrogen bonding from either water or intramolecularly from the enzyme. Similarly, other potentially reactive groups will usually be 'charge neutralized' in the initial state of the enzyme. These reactive groups will normally require 'desolvation' before bond making or breaking can occur. This process is energetically expensive and yet an essential part of the normal activation energy but may be compensated by favourable interactions between the substrate and enzyme.

Attempts to elucidate the importance of solvation have traditionally been investigated by determining the effect of changing the solvent on the reaction rate. Again it is essential to appreciate that any changes in rate result from the *differences* in ground state and transition state energies which the solvent may cause.

The parameter k_{cat}/K_m measures the free energy difference between transition state ES‡ and ground state E and S (equation 13). If γ_S, γ_E and $\gamma_{\ddagger}$ represent the activity coefficients of the substrate, enzyme and transition state, respectively, and are defined relative to a common value of unity in purely aqueous solution, then k_{cat}/K_m for the enzyme-catalyzed reaction in a given solvent mixture is related to its value, $k_{cat}^{\circ}/K_m^{\circ}$, in pure aqueous solution by equation (14). For sparingly soluble substrates γ_S may be obtained from the solubilities, S and S°, measured respectively in the presence and absence of organic solvent (equation 15). The major difficulty is to estimate solvation effects on the enzyme because of the lack of a method for measuring γ_E. This can be overcome if solubility and kinetic data are obtained for *two* substrates. In a given solvent system it then becomes possible to eliminate γ_E because this is the same for both substrates. The effect of the solvent on the ratio of the two transition state activity coefficients is given by equation (16). The right hand side of equation (16) contains only measurable quantities and the transition state ratio for the two substrates 1 and 2, $\gamma_1^{\ddagger}/\gamma_2^{\ddagger}$, may be compared with the ground state ratio, γ_1/γ_2.

$$E + S \xrightarrow{k_{cat}/K_m} ES^{\ddagger} \qquad (13)$$

$$\frac{k_{cat}}{K_m} = \frac{k_{cat}^{\circ}}{K_m^{\circ}} \frac{\gamma_S \gamma_E}{\gamma_{\ddagger}} \qquad (14)$$

$$\gamma_S = S^{\circ}/S \qquad (15)$$

$$\frac{\gamma_1^{\ddagger}}{\gamma_2^{\ddagger}} = \frac{(k_{cat}/K_m)_2}{(k_{cat}/K_m)_1} \frac{S_2}{S_1} \frac{(k_{cat}^{\circ}/K_m^{\circ})_1}{(k_{cat}^{\circ}/K_m^{\circ})_2} \frac{S_1^{\circ}}{S_2^{\circ}} \qquad (16)$$

The rate of the α-chymotrypsin-catalyzed hydrolysis of 4-nitrophenyl acetate and *N*-acetyl-L-tryptophan methyl ester in organic solvent mixtures decreases with increasing amounts of dioxane or propan-2-ol.[39] Measurement of the solubilities of the substrates in the solvent mixtures indicates that the difference in reactivity of the two substrates with solvent composition is largely a ground state effect. In general, the different specificities shown by α-chymotrypsin may be due to ground state solvation effects rather than transition state ones. Desolvation of the substrate may be responsible for the fact that specific substrates bearing an acylamino side chain are more reactive by several powers of ten than their analogues containing a free amino group. This has sometimes been explained in terms of hydrogen-bonding effects, but the differences between the hydrogen bond energies of the amino and acylamino groups with the enzyme are probably inadequate to explain the large rate differences. It seems more likely that the increase in reactivity observed when the amino group is converted into acylamino arises from the relatively weak solvation of the latter group in the ground state than from favourable transition state interactions, though of course the net result will depend upon the sum of these two effects.

It should be emphasized that structural specificity attributed to the desolvation of the substrate in enzyme-catalyzed reactions results from a retardation effect on the less reactive substrates. It has been suggested[40] that a part of the rate enhancement associated with enzyme-catalyzed reactions is due to the preliminary formation of a complex in which the site of reaction is made susceptible to attack by the removal of solvating water molecules. However, since the overall rate of reaction depends on the free energy difference between ground and transition states, any desolvation of the substrate, other than that occurring in comparable nonenzymic reactions and necessary for approach of the reacting groups, will raise the transition state energy and decrease the rate unless the unfavourable energy of desolvation is compensated by favourable binding energy between substrate and enzyme.

5.3.4 BINDING ENERGIES

Although the contribution of a substituent to the binding of a substrate to an enzyme is sometimes estimated from its effect on the Michaelis K_m value, the deduced contribution is often less than the true intrinsic binding energy. This is because K_m values do not necessarily reflect the equilibrium constants for binding to the active site when nonproductive binding or complicated kinetic phenomena occur. Furthermore, much of the binding energy is 'used up' in bringing about the required loss of entropy and inducing any destabilization, relative to their ground states, of the substrate or enzyme.[1,26,32] Another problem is that the contribution of a group R to the binding of the molecule SR to an enzyme cannot be estimated from the free energy of binding SH and the substituent molecule RH (see Chapter 18.8). Entropy differences can make the binding of SR up to 10^8 times more favourable than that estimated from the product of the binding constants for SH and RH.[1,35] Binding energies of substituents are most easily estimated from their effect on the second-order rate constants, k_{cat}/K_m, which reflects the free energy difference between initial states and transition state.[1,26,32,41,42]

Bringing two molecules together is entropically unfavourable, but the significance of this entropic contribution to the binding energy of R can be decreased if the difference in the free energy of activation of SH and SR is measured. This is because the tight binding of the small substituent molecule RH to an enzyme requires the loss of its translational and rotational entropy, which makes an unfavourable contribution to the overall free energy change. However, when R is a substituent in the molecule SR, most of this entropy loss is accounted for in the binding of SH because the loss of entropy upon binding SR will be much the same as that upon binding SH. The total translational and rotational entropy of the large molecule SH will be much the same as that for SR.[36,38] The difference in the free energies of binding SR and SH therefore gives an estimation of the true binding energy of R (equations 17–19). The difference, $\Delta G_{SR} - \Delta G_{SH}$, is very much greater than ΔG_{RH}, the observed free energy of binding RH, because it is free from the unfavourable entropy term accompanying the binding of RH. Consequently, just because $\Delta G_{SR} - \Delta G_{SH}$ may give a very favourable negative free energy change for the intrinsic binding energy of R, it would be *incorrect* to conclude that the favourable binding of SR was due *mainly* to the favourable binding of the substituent R. The binding energy of S in SR has been 'used' to compensate for the large loss of translational and rotational entropy upon binding SH and SR to the enzyme. The maximum increase in free energy from the entropy loss of binding is *ca.* 150 kJ mol^{-1} (a factor of 10^8). If the anchor molecule SH is only loosely bound, then the intrinsic binding energy of R in SR may cause a greater restriction of motion in SR so that $T\Delta S_{SR} > T\Delta S_{SH}$ and there will be no change, or even a decrease, in the observed binding constant. Intrinsic binding energies estimated in this way are, therefore, likely to be lower limits.[43]

$$\Delta G_{SR} - \Delta G_{SH} = \Delta H_{SR} - T\Delta S_{SR} - \Delta H_{SH} + T\Delta S_{SH} \tag{17}$$

$$T\Delta S_{SR} \approx T\Delta S_{SH} \tag{18}$$

$$\Delta G_{SR} - \Delta G_{SH} \approx \Delta H_{SR} - \Delta H_{SH} \approx \Delta H_R \tag{19}$$

If the small molecule R—H binds to the enzyme, an estimation of the factor by which S—R binds more favourably compared with that estimated from the binding of S—H and R—H can be made. A comparison of ΔG_{SH} with $\Delta G_{SR} - \Delta G_{RH}$ provides a lower limit to the intrinsic loss of entropy upon binding R—H[37] and indicates the error of estimating K_{SR} from $K_{SH}K_{RH}$ (equations 20 and 21).

$$\Delta G_{SR} - \Delta G_{SH} - \Delta G_{RH} \approx T\Delta S_{RH} + T\Delta S_{SH} - T\Delta S_{SR} \tag{20}$$

$$(K_{RH}K_{SH})/K_{SR} = \text{antilog}(T\Delta S_{RH}/2.303RT) \tag{21}$$

Assuming, as before, that the intrinsic enthalpy of binding S and R approximates to that of S—R, the free energy difference, $\Delta G_{SR} - (\Delta G_{SH} + \Delta G_{RH})$, or $2.303\,RT\,\log(K_{RH}K_{SH}/K_{SR})$, indicates the difference in entropy loss between binding S—H and R—H separately compared with S—R. If the entropy loss upon binding the large molecule S—H is similar to that for binding S—R, then $K_A K_B / K_{AB}$ approximates the entropy loss of binding the residue R. At a standard state of 1 M, the maximum loss of entropy upon binding is about -150 kJ mol^{-1}, and therefore the equilibrium constant for binding S—R to the enzyme may therefore be up to 10^8 greater than that estimated from the product of the binding constants for S—H and R—H.

The most interesting problem to arise from the estimation of intrinsic binding energies is the apparently large values associated with small substituents. Examples are 9–16 kJ mol^{-1} for CH$_2$, 21 kJ mol^{-1} for SMe, 34 kJ mol^{-1} for OH and 23–38 kJ mol^{-1} for SH. These intermolecular interactions are much greater than those generally observed between solute and solvent or between molecules within a crystal.[41,44] Binding energies have also been determined using site-directed mutagenesis.[18]

The difficulty of interpreting binding energies is illustrated by the following example. The addition of a hydrogen-bonding group to the substrate, for example replacement of Me by CH$_2$OH, may increase binding by several kilojoules per mole. There are several interpretations of this observation, which are outlined in equations (22) to (25).[34] Similar arguments could be applied to replacing amino acid residues in the enzyme. The proton donor substrate SOH will initially be hydrogen bonded to a solvent water molecule. The enzyme's proton acceptor site may also be solvated (equation 22) or, conceivably, unsolvated (equation 23). The latter process would release a water molecule upon complexation, which may form a hydrogen bond with the solvent, thus increasing the number of hydrogen bonds from left to right. However, because of the small size of the water molecule, and therefore its ready accessibility to small cavities in the enzyme, an unsolvated enzyme acceptor site is unlikely.[1] When the OH is replaced by H in the substrate, SH may still displace a water molecule from the enzyme when binding occurs (equation 24). The number of hydrogen bonds then remains the same upon complex formation in both equations (22) and (24). In process (24), however, there must be a decrease in entropy because there are three species on the left hand side and only two on the right hand side. The poorer binding of SH compared with SOH could be attributed to either an unfavourable loss of entropy because of water–water hydrogen bonding in the binding of SH (equation 24) or a decrease in the favourable dispersion interactions between the substrate and enzyme because of the empty space in the complex when H replaces OH.

$$\text{S—OH} \cdots \text{OH}_2 + \text{E} \cdots \text{H}_2\text{O} \rightleftharpoons \text{S—OH} \cdots \text{E} + \text{H}_2\text{O} \cdots \text{H}_2\text{O} \tag{22}$$

$$\text{S—OH} \cdots \text{OH}_2 + \text{E} + \text{H}_2\text{O} \rightleftharpoons \text{S—OH} \cdots \text{E} + \text{H}_2\text{O} \cdots \text{H}_2\text{O} \tag{23}$$

$$\text{S—H} + \text{OH}_2 + \text{E} \cdots \text{H}_2\text{O} \rightleftharpoons [\text{SH E}] + \text{H}_2\text{O} \cdots \text{H}_2\text{O} \tag{24}$$

$$\text{S—H} + \text{E} \cdots \text{H}_2\text{O} \rightleftharpoons [\text{SH E} \cdots \text{H}_2\text{O}] \tag{25}$$

Because of the small size of the water molecule it is also conceivable that water remains bound to the enzyme when SH binds (equation 25). In this case also, there is a similar number of hydrogen bonds, left and right, in equations (22) and (25). Now, the poorer binding of SH compared with SOH could be attributed to the unfavourable entropy change of binding SH (equation 25). In the latter, two species combine to give one, whereas in equation (22), the number of entities remains the same when complexation occurs.

5.3.5 MOLECULAR FIT

Both geometrical compatibility and electrical compatibility are required between the enzyme and the substrate if the maximum interaction energy is to be realized. To determine the magnitude of discrimination between substrates that can be achieved by molecular fit, we need to know how the energy of interaction changes with geometry, the magnitude of the binding energy and the accompanying effect upon entropy changes. If a hydrogen bond occurs between the desired substrate and the enzyme with optimal geometry, we would like to know what effect a less favourable geometry will have upon catalysis with another substrate.

Because of the way in which molecular models are made and reaction mechanisms are 'drawn', the differences in the geometry of the ground state and transition state structures are often exaggerated, whereas the actual distances that atoms move to reach a transition state are not much more than those experienced by normal vibrations. For example, at 298 K mean vibrational amplitudes commonly range from about 0.05 to 0.10 A and bending amplitudes of $\pm 10°$ are ubiquitous and may be even greater in weakly bonded systems. The important factor contributing to the increased binding energy of the transition state to the enzyme around the reactant site is that formation of the transition state is often accompanied by large changes in the electron density surrounding the reacting atoms, which provides the increased favourable interaction between the substrate and enzyme. One consequence of this, of course, is that enzymes can often catalyze different reactions which have similar electron density changes. For example, carboxypeptidase A catalyzes the hydrolysis of amides (8) and the enolization of ketones (9).[45] Both reactions involve an increase in electron density on oxygen, which favourably interacts with the electrophilic zinc.

(8)   (9)

To understand the energetics and specificity of enzyme–substrate interactions it is necessary to know the magnitude and geometry of intermolecular and intramolecular forces. It is of interest to know whether the *intermolecular* force field generated by the enzyme is strong enough to overcome the conformational and geometrical dictates of the *intramolecular* force field of the substrate. Can the enzyme distort the substrate or *vice versa*? How much does it cost in energy to change the ground state arrangement of atoms in the substrate and enzyme?

The forces controlling the distortion of a molecule may be partitioned into the following contributions: (i) bond stretching; (ii) bond angle bending; (iii) torsional effects; (iv) attractive and repulsive nonbonded interactions; (v) electrostatic interactions such as dipole–dipole and polar effects. Unfortunately, there is by no means universal agreement upon the values of the parameters to be used in the quantitative estimation of these effects, and there is a tendency to treat them as adjustable parameters (see Volume 4).[38]

Deformation of bond lengths is very difficult and rarely occurs in 'normal' molecules. Bond angle deformation is fairly easy; for example a 10° change in $\hat{CCC}$ costs about 11 kJ mol^{-1}, and this is the pathway commonly used to relieve nonbonded interaction strain in a molecule. Torsional energy is the 'softest' of all the potential energy terms, and hence distortion of dihedral angles is relatively easy; it generally costs relatively little energy to change conformation of molecules. Probably the most important, but the least understood, energy function is that describing nonbonded interactions and consequently there are a variety of functional forms used to describe this interaction. The calculation of meaningful nonbonded interaction energies is beset by a number of complications: unlike the free atoms, those in molecules do not possess spherical symmetry; the effective dielectric constant of the molecule may influence the transmission of the forces involved, and the calculations apply to the gas phase and in solution the attractive part of the nonbonded interaction would be decreased by the solvent.

A simplified approach to estimate electrostatic interactions is to use partial charges on the individual atoms, obtained from group dipole moments, and to calculate the electrostatic interaction by Coulomb's law (equation 26) as a function of the distance (r, Å) between the partial charges (q) expressed in terms of the electronic charge, in a medium of dielectric constant D. On a qualitative basis, a system of alternative positive and negative partial charges imparts stability to a molecule, while destabilization is associated with adjacent like charges. The use of Coulomb's law should be regarded as a purely empirical procedure, since when two partial charges are not well separated, the solvent molecules and the rest of the solute between and around the two charges do not behave like a continuous medium of constant dielectric constant, and it is also difficult to know where the point dipoles should be located. For two partial charges separated by greater than one width of water layer it has been suggested that the effective dielectric constant approaches that of bulk water, *i.e.* 80; hence electrostatic interactions would be negligible at these distances.

$$E_{el} = 1389q_1q_2/Dr \quad \text{kJ mol}^{-1} \tag{26}$$

Attractive *intermolecular* interactions and the noncovalent forces that determine the geometry of molecular complexes are usually interpreted using classical and semiclassical pictures.[46] The interaction energy ΔE is partitioned as in equation (27), where the individual components have the following physical significance. E_{es}, the electrostatic energy, results from the interaction between charge distribution on one molecule with that of another. This contribution includes the interactions of all permanent charges and multipoles such as dipole–dipole, dipole–quadrupole *etc.* and may be either attractive or repulsive. E_{pol}, the polarization interaction, is the effect of the distortion of the electron distribution of one molecule by another and the higher order coupling resulting from such distortions. This component includes the interactions between all permanent charges and induced multipoles and is always an attractive interaction. E_{ex}, the exchange repulsion, is the interaction caused by exchange of electrons between two molecules. Physically, this is the short range repulsion due to overlap of the electron clouds. E_{ct}, charge transfer or electron delocalization interaction, is the interaction caused by charge transfer from occupied molecular orbitals of one molecule to vacant MOs of another. E_{disp} is the dispersion energy, which is the second-order attraction between fluctuating charges in different molecules and is relatively unimportant for interactions between polar molecules.

$$\Delta E = E_{es} + E_{pol} + E_{ex} + E_{ct} + E_{disp} \tag{27}$$

The factors determining the equilibrium geometry of a complex depend upon its nature but the electrostatic interaction is often the principal factor. For example, in hydrogen bonding, with a small distance separating the proton donor and acceptor, E_{es}, E_{ct} and E_{pol} can all be important attractive components competing against a large E_{ex} repulsion. At longer distances, for the *same complex*, the short range attractive forces, E_{ct} and E_{pol}, are usually unimportant and E_{es} is the only important attraction. However, the importance of an individual component depends on the type of hydrogen bonding, but the electrostatic term is dominant in 'normal' hydrogen bonds between neutral electronegative atoms.[46]

Two aspects of the geometry of hydrogen bonding are important in determining the specificity of enzyme–substrate interactions. Can the intermolecular force field generated by hydrogen bonding override the intramolecular force field determining the geometry of the substrate? How does the energy of intermolecular hydrogen bonding change with perturbations from optimal geometry? The distance between the electronegative atoms in hydrogen bonding is important, but not critical. The $O \cdots O$ and the $O \cdots N$ distances in hydrogen-bonded systems vary between 2.4 and 3.0 Å and 2.65 and 3.15 Å, respectively.[1] The force constant for $O \cdots O$ stretching in ice is 121 kJ mol^{-1} Å^{-1} and it is easier to stretch a hydrogen bond than it is to distort a $C\hat{C}C$ bond angle. A quadratic probably overemphasizes the energy dependence upon distance for increasing the distance from the minimum; theoretical calculations indicate that, for example, $O \cdots O$ distance can vary ± 0.3 Å for less than 5 kJ mol^{-1}.[1]

Although it is commonly thought that 'normal' hydrogen bonds are linear, deviations from linearity are ubiquitous, especially in solids. An examination of the distribution of $X-\hat{H} \cdots Y$ angles shows that deviations by 15° from linearity occur as frequently as strictly linear hydrogen bonds. The $O-\hat{H} \cdots O$ and $N-\hat{H} \cdots O$ angles in intermolecular hydrogen-bonded systems vary from 140 to 180° and from 130 to 180°, respectively. Theoretical calculations show that large changes in angle correspond to only 1 or 2 kJ mol^{-1}; for example $\pm 30°$ for $O-\hat{H} \cdots O{=}C$ costs *ca.* 3 kJ mol^{-1}. Experimental and computational data do not support the hypothesis that the hydrogen bond necessarily lies along the maximum of the electron donor lone pair.[42]

Distortion from the optimum hydrogen bond geometry is easy for changes in orientation and only slightly less facile for changes in distance. Conversely, the intermolecular force field generated by a hydrogen bond seems unlikely to be able to cause major distortions, except for conformational changes, in the geometry of the substrate.

The electrostatic energy dominates the medium to large range interaction between two molecules and it seems reasonable therefore to emphasize its importance. A promising lead in this direction is the use of electrostatic molecular potentials, which simulates the coulombic contribution to the intermolecular interaction by substituting appropriately chosen point charges, which may be fractional, to replace actual neighbouring molecules.[46]

The potential is a function of the electronic distribution and the position of the nuclei in the molecule and is computed from molecular wave functions. It is the basic premise of the use of electrostatic potentials that E_{pol}, E_{ct} and E_{disp} are much less important than electrostatic effects in determining structural features of complex formation. Formally, the electrostatic potential is the value of the interaction between a molecule and a point charge placed at a given distance. Similar chemical groups have portions of space where the shape of the electrostatic potential is reasonably similar. A general property of electrostatic potentials is that different atoms with different local geometries can yield the same value and thus, presumably, the same binding energy. This could be important in the design of enzyme inhibitors and of drugs in general. The combined action of several atoms bound together is not at all apparent from a 'ball and stick' model but is an automatic result of electrostatic potentials.

Considerable electrostatic interaction can arise from the dipole moment of 3.5 D for the peptide unit. In an α helix the peptide dipole moments are aligned nearly parallel to the helix axis, giving rise to a significant electric field. Its dipole runs from the C terminal to the positive end of the N terminal.[47] For example, the potential at 5 Å from the N terminus of an α helix of 10 Å length is *ca.* 0.5 V so that, ignoring the effect of solvent, a negatively charged group at that position involves an attractive energy of *ca.* 50 kJ mol^{-1}.

Ideas about the enzyme being complementary to the transition state structure, particularly with respect to the design of transition state analogues, have tended to be dominated by geometry. Based on the previous discussions, it is apparent that it is the electrostatic interactions between the transition state and the enzyme, rather than geometry, which are crucial in determining the binding energy.

Ideas about the flexibility of enzymes are often dominated by one's preference for space-filling or skeletal models. In order to bring about the maximum loss of entropy of substrates, and provide optional alignment of substrates with catalytic groups at the active site, it is desirable that the enzyme be as rigid as possible. This does not mean that groups within a protein cannot undergo internal rotation; in fact there is much evidence, particularly from NMR studies, that the interior of some proteins is mobile and that some proteins 'breathe', *i.e.* the protein as a whole undergoes some vibronic motion. Part of the protein may be mobile, while another part is not, and these parts may be of diverse sizes for different proteins.

5.3.6 REFERENCES

1. M. I. Page, in 'The Chemistry of Enzyme Action', ed. M. I. Page, Elsevier, Amsterdam, 1984, p. 1.
2. C. Grathwohl and K. Wüthrich, *Biopolymers*, 1976, **15**, 2025.
3. J. N. Spencer, *et al.*, *J. Phys. Chem.*, 1981, **85**, 1236.
4. A. A. Zamyatnin, *Prog. Biophys. Mol. Biol.*, 1972, **24**, 107.
5. C. Chothia, *J. Mol. Biol.*, 1975, **105**, 1.
6. M. H. Klapper, *Biochem. Biophys. Res. Commun.*, 1977, **78**, 1018.
7. J. S. Zigler and J. D. Goosey, *Trends Biochem. Sci.*, 1981, **6**, 133.
8. J. Kay, *Biochem. Soc. Trans.*, 1978, **6**, 789.
9. H. Holzer and P. C. Heinrich, *Annu. Rev. Biochem.*, 1980, **49**, 63.
10. R. L. Baldwin, *Trends Biochem. Sci.*, 1986, **11**, 6.
11. P. S. Kim and R. L. Baldwin, *Annu. Rev. Biochem.*, 1982, **51**, 459; see, however, S. C. Harrison and R. Durbin, *Proc. Natl. Acad. Sci., USA*, 1985, **82**, 4028.
12. G. E. Schulz and R. H. Schirmer, 'Principles of Protein Structure', Springer, New York, 1979, p. 95.
13. C. Chothia and A. M. Lesk, *Trends Biochem. Sci.*, 1985, **10**, 116.
14. C.-L. Tsou, *Trends Biochem. Sci.*, 1986, **11**, 427.
15. M. G. Rossman and P. Argos, *Annu. Rev. Biochem.*, 1981, **50**, 497.
16. T. W. Traut, *Mol. Cell. Biochem.*, 1986, **70**, 3.
17. D. B. Wetlaufer, *Adv. Protein Chem.*, 1981, **34**, 61.
18. J. R. Knowles, *Science (Washington, D.C.)*, 1987, **236**, 1252.
19. K. Wüthrich and G. Wagner, *Trends Biochem. Sci.*, 1984, **9**, 152.
20. A. Warshel, *J. Phys. Chem.*, 1979, **83**, 1640; J. T. Edsall and H. A. McKenzie, *Adv. Biophys.*, 1983, **16**, 53.
21. F. M. Richards, *Annu. Rev. Biophys. Bioeng.*, 1977, **6**, 151.
22. J. Gavezzotti, *J. Am. Chem. Soc.*, 1985, **107**, 962.
23. M. L. Connolly, *Science (Washington, D.C.)*, 1983, **221**, 709; M. L. Connolly, *J. Am. Chem. Soc.*, 1985, **107**, 1118.
24. W. P. Jencks, *Mol. Biol., Biochem. Biophys.*, 1980, **32**, 3.
25. A. R. Fersht and H. Kaethner, *Biochemistry*, 1976, **15**, 3342.
26. W. P. Jencks, *Adv. Enzymol.*, 1975, **43**, 219.
27. A. R. Fersht, *Proc. R. Soc. London, Ser. B*, 1974, **187**, 397.

28. W. J. Albery and J. R. Knowles, *Biochemistry*, 1976, **15**, 5631.
29. T. N. C. Wells and A. R. Fersht, *Biochemistry*, 1986, **25**, 1881.
30. L. Pauling, *Chem. Eng. News*, 1946, **24**, 1375.
31. E. H. Serpersu, D. Shortle and A. S. Mildvan, *Biochemistry*, 1987, **26**, 1289.
32. A. R. Fersht, 'Enzyme Structure and Mechanism', 2nd edn., Freeman, San Francisco, 1984.
33. M. I. Page, in 'Enzyme Mechanisms', ed. M. I. Page and A. Williams, Royal Society of Chemistry, London, 1987.
34. M. I. Page, *J. Mol. Catal.*, 1988, **47**, 241.
35. M. J. S. Dewar and D. M. Storch, *Proc. Natl. Acad. Sci. USA*, 1985, **82**, 2225.
36. M. I. Page and W. P. Jencks, *Proc. Natl. Acad. Sci. USA*, 1971, **68**, 1678.
37. M. I. Page, *Angew. Chem., Int. Ed. Engl.*, 1977, **16**, 449.
38. M. I. Page, *Chem. Soc. Rev.*, 1973, **2**, 295.
39. R. P. Bell, J. E. Critchlow and M. I. Page, *J. Chem. Soc., Perkin Trans 2*, 1974, 66.
40. S. G. Cohen, V. M. Vaidya, and R. M. Schultz, *Proc. Natl. Acad. Sci. USA*, 1970, **66**, 249.
41. W. P. Jencks, *Proc. Natl. Acad. Sci. USA*, 1981, **78**, 4046.
42. M. I. Page, in 'Accuracy in Molecular Processes', ed. T. B. L. Kirkwood, R. F. Rosenberge and D. J. Galas, Chapman and Hall, London, 1986, p. 37.
43. W. P. Jencks and M. I. Page, *Proc. 8th FEBS Meeting Amsterdam*, 1972, **29**, 45.
44. A. R. Fersht, *Trends Biochem. Sci.*, 1987, **12**, 301.
45. T. E. Spratt and E. T. Kaiser, *J. Am. Chem. Soc.*, 1983, **105**, 3679.
46. P. Kollman, in 'The Chemistry of Enzyme Action', ed. M. I. Page, Elsevier, Amsterdam, 1984, p. 55.
47. W. G. T. Hol, P. T. van Duijnen and H. J. C. Berendsen, *Nature (London)*, 1978, **273**, 443.
48. T. E. Creighton, 'Proteins', Freeman, New York, 1984, p. 142.

5.4
Enzyme Inhibition

MICHAEL I. PAGE
Huddersfield Polytechnic, UK

5.4.1	INTRODUCTION	61
5.4.2	REVERSIBLE INHIBITORS	62
5.4.3	MULTISUBSTRATE ANALOGUES	63
5.4.4	TRANSITION STATE ANALOGUES	66
5.4.4.1	*Glycosidase Inhibitors*	68
5.4.4.2	*Serine Protease Inhibitors*	69
5.4.4.3	*Metalloprotease Inhibitors*	71
5.4.4.4	*Aspartic Protease Inhibitors*	72
5.4.4.5	*Slow Binding of Inhibitors*	74
5.4.5	IRREVERSIBLE INHIBITION/INACTIVATION	74
5.4.6	ACTIVE-SITE-DIRECTED IRREVERSIBLE INHIBITORS	76
5.4.7	MECHANISM-BASED IRREVERSIBLE INHIBITORS	79
5.4.8	REFERENCES	84

5.4.1 INTRODUCTION

Enzymes are frequent targets for drug design. The inhibition of an enzyme-catalyzed reaction can make cell growth, division and viability untenable and interrupt major metabolic pathways and hinder the formation of an essential or undesirable metabolite. A desired biological response may be obtainable by only reversibly and partially inhibiting the action of an enzyme. Conversely, inhibition may have to be essentially irreversible to achieve a pharmacological goal. There are some physiological and immunological concerns that make irreversible inhibition of a target enzyme belonging to the host, as opposed to an invading organism, undesirable. However, it makes little difference whether 'irreversible' inhibition is achieved by covalent modification of the enzyme or by the tight binding of a noncovalent inhibitor.

This chapter reviews the important basic concepts of enzyme inhibition and ignores the complications of absorption, transport, metabolism, excretion and toxicological problems associated with the development of an enzyme inhibitor as a successful drug.

The key to understanding enzyme catalysis is the recognition of the importance of the favourable binding interactions between the *nonreacting* parts of the substrate and the enzyme.[1-3] It is generally agreed that this interaction reaches a maximum in the transition state, but it is equally important for catalysis that this stabilization is not expressed in the binding of substrates, intermediates or products.[4,5] These favourable binding energies are either used to compensate for unfavourable interactions and entropy loss upon binding, and therefore do not result in tight binding of substrates, intermediates or products, or they are apparent exclusively in the transition state.

The binding interactions between the nonreacting parts of the substrate and enzyme away from the reaction centre are equally important for enzyme catalysis and for efficient enzyme inhibition. However, whereas for catalysis it is important that these favourable interactions do not result in tight binding of substrate, intermediate or product, for inhibition it would be advantageous if they do result in stabilization of the enzyme–inhibitor complex. These binding energies are equally important for noncovalent, transition state analogue, multisubstrate analogue or mechanism-based inhibitors.

5.4.2 REVERSIBLE INHIBITORS

The kinetics of the reversible inhibition of enzymes is that the enzyme–inhibitor complex (EI) regenerates the inhibitor (I) and enzyme (E; equation 1). This neither excludes bond formation between the enzyme and inhibitor nor indicates that the complex necessarily decomposes rapidly. The covalent bond may be relatively unstable and the enzyme therefore rapidly regenerated. Conversely, a tight noncovalent complex may be relatively stable; for example, if the equilibrium constant for dissociation of EI is 10^{-13} M and the forward rate of association is diffusion controlled at 10^8 M^{-1} s^{-1} then the half-life for EI is 19 h.

$$\text{E} + \text{I} \underset{k_{-1}}{\overset{k_1}{\rightleftharpoons}} \text{EI} \tag{1}$$

In general, however, reversible inhibition is associated with noncovalent bond formation. Consequently a characteristic of such enzyme–inhibitor complex formation is that the onset of inhibition shows no time dependence under normal assay conditions. Furthermore, since the interaction is reversible, inhibition can be reversed by dialysis, gel filtration or dilution to reduce the inhibitor concentration. Under assay conditions the concentration of the inhibitor is usually greater than that of the enzyme and the fraction of free enzyme is given by equation (2), where K_i is the dissociation constant of EI. Therefore, if the concentration of I equals K_i, half of the enzyme is complexed with the inhibitor and dilution of this solution tenfold would give a solution in which only 9% of the enzyme is complexed.

$$[\text{E}]/([\text{E}] + [\text{EI}]) = K_i/(K_i + [\text{I}]) \tag{2}$$

Reversible inhibitors may be classified by their effects on the *kinetic* parameters of an enzyme-catalyzed reaction but there are often a variety of *mechanisms* compatible with the kinetic observations.[6] There is often confusion between the experimental observations and their interpretation. Consequently it is less ambiguous to define inhibitors kinetically rather than mechanistically. For an enzyme-catalyzed reaction exhibiting Michaelis–Menten kinetics, K_m is *defined* as the substrate concentration which gives half the maximum rate at saturation, V_{max}. The Michaelis constant is not necessarily the dissociation constant of the enzyme–substrate complex. Hence the measured or apparent K_m value, determined from, say, a double reciprocal plot, may change in the presence of inhibitor but the true dissociation constant of the enzyme–substrate complex may remain unchanged.

For example, competitive inhibition is characterized kinetically by no change in V_{max} but an increase in the observed K_m because in the presence of inhibitor a greater concentration of substrate is required to reach half of V_{max}. There are several interpretations of this behaviour. A simple kinetic scheme is shown in equation (3) and the initial rate of conversion of the substrate S to product P is given in equation (4). The right half of the denominator is the observed K_m term which therefore changes as the balance between 1 and the concentration of I changes. The dissociation constant K_s does not change. If the concentration of I equals K_i then K_m is doubled from its measured value in the absence of inhibitor. This does not mean that the initial rate is halved, except at low concentrations of substrate, where the enzyme is far from saturated with substrate. Conversely, at high concentrations of substrate, where the enzyme is saturated with substrate, such a concentration of inhibitor would have no effect on the initial rate.

$$\text{EI} \overset{K_i}{\rightleftharpoons} \text{E} + \text{S} \overset{K_s}{\rightleftharpoons} \text{ES} \overset{k_{cat}}{\longrightarrow} \text{P} \tag{3}$$

$$v = \frac{k_{cat}[\text{E}][\text{S}]}{[\text{S}] + K_s\{1 + ([\text{I}]/K_i)\}} = \frac{V_{max}[\text{S}]}{[\text{S}] + K_s\{1 + ([\text{I}]/K_i)\}} \tag{4}$$

A simple mechanism compatible with the kinetics of competitive inhibition is the binding of the substrate and inhibitor at the *same* site in the enzyme, *i.e.* binding of substrate and inhibitor is mutually exclusive. However, it is equally easy to visualize a situation where binding of the inhibitor at a site other than the active site causes a conformational or other change in the enzyme, which in turn prohibits the normal binding of the substrate. The first mechanism has given rise to the substrate analogy for competitive inhibitors which bind to the active site because they are similar in shape and electron distribution to the substrate. However, there are many cases of competitive inhibition where the inhibitor bears no structural resemblance to the substrate. Furthermore, there are relatively few cases where binding of the substrate and inhibitor has been shown unambiguously to occur at the same site.

Salicylate is a competitive inhibitor of NADH in various dehydrogenases[7] and of ATP in various kinases.[8] In these cases, X-ray crystallographic studies have shown that salicylate binds to the adenine site in liver alcohol dehydrogenase and adenylate kinase.[8]

The classic example of a competitive inhibitor which is a substrate analogue is sulfanilamide, which forms a tight complex with the enzyme 7,8-dihydropteroate synthetase with a dissociation constant of *ca.* 1×10^{-7} M.[9] One of the substrates for this folate synthesis is 4-aminobenzoic acid, which has a structural resemblance to sulfonamides (1). There is a significant correlation between the logarithm of the minimum inhibitory concentration and the pK_a of the sulfonamide and the chemical shift of the amino group protons of the precursor anilines. The anion of the sulfonamide is the active form of the inhibitor and these correlations are indicative of the importance of the electron density at the sulfonamide group (2) and its interaction with the enzyme (see Chapter 7.1).[9] However, under certain conditions, the sulfonamides act as dead-end substrates.

$$\text{H}_2\text{N}\!-\!\!\underset{}{\bigcirc}\!\!-\!\!\overset{\displaystyle O}{\underset{\displaystyle O}{\overset{\|}{\underset{\|}{S}}}}\!-\!\text{NHAr} \qquad\qquad \text{H}_2\text{N}\!-\!\!\underset{}{\bigcirc}\!\!-\!\!\overset{\displaystyle O}{\underset{\displaystyle O}{\overset{\|}{\underset{\|}{S}}}}\!-\!\overset{-}{\text{N}}\text{Ar}$$

(1)   (2)

Sulfonamides also inhibit the interconversion of CO_2 and HCO_3^- catalyzed by the zinc metalloenzyme carbonic anhydrase[10] and are used clinically as diuretics. One of the isozymes is extraordinarily efficient, showing k_{cat}/K_m near the diffusion limit. The mechanism is thought to involve direct nucleophilic attack of zinc-bound hydroxide on carbon dioxide to give a zinc-bound bicarbonate ion. Displacement of bicarbonate gives a zinc-bound water, which transfers a proton to buffer by an intermolecular process or intramolecularly to a neighbouring histidine to regenerate the catalyst.[11]

The binding of inhibitors such as sulfonamides and anions takes place at the zinc of carbonic anhydrase. The sulfonamide binds as its anion (2) and is coordinated to the metal by the sulfonamide nitrogen.[10] This is another interesting case of an inhibitor showing some resemblance to substrate/product structure (the bicarbonate anion), but obviously the inclusion of the aromatic residue would be difficult to predict *a priori*. Simple anions such as NCO^-, N_3^- and I^- that inhibit the carbon dioxide hydration activity of carbonic anhydrase bind to the zinc ion either by displacing a coordinated water or by expansion of the coordination sphere.[12] The binding of these small anions, however, is weak and is formally competitive with hydroxide ion for the enzyme in its high pH form.[12] Below pH 7, inhibition appears to be noncompetitive since the anions only affect k_{cat} and not K_m. At pH 9, inhibition appears to become uncompetitive since k_{cat} and K_m are affected to the same extent, leaving the ratio k_{cat}/K_m unchanged.[13] This unusual mode of inhibition is usually attributed to the formation of a ternary complex in which the inhibitor only binds in the presence of the substrate. This does not necessarily mean that substrate and inhibitor bind simultaneously to the active site, but could be a consequence of rate-limiting deprotonation of zinc-bound water and reflect diverse kinetic pathways of inhibitor binding.[14]

The binding energy of substrates to enzymes generally comprises contributions from both the 'nonreacting' or recognition sites and the catalytic site. The latter undergo electrical and geometrical changes in structure on going to the transition state. Although these interactions between the substrate and enzyme are important, and give rise to the development of transition state analogues as inhibitors, the contribution of favourable interactions from the 'nonreacting' subsites can be enormous.[1] It is these interactions which are often used to compensate for unfavourable entropy, solvation and charge effects.[1, 2, 3] The inclusion of these recognition sites in a molecule which does not require the unfavourable energy changes necessary for substrate catalysis should give rise to good inhibitors.

5.4.3 MULTISUBSTRATE ANALOGUES

Some enzymes catalyze reactions between two substrates and proceed by the formation of a ternary complex in which both substrates are bound to the enzyme. Compounds which are structural analogues of both substrates can be effective inhibitors with K_i values much lower than the dissociation constants of either substrate from the ternary complex and also lower than the product of both dissociation constants of the two substrates.[15] A similar situation obviously exists for analogues of multiproduct reactions.

The bringing together of molecules is entropically unfavourable.[1-3] The free energy of binding two substrates or products to an enzyme has a large unfavourable contribution from the negative entropy change accompanying the loss of translational and rotational degrees of freedom. If the favourable binding energy between the enzyme and two substrates or two products is available to a bisubstrate or biproduct analogue then such compounds could bind up to 10^8 M tighter than suggested by the product of the two separate binding constants.[1-3] It is possible to express the efficiency of binding multisubstrate analogues using the effective molarity concept used for intramolecular reactions.[16] The binding of two substrates/products X and Y (equation 5) and an analogue inhibitor X–Y, in which the two fragments are covalently linked (equation 6), may be described by the free energy changes ΔG_s and ΔG_i respectively (equations 7 and 8). This simple treatment assumes that the enthalpy changes may be partitioned into group contributions and that the interaction between the enzyme and X and Y is similar whether X and Y are separate or covalently linked. The large difference in free energy between these two systems comes from the entropy change. The dissociation of the ternary complex in equation (5) generates three molecules, whereas the binary complex in equation (6) gives only two. The largest contribution to this entropy change is the gain of translational and rotational freedom but the entropies of these motions are not very dependent on the mass and size of the molecule.[16] Consequently the maximum difference in the free energies of dissociation of the ternary and binary complexes based only on entropy differences is $150\,kJ\,mol^{-1}$, a factor of 10^8 M.[17] The ratio of the dissociation constants K_s/K_i has units of concentration and is a measure of the effectiveness of the bisubstrate analogue.

$$X + Y + Enz \underset{K_s}{\rightleftharpoons} X \cdot Y \cdot Enz \tag{5}$$

$$X—Y + Enz \underset{K_i}{\rightleftharpoons} X—Y \cdot Enz \tag{6}$$

$$\Delta G_s = \Delta H_{X,E} + \Delta H_{Y,E} - T\Delta S_{X,E} - T\Delta S_{Y,E} \tag{7}$$

$$\Delta G_i = \Delta H_{XY,E} - T\Delta S_{XY,E} \tag{8}$$

$$\Delta G_s - \Delta G_i \approx T\Delta S_{X,E} \approx 150\,kJ\,mol^{-1} \tag{9}$$

A design problem based on this hypothesis is that by definition the two fragments are on either side of the reacting site. It is therefore difficult to synthesize a bisubstrate or biproduct analogue

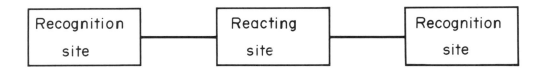

which has the correct geometrical relationship between the two subsites. The maximum binding energy between the enzyme and substrate is expressed in the transition state but it may often be that only one side of the bisubstrate or biproduct inhibitor binds correctly to the enzyme because geometrical constraints may prevent both sides from binding simultaneously. Consequently, the examples of this type of inhibitor do not often show large ratios of $K_X K_Y/K_{XY}$, which are generally less than 100 (expressed as $K_m^X K_m^Y/K_i$ in Table 1).

The compound *N*-phosphonoacetyl-L-aspartate (**3**) is similar in structure to a combination of the two substrates of aspartate transcarbamylase, L-aspartate (**4**) and carbamoyl phosphate (**5**). It is one of the most potent reversible inhibitors of this enzyme, showing a dissociation constant of 3×10^{-8} M. However, this tight binding appears to owe little to entropic factors because the dissociation constant is similar to the products of those for carbamoyl phosphate (2.7×10^{-5} M) and succinate (9×10^{-4} M) or aspartate (1.2×10^{-4} M).[18]

The potent inhibition of adenylate kinase by P^1-(adenosine-5′)-P^5-(adenosine-5′)-pentaphosphate (**6**)[19,20] could result from there being two nucleotide binding sites on the enzyme and is consistent with an associative mechanism for phosphoryl transfer.[23] Both adenosine residues of the multisubstrate analogue are correctly bound to the sites normally occupied by the substrates ATP and

Table 1 Multisubstrate Analogue Inhibitors

Enzyme	Substrates	K_m (M)	Inhibitor	K_i (M)	$K_m^x K_m^y / K_i$ (M)	Ref.
Aspartate transcarbamylase	Aspartate (4) Carbamyl phosphate (5)	1.2×10^{-4} 2.7×10^{-5}	N-Phosphonacetyl L-Aspartate (3)	2.7×10^{-8}	11	18
Adenylate kinase	Mg ATP AMP	1.0×10^{-4} 6.3×10^{-4}	P^1-(Adenosine-5')-P^5-adenosine-5'-pentaphosphate (6)	2.5×10^{-9}	25	19
Lactate dehydrogenase	NADH Pyruvate	2.4×10^{-5} 1.4×10^{-4}	$NAD\text{-}CH_2COCO_2^-$	$<1 \times 10^{-9}$	>3	22
Carnitine acetyl transferase	Acetyl CoA Carnitine	3.4×10^{-5} 1.2×10^{-4}	$Me_3\overset{+}{N}CH_2CH(CH_2CO_2^-)_2$ $^-OCOCH_2SCoA$	$<1.2 \times 10^{-8}$	>0.3	21
Spermidine acetyl transferase	Acetyl CoA Spermidine	1.5×10^{-6} 1.0×10^{-3}	$RNHCOCH_2SCoA$	$<1 \times 10^{-8}$	>0.2	25
Pyridoxamine-pyruvate transaminase	Pyridoxamine Alanine	3.1×10^{-5} 2.0×10^{-3}	Pyridoxylalanine	1.8×10^{-7}	0.3	97

$$\underset{\text{CO}_2^-}{\overset{\text{O}}{\underset{|}{\overset{\|}{\text{C}}}}}-\text{CH}_2\text{PO}_3^{2-}$$

$$^-\text{O}_2\text{CCH}_2-\underset{|}{\overset{|}{\text{CH}}}$$

$$\overset{\text{NH}}{\underset{|}{\text{NH}}}$$

$$^-\text{O}_2\text{CCH}_2-\underset{\text{CO}_2^-}{\overset{|}{\underset{|}{\text{CH}}}}$$

(3)

$$\overset{\text{NH}_2}{\underset{|}{}}$$
$$^-\text{O}_2\text{CCH}_2-\underset{\text{CO}_2^-}{\overset{|}{\underset{|}{\text{CH}}}}$$

(4)

$$\text{H}_2\text{N}\overset{\overset{\text{O}}{\|}}{\text{C}}-\text{OPO}_3^{2-}$$

(5)

$$\text{ADP}-\text{O}\cdots\overset{\overset{\text{O}}{\|}}{\text{P}}\cdots\text{O}-\text{AMP}$$

(6)

$$\text{ADP}-\text{O}---\overset{\overset{\text{O}}{\|}}{\text{P}}---\text{O}-\text{AMP}$$

(7)

AMP. The analogue contains an extra phosphoryl residue compared with the combined substrates but this is presumably required to accommodate the longer axial bonds and the linear geometry of the two substituents in the pentacoordinate intermediate formed in the normal reaction pathway (7).

These two examples highlight the difficulties of distinguishing between the effectiveness of an inhibitor as a multisubstrate analogue or as a transition state analogue.[3,24] However, as has been emphasized continually, much of the binding energy between the inhibitor and enzyme comes from interactions at subsites away from the reactive site and even an ideal transition state analogue inhibitor is unlikely to owe its high affinity to those interactions solely related to the reactive centre.

5.4.4 TRANSITION STATE ANALOGUES

The interaction energy between a substrate and enzyme is greater in the transition state than in the initial state. This differential binding has led to the suggestion that compounds resembling the transition state structure in geometry and electrical charge distribution should bind tightly to the enzyme.[24,26,27] Examples of transition state analogues acting as enzyme inhibitors, reported to date, bind 10^2 to 10^5 times more tightly than substrates of the same enzyme. However, although the concept of such analogues is a very useful tool, it should not dominate, too rigorously, the design of potential inhibitors. Firstly, the specificity of most enzyme-catalyzed reactions is such that there is rarely a *unique* transition state for all reactions catalyzed. There will be changes in transition state structure or even changes in rate-limiting step with changing substrate structure. Secondly, the stabilization of transition states relative to ground states in reactions is accompanied by electron density changes as a result of bond making and breaking. An analogue inhibitor that already possesses the geometrical and electrical features of the substrate in the transition state may be substantially more stabilized in water compared with the substrate in the initial state.

There is a similarity in the perceptions of enzyme specificity and the binding of inhibitors. Unfortunately, the specificity of enzymes is an ambiguous term. It is well known that some enzymes can not only catalyze a particular reaction of a variety of class types but also different types of reactions.[28] This degree of flexibility must also be reflected in a variety of inhibitor structures.

A problem with the different perceptions of specificity is that they indirectly influence our perception of enzymes. Transition state stabilization is often interpreted in terms of the enzyme stabilizing a transition state structure of the substrate. The enzyme is sometimes seen as a rigid immutable state with the correct distribution of atoms to provide the perfect complementarity in shape and charge to the 'transition state' of the *substrate*. The transition state represents the combined state of enzyme and substrate. Instead of describing the enzyme as stabilizing the substrate one could equally well describe the substrate as stabilizing the enzyme.[1,2,17] The free energy of activation, measured by k_{cat}/K_m, includes the energy difference between the isolated enzyme in its initial state and the enzyme substrate complex in the transition state. The efficiency of enzyme catalysis depends on differences (see Chapter 5.3). Although there may be a positive charge on the enzyme which is mutually stabilized by the generation of a negative charge on the substrate in the transition state, this charge will almost certainly be 'neutralized' in the isolated enzyme by a neighbouring part of the protein or by the 'solvation shell' provided by the rest of the enzyme and water. The rate of reaction is controlled by the *difference* in free energy provided by these stabilizations in the initial and transition states.[28] Similarly, the dissociation constant of an enzyme inhibitor complex is the result of free energy *differences* between initial and complex states.

For example, an enzyme-catalyzed reaction of a neutral substrate, A—B, may involve protein stabilization of positive and negative charges generated in the transition state of the substrate (equation 10). A transition state analogue, $^+$X—Y$^-$, may therefore be synthesized carrying the appropriate charges in anticipation that this should bind tightly to the enzyme (equation 11). The solvation of the neutral substrate, A—B, could be very weak and therefore the free energy *change* of interaction between the charged transition state and the enzyme could be enormous. Conversely, the charged inhibitor will be highly solvated and consequently the free energy of transfer from solvent to enzyme is a much less favourable process than that accompanying the *reaction* of the substrate (equation 10). It is therefore unlikely that a perfect noncovalent transition state analogue inhibitor can be made which takes maximum advantage of the active site binding energy in the transition state of the enzyme-catalyzed reaction. Covalent bond formation between the inhibitor and enzyme may, however, generate an entity with an electron density distribution which is different from the initial state but resembles the transition state. Whether such inhibitors should be called transition state analogues or mechanism-based inhibitors is a matter of nomenclature, not understanding.

$$\text{Substrate (A—B)}_{\text{solv}} + \begin{smallmatrix}+\\-\end{smallmatrix}\overline{}\begin{smallmatrix}-\\+\end{smallmatrix}\text{ Enz} \longrightarrow \overset{+}{\text{A}}\text{------}\overset{-}{\text{B}}\ \ \text{Enz}^{\neq} \qquad (10)$$

$$\text{Inhibitor }(\overset{+}{\text{X}}\text{—}\overset{-}{\text{Y}})_{\text{solv}} + \begin{smallmatrix}+\\-\end{smallmatrix}\overline{}\begin{smallmatrix}-\\+\end{smallmatrix}\text{ Enz} \longrightarrow \overset{+}{\text{X}}\text{———}\overset{-}{\text{Y}}\ \ \text{Enz} \qquad (11)$$

Enzymes must stabilize all intermediate states between reactants and products to be successful catalysts.[1,2] Stabilization of the transition state alone is not a sufficient criterion for catalysis. However, over-stabilization of intermediates, of course, leads to inhibition. There must therefore be a fine balance of interactions between substrate and enzyme along the whole reaction coordinate. If these interactions were not available, catalysis would not occur and changes in substrate structure, leading to changes in rate-limiting step and transition state structure, would give enzymes which were highly specific for one substrate only. This is rarely the case.

In nonenzyme-catalyzed reactions the concept of changing transition state structure with changes in substituents or reaction conditions is well accepted. An extreme occurs when there is a change in rate-limiting step from breakdown to formation of an intermediate, *e.g.* k_2 to k_1 in equation (12). The intermediate may have a very different charge distribution and geometry from reactants and products. However, even within *one* of these steps, changes in transition structure are discernible, reflecting greater or lesser degrees of bond making or breaking. These different transition state structures may have different charge distributions and geometries. A similar situation exists for enzyme-catalyzed reactions. For example, not only may the α-chymotrypsin-catalyzed hydrolysis of amides proceed by rate-limiting formation or breakdown of the acyl enzyme (**8**) and of the tetrahedral intermediates preceding and following this (**9** and **10**), but also the transition structure within one of these steps (*e.g.* **11**) may vary in charge and geometry with substituent changes.

$$\text{A—B} + \text{C} \underset{k_{-1}}{\overset{k_1}{\rightleftharpoons}} \text{A—B—C} \overset{k_2}{\longrightarrow} \text{A} + \text{B—C} \qquad (12)$$

(**8**)   (**9**)   (**10**)   (**11**)

Changes in charge density of transition states, relative to initial stages, may be more important than geometrical ones.[2] Chemists often exaggerate, in drawings, the differences in geometry of initial state and transition state structures. However, the actual distance that atoms move to reach a transition state is often not significantly much more than that experienced during normal molecular vibrations.[2] Converting a double bond to a single one increases the bond length by about 0.2 Å. Converting three-coordinate carbon to four-coordinate carbon decreases the bond angle by about

$10°$. The distance between the carbonyl oxygen and the nitrogen in amides increases by 0.07 Å when amides are converted to tetrahedral intermediates. At 298 K mean vibrational amplitudes commonly range from about 0.05 to 0.1 Å—and bending amplitudes of $\pm 10°$ are not rare.[2]

The substitution of one group by another in a substrate may itself cause changes in bond angles, bond lengths and torsional angles which could alter interactions elsewhere in the substrate with the enzyme as well as affecting initial state energies. This is one reason why binding energies of substituents measured by their effect on k_{cat}/K_m may vary from enzyme to enzyme.[2]

The transition states of an enzyme-catalyzed reaction of a series of substrates may change because: (i) substituents may directly affect the ease of bond making and breaking and hence cause different amounts of charge development (Hammond or anti-Hammond effects); (ii) substituents, although not directly affecting chemical reactivity, indirectly use their binding energy to facilitate the ease of bond making or breaking; and (iii) changing solvents, pH, ionic strength and temperature may significantly alter initial state and transition state energies.

The effect of changing transition state structures upon k_{cat}/K_m will reflect the optimization of interactions between the substrate and enzyme. If this interaction is not optimal, k_{cat}/K_m will be reduced but transition state stabilization in enzyme catalysis is still of major importance. The variable geometry and charge distribution in transition states indicates that many enzymes may bind transition state analogues as effective inhibitors even though they have a variety of structures. This is a reflection of the flexibility of intermolecular interactions, particularly electrostatic and hydrogen bonding.[2]

There is sometimes an *ad hoc* rationalization of enzyme inhibition using the transition state analogy, which is ambiguous if the mechanism itself is doubtful. For example, phosphoryl transfer can occur by a dissociative mechanism to give monomeric metaphosphate (PO_3^-) or by an associative pathway involving pentacoordinate intermediates.[29] Both mechanisms involve a phosphorus coordinated to three electron rich oxygens. The inhibition of ATPases[30] and phosphatases[31] by vanadate, molybdate and tungstate ions has been rationalized in terms of resemblance of these ions to the transition state for associative phosphoryl transfer. Similarly, some kinases form tight complexes with ADP, nitrate ion and unphosphorylated cosubstrate.[32] However, it is very difficult to dismiss the possibility that the planar nitrate ion is a mimic of the metaphosphate intermediate, consistent with the dissociative pathway. Sections 5.4.1 to 5.4.4 give some general examples of enzyme inhibition; some of these areas are discussed in more detail later in this volume.

5.4.4.1 Glycosidase Inhibitors

The substrates of glycosyl-transferring enzymes pass through glycosyl-cation-like transition states (**12**).[33] The original suggestion[34] that the lysozyme-catalyzed hydrolysis of oligosaccharides proceeds by inducing a conformational change at site D, so that the glycoside residue is planar at C-5, O-5, C-1 and C-2 to resemble the glycosyl cation (**12**), has been disputed by recent evaluations which indicate that the saccharide residue bound in the D site is not distorted.[35] The general mechanism for the action of glycosidases is outlined in Scheme 1.[33] Two features distinguish the glycosyl cation from its parent glycoside: O-5 and C-1 carry between them a unit positive charge and C-5, O-5, C-1 and C-2 are coplanar (**12**). Saccharide derivatives which possess these features *and* have the appropriate hydroxylation pattern should therefore bind tightly to glycosyl-transferring enzymes as transition state analogues. Aldonolactones (**13**) bind tightly to glycosidases with K_i values two to three orders of magnitude lower than K_s values for substrates.[36] This is presumably attributable to the positive charge on the ring oxygen of the lactone and the partial planarity of the system (**13**). The canonical form (**14**) indicates the charge distribution in lactams, and suitably substituted lactams show K_i values in the nanomolar region.[37] Carbinolamine derivatives (**15**) are in equilibrium with the protonated imine (**16**), another analogue of the glycosyl cation. Suitably hydroxylated derivatives of (**15**) are potent inhibitors of β-mannosidases.[38]

Protonated secondary amines (**17**) are also potential transition state analogues and suitably hydroxylated derivatives are weak to potent inhibitors.[38,39] However, at pH values where the

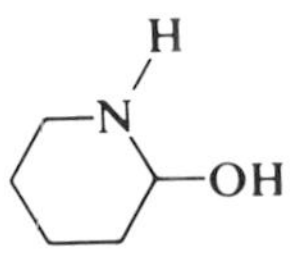

(15)

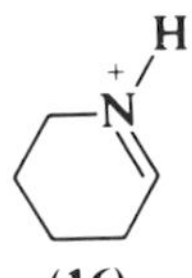

(16)

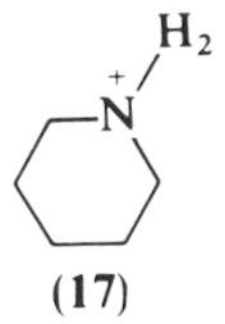

(17)

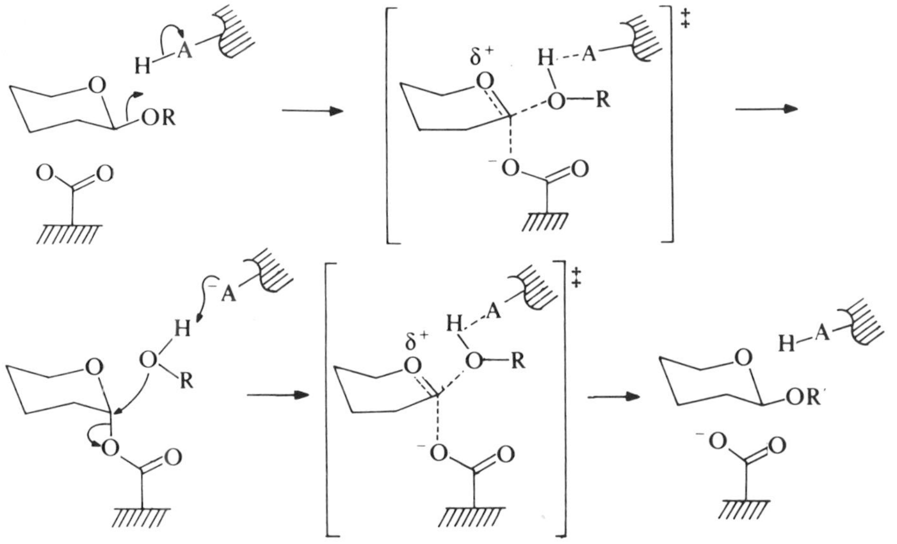

Scheme 1

inhibitor is fully protonated, the binding *increases* with pH, indicating that the EI complex has one proton less than the free enzyme and protonated inhibitor.[39,40] Glucosylamines are similarly good inhibitors of glycosyl enzymes.[41]

5.4.4.2 Serine Protease Inhibitors

Serine proteases catalyze the hydrolysis of amides and esters by the intermediate formation of an acyl enzyme, the formation and breakdown of which is thought, in turn, to involve a tetrahedral intermediate (Scheme 2).[42] X-Ray crystallographic studies have been performed on enzyme–product complexes, on the acyl–enzyme[44] and on enzyme–polypeptide inhibitors.[45] There appears to be no distortion of the substrate on binding; although some polypeptide inhibitors do resemble the tetrahedral intermediate, this distortion of the peptide is present in the unbound inhibitor.[46] The tetrahedral intermediate does not accumulate during the course of the reaction, and the demonstration of its existence is based largely on circumstantial evidence derived from the analogy of nonenzyme-catalyzed reactions of esters and amides.

$$\text{Enz—OH} + \text{RCONHR}' \rightleftharpoons \text{Enz—OH}\cdot\text{RCONHR}' \overset{-\text{H}^+}{\rightleftharpoons} \text{Enz—O—}\underset{\text{NHR}'}{\overset{\text{O}^-}{\text{C}}}\text{—R} \overset{\text{H}^+}{\rightleftharpoons}$$

$$\text{Enz—O—COR} + \text{R}'\text{NH}_2 \overset{\text{OH}^-}{\rightleftharpoons} \text{EnzO—}\underset{\text{OH}}{\overset{\text{O}^-}{\text{C}}}\text{—R} \overset{\text{H}^+}{\rightleftharpoons} \text{EnzOH}\cdot\text{RCO}_2\text{H} \rightleftharpoons \text{RCO}_2\text{H} + \text{EnzOH}$$

Scheme 2

The inhibition of many serine proteases is attributed to an analogy of the inhibitor structure with the tetrahedral intermediate (**9**). For example, there are several inhibitors which are substrate-related aldehydes but there are a variety of mechanisms which could account for this inhibition. The free aldehyde (18) or its hydrate (19) may be noncovalently bound to the enzyme. Alternatively, the enzyme's serine hydroxyl group may form a covalent bond with the aldehyde so that it is bound as a hemiacetal (20). In the case of papain, elastase and chymotrypsin, NMR, secondary deuterium and solvent isotope effects and X-ray diffraction studies indicate that binding is covalent.[47] However, leucine aminopeptidase appears to bind inhibitory aldehydes noncovalently as hydrates, based on a comparison of leucinal with bestatin, which resembles a *gem*-diol and binds to the enzyme without exchange of the hydroxyl oxygen atoms.[25]

(18)   (19)   (20)

Chymostatin (21) forms a hemiacetal adduct with the catalytic Ser-195 residue of *Streptomyces griseus* protease A (22).[48] The role of the active site serine in inhibition is indicated by the 250-fold difference between K_i values for the binding of elastatinal with intact porcine pancreatic elastase and the chemically modified congener, anhydroelastase, in which Ser-195 has been converted into dehydroalanine.[49] Pyruvate derivatives are also powerful inhibitors of serine proteases but the carbonyl group is not fully tetrahedral in the complex adduct formed with Ser-195 in trypsin and 4-amidinophenyl pyruvate.[50] The stabilization of the similar hemiketal complex formed by thrombin and 4-amidinophenyl pyruvate is enhanced by a hydrogen bond between a carboxylate oxygen of the inhibitor molecule and the nitrogen of His-57.[51] ^{13}C NMR studies[52] of α-chymotrypsin complexed with the aldehyde inhibitor, N-acetyl-PheH, show two signals for the hemiacetal at pH 7, which could arise from slow interconverting conformations of the complex involving the protonation of the hemiacetal anion.[53]

(21)   (22)

Although peptidyl methyl ketones are not as electron deficient as aldehydes, some exceptionally powerful inhibitors of this type are known. For example, the acetyl-Ala-Ala-Ala-Phe-Me complex with thermitase shows a K_i of 4×10^{-8} M.[54]

Much more electrophilic are the chloromethyl ketones, which also appear to bind either as *gem*-diols[25] or as hemiketals.[55] These compounds also irreversibly alkylate histidine residues and are considered further under irreversible inhibitors. Peptidyl trifluoromethyl ketones are generally more potent inhibitors than the corresponding aldehyde[56] and sometimes show dissociation constants in the nanomolar region.[57]

Boronic acids (23) readily add hydroxide ion to form tetrahedral adducts (24), which are also reminiscent of the tetrahedral intermediates formed in serine protease-catalyzed reactions. Boronic acid inhibitors are thought to bind covalently by serine hydroxyl addition (25).

(23)   (24)   (25)   (26)

Difluoroborane analogues of amino acids (26), which are convenient precursors of boronic acids, owing to their hydrolysis in water, competitively inhibit serine proteases. The appropriate boronic acids inhibit chymotrypsin ($K_i = 3 \times 10^{-7}$ to 3×10^{-6} M) and porcine pancreatic elastase ($K_i = 1 \times 10^{-7}$ to 3×10^{-6} M).[58,59] Peptide substrate structures which have the boronic acid moiety on their C-terminal end are much more effective inhibitors.[60] The human leucocyte elastase inhibitor, MeOSuc-Ala-Ala-Pro-Boro-ValOH ($K_i = 6 \times 10^{-10}$ M) is as potent as the 70-residue polypeptide inhibitor eglin. This peptide boronic acid is among the first examples of a reversible synthetic protease inhibitor preventing emphysema in an animal model.[61]

5.4.4.3 Metalloprotease Inhibitors

Metalloproteases are invariably zinc enzymes and comprise endoproteases such as thermolysin and collagenases, and exoproteases such as aminopeptidases, dipeptidyl carboxypeptidases such as angiotensin converting enzyme, and the carboxypeptidases.[62] The zinc ion is usually coordinated to two histidine residues, one glutamate residue and water arranged in a distorted tetrahedron. Despite extensive studies, many aspects of the mechanism remain unknown and the general features are based mainly on the studies with carboxypeptidase A. This has not prevented the rationalization of many inhibitors as being transition state analogues.

The activity of carboxypeptidase depends on the ionization state of two groups of pK_a 6.2, due to zinc-bound water or glutamate, and 9.0, attributed to tyrosine.[62] The general features of the mechanism involve general base catalysis by glutamate and electrophilic catalysis by the zinc ion (27a), although other roles have been suggested for the zinc.[63] Transition state analogue inhibitors are thus usually based on the presumed tetrahedral intermediate (27b).

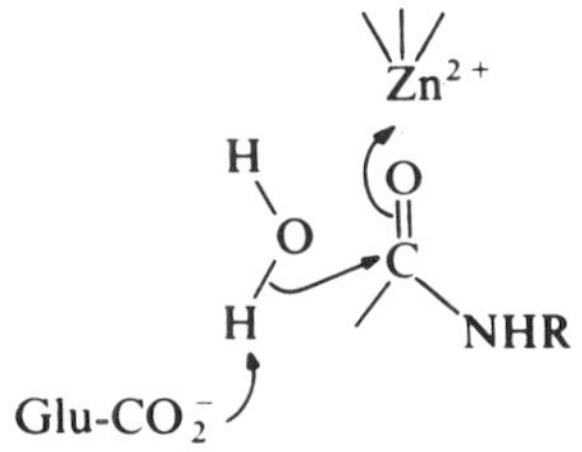

(27a) 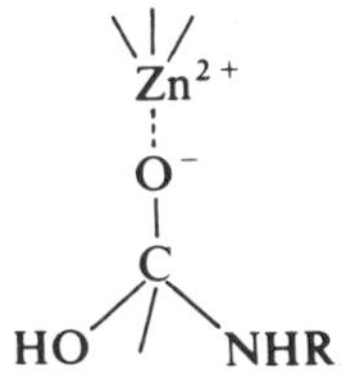　　　(27b)

Phosphoramidon (28) is a naturally occurring potent inhibitor of zinc proteases such as thermolysin,[64] enkephalinase[65] and elastase.[66] The rationalization of activity is based on an analogy of the structure with the tetrahedral intermediate (27b). Other phosphorus-containing amino acid derivatives are phosphoramidates[67] and phosphonamidates.[68] The K_i values for phosphonamidates correlate with k_{cat}/K_m values for the corresponding substrates.[69] The zinc ion in the thermolysin–phosphoramidon complex is four coordinate and is bound to the phosphoryl oxygen.[70]

Although the phosphonate ester (28) binds nearly 10^3-fold less strongly to thermolysin than its NH analogue (29),[71] the mode of binding of both inhibitors is similar.[72] The 17 kJ mol^{-1} of extra binding energy is attributed to the formation of a hydrogen bond between the P—NH proton and the carbonyl oxygen of Ala-133. Part of the difference in the energetics of binding (28) and (29) is attributed to different solvation energies.[73]

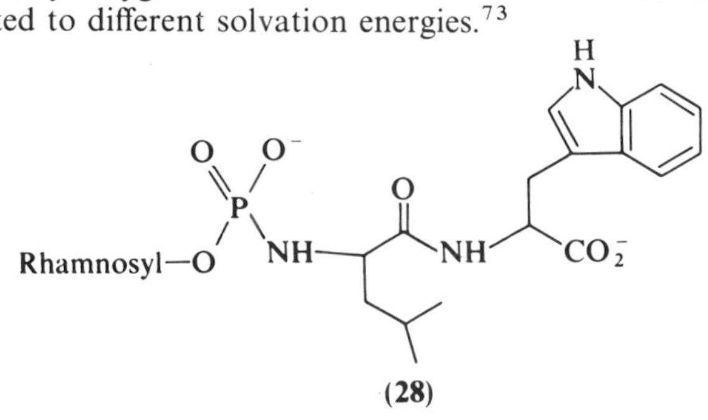

(28) 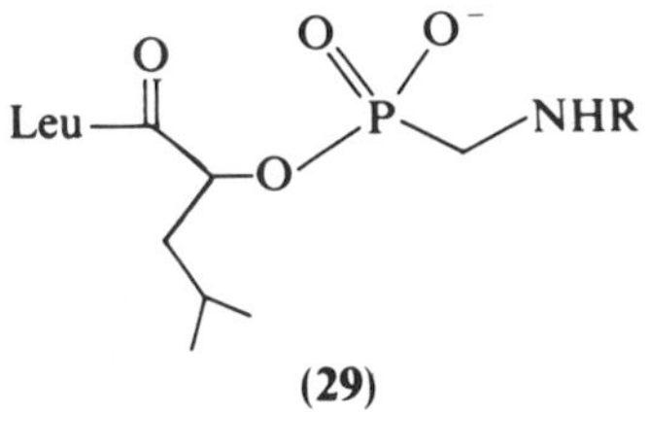　　　(29)

The kinetics of inhibition of angiotensin converting enzyme by phosphorus-containing inhibitors is complex.[74] The apparent K_i values are substrate dependent and chloride ion concentration dependent[75] and so cannot be interpreted literally as true dissociation constants.

Enzymes

The zinc ion in metalloproteases is an electron acceptor and coordinates well to electron rich centres. Naturally, attention is focused on inhibitors binding to this site (**30–33**) as analogues of the presumed intermediate (**27**) but it is important to recognize that this interaction is but one of many and is not solely responsible for binding. Nonetheless, many inhibitors of metalloproteases have been synthesized where this interaction is emphasized.

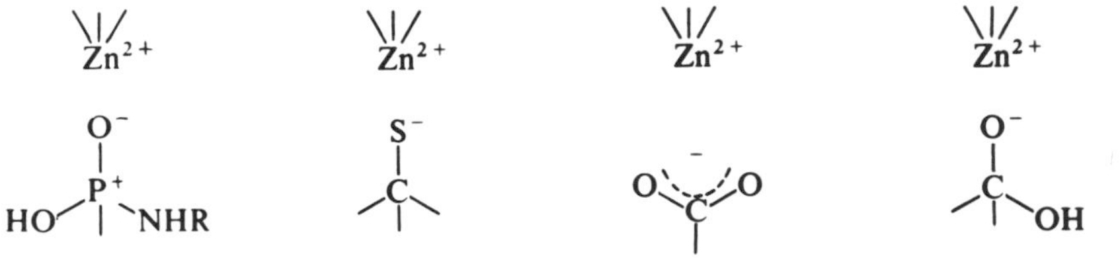

(30)   (31)   (32)   (33)

Captopril (**34**)[76] and enalaprilat (**35**)[77] are active in the treatment of hypertension because they are potent inhibitors of angiotensin converting enzyme. In addition to the coordination of the thiol of (**34**) or the carboxylate of (**35**) to zinc, other binding interactions are important[78] and alternative modes of binding may even occur.[79] The apo-enzyme of angiotensin converting enzyme binds 2×10^4 times less strongly to enalaprilat (**35**),[80] but this does not necessarily indicate the strength of metal ion coordination. The carboxylate of a carboxyalkyl dipeptide acts as a bidentate ligand for the zinc ion of thermolysin.[81]

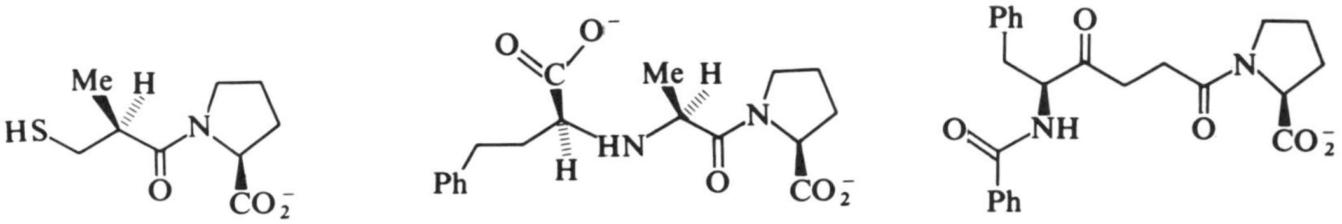

(34)   (35)   (36)

Similar to serine proteases, there are many carbonyl-containing peptide inhibitors of metalloproteases. For carboxypeptidase A, the *gem*-diol product is the bound form of the inhibitor and sometimes, but not always, both hydroxyl groups of the hydrate point to the zinc ion,[82] in a manner analogous to the bidentate behaviour of some carboxylate inhibitors. The zinc ion takes up different positions in the various enzyme–inhibitor complexes,[82] but it is not generally known whether it is the neutral diol or its monoanion which is bound. Leucine aminopeptidase actually catalyzes water addition to the carbonyl group.[83] The slow-binding, potent inhibitor of angiotensin converting enzyme, ketoace (**36**), exists at less than 3% as its hydrate in aqueous solution, although it is thought to bind in this form to the enzyme.[84]

5.4.4.4 Aspartic Protease Inhibitors

Aspartic proteases include the important enzymes pepsin, chymosin (previously called rennin), renin, cathepsin D and, possibly, retroviral proteases.[85] Renin shows no general protease activity but is involved in blood pressure regulation and catalyzes the cleavage of the decapeptide angiotensin I from the N-terminal end of the prohormone angiotensinogen. The optimum pH for the catalytic action of aspartic proteases is 2–4 and is controlled by the ionization of two aspartate residues of pK_a 1.4 and 4.5. This group of enzymes do not catalyze the hydrolysis of esters except as depsipeptide analogues.[86]

There is no general consensus on the mechanism of action of aspartate proteases which, again, widens the scope for the interpretation of inhibitors as transition state analogues. The nucleophilic mechanism (**37**) generates an anhydride intermediate, whilst no covalent intermediate is formed in the general base–general acid mechanism (**38**). Both mechanisms involve the formation of tetrahedral intermediates with two oxygens and nitrogen attached to the central carbon and it is not possible to use transition state analogue inhibitors to distinguish between them. Furthermore, there is a tendency to focus on the formation of the tetrahedral intermediate as the rate-limiting step and hence as the guide to the design of transition state analogues. However, it has been emphasized that

C—N bond cleavage is often rate limiting in the hydrolysis of amides.[63] This reflects a general problem with a literal interpretation of transition state analogues, *i.e.* the tendency to concentrate on the bond-making and bond-breaking sites of the substrate, whereas transition state stabilization in catalysis means the expression of the maximum binding energy between the *whole* substrate and enzyme. It is often the binding energy of the 'nonreacting' parts of the substrate analogue which is responsible for the tight binding of so-called transition state analogue inhibitors.[2,3] Furthermore, efficient catalysis depends on *all* steps of the reaction being facilitated so it probably makes little difference which actual transition state is chosen for mimicry. Just as important as the active site part is the 'nonreacting' part in the design of effective inhibitors. An extreme case, of course, is reached when the rate-limiting step of the enzyme-catalyzed reaction is diffusion controlled encounter or product release.

(37)   (38)

A competitive and tight binding inhibitor of most aspartic proteases is the naturally occurring peptide pepstatin (**39**), which contains the unusual amino acid statine, (3*S*,4*S*)-4-amino-3-hydroxy-6-methylheptanoic acid, and in which a peptide link is replaced by a hydroxymethylene.[87] It is difficult to imagine that the unionized hydroxyl group of an alcohol (**41**) is a good mimic of the oxyanion of the tetrahedral intermediates (**40**) and yet the (3*S*)-hydroxyl group is needed for maximal inhibition. Pepstatin analogues lacking a *pro-S* C-3 hydroxyl group and those with a (3*R*) hydroxyl are much (10^3 to 10^4 fold) weaker inhibitors of aspartic proteases.[88] Furthermore, the C-3 hydroxyl group of pepstatin is within hydrogen-bonding distance of the catalytic aspartate residues.[89] In the native enzyme, the aspartate residues are hydrogen bonded to a water molecule and it has been suggested that binding of the inhibitor causes displacement of this water molecule by the C-3 hydroxyl group.[90] Consequently there is no net change in the number of hydrogen bonds and the binding energy results from the entropic advantage of a water molecule being transferred to bulk solvent.

(39)   (40) Tetrahedral intermediate   (41) Statine analogue

The structure of statine is interesting in that it is either one atom too short to be isosteric with a dipeptide or two atoms too long to be isosteric with a normal α-amino acid. From the X-ray data for a pepstatin–aspartic protease complex it appears that statine is more like a dipeptide analogue in a restricted conformation.[91] A similar situation is observed with the inhibition of adenosine deaminase by (8*R*)-deoxycoformycin (**44**; $K_i = 10^{-12}$ M). The enzyme catalyzes the hydrolysis of many purine ribonucleosides by general base-catalyzed addition of water to give the carbinolamine tetrahedral intermediate (**45**).[92] The inhibitor (**44**) has one more atom in the ring than expected to be isosteric with the natural substrate.

An α,α-difluoroketone residue of 2,2-difluorostatone located within a suitable peptide chain gives a very strong inhibitor of porcine pepsin with a dissociation constant in the subnanomolar range for

(42) Ketone hydrate analogue　　　**(43) Phosphinic acid analogue**

(44)　　**(45)**

the inhibitor–enzyme complex.[93] The hydrate of the ketone is thought to be the active inhibitor. Oxidation of the alcohol residue of statine to the corresponding statone-containing peptide gives inhibitors which bind about 200-fold less strongly to the enzyme.[94] Nonetheless, pepsin catalyzes the addition of water to the ketone residue and it may be the *gem*-diol forms of even unactivated ketones **(42)** which bind to the enzyme.[95]

One of the most potent inhibitors of pepsin is the phosphinic acid derived Iva-Val-Sta^P-Ala-Iaa, which shows a K_i of 7×10^{-11} M but is extremely slow binding. Based on comparisons with other derivatives, it is thought that the uncharged phosphinate **(43)** is bound to the enzyme.[96]

5.4.4.5 Slow Binding of Inhibitors

The regulation of protease activity is mainly mediated by naturally occurring inhibitors, which often exhibit slow binding characteristics and several identifiable enzyme–inhibitor complexes (equations 13 and 14). The slow onset of inhibition is sometimes observed with other inhibitors, but this in itself is not directly informative and is not necessarily related to the mechanism of inhibition. Furthermore, even if saturation kinetics is observed, this does not necessarily imply that EI_2 is derived directly from EI_1, since EI_2 may simply represent an alternative mode of binding. In the simple process (equation 14), the first complex, EI_1, is formed rapidly and reversibly, whereas EI_2 is formed slowly and, maybe, effectively irreversibly, *i.e.* $k_2 \gg k_{-2}$. The second-order rate constants for inhibition are then given by $k_1 k_2/k_{-1} = k_2/K_1$, where K_1 is the dissociation constant k_{-1}/k_1 of the first-formed complex, EI_1, and many inhibitors show values of less than 10^5 M^{-1} s^{-1}.[98] Low values for second-order rate constants can result from low values of k_2 or high values of K_1. For example, the phosphinic acid peptide inhibitor of pepsin **(43)** rapidly forms EI_1 with $K_1 = 7 \times 10^{-9}$ M followed by a slow ($k_2 = 1 \times 10^{-4}$ s^{-1}, $t_{\frac{1}{2}} = 115$ min) conversion to the final enzyme–inhibitor complex, EI_2.[96] The slow dissociation of the complex is not significant with respect to mechanism and is not necessarily associated with slow binding because it may simply represent a small dissociation constant.

$$E + I \underset{k_{\text{off}}}{\overset{k_{\text{on}}}{\rightleftharpoons}} (EI)_{\text{final}} \tag{13}$$

$$E + I \underset{k_{-1}}{\overset{k_1}{\rightleftharpoons}} EI_1 \underset{k_{-2}}{\overset{k_2}{\rightleftharpoons}} EI_2 \tag{14}$$

An enzyme–inhibitor complex may form slowly because a chemical transformation (bond making/breaking) is required. For example, the inhibition of β-galactosidase by D-galactal is due to the formation of a relatively stable deoxygalactosyl–enzyme intermediate.[99] But this is not always the case; the binding of methotrexate to dihydrofolate reductase is slow, which may be attributable to the 'upside down' binding of the inhibitor relative to the substrate.[100] Binding could be slow if the enzyme–inhibitor complex has to undergo a slow conformational change,[101] although the structure of the protein in the complex is often little different from that in the native, uncomplexed state. It has been suggested that the release of tightly bound water molecules may be responsible for slow binding[30] but, generally, enzyme-bound water is exchanged rapidly with bulk solvent.

Slow binding is certainly not a characteristic of transition state analogue inhibitors. In view of the emphasis in this chapter on the contribution of binding at sites away from the active site and the smallness of geometrical changes in the substrate on going to the transition state, it does not seem constructive to examine the problems of geometrical fit of so-called transition state analogues to an enzyme complementary in shape to the substrate.

5.4.5 IRREVERSIBLE INHIBITION/INACTIVATION

Irreversible inhibition of enzymes is not reversible by dialysis, gel filtration or dilution. This definition does not preclude noncovalent, but tightly bound, inhibitors which dissociate slowly.

Inhibitors which form a covalent bond with the enzyme or are covalently modified by a chemical reaction, which may or may not be catalyzed by the enzyme, will usually show time dependent inhibition. The reaction between inhibitor and enzyme may be nonspecific and the rate of inhibition may show second-order behaviour but, if the concentration of inhibitor is much higher than that of the enzyme, pseudo-first-order kinetics will be observed. Specific inhibitors form a complex with the enzyme before chemical reaction (equation 15) and show a saturation phenomenon on the rate with high concentrations of inhibitor. This behaviour is characteristic of active-site-directed inhibitors[102] and mechanism-based or suicide inhibitors.[103] Similar to the parameter K_m for enzyme catalysis, K_i, the concentration of inhibitor that gives half the maximal rate of inhibition, is not necessarily the dissociation constant of EI (equation 16), and, in any case, is not a measure of the total binding energy between inhibitor and enzyme. Some of this favourable interaction may be used to lower the activation energy of the chemical steps involved in covalent bond formation between the enzyme and inhibitor.[1-3] It may be more appropriate to use the words 'inactivator' and 'inactivation' to describe the irreversible chemical reaction between a compound and an enzyme.

$$E + I \underset{k_{-1}}{\overset{k_1}{\rightleftharpoons}} EI \xrightarrow{k_2} E{-}I_2 \tag{15}$$

$$\text{Rate} = \frac{k_2[E][I]}{[I] + \{(k_{-1} + k_2)/k_1\}} \tag{16}$$

Inhibitors which form an initial noncovalent complex with the active site of the enzyme are less effective in the presence of substrate (equation 17). The rate equation is then of the same form as simple competitive inhibition, and with excess inhibitor, pseudo-first-order kinetics for inhibition will be observed (equation 18). With some inhibitors, such as mechanism-based ones, the enzyme may actually turnover to produce a product before being completely inactivated (equation 19). The formation of product and inactivation of the enzyme occur concurrently. The ratio of the rates of these processes, the partition ratio k_3/k_4, remains constant and independent of inhibitor concentration throughout the reaction if inhibitor is in excess of enzyme concentration. If the partition ratio k_3/k_4 is small and less than the ratio of inhibitor to enzyme, all of the enzyme will eventually become inactivated. On the other hand, if the ratio k_3/k_4 is large and greater than [I]/[E], only some of the enzyme will be inactivated by the time all of the inhibitor has been consumed.

$$\text{Products} + E \underset{s}{\overset{k_4}{\longleftarrow}} ES \underset{k_{-1}}{\overset{K_s}{\rightleftharpoons}} E + I \underset{k_{-1}}{\overset{k_1}{\rightleftharpoons}} EI \xrightarrow{k_2} E{-}I_2 \tag{17}$$

$$k_{\text{obs}} = \frac{k_2[E][I]}{[I] + K_i(1 + [S]/K_s)} \tag{18}$$

$$E + I \underset{k_{-1}}{\overset{k_1}{\rightleftharpoons}} EI \xrightarrow{k_2} EX \overset{k_3}{\underset{k_4}{\diagdown}} \begin{array}{l} E + P \\ \\ EY_i \end{array} \tag{19}$$

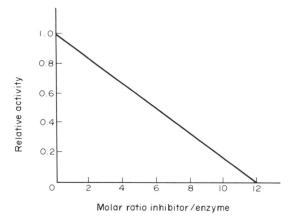

Figure 1 A plot of relative enzyme activity *versus* the molar ratio of inhibitor to enzyme. Extrapolation to zero activity indicates the number of molecules required to inactivate the enzyme, 12 in this case

When the rate of inactivation is followed by withdrawing aliquots from the enzyme–inhibitor solution at various times and activity assayed by addition to a normal substrate, the relative activity is a linear function of the molar ratio of inhibitor to enzyme. Extrapolation to zero activity indicates the number of inhibitor molecules required to inactivate one molecule of enzyme (Figure 1). The slope of such a graph is $(1 + k_3/k_4)$, from which the partition ratio is easily calculated.[104]

5.4.6 ACTIVE-SITE-DIRECTED IRREVERSIBLE INHIBITORS

Active-site-directed irreversible inhibitors (ASDIIs) have chemical reactive functionalities bound to a recognition site characteristic of the substrate. Binding of the inhibitor to the enzyme hopefully occurs at the active site because of the similarity to the substrate structure. Just as in normal enzyme catalysis, some of this binding energy is then used to lower the activation energy between the chemically reactive group of the inhibitor and an amino acid residue on the enzyme (see diagram below). One characteristic of such active-site-directed inhibitors is, therefore, that they should react faster than an analogous nonenzymic reaction. For example, organophosphorus compounds with good leaving groups inhibit serine proteases almost irreversibly by phosphorylation or phosphonylation of the active site serine to give a stable phosphoryl or phosphonyl enzyme. The inhibitors are not simply reactive phosphorylating agents because the esters and fluorides used are relatively stable towards nucleophiles. The rates of phosphonylation of serine proteases and acetylcholinesterase are 10^8–10^{12} fold faster than the analogous reaction with water.[105] A significant amount of the binding energy of favourable interactions between the enzyme and substrate used for catalysis must also be used for catalyzing the reaction with the inhibitor.

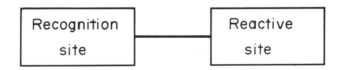

The particular case of the phosphonylation of serine enzymes is of general interest because the tetrahedral structure of the phosphonate ester inhibitor does not resemble the trigonal reactant structure of carboxylic acid esters and amides, nor does the trigonal bipyramidal intermediate expected for displacement at phosphorus resemble the tetrahedral intermediate in acyl transfer, although the product phosphonyl enzyme does have this geometry. This is a good example of the problem identified earlier of relating transition state stabilization to only the local geometrical and charge arrangements of the bond-making and bond-breaking site. There is an obvious degree of flexibility around the active site which can accommodate different geometrical arrangements.

The rate of alkylation of the thiol group of cysteine proteases by peptidyl halomethyl ketones is up to 10^{11}-fold faster than a simple model reaction.[106] Halomethyl ketones are too reactive to be selective enzyme inhibitors and although the chemical reactivity of fluoromethyl ketones is 500-fold less than the corresponding chloro derivative, the difference in the rate of their enzyme-catalyzed reactions is only a factor of 2.[107] Selectivity may be introduced by using the less reactive α-diazomethyl ketones,[108] which, for example as pyroglutamyl derivatives, specifically inactivate pyroglutamyl-peptide hydrolase.[109]

The majority of the functional groups in enzymes are nucleophiles and, consequently, the majority of reactive groups incorporated into active-site-directed irreversible inhibitors are electrophiles (Table 2). The process of covalent bond formation is usually one of alkylation, acylation, phosphorylation or sulfurylation to form a relatively stable derivative.

A simple general procedure of generating or accentuating an electrophilic centre within a substrate is to add a proton. Hence enzymes with suitably placed acidic groups may increase the electrophilicity of the inhibitor so that it is readily attacked by a nucleophilic group on the enzyme. This covalent modification of the enzyme may lead to irreversible inhibition.

The opening of epoxides is acid catalyzed and therefore the incorporation of an epoxide into a suitably recognizable substrate of an enzyme using general acid catalysis can potentially yield an inhibitor. For example, β-glucosidases and β-galactosidases are inhibited by cyclohexene polyol epoxides of a suitable configuration.[110] This is presumably due to the inactivation mechanism shown in (**46**), which is supported by the sigmoidal dependence of the inactivation rate.[111] The alkylated enzyme, in the form of an ester, is relatively stable. Within the active site there may be

Table 2 Typical Reactive Groups of Active-site-directed Irreversible Inhibitors

Enzyme nucleophile	*Electrophile*	*Products*
Amino groups Enz—NH$_2$	Anhydrides RCO—O—COR	Amides R'CONH—Enz
	Imidates $RO—C{\overset{\nearrow NH}{\underset{\searrow R'}{}}}$	Amidines $Enz—NH—C{\overset{\nearrow NH}{\underset{\searrow R''}{}}}$
	Isoureas $RO—C{\overset{\nearrow NH}{\underset{\searrow NHR}{}}}$	Guanidines $Enz—NH—C{\overset{\nearrow NH}{\underset{\searrow NHR}{}}}$
	Ketones $>C{=}O$	Imines $>C{=}N—Enz$
	Imines $>C{=}N—R$	Imines $—C{=}N—Enz$
	Arenesulfonyl halides ArSO$_2$X	Arenesulfonamides ArSO$_2$NH—Enz
	Cyanates R—N$=$C$=$O	Ureas R—NH—CO—NH—Enz
	Epoxides (epoxide ring)—R	Amino alcohols RNHCH$_2$CH(OH)Enz
Carboxyl groups Enz—CO$_2$H	Carbodiimides + amines RN$=$C$=$NR + RNH$_2$	Amides Enz—CONHR
	Epoxides (epoxide ring)—R	Hydroxy esters Enz—CO$_2$CH$_2$CH(OH)R
	α-Diazoamides N$_2$CHCONHR	Amido esters Enz—CO$_2$CH$_2$CONHR
	α-Haloacetates XCH$_2$CO$_2^-$	Half-esters Enz—CO$_2$CH$_2$CO$_2^-$
Hydroxy groups Enz—OH	Carbamates ROCONHR	Carbamates Enz—OCONHR
	Phosphates (RO)$_2$PO(X)	Phosphates Enz—O—PO(OR)$_2$
Imidazole group Enz (imidazole)—H	α-Halo ketones XCH$_2$COR	Alkylimidazole Enz (imidazole)—N—CH$_2$COR
	Pyrocarbonates (ROCO)$_2$O	Carbamate Enz (imidazole)—N—$\overset{O}{\overset{\|}{C}}$—OR
Thiol groups Enz—SH	Epoxides CH$_2$—C(epoxide)$\overset{H}{\underset{R}{}}$	Hydroxy sulfides Enz—S—CH$_2$CH(OH)R
	α,β-Unsaturated carbonyls RCH$=$CH—C$=$O	Thioesters Enz—S—CHR—CH$_2$—CO—
	(maleimide) N—Et	Enz—S—(succinimide) N—Et
	α-Halocarbonyls XCH$_2$CO$_2^-$ XCH$_2$CO$_2$R	Sulfides Enz—S—CH$_2$CO$_2^-$ Enz—S—CH$_2$CO$_2$R

several nucleophilic groups and the residue alkylated is not necessarily one required for the reaction of the normal substrate. For example, a suitably substituted epoxide alkylates a glutamate carboxyl group of glucose-6-phosphate isomerase which is not the one postulated in the catalyzed reaction of glucose 6-phosphate.[112]

(46)   **(47)**   **(48)**

Other alkylating agents can be generated *in situ* using the proton-donating ability of an acidic group on the enzyme. Upon protonation α-diazocarbonyl compounds (**47**) give highly reactive diazonium salts (**48**) which readily undergo nucleophilic substitution by a suitably placed group in the enzyme.

Triazenes are also potential sources of carbonium ions through the intermediate formation of the diazonium ion. Thus the triazene (**49**) inhibits β-galactosidase by alkylation of the active site glutamate,[113] although it appears that the anilide may be expelled by an unprotonated route.[114]

(49)   **(50)**   **(51)**

The classic alkylating agents are α-halomethyl ketones, which, when attached to a fragment of the natural substrate structure, are powerful enzyme inhibitors. The serine proteases use a histidine imidazole to act as a general base catalyst in the nucleophilic attack of the serine hydroxyl on the amide carbonyl to form a tetrahedral intermediate (Scheme 2). It is the histidine imidazole which becomes attached to the methylene carbon in α-halomethyl ketone inhibitors of serine proteases.[115] However, nucleophilic displacement of halide by imidazole does not take place directly. α-Halomethyl ketone derivatives of specific substrates of serine proteases form at least two reversible complexes with the enzyme before irreversible alkylation occurs.[116] Nucleophilic attack of the serine on the ketone (**50**) generates a hemiketal which is stable relative to the Michaelis complex for inhibitors derived from specific substrates. The second-order rate constant, k_i/K_i, corresponds to rate-limiting hemiketal formation for specific chloromethyl ketones (**50**) but rate-limiting alkylation for nonspecific derivatives (**51**). The resulting attachment of the enzyme by two covalent bonds to the inhibitor has been confirmed by an X-ray diffraction study of the complex formed from proteinase K with Z-Ala-Ala-CH$_2$Cl.[117] Furthermore the tetrahedral nature of the inhibitor's carbonyl carbon in the final complex is apparent from ^{13}C NMR observations.[118] In this sense, α-halomethyl ketones are 'second-trap' inhibitors—the serine hydroxyl is trapped by the ketone and the imidazole by the alkyl halide. In the next section, several examples of 'second-trap' inhibitors are described where the second trap is not exposed until a reaction has occurred with the enzyme.

The alkyl halide residue in the hemiketal (**51**) is not particularly electrophilic (less so, in fact, that in the free α-halomethyl ketone) and it is possible that epoxide formation occurs by oxygen anion displacement of chloride prior to attack by imidazole. In fact, the pK_a of the hydroxyl group of the hemiketal appears to be less than that of the imidazolium ion, which is presumably due to stabilization of the oxyanion by backbone peptides.[119]

The simplest active site directed inhibitors are those that progress through the normal catalytic mechanism to generate stable covalent intermediates which only slowly turn over. This can usually be achieved by steric and electronic modification of the substrate to prevent hydrolysis or other reaction of the intermediate. For example, the simplest inhibitors of serine proteases (Scheme 2) are those substrates which exhibit fast acylation of the substrate and slow deacylation of the acyl enzyme intermediate.

Acylation reactions of the serine proteases with arenesulfonyl fluorides,[120] phosphorofluoridates[121] and phosphonates[122] can produce not only effective inhibitors but also methods of titrating the enzyme. Such inhibitors may be considered also as mechanism based, and further examples are considered in the next section, but the general requirement for effective inhibition is simply a stable acylated enzyme. α,β-Unsaturated esters are less susceptible to hydrolysis and enzymes acylated with these derivatives are temporarily inhibited. For example, 4-amidinophenyl esters of cinnamic acid inhibit vitamin K dependent serine proteases.[123] A similar situation exists for the slowly deacylating 4-guanidinobenzoyl enzymes (52)[124] and carbamoyl enzymes (Table 2). Peptidyl carbamates and thiocarbamates are effective inhibitors of elastase and it is possible to introduce specificity by controlling substituents.[125] Stable acyl enzymes are also formed using peptides of α-aza amino acids, in which the α-methine of an α-amino acid is replaced by a nitrogen.[126]

N-Acylsaccharins (53) and *N*-acylbenzoisothiazolone (54) derivatives are powerful acylating agents for human leucocyte elastase but generate acyl enzymes which are only slowly deacylated ($t_{\frac{1}{2}} \approx 11$ d).[127]

(52)   (53)   (54)

Steric stabilization of the acyl enzyme intermediate is shown by the inhibition of elastase by the substituted benzoxazinones (55). Deacylation of the acyl enzyme (56) is retarded up to 10^3-fold by bulky 5-alkyl substituents compared with the unsubstituted derivatives.[128]

(55)   (56)   (57)

Another example of a stable covalent intermediate inhibiting enzyme turnover is found with the enzymes which catalyze decarboxylation by forming substrate-bound imine intermediates. Aceto-acetate decarboxylase catalyzes the decarboxylation of acetoacetate *via* the intermediate formation of an iminium ion. A variety of β-diketones are very effective inhibitors of this enzyme; for example, acetopyruvate is a competitive inhibitor with $K_i = 10^{-7}$M and is not reducible by borohydride when bound to the enzyme. Inhibition is attributed to formation of a relatively stable enamine (57).[129]

5.4.7 MECHANISM-BASED IRREVERSIBLE INHIBITORS

Several inhibitors have already been described which owe their activity to the catalytic mechanism of the enzyme; sometimes this has involved simply noncovalent interactions with particular protein residues used for binding the substrate, whilst in other cases actual covalent bond formation occurs between substrate and enzyme. There are inevitably areas of overlap in the attempts to classify the mechanisms of enzyme inhibition. The design of mechanism-based irreversible inhibitors[130] uses knowledge of the chemical, *i.e.* bond making and breaking, mechanism used by the enzyme to catalyze a reaction as well as the binding energy of the nonreacting parts of the substrate and enzyme. These inhibitors, which are also called suicide, k_{cat} and enzyme-activated inhibitors, act by generating a chemically reactive electrophilic group at the active site, which in turn irreversibly reacts with a nucleophilic group on the enzyme. The advantage of this class of inhibitors is that they contain relatively unreactive groups until they are transformed specifically by the enzyme. The kinetics of inhibition of these suicide inhibitors were described earlier (equation 19).

There are a variety of electrophilic centres which are unmasked by the enzyme. Often an α,β-unsaturated carbonyl or imine (58) is generated, which undergoes a Michael addition reaction with a nucleophilic group on the enzyme (equation 20). The formation of the α,β-unsaturated systems often rely on the effective reverse of the Michael addition reaction, *i.e.* generation of a carbanion type intermediates (resonance stabilized by X) to activate elimination or protonation of an adjacent unsaturated centre (**59–61**).

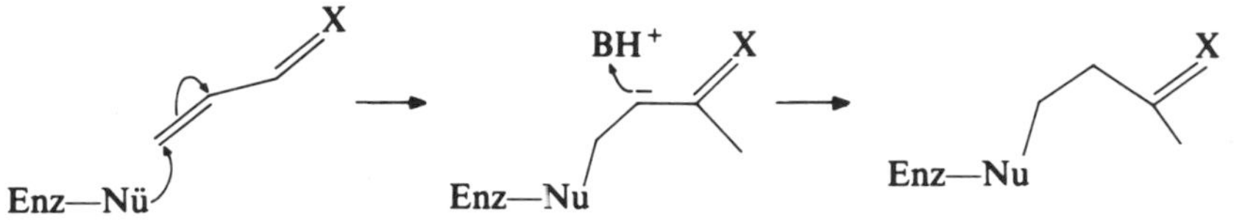

(20)

(58) X = 0, S, NR

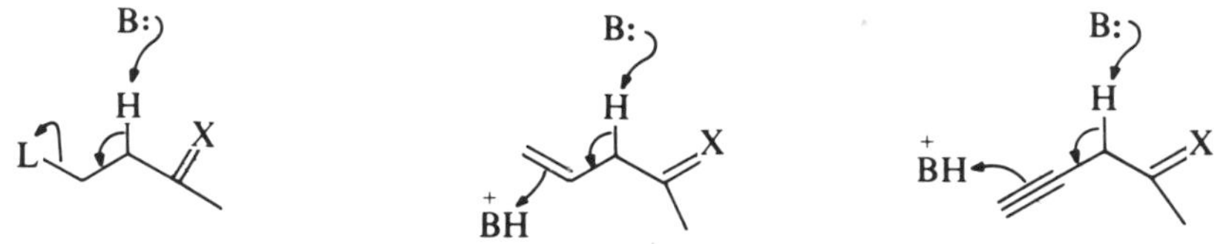

(59)   (60)   (61)

In addition to using the enzyme to generate potent electrophilic centres, suicide inhibitors can use the enzyme mechanism to unmask a group to stabilize a normally labile enzyme–substrate intermediate. For example, the mechanism of the serine protease catalyzed hydrolysis of peptides involves the formation of an acyl enzyme intermediate which is normally rapidly hydrolyzed (Scheme 2). Electron-releasing substituents in benzoyl-α-chymotrypsins stabilize the acyl enzyme towards hydrolysis. For example, 4-aminobenzoyl-α-chymotrypsin is sufficiently stable to inactivate the enzyme with a half-life for deacylation of 23 h.[131] A suicide inhibitor of serine proteases could thus be designed by ensuring that the electron-releasing properties of the amino group are not unmasked until after acylation of the enzyme has occurred. The initial acyl enzyme (**63**) formed from the reaction of thrombin with the anhydride (**62**) is readily decarboxylated to give an *o*-amino-benzoyl enzyme which is sufficiently stable to inhibit turnover of the enzyme.[132]

Enol esters (**64**) and lactone derivatives are potentially useful mechanism-based inhibitors for proteases and esterases because hydrolysis of the ester generates an enol-derived carbonyl group from which a feast of chemical activity can occur.

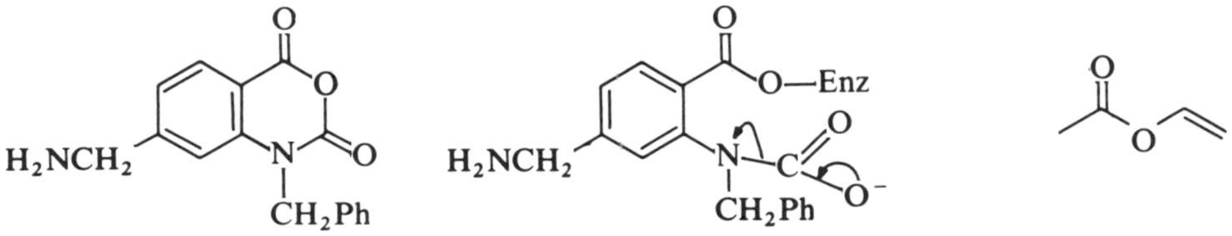

(62)   (63)   (64)

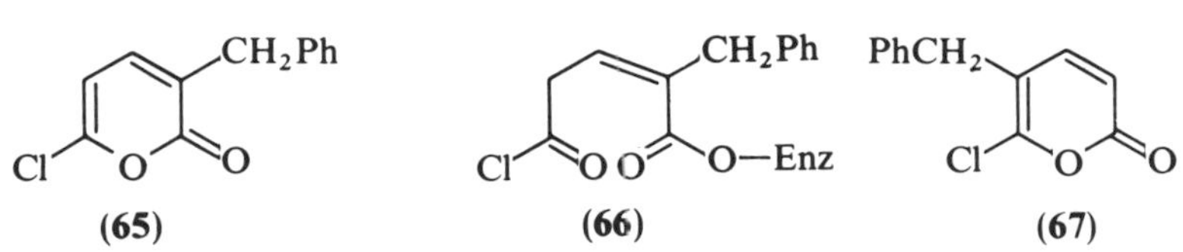

(65)   (66)   (67)

Substituted 6-chloro-2-pyrones such as (**65**) are good inhibitors of α-chymotrypsin, but although acylation initially produces the electrophilic acid chloride (**66**), this is rapidly hydrolyzed and inhibition is due to the slow rate of hydrolysis of the corresponding acyl enzyme with a free carboxylate group. A strong salt bridge between the carboxylate group and the protein is thought to prevent access of water and, therefore, hydrolysis of the acyl enzyme.[133]

Interestingly, inhibition of α-chymotrypsin by 5-benzyl-6-chloro-2-pyrone (**67**) is due to alkylation of the active site serine by the electron deficient centre at C-6. Presumably nucleophilic attack and displacement of chloride occurs at this carbon in preference to attack at the acyl carbon because of positioning controlled by the binding of the benzyl group in the enzyme's hydrophobic pocket.[134]

The unmasking of an alkylation site is provided by 3-alkoxy-7-amino-4-chloroisocoumarins (**68**) in which the electron rich sp^2 halogen bearing carbon is initially resistant to nucleophilic attack.

However, acylation of several serine proteases by (68) gives an acyl enzyme with an active 4-aminobenzyl chloride residue which can alkylate a nucleophilic group at the active site to give an irreversibly inactivated enzyme but the alkylated residue is not histidine.[135]

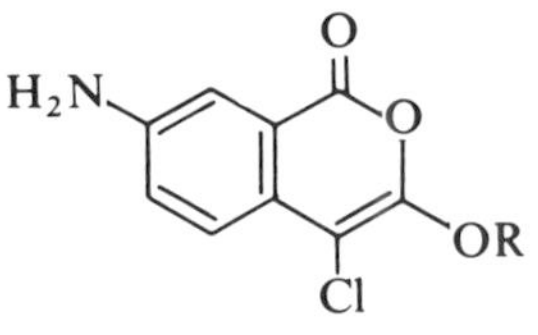

(68) 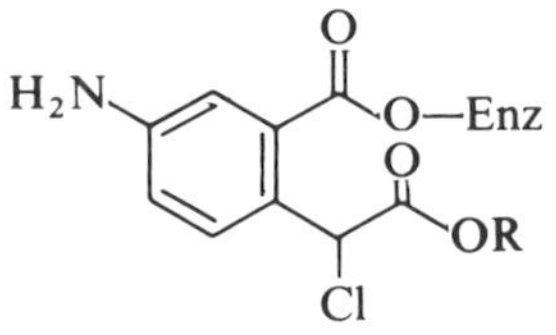   (69) 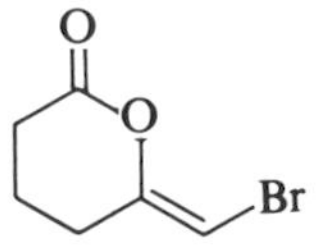   (70)

A similar second trap alkylating site is provided by the haloenol lactones such as (70), which upon ring opening by a serine protease generate an acyl enzyme with an exposed α-halomethyl ketone residue.[136]

Inhibition of α-chymotrypsin by (71) may be due to acylation, generating a β-chloro aldehyde which, after elimination, gives an α,β-unsaturated aldehyde as a potent electrophilic entity.[137]

Ring opening of the lactone of the dichloroisocoumarin (72) to give an acyl enzyme could generate either an acid chloride or a ketene as an acylating agent. The diacylated enzyme is stable and the half-life for deacylation is about 10 h for several serine enzymes.[138]

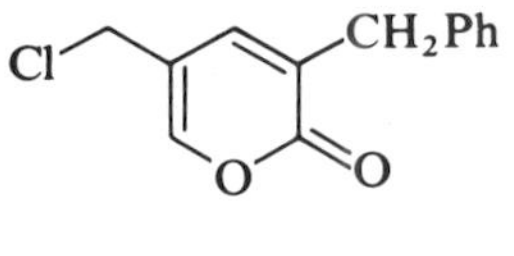

(71) 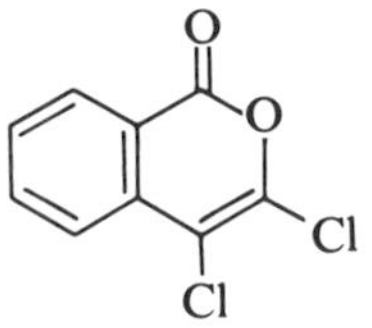   (72) 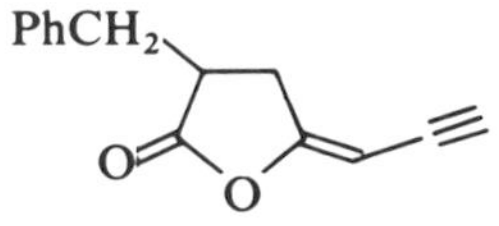   (73)

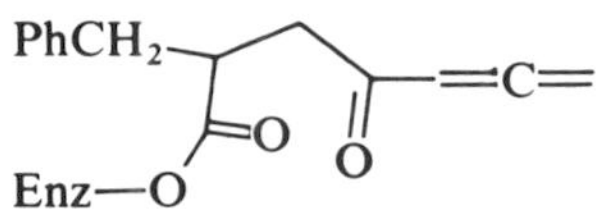

(74) 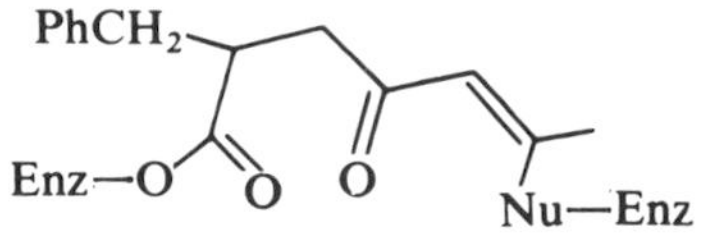   (75)

Acylation of the ynenol lactone (73) by serine proteases generates an acyl enzyme with an exposed electrophilic allenone (74) as a second trap, which readily alkylates a nucleophilic group on the enzyme (75). The lactone derivative (73) is an effective suicide inhibitor of elastases and trypsin and shows a partition ratio r of 1.6 (equation 19). Nearly every molecule of (73) acylated by the enzyme leads to inhibition.[139]

Cephalosporins are well-known inhibitors of the transpeptidase enzymes required for bacterial cell wall production. However, the 7-α-chloro sulfon cephalosporin (76) is a potent inhibitor of porcine pancreatic elastase. The inactivation is time dependent and irreversible and results from alkylation of His-57 of elastase.[140]

There are many mechanism-based inhibitors of β-lactamase, the enzyme responsible for bacterial resistance to penicillins and other β-lactam antibiotics.[141] A generalized scheme for the mechanism of inhibition by β-lactam derivatives such as clavulanic acid and penam sulfones is that after formation of the initial acyl enzyme (77) an electrophilic iminium ion (78) or α,β-unsaturated ester (79) is produced which reacts with a nucleophilic group on the enzyme. The 6-β-halopenicillins are also potent inhibitors of β-lactamase and their mechanism of inhibition is also thought to be of the suicide type. The initial acyl enzyme undergoes displacement of halide by sulfur in the intact thiazolidine[142] or by the sulfide ion in the ring-opened iminium ion (80).[143] Both mechanisms cannot occur, because of geometrical constraints, until the β-lactam ring has opened and in this sense the 6-halopenicillins are mechanism based. It is interesting to note that the binding of the 3-carboxylate residue attached to the thiazolidine, which is so important for recognition and activity of the β-lactam antibiotics, must presumably change or be lost when displacement of halide (80) takes place. Inhibition is attributed to the formation of a relatively stable acyl enzyme because it is a vinylogous urethane.[144]

For the same chemical reasons that imines and iminium ions are so commonly involved in enzyme-catalyzed reactions (*i.e.* compared with ketones, increased susceptibility to nucleophilic

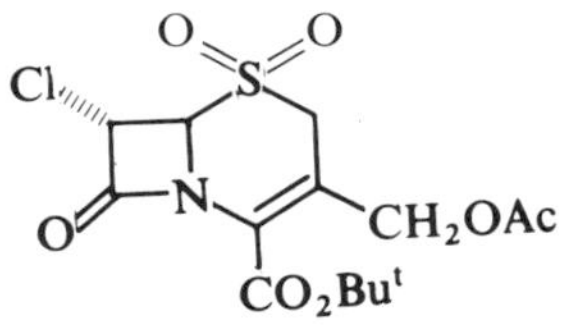

(76) 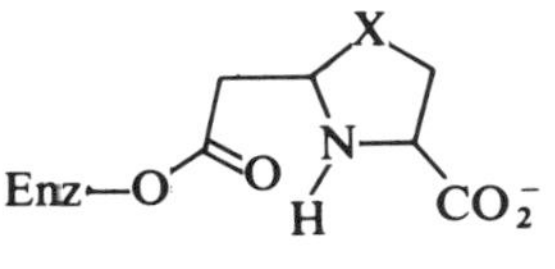   (77) 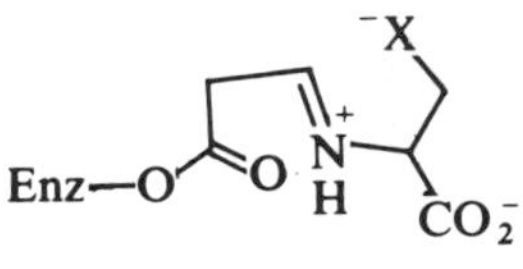   (78)

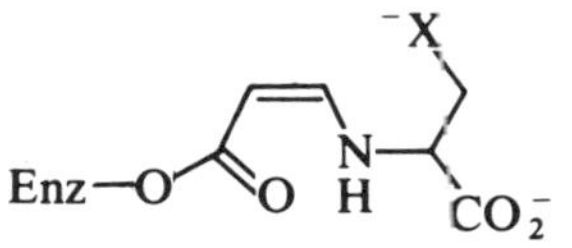

(79) 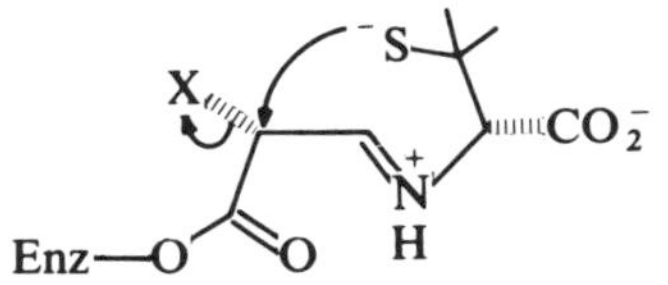   (80)

attack and lability of α-methine hydrogens), they are frequently formed intermediates in the suicide inhibition of enzymes.

The binding of the ketone (81) to aldolase by formation of the imine, followed by elimination of hydrogen fluoride, leads to suicide inhibition of the enzyme, presumably by formation of a conjugated iminium ion which alkylates a cysteine residue in the enzyme (82).

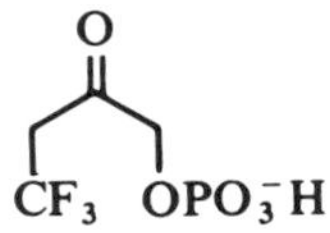

(81) 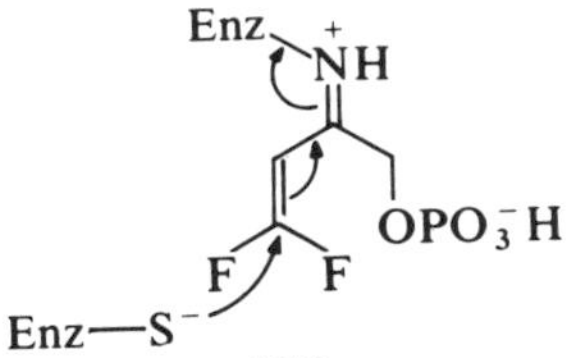   (82)

Elimination of HF is also a key step in the use of β-fluoro-D-alanine as an antibacterial inhibitor of pyridoxal phosphate dependent alanine racemases. These bacterial enzymes convert L-alanine into the D enantiomer required for the synthesis of the mucopeptide layer of bacterial cell walls. Normal iminium formation occurs between the fluoro inhibitor and pyridoxal phosphate but subsequent carbanion generation provides the driving force for liberation of fluoride (83) to give an α,β-unsaturated iminoacrylate intermediate, which may be covalently captured by a nucleophilic group on the enzyme (84).[146] In general, monohaloalanines are good mechanism-based inhibitors of transaminases, decarboxylases and other enzymes using pyridoxal phosphate that involve the generation of a carbanion at the α-carbon of amino acids (83) (see also Chapter 6.6).[147]

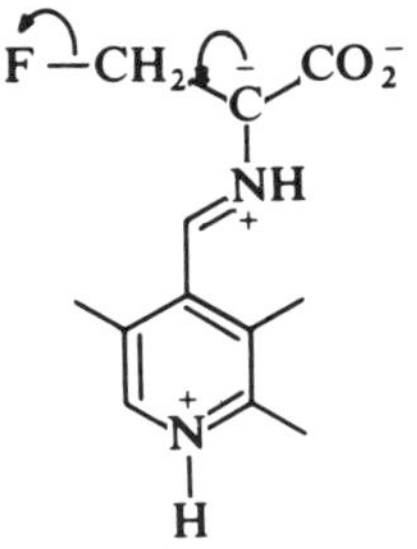

(83) 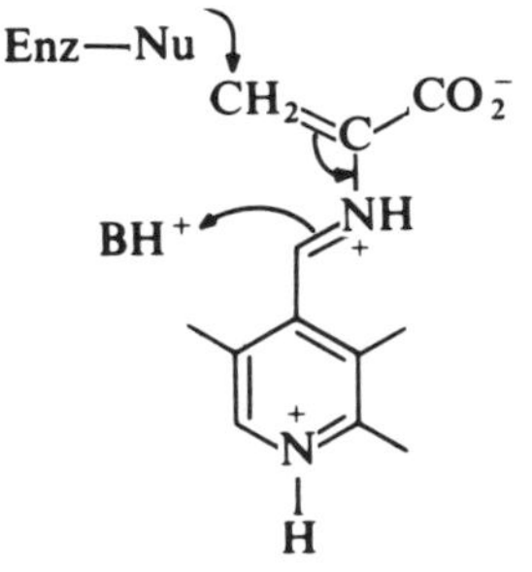   (84)

The *in situ* production of α-alkynic carbanions causes enzymes to unwittingly convert unreactive alkynes to reactive electrophilic conjugated allenes (85), which are potent alkylating agents (86). Thus, propargylglycine (propargyl = HC≡CCH₂—) is an irreversible inhibitor of many pyridoxal phosphate enzymes catalyzing β and γ replacement–elimination reactions.

Mechanism-based inhibitors do not necessarily have to form a covalent bond with a residue on the enzyme. Inactivation can occur by preventing a tightly bound coenzyme from turning over. For example, the fungal metabolite gabaculine (87), a good structural analogue of the natural substrate γ-aminobutyric acid (GABA; 88), completely inactivates γ-aminobutyric acid transaminase (GABA-T) by aromatizing the pyridoxal-bound inhibitor (89).[149]

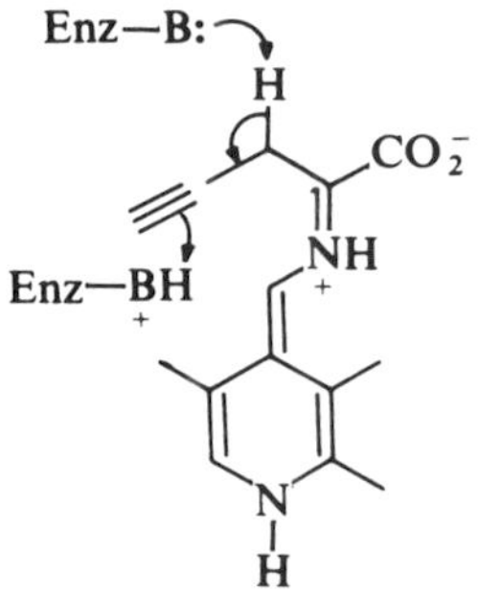

(85)

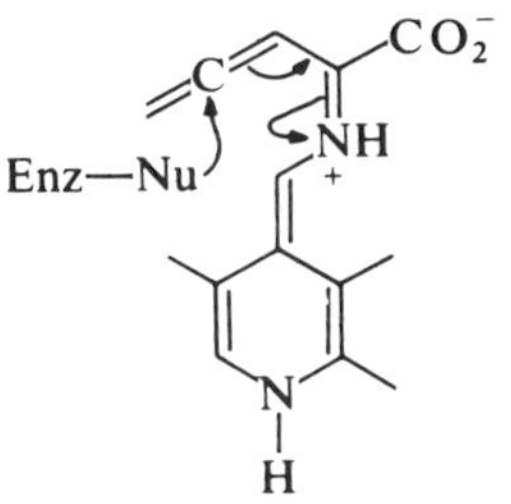

(86)

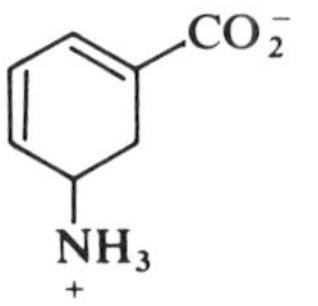

(87)

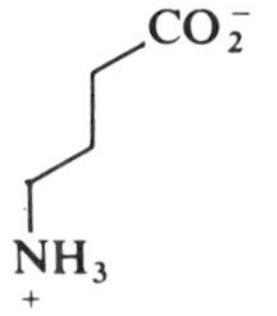

(88)

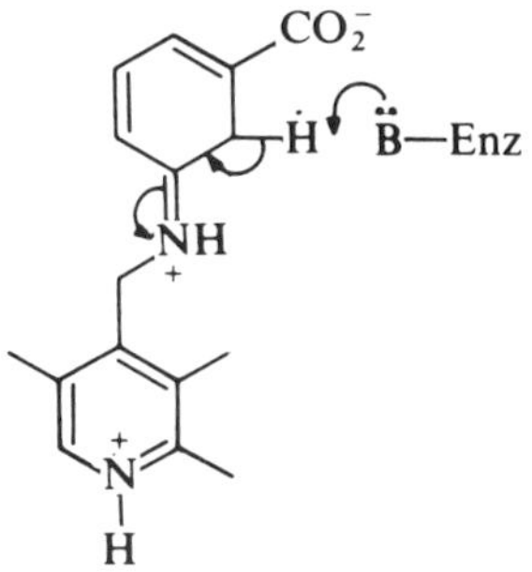

(89)

Many other enzyme-catalyzed reactions involve the generation of carbanion-like intermediates either by simple proton abstraction or hydride, or other nucleophilic, addition. Suitably placed leaving groups or unsaturated sites capable of protonation can, in turn, produce electrophilic centres to inhibit the enzyme irreversibly.

Flavin-dependent enzymes which catalyze the oxidation of substrates with activated methine hydrogens (90) do so by proton abstraction to form an intermediate carbanion.[150] The carbanion can be oxidized by either transfer of hydride or addition of the carbanion to N-5 or C-4a of the isoalloxazine of the flavin (91). Consequently, substrates containing a suitably placed alkyne residue are effective inhibitors of flavoenzymes, presumably because they can produce an allene at the active site. For example, 2-hydroxy-but-3-ynoate (92) inactivates L-lactate oxidase, glycollate oxidase and D-lactate dehydrogenase.[151]

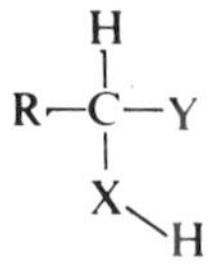

(90)

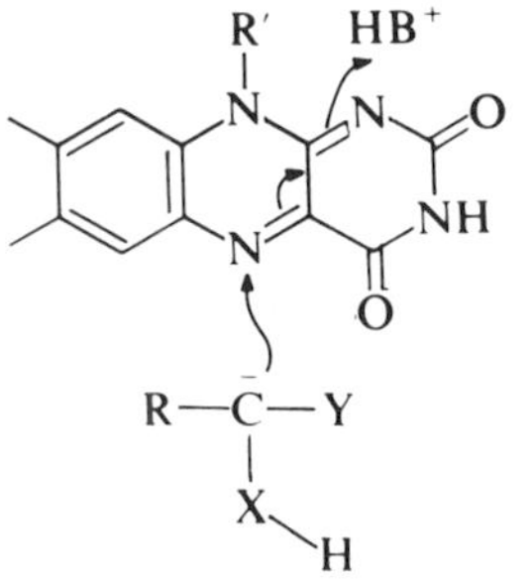

(91)

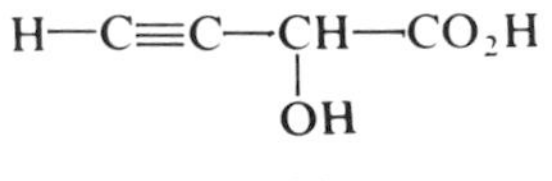

$$H-C\equiv C-\underset{\underset{OH}{|}}{CH}-CO_2H$$

(92)

Flavin-linked enzymes also catalyze the oxidation of ketones to esters in Baeyer–Villiger-type reactions. In biological systems peroxide is bound to flavin, which then adds to the ketone. Thiolactones are suicide inhibitors of the flavoenzyme cyclohexanone oxygenase.[152] The electrophile responsible for inactivation is thought to be an acylated sulfenic acid (93) resulting from the normal 1,2 shift in the breakdown of the tetrahedral intermediate (94). Nucleophilic attack of the acyl centre of (93) liberates the electrophilic sulfenic acid group as a second trap, capable of capturing a nearby enzyme nucleophile.[152]

For the very reasons that coenzymes are so versatile and possess chemistry transferable from one system to another, the design of highly specific mechanism-based inactivators of coenzyme-linked

(93)   (94)   (95)

enzymes is very difficult. The general assumption is that suicide inhibitors are highly selective but, as can be seen from the above examples, similar general principles often apply to a variety of systems. Alkynic GABA (95) is, as expected, a mechanism-based inhibitor of GABA transaminase[153] but it also inactivates glutamate decarboxylase[154] and ornithine δ-aminotransferase.[155]

5.4.8   REFERENCES

1. M. I. Page, in 'Comprehensive Medicinal Chemistry', ed. C. Hansch, P. G. Sammes and J. B. Taylor, Pergamon Press, Oxford, 1989, vol. 2, chap. 5.3.
2. M. I. Page, in 'The Chemistry of Enzyme Action', ed. M. I. Page, Elsevier, Amsterdam, 1984, p. 1.
3. W. P. Jencks, *Adv. Enzymol.*, 1975, **43**, 219.
4. W. P. Jencks, *Mol. Biol., Biochem. Biophys.* 1980, **32**, 3.
5. W. J. Albery and J. R. Knowles, *Biochemistry*, 1976, **15**, 5631.
6. C. Frieden, *J. Biol. Chem.*, 1964, **239**, 3522; I. H. Segal, 'Enzyme Kinetics', Wiley, New York, 1975, p. 27.
7. P. D. Dawkins, B. J. Gould, J. A. Sturman and M. J. H. Smith, *J. Pharm. Pharmacol.*, 1967, **19**, 355.
8. N. C. Price, *Biochem. J.*, 1979, **177**, 603.
9. J. K. Seydel and K. J. Schaper, in 'Enzyme Inhibitors as Drugs', ed. M. Sandler, Macmillan, London, 1980, p. 53.
10. J. L. Evelhoch, D. F. Bocian and J. L. Sudmeier, *Biochemistry*, 1981, **20**, 4951.
11. Y. Pocker, N. Janjic and C. H. Miao, in 'Zinc Enzymes', ed. I. Bertini, C. Luchinat, W. Maret and M. Zepperzauer, Birkhauser, Boston, 1986, p. 341.
12. S. Lindskog, in 'Zinc Enzymes', eds. I. Bertini, C. Luchinat, W. Maret and M. Zepperzauer, Birkhauser, Boston, 1986, p. 307; T. H. Maren and G. Sanyl, *Annu. Rev. Pharmacol. Toxicol.*, 1983, **23**, 439.
13. L. Tibell, C. Forsman, I. Simonsson and S. Lindskog, *Biochim. Biophys. Acta*, 1984, **789**, 302; Y. Pocker and T. L. Deits, *J. Am. Chem. Soc.*, 1982, **104**, 2424.
14. L. Tibell, C. Forsman, I. Simonsson and S. Lindskog, *Biochim. Biophys. Acta*, 1985, **829**, 202.
15. L. D. Byers, *J. Theor. Biol.*, 1978, **74**, 501.
16. M. I. Page, *Chem. Soc. Rev.*, 1973, **2**, 295.
17. M. I. Page, in 'Enzyme Catalysis', ed. M. I. Page and A. Williams, Royal Society of Chemistry, London, 1987, p. 1.
18. K. D. Collins and G. R. Stark, *J. Biol. Chem.*, 1971, **246**, 6599.
19. G. E. Lienhard and I. I. Secemski, *J. Biol. Chem.*, 1973, **248**, 1121.
20. R. Wolfenden, *Annu. Rev. Biophys. Bioeng.*, 1976, **5**, 271.
21. J. F. A. Chase and P. K. Tubbs, *Biochem. J.*, 1969, **111**, 225.
22. L. J. Arnold and N. O. Kaplan, *J. Biol. Chem.*, 1974, **249**, 652.
23. P. M. Cullis, in 'Enzyme Mechanisms', ed. M. I. Page and A. Williams, Royal Society of Chemistry, London, 1987, p. 178.
24. R. Wolfenden, *Acc. Chem. Res.*, 1972, **5**, 10.
25. P. M. Cullis, R. Wolfenden, L. S. Cousens and B. M. Alberts, *J. Biol. Chem.*, 1982, **257**, 12 165.
26. G. E. Lienhard, *Science (Washington, D.C.)* 1973, **180**, 149.
27. L. Pauling, *Chem. Eng. News*, 1946, **24**, 1375; *Am. Sci.*, 1948, **36**, 51.
28. M. I. Page, *J. Mol. Catal.*, 1988, **47**, 241.
29. P. M. Cullis, in 'Enzyme Mechanisms', ed. M. I. Page and A. Williams, Royal Society of Chemistry, London, 1987, p. 178.
30. L. C. Shantley, L. Josephson, R. Warner, M. Yanagisawa, C. Lechene and G. Guidotti, *J. Biol. Chem.*, 1977, **252**, 7421.
31. L. E. Seargeant and R. A. Stinson, *Biochem. J.*, 1979, **181**, 247.
32. D. H. Buttlaire and M. Cohn, *J. Biol. Chem.*, 1974, **249**, 5733.
33. M. L. Sinnott, in 'Enzyme Mechanisms', ed. M. I. Page and A. Williams, Royal Society of Chemistry, London, 1989, p. 259.
34. C. C. F. Blake, G. A. Mair, A. C. T. North, D. C. Phillips and V. R. Sarma, *Proc. R. Soc. London, Ser. B.*, 1967, **167**, 365.
35. M. Schindler, Y. Assaf, N. Sharon and D. M. Chipman, *Biochemistry*, 1977, **16**, 423; A. Warshel and M. Levitt, *J. Mol. Biol.*, 1976, **103**, 227; A. Warshel, *Proc. Natl. Acad. Sci. USA*, 1984, **81**, 444; M. R. Pincus and H. A. Scheraga, *Macromolecules*, 1979, **12**, 633.
36. P. Lalégerie, G. Legler and J. M. Yon, *Biochimie*, 1982, **64**, 977; D. H. Leaback, *Biochem. Biophys. Res. Commun.*, 1968, **32**, 1025.
37. T. Niwa, T. Tsuruoka, S. Inouye, Y. Naito, T. Koeda and T. Niida, *J. Biochem.*, 1972, **72**, 207.
38. G. Legler and E. Juelich, *Carbohydrate Res.*, 1984, **128**, 61.
39. I. C. di Bello, P. Dorling, L. Fellows and B. Winchester, *FEBS Lett.*, 1984, **176**, 61; M. P. Dale, H. E. Ensley, K. Kern, K. A. R. Sastry and L. D. Byers, *Biochemistry*, 1985, **24**, 3530; P. J. Card and W. D. Hitz, *J. Org. Chem.*, 1985, **50**, 891.
40. G. W. Fleet, *Tetrahedron Lett.*, 1985, **26**, 5073.
41. G. Legler and M. Herrchen, *Carbohydrate Res.*, 1983, **116**, 95.
42. J. Fastrez and A. R. Fersht, *Biochemistry*, 1973, **12**, 2025.
43. T. A. Steitz, R. Henderson and D. M. Blow, *J. Mol. Biol.*, 1969, **46**, 337.
44. R. Henderson, *J. Mol. Biol.*, 1970, **54**, 341.

45. A. Rühlmann, D. Kukla, P. Schwager, K. Bartels and R. Huber, *J. Mol. Biol.*, 1974, **77**, 417.
46. J. Deisenhofer and W. Steigemann, *Acta Crystallogr., Ser. B*, 1975, **31**, 238.
47. A. Srinivasan, V. Amarnath, A. D. Broom, F. C. Zou and Y. C. Cheng, *J. Med. Chem.*, 1984, **27**, 1710; R. L. Stein and A. M. Strimpler, *Biochemistry*, 1987, **26**, 2611.
48. L. T. J. Delbaere and G. D. Brayer, *J. Mol. Biol.*, 1985, **183**, 89.
49. H. R. Williams, T.-Y. Lin, M. A. Navia, J. P. Springer and K. Hoogsteen, *Biochem. J.*, 1987, **242**, 267.
50. J. Walter and W. Bode, *Hoppe-Seyler's Z. Physiol. Chem.*, 1983, **364**, 949.
51. K. Tanizawa, Y. Kanaoka, J. D. Wos and W. B. Lawson, *Hoppe-Seyler's Z. Physiol. Chem.*, 1985, **366**, 871.
52. D. O. Shah, K. Lai and D. G. Gorenstein, *J. Am. Chem. Soc.*, 1984, **106**, 4272.
53. N. E. Mackenzie, J. P. Malthouse and A. I. Scott, *Science (Washington, D.C.)*, 1984, **255**, 883.
54. D. Broemme and S. Fittkau, *Biomed. Biochim. Acta*, 1985, **44**, 1089.
55. S. S. Al-Hassan, R. J. Kulick, D. B. Livingstone, C. J. Suckling and H. C. S. Wood, *J. Chem. Soc., Perkin Trans 2*, 1980, 2645.
56. B. Imperiali and R. H. Abeles, *Biochemistry*, 1986, **25**, 3760; R. P. Dunlap, P. J. Stone and R. H. Abeles, *Biochem. Biophys. Res. Commun.*, 1987, **145**, 509.
57. R. L. Stein, A. M. Strimpler, P. D. Edwards, J. J. Lewis, R. C. Mauger, J. A. Schwartz, M. M. Stein, D. A. Trainor, R. A. Wildonger and M. A. Zottola, *Biochemistry*, 1987, **26**, 2682.
58. D. H. Kinder and J. A. Katzenellenbogen, *J. Med. Chem.*, 1985, **28**, 1917.
59. P. Amiri, R. Lindquist, D. S. Matteson and K. M. Sadhu, *Arch. Biochem. Biophys.*, 1984, **234**, 531.
60. C. A. Kettner and A. B. Shenvi, *J. Biol. Chem.*, 1984, **259**, 15 106.
61. A. Baici and U. Seemueller, *Biochem. J.*, 1984, **218**, 829; N. T. Soskel, S. Watanabe, R. Hardie, A. B. Shenvi, J. A. Punt and C. Kettner, *Annu. Rev. Respir. Dis.*, 1986, **133**, 635, 639.
62. D. S. Auld, in 'Enzyme Mechanisms', ed. M. I. Page and A. Williams, Royal Society of Chemistry, London, 1987, p. 240.
63. N. P. Gensmantel, P. Proctor and M. I. Page, *J. Chem. Soc., Perkin Trans. 2*, 1980, 1725.
64. H. Suda, T. Aoyagi, T. Takeuchi and H. Umezawa, *J. Antibiot.*, 1973, **26**, 621.
65. R. L. Elliot, N. Marks, M. J. Berg and P. S. Portoghese, *J. Med. Chem.*, 1985, **28**, 1208.
66. L. Poncz, T. A. Gerken, D. G. Dearborn, D. Grobelny and R. E. Galardy, *Biochemistry*, 1984, **23**, 2766.
67. R. E. Galardy and D. Grobelny, *J. Med. Chem.*, 1985, **28**, 1422; C. M. Kam, N. Nishino and J. C. Powers, *Biochemistry*, 1979, **18**, 3032.
68. K. Yamauchi, S. Ohtsuki and M. Konshita, *Biochim. Biophys. Acta*, 1985, **827**, 275; R. E. Galardy, V. Kontoyiannidou-Ostrem and Z. P. Kortylewicz, *Biochemistry*, 1983, **22**, 1990; N. E. Jacobsen and P. A. Bartlett, *J. Am. Chem. Soc.*, 1981, **103**, 654.
69. P. A. Bartlett, *Stud. Org. Chem.*, 1985, **20**, 439; P. A. Bartlett and C. K. Marlowe, *Biochemistry*, 1983, **22**, 4618.
70. D. E. Tonrud, A. F. Monzigo and B. W. Matthews, *Eur. J. Biochem.*, 1986, **157**, 261.
71. P. A. Bartlett and C. K. Marlowe, *Science (Washington, D.C.)*, 1987, **235**, 569.
72. D. E. Tonrud, H. M. Holden and B. W. Matthews, *Science (Washington, D.C.)*, 1987, **235**, 571.
73. B. A. Bash, U. C. Singh, F. K. Brown, R. Langridge and P. A. Hollman, *Science (Washington, D.C.)*, 1987, **235**, 574.
74. U. B. Goli and R. E. Galardy, *Biochemistry*, 1986, **25**, 7136.
75. R. Shapiro and J. F. Riordan, *Biochemistry*, 1984, **23**, 5225, 5234.
76. M. A. Ondetti, B. Rubin and D. W. Cushman, *Science (Washington, D.C.)* 1977, **196**, 441; E. W. Petrillo and M. A. Ondetti, *Med. Res. Rev.*, 1982, **2**, 1.
77. A. A. Patchett and E. H. Cordes, *Adv. Enzymol.*, 1985, **57**, 1; J. M. Wyvratt and A. A. Patchett, *Med. Res. Rev.*, 1985, **5**, 483.
78. W. H. Parsons, J. L. Davidson, D. Taub, S. D. Aster, E. D. Thorsett, A. A. Patchett, E. H. Ulm and B. I. Lamont, *Biochem. Biophys. Res. Commun.*, 1983, **117**, 108.
79. S. Klutchko, C. J. Blankley, R. W. Fleming, J. M. Minkley, A. E. Werner, I. Nordin, A. Holmes, M. Hoefle, L. Milton, D. M. Cohen, A. D. Essenberg and H. D. Kaplan, *J. Med. Chem.*, 1986, **29**, 1953.
80. H. G. Bull, N. A. Thornberry, M. H. J. Cordes, A. A. Patchett and E. H. Cordes, *J. Biol. Chem.*, 1985, **260**, 2952.
81. A. F. Monzingo and B. W. Matthews, *Biochemistry*, 1984, **23**, 5724.
82. D. W. Christianson and W. N. Lipscomb, *J. Am. Chem. Soc.*, 1986, **108**, 4998; 1985, **107**, 8281; *Proc. Natl. Acad. Sci. USA*, 1985, **82**, 6840.
83. L. Frick and R. Wolfenden, *Biochim. Biophys. Acta*, 1985, **829**, 311.
84. D. Grobelny and R. E. Galardy, *Biochemistry*, 1986, 25, 1072; R. G. Almquist, W. R. Chao, M. E. Ellis and H. L. Johnson, *J. Med. Chem.*, 1980, **23**, 1392.
85. L. H. Pearl and W. R. Taylor, *Nature (London)*, 1987, **329**, 351.
86. G. Fischer, in 'Enzyme Mechanisms', ed. M. I. Page and A. Williams, Royal Society of Chemistry, London, 1987, p. 229.
87. H. Umezawa, T. Aoyagi, H. Morishima, M. Matzusaki, H. Hamada and T. Takeuchi, *J. Antibiot.*, 1970, **23**, 259.
88. D. H. Rich, E. T. O. Sun and E. Ulm, *J. Med. Chem.*, 1980, **23**, 27.
89. R. Bott, E. Subramanian and D. R. Davies, *Biochemistry*, 1982, **21**, 6956.
90. D. H. Rich, *J. Med. Chem.*, 1985, **28**, 263.
91. J. Boger, in 'Peptides Structure and Function', ed. V. J. Hruby and D. H. Rich, Pierce Chemical Co., Rockford, 1983, p. 569.
92. L. Frick, R. Wolfenden, E. Smal and D. C. Baker, *Biochemistry*, 1986, **25**, 1616.
93. M. H. Gelb, J. P. Svaren and R. H. Abeles, *Biochemistry*, 1985, **24**, 1813.
94. S. Thaisrivongs, D. T. Pals, D. W. Harris, W. M. Kati and S. R. Turner, *J. Med. Chem.*, 1986, **29**, 2088.
95. P. G. Schmidt, M. W. Holladay, F. G. Salituro and D. H. Rich, *Biochem. Biophys. Res. Commun.*, 1985, **129**, 597; D. H. Rich, M. S. Bernatowicz and P. G. Schmidt, *J. Am. Chem. Soc.*, 1982, **104**, 3535.
96. P. A. Bartlett and W. R. Kezer, *J. Am. Chem. Soc.*, 1984, **106**, 4282; P. A. Bartlett, J. E. Hanson, F. Acher and P. P. Giannousis, in 'Biophosphates and Their Analogues', ed. K. S. Bruzik and W. J. Stec, Elsevier, Amsterdam, 1987, p. 429.
97. W. B. Dempsey and E. E. Snell, *Biochemistry*, 1963, **2**, 1414.
98. R. Wolfenden and L. Frick, in 'Enzyme Mechanisms', ed. M. I. Page and A. Williams, Royal Society of Chemistry, London, 1987, p. 97.

99. D. F. Wentworth and R. Wolfenden, *Biochemistry*, 1974, **13**, 4715.
100. J. W. Williams, J. F. Morrison and R. G. Duggleby, *Biochemistry*, 1979, **18**, 2567; D. A. Mathews, R. A. Alden, J. T. Bolin, S. T. Freer, R. Hamlin, N. Xuong, J. Kraut, M. Poe, M. Williams and K. Hoogsteen, *Science (Washington, D.C.)*, 1977, **197**, 452.
101. C. Frieden, L. C. Kurz and H. R. Gilbert, *Biochemistry*, 1980, **19**, 5303.
102. B. R. Baker, 'Design of Active Site Directed Irreversible Inhibitors', Wiley, New York, 1967.
103. R. R. Rando, *Science (Washington, D.C.)*, 1974, **185**, 320; R. H. Abeles and A. L. Maycock, *Acc. Chem. Res.*, 1976, **9**, 313; N. Seiler, M. J. Jung and J. Koch-Weser, 'Enzyme Activated Irreversible Inhibitors', Elsevier, Amsterdam, 1978.
104. S. G. Waley, *Biochem. J.*, 1980, **185**, 771.
105. I. M. Kovach, J. H.-A. Huber and R. L. Schowen, *J. Am. Chem. Soc.*, 1988, **110**, 590; I. M. Kovach, M. Larson and R. L. Schowen, *J. Am. Chem. Soc.*, 1986, **108**, 5490.
106. C. Kattner and E. Shaw, *Thromb. Res.*, 1979, **14**, 969.
107. H. Angeliker, P. Wikstrom, P. Rauber and E. Shaw, *Biochem. J.*, 1987, **241**, 871.
108. G. D. J. Green and E. Shaw, *J. Biol. Chem.*, 1981, **256**, 1923.
109. T. C. Friedman and S. Wilk, *J. Neurochem.*, 1986, **46**, 1231.
110. P. Lalégerie, G. Legler and J. M. Yon, *Biochemie*, 1982, **64**, 977; M. Herrchen and G. Legler, *Eur. J. Biochem.*, 1984, **138**, 527.
111. G. Leggler, *Hoppe Seyler's Z. Physiol. Chem.*, 1970, **351**, 25.
112. E. L. O'Connell and I. A. Rose, *J. Biol. Chem.*, 1973, **248**, 2225.
113. A. V. Fowler and P. J. Smith, *J. Biol. Chem.*, 1983, **258**, 10 204.
114. M. L. Sinnott, G. T. Tzotzos and S. E. Marshall, *J. Chem. Soc., Perkin Trans. 2*, 1982, 1665.
115. H. Angeliker, P. Wikstrom, P. Rauber and E. Shaw, *Biochem. J.*, 1987, **241**, 871.
116. J. S. McMurray and D. F. Dyckes, *Biochemistry*, 1986, **25**, 2298; R. L. Stein and D. A. Trainor, *Biochemistry*, 1986, **25**, 5414.
117. C. Betzel, G. P. Pal, D.-D. Jany and W. Saenger, *FEBS Lett.*, 1986, **197**, 105.
118. J. G. P. Malthouse, W. U. Primrose, N. E. Mackenzie and A. I. Scott, *Biochemistry*, 1985, **24**, 3478; A. I. Scott, N. E. Mackenzie, J. G. P. Malthouse, W. U. Primrose, P. E. Fagerness, A. Brisson, L. Z. Qi, W. Bode, C. M. Carter and Y. J. Jang, *Tetrahedron*, 1986, **42**, 3269.
119. W. U. Primrose, A. I. Scott, N. E. McKenzie and J. P. G. Malthouse, *Biochem. J.*, 1985, **231**, 677.
120. M. E. Ando, J. T. Gerig and K. F. S. Luk, *Biochemistry*, 1986, **25**, 4772; J. C. Powers *et al.*, *Biochemistry*, 1985, **24**, 2048.
121. J. Krant, *Annu. Rev. Biochem.*, 1977, **46**, 331.
122. L. A. Lambden and P. A. Bartlett, *Biochem. Biophys. Res. Commun.*, 1983, **112**, 1085.
123. A. D. Turner, D. M. Monroe, H. R. Roberts, N. A. Porter and S. V. Pizzo, *Biochemistry*, 1986, **25**, 4929.
124. J. M. Kaminski, J. Bauer, S. R. Mack, R. A. Anderson, Jr., D. P. Waller and L. J. D. Zaneveld, *J. Med. Chem.*, 1986, **29**, 514.
125. G. A. Digenis, B. J. Agha, K. Tsuji, M. Kato and M. Shinogi, *J. Med. Chem.*, 1986, **29**, 1468.
126. J. C. Powers, R. Boone, D. L. Carroll, B. F. Gupton, C.-M. Kam, N. Nishino, M. Sakomoto and P. M. Tuhy, *J. Biol. Chem.*, 1984, **259**, 4288.
127. M. Zimmerman, H. Morman, D. Mulvey, H. Jones, R. Frankshun and B. M. Ashe, *J. Biol. Chem.*, 1980, **255**, 9848.
128. L. Hedstrom, A. R. Moorman, J. Dobbs and R. H. Abeles, *Biochemistry*, 1984, **23**, 1753; A. Krantz, R. W. Spencer, T. F. Tam, E. Thomas and L. J. Copp, *J. Med. Chem.*, 1987, **30**, 589.
129. W. Tagaki, J. P. Guthrie and F. Westheimer, *Biochemistry*, 1968, **7**, 905.
130. K. Bloch, *Acc. Chem. Res.*, 1969, **2**, 193; R. R. Rando, *Science (Washington, D.C.)*, 1974, **185**, 320; R. R. Rando, *Acc. Chem. Res.*, 1975, **8**, 281; R. Abeles and A. L. Maycock, *Acc. Chem. Res.*, 1976, **9**, 313; B. W. Metcalf, *Med. Chem. Adv.*, 1981, 397; C. Walsh, *Tetrahedron*, 1982, **38**, 871.
131. M. Caplow and W. P. Jencks, *Biochemistry*, 1962, **1**, 883.
132. A. R. Moorman and R. H. Abeles, *J. Am. Chem. Soc.*, 1982, **104**, 6785; M. Gelb and R. H. Abeles, *J. Med. Chem.*, 1986, **29**, 585.
133. D. Ringe, J. M. Mottonen, M. H. Gelb and R. H. Abeles, *Biochemistry*, 1986, **25**, 5633.
134. D. Ringe, B. A. Seaton, M. H. Gelb and R. H. Abeles, *Biochemistry*, 1985, **24**, 64.
135. J. W. Harper and J. C. Powers, *Biochemistry*, 1985, **24**, 7200; E. F. Mayer, L. G. Presta and R. Radakrishnan, *J. Am. Chem. Soc.*, 1985, **107**, 4091.
136. S. B. Daniels and J. A. Katzenellenbogen, *Biochemistry*, 1986, **25**, 1436; M. J. Sofia and J. A. Katzenellenbogen, *J. Med. Chem.*, 1986, **29**, 230.
137. W. A. Boulanger and J. A. Katzenellenbogen, *J. Med. Chem.*, 1986, **29**, 1483.
138. J. W. Harper, K. Hemmi and J. C. Powers, *Biochemistry*, 1985, **24**, 1831.
139. R. W. Spencer, T. F. Tam, E. M. Thomas, V. J. Robinson and A. Krantz, *J. Am. Chem. Soc.*, 1986, **108**, 5589; L. J. Copp, A. Krantz and R. W. Spencer, *Biochemistry*, 1987, **26**, 169.
140. M. A. Navia, J. P. Springer, T.-Y. Lin, H. R. Williams, R. A. Firestone, J. M. Pisano, J. B. Doherty, P. E. Finke and K. Hoogsteen, *Nature (London)*, 1987, **327**, 79.
141. J. R. Knowles, *Acc. Chem. Res.*, 1985, **18**, 97; S. J. Cartwright and S. G. Waley, *Med. Res. Rev.*, 1983, **3**, 341.
142. F. De Maester, J.-M. Frère, G. Dive and J. Lamotte-Brasseur, *Biochem. J.*, in press.
143. V. Knott-Hunziker, B. Orlek, P. G. Sammes and S. G. Waley, *Biochem. J.*, 1979, **147**, 365; M. Loosemore, S. Cohen and R. Pratt, *Biochemistry*, 1980, **19**, 3990.
144. B. S. Orlek, P. G. Sammes, V. Knott-Hunziker and S. G. Waley, *J. Chem. Soc., Perkin Trans. 1*, 1980, 2322.
145. A. Magnien, B. Le Clef and J. F. Biellmann, *Biochemistry*, 1984, **23**, 6858.
146. M. C. Willingham, A. V. Rutherford, M. G. Gallo, J. Wehland, R. B. Dickson, R. Schegel and I. H. Pastan, *J. Histochem. Cytochem.*, 1981, **29**, 1003.
147. P. Bey, *Chem. Ind.*, 1981, 139.
148. P. Marcotte and C. Walsh, *Biochim. Biophys. Res. Commun.*, 1975, **62**, 677.
149. G. Burnett, K. Yonaha, S. Toyama, K. Soda and C. Walsh, *J. Biol. Chem.*, 1980, **255**, 428.
150. R. Wrigglesworth, in 'Enzyme Mechanisms', ed. M. I. Page and A. Williams, Royal Society of Chemistry, London, 1987, p. 506.

151. S. Ghisla, H. Ogata, V. Massey, A. Schonbrunn, R. H. Abeles and C. T. Walsh, *Biochemistry*, 1976, **15**, 1791; T. H. Cromartie and C. T. Walsh, *Biochemistry*, 1975, **14**, 3482.
152. J. A. Latham and C. A. Walsh, *J. Am. Chem. Soc.*, 1987, **109**, 3421.
153. B. W. Metcalf, B. Lippert and P. Casara, in 'Enzyme Activated Irreversible Inhibitors', ed. N. Seiler, M. J. Jung and J. Koch-Weser, Elsevier, Amsterdam, 1978, p. 123.
154. M. J. Jung, in 'Enzyme Activated Irreversible Inhibitors', ed. N. Seiler, M. J. Jung and J. Koch-Weser, Elsevier, Amsterdam, 1978, p. 135.
155. J. F. Rumigny, M. Maitre, M. Racasens, J. M. Blinderman and P. Mandel, *Biochem. Pharmacol.*, 1981, **30**, 305.

5.5

Resistance and Tolerance to Antimicrobial Drugs

LOUIS B. QUESNEL

University of Manchester, UK

5.5.1	THE ANTIBIOTIC ERA AND THE RISE OF ANTIBIOTIC RESISTANCE	90
5.5.1.1	*The Interpretation of Resistance*	90
5.5.1.2	*The Prevalence of Resistance in the Environment, Animals and Man*	91
5.5.2	THE IMPACT OF RESISTANCE ON THE TREATMENT OF DISEASE	92
5.5.3	CELL ENVELOPE IN RELATION TO SUSCEPTIBILITY AND RESISTANCE	95
5.5.3.1	*Differences Between Bacteria, Fungi, Protozoa and Viruses Affecting Susceptibility*	95
5.5.3.2	*Intrinsic Resistance, Crypticity and Tolerance*	97
5.5.3.2.1	*Intrinsic resistance*	97
5.5.3.2.2	*Crypticity*	97
5.5.3.2.3	*Tolerance*	98
5.5.4	THE MECHANISMS OF RESISTANCE IN BACTERIA	98
5.5.4.1	*Alteration of the Drug Target*	98
5.5.4.1.1	*Mutations in envelope components*	98
5.5.4.1.2	*Mutations in ribosomal components*	100
5.5.4.1.3	*Mutations affecting nucleic acid synthesis*	101
5.5.4.2	*Modification of the Drug: Detoxification*	102
5.5.4.2.1	*β-Lactams*	102
5.5.4.2.2	*Chloramphenicol*	106
5.5.4.2.3	*Aminoglycosides and aminocyclitols*	107
5.5.4.2.4	*Miscellaneous modifying enzymes*	109
5.5.4.3	*Altered Permeation and Transport*	110
5.5.4.3.1	*Altered permeability*	110
5.5.4.3.2	*Uptake: modifications of transport systems for D-cycloserine, phosphonomycin (fosfomycin), aminoglycosides*	111
5.5.4.3.3	*Efflux enhancement: tetracyclines*	111
5.5.4.4	*Metabolic By-pass Mechanisms*	113
5.5.4.5	*Overproduction of Antagonistic Metabolites and Target Enzymes*	113
5.5.5	RESISTANCE TO DRUGS ACTING AGAINST FUNGI	114
5.5.5.1	*Polyenes*	114
5.5.5.2	*Imidazoles*	114
5.5.5.3	*Flucytosine (5-fluorocytosine)*	115
5.5.6	DRUG DEPENDENCE	115
5.5.7	GENE MOBILITY AND THE SPREAD OF RESISTANCE	116
5.5.7.1	*Plasmids*	116
5.5.7.2	*Transposons*	117
5.5.7.3	*Other systems of Gene Transfer*	118
5.5.7.3.1	*Transformation*	118
5.5.7.3.2	*Transduction*	118
5.5.7.3.3	*Phage-meditated conjugation*	118
5.5.7.3.4	*Sex pheromone-induced conjugation*	118
5.5.8	REFERENCES	119

5.5.1   THE ANTIBIOTIC ERA AND THE RISE OF ANTIBIOTIC RESISTANCE

The use of antibiotics in chemotherapy is, probably, the single most important life-saving procedure to be introduced in the long history of medical treatment. No scourges of mankind have been greater or swifter in their spread than the pandemics of infectious microbial disease that swept through a defenceless and ignorant world before the antibiotic era. In a mere four months between June and September 1831, cholera raged through Hungary infecting more than a quarter of a million people and killing over 100 000. In Cairo and Alexandria, 30 000 people were believed to have died in a single day.

The first breakthroughs in antimicrobial chemotherapy came with the use of quinine from cinchona bark extract to treat malaria and emetine from the roots of the ipecacuanha tree to treat amoebic dysentery. Here specific chemicals had been identified as the curative agents of long-established or tribal remedies, but the application of modern scientific method to the search for the magic bullets which would 'seek out' and destroy the agents of disease really began with the work of Ehrlich who had clearly grasped the concept of selective toxicity and first coined the word 'chemotherapy'. Salvarsan (1909) and neosalvarsan (1912), the two arsenicals he introduced for the treatment of syphilis, were miracle drugs of their day, and with Fleming's discovery of penicillin, reported to the Medical Research Club in 1928, and Domagk's discovery of Prontosil (1935) from which the Trefuels (1935) identified the sulfanilamide 'moiety' as the antibacterial molecule, the new antibiotic era had dawned. Although Fleming was unable to identify and purify the active compound later deduced by Abraham and his Oxford group in 1943, there is no question that he realized its importance as a potential chemotherapeutic agent.[1]

Since these early beginnings, many thousands of antibiotics have been either discovered as natural products of organisms or invented in the chemical laboratory *de novo* or as modifications of the natural compound. In this contribution an antibiotic is described as an organic compound acting at very low concentrations to inhibit the growth or kill microorganisms, without deference to their origin.

5.5.1.1   The Interpretation of Resistance

While an antibiotic, by definition, must be able to inhibit growth or kill, for clinical application it must achieve therapy without significant damage to the host. Of the thousands in existence only a few dozens find daily clinical application and no one drug provides therapy for all infections. Those organisms which respond are said to be susceptible, while those that are not eliminated by clinically achievable levels of antibiotic are considered resistant. However, clinical efficacy depends on a range of interactions between the antibiotic and the host as well as between the antibiotic and the parasite. The pharmacological and pharmacokinetic attributes of an administered drug will determine how much drug is available at the site of infection and for how long, and these will differ with different drugs under different circumstances of administration route, dosage, drug distribution, metabolism and excretion (see Volume 5). In addition, the immunological status of the host is of major significance as has become all too obvious in the context of the AID syndrome.

The results of *in vitro* sensitivity tests give an indication of the likely performance and efficacy of the drug *in vivo*, but only a wide experience of the use of the drug and the correlation of clinical outcome with laboratory tests can give a reliable evaluation of susceptibility. In practical terms the response of the organism isolated from the infection is compared with that of organisms of known susceptibility status. If the clinically isolated organism is inhibited (or killed) by the same or lower concentrations of antibiotic required to inhibit the control organism under the same test conditions, then the pathogen may be regarded as susceptible. If relatively higher concentrations of drug are required to produce the same *in vitro* response, the organism will be regarded as resistant unless proved otherwise by subsequent clinical experience. Depending on how great the difference and how wide the tolerance of the host for the antibiotic, categories of 'moderate' or 'intermediate' resistance may be defined.

It is important, in these comparisons, that control organisms should be of the same type, responding in the same way to factors affecting drug action, such as pH, divalent cations, oxygen availability, *etc.*

In the context of the patient it should always be remembered that the situation is a dynamic one, the antibiotic level continuously changing while the population density of the invading organism will be increasing when the 'available' drug is at a concentration below the minimum inhibitory concentration (MIC) and decreasing when it is above the MIC. Because the concentration of drug in different body compartments may differ significantly, it has been suggested that attainable safe serum levels should be at least fourfold greater than the MIC for the parasite.

Resistance to an antibiotic may arise abruptly by mutation of the organism, thereby conferring the ability to survive very high drug concentrations, even several orders of magnitude greater. On the other hand, it may arise more insidiously through a succession of small but cumulative mutational changes, ultimately resulting in an increase of therapeutic failure which renders the drug useless. It is also possible to recognize, in a bacterial population descended from a single cell, a distribution of response to the drug such that some members of the population are more resistant, and the greater the total population the greater will be the number of these more resistant cells. When selected and regrown into new populations, similar distributions are obtained, so these are 'phenotypically more resistant' members of a population.

5.5.1.2 The Prevalence of Resistance in the Environment, Animals and Man

While it is probable that new resistance genes are continually evolving, some genes for resistance have existed for a very long time, among which are the genes for resistance found in the antibiotic-producing organisms themselves. Although it is not provable, antibiotic-synthesizing organisms may have existed for many millenia, having evolved genes which protect them against their own antibiotic products of secondary metabolism. Such would seem to be the case, for example, in streptomycin-producing strains of *Streptomyces griseus* which possess the enzyme aminoglycoside 6-phosphotransferase, which converts streptomycin to its 'inactive' 6-phosphoryl derivative. This enzyme is induced by the presence of streptomycin so that the producer strain attains high levels during synthesis, [2,3] which also protect the organism against streptomycin re-entering the producer from the external milieu.[4] Conversely, *S. griseus* strains not able to produce the enzyme do not synthesize streptomycin.[5] Similarly, the rifampicin producer *Nocardia mediterranei* produces a resistant RNA polymerase so that it cannot be self-intoxicated,[6] and *Micromonospora purpurea* possesses ribosomes which are resistant to gentamicin, which it produces.[7] *Streptomyces azureus* strains producing thiostrepton are preserved from its action by the functioning of a methylase enzyme modifying the RNA of the producer.[8] Other mechanisms of self-protection have been reviewed by Cundliffe.[9]

Genes conferring resistance have probably been spread from producer organisms to non-producers *via* the agencies of transposons and plasmids (see Section 5.5.7 below) so that antibiotic resistance genes have been present in potential pathogens even before the advent of antibiotic chemotherapy. This has been confirmed by the examination of cultures lyophilized before the clinical use of antibiotics.[10] These strains have been found to possess not only resistance genes, but plasmids which provide the vectors for their transmission and spread.

However, in the absence of selective pressure imposed by the presence of drugs in the environment, the incidence of resistance was extremely low (less than 1% of *Staphylococcus aureus* resistant to penicillin) and in 1941 nearly all pathogenic strains were sensitive to antibiotics when they were first used. Since then the abundant and continued use of antibiotics throughout the world has led to the very rapid selection and spread of resistant strains and the dissemination of resistant genes. Both the variety and quantity of antibiotics manufactured and used continue to increase and the production of antibiotics in the USA rose exponentially between 1950 and 1978, doubling about every five years to over 11 000 tonnes in 1978, about half of which was incorporated in animal feeds and feed additives.[11] Other developed countries also have substantial antibiotic industries; the burden of these drugs in the clinical and natural environments is now considerable, and the incidence of resistant bacteria is directly related to the intensity of antibiotic usage.[12] Furthermore, their use continues unrestricted in many countries and, even where restricted, abuse is often difficult to monitor or detect, and the problems of the spread of drug resistance can be expected to increase for the foreseeable future.

Although the susceptibility of some animal pathogens, such as the streptococci of bovine mastitis, has not changed after many years of antibiotic use, their prevalence as causative organisms has declined and most cases of mastitis are now caused by penicillin-resistant staphylococci or penicillin-tolerant bacteria such as *Escherichia coli*.[13] Not long after the introduction of penicillin therapy for mastitis, resistant strains of *Staphylococcus aureus* producing β-lactamases appeared and increased in frequency, such that since 1961 about 70% of all isolates are penicillin resistant.[14]

While the incidence of resistance to antibacterial agents among the enteric bacteria of farm animals varies both among and between the different genera and species,[15] the incidence of resistance among salmonellas in the UK and other countries has been high and frequently associated with the spread of disease from animals to man.[16] The occurrence and spread of strains of *Salmonella typhimurium* of phage types 193, 204, 204a and 204c in cattle and man in the UK has been well documented.[17-20] By 1985, 77% of strains from cattle were drug resistant and 84% of the resistant strains were multiresistant with increasingly more complex plasmid-borne drug-resistance patterns. The sulfonamide and tetracycline resistance of phage type 204 (1974) was largely replaced by 1980 by strains of phage type 204c resistant to ampicillin (A), chloramphenicol (C), kanamycin (K), streptomycin (S), sulfonamides (Su), tetracyclines (T) and trimethoprim (Tm); by 1985, 25% were gentamicin resistant as well.[20] Infections in man followed infections in the animals. Some investigators[21] find the evidence of transfection from animals to man less convincing; nonetheless, other resistant pathogens such as *Campylobacter jejuni*[22] and enteropathogenic *E. coli*, as well as resistant commensal coliforms and other Gram-negative bacilli, are passed from animals to man *via* foods or by direct contact with animals.[23] In an experiment with human volunteers, Linton *et al.*[24] showed conclusively that *E. coli* with R-plasmids of animal origin could be established in the gut of humans consequent upon the normal handling and preparation of commercially obtained chicken carcasses.

While the selective pressure of antibiotics will ensure the maintenance of resistant strains in animals or man, the removal of antibiotics does not necessarily lead to the disappearance of resistant strains. In a long-term study of drug resistance in coliforms isolated from pigs, Langlois *et al.*[25] showed that in spite of an initial drop in the incidence of ampicillin, streptomycin, sulfonamide and tetracycline resistance during the first year after total withdrawal of all antibiotics, resistance levels persisted at relatively high frequencies over a period of ten and a half years in the absence of antibiotics.

For further information the reviews of Linton,[12] Hinton *et al.*[15] and Feinman[26] should be consulted.

5.5.2 THE IMPACT OF RESISTANCE ON THE TREATMENT OF DISEASE

As mentioned earlier, not all bacteria will be clinically susceptible to any one drug and a bacterial species sensitive to one may be resistant to other antibiotics. This obviously implies the necessity to match the antibiotic to the infection and Table 1 gives the spectrum of response of some commonly

Table 1 Susceptibility of Pathogenic Microbes to Commonly Used Antimicrobials[a]

Organism	Penicillins	Cephalosporins	Aminoglycosides	Tetracyclines	Macrolides	Chloramphenicol	Sulfonamides	Trimethoprim	Metronidazole	Polymyxins	4-Quinolones	Imidazoles	Polyenes
Bacteria													
Gram-positive													
Staph. aureus	±	+	(+)	(+)	(+)	+	(+)	+	−	−	+	−	−
Str. pyogenes	+	+	−	(+)	+	+	(+)	+	−	−	+	−	−
Str. faecalis	+	−	−	(+)	+	+	(+)	(+)	−	−	+	−	−
Other streptococci	+	+	−	(+)	+	+	(+)	+	−	−	+	−	−
Clostridium spp.	+	+	−	+	+	+	(+)		+	−	(+)		−
Gram-negative													
E. coli	±	±	(+)	(+)	−	(+)	(+)	(+)	−	+	+	−	−
Other enterobacteria	±	±	(+)	(+)	−	(+)	(+)	(+)	−	+	+	−	−
Ps. aeruginosa	±	±	±	−	−	−	−	−	−	+	(+)	−	−
H. influenzae	±	±	−	(+)	+	+	(+)	(+)	−	−	+	−	−
Neisseria spp.	±	+	−	(+)	+	+	(+)	−	−	−	+	−	−
Bacteroides spp.	−	±	−	(+)	+	+	(+)	−	+	−	(+)	−	−
Others													
Mycobacterium spp.	−	−	±	−	−	−	−	−	−	−	(+)	−	−
Chlamydia	−	−	−	+	+	+	+	−	−	−	+	−	−
Mycoplasma	−	−	−	+	+	+	−	−	−	−	+	−	−
Fungi	−	−	−	−	−	−	−	−	−	−	−	±	±

[a] + = usually sensitive; − = usually resistant; (+) = variable response among strains; ± = sensitivity dependent on strain and specific drug.

encountered pathogens to a range of antibiotics. The development of drug resistance will limit the choice of therapeutic agent and may impose on the physician the use of a less effective drug, or a more expensive one, frequently with greater toxicity or undesirable side effects.

Natural insensitivity or intrinsic resistance is determined by the genetic make-up of a strain or species and is persistent so that, for example, a drug unable to penetrate to its target within the cell because of impermeability will remain ineffective unless a mutation occurs to reduce the barrier to drug entry. Where the drug is actively transported into the cell, mutation of transport-related genes which enhance uptake may actually convert strains to a supersensitive state, as found with *unc* mutants to enhanced aminoglycoside susceptibility.[27] Such mutations would obviously be of no adverse clinical significance in the disease process.

Resistance of clinical significance develops in one of two distinct ways: (a) by mutation of the individual genotype or (b) by the acquisition of new genetic elements from another organism. In the former case the mutational event may confer single-step high-level resistance (*e.g.* to streptomycin) or by a succession of events lead to increasingly higher levels of resistance (*e.g.* to penicillin or gentamicin). The spontaneous mutation rate varies for each resistance gene but, by definition, the greater the microbial population, the higher the probability of finding a resistant mutant. If the mutation rate for a particular 'resistance gene' is 10^{-6}, then a population greater than a million will almost certainly contain one or more individuals with the mutant gene and those few individuals can be selected by application of an appropriate concentration of the drug, killing all the sensitive organisms and permitting their replacement by a population totally resistant to that drug concentration, the situation which obtains when therapy is initiated late against a well-established infecting population.

In theory, the use of combination therapy would prevent such a situation since the probability of mutation to two drugs simultaneously is the product of the probabilities of each singly. Unfortunately, the mobility of genes by transposition, and their accumulation on transmissible genetic elements, has made such treatment likely to fail since resistance to both drugs (and several others) may be simultaneously carried (see Section 5.5.7 below).

Resistance to streptomycin, novobiocin, rifampicin and fusidic acid can all arise by mutation in *Staphylococcus aureus* during therapy and may have fatal consequences.[28] Generally, however, clinically important drug resistance results from the spread of R-plasmids carrying multiple resistance to antibiotics and the problems they have generated are serious and widespread.

An outbreak of dysentery started in 1969 in Guatemala[29] and spread rapidly through Central America, causing thousands of deaths. The causative *Shigella dysenteriae* type 1 carried plasmid-determined resistance to sulfonamides, chloramphenicol, streptomycin and tetracyclines[30] (CSSuT), which rendered these drugs useless and an ampicillin resistance plasmid was later acquired.[31]

A widespread and long-lasting outbreak due to multiply resistant *Sh. dysenteriae* type 1 also occurred in Bangladesh, India and Sri Lanka[32,33] in the early 1970s and the strain carried an incompatibility Group B plasmid coding for C and T resistance and an independent non-conjugative SSu plasmid.[34] Since 1979 another extensive outbreak of multiply resistant *Sh. dysenteriae* type 1 has spread from Zaire and Somalia to Rwanda and Burundi,[35] carrying two R-plasmids (Inc group X) coding for ACT and SSu (non-conjugative). By 1981 it had acquired an additional plasmid (Inc group I) coding for ACSSuTTm.

In a survey of resistance to 12 antibiotics in strains of *Sh. dysenteriae*, *Sh. flexneri* and *Sh. boydii* from England and Wales in 1974–1978, 80% of strains were found resistant to two or more drugs; Su resistance was commonest.[36] In *Sh. sonnei*, multiple resistance fell from 38% in 1972 to 8% in 1977;[36] however a survey of all four species in Manchester (UK) during 1981–1984 found 85% of strains resistant to two or more of 10 drugs tested and 12% to six or seven drugs; 86% of 386 *Sh. sonnei* isolates were resistant to two or more drugs, 61% being resistant to ampicillin.[37]

Since 1970, multiply resistant salmonellas have caused severe infection with high incidences of septicaemia and mortality in paediatric units in many countries. Serotypes included *S. typhimurium* phage type 208 in the Middle East,[38] and later a phage type 66/122 in Asia and the Middle East since 1978.[39] The majority of isolates were ACKSSuT resistant and, in addition, frequently Tm and gentamicin (Gm). Other serotypes which have caused serious outbreaks were *S. wein* in North Africa, Southern Europe and the Middle East,[40] *S. saint-paul* in South America, *S. johannesburg* in Hong Kong,[41,42] *S. newport* in New Delhi,[43] *S. oranienburg* in Brazil[44] and *S. ordonez* in Senegal.[45]

Salmonella typhi, for long chloramphenicol-sensitive almost everywhere, emerged in C resistant form and caused an epidemic of major proportions with many fatalities in Mexico in 1972; the strain carried a CSSuT R-plasmid.[46] Since then, chloramphenicol-resistant *S. typhi* have been found associated with outbreaks in several countries.[47]

Other organisms readily treated in the past have now developed resistances that demand the use of alternative therapies. Strains of *Neisseria gonorrhoeae* readily eliminated by benzylpenicillin exhibited gradual small increases in resistance, from less than 0.01 mg L^{-1} but rarely to greater than 2 mg L^{-1}, and remained controllable by adequate doses of penicillin until the abrupt appearance in 1976 of strains producing β-lactamase in both the United States[48] and the UK,[49] with origins in the Far East and West Africa. The β-lactamases are TEM-1 like and several different plasmids have been found. They have now been disseminated through Europe and therapy with spectinomycin or one of the newer β-lactams such as cefuroxime, cefoxitin or cefamandole has been used.

Similarly, gonococcal strains were initially sensitive to less than 0.2 mg L^{-1} of tetracycline, but have developed several mechanisms increasing resistance. Mutations involve the ribosome or outer membrane proteins. Mutations in *mtr* controlling a 52 kDa outer membrane protein limit drug uptake and enhance resistance to acridine orange, rifampicin, penicillin and chloramphenicol as well as tetracycline. Even more alarmingly, high-level tetracycline resistant *N. gonorrhoeae* are now becoming widespread in the U.S.A, Europe and Asia, with strains carrying and transferring resistance to penicillin (β-lactamase) and *tet* M tetracycline resistance (MICs 24–32 mg L^{-1}).

Table 2 Therapies Under Threat or Unavailing Because of Resistance

Infection	Under threat	Unavailing
Meningococcal meningitis and septicaemia		Sulfonamides (initially)
H. influenzae infections	Chloramphenicol	Ampicillin (initially)
Urinary infections	Ampicillin, trimethoprim	
Gonorrhoea	High-dose penicillin	Low-dose penicillin
Pneumococcal (including meningitis)	Penicillin	
Salmonellosis (including typhoid)	Chloramphenicol, cotrimoxazole, amoxycillin	
Cholera		Tetracycline
Methicillin-resistant staphylococcal infection	Many antibiotics	
Shigellosis		Sulfonamides, ampicillin
Hospital coliform infection	Many antibiotics	Ampicillin

Table 3 Antibiotic Resistance of Bacteria Isolated in Hospitals in 17 Locations in W. Germany, Three in Switzerland and Three in Austria[53]

Bacteria[a]	ESC	PRI	PRM	ENT	SER	KLE	CIT	SAL	PSA	SAU	SFA
Number of strains	834	178	345	236	57	411	66	49	360	621	427
Antimicrobials					*Percentage of resistant strains*						
Ampicillin	22	94	13	94	91	99	91	6			<1
Penicillin G										66	
Mezlocillin	9	9	2	19	23	20	15	6			<1
Azlocillin									15		
Oxacillin										2	
Cefazolin	1	87	2	75	95	11	32				
Cefoxitin	1	1	0	79	39	3	50	0			
Gentamicin	<1	6	2	2	14	5	2	0	11	6	
Tobramycin	<1	2	<1	1	21	4	1	0	5	7	
Amikacin	0	0	0	0	0	0	0	0	0	0	
Kanamycin	11	5	13	9	18	11	11	4		11	
Streptomycin	33	28	26	15	18	22	24	6		10	
Nalidixic acid	2	5	3	2	2	8	2	0			
Sulfamethoxazole	34	7	19	12	26	25	8	6		8	96
Trimethoprim	15	14	19	10	37	17	9	0		6	5
Cotrimoxazole	10	7	10	4	14	12	6	0		3	3
Tetracycline	28	62	98	6	79	17	12	6		14	58
Chloramphenicol	16	18	21	8	77	20	11	4		4	24
Erythromycin										8	23
Lincomycin					·					2	
Vancomycin										0	0
Fusidic acid										2	

[a] ESC, *E. coli*; PRI, *Proteus* other spp.; PRM, *Pr. mirabilis*; ENT, *Enterobacter* spp.; SER, *Serratia* spp.; KLE, *Klebsiella* spp.; CIT, *Citrobacter* spp.; SAL, *Salmonella* spp.; PSA, *Pseudomonas aeruginosa*; SAU, *Staph. aureus*; SFA, *Str. faecalis*.

The *tet* M mediated resistance, first reported in *Streptococcus faecalis*, is now spreading among *N. meningitidis*, *Kingella denitrificans* and *Eikenella corrodens*.[49a]

Nosocomial Gram-negative infections have now become a major problem because of the widespread prevalence of resistance to many drugs and the readiness with which these resistances spread. The organisms that have come into prominence are *Klebsiella*, *Pseudomonas*, *Serratia*, *Providencia*, *Proteus* and *Enterobacter* and among Gram-positive bacteria the methicillin-resistant staphylococci have caused serious problems in some countries.[50]

Gram-negative septicaemias, pneumonias and urinary tract infections have become difficult to eradicate because of high resistance levels and the multiplicity of drugs rendered ineffective. In a study of gentamicin-resistant *Klebsiella*, 89% of strains were resistant to 10 or more antibiotics.[51]

Pseudomonas aeruginosa in particular has a high degree of intrinsic resistance to many antibiotics, and the problem is aggravated by the fact that many different R-factors have been associated with this organism. More than 10 different β-lactamases are found and as many as three in the same strain.[52] Several antibiotics with enhanced 'antipseudomonas' activity have now been introduced, among which are the β-lactams azlocillin, piperacillin, cefsulodin, imipenem and aztreonam, and the aminoglycosides sisomicin and amikacin.

Table 2 lists some of the therapies under threat or unavailable as first-line drugs, and Table 3 gives the results of a survey of hospital isolates from 17 locations in West Germany, Switzerland and Austria showing prevalence of resistance to a number of antibiotics.[53] International surveys from centres in the USA, South America, Europe, South Africa, Australia and the Philippines[54] have shown highly variable levels of resistance between organisms and antibiotics. *E. coli* ampicillin resistance ranged from 16 to 73% and for *Proteus mirabilis* from 3 to 56%. For chloramphenicol resistance, *E. coli* ranged from 2 to 48%, *Klebsiella pneumoniae* from 5 to 52% and *Serratia marcescens* from 8 to 67%. Gentamicin resistance in *K. pneumoniae* ranged from 5 to 54%. Clearly, a knowledge of the likelihood of failure of a proposed therapy will depend on an awareness of the prevailing level of resistance, and this knowledge guides the rational choice of antibiotics in the individual situation.

5.5.3 CELL ENVELOPE IN RELATION TO SUSCEPTIBILITY AND RESISTANCE

The targets of antibiotic action may be found in the wall, the cytoplasmic membrane, the ribosome, nucleic acids or cellular organelles and substructures as well as particular metabolic reactions. The ease and rapidity with which the drug arrives at its target may greatly affect the ability of the cell to survive the attack. For some antibiotics the passage will be easy; for others there are several barriers depending on the type of cell and the physicochemical properties of the antibiotic.

5.5.3.1 Differences Between Bacteria, Fungi, Protozoa and Viruses Affecting Susceptibility

The envelopes of different microorganisms differ considerably. Gram-negative organisms have more complex structures than Gram-positive ones and their walls differ from those of fungi. Their cytoplasmic membranes also differ (Figure 1).

Both Gram-positive and Gram-negative cells may possess a capsule or slime layer on their external surface. These layers are exopolysaccharides more or less loosely associated with the wall and occur in many pathogens. In mucoid strains of *Pseudomonas aeruginosa* they are heteropolymers of varying proportions of D-mannuronosyl and L-guluronosyl moieties.[55] It is not certain to what extent they inhibit the diffusion of antibiotics, but there is some evidence that these anionic layers retard the entry of cationic antibiotics such as the aminoglycosides.[56]

The Gram-negative outer membrane (exterior to the peptidoglycan) is composed of lipopolysaccharides (LPS), phospholipids, lipoproteins and proteins. The outer leaflet of the bilayer contains all the LPS, which is negatively charged and highly variable in its attached sugar moieties, and is responsible for the antigenic variability on which serotyping schemes are based. The hydrophilic external polysaccharide of LPS is covalently linked *via* a phosphorylated core to the 'lipid A' moiety which is hydrophobically anchored in the lipid layer. The inner leaf contains the majority of the phospholipid and the embedded Braun lipoprotein links the outer membrane to the peptidoglycan (usually covalently).[57] Traversing both leaves and breaching the hydrophobic barrier posed by the lipid bilayer are the major outer membrane proteins (Omp) or porins, which provide aqueous channels for the access of hydrophilic molecules. For details consult the excellent review by Nikaido and Vaara.[58]

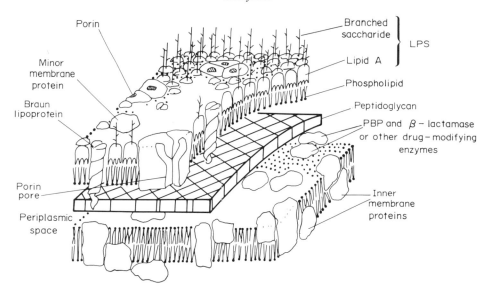

Figure 1 Diagram of the Gram-negative bacterial cell envelope

Both *Escherichia coli* and *Salmonella typhimurium* produce multiple species of porins, in *E. coli* coded by *Omp*F and *Omp*C genes, and under phosphate starvation a *pho*E gene is derepressed for phosphate transport. In addition to OmpF and OmpC proteins, salmonellas possess an OmpD.[59] The porins form trimeric clusters to give pores of diameter 1.2 nm for OmpF (porin 1a) and 1.1 nm for OmpC (porin 1b).[58] Permeability discrimination is on the basis of gross physicochemical properties of molecular size, hydrophobicity and charge, with a cut-off for diffusion of about 600 Da. Table 4 gives partition coefficients and molecular weights of some antibiotics.

Beneath the outer membrane of the Gram-negative cell (which is absent from Gram-positive cells) is a thin layer of peptidoglycan comprising cross-linked polymers of *N*-acetylglucosaminyl-*N*-acetylmuramyl peptide, covalently cross-linked both glycosidically and by transpeptide bonds. Four different chemotypes are found.[60] The Gram-positives contain, in addition, negatively charged, hydrophilic, covalently linked polymers of teichoic and teichuronic acid and the peptidoglycan is

Table 4 Factors Affecting Accumulation of Antibiotics

Antibiotic	M(Da)	Partition coefficient[a]	OM[b] passage	IM[c] passage
Novobiocin	613	>20	fat soluble	fat soluble
Chloramphenicol	323	12.4	fat soluble	fat soluble
Rifamycin SV	698	8.8	(fat soluble)[d]	?ED[e]-transport
Nalidixic acid	232	3.16	(fat soluble)	fat soluble
Minocycline	494	1.2	(fat soluble)	ED-transport
Erythromycin	734	0.79	(fat soluble)	?ED-transport
Climdamycin	390	0.7	diffusion[f]	?ED-transport
Nafcillin	414	0.31	diffusion	not needed
Chlortetracycline	479	0.31	diffusion	ED-transport
Bacitracin	1411	0.12	amphipathic	not needed
Oxacillin	418	0.07	diffusion	not needed
Tetracycline	444	0.07	diffusion	ED-transport
Polymixin B	1200	<0.05	amphipathic	not needed
Penicillin G	334	0.02	diffusion	not needed
Cloxacillin	458	0.02	diffusion	not needed
Neomycin	909	<0.01	diffusion	ED-transport
Vancomycin	1450	<0.01	Gm positive[g]	not needed
D-Cycloserine	102	<0.01	diffusion	ED-transport
Cephalothin	418	<0.01	diffusion	not needed
Carbenicillin	433	<0.01	diffusion	not needed

[a] In 1:1 octanol:sodium phosphate, 0.05 M, pH 7, 24 °C. [b] OM, outer membrane. [c] IM, inner membrane. [d] Brackets denote diminishing hydrophobicity. [e] ED, energy dependent. [f] Entrance primarily through porin pores. [g] Action primarily on OM-free cells.

considerably denser and a significant proportion of the total dry weight. In neither cell type is the peptidoglycan likely to be a barrier to antibiotic entry (other than by weak charge effects) since the exclusion molecular size is about 10^5 Da.

The cytoplasmic membrane is the final and most serious barrier to entry. In both types of bacteria it is a lipid bilayer primarily of phosphatidylethanolamine, phosphatidylglycerol and candiolipin, but variations are found. Embedded in or attached to the membrane are numerous proteins such as the enzymes assembling peptidoglycan, enzymes of the respiratory chain, transport proteins, *etc.* Many antibiotics exert their action either in or outside the cytoplasmic membrane and only those which have targets within the cytoplasm must find means of penetration to the cell interior (Table 4).

The fungal cell envelope is a complex structure in which the extension zone (hyphal tip) is simpler and flexible, becoming rigid and thickened by additional components as it matures. The tips of all hyphae examined are bounded by a wall comprising chitin or cellulose microfibrils embedded in an amorphous matrix of protein. Glucans and glycoprotein may also be incorporated. The secondary (older) wall may consist of several layers; in *Neurospora crassa*, for example, the inner layer of chitin is overlaid by a proteinaceous layer, outside which is a glycoprotein reticulum overlaid by a layer of mixed α- and β-glucans.[61]

It is not clear how significant a barrier the fungal cell wall is. There is some evidence that the cut-off size is about 800 Da in *Saccharomyces cerevisiae*;[62] nevertheless, polyenes with molecular weights of about 1000 readily pass into the plasma membrane.

The membrane of fungi may contain high proportions of carbohydrate and, in addition to phospholipids, sterols such as cholesterol and ergosterol are found. The relative proportions of the components vary with growth phase and growth form whether yeast like or mycelial.[63] These variations may affect the response of the organism to antimycotic drugs.

The structures of viruses have been reviewed recently.[64] Some possess lipid bilayer membranes ('enveloped' viruses), acquired on budding through the host cell membrane, reflecting the composition of those membranes, and usually containing one or more surface glycoproteins; others are without membranes. The possession of membranes may affect the sensitivity of viruses to antibiotic compounds, but most useful antivirals act against stages of virus replication in which biochemical differences from normal host cell metabolism permit the exploitation of specific 'attack' strategies.

5.5.3.2 Intrinsic Resistance, Crypticity and Tolerance

5.5.3.2.1 *Intrinsic resistance*

Intrinsic resistance is the resistance common to the majority of isolated strains. Genetically, intrinsic resistance is associated with chromosomal determinants—the 'normal' genomic constituents of the cell.[65] The gradual shift in resistance seen with some species such as gonococci and pneumococci already mentioned are shifts in intrinsic (non-enzyme) factors. The barrier functions of the envelope layers of the cell described in the previous section also represent aspects of the intrinsic resistance of organisms.

5.5.3.2.2 *Crypticity*

The outer membrane of the Gram-negative bacterium provides a significant barrier to antibiotic permeation and may greatly affect the expression of resistance by β-lactamase-producing cells, where these hydrolytic enzymes lie in the periplasmic space between inner and outer membranes. The rate at which molecules enter and the rate constants determining hydrolysis of the β-lactam substrate (see Section 5.5.4.2.1) may govern the outcome of interaction between drug and cell. Although the exclusion limit of porins may be *ca.* 600 Da, diffusion is affected in proportion to molecular size, especially nearer the upper entry limit, so that, for example, lactose (342 Da) penetrates 300 times more slowly than glycerol (92 Da). However, many β-lactams have similar molecular weights but different permeability coefficients because of the differences in hydrophobicity resulting from side-chain substitutions. There is an inverse linear relationship between permeability coefficient and hydrophobicity such that a tenfold increase in octanol/water partition coefficient (hydrophobicity) of the unionized drug reduces permeability about five- to six-fold.[66] Cephacetrile and cefazolin, which are of low hydrophobicity, penetrate about 10 times more rapidly than cephalothin and cephaloram. Other factors being equal we would, therefore, expect a β-lactamase Gram-negative producer strain to be better able to defend itself against the latter than the former.

The quantification of this advantage is not easy, but a means of estimating it has been suggested by Richmond and Sykes,[67] who define the 'crypticity factor' as the ratio of β-lactamase activity of broken bacterial cells to the β-lactamase activity of unbroken bacteria. In other words, the concentration of available substrate (antibiotic) is rate limiting in the 'unbroken' system, but it is not in the 'broken cell suspension'; hence the relative efficiency of the 'barrier' is indicated. Cephaloridine, which penetrates *E. coli* rapidly, will 'overcome' a cell with relatively large amounts of β-lactamase, but ampicillin, which penetrates poorly, is hydrolyzed as it enters and prevented from accumulating in lethal concentration at the target site (PBPs—see Section 5.5.4.1.1). It is clear, however, that the concentration of β-lactamase in the periplasmic space will be important and the inducibility of the β-lactamase will also play a part in the outcome. The crypticity factor will, therefore, differ between induced and non-induced cells. These phenomena have been considered in detail in a recent symposium.[67a]

5.5.3.2.3 *Tolerance*

Pneumococcal mutants which were 'penicillin-tolerant' were first isolated in 1970.[68] Although they contained normal penicillin-binding proteins (PBPs), which were as accessible to the drug as those of 'sensitive' cells, inhibition of these proteins did not initiate lysis and the cells lost viability only slowly. Antibiotic inhibition of PBPs, therefore, does not automatically lead to lysis and death.

The ability to tolerate penicillin has been found to be due to a defect in the autolysin system responsible for the degradation of preformed peptidoglycan, which is necessary before the insertion of peptidoglycan units augmenting the cell wall and permitting growth of small cells into larger ones. It has been found that the 'excess' autolytic activity of wild-type sensitive pneumococci is not essential for normal growth and division, and Tomasz[69] has argued that penicillin treatment was unable to 'trigger' autolytic activity in these mutants.

5.5.4 THE MECHANISMS OF RESISTANCE IN BACTERIA

Almost as soon as antibiotics were isolated, mechanisms of resistance to them were also found and by 1940 the existence of penicillinase had already been recorded. Although the systematic elucidation of the numerous strategies of resistance we now know was to follow much later, seven possible mechanisms of resistance were forecast as early as 1952.[70]

In their 1978 review, Davies and Smith[71] considered plasmid-determined antibiotic resistance under the following headings: (i) alteration of the target site; (ii) blocking the transport of the antibiotic into the cell, whether specific or active mechanisms were involved; (iii) detoxification or inactivation of the antibiotic; (iv) replacement of the inhibited metabolic step, *i.e.* a by-pass mechanism; (v) increasing the level of enzyme inhibited by the drug; (vi) production of a metabolite antagonizing the inhibitory effect; and (vii) decreasing the cells' metabolic requirement for the pathway or reaction inhibited by the drug.

At the time, Davis and Maas made their 1952 predictions (now largely fulfilled), the possibility that organisms could develop resistance to several drugs simultaneously was not considered and the first reports of conjugal transfer of multiple drug resistance in 1959 (reported by Watanabe 1960[72]) surprised and alarmed microbiologists. Several of the resistance mechanisms can thereby be simultaneously transferred.

It must be admitted that, to date, no examples have been found of certain resistance categories. They may yet be found. Also, some resistance mechanisms result from single gene mutations, for example (i), (ii) and (v) above; others (iii, iv and vi) may require multiple gene mutations.

5.5.4.1 Alteration of the Drug Target

5.5.4.1.1 *Mutations in envelope components*

As mentioned in Section 5.5.3.1, the (inner) cytoplasmic membrane of bacterial cells is associated with enzymes responsible for the assembly of the wall peptidoglycan. The use of radiolabelled benzylpenicillin has permitted the isolation and identification of several proteins which bind penicillin (PBPs) and whose enzymic activity is related to wall synthesis. The numbers of these PBPs varies between species but can be separated into two broad categories: the transpeptidases and the

DD-carboxypeptidases.[73] Transpeptidases also perform endopeptidic cleavage of existing peptide bonds and it now seems that in *E. coli*, and at least some others, the transpeptidases are also transglycosylases.[74]

The PBPs can be separated electrophoretically and are numbered from 1 in descending order of molecular weight. Seven are found in *E. coli*, 1a and 1b of *M* about 92 kDa and 90 kDa being the largest and PBP6 at 40 kDa being the smallest. *Enterobacter cloacae, Klebsiella aerogenes, Proteus rettgeri, Acinetobacter calcoaceticus* and *Pseudomonas aeruginosa* all have six or seven PBPs, of differing molecular weights, but *Neisseria gonorrhoeae* has only three. Among Gram-positive organisms, *Staphylococcus aureus, Micrococcus luteus* and *Bacillus stearothermophilus* have four, *Streptococcus pneumoniae, Bacillus subtilis* and *B. megaterium* have five, *Clostridium perfringens* has six, but *Bacillus cereus* only three.

Binding of a β-lactam to a PBP does not necessarily mean that inhibition of its enzyme activity will lead to death. Mutants lacking some PBPs do not suffer adversely and PBP 4, 5 and 6 of *E. coli* are not lethal targets, but PBPs 1, 2 and 3 are. Interestingly, the requirement for 1a can be substituted by 1b and *vice versa* but in double mutants in which both are non-functional the result is lysis, *e.g.* at the restrictive temperature for *ts* mutants.[73] PBPs 1, 2 and 3 of *E. coli* are transpeptidase/transglycosylase enzymes, and DD-carboxypeptidase activity has been shown for PBP 4, 5 and 6. DD-Carboxypeptidase activity has also been shown for PBP5 of *B. megaterium* and *B. subtilis*, PBP4 of *Staph. aureus*, PBP6 of *Streptococcus faecium* and PBP3 of *Str. pneumoniae*.

The development of resistance by PBP target alterations may have three different causes. These involve decrease of affinity for a particular PBP, an alteration of the quantity of PBP produced or increased rate of breakdown of the PBP–inhibitor complex, and more than one of these properties may be exhibited simultaneously. In some cases PBPs disappear and new PBPs (different electrophoretically) appear.[75] A compilation of PBP changes in resistant strains is given in Table 5. Perhaps the most clear-cut example of PBP involvement in resistance is provided by the finding that mecillinam-resistant *E. coli* possess a PBP2 of greatly lowered affinity. The causality is unequivocal since mecillinam binds specifically and only to PBP2.

A series of 10 increasingly more resistant mutants of *Cl. perfringens* showed affinity changes only in PBP1, eventually conferring a fiftyfold increase in MIC to 3.0 mg L^{-1}, which correlated with a ninefold increase in concentration required to half-saturate the PBP1 enzyme, and this occurred without change in quantities of PBP synthesized.

On the other hand, cloxacillin resistance in a strain of *B. megaterium* was associated with changes in the relative amounts of PBP1 and 3. Similarly, in *Str. faecium* the affinities of PBPs did not change but the resistant strain contained much more PBP5 than the sensitive strain.

The situation in methicillin-resistant (MR) staphylococci is varied and complex.[83a] One MR strain isolated in Yugoslavia exhibited a thousandfold decrease in affinity of PBP3 for methicillin with no significant change for 1, 2 and 4 and this was paralleled by a thousandfold greater MIC.[84] In another MR staphylococcus, methicillin resistance was heterogeneously expressed, the population being almost 'wild-type sensitive' when grown at 37 °C in low salt concentration, but highly resistant

Table 5 Some Resistances Due to Altered Penicillin Binding Proteins (PBP)

Organism	Resistance	Target	Alteration	Ref.
E. coli	Mecillinam	PBP2	Affinity lowered	76
N. gonorrhoeae	Benzylpenicillin	PBP1, 2	Affinity lowered	81
Cl. perfringens	Benzylpenicillin	PBP1	Affinity lowered	77
B. subtilis	Cloxacillin	PBP2	Affinity lowered	78
		PBP1	Lost	
B. megaterium	Cloxacillin	PBP1	Reduced	79
		PBP3	Increased	
Str. pneumoniae	Benzylpenicillin	PBP2	Affinity lowered	82, 83
		Others	Lost or novel	
Staph. aureus	Methicillin	PBP3	Affinity lowered	84
	Methicillin	PBP1, 2, 3	Affinity lowered	88
	Methicillin (variable)	PBP3	Lost	86, 87
		PBP2′	Low affinity, increased, labile PBP2′-drug complex	
		PBP3	Lost	85
	Cephradine	PBP2′	Novel	

when cells are grown at 30 °C in the presence of 5% NaCl. In the resistant strain under the latter conditions the affinity of PBP3 for several penicillins and cephalosporins was greatly decreased, but was accompanied by a substantial increase in the amount of PBP produced, and since its electrophoretic mobility was also changed it was renamed PBP2'. Grown under the sensitive conditions (37 °C, low NaCl) the normal PBP3 was expressed and no PBP2' was observed, implying a temperature sensitive protein. In addition, the half-life of the covalent PBP2'–benzylpenicillin complex was only 2–3 min at 30 °C, while for the isogenic sensitive strain it was 30–40 min for PBP2 and 120 min for PBP3, and its affinity is about a thousandfold lower than either.[75] The genetics of resistance in *S. aureus* has recently been reviewed,[89] and the structure and function of PBPs are being intensively researched.[67a]

With the increasingly serious problem of treating infections due to multiply resistant (MR) staphylococci,[83a] vancomycin was re-introduced as the drug of choice for serious conditions, such as endocarditis cases. The unusual mode of action of vancomycin, in 'shielding' the murein subunits which act as the 'substrate' in the enzymatic extension of the cell wall, suggested that development of resistance was unlikely,[74] yet vancomycin resistance has been reported in a strain of *Staph. haemolyticus* from a dialysis patient with peritonitis.[89a] How widespread this phenomenon becomes remains to be seen but, already, transferable vancomycin and teicoplanin resistance in *Enterococcus faecalis* has been reported.[89b]

5.5.4.1.2 *Mutations in ribosomal components*

Bacterial ribosomes are complex structures assembled from RNA and proteins. The smaller (30S) subunit has one molecule of 15S RNA and 21 proteins and during protein synthesis it is closely associated with a larger (50S) subunit of two RNAs (23S and 5S) and 32 proteins to form a 70S unit moving along a messenger RNA molecule with the aid of translocation factors. The topography of ribosomes is not yet understood and relationships between RNA and proteins, and protein and protein, can seldom be adduced from study of the interactions between antibiotic and ribosome, but they are undoubtedly subtle as well as fundamental. For example, streptomycin resistance is certainly determined by a mutation in the *rps*L gene controlling the small subunit protein S12, yet Sm does not bind to S12 protein by itself, implying that its situation, and probably conformation, within the intact ribosome determines the ability of the ribosome to interact with streptomycin. While other aminoglycoside resistances such as gentamicin, neomycin and neamine are determined by ribosomal protein changes, other antibiotics such as the aminocyclitol spectinomycin, and the very dissimilar molecules of erythromycin and thiostrepton, also have resistances based on protein modifications. Alterations in the RNA determine resistance to kasugamycin (an aminoglycoside), erythromycin, thiostrepton and viomycin. The specific identification of S12 as the target protein of Sm resistance in *E. coli* was achieved by dissociating RNA and proteins from both sensitive and resistant ribosomes, followed by reassembly of combinations of 'sensitive' and 'resistant' components into hybrid ribosomes and testing for protein synthesizing capacity in the presence of antibiotic.[90] The 'omission' of 'sensitive S12' could be shown to restore functions lost when it was included in the presence of Sm during synthesis, while substitution by 'resistant S12' would also restore protein-synthesizing ability. In fact the expression of resistance, sensitivity or dependence were all allelic manifestations of the *rps*L gene (*str*A).

Genetic analysis has revealed that mutation to high-level resistance may result from mutation at two distinct sites: replacement of Lys by Thr, Arg or AspNH$_2$ at position 42, or replacement of Lys by Arg at position 87.[91]

Similar reconstitutions of hybrid 30S subunits from components of spectinomycin (Sp) sensitive and spectinomycin resistant ribosomes have established S5 as the target of this antibiotic. Like Sm resistance, high-level single-step Sp resistance is possible, but is not cross-resistant with Sm. The amino acid substitutions are found in positions 19–21 of the *rps*E (*spc*A) gene product.[92]

While streptomycin high-level resistance (as high as 10 000 mg L^{-1}) is expressed at a drug ribosome ratio of 1:1, the 2-deoxystreptamine antibiotics such as gentamicin (Gm) and neomycin (Nm) show step-wise increases in drug resistance correlated with binding to more than one ribosome site. In some Gm-resistant mutants the large subunit protein L6 is abnormal.[93] In Nm-resistant mutants, alterations have been detected in S5 and S12.[94] Neamine-resistant mutants have altered S17.[95] Kasugamycin does not possess a 2-deoxystreptamine moiety but exhibits resistance related to a change in S2 (*rps*B, *ksg*C). Erythromycin resistance has been associated with mutation in L4 of *E. coli*, and probably also with L22,[96] but most erythromycin-resistant strains have a RNA-based mechanism of resistance (see below).

The peptidyl transferase enzyme (L16) responsible for amino acid transpeptidation during protein synthesis is fixed in the large ribosomal subunit, and chloramphenicol-resistant mutants having altered proteins of the 50S subunit with decreased Cm-binding have been identified in *Bacillus subtilis*.[97]

Thiostrepton has no significant action on Gram-negative bacteria, but resistance in *B. megaterium* and *B. subtilis* results from the loss of L11 protein in these mutants. These Gram-positive ribosomal proteins are serologically homologous with L11 of *E. coli* and their loss causes reduced GTP hydrolysis in the presence of elongation factor EF-G.[98]

While no mutants to fusidic acid resistance have revealed changes in ribosomal components, a *fus* gene controlling fusidic acid resistance in *E. coli* was found to code for the elongation factor EF-G. In the presence of the drug a stable complex of ribosome, EF-G and GDP is formed, which prevents the repetition of cycles of aminoacyl–tRNA binding and GTP hydrolysis required to drive the translocation of ribosomes during protein synthesis. The mutant strain possessed an altered EF-G which allowed GTP hydrolysis to continue.[99]

Resistance to several antibiotics is mediated by alterations in the methylation of ribosomal RNA. In kasugamycin resistance the mutation in a RNA dimethyl transferase coded by the *ksg*A gene mapping at 1 min on the *E. coli* genome leads to the non-appearance of methyl groups on two adenine residues in positions AACCUG of the 16S RNA of the 30S subunit[100] located near the Shine–Dalgarno sequence of bases involved in mRNA recognition.

Within a few years of the introduction of the macrolide erythromycin, the incidence of resistance in *Staphylococcus* spp. rose rapidly to a plateau of 40%. Resistance to erythromycin was usually combined with resistance to lincosamides and streptogramin B, the MLS$_B$ resistance phenotype. MLS$_B$ resistance is the result of post-transcriptional N^6-methylation of a specific adenine of the 23S-RNA of the 50S subunit. Seven classes of methylase have so far been found, coded by genes *erm*A to G.[100a, 100b] The synthesis of rRNA methylase may be either constitutive or induced by different subsets of MLS antibiotics; the mechanism of induction has been largely elucidated.[101] Alterations of ribosomal proteins have also been found which give high level resistance to lincosamide by changes in either protein S7 or L15.[101a]

5.5.4.1.3 *Mutations affecting nucleic acid synthesis*

The 4-quinolone antibacterials are a group of synthetic compounds whose activity is related to the function of bacterial topoisomerase II (DNA gyrase), an enzyme which introduces negative super-helical twists into double-stranded DNA during several procedures essential for the growth and survival of the organism. It functions during DNA replication and repair, and in the transcription of some operons.[102]

Nalidixic acid, the first of the 4-quinolones, although a very useful clinical antibacterial, is among the least effective of the group and newer derivatives, such as ciprofloxacin, ofloxacin and norfloxacin, hold great promise for the treatment of otherwise intractable disease. The new 4-quinolones show potent antibacterial activity against both Gram-positive and Gram-negative organisms, and have the great advantage that resistance has never been found on a transposon or plasmid. In fact, R-plasmids often increase the sensitivity of the organism to these drugs which, in addition, may cause elimination of plasmids from strains harbouring them.[103] Mutations to nalidixic acid are readily found, but species exhibiting resistance have a reduced ability to donate, accept and maintain plasmids.

Several chromosomal mutations affect the action of 4-quinolones. The *nal*A (*gyr*A) mutation mapping at 48 min in *E. coli* alters the two α-subunits of the DNA gyrase molecule which acts in concert with the two β-subunits coded for by *gyr*B which maps at 82 min, and is the mapsite for both *nal*C and *nal*D mutants.[104] The *nal*B mutation (at 57 min) reduces permeability to some 4-quinolones, conferring a moderate level of resistance. Smith[105] investigated the effect of each of these types of mutation in *E. coli* on the response to a range of 4-quinolones and found for the *nal*A mutant a tenfold to twenty-fivefold or greater resistance to most 4-quinolones. The drugs most active against the wild type were also relatively more active against the mutant which showed only five- to seven-fold increase in MIC. Surprisingly, these drugs—ciprofloxacin, ofloxacin and norfloxacin—were more active against the wild-type parent than against the *nal*C (*gyr*B) mutant and suggested this activity was related to the piperazine moiety possessed by these drugs. The *nal*D mutants were two- to ten-fold more resistant than the sensitive parent. The *nal*B mutation caused moderate resistance to some of the 4-quinolones by reducing permeability to their entry.

More recently, a range of norfloxacin-resistant mutants of *E. coli* K12 have been examined[106] and *nor*A, B and C and *nfx*A and B genes implied. The *nor*A and *nfx*A genes were alleles of *gyr*A, while *nor*B, *nor*C and *nfx*B determined outer membrane permeability resistance to norfloxacin resulting from decrease in OmpF porin protein. In *Pseudomonas aeruginosa* a mutation in *nfx*B decreased cell permeability, resulting in four- to sixteen-fold increase in MIC of the newer quinolones but only a two-fold increase in MIC of nalidixic acid and pipemidic acid. In this mutant a new 54 000 Da outer membrane protein was identified.[107]

The *gyr*B locus is also able to mutate to give resistance to two other antibiotics, coumermycin and novobiocin, and the use of these drugs and organisms mutant in *gyr*A and B have helped to shed light on the action of gyrase. It appears that the α-subunits introduce nicks in each DNA strand, staggered by four base pairs, these being necessary to allow negative supercoiling to occur. The β-subunits catalyze the hydrolysis of ATP driving the supercoiling and the α-subunits subsequently reseal the nicked strands. Novobiocin has been shown to prevent the binding of ATP to the GyrB protein.[108]

The DNA-dependent RNA polymerase of bacteria is also a specific target for antibiotic action and this enzyme is 'inhibited' by streptovaricins and rifamycins. RNA polymerase from *E. coli* contains four polypeptide chains working in conjunction with a protein which enables it to recognize and bind to the correct promoter sites for initiation of transcription. The core enzyme consists of two α-, a β- and a β'-subunit. In rifampicin-resistant organisms the only modified polymerase polypeptide is the β-subunit.[109] Reconstruction of hybrid molecules from resistant and sensitive subunits prove that the β-subunit is responsible for resistance although it does not, by itself, bind rifampicin. A binding conformation can be established by a $2\alpha + \beta$ complex or the core enzyme and, when drug-bound by rifampicin, the initiation of the RNA polymerase action is prevented. If drug is added after the polymerization has started, there is no inhibitory effect. Streptovaricin acts in a similar fashion but streptolydigin acts to inhibit the polymerization directly. Although the streptolydigin target is also the β-subunit, binding is at a different site and rifampicin-resistant mutants remain sensitive to streptolydigin.[110]

5.5.4.2   Modification of the Drug: Detoxification

Enzymatic modification of antibiotic molecules may be achieved in two broadly different ways. Many enzymes can act directly upon the antibiotic and catalyze the excision of side groups, or cleave ring structures, *i.e.* bonds are broken. Others act in conjunction with cofactors (or cosubstrates), serving as donors to provide groups which are added to the antibiotic molecule by the formation of new bonds. These reactions furnish mechanisms of resistance to many of the most popular and clinically active antibiotics, and genes for these enzymes may be either chromosomally located or exist on highly mobile plasmids and transposons.

5.5.4.2.1   β-Lactams

A wide range of β-lactam structures has now been investigated, all containing the four-membered β-lactam ring, but the associated ring structure may be five-membered or six-membered and the 1-position in this ring may be sulfur (as in penicillin), oxygen (as in moxalactam) or carbon (as in imipenem); in addition, the ring may be saturated or unsaturated. In the monobactams (aztreonam) only the β-lactam ring is found, in association with a stabilizing acidic group. The β-lactam antibiotics are considered in detail in Chapters 9.2 and 9.3 of this volume.

The importance of β-lactam antibiotics as therapeutic agents depends on their ability to bind to proteins involved in the final stages of cell-wall synthesis. These proteins are found attached to the cytoplasmic membrane or in the periplasmic space between inner and outer membrane of Gram-negative organisms and are the penicillin-binding proteins (PBP) described in Section 5.5.4.1.1. They function as transpeptidases, transglycosylases and DD-carboxypeptidases, and, for some, endopeptidase activity has also been shown. There are no clear examples in which inhibition of DD-carboxypeptidase or endopeptidase action has been shown to lead to the death of the cell, but in many cases lethality has been shown to depend upon transpeptidase inhibition.[74]

β-Lactam antibiotic-binding to PBPs may be eliminated or diminished by the action of three classes of enzymes found in bacterial cells: (i) β-lactamases, which open the β-lactam ring; (ii) acylases (amidases) which excise the side chains attached through position 6 (or 7); and (iii) esterases, which eliminate ester-linked groups of the substituent at position 3 in (**1**) and (**2**). The important reaction in clinical drug resistance is the hydrolysis of the cyclic amide bond by β-lactamases. While acylases are widespread throughout the microbial world, most are not synthesized or have little activity at 37 °C.

(1) Penicillins   **(2)** Cephalosporins

β-Lactamases are produced by Gram-positive and by Gram-negative species of bacteria, by actinomycetes,[111] yeasts[112] and blue-green algae.[113] Shortly after the introduction of penicillin, β-lactamase-producing *Staphylococcus aureus* became common in hospitals and by the mid-1950s their incidence was as high as 80% in some.[114] It became necessary to find new β-lactams which would resist β-lactamase action and methicillin, nafcillin, oxacillin, cloxacillin, dicloxacillin and flucloxacillin were soon produced by modification of the penicillin side chain. Enzymes capable of hydrolyzing these have also been found, mainly in Gram-negative bacteria.

(i) Mechanism of action

'For the PBPs that have been studied in detail, it has been demonstrated unequivocally that β-lactam antibiotics bind and acylate the catalytic site'.[115] The reaction of PBP with a substrate occurs in three steps. Initially the PBP (E) and substrate (S) form a non-covalent complex (E · S), which may revert to the original unchanged reactants or may proceed to a covalently bound acyl–enzyme intermediate (E–P) with the release of the terminal D-alanine of the pentapeptide side chain of the peptidoglycan polymer unit during normal wall synthesis (Chapter 9.1). Reaction of E–P with H_2O yields the hydroxylated product, P–OH, representing carboxypeptidase action; reaction with RNH_2 (adjacent peptide chain) generates a peptide bond and the linking of two peptidoglycan polymer subunits by transpeptidase action, as represented in the generalized equation (1). Each type of reaction regenerates free enzyme.

$$E + S \underset{k_{-1}}{\overset{k_1}{\rightleftharpoons}} E \cdot S \xrightarrow{k_2} E{-}P \tag{1}$$

A similar scheme can be drawn to represent the reaction of PBP with β-lactam (equation 2), where the substrate I represents the antibiotic. In this case the unstable intermediate proceeds to form a covalently bound penicilloyl–PBP complex through the C atom of the cyclic amide bond in the β-lactam ring, with concomitant breaking of the bond. In broad terms, the better the antibiotic the lower the rate constant k_3, so that the PBP remains bound in the penicilloyl complex preventing normal wall-synthesizing function. If k_3 is relatively large then the reaction of E-1′ with H_2O leads to hydrolysis of the β-lactam to the (degraded) penicilloic acid form of the antibiotic, thus releasing the enzyme for normal function or for further hydrolysis of drug. In this reaction the PBP is behaving with a β-lactamase activity. We can, therefore, describe β-lactamase activity by making E in equation (2) the β-lactamase and raising the value of k_3.

Examination of the amino acid sequences of class A staphylococcal β-lactamase and PBP carboxypeptidases has shown a high degree of conservation, especially in the region of the active serine at position 36 which could be aligned with the active serine in the β-lactamase.[116] This points to the probable common evolution of PBPs and β-lactamases. An interesting 'parallel' has been provided by Hedge and Spratt,[117] who have found that high-level resistance to cephalosporins (including third generation drugs) could be produced by mutation of only four amino acids in PBP3 of *E. coli*.

The scheme above is a general one, descriptive of the mechanism for active-site serine enzymes; the mechanism for zinc metalloenzymes, *e.g. B. cereus* β-lactamase II, is different and involves a branched pathway and general-base catalysis rather than covalent, nucleophilic catalysis.[118]

(ii) Classification of β-lactamases

The majority of β-lactamases found in Gram-positive bacteria are inducible and extracellularly produced,[119] while most Gram-negative β-lactamases are membrane-bound and restricted to the periplasmic space, and may be inducible or constitutive.[67] In theory at least the absence of an outer membrane in Gram-positive organisms means that large amounts of enzyme are required to protect the organism against the 'immediate' attack posed by the arrival of antibiotic in the cell's environment. Also, since it would be wasteful for the cell to produce large amounts of protein continuously, an induction system is required. Conversely, the protection afforded by the Gram-negative outer membrane and the slower approach of drug to its target allow Gram-negatives more easily to protect themselves and with lower amounts of enzyme.

Gram-positive β-lactamases

Four serologically different β-lactamases (A–D) have been described in *Staphylococcus aureus*, which are usually inducible and active primarily against the penicillins, but are inhibited by the isoxazoyl penicillins and methicillin. Their R-plasmids are spread among different strains by means of transducing phages.[120] β-Lactamases have also been found in streptococci. Wild-type strains of *B. cereus* produce two β-lactamases which do not cross-react immunologically and are coded by separate genes. The action of β-lactamase I is similar to that of other Gram-positive enzymes and has a similar spectrum of hydrolysis, while β-lactamase II is a zinc metalloenzyme with a wider spectrum of hydrolysis.

Bacillus licheniformis produces two 'versions' of the enzyme, one of which is membrane bound and terminates in a phosphatidylserine.[121] This polypeptide is released to give the extracellular enzyme by cleavage at amino acid 24 from the N-terminal (fixed) end. Its spectrum is similar to other Gram-positive β-lactamases.

Gram-negative β-lactamases

An increasing number and variety of β-lactamases have been identified and their classification has become progressively more difficult. Several schemes have been proposed. Richmond and Sykes[67] proposed five classes. Class I are primarily cephalosporinases (includes most of the chromosomally determined enzymes); Class II are primarily penicillinases; Class III are almost equally active against both drug types, but are sensitive to cloxacillin and resistant to *p*-chloromercuribenzoate (pCMB); Class IV are enzymes of similar substrate profile to Class III, but are resistant to cloxacillin and sensitive to pCMB; Class V enzymes are primarily penicillinases that also hydrolyze cloxacillin, but are resistant to pCMB.

More recently, the enzymes have been grouped into fewer classes based upon substrate profile, the action of inhibitors, molecular weight and isoelectric point (p*I*). For substrate profiles the rates of hydrolysis of saturating concentrations of a selection of penicillins and cephalosporins are compared with the rate of hydrolysis of benzylpenicillin taken as 100%. The substrates usually selected are given in Table 6. The inhibitors frequently tested are cloxacillin, pCMB, sodium chloride and clavulanate (see Chapter 9.3). Matthew[122] collected data on the plasmid-mediated β-lactamases of Gram-negative bacteria and grouped the enzymes into three broad categories: the oxacillinases (OXA), which are able to hydrolyze the isoxazoyl β-lactams and methicillin, the carbenicillinases found in *Pseudomonas* (PSE), and a more general class of enzymes which were relatively highly active against cephaloridine as well as penicillin and ampicillin but did not readily hydrolyze the isoxazoyl β-lactams or methicillin. This group includes TEM-1 and TEM-2, SHV-1 and HMS-1. Since 1979, many more plasmid-borne β-lactamases have been discovered in Gram-negative organisms[123] and newer data have been accumulated and presented by Amyes.[124] For further information the reviews by Sykes and Matthew,[125] Sykes and Bush[126] and Bush and Sykes[127] should be consulted.

Of the plasmid-mediated enzymes listed in Table 6, the TEM-type enzymes are the most widely distributed and frequently encountered. They are found in species of *Alkalescens, Caulobacter, Citrobacter, Clostridium, Enterobacter, Escherichia, Haemophilus, Klebsiella, Neisseria, Proteus, Providencia, Pseudomonas, Salmonella, Serratia, Shigella, Vibrio* and *Yersinia*.[122] TEM-1 and TEM-2 differ by a single amino acid, lysine replacing glutamine at residue 14 in TEM-2.[126]

Table 6 Properties of Some Plasmid Encoded β-Lactamases

	Relative rates of hydrolysis[a]							Inhibitors			M(kDa)	Isoelectric point
	Amp[b]	Carb	Oxa	Meth	Clox	Cer	Ctx	Clox	pCMB	Clav		
Broad spectrum												
TEM-1	106	10	5	0	0	76	<1	+	−	+	22	5.4
TEM-2	107	10	5	0	0	74	0	+	−	+	23.5	5.6
SHV-1	212	8	0	<2	<2	56	0	+	+/−	+	17	7.6
SHV-2	145	ND	0	ND	ND	32	4	+	+/−	+	17	7.6
HMS-1	253	14	<2	<2	<2	183	ND	+	+	−	21	5.2
TLE-1	67	13	4	5	6	52	6	+	ND	+	19.8	5.55
TLE-2	140	13	ND	<0.1	<0.1	NM	ND	+	ND	+	19	6.5
ROB-1	186	25	6	ND	ND	24	ND	ND	ND	ND	ND	8.1
LCR-1	145	4	ND	20	3	55	ND	+	−	+	44	6.5(5.85)
NPS-1	223	18	40	<0.1	ND	3	<1	−	ND	ND	25	6.5
BRO-1	103	78	ND	131	5	23	ND	+	ND	+	ND	5.6
OH10-1	140	11	<0.5	<0.5	<0.5	79	ND	+	ND	+	22	7.0
Oxacillinases												
OXA-1	382	30	197	332	190	30	22	−	+/−	ND	23.3	7.4
OXA-2	179	15	646	23	200	37	1	−	−	−	44.6	7.45, 7.7
OXA-3	178	10	336	29	350	44	0	−	−	ND	41.2	7.1
OXA-4	438	39	220	711	64	194	63	+/−	ND	−	23	7.5
OXA-5	188	40	210	109	258	89	49	+	ND	−	27	7.62
OXA-6	596	46	1048	585	301	149	28	+	ND	−	40	7.8
OXA-7	545	48	702	424	494	136	31	−	ND	−	25	7.65
Carbenicillinases												
PSE-1	90	97	<2	<2	<2	18	27	−	+	−	28.5	5.7
PSE-2	267	121	317	803	371	32	16	+	+	−	12.4	6.1
PSE-3	101	253	ND	ND	3	10	<1	+	−	−	12	6.9
PSE-4	88	150	8	16	<2	40	1	−	−	−	32	5.3
CARB-4	130	79	1	<1	<1	18	ND	−	+	−	22	4.3
SAR-1	63	122	<0.1	<0.1	<0.1	21	<0.1	+	+	+	33.7	4.3
AER-1	38	98	0.9	0.3	0	26	20	ND	ND	ND	22	5.9
Cephalosporinase												
CEP-2	ND	48	ND	ND	ND	108	ND	−	−	ND	36.2	8.1

[a] Benzylpenicillin = 100%; [b] Amp, ampicillin; Carb, carbenicillin; Oxa, oxacillin; Meth, methicillin: Clox, cloxacillin; Cer, cephaloridine; Ctx, cefotaxime; pCMB, p-chloromercuribenzoic acid; Clav, clavulanic acid; ND, not done; NM, not measurable.

Unlike most of the other enzymes, SHV-1 hydrolysis of benzylpenicillin is unaffected by 0.5 mM pCMB, whereas the same concentration of pCMB completely inhibits the hydrolysis of cephaloridine. SHV-1 is frequently associated with *Klebsiella* strains that have been implicated in serious hospital cross-infection.

The OXA-type enzymes are relatively rare compared to the TEM-type and unlike them have the unfortunate ability to hydrolyze the 'β-lactamase stable' antibiotics, methicillin and the isoxazoyl penicillins. Although resistant to pCMB, they are susceptible to inhibition by chloride ions. Oxacillinases have been found in *Bordetella*, *Escherichia*, *Klebsiella*, *Proteus*, *Providencia*, *Salmonella*, *Pseudomonas*, *Shigella* and *Bacteroides*.

While most plasmid-mediated β-lactamases are reported to be constitutive, the chromosomally mediated enzymes may be inducible or constitutive. The majority fall into the Class I of Richmond and Sykes,[67] and some of these are species or even sub-species specific, *e.g.* in *Yersinia enterocolitica*.[128] Organisms producing Class I enzymes are widespread and of great clinical importance. They may be divided broadly into inducible and constitutive β-lactamases. The former are found in *Acinetobacter*, *Citrobacter*, *Enterobacter*, indole-positive *Proteus*, *Providencia*, *Pseudomonas*, *Serratia* and *Yersinia*.

The constitutive chromosomally mediated β-lactamases are found in *E. coli*, *Bacteroides*, *Salmonella*, *Shigella*, *Enterobacter*, *Proteus*, *Yersinia*, *Aeromonas*, *Aerobacter* and *Klebsiella*.

Most strains of *Y. enterocolitica* produce two distinct β-lactamases, differing in substrate profile, molecular weight, inducibility, resistance to pCMB and isoelectric point. Most *Bacteroides fragilis* and *Legionella pneumophila* are also β-lactamase producers. Isoelectric focusing has revealed a family of β-lactamases in the genus *Klebsiella* and these enzymes have a much broader range of substrate corresponding to the Richmond and Sykes[67] Class IV enzymes.

Organisms are increasingly found in which more than one β-lactamase is present. Some contain both a constitutive chromosomal enzyme and a plasmid-mediated β-lactamase. Others may possess two or even three plasmid-determined β-lactamases; a strain of *P. aeruginosa* produced TEM-1, OXA-2 and PSE-1 enzymes.[123] It cannot be assumed, however, that the presence in a cell of genes for two different β-lactamases will lead to the additive expression of substrate-hydrolyzing ability. A strain of *K. pneumoniae* which carried only K1 β-lactamase was resistant to aztreonam, cefotaxime and cefuroxime, while a strain of *E. coli* carrying only the TEM-1 enzyme was susceptible to these antibiotics. However, when both genes were present in *K. pneumoniae* the total amount of β-lactamase activity was greatly reduced and the organism had a susceptibility pattern equivalent to the *E. coli*, not the resistance pattern that would be expected.[127]

5.5.4.2.2 *Chloramphenicol*

Only one of the four possible stereoisomers of chloramphenicol—the D-*threo* isomer—acts specifically to inhibit the peptidyl transferase centre located on the 50S ribosomal subunit and responsible for peptide bond formation during polypeptide chain elongation. A similar specificity exists for interaction with the enzyme responsible for enzymatic chloramphenicol resistance, chloramphenicol acetyltransferase (CAT).

Chloramphenicol is a broad-spectrum antibiotic, almost equally active against most wild-type Gram-positive and Gram-negative bacteria and rikettsiae. Chloramphenicol is not ionized at physiological pH and, while amphiphilic in character, the hydrophobic moieties enable it to permeate membranes and it readily crosses the blood–brain barrier, making it a useful antibiotic in the treatment of meningitis. Resistance is found in Gram-negative bacteria as well as *Staph. aureus* and streptococci, and in most clinical isolates is plasmid-borne and determined by chloramphenicol acetyltransferase, although other mechanisms of resistance exist.[130]

Chloramphenicol acetyltransferase reacts with chloramphenicol and acetyl-CoA to yield the inactivated product 1,3-diacetoxychloramphenicol. The reaction occurs in three steps: firstly, the C-3 hydroxyl group reacts to give 3-acetoxychloramphenicol, which then undergoes a non-enzymic rearrangement to give 1-acetoxychloramphenicol, thus freeing the C-3 position for a second catalytic addition of an acetyl group to give the diacetoxy end product (Scheme 1).[129] The first reaction is rapid and the formation of 3-acetoxychloramphenicol totally inactivates the antibiotic. The reaction forming 1,3-diacetoxychloramphenicol is at least one-hundredfold slower.

A considerable homology occurs among CAT enzymes and a histidine residue is a critical participant in the catalytic reaction. The active enzyme in all cases examined consists of a tetramer of identical polypeptide units 220 amino acids long with a gross molecular weight of about 100000.

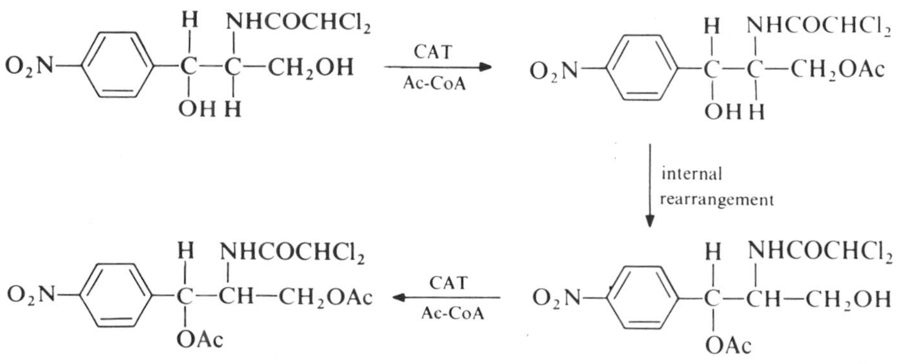

Scheme 1

Detailed examination has revealed two series of related enzymes and a number of variants. CAT types I, II and III are plasmid specified and found in Gram-negative bacteria. A series of four related plasmid-borne enzymes, types A, B, C and D, have been found in *Staphylococcus* species. None of the CAT A–D enzymes are inhibited by 5,5'-dithio-bis-2-nitrobenzoate (DTNB), while all the variants (except type I) are inhibited to greater or lesser extents.[129] CAT resistance has been found in many strains: variant enzymes have been characterized from *Proteus mirabilis*, *Haemophilus influenzae*, *Agrobacterium tumefaciens*, *Bacteroides fragilis*, *Flavobacterium* spp., *Streptomyces acrimycini*, *Streptococcus pneumoniae*, *Strep. agalactiae*, *Strep. faecalis* and *Clostridium perfringens*.

Strains lacking plasmids are inhibited by concentrations of about 2–4 mg L^{-1} but chromosomal mutants to higher levels of resistance are readily obtained. These mutants are resistant as a result of decreased permeability to the drug[131] and no cases of selection for CAT-producing organisms are known. Permeability mutants have been found in *E. coli*, *Shigella*, *Salmonella* and *Ps. aeruginosa*. In *E. coli* strains with mutant OmpF porin there is increased resistance to chloramphenicol as well as other antibiotics, implying normal penetration *via* the porin pores.

The Gram-negative organisms with CAT-coding plasmids produce the enzyme constitutively,[130] but they are usually inducible in Gram-positive organisms. Since chloramphenicol itself prevents protein synthesis, and protein synthesis is necessary for the production of CAT,[132] the problem of induction by an inhibitor of synthesis is an interesting one. Rapid induction occurs only when the concentration of chloramphenicol is less than 100 mM and it has been suggested that translation of CAT mRNA only occurs when an inducer is bound to the ribosome, giving it an enhanced affinity for a GGAGG sequence upstream of the translated sequence and allowing the ribosome to read through into CAT synthesis.[133]

In constitutive Gram-negative CAT producers, the level of enzyme produced is at least fivefold greater than in the inducible Gram-positive systems. However, most of the enteric plasmid-borne resistance is carried on low copy number plasmids, while the smaller staphylococcal CAT plasmids have high copy numbers (20–40 per genome) and this gene dosage effect leads to an enhanced production of the inducible enzyme.

5.5.4.2.3 *Aminoglycosides and aminocyclitols*

This group contains a number of clinically important antibiotics and some useful in veterinary medicine and agriculture (see also Chapter 10.5). The more important belong to the subgroup of disubstituted deoxystreptamines, among which are gentamicin, kanamycin, amikacin, netilmicin, sisomicin, tobramycin, dibekacin, neomycin and paromomycin; others are aminocyclitol-based such as streptomycin and bluensomycin, while other useful members are kasugamycin, spectinomycin, apramycin and fortimicin A.

Three different classes of enzymes have been found to modify and inactivate aminoglycoside antibiotics by transfer reactions involving cofactors. The cofactors are nucleoside triphosphates (especially ATP) or acetyl-CoA. The former participate in phosphorylating reactions (phosphotransferase) and adenylylating or nucleotidylating (nucleotidyltransferase) reactions, both of hydroxyl groups. The acetyltransferase enzymes acetylate NH$_2$ groups and are denoted by the code AAC, the phosphorylating enzymes by APH, and the nucleotidyl- or adenylyl-transferases by ANT or AAD. The site of the modified group is denoted by the ring position number in brackets and a subsequent roman numeral differentiates between enzymes with differing modification profiles. The

Table 7 Enzymes Modifying Aminoglycosides and Aminocyclitols

Enzyme	Antibiotics modified
Phosphotransferases	
APH(3')-I	Nm, Pm, Rm, Lv(5'') Km
APH(3')-II	Nm, Pm, Rm, Bt, Km
APH(3')-III	Rm, Bt, Lv(5''), Km
APH(3')	Km, Ak, Dbk
APH(5'')	Rm
APH(2'')	Km, Gm, Tm, Dbk, Ss
APH(3'')	Sm
APH(6)	Sm
Adenylyltransferases	
AAD(2'')	Km, Gm, Dbk
AAD(4')	Nm, Pm, Bt, Km, Tm, Ak
AAD(4', 4'')	Nm, Rm, Bt, Km, Tm, Gm
AAD(3'')	Sm, Sp
AAD(6)	Sm
Acetyltransferases	
AAC(6')-I	Km, Km-B
AAC(6')-II	Km, Km-B, Gm-C_{1a}, -C_2
AAC(6')-III	Km, Km-B, Gm-C_{1a}, -C_2, Dbk
AAC(6')-IV	Ak
AAC(3')-I	Gm, Ss
AAC(3')-II	Km-B, Gm, Tm, Ss
AAC(3')-III	Nm, Pm, Km-A, -B, Gm, Tm, Ss
AAC(3')-IV	Apm, Nm, Rm, Bt, Km-A, Gm, Tm, Ak, Ntl
AAC(2')	Nm, Rm, Bt, Lv(5''), Km-B, Gm, Dbk

[a] Ak, amikacin; Apm, apramycin; Bt, butirosin; Dbk, dibekacin; Gm, gentamicin; Km, kanamycin; Lv, lividomycin; Nm, neomycin; Ntl, netilmicin; Pm, paromomycin; Rm, ribostamycin; Sm, streptomycin; Sp, spectinomycin; Ss, sisomicin; Tm, tobramycin.

positions on the cyclitol ring (streptidine, bluensidine, deoxystreptamine) are numbered 1–6, those of the 4-substituent ring from 1' to 6' and those at the 5- or 6-position of deoxystreptamine 1''–6''.

Table 7 gives a list of aminoglycoside–aminocyclitol modifying enzymes, showing that some enzymes modify the molecule at two different positions determined by spatial configuration. The enzyme AAD(4') will also modify 4''-hydroxyl groups if they are equatorial as in kanamycin, dibekacin and tobramycin, but not when they are polar as in gentamicin. Some enzymes are specific for a single antibiotic, *e.g.* APH(3''), APH(6) and AAD(6) modify only streptomycin; others, *e.g.* AAC(3')-IV, modify a wide range of drugs. A single aminoglycoside may be modified at different sites by enzymes of all these classes as shown in (3).

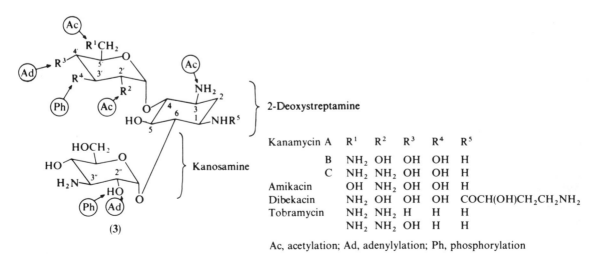

Kanamycin A	R^1	R^2	R^3	R^4	R^5
B	NH_2	OH	OH	OH	H
C	NH_2	NH_2	OH	OH	H
Amikacin	OH	NH_2	OH	OH	H
Dibekacin	NH_2	OH	OH	OH	$COCH(OH)CH_2CH_2NH_2$
Tobramycin	NH_2	NH_2	H	H	H
	NH_2	NH_2	OH	H	H

Ac, acetylation; Ad, adenylylation; Ph, phosphorylation

Positions possibly available for aminoglycoside modification

As mentioned above, modification occurs at either OH or NH_2 groups and the likelihood of modification depends on the position of the group on the drug molecule. Conversely, if the drug is modified so that the vulnerable group is removed or replaced by another non-modifiable group, the drug will become resistant to enzyme action at those locations. Knowledge of these vulnerable groups has led to the development of semisynthetic aminoglycosides with enhanced resistance to modification and sometimes with enhanced antibacterial activity as well.[134]

While the possession of aminoglycoside-inactivating enzymes has been found to be chromosomally determined, *e.g.* AAC(6') in *Serratia marcescens*,[135] the vast majority of these resistance-conferring enzymes have been plasmid associated and many occur on transposons.[136] A single plasmid can specify more than one enzyme activity and plasmids can be transferred between strains by conjugation, transformation or transduction.[136]

In an extensive survey[137] of aminoglycoside resistance mechanisms, APH(3') was the most commonly encountered enzyme, followed by AAC(6') and AAD(2''). In other surveys, similar results have been found, with AAD(2''), AAC(6'), APH(3') and AA6(3) the commonest enzymes among Gram-negatives, while some of these revealed production of two different enzymes by the same strain giving Gm and Nm/Km resistance, while other strains possessed two different Gm-modifying enzymes simultaneously, such as AAD(2'') and AAC(3)-I in *E. coli* and *Citrobacter freundii*, AAD (2'') and AAC(6') by *Ps. aeruginosa* and AAC(3)-I and AAC(6') by *Serratia*.[138] A newly discovered APH(3') in *Acinetobacter baumanii* in a clinical isolate was found to confer resistance to kanamycin and related aminoglycosides. The gene, located on a self-transmissible plasmid, also phosphorylated amikacin with high efficiency, *in vitro*, and conferred high level resistance to this drug.[138a]

The enzyme-determined resistance of a bacterium to a particular aminoglycoside will depend on the affinity of the enzyme for the substrate since this will ultimately determine the rate at which drug accumulates within the cell. If the K_m is low then high levels of resistance will usually be found, whereas a high K_m will allow the accumulation of unmodified drug and lower the resistance threshold of the cell, since more drug will be able to combine with its ribosomal target.

Against a range of pathogens—*Staph. aureus, Enterobacter, E. coli, K. pneumoniae, Pseudomonas, Pr. mirabilis*, indole-positive *Proteus, Salmonella* and *Serratia*—many of the better aminoglycosides have MIC values in the range 0.5 to about 4 mg L^{-1} for the majority of strains in the presence of the kanamycins, gentamicins, amikacin, tobramycin, dibekacin, butakacin, sisomicin, verdamicin and netilmicin. *Pseudomonas* is a troublesome exception, being inherently resistant to the kanamycins, gentamicin B, paromomycin, streptomycin and neomycin. The presence of modifying enzymes able to use the drug as substrate increases the MIC value to more than 64 mg L^{-1} in most cases,[134] a level which excludes these drugs from clinical use because of the nephrotoxic[139] and ototoxic[140] effects associated with many aminoglycosides.

In the case of some of the enzymes, very high levels of resistance are attained, *e.g.* an APH(3')(5'')-III giving Km/Nm MICs of 250–2000 mg L^{-1} in *Staph. aureus* conferred levels of 65 000 mg L^{-1} on *Str. faecalis*.[141]

The precise cause of the death of cells treated with aminoglycosides is much disputed[138] and similar doubts exist over the true cause of resistance due to modifying enzymes. Modified aminoglycosides do not bind to ribosomes and are ineffective as inhibitors of protein synthesis *in vitro*[142] and many consider this a sufficient explanation of resistance. Others argue that the modification of the drugs prevents their transport across the membrane and into the cell.[136, 143] Further information is given in Section 5.5.4.3.2 below.

5.5.4.2.4 *Miscellaneous modifying enzymes*

Enzymes capable of modifying active antibiotic molecules continue to be discovered and the clinical significance, if any, of these remains to be determined. Wiley *et al.*[144] have obtained positive evidence for the phosphorylation of macrolides by extracts of *Streptomyces coelicolor*, which modified the OH group of the amino sugar in erythromycin A, oleandomycin, tylosin, spiramycin and leucomycin A_3.

Arthur *et al.*[145] studied the production of erythromycin esterase in 52 strains of highly resistant enterobacteria and found, by hybridization probes, the existence of either type I enzyme (*ere*A, 23 strains) or type II (*ere*B, 26 strains), while six strains possessed both enzymes. In 18 strains they found both an esterase (I or II) and a methylase (see Section 5.5.4.1.2).

High level lincosamide resistance has been found in strains of *Staph. haemolyticus* and *Staph. aureus* which was mediated by enzymic nucleotidylation of the 4-position. The genes, designated respectively *lin*A and *lin*A′, were found on small plasmids (2.5 and 2.6 kb) and the 91% sequence homology indicated a close relationship between the two enzymes.[101b]

Dabbs[146] has demonstrated fusidic acid resistance in *Rhodococcus erythropolis*, which was due to an inducible extracellular enzyme inactivating the drug molecule, and Kobayashi *et al.*[147] have found a streptothricin acetyltransferase in *Streptomyces lavendulae*.

5.5.4.3 Altered Permeation and Transport

5.5.4.3.1 *Altered permeability*

Reference has already been made to the barrier functions of the outer layers of the microbial cell envelope and their possible effect on antibiotic resistance. These properties are aspects of the intrinsic resistance of certain species or types of cells. In Gram-negative strains the outer membrane provides a barrier to penetration dependent on the size, hydrophobicity and charge of the antibiotic molecule. Because of the membrane structural differences found, strains of *Haemophilus* and *Neisseria* are usually more antibiotic sensitive than *E. coli*, and strains of indole-positive *Proteus* and *Pseudomonas aeruginosa* are even less easily penetrated and more resistant.

Because of the widespread and varied β-lactam resistance now often found in association with many other resistances, vancomycin (to which mutation endowing resistance is rare) has been used as an alternative therapy. However, because of its size and hydrophilic character it is usually excluded by the Gram-negative outer membrane so that these organisms are usually resistant. Gram-negative bacteria are also intrinsically resistant to the macrolides and lincosamides; *E. coli*, *Serratia*, *Klebsiella*, *Proteus* and *Enterobacter* are all resistant but *Haemophilus*, *Neisseria* and *Campylobacter foetus* are susceptible.[148] Rifamycins are hydrophobic antibiotics primarily active against Gram-positive organisms and most species of *Mycobacteria*. Gram-negative species are more resistant with the exception of *Haemophilus*, *Neisseria* and *Legionella*.[149] Although intrinsic resistance is a generally stable property, mutation may cause otherwise sensitive strains to become resistant by means of permeability changes. Some strains may lose the ability to produce one or other porin. Cefoxitin-resistant *Pr. mirabilis* produced only one 40 kDa porin, while resistant *Pr. vulgaris Providencia rettgeri*, *Providencia alcalifaciens* and *Morganella morganii* produced only a 37 kDa porin, leading to increases in MIC from two- to thirtyfold greater for cephalosporins.[150] *Pseudomonas aeruginosa* resistant to imipenem could be selected relatively easily and were found to be the result of reduced production of a 46 kDa outer membrane protein.[151] The LPS production was unchanged. However, alterations in the structure of the LPS in strains of *E. coli* and *Salmonella typhimurium* were found to cause polymyxin resistance. Nuclear magnetic resonance studies suggested that esterification of the core lipid A phosphates had reduced the anionic charge of the outer membrane.[152] Mutation of the LPS also affects sensitivity to tetracyclines.[153]

A recently isolated *Salmonella paratyphi* A spontaneous mutant to ampicillin resistance was cross-resistant to all β-lactams, except imipenem, and to aminoglycosides, chloramphenicol, tetracycline, trimethoprim and quinolones. There were quantitative reductions in three outer membrane proteins and quantitative and qualitative modifications of the LPS. The strain also lost invasiveness into HeLa cells; a remarkable example of the pleiotropic effects of (presumably) a single mutational event.[153a]

Other membrane changes and intrinsic factors in relation to antibiotics have been reviewed by Godfrey and Bryan,[56] Parr and Bryan[154] and Collatz and Gutmann.[154a]

In addition to the properties of the cell which affect permeation by molecules are the numerous physicochemical factors which may play a significant role. Many antibiotics are charged molecules; the aminoglycosides, polymyxins, octapeptins and biguanides, for example, are polycationic and readily interact with the anionic outer surfaces of the bacterial cell. It is not surprising to find that other cations may act competitively against these drugs, causing a significant shift in MIC. In the case of aminoglycosides, cations antagonize activity in the order spermidine^{3+}, Mg^{2+}, Ca^{2+}, putrescine.

Bacteria are also more resistant to aminoglycosides at low pH, and conversely the drugs are more active at higher pH. The external pH affects the $\Delta\psi$ (electrical potential) component of the proton motive force which is to a large extent responsible for driving the uptake of aminoglycosides. However, it is also likely that the ionization state of the antibiotic amino groups is important, quite independently of the effect of $\Delta\psi$.[149]

5.5.4.3.2 Uptake: modifications of transport systems for D-cycloserine, phosphonomycin (fosfomycin), aminoglycosides

D-Cycloserine is a broad-spectrum 'second-line' antibiotic which is a competitive inhibitor of alanine racemase and D-alanine:D-alanine synthetase, enzymes involved in peptidoglycan synthesis (see Chapter 9.2). These stages of wall synthesis take place inside the cytoplasm and the drug must penetrate the cytoplasmic membrane to reach its targets. It does so by using the D-alanine transport system and will inhibit D-alanine uptake in the wild-type organism. D-Alanine and glycine share a common transport system in *E. coli* different from the L-alanine system and is important in the resistance of cells to D-cycloserine. In first-step-resistant mutants the transport system was found to have lost the high-affinity component and in a second-step mutant the low-affinity component of the D-alanine/glycine system and corresponding ability to transport D-cycloserine. In a multistep mutant, 90% transport ability was lost to give an eightfold increase in D-cycloserine resistance. L-Alanine could still be transported with 75% efficiency.[155] Resistant mutants of *Str. faecalis* also show similar defects in the transport of D-cycloserine.[156]

Phosphonomycin inhibits the enzyme UDP-GlcNAc-3-enolpyruvyl transferase transferring an enolpyruvate moiety to *N*-acetylglucosamine in the synthesis of muramic acid (see Chapter 9.2). For entry to the cell it utilizes the α-glycerophosphate transport system and as a secondary portal the hexose phosphate system when induced by the presence of glucose or fructose 6P; the former is repressed by glucose and inhibited by phosphate so that the presence of these causes resistance.[157] Because of the double entry system, resistance does not arise *in vivo* and phosphonomycin expresses broad-spectrum chemotherapeutic activity.[158]

Extensive studies have been carried out on the mechanism of transport of aminoglycosides and largely elucidated by Bryan and his colleagues.[136,149] The uptake of streptomycin occurs in three stages: an initial energy independent, immediate, electrostatic uptake is followed by an energy dependent phase I (EDP-I), which is relatively slow and lasts up to 15 min, followed by a much more rapid EDP-II. The EDP-I is interpreted to encompass the transfer of drug molecules across the membrane by a transporter molecule, a (possibly quinone) component of the membrane to the inner surface of the membrane from which they are removed by ribosome binding during EDP-II.

Resistant bacteria producing aminoglycoside-modifying enzymes do not cause extensive drug inactivation in the environment of treated cells and it has been suggested that resistance is the result of inhibited transport consequent upon modification during transport.

Because of energy-requirements of the transport process and its linkage with the respiratory chain, mutations in these components alter the aminoglycoside response of the affected cells. *E. coli* cells deficient in respiratory ubiquinone and menaquinone were incapable of transporting gentamicin and dihydrostreptomycin, and were resistant. Similarly anaerobes are resistant although their ribosomes can bind streptomycin, and facultative organisms grown anaerobically in the absence of nitrate (as electron acceptor) cannot generate a sufficient proton motive force for the uptake of drug,[159] but the PMF may be equivalent under anaerobic and aerobic growth and no single hypothesis fits all observations.[159a]

Conversely, those circumstances which enhance uptake of drug lead to the reduction of resistance or the formation of supersensitive mutants. Such mutants are supersensitive when the α-subunit of the F_1 component of ATPase is defective (*unc*A), when there is a defect in the F_0 component (*unc*B) or when there is increased nitrate reductase activity, leading to increased terminal electron transport in *Ps. aeruginosa*.[149] Where there is a decreased nitrite reductase activity (*ag1*C), *Pseudomonas* is resistant due to diminished aminoglycoside transport.[160] Mutation of other transport-related genes, adenyl cyclase (*cya*) and cAMP receptor protein (*crp*) in *E. coli* leads to reduced synthesis of components of the electron transport system and the expression of a resistance phenotype.[149]

5.5.4.3.3 Efflux enhancement: tetracyclines

The seven commonly used tetracyclines form an important group of antibiotics of similar mode of action but differing in lipophilic character, affecting both intestinal absorption and their ability to penetrate bacterial cells. The most lipophilic is minocycline followed by doxycycline, methacycline, chlortetracycline, demeclocycline, oxytetracycline and tetracycline. They have a broad spectrum of activity against Gram-positive and -negative organisms as well as *Mycoplasma*, *Chlamydia* and *Rickettsia*. They are bacteristatic inhibitors of protein synthesis binding to the 30S subunit of the ribosome, a 1:1 binding ratio being sufficient for inhibition.[161]

Unfortunately, resistance to tetracyclines is common and widespread among bacteria and is a great drawback to therapeutic use. Resistance may be chromosomally determined or carried on plasmids and is a component of several transposons.[149] Tetracycline resistance was a property of the first plasmids to be discovered, in company with sulfonamide, chloramphenicol and streptomycin resistance.[6] It is the most commonly found resistance in bacterial species, occurring in staphylococci, *Clostridium*, *Bacteroides*, streptococci, *Neisseria*, *Haemophilus*, *Campylobacter*, *Corynebacterium*, *Vibrio cholerae*, *Mycoplasma*, *Ureaplasma*, *Citrobacter*, *Proteus*, *Enterobacter*, *Escherichia*, *Klebsiella*, *Providencia*, *Pseudomonas*, *Salmonella*, *Shigella*, *Serratia* and *Yersinia*.[162]

The entry of tetracyclines into the bacterial cytoplasm involves energy-assisted transport but the mechanism differs from that found for aminoglycosides. Two uptake systems were revealed when cells were studied in a rich growth environment: an early, rapid energy independent phase was followed by a slower energy dependent accumulation.[163] The latter is sensitive to inhibitors of electron transport and energy coupling (*e.g.* DNP) and to uncoupling mutations. If minimal medium was used, the accumulation went to 'saturation' in a single step.[162] The lipophilic tetracycline, minocycline, was more rapidly accumulated in *E. coli* by active transport but was not more efficient than tetracycline, *in vivo*, as an inhibitor of cell growth. Unlike the aminoglycosides they are active against anaerobes, so the energy requirement for transport is different.

Using everted membrane vesicles, McMurry *et al.*[164] found that tetracycline and minocycline could be actively *accumulated*, thereby finding proof of an active efflux mechanism in normal cells. The net accumulation of drug upon which sensitivity depends will be determined by the relative rates of influx and efflux. At the MIC, about 0.03 nmol of drug is associated with 10^8 cells despite net efflux kinetics,[162] the internal drug being in great measure bound to membrane and ribosomes. DNA–DNA hybridization methods have revealed four different classes of tetracycline-resistant determinants on 25 different plasmids in Gram-negative bacteria. These were denoted A–D. In a survey of 225 lactose fermenters, 73% were of class B, with less frequent occurrence of A and C, while no class D example was found.[165] A Class E enzyme has recently been found in Enterobacteriaceae.[165a] In addition to the Tet A–D determinants, others were found in a variety of different Gram-negative organisms.[162]

Among Gram-positive organisms three determinants different from any Gram-negative ones were found in streptococci and given gene designations *tet*L, *tet*M and *tet*N.[162] From the sequencing studies so far performed it has been concluded that the Gram-positive and Gram-negative resistance determining proteins have evolved separately.[166]

A plasmid-borne staphylococcal Tc[r] gene recently sequenced had only 69% homology with *tet*L and has been designated *tet*K, while another on a self-transmissible plasmid in *Compylobacter coli* has a sequence homology suggesting a Gram-positive origin. This gene, designated *tet*O, has also been found in *Enterococcus* and *Streptococcus* spp.[166a] In addition, a transposon-mediated Tc[r] determinant in *B. fragilis* specifies both the efflux of tetracycline and the chemical degradation of the drug. Only intact drug was effluxed.[166b]

Resistance to tetracycline in both Gram-positive and Gram-negative organisms is inducible and is under the control of regulatory genes. The control was found to be a negative induction system in which the drug acting as inducer caused de-repression, leading to the appearance of the Tet protein which could then be identified as a component of the cytoplasmic membrane. Induction could even be performed in R+ extracts *in vitro*.[162] Constitutive resistance is also known, especially in streptococci, but also in *Bacteroides* and *Clostridium* spp.

While the different plasmid-borne resistance determinants all lead to the appearance of new membrane-bound proteins associated with energy-dependent efflux of drug, the proteins differ in their efflux profile; thus TetA (RP1) extrudes tetracycline, oxytetracycline and chlortetracycline, TetB (R100) extrudes, in addition, demethylchlortetracycline, 6-demethyl-6-deoxy-5-hydroxy-6-methylenetetracycline, doxycycline and minocycline; TetC (pSC101) extrudes all but doxycycline and minocycline.[167] Genetic studies of the determinants on transposons Tn*10* (TetB, R100) and Tn*1721* (TetA, RP1) have shown that there are two regions, one encoding the Tet protein and the other a repressor protein. The regions overlap and have two repressor-sensitive operator regions controlling overlapping promotors, dictating that transcription of the two products occurs in opposing directions.[166] The Tet proteins have hydrophilic and lipophilic domains, the latter being embedded in the membrane while the former probably provide the sites for proton and tetracycline binding in an electrically neutral antiport system of efflux driven by the proton motive force.[168]

When the gene dosage of the TetA determinant in transposon Tn*1721* was increased by the generation of multiple tandem repeats, the efflux of drug did not parallel the increase in gene copy number in the range 1–36 per chromosome.[169] This was accompanied by an increase in the osmotic lability of the cells and the results were interpreted to mean that the insertion of excessive Tet proteins into the membrane led to membrane damage, and that only a limited number of functional insertion sites were available in the cell membrane. When the Tet B determinant was cloned into plasmid vectors with different copy numbers, the increase in gene copies led to a decrease in the expression of induced resistance and concomitant reduced efflux.[167] A detailed explanation of these phenomena has not yet been found.

5.5.4.4 Metabolic By-pass Mechanisms

All living organisms need tetrahydrofolic acid (THF) as a carrier of 1-C fragments for the synthesis of several amino acids and nucleotides for RNA and DNA synthesis. It is also required as a cofactor in the reaction modifying methionyl tRNA to give *N*-formylmethionyl tRNA. Many bacteria are able to synthesize THF from basic components, while mammals must have folic acid as an ingredient of their diet. Folic acid is reduced by mammalian dihydrofolate reductase to tetrahydrofolic acid. In competent bacteria, 2-amino-4-hydroxy-6-hydroxymethylpteridine is condensed with *p*-aminobenzoic acid (PABA) to form dihydropteroic acid (DHPA), catalyzed by the enzyme dihydropteroic acid synthase. The DHPA then reacts with glutamic acid in the presence of dihydrofolate synthase to give dihydrofolate (DHF), which in turn is reduced by bacterial dihydrofolate reductase to form THF. The dihydropteroic acid synthase is the target molecule inhibited by sulfonamides competing with PABA for the enzyme active site, while dihydrofolate reductase action is inhibited by trimethoprim and tetroxoprim. The clinical success of sulfonamides and trimethoprim depends on two essential prerequisites: (i) the inability of pathogens (which make their own DHF) to accumulate folic acid from their environment; and (ii) the difference between mammalian and bacterial DHF-reductases, which give the bacterial enzyme a high affinity for trimethoprim, while the human enzyme has almost none (see Chapters 7.1 and 7.2).

Sulfonamide resistance was one of the group of four first discovered to be plasmid borne,[6] and the mechanism for this is the existence of dihydropteroate synthase enzymes which differ from those found in the sensitive cells. The resistant enzyme in *E. coli* was about three orders of magnitude more resistant to sulfonamide inhibition than was the corresponding sensitive chromosomal enzyme,[170] which correlated with the increased MIC of similar magnitude. The plasmid-coded enzyme was of lower molecular weight (45 kDa *versus* 49 kDa) and was more heat sensitive. Further work has shown that at least two different resistant DHPA synthase enzymes exist which confer clinical sulfonamide resistance.[171]

Similarly, resistance to trimethoprim is determined by possession of alternative (additional) dihydrofolate reductase (DHFR) enzymes, the first of which was discovered about 10 years prior to the isolation of the resistant DHPA synthases. A number of different resistant enzymes are now recognized on the basis of drug concentration for 50% inhibition, K_m value, isoelectric point and molecular weight. Amyes[172] lists six different plasmid-borne enzymes in four different class types. The gene for the type 1a enzyme is located (with a streptomycin/spectinomycin resistance gene) on the transposon Tn*7*, and is widespread among the enterobacteria. In a survey of hospital strains of urinary tract bacteria, 25% had resistance to over 1000 mg L^{-1} of trimethoprim and about half of these were unable to transfer their resistance, implying that the Tn*7* transposon had migrated to the chromosome.[172] Both type Ia and Ib enzymes are 50% inhibited by relatively low concentrations of trimethoprim and methotrexate (but still four and three orders of magnitude different), while for the type IIa and IIb enzymes they are six and five orders of magnitude greater than for the wild-type chromosomal enzyme. The type II enzymes, which also confer high level resistance and are transposon coded, are heat resistant and produced in smaller quantities than the type I.[173]

Type III and type IV DHFR enzymes confer only moderate levels of resistance, with MIC values of 64–160 mg L^{-1}. The type IV DHFR found in India was inducible.[174] A fifth type, conferring high-level resistance, was recently isolated from Enterobacteria in Sri Lanka,[174a] while a sixth type with high-level resistance (>1000 mg L^{-1}) isolated from *Vibrio cholerae* hybridized only weakly with type I and not with a *dfr*II-specific probe.[174b]

5.5.4.5 Overproduction of Antagonistic Metabolites and Target Enzymes

The expression of resistance by an organism is the outcome of numerous properties and interactions of the drug and the structure and metabolism of the organism. Where a drug and metabolite compete for the same enzyme site, the affinity of each substrate, their relative concentrations and the stability of the reaction complexes formed are of fundamental importance. The hyperproduction of PABA by *Neisseria*[175] and *Staphylococcus*[176] have been the cause of resistance, but strains expressing this mechanism of sulfonamide resistance have never been shown to be a serious clinical problem.

Mutants of *Streptococcus* have been found in which the enzyme targets of D-cycloserine have been 'overproduced'; some strains produced up to five times more D-alanyl-D-alanine synthetase or up to eight times more alanine racemase.[177]

A somewhat similar situation results from the mutation of regulatory genes controlling the production of β-lactamases by *Enterobacter*, *Citrobacter*, *Serratia*, indole-positive *Proteus*, *Providencia* and *Ps. aeruginosa*. Spontaneous mutation to a stably derepressed state occurred at a frequency of 10^{-6} to 10^{-7}, leading to production of large amounts of these chromosomally mediated cephalosporinases, resulting in resistance to cefamandole, cefuroxime, moxalactam, cefotaxime, ceftriaxone and ceftazidime, as well as to broad-spectrum penicillins such as carbenicillin, ticarcillin, mezlocillin, piperacillin, azlocillin and even the new monobactams, even though many of these drugs were poor substrates for the enzymes.[178]

5.5.5   RESISTANCE TO DRUGS ACTING AGAINST FUNGI

Of the many thousands of fungal species, only a handful are important pathogens of man, but as a consequence of the metabolic similarity between fungi and man and the paucity of highly efficient antibiotics, deep-seated infections are particularly difficult to eradicate. The pathogenic fungi fall into four morphological groups: (i) the true yeasts (*Cryptococcus neoformans*); (ii) yeast-like fungi with a pseudomycelial stage (*Candida albicans*); (iii) fungi producing a true mycelium (*Aspergillus fumigatus*, *Trichophyton*, *Epidermophyton floccosum*, *Microsporon*); and (iv) dimorphic fungi growing as yeast or mycelium depending on cultural conditions (*Histoplasma capsulatum*, *Coccidioides immitis*, *Blastomyces dermatitidis*). The drugs available for their treatment fall mainly into the polyene and imidazole groups, griseofulvin, 5-fluorocytosine, tolnaftate and benzoic acid. Most compounds inhibitory to fungi are too toxic for systemic use and only 5-fluorocytosine is well tolerated by humans,[179] but antifungal triazoles and imidazoles under development as orally active antimycotic agents[180] hold prospects for the future. Although amphotericin B, miconazole and ketoconazole are used systemically, they have serious side effects and there is a small margin between relatively safe serum levels and fungistatic drug concentrations.[181]

5.5.5.1   Polyenes

Among the polyenes, a group of over 200 compounds, characteristically with opposing hydrophilic and hydrophobic faces, only amphotericin B, candicidin, natamycin and nystatin are frequently used clinically. The biochemical basis of their action is their ability to penetrate into and accumulate within the membrane, a property determined by the presence of sterols and their amphipathic structure. Amphotericin B is specifically 'bound' by ergosterol and it is thought that eight molecules cluster to form a hydrophilic pore through which cytoplasmic components leak from the cell.[182] The effect is bactericidal.

Resistance to polyene antibiotics is not a serious clinical problem, although resistant yeasts have been found and the necessity for long-term therapy in immune deficient patients may lead to further problems. Examination of resistant mutants reveals either qualitative or quantitative changes in sterol composition of the membranes, reduced amounts of ergosterol being the cause of amphotericin B resistance.[183]

5.5.5.2   Imidazoles

All the useful imidazoles (*e.g.* clotrimazole, econazole, isoconazole, butaconazole, miconazole, ketoconazole, *etc.*) have similar effects on sensitive cells, although the biochemical mechanisms of inhibition have not been resolved in detail.[184] A variety of effects are observed depending upon the concentration of drug used. Vanden Bossche *et al.*[185] proposed the following. (a) Drug concentration > 10 nmol L^{-1}: interference with the fatty acid desaturase system leading to more saturated acids (mainly palmitic). (b) Drug concentration > 10 μmol L^{-1}: direct interaction of some (*e.g.* miconazole) with lipids, causing membrane disorganization. (c) Drug concentration > 0.1 mmol L^{-1}: interference with the lanosterol 14-α-demethylase system, which is cytochrome

P-450 dependent; decreased availability of ergosterol and accumulation of 14-α-methyl sterols (see also Chapter 7.4).[185] However, the MIC values can vary as much as one-thousandfold for miconazole and ketoconazole against a range of fungi,[186] a consideration of some importance since systemic use of ketoconazole can lead to liver damage.[187]

Very few reports of 'clinically developed' resistance have been published and mutants are not readily obtained in the laboratory. Convincing explanations of resistance are rare. Sterol biosynthesis was less sensitive to ketoconazole in two strains of *C. albicans* isolated from patients who had relapsed after treatment, but the cell-free system was sensitive and the whole cells failed to accumulate a radiolabelled triazole. Resistance was attributed to a change in the membrane, making it impermeable to the drug.[188] In an examination of 10 strains of miconazole-resistant *C. glabrata*, two types of mutants were found. In the first group the MIC was raised threefold, but the second group appeared to have similar MICs to the parental strain. In both groups, and in the sensitive strain, oxygen uptake was inhibited 88% at the MIC; however, spheroplasts of resistant and sensitive strains gave the same oxygen uptake correlation with drug concentration, ruling out the cell wall as a barrier to drug penetration.[181]

5.5.5.3 Flucytosine (5-Fluorocytosine)

Flucytosine is effective in treating candidosis and cryptococcosis but serious resistance problems have arisen during therapy. In sensitive fungi, flucytosine is metabolized to 5-fluorodeoxyuridine monophosphate, which blocks thymidylate synthetase, and there is extensive replacement of uracil by 5-fluorouracil in fungal RNA. For these reactions to occur in the cell the drug must be transported and deaminated to 5-fluorouracil by cytosine deaminase.[186] Unfortunately the poor uptake of 5-fluorouracil by fungi prevents its use as a drug.

The useful antifungal spectrum of flucytosine is limited to species of *Candida*, *Cryptococcus*, *Aspergillus*, *Phialophora verrucosa*, *Rhinocladiella pedrosoi*, *Exophiala dermatitidis* and *Cladosporium carrionii*,[186] and it acts synergistically with amphotericin B against these organisms. Surveys of clinical isolates of *Candida albicans* found less than 10% highly resistant to growth inhibition, but 35–40% were partially resistant. The partially resistant strains were heterozygous for resistance and segregated to give sensitive and highly resistant strains.[181] In the highly resistant segregants the majority of strains are deficient in UMP pyrophosphorylase.[189] Five flucytosine resistance mechanisms have been revealed by UV-induced mutation in *Saccharomyces cerevisiae*, viz. loss of cytosine permease, loss of cytosine deaminase, loss of UMP pyrophosphorylase, loss of feedback regulation of aspartate transcarbamylase, and stimulation of orotidylate pyrophosphorylase and decarboxylase. It is likely that these mechanisms also occur in *C. albicans*, and since cytosine permease and cytosine deaminase are not essential for growth of *C. albicans*, their loss conferring resistance should still allow efficient growth. In *C. glabrata*, mutants lacking cytosine permease have been found.[181]

(i) Griseofulvin

Although resistance to this antibiotic has often been shown to be the cause of treatment failure with griseofulvin, no clear indications of the biochemical mechanism of resistance have been published.[186]

5.5.6 DRUG DEPENDENCE

Dependence is a special category of drug resistance found in Gram-positive and Gram-negative bacteria in which the result of mutation confers not only the ability to grow in high concentrations of the drug but an absolute requirement for it if growth is to be logarithmic. It occurs most commonly with the aminoglycoside antibiotics and was recognized as early as 1951.[190] In the absence of the required antibiotic protein, synthesis and growth are arithmetic for several hours before the culture loses viability. The rate of protein synthesis in drug-limited cultures is dependent upon drug concentration up to a saturating value.[191] Momose and Gorini[192] distinguished four classes of aminoglycoside-dependent mutants in *E. coli*: (i) drug^D were dependent either on streptomycin (Sm), paromomycin (Pm) or ethanol; (ii) Sm^{D}Et^D were dependent on (Sm) or Et; (iii) Sm^D were dependent on Sm only; and (iv) Pm^{D}Sm^R were dependent on Pm but resistant to Sm. All these phenotypes mapped in the *rps*L gene determining the structure of ribosomal protein S12. Other Sm^D mutants have alterations in both S8 and S12 ribosomal proteins,[193] while a spectinomycin-dependent mutant had an altered S5 (conferring resistance) and an altered 50S subunit protein (which conferred dependence).[194] In kasugamycin dependence, more than one locus was mutated and changes were found in S8, S9, S13, S18, L11 or L14.[195]

When drug[D] mutants were grown on Sm medium and transferred directly to Pm medium or *vice versa*, they died. However, if washed in buffer before regrowth on the second medium, they survived. If both Pm and Sm were present simultaneously at concentrations above the MIC of the parent organism, they also died.[196] The aminoglycosides could be divided into two groups in which combinations of drugs within a group permitted logarithmic growth, while combinations from both groups were lethal. It was concluded that the ribosome possessed two adjacent binding sites, one for each group of drugs. If both sites were occupied simultaneously, the cells died; if only one site was occupied, protein synthesis was possible; and if neither site was occupied, protein synthesis of active enzymes was not possible.[197]

It has been proposed that drug-dependent organisms might be used as living vaccines of pathogenic organisms, administered with the required drug, since the population could easily be killed by withdrawing the drug. However, dependent mutants revert to independence at high frequency, most of the revertants showing a second (suppressor) mutation in S4 associated with 'ribosomal ambiguity' (*ram*)[198] and the fidelity of mRNA translation.

5.5.7 GENE MOBILITY AND THE SPREAD OF RESISTANCE

5.5.7.1 Plasmids

While genes capable of conferring drug resistance existed long before the antibiotic era and are intrinsic sources of protection for producer organisms, the real problems of resistance in clinical situations have arisen through the mobilization of these genes by plasmids and transposons. Plasmids have been found in nearly all species where specifically sought and determine a wide range of properties other than antibiotic resistance, such as bacteriophage resistance, resistance to heavy metal ions and disinfectants, bacteriocin production, siderophore production, adhesion and colonization factors, enterotoxin production, haemolysin and metabolic abilities of various kinds. Those carrying resistance determinants are called R-plasmids.

Plasmids are extrachromosomal, non-essential (except in certain environments) genetic elements comprising circular double-stranded DNA molecules ranging in size from a few kilobase pairs (kbp) to several hundred kbp. A single bacterium may carry one or more different plasmids. Such plasmids must be compatible to be stably maintained and plasmids which cannot stably coexist in the same strain are said to belong to the same incompatibility group (Inc). A single type of plasmid may be maintained stably at a copy number relative to the chromosome from approximately 1 to about 40 copies. In general, large plasmids have low copy numbers and small plasmids may have high copy numbers (Table 8).

Table 8 Examples of R-plasmids

Plasmid	Phenotype[a]	M (MDa)	Organisms
R16	Tra⁺, Ap, Cm, Sm, Su, Tc	69	*E. coli, Salmonella, Shigella*
R1	Tra⁺, Ap, Cm, Km, Sm, Su, Sp	62	*E. coli, Proteus, Salmonella, Providencia, Serratia, Klebsiella*
R391	Tra⁺, Km, Nm, Hg		*E. coli, Proteus*
R69	Tra⁺, Ap, Km, Nm, Pm, Tc		*E. coli, Sal. paratyphi* B, *Serratia, Klebsiella, Proteus*
R300B	Tra⁻, Sm, Su	5.7 (11 copies)	*E. coli, Sal. typhi, Salmonella* spp. *Proteus, Pseudomonas, Providencia*
PM41	Tra⁺, Cb, Gm, Km, Sm, Tm	57	*Pseudomonas*
R477	Tra⁺, Cm, Km, Sm, Su, Tc	173	*Serratia marcescens, E. coli*
pUB101	Tra⁻, Pc, Cd, Fa	14.6	*Staph. aureus*
Ero	Em, Ln, Vm	17.6 (6 copies)	*Str. faecalis*

[a] Ap, ampicillin; Cb, carbenicillin; Cd, cadmium; Cm, chloramphenicol; Em, erythromycin; Fa, fusidic acid; Gm, gentamicin; Hg, mercury; Km, kanamycin; Ln, lincomycin Nm, neomycin; Pc, penicillin; Pm, paromomycin; Sm, streptomycin; Sp, spectinomycin; Su, sulfonamide; Tc, tetracycline; Tm, tobramycin; Tra, transfer; Vm, vernamycin B.

Some plasmids possess all the genes necessary for the production of components of the 'mating' and transfer system, the *tra* genes, and are capable of initiating their own transfer by means of conjugation with another organism of the same or different species. These are the conjugative (autotransmissible) plasmids. Small plasmids (< 30 kbp) are usually non-autotransmissible but may be 'mobilized' by the larger conjugative plasmids in the same cell for their transfer to a recipient cell. Even a non-mobilizable plasmid (lacking transfer origin and mobilization functions) can be transferred 'cointegrated' with a conjugative plasmid.[199]

When a plasmid is transferred to a recipient bacterium it may not be able to establish itself within the host in a permanent replicative state; a plasmid, therefore, will only become permanently established in bacteria within the host range of the plasmid. If the transfer range is wider than the host range then the plasmid may become a suicide vector, enabling plasmid genes to be integrated into the host chromosome before the plasmid DNA is 'segregated out' on subsequent cell division.

There are therefore two aspects to the spread of genes by plasmids: the necessity that the plasmid acts as a replicon (makes copies for transfer to offspring) and that it is compatible with, and stably maintained within, the host strain. These properties determine the vertical transfer of the plasmid. The conjugative, mobilizable and host range properties determine the horizontal transfer of the plasmid and its genes. Usually Gram-positive plasmids can be horizontally transferred only among Gram-positive organisms and similarly Gram-negative plasmids among Gram-negative organisms, but the finding of a typical Gram-positive Gm-inactivating enzyme in *Campylobacter* suggests that the barrier may not be absolute.[200]

5.5.7.2 Transposons

Transposons are sequences of DNA that can transfer themselves from one replicon (DNA donor molecule) to another (recipient) replicon with little regard for sequence homology. They are not independent replicons and must form part of a replicon sequence in order to replicate. Most of the known antibiotic resistance determinants occur on one or other transposon (Table 9).

There are three types of transposons.[201] (i) Composite transposons consisting of one or more resistance genes bounded by insertion sequences (IS) 7–15 kbp in length and usually in inverse orientation. The transposition functions are determined by one or other IS. In transposon Tn*5*, which codes APH(3′) enzyme, the left side IS (IS50L) contains a protein synthesis terminating codon, which also creates a strong promoter for the resistance gene. The right side IS (IS50R) codes for a transposase and the transposase regulatory protein. (ii) Transposons similar to Tn*3*. Tn*3* has invert repeat sequences of 38 bp at each end. It codes for two genes essential for transposition. Gene *tnp*A codes for a transposase enzyme recognizing the IR sequences and generates a cointegrate structure of donor and target DNA molecules. The *tnp*R gene product is a resolvase enzyme, which resolves the cointegrate at a specific (*res*) site[202] between *tnp*A and *tnp*R, acting also to regulate the synthesis of both resolvase and transposase. To the right of *tnp*R is the β-lactamase (*bla*) gene coding for TEM-1 enzyme (Section 5.5.4.2.1). Similar transposons are Tn*1* and Tn*2*, coding for TEM-2 and TEM-1 respectively; Tn*21*, determining resistance to Sm, Sp, Su and Hg; and Tn*1721*, coding for class A Tet protein (Section 5.5.4.3.3). (iii) Conjugative transposons. These are widespread in streptococci and have played a major role in the spread of multiple antibiotic resistance in this genus.[203] Strains exhibiting conjugative transfer of resistance (*e.g.* Tc, MLS) were carefully examined for plasmids but none were found, indicating a chromosomal location for these transposons which were nonetheless

Table 9 Some Antibiotic Resistance Transposons

Transposon	Resistance phenotype[a]	Resistance mechanisms	Size (kbp)	Original source
Tn*1*	Ap	TEM-2	5.0	RP4 *Pseudomonas*
Tn*3*	Ap	TEM-1	5.0	R1-19 *Salmonella*
Tn*4*	Ap, Sm, Su, Hg	TEM-1	24.0	R1-19 *Salmonella*
Tn*5*	Km	APH(3′)II	5.1	JR67 *Klebsiella*
Tn*9*	Cm	CAT(type I)	2.7	R10 *Shigella*
Tn*10*	Tc (class B)	Efflux	9.3	R100 *Shigella*
Tn*551*	MLS	Adenine N^6-methylase	5.2	PI258 *Staphylococcus*
Tn*1401*	Ap, Sm, Sp, Su, Hg	PSE-1	12.0	*Pseudomonas*

[a] See Table 8 for abbreviations of resistance.

able to cause conjugation and gene transfer between different streptococci. *Str. pneumoniae, Str. faecalis, Str. pyogenes* and *Str. agalactiae* have all been implicated. The gene transfer occurs between cells in close contact (filter matings in the laboratory) and is resistant to the action of DNAase. It is also probable that clindamycin or Tc resistance may be transferred in *Bacteroides* by a similar mechanism.[204,205]

The fact that transposition occurs independently of the normal bacterial mechanism for recombining homologous DNA molecules involving the *rec*A gene product, gives transposition an increased probability of occurrence and flexibility in the transfer of genes between molecules and cells. Bacteria have, in consequence, accumulated different resistance genes by tandem insertions of transposons into non-essential plasmid sites.[206] The continued selective pressure of heavy antibiotic use, especially in hospitals, has ensured the progressive acquisition of drug resistance genes on individual plasmids, and encouraged the accumulation of different R-plasmids in the same pathogenic bacterial cell, often passed on from harmless commensal organisms.

5.5.7.3 Other Systems of Gene Transfer

5.5.7.3.1 *Transformation*

In Gram-positive cells, double-stranded (ds)DNA (probably from cell lysis of the donor) binds reversibly to specific surface proteins of competent recipient cells. The dsDNA then becomes irreversibly bound, suffering at the same time a number of single-strand nicks at 6–8 kbp intervals. Double-strand breaks then appear and one of the strands is completely hydrolyzed by an exterior endonuclease, while the resulting single strand becomes coated with a polypeptide which allows the protein–DNA complex to enter the cell. Here the ssDNA is incorporated into the genome to give a heteroduplex region, which leads on subsequent division and segregation to the formation of mutant offspring. Gram-negative cells have a similar mechanism but the adsorbed dsDNA is not degraded to ssDNA for entry into the cell; however, only a single strand of the transforming DNA is incorporated into the genome, displacing the endogenote strand at the point of incorporation. The non-incorporated single strands are then degraded. This system probably does not play a significant role in resistance gene transfer in natural environments, although the phenomenon is much utilized in genetic experimentation.

5.5.7.3.2 *Transduction*

Transduction is mediated by bacteriophage and does play a significant part in the spread of resistance genes, especially among staphylococci and streptococci. In generalized transduction the bacterial genome is cut by nuclease into a series of dsDNA fragments and these are packaged into phage capsids instead of an appropriate aliquot of phage DNA. When these particles are released and adsorb to another cell within their host range, the bacterial dsDNA is injected into the recipient and may recombine with its genome. If recombination fails to occur, the recipient is termed an abortive transductant. If the donor cells contain plasmid molecules of appropriate size, they may be incorporated into the capsids so that the plasmid is transferred and is capable of existing in the new host (if compatible) as an independent replicon.

In specialized transduction a specific small segment of donor genome, adjacent to the integrated prophage DNA, is erroneously excised with (most of) the phage DNA when the temperate phage is induced to reproduce new phage particles. This phenomenon is not important in the transfer of resistance genes.

5.5.7.3.3 *Phage-mediated conjugation*

In this system the presence of a bacteriophage in either the donor or recipient enables the transfer of plasmids at high frequency, without death of the donor or the lysis of the recipient. In these cases the lytic functions of the phage are not induced and up to one-tenth of the recipients in the culture may receive plasmids.[28]

5.5.7.3.4 *Sex pheromone-induced conjugation*

Perhaps the most cunning system devised by bacteria for the facilitation of gene transfer is the excretion by certain recipient strains of streptococci (and probably also by staphylococci) of small peptide sex pheromones which induce donors harbouring certain conjugative plasmids to synthesize an adhesin or 'aggregation substance' to promote mating. The adhesion is a protein antigen coating the donor and enabling clumping either of donors or donors and recipients, and is also called 'clumping inducing agent' (CIA). When the mating aggregate is formed and the recipient has received a plasmid from the donor, the recipient ceases to produce that plasmid-related pheromone, but may continue to produce pheromones attractive to carriers of different plasmids.[207,208]

5.5.8 REFERENCES

1. A. Fleming, *Br. J. Exp. Pathol.*, 1929, **10**, 226.
2. A. L. Miller and J. B. Walker, *J. Bacteriol.*, 1969, **99**, 401.
3. O. Nimi, G. Ito, S. Sueda and R. Nomi, *Agric. Biol. Chem.*, 1971, **35**, 848.
4. R. Cella and L. C. Vining, *Can. J. Microbiol.*, 1975, **21**, 463.
5. J. M. Piwowarski and P. D. Shaw, *Antimicrob. Agents Chemother.*, 1979, **16**, 176.
6. S. Watanabe and K. Tanaka, *Biochem. Biophys. Res. Commun.*, 1976, **72**, 522.
7. W. Piendl and A. Böck, *Antimicrob. Agents Chemother.*, 1982, **22**, 231.
8. C. J. Thompson, R. H. Skinner, J. Thompson, J. M. Ward, D. A. Hopwood and E. Cundliffe, *J. Bacteriol.*, 1982, **151**, 678.
9. E. Cundliffe, *Br. Med. Bull.*, 1984, **40**, 61.
10. D. H. Smith, *J. Bacteriol.*, 1967, **94**, 2071.
11. Report, 'The Effects on Human Health of Subtherapeutic Use of Antimicrobials in Animal Feeds', *National Academy of Sciences, Washington*, DC, 1980, vol. 8, p. 376.
12. A. H. Linton, *Br. Med. Bull.*, 1984, **40**, 91.
13. A. H. Linton and T. C. Robinson, *Br. Vet. J.*, 1984, **140** (4), 368.
14. J. K. L. Pearson, in 'Antibiotics and Antibiosis in Agriculture', ed. M. Woodbine, Butterworths, London, 1977, p. 217.
15. M. Hinton, A. Kaukas and A. H. Linton, *J. Appl. Bacteriol.*, 1986, **61**, Suppl., 77S.
16. Report, 'Animal Salmonellosis', Communicable Disease Report, Public Health Laboratory Service, 1984, Edition 84/49.
17. E. J. Threlfall, L. R. Ward and B. Rowe, *Vet. Rec.*, 1978, **103**, 438.
18. B. Rowe, E. J. Threlfall, L. R. Ward and A. S. Ashley, *Vet. Rec.*, 1979, **105**, 468.
19. E. J. Threlfall, L. R. Ward, A. S. Ashley and B. Rowe, *Br. Med. J.*, 1980, **280**, 1210.
20. E. J. Threlfall, L. R. Ward and B. Rowe, *J. Antimicrob. Chemother.*, 1986, Suppl. C, 175.
21. C. E. Cherubin, *Rev. Infect. Dis.*, 1981, **3**, 1105.
22. M. B. Skirrow, *J. Hyg.*, 1982, **89**, 175.
23. A. H. Linton, *J. Antimicrob. Chemother.*, 1986, Suppl. C, 189.
24. A. H. Linton, K. Howe, C. L. Hartley, H. M. Clements and M. H. Richmond, *J. Appl. Bacteriol.*, 1977, **42**, 365.
25. B. E. Langlois, G. L. Cromwell, T. S. Stahly, K. A. Dawson and V. W. Hays, *Appl. Environ. Microbiol.*, 1983, **46**, 1433.
26. S. E. Feinman, in 'Handbook Series in Zoonoses', ed. J. H. Steele and G. W. Beran, CRC Press, Boca Raton, FL, 1984, vol. 1, p. 151.
27. G. Turnock, S. J. Erickson, B. A. Sckrell and B. Birch, *J. Gen. Microbiol.*, 1972, **70**, 507.
28. R. W. Lacey, *Br. Med. Bull.*, 1984, **40**, 77.
29. E. J. Gangarosa, D. R. Perera, L. J. Mata, C. M. Morrison, G. Guzman and L. B. Reller, *J. Infect. Dis.*, 1970, **122**, 181.
30. W. E. Farrar and M. Eidson, *J. Infect. Dis.*, 1971, **124**, 327.
31. J. H. Crosa, J. Olarte, L. J. Mata, L. K. Luttropp and M. E. Peñaranda, *Antimicrob. Agents Chemother.*, 1977, **11**, 553.
32. M. M. Rahaman, F. Huq, C. R. Dey, A. K. M. Kibirya and G. Curlin, *Lancet 1*, 1974, 406.
33. M. M. Rahaman, M. M. Khan, K. Aziz, M. S. Isham and A. K. M. Kibirya, *J. Infect. Dis.*, 1975, **132**, 15.
34. J. A. Frost, G. A. Willshaw, E. A. Barclay, B. Rowe, P. Lemmens and J. Vandepitte, *J. Hyg.*, 1985, **94** (2), 163.
35. B. Rowe and E. J. Threlfall, *Br. Med. Bull.*, 1984, **40**, 58.
36. R. J. Gross, L. V. Thomas and B. Rowe, *Br. Med. J.*, 1979, **2**, 744.
37. J. C. L. Mwansa, Ph.D. Thesis, University of Manchester, 1986.
38. E. S. Anderson, E. J. Threlfall, J. M. Carr, M. M. McConnell and H. R. Smith, *J. Hyg.*, 1977, **79**, 425.
39. B. Rowe, E. J. Threlfall, J. A. Frost and L. R. Ward, *Lancet 1*, 1980, 1070.
40. S. Le Minor, *Med. Mal. Infect.*, 1972, **2**, 441.
41. P. Y. Chau, W. T. Wong and Y. P. Fok, *J. Hyg.*, 1978, **81**, 343.
42. P. Y. Chau, J. Ling, E. J. Threlfall and S. W. K. Im, *J. Gen. Microbiol.*, 1982, **128**, 239.
43. K. B. Sharma, K. Bheem Bhat, P. Asha, N. Diwan and S. Vaze, *Indian J. Med. Res.*, 1979, **69**, 720.
44. L. Le Minor, C. Coynault and G. Pessoa, *Ann. Microbiol. (Paris)*, 1974, **125A**, 261.
45. H. Sarrat, L. Le Minor, S. Le Minor, C. Lafaix and L. Maydat, *Pathol. Biol.*, 1972, **20**, 577.
46. E. S. Anderson, *J. Hyg.*, 1975, **74**, 289.
47. C. Herzog, *Acta Trop.*, 1980, **37**, 275.
48. L. P. Elwell, M. Roberts, L. W. Mayer and S. Falkow, *Antimicrob. Agents Chemother.*, 1977, **11**, 528.
49. A. Percival, T. E. Corkhill, O. P. Arya, J. Rowlands, C. D. Alergent, E. Rees and E. H. Annels, *Lancet 2*, 1976, 1379.
49a. J. Heritage and P. M. Hawkey, *J. Antimicrob. Chemother.*, 1988, **22**, 575.
50. R. P. Wenzel, *Ann. Intern. Med.*, 1982, **97**, 440.
51. M. W. Casewell, in 'Recent Advances in Infection', ed. D. S. Reeves and A. M. Geddes, Churchill Livingstone, Edinburgh, 1982, p. 31.
52. A. A. Medeiros, R. W. Hedges and G. A. Jacoby, *J. Bacteriol.*, 1982, **149**, 700.
53. M. Kresken and B. Wiedemann, *J. Antimicrob. Chemother.*, 1986, **18**, Suppl. C, 235.
54. T. F. O'Brien and The International Survey of Antibiotic Resistance Group, *J. Antimicrob. Chemother.*, 1986, **18**, Suppl. C, 243.
55. B. K. Pugashetti, H. M. Metzger, Jr., L. Vadas and D. S. Feingold, *J. Clin. Microbiol.*, 1982, **16**, 686.
56. A. J. Godfrey and L. E. Bryan, in 'Antimicrobial Drug Resistance', ed. L. E. Bryan, Academic Press, New York, 1984, p. 113.

57. V. Braun, *Biochim. Biophys. Acta*, 1975, **415**, 335.
58. H. Nikaido and M. Vaara, *Microbiol. Rev.*, 1985, **49**, 1.
59. H. Nikaido, in 'Microbiology—1984', ed. L. Leive and D. Schlessinger, American Society for Microbiology, Washington, DC, 1984, p. 381.
60. J. Baddiley, in 'β-Lactam Antibiotics—Mode of Action, New Developments, Future Prospects', ed. M. R. J. Salton and G. D. Shockman, Academic Press, New York, 1981, p. 13.
61. G. W. Gooday and A. P. J. Trinci, in 'The Eukaryotic Microbial Cell', ed. G. W. Gooday, D. Lloyd and A. P. J. Trinci, Cambridge University Press, Cambridge, 1980, p. 207.
62. R. Scherrer, L. Louden and P. Gerhardt, *J. Bacteriol.*, 1974, **118**, 534.
63. D. Kerridge, in 'The Eukaryotic Microbial Cell', ed. G. W. Gooday, D. Lloyd and A. P. J. Trinci, Cambridge University Press, Cambridge, 1980, p. 103.
64. S. C. Harrison, in 'The Microbe—1984 Part 1: Viruses', ed. B. W. J. Mahy and J. R. Pattison, Cambridge University Press, Cambridge, 1984, p. 29.
65. H. Nikaido, *Biochim. Biophys. Acta*, 1976, **433**, 118.
66. H. Nikaido, in 'β-Lactam Antibiotics, Mode of Action, New Developments, Future Prospects', ed. M. R. J. Salton and G. D. Shockman, Academic Press, New York, 1981, p. 249.
67. M. H. Richmond and R. B. Sykes, *Adv. Microb. Physiol.*, 1973, **9**, 31.
67a. P. Actor, L. Daneo-Moore, M. L. Higgins, M. R. S. Salton and G. D. Shockman (ed.), 'Antibiotic Inhibition of Bacterial Cell Surface Assembly and Function', American Society for Microbiology, Washington, 1984.
68. A. Tomasz, A. Albino and E. Zanati, *Nature (London)*, 1970, **227**, 138.
69. A. Tomasz, in 'β-Lactam Antibiotics, Mode of Action, New Developments and Future Prospects', ed. M. R. J. Salton and G. D. Shockman, Academic Press, New York, 1981, p. 227.
70. B. D. Davis and W. K. Mass, *Proc. Natl. Acad. Sci. USA*, 1952, **38**, 775.
71. J. Davies and D. I. Smith, *Annu. Rev. Microbiol.*, 1978, 32, 468.
72. T. Watanabe and T. Fukasawa, *Biochem. Biophys. Res. Commun.*, 1960, **3**, 660.
73. B. G. Spratt, *J. Gen. Microbiol.*, 1983, **129**, 1247.
74. P. E. Reynolds, in 'The Scientific Basis of Antimicrobial Chemotherapy', ed. D. Greenwood and F. O'Grady, Cambridge University Press, Cambridge, 1985, p. 13.
75. P. E. Reynolds, *Br. Med. Bull.*, 1984, **40**, 3.
76. B. G. Spratt, *Nature (London)*, 1978, **274**, 713.
77. R. Williamson, in 'The Target of Penicillin', ed. R. Hakenbeck, J.-V. Holte and H. Labischinski, de Gruyter, Berlin, 1983, p. 487.
78. C. E. Buchanan and J. L. Strominger, *Proc. Natl. Acad. Sci. USA*, 1976, **73**, 1816.
79. A. F. Giles and P. E. Reynolds, *Nature (London)*, 1979, **280**, 167.
80. R. Fontana, P. Canepari and G. Salta, in 'The Target of Penicillin', ed. R. Hakenbeck, J.-V. Holte and H. Labischinski, de Gruyter, Berlin, 1983, p. 531.
81. T. J. Dougherty, A. E. Koller and A. Tomasz, *Antimicrob. Agents Chemother.*, 1980, **18**, 730.
82. R. Hakenbeck, M. Tarpay and A. Tomasz, *Antimicrob. Agents Chemother.*, 1980, **17**, 364.
83. S. Zighelboim and A. Tomasz, *Antimicrob. Agents Chemother.*, 1980, **17**, 434.
83a. H. F. Chambers, *Clin. Microbiol. Rev.*, 1988, **1**, 173.
84. M. V. Hayes, N. A. C. Curtis, A. W. Wyke and J. B. Ward, *FEMS Microbiol. Lett.*, 1981, **10**, 119.
85. N. H. Georgopapadakou, S. A. Smith and D. P. Bonner, *Antimicrob. Agents Chemother.*, 1982, **22**, 172.
86. D. F. J. Brown and P. E. Reynolds, *FEBS Lett.*, 1980, **122**, 275.
87. D. F. J. Brown and P. E. Reynolds, in 'The Target of Penicillin', ed. R. Hakenbeck, J.-V. Holte and H. Labischinski, de Gruyter, Berlin, 1983, p. 537.
88. B. Hartman and A. Tomasz, *Antimicrob. Agents Chemother.*, 1981, **19**, 726.
89. B. R. Lyon and R. Skurray, *Microbiol. Rev.*, 1987, **51**, 88.
89a. R. S. Schwalbe, J. T. Stapleton and P. H. Gilligan, *New Engl. J. Med.*, 1988, **316**, 927.
89b. R. Leclercq, E. Derlot, M. Weber, J. Duval and P. Courvalin, *Antimicrob. Agents Chemother.*, 1989, **33**, 10.
90. M. Ozaki, S. Mizushima and M. Nomura, *Nature (London)*, 1969, **222**, 333.
91. G. Funatsu and H. G. Wittmann, *J. Mol. Biol.*, 1972, **68**, 547.
92. B. Wittmann-Liebold and B. Greuer, *FEBS Lett.*, 1978, **95**, 91.
93. P. Buckel, A. Buchberger, A. Böck and H. G. Wittmann, *Mol. Gen. Genet.*, 1977, **158**, 47.
94. M. de Wilde, T. Cabezón, R, Villarroel, A. Herzog and A. Bollen, *Mol. Gen. Genet.*, 1975, **142**, 19.
95. M. Cannon, T. Cabezón and A. Bollen, *Mol. Gen. Genet.*, 1974, **130**, 321.
96. H. G. Wittmann, G. Stöffler, D. Apirion, *et al.*, *Mol. Gen. Genet.*, 1973, **127**, 175.
97. S. Osawa, R. Takata, K. Tanaka and M. Tomaki, *Mol. Gen. Genet.*, 1973, **127**, 163.
98. M. J. R. Stark, E. Cundliffe, J. Dijk and G. Stöffler, *Mol. Gen. Genet.*, 1980, **180**, 11.
99. N. Tanaka, G. Kawano and T. Kinoshita, *Biochem. Biophys. Res. Commun.*, 1971, **42**, 564.
100. T. L. Helsen, J. E. Davies and J. E. Dahlberg, *Nature (London), New Biol.*, 1971, **233**, 12.
100a. C. Mabilat and P. Courvalin, *Ann. Inst. Pasteur Microbiol.*, 1988, **139**, 677.
100b. M. Arthur, A. Brisson-Noel and P. Courvalin, *J. Antimicrob. Chemother.*, 1987, **20**, 783.
101. B. Weisblum, *Br. Med. Bull.*, 1984, **40**, 47.
101a. L. M. Quiros, S. Fidalgo, F. J. Mendez, C. Hardisson and J. A. Salas, *Antimicrob. Agents Chemother.*, 1988, **32**, 420.
102. M. Gellert, *Annu. Rev. Biochem.*, 1981, **50**, 879.
103. G. C. Crumplin and J. T. Smith, *J. Antimicrob. Chemother.*, 1981, **7**, 379.
104. S. Inoue, T. Ohue, J. Yamagishi, S. Nakamura and M. Shimizu, *Antimicrob. Agents Chemother.*, 1978, **14**, 240.
105. J. T. Smith, in 'The Scientific Basis of Antimicrobial Chemotherapy', ed. D. Greenwood and F. O'Grady, Cambridge University Press, Cambridge, 1985, p. 69.
106. K. Hirai, H. Aoyama, S. Suzue, T. Irikura, S. Iyobe and S. Mitsuhashi, *Antimicrob. Agents Chemother.*, 1986, **30**, 248.
107. K. Hirai, S. Suzue, T. Irikura, S. Iyobe and S. Mitsuhashi, *Antimicrob. Agents Chemother.*, 1987, **31**, 582.
108. K. Mizuuchi, M. H. O'Dea and M. Gellert, *Proc. Natl. Acad. Sci. USA*, 1978, **75**, 5960.
109. D. Rabussay and W. Zillig, *FEBS Lett.*, 1969, **5**, 104.

110. G. Cassani, R. R. Burgess, H. M. Goodman and L. Gold, *Nature (London), New Biol.*, 1971, **230**, 197.
111. H. Ogawara, *Antimicrob. Agents Chemother.*, 1975, **8**, 402.
112. R. J. Mehta and C. H. Nash, *J. Antibiot.*, 1978, **31**, 239.
113. D. J. Kushner and C. Breull, *Arch. Microbiol.*, 1977, **112**, 219.
114. M. Ridley, D. Barrie, R. Lynn and K. C. Stead, *Lancet 1*, 1970, 230.
115. D. J. Waxman and J. L. Strominger, *Annu. Rev. Biochem.*, 1983, **52**, 825.
116. D. J. Waxman and J. L. Strominger, *J. Biol. Chem.*, 1980, **255**, 3964.
117. P. J. Hedge and B. G. Spratt, *Nature (London)*, 1985, **318**, 478.
118. R. Bicknell, S. J. Cartwright, E. L. Emanuel, G. C. Knight and S. G. Waley, in 'Recent Advances in the Chemistry of β-Lactam Antibiotics', ed. A. G. Brown and S. M. Roberts, The Royal Society of Chemistry, London, 1985, No. 3, p. 280.
119. N. Citri and M. R. Pollock, *Adv. Enzymol.*, 1966, **28**, 237.
120. K. G. H. Dyke, in 'β-Lactamases', ed. J. M. T. Hamilton-Miller and J. T. Smith, Academic Press, New York, 1979, p. 291.
121. S. Yamamoto and J. O. Lampen, *Proc. Natl. Acad. Sci. USA*, 1976, **73**, 1457.
122. M. Matthew, *J. Antimicrob. Chemother.*, 1979, **5**, 349.
123. A. A. Medeiros, *Br. Med. Bull.*, 1984, **40**, 18.
124. S. G. B. Amyes, in 'Proceedings of the 3rd International Congress of Clinical Microbiology', ed. D. M. Livermore, Theracom, Winchester, 1988, p. 31.
125. R. B. Sykes and M. Matthew, *J. Antimicrob. Chemother.*, 1976, **2**, 115.
126. R. B. Sykes and K. Bush, in 'Chemistry and Biology of β-Lactam Antibiotics', ed. R. B. Morin and M. Gorman, Academic Press, New York, 1982, vol. 3, p. 155.
127. K. Bush and R. B. Sykes, in 'Antimicrobial Drug Resistance', ed. L. E. Bryan, Academic Press, New York, 1984, p. 1.
128. M. Matthew, G. Cornelis and G. Wauters, *J. Gen. Microbiol.*, 1977, **102**, 55.
129. W. V. Shaw, *CRC Crit. Rev. Biochem.*, 1983, **14**, 1.
130. W. V. Shaw, *Br. Med. Bull.*, 1984, **40**, 36.
131. S. Mitsuhashi, K. Kawabe, A. Fuse and A. Iyobe, in 'Microbial Drug Resistance', ed. S. Mitsuhashi and H. Hosimoto, University of Tokyo Press, Tokyo, 1975, p. 515.
132. E. Winshell and W. V. Shaw, *J. Bacteriol.*, 1969, **98**, 1248.
133. E. J. Duvall, D. M. Williams, P. S. Lovett, C. Rudolph, N. Vasantha and M. Guyer, *Gene*, 1983, **24**, 171.
134. T. L. Nagabhushan, G. H. Miller and M. J. Weinstein, in 'The Aminoglycosides, Microbiology, Clinical Use and Toxicology', ed. A. Whelton and H. C. Neu, Dekker, New York, 1982, p. 3.
135. J. F. John, W. F. McNeill, K. E. Price and P. A. Kresel, *Antimicrob. Agents Chemother.*, 1982, **21**, 587.
136. L. E. Bryan, in 'Antimicrobial Drug Resistance', ed. L. E. Bryan, Academic Press, New York, 1984, p. 241.
137. K. E. Price, P. A. Kresel, L. A. Farchione, S. B. Siskin and S. A. Karpow, *J. Antimicrob. Chemother.*, 1981, **8**, Suppl. A, 89.
138. I. Phillips and K. Shannon, *Br. Med. Bull.*, 1984, **40**, 28.
138a. T. Lambert, G. Gerbaud and P. Courvalin, *Antimicrob. Agents Chemother.*, 1988, **32**, 15.
139. G. B. Appel, in 'The Aminoglycosides, Microbiology, Clinical Use and Toxicology', ed. A. Whelton and H. C. Neu, Dekker, New York, 1982, p. 269.
140. R. E. Brumfitt and K. E. Fox, in 'The Aminoglycosides, Microbiology, Clinical Use and Toxicology', ed. A. Whelton and H. C. Neu, Dekker, New York, 1982, p. 269.
141. P. Courvalin and C. Carlier, *J. Antimicrob. Chemother.*, 1981, **8**, Suppl. A, 57.
142. T. Yamada, D. Tipper and J. Davies, *Nature (London)*, 1968, **219**, 288.
143. J. Davies and D. I. Smith, *Annu. Rev. Microbiol.*, 1978, **32**, 469.
144. P. F. Wiley, L. Backynskyj, L. A. Dolak, J. I. Ciadella and V. P. Marshall, *J. Antibiot.*, 1987, **40**, 195.
145. M. Arthur, A. Andremont and P. Courvalin, *Antimicrob. Agents Chemother.*, 1987, **31**, 404.
146. E. R. Dabbs, *FEMS Microbiol. Lett.*, 1987, **40**, 135.
147. T. Kobayashi, T. Uozumi and T. Beppu, *J. Antibiot*, 1986, **39**, 688.
148. D. W. Garrison, R. M. de Haan and J. B. Lawson, *Antimicrob. Agents Chemother.*, 1967, 397.
149. L. E. Bryan, in 'Bacterial Resistance and Susceptibility to Chemotherapeutic Agents', Cambridge University Press, Cambridge, 1982.
150. J. Mitsuyama, R. Hiruma, A. Yamaguchi and T. Sawai, *Antimicrob. Agents Chemother.*, 1987, **31**, 379.
151. K.-H. Büscher, W. Cullmann, W. Dick and W. Opferkuch, *Antimicrob. Agents Chemother.*, 1987, **31**, 703.
152. A. A. Peterson, S. W. Fesik and E. J. McGroarty, *Antimicrob. Agents Chemother.*, 1987, **31**, 230.
153. L. Leive, S. Telesetsky, W. G. Coleman, Jr. and D. Carr, *Antimicrob. Agents Chemother.*, 1984, **25**, 539.
153a. L. Gutmann, D. Billot-Klein, R. Williamson, F. W. Goldstein, J. Mounier, J. F. Acar and E. Collatz, *Antimicrob. Agents Chemother.*, 1988, **32**, 195.
154. T. R. Parr, Jr. and L. E. Bryan, in 'Antimicrobial Drug Resistance', ed. L. E. Bryan, Academic Press, New York, 1984, p. 81.
154a. E. Collatz and L. Gutmann, in 'Antimicrobial Agents Annual—2', ed. P. K. Peterson and J. Verhoef, Elsevier, Amsterdam, 1987.
155. R. J. Wargel, C. A. Shadur and F. C. Neuhaus, *J. Bacteriol.*, 1971, **105**, 1028.
156. I. Chopra and P. R. Ball, *Adv. Microb. Physiol.*, 1982, **23**, 183.
157. D. Hendlin, E. O. Stapley, M. Jackson, H. Wallick, A. K. Miller, F. J. Wolf, T. W. Miller, L. Chaiet, F. M. Kahan, E. L. Foltz, H. B. Woodruff, J. M. Mata, S. Hernandez and S. Mochales, *Science (Washington, D.C.)*, 1969, **166**, 122.
158. H. B. Woodruff, J. M. Mata, S. Hernandez, A. Rodriguez, E. O. Stapley, H. Wallick, A. K. Miller and D. Hendlin, *Chemotherapy*, 1977, **23**, Suppl. 1, 1.
159. L. E. Bryan and S. Kwan, *J. Antimicrob. Chemother.*, 1981, **8**, Suppl. D, 1.
159a. D. Schlessinger, *Clin. Microbiol. Rev.*, 1988, **1**, 54.
160. L. E. Bryan, T. Nigas, B. W. Holloway and C. Crowther, *Antimicrob. Agents Chemother.*, 1980, **17**, 71.
161. T. R. Tritton, *Biochemistry*, 1977, **16**, 4133.
162. S. B. Levy, in 'Antimicrobial Drug Resistance', ed. L. E. Bryan, Academic Press, New York, 1984, p. 191.
162a. V. Burdett, *J. Bacteriol.*, 1986, **165**, 564.
163. L. M. McMurray and S. B. Levy, *Antimicrob. Agents Chemother.*, 1978, **14**, 201.
164. L. M. McMurray, D. Aronson and S. B. Levy, *Antimicrob. Agents Chemother.*, 1983, **24**, 544.
165. B. Marshall, C. Tachibana and S. B. Levy, *Antimicrob. Agents Chemother.*, 1983, **24**, 835.

165a. B. Marshall, S. Morrissey, P. Flynn and S. B. Levy, *Gene*, 1987, **50**, 111.
166. I. Chopra, *J. Antimicrob. Chemother.*, 1986, **18**, Suppl. C, 51.
166a. R. Zilhao, B. Papadopoulou and P. Courvalin, *Antimicrob. Agents Chemother.*, 1988, **32**, 1793.
166b. B. H. Park and S. B. Levy, *Antimicrob. Agents Chemother.*, 1988, **32**. 1797.
167. I. Chopra, *Br. Med. Bull.*, 1984, **40**, 11.
168. I. Chopra, in 'Handbook of Experimental Pharmacology', ed. J. J. Hlavka and J. H. Boothe, Springer-Verlag, Heidelberg, 1985, vol. 78, p. 317.
169. K. Wiebauer, S. Schraml, S. W. Shales and R. Schmitt, *J. Bacteriol.*, 1981, **147**, 851.
170. E. M. Wise, Jr. and M. M. Abou-Donia, *Proc. Natl. Acad. Sci. USA*, 1975, **72**, 2621.
171. G. Swedberg and O. Sköld, *J. Bacteriol.*, 1983, **153**, 1228.
172. S. G. B. Amyes, *J. Antimicrob. Chemother.*, 1986, **18**, Suppl. C, 215.
173. J. T. Smith and S. G. B. Amyes, *Br. Med. Bull.*, 1984, **40**, 42.
174. H. K. Young and S. G. B. Amyes, *J. Biol. Chem.*, 1986, **261**, 2503.
174a. L. Sundstrom, T. Vinayagamoorthy and O. Skold, *Antimicrob. Agents Chemother.*, 1987, **31**, 60.
174b. M. Ouelette, G. Gerbaud and P. Courvalin, *Ann. Inst. Pasteur Microbiol.*, 1988, **139**, 105.
175. D. Ivler, J. M. Leedom, L. D. Thrupp, P. Wehrle, B. Portnoy and A. C. Matthies, Jr., *Antimicrob. Agents/Chemother.*, 1965, p. 444
176. P. J. White and D. D. Woods, *J. Gen. Microbiol.*, 1965, **40**, 243.
177. R. H. Reitz, H. D. Slade and F. C. Neuhaus, *Biochemistry*, 1967, **6**, 2561.
178. C. C. Sanders, in 'Microbiology—1984', ed. L. Leive and D. Schlessinger, American Society for Microbiology, Washington, DC, 1984, p. 391.
179. H. J. Scholer, in 'Antifungal Chemotherapy', ed. D. C. E. Speller, Wiley, New York, 1980, p. 35.
180. M. Borgers, in 'The Scientific Basis of Antimicrobial Chemotherapy', ed. D. Greenwood and F. O'Grady, Cambridge University Press, Cambridge, 1985, p. 133.
181. D. Kerridge and R. O. Nicholas, *J. Antimicrob. Chemother.*, 1986, **18**, Suppl. B, 39.
182. E. F. Gale, E. Cundliffe, P. E. Reynolds, M. H. Richmond and M. J. Waring, in 'The Molecular Basis of Antibiotic Action', 2nd edn., Wiley, New York, 1981, p. 175.
183. A. M. Pierce, H. D. Pierce, A. M. Unrau and A. C Oelschlager, *Can. J. Biochem.*, 1978, **56**, 135.
184. D. Kerridge, *Adv. Microb. Physiol.*, 1986, 1.
185. H. Vanden Bossche, W. Lauwers, G. Willemsens, P Marichal, F. Cornelissen and W. Cools, *Pestic. Sci.*, 1984, **15**, 188.
186. H. J. Scholer and A. Polak, in 'Antimicrobial Drug Resistance', ed. L. E. Bryan, Academic Press, New York, 1984, p. 394.
187. R. J. Hay, *Br. Med. J.*, 1985, **290**, 260.
188. J. F. Ryley, R. G. Wilson and K. J. Barrett-Bee, *J. Med. Vet. Mycol.*, 1984, **22**, 53.
189. W. L. Whelan and D. Kerridge, *Antimicrob. Agents Chemother.*, 1984, **26**, 570.
190. G. Bertani, *Genetics*, 1951, **36**, 598.
191. C. R. Spotts, *J. Gen. Microbiol.*, 1962, **28**, 347.
192. H. Momose and L. Gorini, *Genetics*, 1970, **67**, 19.
193. E. R. Dabbs and H. G. Wittmann, *Mol. Gen. Genet.*, 1976, **149**, 303.
194. E. R. Dabbs, *Mol. Gen. Genet.*, 1977, **151**, 261.
195. E. R. Dabbs, *J. Bacteriol.*, 1978, **136**, 994.
196. L. B. Quesnel and H. S. N. Hussain, *Microbios*, 1971, **4**, 33.
197. I. A. Nami and L. B. Quesnel, *IRCS Med. Sci.: Libr. Compend.*, 1983, **11**, 719.
198. W. Piepersberg, A. Bock and H. G. Wittmann, *Mol. Gen. Genet.*, 1975, **140**, 91.
199. K. N. Timmis, M. I. Gonzalez-Carrero, T. Sekizaki and F. Rojo, *J. Antimicrob. Chemother.*, 1986, **18**, Suppl. C, p. 1.
200. P. Trieu-Cuot, G. Gerbaud, T. Lambert and P. Courvalin, *EMBO J.*, 1985, **4**, 3583.
201. G. A. Jacoby, in 'The Scientific Basis of Antimicrobial Chemotherapy', ed. D. Greenwood and F. O'Grady, Cambridge University Press, Cambridge, 1985, p. 185.
202. F. Heffron, in 'Mobile Genetic Elements', ed. J. A. Shapiro, Academic Press, New York, 1985, p. 233.
203. D. B. Clewell and C. Ganron-Burke, *Annu. Rev. Microbiol.*, 1986, **40**, 635.
204. T. D. Mays, C. J. Smith, R. A. Welch, C. Delfini and F. L. Macrina, *Antimicrob. Agents Chemother.*, 1982, **21**, 110.
205. A. Rashtchian, G. R. Dubes and S. J. Booth, *J. Bacteriol.*, 1982, **150**, 141.
206. J. R. Saunders, *Br. Med. Bull.*, 1984, **40**, 54.
207. D. B. Clewell, *Microbiol. Rev.*, 1981, **45**, 409.
208. D. B. Clewell, B. A. White, Y. Ike and F. Y. An, in 'Microbial Development', ed. R. Losick and L. Shapiro, Cold Spring Harbor Press, Cold Spring Harbor, 1984, p. 133.

6.1

Oxygenases

LAWRENCE I. KRUSE
Smith Kline & French Research Ltd, Welwyn, UK

6.1.1	INTRODUCTION TO ENZYME PROSTHETIC GROUPS WHICH ACTIVATE OXYGEN	123
	6.1.1.1 Flavin	123
	6.1.1.2 Hemoprotein Cofactor	124
	6.1.1.3 Copper	125
6.1.2	AMINE OXIDASES	125
	6.1.2.1 Monoamine Oxidase	125
	6.1.2.2 Polyamine Oxidase	128
6.1.3	DOPAMINE β-HYDROXYLASE	128
	6.1.3.1 Mechanism-based Inactivators	129
	6.1.3.2 Multisubstrate Inhibitors	131
6.1.4	CYTOCHROME P-450	133
	6.1.4.1 Reactions Catalyzed by Cytochrome P-450	133
	6.1.4.2 Xenobiotic Metabolizing P-450 Isozymes	134
	6.1.4.3 Mechanisms of P-450 Inhibition	134
	6.1.4.4 Substrate Specific P-450 Isozymes	137
	6.1.4.4.1 Aromatase	137
	6.1.4.4.2 14α-Demethylase	138
	6.1.4.4.3 Steroid side-chain cleavage	141
6.1.5	XANTHINE OXIDASE	143
6.1.6	REFERENCES	143

6.1.1 INTRODUCTION TO ENZYME PROSTHETIC GROUPS WHICH ACTIVATE OXYGEN

6.1.1.1 Flavin

Flavin (1), the heterocyclic base contained in vitamin B_2, is found in numerous oxygenases, monooxygenases, and other redox active enzymes. Critical to the biochemical activity of this cofactor is its capacity to react as either a single- or two-electron donor/acceptor. This reactivity is based on the stability of the half-reduced radical anion form (2) of the cofactor, as illustrated in equation (1). The reduction/oxidation of the flavin cofactor according to this equation is readily reversible. Two chemical properties of the flavin cofactor confer special reactivities. First, the redox potential for equation (1), -0.18 V, makes both oxidized and reduced forms readily accessible under physiological conditions. Second, the reduced flavin readily reacts with molecular oxygen to yield H_2O_2 and oxidized flavin cofactor (1). This mode of reactivity is particularly important in the flavin-dependent amine oxidases.

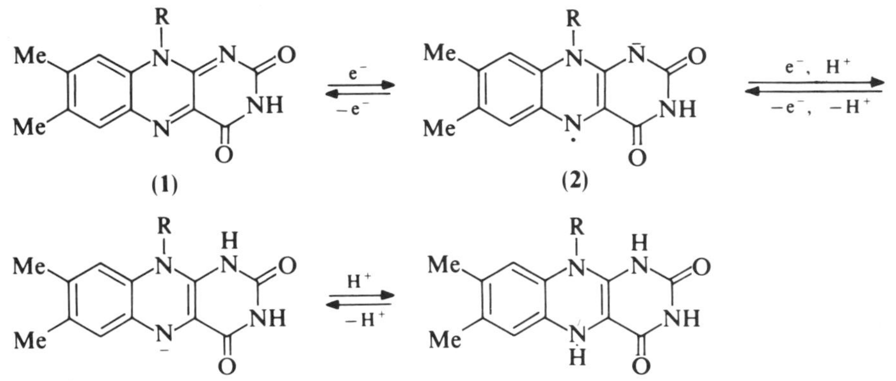

(1)

6.1.1.2 Hemoprotein Cofactor

A majority of iron-dependent mixed function oxidases have an iron–protoporphyrin IX cofactor (**3**) as a tightly bound prosthetic group. The tetrapyrrole frame of the protoporphyrin provides a square-planar nitrogen chelate for the iron atom while the protein envelope makes one axial ligand available to the iron atom. In the P-450 family of mixed function oxidases the final axial ligand is a solvent water molecule until the binding of the organic substrate, which leads to displacement of the front-side water ligand.

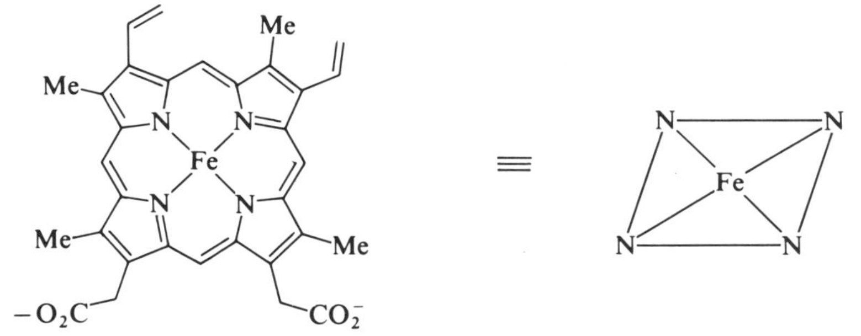

(3)

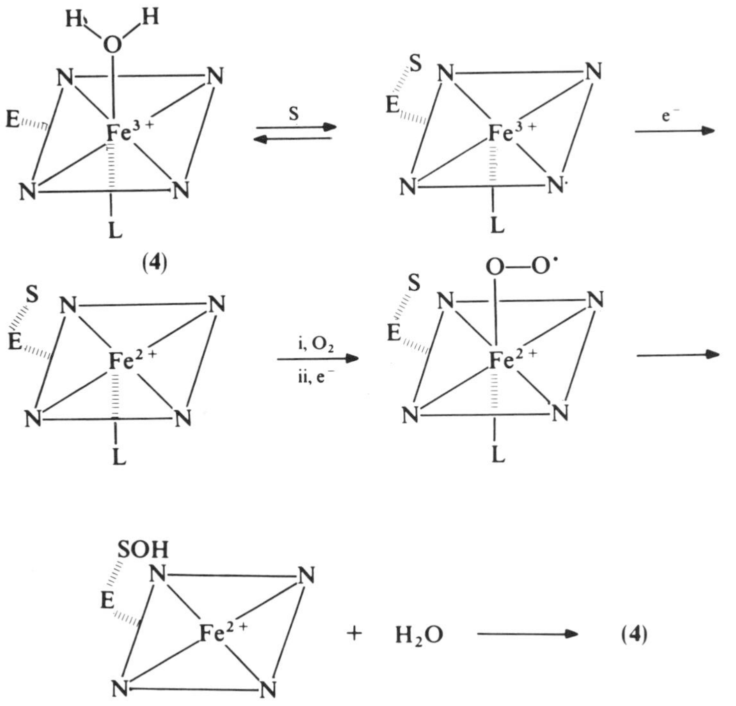

(2)

During catalysis the iron undergoes a rather complex redox cycle (equation 2). Initially, the resting enzyme–heme complex (4) binds organic substrate. Addition of an electron generates the reduced 'high-spin' Fe^{2+} oxidation state which then reacts with molecular oxygen and a second electron to generate a highly reactive oxygen species generally believed to be Fe^{5+}=O (or a resonance equivalent). It is this iron–oxene which catalyzes substrate oxidation. The high iron oxidation state is no doubt stabilized by the very electron rich protoporphyrin ring system. Release of water and product regenerates the resting enzyme (4).

6.1.1.3 Copper

Several dozen copper-containing oxidases have been identified. Without exception, far less is known about the active site arrangement, copper ligands and chemical mechanism of the copper-containing oxidases, than is known for the iron–protoporphyrin oxidases. The very positive redox potentials for copper proteins ($+0.2$ to $+0.65$ V) make obligatory the use of molecular oxygen as cosubstrate. The actual inorganic chemistry of oxygen activation by copper proteins may involve a catalytic cycle similar to that in the iron-containing systems of equation (2), or the intermediacy of powerful oxidizing species such as hydroxyl radicals.

6.1.2 AMINE OXIDASES

6.1.2.1 Monoamine Oxidase

Monoamine oxidase (MAO, EC 1.4.3.4) is a family of flavin-dependent enzymes which catalyze the oxidative deamination of biogenic amines to the corresponding aldehyde according to the stoichiometry of equation (3).[1] Use of isotopically labelled molecular oxygen demonstrates this cosubstrate serves as a two-electron oxidizing species and it is not cleaved to H_2O or incorporated into the aldehyde products. The chemical mechanism of monoamine oxidation has been shown to occur *via* a formal dehydration of amine to imine or iminium species concomitant with the reduction of flavin cofactor.[2,3] Subsequent hydrolysis of the imine intermediate generates aldehyde and amine products. Reoxidation of reduced flavin cofactor by dioxygen produces an equivalent of H_2O_2 while regenerating the catalytically competent, oxidized form of the enzyme. A reasonable mechanism by which MAO cleaves the chemically unreactive α-hydrogen of amines has been advanced by Silverman (equation 4).[4] In this proposed mechanism the flavin cofactor acts twice as a one-electron recipient. Initial, slow electron transfer from amine nitrogen to flavin generates a nitrogen-centred radical cation in which the α-hydrogen atom is labilized and rapidly lost (as a proton) to yield a carbon-based radical stabilized by the adjacent amine electron lone-pair. A subsequent single-electron transfer from this radical intermediate to the flavin radical produces the iminium intermediate and the fully reduced flavin cofactor. The iminium intermediate undergoes hydrolysis, and reoxidation of flavin cofactor by molecular oxygen completes the catalytic cycle.

$$R^1CH_2NR_2^2 + {}^{18}O_2 + H_2O \rightarrow R^1CHO + H_2{}^{18}O_2 + HNR_2^2 \qquad (3)$$

$$\underset{\substack{H \\ |}}{R^1-CH-\ddot{N}R_2^2} \xrightarrow[\text{slow}]{Fl_{ox} + H^+ \; FlH\cdot} \underset{\substack{H \\ |}}{R^1-\overset{\cdot+}{CH}-NR_2^2} \xrightarrow[\text{fast}]{-H^+} \underset{\substack{H \\ |}}{R^1-\underset{\cdot}{C}-\ddot{N}R_2^2}$$

$$\xrightarrow{FlH\cdot + H^+ \; FlH_2} R^1-CH{=}\overset{+}{N}R_2^2 \xrightarrow{H_2O} R^1-CHO + HNR_2^2 \qquad (4)$$

$$FlH_2 \xrightarrow{O_2} Fl_{ox} + H_2O_2$$

Monoamine oxidases have been categorized into two functional classes, Type A and Type B, based upon substrate specificities and selective inhibition. MAO Type A selectively deaminates 5-hydroxytryptamine and norepinephrine, whereas preferred substrates for Type B are benzyl-amines and phenethylamines. Type A is selectively inhibited by clorgyline (5a), whereas Type B is selectively inhibited by L-deprenyl (5b) and MDL 72145 (6). Monoamine oxidase inhibitors which enter the central nervous system are effective as antidepressants although they have received only limited clinical utility as a result of peripheral side effects. Serious hypertensive crises have been observed in MAO-inhibited patients who ingest tyramine-containing substances.[5] It is likely, based

(5a)   (5b)   (5c)   (6)

upon substrate specificities, that inhibition of the Type B form results in these untoward cardio-vascular effects.

A large number of different MAO inhibitors are known; these comprise three major classes. Substituted allyl-, propargyl- and cyclopropyl-amines are well-characterized classes of MAO inhibitors that produce time-dependent inactivation of enzyme which results from a mechanism-based event. Another class of irreversible MAO inhibitor, the hydrazines, are also mechanism-based inactivators, but the chemical mechanism by which these inactivate is less well understood. Finally, a number of heterocyclic secondary and tertiary amines are reversible MAO inhibitors of varying potency.

Propargylic or allylic amines such as pargyline (5c), clorgyline (5b) or deprenyl (5a) are capable of discriminating between MAO Types A or B, but are likely to cause mechanism-based inactivation by a common chemical event.[6-8] Oxidation of these propargylic amines by MAO leads initially to an iminium intermediate and, concomitantly, to the reduced form of the flavin cofactor as shown in equation (5). Electrophilic attack of the propenyl-substituted iminium cation upon the flavin leads to a covalent trapping of the cofactor as the N-5 adduct, which produces irreversible inactivation.

(5)

The inactivation of MAO by the antidepressant *trans*-2-phenylcyclopropylamine (7) has been shown to follow a different course in that the protein, rather than the flavin cofactor, undergoes covalent modification according to equation (6). Evidence has been presented in support of an enzymatic thiol group as the nucleophile X in equation (6).[9]

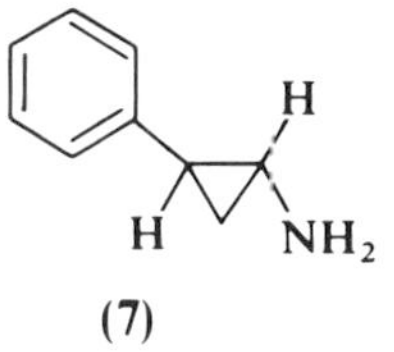

(7)

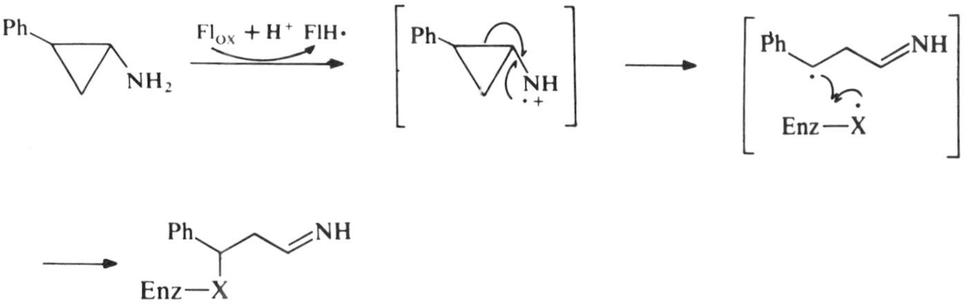

(6)

The substituted allylamine (6) has been shown to be a selective mechanism-based inhibitor of MAO Type B.[10] Recent studies of structure–activity relationships have led to an understanding of what structural features impart selectivity for Type A or Type B MAO for phenylallylamines related to (6).[11] Of interest are observations that the α-carboxy analogue (8) of appropriately substituted allylamines is decarboxylated *in vivo* to the corresponding allylamine by DOPA decarboxylase.[12] Since (8) is an aromatic α-amino acid, this compound will have access to the central nervous system *via* the aromatic amino acid transport system. It is hoped that coadministration of (8) and a *peripheral* DOPA decarboxylase inhibitor will lead to a selective decarboxylation to the active allylamine in the central nervous system, thereby avoiding the hypertensive crisis which can result from tyramine ingestion by patients on MAO inhibitor therapy.[13]

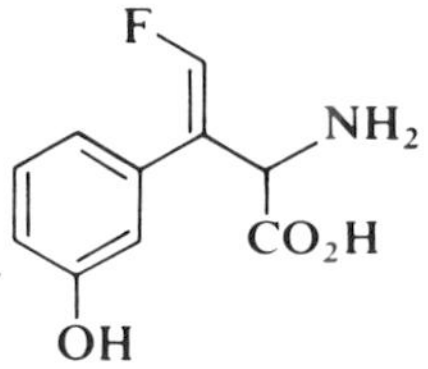

(8) $R{-}NHNH_2$ (9) $R{-}\overset{O}{\overset{\|}{C}}{-}NHNH_2$ (10) 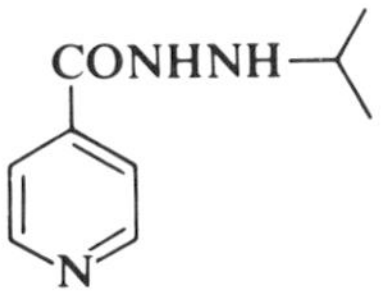 (11)

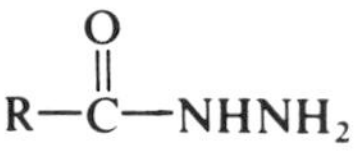

A large variety of substituted hydrazines (9) and hydrazides (10) have been shown to inhibit MAO *in vivo*. Extensive reviews of hydrazine derivatives as MAO inhibitors have appeared. One hydrazide in particular, iproniazide (11), has shown clinical utility. The inhibition of MAO by hydrazine derivatives is believed to occur *via* several pathways. In the case of phenylhydrazine (9; R = Ph), evidence has been presented[14] in support of an initial oxidation to a phenylazo compound followed by loss of N_2 and covalent modification of the flavin by phenyl substitution at the C-4a position, as shown in equation (7).

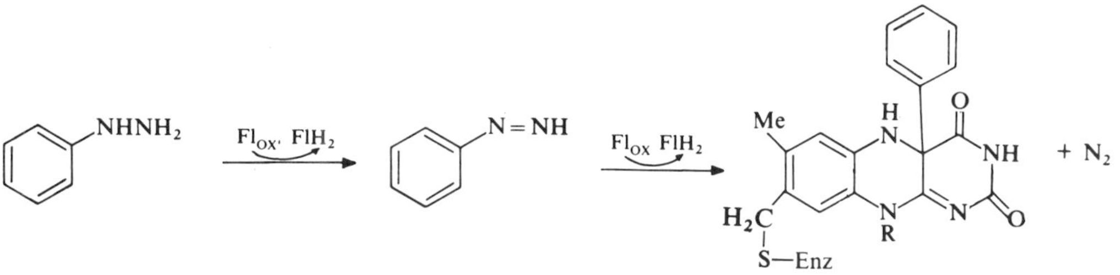

(7)

The harmala alkaloids harmine (12a) and harmaline (12b) are very potent, reversible inhibitors of Type A MAO and a wide range of related substituted amines and (hetero)arylalkylamines have been characterized as MAO inhibitors of varying potencies and specificities for Type A or B.[15]

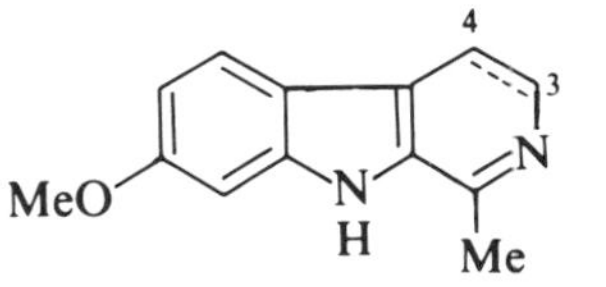

(12a) 3,4-CH=CH
(12b) 3,4-CH$_2$CH$_2$

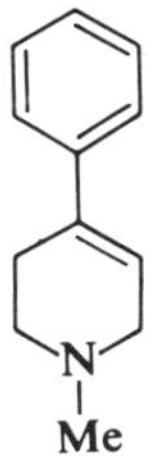

(13) 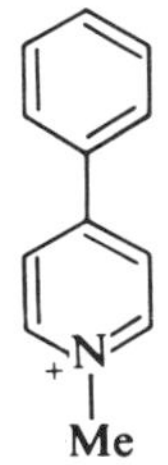   **(14)**

The tetrahydropiperidine derivative (13) has achieved considerable notoriety as a selective neurotoxin that induces Parkinson-like symptoms which result from lesions of the dopaminergic neurons in the central nervous system. These actions have been suggested to result from the oxidized pyridinium active metabolite (14), which is produced *in vivo* by the action of MAO upon (13). Support for this can be drawn from experiments which show MAO inhibitors to have a protective effect against (13).[16]

6.1.2.2 Polyamine Oxidase

Polyamine oxidase (PAO, EC 1.4.3.4) is another flavin dependent oxidase which is responsible for the metabolic oxidation of polyamines such as putrescine, spermine and spermidine. Some allenyl- and bis(allenyl)diamines (15a–c) have been reported recently as specific mechanism-based inhibitors of PAO.[17] The *in vivo* activity and therapeutic evaluation of PAO inhibitors awaits further disclosure.

$$R^1NH(CH_2)_4NHR^2$$

(15a) R^1 = CH$_2$=C=CHCH$_2$—, R^2 = H
(15b) R^1 = CH$_2$=C=CHCH$_2$—, R^2 = Me
(15c) R^1 = CH$_2$=C=CHCH$_2$—, R^2 = CH$_2$=C=CHCH$_2$—,

6.1.3 DOPAMINE β-HYDROXYLASE

Dopamine β-hydroxylase (DBH, EC 1.14.17.1) is a copper-containing, glycosylated, tetrameric monooxygenase of molecular weight 290 000.[18,19] The enzyme is localized to the chromaffin vesicles which are found in the adrenal medullary complex and the terminii of noradrenergic neurons. DBH catalyzes the benzylic hydroxylation of dopamine to norepinephrine according to equation (8) where, under physiological conditions, ascorbic acid provides the two electrons required for catalysis. A 2:1 copper atom:subunit stoichiometry leads to maximum catalytic activity with dopamine substrate,[20] a finding which is entirely consistent with a net two-electron catalytic process and the long known copper redox change from Cu^{2+} to Cu$^+$ which accompanies catalysis.[18] The carbohydrate content of DBH is variable and comprises approximately 6–10% of the total molecular weight. This carbohydrate has little if any role in chemical catalysis since studies with DBH which has been deglycosylated have shown the deglycoenzyme to have kinetic parameters identical to native enzyme.[21] Until very recently, the primary structure of DBH was unknown and different enzyme preparations showed considerable variation in amino acid composition.[18] A protein sequence for human enzyme has been predicted from the gene which has now been sequenced and cloned.[22]

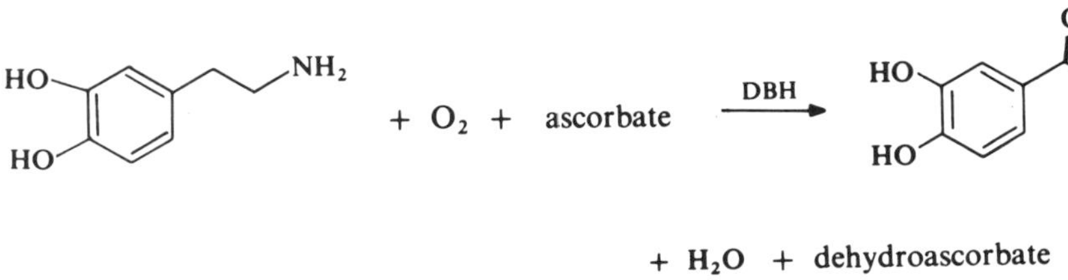

$$+ H_2O + \text{dehydroascorbate}$$

(8)

The order of substrate addition to DBH has been shown to be exceedingly complex and to be further complicated by pH and anion-dependent 'activations' which result from changes in the order of dopamine and oxygen substrate addition.[23] DBH has been recognized as the key enzyme in the catecholamine biosynthetic cascade in that the endogenous ratio of dopamine and norepinephrine is directly related to catalysis by DBH. Dopamine is a vasodilator diuretic and has been used as intravenous therapy in seriously ill congestive heart failure patients. Norepinephrine, the product of the DBH-catalyzed transformation, is a vasoconstrictor substance whose primary effect, in normotensive individuals, is the maintenance of cardiovascular tone. Selective inhibitors of DBH are therefore expected to act as modulators of the catecholaminergic neuronal system which will have therapeutic potential as antihypertensives, antianginals, and perhaps have efficacy in the treatment of congestive heart failure.

DBH cannot be classified into any of the three simple categories of copper proteins, based upon magnetic and spectroscopic properties of the prosthetic metal.[19] In this regard, as well as from the perspective of the biochemical redox transformation, DBH appears unique. Until a recent study of the Cu^+ and Cu^{2+} forms of DBH by EXAFS (*Extended X-ray Absorption Fine Structure*),[24] virtually nothing was known about the copper ligand environments during catalytic turnover or upon substrate binding. This study has shown the resting (Cu^{2+}) copper environment to be essentially a square-planar complex with four nitrogen ligands (presumably imidazoles) at 1.98 Å. Upon one-electron reduction the copper atom environment undergoes a dramatic reorganization with a reduction in the nitrogen ligand number. Concomitant with this change is the addition of a sulfur ligand at 2.30 Å. The reduced, catalytically active form of enzyme is best described as having an average copper ligand environment with two nitrogen ligands at 1.94 Å and one sulfur at 2.30 Å. This change in ligand environment could explain the very favourable redox potential of $+0.35$ V for $Cu^{2+} \rightarrow Cu^+$.

6.1.3.1 Mechanism-based Inactivators

The substantial therapeutic potential of a DBH inhibitor has been widely recognized. In spite of the complexity of the DBH active site and ambiguities regarding the actual chemical mechanism of catalysis, this enzyme has been studied by several research groups whose interest is the design of mechanism-based inhibitors. A large number of simple substrate-like inhibitors, bearing latent electrophilic or reactive functional groups, has been reported as time dependent inactivators of DBH. Representative examples include alkenes, alkynes, cyclopropanes, hydrazines and nitriles. Complex mechanisms[25] have been proposed for the inactivation of DBH by these alternative substrates (Table 1). However, these chemical mechanisms remain speculative, since no structures for covalently modified enzyme active site residues have been reported. Even though a large number of different classes of mechanism-based inhibitors of DBH have been described, the kinetic data reported in Table 1 for representatives of these show the majority to be of mechanistic interest but of little genuine therapeutic utility. Partition ratios are very large for some inhibitors, while for others the affinity for enzyme (K_I) is in the millimolar range, a concentration which is probably unachievable *in vivo*. A comparison of the therapeutically relevant k_{inact}/K_I ratio, which describes both affinity for enzyme and rate of inactivation, shows only inhibitors (**16–18**; see Table 1) to have the requisite affinity and inactivation rates. Inhibitor (**18**) also shows a very low partition ratio. The high affinity for DBH shown by (**17**) ($K_I = 35 \, \mu M$) and (**18**) ($K_I = 57 \, \mu M$) relative to dopamine substrate [K_m(app) ≈ 1 mM] is notable. The improved affinity which results from substituting an alkyne into the benzylic methylene of the substrate, as in (**18**), or from substituting a thiophene for the catechol ring in dopamine, is most striking. In this context it is worth noting that the (*S*) enantiomer of (**18**) shows even higher affinity for DBH ($K_{is} = 16 \, \mu M$) than the inactivator (*R*)-(**18**), but no time-dependent inactivation occurs with this isomer.[33] A benzylic hydrogen atom with the correct configuration for abstraction by DBH appears to be a requirement for inactivation by these β-ethynyltyramines.

The chemical process by which the alternative substrates in Table 1 inactivate DBH remains entirely unclear. While the inhibitors in Table 1 were prepared by analogy with known inhibitors of other mixed function oxidases and contain latent electrophilic groups, these functional groups are not required for mechanism-based inactivation of DBH, as a recent study reports. Mechanism-based inactivation of DBH appears to be a generic property of alternative substrates which lack the basic side-chain nitrogen present in phenethylamine substrates. Thus, *p*-cresol (**19**; see Table 2) and simple analogues, which lack a basic nitrogen, show effective mechanism-based inhibition of DBH

Table 1 Mechanism-based Inhibitors of DBH

Compound	k_{inact} (min^{-1})	K_I (mM)	k_{inact}/K_I (M^{-1}min^{-1})	Partition ratio	Ref.
	0.05 (37 °C; 0.22 mM O$_2$)	1.9	26.3	—	26
	1.8 (25 °C; 1.2 mM O$_2$)	12.0	150	100:1	27
	0.07 (37 °C; 0.22 mM O$_2$)	0.021	3330	—	28
(16)	0.81 (37 °C; 0.24 mM O$_2$)	0.52	1560	1000:1	29
	0.2 (25 °C; 0.25 mM O$_2$)	3.6	56	40:1	30
(17)	0.12 (37 °C; 0.22 mM O$_2$)	0.035	3430	—	28
	0.0066 (35 °C; X O$_2$)	5.0	1.32	—	31
	0.066 (35 °C; 0.75 mM O$_2$)	7.9	0.84	—	32
(18)	0.184 (37 °C; 0.24 mM O$_2$)	0.057	3230	<5:1	33

(Table 2).[34] In the case of (19), covalent labelling of DBH has been demonstrated by [³H]-incorporation experiments. The label is associated with a single tyrosine residue in a 26 amino acid tryptic peptide.[35] Covalent labelling of the tyrosine residue occurs by insertion of a *p*-cresol-derived hydroxybenzyl radical into the aromatic side chain of this amino acid.[21] This result, together with the diversity of structure in Table 1, and others reported, underscores the complexity of the mechanism-based inactivation of DBH by alternative substrates and cautions against drawing elaborate mechanistic conclusions from simple kinetic results.

Table 2 Substituted Hydroxyphenyl Inhibitors of DBH[a]

Compound	K_{is} (mM)	K_I (mM)	k_{inact} (min^{-1})	k_{inact}/K_I (M^{-1} min^{-1})	Partition ratio
(19)	0.18	2.0	0.3	150	1300:1
	2.6	12.8	0.3	23	—
	3.0	1.5	0.033	22	—
	—	1.1	0.11	100	—
	—	10.4	0.026	2.5	—
	1.7	—	0	—	—

[a] 37 °C, 0.24 mM O_2.

6.1.3.2 Multisubstrate Inhibitors

An alternate approach to the inhibition of DBH led to the design of 'multisubstrate' inhibitors of which compound (**20**) is prototypical.[36–38] These inhibitors bind the enzyme with K_i values in the nanomolar range, fully 10^6-fold more tightly than either oxygen or tyramine substrates. The design of multisubstrate inhibitors is illustrated schematically in Figure 1. Here the simultaneous binding of substrates A and B to a model enzyme occurs with dissociation constants (approximated by K_m) of K_mA and K_mB prior to release of products A' and B'. It has been shown experimentally for several enzymes that by joining substrates A and B with a suitable covalent linkage, an inhibitor of greatly

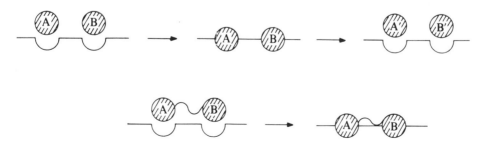

Figure 1 Multisubstrate enzyme inhibitors

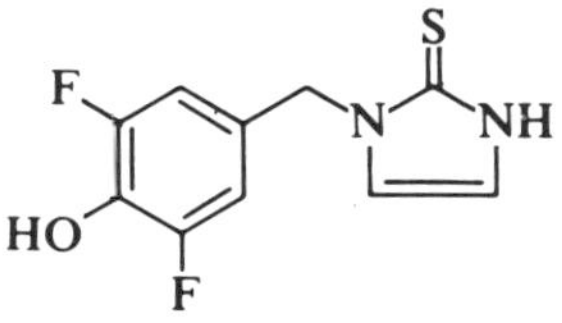

(20)

enhanced binding can result.[39] As an empirical observation, the dissociation constant K_i of this inhibitor is approximated by equation (9) where the entropic advantage of binding two enzymatic sites by a single molecule is translated into a 10^2–10^3-fold increase in the affinity of inhibitor for enzyme. In the design of multisubstrate inhibitors of DBH, it was hypothesized that the oxygen and tyramine substrate bound enzyme simultaneously and in a close spacial arrangement at the enzyme active site. While the covalent attachment of a tyramine structure to *oxygen* substrate was obviously precluded by the structure of molecular oxygen, the substitution of a soft, sulfur-containing bidentate ligand in place of the oxygen (Figure 2) substrate in the multisubstrate inhibitors led to compounds which satisfied all the criteria expected for multisubstrate inhibitors: potency, structural resemblance to the natural substrates and (under the appropriate conditions) competitive binding *vs.* these substrates.[36] A large number of these inhibitors with different phenyl substituents was prepared and subjected to a modified Hansch analysis.[38] This demonstrated a correlation of DBH-inhibitory potency with a few key structural parameters including bulk, lipophilicity, π-density and, not surprisingly, the presence of a 4-hydroxyl group on the phenethylamine mimic portion of inhibitor. Regression analysis indicated an excellent correlation ($r = 0.91$) between predicted and observed inhibitor potency (equation 10).

$$K_i \approx (K_mA)(K_mB)(10^{-2}\text{–}10^{-3}) \tag{9}$$

$$-\log IC_{50} = 1.28(\pm 0.22)\,I(4\text{-OH}) + 0.65(\pm 0.16)\,\pi_{345} - 0.14(\pm 0.02)MR_{345} + 1.42(\pm 0.33)F_{345} - 1.26 \tag{10}$$

$$n = 25, \quad r = 0.91, \quad F = 22.9, \quad S = 0.44$$

Certain of these multisubstrate inhibitors of DBH have been evaluated *in vivo* for their biochemical and pharmacological effects. Following oral administration of these inhibitors to spontaneously hypertensive rats, significant increases occur in vascular dopamine levels and the vascular norepinephrine levels decrease, as expected for the inhibition of DBH.[40,41] These biochemical effects are coincident with a decrease in mean arterial blood pressure.

Similar antihypertensive effects have been demonstrated for DBH inhibitors in the DOCA rat model of hypertension and in the normotensive rat. DBH inhibitors have been shown to prevent the cardiac hypertrophy which occurs in the developing SHR[41] and they show significant diuretic effects in saline-loaded SHR or normotensive rats. It is anticipated that one of these inhibitors, presently in preclinical evaluation, will fulfill the promise of a novel cardiovascular agent, which is efficacious as an antihypertensive, diuretic vasodilator with additional potential for efficacy in the treatment of congestive heart failure.

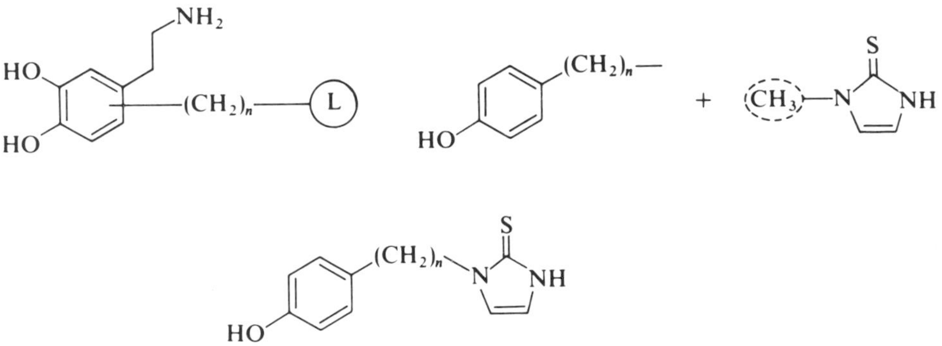

$$n = 0\text{–}5$$
$$n = 1, 3 \text{ optimum}$$

Figure 2 Multisubstrate inhibitors of DBH

6.1.4 CYTOCHROME P-450

6.1.4.1 Reactions Catalyzed by Cytochrome P-450

A diverse family of mixed function oxidases is categorized by the descriptor 'cytochrome P-450'.[42] These oxidases, which are present in all living organisms, are critically dependent upon the Fe/heme cofactor. Cytochrome P-450 oxidase catalysis can be described in a generic way by equation (11). However, this simple equation belies the chemical diversity found in the P-450 mediated oxidations, since the generalized substrate S shows a wide range of structural variation. Thus, oxygen insertion into unactivated C—H bonds, as in camphor hydroxylase (equation 12),[43–45] leads to the conversion of hydrophobic substrates into more hydrophilic products. If the C—H bond is adjacent to an oxygen or nitrogen heteroatom (equation 13, X = O, N), subsequent cleavage to a dealkylated product can occur.[46–49]

$$S + NADPH + O_2 \xrightarrow{P-450} S{-}O + H_2O + NADP^+ \tag{11}$$

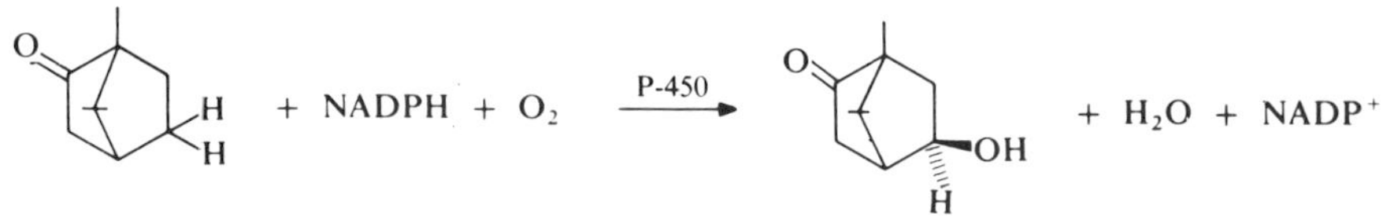

(12)

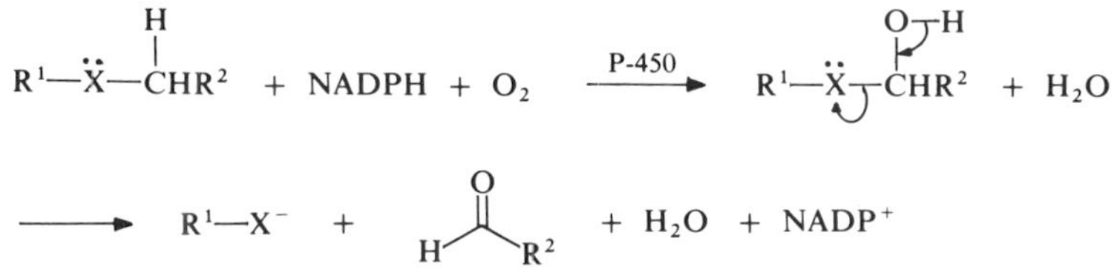

(13)

If the heteroatom X in equation (13) is nitrogen or sulfur the oxidation can also follow a mechanistically different pathway but with an identical net stoichiometry, as shown in equation (14). With substrates that contain nitrogen or sulfur heteroatoms, an alternative mode of catalysis, oxidation at the heteroatom, occurs to give the *N*-oxide, sulfoxide or sulfone as indicated by equation (15).[50–52] Not only are the electron lone-pairs on nitrogen or sulfur subject to this type of direct oxidation, those present in alkenes and alkynes are also oxidized, as shown in equation (16).[53–58] The highly strained intermediate oxirene product which results from oxidation of an alkyne can undergo a variety of further reactions, including the hydride migration/hydration sequence shown in equation (16).[59–65]

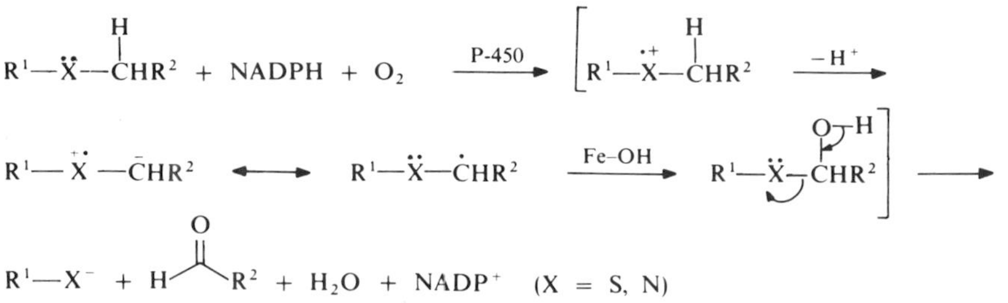

(14)

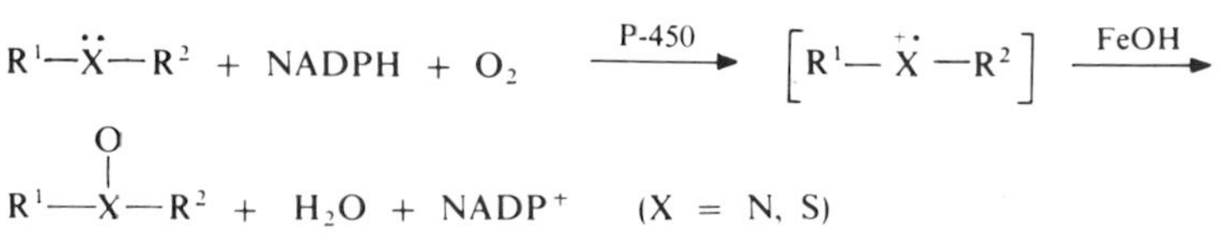

(15)

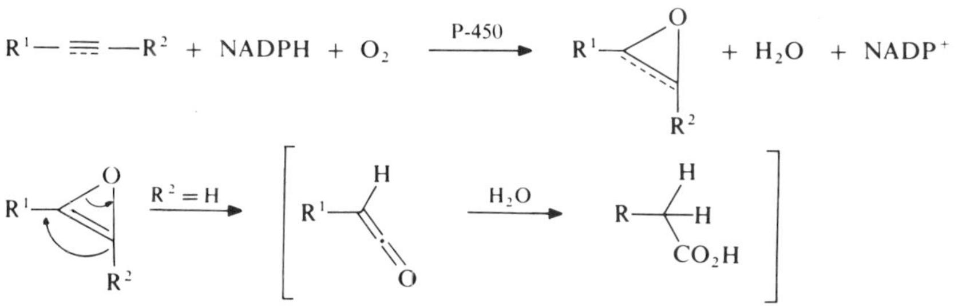

(16)

One of the most fascinating chemical transformations carried out by a P-450 system is the oxidative deformylation of a steroidal aldehyde to yield an alkene (equation 17).[66] The aldehyde group is removed in the formate oxidation state with half of the dioxygen substrate appearing in formate product. Another complex cleavage transformation occurs in the cytochrome P-450 mediated steroid diol cleavage reaction wherein the alkyl side chain is removed during steroid biosynthesis, as shown in equation (18).[67,68] Other electron-rich or readily oxidized functional groups can also serve as cytochrome P-450 substrates. In a mechanistic variation of equations (14, 15), substituted hydrazines are substrates for P-450. These undergo formal dehydrogenation as illustrated in equation (19).[69] The exceedingly rich chemistry exhibited by various P-450 types is a reflection of the fundamental importance of this class of enzymes to cells.

$$R^1-NH-NH-R^2 + NADPH + O_2 \xrightarrow{P-450} R^1-N=N-R^2 + 2H_2O + NADP^+ \qquad (19)$$

6.1.4.2 Xenobiotic Metabolizing P-450 Isozymes

The P-450 enzymes present in higher organisms may be classified into two broad categories: the xenobiotic metabolizing systems and the specific biotransformation type. The xenobiotic metabolizing P-450 systems occur predominantly in the endoplasmic reticulum of liver cells. These P-450 isozymes share several common characteristics. They are relatively nonspecific with respect to preferred substrate, but each of the several dozen isotypes prefers a common substrate class, *i.e.* alkyl ether, alkylamine, aromatic hydrocarbon, *etc*. This catalysis corresponds to the stoichiometry of equations (12)–(16). As might be expected for enzymes with relatively low substrate specificity, these isozymes have high (greater than micromolar) K_m constants for substrate and are readily induced *in vivo* by the presence of the appropriate substrate class.[70,71] Levels of hepatic enzyme induction can be rather dramatic, since P-450 isozymes can constitute as much as 15% of the total cellular protein. The hepatic P-450 systems appear to provide a general detoxifying/metabolizing function which is rapidly amplified in response to the presence of xenobiotics in the body. Practising medicinal chemists need to be cognizant of the xenobiotic-metabolizing P-450 systems for two reasons. Firstly, the *in vivo* life-time of administered drugs will depend in part upon their activity as substrate for the hepatic P-450 enzymes. Secondly, a number of drug substances are capable of inhibiting the hepatic P-450 systems, leading to changes in the metabolism of *other* drugs. The resulting 'drug interactions' can have serious side effects *in vivo*.

The second general category of P-450 systems is comprised of those enzymes which catalyze a specific biotransformation. These show a high degree of substrate specificity and, as expected, a low K_m for the preferred substrate. In man the specific P-450 isozymes are found primarily in the adrenal cortex and gonads, where steroid biosynthesis occurs. Certain of the steroid biosynthetic P-450 enzymes catalyze the simple hydrocarbon oxidation shown in equation (12), whereas others carry out the much more complex transformations indicated in equations (17) and (18).

6.1.4.3 Mechanisms of P-450 Inhibition

Given the rich chemistry carried out by the various cytochrome P-450 systems (equations 11–19), it is not surprising that a number of appropriately substituted substrate analogues are capable of either reversible or irreversible inhibition. The simplest class of reversible P-450 inhibitor is an

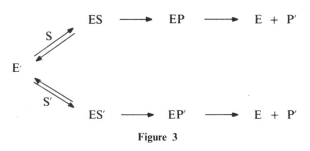

Figure 3

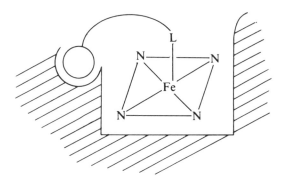

Figure 4

alternative substrate (Figure 3, S′) which is capable of binding P-450 in competition with another substrate (S). Because S′ competes with S for the enzyme, at suitable concentrations of S′ the catalysis of S to P will be inhibited by simple mass action effects. A second simple class of P-450 inhibitor can be described as an iron ligand attached to an organic functional group. The mechanism of action of this large class of P-450 inhibitors appears to involve simultaneous binding of the heme iron and the enzymatic pocket which holds substrate during catalytic turnover (Figure 4). The mechanisms of action for some of these simple inhibitors of P-450 have recently been clarified by X-ray crystallographic studies on inhibitor bound to P-450.[45] While a large number of compounds inhibit P-450 in this fashion, several of the more selective classes have found therapeutic utility (see Section 6.1.4.4.2). In general, this type of P-450 inhibition is most often found in the form of unexpected (and undesirable) drug interactions which result from inhibition of xenobiotic metabolism by another drug.

A somewhat more complex mode of P-450 inhibition is found in a structurally diverse class of alternative substrates that bind and undergo catalysis by P-450 to yield products which are much more tightly bound than the initial substrate. The mode of binding is thought to occur *via* interaction with a substrate binding pocket as well as with the heme Fe as shown in Figure 4. However, the inhibitory complex is considerably tighter, a result of a very strong interaction between the metabolite and the heme Fe. The various substrate classes which inhibit P-450 by this mechanism include amphetamines,[72] selected hydrazines,[73] methylenedioxybenzenes,[74,75] halocarbons,[76] and macrolide antibiotics.[77,78] The inhibitory metabolites formed from these substrates are best described as iron–'nitrone' (equation 20), iron–'nitrene' (equation 21), and iron–'carbene' (equation 22) complexes.

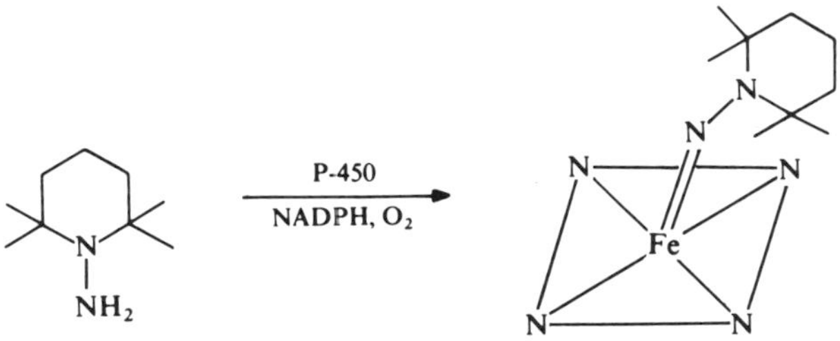

(21)

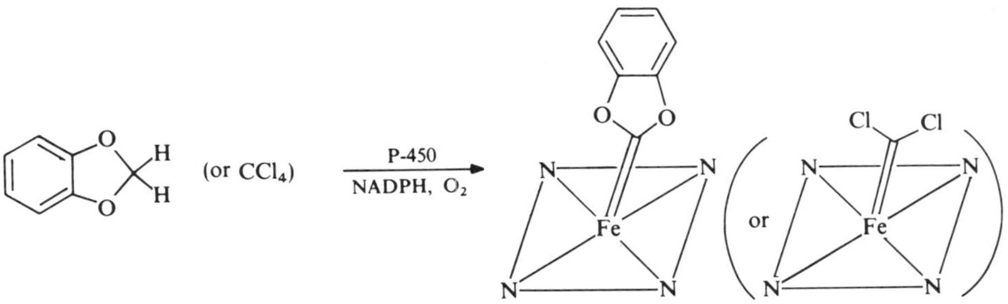

(22)

Substrates for P-450 catalyzed oxidations which incorporate certain classes of functional groups are capable of irreversible, mechanism-based inhibition of P-450. Appropriately substituted alkenes are capable of a selective, irreversible inactivation of P-450 isotypes. Allylisopropylacetamide (21) is a simple alkene that inactivates P-450 [in addition to undergoing normal substrate turnover according to equation (16)].[79] A wide range of alkynic inhibitors cause mechanism-based destruction of P-450 isotypes (inactivation by related allenes has also been demonstrated).[80] In the case of acetylene (22), this has been unambiguously shown to involve *N*-alkylation of the heme prosthetic group (equation 23).[59,60] Other functional groups which lead to irreversible inactivation of P-450 are the 1-aminobenzo-1,2,3-triazole (23),[81] and certain cyclopropylamine derivatives such as (24).[82,83] The amine (23) is believed to react initially *via* equation (19) to produce an azamine intermediate. This is then capable of collapse to the highly reactive benzyne *via* loss of two moles of nitrogen. Insertion of the benzyne into the heme prosthetic group leads to loss of catalytic activity, as illustrated by equation (24). Oxidation of the cyclopropylamine (24) by P-450 leads to a radical-induced ring opening which yields a rearranged radical that ultimately inserts into the protein, as in equation (25).

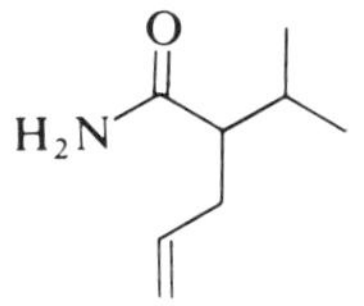

(21)

(22)

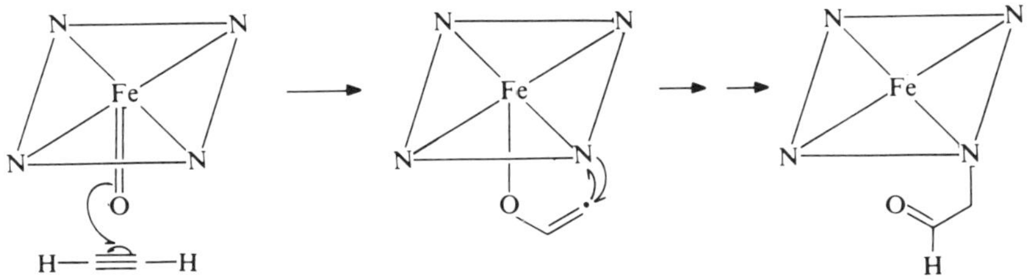

(23)

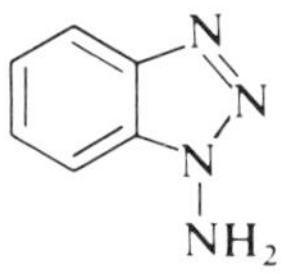

(23)

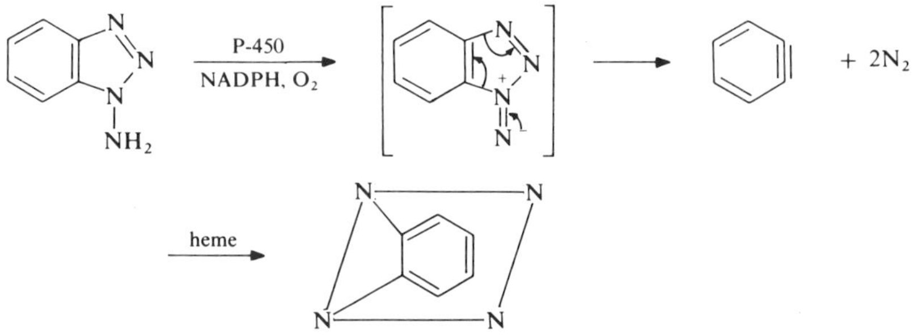

(24)

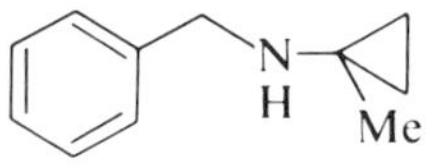

(**24**)

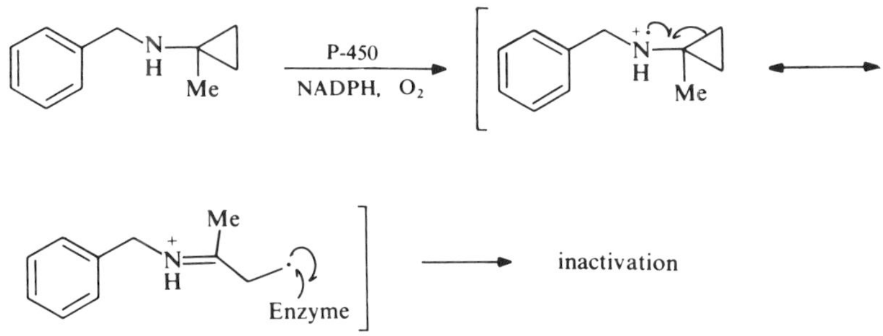

(25)

6.1.4.4 Substrate Specific P-450 Isozymes

6.1.4.4.1 Aromatase

Aromatase is a substrate-specific, microsomal P-450 enzyme which catalyzes the aromatization of A-ring steroids to produce estrone (equation 26).[84-87] Catalysis has been shown to proceed in three steps *via* 19-hydroxyandrostenedione and 19-oxoandrostenedione. These initial oxidations proceed twice *via* equation (12) to yield an intermediate 19-diol which spontaneously dehydrates to the carbaldehyde. The final oxidative aromatization of the steroid A-ring appears to be a biochemical transformation, unique to aromatase wherein the angular 19-methyl group is expelled in the formic acid oxidation state. Several chemical mechanisms for this unusual transformation have been proposed, although only the most recent proposal satisfies all of the experimental data and is chemically feasible. In this proposal, radical formation at C-1 assists fragmentation and removal of the 19-carbonyl group (Figure 5).[66]

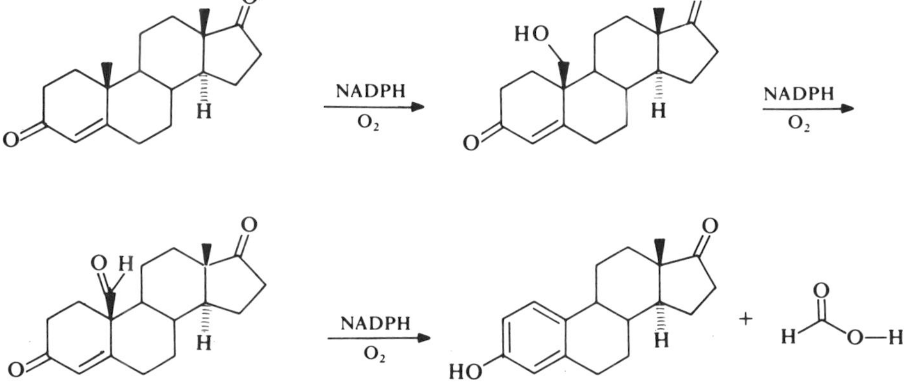

(26)

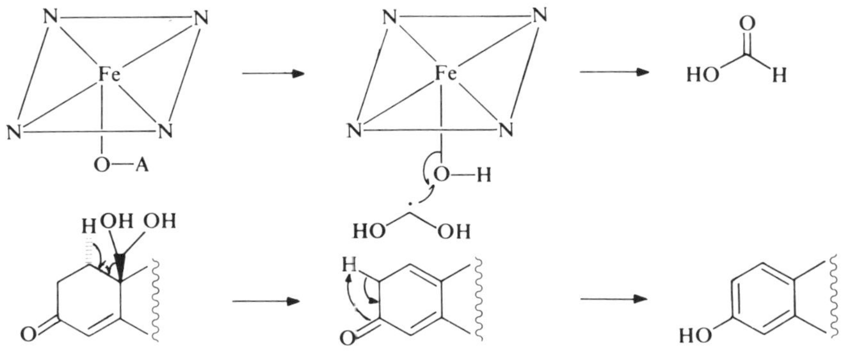

Figure 5

Aromatase is of considerable therapeutic interest since estrogen production has been associated with gynecomastia, endometriosis, and endometrial and breast cancers.[85,86] A wide range of both reversible and irreversible aromatase inhibitors has been reported (Table 3). In addition to those inhibitors which are substrate analogues, compounds such as aminoglutethimide (25) have been shown to inhibit aromatase. Compound (25) is being used with increasing frequency in treating estrogen-dependent breast cancers.[88]

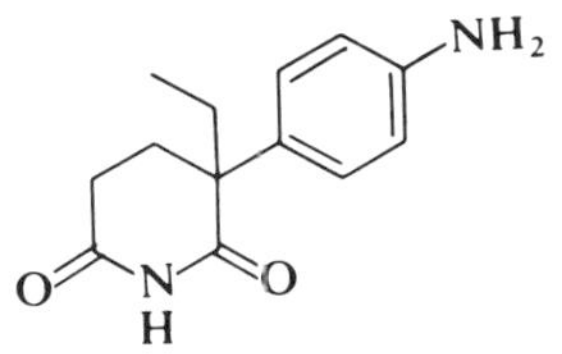

(25)

The aromatase inhibitors depicted in Table 3 show several different modes of inhibition. For the simple substrate analogues (26) which bear additional functional groups at the steroid 10-position, both reversible and time-dependent, irreversible inhibition is observed. More than a 10^3-fold variation in affinity for enzyme is shown by these different inhibitors. The (R)- and (S)-oxirane inhibitors show potent time-dependent inhibition, with the (R) isomer exhibiting both higher affinity and a more rapid inactivation. Interestingly, the closely related thiiranes show only reversible (but potent) inhibition. The 19-azidomethyl and 19-methylthiomethyl analogues show potent, reversible inhibition of aromatase; in the thioether inhibitor, spectroscopic changes upon inhibitor binding have been used to argue for ligation of the thioether with the heme iron, as depicted schematically in Figure 4. The corresponding 19-thiomethyl inhibitor shows somewhat reduced affinity for enzyme with a very slow time-dependent inactivation. The 10-allenyl and 10-propenyl substituted inhibitors show both high affinity and a facile time-dependent inactivation. The mechanism of this inactivation has been suggested to involve chemistry related to the inactivation of P-450 by simple acetylene (equation 23). A second series of steroidal substrates with B-ring modifications, compounds (27)–(29), show time-dependent irreversible inactivation of aromatase. The chemical mechanism by which these inhibitors act remains undefined. In spite of the clinical success of drugs such as aminogluteth- imide and the discovery of the many effective aromatase inhibitors shown in Table 3, the search for novel, effective aromatase inhibitors continues. This will undoubtedly remain an area of consider- able interest for some time to come.

6.1.4.4.2 14α-Demethylase

An early enzymatic step in the biosynthesis of both mammalian and fungal cell membrane sterols is the demethylation of the sterol lanosterol (30) to the heteronuclear $\Delta^{8,14}$-diene (31) (Figure 6).[99] The stoichiometry of this P-450 mediated oxidation is identical to that observed in the aromatase reaction. The 14α-demethylase enzymes present in both mammalian and fungal systems have become important targets for the design of inhibitors. A fungal-selective 14α-demethylase inhibitor is expected to act as an antifungal agent while an inhibitor of the mammalian enzyme might act as a hypocholesterolemic agent.

Table 3 Inhibitors of Aromatase

Structure	K_i (nM)	$t_{1/2}$ (min)	Ref.	Structure	K_i (nM)	$t_{1/2}$ (min)	Ref.

(26)

Structure	K_i (nM)	$t_{1/2}$ (min)	Ref.	Structure	K_i (nM)	$t_{1/2}$ (min)	Ref.
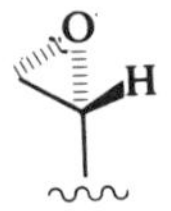 R = (epoxide, O above)	7	1.6	89	R = —CH$_2$—SH	34	462	93
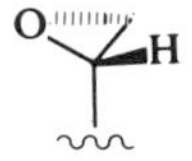 = (epoxide, O behind)	75	20	89	= —CH=C=CH$_2$	14	24	94
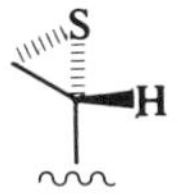 = (thiirane, S front)	1	—	90, 91	= —CH$_2$—C≡CH	10	7	94
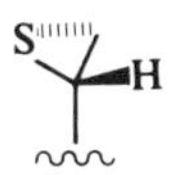 = (thiirane, S behind)	75	—	90, 91	= —SH 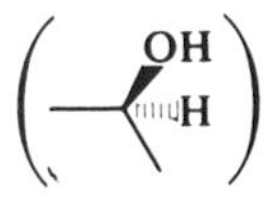 (—CH(OH)H)	106	281	93
= —CH$_2$—N=$\overset{+}{N}$=$\bar{N}$	5	—	92	= —O—OH	330	~1.5	95
= —CH$_2$—SCH$_3$	1	—	92	= —OH	1500	—	95

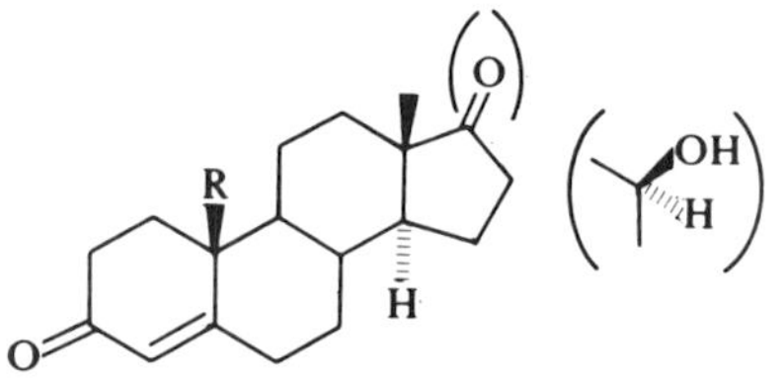

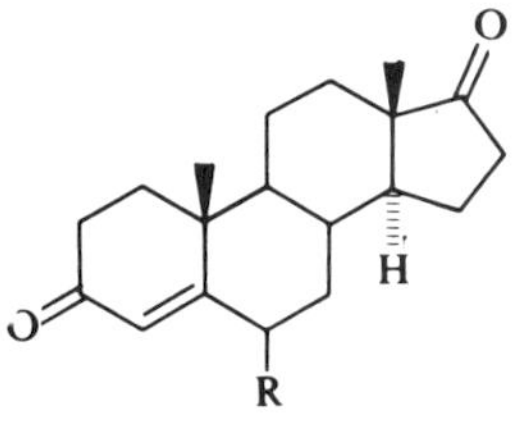

(27)

Structure	K_i (nM)	$t_{1/2}$ (min)	Ref.	Structure	K_i (nM)	$t_{1/2}$ (min)	Ref.
R = α-Br	3.4	—	96	R = α-O—O—H	167	~90	97
= β-Br	800	28	96	= β-O—O—H	163	~90	97

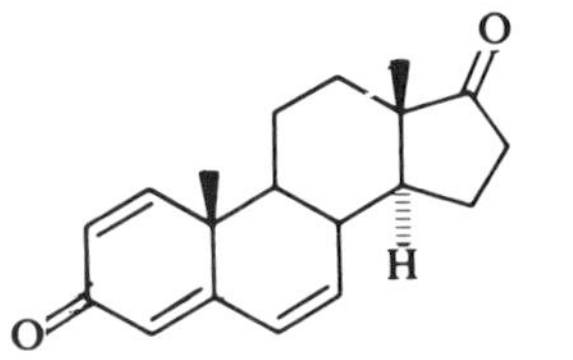

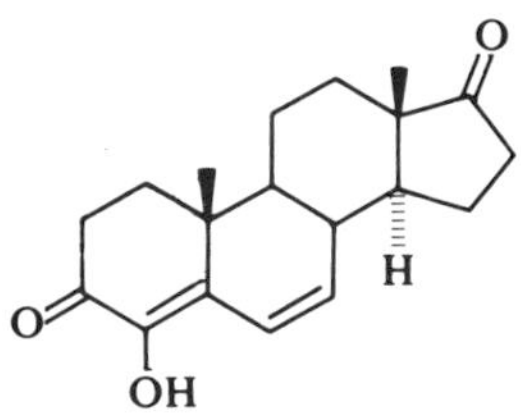

(28)

(29)

	K_i (nM)	$t_{1/2}$ (min)	Ref.		K_i (nM)	$t_{1/2}$ (min)	Ref.
	—	2.5	98		—	6.8	98

Considerable effort has been invested in the design of fungal-selective 14α-demethylase inhibitors and this has resulted in useful, orally active antifungal agents which are effective against both topical (or vaginal) and systemic fungal infections.[100] The recognition of 14α-demethylase as an antifungal

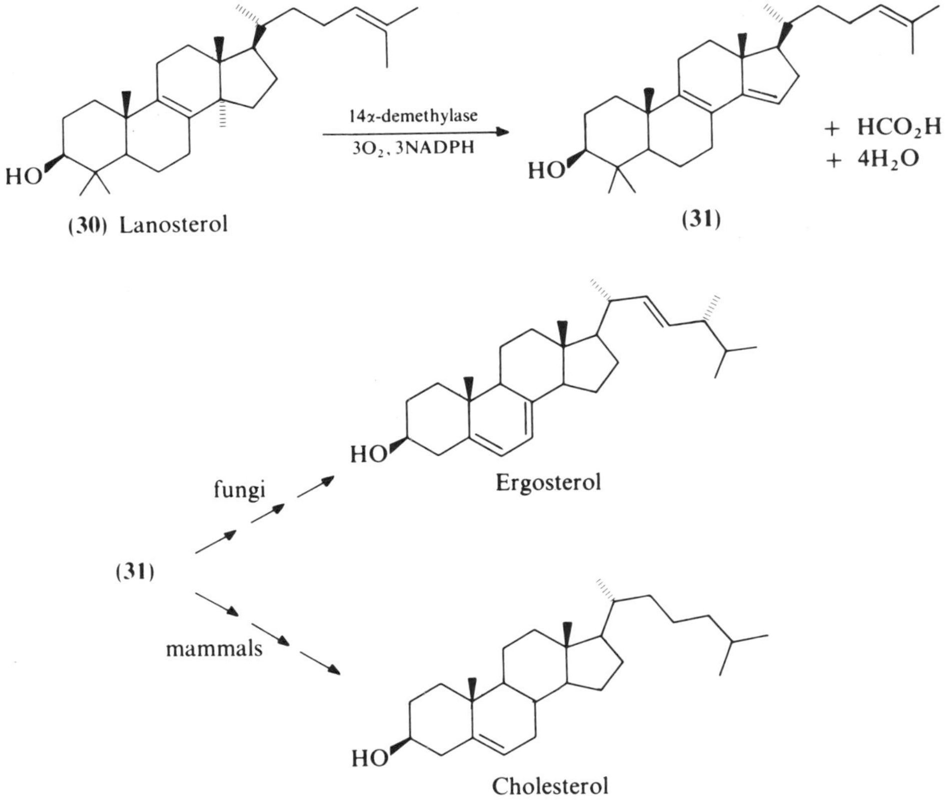

$$(30)\ \text{Lanosterol} \xrightarrow[3O_2,\ 3NADPH]{14\alpha\text{-demethylase}} (31) + HCO_2H + 4H_2O$$

Figure 6

target occurred rather fortuitously with the observation that *N*-tritylimidazole (**32**) and the ring-chlorinated analogue clotrimizole (**33**)[101] possessed antifungal activity. Subsequent work with classical structure–activity relationships led to the discovery that the imidazoles miconazole (**34**)[102] and ketoconazole (**35**)[103–105] had improved antifungal activity in man, coupled with a somewhat improved selectivity for the fungal enzyme relative to the mammalian 14α-demethylase. Despite these advantages, the high doses of (**35**), and the therapeutic time course which were required for antiinfective action, led to some inhibition of the mammalian enzyme and side effects such as gynecomastia.[106,107] Further research led to the identification of itraconazole (**36**)[108,109] as a somewhat more selective, safe, well-absorbed antifungal agent. Itraconazole is highly active *in vitro* against yeasts, aspergilli, dermatophytes, and other fungi. Improved *in vivo* activity has also been observed and has been attributed to the enhanced pharmacokinetic properties of this triazole.

A wide range of structural relatives to (**32**)–(**36**) has been prepared and evaluated for antifungal activity. Extensive reviews summarize the SAR for these P-450 inhibitors.[110,111] Unfortunately, most of these compounds have highly crystalline and water-insoluble properties; these structural characteristics impart poor oral bioavailability, less than ideal systemic distribution, and a proclivity toward inhibiting hepatic or other mammalian P-450 isozymes. Perhaps the most important innovation in this approach to the inhibition of fungal 14α-demethylase was the observation that fluconazole (**37**) was 10–100 times more active than ketoconazole and was water-soluble, orally absorbed, was 100% bioavailable in man and, in addition, had an extremely long plasma half-life.[112,113] This fortunate combination of physicochemical properties and potency has led to excellent activity against vaginal and systemic candidiasis and a variety of dermatophytes. No side effects have so far been reported.

It is generally believed that compounds such as (**32**)–(**37**) inhibit the fungal 14α-demethylase by binding in a fashion which is illustrated in Figure 4.[114] An imidazole or triazole heterocycle is a common structural feature of these inhibitors, together with a lipophilic aromatic or aralkyl group. Presumably the varying degree of selectivity between mammalian and fungal enzymes as well as potency result from binding affinities of the lipophilic portion of inhibitor. The studies of structure–activity relationships which have been carried out appear to have afforded certain

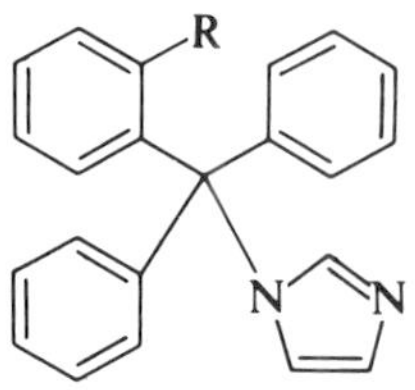

(32) R = H
(33) R = Cl

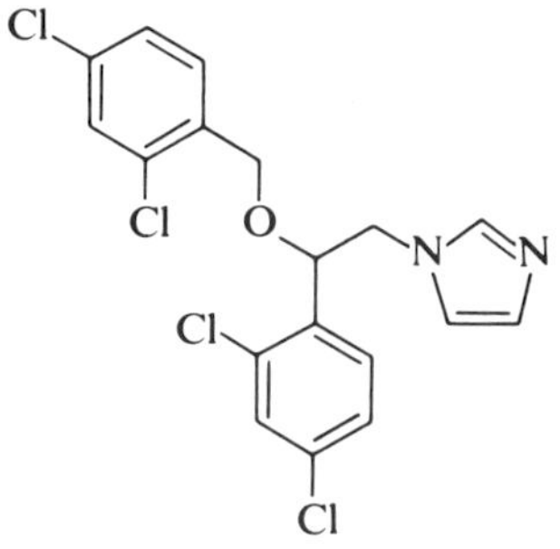

(34)

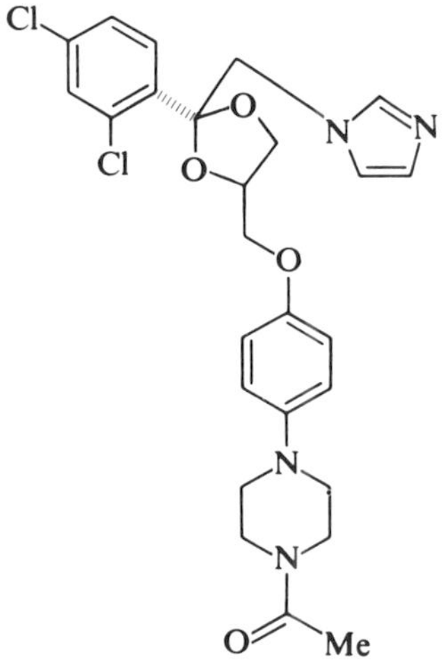

(35)

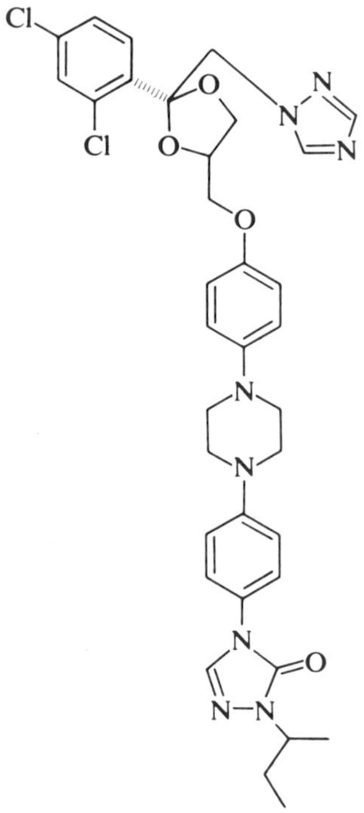

(36)

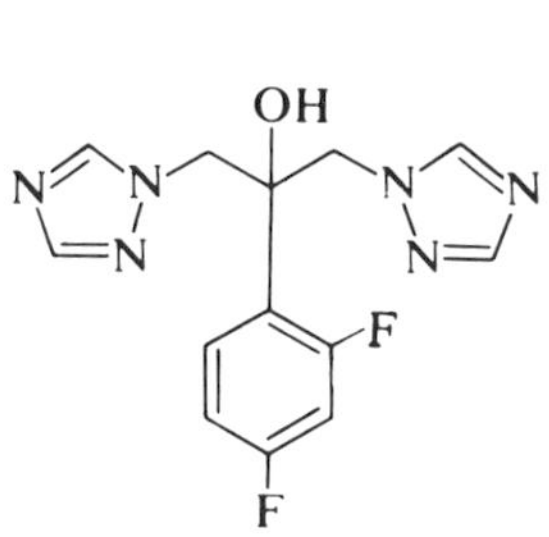

(37)

elements of species selectivity, potency and bioavailability and these are reflected in enhanced therapeutic utility.

The inhibition of mammalian P-450 isozymes by antifungal agents such as (32)–(36) produces varying degrees of side effects. The design of *selective* reversible or time-dependent inhibitors of the mammalian 14α-demethylase may lead to novel hypocholesterolemic therapy. Lowering of blood cholesterol levels is an important target of several research programmes, since high cholesterol levels are associated with increased risk of heart disease. While it is unlikely that inhibition of 14α-demethylase (Figure 6) will lead directly to lowered plasma cholesterol, it is likely that the accumulated sterol biosynthetic intermediates will undergo further catabolic oxidation to oxysteroids. These have been shown to down-regulate the activity of 3-hydroxy-3-methylglutaryl coenzyme A reductase, with a resultant overall decrease of sterol biosynthesis. One anecdotal clinical observation of cholesterol lowering has been observed to result from low-dose ketoconazole antifungal therapy.[115] This is postulated to result from the accumulation of oxygenated sterol intermediates. While the feedback regulation of early biosynthetic enzymes by accumulating intermediate sterol metabolites could produce rather complex pharmacological scenarios, the therapeutic importance of this approach has yet to be fully explored.

6.1.4.4.3 Steroid side-chain cleavage

One of the most interesting cytochrome P-450 mediated events is the sterol side-chain cleavage reaction [equations (18) and (27)] which occurs during steroid biosynthesis. Several novel inhibitors of this enzyme have been described. The diyne-diol (**38**)[116] shows mechanism-based inactivation which occurs with a very low (6:1) partition ratio. It has been suggested that attempted side-chain cleavage results in production of an oxirene intermediate at the 22,23 side-chain position and that this oxirene traps the enzyme by alkylation of an active-site amino acid residue, by a chemical mechanism different from that shown in equation (23), where the heme is modified. A second mechanism-based inhibitor of the side-chain cleavage P-450 has been reported.[117] The novel silicon containing sterol (**39**) leads to very efficient time-dependent inactivation and a low, <6:1, partition ratio. A chemical mechanism involving silylation of enzyme (Figure 7) was suggested to account for the inactivation. A final example of mechanism-based inhibition of the side-chain cleavage P-450 was observed with thiasterol (**40**).[118,119] This thioether has been suggested to undergo P-450 mediated oxidation to the corresponding sulfoxide which then binds very tightly, but not covalently to the heme iron. Spectroscopic evidence (EXAFS) supported a ligation of the heme iron by the sulfoxide oxygen. Inactivation with (**40**) thus follows the general mechanistic schemes outlined by equations (22)–(24).

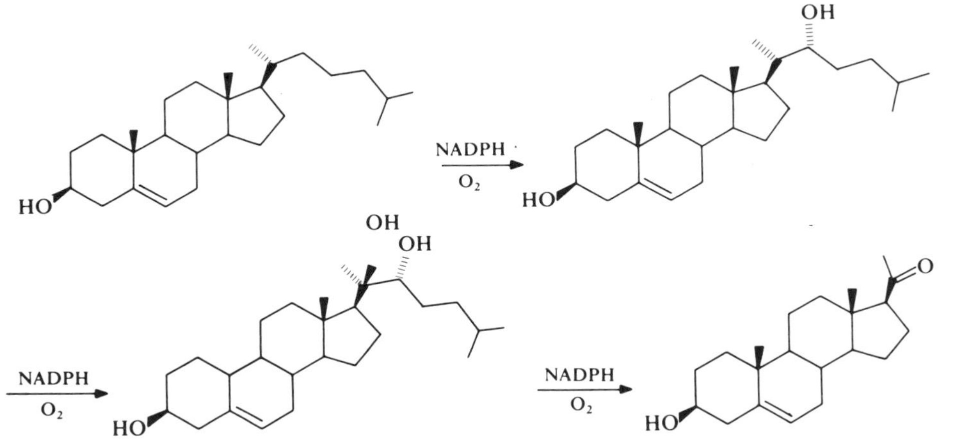

(27)

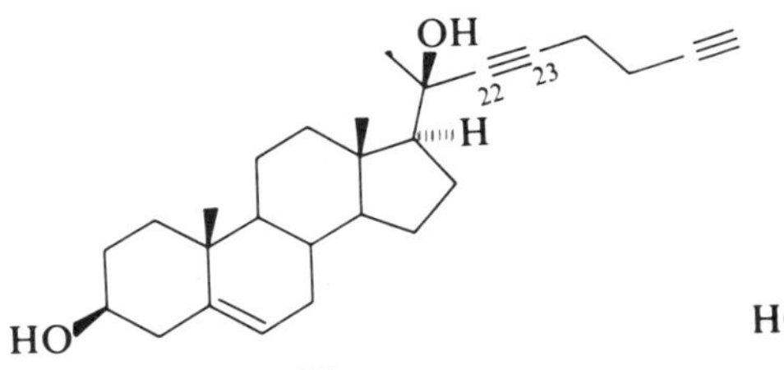

(**38**) 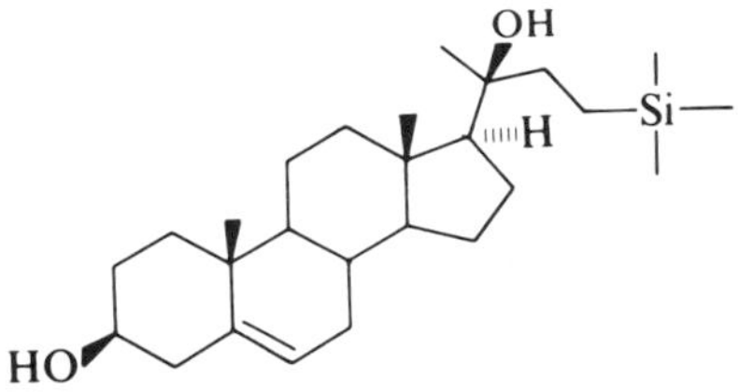   (**39**)

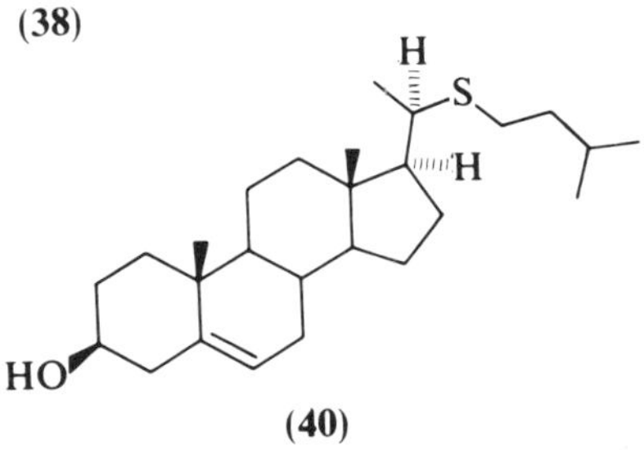

(**40**)

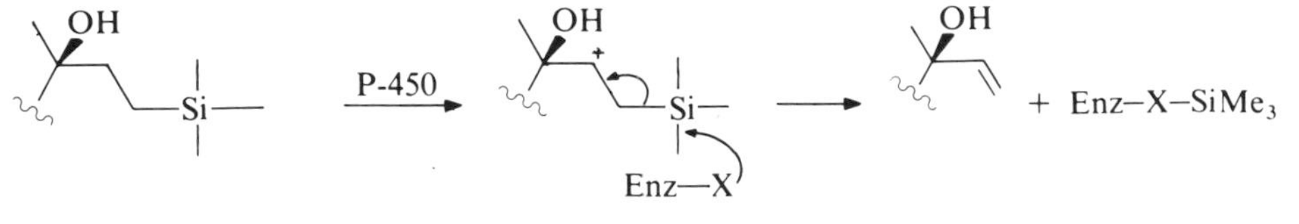

Figure 7

6.1.5 XANTHINE OXIDASE

Xanthine oxidase is a high molecular weight ($\sim 280\,000$ Da) Mo-containing mixed-function oxidase which catalyzes the catabolism of hypoxanthine to xanthine and finally to uric acid, according to equation (28).[120,121] The xanthine oxidase present in milk has been purified and shown to be a dimer.[122,123] The active-site structure of xanthine oxidase is extraordinarily complex: one atom of Mo, four atoms of Fe, four atoms of inorganic (acid-labile) sulfide and a molecule of flavin cofactor are present in each subunit! Despite considerable efforts and prolonged study, the chemical mechanism of xanthine oxidation is still poorly understood.

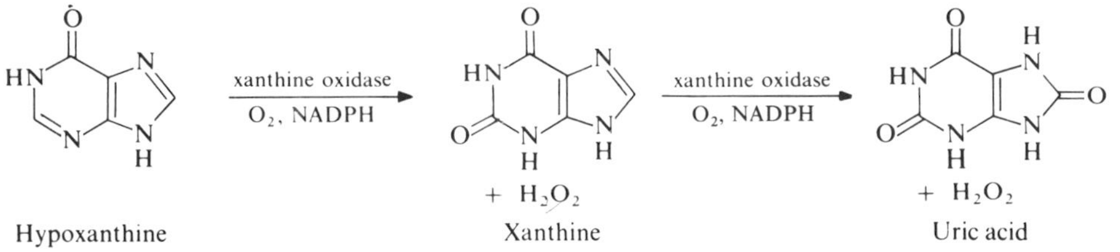

$$\text{(28)}$$

Hypoxanthine Xanthine Uric acid

Xanthine oxidase is important in several disease states. In normal individuals the uric acid produced according to equation (28) is eliminated in the urine. In individuals who suffer from gout, crystals of the uric acid calcium salt are formed in joints, leading to the inflammation and severe pain associated with this disease state. Allopurinol (41) has shown clinical efficacy in the treatment of gout. This compound is a very potent inhibitor of xanthine oxidase. Inhibition occurs by formation of a very tightly bound (but noncovalent) complex which involves an initial hydroxylation of (36) to alloxanthine (42), in the normal catalytic cycle.[124] In the subsequent inactivation event, (42) binds xanthine oxidase reversibly followed by a slow, essentially irreversible conversion of the bound complex to a tightly bound complex. The formation of the tight-binding complex is associated with a reduction of Mo^{6+} cofactor to the Mo^{4+} species.[125] Although potentially reversible, the very slow off-rate ($k \approx 0.002$ min^{-1}) makes inhibition effectively irreversible under physiological conditions.

(41)   (42)

The role of xanthine oxidase in the tissue injury which occurs in reperfusion following ischemia may be far more important than in gout. During ischemia, significant catabolism of ATP to hypoxanthine occurs (Scheme 1).[126-128] Subsequent reperfusion is accompanied by a burst of xanthine oxidase activity and a consequent 'pulse' of H_2O_2 and O_2^- which are the suspected mediators of tissue damage. The actual chemical mechanisms by which the reduced oxygen products of xanthine oxidase produce tissue damage is unknown. However, it appears that conversion to hydroxyl radical (HO·) may occur and that this is the species which causes tissue damage. Interestingly, allopurinol shows some efficacy in the treatment of reperfusion injury.[129,130] It remains to be determined whether more effective inhibitors of this 'oxidative pulse' can be designed and shown to be useful in the prevention of reperfusion tissue damage.

$$\text{ATP} \rightarrow \text{ADP} \rightarrow \text{AMP} \rightarrow \text{Adenosine} \rightarrow \text{Inosine} \rightarrow \text{Hypoxanthine}$$

Scheme 1 Ischemia

6.1.6 REFERENCES

1. T. P. Singer, R. W. Von Korff and D. L. Murphy (ed.), 'Monoamine Oxidase: Structure, Function and Altered Functions', Academic Press, New York, 1979.
2. D. R. Patek, H. Y. K. Chuang and L. Hellerman, *Fed. Proc., Fed. Am. Soc. Exp. Biol.*, 1972, **31**, 420.
3. K. T. Yasunobu and B. Gomes, *Methods Enzymol.*, 1971, **17B**, 709.
4. R. B. Silverman, S. J. Hoffman and W. B. Catus, III, *J. Am. Chem. Soc.*, 1980, **102**, 7126.
5. M. B. H. Youdim and J. P. M. Finberg, *Mod. Prob. Pharmacopsychiat.*, 1983, **5**, 393.

6. A. Maycock, R. H. Abeles, J. I. Salach and T. P. Singer, *Biochemistry*, 1976, **15**, 114.
7. H. Y. K. Chuang, D. R. Patek and L. Hellerman, *J. Biol. Chem.*, 1974, **249**, 2381.
8. T. P. Singer, In 'Enzyme Inhibitors', ed. V. Brodbeck, Verlag Chemie, Weinheim, 1980, p. 7.
9. R. B. Silverman, *J. Biol. Chem.*, 1983, **258**, 14766.
10. P. Bey, J. Fozard, J. M. Lacoste, I. A. McDonald, M. Zreika and M. G. Palfreyman, *J. Med. Chem.*, 1984, **27**, 9.
11. I. A. McDonald, J. M. Lacoste, P. Bey, M. G. Palfreyman and M. Zreika, *J. Med. Chem.*, 1985, **28**, 186.
12. I. A. McDonald, J. M. Lacoste, P. Bey, J. Wagner, M. Zreika and M. G. Palfreyman, *Bioorg. Chem.*, 1986, **14**, 103.
13. M. G. Palfreyman, I. A. McDonald, J. R. Fozard, Y. Mely, A. J. Sleight, M. Zreika, J. Wagner, P. Bey and P. J. Lewis, *J. Neurochem.*, 1985, **45**, 1850.
14. A. Pletscher, K.F. Grey and P. Zeller, in 'Progress in Drug Research', ed. E. Zucker, Birkhäuser, Basel, 1960, p. 417.
15. S. Udenfriend, B. Witkop, R. G. Redfield and H. Weissbach, *Biochem. Pharmacol.*, 1958, **1**, 60.
16. R. E. Heikkila, L. Manzino, F. S. Cabbat and R. C. Duvoisin, *Nature (London)*, 1984, 467.
17. P. Bey, F. N. Bolkenius, N. Seiler and P. Casara, *J. Med. Chem.*, 1985, **28**, 1.
18. R. C. Rosenberg and W. Lovenberg, in 'Essays in Neurochemistry and Neuropharmacology', ed. M. B. H. Youdim, W. Lovenberg, D. F. Sharman and J. R. Lagnado, Wiley, New York, 1980, vol. 4, p. 163.
19. L. C. Stewart and J. P. Klinman, *Ann. Rev. Biochem.*, 1989, in press.
20. J. P. Klinman, M. Krueger, M. Brenner and D. E. Edmondson, *J. Biol. Chem.*, 1984, **259**, 3399.
21. W. E. DeWolf, Jr. and L. I. Kruse, unpublished observation.
22. A. Lamouroux, A. Vigny, N. F. Biguet, M. C. Darmon, R. Franck, J.-P. Henry and J. Mallet, *EMBO J.*, 1987, **6**, 3931.
23. N. G. Ahn and J. P. Klinman, *Biochemistry*, 1983, **22**, 3096.
24. R. A. Scott, R. J. Sullivan, W. E. DeWolf, Jr., R. E. Dolle and L. I. Kruse, *Biochemistry*, 1989, in press.
25. P. F. Fitzpatrick and J. J. Villafranca, *Arch. Biochem. Biophys.*, 1987, **257**, 231.
26. J. M. Baldoni and J. J. Villafranca, *J. Biol. Chem.*, 1980, **255**, 8987.
27. P. F. Fitzpatrick and J. J. Villafranca, *J. Am. Chem. Soc.*, 1985, **107**, 5022.
28. T. M. Bargar, R. J. Broersma, L. C. Creemer, J. R. McCarthy, J.-M. Hornsperger, M. G. Palfreyman, J. Wagner and M. J. Jung, *J. Med. Chem.*, 1986, **29**, 315.
29. S. R. Padgette, K. Wimalasena, H. H. Herman, S. R. Sirimane and S. W. May, *Biochemistry*, 1985, **24**, 5826.
30. B. Rajashekhar, P. F. Fitzpatrick, G. Colombo and J. J. Villafranca, *J. Biol. Chem.*, 1984, **259**, 6925.
31. J. B. Mangold and J. P. Klinman, *J. Biol. Chem.*, 1984, **259**, 7772.
32. M. J. Bossard and J. P. Klinman, *J. Biol. Chem.*, 1986, **261**, 16421.
33. L. I. Kruse, C. Kaiser, W. E. DeWolf, Jr., P. A. Chambers, P. J. Goodhart, M. Ezekiel and E. H. Ohlstein, *J. Med. Chem.*, 1988, **31**, 704.
34. P. J. Goodhart, W. E. DeWolf, Jr. and L. I. Kruse, *Biochemistry*, 1987, **26**, 2576.
35. W. E. DeWolf, Jr., P. J. Goodhart, A. Varrichio, R. G. L. Shorr and L. I. Kruse, *Fed. Proc., Fed. Am. Soc. Exp. Biol.*, 1987, **46**, 2229.
36. L. I. Kruse, W. E. DeWolf, Jr., P. A. Chambers and P. J. Goodhart, *Biochemistry*, 1986, **25**, 7271.
37. L. I. Kruse, C. Kaiser, W. E. DeWolf, Jr., J. S. Frazee, E. Garvey, E. L. Hilbert, W. A. Faulkner, K. E. Flaim, J. L. Sawyer and B. A. Berkowitz, *J. Med. Chem.*, 1986, **29**, 2465.
38. L. I. Kruse, C. Kaiser, W. E. DeWolf, Jr., J. S. Frazee, S. T. Ross, J. Wawro, M. Wise, K. E. Flaim, J. L. Sawyer, R. W. Erickson, M. Ezekiel, E. H. Ohlstein and B. A. Berkowitz, *J. Med. Chem.*, 1987, **30**, 486.
39. L. D. Byers, *J. Theor. Biol.*, 1978, **74**, 501.
40. L. I. Kruse, C. Kaiser, W. E. DeWolf, Jr., J. S. Frazee, R. W. Erickson, M. Ezekiel, E. H. Ohlstein, R. R. Ruffolo, Jr. and B. A. Berkowitz, *J. Med. Chem.*, 1986, **29**, 887.
41. E. H. Ohlstein, L. I. Kruse, M. Ezekiel, S. S. Sherman, R. Erickson, W. E. DeWolf, Jr. and B. A. Berkowitz, *J. Pharmacol. Exp. Ther.*, 1987, **241**, 554.
42. P. R. Ortiz de Montellano (ed.), 'Cytochrome P-450', Plenum Press, New York, 1986.
43. T. L. Poulos, B. C. Finzel, I. C. Gunsalus, G. C. Wagner and J. Kraut, *J. Biol. Chem.*, 1985, **260**, 16122.
44. T. L. Poulos, B. C. Finzel and A. J. Howard, *Biochemistry*, 1986, **25**, 5314.
45. T. L. Poulos and A. J. Howard, *Biochemistry*, 1987, **26**, 8165.
46. G. T. Miwa, J. S. Walsh and A. Y. H. Lu, *J. Biol. Chem.*, 1984, **259**, 3000.
47. M. M. Abdel-Monem, *J. Med. Chem.*, 1975, **18**, 427.
48. G. T. Miwa, W. A. Garland, B. J. Hodshon, A. Y. H. Lu and D. B. Northrup, *J. Biol. Chem.*, 1980, **255**, 6049.
49. G. T. Miwa, J. S. Walsh, G. L. Kedderis and P. F. Hollenberg, *J. Biol. Chem.*, 1983, **258**, 14445.
50. Y. Watanabe, T. Iyanagi and S. Oae, *Tetrahedron Lett.*, 1980, **21**, 3685.
51. Y. Watanabe, T. Numata, T. Iyanagi and S. Oae, *Bull. Chem. Soc. Jpn.*, 1981, **54**, 1163.
52. Y. Watanabe, T. Iyanagi and S. Oae, *Tetrahedron Lett.*, 1982, **23**, 533.
53. T. Watanabe and K. Akamatsu, *Biochem. Pharmacol.*, 1974, **23**, 1079.
54. K. L. Kunze, B. L. K. Mangold, C. Wheeler, H. S. Beilan and P. R. Ortiz de Montellano, *J. Biol. Chem.*, 1983, **258**, 4202.
55. D. C. Liebler and F. P. Guengerich, *Biochemistry*, 1983, **22**, 5482.
56. J. T. Groves and Y. Watanabe, *J. Am. Chem. Soc.*, 1986, **108**, 507.
57. T. G. Traylor, Y. Iamamoto and T. Nakano, *J. Am. Chem. Soc.*, 1986, **108**, 3529.
58. S. E. Creager, S. A. Raybuck and R. W. Murray, *J. Am. Chem. Soc.*, 1986, **108**, 4225.
59. P. R. Ortiz de Montellano, K. L. Kunze, H. S. Beilan and C. Wheeler, *Biochemistry*, 1982, **21**, 1331.
60. P. R. Ortiz de Montellano and E. A. Komives, *J. Biol. Chem.*, 1985, **260**, 3330.
61. P. R. Ortiz de Montellano and N. O. Reich, *J. Biol. Chem.*, 1984, **259**, 4136.
62. C. A. Jacob and P. R. Ortiz de Montellano, *Biochemistry*, 1986, **25**, 4705.
63. P. R. Ortiz de Montellano and K. L. Kunze, *Arch. Biochem. Biophys.*, 1981, **209**, 710.
64. P. R. Ortiz de Montellano, in 'Bioactivation of Foreign Compounds', ed. M. W. Anders, Academic Press, New York, 1985, p. 121.
65. R. E. McMahon, J. C. Turner, G. W. Turner and H. R. Sullivan, *Biochem. Biophys. Res. Commun.*, 1981, **99**, 662.
66. D. D. Benson, H. L. Carrell and D. F. Covey, *Biochemistry*, 1987, **26**, 7833.
67. S. Burstein and M. Gut, *Steroids*, 1976, **28**, 115.
68. G. Constantopoulos, P. S. Satoh and T. T. Tchen, *Biochem. Biophys. Res. Commun.*, 1962, **8**, 50.

69. S. F. Muakkassah, W. R. Bidlack and W. C. T. Yang, *Biochem. Pharmacol.*, 1981, **30**, 1651.
70. F. P. Guengerich, G. A. Dannan, S. T. Wright, M. V. Martin and L. S. Kaminsky, *Biochemistry*, 1982, **21**, 6019.
71. D. E. Ryan, S. Iida, A. W. Wood, P. E. Thomas, C. S. Lieber and W. Levin, *J. Biol. Chem.*, 1984, **259**, 1239.
72. E. H. Jeffrey and G. J. Mannering, *Mol. Pharmacol.*, 1983, **23**, 748.
73. S. J. Moloney, B. J. Snider and R. A. Prough, *Xenobiotica*, 1984, **14**, 803.
74. D. Mansuy, *Rev. Biochem. Toxicol.*, 1981, **3**, 283.
75. C. F. Wilkinson, M. Murray and C. B. Marcus, *Rev. Biochem. Toxicol.*, 1984, **6**, 27.
76. D. Mansuy, J. Large, J. Chattard, J. Bartoli, B. Chervier and R. Weiss, *Angew. Chem., Int. Ed. Engl.*, 1978, **17**, 781.
77. D. Pressayre, D. Larrey, J. Vitaux, P. Breil, J. Belghiti and J.-P. Benhamou, *Biochem. Pharmacol.*, 1982, **31**, 1699.
78. M. Delaforge, M. Jaouen and D. Mansuy, *Biochem. Pharmacol.*, 1983, **32**, 2309.
79. P. R. Ortiz de Montellano and M. A. Correira, *Ann. Rev. Pharmacol.*, 1983, **23**, 481.
80. P. R. Ortiz de Montellano and K. L. Kunze, *J. Biol. Chem.*, 1980, **255**, 5578.
81. P. R. Ortiz de Montellano, J. Mathews and K. L. Langry, *Tetrahedron*, 1983, **40**, 511.
82. R. P. Hanzlik and R. H. Tullman, *J. Am. Chem. Soc.*, 1982, **104**, 2048.
83. T. L. MacDonald, K. Zirvi, L. T. Burka, P. Peyman and F. P. Guengerich, *J. Am. Chem. Soc.*, 1982, **104**, 2050.
84. J. Fishman and M. S. Raju, *J. Biol. Chem.*, 1981, **256**, 4472.
85. A. M. H. Brodie, *Cancer Res. (Suppl.)*, 1982, **42**, 3312s.
86. A. M. H. Brodie, *Biochem. Pharmacol.*, 1985, **34**, 3213.
87. E. Capsi, T. Arunachalam and P. A. Nelson, *J. Am. Chem. Soc.*, 1986, **108**, 1847.
88. A. F. Harris, T. J. Powles, I. E. Smith, R. C. Coombes, H. T. Ford, J. C. Gazet, C. L. Harmer, M. Morgan, H. White and C. A. Parsons, *Eur. J. Cancer Clin. Oncol.*, 1983, **19**, 11.
89. M.-J. Shih, M. H. Carrell, H. L. Carrell, C. L. Wright, J. O. Johnson and C. H. Robinson, *J. Chem. Soc., Chem. Commun.*, 1987, 213.
90. W. E. Childers and C. H. Robinson, *J. Chem. Soc., Chem. Commun.*, 1987, 320.
91. J. T. Kellis, Jr., W. E. Childers, C. H. Robinson and L. E. Vickery, *J. Biol. Chem.*, 1987, **262**, 4421.
92. J. N. Wright, M. R. Calder and M. Akhtar, *J. Chem. Soc., Chem. Commun.*, 1985, 1733.
93. P. J. Bednarski, D. J. Porubek and S. D. Nelson, *J. Med. Chem.*, 1985, **28**, 775.
94. B. W. Metcalf, C. L. Wright, J. P. Burkhart and J. O. Johnson, *J. Am. Chem. Soc.*, 1981, **103**, 3221.
95. D. F. Covey, W. F. Hood, D. D. Beusen and H. L. Carrell, *Biochemistry*, 1984, **23**, 5398.
96. Y. Osawa, Y. Osawa and M. J. Coon, *Endocrinology*, 1987, **121**, 1010.
97. L. Tan and A. Petit, *Biochem. Biophys. Res. Commun.*, 1985, **128**, 613.
98. P. A. Marsh, H. J. Brodie, W. Garrett, C.-H. Tsai-Morris and A. M. H. Brodie, *J. Med. Chem.*, 1985, **28**, 788.
99. Y. Aoyama and Y. Yoshida, *Biochem. Biophys. Res. Commun.*, 1978, **85**, 26.
100. K. Richardson and M. S. Marriott, *Annu. Rep. Med. Chem.*, 1987, **22**, 159.
101. D. C. Plempel *et al.*, *Antimicrol. Agents Chemother.*, 1969, **00**, 271.
102. D. A. Stevens, *Drugs*, 1983, **26**, 347.
103. J. Symoens and G. Cauwenbergh, *Prog. Drug Res.*, 1983, **27**, 63.
104. H. B. Levine, 'Ketoconazole in the Management of Fungal Disease', ADIS Press, Australia, 1982.
105. H. Koch, *Pharm. Int.*, 1983, **4**, 151.
106. A. Pont, P. L. Williams, S. Azhar, R. E. Reitz, C. Bochra, E. R. Smith and D. A. Stevens, *Arch. Int. Med.*, 1982, **142**, 2137.
107. D. S. Grosso, T. W. Boyden, R. W. Parmenter, R. C. Rumbaugh and H. D. Colby, *Life Sci.*, 1977, **20**, 1017.
108. M. Borgers, in 'The Scientific Basis of Antimicrobial Chemotherapy', ed. D. Greenwood and F. O'Grady, Cambridge University Press, Cambridge, 1985, p. 133.
109. Anon., *Drugs Future*, 1986, **11**, 335.
110. G. J. Ellames, in 'Topics in Antibiotic Chemistry', ed. P. G. Sammes, Ellis Horwood, Chichester, 1982, vol. 6, p. 14.
111. L. Zirngibl, *Prog. Drug Res.*, 1983, **27**, 253.
112. P. F. Troke, R. J. Andrews, K. W. Brammer, M. S. Marriott and K. Richardson, *Antimicrob. Agents Chemother.*, 1985, **28**, 815.
113. M. S. Marriott, M. J. Humphrey and M. H. Tarbit, in 'Recent Advances in Chemotherapy', ed. J. Ishigami, University of Tokyo Press, 1985, p. 1937.
114. I. Schuster, *Xenobiotica*, 1985, **15**, 529.
115. A. Gupta, R. C. Sexton and H. Rudney, *J. Biol. Chem.*, 1986, **261**, 8348.
116. A. Nagahisa, R. W. Spencer and W. H. Orme-Johnson, *J. Biol. Chem.*, 1983, **258**, 6721.
117. A. Nagahisa, W. H. Orme-Johnson and S. R. Wilson, *J. Am. Chem. Soc.*, 1984, **106**, 1166.
118. R. J. Kruger, A. Nagahisa, M. Gut, S. R. Wilson and W. H. Orme-Johnson, *J. Biol. Chem.*, 1984, **259**, 852.
119. A. Nagahisa, R. W. Spencer and W. H. Orme-Johnson, *J. Biol. Chem.*, 1985, **260**, 852.
120. V. Massey, in 'Iron-Sulfur Proteins', ed. W. Lovenberg, Academic Press, New York, vol. 1, p. 301.
121. R. C. Bray, in 'The Enzymes', 3rd edn., ed. P. Boyer, Academic Press, New York, 1975, vol. 12, p. 300.
122. L. I. Hart, M. A. McGartoll, H. R. Chapman and R. C. Bray, *Biochem. J.*, 1970, **116**, 851.
123. R. C. Bray, in 'Flavins and Flavoproteins', ed. V. Massey and C. H. Williams, Elsevier, Amsterdam, 1982, p. 775.
124. V. Massey, H. Komai, G. Palmer and G. B. Elion, *J. Biol. Chem.*, 1970, **245**, 2837.
125. T. R. Hawkes, G. N. George and R. C. Bray, *Biochem. J.*, 1984, **218**, 961.
126. S. Imai, A. L. Riley and R. M. Berne, *Circ. Res.*, 1964, **15**, 443.
127. M. R. Buhl, C. Kemp and E. Kemp, *Transplantation*, 1976, **21**, 460.
128. O. D. Saugstad, H. Schrader and A. O. Aasen, *Brain Res.*, 1976, **112**, 188.
129. S. W. Werns, M. J. Shea, S. E. Mitsos, R. C. Dysko, J. C. Fantone, M. A. Schork, G. D. Abrams, B. Pitt and B. R. Lucchesi, *Circulation*, 1986, **73**, 518.
130. S. Akizuki, S. Yoshida, D. E. Chambers, L. J. Eddy, L. F. Parmley, D. M. Yellon and J. M. Downey, *Cardiovasc. Res.*, 1985, **19**, 686.

6.2

The Arachidonic Acid Cascade

GERALD A. HIGGS, E. ANNIE HIGGS and SALVADOR MONCADA, FRS
Wellcome Research Laboratories, Beckenham, Kent, UK

6.2.1   INTRODUCTION   147

6.2.2   CYCLOOXYGENASE PATHWAY OF ARACHIDONIC ACID METABOLISM   149

   6.2.2.1   Biosynthesis of Cyclooxygenase Products   149
   6.2.2.2   Biological Properties of Cyclooxygenase Products   150
      6.2.2.2.1   The stable prostaglandins   150
      6.2.2.2.2   Prostacyclin   152
      6.2.2.2.3   Thromboxane A$_2$ 155

6.2.3   LIPOXYGENASE PATHWAY OF ARACHIDONIC ACID METABOLISM   156

   6.2.3.1   Biosynthesis of Lipoxygenase Products   156
   6.2.3.2   Biological Properties of Lipoxygenase Products and their Role in Disease   158
      6.2.3.2.1   Lipoxygenase products in diseases of the airway   158
      6.2.3.2.2   Lipoxygenase products in inflammatory diseases   158
      6.2.3.2.3   Lipoxygenase products and the vasculature   159

6.2.4   OTHER PATHWAYS OF ARACHIDONIC ACID METABOLISM   160

6.2.5   INHIBITORS OF ARACHIDONIC ACID METABOLISM   160

   6.2.5.1   Aspirin-like Drugs   160
      6.2.5.1.1   Anti-inflammatory actions   160
      6.2.5.1.2   Antithrombotic actions   162
   6.2.5.2   Thromboxane Synthase Inhibitors and Thromboxane Antagonists   163
   6.2.5.3   Lipoxygenase Inhibitors   163
      6.2.5.3.1   Mechanism of action   163
      6.2.5.3.2   Dual inhibitors of cyclooxygenase and lipoxygenase   164
      6.2.5.3.3   Selective lipoxygenase inhibitors   166
      6.2.5.3.4   The effect of lipoxygenase inhibitors on bronchial anaphylaxis   166
      6.2.5.3.5   The effect of lipoxygenase inhibitors on inflammation   167

6.2.6   DIETARY MANIPULATION OF ARACHIDONIC ACID METABOLISM   168

6.2.7   SUMMARY   170

6.2.8   REFERENCES   170

6.2.1   INTRODUCTION

Arachidonic acid is a 20-carbon polyunsaturated fatty acid which belongs to the group of essential fatty acids first identified in the 1930s. At about the same time, Von Euler[1] detected in semen an activity which contracted uterine smooth muscle, and which he named 'prostaglandin' since the prostate gland was thought to be a major source of this activity.

More than 20 years later it was shown that 'prostaglandin' was, in fact, a family of compounds, of which two prostaglandins (PGs) were isolated in crystalline form and their structures were elucidated in the early 1960s.[2] When the general structure of the prostaglandins was realized, their kinship with essential fatty acids was recognized.

At first it was believed that the stable PGs, E$_2$ and F$_{2\alpha}$, were the most important prostaglandins. However, since 1973, several discoveries have caused a radical shift in emphasis away from PGEs

147

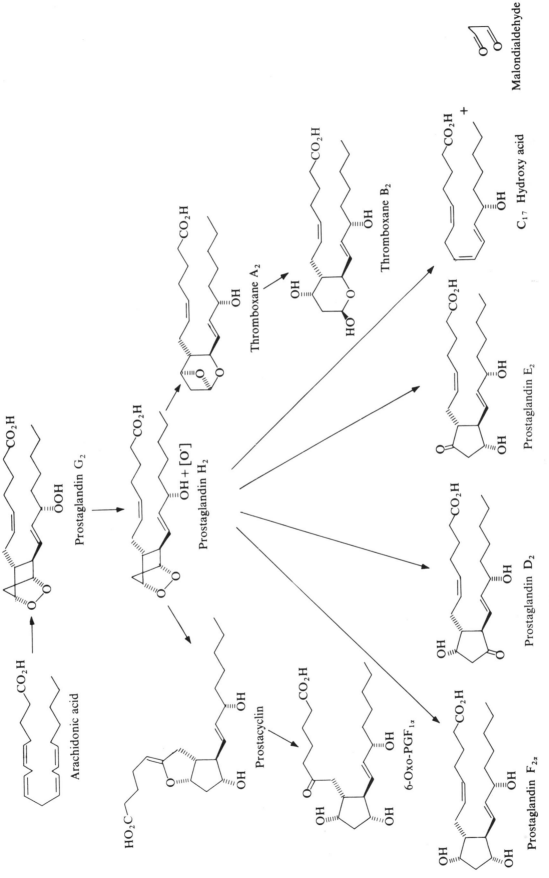

Figure 1 The cyclooxygenase pathway of arachidonic acid metabolism

and PGFs. The first was the isolation and identification of two unstable cyclic endoperoxides, prostaglandin G_2 (PGG$_2$) and prostaglandin H_2 (PGH$_2$) (Figure 1).[3] Later came the elucidation of the structure of thromboxane A_2 (TXA$_2$) and that of its degradation product, thromboxane B_2 (TXB$_2$),[4] and then the discovery of prostacyclin (PGI$_2$).[5] These findings, coupled with the recognition of a different enzymatic pathway (a lipoxygenase), which converts arachidonic acid to hydroxy acids such as 12-hydroxyeicosatetraenoic acid (HETE), led to the realization that the 'classically known' stable prostaglandins constitute only a fraction of the active products of arachidonic acid metabolism.

Recently, the products of a pathway initiated by the action of a 5-lipoxygenase enzyme have been characterized. These compounds have been named leukotrienes (LTs) because of their initial discovery in leukocytes and their conjugated triene structure.[6] Other lipoxygenase products of arachidonic acid, including lipoxins and trihydroxyeicosatetraenoic acids, have also been identified. Thus it is now known that there are a number of enzymes, in most cells of the body, that catalyze the conversion of arachidonic acid to oxygenated products, known collectively as eicosanoids. These arachidonic acid metabolites have potent and sometimes opposing biological and physiological activities. Their presence or absence may contribute to the pathogenesis of a number of clinical conditions, including inflammation, asthma, thrombosis and atherosclerosis.

6.2.2 CYCLOOXYGENASE PATHWAY OF ARACHIDONIC ACID METABOLISM

6.2.2.1 Biosynthesis of Cyclooxygenase Products

Arachidonic acid is the most abundant precursor of eicosanoids in mammalian tissue and it is either derived from dietary linoleic acid (octadecadienoic acid) or is ingested as a constituent of meat. Arachidonate is then esterified as a component of the phospholipids of cell membranes or is found in ester linkage in other complex lipids. The concentration of free arachidonic acid is low, and the biosynthesis of the eicosanoids (which are not stored as such) depends primarily upon its release from membrane phospholipids by various acyl hydrolases. These enzymes may be activated by a number of chemical stimuli, such as bradykinin, angiotensin II and noradrenaline, as well as by immunological, mechanical or electrical stimulation.

Once released, arachidonic acid is metabolized by two types of enzyme, one of which is cyclooxygenase. This enzyme seems to be present in all cell types except erythrocytes and is located on cellular membranes in the microsomal fraction. It has been identified as a glycoprotein containing mannose and N-acetylglucosamine, and there is evidence that the native enzyme exists as a dimer of about 130 000 Da whereas the purified enzyme migrates on polyacrylamide gel electrophoresis as a single protein of 69 000 Da.[7] Heme iron is necessary for the enzyme's activity and hematin has been found bound to it.[8] Cyclooxygenase, which is sometimes called 'prostaglandin synthetase', has a markedly greater affinity for C_{20} unsaturated fatty acids than C_{19} or C_{21}.[9] Similarly, there is a greater affinity for C20:4ω6* than C20:3ω6 or C20:5ω3.

Cyclooxygenase catalyzes the incorporation of molecular oxygen into arachidonic acid, leading to peroxidation at C-11 and C-15, followed by ring closure between C-8 and C-12. The products of this reaction are the cyclic endoperoxides PGG$_2$ and PGH$_2$, which are chemically unstable (*i.e.* half lives of 5 min at 37 °C and pH 7.5) and are isomerized enzymatically or nonenzymatically into different products (Figure 1). The major products of endoperoxide metabolism are the stable prostaglandins PGD$_2$, PGE$_2$ and PGF$_{2\alpha}$, thromboxane A_2 and prostacyclin. The metabolism of endoperoxides is determined by the tissues in which they are generated so that, for example, the major cyclooxygenase metabolite of arachidonic acid in the platelet is TXA$_2$, whereas in the vascular endothelium it is prostacyclin.

The development of highly sensitive techniques of analytical biochemistry permitted Bergstrom and his colleagues to elucidate the structure of two so-called primary prostaglandins.[2] These were designated prostaglandin E and F according to their partition between *E*ther and phosphate (Swedish — *F*osfat) buffer. Since then, subsequent prostaglandins have been named on an alphabetical basis and to date there are subdivisions from PGA to PGJ. Prostaglandins E and F were shown to be 20-carbon cyclized fatty acids and it was their structural similarity to the essential polyunsaturated fatty acids that led to the realization that arachidonic acid and other 20-carbon fatty acids are the precursors of prostaglandins.[10,11]

* This trivial nomenclature indicates the length of the carbon chain (C_{20}), the number of skipped cisoidal alkenic bonds (4) and where these start (carbon 6, counting from the methyl terminus). Thus this acid (arachidonic acid) is eicosa-5c,8c,11c,14c-tetraenoic acid. C20:5ω3 indicates eicosa-5c,8c,11c,14c,17c-pentaenoic acid, *etc.*

Prostaglandins derived from arachidonic acid (C20:4ω6) contain two double bonds (see Figure 1); hence PGD_2, PGE_2, *etc.* Prostaglandins derived from eicosatrienoic acid (C20:3ω6) or eicosapenta-enoic acid (C20:5ω3) contain one or three double bonds and are denoted PGD_1, PGE_1, or PGD_3, PGE_3, respectively.

The primary prostaglandins can be formed by nonenzymic degradation of the endoperoxides but the formation of PGE_2 and PGD_2 can also be catalyzed by isomerases.[12,13] There is little evidence for a PGF isomerase and it is most likely that $PGF_{2\alpha}$ is formed by reductive cleavage of the endoperoxide in the presence of reducing agents.

The endoperoxide PGH_2 is also metabolized into two unstable and highly biologically active compounds with structures that differ from those of the primary prostaglandins. One of these is thromboxane A_2 (TXA_2), formed by an enzyme, thromboxane synthase, first isolated from human and equine platelets. This enzyme has been solubilized and separated from the cyclooxygenase and there are detailed studies of human and bovine platelet thromboxane synthase. Separation of platelet membrane fractions suggests that the biosynthesis of thromboxane from arachidonic acid is associated with intracellular membrane components of the dense tubular system.[14] TXA_2 has a very short chemical half-life ($t_{1/2} = 30$ s at 37 °C and pH 7.5) and it breaks down nonenzymatically into the stable thromboxane B_2 (TXB_2) (see Figure 1). Thromboxane production has been reported in many tissues including lung, kidney, spleen and some vascular tissues,[15] but the presence of thromboxane synthase in circulating platelets and leukocytes indicates that at least some of this activity may be due to contaminating blood.

The other route of metabolism of PGH_2 is to prostacyclin, yet another unstable compound ($t_{1/2} = 3$ min at 37 °C and pH 7.5) formed by an enzyme, prostacyclin synthase, first discovered in vascular tissue. This enzyme has been purified using monoclonal antibodies and is a heme-containing protein of about 52 000 Da.[16] Prostacyclin production proceeds by polarization of the 9–11 endoperoxide and the formation of a 6,9-epoxy derivative, giving rise to a bicyclic structure (Figure 1). This is accompanied by loss of hydrogen from C-6. Prostacyclin is hydrolyzed nonenzymatically to a stable compound, 6-oxo-$PGF_{1\alpha}$ (6-keto-$PGF_{1\alpha}$). Prostacyclin synthase has been detected in many tissues other than the vasculature; these include fibroblasts, smooth muscle cells, macrophages, white cells, lung, heart, kidney, spleen, uterus and gut.[15]

6.2.2.2 Biological Properties of Cyclooxygenase Products

6.2.2.2.1 *The stable prostaglandins*

(i) *Biological actions*

The prostaglandins display a broad spectrum of biological actions, the more important of which are described below.

In most species and in most vascular beds, PGEs and PGAs are potent vasodilators. Responses to $PGF_{2\alpha}$ show species variation, but vasodilatation has been observed following injection into the human brachial artery of $PGF_{2\alpha}$ and PGs A_1, B_1, E_1 and E_2.[17] Dilatation in response to prostaglandins seemingly involves arterioles, precapillaries, sphincters and postcapillary venules. PGEs are not universally vasodilatory; constrictor effects have been noted at selected sites. Superficial veins of the hand are contracted by $PGF_{2\alpha}$, but not by PGEs. The behaviour of other large capacitance veins in various animals is similar.[18]

Cardiac output is generally increased by PGs E, F and A. Weak, direct inotropic effects (increased cardiac output) have been noted in various isolated preparations. In the intact animal, however, increased force of contraction as well as increased heart rate is largely a reflex consequence of fall in total peripheral resistance. Systemic blood pressure generally falls in response to PGs E and A, and blood flow to most organs, including the heart and kidney, is increased. These effects are particularly striking in some patients with hypertensive disease.[19]

The prostaglandins exert powerful actions on platelets. Some of them, like PGE_1 and PGD_2, are inhibitors of the aggregation of human platelets *in vitro* at concentrations around 0.1 µM. PGE_2 exerts variable effects on platelets; it is a potentiator of some forms of aggregation at low concentrations (below 1 µM) and an inhibitor at higher concentrations.

PGA_2, PGE_1 and PGE_2 induce erythropoiesis by stimulating the release of erythropoietin from the renal cortex.[20] Moreover, PGE_1 and PGE_2 produce variable effects on the fragility of red cells; at very low concentrations (10–100 pM) they decrease fragility, while at higher concentrations (1 nM) they increase it.[21]

In general, PGFs contract and PGEs relax bronchial and tracheal muscle from various species, including man. Asthmatic individuals are particularly sensitive, and $PGF_{2\alpha}$ can cause intense bronchospasm. In contrast, both PGE_1 and PGE_2 are potent bronchodilators when given to such patients by aerosol.[22]

Strips of nonpregnant human uterus are contracted by PGFs but relaxed by PGs E, A, and B. The contractile response is most prominent before menstruation, whereas relaxation is greatest at mid-cycle. Uterine strips from pregnant women are uniformly contracted by PGFs and by low concentrations of PGE_2; high concentrations of PGE_2 produce relaxation. The intravenous infusion of PGE_2 or $PGF_{2\alpha}$ to pregnant human females produces a dose-dependent increase in the frequency and intensity of uterine contraction. Uterine responsiveness to prostaglandins increases somewhat as pregnancy progresses, but far less than does that to oxytocin. The roles of prostaglandins in reproductive processes have been reviewed.[23]

PGEs and PGAs inhibit gastric acid secretion stimulated by feeding, histamine or gastrin. Volume of secretion, acidity and content of pepsin are all reduced, probably by an action exerted directly on the secretory cells. In addition, these prostaglandins are vasodilators in the gastric mucosa. Mucus secretion in the stomach and small intestine is increased by prostaglandins, and there is substantial movement of water and electrolytes into the intestinal lumen. The role of prostaglandins as regulators of gastrointestinal function has been reviewed.[24]

Infusions of PGE_2 directly into the renal arteries of dogs increase renal blood flow and provoke diuresis, natriuresis and kaliuresis; there is little change in the rate of glomerular filtration unless renal vasoconstriction is present. PGEs inhibit water reabsorption induced by antidiuretic hormone in the toad bladder and in rabbit collecting tubules. PGE_2 also inhibits chloride reabsorption in the thick ascending limb of the loop of Henle in the rabbit. In addition, PGE_2 and PGD_2 can cause the release of renin from the renal cortex.[25]

(ii) Pathological implications

Prostaglandins such as PGE_2 have been associated with the pathogenesis of hypertension, since it induces vasodilatation by a direct action on vascular smooth muscle, reduces the action of vasoconstrictor substances such as angiotensin and noradrenaline, and releases renin from the kidney.[26] Moreover, it has been suggested that some of the antihypertensive effects of angiotensin-converting enzyme inhibitors may be due to an increase in kinins, leading to an increased release of vasodilator prostaglandins.[27]

Increased generation of prostaglandins has been reported in animals with spontaneous or experimentally induced hypertension. Whether this represents a defensive response against the ensuing increase in blood pressure or is part of its pathogenesis has not been elucidated. This fact, together with the conflicting reports on the effects of indomethacin and other cyclooxygenase inhibitors, which increase blood pressure in some animal models of hypertension but are inactive in others,[26] indicate that more work is needed before the precise role of vasodilator prostaglandins in the genesis and development of hypertension is finally clarified.

Inflammation is one of the few conditions in which PGE_2 is a major product of cyclooxygenase and high levels of PGE_2 have been detected in many human inflammatory diseases (Table 1).[28,29] It is the predominant eicosanoid detected in inflammatory conditions ranging from experimental acute edema and sunburn through to chronic arthritis in man. It is a potent dilator of vascular smooth muscle, accounting for the characteristic vasodilatation and erythema (redness) seen in acute inflammation. The effect of vasodilatation is to increase the flow of blood through inflamed tissues and this augments the extravasation of fluid (edema) caused by agents which increase vascular permeability, such as bradykinin and histamine. Prostaglandin E_2 also acts synergistically with other mediators to produce inflammatory pain. Without having any direct pain-producing activity, PGE_2 sensitizes receptors on afferent nerve endings to the actions of bradykinin and histamine. Thirdly, PGE_2 is a potent pyretic agent and its production in bacterial and viral infections contributes to the fever associated with these diseases. The pyrexia induced by endogenous pyrogen, now known to be interleukin-1, is reduced by cyclooxygenase inhibitors and is thought to be principally mediated by PGE_2.[30]

In addition to producing pain and swelling, PGE_2 has been implicated in the tissue destruction found in the joints in rheumatoid arthritis.[31] A number of *in vitro* studies have been used as the basis for the claim that PGE_2 will induce damage to cartilage and bone. There is, however, no clinical evidence that cyclooxygenase inhibitors, such as the nonsteroid anti-inflammatory drugs often used in the treatment of arthritis, inhibit the process of tissue destruction.[31]

Table 1 Inflammatory Diseases in Man in which Elevated Levels of Eicosanoids have been Detected in Inflamed Tissues[a]

Disease	Eicosanoids detected
Allergic contact eczema	PGE, PGF
Rheumatoid arthritis	PGE_2, PGE_1, $PGF_{2\alpha}$, $PGF_{1\alpha}$, TXB_2, 6-keto-$PGF_{1\alpha}$, LTB_4, LTC_4, 5-HETE
Osteoarthritis and other chronic inflammatory joint disorders	PGE_2, $PGF_{2\alpha}$, TXB_2, 6-keto-$PGF_{1\alpha}$, LTB_4
Psoriasis	PGE_2, $PGF_{2\alpha}$, 12-HETE, LTB_4
Gout	LTB_4
Inflammatory bowel disease	PGE_2, $PGF_{2\alpha}$, 6-keto-$PGF_{1\alpha}$, TXB_2, 12-HETE, 5-HETE, 15-HETE, LTB_4

[a] For reviews, see refs. 28 and 29.

Many other cyclooxygenase products have been detected in inflammatory lesions. These include $PGF_{2\alpha}$, PGD_2, prostacyclin and TXB_2, but usually they are present at less than a quarter of the concentrations of PGE_2 and it is possible that the process of inflammation actually directs the enzymic pathway towards the production of PGE_2.

6.2.2.2.2 *Prostacyclin*

(i) *Biological actions*

Prostacyclin is the main product of arachidonic acid in all vascular tissue so far tested, including those of man. The ability of the vessel wall to synthesize prostacyclin is greatest at the intimal surface and progressively decreases towards the adventitia. Cultures of cells from vessel walls show that endothelial cells have the greatest capacity to produce prostacyclin.[32]

Prostacylin relaxes isolated vascular strips and is a strong hypotensive agent through vasodilatation of all vascular beds studied, including the pulmonary and cerebral circulations.[33] It has been suggested that prostacyclin generation participates in, or accounts for, functional hyperemia.[34]

Prostacyclin is the most potent endogenous inhibitor of platelet aggregation yet discovered. This effect is short-lasting *in vivo*, disappearing within 30 minutes of cessation of intravenous administration. Prostacyclin disperses platelet aggregates *in vitro* and in the circulation of man. Moreover, it inhibits thrombus formation in arterioles of the hamster cheek pouch, the carotid artery of the rabbit and the coronary artery of the dog, protects against sudden death (thought to be due to platelet clumping) induced by intravenous arachidonic acid in rabbits, and inhibits platelet aggregation in pial venules of the mouse when applied locally.[32]

Prostacyclin inhibits platelet aggregation by stimulating adenylate cyclase, leading to an increase in cAMP level in the platelets. In this respect, prostacyclin is much more potent than either PGE_1 or PGD_2 and its effect is longer lasting. Increased cAMP in platelets leads to enhancement of Ca^{2+} sequestration in membranes, together with inhibition of phospholipase and cyclooxygenase. Thus, by inhibiting several steps in the activation of the arachidonic acid metabolic cascade, prostacyclin exerts an overall control of platelet aggregability.[32]

Prostacyclin increases cAMP levels in cells other than platelets,[35] raising the possibility that, in these cells also, a balance with the thromboxane system exerts a similar homeostatic control of cell behaviour. Thus, the prostacyclin/TXA_2 system may have wider biological significance in cell regulation. An example is that prostacyclin inhibits white cell adherence to the vessel wall,[36] to nylon fibres and to endothelial monolayers *in vitro*.[37] Prostacyclin increases cAMP in the endothelial cell itself, suggesting a possible negative feedback control for prostacyclin production by the endothelium.[38]

Prostacyclin inhibits platelet aggregation (platelet–platelet interaction) at much lower concentrations than those needed to inhibit adhesion (platelet–collagen interaction) (Figure 2).[39] Thus, prostacyclin may permit platelets to stick to damaged vascular tissue and to interact with it, so allowing them to participate in the repair process, while at the same time preventing or limiting thrombus formation. In addition, platelets adhering to a site where prostacyclin synthase is present could well feed the enzyme with endoperoxide, thereby producing prostacyclin and preventing other platelets from clumping on to the adhering platelets, limiting the cells to a monolayer.

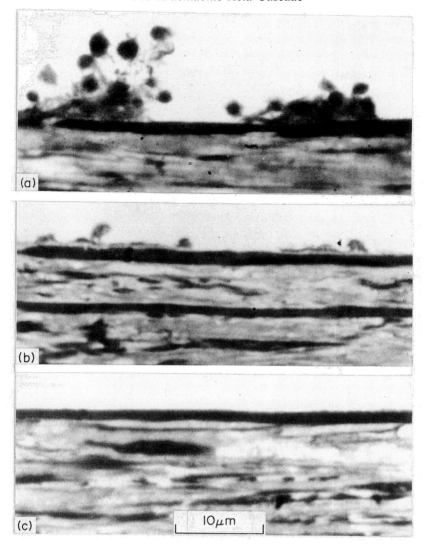

Figure 2 Light micrograph showing the effect of prostacyclin on platelet aggregation and adhesion to the endothelium-denuded rabbit aorta: (a) when the subendothelium is perfused with human blood, platelet thrombi are visible; (b) when prostacyclin (2 ng mL^{-1}) is added to the blood, adhesions but no thrombi are visible; (c) treatment of the blood with prostacyclin (100 ng mL^{-1}) prevents both adhesion and aggregation of platelets[39]

Prostacyclin interacts with endothelium-derived relaxing factor (EDRF), the labile humoral agent released by vascular endothelium which is responsible for the vascular relaxant properties of some vasodilators.[40] Endothelium-derived relaxing factor, the chemical nature of which has now been identified as nitric oxide (NO), is also a potent inhibitor of platelet aggregation and adhesion.

The inhibitory action of EDRF on platelet aggregation can be clearly differentiated from that of prostacyclin in that it is mediated *via* stimulation of guanylate cyclase. The anti-aggregating activity of both EDRF and authentic NO have been shown to be potentiated by subthreshold concentrations of prostacyclin.[41] Similarly, subthreshold concentrations of NO and EDRF potentiate the anti-aggregating activity of prostacyclin. Synergy also occurs between prostacyclin, EDRF and NO on platelet disaggregation. In addition, prostacyclin and NO released from vascular endothelial cells by bradykinin synergize with each other to inhibit platelet aggregation. This anti-aggregating activity is abolished by treatment of the cells with indomethacin and partially reversed by treatment with hemoglobin, a known inhibitor of NO.[41]

These findings suggest that prostacyclin and NO may regulate platelet–vessel wall interactions *in vivo*, at far smaller concentrations than those detectable by nonbiological means. Prostacyclin may indeed have a physiological homeostatic role in controlling platelet aggregability if it acts on a background of NO close to the endothelial surface. Unlike platelet aggregation and disaggregation, potentiation between NO and prostacyclin as inhibitors of adhesion was not observed.

In addition to its well-known vasodilator and anti-aggregating actions, prostacyclin shares with other prostaglandins a 'cytoprotective activity'. This term arose from studies, mainly in the rat, of the effects of prostaglandins on the gastric mucosa subjected to a chemical or physical trauma.[24] Prostacyclin also shares this activity and it is suggested that this other property of prostacyclin may be important in explaining certain of its therapeutic effects. For instance, in some models of myocardial infarction, prostacyclin reduces infarct size and arrhythmias and also decreases oxygen demand and the release of enzymes from infarcted areas. In sheep, prostacyclin protects the lungs against injury induced by endotoxin. A beneficial effect of prostacyclin has also been reported in endotoxin shock in the dog and in the cat, where it improves splanchnic blood flow and reduces the formation and release of lysosomal hydrolases (cathepsin D). The effects of hypoxic damage in the cat isolated perfused.liver are also substantially reduced by prostacyclin. Canine livers can be preserved *ex vivo* for up to 48 h and then successfully transplanted, using a combination of refrigeration, Sacks' solution and prostacyclin.[32]

All of these effects could be related to the cytoprotective activity, another example of which occurs in platelets when addition of prostacyclin during their separation from blood and subsequent washing gives rise to a substantial improvement in their subsequent functionality *in vitro*. In addition, the *in vitro* survival time of platelets prepared with the addition of prostacyclin is increased from 4 to 8 h so that they remain functional for more than 72 h.[42] This effect is not accompanied by a prolonged increase in cAMP level in platelets, thus separating it from the classical anti-aggregating effect.[43] Interestingly, a prostacyclin analogue shows a dissociation between anti-aggregating and cytoprotective effects in a model of acute myocardial ischemia.[44] All these results suggest that some of the therapeutic effects of prostacyclin might be related to the cytoprotective property and point to even wider indications for prostacyclin in cell or tissue preservation *in vivo* and *in vitro*.

High doses of prostacyclin infused intravenously in dogs induce significant fibrinolytic activity in plasma.[45] Studies in man suggest that the induction of fibrinolytic activity is short lasting, reaching a peak three hours after the start of the infusion and disappearing at 24 h in spite of continuation of the infusion.[46] The underlying mechanism of this fibrinolytic effect is not yet fully understood but experiments *in vitro*[47] suggest that prostacyclin and PGE_1 induce the release of a plasminogen activator, probably through a cAMP-mediated effect.

Prostacyclin has recently been shown to inhibit the release of mitogenic activity ('growth factors') from stimulated human platelets.[48] Such growth factors are thought to mediate the progression of atherosclerosis by promoting smooth muscle cell and fibroblast proliferation.

(ii) Pathological implications

A number of diseases are associated with an imbalance in the prostacyclin/TXA_2 system.[49, 50] A decrease in prostacyclin formation by atherosclerotic vascular tissue has been demonstrated both in experimental animals and in man. Moreover, smooth muscle cells obtained from atherosclerotic lesions and cultured *in vitro* consistently produce less prostacyclin than normal vascular smooth muscle cells. These findings probably reflect the increased generation by atherosclerotic vessels of lipoxygenase products, such as 15-HETE, which are selective inhibitors of prostacyclin formation.[51]

Platelets from rabbits made atherosclerotic by dietary manipulation, and from patients who have survived myocardial infarction, are abnormally sensitive to aggregating agents and produce more TXA_2 than controls. Interestingly, cholesterol stimulates TXA_2 production by platelets *in vitro*. Arteries from rabbits made atherosclerotic by cholesterol feeding, as well as some atheromatous coronary arteries from autopsy cases, produce more TXB_2 than 6-oxo-$PGF_{1\alpha}$, suggesting that vascular production of TXA_2 may be a contributory factor in atherosclerosis.[50]

Prostaglandin production by vascular tissue and platelets is also altered in diabetes. Vascular prostacyclin production is depressed in diabetic animals and man, an effect reversed by insulin or islet tissue transplantation. Thrombotic thrombocytopenic purpura (TTP), like diabetes, is associated with formation of microvascular thromboemboli, and a deficiency in prostacyclin production may contribute to the increased platelet consumption that occurs in TTP.

Prostacyclin production is significantly lower in umbilical and placental vessels from pre-eclamptic patients than in those from normally pregnant women and thromboxane levels are elevated. A reduced vascular capacity for prostacyclin production has also been observed in patients with systemic lupus erythematosus.[52] Patients with chronic glomerular disease excrete less urinary 6-oxo-$PGF_{1\alpha}$ than do healthy control subjects.[53] Cigarette smoking is a well-recognized risk factor in atherosclerosis and is known to increase platelet reactivity and to reduce both basal and stimulated levels of urinary 6-oxo-$PGF_{1\alpha}$. Prostacyclin formation by kidney microsomes and rabbit

Table 2 Clinical Conditions in which Prostacyclin has been Used

Extra-corporeal circulation: renal dialysis, heart/lung bypass
Pulmonary hypertension
Septic shock
Neonatal post-operative hypertension
Pre-eclampsia
Transplant organ preservation
Acute coronary artery re-occlusion
Adult respiratory distress syndrome

isolated hearts is reduced by nicotine. Umbilical arteries from mothers who smoke produce significantly less prostacyclin than those from nonsmoking control subjects.[50]

Continuously high estrogen levels have also been associated with cardiovascular disease. Estrogen administration has been shown to reduce prostacyclin production in the uterus and in the vasculature of experimental animals. In addition, spontaneous *ex vivo* platelet aggregation occurs in oral contraceptive users but not in female nonusers or men.[50]

Prostacyclin is available as a stable freeze-dried preparation (Epoprostenol; Flolan) for administration to man in a number of cardiothrombotic indications (Table 2). The use of prostacyclin in several other clinical conditions is now being investigated. In addition to its use in extracorporeal circulation systems such as cardiopulmonary bypass operations, renal dialysis and charcoal hemoperfusion, prostacyclin has been shown to be effective in the treatment of atherosclerotic peripheral vascular disease and Raynaud's syndrome.[54] It is also being tested for the treatment of other thrombotic conditions such as stroke and myocardial injury in patients with acute myocardial infarction.[55] Prostacyclin has been used in patients with pulmonary hypertension and has been found to have pulmonary hemodynamic effects similar to those of hydralazine[56] and nifedipine.[57] Prostacyclin may have a therapeutic use in testing pulmonary vasoreactivity in such patients and, because of its prompt, brief action, may provide greater patient safety. Extensive reviews of the clinical applications of prostacyclin are available.[54]

What is becoming clear is that the efficacy of prostacyclin, particularly its long-lasting effects in certain clinical conditions like peripheral vascular disease and Raynaud's syndrome, cannot be explained solely in terms of its short-lasting vasodilator and anti-aggregatory actions. Consequently, interest is being focused on the other activities of prostacyclin such as cytoprotection, fibrinolysis and stimulation of cholesterol metabolism.

6.2.2.2.3 Thromboxane A_2

(i) Biological actions

Thromboxane A_2, a powerful vasoconstrictor and promoter of platelet aggregation,[4] is released from platelets during aggregation and from guinea-pig lungs perfused with arachidonic acid. Its stable degradation product, TXB_2, has considerably diminished biological activity. TXA_2 was first observed in 1969 by Piper and Vane as a labile substance found in the effluent of guinea-pig isolated perfused lungs during anaphylactic challenge.[58] Because of its biological action it was originally known as 'rabbit aorta contracting substance'.

TXA_2 also contracts rabbit coeliac and mesenteric artery, human umbilical artery, guinea-pig trachea, bovine and pig coronary artery and lamb ductus arteriosus and has a coronary vasospastic action in the guinea-pig isolated heart. In addition, it contracts helically cut strips prepared from bovine cerebral conductance arteries and isolated segments of human basilar artery. TXA_2 has been shown to cause an immediate and short-lasting vasoconstriction in the perfused canine mesenteric and femoral vascular beds. TXA_2 is a potent stimulator of platelet aggregation. In contrast to prostacyclin, TXA_2 and PGH_2 can reduce platelet intracellular cAMP levels but only when these levels are already elevated by an adenylate cyclase stimulant such as PGE_1 or prostacyclin. The induction of platelet aggregation by these compounds may not necessarily be directly linked to inhibition of adenylate cyclase, since basal levels of cAMP are not lowered by TXA_2 or PGH_2. Furthermore, an adenylate cyclase inhibitor (SQ22536), which can lower basal concentrations of cAMP as well as reduce those stimulated by PGE_1, fails to induce platelet aggregation. However, the ability of TXA_2 to modulate the anti-aggregating actions of prostacyclin by attenuating its actions on intracellular cAMP levels may be an important interactive mechanism between these two agents under physiological and pathophysiological conditions. Prostacyclin and TXA_2 also have opposing effects on cAMP levels in B16 melanoma cells.

In addition to their opposing effects on platelet function and the cardiovascular system, prostacyclin and thromboxane A_2 have counteracting effects on the bronchopulmonary system and the stomach. In the gastric mucosa, TXA_2 is a potent ulcerogen whereas prostacyclin can inhibit gastric damage. Likewise, in myocardial and hepatic tissue, TXA_2 can exert a cytolytic action, whereas prostacyclin can protect these tissues from damage, perhaps reflecting again the opposite poles of the same biological mechanisms, being in this case cellular integrity. The possible interactions between TXA_2 and prostacyclin in the modulation of smooth muscle tone and motility in the gastrointestinal and reproductive tracts are less clear, since in general both prostanoids are only weak spasmogens on non-vascular tissue. Reviews of the biological properties of TXA_2 have been published.[15,59]

(ii) Pathological implications

Elevated TXA_2 production has been demonstrated in a number of diseases in which there is a tendency for thrombosis to develop.[15,49,50] Platelets from patients with arterial thrombosis, deep-vein thrombosis, or recurrent venous thrombosis produce more prostaglandin endoperoxides and TXA_2 than normal and have a shortened survival time.[50]

The platelets of both diabetic animals and patients produce more TXB_2 than platelets from normal subjects, which is attributed to increased release of platelet arachidonate.[60] An elevated number of circulating platelet aggregates and greater platelet hyperaggregability have been reported in diabetic patients of all types, from preclinical to those with vascular complications.[61] Platelet hyperactivity and increased synthesis of prostaglandin endoperoxides have been demonstrated in the infants of diabetic mothers, suggesting an underlying genetic predisposition.[62] Intravascular platelet aggregation, elevated amounts of the vasoconstrictor TXA_2, and reduced prostacyclin levels all may contribute to the increased incidence and severity of retinopathy and coronary heart disease associated with diabetes. The strong relationship between elevated TXA_2 and decreased prostacyclin synthesis has been demonstrated by the fact that treatment of diabetic animals with either a thromboxane synthase inhibitor or with a compound which promotes prostacyclin formation restored both thromboxane and prostacyclin synthesis to normal.[63] Increased urinary excretion of TXB_2 and decreased excretion of PGE_2 have been described in rats in which diabetes was induced with streptozotocin.[64]

Platelets from patients who have survived myocardial infarction are abnormally sensitive to aggregating agents, produce more TXA_2 and have a shorter bleeding time than control subjects.[65] Elevated TXB_2 levels have been demonstrated in the blood of patients with Prinzmetal's angina.[66] Studies of TXB_2 levels in coronary sinus blood of patients with unstable angina showed that local TXA_2 release is associated with recent episodes of angina, but the authors were unable to distinguish whether the release was cause or effect.[67] Elevated thromboxane levels have also been described in septic shock, incipient renal graft rejection, and vasoconstriction of the hydronephrotic kidney.[50]

6.2.3 LIPOXYGENASE PATHWAY OF ARACHIDONIC ACID METABOLISM

6.2.3.1 Biosynthesis of Lipoxygenase Products

Arachidonic acid liberated from membrane phospholipids may also be metabolized by a group of enzymes known as lipoxygenases. These enzymes are common in plants but were not detected in mammalian tissues until the discovery of 12-lipoxygenase in platelets[68] and 5-lipoxygenase in leukocytes.[69] Lipoxygenase activity has been found in the cytoplasmic and particulate fractions of platelets from a number of species, including man.

Lipoxygenase enzymes catalyze the peroxidation of fatty acids but, unlike the reaction catalyzed by cyclooxygenase, ring formation does not follow this peroxidation step. Instead, this reaction gives rise to straight-chain hydroperoxy acids (HPETEs) which can then be converted to hydroxy acids (HETEs) (Figure 3).

In the absence of cyclooxygenase, peroxidation at C-11 or C-15 leads to the production of 11-HETE or 15-HETE, but in 1974 a separate lipoxygenase pathway was discovered in platelets which produces 12-HETE.[68] Most is known about the soluble lipoxygenase which resolves on gel filtration into two components of 100 000 and 160 000 Da.[70] The smaller molecule produces mainly 12-HPETE whilst the larger molecule produces both 12-HPETE and 12-HETE. This has prompted the suggestion that the larger component may represent a complex between a 12-lipoxygenase and a

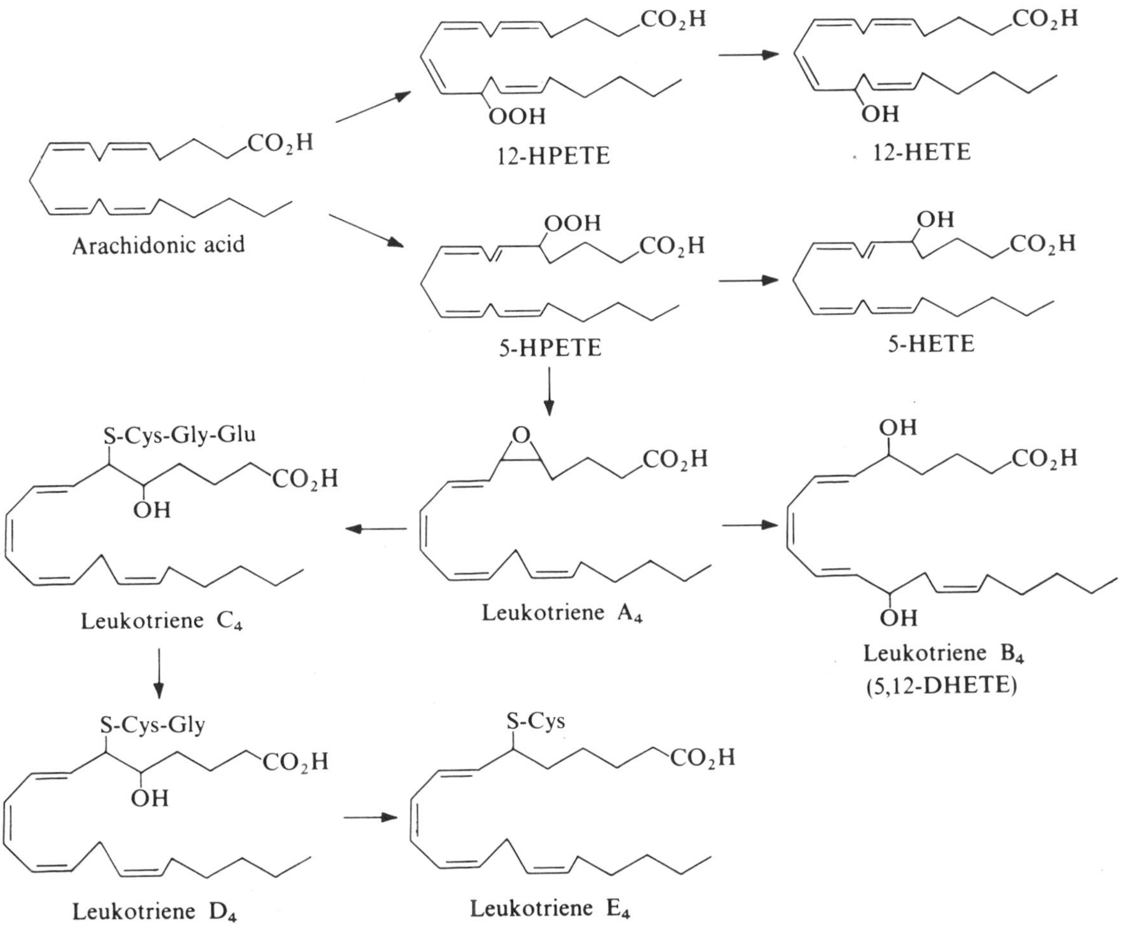

Figure 3 The 12- and 5-lipoxygenase pathways of arachidonic acid metabolism

peroxidase. Platelet lipoxygenase has a higher affinity for arachidonic acid and other 20-carbon unsaturated fatty acids than for linoleic acid, the preferred substrate of plant enzymes. Lipoxygenases catalyze the abstraction of a hydrogen radical from the methylene group of a pentadiene in the fatty acid chain and this step resembles the initiation of the cyclooxygenase pathway. The activated pentadiene is then vulnerable to oxygen attack across either of the double bonds. In common with plant enzymes, mammalian lipoxygenases require iron.[71] Although the nature of the peroxidase is not well characterized, there is evidence that reduced glutathione is an important cofactor.[72]

In 1976, polymorphonuclear leukocytes (PMNs) were found to contain a 5-lipoxygenase.[69] As well as producing 5-HETE, the leukocyte 5-lipoxygenase initiates the formation of a family of compounds which have in common a conjugated triene structure and, because of their source, are called the leukotrienes.[73] 5-HPETE is converted to the 5,6-epoxide of arachidonic acid (leukotriene A_4; LTA$_4$; Figure 3) which can be converted to the 5,12-dihydroxy acid, LTB$_4$. Alternatively, the addition of glutathione to the epoxide by glutathione-S-transferase results in the formation of LTC$_4$. The removal of glutamate from LTC$_4$ by α-glutamyl transpeptidase gives LTD$_4$, which is further metabolized to LTE$_4$ with the loss of glycine. The addition of glutamate to LTE$_4$, giving cysteine-glutamate at C-6, results in LTF$_4$.[74]

The leukotrienes do not contain a cyclopentane ring or an oxane ring and so they retain the same number of double bonds as their precursor fatty acids. This explains the numerical suffix given to the leukotrienes, so that LTA$_3$ is derived from eicosatrienoic acid, LTA$_4$ from arachidonic acid (eicosatetraenoic acid) and LTA$_5$ from eicosapentaenoic acid.

The substrate specificity of 5-lipoxygenase is different from the platelet enzyme, requiring double bonds at the 5, 8 and 11 positions and not being so dependent upon chain length.[75] Also, the 5-lipoxygenase differs from the 12-lipoxygenase in its calcium dependence. There is some evidence that calcium interacts with inactive monomers of the 5-lipoxygenase (M 90 000 Da) to give an activated dimer (M 180 000 Da).[76] The conversion of LTA$_4$ to LTB$_4$ is enzymic and an epoxide hydrolase has been identified in the cytoplasm.[77] This is in contrast to the enzymes which convert LTA_4 to the peptido-leukotrienes, glutathione-S-transferase and α-glutamyl transpeptidase, which are particulate.[78]

In addition to platelets and leukocytes, lipoxygenase activity has been detected in lungs, skin, ocular tissue and blood vessels.

6.2.3.2 Biological Properties of Lipoxygenase Products and their Role in Disease

6.2.3.2.1 *Lipoxygenase products in diseases of the airway*

In 1938, after challenging perfused lungs *in vitro* with snake venom, Feldberg and Kellaway observed that there was a release of a biologically active principle that induced a slow, long-lasting contraction of smooth muscle.[79] This activity was named 'slow reacting substance of anaphylaxis' (SRS-A) and became a focus of research interest as it was believed to be an important mediator of human asthma. SRS-A is released concomitantly with prostaglandins and TXA_2 during anaphylaxis.[58] Moreover, experiments showing that SRS-A release is enhanced in the presence of inhibitors of prostaglandin synthesis prompted the suggestion that prostaglandins and SRS-A may be formed from the same precursor.[80] This was shown to be the case, with identification of a peptido-leukotriene that was generated from arachidonic acid by 5-lipoxygenase and that possessed slow-reacting properties on bronchial smooth muscle. SRS-A is now known to be a mixture of leukotrienes C_4, D_4 and E_4.[73]

The peptido-leukotrienes are two to three orders of magnitude more potent than histamine as bronchoconstrictors; however, there is considerable interspecies variation, with human and guinea-pig respiratory smooth muscle being more sensitive to the leukotrienes than rat, cat or dog airways.[81] Leukotrienes contract isolated preparations of trachea, bronchial and parenchymal smooth muscle, but experiments *in vivo* indicate that they have a selective action on small airways. LTC_4 and LTD_4 induce a preferential reduction in lung compliance and are relatively less effective in reducing specific airway conductance.[82]

When administered to human volunteers, LTC_4 and LTD_4 caused coughing, bronchoconstriction, wheezing, tightness of the chest and a reduction in expiratory maximum airflow rate.[83] Furthermore, peptido-leukotrienes have been detected in the sputum from asthmatics[84] and in nasal washes from allergic patients following antigen challenge.[85] These findings have strengthened the opinion that leukotrienes are important mediators of respiratory pathology.

6.2.3.2.2 *Lipoxygenase products in inflammatory diseases*

Lipoxygenase activity can be stimulated by chemical, mechanical or immunological challenge and is elevated in inflammatory responses in experimental animals and man. The first indication that lipoxygenase activation occurs in inflammation came from the observation that 12-HETE was present in the involved epidermis of patients with psoriasis.[86] LTB_4 has now been detected in fluid from involved tissue in rheumatoid arthritis, gout, psoriasis and ulcerative colitis. There is also one report of LTC_4 being present in synovial fluids from arthritic patients.[28, 87]

Some experimental evidence suggests that leukotrienes may contribute to the vascular changes in acute inflammation. In combination with inflammatory mediators such as bradykinin or prostaglandin E_2, LTB_4 increases vascular permeability in rabbit skin (Figure 4);[88] furthermore, this effect is dependent upon circulating neutrophils. LTC_4 and LTD_4 are vasoconstrictors, but they also cause plasma leakage in guinea-pig and rat skin; they have little activity in rabbit skin. In the microcirculation of the hamster cheek pouch, LTC_4 and LTD_4 are approximately 1000 times more potent than histamine in inducing macromolecular leakage from postcapillary venules. However, unlike LTB_4, this activity is not dependent on circulating neutrophils. In human skin, LTB_4, LTC_4 and LTD_4 cause transient wheal and flare responses either by a direct action or through the release of endogenous mediators.[28, 87]

There are conflicting reports on the role of leukotrienes in inflammatory pain. Injections of LTB_4 into rat paws induced a prolonged reduction in the nociceptive pressure threshold and LTB_4 was approximately equipotent with bradykinin in this activity. In another model, though, LTB_4, LTC_4 and LTD_4 antagonized bradykinin-induced algesia in the rabbit ear. High doses of LTB_4 and LTC_4 also inhibited intradental nerve excitability.[87]

It is likely, however, that the major contribution of LTB_4 and 12-HETE to inflammation is through an effect on leukocytes. Turner *et al.*[89] found that platelet lipoxygenase products were

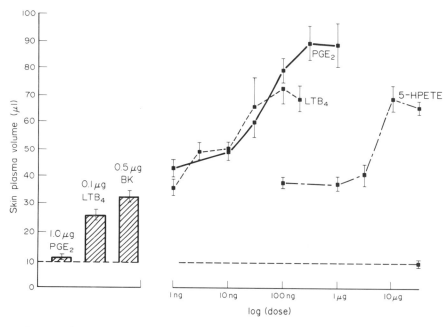

Figure 4 Synergy between eicosanoids and bradykinin in producing plasma leakage in rabbit skin. The hatched columns show the effects of intradermal injections of prostaglandin E_2 (PGE$_2$), leukotriene B_4 (LTB$_4$) or bradykinin (BK) on plasma leakage in rabbit skin. The dose response curves show the effects of combining increasing doses of PGE$_2$, LTB$_4$ or the leukotriene precursor 5-HPETE with 0.5 μg BK. Data from ref. 88

Table 3 The Chemokinetic Activity of Arachidonate Lipoxygenase Products for Human Neutrophils *in vitro*[91]

Lipoxygenase product	Concentration for maximal chemokinetic activity ($\mu g\ mL^{-1}$)
LTB$_4$ (5,12-di-HETE)	0.001
5-HPETE	0.32
5-HETE	0.32
11-HPETE	1.0–3.2
11-HETE	3.2–10.0
12-HPETE	1.0
12-HETE	3.2
15-HPETE	> 10.0
15-HETE	> 10.0

chemotactic for PMNs. The leukocyte product 5-HETE is more potent than the platelet product 12-HETE, and the 5,12-dihydroxy acid LTB$_4$ is even more potent (Table 3).[90,91] It seems, therefore, that 5-lipoxygenase activity in migrating leukocytes represents a local control mechanism to amplify the recruitment of inflammatory cells to damaged tissues.

6.2.3.2.3 Lipoxygenase products and the vasculature

The presence of a 5-lipoxygenase system, and the production of leukotriene-like material, has been demonstrated in vascular tissue.[92] Leukotrienes C_4 and D_4 are powerful vasoconstrictors in the coronary and other vascular beds in several species, as well as in the microvasculature, and it has been proposed that they may contribute to myocardial ischemia and angina.[93] It is interesting that the production of lipoxygenase products is greatest in the adventitia and decreases towards the intima,[92] following a reverse pattern of production to that of prostacyclin. The interaction between products of the two eicosanoid pathways in the vessel wall has yet to be investigated; however, it is possible that generation *in vivo* of lipoxygenase products in the vascular wall may contribute to pathological changes in blood vessel tone. Indeed, lipoxygenase products, such as 15-HPETE, are inhibitors of prostacyclin formation.[51]

6.2.4 OTHER PATHWAYS OF ARACHIDONIC ACID METABOLISM

A new series of arachidonic acid products has recently been discovered in human neutrophils.[94] These compounds, which contain a conjugated tetraene structure, arise from the interactions of 5- and 15-lipoxygenases and other reactions and have been given the name lipoxins. Formation of lipoxins has also been demonstrated in eosinophil-rich granulocyte suspensions from blood of patients with hypereosinophilia syndrome.[95]

Lipoxins possess biological activity.[96] One of the series, lipoxin A, is a potent releaser of lysosomal enzymes from human neutrophils. This action is accompanied by superoxide anion generation but not by neutrophil aggregation. Lipoxin A also causes chemotaxis, contracts guinea-pig lung strips and induces arteriolar dilatation. In addition, lipoxin A is a potent activator of protein kinase C, suggesting a possible role as a regulator of specific cellular responses. The biological significance of this group of arachidonic acid metabolites is, however, unclear as their formation *in vivo* has yet to be demonstrated.

An additional pathway of arachidonic acid metabolism has been demonstrated in the kidney and in rat pituitary microsomes where epoxyeicosatrienoic acids (EETs) are formed *via* an NADPH-dependent cytochrome P-450 enzyme.[97] These epoxides have been shown to stimulate release of prolactin,[98] oxytocin and arginine vasopression[99] from rat pituitary cells, to stimulate the release of somatostatin from microsomal fractions from the rat hypothalamus,[100] and to stimulate glucagon and insulin release from rat isolated pancreatic islets.[101] They also inhibit vasopressin-stimulated osmotic water flow in the toad bladder.[102] 5,6-EET, but not 8,9-, 11,12- or 14,15-EET, also reduces vascular resistance in the rat tail artery.[103]

As with the lipoxins, the biological significance of the epoxy metabolites of arachidonic acid will not be apparent until they have been demonstrated *in vivo*.

6.2.5 INHIBITORS OF ARACHIDONIC ACID METABOLISM

6.2.5.1 Aspirin-like Drugs

6.2.5.1.1 *Anti-inflammatory actions*

A key step in the realization of the importance of arachidonic acid metabolism was the discovery in 1971 that the nonsteroid anti-inflammatory drugs, of which aspirin is the prototype, selectively prevent the biosynthesis of prostaglandins.[104-106]

Inhibition of prostaglandin synthesis by aspirin-like anti-inflammatory drugs has been demonstrated in a wide variety of cell types and tissues, ranging from whole animals and man to microsomal enzyme preparations. Cyclooxygenase is inhibited by preventing the abstraction of hydrogen from C-13 and, therefore, blocking peroxidation at C-11 and C-15. This action is highly specific, for similar abstraction and peroxidation reactions at other points in the fatty acid molecule are not inhibited. Furthermore, these drugs do not prevent the generation of prostaglandins from cyclic endoperoxides.

Within two years of the discovery that nonsteroid anti-inflammatory drugs inhibit prostaglandin synthesis, several classes of inhibitors had been identified[107] and this list has now been augmented by a remarkable variety of compounds shown selectively to inhibit cyclooxygenase. This diverse group of chemicals has been broadly classified to exhibit three types of inhibition: reversible competitive, irreversible and reversible noncompetitive.[108] Examples of reversible competitive inhibitors are fatty acids, closely related to the substrate, which have a comparable affinity for the enzyme but are not converted to oxygenated products. The anti-inflammatory drug ibuprofen (Figure 5) has a binding affinity for cyclooxygenase similar to that of the substrate arachidonic acid and this explains why ibuprofen inhibits the enzyme.[109] Aspirin (Figure 5) itself is an irreversible inhibitor which covalently acetylates cyclooxygenase, probably at a lysine residue in the active site of the enzyme.[110] This mechanism does not, however, explain the action of other inhibitors such as indomethacin and flurbiprofen (Figure 5). The time-dependent inactivation of cyclooxygenase by these inhibitors is associated with their carboxylic acid groups, and the importance of an aryl halogen has also been recognized.[110]

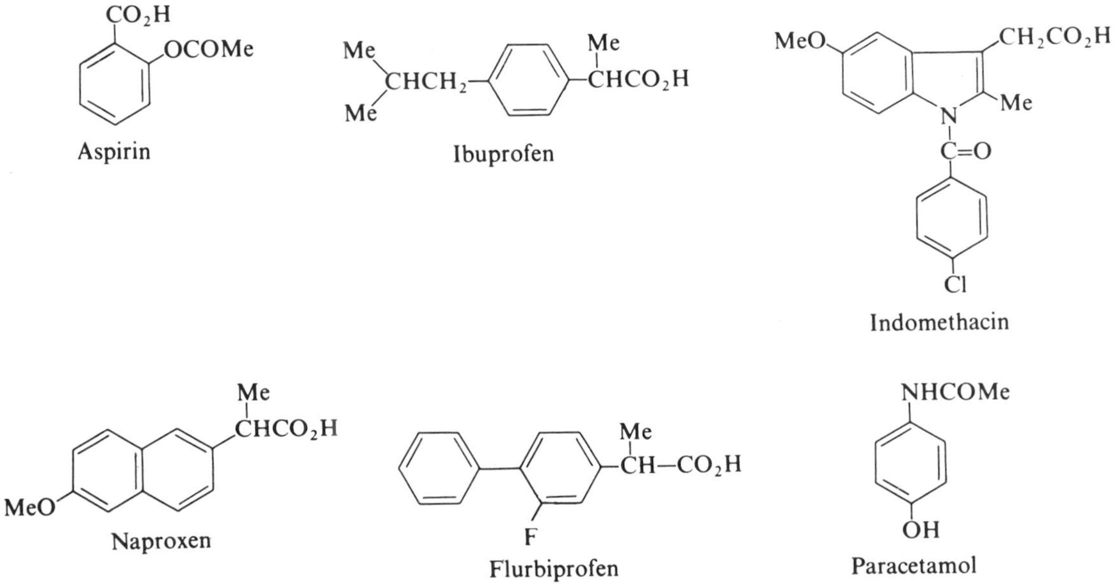

Figure 5 Drugs which selectively inhibit cyclooxygenase

The reversible noncompetitive inhibitors have antioxidant or radical trapping properties. It has been proposed that cyclooxygenase activity is sustained by a continual presence of lipid peroxide that induces a free-radical chain reaction; this is blocked by the addition of radical scavengers or antioxidants. Paracetamol (Figure 5) probably works by this mechanism.[111]

In man, aspirin blocks cyclooxygenase activity in platelets within an hour of oral administration,[105] and this observation has been confirmed in several species. Because aspirin irreversibly acetylates the enzyme and platelets are unable to generate new enzyme, inhibition of platelet cyclooxygenase lasts for the lifetime of the cell. This results in effects on platelet function for several days after a single dose of aspirin.

Aspirin is significantly more potent in inhibiting cyclooxygenase *in vitro* than salicylate,[104] but the anti-inflammatory potency of the two drugs is similar, as is their ability to reduce urinary output of prostaglandin metabolites.[112] This can be explained by aspirin acting as a pro-drug for salicylate, which is, after all, why aspirin was originally synthesized. Aspirin acts directly on blood cells such as platelets and leukocytes but has little effect on peripheral tissues as most of the drug is hydrolyzed to salicylate before leaving the hepatic portal circulation.[113] Following large oral doses of aspirin, the

Table 4 Rank Order of Potency of Aspirin-like Drugs in Inhibiting Edema and Prostaglandin E_2 Synthesis in an *in vivo* Model of Experimental Inflammation[a]

| | Relative potency in vivo | |
| | *Inhibition of* | |
Nonsteroid anti-inflammatory drug	*PGE_2 synthesis*	*Inhibition of edema*
Sodium salicylate	0.02	0.10
Aspirin	0.07	0.11
Fenclofenac	0.05	0.22
Alclofenac	0.35	0.40
Benoxaprofen	1.2	0.40
Phenylbutazone	0.39	0.73
Ibuprofen	0.20	0.85
Naproxen	1.0	1.0
Sulindac	0.7	4.8
Diclofenac	5.5	5.4
Indomethacin	7.3	5.4
Ketoprofen	64.7	8.5
Flurbiprofen	34.4	25.9

[a] Each value is relative to Naproxen which has been arbitrarily set at 1.0. For methodology see ref. 115.

drug can be detected in the plasma of peripheral blood and even in the peripheral tissues, but aspirin concentrations are rapidly exceeded (50–100-fold) by salicylate concentrations.[114] After 1–2 hours, aspirin is undetectable in the periphery but salicylate persists for more than 6 hours. During this time the synthesis of prostaglandins at a peripheral site of inflammation is reduced and this correlates with anti-inflammatory activity. It is likely, therefore, that the anti-inflammatory activity of orally dosed aspirin or salicylate is due to the inhibition by salicylate of prostaglandin production in the inflamed tissues.[114]

There is a good correlation between the relative potencies of aspirin-like drugs in reducing prostaglandin concentrations in inflammatory exudates and their inhibition of inflammatory edema (Table 4).[115] Furthermore, in groups of patients with arthritis receiving aspirin-like drugs, the mean concentration of cyclooxygenase products in synovial fluids is only one tenth of the concentrations in fluids from untreated patients[116] and this is associated with symptomatic relief.

6.2.5.1.2 Antithrombotic actions

Arachidonic acid metabolism gives rise to prostacyclin and TXA_2, which have opposing effects on blood vessels and platelets. The balance between these products is altered in a number of pathological conditions of the vasculature, suggesting that compounds that can modify their synthesis may have an antithrombotic potential.

Inhibition of TXA_2 synthesis may be achieved by inhibiting the cyclooxygenase enzyme with aspirin-like drugs. However, because aspirin prevents the production of the endoperoxide intermediates from which all the cyclooxygenase products of arachidonic acid are formed, it can prevent not only production of TXA_2 in platelets, but also production of prostacyclin in vascular tissue. Certain factors, however, indicate that aspirin may inhibit TXA_2 production at doses that do not impair prostacyclin production. First, inhibition of the vascular cyclooxygenase, unlike the platelet cyclooxygenase which is inhibited by aspirin for the entire lifespan of the platelet, persists for a much shorter period because of the generation of new enzyme. Second, the platelet cyclooxygenase appears to be more sensitive *in vitro* and *in vivo* than the vascular cyclooxygenase to the inhibitory action of aspirin.[117] Third, studies using a deuterated aspirin analogue have shown that platelets passing through the gut capillaries while an oral dose of aspirin is undergoing presystemic hydrolysis can be exposed to higher concentrations of aspirin than platelets present in the peripheral circulation.[113] Thus, serum thromboxane levels are significantly reduced by orally administered aspirin prior to the drug's appearance in the systemic circulation where it can inhibit prostacyclin synthesis by systemic vascular endothelium.

During the past few years, attempts have been made to titrate aspirin dosage down to find one that will achieve inhibition of TXA_2 formation in man without affecting prostacyclin production. This has been done by using doses as low as 20 mg day^{-1}, or by 'slow release' administration of very low doses. Recent reports suggest that this may be difficult to achieve in practice, although the hypothesis is still being tested.[50]

A number of controlled long-term clinical trials in patients with transient ischemic attacks (TIA) have shown aspirin to reduce the risk of further TIA, cerebral infarction and mortality.[118,119] Low doses of aspirin have been effective in preventing thrombosis in patients undergoing long-term hemodialysis[120] or aortocoronary bypass[121] and in reducing thrombotic complications associated with artificial heart valves.[122] Long-term aspirin treatment in patients with unstable angina significantly reduces the risk of acute myocardial infarction or death.[123] On the other hand, results from studies in which aspirin was administered on a long-term basis to patients who had had myocardial infarction failed to show a benefit from aspirin.[124,125] Some trials did show a trend towards a reduction in mortality rate with aspirin treatment, but the results were not significant. A preliminary report has recently been published of a trial in which 22 000 male physicians were given buffered aspirin (325 mg) and/or placebo on alternate days. After almost five years the total number of myocardial infarctions among those subjects taking aspirin was reduced by nearly half. In contrast, the incidence of strokes was slightly, but not significantly, greater among aspirin takers. When all important vascular events (nonfatal myocardial infarction, stroke and cardiovascular death) were combined, there was a significant 23% reduction in risk among those assigned to the aspirin group.[126]

The problem with aspirin as a potential antithrombotic agent, however, lies not in the fact that the ideal regimen has not been found but in the fact that platelet aggregation is a complex mechanism in which several pathways are involved, only one of them depending on the release of TXA_2. ADP and platelet-activating factor, both of which are released by activated platelets, can cause platelet aggregation independently of TXA_2. Since we do not know which of the pathways is determining aggregation in the different pathophysiological entities that we aim to prevent or treat, then the use of aspirin as a comprehensive antithrombotic agent is not justified.

6.2.5.2 Thromboxane Synthase Inhibitors and Thromboxane Antagonists

Theoretically, a selective inhibitor of thromboxane synthase should prove to be a superior antithrombotic agent to aspirin as it would allow continued prostacyclin formation by vessel walls or other cells, either from their own endoperoxides or from those released from platelets. The major types of selective thromboxane synthase inhibitor that have now been developed comprise imidazole and pyridine derivatives and structural analogues of endoperoxides and TXA_2.[127] Selective inhibition of thromboxane synthesis has been demonstrated *in vitro* and *in vivo*, in some cases accompanied by elevated serum levels of 6-oxo-$PGF_{1\alpha}$.[128]

Studies of the effect of dazoxiben, an imidazole analogue, on aggregation of human platelets *ex vivo* showed that in some individuals, termed 'non-responders', platelet aggregation in response to arachidonic acid is unaffected, even after total inhibition of TXA_2 synthesis.[129] Similar results have been reported with other TXA_2 synthase inhibitors. This may be explained by diversion of cyclic endoperoxide metabolism after TXA_2 synthase inhibition towards other prostaglandins that may themselves modulate platelet function. Thus, the overall balance between pro-aggregatory cyclic endoperoxides and PGE_2 and the anti-aggregatory prostacyclin and PGD_2 may determine whether or not platelet aggregation is inhibited in the absence of TXA_2 production. Such a balance could obviously vary between individuals, thus determining whether or not the individual is a 'responder' to TXA_2 synthase inhibitors.

Thromboxane synthase inhibitors have been tested clinically in patients with Raynaud's syndrome and with peripheral vascular disease. Some beneficial effects were observed, but the results were not conclusive. OKY-1581, a pyridine derivative, was effective in reducing the severity of myocardial infarction in 11 patients and dazmegrel (UK 38485) has been reported to inhibit microalbuminuria in a small group of diabetic patients.[127] In general, the effects of selective thromboxane synthase inhibitors in the clinic have been disappointing; it is possible, however, that compounds with a longer duration of action are required to test the hypothesis that this type of drug is efficacious in human disease.

The combination of a thromboxane synthase inhibitor with a TXA_2 antagonist which will also block endoperoxide activity may prove to be clinically useful.[130] An ideal compound should inhibit thromboxane formation and prevent the pro-aggregatory effects of the endoperoxides without affecting cyclooxygenase, prostacyclin synthase and PGD_2 isomerase; nor should it antagonize the anti-aggregatory effects of prostacyclin or PGD_2.

TXA_2-receptor antagonists have also been synthesized, such as 9,11-azaprosta-5,13-dienoic acid (azo analogue 1), which is not only a PGH_2/TXA_2-receptor antagonist but also inhibits TXA_2 synthase. This compound prevented platelet aggregation induced by arachidonic acid, ADP (second phase) or the endoperoxide analogue U46619 in human platelet-rich plasma from subjects who did not respond to an imidazole derivative, dazoxiben.[129] The thromboxane-receptor antagonist BM 13.177, when given orally to human volunteers, has been shown to inhibit platelet aggregation induced by arachidonic acid, low-dose collagen and two stable endoperoxide analogues. The bleeding time was slightly prolonged.[131] Synergism between BM 13.177 and dazoxiben on platelets following administration to man has been reported.[132] BM 13.177 is currently being investigated in the clinic. Preliminary data show it to reduce elevated levels of platelet factor 4 and β-thromboglobulin, two indices of platelet activation, in atherosclerotic patients.

6.2.5.3 Lipoxygenase Inhibitors

6.2.5.3.1 *Mechanism of action*

Lipoxygenase activity can result in the formation of at least six different HPETEs. The aspirin-like drugs prevent abstraction of hydrogen at C-13 and, therefore, inhibit the production of 11-HPETE and 15-HPETE as well as the cyclized prostaglandins.[70] However, these drugs have little or no inhibitory action on lipoxygenase reactions at other points in the molecule. For example, concentrations of indomethacin which completely block cyclooxygenase, increase production of 5-HETE in leukocytes.[134]

In the same study in which lipoxygenase activity was first revealed in platelets, Hamberg and Samuelsson[68] also showed that the enzyme was inhibited by the alkynic analogue of arachidonic acid, ETYA. This compound, which has four triple bonds in place of the four double bonds in arachidonic acid, inhibited cyclooxygenase as well. In both reactions, ETYA competes with arachidonic acid for the enzymes; however, with ETYA, hydrogen abstraction cannot occur and so the peroxidation that is normally catalyzed by cyclooxygenase or lipoxygenase is blocked. Selective inhibition of 5-, 12- or 15-lipoxygenase has also been described using 5,6-, 11,12- or 14,15-dehydroarachidonic acids, respectively.[87] There are no reports that these compounds act as lipoxygenase inhibitors *in vivo*; in fact, in one study, ETYA did not prevent prostaglandin production and did not have anti-inflammatory activity after oral administration to rats.[135] This can almost certainly be explained by rapid metabolism of the compound *in vivo*.

A radical scavenger, nordihydroguaiaretic acid (NDGA), selectively inhibits lipoxygenase *in vitro*.[136] Other substances with potent activity against lipoxygenase *in vitro* include baicalein, esculetin and caffeic acid.[87] However, there is no evidence that any of these compounds act specifically to inhibit lipoxygenase *in vivo*. As a result, any predictions concerning the value of lipoxygenase inhibitors as therapeutic agents, based on the activity or lack of activity of these compounds in experimental models of disease, must be treated with caution.

The anti-inflammatory drug benoxaprofen inhibits the ionophore-induced synthesis of LTB_4 by isolated leukocytes,[137] and this activity has been linked with the therapeutic effects of the drug in human psoriasis.[138] In an experimental model of inflammation, though, benoxaprofen had no effect on LTB_4 levels in inflammatory exudates; instead, it acted as a selective inhibitor of prostaglandin production.[139] More detailed investigations of the *in vitro* properties of benoxaprofen revealed that the compound had no activity at all against a semipurified lipoxygenase, but only prevented ionophore stimulation of LTB_4 synthesis in whole leukocytes. Furthermore, the compound was considerably less active in preventing LTB_4 synthesis by stimuli other than the calcium ionophore.[87] This series of findings emphasizes the need for specific measurements both *in vitro* and *in vivo* before a therapeutic activity can be attributed to a specific mechanism of action. Conversely, there is a danger that credible targets for novel therapy may be ruled out because the compounds used to test the hypothesis, while having potent activity *in vitro*, may simply not be bioavailable.

6.2.5.3.2 *Dual inhibitors of cyclooxygenase and lipoxygenase*

One of the few compounds that has been shown to inhibit leukotriene production *in vivo* is the phenylpyrazoline BW755C (Figure 6).[140] This compound is a powerful antioxidant that inhibits cyclooxygenase, as well as 5-, 11-, 12- and 15-lipoxygenases.[87] Following oral administration of BW755C to rats, there is a dose-dependent reduction in the concentration of LTB_4, PGE_2 and TXB_2 in acute inflammatory exudates. The compounds timegadine and CBS-1108 also act as dual inhibitors *in vitro* and have a similar profile of anti-inflammatory activity to BW755C. However, it is not known if these compounds act as dual inhibitors *in vivo*. The benzoquinone AA-861 is a selective inhibitor of 5-lipoxygenase *in vitro* and local administration of AA-861 in a model of reversed passive Arthus pleurisy in the rat caused a reduction in the synthesis of LTD_4. Nothing, though, is known about the selectivity of this effect or whether AA-861 inhibits leukotriene synthesis following systemic administration.

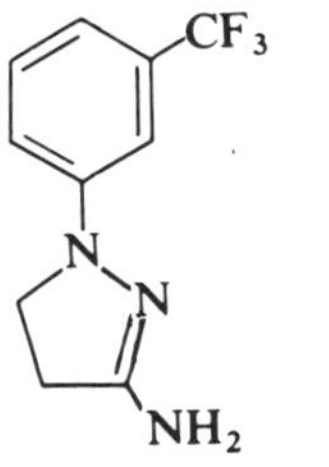

BW755C

BW A4C

3-Amino-1-[*m*-(trifluoromethyl)phenyl]-2-pyrazoline

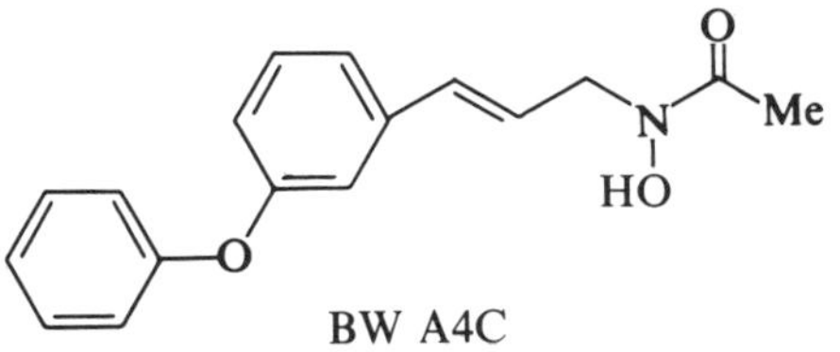

N-(3-Phenoxycinnamyl)acetohydroxamic acid

Figure 6 Experimental compounds which inhibit lipoxygenase. BW755C inhibits cyclooxygenase as well as 5-, 11-, 12- and 15-lipoxygenase. BW A4C selectively inhibits 5-lipoxygenase

Dual inhibition of cyclooxygenase and lipoxygenase results in greater anti-inflammatory activity than that achieved by selective cyclooxygenase inhibitors.[140] For example, BW755C has similar antiedema, analgesic and antipyretic activity to indomethacin, but is more effective in reducing leukocyte accumulation at therapeutic doses.[141] In addition, BW755C is thought to limit inflammatory damage and necrosis in a model of foreign body rejection by preventing the accumulation of phagocytic leukocytes.[142]

BW755C has also been shown to reduce the area of damaged tissue in experimental models of myocardial infarction.[93] Infarcted tissue converts arachidonic acid primarily to the lipoxygenase product 12-HETE, with smaller quantities of 15-HETE also formed.[143] The formation of these lipoxygenase products is attributed to the invading leukocytes since the formation of 12- and 15-HETE is proportional to the degree of cell invasion. The PMNs appear to play a central role in enhancing the myocardial injury observed in a model of myocardial ischemia in the dog.[143] Activation of peripheral blood PMNs has also been reported in patients with myocardial infarction.[144] The effects of BW755C and indomethacin have been compared in a model of myocardial infarction produced by 60 min of coronary occlusion followed by 5 hours reperfusion in anesthetized beagles.[145] BW755C attenuated leukocyte infiltration into the myocardium, abolished 12-HETE formation and significantly reduced infarct size, whereas indomethacin had no effect on these parameters. BW755C also reduced the incidence of ventricular arrhythmias in this model.[143] These effects of BW755C are observed when the drug is given after the period of occlusion and are independent of any hemodynamic effect, thus ruling out any effect of BW755C on oxygen consumption or delivery and indicating that the drug is acting on the cellular infiltration which is known to occur in the later phases of myocardial infarction.[93]

Because BW755C is a dual inhibitor, it is difficult to interpret which of its activities are due to inhibition of lipoxygenase. There are some clues, though, generated in experiments where selective cyclooxygenase inhibitors are relatively inactive. In isolated airway smooth muscle from sensitized animals, BW755C blocks antigen-induced contractions, whereas cyclooxygenase inhibitors either are inactive or enhance constriction.[146] Similarly, in a naturally occurring IgE-mediated bronchospasm in monkeys (which closely resembles human allergic asthma), BW755C blocks antigen-induced increases in both pulmonary resistance and dynamic compliance and significantly attenuates other changes in pulmonary function (Figure 7).[147] In these animals the greatest effect of exogenous SRS-A is on pulmonary resistance and compliance[148] and, therefore, it seems likely that the therapeutic benefit derived from BW755C treatment is due to the inhibition of leukotriene synthesis by 5-lipoxygenase. These experiments give the strongest support to the view that selective 5-lipoxygenase inhibitors would be useful antiasthmatic drugs.

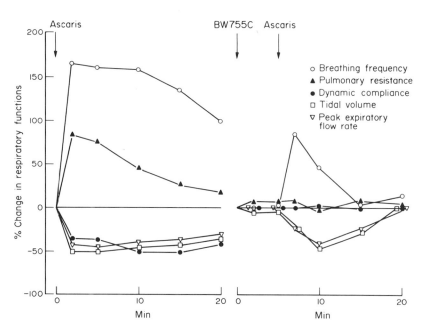

Figure 7 The effect of BW755C on ascaris-induced changes in respiratory functions in a rhesus monkey. The curves on the left of the figure show the changes induced by ascaris challenge in an untreated animal. The curves on the right show the effects of challenging the same animal 5 min after aerosol treatment with BW755C (10 mg mL^{-1}). Data taken from ref. 147

6.2.5.3.3 *Selective lipoxygenase inhibitors*

Working on the hypothesis that iron plays a key role in lipoxygenase catalysis,[149] Corey and his colleagues synthesized some amide analogues of arachidonic acid in which strong coordination to iron is possible.[150] They found that analogues containing a hydroxamic acid moiety are potent and selective inhibitors of leukotriene biosynthesis.

A novel series of acetohydroxamic acid lipoxygenase inhibitors has now been reported.[151] Phenoxycinnamylacetohydroxamic acid (BW A4C; see Figure 6) inhibited the synthesis of LTB_4 at lower concentrations than were required to inhibit the synthesis of TXB_2, whereas BW755C inhibited both products equally. The IC_{50} value against lipoxygenase for BW A4C was 0.1 μM. Concentrations required to inhibit cyclooxygenase, 12-lipoxygenase or 15-lipoxygenase were up to 24 times greater. In addition, BW A4C inhibited the synthesis of $[^{14}C]$-5-HETE from $[^{14}C]$-arachidonic acid and the calcium-dependent synthesis of LTB_4 from 5-HPETE. This therefore suggests that this compound inhibits 5-lipoxygenase and possibly LTA_4 synthase. In whole leukocytes, BW A4C selectively inhibits LTB_4 synthesis induced by the calcium ionophore A23187 and by zymosan.[87]

A single oral dose of the dual inhibitor BW755C (50 mg kg^{-1}) reduced the *ex vivo* synthesis of both LTB_4 and TXB_4 in whole blood for up to 6 h, but nafazatrom produced only a modest reduction in LTB_4 for about 1 h and NDGA had no significant effect on either LTB_4 or TXB_2 synthesis (Figure 8).[152] In contrast, BW A4C caused a prolonged and selective inhibition of ionophore-stimulated LTB_4 production *ex vivo* (Figure 8). The inhibition of LTB_4 synthesis by BW A4C was dose-dependent (ED_{50} at 6 h = 9.0) and the degree of inhibition correlated with the plasma concentrations of unchanged drug determined by high-performance liquid chromatography.[152]

6.2.5.3.4 *The effect of lipoxygenase inhibitors on bronchial anaphylaxis*

The potent bronchoconstrictor properties of the peptido-leukotrienes indicate that selective 5-lipoxygenase inhibitors would be useful in treating asthma. The acetohydroxamic acids have been

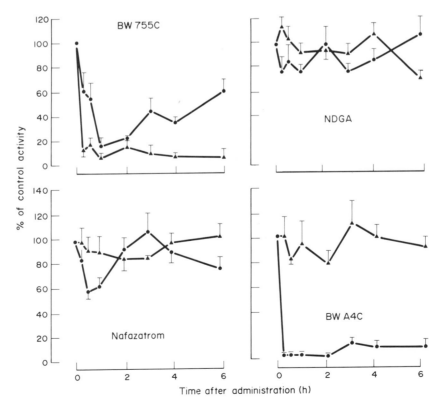

Figure 8 The effects of various lipoxygenase inhibitors on the ionophore-stimulated synthesis of thromboxane B_2 (▲) and leukotriene B_4 (●) in rat blood *ex vivo*. Each value is the mean ($\pm$s.e. mean) from five animals dosed orally with 50 mg kg^{-1} drug at time 0. Data taken from ref. 152

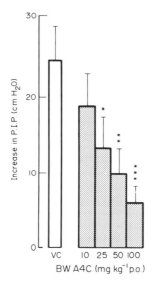

Figure 9 The effect of BW A4C (10–100 mg kg^{-1} p.o.) on 'leukotriene-dependent' anaphylactic bronchospasm in anesthetized pump-ventilated guinea-pigs sensitized to ovalbumin and challenged by aerosol. The open column shows the increase in pulmonary inflation pressure (P.I.P.) in animals receiving vehicle alone (VC). Data taken from ref. 153

tested in anesthetized, pump-ventilated guinea-pigs actively sensitized to ovalbumin and challenged with aerosolized antigen.[153] Aerosol challenge provoked a rapidly developing and intense bronchoconstriction measured by a prolonged increase in pulmonary inflation pressure (P.I.P.). Pretreatment with a combination of mepyramine (2 mg kg^{-1}, i.v.) and indomethacin (10 mg kg^{-1}, i.v.) decreased the maximum rise in P.I.P. by approximately 45%. The residual histamine and cyclooxygenase-independent bronchoconstriction was markedly reduced (65%) by additional pretreatment with a leukotriene antagonist, FPL 55712 (10 mg kg^{-1}, i.v.). This indicates that there is a significant 'leukotriene-dependent' component of anaphylactic bronchospasm in this model.

Oral administration of BW A4C caused a dose-dependent reduction in the leukotriene-dependent bronchospasm up to a maximum effect (60–75%; Figure 9) that was equivalent to the maximum effect obtained with FPL 55712. In animals pretreated with mepyramine alone, the lipoxygenase inhibitors did not reduce the response to antigen, nor did they modify histamine-induced bronchospasm. The duration of action of BW A4C correlated with the plasma concentrations of unchanged drug. These results indicate that BW A4C attenuated anaphylactic bronchospasm by selectively inhibiting antigen-induced leukotriene synthesis by arachidonate 5-lipoxygenase.

6.2.5.3.5 *The effect of lipoxygenase inhibitors on inflammation*

The acetohydroxamic acids have been tested in a series of animal models of acute inflammation.[154] The concentrations of LTB$_4$ found in 6 h inflammatory exudates, induced in rats by the subcutaneous implantation of carrageenin-soaked polyester sponges, were reduced dose-dependently by BW A4C (ED$_{50}$ = 2.6 mg kg^{-1}, p.o.) with little or no effect on PGE$_2$ concentrations in the exudates (ED$_{50}$ > 100 mg kg^{-1}). Furthermore, this effect was accompanied by a reduction in the numbers of leukocytes accumulating in the exudate. It is also possible that inhibition of leukocyte migration was due to inhibition of the synthesis of the chemotactic agent LTB$_4$.

Doses of BW A4C up to 200 mg kg^{-1} p.o. had little or no effect on carrageenin-induced edema and hyperalgesia in rat paws or on phenylbenzoquinone-induced writhing in mice (Table 5).[154] In contrast, yeast-induced pyrexia in rats was reduced by BW A4C (ED$_{50}$ = 32 mg kg^{-1}). However, there was no direct correlation between the antipyretic effects of BW A4C and the inhibition of lipoxygenase. Therefore it cannot be discounted that the antipyretic effect may be due to a property of BW A4C other than inhibition of lipoxygenase. From these experiments it would appear that cyclooxygenase products are the major mediators of vascular and pain responses in acute inflammation and that lipoxygenase products are relatively unimportant in these models.

Acetohydroxamic acids such as BW A4C are, therefore, potent and selective inhibitors of 5-lipoxygenase *in vivo*. They attenuate leukotriene-dependent anaphylactic bronchospasm, the accumulation of inflammatory leukocytes and the development of fever in experimental models. It now remains to be determined whether these compounds have any therapeutic value in man.

Table 5 Effects of Selective a Cyclooxygenase Inhibitor Indomethacin (indo), a Dual Inhibitor (BW755C) and a Selective Lipoxygenase Inhibitor (BW A4C) in Models of Acute Inflammation[154]

	ED_{50} (mg kg^{-1} p.o.)		
Inflammatory response	Indo	BW755C	BW A4C
Carrageenin-induced edema (rat)	5.7	17.0	> 200
Carrageenin-induced hyperalgesia (rat)	3.5	31.1	233
PBQ-induced writhing (mouse)	1.3	39.8	> 200
Yeast-induced hyperthermia (rat, 1 h)	1.56	4.15	37.0

6.2.6 DIETARY MANIPULATION OF ARACHIDONIC ACID METABOLISM

Dietary manipulation of fatty acid precursors has been suggested as a means of achieving an antithrombotic effect. In the middle of the 1970s, enrichment of the diet with dihomo-γ-linolenic acid (C20:3ω6), the precursor of monoenoic prostaglandins, was suggested since PGE$_1$ is anti-aggregatory (Table 6).[155] However, it became clear that feeding rabbits with sufficient dihomo-γ-linolenic acid to elevate the tissue content did not alter the platelet sensitivity to ADP.[156] The discovery of prostacyclin made it apparent that this was not the most rational approach to dietary manipulation, since prostaglandin endoperoxides of the '1' series cannot give rise to an anti-aggregatory prostacyclin-type compound and there is little or no evidence that tissues form PGE$_1$ in animals or man.

Eicosapentaenoic acid (EPA; C20:5ω3), which is found in fish oil, is a polyunsaturated fatty acid like arachidonic acid (C20:4ω3) but has a higher degree of unsaturation. It gives rise to prostaglandins of the '3' series and when incubated with vascular tissue leads to the release of an anti-aggregating substance.[157] Synthetic Δ^{17}-prostacyclin, or PGI$_3$, is as potent an anti-aggregating agent as prostacyclin (Table 6). In contrast, TXA$_3$ has a weaker pro-aggregating activity than TXA$_2$.[157]

EPA inhibits platelet aggregation in platelet-rich plasma stimulated by ADP, collagen, arachidonic acid, and a synthetic analogue of PGH$_2$.[157] EPA also inhibits aggregation in aspirin- and imidazole-treated platelets and inhibits thrombin-induced aggregation.[35] It is clear, therefore, that both prostaglandin-dependent and -independent pathways of platelet aggregation are inhibited by EPA *in vitro*. *In vivo*, EPA is incorporated into platelet phospholipids, to some extent replacing arachidonic acid. Its antithrombotic effect may be exerted by its competition with released arachidonic acid for cyclooxygenase, leading to reduced synthesis of TXA$_2$,[158] or by its conversion to the less pro-aggregatory PGH$_3$ and TXA$_3$.[157]

Table 6 Structures of Dihomo-γ-linolenic Acid (top), Arachidonic Acid (middle) and Eicosapentaenoic Acid (bottom) and the Effect on Platelet Aggregation of their Metabolites Produced in either Platelets or the Blood Vessel Wall

Pro-aggregating	Platelets	Precursor	Vessel wall	Anti-aggregating
No	TXA$_1$		No PGI type compound	No
Yes	TXA$_2$		PGI$_2$	Yes
No	TXA$_3$		PGI$_3$	Yes

In vitro production of PGI_3 from EPA has been demonstrated in human umbilical veins[159] and in rabbit renal cortex microsomes.[160] More significantly, PGI_3 has been demonstrated to be formed *in vivo* in man after ingesting cod liver oil or dietary fish.[161] Excretion of the major urinary metabolite of PGI_3 was not detectable in control conditions, but was clearly measurable after intake of cod liver oil or mackerel and increased with increasing EPA incorporation into plasma phospholipids, reaching levels one-half of those measured for the major urinary metabolite of prostacyclin. Interestingly, the urinary levels of the metabolite of prostacyclin were, if anything, also increased on the fish diet. The *in vivo* formation of TXA_3 by collagen-stimulated platelets from human subjects taking cod liver oil has also been demonstrated. These platelets showed reduced sensitivity to the pro-aggregatory effects of collagen and reduced formation of TXA_2.[162]

In addition to conversion into TXA_3, PGH_3 undergoes a rapid spontaneous degradation to a mixture of PGE_3 and PGD_3 in human platelets *in vitro* and, since PGD_3 increases platelet cyclic AMP, it can inhibit platelet aggregation.[157, 163] Moreover, PGE_2 is an antagonist of the action of PGD_2, but PGE_3 is not an antagonist of PGD_3.[157] Overall, therefore, the final mixture of products derived from EPA may be more anti-aggregatory than those derived from arachidonic acid and this could add to the potential anti-aggregatory effect of EPA. The antiplatelet effect of EPA is unlikely to be due to a single mechanism of action but is more likely to be the result of a combination of several mechanisms. EPA has a hypolipidemic effect on triglycerides, low density lipoprotein (LDL), very low density lipoprotein (VLDL) and cholesterol; high levels of these lipids are known risk factors in coronary disease.

Evidence from epidemiological studies shows a low incidence of thrombotic disorders and ischemic heart disease in populations, such as Eskimos and Japanese, that eat a diet rich in fish.[164] High dietary consumption of fish is associated with prolonged cutaneous bleeding time, reduced *ex vivo* platelet aggregability, low blood viscosity, low concentrations of cholesterol, tryglyceride, LDL and VLDL and elevated EPA levels in plasma cholesteryl esters, triglycerides and phospholipids. In a 20-year study in Dutch men it has been shown that as little as one or two fish dishes per week may be of preventative value in relation to coronary heart disease.[165]

There are numerous reports of feeding fish oil, fish meat or purified EPA to humans.[164] Many of the studies used fish oil or fish meat as a source of EPA and, therefore, it is important to recognize that they might not only be showing the effects of EPA but also those of other polyunsaturated fatty acids, among them docosahexaenoic acid (DHA), which is also present in considerable concentrations. The administration periods reported in these studies vary from one week to three years.

Dietary supplementation with EPA in one form or another leads to an increase in the ratio of EPA : arachidonic acid in plasma, platelets and red blood cells. This is accompanied by changes in hemostatic parameters consistent with an antithrombotic effect, so that platelet aggregation to collagen and ADP is reduced, as is TXB_2 formation during platelet aggregation. Plasma levels of cholesterol, triglyceride, β-thromboglobulin and plasma antithrombin III are also reduced. In addition, in some studies there is an increase in cutaneous bleeding time and in red cell deformability together with a decrease in systemic blood pressure, platelet count, whole blood viscosity and platelet retention in glass bead columns.[164] All the changes described tend to disappear within 4–6 weeks of stopping the supplementation of the diet.

There have been several studies on the effect of EPA on patients with hyperlipidemia.[166–168] All report a fall in plasma triglyceride levels and in some, but not all studies, a reduction in bleeding time was observed. In a recent study, in which patients with mild essential hypertension were fed a diet supplemented with EPA (approx. 2.2 g daily) in the form of mackerel, significant decreases in serum triglycerides, total and LDL cholesterol, blood pressure and TXB_2 levels were found, while HDL cholesterol was elevated.[169] When the dose of EPA was reduced (approx. 0.47 g daily) the blood pressure effects were maintained, although the other parameters returned to normal.

EPA has also been shown to cause a significant increase, both *in vitro* and *in vivo*, in the formation of LTB_5,[170, 171] which is less active biologically than LTB_4, and a decrease in the synthesis of LTB_4 by stimulated leukocytes. In addition, dietary supplementation with fish oil increases the EPA content in neutrophils and monocytes.[172] The release of arachidonic acid and formation of its leukotriene metabolites was reduced in both cell types, and the response of neutrophils to LTB_4 (as measured by chemotaxis and endothelial cell adherence) was greatly reduced or abolished. Furthermore, carrageenin-induced edema in rat paws was significantly reduced in animals fed an EPA-rich diet.[173] Thus, dietary manipulation with EPA could influence the course of acute and chronic inflammatory conditions; indeed, populations with a high dietary intake of EPA have a reduced incidence of inflammatory disease.[174]

There is now considerable evidence to support the suggestion that an EPA-rich diet will lead to changes in hemostatic and other parameters consistent with an overall antithrombotic effect, as well as having a beneficial effect on inflammatory conditions. However, it still remains to be demonstrated whether the observed changes translate themselves into clear-cut beneficial clinical effects in well-conducted double-blind studies. In addition, it has not yet been determined whether all the observations are linked with EPA or if other long-chain polyunsaturated fatty acids in fish oils also contribute to the observed effects.

6.2.7 SUMMARY

The study of arachidonic acid metabolism over the last 30 years has advanced our understanding of physiological and pathological processes, helped to elucidate the mechanism of action of established drugs such as aspirin, and revealed new targets for the development of novel therapies. Prostaglandins, thromboxanes and leukotrienes are now established as groups of lipids with the properties of local hormones involved in the modulation of cardiovascular homeostasis, bronchial anaphylaxis, thrombosis and the inflammatory response. It is widely accepted that the mechanism of action of the large group of nonsteroid anti-inflammatory drugs is through the selective inhibition of prostaglandin synthesis by the enzyme cyclooxygenase. The discovery of eicosanoids with potent vasodilator and platelet-inhibiting properties has led to the development of drugs such as prostacyclin for use in a number of cardiothrombotic conditions. An understanding of the enzymes involved in the biosynthesis of these agents has focused attention on their inhibition. Inhibitors of the lipoxygenase enzymes and antagonists of leukotrienes are now being developed and it remains to be seen if these compounds will have the predicted therapeutic value in the treatment of human bronchial asthma and inflammation.

6.2.8 REFERENCES

1. U. S. Von Euler, *J. Physiol.*, 1937, **88**, 213.
2. S. Bergström, R. Ryhage, B. Samuelsson and J. Sjövall, *J. Biol. Chem.*, 1963, **238**, 3555.
3. R. J. Flower, in 'Handbook of Experimental Pharmacology', ed. J. R. Vane and S. H. Ferreira, Springer-Verlag, Berlin, 1978, vol. 50, p. 374.
4. M. Hamberg, J. Svensson and B. Samuelsson, *Proc. Natl. Acad. Sci. U.S.A.*, 1975, **72**, 2994.
5. S. Moncada, R. Gryglewski, S. Bunting and J.R. Vane, *Nature (London)*, 1976, **263**, 663.
6. B. Samuelsson, *Science (Washington, DC)*, 1983, **220**, 568.
7. F. J. Van der Ouderaa, M. Buytenhek, D. H. Nugteren and D. A. Van Dorp, *Biochim. Biophys. Acta*, 1977, **487**, 315.
8. F. J. Van der Ouderaa, M. Buytenhek, F. J. Slikkerveer and D. A. Van Dorp, *Biochim. Biophys. Acta*, 1979, **572**, 29.
9. D. A. Van Dorp, *Progr. Biochem. Pharmacol.*, 1967, **3**, 71.
10. S. Bergstrom, H. Danielsson and B. Samuelsson, *Biochim. Biophys. Acta*, 1964, **90**, 207.
11. D. A. Van Dorp, R. K. Beerthuis, D. H. Nugteren and H. Vonkeman, *Biochim. Biophys. Acta*, 1964, **90**, 204.
12. N. Ogino, T. Miyamoto, S. Yamamoto and O. Hayaishi, *J. Biol. Chem.*, 1977, **252**, 890.
13. E. Christ-Hazelhof and D. H. Nugteren, *Biochim. Biophys. Acta*, 1979, **572**, 43.
14. S. Bunting, S. Moncada and J. R. Vane, *Br. Med. Bull.*, 1983, **39**, 271.
15. S. Moncada and J. R. Vane, *Pharmacol. Rev.*, 1979, **30**, 293.
16. W. L. Smith, D. L. DeWitt and J. S. Day, *Adv. Prostaglandin Thromboxane Leuk. Res.*, 1983, **11**, 87.
17. B. F. Robinson, J. G. Collier, S. M. M. Karim and K. Somers, *Clin. Sci.*, 1973, **44**, 367.
18. J. Nakano, In 'The Prostaglandins: Pharmacological and Therapeutic Advances', ed. M. F. Cuthbert, Lippincott, Philadelphia, 1973, p. 23.
19. J. B. Lee, *Arch. Intern. Med.*, 1974, **133**, 56.
20. J. W. Fisher and D. M. Gross, in 'Prostaglandins in Hematology', ed. M. Silver, B. J. Smith and J. J. Kocsis, Spectrum Publications, New York, 1977, p. 159.
21. H. Rasmussen and W. Lake, in 'Prostaglandins in Hematology', ed. M. Silver, B. J. Smith and J. J. Kocsis, Spectrum Publications, New York, 1977, p. 187.
22. M. F. Cuthbert, in 'The Prostaglandins: Pharmacological and Therapeutic Advances', ed. M. F. Cuthbert, Lippincott, Philadelphia, 1973, p. 253.
23. V. J. Goldberg and P. W. Ramwell, *Physiol. Rev.*, 1975, **55**, 325.
24. B. J. R. Whittle and J. R. Vane, in 'Physiology of the Gastrointestinal Tract', 2nd edn., ed. L. R. Johnson, Raven Press, New York, 1987, vol. 1, p. 143.
25. S. Moncada, R. J. Flower and J. R. Vane, in Goodman and Gilman's 'The Pharmacological Basis of Therapeutics', 7th edn., ed. A. G. Gilman, L. S. Goodman, T. W. Rall and F. Murad, Macmillan, New York, 1985, p. 660.
26. J. P. Codde and L. J. Beilin, *J. Hypertens.*, 1986, **4**, 675.
27. K. Mullane, S. Moncada and J. R. Vane, in 'Prostaglandins and the Kidney', ed. M. J. Dunn, C. Patrono and G. A. Cinotti, Plenum, New York, 1982, p. 213.
28. G. A. Higgs, S. Moncada and J. R. Vane, *Ann. Clin. Res.*, 1984, **16**, 287.
29. D. S. Rampton and C. J. Hawkey, *Gut*, 1984, **25**, 1399.
30. H. A. Bernheim, T. M. Gilbert and J. T. Stitt, *J. Physiol.*, 1980, **301**, 69.
31. B. Henderson, E. R. Pettipher and G. A. Higgs, *Br. Med. Bull.*, 1987, **43**, 415.
32. S. Moncada, in 'Prostaglandins: Research and Clinical Update', ed. G. L. Longenecker and S. W. Schaffer, Alpha Editions, Minneapolis, 1985, p. 1.

33. G. J. Dusting, K. M. Mullane and S. Moncada, in 'Handbook of Hypertension', ed. A. Zanchetti and R. C. Tarazi, Elsevier, Amsterdam, 1986, vol. 7, p. 408.
34. B. J. R. Whittle, in 'Gastrointestinal Mucosal Blood Flow', Churchill Livingstone, Edinburgh, 1980, p. 180.
35. S. Moncada, *Br. J. Pharmacol.*, 1982, **76**, 3.
36. G. A. Higgs, in 'Cardiovascular Pharmacology of the Prostaglandins', Raven Press, New York, 1982, p. 315.
37. L. A. Boxer, J. M. Allen, M. Schmidt, M. Yoder and R. L. Baehner, *J. Lab. Clin. Med.*, 1980, **95**, 672.
38. A. I. Schafer, M. A. Gimbrone Jr. and R. I. Handin, *Biochem. Biophys. Res. Commun.*, 1980, **96**, 1640.
39. E. A. Higgs, S. Moncada, J. R. Vane, J. P. Caen, H. Michel and G. Tobelem, *Prostaglandins*, 1978, **16**, 17.
40. S. Moncada, R. M. J. Palmer and E. A. Higgs, in 'Thrombosis and Haemostasis', ed. M. Verstraete, J. Vermylen, H. R. Lijnen and J. Arnout, Leuven University Press, Belgium, 1987, p. 597.
41. M. W. Radomski, R. M. J. Palmer and S. Moncada, *Br. J. Pharmacol.*, 1987, **92**, 639.
42. M. Radomski and S. Moncada, *Thromb. Res.*, 1983, **30**, 383.
43. G. J. Blackwell, M. Radomski, J. R. Vargas and S. Moncada, *Biochim. Biophys. Acta*, 1982, **718**, 60.
44. K. Schror, R. Ohlendorf and H. Darius, *J. Pharmacol. Exp. Ther.*, 1981, **219**, 243.
45. T. Utsunomiya, M. M. Krausz, C. R. Valeri, D. Shepro and H. B. Hechtman, *Surgery*, 1980, **88**, 25.
46. A. Szczeklik, M. Kopec, K. Sladek, J. Musial, J. Chmielewska, E. Teisseyre, G. Dudek-Wojciechowska and M. Palester-Chlebowczyk, *Thromb. Res.*, 1983, **29**, 655.
47. D. J. Crutchley, L. B. Conanan and J. R. Maynard, *J. Pharmacol. Exp. Ther.*, 1982, **222**, 544.
48. A. L. Willis, D. L. Smith, C. Vigo and A. F. Kluge, *Lancet*, 1986, **ii**, 682.
49. S. Moncada, in 'Pharmacological Control of Hyperlipidaemia', Prous, Barcelona, 1986, p. 439.
50. S. Moncada and E. A. Higgs, *Clin. Haematol.*, 1986, **15**, 273.
51. S. Moncada, R. J. Gryglewski, S. Bunting and J. R. Vane, *Prostaglandins*, 1976, **12**, 715.
52. L. O. Carreras, G. Defreyn, S. J. Machin, J. Vermylen, R. Deman, B. Spitz and A. Van Assche, *Lancet*, 1981, **i**, 244.
53. C. Patrono, G. Ciabattoni, G. Remuzzi, E. Gotti, S. Bombardieri, O. DiMunno, G. Tartarelli, G. A. Cinotti, B. M. Simonetti and A. Pierucci, *J. Clin. Invest.*, 1985, **76**, 1011.
54. S. Moncada, in 'Advanced Medicine', ed. D. R. Triger, Ballière Tindall, London, 1980, vol. 22, p. 323.
55. P. Henriksson, O. Edhag and A. Wennmalm, *Br. Heart J.*, 1985, **53**, 173.
56. B. M. Groves, L. J. Rubin, M. F. Frosolono, A. E. Cato and J. R. Reeves, *Am. Heart J.*, 1985, **110**, 1200.
57. R. J. Barst, *Chest*, 1986, **89**, 497.
58. P. J. Piper and J. R. Vane, *Nature (London)*, 1969, **223**, 29.
59. B. J. R. Whittle and S. Moncada, *Br. Med. Bull.*, 1983, **39**, 232.
60. H. E. Harrison, A. H. Reece and M. Johnson, *Life Sci.*, 1978, **23**, 351.
61. G. F. Gensini, R. Abbate, S. Favilla and G. G. Neri Serneri, *Thromb. Haem.*, 1979, **42**, 983.
62. M. J. Stuart, H. Elrad, J. E. Graeber, D. O. Hakanson, S. G. Sunderji and M. K. Barvinchak, *J. Lab. Clin. Med.*, 1979, **94**, 12.
63. M. Johnson, A. H. Reece and H. E. Harrison, in 'Prostaglandins and Thromboxanes', ed. W. Forster, VEB Gustav Fischer Verlag, Jena, DDR, 1980, p. 79.
64. J. Quilley, *Fed. Proc., Fed. Am. Soc. Exp. Biol.*, 1984, **43**, 1040.
65. P. C. Milner and J. F. Martin, *Brit. Med. J.*, 1985, **290**, 1767.
66. R. I. Lewy, J. B. Smith, M. J. Silver, J. Saia, P. Walinsky and L. Wiener, *Prostaglandins Med.*, 1979, **2**, 243.
67. P. D. Hirsh, L. D. Hillis, W. D. Campbell, B. G. Firth and J. T. Willerson, *New Engl. J. Med.*, 1981, **304**, 685.
68. M. Hamberg and B. Samuelsson, *Proc. Natl. Acad. Sci. USA*, 1974, **71**, 3400.
69. P. Borgeat, M. Hamberg and B Samuelsson, *J. Biol. Chem.*, 1976, **251**, 7816.
70. M. I. Siegel, R. T. McConnell, N. A. Porter and P. Cuatrecasas, *Proc. Natl. Acad. Sci. USA*, 1980, **77**, 308.
71. D. Aharony, J. B. Smith and M. J. Silver, *Prostaglandins Med.*, 1981, **6**, 237.
72. W. C. Chang, J. Nakao, H. Orimo and S. I. Murota, *Biochem. J.*, 1982, **202**, 771.
73. R. C. Murphy, S. Hammarström and B. Samuelsson, *Proc. Natl. Acad. Sci. USA*, 1979, **76**, 4275.
74. B. Samuelsson, *Science (Washington, DC)*, 1983, **220**, 568.
75. B. A. Jakschik, A. R. Sams, H. Sprecher and P. Needleman, *Prostaglandins*, 1980, **20**, 401.
76. C. W. Parker and S. Aykent, *Biochem. Biophys. Res. Commun.*, 1982, **109**, 1011.
77. A. L. Maycock, M. S. Anderson, D. M. DeSousa and F. A. Kuehl, *J. Biol. Chem.*, 1982, **257**, 13911.
78. B. A. Jakschik and C. G. Kuo, *Adv. Prostaglandin Thromboxane Leuk. Res.*, 1983, **11**, 141.
79. W. Feldberg and C. H. Kellaway, *J. Physiol.*, 1938, **94**, 187.
80. J. L. Walker, in 'Advances in Biosciences', ed. S. Bergstrom and S. Bernhard, Pergamon, New York, 1973, vol. 9, p. 235.
81. P. J. Piper, in 'Development of Anti-asthmatic Drugs', ed. D. R. Buckle and H. Smith, Butterworths, London, 1984, p. 55.
82. J. M. Drazen, K. F. Austen, R. A. Lewis, D. A. Clark, G. Goto, A. Marfat and E. J. Corey, *Proc. Natl. Acad. Sci. USA*, 1980, **77**, 4354.
83. M. C. Holroyde, R. E. C. Altounyan, M. Cole, M. Dixon and E. V. Elliott, *Lancet*, 1981, **ii**, 17.
84. J. T. Zakrzewski, N. C. Barnes, P. J. Piper and J. F. Costello, *Br. J. Pharmacol.*, 1985, **19**, 574P.
85. P. S. Creticos, S. P. Peters, N. F. Adkinson, Jr., R. M. Naclerio, E. C. Hayes, P. S. Norman and L. M. Lichtenstein, *New Engl. J. Med.*, 1984, **310**, 1626.
86. S. Hammarström, M. Hamberg, B. Samuelsson, E. A. Duell, M. Stawiski and J. J. Voorhees, *Proc. Natl. Acad. Sci. USA*, 1975, **72**, 5130.
87. P. Bhattacherjee, N. K. Boughton-Smith, R. L. Follenfant, L. G. Garland, G. A. Higgs, H. F. Hodson, P. J. Islip, W. P. Jackson, S. Moncada, A. N. Payne, R. W. Randall, C. H. Reynolds, J. A. Salmon, J. E. Tateson and B. J. R. Whittle, *Ann. N.Y. Acad. Sci.*, 1988, **524**, 307.
88. G. A. Higgs, J. A. Salmon and J. A. Spayne, *Br. J. Pharmacol.*, 1981, **74**, 429.
89. S. R. Turner, J. A. Tainer and W. S. Lynn, *Nature (London)*, 1975, **257**, 680.
90. A. W. Ford-Hutchinson, M. A. Bray, M. V. Doig, M. E. Shipley and M. J. H. Smith, *Nature (London)*, 1980, **286**, 264.
91. R. J. Palmer, R. Stepney, G. A. Higgs and K. E. Eakins, *Prostaglandins*, 1980, **20**, 411.
92. P. J. Piper and S. A. Galton, *Prostaglandins*, 1984, **28**, 905.
93. S. Moncada, in 'The Leukotrienes: Their Biological Significance', ed. P. J. Piper, Raven Press, New York, 1986, p. 99.

94. C. N. Serhan, M. Hamberg and B. Samuelsson, *Biochem. Biophys. Res. Commun.*, 1984, **118**, 943.
95. C. N. Serhan, U. Hirsch, J. Palmblad and B. Samuelsson, *FEBS Lett.*, 1987, **217**, 242.
96. B. Samuelsson, S. E. Dahlen, J. A. Lindgren, C. A. Ronzer and C. N. Serhan, *Science (Washington, DC)*, 1987, **237**, 1171.
97. J. Capdevila, L. Marnett, N. Chacos, K. A. Prough and R. W. Estabrook, *Proc. Natl. Acad. Sci. USA*, 1982, **79**, 767.
98. P. L. Canonico, A. M. Judd, K. Koike, C. A. Valdenegro and R. M. Macleod, *Endocrinology*, 1985, **116**, 218.
99. A. Negro-Vilar, G. D. Snyder, J. R. Falck, S. Manna, N. Chacos and J. Capdevila, *Endocrinology*, 1985, **116**, 2663.
100. J. Capdevila, N. Chacos, J. R. Falck, S. Manna, A. Negro-Vilar and S. R. Ojeda, *Endocrinology*, 1983, **113**, 421.
101. J. R. Falck, S. Manna, J. Moltz, N. Chacos and J. Capdevila, *Biochem. Biophys. Res. Commun.*, 1983, **114**, 743.
102. D. Schlondorff, E. Petty, J. A. Oates, M. Jacoby and S. D. Levine, *Am. J. Physiol.*, 1987, **253**, F464.
103. M. A. Carroll, M. Schwartzmann, J. Capdevila, J. R. Falck and J. C. McGiff, *Eur. J. Pharmacol.*, 1987, **138**, 281.
104. J. R. Vane, *Nature (London), New Biol.*, 1971, **231**, 232.
105. J. B. Smith and A. L. Willis, *Nature (London), New Biol.*, 1971, **231**, 235.
106. S. H. Ferreira, S. Moncada and J. R. Vane, *Nature (London), New Biol.*, 1971, **231**, 237.
107. R. J. Flower, *Pharm Rev.*, 1974, **26**, 33.
108. W. E. M. Lands, *Trends Pharmacol.*, 1981, **2**, 78.
109. L. H. Rome and W. E. M. Lands, *Proc. Natl. Acad. Sci. USA*, 1975, **72**, 4863.
110. G. J. Roth and P. W. Majerus, *J. Clin. Invest.*, 1975, **56**, 624.
111. W. E. M. Lands, H. W. Cook and L. H. Rome, *Adv. Prostaglandin Thromboxane Res.*, 1976, **1**, 7.
112. M. Hamberg, *Biochem. Biophys. Res. Commun.*, 1972, **49**, 720.
113. A. K. Pedersen and G. A. Fitzgerald, *New Eng. J. Med.*, 1984, **311**, 1206.
114. G. A. Higgs, J. A. Salmon, B. Henderson and J. R. Vane, *Proc. Natl. Acad. Sci. USA*, 1987, **84**, 1417.
115. G. A. Higgs, K. E. Eakins, K. G. Mugridge, S. Moncada and J. R. Vane, *Eur. J. Pharmacol.*, 1980, **66**, 81.
116. G. A. Higgs, J. R. Vane, F. D. Hart and J. A. Wojtulewski, in 'Prostaglandin Synthetase Inhibitors', ed. J. H. Robinson and J. R. Vane, Raven Press, New York, 1974, p. 165.
117. J. W. Burch, N. L. Baenziger, N. Stanford and P. W. Majerus, *Proc. Natl. Acad. Sci. USA*, 1978, **75**, 5181.
118. The Canadian Co-operative Study Group, *New Engl. J. Med.*, 1978, **299**, 53.
119. W. S. Fields, *Am. J. Med.*, 1983, **74**, 61.
120. H. R. Harter, J. W. Burch, P. W. Majerus, N. Stanford, J. A. Delmez, C. B. Anderson and C. A. Weerts, *New Engl. J. Med.*, 1979, **301**, 577.
121. R. L. Lorenz, M. Weber, J. Kotzur, K. Theisen, C. V. Schacky, W. Meister, B. Reichardt and P. C. Weber, *Lancet*, 1984, **i**, 1261.
122. J. Dale, E. Myhre, O. Storstein, H. Stormorken and L. Efskind, *Am. Heart J.*, 1977, **94**, 101.
123. H. D. Lewis Jr., J. W. Davis, D. G. Archibald, W. E. Steinke, T. C. Smitherman, J. E. Doherty, III, H. W. Schnaper, M. M. LeWinter, E. Linares, J. M. Pouget, S. C. Sabharwal, E. Chesler and H. Demots, *New Engl. J. Med.*, 1983, **309**, 396.
124. G. S. May, K. A. Eberlein, C. D. Furberg, E. R. Passamain and D. S. De Mets, *Prog. Cardiovasc. Dis.*, 1982, **24**, 331.
125. P. C. Elwood, *Am. J. Med.*, 1983, **74**, 50.
126. The Steering Committee of the Physicians' Health Study Research Group, *New Engl. J. Med.*, 1988, **318**, 262.
127. J. A. Salmon, *Adv. Drug Res.*, 1986, **15**, 111.
128. J. Vermylen, G. Defreyn, L. O. Carreras, S. J. Machin, J. Van Schaeren and M. Verstraete, *Lancet*, 1981, **i**, 1073.
129. V. Bertele, A. Falanga, M. Tomasiak, C. Chiabrando, C. Cerletti and G. De Gaetano, *Blood*, 1984, **63**, 1460.
130. P. Gresele, E. Van Houtte, J. Arnout, M. Deckmyn and J. Vermylen, *Thromb. Haem.*, 1984, **52**, 364.
131. P. Gresele, M. Deckmyn, J. Arnout, J. Lemmens, W. Janssens and J. Vermylen, *Lancet*, 1984, **i**, 991.
132. G. De Gaetano, C. Cerletti, E. Dejana and J. Vermylen, *Drugs*, 1986, **31**, 517.
133. H. Reiss, E. Hiller, B. Reinhardt and C. Branning, *Thromb. Res.*, 1984, **35**, 371.
134. R. W. Randall, K. E. Eakins, G. A. Higgs, J. A. Salmon and J. E. Tateson, *Agents Actions*, 1980, **10**, 553.
135. G. A. Higgs and K. G. Mugridge, *Br. J. Pharmacol.*, 1982, **76**, 284P.
136. M. Hamberg, *Biochim. Biophys. Acta*, 1976, **431**, 651.
137. J. Harvey, H. Parish, P. P. K. Ho, J. R. Boot and W. Dawson, *J. Pharm. Pharmacol.*, 1983, **35**, 44.
138. B. R. Allen and S. M. Littlewood, *Br. Med. J.*, 1982. **285**, 1241.
139. J. A. Salmon, G. A. Higgs, L. Tilling, S. Moncada and J. R. Vane, *Lancet*, 1984, **ii**, 848.
140. G. A. Higgs, R. J. Flower and J. R. Vane, *Biochem. Pharmacol.*, 1979, **28**, 1959.
141. G. A. Higgs, K. E. Eakins, K. G. Mugridge, S. Moncada and J. R. Vane, *Eur. J. Pharmacol.*, 1980, **66**, 81.
142. G. A. Higgs, K. G. Mugridge, S. Moncada and J. R. Vane, *Proc. Natl. Acad. Sci. USA*, 1984, **81**, 2890.
143. K. M. Mullane, N. Read, J. A. Salmon and S. Moncada, *J. Pharm. Exp. Ther.*, 1984, **228**, 510.
144. L. B. Lauter, M. R. El Kharib, J. A. Rising and E. Robin, *Ann. Intern. Med.*, 1973, **79**, 59.
145. K. M. Mullane and S. Moncada, *Prostaglandins*, 1982, **24**, 255.
146. B. J. Everitt, J. A. Bentley, W. P. Spiegel and N. A. Porter, *Pharmacologist*, 1979, **21**, 153.
147. R. Patterson, J. L. Pruzansky and K. E. Harris, *J. Allergy Clin. Immunol.*, 1981, **67**, 444.
148. R. Patterson, R. Orange and K. E. Harris, *J. Allergy Clin. Immunol.*, 1978, **62**, 371.
149. C. A. Appleby, B. A. Wittenberg and J. B. Wittenberg, *J. Biol. Chem.*, 1973, **249**, 3183.
150. E. J. Corey, J. R. Cashman, S. S. Kantner and S. W. Wright, *J. Am. Chem. Soc.*, 1984, **106**, 1503.
151. W. P. Jackson, P. J. Islip, G. Kneen, A. Pugh and P. J. Wates, *J. Med. Chem.*, 1988, **31**, 499.
152. J. E. Tateson, R. W. Randall, C. H. Reynolds, W. P. Jackson, P. Bhattacherjee, J. A. Salmon and L. G. Garland, *Br. J. Pharmacol.*, 1988, **94**, 528.
153. A. N. Payne, L. G. Garland, I. W. Lees and J. A. Salmon, *Br. J. Pharmacol.*, 1988, **94**, 540.
154. G. A. Higgs, R. L. Follenfant and L. G. Garland, *Br. J. Pharmacol.*, 1988, **94**, 547.
155. A. L. Willis, K. Comai, D. C. Kuhn and J. Paulsrud, *Prostaglandins*, 1974, **8**, 509.
156. O. Oelz, H. W. Seyberth, H. R. Knapp, B. J. Sweetman and J. A. Oates, *Biochim. Biophys. Acta*, 1976, **431**, 268.
157. R. J. Gryglewski, J. A. Salmon, F. B. Ubatuba, B. C. Weatherly, S. Moncada and J. R. Vane, *Prostaglandins*, 1979, **18**, 453.
158. B. R. Culp, B. G. Titus and W. E. M. Lands, *Prostaglandins Med.*, 1979, **3**, 269.
159. J. Dyerberg, K. A. Jorgensen and T. Arnfred, *Prostaglandins*, 1981, **22**, 857.

160. R. Lorenz, U. Spengler, S. Fischer, J. Duhm and P. C. Weber, *Circulation*, 1983, **67**, 504.
161. S. Fischer and P. C. Weber, *Nature (London)*, 1984, **307**, 165.
162. S. Fischer and P. C. Weber, *Biochem. Biophys. Res. Commun.*, 1983, **116**, 1091.
163. P. Needleman, A. Raz, M. Minkes, J. A. Ferrendelli and H. Sprecher, *Proc. Natl. Acad. Sci. USA*, 1979, **76**, 944.
164. E. A. Higgs, S. Moncada and J. R. Vane, *Prog. Lipid Res.*, 1986, **25**, 5.
165. D. Kromhout, E. B. Bosschieter and C. De Lezenne Coulander, *New Engl. J. Med.*, 1985, **312**, 1205.
166. L. A. Simons, J. B. Hickie and S. Balasubramaniam, *Atherosclerosis*, 1985, **54**, 75.
167. R. Saynor, D. Verel and I. Gillott, *Atherosclerosis*, 1984, **50**, 3.
168. B. E. Phillipson, D. W. Rothrock, W. E. Connor, W. S. Harris and D. R. Illingworth, *New Engl. J. Med.*, 1985, **312**, 1210.
169. P. Singer, I. Berger, K. Luck, C. Taube, E. Naumann and W. Goedicke, *Atherosclerosis*, 1986, **62**, 259.
170. S. M. Prescott, *J. Biol. Chem.*, 1984, **259**, 7615.
171. T. Terano, J. A. Salmon and S. Moncada, *Biochem. Pharmacol.*, 1984, **33**, 3071.
172. T. H. Lee, R. L. Hoover, J. D. Williams, R. I. Sperling, J. Ravalese, B. W. Spur, D. R. Robinson, E. J. Corey, R. A. Lewis and K. F. Austen, *New Engl. J. Med.*, 1985, **312**, 1217.
173. T. Terano, J. A. Salmon, G. A. Higgs and S. Moncada, *Biochem. Pharmacol.*, 1986, **35**, 779.
174. N. Kromann and A. Green, *Acta Med. Scand.*, 1980, **208**, 401.

6.3

Agents Acting on Passive Ion Transport

PETER M. MAY

Murdoch University, Perth, Australia

6.3.1 METAL IONS IN BIOLOGICAL SYSTEMS	175
6.3.1.1 The Physiological Role of Metal Ions	176
6.3.1.2 Metal Ions as Drug Targets	177
6.3.1.2.1 Chemical selectivity	177
6.3.1.2.2 Biochemical competition	178
6.3.1.2.3 Physiological transport	179
6.3.1.2.4 Metal binding in medicine	180
6.3.2 METAL BINDING BY PHARMACEUTICALS	181
6.3.2.1 Chelating Therapeuticals	181
6.3.2.1.1 Essential metals	182
6.3.2.1.2 Non-essential metals	184
6.3.2.1.3 Synergistic chelation therapy	185
6.3.2.2 Cytotoxic Agents	186
6.3.2.2.1 Anticancer compounds	186
6.3.2.2.2 The ionophores	188
6.3.2.2.3 The tetracyclines	189
6.3.3 FUTURE DEVELOPMENTS	190
6.3.4 REFERENCES	190

6.3.1 METAL IONS IN BIOLOGICAL SYSTEMS

Two interesting paradoxes are associated with the subject of this chapter. First, the metal-binding agents to be discussed are often used routinely in the clinic but, in spite of their major practical applications, they are commonly regarded as peripheral to the mainstream of medicinal chemistry. Second, the great potential which metal binding in medicine undoubtedly has for making systematic progress against a wide variety of diseases has not been realized, even though many of the underlying chemical principles have been established for decades.

Both of these paradoxes arise because the agents being considered interact with metal ions, most commonly the transition metal ions. Whilst the importance of metal ions is well recognized, the medical implications of inorganic rather than organic chemistry tend to be seen as an esoteric aspect of biochemistry. Of course, this view reflects the way chemical disciplines are studied (much more than it represents any fundamental biological division) but it does raise some worrying barriers to comprehension.

The second of the paradoxes noted above also has worrying implications for those interested in drug design. Metal ions are employed in biology[1] for a number of their chemical characteristics (such as their reversible binding and readily interconverted redox states), presumably because they can be used for these attributes more easily and cleanly than organic alternatives. Although the basic chemistry involved is now very well understood at the level of individual reactions, the same cannot be said for the overall behaviour of either metal ions or metal-binding agents present in a biological milieu. The labile nature of many of the reactions is, perhaps, the most obvious reason: it contrasts sharply with normal organic experience, making it particularly difficult for experimentalists to interpret their observations correctly and to avoid artefacts.

The main object of this chapter is to summarize what is known about the action of metal-binding agents *in vivo*. This will indicate how metal binding impacts on the design and testing of drugs. To put this into perspective, it is first necessary to discuss the general role played by metal ions in biological systems.

6.3.1.1 The Physiological Role of Metal Ions

Historically, the physiological role of the trace metals has been established in two main ways: (i) by depriving laboratory animals of them until signs of ill health appeared (ensuring these were reversible by the sole addition of the element to the deficient diet), and (ii) studying deficiency syndromes in humans, usually as caused by defects in the metabolism of the element in question. With these unsubtle and limited techniques, a great deal of knowledge has been accumulated.[2]

Figure 1 shows that some 30 elements have thus been found essential to at least one living species;[3] about 20 have been associated with deficiency symptoms in human beings and some 13 are needed by every known form of life. Of special interest in the present context is the fact that the transition series metals, which number nearly one third of the 20 elements required by humans, account for only some 10 g of a (standard) 70 kg person. Moreover, iron accounts for rather more than 50% of this amount so that the others are only present in very small quantities indeed. In spite of this, anyone deprived of copper, zinc, manganese, *etc.* for particularly long periods suffers a deterioration in health which, ultimately, can be fatal.

Coupled with such low levels of abundance, this obligatory need for some trace elements clearly points to their general biochemical function—although they do have structural and regulatory roles as well, they act mainly as catalysts.[4-6] Very frequently, their catalytic activity is achieved in association with enzymes. Sometimes the metal ion is firmly imbedded in the protein structure, forming a metalloenzyme with a small but invariant number of metal atoms per molecule, which loses its activity if the metal is removed, for example by dialysis. On other occasions, a cation is required as a cofactor but is not an integral part of the enzyme molecule. Metal involvement of both sorts has been proven as essential in the function of about one fifth of all known enzymes; probably many more depend on metals in a less clear-cut but nevertheless significant way.

In fact, all living systems show a propensity to accumulate the transition metals whether they be physiologically essential or not.[7] For example, the minute traces of radioactive metals such as plutonium in seawater are concentrated so efficiently by certain crustaceans that, even though the animal can have no biological need for them, its intestines can become (at least in some countries) officially classifiable as a radioactive source! As is widely appreciated by the general public, the assimilation of many of these 'heavy metals' is potentially harmful to health. Lead, cadmium and mercury are especially notorious (see, for example, refs. 8 and 9). They interfere with enzymes and other biomolecules, often by binding to sulfhydryl groups.[10] However, it is also important to realize that even the essential metals may accumulate beyond desirable levels and that all metals are toxic when they reach sufficiently high concentrations.[2] As Paracelsus said, 'All substances are poisons: there is none which is not a poison. The right dose differentiates a poison and a remedy' (cited in ref. 11). As a mildly amusing but effective illustration, note that in addition to being notoriously toxic, arsenic also appears[12] in Figure 1 as an essential element.

Strictly speaking, it is the balance between various metals rather than just their absolute concentrations that is overridingly important. It is, for example, often possible to offset the ill effects

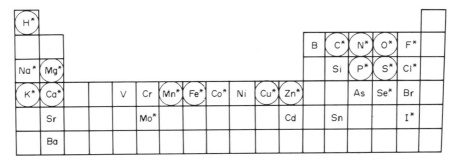

Figure 1 Biologically essential elements. The elements shown are known to be essential to at least one species. A circle denotes that the element appears to be essential to *every* known form of life. An asterisk marks elements associated with deficiency symptoms in humans

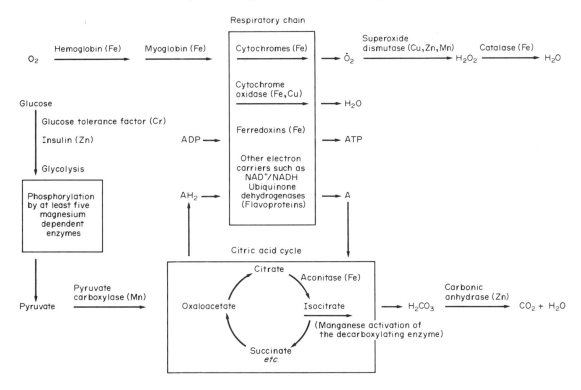

Figure 2 Schematic representation of the role of metals in the production of metabolic energy

of an imbalance in one metal by altering the levels of another (see, for instance, ref. 13). This reflects interactions *in vivo* that occur through competition for binding sites on enzymes and/or metal storage proteins. Much has been published on this matter but few reports based on experiments with animals elucidate what is happening at the molecular level. (For example, it is always difficult to exclude the possibility that the primary effect of the one metal is not simply to exclude the uptake of the other, either during gastrointestinal uptake or assimilation through the cell membrane.[14]) This chapter will be almost exclusively concerned with organic agents acting on ion transport (rather than with the effects of other metal ions).

The concept that proper biological function and, hence, good health depends on optimum *in vivo* concentrations of metal ions is central to everything discussed in this chapter. Metal-binding agents exert many of their pharmacological properties by altering the levels of metal ions *in vivo* relative to their overall optimum concentrations. In essence, they are able to activate and inactivate enzymes by regulating the availability of metal ion cofactors and/or inhibitors.

It is impossible here to discuss comprehensively the particular physiological functions of metal ions. This should not obscure an important point: they are involved in every major arena of biochemistry.[15] Metalloenzymes participate in the synthesis and in the degradation of proteins, nucleic acids, carbohydrates and lipids.[15, 16] As a single illustration of this, Figure 2 shows just how intimately the transition metals interweave themselves throughout the realm of bioenergetics. They appear at many stages in the biochemical reaction of molecular oxygen and glucose (taken as a representative of dietary carbohydrates) to form carbon dioxide and water. Their impressive generality and ubiquity form the basis of the hope mentioned at the start of this chapter, namely that, by adroit manipulation of metal concentrations *in vivo*, progress can be made in controlling a host of disease states.

6.3.1.2 Metal Ions as Drug Targets

6.3.1.2.1 *Chemical selectivity*

Broad chemical principles for predicting the relative strength of the binding between a given metal ion and a given ligand are presented in most textbooks on advanced inorganic chemistry (*e.g.* ref. 17).

There are distinct preferences shown between certain types of metal ion and particular electron-donating functional groups which permit various generalizations to be made about the likely compatibility of metal ion and ligand. In essence, both metals and ligands are divided into two categories on the basis of their known chemical properties—bonding interactions are likely to be favourable if the acceptor and the donor both belong to the same group, *i.e.* are alike in their electronic characteristics. Metal ions and donor atoms that are mostly of an ionic nature bind together more strongly than they would with partners which are mostly covalent and *vice versa*. This forms the basis of the so-called HSAB (*H*ard and *S*oft *A*cids and *B*ases) approach.[18,19]

Other chemical criteria for enhancing the selectivity of interactions between metal ions and ligands are also well established.[20] These include the effects of certain types of donor atom to stabilize electrons in *d* orbitals. (Donor atoms are arranged in an order, known as the 'spectrochemical series', which indicates how likely they are to stabilize metal ions with certain *d* orbital configurations, most notably Cu^{II} and Ni^{II}.) Ligand molecules with structures that fit around a particular ion most snugly can also give rise to very pronounced selectivity as shown by some of the ionophore compounds (*vide infra*).

Good metal–ligand selectivity also depends on the context in which binding takes place. The multicomponent nature of biological fluids is an obvious factor to be considered but one which is nevertheless often overlooked or underestimated. The lability of many metal–ligand interactions of interest means that they must be regarded as participating in a system of equilibria that manifests itself in the biological fluid.[21] Generally speaking there will be hundreds of possible components and many thousands of reactions to consider.[22] It is much more difficult to predict the outcome of this competitive environment than many researchers blithely imagine: implicit guesses based on the magnitude of the equilibrium constant of the 1:1 metal–ligand reaction frequently appear in the literature. Typically, no consideration is given to the generally dominant effects of ligand binding by protons and/or other metal ions. The formation of ternary complexes is another confounding possibility which tends to be ignored.

The only reliable way to ascertain the outcome of the equilibrium competition and, hence, to identify which chemical species are most stable in the multicomponent mixture is to perform a computer simulation using thermodynamic parameters individually determined for each of the relevant reactions.[22] This information lies well outside the present capabilities of all known analytical techniques. Of course, computer simulations cannot be thought to tell exactly what is happening in the biological fluid: apart from the practical difficulties in properly performing such calculations for mixtures of such complexity (as discussed below), there are important effects to consider of reaction kinetics and of the transport between biological compartments. But such calculations can and should be used, rather than making rash chemical judgements.

6.3.1.2.2 *Biochemical competition*

Metal ions in biological fluids can exist either as complexes or as the so-called free ions (meaning not associated with anything but water molecules). The complexes span a wide range of sizes but, just for convenience, they tend to be classified as being of high or of low molecular weight, depending on whether or not the metal ion is bound to a macromolecule, typically a protein. Low molecular weight, naturally occurring ligands such as amino acids generally form complexes that are labile; high molecular weight complexes divide themselves, fairly distinctly, into both the labile and the inert categories.

Metal ions such as sodium and potassium occur in solution almost entirely uncomplexed, *i.e.* as the free ions[23] (the only important exceptions being provided by their ionophore complexes mentioned below). They act as charge carriers and participate in very fast biochemical reactions. The other 'bulk' ions, calcium and magnesium, are also present to some degree as free ions; however, their high molecular weight complexes tend to predominate. They act largely as biological triggers and they participate in moderately fast reactions.[24] In sharp contrast, the transition metals are found to exist virtually entirely as complexes. Their free ion concentrations are almost never more than a diminishingly small fraction of their total amount. A very elegant assay published recently[25] reveals the free ion concentration for Zn^{II} in plasma to be about 2×10^{-9} M. The free ion concentrations of Cu^{II} and Fe^{III} in blood plasma must be considerably lower than this and have been estimated[22] to be about 10^{-18} M and 10^{-23} M respectively. Even the low molecular weight complexes are present only in very small concentrations—a figure for Cu^{II} of 10^{-12} M, say, compared with a total concentration in plasma of about 10^{-6} M.

The extraordinarily low concentrations of the free transition metal ions, particularly of Cu^{II} and Fe^{III}, mean that these species are themselves not able to participate mechanistically in the metabolic processes of these elements. Many reported experiments are rendered valueless by the use of unrealistically high free ion concentrations.[25] Reactions which have kinetic requirements that exclude macromolecular complexes (such as passive diffusion through biological membranes) must thus be mediated by low molecular weight complexes rather than the free ions. These complexes also play an important physiological role in facilitating exchange of the metal ion between macromolecular binding sites.

Given that most compounds with electron donor groups (*i.e.* the vast majority of organic compounds) will tend to interact with metal ions in a well-established order when introduced to metal ions in a test tube, there is a natural inclination to assume that this order will also manifest itself between agents in a biological fluid. For example, it is generally accepted that Fe^{III} is much more likely to form strong complexes than Mn^{II}, so it may seem reasonable to think that administered chelating agents will almost inevitably target *in vivo* Fe^{III} in preference to Mn^{II}. Indeed, the *in vitro* selectivity makes it difficult to see how Mn^{II}, Zn^{II}, Pb^{II} and even Cu^{II} can ever be sequestered by a chelator in a biological fluid that contains roughly equivalent concentrations of Fe^{III}. This kind of thinking created a major conceptual difficulty for early research with metalloenzymes because it was impossible to understand how Zn^{II}-binding enzymes were not immediately substituted with the more strongly bound, and almost as abundant, Cu^{II} ion. The answer, of course, lies in the generally staggered levels of the free ions present in most biological fluids: Na^{I}, $K^{I} > Ca^{II}$, $Mg^{II} > Zn^{II}$, $Mn^{II} > Cu^{II} > Fe^{III}$. However, what is not sufficiently appreciated, even today, is that this reversal of metal-binding order does not require a particularly delicate biological balancing act.[14]. It is a natural consequence of the binding of these metals by a multicomponent soup of organic compounds. To achieve metal ion selectivity, therefore, one need only find a chemical structure that has just a slight preference for the target ion compared with most other compounds. Nature sometimes overrides this inherent arrangement of free metal ion concentrations by maintaining different overall levels in different compartments (*e.g.* the sodium/potassium and calcium/magnesium ratios are reversed inside and outside of cells) but this can only be achieved at some metabolic cost.

The properties of all agents which are potentially capable of complexing metal ions *in vivo* must be assessed in the context of the competitive metal-binding environment described above. In general, the situation can be summarized as follows: the alkali metal ions are present in high concentrations but they are immune to binding unless the agent has very unusual and selective abilities; the transition metal ions are present in lower concentrations and tend to be strongly bound to endogenous ligands, mainly proteins, which make sequestration difficult both kinetically and thermodynamically; the alkaline-earth metal ions take up a position intermediate between these two extremes.

6.3.1.2.3 *Physiological transport*

The electrical charge on metal ions in solution constitutes an obvious impediment to their transport across biological membranes. With the alkali and alkaline-earth metals, large fluxes are required to maintain essential concentration gradients and, consequently, specific mechanisms involving active expenditure of metabolic energy have evolved to perform this task. These processes are dealt with in detail in other chapters. They contrast profoundly with the general picture which is emerging for the transition metal ions, where the mechanisms tend to be passive and are frequently non-specific. Most data in this regard have been obtained from studies on iron, copper and zinc, so the following account concentrates on these elements. Once again, it is impossible to consider individually the large number of contributions which have recently been made in this area and for which there are some excellent, comprehensive reviews;[26, 27] space permits only a general summary of those aspects which seem common to the above-mentioned transition metals.

Uptake from the gastrointestinal tract generally occurs predominantly in the duodenum. Transition metal ions released by food digestion in the stomach are initially kept in solution due to the prevailing acidity but, if the ions are not rapidly taken up into mucosal cells, they are lost by precipitation as the pH rises. This is the first of two barriers which act to limit assimilation of both essential and toxic trace metals. The second barrier operates within the mucosal cell itself; if the transfer of the metal into plasma can be delayed or prevented by intracellular binding, the metal will be lost through epithelial sloughing. There have been dozens of reports regarding the action of endogenously secreted 'factors', which are supposed to regulate assimilation of transition metal ions during the absorption stage (see, for example, refs. 28 and 29). However, none of these claims has withstood the tests of time and independent scrutiny. It seems that basic chemical considerations account well enough for what is observed: absorption is promoted by agents which keep the ion in solution and can either deliver it to the membrane or form a complex which can readily diffuse into the mucosal cell.[30] Dietary or administered agents which produce electrically neutral and/or lipophilic complexes help; high molecular weight ligands that bind metals strongly and cannot be absorbed (like phytate) hinder.

In plasma, special transport proteins carry the metal ion from the serosal interface to its primary metabolic destination, usually the liver. Cu^{II}, Zn^{II} and most toxic transition metals are bound and transported by albumin. The majority of these interactions are only partly or non-specific but a special binding site for Cu^{II} exists at the N terminus of the protein of most mammalian species including *Homo sapiens*. Iron, on the other hand, has its own special transport protein called transferrin.[31] Whilst being extraordinarily specific in its affinity for iron, transferrin is also known to carry a few other metal ions, such as plutonium. The ability of chelating agents to remove metal ions whilst they are being transported in plasma (*i.e.* before they become deposited in the tissues) is a major factor determining their effectiveness.

In cells, the metal ions which are not immediately required for metalloenzyme synthesis or other metabolic purposes are retained in special storage proteins. Metallothionein holds zinc and copper, whilst iron is deposited in ferritin.[32] Synthesis of both proteins is induced by increasing intracellular levels of their respective free metal ions. They both have unusually high metal contents (one molecule of ferritin, for example, typically contains thousands of iron atoms). They are also remarkably similar in the way their metal ions are released: whilst they will give up metal ions to chelating agent competitors, major turnover is achieved by normal protein decay processes taking place within the cell. Presumably, this cycle of protein synthesis and degradation establishes a steady state that effectively buffers intracellular metal ion levels.

The transition metals are mainly excreted *via* the faeces, with the bile as the major pathway (except in the case of zinc, which is excreted across the intestinal wall).[33] Recently, the way in which the essential transition metal ions are transported into and out of hepatocytes has become much clearer.[34-37] The evidence for a dynamic equilibrium between a labile, intracellular 'pool' of the metal and the metal-binding proteins in plasma has crystallized, at least for iron, copper and zinc. The first convincing suggestions of this came from studies of chelating agents with the ability to sequester a dialyzable, low molecular weight fraction of the intracellular pool of iron in Chang cells.[38] However, the concept has since been amply confirmed by quantitative analysis of the concentrations[39] and of the transport kinetics (*vide infra*).

In general, the hepatocyte transport processes are saturable and they are bidirectional (*i.e.* they have the same parameters for both uptake and efflux). With copper, the exchange has been shown to be effected by a low molecular weight complex, almost certainly the electrically neutral bishistidinato species.[40] Interestingly, the histidine does not itself pass through the membrane; this implies that the neutrality of the complex merely permits it to enter/approach the membrane, not to diffuse through it. The kinetic information suggests that the mechanism is one of passive facilitated transport.[41,42] Iron, on the other hand, enters hepatocytes bound to its transport protein. After transferrin has attached itself to specific receptors on the membrane surface, it is endocytosed into a vesicle where the iron is released.[43] The transferrin is subsequently returned, undegraded, to the extracellular medium.[44] With both iron and zinc, there is an unsaturable component to uptake (in addition to the saturable component mentioned above), which seems likely to be of significance under conditions of metal ion overload.[45]

6.3.1.2.4 *Metal binding in medicine*

In accordance with their important and ubiquitous biochemical roles, metals are involved in a strikingly diverse number of clinical conditions, as listed in Table 1.

About one third of the items shown are symptoms associated with overt toxicity. Industrial exposure accounts for most of these but some, such as hemochromatosis and Wilson's disease, are due to physiological defects; others, like transfusional siderosis arise as secondary complications to excessive intakes. There are few items in Table 1 due to overt metal deficiencies. Apart from iron, most trace elements are required in such small quantities and are so well absorbed that it is difficult to induce deficiency. Even in chronic malnutrition other nutrients tend to be the limiting factors.

In a few cases, overt deficiency symptoms do arise. As a rule, this either occurs very rarely (*e.g.* in the dwarfism and hypergonadism associated with deficient absorption of dietary zinc) or it is caused by rather special circumstances (as in cases where infusions for total parenteral nutrition must be

Table 1 Clinical Conditions Associated with Metals

Condition	Metal	Condition	Metal
Acne	Zn	Hemochromatosis	Fe
Acrodermatitis enteropathica	Zn	Hemolysis	Cu
Alcoholism	Fe, Zn	Hypergonadism	Zn
Alzheimer's disease	Al?	Hypertension	Cd, Zn?
Anaemia	Fe	Immune deficiency	Cu, Zn
Aspermia	Zn	Kidney failure	Cd, Hg, Pb
Bantu siderosis	Fe	Leishmaniasis	Sb
Behavioural disorder	Pb	Liver necrosis	As, Cd, Cu, Fe
Cancer	Cu, Fe, Zn	Manic depression	Li
Cardiovascular disease	Cd, Fe, Mg, Pb	Menke's syndrome	Cu
Celiac disease	Zn	Neurological disorder	As, Cu, Mn, Pb
Cirrhosis	Cu	Osteomalacia	Al, Cd
Crohn's disease	Zn	Parakeratosis	Zn
Congenital abnormality	Mn, Zn	Psoriasis	Zn
Cystic fibrosis	Mn?, Zn?	Pulmonary collapse	Cd, Hg, Mn, Zn
Dermatitis	Ag, Au, Cr, Ni, Zn	Rheumatoid arthritis	Cu, Fe
Diabetes	Zn	Taste acuity loss	Zn
Down's syndrome	Cu?, Zn?	Transfusional siderosis	Fe
Dwarfism	Zn	Ulcers	Bi, Cu, Zn
Encephalopathy	Al, Pb	Wilson's disease	Cu
Gastroenteritis	As, Cd, Cr, Fe, Hg	Wound healing impairment	Zn
Gout	Pb		

supplemented with copper, zinc and chromium[46]). However, there are two important exceptions to this rule. Marginal zinc deficiency (for example, leading to acrodermatitis enteropathica, loss of taste acuity and slow wound healing) is one. The other, even more important, has already been noted: iron deficiency is a common cause of anaemia, especially in women living on refined western diets.

This leaves the majority of items in Table 1 representing less obvious connections between the metal and the medical condition. Sometimes there is no doubt about the link (*e.g.* with zinc through insulin in diabetes) but often it is fairly obscure. Nevertheless, the general impression of a comprehensive medical involvement with metals is overwhelming.

It is noteworthy that, in spite of this widespread role for metals in medicine, there are relatively few therapeuticals which are specifically thought of as metal binding and even fewer which actually contain a metal in the form that they are administered. Of the latter, the various gold-containing agents used in the treatment of rheumatoid arthritis[47,48] and the radionuclide-containing contrast agents used, for example, in imaging tumours[49,50] deserve to be mentioned specifically.

The importance of free radical reactions in tissue injury and the initiation of disease has received considerable attention in recent times.[51-53] Metals, especially iron and copper, are manifestly involved in both the generation and biochemical control of these dangerous species.[54] This reflects their chemical tendency to engage in rapid electron exchange reactions. However, since they can be responsible for the problem as well as being a potential remedy, it has not been easy to understand their physiological effects. This ambivalence is very evident in the confusion surrounding possible roles for metals in both the cause and the cure of diseases such as cancer and rheumatoid arthritis. Some noble metals have been advocated for treatment of the latter disease; in particular gold compounds are prescribed for the treatment of arthritis and rheumatism. A possible role is their quenching of singlet oxygen to prevent further tissue damage.[108]

6.3.2 METAL BINDING BY PHARMACEUTICALS

6.3.2.1 Chelating Therapeuticals

A number of substantial reviews on chelation therapy have appeared recently.[20,55,56] Table 2 shows the drugs of choice presently used to treat various kinds of metal poisoning.

The agents used for clinical purposes provide the most clear-cut information about the behaviour of chelating compounds *in vivo*. This is because they have been long established as pharmaceuticals and, consequently, their mechanisms of action have been intensively studied. Also, since they bind

Table 2 Clinical Chelating Agents

Rank (approx.)	Metal	Drug of Choice
1	Fe	Desferrioxamine
2	Cu	Penicillamine, triethylenetetramine
3	Pb	BAL, EDTA
4	Au	BAL
5	Al	Desferrioxamine
6	Cd	None available
7	Ni	Diethyl dithiocarbamate
8	Hg	BAL
9	As	BAL
10	Pu	DTPA

strongly and tend to exhibit their effects in a direct way, they make it much easier to interpret the results of experiments compared, say, with other compounds discussed later in this chapter.

Two features characterize all of the chelating therapeuticals in current clinical use: (i) they are very powerful metal-binding agents (sometimes, but not always, showing great selectivity for their target ion); and (ii) their therapeutic properties are closely related to, and often limited by, their ability to cross biological membranes. These facts have inclined many researchers to believe that by enhancing selectivity and modifying the lipophilicity of compounds, improved agents would result. The striking lack of progress which has been made over the past 30 years, in spite of many efforts to introduce new chelating agents into the clinic, argues convincingly that this view is far too simplistic.

It is now clear[20] that the design of better chelating agents requires a broad understanding of the whole sequence of mechanistic steps that are involved in the way a toxic metal is sequestered from deposits within the body and in the way it may subsequently be excreted. Ideally, this involves transport of the agent from the blood stream into the target tissue, to maximize mobilization of the metal from intracellular binding sites, followed by the return of the resulting complex into the blood and thence into the urine for excretion. Alternatively, if the toxic metal is mainly deposited in the liver, the complex must be well suited to carry the metal ion into the enterohepatic circulation for removal in the faeces. In either case, the effectiveness of the agent requires a combination of properties enabling it to surmount a variety of physiological hurdles.

In addition to the evident difficulty, encountered with all drug design, of finding compounds that must satisfy a multiplicity of requirements, the search for new chelating therapeuticals is hindered by another: many of the chemical properties one would select to achieve particular aspects of ideal chelating behaviour tend to be mutually exclusive.[20] For example, only compounds which are reasonably lipophilic are able to penetrate the various tissues in which toxic metals accumulate, but increasing lipophilicity is strongly correlated with increasing toxicity and it also tends to facilitate redistribution of the toxic metal into other tissues rather than to promote excretion. This inherent contradiction between desirable characteristics even extends to the quest for strong metal binding: coordinating functional groups tend to make the compound polar and, hence, tend to confine it to the blood stream; moreover, the more powerful the chelator, the more likely it is to interact adversely with the function of essential metal ions.

Many of the limitations with current chelation therapies can be attributed to such problems. Only by understanding how they arise and working out how they can be resolved will it be possible to achieve renewed improvements in this kind of therapy.

6.3.2.1.1 *Essential metals*

Only two essential metals give rise to toxicity symptoms sufficiently frequently to discuss here. These are iron and copper.

(i) *Iron*

Chelation therapy for iron is a clinical concern throughout the western world, both for acute and chronic poisoning.[57] Acute incidents are mainly associated with accidental overdoses in young children who get hold of iron supplement tablets from the family medicine cupboard. Over 2000 cases (in which large numbers of the brightly coloured, sweet-like pills had been consumed) used to be reported annually in the USA. Mortality rates were typically as high as 45%.[58] Fortunately, the development of child-proof bottles and of effective chelation treatments has dramatically improved this appalling statistic.

Chelation therapy for chronic iron overload is also a problem that is mainly confined to the so-called developed countries. Although iron overload conditions do occur elsewhere, particularly in Africa, they can usually be treated very effectively by phlebotomy.[59] However, this approach is not possible in the case of β-thalassaemia, a hereditary disease associated with defective development of hemoglobin. Poisoning by iron occurs as a secondary consequence of the long term blood transfusions required for the treatment of these patients. Iron accumulates from the turnover of administered hemoglobin because, uniquely amongst trace elements, there is no normal physiological mechanism for its excretion.[57] Left untreated, iron deposition causes heart and/or liver failure and patients die in their early teens. However, through early and continuous chelation treatment, levels of iron can be held down and the chances of the patients surviving at least into adulthood are considerably enhanced.[60]

By far the most widely used agent for iron decorporation[61] is desferrioxamine (**1**). This compound is produced naturally by certain bacteria for the purpose of sequestering the metal from the very insoluble hydroxy precipitates which are formed under normal physiological conditions of pH. Desferrioxamine is capable of penetrating the various tissues in the body, especially the liver, where iron is deposited. It is thus effective in promoting iron excretion both in the faeces and the urine. However, it is degraded rapidly *in vivo* and it is unable to remove iron from transferrin for kinetic (not thermodynamic) reasons. Together, these limitations make the iron chelator much less effective than it otherwise might be.

$$H_2N-(CH_2)_5-\underset{\underset{OH}{|}}{N}-\underset{\underset{O}{\|}}{C}-(CH_2)_2-\underset{\underset{OH}{|}}{N}-\underset{\underset{O}{\|}}{C}-(CH_2)_5-\underset{\underset{OH}{|}}{N}-\underset{\underset{O}{\|}}{C}-(CH_2)_2-\underset{\underset{OH}{|}}{N}-\underset{\underset{O}{\|}}{C}-(CH_2)_5-\underset{\underset{OH}{|}}{N}-\underset{\underset{O}{\|}}{C}-Me$$

(1)

$$\underset{HO_2CCH_2}{\overset{HO_2CCH_2}{>}}N-CH_2CH_2-\underset{\underset{CH_2CO_2H}{|}}{N}-CH_2CH_2-N\underset{CH_2CO_2H}{\overset{CH_2CO_2H}{<}}$$

(2)

The compound DTPA (diethylenetriaminepentaacetic acid; **2**) has also been employed to treat iron toxicity in humans [62] but interest in it is mainly confined to the conclusions which can be drawn from certain experiments with animals. Since the compound is exceedingly hydrophilic, it is strictly confined to the blood plasma compartment. Thus, the urinary iron excretion it promotes must arise from a successful competition for the metal returning to plasma (from organs such as the liver) prior to the metal's sequestration by transferrin. It can be inferred that, as an even more powerful iron-binding agent, desferrioxamine would be capable of acquiring the metal in a similar way. Accordingly, it seems certain that this represents the source of the urinary iron which the latter agent also induces.

(ii) Copper

The other essential element associated with toxicity in the clinic is copper.[63] As with iron, instances of both acute and chronic poisoning are well documented in the literature. Acute cases have been mainly associated with ingestion of copper sulfate as a means of committing suicide[64] and with hemodialysis.[65] Chronic poisoning arises through a physiological defect that gives rise to the hereditary condition called Wilson's disease. The exact nature of the physiological malfunction causing Wilson's disease is unknown but it seems to be linked to an abnormality of copper metabolism in the liver. This leads to an inadequate excretion of the metal[66] and, in many cases, to deficient synthesis of the copper metalloprotein, caeruloplasmin. Accordingly, copper levels increase inexorably until individuals suffering from the disease die from the failure of vital organs, usually the liver. So high can internal concentrations of the metal become, that elemental copper is deposited as visible brown rings around the corneas of the eyes, giving rise to one of the most characteristic symptoms of the disease.

Nowadays there are two chelating agents used for copper decorporation in the clinic, penicillamine (3; $Me_2C(SH)CH(NH_2)CO_2H$) and triethylenetetramine (4; $NH_2CH_2CH_2NHCH_2CH_2$-$NHCH_2CH_2NH_2$). Both were introduced by Walshe,[67,68] a medical practitioner who with insight, persistence and some good fortune has radically transformed the prognosis of Wilson's disease almost single-handedly. He first tried penicillamine in 1952, apparently because it was thought likely to complex copper strongly and because, being an amino acid, he believed it would not be toxic. In fact, it is now known that this compound does not complex copper, at least not in a conventional way, because it reduces the metal from the Cu^{II} to the Cu^I state. Also, serious limitations due to toxicity were initially encountered, although these have now been much reduced by routine administration of the pure D isomer.[69] Drug tolerance nevertheless remains something of a problem for a small percentage of Wilson's disease patients.

For this reason, Walshe sought an alternative to penicillamine and, in 1972, introduced triethylenetetramine.[68] This proved to be almost as effective as penicillamine itself. It has thus become the standard treatment for Wilson's disease in those sufferers who develop an allergic response to penicillamine. These chelating agents are capable of extending the lives of patients by as much as 20 years or more. Sadly, many of those who die from this disease are diagnosed too late or not at all.[70]

6.3.2.1.2 *Non-essential metals*

Heavy metal poisoning is nowadays of much less clinical concern than it used to be. Greater public awareness of the potential danger and better industrial working conditions are mainly responsible for this improvement. Consequently, much of the literature in this area is concentrated on a few major incidents, attributable to particular circumstances, rather than any on-going cause for concern. For example, over the past couple of decades, mercury poisoning[71] has gained considerable notoriety from just two tragic episodes: about 50 fatalities in Minimata, Japan from industrial pollution affecting various seafoods and several hundred casualties in Iraq, as a result of contamination of wheat by certain mercury-containing fungicides. Similarly, localized cadmium pollution was responsible for the repeated outbreaks of Japanese 'itai-itai' disease, an excruciating condition characterized by extreme bone brittleness.[72]

In sharp contrast to the impression which may be created by the attention to such episodes, chronic rather than acute exposure to heavy metals is a more justifiable reason for concern.[73] Cadmium assimilation is increasingly being found responsible for increased mortality, hypertension and renal dysfunction in those living in industrialized environments, not only from exposure in the workplace[74] but also from more insidious sources such as cigarette smoking.[75] The same seems likely to be true of mercury poisoning, where careless or uninformed handling of the element (*e.g.* in laboratories or by school children playing with the contents of broken thermometers) represents a hazard that, whilst difficult to assess, may well be of more significance than is widely appreciated.

Of all the heavy metals, lead is the one requiring clinical treatment by far the most frequently. Lead from petrol, from some water pipes and, in peculiar circumstances, from old paint affects a strikingly large number of people, especially children.[76] Diminished heme synthesis and damage to the central nervous system are responsible for the commonest acute symptoms. As many as a few percent of those living in highly industrialized societies accumulate sufficient levels of lead to affect their health in a significant way. The detrimental impact of lead on early cognitive development is a controversial but nevertheless worrying issue.[77,78] Fortunately, all of the main sources of lead[79] are now under some control and their impact should decline substantially. However, it remains to be seen whether legal measures to limit the amount of lead emitted in vehicle exhaust fumes are going to solve, within a reasonable timescale, what has become a global problem.

None of the chelating agents available to treat chronic poisoning by these heavy metals is truly satisfactory. The most outstanding of those in clinical use has been known for nearly half a century; yet in several instances of metal poisoning it remains to be superseded by anything better.[20] This is BAL (2,3-dimercaptopropanol; $HSCH_2CH(SH)CH_2OH$; 5), the first chelating therapeutical and the one still recommended for treating lead, arsenic, gold and mercury toxicity. It is also one of, if not *the*, best of agents for mobilizing deposits of cadmium (but it is not recommended in cadmium poisoning since it may aggravate this metal's nephrotoxicity).

There are two reasons for the striking effectiveness of BAL compared with other compounds. First, the two sulfhydryl groups confer on it a very powerful affinity for so-called 'HSAB soft' metal ions such as mercury and cadmium. Secondly, both the agent itself and the complexes it forms with these metal ions are likely to exist as electrically neutral species in solution, hence being best capable of passively diffusing through biological membranes. This enables the chelating agent to access toxic metal deposits within organs such as the liver. But, as pointed out above, it also leads to a number of clinical disadvantages. BAL is an extremely noxious drug, being both inherently toxic and responsible for possible redistributions of heavy metals to even more sensitive organs than are initially affected. Moreover, the agent also complexes zinc *in vivo*, which would tend to counteract the protective effects of this element against poisoning by other heavy metals. Less toxic, water soluble chelating agents related chemically to BAL have been extensively tested in animals[80] but are not yet recommended for use in people.

Two other non-essential metals, responsible for much of the current research in chelation therapy, need to be mentioned in this section. These are aluminum and plutonium.

Aluminum poisoning is a relative newcomer.[81] It has, however, attracted much medical concern because it is associated with both an osteomalacia and an encephalopathy in patients with renal failure undergoing long term dialysis.[82] There has been considerable debate about the relative importance of two possible sources of the metal, namely the water supply used for dialysis and the aluminum-based drugs given to these patients to control hyperphosphataemia.[83] Poisoning by intravenous administration of aluminum-contaminated albumin solutions has been frequently reported in recent years (*e.g.* ref. 84). A strong tendency to hydrolyze, together with slow ligand exchange rates,[85] makes Al^{III} a difficult target for chelators.

In contrast to aluminum, poisoning by plutonium and other radionuclides has been actively researched since the start of the atomic age. Very many projects and huge amounts of money have been devoted to finding ways of decorporating radionuclides generally. This is in spite of there being only a handful of people who have ever been sufficiently contaminated to warrant chelation therapy.

Interestingly, the chelating agents of choice for treating both aluminum and plutonium poisoning are both associated with removal of iron. These are desferrioxamine and DTPA, respectively. Desferrioxamine also removes plutonium and is especially effective in combination with DTPA.[86] A degree of chemical similarity between iron and both aluminum and plutonium is no doubt partly responsible for the affinity these agents have for all three metal ions. However, another factor, probably of as much significance, is the interaction which occurs on the binding sites of endogenous ligands—iron mobilization by desferrioxamine and DTPA is likely to release simultaneously any aluminum or plutonium that has become involved in the normal metabolic processes for handling iron (such as by storage in ferritin).

6.3.2.1.3 *Synergistic chelation therapy*

Perhaps the most promising way to overcome the difficulties arising from the mutual exclusiveness of ideal properties, outlined above, is to administer not one but a combination of agents. The idea has been termed 'synergistic chelation therapy' to emphasize that two or more compounds may have an effect even greater than their sum.[87] This occurs because they are able, individually, to achieve different pharmacological objectives. In essence, the concept is that it should be easier to meet a sequence of demanding chemical goals if they do not all have to be accomplished by the molecular properties of a single compound.[20]

Whilst the theoretical potential of synergistic chelation therapy is considerable, it has only occasionally been realized in practice. However, where it has been sensibly tried, the evidence supporting the idea is promising.

In the treatment of lead poisoning, for example, the use of BAL in conjunction with EDTA (6) gives markedly better results than when either agent is administered on its own.[88] This synergism is undoubtedly due to the ability of BAL to access intracellular lead, whilst EDTA sequesters the metal ion in plasma. The effectiveness of the EDTA is greatly enhanced by BAL shuttling lead from cells

$$HO_2CCH_2 \diagdown \qquad \diagup CH_2CO_2H$$
$$NCH_2CH_2N$$
$$HO_2CCH_2 \diagup \qquad \diagdown CH_2CO_2H$$

(6)

into the plasma. Being confined to the plasma compartment, EDTA would otherwise have to rely on a kind of leaching process by endogenous ligands. Conversely, by chelating lead that enters the plasma, EDTA reduces the neurological problems associated with BAL administered on its own (due to the effects of a lead redistribution *in vivo*).

Another example of the potential which synergistic chelation therapy offers concerns iron decorporation. It is known that desferrioxamine is thermodynamically capable of removing iron from transferrin in plasma yet the transfer does not take place (at a significant rate). Since the major factor limiting iron removal by desferrioxamine is the short half-life of the agent *in vivo*, a very dramatic improvement in iron chelation therapy would be made if some way to overcome this kinetic obstruction could be found. One hope is that a small iron-binding molecule might be able to extract the iron from within the protein structure and transfer it to desferrioxamine under the thermodynamic driving force provided by the latter. Some experiments using the chelator NTA, nitrilotriacetic acid, have yielded auspicious results.[89]

6.3.2.2 Cytotoxic Agents

The bacteriocidal action of chelates and chelating agents has been a focus of medical interest for many years.[90] Studies of compounds such as 8-hydroxyquinoline and 1,10-phenanthroline, whose antimalarial activity is well understood in terms of complex formation *in vivo*, have provided the basis of much of our present understanding.

Metal binding can lead to cytotoxicity by one of three mechanisms:[91] (i) it can develop or enhance the inherent toxic properties of a metal; (ii) it can deplete the intracellular concentration of an essential metal; and (iii) not so obviously, it can facilitate uptake of an agent that is toxic but not metal binding and that cannot otherwise reach and/or penetrate the target cell. These mechanisms are clearly illustrated by the three main groups of clinical antibiotic whose therapeutic properties are known to reflect metal binding *in vivo*.

6.3.2.2.1 Anticancer compounds

Of all the agents discussed in this chapter none has had greater clinical impact in recent years than a group of anticancer compounds whose biological activity depends on the formation of a metal complex. These compounds fall into distinct subgroups on the basis of chemical structure and differences in pharmacological properties as listed in Table 3.

To offset any contrary impression which the following discussion may create, it is important to stress these differences: the various subgroups have distinct properties and are used for treating different types of cancer; indeed, a number of compounds shown in Table 3 are not even in clinical use. Neither is there complete consensus as to the mechanism by which these agents exert their cytotoxic effects. These aspects are well covered in the very large literature which exists about these compounds but they fall largely outside the scope of this review. Here, attention is given only to those similarities and associations which do occur between the compounds, particularly as it relates to the role of metal binding in their mode of action.

One of the most striking characteristics which these agents share is, simply, their marked effectiveness. Another is that their application in the clinic is either limited or prevented by side effects. Nevertheless, they have collectively provided clinicians with a battery of anticancer therapeuticals which, over the past decade or so, have transformed the prognosis of patients with certain cancers in a truly remarkable fashion. In particular, cisplatin (**7**) is the agent of choice for testicular

Table 3 Metal-binding Anticancer Compounds

Noble metal complexes	Aminoquinones	Quinones Anthracyclines	Glycopeptides	Bisdioxopiperazines
Cisplatin	Streptonigrin	Daunomycin	Bleomycin	Razoxane
	Mitomycin	Adriamycin	Tallysomycin	
	Porfiromycin		Phleomycin	
	Actinomycin		Peplomycin	
	Rifamycin			
	Geldanomycin			

and ovarian cancers; the anthracyclines (daunomycin and adriamycin; **8**), are widely used to deal with solid tumours in the breast, ovary and lung; and the bleomycins have provided a breakthrough in the treatment of Hodgkin's lymphoma as well as being active against some carcinomas of the skin. This general efficacy depends on the ability, needed by all good anticancer agents, to kill cells selectively, especially those undergoing replication. Excluding, for the time being, the bisdioxopiperazines (**9**), whose mechanism of action is not known, this reflects a critical property which all these compounds share, namely the capacity to interact with, and to impair, DNA.

Since the discovery of cisplatin by Rosenberg *et al.* in 1965,[92] many related inorganic complexes have been tested for antitumour activity.[93] Very impressive clinical results have been achieved. The metal ion coordinates to specific donor groups on adjacent guanosine residues of the nucleic acid, forming a ternary complex that bridges cross-strands of the macromolecule. The square planar geometry and *cis* configuration of the metal ion are essential structural characteristics. Expulsion of labile chlorides (or other good leaving groups) is also a critical step in the mechanism of action. A second generation of platinum drugs with carboxylate ligands is now being researched. It is supposed that their greater biological activity is due to secondary chemical and pharmacological properties rather than being the result of a different kind of cytotoxic mechanism.

The role of metal ions in the mechanisms of action of the quinone-containing compounds has recently become increasingly well understood,[94] although the details are by no means entirely certain as yet. There is no doubt that the drugs intercalate themselves into the DNA helix and that the damage they cause involves some kind of free radical reaction. Free radical scavengers (such as α-tocopherol) and chelators (such as desferrioxamine) significantly reduce DNA degradation. Although the involvement of hydroxyl radicals cannot be ruled out, superoxide, in particular, is implicated:[95] superoxide dismutase levels have been observed to rise appreciably in some experiments. At limiting concentrations of these drugs, the number of lesions they cause exceeds the number of molecules which are present, indicating that they act as catalysts. Finally, there is considerable evidence that the presence of both molecular oxygen and a transition metal ion is necessary for the DNA strand-scissioning which characterizes these quinone-containing drugs.

Hydroxy- and amino-quinones are powerful chelators of transition metal ions, especially iron and copper, and it is well known that complexes of these redoxing metals can promote the formation of free radicals. For example, the Fenton reaction (equation 1) is often cited in this context (although frequently overlooked is the fact that most experiments based on this reaction have employed unrealistically high levels of reactants and so they cannot be used to infer much of physiological consequence). More to the point, it seems that the one-electron reduction of molecular oxygen by Fe^{II} to form Fe^{III} may be mediated by the formation of an iron–quinone complex.[96] The redoxing capabilities of the quinone/semiquinone no doubt facilitate this and there is some experimental indication that the catalytic cycle is driven by the oxidation of intracellular thiols, possibly glutathione.

$$Fe^{2+} + H_2O_2 \rightarrow Fe^{3+} + OH\cdot + OH^- \tag{1}$$

Against this background, the suggestion that DNA cleavage results from the formation of a ternary (iron) complex which holds the drug in close proximity to the macromolecule makes an attractive hypothesis. Support for the idea also comes from the selectivity of the interactions, which clearly favour attack at particular sites on the DNA. This would be expected from a mechanism involving coordination; adventitious generation of free radicals would be more random in its effects. Some elegant model studies have been carried out in this area.[108]

A final point to mention here concerns the interesting and clinically important link that exists between the anthracyclines and Razoxane (ICRF 159; **10**). Razoxane is the bisdioxopiperazine actually employed in therapy. This and many similar compounds were developed by Creighton at the Imperial Cancer Research Fund in the hope that they would act against malignant growth by chelating essential metal ions intracellularly.[97] The administered compounds are known to diffuse through cell membranes and then to hydrolyze into EDTA-like molecules (10) as was originally intended. However, in spite of many years of investigation into their mechanism of action, this remains unclear.[98] Although metal ions are likely targets (because, for example, it is known that the drug does not interact with an asymmetric structure), this is still unproven. If, indeed, the drug does target a metal ion, there are few clues as to its identity (see, for example, ref. 99). However, the drugs have a marked ability to reduce the dose dependent cardiotoxicity of the anthracycline drugs (see, for example, ref. 100). The obvious conclusion which can be drawn from this, and from some other relevant evidence, is that competition for intracellular iron by the hydrolysis product of Razoxane restricts the peroxidation of lipids caused by the anthracycline–iron system.

$$HO_2CCH_2 \diagdown \qquad \qquad CH_2CO_2H$$
$$\diagup NCHCH_2N \diagdown$$
$$HO_2CCH_2 \diagup \quad \overset{|}{R} \qquad CH_2CONH_2$$

(10)

6.3.2.2.2 *The ionophores*

The ionophores are a very numerous family of naturally-occurring antibiotics that function by altering the permeability of cell membranes to metal ions.[101] They are synthesized by a variety of microorganisms, most notably of the genus *Streptomyces*. Their discovery was due to the observation that they uncouple oxidative phosphorylation in the mitochondria of rat liver (by inhibition of the enzyme ATPase). The transport of ions across membranes discharges concentration and potential gradients, which the cell strives to maintain. Ultimately, the cell dies because its metabolic capacity is overwhelmed.

These ionophores differ markedly from the other compounds discussed in this chapter in that they are avid binders of metal ions like potassium and sodium. Such ions are not at all strongly coordinated by most chelators. The agents have special molecular structures/configurations, with oxygen-dominated donor groups, designed to encapsulate ions with high charge/radius ratios. The electrostatic nature of the ion is thus shielded by the ionophore, permitting it to enter and to diffuse through both natural and synthetic lipid membranes.

Since the size of the cavity created by the ionophore critically determines the radius of ions which will fit, these compounds exhibit an extraordinary degree of selectivity. As an extreme example of this, valinomycin discriminates in favour of potassium ions over those of sodium by a factor of 10 000.[102] It is important to note, however, that it is the radius of the hydrated ion (rather than the true ionic radius) which is the determining factor: this causes ions with low atomic number to be the ones disadvantaged by large size since, having the highest intrinsic charge/radius ratio, they become surrounded by water molecules to the greatest extent.

The ionophores divide into several classes according to chemical make-up and mechanism of action.[101] Figure 3 shows the categories to which each of the best-known compounds belongs. Most common are the carrier-type, meaning that the ionophore wraps itself around the metal ion to catalyze its diffusion through the cell membrane. Such carriers are further subdivided amongst the neutral and the carboxylate groups. The former are cyclic peptides which, being in themselves electrically neutral molecules, form complexes which are (positively) charged. The latter are named for the carboxylate functions which characteristically counteract the charge on the metal ion to form transportable species which are neutral. Accordingly, and somewhat paradoxically, the neutral carriers have a direct effect on cell membrane potentials, whereas the carboxylate carriers do not.

As opposed to both of these carrier-type ionophores, there is another important class of compounds which act by forming channels through the membrane. In other words, unlike the carriers, these ionophores do not codiffuse with the ion but, instead, create a pore through which the ion can pass. The channel-forming ionophores are linear rather than cyclic peptides. They exhibit higher transport rates than the carrier-type compounds. As one might expect, their effectiveness is not as directly dependent on membrane thickness (although, if the membrane is too thick, the stability of the channel is lost).

Despite their seeming suitability, the ionophores have not proved useful in the decorporation of heavy metals. Whilst they have some very important applications (for example, as antibiotics and as

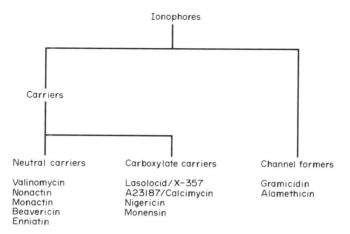

Figure 3   Classification of ionophores

exchangers in ion selective electrodes), oft-repeated justifications of research with these compounds as potential heavy metal chelating agents now ring hollow. It should be clear that simply playing molecular roulette with these structures is not going to achieve the quantum jump necessary to produce new clinical heavy metal chelating agents.

6.3.2.2.3   *The tetracyclines*

This is another numerous family of broad spectrum antibiotics. They have been of great clinical usefulness for over 30 years. Once again, concern is given here only to what is known about their interactions with metal ions. Although this information is limited, and in some respects controversial, it seems certain that the avid metal-binding capabilities of tetracycline (**11**) do play an important role in determining its pharmacological properties.[103] Their relative abilities to coordinate metal ions have been extensively investigated by Berthon and coworkers to determine whether they can account for some of the more subtle therapeutic differences observed between various members of the tetracycline family (see, for example, ref. 104 and previous parts of that series).

$$HO \quad Me \quad HN^+Me_2$$

(structure of tetracycline)

(11)

The tetracyclines act by inhibiting protein synthesis at the ribosome. This happens in a manner that is not entirely understood but which may involve a metal complex (with Mg^{2+} being the most likely candidate). Apart from this, metal ions are now clearly implicated in affecting gastrointestinal uptake of the drugs and in causing some of their most troublesome side effects. Interference with mineralization occurs in several guises, most notably in the depression of long-bone growth rates in premature children, the hypoplasia of tooth crowns and the discolouration of tooth enamel. Calcium depletion is clearly the main factor responsible for these side effects but other metal ions may also be involved, *e.g.* binding of Fe^{II} has been suggested as affecting collagen synthesis by inhibition of protocollagen hydroxylase.

In blood plasma the drugs exist almost entirely in the form of their calcium and (to a lesser extent) magnesium complexes.[103,105] It is likely that the magnesium complexes will predominate intracellularly, where the magnesium:calcium ratio is much higher than it is in plasma.

6.3.3 FUTURE DEVELOPMENTS

It is salutary to realize that very few of the drugs discussed in this chapter are the products of goal-directed design and synthesis. Ironically, the first deliberate efforts to produce a chelating therapeutical led to the development of BAL but this is the exception which proves the rule. Almost all of the remaining compounds were discovered just by testing biologically synthesized substances for their therapeutic effects. It is noteworthy that this can be said of all those agents like desferrioxamine, valinomycin and penicillamine which have particularly remarkable chemical properties.

The lesson to be learned from this observation goes beyond appreciating nature's chemical prowess. We must understand why efforts to develop new metal-binding agents have been so unproductive. As noted at the start of this chapter, progress has been lacking even though the basic chemical principles are all well understood. One might reasonably think that, with such knowledge, systematic advances in chelating drug design should follow. Yet, there is little to suggest that metal-binding agents can nowadays be developed for chosen medical purposes without luck still being an essential and major ingredient. Even well-funded and multigroup projects have been depressingly unsuccessful, for example that mounted in the 1970s by the USA to produce a better iron chelator for treating Cooley's anaemia.[106]

If the problem does not lie in understanding the individual chemical attributes required for therapeutic efficacy, it must be that finding the correct combination of properties is presently beyond our intellectual reach. The lack of progress in exploiting the concept of synergistic chelation therapy shows that considering only one kind of effect at a time is insufficient. Many chemical aspects of physiological systems are relevant to the action of a metal-binding drug and, clearly, all must be taken into account simultaneously to understand its behaviour.

Computer simulation models have the potential to deliver this comprehensive understanding. However, at present, most such calculations are confined to static equilibrium systems of single biofluids. Some attempts to include kinetic effects have been made (see, for example, ref. 107) but these primitive moves need to be taken much further. More attention must be given to the effects of redox equilibria and of solubility. Metal binding by proteins must be dealt with in a much more sophisticated way than it is at the moment. Perhaps most importantly of all, future models must describe the multicompartmental nature of biological systems, thus providing a dynamic view of metal binding *in vivo*, which has to date been so noticeably lacking.

There is, in principle, no reason why better models such as these cannot be developed. The passive nature of most transition metal interactions *in vivo* and the passive nature of the transition metal ion transport processes greatly simplify the task. Some important physiological parameters remain to be determined and a substantial programming effort will be required but neither of these constitute insurmountable obstacles. Accordingly, it is only a matter of time before a comprehensive simulator is developed which encompasses all the important processes affecting metal-binding therapeuticals. It is more than likely that models with very impressive predictive powers will appear before the turn of the century. These will enable researchers to consider, for the first time, the *in vivo* actions of metal-binding agents in their entirety, taking into proper account all that is known about their chemistry. Then, and only then, will systematic progress in the design of metal-binding drugs commence.

6.3.4 REFERENCES

1. R. Österberg, *Nature (London)*, 1974, **249**, 382.
2. E. J. Underwood, 'Trace Elements in Human and Animal Nutrition', Academic Press, New York, 1977.
3. F. A. Cotton and G. Wilkinson, 'Advanced Inorganic Chemistry', 4th edn., Wiley, New York, 1980, p. 1311.
4. B. L. Vallee and R. J. P. Williams, *Chem. Br.*, 1968, **4**. 397.
5. B. L. Vallee and R. J. P. Williams, *Proc. Natl. Acad. Sci. USA*, 1968, **59**, 498.
6. R. J. P. Williams, *Pure Appl. Chem.*, 1982, **54**, 1889.
7. R. Albert, M. Berlin, J. Finklea, L. Friberg, R. A. Goyer, R. Henderson, S. Hernberg, G. Kazantzis, R. A. Kehoe, A. C. Kolbye, L. Magos, J. K. Miettinen, G. F. Nordberg, T. Norseth, E. A. Pfitzer, M. Piscator, S. I. Shibko, A. Singerman, K. Tsuchiya and J. Vostal, *Environ. Physiol. Biochem.*, 1973, **3**, 65.
8. D. Gloag, *Br. Med. J.*, 1981, **282**, 41.
9. D. Gloag, *Br. Med. J.*, 1981, **282**, 879.
10. R. B. Martin, *Met. Ions Biol. Syst.*, 1986, **20**, 21.
11. L. W. Oberley, in 'Superoxide Dismutase', ed. L. W. Oberley, CRC Press, Boca Raton, FL, 1985, p. 2.
12. F. H. Nielsen, *Annu. Rev. Nutr.*, 1984, **4**, 21.
13. G. J. Brewer, G. M. Hill, R. D. Dick, T. T. Nostrant, J. S. Sams, J. J. Wells and A. S. Prasad, *J. Lab. Clin. Med.*, 1987, **109**, 526.

14. I. Bremner and P. M. May, in 'Zinc in Human Biology', ed. C. F. Mills, Springer-Verlag, Berlin, 1989, p. 95.
15. M. N. Hughes, 'The Inorganic Chemistry of Biological Processes,' Wiley, Chichester, 1972, p. 1.
16. A. S. Mildvan and L. A. Loeb, *CRC Crit. Rev. Biochem.*, 1979, **6**, 219.
17. J. E. Huheey, 'Inorganic Chemistry. Principles of Structure and Reactivity', Harper and Row, London, 1975.
18. R. G. Pearson, *J. Am. Chem. Soc.*, 1963, **85**, 3533.
19. R. G. Pearson, *J. Chem. Educ.*, 1987, **64**, 561.
20. P. M. May and R. A. Bulman, *Prog. Med. Chem.*, 1983, **20**, 225.
21. D. D. Perrin, *Nature (London)*, 1965, **206**, 170.
22. P. M. May, P. W. Linder and D. R. Williams, *J. Chem. Soc., Dalton Trans.*, 1977, 588.
23. R. J. P. Williams, *Chem. Soc. Rev.*, 1970, **24**, 331.
24. R. J. P. Williams, *Chem. Soc. Rev.*, 1980, **9**, 325.
25. G. R. Magneson, J. M. Puvathingal and W. J. Ray, Jr., *J. Biol. Chem.*, 1987, **262**, 11 140.
26. R. W. Charlton and T. H. Bothwell, *Annu. Rev. Med.*, 1983, **34**, 55.
27. R. J. Cousins, *Physiol. Rev.*, 1985, **65**, 238.
28. Anon., *Nutr. Rev.*, 1986, **44**, 181.
29. M. K. Song, *Comput. Biochem. Physiol.*, 1987, **87A**, 223.
30. E. R. Morris, *Fed. Proc., Fed. Am. Soc. Exp. Biol.*, 1983, **42**, 1716.
31. H. A. Huebers and C. A. Finch, *Physiol. Rev.*, 1987, **67**, 520.
32. E. C. Theil, *Annu. Rev. Biochem.*, 1987, **56**, 289.
33. C. D. Klaasen and J. B. Watkins, III, *Pharmacol. Rev.*, 1984, **36**, 1.
34. M. J. Ettinger, H. M. Darwish and R. C. Schmitt, *Fed. Proc., Fed. Am. Soc. Exp. Biol.*, 1986, **45**, 2800.
35. S. E. Pattison and R. J. Cousins, *Fed. Proc., Fed. Am. Soc. Exp. Biol.*, 1986, **45**, 2805.
36. E. H. Morgan and E. Baker, *Fed. Proc., Fed. Am. Soc. Exp. Biol.*, 1986, **45**, 2810.
37. V. L. Schramm and M. Brandt, *Fed. Proc., Fed. Am. Soc. Exp. Biol.*, 1986, **45**, 2817.
38. A. Jacobs, *Blood*, 1977, **50**, 433.
39. M. Mulligan, B. Althaus and M. C. Linder, *Int. J. Biochem.*, 1986, **18**, 791.
40. H. M. Darwish, J. C. Cheney, R. C. Schmitt and M. J. Ettinger, *Am. J. Physiol.*, 1984, **246**, G72.
41. R. C. Schmitt, H. M. Darwish, J. C. Cheney and M. J. Ettinger, *Am. J. Physiol.*, 1983, **244**, G183.
42. H. M. Darwish, R. C. Schmitt, J. C. Cheney and M. J. Ettinger, *Am. J. Physiol.*, 1984, **246**, G48.
43. M. A. Page, E. Baker and E. H. Morgan, *Am. J. Physiol.*, 1984, **246**, G26.
44. E. Baker, M. A. Page and E. H. Morgan, *Am. J. Physiol.*, 1985, **248**, G93.
45. E. H. Morgan, *Mol. Aspects Med.*, 1981, **4**, 1.
46. D. Rudman and P. J. Williams, *Nutr. Rev.*, 1985, **43**, 1.
47. I. L. Bonta, M. J. Parnham, J. E. Vincent and P. C. Bragt, *Prog. Med. Chem.*, 1980, **17**, 185.
48. J. H. Leibfarth and R. H. Persellin, *Agents Actions*, 1981, **11**, 458.
49. P. P. Dendy, P. F. Sharp, W. I. Keyes and J. R. Mallard, *Br. Med. Bull.*, 1980, **36**, 223.
50. R. L. Hayes and K. F. Hubner, *Met. Ions Biol. Syst.*, 1983, **16**, 279.
51. T. F. Slater, *Biochem. J.*, 1984, **222**, 1.
52. W. A. Pryor, *Annu. Rev. Physiol.*, 1986, **48**, 657.
53. P. J. Simpson and B. R. Lucchesi, *J. Lab. Clin. Med.*, 1987, **110**, 13.
54. B. Halliwell and J. M. C. Gutterdge, *Biochem. J.*, 1984, **219**, 1.
55. M. M. Jones, *Met. Ions Biol. Syst.*, 1983, **16**, 47.
56. J. Aaseth, *Hum. Toxicol.*, 1983, **2**, 257.
57. C. A. Finch and H. Huebers, *N. Engl. J. Med.*, 1982, **306**, 1520.
58. W. F. Westlin, *Clin. Pediatr.*, 1966, **5**, 531.
59. A. Bomford and R. Williams, *Q. J. Med.*, 1976, **45**, 611.
60. B. Modell, *Prog. Hematol.*, 1979, **11**, 267.
61. A. V. Hoffbrand, in 'Iron in Biochemistry and Medicine, II', ed. A. Jacobs and M. Worwood, Academic Press, London, 1980, p. 499.
62. D. G. D. Barr and D. K. B. Fraser, *Br. Med. J.*, 1968, **1**, 737.
63. I. Sternlieb, *Gastroenterology*, 1980, **78**, 1615.
64. H. K. Chuttani, P. S. Gupta, S. Gulati and D. N. Gupta, *Am. J. Med.*, 1965, **39**, 849.
65. W. J. Klein, E. N. Metz and A. R. Price, *Arch. Intern. Med.*, 1972, **129**, 578.
66. D. Frommer, *Gut*, 1974, **15**, 125.
67. J. M. Walshe, *Am. J. Med.*, 1956, **21**, 487.
68. J. M. Walshe, *Lancet*, 1969, **2**, 1401.
69. J. M. Walshe, *Ann. Intern. Med.*, 1960, **53**, 1090.
70. D. M. Danks, *Lancet*, 1982, **2**, 435.
71. F. Bakir, S. F. Damluji, L. Amin-Zaki, M. Murtadha, N. Y. Al-Rawi, S. Tikriti, H. I. Dhahir, T. W. Clarkson, J. C. Smith and R. A. Doherty, *Science (Washington, D.C.)*, 1973, **181**, 230.
72. Department of the Environment Central Directorate on Environmental Pollution, 'Cadmium in the Environment and its Significance to Man', Pollution Paper No. 17, HMSO, London, 1980.
73. R. C. Schnell, *Fundam. Appl. Toxicol.*, 1981, **1**, 347.
74. B. G. Armstrong and G. Kazantis, *Lancet*, 1983, **1**, 1425.
75. D. B. Louria, M. M. Joselow and A. A. Browder, *Ann. Intern. Med.*, 1972, **76**, 307.
76. R. L. Boeckx, *Anal. Chem.*, 1986, **58**, 274A.
77. J. M. Davis and D. J. Svendsgaard, *Nature (London)*, 1987, **329**, 297.
78. D. Bellinger, A. Leviton, C. Waternaux, H. L. Needleman and M. Rabinowitz, *N. Engl. J. Med.*, 1987, **316**, 1037.
79. Report of DHSS Working Party on Lead in the Environment, 'Lead and Health', HMSO, London, 1980.
80. H. V. Aposhian, *Annu. Rev. Pharmacol. Toxicol.*, 1983, **23**, 193.
81. H. G. Nebeker and J. W. Coburn, *Annu. Rev. Med.*, 1986, **37**, 79.
82. M. R. Wills and J. Savory, *Lancet*, 1983, **2**, 29.
83. S. P. Andreoli, J. M. Bergstein and D. J. Sherrard, *N. Engl. J. Med.*, 1984, **310**, 1079.

84. D. Maharaj, G. S. Fell, B. F. Boyce, J. P. Ng, G. D. Smith, J. M. Boulton-Jones, R. L. C. Cumming and J. F. Davidson, *Br. Med. J.*, 1987, **295**, 693.
85. R. B. Martin, *Clin. Chem.*, 1986, **32**, 1797.
86. V. Volf, A. Seidel and K. Takada, *Health Phys.*, 1977, **32**, 155.
87. P. M. May and D. R. Williams, *Nature (London)*, 1979, **278**, 581.
88. J. J. Chisolm, Jr., *J. Pediatr.*, 1968, **73**, 1.
89. S. Pollack and S. Ruocco, *Blood*, 1981, **57**, 1117.
90. A. Albert, 'Selective Toxicity', 5th edn., Chapman and Hall, London, 1973, p. 334.
91. J. B. Neilands and J. R. Valenta, *Met. Ions Biol. Syst.*, 1985, **19**, 313.
92. B. Rosenberg, L. Van Camp and T. Krigas, *Nature (London)*, 1965, **205**, 698.
93. M. J. Cleare, *Coord. Chem. Rev.*, 1974, **12**, 349.
94. N. R. Bachur, S. L. Gordon and M. V. Gee, *Cancer Res.*, 1978, **38**, 1745.
95. L. W. Oberley and G. R. Buettner, *Cancer Res.*, 1979, **39**, 1141.
96. C. Myers, L. Gianni, J. Zweier, J. Muindi, B. K. Sinha and H. Eliot, *Fed. Proc., Fed. Am. Soc. Exp. Biol.*, 1986, **45**, 2792.
97. A. M. Creighton, K. Hellman and S. Whitecross, *Nature (London)*, 1969, **222**, 384.
98. E. H. Herman, D. T. Witiak, K. Hellmann and V. S. Waravdekar, *Adv. Pharmacol. Chemother.*, 1982, **19**, 249.
99. Z.-X. Huang, P. M. May, K. M. Quinlan, D. R. Williams and A. M. Creighton, *Agents Actions*, 1982, **12**, 536.
100. E. H. Herman and V. J. Ferrans, *Drugs Exp. Clin. Res.*, 1983, **9**, 483.
101. B. C. Pressman, *Met. Ions Biol. Syst.*, 1985, **19**, 1.
102. K. R. K. Easwaran, *Met. Ions Biol. Syst.*, 1985, **19**, 109.
103. M. Brion, G. Berthon and J.-B. Fourtillan, *Inorg. Chim. Acta*, 1981, **55**, 47.
104. L. Lambs, B. Decok-LeRévérend, H. Kozlowski and G. Berthon, *Inorg. Chem.*, 1988, **27**, 3001.
105. G. Berthon, M. Brion and L. Lambs, *J. Inorg. Biochem.*, 1983, **19**, 1.
106. W. F. Anderson, in 'Symposium on Development of Iron Chelators for Clinical Use', ed. W. F. Anderson and M. C. Hiller, DHEW Publication No. NIH 76-994, 1976, p. 1.
107. M. M. Jones and P. M. May, *Inorg. Chim. Acta*, 1987, **138**, 67.
108. E. J. Corey, M. M. Mehrotra and A. U. Khan, *Science (Washington, D.C.)*, 1987, **236**, 68.

6.4

Agents Acting on Active Ion Transport

JOHN L. SUSCHITZKY and EDWARD WELLS
Fisons Pharmaceutical Division, Loughborough, UK

6.4.1 INTRODUCTION	193
6.4.1.1 Reaction Mechanism of the Na$^+$,K$^+$- and H$^+$,K$^+$-ATPases	194
6.4.2 THE H$^+$,K$^+$-ATPase	194
6.4.2.1 Rôle in Acid Secretion	194
6.4.2.2 Structure	195
6.4.2.3 Chemical Treatments of Peptic Ulcers	195
6.4.2.3.1 Cytoprotective agents	195
6.4.2.3.2 Antibacterial agents against Campylobacter pylori	195
6.4.2.3.3 New agents which are more effective inhibitors of gastric acid secretion	195
6.4.2.4 Biological Screening of H$^+$,K$^+$-ATPase Inhibitors	196
6.4.2.5 Competitive Inhibitors of the H$^+$,K$^+$-ATPase	196
6.4.2.5.1 SCH 28080	196
6.4.2.5.2 The mechanism of action of the imidazopyridines	197
6.4.2.5.3 Structure–activity relationships	197
6.4.2.6 Non-competitive Inhibitors of the H$^+$,K$^+$-ATPase	198
6.4.2.6.1 Mode of action of pyridylmethylbenzimidazole sulfoxides	198
6.4.2.6.2 The nature of the chemical rearrangement	199
6.4.2.6.3 Structure–activity relationships	201
6.4.2.6.4 Compounds structurally distinct from the pyridylmethylbenzimidazole sulfoxides	204
6.4.2.6.4 Omeprazole	204
6.4.2.7 Other Putative Proton Pump Blockers	205
6.4.3 THE Na$^+$,K$^+$-ATPase	205
6.4.3.1 Structure and function	205
6.4.3.2 Inotropic Action of the Cardiac Glycosides	206
6.4.3.3 Endogenous Digitalis-like Factors	206
6.4.3.4 Clinical Applications	206
6.4.3.5 The Relationship Between Structure and Activity of the Cardenolide Derivatives	207
6.4.3.5.1 The steroid nucleus	207
6.4.3.5.2 The lactone side chain	208
6.4.3.5.3 The substituent at C-3 (sugar residue)	209
6.4.3.5.4 Hydrophobic interactions	209
6.4.3.5.5 Computational studies	209
6.4.3.6 Other (Non-steroidal) Inhibitors	210
6.4.4 REFERENCES	210

6.4.1 INTRODUCTION

This chapter will describe the ATP dependent ion pumps for protons and potassium ions, and sodium and potassium ions. These are controlled by H$^+$,K$^+$-ATPase and Na$^+$,K$^+$-ATPase enzymes; it will also cover recent work on the inhibitors of these enzymes. The H$^+$,K$^+$-ATPase is responsible for gastric acid secretion and its inhibition presents an opportunity for peptic ulcer therapy. The Na$^+$,K$^+$-ATPase is the site of action of the cardiac glycosides. Inhibition, by mechanisms which will be discussed, influences intracellular Ca^{2+} levels, resulting in positive inotropy (increased force of contraction), which is the basis for their use in congestive heart failure.

These proteins are functionally related to a broader group of cation pumps, which include the Ca^{2+}-ATPases of both the sarcoplasmic reticulum[1] and the distinct Ca^{2+}/calmodulin-activated enzyme in the plasma membrane.[2] All these enzymes show a similar catalytic cycle involving an aspartyl phosphate intermediate as outlined below. Although interference with the Ca^{2+}-ATPases may have therapeutic applications, they are not considered further here because of the lack of reported medicinal chemistry.

6.4.1.1 Reaction Mechanism of the Na^+,K^+- and H^+,K^+-ATPases

A simplified general mechanism of the reaction pathway is shown in Scheme 1.[3,4] The enzyme exists in two major conformations, E_1 and E_2, where the ion-binding sites are available at the cytosolic and extracellular surfaces of the membrane respectively. Following binding of an ion, the binding site changes to an occluded form (shown by parentheses in Scheme 1) in which it probably cannot exchange with ions in solution on either side of the membrane. Completion of the translocation step is accompanied by an E_1–E_2 transition. Energy is derived from the hydrolysis of ATP, part of which is retained in the formation of a phosphoprotein intermediate whereby the γ-phosphate is transferred to an aspartate residue. Dephosphorylation is stimulated by binding of K^+ at the extracellular surface.

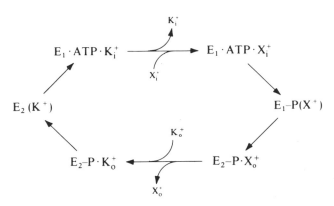

$X^+ = Na^+$ for Na^+,K^+-ATPase and H^+ for H^+,K^+-ATPase; subscript i = intracellular, subscript o = extracellular; occluded binding sites are in parentheses

Scheme 1   Generalized reaction pathway

6.4.2 THE H^+,K^+-ATPase

The H^+,K^+-ATPase is localized in the specialized acid-secreting parietal cell of the gastric mucosa and produces a transmembrane proton gradient in excess of 10^6. H^+ is pumped in electroneutral exchange for K^+.

6.4.2.1 Rôle in Acid Secretion

In the resting cell, the pump is localized on an intracellular smooth membrane structure (tubulovesicle system). Upon stimulation, however, this network fuses to become part of an expanded canaliculus which connects with the extracellular space. These morphological changes are accompanied by an increase in oxygen consumption and secretion of acid into the canaliculus.[4]

The major stimulus of acid secretion at the parietal cell is by histamine acting at the H_2 receptor, which induces an increase in intracellular cAMP levels. Gastrin and acetylcholine also have specific receptors on the parietal cell, stimulation of which results respectively in a transient mobilization of Ca^{2+} from intracellular stores and a more prolonged rise in intracellular calcium from outside the cell.[4,5] Involvement of different second messengers allows for synergistic interactions between stimuli.

The intracellular mechanisms involved in activation of the parietal cell have not been fully defined but are associated with an increased rate of ATP production and modification of the cytoskeleton to bring about the morphological changes. The proton pump appears to be stimulated by an increase in the availability of the K^+ counterion at the luminal (extracellular) surface, as a result of an increased permeability of the membrane to KCl.[4,6,7] Carbonic anhydrase, also associated with the canalicular membrane, converts cytosolic OH^- to HCO_3^-. A HCO_3^-/Cl^- exchanger in the basolateral membrane serves the dual purpose of providing Cl^- for HCl secretion and preventing a rise in cytosolic pH consequent to proton extrusion.[4] It should be noted that the above description of the 'ion balance sheet' is somewhat simplified.

6.4.2.2 Structure

The catalytic functions are all associated with a single polypeptide chain, the amino acid sequence of which has been deduced from cDNA clones of the rat enzyme.[8] It contains 1033 amino acid residues ($M_r = 114\,000$) and hydropathy plots suggest the existence of eight transmembrane α-helical regions. Similarities with the Na^+,K^+-ATPase will be discussed later in the chapter.

The chemical nature of the groups involved in proton translocation and the mechanism by which the proton is released into such a unique acid environment are unresolved.

6.4.2.3 Chemical Treatments of Peptic Ulcers

In the 1970s and 1980s the principal chemical treatment of peptic ulcers has been by the use of antagonists of the H_2-histamine receptors (see Volume 3, Chapter 12.5). Three approaches to the treatment of peptic ulcer and other related 'gastric acid dependent' diseases are now emerging to challenge the dominance of these H_2-receptor blockers.

6.4.2.3.1 *Cytoprotective agents*

Cytoprotection is a catch-all expression, which includes phenomena such as improvement in the quantity and quality of mucus production, increase in mucosal blood flow and stimulation of sodium bicarbonate secretion. The early claims that prostaglandins afford 'cytoprotection' do not appear to have been substantiated in man.[9] At best, their ulcer-healing properties approach those of the H_2 antagonists and appear to be a consequence only of their antisecretory activity. However, certain agents, notably colloidal bismuth subcitrate (CBS, De Nol)[10] and sucralfate[11] may owe their clinical efficacy, at least in part, to the enhancement of the defensive forces which help to protect the gastric mucosa.

6.4.2.3.2 *Antibacterial agents against* **Campylobacter pylori**

A strong correlation between the occurrence of *C. pylori* and the presence of gastritis and duodenitis in man has been observed.[12] It is not clear, however, whether the bacterium is cause or effect. Interest in this area has been spurred by the finding that CBS not only heals ulcers as effectively as the antisecretory agents but also appears to delay their recurrence after therapy has been discontinued. This appears to be associated with the removal of the bacterium.[13]

6.4.2.3.3 *New agents which are more effective inhibitors of gastric acid secretion*

Of the three approaches, only the new highly efficacious antisecretory agents so far appear to show a real advance in peptic ulcer therapy and potential as drugs for the 1990s.

In order to discover such agents, we need to understand the physiological processes which lead to acid production. There are many endogenous secretagogues; some of their actions with respect to cellular events are shown in Figure 1. Two cells are principally involved: a histamine-secreting cell which, in man, appears to resemble a mast cell but in some animals, notably the rat, is an enterochromaffin-like (ECL) cell. Secondly, the oxyntic or parietal cell is responsible for acid production. The 'principal pathway' for acid production can probably be considered to be as follows. Under the influence of various stimuli, especially the vagus nerve (anticipation of food *etc.*) gastrin is released from the G cells which occur in the antrum of the stomach. This hormone can

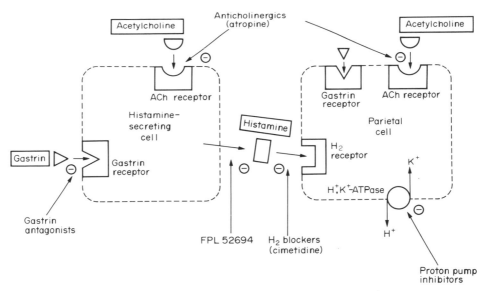

Figure 1 Approaches to the inhibition of gastric acid secretion

stimulate a receptor on the 'mast' cell, leading to the release of histamine. The histamine can occupy the H_2 receptor of the parietal cell, resulting in its stimulation. A number of events connect this to the H^+,K^+-ATPase enzyme, which is activated to produce acid (see earlier).

Additionally, there are muscarinic receptors on both mast and parietal cells, stimulation of which can give rise to acid production and there are gastrin receptors on the parietal cell. Muscarinic antagonists are moderately effective inhibitors of gastric acid secretion. However, the selective M_1 blockers (notably pirenzepine) probably owe their efficacy to ganglion blockade.[14]

Inhibitors of the various secretagogues all appear to be able to reduce gastric acid secretion. However, the efficacy and potency of drugs which act on the histamine-secreting cell, such as gastrin antagonists (proglumide,[15] RS-2039,[16] benzotript[17]) and inhibitors of histamine release (*e.g.* the chromone FPL 52694)[18] are generally poor. Only agents which operate on the parietal cell, notably the H_2 antagonists, appear to be capable of achieving useful therapeutic effects. The efficacy of drugs appears to improve the closer their interference is to the final event along the secretory pathway. The identification of drugs which inhibit the enzyme responsible for acid production is therefore an attractive proposition.

6.4.2.4 Biological Screening of H^+,K^+-ATPase Inhibitors

The H^+,K^+-ATPase can readily be assayed in membrane fragments or leaky vesicles by the measurement of K^+-stimulated ATPase activity (or the K^+-stimulated *p*-nitrophenol phosphatase part reaction). However, for the assay of inhibitors which accumulate in the acid space, with or without subsequent acid-catalyzed activation, the use of tight vesicles which, upon stimulation with K^+ and valinomycin generate a low intraluminal pH, are more representative of the activity *in vivo*.[19] Acid transport is measured by accumulation of weak bases such as [^{14}C]aminopyrine which become trapped in the lumen by protonation. The isolated rabbit gastric gland is also used.[20] This is essentially the system depicted in Figure 1. Proton pump inhibitors can be evaluated against secretion stimulated by K^+ or dibutyryl-cAMP. Luminal acidification is measured by [^{14}C]aminopyrine uptake.

The Heidenhain pouch or gastric fistula dog are most commonly used for the assessment of proton pump blockers *in vivo*. Drugs may be administered orally, intravenously or intraduodenally. Various rat models have also been used including the gastric fistula, Shay (pyloris ligated) and Ghosh and Schild preparations. So far no reliable tissue culture assays have been developed.

6.4.2.5 Competitive Inhibitors of the H^+,K^+-ATPase

6.4.2.5.1 *SCH 28080*

In the early 1980s reports appeared from the Schering-Plough Corporation of a potent antisecretory agent which did *not* possess anticholinergic or H_2-antagonist properties.[21,22] This imidazo[1,2-

a]pyridine derivative (**1**; SCH 28080) displayed inhibition of acid secretion *in vitro*,[23] in animal models[21] and in man[24] and also interestingly demonstrated cytoprotection of the gastric mucosa,[21,25] an activity which is probably independent of antisecretory activity. Unfortunately, progression of SCH 28080 had to be discontinued due to toxicological problems (hepatotoxicity and hyperplasia of the gastric mucosa).[25,26] This compound was 50 times less potent by the oral (ED_{50} = 4.4 mg kg^{-1}) than the intravenous route (ED_{50} = 0.09 mg kg^{-1}) against histamine-stimulated acid secretion in the Heidenhain Pouch dog, which appears to be a consequence of its metabolism (first pass effect) and *not* poor absorption.[27]

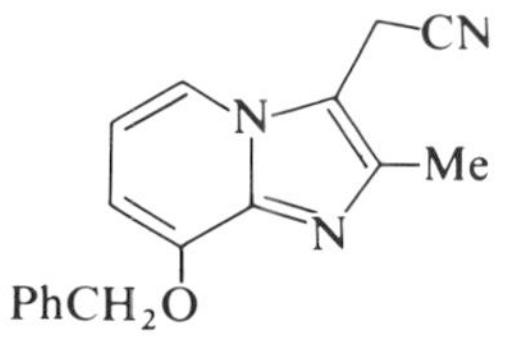

(**1**) SCH 28080

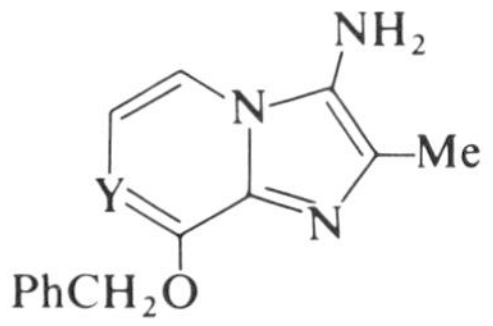

(**2a**). Y = N; SCH 32651
(**2b**) Y = CH

Subsequently, another imidazopyridine structure (**2a**; SCH 32651) with potential as an antiulcer agent has been described. Its antisecretory potency *in vitro*[28] is, however, considerably less than that of SCH 28080, though it does again possess gastric cytoprotective properties[29] and is in fact of comparable potency *in vivo* in the dog.[27]

6.4.2.5.2 *The mechanism of action of the imidazopyridines*

Studies with purified H$^+$,K$^+$-ATPase showed that inhibition of this enzyme by SCH 28080 occurred in a competitive manner with respect to K$^+$ ions. The compound was virtually inactive against the Na$^+$,K$^+$-ATPase. It was concluded that SCH 28080 acts with high selectivity on the H$^+$,K$^+$-ATPase by a competitive inhibition at the high affinity luminal K$^+$ site.[30]

Further, it was shown that the potency of inhibition of K$^+$-stimulated H$^+$,K$^+$-ATPase increased with decreasing pH and is therefore related to the concentration of the conjugate acid of SCH 28080 (a weak base with pK_a 5.6). The quaternary salt derivative (**3**) which is methylated on the N-1 atom, was also active on the enzyme but did not show a pH dependence.[31] These results strongly suggest that the inhibiting species is the protonated drug, which would be the predominant form present at its site of action in the canaliculus of a secreting parietal cell.

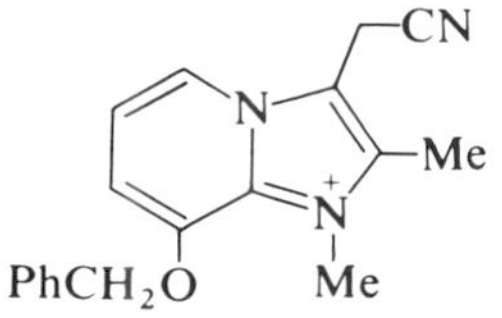

(**3**)

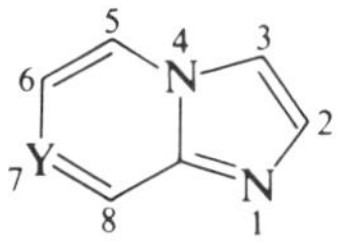

(**4a**) Y = CH
(**4b**) Y = N

6.4.2.5.3 *Structure–activity relationships*

Eighty-one imidazo[1,2-*a*]pyridine derivatives of (**4a**) related to SCH 28080 were synthesized and studied for their antisecretory activity *in vivo* (Shay rat and histamine-stimulated Heidenhain Pouch dog) and also for cytoprotective activity (ethanol lesion test in the rat).[32] For the dog data, the following structure–activity relationships were found: (i) a small alkyl group at C-2 (methyl or ethyl) favoured activity; (ii) a cyanomethyl group at C-3 was a requirement for maintaining both antisecretory and cytoprotective activity; and (iii) activity in the 8-position was maximized with benzyloxy, 3-thienylmethoxy or phenylmethylamino substituents.

Subsequently, in an attempt to find a successor to SCH 28080, a number of 3-amino (and related) analogues of imidazo[1,2-*a*]pyridine (**4a**) and imidazo[1,2-*a*]pyrazine (**4b**) were synthesized.[27] The position of protonation was considered to be important in determining their activity. Both ring systems (when substituted by a 3-NH$_2$ function as in **2a** and **2b**) were shown by X-ray analysis to be protonated at N-1, which was in accord with calculated heats of formation for the various protonated products, but at variance in the case of (**2b**) with ground state N atom charge densities calculated using MINDO/3 or MINDO/2 computations.[27]

In an attempt to throw light on the part played by metabolites on activity and toxicity, studies were carried out using [13]C- and ^{14}C-labelled (**1**) and the corresponding 3-amino derivative (**2b**). The metabolites of (**1**) were shown to be the thiocyanate anion and a small amount of the debenzylated (8-hydroxy) compound, the former was said to result from production of cyanide *via* oxidative decyanation of the cyanomethyl group, followed by detoxification by the enzyme rhodanese.[27] Interestingly, thiocyanate was also shown to be a major metabolite of compound (**2b**)!

Structure–activity relationships established that the only acceptable alternative to the CH_2CN substituent at C-3 appeared to be NH_2. Introduction of a ring N atom at position 7, as in (**2a**), gave rise to a retention of activity. The relationships between structure, activity and toxicity (hyperplasia of the glandular mucosa) were also studied.[26] This work established SCH 32651 (**2a**) as the compound with the best biological profile.

Surprisingly little work has been reported on these interesting reversible inhibitors of the H^+,K^+-ATPase enzyme. Highly efficacious drugs could emerge from such research with certain advantages over compounds which non-competitively inhibit the enzyme, *viz.* the sulfoxides described in the following sections.

6.4.2.6   Non-competitive Inhibitors of the H^+,K^+-ATPase

In the early 1970s, researchers at AB Hässle (Sweden) initiated a chemical programme to optimize the antisecretory activity of analogues of the pyridylthioacetamide (**5**; CMN 131). This led to the discovery of a class of extremely efficacious inhibitors of gastric acid secretion with a novel mode of action, of which the pyridylmethylbenzimidazole sulfoxide, timoprazole (**6a**), is the archetypal structure.[33] Fortuitously coinciding with this work was the discovery of the H^+,K^+-ATPase enzyme,[34] which enabled the demonstration that compounds related to timoprazole were non-competitive inhibitors of the enzyme and provided a rational approach to the design of more potent analogues. This led to the synthesis of picoprazole (**6b**) and omeprazole (**6c**), a new drug for the treatment of peptic ulcer and related diseases (see later). This work also helped to generate an understanding of the way in which the enzyme operates.

(**5**) CMN 131   (**6b**) Picoprazole   (**6c**) Omeprazole

6.4.2.6.1   *Mode of action of pyridylmethylbenzimidazole sulfoxides*

On studying the mechanism of action of these inhibitors of the H^+,K^+-ATPase ('proton pump blockers') several salient features of their action became apparent:[35] (i) the weak basicity of the compounds ($pK_a \approx 4$) allows them to accumulate in the acid space adjacent to their site of action (the secretory canaliculus of the parietal cell); (ii) the sulfoxides themselves have no intrinsic activity, but under the influence of acid undergo a chemical rearrangement to an active species; and (iii) the active species is thiophilic in nature and covalently binds to thiol functions (cysteinyl residues) generating disulfide bridges to the enzyme, thereby causing its inactivation.

These features account for the long duration of action (days) of this class of compound, which has been demonstrated in animals and in man and which persists long after the parent drug has been eliminated. This unique mode of drug action shows some analogy with suicide inhibitors. Their specificity of action, however, is dependent on chemical stability at normal physiological pH rather than a stereoelectronic relationship with the enzyme.

Experiments using labelled omeprazole showed a linear relationship between covalent attachment to the enzyme and inhibition; the number of molecules of inhibitor approached two at 100% inhibition.[36]

6.4.2.6.2 *The nature of the chemical rearrangement*

Several groups, including workers at Byk Gulden collaborating with SK&F and the original group at Hässle, have published their findings concerning the mechanism by which the sulfoxides inhibit the H^+,K^+-ATPase enzyme.[37,38,39] The chemical transformations leading to the inactivation of the enzyme have also been elucidated.[40] Timoprazole and *t*-butylthiol (as a 'model' for the enzyme) were reacted together in the presence of HCl (to 'mimic' the gastric conditions) in aqueous methanol overnight. On cooling the mixture, large crystals were obtained, which were suitable for X-ray analysis. The structure of the product was shown to be the disulfide (7) and strongly suggested that the sulfoxide had undergone an acid-catalyzed Smiles rearrangement to the spiro compound (8). On ring opening, the reactive sulfenic acid (9) would be formed, which was considered to be the thiophilic species responsible for the inactivation of the enzyme. These events are summarized in Scheme 2.

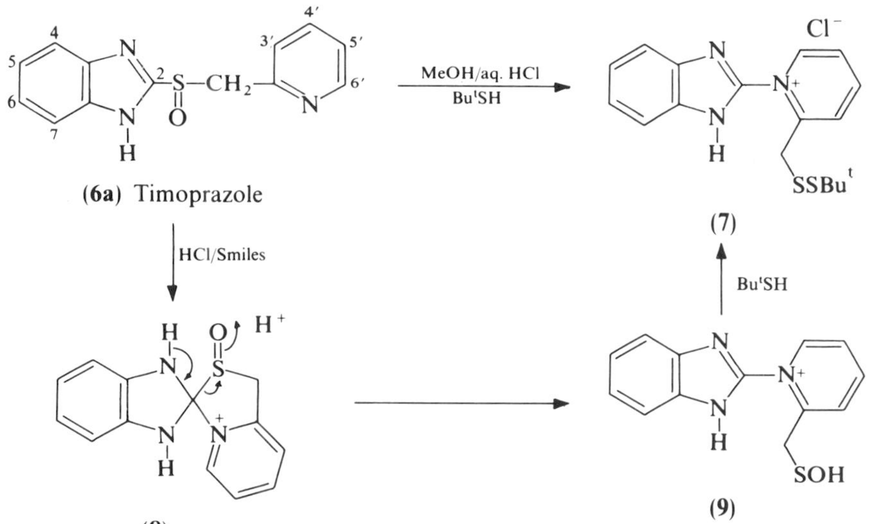

Scheme 2

The interesting feature of the chemistry of timoprazole and its congeners is their ability to undergo a Smiles rearrangement under acid (rather than the normal base) catalysis. A precedent for such a reaction is, however, known.[41]

By use of large counterions (PF_6^- or BF_4^-) sulfenamide analogues produced by intramolecular loss of one water molecule from the sulfenic acids of timoprazole derivatives were isolated.[37,38] These relatively stable molecules were proposed to be the active species (see Scheme 3).

The parent radiolabelled sulfoxide can be activated with acid and subsequent incubation of the mixture with the enzyme results in rapid incorporation of omeprazole into the enzyme.[42] Also, the sulfenamide itself produces a rapid block of the enzyme ($IC_{50} = 0.6 \mu M$).

Another interesting feature of the compounds (see Scheme 3) is that in the presence of an excess of thiol, the disulfide is converted *via* a second (but base-catalyzed) Smiles rearrangement into the sulfide which corresponds to the original sulfoxide. Sulfides of this type are known to be oxidized by the liver to the parent sulfoxide,[37,43] which raises the intriguing possibility of 'catalytic' drug action in which cycling occurs as shown in Scheme 3 for omeprazole. (Note also the formation of regioisomers.) In practice, only a fraction of the total dose completes the cycle because of extensive metabolism, largely to the sulfone. Further, the possibility exists that endogenous thiols or a free thiol group of the enzyme itself may regenerate the enzyme by displacing the drug from the binding site.[44] It is probably more likely, however, that recovery of activity requires *de novo* synthesis of the enzyme,[37,45] which is consistent with the long duration of action of the drug.

The kinetics of the inactivation process have also been studied; the rate of disappearance of sulfoxide increases with decreasing pH but is independent of the concentration of thiol trap and the steps leading to the production of sulfenamide are all reversible.[37–39a]

One further feature of the chemistry of the timoprazole analogues may be of biological significance. The inactivating species (sulfenic acid or sulfenamide) is a quaternary ammonium salt and is

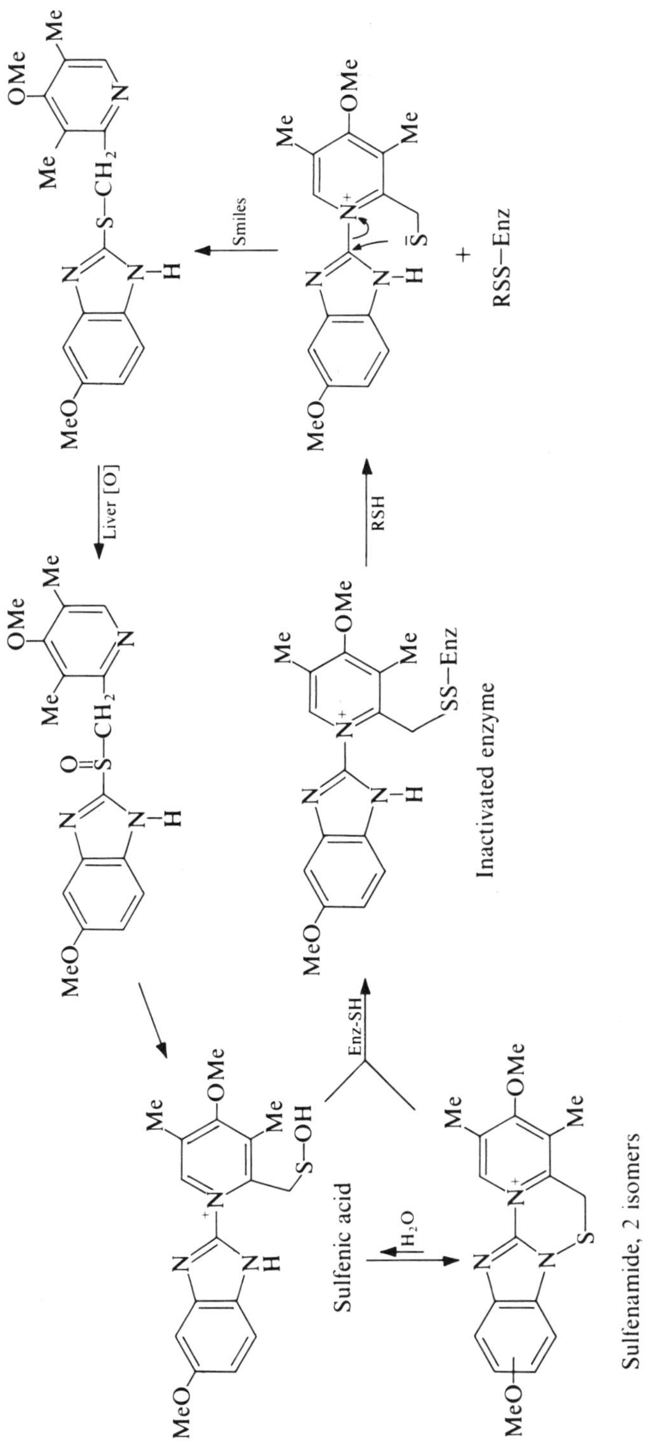

Scheme 3   The omeprazole cycle

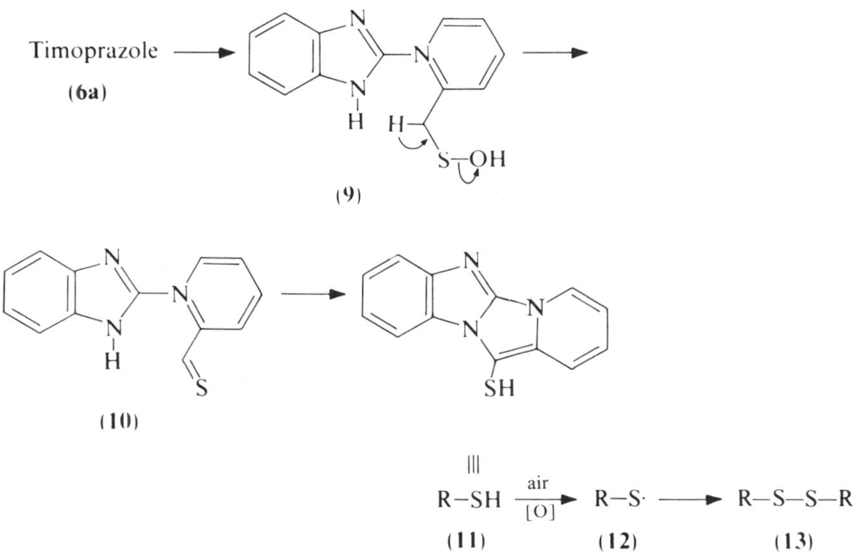

Scheme 4

therefore unlikely to move out of the acid space by crossing cell membranes, thus greatly reducing the possibility of binding to proteins at other sites.

The chemistry of these compounds is further complicated by other reactions which can take place in the absence of a sulfhydryl trap. On standing at low concentrations in acid, highly coloured (red and mauve) products are formed *via* the thioaldehyde (10) which are due to the thiol (11), the corresponding radical (12) and disulfide (13; shown for timoprazole in Scheme 4).[35,39a,46]

At high concentrations, the intermediate sulfenic acids can give rise to disulfides (14a) and the corresponding thiosulfinates (14b).[39b] These products are likely to be artefacts of high concentration test-tube experiments and probably do not occur *in vivo*.

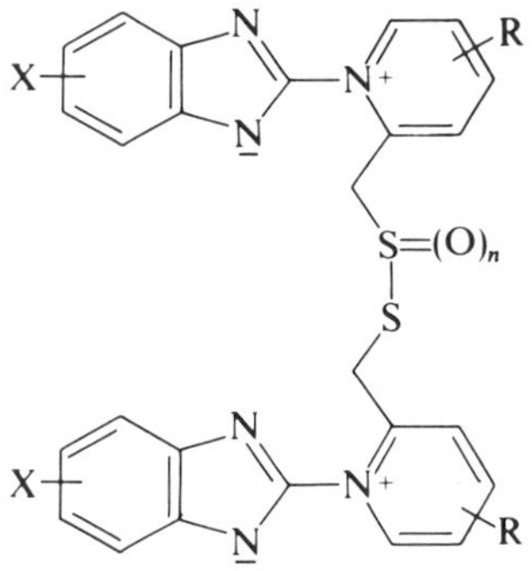

(14a) $n = 0$
(14b) $n = 1$

6.4.2.6.3 Structure–activity relationships

The pyridylmethylbenzimidazole sulfoxides can be considered to possess three structural elements: the pyridine ring, the benzimidazole ring system and the linking chain. Structure–activity relationships of compounds can be largely, but not completely, rationalized in terms of their ability to undergo the acid-catalyzed Smiles rearrangement. In the following discussion, the numbering system employed is shown on the structure of timoprazole (6a).

(i) The linking chain

Workers at Hässle have demonstrated that (as expected) replacement of the SOCH$_2$ linking chain by a variety of other groups (*e.g.* SCH$_2$, SO$_2$CH$_2$, SCH$_2$CH$_2$ and various carbon- and oxygen-containing chains) leads to total loss of activity *in vitro*.[33] Only SOCH(Me) gave significant activity. More surprisingly, we have shown that extending the length of the chain ($SOCH_2CH_2$) gives rise to an inactive and acid stable compound,[40] even though a Smiles rearrangement to a sulfenic acid *via* a six-membered spiro intermediate is theoretically possible in this case. The activity *in vivo* shown by the compound containing the SCH_2 connecting chain is due to metabolic transformation to the sulfoxide.[37]

As predicted from mechanistic considerations, the 3-pyridyl and 4-pyridyl isomers of timoprazole are completely inactive.

(ii) Substitution in the pyridine ring

A great dependence of biological activity on pyridine ring substitution is in most cases due to the effect of the substituents on the pK_a of the ring nitrogen atom. This has been ascribed to the development of a positive charge on the pyridine ring in the rate-determining step of the chemical activation process,[45] but actually correlates well with the *nulceophilicity* (rather than basicity) of the nitrogen atom, a reflection of the ease of spiro intermediate formation. For example, substitution in the 6'-position of the ring results in loss of activity resulting from a disfavoured steric interaction in the transition state for spiro ring formation. When significant steric effects are absent, a pK_a value of $\geqslant 4$ is probably optimal for activity. Weaker bases such as those encountered in timoprazole and the 4'-CO_2Me derivative generally show greatly reduced activity. Even the 4'-methyl compound ($pK_a = 3.8$, $pIC_{50} = 4.6$) is 100 times less active than the 4'-alkoxy analogues such as 4'-OPr^i ($pK_a = 4.43$, $pIC_{50} = 6.5$).[33] In the case of omeprazole ($pK_a = 4.0$, $pIC_{50} = 6.3$) the 4'-OMe substituent has little effect on pK_a as it is bent out of plane by the two flanking methyl groups (all pIC_{50} values were determined using rabbit gastric gland).

Conformational effects probably play a part in the activity of compounds which possess a substituent at C-3' (notably picoprazole and omeprazole) to facilitate the Smiles rearrangement.[47] In such cases, the pyridine ring is twisted into a conformation that favours attack at the 2-position of the benzimidazole ring. This also applies to the anilino compounds (Table 1, entry 3).

(iii) Substituents in the benzimidazole ring

Effects of substitution in the benzimidazole ring on potency are not as dramatic as in the pyridine ring, but can serve in the fine tuning to optimize activity. It is perhaps surprising that activity does *not* appear to correlate with increasing electrophilicity of the ring system. In fact the reverse is generally true; activity in the rabbit gastric gland is generally favoured by electron-donating substituents in this ring system. Two explanations for this phenomenon have been offered. Firstly, the workers at Hässle have identified two processes by which the sulfoxides can rearrange: the desired acid-catalyzed reaction and an unwanted non-catalyzed process (again producing the sulfenic acid), which becomes more important at elevated pH. Introduction of electron-withdrawing substituents (*e.g.* 5-NO_2, 5-MeSO, 5-CF_3) results in the enhancement of the uncatalyzed process, leading to decreased potency on the enzyme as well as the highly undesirable property of reduced stability at neutral pH. By measuring the pH–rate profile of compounds, the acid-catalyzed process was deduced for each compound and hence the amount of 'active blocker' produced at any pH.[45] The favourable effects of electron-donating substituents may also be explained in terms of the need for protonation of the benzimidazole (rather than the pyridine) ring in the rate-determining step. By using standard MNDO methodology, we have studied the model system (15), structurally distinct from timoprazole, but able to undergo an analogous acid-catalyzed Smiles rearrangement. The variation in energy was computed as the pyridine N atom was moved towards the imidazole 2-position. An energy minimum could only be obtained (corresponding to bond formation) when the imidazole N atom was protonated.[48]

(15)

Table 1 Sulfoxide Inhibitors of H^+, K^+-ATPase

Entry	Parent structure	Spiro ring size	Acid-catlyzed Smiles?	Biological activity	Ref.
1		5	Yes	+ +	33
2		5	No	−	40
3		6	Yes	+ +	52
4		5	Slow	+	52
5		5	Very fast	+	52
6		6	Yes	+ + +	48

For the pyridylmethylbenzimidazole sulfoxides, three classes of compounds were identified in terms of potency.[33] The most active family had lipophilic electron-releasing substituents. This group of compounds was also generally potent *in vivo* (gastric fistula dog), although more detailed correlations between *in vitro* and *in vivo* activity were not good.

Finally, *N*-acylated benzimidazoles may be used as prodrugs for the corresponding deacylated compounds, but *N*-alkylbenzimidazoles are essentially inactive.[33]

(iv) Stereochemical and hydrophobic dependence on activity

There is no evidence supporting a stereochemical requirement for activity. This is not surprising in the light of the covalent nature of the drug–receptor interaction.

However, a hydrophobic factor does play a part since the introduction of polar substituents (*e.g.* an acetamido-substituted pyridine or benzimidazole ring) leads to greatly reduced activity *in vitro*. We have also shown that the introduction of polar groups into the benzimidazole ring of anilino compounds (see below) is detrimental to activity.[40] These effects may be due to a hydrophobic pocket surrounding the enzymic sulfhydryl function which is bound by the drug.

6.4.2.6.4 *Compounds structurally distinct from the pyridylmethylbenzimidazole sulfoxides*

Although more than 40 companies have now patented in this area, very few have claimed sulfoxides which do not contain the timoprazole template. As a consequence, relatively few compounds have reached the clinic, presumably because of patent clashes and because advantages over omeprazole could not be found. Of the pyridylmethylbenzimidazole sulfoxides which have been taken into a development programme, only those from Upjohn,[49] Byk Gulden,[43] Hoffmann-La Roche[50] and Takeda[51] have been described in any detail.

A number of systems have been examined for their ability to undergo acid-catalyzed Smiles rearrangements with consequent inhibition of gastric acid secretion (Table 1, entries 2–6).[40] Because of the loss of activity (identified by workers at Hässle)[33] associated with the replacement of the benzimidazole ring system by other heterocycles, we initially chose to retain the benzimidazole 2-sulfoxide group to examine the effects of the replacement of the 2-pyridylmethyl moiety by other nucleophilic termini. Perhaps the simplest structure, which retains a weakly nucleophilic tertiary nitrogen as well as the potential to form a 5-membered spiro intermediate, is the ethylmethylanilino compound (Table 1, entry 2) but this structure showed no propensity for undergoing an acid-catalyzed rearrangement. Following this, we studied a number of anilino derivatives (Table 1, entries 3–5),[52] which all showed acid reactivity and biological activity. An optimal profile was displayed by analogues of the *N,N*-dimethyltoluidine compound (Table 1, entry 3) where a 6-membered spiro intermediate would be involved. This parent structure (NC 1300) has been taken into a development programme by Nippon Chemiphar.[49] More recently, analogues of the phenylpyridyl system (Table 1, entry 6) have been demonstrated to possess potent antisecretory properties.[48]

6.4.2.6.5 *Omeprazole*

(i) Clinical Profile

Omeprazole (Losec) is the first therapeutically proven proton pump blocker and is clearly emerging as a significant advance in the treatment of peptic ulcer and related diseases. Its high efficacy in inhibiting gastric acid further supports the adage 'no acid — no ulcer'. At the present time product licences have been granted in some 20 countries, including France and USA, and it is marketed in Sweden, UK, the Netherlands and several other countries. Medical opinion is showing increasing support for the drug.[53,54]

Peptic ulcer treatment is associated with faster relief of pain and a considerable decrease in healing time compared with H_2-receptor antagonists.[55] Healing has been demonstrated in patients whose ulcers are resistant to treatment with cimetidine or ranitidine[56] and superiority over ranitidine has been shown in patients suffering from reflux oesophagitis.[57] Omeprazole has also been shown to give considerable relief to patients with Zollinger–Ellison syndrome, a condition in which a pancreatic tumour results in continuous gastrin production, giving rise to unremitting gastric acid secretion.[58]

A comprehensive bibliography of omeprazole is given in ref. 59.

(ii) Disadvantages

The following limitations of omeprazole should be considered by the medicinal chemist designing new proton pump blockers with advantages over omeprazole.

(1) Long term treatment of rats with the drug has resulted in enterochromaffin-like cell hyperplasia and carcinoid syndrome.[58] A detailed discussion of the implications of this finding is beyond the scope of this review. However, it is probably a consequence of the high level, long term inhibition of gastric acid secretion and not a direct toxicological effect of the drug.

(2) The sensitization of personnel handling the drug has been reported,[60] but the phenomenon has not been recorded in patients.

(3) Early reports that the drug gives rise to hepatotoxicity[61] have now been largely discounted.[62]

(4) Omeprazole inhibits liver microsomal mixed function oxidase, which may affect the hepatic degradation of concomitantly administered drugs.[58]

(5) The main disadvantage of omeprazole is its ready activation at mildly acidic pH. The ideal proton pump inhibitor should be stable at pH 5, the lowest value found outside the gastric acid space. Although this has apparently not led to any long term toxicological problems for omeprazole, the drug has to be administered as an enterically coated formulation to prevent its destruction in the stomach. (Like most other antisecretory drugs, omeprazole must be absorbed in order to exert its therapeutic effects, and does not show significant topical action.)

The variation in the extent of the inhibitory effects on acid secretion shown by omeprazole at lower doses[63] may also be a consequence of its low stability at higher pH values.

6.4.2.7 Other Putative Proton Pump Blockers

Other structures which have been claimed to possess inhibitory action on the H^+,K^+-ATPase enzyme include allyl isothiocyanate,[64] trifluoperazine,[65] nolinium bromide,[66] RP 40749,[67] and fenoctimine.[68] Biochemical studies and the quaternary nature of its structure suggest that nolinium bromide may share with the imidazopyridines (*e.g.* SCH 28080) an ability to antagonize the K^+ site of the enzyme.

6.4.3 THE Na^+,K^+-ATPase

6.4.3.1 Structure and Function

The Na^+,K^+-ATPase is a ubiquitous integral plasma membrane protein which pumps Na^+ out of and K^+ into the cell against the electrochemical gradients, the necessary energy being derived from the hydrolysis of ATP. The process is electrogenic, the stoichiometry being $3Na^+:2K^+$. The enzyme thus maintains the Na^+ electrochemical gradient, which provides energy for a wide varity of processes such as action potential generation in excitable cells, and coupled transport of ions and metabolites.[3,69]

The enzyme is an $\alpha\beta$ heterodimer (see Chapter 5.1). The β subunit is a glycoprotein which cDNA analysis has shown to comprise 302 amino acids. Hydrophobicity analysis suggests that there is a single α-helical transmembrane domain located close to the N terminus,[70] the polypeptide having a predominantly extracellular location. The α subunit possesses all the catalytic functions and contains the cardiac glycoside-binding site. The primary sequences deduced from the cDNA from a variety of species contain 1016 or 1022 amino acids.[71-73] There is 90% homology between species. Hydrophobicity analysis, supported by other biochemical studies, has led to the suggestion that the enzyme has 6,[73] 7[74] or 8[72] transmembrane α helices, although the positions of the first five are agreed.

There is also a 62% homology with the rat H^+,K^+-ATPase[8] and this is particularly strong between the fourth and fifth transmembrane helices where the ATP-binding site and the phosphorylated asparatate are located.[8,72] The cardiac glycosides can bind strongly to the enzyme and the cardiac glycoside-binding site is formed, at least in part, by the extracellular loop between the third and fourth transmembrane helices.[69,72]

The Na^+,K^+-ATPase exists as at least two isozymes with different sensitivities to inhibition by cardiac glycosides. These were first demonstrated as distinct biochemical entities in brain tissues. The $\alpha(+)$ isozyme of the axolemma shows a 300-fold greater sensitivity to ouabain than the α isozyme found in astrocytes, which appears to be the same as the isozyme isolated from kidney.[75]

More recently, both isozymes have been identified in rat heart, the ouabain sensitive form comprising 20% of the total enzyme in neonates but declining during maturation.[76] Interestingly, a model of cardiac hypertrophy in the rat re-expresses the ouabain sensitive form.[77] Canine ventricle also possesses a 34:66 distribution of isozymes with IC_{50} values for inhibition by ouabain of 2 and 300 nM respectively.[78] The discovery of isozymes is consistent with earlier studies which resolved two ouabain-binding sites in rat heart.[79-81]

6.4.3.2 Inotropic Action of the Cardiac Glycosides

Although the mechanism of the inotropy produced by the cardiac glycosides is widely believed to be due to inhibition of the Na^+,K^+-ATPase,[82] this has been questioned because of the discrepancies in some preparations where lower levels of ouabain were required to produce inotropy than were necessary to inhibit the pump in membrane preparations.[79,80] The existence of the high affinity isozyme in cardiac muscle, which has been reported to be inhibited by concentrations of ouabain at which the onset of inotropy is first observed, may go some way to explain these discrepancies.

The mechanism by which inhibition of the Na^+,K^+-ATPase produces positive inotropy is now believed to depend on the presence of a second protein, the Na^+/Ca^{2+} exchanger, in the plasma membrane.[83] This exchange process is electrogenic with a stoichiometry of $3Na^+:1Ca^{2+}$ and is driven by the Na^+ electrochemical gradient. At resting membrane potentials, the exchanger acts to extrude Ca^{2+} from the cell. Because of the stoichiometry, an increase in $[Na^+]_i$ due to inhibition of the Na^+,K^+-ATPase can give rise to a proportionally much larger increase in $[Ca^{2+}]_i$.[83] The increase in Ca^{2+} available to the contractile elements gives rise to the increased force of contraction. Cardiac glycosides also have stimulatory effects on autonomic transmission,[84] which may contribute to the inotropy[83] as well as other therapeutic and toxic effects.[85,86] It is thought that these neuronal effects are also due to Na^+,K^+-ATPase inhibition.

6.4.3.3 Endogenous Digitalis-like Factors

Identification of the Na^+,K^+-ATPase as the receptor for the cardiac glycosides has excited interest in the possible existence of an endogenous ligand. The report that the plasma of individuals with essential hypertension contains a factor that inhibits the Na^+,K^+-ATPase[87] has stimulated a search for a digitalis-like factor with natriuretic and hypertensive properties. This has produced a literature (for reviews see refs. 88 and 89) too large and diverse to discuss here. This factor is variously assayed by inhibition of Na^+,K^+-ATPase, by inhibition of $^{86}Rb^+$ uptake or by cross-reactivity with antidigoxin antibodies. The differences in the reported profile of the factor observed in a range of models suggest that not all groups are investigating the same material. Full characterization is awaited and, until this occurs, reservations will remain.[88]

6.4.3.4 Clinical Applications

The design and selection of drugs have generally been on the basis of their activity as Na^+,K^+-ATPase inhibitors and as positive inotropes in pharmacological models. However, clinically, cardiac glycosides have two principal applications. They are useful in the treatment of atrial fibrillation. They control the ventricular response by increasing the refractoriness of the atrioventricular node.[86,90,91] However, their use to provide long term inotropic support in heart failure patients in sinus rhythm is more controversial[86,90,91] and there is debate as to whether the inotropic effect is only useful in the short term.[91] Nevertheless, increased cardiac performance is observed[91] and these drugs are still widely used. Also, because of the low therapeutic index, clinical benefit has to be weighed against the high incidence of side effects. These include arrhythmias, gastrointestinal disturbances and CNS effects, particularly emesis and visual disturbance.[84,85]

The most frequently used pure compounds are digitoxin and digoxin which are the tris-2,6-dideoxyribohexose derivatives of (16) and (17) respectively (R = tridigitoxose). Digitoxin is better absorbed and is metabolized by the liver, whereas digoxin is excreted unchanged by the kidneys, giving rise to problems of dose management in patients with renal impairment.[84] Some compounds, such as gitaloxin, are claimed to have therapeutic advantages.[92] It has been suggested that pharmacokinetic properties, particularly with respect to the response of the CNS, may account for the differences in therapeutic index.[84]

6.4.3.5 The Relationship Between Structure and Activity of the Cardenolide Derivatives

Several hundred digitalis-like structures have been isolated from natural sources and shown to possess cardiotonic properties (see, for example, Steidle[93]). Their detailed description is obviously beyond the scope of this brief review, which concentrates on those molecules which have been part of structure–activity studies.

In this discussion, the parent structure is considered to be $5\beta,14\beta$-androstane-$3\beta,14$-diol to which an unsaturated lactone side chain is attached at C-17 to give the basic butenolide structure of digitoxigenin (**16**). The glycosylated derivatives have sugar residues attached to the C-3 alcoholic function. Substitution at C-12 and C-16 by OH gives rise to the digoxigenin (**17**) and gitoxigenin (**18**) series respectively. The strophanthidin derivatives (*e.g.* **19**) and ouabain (**20**) are hydroxylated at C-5 and the angular methyl group (R^{19}) is oxidized.

The reliability of structure–activity information in this field is limited by a number of factors and has led to some confusion and contradictory statements in the literature. Firstly, the earlier investigations dealt with impure plant extracts, often involving complex mixtures of cardiac glycosides. Secondly, the sensitivity of the Na^+,K^+-ATPase enzyme can vary greatly between different tissues of one species, from species to species,[94,95] and even in one animal according to season[96] (see also the earlier discussion of isozymes). The ranking of potencies of substrates can be affected by these discrepancies. Further, the inotropic potency (both *in vitro* and *in vivo*) often correlates poorly with activity on the isolated enzyme.[92,97] This has led to speculation that the cardenolides may owe a significant part of their action to an ability to stimulate the release of catecholamines.[98]

However, there have been several useful investigations into the structure–activity relationships of the cardiac glycosides utilizing all the modern approaches normally employed by the medicinal chemist.[99]

The minimum structure required to inhibit the Na^+,K^+-ATPase enzyme is the $5\beta,14\beta$-androstane-$3\beta,14$-diol unit.[99-101] However, the glycoside residue and butenolide side chain can substantially increase activity by binding to auxiliary sites on the enzyme. Replacement with a C/D *trans* flat ring junction prevents recognition by the enzyme.[99,100]

The inhibition of the Na^+,K^+-ATPase enzyme by some other steroids and related molecules such as cassaine, canrenone, the diguanyl hydrazone derivatives of glucocorticoids and progestin chlormadinol acetate was discounted by Schönfeld *et al.* because of their lack of potency and selectivity of action.[100] However, certain cholestane derivatives were shown by Kamernitskii *et al.* to possess activity on the enzyme at the micromolar level.[102]

6.4.3.5.1 *The steroid nucleus*

Most of the structure–activity investigations associated with the androstane unit have concentrated on functionalization on, or close to, ring D, related to the digoxin (**17**) and gitoxin (**18**) series, though it is clear that ouabain (**20**) and related structures owe their potency, in part, to the oxidized positions of rings, A, B and C. However, Shimada *et al.*[103] demonstrated that oxygenation at C-2 and C-11 normally leads to an increase in activity. The former group included the elaeodendrosides in which the sugar moiety is doubly linked to ring A at C-2 and C-3, of which the diosphenol derivative (**21**) was the most potent ($IC_{50} = 0.4$ μM) of 43 compounds tested. Introduction of a polar substituent, such as a sugar residue or carboxylate, into the C-19 angular methyl group led to loss of activity.[101]

Gitoxigenin (**18**) is less active than digitoxigenin (**16**), but the 16β-esters (formate (**22**) and acetate) are more potent inhibitors of the Na^+,K^+-ATPase.[104,105] This also holds true for the 3β-formate, acetate and monodigitoxoside series.[104] The underlying reason for this well-established phenomenon has been the subject of much study and discussion. It has been suggested that for the 16-hydroxy derivatives, an unfavourable hydrogen bond with the lactone carbonyl group prevents the butenolide ring from binding to the enzyme.[105] Griffin *et al.* have synthesized a number of gitoxigenin analogues[106] and studied their X-ray structures,[104] which indicated that intramolecular hydrogen bonding normally occurred between the 16-OH and 14-OH substituents, though in some compounds an intermolecular hydrogen bond was formed with the 16-OH function. In either case, the ring D conformation was altered, resulting in an orientation of the lactone ring in which the carbonyl oxygen pointed away from the putative binding site on the enzyme. On esterification of gitoxigenin, the desired conformation of the lactone ring is restored. However, *increased* activity of gitaloxigenin (**22**) over digitoxigenin (**16**) can be attributed to an additional interaction of the 16-OCOR function with the enzyme.[104] De Pover and Godfraind[105] have suggested that the site may

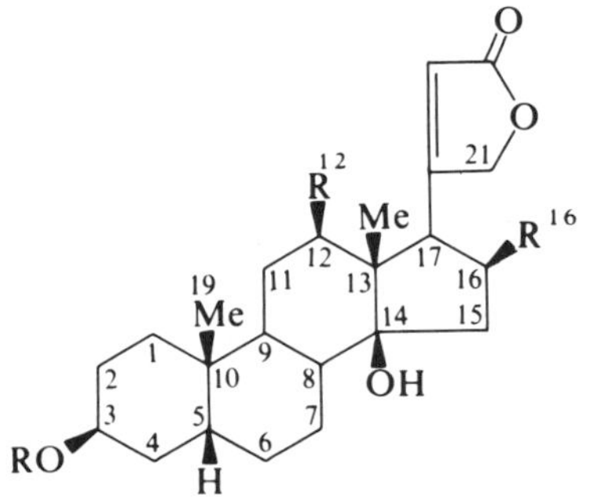

	R	R^{16}	R^{12}			R^{19}	X	
(16)	H	H	H	Digitoxigenin	(19)	CHO	H	Strophanthidin (α-L-rhamnoside)
(17)	H	H	OH	Digoxigenin	(20)	CH$_2$OH	OH	Ouabain
(18)	H	OH	H	Gitoxigenin				
(22)	H	OCHO	H	Gitaloxigenin				

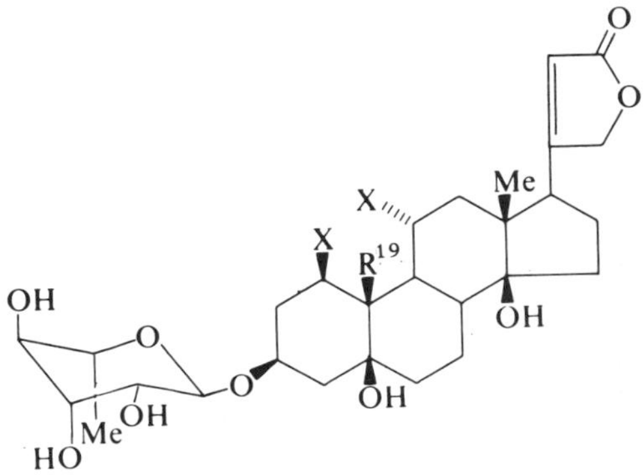

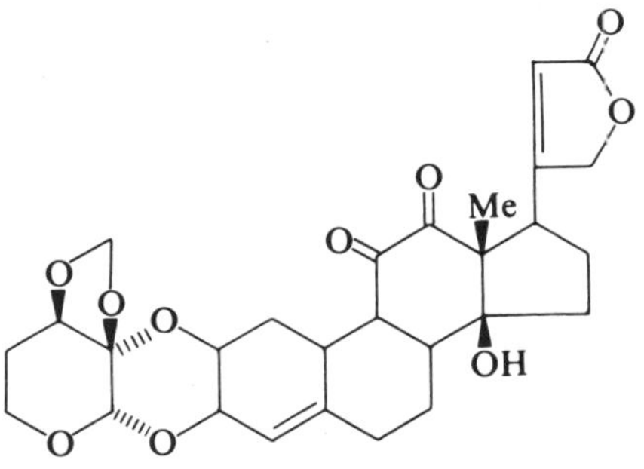

(21)

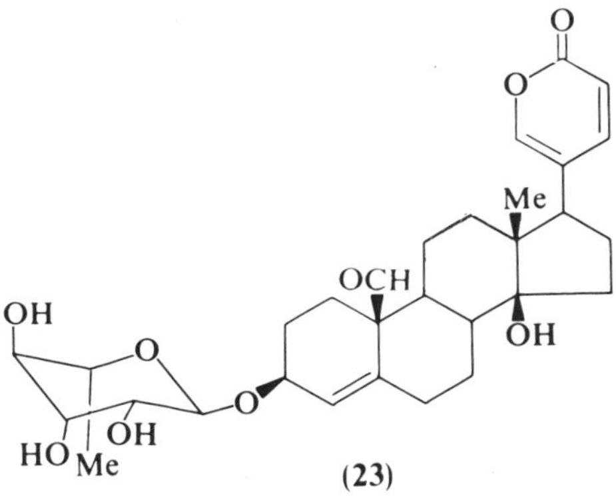

(23)

be the same one that normally binds the lactone carbonyl. Activity of the 16-esters decreases in the order formate (OCHO) > acetate (OCOMe) > carbonate (OCO$_2$Me).

6.4.3.5.2 *The lactone side chain*

The electronic properties of the butenolide moiety are clearly important for activity. For example, reduction of the double bond to the corresponding butanolide results in considerable lowering of activity.[107] A change in the stereochemistry of attachment of the 17-substituent (demonstrated for 17-α-strophanthidin derivatives) also results in loss of activity.[101] Biological activity is retained or slightly enhanced if the butenolide is replaced by a pentadienolide function as in the bufadienolide rhamnosides,[103] *e.g.* (23). Other beneficial C-17 substitutions have often been associated with carbonyl-containing substituents (see, for example, ref. 102).

Using X-ray data, From *et al.*[108] demonstrated a good correlation between potency on the enzyme and the change in position of the carbonyl oxygen atom relative to that found in digitoxigenin. Explanations of the type of interaction involved between the enone system of the butenolide and the receptor have included a combination of hydrogen bonding and π-complex formation[109] and even covalent binding *via* a Michael reaction.[110] However, the butenolide can be replaced by a furan or pyridine moiety, which gives rise to a retention of the free energy of binding to the Na$^+$,K$^+$-ATPase enzyme.[99,100] Common to all these side chains is a negative electrostatic potential well in the vicinity of the ring heteroatoms, which may be the site of hydrogen acceptor bonding.

6.4.3.5.3 The substituent at C-3 (sugar residue)

The binding energy of the aglycone digitoxigenin with the human cardiac muscle ATPase increases on glycosylation in the order arabinofuranose < tridigitoxose < 3'-*O*-methylrhamnose < 2',3'-*O*-isopropylidenerhamnose < glucose < digitoxose < rhamnose < 4'-deoxy-4'-aminorhamnose. In view of the large variation in the stereoelectronic properties of these sugar moieties, the range of interaction energies observed (-2.8 to -8.6 kJ mol^{-1}) is surprisingly small, indicating a remarkable degree of conformational adaptability of the sugar-binding subsite.[99,100] The second and third sugar residues of di- and tri-digitoxides make a small detracting contribution to enzyme binding, which demonstrates that the auxiliary sugar-binding site is subject to a small steric constraint.[100,111]

Glycosylation of different steroid nuclei (digoxigenin, 16-epigitoxigenin and gitoxigenin) gave in each case a very similar benefit in interaction energy.[100]

The presence of an anionic binding site on the enzyme is suggested by the energy increase caused by a 4'-amino group in the rhamnose side chain and also that this energy is decreased on sulfonation of the 3β-hydroxy group of digitoxigenin. Acylation or benzoylation of the 3-hydroxy function normally leads to little change in binding energy[100] or inhibitory activity.[101,106] However, acylation of the sugar residue hydroxyls led to considerable decrease in activity (see Section 6.5.3.5.4).[101]

6.4.3.5.4 Hydrophobic interactions

The relationship between lipophilicity and inhibitory action on guinea-pig heart Na$^+$,K$^+$-ATPase (ID$_{50}$ values) was investigated by Dzimiri *et al.*[107] The lipophilicities of 32 cardenolide derivatives were measured and generally decreased in the order digitoxigenin > gitoxigenin > digoxigenin > strophanthidin and an *overall* correlation between log P and ID$_{50}$ was naturally not seen across all structural types. The following structural features *not* associated with hydrophobicity were noted.

(i) The relatively poor activity of gitoxin was associated with unfavourable interactions of the 16-OH group with the receptor (in contrast to the 12-OH analogue, digoxin, which is less lipophilic but more potent). In particular the benefit to activity of acylation of the former but not the latter derivative was noted (see earlier). (ii) Derivatization of the sugar alcohol groups was detrimental to activity, suggesting that hydrogen bonding between glycoside and receptor may take place. (iii) Reduction of the double bond in the butenolide fragment resulted in almost complete loss of activity.

After allowing for these over-riding stereoelectronic effects, a moderate correlation between log P and potency on the enzyme was indeed seen, and the data suggest that an important hydrophobic interaction occurs, particularly with ring D of the steroid, and a weaker one with the sugar moiety.

In contrast, Repke[99] claimed that hydrophobic binding does not appear to play a dominant rôle. The two derivatives digitoxin and ouabain, at opposite ends of the lipophilicity spectrum, gave the same pattern of dissociation kinetics in pig cardiac muscle and brain cortex. Furthermore, hydrophobic bonding would demand negative enthalpy and entropy changes on binding with the enzyme, which is in contradiction to found values.

6.4.3.5.5 Computational studies

The use of computer graphics and computational techniques to study the cardiac glycosides has been reviewed by Fullerton *et al.*[112] Using molecular superposition programs, the striking correlation between the C-17 side group carbonyl oxygen position and biological activity was confirmed; for each 2.2 Å shift in oxygen position relative to digitoxigenin, activity changed by one order of magnitude. The energy barriers of rotation of the C-17 group[112] and the C(3)–glycoside bond[113] have also been studied using molecular mechanics calculations.

Dittrich *et al.*[114] have studied the relationship between the dipole moment of the cardiotonic steroids and their biological activity. Dipole moments computed for the steroids for low energy conformations (which had either C(14)–C(22) *or* C(14)–C(21) opposition) correlated linearly with the standard Gibbs energy of their interaction with the Na$^+$,K$^+$-ATPase.

The dipole component was found by means of regression analysis to correspond to the electrical field which runs parallel to the line connecting C-6 and C-9 of a given steroid. This field encompasses the whole digitalis molecule and can account for the variation in activity as the orientation of the butenolide chain changes.

6.4.3.6 Other (Non-steroidal) Inhibitors

A number of structures, unrelated to the cardiac glycosides, have been demonstrated to inhibit the Na^+,K^+-ATPase. The inotropic action of the indole derivative DPI 201-106 has recently been attributed, in part, to its ability to inhibit the enzyme.[115] Phenothiazines[116] inhibit the enzyme and this correlates with antipsychotic activity, but may involve a non-specific interaction resulting from their cationic detergent properties or through formation of a semiquinone radical. Derivatives of *n*-alkylpiperidines and nicotinamide inhibited the enzyme and a correlation with alkyl chain length was seen.[117] The biphasic inhibition kinetics supported the hypothesis that two inhibitory sites were involved; a high affinity (inhibition of Na^+-stimulated phosphorylation) and a low affinity binding to the K^+-stimulated dephosphorylation reaction but again a detergent-like action may have been responsible. Other claimed inhibitors include styrene and its oxide[118] and a number of other compounds cited by Shlevin.[94] Generally, inhibitory activity with all these agents occurred above the micromolar level and was probably 'non-specific' in nature. In any case, none of them has been the subject of a systematic medicinal chemical investigation and are, therefore, not considered here.

ACKNOWLEDGEMENTS

The authors would like to express their gratitude to Professor George Sachs (CURE, Wadsworth VA Medical Centre, Los Angeles, CA), Professor Hans Suschitzky (University of Salford) and Dr Anthony Ingall, Dr David Hall and Mr Henryk Radziwonik (Fisons R & D, Loughborough) for their help in the preparation of this review.

6.4.4 REFERENCES

1. D. H. MacLennan, C. J. Brandl, B. Korczak and N. M. Green, *Nature (London)*, 1985, **316**, 696.
2. E. Carafoli and M. Zurini, *Biochim. Biophys. Acta*, 1982, **683**, 279.
3. P. L. Jørgensen, *Kidney Int.*, 1986, **29**, 10.
4. G. Sachs, *Scand. J. Gastroenterol.*, 1986, **21** (suppl. 118), 1.
5. S. Muallem and G. Sachs, *Biochim. Biophys. Acta*, 1984, **805**, 181.
6. J. M. Wolosin and J. G. Forte, *J. Membr. Biol.*, 1983, **71**, 195.
7. J. Cuppoletti and G. Sachs, *J. Biol. Chem.*, 1984, **259**, 14952.
8. G. E. Shull and J. B. Lingrel, *J. Biol. Chem.*, 1986, **261**, 16788.
9. C. J. Hawkey and R. P. Walt, *Lancet*, 1986, **2**, 1084.
10. G. N. J. Tytgat, *Digestion* (suppl.), 1987, **37** (suppl. 2), 1.
11. R. Nagashima, *J. Clin. Gastroenterol.*, 1981, **3** (suppl. 2), 103.
12. D. M. Jones, A. M. Lessells and J. Eldridge, *J. Clin. Pathol.*, 1984, **37**, 1002.
13. J. G. Coghlan, D. Gilligan, H. Humphries, D. McKenna, D. Dooley, E. Sweeney, C. Keane and C. O'Morain, *Lancet*, 1987, **2**, 1109.
14. M. Feldman, *Gastroenterology*, 1984, **86**, 361.
15. C. J. Loewe, J. R. Grider, J. Gardiner and Z. R. Vlahcevic, *Gastroenterology*, 1985, **89**, 746.
16. S. Kobayashi, M. Miyamoto, Y. Shumada, K. Endo, F. Asai and T. Ito, *Jpn. J. Pharmacol.*, 1984, **36** (suppl.), 89P.
17. *Drugs of the Future*, 1983, **8**, 487.
18. (a) A. K. Nicol, M. Thomas and J. Wilson, *J. Pharm. Pharmacol.*, 1981, **33**, 554; (b) H.-J. Reimann, U. Schmidt, B. Ultsch, T. J. Sullivan and P. Wendt, *Gut*, 1984, **25**, 1221.
19. W. B. Im, J. C. Sih, D. P. Blakeman and J. P. McGrath, *J. Biol. Chem.*, 1985, **260**, 4591.
20. T. Berglindh and K. J. Obrink, *Acta Physiol. Scand.*, 1976, **96**, 150.
21. J. F. Long, M. Steinberg and M. Derelanko, *Gastroenterology*, 1981, **80**, 1216.
22. *Drugs of the Future*, 1982, **7**, 755.
23. P. J. S. Chiu, C. Casciano, G. Tetzloff, J. F. Long and A. Barnett, *J. Pharmacol. Exp. Ther.*, 1983, **226**, 121.
24. M. D. Ene, T. Khan-Daneshmend and C. J. C. Roberts, *Br. J. Pharmacol.*, 1982, **76**, 389.
25. J. F. Long, P. J. S. Chiu, M. J. Derelanko and M. Steinberg, *J. Pharmacol. Exp. Ther.*, 1983, **226**, 114.
26. J. J. Kaminski, D. G. Perkins, J. D. Frantz, D. M. Solomon, A. J. Elliott, P. J. S. Chiu and J. F. Long, *J. Med. Chem.*, 1987, **30**, 2047.
27. J. J. Kaminski, J. M. Hilbert, B. N. Pramanik, D. M. Solomon, D. J. Conn, R. K. Rizvi, A. J. Elliott, H. Guzik, R. G. Lovey, M. S. Domalski, S. Wong, C. Puchalski, E. H. Gold, J. F. Long, P. J. S. Chiu and A. T. McPhail, *J. Med. Chem.*, 1987, **30**, 2031.
28. C. K. Scott, E. Sundell and L. Castrovilly, *Biochem. Pharmacol.*, 1987, **36**, 97.
29. P. J. S. Chiu, A. Barnett, C. Gerhart, M. Policelli and J. Kaminski, *Arch. Int. Pharmacodyn. Ther.*, 1984, **270**, 128.
30. W. Beil, I. Hackbarth and K.-Fr. Sewing, *Br. J. Pharmacol.*, 1986, **88**, 19.
31. D. J. Keeling, S. M. Laing, J. Senn-Bilfinger and A. H. Underwood, *Biochem. Soc. Trans.*, 1987, **15**, 694.
32. J. J. Kaminski, J. A. Bristol, C. Puchalski, R. G. Lovey, A. J. Elliott, H. Guzik, D. M. Solomon, D. J. Conn, M. S. Domalski, S. Wong, E. H. Gold, J. F. Long, P. J. S. Chiu, M. Steinberg and A. T. McPhail, *J. Med. Chem.*, 1985, **28**, 876.
33. A. Brändström, P. Lindberg and U. Junggren, *Scand. J. Gastroenterol.*, 1985, **20** (suppl. 108), 15.
34. A. L. Ganser and J. G. Forte, *Biochim. Biophys. Acta*, 1973, **307**, 169.
35. B. Wallmark, A. Brändström and H. Larsson, *Biochim. Biophys. Acta*, 1984, **778**, 549.

36. P. Lorentzon, R. Jackson, B. Wallmark and G. Sachs, *Biochim. Biophys. Acta*, 1987, **897**, 41.
37. P. Lindberg, P. Nordberg, T. Alminger, A. Brändström and B. Wallmark, *J. Med. Chem.*, 1986, **29**, 1327.
38. V. Figala, K. Klemm, B. Kohl, U. Krüger, G. Rainer, H. Schaefer, J. Senn-Bilfinger and E. Sturm, *J. Chem. Soc., Chem. Commun.*, 1986, 125.
39. (a) E. Sturm, U. Krüger, J. Senn-Bilfinger, V. Figala, K. Klemm, B. Kohl, G. Rainer, H. Schaefer, T. J. Blake, D. W. Darkin, R. J. Ife, C. A. Leach, R. C. Mitchell, E. S. Pepper, C. J. Salter, N. J. Viney, G. Huttner and L. Zsolnai, *J. Org. Chem.*, 1987, **52**, 4573; (b) 1987, **52**, 4582.
40. Fisons plc, Pharmaceutical Division, Research and Development Laboratories, UK, unpublished results.
41. Y. Maki, K. Yamane and M. Sato, *Yakugaku Zasshi*, 1966, **86**, 50 (*Chem. Abstr.*, 1966, **64**, 11 165).
42. D. J. Keeling, C. Fallowfield, K. J. Milliner, S. K. Tingley, R. J. Ife and A. H. Underwood, *Biochem. Pharmacol.*, 1985, **34**, 2967.
43. W. Beil, M. Eltze, K. Heintze, K. Klemm, R. Riedel, C. Schudt, K.-Fr. Sewing and A. Simon, *Br. J. Pharmacol.*, 1986, **88**, 389.
44. W. B. Im, D. P. Blakeman and G. Sachs, *Biochim. Biophys. Acta*, 1985, **845**, 54.
45. P. Lindberg, A. Brändström and B. Wallmark, *Trends Pharmacol. Sci.*, 1987, **8**, 399.
46. G. Rackur, M. Bickel, H.-W. Fehlhaber, A. Herling, V. Hitzel, H.-J. Long, M. Rösner and R. Weyer, *Biochem. Biophys. Res. Commun.*, 1985, **128**, 477.
47. P. Lindberg, AB Hässle Research Laboratories, Sweden, personal communication.
48. P. A. Cage, D. Cox, H. A. Dowlatshahi, N. P. Gensmantel, A. H. Ingall, S. J. King, D. P. Marriott, V. Samani and J. L. Suschitzky, poster presented to Medicinal Chemistry Conference, Cambridge, UK, September 1987.
49. *Drugs of the Future*, 1987, **12**, 34.
50. K. Sigrist-Nelson, R. K. M. Müller and A. E. Fischli, *FEBS Lett.*, 1986, **197**, 187.
51. H. Satoh, N. Inatomi, H. Nagaya, I. Inada, A. Nohara and Y. Maki, *Jpn. J. Pharmacol.*, 1986, **40** (suppl.), 226P.
52. A. H. Ingall, D. Cox, J. L. Suschitzky and P. A. Cage, poster presented to IXth International Symposium on Medicinal Chemistry, Berlin, 1986.
53. Editorial, *Lancet*, 1987, **2**, 1187.
54. D. G. Weir, *Br. Med. J.*, 1988, **296**, 195.
55. (a) K. D. Bordhan, G. B. Porro, K. Bose, R. F. C. Hinchcliffe, M. Lazzaroni and P. Morris, *Gut*, 1985, **26**, A557; (b) K. Lauritsen, S. J. Rune, P. Bytzer, H. Kelbaek, K. G. Jensen, J. Rask-Madsen, F. Bendtsen, J. Linde, M. Højlund, H. H. Andersen, K. Møllmann, V. R. Nissen, L. Ovesen, P. Schlichting, U. Tage-Jensen and H. R. Wulff, *N. Engl. J. Med.*, 1985, **312**, 958.
56. G. N. J. Tytgat, C. B. H. W. Lamers, W. Hameeteman, J. M. B. J. Jansen and J. A. Wilson, *Aliment. Pharmacol. Ther.*, 1987, **1**, 31.
57. T. Havelund, L. S. Laursen, E. Skoubo-Kristensen, B. N. Andersen, S. A. Pedersen, K. B. Jensen, C. Fenger, F. Hanberg-Sørensen and K. Lauritsen, *Br. Med. J.*, 1988, **296**, 89.
58. S. P. Clissold and D. M. Compoli-Richards, *Drugs*, 1986, **32**, 15.
59. Scrip's New Product Review No. 18, 'Omeprazole', 1987.
60. B. Meding, *Contact Dermatitis*, 1986, **15**, 36.
61. S. Gustavsson, H. Adami, L. Lööf, H. A. A. Nyberg and O. Nyrén, *Lancet*, 1983, **2**, 124.
62. L. Lööf, H.-O. Adami, S. Gustavsson, A. Nyberg, O. Nyrén and P. Lundborg, *Lancet*, 1984, **1**, 1347.
63. C. W. Howden, J. K. Derodra, D. W. Burget, C. Silletti, M. Dickey and R. H. Hunt, *Gastroenterology*, 1986, **90**, 1466.
64. N. Takeguchi, Y. Nishimura, T. Watanabe, Y. Mori and M. Morii, *Biochem. Biophys. Res. Commun.*, 1983, **112**, 464.
65. W. B. Im, D. P. Blakeman, J. Mendlein and G. Sachs, *Biochim. Biophys. Acta*, 1984, **770**, 65.
66. J. Nandi, M. V. Wright and T. K. Ray, *Gastroenterology*, 1983, **85**, 938.
67. *Drugs of the Future*, 1983, **8**, 1028.
68. W. W. Reenstra, B. Shortridge and J. G. Forte, *Biochem. Pharmacol.*, 1985, **34**, 2331.
69. B. C. Rossier, K. Geering and J. P. Kraehenbuhl, *Trends Biochem. Sci.*, 1987, **12**, 483.
70. G. E. Shull, L. K. Lane and J. B. Lingrel, *Nature (London)*, 1986, **321**, 429.
71. Y. A. Ovchinnikov, N. N. Modyanov, N. E. Broude, K. E. Petrukhin, A. V. Grishin, N. M. Arzamazova, N. A. Aldanova, G. S. Monastyrskaya and E. D. Sverdlov, *FEBS Lett.*, 1986, **201**, 237.
72. G. E. Shull, A. Schwartz and J. B. Lingrel, *Nature (London)*, 1985, **316**, 691.
73. K. Kawakami, S. Noguchi, M. Noda, H. Takahashi, T. Ohta, M. Kawamura, H. Nojima, K. Nagano, T. Hirose, S. Inayama, H. Hayashida, T. Miyata and S. Numa, *Nature (London)*, 1985, **316**, 733.
74. Y. A. Ovchinnikov, N. M. Luneva, E. A. Arystarkhova, N. M. Gevondyan, N. M. Arzamazova, A. T. Kozhich, V. A. Nesmeyanov and N. N. Modyanov, *FEBS Lett.*, 1988, **227**, 230.
75. K. J. Sweadner, *J. Biol. Chem.*, 1985, **260**, 11 508.
76. K. J. Sweadner and S. K. Farshi, *Proc. Natl. Acad. Sci. USA*, 1987, **84**, 8404.
77. D. Charlemagne, J.-M. Maixent, M. Preteseille and L. G. Lelievre, *J. Biol. Chem.*, 1986, **261**, 185.
78. J.-M. Maixent, D. Charlemagne, B. de la Chapelle and L. G. Lelievre, *J. Biol. Chem.*, 1987, **262**, 6842.
79. E. Erdmann, G. Philipp and H. Scholz, *Biochem. Pharmacol.*, 1980, **29**, 3219.
80. R. J. Adams, A. Schwartz, G. Grupp, I. Grupp, S.-W. Lee, E. T. Wallick, T. Powell, V. W. Twist and P. Gathiram, *Nature (London)*, 1982, **296**, 167.
81. F. Noel and T. Godfraind, *Biochem. Pharmacol.*, 1984, **33**, 47.
82. T. Akera, *Science (Washington, D.C.)*, 1977, **198**, 569.
83. M. P. Blaustein, *Trends Pharmacol. Sci.*, 1985, **6**, 289.
84. C. M. Lathers, L. J. Lipka and H. A. Klions, *J. Clin. Pharmacol.*, 1985, **25**, 501.
85. J. C. Somberg, *J. Clin. Pharmacol.*, 1985, **25**, 529.
86. B. Levitt and D. L. Keefe, *J. Clin. Pharmacol.*, 1985, **25**, 507.
87. J. M. Hamlyn, R. Ringel, J. Shaeffer, P. D. Levinson, B. P. Hamilton, A. A. Kowarski and M. P. Blaustein, *Nature (London)*, 1982, **300**, 650.
88. M. R. Wilkins, *Trends Pharmacol. Sci.*, 1985, **6**, 286.
89. F. J. Haddy and M. B. Pamnani, *Fed. Proc., Fed. Am. Soc. Exp. Biol.*, 1985, **44**, 2789.
90. J. Somberg, D. Keefe and B. Levitt, *J. Clin. Pharmacol.*, 1985, **25**, 507.

91. M. D. Cheitlin, *Cardiology*, 1987, **74**, 376.
92. B. G. Woodcock and N. Rietbrock, *Trends Pharmacol. Sci.*, 1985, **6**, 273.
93. W. Steidle, in 'Wandlungen in der Therapie der Herzinsuffizienz', ed. N. Rietbrock, B. Schneiders and J. Schuster, Vieweg, Weisbaden, 1983, p. 7.
94. H. H. Shlevin, *Drug Dev. Res.*, 1984, **4**, 275.
95. S. L. Bonting, N. M. Hawkins and M. R. Canady, *Biochem. Pharmacol.*, 1964, **13**, 13.
96. J. S. Charnock, W. F. Dryden, P. A. Lauzon and C. Skoog, *Comp. Biochem. Physiol. B*, 1980, **65**, 675.
97. K. R. H. Repke and W. Schönfeld, *Trends Pharmacol. Sci.*, 1984, **5**, 393.
98. T. J. Hougen, N. Spicer and T. W. Smith, *J. Clin. Invest.*, 1981, **68**, 1207.
99. K. Repke, *Trends Pharmacol. Sci.*, 1985, **6**, 275.
100. W. Schönfeld, J. Weiland, C. Lindig, M. Masnyk, M. M. Kabat, A. Kurek, J. Wicha and K. R. H. Repke, *Naunyn-Schmiedebergs Arch. Pharmacol.*, 1985, **329**, 414.
101. N. M. Mirsalikhova, *Khim. Prir. Soed.*, 1983, 320.
102. A. V. Kamernitskii, I. G. Reshetova, N. M. Mirsalikhova, V. G. Levi and E. I. Chernoburova, *Biokhimia (Moscow)*, 1984, **49**, 263.
103. K. Shimada, N. Ishii, K. Ohishi, J. S. Ro and T. Nambara, *J. Pharmacobio-Dyn.*, 1986, **9**, 755.
104. J. F. Griffin, D. C. Rohrer, K. Ahmed, A. H. L. From, T. Hashimoto, H. Rathore and D. S. Fullerton, *Mol. Pharmacol.*, 1986, **29**, 270.
105. A. DePover and T. Godfraind, *Naunyn-Schmiedebergs Arch. Pharmacol.*, 1982, **321**, 135.
106. T. Hashimoto, H. Rathore, D. Satoh, G. Hong, J. F. Griffin, A. H. L. From, K. Ahmed and D. S. Fullerton, *J. Med. Chem.*, 1986, **29**, 997.
107. N. Dzimiri, U. Fricke and W. Klaus, *Br. J. Pharmacol.*, 1987, **91**, 31.
108. A. H. L. From, D. S. Fullerton, T. Deffo, E. Kitatsuji, D. C. Rohrer and K. Ahmed, *J. Mol. Cell. Cardiol.*, 1984, **16**, 835.
109. W. E. Wilson, W. I. Siuitz and L. T. Hana, *Mol. Pharmacol.*, 1970, **6**, 449.
110. D. S. Fullerton, M. C. Pankaskie, K. Ahmed and A. H. L. From, *J. Med Chem.*, 1976, **19**, 1330.
111. E. Erdman and W. Schener, *Naunyn-Schmiedebergs Arch. Pharmacol.*, 1974, **283**, 335.
112. D. S. Fullerton, J. F. Griffin, D. C. Rohrer, A. H. L. From and K. Ahmed, *Trends Pharmacol. Sci.*, 1985, **6**, 279.
113. D. C. Rohrer, M. Kihora, T. Deffo, H. Rathore, K. Ahmed, A. H. L. From and D. S. Fullerton, *J. Am. Chem. Soc.*, 1984, **106**, 8269.
114. F. Dittrich, P. Berlin, K. Köpke and K. R. H. Repke, *Curr. Top. Membr. Trans.*, 1983, **19**, 251.
115. A. J. Kaumann, D. E. Richards and D. A. Russell, *Br. J. Pharmacol.*, 1987, **91**, 3.
116. P. Palatini, *Gen. Pharmacol.*, 1978, **9**, 215.
117. A. A. Abdelfattah and R. B. Koch, *Biochem. Pharmacol.*, 1981, **30**, 3195.
118. A. P. Singh and S. P. Srivastava, *J. Environ. Pathol. Toxicol. Oncol.*, 1986, **6**, 29.

6.5

Pyridoxal Dependent Systems

DAVID GANI

University of Southampton, UK

6.5.1	INTRODUCTION TO COENZYME B$_6$	214
6.5.2	STEREOCHEMICAL, MECHANISTIC AND STRUCTURAL FEATURES	214
	6.5.2.1 General Characteristics	214
	6.5.2.2 Transaminases	215
	6.5.2.2.1 L-α-Amino acid transaminases	215
	6.5.2.2.2 ω-Amino acid transaminases	220
	6.5.2.2.3 D-Amino acid transaminases	222
	6.5.2.3 Decarboxylases	222
	6.5.2.3.1 General features	222
	6.5.2.3.2 Glutamate decarboxylase	225
	6.5.2.3.3 Ornithine decarboxylase	226
	6.5.2.3.4 Histidine decarboxylase	227
	6.5.2.3.5 Aromatic amino acid decarboxylases	227
	6.5.2.4 Amino Acid Racemases	228
	6.5.2.4.1 General features	228
	6.5.2.4.2 Alanine racemase	229
	6.5.2.4.3 Other racemases	229
	6.5.2.5 Other PLP-dependent Systems	230
	6.5.2.5.1 Serine hydroxymethyltransferase	230
	6.5.2.5.2 Reactions of amino acids at C-3	233
	6.5.2.5.3 Reactions of amino acids at C-4	236
	6.5.2.5.4 Glycogen phosphorylase	239
6.5.3	INHIBITORS	239
	6.5.3.1 Introduction	239
	6.5.3.2 'Suicide' Inhibitors (Mechanism Activated Inhibitors)	240
	6.5.3.3 Transaminase Inhibitors	240
	6.5.3.3.1 Inhibitors for L-α-amino acid transaminases	240
	6.5.3.3.2 Inhibitors for γ-aminobutyrate transaminase (GABA-T)	242
	6.5.3.3.3 Inhibitors for ω-ornithine aminotransferase	243
	6.5.3.3.4 Inhibitors for D-amino acid transaminase	244
	6.5.3.4 Inhibitors for Decarboxylases	244
	6.5.3.4.1 Glutamate decarboxylase inhibitors	244
	6.5.3.4.2 Ornithine decarboxylase inhibitors	245
	6.5.3.4.3 Histidine decarboxylase inhibitors	245
	6.5.3.4.4 Aromatic amino acid decarboxylase inhibitors	246
	6.5.3.5 Inhibitors for Racemase Enzymes	246
	6.5.3.6 Inhibitors for Other PLP-dependent Enzymes	247
	6.5.3.6.1 Serine hydroxymethyltransferase inhibitors	247
	6.5.3.6.2 Inhibitors for enzymes catalyzing β-elimination or β-replacement reactions	247
	6.5.3.6.3 Inhibitors for PLP-dependent enzymes catalyzing reaction at C-4	248
6.5.4	OVERVIEW	248
6.5.5	REFERENCES	250

6.5.1 INTRODUCTION TO COENZYME B₆

Pyridoxal 5′-phosphate (coenzyme B_6) serves as a cofactor for a myriad of enzyme-catalyzed reactions. The coenzyme was originally referred to as codecarboxylase as it was observed that the resolved inactive apoenzyme of L-lysine decarboxylase was reactivated upon treatment with codecarboxylase.[1] The connection between codecarboxylase and vitamin B_6, a factor present in unrefined foodstuffs that cured a florid acrodynia in rats fed on defined diets, was not realized until Bellamy and Gunsalus published their work on tyrosine decarboxylase.[2] Following on from work by Gale,[3] which had revealed that the production of active enzyme in *Streptococcus faecalis* depended upon the inclusion of unidentified factors in addition to tyrosine in media, Bellamy and Gunsalus showed that media containing larger amounts of pyridoxine and nicotinic acid than were required for optimum growth gave high yields of active enzyme. Slightly later, in the mid-1940s, Snell identified pyridoxal phosphate as the coenzyme for transaminases.[4] Over the following 10 years it was established that the coenzyme was also involved in the tryptophanase, D-serine dehydratase and cysteine desulfhydrase reactions. It is now known that pyridoxal phosphate serves as a coenzyme for hundreds of enzyme-mediated reactions involving amino acids and, in a different manner, as a coenzyme for the phosphorylase reaction.

Over the past 20 years the mechanisms by which pyridoxal-dependent enzymes operate have been subjected to investigation and a great deal has been learned; not least, that the enzymes are diverse in many ways. It is only recently that detailed 3-D structural information has become available, but for only one class of enzyme, the transaminases. Thus the basis of our understanding of pyridoxal-dependent enzymes in general rests largely upon stereochemical, kinetic and primary structural information. Not surprisingly, this important cofactor has attracted the attention of the medicinal chemist for the rational design of new inhibitors with pharmaceutical potential.[5-10]

6.5.2 STEREOCHEMICAL, MECHANISTIC AND STRUCTURAL FEATURES

6.5.2.1 General Characteristics

Pyridoxal 5′-phosphate (PLP, **1**) is probably nature's most versatile coenzyme. It is intimately associated with amino acid metabolism and plays a central role in linking the carbon and nitrogen cycles. In all reactions involving PLP, including phosphorylase, a protein-bound lysine residue and the C-4 formyl group of PLP form a holoenzyme aldimine (**2**). In the phosphorylase reaction the aldimine anchors the PLP to the protein and the phosphate ester participates in the catalytic process (see Section 6.5.2.5.4). In all other reactions the condensation of an amino acid with the internal aldimine leads to the formation of a Schiff's base substrate aldimine (**3**). In some systems this transaldimination step leads to a large change in the conformation of the coenzyme complex at the active site of the enzyme; the relevance of this conformational change to the steric courses of PLP-dependent enzymic reactions is considered in detail later in this section. In the presence of the appropriate enzyme, all of the bonds connected to C^α of the imino acid moiety can be cleaved.

(1) PLP   (2)

On formation of the enzyme–substrate Schiff's base complex, rotation about the $N—C^\alpha$ bond is fixed by the enzyme so that the bond to be cleaved at C^α is held orthogonal to the plane of the pyridinium ring.[11] In this conformation, maximum orbital overlap is achieved between the developing negative charge and the conjugated electron-deficient π-electron system (Scheme 1). The cleavage is further facilitated by an enzyme-bound base, usually the ε-amino group of the lysyl residue, which stabilizes the positive charge as it forms on the electrofuge. Following these general, initial events, the exact course of the reaction varies depending upon which of the bonds at C^α is to be cleaved.

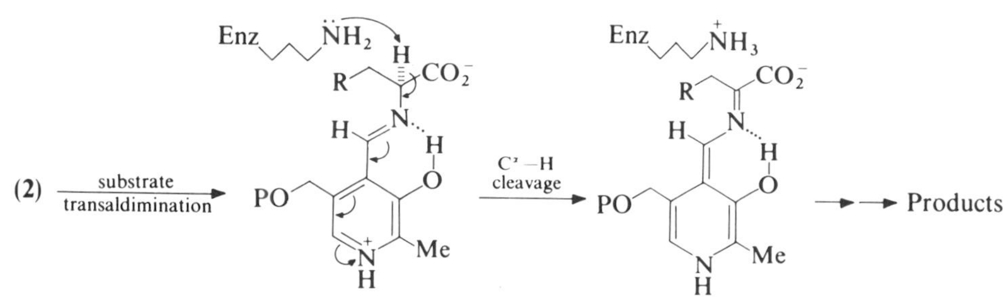

Scheme 1

Reviews of earlier work on PLP-dependent reactions are available.[7-13]

6.5.2.2 Transaminases

6.5.2.2.1 *L-α-Amino acid transaminases*

L-Glutamic acid is almost unique amongst amino acids in that the C^α—N bond is formed directly from ammonia and 2-oxoglutarate. Subsequent reduction of the transient imine with NADPH is reversibly catalyzed by glutamate dehydrogenase and results in the production of L-glutamic acid and NADP. Other amino acids are then biosynthesized from the corresponding α-oxo acids and L-glutamic acid through the action of PLP-dependent transaminases. The reaction sequence of the general L-amino acid transamination mechanism is outlined in Scheme 2, where proton transfer from C^α of the aldimine (**3**) to the ketimine (**4**) occurs on the C-4′ *si* face of the coenzyme. The products are pyridoxamine 5′-phosphate (PMP, **5**) and 2-oxoglutaric acid. The second half of the transamination involves reaction of the PMP with a new α-oxo acid (*e.g.* pyruvic acid) to give a new ketimine (**6**). Suprafacial proton transfer of C-4′ hydrogen atoms to C^α gives the aldimine (**7**), which undergoes transaldimination to give the new L-amino acid (alanine) and the internal PLP-aldimine (**2**) (Scheme 2). The overall reaction as illustrated for glutamic-pyruvic transaminase is thus: L-glutamate and pyruvate give 2-oxoglutarate and L-alanine. It is interesting to note that glycine exchanges the 2-*pro-R* hydrogen and L-amino acids exchange H^α with solvent hydrogen when no alternative oxo acids are available. D-Amino acid transaminases are also known (see Section 6.5.2.2.3).

The PLP-dependent transaminases are a well studied group of enzymes. In particular, aspartate aminotransferase (AAT) has received much attention and X-ray structures of the holoenzyme–substrate complex have now been published (see below). Indeed, most current ideas concerning PLP-dependent enzymic catalysis in general stem from studies of AAT. The configuration of the C-4′—N double bond of the aldimine–holoenzyme complex (**3**) has been assumed to be *trans* owing to the unfavourable interactions that would occur with the ring substituent in the *cis* coplanar conformation. On the basis of this assumption, both Arigoni[14] and Dunathan[15] independently conducted experiments with aspartate transaminase to define the site of protonation at C-4′. Both groups determined that solvent hydrogen was introduced from the C-4′ *si* face of the coenzyme to give samples of C-4′ labelled (4′*S*)-pyridoxamine when the reaction was conducted in labelled water. Using this system as a stereochemical assay of C-4′ it was shown that glutamic-pyruvic transaminase,[16] pyridoxamine-pyruvic transaminase[17] and dialkylamino acid transaminase[18] all labilize the 4′-*pro-S* hydrogen of pyridoxamine. Thus it was then only necessary to determine the mode of proton transfer from C^α to C-4′, as suprafacial *versus* antarafacial, to define the stereochemistry of the C^α—N double bond of the ketimine. Direct internal hydrogen transfer from C^α to C-4′ was demonstrated for abortive transamination reactions and for the reaction catalyzed by pyridoxamine-pyruvic transaminase. These observations are in accord with a suprafacial single-base mechanism. Hence H^α and the 4′-*pro-S* proton must be disposed on the same face of the coenzyme, and so the C-4′—N double bond of the aldimine must be *trans*. Arigoni came to the same conclusions by reducing an equilibrium mixture of AAT, aspartate and oxaloacetate with sodium borotritide. Degradation of the phosphopyridoxylaspartic acid to aspartate and pyridoxamine, followed by stereochemical assay of the tritiated positions, showed that (2*S*)-[2-³H$_1$]aspartate and (4′*S*)-[4′-³H$_1$]pyridoxamine were the predominant products, thus providing further evidence for a *trans* aldimine double bond.

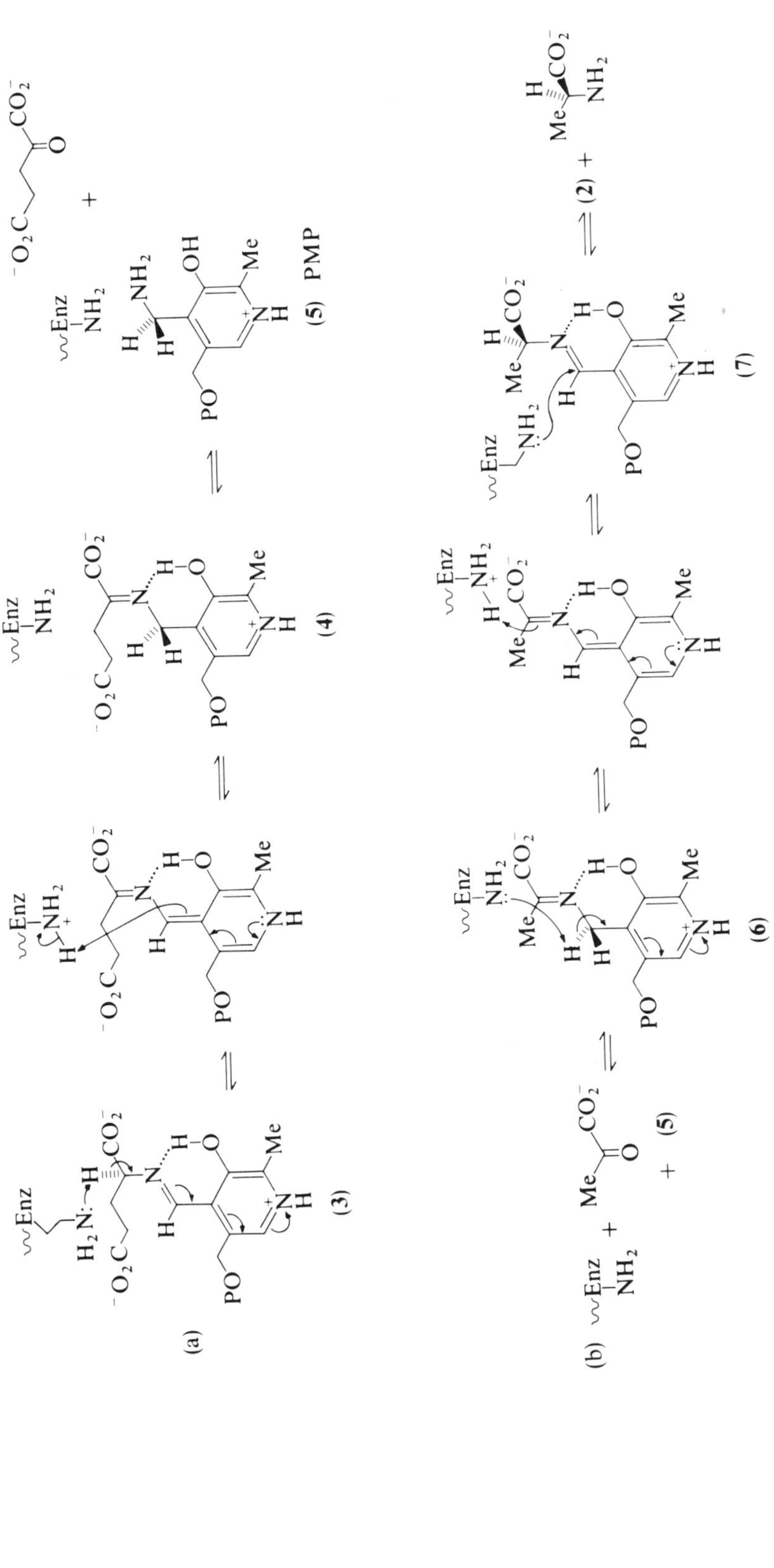

Scheme 2

The results also defined the configuration of the C^α—N double bond in the tautomeric ketimine in the AAT reaction.

Reductions with labelled sodium borohydride on other holoenzyme–substrate complexes have also shown that the exposed face of the coenzyme at which proton transfer occurs, the C-4' *si* face, is also more accessible to hydride. However, in the absence of substrate, reduction primarily occurs from the opposite C-4' *re* face of the coenzyme to give (4'R)-[4'-^{3}H]pyridoxyllysine, as has been demonstrated for both aspartate and alanine transaminase.[19] These results suggest that a significant change in conformation occurs during transamination with respect to the accessibility of the C-4' *re* face of the coenzyme to the solvent. Recent X-ray studies of aspartate transaminase[20–24] have confirmed these suggestions and it is now possible to envisage the three-dimensional catalytic function of the protein for the entire transamination process.

Aspartate transaminase from chicken heart mitochondria is composed of two identical subunits ($M_r \sim 45\,000$) which consist of two domains. PLP is bound to the larger domain in a pocket near the subunit interface. It is now evident that the proximal and distal carboxylate groups of dicarboxylic acid substrates (*viz.* aspartate, glutamate, oxaloacetate and α-ketoglutarate) are bound by two arginine residues (386 and 292 from adjacent subunits) and that substrate specificity is largely determined by these interactions. The mode of substrate binding not only ensures efficient catalysis but also causes a bulk movement in the smaller domain that closes the active-site pocket and moves Arg-386 3 Å closer to the coenzyme. The transaldimination of the ε-amino group of Lys-258 by

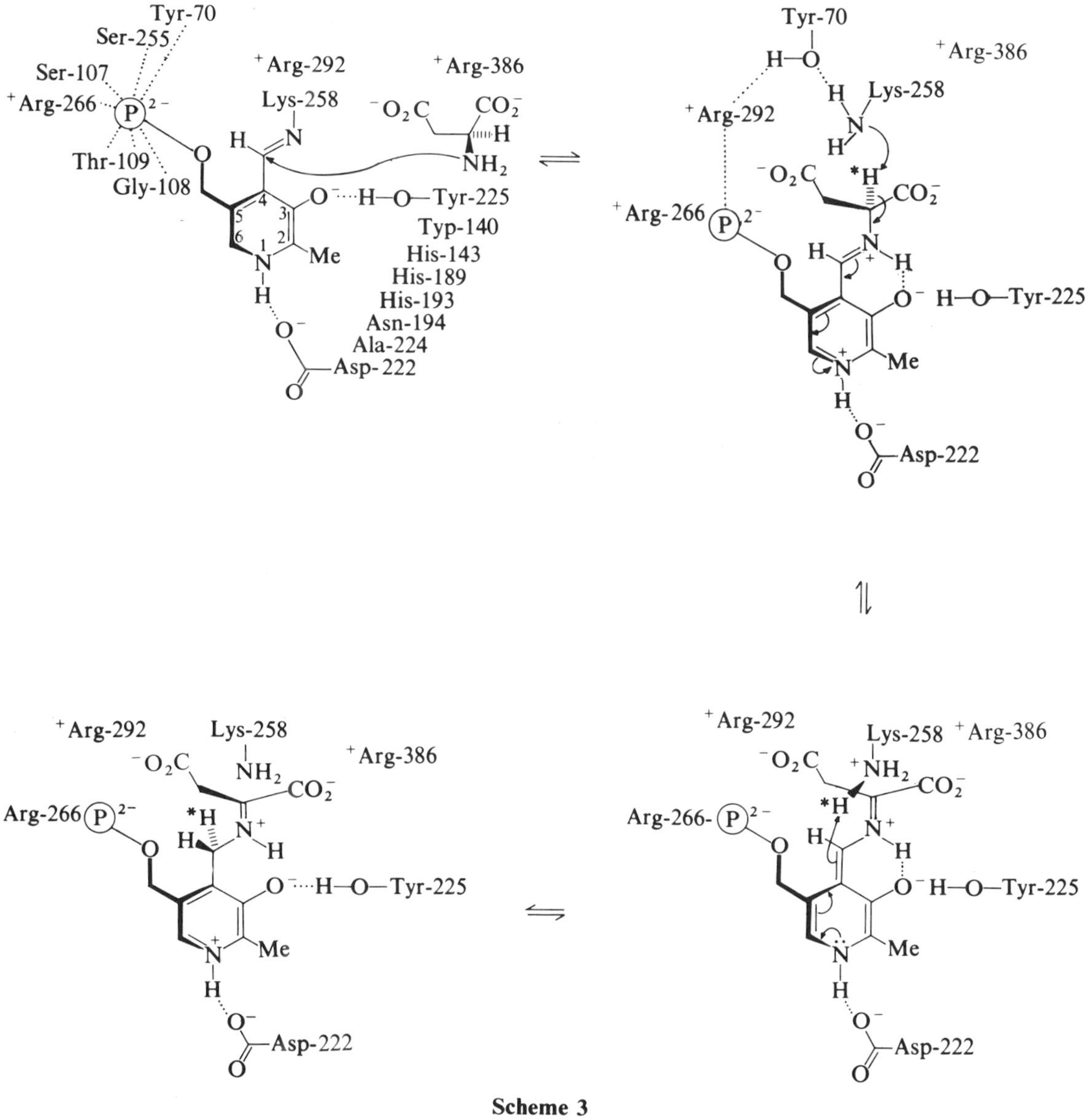

Scheme 3

Table 1 Comparison of Amino Acid Sequence for Selected Regions of Aspartate Aminotransferase Isoenzymes

Isoenzyme source	70	108	140	190	222	258	266	292	360	386
	→	→	→	→	→	→	→	→	→	→
Human (m)[a,35]	EYL[b]	S GTG	TWGNH	LHACAHNPTG	FFDMAYQGF	QSYAKN	ERV	I RP	MFC	GRI
Chicken (m)[31]	EYL	S GTG	SWGNH	LHACAHNPTG	YFDMAYQGF	QSYAKN	ERA	I RP	MFC	GRI
Pig (m)[33]	EYL	S GTG	SWGNH	LHACAHNPTG	FFDMAYQGF	QSYAKN	ERV	I RP	MFC	GRI
Rat (m)[34]	EYL	S GTG	SWGNH	LHACAHNPTG	FFDMAYQGF	QSYAKN	ERV	I RP	MFC	GRI
Chicken (c)[36]	EYL	GGTG	TWE NH	LHACAHNPTG	FFDS AYQGF	QSFSKN	ERV	VRT	MFS	GRI
Pig (c)[37]	EYL	GGTG	TWE NH	LHACAHNPTG	FFDS AYQGF	QSFSKN	ERV	VRV	MFS	GRI
E. coli[38,39]	NYL	GGTG	SWPNH	FHGCCHNPTG	LFDF AYQGF	S SYSKN	ERV	I RA	DFS	GRV
	*[c]	***	*	* **	*	* *	*	*	*	*

[a] m = mitochondrial; c = cytosolic.
[b] One letter codes: Gly, G; Ala, A; Val, V; Leu, L; Ile, I; Phe, F; Tyr, Y; Trp, W; Ser, S; Thr, T; Cys, C; Met, M; Asn, N; Gln, Q; Asp, D; Glu, E; Lys, K; Arg, R; His, H; Pro, P.
[c] Amino acid residues involved in binding coenzyme or substrate (see Scheme 3).

aspartic acid to form the external aldimine, which is thought to occur *via* the intermediacy of a geminal diamine,[25,26] causes the coenzyme to tilt by 30° (Scheme 3). The released ε-amino group then serves as the enzyme-bound proton carrier for the 1,3-prototropic shifts accompanying transamination. The ε-amino group of Lys-258 initially hydrogen bonds to the phenolic group of Tyr-70. The ε-amino group then abstracts the C^α hydrogen. This step is at least partially rate-limiting.[27,28] The hydrogen atom is then transferred to C-4′ without substantial exchange with solvent hydrogen to give the ketimine. After formation and hydrolysis of the initial ketimine, the coenzyme rotates back by 15°.

The coenzyme is held in place by several amino acid residues. The protonated pyridinium ring is hydrogen bonded to Asp-222, which forms a second hydrogen bond to His-143. The 2-methyl group is located in a pocket formed by eight amino acid residues and the 3′-oxygen atom is hydrogen bonded to the phenolic OH of Tyr-225. X-ray data also indicate that an O-3′ cisoid ε-lysine aldimine conformer exists in the absence of substrate and this in turn lends weight to a C-5—C-5′ rotation mechanism to facilitate the change in conformation during transaldimination. The 5′-phosphate ester group of the coenzyme appears to be bound by seven or eight hydrogen bonds as the dianion. The guanidinium group of Arg-266 forms two of these hydrogen bonds and largely offsets the double negative charge (Scheme 3). It is also worth noting that the covalent bonds between the phosphate group and the pyridine ring are strained. All of these results are in accord with stereochemical observations and it is thus evident that the dynamic model of transaminase catalysis based on solution properties complements the X-ray diffraction model.[29]

A particularly puzzling result, however, came from the laboratory of Martinez-Carrion.[30] These workers showed that aspartate aminotransferase holoenzyme carbamylated at Lys-258 in the presence of aspartate was reduced by borohydride from the C-4′ *re* face of the coenzyme rather than at the *si* face. However, single-turnover half-reaction of the modified enzyme in deuterium oxide gave (4′S)-[4′-$^2H_1$]pyridoxamine as a result of protonation on the 4′-*si* face of the coenzyme, the same stereochemical course as observed for the native enzyme. From these results, Martinez-Carrion suggested that either Lys-258 is not the proton acceptor/donor or that the change in accessibility of C-4′ to borohydride is not simply related to the conformational requirements of the catalytic process. It should be noted, however, that these findings could merely reflect the increased steric bulk of the carbamylated lysine residue on the 4′-*si* face of the coenzyme, which could allow preferential C-4′ *re* face reduction.

The primary structure of chicken heart mitochondrial enzyme was determined by Graf-Hausner *et al.*[31] The sequences of several other aspartate aminotransferase isoenzymes have been determined[32–40] and these show almost 100% homology for the regions of the protein which correspond to substrate or coenzyme binding residues of the chicken mitochondrial enzyme for which X-ray data are available. Table 1 illustrates this high degree of homology.

Owing to their strikingly similar stereochemical and mechanistic properties, it has often been suggested that PLP-dependent enzymes have evolved from a common progenitor. Indeed, comparison of some seemingly unrelated PLP-dependent enzymes, for example porcine aspartate aminotransferase and *E. coli* D-serine dehydratase, revealed significant sequence homology when aligned optimally.[41] However, two recent papers describing the primary structure of rat liver tyrosine[42] and *E. coli* branched-chain amino acid aminotransferase,[43] enzymes which also use L-glutamate as an amino group donor, do not show significant homology with other transaminases. The primary structures of rat ornithine aminotransferase[44] and aromatic amino acid transaminase, as deduced from the nucleotide sequence[45] of the *Tyr* B gene of *E. coli*, do show sequence homology when compared with aspartate aminotransferase. It is interesting to note that the deduced primary structure of tyrosine[42] (EC 2.6.1.5) and aromatic amino acid transaminase[45] (EC 2.6.1.57) do not show similar extents of homology with aspartate aminotransferase.

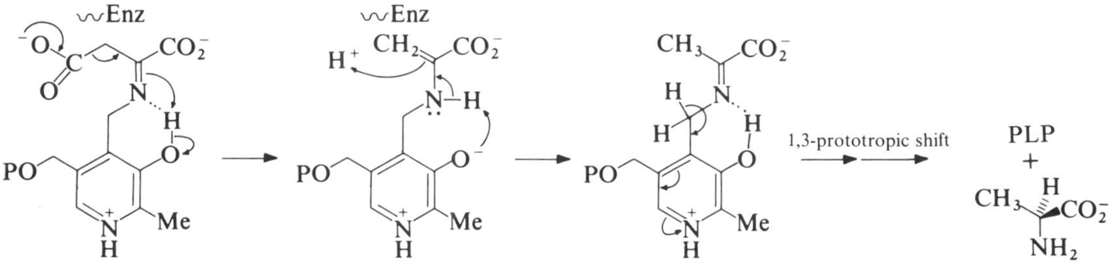

Scheme 4

Very recently, *E. coli* aspartate aminotransferase has been subjected to a site-specific mutagenesis.[46,47] When the amino acid residue Arg-292 responsible for binding to the β-carboxyl group of the substrate, aspartic acid, was altered to Asp-292 the mutant protein was able, albeit slowly, to transaminate ornithine, lysine and arginine. This is a conceptually rewarding result which clearly illustrates the power of these modern combined X-ray–computer graphic–mutagenesis–enzyme kinetic techniques in testing mechanistic and structural ideas. When the active-site lysine was altered to alanine (Lys-258 → Ala-258), the mutant PMP–holoenzyme acted as a decarboxylase for oxaloacetic acid.[47] Presumably decarboxylation occurs at the β-carbon to give L-alanine (Scheme 4).

6.5.2.2.2 ω-*Amino acid transaminases*

γ-Aminobutyric acid (GABA), and the ω-amino groups of ornithine and lysine, are converted to the corresponding aldehydes by a small group of related PLP-dependent transaminases.[48,49] The mechanisms of these reactions are similar to those of α-amino acid transaminases (Schemes 5 and 6). However, as the substrates are prochiral rather than chiral, stereospecifically labelled amines were required to investigate the stereochemical courses of the reactions. Using the known retentive mode of the decarboxylation of L-glutamate,[50] (4R)- and (4S)-[4-^{3}H$_1$]GABA were prepared.[51] When each labelled sample of GABA was converted into succinate semialdehyde with bacterial GABA transaminase, tritium was only retained by the (4R) isomer.[51] A similar result was obtained using the mammalian brain enzyme.[52] On the other hand, when ω-amino acid pyruvic transaminase from *Pseudomonas sp.* F.126 was investigated, the opposite stereochemical preference was observed.[51] This latter enzyme thus catalyzed removal of 4-*pro-R* hydrogen from the substrate. In order to explain these results, two situations can be envisaged. Either the enzymes could act upon opposite faces of the coenzyme with the distal carboxyl group of the substrate bound in similar conformations, or the same face of the coenzyme could be used if the substrate were bound in totally different modes (Scheme 7). Recent work has revealed that both lysine[53,54] and ornithine transaminase[55] catalyze the stereospecific removal of *pro-S* hydrogen from the terminal C atoms of their respective substrates.

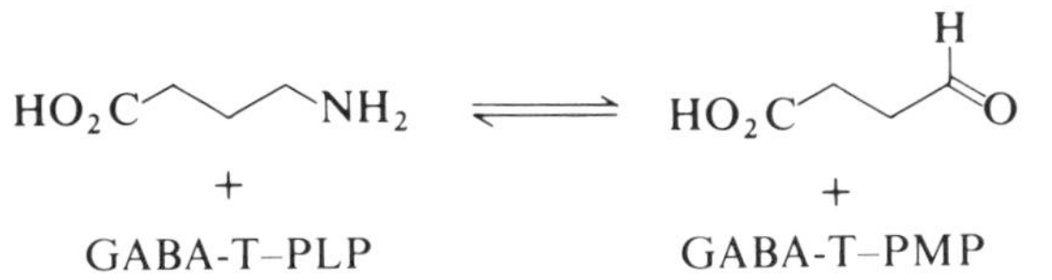

Scheme 5

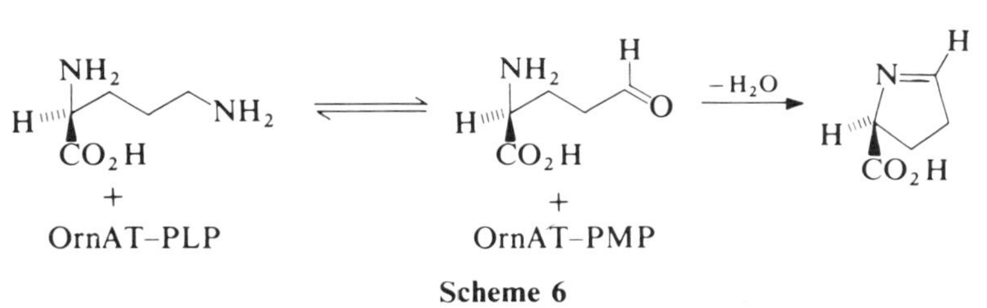

Scheme 6

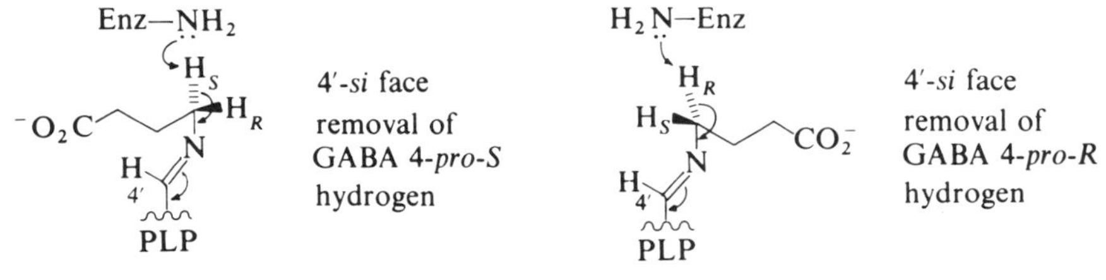

Scheme 7

Mammalian γ-aminobutyric acid transaminase (GABA-T) is worthy of special mention.[56] GABA-T is a key enzyme involved in regulating the GABA-ergic system (Scheme 8) in the mammalian central nervous system.[57] Studies of GABA-T have been stimulated by the finding that the cerebral γ-aminobutyric acid concentrations can be altered by controlling GABA-T activity. γ-Aminobutyric acid is the major inhibitory neurotransmitter and there is much evidence to suggest that high cerebral concentrations prevent convulsions.[58] Thus the inhibition of GABA-T, the first enzyme on the catabolic pathway, may be important in anticonvulsive therapy.[59] Very recently, work on the GABA receptor has given clues as to the 3-D structure of the receptor trans-membrane polypeptide.[60] There are some similarities with the β-adreno-type receptors[61,62] (see also Volume 3, Part 12).

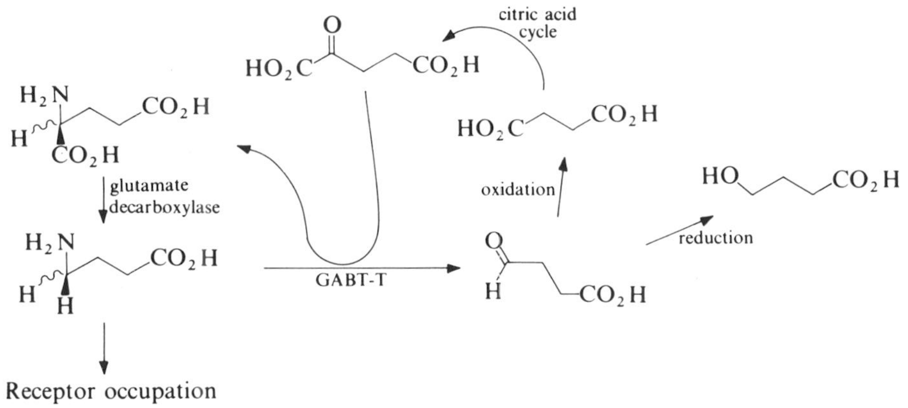

Scheme 8

GABA-T is a mitochondrial enzyme which has been purified to homogeneity from a variety of mammalian brain sources.[63,64] The enzyme is a dimer (M_r 108 000) of identical subunits. The intact protein contains no disulfide bridges and shows a pH optimum at 8.5. The K_m for γ-aminobutyric acid varies with the source, but is of the order 1 mM for the human enzyme and 4 mM for the rat enzyme.

Judged by the comparison of their amino acid compositions, these two enzymes show much less homology than mitochondrial aspartate aminotransferases from the same species. GABA-T from some species appears to bind only one mole of PLP per dimer.[65] Kim and Churchich have provided evidence, based upon experiments conducted with phthalaldehyde modified enzyme, that two active-site lysine residues bind the carboxylate groups of dicarboxylic acids.[66] β-Alanine is a good substrate for GABA-T but side reactions occur to give inactivated enzyme.[67] In view of the importance of controlling GABA-T activity in anticonvulsive therapy, much effort has been expended on the synthesis and development of GABA-T inhibitors. Their mode of action and therapeutic potential is considered below (Section 6.5.3.3.2).

ω-Ornithine aminotransferase is also worthy of special mention. This enzyme not only interconnects the citric acid cycle and the urea cycle but also couples ornithine to proline metabolism (Scheme 9).

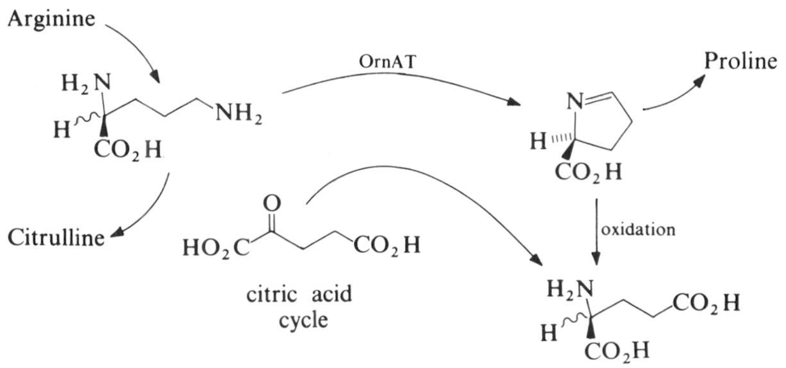

Scheme 9

The enzyme has been purified from a number of sources including rat liver, kidney and intestine. The enzymes from all three tissue-types are chemically, immunochemically and kinetically identical although there may be some differences in the primary structures.[48] The liver enzyme is an oligomer, probably a tetramer (M_r 170 000) of identical subunits (M_r 43 000).[68] Evidence has been presented to suggest that each subunit binds to 1 mol of PLP.[69] Very recently the enzyme has been reported to exist as a hexamer.[70] The primary structures of the human[71] and rat[72] enzymes have been investigated.

6.5.2.2.3 D-*Amino acid transaminases*

D-Amino acid transaminases are present in many microorganisms and plants.[73] Although alanine racemase is ubiquitous in all bacteria (see Section 6.5.2.4), glutamate racemase is quite rare.[74] The biosynthesis of D-glutamate thus often involves the transfer of ammonia from D-alanine to α-ketoglutarate, the reaction catalyzed by D-amino acid transaminase. The enzyme from *B. sphaericus* has been purified to homogeneity and crystallized.[75] The enzyme has a molecular weight of ∼ 60 000 amu and is composed of two identical subunits. Two moles of PLP bind to the intact dimer.[76] One mole of PLP binds to an active-site lysine as an aldimine (λ_{max} 415 nm) and participates in catalysis. Treatment of the holoenzyme with phenylhydrazine gives a catalytically inactive semiapoenzyme, of which the remaining PLP adduct, possibly a *gem*-diamine, absorbs at 330 nm and is not susceptible to $NaBH_4$ reduction.

Clearly the substrates for D-amino acid transaminases are enantiomers of the more common L-type transaminases. Comparison of the actual substrate holoenzyme binding modes would be interesting and would reveal which face of the coenzyme with respect to C-4′ was involved in proton transfers during transamination.

D-Amino acid transaminase has a wide substrate specificity and is able to transaminate almost 20 different D-amino acid substrates. The enzyme contains eight cysteinyl residues per molecule and the reaction of these with thiol-attacking reagents has been used to assess their environment in the absence and presence of substrates.[77]

6.5.2.3 Decarboxylases

6.5.2.3.1 *General features*

The conversion of amino acids to amines is one of the most important physiological roles of PLP. The formation of dopamine, histamine, γ-aminobutyric acid (GABA) and putrescine all depend upon PLP-dependent decarboxylation of their respective parent amino acids. The biosynthesis of the higher polyamines, spermine and spermidine, from putrescine involves the pyruvate-dependent decarboxylation of S-adenosylmethionine (SAM). Recently, studies of both of the enzymes involved in this biosynthetic path (*viz.* ornithine and SAM decarboxylase) have been intensified in order to procure inhibitors as there is now good evidence to show that polyamines are required for cellular replication and growth (see Section 6.5.3.4).

The stereochemical and mechanistic features of enzymic decarboxylation are outlined in Scheme 10. The N—C$^\alpha$ bond is disposed so that the carboxyl group of an L-amino acid is perpendicular to the conjugated π-electron system, presumably on the 4′-*si* face of the coenzyme. The retentive mode of decarboxylation has been shown to occur for glutamate,[50,78,79] histidine,[80,81] tyrosine,[82] lysine,[83] ornithine,[84,85] arginine[84,85] and methionine decarboxylase.[86] Only two decarboxylases show alternative stereochemical courses, the aminomalonate decarboxylase activity of serine hydroxy-methyltransferase (SHMT),[87] which produces racemic labelled glycine (see Section 6.5.2.2.1), and α,ω-*meso*-diaminopimelic acid decarboxylase from both prokaryotic[88] and eukaryotic[89] sources. α,ω-*meso*-Diaminopimelic acid decarboxylase catalyzes the conversion of α,ω-*meso*-diaminopimelic acid to L-lysine with inversion of configuration. These latter systems are thus the only *bona fide* examples of an enzymic decarboxylation which proceeds with inversion of configuration and also the only known PLP-dependent enzymes which decarboxylate a D-amino acid centre. Only α,ω-*meso*-(2R,6S)-diaminopimelic acid is a substrate for the enzyme from *Bacillus sphaericus* and both the (2R,6R) and (2S,6S) isomers fail to react.[90] These observations tend to indicate that the distal binding group for the ε-carbon centre requires the (S) configuration.

The ability of the enzyme to decarboxylate a D-amino acid centre with inversion of stereochemistry, although unique, is consistent with current ideas regarding PLP-dependent catalysis. In all of

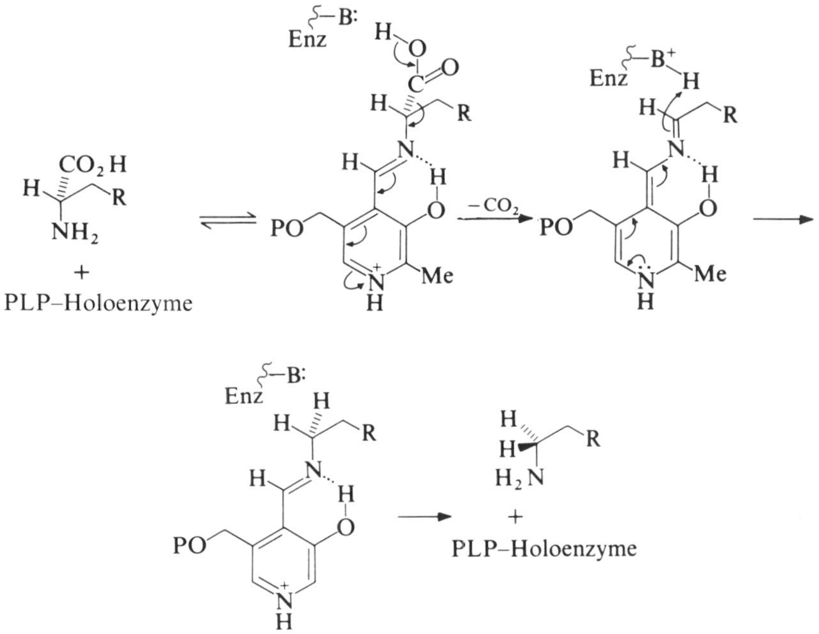

Scheme 10

the 10 or so cases studied, PLP binds to the protein in the presence of substrate with the C-4′-*si* face of the coenzyme exposed to the solvent. Binding of the D-amino acid α-centre of *meso*-diaminopimelate at the active site of the enzyme, with an α-hydrogen atom and alkyl side chain occupying similar positions to those normally occupied in L-amino acid decarboxylases, would dispose the carboxyl group on the C-4′-*re* face of the coenzyme. In this conformation the carboxyl group is also perpendicular to the π-electron system and thus decarboxylation of the aldimine would be stereoelectronically assisted, according to the Dunathan postulate. The resulting ketimine is identical to the product of L-amino acid decarboxylation and would be protonated in the usual manner from the C-4′-*si* face of the coenzyme (Scheme 10). This sequence leads to decarboxylation of a D-amino acid centre with inversion of configuration.

Aspartate β-decarboxylase catalyzes α-decarboxylation of aminomalonic acid[91] and β-decarboxylation of aspartic acid.[92] The enzyme also catalyzes desulfination of L-cysteine sulfinate, β-elimination of L-3-chloroalanine, and transamination between several L-amino acids and α-ketoglutarate, pyruvate and oxaloacetate.[93] α-Decarboxylation of aminomalonic acid occurs with retention of configuration at C^α but the enzyme removes the *pro-R* carboxyl group, thus acting as a D-amino acid decarboxylase.[94] Clearly the aminomalonate either binds at the active site with the *pro-R* carboxyl group disposed on the C-4′-*re* face of the coenzyme, or in the opposite conformation with the *pro-R* carboxyl group disposed on the C-4′-*si* face (Scheme 11).

The former arrangement, originally put forward by O'Leary and Floss and coworkers, requires that both decarboxylation and reprotonation occur from the C-4′-*re* face of the coenzyme. This suggestion arose after an elegant study of the stereochemical course of the β-decarboxylation of L-aspartic acid had indicated that a second enzyme-bound base may cover the C-4′-*re* face of the

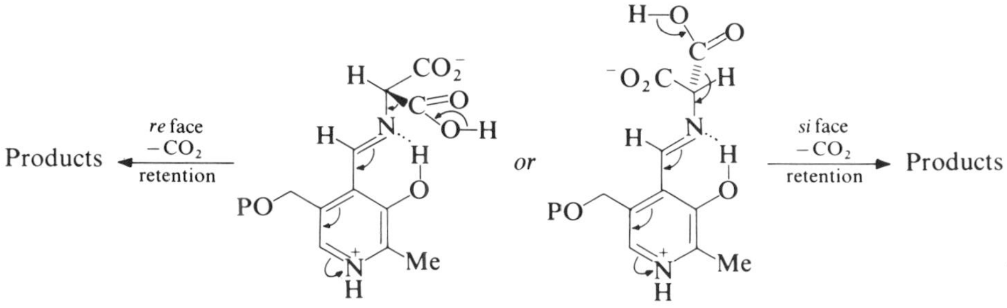

Scheme 11

coenzyme.[95] Such a two-base system would facilitate α-decarboxylation on the C-4'-*re* face. A similar two-base system has also been proposed to account for the action of PLP-dependent racemases (see Section 6.5.2.4.1). However, in order to assess the mode of binding of aminomalonate to aspartate β-decarboxylase it is first necessary to follow the arguments of Floss relating to a two-base active site.[95] Incubation of C-3 chirally labelled L-aspartic acid with the enzyme in water containing a further label gave samples of L-alanine containing chiral methyl groups. Stereochemical assay revealed that β-decarboxylation had occurred predominantly with inversion of configuration at C-3 of aspartate. The racemization that accompanied the decarboxylation could be reduced by reducing the incubation times and it was thus suggested that enzyme-catalyzed exchange of the methyl protons of alanine *after* decarboxylation was largely, if not solely, responsible. The abortive transamination–decarboxylation that accompanies the normal decarboxylation was also studied. Incubation of the enzyme with C$^{\alpha}$-tritiated L-aspartic acid led to 17% incorporation of label at C-4' *pro-S* position of pyridoxamine *via* internal suprafacial transfer on the C-4'-*si* face of the aldimine (8), but only to the extent of 1% at the combined C-2 and C-3 centres of alanine after normal β-decarboxylation. Although it was noted that the stereochemical features of the reaction could be accounted for by a single active-site base (*cf.* tryptophan synthase, see Section 6.5.2.4) covering the C-4'-*si* face, a higher proportion of tritium transfer from C-2 to C-3 would be expected. However, before accepting a two-base system, the possibility that transfer of hydrogen from C-2 of aspartate to C-4' of the coenzyme occurs in both the normal β-decarboxylation and abortive transamination, before further reaction, should be considered. β-Decarboxylation would then yield the enamine (9) which would be protonated from the C-4'-*si* face of the coenzyme by the enzyme-bound base to give the ketimine (10) with inversion of configuration (Scheme 12). As most of the label present at C-2 of an aspartate isotoper would have been already transferred to C-4' of the coenzyme, the remaining label associated with the base (ε-NH$_2$ of lysine) would not be expected to result in a large enrichment

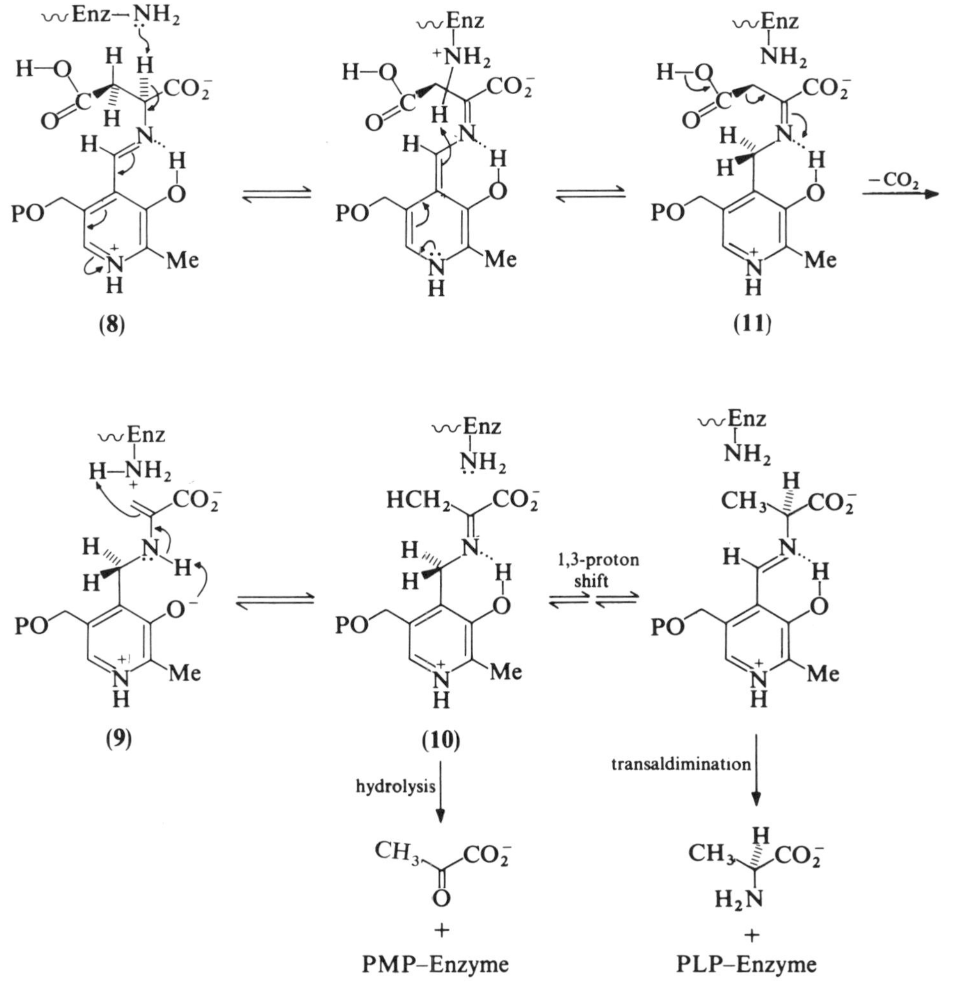

Scheme 12

at C-3 of the ketimine. Also, solvent exchange would become increasingly important in the latter stages of the reaction and would further reduce the amount of tritium available for incorporation at C-3. Solvent exchange must account for the small enrichment observed on the back-transfer of 4'-H atoms to C-2 as otherwise both one- and two-base mechanisms would predict significant incorporation at C-2. A possible mechanism for the dual reaction pathways of L-aspartic acid with aspartate β-decarboxylase is depicted in Scheme 12 for a single-base system. The two-reaction pathways differ only after β-decarboxylation and protonation have occurred to give the ketimine (**10**). The identical mechanism has been recently proposed by Rosenberg and O'Leary[96] to account for results which showed that α- and β-hydrogen/solvent hydrogen exchange are faster than the decarboxylation of L-aspartate and that decarboxylation is not entirely rate-limiting.

In the light of the above analysis it is thus probable that aminomalonate undergoes α-decarboxylation with retention of configuration on the C-4'-*si* face of the coenzyme since an alternative arrangement would require a two-base system (Scheme 11). It is also likely that D-aspartate binds at the active site of the *N*-methyl-PLP–apoenzyme complex[97] with the α- and β-carboxyl groups in positions similar to the natural holoenzyme with L-aspartate. Stereoelectronically assisted labilization of the α-hydrogen atom (disposed on the C-4'-*re* face of the coenzyme) would give a ketimine similar to that derived from L-aspartate, *cf.* structure (**11**). β-Decarboxylation and subsequent protonation from the C-4'-*si* face would give, after hydrolysis, the observed product L-alanine.

Kynureninase catalyzes the conversion of kynurenine to L-alanine and anthranilic acid (Scheme 13). The reaction was originally thought to follow an α,β-elimination–β-replacement mechanism but now there is evidence to suggest that the mechanism is similar to that for the β-decarboxylation of L-aspartate. Cleavage of the β,γ-bond would be facilitated through nucleophilic attack at the β-carbonyl group (Scheme 13). Interestingly, the reaction occurs with retention of configuration at C-3.[98]

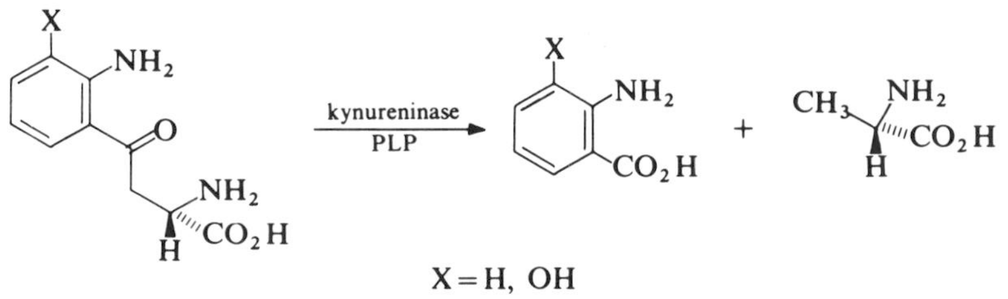

X = H, OH

Scheme 13

6.5.2.3.2 Glutamate decarboxylase

Glutamate decarboxylase (GAD) is an ubiquitous enzyme and has been isolated and purified from bacteria, plants and mammals.[99] The *E. coli* and mammalian enzymes have been studied most. The enzymes differ in many respects. GADs from different tissues within the same species are different,[100] and there are several in the mammalian brain.[101] The *E. coli* plant and mammalian enzymes decarboxylate L-glutamic acid with retention of configuration of C-2 (see Section 6.5.2.3.1). However, the *E. coli* and rat brain enzyme show opposite sensitivities to the two enantiomers of the inhibitor 4-aminohex-5-ynoic acid (see Section 6.5.3.4.1). The mammalian brain enzyme is directly responsible for the biosynthesis of GABA in the GABA-ergic system (Scheme 8). The enzyme from mouse brain synaptic terminals[102] is a homodimer (M_r 85 000) whereas the bacterial enzyme is a hexamer (M_r 300 000).[103] Bovine brain L-cysteic acid/cysteine sulfinic acid decarboxylase, a key enzyme in taurine biosynthesis which copurifies with GAD, is distinct from GAD.[104] Recently a preparation from whole rat brain of M_r 120 000 has been reported.[105] The protein is a dimer of unequal subunits (M_r 40 000 and 80 000).

Purification, isolation and assays procedures for brain[106] and bacterial[107] GAD have been reviewed. The rat brain enzyme has been cloned[108] and Kaufman *et al.* have shown that a brain enzyme fusion protein is able to convert substrate to γ-aminobutyric acid.[109]

Recent studies by Choi and Churchich have probed the side reactions catalyzed by homogeneous porcine brain glutamate decarboxylase.[110] The enzyme catalyzes a slow abortive decarboxylation–transamination reaction with L-glutamate as substrate to give succinic semialdehyde in a similar manner to the *E. coli* enzyme. Incubation of the resolved brain enzyme with phosphopyridoxylethanolamine phosphate gave active holoenzyme. A possible mechanism for this reaction is given in Scheme 14.

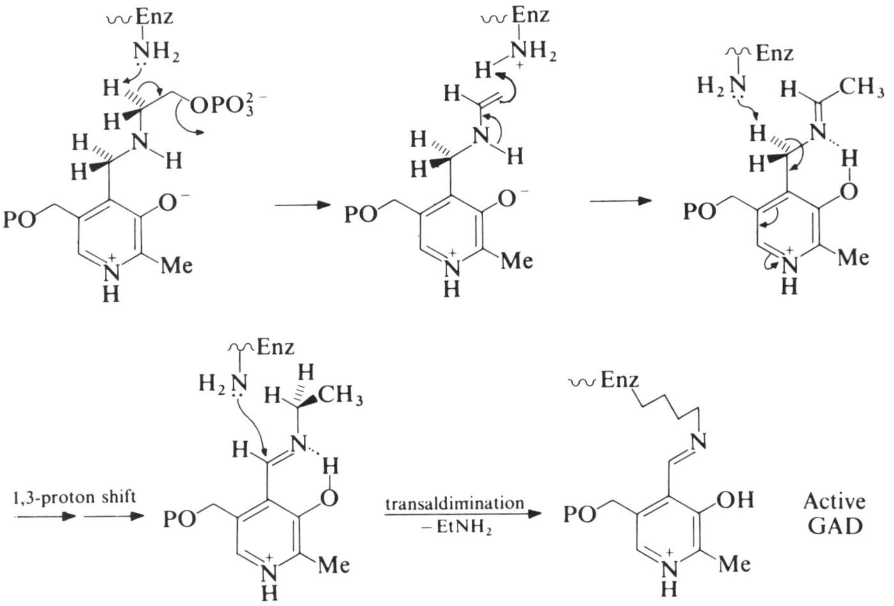

Scheme 14

The mechanisms of several nonphysiological reactions of the decarboxylase have also been investigated (see Section 6.5.3.4.1).

6.5.2.3.3 *Ornithine decarboxylase*

Ornithine decarboxylase (ODC) is the first and rate-limiting enzyme in the biosynthetic pathway leading to putrescine and higher polyamines (Scheme 15).[111] The activity of the enzyme *in vivo* increases dramatically in response to cellular stimulation which promotes regeneration and replication. Indeed, cellular levels of the enzyme are high during growth and very low in growth-arrested cells. The enzyme has the shortest half-life of any known mammalian enzyme, 8–30 min, and an extremely short induction time. The association of high polyamine levels with rapid cellular proliferation, and protein biosynthesis, led to the idea that polyamines may be required for RNA/DNA biosynthesis. Hence the enzyme has been identified as a target for cancer chemotherapy.[111]

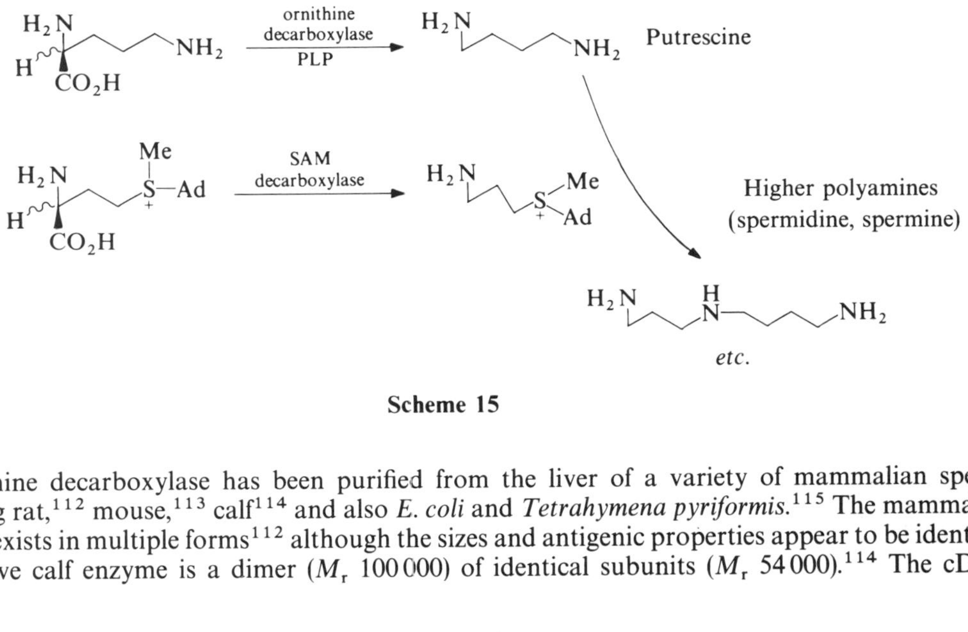

Scheme 15

Ornithine decarboxylase has been purified from the liver of a variety of mammalian species including rat,[112] mouse,[113] calf[114] and also *E. coli* and *Tetrahymena pyriformis*.[115] The mammalian enzyme exists in multiple forms[112] although the sizes and antigenic properties appear to be identical. The active calf enzyme is a dimer (M_r 100 000) of identical subunits (M_r 54 000).[114] The cDNA deduced sequence of the murine enzyme has been reported and on this basis the murine subunits are 51 100 amu.[116,117] The active site peptide of the bacterial enzyme[118] does not appear to show homology with the 'likely' active-site region of the murine enzyme. For a review of active-site peptide sequences, see Morino and Nagashima.[119]

In view of the importance of ornithine decarboxylase in cellular replication, many inhibitors for the enzyme have been designed (see Section 6.5.3.4.2).

6.5.2.3.4 Histidine decarboxylase

Histidine decarboxylase (HDC) catalyzes the decarboxylation of histidine to give histamine (Scheme 16). In mammals, histamine is important as the major receptor agonist.[120] Overproduction of histamine is associated with many biological responses including gastric secretion, secretion of nasal mucosa, tears and skin rashes. The development of antihistamines, antagonists for H_1 and H_2 receptors, has been and still is an important area in medicinal chemistry (see Volume 3, Chapters 12.1 and 12.2). An alternative and more recent approach to the problem has been to evaluate methods for controlling the biosynthesis of histamine *in vivo* through the use of specific histidine decarboxylase inhibitors (Section 6.5.3.4.3); hence there has been a great deal of interest in the enzyme.

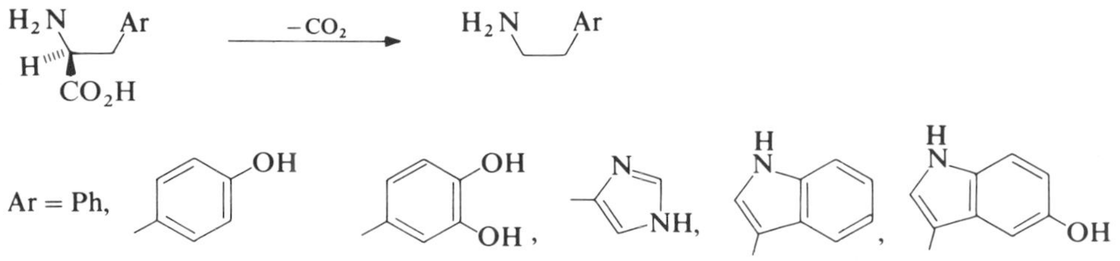

Scheme 16

Mammalian histidine decarboxylase has been purified from fetal rats.[121,122] The activity of the protein is extremely low, much lower than that of the bacterial pyruvate-dependent enzymes.[123–126] The PLP-dependent mammalian enzyme has been extremely difficult to study due to its low abundance and lability, but recently Snell and coworkers have reported on the purification and characterization of a bacterial PLP-dependent histidine decarboxylase from *Morganella morganii*.[126] The enzyme is a tetramer (M_r 170 000) composed of identical subunits (M_r 43 000) and is approximately 70 times more active than the dimeric mammalian enzymes (M_r 110 000).

The enzyme is inactivated by (S)-2-fluoromethylhistidine (see also Section 6.5.3.4.3). Using the tritiated compound it was established that a nonactive-site peptide serine residue becomes attached to the inhibitor during inactivation.[127] Subsequent work, including sequencing the gene for the enzyme, revealed that 377 amino acids were encoded. Comparison of the sequence with tryptic peptides containing the active-site lysine and inactivator sensitive serine residues indicated that these occurred at positions 232 and 322 respectively.[128] Clearly, although separated by 90 residues, Lys-232 and Ser-322 are close in 3-D space.

6.5.2.3.5 Aromatic amino acid decarboxylases

The aromatic amino acid decarboxylases catalyze the formation of many pharmacologically important amines, including catecholamines and indolamines (Scheme 16). The enzymes are widely distributed in bacteria, plants and animals. The mammalian broad specificity aromatic amino acid decarboxylase, DOPA decarboxylase (DDC), is able to convert many aromatic substrates to the corresponding amines including phenylalanine to phenylethylamine, tyrosine to tyramine, tryptophan to tryptamine, 5-hydroxytryptophan to serotonin, and histidine to histamine.[129] This last activity is distinct from the specific mammalian histidine decarboxylase (Section 6.5.2.3.4), although antirat HDC antibodies recognise guinea pig DDC but not rat DDC.[130] DOPA decarboxylase from porcine kidney is a dimeric enzyme (M_r 100 000) composed of nonidentical subunits.[131] The enzyme contains 1 mol of PLP per mol of protein. Recent studies by Borri Voltattorni and coworkers using diethyl pyrocarbonate, phenylglyoxal and butanedione have indicated that the enzyme contains an essential histidine residue[132] and an essential arginine residue.[133] The arginine residue, unlike in many other PLP-dependent enzymes, does not appear to be involved in binding to the anionic phosphate ester group of the coenzyme or the carboxylate group of the substrate.[133]

Sodium borotritide reduction of the porcine kidney holoenzyme gives a labelled ε-*N*-pyridoxyl-lysine adduct.[134] Reduction occurs from the 4'-*re* face of the internal aldimine.

The enzyme from rat liver is also a dimer of M_r 100 000.[135] However, unlike the porcine kidney enzyme, the subunits appear to be identical. Monoclonal antibody for the rat enzyme immunoprecipitates rat and guinea pig liver DOPA decarboxylase to the same extent but does not affect the rabbit enzyme. The primary sequence is known for Drosophila DDC.[136] The active-site regions of the porcine and Drosophila enzyme are perfectly conserved.

6.5.2.4 Amino Acid Racemases

6.5.2.4.1 *General features*

Amino acid racemases catalyze the racemization of L- and D-amino acids. They are extremely common in prokaryotes and many are involved in the biosynthesis of peptidoglycan in bacterial cell walls.[137] PLP-dependent enzymes in general, with only the exceptions discussed so far, catalyze group transfers on the C-4'-*si* face of the coenzyme. However, the amino acid racemases must allow deprotonation of the substrate aldimine and subsequent protonation of the ketimine to occur from opposite faces of the coenzyme. To achieve racemization, a racemase may operate using two enzyme-bound bases, one on each face of the coenzyme (Scheme 17), or one base as depicted in Scheme 18. The one-base system requires the protein to undergo large changes in conformation so that each face of the coenzyme is accessible. This type of operation has been referred to as the 'swinging door' mechanism.[138] The non-B₆-dependent enzyme proline racemase has been shown to operate using a two-base system and a rather detailed kinetic and mechanistic analysis has been conducted by Knowles and coworkers.[139-144] Gani has reviewed this work.[145] Note that α,ω-diaminopimelate racemase is also non-PLP dependent and operates *via* a two-base mechanism.[146] Adams suggested that alanine racemase from *Pseudomonas* may utilize a two-base system but there is now growing evidence to show that only one base is involved.[147]

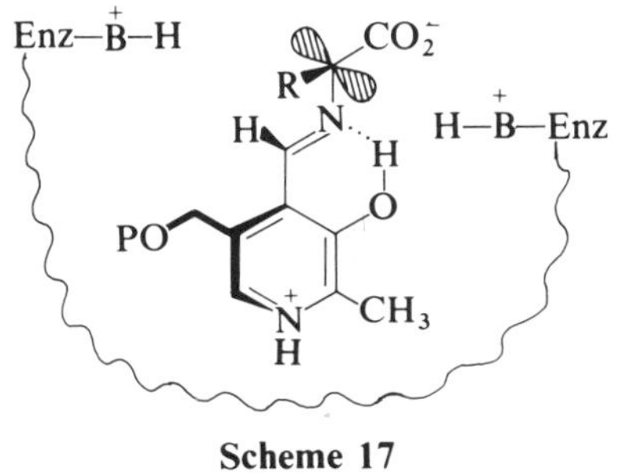

Scheme 17

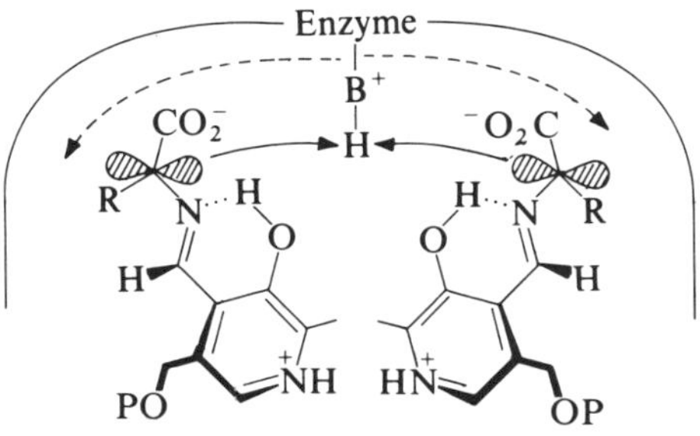

Scheme 18

The 'swinging door' mechanism was originally put forward by Henderson and Johnston[138] following studies of the differential inactivation of alanine racemase from *B. subtilis* with D- and L-β-chloroalanine (see Section 6.5.3.5). These workers showed that the D-isomer was a more powerful inhibitor than the L-isomer and suggested that the asymmetry at the active site would best be rationalized in terms of a single-base swinging-door mechanism where the D-alanine conformation would be most stable. Hence L-alanine, on binding at the active site, would exchange H^{α} and undergo racemization at similar rates, whereas D-alanine would exchange H^{α} at a rate much faster than racemization. Using a variety of other β-substituted alanine inhibitors, results have been obtained to support further the single-base mechanism.[148,149] In general, D-substrate analogues appear to be better irreversible inactivators than the L-antipodes[149] (see Section 6.5.3.5).

A decisive method for distinguishing between a single-base and a twin-base system is to assess the extent of C^{α} hydrogen atom transfer between enantiomers during the racemization under single turnover conditions. If no internal hydrogen transfer occurred, a two-base system would be implied (or very rapid solvent hydrogen exchange), whereas internal hydrogen transfer would be consistent with only a single-base mechanism (Scheme 18). Floss and coworkers[150] showed that for both tyrosine phenollyase (from *Escherichia intermedia*) and amino acid racemase (from *Pseudomonas striata*) the conversion of L-alanine to D-alanine under essentially single turnover conditions occurred with significant internal hydrogen transfer, indicating a single-base mechanism. On the other hand, alanine racemase (from *E. coli*) showed no internal hydrogen transfer. However, in this latter experiment the operation of a single-base system cannot be ruled out.

6.5.2.4.2 Alanine racemase

Alanine racemase is an important enzyme in the biosynthesis of peptidoglycan and the enzyme is a target for the action of many antibacterial agents[151] (see Volume 2, Part 9).

Walsh and coworkers have studied the enzyme from a number of sources in some detail. Alanine racemase from *E. coli* B is a dimer, M_r 100 000, which contains 1 mol of PLP per dimer.[149] The catabolic *Salmonella typhimurium* enzyme, however, is a 39 000 amu monomer which contains 1 molar equivalent of coenzyme.[152]

The biosynthetic enzyme from the same species is also monomeric (M_r 40 000) and binds one mol of coenzyme per mol of protein.[153] The primary structure of both the catabolic enzyme, the *dad* B gene product,[152,155] and the biosynthetic enzyme, the *alv* t gene product,[154] have been determined. The proteins display 43% overall sequence homology and the active-site decapeptides containing the PLP-binding lysine residues are identical.

The broad specificity amino acid racemase from *Pseudomonas striata* contains one mol of PLP per subunit (M_r 42 000) and is probably a homodimer (M_r 110 000).[156] The sequence of the active-site peptide is similar to the *Salmonella typhimurium* enzymes. All of these enzymes are derived from Gram-negative bacteria. Alanine racemase from a Gram-positive bacterium, *Streptococcus faecalis*, is probably a monomer (M_r 42 000) which binds 1 mol of coenzyme per mol of protein.[157]

The enzyme from the thermophilic Gram-positive bacterium *Bacillus stearothermophilus* is, however, a dimer.[158] The enzyme has now been cloned and over-produced. Walsh and coworkers have undertaken a detailed investigation of the inhibition of both Gram-negative[149-156] and Gram-positive bacterial alanine racemase.[157-159] The results are extremely interesting and reveal major differences between the two 'types' of enzyme (see Section 6.5.3.5).

6.5.2.4.3 Other racemases

In addition to alanine and broad substrate specificity amino acid racemase, many other PLP-dependent racemases are known.[160] Glutamate racemase is not present in many strains of bacteria and thus although D-glutamate is an essential precursor in the biosynthesis of peptidoglycan, its formation results from the action of D-amino acid transaminase on α-ketoglutamate (see Section 6.5.2.2.3). Arginine[161] and α-amino-ε-caprolactam[162] racemase are both PLP-dependent, whereas α,ω-diaminopimelic acid and proline racemase do *not* require PLP. Indeed, it now appears that *Pediococcus pentosaceus* L-glutamate racemase is not PLP-dependent.[163] The enzyme has been cloned and is a monomeric protein (M_r 40 000). Broad specificity amino acid racemase from *Aeromonas caviae*[164] is a homodimer (M_r 76 000) which binds to 2 mol of coenzyme per mol of enzyme and is similar to the *Pseudomonas striata* enzyme.[156] However, the two enzymes share no common antigenic determinants.[164] Two reviews on racemases are recommended for fuller details of the earlier work.[147,160]

6.5.2.5 Other PLP-dependent Systems

6.5.2.5.1 *Serine hydroxymethyltransferase*

Serine hydroxymethyltransferase (SHMT) catalyzes the formation of 5,10-methylenetetrahydrofolic acid (**12**) and glycine from tetrahydrofolic acid (THF, **13**) and L-serine and is thus a key enzyme in single carbon (C_1) metabolism at all oxidation levels (see Scheme 19; also see Parts 7 and 10). In the absence of THF, a slower reaction to give glycine and formaldehyde occurs. Indeed, SHMT is remarkable in that the enzyme catalyzes the retro-aldol cleavage of L-allothreonine, L-threonine, L-*erythro*-β-phenylserine, L-*threo*-β-phenylserine and a variety of other *para* and *meta* substituted β-phenylserines to give glycine and the appropriate aldehyde.[165-169] In addition to catalyzing the retro-aldol reaction, serine hydroxymethyltransferase also shows a number of other activities including an ability to decarboxylate aminomalonic acid,[170] transaminate D-alanine[171-173] and catalyze α-hydrogen exchange and aldol condensation with L-amino acids.[174] These additional activities vary with the source of the enzyme and are unlikely to be of any physiological importance. The exceptional range of catalytic activity shown by serine hydroxymethyltransferase has stimulated mechanistic investigations into each of these reactions.

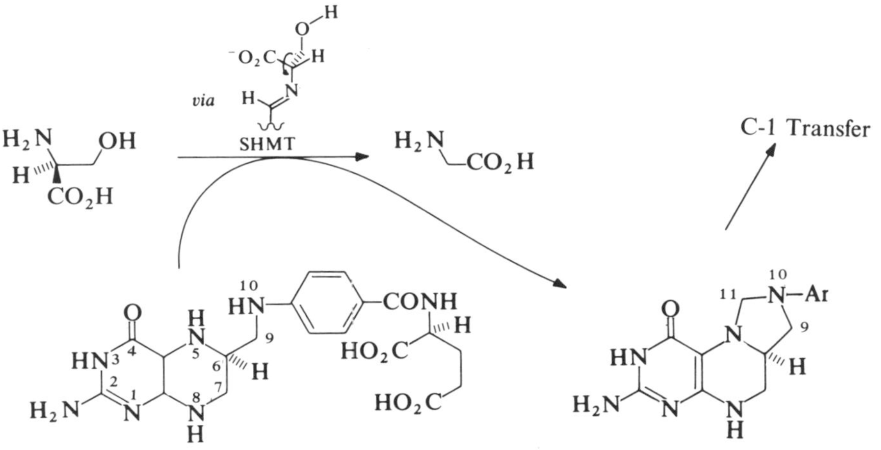

(13) THF   **(12) 5,10-CH$_2$-THF**

Scheme 19

In vivo serine hydroxymethyltransferase catalyzes β-carbon cleavage of L-serine to provide formaldehyde for further reaction with tetrahydrofolic acid to give 5,10-methylenetetrahydrofolic acid (5,10-CH$_2$-THF) (Scheme 19). Benkovic and Floss and coworkers[175] investigated the stereochemical course of formaldehyde transfer using samples of serine stereospecifically labelled with tritium at C-3 and found that the reaction was only partially stereospecific. Then stereochemical assay of the resulting 5,10-CH$_2$-THF was achieved using three enzymes. Oxidation with 5,10-methylenetetrahydrofolate dehydrogenase, now known to be specific for removal of the 11-*pro-R* hydrogen,[176] which transfers hydrogen to the 4-*re* face of NADP; removal of the 4-*pro-S* hydrogen of NADPH using glutamate dehydrogenase; and, finally, degradation of NADP to nicotinamide with NADase. The label distribution between 5,10-methylene-THF and nicotinamide thus reflected the stereospecificity of the hydroxymethyl group transfer. These results complemented the findings of Biellmann and Schuber,[177] who showed that SHMT-catalyzed condensation of samples of chirally tritiated 5,10-methylene-THF with glycine gave samples of L-serine which were partially racemic at C-3. SHMT thus showed incomplete stereochemical control with regard to the methylene hydrogen atoms for both the forward and reverse reactions to the extent that approximately 50% of the product was epimerized in each case.

To account for these observations, two mechanistic schemes were put forward. The first was that free formaldehyde is transiently formed at the active site and that rotation of this intermediate, prior to reaction with THF, would allow some racemization (Scheme 20).[166] The alternative explanation was that when serine is bound at the active site, the hydroxymethyl group is able to react in two distinct conformations to give formaldehyde.[175] Rotation of the formaldehyde and reaction with THF would then occur in a manner dependent upon the original distribution of serine conformers to give partially epimerized products (Scheme 21). The major argument in favour of the second proposal is that serine hydroxymethyltransferase catalyzes cleavage of the (2S) diastereomers of *allo*-threonine and threonine. However, although this observation clearly indicates that the active site is flexible with regard to binding substrates containing β-substituents (indeed, β-phenylserines also react), a crucial binding group, the hydroxyl group of both of the serine conformers, would need to interact with the enzyme to almost equal extents.

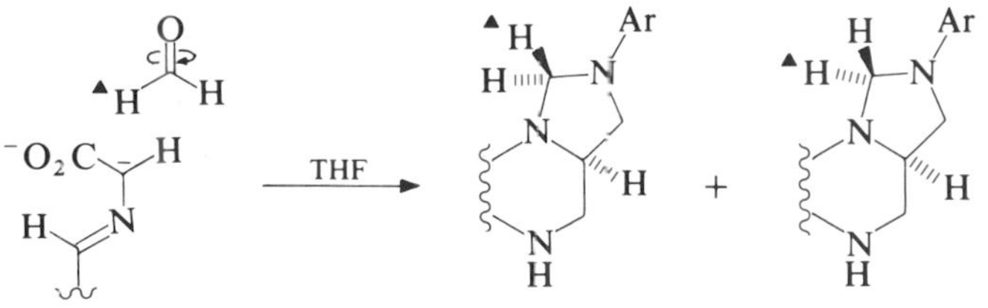

Scheme 20

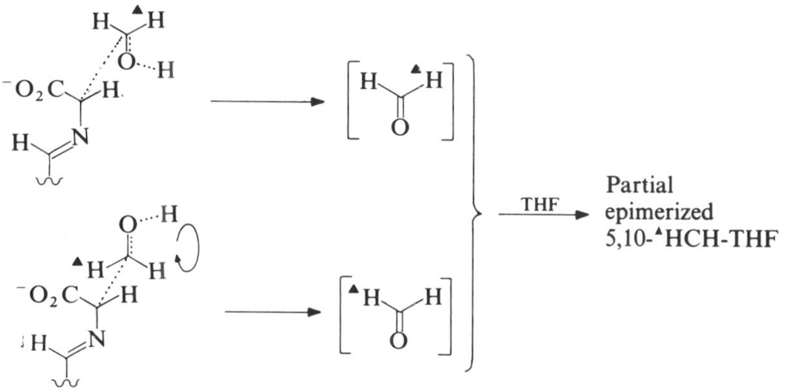

Scheme 21

A disadvantage in both proposals is that THF serves no direct catalytic function. In view of the large rate enhancement observed for the reaction of glycine with formaldehyde in the presence of THF, it seems reasonable that the correct mechanism should involve THF directly. Such a mechanism can be written (Scheme 22) where the aminol (14) is formed directly *via* nucleophilic displacement of the stabilized glycyl anion (15) by N-5 of THF. As all PLP-dependent enzymes studied to date, including SHMT,[171] present the C-4'-*si* face of substrate–coenzyme complex to the solvent (see Section 6.5.2.3.1), it is likely that THF also approaches from this face. Dehydration of the aminol need not be under enzymic control and thus could occur for both rotomeric forms of the aminol (16) to give the imines. Ring closure would then yield the epimeric 5,10-methylene THF samples.

Benkovic has recently assigned the ^{1}H NMR spectrum of 5,10-methylene-THF using nuclear Overhauser effect techniques and has determined the absolute stereochemistry of samples chirally labelled with deuterium at C-11.[176] From these studies it is apparent that the predominant products from (2S,3S)-[3-^{2}H$_1$]- and (2S,3R)-[3-^{2}H$_1$]-serine are (6R,11S)-[11-^{2}H$_1$]- and (6R,11R)-[11-^{2}H$_1$]-5,10-methylene-THF, respectively. These results are important in rationalizing the observed stereochemical course of many reactions involving 5,10-methylene-THF in the thymidylate synthase reaction.[178]

SHMT catalyzes a number of other reactions (as well as other glycine aldol/retroaldol reactions) including transamination of D-alanine,[171–173] decarboxylation of aminomalonic acid,[170] and α-hydrogen exchange and aldol reaction of L-amino acids.[174] α-Hydrogen exchange and transamination of D-alanine could be predicted on stereoelectronic grounds as the hydrogen atom occupies the same configurational position as is occupied by the hydroxymethyl group of L-serine, or the 2-*pro-S* hydrogen atom of glycine which also exchanges.[166] During the transamination the aldimine complex is deprotonated–protonated at both C$^\alpha$ and C-4' from the C-4'-*si* face of the coenzyme.[171] L-Amino acid transaminases (*e.g.* glutamic pyruvic transaminase) also protonate the aldimine from the C-4'-*si* face but the 2-*pro-R* hydrogen of glycine exchanges with solvent

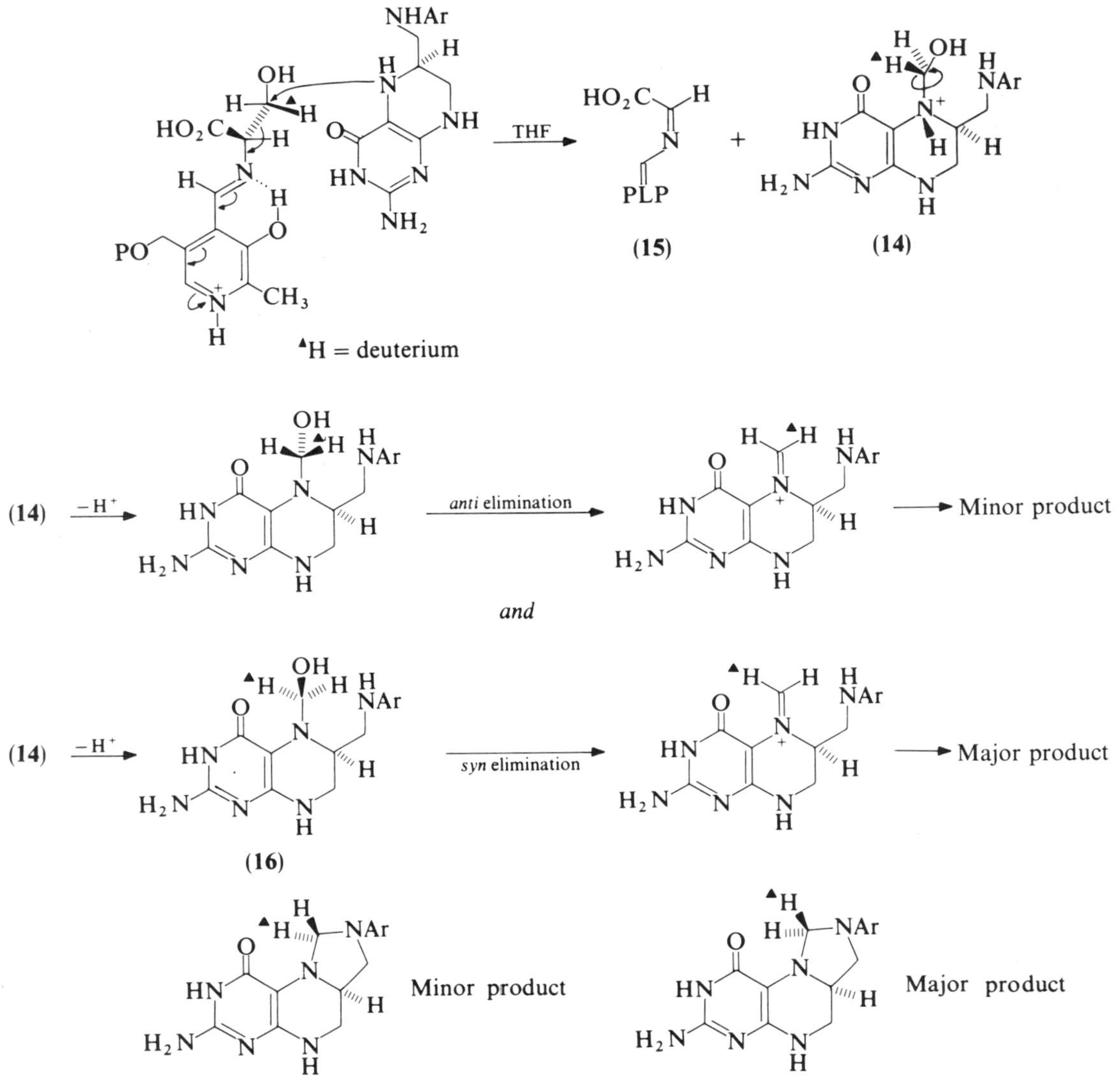

Scheme 22

This comparison directly reveals the differences in substrate binding groups at the active sites of the two enzymes.

The nonstereospecific SHMT catalyzed decarboxylation of labelled aminomalonate to give glycine was initially puzzling. However, since that time it has become apparent that the holoenzyme can bind L-amino acids and catalyze exchange of the α-H atoms.[179] The authors of this work suggest that the ability of L-amino acids to form quinonoid intermediates and exchange α-hydrogen atoms with the solvent provide examples of a break in the concept of stereoelectronically assisted bond cleavage as originally proposed by Dunathan[11,12] (see Section 6.5.2.1). Despite this suggestion, all of the observed stereochemical features of these 'unusual' reactions could be rationalized if the enzyme pockets on the C-4'-*re* face of the coenzyme (the protein-covered side) were flexible. Indeed, in view of the vast number of substrates for SHMT, the active-site region must be flexible. Scheme 23 depicts the mode of binding of D- and L-alanine to SHMT–holoenzyme compared with that of L-serine. It should be noted that α-methylserine is also a substrate for a SHMT catalyzed retro-aldol reaction.[179] With the knowledge that the enzyme can bind both D- and L-amino acids to form quinonoid intermediates, it seems likely that observed nonstereospecific decarboxylation of chiral aminomalonate (Scheme 24) is best accounted for by the substrate binding in two distinct conformations as for D- and L-amino acids. Decarboxylation could thus occur stereospecifically with, for example, retention of configuration for each conformer to give an apparently racemic product.

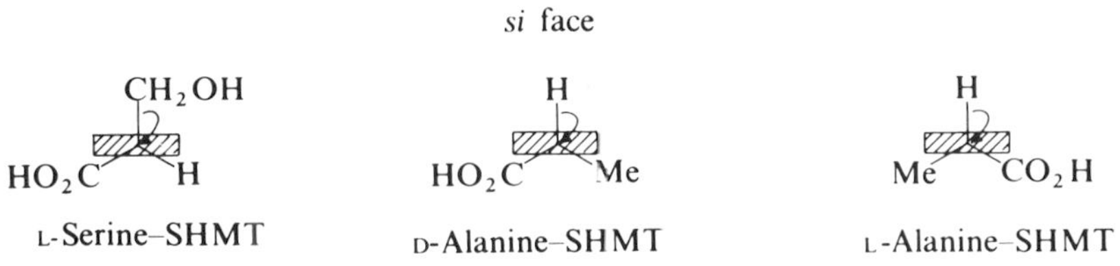

si face

CH$_2$OH   H   H

HO$_2$C H   HO$_2$C Me   Me CO$_2$H

L-Serine–SHMT   D-Alanine–SHMT   L-Alanine–SHMT

Scheme 23

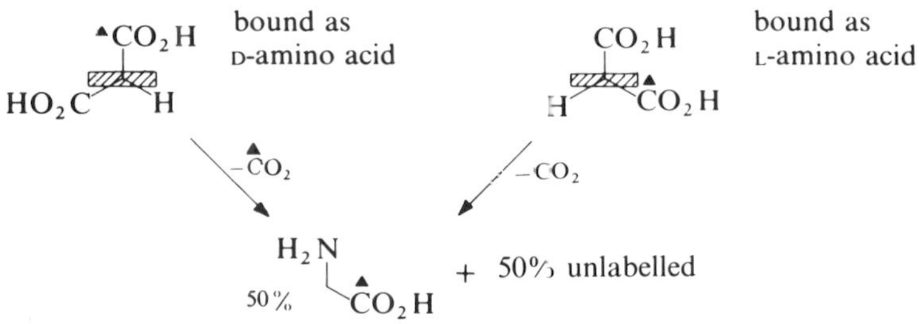

CO$_2$H   bound as   CO$_2$H   bound as

HO$_2$C H D-amino acid   H CO$_2$H L-amino acid

$-CO_2$ $-CO_2$

H$_2$N

50%   CO$_2$H   + 50% unlabelled

Scheme 24

Serine hydroxymethyltransferase has been purified from a variety of sources including rat,[170] sheep[180] and rabbit liver[181] and *E. coli*.[182] Both cytosolic and mitochondrial enzyme from rabbit liver are each tetramers (M_r 215 000) composed of identical subunits. The mitochondrial enzyme is slightly larger by 1000 Da per subunit than the cytosolic enzyme.[183] Addition of glycine to either enzyme results in the formation of holoenzyme–substrate complexes absorbing at 343, 425 and 495 nm. The relative concentration of each complex is not identical for each enzyme. The affinity of each enzyme for THF and amino acid substrate are almost identical, except for glycine where the mitochondrial enzyme has a greater affinity.

Structural studies also suggest that the two enzymes are very similar. Both enzymes have isoelectric points of 7.2 and copurify through six purification steps. Also antibodies raised from cytosolic enzyme cross react with mitochondrial enzyme, although no reaction occurs between antibodies for mitochondrial enzyme and cytosolic enzyme. From the above data it is clear that the enzyme forms are related. Indeed, studies have indicated that there is a high degree of homology between the proteins.[183,184] In these studies, each of eight cysteine-containing peptides from the cytosolic protein and six cysteine-containing peptides from the mitochondrial protein were isolated and sequenced. Four peptides from each group showed a high degree of homology. Comparison of 60 tryptic and chymotryptic peptides from the cytosolic enzyme and 20 from the mitochondrial enzyme revealed that 12 peptides were homologous. Interestingly, the active-site peptides of both forms of the protein from rabbit liver[183] and from *E. coli* are identical.[185] These peptides contain the lysine residue which forms an internal aldimine with PLP.[184] It should be noted that the original assignment of the active-site peptide for rabbit liver cytosolic SHMT was incorrect when first published and contained only three consecutive threonine residues instead of four. The complete amino acid sequence of the cytosolic rabbit enzyme[186] has been reported. The complete amino acid sequence of *E. coli* SHMT was proposed on the basis of the nucleotide sequence of the *E. coli gly* A gene,[185] before the enzyme had been purified.

Schirch has recently purified the enzyme from *E. coli* GS245 and has determined that the protein exists as a dimer (M_r 95 000), in contrast to the tetrameric mammalian enzymes.[182] However, in all other respects the *E. coli* enzyme is very similar to the mammalian liver isoenzymes. Hopkins and Schirch have altered His-228 in the *E. coli* enzyme to an asparagine residue using site specific mutagenesis.[187] The resultant mutant enzyme was 25% active compared with the wild type. The His-228 is believed to be involved in binding the substrate or coenzyme but not, apparently, in catalyzing the reaction. SHMT has also been purified from mung bean.[188] Schirch has reviewed the mechanism of SHMT catalyzed reactions in work published up to 1982.[189]

6.5.2.5.2 *Reactions of amino acids at C-3*

A large group of PLP-dependent enzymes catalyze the elimination of C-3 substituents to give the aminoacrylate intermediate.[190-192] Some of these enzymes then catalyze Michael addition to the intermediate (Scheme 25, route A) to give new amino acids, whereas others catalyze hydrolysis to give α-keto acid and ammonia (route B). Following labilization of H^α to give the ketimine, presumably from the C-4'-*si* face of the coenzyme, the leaving group (X) is dispelled to give the acrylate derivative. To allow efficient elimination the leaving group should be disposed perpendicularly to the π-electron system on either the 4'-*si* face or the 4'-*re* face of the coenzyme, thus leading to syn- and anti-periplanar elimination, respectively. For β-replacement reactions the trajectory of the new (incoming) nucleophile would then define the stereochemical course of the reaction with respect to C-3.

The steric courses of several β-replacement and α,β-elimination–deamination reactions have now been determined and the results are presented in Scheme 25. To date, all of these reactions have been shown to occur in a retentive mode. Much of the work on α,β-elimination and β-replacement reactions has focused on the tryptophan synthase. The complete enzyme exists as an $\alpha_2\beta_2$ tetramer and is responsible for the biosynthesis of tryptophan from L-serine and indole-3-glycerol phosphate. The α-subunit converts indole-3-glycerol phosphate to protein-bound indole while the PLP-dependent β-subunits promote the dehydration of serine for subsequent β-replacement by the indole.[193] The β_2 subunit also catalyzes the conversion of L-serine to pyruvate and ammonia but only in the absence of α-protein.[190] This reaction has been used to probe the mechanism of the analogous *bona fide* α,β-elimination reactions.

The reaction of the β_2-protein with C-3 chirally tritiated L-serine in ^2H_2O generated pyruvate with an apparently racemic methyl group.[194] Further studies revealed that solvent deuterium had not been incorporated at C-3 and that the third hydrogen originated from C-2 of serine. In the latter experiment, Floss and coworkers showed that samples of L-serine chirally labelled at C-3 with deuterium and tritium but unlabelled at C-2 gave samples of pyruvate with retention of configuration at C-3 (Scheme 25). These incubations were conducted in deuterium oxide and thus the C^α position contained the only available protium. On the basis of these results, Floss suggested that the 1,3-prototopic shift must occur suprafacially and thus by means of a single enzyme-bound base. Also, as the hydroxyl group of serine is replaced by hydrogen in a retentive mode, the elimination must occur in a synperiplanar fashion.[194]

To determine which face of the coenzyme was involved in these reactions, the β_2-protein was incubated with unlabelled L-serine in deuterium oxide in the presence of mercaptoethanol. This abortive transamination yielded PMP which contained 25% of label exclusively in the 4'-*pro-S* position. The low incorporation implied that 75% of the hydrogen transferred to C-4' was derived from the holoenzyme–substrate complex, presumably from the α-position of the substrate. These results are consistent with a 1,3-suprafacial prototropic shift on the C-4'-*si* face of the coenzyme in a conformation where the C^α—C^β and C-4'—N bonds are eclipsed. An enzyme-bound base must therefore be situated on the *si* face in close proximity to C-4', C^α and C^β. Also, as the hydrogen transfers from C^α to both C-4' and C^β are not appreciably affected by exchange of label with the solvent, it is likely that the imidazole ring nitrogen of a histidine residue rather than a polyprotic amino group acts as the base.[9] Further information regarding the conformation of substrate binding at the active site has also come from the Floss laboratory.[195] Borohydride reduction of [4'-^{3}H]holoenzyme $\alpha_2\beta_2$ complex in the presence of L-serine gave pyridoxylalanine *via* C-4'-*si* face reduction. However, in the absence of substrate the reduction gave pyridoxyllysine. Analysis revealed that the internal aldimine had been reduced from the C-4'-*re* face of the coenzyme. These findings[195] are analogous to results obtained from the reduction of transaminases and thus tryptophan synthase also undergoes significant conformational modification on binding substrate.[9] In a further experiment the unlabelled holoenzyme was reduced with borotritide in the presence of the inhibitor indolepropanol phosphate with (2S,3S)-[2-^{2}H,3-^{2}H]serine as substrate. Examination of the resulting pyridoxylalanine revealed that incorporation of tritium into the C-4', C-2 and C-3 positions had occurred to the extent of 50%, 20% and 30%, respectively. It was also evident that the methyl group of the product was of the (S) configuration and thus reduction had occurred at C^β from the same (C-4'-*si*) face of the coenzyme as for C-4' reduction.[195] It is also likely, but not proven, that reduction at C^α also occurs from the C-4'-*si* face.

Recently a new single-turnover, half-transamination catalyzed by the $\alpha_2\beta_2$ complex has been reported.[196] The reaction involved transfer of the 4'-*pro-S* hydrogen of PMP to indole-3-pyruvic acid. Other work in the area has focused on the enzyme-catalyzed isomerization of 5-fluoro- and (3S)-2,3-dihydro-5-fluoro-L-tryptophan,[197] the β-elimination reaction of L-tryptophan[198] inhibition by 2,3-dihydro-L-tryptophan[199] and the hydrogen exchange kinetics of H^α-substrate solvent hydrogen exchange.[200]

Tryptophanase shows very similar stereochemical properties to tryptophan synthase.[201,202] However, the enzyme-bound base in tryptophanase appears to be polyprotic, unlike the imidazole group of histidine in the synthase.[9] Another difference is that the C^α hydrogen of L-tryptophan is

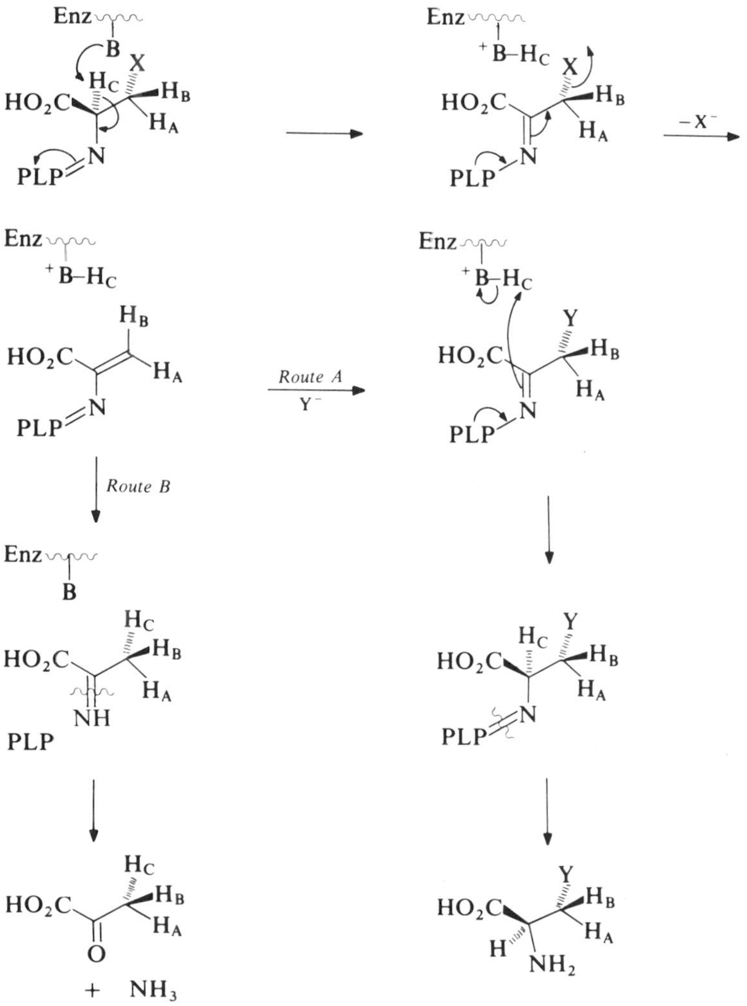

Route A X	Y	Enzyme	*Route B* X	Enzyme
AcO—	HS—	Cysteine synthase	HO—	Tryptophan synthase β_2-protein
HS—	NC—	β-Cyanoalanine synthase		Tyrosine phenol-lyase
				Tryptophanase
				D-Serine dehydrase
HO—	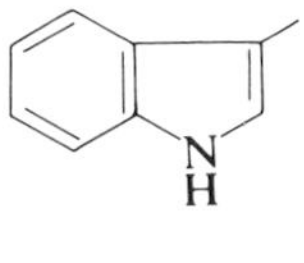	Tryptophan synthase Tryptophan synthase β_2-protein	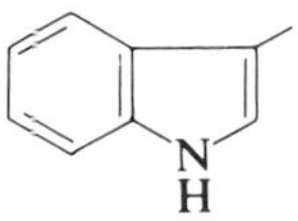	Tryptophanase
HO—	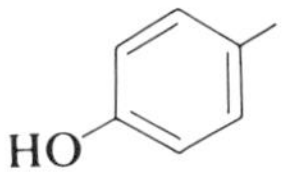	Tryptophanase Tyrosine phenol-lyase	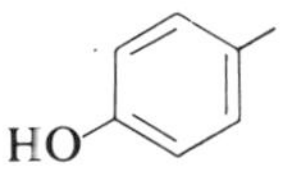	Tyrosine phenol-lyase
HO—	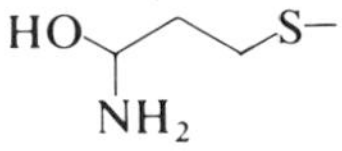	Cystathione β-synthase	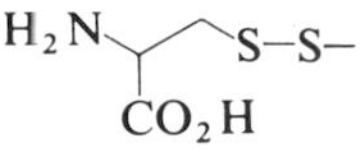	*S*-Alkylcysteine lyase

Scheme 25

initially transferred to C-3 of the indole to protonate the leaving group and hence facilitate elimination. Consequently H^α transfer from tryptophan or serine to the pyruvate methyl group (Scheme 25, route B) occurs to only a limited extent and most of the hydrogen is derived from the solvent.[201]

The purification and characterization of a new PLP-dependent enzyme from pig liver was recently described by Soda and coworkers.[203] The enzyme, selenocysteine β-lyase, exclusively converts L-selenocysteine into L-alanine and hydrogen selenide. L-Cysteine is not a substrate for the enzyme, although it does competitively inhibit the physiological reaction. The enzyme has been detected in several species and in mammals occurs in several major organs.[204] Selenium now appears to be an essential nutrient for mammals, birds and some bacteria and the role of selenium in biochemical processes has been reviewed by Stadtman.[205]

Soda has examined the selenocysteine lyase reaction in some detail[203,206] and has provided evidence to show that the true enzymic products are L-alanine and elemental selenium (Scheme 26). Nonenzymic reduction by endogenous reducing agents must therefore account for the observed formation of hydrogen selenide. Incubation of the enzyme with C^{α}-deuteriated substrate showed an isotope effect of 2.4, indicating that transfer of the H^{α} occurs in a rate-limiting step. Furthermore, incubation of the protio substrate in deuterium oxide gave [2-^{1}H,3-^{2}H$_1$]- and [2-^{1}H,3-^{2}H$_2$]-L-alanine in the ratio 1:1 as judged by ^{13}C and ^{1}H NMR. This latter result indicates that while H^{α} is transferred to the enzyme-bound base and then back to C^{α} of the product internally, solvent hydrogen replaces selenium at C^{β} and that in 50% of catalytic turnovers an additional hydrogen atom from C^{β} is exchanged with solvent. From all of these observations it is apparent that selenium leaves as an electrofuge in a reaction which closely parallels the aspartate β-decarboxylase reaction (Scheme 12). It is interesting to note that the β-decarboxylase also catalyzes desulfination of L-cysteine sulfinate to L-alanine and sulfur dioxide.[94] The lyase has now been purified from *Citrobacter freundii*.[207] Although the catalytic properties are similar to the mammalian enzyme,[203] the amino acid compositions and subunit structures of the enzymes are different.

Scheme 26

6.5.2.5.3 *Reactions of amino acids at C-4*

Several PLP-dependent enzymes catalyze β,γ-elimination and γ-replacement reactions.[208] Mechanistically, the first steps of β,γ-elimination resemble those of aspartate β-decarboxylase reactions (Scheme 12) in that after initial labilization of H^{α} to form the quinoid, a C^{β} substituent (in this case a proton) is removed to give a stabilized C^{β} carbanion. Subsequent elimination of a nucleofuge from C^{γ} then generates an enzyme-bound vinylglycine intermediate (17). The intermediate is able to undergo a variety of reactions at both C^{β} or C^{γ}; these are depicted in Scheme 27.

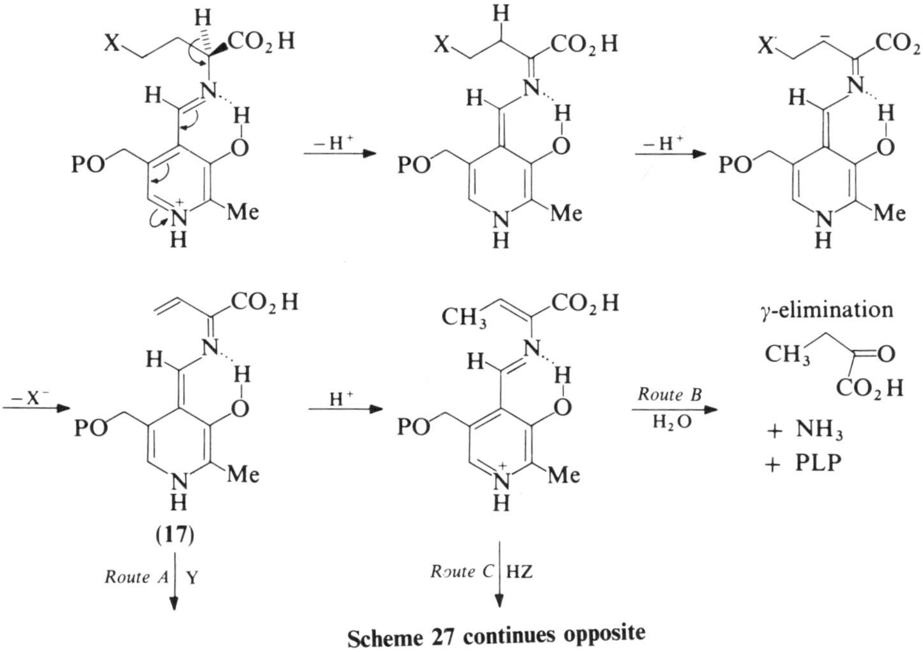

Scheme 27 continues opposite

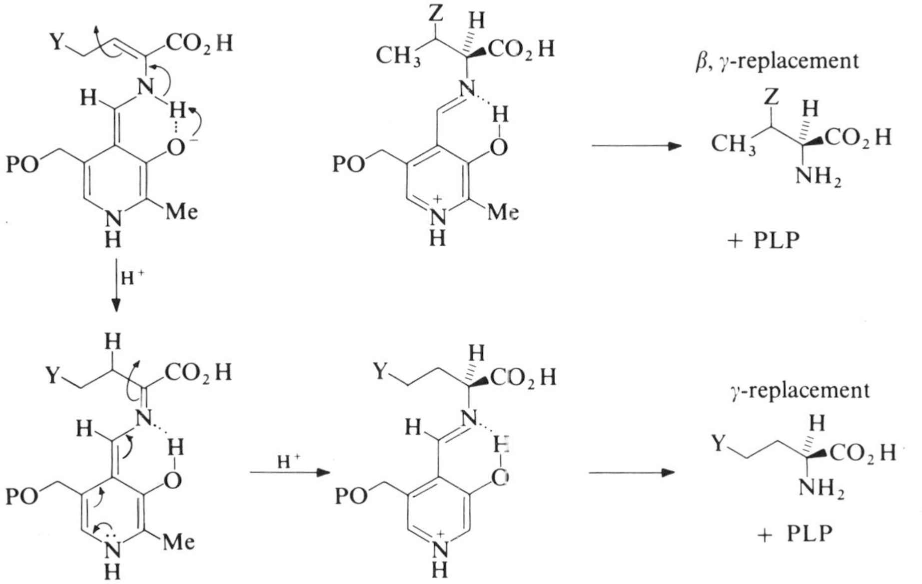

γ-Replacement (Route A)

X	Y	Enzyme	Product
RO—	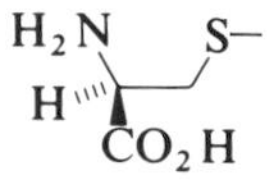 (H₂N, S—, H, CO₂H amino acid)	Cystathionine γ-synthase	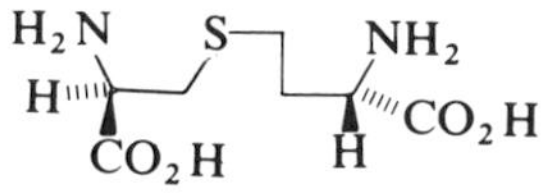 Cystathionine

R = succinyl, acetyl,
phosphoryl, malonyl

γ-Elimination–deamination (Route B)

X		Enzyme	Product
HO_2C CO_2—		Cystathione γ-synthase	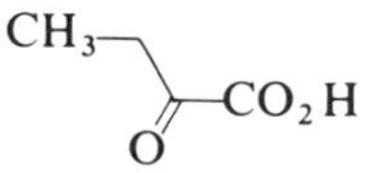 CH_3 —CO₂H + NH_3
H_2N S— H CO₂H		Cystathionase	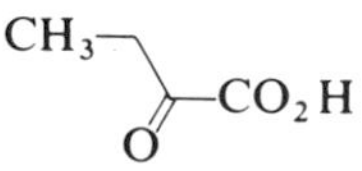 CH_3 —CO₂H + NH_3
HO—		Cystathionase (Homoserine dehydrase)	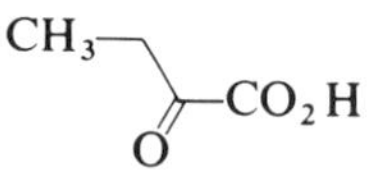 CH_3 —CO₂H + NH_3

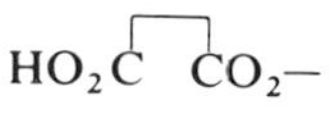

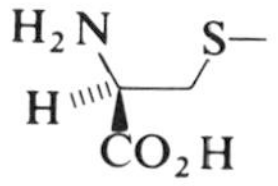

β,γ-Replacement (Route C)

X	Z	Enzyme	Product
$^{2-}O_3P$—O—	HO—	Threonine synthase	L-Threonine

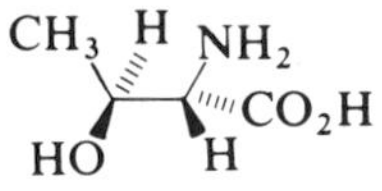

Scheme 27

Most of the mechanistic information regarding reaction at C-4 has been acquired from studies conducted with γ-cystathionine synthase from *Salmonella*.[209] The normal physiological reaction (Scheme 28) involves γ-replacement of an ester group by the thiol group of L-cysteine to give the methionine precursor cystathione. In the absence of cysteine, however, *O*-succinylhomoserine is converted to α-ketobutyrate and ammonia. In order to probe the stereochemical course of this reaction, Posner and Flavin[210] incubated the substrate with the enzyme in deuterium oxide. Analysis of the resulting α-ketobutyrate revealed that incorporation of deuterium had occurred at both C-3 and C-4 to the extent of 1.0 and 0.2 atom equivalents, respectively. Degradation of this sample to propionate followed by comparison with authentic chirally deuteriated propionate by ORD showed that the enzymic product contained deuterium in the 3-*pro-S* position.[210] α-Ketobutyrate derived from the incubation of homoserine dehydrase (γ-cystathionase) with homoserine in deuterium oxide also possessed (3S) stereochemistry.[211] Further experiments with the synthase demonstrated that the enzyme catalyzed solvent hydrogen exchange at C-2 and C-3 of cystathionine and homoserine and that exchange was rapid for only one of the C-3 methylene hydrogens. Using ^{1}H NMR techniques, Fuganti and Coggiola[212] determined that the 3-*pro-R* hydrogen was most rapidly exchanged and thus that C-3 deprotonation and subsequent reprotonation occurred in a retentive mode. (Note that the 3-*pro-R* hydrogen of substrate and the 3-*pro-S* hydrogen of α-ketobutyrate occupy spatially equivalent positions). The low incorporation of deuterium into C-4 of α-ketobutyrate (0.2 atom equivalents) was accounted for by showing that protium from both C-2 and C-3 hydrogen atom positions of the substrate were transferred to C-4. These results suggest that all of the deprotonation and protonation steps occurring at C-2, C-3 and C-4 are promoted by a single enzyme-bound base in a suprafacial manner.

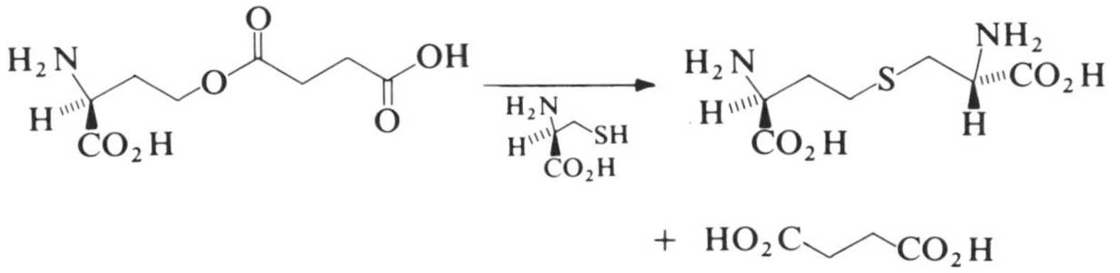

Scheme 28

The steric course of events at C-4 were deduced by Walsh and coworkers[213] using vinylglycine. Vinylglycine is an excellent alternative substrate to *O*-succinylhomoserine and is able to react with holoenzyme to give the same enzyme-bound vinylglycine intermediate (**17**) as is derived from the true substrate. When samples of (*Z*)-DL-[4-^{2}H$_1$]- and (*E*)-DL-[3-^{2}H$_1$,4-^{2}H$_1$]-vinylglycine were incubated with the enzyme in the presence of L-cysteine and the resulting cystathione samples converted to homoserine, for assessment of configuration at C-4, it was apparent that nucleophilic attack by sulfur had occurred at the *si* face of the vinylglycine intermediate (**17**) with respect to C-3 (Scheme 27). Further experiments with these samples of homoserine involved re-incubation of the *O*-succinyl derivatives as before to give new samples of cystathionine. Comparison of these new samples with the original cystathionine samples derived from deuteriated vinylglycine revealed that the succinate ester group was replaced by sulfur with retention of configuration at C-4.[214] Walsh also showed that protonation at C-4 in the γ-elimination reaction occurs in a retentive mode. It is thus apparent that all events in the β,γ-elimination and γ-replacement reactions catalyzed by cystathionine synthase occur on one face of the coenzyme–substrate complex, the *re* face with respect to C-2, in a conformation where the C-2—N and C-3—C-4 bonds are cisoid (eclipsed).

In contrast to other enzymes acting at C-4, threonine synthase removes the 3-*pro-S* hydrogen from the physiological substrate phosphohomoserine (Scheme 27, route C).[215] Furthermore, incubations conducted in deuterium oxide lead to the incorporation of two atoms of deuterium, one at C-2 and one at C-4 of the product. Thus no significant internal proton transfer is apparent. The 3-*pro-S* hydrogen of the substrate is replaced by a hydroxyl group during the conversion and as the absolute configuration of L-threonine is (2S,3R) it is evident that the process occurs in a retentive mode. Clearly the mechanism of this β,γ-replacement reaction calls for an entirely different conformation at the active site compared with β-replacement systems. Two possibilities, either a transoid quinoid intermediate or a two-base system, however, would adequately explain the differences in the stereochemical courses of the two types of reaction. Interestingly, the retentive mode of proton replacement corresponds to, in a reverse sense, the stereochemical course of hydroxyl group elimination and subsequent protonation of the aminocrotonate intermediate in the α,β-elimination of D- and L-threonine as catalyzed by D-threonine dehydrase and L-threonine dehydrase from *Serratia marcescens*.[216,217] In the latter case the proton is added from the C-3-*re* face.

1-Aminocyclopropane-1-carboxylate (ACPC) deaminase catalyzes the fragmentation of ACPC to give α-ketobutyrate and ammonia (Scheme 29). The enzyme from *Pseudomonas* spp. has been studied in a series of papers by Walsh and coworkers.[218-220] The enzyme is able to utilize D-vinylglycine as a substrate and thus resembles many others catalyzing γ-elimination reactions.[213] The stereochemical and mechanistic features of the catalyzed reaction have been reviewed recently.[145a]

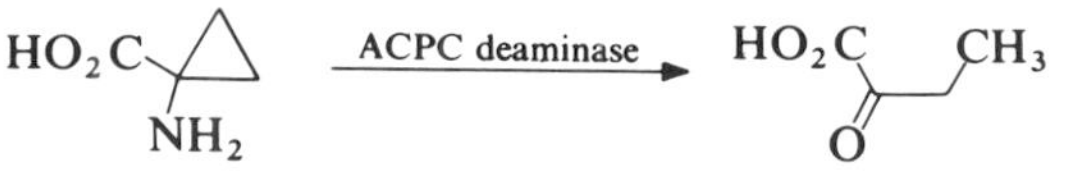

Scheme 29

6.5.2.5.4 *Glycogen phosphorylase*

PLP is a coenzyme for glycogen phosphorylase.[221] The enzyme is responsible for the control of glycogen degradation and catalyzes the sequential phosphorylation of glycogen from the non-reducing terminus of α(1-4) linked glucose residues. The rabbit skeletal muscle enzyme exists in two forms: phosphorylase b, a dimer (inactive) and phosphorylase a, a tetramer (active). *In vivo* the two forms are interconverted by specific enzymes which are in turn activated by cAMP. Glycogen phosphorylase contains one mole of PLP per monomer and forms an aldimine with Lys-679. The activity of the enzyme is not destroyed by treatment with borohydride, as would be expected if PLP functioned in its usual role. It was originally thought that PLP was merely involved in stabilizing the structural conformation of the protein. However, from several recent studies[222-226] it is evident that PLP is actually involved in the catalytic process and that the 5'-phosphate group is the site where catalysis occurs. Several mechanistic schemes have been suggested in order to account for the catalytic activity of the coenzyme and it has been proposed that the phosphate ester dianion acts as a nucleophile,[222] as an acid (monoanion)/base (dianion), and as an electrophile (at phosphorus). Research in the area has been reviewed recently.[221,227]

6.5.3 INHIBITORS

6.5.3.1 Introduction

In general, PLP-dependent enzymes are inhibited by carbonyl-attacking reagents, for example, hydroxylamines and hydrazines, and by substrate analogues which cannot undergo reaction but which contain essential active-site binding groups. In the absence of endogenous coenzyme, many PLP-dependent enzymes are inhibited by the production of PMP in abortive transaminase side reactions. In inhibitor design, some effort has been directed towards the synthesis of substrate analogues containing carbonyl-attacking groups, for example, hydrazino acids,[228] and also reaction-blocked substrates, for example, Cα-methyl amino acids.[229] However, most effort has focused upon the design of mechanism-based (k_{cat}) inhibitors.[230-232] These types of compound are non-reactive until activated by the holoenzyme. At the active site of the target enzyme their latent reactivity promotes side reaction(s) with the holoenzyme complex which occasionally result in the formation of a covalently bound inhibitor–coenzyme–protein complex which is unable to return to an active form. More often, the side reactions catalyzed by the enzyme do not result in an irreversible binding ('killing') event. Thus most suicide inhibitors form intermediates which then partition between subsequent safe and 'killing' reactions in a manner characteristic of the holoenzyme suicide substrate reactive intermediate, independent of its actual mode of formation.

The kinetic model of suicide inhibition is depicted in Scheme 30; also see Section 6.5.5.4. From the kinetic model it is evident that for the reaction of a given enzyme with inhibitors which lead to formation of intermediates that cause inactivation frequently, the value of k_{+3}/k_{-3} is small and will be most specific at a given concentration. These 'suicide inhibitors' thus offer great potential as selective or even specific enzyme inhibitors. Also, as many possible inhibitor starting structures may be used to create the same reactive enzyme bound intermediate, the structure can be optimized in consideration of potential therapeutic requirements such as lipophilicity and metabolic stability. The challenge to medicinal chemists is, therefore, to use a knowledge of the mechanism of the reaction catalyzed by the target enzyme together with information regarding the position of key binding groups and potentially reactive species at the active site, to design inhibitors that are highly specific for the target enzyme.

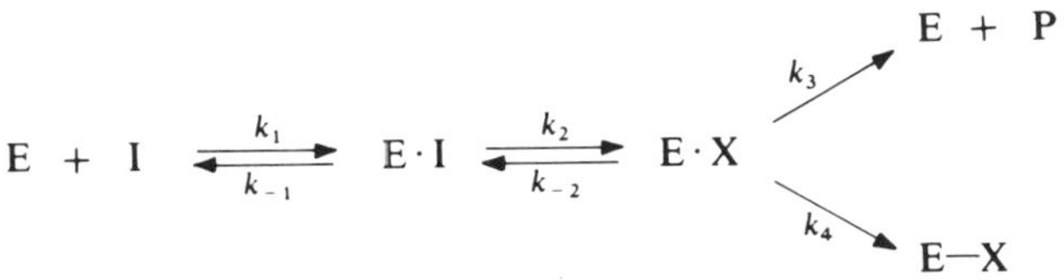

$$E + I \xrightleftharpoons[k_{-1}]{k_1} E \cdot I \xrightleftharpoons[k_{-2}]{k_2} E \cdot X \begin{array}{c} \xrightarrow{k_3} E + P \\ \xrightarrow{k_4} E{-}X \end{array}$$

Scheme 30

6.5.3.2 'Suicide' Inhibitors (Mechanism Activated Inhibitors)

The design and action of 'suicide' substrates has been an area of great interest over the past decade and PLP-dependent systems have been the focus of much of this interest.[230-232] Many types of system have been devised to trap active-site nucleophiles, although most of these systems share common features. For these reasons, suicide inhibitors for given target enzymes are grouped together according to their inhibitory properties, rather than by their mode of inhibition.

6.5.3.3 Transaminase Inhibitors

6.5.3.3.1 *Inhibitors for L-α-amino acid transaminases*

Aspartate aminotransferase (AAT) is not a system of choice for specific inhibition due to the ubiquitous nature of the enzyme and the highly conserved chemical and structural properties of the enzyme from a wide range of species. Recently, some studies have revealed that significant differences do exist between the mitochondrial and cytosolic mammalian enzymes in spite of their similarities and thus in the future, when bacterial isoenzymes are better characterized, it may be possible selectively to inhibit one type in the presence of the other types. However, much more fundamental information is required before such a strategy is viable in therapy. Nevertheless, AAT is the best studied of all PLP-dependent enzymes (see Section 6.5.2.2.1), and the modes of inactivation of the enzyme by many different types of inhibitor have been investigated.

AAT is inhibited by D- and L-hydrazinosuccinic acid,[228] 2-methylaspartic acid[229] and maleic acid.[233] The last two compounds have been co-crystallized with the enzyme for X-ray diffraction analysis and so their binding modes are well defined.[23] AAT is also inhibited by a wide range of suicide substrates including L-serine-O-sulfate,[234] 3-methenylaspartic acid[235] and (2R,3S)-3-chloroaspartic acid.[236] Interestingly, neither diastereomer of L-3-chloroglutamic acid inhibit the enzyme although both compounds are substrates.[237]

The mechanism of inhibition by L-serine-O-sulfate has been studied in some detail by Metzler and coworkers.[238] Originally it was assumed that after H^α abstraction and elimination of the 3-leaving group, the resulting Michael acceptor reacted directly with an active-site bound nucleophile, possibly the ε-amino group of Lys-258, to give the inactivated complex (see Scheme 31).

This classical mechanism was ruled out as the mode of inhibition of L-serine-O-sulfate, however when Metzler showed that the coenzyme adduct isolated from the inactivated complex was the product of an enamine reaction with the internal aldimine, the Schnackerz adduct (**19**) (Scheme 32). Thus the suicide inhibitor was converted to the 'free' nucleophilic aminoacrylate. This inactivation mechanism is extremely interesting and clearly indicates that the most obvious chemical mechanism does not always apply.

The formation of similar adducts upon inhibition of alanine racemase and glutamate decarboxylase with β-nucleofuge substituted alanines shows that the reaction is quite general. However, for inhibitors containing further functionality, alternative inhibition modes may prevail, for example inhibition by the two aspartic acid derivatives[235,236] mentioned above and also by the bacterial toxin (E)-3-methoxyvinylglycine (**20**),[239] which is thought to involve an addition–elimination reaction (Scheme 33).

For 3-chloroaspartic acid,[236] where in contrast to serine-O-sulfate the leaving group is not a binding group, a strong Arg-292–substrate β-carboxylate interaction would need to be overcome in order to allow an enamine inhibition mechanism;[238] thus a Michael reaction cannot be discounted (Scheme 31). There is evidence to suggest that intermediates derived from both β-methenylaspartic

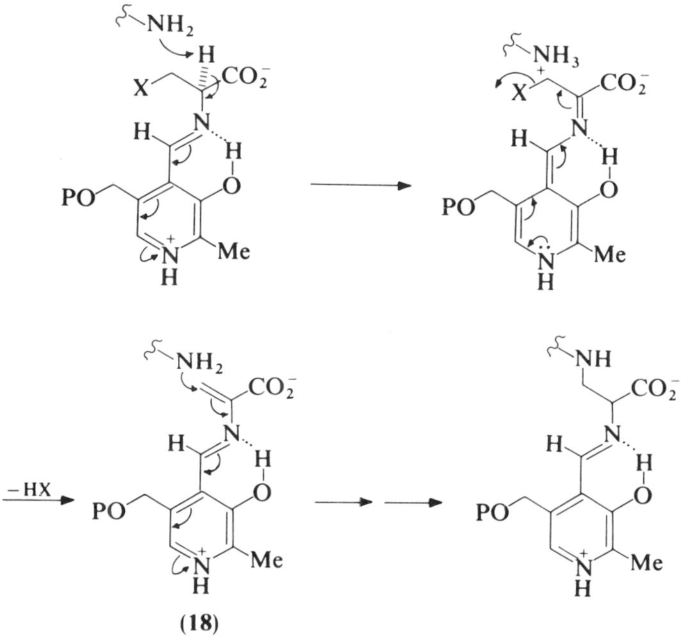

Scheme 31

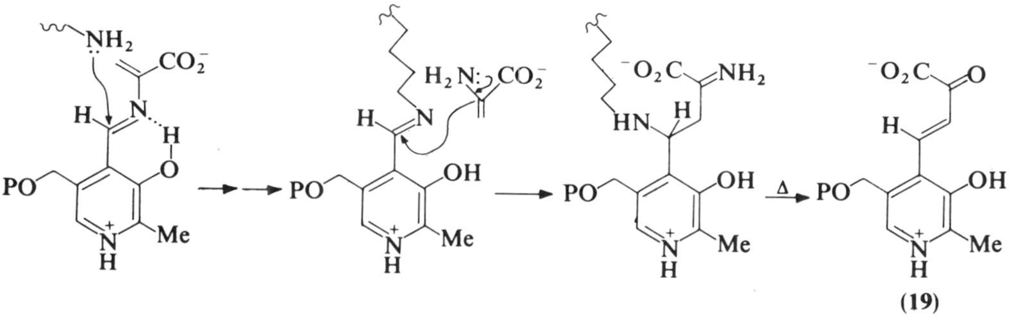

Scheme 32

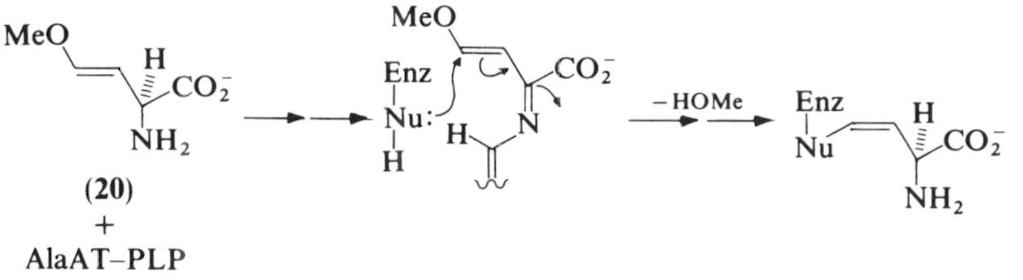

Scheme 33

acid and the toxin (E)-3-methoxyvinylglycine[239] undergo Michael reaction during the course of enzyme inactivation. L-Vinylglycine also inactivates AAT[240,241] but does not inactivate the other common mammalian α-amino acid transaminase, L-alanine transaminase, AlaAT. AlaAT is inhibited, however, by (E)-3-methoxyvinylglycine[239] and 3-methenylaspartic acid.[235] Both enzymes are inhibited by L-3-chloroalanine[242,243] and L-propargylglycine (21),[244-246] an alkyne-containing natural product from *Streptomyces*. AlaAT is inhibited by cycloserine (22) very efficiently[247] while AAT is inhibited to varying degrees by α- and γ-cycloglutamic acids.[248-250] α-Cycloglutamate (23)[248,250] is believed to acylate an active-site bound nucleophile to give a stable inactivated complex (Scheme 34), while γ-cycloglutamate is thought to form an oxime with 3-aminooxyglutamate produced by ring opening of the isoxazolidane ring.

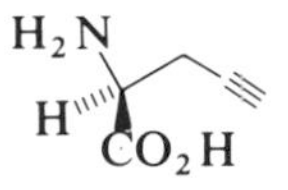

(21)

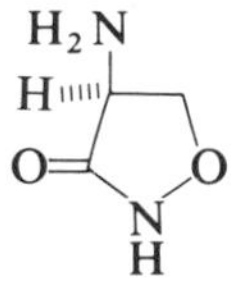

(22)

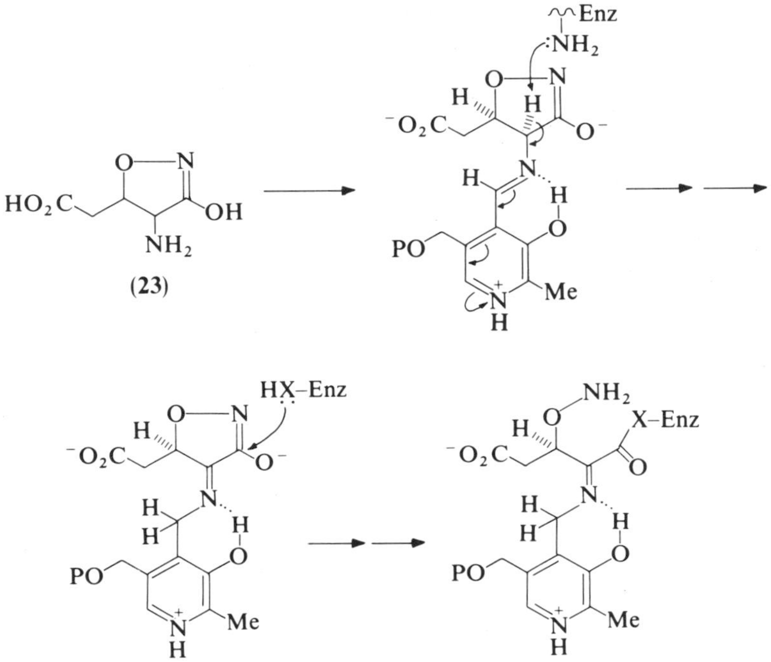

(23)

Scheme 34

6.5.3.3.2 *Inhibitors for γ-aminobutyrate transaminase (GABA-T)*

The inhibition of GABA-T, the key enzyme controlling the catabolism of GABA, is an important area in medicinal chemistry owing to the well established relationship between low brain GABA levels and convulsions (see Section 6.5.2.2.2). The area is particularly exciting because the key biosynthetic and catabolic enzymes, glutamate decarboxylase and GABA-T respectively, share many common features in that they are both PLP dependent and both bind GABA. A special degree of specificity in inhibition is thus called for, and the price of failure could be severe and result in the efficient inhibition of the decarboxylase. This would lead to exactly the wrong situation: depletion of brain GABA and the induction of convulsions. For an overview of the pharmacology of the GABA-ergic system, see Iversen[59] and Olsen and Venter.[251] GABA-T is inhibited reversibly by various carbonyl-attacking reagents including hydroxylamine, aminoxyacetic acid, hydrazinopropionic acid, hydrazines and carbazides. Various GABA analogues also inhibit, including β-chloro- and 3-phenyl-GABA and 4-aminocrotonic acid.[252] The enzyme is inactivated irreversibly by ethanolamine-*O*-sulfate[252] and *in vivo* the compound leads to an increase in brain GABA levels.[253] γ-Ethynyl-GABA[254] and γ-vinyl-GABA[255] are also powerful inhibitors and have been investigated for their pharmacology in animals.[256,257] L-Ethynyl-GABA is less useful *in vivo* as the compound inactivates brain glutamate decarboxylase (see Section 6.5.3.4.1),[52] and ornithine transaminase,[259] which is on the biosynthetic pathway to L-glutamate. Indeed, although both γ-ethynyl-GABA and γ-vinyl-GABA give peak brain GABA concentrations 4 hours after administration, the elevated level is sustained for much longer by γ-vinyl-GABA.[258]

Various halogenated GABA analogues have been prepared and some of these inactivate the enzyme. 5-Fluoromethyl-[260] and 5-trifluoromethyl-GABA[261] inactivate the enzyme whereas 3-chloro-[262] and 3-fluoro-GABA[263] do not. Likewise, 4-amino-2-butenoic acid does not inactivate the enzyme. Mammalian GABA-T is inhibited by the Streptomyces natural product gabaculine **(24)** *via* an interesting aromatization mechanism to give an aromatic PMP adduct (Scheme 35).[264,265]

However, gabaculine is not specific for GABA-T and inactivates a number of other enzymes (K_i 10^{-2} to 10^{-4} M) including D-amino acid transaminase,[266] AlaAT and AAT.[267] Isogabaculine (**25**) also inactivates GABA-T.

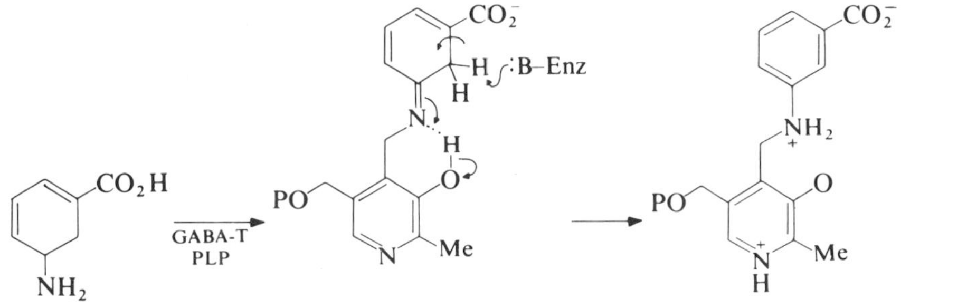

(24)   **Scheme 35**

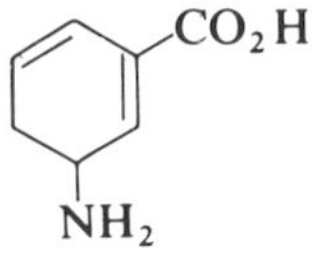

(25)

The mechanism of the inhibition of porcine brain GABA-T by fluoromethyl-GABA has been investigated by Silverman and coworkers.[268] The inactive complex contains a new C—C double bond linking the coenzyme at C-4' to the inhibitor at C-5 and is thus the product of an enamine reaction. The inactivation mechanism, which is similar to that reported by Metzler and coworkers,[238] is the first example for a γ-amino acid. The synthesis and activity of 4-amino-2-(substituted methyl)-2-butenoic (substituent = F, Cl, OH),[269] 4-amino-3-arylbutenoic and 4-amino-3-arylbut-2-enoic acids[270] as GABA-T inhibitors have been reported by Silverman recently. Bright and coworkers have shown that 3-nitro-1-propanamine is a good substrate for porcine brain GABA-T but an inactivator for the *Pseudomonas fluorescens* enzyme.[271] The homologue 4-nitro-1-butanamine inactivates both enzymes. 5-Nitro-L-norvaline (**26**) also inactivates GABA-T but coenzyme-attacking inhibitors are in general less specific and AAT and AlaAT are also inhibited (Scheme 36).[272]

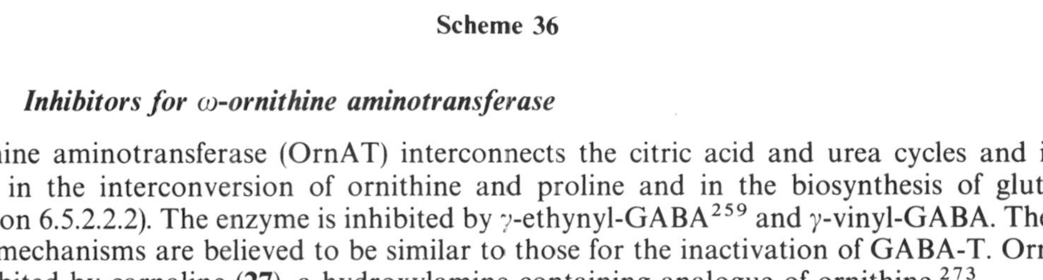

(26)

Nitronate anion

Scheme 36

6.5.3.3.3 *Inhibitors for ω-ornithine aminotransferase*

Ornithine aminotransferase (OrnAT) interconnects the citric acid and urea cycles and is also involved in the interconversion of ornithine and proline and in the biosynthesis of glutamate (see Section 6.5.2.2.2). The enzyme is inhibited by γ-ethynyl-GABA[259] and γ-vinyl-GABA. The inactivation mechanisms are believed to be similar to those for the inactivation of GABA-T. OrnAT is also inhibited by carnaline (**27**), a hydroxylamine-containing analogue of ornithine.[273]

(27)   **(28)**

6.5.3.3.4 *Inhibitors for D-amino acid transaminase*

As outlined earlier, D-amino acid transaminase can be important in the biosynthesis of peptidoglycan and specific inhibitors may be of use as antibacterial agents. In this context the key reaction is the transamination of α-ketoglutarate to D-glutamate using D-alanine. D-3-Chloroalanine[274] and D-3-fluoroalanine[266] are good suicide inhibitors of D-amino acid transaminase. D-Cycloserine and both D- and L-penicillamine (28) function as competitive inhibitors against D-alanine.[275] Other inhibitors for the enzyme are D-vinylglycine,[276] D-propargylglycine[230] and gabaculine.[266] Many of these compounds are effective inhibitors of alanine racemase,[277] which is usually the enzyme responsible for supplying D-alanine for both peptidoglycan biosynthesis and the transamination of α-ketoglutarate to D-glutamate (see Section 6.5.3.5).

6.5.3.4 Inhibitors for Decarboxylases

The inhibition of PLP-dependent decarboxylases is a particularly important area owing to the role of the enzymes in the biosynthesis of many pharmacologically active amines (see Section 6.5.2.3).

6.5.3.4.1 *Glutamate decarboxylase inhibitors*

E. coli glutamate decarboxylase is inhibited by L-serine-*O*-sulfate to give an inactive complex[278] similar to that derived from AAT.[238] The stereochemical mode of the killing reaction and the mode of substrate binding at the active site of the decarboxylase are especially interesting because decarboxylation does not occur during the inactivation process even though the suicide substrate strongly resembles L-glutamic acid. Some clues as to the conformation of substrate binding have come from Marquet's laboratory, where it was shown that the bacterial enzyme catalyzes the decarboxylation of (2R,3R)-3-fluoroglutamic acid to the corresponding 3-fluoro-GABA and (2R,3S)-3-fluoroglutamic acid to succinate semialdehyde.[279] Neither substrate inactivates the enzyme, but the elimination of HF from only the (2R,3S) diastereomer confirmed that the side chain was restricted in one of two conformations with the F atom *anti* or *syn* to the stabilized carbanion.

The stereochemistry of the inactivation of *E. coli* and mammalian brain L-glutamate decarboxylase by 4-aminohex-5-ynoic acid (ethynyl-GABA, 29) has been investigated by Bouclier *et al.*[52] and Danzin *et al.*[280] The bacterial enzyme is inhibited by the (*R*) enantiomer while the rat brain enzyme is inhibited by the (*S*) enantiomer. In the physiological reaction both enzymes catalyze decarboxylation in a retention mode and thus inhibition must be initiated *via* different mechanisms. It has been suggested that inhibition of the bacterial enzyme is a direct consequence of microscopic reversibility in as much as the proton of the (*R*) enantiomer of the inhibitor should occupy the spatial position as the 4-*pro-R* hydrogen of the product GABA (Scheme 37). However, to account for inhibition of the

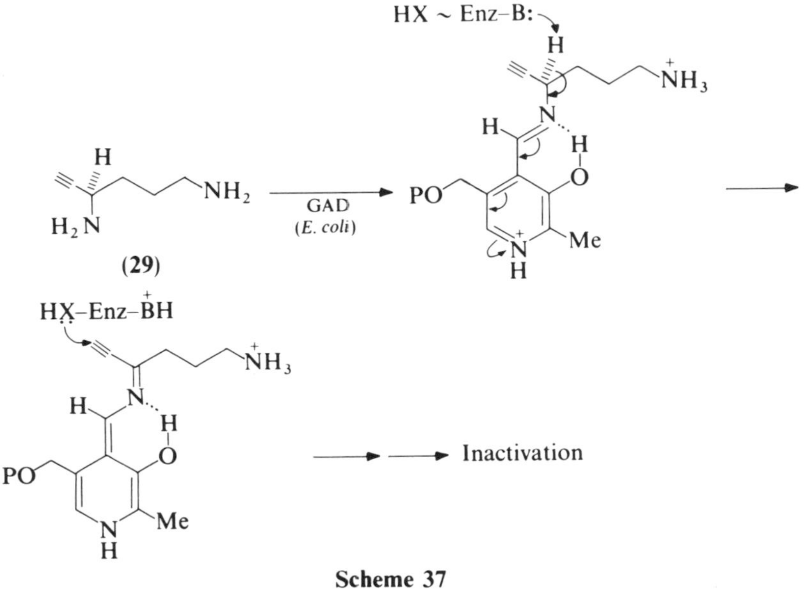

Scheme 37

mammalian enzyme by (S)-4-aminohex-5-ynoic acid a mechanism similar to the first stages of transamination was invoked.[232,281,282] It should be noted that such a process would occur on the $4'$-*re* face of the coenzyme if the inhibitor were to adopt a binding conformation similar to that of the physiological substrate. It is difficult to imagine why the bacterial enzyme is not inhibited by the (S) enantiomer since inhibition by (2S)-serine-O-sulfate involves removal of the α-proton on, presumably, the $4'$-*re* face of the coenzyme, rather than decarboxylation.[278]

Recently, inactivation of the mammalian and bacterial enzymes by α-chloromethyl-, α-fluoromethyl- and α-difluoromethyl-glutamic acid (30)[282] has been investigated. Clearly much further work in the area is required before all of these observations can be rationalized.

(30) (31) (32)

6.5.3.4.2 Ornithine decarboxylase inhibitors

Owing to the growing interest in understanding the role of putrescine and the higher polyamines spermine and spermidine in cellular growth (see Section 6.5.2.2.3), many inhibitors for ornithine decarboxylase have been designed. α-Ethynylornithine (31)[283] and 5-hexyne-1,4-diamine (ethynylputrescine, 32)[284] both inactivate the enzyme, presumably *via* a similar mechanism in which decarboxylation or proton abstraction initiates the inactivation process. Whereas α-ethynyl- and α-vinyl-ornithine are time-dependent irreversible inhibitors,[283] incorporation of a double bond into the side chain to give β,γ-dehydroornithine (33) yields a very potent competitive inhibitor.[285]

(33) (34)

α-Fluoromethylputrescine is a suicide substrate for the *E. coli* enzyme, whereas the α-difluoro compound inhibits the *P. aeruginosa* enzyme.[286] α-Fluoromethyl-[287] and α,α-difluoromethyl-ornithine[288] inhibit the mammalian enzyme but the inactivation process with the monofluoro compound occurs extremely slowly. Interestingly, α-chloromethylornithine is an irreversible inhibitor for ornithine decarboxylase[288] whereas 3-chloroornithine acts as a competitive inhibitor.[289] α-Cyanomethylornithine also slowly inactivates the enzyme.[288]

Bey *et al.* have described compounds containing two potential inactivating functionalities, an alkenic group and an α-halomethyl group. α-Fluoromethyldehydroornithine (34) and α-fluoromethyldehydroputrescine have been tested as ornithine decarboxylase inhibitors.[290] The dehydroornithine (34) is more effective as an inhibitor than the saturated analogue and its K_i is 30 times lower. Note that β,γ-unsaturation in ornithine leads to very tight binding.[285] Casara *et al.* have recently reported that the (1R,4R) isomer of 1-methyl-5-hexyne-1,4-diamine (the methyl analogue) of ethynylputrescine is the most potent inhibitor of the four possible stereoisomers.[291] Heby has reviewed the applications of inhibitors of chemotherapeutic potential targetted at ornithine decarboxylase.[111] Other inhibitors of polyamine biosynthesis which act upon the pyruvate dependent enzyme S-adenosylmethionine decarboxylase will not be described here.

6.5.3.4.3 Histidine decarboxylase inhibitors

Histidine decarboxylase catalyzes the formation of histamine (see Section 6.5.2.3.4) and is thus a target for potential inhibitors. Histidine decarboxylase is inactivated by similar types of compound as described for glutamate and ornithine decarboxylase, although the area has been less active due to limitations imposed by the purity and availability of the PLP-dependent enzyme. α-Chloromethyl-histidine is a selective time-dependent inhibitor of mammalian histidine decarboxylase;[292] the α-fluoromethyl compound also inhibits.[287] The mechanism of inhibition of the rat fetal enzyme by (S)-α-fluoromethylhistidine involves decarboxylation and three such events occur for each inactivation event.[293,294] (S)-α-Fluoromethylhistamine is a more potent inhibitor than the (R) enantiomer.

6.5.3.4.4 *Aromatic amino acid decarboxylase inhibitors*

Aromatic amino acid (DOPA) decarboxylase, the second enzyme in the biosynthesis of catechol-amine and indolamine neurotransmitters[129,295] (see Section 6.5.2.3.5), has been subject to intense investigation. In nervous tissue the activity of tyrosine or tryptophan hydroxylase is approximately 100-fold greater than the activity of the decarboxylase and it was thus predicted that the biosynthesis of the biogenic amines would be difficult to block at the decarboxylation step.[296] Indeed, neither α-methylhydrazine-DOPA, a potent competitive inhibitor, nor the suicide substrate α,α-difluoromethyl-DOPA alter endogenous catecholamine levels.[297] However, α-monofluoromethyl-DOPA does decrease these levels *in vivo*.[298] The inhibition mechanism is similar to those for other decarboxylase inactivation processes where a carboxyl group and fluoride ion are expelled.[299] One turnover event leads to complete inactivation. The (R) enantiomer of α-fluoromethyldopamine causes irreversible inhibition, but inactivation by the (S) enantiomer is completely reversible.[299] Aromatic amino acid decarboxylase is also inhibited by α-chloromethyl-DOPA,[289] α-difluoromethyl-DOPA,[300] α-vinyl-DOPA and α-ethynyl-DOPA. Inhibition by the unsaturated compounds is incomplete and is reversible; activity returns upon exhaustive dialysis.[301] Inhibition by the α-allyl-DOPA (35), α-allyltyrosine or α-allylphenylalanine, however, is essentially irreversible.[302]

6.5.3.5 Inhibitors for Racemase Enzymes

PLP-dependent racemase enzymes, in particular alanine racemase, are important in biosynthetic pathways leading to peptidoglycan. Alanine racemase from a variety of sources is inhibited by D- and L-3-chloroalanine. Indeed, it was during studies of the inhibition of the enzyme from *B. subtilis* with the two enantiomers of chloroalanine that the swinging door mechanism was proposed in order to account for the greater potency of the D-isomer.[138] 3-Fluoroalanine and *O*-acetylserine are also inhibitors.[149] Interestingly, the partition ratio, the number of turnover events per inactivation event, is 850:1 for each of the *E. coli* B[149] and *S. typhimurium*[155] catabolic alanine racemase and *P. striata* amino acid racemase[156] with either enantiomer of 3-fluoro- or 3-chloro-alanine. Walsh and coworkers have shown that inhibition involves an aminoacrylate enamine condensation with the internal aldimine of the biosynthetic[153] and catabolic *S. typhimurium*[155] enzymes and *P. striata* amino acid racemase. It is likely that the other enzymes are inactivated in an identical manner (see Section 6.5.2.4). A similar inhibition mechanism also appears to account for inactivation by D-*O*-acetylserine and the *Streptomyces* antibacterial agents *O*-carbamoyl-D-serine (36) and D-cyclo-serine (37).[149]

(35)   (36)   (37)   (38)

L-Enantiomers of *O*-carbamoyl- and *O*-acetyl-serine are not irreversible inhibitors of the *E. coli* B enzyme. The enzyme is inhibited by 3,3-difluoroalanine. The rate of inactivation by the D-enantiomer is 14-fold faster than for the L-enantiomer.[303] Inactivation, which produces 3-fluoropyruvate, is reversible on dilution in both cases, presumably *via* further fluoride loss and generation of a labile 3-imino-2-aminomalonate–enzyme adduct. Trifluoroalanine also inactivates the *E. coli* B racemase;[303] the inactivated complex may resemble the adduct formed from γ-cystathionase[304] (see Section 6.5.3.6.3). L-Ethynylglycine inhibits alanine racemase although vinyl-glycine does not.[277]

In Section 6.5.2.4.2 it was noted that very recently alanine racemase from the Gram-positive bacterium *Bacillus stearothermophilis* had been cloned, purified and characterized.[156] The racemase shows different properties compared with the Gram-negative bacterial racemases with respect to inhibition by the phosphonate analogue of alanine, 1-aminoethylphosphonate (38).[159] While the Gram-positive enzyme is sensitive to time-dependent inhibition by (R)- and (S)-1-aminoethyl-phosphonate, the Gram negative enzymes are not. The inactivation mechanism is thought to involve the initial reversible formation of a weak complex (competitive) that slowly isomerizes to a stoichiometric complex which dissociates extremely slowly.[159] The complex is not reducible by borohydride and shows a nonperturbed fluorescence spectrum for the bound coenzyme.

6.5.3.6 Inhibitors for Other PLP-dependent Enzymes

6.5.3.6.1 *Serine hydroxymethyltransferase inhibitors*

Serine hydroxymethyltransferase and its role in C_1 transfer and mode of action has been discussed at length in Section 6.5.2.5.1. The enzyme catalyzes a number of reactions in addition to the retro-aldol cleavage of serine. Mammalian liver SHMT is inactivated by (*S*)-3-fluoroalanine about once in every 50 turnovers.[305] The usual products are pyruvate, ammonium and fluoride ion. The inactivation involves the formation of an internal thioether linkage between the aminoacrylic–PLP intermediate and a cysteine residue (Scheme 38). In the presence of tetrahydrofolic acid the rates of deamination and inactivation are increased 200-fold. It is unlikely that inhibitors such as 3-fluoroalanine will have potential as selective *in vivo* inhibitors of SHMT.

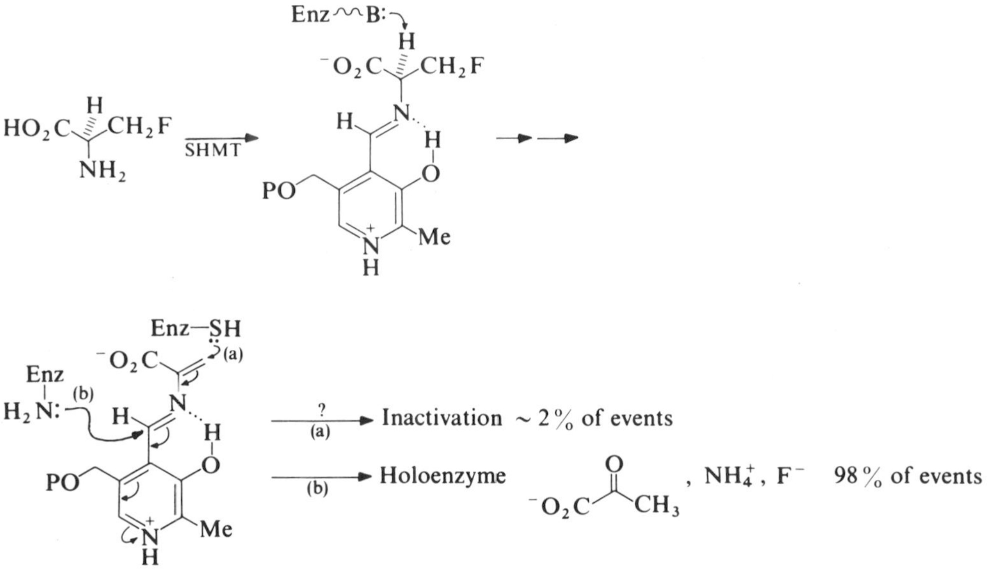

Scheme 38

6.5.3.6.2 *Inhibitors for enzymes catalyzing β-elimination or β-replacement reactions*

β-Elimination is well documented as part of a physiological reaction (see Section 6.5.2.5.2) or as part of a nonphysiological or abortive processes for many PLP-dependent enzymes. Enzymes which have evolved to catalyze α,β-elimination are not so susceptible to inhibition initiated by many of the typical suicide substrates containing halogen, alkynic or alkenic functionalities.[192] However, occasionally an abortive reaction occurs even with a physiological substrate. For example, threonine deaminase is inactivated by L-serine once in every 105 turnovers;[306] 3-chloroalanine has a similar effect.[307] However, L-threonine does not cause inactivation events, presumably because the intermediate aminocrotonate is less reactive whichever inactivation mechanism applies: classical Michael reaction *versus* enamine condensation to give the Schnackerz-type product (**19**) (Schemes 31 and 32).

E. coli tryptophanase is inactivated by trifluoroalanine.[308] It has been suggested that after the formation of the 3,3-difluoro-2-aminoacrylate–PLP-aldimine adduct at the active site of the enzyme, the species would be more susceptible to attack at C-3 by an enzyme-bound nucleophile than to hydrolysis at the aldimine carbon atom. The observation that each turnover event leads to inactivation is certainly consistent with this hypothesis. A similar situation occurs during the inactivation of β-[309] and γ-cystathionase.[304] β-Cystathionase (see Section 6.5.2.5.2) is also inhibited irreversibly by the fungal toxin rhizobiotoxin (**39**).[310] This compound resembles cystathionine and is presumably activated by initial α-proton abstraction (Scheme 39). A number of possible mechanistic courses could explain the formation of an inactivated complex, including an addition–elimination reaction similar to that proposed for (*E*)-methoxyvinylglycine with AlaAT (Scheme 33), or the formation of a very high binding affinity PLP-aminocrotonate ether. Tryptophan synthase (Section 6.5.2.5.2) is also inhibited by trifluoroalanine.[309] This enzyme has been studied in more detail than other PLP-dependent enzymes active at C-3 and it is known that both difluoro- and dichloro-alanine also inhibit. The *E. coli* enzyme reacts with cyanoglycine to give an inactivated complex

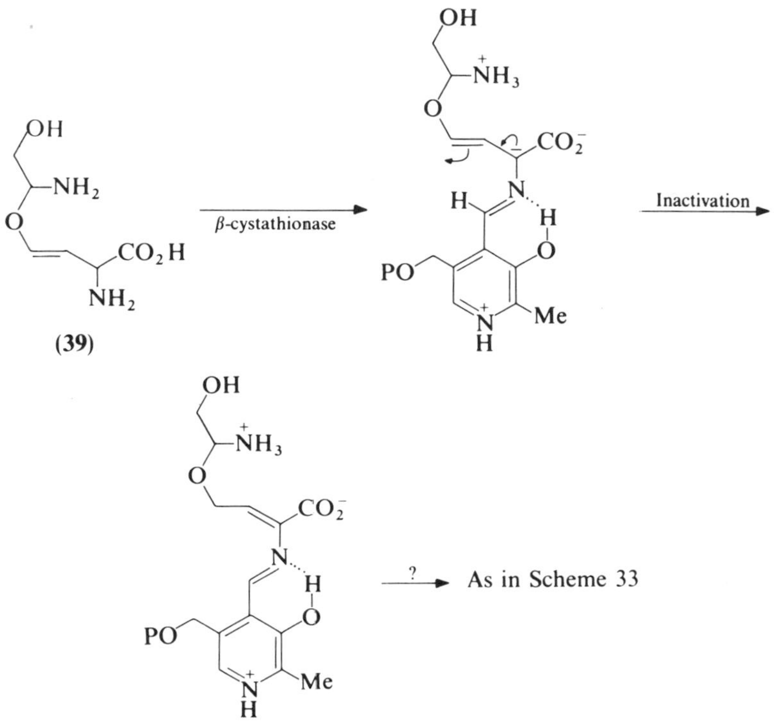

Scheme 39

which reactivates on dialysis. The inactivated coenzyme adduct is probably a stable α-aminomalo-mononitrile anion.[311]

6.5.3.6.3 *Inhibitors for PLP-dependent enzymes catalyzing reaction at C-4*

Enzymes catalyzing reaction at the γ-position of amino acids have been described in Section 6.5.2.5.3. For many of these enzymes, vinylglycine, which is a suicide substrate for other PLP-dependent systems, is a true physiological substrate in as much as its coenzyme-bound intermediates are generated or utilized during normal physiological reactions. However, suicide inhibitors for the α-synthetases and lyases are known. (*E*)-Aminoethoxyvinylglycine (**40**) inhibits 1-amino-1-carboxy-cyclopropane synthetase.[277,312] Propargylglycine inhibits cystathione γ-synthetase.[313] The weak antibacterial activity of the compound may be due to its ability to inhibit the *Salmonella* enzyme with a partition ratio of 4:1. Cystathionine and methionine γ-synthetase are inhibited by allyl sulfoxides.[314] The killing species are the products of 2,3-sigmatropic rearrangement, allylsulfenates (Scheme 40).

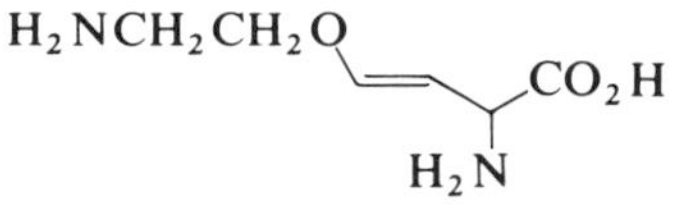

(40)

The action of these compounds has been discussed by Walsh.[230] Finally, the inhibition of α-cystathionase by trifluoroalanine has been mentioned.[304] The proposed mechanism is outlined in Scheme 41.

6.5.4 OVERVIEW

There is a myriad of literature available containing information on hundreds of different types of PLP-dependent enzyme. Clearly we know a little about a lot of PLP-dependent enzymes.

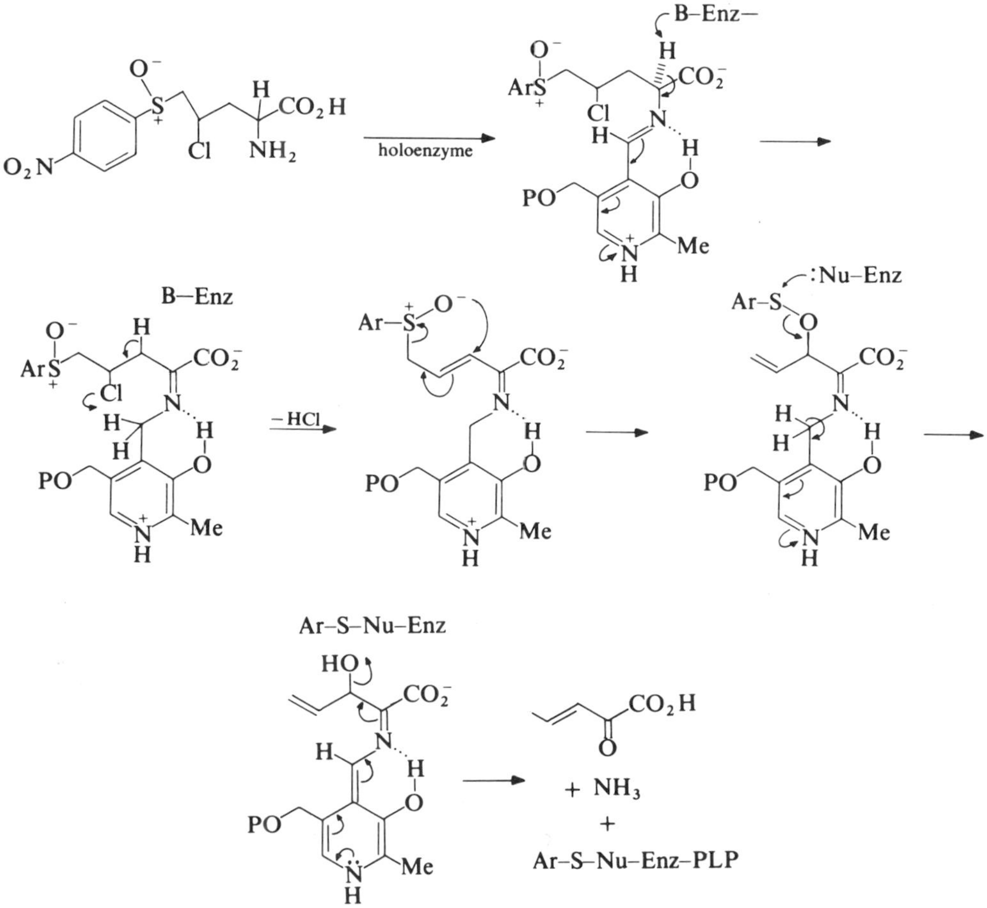

Scheme 40

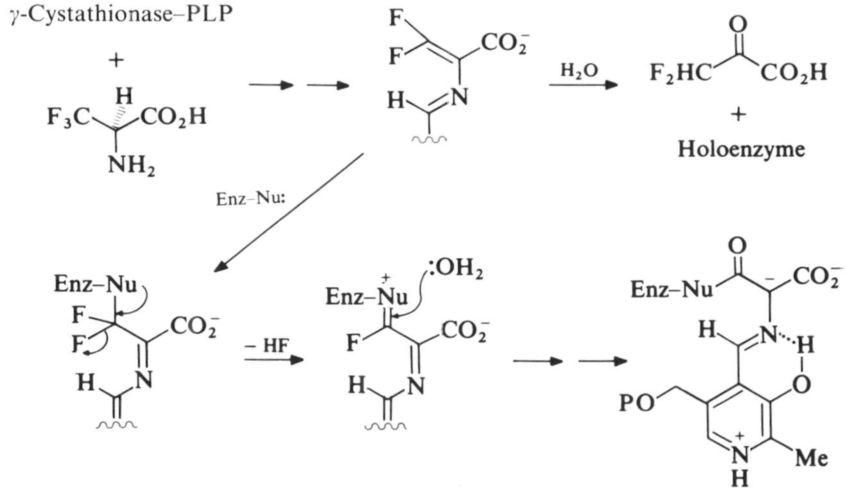

Scheme 41

Unfortunately we do not know much about any single class of enzymes with maybe the exception of the α-amino acid α-transaminases and specifically AAT. From the viewpoint of a medicinal chemist, AAT is unlikely to be a useful target enzyme but much can be learned about specific drug design in general from these relatively well studied systems. Given that studies of the mechanism of enzymes using techniques which have revolutionalized the area have not yet filtered through to the literature, the coming few years will tell whether a better understanding of enzyme structure and mechanism leads to the efficient rational design of new inhibitors and drugs. Considering the importance of the pyridoxal area and its well established role as a major testing ground for mechanism-based inhibitor design, the next few years are likely to be particularly exciting.

6.5.5 REFERENCES

1. E. F. Gale and H. M. R. Epps, *Biochem. J.*, 1944, **38**, 232.
2. W. D. Bellamy and I. C. Gunsalus, *J. Bacteriol.*, 1943, **46**, 573.
3. E. F. Gale, *Biochem. J.*, 1940, **34**, 846.
4. E. F. Snell, *J. Am. Chem. Soc.*, 1945, **67**, 194.
5. D. Dolphin, R. Poulson and O. Avramovic (eds.), 'Vitamin B$_6$; Pyridoxal Phosphate, Chemical, Biochemical and Medical Aspects', Wiley, New York, 1986.
6. P. Christen and D. E. Metzler (eds.), 'Transaminases', Wiley, New York, 1985.
7. A. E. Evangelopolous (ed.), 'Chemical and Biological Aspects of Vitamin B$_6$ Catalysis', Parts A and B, Liss, New York, 1984.
8. M. Akhtar, V. C. Emery and J. A. Robinson, in 'The Physical and Organic Basis of Biochemistry', ed. M. I. Page, Elsevier, Amsterdam, 1984, p. 303.
9. H. G. Floss and J. C. Vederas, in 'Stereochemistry', ed. C. Tamm, Elsevier, Amsterdam, 1982, p. 161.
10. A. E. Martell, *Adv. Enzymol.*, 1982, **53**, 163.
11. H. C. Dunathan, *Proc. Natl. Acad. Sci. USA*, 1966, **55**, 712.
12. H. C. Dunathan, *Adv. Enzymol.*, 1971, **35**, 79.
13. L. Davis and D. E. Metzler, in 'The Enzymes', ed. P. D. Boyer, Academic Press, New York, 1972, vol. 7, p. 33.
14. P. Besmer and D. Arigoni, *Chimia*, 1969, **23**, 190.
15. H. C. Dunathan, L. Davis, P. G. Kury and M. Kaplan, *Biochemistry*, 1968, **7**, 4532.
16. D. Arigoni, as cited in ref. 9.
17. J. E. Ayling, H. C. Dunathan and E. E. Snell, *Biochemistry*, 1968, **7**, 4537.
18. G. B. Bailey, T. Kusamram and K. Vuttivej, *Fed. Proc., Fed. Am. Soc. Exp. Biol.*, 1970, **29**, 857.
19. E. Austermuhle-Bertola, Ph.D. Dissertation, 1973, ETH Zurich.
20. G. C. Ford, G. Eichele and J. N. Jansonius, *Proc. Natl. Acad. Sci. USA*, 1980, **77**, 2559.
21. A. Arnone, P. H. Roger, C. C. Hyde, P. D. Briley, C. M. Metzler and D. E. Metzler, in ref. 6, p. 138.
22. J. N. Jansonius, G. Eichele, G. C. Ford, D. Picot, C. Thaller and M. G. Vincent, in ref. 6, p. 110.
23. J. F. Kirsch, G. Eichele, G. C. Ford, M. G. Vincent, J. N. Jansonius, H. Gehring and P. Christen, *J. Mol. Biol.*, 1984, **174**, 497.
24. V. V. Borisov, S. N. Bomsova, G. S. Kachalova, N. I. Sosfenov and B. K. Vainshtein, in ref. 6, p. 155.
25. F. S. Furbish, M. L. Fonda and D. E. Metzler, *Biochemistry*, 1969, **8**, 5169.
26. C. S. Frederiuk and J. A. Shafer, *J. Biol. Chem.*, 1983, **258**, 5372.
27. S.-M. Fang, H. J. Rhodes and M. I. Blake, *Biochim. Biophys. Acta*, 1970, **212**, 281.
28. J. F. Kirsch and D. A. Julin, *Fed. Proc., Fed. Am. Soc. Exp. Biol.*, 1982, **41**, 628.
29. A. Arnone, P. Christen, J. N. Jansonius and D. E. Metzler, in ref. 6, p. 326.
30. S. W. Zito and M. Martinez-Carrion, *J. Biol. Chem.*, 1980, **255**, 8645.
31. U. Graf-Hausner, K. J. Wilson and P. Christen, *J. Biol. Chem.*, 1983, **258**, 8813.
32. D. Barra, F. Martini, G. Montarani, S. Doonan and F. Bossa, *FEBS Lett.*, 1979, **108**, 103.
33. D. Barra, F. Bossa, S. Doonan, H. M. A. Fahmy, G. J. Hughes, F. Martini, R. Petruzzelli and B. Wittmann-Liebold, *Eur. J. Biochem.*, 1980, **108**, 405; H. Kagamiyama, R. Sakakibara, S. Tanase, Y. Morino and H. Wade, *J. Biol. Chem.*, 1980, **255**, 6155.
34. Q. K. Huynh, R. Sakakibara, T. Watanabe and H. Wada, *Biochem. Biophys. Res. Commun.*, 1980, **97**, 474; J. R. Mattingley Jr, F. J. Rodriguez-Berrocal, J. Gordon, A. Iriarte, M. Martinez-Carrion, *Biochem. Biophys. Res. Commun.*, 1987, **149**, 859.
35. F. Martini, S. Angelaccio, D. Barra, S. Pascarella, B. Maras, S. Doonan and F. Bossa, *Biochim. Biophys. Acta*, 1985, **832**, 46.
36. S. V. Shlyapnikov, A. N. Myasnikov, E. S. Severin, M. A. Myagkova, Y. M. Torchinsky and A. E. Braunstein, *FEBS Lett.*, 1979, **106**, 385.
37. S. Doonan, H. J. Doonan, R. Hanford, C. A. Vernon, J. M. Walker, L. P. S. Airoldi, F. Bossa, D. Barra, M. Carloni, P. Fasella and F. Riva, *Biochem. J.*, 1975, **149**, 497.
38. K. Kondo, S. Wakabayashi, T. Yagi and H. Kagamiyama, *Biochem. Biophys. Res. Commun.*, 1984, **122**, 62.
39. S. Kuramitsu, S. Okumo, T. Ogawa, H. Ogawa and H. Kagamiyama, *J. Biochem. (Tokyo)*, 1985, **97**, 1259.
40. S. Doonan, F. Martini, S. Angelaccio, S. Pascarella, D. Barra and F. Bossa, *J. Mol. Evol.*, 1986, **23**, 328.
41. E. Schiltz and W. Schmitt, *FEBS Lett.*, 1981, **134**, 57.
42. T. Grange, C. Guenet, J. G. Dietrich, S. Chasserot, M. Fromont, N. Befort, J. Jami, G. Beck and R. Pictet, *J. Mol. Biol.*, 1985, **184**, 347.
43. S. Kuramitsu, T. Ogawa, H. Ogawa and H. Kagamiyama, *J. Biochem. (Tokyo)*, 1985, **97**, 993.
44. M. M. Mueckler and H. C. Pitot, *J. Biol. Chem.*, 1985, **260**, 12 993.
45. S. Kuramitsu, K. Inoue, T. Ogawa, H. Ogawa and H. Kagamiyama, *Biochem. Biophys. Res. Commun.*, 1985, **133**, 134.
46. C. N. Cronin, B. A. Malcolm and J. F. Kirsch, *J. Am. Chem. Soc.*, 1987, **109**, 2222.
47. D. Ringe and D. L. Smith, ACS Meeting, New Orleans, 1987, Abstr. No. 87; J. R. Knowles, *Science (Washington, DC)*, 1987, **236**, 1252.
48. R. A. John and L. J. Fowler, in ref. 6, p. 413.
49. K. Soda, in ref. 6, p. 421.
50. H. Yamada and M. H. O'Leary, *Biochemistry*, 1978, **17**, 669.

51. G. Burnett, C. T. Walsh, K. Yonaka, S. Toyama and K. Soda, *J. Chem. Soc., Chem. Commun.*, 1979, 826.
52. M. Bouclier, M. J. Jung and B. Lippert, *Eur. J. Biochem.*, 1979, **98**, 363.
53. D. J. Aberhart and D. J. Russell, *J. Am. Chem. Soc.*, 1984, **106**, 4902.
54. D. J. Aberhart and D. J. Russell, *J. Am. Chem. Soc.*, 1984, **106**, 4907.
55. K. Tanizawa, T. Yoshimura, Y. Asada, S. Sawada, H. Misono and K. Soda, *Biochemistry*, 1982, **21**, 1104.
56. P. Mandel and F. V. Defeudis (eds.), 'GABA-Biochemistry and CNS Functions', Plenum Press, New York, 1979.
57. R. W. Olsen, *Annu. Rev. Pharmacol. Toxicol.*, 1982, **22**, 245.
58. D. R. Curtis and G. A. R. Johnston, *Ergebn. Physiol.*, 1974, **69**, 98.
59. L. L. Iverson, in 'Biochemical Psychopharmacology of GABA', ed. M. A. Lipton, A. D. Marcio and K. F. Killam, Raven Press, New York, 1978, p. 25.
60. P. R. Schofield, M. G. Darlison, N. Fujita, D. R. Burt, F. A. Stevenson, H. Rodriguez, L. M. Rhee, J. Ramachandran, V. Reale, T. A. Glencorse, P. H. Seeburg and E. A. Barnard, *Nature (London)*, 1987, **328**, 221.
61. H. G. Dohlman, M. G. Caron and R. F. Lefkowitz, *Biochemistry*, 1987, **26**, 2657.
62. R. A. F. Dixon, I. S. Sigal, M. R. Candelore, R. B. Register, W. Scattergood, E. Rands and C. D. Strader, *EMBO J.*, 1987, **6**, 3269.
63. M. Maitre, L. Ossola and P. Mandel, in ref. 56, p. 3.
64. T. Beeler and J. E. Churchich, *Eur. J. Biochem.*, 1978, **85**, 365.
65. J. E. Churchich and U. Moses, *J. Biol. Chem.*, 1981, **256**, 1101.
66. D. S. Kim and J. E. Churchich, *Biochem. Biophys. Res. Commun.*, 1981, **99**, 1333.
67. L. J. Fowler and R. A. John, as cited in ref. 48.
68. Y. Sanada, T. Shiotani, E. Okuno and N. Katanuma, *Eur. J. Biochem.*, 1976, **69**, 507.
69. C. C. Kalita, J. D. Kerman and H. J. Strecker, *Biochim. Biophys. Acta*, 1976, **429**, 780.
70. Z. Markovic-Houstey, M. Kania, A. Lustig, M. G. Vincent, J. N. Jansonius and R. A. John, *Eur. J. Biochem.*, 1987, **162**, 345.
71. G. Inana, S. Totsuka, M. Redmond, T. Dougherty, J. Nagle. T. Shiono, T. Ohura, F. Kominami and N. Katunuma, *Proc. Natl. Acad. Sci. USA*, 1986, **83**, 1203.
72. M. Simmaco, R. A. John, D. Barra and F. Bossa, *FEBS Lett.*, 1986, **199**, 39.
73. J. J. Corrigan, *Science (Washington, DC)*, 1969, **164**, 142.
74. J. I. Durham, P. W. Morgan, J. W. Prescott and C. M. Lyman, *Phytochemistry*, 1973, **12**, 2123.
75. K. Soda, K. Yonaha, H. Misono and M. Osugi, *FEBS Lett.*, 1974, **46**, 359.
76. K. Yonaha, H. Misono and K. Soda, *FEBS Lett.*, 1975, **55**, 265.
77. T. S. Soper, H. Veno and J. M. Manning, *Arch. Biochem. Biophys.*, 1985, **240**, 1.
78. D. Voss, J. Gerdee and E. Leistner, *Phytochemistry*, 1985, **24**, 1471.
79. E. Santaniello, M. G. Kienle, A. Manzocchi and E. Bosisio, *J. Chem. Soc., Perkin Trans. 1*, 1979, 1677.
80. A. R. Battersby, R. Joyeau and J. Staunton, *FEBS Lett.*, 1979, **107**, 231.
81. A. R. Battersby, M. Nicoletti, J. Staunton and R. Vleggaar, *J. Chem. Soc., Perkin Trans. 1*, 1980, 43.
82. A. R. Battersby, E. J. T. Chrystal and J. Staunton, *J. Chem. Soc., Perkin Trans. 1*, 1980, 31.
83. A. R. Battersby, R. Murphy and J. Staunton, *J. Chem. Soc., Perkin Trans. 1*, 1982, 449.
84. G. R. Orr and S. J. Gould, *Tetrahedron Lett.*, 1982, **23**, 3139.
85. J. C. Richards and I. D. Spenser, *Can. J. Chem.*, 1982, **60**, 2810.
86. D. E. Stevenson, M. Akhtar and D. Gani, *Tetrahedron Lett.*, 1986, **27**, 5661.
87. J. W. Thanassi and J. S. Fruton, *Biochemistry*, 1961, **1**, 975.
88. Y. Asada, K. Tanizaka, S. Sawada, T. Suzuki, H. Misano and K. Soda, *Biochemistry*, 1981, **20**, 6881.
89. J. G. Kelland, M. M. Palcic, M. A. Pickard and J. C. Vederas, *Biochemistry*, 1985, **24**, 3263.
90. Y. Asuda, K. Tanizawa, Y. Kawabata, H. Misono and K. Soda, *Agric. Biol. Chem.*, 1981, **45**, 1513.
91. A. G. Palekar, S. S. Tate and A. Meister, *Biochemistry*, 1970, **9**, 2310.
92. S. S. Tate and A. Meister, *Biochemistry*, 1969, **8**, 1660.
93. S. S. Tate and A. Meister, *Adv. Enzymol.*, 1971, **35**, 503.
94. A. G. Palekar, S. S. Tate and A. Meister, *Biochemistry*, 1971, **10**, 2180.
95. C.-C. Chang, A. Laghai, M. H. O'Leary and H. G. Floss, *J. Biol. Chem.*, 1982, **257**, 3564.
96. R. M. Rosenberg and M. H. O'Leary, *Biochemistry*, 1985, **24**, 1598.
97. S. S. Tate and A. Meister, *Biochemistry*, 1969, **8**, 1056.
98. M. M. Palcic, M. Antoun, K. Tanizawa, K. Soda and H. G. Floss, *J. Biol. Chem.*, 1985, **260**, 5248.
99. P. Y. Sze, in ref. 56, p. 59.
100. J. Y. Wu, *J. Neurochem.*, 1977, **28**, 1359.
101. L. A. Denner and J. Y. Wu, *J. Neurochem.*, 1985, **44**, 957.
102. J. Y. Wu, T. Matsuda and E. Roberts, *J. Biol. Chem.*, 1973, **248**, 3029.
103. P. H. Strausbauch and E. H. Fischer, *Biochemistry*, 1970, **9**, 226.
104. J. Y. Wu, *Proc. Natl. Acad. Sci. USA*, 1982, **79**, 4270.
105. L. A. Denner, S. C. Wei, H. S. Lin, C. T. Lin and J. Y. Wu, *Proc. Natl. Acad. Sci. USA*, 1987, **84**, 668.
106. J.-Y. Wu, L. Denner, C. T. Lin and G. Song, *Methods Enzymol.*, 1985, **113**, 3.
107. M. L. Fonda, *Methods Enzymol.*, 1985, **113**, 11.
108. J. F. Julien, F. Legay, S. Dumas, M. Tappaz and J. Mallet, *Neurosci. Lett.*, 1987, **73**, 173.
109. D. L. Kaufman, J. F. McGinnis, N. R. Krieger and A. J. Tobin, *Science (Washington, DC)*, 1986, **232**, 1138.
110. S. Y. Choi and J. E. Churchich, *Eur. J. Biochem.*, 1986. **160**, 515.
111. O. Heby, *Adv. Enzyme Regul.*, 1985, **24**, 103.
112. M. F. Obenrader and W. F. Prouty, *J. Biol. Chem.*, 1977, **252**, 2860.
113. J. E. Seely, H. Poso and A. E. Pegg, *Biochemistry*, 1982, **21**, 3394.
114. M. K. Haddox and D. H. Russell, *Biochemistry*, 1981, **20**, 6721.
115. T. S. Sklaviadis, J. G. Georgatsos and D. A. Kyriakidis, *Biochim. Biophys. Acta*, 1985, **831**, 288.
116. M. Gupta and P. Coffino, *J. Biol. Chem.*, 1985, **260**, 2941.
117. C. Kahana and D. Nathans, *Proc. Natl. Acad. Sci. USA*, 1985, **82**, 1673.
118. D. Applebaum, D. L. Sabo, E. H. Fischer and D. R. Morris, *Biochemistry*, 1975, **14**, 3675.

119. Y. Morino and F. Nagashima, *Methods Enzymol.*, 1984, **106**, 116.
120. 'International Encyclopedia of Pharmacology and Therapeutics', Section 74, 'Histamines and Antihistamines', ed. M. Schachter, Pergamon, Oxford, 1973.
121. L. Hammar and S. Hjerten, *Agents Actions*, 1980, **10**, 93.
122. Y. Taguchi, T. Watanabe, H. Kubota, H. Hayashi and H. Wada, *J. Biol. Chem.*, 1984, **259**, 5214.
123. P. A. Recsei, W. M. Moore and E. E. Snell, *J. Biol. Chem.*, 1983, **258**, 439.
124. P. A. Recsei and E. E. Snell, *Annu. Rev. Biochem.*, 1984, **53**, 357.
125. Q. K. Huynh and E. E. Snell, *J. Biol. Chem.*, 1985, **260**, 2798.
126. S. Tanase, B. M. Guirard and E. E. Snell, *J. Biol. Chem.*, 1985, **260**, 6738.
127. H. Hayashi, S. Tanase and E. E. Snell, *J. Biol. Chem.*, 1986, **261**, 11003.
128. G. L. Vaaler, M. A. Brasch and E. E. Snell, *J. Biol. Chem.*, 1986, **261**, 11010.
129. M. J. Jung, *Bioorg. Chem.*, 1986, **14**, 429.
130. M. Ando-Yamamato, H. Hayashi, Y. Taguchi, H. Fukui, T. Watanabe and H. Wada, *Biochem. Biophys. Res. Commun.*, 1986, **141**, 306.
131. C. Borri Voltattorni, A. Minelli, C. Cirotto, D. Barra and C. Turano, *Arch. Biochem. Biophys.*, 1982, **217**, 58.
132. P. Dominici, B. Tancini and C. Borri Voltattorni, *J. Biol. Chem.*, 1985, **260**, 10583.
133. B. Tancini, P. Dominici, D. Barra and C. Borri Voltattorni, *Arch. Biochem. Biophys.*, 1985, **238**, 565.
134. P. Dominici, B. Tancini and C. Borri Voltattorni, *Arch. Biochem. Biophys.*, 1986, **251**, 762.
135. M. Ando-Yamamoto, H. Hayashi, T. Sugiyama, H. Fukui, T. Watanabe and H. Wada, *J. Biochem.*, 1987, **101**, 405.
136. D. D. Eveleth, R. D. Gietz, C. A. Spencer, F. E. Nargang, R. B. Hodgetts and J. L. Marsh, *EMBO J.*, 1986, **5**, 2663.
137. J. T. Park, *Symp. Soc. Gen. Microbiol.*, 1958, **8**, 49.
138. L. L. Henderson and R. B. Johnston, *Biochem. Biophys. Res. Commun.*, 1976, **68**, 793.
139. L. M. Fisher, W. J. Albery and J. R. Knowles, *Biochemistry*, 1986, **25**, 2529.
140. L. M. Fisher, W. J. Albery and J. R. Knowles, *Biochemistry*, 1986, **25**, 2538.
141. L. M. Fisher, J. G. Belasco, T. W. Bruice, W. J. Albery and J. R. Knowles, *Biochemistry*, 1986, **25**, 2543.
142. J. G. Belasco, W. J. Albery and J. R. Knowles, *Biochemistry*, 1986, **25**, 2552.
143. J. G. Belasco, T. W. Bruice, W. J. Albery and J. R. Knowles, *Biochemistry*, 1986, **25**, 2558.
144. J. G. Belasco, T. W. Bruice, L. M. Fisher, W. J. Albery and J. R. Knowles, *Biochemistry*, 1986, **25**, 2564.
145. D. Gani, *Annu. Rep. Prog. Chem., Sect. B*, 1985, **82**, 287; 1986, **83**, 303.
146. J. S. Wiseman and J. S. Nichols, *J. Biol. Chem.*, 1984, **259**, 8907.
147. E. Adams, *Adv. Enzymol.*, 1976, **44**, 69.
148. R. B. Johnston, E. C. Schreiber, M. P. Davis, L. Jillson, W. T. Sorrell and M. E. Kirker, in ref. 7, p. 339.
149. E. Wang and C. T. Walsh, *Biochemistry*, 1978, **17**, 1313.
150. S. Shen, H. G. Floss, H. Kumagai, H. Yamada, N. Esaki, K. Soda, S. A. Wasserman and C. T. Walsh, *J. Chem. Soc., Chem. Commun.*, 1983, 82.
151. C. T. Walsh, R. Badet, E. Daub, N. Esaki and N. Galakatos, *Spec. Publ. Roy. Soc. Chem.*, 1986, **55**, 193.
152. S. A. Wasserman, E. Daub, P. Grisafi, D. Botstein and C. T. Walsh, *Biochemistry*, 1984, **23**, 5182.
153. N. Esaki and C. T. Walsh, *Biochemistry*, 1986, **25**, 3261.
154. N. G. Galakatos, E. Daub, D. Botstein and C. T. Walsh, *Biochemistry*, 1986, **25**, 3255.
155. B. Badet, D. Roise and C. T. Walsh, *Biochemistry*, 1984, **23**, 5188.
156. D. Roise, K. Soda, T. Yagi and C. T. Walsh, *Biochemistry*, 1984, **23**, 5195.
157. B. Badet and C. T. Walsh, *Biochemistry*, 1984, **24**, 1333.
158. K. Inagaki, K. Tanizawa, B. Badet, C. T. Walsh, H. Tanaka and K. Soda, *Biochemistry*, 1986, **25**, 3268.
159. B. Badet, K. Inagaki, K. Soda and C. T. Walsh, *Biochemistry*, 1986, **25**, 3275.
160. K. Soda, H. Tanaka and K. Tanizawa, in ref. 5, p. 223.
161. T. Yorifuji, K. Ogata and K. Soda, *J. Biol. Chem.*, 1971, **246**, 5085.
162. S. A. Ahmed, N. Esaki and K. Soda, *FEBS Lett.*, 1982, **150**, 370.
163. N. Nakajima, K. Tanizawa, H. Tanaka and K. Soda, *Agric. Biol. Chem.*, 1986, **50**, 2823.
164. K. Inagaki, K. Tanizawa, H. Tanaka and K. Soda, *Agric. Biol. Chem.*, 1987, **51**, 173.
165. L. Schirch and T. Gross, *J. Biol. Chem.*, 1968, **243**, 5651.
166. M. Akhtar, H. A. El-Obeid and P. M. Jordan, *Biochem. J.*, 1975, **145**, 159.
167. R. J. Ulevitch and R. G. Kallen, *Biochemistry*, 1977, **16**, 5342.
168. R. J. Ulevitch and R. G. Kallen, *Biochemistry*, 1977, **16**, 5350.
169. R. J. Ulevitch and R. G. Kallen, *Biochemistry*, 1977, **16**, 5355.
170. A. G. Palekar, S. S. Tate and A. Meister, *J. Biol. Chem.*, 1973, **248**, 1158.
171. J. G. Voet, D. M. Hindenlang, T. J. J. Blanck, R. J. Ulevitch, R. G. Kallen and H. C. Dunathan, *J. Biol. Chem.*, 1973, **248**, 841.
172. L. Schirch and W. T. Jenkins, *J. Biol. Chem.*, 1964, **239**, 3797.
173. L. Schirch and W. T. Jenkins, *J. Biol. Chem.*, 1964, **239**, 3801.
174. J. Hansen and L. Davis, *Biochim. Biophys. Acta*, 1979, **568**, 321.
175. C. M. Tatum, P. A. Benkovic, S. J. Benkovic, R. Potts, E. Schleicher and H. G. Floss, *Biochemistry*, 1977, **16**, 1093.
176. L. J. Slieker and S. J. Benkovic, *J. Am. Chem. Soc.*, 1984, **106**, 1833.
177. J. R. Biellmann and F. Schuber, *Biochem. Biophys. Res. Commun.*, 1967, **27**, 517; *Bull. Soc. Chim. Biol.*, 1970, **52**, 211.
178. C. Tatum, J. Vederas, E. Schleicher, S. J. Benkovic and H. G. Floss, *J. Chem. Soc., Chem. Commun.*, 1977, 218.
179. L. Schirch and A. Diller, *J. Biol. Chem.*, 1971, **246**, 3961.
180. W.-M. Ching and R. G. Kallen, *Biochemistry*, 1979, **18**, 821.
181. L. Schirch and D. Peterson, *J. Biol. Chem.*, 1980, **255**, 7801.
182. V. Schirch, S. Hopkins, E. Villar and S. Angelaccio, *J. Bacteriol.*, 1985, **163**, 1.
183. L. Schirch, F. Gavilanes, D. Peterson, B. Bullis, D. Barra and F. Bossa, in ref. 7, p. 301.
184. D. Barra, F. Martini, S. Angelaccio, F. Bossa, F. Gavilanes, D. Peterson, B. Bullis and L. Schirch, *Biochem. Biophys. Res. Commun.*, 1983, **116**, 1007.
185. M. D. Plamann, L. T. Stauffer, M. W. Urbanowski and G. V. Stauffer, *Nucleic Acids Res.*, 1983, **11**, 2065.
186. F. Martini, S. Angelaccio, S. Pascarella, D. Barra, F. Bossa and V. Schirch, *J. Biol. Chem.*, 1987, **262**, 5499.

187. S. Hopkins and V. Schirch, *J. Biol. Chem.*, 1986, **261**, 3363.
188. D. N. Rao and N. A. Rao, *Plant Physiol.*, 1982, **69**, 11.
189. L. Schirch, *Adv. Enzymol.*, 1982, **53**, 83.
190. E. W. Miles, *Adv. Enzymol.*, 1979, **49**, 127.
191. A. E. Braunstein and E. V. Goryachenkova, *Adv. Enzymol.*, 1984, **56**, 1.
192. E. W. Miles, in ref. 5, p. 253.
193. P. Bartholmes, in ref. 7, p. 309.
194. M. D. Tsai, E. Schleicher, R. Potts, G. E. Skye and H. G. Floss, *J. Biol. Chem.*, 1978, **253**, 5344.
195. E. W. Miles, D. R. Houck and H. G. Floss, *J. Biol. Chem.*, 1982, **257**, 14 203.
196. E. W. Miles, *Biochemistry*, 1987, **26**, 597.
197. E. W. Miles, R. S. Phillips, H. J. C. Yeh and L. A. Cohen, *Biochemistry*, 1986, **25**, 4240.
198. S. A. Ahmed, B. Martin and E. W. Miles, *Biochemistry*, 1986, **25**, 4233.
199. R. S. Phillips, E. W. Miles and L. A. Cohen, *J. Biol. Chem.*, 1985, **260**, 14 665.
200. P. D. Turner, H. C. Loughrey and C. J. Bailey, *Biochim. Biophys. Acta*, 1985, **832**, 280.
201. J. C. Vederas, E. Schleicher, M.-D. Tsai and H. G. Floss, *J. Biol. Chem.*, 1978, **253**, 5350.
202. E. Schleicher, K. Mascaro, R. Potts, D. R. Mann and H. G. Floss, *J. Am. Chem. Soc.*, 1976, **98**, 1043.
203. N. Esaki, T. Nakamura, H. Tanaka and K. Soda, *J. Biol. Chem.*, 1982, **257**, 4386.
204. K. Soda, N. Esaki, T. Nakamur, N. Karai, P. Chocat and H. Tanaka, in ref. 7, p. 319.
205. T. C. Stadtman, *Adv. Enzymol.*, 1979, **48**, 1.
206. N. Esaki, N. Karai, H. Tanaka and K. Soda, in ref. 7, p. 329.
207. P. Chocat, N. Esaki, K. Tanizawa, K. Nakamura, H. Tanaka and K. Soda, *J. Bacteriol.*, 1985, **163**, 669.
208. J. E. Churchich, in ref. 5, p. 311.
209. M. M. Kaplan and M. Flavin, *J. Biol. Chem.*, 1966, **241**, 5781.
210. B. I. Posner and M. Flavin, *J. Biol. Chem.*, 1972, **247**, 6412.
211. M. Krongelb, T. A. Smith and R. H. Abeles, *Biochim. Biophys. Acta*, 1968, **167**, 473.
212. C. Fuganti and D. Coggiola, *Experientia*, 1977, **33**, 847.
213. M. Johnston, P. Marcotte, J. Donovan and C. T. Walsh, *Biochemistry*, 1979, **18**, 1729.
214. M. N. T. Chang and C. T. Walsh, *J. Am. Chem. Soc.*, 1981, **103**, 4921.
215. C. Fuganti, *J. Chem. Soc., Chem. Commun.*, 1979, 337.
216. S. Komatsubara, M. Kisumi, I. Chibata, M. M. V. Gregorio, U. S. Muller and D. H. G. Crout, *J. Chem. Soc., Chem. Commun.*, 1977, 839.
217. D. H. G. Crout, M. M. V. Gregorio, U. S. Muller, S. Komatsubara, M. Kisumi and I. Chibata, *Eur. J. Biochem.*, 1980, **106**, 97.
218. C. T. Walsh, R. A. Pascal, M. Johnston, R. Raines, D. Dikshit, A. Krantz and M. Honma, *Biochemistry*, 1981, **20**, 7509.
219. R. K. Hill, S. R. Prakash, R. Wiesendanger, W. Angst, B. Martinoni, D. Arigoni, H. W. Liu and C. T. Walsh, *J. Am. Chem. Soc.*, 1984, **106**, 795.
220. H. W. Liu, R. Auchus and C. T. Walsh, *J. Am. Chem. Soc.*, 1984, **106**, 5335.
221. N. B. Madsen and S. G. Withers, in ref. 5, p. 355, and ref. 7, p. 117.
222. L. N. Johnson, J. A. Jenkins, K. S. Wilson, E. A. Stura and G. Zanotti, *J. Mol. Biol.*, 1980, **140**, 565.
223. S. Shimonura and T. Fukui, *Biochemistry*, 1978, **17**, 5359.
224. M. Takagi, T. Fukui and S. Shimomura, *Proc. Natl. Acad. Sci. USA*, 1982, **79**, 3716.
225. H. W. Klein and E. J. M. Helmreich, *FEBS Lett.*, 1979, **108**, 209.
226. H. W. Klein, D. Palm and E. J. M. Helmreich, *Biochemistry*, 1982, **21**, 6675.
227. M. S. P. Sansom, E. A. Stura, Y. S. Babu, P. McLaughlin and L. N. Johnson, in ref. 7, p. 127.
228. R. Yamada, Y. Wakabayashi, A. Iwashima and T. Hasegawa, *Biochim. Biophys. Acta*, 1986, **871**, 279.
229. G. G. Hammes and J. L. Haslam, *Biochemistry*, 1968, **7**, 1519.
230. C. T. Walsh, in ref. 5, p. 43.
231. M. Johnston and C. T. Walsh, in 'Molecular Basis for Drug Action', ed. T. Singer and R. Ondarza, Elsevier, Amsterdam, 1981, p. 167.
232. C. Danzin and M. J. Jung, in ref. 7, p. 377.
233. C. M. Michuda and M. Martinez-Carrion, *J. Biol. Chem.*, 1970, **245**, 262.
234. R. A. John and P. Fasella, *Biochemistry*, 1969, **8**, 4477.
235. A. J. L. Cooper, S. M. Fitzpatrick, C. Kaufman and P. Dowd, *J. Am. Chem. Soc.*, 1982, **104**, 332.
236. N. P. Botting, M. Akhtar, D. M. Smith and D. Gani, 1988, unpublished results.
237. J. M. Manning, R. M. Khomutov and P. Fasella, *Eur. J. Biochem.* 1968, **5**, 199.
238. H. Ueno, J. J. Likos and D. E. Metzler, *Biochemistry*, 1982, **21**, 4387.
239. R. R. Rando, N. Relyea and L. Cheng, *J. Biol. Chem.*, 1976, **251**, 3306.
240. R. R. Rando, *Biochemistry*, 1974, **13**, 3859.
241. H. Gehring, R. R. Rando and P. Christen, *Biochemistry*, 1977, **16**, 4832.
242. Y. Morino and M. Okamoto, *Biochem. Biophys. Res. Commun.*, 1973, **50**, 1061.
243. Y. Morino, A. M. Osman and M. Okamoto, *J. Biol. Chem.*, 1974, **249**, 6684.
244. S. Tanase and Y. Morino, *Biochem. Biophys. Res. Commun.*, 1976, **68**, 1301.
245. G. Burnett, P. Marcotte and C. T. Walsh, *J. Biol. Chem.*, 1980, **255**, 3487.
246. G. Burnett, K. Yonaha, S. Toyama, K. Soda and C. T. Walsh, *J. Biol. Chem.*, 1980, **255**, 428.
247. R. M. Khomutov, E. S. Severin, G. K. Kovaleva, N. N. Gulyaev, N. V. Gnuchev and L. P. Sastchenko, in 'Pyridoxal Catalysis: Enzymes and Model Systems', ed. E. E. Snell, A. E. Braunstein, E. S. Severin and Y. M. Torchinsky, Wiley, New York, 1968, p. 631.
248. G. K. Kovaleva, E. S. Severin, P. Fasella and R. M. Khomutov, *Biokhimiya*, 1973, **38**, 365.
249. G. K. Kovaleva and E. S. Severin, *Biokhimiya*, 1972, **37**, 1282.
250. A. E. Braunstein, in 'The Enzymes', 3rd edn., ed. P. D. Boyer, Academic Press, New York, p. 379.
251. R. W. Olsen and J. C. Venter (eds.), 'Benzodiazepine/GABA Receptors and Chloride Channels: Structural and Functional Properties', Liss, New York, 1986.
252. L. J. Fowler and R. A. John, *Biochem. J.*, 1972, **130**, 569.

253. M. J. Jung and B. W. Metcalf, *Biochem. Biophys. Res. Commun.*, 1975, **67**, 301.
254. B. Lippert, B. W. Metcalf, M. J. Jung and P. Casara, *Eur. J. Biochem.*, 1977, **74**, 441.
255. P. J. Schechter and Y. Tranier, in 'Enzyme Activated Irreversible Inhibitors', ed. N. Seiler, M. H. Jung and J. Koch-Weser, Elsevier, Amsterdam, 1978, p. 149.
256. M. Palfreyman, P. J. Schechter, W. Buckett, G. P. Tell and J. Koch-Weser, *Biochem. Pharmacol.*, 1981, **30**, 817.
257. A. Sjoerdsma, *Clin. Pharmacol. Ther.*, 1981, **30**, 3.
258. P. J. Schechter, Y. Tranier and J. Grove, in ref. 56, p. 43.
259. R. A. John, E. D. Jones and L. J. Fowler, *Biochem. J.*, 1979, **177**, 721.
260. R. B. Silverman and M. A. Levy, *Biochemistry*, 1981, **20**, 1197.
261. T. S. Soper and J. M. Manning, in ref. 6, p. 266.
262. N. T. Buu and N. M. Val Gelder, *Br. J. Pharmacol.*, 1974, **52**, 401.
263. R. B. Silverman and M. A. Levy, *J. Biol. Chem.*, 1981, **256**, 11 565.
264. R. Rando and F. W. Bangenter, *J. Am. Chem. Soc.*, 1976, **98**, 6762.
265. R. Rando, *Biochemistry*, 1977, **16**, 4604.
266. T. S. Soper and J. M. Manning, *J. Biol. Chem.*, 1981, **256**, 4263.
267. T. S. Soper and J. M. Manning, *J. Biol. Chem.*, 1982, **257**, 13 930.
268. R. B. Silverman and B. J. Invergo, *Biochemistry*, 1986, **25**, 6817.
269. R. B. Silverman, S. C. Durkee and B. J. Invergo, *J. Med. Chem.*, 1986, **29**, 764.
270. R. B. Silverman, B. J. Invergo, M. A. Levy and C. R. Andrew, *J. Biol. Chem.*, 1987, **262**, 3192.
271. T. A. Alston, D. J. T. Porter and H. J. Bright, *J. Enzyme Inhib.*, 1987, **1**, 215.
272. T. A. Alston and H. J. Bright, *FEBS Lett.*, 1981, **126**, 269.
273. K. Kito, Y. Sanada and N. Kutunama, *J. Biochem.*, 1978, **83**, 201.
274. T. S. Soper, W. M. Jones, B. Lerner, M. Trop and J. M. Manning, *J. Biol. Chem.*, 1977, **252**, 3170.
275. K. Soda and N. Esaki, in ref. 6, p. 463.
276. T. S. Soper, J. M. Manning, P. A. Marcotte and C. T. Walsh, *J. Biol. Chem.*, 1977, **252**, 1571.
277. R. A. John, in 'Enzyme Inhibitors as Drugs', ed. M. Sandler, Macmillan Press, London, 1980, p. 73.
278. J. J. Likos, H. Ueno, R. W. Feldhaus and D. E. Metzler, *Biochemistry*, 1982, **21**, 4377.
279. A. Vidal-Cros, M. Gaudry and A. Marquet, *Biochem. J.*, 1985, **229**, 675.
280. C. Danzin, N. Claverie and M. J. Jung, *Biochem. Pharmacol.*, 1984, **33**, 1741.
281. M. J. Jung, J. Koch-Weser and A. Sjoerdsma, in ref. 277, p. 95.
282. M. J. Jung and J. Koch-Weser, in ref. 231, p. 135.
283. C. Danzin, P. Casara, N. Claverie and B. W. Metcalf, *J. Med. Chem.*, 1981, **24**, 16.
284. B. W. Metcalf, P. Bey, C. Danzin, M. Jung, P. Casara and J. Vevert, *J. Am. Chem. Soc.*, 1978, **100**, 2551.
285. N. Relyea and R. R. Rando, *Biochem. Biophys. Res. Commun.*, 1975, **67**, 392.
286. C. Danzin, P. Bey, D. Schirlin and N. Claverie, *Biochem. Pharmacol.*, 1982, **31**, 3871.
287. J. Kollonitsch, A. A. Patchett, S. Marburg, A. L. Maycock, L. M. Perkins, G. A. Doldouras, D. E. Duggan and S. D. Aster, *Nature (London)*, 1978, **274**, 906.
288. B. W. Metcalf, P. Bey, C. Danzin, M. J. Jung, P. Casara and J. P. Vevert, *J. Am. Chem. Soc.*, 1978, **100**, 2551.
289. P. Bey, in ref. 255, p. 27.
290. P. Bey, F. Gerhart, V. Van Dorsselaer and C. Danzin, *J. Med. Chem.*, 1983, **26**, 1551.
291. P. Casara, C. Danzin, B. Metcalf and M. Jung, *J. Chem. Soc., Perkin Trans. 1*, 1985, 2201.
292. B. Lippert, P. Bey, V. Van Dorsselaer, J. P. Vevert, C. Danzin, G. Ribereau-Gayon and M. J. Jung, *Agents Actions*, 1979, **9**, 38.
293. H. Kubota, H. Hayashi, T. Watanabe, Y. Taguchi and H. Wada, *Biochem. Pharmacol.*, 1984, **33**, 983.
294. H. Wada, T. Watanabe, K. Maeyama, Y. Taguchi and H. Hayashi, in ref. 7, p. 245.
295. K. Srinivasan and J. Awapara, *Biochim. Biophys. Acta*, 1978, **526**, 597.
296. M. Levitt, S. Spector, A. Sjoerdsma and S. Udenfriend, *J. Pharmacol. Exp. Ther.*, 1965, **148**, 1.
297. M. Palfreyman, C. Danzin, M. J. Jung, J. R. Fozard, J. Wagner, J. K. Woodward, M. Aubry, R. C. Dage and J. Koch-Weser, in ref. 255, p. 221.
298. M. J. Jung, M. G. Palfreyman, J. Wagner, P. Bey, G. Ribereau-Gayon, M. Zraika and J. Koch-Weser, *Life. Sci.*, 1979, **24**, 1037.
299. A. L. Maycock, S. D. Aster and A. A. Patchett, *Biochemistry*, 1980, **19**, 709.
300. M. J. Jung, M. G. Palfreyman, G. Ribereau-Gayon, P. Bey, B. Metcalf, J. Koch-Weser and A. Sjoerdsma, in 'Drug Action and Design: Mechanism-Based Enzyme Inhibitors', ed. T. Kalman, Elsevier, Amsterdam, 1979, p. 131.
301. A. Maycock, S. Aster and A. Patchett, in ref. 255, p. 221.
302. A. L. Castelhano, D. H. Pliura, G. J. Taylor, K. C. Hsieh and A. Krantz, *J. Am. Chem. Soc.*, 1984, **106**, 2734.
303. E. A. Wang and C. T. Walsh, *Biochemistry*, 1981, **20**, 7539.
304. R. B. Silverman and R. H. Abeles, *Biochemistry*, 1977, **16**, 5515.
305. E. A. Wang, R. Kallen and C. T. Walsh, *J. Biol. Chem.*, 1981, **256**, 6917.
306. A. T. Phillips, *Biochim. Biophys. Acta*, 1968, **151**, 523.
307. H. Yoshida, K. Hanada, H. Ohsawa, H. Kumagai and H. Yamada, *Agric. Biol. Chem.*, 1982, **46**, 1035.
308. E. Groman, Y. Z. Huang, T. Watanabe and E. E. Snell, *Proc. Natl. Acad. Sci. USA*, 1972, **69**, 3297.
309. R. B. Silverman and R. Abeles, *Biochemistry*, 1976, **15**, 4718.
310. J. Giovanelli, L. D. Owens and S. H. Mudd, *Biochem. Biophys. Acta*, 1971, **227**, 671.
311. E. W. Miles, *Biochem. Biophys. Res. Commun.*, 1975, **64**, 248.
312. J. Baker, M. Lieberman and J. Anderson, *Plant Physiol.*, 1978, **61**, 886.
313. M. Johnston, D. Jankowski, P. Marcotte, H. Tanaka, N. Esaki, K. Soda and C. T. Walsh, *Biochemistry*, 1979, **18**, 4690.
314. M. Johnston, R. Raines, C. T. Walsh and R. A. Firestone, *J. Am. Chem. Soc.*, 1980, **102**, 4241.

7.1

Sulfonamides and Sulfones

PETER G. SAMMES
Brunel University of West London, Uxbridge, UK

7.1.1	INTRODUCTION	255
7.1.2	CHEMICAL STRUCTURE OF THE SULFONAMIDE GROUP	257
7.1.3	ANTIBACTERIAL ACTION OF SULFONAMIDES	258
7.1.4	STRUCTURE–ACTIVITY STUDIES ON SULFONAMIDES	260
7.1.4.1	*General Considerations*	260
7.1.4.2	*Whole Cell Inhibitory Activities*	262
7.1.4.3	*Cell Free Studies*	264
7.1.5	RESISTANCE	264
7.1.6	PHARMACOKINETICS, METABOLISM AND SIDE EFFECTS OF SULFONAMIDES	266
7.1.7	OTHER CHEMOTHERAPEUTIC APPLICATIONS OF SULFONAMIDES	266
7.1.8	REFERENCES	269

7.1.1 INTRODUCTION

The historic and practical importance of this group of drugs, currently over 30 agents are used clinically, together with the detailed knowledge of their antibacterial properties demands that they be treated separately from other groups of enzyme inhibitors. This short review concentrates on their mode of action as antiinfective drugs and some of the studies carried out to establish structure–activity relationships; brief mention is made of the applications of these compounds in other chemotherapeutic areas such as their use as diuretics and as hypoglycaemic agents.

The initial discovery of the sulfonamides was largely the result of serendipity. At the turn of the century a wide variety of azo dyes were in use and continuous efforts were being made to improve the properties of these very useful compounds. It became popular to introduce the sulfonamide group (**1**), which often gave the dyes extra light stability, greater water solubility during application and a greater fixation to certain fibres, particularly proteins such as wool. Of note were observations such as that made in 1932 that the dye prontosil (**2**) possessed effective bactericidal properties *in vivo* in small animals, for example the protection of mice against streptococcal infections.[1] This observation stimulated further studies and shortly afterwards, in 1933, prontosil was used clinically in a case of staphylococcal septicaemia.[2]

(**1**) R^1 = aryl or heteroaryl group; R^2 = H, alkyl, aryl or acyl group

(**2**)

During clinical studies it was confirmed that, whereas prontosil was active *in vivo*, *in vitro* assays proved ineffective. Attempts to modify the molecule showed that variations in the nonsulfonamide-containing portion of the azo dye had little effect on *in vivo* potency, only varying degrees of effectiveness being observed. In contrast variations in the sulfonamide-containing moiety were of consequence, different potencies and ranges of activity being observed.[3] The subsequent finding that prontosil was metabolized *in vivo* to give sulfanilamide (3) as the major product,[4] coupled with the observation that the latter exhibited bacteriostatic properties, both *in vivo* and *in vitro*,[5] suggested that prontosil itself was not the active drug species but that it acted as a prodrug, being reduced *in vivo* to produce the active metabolite sulfanilamide (3). Rapid progress in the clinical use of the latter and a range of related compounds became possible, aided by the facts that sulfanilamide was relatively easy and cheap to manufacture and that, since it was a well-known compound, first prepared in 1909,[6] it was not subject to patent protection as a chemotherapeutic and antibacterial agent.

(3)　　　　　　　　　　　　(4)

The sulfonamides, predating the penicillins, were thus the first group of chemical compounds to be widely used as specific antibacterial agents, in turn helping to spawn a host of new companies willing to research into the scope and limitations of this group of compounds. Indeed the modern pharmaceutical industry owes much to the discovery of the sulfonamides. They remain important to this day. One of the most famous, named M&B 693 (4; sulfapyridine) was prescribed to Winston Churchill in 1943, during the Second World War, when he contracted pneumonia during a trip to Africa. The drug effected a cure of the infection, which previously was often fatal.[7]

Development of the sulfonamides led for the first time to the design of evaluated clinical trials and these, in turn, resulted in important observations on their side effects. Some of these proved to give unexpected benefits, eventually harnessed as new drug leads and the application of various sulfonamide derivatives in other important chemotherapeutic areas, such as their use as diuretics,[8] as orally active antidiabetics[9] (particularly in maturity onset diabetes) and as carbonic anhydrase inhibitors.[10]

The clinical use of the sulfonamides was thus of historical importance in helping to lay the foundations for the whole of medicinal chemistry on which this publication is based. The need to assay the level of sulfonamides in tissue fluids, in order to investigate drug distribution, was studied by various people, including Bratton and Marshall,[11] who developed a simple but precise assay, based, not surprisingly, on the colour reaction involved in the diazotization of sulfanilamide (*i.e.* making azo dyes akin to prontosil!). This assay allowed the introduction of proper dosage regimes related to drug absorption, half-lives, distribution and excretion factors (see also Chapters 23.1–23.3). The study of residence times and distribution studies led to the use of pharmacokinetic studies as an essential part of drug development programmes.

Alongside the clinical studies, fundamental work on the mode of action of the sulfonamides commenced. The first clue on this was the observation by Woods[12] that the *in vitro* antibacterial action of the sulfonamides could be reversed by addition of *p*-aminobenzoic acid (5; PABA). This observation suggested that the sulfonamides acted as a competitive antagonist to the naturally occurring PABA, known to be an essential growth factor for some microorganisms, *i.e.* that they are acting as antimetabolites. The concept of antimetabolites was developed by Fildes[13] and has subsequently been applied to several other areas of chemotherapy, notably in the nucleotide area (see Chapter 7.3). The early interest in the sulfonamides has been maintained to the present and the studies on separating out the beneficial side effects from the antibacterial properties have helped to establish methods for developing new leads from drugs as well as protocols for toxicological studies. Finally, because of their relatively simple structure many variations and analogues have been synthesized and these in turn have helped to establish the principles for optimizing drug design (see also Chapters 21.1–21.3). In this chapter a survey of the studies used to establish the mode of action of the sulfonamides as antibacterial agents and some aspects of structure–activity relationships will be presented. Many reviews on this area have been written and the reader is also referred to some of these.[14]

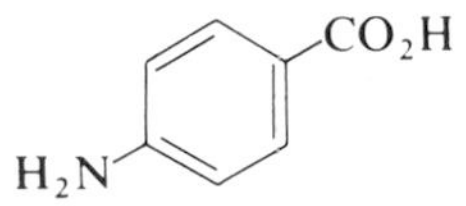

(5)

7.1.2   CHEMICAL STRUCTURE OF THE SULFONAMIDE GROUP

The availability of many crystalline derivatives of sulfonamide drugs has given scope for several structural studies. The sulfonamide group is strongly electron attracting, leading to a high electron density on the sulfone oxygen atoms and enhancing the acidity of the proton attached to nitrogen; *N*-unsubstituted or monosubstituted sulfonamides behave as acids and readily form salts. Structurally the atoms linked to the central hexavalent sulfur atom are arranged tetrahedrally, as exemplified by the structures of the drugs hydrochlorothiazide (6)[15] and torasemide (7),[16] which behave as diuretics,[17] and sulthiame (8),[18] a carbonic anhydrase inhibitor (see Figure 1).[19] It will be noted that, in each case, the sulfonamide group adopts a conformation in which the nitrogen atom prefers to lie orthogonal to the plane of the aromatic ring, unless strong steric influences are present such as bulky or hydrogen-bonding *ortho* substituents or the sulfonamide group is held in a rigid ring. Fewer studies seemed to have been made of salts of the sulfonamides.

The pK_a values of sulfonamides bearing a N—H bond are generally in the range 4–9. Because of this, the simple *N*-unsubstituted sulfonamide group has often been compared to a carboxyl group, *i.e.* a carboxyl isostere; the structures of sulfanilamide (3) and PABA (5) are very similar. Thus in the majority of biological situations the sulfonamides are likely to be present as salts, although for transport through membranes *etc.*, the unionized form is likely to be more easily transported, hence the pK_a value is important in helping to determine the penetration of these agents into cells (see also the discussion below). In sulfonamides in which the nitrogen is further conjugated, such as sulfaguanidine (9)[20] and sulfapyridine (4),[21] the doubly bonded nitrogen is the preferred tautomer.[22]

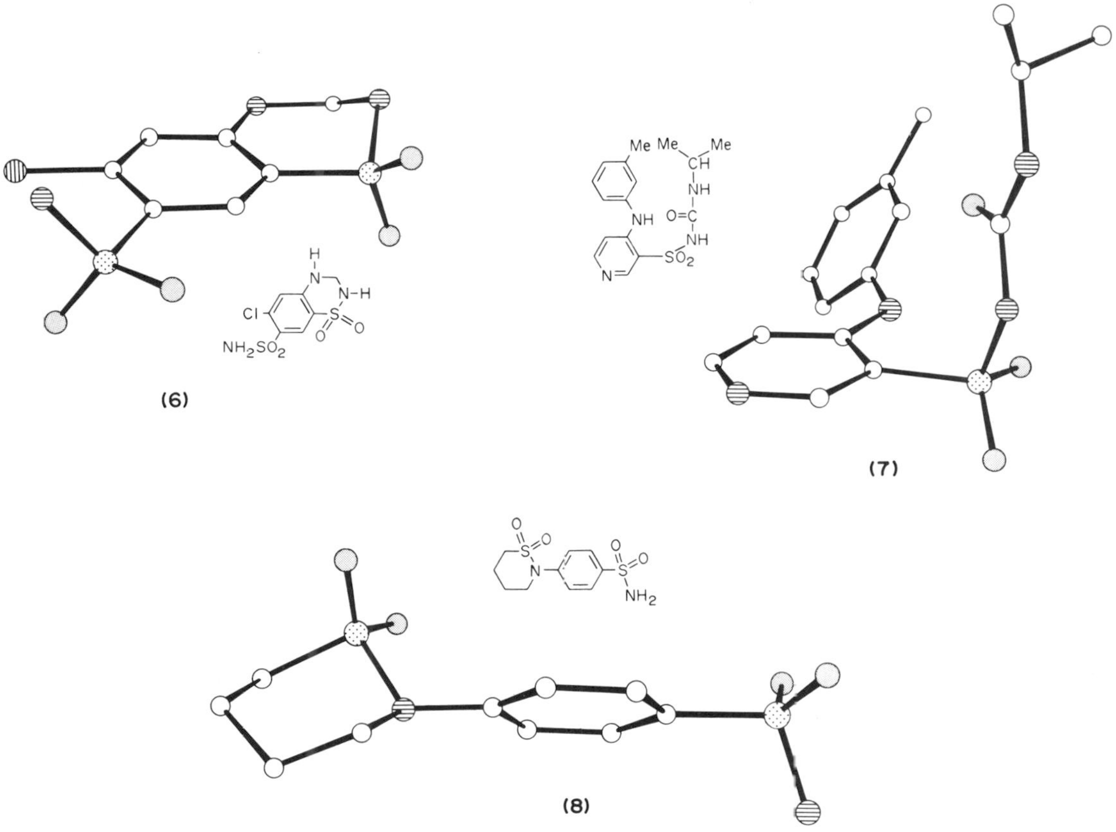

(6)

(7)

(8)

Figure 1

Finally, in any intermolecular interactions account must be given of the large dipole moment associated with the sulfonamide group and the surface charges, particularly those placed on the oxygen atoms. During attempts to find active variations in the sulfonamide structure it was found that some diaryl sulfones also acted as antagonists of PABA, for example the agent dapsone (10),[23] which is still particularly useful in mycobacterial infections such as leprosy.[24]

Structure–activity studies have demonstrated that these bind in a similar manner to sulfonamides, substituents reinforcing negative charge on the oxygen atoms also enhancing activity of the derivatives (see below).

7.1.3   ANTIBACTERIAL ACTION OF SULFONAMIDES

The sulfonamides inhibit growing bacterial organisms. The finding that sulfanilamide antagonizes the role of PABA focused early investigations on the precise role of the latter.[5] Of importance was the finding that PABA is an essential growth ingredient for a variety of microorganisms and that PABA itself was a cofactor involved in the biosynthesis of single carbon units (the C_1 pool) *via* participation in folic acid (11).[25] If supplies of folic acid are limited, other C_1 sources can be substituted into the C_1 pool, such as methionine and glycine. Alternatively, in the absence of normal amounts of folic acid, maintenance of growth can be sustained by supplementation with nucleic acid bases such as adenine, guanine and thymine, since a key role of folic acid is donation of C_1 units in the biosynthesis of the latter.

Determination of the structure of folic acid (11) showed PABA incorporation and subsequent studies have fully elucidated the steps involved in the biosynthesis of folic acid and its related reduced partners.[26] The biosynthesis is outlined in Scheme 1; PABA is itself normally metabolized *via* the chorismic acid route.[27] In the biosynthesis of folic acid, the dihydropteridine derivative (12) is either coupled with PABA, in an ATP-activated process, leading to the dihydropteroic acid (13) and this in turn conjugated with glutamic acid to give dihydrofolic acid (14) or the latter is formed directly from the pteridine (12) and the PABA–glutamic acid conjugate (15), again *via* an ATP-activated process. Experiments suggest that the same enzyme can be involved in the incorporation of either PABA (5) or its glutamic acid conjugate (15).[28] The dihydrofolic acid (14) can undergo either reversible oxidation to folic acid (11) or reversible reduction to tetrahydrofolic acid (16). It is the enzyme involved in the latter step (dihydrofolate reductase) that is inhibited by agents such as methotrexate and trimethoprim (see Chapter 7.2). Tetrahydrofolic acid acts as the one-carbon carrier, involving methylation, hydroxymethylation and formylation at the 5- and 10-positions (Scheme 2; see structures 17) and is involved in the enzyme-catalyzed transfer of these one-carbon units to various substrates; an example is given in Scheme 3. Many studies have been made to elucidate the role of the sulfonamides in the disturbance of the normal biosynthetic steps. Indirect evidence that the sulfonamides are incorporated into covalently bound analogues of the folic acid type, rather than acting as passive enzyme inhibitors, was reported by Brown[29] in the 1960s. He showed that when sulfathiazole (18) was preincubated with small quantities of pteridine substrates in cell free bacterial extracts before addition of PABA, a much higher degree of inhibition was found than when all the substrates were added simultaneously, suggesting that in the former case the sulfonamide reacts enzymically with the pteridine substrate during the preincubation, so that when PABA is added there is no pteridine left for formation of the folic acids; as expected, addition of more pteridine allows synthesis of folic acid to proceed. Subsequently, the new conjugates, of the general type (19), have been isolated and identified.[30] A similar incorporation is observed for the diaryl sulfones.[31]

The relationship between sulfonamide and PABA was strictly competitive when the incubation was carried out with all substrates present at the same time and with normal concentrations of pteridines; under these conditions the inhibitory effect of the sulfonamide can be reversed by addition of more PABA.

These experiments show that sulfonamides can compete with PABA not only as competitive binding agents on the particular enzyme involved in the biosynthesis of the folic acids, but also by

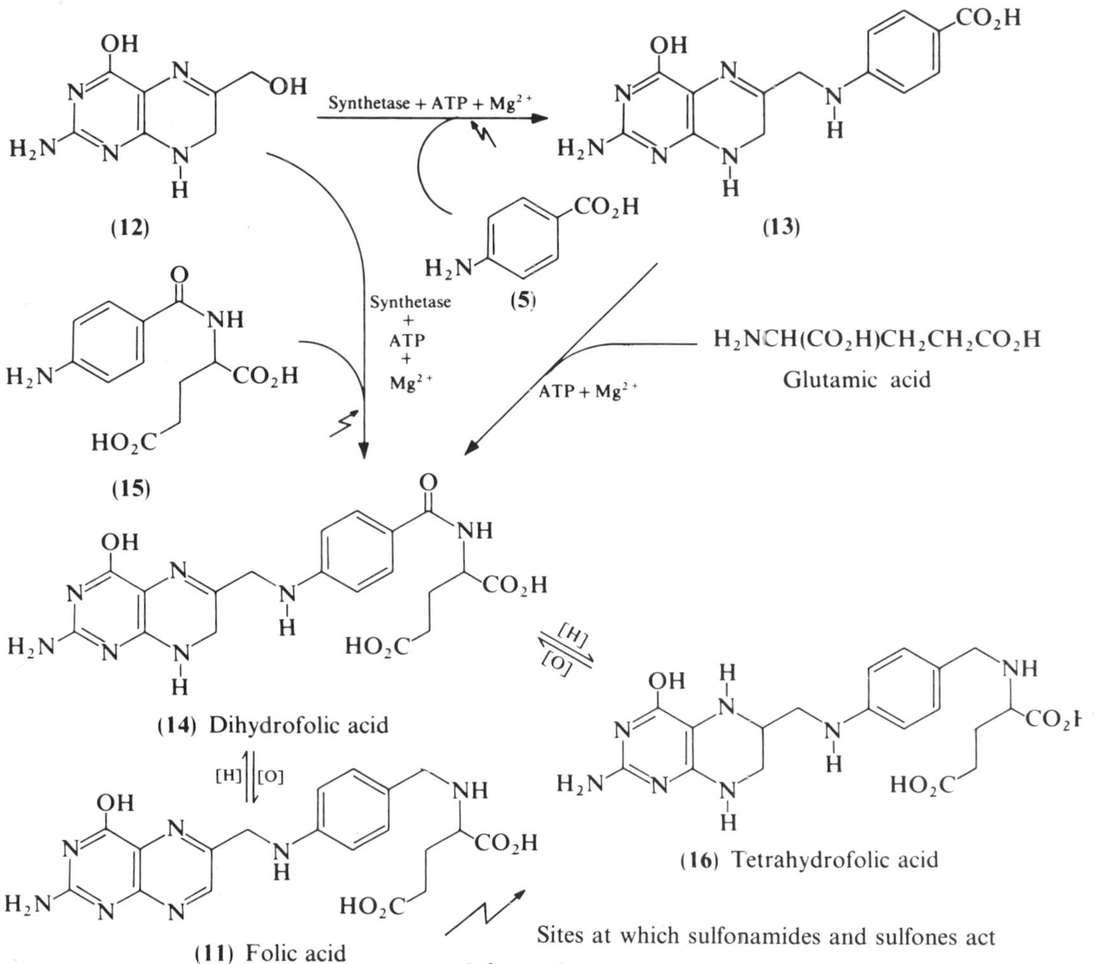

(12)

Synthetase + ATP + Mg²⁺

(13)

H_2N

CO_2H

(5)

$H_2NCH(CO_2H)CH_2CH_2CO_2H$
Glutamic acid

Synthetase
+
ATP
+
Mg²⁺

ATP + Mg²⁺

(15)

(14) Dihydrofolic acid

$\frac{[H]}{[O]}$

(16) Tetrahydrofolic acid

$[H] \parallel [O]$

(11) Folic acid

Sites at which sulfonamides and sulfones act

Scheme 1

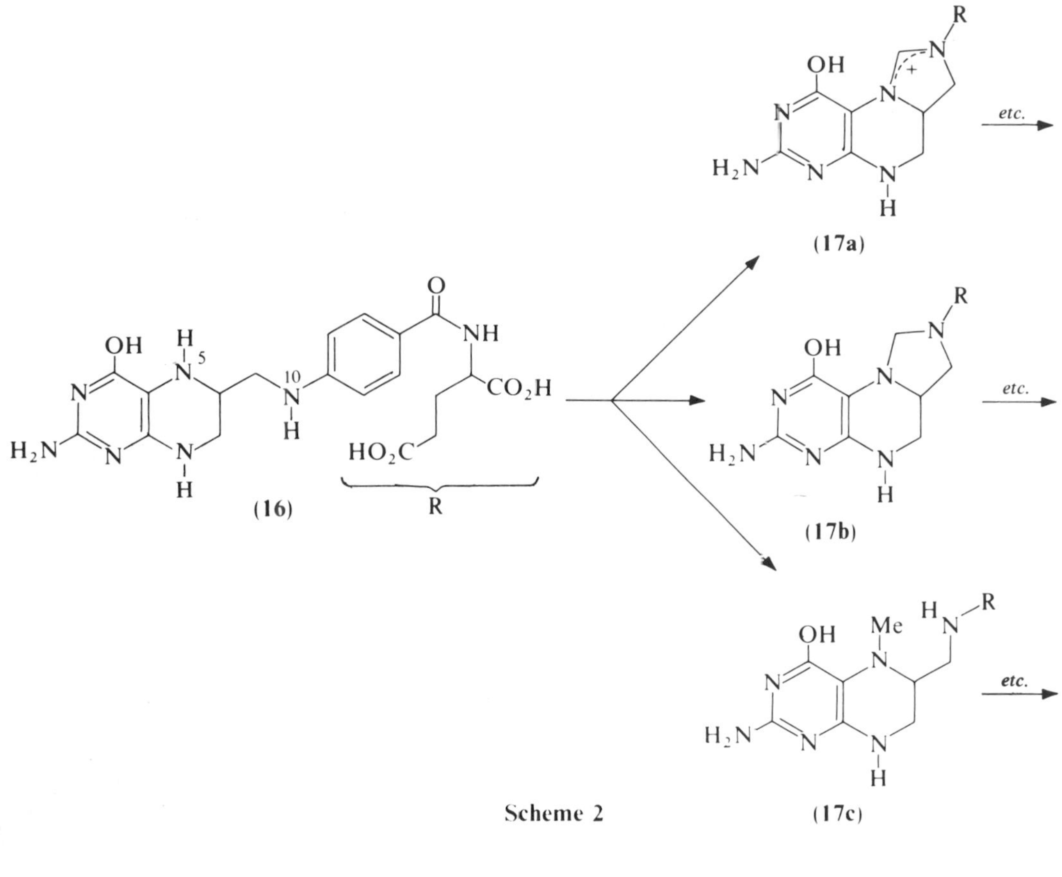

(17a)

etc.

(16)

R

(17b)

etc.

(17c)

etc.

Scheme 2

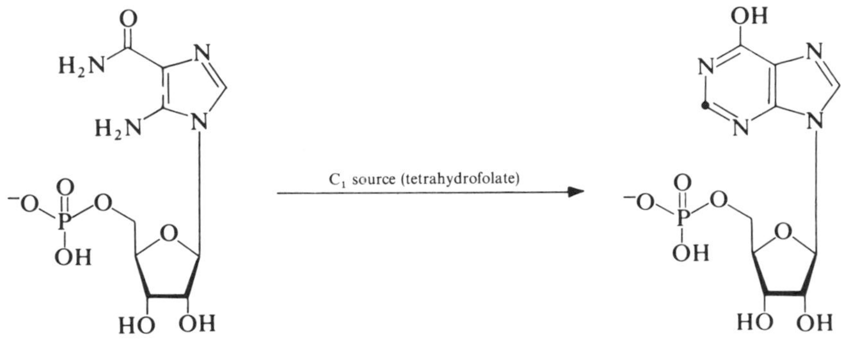

Scheme 3

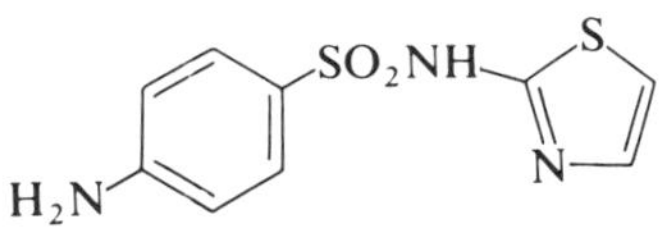

(18) 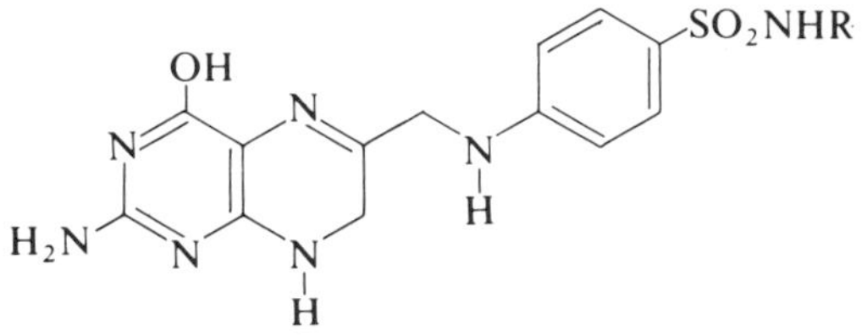   **(19)**

substituting for folic acid production by formation of the sulfonamide analogues; the latter have been shown to be inactive as C_1 carriers. There remains the question as to what is the rate-determining step in the interference by the sulfonamides of the biosynthesis of the folic acids. If incorporation of the sulfonamides into the analogues (**19**) were fast, the rate-determining step in their inhibitory role would then occur later in the sequence of events. Studies showed that although incorporation of the sulfonamides occurs, *i.e.* can compete with the rate of incorporation of PABA, the sulfonamides acted as competitive antagonists, binding strongly to the enzymes responsible for the activation of the starting pteridine (**12**) and of the coupling of this to PABA (see Scheme 1). Thus the sulfonamides appear to act as antibacterial agents mainly by stopping the formation of folic acid and hence turning off the normal supply to the C_1 metabolic pool. This action explains why the sulfonamides only act against growing microorganisms and why the onset of bacteriostasis is preceded by a lag phase, during which stores of PABA and folic acid are being used up; the length of the lag phase depends on the level of these stores. Considerable effort has been expended on fully characterizing the enzymes involved in the susceptible steps of folic acid synthesis in order to gain a better understanding of the molecular interactions of the natural and unnatural substrates with the enzymes.[32] Such information is vital in order to study the relative sensitivity of the synthetase enzymes from different microorganisms and the mechanism whereby some microorganisms can become resistant to these drugs (see also below).

It is interesting to note that despite their structural dissimilarity, the mechanism of action of 4,4′-diaminodiphenyl sulfone (**10**; dapsone) and related sulfones is similar to that of the sulfonamides; as expected, their *in vitro* inhibitory action against microorganisms is also reversed by the addition of PABA.

The sulfonamides are only active in organisms where *de novo* folic acid synthesis is crucial for survival. The higher organisms, such as man, do not synthesize their own folic acid and their requirements for folic acid are assimilated in their diets — for example folic acid is produced by plants. *De novo* folic acid synthesis also takes place in a wide range of microorganisms, including bacteria, protozoa and yeasts and this range explains both the broad spectrum of activity of the sulfonamides and their selectivity. It appears that these microorganisms do not readily extract folic acid from the host organism, such as that occurring in the blood stream of man, and that the concentrations of other folic acid precursors, such as PABA, are too low *in vivo* to reverse the antagonism of the sulfonamides.

7.1.4 STRUCTURE–ACTIVITY STUDIES ON SULFONAMIDES

7.1.4.1 General Considerations

Because of their relatively simple structure and their chemotherapeutic importance, a great many analogues of sulfanilide are known. This has enabled chemists to make detailed studies on structure–activity relationships within this group and they have become a model for relating chemical structure to biological activity.[33]

That a chemical has potency against a target enzyme is essential, but not necessarily sufficient, in order to develop a useful chemotherapeutic agent. In order to serve as the latter a compound must also have the right physicochemical properties to reach the target site (for the sulfonamides the pathogen), must not be involved in any distractions on the way, such as strong binding to other biological sites, for example proteins, must reach active concentrations at the target (lethal ones in the target microorganisms), and not have any adverse side effects on the host.

In order to unravel this complex set of parameters the problem is often divided into parts: an examination of the efficacy of a potential drug in cell free systems, followed by studies on intact cells and then studies in animals and, finally, clinical studies in man. An important aim of studies on structure–activity relationships (SAR) is to try to establish *quantitative* expressions (QSAR) relating chemical structure to physicochemical properties of a group of drugs. The overall relationship between dosing and effect may be divided into three separate steps, the pharmaceutical, pharmacokinetic and pharmacodynamic processes (Figure 2). These steps are more fully considered in Volume 5. In general two types of physicochemical conditions are necessary for biological action: (i) chemical properties necessary to induce a physicochemical or chemical reaction at the receptor site; and (ii) physicochemical properties which determine the pharmacokinetics of the drug, *i.e.* absorption, distribution, metabolism, binding to nonspecific binding sites and excretion with respect to the biological system involved.

Optima in chemical properties with respect to biological activity at the specific target do not necessarily coincide with the physicochemical optima that determine pharmacokinetic parameters. There are many examples where the chemical requirements for each are opposed or related in a complex manner. As a consequence the rational design of a new drug is almost always a compromise. To deduce the QSAR of a drug lead it is normal to first consider the chemical and physicochemical properties with respect to the biological response at the site of action based on real constants and then to correlate these to the optimum conditions for the pharmacokinetic properties of the drug; pharmacokinetic parameters often have a wider spread than the former. The parameters to be considered may be considered in two groups: (i) those that describe the physical properties of the compounds, such as water solubility, surface tension, partition coefficient, vapour pressure, molal activity, parachor and R_f or R_m values,[34] *i.e.* the 'unpolar parameters'; and (ii) parameters related to chemical reactivity, such as the Hammett or Taft constants, electronic data gained from spectroscopic measurements, rate constants, quantum chemical data, steric factors, *i.e.* the 'polar parameters'.

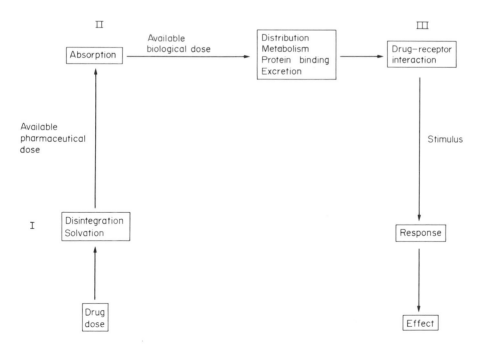

Figure 2 The overall relationship between dosing and effect: I, pharmaceutical processes; II, pharmacokinetic processes; III, pharmacodynamic processes

Guidelines for the study and interrelationships between these parameters have been published[35] and several approaches have been proposed.[36] The most widely used approach is the linear free energy related model, the Hansch approach,[37] where the variance in biological effect (ΔBE) is explained by the variance of certain linear free energy related substituent constants which describe the change in lipophilic/hydrophilic ($\Delta L/\Delta H$), electronic (ΔE), steric (ΔE_s), or other properties induced by varying the chemical substituents in the main structure under consideration.

$$\Delta BE = \log(c_1/c_2) = f(\Delta L/\Delta H; \Delta E; \Delta E_s, \text{etc.})$$

The variance in electronic properties can be expressed using parameters such as the Hammett constant (σ), pK_a values, or spectroscopic data, like chemical shifts from NMR spectroscopy, intensity and wavelength changes of signals from IR or UV spectra, *etc.* Lipophilicity refers to partition coefficients, log P values or the π substituent constant, or R_m values from reverse phase chromatography. The steric influence of the substituent can be expressed by the Taft steric constant E_s, the van der Waals volume, or the molar volume according to Bondi and Exner. In a stepwise, linear, multiregressional analysis the physicochemical parameters which have the most statistical significance for the explanation of the observed change in biological effect undergo evaluation as follows

$$\log \Delta BE = a + b\pi$$

$$\log \Delta BE = a + c\sigma$$

$$\log \Delta BE = a + dE_s$$

$$\log \Delta BE = a + b\pi + c\sigma$$

etc.

In order to be able to cope with the multifactorial situation confronting dosage of a drug in a whole animal, simple biological models are initially selected in order to reduce errors in measurements. It is for good reasons that most QSAR techniques are applied to descriptions of biological activity of enzyme inhibitors[38] since these are generally the easiest to cope with.

The discussion on sulfonamides below is limited to *in vitro* studies on bacteria in a culture medium and to isolated enzyme proteins, in which case, as indicated above, the possible rate-determining steps may be reduced to: (a) passage through the bacterial cell wall; (b) binding to the specific receptor protein; and (c) reaction with the enzyme proteins or with a substrate to form a product after irreversible interaction with the enzyme protein.

7.1.4.2 Whole Cell Inhibitory Activities

Early studies by Bell and Roblin[39] on the action of a wide range of sulfonamides against whole cells of *E. coli* showed that a parabolic relationship was followed when pK_a was plotted against log $(1/MIC)$ (MIC = minimal inhibitory growth concentrations), with the maximum activity occurring around $pK_a = 7$, the physiological pH region. Increasing the pH of the growth medium *increased* the relative activity of a particular sulfonamide up to the point at which the degree of ionization of the sulfonamide reached 50%.[40] This effect was explained on the basis that penetration of the antibiotic into the cell occurred in the unionized form and that the active form, once inside the cell, was the ionized form.[41]

More recently Seydel *et al.* have made studies on a series of substituted N^1-phenylsulfanilamides (**20**) and N^1-pyridylsulfanilamides (**21**)[42] using serial dilution techniques against growing *E. coli* bacteria.

(**20**)   (**21**)

Minimal inhibitory growth concentrations (MIC) were measured for the various derivatives, since it had been shown that MIC values paralleled activity rate constants obtained from kinetic studies on bacterial growth.[43] They showed that, for sulfonamides with $pK_a > 6.5$, log MIC gives a linear relationship with the acidity of the N(1)—H proton, as measured by pK_a or NMR studies (Figure 3).[44] The following relationship was observed in a study of 18 derivatives

$$\log(MIC)_{o,m,p} = (0.66 \pm 0.07)pK_a - 4.76$$

$$n = 18, r = 0.92, s = 0.17, F = 96$$

Deviations occurred for some *ortho*-substituted anilines (**20**; R = *o* substituent). Although some of these deviations could be ascribed to not taking into account intramolecular hydrogen-bonding effects, even when these were considered some deviants remained, all being more active than predicted by a constant factor. Since these compounds all contained an *ortho* substituent, a steric factor was assumed and was represented by an indicator factor I (value 1 when present, 0 when absent), to give the statistically better regression curve

$$\log(MIC)_{o,m,p} = (0.68 \pm 0.06)pK_a - (0.24 \pm 0.08)I - 4.78$$

$$n = 18, r = 0.95, s = 0.14, F = 67$$

This new relationship tells us something about the interaction beween the sulfanilamides and the active site: possibly the *ortho*-substituted derivatives adopt a slightly different conformation than the series without such substituents and *that this modified conformation is a better fit*.[45] Furthermore, kinetic studies showed that these derivatives are more efficiently incorporated to form folic acid analogues, thus preventing residual folic acid formation. This deduction illustrates the power of simple QSAR relationships to help detect events at the molecular recognition level. Not surprisingly, other regression curves, using electronic parameters such as the Hammett substituent constant σ, also give good correlations.

Studies of dose–response curves, focusing on the fractional inhibition of folate synthesis against inhibitory concentrations, confirmed the prime activity of the sulfonamides acting in this manner, *i.e.* as direct competitive antagonists and not as noncompetitive antagonists and also enabled measurement of 50% inhibition concentrations. These gave a 25-fold variation in inhibitory activity for the range of 50 derivatives studied. In order to check whether or not these differences were due to permeability differences through the cell wall, repeated studies were made using cell free systems. In this case, most derivatives gave a similar relationship, showing that for the majority of derivatives, the variation in activity is not due to rate-controlling permeability differences and that it must be associated with the action of the sulfonamides at the enzyme level. However, inspection of the set of results from the whole cell studies showed that those few derivatives that did not fit on the straight line were all compounds with pK_a values of less than 6, that is, compounds that are more than 90% ionized at physiological pH. For these the rate-limiting step must be permeation across the cell wall, since passage of fully ionized species does not occur and, since the fraction unionized is so low, transport into the cell virtually ceases. All these derivatives gave high activity in the cell free experiments. Of interest was the early finding that even sulfanilic acid (**3**; pK_a 3.7) is active in cell free systems but, because of its complete ionization, is totally inactive in whole cell assays.[46]

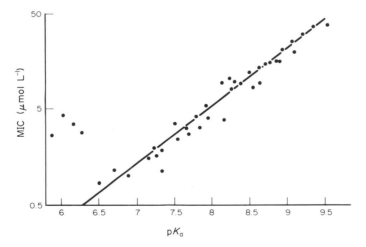

Figure 3 Typical correlation curve for sulfonamide activity *versus* pK_a (data provided by Professor J. K. Seydel)

Attention was then directed at explaining the 25-fold variation in inhibitory activity against the folic acid synthesizing systems observed with this set of derivatives. Studies on the rate of formation of incorporated sulfanilide into the folic acid analogues (*e.g.* see **19** above) occurred at the same rate, despite the large variance in their inhibitory power, and only the natural substrate, PABA, showed a faster rate of uptake. A similar set of results were obtained when using substituted PABA derivatives which were all totally ionized under the experimental conditions. For the PABA derivatives the substituents did not unduly influence the rate of incorporation except when bulky *ortho* substituents were present, which inhibited incorporation. It may be concluded that the rate-limiting step in folate inhibition by both the sulfonamides and sulfones is not the rate of formation of the dihydropteroic acid analogue, although this is important to help divert stores of precursors from forming folic acid, but must be due to the different affinities of the inhibitory agents for the enzyme dihydropteroic acid synthetase compared with the affinity of the natural substrate PABA.

7.1.4.3 Cell Free Studies

There remained the question of confirming whether or not the sulfonamides act on the enzyme system in the ionized or unionized form. From the cell free experiments it was shown that the most active agents are those that are ionized. However, as shown by the activity of the diaryl sulfones, which cannot ionize, the unionized forms are also active but less so than the ionized forms. Competition experiments, designed using PABA to reverse the inhibitory effect of a range of sulfonamides, showed that the concentration of PABA (K) needed to reverse the inhibitory effect of the sulfonamides on folate production to 50% of that obtained in the absence of the inhibitors, could be related by the regression curve

$$K_{PABA} = (5.76 \pm 0.53)[SA^-] + (0.19 \pm 0.0350)[SA] + (41.9 \pm 18.5)I - 2.36$$

$$n = 23, r = 0.97, s = 23.2, F = 107$$

where $[SA^-]$ is the concentration of ionized sulfonamide and $[SA]$ the concentration of unionized sulfonamide.[33] This shows that the ionized sulfonamides are about 30 times more active than the unionized forms.

Another interesting correlation was obtained by comparing the activity of a range of sulfones and sulfonamides on the mycobacterium *M. smegmatis*, related to *M. leprae*, the causative agent of leprosy. It was found that intrinsic activity mainly depends on the negative charge density on the sulfone group, although equations correlating activity with an electronic effect descriptor have opposite signs of the regression coefficients. The explanation for this paradoxical observation is that an increase in the negative charge at the SO_2 group is produced by electron-releasing groups for sulfones, whereas, for sulfonamides, ionization increases with electron attraction by the substituent.[47] Molecular modelling of diaryl sulfones has been carried out.[48]

Studies have also been carried out to see whether any other parameters are of significance in fitting observed activities to a regression curve. Several studies on hydrophobic effects have shown these to be relatively unimportant for intrinsic activity at the enzyme but of some relevance with respect to optimizing transmembrane transport into the cells. Thus measurements of the lipophilicity of the unionized sulfonamides are of value in determining the transport properties of the drugs and these in turn influence their pharmacokinetics, including parameters such as biological half-lives. Longer acting sulfonamides generally have higher lipophilicities.[49] Some of the current sulfonamides in clinical use are listed in Table 1.

7.1.5 RESISTANCE

As with many antibiotics, parasites are often able to learn to resist the action of drugs such as the sulfonamides, using random mutations, selection or the transfer of nuclear resistance factors from other resistant organisms. Three principle methods can be used: altering the permeability of the cell to the sulfonamides so that less is transported in, hence requiring increased dosage of the drug to reach lethal concentrations in the pathogen; increasing the production of PABA so that more effective competition with the antagonist occurs; and induction of changes into the target enzyme so that selectivity for the natural substrate is increased. Examples of all three types of resistance, and some combinations of effects have been recorded.[50] Naturally resistant strains of some organisms have been discovered, for example of *E. coli*, and it has been shown that these contain an altered dihydropteroate synthetase enzyme.[51]

Table 1 Typical Sulfonamides

$$H_2N \text{—} \langle\text{benzene ring}\rangle \text{—} SO_2NHR$$

Generic name		R	pK_a	Comments
Sulfaguanidine	**(9)**	NH, C, NH₂	Basic	Poorly absorbed, locally acting, used in gastro-instestinal infections
Sulfathiazole	**(18)**	(thiazole ring)	7.25	Well absorbed, rapidly excreted
Sulfacetamide		—COME	5.4	Well absorbed, used in eye infections
Sulfamethoxazole		(isoxazole ring, Me)	6.0	Well absorbed, moderate excretion
Sulfadiazine		(pyrimidine ring)	6.52	Well absorbed, moderate excretion
Sulfamethoxine		(dimethoxy pyrimidine ring, OMe, MeO)	6.1	Well absorbed, slowly excreted, used in dysentery, *etc.*

Efforts to find out the exact mechanism for a particular resistant organism are essential in order to determine the best way of overcoming the problem. An example is on studies with resistant strains of *Mycobacterium lufu*, related to the leprosy pathogen *M. leprae*. Dapsone resistant strains have been isolated and studied. The resistance was shown not to be due to a change in the target enzyme; cell free assays showed the target enzymes from the resistant organisms to be just as susceptible to the drug as normal strains. Since altered permeability of the drug into the cell was also discounted, it was concluded, in this case, that the most likely cause of resistance was gene amplification of the enzyme, *i.e.* that far more copies of the enzyme were being produced than in the normal strain so that saturation of the enzyme by the drug becomes far more difficult and folic acid synthesis is maintained.[50] As a result, attempts to overcome this type of resistance by structural modification to the dapsone molecule would be unrealistic.

Resistance to sulfonamides is also observed in the treatment of the plasmodia, the causative agents of malaria. Although these are eukaryotic organisms, the sulfonamides act in a similar way to that observed on prokaryotes.[52] In these parasites resistance is by means of a by-pass mechanism, *viz.* whereas they normally produce their own folic acid, in the presence of high concentrations of the antagonists, such as sulfadimethoxine **(22)**, resistant forms can survive by using reduced forms of folic acid available from the host erythrocytes.[53]

Not surprisingly, since the modes of action of the sulfonamides as antibacterial agents revolve around the same basic mechanism, different sulfonamides usually show cross-resistance, *i.e.* resistance against one member of the family is rapidly followed by resistance to other members of these antibiotics.

Because of the problems of resistance using sulfonamides on their own and because of the intrinsically more powerful activity of the β-lactam antibiotics and other agents, sulfonamides are not usually prescribed on their own but as mixtures with other antibiotic agents. In particular use is made of mixtures of the sulfonamides with inhibitors of the dihydrofolic acid reductase enzyme, such as trimethoprim **(23)**, pyrimethamine **(24)** or methotrexate **(25**; see Chapter 7.2). These mixtures are found to act synergistically,[54] since they each act on enzymes sequentially involved in a common biosynthetic pathway (see Scheme 1). Provided one selects agents that are pharmacokinetically

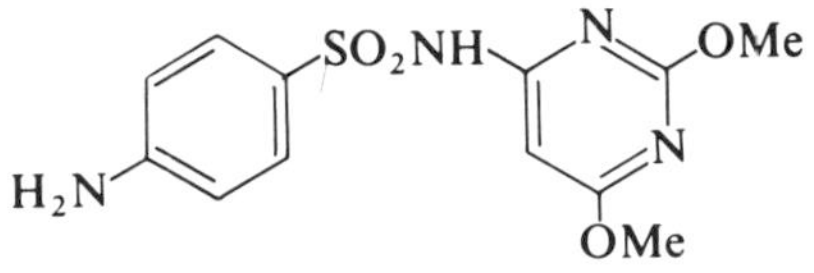

(22)

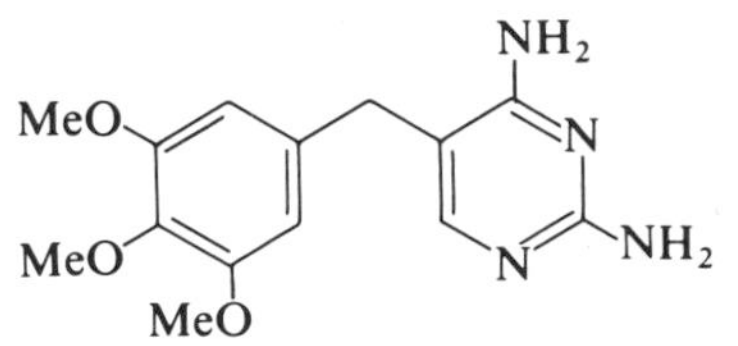

(23)

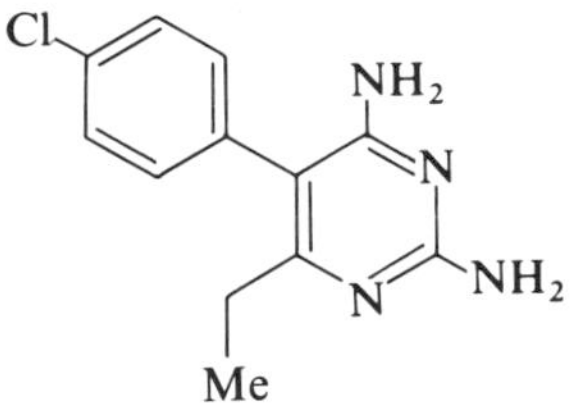

(24)

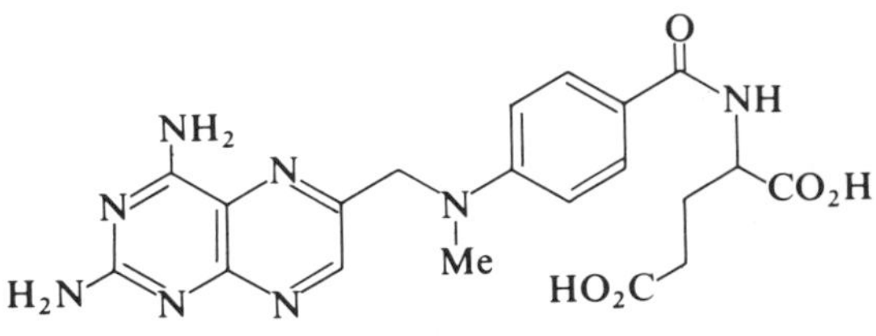

(25)

similar, *i.e.* that can arrive at the target at the same time and have similar half-lives *in vivo*, they have therapeutic advantages.[55] Furthermore the chances of resistance occurring against such combinations of drugs is greatly reduced.

7.1.6 PHARMACOKINETICS, METABOLISM AND SIDE EFFECTS OF SULFONAMIDES

One of the reasons why so many members of the sulfonamide family have been prepared followed from the finding that there were great variations in their pharmacokinetics. After oral intake absorption from the gut occurs for those derivatives not having extra ionizable groups; the latter, not being absorbed, stay in the gut and are sometimes used to treat gastrointestinal infections. Absorption of the former takes place *via* the neutral (unionized) form and they are distributed fairly evenly throughout the body. The more water soluble forms tend to be excreted fairly rapidly, useful for treating urinary infections, whilst the others are useful for the treatment of systemic infections. Conjugate formation can occur, mainly in the liver, and consists mainly of *N*-acetylation and glucuronidation. As would be expected, variation of structural features affects both metabolic fates and rates of excretion and a wide range of half-life variations can be tailored into these molecules, from periods as short as to 2 h over 6 d. Attempts to model lifetimes from structural features have been made.[56]

As for many drugs general protein binding occurs with the sulfonamides, effectively lowering the local free concentration of the drug and hence its ability to inactivate its target enzyme, as well as preventing the metabolism of the drug itself. Since protein binding of the sulfonamides is reversible, besides lowering the drug concentration, it also has the effect of increasing its half-life in the body. In general protein binding seems to be favoured by increased lipophilicity.

A problem encountered with some of the older sulfonamides was their tendency to crystallize out at the lower pH values associated with urine formation in the kidneys (crystalluria). This problem has largely been overcome with derivatives that are more water soluble and that build up adequate levels in the target pathogen at lower host concentrations. Blood dyscrasias occasionally occur, which prevent further use of the drugs and some patients suffer from allergic responses such as hypersensitivity and urticaria but these side effects are uncommon. By comparison with many groups of antibiotics the sulfonamides are relatively safe agents.

7.1.7 OTHER CHEMOTHERAPEUTIC APPLICATIONS OF SULFONAMIDES

Clinical observations on the earlier sulfonamides showed that some members had diuretic properties. Diuresis depends on water transport rates through the kidneys and this in turn is related to ionic balances, such as the sodium and chloride balance. For example, sodium uptake by the kidneys is an active process needing ATP and results in retention of water, which is simultaneously absorbed by osmosis. Blocking of these sodium channels leads to a net loss of water.

The earlier sulfonamides, particularly those not having substituents on the amide nitrogen, were shown not to behave as sodium channel blocking agents, in contrast to the agents such as triamterene (26) and amiloride (27) or chloride channel blocking agents, such as furosemide (28). Investigations of their action showed that the sulfonamides act as potent inhibitors of an enzyme system of importance in controlling water retention, the zinc metalloenzyme carbonic anhydrase.

(26) (27) (28)

Carbonic anhydrase, which normally exists in at least two isoenzymic forms, is responsible for catalyzing the hydrolysis of carbon dioxide to form carbonic acid, as well as the reverse process, and is found in the kidneys as well as the pancreas and other tissues. Under physiological conditions the carbonic acid formed immediately ionizes to form the biocarbonate ion. Although the hydration of carbon dioxide in the absence of the enzyme is already quite a fast process, nature has evidently found that it is still not quick enough![57]

The carbonic anhydrase enzymes have been fully studied and the zinc acts as a Lewis acid towards water, thus releasing a proton, which is eventually donated to the carbon dioxide molecule. The active sulfonamides, which must have the amide nitrogen unsubstituted, slip into the active site and chelate to the zinc atom, displacing the normally associated water ligand and preventing ingress of the carbon dioxide molecule.[58] The sulfonamides of the type $ArSO_2NH_2$ (29; Ar = aryl group) thus prevent both the hydration and dehydration of carbon dioxide. Because the enzyme is present in large quantities, inhibition must be largely complete before the effect of the blockade is realized. Normally bicarbonate ion is excreted from the glomerulus to the proximal tubule, where sodium ions are reabsorbed in an active process. At the same time hydrogen ions are released into the proximal tubule and these react with the bicarbonate ions to form carbonic acid which, under the influence of the enzyme situated in the membrane of the tubule, is converted into carbon dioxide and water. The carbon dioxide diffuses through the membrane, where more carbonic anhydrase catalyzes the reverse, hydration procedure. The bicarbonate ion thus formed produces a salt with the sodium ions, whilst the hydrogen ion is transported into the lumen to compensate for the migration of the sodium ion. The net process is the active, catalyzed reabsorption of sodium bicarbonate from the lumen. Water is also reabsorbed during this process to help maintain osmotic pressure.

Blocking of the enzyme results in retention of the bicarbonate ions in the tubule, thus increasing the pH, thus restricting reabsorption of sodium ions and the associated water. Under blockade only the passive (noncatalyzed) equilibration of carbon dioxide with carbonic acid occurs and this is relatively slow. The overall effect of the sulfonamides as diuretics is considerably offset by the fact that after passage through the proximal tubule the filtrate passes into the descending and ascending limbs of the loop of Henle (Figure 4), where reabsorption occurs by different, nonbicarbonate dependent processes, *viz.* absorption of sodium and chloride ions, and these processes can compensate for the loss of water in the proximal tubule by more efficient reabsorption processes. In practice, combinations of sulfonamides with the sodium or chloride blocking agents, such as triampterene, are prescribed. These agents are of considerable value in the treatment of oedema (water retention) and general excess of fluid in tissues, which causes stress on the heart, and hence they are often prescribed to patients with heart diseases.

One further application of the sulfonamide carbonic anhydrase inhibitors is in the alleviation of the effects of the eye disease, glaucoma. The carbonic anhydrase enzymes are present in the membrane of the cells separating the aqueous humour from the blood stream. In glaucoma an excess of water builds up in the aqueous humour of the eyeball and this in turn exerts pressure on the iris and the optic nerve. Although surgery is usually required to remove the obstruction causing the condition, the sulfonamides can help to alleviate the pressure, and hence permanent damage to the eye, before surgery is needed. The mechanism of action is slightly different from that occurring in the kidney. Bicarbonate ion is normally secreted into the aqueous humour, with its concomitant water,

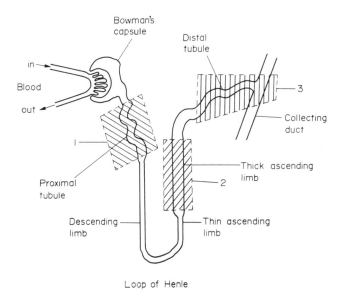

Figure 4 Schematic representation of a kidney nephron: 1, site of action of carbonic anhydrase antagonists; 2, site of action of chloride channel blockers; 3, site of action of sodium channel blockers

and the effect of the carbonic anhydrase inhibitors is to block this passage, resulting in a marked reduction in the concentration of the ion in the aqueous humour and hence the intraocular pressure.[59] The sulfonamide methazolamide (**30**) is particularly useful in the treatment of glaucoma; it is rather lipophilic and tends to concentrate in the tissue of the eyes, thus achieving the concentrations needed to have a marked effect on the carbonic anhydrase enzymes located there.

(**30**)   (**31**)

(**32**)

There is one other area where sulfonamide derivatives have found clinical applications and that is in the treatment of noninsulin dependent diabetes, such as some forms of 'maturity onset diabetes'. The sulfonamides used contain the sulfonylurea function, such as tolbutamide (**31**) and glipizide (**32**). Studies on these agents show that they do not primarily act as enzyme inhibitors. Their effect in lowering levels of glucose circulating in patients suffering from noninsulin dependent diabetes appears to rely on two separate actions. They act on the cells of the pancreas to help stimulate secretion of the natural glucose-controlling agent insulin and they potentiate the action of insulin itself. The former effect is caused by the sulfonamide binding to the membrane of the β cells of the pancreas, those involved in insulin secretion, causing a depolarization and increased calcium ion influx; since calcium ions are needed for the release of insulin, this release is stimulated.[60]

It is found that chronic dosage of the sulfonylureas in treating this dysfunction often leads to tolerance and over time the beneficial effects of these drugs subside.

ACKNOWLEDGEMENT

The author wishes to thank Professor J. K. Seydel for providing much of the information needed to write this short review.

7.1.8 REFERENCES

1. G. Domagk, *Dtsch. Med. Wochenschr.*, 1935, **61**, 250.
2. Foerster, *Zentralbl. Huetten Walzwerke*, 1933, **45**, 549; H. Hörlein, *Proc. Roy. Soc. Med.*, 1936, **29**, 313.
3. J. Trefouel, J. Trefouel, F. Nitti and D. Bovet, *CR Seances Soc. Biol. Ses Fil.*, 1935, **120**, 756.
4. A. T. Fuller, *Lancet*, 1937, **1**, 194.
5. G. A. H. Buttle, W. H. Grey and D. Stephenson, *Lancet*, 1936, **1**, 1286.
6. P. Gelmo, *J. Prakt. Chem.*, 1908, **27**, 369.
7. W. S. Churchill, 'The Second World War', 2nd edn., Cassell, London, 1962, vol. 4, p. 582.
8. K. H. Bayer and J. E. Baer, *Pharmacol. Rev.*, 1961, **13**, 517; O. B. T. Nielsen, H. Bruun, C. Bretting and P. W. Feit, *J. Med. Chem.*, 1975, **18**, 41.
9. A. Loubatieres, in 'Oral Hypoglycemic Agents', ed. G. D. Campbell, Academic Press, New York, 1969, p. 1.
10. T. H. Maren, *Drug. Dev. Res.*, 1987, **10**, 255; P. G. Benedetti, M. C. Menziani and C. Frassineti, *Quant. Struct. Act. Relat. Pharmacol. Chem. Biol.*, 1985, **4**, 23.
11. A. C. Bratton and E. K. Marshall, Jr., *J. Biol. Chem.*, 1939, **128**, 537.
12. D. D. Woods, *Br. J. Exp. Pathol.*, 1940, **21**, 74.
13. P. Fildes, *Lancet*, 1940, **1**, 955.
14. N. Anand, in 'Burger's Medicinal Chemistry', 4th edn., ed. M. E. Wolff, Wiley-Interscience, New York, vol. 2, p. 1; R. C. Allen, *Chem. Pharmacol. Drugs*, 1983, **2**, 49; J. K. Seydel, *Handb. Exp. Pharmacol.*, 1983, **64**, 25; N. Anand, *Handb. Exp. Pharmacol.*, 1983, **64**, 25.
15. L. Dupont and O. Dideberg, *Acta Crystallogr., Ser. B*, 1972, **28**, 2340.
16. L. Dupont, J. Lamotte, H. Campsteyn and M. Vermiere, *Acta Crystallogr., Ser. B.*, 1978, **34**, 1304.
17. J. D. Crawford and G. C. Kennedy, *Nature (London)*, 1959, **183**, 891, J. E. Delarge, C. L. Lapiere and A. H. Georges, *Chem. Abstr.*, 1976, **84**, 59 218.
18. A. Camerman and N. Camerman, *Can. J. Chem.*, 1975, **53**, 2194.
19. H. Tanimukai, M. Inoui, S. Hariguchi and Z. Kaneko, *Biochem. Pharmacol.*, 1965, **14**, 961.
20. L. C. Leitch, B. E. Baker and L. Brickman, *Can. J. Res., Sect. B*, 1945, **23**, 139.
21. L. E. H. Whitby, *Lancet*, 1938, **2**, 1210.
22. A. Rastelli, P. G. DeBenedetti, A. Albasini and P. G. Pecorari, *J. Chem. Soc., Perkin Trans. 2*, 1975, 522.
23. N. Rist, *Nature (London)*, 1940, **146**, 838.
24. M. Hooper, *Chem. Soc. Rev.*, 1987, **16**, 437.
25. A. D. Welch and C. A. Nichol, *Annu. Rev. Biochem.*, 1952, **21**, 633; D. D. Woods in 'Chemistry and Biology of Pteridines', Ciba Foundation Symposium Proceedings, Little, Brown, Boston, 1954, p. 220.
26. R. L. Blakley, 'The Biochemistry of Folic Acid and Related Pteridines', Ciba Foundation Proceedings, Little, Brown, Boston, 1954, p. 220.
27. E. Haslam, 'The Shikimate Pathway', Butterworths, London, 1974.
28. T. Shiota, M. N. Disraely and M. P. McCann, *J. Biol. Chem.*, 1964, **239**, 326.
29. G. M. Brown, *J. Biol. Chem.*, 1962, **237**, 536.
30. L. Bock, G. H. Miller, K.-J. Schaper and J. K. Seydel, *J. Med. Chem.*, 1974, **17**, 23; L. Bock, W. Butte, M. Richter and J. K. Seydel, *Anal. Biochem.*, 1978, **86**, 238.
31. J. K. Seydel, M. Richter and E. Wempe, *Int. J. Lepr. Other Mycobact. Dis.*, 1980, **48**, 18.
32. For a recent review see L. C. Allen, *ACS Symp. Ser.*, 1988, **363**, 91.
33. J. K. Seydel and K.-J. Schaper, in 'Enzyme Inhibitors as Drugs', ed. M. Sandler, Biological Council Publications, London, 1979, p. 53.
34. C. B. C. Boyce and B. V. Milborrow, *Nature (London)*, 1965, **208**, 537.
35. C. Hansch, in 'Drug Design', ed. E. J. Ariens, Academic Press, New York, 1971, vol. 1, p. 270.
36. J. C. Deardon (ed.), 'Quantitative Approaches in Drug Design', Elsevier, Amsterdam, 1983, p. 253.
37. C. Hansch, P. P. Maloney, T. Fujita and R. M. Muir, *Nature (London)*, 1962, **194**, 178.
38. J. K. Seydel and K.-J. Schaper, 'Chemische Struktur und Biologische Activitat von Wirkstoffen, Methoden der Quantitativen Struktur-Wirkung Analyse', Verlag Chemie, Weinheim, 1979.
39. P. H. Bell and R. O. Roblin, Jr., *J. Am. Chem. Soc.*, 1942, **64**, 2905.
40. P. B. Cowles, *Yale J. Biol. Med.*, 1942, **14**, 599.
41. A. H. Brueckner, *Yale J. Biol. Med.*, 1943, **15**, 813.
42. J. K. Seydel, *J. Med. Chem.*, 1971, **14**, 271.
43. E. R. Garrett, J. M. Mielck, J. K. Seydel and H. J. Kessler, *J. Med. Chem.*, 1969, **12**, 740.
44. J. K. Seydel, in 'Drug Design', ed. E. J. Ariens, Academic Press, New York, 1971, vol. 1, p. 343.
45. J. K. Seydel, in 'Topics in Infectious Diseases', ed. J. Drews and F. E. Hahn, Springer-Verlag, Vienna, 1974, vol. 1, p. 25.
46. G. M. Brown, *J. Biol Chem.*, 1962, **237**, 536.
47. E. A. Coates, H.-P. Cordes, V. M. Kulkarni, M. Richter, K.-J. Schaper, M. Wiese and J. K. Seydel, *Quant. Struct. Act. Relat.*, 1985, **4**, 99.
48. A. J. Hopfinger, R. L. Lopez de Compadre, M. G. Koehler, S. Emery and J. K. Seydel, *Quant. Struct. Act. Relat.*, 1987, **6**, 111.
49. J. Rieder, *Arzneim.-Forsch.*, 1963, **13**, 95.
50. For an earlier review see A. Bishop, *Biol. Rev.*, 1959, **34**, 445.
51. E. M. Wise, Jr. and M. M. Abou-Donia, *Proc. Natl. Acad. Sci., USA*, 1975, **72**, 2621.
52. M. Wiese, J. K. Seydel, H. Pieper, G. Kruger, K. R. Knoll and J. Keek, *Quant. Struct. Act. Relat.*, 1987, **6**, 164.

53. A. Bishop, in 'Drugs, Parasites and Hosts', ed. L. G. Goodwin and R. Nimmo-Smith, Little, Brown, Boston, 1962, p. 98.
54. S. R. M. Bushby and G. H. Hitchings, *Br. J. Pharmacol. Chemother.*, 1968, **33**, 72; M. R. Moody and V. M. Young, *Antimicrob. Agents Chemother.*, 1975, **7**, 836.
55. S. Ramachandran, J. T. Godfrey and N. D. W. Lionel, *J. Trop. Med. Hyg.*, 1978, **81**, 36.
56. T. Fujita, in 'Biological Correlations — the Hansch Approach', ed. R. F. Gould, American Chemical Society, Washington, DC, 1972, p. 80.
57. T. H. Maren, *Ann. N. Y. Acad. Sci.*, 1984, **429**, 568.
58. J. D. Coleman, *Annu. Rev. Pharmacol.*, 1975, **15**, 221; A. Vedani and E. F. Meyer, *J. Pharm. Sci.*, 1984, **73**, 352.
59. M. G. Garrino, H. P. Meissner and J. C. Henquin, *Eur. J. Pharmacol.*, 1986, **124**, 309.

7.2

Reductases

JOHN J. McCORMACK

University of Vermont, Burlington, VT, USA and National Institutes of Health, Bethesda, MD, USA

7.2.1 INTRODUCTION   271

7.2.2 BRIEF OVERVIEW OF FOLATE BIOCHEMISTRY   272

7.2.3 DIHYDROFOLATE REDUCTASE   272

    7.2.3.1 *Enzyme Structure*   274
    7.2.3.2 *Classical DHFR Inhibitors*   275
    7.2.3.3 *Quantitative Aspects of Drug–Enzyme Interactions*   277
    7.2.3.4 *Nonclassical Inhibitors of DHFR*   282
    7.2.3.5 *Synthesis of Nonclassical DHFR Inhibitors*   288
    7.2.3.6 *Clinical Uses and Major Toxicities of DHFR Inhibitors* 290

7.2.4 DIHYDROPTERIN REDUCTASE   290

7.2.5 ALDOSE REDUCTASE   292

7.2.6 FUMARATE REDUCTASE   294

7.2.7 TRYPANOTHIONE REDUCTASE AND GLUTATHIONE REDUCTASE   294

7.2.8 REFERENCES   295

7.2.1 INTRODUCTION

The major emphasis in this chapter will be placed on *dihydrofolate reductases*, because of the medical importance of inhibitors of these in the management of neoplastic, bacterial and protozoal diseases. The rich store of information that exists on the biochemistry of dihydrofolate reductases and the chemical determinants of interactions of drugs with these enzymes further establishes dihydrofolate reductases as excellent models for exemplifying enzymes as molecular targets (receptors) for design and development of drugs. It is appropriate, in addition, to summarize pertinent information on other reductases that can serve as targets for drug action. Attention will be given to dihydropteridine reductase, aldose reductase (sugar alcohol dehydrogenase), fumarate reductase (succinate dehydrogenase) and glutathione reductase, and potential medical applications of inhibitors of these enzymes will be discussed. The enzymes involved in conversion of ribonucleotides to deoxyribonucleotides, commonly termed ribonucleotide reductases, will not be considered in this chapter; they are discussed in the context of compounds that interfere with purine and pyrimidine metabolism at the nucleotide level (Chapter 7.3).

Compounds that inhibit dihydrofolate reductase (DHFR) have assumed an important role in clinical medicine, as exemplified by the use of methotrexate (**1b**) in neoplastic diseases[1] and rheumatoid arthritis,[2] of trimethoprim (**2a**) in bacterial diseases[3] and of pyrimethamine in protozoal diseases.[4] Recent investigations have provided evidence for antineoplastic[5] and antiprotozoal[6] activity for the potent DHFR inhibitor trimetrexate (**2b**). In a very real sense, DHFR acts as a receptor for mediating the effects of these drugs in biological systems, and a discussion of the biochemistry and medicinal chemistry of DHFRs from diverse sources provides a valuable perspective on using information from studies of drug–receptor interactions in drug design and

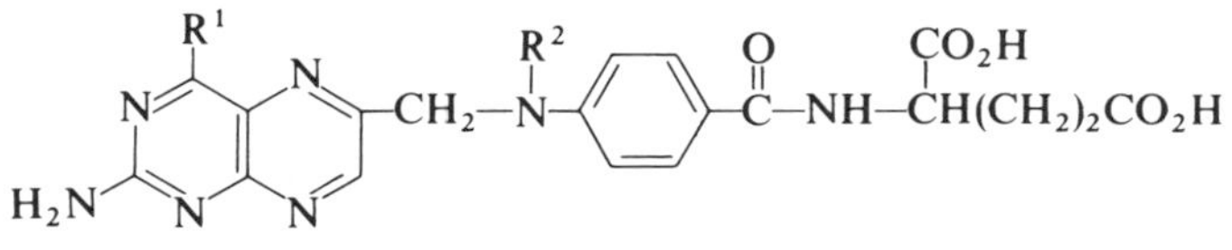

(1a)   Folic acid; $R^1 = OH$, $R^2 = H$

(1b)   Methotrexate; $R^1 = NH_2$, $R^2 = Me$

Figure 1

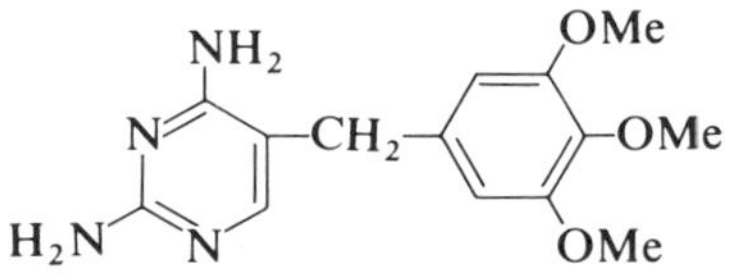

(2a)   Trimethoprim 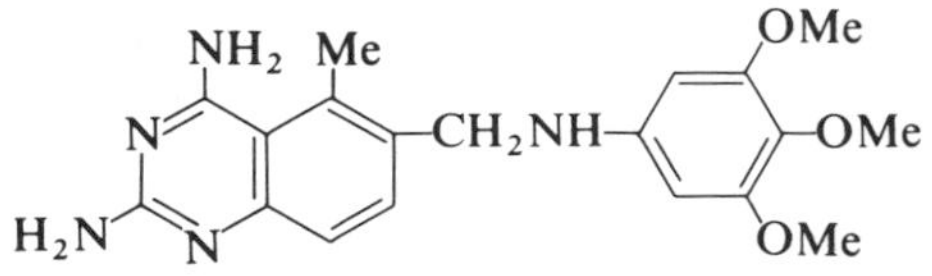   (2b)   Trimetrexate

Figure 2

development. It is appropriate, at the outset, to differentiate between a 'classical' and a 'nonclassical' folate antagonist. Those DHFR inhibitors classified as classical folate antagonists are characterized by a *p*-aminobenzoylglutamic acid side chain in the molecule and thus closely resemble folic acid itself. This relationship is exemplified in the comparison of the structures of methotrexate (1b) and folic acid (1a) shown in Figure 1. A classical antimetabolite is one in which a simple, usually isosteric alteration of a normal metabolite results in a compound with biological activity antagonistic to that of the metabolite. The term nonclassical is applied to antimetabolites in which substantial structural changes are made in relation to the metabolite.[7]

Compounds classified as nonclassical inhibitors of DHFR do not possess a *p*-aminobenzoylglutamic acid side chain but rather have a side chain that is relatively lipophilic. Representative nonclassical folate antagonists are trimethoprim (2a) and trimetrexate (2b), the structures of which are shown in Figure 2.

7.2.2   BRIEF OVERVIEW OF FOLATE BIOCHEMISTRY

Folic acid (1a; pteroylglutamic acid; Figure 1) is a water soluble vitamin that plays a crucial role in cellular economy. The vitamin is a 2-amino-4-oxopteridine with a side chain incorporating both *p*-aminobenzoic acid and glutamic acid. Folic acid must be reduced to the tetrahydro form, which is the active 'acceptor' of single-carbon units that are subsequently transferred enzymatically from an appropriate cofactor to precursor molecules that lead to the synthesis of purine and pyrimidine (*e.g.* thymine) components of nucleic acids, and of the important amino acid, methionine. Excellent reviews of the complex biochemistry of folate derivatives have been published (see also Chapter 7.1).[8] Structures of several tetrahydrofolate derivatives bearing single-carbon substituents at different oxidation states are shown in Figure 3. It is important to note that the biochemistry of folate and derived coenzymes is substantially complicated by the ability of folates to be converted to polyglutamated derivatives with additional glutamic acid residues linked by amide bonds involving the γ-carboxyl group of the preceding glutamate residue. Commentary on this aspect of folate/antifolate action will be made when discussing methotrexate but it is pertinent in this introductory section to stress that observations concerning greater intracellular retention of polyglutamates, because of their greater electrostatic charge and more effective interaction of folate polyglutamates with several enzymes, serve to indicate the biological importance of polyglutamate formation.[9]

7.2.3   DIHYDROFOLATE REDUCTASE

Dihydrofolate reductase (DHFR; tetrahydrofolate dehydrogenase; 5,6,7,8-tetrahydrofolate-$NADP^+$ oxidoreductase; EC 1.5.1.3) mediates the reaction shown below (Figure 4). Reductases of this type have been isolated from a wide variety of biological sources and represent early and, consequently, well-studied examples of discrete biochemical targets that are accurately classified as

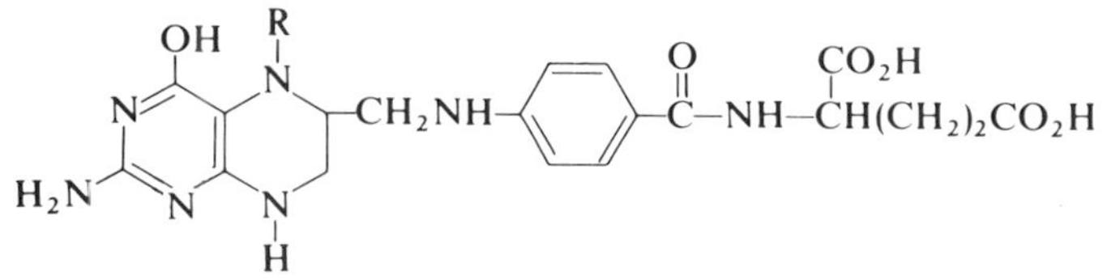

(3a) 5-Methyltetrahydrofolate; R = Me

(3b) 5-Formyltetrahydrofolate (leucovorin); R = CHO

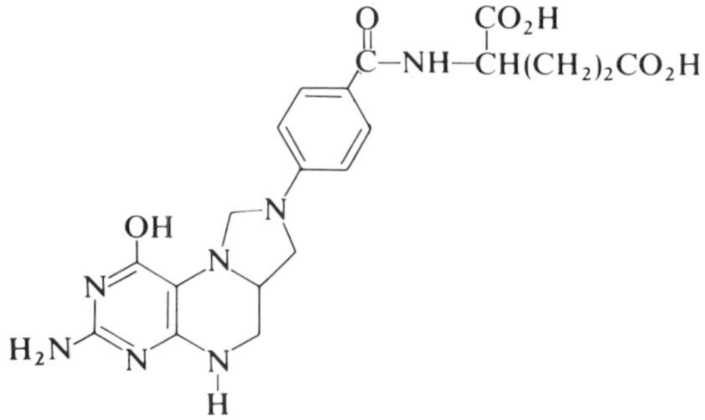

(3c) 5,10-Methylenetetrahydrofolate

Figure 3

drug receptors. It is entirely appropriate therefore, that Freisheim and Matthews began an excellent comprehensive review of comparative aspects of the biochemistry of DHFR with a description of this enzyme as 'an intracellular receptor for folate antagonist drugs'.[10] 7,8-Dihydrofolate is much more efficiently reduced by DHFR from most sources than is folate itself under conventional assay conditions (*e.g.* pH 7.0). Polyglutamate derivatives of dihydrofolate are reduced by DHFR and, in the case of an enzyme from human cells, the polyglutamates appear to be somewhat more efficient substrates.[11] NADPH is a better cofactor (cosubstrate) than is NADH for the reaction catalyzed by DHFR. It is noteworthy that analogs of NADPH have been reported to substitute for NADPH in reductions mediated by DHFR from a bacterial source.[12] The kinetic mechanism of the DHFR reaction seems relatively uncomplicated in most cases. Enzymatic reduction involves a random 'bi-bi' process in which either the substrate (dihydrofolate) or the cofactor (NADPH) form a binary complex with the enzyme with subsequent rapid binding of the second small molecule component to form the ternary complex. The transfer of a hydride ion from NADPH to dihydrofolate occurs stereospecifically and is thought to be facilitated by protonation of N-5 of the substrate molecule.[10] Despite much elegant chemical and biochemical work on DHFR, questions remain concerning the details of the mechanism of enzyme-catalyzed reduction. One possible mechanism, proposed by Filman and his colleagues,[13] for DHFR purified from *Lactobacillus casei* involves a major contribution by the enzyme to stabilization of an isomer of oxidized nicotinamide with carbonium ion character. The formation of such an isomer would be expected to facilitate hydride ion transfer (Figure 5). The mechanism of reduction appears to involve protonation of a single group (pK_a 8.9) for the reaction mediated by the enzyme from *Escherichia coli*.[14]

K_m values for dihydrofolate estimated for DHFR from avian and mammalian liver vary considerably; for example, the K_m for chicken liver DHFR was reported to be 0.1 μM, while that for beef liver was approximately 50-fold higher.[10] Estimated K_m values for typical bacterial reductases are in the range 0.5–1 μM but K_m values for dihydrofolate may be increased more than 10-fold for DHFR isolated from bacteria resistant to methotrexate or other folate antagonists.[10] Ferone has

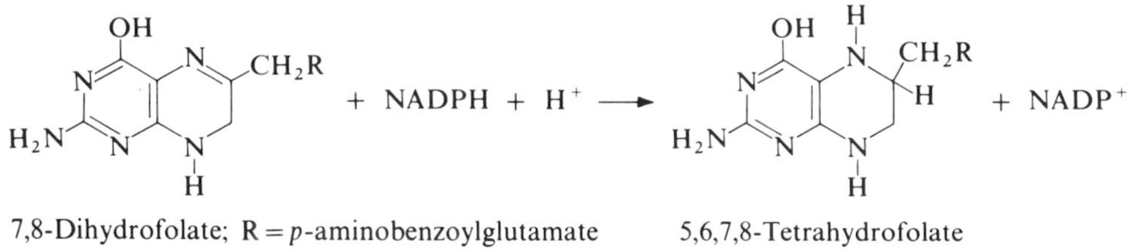

7,8-Dihydrofolate; R = *p*-aminobenzoylglutamate   5,6,7,8-Tetrahydrofolate

Figure 4

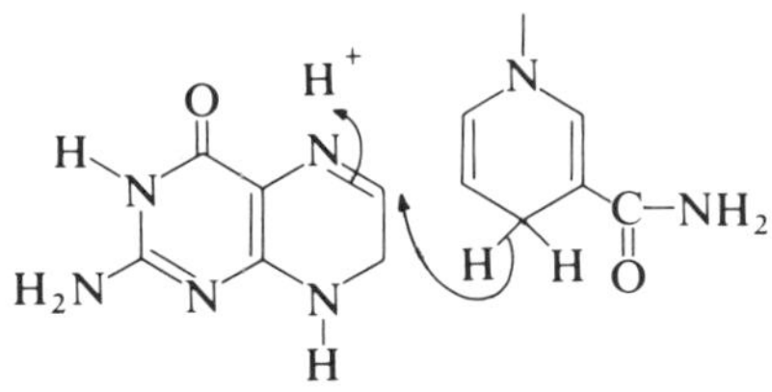

Only dihydropyridine portion of cofactor and dihydropterin portion of substrate (keto form) are shown.

Figure 5

observed a comparable alteration in the K_m for dihydrofolate in a protozoal DHFR obtained from organisms resistant to the 2,4-diaminopyrimidine pyrimethamine; the K_m for dihydrofolate for the enzyme obtained from the drug sensitive organisms was 4.2 μM, while that for the enzyme obtained from the drug resistant organisms was 51.9 μM.[15] K_m values for NADPH also vary considerably from source to source. The K_m for NADPH estimated for DHFR from chicken liver was 1.8 μM, while that for the enzyme from beef liver was 15 μM. NADPH K_m values for DHFR from bacteria generally are in the range 1–10 μM. K_m values for dihydrofolate for DHFR from trypanosomal species ranged from 3 μM to 10 μM and the corresponding values for NADPH were reported to be approximately 10 μM;[16] similar estimates of kinetic parameters were made for DHFR isolated from schistosomes and filariae.[17] Blakley's excellent review[8b] provides kinetic information for DHFRs from a wide variety of sources as well as a valuable perspective on the potential problems in comparing values obtained under differing, possibly nonoptimal conditions.

7.2.3.1 Enzyme Structure

DHFRs isolated from mammalian, avian and bacterial sources are relatively small proteins (M approximately 20 000 Da) and this fact has facilitated structural studies of these enzymes. Early observations of large molecular weights for DHFR obtained from protozoa[18,19] have been extended in recent studies that have revealed a bifunctionality of DHFR isolated from protozoa, with the enzyme complex capable of mediating both reduction of dihydrofolate and the single-carbon unit transfer that converts deoxyuridylate to thymidylate required for DNA synthesis.[20–22] X-ray crystallographic studies[23] have shown substantial similarities in the structures of DHFR from avian and bacterial sources. Such structural information has stimulated investigations aimed at ascertaining the mechanism(s) of catalysis mediated by DHFR and additional details on such investigations can be found in comprehensive reviews prepared by Blakley[8b] and by Freisheim and Matthews.[10] The amino acid sequences of DHFR from beef liver and from murine leukemia cells are remarkably similar, with 85% of the residues identical. Amino acids associated with binding of the substrate and cofactor and of representative inhibitors, such as methotrexate, to DHFR from both mammalian and bacterial sources are identical or closely similar, indicating a high degree of functional homology at critical binding sites of DHFR. Molecular biological studies of DHFR using techniques that permit specified changes in the protein have shown that changes in amino acids at the active site of DHFR from *E. coli* profoundly decrease catalytic activity; thus replacement of an aspartic acid residue by asparagine results in a protein that reduces dihydrofolate less rapidly than the native protein.[24]

Use of affinity labels that form covalent bonds with residues in the active site of DHFR has provided valuable information on the differences between DHFRs from different sources. The compound containing a reactive iodoacetyl function, depicted in Figure 6, reacted rapidly with bacterial DHFR to yield a modified histidine residue, while reaction with enzymes from vertebrates involved modification of a cysteine residue.[25] Analogous irreversible inactivation of DHFR from mouse leukemia cells and fungal cells (*Candida albicans*) has been reported by Rosowsky and his colleagues for bromoacetyl but not chloroacetyl compounds related to methotrexate.[26] Probes of the hydrophobic cavities of DHFR can be obtained by incorporating a spin-labeled group into inhibitor molecules and analyzing enzyme–inhibitor interactions by electron spin resonance spectroscopy. Such probes are capable of providing valuable quantitative information as well as qualitative information of modes of binding of different inhibitors. For example, the TEMPO function of the nonclassical pteridine inhibitor (**6a**; Figure 6) undergoes strong hydrophobic interaction with bacterial DHFR while the TEMPO function of the potent classical inhibitor (**6b**; Figure 6) binds

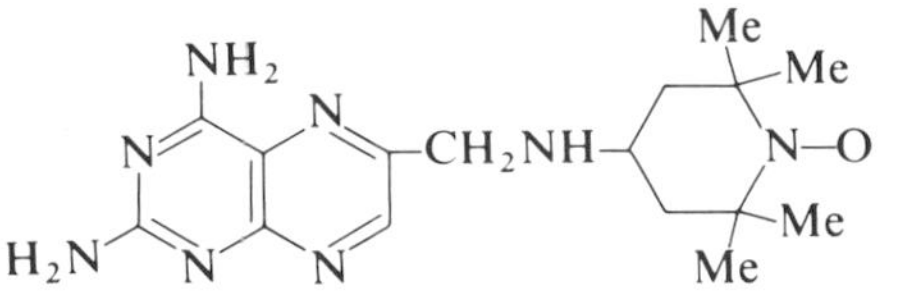

(6a)

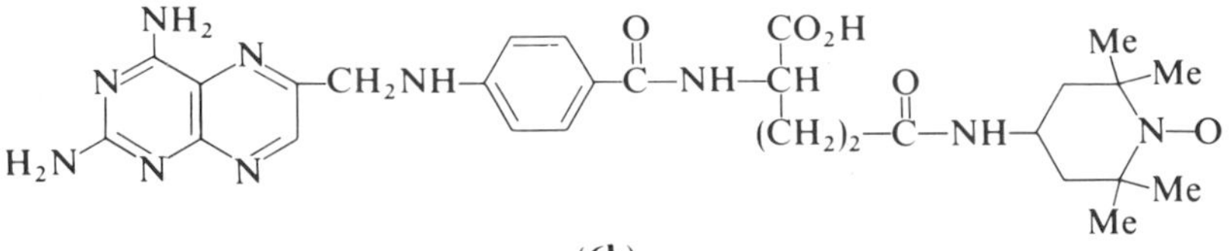

(6b)

Figure 6

weakly to different residues outside the primary binding region of the enzyme.[27] Use of X-ray crystallographic data, together with quantitative structure–activity relationships and computer graphics, is providing increasingly valuable insights into the molecular structure of DHFR and the chemical processes that govern interactions of the enzyme with inhibitors.[28]

7.2.3.2 Classical DHFR Inhibitors

Replacement of the 4-oxo function of folic acid by an amino group resulted in the class of compounds now designated as classical folate antagonists. Methotrexate is the DHFR inhibitor with the longest period of application in the treatment of human disease. The sequence of investigations leading to the introduction of methotrexate into clinical medicine has been presented by Jukes.[29] Important aspects of this sequence include the observations of temporary remissions in children with acute leukemia treated with aminopterin,[30] which differs from methotrexate in lacking a 10-methyl group, observations by Burchenal and his colleagues of antileukemic effects of methotrexate in mice and pediatric patients,[31] and observation by Hertz and his colleagues of the dramatic efficacy of methotrexate in treatment of some patients with choriocarcinoma, a gestational trophoblastic tumor, derived from placental tissue.[32] It should be pointed out that while the antileukemic activity of aminopterin was recognized before that of methotrexate, the toxicity of aminopterin was substantially greater than that observed for methotrexate and thus the latter compound was preferable for extensive clinical development. The observation by Nichol and Welch that aminopterin inhibited the conversion of folic to leucovorin (3b; citrovorum factor; 5-formyl-5,6,7,8-tetrahydrofolate) focused attention on reduction of folate derivatives as a primary mechanism of action of folate antagonists.[33] Inhibition, by aminopterin, of enzymatic reduction of folate and dihydrofolate by an enzyme obtained from chicken liver was demonstrated first by Futterman and Silverman[34] and comparable observations were soon reported for methotrexate.[35]

The clinically useful properties of methotrexate have provided impetus for the synthesis of a variety of analogs with modifications in the heterocyclic ring, the aromatic ring of the side chain and the glutamic acid residue among those reported. A compilation of analogs of methotrexate as well as summaries of methods used to synthesize methotrexate and related compounds has been contributed by Rahman and Chhabra.[36] Synthetic approaches to methotrexate have varied considerably and overall reaction sequences proceed with sharply different levels of efficiency.

The initial approach to the synthesis of methotrexate is exemplified by the method of Seeger and her colleagues[37] involving a 'one-pot' condensation of 2,4,5,6-tetraaminopyrimidine (7a) with dibromopropionaldehyde and *p*-methylaminobenzoylglutamic acid in aqueous solution (Figure 7). This procedure results in a very complex reaction mixture with a poor yield (5% or lower) of the desired material.

A much more efficient and elegant approach to the synthesis of methotrexate and analogous compounds was devised by Taylor and his colleagues;[38] this approach (Figure 7) uses a pyrazine precursor rather than a pyrimidine precursor for cyclization and involves reaction of 2-amino-3-cyano-6-chloromethylpyrazine *N*-oxide (7b) with a diester of *p*-methylaminobenzoylglutamic acid, deoxygenation of the 'conjugated' pyrazine *N*-oxide (7c) with triethyl phosphite and subsequent cyclization with guanidine and hydrolysis of the ester. A third example of a synthetic approach to classical DHFR inhibitors is that described by DeGraw and his colleagues[39] for the synthesis of

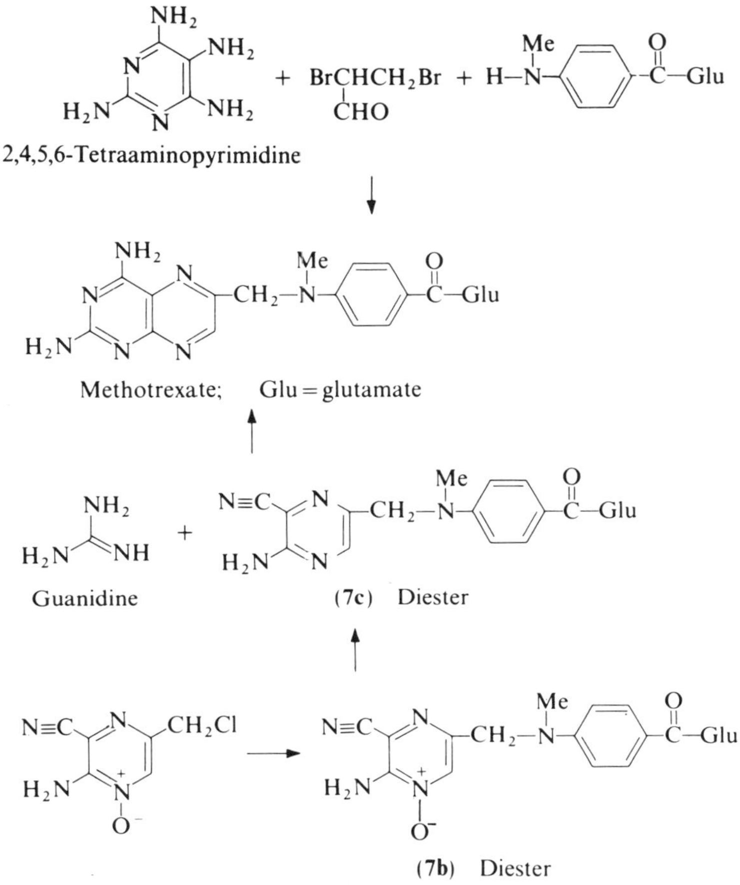
(7a) 2,4,5,6-Tetraaminopyrimidine
BrCHCH2Br
CHO
Methotrexate; Glu = glutamate
Guanidine
(7c) Diester
(7b) Diester

Figure 7

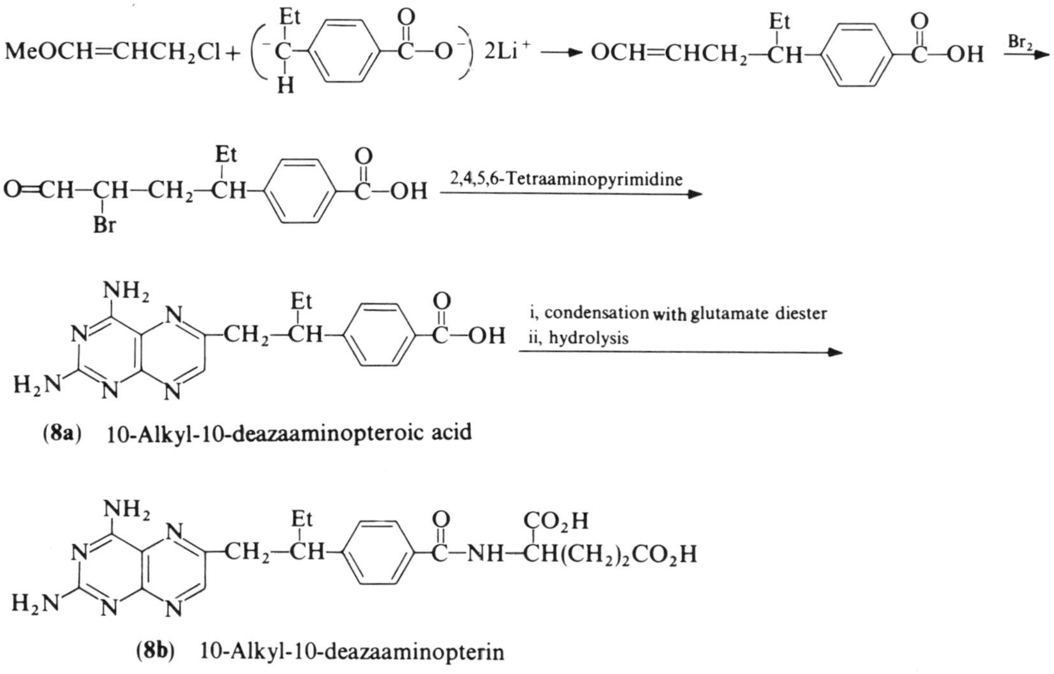
MeOCH=CHCH2Cl+
2Li+
OCH=CHCH2
Br2
2,4,5,6-Tetraaminopyrimidine
i, condensation with glutamate diester
ii, hydrolysis
(8a) 10-Alkyl-10-deazaaminopteroic acid
(8b) 10-Alkyl-10-deazaaminopterin

Figure 8

compounds ('deazaaminopterins') in which the N^{10}-methylamino group of methotrexate is replaced by a methylene unit (Figure 8). The dianion of an appropriate alkylbenzoic acid is alkylated with 3-methoxyallyl chloride and the resulting enol ether carboxylate is brominated to yield a bromoaldehyde which, in turn, is condensed with 2,4,5,6-tetraaminopyrimidine, affording the aminopteroic acid. The aminopteroic acid (**8a**) is coupled with diethylglutamic acid and the product hydrolyzed in alkali to yield the desired 10-deazaaminopterin derivative (**8b**). A somewhat different approach was taken to synthesize the individual diastereoisomers of 10-alkyl-10-deazaaminopterins.[40]

7.2.3.3 Quantitative Aspects of Drug–Enzyme Interactions

Analysis of the interaction between methotrexate and DHFR is complicated by several factors. First, the inhibition observed, while formally competitive, can exhibit, under appropriate conditions, tight-binding ('stoichiometric'[41]) characteristics that make the interaction effectively 'noncompetitive'. Second, the affinity of methotrexate for DHFR may be increased considerably in the presence of the reduced pyridine nucleotide cofactor (NADPH). Representative estimates of the inhibition constant (K_i) for binding of methotrexate to mammalian DHFRs in the range 5–7 pM were reported by Jackson and his colleagues.[42] K_i values for methotrexate inhibition of DHFR from other sources[43] are summarized in Table 1.

The values listed in Table 1 were selected to exemplify the high affinity of methotrexate for this target enzyme and are not intended as a comprehensive list of constants for these drug–enzyme interactions. It is important to note that the values tabulated relate to dissociation of methotrexate from the ternary complex and thus include the contribution of NADPH to drug binding.

The binding of methotrexate to DHFR is not a simple interaction and appears to involve secondary processes such as comparatively slow isomerization of the complex formed initially. Blakley and Cocco have speculated that isomerization 'might involve relatively subtle rearrangements of side chains, bound water and inhibitor in the active site without any notable change in the folding of the backbone' of DHFR.[44] Of considerable importance are observations by Stone and Morrison that there are substantial differences in the mode of interaction of various classes of DHFR inhibitors with DHFR from a bacterial (*E. coli*) and a vertebrate (chicken liver) source[45] but that the slow tight-binding type of inhibition observed for bacterial DHFR also is demonstrable for the enzyme isolated from chicken liver.

Especially noteworthy is the demonstration that there is selectivity in the enhancement of diaminoheterocyclic binding to DHFR produced by the cofactor NADPH. Thus, the 4,6-diaminodihydrotriazine analog (**9a**; Figure 9) binds nearly 50 000 times better to chicken liver DHFR in a ternary complex (enzyme–NADPH–inhibitor) than in the binary complex formed in the absence of NADPH. On the other hand, a 2,4-diaminopyrimidine derivative, pyrimethamine (**9b**; Figure 9) shows relatively little difference (six-fold) in binding to this enzyme in the binary or ternary complex. Stone and Morrison have commented that the ability of a compound to strongly inhibit DHFR

Table 1 Representative Estimates of Dissociation Constants for Methotrexate from DHFR Complexes

Enzyme source	K_i (nM)
Mouse liver	0.004
Rat liver	0.004
Mouse L 1210 leukemia	0.005
Human lymphoblasts	0.007
L. casei	0.190
E. coli	1.000

(**9a**) R = phenylthiomethyl   (**9b**)

Figure 9

from a vertebrate source but not a bacterial source may be due to 'favorable Van der Waals interactions that occur . . . within the active site'. In this connection it is well to recognize that Baker, over 20 years ago, focused sharply on hydrophobic bonding and van der Waals forces as determinants of the selectivity of interactions of drugs with enzymes such as DHFR.[46]

Proton magnetic resonance studies at high field (500 MHz) have provided strong evidence for conformational changes produced when NADPH complexes with bacterial DHFR.[47] Careful studies of this type provide information that, when considered in the light of X-ray crystallographic data on DHFR (especially as a ternary complex with cofactor and prototype inhibitors), permits the assignment of resonances for critical amino acid components of the protein and assessment of the interactions of DHFR inhibitors with amino acid residues.[48] Such investigations afford an invaluable basis for more incisive and rational design of inhibitors of DHFR more selective than those currently available for treatment of cancer and other diseases (*e.g.* rheumatoid arthritis) in which the target cells do not differ greatly in biochemical profiles from other cells in the body.

Formation of polyglutamate derivatives, catalyzed by the enzyme folyl polyglutamate synthetase, occurs not only with folic acid and its metabolites but also with methotrexate and other folate antagonists.[49] The polyglutamates formed from methotrexate are potent inhibitors of DHFR and therefore must be considered when evaluating the effects of methotrexate treatment.[50,51]

A classical DHFR inhibitor of potential clinical importance has been developed by DeGraw, Sirotnak and their colleagues.[39] This compound, 10-ethyl-10-deazaaminopterin (10-EDAM; **8b**), has shown excellent activity in preclinical antitumor evaluation systems[52] and may have significantly improved selectivity (lower host toxicity) compared to methotrexate, in part due to more rapid clearance from normal host tissue susceptible to toxic effects. The K_i estimated for 10-EDAM for DHFR isolated from mouse leukemia cells was 2.8 pM, essentially identical to that estimated concomitantly for methotrexate (3.2 pM).[39] 10-EDAM also inhibited DHFR from *L. casei* to an extent comparable to that observed with methotrexate. 10-EDAM has been reported to produce a relatively high response rate of 32% in patients with non-small-cell lung cancer participating in a phase II trial.[53] The provocative clinical activity of 10-EDAM is of considerable importance, especially in view of the comparatively low incidence of leucopenia and thrombocytopenia observed in the study. Degraw and his colleagues have also synthesized the diastereo isomers of 10-EDAM and found relatively little difference in the biochemical and biological activities for the stereomers.[40]

Efforts have been made to improve the selectivity of antitumor action of methotrexate by conjugating the drug with antitumor antibodies, but it seems clear that extensive work, both chemical and biological, is required in order to achieve a clinically useful preparation that will bind specifically to tumor cell surfaces, undergo internalization (endocytosis) and hydrolytic liberation of the active drug.[54]

Membrane transport of methotrexate and related classical folate antagonists is an important determinant of the biological activity of such compounds and this aspect of the pharmacology of methotrexate and other folate derivatives has been the subject of numerous investigations.[55] The entry of methotrexate, which exists as a highly polar dianion at physiological pH, into mouse leukemia cells requires energy and the active process involved also serves to transport folate metabolites such as 5-methyltetrahydrofolate. The transport of methotrexate into mouse leukemia cells has been shown to vary with the phase of cell growth;[56] the velocity of influx was faster during the logarithmic phase of growth compared with that estimated during the stationary phase. Altered transport of methotrexate has been identified as a major contributing factor to drug resistance in several cell lines and, as discussed later in this chapter, design of newer inhibitors of DHFR has involved strategies to circumvent such transport problems. Accumulation of methotrexate in cells is determined not only by an active influx process but also by an efflux process that, like the influx process, is susceptible to inhibition by probenecid (**10a**; Figure 10), a compound widely employed to inhibit cellular transport of organic anions.[57] Many analogs of methotrexate have been synthesized and evaluated as inhibitors of DHFR and as antitumor agents.[58] Such investigations have revealed a number of compounds which are potent inhibitors of DHFR which did not appear sufficiently promising to warrant further development when they were evaluated initially. One wonders whether such compounds may merit reconsideration as candidates for preclinical evaluation in newer test systems that may prove more useful in selecting compounds for development as clinical agents.[59] In any event a discussion of structure–activity relationships of selected classical folate antagonists is warranted in order to provide a summary of salient structural features that govern interactions of compounds with DHFR. At the outset, it is important to stress that the 2,4-diaminopyrimidinyl moiety of methotrexate and related compounds is essential for potent inhibition of DHFR from virtually all sources studied. This is exemplified by the dramatic reduction in potency as a DHFR inhibitor when the 2,4-diaminopteridine derivative aminopterin is compared with folic acid, the corresponding 2-amino-4-oxopteridine; for inhibition of a bacterial enzyme folic acid is 10 000-fold less potent that aminopterin[60] and for inhibition of a protozoal (trypanosomal) enzyme a similar difference was observed.[61] Alkylation of amino groups of 2,4-diaminofolate analogs also sharply reduced inhibitory potency.[62] Replacement of the 4-amino group of methotrexate by a thio function produces a 10 000-fold reduction in inhibitory potency evaluated against DHFRs from mammalian, avian, bacterial and protozoal sources.[63,64] A hydrazino substituent in position 4 also produces a substantial (over 1000-fold) loss of inhibitory activity, compared with methotrexate, against DHFR obtained from pigeon liver.[64]

Inhibition of DHFR by methotrexate is moderately stereospecific, since the isomer with L-glutamate in the side chain is approximately 10-fold more potent than that containing D-glutamate.[65] Replacement of the amide carbonyl group of aminopterin by a methylene function results in a 10–20-fold decrease in binding to DHFR from *L. casei* and L 1210 mouse leukemia cells.[66] Protonation of the additional basic nitrogen center in the analog (**10b**; Figure 10) and consequent interference with electrostatic determinants of binding to the enzyme was considered to be a major contributory factor to the lowered binding to DHFR.

Although the data obtained are for nonclassical folate antagonists, it seems useful to mention here interesting observations made by Hynes and his colleagues[67] concerning the relative potencies of 2,4-diamino-, 2-amino-4-oxo- and 2-amino-4-thio-quinazolines as inhibitors of DHFR from rat liver and the bacterium *Streptococcus faecium*. These results are summarized in Table 2.

While the 2-amino-4-thioquinazoline analog was a much less potent inhibitor of DHFR from both sources studied than the 2,4-diaminoquinazoline, the loss of activity that accompanied conversion of the 4-amino group to a 4-oxo function was considerably less than anticipated, especially for the rat liver enzyme. Extensive chemical, biochemical and pharmacological investigations have been carried out on the effects produced by removing nitrogen atoms from the pteridine nucleus of prototype classical folate antagonists such as methotrexate. Replacement of the nitrogen atom at position 3 of methotrexate gives compound (**11a**; Figure 11), which produced 50% inhibition of pigeon liver DHFR at a concentration of 63 nM, roughly five-fold higher than the concentration of methotrexate required to produce equivalent inhibition of the enzyme.[68] Compound (**11b**), lacking the 1-nitrogen of methotrexate produced 50% inhibition of DHFR at a concentration of 4700 nM, approximately 400-fold higher than for methotrexate and 75-fold higher than for the 3-deaza analog.[68] 5-Deazaaminopterin and 5-deazamethotrexate inhibit DHFR with a potency comparable to that of methotrexate and showed efficacy similar to that of methotrexate against L 1210 leukemia in mice.[69] Removal of both nitrogens from the pyrazine portion of aminopterin, yielding the quinazoline analog (5,8-deazaaminopterin; **11c**) did not substantially lower potency against DHFR.[58] The 2,4-diamino-5,8-deaza analog of aminopterin with a chloro

(**10a**) Probenecid

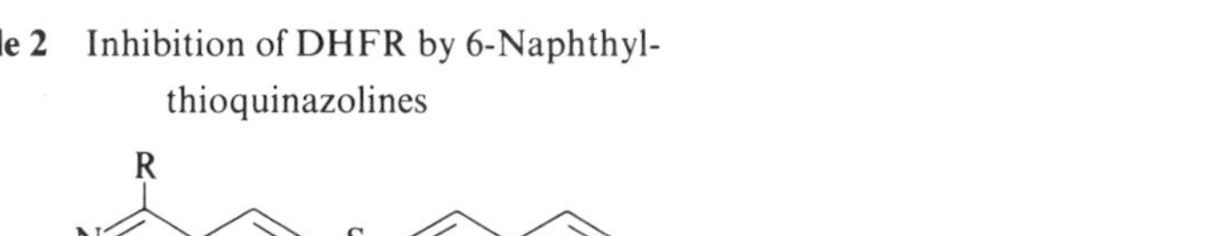

(**10b**)

Figure 10

Table 2 Inhibition of DHFR by 6-Naphthyl-thioquinazolines

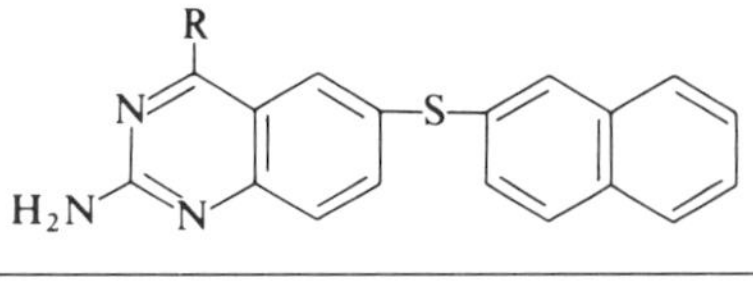

| R | IC$_{50}$ (nM) | |
	Rat liver	*S. faecium*
NH$_2$	7	7
OH	45	180
SH	230	800

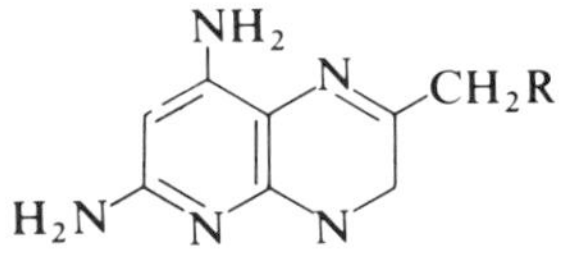

(11a) 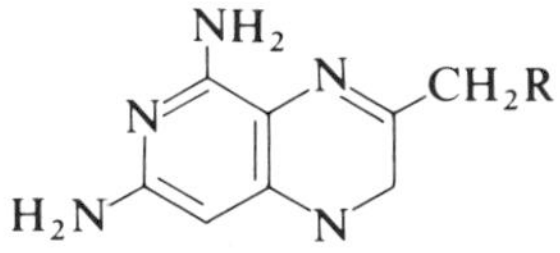   (11b) 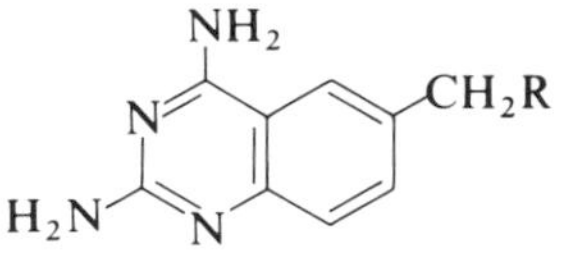   (11c)

R = p-Amino (or methylamino)benzoylglutamic acid

Figure 11

substituent in position 5 was found to inhibit the growth of some human gastrointestinal adeno-carcinoma cells *in vitro* at concentrations as much as 10 times lower than that required for equivalent inhibition by methotrexate;[70] this observation seems worth pursuing in relevant *in vivo* models. The 7-methyl analogs of both methotrexate and its 3',5'-dichloro derivative were synthesized by Rosowsky and Chen and found to be more than 1000-fold less active than the 'parent' compounds as inhibitors of DHFR from mouse leukemia and bacterial cells; the detrimental effect of the 7-methyl group for binding to the enzyme has been attributed to steric interference with binding of the side chain and, possibly, of the heterocyclic ring to the enzyme.[71] Oxidation of methotrexate and dichloromethotrexate occurs *in vivo*, resulting in 7-oxo derivatives; the enzyme aldehyde oxidase catalyzes this reaction and the oxidation of a variety of other heterocyclic compounds.[72,73] The 7-oxo metabolites are several hundred times less active than are the precursors as inhibitors of DHFR.[72] The 3',5'-dichloro derivative of methotrexate is readily obtainable by chlorination of methotrexate under appropriate reaction conditions;[74] chlorination of methotrexate in the benzene ring does not adversely alter binding to DHFR. Although dichloromethotrexate has been evaluated previously in the clinic and considered to offer no prominent advantages over methotrexate, it deserves mention that aspects of the clinical pharmacology of this drug are still being elucidated.[75] One reason for the continued interest in dichloromethotrexate is the relatively rapid hepatic clearance of this drug that may permit its use with drugs such as cisplatin which are nephrotoxic and thus dangerous to use with methotrexate, which itself may produce renal toxicity related to its high renal clearance.[76] In addition renal impairment produced by cisplatin would be expected to exacerbate the toxicity of methotrexate but not of the dichloro derivative because the latter compound is cleared primarily by the liver.

The *p*-aminobenzoylglutamic acid side chain *per se* is not essential for potent binding to DHFR, as evidenced by high activity of *nonclassical* DHFR inhibitors discussed later in this chapter. Nevertheless, for classical inhibitors, this side chain contributes substantially to interaction with DHFR, as demonstrated by the 60–70-fold enhancement in inhibitory potency for methotrexate compared to the pteroic acid derivative that lacks the glutamic acid moiety of the side chain.[68] 2,4-Diaminopteridine derivatives that lack the side chain are very poor inhibitors of DHFR; thus, 2,4-diamino-6-methylpteridine is nearly 20 000 times less active than methotrexate (Table 3).[68]

Additional modifications of the structure of methotrexate include replacement of the glutamic acid moiety with other amino acids, alteration of the bridging function (carbon 9–nitrogen 10) that links the diaminoheterocyclic nucleus and the benzoylglutamic acid function, and formation of esters and amides in the glutamic acid residue. The promising activity of the 10-ethyl-10-deazaaminopterin analog has been mentioned previously in this chapter and Montgomery and Piper[58] and Rahman and Chhabra[36] have provided useful summaries of the pertinent medicinal chemistry of a variety of other classical folate antagonists. Since there is little evidence for superiority

Table 3 Inhibition of DHFR by 2,4-Diaminopteridines

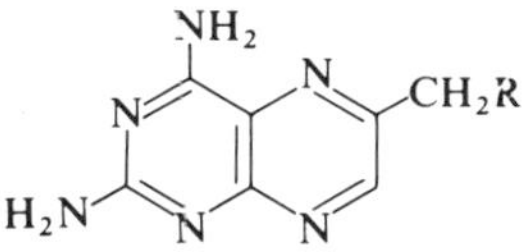

R	IC_{50} (nM)
H	21 000
p-Methylaminobenzoic acid	88
p-Methylaminobenzoylglutamic acid	1.3

of most of the analogs over methotrexate, only a few representative examples will be selected to further illustrate structure–activity relationships in this series. It must be borne in mind that much creative chemical work has characterized the investigations aimed at synthesis of methotrexate analogs and new information on the cellular pharmacokinetics and pharmacodynamics of folate antagonists in cancer cells may refocus attention on some of these compounds as candidates for drug development.

Esters of methotrexate have been studied extensively as potential drugs, initially because of the expectation that the esters, being more hydrophobic than methotrexate, would overcome cellular transport barriers that exist because of the ionization of the carboxylate functions of methotrexate. The pharmacology of such esters is complicated by esterase-mediated hydrolysis that occurs at variable rates in different animal species.[77]

Methotrexate diesters retain the ability to inhibit DHFR, albeit at a level less than methotrexate itself. Inhibition, in such diesters, is related to the length of the alkyl substituent on the ester function and this is exemplified by the 100-fold increase in potency for the dibutyl ester compared to the dimethyl ester.[78] Monoesters of methotrexate at both the α- and γ-carboxyl groups have been synthesized and evaluated biochemically and biologically.[79] Table 4 summarizes the results of studies of the butyl esters of methotrexate as inhibitors of DHFR from rabbit liver. The two monobutyl esters exhibited identical inhibitory activity in this system and this activity was comparable (only two-fold difference) to that of methotrexate; on the other hand the dibutyl ester was nearly 20-fold less potent than methotrexate. A later study using DHFR from L 1210 cells found the α-butyl ester to be 10-fold less active than the γ-butyl ester, which gave an IC_{50} of 5.4 nM, closely similar to that of methotrexate (3.3 nM) determined under identical reaction conditions.[80]

The γ-t-butyl ester of methotrexate is closely similar to methotrexate in terms of potency as an inhibitor of DHFR purified from mouse leukemia (L 1210) cells with 50% inhibition observed at a concentration of 55 nM with the ester and 67 nM with methotrexate.[81] The possibility exists that the higher susceptibility of the t-butyl ester of methotrexate to oxidation by aldehyde oxidase can be exploited for better selectivity in treatment of some cancers.[82] When the carboxyl functions of methotrexate are converted to amides, a profound difference in inhibitory activity against L 1210 DHFR was observed between α- and γ-amides. The γ-amide inhibits DHFR slightly better than methotrexate but the α-amide is approximately 100 000-fold less potent;[58] this enormous difference in the activities of two structurally related compounds is observed also when the activities of α- and γ-amino acid conjugates are compared. The vastly different comparative DHFR inhibitory potencies observed for the esters in relation to that observed for the amides clearly indicates substantially different modes of binding to the enzyme.

Rosowsky and his colleagues have extended their investigations of structure–activity relationships in the classical antifolate series to the synthesis and evaluation of aminopterin analogs in which the glutamic acid amide function is attached at *ortho* and *meta* positions of the benzene ring rather than at the *para* position. The *ortho* and *meta* isomers (**12a**, **12b**; Figure 12) both proved to be extremely weak inhibitors of L 1210 DHFR, with inhibitory potencies approximately 100 000-fold lower than that for aminopterin itself.[83] The *ortho* and *meta* positional isomers were also found to be inactive as substrates for mouse liver folyl polyglutamate synthetase, an additional pronounced difference from the *para* isomer. Rosowsky's group also has reported an interesting activity profile for a compound in which the propionic acid moiety of methotrexate was moved from the α-carbon to the adjacent nitrogen (**12c**; Figure 12). This compound inhibited the growth of L 1210 cells at a concentration (IC_{50} of 12 000 nM) much higher than that observed for methotrexate (IC_{50} of 5 nM). The estimated

Table 4 Inhibition of DHFR by Methotrexate Esters

R^1	R^2	IC_{50} (nM)
H	H (methotrexate)	9
Bu	H	17
H	Bu	17
Bu	Bu	140

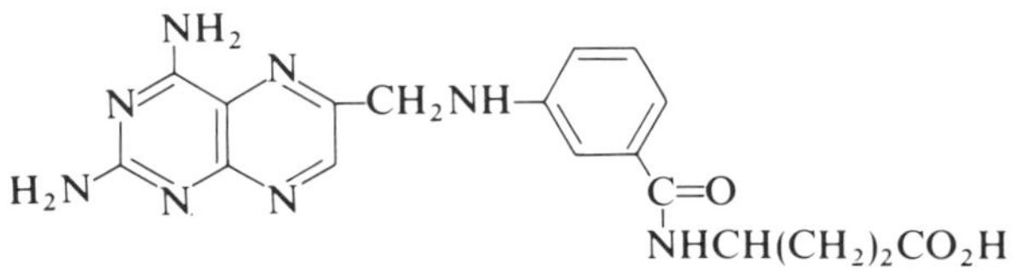

(12a)

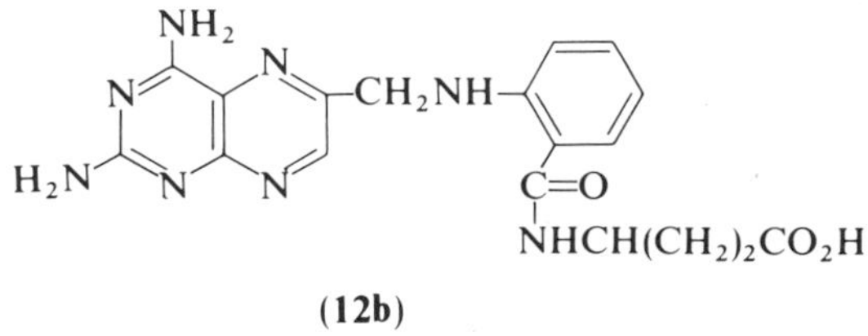

(12b)

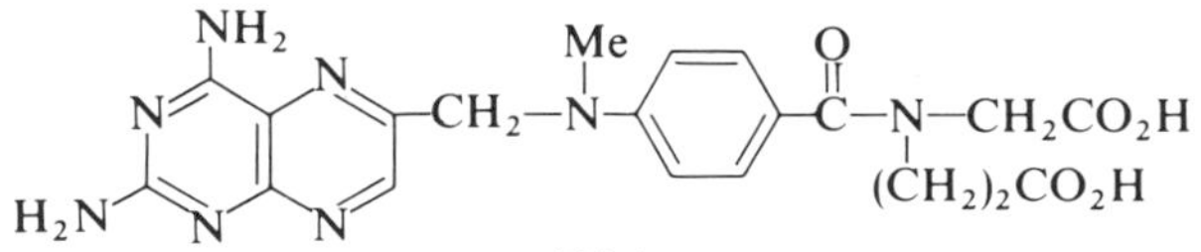

(12c)

Figure 12

Table 5 Representative Interactions Between *E. coli* DHFR and Methotrexate

Methotrexate function	Interaction type	DHFR amino acid
N-1 (base)	Ionic	Asp-27 (acid)
α-Carboxyl	H bonds	Arg-57
2-NH$_2$	H bond	Thr-43
4-NH$_2$	H bond	Ile-5 (carbonyl)
Pteridine ring	Hydrophobic	Ile-5, Ala-7, Phe-31, Ile-94
Benzene ring	Hydrophobic	Leu-28, Ile-50, Leu-54, Ile-94

IC$_{50}$ (38 nM) for inhibition of DHFR differed relatively little from that of methotrexate but unusual curvilinearity in the plot of enzyme inhibition *versus* concentration of inhibitor made it difficult to assess the potency of the compound.[83] The unusual nature of the concentration–inhibition curve for this compound may make it an interesting probe for structural studies of DHFR, especially if the curvilinearity is different with enzymes from different sources.

The final compound to be mentioned in this discussion of classical folate antagonists is the isomer of methotrexate in which the *p*-methylaminobenzoylglutamic acid function is attached at position 7 rather than at position 6. The 7-isomer was much less effective than methotrexate in treatment of L 1210 leukemia in mice; unfortunately DHFR inhibitory activity does not appear to have been evaluated for this compound.[84]

Roth[85] has summarized concisely (Table 5) some of the important interactions deduced from X-ray crystallographic data concerning the interaction of methotrexate with DHFR from *E. coli* (binary complex). Ionic bonding, hydrogen bonding and hydrophobic bonding are all indicated as contributing to the stabilization of the drug–receptor complex. It is noteworthy that, in the complex with the substrate dihydrofolate, interactions with DHFR are comparable except that the orientation of the pteridine ring is altered by a 180° rotation so that the association with aspartate 27 occurs at N-5, which is most basic nitrogen in the 7,8-dihydropteridine ring, while for methotrexate the most basic nitrogen is at N-1. This 'flip' in the orientation of the pteridine portion of the molecule permits dihydrofolate to interact with the aspartate residue that is thought to participate as a proton donor in the reduction of the 5,6-double bond catalyzed by the enzyme.

7.2.3.4 Nonclassical Inhibitors of DHFR

It is essential to begin a discussion of nonclassical inhibitors of DHFR with a clear statement of the importance of the investigations of Hitchings and his colleagues[86] and Baker and his colleagues[46] in providing crucial structure–activity relationships that have permitted insights into the comparative biochemistry of DHFR and that deepened our understanding of chemical determinants of drug action. Initial evidence that compounds which we now classify as nonclassical DHFR inhibitors act by interfering with folate metabolism was provided by Daniel and Norris, in 1947, who found that antibacterial effects of 2,4-diaminopteridines such as (**13**; Figure 13)[87] were reversed by folic acid. Similar observations were made by Hitchings and his colleagues in studies of other 2,4-diaminopyrimidines such as (**13b**).[88] The medical importance of nonclassical inhibitors of DHFR is exemplified by the established utility of trimethoprim (**13b**) in bacterial[3] and protozoal[89] diseases, of pyrimethamine (**13c**) in protozoal diseases and the investigational use of trimetrexate (**13d**) and piritrexem (**13e**) in neoplastic diseases,[5] protozoal diseases[6] and psoriasis.[90] Trimethoprim, a 2,4-diaminopyrimidine with a trimethoxybenzyl substituent at position 5 of the heterocyclic ring, is an important compound, not only because of its substantial clinical activity but also because of its relatively early identification by Burchall and Hitchings as a prototypic highly selective inhibitor of DHFR.[91] The selectivity of trimethoprim is summarized in Table 6 in comparison with that of two other folate antagonists in clinical use. Inspection of the table reveals that there is relatively little difference in the susceptibilities of the four enzymes listed to inhibition by the classical inhibitor methotrexate. It is recognized that there are pitfalls in comparing IC_{50} values for a given drug as an inhibitor of enzymes from different sources, in the absence of definitive data on the kinetics of inhibition as well as of the reaction itself.[101] Nevertheless, even at a relatively phenomenological level, the consistency of susceptibility of DHFRs from most sources studied to inhibition by methotrexate is remarkable. The pattern of sensitivities of DHFRs to the 5-benzylpyrimidine trimethoprim differs sharply from that of methotrexate; thus the mammalian enzyme is approximately 50 000-fold less sensitive to inhibition than is the bacterial enzyme obtained from *E. coli*. The 5-phenylpyrimidine pyrimethamine inhibits DHFR from the malaria parasite *Plasmodium berghei* at a concentration more than 1000-fold lower than that required to inhibit the enzyme from mammalian liver. The relative selectivities of trimethoprim and pyrimethamine clearly are of importance in determining the clinical utility of these compounds in bacterial and protozoal diseases. DHFR isolated from another parasitic protozoan, *Trypanosoma equiperdum*, exhibited still another pattern of sensitivities to the inhibitors studied, especially with regard to the relatively small difference in the concentrations of trimethoprim and pyrimethamine required to produce equivalent enzyme inhibition. This pattern is similar for DHFR isolated from other trypanosomal species as well.[16]

In general the chemical determinants of the potencies of nonclassical DHFR inhibitors are similar to those established for classical inhibitors. Thus 2,4-diaminopyrimidine derivatives are bound much more tightly to the enzyme than are the corresponding 2-amino-4-oxo compounds, as indicated in Figure 14.[93] The importance of the 2,4-diaminopyrimidinyl moiety is further exemplified by observations that 2,4-diamino-6,7-diphenylpteridine (**14c**) is a moderately potent inhibitor

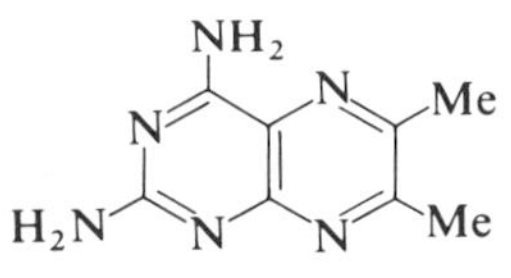

(13a) 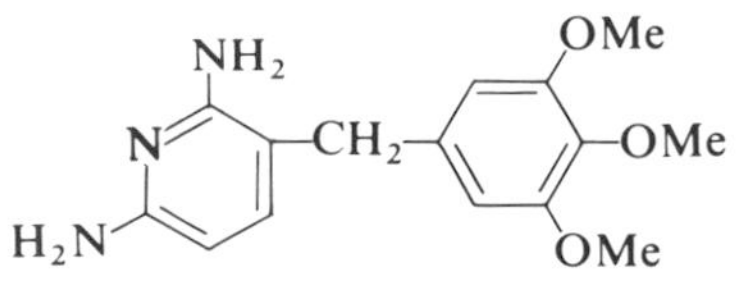 (13b) 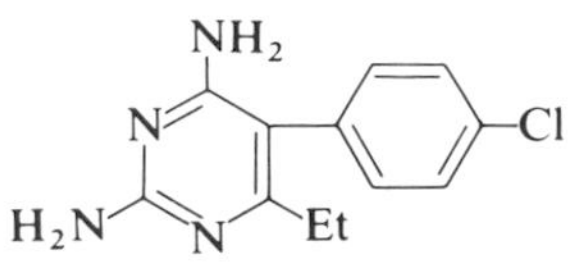 (13c)

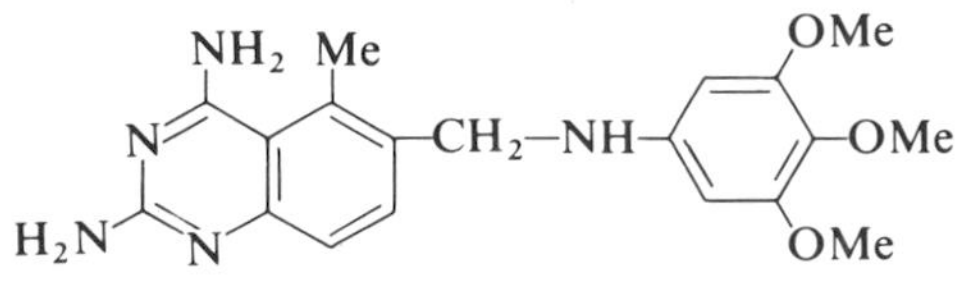

(13d) 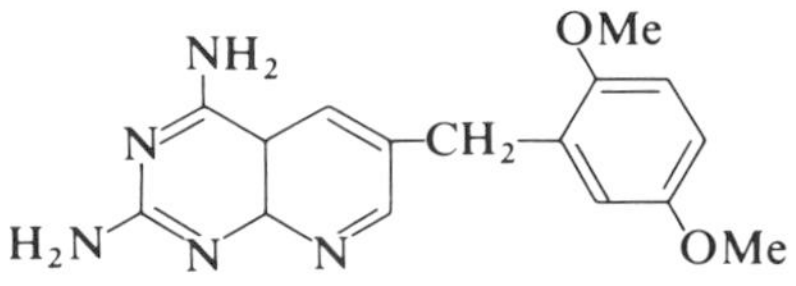 (13e)

Figure 13

Table 6 Inhibition of DHFRs from Various Sources

Compound	IC_{50} (nM)			
	Rat liver	E. coli	T. equiperdum	P. berghei
Methotrexate	2	1	0.2	0.7
Pyrimethamine	700	2500	200	0.5
Trimethoprim	260 000	5	1000	70

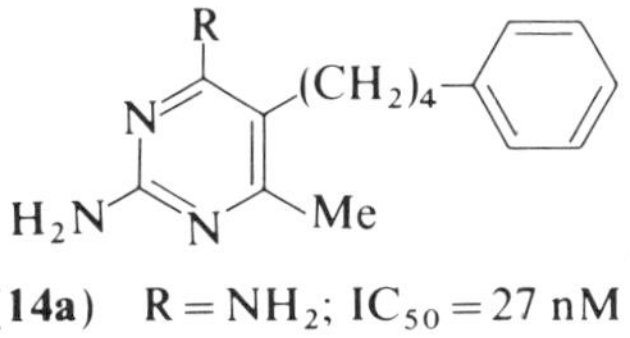

(14a)   R = NH$_2$; IC$_{50}$ = 27 nM

(14b)   R = OH;   IC$_{50}$ = 30 000 nM

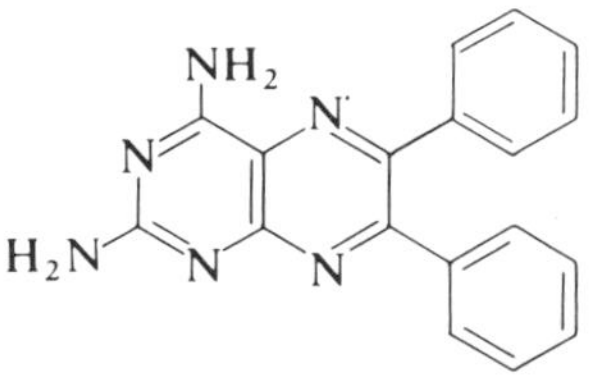

(14c)

Figure 14

Table 7 Inhibition of DHFR by Pyrimidine Derivatives

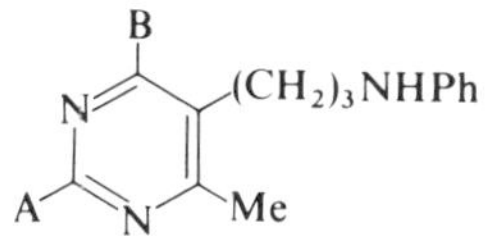

A	B	IC$_{50}$ (μM)
NH$_2$	NH$_2$	2
NH$_2$	H	950
H	NH$_2$	140
NH$_2$	NMe$_2$	58

(IC$_{50}$ = 400 nM) of a protozoal DHFR and that removal of either the 2- or the 4-amino group abolishes activity.[94] Similar observations for inhibition of DHFR from pigeon liver by 2,4-diamino, 2-amino and 4-amino derivatives of 5-(3-anilinopropyl)-6-methylpyrimidine have been reported by Baker and Ho[95] as summarized in Table 7, although it should be noted that the decrement in binding of only 30-fold when the 4-amino function is replaced by a dimethylamino function is unexpectedly small.

The proposed mode of binding of trimethoprim to bacterial DHFR shows important similarities to the binding of methotrexate, such as the ionic interaction of N-1 with Asp-27 and the major contributions of hydrophobic interactions to stabilization of the drug–DHFR complex.[96] In several instances it has been possible to show that binding of NADHP to DHFR enhances the binding of an inhibitor molecule to a very large degree. This enhancement of binding of one ligand in the presence of a second ligand is known as positive cooperativity. Such cooperativity is exemplified by the pronounced (more than 200-fold) enhancement of trimethoprim binding to *E. coli* DHFR in the presence of NADPH.[97] Baccanari and his colleagues have reported some interesting observations about cooperativity with DHFRs obtained from fungi and bacteria.[98] DHFR from *Candida albicans* was 3000-fold less sensitive to inhibition by trimethoprim than was the enzyme from *E. coli* and, in addition, the binding of the substrate, dihydrofolate, was diminished in the presence of NADPH (*i.e. negative* cooperativity). These investigators also found that the positive cooperativity for trimethoprim binding to a bacterial DHFR was 100–500-fold higher than observed for mammalian enzymes. The importance of hydrophobic interactions for binding of 2,4-diaminoheterocyclic compounds to DHFR from various sources has been well documented.[46,99]

A useful illustration of the contribution of hydrophobic interactions to enzyme inhibitory potency is provided by comparison of the inhibitory activity of compound (15a; *n* = 4) with that of compound (15a; *n* = 10) evaluated against DHFR from rat liver. The 2,4-diaminocycloalka[*g*]pteridine with four methylene groups attached to the pyrazine portion of the molecule is 1000-fold less potent than the compound with 10 methylene groups. Comparable observations were made for DHFRs isolated from L 1210 leukemia cells and the protozoan *Trypanosoma cruzi*.[100] Taira and Benkovic[101] provided additional evidence for the importance of hydrophobic interactions in determining the efficacy of inhibitor binding to DHFR. These investigators have shown, using site-directed mutagenesis, that binding of the weak DHFR inhibitor 2,4-diamino-6,7-dimethylpteridine is approximately 200-fold less effective, evaluated against the enzyme from *E. coli*, when an aromatic amino acid, phenylalanine, is replaced by valine at position 31 of the protein sequence. A substantial

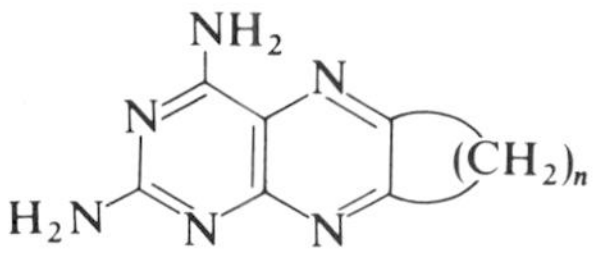

(15a) 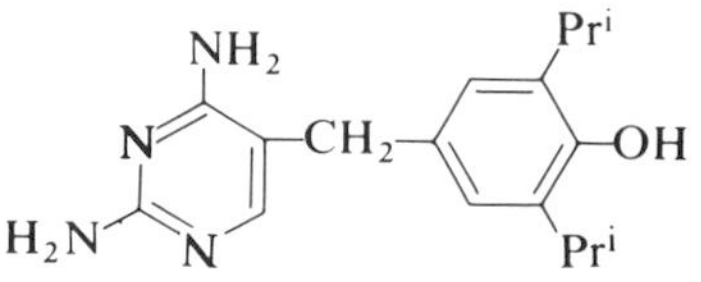 **(15b)**

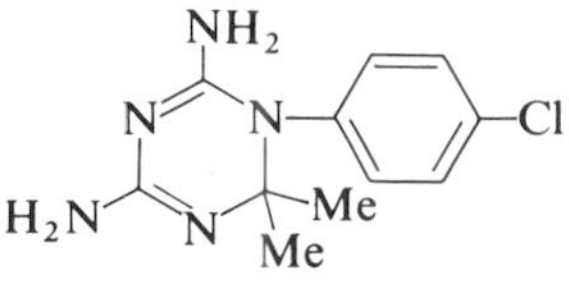

(15c) 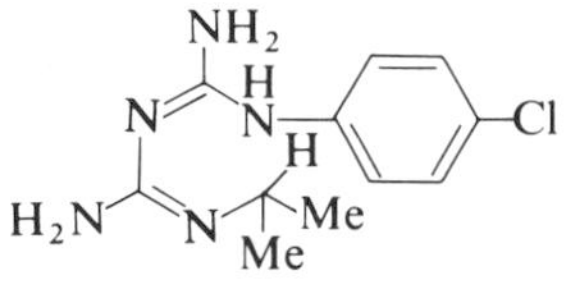 **(15d)** Proguanil

Figure 15

decrease (140-fold) in the binding of methotrexate to *E. coli* DHFR was observed when phenylalanine was replaced by valine, indicating the prominent contribution of hydrophobic interactions as well as ionic ones to the interaction of methotrexate and related compounds to DHFR.

2,4-Diamino-5-(4-hydroxy-3,5-diisopropylbenzyl)pyrimidine (**15b**) has been reported to be a potent ($IC_{50} = 29$ nM) inhibitor of DHFR from the gonococcus *Neisseria gonorrhoeae*, with much less inhibitory activity ($IC_{50} = 7200$ nM) against DHFR from rat liver; this compound was 15-fold more potent than trimethoprim against the gonococcal reductase and also more potent (20–30-fold) than trimethoprim in suppressing the growth of *N. gonorrhoeae in vitro.*[102] One of the difficulties in drug development is illustrated in the report by Roth and her colleagues[111] dealing with compound (**15b**) and related agents. The profile of activity for (**15b**) appeared to be promising, based on *in vitro* information, and pharmacokinetic studies were initiated to further define the preclinical pharmacological properties of the drug. Unfortunately the extensive and rapid metabolism in animals involving oxidation of a methyl group to a carboxyl group resulted in a loss of priority for the compound as a candidate for development.

4,6-Diamino-1,2-dihydro-2,2-dialkyl-1-phenyl-*s*-triazines (**15c**) have been known for many years to be potent folate antagonists[77] and inhibitors of DHFR. The antimalarial agent chlorguanide (**15d**; proguanil) owes its characteristic activity to metabolic conversion to the dihydrotriazine by mammals.[103] Quantitative structure–activity relationships (QSAR) have been established for nonclassical DHFRs from several sources and this work can be exemplified by the contributions of Hansch and his colleagues.[99] QSAR for 4,6-diaminodihydrotriazines as inhibitors of DHFR from the protozoal parasite *Leishmania major* can be represented by the following equation:[104]

$$\log 1/K_i = 0.65\pi_3' - 1.22\log(\beta 10^{\pi'3} + 1) - 1.12I_{OR} + 0.58MR_y + 5.05.$$

In this equation K_i is the inhibition constant for a compound of general structure (**15c**), π_3' is an expression of the hydrophobicity of the 3-substituent in the phenyl ring, I_{OR} is a so-called indicator variable that serves to express the presence or absence of a given function or structural feature in a given molecule and in this case represents the presence of an alkoxy function in the phenyl ring, MR_y (molar refractivity) represents steric effects and polarizability of a substituent in this subset of dihydrotriazines. Of considerable interest is the observation that there is a very good correlation between values determined for inhibition of leishmanial DHFR and those determined for inhibition of growth of the parasites *in vitro*. There are indications that dihydrotriazines with large hydrophobic substituents have modest selectivity for inhibition of DHFR from *L. major* relative to inhibition of human DHFR. It is instructive to compare the QSAR expression for inhibition of leishmanial DHFR by dihydrotriazines to that for inhibition of human DHFR isolated from lymphoblastoid cells shown below.[105]

$$\log 1/K_i = 1.07\pi_3' - 1.10\log(\beta 10^{\pi_3} + 1) + 0.50I + 0.82\sigma + 6.07$$

Prominent differences in the two equations include the inclusion or not of expressions for molar refractivity (MR) and electronic effects (σ) of substituents. These differences indicate that electronic influences of the 3-substituent are relatively inconsequential in determining inhibitor binding to *L. major* DHFR but that this enzyme is sensitive to the bulk and polarizability of the substituent. A useful perspective on the QSAR approach, modified by X-ray crystallographic data, in medicinal chemistry has been presented by Hansch and his colleagues:[106]

'The advantage of QSAR is that by probing an enzyme with a well-designed set of congeners, one can obtain information about the active site region in terms of its hydrophobic, steric and electronic requirements for ligand interaction . . . . Practicing QSAR is somewhat like a blind man defining an object or a room by exploring it with his hands: eventually a good image can be developed. The use of crystallography is somewhat like giving sight to the blind man: one can immediately 'see' what sort of ligands it would be foolish to make from, say, the steric or hydrophobic point of view and which ligands might be good bets to test.'

Roth[107] has contributed a clear, concise account of the use of X-ray crystal structures and a molecular mechanics computer program in the design of DHFR inhibitors more potent than the prototype compound trimethoprim. Of particular interest and importance is the demonstration that ionic and hydrophobic sites on the enzyme can be exploited in drug design. Two nonclassical DHFR inhibitors are currently under clinical investigation. These compounds are the 2,4-diaminoquinazoline derivative trimetrexate (**13d**) and the 2,4-diaminopyrido[2,3-*d*]pyrimidine derivative piritrexem (**13e**). Elslager and his colleagues described the potent antiprotozoal and antibacterial activity of 2,4-diaminoquinazolines and related pyridopyrimidine derivatives;[108] several such compounds were found to inhibit DHFR from L 1210 cells and rat liver in a range of ($IC_{50} = 10^{-8}$–10^{-9}) comparable to that observed for methotrexate.[109] Striking structure–activity relationships can be demonstrated in this series of compounds with regard to inhibition of DHFR. For example insertion of a chloro substituent at position 5 of the quinazoline nucleus (Figure 16) results in an increase of nearly 10 000-fold in inhibitory potency for DHFR obtained from *L. casei* and of approximately 1000-fold for the enzyme from *T. cruzi*. Increases in inhibitory potency were also noted for DHFRs from four sources: rat liver, L 1210 cells, *L. casei* and *T. cruzi*, for trimetrexate compared with the analogous compound lacking the 5-methyl group. The high potency of several 2,4-diaminoquinazolines as inhibitors of DHFRs from neoplastic and protozoal cells led to the suggestion that such compounds, or appropriately substituted analogs, might ultimately find application in the treatment of neoplastic or protozoal diseases.[63]

Trimetrexate serves as the prototype quinazoline inhibitor of DHFR because of the relatively extensive preclinical and clinical information available about this drug. Sharper definition of the potential promise of trimetrexate as an antineoplastic agent was provided by observations by Bertino and his colleagues that trimetrexate inhibited strongly the growth of human cancer cells *in vitro*, was more efficiently transported into cells than methotrexate, and showed substantial activity against experimental neoplasms in mice, particularly in neoplasms that were relatively insensitive to methotrexate.[110] The increased efficiency of trimetrexate transport over that of methotrexate appears attributable to the lack, in trimetrexate, of the glutamic acid chain with its two anionic centers. It is of importance to note that trimetrexate is capable of inhibiting the growth of human cell lines resistant to methotrexate because of impaired drug transport.[111,112] Impressive synergistic activity has been reported for trimetrexate and several established anticancer drugs (adriamycin, cyclophosphamide, 6-thioguanine[113]) in P 388 leukemia in mice. In the case of adriamycin, combination with trimetrexate produced a cell kill roughly three logs greater than could be attained with either drug alone. A high degree of synergism was also found for trimetrexate and cyclophosphamide and such observations indicate the desirability of future clinical trials of trimetrexate in combination with other anticancer drugs.

Several phase I trials have been completed with trimetrexate administered in a variety of regimens,[114] ranging from a rapid bolus injection to continuous infusion protocols. A study at the University of Vermont employed a regimen in which trimetrexate, as the glucuronate salt, was given as a short (10 min) intravenous infusion daily for 5 d, with treatment repeated at intervals of approximately one month. A representative clinical toxicological profile was observed in this study; for example the primary dose-limiting toxicity was myelosuppression, with both leucopenia and thrombocytopenia ascribed to drug treatment.[115] Thrombocytopenia was the primary toxicity

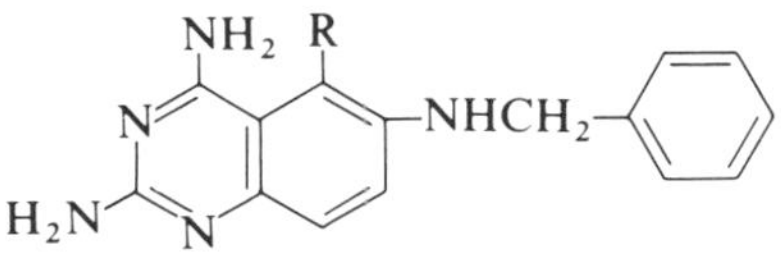

(**16a**)   R = H; $IC_{50} = 28\,000$ nM (against *L. casei*)

(**16b**)   R = Cl; $IC_{50} = 4.5$ nM (against *L. casei*)

Figure 16

observed when trimetrexate was administered as nine consecutive daily bolus injections, a regimen chosen on the basis of optimal preclinical efficacy studies.[116]

Types of nonhematological toxicity noted by Stewart and his colleagues included gastrointestinal problems (nausea, vomiting and diarrhea), oral mucositis, skin rash and somnolence. Elimination of trimetrexate in this study was found to be described by either a biexponential equation or a triexponential equation. 'Distribution' (α) half-lives were under 30 min, and the terminal elimination half-lives were in the range 2–9 h (β) for the two-compartment model. Terminal elimination half-lives (τ) for the three-compartment model were in the range 5–14 h, except for one patient for whom a terminal elimination half-life of over 30 h was estimated.

In addition to standard assays employing HPLC,[117] methods are available for determining trimetrexate concentrations in biological samples by exploiting the high potency of trimetrexate as an inhibitor of DHFR.[118,119] Very early in the course of clinical investigation of trimetrexate it was noted that concentrations of the drug measured by the biochemically based assay (enzyme inhibition) were substantially higher, particularly at time points several hours after drug administration, than those estimated by HPLC assay.[114] This discrepancy has been explained by the finding that trimetrexate is converted metabolically to compounds that retain the ability to inhibit DHFR and that, accordingly, are 'seen' as trimetrexate equivalents in the enzyme inhibition assay. The susceptibility of trimetrexate to oxidative demethylation was reported first by Heusner and Franklin[120] and the sensitivity of the oxidative process to inhibition by cimetidine has been noted by these workers, using rat hepatic microsomes, and by Webster and her colleagues, using the isolated perfused rat liver.[121] Relatively little trimetrexate was excreted unchanged by the kidney in most patients participating in phase I trials but it should be noted that, in a patient with low plasma clearance, presumably reflecting a defect in hepatic metabolism of the drug, a large amount of the drug could be accounted for, unchanged, in the urine.[115] One major metabolite of trimetrexate has been identified in mice and in human patients;[122] this metabolite is formed by oxidative demethylation at position 4 of the trimethoxyphenyl substituent and subsequent, apparently rapid, conjugation of the phenol to form a glucuronide. The glucuronide and other polar metabolites of trimetrexate are thought to contribute little to the toxicity or efficacy of the drug, even though they are capable of strongly inhibiting DHFR, because they are not transported into cells. It must be emphasized, however, that the activity of metabolites of trimetrexate is not a settled question and considerable investigation remains to be done to clarify this point adequately. Concise reviews of the preclinical and clinical pharmacology of trimetrexate are available[114,123] and are excellent sources for additional information about this drug. There has been great enthusiasm generated for trimetrexate as an anticancer agent because of provocative, albeit generally minor, responses to treatment in phase I trials.[114,115] Significant activity has been reported for treatment with trimetrexate in patients with breast cancer, head and neck cancer and non-small-cell lung cancer[5,124,125] in patients participating in phase II trials but the rates of response are low (approximately 20%) and some studies have been negative.[126,127] It is far from clear that the schedules currently used for administration of trimetrexate are appropriate for ultimate clinical use.[124]

Balis and his colleagues have studied the clinical pharmacology of trimetrexate in pediatric patients[129] and found that the dose-limiting toxicities for the drug given as an intravenous bolus once a week for three weeks, were both hematological (granulocytopenia and thrombocytopenia) and nonhematological (mucositis, diffuse pruritic rash). Terminal elimination half-lives estimated in this study varied considerably, with the range of estimates from 2 h to 20 h, and the mean for all patients approximately 8 h. It is interesting to note that pharmacokinetic studies in adult patients also revealed a wide range of elimination half-lives (2 to >30 h). Balis and his colleagues also observed the presence in the urine of pediatric patients of metabolic products of trimetrexate, one of which appears to be a glucuronide; these investigators also reported that trimetrexate does not penetrate the central nervous system efficiently, in agreement with prior animal studies. The poor transport of the drug into the central nervous system was considered to be related to high plasma protein binding of the drug. Kamen and his colleagues have completed a pediatric phase I investigation of trimetrexate in which the drug was administered as a 30–45 min infusion daily for 5 d, with the course of treatment repeated every 21 d.[130] In this study, for leukemia patients, the dose-limiting toxicity was mucositis. These investigators considered trimetrexate elimination from plasma to be best represented as a biexponential process with half-lives (α, β respectively) of 20 min and 5–6 h.

Patients with acquired immunodeficiency syndrome (AIDS) are highly susceptible to a variety of opportunistic infections including pneumonia due to *Pneumocystis carinii* and focal encephalitis due to *Toxoplasma gondii*.[131] Allegra and his colleagues[132] have shown that trimetrexate is 1000 times more potent than trimethoprim as an inhibitor of DHFR from *P. carinii*, and comparable differences in inhibitory potency were observed[133] for inhibition of *T. gondii* DHFR by trimetrexate relative to the inhibitory potencies of trimethoprim and pyrimethamine. Such findings are of considerable importance because combinations of trimethoprim or pyrimethamine with a sulfonamide, standard therapy for infections with *P. carinii* or *T. gondii*, are poorly tolerated by AIDS patients. Although trimetrexate is not a selective inhibitor of DHFR from *P. carinii* or *T. gondii* the drug can be given with the tetrahydrofolate derivative leucovorin, which is transported into mammalian cells but not into the parasitic cells, thus affording a pharmacokinetically determined selectivity. Clinical investigations have been carried out with trimetrexate and leucovorin in AIDS patients with *P. carinii* pneumonia and the drug combination was found to be effective and reasonably well tolerated. Intensive studies are in progress to define the utility of trimetrexate and related DHFR inhibitors in the treatment of infections that are common in AIDS patients. It is conceivable that other nonclassical inhibitors of DHFR[77,134] may emerge as superior alternatives to trimetrexate for use in such infections.

2,4-Diaminopyrido[2,3-*d*]pyrimidines (5-deazapteridines), such as aspiritrexem, are potent inhibitors of DHFR from several sources.[77] Pyridopyrimidine derivatives inhibit mammalian DHFR approximately as effectively as do comparable quinazolines. Piritrexem, substituted at position 6 with a dimethoxybenzyl group and at position 5 with a methyl group, is a potent inhibitor of DHFR isolated from human leukemic cells; its IC_{50}, in this system, of 5 nM is essentially identical with that determined for methotrexate.[135] Piritrexem also exhibits cytotoxicity for cancer cells in culture that is closely comparable to that observed for methotrexate; for example, the IC_{50} values for piritrexem and methotrexate, respectively are 22 and 16 nM for inhibition of the growth of L 1210 cells. The activity of piritrexem in preclinical anticancer evaluation models[136] was sufficient to warrant initiation of broad clinical trials that are still in progress.[137] The activity described for methotrexate in the treatment of psoriasis has suggested the possible role of nonclassical inhibitors of DHFR in this condition; in this connection it is of interest to note that piritrexem accumulates in rat skin to a moderate extent, with concentration of the drug in skin being 16-fold higher than in plasma several hours after intravenous administration of the drug.[136] It is important to clarify a point pertinent to the classification of compounds such as piritrexem and trimetrexate as 'lipid soluble' DHFR inhibitors. These compounds are certainly more lipid soluble than the dianionic methotrexate in conventional physiological media, and this accounts, to a large extent, for the facile entry of these compounds into cells that are poorly permeable to methotrexate. However the concentrations of both piritrexem and trimetrexate in brain tissue of animals was much lower than anticipated. Phase I evaluations of piritrexem revealed that the major dose-limiting toxicity for the drug is hematological with leucopenia, granulocytopenia and thrombocytopenia reported;[137] a concentration dependent thrombophlebitis was also described when the drug was administered intravenously. Other toxicities reported included oral mucositis and skin rash. Laszlo and his colleagues also investigated the clinical toxicity of piritrexem after oral administration and found that the toxicological profile resembled that determined for intravenous administration except for a higher incidence of gastrointestinal toxicity, manifested as nausea and vomiting.[137] The gastrointestinal problems could be mitigated by giving the oral dose twice daily. A phase II trial of piritrexem, administered orally, has indicated that the drug is only marginally active in patients with non-small-cell lung cancer.[138] Clinical pharmacokinetic studies with piritrexem have been carried out and an elimination half-life in the range 3–5 h has been reported;[156] moderate liver dysfunction has been observed to be associated with a substantial prolongation of elimination (half-life = 14 h), consistent with expectation based on preclinical studies demonstrating extensive metabolic alteration as a primary mode of piritrexem clearance.[136]

7.2.3.5 Synthesis of Nonclassical DHFR Inhibitors

Two approaches to the synthesis of 2,4-diamino-5-benzylpyrimidines related to trimethoprim have been described by Roth and her colleagues.[102] One approach involves the acid-catalyzed condensation of a 2,6-dialkylphenol with 2,4-diamino-5-hydroxymethylpyrimidine to yield the desired 2,4-diamino-5-(3,5-dialkyl-4-hydroxybenzyl)pyrimidine. A second approach involves base-mediated condensation of a phenolic Mannich base with uracil to yield 5-(3,5-dialkyl-4-hydroxybenzyl)uracil, which is chlorinated and the chloro compound subsequently aminated to yield the candidate DHFR inhibitor (Figure 17).

A sequence reported by Russell and Hitchings[139] is representative of approaches to the synthesis of 5-arylpyrimidines of the pyrimethamine type. The initial reaction of this sequence is the formation of the cyano ketone from the condensation of phenylacetonitrile and ethyl propionate in the

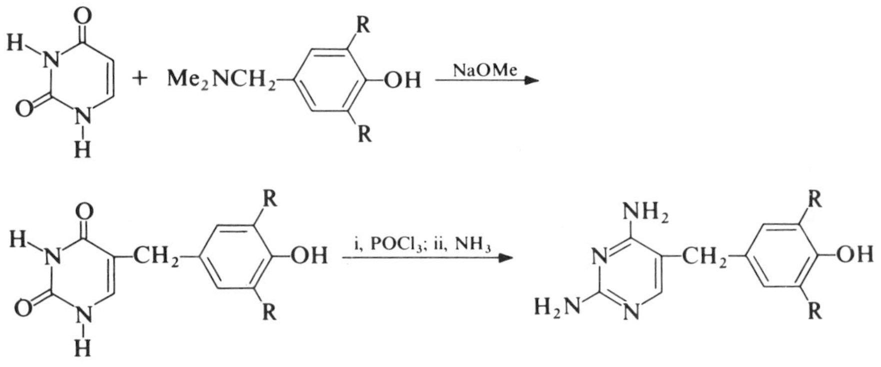

Figure 17

R = Bu^t

presence of sodium ethoxide; treatment of the cyano ketone with diazomethane results in the formation of the enol ether which is condensed with guanidine to yield pyrimethamine (**18a**; Figure 18). The method described by Modest[140] has been used extensively to synthesize a wide variety of dihydrotriazines that act as folate antagonists. In this method (Figure 18) a substituted aniline hydrochloride, dicyandiamide, and acetone are reacted together with heating and, luckily, the product of the reaction, such as (**18b**), often crystallizes from the reaction mixture. 2,4-Diamino-5-methyl-6-(3,4,5-trimethoxyanilino)methylquinazoline (trimetrexate), as indicated earlier, is of considerable potential interest as an antiprotozoal and antineoplastic agent. Key steps in the synthesis of trimetrexate[141] are outlined in Figure 19. Condensation at the chloro and cyano functions of (**19a**) with guanidine results in ring closure to give the 2,4-diamino-6-nitroquinazoline (**19b**). The nitroquinazoline is reduced to the 6-amino derivative, which is then converted to the nitrile by a

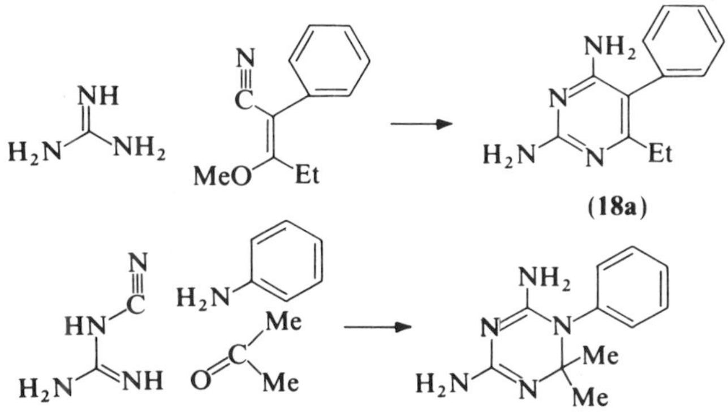

Figure 18 (**18b**)

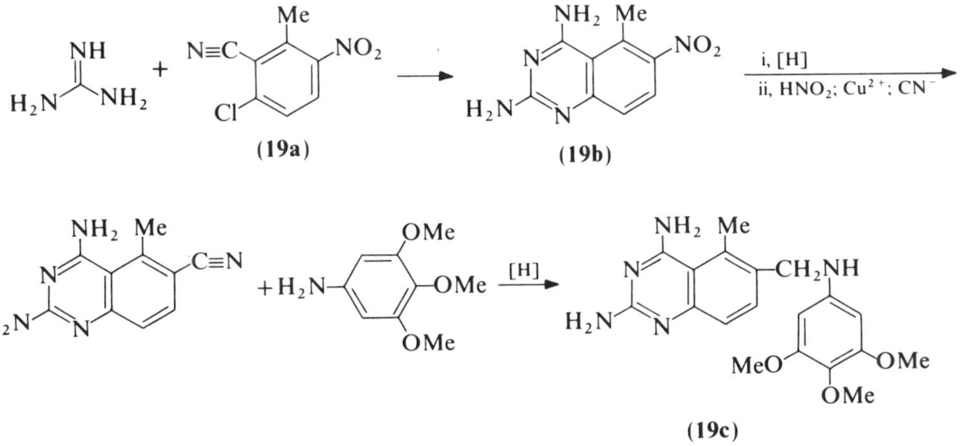

Figure 19

Sandmeyer-type reaction (diazotization and treatment with copper cyanide). Reduction of the nitrile in the presence of the trimethoxyaniline affords trimetrexate (**19c**) in a yield of approximately 30%.

7.2.3.6 Clinical Uses and Major Toxicities of DHFR Inhibitors

Methotrexate has established value in the treatment of acute lymphocytic leukemia in children, and in the treatment of patients with gestational trophoblastic tumors such as choriocarcinoma. Several other neoplastic diseases, such as breast cancer, cancer of the head and neck, and Burkitt's lymphoma, are treated with regimens that include methotrexate. High does of methotrexate, with 'rescue' from toxicity to normal cells achieved by use of leucovorin, have been found to produce beneficial effects in osteogenic sarcoma, although there is no universal agreement about the value of high dose methotrexate regimens.[142,143] Methotrexate also has well-documented activity in psoriasis that is poorly responsive to other more conventional therapeutic modalities. Numerous reports have appeared concerning the effectiveness of methotrexate, presumably functioning as an immunosuppressive agent, in severe rheumatoid arthritis.[144] Toxic effects of methotrexate are produced on the bone marrow, oral mucosa and the gastrointestinal tract; kidney and liver damage also are recognized potential hazards of treatment with methotrexate. Concomitant use of acidic drugs such as aspirin or sulfonamides can exacerbate methotrexate toxicity by displacing the drug from binding sites on plasma proteins. The pharmacokinetic pattern for methotrexate in humans is complex owing, in part, to intracellular conversion to polyglutamate derivatives that may contribute importantly to biological activity. The elimination of methotrexate from plasma is a triexponential process with representative half-lives of 30 min (α), 3 h (β) and 10 h (γ).[145]

Pyrimethamine (Daraprim) is administered orally, usually combined with a sulfonamide or sulfone, for the prophylaxis and treatment of falciparum malaria. Pyrimethamine and a sulfonamide are also used in the treatment of toxoplasmosis. The pyrimethamine regimen for toxoplasmosis is more intensive than that for malaria and leucovorin is administered to patients being treated with pyrimethamine for toxoplasmosis to reduce the likelihood of host toxocity with no effect on antiprotozoal efficacy. Potential toxic effects of pyrimethamine reflect interference with folate metabolism and include hematological toxicity (megaloblastic anemia, leukopenia and thrombocytopenia). Pyrimethamine elimination is very slow in humans with elimination half-lives of 3–5 d reported.[146]

Trimethoprim, in combination with an appropriate sulfonamide, is active against a wide range of bacteria and other microorganisms. Trimethoprim itself has significant antibacterial activity but the combination formulation of trimethoprim with sulfamethoxazole (Bactrim, Septra) is the prototype for use in the chemotherapy of bacterial diseases. Representative established clinical applications include the treatment of urinary tract infections, acute otitis media, enteritis due to *Shigellae* and bronchial infections. Trimethoprim and sulfamethoxazole is an effective treatment for pneumonia caused by *Pneumocystis carinii*. Trimethoprim is cleared moderately efficiently in humans with elimination half-lives in the range 10–12 h.[147] The toxicity of trimethoprim–sulfonamide combinations frequently seems due primarily to the sulfonamide component, although folate deficiency manifestations can be observed to be produced by trimethoprim in patients with marginal folate status. The incidence of untoward reactions to trimethoprim–sulfonamide treatment has, as mentioned earlier, been alarmingly high in patients with AIDS.

7.2.4 DIHYDROPTERIN REDUCTASE

Dihydropterin reductase (EC 1.6.99.7; NADPH:6,7-dihydropteridine reductase; DHPR) from most sources is an enzyme of moderate size ($M \approx 50\,000$ Da) that is dimeric. The enzyme is also called dihydropteridine reductase but the dihydropterin reductase designation seems more appropriate, since representative naturally occurring substrates are all reduced pterins (2-amino-4-oxopteridines). DHPR catalyzes the reaction shown in Figure 20. The enzymatic reaction is distinctive in that the actual substrate is not a conventional 7,8-dihydropterin but rather a quinonoid derivative. A succinct review of DHPR and of phenylalanine hydroxylase, which requires a tetrahydropterin cofactor provided by DHPR, has been provided by Shiman[148] and a more comprehensive review of DHPR has been contributed by Armarego and his colleagues.[149] The natural tetrahydropterin cofactor is tetrahydrobiopterin and it plays a critical role in hydroxylation reactions mediated by aromatic amino acid hydroxylases that subserve the continued synthesis of biogenic amines such as dopamine, norepinephrine and 5-hydroxytryptamine. The substrate specificity of DHPR is not particularly narrow since quinonoid pterins with several different substituents in positions 6 and/or 7 will be reduced and even appropriately substituted pyrimidines[150] can serve as substrates. Estimated K_m values for dihydropterin substrates, with methyl substituents in positions 6 and 7, are in the range of 10–100 μM with values for dihydrobiopterin being somewhat lower. NADH and NADPH both will serve as cofactors (cosubstrates) for DHPR from most sources. Although NADH often appears to be a more effective cofactor, kinetically, for the reduction of quinonoid dihydropterins, Shiman has emphasized that more favorable concentrations of NADPH *in vivo* may make this compound an important physiological cofactor in biological systems.[148] The point made by Shiman deserves careful attention, not only for DHPR, but also for other enzyme systems, because optimization of biochemical reactions in easily manipulated reaction mixtures *in vitro* may not properly reflect conditions *in vivo* and this could lead to misleading conclusions about biochemical processes *in situ*. There is insufficient structural information available that permits commentary on molecular details of the catalytic mechanism for the reduction, although there is evidence that the enzyme binds the reducing agent (*e.g.* NADH) first, with subsequent addition of the dihydropterin. The hydride transfer from NADH occurs in a stereoselective manner. A model for the reduction involving protonation at N-3 of the quinonoid substrate, facilitating hydride transfer to N-5, has been proposed.[149] Deficiency or inhibition of DHPR can lead to neurological dysfunction owing to the role of tetrahydropterin cofactors in the synthesis of biogenic amine neurotransmitters.[150] Neurotoxicity associated with excessive aluminum concentrations in serum may be due, at least in part, to inhibition of DHPR by aluminum.[151] DHPR differs from other reductases discussed in this chapter in that drugs of clinical relevance are likely to be *substrates* for the enzyme rather than *inhibitors*. Nevertheless representative studies of inhibitors of DHPR deserve mention because they elaborate structure–activity relationships for the enzyme. Quinonoid 6,6,8-trimethyl-7,8-dihydro(6)pterin (**20a**; Figure 20) was found to be inactive as a substrate for DHPR but it was capable of inhibiting the enzyme, albeit at a relatively high concentration (IC$_{50}$ = 200 μM).[152] The lack of substrate activity was considered to be due to steric effects of the 8-methyl group, since the compound lacking this group exhibits substrate activity. Gessner and his colleagues have reported that hydroxylated derivatives of 4-phenylpyridine, 4-phenyl-1,2,3,6-tetrahydropyridine and 4-phenylpiperidine are moderately potent inhibitors of DHPR[153] from human liver and from rat brain. The inhibition seems to be of the noncompetitive type and reduction of the heterocyclic nucleus led to a diminution in enzyme inhibitory activity as exemplified by the much lower IC$_{50}$ (0.38 μM) for the pyridine derivative (**21a**; Figure 21) compared to that (44 μM) for the piperidine (**21b**; Figure 21) evaluated against human liver DHPR. DHFR inhibitors such as aminopterin and 2,4-diamino-6,7-dimethylpteridine inhibit DHPR from mammalian sources but there seems to be no correlation between abilities to inhibit these enzymes. Several quinones, including those generated by oxidation of catecholamines and other catechol derivatives, can inhibit DHPR. It is important to consider the possibility that the activity of inhibitors with catechol substituents, as inhibitors of DHPR, is due to conversion of the catechol to a quinone.[149] Webber and her colleagues[154] have investigated a photoaffinity probe for DHPR; the probe consists of an arylazido-β-alanyl function attached to NADH. The probe appears to react

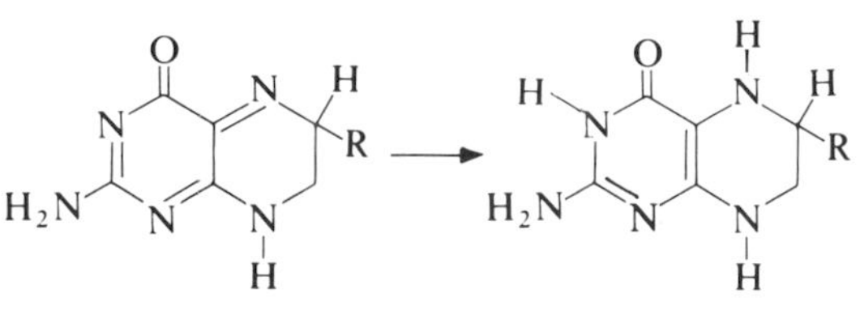

R = CH(OH)CH(OH)Me

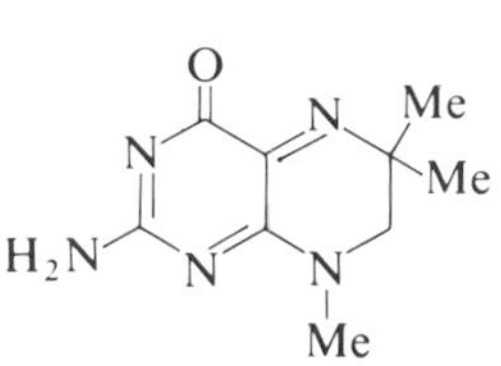

Figure 20

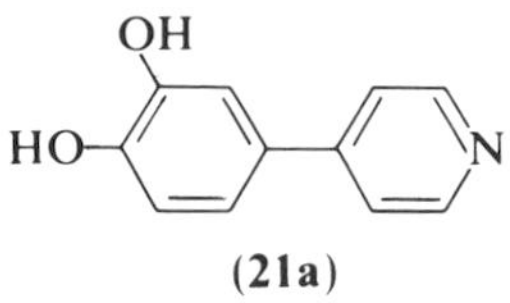

(**21a**) 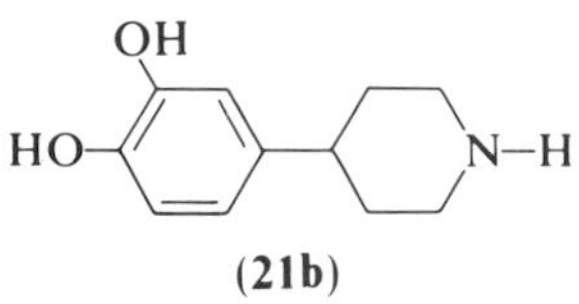   (**21b**)

Figure 21

specifically with DHPR and is a relatively potent inhibitor of the enzyme ($K_i = 0.1–0.4$ μM) that inhibits in a formally competitive manner with respect to the pyridine nucleotide cofactor. The amino acid sequence has been deduced for DHPR from mammalian liver and the enzymes from human and rat liver show a high degree of homology.[155] Crystals of DHPR suitable for three-dimensional structure determination have been described but no detailed structure has yet been reported.[156]

7.2.5   ALDOSE REDUCTASE

Aldose reductase (alditol:NADP$^+$ oxidoreductase; EC 1.1.1.21) is also called sugar alcohol dehydrogenase. This enzyme catalyzes the reduction of glucose to sorbitol (Figure 22); a second enzyme, sorbitol dehydrogenase, oxidizes sorbitol to fructose. This two-enzyme sequence is known as the polyol (sorbitol) pathway. Aldose reductase (AR) has a broad substrate specificity and many other sugars are reduced by the enzyme, as are other aldehydic compounds. NADPH is the cosubstrate for the reaction. NADH is not an effective substitute for NADPH. The enzyme exhibits stereospecificity; for example, D-glucose is a much better substrate than is L-glucose. Transfer of the hydride ion from the reduced pyridine nucleotide cofactor is also stereospecific and is comparable to the process observed for other oxidoreductases such as glucose-6-phosphate dehydrogenase.[157] Estimates of the molecular weight of AR range from 25 to 50 kDa. Sugar alcohols produced by AR have been linked with pathological changes in the eye and other organs observed in patients with diabetes and interest is high in assessing the feasibility of minimizing the incidence and severity of such complications by inhibiting AR. The level of interest in this area of research is exemplified by several informative reviews that summarize salient aspects of the medicinal chemistry, biochemistry and pharmacology of AR inhibition.[158–160] Larson and his colleagues[161] have reviewed extensively progress in the medicinal chemistry of AR inhibitors. It seems important, at the outset, to mention the caveat emphasized by these investigators that 'attempts at defining *the* aldose reductase inhibitor binding site are seriously compromised by the lack of evidence that the structurally unrelated inhibitors considered in the analysis bind in a similar fashion to the enzyme'.

Information on the structure of AR is very fragmentary but it appears that cationic sites, consisting of protonated basic amino acids, are responsible for binding of the reduced pyridine nucleotide cofactor and that there is substantial hydrophobic character in the substrate–binding region. A wide variety of compounds has been evaluated for inhibitory activity against AR; emphasis is placed on investigation of the enzyme from the ocular lens but it should be recognized that AR is present in other tissues that are damaged in diabetic patients and that the sensitivities of AR in different tissues to a given inhibitor may be quite different. The prototype AR inhibitor is sorbinil, a spirohydantoin, with an imidazolidinedione (hydantoin) linked (spiro) to a substituted chroman ring (22). Inhibition of AR by sorbinil is stereoselective as evidenced by the 30–100-fold differences in inhibitory potencies between the (R) and (S) forms of the drug. Kador, Kinoshita and Sharpless have stated that sorbinil 'may be considered a benchmark by which the potency and effectiveness of other aldose reductase inhibitors are measured'.[162] Sorbinil does not appear to compete either with the sugar (aldehyde) substrate or with NADPH for binding to AR since inhibition observed is formally noncompetitive.[163] Representatives (Figure 23) of other structural types of compounds that inhibit AR are the rhodanine-3-acetic acid derivative (epalrestat; ONO-2235; 23a), the phthalazineacetic

Glucose   +   NADPH $\rightleftharpoons$ Sorbitol   +   NADP$^+$
(Aldehyde)

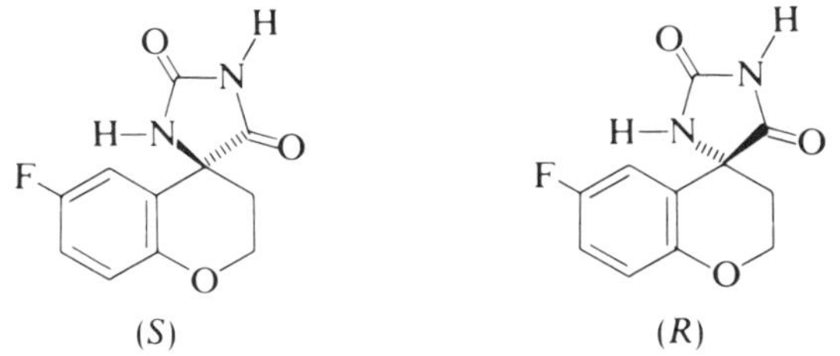

(S)   (R)

(22)   Sorbinil

Figure 22

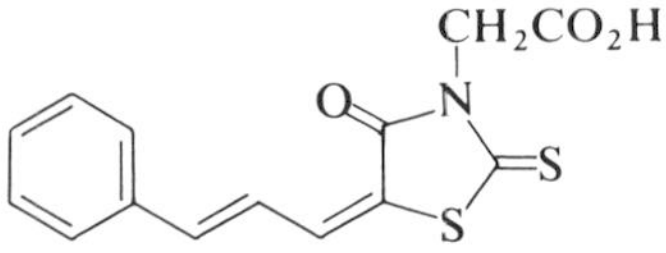

(23a) Epalrestat 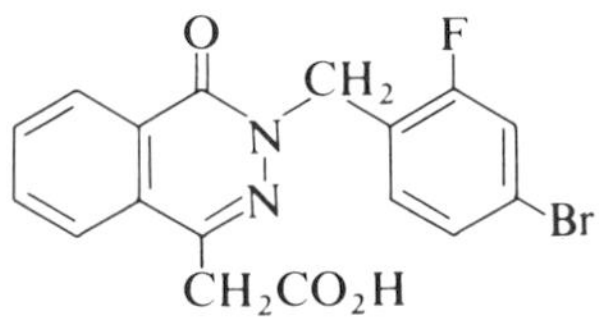   (23b) Statil 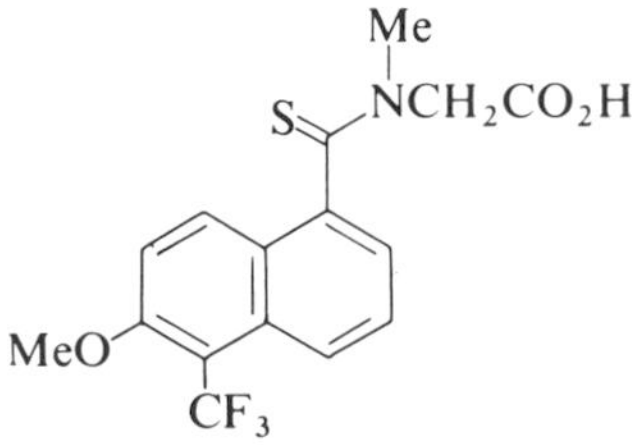   (23c) Tolrestat

Figure 23

acid derivative (statil; ICI 128 436; **23b**) and the naphthalene derivative tolrestat (AY-27 773; **23c**). Numerous other compounds are capable of inhibiting AR and these include acidic compounds related to the nonsteroidal antiinflammatory agents and to mast cell 'stabilizers' such as disodium cromoglycate. Naturally occurring complex phenols (flavonoids) also inhibit AR but it should be stressed that flavonoids also have a multiplicity of other effects.[164]

Kador and his colleagues have offered some useful generalizations concerning structure–activity relationships for the inhibition of AR.[162] Thus the interaction between enzyme and inhibitor is thought to involve a nucleophilic center in the enzyme and a carbonyl function of the inhibitor. Other characteristics of potent inhibitors of AR include proper stereochemistry, a considerable degree of molecular planarity, and capability to participate in hydrophobic interactions. While hydrogen bonding capability near hydrophobic interaction sites has been mentioned as a determinant of optimal binding for many AR inhibitors, the hydrophobic interactions are thought to be of primary importance for sorbinil binding to AR.[158] A series of N-benzenesulfonylamino acids have been evaluated as inhibitors of AR obtained from rat lens.[165] None of these compounds proved to be more potent than sorbinil; a discussion of the structural determinants of inhibitory activity in this series of AR inhibitors included consideration of a possible protonated base (*e.g.* guanidinium) site on AR that binds carboxylate functions or comparable anionic groups in addition to the determinants of effective enzyme–inhibitor interactions described by Kador and his colleagues. While the proposed receptor site model has substantial heuristic value, there are several limitations to its general applications.

The excellent account of the development of Ar inhibitors by Larson and his colleagues summarizes the chronology of prominent findings in this area.[161] For example, early studies identified tetramethyleneglutaric acid ($IC_{50} = 1$–$10 \mu M$) as a relatively potent inhibitor of AR obtained from bovine lens. Observations of the activity of cyclic urea derivatives, especially hydantoins, led to the identification of sorbinil and related compounds as potent inhibitors of AR that also were effective *in vivo*. Larson and his colleagues emphasize that *in vitro* potency frequently does not correlate with *in vivo* potency and suggest that pharmacokinetic factors such as protein binding and cellular transport play an important role in determining activity *in vivo*. Many interesting structure–activity relationships have been established for inhibition of AR by compounds related to sorbinil. For example (Figure 24) inhibition of AR is produced by the chroman (**24a**) and thiachroman (**24b**) derivatives at identical concentrations ($IC_{50} = 5 \mu M$) but expansion of the ring by replacing the oxygen or sulfur atom by two methylene units (to **24c**) results in a pronounced decrease in inhibitory activity. Substitution at the 6-position of the chroman molecule with a halogen atom results in a 5–50-fold increase in potency evaluated against bovine lens AR. Tolrestat (AY-27773; **23c**) is synthesized from 6-methoxy-5-trifluoromethylnaphthalene-1-carboxylic acid by a reaction sequence involving formation of the acyl chloride by treatment with thionyl chloride, amide formation by treatment of the acyl chloride with the methyl ester of N-methylglycine (sarcosine),

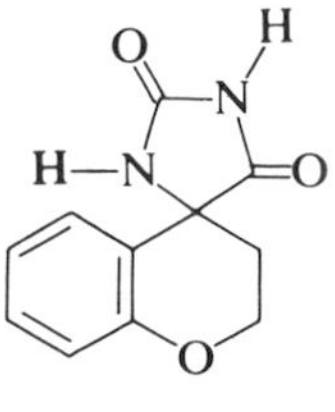

(24a) 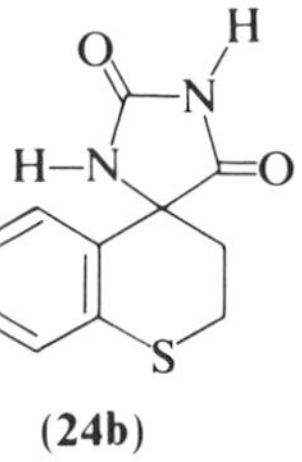   (24b) 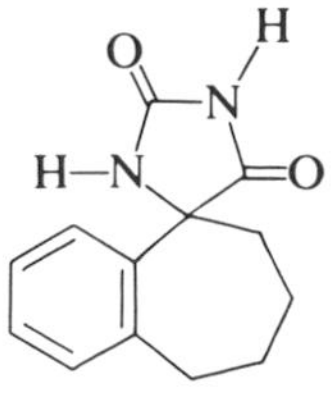   (24c)

Figure 24

conversion of the amide to a thioamide by thiation with phosphorus pentasulfide, and, finally, alkaline hydrolysis of the ester.[166] Tolrestat is a very potent inhibitor of AR with an IC_{50} (*ca.* 30 nM) approximately 50-fold lower than that determined for sorbinil under identical conditions. Tolrestat is currently under clinical investigation. Another compound representative of those in clinical trials is the phthalazinecarboxylic acid statil (ICI 128 436; **23b**). This compound inhibits AR in the 20–100 nM range; the *N*-benzyl group plays a significant role in the activity of this compound as judged by the sharp losses in activity when this group is replaced by small alkyl groups.

7.2.6 FUMARATE REDUCTASE

Certain trypanosomal species and helminths possess a mitochondrial membrane fumarate reductase system (FR) that catalyzes hydride transfer from NADH to fumarate yielding succinate; the K_m for NADH in this system is 60 μM.[167] This enzyme system has been reported, in *T. brucei*, to be capable of generating hydrogen peroxide and superoxide anion and fumarate inhibits the formation of these reactive oxygen species. Turrens has commented on the possibility of developing an antitrypanosomal drug that inhibits the binding of fumarate to the reductase, with consequent enhanced generation of cytotoxic oxygen species.[167] FR has been identified as a possible enzyme locus for action of anthelminthic drugs also.[168] K_m values for fumarate in the FR system from *Trichinella* isolates range from 2 to 7 mM; while drugs such as mebendazole, thiabendazole and ivermectin inhibit FR from *Trichinella*, high concentrations of the drugs are required (0.6–1 mM) and there is uncertainty about the contribution of this biochemical effect to the characteristic pharmacological action of these drugs.[169] Although FR activity has been reported to be diminished in strains of helminths resistant to drugs such as thiabendazole, this phenomenon is not invariably associated with resistance to drugs of the benzimidazole class.[170]

7.2.7 TRYPANOTHIONE REDUCTASE AND GLUTATHIONE REDUCTASE

Trypanothione reductase (TR) is a flavoprotein obtained from trypanosomes and related protozoa that catalyzes the NADPH dependent reduction of the macrocyclic disulfide trypanothione (Figure 25). Trypanothione is a glutathione–spermidine conjugate, N^1,N^8-bis(glutathionyl)spermidine.[171] TR appears to serve a protective function in trypanosomatids analogous to that of glutathione (see Figure 26) in mammalian cells and its occurrence in the parasites but not host cells may be exploitable in the design of selective chemotherapeutic agents. Maintenance of trypanothione in the reduced (thiol) form is important in protecting trypanosomatids from damage by oxygen-derived radical species. Shames and his colleagues[172] have determined the primary sequence of amino acids in TR obtained from *T. congolense*, and found nearly 90% homology with the comparable enzyme isolated from the related protozoan *Crithidia fasciculata* and substantial homology (> 50%) with human glutathione reductase. The poor affinity of TR for glutathione has been attributed, in part, to absence of basic residues (arginines) that contribute to glutathione binding to mammalian glutathione reductase. Henderson and his colleagues[173] have described a group of 'subversive' substrates for TR, so called because they are subject to one-electron reduction mediated by this enzyme, with subsequent reoxidation to produce the toxic superoxide radical; in

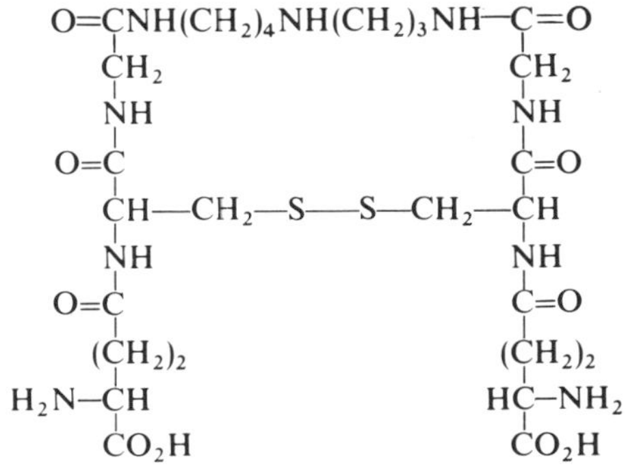

O=CNH(CH$_2$)$_4$NH(CH$_2$)$_3$NH—C=O
CH$_2$ CH$_2$
NH   NH
O=C   C=O
CH—CH$_2$—S——S—CH$_2$—CH
NH   NH
O=C   C=O
(CH$_2$)$_2$ (CH$_2$)$_2$
H$_2$N—CH   HC—NH$_2$
CO$_2$H   CO$_2$H

Trypanothione

Figure 25

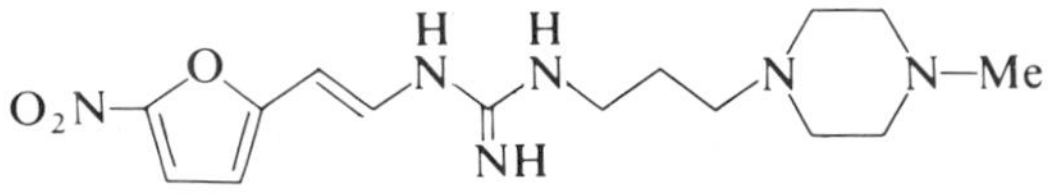

(26a)

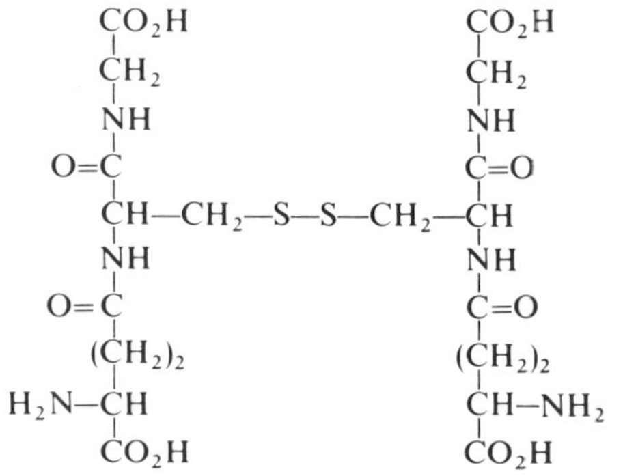

(26b) Oxidized glutathione

Figure 26

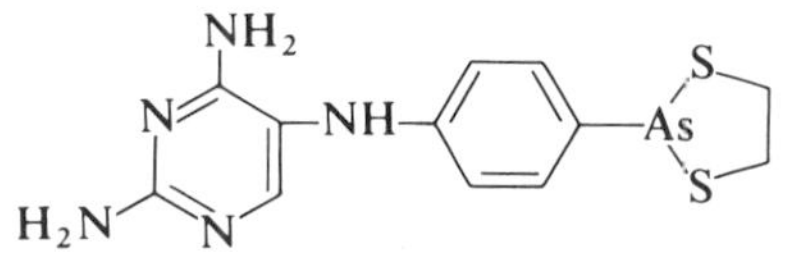

Figure 27

this way, the compounds subvert the normal protective role of TR. In addition, such compounds inhibit TR and thus limit the generation of the protective thiol species that mitigates oxidant stress. The trypanocidal action of these compounds, exemplified by structure (**26a**), correlates well with ability to inhibit TR. For the nitrofuran derivative shown, 50% inhibition of TR activity is observed at a concentration of 10 μM when the substrate, trypanothione, is present in the reaction mixture at 250 μM. Henderson and his colleagues suggest that new selective trypanocidal drugs can be designed using their observations as a point of departure. Glutathione reductase (EC 1.6.4.2; glutathione:NADPH reductase) has been identified as a possible target for selective antifilarial drug action because of observations that the drug melarsen oxide (Figure 27) inhibits the enzyme from *Litomosoides carnii* approximately 25-fold more effectively ($IC_{50} = 9$ μM) than the comparable enzyme from human erythrocytes. Neither the K_m for oxidized glutathione (65–90 μM) nor the K_m for NADPH (8.5–9.0 μM) differed significantly for the enzyme from the two sources.[174]

7.2.8 REFERENCES

1. J. Jolivet, K. H. Cowan, G. A. Curt, N. J. Clendenium and B. A. Chabner, *N. Engl. J. Med.*, 1983, **309**, 1094.
2. M. E. Weinblatt, J. S. Coblyn, D. A. Fox, P. A. Fraser, D. E. Holdsworth, D. N. Glass and D. E. Trentham, *N. Engl. J. Med.*, 1985, **312**, 818.
3. E. Green and C. H. Demos, in 'Folate Antagonists as Therapeutic Agents', ed. F. M. Sirotnak, J. J. Burchall, W. B. Ensminger and J. A. Montgomery, Academic Press, Orlando, FL, 1984, vol. 2, p. 192.
4. I. M. Rollo, in 'Handbook of Experimental Pharmacology', ed. G. H. Hitchings, Springer Verlag, Berlin, 1983, vol. 64, p. 293.
5. J. Maroun, *Semin. Oncol.*, 1988, **15** (Suppl. 2), 17.
6. C. J. Allegra, B. A. Chabner, C. U. Tuazon, D. Ogata-Arakaki, B. Baird, J. C. Drake, J. T. Simmons, E. E. Lack, J. H. Shelhamer, F. Balis, R. Walker, J. A. Kovacs, H. C. Lane and H. Masur. *N. Engl. J. Med.*, 1987, **317**, 978.
7. B. R. Baker, *Ann. N.Y. Acad. Sci.*, 1971, **186**, 214.
8. (a) R. L. Kisliuk, in 'Folate Antagonists as Therapeutic Agents', ed. F. M. Sirotnak, J. J. Burchall, W. B. Ensminger and J. A. Montgomery, Academic Press, Orlando, FL, 1984, vol. 1, p. 2; (b) R. L. Blakley, in 'Folates and Pterins', ed. R. L. Blakley and S. J. Benkovic, Wiley, New York, 1984, vol. 1, p. 191.
9. (a) C. J. Allegra, R. L. Fine, J. C. Drake and B. A. Chabner, *J. Biol. Chem.*, 1986, **261**, 6478; (b) R. G. Matthews, C. Ghose, J. M. Green, K. D. Matthews and R. B. Dunlap, *Adv. Enzym. Regul.*, 1987, **26**, 157.

10. J. H. Freisheim and D. A. Mathews, in 'Folate Antagonists as Therapeutic Agents', ed. F. M. Sirotnak, J. J. Burchall, W. B. Ensminger and J. A. Montgomery, Academic Press, Orlando, FL, 1984, vol. 1, p. 70.
11. J. K. Coward, K. N. Parameswaran, A. R. Cashmore and J. R. Bertino, *Biochemistry*, 1974, **13**, 3899.
12. T. J. Williams, T. K. Lee and R. B. Dunlap, *Arch. Biochem. Biophys.*, 1977, **181**, 569.
13. D. J. Filman, J. T. Bolin, D. A. Mathews and J. Kraut, *J. Biol. Chem.*, 1982, **257**, 13 663.
14. J. F. Morrison and S. R. Stone, *Biochemistry*, 1988, **27**, 5499.
15. R. Ferone, *J. Biol. Chem.*, 1970, **245**, 850.
16. J. J. Jaffe, J. J. McCormack and W. E. Gutteridge, *Exp. Parasitol.*, 1969, **25**, 311.
17. J. J. Jaffe, J. J. McCormack and E. Meymarian, *Biochem. Pharmacol.*, 1972, **21**, 719.
18. R. Ferone, J. J. Burchall and G. H. Hitchings, *Mol. Pharmacol.*, 1969, **5**, 49.
19. W. E. Gutteridge, J. J. Jaffe and J. J. McCormack, Jr., *Biochim. Biophys. Acta*, 1969, **191**, 753.
20. R. Ferone and S. Roland, *Proc. Natl. Acad. Sci. USA*, 1980, **77**, 5802.
21. C. E. Garrett, J. A. Coderre, T. D. Meek, E. P. Garvey, D. M. Claman, S. M. Beverley and D. V. Santi, *Mol. Biochem. Parasitol.*, 1984, **11**, 257.
22. T. D. Meek, E. P. Garvey and D. V. Santi, *Biochemistry*, 1985, **24**, 678.
23. (a) D. A. Matthews, J. T. Bolin, J. M. Burridge, D. J. Filman, K. W. Volz, B. T. Kaufman, C. R. Beddell, J. N. Champness, D. K. Stammers and J. Kraut, *J. Biol. Chem.*, 1985, **260**, 381; (b) D. A. Matthews, J. T. Bolin, J. M. Burridge, D. J. Filman, K. W. Volz and J. Kraut, *J. Biol. Chem.*, 1985, **260**, 392.
24. J. J. McCormack, unpublished results.
25. T. J. Delcamp, A. Rosowsky, P. L. Smith, J. E. Wright and J. H. Freisheim, in 'Pteridines and Folic Acid Derivatives', ed. B. A. Cooper and V. M. Whitehead, Walter de Gruyter, Berlin, 1986, p. 807.
26. A. Rosowsky, V. C. Solan, R. A. Forsch, T. J. Delcamp, D. P. Baccanari and J. H. Freisheim, *J. Med. Chem.*, 1987, **30**, 1463.
27. J. J. McCormack, unpublished results.
28. C. Hansch, R.-L. Li, J. M. Blaney and R. Langridge, *J. Med. Chem.*, 1982, **25**, 777.
29. T. H. Jukes, *Cancer Res.*, 1987, **47**, 5528.
30. S. K. Farber, R. D. Diamond, R. F. Mercer, R. F. Sylvester, Jr. and J. A. Wolff, *N. Engl. J. Med.*, 1948, **238**, 787.
31. J. H. Burchenal, J. R. Burchenal, M. N. Kushida, S. F. Johnson and B. S. Williams, *Cancer*, 1949, **2**, 113.
32. R. Hertz, G. T. Ross and M. B. Lipsett, *Ann. N.Y. Acad. Sci.*, 1964, **114**, 881.
33. C. A. Nichol and A. D. Welch, *Proc. Soc. Exp. Biol. Med.*, 1950, **74**, 403.
34. (a) S. Futterman and M. Silverman, *J. Biol. Chem.*, 1957, **224**, 31; (b) S. Futterman, *J. Biol. Chem.*, 1957, **228**, 1031.
35. (a) S. F. Zakrzewski and C. A. Nichol, *Biochim. Biophys. Acta*, 1958, **27**, 425; (b) M. J. Osborn, M. Freeman and F. M. Huennekens, *Proc. Soc. Exp. Biol. Med.*, 1958, **97**, 429.
36. L. K. A. Rahman and S. R. Chhabra, *Med. Res. Rev.*, 1988, **8**, 95.
37. D. R. Seeger, D. B. Cosulich, J. M. Smith, Jr. and M. E. Hultquist, *J. Am. Chem. Soc.*, 1949, **71**, 1753.
38. E. C. Taylor, K. L. Perlman, Y.-H. Kim, I. P. Sword and P. A. Jacobi, *J. Am. Chem. Soc.*, 1973, **95**, 6413.
39. J. I. DeGraw, V. H. Brown, H. Tagawa, R. L. Kisliuk, Y. Gaumont and F. M. Sirotnak, *J. Med. Chem.*, 1982, **25**, 1227.
40. J. I. DeGraw, P. H. Christie, H. Tagawa, R. L. Kisliuk, Y. Gaumont, F. A. Schmid and F. M. Sirotnak, *J. Med. Chem.*, 1986, **29**, 1056.
41. W. C. Werkheiser, *J. Biol. Chem.*, 1961, **236**, 888.
42. R. C. Jackson, L. I. Hart and K. R. Harrap, *Cancer Res.*, 1976, **36**, 1991.
43. S. Cha, S. Y. R. Kim, S. G. Kornstein, P. W. Kantoff, K. H. Kim and F. N. M. Naguib, *Biochem. Pharmacol.*, 1981, **30**, 1507.
44. R. L. Blakley and L. Cocco, *Biochemistry*, 1988, **24**, 4772.
45. S. R. Stone and J. F. Morrison, *Biochim. Biophys. Acta*, 1986, **869**, 275.
46. B. R. Baker, 'Design of Active-Site-Directed Irreversible Enzyme Inhibitors', Wiley, New York, 1967.
47. *J. Mol. Biol.*, 1981, **188**, 81.
48. B. Birdsall, J. Feeney, C. Pascual, G. C. K. Roberts, I. Kompis, R. L. Then, K. Muller and A. Kroen. *J. Med. Chem.*, 1984, **27**, 1672.
49. T. B. Johnson, M. G. Nair and J. Galivan, *Cancer Res.*, 1988, **48**, 2426.
50. J. Jolivet, R. Schilsky, B. Bailey, J. C. Drake and B. A. Chabner, *J. Clin. Invest.*, 1982, **70**, 351.
51. B. A. Chabner, C. J. Allegra, G. A. Curt, N. J. Clendenium, J. Baram, S. Koiz, J. C. Drake and J. Jolivet, *J. Clin. Invest.*, 1985, **76**, 907.
52. F. A. Schmid, F. M. Sirotnak, G. M. Otter and J. I. DeGraw, *Cancer Treat. Rep.*, 1987, **71**, 727.
53. K. Y. Shum, M. G. Kris, R. G. Gralla, M. T. Burke, L. D. Marks and R. T. Heelan, *J. Clin. Oncol.*, 1988, **6**, 446.
54. N. Endo, Y. Takeda, N. Umemoto, K. Kishida, K. Watanabe, M. Saito, Y. Kato and T. Hara, *Cancer Res.*, 1988, **48**, 3330.
55. M. Dembo and F. M. Sirotnak, in 'Folate Antagonists as Therapeutic Agents', ed. F. M. Sirotnak, J. J. Burchall, W. B. Ensminger and J. A. Montgomery, Academic Press, Orlando, FL, 1984, p. 173.
56. P. L. Chello, F. M. Sirotnak and D. M. Dorick, *Mol. Pharmacol.*, 1980, **18**, 274.
57. F. M. Sirotnak, D. M. Moccio, C. H. Hancock and C. W. Young, *Cancer Res.*, 1981, **41**, 3944.
58. J. A. Montgomery and J. R. Piper, in 'Folate Antagonists as Therapeutic Agents', ed. F. M. Sirotnak, J. J. Burchall, W. B. Ensminger and J. A. Montgomery, Academic Press, Orlando, FL, 1984, p. 219.
59. A. Monks, D. Scudiero, R. Shoemaker, K. Paull, D. Fine, J. Mayo and M. Boyd, *Proc. Am. Assoc. Cancer Res.*, 1988, **29**, 488.
60. R. L. Blakley and B. M. McDougall, *J. Biol. Chem.*, 1961, **236**, 1163.
61. J. J. Jaffe and J. J. McCormack, *Mol. Pharmacol.*, 1967, **3**, 662.
62. B. R. Baker, 'Design of Active-Site-Directed Irreversible Enzyme Inhibitors', Wiley, New York, 1967, p. 199.
63. J. J. McCormack, in 'Chemistry and Biology of Pteridines', ed. W. Pfleiderer, W. de Gruyter, New York, 1975, p. 131.
64. R. D. Elliott, C. Temple, Jr., J. L. Frye and J. A. Montgomery, *J. Med. Chem.*, 1975, **18**, 492.
65. S. M. Cramer, J. H. Schornagel, K. K. Kalghatgi, J. R. Bertino and C. Horvath, *Cancer Res.*, 1984, **44**, 1843.
66. A. Rosowsky and R. Forsch, *J. Med. Chem.*, 1982, **25**, 1454.

67. J. B. Hynes, W. T. Ashton, D. Bryansmith and J. H. Freisheim, *J. Biol. Chem.*, 1974, **17**, 1023.
68. J. A. Montgomery, R. D. Elliott, S. L. Straight and C. Temple, Jr., *Ann. N.Y. Acad. Sci.*, 1971, **186**, 227.
69. J. R. Piper, G. S. McCaleb, J. A. Montgomery, R. L. Kisliuk, Y. Gaumont and F. M. Sirotnak, *J. Med. Chem.*, 1986, **29**, 1080.
70. J. B. Hynes, A. Kumar, A. Tomažič and W. L. Washtien, *J. Med. Chem.*, 1987, **30**, 1515.
71. A. Rosowsky and K. K. N. Chen, *J. Med. Chem.*, 1974, **17**, 1308.
72. D. G. Johns and D. M. Valerino, *Ann. N.Y. Acad. Sci.*, 1971, **186**, 378.
73. C. N. Hodnett, J. J. McCormack and J. A. Sabean, *J. Pharm. Sci.*, 1976, **65**, 1150.
74. R. B. Angier and W. V. Curran, *J. Am. Chem. Soc.*, 1959, **81**, 2814.
75. A. Hantel, E. K. Rowinsky, G. B. Dole, D. S. Ettinger, L. B. Grochow, W. P. McGuire and R. C. Donehower, *Proc. Am. Assoc. Cancer Res.*, 1987, **28**, 201.
76. R. B. Natale, R. H. Wheeler and J. A. Roberts, *NCI Monographs*, 1987, No. 5, 219.
77. J. J. McCormack *Med. Res. Rev.*, 1981, **1**, 303.
78. D. G. Johns, D. Farquhar, B. A. Chabner and J. J. McCormack, *Biochem. Soc. Trans.*, 1974, **2**, 602.
79. A. Rosowsky, G. P. Beardsley, W. D. Ensminger, H. Lazarus and C.-S. Yu, *J. Med. Chem.*, 1978, **21**, 380.
80. A. Rosowsky, R. A. Forsch, C.-S. Yu, H. Lazarus and G. P. Beardsley, *J. Med. Chem.*, 1984, **27**, 605.
81. A. Rosowsky, J. H. Freisheim, H. Bader, R. A. Forsch, S. S. Susten, C. A. Cucchi and E. Frei, III, *J. Med. Chem.*, 1985, **28**, 660.
82. J. E. Wright, A. Rosowsky, D. J. Waxman, D. Trites, C. A. Cucchi, J. Flatow and E. Frei, *Biochem. Pharmacol.*, 1987, **36**, 2209.
83. A. Rosowsky, H. Bader, R. A. Forsch, R. G. Moran and J. H. Freisheim, *J. Med. Chem.*, 1988, **31**, 763.
84. D. C. Suster, L. V. Feyns, G. Ciustea, G. Botez, V. Dobre, R. Bick and I. Niculescu-Duvâz, *J. Med. Chem.*, 1974, **17**, 758.
85. B. Roth, *Handb. Exp. Pharmacol.*, 1983, **64**, 107.
86. G. H. Hitchings and B. Roth, in 'Enzyme Inhibitors as Drugs', ed. M. Sandler, University Park Press, Baltimore, MD, 1980, p. 263.
87. L. J. Daniel and L. C. Norris, *J. Biol. Chem.*, 1947, **170**, 747.
88. G. H. Hitchings, G. B. Elion, H. Vanderwerff and E. A. Falco, *J. Biol. Chem.*, 1948, **174**, 765.
89. W. T. Hughes, in 'Folate Antagonists as Therapeutic Agents', ed. F. M. Sirotnak, J. J. Burchall, W. B. Ensminger and J. A. Montgomery, Academic Press, Orlando, FL, 1984, p. 251.
90. J. J. McCormack, unpublished results.
91. J. J. Burchall and G. H. Hitchings, *Mol. Pharmacol.*, 1965, **1**, 126.
92. S. W. Dietrich, J. W. Blaney, M. A. Reynolds, P. Y. C. Jow and C. Hansch, *J. Med. Chem.*, 1980, **23**, 1205.
93. B. R. Baker and B.-T. Ho, *J. Pharm. Sci.*, 1965, **54**, 1261.
94. J. J. McCormack and J. J. Jaffe, *J. Med. Chem.*, 1969, **12**, 662.
95. B. R. Baker and B.-T. Ho, *J. Pharm. Sci.*, 1964, **53**, 1457.
96. D. J. Baker, C. R. Beddell, J. N. Champness, P. J. Goodford, F. E. A. Norrington, D. R. Smith and D. K. Stammers, *FEBS Lett.*, 1981, **126**, 49.
97. D. P. Baccanari, S. Daluge and R. W. King, *Biochemistry*, 1982, **21**, 5068.
98. D. P. Baccanari, R. L. Tansik and G. H. Hitchings, *Adv. Enzym. Regul.* 1987, **26**, 3.
99. J. M. Blaney, C. Hansch, C. Silipo and A. Vittoria, *Chem. Rev.*, 1984, **84**, 333.
100. E. C. Taylor, J. V. Berrier, A. J. Cocuzza, R. Kobylecki and J. J. McCormack, *J. Med. Chem.*, 1977, **20**, 1215.
101. K. Taira and S. J. Benkovic, *J. Med. Chem.*, 1988, **31**, 129.
102. B. Roth, D. P. Baccanari, C. W. Sigel, J. P. Hubbell, J. Eaddy, J. C. Kao, M. E. Grace and B. S. Rauckman, *J. Med. Chem.*, 1988, **31**, 122.
103. H. C. Carrington, A. F. Crowther, D. G. Davey, A. A. Levi and F. L. Rose, *Nature (London)*, 1951, **168**, 1080.
104. R. G. Booth, C. D. Selassie, C. Hansch and D. V. Santi, *J. Med. Chem.*, 1987, **30**, 1218.
105. B. A. Hathaway, Z.-R. Guo, C. Hansch, T. J. Delcamp, S. S. Susten and J. H. Freisheim, *J. Med. Chem.*, 1984, **27**, 144.
106. C. Hansch, B. A. Hathaway, Z.-R. Guo, C. D. Selassie, S. W. Dietrich, J. M. Blaney, R. Langridge, K. W. Volz and B. T. Kaufman, *J. Med. Chem.*, 1984, **27**, 129.
107. B. Roth, *Fed. Proc., Fed. Am. Soc. Exp. Biol.*, 1986, **45**, 2765.
108. E. F. Elslager, *Prog. Drug Res.*, 1974, **18**, 99.
109. W. E. Richter, Jr. and J. J. McCormack, *J. Med. Chem.*, 1974, **17**, 943.
110. J. R. Bertino, W. L. Sawicki, B. A. Moroson, A. R. Cashmore and E. F. Elslager, *Biochem. Pharmacol.*, 1979, **28**, 1983.
111. H. Diddens, D. Niethammer and R. C. Jackson, *Cancer Res.*, 1983, **43**, 5286.
112. T. Ohnuma, R. J. Lo, K. J. Scanlon, B. A. Kamen, T. Ohnoshi, S. R. Wolman and J. F. Holland, *Cancer Res.*, 1985, **45**, 1815.
113. W. L. Leopold, D. J. Dykes and D. P. Griswold, Jr., *NCI Monographs*, 1987, No. 5, 99.
114. J. T. Lin and J. R. Bertino, *J. Clin. Oncol.*, 1987, **5**, 2032.
115. J. A. Stewart, J. J. McCormack, W. Tong, J. B. Low, J. D. Roberts, A. Blow, L. R. Whitfield, L. D. Haugh, W. R. Grove, A. J. Grillo-Lopez and R. J. DeLap, *Cancer Res.*, 1988, **48**, 5029.
116. J. Jolivet, L. Landry, M.-F. Pinard, J. J. McCormack, W. P. Tong and E. Eisenhauer, *Cancer Chemother. Pharmacol.*, 1987, **20**, 169.
117. C. A. Ackerly, J. Hartshorn, W. P. Tong and J. J. McCormack, *J. Liq. Chromatogr.*, 1985, **8**, 125.
118. J. J. Heusner and J. J. McCormack, *J. Pharm. Sci.*, 1981, **70**, 827.
119. F. M. Balis, C. M. Lester and D. G. Poplack, *Cancer Res.*, 1986, **46**, 169.
120. J. J. Heusner and M. R. Franklin, *Pharmacology*, 1985, **30**, 266.
121. L. K. Webster, W. P. Tong and J. J. McCormack, *J. Pharm. Pharmacol.*, 1987, **39**, 942.
122. J. J. McCormack, L. K. Webster, W. P. Tong, L. A. Mathews and J. A. Stewart, *Proc. Am. Assoc. Cancer Res.*, 1986, **27**, 256.
123. P. J. O'Dwyer, R. J. DeLap, S. A. King, A. J. Grillo-Lopez, D. F. Hoth and B. Leyland-Jones, *NCI Monographs*, 1987, No. 5, 105.
124. Warner-Lambert/Parke-Davis North American Phase 2 Study Group, *Proc. Am. Soc. Clin. Oncol.*, 1987, **6**, 54.

125. Warner-Lambert/Parke-Davis North American Phase 2 Study Group, *Proc. Am. Soc. Clin. Oncol.*, 1987, **6**, 131.
126. M. E. Costanza, A. H. Korzun, M. A. Rice and W. C. Wood, *Proc. Am. Soc. Clin. Oncol.*, 1988, **7**, 30.
127. J. M. Baselga, M. G. Kris, R. J. Gralla, E. Cheng, L. M. Potanovich, R. W. D'Acquisto, M. P. Fanucchi and R. T. Heelan, *Proc. Am. Soc. Clin. Oncol.*, 1988, **7**, 220.
128. R. C. Donehower, *J. Natl. Cancer Inst.*, 1988, **80**, 1268.
129. F. M. Balis, R. Patel, E. Luks, K. M. Doherty, J. S. Holcenberg, C. Tan, G. H. Reaman, J. Belasco, L. J. Ettinger, S. Zimm and D. G. Poplack, *Cancer Res.*, 1987, **47**, 4973.
130. B. A. Kamen, A. Streckfuss, J. Sanders, W. P. Tong and J. J. McCormack, *J. Natl. Cancer Inst.*, 1988, submitted.
131. A. E. Glatt, K. Chirgwin and S. H. Landesman, *N. Engl. J. Med.*, 1988, **318**, 1439.
132. C. J. Allegra, J. A. Kovacs, J. C. Drake, J. C. Swan, B. A. Chabner and H. Masur, *J. Exp. Med.*, 1987, **165**, 926.
133. C. J. Allegra, J. A. Kovacs, J. C. Drake, J. C. Swan, B. A. Chabner and H. Masur, *J. Clin. Invest.*, 1987, **79**, 478.
134. L. E. Werbel, in 'Folate Antagonists as Therapeutic Agents', ed. F. M. Sirotnak, J. J. Burchall, W. B. Ensminger and J. H. Freisheim, Academic Press, Orlando, FL, 1984, p. 261.
135. E. M. Grivsky, S. Lee, C. W. Sigel, D. S. Duch and C. A. Nichol, *J. Med. Chem.*, 1980, **33**, 327.
136. C. W. Sigel, A. W. Macklin, J. L. Woolley, Jr., N. W. Johnson, M. A. Collier, M. R. Blum, N. J. Clendennin, B. J. M. Everitt, G. Grebe, A. Mackars, R. G. Foss, D. S. Duch, S. W. Bowers and C. A. Nichol, *NCI Monographs*, 1987, No. 5, 111.
137. J. Laszlo, W. D. Brenckman, Jr., E. Morgan, N. J. Clendennin, T. Williams, V. Currie and C. Young, *NCI Monographs*, 1987, No. 5, 121.
138. M. G. Kris, R. J. Gralla, M. T. Burke, L. D. Berkowitz, L. D. Marks, D. P. Kelsen and R. T. Heelan, *Cancer Treat. Rep.*, 1987, **71**, 763.
139. P. B. Russell and G. H. Hitchings, *J. Am. Chem. Soc.*, 1951, **73**, 3763.
140. E. J. Modest, *J. Org. Chem.*, 1956, **21**, 1.
141. E. F. Elslager, J. L. Johnson and L. M. Werbel, *J. Med. Chem.*, 1983, **26**, 1753.
142. J. H. Edmonson, *NCI Monographs*, 1987, No. 5, 67.
143. B. A. Kamen and N. J. Winick, *Biochem. Pharmacol.*, 1988, **37**, 2713; J. L. Grem, S. A. King, R. E. Wittes and B. Leyland-Jones, *J. Natl. Cancer Inst.*, 1988, **80**, 626.
144. W. S. Wilke and A. H. Mackenzie, *Drugs*, 1986, **32**, 103.
145. W. B. Ensminger, in 'Folate Antagonists as Therapeutic Agents', ed. F. M. Sirotnak, J. J. Burchall, W. B. Ensminger and J. H. Freisheim, Academic Press, Orlando, FL, 1984, vol. 2, p. 133; B. A. Chabner, in 'Pharmacological Principles of Cancer Treatment', ed. B. A. Chabner, Saunders, Philadelphia, 1982, p. 229.
146. J. C. Cavallito, C. A. Nichol, W. D. Brenckman, Jr., R. L. DeAngelis, D. R. Stickney, W. S. Simmons and C. W. Sigel, *Drug Metab. Dispos.*, 1978, **6**, 329.
147. B. Odlund, P. Hartvig, K. E. Fjellstrom, B. Lindstrom and S. Bengtsson, *Eur. J. Clin. Pharmacol.*, 1984, **26**, 393.
148. R. Shiman, in 'Folates and Pterins', ed. R. L. Blakley and S. J. Benkovic, Wiley, New York, vol. 2, 1984, p. 232.
149. W. L. F. Armarego, D. Randles and P. Waring, *Med. Res. Rev.*, 1984, **4**, 267.
150. S. W. Bailey and J. E. Ayling, *J. Biol. Chem.*, 1980, **255**, 7774.
151. P. Altmann, F. A1-Salihi, K. Butter, P. Cutler, J. Blair, R. Leeming, J. Cunningham and F. Marsh, *N. Engl. J. Med.*, 1987, **317**, 80.
152. D. Randles and W. L. F. Armarego, *Eur. J. Biochem.*, 1985, **146**, 467.
153. W. Gessner, A. Brossi, R.-S. Shen and C. W. Abell, *J. Med. Chem.*, 1985, **28**, 311.
154. S. Webber, K. Baumgartner, J. N. R. Blair and J. M. Whiteley, *Biochem. Pharmacol.*, 1988, **37**, 2869.
155. M. Shahbaz, J. A. Hoch, K. A. Trach, J. A. Hural, S. Webber and J. M. Whiteley, *J. Biol. Chem.*, 1987, **262**, 16 412.
156. D. A. Matthews, S. Webber and J. M. Whiteley, *J. Biol. Chem.*, 1986, **261**, 3891.
157. H. B. Feldman, P. A. Szczepanik, P. Havre, R. J. M. Corrall, L. C. Yu, H. M. Rodman, B. A. Rosner, P. D. Klein and B. R. Landau, *Biochim. Biophys. Acta*, 1977, **480**, 14.
158. C. A. Lipinski and N. J. Hutson, *Annu. Rep. Med. Chem.*, 1984, **19**, 169.
159. P.F. Kador, *Med. Res. Rev.*, 1988, **8**, 325.
160. J. D. Ward and P. Benfield, *Drugs*, 1986, **32**, Suppl. 2.
161. E. R. Larson, C. A. Lipinski and R. Sarges, *Med. Res. Rev.*, 1988, **8**, 159.
162. P. F. Kador, J. H. Kinoshita and N. E. Sharpless, *J. Med. Chem.*, 1985, **28**, 841.
163. P. F. Kador and N. E. Sharpless, *Mol. Pharmacol.*, 1983, **24**, 521.
164. B. Havsteen, *Biochem. Pharmacol.*, 1983, **32**, 1141.
165. J. De Ruiter, A. N. Brubaker, M. A. Garner, J. M. Barksdale and C. A. Mayfield, *J. Pharm. Sci.*, 1987, **76**, 149.
166. K. Sestanj, F. Bellini, S. Fung, N. Abraham, A. Treasurywala, L. Humber, N. Simard-Duquesne and D. Dvornik, *J. Med. Chem.*, 1984, **27**, 255.
167. J. F. Turrens, *Mol. Biochem. Parasitol.*, 1987, **25**, 55.
168. J. C. W. Comley and J. D. Wright, *Int. J. Parasitol.*, 1981, **11**, 79.
169. F. Rodriquez-Caabeiro, A. Griado-Fornelo and A. Jimenez-Gonzalez, *Parasitology*, 1985, **91**, 577.
170. C. Bryant and E.-M. Bennet, *Mol. Biochem. Parasitol.*, 1983, **7**, 281.
171. A. H. Fairlamb, P. Blackburn, P. Ulrich, B. T. Chait and A. Cerami, *Science (Washington, D.C.)*, 1985, **227**, 1485.
172. S. L. Shames, B. E. Kimmel, O. P. Peoples, N. Agabian and C. Walsh, *Biochemistry*, 1988, **27**, 5014.
173. G. B. Henderson, P. Ulrich, A. H. Fairlamb, I. Rosenberg, M. Pereira, M. Sela and A. Cerami, *Proc. Natl. Acad. Sci. USA*, 1988, **85**, 5374.
174. K. K. Bharghava, N. L. Trang, A. Cerami and J. W. Eaton, *Mol. Biochem. Parasitol.*, 1983, **9**, 29.

7.3

Purine and Pyrimidine Targets

JOHN B. HOBBS

Formerly of the City University, London, UK

7.3.1   INTRODUCTION   299

7.3.2   THE METABOLISM OF PURINES AND PYRIMIDINES   300

  7.3.2.1   Purine Metabolism   300
  7.3.2.2   Pyrimidine Metabolism   303

7.3.3   ALLOPURINOL AND XANTHINE OXIDASE   305

7.3.4   ANTINEOPLASTIC AGENTS   306

  7.3.4.1   Purine Antagonists   306
    7.3.4.1.1   Thiopurines   306
    7.3.4.1.2   Azapurines   308
    7.3.4.1.3   Ara-adenosine   309
    7.3.4.1.4   Other agents   310
  7.3.4.2   Pyrimidine Antagonists   311
    7.3.4.2.1   Azapyrimidines   311
    7.3.4.2.2   Fluoropyrimidines   312
    7.3.4.2.3   Ara-cytidine and its derivatives   313
    7.3.4.2.4   Other agents   314

7.3.5   TARGETS FOR ANTIVIRAL CHEMOTHERAPY   314

  7.3.5.1   Pyrimidine Nucleoside Antiviral Agents   315
    7.3.5.1.1   Halogenated pyrimidine 2'-deoxynucleosides   315
    7.3.5.1.2   5-Alkyl and other 5-substituted pyrimidine 2'-deoxynucleosides   316
    7.3.5.1.3   Pyrimidine arabinonucleosides and related compounds   317
    7.3.5.1.4   Other agents   318
  7.3.5.2   Purine Nucleoside Antiviral Agents   319
    7.3.5.2.1   Ara-adenosine and related compounds   319
    7.3.5.2.2   Agents containing modified and 'open ring' sugars   320
    7.3.5.2.3   Other agents   321
  7.3.5.3   Azole Nucleosides and Related Antiviral Compounds   322

7.3.6   ANTIPARASITIC PURINE AND PYRIMIDINE-RELATED AGENTS   323

  7.3.6.1   Allopurinol and Other Hypoxanthine Analogues   323
  7.3.6.2   Purine Derivatives   325
  7.3.6.3   Pyrimidine Derivatives and Analogues   325

7.3.7   ANTIFUNGAL AGENTS   326

7.3.8   SUMMARY—FUTURE DIRECTIONS   326

7.3.9   REFERENCES   327

7.3.1   INTRODUCTION

This chapter will consider the processes involved in the anabolism and catabolism of purines and pyrimidines, and the ways in which they are elaborated to form more highly organized species such as the nucleic acids. The enzymes catalyzing all these processes are potential targets for inhibitors, and in many cases these inhibitors are analogues of the naturally occurring purines and pyrimidines, differing from them either in the pattern of substitution on the heterocycle or in a modification of the ring system of the corresponding heterocycle. The resultant interruption in the primary metabolic processes consequent on the action of these inhibitory agents qualifies these agents to be described as antimetabolites; they are utilized widely for the control of neoplastic viral, parasitic and fungal conditions. The subject matter of the chapter will be concentrated in these areas. Use of purine antagonists for treatment of immune response diseases is described in Chapter 8.4. A number of efficacious and well-known drugs are coincidentally derivatives of pyrimidine, such as the barbiturates, and certain sulfonamides (sulfadiazine, sulfadimethoxine, *etc.*) but they do not express their activities primarily as antimetabolites affecting processes of purine and pyrimidine metabolism and thus will not be dealt with in this chapter.

7.3.2 THE METABOLISM OF PURINES AND PYRIMIDINES

7.3.2.1 Purine Metabolism

Antimetabolites interfere with the formation or utilization of the cell's natural metabolites. Purine antimetabolites are for the most part rendered effective *via* initial conversion to nucleosides and nucleotides, which are the true inhibitory agents. To appreciate the reasons for their activity, it is first

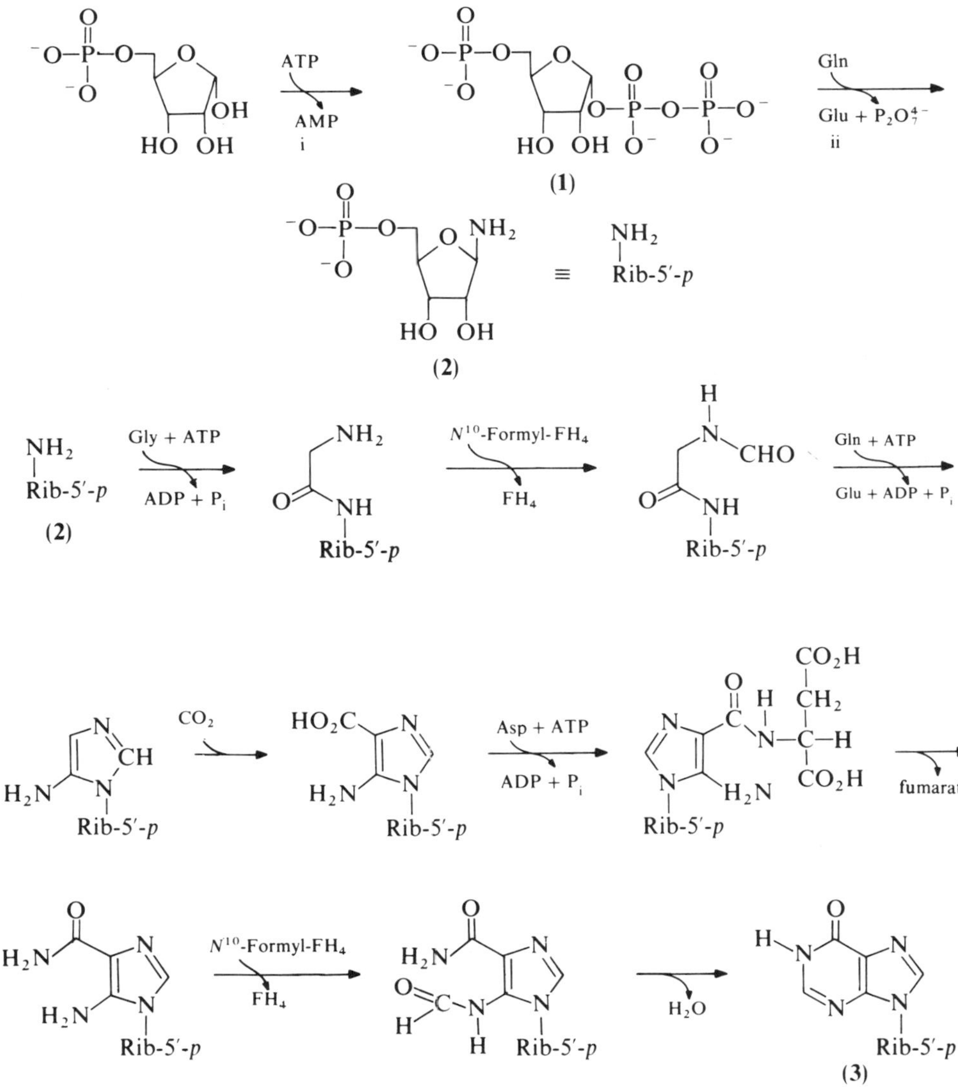

Figure 1 *De novo* biosynthesis of IMP (**3**). Enzymes: i, 5-phosphoribosyl-1-pyrophosphate; ii, phosphoribosylpyrophosphate amidotransferase; the other enzymes in the sequence are less significantly affected by the action of purine antimetabolic drugs

necessary to be acquainted with the normal cellular pathways of purine nucleotide anabolism and interconversion.

The pathway of the *de novo* synthesis of inosinic acid, largely elucidated by Buchanan and his colleagues,[1] is shown in Figure 1. It is seen that D-ribofuranosyl 5-phosphate is pyrophosphorylated to 5-phosphoribosyl 1-pyrophosphate (PRPP; **1**), which is subsequently aminated with displacement of pyrophosphate, forming 5-phosphoribosylamine (**2**). Both enzymes catalyzing these steps are subject to feedback inhibition, and it is these reactions which are principally affected by purine and purine analogue antimetabolites. The reactions of the pathway involving tetrahydrofolate derivatives are affected by folate antagonists, and those involving amino transfer from glutamine are inhibited by azaserine and 6-diazo-5-oxo-L-norleucine, but these will not be dealt with further.

Once formed, inosinic acid (IMP; **3**) is converted into adenylic acid (AMP; **4**) and guanylic acid (GMP; **5**). These purine nucleotide interconversions are outlined in Figure 2. The pathway to AMP involves the formation of adenylosuccinate as intermediate, while that to GMP involves initial oxidation of (**3**) to xanthylic acid (**6**), which is then aminated to form GMP.

Another means of forming ribonucleotides is *via* certain 'salvage pathways'.[2] Adenine, hypoxanthine and guanine, either supplied exogenously or formed intracellularly during the breakdown of nucleosides and nucleotides, can displace pyrophosphate from PRPP (**1**) directly to form AMP, IMP and GMP respectively, in reactions catalyzed by adenine phosphoribosyltransferase (for AMP) or hypoxanthine-guanine phosphoribosyltransferase (HGPRT) (for GMP, IMP). These enzymes are not only important for their role in the economy of nitrogen metabolism, but also because they accept a substantial number of purine antimetabolite drugs as substrates, potentiating them by converting them to the corresponding nucleotides.[3] In an alternative salvage pathway, catalyzed by purine nucleoside phosphorylase, hypoxanthine and guanine can displace phosphate from ribose 1-phosphate or deoxyribose 1-phosphate to afford the corresponding nucleosides, which are then phosphorylated by nucleoside kinases.

The purine nucleosides, and many of their analogues, are converted to their 5'-monophosphates by the purine nucleoside kinases *via* phosphoryl transfer from ATP. Adenosine kinase has been well characterized, inosine and guanosine kinases less so. The monophosphates are in turn converted to the 5'-diphosphates by the nucleoside monophosphokinases, and the nucleoside 5'-diphosphates to the 5'-triphosphates by the nucleoside diphosphokinases. Again phosphoryl transfer from ATP (which is constantly regenerated by oxidative phosphorylation) is utilized in each process. Adenylate kinase of muscle (myokinase) is the best-known example of a nucleoside monophosphokinase. It is due to the agency of these enzymes that nucleosides derived from purine antimetabolites reach the levels of 5'-di- and tri-phosphates at which their activity is often expressed.

To effect the synthesis of DNA, or indeed to become incorporated in DNA chains, a ribonucleotide must first be reduced to a 2'-deoxyribonucleotide.[4] This process is well established in microorganisms, where it is catalyzed principally at the diphosphate level by the enzyme ribonucleotide

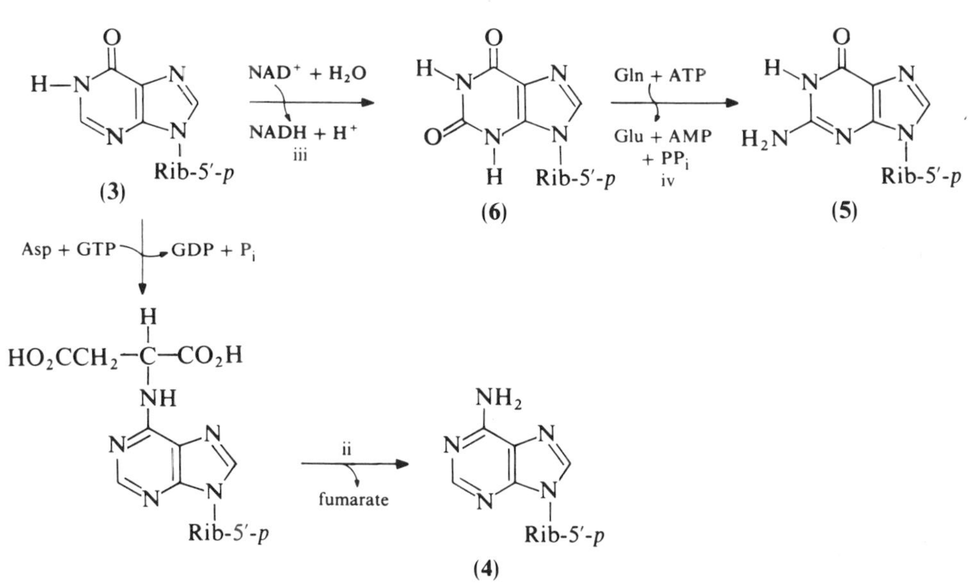

Figure 2 Biosynthesis of AMP (**4**) and GMP (**5**) from IMP (**3**). Enzymes: i, adenylosuccinate synthetase; ii, adenylosuccinate lyase; iii, IMP dehydrogenase; iv, GMP synthetase

reductase, a member of the nucleotide reductases. Thus, ADP and GDP are converted to dADP and dGDP respectively. The mechanism of the process is not yet fully elucidated, but appears to involve free radical intermediates. The role of nucleotide reductases in the cells of higher organisms is less clear. While purine analogues have been shown to become incorporated into DNA, routes to the 2′-deoxynucleoside triphosphates which do not involve a nucleotide reductase are possible.

Methylases transfer methyl groups from *S*-adenosylmethionine to purine (and pyrimidine) bases, generally at the level of the preformed nucleic acids,[5] although in some organisms methylation occurs at the ribonucleoside phosphate level. An *S*-adenosylmethionine-linked sulfhydryl trans-methylase system, found in many mammalian tissues, methylates sulfhydryl functions in purine analogues containing this group.[6] Demethylases occur mainly in liver, associated with the micro-somes, performing an oxidative function to oxygenate methyl groups, which are then lost as formaldehyde. Such demethylation may lead to the potentiation of a prodrug.

The anabolic processes considered hitherto are usually involved when a purine antimetabolite is metabolized prior to expressing its activity. The duration of that activity will be limited by the catabolic processes which now follow. The ribonucleases and deoxyribonucleases degrade RNA and DNA, respectively, and are unlikely to be significant unless the drug's activity is expressed *via* incorporation into these polynucleotides, but phosphatases and nucleotidases of low specificity are found in many tissues, and dephosphorylate nucleoside monophosphates (and sometimes di- and tri-phosphates) hydrolytically to the parent nucleosides. If a drug's principal activity is expressed at the nucleotide level, this hydrolysis reaction can clearly limit the activity.

The nucleoside phosphorylases mentioned briefly above catalyze the reversible cleavage of purine nucleosides to the free bases and ribose 1-phosphate.[7] *In vivo*, the equilibrium lies on the side of nucleoside degradation, and these enzymes (which are found in most cells) degrade many purine ribonucleoside analogues in this manner, to release the base. The hydrolytic cleavage of nucleosides into their base and sugar components by hydrolases has been observed in bacterial cells, but not in mammals.

Adenosine deaminase deaminates adenosine to form inosine. The enzyme is found in all types of cells, and the process constitutes a major step in nucleoside catabolism. Many adenosine analogues are deaminated by this enzyme, which also catalyzes hydrolytic removal of methylamino, hydrazino, hydroxylamino, chloro and methoxy groups from the 6-position of purine ribonucleosides.[8] Since

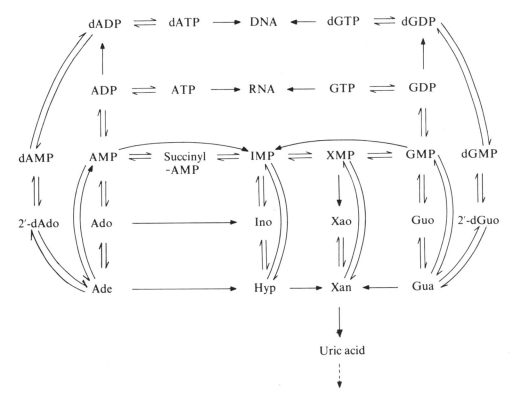

Figure 3 Cellular transformations of the naturally occurring purines and their nucleosides and nucleotides. The standard IUPAC abbreviations are used for all species. IMP is formed by *de novo* biosynthesis (see Figure 1)

inosine and its analogues are largely inactive, the activity of this enzyme limits the applicability of many adenosine analogues. Adenine deaminase has been described in microorganisms, converting adenine to hypoxanthine, and guanine deaminase has been found in animal tissues as well as microorganisms, converting guanine to xanthine. These deaminases show comparable hydrolytic activity with certain analogues of adenine and guanine, respectively. Adenylate deaminase is a well-characterized enzyme, which converts AMP to IMP, but its wider substrate specificity is uncertain.

Xanthine oxidase[9] oxidizes hypoxanthine to xanthine, and xanthine to uric acid. If the species concerned possesses uricase, allantoinase and allantoicase, further oxidation to allantoin, allantoic acid, glyoxylic acid and urea may take place; in humans uric acid is the product excreted as a result of purine oxidation. As well as hypoxanthine, many purine analogues are substrates for xanthine oxidase, and the resultant oxidation eliminates their possible utility as antimetabolites. Consequently, inhibitors of xanthine oxidase may be useful adjuvants (synergists) in therapy using purine drugs, although 9-substituted purines are in general not substrates for xanthine oxidase.

The reader should now find that the enzymic processes summarized in the foregoing pages account for practically all the purine metabolic transformations occurring in the cell, as depicted in Figure 3.

7.3.2.2 Pyrimidine Metabolism

The pathway by which pyrimidine nucleotides are synthesized *de novo*[10] in the cell is depicted in Figure 4. In the committing reaction, carbamoyl phosphate condenses with aspartic acid in a process catalyzed by the celebrated allosteric enzyme aspartate transcarbamoylase. Condensation with elimination of water followed by dehydrogenation affords orotic acid (**7**), which displaces pyrophosphate from PRPP under the influence of orotate phosphoribosyltransferase (*cf.* the purine salvage pathway, above) to afford orotidylic acid (**8**). Decarboxylation then gives uridylic acid (UMP; **9**). This key intermediate is converted into its related pyrimidine nucleotides *via* the pathways depicted in Figure 5.

Nucleoside mono- and di-phosphokinases convert UMP to UDP and UTP respectively. Ribonucleotide reductase then converts UDP to the 2'-deoxynucleotide dUDP, after which hydrolysis by a phosphatase yields dUMP. In contrast, UTP is aminated *via* transfer of amide nitrogen from glutamine, giving cytidine 5'-triphosphate (CTP; **10**), which together with UTP is a pyrimidine nucleotide substrate for DNA-directed RNA polymerase in the biosynthesis of RNA. However, CTP is also hydrolyzed to CDP, which is a substrate for ribonucleotide reductase, forming dCDP. This may in turn be hydrolyzed to dCMP by a phosphatase, after which a deaminase converts dCMP to dUMP, thus establishing a second pathway to dUMP. The dCDP is also converted to dCTP by nucleoside diphosphokinase. In a highly important reaction, dUMP is converted to 2'-deoxythymidylic acid (dTMP; **11**; usually just styled, inconsistently, 'thymidylic acid') by transfer of a methylene group from N^5,N^{10}-methylenetetrahydrofolate by the enzyme thymidylate synthetase.[11] This is a critical reaction: cells deprived of the base thymine eventually suffer 'thymineless death' since they

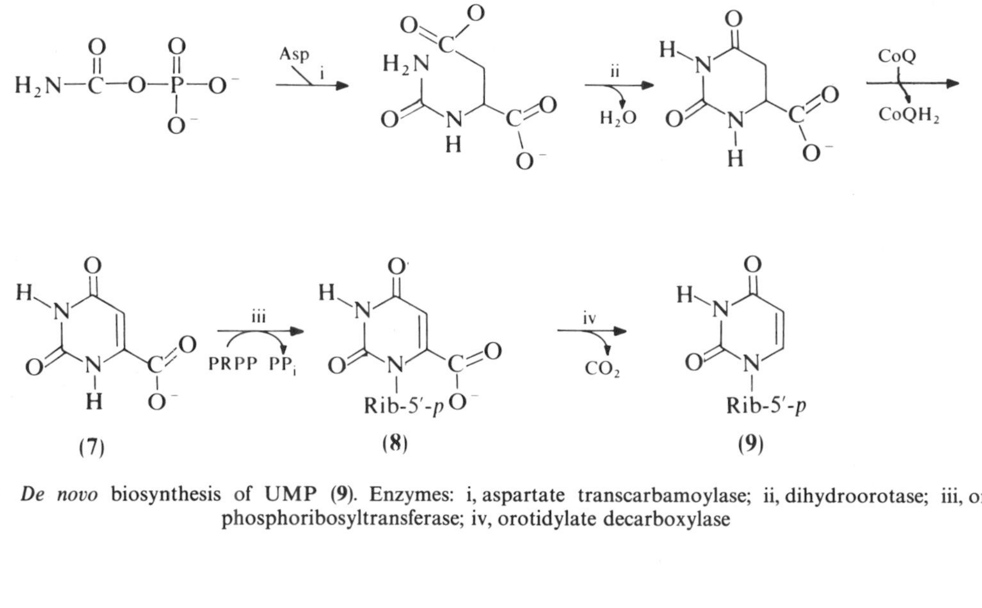

Figure 4 *De novo* biosynthesis of UMP (**9**). Enzymes: i, aspartate transcarbamoylase; ii, dihydroorotase; iii, orotate phosphoribosyltransferase; iv, orotidylate decarboxylase

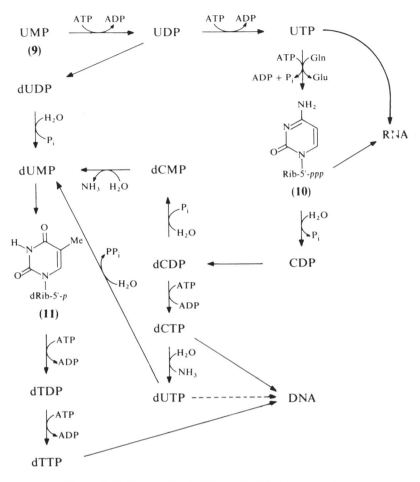

Figure 5   Pathways of pyrimidine nucleotide interconversion

can no longer form DNA, and thus this reaction to form dTMP has become an important target enzyme in tumour therapy. Once again nucleotide phosphokinases convert dTMP to dTDP and dTTP, and the latter, together with dCTP, form the principal pyrimidine nucleotide substrates for DNA polymerase. However, a deaminase also converts dCTP to dUTP, and although a pyrophosphatase enzyme acts to convert dUTP to dUMP, dUTP is a substrate for DNA polymerases, and consequently deoxyuridine is occasionally misincorporated into the DNA chain. If this occurs, the error is corrected by removal of uracil from the polynucleotide by breakage of the glycosidic bond by the enzyme uracil *N*-glycosylase. This removal of error bases from DNA (the uracil could also arise *via* spontaneous deamination of cytosine base) serves to maintain the integrity of the DNA.[12]

Uridine can be phosphorolyzed by uridine phosphorylase to afford uracil and ribose 1-phosphate; this reaction is reversible. Cytidine is not phosphorolyzed directly, but is first deaminated to uridine by cytidine deaminase. Uracil can participate in a 'salvage pathway', reacting directly with PRPP to form UMP, a process which has been reported only in microbial cells. Thymine can also participate in a salvage pathway, reacting with deoxyribose 1-phosphate to afford thymidine (a reversible reaction catalyzed by thymidine phosphorylase), which can then be phosphorylated to dTMP *via* phosphoryl transfer from ATP by thymidine kinase. This latter enzyme is sometimes specified by viral genomes and in some cases is of central importance in potentiating antiviral drugs.[13]

Cytosine is deaminated to uracil by cytosine deaminase. The deaminases are primary causes of the loss of drug activity of aminated purine and pyrimidine analogues, a point which will be reconsidered later. Uracil is degraded *via* reduction to 5,6-dihydrouracil, which is then hydrolyzed to β-ureidopropionic acid and then to β-alanine. Thymine is degraded *via* an exactly analogous pathway. Neither pathway is particularly significant in terms of the consequences of its malfunction, or in the useful control which can be obtained over it by drug action, when compared to purine degradation and its control by allopurinol.

7.3.3 ALLOPURINOL AND XANTHINE OXIDASE

As described briefly above, the oxidation of hypoxanthine (**12**) to xanthine (**13**) and of xanthine to uric acid (**14**) is catalyzed by the molybdenum-containing flavoprotein xanthine oxidase.[9,14] The enzyme from liver is a dimer of $M \approx 300\,000$, each subunit containing one FAD residue and one molybdopterin molecule as well as two Fe_2S_2 centres. The reaction catalyzed has the general form

$$AH + H_2O + X \longrightarrow AOH + XH_2$$

where AH is the reduced substrate and X an electron-accepting substrate, in this case oxygen. The reduction of oxygen by a single electron from the reduced substrate results in the formation of superoxide anion. During the initial processes of the reaction, the Mo^{VI} centre is believed to be reduced to the Mo^{IV} state in the same stage which introduces a nucleophilic atom X on the purine nucleus (**15a** → **15b**).[15] The atom X has been suggested to be the terminal oxo group present in the oxidized enzyme.[16] Either direct hydrolysis of (**15b**) or (at high pH) loss of the thiol proton followed by hydrolysis then affords the oxopurine product together with Mo^{IV} states of the enzyme, which give rise to the various EPR signals characteristic of Mo^V species which have been observed for the enzyme. Despite extensive investigation of the enzyme using inhibitors, flash photolysis, EPR, EXAFS and other spectroscopic studies, no unchallenged chemical mechanism for this complex reaction has yet been formulated.[17]

Uric acid possesses an acidic proton, $pK_a = 5.4$, at the N-9 position, and consequently exists in body fluids mainly as monosodium urate, the plasma concentrations in adults being *ca.* 4.1 mg dL^{-1} for females and *ca.* 5.0 mg dL^{-1} for males. The solubility of monosodium urate in plasma is *ca* 6.4 mg dL^{-1}. If this level is exceeded, due possibly to defects in renal clearance, or overproduction of urate due to an overactive PRPP synthetase or a defective salvage pathway (as in the macabre Lesch–Nyhan syndrome), hyperuricaemia results. Crystals of monosodium urate form in leukocytes in synovial fluid to give the excruciating pain of the acute arthritic condition termed gout (see Section 8.4.4.1).[18]

Allopurinol (**16**, 1*H*-pyrazolo[3,4-*d*]pyrimidin-4-ol), marketed under a variety of names, such as Zyloric, affords a very effective method of controlling serum urate concentrations. It is oxidized by xanthine oxidase to alloxanthine (**17**), which binds very tightly to the Mo^{IV} state of the enzyme to prevent further activity,[19] thus providing an example of 'suicide inactivation'. The complex between the reduced enzyme and alloxanthine is formed slowly from a rapid preequilibrium, and an overall K_i value of 3×10^{-8} M has been calculated.[20] The complex may be oxidized by potassium ferricyanide to release alloxanthine and reactivate the enzyme, but it is stable in air for hours, and under physiological conditions the inactivation is thus highly effective. The levels of uric acid in urine and plasma drop in consequence and patients receiving the drug excrete purines predominantly in the form of xanthine. The reutilization of hypoxanthine for nucleotide synthesis *via* the salvage pathway is increased, the resultant drop in PRPP levels tending to suppress purine nucleotide biosynthesis. The decrease in serum urate levels leads to alleviation of the symptoms of gout, and also to decreased incidence of urinary calculi in susceptible patients.

This suppression of oxidative purine elimination together with stimulation of the salvage pathway means that allopurinol enhances the action of certain antineoplastic purine drugs such as 6-mercaptopurine (1,7-dihydropurine-6-thione), and its prodrugs such as azathioprine, and may consequently be given as an adjuvant in conditions in which these drugs are employed.[21]

Xanthine oxidase gives rise both to superoxide ion or to hydrogen peroxide as the by-products of reaction, depending on the state of reduction of the enzyme.[14, 22] Protonation of superoxide, O_2^-, affords the hydroperoxyl radical, which undergoes spontaneous dismutation to yield hydrogen peroxide and oxygen. Iron(III) ions can also oxidize superoxide to oxygen

$$O_2^- + Fe^{3+} \longrightarrow O_2 + Fe^{2+}$$

and the resultant iron(II) ions can reduce hydrogen peroxide to afford the highly deleterious hydroxyl radical[23]

$$Fe^{2+} + H_2O_2 \longrightarrow Fe^{3+} + {}^-OH + \cdot OH$$

In consequence, superoxide requires to be scavenged as soon as it is formed, to prevent the formation of dangerous mutagens. This is performed by superoxide dismutases,[24] which catalyze the dismutation reaction

$$O_2^- + O_2^- + 2H^+ \longrightarrow H_2O_2 + O_2$$

In eukaryotes, the cytosolic dismutase contains one atom each of copper and zinc in each of its two identical subunits, while the mitochondrial dismutase contains two manganese atoms per molecule.

The importance of the function of superoxide dismutases has stimulated a search for simpler species with comparable activity, and a number of copper complexes which mimic the enzymic activity, such as copper penicillinamine, have been identified.[25] The hydrogen peroxide formed is converted to oxygen and water by the heme protein catalase.[26]

7.3.4 ANTINEOPLASTIC AGENTS[27]

Neoplasm means 'new growth', and it is characteristic of tumour cells that they grow and divide more rapidly than the parent tissue from which they have differentiated. However, their biochemistry will in essence be the same as that of the parent tissue. In the search for antineoplastic agents, the medicinal chemist seeks an agent which kills more of the tumour cells than the host cells, due to the neoplastic tissue taking up more of the drug. It is therefore not surprising that the greatest successes have been registered against the more rapidly growing cancers, such as leukaemias and lymphomas, since in these conditions the cells are constantly dividing, and are more sensitive to antimetabolites of the types described in this section. It must be realized that not all cytotoxic drugs will be useful antitumour agents. Ideally, one seeks an agent which is selectively cytotoxic for the tumour cell population, and current developments may eventually lead to higher degrees of selectivity against tumour cells than has previously been the case. For the most part, successful antitumour agents either inhibit the formation or function of the nucleic acids, or they prevent cell division. As will be seen, the antimetabolites considered in this section, and indeed in this chapter, belong to the former class.

7.3.4.1 Purine Antagonists[28, 29]

7.3.4.1.1 Thiopurines

Of the many purine antagonists which have been evaluated for antineoplastic activity, 6-mercaptopurine (MP; 1,7-dihydropurine-6-thione; **18**) and 6-thioguanine (TG; **19**) are the most active against a range of experimental tumours, and have been widely used to treat cancer in humans. To display activity, MP is first converted to the nucleotide (**20**) by reaction with PRPP catalyzed by HGPRT, an enzyme which it inhibits competitively. Cell lines which are resistant to the cytotoxic effect of MP lack (or have lost) the ability to perform this conversion.[30] The nucleotide (20) notably inhibits phosphoribosyl-pyrophosphate amidotransferase (Figure 1), thus blocking *de novo* purine nucleotide synthesis at an early stage. However, it is also methylated intracellularly to 6-(methylthio)purine ribonucleotide (21), which is a more powerful inhibitor of the same enzyme.[31] While the nucleoside (22) is taken up by cells, it is not normally phosphorylated, but rather cleaved by purine nucleoside phosphorylase to give MP (18).[32] While MP can be methylated *in vivo* to (23), this species is not a substrate for the purine phosphoribosyltransferases. However, the nucleoside (24) is taken up by cells, is not significantly cleaved by purine nucleoside phosphorylase, and is readily phosphorylated by adenosine kinase to give (21).[33] Consequently there are two possible routes to (21), one starting from MP and the other from (24), and these two agents are therapeutically synergistic against L1210 (mouse leukaemia) cells, although not against resistant cell lines in which one or the other path to (21) is blocked.[34]

(18)　(19)　(20)　R = PO$_3$H$_2$
(21)　R = PO$_3$H$_2$　(22)　R = H
(24)　R = H　(23)

While (23) is not converted to (21) in the same way that (18) is converted to (20), it displays antitumour activity *in vivo*, which must result from demethylation to (18) followed by conversion to (20) and (21). The demethylation occurs in the liver, and probably represents the pathway by which a number of 6-(alkylthio)purines and their derivatives express their cytotoxic activity.[35] The drug azathioprine (Imuran; 25) displays cytotoxic and immunosuppressive properties (see also Chapter 8.4), and is cleaved mainly in the liver, probably by sulfhydryl compounds, to release MP.[36] It is used to treat acute leukaemia (see Section 8.4.3). Its toxicity is dose dependent and may be enhanced if given together with allopurinol.

The main degradative pathway of (20) involves dephosphorylation to (22), followed by loss of the sugar to give (18), which can be oxidized by xanthine oxidase to 8-oxy-6-mercaptopurine and 6-thiouric acid. This process is inhibited by allopurinol, but its effectiveness as an adjuvant in increasing the toxicity of MP is not unequivocal.[37] In contrast, (23) is oxidized to 6-methylthio-8-oxopurine by hepatic aldehyde oxidase. Subsequent oxidation leads to the release of inorganic sulfate.

6-Thioguanine (TG; 19) is also a good substrate for HGPRT, being converted to 6-thioguanylic acid (26), which, in contrast to (20), is further converted to the 5'-diphosphate (27) and triphosphate (28). The triphosphate is a substrate for polymerases, leading to the incorporation of units of (26) into RNA,[38] and TG is also incorporated into DNA as (29).[38] For this to occur, the 2'-deoxynucleoside 5'-triphosphate (30) must first be formed, probably *via* reduction of the intermediate diphosphate (31) by ribonucleotide reductase. The incorporation of TG into DNA is thought to be largely responsible for its antitumour activity, although other metabolic effects due to (26) may be involved.[39] The incorporation of MP into DNA has been reported, but appears to be less significant in terms of expression of cytotoxicity than that of TG. Certainly the conversion of TG to nucleotides is much more significant than that of MP, and the development of resistance to TG by tumours has been attributed to increased breakdown of the nucleotides by phosphatases.[40] A delayed cytotoxic effect has been observed for MP, however, which is distinct from that due to its effect on purine nucleotide synthesis, and this has been ascribed to its conversion to TG nucleotides which are then incorporated into DNA and RNA.[41] Like MP, TG, after deamination to 6-thioxanthine, is oxidized to 6-thiouric acid, and subsequently gives rise to inorganic sulfate. S-Methylation of TG to 2-amino-6-(methylthio)purine, which is subsequently deaminated to 2-hydroxy-6-(methylthio)purine, is also a major pathway in man. The S-heterocyclic derivative of TG, Guaneran (32), the analogue of azathiopurine (Imuran), has also been prepared as a protected prodrug of TG, and is similarly split nonenzymically by sulfhydryl groups to release TG.[36]

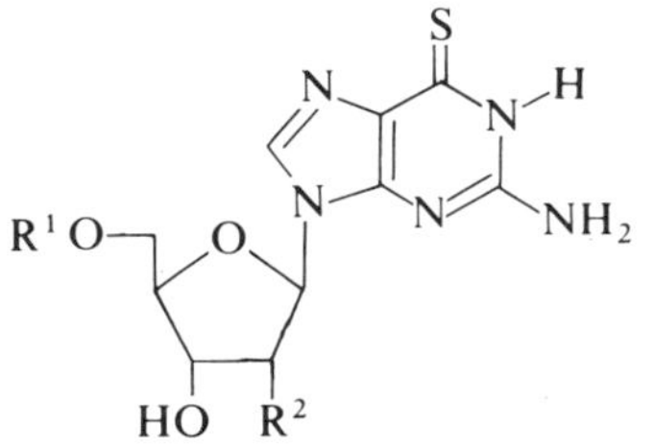

(26)	$R^1 = PO_3H_2$,	$R^2 = OH$
(27)	$R^1 = P_2O_6H_3$,	$R^2 = OH$
(28)	$R^1 = P_3O_9H_4$,	$R^2 = OH$
(29)	$R^1 = PO_3H_2$,	$R^2 = H$
(30)	$R^1 = P_3O_9H_4$,	$R^2 = H$
(31)	$R^1 = P_2O_6H_3$,	$R^2 = H$

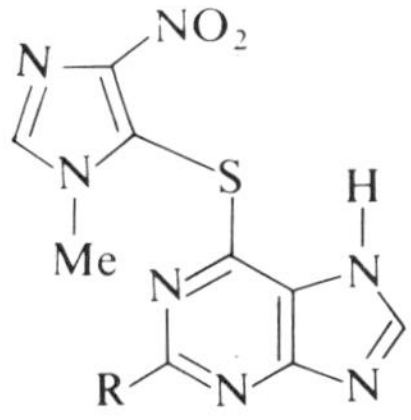

(25) R = H
(32) R = NH₂

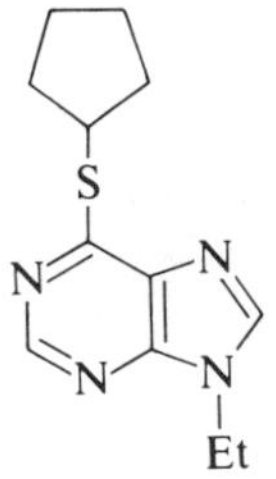

(33)

It is interesting that, while work with leukaemic cell lines incubated with MP showed that (20) was formed as the major intracellular metabolite, in erythrocytes of patients receiving MP by intravenous infusion the major intracellular metabolite was the TG-derived nucleotide (26).[42]

Other cytotoxic thiopurines with antitumour activity include 9-alkyl, 1-alkyl and 3-alkyl derivatives of MP.[43] The mode of action of these compounds does not seem to have been investigated fully. 9-Ethyl-6-mercaptopurine was about as active as MP in a limited clinical trial against chronic leukaemia.[44] 6-(Cyclopentylthio)-9-ethylpurine (33), a derivative of MP designed for lipid solubility, is reportedly more effective than MP at crossing the blood–brain barrier to kill intracerebral tumour cells.[45]

7.3.4.1.2 *Azapurines*

8-Azaguanine (34) was found to possess antitumour activity as long ago as 1949,[46] and was the first purine analogue shown to be incorporated into polynucleotides.[47] It is a substrate for HGPRT, thus becoming converted *in vivo* to 8-azaguanosine 5'-monophosphate and thence to its 5'-di- and tri-phosphates. Its primary metabolic effect is on protein synthesis, and its mode of action is believed to result from its incorporation into RNA as 8-azaguanylic acid.[48] 8-Azahypoxanthine (35) is less cytotoxic, but its ribonucleoside, 8-azainosine (36), shows substantial activity against several tumour cell lines.[49] It requires to be converted to its monophosphate in order to express its activity, a process which may occur *via* the agency of adenosine kinase or *via* cleavage to (35) by a nucleoside phosphorylase followed by phosphoribosylation as for (34) above. 8-Azaadenosine may be converted to its 5'-monophosphate *via* the same routes as (36), or it may be converted to (36) *via* adenosine deaminase and then to the monophosphate of (36), 8-azainosinic acid. This species becomes converted to 8-azaguanylic acid (*cf.* Figure 2) and hence (36) becomes incorporated into polynucleotides as 8-azaguanylic acid, while (37) becomes incorporated partly as 8-azaguanylic acid

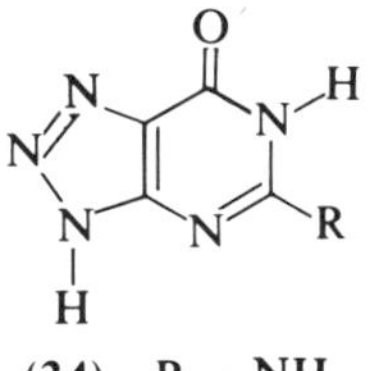

(34) R = NH₂
(35) R = H

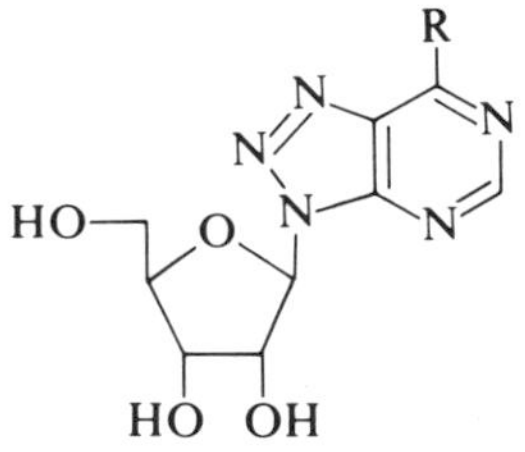

(36) R = OH
(37) R = NH₂

and partly as 8-azaadenylic acid, depending on the cell line involved.[50] Thus all three 8-azapurines cause the same base analogue to be introduced in RNA. It is noteworthy that (37) is also a strong inhibitor of *de novo* purine nucleotide synthesis.

8-Aza-6-thioinosine (38) is cytotoxic to human epidermoid carcinoma cells in culture, but it rearranges rapidly under the conditions of the cytotoxicity tests to the [1,2,3]thiadiazolo[5,4-*d*]-pyrimidin-7-amine (39), which displays 10 times its cytotoxicity. Both compounds are phosphorylated rapidly by adenosine kinase, and it is difficult to study the activity of (38) separately from that of its rearrangement product. The situation is further complicated by the finding that the mono-phosphate of (39), the *β*-anomer, can rearrange *in vivo* to form the *α*-anomer, although it is the *β*-anomer which is responsible for cytotoxicity. Neither (38) nor (39) seems to be able to give rise to 8-azaguanine or 8-azahypoxanthine or their conjugates. Instead, they selectively reduce the guanine nucleotide pools and inhibit DNA and RNA synthesis without directly affecting protein synthesis.[51]

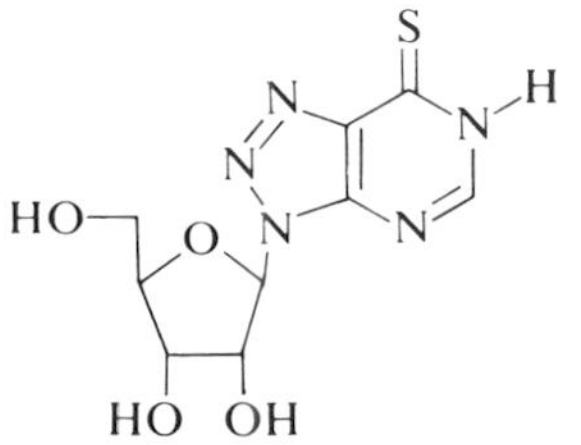

(38) 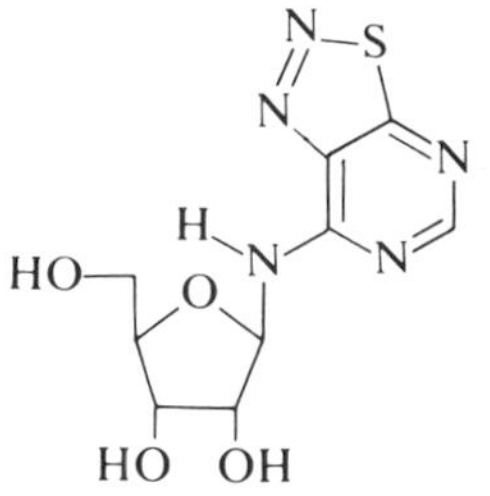   (39)

Among other 8-azapurine analogues, 8-aza-6-thioguanine displays similar cytotoxicity to 8-azaguanine, though its ribonucleoside is inactive. It is probably potentiated *via* the action of HGPRT.[52] 8-Aza-*S*-methyl-6-thioinosine is highly cytotoxic, and expresses its activity following conversion to its nucleotides *via* adenosine kinase.[53] O^6-Methyl-8-azainosine, O^6-ethyl-8-azainosine and O^6-methyl-8-azaguanosine all display cytotoxicity, which appears in the case of the O^6-methyl compounds to be wholly or partly due to their conversion to 8-azainosine by the agency of adenosine deaminase. In the presence of 2'-deoxycoformycin (40), a powerful inhibitor of this enzyme, their cytotoxicity is lessened considerably. Adenosine kinase can act to potentiate the O^6-alkylated-8-azainosine derivatives, but not O^6-methyl-8-azaguanosine.[28]

2-Azaadenosine (41) is a substrate for both adenosine kinase, and for adenosine deaminase, which converts it to 2-azainosine (42), which is cleaved to 2-azahypoxanthine. The cytotoxicity of (41) is expressed both *via* its phosphorylation and *via* its conversion to 2-azahypoxanthine. It shows moderate activity against L1210 leukaemia cells.[54]

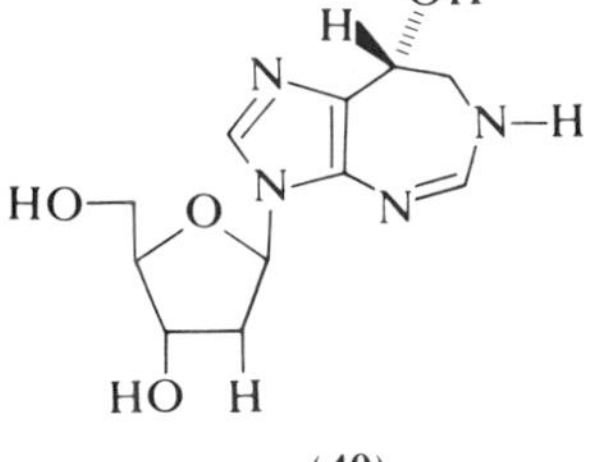

(40)

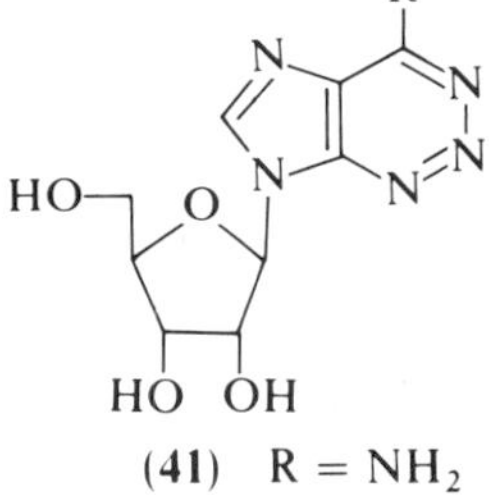

(41) R = NH$_2$

(42) R = OH

7.3.4.1.3 Ara-*adenosine*

9-*β*-D-Arabinofuranosyladenine (*ara*-adenosine; *ara*-A; Vidarabine; 43) is a strong inhibitor of growth of certain ascites tumours[55] and human leukaemias.[56] It suffers from the considerable disadvantage that it is rapidly deaminated *in vivo* to the inactive *ara*-inosine, but coadministration of known inhibitors of adenosine deaminase (see Section 8.4.2.1) such as pentostatin (40) suppress this reaction,[57] permitting *ara*-A to be phosphorylated to its monophosphate by deoxycytidine kinase or adenosine kinase (for which it is a poor substrate) and subsequently to its triphosphate which is a strong noncompetitive inhibitor of DNA polymerase.[58] In combination with (40), *ara*-A has afforded high cure rates of murine leukaemias. 2-Fluoro-*ara*-adenosine (**44**) is resistant to adenosine deaminase,[59] and is a substrate for deoxycytidine kinase, being subsequently converted to its diphosphate and triphosphate. It acts as a good inhibitor of ribonucleotide reductase and DNA polymerase,[60] and is active, for instance, against P388 and L1210 leukaemia resistant to *ara*-A. Among other analogues of *ara*-A which have been synthesized and tested, 2'-azido-2'-deoxy-*ara*-adenosine (**45**) has similar cytotoxicity to *ara*-A against L1210 cells, while 2'-amino-2'deoxy-*ara*-adenosine (**46**) is less active.[61]

(43)	$R^1 = H,$	$R^2 = OH$
(44)	$R^1 = F,$	$R^2 = OH$
(45)	$R^1 = H,$	$R^2 = N_3$
(46)	$R^1 = H,$	$R^2 = NH_2$

(47)	$X = CH,$	$R^1 = OH,$	$R^2 = H$
(48)	$X = N,$	$R^1 = OH,$	$R^2 = H$
(49)	$X = N,$	$R^1 = H,$	$R^2 = OH$

(50)	$R = H$
(51)	$R = CONH_2$
(52)	$R = CN$

(53)

7.3.4.1.4 Other agents

Many other purine analogues have been investigated for their antitumour properties, and brief mention of a few must suffice. Aristeromycin (**47**), the analogue of adenosine containing a cyclopentane ring in place of the ribofuranose moiety, is highly cytotoxic to L1210 cells in culture, but ineffective against L1210 leukaemia *in vivo*.[62] The corresponding 8-azaadenine derivative (**48**), however, does display antileukaemic activity in mice,[63] and so also does the carbocyclic 'xylofuranosyl'-8-azaadenine (**49**).[64] 2'-Deoxy-2-fluoroadenosine, 2-bromo-2'-deoxyadenosine and 2-chloro-2'-deoxyadenosine are all highly cytotoxic and active against leukaemia L1210 *in vivo*.[65] All appear to be activated by deoxycytidine kinase and probably act in a similar way to 2-fluoro-*ara*-adenosine. 2-Chloro-2'-deoxyadenosine, which, like its bromo congener, is resistant to adenosine deaminase, showed useful activity in a phase I trial in patients with advanced leukaemia. The corresponding ribonucleosides have little if any activity, probably because they are poor substrates for adenosine kinase. Certain deazapurine derivatives have been found to be active antimetabolites, although with limited applicability as antineoplastic agents. 3-Deazaadenosine is converted *in vivo* to 3-deazaadenosylhomocysteine, which is a strong inhibitor of the methyltransferases which utilize *S*-adenosylmethionine as the methyl donor.[66] 3-Deazaguanine has been found to be active against murine L1210 leukaemia *in vivo*.[67] 9-Deazaadenosine is markedly cytotoxic to a number of human and murine leukaemia cell lines.[68] 2-Amino-6-chloro-1-deazapurine is also similarly active *in vitro* and *in vivo*. It is converted to its ribonucleotide in L1210 cells, and inhibits *de novo* purine nucleotide biosynthesis and also IMP dehydrogenase.[69] Among purine nucleoside antibiotics, the 7-deaza-purine series tubercidin (**50**), sangivamycin (**51**) and toyocamycin (**52**) all display antineoplastic activity.[70] Tubercidin 5'-triphosphate is a powerful inhibitor of the synthesis of DNA and RNA generally, the suppression of mRNA synthesis being particularly important. Another nucleoside antiobiotic, neplanocin (53), also shows significant *in vivo* activity against L1210 cells.[71]

7.3.4.2 Pyrimidine Antagonists

7.3.4.2.1 *Azapyrimidines*

5-Azacytidine (Azacitidine; 54), the most active of the azapyrimidines, shows substantial activity in the treatment of murine and human leukaemias, but little useful activity against solid tumours.[72] It is phosphorylated *in vivo* to the mono-, di- and tri-phosphate levels, and inhibits nucleic acid synthesis and function. In calf thymus nuclei, the synthesis of mRNA is inhibited, while the incorporation of 5-azacytidine into tRNA decreases amino acid acceptor activity for most of the amino acids, further suppressing protein synthesis.[73] In *Escherichia coli* it becomes incorporated, as the deoxynucleotide, into DNA, and may express its mutagenic and other activities by this route also. Although the 1,3,5-triazine ring is unstable in solution in the nucleotides, it seems to be stabilized upon incorporation into polynucleotides.[74]

5-Azauracil (55) and 5-azaorotic acid (56) are competitive inhibitors of orotate phosphoribosyl-transferase[75] (Figure 4) and show activity against adenocarcinoma 755, but not against L1210 leukaemia. 5-Azauracil may also exert its activity, in part, *via* hydrolysis to *N*-formylbiuret, which inhibits dihydroorotase. However, 5-azauracil is also converted *in vivo* to 5-azauridine and 5-azauridylic acid, and 5-azaorotic acid to 5-azaorotidylic acid, and both these agents are additionally inhibitors of orotidine-5'-phosphate decarboxylase. In contrast to (54), therefore, (55) and (56) primarily suppress *de novo* pyrimidine biosynthesis.[76]

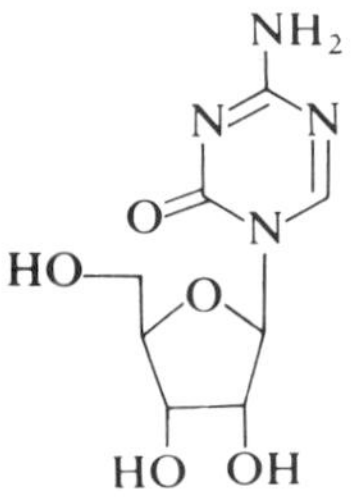

(54)

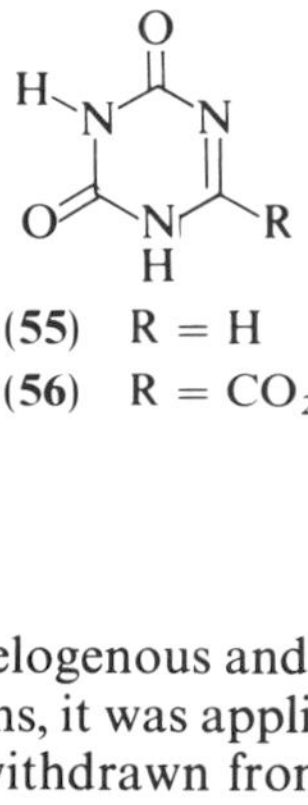

(55) R = H
(56) R = CO$_2$H

6-Azauridine (57) was formerly used to treat chronic myelogenous and acute leukaemias,[77] as well as psoriasis and mycosal infections. For the latter conditions, it was applied as its prodrug 2',3',5'-tri-*O*-acetyl-6-azauridine (Azaribine; 58), but this has been withdrawn from the US market following reports of thrombosis. 6-Azauridine is phosphorylated *in vivo* to its 5'-monophosphate, which is a powerful inhibitor of orotidine-5'-phosphate decarboxylase,[78] and despite inhibitory activity against a number of other enzymes, this inhibition of orotidylic acid metabolism is primarily responsible for its cytotoxic effects. It is not now regarded as having proven clinical utility.

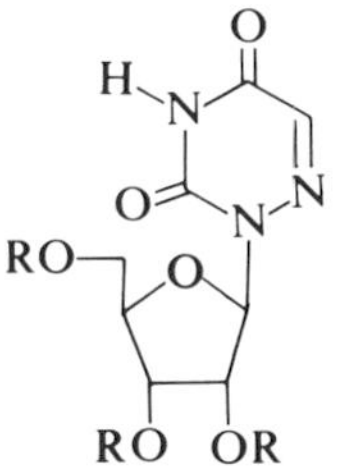

(57) R = H
(58) R = MeCO

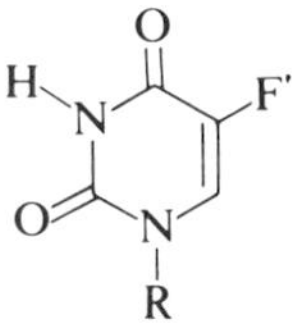

(59) R = H
(60) R = 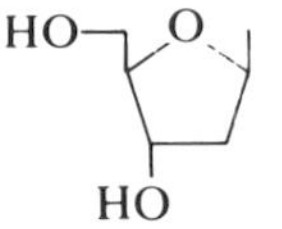 HO—[deoxyribose]

(62) R = [tetrahydrofuranyl]

(61)

(63)

7.3.4.2.2   Fluoropyrimidines[79]

Of the various halogenated pyrimidines which have been prepared, only the fluorinated pyrimidines and their nucleosides have useful antitumour activity, and of these the most significant are 5-fluorouracil (FU; 59) and its 2'-deoxyribonucleoside (FUdR; Floxuridine; 60). Both of these drugs have found wide uses in patients suffering from metastatic cancer, and have been administered both singly and in combination therapy. Both are converted to 5-fluoro-2'-deoxyuridylic acid, by phosphorylation (in the case of 60), or *via* reaction with deoxyribose 1-phosphate catalyzed by thymidine phosphorylase, or a longer route involving 5-fluorouridine (for 59). 5-Fluoro-2'-deoxyuridylic acid (FdUMP) is a powerful inhibitor of thymidylate synthetase[80] and, by suppressing the formation of dTMP, it causes 'thymineless death' in neoplasms. The structure of the ternary covalent complex believed to be formed by reaction between the enzyme, FdUMP and N^5,N^{10}-methylenetetrahydrofolate is depicted in (61). There is now good evidence that the methylene bridge is bonded to N-5 as shown, rather than to N-10 of tetrahydrofolate.[81] When dUMP is the substrate, the proton which replaces the fluorine atom in (61) is abstracted by a basic group on the enzyme, and elimination of the tetrahydrofolate moiety ensues. Reduction of the resultant 5-methylene group by hydride transfer from tetrahydrofolate and subsequent expulsion of the sulfhydryl group of the enzyme affords dTMP. The presence of fluorine, which cannot be abstracted by base, halts the process. In contrast, the 5'-monophosphates of 5-chloro-, 5-bromo- and 5-iodo-2'-deoxyuridine, in which the halogen atom is bulkier and less electronegative, have little or no useful inhibitory activity for the enzyme. Preliminary QSAR studies of the enzyme from *Lactobacillus casei*[82] led to the conclusion that electron-withdrawing groups at the 5-position of 2'-deoxyuridylate species enhanced the affinity of the analogues for the enzyme, probably primarily by altering the electron distribution in the heterocycle, but also by increasing the acidity of the proton at N-3 of the ring. Larger substituents at the 5-position are detrimental to binding. The structure of the enzyme has recently been disclosed[83] and this information should permit the more precise design of inhibitors.[84]

5-Fluorouridine is highly cytotoxic but less effective against tumours, possibly because the reaction sequence needed to convert it to FdUMP is inefficient. 5-Fluorocytidine, 5-fluoro-2'-deoxycytidine and 5-fluoroorotic acid probably owe much of their antitumour activity to the ability to be converted to FdUMP *in vivo*.[85] 5-Fluoro-*ara*-uridine exhibits antitumour activity, albeit less than that of FUdR, but 5-fluoro-2'-deoxy-*lyxo*-uridine does not.[86]

A considerable number of prodrugs of 5-fluorouracil have been prepared, one of the most widely researched being 5-fluoro-1-(tetrahydro-2-furyl)uracil (Ftorafur; Tegafur; 62).[87] The corresponding 1,3-bis(tetrahydro-2-furyl) derivative has also been described, and, like Ftorafur, it is active against a range of solid tumours.[88] These agents are cleaved to release FU, possibly by thymidine phosphorylase or *via* hydroxylation at the 2'- and 3'-carbon atoms by microsomal monooxygenases. Other prodrugs of FU showing good antitumour activity include its 1-*t*-butylcarbamoyl,[89] 1-hexylcarbamoyl[90] and 1-methoxycarbonylmethylcarbamoyl[91] derivatives, 5'-deoxy-5-fluorouridine[92] and methyl 1-(5-fluoro-1*H*-2-oxopyrimidin-4-yl)-β-D-glucopyranuronate (63).[93] All of these contain in essence the FU moiety, ready for release *via* oxidative or hydrolytic means. 5-Fluoropyrimidin-4(1*H*)-one (64), with activity against tumours in experimental animals, is a different type of prodrug, being oxidized to FU by xanthine oxidase.[94]

A number of prodrugs of FUdR have also been prepared, a notably potent member being 5-bromo-6-methoxydihydro-5-fluoro-2'-deoxyuridine (65).[95] Finally, a different species of fluoro-pyrimidine derivative, which as its 5′-monophosphate inhibits thymidylate synthetase, is 5-trifluoromethyl-2′-deoxyuridine (Trifluridine; **66**). It possesses a higher therapeutic index against adenocarcinoma 755 in mice than FUdR.[96]

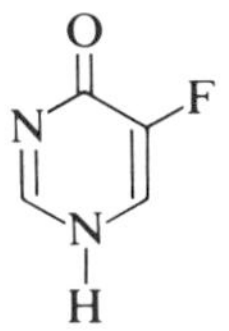

(64) 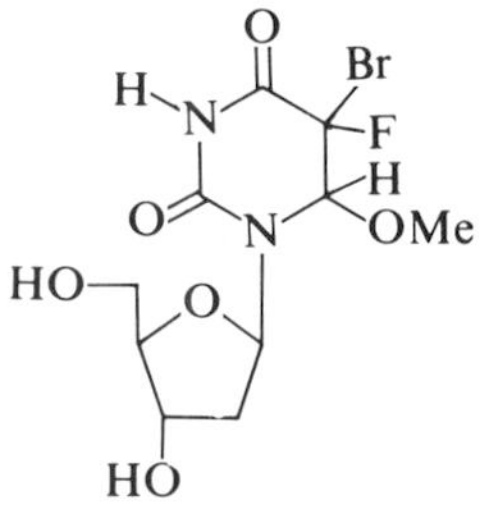   (65) 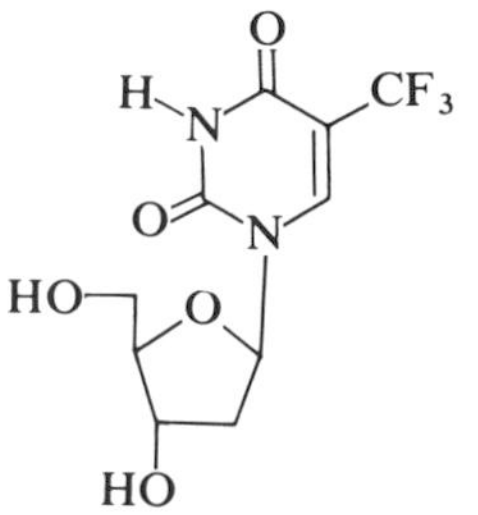   (66)

7.3.4.2.3 Ara-*cytidine and its derivatives*

1-β-D-Arabinofuranosylcytosine (*ara*-cytidine; Cytarabine, **67**; *ara*-C) is highly active against both mouse and human leukaemias,[97] and has found wide application in their treatment. The drug is active in its phosphorylated forms,[98] and *ara*-CTP is a powerful inhibitor of mammalian DNA polymerase,[99] and to some extent also a substrate for the polymerases, since *ara*-C has been shown to be incorporated into both DNA and RNA in human leukocytes. In addition, *ara*-CDP is a moderate inhibitor of a ribonucleotide reductase from human tumour cells, which may account for the observation that *ara*-C blocks the conversion of uridine to dCMP, but not to CMP.

Ara-C possesses the disadvantage that it is rather readily deaminated to the inactive *ara*-U by cytidine deaminase,[100] a finding which has prompted a search for cytidine deaminase inhibitors for coadministration with the drug, and also for prodrugs, slow release forms or analogues of *ara*-C which resist the deaminase. Tetrahydrouridine [1-(β-D-ribofuranosyl)-3,4,5,6-tetrahydro-2-oxopyrimidin-4-ol][101] has been found to be an efficient inhibitor of cytidine deaminase and is probably the one most commonly used, although the 1-β-D-ribofuranosides of 2-oxopyrimidine, 2-oxo-1,3-diazepine, and 3,4-dihydro-4-hydroxymethyl-2-oxopyrimidine are also effective.[102] 2′-Azido- and 2′-amino-2′-deoxy-*ara*-C (Cytarazid; **68** and Cytaramin; **69**, respectively) are effective against L1210 leukaemia in mice, probably because they are resistant to the deaminase.[103] The same applies to 2′-*O*-nitro-*ara*-C.[104] 2,2′-Anhydro-1-β-D-arabinofuranosylcytosine (anhydro-*ara*-C; **70**) is another prodrug, undergoing slow autohydrolysis to *ara*-C,[105] and the 3′,5′-di-*O*-acyl and 3′-*O*-acyl-5′-phosphate derivatives of (**70**) also act as prodrugs and are active against L1210 leukaemia.[106] So, too, are 5-aza-*ara*-C[107] and the carbocyclic analogue of *ara*-C.[108] The

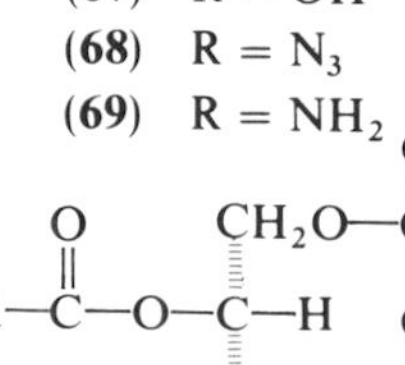

(67) R = OH
(68) R = N₃
(69) R = NH₂

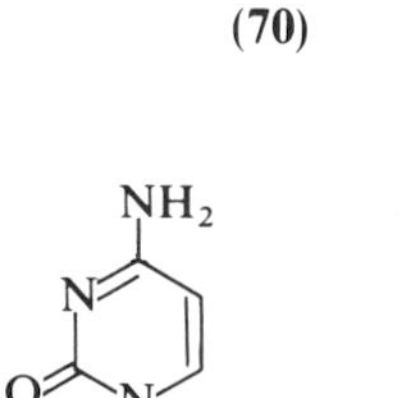

(70)

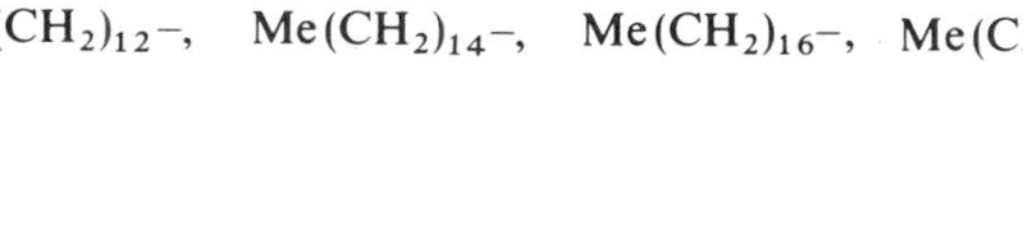

(71) R = Me(CH₂)₁₂–,   Me(CH₂)₁₄–,   Me(CH₂)₁₆–,   Me(CH₂)₇CH=CH(CH₂)₇–

conjugation of *ara*-CMP with corticosteroids such as cortisol and cortisone has afforded derivatives with higher activity against L1210 leukaemia than *ara*-C itself.[109] N^4-Palmitoyl-*ara*-C is resistant to the deaminase and is hydrolyzed slowly to afford adequately cytotoxic levels of *ara*-C.[110] Several phospholipid conjugates of *ara*-CMP (71) have been described[111] and found to be more effective than *ara*-C, possibly because they release *ara*-CMP directly. They are also resistant to cytidine deaminase. 5-Fluoro-*ara*-C also shows substantial activity against murine leukaemias, and is thought to act more as an analogue of *ara*-C than of 5-fluorocytidine, blocking the reduction of CDP to dCDP by ribonucleotide reductase.[112]

Ara-cytidine is frequently used in combination with vincristine or vinblastine, inhibitors of mitosis, in the treatment of human leukaemias. It is possibly surprising that it shows little activity against solid tumours.

7.3.4.2.4   Other agents

A number of other pyrimidine antagonists displaying antitumour activity, in which the base is conjugated to a modified sugar ring, have been reported. Although *ara*-uridine shows no useful activity, *ara*-thymidine and 5-bromo- and 5-iodo-*ara*-uridine inhibit the growth of sarcoma 180 and L1210 cells in culture.[113] Other thymidine analogues with similar activity include 5-azidomethyl-, 5-aminomethyl- and 5-hydroxymethyl-2′ -deoxyuridine.[114] 3′-Amino-3′-deoxythymidine[115] and 3′-amino-2′,3′-dideoxycytidine[116] also possess strong activity against L1210 leukaemia. 2′-Deoxy-2′-fluoro-5-methyl-1-β-D-arabinofuranosyluracil (FMAU; 72) is highly active against *ara*-C resistant L1210 and P815 cell lines both *in vitro* and *in vivo*.[117]

2-β-D-Ribofuranosylthiazole-4-carboxamide (Tiazofurin; 73) has aroused much interest recently for its activity against solid tumours such as lung carcinoma. It is metabolized to an analogue of NAD^+ in which the thiazole-4-carboxamide moiety replaces the nicotinamide ring. However, it also depresses the synthesis of DNA and RNA, and thus merits inclusion as an antagonist of normal purine and pyrimidine metabolism.[118]

(72)   (73)

7.3.5   TARGETS FOR ANTIVIRAL CHEMOTHERAPY[13]

Viral chemotherapy presents a quite different problem from tumour chemotherapy. A virus[119] consists of a core of nuclear material, DNA or RNA, containing specific viral genes, which may be associated with 'core proteins', and which is surrounded by a protective protein coat. The coat may have functional protein appendages, and there may be present an 'envelope', rich in lipoprotein and glycoprotein. The virus possesses no energy-producing or protein-synthesizing machinery of its own. In order to reproduce, it must therefore become adsorbed to a host cell, be transported across the cell membrane, and uncoat in order that its viral genes may be expressed. The genome may require to be transported to the cytoplasm or the nucleus. The nuclear material must then be replicated, and the viral genes also transcribed into RNA (assuming that we are dealing with a DNA virus) and translated into virally specified protein. In addition to the coat protein, viral enzymes will be produced in this process, and these are vital for suborning the host cell's internal biochemistry to serve the requirements of the virus. Before the viral proteins can be made, a process of maturation of the viral mRNA molecules, involving guanylylation and methylation to produce a 'cap' structure containing 7-methylguanosine 5′-phosphate in a 5′–5′-pyrophosphate structure at the 5′-end of the mRNA, must occur. Once the synthesis of the components needed to make fresh viral particles has been completed, maturation occurs, in which the nucleic acid and protein components are assembled to form the complete viral particle or virion, and finally this is released from the cell as a new, infective virus.

RNA viruses require that their nucleic acid be replicated also, in order that new viral particles can form. Some RNA viruses code for an enzyme 'transcriptase' or 'replicase' (RNA-directed RNA polymerase), which replicates RNA strands directly without involving DNA,[120] while others encode an enzyme 'reverse transcriptase' (RNA-directed DNA polymerase), which transcribes the RNA sequence into DNA — the reverse of the usual sequence — which can subsequently become integrated into the host cell chromosome and is transcribed to form copies of the viral RNA.[121]

Purine and pyrimidine antimetabolites which are active against viruses are thus most likely to be effective by inhibiting specifically the viral enzymes which are required to replicate the viral nucleic acids, or inhibiting the viral enzymes responsible for transcription and capping of mRNA, without inhibiting the corresponding enzymes of the host cell. Alternatively, incorporation of an antimetabolite into a viral nucleic acid causing the cessation of strand synthesis, or otherwise disturbing the normal function of the nucleic acid in replication or in directing protein synthesis, offers another way of preventing the replication of functional virus.[13] Again, it is important that the integrity of the nuclear material of the host cell should not be compromised. One therefore seeks to exploit the differences between the viral enzymes and host cell enzymes to the discomfiture of the virus.

By way of illustration, some of the more widely investigated purine and pyrimidine antiviral agents will now be described, together with their mode of action.

7.3.5.1 Pyrimidine Nucleoside Antiviral Agents[122]

7.3.5.1.1 Halogenated pyrimidine 2'-deoxynucleosides

5-Iodo-2'-deoxyuridine (Idoxuridine; **74**)[123] was the first drug approved for clinical use in treatment of herpes keratitis, a sight-threatening eye infection. It is phosphorylated by both viral and cellular thymidine kinases to the 5'-monophosphate.[124] Since herpes-infected cells contain high levels of thymidine kinase,[125] the drug attains higher levels of its effective phosphorylated forms in infected cells. The primary reason for the antiviral activity of (**74**) lies in its incorporation into viral DNA.[126] This will also occur in uninfected cells exposed to the drug, hence it cannot be administered systemically. The resulting abnormal viral DNA results in the formation of defective viral subunits which cannot function correctly in the replicative process. Changes in RNA structure and the depression of viral protein synthesis have been reported, possibly related to the mutagenic activity of the drug. In both 5-iodo- and 5-bromo-deoxyuridine, the enol tautomer of the uracil ring is believed to be stabilized with respect to the proportion occurring in thymidine, permitting mispairing with guanine to occur and leading to 'transitional' mutants. This view has been challenged, however.[127]

Iododeoxyuridine is principally effective against DNA viruses and oncogenic RNA viruses. In addition to its use topically against herpes keratitis, it has been applied in DMSO solution for treatment of herpes zoster, genital herpes and vaccinia infections, with varying success.[128] 5'-Amino-5-iodo-2',5'-dideoxyuridine (Aminoidoxuridine; **75**) also displays good antiherpes activity and is less cytotoxic than (**74**).[124,129] The phosphorylation of (**75**) is catalyzed specifically by virus-induced thymidine kinase, and thus (**75**) is only potentiated in virus-infected cells. It inhibits the synthesis of DNA in infected cells. The 5-bromo-, 5-chloro- and 5-trifluoro-methyl derivatives of 5'-amino-2',5'-dideoxyuridine also exhibit antiviral activity.[130]

(**74**)   R = OH

(**75**)   R = NH₂

(**76**)   X = I

(**77**)   X = Br

5-Iodo-2'-deoxycytidine (**76**) and 5-bromo-2'-deoxycytidine (**77**) also show useful activity against herpes viruses and vaccinia, with less cytotoxicity than the uridine compounds and more selective inhibition of viral replication. This selectivity is thought to be due to specific phosphorylation of (**76**) and (**77**) by a viral kinase, once again.[131] However, human tissues contain a highly active cytidine deaminase, which would convert (**76**) to the cytotoxic (**74**) in uninfected cells, and this has precluded its wider exploitation.

5-Trifluoromethyl-2'-deoxyuridine (Trifluridine; **66**) is also used for the treatment of herpetic keratitis, particularly in patients unresponsive to other agents.[132] It resembles idoxuridine (**74**) in being phosphorylated by both cellular and viral kinases, and being incorporated into cellular and viral DNA, thus interfering with mRNA synthesis and the formation of late viral proteins.[133] The nucleotides of (**66**) also inhibit viral and cellular DNA polymerases and thymidylate synthetase (Section 7.3.4.2.2) conferring the cytotoxicity which has restricted its use as an antiviral drug to topical application.[132]

7.3.5.1.2 *5-Alkyl and other 5-substituted pyrimidine 2'-deoxynucleosides*

A very considerable number of these compounds have been prepared and tested for their antiviral activity. 5-Ethyl-2'-deoxyuridine, the first active member of the 5-alkyl-2'-deoxyuridines to be recognized, is highly active against vaccinia and certain herpes viruses.[134] It is readily phosphorylated by herpes viral thymidine kinases, converted to the 5'-triphosphate, and incorporated into DNA. However, this process also occurs in uninfected cells. The exact mode of action of 5-ethyl-2'-deoxyuridine has not been clearly established. It inhibits DNA synthesis, but not at concentrations consistent with its antiviral effect. It is not mutagenic or immunosuppressive and is relatively nontoxic, and is presumably effective by virtue of its incorporation into DNA. 5-Propyl-[135] and 5-isopropyl-2'-deoxyuridine show broadly similar activity and properties. 5-Vinyl- and 5-ethynyl-2'-deoxyuridine[136] are potent inhibitors of HSV-1 and HSV-2 replication (HSV = herpes simplex virus), but are nonselective in their activity, inhibiting cell metabolism and viral replication at similar concentrations. 5-(1-Propynyl)-2'-deoxyuridine shows similar properties but in a series of 5-(1-alkynyl) derivatives the antiviral activity decreased with increasing length of the carbon chain.[137] It is thought that the antiviral properties of these compounds which bear a double or triple carbon–carbon bond attached at C-5 of 2'-deoxyuridine may be due to inhibition of thymidylate synthetase by the 5'-monophosphates.

Certain derivatives of 5-vinyl-2'-deoxyuridine possess quite remarkable potency, particularly (*E*)-5-(2-bromovinyl)-2'-deoxyuridine (BVDU; **78**) and (*E*)-5-(2-iodovinyl)-2'-deoxyuridine (IVDU; **79**).[136] The analogous propenyl (**80**),[138] 3,3,3-trifluoropropenyl (**81**)[139] and cyanovinyl (**82**)[139] derivatives are much less potent by comparison, and longer chain derivatives virtually inactive. Structure–activity correlation indicates that optimum inhibition of HSV-1 is achieved when the 5-substituent is unsaturated, conjugated with the pyrimidine ring, has *E* stereochemistry, includes an electronegative function, is hydrophobic, unbranched and not more than four carbon atoms in length.[139] BVDU[140] is highly active against HSV-1 (though less so against HSV-2), varicella zoster (VZV), pseudorabies virus and various other herpes viruses. At the concentrations at which it inhibits HSV-1 DNA synthesis, it does not affect normal cell metabolism. It is phosphorylated by the viral thymidine kinase successively to its mono- and di-phosphate and thence converted to its triphosphate, which inhibits HSV-1-induced DNA polymerase very powerfully. It also becomes incorporated into viral and cellular DNA, suppressing its function for RNA synthesis. The specificity of the viral kinase for BVDU restricts these effects to the virus-infected cell. The relative lack of potency against HSV-2 is due to the apparent inability of HSV-2 thymidine kinase to convert the 5'-monophosphate of BVDU to the 5'-diphosphate.

(**78**)   R = Br
(**79**)   R = I
(**80**)   R = Me
(**81**)   R = CF_3
(**82**)   R = CN

BVDU has proved effective in the topical and systemic treatment of a number of herpes viral infections in animals, and in certain clinical conditions in humans. Unfortunately, it is a substrate for cellular deoxyuridine phosphorylase, which, by removing the base, destroys its activity. To circumvent this defect, analogues containing the carbocyclic (as opposed to the ribose) ring have been synthesized. These are resistant to the enzyme, and also show promising antiviral activity.[141]

Many other 5-substituted 2'-deoxyuridine species have been tested for antiviral activity. 5-Formyl-[142] and 5-nitro-2'-deoxyuridine[143] possess similar activity to 5-vinyl- and 5-ethynyl-2'-deoxyuridine and are thought to exert their activity *via* inhibition of thymidylate synthetase. 5-Methylamino-,[144] 5-cyanomethoxy-,[145] 5-propynyloxy-,[146] 5-methoxy-,[147] 5-methylthio-[148] and 5-methylthiomethyl-2'-deoxyuridine,[142] to name but a few, exhibit selective but moderate antiherpes activity in *in vitro* systems. In most cases their mode of action has not been fully elucidated.

In general, if a 5-substituted 2'-deoxyuridine species exhibits strong antiviral activity, the correspondingly 5-substituted 2'-deoxycytidine exhibits comparable potency but higher selectivity. There are several reasons for this: if the analogue expresses its activity in the form of the 5-substituted-2'-deoxyuridine nucleotide, *i.e.* following deamination, the deamination may take place at the nucleoside or nucleotide level. In herpes-virus-infected cells, the latter course is followed, thus requiring the agency of thymidine (or deoxycytidine) kinase and deoxycytidylate deaminase, two virus-specified enzymes. If deamination does not occur, or is prevented by the addition of an inhibitor such as tetrahydrouridine, the 5-substituted-2'-deoxycytidylate may be incorporated into DNA and affect the methylation of DNA, which is vital in the maturation process. In addition, the 5-substituted-2'-deoxycytidine nucleotides are not substrates for deoxyuridine phosphorylase and do not inhibit host thymidylate synthetase or ribonucleotide reductase, properties which are associated with the cytotoxicity of the corresponding 2'-deoxyuridine species.

The 5-substituted-2'-deoxycytidine species may thus offer certain advantages in antiviral therapy over the corresponding 2'-deoxyuridine analogues.[149] However, they are generally more difficult of synthetic access, and this, together with a shift of focus towards other agents described below, may account for the lack of complete evaluation of these compounds.

7.3.5.1.3 *Pyrimidine arabinonucleosides and related compounds*

The antiviral activity of the arabinofuranosyluridine nucleosides is largely confined to herpes viruses. While *ara*-uridine (*ara*-U; **83**) is ineffective, *ara*-thymidine (*ara*-T; **84**) is an effective inhibitor of HSV-1, HSV-2, VZV and Epstein–Barr virus.[136,150] Its antiviral selectivity is due to its preferential phosphorylation by virus-induced thymidine kinase, since the 5'-triphosphate inhibits both viral and cellular DNA polymerases with similar efficiency, and is also a substrate for these enzymes, permitting *ara*-T to become incorporated in DNA.[151] Little toxicity due to *ara*-T has been observed in uninfected cells. *Ara*-T has been shown to be effective against herpes virus infections in rodents. 5-Ethyl-*ara*-U (**85**) shows similar activity to *ara*-T against HSV-1 and is even less cytotoxic in uninfected cells.[152] Its mode of action is thought to be identical. In fact, most of the 5-substituents which confer antiherpes viral activity in the 2'-deoxyuridine series are also associated with activity in the *ara*-uridine series. 5-Nitro-*ara*-U is an exception, being almost devoid of antiviral activity.[153] (*E*)-5-(2-Bromovinyl)-*ara*-U (**86**) is extremely potent against VZV.[154] It is metabolized and exhibits activity in a similar way to *ara*-T but, unlike BVDU, it is not incorporated into the DNA of infected cells.

(**83**) R = H
(**84**) R = Me
(**85**) R = Et
(**86**) R = CH $\overset{E}{=}$ CHBr

Ara-cytidine (**67**) has been described previously as an antineoplastic agent (Section 7.3.4.2.3). Despite showing strong antiviral activity, its high cytotoxicity precludes its use as an antiviral agent. The same is true of 5-fluoro-*ara*-C. Certain 5-substituted-2'-deoxy-2'-haloarabinofuranosylpyrimidines do exhibit high and selective antiherpes activity, however, the most potent and thoroughly researched being 2'-fluoro-2'-deoxy-5-iodo-*ara*-C (FIAC; **87**)[155,156] and 2'-fluoro-2'-deoxy-5-methyl-*ara*-U (FMAU; **88**).[156] Both are active against HSV-1, HSV-2, VZV and cytomegalovirus, for the last at unusually low concentrations compared with other compounds. These agents are phosphorylated selectively by the viral thymidine kinase (deoxycytidine kinase), and subsequently converted to the 5'-triphosphates. The FIAC triphosphate inhibits HSV-1 and HSV-2 DNA polymerases more strongly than cellular polymerases, and appears to be a substrate only for the viral polymerase.[157] Metabolic studies indicate that FIAC is converted mainly to its deaminated derivative FIAU (**89**) and also, *via* deiodination, to FAU (**90**) and FMAU (**88**).[158] Thus its metabolites are also potent antiherpetic agents. FMAU has proved effective in the systemic treatment of HSV-1 and HSV-2 infections in mice, and has been found to inhibit the growth of murine and human leukaemic cell lines, including some which are resistant to *ara*-C.[159]

(**87**) X = F
(**91**) X = Cl

(**88**) R = Me
(**89**) R = I
(**90**) R = H

2'-Chloro-2'-deoxy-5-iodo-*ara*-C (**91**) and 2'-chloro-2'-deoxy-5-methyl-*ara*-C, while less active than FIAC, also exhibit antiherpetic activity with greater effectiveness against HSV-2 than HSV-1. The 2'-halo-2'-deoxyribofuranosyl analogues of (**87**)–(**91**) show little activity in comparison with the compounds in the *ara* configuration.[160] The differences in activity between the two series are presumably largely the consequence of conformational differences altering the ability of the compounds to act as substrates for enzymes, although differences in chemical reactivity and polarity may play a part.

7.3.5.1.4 Other agents

6-Azauridine (**57**)[161] exhibits an unusual spectrum of antiviral activity against both DNA and RNA viruses. Its metabolism and mode of action have been described previously (Section 7.3.4.2.1). Its antiviral specificity is thought to be due to increased uridine kinase specificity in virus-infected cells affording 6-aza-UMP. Studies of 6-azauridine have tended to concentrate on DNA viruses such as pox and herpes viruses, although the drug is not highly potent against them. Reports of immunosuppression and teratogenicity have curtailed its use, but the drug cannot be regarded as having been fully evaluated. 5,6-Dihydro-5-azathymidine (**92**), a nucleoside antibiotic, is active *in vitro* against HSV-1 and HSV-2 and VZV, but its spectrum of activity is restricted to herpes viruses.[162] Activity against HSV-1 and HSV-2 infections has also been demonstrated in experimental animals.[163] The mechanism of action has not been resolved, although inhibition of thymidine phosphorylation in the infected cell has been suggested.[162] Among related aglycones, 6-azauracil has a mild inhibitory effect on vaccinia virus *in vivo*, and several alkyl derivatives of 2-thio- and 4-thio-6-azauracil inhibit DNA viruses *in vitro*.[164]

3-Deazauridine (**93**) and 3-deazacytidine are active against a number of RNA viruses *in vitro*,[165] but their antiviral efficacy *in vivo* is in doubt,[166] and they are of greater interest as potential antitumour compounds. 3-Deazauridine is metabolized to its 5'-triphosphate, which inhibits CTP synthetase (Figure 5),[167] although it is not incorporated into RNA or DNA. While this provides a rationale for the antitumour activity, the reasons for the observed antiviral activity are not clear.

The threatened pandemic of AIDS (acquired immunodeficiency syndrome) has spurred considerable effort to find agents active against human immunodeficiency virus (HIV) and sugar-modified analogues of nucleosides, in particular have been described as active. HIV is a retrovirus, belonging

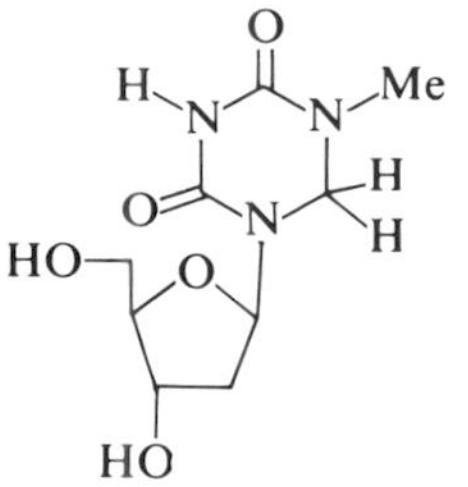

(92)

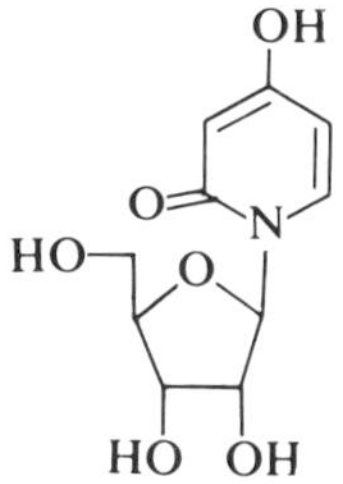

(93)

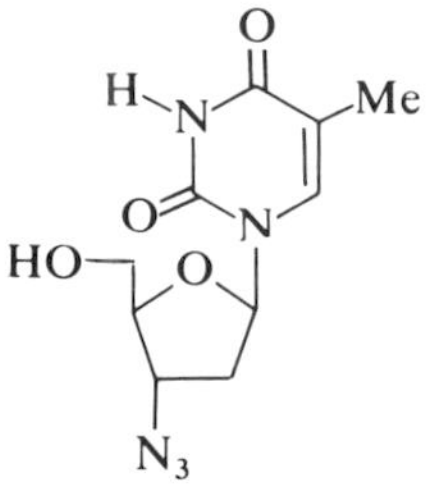

(94)

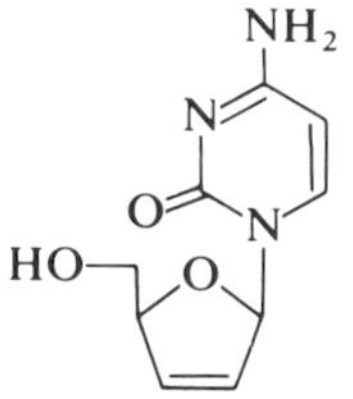

(95)

to the class of RNA viruses which encodes a reverse transcriptase necessary for the eventual integration of its genetic information into the host cell genome in the form of DNA.[168] 3'-Azido-2',3'-dideoxythymidine (AZT; Retrovir; **94**)[169], 2',3'-dideoxy-2',3'-didehydrocytidine (**95**) and 2',3'-dideoxycytidine[170] (in which the double bond in the sugar ring of (**95**) is reduced)[170] all show activity against HIV *in vitro* and AZT has been licensed for use in patients with AIDS. The azido group in AZT confers extra lipophilicity, and it crosses cell membranes rather readily by diffusion.[171] These and similar nucleosides, on conversion to their 5'-triphosphates, are strong inhibitors of HIV reverse transcriptase.[172] 2',3'-Dideoxynucleosides have been found to show broad spectrum activity against retroviruses.[173]

7.3.5.2 Purine Nucleoside Antiviral Agents[174]

7.3.5.2.1 Ara-*adenosine and related compounds*

Ara-adenosine (Vidarabine; **43**)[175] has already been described as an antitumour agent (Section 7.3.4.1.3). It is also active *in vitro* and *in vivo* against a number of DNA viruses, and also some retroviruses (RNA tumour viruses; see above), at concentrations which do not produce apparent cytotoxicity. It is used clinically for topical therapy of herpes keratitis and systemically for herpes encephalitis, and is also valuable for treating VZV in immunocompromised patients. As indicated previously, *ara*-A is converted to its mono-, di- and tri-phosphates *in vivo*. No differences in the rates of phosphorylation or deamination (see below) have been observed between HSV-1-infected and uninfected cells.[176] *Ara*-A expresses its activity primarily through the inhibition of DNA synthesis, and the level of intracellular *ara*-ATP shows good correlation with the suppression of DNA synthesis and the antiviral activity observed.[176,177] Cellular and HSV-1 and HSV-2 DNA polymerases are inhibited by *ara*-ATP, the K_i values for the viral enzymes being rather lower than those for the cellular enzymes.[178] This may be the basis of the selective antiviral activity, although this view has been challenged.[174] Also, *ara*-AMP becomes incorporated at low levels into the HSV genome, due to *ara*-ATP acting as a substrate for the polymerase.[179] Incorporation may slow the rate of DNA synthesis by distortion of the primer terminus or affect the function of the DNA by introducing helical distortion. In addition, *ara*-ATP inhibits ribonucleotide reductase[180] and the polyadenylylation of mRNA,[181] as well as RNA-dependent RNA polymerase (replicase) and terminal deoxynucleotidyl transferase.[182] Moreover, *ara*-A itself inhibits *S*-adenosylhomocysteinase, suppressing biological methylation, including the methylation of mRNA.[183] There are thus a number of activities which could contribute to the antiviral potency and/or the observed cytotoxicity of *ara*-A, and, while the suppression of DNA synthesis is regarded as its principal function, a satisfactory explanation for the selective antiviral activity is still sought.

Ara-A is deaminated *in vivo* to 9-β-D-arabinofuranosylhypoxanthine (*ara*-H; **96**) by adenosine deaminase.[184] *Ara*-H is also active against DNA viruses both *in vitro* and *in vivo*, although generally less so than *ara*-A.[185] Its mode of action, like that of *ara*-A, appears to involve conversion to the 5'-triphosphate, *ara*-ITP, which inhibits mammalian DNA polymerases α and β as a competitor with substrate dATP,[186] and HSV-1 DNA polymerase as a competitor with dGTP.[187] There are also some grounds for thinking that *ara*-H can be converted *in vivo*, via *ara*- IMP, to nucleotides of *ara*-A[188] and *ara*-G,[180,186] following the pathways depicted in Figure 2, and thus that *ara*-H may express part of its activity by this route.

In order to suppress its deamination, studies of the administration of *ara*-A in combination with an inhibitor of adenosine deaminase such as 2'-deoxycoformycin (**40**) or the corresponding ribonucleoside coformycin have been performed.[189] The antiviral activity of *ara*-A was found to increase, but so, too, did the cytotoxic effects of the drug.[190] A number of deaminase resistant analogues of *ara*-A have therefore been prepared and tested, including 2-fluoro- (**44**)[191] and 2'-azido-*ara*-A (**45**),[192] and carbocyclic *ara*-A (cyclaridine; **97**).[193] All show antiviral activity, being converted to their respective 5'-triphosphates, which presumably inhibit DNA polymerase. Cyclaridine and its 5'-methoxyacetate have shown useful activity in trials against genital herpes and herpes encephalitis.[194] 2-Amino-*ara*-A also displays substantial antiviral activity, as does its deamination product, *ara*-G.[195] Both appear to be less potent *in vitro* than *ara*-A, but less cytotoxic and consequently more selective, and *ara*-G has shown activity *in vivo* in animals.

(96)   (97)

Besides susceptibility to deamination, two further disadvantages of *ara*-A are its poor ability to penetrate skin and mucous membranes and its low solubility in water. A number of *O*-acyl ester prodrugs have been prepared with a view to enhancing topical penetration, and the 2',3'-di-*O*-acetate[196] and the 5'-*O*-valerate of *ara*-A proved active against HSV-2 genital herpes in female guinea pigs, although the latter compound was ineffective in mice.[197] However, the 5'-*O*-valerate is an inhibitor, but not a substrate, of adenosine deaminase.[198] The 5'-monophosphate of *ara*-A has been studied as a water soluble formulation for systemic administration. *Ara*-AMP is not readily deaminated, and penetrates cells intact.[199] It appears to be as effective as *ara*-A in treating patients with herpes encephalitis and herpes zoster,[200] but was ineffective as a cream in treating genital herpes.[201] The 5'-monophosphate of *ara*-H, and the 5'-*O*-methylphosphates and 3',5'-cyclic phosphates of *ara*-A and *ara*-H, have also been prepared and tested as water-soluble prodrugs, and show significant antiviral activity.[129]

7.3.5.2.2 *Agents containing modified and 'open ring' sugars*

In addition to *ara*-A, certain other purine nucleosides containing modified sugar rings have been reported as possessing significant antiviral activity. Both 2'- and 3'-*C*-methyl-adenosine show activity against vaccinia virus infections,[202] and 9-β-D-xylofuranosylguanine shows moderate antiherpes virus activity *in vivo*.[203] A number of 6-substituted analogues of the carbocyclic purine and 8-azapurine nucleosides (**47**) and (**48**) show *in vitro* activity against herpes and vaccinia viruses, with little cytotoxicity.[204] The most remarkable agents developed in recent years, however, are the 'open ring' nucleoside analogues.

Acyclovir (9-(2-hydroxyethoxymethyl)guanine; acycloguanosine; Zovirax) can be drawn (**98**) as representing a part-structure of guanosine (or 2'-deoxyguanosine). It displays remarkable and specific potency against herpes virus infections, and extremely low toxicity.[205] It has been licensed in the USA for topical or intravenous treatment of primary genital herpes[206] and for primary or recurrent mucosal or cutaneous HSV-1 and HSV-2 infections in patients who are immunosuppressed.[207] Oral administration of the drug for treatment of these conditions proved similarly effective. Clinical trials also show promise for the treatment of herpes keratitis and VZV,[207] and effective treatment of recurrent herpes labialis by topical application of acyclovir cream has been reported.[208]

Studies in cell culture show that acyclovir is a particularly potent inhibitor of the replication of HSV-1, less potent against HSV-2, varicella zoster and Epstein–Barr virus, and least potent against cytomegalovirus.[207,209] Examination of the mode of action of the drug permits these differences to be rationalized. Acyclovir is converted to its 5′-monophosphate specifically by the thymidine (deoxycytidine) kinase induced by the herpes viruses, being only a poor substrate for cellular thymidine and purine kinases.[210] The 5′-monophosphate is then converted to the diphosphate by cellular guanylate kinase,[211] and thence to its active 5′-triphosphate (acyclo-GTP) by further cellular kinases. Acyclo-GTP is a much more powerful inhibitor of HSV-1 DNA polymerase, competing with dGTP, than it is of mammalian DNA polymerases α and β.[212,213] Indeed, all the human herpes viruses sensitive to acyclovir induce DNA polymerases, which are inhibited by acyclo-GTP. The sensitivity to acyclovir seems to reflect the efficiency of the formation of acyclo-GTP in the infected cell, which in turn reflects the efficiency of the phosphorylation by the viral kinase, and human cytomegalovirus, unlike HSV-1 and HSV-2, does not induce a viral kinase.[212] The specificity and activity of acyclovir are thus dictated by the viral kinase and DNA polymerase.

In addition, acyclo-GTP is a better substrate for the viral polymerases than mammalian cell polymerases. Incorporation of acyclo-GMP at the 3′-terminus of a DNA chain results in chain termination,[214] and studies with HSV-1 DNA polymerase show that the 3′-exonuclease activity of the enzyme cannot remove the misincorporated acyclo-GMP.[213] Moreover, the prematurely terminated chain inhibits the polymerase function of the enzyme, thus possibly reinforcing the antiviral activity. The overall effect of the inhibition of HSV-1 DNA synthesis leads to a reduction of synthesis of herpes proteins and loss of infectivity.

Not surprisingly, much research has been invested with a view to obtaining prodrugs and analogues of acyclovir. 2,6-Diamino-9-(2-hydroxyethoxymethyl)purine is metabolically deaminated to acyclovir by adenosine deaminase[215] and is reportedly absorbed better from the gut.[216] 9-(1,3-Dihydroxy-2-propoxymethyl)guanine (DHPG; 2′-nor-2′-deoxyguanosine; **99**)[217,218] has antiviral activity *in vitro* which is equal or superior to that of acyclovir for all human herpes viruses including cytomegalovirus.[219] It exhibits very low cytotoxicity, and is reportedly superior to acyclovir in animals infected with HSV-1 and HSV-2. It is phosphorylated more efficiently than acyclovir by both viral thymidine (deoxycytidine) kinase and cellular guanylate kinase. (R)-9-(3,4-Dihydroxybutyl)guanine (DHBG; **100**) is also a selective antiherpetic agent active against HSV-1 and HSV-2 *in vitro* and in animal model studies.[220] Like acyclovir, it is selectively phosphorylated by the viral kinase and inhibits DNA synthesis selectively in HSV-1-infected cells. The (S) enantiomer is less active.

(**98**)   R = H
(**99**)   R = CH₂OH

(**100**)

(**101**)

(S)-9-(2,3-Dihydroxypropyl)adenine (DHPA; **101**) is active against both DNA and RNA viruses, including herpes, vaccinia, vesicular stomatitis and measles virus at concentrations which do not affect cell viability or DNA, RNA and protein synthesis.[221] However, it is reportedly embryotoxic in chicks and mice[222] and cytotoxic to lymphocytes.[223] It is thought to act *via* suppression of methylation of viral mRNA,[224] due to the inhibition of S-adenosylhomocysteinase.[225]

Many other such 'open ring' aliphatic nucleosides have been prepared and evaluated recently: the above examples represent only a few of the more interesting and active members in this class.

7.3.5.2.3   *Other agents*

3-Deazaguanine, 3-deazaguanosine (**101**) and 3-deazaguanylic acid are active against a broad spectrum of DNA and RNA viruses *in vitro*, and against influenza A and B, parainfluenza, herpes and Friend leukaemia virus infections in laboratory animals.[226] These compounds inhibit HGPRT and IMP dehydrogenase,[227] the antiviral activity being associated with the latter function, leading to suppression of viral nucleic acid synthesis. 3-Deazaadenosine (**102**) and its carbocyclic analogue (3-deazaaristeromycin; **103**) inhibit a number of DNA and RNA viruses, with low associated cytotoxicity.[228-230] Both agents are thought to inhibit S-adenosylhomocysteinase.[231] While it has been speculated that this inhibition leads to undermethylation of viral mRNA in the case of (**102**),[229] in a study using (**103**) and involving vaccinia virus, methylation of viral mRNA was not suppressed.[232] Higher potency against vesicular stomatitis, parainfluenza, measles and vaccinia viruses has been observed for (**103**) than for (**102**), DHPA, or ribavirin (see below).

(**102**)   X = O
(**103**)   X = CH$_2$

7.3.5.3   Azole Nucleosides and Related Antiviral Compounds

1-β-D-Ribofuranosyl-1,2,4-triazole-3-carboxamide (Ribavirin; Virazole; **104**)[223] is a broad spectrum antiviral inhibiting the replication of a wide range of DNA and RNA viruses *in vitro* and *in vivo*, the most sensitive being HSV-1 and HSV-2, vaccinia, influenza, parainfluenza, measles, rhino and some tumour viruses.[234] While oral administration of ribavirin is only marginally effective in alleviating influenza A and B symptoms in humans, good results against these and respiratory syncytial virus infections were reported for ribavirin administered by inhalation as an aerosol.[235] The drug exhibits little toxicity in the dose range at which it is effective *in vitro* and *in vivo* as an antiviral,[234] but has been found to be teratogenic in mice, apparently inhibiting embryonic DNA synthesis.[236]

The precise reasons for the selective antiviral effects of ribavirin are unknown. It is readily phosphorylated by cellular adenosine or deoxyadenosine kinase to its 5'-monophosphate, which is a strong inhibitor of IMP dehydrogenase.[237] However, this does not seem sufficient to account for the activity against RNA viruses: tiazofurin (**73**) is also a strong inhibitor of this enzyme, but has little antiviral effect. Ribavirin is further phosphorylated to its 5'-triphosphate, which has been reported to inhibit influenza virus RNA polymerase selectively in a cell free assay, but not inhibit cellular RNA polymerase.[238] The 5'-triphosphate also shows selective inhibition of HSV DNA synthesis relative to cellular DNA synthesis in infected cells,[239] possibly due to its inhibiting effect on HSV-1 DNA polymerase. Finally, the 5'-triphosphate has been shown to inhibit vaccinia virus mRNA guanylyltransferase, thus blocking the 'capping' of mRNA,[240] and this may account for the activity of ribavirin against both DNA and RNA viruses and provide a rationale for antiviral selectivity.

The 2',3',5'-tri-O-acetate of ribavirin has been prepared as a lipophilic prodrug and appears to be able to cross the blood–brain barrier, showing activity against intracerebral viral infections in experimental animals.[241] Many analogues of ribavirin have been synthesized,[129, 233] including the carboxamidine (**105**), which is active against both RNA and DNA viruses, and thiocarboxamide (**106**), which is active only against DNA viruses. Structure–activity analysis shows that the carboxamide group, the 1,2,4-triazole ring, and the β-D-ribofuranosyl moiety are essential for antiviral activity. The heterocycle of ribavirin, 1,2,4-triazole-3-carboxamide, also displays antiviral properties, being converted enzymatically to ribavirin *in vivo*.[242] 5-Fluoro-1-β-D-ribofuranosylimidazole-4-carboxamide (**107**) and other 5-halogenated imidazole-4-carboxamides of this type show broad spectrum antiviral activity, albeit less than that of ribavirin.[243]

Selenazole (2-β-D-ribofuranosylselenazole-4-carboxamide),[244] an analogue of tiazofurin (**73**) in which the sulfur atom is replaced by selenium, has broad spectrum antiviral activity *in vitro* against both DNA and RNA viruses which is superior to that of ribavirin.[245] It is particularly active against certain families of RNA viruses. It is, however, markedly cytotoxic towards certain cell lines in culture, and showed antitumour activity against Lewis lung carcinoma in mice.[244] 4-Hydroxy-3-β-D-ribofuranosylpyrazole-5-carboxamide (pyrazofurin; **108**) likewise possesses broad spectrum

(104)	X = O,	Y = N
(105)	X = NH,	Y = N
(106)	X = S,	Y = N
(107)	X = O,	Y = CF

(108)

antiviral activity against both DNA and RNA viruses *in vitro*, but also exhibits greater cytotoxicity than ribavirin and is of greater interest as an antitumour agent.[246]

7.3.6 ANTIPARASITIC PURINE AND PYRIMIDINE-RELATED AGENTS

The chemotherapy of parasitic infection poses a different set of problems to those found in antitumour and antiviral drug development. Here one attempts to eliminate a parasite which possesses its own specific biochemistry which may differ from that of the host cell.[247] The parasitic protozoa cause a number of human and animal diseases including trypanosomiasis (African and South American), leishmaniasis, coccidiosis, toxoplasmosis and malaria. It has only recently been appreciated that the nucleotide metabolism in these protozoan parasites may differ markedly from that of the host mammalian cells, thereby offering opportunities for therapy.[248] For instance, they appear incapable of *de novo* purine synthesis, and are thus dependent upon the host for the supply of purine bases. However, they can effect all or part of the synthesis of pyrimidines *de novo*, and some are totally dependent on *de novo* synthesis. In addition, most, if not all, of the protozoan parasites seem to combine the enzyme activities of dihydrofolate reductase and thymidylate synthetase in a single protein. These, and other protozoan enzymes may offer targets for chemotherapeutic control, and the individual biochemistries may be exploited for the potentiation of an agent which ultimately proves fatal to the protozoon. Some examples are described below.

7.3.6.1 Allopurinol and Other Hypoxanthine Analogues

As indicated above, the parasitic protozoa require an external supply of purine bases for their metabolism, and hypoxanthine (11) appears to be the main source for *Plasmodia*, *Leishmania*, *Crithidia fasciculata*, *Toxoplasma gondii*, *Eimeria tenella* and some *Trypanosoma*.[248] Allopurinol (16) is metabolized in a similar way to hypoxanthine in a number of these species, resulting in antiparasitic activity.[249] *Leishmania* promastigotes (a flagellated extracellular form) contain high levels of HGPRT, converting allopurinol to its ribonucleotide (109), and simultaneously lack xanthine oxidase to oxidize allopurinol to alloxanthine. The allopurinol ribonucleotide is then converted *via* the parasites' adenylosuccinate synthetase and adenylosuccinate lyase to the AMP analogue (110), a process which is not catalyzed by the mammalian enzyme.[250] Subsequently, (110) is converted to the 5'-triphosphate, which is incorporated into RNA, causing defective function.[249] Allopurinol ribonucleoside (111) is more effective than allopurinol in inhibiting *in vitro* growth of some *Leishmania* promastigotes.[251] It is converted to (109) *via* a nucleoside phosphotransferase in the parasite.

Leishmania also readily convert (112) into (110), and thence *via* adenylate deaminase to (109).[249] The effectiveness of these pyrazolopyrimidine drugs thus still depends upon the activity of adenylosuccinate synthetase forming (110) for eventual incorporation into RNA. In tests on *Leishmania* promastigotes of various isolates, (112) was found more effective than allopurinol,[252] although allopurinol was found effective orally against *Leishmania major* and *Leishmania mexicana amazonensis* in mice,[253] and gave good cure rates in antimony resistant visceral leishmaniasis in humans.[254] Against members of the genus *Trypanosoma*, allopurinol and (112) have shown varying activity. *T. rangeli*, for instance, does not convert (109) to (110) and is insensitive to allopurinol and only slightly sensitive to (112),[255] while some strains of *T. cruzi* were sensitive to allopurinol and (112), both *in vitro* and *in vivo*.[256] However, results in *T. cruzi* were highly dependent upon the strain tested. *C. fasciculata* and *E. tenella* are also sensitive to allopurinol *in vitro*.[257] *L. tropica* is much more sensitive *in vitro* to the 3-bromo analogue of allopurinol ribonucleoside (**113**) than to (**111**),[258] and its promastigotes are sensitive to 3′-deoxyinosine and to a number of derivatives of tubercidin (**50**).[259] 3′-Deoxyinosine is phosphorylated and aminated by *Leishmania* to afford nucleotides of 3′-deoxyadenosine (cordycepin). 7-Deazainosine, 9-deazainosine, formycin B (**114**) and a number of other inosine analogues have been tested against *L. tropica*, *L. donovani*, *T. cruzi* and *T. gambiensi*.[260] In general these showed strong activity with little associated toxicity for mammalian cells.

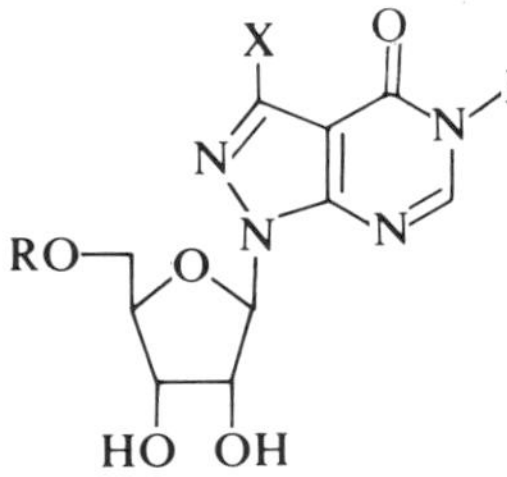

(**109**)   R = PO$_3$H$_2$,   X = H
(**111**)   R = H,   X = H
(**113**)   R = H,   X = Br

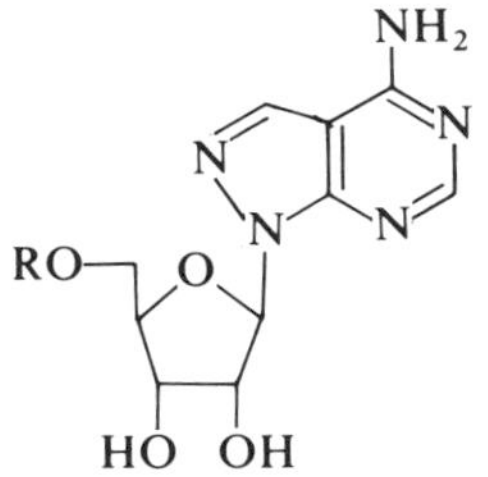

(**110**)   R = PO$_3$H$_2$

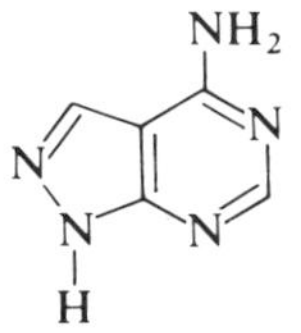

(**112**)

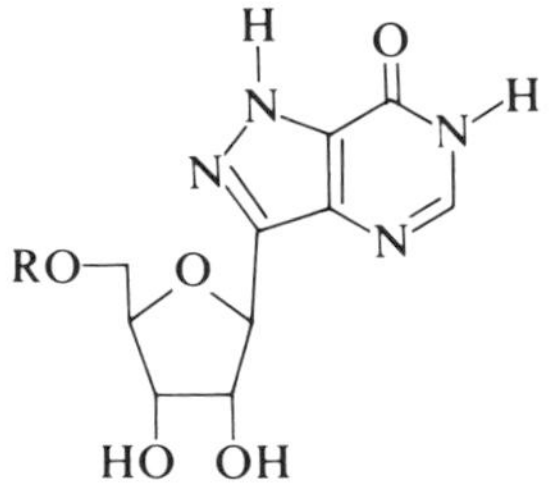

(**114**)   R = H
(**115**)   R = PO$_3$H$_2$

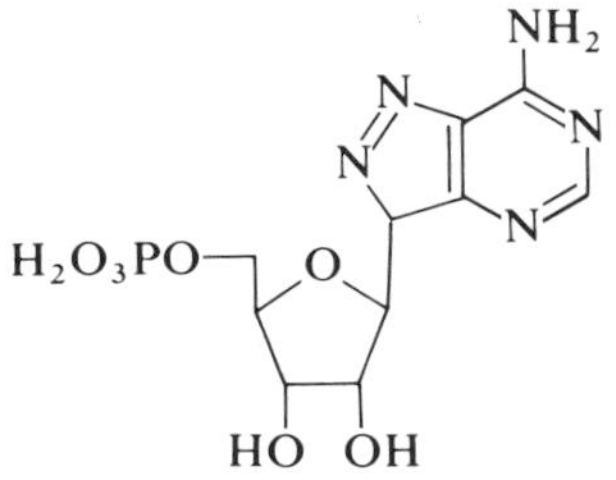

(**116**)

Formycin B, which is not phosphorylated or cleaved in mammalian cells, has been studied particularly thoroughly. It is active against promastigotes and amastigotes (the rounded intracellular form) of a number of *Leishmania* species both *in vitro* and *in vivo*.[261] The parasites convert (**114**) to its 5′-monophosphate (**115**), which is a substrate for the parasitic and host cell adenylosuccinate synthetases, forming the 5′-monophosphate of formycin A (**116**),[262] which is in turn converted to the 5′-triphosphate and incorporated into cellular RNA. The consequence appears to be suppression of protein synthesis, and this has been demonstrated for some species.[263] In *in vitro* testing against *L. tropica* in macrophages, formycins A and B and their nucleotides displayed much higher activity than allopurinol.[264] Formycin B is also effective against *T. cruzi*, in which it is metabolized in the same way as in *Leishmania*.[265]

Thiopurinol, the sulfur analogue of allopurinol, and its ribonucleoside (**117**) also show antiparasitic activity.[266] Thiopurinol shows similar antileishmanial activity to allopurinol, while the ribonucleosides show greater activity. Thiopurinol and (**117**) are converted to the 5′-monophosphate (**118**) by the same routes by which allopurinol and (**111**) are converted to (**109**). The nucleotide (**118**),

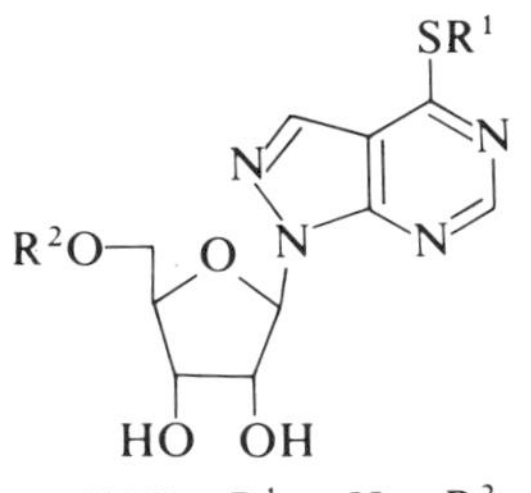

(**117**)   R^1 = H,   R^2 = H
(**118**)   R^1 = H,   R^2 = PO$_3$H$_2$
(**119**)   R^1 = Et,   R^2 = H
(**120**)   R^1 = CH$_2$CH $\overset{E}{=}$ CHPh,   R^2 = H

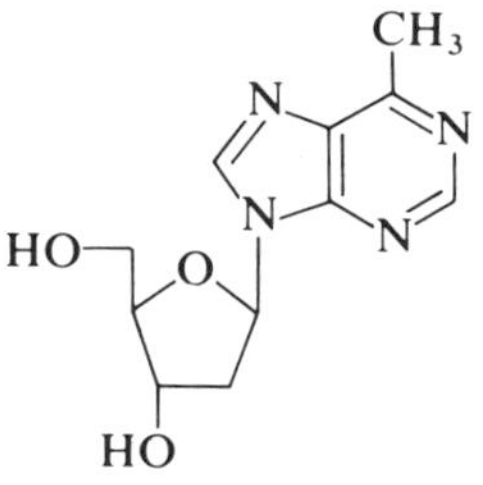

(**121**)

like (109) and (115), inhibits leishmanial GMP reductase, an enzyme which catalyzes the reductive deamination of GMP to IMP, more powerfully than it inhibits the corresponding enzyme in human erythrocytes.[267] The nucleotide (118) also inhibits the reaction of IMP with adenylosuccinate synthetase, blocking the synthesis of AMP.[268] Some strains of *T. cruzi* are sensitive to thiopurinol and (117), while others are not.[249] Against *Eimeria tenella* (an intestinal sporozoon causing coccidiosis) thiopurinol showed no activity in chicks *in vivo*, while (117) showed some activity,[269] and the *S*-alkylated derivatives (119) and (120) showed rather more.[270] Against other *Eimeria* species, (120) showed a broad spectrum of activity but it was ineffective against *E. acervulina*, despite being a substrate for adenosine kinase from this organism.[271]

7.3.6.2 Purine Derivatives

N^6-Methyladenine inhibits the growth of *Leishmania* and *T. cruzi in vitro*. The mechanism of inhibition is obscure, but the degree of inhibition correlates with the sensitivity of guanine deaminase to the drug.[272] 9-β-D-(2-Deoxyribofuranosyl)-6-methylpurine (121) does not inhibit promastigotes of *L. donovani* and *L. mexicana* directly, but kills mouse macrophages infected with their amastigotes, while noninfected cells remain intact. The *Leishmania* secrete a 2'-deoxyribonucleosidase which releases 6-methylpurine from (121), which is toxic to the macrophage.[273] 4-Amino-5-imidazolecarboxamide, an intermediate (in the form of its ribonucleotide, see Figure 1) in purine biosynthesis, shows *in vitro* activity against several *Leishmania* species, and has been found to inhibit the guanine aminohydrolase activity in their promastigotes.[274]

Plasmodium falciparum, a malaria parasite, requires a supply of guanine nucleotides in order to multiply, and has been controlled in erythrocyte culture by the administration of bredinin (122), which inhibits the conversion of IMP to GMP.[275] Erythrocytes infected with *P. falciparum* show heightened adenosine deaminase activity,[276] which can be inhibited by 2'-deoxycoformycin (40), and *in vivo* treatment with (40) reduced parasitaemia in *P. knowlesi*-infected rhesus monkeys.[277]

Arprinocid (9-(2-chloro-6-fluorobenzyl)adenine) is a powerful anticoccidial agent, inhibiting the growth of *E. tenella* in cell culture.[278] It is partially converted *in vivo* by microsomal oxidation to its 1-oxide, which has much greater potency than the parent drug.[279] Its mode of action is to block the transport of hypoxanthine into the parasites.[280]

7.3.6.3 Pyrimidine Derivatives and Analogues

Since most, if not all, parasitic protozoa are capable of incorporating bicarbonate and orotate into the pyrimidine rings of nucleic acids, they must be able to perform at least part of the *de novo* biosynthesis of pyrimidines (Figure 4).[247] *Plasmodium*, for instance, is totally dependent on *de novo* synthesis, lacking a salvage pathway for pyrimidines. Thus, inhibition of enzymes of the pyrimidine synthetic pathway offers an opportunity for control of these parasites. For instance, orotate phosphoribosyltransferase from *P. berghei* is inhibited by 5-azaorotate (56), 5-azauracil (55) and 6-azauracil (123) and, less powerfully, by alloxanthine and allopurinol.[281] *Toxoplasma gondii*, by contrast, can salvage uracil,[282] although it can probably survive on *de novo* pyrimidine synthesis if the salvage pathway is blocked. For this species, simultaneous inhibition of both pathways would be required. A series of N^1-arylated-6-azauracil derivatives has been found to possess strong anticoccidial activity, with (124) possessing particular broad spectrum potency, some 4000 times stronger than 6-azauracil itself.[283] A similarly substituted series of 1-phenyluracils also exhibited anticoccidial activity.[284] It seems unlikely that orotidylate decarboxylase, the presumed enzyme target for

(122)

(123) R = H
(124) R = Cl—⟨benzene⟩—S—⟨xylene⟩—

X = Cl or Me

6-azauridylate formed from (123), can be the target for the 1-arylated uracils and 6-azauracils.[285] Probably one or more salvage enzymes, the uracil or orotate phosphoribosyltransferases, are inhibited.

Compounds entrapped in liposomes show enhanced activity against visceral leishmaniasis (*L. donovani*). 5-Fluorocytosine has been shown to be an active antileishmanial drug when administered in this way.[286]

7.3.7 ANTIFUNGAL AGENTS

5-Fluorocytosine (Flucytosine) is an effective but narrow spectrum antifungal agent, applied systemically for the treatment of yeast (*Candida*, *Cryptococcus* and *Torulopsis*) infections, against which it shows high activity. Studies using *Candida albicans* show that it possesses both a specific permease and a specific deaminase which animal cells lack, permitting the drug to be taken up and converted to 5-fluorouracil (59).[287] This is in turn metabolized to 5-fluoro-2'-deoxyuridylate, which irreversibly inhibits thymidylate synthetase[288] and thus DNA synthesis[289] (Section 7.3.4.2.2). In addition, 5-fluorouracil becomes incorporated into RNA, and for *C. albicans* the degree of inhibition of growth by 5-fluorocytosine correlates directly with this degree of incorporation of 5-fluorouracil.[287] There is thus a dual mechanism of action, and the relative contribution of each mechanism to the overall inhibition of growth in fungi and yeasts is not clear. 5-Fluorouracil itself cannot be used because it penetrates the fungi only poorly, and has high cytotoxicity, as discussed earlier. Amphotericin B markedly enhances the fungicidal properties of 5-fluorocytosine, the polyene antibiotic presumably damaging the fungal cell membrane to permit increased uptake of the drug. The clinical pharmacokinetics of 5-fluorocytosine and amphotericin B as antifungals have been reviewed[290] and their use in combination for the therapy of chronic cryptococcal meningitis has been described.[291]

7.3.8 SUMMARY — FUTURE DIRECTIONS

The dramatic discoveries and developments in the areas of monoclonal antibodies, nucleic acid sequencing, oligonucleotide synthesis, and genetic engineering made during the 1970s seem likely to provide the greatest long term influence in future developments of chemotherapy. The ability to clone a particular gene *ad libitum* should mean that large quantities of enzymes and cellular proteins can be produced selectively, and crystallization and structural determination using X-ray diffraction methods should permit more precise and specific enzymes inhibitors to be synthesized, complementing the data provided by QSAR analysis together with NMR data and computer graphics investigations. The principal thrust in the future development of purine and pyrimidine antagonists for the treatment of neoplasia and viral conditions is likely to lie in this direction.[292] Monoclonal antibodies, loaded with cytotoxic antimetabolites, which are specific to antigens on the surfaces of tumour cells may provide much more precisely directed 'magic bullets' than has hitherto been the case.

There has been interest for some time in the preparation of prodrugs of inhibitory nucleotides, as opposed to nucleosides or purine or pyrimidine bases. As we have seen, most of the purine or pyrimidine antagonists are active at the nucleotide level, and the introduction into the cell of a masked nucleotide, which is readily unmasked under physiological conditions to liberate the active nucleotide is a desirable aim, and obviates the requirement for phosphorylation of the nucleoside by a cellular or viral kinase. While some imaginative forays in this direction have afforded disappointing results,[293] others provide positive promise for the future.[294]

While the present impetus in the search for new antiviral drugs seems likely to continue to throw up new broad spectrum antivirals (interest in phosphonate derivatives of 'open chain' nucleoside analogues has recently become intense),[295] it is likely that future treatment of viral, and possibly neoplastic, conditions will seek to disrupt or eliminate the function of the nucleic acids, rather than blocking metabolism at the nucleoside or nucleotide level. For instance, the unusual 2'–5'-linked oligoadenylate '2–5A' (pppA(2')pA(2')pA), which is produced in cells as part of their response to interferon, potentiates a latent endoribonuclease (RNase L), which then digests mRNA to suppress protein synthesis.[296] In consequence it has a marked antiviral effect, and since it appears that conjugation with poly(L-lysine) permits the '2–5A' to pass the cell membrane to activate RNase L intracellularly,[297] the use of '2–5A' and related analogues[298] in this way may offer new agents for broad spectrum antiviral therapy. Other work of note utilizes oligonucleotide sequences complementary to sequences governing the expression of viral genes, which hybridize to the complementary sequence to suppress the expression of the genes.[299] The use of oligonucleotides conjugated with poly(L-lysine)[300] or analogous (electrically neutral) oligonucleoside methylphosphonates[301] shows promise.

Some antitumour agents (bleomycin,[302] neocarzinostatin,[303] *etc.*) act by cleaving DNA, and much effort has been invested to produce DNA cleavage agents, often consisting of a hydroxyl-radical-generating system linked covalently to an oligonucleotide, which are 'complementarily addressed' to hybridize with and subsequently cleave a complementary sequence of DNA.[304] Other agents designed for the same purpose are nonnucleotidic, but bind selectively to a particular DNA sequence prior to cleaving the nucleic acid strand.[305] The development and refinement of such agents offers tantalizing prospects for future antitumour therapy. Of course, many other agents of proven efficacy act by disrupting nucleic acid function as described in Part 10.

The future development of antiparasitic purine and pyrimidine antagonists is likely to be attendant on a fuller appreciation of the biochemistry of each individual organism, and the way in which it differs from that of the host cell. Once again, more precise knowledge of the structures of the parasitic enzymes, QSAR and computer graphics may hold the key to the development of more effective drugs. The effectiveness of Flucytosine in treating fungal infections depends, as described above, upon its uptake by a specific permease, and future progress here may attend the further exploitation of the permease or, more likely, the development of efficacious drugs which can pass the membrane by simple diffusion, without reliance upon the permease.

Finally the reader is reminded that this chapter has been restricted essentially to purine and pyrimidine antagonists: there are many effective antitumour, antiviral, antiparasitic and antifungal agents which do not belong to this class. However, the purine and pyrimidine antagonists, considered as a whole, provide a satisfying array of examples of the uses of perturbation and inhibition of normal biochemical processes for the purpose of combating disease.

7.3.9 REFERENCES

1. J. M. Buchanan, 'The Nucleic Acids', ed. E. Chargaff and J. N. Davidson, Academic Press, New York, 1960, vol. 3, p. 303.
2. A. W. Murray, D. C. Elliott and M. R. Atkinson, *Prog. Nucleic Acid Res. Mol. Biol.*, 1970, **10**, 87.
3. J. A. Montgomery, *Prog. Med. Chem.*, 1970, **7**, 69: a most valuable review pertinent to this whole section.
4. L. Thelander and P. Reichard, *Annu. Rev. Biochem.*, 1979, **48**, 133.
5. R. H. Burdon and R. L. P. Adams, *Trends Biochem. Sci.*, 1980, **5**, 294 and refs. therein.
6. C. N. Remy, *J. Biol. Chem.*, 1963, **238**, 1078.
7. T. A. Krenitsky, *Mol. Pharmacol.*, 1967, **3**, 526.
8. See, for example, R. Wolfenden, *J. Am. Chem. Soc.*, 1966, **88**, 3157; B. M. Chassy and R. J. Suhadolnik, *J. Biol. Chem.*, 1967, **242**, 3655.
9. R. C. Bray, in 'The Enzymes', ed. P. D. Boyer, Academic Press, New York, 1975, vol. 12, p. 299.
10. M. E. Jones, *Annu. Rev. Biochem.*, 1980, **49**, 253; P. A. Hoffee and M. E. Jones, *Methods Enzymol.*, 1978, **51**.
11. M. Friedkin, *Adv. Enzymol. Relat. Areas Mol. Biol.*, 1973, **38**, 235.
12. P. C. Hanawalt, P. K. Cooper, A. K. Ganesan and C. A. Smith, *Annu. Rev. Biochem.*, 1979, **48**, 783.
13. See, for example, W. H. Prusoff, T.-S. Lin, W. R. Mancini, M. J. Otto, S. A. Siegel and J. J. Lee, in 'Topics in Antiviral Chemotherapy', ed. R. T. Walker and E. De Clercq, Plenum Press, New York, 1984, p. 1.
14. R. Hill and V. Massey, in 'Molybdenum Enzymes', ed. T. G. Spiro, Wiley-Interscience, New York, 1985, p. 443.
15. R. C. Bray, in 'Biological Magnetic Resonance', ed. J. Reuben and L. J. Berliner, Plenum Press, New York, 1980, vol. 2, p. 45.
16. R. C. Bray and S. Gutteridge, *Biochemistry*, 1982, **21**, 5992.
17. Ref. 14, p. 498.
18. J. B. Wyngaarden and W. N. Kelley, in 'The Metabolic Basis of Inherited Disease', ed. J. B. Stanbury, J. B. Wyngaarden and D. S. Fredrickson, 4th edn., McGraw-Hill, New York, 1978.
19. V. Massey, H. Komai, G. Palmer and G. B. Elion, *J. Biol. Chem.*, 1970, **245**, 2837.
20. J. W. Williams and R. C. Bray, *Biochem. J.*, 1981, **195**, 753.
21. G. B. Elion, S. Callahan, H. Nathan, S. Bieber, R. W. Rundles and G. H. Hitchings, *Biochem. Pharmacol.*, 1963, **12**, 85.
22. J. S. Olson, D. P. Ballou, G. Palmer and V. Massey, *J. Biol. Chem.*, 1974, **249**, 4363.
23. See, for example, A. C. Mello Filho and R. Meneghini, *Biochim. Biophys. Acta*, 1984, **781**, 56.
24. I. Fridovich, *Annu. Rev. Biochem.*, 1975, **44**, 147; 'Developments in Biochemistry', ed. J. V. Bannister and H. A. O. Hill, Elsevier North Holland, New York, 1980, vol. 11A.
25. E. Lengfelder and U. Weser, *Clin. Respir. Physiol.*, 1981, **17**, 73.
26. A. Diesseroth and A. L. Dounce, *Physiol. Rev.*, 1970, **50**, 319.
27. J. A. Montgomery, T. P. Johnston and Y. F. Shealy, in 'Burger's Medicinal Chemistry', ed. M. E. Wolff, 4th edn., Wiley, New York, 1980, part 2, p. 595.
28. J. A. Montgomery, *Med. Res. Rev.*, 1982, **2**, 271.
29. R. K. Robins and G. R. Revankar, *Med. Res. Rev.*, 1985, **5**, 273.
30. R. W. Brockman, in 'Advances in Cancer Research', ed. A. Haddow and S. Weinhouse, Academic Press, New York, 1963, vol. 7, p. 129.

31. R. J. McCollister, W. R. Gilbert, Jr., D. M. Ashton and J. B. Wyngaarden, *J. Biol. Chem.*, 1964, **239**, 1560; D. L. Hill and L. L. Bennett, Jr., *Biochemistry*, 1969, **8**, 122.
32. J. A. Montgomery and R. F. Struck, *Prog. Drug. Res.*, 1973, **17**, 322.
33. H. P. Schnebli, D. L. Hill and L. L. Bennett, Jr., *J. Biol. Chem.*, 1967, **242**, 1997.
34. F. M. Schabel, Jr., W. R. Laster, Jr. and H. E. Skipper, *Cancer Chemother. Rep.*, 1967, **51**, 11.
35. F. M. Schabel, Jr., J. A. Montgomery, H. E. Skipper, W. R. Laster, Jr. and J. R. Thomson, *Cancer Res.*, 1961, **21**, 690.
36. G. B. Elion, *Fed. Proc., Fed. Am. Soc. Exp. Biol.*, 1967, **26**, 898; K. G. Vanscoik, C. A. Johnson and W. R. Porter, *Drug Metab. Rev.*, 1985, **16**, 157.
37. A. S. Levine, H. L. Sharp, J. Mitchell, W. Krivit and M. E. Nesbit, *Cancer Chemother. Rep.*, 1969, **53**, 53; J. J. Coffey, C. A. White, A. B. Lesk, W. I. Rogers and A. A. Serpick, *Cancer Res.*, 1972, **32**, 1283.
38. G. A. LePage, *Cancer Res.*, 1963, **23**, 1202; G. A. LePage and M. Jones, *Cancer Res.*, 1961, **21**, 1590.
39. R. P. Miech, R. York and R. E. Parks, Jr., *Mol. Pharmacol.*, 1969, **5**, 30.
40. M. K. Wolpert, S. P. Damle, J. E. Brown, E. Sznycer, K. C. Agrawal and A. C. Sartorelli, *Cancer Res.*, 1971, **31**, 1620, see also M. Rosman, M. H. Lee, W. A. Creasey and A. C. Sartorelli, *Cancer Res.*, 1974, **34**, 1952.
41. J. P. Scannell and G. H. Hitchings, *Proc. Soc. Exp. Biol. Med.*, 1966, **122**, 627; D. M. Tidd and A. R. P. Paterson, *Cancer Res.*, 1974, **34**, 738.
42. S. Zimm, L. J. Ettinger, J. S. Holcenberg, B. A. Kamen, T. J. Vietti, J. Belasco, N. Cogliano-Shutta, F. Balis, L. E. Lavi, J. M. Collins and D. P. Poplack, *Cancer Res.*, 1985, **45**, 1869.
43. J. A. Montgomery, H. J. Thomas and K. Hewson, *J. Med. Chem.*, 1972, **15**, 1189.
44. C. B. Johnson, W. B. Frommeyer, Jr., W. J. Hammack and C. E. Butterworth, Jr., *Cancer Chemother. Rep.*, 1962, **20**, 137.
45. C. Temple, Jr., C. L. Kussner and J. A. Montgomery, *J. Med. Chem.*, 1968, **11**, 41.
46. G. W. Kidder, V. C. Dewey, R. E. Parks, Jr. and G. L. Woodside, *Science (Washington, D.C.)*, 1949, **109**, 511.
47. J. H. Mitchell, H. E. Skipper and L. L. Bennett, *Cancer Res.*, 1950, **10**, 647.
48. H. G. Mandel, *Fed. Proc., Fed. Am. Soc. Exp. Biol.*, 1967, **26**, 905.
49. L. L. Bennett, Jr., M. H. Vail, P. W. Allan and W. R. Laster, Jr., *Cancer Res.*, 1973, **33**, 465.
50. L. L. Bennett, Jr. and P. W. Allan, *Cancer Res.*, 1976, **36**, 3917.
51. L. L. Bennett, Jr., J. A. Montgomery, R. W. Brockman and Y. F. Shealy, in 'Advances in Enzyme Regulation', ed. G. Weber, Pergamon Press, New York, 1978, vol. 16, p. 255; L. L. Bennett, Jr., L. M. Rose, P. W. Allan, D. Smithers, D. J. Adamson, R. D. Elliott and J. A. Montgomery, *Mol. Pharmacol.*, 1979, **16**, 981.
52. R. D. Elliott and J. A. Montgomery, *J. Med. Chem.*, 1976, **19**, 1186.
53. R. D. Elliott and J. A. Montgomery, *J. Med. Chem.*, 1977, **20**, 116.
54. J. A. Montgomery, R. D. Elliott and H. J. Thomas, *Ann. N.Y. Acad. Sci.*, 1975, **255**, 292.
55. F. M. Schabel, Jr., *Chemotherapy*, 1968, **13**, 21.
56. G. P. Bodey, J. Gottlieb, K. B. McCredie and E. J. Freireich, *Proc. Am. Assoc. Cancer Res.*, 1974, **15**, 129.
57. G. A. LePage, L. S. Worth and A. P. Kimball, *Cancer Res.*, 1976, **36**, 1481.
58. J. L. York and G. A. LePage, *Can. J. Biochem.*, 1966, **44**, 19.
59. J. A. Montgomery and K. Hewson, *J. Med. Chem.*, 1969, **12**, 498.
60. L. White, S. C. Shaddix, Y.-C. Cheng, R. W. Brockman and L. L. Bennett, Jr., *Proc. Am. Assoc. Cancer Res.*, 1981, **22**, 33.
61. S. H. Lee, F. M. Unger, R. Christian and A. C. Sartorelli, *Biochem. Pharmacol.*, 1979, **28**, 1267.
62. Y. F. Shealy, M. C. Thorpe, W. C. Coburn, Jr. and J. D. Clayton, *Chem. Pharm. Bull.*, 1980, **28**, 3114.
63. Y. F. Shealy and J. D. Clayton, *J. Pharm. Sci.*, 1973, **62**, 858.
64. R. Vince, J. Brownell and S. Daluge, *J. Med. Chem.*, 1984, **27**, 1358.
65. D. A. Carson, D. B. Wasson, J. Kaye, B. Ullman, D. W. Martin, Jr., R. K. Robins and J. A. Montgomery, *Proc. Natl. Acad. Sci. USA*, 1980, **77**, 6865.
66. P. K. Chiang, H. H. Richards and G. L. Cantoni, *Mol. Pharmacol.*, 1977, **13**, 939.
67. T. A. Khwaja, *Cancer Treat. Rep.*, 1982, **66**, 1853.
68. M. I. Lim and R. S. Klein, *Tetrahedron Lett.*, 1981, **22**, 25.
69. L. L. Bennett, Jr., D. Smithers, N. F. DuBois, J. E. Hughes and R. W. Brockman, *Proc. Am. Assoc. Cancer Res.*, 1981, **22**, 27.
70. P. S. Ritch and R. I. Glazer in 'Developments in Cancer Chemotherapy', ed. R. I. Glazer, CRC Press, Boca Raton, FL, 1984.
71. M. Hayashi, S. Yaginuma, H. Yoshioka and K. Nakatsu, *J. Antibiot.*, 1981, **34**, 675.
72. R. B. Livingston and S. K. Carter, 'Single Agents in Cancer Chemotherapy', IFI/Plenum, New York, 1970.
73. See F. Kalousek, K. Raska, Jr., M. Turovcik and F. Sorm, *Collect. Czech. Chem. Commun.*, 1966, **31**, 1421, and refs. therein.
74. W. S. Zielinski and M. Sprinzl, *Nucleic Acids Res.*, 1984, **12**, 5025.
75. R. E. Handschumacher, *Cancer Res.*, 1963, **23**, 634.
76. See T. W. Kensler and D. A. Cooney, *Adv. Pharmacol. Chemother.*, 1981, **18**, 273 for a review of chemotherapeutic inhibitors of enzymes of the *de novo* pyrimidine pathway.
77. K. Hernandez, D. Pinkel, S. Lee and L. Leone, *Cancer Chemother. Rep.*, 1969, **53**, 203.
78. R. E. Handschumacher, *J. Biol. Chem.*, 1960, **235**, 764; T. W. Traut and M. E. Jones, *J. Biol. Chem.*, 1977, **252**, 8374.
79. C. Heidelberger, P. V. Danenberg and R. G. Moran, *Adv. Enzymol. Relat. Areas Mol. Biol.*, 1983, **54**, 57.
80. P. V. Danenberg, *Biochim. Biophys. Acta*, 1977, **473**, 73.
81. A. M. Pellino and P. V. Danenberg, *J. Biol. Chem.*, 1985, **260**, 10996; see also J. S. Park, C. T.-C. Chang and M. P. Mertes, *J. Med. Chem.*, 1979, **22**, 1134.
82. Y. Wataya, D. V. Santi and C. Hansch, *J. Med. Chem.*, 1977, **20**, 1469; see also M. Yashimoto and C. Hansch, *J. Med. Chem.*, 1976, **19**, 71.
83. L. W. Hardy, J. S. Finer-Moore, W. R. Montfort, M. O. Jones, D. V. Santi and R. M. Stroud, *Science (Washington, D.C.)*, 1987, **235**, 448.
84. D. V. Santi, *J. Med. Chem*, 1980, **23**, 103.
85. J. Lichenstein, H. D. Barner and S. S. Cohen, *J. Biol. Chem.*, 1960, **235**, 457.
86. J. J. Fox and N. C. Miller, *J. Org. Chem.*, 1963, **28**, 936.
87. S. Hillers, R. A. Zhuk, M. Lidaks and A. Zidermane, *Br. Pat.*, 1168391 (*Chem. Abstr.*, 1970, **72**, 43715).

88. M. Yasumoto, I. Yamawaki, T. Marunaka and S. Hashimoto, *J. Med. Chem.*, 1978, **21**, 738.
89. A. Hoshi, M. Igo, A. Nakamura, T. Inomata and K. Kuretani, *Chem. Pharm. Bull.*, 1978, **26**, 161.
90. T. Tsuruo, H. Ida, K. Naganuma, S. Tsukagoshi and Y. Sakurai, *Cancer Chemother. Pharmacol.*, 1980, **4**, 83.
91. M. Igo, A. Hoshi, M. Inomata, N. Audo and K. Kuretani, *J. Pharmacobio-Dyn.*, 1981, **4**, 203.
92. W. Bollag and H. R. Hartmann, *Eur. J. Cancer*, 1980, **16**, 427.
93. M. Arakawa, F. Shimizu, K. Sasagawa, T. Inomata and K. Shinkai, *Gann*, 1981, **72**, 220.
94. D. G. Johns, A. C. Sartorelli, J. R. Bertino, A. T. Iannotti, B. A. Booth and A. D. Welch, *Biochem. Pharmacol.*, 1966, **15**, 400.
95. R. Duschinsky, T. Gabriel, W. Tautz, A. Nussbaum, M. Hoffer, E. Grunberg, J. H. Burchenal and J. J. Fox, *J. Med. Chem.*, 1967, **10**, 47.
96. See P. Reyes and C. Heidelberger, *Mol. Pharmacol.*, 1965, **1**, 14, and refs. therein.
97. A. Rossi, in 'Nucleoside Analogues: Chemistry, Biology, and Medical Applications', ed. R. T. Walker, E. De Clercq and F. Eckstein, Plenum Press, New York, 1979, p. 411.
98. D. Kessel, T. C. Hall and I. Wodinsky, *Science (Washington, D.C.)*, 1967, **156**, 1240.
99. J. J. Furth and S. S. Cohen, *Cancer Res.*, 1968, **28**, 2061.
100. W. A. Creasey, R. J. Papac, M. E. Markiw, P. Calabresi and A. D. Welch, *Biochem. Pharmacol.*, 1966, **15**, 1417.
101. D. F. Wentworth and R. Wolfenden, *Methods Enzymol.*, 1978, **151**, 401.
102. C.-H. Kim, V. E. Marquez, D. T. Mao, D. R. Haines and J. J. McCormack, *J. Med. Chem.*, 1986, **29**, 1374.
103. M. Bobek, Y.-C. Cheng and A. Bloch, *J. Med. Chem.*, 1978, **21**, 597.
104. T. L. Chwang, A. Fridland and T. L. Avery, *J. Med. Chem.*, 1983, **26**, 280.
105. A. Hoshi, F. Kanzawa, K. Kuretani, M. Sanoyoshi and Y. Arai, *Gann*, 1971, **62**, 145.
106. K. Kondo, T. Nagura, Y. Arai and I. Inoue, *J. Med. Chem.*, 1979, **22**, 639.
107. J. A. Beisler, M. M. Abbasi and J. S. Driscoll, *J. Med. Chem.*, 1979, **22**, 1230.
108. Y. F. Shealy and C. A. O'Dell, *J. Pharm. Sci.*, 1979, **68**, 668.
109. C. I. Hong, A. Nechaev, A. J. Kirisits, D. J. Buchheit and C. R. West, *J. Med. Chem.*, 1980, **23**, 1343.
110. T. Tsuruo, H. Iida, K. Hori, S. Tsukagoshi and Y. Sakurai, *Cancer Res.*, 1981, **41**, 4484.
111. E. K. Ryu, R. J. Ross, T. Matsushita, M. MacCoss, C. I. Hong and C. R. West, *J. Med. Chem.*, 1982, **25**, 1322.
112. J. H. Burchenal, H. H. Adams, N. S. Newell and J. J. Fox, *Cancer Res.*, 1966, **26**, 370.
113. W. H. Prusoff, R. F. Schinazi and M. S. Chen, *J. Med. Chem.*, 1979, **22**, 1273.
114. G. T. Shiau, R. F. Schinazi, M. S. Chen and W. H. Prusoff, *J. Med. Chem.*, 1980, **23**, 127.
115. T.-S. Lin, P. H. Fischer and W. H. Prusoff, *Biochem. Pharmacol.*, 1982, **31**, 125.
116. T.-S. Lin and W. R. Mancini, *J. Med. Chem.*, 1983, **26**, 544.
117. J. H. Burchenal, T.-C. Chou, L. Lokys, R. S. Smith, K. A. Watanabe, T.-L. Su and J. J. Fox, *Cancer Res.*, 1982, **42**, 2598.
118. P. P. Saunders, R. Kuttan, M. M. Lai and R. K. Robins, *Mol. Pharmacol.*, 1983, **23**, 534 and refs. therein.
119. S. J. Martin, 'The Biochemistry of Viruses', Cambridge University Press, Cambridge, 1978, provides a succinct introduction to the subject.
120. See, for example, A. W. Senear and J. A. Steitz, *J. Biol. Chem.*, 1976, **251**, 1902.
121. H. Temin and D. Baltimore, *Adv. Virus Res.*, 1973, **17**, 129; *Cold Spring Harbor Symp. Quant. Biol.*, 1974, **39**.
122. E. De Clercq, in 'Targets for the Design of Antiviral Agents', ed. R. T. Walker and E. de Clercq, Plenum Press, New York, 1984, p. 203.
123. B. Goz, *Pharmacol. Rev.*, 1978, **29**, 249; W. H. Prusoff and B. Goz, in 'Antineoplastic and Immunosuppressive Agents', Springer Verlag, Berlin, 1975, vol. 2, p. 272.
124. W. H. Prusoff, M. S. Chen, P. H. Fisher, T.-S. Lin, G. T. Shiau, R. F. Schinazi and J. Walker, *Pharmacol. Ther.*, 1979, **7**, 1.
125. S. Kit, *Pharmacol. Ther.*, 1979, **4**, 501.
126. W. H. Prusoff and B. Goz, in 'The Herpesviruses', ed. A. S. Kaplan, Academic Press, New York, 1973, p. 641.
127. D. Shugar, in 'Antimetabolites in Biochemistry, Biology, and Medicine', ed. J. Skoda and P. Langen, Pergamon Press, Oxford, 1979.
128. J. Verbov, *J. Antimicrob. Chemother.*, 1979, **5**, 126; J. D. Parker, *J. Antimicrob. Chemother.*, 1977, **3** (Suppl. A), 131.
129. R. W. Sidwell and J. T. Witkowski, in 'Burger's Medicinal Chemistry', 4th edn., ed. M. E. Wolff, Wiley, New York, 1979, p. 543.
130. T. S. Lin, J. P. Neenan, Y.-C. Cheng, W. H. Prusoff and D. C. Ward, *J. Med. Chem.*, 1976, **19**, 495.
131. M. J. Doberson, M. Jerkofsky and S. Greer, *J. Virol.*, 1976, **20**, 478.
132. C. Heidelberger and D. H. King, *Pharmacol. Ther.*, 1979, **6**, 427.
133. C. Heidelberger, in 'Antineoplastic and Immunosuppressive Agents', Springer Verlag, Berlin, 1975, vol. 2, p. 193.
134. E. De Clercq and D. Shugar, *Biochem. Pharmacol.*, 1975, **24**, 1073.
135. E. De Clercq, J. Descamps and D. Shugar, *Antimicrob. Agents Chemother.*, 1978, **13**, 545.
136. E. De Clercq, J. Descamps, P. De Somer, P. J. Barr, A. S. Jones and R. T. Walker, *Proc. Natl. Acad. Sci. USA*, 1979, **76**, 2947.
137. E. De Clercq, J. Descamps, J. Balzarini, J. Giziewicz, P. J. Barr and M. J. Robins, *J. Med. Chem.*, 1983, **26**, 661.
138. Y.-C. Cheng, S. Grill, J. L. Ruth and D. E. Bergstrom, *Antimicrob. Agents Chemother.*, 1980, **18**, 957.
139. D. E. Bergstrom, J. L. Ruth, P. A. Reddy and E. De Clercq, *J. Med. Chem.*, 1984, **27**, 279.
140. E. De Clercq, *Pure Appl. Chem.*, 1983, **55**, 623.
141. P. Herdewijn, E. De Clercq, J. Balzarini and H. Vanderhaeghe, *J. Med. Chem.*, 1985, **28**, 550.
142. E. De Clercq, J. Descamps, C. L. Schmidt and M. P. Mertes, *Biochem. Pharmacol.*, 1979, **28**, 3249.
143. E. De Clercq, J. Descamps, G.-F. Huang and P. F. Torrence, *Mol. Pharmacol.*, 1978, **14**, 422.
144. T. Y. Shen, J. F. McPherson and B. O. Linn, *J. Med. Chem.*, 1966, **9**, 366.
145. G.-F. Huang, M. Okada, E. De Clercq and P. F. Torrence, *J. Med. Chem.*, 1981, **24**, 390.
146. P. F. Torrence, J. W. Spencer, A. M. Bobst, J. Descamps and E. De Clercq, *J. Med. Chem.*, 1978, **21**, 228.
147. L. A. Babiuk, B. Meldrum, V. A. Gupta and B. T. Rouse, *Antimicrob. Agents Chemother.*, 1975, **8**, 643.
148. R. Hardi, R. G. Hughes, Jr., Y. K. Ho, K. C. Chadha and T. J. Bardos, *Antimicrob. Agents Chemother.*, 1976, **10**, 682.
149. See, for example, M. J. Dobersen, M. Jerkofsky and S. Greer, *J. Virol.*, 1976, **20**, 478; for a brief survey of these compounds and leading refs., see ref. 122.
150. G. Gentry, J. McGowan, J. Barnett, R. Nevins and G. Allen, *Adv. Opthalmol.*, 1979, **38**, 164.

151. W. E. G. Müller, R. K. Zhan, J. Arendes and D. Falke, *J. Gen. Virol.*, 1979, **43**, 261.
152. T. Kulikowski, Z. Zawadzki, D. Shugar, J. Descamps and E. De Clercq, *J. Med. Chem.*, 1979, **22**, 647.
153. P. F. Torrence, G.-F. Huang, M. W. Edwards, B. Bhooshan, J. Descamps and E. De Clercq, *J. Med. Chem.*, 1979, **22**, 316.
154. S. Shigeta, T. Yokota, T. Iwabuchi, M. Baba, K. Konno, M. Ogata and E. De Clercq, *J. Infect. Dis.*, 1983, **147**, 576.
155. C. Lopez, K. A. Watanabe and J. J. Fox, *Antimicrob. Agents Chemother.*, 1980, **17**, 803.
156. K. A. Watanabe, U. Reichman, K. Hirota, C. Lopez and J. J. Fox, *J. Med. Chem.*, 1979, **22**, 21.
157. H. S. Allaudeen, J. Descamps, R. K. Sehgal and J. J. Fox, *J. Biol. Chem.*, 1982, **257**, 11 879.
158. T.-C. Chou, A. Feinberg, A. J. Grant, P. Vidal, U. Reichman, K. A. Watanabe, J. J. Fox and F. S. Philips, *Cancer Res.*, 1981, **41**, 3336.
159. F. S. Philips, A. Feinberg, T.-C. Chou, P. M. Vidal, T.-L. Su, K. A. Watanabe and J. J. Fox, *Cancer Res.*, 1983, **43**, 3619, and refs. therein.
160. K. A. Watanabe, T.-L. Su, R. S. Klein, C. K. Chu, A. Matsuda, M. W. Chun, C. Lopez and J. J. Fox, *J. Med. Chem.*, 1983, **26**, 152.
161. B. Rada and J. Doskocil, *Pharmacol. Ther.*, 1980, **9**, 171.
162. H. E. Renis, *Antimicrob. Agents Chemother.*, 1978, **13**, 613.
163. H. E. Renis and E. E. Eidson, *Antimicrob. Agents Chemother.*, 1979, **15**, 213.
164. F. Smejkal, J. Gut and F. Sorm, *Acta Virol.*, 1962, **6**, 364.
165. G. P. Khare, R. W. Sidwell, J. H. Huffman, R. L. Tolman and R. K. Robins, *Proc. Soc. Exp. Biol. Med.*, 1972, **140**, 880.
166. W. M. Shannon, *Ann. N.Y. Acad. Sci.*, 1977, **284**, 472.
167. R. P. McPartland, M. C. Wang, A. Bloch and H. Weinfeld, *Cancer Res.*, 1974, **34**, 3107.
168. E. De Clercq, *J. Med. Chem.*, 1986, **29**, 1561; *Trends Pharmacol. Sci.*, 1987, **8**, 339; *Proc. R. Belg. Acad. Sci.*, 1988, 166.
169. H. Mitsuya, J. K. Weinhold, P. A. Furman, M. H. St. Clair, S. N. Lehrman, R. C. Gallo, D. Bolognesi, D. W. Barry and S. Broder, *Proc. Natl. Acad. Sci. USA*, 1985, **82**, 7096.
170. J. Balzarini, R. Pauwels, P. Herdewijn, E. De Clercq, D. A. Cooney, G.-J. Kang, M. Dalal, D. G. Johns and S. Broder, *Biochem. Biophys. Res. Commun.*, 1986, **140**, 735.
171. T. P. Zimmermann, W. B. Mahony and K. L. Prus, *J. Biol. Chem.*, 1987, **262**, 5748.
172. Y.-C. Cheng, G. E. Dutschman, K. F. Bastow, M. G. Sarngadharan and R. Y. C. Ting, *J. Biol. Chem.*, 1987, **262**, 2187.
173. J. E. Dahlberg, H. Mitsuya, S. B. Blam, S. Broder and S. A. Aaronson, *Proc. Natl. Acad. Sci. USA*, 1987, **84**, 2469; H. Mitsuya and S. Broder, *Proc. Natl. Acad. Sci. USA*, 1986, **83**, 1911; see also J. Balzarini, M. J. Robins, R. Zou, P. Herdewijn and E. De Clercq, *Biochem. Biophys. Res. Commun.*, 1987, **145**, 277, and refs. therein.
174. J. C. Drach, in 'Targets for the Design of Antiviral Agents', ed. R. T. Walker and E. De Clercq, Plenum Press, New York, 1984, p. 231.
175. D. Pavan-Langston, R. A. Buchanan and C. A. Alford, Jr. (eds.), 'Adenine Arabinoside; An Antiviral Agent', Raven Press, New York, 1975.
176. P. M. Schwartz, J. N. Sandberg, C. Shipman, Jr. and J. C. Drach, *15th Intersci. Conf. Antimicrob. Agents Chemother., Washington, D.C.*, 1975, Abstr. 357.
177. W. Plunkett, S. Chubb, L. Alexander and J. A. Montgomery, *Cancer Res.*, 1980, **40**, 2349.
178. W. E. G. Müller, R. K. Zahn and D. Falke, *Virology*, 1978, **84**, 320; M. Ostrander and Y.-C. Cheng, *Biochim. Biophys. Acta*, 1980, **609**, 232.
179. W. E. G. Müller, R. K. Zahn, K. Bittlingmaier and D. Falke, *Ann. N.Y. Acad. Sci.*, 1977, **284**, 34.
180. S. S. Cohen, *Med. Biol.*, 1976, **54**, 299.
181. K. M. Rose, T. B. Leonard and T. H. Carter, *Mol. Pharmacol.*, 1982, **22**, 517.
182. W. E. G. Müller, R. K. Zahn and J. Arendes, *FEBS Lett.*, 1978, **94**, 47.
183. M. S. Hershfield, *J. Biol. Chem.*, 1979, **254**, 22.
184. S. S. Cohen, *Prog. Nucleic Acids Res., Mol. Biol.*, 1966, **5**, 1; See also S. S. Cohen, in 'Nucleoside Analogues: Chemistry, Biology, and Medical Applications', ed. R. T. Walker, E. De Clercq and F. Eckstein, Plenum Press, New York, 1979, p. 225 for a valuable review of *ara*-nucleoside metabolism and function.
185. F. A. Miller, G. J. Dixon, J. Ehrlich, B. J. Sloan and I. W. McLean, Jr., *Antimicrob. Agents Chemother.*, 1968, 136.
186. W. E. G. Müller, J. Arendes, A. Maidhof, R. K. Zahn and W. Geurtsen, *Chemotherapy*, 1981, **27**, 53.
187. D. Derse, K. F. Bastow and Y.-C. Cheng, *J. Biol. Chem.*, 1982, **257**, 10 251.
188. G. A. LePage, *Can. J. Biochem.*, 1970, **48**, 75.
189. P. M. Schwartz, C. Shipman, Jr. and J. C. Drach, *Antimicrob. Agents Chemother.*, 1976, **10**, 64.
190. W. Plunkett and S. S. Cohen, *Ann. N.Y. Acad. Sci.*, 1977, **284**, 91.
191. R. W. Brockman, Y.-C. Cheng, F. M. Schabel and J. A. Montgomery, *Cancer Res.*, 1980, **40**, 3610.
192. C. M. Cermak-Mörth, R. Christian and F.M. Unger, *Biochem. Pharmacol.*, 1979, **28**, 2105.
193. R. Vince and S. Daluge, *J. Med. Chem.*, 1977, **20**, 612.
194. W. M. Shannon, L. Westbrook, G. Arnett, S. Daluge, H. Lee and R. Vince, *Antimicrob. Agents Chemother.*, 1983, **24**, 538.
195. G. B. Elion, J. L. Rideout, P. de Miranda, P. Collins and D. J. Bauer, *Ann. N.Y. Acad. Sci.*, 1975, **255**, 468.
196. D. C. Baker, T. H. Haskell, S. R. Putt and B. J. Sloan, *J. Med. Chem.*, 1979, **22**, 273.
197. J. T. Richards, E. R. Kern, J. C. Overall, Jr. and L. A. Glasgow, *Antiviral Res.*, 1982, **2**, 27.
198. R. A. Lipper, S. M. Machkovech, J. C. Drach and W. I. Higuchi, *Mol. Pharmacol.*, 1978, **14**, 366.
199. W. Plunkett and S. S. Cohen, *Cancer Res.*, 1975, **35**, 1547.
200. O. Sauer, G. T. Werner, H. Schneider, H. G. Lenard, L. V. Haselberg and H.-J. Nettesheim, *Klin. Paediatr.*, 1979, **191**, 566.
201. V. A. Hatcher, A. E. Friedman-Kien, E. L. Marcus and R. J. Klein, *Antiviral Res.*, 1982, **2**, 283.
202. E. Walton, S. R. Jenkins, R. F. Nutt, F. W. Holly and M. Nemes, *J. Med. Chem.*, 1969, **12**, 306.
203. G. R. Revankar, J. H. Huffmann, R. W. Sidwell, R. L. Tolman, R. K. Robins and L. B. Allen, *J. Med. Chem.*, 1976, **19**, 1026.
204. L. L. Bennett, Jr., W. M. Shannon, P. W. Allan and G. Arnett, *Ann. N.Y. Acad. Sci.*, 1975, **255**, 342.
205. G. B. Elion, *Am. J. Med.*, 1982, **73** (1A), 7; *J. Antimicrob. Chemother.*, 1983, **12**, Suppl. B.
206. L. Corey, A. J. Nahmias, M. E. Guinan, J. K. Benedetti, C. W. Critchlow and K. K. Holmes, *New Engl. J. Med.*, 1982, **306**, 1313.
207. *Am. J. Med.*, 1982, **73**, 1.

208. A. P. Fiddian, J. M. Yeo, R. Stubbings and D. Dean, *Br. Med. J.*, 1983, **286**, 1699.
209. C. S. Crumpacker, L. E. Schnipper, J. A. Zaia and M. J. Levin, *Antimicrob. Agents Chemother.*, 1979, **15**, 642.
210. J. A. Fyfe, P. M. Keller, P. A. Furman, R. L. Miller and G. B. Elion, *J. Biol. Chem.*, 1978, **253**, 8721.
211. W. H. Miller and R. L. Miller, *J. Biol. Chem.*, 1980, **255**, 7204; *Biochem. Pharmacol.*, 1982, **31**, 3879.
212. M. H. St. Clair, P. A. Furman, C. M. Lubbers and G. B. Elion, *Antimicrob. Agents Chemother.*, 1980, **18**, 741.
213. D. Derse, Y.-C. Cheng, P. A. Furman, M. H. St. Clair and G. B. Elion, *J. Biol. Chem.*, 1981, **256**, 11 447.
214. P. V. McGuirt and P. A. Furman, *Am. J. Med.*, 1982, **73**, 67.
215. T. Spector, T. E. Jones and L. M. Beacham, III, *Biochem. Pharmacol.*, 1983, **32**, 2505.
216. H. C. Krasny, S. H. T. Liao and S. S. Good, *Clin. Pharmacol. Ther.*, 1983, **33**, 256.
217. K. O. Smith, K. S. Galloway, W. L. Kennell, K. K. Ogilvie and B. K. Radatus, *Antimicrob. Agents Chemother.*, 1982, **22**, 55; W. T. Ashton, J. D. Karkas, A. K. Field and R. L. Tolman, *Biochem. Biophys. Res. Commun.*, 1982, **108**, 1716.
218. D. F. Smee, J. C. Martin, J. P. H. Verheyden and T. R. Mathews, *Antimicrob. Agents Chemother.*, 1983, **23**, 676.
219. E.-C. Mar, Y.-C. Cheng and E.-S. Huang, *Antimicrob. Agents Chemother.*, 1983, **24**, 518.
220. A. Larsson, B. Öberg, S. Alenius, G.-E. Hagberg, N.-G. Johansson, B. Lindborg and G. Stening, *Antimicrob. Agents Chemother.*, 1983, **23**, 664.
221. E. De Clercq, J. Descamps, P. De Somer and A. Holy, *Science (Washington, D.C.)*, 1978, **200**, 563.
222. R. Jelinek, A. Holy and I. Votruba, *Teratology*, 1981, **24**, 267.
223. J. R. Wingard, A. D. Hess, R. K. Stuart, R. Saral and W. H. Burns, *Antimicrob. Agents Chemother.*, 1983, **23**, 593.
224. E. De Clercq, *Biochem. J.*, 1982, **205**, 1.
225. I. Votruba and A. Holy, *Collect. Czech. Chem. Commun.*, 1980, **45**, 3039.
226. L. B. Allen, J. H. Huffman, P. D. Cook, R. B. Meyer, Jr., R. K. Robins and R. W. Sidwell, *Antimicrob. Agents Chemother.*, 1977, **12**, 114.
227. D. G. Streeter and H. H. P. Koyama, *Biochem. Pharmacol.*, 1976, **25**, 2413.
228. J. P. Bader, N. R. Brown, P. K. Chiang and G. L. Cantoni, *Virology*, 1978, **89**, 494.
229. A. J. Bodner, G. L. Cantoni and P. K. Chiang, *Biochem. Biophys. Res. Commun.*, 1981, **98**, 476.
230. E. De Clercq and J. A. Montgomery, *Antiviral Res.*, 1983, **3**, 17.
231. P. K. Chiang, H. H. Richards and G. L. Cantoni, *Mol. Pharmacol.*, 1977, **13**, 939.
232. P. K. Chiang, R. K. Gordon, N. D. Brown, E. De Clercq and J. A. Montgomery, *Am. Chem. Soc., Div. Med. Chem., 185th Natl. Meet.*, Seattle, 1983, Abstr. 34.
233. R. A. Smith and W. Kirkpatrick (eds), 'Ribavirin, A Broad Spectrum Antiviral Agent', Academic Press, New York, 1980; T. W. Chang and R. C. Heel, *Drugs*, 1981, **22**, 111.
234. R. W. Sidwell, R. K. Robins and I. W. Hillyard, *Pharmacol. Ther.*, 1979, **6**, 123.
235. H. W. McClung, V. Knight, B. E. Gilbert, S. Z. Wilson, J. M. Quarles and G. W. Divine, *J. Am. Med. Assoc.*, 1983, **249**, 2671.
236. D. M. Kochhar, J. D. Penner and T. B. Knudsen, *Toxicol. Appl. Pharmacol.*, 1980, **52**, 99.
237. D. G. Streeter, J. T. Witkowski, G. P. Khare, R. W. Sidwell, R. J. Bauer, R. K. Robins and L. N. Simon, *Proc. Natl. Acad. Sci. USA*, 1973, **70**, 1174.
238. B. Ericksson, E. Helgstrand, K. N. G. Johansson, A. Larsson, A. Misiorny, J. O. Norén, L. Philipson, K. Stenburg, G. Stening, S. Stridh and B. Öberg, *Antimicrob. Agents Chemother.*, 1977, **11**, 946.
239. J. C. Drach, M. A. Thomas, J. W. Barnett, S. H. Smith and C. Shipman, Jr., *Science (Washington, D.C.)*, 1981, **212**, 549.
240. B. B. Goswami, E. Borek, O. K. Sharma, J. Fujitaki and R. A. Smith, *Biochem. Biophys. Res. Commun.*, 1979, **89**, 830.
241. D. F. Smee, R. W. Sidwell, S. M. Clark, B. B. Barnett and R. S. Spendlove, *Antimicrob. Agents Chemother.*, 1981, **20**, 533.
242. J. P. Miller, L. J. Kigwana, D. G. Streeter, R. K. Robins, L. N. Simon and J. Roboz, *Ann. N.Y. Acad. Sci.*, 1977, **284**, 211.
243. P. C. Srivastava, D. G. Streeter, T. R. Mathews, L. B. Allen, R. W. Sidwell and R. K. Robins, *J. Med. Chem.*, 1976, **19**, 1020.
244. P. C. Srivastava and R. K. Robins, *J. Med. Chem.*, 1983, **26**, 445.
245. J. J. Kirsi, J. A. North, P. A. McKernan, B. K. Murray, P. G. Canonico, J. W. Huggins, P. C. Srivastava and R. K. Robins, *Antimicrob. Agents Chemother.*, 1983, **24**, 353.
246. G. E. Gutowski, M. J. Sweeney, D. C. DeLong, R. L. Hamill, K. Gazon and R. W. Dyer, *Ann. N.Y. Acad. Sci.*, 1975, **255**, 544.
247. W. E. Gutteridge and G. H. Coombs, in 'Biochemistry of Parasitic Protozoa', Macmillan, London, 1977, p. 75.
248. C. C. Wang, *Annu. Rep. Med. Chem.*, 1981, **16**, 269.
249. J. J. Marr and R.L. Berens, *Mol. Biochem. Parasitol.*, 1983, **7**, 339.
250. T. Spector, T. E. Jones, S. W. LaFon, D. J. Nelson, R. L. Berens and J. J. Marr, *Biochem. Pharmacol.*, 1984, **33**, 1611.
251. D. J. Nelson, S. W. LaFon, J. V. Tuttle, W. H. Miller, R. L. Miller, T. A. Krenitsky, G. B. Elion, R. L. Berens and J. J. Marr, *J. Biol. Chem.*, 1979, **254**, 11 544.
252. J. L. Avila and M. A. Casanova, *Antimicrob. Agents Chemother.*, 1982, **22**, 380.
253. W. Peters, E. R. Trotter and B. L. Robinson, *Ann. Trop. Med. Parasitol.*, 1980, **74**, 321.
254. P. A. Kager, P. H. Rees, B. T. Wellde, W. T. Hockmeyer and W. H. Lyerly, *Trans. R. Soc. Trop. Med. Hyg.*, 1981, **75**, 556.
255. J. L. Avila, A. Avila and M. A. Casanova, *Mol. Biochem. Parasitol.*, 1981, **4**, 265.
256. D. J. Hupe, *Annu. Rep. Med. Chem.*, 1986, **21**, 247: a valuable summary of recent work in this field, with leading refs.
257. S. B. Stepanova, N. A. Koreshkova, N. N. Guliaev and N. A. Fedorov, *Bull. Exp. Biol. Med.*, 1980, **89**, 590; D. E. Wilman and P. M. Goddard, *J. Med. Chem.*, 1980, **23**, 1052.
258. H. B. Cottam, C. R. Petrie, P. A. McKernan, R. J. Groebel, N. K. Dalley, R. B. Davidson, R. K. Robins and G. R. Revankar, *J. Med. Chem.*, 1984, **27**, 1119.
259. Y. Wataya, O. Hiraoka, Y. Sonobe, A. Yoshioka, A. Matsuda, T. Miyasaka, M. Saneyoshi and T. Ueda, *Nucleic Acids Symp. Ser.*, 1984, **15**, 69; Y. Wataya and O. Hiraoka, *Biochem. Biophys. Res. Commun.*, 1984, **123**, 677.
260. J. J. Marr, R. L. Berens, N. K. Cohn, D. J. Nelson and R. S. Klein, *Antimicrob. Agents Chemother.*, 1984, **25**, 292; P. Rainey, P. A. Nolan, L. B. Townsend, R. K. Robins, J. J. Fox, J. A. Secrist, III and D. V. Santi, *Pharm. Res.*, 1985, 217.
261. D. A. Carson and K.-P. Chang, *Biochem. Biophys. Res. Commun.*, 1981, **100**, 1377; P. Rainey and D. V. Santi, *Proc. Natl. Acad. Sci. USA*, 1983, **80**, 288.
262. D. J. Nelson, S. W. Lafon, T. E. Jones, T. Spector, R. L. Berens and J. J. Marr, *Biochem. Biophys. Res. Commun.*, 1982, **108**, 349.

263. L. L. Nolan, J. D. Berman and L. Giri, *Biochem. Int.*, 1984, **9**, 207.
264. J. D. Berman, L.S. Lee, R. K. Robins and G. R. Revankar, *Antimicrob. Agents Chemother.*, 1983, **24**, 233.
265. P. Rainey, C. E. Garrett and D. V. Santi, *Biochem. Pharmacol.*, 1983, **32**, 749.
266. J. J. Marr, R. L. Berens, D. J. Nelson, T. A. Krenitsky, T. Spector, S. W. LaFon and G. B. Elion, *Biochem. Pharmacol.*, 1982, **31**, 143; *Clin. Res.*, 1981, **29**, 390A.
267. T. Spector and T. E. Jones, *Biochem. Pharmacol.*, 1982, **31**, 3891.
268. J. V. Tuttle and T. A. Krenitsky, *J. Biol. Chem.*, 1980, **255**, 909.
269. T. A. Krenitsky, J. L. Rideout, G. W. Koszalka, R. B. Inmon, E. Y. Chao, G. B. Elion, V. S. Latter and R. B. Williams, *J. Med. Chem.*, 1982, **25**, 32.
270. J. L. Rideout, T. A. Krenitsky, E. Y. Chao, G. B. Elion, R. B. Williams and V. S. Latter, *J. Med. Chem.*, 1983, **26**, 1489.
271. R. L. Miller, D. L. Adamczyk, J. L. Rideout and T. A. Krenitsky, *Mol. Biochem. Parasitol.*, 1982, **6**, 209.
272. L. L. Nolan and G. W. Kidder, *Antimicrob. Agents Chemother.*, 1980, **17**, 567.
273. D. A. Carson and K.-P. Chang, *Life Sci.*, 1981, **29**, 1617.
274. G. W. Kidder and L. L. Nolan, *Mol. Biochem. Parasitol.*, 1981, **3**, 265.
275. H. K. Webster and J. M. Whaun, *J. Clin. Invest.*, 1982, **70**, 461.
276. W. P. Wiesmann, H. K. Webster, C. Lambros, W. N. Kelley and P. E. Daddona, *Prog. Clin. Biol. Res.*, 1984, **165**, 325.
277. H. K. Webster, W. P. Wiesmann and C. S. Pavia, *Adv. Exp. Med. Biol.*, 1984, **165A**, 225.
278. C. C. Wang and P. M. Simashkevich, *Mol. Biochem. Parasitol.*, 1980, **1**, 335.
279. R. S. Slaughter and E. M. Barnes, Jr., *Arch. Biochem. Biophys.*, 1979, **197**, 349.
280. C. C. Wang, R. L. Tolman, P. M. Simashkevich and R. L. Stotish, *Biochem. Pharmacol.*, 1979, **28**, 2249.
281. W. J. O'Sullivan and K. Ketley, *Ann. Trop. Med. Parasitol.*, 1980, **74**, 109.
282. E. R. Pfefferkorn, *Exp. Parasitol.*, 1978, **44**, 26.
283. M. W. Miller, B. L. Mylari, H. L. Howes, Jr., S. K. Figdor, M. J. Lynch, J. E. Lynch, S. K. Gupta, L. R. Chappel and R. C. Koch, *J. Med. Chem.*, 1981, **24**, 1337.
284. M. W. Miller (Pfizer Inc.), *US Pat.* 4 239 888 (1980) (*Chem. Abstr.*, 1981, **94**, 175 157z).
285. M. W. Miller, B. L. Mylari, H. L. Howes, Jr., M. J. Lynch, J. E. Lynch and R. C. Koch, *J. Med. Chem.*, 1979, **22**, 1483.
286. R. R. C. New, M. L. Chance and S. Heath, *J. Antimicrob. Chemother.*, 1981, **8**, 371.
287. A. Polak and H. J. Scholer, *Chemotherapy*, 1975, **21**, 113.
288. R. B. Diasio, J. E. Bennett and C. E. Myers, *Biochem. Pharmacol.*, 1978, **27**, 703.
289. A. Polak and W. H. Wain, *Chemotherapy*, 1977, **23**, 243.
290. T. K. Daneshmend and D. W. Warnock, *Clin. Pharmacokinet.*, 1983, **8**, 17.
291. J. R. Perfect and D.T. Durak, *J. Infect. Dis.*, 1982, **146**, 429.
292. J. A. Montgomery, *Acc. Chem. Res.*, 1986, **10**, 293.
293. D. Farquhar, N. J. Kuttesch, M. G. Wilkerson and T. Winkler, *J. Med. Chem.*, 1983, **26**, 1153; A. S. Jones, C. McGuigan, R. T. Walker, J. Balzarini and E. De Clercq, *J. Chem. Soc., Perkin Trans. 1*, 1984, 1471.
294. C. I. Hong, S.-H. An, D. J. Buchheit, A. Nechaev, A. J. Kirisits, C. R. West and W. E. Berdel, *J. Med. Chem.*, 1986, **29**, 2038.
295. See, for example, E. De Clercq, A. Holy, I. Rosenberg, T. Sukuma, J. Balzarini and P. C. Maudgal, *Nature (London)*, 1986, **323**, 464.
296. B. R. G. Williams and I. M. Kerr, *Trends Biochem. Sci.*, 1980, **5**, 138 and refs. therein.
297. B. Bayard, C. Bisbal and B. Lebleu, *Biochemistry*, 1986, **25**, 3730.
298. P. F. Torrence, J. Imai, J. C. Jamoulle, and K. Lesiak, *Methods Enzymol.*, 1986, **119**, 522.
299. See for example, P. C. Zamecnik, J. Goodchild, Y. Taguchi and P. S. Sarin, *Proc. Natl. Acad. Sci. USA*, 1986, **83**, 4143.
300. M. Lemaitre, B. Bayard and B. Lebleu, *Proc. Natl. Acad. Sci. USA*, 1987, **84**, 648.
301. C. H. Agris, K. R. Blake, P. S. Miller, M. P. Reddy and P. O. P. Ts'o, *Biochemistry*, 1986, **25**, 6268.
302. S. M. Hecht, *Acc. Chem. Res.*, 1986, **19**, 383.
303. L. S. Kappen, T. E. Ellenberger and I. H. Goldberg, *Biochemistry*, 1987, **26**, 384 and refs. therein.
304. G. B. Dreyer and P. B. Dervan, *Proc. Natl. Acad. Sci. USA*, 1985, **82**, 968; B. C. F. Chu and L. E. Orgel, *Proc. Natl. Acad. Sci. USA*, 1985, **82**, 963. For an interesting variant, see B. L. Iverson and P. B. Dervan, *J. Am. Chem. Soc.*, 1987, **109**, 1241.
305. P. B. Dervan, *Science (Washington, D.C.)*, 1986, **232**, 464.

7.4

Therapeutic Consequences of the Inhibition of Sterol Metabolism

JERRY L. ADAMS and BRIAN W. METCALF

Smith Kline & French Laboratories, Philadelphia, PA, USA

7.4.1 INTRODUCTION	333
7.4.2 THE INHIBITION OF CHOLESTEROL BIOSYNTHESIS	334
7.4.2.1 *Cholesterol and Atherosclerosis*	334
7.4.2.1.1 *The role of the LDL receptor in cholesterol homeostasis*	334
7.4.2.1.2 *Regulation of cholesterol biosynthesis via hydroxymethylglutaryl-CoA reductase (HMGCoA reductase)*	335
7.4.2.2 *Inhibition of HMGCoA Reductase*	336
7.4.2.2.1 *Enzymology of HMGCoA reductase*	336
7.4.2.2.2 *Inhibitors of HMGCoA reductase*	337
7.4.2.3 *Genomic Regulation of HMGCoA Reductase*	339
7.4.2.3.1 *Oxysterols as endogenous regulators*	339
7.4.2.3.2 *Lanosterol 14α-demethylase*	340
7.4.2.3.3 *Oxidosqualene cyclase*	342
7.4.2.4 *Inhibition of the Later Steps of Cholesterol Biosynthesis*	343
7.4.2.4.1 *Early hypocholesterolemics*	343
7.4.2.4.2 *The mechanistic basis for the inhibition of enzymes that proceed through carbocationic intermediates*	343
7.4.3 INHIBITION OF STEROID HORMONE BIOSYNTHESIS	347
7.4.3.1 *Inhibition of Cytochrome P-450 Isozymes Involved in Steroid Hormone Biosynthesis*	347
7.4.3.1.1 *Metyrapone*	347
7.4.3.1.2 *Aminoglutethimide*	347
7.4.3.1.3 *Ketoconazole*	348
7.4.3.1.4 *Mechanism of hydroxylation by cytochrome P-450*	348
7.4.3.1.5 *Mechanism of inhibition by reversible inhibitors*	349
7.4.3.1.6 *Mechanism of inhibition by irreversible inhibitors*	349
7.4.3.2 *Selective Inhibition of Androgen Biosynthesis; Steroid 5α-Reductase*	353
7.4.4 INHIBITION OF STEROL METABOLISM IN NONMAMMALIAN ORGANISMS	354
7.4.4.1 *Inhibition of Lanosterol 14α-Demethylase*	354
7.4.4.1.1 *Mode of action of azoles*	356
7.4.4.1.2 *Molecular basis for the action of azoles*	357
7.4.4.1.3 *Enhanced selectivity in the inhibition of lanosterol 14α-demethylase as the basis for antifungal drug design*	357
7.4.4.2 *Inhibition of Squalene Oxidase*	359
7.4.4.2.1 *Basis of antifungal effect*	359
7.4.4.2.2 *Basis of enzyme inhibition*	359
7.4.4.3 *Other Sites of Action*	360
7.4.4.4 *The Polyene Antibiotics*	361
7.4.5 ADDENDUM	361
7.4.6 REFERENCES	362

7.4.1 INTRODUCTION

The biosynthesis of sterols is not limited to mammalian organisms, but instead is a process common to all eukaryotes. The almost universal requirement for sterols is related to the unique role that these compounds serve in modulating the mechanical properties of phospholipid bilayers. Sterols are oriented in bilayers with the polar head group facing the aqueous medium. This allows the lipophilic steroid skeleton and side chain to interact *via* cooperative van der Waals' contacts with the phospholipid. The removal of sterols from phospholipid bilayers results in membranes with a sharper transition temperature and lower stability above the transition temperature. The inclusion of sterols in membranes then serves to broaden the range in which the membrane can retain the fluidity required to function while at the same time serving as structurally rigid anchors to maintain membrane organization.[1]

This universal need for sterols is a characteristic of both animals and plants. The list of organisms requiring sterols includes the fungi as well as many pathogenic parasites. Excluded from a requirement for sterols are bacteria, whose cell wall construction is based upon a rigid polysaccharide skeleton. In many prokaryotes, polyterpenoids are seen to fulfill a function analogous to that of sterols and hence these polyterpenoids may be viewed as the evolutionary precursors to sterols. An apparent exception to this division of sterol biosynthesis are the mycoplasmas. These wall-less, parasitic bacteria are the pathogenic agent causing tuberculosis and normally contain sterols. The sterols found in the mycoplasmas are not biosynthesized but are instead host derived. When these bacteria are grown in the absence of sterols they choose to synthesize polar carotenoids. The ability of this organism to utilize both sterols and carotenoids appears to illustrate the similar role these compounds fulfill.[2] Apart from their essential structural role, sterols also serve as the precursors for hormones in animals. Sterol based hormonal systems are also present in fungi, but are less well characterized.

7.4.2 THE INHIBITION OF CHOLESTEROL BIOSYNTHESIS

7.4.2.1 Cholesterol and Atherosclerosis

In November 1984, a panel of experts making up the NIH-sponsored Consensus Development Conference on Lowering Blood Cholesterol to Prevent Heart Disease agreed that:

(a) There is a causal relationship between blood cholesterol and coronary heart disease. The evidence: nonhuman primates develop atherosclerosis when fed high cholesterol diets. As a corollary, when blood cholesterol levels in Rhesus monkeys are reduced by diet or drugs, atherosclerotic lesions regress.
(b) Reduction of cholesterol levels will help prevent coronary heart disease.
(c) Attempts should be made to lower the blood cholesterol levels in the general population.
(d) There is a need for more effective hypolipidemic agents.

7.4.2.1.1 *The role of the LDL receptor in cholesterol homeostasis*

The central role of the low density lipoprotein (LDL) receptor in cholesterol homeostasis was first appreciated when Brown and Goldstein showed that defects in this receptor were responsible for the disease of familial hypercholesterolemia (FH).[3] FH heterozygotes (one mutant LDL receptor gene inherited) have elevated plasma LDL levels and suffer heart attacks by the age of 35. FH homozygotes (two copies of the mutant gene, one from each parent) have circulating levels of LDL six times higher than normal and are subject to heart attacks by the age of two. Studies of these individuals demonstrate the causal relationship between elevated circulating LDL and atherosclerosis.

In studying the LDL receptor, Brown and Goldstein not only pointed the way to the treatment or prevention of atherosclerosis (lowering circulating LDL levels) but in doing so developed such fundamental concepts in cell biology as 'receptor mediated endocytosis'. The LDL particle has a core composed of esterified cholesterol and binds to its receptor on the cell surface. The receptor is then internalized *via* an indentation in the cell surface known as a coated pit. A change in pH induces the release of LDL and the free receptor moves back to the cell surface to bind more LDL. Inside the cell the LDL is degraded in the lysosome and the esterified cholesterol is hydrolyzed to liberate free cholesterol. The amount of cholesterol liberated in this fashion controls the cell's cholesterol metabolism *via* three processes.

(i) Cholesterol biosynthesis is suppressed by regulation of the enzyme for the rate limiting step, hydroxymethylglutaryl-CoA reductase (HMGCoA reductase).
(ii) The enzyme acylcholesterol acyltransferase (ACAT) is activated. This enzyme catalyzes the formation of cholesterol esters for storage.
(iii) The synthesis of LDL receptors is decreased.

Conversely, when the cell has a need for cholesterol, HMGCoA reductase activity is increased *via* increased enzyme synthesis, and the synthesis of the LDL receptor is activated at the transcriptional level. This coordinated regulation of the LDL receptor suggested an approach for the up-regulation of the receptor in FH homozygotes. The liver requires cholesterol for the synthesis of bile acids, necessary for the emulsification of dietary fats. The bile acids are then reabsorbed from the intestine and taken up again by the liver. Nonabsorbable bile acid sequestrants, polymers which contain positively charged groups which bind bile acids, are able to interrupt this recirculation of bile acids and cause the liver to take up more cholesterol by increasing LDL receptor synthesis. Administration of the bile acid sequestrant, cholestyramine, leads to a decrease in circulating LDL levels which correlates with a decrease in the incidence of heart attacks. As the liver also increases the synthesis of HMGCoA reductase in response to a demand for cholesterol, it was proposed that inhibition of this enzyme would increase the demand for LDL uptake and hence further up-regulate receptor synthesis.[3]

7.4.2.1.2 *Regulation of cholesterol biosynthesis* via *hydroxymethylglutaryl-CoA reductase (HMGCoA reductase)*

The biosynthesis of cholesterol comprises some 26 reaction steps, many of which are now known in detail as a result of the work of Block, Lynen, Popjak and Cornforth.[4] These are summarized in Scheme 1. The enzyme HMGCoA reductase has been shown to be rate limiting in the overall pathway to cholesterol. This conclusion results from a study in rats of the response to dietary cholesterol which suppresses overall hepatic sterol synthetic capacity in parallel with a reduction in HMGCoA reductase activity. Incorporation of radioactivity from [1-^{14}C]acetyl-CoA into HMGCoA and of [2-^{14}C]mevalonate into cholesterol are unaffected.[5] Human and animal cells possess at least two mechanisms for obtaining the cholesterol necessary for the synthesis of membranes, steroid hormones and bile acids: (i) they can synthesize cholesterol *de novo via* the pathway shown in Scheme 1; (ii) they can supply themselves with cholesterol through the receptor mediated endocytosis of cholesterol-containing low density lipoprotein.

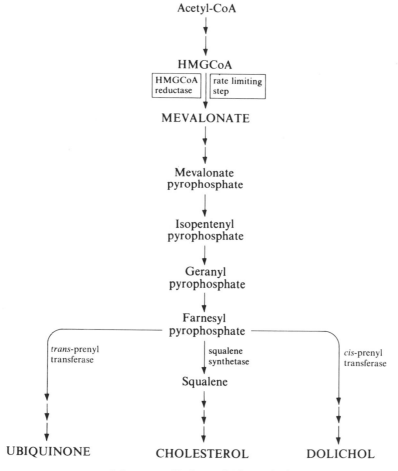

Scheme 1 Cholesterol Biosynthesis

It has been shown using tissue culture cells that both the LDL receptor and HMGCoA reductase are subject to end-product feedback regulation by cholesterol. When cellular cholesterol levels rise, the synthesis of HMGCoA reductase and the synthesis of the LDL receptor are suppressed. On the other hand, when the cells have an increased demand for cholesterol, the production of LDL receptors and of HMGCoA reductase is induced.[6] The suppression of HMGCoA reductase by cholesterol derived from LDL is accomplished by two mechanisms: (i) decreased transcription of the gene;[7] (ii) enhanced degradation of the protein.[8]

Although HMGCoA reductase is the rate limiting step of cholesterol biosynthesis, and is suppressed as a consequence of receptor mediated uptake of cholesterol in the form of LDL leading to a reduced amount of reductase protein, there is a secondary regulatory process involving squalene synthetase (Scheme 1). When squalene synthetase activity is suppressed, there is enough cellular farnesyl pyrophosphate to support the synthesis of the nonsterol products, ubiquinone and dolichol. This diversion of farnesyl pyrophosphate occurs because of the low K_m values of this substrate for the initial enzymes of the alternate pathways.[9]

7.4.2.2 Inhibition of HMGCoA Reductase

7.4.2.2.1 *Enzymology of HMGCoA reductase*

Structural studies of this protein and cloning of its cDNA were made possible through the development of UT-1 cells, a line of Chinese hamster ovary cells that was adapted to growth in the presence of compactin (**1**), a competitive inhibitor of the enzyme. The native HMGCoA reductase from Chinese hamster ovary cells appears to be a dimer of molecular weight 200 000. The hamster enzyme has a subunit molecular weight of 97 000 and consists of 887 amino acids. The protein is divided into two domains. One, the NH$_2$ terminus domain, is extremely hydrophobic. Using computer modelling schemes that are known to predict, with reasonable accuracy, the secondary structures of proteins based on the amino acid sequence, Liscum *et al.* have proposed that the NH$_2$ terminus domain criss-crosses the endoplasmic reticulum seven times. The second domain is the carboxyl terminus which projects into the cytoplasm.[10] A full length cDNA for the human enzyme has recently been isolated and this shares a high degree of homology with the hamster enzyme. The NH$_2$ terminus domain is highly conserved with only 7 substitutions out of 339 amino acids. The carboxyl terminus catalytic domain is also highly conserved with 22 substitutions out of 439 amino acids. The linker region between the two domains has diverged with 32 substitutions out of 110 amino acids.[11] The linker region is subject to facile proteolysis during a freeze–thaw process which leads to isolation of a soluble enzyme of molecular weight 52 000. This proteolytic fragment retains the catalytic activity of the intact protein.[12,13] Procedures for solubilizing HMGCoA reductase in the presence of inhibitors of proteolysis by processes independent of freeze-thawing have recently been developed although it is not known whether the protein prepared in this manner is the intact protein with a subunit molecular weight of 97 000.

(**1**) $R^1 = R^2 = H$
(**2**) $R^1 = Me, R_2 = H$
(**3**) $R^1 = R^2 = Me$

(**4**) $R = H$
(**5**) $R = Me$

The transformations catalyzed by HMGCoA reductase are shown in equation (1). It has been shown with the rat liver enzyme that HMGCoA binds first followed by one molecule of NADPH. Reduction then occurs to afford enzyme-bound hemithioacetal and NADP$^+$ leaves to be replaced by NADPH. Presumably, binding of the reduced cofactor causes the hemithioacetal to collapse to liberate the aldehyde, mevaldate, which is subsequently reduced to mevalonate. In dead-end

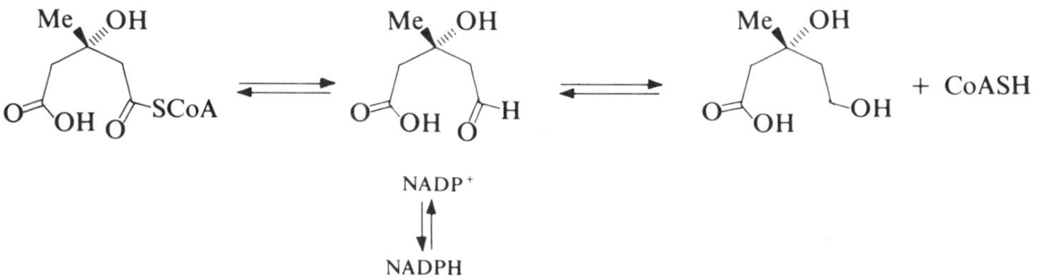

$$\text{(structure)} \rightleftharpoons \text{(structure)} \rightleftharpoons \text{(structure)} + \text{CoASH} \qquad (1)$$

NADP$^+$ ⇅ NADPH

inhibition studies, adenosine 2′-monophospho-5′-diphosphoribose was found to inhibit the enzyme competitively with respect to NADPH, and uncompetitively with respect to 3-hydroxy-3-methylglutaryl-CoA. These kinetic patterns are consistent with the reductase interacting first with 3-hydroxy-3-methylglutaryl-CoA to form a binary complex, which in turn forms a ternary complex with NADPH. In product inhibition studies, the product, mevalonate, is competitive *versus* HMGCoA and noncompetitive *versus* NADPH, suggesting that mevalonate is the last product to be released. The kinetic mechanism for the rat liver enzyme is summarized in Scheme 2.[14] In this scheme the symbols P and Q are used as the sequence of release of the second NADP$^+$ and CoASH is not known.

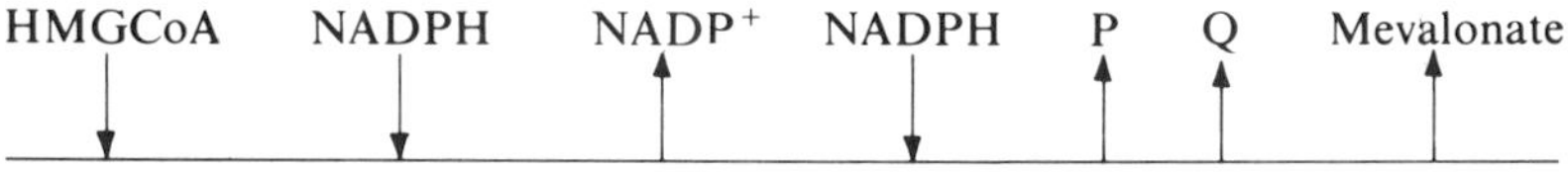

HMGCoA   NADPH   NADP$^+$ NADPH   P   Q   Mevalonate

Scheme 2

Recently, the reverse transformations have been studied with both the freeze–thaw preparation and with the enzyme prepared using a detergent in the presence of protease inhibitors. When [2-^{3}H] mevalonate is incubated with NADP$^+$ and CoASH in the presence of either enzyme preparation, HMGCoA is formed. This process is inhibited by mevinolin (2), a potent inhibitor of the enzymatic reaction in the forward direction. Similarly, when either preparation is incubated with mevaldate, NADP$^+$ and [^{3}H]CoASH, the isolated product is HMGCoA. This half reaction is also inhibited by mevinolin. CoASH plays a role in each half reaction: it is an absolute requirement in the reverse reaction which uses mevaldate as substrate, while in the forward reaction, also using mevaldate as substrate, it was observed to greatly stimulate the formation of mevalonate. In fact, formation of mevalonate in its absence was barely detectable. Analogs of CoASH were less effective than CoASH itself in stimulating this half reaction in the forward direction. Pantothenate and acetyl-CoA each supported the rate of reduction at 1% that of CoASH.[13]

The entire molecule of HMGCoA is not necessary for substrate recognition. The truncated HMGCoA (6) in which the nucleotide portion has been deleted is a substrate ($K_m = 12 \, \mu M$), while the analogous thioether (7) is an inhibitor ($K_i = 15 \, \mu M$).[15] The K_m for HMGCoA is $4.0 \, \mu M$.

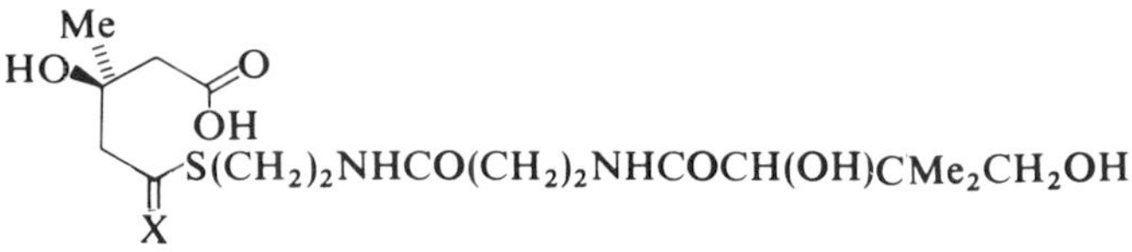

Me
HO
O
OH
S(CH$_2$)$_2$NHCO(CH$_2$)$_2$NHCOCH(OH)CMe$_2$CH$_2$OH
X

(6) X = O
(7) X = 2 H

7.4.2.2.2 *Inhibitors of HMGCoA reductase*

Very potent competitive inhibitors of HMGCoA reductase have become available from natural products isolation programs. Compactin (1) was isolated from cultures of *Penicillium citrinum*[16] and from *P. brevicopactum*,[17] while *Aspergillis terreus* yielded mevinolin (2) which is three times more potent.[18] The dihydro analogs of (1) and (2) [(4) and (5), respectively] have also been isolated[19,20] and have similar activity (Table 1).

Compactin has been shown to lower plasma cholesterol levels in dog, monkey and man.[21] Similarly, mevinolin lowers serum cholesterol in man.[22] At doses of 6.5 mg twice daily for four

Table 1   Inhibition of Rat Liver HMGCoA Reductase

Compound	K_i
(1)	1.4 nM (Alberts *et al.*, 1980)
(2)	0.6 nM (Alberts *et al.*, 1980)
(4)	3.7 nM (Lam *et al.*, 1981)
(5)	$IC_{50} = 2.7$ nM (Alberts-Schonberg *et al.*, 1981)
HMGCoA	$K_m = 4.0\,\mu M$ (Alberts *et al.*, 1980)

weeks, mevinolin lowered total cholesterol 28% and LDL cholesterol 34% in patients with heterozygous familial hypercholesterolemia.[23] Furthermore, on combination with colestipol, a bile acid sequestrant, LDL cholesterol levels decreased by 52%.[24] It is believed that this increased effect is due to up-regulation of the LDL receptor.[25] Mevinolin has been reported to be teratogenic in rats, but the effect is antagonized by the coadministration of mevalonate, suggesting that the effects of enzyme inhibition and not mevinolin itself are causative.[26]

The ester side chain of mevinolin can be removed and replaced by more bulky ones. This approach has led to synvinolin (3), which is reported to increase potency 2.5-fold.[27]

CS-514 (8) was found first as a minor urinary metabolite in dogs and later was obtained by microbial transformation of compactin. This compound is also a powerful competitive inhibitor of HMGCoA reductase with respect to HMGCoA ($K_i = 2.3$ nM), but unlike compactin and mevinolin it appears to be tissue selective in that a more profound inhibition of sterol biosynthesis occurs in liver and intestine than in other organs, including hormone producing ones. Using radiolabelled (8) it has been shown that it is taken up similarly to compactin in hepatocytes, but much less so in spleen cells and mouse L cells. Thus the less potent inhibition of cholesterol synthesis in cells of nonhepatic origin is ascribed to a lower uptake of the drug in these cells.[28]

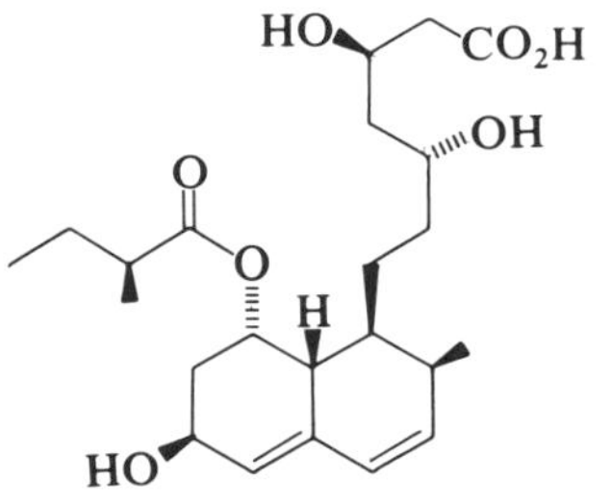

(8)

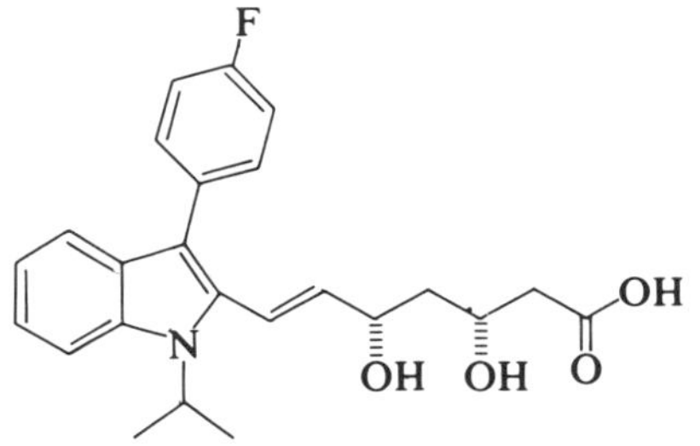

(10)

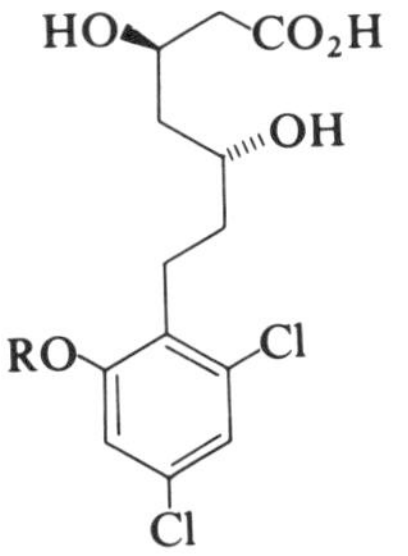

(9a) R = H
(9b) R = *p*-F-benzyl

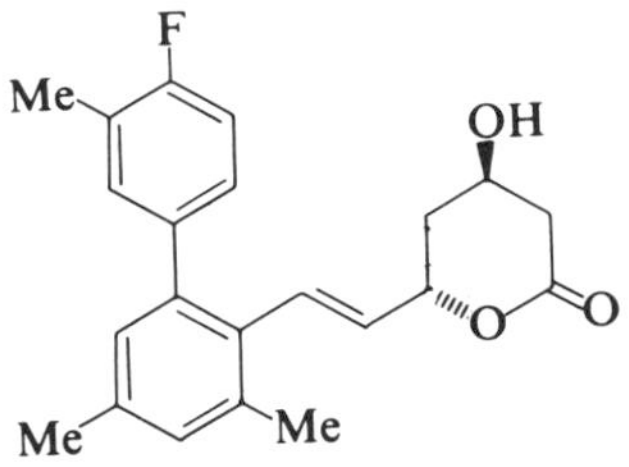

(9c)

Paradoxically, in the livers of animals[29] and in cultured cells[30] treated with these inhibitors, the activity of HMGCoA reductase is increased. This has been shown to result from increased enzyme synthesis and protection of the enzyme against degradation by the competitive inhibitor. The enzyme half-life is increased from 2 hours to 11 hours in the presence of mevinolin.[31] In an extensive structure–activity study, workers at Merck have demonstrated that the synthetically complex bicyclic portion of mevinolin (2) can be replaced by a simpler aromatic fragment as in (9). Compound (9a) has the same K_i as compactin on the yeast enzyme (0.2 nM)[32] but is considerably less active on the rat liver enzyme.[33] On the other hand, compound (9b) has half the inhibitory activity of compactin on the rat liver enzyme. For the rat liver enzyme, among the most potent synthetic compounds yet reported is the biphenyl (**9c**), which is similar in potency to mevinolin (**2**).[34]

Recently, using the yeast enzyme, Nakamura and Abeles[32] have demonstrated that HMGCoA and CoASH prevent the binding of compactin (**1**) to the enzyme. A consideration of the structure of (**1**) suggests that the bicyclic portion takes up the CoASH binding site. NADPH does not prevent the binding of (**1**). On the other hand, HMGCoA but not CoASH prevents the binding of (**9a**), suggesting that the upper portion of (**9a**) takes up the HMGCoA site. It seems that the aromatic portion of (**9a**) takes up yet another binding site which is different from that occupied by CoASH. It is not known whether this situation will pertain with the mammalian enzyme with the more potent (**9b**) and (**9c**).

There is an ambiguity in that the reported chirality at the C-5 position of (**9a**), the secondary alcohol, is the same as that of compactin (**1**) at this position. This does not appear to be reasonable if the bicyclic portion of (**1**) takes up the CoASH binding site while the phenethyl part of (**9a**) does not, unless the part of (**1**) which overlaps the CoASH site is remote from position C-5.

Other reported synthetic inhibitors of HMGCoA reductase include XU 62-320 (**10**) which was synthesized at Sandoz. Compound (**10**) has been described to be more potent than compactin and is in clinical trial.[35]

In September 1987 the F.D.A. approved the commercialization of mevinolin under the trade name of Mevacor.

7.4.2.3 Genomic Regulation of HMGCoA Reductase

As previously discussed, cells are known to regulate cholesterol levels by adjusting their biosynthetic capacity in response to the amount of cholesterol internalized by LDL receptor-mediated endocytosis. It has been demonstrated that it is the free cholesterol released from LDL by this process that suppresses HMGCoA reductase and thereby limits endogenous biosynthesis. There are two mechanisms proposed for the regulation of HMGCoA reductase (as well as ACAT and LDL receptors, see Introduction) at the genomic level by LDL-derived cholesterol: (i) LDL-derived cholesterol is converted to an oxysterol, or (ii) LDL-derived cholesterol indirectly mediates the synthesis of an endogenous regulator, again presumably an oxysterol.

The term oxysterol may be broadly defined as a derivative of cholesterol possessing a 3β-hydroxyl plus additional oxygenation at some remote position, for example at carbons 7, 9, 15, 20, 22, 25 or 32. The appeal for the involvement of a regulatory oxysterol arose from the observation that highly purified preparations of cholesterol when added to cell culture medium were very poor inhibitors of cholesterol synthesis. The initial observations to the contrary were evidently due to impurities, and the pure oxysterols were found to be much more potent than cholesterol in repressing HMGCoA reductase and hence cholesterol synthesis.[36] The role of oxysterols as mediators is further supported by the discovery of an oxysterol binding protein. The binding of oxysterols to this protein has been positively correlated with the repression of HMGCoA reductase.[37] It has been proposed, but as yet not demonstrated, that it is the binding of the oxysterol–protein complex to regulatory sequences on the HMGCoA reductase gene that prevents transcription, by analogy with the process known for steroid hormones.

7.4.2.3.1 Oxysterols as endogenous regulators

The development of the correlation of oxysterol binding affinity and suppression of HMGCoA reductase was aided by the preparation of a large number of oxygenated cholesterol analogs. The majority of these compounds were prepared by Schroepfer and coworkers.[38] Their search for potential endogenous regulators was centered around cholesterol derivatives having additional oxygenation at either C-15 or C-32 since compounds such as these were logical potential intermediates in the 14α-demethylation of lanosterol (see Scheme 3). Many of these compounds were found to be extremely potent inhibitors of HMGCoA reductase synthesis in cell culture and, additionally, some of these compounds blocked the metabolism of lanosterol. One outstanding compound that was uncovered was 5α-cholest-8(14)-en-3β-ol-15-one (**11**). This sterol was equipotent to 25-hydroxycholesterol (the most potent oxysterol known) and did not cause a buildup of lanosterol. In contrast to the results obtained with 25-hydroxycholesterol and 7-ketocholesterol, which lowered cholesterol in cell culture but not in whole animals,[39] (**11**) was an effective agent both in cell culture and whole animals.[40] The oral administration of (**11**) to Rhesus monkeys fed a moderate diet of cholesterol caused a 41% drop in serum cholesterol, a 61% drop in LDL cholesterol, and a 61% increase in HDL cholesterol. The changes produced by (11) in the serum cholesterol profile are all considered desirable for the treatment and prevention of atherosclerosis. The promise of this oxysterol, which is currently undergoing clinical trials, suggests that the regulatory control exerted by oxysterols on cholesterol metabolism can be exploited for the development of novel therapeutic agents to treat hypercholesterolemia. Progress towards this goal will be facilitated by elucidation of the mechanisms through which endogenously produced oxysterols operate.

(11)

The dilemma which remains, if the central role of oxysterols is accepted, is how do these agents arise from LDL-derived cholesterol? As previously stated, one proposal is the direct oxidation of cholesterol to an oxysterol. Researchers have attempted to correlate the potent inhibition of HMGCoA reductase biosynthesis by oxysterols such as 25-hydroxycholesterol and 7-ketocholesterol to a direct oxidative transformation of cholesterol and have yet to demonstrate convincingly that this mechanism is operative. This has led to the suggestion that cholesterol may be undergoing autoxidation to various oxysterols, a nonenzymatic process. Alternative hypotheses have been presented which invoke an indirect role for cholesterol in the control of endogenous oxysterol biosynthesis. These hypotheses are predicated on the ability of cholesterol to alter membrane structure, leading to an increase in the production of oxysterol intermediates that are part of the normal biosynthetic pathway. The enzymes most likely to be involved in the production of endogenous biosynthetic oxysterols are squalene oxidase and lanosterol 14α-demethylase.

7.4.2.3.2 *Lanosterol 14α-demethylase*

The oxidative removal of C-32 (*Chemical Abstracts* refers to this as C-30; however, the older nomenclature is most commonly used) of lanosterol (12) is catalyzed by a cytochrome P-450 dependent monooxygenase. The microsomal rat enzyme prefers the unsaturated side-chain of lanosterol, while dihydrolanosterol (13) is processed ten times more slowly.[41] The reaction requires both NADPH and O_2 and uses cytochrome P-450 reductase to mediate the electron shuttle. The reaction proceeds through two discrete oxygenated intermediates (14) and (15) to yield the diene (16) and formic acid (Scheme 3). The resemblance to aromatase (see Scheme 7) is striking. The reaction is effected by a single protein; the first two oxygenations appear to be typical cytochrome P-450 insertions into C—H bonds to generate sequentially an alcohol (14) and then a diol which collapses to aldehyde (15). The mechanism of the third oxygenation is unknown. In contrast to aromatase, the oxygenated intermediates appear to be more tightly bound than the initial substrate lanosterol (data on yeast enzyme).[42] Gibbons et al.[43] were the first group to report the suppression of HMGCoA reductase in cell culture in response to 32-oxygenated lanostenes, and they suggested, along with Watson and coworkers,[44] the possible role of these compounds as endogenous regulators of cholesterol biosynthesis. In the subsequent discussion the term 32-oxylanostenes will be used to refer to both (14) and (15) possessing either a saturated or unsaturated side chain. The importance of the Δ^{24}-alkene in mediating oxysterol synthesis and/or suppression of HMGCoA reductase is unknown.

Ketoconazole (17), a nonselective inhibitor of cytochrome P-450 enzymes including both mammalian and yeast sterol 14α-demethylase, has recently been demonstrated to lower serum cholesterol effectively in man.[45] Several lines of evidence suggest that this hypocholesterolemic effect may be a result of the inhibition of lanosterol demethylation. The suppression of HMGCoA reductase by (17) in rat intestinal epithelial and Chinese hamster ovary (CHO) cells is biphasic; low concentrations of (17) lower HMGCoA reductase activity while higher concentrations raise HMGCoA reductase levels to values slightly greater than controls.[46,47] Experiments in cultured normal human fibro-

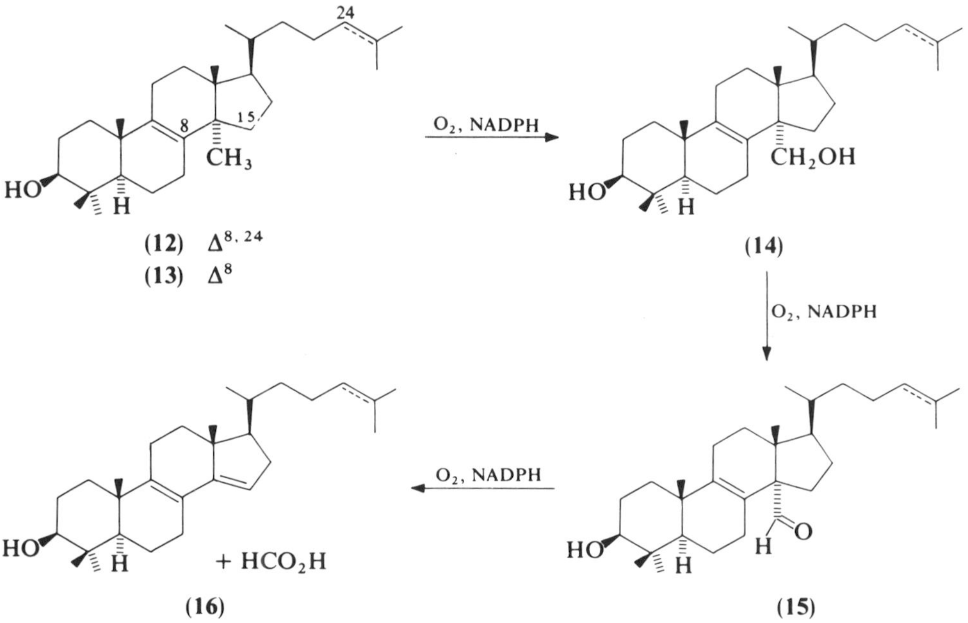

Scheme 3

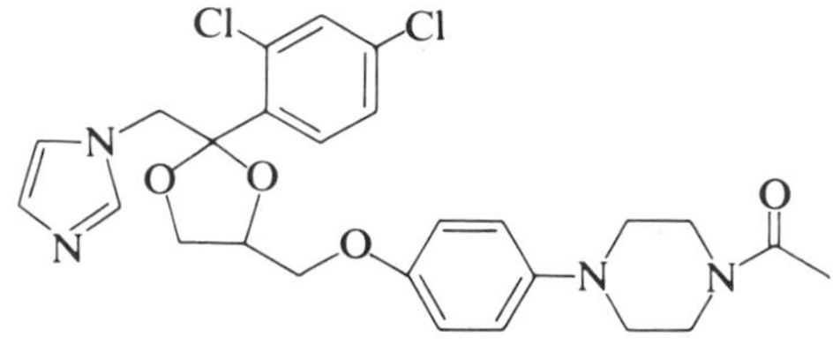

(17)

blasts have correlated the effect of increasing concentrations of (17) with a drop in cholesterol, a rise in lanosterol, and a suppression of HMGCoA reductase.[48] These data appear to demonstrate that the hypocholesterolemic effect arises from the inhibition of sterol 14α-demethylase.

Studies by Trzaskos and coworkers on the effect of (17) on the activity of sterol 14α-demethylase in rat liver hepatic microsomes offer a possible explanation for the above results.[41] These studies show that the levels of oxysterol metabolites (14) and (15) increase at low concentrations of (17) and fall at higher concentrations where the inhibition of sterol 14α-demethylase is complete. Their data suggest that all three substrates are in competition for processing by the enzyme and that as (12) builds up, increasing amounts of the intermediates are released from the enzyme. These workers have further established sterol 14α-demethylase as the site of action for the regulation of HMGCoA reductase by ketoconazole and miconazole in both CHO cells and primary rat hepatocytes by correlating the amount of HMGCoA reductase protein with the production of 32-oxylanostenes at varying drug concentrations.[47,49] The failure of these sterol 14α-demethylase inhibitors to affect HMGCoA reductase levels in 14α-demethylase deficient mutant CHO cells is strong evidence for the role of this enzyme in production of regulatory oxysterols. A comprehensive explanation for the hypocholesterolemic effect of ketoconazole can be proposed. The partial inhibition of sterol 14α-demethylase increases endogenous 32-oxylanostenes, resulting in the suppression of HMGCoA reductase synthesis and hence of *de novo* cholesterol biosynthesis. These data suggest that the inhibition of mammalian sterol 14α-demethylase might be therapeutically beneficial in the treatment of hypercholesterolemia (see Sections 7.4.3.1.6 and 7.4.4.2.2 for approaches to the inhibition of cytochrome P-450 enzymes). Alternatively, the incomplete processing of a lanosterol mimic by the enzyme might allow for the *in vivo* synthesis of oxysterols. Workers at DuPont have reported in preliminary form their attempts to utilize both approaches.[50a]

Evidence for the operation of this control mechanism under normal physiological conditions is limited. The accumulation of 32-oxylanostenes in cell culture upon incubation with high concentrations of mevalonate, conditions which suppress HMGCoA reductase, has been demonstrated.[51] The addition of mevalonate results in high lanosterol concentrations and thus the effect is similar to an incomplete inhibition of demethylation, *i.e.* elevated oxysterol production. However, the physiological stimulus is LDL-derived cholesterol. The observation that the addition of cholesterol to a microsomal preparation of 14α-demethylase increases the proportion of 32-oxylanostenes suggests how LDL-derived cholesterol might modulate oxysterol production.[47] Additional mechanisms must be operative since the effect of LDL on HMGCoA reductase does not depend only on the sterol products derived from mevalonate.[9] This suggest that the ability of high ketoconazole concentrations to block the suppression of HMGCoA reductase by LDL may not be solely due to a total inhibition of 32-oxylanostene production.[46]

7.4.2.3.3 Oxidosqualene cyclase

The recently proposed alternate pathway for the synthesis of endogenous regulatory oxysterols invokes a second epoxidation of squalene oxide to yield squalene dioxide. Squalene dioxide, which has no inherent oxysterol-like properties, is a viable substrate for all of the cholesterol biosynthetic enzymes (see Scheme 5 for normal pathway). Thus cyclization of squalene dioxide affords 24(*S*),25-epoxylanosterol; further processing then leads to 24(*S*),25-epoxycholesterol (Scheme 4). The presence of 24(*S*),25-epoxycholesterol in human liver,[52] and the correlation of its ability to suppress HMGCoA reductase with binding to the oxysterol-binding protein,[53] suggest that this pathway could have physiological relevance. The investigation of this pathway has been facilitated by the discovery of inhibitors of oxidosqualene cyclase. Both 4,4,10β-trimethyl-*trans*-decal-3β-ol (**18**) and U18666A (**19**) have been shown to block cholesterol synthesis in both broken and whole cell preparations with a concomitant rise in squalene mono- and di-oxide.[54,55] These compounds have been examined using a microsomal enzyme preparation from rat liver and found to be potent inhibitors. Similar to the results obtained with ketoconazole, the regulation of HMGCoA reductase by (**19**) is biphasic. The activity of HMGCoA reductase drops at low concentrations of inhibitor and at higher concentrations, at which oxidosqualene cyclase should be completely blocked, no effect or a slight enhancement of HMGCoA reductase is observed.[56] This finding is consistent with an increase in the production of squalene derived oxysterols [24(*S*),25-epoxycholesterol?] at low inhibitor concentrations, while at higher concentrations this process is totally blocked. Preliminary reports indicate that the levels of oxysterols derived from squalene dioxide increase in response to dietary cholesterol (rats) or LDL (CHO mutant cells, cholesterol auxotroph).[57] The importance of this pathway in the regulation of cholesterol under normal physiological conditions is as yet not known. Moreover, the potential for inhibitors of squalene cyclase to exert a therapeutically desirable effect is unknown.

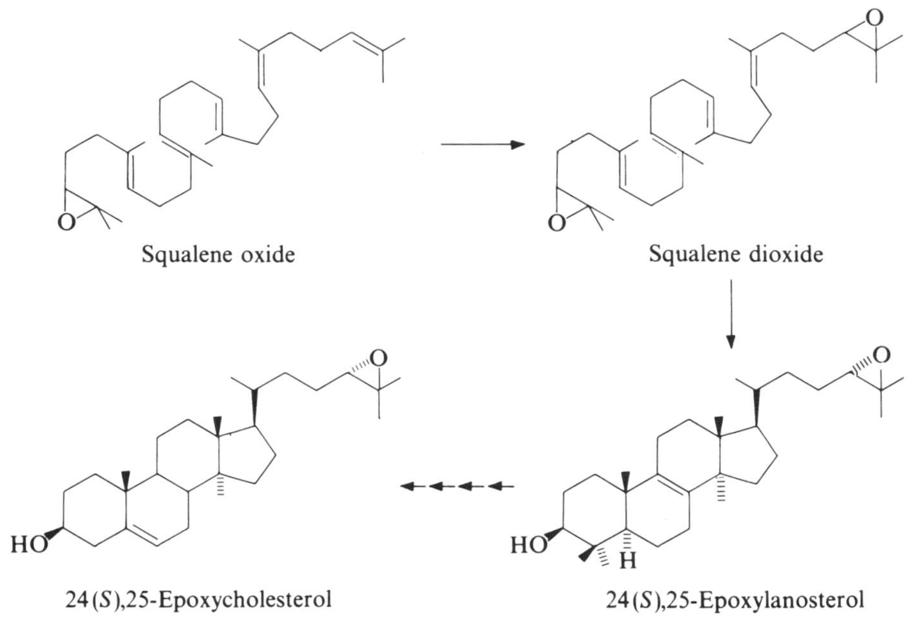

Squalene oxide   Squalene dioxide

24(*S*),25-Epoxycholesterol   24(*S*),25-Epoxylanosterol

Scheme 4

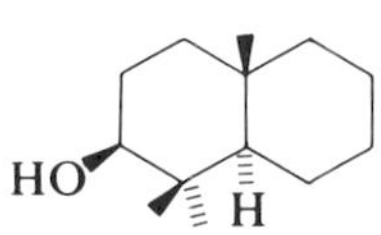

(18) 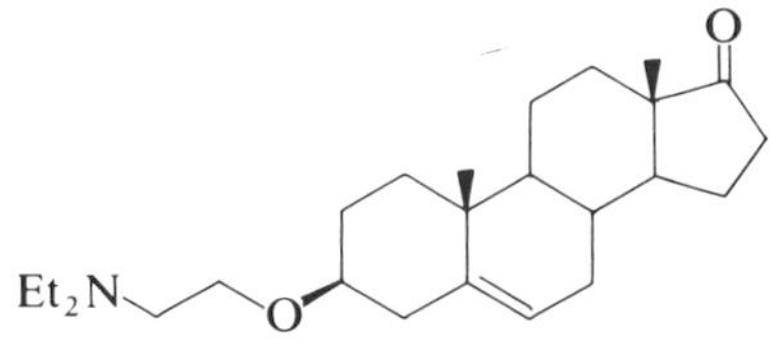   (19)

7.4.2.4 Inhibition of the Later Steps of Cholesterol Biosynthesis

7.4.2.4.1 *Early hypocholesterolemics*

Early attempts to develop agents to lower serum cholesterol resulted in the identification of compounds whose hypercholesterolemic mechanism of action derives from the inhibition of steps late in the biosynthetic pathway. AY-9944 (20), which inhibits $\Delta^{5,7}$-sterol 7-reductase, and triparanol (21), which inhibits sterol 24-reductase, serve to illustrate the dangers of the indiscriminate blockade of this pathway (Scheme 5). The failure of these agents was in part due to buildup of cholesterol precursors to toxic levels.[58] For example, the formation of irreversible cataracts in some of the patients treated with (21) is believed to be associated with the observed elevation of serum desmosterol concentrations.[59,60] Another interesting example is 20,25-diazacholesterol (22), which like (21) blocks the sterol 24-reductase, leading to an increase in desmosterol. Although (22) effects a net decrease in sterol levels, a net increase in the activity of HMGCoA reductase has been noted.[61] Both the suppression[62] and elevation[44] of HMGCoA reductase by (21) has been reported. In all of these examples the actual basis of the hypocholesterolemic effect appears to be more than can be explained by the inhibition of a single enzyme. Additional sites of enzyme inhibtion have been shown; for example, $\Delta^{8,14}$-sterol 14-reductase and oxidosqualene cyclase. These early failures illustrate that the successful intervention in this pathway must be based on a sound understanding of the overall regulation of the system.

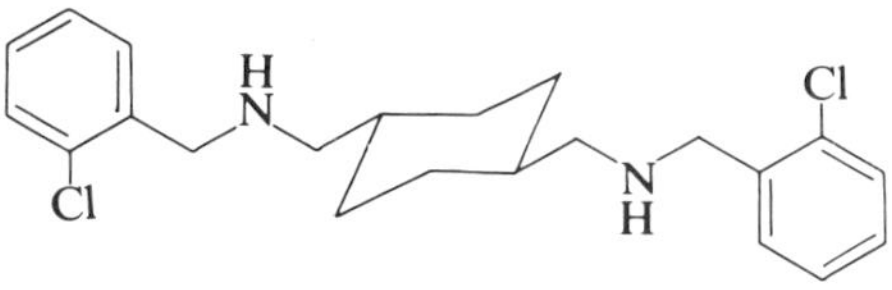

(20)

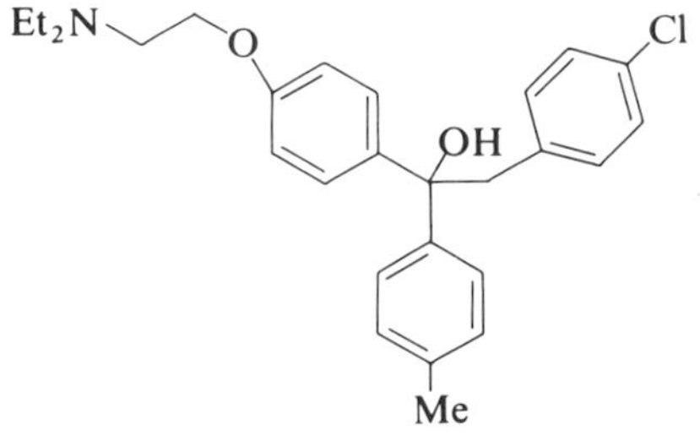

(21)

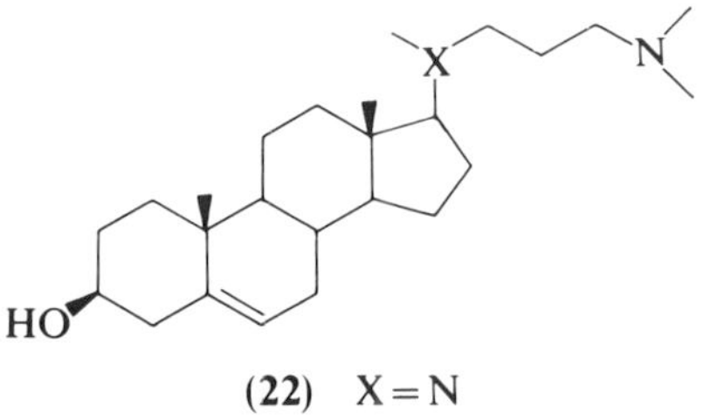

(22) X = N
(23) X = CH

7.4.2.4.2 *The mechanistic basis for the inhibition of enzymes that proceed through carbocationic intermediates*

The mechanistic basis for the inhibition of sterol 24-reductase by the side chain azacholesterols illustrates a general theme for the inhibition of enzymes that generate carbocationic intermediates in transforming substrate to product. Several examples of such enzymes are in the sterol biosynthetic pathway. These include oxidosqualene cyclase, $\Delta^8 \rightarrow \Delta^7$-sterol isomerase and $\Delta^{8,14}$-sterol 14-reductase, which are common to all sterol synthesizing organisms, and Δ^{24}-sterol *S*-methyltransferase which is found only in plants and fungi. The mechanistic studies of these enzymes are consistent

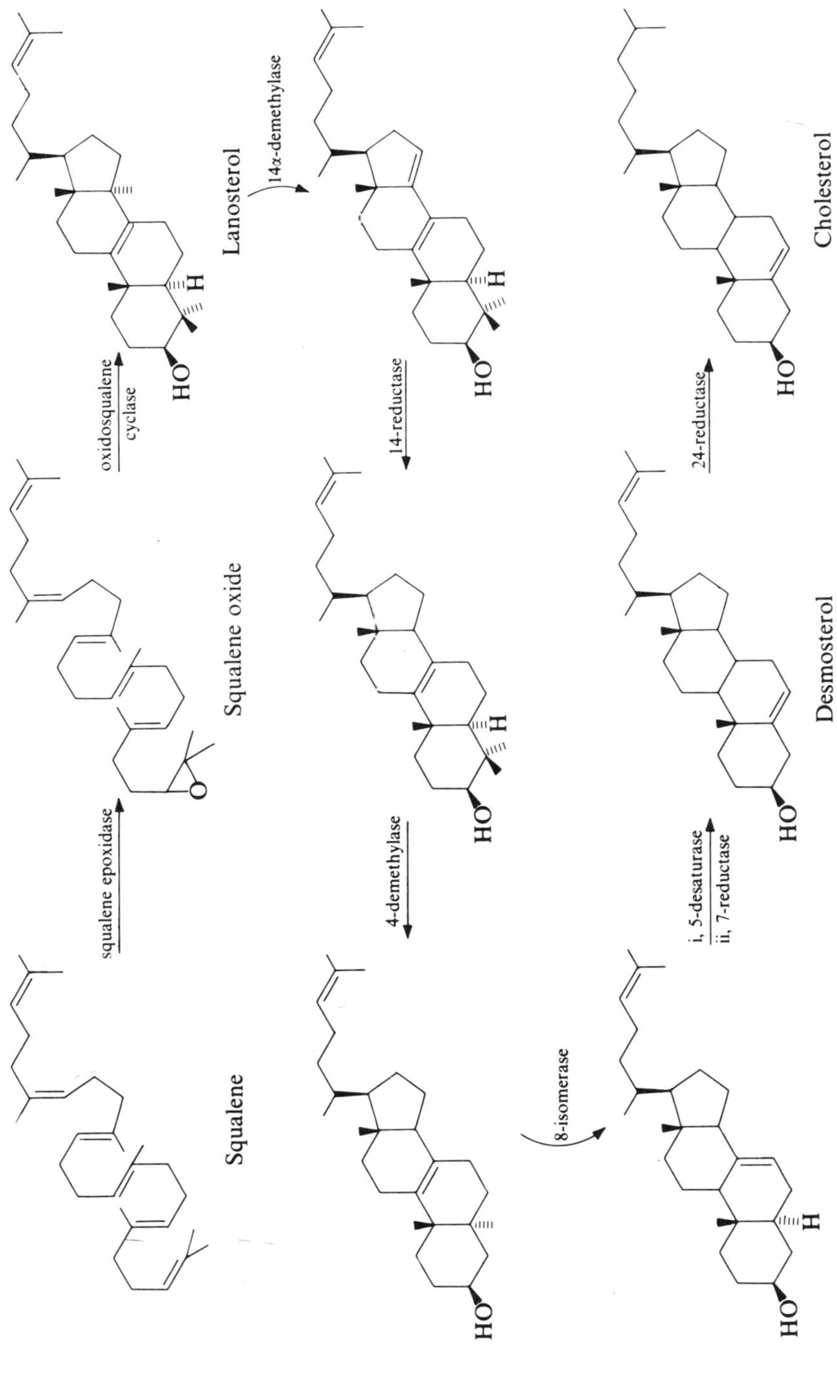

Squalene

Squalene oxide

Lanosterol

squalene epoxidase

oxidosqualene
cyclase

14α-demethylase

14-reductase

4-demethylase

8-isomerase

i, 5-desaturase
ii, 7-reductase

24-reductase

Desmosterol

Cholesterol

Scheme 5

with the initiation of catalysis by protonation of the double bond to yield the more stable carbocation followed by hydride addition from the nucleotide cofactor, in the case of the reductases, or rearrangement and deprotonation as is seen in oxidosqualene cyclase and $\Delta^8 \rightarrow \Delta^7$-sterol isomerase.[63] Compounds which closely mimic this high-energy cationic intermediate (which should be structurally and electronically similar to the actual transition state) should be potent inhibitors of the enzymatic reaction.[64] The structure of the expected intermediate carbocation for the sterol 24-reductase reaction is illustrated in equation (2). The replacement of the positively charged carbon in this structure with a protonated nitrogen, *i.e.* 25-azacholesterol (**23**), affords a similarly charged species which is an imperfect mimic of the transition state due to the change in geometry and the addition of a proton. Nevertheless, (**23**) at nanomolar concentrations inhibits the enzyme, a behavior that is also seen with a number of side chain azasterols including 20,25-diazacholesterol (**22**). The most important feature for potent inhibition is the position of the nitrogen in the side chain and little effect is seen with changes in the steroid nucleus. A logical extension of this reasoning may explain how (**20**) and (**21**) inhibit these steroidal reductases with varying specificity: they both possess a basic nitrogen on a lipophilic framework.

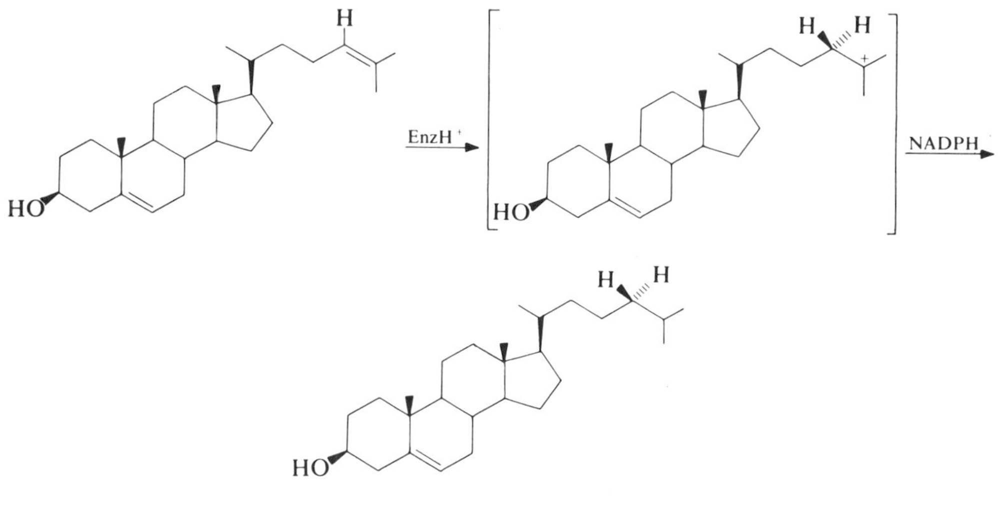

(2)

The inhibition of enzymes which proceed through a carbocationic high-energy intermediate (HEI) using the above outlined principle has recently been systematically investigated.[65] These researchers have examined HEI analogs for oxidosqualene cyclase, Δ^{24}-sterol *S*-methyltransferase and $\Delta^8 \rightarrow \Delta^7$-sterol isomerase. Their results reinforce the validity of the preceding proposal and therefore the suggested principles underlying the enzymatic reaction. The reaction catalyzed by oxidosqualene cyclase is illustrated in equation (3), along with the initial carbocationic 'intermediate'. Although the formation of the steroidal tetracyclic structure is thought to be nearly synchronous, it is probably not concerted and the enzyme must be able to direct some of the steps of the cyclization and the subsequent methyl and hydride migrations by stabilization of carbocationic HEIs.

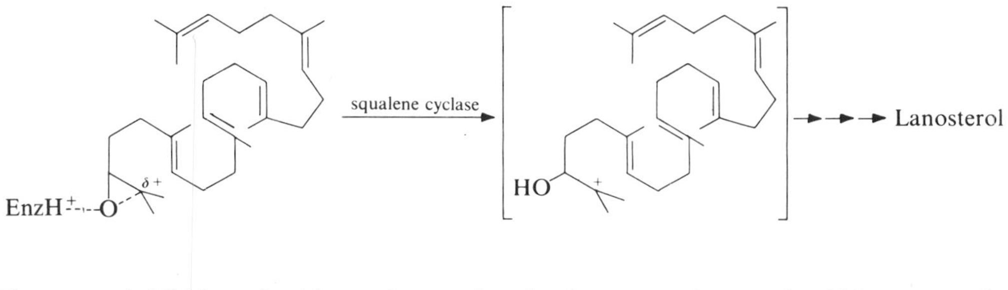

(3)

The potent inhibition of oxidosqualene cyclase by the azasqualene analog (**24**) suggests that it is a reasonable mimic of the initial carbocation formed from cleavage of the epoxide. Furthermore, the similar behavior of its quaternary ammonium salt (**25**) illustrates the importance of the positive charge.[66]

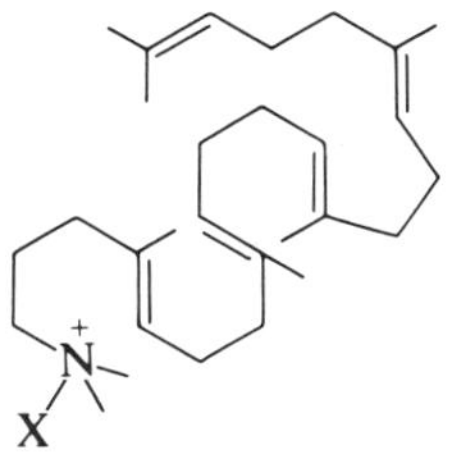

IC$_{50}$ Rat liver
Oxidosqualene cyclase
(μM)

(**24**)   X = H   8.8
(**25**)   X = Me   5.1

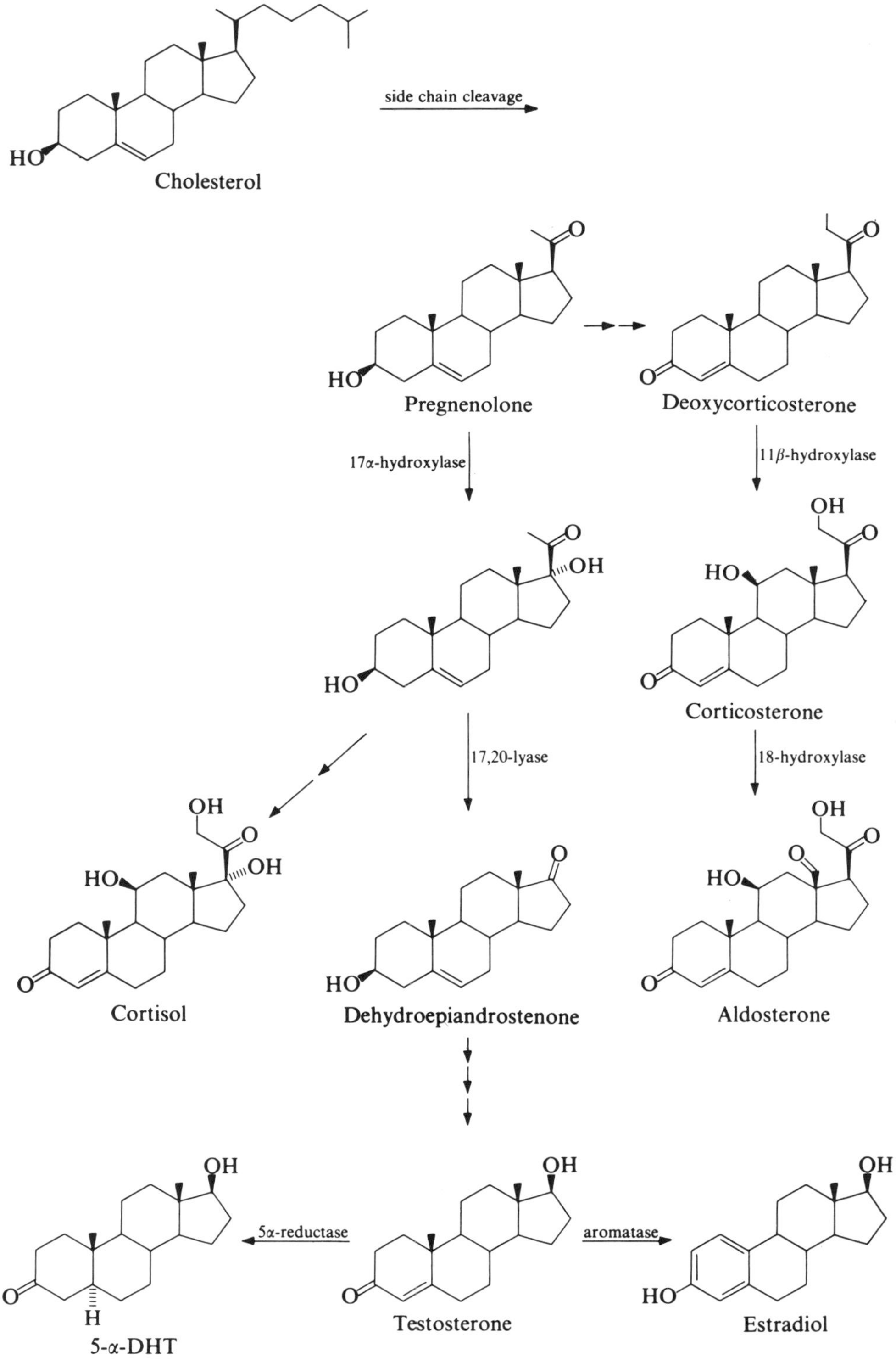

Cholesterol

side chain cleavage

Pregnenolone   Deoxycorticosterone

17α-hydroxylase   11β-hydroxylase

Corticosterone

17,20-lyase   18-hydroxylase

Cortisol   Dehydroepiandrostenone   Aldosterone

5α-reductase   aromatase

5-α-DHT   Testosterone   Estradiol

Scheme 6

7.4.3 INHIBITION OF STEROID HORMONE BIOSYNTHESIS

7.4.3.1 Inhibition of Cytochrome P-450 Isozymes Involved in Steroid Hormone Biosynthesis

Scheme 6 shows in abbreviated form the biosynthesis of the steroid hormones from cholesterol. These pathways involve a number of cytochrome P-450 enzymes. The rate determining step is the conversion of cholesterol to pregnenolone by the cytochrome P-450 enzyme, cholesterol side chain cleavage (cholesterol scc) enzyme. The formation of the glucocorticoid cortisol (hydrocortisone) from 11-deoxycortisol is catalyzed by another cytochrome P-450 enzyme, 11β-hydroxylase. Similarly, the principal mineralocorticoid, aldosterone, results from sequential 11β- and 18-hydroxylations of deoxycorticosterone. The biosynthesis of estrone and estradiol from androstenedione and testosterone respectively is also a cytochrome P-450 dependent process and involves the enzyme aromatase. As many physiologically important pathways involve cytochrome P-450 enzymes, nonspecific inhibitors of cytochrome P-450 dependent processes could be expected to have profound direct effects on steroid hormone biosynthesis.

The inhibition of 11β-hydroxylase which leads to a reduction of cortisol concentrations results in potent indirect effects. Secretion of adrenocorticotropic hormone (ACTH) by the pituitary gland is suppressed by cortisol, but not by cortisol precursors, so agents which inhibit 11β-hydroxylase cause a hypersecretion of ACTH which in turn activates steroid hormone biosynthesis *via* cholesterol scc enzyme in the adrenal cortex. As cholesterol scc is the rate limiting step in adrenal steroid biosynthesis, the overall flux of adrenal steroid synthesis is increased. Given sufficient time, ACTH also induces hyperplasia of the adrenal cortex.

Such steroid hormone ablation can be used to advantage, as will be described in the following section on the clinical utilities of nonspecific cytochrome P-450 inhibitors such as metyrapone (**26**), aminoglutethimide (**27**) and ketoconazole (**17**). The clinical consequences of administration of these agents also indicate that specific inhibitors of the individual isozymes may offer therapeutic advantages.

7.4.3.1.1 Metyrapone

Metyrapone (**26**) has been used clinically both in a diagnostic sense and as a therapeutic.[67] Its diagnostic utility stems from the direct inhibition of 11β-hydroxylase accompanied by the compensatory increase in ACTH secretion from the pituitary. Patients with hypercortisolism (Cushing's syndrome), due to a primary excess of ACTH secretion by the pituitary, respond to metyrapone with a large increase in urinary excretion of 11-deoxycortisol (17-hydroxycorticoids). Patients whose excessive secretion of cortisol is due to an adrenocortical tumor do not excrete increased amounts of 17-hydroxycorticoids on metyrapone administration.

(26) (27) (28)

Because it blocks the biosynthesis of cortisol, metyrapone is clinically effective in the treatment of patients with Cushing's syndrome due to adrenal tumor. Its use in treatment of Cushing's syndrome resulting from hypersecretion of ACTH from the pituitary is limited because such patients respond to cortisol inhibition with increasing secretion of ACTH from the pituitary, which in turn stimulates cortisol production.

The biosynthesis of aldosterone also involves 11β-hydroxylase and is inhibited by metyrapone. Metyrapone, however, is not a reliable treatment for correcting the electrolyte imbalance in patients with hyperaldosteronism because the blockade of 11β-hydroxylase leads to a buildup of the cortisol precursor, 11-deoxycorticosterone, which is itself a mineralocorticoid.

7.4.3.1.2 Aminoglutethimide

Aminoglutethimide (**27**), as a consequence of its ability to block 11β-hydroxylase, has been used successfully to decrease the hypersecretion of cortisol in patients with autonomously functioning adrenal tumors. Its main utility, however, results from aromatase inhibition leading to a decrease in estrogen production. Approximately one third of the cases of breast carcinoma are estrogen dependent and the withdrawal of estrogen following surgical ablation, or treatment with antiestrogens, results in tumor regression.[68] Aminoglutethimide, given in conjunction with replacement glucocorticoid therapy, produces objective disease regression in 32% of unselected patients with metastatic breast carcinoma. When such patients are preselected as having tumors which are estrogen receptor positive, this response rises to 52%. Replacement hydrocortisone must be administered with aminoglutethimide in order to prevent hypersecretion of ACTH. This hypersecretion is a consequence of 11β-hydroxylase inhibition.[69] Aminoglutethimide also inhibits other cytochrome P-450 isozymes. Santen and Misbin[69] have estimated inhibitory concentrations for a number of cytochrome P-450 hydroxylations: aromatase ($0.3\,\mu M$), 18-hydroxylase ($0.2\,\mu M$), cholesterol scc ($3.5\,\mu M$), and 11β-hydroxylase ($120\,\mu M$). Structural modifications of the aminoglutethimide structure have recently led to analogs possessing greater selectivity towards aromatase.[70,71]

7.4.3.1.3 Ketoconazole

Ketoconazole (17) was initially introduced as an orally active antifungal agent (see Section 7.4.4.1). The compound functions as an antifungal by inhibiting the cytochrome P-450 dependent lanosterol 14α-demethylase of fungal sterol biosynthesis. The lack of specificity of ketoconazole was first suggested by the appearance of gynecomastia in patients receiving high doses of the drug. Studies demonstrated that a profound transient drop in testosterone levels occurred upon administration of a single dose of ketoconazole and that chronic treatment resulted in a sustained depression of testosterone levels. Further investigation revealed ketoconazole to inhibit a variety of cytochrome P-450 processes including 17,20-lyase, cholesterol scc enzyme, 11β-hydroxylase, aromatase and mammalian lanosterol 14α-demethylase. The effect of high dose ketoconazole on androgen levels is similar to that of orchiectomy as a result of the inhibition of androgen biosynthesis in both the testis and adrenal. This suggested a utility in the treatment of prostatic cancer, a disease which is known to be androgen dependent. The initial clinical results suggest that ketoconazole may be as effective as other hormonal or surgical treatments in slowing the progression of the tumor. The suppression of corticoid sterol biosynthesis did not in general prove problematic, presumably because the inhibition of 11β-hydroxylase is overridden by an increase in the release of ACTH.[72]

Nonspecific inhibitors of cytochrome P-450 isozymes clearly have clinical utility. More specific biological effects might be attainable by increasing the enzyme selectivity of inhibitors of cytochrome P-450 isozymes. The design of such inhibitors should be facilitated by an understanding of the mechanisms by which cytochrome P-450 functions, as well as an appreciation of the mechanisms of inhibition.

7.4.3.1.4 Mechanism of hydroxylation by cytochrome P-450

The transformations mediated by cytochrome P-450 enzymes involve enzyme complexes where NADPH transfers two electron equivalents to a carrier which acts as a redox two-electron/one-electron switch and in turn transfers electrons, one at a time, to the terminal cytochrome P-450. The catalytic sequence for mitochondrial cytochrome P-450 enzymes such as cholesterol scc enzyme and 11β-hydroxylase involve substrate binding followed by one-electron transfer from adrenodoxin, an iron–sulfur protein. Dioxygen then binds to the reduced cytochrome P-450 to form a dioxygen complex, the cleavage of which is initiated by transfer of a second electron from adrenoredoxin. The cleavage of the dioxygen complex is not understood, but the likely catalytic species is thought to have a single oxygen atom attached to the heme iron. Such a species can be represented by $[FeO]^{3+}$ in which iron is formally in the $+5$ oxidation state. P-450 isozymes from the endoplasmic reticulum differ from their mitochondrial counterparts as they are supported by cytochrome P-450 reductase, a flavoprotein which reduces the cytochrome P-450/substrate complex.

The transfer of the oxygen atom to substrate is probably nonconcerted. This conclusion results from the observation of stereochemical scrambling in the hydroxylation of exodeuterated norbornane with rabbit liver microsomes, which yields both the *exo* and *endo* alcohols with varying numbers of deuterium atoms.[73] Scrambling has also been observed in the hydroxylation of camphor by P-450$_{cam}$. The enzyme removes both the *endo* and *exo* hydrogen, but delivers the hydroxyl to the *exo* position.[74] Using a prochiral substrate, phenylethane, and a purified P-450$_{LM2}$, it has been demonstrated that 23–40% of the hydroxylation events are associated with products bearing deuterium in the opposite configuration to the starting specifically deuterated substrate. These events require the intermediacy of a discrete tricoordinate carbon intermediate which can rotate before radical recombination.[75] Kinetic isotope effect data also argue that the C—H bond is broken in the transition state.[76] Evidence also exists that the epoxidation of alkenes and the oxidation of terminal alkynes occur *via* nonconcerted mechanisms. Thus while terminal alkenes are epoxidized with retention of stereochemistry, heme alkylation also occurs. As the epoxide products do not alkylate the heme, there must be divergence between epoxidation and heme alkylation prior to epoxide formation.[77] The P-450 catalyzed oxidation of terminal alkynes is accompanied by quantitative rearrangement of the alkynic hydrogen to the vicinal carbon, as well as an inactivation process involving alkylation of the prosthetic heme. As in the case with the epoxidation of terminal alkenes, the oxidation of terminal alkynes is also occurring *via* a sequence with divergence at a point prior to transfer of oxygen to the π-bond, consistent with a nonconcerted mechanism.[78]

7.4.3.1.5 Mechanism of inhibition by reversible inhibitors

Inhibitors such as metyrapone, aminoglutethimide and ketoconazole bind reversibly to cytochrome P-450 as a consequence of their ability to coordinate to the prosthetic heme iron, as well as to a lipophilic pocket near the active site. Such binding results in type II difference spectra as the enzyme is shifted from high- to low-spin form. This results in a change in the redox potential of the enzyme so that its reduction by cytochrome P-450 reductase or by adrenodoxin is more difficult. Generally, these inhibitors are aliphatic or aromatic nitrogen containing compounds. While compounds which inhibit by this mechanism are nonspecific inhibitors of all cytochrome P-450 enzymes, recent structural changes have led to more selective compounds. For example, when the aniline ring of aminoglutethimide (**27**) is replaced by pyridine as in (**28**), a more selective aromatase inhibitor is obtained.[79]

High selectivity can also be achieved by incorporating the iron ligand functionality into the structure of the substrate itself. In this way, potent and selective inhibitors of cholesterol scc enzyme have been obtained as illustrated by (22*R*)-22-aminocholesterol (**29**).[80] Compound (**29**) converts cholesterol scc enzyme from predominately high-spin to the low-spin form and has an apparent K_i of 30 nM. Both EPR and UV–VIS spectroscopic studies are consistent with binding of (**29**) in the cholesterol binding site with additional binding of the amino group to the iron(III) atom.

(29)

7.4.3.1.6 Mechanism of inhibition by irreversible inhibitors

The approach of 'suicide' enzyme inhibition[81] has lent itself successfully to the design of specific, irreversible inactivators of cytochrome P-450 enzymes. An inhibitor is described as a 'suicide' inhibitor if the inhibition process is dependent on enzyme turnover. Such inhibitors are therefore substrates for the target enzyme and on catalysis form an overtly active species which combines covalently either with the prosthetic group or with the protein. Often, time dependency of the inactivation process is taken as a preliminary indication of suicide inhibition. Complete characterization, of course, requires evidence of enzyme turnover. The specificity of this class of inhibitors resides in the requirement that they be substrates for the target enzyme.

As mentioned earlier, the epoxidation of terminal alkenes is accompanied by heme alkylation. In the simplest example, administration of ethylene to phenobarbital pretreated rats leads to accumulation of an abnormal porphyrin in the liver. This pigment has been isolated and its structure determined as one of the four possible isomers of *N*-(hydroxyethyl)protoporphyrin IX (**30**). This finding supports the notion that the prosthetic group of cytochrome P-450 is alkylated during transfer of the catalytically activated oxygen to the π-bond of the ethylene.[82] Later studies using [18]O demonstrated that the oxygen atom in the isolated *N*-(hydroxyethyl)protoporphyrin IX derives from molecular oxygen.[83] Vinyl fluoride, vinyl bromide and fluroxene (2,2,2-trifluoroethyl vinyl ether) also alkylate the heme of cytochrome P-450. The same alkylated heme derivative, namely *N*-(2-oxoethyl)protoporphyrin IX (**31**), has been isolated in all cases.

(**30**)   R = CH$_2$OH
(**31**)   R = CHO

The formation of this adduct requires the introduction of oxygen at the heteroatom substituted terminus, followed by addition of the unsubstituted terminus at the heme nitrogen. This regiochemistry of addition is most consistent with a radical intermediate and suggests the mechanism shown in equation (4) for the inactivation of cytochrome P-450 by alkenes.

$$(4)$$

Alkynic compounds also inactivate cytochrome P-450. Processing of acetylene itself leads to the formation of *N*-(2-oxoethyl)protoporphyrin IX (**31**).[84] By analogy with the scheme proposed by Ortiz de Montellano for the inactivation of cytochrome P-450 by alkenes, it was also suggested that inactivation by alkynes proceeds *via* an oxyvinyl radical, as depicted in equation (5). The alkyne function has proven particularly useful in the irreversible inhibition of cytochrome P-450 enzymes as its incorporation into the appropriate substrate structures leads to potent, selective inhibitors of cytochrome P-450 isozymes.

$$(5)$$

(i) Aromatase

The encouraging results obtained with aminoglutethimide in breast cancer suggest that a more specific, more potent inhibitor of aromatase would offer advantages. Aromatase catalyzes the removal of the 19-methyl group from the androgen, androstenedione, to yield the estrogen, estrone. It is accepted that of the three reducing equivalents and three equivalents of oxygen necessary for the complete transformation, two of these are involved in sequential hydroxylations at C-19. On further transformation, the 19-aldehyde collapses to products, estrone and formic acid (Scheme 7). The precise mechanism of this final transformation is controversial.[85–87].

The incorporation of an alkyne group at C-19 in the substrate androstenedione as in compound (**32**) has led to a potent irreversible inhibitor of aromatase,[88–90] which has no inhibitory effect *in vitro* on 11β-hydroxylase, 18-hydroxylase and cholesterol scc enzyme.[91] The alkyne (**32**) has been

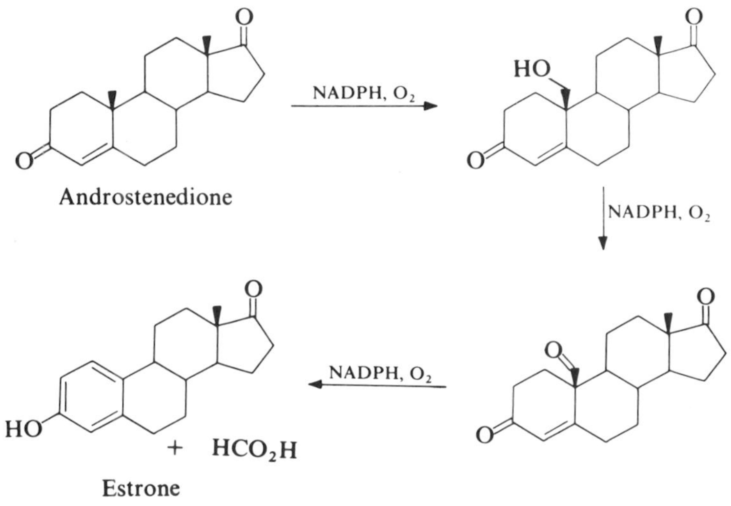

Androstenedione

Estrone + HCO$_2$H

Scheme 7

demonstrated to inhibit aromatase activity *in vivo*.[92] Other mechanism based inhibitors designed for aromatase include the allene (**33**),[89] and the 19,19-difluoro steroid (**34**).[93] Presumably the allene (**33**) inactivates aromatase as a consequence of oxygen addition to generate an allylic radical which could be intercepted by either the heme prosthetic group, or by the protein. It was proposed that the difluoro steroid (**34**) would be hydroxylated to generate, after loss of HF, an acyl fluoride which could acylate some nucleophilic residue in the active site. The increased interest in irreversible, mechanism-based inhibitors of aromatase has led to a reexamination of the kinetics of inhibition of previously known competitive inhibitors. It now appears that the presence of a double bond at C-1 leads to suicide inhibition. Thus, incorporation of this feature into the substrate, androstenedione, to afford androsta-1,3-diene-3,17-dione (**35**) gives rise to a time-dependent inactivation process and irreversible inhibition of the enzyme, the K_i for (**35**) being 0.32 μM.[94] Similarly, the incorporation of a double bond at C-6 into androstenedione as in compound (**36**) affords an irreversible inhibitor with a K_i of 0.18 μM.[95] The likely mechanism of inactivation is not apparent.

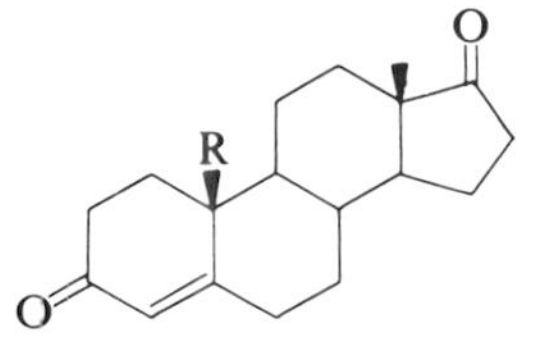

(**32**) R = CH$_2$—C≡C—H
(**33**) R = CH=C=CH$_2$
(**34**) R = CHF$_2$
(**38**) R = CH$_2$SH
(**39**) R = CH$_2$SCH$_3$

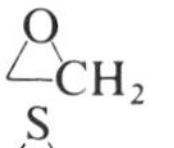

(**40**) R = (epoxide with CH$_2$)

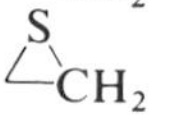

(**41**) R = (thiirane with CH$_2$)

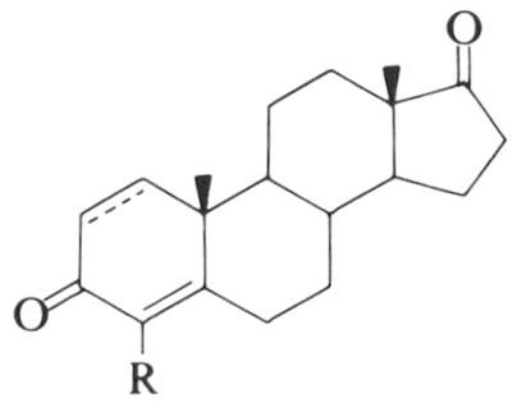

(**35**) R = H, $\Delta^{1,4}$
(**37**) R = OH, Δ^4

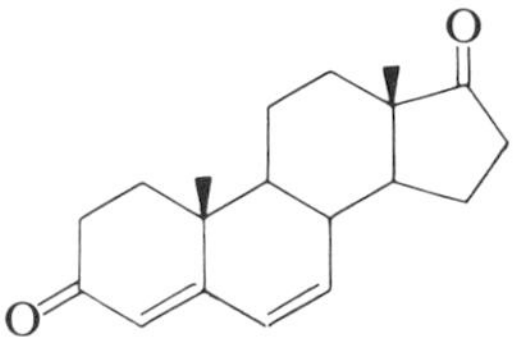

(**36**)

4-Hydroxyandrostenedione (**37**) is also a suicide inhibitor of aromatase[96] and is presently undergoing clinical trials in breast cancer patients, where complete or partial tumor regression has been observed after 4 months in 30% of patients. These patients had relapsed from previous therapy.[97] The mechanism by which (**37**) inhibits aromatase is not apparent, although an attempt has been made to rationalize the observed NADPH-dependent, time-dependent irreversible inhibition.[98]

19-Mercaptoandrostenedione (**38**) has also been described as a suicide inhibitor of aromatase, the inactivation process being time and NADPH dependent. It was suggested that the thiol group may be oxidized by aromatase to afford an electrophilic sulfenic acid.[99]

Potent inhibition of aromatase can also be realized in a competitive sense. For example, 19-methylthioandrostenedione (**39**) was shown to inhibit the enzyme *via* coordination of the steroidal sulfur atom to the heme iron. The K_i was found to be 1 nM.[100] Similarly, 10-oxirane (**40**) and 10-thiirane (**41**) substituted androstenediones have high affinity for aromatase. The (19*R*) isomers were more potent than the (19*S*) isomers, and (**40**) and (**41**) exhibited type II difference spectra, consistent with their binding to the heme iron.[101]

Another development is the identification of a site of bulk tolerance and lipophilicity at the aromatase active site which is exploited in the design of 7α-(4'-aminophenylthio)androst-4-ene-3,17-dione (**42**) which is a competitive inhibitor with a K_i of 0.018 μM.[102] The incorporation of a double bond at C-1 into (**42**), as in compound (**43**), once again leads to a suicide inhibitor. Compound (**43**) is among the most potent inhibitors of this type reported for aromatase.[103]

(**42**) Δ⁴
(**43**) Δ¹·⁴

(ii) Other Cytochrome P-450 Isozymes Inactivated by Alkynic Substrate Analogs

The concept of inactivating cytochrome P-450 enzymes by substrate analogs bearing alkyne functionalities has been extrapolated to steroid 18-hydroxylase, which is inactivated in a time-dependent manner by the 18-ethynyl steroid (**44**).[104] The 11β-hydroxy group was found to be necessary for time-dependent inhibition. Similarly, a series of alkynic steroids, exemplified by structure (**45**), has been reported to inactivate cholesterol scc enzyme irreversibly. These compounds exhibited type II difference spectra with purified cholesterol scc enzyme. Incubation in the presence of electron donors and oxygen led to a time-dependent decrease of absorption in the Soret region which was found to be dependent on adrenodoxin, adrenodoxin reductase, NADPH and oxygen. These results are consistent with a direct association with turnover of the alkynic steroids.[105]

(**44**)   (**45**)

A conceptually different approach has been developed for the inactivation of cytochrome P-450 enzymes which relies on the fragmentation of silicon species in the presence of a β-carbocation. The trimethylsilyl steroid (**46**) has been found to be a time-dependent inhibitor of cholesterol scc enzyme. As with the alkynic steroid (**44**), the addition of (**46**) to the purified enzyme produced a type II difference spectrum, indicating that the hydroxyl function binds to the heme iron. In the presence of electron donors and oxygen, (**46**) led to time-dependent inhibition. The fragmentation process illustrated was proposed to account for the time dependency.[106]

(iii) Spironolactone and Cytochrome P-450

Aldosterone is the most potent, endogenous mineralocorticoid. This class of steroids acts by augmenting the readsorption of sodium and chloride, and increasing the excretion of potassium.

(46)

Excess mineralocorticoid activity therefore leads to an expansion of extracellular volume and hypokalemia. Spironolactone (47) is an aldosterone antagonist which competes with aldosterone for the mineralocorticoid receptor. The administration of spironolactone, however, is known to cause a depletion of microsomal cytochrome P-450 in tissues which have a high activity of steroid 17α-hydroxylase. The sulfur atom at position 7α is required for the loss of adrenal and testicular cytochrome P-450. Deacetylated spironolactone (48) is also able to destroy cytochrome P-450. *In vitro* studies demonstrate that the destruction of microsomal cytochrome P-450 is NADPH dependent.[107] The mechanism of inactivation would therefore seem to be of the suicide type. The destruction of adrenal cytochrome P-450 by (48) was concurrent with a decrease in the activity of steroid 17α-hydroxylase. Studies with radiolabelled (48) suggest that during the loss of cytochrome P-450 by thio steroids, the sulfur atom of the thio group, after activation by the cytochrome P-450, is eliminated from the steroid moiety and binds covalently to the cytochrome P-450 apoenzymes. This leads to loss of steroid hydroxylase activity.[108]

(47) R = COMe
(48) R = H

7.4.3.2 Selective Inhibition of Androgen Biosynthesis; Steroid 5α-Reductase

Considerable evidence has now accrued that, in some tissues, testosterone is the active androgen, while in others it functions as a prohormone and its metabolite, 5α-dihydrotestosterone (DHT), is the actual hormone. The prostate in particular appears to be under 5α-dihydrotestosterone stimulation while other androgenic functions (*e.g.* muscle development, spermatogenesis) are supported by testosterone. These conclusions are substantially supported by studies of males who are genetically deficient in steroid 5α-reductase. Post-puberty the subjects have been found to have tiny prostates which can be induced to enlarge on DHT administration.[109]

DHT is also viewed as an important androgenic stimulus for prostatic cancer cells.[110] Current treatments for prostatic cancer include blockade of testosterone biosynthesis and androgen receptor antagonists. Diethylstilbestrol (DES), an estrogen agonist, inhibits secretion of pituitary LH, a hormone which stimulates gonadal steroidogenesis. DES treatment reduces testosterone production to castrate levels, but is associated with side effects such as cardiovascular complications and feminization. Under chronic conditions, LHRH agonists disrupt regulatory feedback loops and inhibit pituitary release of LH and FSH. These agents are also less than ideal owing to initial 'tumor flare' which occurs before feedback suppression of pituitary hormone release occurs. Specific inhibitors of steroid 5α-reductase may offer an alternative not associated with such side effects; the functions of testosterone should be maintained, while the trophic support to the prostate provided by DHT should be removed. The aza steroid MK-906 (49) has been reported to be a potent inhibitor of steroid 5α-reductase, and is undergoing clinical trial in patients with benign prostatic hypertrophy.[111] Presumably, (49) binds tightly to the enzyme because it mimics the putative enzyme bound enolate intermediate in the conversion of testosterone to DHT (equation 6).

(49)

$$E^+ \quad O \quad \text{NADPH} \qquad EO \qquad \qquad O \qquad \qquad (6)$$

Testosterone   Dihydrotestosterone

7.4.4 INHIBITION OF STEROL METABOLISM IN NONMAMMALIAN ORGANISMS

The major therapeutic application for the inhibition of sterol metabolism in nonmammalian organisms is in the area of antifungal chemotherapy. Since the goal is to kill the pathogenic organism, any means of interrupting this pathway is allowable in principle. Therefore the potential opportunities for the interruption of sterol biosynthesis in nonmammalian systems are not limited by the same considerations as those applied to man. Both man and the fungi, which are also eukaryotes, synthesize lanosterol and further process it to a structural sterol. In fungi, ergosterol serves a structural role similar to that played by cholesterol in mammals. Because of the close structural similarity of the end products (cholesterol and ergosterol), most of the enzymes in the biosynthetic pathway leading to ergosterol in fungi have mammalian counterparts (compare Schemes 5 and 8). The key to exploiting fungal sterol biosynthesis as a target for antifungal chemotherapy is to inhibit selectively the fungal enzyme; a goal that, as we will see, has not proven trivial.

Several of the pathogenic parasites which cause human disease also synthesize sterols, and reports suggesting that sterol biosynthesis inhibitors may be useful in the treatment of parasitic diseases have appeared. The major sterol of several *Leishmania* species is ergosta-5,7,24(28)-trien-3β-ol, which is apparently biosynthesized by a route similar to that producing ergosterol in fungi. Likewise, *Trypanosoma cruzi* epimastigotes contain an ergosterol-like sterol. Ketoconazole (17) inhibits the multiplication of both *L. mexicana mexicana* promastigotes and *T. cruzi* epimastigotes. This inhibition is accompanied by alterations in the pattern of sterol biosynthesis with an accumulation of 14α-methylsterols (see Section 7.4.4.1.1).[112,113] Ketoconazole is also effective against cutaneous and mucotaneous leshmaniasis in man. These observations suggest that the inhibition of sterol biosynthesis may offer opportunities for novel therapeutic agents in the treatment of parasitic diseases.

7.4.4.1 Inhibition of Lanosterol 14α-Demethylase

The major thrust of medicinal chemists in the area of antifungal chemotherapy for the last two decades has been centered on a group of heterocyclic compounds generally referred to as the 'azoles'. The initial members of this class were clotrimazole (50) (Bayer AG, 1969) and miconazole (51) (Janssen Pharmaceutica, 1969). These agents have been used effectively as topical agents in the treatment of a variety of dermatomycoses including vaginal candidiasis. The systemic administration of these compounds has proven effective in the treatment of some deep-seated mycosis, but these agents are not ideal due to their poor bioavailability. Moreover, the initial azoles showed numerous side effects including gastric irritation, vomiting and diarrhoea plus more serious neurological and

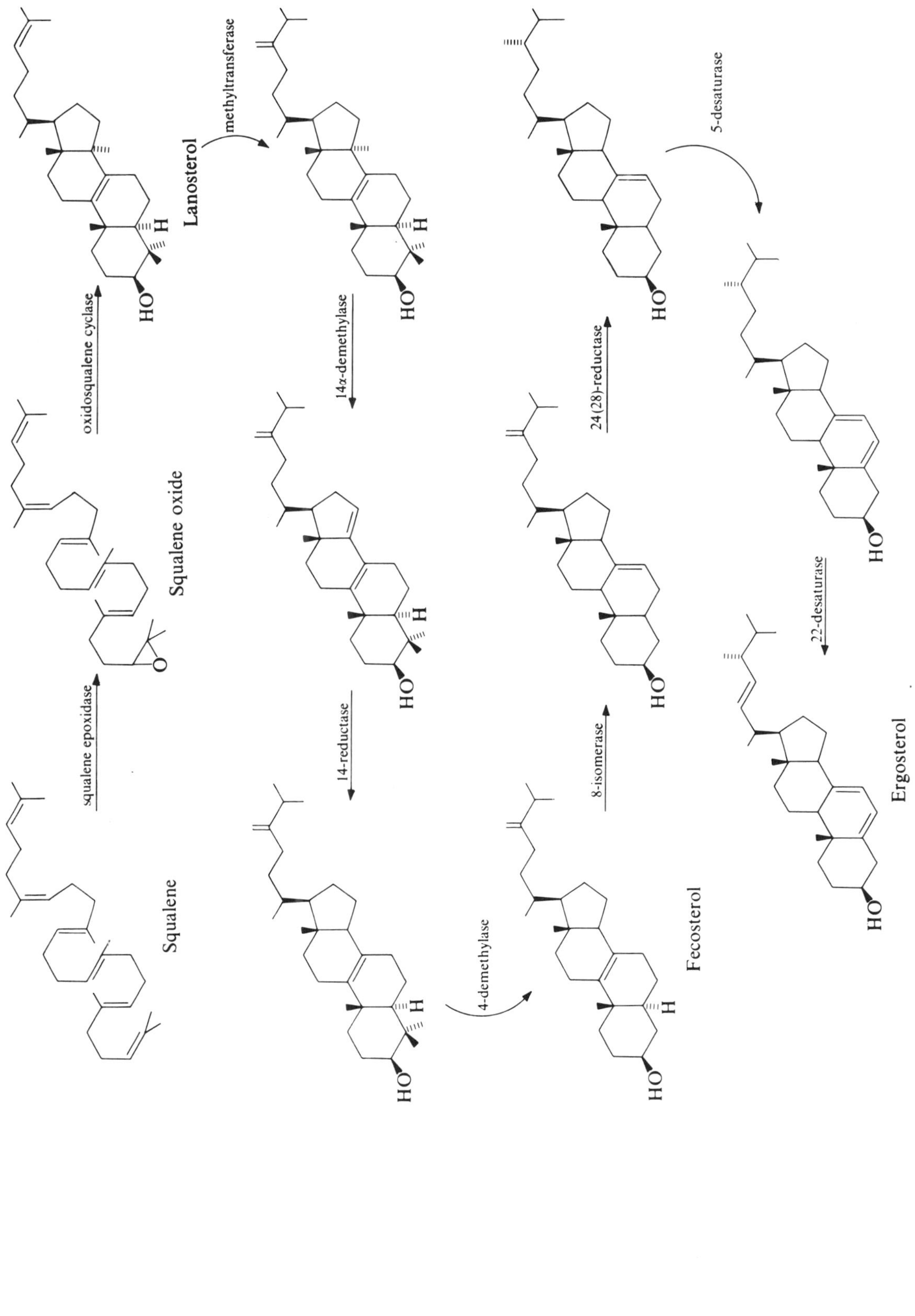

Squalene

Squalene oxide

Lanosterol

Fecosterol

Ergosterol

squalene epoxidase

oxidosqualene cyclase

methyltransferase

14α-demethylase

14-reductase

4-demethylase

8-isomerase

24(28)-reductase

5-desaturase

22-desaturase

The order of these steps is organism dependent

Scheme 8

hematological effects. The introduction of a safe orally active azole, ketoconazole (**17**), by Janssen in 1979 was hailed as a major breakthrough in the treatment of systemic mycosis.[114] Ketoconazole has proven to be of clinical importance for the treatment of mycosis and has undoubtedly been a major stimulus for the development of this class of compounds. It has also served as an important research tool for elucidating the mechanism of action of the azoles.

(**50**)   (**51**)

7.4.4.1.1 *Mode of action of azoles*

By 1974 it was realized that the azoles produced marked alterations in the membrane structures of *Candida albicans* as evidenced by electron microscopy. As ergosterol is the primary membrane sterol in fungi (Scheme 8 shows the sterol biosynthetic pathways in fungi), its accumulation in drug treated *C. albicans* was studied and the sterol was found to be depleted.[115] These observations led to the hypothesis that the azoles are fungistatic at low doses because they block the biosynthesis of ergosterol. Consistent with this proposal is the observed delay in the appearance of effects on growth. The delay would represent the time necessary to deplete preexisting cellular levels. Depletion of ergosterol concentrations occurs concomitantly with accumulation of 14α-methyl sterols, thus suggesting that lanosterol 14α-demethylase is the site of inhibition.[116]

Accumulation of bulky 14α-methylsterols in the membrane is likely to be responsible for growth inhibition. It has been argued that demethylation at C-4 and C-14 of lanosterol, which results in a planar α face, renders the molecule more effective in interacting with hydrophobic phospholipids present in the lipid bilayer of membranes. The bulky 14α-methyl group destroys the planarity of the α face of the sterol and hence the integrity of the membrane.[117] Enzymatic removal of the 14α-methyl group is catalyzed by lanosterol 14α-demethylase, an oxygen requiring cytochrome P-450 isozyme. Organisms growing under strict anaerobic conditions, as a consequence of the oxygen requirement of sterol 14α-demethylase and of squalene oxidase, another enzyme involved in ergosterol biosynthesis, should not be able to demethylate lanosterol and thereby produce ergosterol. No growth of *S. cerevisiae* was observed when incubated anaerobically in the absence of ergosterol, but in the presence of ergosterol the yeast increased in numbers until a stationary phase was reached.[118] Lanosterol supplementation was not capable of supporting growth.

The antifungal mechanism of action of azoles is more complex, however, as the concentrations necessary to deplete ergosterol concentrations resulting in fungistatic activity are considerably lower than those necessary to have a fungicidal effect. Miconazole (**51**) has at least two distinct antifungal mechanisms: sterol synthesis inhibition and, at higher concentrations, a direct membrane effect. On the other hand, ketoconazole has been reported to have only a sterol synthesis inhibition component in its antifungal action.[119] As ergosterol depletion clearly has an effect on membrane function, other membrane bound enzymes in azole treated yeast have been studied and cytochrome *c* oxidase (miconazole and ketoconazole)[120] and ATPase (miconazole)[121] have been found to be inhibited. The discovery of the multiple and varying effects produced by the azoles has led to some confusion and considerable debate as to the underlying mechanism of antifungal activity.[122] The arguments relating to additional mechanisms of action depending upon the azole under discussion are not sufficient to weaken the hypothesis of lanosterol 14α-demethylase inhibition as the primary event for the expression of antifungal activity. Within this class there are no exceptions with the compounds examined to date: the sensitivity of the organism can be demonstrated by directly assaying the enzyme, and this in turn correlated both *in vitro* and *in vivo* with a buildup of 14α-alkylated sterols. This hypothesis has served as an important tool for the discovery of azoles having improved therapeutic profiles (see Section 7.4.4.1.3).

What still remains unclear is how this pertubation of fungal membrane sterol composition, and hence membrane structure and function, prevents fungal growth. A key finding in this regard is the uncoordinated activation of chitin synthetase caused by ketoconazole.[116] Chitin is a major component of the primary septum, and as such its regulation is important in the separation of the bud and mother cell. The excessive and inappropriate deposition of chitin may impede the degradation of this septum and thereby prevent budding. If one considers the greater structural importance of chitin in the cell wall of hyphae when compared to the yeast, the observed enhanced senstivity to ketoconazole of the hyphal form of *C. albicans* relative to the yeast lends support to this proposal. It is possible that this more selective mechanism of action is responsible for the improved therapeutic index of ketoconazole.

7.4.4.1.2 *Molecular basis for the action of azoles*

Lanosterol 14α-demethylase is a cytochrome P-450 isozyme,[123] and as such its sensitivity to different substrates and inhibitors can be readily assessed spectrally. At least three types of spectral changes are induced by ligands bound to the cytochrome P-450. Pertinent to this discussion is that, in general, substrate binding is characterized by a difference spectrum with an absorption maximum at about 390 nm and a minimum at about 420 nm (type I). Nitrogenous bases elicit a spectral change with a maximum about 425 nm and a minimum about 390 nm (type II). Lanosterol has been shown to induce a type I spectral change with yeast lanosterol 14α-demethylase which has been solubilized with detergent and purified to gel electrophoretic homogeneity; the derived K_d is 6.9 μM.[124] On the other hand, ketoconazole (**17**) and other azoles cause type II spectral changes using yeast.[116] That (**17**) can induce such spectral changes confirms that it binds to the sterol 14α-demethylase and suggests that the inhibitory mechanism may involve complexation of the nitrogen of the imidazole ring with the iron of the cytochrome P-450. Similarly, diniconazole (**52**), a triazole containing antifungal, induces a type II spectral change on purified yeast lanosterol 14α-demethylase, providing direct evidence for the binding of this antifungal to the cytochrome iron. Interestingly, the binding of *R*(−)-diniconazole to the enzyme inhibits the formation of an iron(II)–CO complex, but *S*(+)-diniconazole does not.[125] The ability of an azole to prevent the formation of the reduced enzyme–CO complex appears to be related to the tenacity or potency of the inhibition .

(**52**)

7.4.4.1.3 *Enhanced selectivity in the inhibition of lanosterol 14α-demethylase as the basis for antifungal drug design*

When rats are infected intravaginally with *C. albicans* and then treated orally with ketoconazole (**17**), a dose-dependent decrease in the incorporation of labelled mevalonate (administered i.p.) into ergosterol/cholesterol (C-14 demethylated sterols) was observed in the liver and in the reisolated fungus. This decrease coincided with an increase of radioactivity incorporated into C-14 methylated sterols (lanosterol). This experiment was one of the first indications of a theme that was to become a common facet of the biological activity of ketoconazole: that is, the lack of sufficiently high specificity for the target enzyme. As discussed earlier (Section 7.4.2.3.2), a cytochrome P-450 14α-demethylase system is also used in mammalian organisms in the transformation of lanosterol to cholesterol. These enzymes appear to share many common mechanistic features, among which is their sensitivity to inhibition by ketoconazole. In this experiment almost six times more ketoconazole is needed to obtain a similar accumulation of radioactivity in C-14 methylated sterols in rat liver compared with *C. albicans*.[126] If one examines ketoconazole in cell free preparations, the yeast enzyme appears to be 28 times more sensitive than the liver enzyme. Similarly, type II spectral changes on microsomes demonstrated a greater sensitivity of *S. cerevisiae* compared with the rat liver system.[116] These results demonstrate a difference in the sensitivity of the fungal lanosterol 14α-demethylase compared with the mammalian enzyme and suggest that the lanosterol 14α-demethylase enzymes from these two sources are different.

The lack of discrimination by ketoconazole for the fungal demethylase extends into many, if not all, of the constitutive cytochrome P-450 enzymes of mammalian sterol biosynthesis. Among the various cytochrome P-450 isozymes inhibited at micromolar concentrations by ketoconazole are 11β-hydroxylase, cholesterol side-chain cleavage enzyme, 17,20-lyase and aromatase. The ability of ketoconazole to block sterol metabolism effectively at these additional sites has major clinical consequences (see Section 7.4.3.1.3). The zenobiotic metabolizing cytochrome P-450 enzymes of rat liver, in contrast, are less potently effected by ketoconazole. Still, occasional liver toxicity is seen with ketoconazole, and it is reasonable to speculate that this is a result of interference with cytochrome P-450 metabolism in the liver. This lack of specificity is not surprising given the mechanism of action of these compounds, which involves complexation of the iron of the cytochrome by the heterocyclic nitrogen atom thereby blocking the obligatory reduction of the cytochrome by the cytochrome P-450 reductase, as well as substrate binding.[124] In principle, this type of interaction is available to all iron–heme proteins.

Ketoconazole has served to demonstrate the potential promise of an orally effective azole to treat a variety of fungal infections. The elucidation of the azoles' major mechanism of action has allowed researchers to further narrow their focus for the discovery of new compounds to a search for agents that specifically inhibit fungal lanosterol 14α-demethylase in order to minimize undesirable side effects. Itraconazole (53) is an example of the application of this strategy. The Janssen group was able to improve the specificity of ketoconazole by changing the heme iron ligating functionality from imidazole to triazole and by further manipulation of more remote binding sites. This increased selectivity for the fungal demethylase to 85 to 1 over the mammalian enzyme and, more importantly, essentially abrogated the inhibition of all other P-450 isozymes.[127] The selectivity of triazole based inhibitors has been most impressively demonstrated by the bis-triazole fluconazole (54). This compound is equipotent to ketoconazole and itraconazole in its inhibition of the fungal enzyme, but shows no affinity for mammalian lanosterol 14α-demethylase![128] The selectivity achieved with both of these inhibitors appears to have translated into improved therapeutic agents and these compounds, as well as others, are expected to replace ketoconazole as the azole of choice.

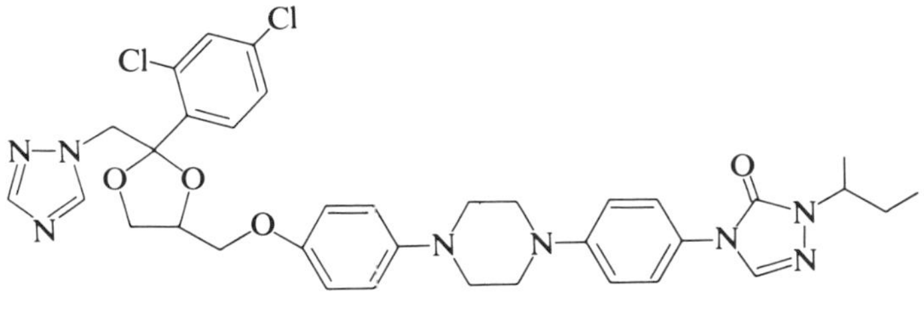

(53)

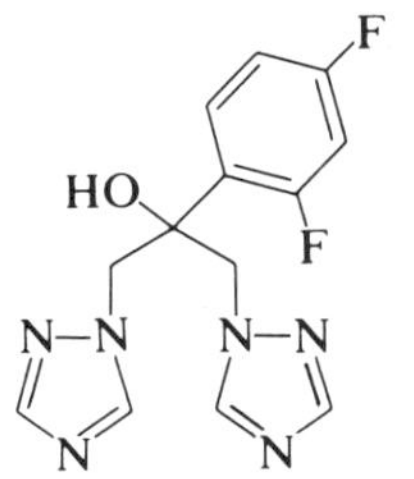

(54)

The successful realization of this strategy demonstrates: (i) that the differences in the binding pockets of P-450 isozymes are significant and exploitable, and (ii) that this strategy could serve as the basis for the design of selective inhibitors of other cytochrome P-450 isozymes. The application of this approach to a mechanistically very closely related enzyme, aromatase, has been reported.[129] An alternative approach would be to synthesize analogs of the substrate which incorporate latently reactive groups at the carbon undergoing oxidative removal. This approach has been applied to aromatase (see Section 7.4.3.1.6) and recently to the inhibition of both mammalian and fungal lanosterol 14α-demethylase.[50,130] The advantage of a substrate analog approach is that specificity is inherently a part of the design process.

7.4.4.2 Inhibition of Squalene Oxidase

7.4.4.2.1 Basis of antifungal effect

The allylamine antimycotics, which were first disclosed in 1979, are representatives of a class of ergosterol biosynthesis inhibitors distinct from the azoles. The utility of these compounds, as exemplified by naftifine (55), is primarily as topical agents. The accumulation of squalene in the presence of (55) in both whole and broken cell preparations of *C. albicans* locates the site of sterol biosynthesis inhibition as squalene oxidase. This result has been confirmed using particulate enzyme preparations from two *Candida* species for both (55) and a more potent allylamine, terbinafine (56) (Table 2).[131] The correlation of the *in vitro* inhibition of the enzyme, the inhibition of sterol biosynthesis in whole cells, and the antifungal potency (MIC) is consistent with a causal relationship between squalene oxidase inhibition and antifungal activity.[132] The relative importance of ergosterol deficiency *versus* squalene accumulation on the fungicidal activity of the allylamines appears to be organism specific. The growth of the more sensitive dermatophytes is arrested quickly at drug concentrations far below that needed for a total inhibition of ergosterol biosynthesis. *Candida* species require higher drug concentrations and longer incubations before their growth is halted. Under these conditions, ergosterol is significantly depleted and squalene accumulates in high concentrations. This suggests that dermatophyte membrane function is easily disrupted by squalene whereas that of *Candida* is not. How squalene affects the organism is not known. One suggestion is that squalene acts to increase membrane permeability. Alternatively, high concentrations of squalene may parallel the effects of lanosterol accumulation (Section 7.4.4.1.1), causing a deregulation of key enzymes. The thiocarbamate drugs, for example tolnaftate (57), are selectively active against dermatophytes. They are also effective inhibitors of squalene oxidase. This activity has been proposed as the basis of the antifungal activity for these agents, whose mode of action had previously not been adequately explained.[133]

(55)   (56)   (57)

Table 2 Comparison of Antinfungal and Squalene Oxidase Activity of Allylamines[a]

| Organism | Naftifine (55) | | Terfinabine (56) | | |
	MIC (mg L^{-1})	IC$_{50}$	K_i (μM)	MIC (mg L^{-1})	IC$_{50}$
C. albicans	50	0.350	0.03	3.1	0.008
C. parapsilosis	1.6	0.23	0.04	0.4	0.006
T. mentagrophytes	0.05	0.006	—	0.003	0.002
Rat liver	—	41	77	—	27

[a] MIC, minimum inhibitory concentration; IC$_{50}$ and K_i are for squalene oxidase inhibition on microsomal preparations.

7.4.4.2.2 Basis of enzyme inhibition

The epoxidation of squalene by squalene oxidase is the first oxygenase-catalyzed reaction in the later stages of ergosterol biosynthesis (see Scheme 8). The enzyme from rat liver has been purified, allowing for the partial characterization of the protein.[134] The reconstituted enzyme requires NADPH-cytochrome P-450 reductase, NADPH, FAD, and molecular oxygen for full activity. Both its absorption spectrum and its lack of inhibition by standard P-450 inhibitors plus its chromatographic behavior suggest that this is not a cytochrome P-450 enzyme. Similar conclusions have been reached with the particulate enzyme from *C. albicans*.[135] One notable difference between the two enzymes is the preference by the *Candida* enzyme for NADH over NADPH. The inhibition of these enzymes by (54) and (55) is, however, substantially different. Terbinafine (54) is a weak competitive inhibitor of the rat liver epoxidase ($K_i = 77\,\mu M$) and a potent noncompetitive inhibitor of the *C. albicans* enzyme ($K_i = 0.03\,\mu M$). The basis of this potent inhibition is unknown. The presence of an allylamine functionality might suggest that they are acting as irreversible inhibitors, as has been demonstrated for pargyline, a propargylic amine, which irreversibly inhibits monoamine oxidase. However, the inhibition in this case is reversible. The very close structural similarity of the allylamines and the thiocarbamates suggests that these compounds interact in a similar manner with enzymes. Based on this, and the affinity of both sulfur and nitrogen for iron, one can speculate that the inhibition is the result of an interaction of these moieties at the active site (this assumes that the protein does indeed contain iron). If so, this affinity for iron does not carry over to cytochrome P-450 enzymes. These compounds do not appreciably inhibit either the liver microsomal or the sterol constitutive cytochrome P-450 enzymes, a property that distinguishes this class from the azole antifungals.[136] Indeed, the animal toxicology and initial clinical data with terbinafine given orally show it to be remarkably nontoxic. This apparent advantage is offset by its ineffectiveness in treating serious systemic mycosis, and it appears that these agents will only be useful in the treatment of superficial infections.[137]

7.4.4.3 Other Sites of Action

There exists a large number of antifungal agents whose mechanism of action appears to be related to their ability to inhibit fungal sterol biosynthetic enzymes other than lanosterol 14α-demethylase and squalene oxidase. None of these agents has proven useful in the treatment of human mycosis. Aside from any overt toxicity that these compounds might possess, their primary reason for failure can be attributed to their lack of specificity for fungal over mammalian sterol biosynthesis. This is, of course, not a problem if the compound is used as an agricultural fungicide, and ergosterol biosynthesis inhibitors, referred to as EBI fungicides, are a major product in the treatment of crop fungal diseases.[138] Additional sites of action besides those already discussed are the $\Delta^8 \rightarrow \Delta^7$-sterol isomerase, $\Delta^{8,14}$-sterol 14-reductase, $\Delta^{24(28)}$-sterol reductase and Δ^{24}-sterol *S*-methyltransferase. These enzymatic reactions all can be envisioned to proceed through carbocationic high-energy intermediates (HEI) and the activity of most of the inhibitors can be explained based on the previously outlined principles (Section 7.4.2.4.2).

The tertiary amine triparanol (21) is moderately fungicidal. This compound, which was previously discussed as an inhibitor of sterol 24-reductase (Section 7.4.2.4), is also an effective inhibitor of the Δ^{24}-sterol *S*-methyltransferase.[139] Since both of these enzymatic reactions are believed to be initiated by electrophilic attack at C-24 to yield as an initial enzyme-bound intermediate, a carbocation at C-25, the inhibition by (21) of the two enzymes is consistent with the compound's ability to mimic this intermediate. Furthermore, the side chain azacholesterols (22) and (23) inhibit both enzymes and inhibit fungal growth.[140] The morpholine EBIs [for example, tridemorph (58)], which are of agricultural importance, inhibit both the fungal $\Delta^8 \rightarrow \Delta^7$-sterol isomerase and sterol 24-reductase.[141] A more interesting example is the 15-aza steroid (59), a natural product isolated from *Geotrichum flavobrunneum*. This compound is a potent antifungal, showing good activity against *C. albicans*. The imine (59) inhibits Δ^{24}-sterol *S*-methyltransferase ($K_i = 43\,\mu M$), $\Delta^{24(28)}$-sterol reductase ($K_i = 17\,\mu M$) and $\Delta^{8,14}$-sterol 14-reductase ($K_i = 2\,nM$).[142] The close structural similarity of (59) to the 8,14-diene moiety of the substrate for the $\Delta^{8,14}$-sterol 14-reductase would appear to provide an adequate explanation for the observed potent inhibition. The protonation of the imine, which may or may not be enzyme catalyzed, would afford a close mimic of the presumed HEI. However, the noncompetitive inhibition pattern indicates that (59) does not bind to the same enzyme form as the normal substrate. Compactin (1), which was initially isolated as an antifungal

$C_{13}H_{27}-N$

(58)

(59)

antibiotic, provides a final example. The compound potently inhibits HMGCoA reductase, the rate limiting enzyme in the pathway to ergosterol and cholesterol, from both fungal and mammalian sources, and hence is both an antifungal and a hypocholesterolemic. This latter activity has overshadowed any interest in compactin as an antifungal (see Section 7.4.2.2.2).

7.4.4.4 The Polyene Antibiotics

The polyene antibiotics are included in this review, not because they inhibit sterol biosynthesis, but because their effectiveness as therapeutic agents depends on the presence of sterols. Although the true mechanism of antifungal action of the polyenes remains controversial, their interactions with sterols and sterol containing membranes is not. Two observations strongly suggest that the interaction with sterols is the basis of their antifungal action: (i) organisms that contain sterols are susceptible to polyenes (yeast and mammalian cells) and organisms that do not contain sterols are resistant (bacteria); (ii) sterols added to the medium protect susceptible cells from polyenes.

The study of sterol–polyene complexes in synthetic membrane preparations has served to model their possible mode of action. The clinically more useful polyene, amphotericin B (60), was found to prefer ergosterol containing membranes over those containing cholesterol. The polyene filipin (61) displays a preference for cholesterol containing membranes. This physicochemical behavior is reflected in the *in vivo* activity; filipin is a less potent antifungal and is too toxic to mammalian cells to be clinically useful. The sterol–polyene interaction, which can be demonstrated to cause a leakage of cellular constituents, is believed to result in the formation of pores or channels and eventual cell lysis.[143]

The polyenes, which remain the most useful drugs for the treatment of life-threatening mycosis, and the azoles both rely upon an interference with sterols (function or metabolism) to exert their antifungal effect. The sterols of fungi are found in the cell membrane or plasmalemma, which is analogous to the mammalian plasma membrane. The plasmalemma is surrounded by a cell wall which serves to impart structural rigidity and in some instances acts as a barrier for the internalization of exogenous materials. The fungal cell wall is composed of three different classes of polysaccharides: mannoproteins, glucan, and chitin. The latter two constituents, which are not found in mammalian cells, would appear to be ideal targets for antifungal directed chemotherapy. Inhibitors of both chitin synthetase (the polyoxins and nikkomycins, which are useful agricultural fungicides) and glucan synthetase (echinocandin B) are known. Ironically, inhibitors of cell wall biosynthesis have as yet not proven useful in the treatment of human mycosis.[144]

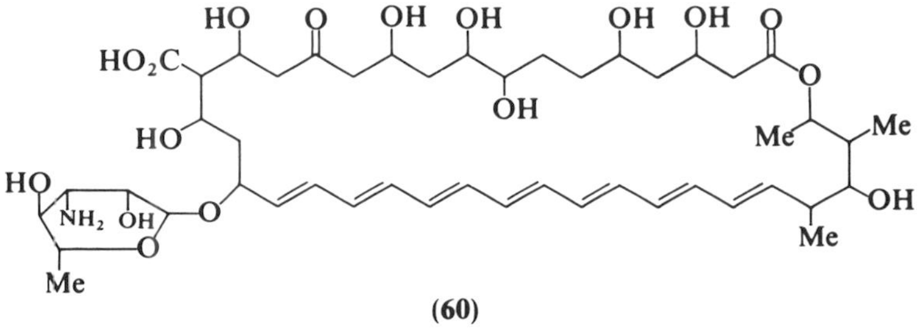

(60)

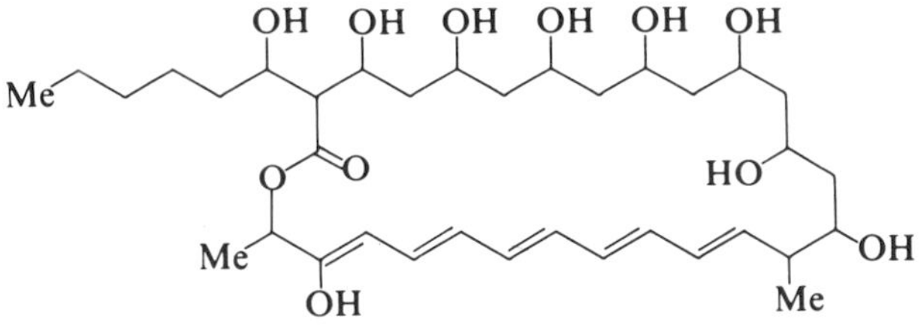

(61)

7.4.5 ADDENDUM

Brown and Goldstein have continued their investigation on the regulation of cholesterol biosynthesis. They have identified a common regulatory element in the genes coding for HMGCoA reductase, HMGCoA synthase and the LDL receptor which represses transcription in the presence of oxysterols.[145] An oxysterol binding protein from rabbit liver has been cloned and expressed. This protein, and similar proteins previously identified in mouse and hamster liver cytosol, are believed to mediate the effect of oxysterols on gene regulation.[146] The binding of the protein to DNA has yet to be demonstrated.

A new class of potent and selective inhibitors of steroid 5α-reductase has recently been reported.[147] These inhibitors, which are 3-androstene-3-carboxylic acids, were designed to bind as enolate mimics (see equation 6, p. 354). The carboxylic acid (62), which has a $K_{i,app}$ of 30–36 nM *versus* the human enzyme, shows uncompetitive binding behavior *versus* testosterone. This is in contrast to the 4-azasteroids, such as (49), which show competitive binding with respect to testosterone, and implies binding of the steroidal carboxylates to the enzyme–NADP$^+$ complex. Because testosterone does not compete with these compounds for binding to the enzyme, they may show some therapeutic advantage.

(62)

7.4.6   REFERENCES

1. R. A. Demel and B. de Kruyff, *Biochim. Biophys. Acta*, 1976, **457**, 109.
2. G. Ourisson and M. Rohmer, in 'Current Topics in Membrane and Transport', ed. S. Razin and S. Rottem, Academic Press, New York, 1982, vol. 17, p. 153.
3. M. S. Brown and J. L. Goldstein, *Angew. Chem., Int. Ed. Engl.*, 1986, **25**, 583.
4. S. F. Gibbons, K. A. Mitropoulos and N. B. Myant, 'Biochemistry of Cholesterol', Elsevier, Amsterdam, 1982, p. 131.
5. D. J. Shapiro and V. W. Rodwell, *J. Biol. Chem.*, 1971, **246**, 3210.
6. J. L. Goldstein and M. S. Brown, *J. Lipid Res.*, 1984, **25**, 1450.
7. K. L. Luskey, J. R. Faust, D. J. Chin, M. S. Brown and J. L. Goldstein, *J. Biol. Chem.*, 1983, **258**, 8462.
8. J. R. Faust, K. L. Luskey, D. J. Chin, J. L. Goldstein and M. S. Brown, *Proc. Natl. Acad. Sci. USA*, 1982, **79**, 5205.
9. M. S. Brown and J. L. Goldstein, *J. Lipid Res.*, 1980, **21**, 505.
10. L. Liscum, J. Finer-Moore, R. M. Stroud, K. L. Luskey, M. S. Brown and J. L. Goldstein, *J. Biol. Chem.*, 1985, **260**, 522.
11. K. L. Luskey and B. Stevens, *J. Biol. Chem.*, 1985, **260**, 10271.
12. D. A. Kleinsek, R. E. Dugan, T. A. Baker and J. W. Porter, *Methods Enzymol.*, 1981, **71**, 462.
13. D. G. Sherban, P. J. Kennelly, K. G. Brandt and V. W. Rodwell, *J. Biol. Chem.*, 1985, **260**, 12579.
14. K. Tanzawa and A. Endo, *Eur. J. Biochem.*, 1979, **98**, 195.
15. T.-G. Nguyen, K. Gerbing and H. Eggerer, *Hoppe-Seyler's Z. Physiol. Chem.*, 1984, **365**, 1.
16. A. Endo, M. Kuroda and K. Tanzawa, *FEBS Lett.*, 1976, **72**, 323.
17. A. G. Brown, T. C. Smale, T. J. King, R. Hasenkemp and R. Thompson, *J. Chem. Soc., Perkin Trans. 1*, 1976, 1165.
18. A. W. Alberts, J. Chen, G. Kuron, V. Hunt, J. Huff, C. Hoffman, J. Rothrock, M. Lopez, H. Joshua, E. Harris, A. Patchett, R. Monaghan, S. Currie, E. Shapley, G. Albers-Schonberg, O. Hensens, J. Hirshfield, K. Hoogstein, J. Liesch and J. Springer, *Proc. Natl. Acad. Sci. USA*, 1980, **77**, 3957.
19. Y. K. Lam, V. P. Gullo, R. T. Goegelman, D. Jorn, L. Huang, C. DeRiso, R. L. Monaghan and I. Putter, *J. Antibiot.*, 1981, **34**, 614.
20. G. Albers-Schonberg, H. Joshua, M. B. Lopez, O. D. Hensens, J. P. Springer, J. Chen, S. Ostrove, C. H. Hoffman, A. W. Alberts and A. A. Patchett, *J. Antibiot.*, 1981, **34**, 507.
21. A. Endo, *Trends Biochem. Sci.*, 1981, **6**, 10.
22. J. A. Tobert, G. D. Bell, J. Birtwell, I. James, W. R. Kukovetz, J. S. Pryor, A. Buntinx, I. B. Holmes, Y.-S. Chao and J. A. Bolognese, *J. Clin. Invest.*, 1981, **69**, 913.
23. D. R. Illingworth and G. J. Sexton, *J. Clin. Invest.*, 1984, **74**, 1972.
24. D. R. Illingworth, *Ann. Intern. Med.*, 1984, **101**, 598.
25. D. W. Bilheimer, S. M. Grundy, M. S. Brown and J. L. Goldstein, *Proc. Natl. Acad. Sci. USA*, 1983, **80**, 4124.
26. D. H. Minsker, J. S. MacDonald, R. T. Robertson and D. L. Bokelman, *Teratology*, 1983, **28**, 449.
27. W. F. Hoffman, A. W. Alberts, P. S. Anderson, J. S. Chen, R. L. Smith and A. K. Willard, *J. Med. Chem.*, 1986, **29**, 849.
28. Y. Tsujita, M. Kuroda, Y. Shimade, K. Tanzawa, M. Arai, I. Kaneko, M. Tanake, H. Masuda, C. Tarumi, Y. Watanabe and S. Fujii, *Biochim. Biophys. Acta*, 1986, **877**, 50.
29. T. Kita, M. S. Brown and J. L. Goldstein, *J. Clin. Invest.*, 1980, **66**, 1094.
30. M. Sinensky and J. Logel, *J. Biol. Chem.*, 1983, **258**, 8547.
31. M. S. Brown, J. R. Faust, J. L. Goldstein, I. Kaneko and A. Endo, *J. Biol. Chem.*, 1978, **253**, 1121.
32. C. E. Nakamura and R. H. Abeles, *Biochemistry*, 1985, **24**, 1364.

33. W. F. Hoffman, A. W. Alberts, E. J. Cragoe, Jr., A. A. Deana, B. E. Evans, J. L. Gilfillan, N. P. Gould, J. W. Huff, F. C. Novello, J. D. Prugh, K. E. Rittle, R. L. Smith, G. E. Stokker and A. K. Willard, *J. Med. Chem.*, 1986, **29**, 159.

34. G. E. Stokker, A. W. Alberts, R. S. Anderson, E. J. Cragoe Jr., A. A. Deana, J. L. Gilfillan, J. Hirshfield, W. J. Holtz, W. F. Hoffman, J. W. Huff, T. J. Lee, F. C. Novello, J. D. Prugh, C. S. Roney, R. L. Smith and A. K. Willard, *J. Med. Chem.*, 1986, **29**, 170.

35. F. G. Kathawala, T. Scallen, R. G. Engstrom, D. B. Weinstein, H. Schuster, R. Stabler, J. Kratunis, J. R. Wareing, C. F. Jewell, L. Widler and S. Wattanasin, presented at the 194th American Chemical Society National Meeting, Division of Medicinal Chemistry, 1987, Abstract 79.

36. A. A. Kandutsch, H. W. Chen and H.-J. Heiniger, *Science*, 1978, **201**, 498.

37. F. R. Taylor, S. E. Saucier, E. P. Shown, E. J. Parish and A. A. Kandutsch, *J. Biol. Chem.*, 1984, **259**, 12 382.

38. F. D. Pinkerton, A. Izumi, C. M. Anderson, L. R. Miller, A. Kisic and G. J. Schroepfer, Jr., *J. Biol. Chem.*, 1982, **257**, 1929.

39. A. A. Kandutsch, H.-J. Heiniger and H. W. Chen, *Biochim. Biophys. Acta*, 1977, **486**, 260.

40. G. J. Schroepfer, Jr., B. C. Sherrill, K.-S. Wang, W. K. Wilson, A. Kisic and T. B. Clarkson, *Proc. Natl. Acad. Sci. USA*, 1984, **81**, 6861.

41. J. M. Trzaskos, R. T. Fischer and M. F. Favata, *J. Biol. Chem.*, 1986, **261**, 16 937.

42. Y. Aoyama, Y. Yoshida, Y. Sonoda and Y. Sato, *J. Biol. Chem.*, 1987, **262**, 1239.

43. G. F. Gibbons, C. R. Pullinger, H. W. Chen, W. K. Cavenee and A. A. Kandutsch, *J. Biol. Chem.*, 1980, **255**, 395.

44. C. Havel, E. Hansbury, T. J. Scallen and J. A. Watson, *J. Biol. Chem.*, 1979, **254**, 9573.

45. F. B. Kraemer and A. Pont, *Am. J. Med.*, 1986, **80**, 616.

46. A. Gupta, R. C. Sexton and H. Rudney, *J. Biol. Chem.*, 1986, **261**, 8348.

47. M. F. Favata, J. M. Trzaskos, H. W. Chen, R. T. Fischer and R. S. Greenberg, *J. Biol. Chem.*, 1987, **262**, 12 254.

48. F. B. Kraemer and S. Spilman, *J. Pharmacol. Exp. Ther.*, 1986, **238**, 905.

49. J. M. Trzaskos, M. F. Favata, R. T. Fischer and S. H. Stam, *J. Biol. Chem.*, 1987, **262**, 12261.

50. (a) S. S. Ko, H. Chen, M. F. Favata, R. T. Fischer, J. L. Gaylor, P. R. Johnson, R. L. Magolda, S. H. Stam and J. M. Trzaskos, presented at the 194th American Chemical Society National Meeting, Division of Medicinal Chemistry, 1987, Abstracts 77 and 78; (b) L. L. Frye and C. H. Robinson, *J. Chem. Soc., Chem. Commun.*, 1988, 129.

51. S. E. Saucier, A. A. Kandutsch, S. Phirwa and T. A. Spencer, *J. Biol. Chem.*, 1987, **262**, 14 056.

52. T. A. Spencer, A. K. Gayen, S. Phirwa, J. A. Nelson, F. R. Taylor, A. A. Kandutsch and S. K. Erikson, *J. Biol. Chem.*, 1985, **260**, 13 391.

53. F. R. Taylor, A. A. Kandutsch, A. K. Gayen, J. A. Nelson, S. S. Nelson, S. Phirwa and T. A. Spencer, *J. Biol. Chem.*, 1986, **261**, 15 039.

54. J. A. Nelson, M. R. Czany, T. A. Spencer, J. S. Limanek, K. R. McCrae and T. Y. Chang, *J. Am. Chem. Soc.*, 1978, **100**, 4900.

55. R. C. Sexton, S. R. Panini, F. Azran and H. Rudney, *Biochemistry*, 1983, **22**, 5687.

56. S. R. Panini, R. C. Sexton and H. Rudney, *J. Biol. Chem.*, 1984, **259**, 7767.

57. S. R. Panini, R. C. Sexton, A. K. Gupta, E. J. Parish, S. Chitrakorn and H. Rudney, *J. Lipid Res.*, 1986, **27**, 1190.

58. R. Fears, in '3-Hydroxy-3-Methylglutaryl-Coenzyme A Reductase', ed. R. Sabine, CRC Press, Boca Raton, FL, 1983, p. 189.

59. R. C. Laughlin and T. F. Carey, *J. Am. Med. Assoc.*, 1962, **181**, 339.

60. D. Steinberg, J. Avagin and E. B. Feigelson, *J. Clin. Invest.*, 1961, **4**, 884.

61. R. Langdon, S. El-Masry and R. E. Counsell, *J. Lipid Res.*, 1977, **18**, 24.

62. A. Boogaard, M. Griffioen and L. H. Cohen, *Biochem. J.*, 1987, **241**, 345.

63. M. Akhtar, D. C. Wilton, I. A. Watkinson and A. D. Rahimtula, *Proc. R. Soc. London, Ser. B*, 1972, **180**, 167.

64. R. Wolfenden, *Annu. Rev. Biophys. Bioeng.*, 1976, **5**, 271.

65. A. Rahier, M. Taton, P. Bouvier-Nav'e, P. Schmitt, P. Benveniste, F. Schuber, A. S. Narula, L. Cattel, C. Anding and P. Place, *Lipids*, 1986, **21**, 52.

66. L. Cattel, M. Ceruti, F. Viola, L. Delprino, G. Balliano, A. Duraittii and P. Bouvier-Nav'e, *Lipids*, 1986, **21**, 31.

67. D. N. North, *Ann. Intern. Med.*, 1978, **89**, 128.

68. W. L. McGuire, *Sel. Top. Clin. Sci. Annu. Rev. Med.*, 1975, **26**, 353.

69. R. J. Santen and R. I. Misbin, *Pharmacotherapy*, 1981, **1**, 95.

70. P. J. Nicholls, M. J. Daly and H. J. Smith, *Breast Cancer Res. Treat.*, 1986, **7**, S55.

71. A. B. Foster, M. Jarman, C. S. Leung, M. G. Rowlands and G. N. Taylor, *J. Med. Chem.*, 1983, **26**, 50.

72. W. K. Amery, R. de Coster and I. Caers, *Drug Dev. Res.*, 1986, **8**, 299.

73. J. T. Groves, G. S. McClusky, R. E. White and M. J. Coon, *Biochem. Biophys. Res. Commun.*, 1978, **81**, 154.

74. M. H. Gelb, D. C. Heimbrook, P. Malkonen and S. G. Sliger, *Biochemistry*, 1982, **21**, 370.

75. R. E. White, J. P. Miller, L. V. Favreau and A. Bhattacharyya, *J. Am. Chem. Soc.*, 1986, **108**, 6024.

76. P. R. Ortiz de Montellano, in 'Cytochrome P-450', ed. P. R. Ortiz de Montellano, Plenum Press, New York, 1986, p. 217.

77. P. R. Ortiz de Montellano and M. A. Correia, *Annu. Rev. Pharmacol. Toxicol.*, 1983, **23**, 481.

78. P. R. Ortiz de Montellano and E. A. Komives, *J. Biol. Chem.*, 1985, **260**, 3330.

79. A. B. Foster, M. Jarman, C.-S. Leung, M. G. Rowlands, G. N. Taylor, R. G. Plevey and P. Sampson, *J. Med. Chem.*, 1985, **28**, 200.

80. A. Nagahisa, T. Foo, M. Gut and W. H. Orme-Johnson, *J. Biol. Chem.*, 1985, **260**, 846.

81. C. Walsh, *Tetrahedron*, 1982, **38**, 871.

82. P. R. Ortiz de Montellano, H. S. Beilan, K. L. Kunze and B. A. Mico, *J. Biol. Chem.*, 1981, **256**, 4395.

83. P. R. Ortiz de Montellano, B. L. K. Mangold, C. Wheeler, K. L. Kunze and N. O. Reich, *J. Biol. Chem.*, 1983, **258**, 4208.

84. P. R. Ortiz de Montellano, K. L. Kunz, H. S. Beilan and C. Wheeler, *Biochemistry*, 1982, **21**, 1331.

85. J. Fishman and J. Goto, *J. Biol. Chem.*, 1981, **256**, 4466.

86. M. Akhtar, M. R. Calder, D. L. Corina and J. N. Wright, *Biochem. J.*, 1982, **201**, 569.

87. E. Caspi, J. Wicha, T. Aruncachalam, P. Nelson and G. Spiteller, *J. Am. Chem. Soc.*, 1984, **106**, 7282.

88. D. F. Covey, W. F. Hood and V. D. Parikh, *J. Biol. Chem.*, 1981, **256**, 1076.

89. B. W. Metcalf, C. L. Wright, J. P. Burkhart and J. O. Johnston, *J. Am. Chem. Soc.*, 1981, **103**, 3221.

90. P. A. Marcotte and C. H. Robinson, *Steroids*, 1982, **39**, 325.

91. J. O. Johnston, C. L. Wright and B. W. Metcalf, *Endocrinology*, 1984, **115**, 776.

92. J. O. Johnston and B. W. Metcalf, in 'Novel Approaches to Cancer Chemotherapy', ed. P. Sunkara, Academic Press, New York, 1984, p. 307.
93. P. A. Marcotte and C. H. Robinson, *Biochemistry*, 1982, **21**, 2773.
94. D. F. Covey and W. F. Hood, *Cancer Res., Suppl.*, 1982, **42**, 3327.
95. D. F. Covey and W. F. Hood, *Endocrinology*, 1981, **108**, 1597.
96. A. M. H. Brodie, W. M. Garrett, J. R. Henderson, C. Tsai-Morris, P. A. Marcotte and C. H. Robinson, *Steroids*, 1981, **38**, 693.
97. A. M. H. Brodie, L.-Y. Wing, P. Goss, M. Dowsett and R. C. Coombes, *J. Steroid Biochem.*, 1986, **24**, 91.
98. D. F. Covey and W. F. Hood, *Cancer Res., Suppl.*, 1982, **42**, 3327.
99. P. J. Bednarski, D. J. Porubek and S. D. Nelson, *J. Med. Chem.*, 1985, **28**, 775.
100. J. N. Wright, M. R. Calder and M. Akhtar, *J. Chem. Soc., Chem. Commun.*, 1985, 1733.
101. J. T. Kellis, W. E. Childers, C. H. Robinson and L. E. Vickery, *J. Biol. Chem.*, 1987, **262**, 4421.
102. R. W. Brueggemeier, E. E. Floyd and R. E. Council, *J. Med. Chem.*, 1978, **21**, 1007.
103. C. E. Snider and R. W. Brueggemeier, *J. Biol. Chem.*, 1987, **262**, 8685.
104. J. O. Johnston, C. L. Wright, G. W. Holbert and B. W. Metcalf, *Endocrinology*, 1985, **116**, 1415.
105. A. Nagahisa, R. W. Spencer and W. H. Orme-Johnson, *J. Biol. Chem.*, 1983, **258**, 6721.
106. A. Nagahisa, W. H. Orme-Johnson and S. R. Wilson, *J. Am. Chem. Soc.*, 1984, **106**, 1166.
107. R. H. Menard, T. M. Guenthner, H. Kon and J. R. Gillette, *J. Biol. Chem.*, 1979, **254**, 1726.
108. R. H. Menard, T. M. Guenthner, A. M. Taburet, H. Kon, L. R. Pohl, J. R. Gillette, H. V. Gelboin and W. F. Trager, *Mol. Pharmacol.*, 1979, **16**, 997.
109. J. Imperato-McGinley and T. Gautier, *Trends Genet.*, 1986, 130.
110. V. Petrow, *Prostate*, 1986, **9**, 343.
111. T. Liang, M. A. Cascieri, A. H. Cheung, G. F. Reynolds and G. H. Rasmusson, *Endocrinology*, 1985, **117**, 571.
112. L. J. Goad, G. G. Holz, Jr. and D. H. Beach, *Mol. Biochem. Parasitol.*, 1985, **15**, 257.
113. D. H. Beach, L. J. Goad and G. G. Holz, Jr., *Biochem. Biophys. Res. Commun.*, 1986, **136**, 851.
114. H. E. Jones, *Arch. Dermatol.*, 1982, **118**, 217.
115. H. Vanden Bossche, G. Willemsens, W. Cools, W. F. J. Lauwers and L. LeJeune, *Chem. Biol. Interact.*, 1978, **21**, 59.
116. H. Vanden Bossche, G. Willemsens, P. Marichal, W. Cools and W. Lauwers, *Symp. Br. Mycol. Soc.*, 1983, **9**, 321.
117. C. E. Dahl, J. S. Dahl and K. Bloch, *Biochemistry*, 1980, **19**, 1462.
118. W. R. Nes, J. H. Adler, B. C. Sekula and K. Krevis, *Biochem. Biophys. Res. Commun.*, 1976, **71**, 1296.
119. I. J. Sud and D. S. Feingold, *Antimicrob. Agents Chemother.*, 1981, **20**, 71.
120. M. L. Shigematsu, J. Uno and T. Arai, *Antimicrob. Agents Chemother.*, 1982, **21**, 919.
121. F. Portillo and C. Gancedo, *Eur. J. Biochem.*, 1984, **143**, 273.
122. W. H. Beggs, F. A. Andrews and G. A. Sarosi, *Life Sci.* 1981, **28**, 111.
123. Y. Yoshida and Y. Aoyama, *J. Biol. Chem.*, 1984, **259**, 1655.
124. Y. Aoyama, Y. Yoshida, S. Hata, T. Nishino and H. Katsuki, *Biochem. Biophys. Res. Commun.*, 1983, **115**, 642.
125. Y. Yoshida, Y. Aoyama, H. Takano and T. Kato, *Biochem. Biophys. Res. Commun.*, 1986, **137**, 513.
126. H. Vanden Bossche, G. Willemsens, W. Cools, F. Cornelissen, W. F. Lauwers and J. M. Van Cutsem, *Antimicrob. Agents Chemother.*, 1980, **17**, 922.
127. H. Vanden Bossche, D. Bellens, W. Cools, J. Gorrens, P. Marichal, H. Verhoeven, G. Willemsens, R. De Coster, W. Lauwers and L. Jeune, *Drug Dev. Res.*, 1986, **8**, 287.
128. M. S. Marriott, G. W. Pye, K. Richardson and P. F. Troke, in '*In vitro* and *In vivo* Evaluation of Antifungal Agents', ed. K. Iwata and H. Vanden Bossche, Elsevier, Amsterdam, 1986, p. 143.
129. H. M. Taylor, C. D. Jones, J. D. Davenport, K. S. Hirsch, T. J. Kress and D. Weaver, *J. Med. Chem.*, 1987, **30**, 1359.
130. A. B. Cooper, J. J. Wright, A. K. Ganguly, J. Desai, D. Loebenberg, R. Parmegiani, D. S. Feingold and I. J. Sud, *J. Chem. Soc., Chem. Commun.*, 1989, 898.
131. N. S. Ryder and M.-C. Dupont, *Biochem. J.*, 1985, **230**, 765.
132. N. Ryder, in '*In vitro* and *In vivo* Evaluation of Antifungal Agents', ed. K. Iwata and H. Vanden Bossche, Elsevier, Amsterdam, 1986, p. 89.
133. N. S. Ryder, I. Frank and M.-C. DuPont, *Antimicrob. Agents Chemother.*, 1986, **29**, 858.
134. T. Ono, K. Nakazono and H. Kosaka, *Biochim. Biophys. Acta*, 1982, **709**, 84.
135. N. S. Ryder and M.-C. DuPont, *Biochem. Biophys. Acta*, 1984, **794**, 466.
136. I. Schuster, *Xenobiotica*, 1985, **15**, 529.
137. A. Stutz, *Angew. Chem., Int. Ed. Engl.*, 1987, **26**, 320.
138. B. C. Baldwin, *Biochem. Soc. Trans.*, 1983, **11**, 659.
139. H. C. Malhotra and W. R. Nes, *J. Biol. Chem.*, 1971, **246**, 4934.
140. A. C. Oehlschlager, R. H. Angus, A. M. Pierce, H. D. Pierce, Jr. and R. Srivasan, *Biochemistry*, 1984, **23**, 3582.
141. R. I. Baloch and E. I. Mercer, *Phytochemistry*, 1987, **26**, 663.
142. L. W. Parks and R. J. Rodriguez, *Biochem. Soc. Trans.*, 1983, **11**, 656.
143. G. Medoff and G. A. Kobayashi, 'Antifungal Chemotherapy', ed. D. C. E. Speller, Wiley, New York, 1980, p. 3.
144. J. F. Ryley, R. G. Wilson, M. B. Gravestock and J. P. Poyser, *Adv. Pharmacol. Chemother.*, 1981, **18**, 49.
145. J. E. Metherall, J. L. Goldstein, K. L. Luskey and M. S. Brown, *J. Biol. Chem.*, 1989, **264**, 15 634 and references therein.
146. P. A. Dawson, N. D. Ridgway, C. A. Slaughter, M. S. Brown and J. L. Goldstein, *J. Biol. Chem.*, 1989, **264**, 16 798.
147. B. W. Metcalf, D. A. Holt, M. A. Levy, J. M. Erb, J. I. Heaslip, M. Brandt and H.-J. Oh, *Bioorg. Chem.*, 1989, **17**, 372.

8.1

Hydrolases

ALAN D. ELBEIN

University of Texas Health Science Center, San Antonio, TX, USA

8.1.1	INTRODUCTION	365
8.1.2	GLYCOSIDASES	365
	8.1.2.1 General Comments	365
	8.1.2.2 Neutral Glycosidases in Glycoprotein Processing	366
	8.1.2.3 Chemistry of Naturally Occurring Glycosidase Inhibitors	369
	8.1.2.3.1 Swainsonine	369
	8.1.2.3.2 Castanospermine	370
	8.1.2.3.3 Related polyhydroxy alkaloids	370
	8.1.2.4 Biochemistry of Naturally Occurring Glycosidase Inhibitors	371
	8.1.2.4.1 Inhibition of mannosidases by swainsonine	371
	8.1.2.4.2 Inhibition of glycoprotein processing	372
	8.1.2.4.3 Effects of swainsonine on animals	374
	8.1.2.4.4 Chemical synthesis of swainsonine	375
	8.1.2.4.5 Castanospermine inhibition of α- and β-glucosidases	376
	8.1.2.4.6 Effect of castanospermine and deoxynojirimycin on glycoprotein processing	376
	8.1.2.4.7 Deoxymannojirimycin	378
	8.1.2.4.8 Pyrrolidine alkaloids	379
	8.1.2.4.9 Chemically produced glycosidase and processing inhibitors	379
8.1.3	ACETYLCHOLINESTERASE	381
8.1.4	SUMMARY	386
8.1.5	REFERENCES	386

8.1.1   INTRODUCTION

Hydrolases are enzymes that catalyze the cleavage of bonds between carbon and some other atom with the addition of water across the bond. This group of enzymes includes neutral and acidic glycosidases, esterases such as acetylcholinesterases, peptidases, lipases, phospholipases, phosphatases, and so on. Since there is such a great diversity in the various types of reactions that are catalyzed by hydrolases, it would be virtually impossible to cover all of the different classes of enzymes and their inhibitors in this one chapter. Thus, the author has chosen to focus most of this chapter on two groups of enzymes which have been well studied (*i.e.* glycosidases and acetylcholinesterase), and for which a number of very useful and powerful inhibitors are known.

8.1.2   GLYCOSIDASES

8.1.2.1   General Comments

Glycosidases catalyze the hydrolysis of glycosidic bonds in simple glycosides, in oligosaccharides and polysaccharides, and in complex carbohydrates such as glycolipids and glycoproteins. The action of these enzymes results in the liberation of monosaccharides or oligosaccharides of lower molecular weight than the starting material. These enzymes can be broadly classified into two groups as follows: (i) exoglycosidases, which attack glycosidic linkages at the nonreducing ends of the saccharide chains to liberate monosaccharide units, and (ii) endoglycosidases, which act on glycosidic linkages within the saccharide chains to produce smaller oligosaccharides. A large number of different glycosidases have been described that show specificity for the sugar residue to be hydrolyzed as well as for the anomeric configuration.[1] However, in many cases, the specificity for the glycosidic linkage is not very rigid and the enzyme will attack the appropriate sugar with the proper anomeric configuration, regardless of the linkage. On the other hand, some of the recently described neutral glycosidases that play a critical role in glycoprotein processing are quite specific for the glycosidic linkage as well as for the specific sugar and the anomeric configuration.

Glycosidases are widespread in nature and are probably present in all organisms. Such great diversity is likely to be a reflection of their vital role in catabolism and turnover, as well as in the modification of various complex carbohydrates. Thus, extracellular glycosidases, such as those found in the intestinal tracts of animals or those secreted by microorganisms, degrade larger molecules to prepare them for uptake and utilization by the organism.[2] There are also a number of intracellular glycosidases that are located in the lysosomes of eukaryotic cells. Such enzymes also degrade complex carbohydrates within the cell and are involved with the turnover of various cellular components such as glycolipids, glycoproteins or proteoglycans.[3]

Glycosidases may also have other critical functions in metabolism. For example, during the growth of organisms that have rigid cell walls (*i.e.* plants, fungi and bacteria), new cell wall material must be inserted at various growing points in order to allow for expansion of the existing wall. Glycosidases are believed to be involved in the loosening of the old wall material.[4] In the biosynthesis of the oligosaccharide portion of the N-linked glycoproteins, glycosidases play key roles in the removal of specific sugars to produce different types of oligosaccharide structures (see below for the details of these reactions).

It is interesting to point out that although glycosidases were among the first enzymes to be studied,[5,6] we know considerably less about structures, mechanisms and cellular roles of this widely distributed group of enzymes than about many other hydrolases. Thus, lysozyme, an endoglycanase that attacks the β-1,4 glycosidic bonds linking together GlcNAc and muramic acid, is the only glycosidase with which detailed structural analysis of the complex between enzyme and substrate has been carried out.

Glycosidase-catalyzed hydrolysis resembles the acid-catalyzed hydrolysis of simple alkyl or arylglycosides in that cleavage of the C(1)–O, *i.e.* the glycosyl bond, occurs. These reactions can therefore be considered as nucleophilic substitutions at the C-1 of the glycoside.[6-8] The first step in the reaction is always considered to be the protonation of the anomeric oxygen by an AH group on the enzyme (for example, see Figure 6). This reaction leads to the release of the aglycon group, while the sugar remains bound to the enzyme. The resulting product may either be a covalent intermediate or a glycosyl ion intermediate that is stabilized by the enzyme; these two possibilities would correspond to the ES form of the enzyme. The second step in the reaction, which would correspond to the decomposition of the ES intermediate, is the base-catalyzed reaction by the A group on the enzyme. The reaction is completed by the stereospecific addition of a hydroxyl group to the oxocarbonium ion. Water of course is the usual hydroxyl donor, but other hydroxylic compounds, including the hydroxyl groups of other sugars, could also act as donors. Glycosidases can be divided into two classes depending on whether the hydrolysis proceeds with retention of the anomeric configuration or inversion of this configuration.[9]

8.1.2.2 Neutral Glycosidases in Glycoprotein Processing

The N-linked or asparagine-linked glycoproteins are commonly found in eukaryotic cells, both as cells surface or membrane proteins, and also as secreted proteins.[10] The oligosaccharide portion of these glycoproteins may be either of the high mannose, the complex or the hybrid type of structure, as shown in Figure 1. All of these structures contain the same basic core region, which is a pentasaccharide composed of a branched trimannose unit linked to an N,N'-diacetylchitobiose (this core region is shown within the box). The GlcNAc at the reducing end of the chitobiose is in turn attached to the amide nitrogen of an asparagine in the protein. In the high mannose structure of Figure 1, this pentasaccharide core region is further elongated by six α-linked mannose residues, although some of the high mannose structures may have fewer than nine total mannoses. On the other hand, in the complex types of oligosaccharides, the pentasaccharide core is elongated by the trisaccharide, sialic acid–galactose–GlcNAc. The complex oligosaccharides may have two (biantennary), three (triantennary) or four (tetrantennary) of these trisaccharide units attached to the core structure. Finally, the hybrid types of oligosaccharides, which seem to be relatively rare in nature

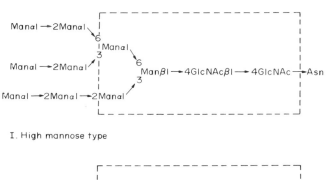

I. High mannose type

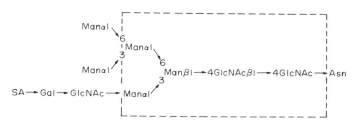

II. Complex type

III. Hybrid type

Figure 1 Generalized structures of the high mannose (upper), complex (middle) and hybrid (lower) types of oligosaccharides found on the N-linked glycoproteins

and have only been found recently in some animal cells, are combinations of the high mannose and complex chains and may have several mannoses linked to the 6-branch, as well as one or two of the sialic acid–galactose–GlcNAc trisaccharides on the 3-branch.[11,12]

The biosynthesis of these N-linked oligosaccharides involves two series of reactions that are outlined in Figures 2 and 3. The first series of reactions gives rise to a common intermediate, which is the oligosaccharide donor for all of the N-linked oligosaccharides, while the second series of reactions modifies or trims this initial oligosaccharide structure to produce the various high mannose or complex structures. The series of reactions shown in Figure 2 is commonly referred to as the dolichol cycle, because dolichyl phosphate serves as a carrier of the sugars.[13,14] The sequence of reactions is initiated by the transfer of GlcNAc-1-P, from UDP-GlcNAc, to dolichyl-P to form the first lipid intermediate, GlcNAc-PP-dolichol.[15-17] A second GlcNAc is then added, also from UDP-GlcNAc, to form the N,N'-diacetylchitobiosyl-PP–dolichol.[18,19] This lipid-linked disaccharide is then the acceptor of mannose and glucose units, as indicated in Figure 2, to form the final lipid-linked oligosaccharide, $Glc_3Man_9(GlcNAc)_2$–pyrophosphoryl–dolichol.[20-22] This large oligosaccharide is finally transferred to protein, while the polypeptide chain is being synthesized on membrane-bound polysomes.[23-25] The enzymes involved in the dolichol cycle are all membrane-bound glycosyltransferases and these reactions appear to occur in the endoplasmic reticulum.[26-28]

Once the oligosaccharide has been transferred to the protein, the oligosaccharide undergoes a series of modification reactions that are catalyzed by a number of specific and unusual glycosidases. Although several of these glycosidases have been highly purified, their mechanism of action has not been studied in detail and it is therefore not known whether the catalytic event is similar in these enzymes to that in other so-called acylglycosidases. However, since some of the inhibitors described here can act on both types of glycosidases, it seems likely that the mechanism of enzyme action may be similar at least in some cases.

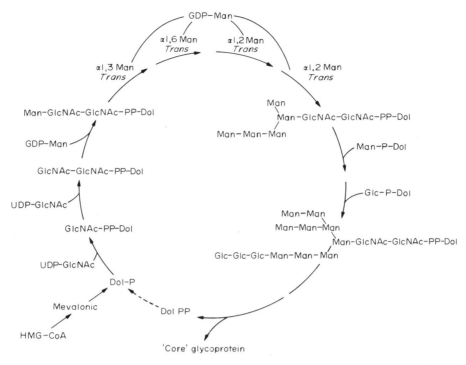

Figure 2. Initial pathway leading to the formation of the precursor for all of the N-linked oligosaccharides. This pathway involves the formation of lipid-linked saccharide intermediates, resulting in the formation of the common intermediate $Glc_3Man_9(GlcNAc)_2$–pyrophosphoryl–dolichol. This pathway is frequently called the dolichol cycle

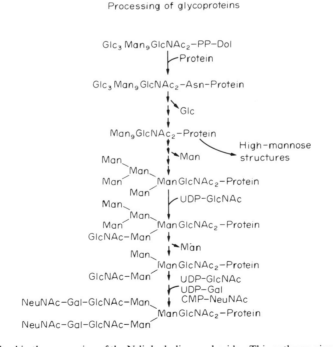

Figure 3 Reactions involved in the processing of the N-linked oligosaccharides. This pathway gives rise to the various kinds of oligosaccharides associated with the asparagine-linked glycoproteins

The initial processing reactions begin in the endoplasmic reticulum and continue as the protein is transported through the Golgi apparatus to its final destination.[29,30] These processing reactions involved in glycoprotein biosynthesis are outlined in Figure 3. Very soon after attachment of the oligosaccharide to protein, the glucose residues are removed from the glycoprotein by two different membrane-bound glucosidases.[31-33] Glucosidase I removed the terminal α-1,2-linked glucose,[34,35] while glucosidase II removes the next two α-1,3-linked glucoses.[36,37] Several of the plant alkaloids discussed below, as well as one of the epoxide derivatives, are relatively good inhibitors of these enzymes.

Removal of all three glucose residues from these oligosaccharides gives rise to glycoproteins having $Man_9(GlcNAc)_2$ structures. This glycoprotein may remain as a high mannose structure such as that found in various viruses and cells, or the oligosaccharides may be further processed to produce the hybrid and complex types of oligosaccharides. These additional processing reactions occur in the endoplasmic reticulum and in the Golgi apparatus, and involve the removal of all four of the α-1,2-linked mannose residues. Thus, specific α-1,2-mannosidases (mannosidase IA/B)[38] have been isolated and purified from the Golgi apparatus of hepatocytes,[39,40] and recently another α-1,2-mannosidase was isolated and partially purified from the endoplasmic reticulum.[41] As a result of the action of these mannosidases, a $Man_5(GlcNAc)_2$–protein results. A GlcNAc is then added from UDP-GlcNAc to the α-3-mannose branch of this $Man_5(GlcNAc)_2$–protein, and this GlcNAc apparently represents an essential signal for another Golgi mannosidase, called mannosidase II, to remove the 3-linked and 6-linked mannoses from the 6-mannose branch.[42-44] After removal of these final two mannoses to give a $GlcNAc-Man_3(GlcNAc)_2$–protein, the remaining sugars of the complex types of oligosaccharides, *i.e.* GlcNAc, galactose, sialic acid and fucose, are added by sequential transfer from their sugar nucleotides to produce the various complex glycoproteins.[45,46] However, all of the details of these additions are not known. It is also not clear why some oligosaccharides (*i.e.* glycoproteins) remain as high mannose types, whereas others become complex chains, or why some chains are biantennary, others are triantennary, and so on.

8.1.2.3 Chemistry of Naturally Occurring Glycosidase Inhibitors

8.1.2.3.1 *Swainsonine*

This indolizidine alkaloid was first isolated from the leaves of the Australian plant *Swainsona canescens*.[47] The leaves of this plant are toxic to animals, causing severe neurological and skeletal difficulties that ultimately lead to death. In the south west of the USA, this disease is called 'locoism' and results from eating plants of the genus *Astragalus*. The toxic principle in these leaves has been identified as swainsonine. Swainsonine is very soluble in water and in hydroxylic solvents, moderately soluble in polar, nonhydroxylic solvents such as acetone and chloroform, and insoluble in nonpolar solvents. Thus, the alkaloid is extracted from the plant material with acetone, ethyl acetate or methanol, and is then purified by chromatography on an AG-50W-X8 (NH_4^+) column. The fractions eluting with dilute ammonium hydroxide are monitored for α-mannosidase inhibitory activity against jack bean α-mannosidase. Extraction from the residue of the active column fraction with ammonia-saturated chloroform and crystallization from the same solvent gives swainsonine as white needles.[47,48] Swainsonine has also been isolated from various *Astragalus* species (locoweed) that grow in the south west of the USA,[49,50] and from the fungus *Rhizoctonia leguminicola*.[51,52] The yield of swainsonine from *Swainsona canescens* was about 0.001%, and from *Astragalus lentiginosus* it was about 0.003%.

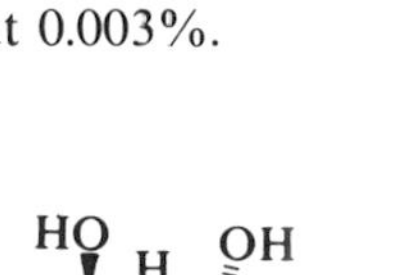

Swainsonine 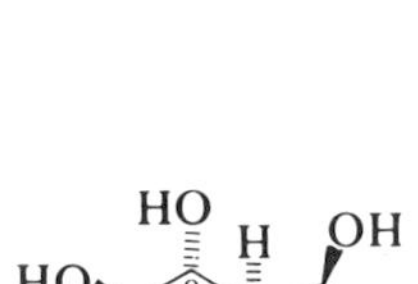 · Castanospermine 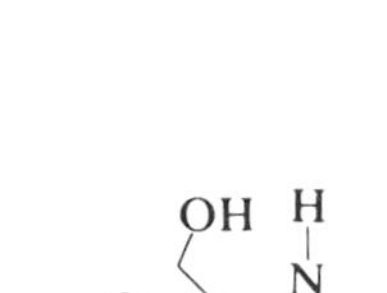 · Deoxynojirimycin 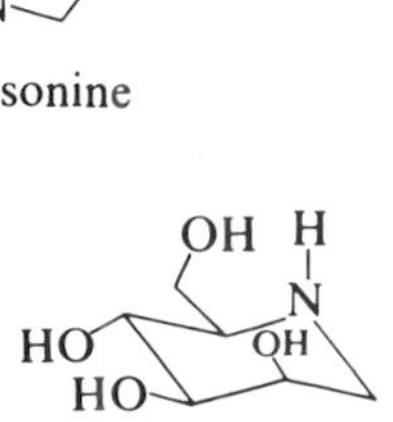 · Deoxymannojirimycin 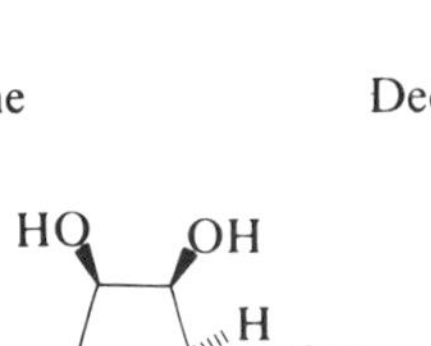 · 1,4-Dideoxy-1,4-iminomannitol

Figure 4 Structures of the various inhibitors of glycosidases and glycoprotein-processing enzymes

The structure of swainsonine was determined primarily by spectroscopic methods and is shown in Figure 4. However, the formation of certain derivatives, such as the diacetate and triacetate and the acetonide, provided confirmatory evidence for the assigned structure.[53, 54] In addition, high resolution mass spectrometry, ^{1}H and ^{13}C NMR spectroscopy and X-ray diffraction of the diacetate derivative have provided evidence for the assigned structure.[55]

8.1.2.3.2 *Castanospermine*

The Moreton Bay chestnut or black bean (*Castanospermum australe*) is a large leguminous tree found in rain forests and along creek banks in eastern Australia. It has also been introduced as a landscape tree in southern California. Although aborigines ate the large chestnut-like seeds after soaking them in water and roasting them, early European settlers found that consumption of raw or roasted seeds produced severe gastrointestinal pain.[56] Both cattle and horses ingest the seeds when available, in spite of the often fatal gastroenteritis that develops. Early reports indicated that the toxin was a saponin, but recently the alkaloid castanospermine was isolated from these seeds[61] and shown to be a potent inhibitor of almond emulsin β-glucosidase as well as various α-glucosidases.[57] In addition, castanospermine is a strong inhibitor of intestinal sucrase and maltase.[58−60] This latter inhibition may well account for the toxicity of these seeds.

Like swainsonine, castanospermine is very soluble in water and in hydroxylic solvents, but is insoluble in nonpolar solvents. Extraction of ground seeds of *Castanospermum australe* with 75% ethanol, water or methanol, and purification of the water soluble portion of the extract by ion exchange chromatography gave a fraction containing basic materials. The castanospermine could be crystallized from this basic material in aqueous ethanol or methanol as white prisms with a melting point of 212–215 °C (dec.) and a rotation of $[\alpha]^{24} = +79.2°$. From mature seeds, castanospermine was isolated in 0.3% yield.[61]

The structural determination of castanospermine was primarily based on spectroscopic methods. However, treatment with acetic anhydride in pyridine gave a tetraacetate derivative, while treatment with iodomethane gave the methiodide, indicating that the compound was a tetrahydroxylated tertiary base alkaloid. A variety of other analytical data, as well as ^{13}C NMR and ^{1}H NMR and high resolution EI mass spectrometry defined the structure of this alkaloid. An X-ray crystal structure determination of castanospermine confirmed the structural assignment and showed that the six-membered ring adopts a chair conformation, whereas the five-membered ring has a pseudoenvelope conformation.[61]

8.1.2.3.3 *Related polyhydroxy alkaloids*

Several naturally occurring polyhydroxy alkaloids are known that are somewhat similar in structure and in inhibitory activity to the simple indolizidine alkaloids like swainsonine and castanospermine. These alkaloids are nitrogen analogs of pyranose or furanose sugars and include nojirimycin and deoxynojirimycin, as well as deoxymannojirimycin and the pyrrolidine alkaloid 2,5-dihydroxymethyl-3,4-dihydroxypyrrolidine. The former two alkaloids have been known for some time and have been studied extensively as glucosidase inhibitors, but the latter two compounds were discovered more recently. These compounds are of considerable interest because of the potential information they may provide as to the structural requirements for glycosidase inhibition, as well as being valuable tools in their own right.

Nojirimycin occurs in the fermentation broths of several strains of *Streptomyces*, including *S. nojiriensis*, from which its name is derived.[62] The compound is quite unstable under neutral or acidic conditions. but may be isolated as the stable bisulfite adduct from which it can be recovered in the pure state by alkaline hydrolysis followed by ion exchange chromatography and crystallization at low temperature. The structure of nojirimycin was determined by chemical methods in combination with mass spectrometry and NMR analysis. A particularly significant transformation was its reduction, either catalytically or with sodium borohydride, to deoxynojirimycin, the structure of which was established by 220 MHz ^{1}H NMR spectroscopy.[63] The D configuration for nojirimycin was established by ORD, and the alkaloid is therefore 5-amino-5-deoxy-α-D-glucopyranose. Deoxynojirimycin, initially obtained as a reduction product of nojirimycin, was subsequently isolated from a number of *Bacillus* species,[63] and has also been obtained from mulberry plants.[64]

Deoxymannojirimycin, the mannose analog of deoxynojirimycin, was synthesized in six steps from D-mannose.[65] The compound was also isolated in 0.24% yield from the seeds of the legume

Lonchocarpus sericeus.[66, 67] The structure was elucidated by EI and CI mass spectra in combination with ^{13}C and ^{1}H NMR spectroscopy. The pyrrolidine alkaloid 2,5-dihydroxymethyl-3,4-dihydroxy-pyrrolidine was isolated in 0.1% yield from fresh leaves of *Derris elliptica*[68] and more recently from the seeds of *Lonchocarpus costariensis,*[69] both of which are members of the Leguminosae. The compound was rapidly degraded on treatment with periodic acid, demonstrating the presence of vicinal hydroxyl groups. The ^{1}H NMR spectrum of the hydrochloride derivative showed that the compound contained four hydroxyl groups and eight nonexchangeable hydrogen atoms. The ^{13}C NMR spectrum of the free base showed only three resonances which corresponded to one methylene and two methine groups, all of which are shifted to low field by attachment to either OH or NH groups, indicating a remarkable symmetry in the molecule. Mass spectral data supported the imino alcohol structure,[70] and this conclusion was further confirmed by 300 MHz NMR studies. The optical activity of the compound eliminated the *meso* forms from consideration, and the structure is therefore (2R,5R)-dihydroxymethyl-(3R,4R)-dihydroxypyrrolidine or its enantiomer.

Various polyhydroxy alkaloids have now been synthesized chemically and tested as inhibitors of glycosidases. Thus the 1,5-dideoxy-1,5-iminohexitols are analogs of pyranose sugars in which the oxygen atom in the ring is replaced by a nitrogen, and the anomeric hydroxyl group is replaced by a hydrogen. These compounds have been found to be reasonably specific inhibitors of the hydrolysis of the corresponding glycopyranosides catalyzed by the specific glycosidase. For example, the synthetic analog 1,5-dideoxy-1,5-imino-L-fucitol is a potent inhibitor of α-L-fucosidase,[71] whereas the analog 1,5-dideoxy-1,5-imino-D-galactitol is a good inhibitor of coffee bean α-galactosidase (K_i = 1.6 nM, or 50% inhibition at 400 nM).[72] This α-galactosidase inhibitor was also reported to inhibit intestinal lactase.[73] 2-Acetamido-1,5-imino-1,2,5-trideoxy-D-glucitol was found to be an effective inhibitor of a number of β-N-acetylhexosaminidases.[74, 75]

Perhaps more surprising than the fact that the above compounds are potent and reasonably specific inhibitors is the finding that furanose derivatives such as 1,4-dideoxy-1,4-imino-D-mannitol, the azafuranose derivative of D-mannose, is also a potent inhibitor of the hydrolysis of mannopyranosides as catalyzed by the enzyme jack bean α-mannosidase.[76] In addition, some 1,4-iminopentitols have been found to be powerful glycosidase inhibitors; for example 1,4-dideoxy-1,4-imino-D-lyxitol is an α-galactosidase inhibitor, while 1,4-dideoxy-1,4-imino-D-arabinitol is a powerful α-glucosdiase inhibitor (50% inhibition at 2×10^{-7} M).[77] Thus, the potential for synthesizing a great variety of inhibitors for lysosomal as well as processing glycosidases exists, and such compounds should be of great value for studying the role of these enzymes as well as their mechanism of catalysis and substrate specificities.

8.1.2.4 Biochemistry of Naturally Occurring Glycosidase Inhibitors

8.1.2.4.1 *Inhibition of mannosidases by swainsonine*

The first studies showing a biological role for any of the indolizidine alkaloids were those done with swainsonine. Early observations indicated that animals that ate the leaves of *Swainsona canescens*, a plant that produces swainsonine, exhibited symptoms similar to those of individuals suffering from the lysosomal storage disease α-mannosidosis.[78] Since this disease is due to the absence of lysosomal α-mannosidase, Dorling *et al.* examined the effects of swainsonine on the lysosomal enzyme and showed that the alkaloid was a potent inhibitor of this mannosidase.

Mammalian tissues generally contain three structurally and genetically distinct types of α-mannosidases, with different locations in the cell and with different optimum pH for activity. These have been designated as the acidic or lysosomal α-mannosidase, the intermediate or Golgi α-mannosidase, and the neutral or cytosolic α-mannosidase. The acidic enzyme has a pH optimum of 4–4.5 and is found in the lysosomes. This mannosidase cleaves terminal mannose units from the nonreducing ends, regardless of whether the glycosidic linkage is α-1,2, α-1,3, or α-1,6. The function of this enzyme is in the degradation or turnover of oligosaccharides of N-linked glycoproteins. There are at least two different Golgi α-mannosidases, both of which have pH optima around 5.5–6.0, or even a little higher. As shown in Figure 3, these mannosidases are involved in the processing of the oligosaccharide chains of glycoproteins. There are several enzymes, referred to as mannosidase IA/B, that have specificity for α-1,2-mannosyl linkages.[79-81] The other Golgi mannosidase that participates in processing is called mannosidase II.[82] This enzyme cleaves the α-1,6- and α-1,3-linked mannosyl residues from the GlcNAc–Man$_5$(GlcNAc)$_2$–protein. Recently, another α-mannosidase was found in the endoplasmic reticulum.[83] Although this enzyme has not been purified as yet, it does cleave α-1,2-mannosyl linkages, and it has a pH optimum of about 6.5. Its precise role is not known, but it is believed to be involved in the early stages of glycoprotein trimming.[84] Finally, an α-mannosidase with a pH optimum of 6.5 is located in the cytoplasm. The function of this enzyme is not known, but it may arise by proteolysis of the above endoplasmic reticulum α-mannosidase.

Dorling and coworkers found that the plant alkaloid swainsonine at a concentration of 20 μM completely inhibited the acidic α-mannosidase (*i.e.* pH 4.0) from all tissues examined, as well as the acidic α-mannosidase from the liver of the lamprey eel and the acidic α-mannosidase from jack bean seeds. On the other hand, a number of glycosidases from mouse liver (*i.e.* α-glucosidase, β-galactosidase, β-hexosaminidase and β-glucuronidase) were not inhibited by swainsonine, even when tested at a 10-fold higher concentration than that needed to completely inhibit the α-mannosidase.[85] These workers point out that the pK_a of swainsonine is 7.4 and therefore at a pH of 4.0 the alkaloid would be fully ionized. Using molecular models they observed that the relative positions of the cationic centers and the three hydroxyl groups are similar to that of the hypothetical mannosyl ion which may serve as an intermediate. This arrangement may account for the apparent specificity of swainsonine for α-mannosidase. Dorling and colleagues have presented evidence to demonstrate that swainsonine is a reversible site-directed inhibitor of lysosomal α-mannosidase.[85]

In another study, swainsonine was tested at a number of concentrations against a variety of commercially available glycosidases in order to determine the specificity of this alkaloid. Even at 1 μg mL^{-1} (about 50 times the amount needed for 50% inhibition of jack bean α-mannosidase), swainsonine had no inhibitory effect on β-mannosidase, α- or β-glucosidase, α- or β-galactosidase, β-*N*-acetylhexosaminidase or α-L-fucosidase.[86] A number of kinetic studies were done to determine the type of inhibition by swainsonine on the jack bean α-mannosidase. Lineweaver–Burk plots of substrate concentration (*i.e.* p-nitrophenyl-α-D-mannopyranoside) *versus* velocity, in the presence of various amounts of swainsonine, showed considerable curvature at high substrate concentrations, suggesting that swainsonine may be a competitive inhibitor that binds tightly to the enzyme and is only slowly displaced. Periodate oxidation of swainsonine completely destroyed the inhibitory activity of this alkaloid, in keeping with the presence of and requirement for vicinal hydroxyl groups.[87]

Studies on the effects of swainsonine on the processing of the N-linked oligosaccharides have shown that this alkaloid alters the normal processing pathway and blocks or prevents the formation of complex types of oligosaccharides (see next section for more details).[88-90] As discussed previously, there are several different mannosidases involved in these processing reactions. Thus, mannosidase I (A and B) is specific for the α-1,2-linked mannose residues and acts on oligosaccharides from Man$_9$(GlcNAc)$_2$ down to Man$_5$(GlcNAc)$_2$, whereas mannosidase II cleaves the α-1,3- and α-1,6-linked mannoses from the GlcNAc-Man$_5$(GlcNAc)$_2$ to complete the mannose-trimming reactions. Swainsonine did not inhibit the Golgi mannosidase I, but mannosidase II was strongly inhibited by this alkaloid.[91] In addition, Tulsiani *et al.* examined the effect of swainsonine on the processing of mannose-labeled oligosaccharides ranging in size from [^{3}H]Man$_9$(GlcNAc)$_2$ to [^{3}H]Man$_5$(GlcNAc)$_2$ by rat liver Golgi fractions.[92] Swainsonine had no effect on the yield of [^{3}H]Man$_5$(GlcNAc)$_2$ or of free [^{3}H]mannose produced by these enzyme preparations, a further indication that α-1,2-mannosidases (*i.e.* mannosidase I) were not inhibited. The processing experiments were also done in the presence of UDP-GlcNAc to promote the addition of GlcNAc to the Man$_5$(GlcNAc)$_2$. In the absence of swainsonine, the Golgi preparations converted the oligosaccharides mainly to the GlcNAc-Man$_3$(GlcNAc)$_2$ and GlcNAc$_2$-Man$_3$(GlcNAc)$_2$ species, whereas in the presence of the alkaloid the product was GlcNAc-Man$_5$(GlcNAc)$_2$. Thus the processing mannosidase II is the site of inhibition of swainsonine in this pathway. Characterization of the oligosaccharides of the N-linked glycoproteins produced in cultured mammalian cells in the presence of swainsonine also indicates that mannosidase II is the site of inhibition *in vivo* (see below for details of these studies).

8.1.2.4.2 *Inhibition of glycoprotein processing*

Since swainsonine is a potent inhibitor of various α-mannosidases, it was tested as an inhibitor of the biosynthesis of N-linked glycoproteins. These studies were performed with animal cells growing in culture in order to see whether swainsonine could permeate these cells, and, if so, whether it would affect the processing pathway. When animal cells such as Madin–Darby canine kidney (MDCK) cells, Chinese hamster ovary (CHO) cells, fibroblasts, *etc.* are grown in the presence of [2-^{3}H]mannose or [6-^{3}H]glucosamine, these radioisotopes are incorporated into N-linked glycoproteins that may have high mannose, hybrid or complex types of oligosaccharides. While the amount of each type of oligosaccharide varies with the particular protein being considered, as well as with the cell type and the conditions of growth, animal cells generally have more complex types of oligosaccharides in their cell surface or membrane glycoproteins. One can easily distinguish the complex structures from the high mannose or hybrid types by digesting the entire glycoprotein with the proteolytic enzyme pronase, which cleaves the protein down to individual amino acids. The oligosaccharides are released as glycopeptides that contain one or two amino acids and the intact carbohydrate chain. The glycopeptides can then be separated by gel filtration on the basis of their size — the complex types of oligosaccharides are larger in size and elute earliest, whereas the high mannose types are smaller and elute later. These glycopeptides can be further analyzed by treating them with the enzyme endoglucosaminidase H, which cleaves high mannose and hybrid structures between the two internal *N*-acetylglucosamine residues. This enzyme, however, will not hydrolyze this glycosidic bond in complex types of oligosaccharides.[93] Thus, after digestion with endo H, the high mannose and hybrid chains show an altered and slower migration on the Biogel columns because they have lost the GlcNAc–peptide portion of the molecule. However, the complex chains continue to migrate in the same position either with or without treatment with endo H since this enzyme does not cleave these glycopeptides.

When MDCK cells or CHO cells were incubated for several hours in the presence of swainsonine, and then labeled with [2-³H]mannose or [6-³H]glucosamine, there was a dramatic decrease in the amount of label in the complex types of glycopeptides and a substantial increase in the amount of label in the high mannose (or hybrid) types of glycopeptides. These changes were monitored by determining the amount of radioactivity that became susceptible to digestion by endoglucosaminidase H with increasing amounts of swainsonine. These experiments demonstrated that swainsonine was inhibiting one or more steps in the processing pathway, but the exact site of action was not determined in these studies.[88]

The effect of swainsonine on the biosynthesis and processing of several viral glycoproteins has also been studied. Influenza virus is an enveloped virus that has a major coat protein called the hemagglutinin. This protein is a glycoprotein that contains seven or eight N-linked oligosaccharide chains, of which about four or five are of the complex type and about two or three of the high mannose structure.[94] Another virus that produces an envelope containing N-linked glycoproteins is vesicular stomatitis virus, usually referred to as VSV. In this case the glycoprotein is called the G protein and has two N-linked oligosaccharides, both of which are of the biantennary, complex structure.[95]

Influenza virus was grown in MDCK cells in the presence of various amounts of swainsonine, and the viral glycoproteins were labeled by the addition of [2-³H]mannose or [6-³H]glucosamine. At concentrations of swainsonine of 1 μg mL^{-1} or higher, more than 90% of the viral glycoprotein was susceptible to the action of endoglucosaminidase H, whereas in the normal (*i.e.* control) virus more than 70% of the N-linked oligosaccharides are resistant to the action of this enzyme. These results suggested that swainsonine was causing the oligosaccharide structure to be changed from complex types to high mannose or hybrid types of oligosaccharides.[89] Essentially the same results were obtained when VSV was grown in baby hamster kidney cells in the presence of swainsonine. In these experiments, a complete change in oligosaccharide structure from endoglucosaminidase H resistant to endoglucosaminidase H sensitive occurred at concentrations of swainsonine of 50 ng mL^{-1} of culture medium.[96] The major oligosaccharide found in the VSV glycoproteins produced in the presence of swainsonine was characterized by a variety of enzymatic and chemical procedures and shown to be a hybrid type of oligosaccharide similar to that shown in Figure 1. With the VSV system, swainsonine had to be added to the medium within the first 4 or 5 h after infection in order to cause the observed changes in structure from complex types of oligosaccharides to hybrid structures. These 4 or 5 h probably represent the time when most of the G protein is synthesized and processed. Interestingly enough, the effects of swainsonine were reversible as long as the alkaloid was removed before most of the G protein was synthesized (*i.e.* during the first 1 or 2 h) and as long as the amount of swainsonine used was not too high. Thus if the swainsonine-containing medium was replaced with fresh medium during these first few hours, the cells could revert to the production of normal, complex oligosaccharides. In these experiments with virus-infected cells, swainsonine did not appear to affect the production of the virus; that is the same number of viral particles were observed in the medium of swainsonine-grown and control cells. Also, the virus produced in the presence of swainsonine appeared to retain its infectivity. These studies indicate that the presence of complex types of oligosaccharides on the enveloped glycoproteins are not necessary for the production and release of virus from the cells, nor for the infectivity of the virus towards other cultured cells.

Swainsonine also caused the production of hybrid types of glycoproteins in cultured fibroblasts. Thus, in the presence of 10 μg mL^{-1} of swainsonine, most of the complex types of glycoproteins were replaced with hybrid-containing structures. The major oligosaccharide was found to have the same hybrid type of structure as that shown in Figure 1. This same hybrid type of oligosaccharide could be produced when rat liver Golgi fractions were incubated in the presence of [³H]Man₅GlcNAc,

UDP-GlcNAc, UDP-galactose, CMP-NeuAc and swainsonine.[97] Swainsonine was also used to study the processing of the N-linked oligosaccharides of α-1-antitrypsin by rat hepatocytes. In the control (normal) hepatocytes, the newly synthesized α-1-antitrypsin was found in the cells as a glycoprotein of apparent molecular weight of 49 000, and having high mannose types of oligosaccharides. However, in the medium, a secreted form of α-1-antitrypsin was identified that had an apparent molecular weight of about 54 000 and oligosaccharides of the complex type. Because pulse chase studies indicated a precursor–product relationship, it seems likely that the high mannose type of glycoprotein is processed to complex oligosaccharides during the transport of the antitrypsin and/or secretion from the cells. When these hepatocytes were treated with swainsonine, the intracellular form of the α-1-antitrypsin was indistinguishable from that of control cells and had an apparent molecular weight of 49 000. However, the α-1-antitrypsin secreted from the cells was different from that of the control cells. Thus the control cell α-1-antitrypsin had a molecular weight of about 54 000 and was resistant to the action of endoglucosaminidase H, whereas the swainsonine-induced protein had a molecular weight of about 51 000 and was sensitive to endoglucosaminidase H. In addition, it had a higher content of (*i.e.* incorporation of) radioactive mannose, but contained only half as much [³H]galactose and about the same amount of [³H]fucose as the controls. Thus, swainsonine did alter the carbohydrate structure of the α-1-antitrypsin, but this alteration apparently did not impair the secretion of the molecule.[90]

Fibronectin is an interesting and important glycoprotein that is secreted by fibroblasts and is necessary for the adhesion of some cells to the substratum. This protein contains about 4 to 10% carbohydrate, consisting principally of complex types of oligosaccharides that have the biantennary structure.[98] In swainsonine-treated fibroblasts, fibronectin was secreted into the medium just as it is in control cells, but, in this case, the oligosaccharide chains were susceptible to the action of endoglucosaminidase H. However, in the normal fibronectin, the oligosaccharides are resistant to endoglucosaminidase H. The structure of the oligosaccharide, produced in the presence of swainsonine and released by endoglucosaminidase H, was analyzed using a variety of glycosidases in conjunction with gel filtration analysis of the products. The major structure of the original oligosaccharide on the protein was deduced to be as follows: Gal-β-GlcNAc-β-Man-α-[Man-α-(Man-α-)Man]-Man-β-GlcNAc-β-(α-Fuc-)GlcNAc. Other studies showed that the synthesis and secretion of fibronectin were not affected by its change in glycosylation. The results also imply that swainsonine inhibits the Golgi mannosidase II in intact cells, and that removal of α-1,3- and α-1,6-linked mannoses by that enzyme is not a prerequisite for the addition of galactose to the outer, nonreducing GlcNAc, or the fucose to the inner reducing GlcNAc. Taken as a whole, these studies indicate that significant changes in oligosaccharide structure have little effect upon the routing and cellular release of a typical N-asparagine-linked glycoprotein,[99] at least in the cultured cells.

Swainsonine also did not affect the insertion or function of the insulin receptor,[100] the epidermal growth factor receptor[101] or the receptor for asialoglycoproteins.[102] The alkaloid did, however, block the receptor-mediated uptake of mannose-terminated glycoproteins by macrophages. This inhibition appeared to be due to the formation of hybrid chains on the glycoproteins present on the macrophage surface, which could then bind to and tie up the mannose receptors.[103]

Some glycoprotein functions are affected by swainsonine. The stimulation of resorptive cells by glucocorticoid probably involves the attachment of osteoclasts and other cells to bone and this appears to be blocked by swainsonine.[104] This alkaloid also reduced the interaction of *Trypanosoma cruzi* with peritoneal macrophages when either host or parasite was treated with the drug.[105] There was a dramatic decline in the ability of B16-F10 melanoma cells to colonize the lungs of experimental animals in the presence of swainsonine.[106] Lymphocytes can be stimulated to proliferate by treating them with concanavalin A, and this stimulation is suppressed by an immunosuppressive factor isolated from the serum of mice bearing a tumor called sarcoma 180. The suppression caused by this factor is overcome by swainsonine, suggesting that this alkaloid might be useful in immunosuppressive diseases.[107] The turnover or degradation of endocytosed glycoproteins is inhibited by swainsonine, leading to an accumulation of the glycoproteins in the lysosomes. The block appears to be an inhibition of lysosomal α-mannosidase (and perhaps other mannosidases) and suggests that the oligosaccharide portion of the glycoprotein must be degraded before the protein is attacked.[108] The above studies suggest some important roles for swainsonine as well as the need for additional experimentation on these systems and others.

8.1.2.4.3 *Effects of swainsonine on animals*

Swainsonine also has dramatic and disastrous effects on animals. Thus, ingestion by grazing livestock of certain leguminous plants of the genera *Astragalus* and *Oxytropis*, as well as the

Australian genus *Swainsona*, induces a chronic neurological disease that resembles the human genetic storage disease α-mannosidosis.[109] Livestock must consume locoweed for an extended period of time before the symptoms of locoweed poisoning become evident. If the locoweed grazing is stopped before the animal becomes too emaciated, the animal will recover but will continue to show neurological difficulties when stressed.[110] Locoweeds can also cause abortion and birth defects if ingested, and swainsonine appears to be secreted in the milk of afflicted animals.[111] Therefore, early detection of the disease is essential. Since humans with various storage diseases frequently secrete in their urine products that result from the disease, the urine of animals fed locoweed or swainsonine was examined to determine whether oligosaccharide accumulation products occurred. The profiles obtained from these animals either by TLC or by HPLC demonstrated that these treated animals did excrete high mannose types of oligosaccharides analogous to individuals suffering from α-mannosidosis.

Rats, sheep and goats treated with swainsonine excrete 'high mannose' types of oligosaccharides in their urine. The major oligosaccharide in the urine of rats and guinea pigs is a $Man_5GlcNAc$, whereas sheep excrete a mixture of oligosaccharides of composition $Man_{2-5}(GlcNAc)_2$ and $Man_{3-5}(GlcNAc)$. The presence of these oligosaccharides suggests that the Golgi mannosidase II as well as lysosomal α-mannosidase is inhibited by swainsonine, resulting in the storage of abnormally processed asparagine-linked glycans from glycoproteins. Although the defect in glycoprotein processing appeared to have little effect on the health of the intoxicated animal, the accompanying lysosomal storage produces a diseased state.[112] In sheep, changes in the patterns of urinary oligosaccharides could be correlated with the onset of visible symptoms, which occurred approximately five weeks after the typical urinary sugars were observed.[113] In locoweed-fed sheep, certain oligosaccharides appeared early in the urine and then rapidly declined, while others appeared only at later times.[114] It seems likely that rapid screening by TLC and by HPLC, as well as a knowledge of the particular oligosaccharide that would be expected in the urine, should make early detection of this disease quite feasible. It should again be pointed out that all of these symptoms appear to be due to the inhibition of the lysosomal and perhaps the processing α-mannosidases in these animals.

Swainsonine was administered to rats by including it in the drinking water, and its effects on tissue enzyme levels were determined. The activity of Golgi mannosidase II was markedly decreased to 22% of control values, but changes in other Golgi enzymes did not occur. However, unexpected and unusual changes occurred in the lysosomal enzymes. In liver, the acid mannosidase increased markedly instead of decreasing as might be expected for a mannosidase inhibitor, and one which causes a mannosidosis-like condition. Also the principal change in brain was an increase in lysosomal α-mannosidase levels. In plasma, most lysosomal enzyme levels increased. These results indicate that pathological effects of swainsonine are not solely attributable to its inhibition of lysosomal α-mannosidase and may also be caused by its inhibition of the glycoprotein-processing glycosidases.[115] Using the pig as an experimental model, the effect of swainsonine and locoweed on the animals was compared directly in order to determine whether swainsonine was responsible for the symptoms of locoism.[116] Both treatments increased lysosomal acid glycosidase activities in most tissues, decreased liver Golgi mannosidase II levels, increased plasma hydrolase levels, and greatly increased the tissue oligosaccharide levels, particularly of $Man_5(GlcNAc)_2$ and $Man_4(GlcNAc)_2$. These results support the idea that swainsonine is truly the agent in locoweed that is responsible for the changes in enzyme levels and in the accumulation of oligosaccharides. In addition, the behavior of the animals was the same, regardless of whether they received locoweed or swainsonine.[116]

8.1.2.4.4 *Chemical synthesis of swainsonine*

A number of laboratories have now accomplished the chemical synthesis of swainsonine. In most cases, the synthesis started with a derivative of glucose or mannose, such as methyl 3-amino-3-deoxy-α-D-mannopyranoside,[117,118] or methyl 4,8-imino-2,3-*O*-isopropylidine-4,6,7,8-tetradeoxy-α-D-mannooctapyranoside.[119] The synthesis of swainsonine was also accomplished from the four-carbon precursor, *trans*-1,4-dichloro-2-butene.[121] Utilizing this synthesis, two other isomers of swainsonine, referred to as 'Glc-swainsonine' [1*S*,2*S*,8*R*,8α*R*-trihydroxyindolizidine] and 'Ido-swainsonine' [1*S*,2*S*,8*S*,8α*R*-trihydroxyindolizidine] were produced. These isomers afforded an opportunity to determine how the chirality of these molecules affected their ability to inhibit various glycosidases. Interestingly enough, the 'Glc-swainsonine' did not inhibit jack bean α-mannosidase, but instead inhibited the fungal α-glucosidase. However, this isomer still inhibited the glycoprotein-processing α-mannosidase II, rather than glucosidase I or glucosidase II. On the other hand, it was much less effective as an inhibitor than was swainsonine.[121] These results indicate that the indolizidine ring structure is not directly analogous to the pyranose ring of the sugar substrates and therefore the chirality of the hydroxyl groups may not be the only factor in the specificity of the inhibition. One of the major advantages of being able to chemically synthesize the various inhibitors is that modifications in the synthesis may produce a number of isomers. Comparisons of the biological activity of these isomers can provide a great deal of information about the requirements for inhibition of specific glycosidases. Several other isomers of swainsonine have also been synthesized, but as yet nothing is known about their biological activity.[122]

8.1.2.4.5 *Castanospermine inhibition of α- and β-glucosidases*

Since castanospermine was known to be a tetrahydroxyindolizidine alkaloid with a structure somewhat related to that of swainsonine, it was tested against a variety of commercially available glycosidases to see whether it would inhibit any of these enzymes. Interestingly enough, castanospermine proved to be a potent inhibitor of almond emulsin β-glucosidase, showing almost complete inhibition of this enzyme at 50 μg mL^{-1}. However, even at levels of 250 μg mL^{-1}, this alkaloid did not inhibit yeast α-glucosidase, α- or β-galactosidase, α-mannosidase, β-glucuronidase, β-*N*-acetylhexosaminidase or α-L-fucosidase. Some inhibition of β-xylosidase was also seen in the presence of castanospermine, but the inhibitor concentration needed for this enzyme was considerably higher than that necessary for the β-glucosidase. The inhibition of β-glucosidase was observed throughout a time course and the inhibition with respect to the substrate *p*-nitrophenyl-β-D-glucopyranoside was of the mixed type. Castanospermine also inhibited β-glucosidase and β-glucocerebrosidase in fibroblast extracts, but in these extracts this alkaloid also inhibited the lysosomal α-glucosidase.[57] In later studies, it was found that castanospermine is in fact an effective inhibitor of many α-glucosidases (see below), but for some unknown reason the yeast α-glucosidase is resistant to this inhibitor.

The mechanism of inhibition of castanospermine on the fungal amyloglucosidase (an *exo*-1,4-α-glucosidase) and on almond emulsin β-glucosidase was studied.[58] Castanospermine proved to be a competitive inhibitor of amyloglucosidase, at both pH 4.5 and 6.0, when assayed with the *p*-nitrophenyl-α-glucoside as substrate. It was also a competitive inhibitor of the β-glucosidase at pH 6.5, but previous studies had indicated that at pH 4.5, the inhibition was of the mixed type. It was found in these experiments that the pH of the incubation mixture had a marked effect on the extent of inhibition. Thus, in all cases, castanospermine was a much better inhibitor at pH 6.0 to 6.5 than it was at lower pH values. The pK_a for castanospermine in water has been found to be 6.09, suggesting that this alkaloid is a much better inhibitor when it is in the unprotonated form. Other evidence to support this idea is the finding that the *N*-oxide of castanospermine is still a competitive inhibitor of amyloglucosidase, but it is 50 to 100 times less effective than is castanospermine. In addition, the activity of the *N*-oxide is not affected by changing the pH of the incubation mixture.[129]

Castanospermine is also a potent, time dependent inhibitor of the purified sucrase–isomaltase complex from rat small intestine. First-order kinetics for the inactivation of sucrase and isomaltase were observed with this alkaloid. Protection studies indicated that castanospermine competed for the glucosyl subsite with the substrates sucrose and isomaltose. The second-order rate constants (K_i) for the association reaction between castanospermine and the protein complex were calculated to be 6.5×10^3 and 0.3×10^3 M^{-1} s^{-1} for sucrase and isomaltase respectively. Only barely detectable reactivation of the inhibited isomaltase could be detected over a 24 h period, but the inhibited sucrase could be reactivated to the extent of 30% in 24 h. The authors suggest that castanospermine functions as a transition state analog that binds extremely tightly to sucrase and isomaltase.[59] Castanospermine was tested against a number of intestinal glycohydrolases and found to be most effective against sucrase with an IC$_{50}$ of 1.1×10^{-7} M. When administered to animals, a significant effect was seen at doses of less than 1 mg kg^{-1} in both normal and streptozotocin-treated rats, in terms of preventing the hyperglycemic response to an oral sucrose challenge.[60]

8.1.2.4.6 *Effect of castanospermine and deoxynojirimycin on glycoprotein processing*

Since castanospermine inhibits α- and β-glucosidases, and since several of the enzymes in the processing pathway are α-glucosidases, this alkaloid was tested as an inhibitor of the trimming of the oligosaccharide chains of the influenza viral hemagglutinin. When influenza virus was raised in MDCK cells in the presence of castanospermine at 10 μg mL^{-1} or higher, 80 to 90% of the viral glycopeptides became susceptible to the action of endoglucosaminidase H, whereas in the normal virus, 70% of the oligosaccharides are resistant to this enzyme. The major oligosaccharide released by endoglucosaminidase H from the castanospermine-treated virus migrated like a hexose$_{10}$GlcNAc upon gel filtration on a calibrated column of Biogel P-4. This oligosaccharide was characterized as a Glc$_3$Man$_7$GlcNAc on the basis of various enzymatic treatments followed by analysis of the products, as well as by methylation analysis of the [2-^{3}H]mannose-labeled and [6-^{3}H]galactose-labeled oligosaccharide. The presence of three glucoses in this oligosaccharide was also confirmed by periodate oxidation studies of the glucose-labeled oligosaccharide. Castanospermine did not inhibit the incorporation of [^{3}H]leucine or [^{14}C]alanine into protein in MDCK cells, even at levels as high as 500 μg mL^{-1}. In addition, castanospermine did not inhibit the production of influenza virus, nor did it alter the infectivity of virus raised in its presence. Castanospermine did, however, alter the carbohydrate structures of molecules on the surfaces of various cells. Thus, MDCK cells grown for several days in the presence of castanospermine were able to bind almost twice as much [^{3}H]concanavalin A as were control cells.[123] These studies indicated that castanospermine inhibited glucosidase I, the first enzyme in the processing pathway, leading to the formation of oligosaccharides still retaining the three glucose residues. Cell free studies with microsomes from rat liver, and recent studies with a partially purified glucosidase I from mung bean seedlings,[124] have shown that this alkaloid is an effective inhibitor of glucosidase I.

Castanospermine also inhibited glycoprotein processing in suspension-cultured soybean cells.[125] In these studies, soybean cells were pulse-labeled with [2-^{3}H]mannose for 30 min and then chased for varying times in unlabeled medium. In the control cells, the initial glycopeptides isolated at the beginning of the chase contained oligosaccharides having mostly Glc$_3$Man$_9$(GlcNAc)$_2$ structures, and these were trimmed to Man$_9$(GlcNAc)$_2$ and then to Man$_7$(GlcNAc)$_2$ with increasing times of chase. On the other hand, when the chase was done in the presence of castanospermine, no trimming of glucose residues occurred, although some mannose residues were apparently still removed. The major oligosaccharide in the glycopeptides of castanospermine-incubated cells after a 90 min chase was the Glc$_3$Man$_7$(GlcNAc)$_2$ structure. Thus, in plants as well as in animals, castanospermine appears to inhibit glucosidase I.[125]

In a study on the synthesis and transport of the E-2 glycoprotein of the coronavirus mouse hepatitis virus A59, a number of glycosylation inhibitors and processing inhibitors were tested including tunicamycin, swainsonine, castanospermine, *N*-deoxynojirimycin and deoxymannojirimycin. These studies demonstrated that treatment of MHV A59-infected cells with tunicamycin, or with processing inhibitors of N-linked glycans, influenced the transport properties as well as the antigenic properties of the spike glycoprotein E-2 and indirectly affected the growth of MHV A59. The processing inhibitors that act earliest in the pathway, such as castanospermine and deoxynojirimycin, had strong effects, whereas swainsonine and deoxymannojirimycin had little or no effect. Thus, the presence of glucose on the N-linked glycans strongly affected transport of E-2. On the other hand, the E-2 synthesized under these conditions was still acylated but accumulated in a compartment distinct from the Golgi. Other processing inhibitors that act at later steps in the processing pathway, *i.e.* swainsonine and deoxymannojirimycin, had little or no effect on the transport or antigenic properties of E-2.[126]

Swainsonine and castanospermine were also used to study the biosynthesis and transport of the insulin receptor in IM-9 lymphocytes. In the presence of swainsonine, the insulin receptor had a slightly lower molecular weight and contained hybrid rather than complex types of oligosaccharides. Nevertheless, the receptor was still autophosphorylated in the presence of hormone and still had the same affinity for insulin. But in the presence of castanospermine or deoxynojirimycin, cultured IM-9 lymphocytes had a 50% reduction in surface insulin receptors as demonstrated by ligand binding and affinity crosslinking with [^{125}I]insulin and lactoperoxidase/Na^{125}I-labeling studies. This reduction in insulin receptor levels was not due to changes in degradation rates but was caused by the fact that the lack of glucose removal delayed the processing of this precursor, which probably accounts for the reduction in cell surface receptors.[127]

In addition to the oligosaccharide-processing reactions shown in Figure 3, other posttranslational modifications may occur on N-linked glycoproteins such as addition of palmitic acid, addition of sulfate residues, fucosylation and so on. Castanospermine and other processing inhibitors were used to determine what oligosaccharide structure (if any) was necessary for these modifications to occur. For example, many animal glycoproteins have an α-1,6-linked fucose attached to the innermost GlcNAc, and biosynthetic studies on this fucosyl transferase have indicated that oligosaccharide acceptors for this fucose must have a GlcNAc attached to the 3-linked mannose unit. Thus the GlcNAc-β-1,2-Man-α-1,3[Man-α-1,6(Man-α-1,3)]Man-α-1,6-Man-β-1,4-GlcNAc-β-1,4-GlcNAc structure was an excellent acceptor for this fucose with the isolated enzyme. In keeping with this data, several studies have indicated that when various animal cells are grown in culture in the presence of deoxynojirimycin or castanospermine, the N-linked glycoproteins do not contain fucose, presumably because these oligosaccharides remain as glucose-containing high mannose structures.[90,99,128] However, when the cells are incubated with swainsonine, they are able to add fucose to the hybrid structures, probably because these oligosaccharides do have the necessary GlcNAc signal. Similar results were obtained with regard to sulfation of the N-linked structures.[129] However, neither castanospermine nor swainsonine had any effect on the palmitylation of the α-subunit of the sodium channel in primary cultures of rat brain neurons.[130]

Some very recent studies with castanospermine have suggested that this alkaloid inhibits the growth of the AIDS virus (or other related enveloped retroviruses) in cell culture. Thus, the alkaloid inhibited syncytium formation induced by the envelope glycoprotein of the human immunodeficiency virus and inhibited viral replication. The decrease in syncytium formation in the presence of castanospermine was attributed to inhibition of processing of the envelope precursor protein gp 160, with resultant decreased cell surface expression of the mature envelope glycoprotein gp 120. The alkaloid might also cause defects in steps involved in membrane fusion after binding of CD4 antigen. Also modification in the structure of the envelope glycoprotein could affect the ability of the virus to enter the cells.[131–133] Deoxynojirimycin and castanospermine have also been found to affect other viral systems. Both of these compounds inhibited the formation of Sindbis virus in BHK cells and in several other cell lines. Growth of Sindbis was inhibited to a much greater extent at 37 °C than at 30 °C.[134] Studies with the San Juan strain of VSV indicated that formation of the virus became temperature sensitive when glucose residues were retained on the oligosaccharide chains of the G protein, as happens in the presence of these two inhibitors.[135] However, in another study,[136] deoxynojirimycin was reported to have no effect on the surface expression of VSV G protein, influenza hemagglutinin and class I histocompatability antigen.

Castanospermine was injected into rats to determine its effects on glycogen metabolism. The alkaloid caused a marked decrease in the activity of the liver α-glucosidase, and, after injections of alkaloid over a three or four day period, glycogen disappeared from the cytoplasm of hepatocytes and accumulated in the lysosomes.[137] This alkaloid was shown to be a potent inhibitor of rat hepatic lysosomal α-glucosidase *in vitro*. The inhibition was time dependent, indicating a slow interaction between inhibitor and enzyme. This slow interaction probably occurred at the active site of the enzyme and, once it occurred, the inhibitor apparently bound very tightly. Attempts to displace the inhibitor or reverse the inhibition by dialysis or substrate displacement were unsuccessful.[138] Castanospermine is also a potent inhibitor of intestinal sucrase and maltase, and these enzymatic activities are markedly reduced in animals treated with castanospermine.[59,60] *In vitro*, castanospermine was most effective against sucrase with an IC_{50} of 1.1×10^{-7} M.[59] *In vivo* in rats,[139] a significant effect of this alkaloid was seen at doses less than 1 mg kg^{-1} in both normal and streptozotocin-treated animals. The castanospermine was a very potent long-acting inhibitor of the glycemic response to an oral sucrose load in rats, explicable by its effects on the intestinal α-glucosidases.

The seeds of the Moreton Bay chestnut also contain another indolizidine alkaloid that was isolated and characterized as 6-epicastanospermine. This isomer might be expected to be an inhibitor of α-mannosidase since the 6-hydroxyl group of the alkaloid should be analogous to the 2-hydroxyl of glucose. Instead, the 6-epicastanospermine differed only from castanospermine in that it no longer inhibited β-glucosidase and was a weaker inhibitor of α-glucosidase. It did not inhibit other glycosidases, including β-mannosidase.[140] Thus, at this state of our knowledge, we cannot predict whether these compounds will be inhibitors of glycosidases nor which glycosidases will be sites of inhibition. We need more inhibitors with varied structures before we can link chirality of the inhibitor to site of action on specific glycosidases.

8.1.2.4.7 *Deoxymannojirimycin*

Since glucosidase I is inhibited by deoxynojirimycin, the 2-epimer of this inhibitor was synthesized chemically[65] and tested against the processing glycosidases. This synthetic compound, called deoxymannojirimycin, was found to be a potent inhibitor of mannosidase IA/B, and caused the accumulation of high mannose oligosaccharides, of the Man$_{8-9}$(GlcNAc)$_2$ type, on the glycoproteins of cultured cells.[141,142] Deoxymannojirimycin did not inhibit the secretion of IgD or IgM in cultured cells, but deoxynojirimycin did inhibit the secretion of these proteins.[143] Deoxymannojirimycin was also without effect on the formation and appearance of the G protein of VSV, as well as the hemagglutinin of influenza virus and the HLA A, B and C antigens.[143]

This inhibitor was used as a tool to study the recycling of membrane glycoproteins in order to see whether oligosaccharides could be processed or modified during endocytosis. Thus, membrane glycoproteins were synthesized in cell cultures in the presence of deoxymannojirimycin and labeled by the addition of $[2\text{-}^3\text{H}]$mannose. The label and the inhibitor were then removed by changing the medium, and the alterations in the structures of the oligosaccharides of several membrane glycoproteins was followed with time. The initial oligosaccharide structure, before the start of the chase, was a $\text{Man}_{8-9}(\text{GlcNAc})_2$, in keeping with the fact that deoxymannojirimycin inhibits mannosidase I. However, during the chase, a small but apparently significant amount of the oligosaccharides of transferrin and other membrane glycoproteins were modified, indicating that some of these glycoproteins did recycle through the mannosidase I compartment during endocytosis.[144] Deoxymannojirimycin was also used to determine the role of the ER mannosidase in the processing of the oligosaccharide chains of the endoplasmic reticulum enzyme HMG CoA reductase using UT-1 cells. These cells are a compactin resistant cell line that has amplified amounts of this important enzyme. In the absence of inhibitor, the predominant oligosaccharides on the reductase are single isomers of $\text{Man}_6(\text{GlcNAc})_2$ and $\text{Man}_8(\text{GlcNAc})_2$. However, in the presence of the inhibitor, the $\text{Man}_8(\text{GlcNAc})_2$ accumulated, indicating that the ER enzyme was responsible for the initial trimming.[145] However, not all hepatocyte glycoproteins were found to be substrates for the ER mannosidase.

8.1.2.4.8 *Pyrrolidine alkaloids*

The pyrrolidine alkaloid 2,5-dihydroxymethyl-3,4-dihydroxypyrrolidine (DMDP) was isolated from the plants *Lonchocarpus sericeus* and *Derris elliptica*,[68] and tested as an inhibitor of various glycosidases.[69] The compound inhibited almond emulsin β-glucosidase ($K_i = 7 \times 10^{-6}$ M), yeast α-glucosidase ($K_i = 6 \times 10^{-6}$ M), and insect trehalase ($K_i = 5.5 \times 10^{-5}$ M). DMDP was also found to inhibit the processing of the N-linked oligosaccharides of the hemagglutinin of the influenza virus when MDCK cells were used as the host. Thus, in the presence of 250 $\mu\text{g mL}^{-1}$ of alkaloid, more than 80% of the oligosaccharide chains became susceptible to the action of endoglucosaminidase H, indicating that they were not processed to complex structures. These oligosaccharides were characterized as having $\text{Glc}_3\text{Man}_{7-9}(\text{GlcNAc})$ structures. These experiments indicated that DMDP inhibited glucosidase I,[146] and this site of action was confirmed by *in vitro* studies.[124] However, in other studies with IEC-6 intestinal cells, 5 mM DMDP also inhibited complex chain formation by 80% or so, but in this case there was an increase in high mannose ($\text{Man}_{7-9}\text{GlcNAc}$) as well as glucose-containing oligosaccharides.[147] These workers suggested that DMDP did not primarily inhibit glucosidase I in these cells. In a study examining the synthesis of the oncogenic protein V-erb B, a membrane glycoprotein, DMDP caused the same alterations in structure as did deoxynojirimycin, indicating that glucosidase I was the site of inhibition of both alkaloids.[148]

Although DMDP is not as effective as an inhibitor as the other glucosidase inhibitors (*i.e.* deoxynojirimycin or castanospermine), it is interesting and informative in several respects. First of all, the fact that it acts as an inhibitor indicates that a six-membered ring structure (*i.e.* pyranose ring) is not absolutely necessary for a compound to be an inhibitor of glycosidases, although six-membered rings may be more effective than five-membered ring structures. Secondly, these studies indicate that a nitrogen (*i.e.* NH) in the ring and several hydroxyl groups, in the proper orientation, are necessary for activity. Recently, DMDP was synthesized chemically and the synthetic route for this synthesis should allow for various isomers to be produced.[149] Such compounds should be valuable additions to the already existing glycosidase inhibitors.

8.1.2.4.9 *Chemically produced glycosidase and processing inhibitors*

1,4-Dideoxy-1,4-diamino-D-mannitol (DIM) was chemically synthesized from benzyl-α-D-mannopyranoside and found to be a competitive inhibitor of jack bean α-mannosidase.[76] This compound was also shown to inhibit the formation of complex types of oligosaccharides in the influenza viral hemagglutinin.[160] The oligosaccharides that accumulated in the presence of DIM were mostly of the $\text{Man}_9\text{GlcNAc}$ structure, suggesting that DIM was inhibiting mannosidase I. In keeping with this observation, DIM was also found to inhibit the *in vitro* action of rat liver mannosidase I using the $[^3\text{H}]$mannose-labeled $\text{Man}_9\text{GlcNAc}$ as substrate in these reactions.[150] Althugh DIM is not as effective as an inhibitor of mannosidase I as swainsonine is on mannosidase II, it can be synthesized in good yield and in a relatively straightforward synthesis. Therefore, DIM and its derivatives or

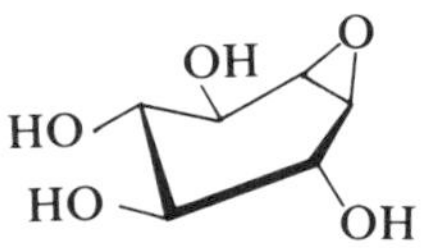

Conduritol epoxide

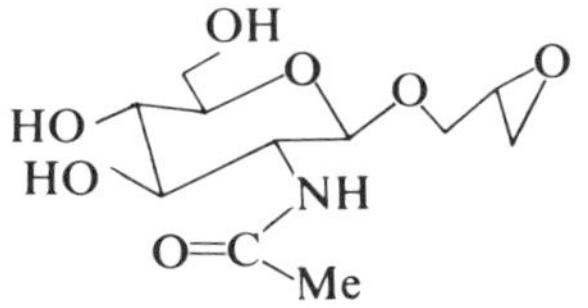

2′,3′-Epoxypropyl-β-glycoside of *N*-acetylglucosamine

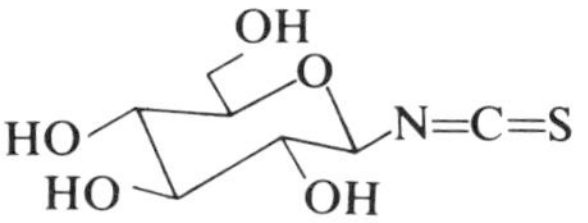

Isothiocyanate derivative

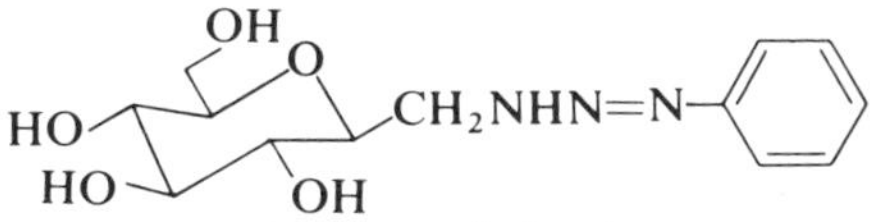

Triazene derivatives

Figure 5 Structures of conduritol epoxide and other chemically synthesized inhibitors

related isomers should be valuable compounds for studies in animals, provided they are not toxic. Such inhibitors could be used to produce animal models of the various lysosomal storage diseases in which various glycosidases are missing.

Conduritol epoxide and other conduritol derivatives (see Figure 5 for structures) are also inhibitors of glycosidases and have been used in various studies.[151] Thus, a covalent, active-site-directed inhibition of various β-D-glucosidases was first obtained by Legler,[9] using bromoinositol derivatives. On the basis of the time dependence of the inhibition, the author suggested that the actual inhibitor was the epoxide that was produced from the bromo derivative by *trans* elimination of HBr. Thus, 1,2-anhydromyoinositol was prepared from conduritol B and found to inhibit the enzyme at least 10^3 times more effectively.[152]

The reaction of epoxides with various nucleophiles is well known and appears to require acid catalysis except in those cases where strong nucleophiles such as alkoxide, thiolate or thiosulfate ions are used.[153] Epoxides in the inositol series are usually stable and rather unreactive compounds and therefore acid catalysis is especially important for their reaction. The reason that they are highly specific inhibitors for glycosidases is apparently because they react only when the binding site contains a suitably oriented acidic group for catalysis in addition to a nucleophilic group in the correct position. These two groups are also believed to be involved in the catalysis of glycoside hydrolysis. Of course, the orientation of the hydroxyl groups in the cyclohexane ring must correspond to those of the sugar for which the enzyme shows specificity.

When conduritol B epoxide is synthesized chemically, one obtains a racemic mixture of two derivatives, *i.e.* the 1-D-1,2-anhydromyoinositol and the 1-L-1,2-anhydromyoinositol. The glucosidases discriminate between these two derivatives since only 1-D-1,2-anhydromyoinositol is an inhibitor of β-glucosidases, while the L enantiomer is not an inhibitor of either α- or β-glucosidases. Reaction of the D enantiomer with β-glucosidases gives rise to (+)-chiroinositol after release from the enzyme with hydroxylamine, whereas reaction with α-glucosidases and release with hydroxylamine gives scylloinositol.[154] On the basis of these data, Legler has proposed the scheme shown in Figure 6 as a possible mechanism for the reaction of conduritol epoxide with β-glucosidase. The reaction with α-glucosidase would be similar but would give a different enantiomer.[9]

This inhibitor (*i.e.* conduritol epoxide) administered to rats produced a model of Gaucher's disease by inhibiting the β-glucosidase of brain, liver and spleen and caused an increase in the levels of glucosylceramide in tissues.[155] Bromoconduritol (6-bromo-3,4,5-trihydroxycyclohex-1-ene) is an active-site-directed covalent inhibitor of glycosides that was also synthesized chemically.[156] In virus-infected cells, bromoconduritol caused the accumulation of endoglucosaminidase H sensitive oligosaccharides that had the structures $GlcMan_9GlcNAc$, $GlcMan_8GlcNAc$ and

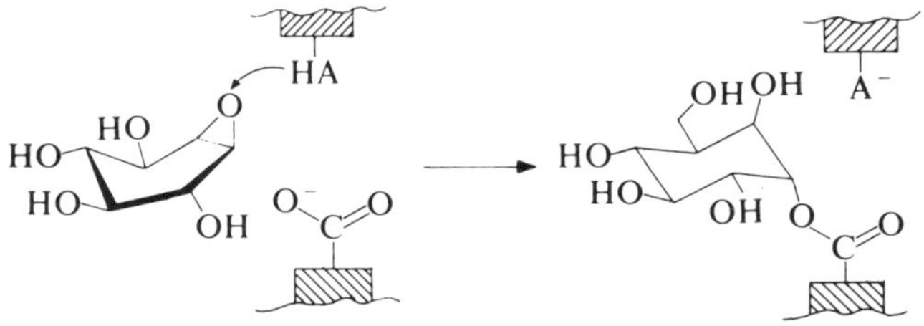

Figure 6 Proposed mechanism involving acid-catalyzed inhibition of β-glucosidase by conduritol epoxide

GlcMan$_7$GlcNAc. This led the authors to suggest that this compound inhibited glucosidase II.[157] However, glucosidase II is believed to remove the two innermost α-1,3-linked glucoses. Thus, inhibition of this enzyme would be expected to give rise to a Glc$_2$Man$_9$GlcNAc. Thus, it is not clear what the specific site of action of this inhibitor is. One problem with bromoconduritol is that it has a half-life of only 15 min in water and is therefore difficult to use with cultured cells. It is not clear whether its degradation products are toxic to cells.

Glycosylmethyl-*p*-nitrophenyltriazenes (Figure 5) irreversibly inactivate lysosomal glycosidases such as β-galactosidase and β-glucosidase *in vitro*.[158] The mannopyranosylmethyl-*p*-nitrophenyl-triazene was tested to determine its effect on the biosynthesis and degradation of α-1-acid glycoprotein in hepatocytes. This compound prevented the conversion of high mannose chains to complex oligosaccharides and gave mostly oligosaccharides of the Man$_{7-9}$GlcNAc types. These data suggested that the inhibitor acted at the mannosidase I stage. This compound also blocked the hydrolysis of endocytosed α-1-acid glycoprotein chains probably by inhibiting lysosomal α-manno-sidase.[159]

8.1.3 ACETYLCHOLINESTERASE

A nerve impulse is an electrical signal that is caused by a flow of ions across the plasma membrane of a neuron or other susceptible cell. In the case of most cells, including the neuron, the interior of the cell is high in K$^+$ and low in Na$^+$. These ion concentrations are generated by means of a specific transport system called the Na$^+$–K$^+$ ATPase.[160–162] The active transport of Na$^+$ and K$^+$ is of great importance as demonstrated by the fact that more than one third of the ATP utilized by a resting animal is used to pump Na$^+$ and K$^+$.

The Na$^+$–K$^+$ ATPase of human erythrocytes, as well as those from dog and sheep kidney, have been purified and their properties have been studied in considerable detail.[163, 164] The enzyme is a 270 kDa tetrameric protein that has a composition of α_2-β_2. The large α-subunit has a molecular weight of 95 kDa and contains a site that is involved in the hydrolysis of ATP and another site that binds the cardiotonic steroid inhibitors such as digitoxigenin and oubain. These inhibitors have profound effects on the physiology of the heart since they inhibit the dephosphorylation of the ATPase. In terms of the orientation or the topology of these different sites, the ATP hydrolysis site is located on the cytosolic side of the membrane, whereas the binding site for the cardiotonic steroids is on the external side. The smaller β-subunit has a molecular weight of 40 kDa and contains covalently bound carbohydrate. Crosslinks can be formed between the α-subunits, or between an α-subunit and a β-subunit, but not between two β-subunits. These data indicate that the α-subunits are oriented in close proximity to each other, but the β-subunits are far apart. This has led to a model of how this protein is oriented in the membrane as a tetramer as β-α-α-β.[164, 165]

In the resting state, the membrane of the nerve axon is more permeable to K$^+$ than to Na$^+$, and, as a result, the membrane potential is largely determined by the ratio of internal to external K$^+$. The membrane potential of -60 mV found in the unstimulated axon is very close to the level (-75 mV) that would be expected for a membrane that was only permeable to K$^+$. A nerve impulse or action potential is generated when the membrane potential is depolarized beyond a critical threshold value, such as from -60 mV to -40 mV. The membrane potential becomes positive within a millisecond and attains a value of $+30$ mV before it becomes negative again. The amplified depolarization is then propagated along the nerve.[167, 168]

Many years ago, Hodgkin and Huxley[166] were able to show that this action potential resulted from large and transient changes in the permeability of the axon membrane to Na$^+$ and K$^+$. The first change is in the conductance of the membrane to Na$^+$. The depolarization of the membrane beyond the threshold value then leads to an opening of the Na$^+$ channels. As a result, sodium ions begin to flow into the cells as a consequence of the large electrochemical gradients across the membrane. As more and more Na$^+$ enters the cell, it causes further depolarization and the opening of more channels.[169] The result, as indicated above, is a very rapid and large change in the membrane potential from about -60 mV to about $+30$ mV, which is the Na$^+$ equilibrium potential. Then, the Na$^+$ channels spontaneously close and the K$^+$ gates open, allowing K$^+$ to flow outward, and returning the membrane potential to -75 mV. This value is the K$^+$ equilibrium potential. Actually, only a very small proportion of the Na$^+$ and K$^+$ in the nerve cell actually flows across the membrane during this action potential. Nevertheless, the action potential is a very effective means of sending signals over considerable distances within the organism.

A synapse is a nerve junction where two nerves come together and can interact by the passage of a signal from one to the other (Figure 7). The presynaptic membrane of the transmitting nerve is

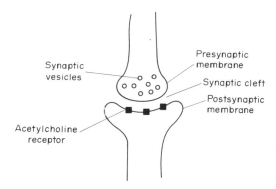

Figure 7 Schematic representation of a synapse showing the pre- and post-synaptic membranes. Within the presynaptic neuron are shown vesicles containing acetylcholine, while on the postsynaptic membrane are acetylcholine receptors

separated from the postsynaptic membrane of the receiving nerve by a gap of about 500 Å, which is referred to as the synaptic cleft.[170] The nerve impulse is transmitted or communicated across the synapse by a chemical transmitter such as acetylcholine or epinephrine, *etc.* This neurotransmitter is also the transmitter that transmits signals between the nerve and striated muscle (*i.e.* at motor end plates).

A cholinergic synapse is one that uses acetylcholine as the neurotransmitter. In this case, the presynaptic axon is filled with vesicles that contain acetylcholine. A nerve impulse causes the fusion of vesicles with the plasma membrane and the release of acetylcholine into the synaptic cleft. The acetylcholine diffuses to the postsynaptic membrane, where it combines with acetylcholine receptors present in this membrane.[171,172] The interaction of acetylcholine with its receptor causes a significant conformational change in the receptor that results in a large change in the permeability properties of the membrane. As a result, a channel is opened that is equally permeable to Na^+ and K^+, but since the electrochemical gradient across the membrane is much steeper for Na^+, the flow of Na^+ is much greater. The neurotransmitter thus causes a depolarization of the membrane which is then propagated along the electrically excitable membrane of the second nerve. Finally, acetylcholine is hydrolyzed by acetylcholinesterase and the polarization of the postsynaptic membrane is restored.

The synthesis of acetylcholine occurs near the presynaptic ends of the axons by the transfer of acetate from acetyl-CoA to choline. The enzyme that catalyzes this reaction is called choline acetyltransferase.[173] Some of this newly synthesized acetylcholine is taken up into vesicles, whereas the remainder stays in the cytoplasm. A cholinergic synaptic vesicle that is about 400 Å in diameter has been estimated to contain about 10 000 molecules of acetylcholine.

The depolarizing signal must be turned off in order to restore the excitability of the membrane. The signal is destroyed by the hydrolysis of acetylcholine by the enzyme acetylcholinesterase. The possibility that an enzyme catalyzing the hydrolysis of acetylcholine, or other choline esters, was associated with nerve fibers and muscle tissue was postulated by Dale in 1914, based on a number of pharmacological experiments.[174] The actual demonstration of such an enzyme in nerve and muscle tissue was first described by Nachmansohn and Lederer and shown to be associated with the particulate fraction of the cell.[175] In the meantime, enzymes that could catalyze the hydrolysis of choline esters of various carboxylic acids were found in many other tissues.[176] The general reaction catalyzed by these enzymes (cholinesterases) is as follows (where R represents the acyl group with its carboxyl function and X represents the amino group with its alcohol function)

$$RX + H_2O \longrightarrow ROH + HX$$

The difference between those enzymes referred to by the general designation 'cholinesterase' and those more specifically referred to as 'acetylcholinesterase' is related to the specificity of the latter group of enzymes. For example, the turnover number for acetylcholinesterase is sharply decreased when the acyl group of the substrate is larger than propionyl,[177] but is relatively insensitive to variations in the alcohol group among either charged or neutral substrates.[178] On the other hand, the turnover number of carboxylic acid ester substrates for the cholinesterases is much less sensitive to the size of the acyl group, but considerably more sensitive to variations in the charge and/or structure of the alcohol group.[179] Also the group of acetylcholinesterases from different tissue sources are much more similar to each other with regard to variation in substrate and inhibitor specificity than are the group of enzymes referred to as cholinesterases.[180]

Since acetylcholinesterase activity was found to be several fold higher in muscle endplates than it is in other regions of the muscle, and since the electric organs of fishes are thought to be phylogenetically modified skeletal muscle, Nachmansohn measured the activity of the enzyme in the electric organs of the ray *Torpedo marmorate* and of the eel *Electrophorus electricus*. He found that these organs did in fact have very high concentrations of acetylcholinesterase and he was able to obtain the first soluble extract of the enzyme from these sources.[181] As a result of those pioneering studies, the electric eel has been a favored source of acetylcholinesterase and many of the studies in this field have used enzyme from that source.

When acetylcholinesterase is solubilized from the electric organ of *Electrophorus*, and the enzyme subjected to sedimentation analysis in sucrose gradients, three forms of the enzyme are detected that have been referred to as A_4, A_8 and A_{12}. These forms contain one, two and three tetrameric assemblies of catalytic subunits, and a rod-like tail or structural element. This tail gives the assembly an asymmetric character and can be clearly seen in the electron microscope.[182] These forms were converted into globular forms by either proteolytic treatment or by sonication. The globular forms were designated as G_1, G_2 and G_4. The asymmetric forms noted above, but not the globular forms, were found to contain hydroxyproline and hydroxylysine and to be sensitive to digestion by protease free collagenase, indicating that the tail-like structure was probably collagen.[183] Other reports have indicated that acetylcholinesterase is found in the synaptic cleft of neurons, where it is bound to a network of collagen and proteoglycans derived from the postsynaptic cell. This enzyme was reported to have a molecular weight of about 260 kDa. The most complex structure, A_{12}, appears to be composed of six dimers and the tail structure. Each dimer is composed of two peptides of 80 000 Da, which are held together by disulfide bridges. One half of the dimers are bridged to the tail, also by disulfide bonds.[184] Apparently there is some dispute as to whether the tail structure is really an integral part of this enzyme.[185] The characterization of the physical properties of the mammalian enzyme has progressed more slowly, probably because of the limited amounts of enzyme available.

The hydrolysis of acetylcholine by acetylcholinesterase proceeds by the transfer of the acyl group, R (see above reaction), to form an acyl enzyme, which is then hydrolyzed.[186] Kinetic studies have distinguished two subsites at the active center of the enzyme,[187] and these subsites have been incorporated into a model of the enzyme proposed by Nachmansohn and Wilson as shown in Figure 8.[177] Cationic substrates and inhibitors apparently bind at a negatively charged anionic subsite, whereas the ester subsite contains an active serine residue which becomes acylated during the catalytic event (see Figure 9 for reaction). The model in Figure 8 shows two distinct but topologically adjacent subsites at the active site of the acetylcholinesterase, and is apparently in accord with the data obtained from kinetic and thermodynamic measurements. On the basis of studies suggesting an additional intermediate in the acylation reaction, the model has been expanded and presented in more detail to suggest an induced-fit complex.[188] The catalytic mechanism of acetylcholinesterase resembles that of the well-known protease chymotrypsin. As shown in Figure 9, acetylcholine reacts with a specific serine residue at the active site (subsite 2) of the enzyme to form a covalent acetyl enzyme intermediate, and free choline is released. The acetyl enzyme intermediate then reacts with water to form acetate and regenerate the free enzyme.

As discussed below in much more detail, several other classes of acylating agents, in addition to carboxylic acid esters, show reactivity towards the acetylcholinesterase catalytic site. These include esters and acyl halides of substituted phosphoric, carbamic and sulfonic acids.[189] The acyl enzymes formed by these compounds are hydrolyzed very slowly and thus these compounds may be potent

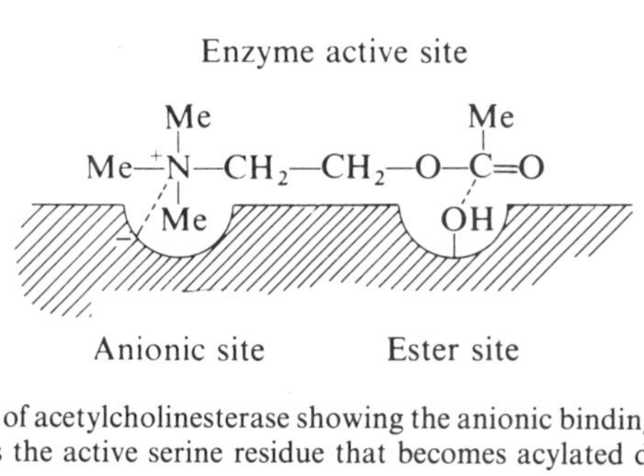

Enzyme active site

$$Me-\overset{+}{\underset{\underset{Me}{|}}{N}}-CH_2-CH_2-O-\overset{\overset{Me}{|}}{C}=O$$

Anionic site   Ester site

Figure 8 A model for the active site of acetylcholinesterase showing the anionic binding site and the site that binds the esters. At the ester site is the active serine residue that becomes acylated during the reaction

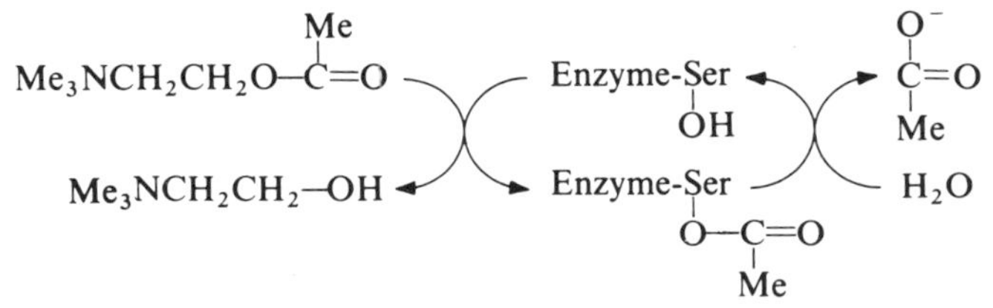

Figure 9 The reaction catalyzed by acetylcholinesterase, whereby acetylcholine is hydrolyzed to give free choline and acetate. The intermediate in this reaction is the covalently modified enzyme which becomes acylated during reaction

inhibitors of the enzyme. Organophosphorylated acetylcholinesterase may be dephosphorylated with nucleophilic agents with concomitant reactivation of the enzyme.[190]

The most potent and well-characterized inhibitors of cholinesterases fall into three classes as follows: organophosphorus inhibitors, carbamates and organosulfonates.[191] Generally all of these inhibitors are regarded as irreversible. The structures of some representative compounds of each class are shown in Figure 10. The mechanism of the reaction of a typical organophosphorus compound such as diisopropyl phosphofluoridate (DIPF) with the acetylcholinesterase is apparently similar to that with the proteolytic enzyme chymotrypsin. In the latter case, reaction with DIPF produces a time dependent irreversible inactivation of chymotrypsin, indicating covalent modification.[192] The authors used [^{32}P]DIPF and found that stoichiometric amounts of radioactivity were incorporated into the enzyme. Since the kinetics of labeling with ^{32}P closely followed the inactiv-

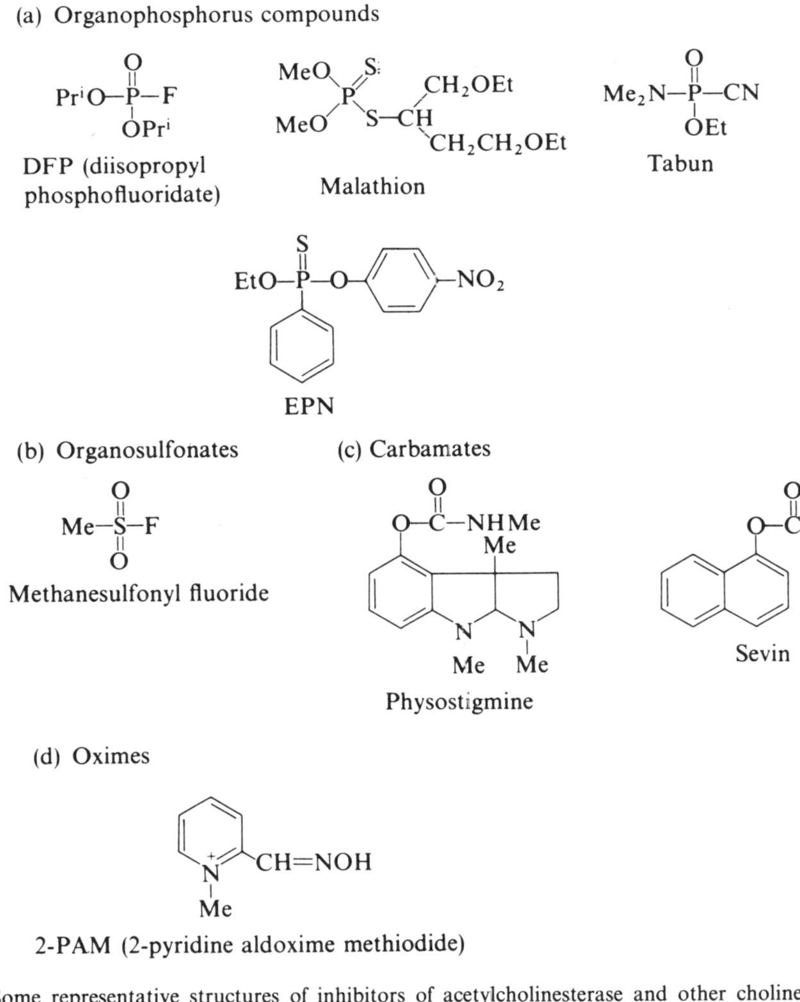

Figure 10 Some representative structures of inhibitors of acetylcholinesterase and other cholinesterases. Included are organophosphorus compounds, organosulfonates and carbamates. The oxime, PAM, is apparently a weak inhibitor of the enzyme, but is a potent antidote for organophosphorus poisoning

ation of the enzymatic activity, the authors suggested that covalent modification was destroying the catalytic activity. More recent studies have demonstrated that the organophosphorus compounds become covalently attached to the serine residues at the active site through a phosphomonoester bond.

Since some of the organic fluorophosphates such as DIPF, tetraethyl alkyl pyrophosphate (TEPP) or isopropoxymethyl fluorophosphonate (Sarin) are chemical warfare agents (nerve gases), whereas others are insecticides, a great deal of effort has been spent towards finding antidotes that can reverse their effect. The toxicity of these phosphorus-containing compounds appears to be mostly due to their remarkable facility to inhibit acetylcholinesterase.[193,194]

Early studies on the reversal of the inhibition indicated that nucleophilic agents such as hydroxylamine could reactivate the DIPF-inhibited acetylcholinesterase. Hydroxylamine is toxic to animals in the concentrations that are necessary to reactivate a DIPF-inactivated enzyme, and, therefore, derivatives of hydroxylamine were synthesized that would hopefully be more active. Because the acetylcholinesterase has an anionic binding site, a derivative with a quaternary nitrogen appeared to be a reasonable choice. A very effective 'antidote' proved to be 2-pyridine aldoxime methiodide (PAM).[195] In this compound (Figure 10), the oxime functional group and the quaternary structure are so arranged that the compound is bound to the inhibited enzyme in an orientation that directs the oxime function against the nucleophilic phosphorus atom.[195] PAM is thus a very effective antidote for many of these organophosphorus compounds. In fact, 10^{-6} M PAM is as effective as 1 M hydroxylamine. Interestingly enough, PAM is more effective when used in combination with atropine.[196]

Since PAM is very poorly soluble in lipid solvents, it appeared likely that its penetration into tissue was restricted to the peripheral nervous system and this expectation seems realistic with the finding that this compound shows little tendency to reactivate brain acetylcholinesterase *in vivo*.[197] However, it shows substantial reactivation of the muscle enzyme. The effectiveness of PAM as an antidote is probably due to the fact that the primary damage done by these poisons is to the peripheral nervous system. Another derivative of PAM, called PAD (2-pyridine aldoxime dodec-iodide) is much more soluble in lipid solvents than is PAM. When the two drugs (PAM and PAD) were used in combination on Sarin-poisoned mice, the survival rate was greatly increased over that seen with PAM alone. These data suggested that PAD was able to penetrate the brain and other areas that PAM could not affect.[198]

Thus, a variety of oxime derivatives can reactivate the acetylcholinesterase from the effects of organophosphorus compounds. The ability of the oximes to displace the organophosphate from the enzyme decreases with time of incubation, presumably because the phosphorylated complex undergoes an additional reaction in which one of the alkyl groups of the organophosphate moiety is eliminated.[199] Berends *et al.*[200] were able to show that the amount of cholinesterase that could not be reactivated following DIPF poisoning was proportional to the amount of monoisopropyl phosphate bound to the enzyme and inversely proportional to the amount of diisopropyl phosphate bound to the enzyme. Thus the cleavage of an alkoxy group from the organophosphate moiety of the cholinesterase–inhibitor complex renders it resistant to removal by the oximes. The group released from the Sarin–cholinesterase complex was identified as the isopropyl group by using tritium-labeled Sarin.[199]

The organosulfonates are analogous to the organophosphates, although they are much weaker cholinesterase inhibitors. Several laboratories have found some advantages in using compounds such as methanesulfonyl fluoride for inhibition studies, since the inhibition is irreversible and the sulfonyl group is small and minimizes the possibility of steric hindrance.[201] The selenophosphorus compounds are less stable than the corresponding sulfur compounds and they are also more toxic.

In addition to organic phosphates and organic sulfonates, various other compounds can serve as inhibitors of acetylcholinesterase. For example, carbamoyl compounds can also react with the active serine to produce a carbamoyl–enzyme complex. Physostigmine (Figure 10) is an alkaloid that is derived from the calabar bean. This plant was once used in witchcraft trials as an ordeal poison. Neostigmine is another inhibitor of acetylcholinesterase that functions in the same way. It binds to the enzyme so that its positively charged trimethylammonium group is at the anionic site of the enzyme, while the carbamoyl group is adjacent to the active serine.[202,203] The enzyme becomes carbamoylated and the alcohol is released.[204] The carbamoyl–enzyme intermediate is subsequently hydrolyzed at a very slow rate in contrast to the acyl–enzyme intermediate. Interestingly enough, it was found that cats treated with a low dose of eserine, which gives mild signs of poisoning, were able to withstand several times the LD_{50} of DFP, administered a few minutes later. On the other hand, the toxicity of carbamates administered after the DFP appears to be additive. The details of these effects are apparently not understood at this time.

Although there are a number of other compounds that have been reported to inhibit cholinesterases, these appear to be less specific types of compounds and include inorganic compounds such as NaCl or NaF, monoquaternary compounds such as choline, and a variety of miscellaneous compounds such as glycine, glutamic acid, creatine, creatinine, serotonin, morphine, and so on. While these are all of interest, the number of compounds is too great to cover in this limited review.

8.1.4 SUMMARY

As indicated in the introduction, a review on hydrolases could cover a great diversity of enzymes, including glycosidases, esterases, peptidases, lipases, phosphatases, *etc.* Such a review on inhibitors of these enzymes could cover several volumes and still not be complete. Thus this review has focused on inhibitors of glycosidases and inhibitors of cholinesterases. These two groups of enzymes have been extensively studied and we now have some powerful tools to inhibit these activities and learn more about the mechanism of catalytic activity and on how inhibition of these enzymes affects the physiological state of the organism. Once some of the above enzymes become crystallized and we can study the active site of the protein, we may be able to design even better and more specific inhibitors.

8.1.5 REFERENCES

1. V. Ginsburg and E. F. Neufeld, 'Methods in Enzymology', Academic Press, New York, 1966.
2. W. W. Shreeve, 'Physiological Chemistry of Carbohydrates in Mammals', Saunders, Philadelphia, 1974.
3. H. G. Hers and F. Van Hoof, 'Lysosomes and Storage Diseases', Academic Press, New York, 1973.
4. J. J. Marshall, 'Mechanisms of Saccharide Polymerization and Depolymerization', Academic Press, New York, 1980.
5. J. Robiquet and C. Boutron, *Ann. Chim. Phys.*, 1980, **44**, 366.
6. J. Liebig and F. Wohler, *Ann. Pharm.*, 1837, **21**, 96.
7. D. E. Koshland, in 'The Enzymes', ed. P. D. Boyer, Academic Press, New York, 1959, vol. 1, p. 305.
8. B. Capon, *Chem. Rev.*, 1969, **69**, 407.
9. P. Lalegerie, G. Legler and J. M. Yon, *Biochimie*, 1982, **64**, 977.
10. A. Kobata, in 'Biology of Carbohydrates', ed. V. Ginsburg and P. W. Robbins, Wiley Interscience, New York, 1984, vol. 2, p. 87.
11. J. F. G. Vliegenthart, L. Dorland and H. Halbeek, *Adv. Carbohydr. Chem. Biochem.*, 1983, **41**, 209.
12. R. Kornfeld and S. Kornfeld, *Annu. Rev. Biochem.*, 1976, **45**, 217.
13. P. V. Wagh and O. P. Bahl, *CRC Crit. Rev. Biochem.*, 1981, **10**, 307.
14. A. D. Elbein, *Annu. Rev. Plant Physiol.*, 1979, **30**, 239.
15. J. A. Levy, H. Carminatti, A. I. Cantarella, N. H. Behrens, L. F. Leloir and E. Tabora, *Biochem. Biophys. Res. Commun.*, 1974, **60**, 118.
16. A. Heifetz and A. D. Elbein, *J. Biol. Chem.*, 1977, **252**, 3057.
17. R. K. Keller. D. Y. Boon and F. C. Crum, *Biochemistry*, 1979, **18**, 3946.
18. C. B. Sharma, L. Lehle and W. Tanner, *Eur. J. Biochem.*, 1982, **126**, 319.
19. G. P. Kaushal and A. D. Elbein, *Plant Physiol.*, 1986, **81**, 1086.
20. E. Li, I. Tabas and S. Kornfeld, *J. Biol. Chem.*, 1978, **253**, 7762.
21. T. Liu, B. Stetson, S. I. Turco, S. C. Hubbard and P. W. Robbins, *J. Biol. Chem.*, 1979, **254**, 4554.
22. H. Hori, D. W. James, Jr. and A. D. Elbein, *Arch. Biochem. Biophys.*, 1982, **215**, 12.
23. M. L. Kiely, G. S. McKnight and R. T. Schimke, *J. Biol. Chem.*, 1976, **251**, 5490.
24. V. R. Lingappa, J. R. Lingappa, R. Prasad, K. E. Ebner and G. Blobel, *Proc. Natl. Acad. Sci. USA*, 1978, **75**, 2338.
25. J. E. Rothman and H. F. Lodish, *Nature (London)*, 1977, **269**, 775.
26. U. Czichi and W. J. Lennarz, *J. Biol. Chem.*, 1977, **252**, 7901.
27. I. Vargas and H. Carminatti, *Mol. Cell. Biochem.*, 1977, **16**, 171.
28. A. J. Parodi and J. Martin-Barrientos, *Biochim. Biophys. Acta*, 1977, **500**, 80.
29. S. C. Hubbard and R. J. Ivatt, *Annu. Rev. Biochem.*, 1981, **50**, 555.
30. L. A. Hunt, J. R. Etchinson and D. F. Summers, *Proc. Natl. Acad. Sci. USA*, 1978, **75**, 754.
31. R. A. Ugalde, R. J. Staneloni and L. F. Leloir, *FEBS Lett.*, 1978, **91**, 209.
32. W. W. Chen and W. J. Lennarz, *J. Biol. Chem.*, 1978, **253**, 5780.
33. J. M. Michael and S. Kornfeld, *Arch. Biochem. Biophys.*, 1980, **199**, 249.
34. J. J. Elting, W. W. Chen and W. J. Lennarz, *J. Biol. Chem.*, 1980, **255**, 2325.
35. R. D. Kilker, Jr., B. Saunier, J. S. Tkacz and A. Hercovics, *J. Biol. Chem.*, 1981, **256**, 5299.
36. L. S. Grinna and P. W. Robbins, *J. Biol. Chem.*, 1980, **255**, 2255.
37. D. M. Burns and O. Touster, *J. Biol. Chem.*, 1982, **257**, 9991.
38. D. J. Opheim and O. Touster, *J. Biol. Chem.*, 1978, **253**, 1017.
39. D. R. P. Tulsiani, D. J. Opheim and O. Touster, *J. Biol. Chem.*, 1977, **252**, 3227.
40. W. T. Forsee and J. S. Schutzbach, *J. Biol. Chem.*, 1981, **256**, 6577.
41. J. Bischoff and R. Kornfeld, *J. Biol. Chem.*, 1978, **253**, 7907.
42. I. Tabas and S. Kornfeld, *J. Biol. Chem.*, 1978, **253**, 7779.
43. H. Schachter, P. Stanley and S. Narasimhan, *J. Biol. Chem.*, 1977, **252**, 3926.
44. N. Harpaz and H. Schachter, *J. Biol. Chem.*, 1980, **255**, 4885.

45. H. Schachter and S. Roseman, in 'The Biochemistry of Glycoproteins and Proteoglycans', ed. W. J. Lennarz, Plenum Press, New York, 1980, p. 85.
46. T. A. Beyer, J. E. Sadler, J. I. Rearick, J. C. Paulson and R. L. Hill, *Adv. Enzymol.*, 1981, **52**, 23.
47. S. M. Colegate, P. R. Dorling and C. R. Huxtable, *Aust. J. Chem.*, 1979, **32**, 2257.
48. R. J. Molyneux, unpublished results.
49. R. J. Molyneux and L. F. James, *Science (Washington, D.C.)*, 1982, **216**, 190.
50. D. Davis, P. Schwarz, T. Hernandez, M. Mitchell, B. Warnock and A. D. Elbein, *Plant. Physiol.*, 1984, **76**, 972.
51. H. P. Broquist, *Annu. Rev. Nutr.*, 1985, **5**, 391.
52. M. J. Schneider, F. S. Ungemach, H. P. Broquist and T. M. Harris, *Tetrahedron*, 1983, **39**, 29.
53. R. J. Molyneux, L. F. James and K. E. Panter, in 'Plant Toxicology', ed. A. A. Seawright, M. P. Hegarty, R. F. Keeler and L. J. James, Queensland Poisonous Plants Committee, 1986, p. 266.
54. B. W. Skeleton and A. H. White, *Aust. J. Chem.*, 1980, **33**, 435.
55. A. J. Aasen, C. C. Culvenor and L. W. Smith, *J. Org. Chem.*, 1969, **34**, 4137.
56. S. L. Everist, 'Poisonous Plants of Australia', Angus and Robertson, Sydney, 1981.
57. R. Saul, J. P. Chambers, R. J. Molyneux and A. D. Elbein, *Arch. Biochem. Biophys.*, 1983, **221**, 593.
58. R. Saul, R. J. Molyneux and A. D. Elbein, *Arch. Biochem. Biophys.*, 1984, **230**, 668.
59. B. L. Rhinehart, K. M. Robinson, A. J. Payne, M. E. Wheatley, J. L. Fisher, P. S. Liu and W. Cheng, *Life Sci.*, 1987, **41**, 2325.
60. C. Danzin and A. Ehrhard, *Arch. Biochem. Biophys.*, 1987, **257**, 472.
61. L. D. Hohenschultz, E. A. Bell, P. A. Jewess, D. P. Leworthy, R. J. Pryce, E. Arnold and J. Clardy, *Phytochemistry*, 1981, **20**, 811.
62. D. D. Schmidt, W. Frommer, L. Muller and E. Truscheit, *Naturwissenschaften*, 1979, **66**, 584.
63. S. Inouye, T. Tsuruoka, T. Ito and T. Niida, *Tetrahedron*, 1968, **24**, 2125.
64. H. Murai, K. Ohata, Y. Enomoto, Y. Yoshikuni, T. Kono and M. Yagi, *Ger. Pat.* 2 656 602; *Chem. Abstr.*, 1977, **87**, 14 127.
65. G. Legler and E. Julich, *Carbohydr. Res.*, 1984, **128**, 61.
66. G. W. J. Fleet, M. J. Gough and T. K. M. Shing, *Tetrahedron Lett.*, 1984, **25**, 4029.
67. L. E. Fellows, E. A. Bell, D. G. Lynn, F. Pilkiewicz, I. Miura and K. Nakanishi, *J. Chem. Soc., Chem. Commun.*, 1979, 977.
68. A. Welter, J. Jadot, G. Dardenne, M. Marlier and J. Casimir, *Phytochemistry*, 1976, **15**, 747.
69. S. V. Evans, L. Fellows and E. A. Bell, *Phytochemistry*, 1985, **22**, 768.
70. R. J. Nash, S. Evans, L. Fellows and E. A. Bell, in 'Plant Toxicology', ed. A. A. Seawright, M. P. Hegarty, R. F. Keeler and L. F. James, Queensland Poisonous Plants Committee, Brisbane, 1985.
71. G. W. J. Fleet, A. N. Shaw, S. V. Evans and L. F. Fellows, *J. Chem. Soc., Chem. Commun.*, 1985, 841.
72. R. C. Bernotas, M. A. Pezzone and B. Ganem, *Carbohydr. Res.*, 1987, **167**, 305.
73. H. Paulsen, Y. Hayauchi and V. Sinnwell, *Chem. Ber.*, 1980, **113**, 2601.
74. R. C. Bernotas and B. Ganem, *Carbohydr. Res.*, 1987, **167**, 312.
75. G. W. J. Fleet, P. W. Smith, R. J. Nash, L. F. Fellows, R. B. Parekh and T. W. Rademacher, *Chem. Lett.*, 1986, 1051.
76. G. W. J. Fleet, P. W. Smith, S. V. Evans and L. E. Fellows, *J. Chem. Soc., Chem. Commun.*, 1984, 1240.
77. G. W. J. Fleet, S. J. Nicolas, P. W. Smith, S. V. Evans, L. F. Fellows and R. J. Nash, *Tetrahedron Lett.*, 1985, **26**, 3127.
78. P. R. Dorling, C. R. Huxtable and P. Vogel, *Neuropathol. Appl. Neurobiol.*, 1978, **4**, 285.
79. D. J. Opheim and O. Touster, *J. Biol. Chem.*, 1978, **253**, 1017.
80. I. Tabas and S. Kornfeld, *J. Biol. Chem.*, 1979, **254**, 11 655.
81. T. Szumilo, G. P. Kaushal and A. D. Elbein, *Plant Physiol.*, 1986, **81**, 383.
82. S. Narasimhan, P. Stanley and H. Schachter, *J. Biol. Chem.*, 1977, **252**, 3926.
83. J. Bischoff and R. Kornfeld, *J. Biol. Chem.*, 1983, **258**, 7907.
84. J. Bischoff, L. Liscum and R. Kornfeld, *J. Biol. Chem.*, 1986, **261**, 4766.
85. P. R. Dorling, C. R. Huxtable and S. M. Colegate, *Biochem. J.*, 1980, **191**, 649.
86. M. S. Kang and A. D. Elbein, *Plant Physiol.*, 1983, **71**, 551.
87. B. Winchester, *Biochem. Soc. Trans.*, 1984, **12**, 552.
88. A. D. Elbein, R. Solf, K. Vosbeck and P. R. Dorling, *Proc. Natl. Acad. Sci. USA*, 1981, **78**, 7393.
89. A. D. Elbein, P. R. Dorling, K. Vosbeck and M. Horisberger, *J. Biol. Chem.*, 1982, **257**, 1573.
90. V. Gross, T. A. Tran-Thi, K. Vosbeck and P. C. Heinrich, *J. Biol. Chem.*, 1983, **258**, 4032.
91. D. R. P. Tulsiani, T. M. Harris and O. Touster, *J. Biol. Chem.*, 1982, **257**, 7936.
92. D. R. P. Tulsiani, H. P. Broquist and O. Touster, *Arch. Biochem. Biophys.*, 1985, **236**, 427.
93. A. L. Tarentino and F. Maley, *J. Biol. Chem.*, 1974, **249**, 811.
94. A. Matsumoto, H. Yoshima and A. Kobata, *Biochemistry*, 1983, **22**, 188.
95. C. L. Reading, E. E. Penhoet and C. E. Ballou, *J. Biol. Chem.*, 1978, **253**, 5600.
96. M. S. Kang and A. D. Elbein, *J. Virol.*, 1983, **46**, 60.
97. D. R. P. Tulsiani and O. Touster, *J. Biol. Chem.*, 1983, **258**, 7578.
98. M. Fukuda and S. Hakomori, *J. Biol. Chem.*, 1979, **254**, 5442.
99. R. G. Arumugham and M. L. Tanzer, *J. Biol. Chem.*, 1983, **258**, 11 883.
100. V. Duronio, S. Jacobs and P. Cuatrecases, *J. Biol. Chem.*, 1986, **261**, 970.
101. A. M. Soderquist and G. Carpenter, *J. Biol. Chem.*, 1984, **259**, 12 586.
102. P. P. Breitfeld, D. Rup and A. L. Schwartz, *J. Biol. Chem.*, 1984, **259**, 10 414.
103. K. N. Chung, V. L. Shepard and P. Stahl, *J. Biol. Chem.*, 1984, **259**, 14 637.
104. Z. Bar-Scavit, A. J. Kahn, L. E. Pegg, K. R. Stone and S. L. Teitelbaum, *J. Clin. Invest.*, 1984, **73**, 1277.
105. F. Villata and F. Kierszenbaum, *Mol. Biochem. Parasitol.*, 1985, **16**, 1.
106. M. J. Humphries, K. Matsumoto, S. L. White and K. Olden, *Proc. Natl. Acad. Sci. USA*, 1986, **83**, 1752.
107. M. Hino, O. Nakayama, Y. Tsurami, K. Adachi and T. Shibata, *J. Antibiot.*, 1985, **38**, 926.
108. J. R. Winkler and H. L. Segal, *J. Biol. Chem.*, 1984, **259**, 1958.
109. C. R. Huxtable and P. R. Dorling, *Am. J. Pathol.*, 1982, **107**, 124.
110. K. R. Van Kampen, R. W. Knes and L. F. James, in 'Effects of Poisonous Plants on Livestock', ed. R. F. Keeler, K. R. Van Kampen and L. F. James, Academic Press, New York, 1978, p. 465.

111. L. F. James and W. J. Hartley, *Am. J. Vet. Res.*, 1977, **38**, 1263.
112. D. Abraham, W. F. Blakemore, R. D. Jolly, R. Sidebotham and B. Winchester, *Biochem. J.*, 1983, **215**, 573.
113. S. Sadeh, C. D. Warren, P. F. Daniel, B. Bugge, L. F. James and R. W. Jeanloz, *FEBS Lett.*, 1983, **163**, 104.
114. C. D. Warren, S. Sadeh, P. F. Daniel, B. Bugge, L. F. James and R. W. Jeanloz, *FEBS Lett.*, 1983, **163**, 99.
115. D. R. P. Tulsiani and O. Touster, *Arch. Biochem. Biophys.*, 1983, **224**, 594.
116. D. R. P. Tulsiani, H. P. Broquist, L. F. James and O. Touster, *Arch. Biochem. Biophys.*, 1984, **232**, 76.
117. T. Suami, K. I. Tadano and Y. Iimura, *Carbohydr. Res.*, 1985, **136**, 67.
118. M. H. Ali, L. Hough and A. C. Richardson, *J. Chem. Soc., Chem. Commun.*, 1984, 447.
119. N. Yasuda, H. Tsutsumi and T. Takaya, *Chem. Lett.*, 1984, 1201.
120. C. E. Adams, F. J. Walker and K. B. Sharpless, *J. Org. Chem.*, 1985, **50**, 420.
121. A. D. Elbein, T. Szumilo, B. A. Sanford, K. B. Sharpless and C. E. Adams, *Biochemistry*, 1987, **26**, 2502.
122. N. Yasuda, H. Tsutsumi and T. Takaya, *Chem. Lett.*, 1985, 31.
123. Y. T. Pan, H. Hori, R. Saul, B. A. Sanford, R. J. Molyneux and A. D. Elbein, *Biochemistry*, 1983, **22**, 3975.
124. T. Szumilo, G. P. Kaushal and A. D. Elbein, *Arch. Biochem. Biophys.*, 1986, **247**, 261.
125. H. Hori, Y. T. Pan, R. J. Molyneux and A. D. Elbein, *Arch. Biochem. Biophys.*, 1984, **228**, 525.
126. R. Repp, T. Tamura, C. B. Boschek, H. Wege, R. T. Schwarz and H. Niemann, *J. Biol. Chem.*, 1985, **260**, 15873.
127. R. F. Arakaki, J. A. Hedo, E. Collier and P. Gorden, *J. Biol. Chem.*, 1987, **262**, 11886.
128. P. M. Schwarz and A. D. Elbein, *J. Biol. Chem.*, 1985, **260**, 14452.
129. R. K. Merkle, A. D. Elbein and A. Heifetz, *J. Biol. Chem.*, 1985, **260**, 1083.
130. J. W. Schmidt and W. A. Catterall, *J. Biol. Chem.*, 1987, **262**, 13713.
131. R. A. Gruters, J. J. Neefjes, M. Tersmette, R. E. Y. de Goede, A. Tulp, H. G. Huisman, F. Miedema and H. L. Ploegh, *Nature (London)*, 1987, **330**, 74.
132. P. S. Sunkara, T. L. Bowlin, P. S. Liu and A. Sjoerdsma, *Biochem. Biophys. Res. Commun.*, 1987, **148**, 206.
133. B. D. Walker, M. Kowalski, W. C. Goh, K. Kozarsky, M. Kreiger, C. Rosen, L. Rohrschneider, W. A. Haseltine and J. Sodroski, *Proc. Natl. Acad. Sci. USA*, 1987, **84**, 8120.
134. S. Schlesinger, A. H. Koyama, C. Malfer, S. L. Gee and M. J. Schlesinger, *Virus Res.*, 1985, **2**, 139.
135. S. Schlesinger, C. Malfer and M. J. Schlesinger, *J. Biol. Chem.*, 1984, **259**, 7597.
136. B. Burke, K. Matlin, E. Bause, G. Legler, N. Peyrieras and H. Ploegh, *EMBO J.*, 1984, **3**, 551.
137. R. Saul, J. J. Ghidoni, R. J. Molyneux and A. D. Elbein, *Proc. Natl. Acad. Sci. USA*, 1985, **82**, 93.
138. B. R. Ellmers, B. L. Rhinehart and K. M. Robinson, *Biochem. Pharm.*, 1987, **36**, 2381.
139. Y. T. Pan, J: J. Ghidoni and A. D. Elbein, manuscript in preparation.
140. R. J. Molyneux, J. N. Roitman, G. Dunnheim, T. Szumilo and A. D. Elbein, *Arch. Biochem. Biophys.*, 1986, **251**, 450.
141. U. Fuhrmann, E. Bause, G. Legler and H. Ploegh, *Nature (London)*, 1984, **307**, 755.
142. A. D. Elbein, G. Legler, A. Tlusty, W. McDowell and R. T. Schwarz, *Arch. Biochem. Biophys.*, 1984, **235**, 579.
143. V. Gross, K. Steube, T. A. Tran-thi, W. McDowell, R. T. Schwarz, K. Decker, W. Gere and P. C. Heinric, *Eur. J. Biochem.*, 1985, **150**, 41.
144. M. D. Snider and O. C. Rogers, *J. Cell Biol.*, 1986, **103**, 265.
145. J. Bischoff, L. Liscum and R. Kornfeld, *J. Biol. Chem.*, 1986, **261**, 4766.
146. A. D. Elbein, M. Mitchell, B. A. Sanford, L. E. Fellows and S. V. Evans, *J. Biol. Chem.*, 1984, **259**, 12409.
147. P. A. Romero, P. Friedlander, L. E. Fellows, S. V. Evans and A. Herscovics, *FEBS Lett.*, 1985, **184**, 197.
148. J. A. Schmidt, H. Beug and M. Hayman, *EMBO J.*, 1985, **4**, 105.
149. G. W. J. Fleet and P. W. Smith, *Tetrahedron Lett.*, 1985, **26**, 1469.
150. G. Palamarczyk, M. Mitchell, P. W. Smith, G. W. J. Fleet and A. D. Elbein, *Arch. Biochem. Biophys.*, 1985, **243**, 35.
151. G. Legler, *Methods Enzymol.*, 1977, **46**, 368.
152. M. Nakajima, I. Tomida, N. Kurihara and S. Takei, *Chem. Ber.*, 1959, **92**, 173.
153. W. C. Ross, *J. Chem. Soc.*, 1950, **240**, 2257.
154. H. Braun, G. Legler, J. Dehusses and G. Semenza, *Biochim. Biophys. Acta*, 1977, **483**, 135.
155. J. N. Kanfer, G. Legler, J. Sullivan, S. S. Raghavan and R. A. Mumford, *Biochem. Biophys. Res. Commun.*, 1975, **67**, 85.
156. G. Legler and W. Lotz, *Hoppe Seyler's Z. Physiol. Chem.*, 1973, **354**, 243.
157. R. Datema, P. A. Romero, G. Legler and R. T. Schwarz, *Proc. Natl. Acad. Sci. USA*, 1982, **79**, 6787.
158. O. P. Van Diggelen, H. Galjaard, M. L. Sinnott and P. J. Smith, *Biochem. J.*, 1980, **188**, 337.
159. P. A. Docherty, M. J. Kuranda, N. N. Aronson, Jr., J. N. BeMiller, R. W. Myers and J. A. Bohn, *J. Biol. Chem.*, 1986, **261**, 3457.
160. I. M. Glynn and S. J. D. Karlish, *Annu. Rev. Physiol.*, 1975, **37**, 13.
161. A. S. Hobbs and R. W. Albers, *Annu. Rev. Biophys. Bioeng.*, 1980, **9**, 259.
162. J. D. Robinson and M. S. Flashner, *Biochem. Biophys. Acta*, 1979, **549**, 145.
163. P. L. Jorgensen, *Biochim. Biophys. Acta*, 1974, **356**, 36.
164. J. Kyte, *J. Biol. Chem.*, 1971, **246**, 4157.
165. B. Forbush, III, J. H. Kaplan and J. F. Hoffman, *Biochemistry*, 1978, **17**, 3667.
166. A. L. Hodkin, 'The Conduction of the Nervous Impulse', Thomas, Springfield, IL, 1964.
167. W. A. Catterall, *Annu. Rev. Biochem.*, 1986, **55**, 953.
168. R. D. Keynes, *Sci. Am.*, 1979, **240**, 126.
169. L. Stryer, 'Biochemistry', Freeman, New York, 1988.
170. G. T. Siegel, R. W. Albers, B. W. Agranoff and R. Katzman, 'Basic Neurochemistry', Little, Brown, Boston, MA, 1981.
171. M. S. Briley and J.-P. Changeux, *Int. Rev. Neurobiol.*, 1977, **20**, 31.
172. D. M. Fambrough, *Physiol. Rev.*, 1979, **59**, 165.
173. H. G. Mautner, *CRC Crit. Rev. Biochem.*, 1977, **4**, 341.
174. H. H. Dale, *J. Pharmacol. Exp. Ther.*, 1914, **6**, 147.
175. D. Nachmansohn and E. Lederer, *Bull. Soc. Chim. Biol.*, 1939, **21**, 797.
176. B. Mendel, D. B. Mundell and H. Rudney, *Biochem. J.*, 1943, **37**, 473.
177. D. Nachmansohn and I. B. Wilson, *Adv. Enzymol.*, 1951, **12**, 259.
178. I. B. Wilson, *J. Biol. Chem.*, 1952, **197**, 215.
179. I. B. Wilson, *J. Biol. Chem.*, 1954, **208**, 123.

180. J. A. Cohen and R. A. Oosterbaan, 'Handbook of Experimental Pharmacology', Springer, Berlin, 1963, Suppl. IV.
181. D. Nachmansohn, 'Chemical and Molecular Basis of Nerve Activity', Academic Press, New York, 1959.
182. T. L. Rosenberry, *Adv. Enzymol.*, 1975, **43**, 103.
183. S. Bon and J. Massoulie, *Eur. J. Biochem.*, 1978, **89**, 89.
184. L. Anglister and I. Silman, *J. Mol. Biol.*, 1978, **125**, 293.
185. J. Massoulie and S. Bon, *Annu. Rev. Neurosci.*, 1982, **5**, 57.
186. D. Nachmansohn, in 'The Structure and Function of Muscle', ed. G. H. Bourne, Academic Press, New York, 1973, vol. III.
187. D. M. Krupka and K. J. Laidler, *J. Am. Chem. Soc.*, 1961, **83**, 1454.
188. T. L. Rosenberry, *Proc. Natl. Acad. Sci. USA*, 1975, **72**, 3834.
189. I. B. Wilson, *Ann. N. Y. Acad. Sci.*, 1966, **135**, 177.
190. H. C. Froede and I. B. Wilson in 'The Enzymes', ed. P. D. Boyer, Academic Press, New York, 1971, p. 87.
191. A. G. Karczmar, E. Usdin and J. H. Wills, 'Acetylcholinesterase Agents', Pergamon Press, Oxford, 1970.
192. E. F. Jansen, M.-D. F. Nutting and A. K. Balls, *J. Biol. Chem.*, 1949, **179**, 201.
193. H. O. Michel, B. E. Hackley, L. Berkowitz, G. List, E. B. Hackley, W. Gillilan and M. Parkau, *Arch. Biochem. Biophys.*, 1967, **121**, 29.
194. K.-B. Augustinsson and D. Nachmansohn, *J. Biol. Chem.*, 1949, **179**, 543.
195. I. B. Wilson and S. Ginsburg, *Biochim. Biophys. Acta*, 1955, **18**, 168.
196. H. Kewitz and I. B. Wilson, *Arch. Biochem. Biophys.*, 1956, **60**, 261.
197. I. B. Wilson, *Biochem. Biophys. Acta*, 1958, **27**, 196.
198. H. Kewitz and D. Nachmansohn, *Arch. Biochem. Biophys.*, 1957, **66**, 271.
199. T. E. Smith and E. Usdin, *Biochemistry*, 1966, **5**, 2914.
200. F. Berends, C. H. Posthumus, I. Sluys and F. A. Deierkauf, *Biochem. Biophys. Acta*, 1959, **34**, 576.
201. H. P. Metzger and I. B. Wilson, *Biochem. Biophys. Res. Commun.*, 1967, **28**, 263.
202. T. Mattio, M. McIlhany, E. Giacobini and T. Hallak, *Neuropharmacology*, 1986, **25**, 1167.
203. A. Brossi, B. Schonenberger, O. E. Clark and R. Ray, *FEBS Lett.*, 1986, **201**, 190.
204. J. H. Fellman and T. S. Fujita, *Biochim. Biophys. Acta*, 1964, **89**, 360.

8.2

Peptidase Inhibitors

DANIEL H. RICH

University of Wisconsin, Madison, WI, USA

8.2.1	INTRODUCTION	391
8.2.2	TYPES OF PEPTIDASES	394
8.2.3	SELECTED PEPTIDASE INHIBITORS	400
	8.2.3.1 Angiotensin Converting Enzyme (ACE) Inhibitors	400
	8.2.3.1.1 Teprotide	400
	8.2.3.1.2 Captopril	400
	8.2.3.1.3 P'$_2$ variations of captopril	401
	8.2.3.1.4 N-Carboxyalkyl dipeptides: enalapril and lisinopril	401
	8.2.3.1.5 Enalapril analogs	402
	8.2.3.1.6 Phosphorus-containing analogs: fosfinopril	403
	8.2.3.1.7 Conformationally restricted captopril and enalapril analogs	404
	8.2.3.1.8 Cilazapril	406
	8.2.3.1.9 Substrate-based ketone and alcohol ACE inhibitors	409
	8.2.3.2 Enkephalinase Inhibitors	410
	8.2.3.3 Collagenase Inhibitors	412
	8.2.3.4 Renin Inhibitors	414
	8.2.3.4.1 Substrate analog inhibitors: RIP	414
	8.2.3.4.2 Transition-state analog inhibitors: reduced peptide 'isosteres'	415
	8.2.3.4.3 Statine-containing inhibitors: SCRIP	416
	8.2.3.4.4 Other transition-state analog designs	419
	8.2.3.5 Elastase Inhibitors: Human Elastase (HNE) Inhibitors	422
	8.2.3.5.1 Peptidyl inhibitors	422
	8.2.3.5.2 Heterocyclic covalent inhibitors	426
	8.2.3.6 Aminopeptidase Inhibitors	428
	8.2.3.7 Inhibitors of Cysteine Proteinases	431
	8.2.3.8 Inhibitors of Prolyl Endopeptidases	433
	8.2.3.9 Thermolysin Inhibitors	434
8.2.4	PROSPECTIVES	435
8.2.5	REFERENCES	436

8.2.1 INTRODUCTION

Peptidases[1] are crucial for the normal functioning of biological systems. Some peptidases release peptide hormones and neuromodulators from their inactive precursors[2] or activate other enzymes, *e.g.* clotting factors,[3] while other peptidases terminate biological responses by degrading the message-transmitting peptide.[4] Within the organism, peptidases encounter immense numbers of proteins that might be potential substrates, so it is not surprising that these enzymes are able to select target substrates and are tightly regulated by endogenous proteinase inhibitors.[5] Nevertheless, a number of disease states appear to be caused by excessive proteolytic activity in susceptible tissue brought on by abnormally low concentrations of endogenous proteinase inhibitors.[6] Insufficient proteinase inhibitor may be caused by an environmental factor, *e.g.* cigarette smoke is known to inactivate the major naturally occurring elastase inhibitor, α_1-PI.[7] Hereditary factors may also be a cause of excessive proteolytic activity.[8] However, in most cases, excessive proteolytic activity as a

391

cause of a disease is not clearly established but is implicated by indirect methods, such as the effects exogenously administered peptidase inhibitors have on a physiological system, or from manifestations of conditions which reasonably might be thought to be effected by peptidase activity, such as muscle wasting diseases.[9]

The possible use of synthetic or natural peptidase inhibitors to control peptidase activity and thereby regulate specific disease states has been recognized for a long time, but only in the last 10 years have major successes actually been achieved. The discovery of captopril (**1**; see Table 1)

Table 1 Representative Inhibitors of Angiotensin Converting Enzyme

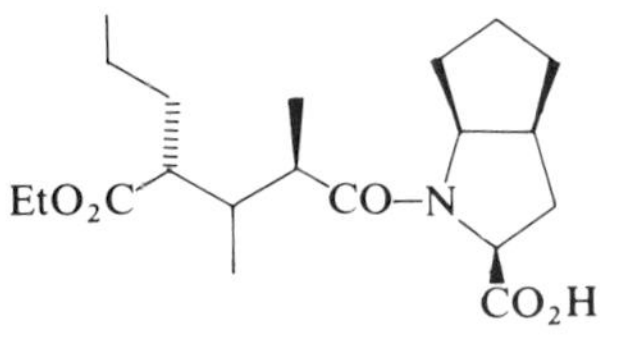

(19)

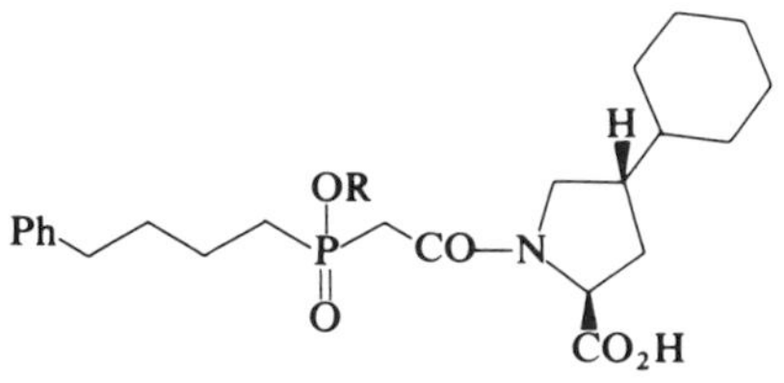

(20) R = OCOEt

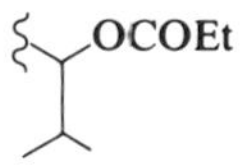

(21) R = H

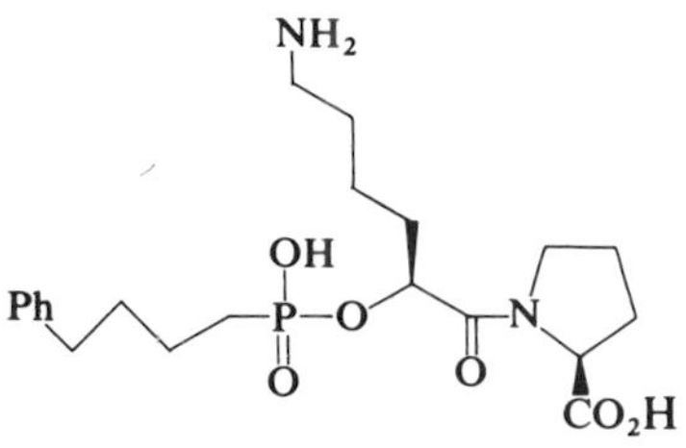

(22)

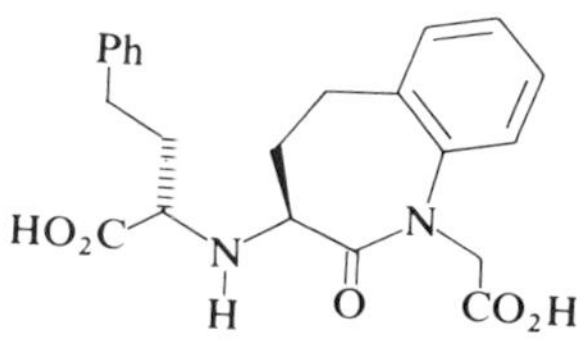

(23)

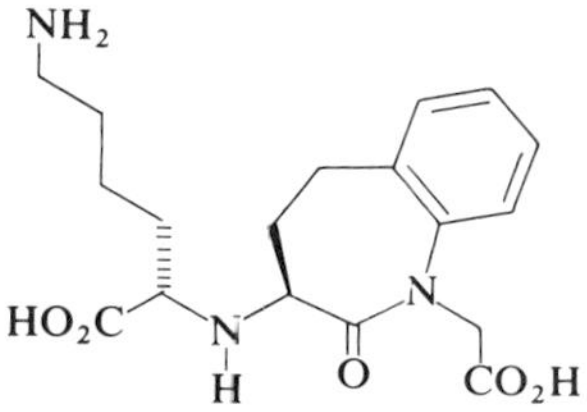

(24)

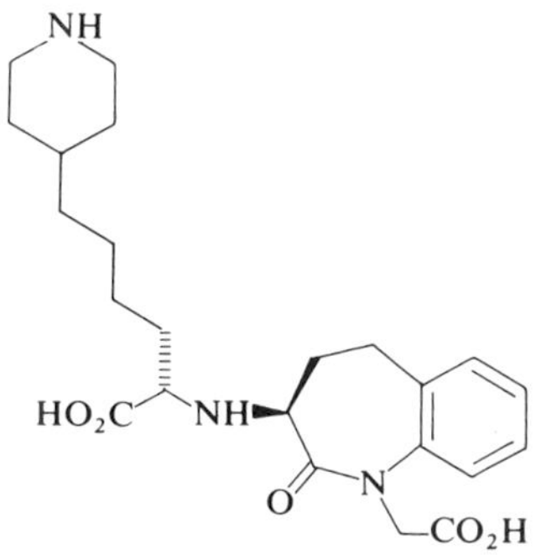

(25)

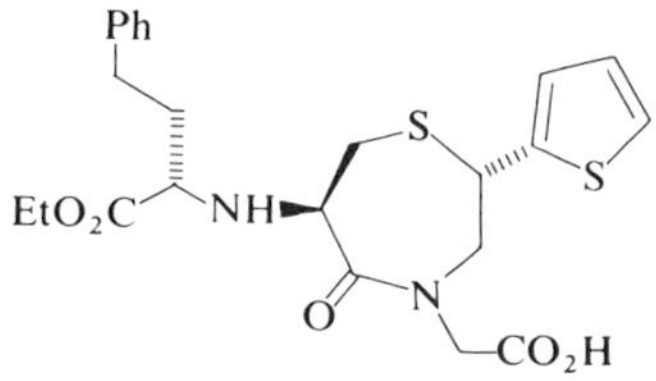

(26)

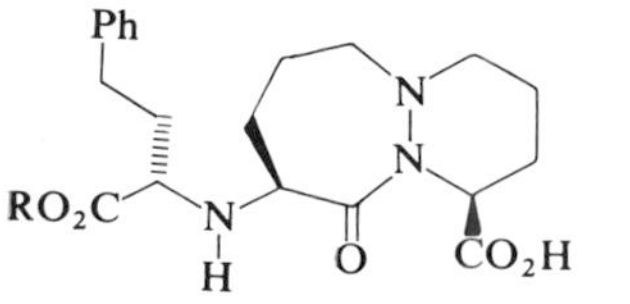

(27) R = Et
(28) R = H

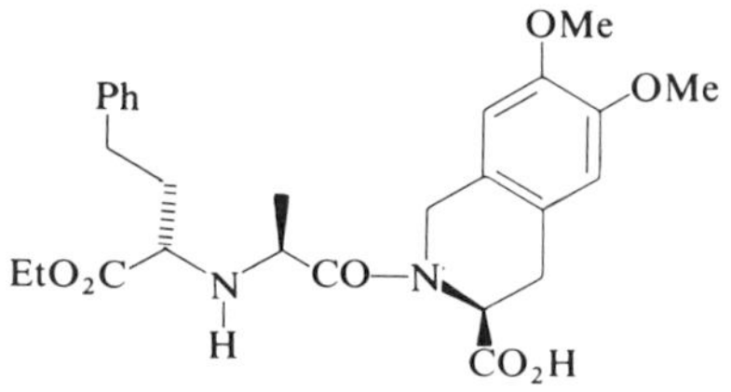

(29)

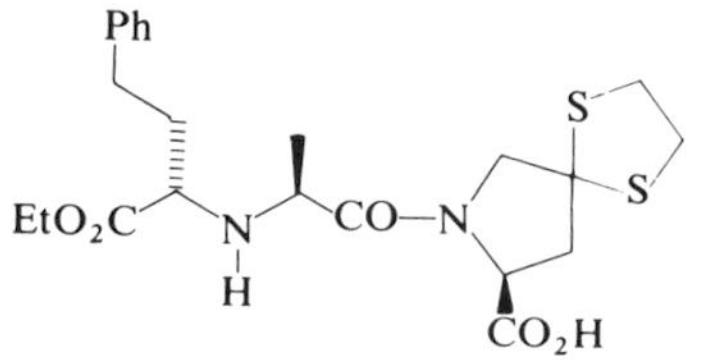

(30)

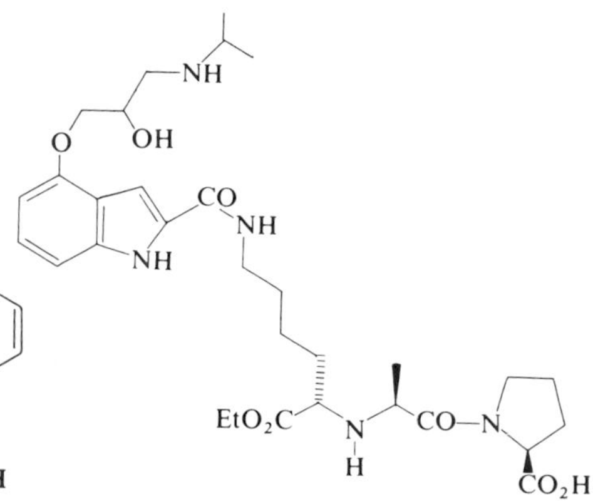

(31)

(32)

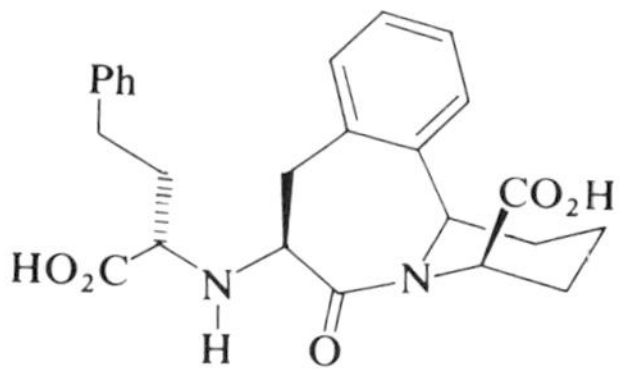

(33)

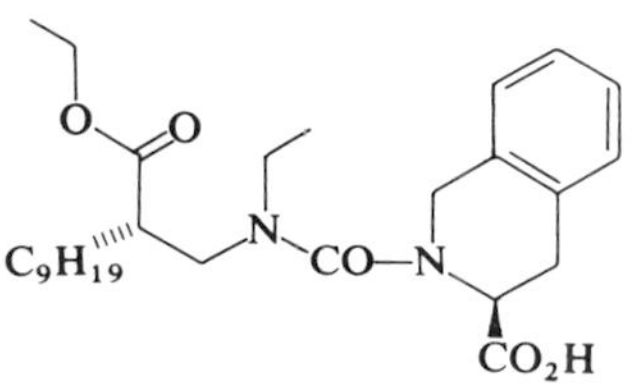

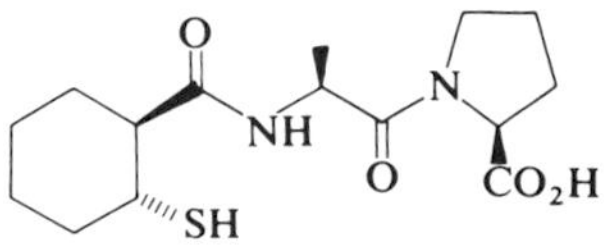

(34)

represents the first major drug developed from a strategy of inhibition of a specific peptidase.[10] Captopril is an orally effective antihypertensive drug that acts by inhibiting a peptidyl dipeptidase (called angiotensin converting enzyme or ACE) for its mechanism of action. ACE is the last peptidase needed for the biosynthesis of angiotensin II, a natural pressor substance (Scheme 1). Inhibition of ACE leads to a lowering of angiotensin II levels *in vivo* and to lowered blood pressure. Enalaprilate (3) and its orally active prodrug, enalapril (2), were discovered[11] shortly after captopril. Many other effective clinical ACE inhibitors and prodrugs thereof, some of which are undergoing active clinical use or evaluation, are shown in Table 1. Several inhibitors of human neutrophil elastase are being developed for the treatment of pulmonary emphysema.[12] Because these compounds have been discovered only recently, clinical trials are not yet complete. If successful, this approach would constitute the first example of a low molecular weight peptidase inhibitor being used to supplement an endogenous proteinase inhibitor.

Low molecular weight peptidase inhibitors can also be used to potentiate natural biological responses by slowing down the degradation of an endogenous biologically active peptide. Several enkephalinase inhibitors have been shown to potentiate the analgesic responses to endogenous or exogenously administered enkephalin.[13] These compounds illustrate how an indirect agonist of an endogenous transmitter can be derived from peptide chemistry. Other research efforts have led to the discovery of renin inhibitors[14] and collagenase inhibitors.[15] The former class of compounds have potential utility as antihypertensive agents because they inhibit the first step in the biosynthesis of angiotensin II (Scheme 1). The collagenase inhibitors are targeted for use in treating inflammatory diseases.

Renin Substrate: Asp–Arg–Val–Tyr–Ile–His–Pro–Phe–His–Leu–Val–Ile–His–protein
Position: 1   2   3   4   5   6   7   8   9   10   11   12   13

Renin ↓

Angiotensin I   Asp–Arg–Val–Tyr–Ile–His–Pro–Phe–His–Leu
                   1   2   3   4   5   6   7   8   9   10

ACE ↓

Angiotensin II   Asp–Arg–Val–Tyr–Ile–His–Pro–Phe + H–His–Leu–OH
                   1   2   3   4   5   6   7   8

Scheme 1   Biosynthesis of angiotensin II by the action of renin and ACE

The development of the aforementioned peptidase inhibitors clearly establishes that inhibition of biosynthetic or metabolizing peptidases is a viable approach to treating disease states, but only a fraction of the potentially important peptidases have been studied. Some of the peptidases that could be inhibited to produce useful therapeutic effects are listed in Table 2.

It is the purpose of this article to identify the therapeutically important peptidase inhibitors and to try to describe the design principles that have been used to develop these inhibitors. As background information, the catalytic mechanisms of peptidases and the basic concepts for binding of substrates and inhibitors to peptidases will be briefly reviewed. This chapter will focus on the applications of modern physical and biochemical methods to the design of five major classes of peptidase inhibitors. Because of limitations in space, it will not be possible to describe all types of compounds that inhibit peptidases. For that purpose, the book 'Proteinase Inhibitors' edited by Barrett and Salvesen provides an excellent and detailed account of proteinase inhibitor history, nomenclature, structures and function.[16]

8.2.2 TYPES OF PEPTIDASES

Four classes of peptidases, each with a distinct catalytic mechanism, have been identified (Table 3). These enzymes are classified according to the most significant functional group or prosthetic group within the active site of the enzyme. Serine proteinases utilize the hydroxyl group in the side chain of a serine residue in the active site (Ser-195 in chymotrypsin) as a nucleophile

Table 2 Potential Targets for New Peptidase Inhibitors

Peptidase	Representative enzymes	Normal function
Serine[a]	Thrombin; plasma kallikrein; Factors VIIa, IXa–XIIa; activated protein C	Blood coagulation
	Factor C1r; Factor C1s; Factor D; Factor B; C3 convertase.	Complement activation
	Trypsin; chymotrypsin; pancreatic elastase; entereokinase plasmin; plasminogen activator	Digestion, fibrinolysis
	Tissue kallikrein; post-protein cleaving enzyme	Hormone metabolism
	Elastase; cathepsin G; mast cell chymases, tryptases	Phagocytosis
	ATP-dependent proteases (intracellular)	Protein turnover
Metallo[a]		
Exopeptidases	Angiotensin converting enzyme; aminopeptidases; carboxypeptidases; renal dipeptidases	Blood pressure regulation
Endopeptidases	Collagenase; endopeptidase 24.11; enkephalinase; IgA proteinase	
	Macrophage elastase	Blood pressure regulation; metabolism of peptides; inactivates carbapenem antibiotics
		Tissue remodeling; peptide metabolism; cleaves hinge region of IgA
		Phagocytosis
Aspartic[b]	Renin	Blood pressure; peptide metabolism; protein turnover
	Cathepsin D	
	HIV protease	HIV replication
Cysteine[b]	Cathepsins B, H, L	Protein turnover; bone resorbtion
	Calcium activated neutral proteases (CANP, calpains)	Bone resorbtion; protein turnover

[a] Adapted from Powers.[17,19] [b] Adapted from Rich.[20,21]

Table 3 Characteristics of Peptidases

Type	Most significant active-site functional groups	Some characteristic irreversible inhibitors	Some selective reversible inhibitors
Serine	Serine (hydroxyl)	Organophosphates	Peptidyl aldehydes
	Histidine (imidazole)	Peptidyl chloromethanes	Peptidyl borates
	Aspartic acid (carboxyl)		
Aspartic	Aspartic acid (carboxyls)	EPNP (epoxide)	Pepstatin
Cysteine	Cysteine (thiol group)	Peptidyl chloromethanes	Peptidyl aldehyde
	Histidine (imidazole)	Peptidyl diazomethanes; E-64 and related analogs	Antipain
Metallo	Zinc ion	Divalent metal chelating reagents;[a]	Phosphoramidon
		8-Hydroxyisoquinoline; EDTA	α-Ketoamides; bestatin; amastatin

[a] Removal of the essential divalent metal is not always an irreversible process, but metallopeptidases are usually inhibited by metal chelating reagents.

(Scheme 2).[17] The hydroxyl group is activated by the close proximity of an imidazole group present on the amino acid histidine (His-57 in chymotrypsin) and attacks the amide bond in the substrate by adding to the carbonyl group to form a tetrahedral intermediate (**35**). A third component of the catalytic system, the side-chain carboxyl group of an aspartic acid [Asp-102 (not shown) in chymotrypsin] apparently is necessary to position properly the His imidazole group and to aid in proton transfer.[18] The formation of the tetrahedral intermediate (**35**) proceeds *via* an anionic oxygen which is stabilized by hydrogen bonding to two ideally placed amide protons, a structure termed the 'oxyanion hole'. The tetrahedral intermediate collapses to form the peptide product (derived from the carboxyl end of the substrate) and an *O*-acyl intermediate (**36**). Hydrolysis of (**36**), which must proceed through a second tetrahedral intermediate (**37**), liberates the second product. All steps of this process are reversible, and under careful controlled experimental conditions the reverse reaction, peptide bond formation, has been observed and used for synthetic purposes.[19]

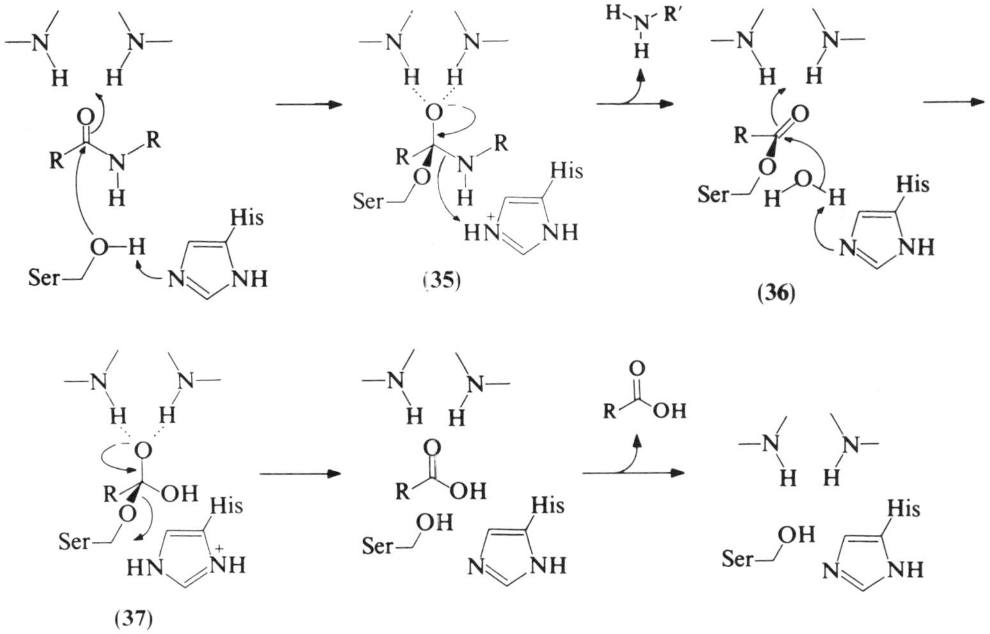

Scheme 2 Schematic representation of amide bond hydrolysis catalyzed by serine peptidases

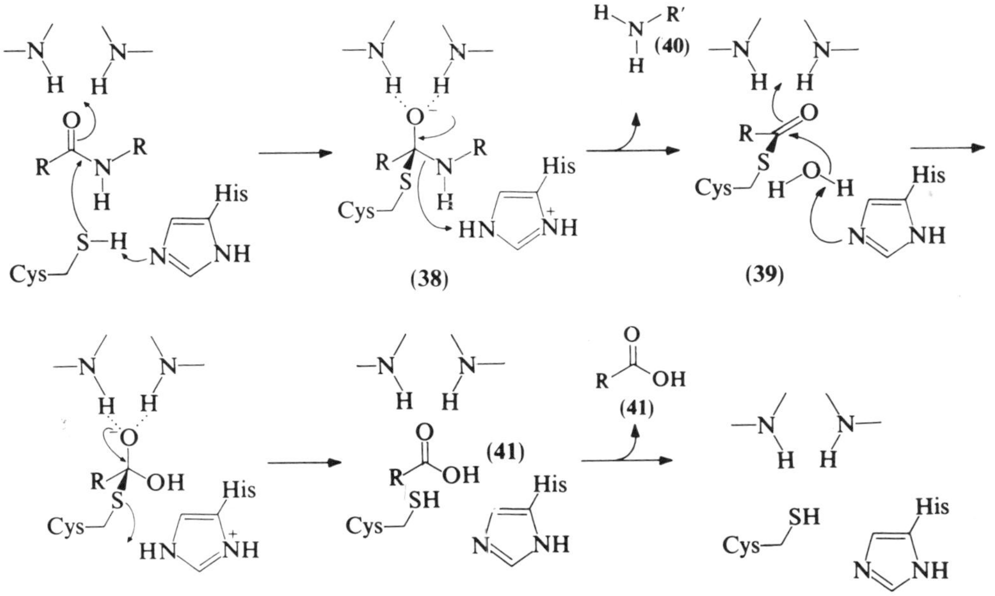

Scheme 3 Schematic representation of amide bond hydrolysis catalyzed by cysteine proteinases

Cysteine proteinases contain a nucleophilic thiol group (Cys-25 in papain) that is deprotonated by the imidazole nitrogen on an adjacent histidine (His-169 in papain) (Scheme 3).[20] An 'oxyanion hole' formed from the amide protons of Cys-25 and Gln-19 may stabilize the formation of the tetrahedral species (**38**). Collapse of this tetrahedral intermediate produces a thiol ester (**39**) and the C-terminal peptide product (**40**). Hydrolysis of (**39**) generates the second product (**41**).

Aspartic and metallo peptidases (Schemes 4 and 5) do not appear to utilize covalent intermediates in their catalytic mechanisms. Instead, the catalytic groups catalyze the addition of water directly to the substrate amide bond. Aspartic proteinases utilize the β-carboxyl groups located on two active-site aspartic acids to catalyze hydrolysis.[21] The catalytic mechanism is still tentative but appears to be consistent with the mechanism shown in Scheme 4. This was proposed on the basis of molecular

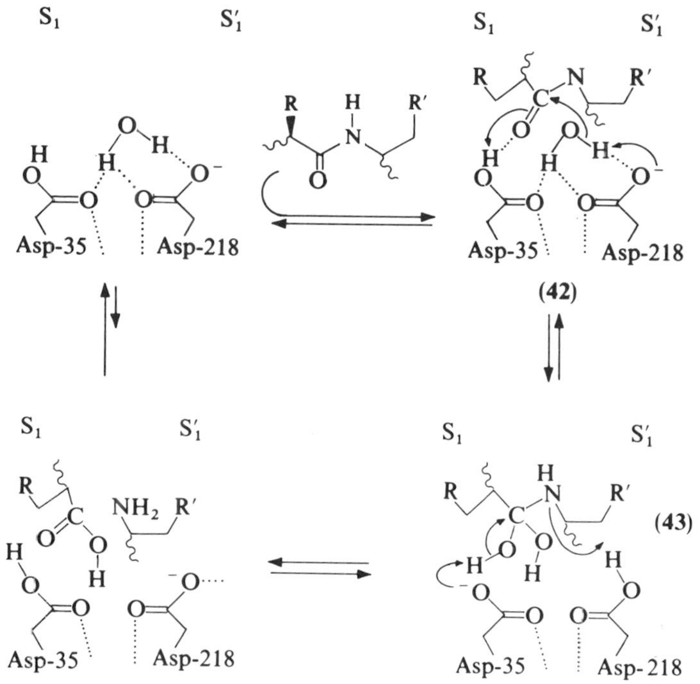

S₁ S′₁ S₁ S′₁

(The scheme shows aspartic proteinase catalytic mechanism with Asp-35 and Asp-218 residues, structures labeled (42) and (43))

$S_1 \qquad S_1'$

$S_1 \qquad S_1'$

Asp-35   Asp-218

(42)

Asp-35   Asp-218

$S_1 \qquad S_1'$

$S_1 \qquad S_1'$

Asp-35   Asp-218

Asp-35   Asp-218

(43)

Scheme 4 Aspartic proteinase catalytic mechanism

modeling studies with the X-ray crystal structure of a reduced amide bond inhibitor bound to *Rhizopus chinensis* aspartic proteinase. A water molecule is hydrogen bonded to the two essential aspartic acid carboxyl groups (Asp-32 and -215, pepsin numbering) in the native enzyme. Upon binding of substrate, the proton shared by the two aspartic acid carboxyl groups protonates the amide carbonyl group in the substrate while the second carboxyl group catalyzes addition of water by a general base mechanism, as shown in (42).[22] Reverse flow of electrons leads to protonation of the departing amine (43) and to product formation. No covalent intermediate is needed for this mechanism although evidence for covalent intermediates obtained from transpeptidation experiments remains to be fully explained.[23]

The well-characterized metallopeptidases can be represented by the zinc-containing peptidase, thermolysin. Our understanding of the catalytic mechanism for this enzyme, and related metallopeptidases, is constantly evolving but much of the evidence supports the mechanism presented in Scheme 5 in which the divalent zinc ion functions to position the substrate water molecule as in (44) and, as a Lewis acid, to activate the amide carbonyl. Upon binding of substrate, additional residues (His-231, Tyr-157) are postulated to facilitate activation of the substrate amide bond (45→46), leading to the formation of the tetrahedral intermediate (47). The carboxylate group on amino acid Glu-143 serves as a general base in this process and possibly as a proton donor to the departing amine.[24] Collapse of the tetrahedral intermediate (47) *via* multiple steps leads to the products.

Competitive and irreversible inhibitors of peptidases are used to identify each new proteinase (see Table 3).[16,25] In general, serine proteinases are irreversibly inhibited by organophosphates or sulfonyl fluorides or competitively inhibited by peptidyl aldehydes. Aspartic proteinases are inhibited by low concentrations of pepstatin. Cysteine proteinases are irreversibly inhibited by reaction with E-64 derivatives (see Section 8.2.3.7),[26] and metalloproteinases are inhibited by phosphoramidon[27] or metal chelating reagents.[24] These tests serve to assign a probable catalytic mechanism to a newly discovered peptidase, but such tests must be interpreted carefully. To date, all well-characterized peptidases can be assigned to one of the aforementioned peptidase classes. However, there is no *a priori* reason that additional catalytic mechanisms have not evolved and will be discovered.[25] Each newly isolated peptidase must be carefully characterized.

Peptidases are also classified according to their substrate specificity, *i.e.* their ability to recognize and hydrolyze amide bonds positioned next to characteristic amino acids or within characteristic sequences of amino acids. To facilitate discussions of peptidase specificity, Schechter and Berger[28] devised the widely accepted convention shown in Figure 1a for describing recognition sites in peptidases. Amino acid sequences are numbered according to their position relative to the amide

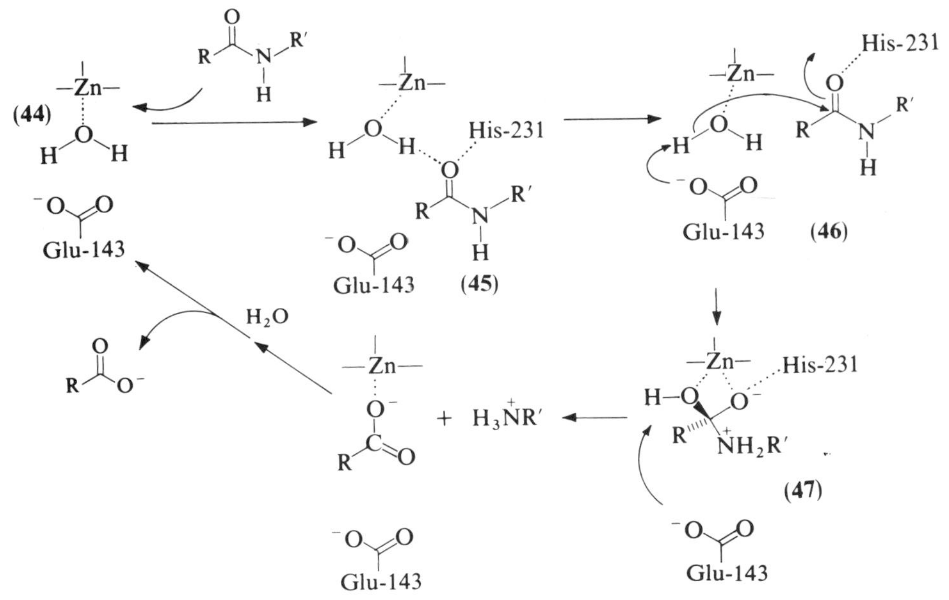

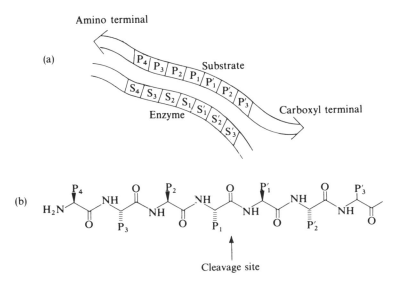

Scheme 5 Schematic representation of amide bond hydrolysis catalyzed by metalloproteinases

bond being cleaved by the enzyme. The cleavage site is designated the P_1-P'_1 site which binds to the enzyme at the corresponding S_1-S'_1 site. Residues extending towards the N-terminus are designated by increasing numbers (P_2, P_3 ... P_n) which bind to the S_2, S_3 ... S_n sites on the enzyme. By convention the N-terminus of a peptide is at the left end of the sequence. Residues extending towards the carboxyl terminus are designated with primes (P'_2, P'_3, *etc.*) which bind to the S'_2, S'_3, *etc.* enzyme sites. The specificity of some peptidases can be limited primarily to one site (*e.g.* trypsin recognizes basic amino acids in the P_1 site) or to extended binding sites of seven or more amino acids (*e.g.* P_4-P'_3 in the case of aspartic proteinases). With extended binding sites it is not unusual for one or two subsites to contribute more to binding than the remaining sites.

Different proteinases may also utilize similar mechanisms for binding substrates and inhibitors. A comparison of the X-ray crystal structures of a number of enzyme–inhibitor complexes reveals that

Figure 1 (a) The active site of papain according to Schechter and Berger.[28] (b) Orientations of side chains in peptides binding in antiparallel β-sheet conformations. Note that in this conformation, the P_1, P_3, P_5, *etc.* side chains are adjacent and on the opposite side of the peptide backbone from the adjacent P_2, P_4, *etc.* side chains

most peptidase inhibitors utilize a portion of an antiparallel β-pleated sheet secondary structure to stabilize binding to the enzyme.[29] By analogy, substrates are expected to bind in a similar fashion. The general form of the antiparallel β-sheet is shown in Figure 2a.[30] This can be viewed as a three-peptide chain structure in which interchain hydrogen bonds between amide protons and amide

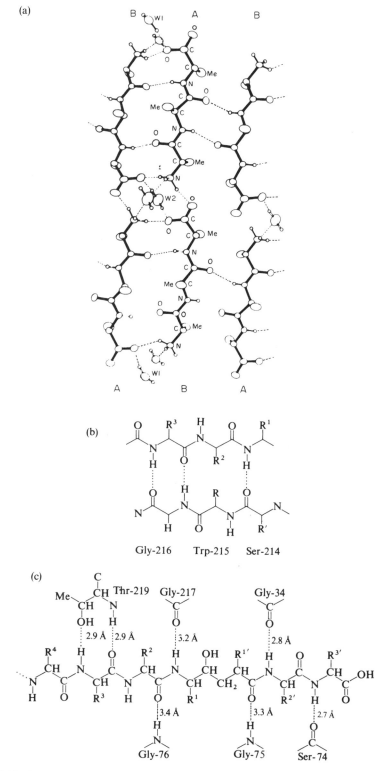

Figure 2 Antiparallel β-pleated sheet structures in proteins and enzyme–inhibitor complexes. (a) Intermolecular hydrogen bonding in L-Ala-L-Ala-L-Ala in an antiparallel sheet.[30] (b,c) Selected examples of antiparallel binding of inhibitors in peptidase active sites: (b) trypsin;[29] (c) aspartic proteinases[22,32]

carbonyls yield a highly ordered and stabilized structure.[31] In the case of inhibitors, the outer two chains are provided by the enzyme while the inner chain is provided by the inhibitor. Some examples of the hydrogen bonding patterns between enzymes and inhibitors are shown in Figures 2b and 2c. Note that some of the possible hydrogen bonds between the inhibitor and protein backbones predicted on the basis of the antiparallel β-sheet structure are not formed and other variations occur, such as hydrogen bonds from the side chains of the enzyme to the inhibitor rather than from the backbone amide bonds.[22,29,32] One interesting feature of antiparallel β-sheet structures is that the side chains follow a regular pattern in which the odd number residues (P_1, P_3, P_5) are on the same side of the peptide backbone and across the peptide backbone from the even numbered side chains (P_2, P_4, P_6) (see Figure 1b). These patterns place the alternating side chains adjacent to each other and may permit the construction of cyclic inhibitors (see Section 8.2.3.4.4) if enzyme atoms do not intrude. In addition, the close proximity of alternating side chains may provide a rationalization for nonadditive effects of inhibitors since enzyme amino acid side chains may be used to form more than one specificity pocket; thus, expansion of one binding site (*e.g.* S_1) to accommodate large ligands may be accomplished by simultaneous contraction of adjacent binding sites (*e.g.* S_1' or S_3).

8.2.3   SELECTED PEPTIDASE INHIBITORS

8.2.3.1   Angiotensin Converting Enzyme (ACE) Inhibitors

ACE [EC 3.4.15.11] catalyzes the conversion of the decapeptide angiotensin I to the octapeptide angiotensin II by specific hydrolysis of the penultimate amide bond between Phe and His (Scheme 1). Angiotensin II elevates blood pressure by constricting smooth muscles in blood vessels and by stimulating the release of aldosterone which causes sodium ion retention.[33] Although early theories of hypertension suggested that angiotensin II might be important only in relatively rare forms of hypertension, *e.g.* malignant hypertension, and thus that ACE inhibitors would have limited utility, we now know that inhibition of ACE in conjunction with diuretics lowers blood pressure in over 80% of patients with essential hypertension. Thus, ACE inhibitors have become an important class of drug for controlling the most commonly encountered forms of hypertension[34] and for controlling congestive heart failure.

8.2.3.1.1   *Teprotide*

The development of ACE inhibitors derives directly from the discovery by Ferreira and his associates of a series of bradykinin potentiating peptides (Bpp) in the venom of the Brazilian snake, *Bothrops jarara*.[35] The bradykinin potentiating properties of these peptides resulted from their inhibition of ACE. Ondetti's group at Squibb determined the structure of teprotide (**48**) and synthesized it.[36] In the clinic, teprotide was found to lower blood pressure in animal models and in humans, thereby establishing ACE as a realistic target for developing antihypertensive agents. However, teprotide itself was not a suitable antihypertensive agent because it was not orally active and lacked the duration of action needed for treatment of this disease over a long period of time. Smaller peptides were found that inhibited ACE, *e.g.* (**49**; $K_i = 1.8$ μM) and (**50**; $K_i = 230$ μM) but these were not sufficiently potent to be used as drugs. Nevertheless, Ondetti and Cushman and their associates reasoned that the dipeptides could be suitable starting points for the discovery of clinically useful drugs if methods could be found to enhance their binding to ACE.

pGlu–Trp–Pro–Arg–Pro–Gln–Ile–Pro–Pro

(**48**)

Val–Trp   Ala–Pro

(**49**)   (**50**)

8.2.3.1.2.   *Captopril*

One of the key ideas in the development of ACE inhibitors was the hypothesis of Ondetti and Cushman that ACE is mechanistically related to the well-characterized metallopeptidase carboxypeptidase A (CPA), and thus that CPA inhibitors could be modified to produce inhibitors of ACE.[10,37] Angiotensin converting enzyme (ACE) is a 130–140 kDa membrane-bound, metalloglycopeptidase which recognizes and cleaves dipeptides from the carboxyl terminus of angiotensin I, bradykinin and other natural peptides and synthetic substrates. The enzyme contains one zinc atom per catalytically active enzyme. Because X-ray crystal structures and active-site data for ACE were lacking, Ondetti and Cushman, reasoning by analogy, hypothesized that the active-site zinc would coordinate to the substrate amide carbonyl to facilitate hydrolysis by a mechanism closely related to the CPA mechanism. A variant of this mechanism is shown in Scheme 5. Carboxypeptidase is related to ACE in that this zinc metallopeptidase removes one amino acid from the carboxy terminus of peptide substrates. Byers and Wolfenden had designed benzylsuccinic acid as a di-product (collected product) inhibitor of CPA,[38] and the Squibb group reasoned that if ACE were sufficiently similar to CPA, an appropriate succinyl *dipeptide* analog (51) might be a good ACE inhibitor.[37] The carboxyalkanoyl derivative (51) was synthesized and found to bind to ACE approximately 100-fold more strongly than (50), which is consistent with the interaction of the carboxy group with the active site zinc ion. Replacement of the carboxy group with the thiol group, a ligand known to chelate tightly to zinc, gave captopril (1) which binds to ACE ($K_i = 1.7$ nM) about 1000 times more tightly than (51).[37]

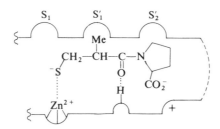

(51)

8.2.3.1.3 P_2' variations of captopril

The relatively nonspecific binding requirements of the S_2' subsite of ACE for both substrates and inhibitors (*e.g.* (49) contains the bulky indole moiety in P_2') has led to many novel ACE inhibitors in which the proline residue in the P_2' inhibitor position has been modified. Thus the thiaproline analog (4),[39] the demethyl, substituted thiaproline analog (5),[40] and the phenylthio substituted prodrug analog (6)[41] (see Table 1) are good inhibitors of ACE. The proline ring system can be modified by benzo fusion, *e.g.* (7),[42] and *N*-substituted glycine derivatives, *e.g.* prodrug (8),[43] also are good ACE inhibitors. Additional prodrug forms of these compounds also have been developed, *e.g.* (9)–(11).[44–47] Structure (12) represents a composite of several modifications.[48] All of these inhibitors can be considered to inhibit ACE by a mechanism closely related to that proposed by the Squibb group (Figure 3) in which the C-terminal carboxyl group binds to a positively charged enzyme group (probably Arg), the amide carbonyl is hydrogen bonded to the enzyme, and the free thiol group is coordinated to the zinc ion in the active site of ACE.

Figure 3 Binding interactions of captopril with the active site of ACE as proposed by Ondetti *et al.*[10]

8.2.3.1.4 *N*-Carboxyalkyl dipeptides: enalapril and lisinopril

An alternative approach to ACE inhibitors was developed by Patchett and colleagues at Merck.[11,49] Beginning with the collected-product idea for inhibiting ACE, they prepared the *N*-carboxyalkyl dipeptide (52), which was expected to mimic more closely the structure of (50) than did the carboxyalkanoylproline derivative (51) by virtue of the retained NH group. However, (52) was found to be a relatively weak ACE inhibitor that was essentially equipotent with (47). Thus replacement of CH_2 with NH did not produce the anticipated better ACE inhibitors in the series of compounds first tried. However, when R substituents were added to the N-carboxyalkyl group, much more potent inhibitors were obtained, *e.g.* (53). The enhanced binding was presumed to result from hydrophobic interactions with the enzyme S_1 subsite and the proper orientation of the —NH— and —CO_2H groups. Systematic extension of these P_1 substituents led to the discovery of the dicarboxylic acid derivative enalaprilat (3), an effective inhibitor of ACE ($K_i = 0.04$ nM) *in vitro.* Enalaprilat, a di-acid, is not well absorbed orally but the ethyl ester derivative enalapril (2), a mono-acid, is a prodrug form that is well absorbed orally and is converted by hydrolysis *in vivo* to the active ACE inhibitor (3).

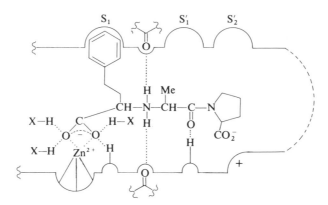

Systematic variations in the structure of enalaprilat demonstrated that the P_1' substituents could be varied considerably. The lysine side chain analog (13; lisinopril) is a good inhibitor of ACE which is well absorbed without needing to be converted into a prodrug.[49,50] [Excellent bioavailability has been shown for another ACE inhibitor which contains a lysine side chain in P_1' (see Section 8.2.3.1.6).]

Patchett and colleagues proposed that enalaprilat binds to ACE as shown in Figure 4.[11] The mechanism is similar to that proposed for captopril, *e.g.* a cationic group electrostatically binds with the C-terminal carboxylate, a donor hydrogen bonds to the carbonyl group of alanine, and the N-carboxyalkyl group binds to the enzyme's zinc ion. The binding of enalaprilat differs mainly in that the phenylethyl side chain binds to the S_1 enzyme subsite (which remains unoccupied in the case of captopril) and a protonated amine hydrogen bonds with two enzyme carbonyl groups.

Figure 4   Binding interactions between enalaprilat and the active site of ACE as proposed by Wyvratt and Patchett[11]

8.2.3.1.5   *Enalapril analogs*

As was shown with the structure–activity relationships for captopril, the S_2' subsite of ACE is capable of accommodating many different substitutions on the inhibitor proline ring system; consequently, numerous enalapril analogs have been synthesized in which the P_2' proline residue has been altered (shown in Table 1 in their prodrug forms). Bicyclic analogs (14)[51] and (15)[52] are effective ACE inhibitors, as are the N-substituted glycine derivatives (16)[53] and the tetrahydroisoquinoline derivatives (17).[54] However, the larger P_2' substituted analogs are not markedly stronger inhibitors of ACE than enalaprilat (3), which has led to the suggestion that the binding of P_2' substituted N-carboxyalkyl dipeptide inhibitors to ACE (Figure 5) differs from the binding of P_2' substituted captopril ACE inhibitors with respect to the positioning of the P_1 group.[11] Note that the P_1 group does not occupy the S_1 pocket in the mechanism proposed in Figure 5. Consistent with this postulate is the fact that variations in the P_1 side chain in the N-carboxyalkyl dipeptide inhibitors which contain larger groups in the P_2' position (*e.g.* 18, 19)[55,56] also result in good ACE inhibitors.

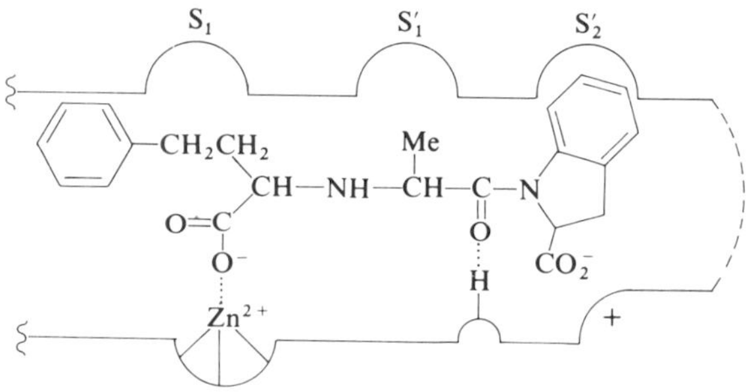

Figure 5 Possible binding interactions between P'$_2$-substituted carboxyalkanoyl inhibitors and the active site of ACE (*cf.* Figure 1(b) for the proximity of S$_1$ and S'$_2$ subsites)

8.2.3.1.6 *Phosphorus-containing analogs: fosfinopril*

Phosphorus-based inhibitor designs were stimulated by the discovery of phosphoramidon [*N*-(α-L-rhamnopyranosyloxyhydroxyphosphinyl)-L-leucyl-L-tryptophan (**54**)], a potent inhibitor of metallopeptidases, *e.g.* thermolysin.[27] Subsequently, the X-ray crystal structure of the phosphoramidon–thermolysin complex revealed that the hydroxyphosphinyl functional group was a monodentate ligand bound to the catalytically active zinc ion; this result led to the proposal that this functional group mimicked the tetrahedral intermediate for amide bond hydrolysis.[57] A subsequent modeling study suggested that the tetrahedral intermediate binds to zinc in a bidentate fashion.[24]

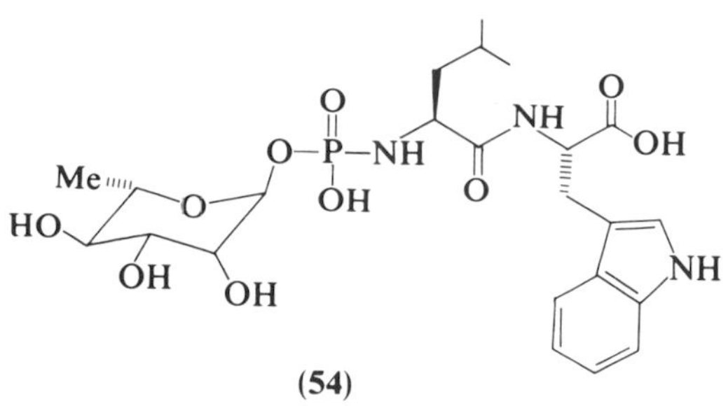

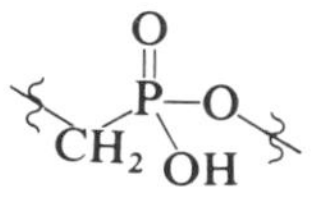

These observations have stimulated the synthesis of numerous ACE inhibitors in which the peptide backbone incorporates different phosphorus moieties, *e.g.* the phosphonates (**55**), the phosphinic acid derivatives (**56**), and phosphonamidate derivatives (**57**). The simplest of the ACE inhibitors that can be designed by this procedure, (**58**), is formed by adding a hydroxyphosphinyl group to the dipeptide.[58] Compound (**58**) was found to be a very potent *in vitro* ACE inhibitor although the P—N bond in this molecule is too unstable in acid for (**58**) to survive oral delivery. Other more acid stable phosphorus derivatives were synthesized and several potent ACE inhibitors have been identified within this group. Surprisingly, the structure–activity relationships for the phosphorus-derived inhibitors suggest that the P—O group binds to zinc in a different orientation than occurs in the inhibitors related to captopril or enalaprilat.[59] Thus in the phosphonate series of ACE inhibitors, the distance between the proline rings and the zinc ligand is shorter relative to the other classes of ACE inhibitors, as illustrated by inhibitor (**21**; see Table 1). The most effective ACE inhibitor is produced when the inhibitor contains a phenylbutyl side chain in the P$_1$ position and hydrophobic substituents in P'$_2$. An unusual double ester prodrug form of (**21**), fosfinopril (**20**), is a particularly effective orally active ACE inhibitor which has a longer duration of action in man than enalapril.[60]

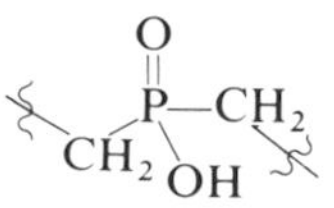

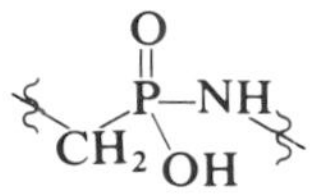

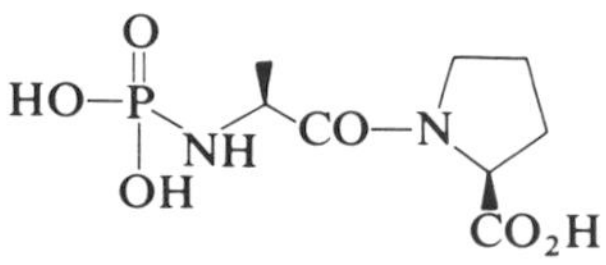

More recently, the Squibb group has described a phosphonyloxyacyl proline derivative, SQ 29852 (**22**) that is an orally active inhibitor of ACE.[61] The aminobutyl side chain in (**22**) is an absolute requirement for full expression of oral activity, and this structure–activity relationship resembles the effect of the lysine side chain in lisinopril (**13**).

8.2.3.1.7 *Conformationally restricted captopril and enalapril analogs*

In the early work to develop both captopril and enalapril, it was noted that introduction of a methyl group α to the carbonyl group in the P'_1 position (corresponding to the methyl group of Ala in the carboxyalkyl series of inhibitors) greatly enhanced the biological activity of these inhibitors. Thus the 2-methyl derivative (**51**) is more effective than structure (**59**), and (**1**) is more potent than structure (**60**) as ACE inhibitors. In principle, the added methyl side-chain substituent could enhance binding by providing a group that hydrophobically binds to the enzyme's S'_1 pocket, or it could force a greater percentage of the analog out of a nonproductive conformation into a conformation favoring biological activity. Since all classes of ACE inhibitors tolerate side chains larger than methyl in the P'_1 position, the enzyme S'_1 subsite must be able to accommodate side chains larger than methyl [see, for example, lisinopril (**13**) and SQ 29852 (**22**)]. Thus the favorable effect of the added methyl group in (**1**) and (**51**) was postulated to be conformational in origin.

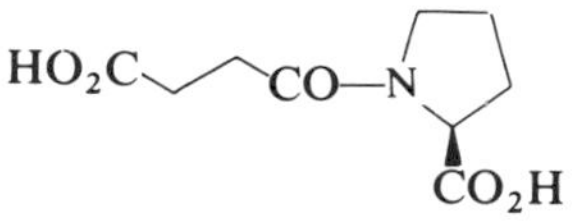

(**59**) 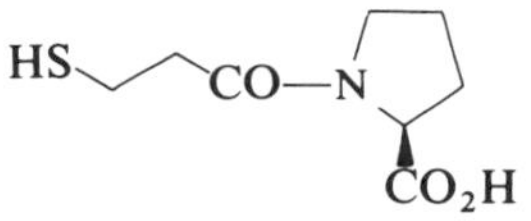   (**60**)

To test this hypothesis, a series of conformationally constrained analogs of captopril and enalapril were synthesized and evaluated as ACE inhibitors.[62] This strategy for identifying the enzyme-bound inhibitor conformation is a variation of the strategy for determining the bioactive conformations of peptide hormones that was first carried out with somatostatin[63] and later with α-melanotropin[64] and enkephalin.[65] At the outset of this work the amide bond between Ala-Pro was assumed, on the basis of molecular mechanics calculations and solution NMR data, to be in the *trans* conformer in the bioactive conformation. Thus the differences between the methylated analogs [(**51**), (**1**)] and unmethylated analogs [(**59**), (**60**)] would be due to the differences in the torsion angle ψ (Figure 6). The methylated derivatives are presumed to adopt a higher concentration of the bioactive conformation.[66]

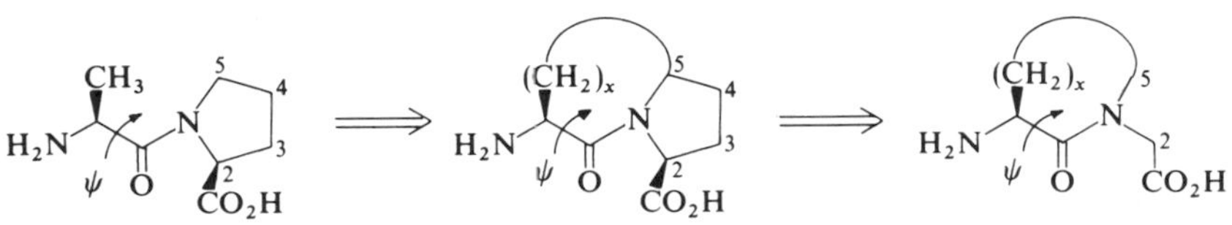

Figure 6   Definition of the torsion angle ψ in monocyclic and bicyclic ACE inhibitors (adapted from Wyvratt and Patchett[11])

The study was carried out in two phases. First, the possible conformations for the dipeptide Ala-Pro were calculated as a function of the torsion angle ψ. Second, cyclic captopril and enalapril analogs were synthesized which bridged the β-carbon of alanine with the δ-carbon of the proline ring (Table 4). In order to limit the synthetic difficulties in the initial phases of this work, the analogs were derived from *N*-alkyl glycine derivatives rather than from proline. By systematically varying the number of the bridging atoms and therefore the size of the resulting ring system, it was possible to constrain the torsion angle ψ within narrow limits. Estimates for ψ in the cyclic portion of the inhibitors were obtained from energy calculations and were compared with known structures of related five- and six-membered ring analogs. The larger ring systems can adopt multiple conformations each with a distinct value of ψ, so that it was necessary to calculate all possible low energy conformations of the seven- and eight-membered lactams by using Still's RINGMAKER/BAKMOD programs.[67] The most potent inhibitors were found to be the seven- through nine-membered lactams; the five- and six-membered rings systems were decidedly inferior inhibitors.

Table 4 Conformationally Restricted Analogs of Enaliprilat

	IC_{50} (nM)	ψ (calc; low energy conformation)
(61)	5300	$-132°$
(62)	430	$-183°$
(63)	19	$166°$
(64)	1.9	$145°$
(65)	8.1	$135°$

The biologically active lactam derivatives (63)–(65) adopt at least one low-energy conformation in which ψ lies between 130–170°. The weaker lactam inhibitors did not have low-energy conformers in which ψ approached 130–170°. For example, in the five- and six-membered lactams (61) and (62), ψ was $-132°$ and $-138°$, respectively. From these data, the Merck group proposed that the conformation of the ACE inhibitors that is bound to the active site of the enzyme is characterized by a *trans* amide bond between Ala-Pro and a ψ angle between 130–170°.[62] In the X-ray crystal structure of enalaprilat (3),[68] ψ is 143°.

Bicyclic inhibitors incorporating both the seven-membered lactam and either the proline (66)[26] or the thiaproline (67) ring system[69] were then synthesized. The expectation was that the more conformationally restricted analogs would be more potent inhibitors of ACE due to a favorable entropic contribution to the free energy of binding. Reintroduction of the proline ring system (66) did not improve the inhibitory potency. However, the replacement of one methylene in (66) by a sulfur atom to give (67) increased inhibitory potency. The Merck group suggested that the added conformational flexibility imparted by the larger ring in thiaproline was important to the binding of the inhibitor relative to the proline analogs. Compound (67) is a remarkably potent inhibitor (K_i = 0.076 nM). Consistent with the proposed bioactive conformation, X-ray crystallography established that $\psi = 168°$ in (67), which was in good agreement with the value calculated (156°) by molecular mechanics methods. Other variations of the seven-membered lactam inhibitors, *e.g.* (23)–(26) (see Table 1),[70–73] are reported to be undergoing clinical evaluations.

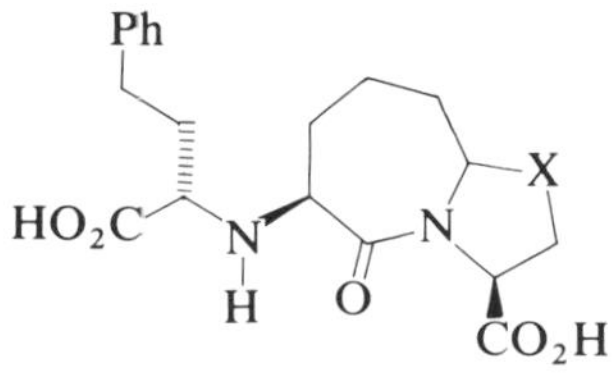

(66) X = CH$_2$
(67) X = S

8.2.3.1.8   Cilazapril

A group at Roche discovered another highly potent bicyclic ACE inhibitor, cilazapril (27; see Table 1) when it applied the principle of conformational restriction to define the bioactive conformation of captopril and enalapril derivatives.[74,75] The Roche approach was initially focused on the question of the relative orientations of the essential amide carbonyl, carboxylate and thiol groups in the bioactive conformation of captopril. The hexahydropyridazine system (Table 5) was chosen so as to restrict the conformations of the thiol and carboxylate groups relative to each other. Bicyclic inhibitors (68) and (69) were tested and the latter compound was found to be the most potent ACE inhibitor in this series, which established that the thiol and carboxyl groups must be approximately *trans* to each other. The thiol group in (69) can still adopt multiple conformations (Figure 7b), but the number of possibilities is limited when compared with the possible conformations of the thiol group in captopril (Figure 7a). These conformations were calculated by assuming that the Ala-Pro amide bond in *trans* in the bioactive conformation, that the carboxylate is axial to the proline and pipicolic rings as it is in the solution and the X-ray conformations, and that the alanine carbonyl group in captopril hydrogen bonds to the enzyme. Under these conditions the permissible orientations of the thiol group are severely limited. The Roche group proposed that these constraints could be accommodated by the enzyme bound conformations of captopril shown in Figure 7c.

Table 5   Conformationally Constrained Bicyclic ACE Inhibitors

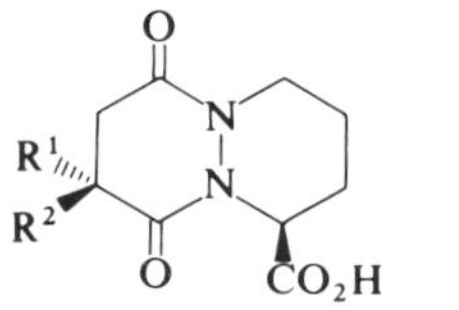

(68) R^1 = CH$_2$SH; R^2 = H
(69) R^1 = H; R^2 = CH$_2$SH

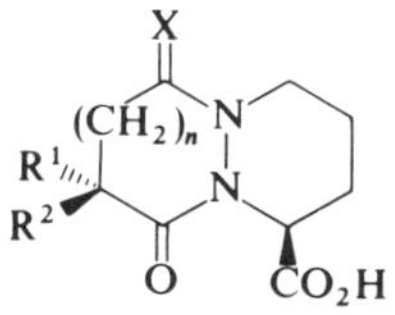

(70)–(77)

| Compound | X = H$_2$ | | | | Compound | X = O | | |
	Ring size	n	ψ (°)	I_{50} (nM)		n	ψ (°)	I_{50} (nM)
(70)	5	0	255	8000	(74)	0	237	—
(71)	6	1	245	28	(75)	1	197	20
(72)	7	2	164	1.6	(76)	2	165	4
(73)	8	3	142	4.5	(77)	3	146	15

The bicyclic structure (69) is not a very potent ACE inhibitor when compared with captopril or enalapril, and the Roche group recognized that this might be caused by a nonoptimal ψ torsion angle imposed by the six-membered lactam in (69).[76] The Roche group synthesized a series of lactam analogs related to enalapril rather than captopril in which the size of the lactam ring system and the torsion angle ψ were systematically varied [Table 5; (70)–(77)]. In two series of compounds the most potent ACE inhibitors (72, 76) were derived from the seven-membered lactams. In each series, ψ is 165–166° in the most potent analogs. The five- and six-membered lactams (70, 71, 74, 75) are much weaker inhibitors and ψ in these compounds lies far outside the 140–170° range indicated for the bioactive conformation of ACE inhibitors. Cilazapril (**27**), the ethyl ester prodrug form of (**28**), is undergoing further development as an antihypertensive drug.[76]

One highly interesting result derived from the studies to determine the bioactive conformations of the *N*-carboxyalkyl dipeptide inhibitors is the discovery that this class of inhibitor adopts different conformations with respect to the torsion angle ψ when it is bound to different metallopeptidases. In

(a)

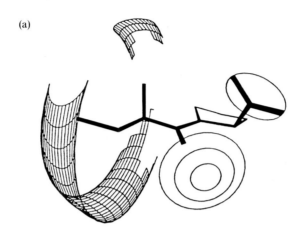

(b)

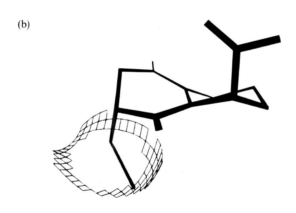

(c)

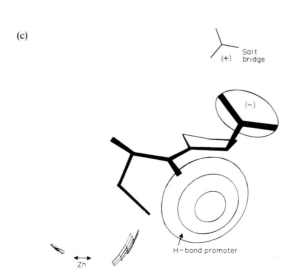

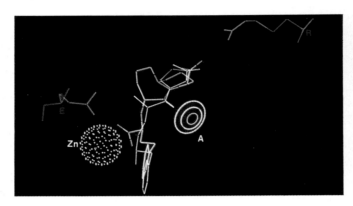

Figure 7 (a) Possible orientations of the thiol group in captopril. (b) Possible location of the thiol group in the bicyclic inhibitor. (c) Postulated enzyme-bound conformation of the thiol group in captopril and related analogs.[75,76] (d) Model of cilazapril and the postulated zinc dipeptidase angiotensin converting enzyme. Observed position of zinc atom (sphere) and the catalytic Glu-143 residue (red) of thermolysin[1] are shown together with a low-energy conformation of cilazapril[2] and substrate H-Phe-Ala-Pro-OH (magenta). Region A (green) shows the limits of the probable location of H-bond donor and R (blue) denotes the postulated position of an arginine residue, which interacts with the terminal carboxy group of cilazapril

the crystal structure of *N*-[1-carboxy-3-phenylpropyl]-L-leucyl-L-tryptophan (**78**) bound to thermolysin, determined by Matthews and coworkers, ψ is 316° in contrast to the 130–170° range preferred by ACE.[77] Calculations of the low-energy conformations of L-Ala-L-Pro indicate two approximately equal low-energy minima lying between $\psi = 90–180°$ and 300–330°.[76] Thus ACE appears to bind inhibitor conformations (and presumably substrate conformations) falling within the first energy well while the closely related metallopeptidase thermolysin binds inhibitors that can adopt conformations falling within the second energy well.

Other attempts to identify the bioactive conformation of the ACE inhibitors by means of comparative molecular modeling and energy minimization techniques have been reported. Andrews and coworkers compared the structures of eight structurally diverse ACE inhibitors and concluded that there is a common inhibitor conformation (with $\psi = 165°$) throughout the series that can bind to the enzyme.[78] The active site of ACE was then 'mapped' by assuming that complementary enzyme groups were needed to bind to the shared low-energy conformations. Ciabatti and coworkers derived a bioactive conformation from the study of a series of captopril analogs designed to constrain the thiol group conformationally as shown in (**79**).[79] In the most active analogs the thiol group is *trans* to the proline carboxyl group and can bind to the ACE zinc atom in a fashion consistent with the Andrews' model.

Another very different class of conformationally restricted ACE inhibitors was devised by Weller *et al.*[141] to define the conformation in the ACE S$_1$ subsite. The captopril analog (**34**) is nearly 10 times more potent than (**1**). Optimally constrained analogs have a dihedral angle of 30–60° between bonds for the zinc-binding and amide groups. Marshall and colleagues carried out the most extensive comparison of the ACE inhibitor structures reported to date.[80] All possible conformations (within 10° increments for each torsion angle) for each ACE inhibitor were generated by using SEARCH procedures. Each conformation was represented by a point in distance space which could be displayed in three dimensions by means of orientational maps (Figure 8a). Proceeding from the assumption that all biologically active ACE inhibitors would bind to the same elements in ACE, two

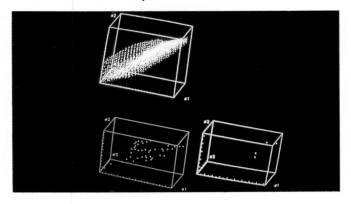

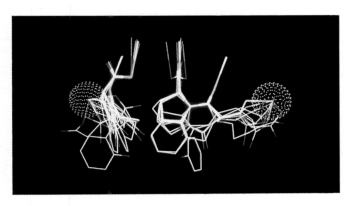

Figure 8 (a) Successive distance orientation maps (OMAPs) for a series of ACE inhibitors. 1. The first OMAP, generated without any constraints on one molecule, Mi. 2. Orientation space common to several ACE inhibitors. 3. Final OMAP representing the column active site patterns for all examined analogues of ACE inhibitors. (b) Orthogonal projection of the conformations of 28 ACE inhibitors that fit the OMAP in (a) (one conformer per structure). The pink dot surface represents the active-site zinc atom

points in distance space were found that are common to all the ACE inhibitors studied. Each of the 28 inhibitors had between 4 and 562 possible conformations that were consistent with the two valid OMAP points. In Figure 8b, 28 examples of superimposed allowed conformations (one each for each class of inhibitor) are shown as they are inferred to bind to the zinc ion (pink), the hydrogen bond donor, and the enzyme cationic group.

8.2.3.1.9 *Substrate-based ketone and alcohol ACE inhibitors*

Several other classes of ACE inhibitors have been reported but have not yet reached clinical evaluation. Almquist reported the synthesis of the ketomethylene derivative (**80**) which is a ketone analog of a substrate-like sequence for ACE.[81] The ketone is an exceptional ACE inhibitor *in vitro* but does not show sufficient oral bioavailability to be used clinically. Almquist's work established that ketomethylene groups could lead to excellent inhibitors of metallopeptidases, a result which inspired chemists to attempt variations of this strategy as a means to develop inhibitors of many peptidases (see Sections 8.2.3.4.3 and 8.2.3.5.1). Gordon and colleagues have described two novel classes of ACE inhibitors.[82] The amino ketone derivatives (**81**) are very potent ACE inhibitors *in vitro*. Surprisingly, the corresponding aminoalcohol derivative (**82**) also is a strong ACE inhibitor.[83]

(**80**)

(**81**)

(**82**)

It has been proposed that both compounds bind to the active site zinc ion in ACE in a bidentate fashion (Figure 9). Additional ACE inhibitors (**29**)–(**33**) are shown in Table 1.[84-87] Compound (**33**) has the lowest K_i (4 pM) for inhibition of ACE yet reported.[88]

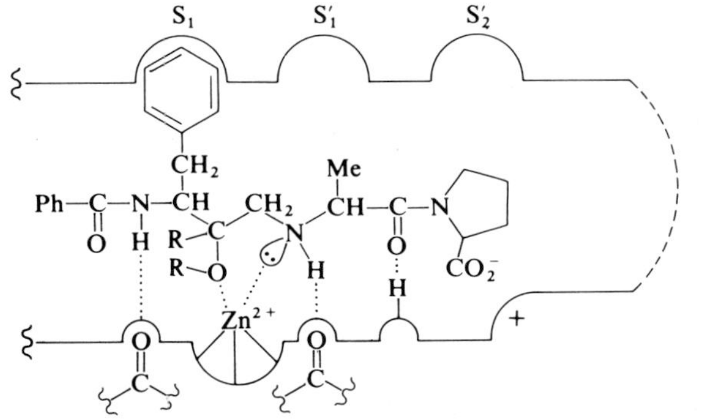

Figure 9   Proposed binding interactions of the amino ketones and amino alcohol derivatives (**81**) and (**82**) with the active site of ACE (adapted from Gordon *et al.*[82,83])

8.2.3.2   Enkephalinase Inhibitors

Neutral endopeptidase [EC 3.4.24.11; NEP] is a membrane-bound zinc metalloendopeptidase located in the plasma membrane of many tissues.[89] In mammalian brain the enzyme has been shown to be involved in the inactivation of the opiod peptides, methionine-enkephalin (**83**; Met-Enk) and leucine-enkephalin (**84**; Leu-Enk) (Figure 10), and thus has been called 'enkephalinase,' but this compound represents only one of many substrates for the enzyme. The biological relevance of NEP as an enkephalin-degrading enzyme is supported by its distribution in rat brain which overlaps that of opiod receptors, its neuronal localization, and by the naloxone-reversible analgesic responses induced by inhibitors such as thiorphan (see below). Inhibitors of enkephalin degradation thus represent a new class of potential analgesic drugs that have attracted attention because of their possible reduced potential for misuse.

Tyr–Gly–Gly–Phe–Met (**83**)

Tyr–Gly–Gly–Phe–Leu (**84**)

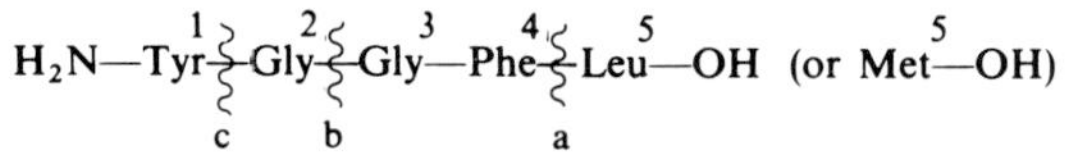

Figure 10   The enzymatic degradation of the enkephalins: (a) endopeptidase 24.11 and angiotensin converting enzyme; (b) aminopeptidase M; (c) dipeptidylaminopeptidase

The design of the enkephalinase inhibitors was based on analogy with the active site model of angiotensin converting enzyme (Figures 3, 4).[90] Various dipeptides were synthesized and evaluated as NEP inhibitors in order to identify the nature of the S_1' and S_2' binding sites of the enzyme. Phe-Ala and Phe-Leu were selected as suitable starting points for further modification and thiol ligands were added to these structures for the purpose of chelating with the enzyme's active-site zinc ion, following the strategies developed originally to prepare ACE inhibitors. Representative examples of these inhibitors are shown in Table 6.

Thiorphan, *N*-[(*R*, *S*)-3-mercapto-2-benzylpropanoyl]glycine (**85**) is a highly potent inhibitor (K_i = 3.5 nM) of 'enkephalinase'.[91] Thiorphan also inhibits ACE although less strongly (K_i = 140 nM). In order to develop more selective compounds, Roques and colleagues studied the effect on

Table 6 Representative Inhibitors of Enkephalinase

modification of the amide bond in (**85**) upon inhibitor potency of both NEP and ACE. Structural modifications of model dipeptides had indicated that N-methylation of the amide bond weakened NEP inhibition without affecting ACE inhibition, suggesting that the amide bond was more important for proper binding to NEP than to ACE. The retro amide bond derivative (**86**) was synthesized and found to retain most of the NEP activity ($K_i = 6$ nM) but almost no ACE inhibition activity.[92] This remarkable specificity is the highest reported to date for peptides designed by replacing an amide bond with a retro-inverso isostere.

Additional enkephalinase inhibitors have been designed by incorporating other possible zinc ligands in place of the thiol group. Mumford *et al.* reported enkephalinase inhibitors, *e.g.* (**87**), derived from the N-carboxyalkyl dipeptide inhibitors,[93] first developed for use as inhibitors of ACE. In this case the important feature is the placement of phenylalanine in the P'_1 position for efficient inhibition of NEP. Schering 32615 (**88**) represents another example of this approach in which the β-alanine at the C-terminus of the inhibitor provides extraordinary selectivity for inhibition of NEP ($K_i = 15$ nM) without inhibiting ACE ($I_{50} > 100\,000$ nM).[94] The di-acid form of (**88**) is not orally active, but the prodrug form (**89**) currently is in safety assessment. Enkephalinase inhibitors derived from phosphonates, *e.g.* Schering 34729 (**90**), have also been developed in which the phosphonate group is thought to coordinate to the zinc ion in the active site of NEP.[95a] A detailed comparison between inhibition of ACE and NEP by over 50 derivatives of mercaptopropanoyl amino acids and related derivatives was reported by Gordon *et al.*[95b]

The enkephalins are degraded *in vivo* by at least three different peptidases at the cleavage sites shown in Figure 10.[96] Since all three enzymes contain zinc in their active sites, the possibility existed that a single, suitably designed chelating inhibitor might inhibit all three. Powers and Nishino had shown that N-hydroxamic acid ligands incorporated into peptide chains could be used to develop potent bidentate inhibitors of zinc metallopeptidases,[97] and Roques, Fournié-Zaluski and colleagues applied this strategy to the synthesis of enkephalinase inhibitors to obtain kelatorphan (**91**).[98] The hydroxamic acid (**91**) is very effective inhibitor of NEP ($K_i = 0.3$ nM) and dipeptidylaminopeptidase (1.0 nM) and to a lesser extent of aminopeptidase-M (720 nM). Both thiorphan and kelatorphan protect enkephalins from degradation *in vitro* and *in vivo*. All inhibitor-induced analgesic responses are antagonized by prior administration of naloxone, demonstrating that the observed effects are due to specific stimulation of opiod receptors. However, very strong nociceptive stimuli are not as effectively inhibited by the enkephalinase inhibitors alone, possibly because other opiod peptides (dynorphans, endorphins, *etc.*) mediate the responses.

The enkephalinase inhibitors have proven to be remarkably effective probes of the physiological functions of the enkephalin systems. A triptiated glycine analog (**92**) of kelatorphan has proven effective for determining the location of the enkephalin degrading enzymes in the central nervous system.[99]

8.2.3.3 Collagenase Inhibitors

Collagenases [EC 3.4.24.7] are zinc metallopeptidases that degrade triple helical collagen under physiological conditions.[100] Vertebrate collagenases preferentially cleave each of the three polypeptide chains of types I, II or III collagen about three-quarters of the distance from the N-termini to produce fragments designated TC ∗ A and TC ∗ B. Prokaryotic collagenases make numerous cleavages thoughout the collagen triple helix.

Inhibitors of collagenase are useful for elucidating the physiology of the associated enzymes[101] and may be of use for treating pathological conditions characterized by excessive proteolysis of collagen, such as occurs in rheumatoid arthritis, corneal ulceration and epidermolysis byllosa.

Collagenases are zinc metalloproteinases.[102] By analogy with the design principles described for the development of inhibitors of the zinc metallopeptidases already described, collagenase inhibitors were expected to be derived from substrate peptides by incorporating groups which would chelate zinc. However, identification of the appropriate substrate sequence on which to begin the analog synthesis was not trivial.

The triple helical structure of collagen had suggested that collagen substrates, and by inference collagenase inhibitors, might have to adopt a defined tertiary structure in order to be recognized by the enzyme, in contrast to other peptidases which more readily recognize conformationally mobile, shorter peptides.[15] However, a detailed computerized search of the protein sequences known to be cleaved by mammalian collagenase led to the discovery that the preferred amino acid sequence for collagenase cleavage (Figure 11) represented a departure from the ordered triple helix, suggesting that collagenase recognized disordered sites in collagen. The substrate sequence was found to span the P_3-P'_3 amino acid residues and is characterized by the absence of proline at the P_2 and P'_2 subsites, *i.e.* the substrate sequence represents a deviation from the regular structure anticipated from the repeat nature of the triple helical polymer. The best synthetic collagenase inhibitors have structures which appear to be consistent with this sequence specificity.

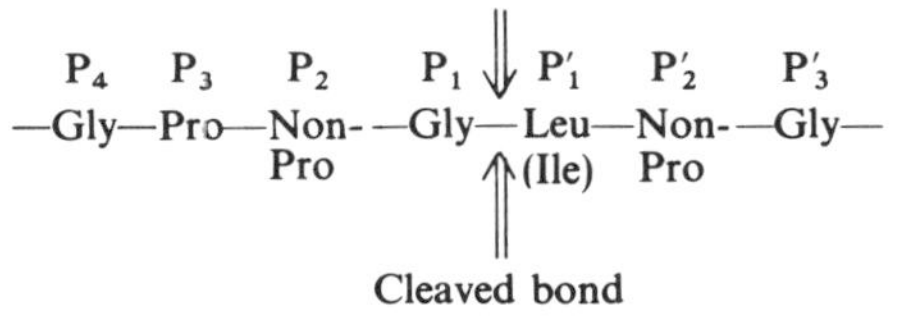

$$P_4 \quad P_3 \quad P_2 \quad P_1 \Downarrow P'_1 \quad P'_2 \quad P'_3$$
$$\text{—Gly—Pro—Non-}\text{—Gly—Leu—Non-}\text{—Gly—}$$
$$\text{Pro} \qquad \Uparrow\text{(Ile)} \quad \text{Pro}$$

Cleaved bond

Figure 11 Preferred amino acid sequence for collagenase substrates[15]

The development of inhibitors of mammalian collagenase has been reviewed by Johnson *et al.*[15] The general strategies for inhibiting metallopeptidases (see Section 8.2.3) were applied to collagenase. Incorporation of the *N*-carboxyalkyl, the hydroxamic acid, and the thiol methylene moieties into a modified Leu-Tyr sequence gave potent collagenase inhibitors shown in Table 7. The importance of hydrophobic sbustituents in the P'_2 position is demonstrated by comparing the potency of the alanine derivative (**94**) with the *O*-methyl tyrosine derivative (**93**) which is five-fold more potent. Mammalian collagenases are reported to be inhibited more effectively by inhibitors that contain leucine or phenylalanine analogs in the P'_1 residue, and Darlak *et al.* found that extension of the peptide chain into the P'_2-P'_3 positions gave even more effective inhibitors of porcine synovial collagenase.[103] Use of (2'-naphthyl)-L-Ala led to a very potent inhibitor (**95**; $K_i = 14$ nM) against this enzyme.[104]

Excellent collagenase inhibitors have been obtained by incorporating the metal-chelating hydroxamic acid functionality into a peptide sequence corresponding to Leu-Leu-Ala-OEt.[15,105] The Roche inhibitor (**96**) has a $K_i \approx 5$ nM. Inhibitors have been derived by incorporating phosphorus ligands into collagenase inhibitors by following the strategies employed to develop ACE inhibitors. Mookhtiar *et al.* synthesized a series of phosphoramidate derivatives in which the amide carbonyl of the substrate was replaced with the PO_2 group.[106] Inhibitor (**97**) of human neutrophil collagenase thus obtained has a $K_i = 14$ µM. Similarly, the *N*-carboxyalkyl inhibitor design has been employed to produce collagenase inhibitors. Compound (**98**) is reported to have an $IC_{50} = 8$ µM against collagenase.[107]

Molecular modeling studies of collagenase substrates have been used to distinguish between active conformations with the P'_1 side chain adjacent to either P'_3 or P'_2 side chains. The latter conformation was proposed to resemble that of the enzyme-bound thermolysin inhibitors. Structure–activity data for the series of compounds related to (**99**) indicates that the side chains R^1,

Table 7 Representative Inhibitors of Collagenases

Compound	R^1	R^2	IC_{50} (μM)
(93)	(benzyl-OMe)	NHMe	0.2
(94)	—Me	NHMe	1.0
(95)	(naphthylmethyl)		0.014
(96)			0.009
(97)			14.0
(98)			8.8
(99)			—
(100)			—
(101)			0.07

Table 7   *(Contd.)*

(102)   0.03

(103)   0.03

R^2, and R^3 point away from the enzyme whereas P'_1 binds to the S'_1 subsite. These results were interpreted to indicate that the P'_1 and P'_2 subsites will be on the opposite side of the inhibitor backbone in an enzyme-bound conformation quite different from that found for inhibitors bound in the active site of thermolysin. Cyclic collagenase inhibitors, which constrain adjoining side chains, *e.g.* (100) in which a lactam joins R^2 and R^1 in (99), have recently been reported. A new series of collagenase inhibitors that incorporate these design principles has also recently been reported.[222] Borkakoti *et al.* prepared substrate-derived phosphinic acid inhibitors, *e.g.* (101), but none exhibited useful pharmacokinetic properties. Systematic modification of the N-terminus and replacement of the C-terminal amino acid ester by an *N*-methyl amide group led to potent, low molecular weight derivatives, *e.g.* (102). Conversion of (102) into cyclic analogs for the reasons cited for (100) led to the discovery of lactam (103), an analog which retained good activity against collagenase but has greatly improved *in vivo* properties.[222] It is likely that additional collagenase inhibitors will be discovered by application of analogous design principles.

8.2.3.4   Renin Inhibitors

The success of the ACE inhibitors as therapeutic agents for treating hypertension and congestive heart failure has increased interest in developing drugs based upon the inhibition of human renin [EC 3.4.99.19].[14,108] The cleavage of angiotensinogen to angiotensin I is the first and rate-controlling step in the biosynthesis of angiotensin II (Scheme 1). Since angiotensinogen is the only known natural substrate for renin, it was thought that inhibition of renin might lead to more selective regulation of blood pressure than produced by ACE inhibitors. At present, renin remains a viable target for developing antihypertensive agents but renin inhibitors have not yet been extensively tested in the clinic and the possibility exists that renin inhibitors will be only equivalent to ACE inhibitors as antihypertensive agents.

8.2.3.4.1   *Substrate analog inhibitors: RIP*

The first successful renin inhibitors were derived from modifications of renin substrates. Skeggs and coworkers had shown that octapeptide (104) was the shortest porcine substrate sequence that could be cleaved by renin,[109] and later the minimal human sequence (105) was determined.[110] Kokubu and coworkers had found that the truncated peptide, Leu-Leu-Val-Tyr-OMe (106) is an inhibitor of renin although only at very high concentrations.[111] Burton and colleagues discovered the first useful inhibitors of renin by synthesizing analogs of substrate sequences in which the amino acids adjacent to the hydrolyzed amide bond were changed.[112] Replacement of the two leucine residues in the intact octapeptide by phenylalanines rendered the peptide resistant to cleavage by renin, a fact not yet explained in molecular terms. By systematic modification of this structure, primarily in order to improve the solubility characteristics of the peptide, they obtained decapeptide

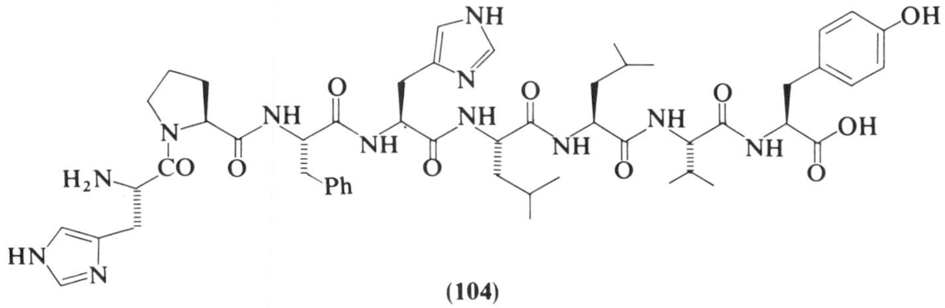

(104)

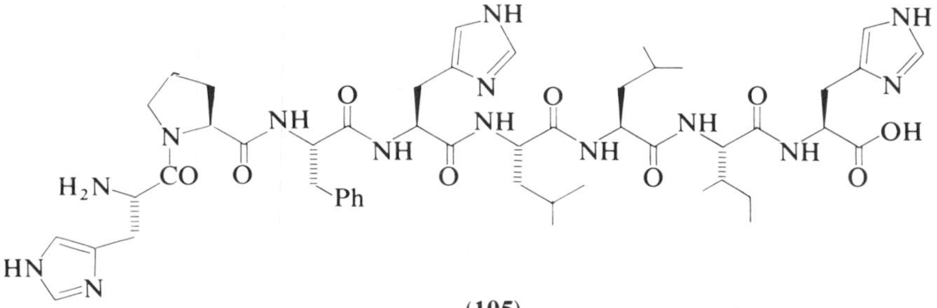

(105)

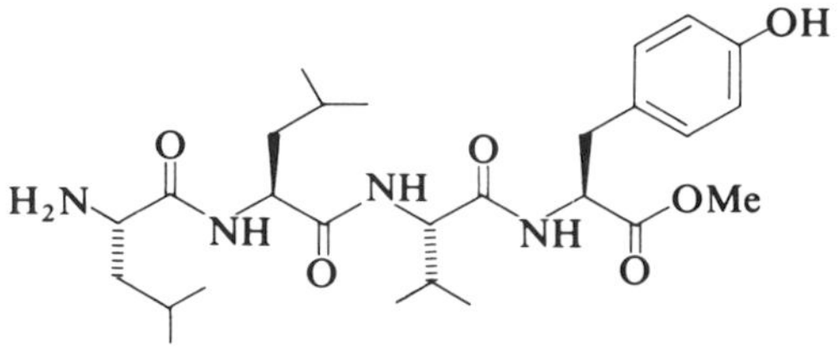

(106)

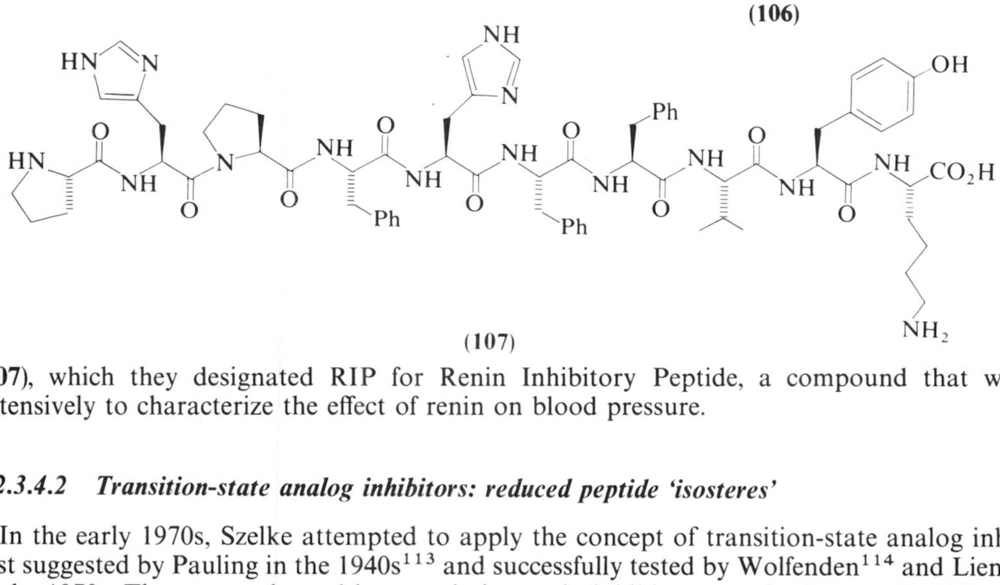

(107)

(107), which they designated RIP for Renin Inhibitory Peptide, a compound that was used extensively to characterize the effect of renin on blood pressure.

8.2.3.4.2 *Transition-state analog inhibitors: reduced peptide 'isosteres'*

In the early 1970s, Szelke attempted to apply the concept of transition-state analog inhibition, first suggested by Pauling in the 1940s[113] and successfully tested by Wolfenden[114] and Lienhard[115] in the 1970s. The approach used here to design renin inhibitors was fundamentally different from that employed to design ACE inhibitors. Reasoning that the planar amide bond in the substrate must be transformed to a tetrahedral species in the transition state for the renin catalyzed reaction (see Scheme 4, Section 8.2.2), Szelke synthesized the reduced amide bond isostere (108) and prepared the corresponding reduced amide bond analog (109) of the Kokubu tetrapeptide (106).[116] The

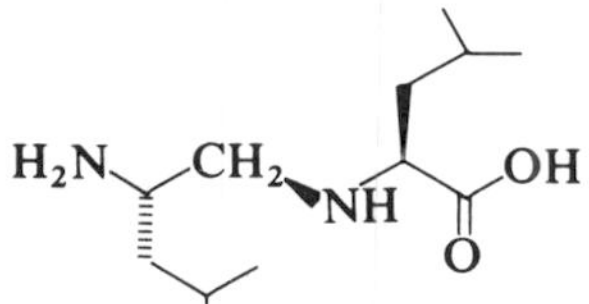

(108)

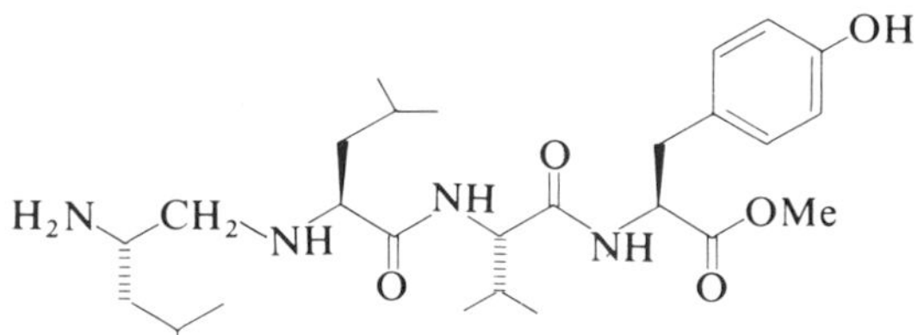

(109)

replacement of the trigonal amide bond by the tetrahedral methylene and nitrogen groups was expected to produce a better inhibitor than the Kokubu peptide since the energy produced by the binding of the inhibitor to the enzyme would not be needed to distort the amide bond. Instead of obtaining stronger renin inhibitors, Szelke found that only relatively weak inhibitors of renin were obtained. However, the basic idea was shown later to be correct when Szelke and coworkers incorporated the reduced peptide 'isostere' into a peptide more closely resembling the decapeptide inhibitor RIP (**107**).[117] This decapeptide, H-142 (**110**), is an excellent renin inhibitor that is some 200 times more potent than RIP.

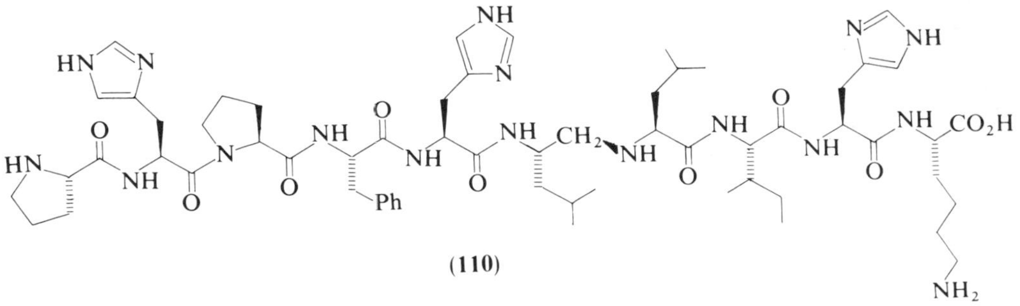

(**110**)

The potency of (**110**) *versus* RIP was remarkable in view of the fact that the reduced amide bond in H-142 only remotely resembles a transition state or tetrahedral intermediate for amide bond hydrolysis catalyzed by renin. Thus although H-142 contains two tetrahedral atoms in place of the planar amide bond, the analog lacks the hydrogen bond accepting hydroxyl groups. To explore whether one or both of the hydroxyl groups were important for binding to renin, Szelke and coworkers introduced a new transition-state mimic, the hydroxyethylene dipeptide 'isostere' (**111**).[118] When (**111**) was used in place of the reduced amide isostere (**108**) in the human renin substrate sequence, the renin inhibitor H-261 (**112**) was obtained which proved to be about 250 times more potent than H-142 (**110**).

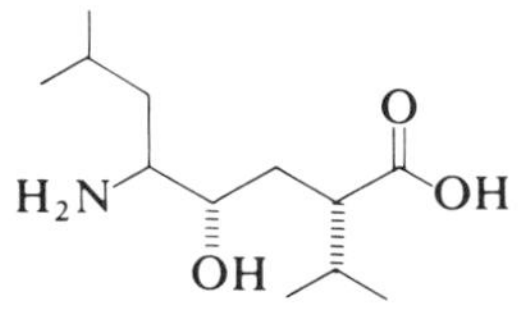

(**111**)

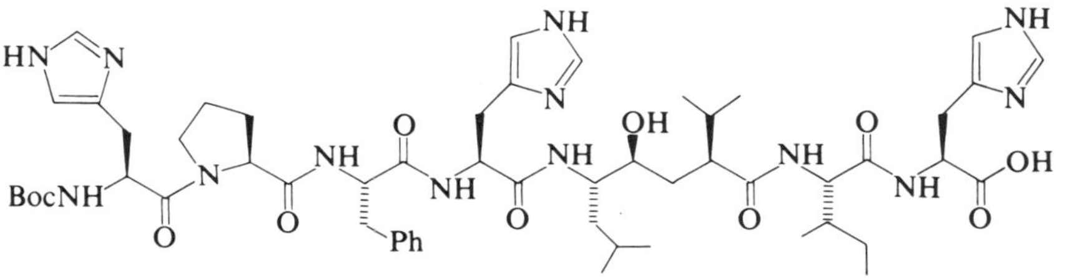

(**112**)

8.2.3.4.3 *Statine-containing inhibitors: SCRIP*[119]

An independent approach to the design of renin inhibitors was derived from pepstatin (**113**), the natural aspartic proteinase inhibitor isolated by Umezawa *et al.*[120] Pepstatin is an exceedingly effective inhibitor of most aspartic proteinases but a relatively poor inhibitor of human renin. That fact, plus the knowledge that the substrate specificity of the various renins is very stringent, suggested that good renin inhibiting analogs of pepstatin might be found if the mechanism of inhibition of aspartic proteinases by this natural product were understood. Systematic study of the interaction of pepstatin first with pepsin and later with renin led to an understanding of the properties of pepstatin needed to inhibit aspartic proteinases efficiently and to the discovery of potent renin inhibitors.

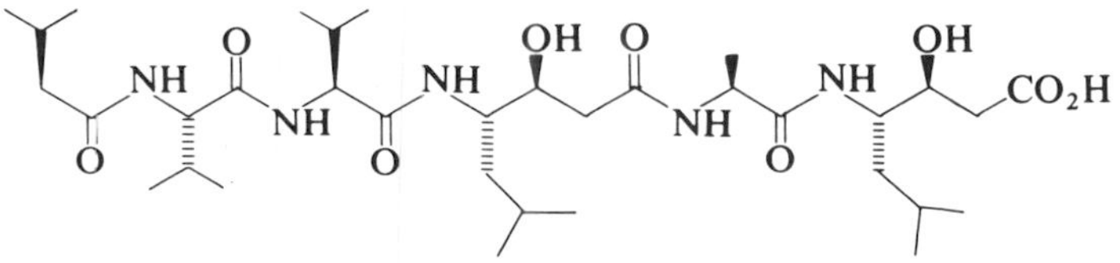

(113) 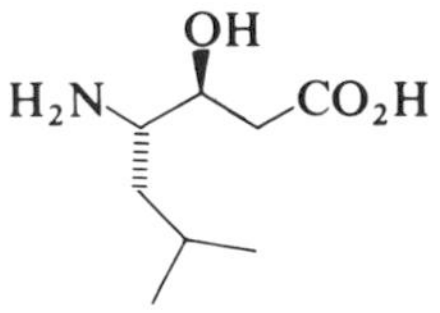 (114)

Pepstatin contains two copies of a unique amino acid, (3S,4S)-3-hydroxy-4-amino-6-methylheptanoic acid, statine (114). The central statine is the important one for inhibition.[21,121] The structure of statine bears some resemblance to the transition state for amide bond hydrolysis, as was suggested by Marshall[122] and Marcinsizyn *et al.*,[123] but the analogy is not perfect especially when compared with the hydroxyethylene isostere (111) (see Section 8.2.3.4.2). Nevertheless, pepstatin is remarkably potent and inhibits several aspartic proteinases with inhibition constants (K_i) of less than 1 nM. Structure–activity studies have shown that the (3S) hydroxyl group in statine is essential for maximal inhibition of several aspartic proteinases, and X-ray crystallographic studies have shown that when the inhibitor binds, this hydroxyl group is hydrogen bonded to the two catalytically active aspartic acid carboxyl groups in the enzyme active site.[124,125] The (3S) statine hydroxyl group occupies the same site on the enzyme that the presumed substrate water occupies in the native enzyme, so substrate water must be displaced when this inhibitor binds to aspartic proteinases. These considerations have led to the suggestion that statine is, in part, a collected-substrate inhibitor of this class of enzyme, in which some of the binding energy is entropic, resulting from release of enzyme-bound substrate water.[121,126]

Molecular modeling studies based on the X-ray crystal structure of pepstatin bound to *R. chinensis* aspartic proteinase in conjunction with the testing of synthetic analogs have led to the concept that statine is an analog of the dipeptidyl tetrahedral intermediate.[127] The functional equivalence is quite remarkable because statine (114) is one main-chain atom shorter and lacks a P'_1 side chain in comparison with the hydroxyethylene isostere (111), yet pepstatin is a remarkably effective inhibitor of most aspartic proteinases. Potent inhibitors of renin, cathepsin D, penicillopepsin, pepsin and other aspartic proteinases have been obtained by replacing the dipeptidyl cleavage site in a substrate sequence with statine.[128] In the case of renin, this has led to the discovery of SCRIP (statine-containing renin inhibitor; 115),[129] and other potent renin inhibitors, *e.g.* (116).[130] Molecular modeling studies based on the X-ray crystal structure of pepstatin bound to *R. chinensis* pepsin suggested that the side chain of statine could be altered to enhance binding to the enzyme's S_1

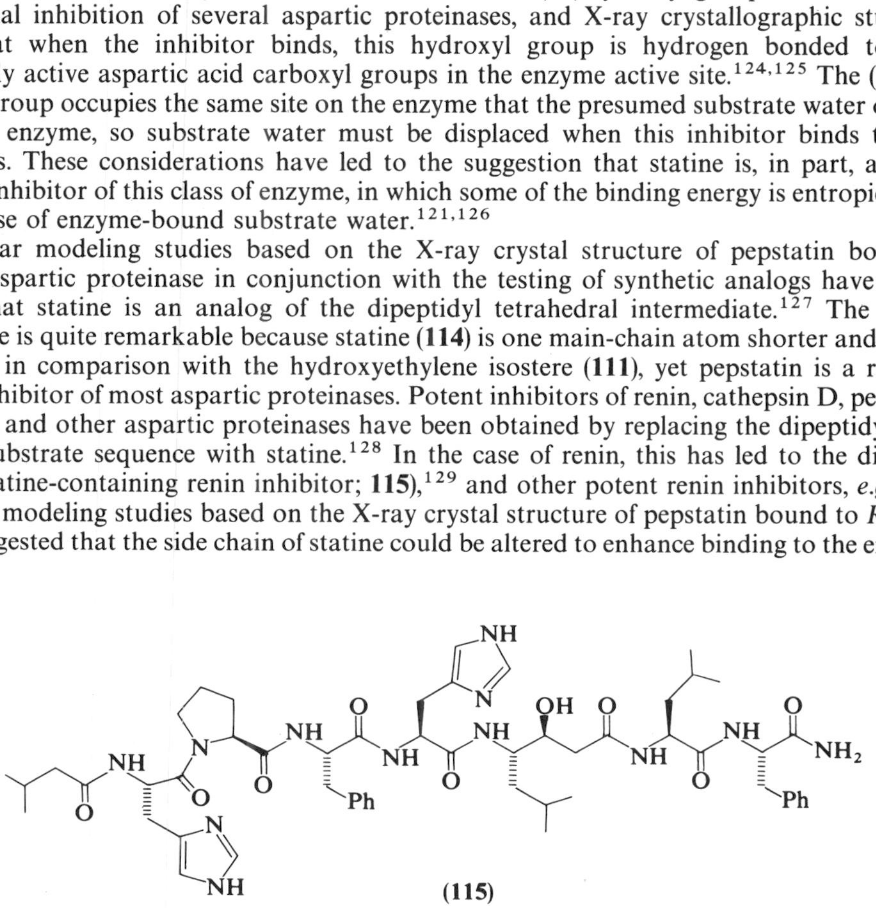

(115)

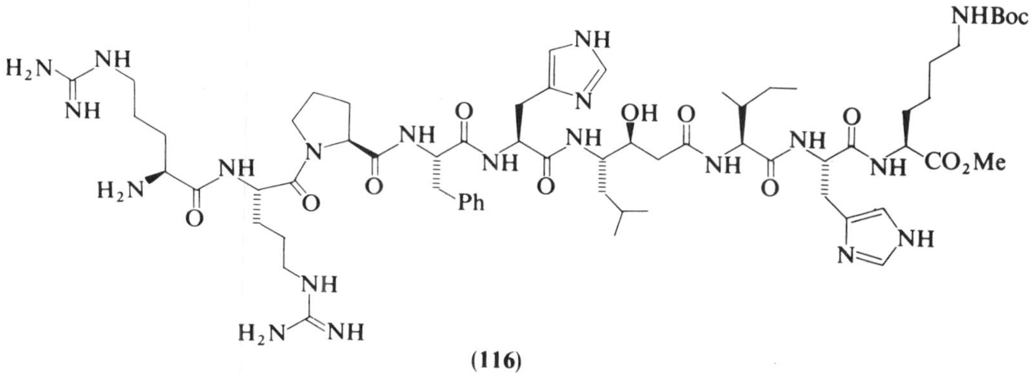

(116)

pocket.[131] The cyclohexyl analog, (3S,4S)-3-hydroxy-4-amino-5-cyclohexylpentanoic acid (ACHPA; **117**), gave inhibitor (**118**) which binds to renin 55–70 times more tightly than analogs with the isobutyl side chain of statine. The cyclohexyl side chain (Cy) has been incorporated into many recently reported dipeptidyl isosteres, *e.g.* the reduced amide isostere (**119**)[132] and into the hydroxyethylene isostere (**120**)[133] that are being used to produce more potent renin inhibitors. The Ciba-Geigy renin inhibitor (**121**) has been reported to have low but significant oral bioavailability and is undergoing clinical evaluation as an antihypertensive agent.[134]

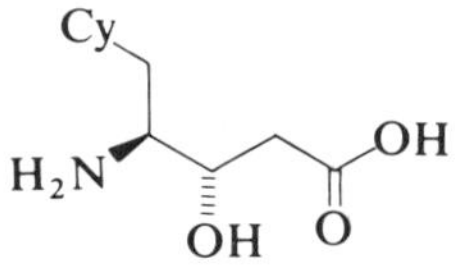

(117)

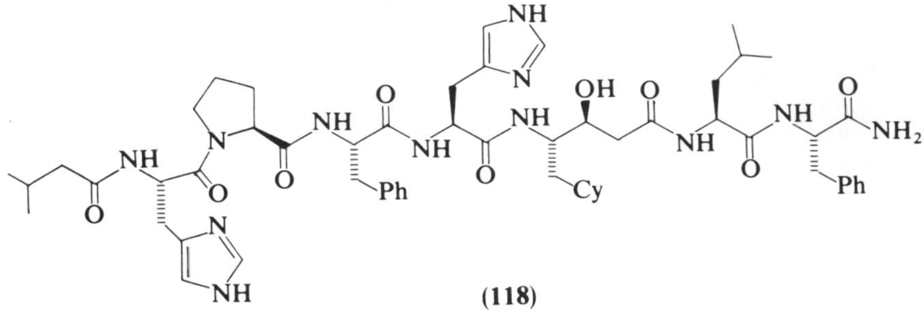

(118)

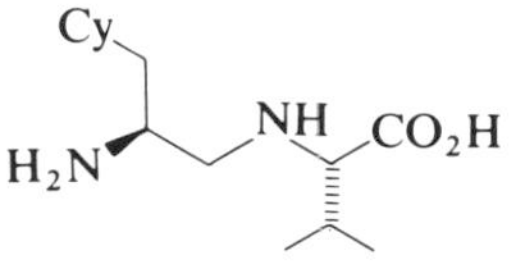

(119)

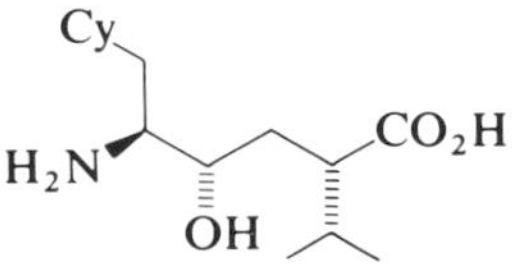

(120)

Many other transition-state analog designs have been incorporated into renin substrate sequences in an attempt to obtain improved renin inhibitors. Peptide aldehydes such as RRM-188 (**122**) are remarkably good renin inhibitors in spite of their small size.[135] The aldehyde functionality is presumably hydrated to the *gem*-diol which mimics the tetrahedral intermediate, and the naphthylalanine derivative at P_3 enhances binding to renin. Peptide ketones derived both from statine, *e.g.* statone (**123**),[136] and from the hydroxyethylene isostere, *e.g.* the ketomethylene group (**124**),[137] have been shown by ^{13}C NMR to form the tetrahedral *gem*-diol when bound to porcine pepsin.[138] These are good inhibitors but are not bound as tightly as the corresponding statine-derived inhibitors, presumably because some of the binding energy is utilized to drive hydration of the ketone. The hydrate can be stabilized in solution by addition of electron withdrawing groups such as di- or tri-fluoroalkyl groups adjacent to the ketone. These di- and tri-fluoromethyl ketone derivatives would be expected to bind more tightly to the enzyme since they hydrate readily in solution prior to binding to the enzyme and, therefore, do not waste binding energy on rehybridization of the trigonal ketone to the *gem*-diol species that is bound. Abeles and Gelb found the difluorostatone peptide

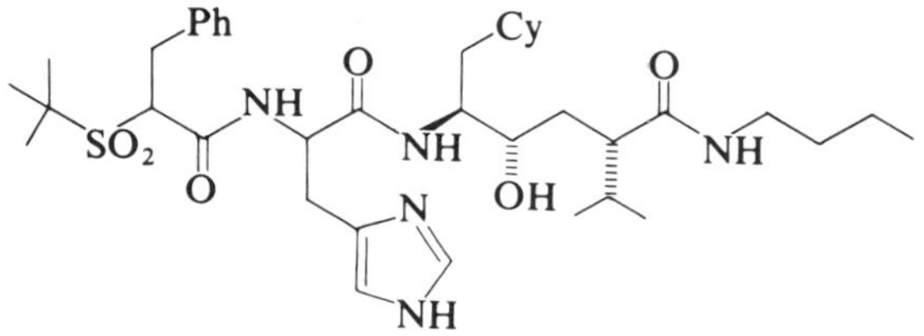

(121)

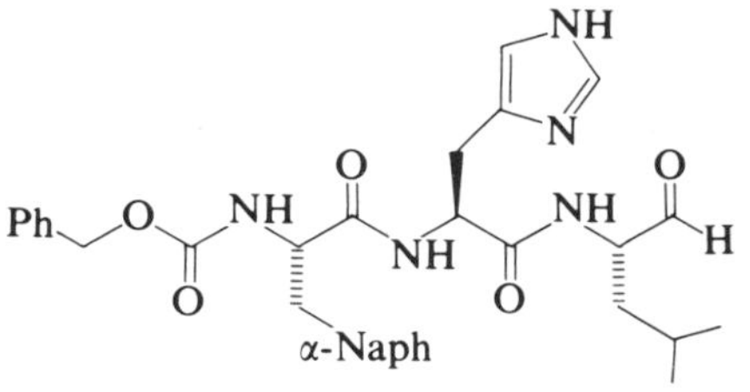

(122)

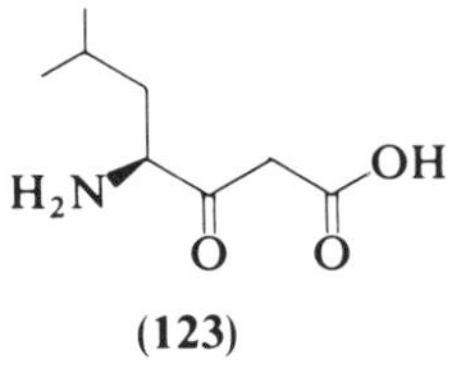

(123)

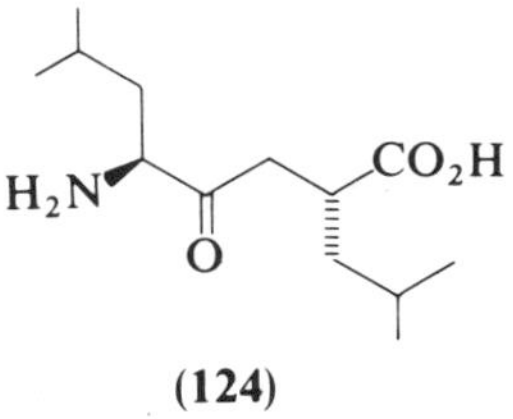

(124)

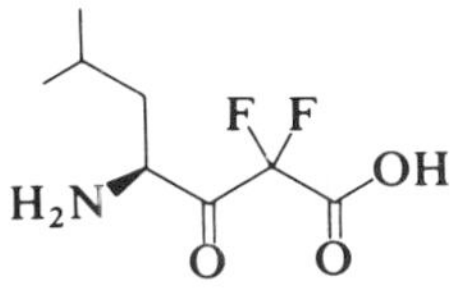

(125)

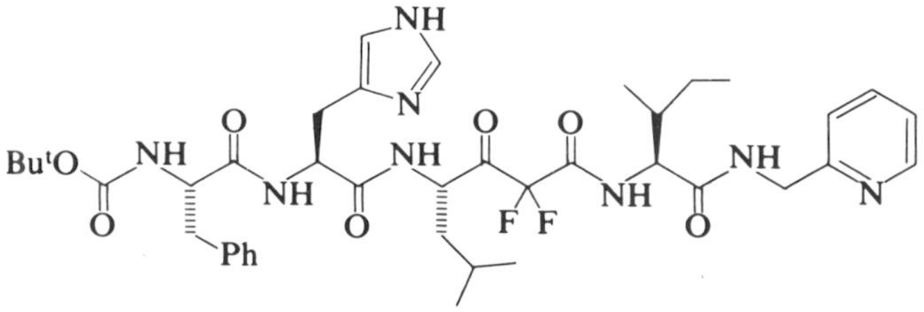

(126)

(125) is a very tight binding inhibitor of porcine pepsin,[139] and Thaisrivongs *et al.* used a similar strategy to develop (126), a very tight binding inhibitor of renin.[140]

8.2.3.4.4 *Other transition-state analog designs*

Numerous variants of the above lead renin inhibitors have been described in the literature. The design of these structures has been motivated by the desire to simplify the structures of the inhibitors in order to obtain low molecular weight, orally active compounds. One fundamental difference between the renin and ACE inhibitors has been the size of the molecule needed to inhibit the enzyme efficiently, a problem anticipated by the fact that the minimal substrate sequence for renin contains eight amino acids whereas the minimal substrate for ACE contains three amino acids. It is difficult to obtain orally active drugs when the molecular weight of the compound exceeds 500. The many amide bonds present in peptide inhibitors cause additional difficulties in that the peptide bond might be hydrolyzed by endogenous peptidases. In addition, the donor–acceptor hydrogen bonding properties of amides impart peculiar distributional properties to these molecules. Thus, many of the isosteres described in the following section no doubt have been guided by the desire to eliminate as many of the amide bonds as possible.

The Abbott group described a series of renin inhibitors designed to eliminate many of the peptide features in the C-terminal region of the compounds while retaining the side chains needed for interaction with the enzyme. Much of this work was guided by molecular modeling techniques in which hypothetical structures were 'docked' into computer models of the active site of renin and energy minimized to see if the structure under consideration would bind favorably.[142] The sulfone derivative (127) corresponds to one such inhibitor in which the P_1' side chain and the P_2' amide NH have been removed. Sulfone (127) is an excellent renin inhibitor in spite of these rather drastic changes. Replacement of an amide group by a sulfide in the ACHPA series led to excellent renin inhibitors, *e.g.* (128). The retroamide derivatives of related compounds, *e.g.* (129), also retain good renin inhibitory potency. The Abbott group devised the unusually effective glycol derivative (130) which retains excellent renin inhibitory activity ($K_i = 0.4$ nM) in spite of the absence of the P_2'-P_3' side chains which would be expected to bind to renin's S_2'-S_3' subsites. This enhanced binding has been rationalized to result from an additional hydrogen bond between the new hydroxyl group and

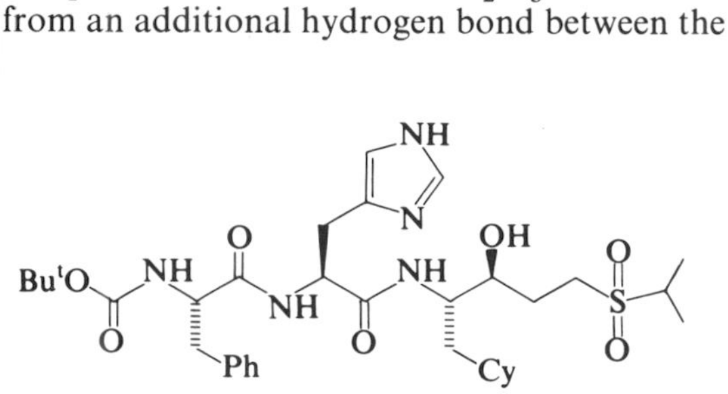

(127)

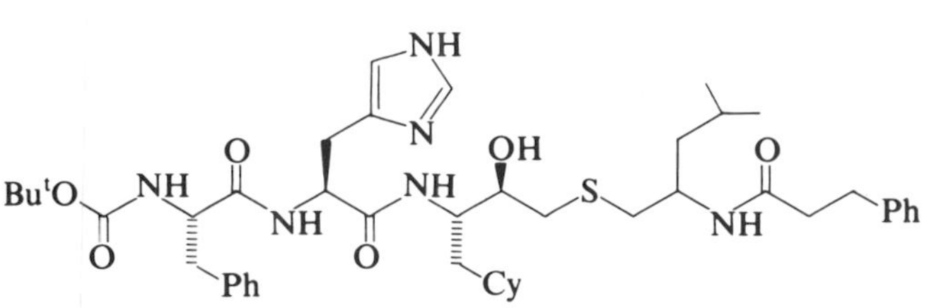

(128)

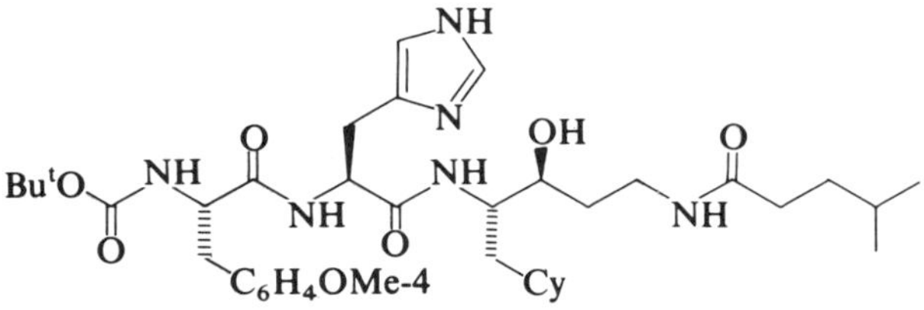

(129) 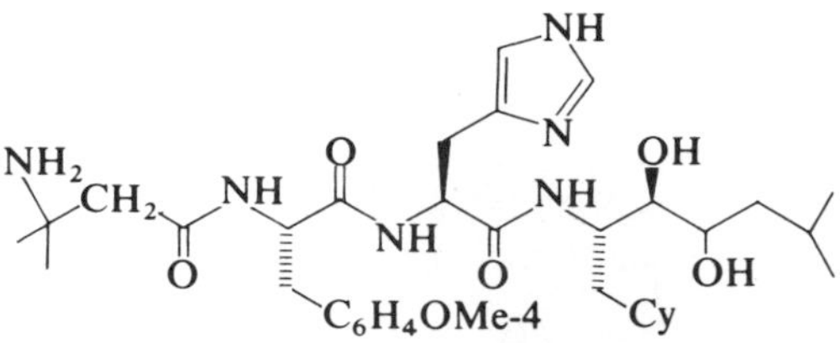   **(130)**

one of the active site carboxyl groups. The C-5 hydroxyl group is expected to bind between the catalytically active aspartic acids by analogy with the pepstatin – *R. chinensis* pepsin crystal structure, while the additional C-4 hydroxyl group is postulated to contribute an additional hydrogen bond to one of the aspartic acids. Consistent with this is the fact that a closely related analog lacking the C-4 hydroxyl group is a significantly weaker inhibitor. Glycol (**130**; A-64662) is undergoing clinical evaluation.

The α-hydroxyester (**131**)[143] and the ACHPA derivative (**132**)[144] are good renin inhibitors, which further indicates that the C-terminal peptide region is not absolutely essential for renin inhibition.

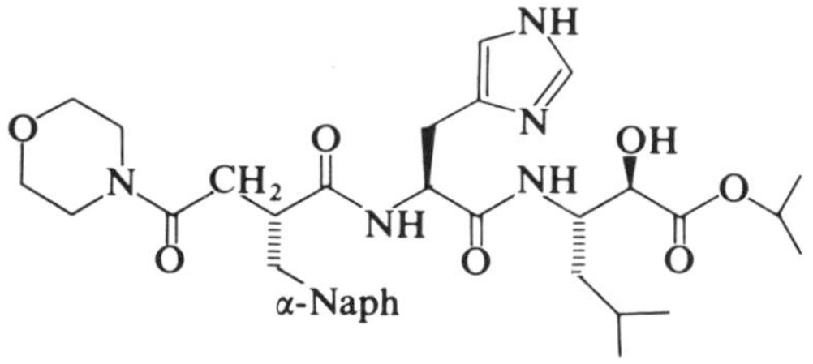

(131)

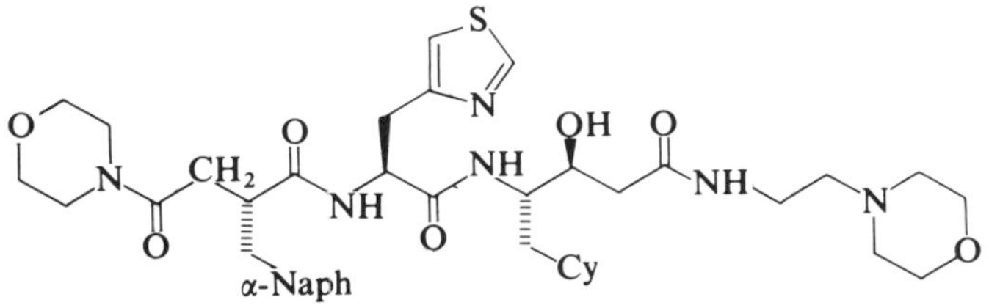

(132)

Many other dipeptide units designed to mimic the tetrahedral intermediate have been tested on the renin system. The phosphinic acid analog of statine (**133**; PSta) produces very potent pepsin inhibitors when incorporated into pepsin substrate sequences.[145] However, when (**133**) is incorporated into renin substrate sequences, the resulting peptide does not appear to inhibit renin effectively, possibly because it is completely ionized at the higher pH needed for renin activity.[14] Substrate sequences containing the aminoalcohol isostere,[83] *e.g.* (**134**), are effective renin inhibitors.[146]

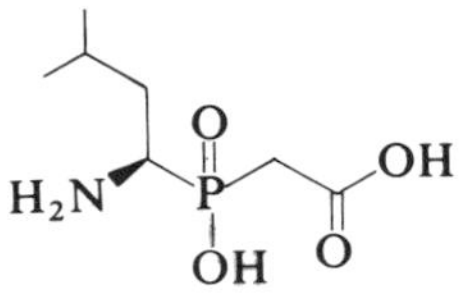

(133) 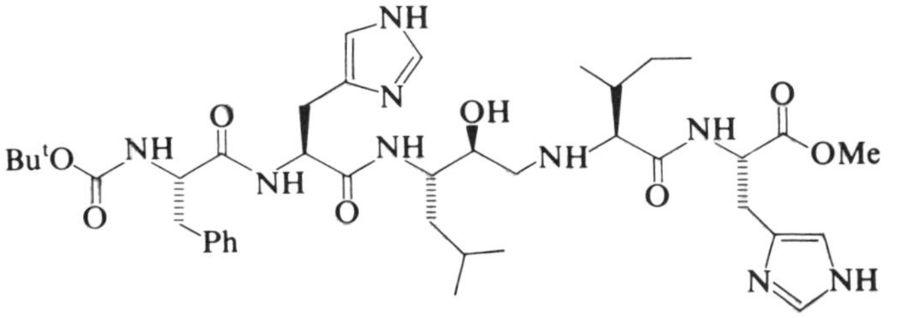   **(134)**

Most renin inhibitors reported to date have an aromatic or bulky substituent in the P_3 position. Removal of the acyl P_3 group dramatically reduces renin inhibitory activity, which explains the negative results Szelke and colleagues reported for the reduced amide analogs (**109**), which lacked both the P_2 and P_3 groups. At least some of the poor oral activity and short duration of action of the renin inhibitors appears to be due to the presence of the amide bond between Phe-His. This bond is cleaved by chymotrypsin *in vitro* and structural modifications that slow chymotrypsin-catalyzed hydrolysis of this bond *in vitro* improve the oral absorbtion of the resulting peptide. For example, the analogs in which the amide bond has been methylated (*N*-MeHis; in **135**)[147] or in which a *p*-methoxy group has been added to phenylalanine [Phe(OMe); in (**129**), (**130**)][148] are less susceptible to cleavage by chymotrypsin and have better bioavailability. Unfortunately, proteolytic cleavage of amide bonds is not the only mode of inactivation; some peptides are thought to be secreted intact into the bile or are not absorbed into the blood stream even though stable to gastric metabolism.

(**135**)

(**136**)

Attempts to prepare conformationally restricted renin inhibitors have been reported. For the most part, these structures are formed by closure of disulfide bonds or lactams that can be readily introduced into the structure by conventional methods of peptide synthesis.[149] Figure 12 illustrates the results obtained when cyclic structure (**136**) is modeled on the enzyme-bound conformation of pepstatin.[119] In general, portions of critical side-chains have been sacrificed for synthetic ease (such as the imidazole group of His in **136**) and, as a result, superactive analogs have not been described. A novel series of cyclic ether derivatives that link the P_2 histidine nitrogen with the P_1 side chain (**137**) were designed on the basis of molecular modeling studies that indicated such ring structures could

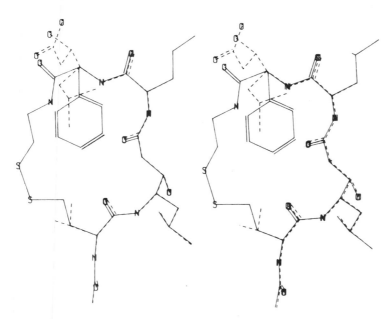

Figure 12 The structure of Ac-Hcy-Sta-Leu-Phe-NH-CH$_2$-CH$_2$-S (solid line) matched on to the Val-Sta-Ala-Sta portion of pepstatin (dashed line) which is in the conformation found in *Rhizopus chinensis* pepsin[119]

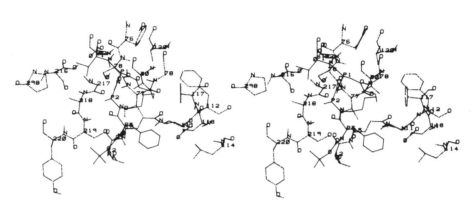

(137)

be formed without destabilizing the bound conformation of the inhibitor and without sterically interacting with the enzyme (Figure 13).[150] For synthetic ease, the P_2 position was changed to an alanine rather than the histidine side chain and the reduced amide isostere (108) was used rather than a hydroxyethylene isostere (111) or its equivalent. Very good inhibitors were obtained relative to the noncyclic parent structure; nevertheless, the entropic factor was not realized and the best inhibitors were only equipotent with the corresponding linear structure.

Figure 13   Stereo representation of the energy minimized structure of the 14-membered cyclic renin inhibitor (136) (solid lines) in the active site of human renin (dotted lines)[150]

8.2.3.5   Elastase Inhibitors: Human Elastase (HNE) Inhibitors

Serine proteinases are classified by their substrate specificity into three types: trypsin-like, chymotrypsin-like and elastase-like.[17] Trypsin cleaves substrates with positively charged amino acid residues (Arg and Lys) in P_1, whereas chymotrypsin prefers substrates with large aliphatic or aromatic groups (*e.g.* Phe, Trp, Leu) at P_1 and elastase prefers smaller aliphatic side chains (*e.g.* Ala, Val) in the P_1 position. Although many serine proteinases are implicated in disease states and thus could serve as sites for therapeutic intervention (Table 2), thus far the only reported clinical candidate low molecular weight inhibitors of serine proteases are elastase inhibitors which have been developed as potential drugs for treating emphysema. This disease is known to be caused by an imbalance between proteases and endogenous antiproteinases in the lung, arising either by a genetic insufficiency of α-1-protease inhibitor (α-1-PI) or by environmental factors, such as oxidants that are present in cigarette smoke. Experimental evidence for this hypothesis has come from induction of emphysema in animals by administration of proteases, primarily elastases.[151] These results thus suggest that synthetic or natural elastase inhibitors could be used to replace the lost elastase inhibitory factors in the lung and thereby prevent or retard the progression of this disease.[12]

8.2.3.5.1   Peptidyl inhibitors

Human elastase [EC 3.4.21.11] is a serine proteinase found in the azurophilic granules of the human polymorphonuclear leukocytes.[152] The enzyme, earlier referred to as human leukocyte elastase (HLE) and more recently as human neutrophil elastase (HNE), acts *via* the catalytic mechanism shown in Scheme 2 (Section 8.2.2). The natural substrate, elastin, is a flexible protein that is highly cross-linked by desmosine and α-isodesmosine and other cross-linking moieties. The enzyme appears to cleave the substrate most efficiently at sites near to the peptide chain cross-linking sites, an observation which Powers *et al.* exploited to produce more efficient synthetic substrates.[153] When structures which more closely resemble the cross-linking units were incorporated into the synthetic substrate MeO-Suc-Ala-Ala-Pro-Val-NHR (138), to obtain MeO-Suc-Lys(Z)-Ala-Pro-Val-NHR (139) where R is a chromophoric leaving group, improved elastase substrates were obtained ($k_{cat}/K_m = 710\,000$ M^{-1} s^{-1} *vs.* 120 000 M^{-1} s^{-1} for (138) with human leukocyte elastase). These peptides have served as the templates for the development of many elastase inhibitors. Initially, these included the corresponding peptidyl chloromethanes (140),[152] azapeptides[154] and trifluoroacetyl peptides.[155] These elastase inhibitors were not selective enough for the target enzyme and were considered to be potentially toxic.

Improved inhibitors, several of which are undergoing clinical evaluation, have been obtained by modifying the lead substrate sequences to resemble transition state analogs or by modifying natural products. The peptide aldehydes, leupeptin (141), chymostatin (142) and elastinal (143),[120] are known to inhibit serine peptidases efficiently. NMR experiments with ^{13}C-labeled aldehyde groups and X-ray crystallography have established that the catalytically active serine hydroxyl group adds to the C-terminal aldehyde to form a hemiacetal derivative,[156] a structure thought to resemble the tetrahedral intermediate for enzyme-catalyzed hydrolysis of the amide bond (35; Scheme 2). However, the natural peptide aldehydes are not selective enough to be used as elastase inhibitors in humans.

In recent years, several synthetic inhibitors have been designed to inhibit elastases by attaching reversibly or irreversibly to the elastase active serine hydroxyl group in the enzyme's active site. The peptide derived inhibitors described below are reversible inhibitors that have been developed by applying the principles of transition-state analog inhibitor theory and the lessons obtained from

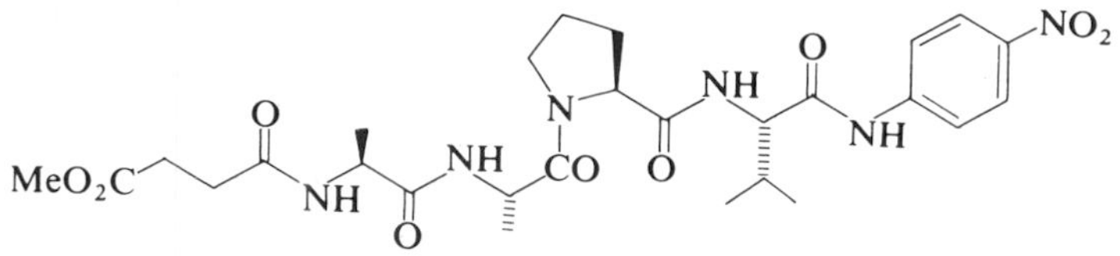

(138)

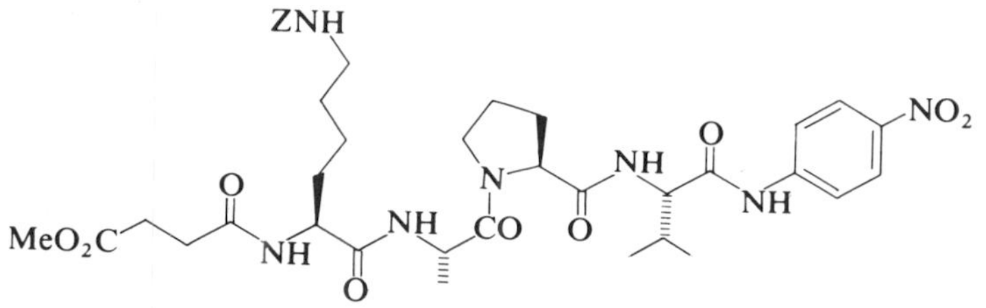

(139)

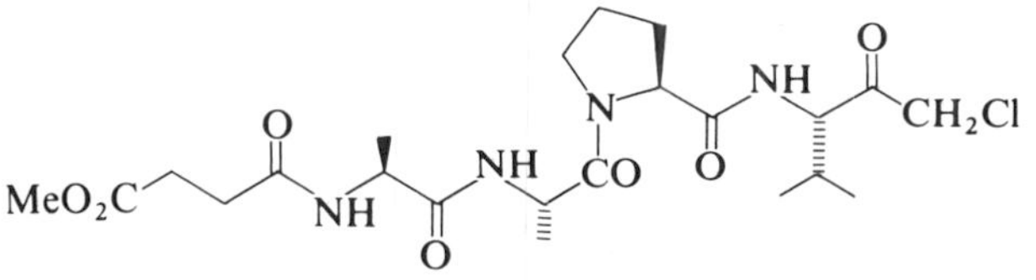

(140)

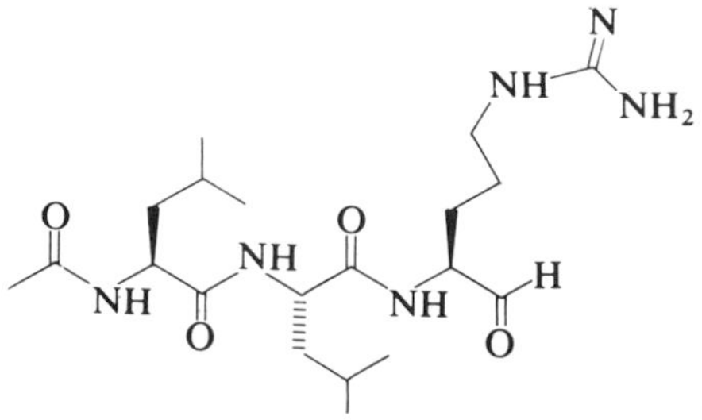

(141)

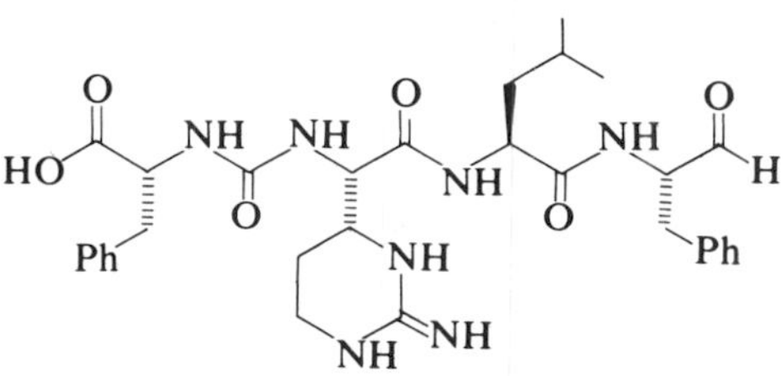

(142)

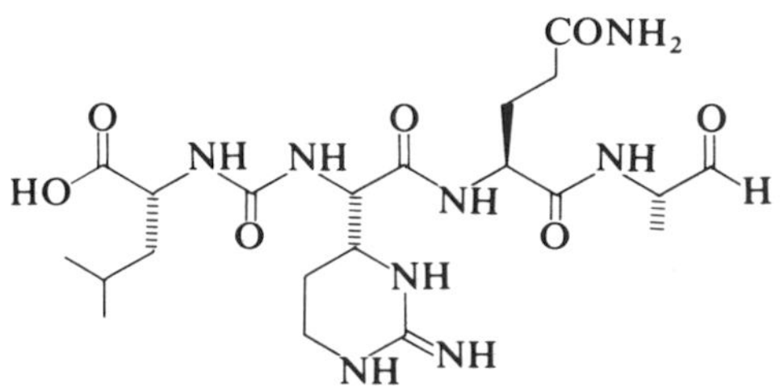

(143)

studies of the natural peptidyl aldehyde inhibitors. The heterocyclic inhibitors (see Section 8.2.3.5.2), in contrast, are covalent inhibitors of these enzymes.

Trainor and colleagues developed potent HNE inhibitors by incorporating the peptide aldehyde group in the HNE substrate (139).[12] Specificity for HNE results from the substrate sequence whereas the aldehyde functional group reacts with the serine hydroxyl group in the enzyme active site to stabilize the complex further. Compound U14804 (144) obtained by this strategy is a very efficient *in vitro* inhibitor of HNE (K_i = 4 nM). A similar strategy was employed by Hassall and colleagues at Roche to develop the aldehydic HNE and porcine pancreatic elastase (PPE) inhibitors (145) and (146).[157,158] These compounds had no overt toxicity and significantly inhibited the development of emphysema-like lung lesions in experimental animal models.

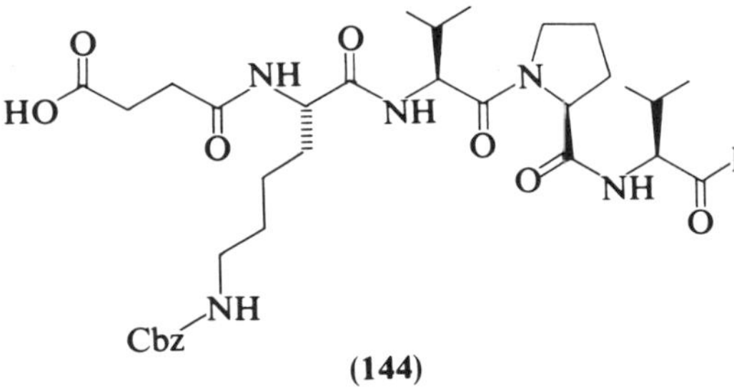

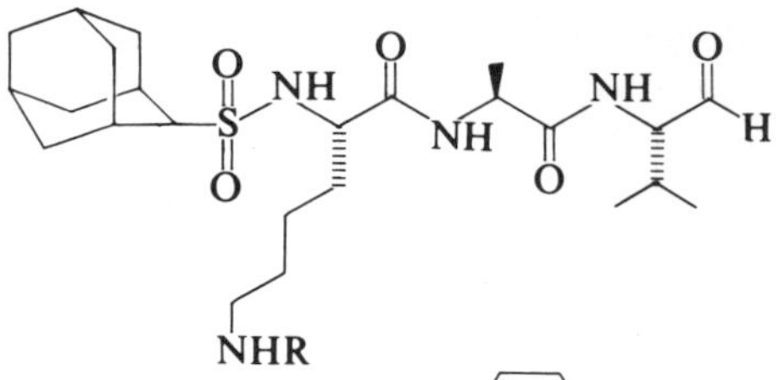

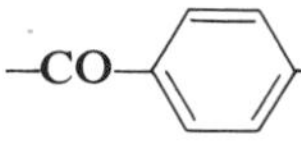

Peptidyl aldehydes have demonstrated *in vitro* HNE and PPE inhibitory activity but these inhibitors lack sufficient duration of *in vivo* activity due to the rapid oxidation of the C-terminal aldehyde to a carboxylic acid. In order to prepare potent, metabolically stable HNE inhibitors, the ICI Pharmaceuticals group replaced the aldehyde group with the metabolically more stable trifluoromethyl ketone functionality (TFMK) (147; Scheme 6).[12] The elastase inhibitor (149) is a very potent inhibitor of HNE (K_i = 0.04 nM). The trifluoromethyl ketone group contains an electrophilic carbonyl group that reacts with the active-site serine hydroxyl group. An X-ray crystallographic analysis of the TFMK (150) in the active site of porcine pancreatic elastase (PPE) shows that the enzyme serine hydroxyl group has reacted with the TFMK carbonyl group to form the hemiacetal derivative (148; Scheme 6).[159] Independently, Abeles and Gelb developed a series of very effective protease and esterase inhibitors that contain trifluoromethyl ketone groups as the reactive moiety.[160,161] These also react with the enzyme active-site serine hydroxyl group or with water to generate transition-state analogs *in situ*. In addition to elastase inhibitor (151; K_i = 1 nM), Abeles and coworkers have reported trifluoromethyl inhibitors of ACE (152)[161] and chymotrypsin (153),[162] and difluoromethylene inhibitors of pepsin (154).[160] Trifluoromethyl ketone inhibitors of acetylcholine esterase are also known.[163]

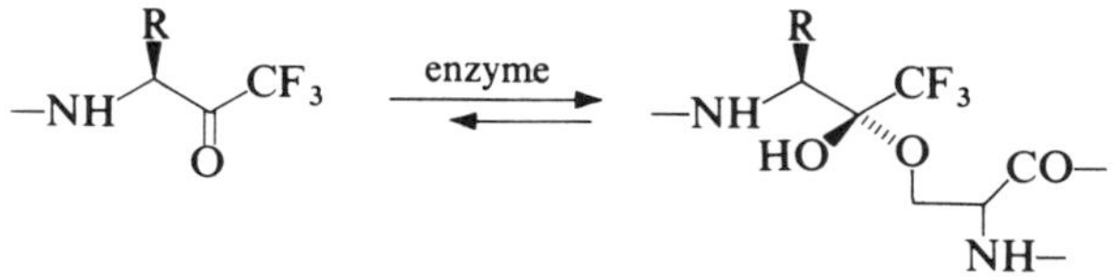

Scheme 6 Reaction of trifluoromethyl ketones with the active-site hydroxyl group of serine peptidases

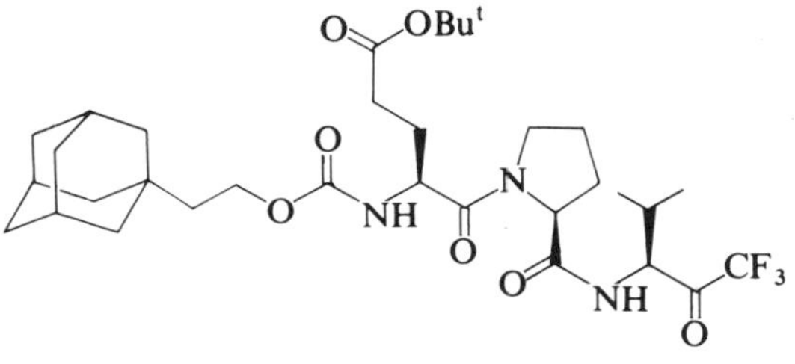

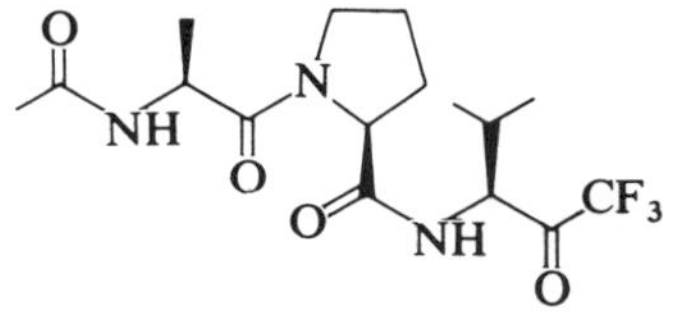

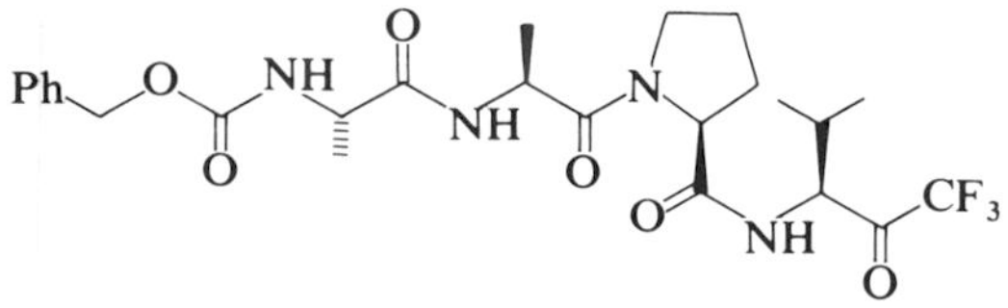

(151)

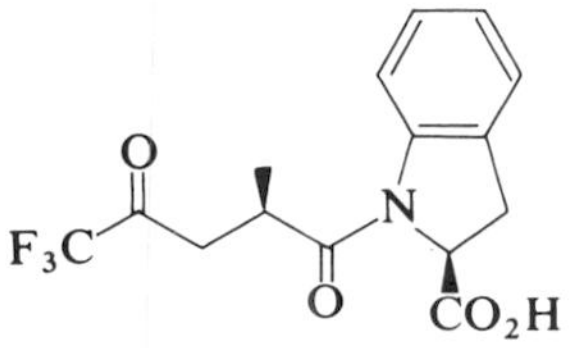

(152)

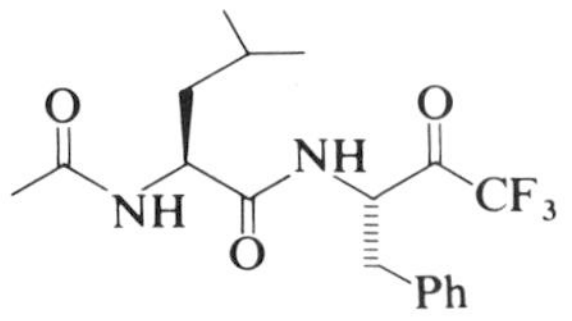

(153)

Boronic acid derivatives have been used as transition-state analog inhibitors of serine proteases since 1971.[164] The mechanism of inhibition of the serine protease subtilisin by this compound was determined by X-ray crystallography, which clearly established that the boronic acid inhibitor (155) added to the active-site serine hydroxyl group.[165] A practical route to α-aminoboronic acids was devised by Matteson in 1981 and enabled this transition-state analog group to be applied to the inhibition of HNE.[166] Kettner and Shenvi developed the peptidyl boronic acid ester derivative (156) and showed that hydrolysis produced the acid (157) which effectively inhibits elastase ($K_i = 0.57$ nM).[167]

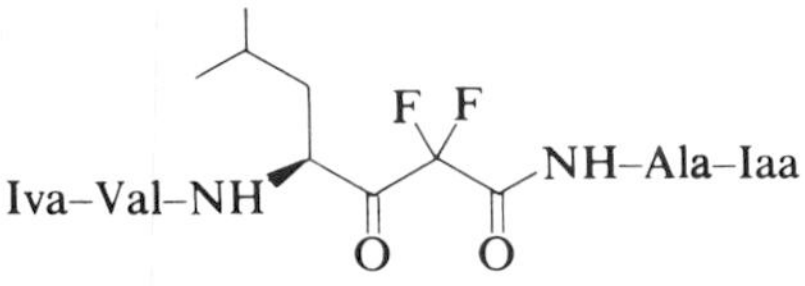

(154)

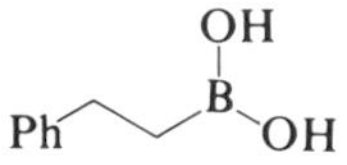

(155)

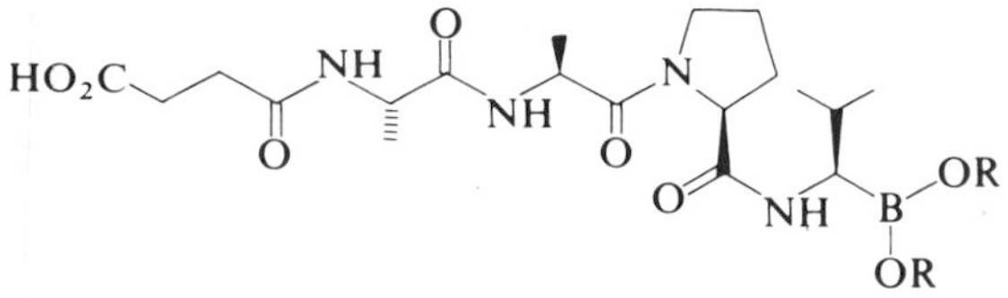

(156) R = diethanolamine ester
(157) R = H

The binding of peptidyl inhibitors (149), (151) and (157) to elastase is limited to interactions with S_1-S_4 enzyme subsites. Little information about the effect of the prime subsites (S_1'-S_3') on the enzyme is available from studies using synthetic substrates, since most substrates in use today are derivatives of active esters or *p*-nitroanilides. However, by analogy with the protein inhibitors of other serine proteinases, or with low molecular weight inhibitors of other peptidases, it is likely that inhibitors which contain side chains that could bind to the primed enzyme subsites (S_2'-S_3', *etc.*) would enhance binding to HNE and other serine peptidases. This hypothesis has been tested by two different approaches. Abeles *et al.* synthesized the chymotrypsin inhibitors (158) and (159), the latter of which

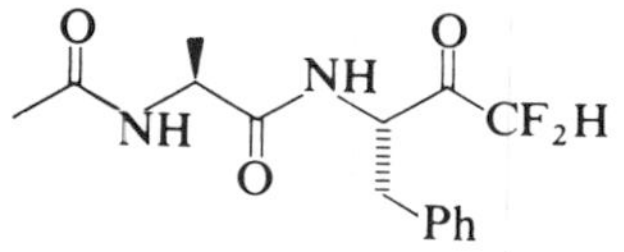

(158)

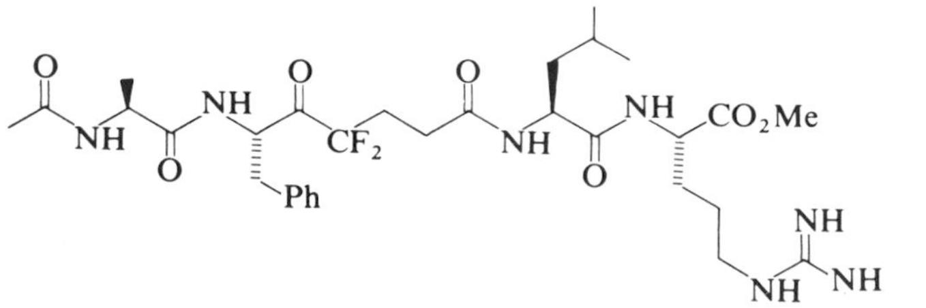

(159)

contains P$'_2$ and P$'_3$ substituents suitable for binding to the S$'_2$-S$'_3$ enzyme subsites.[162] The amino acid side chain was selected on the basis of the work of Laskowski *et al.*, who had examined the effects of substituents in the P$'_2$-P$'_3$ subsites of the serine protease inhibitors, avian ovomucoids.[168] In Laskowski's studies it was found that replacement of Val at P$'_3$ with Arg increased binding to the enzyme by 37-fold. Abeles and colleagues prepared inhibitor (**159**), which contains an Arg residue at P$'_3$, and found that (**159**) binds to chymotrypsin about 250 times more tightly than the uncharged inhibitor (**158**). The other approach, devised by Trainor and colleagues, prepared a series of substituted difluoromethyl ketone elastase inhibitors, *e.g.* (**160**) and (**161**), to see if the C-terminal phenethyl group could bind to the S$'_1$-S$'_3$ subsites of HNE.[169] Although the trifluoromethyl ketone derivatives are intrinsically more active than the difluoromethyl ketone derivatives, the additional hydrophobic interactions of the extended α,α-difluoromethylene-β-diketones makes these compounds a more potent series of HLE inhibitors. Compound (**161**) is a subnanomolar inhibitor of HNE (K_i = 0.4 nM).

(160)   (161)

8.2.3.5.2 *Heterocyclic covalent inhibitors*

The elastase inhibitors described above are competitive, reversible inhibitors of this serine protease by virtue of the facile equilibrium between the hemiacetal intermediate (**148**) and the ketone (**147**) (or corresponding boronic acid derivative). There are a variety of heterocyclic lactones that inhibit HNE by other mechanisms. These compounds are alternative substrates of HNE in which the lactone moiety is opened by the enzymic serine hydroxyl group in an enzyme-catalyzed process. Depending on the structure of the heterocyclic inhibitor, the acyl–enzyme intermediate is either slowly cleaved from the enzyme, thereby reducing the concentration of free enzyme, or it reacts further to inhibit the enzyme irreversibly. Compounds (**162**)–(**165**) are examples of the former class;[170-172] these are alternative substrate inhibitors of HNE and other serine proteinases. In contrast, compounds (**166**) and (**167**)[173,174] are k_{cat} inhibitors of HNE since the intermediate formed can react further with the enzyme to alkylate an active site nucleophile and thereby irreversibly inhibit HNE.[175]

(162)   (163)   (164)

(165)   (166)   (167)

Useful drugs which act by acylating serine proteinases may eventually be developed by modifying structures (**162**)–(**170**), but to date only one class of acylating agent, the neutral cephalosporins, *e.g.* (**168**), has progressed to the point where clinical candidates appear likely. Doherty and colleagues at Merck have reported that derivatives of cephalosporin esters are efficient inhibitors of HNE.[176] These compounds were discovered as a result of early observations that the benzyl ester of the β-lactamase inhibitor, clavulanic acid (**169**),[177] weakly inhibited HNE and other serine proteinases but did not inhibit β-lactamases or transpeptidase, enzymes that are known to require a free carboxylic acid group. Structure–activity studies led to the discovery that neutral cephalosporins, *e.g.* (**168**), can be modified to become potent HLE inhibitors.[176] Cephalosporins bearing substituents in the 7α-position are more potent inhibitors than those with 7β-substituents, presumably since the 7α-derivatives more closely resemble the L-amino acid stereochemistry of mammalian substrates as opposed to the D-Ala-D-Ala substrate in bacteria. Consistent with this is the fact that smaller 7-position substituents are preferred, as would be expected if these groups bind to the S_1 subsite of HLE. The sulfone derivatives are much more potent inhibitors than the corresponding sulfides or sulfoxides. Depending on the structure, the neutral cephalosporins can be alternative substrates or irreversible inhibitors of HLE. The 7-chloro derivative (**170**) is an irreversible inhibitor of HLE (Scheme 7). X-ray crystallographic analysis of the porcine pancreatic elastase–inhibitor complex shows that the imidazole group of His-57 that is needed for catalytic activity has displaced the 3′-substituent (Figure 14).[178] A possible mechanism for (**168**), which rapidly and irreversibly inhibits HNE ($k_{cat}/K_m = 328\,000\ \mathrm{M}^{-1}\mathrm{s}^{-1}$), may stabilize the enzyme–inhibitor complex by a slightly different mechanism involving cleavage of the S-1—C-8 bond, as is postulated to occur with k_{cat} inhibitors of β-lactamases. In either case, HLE inhibitors appear to be new examples of suicide substrates or k_{cat} inhibitors that offer considerable potential for developing drugs which inhibit these and other serine peptidases.

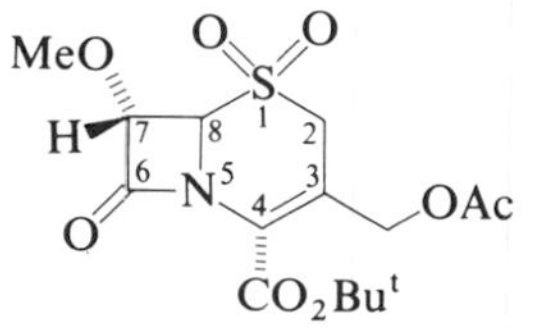

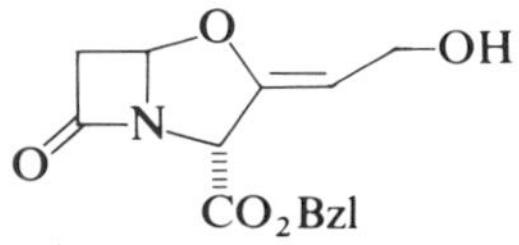

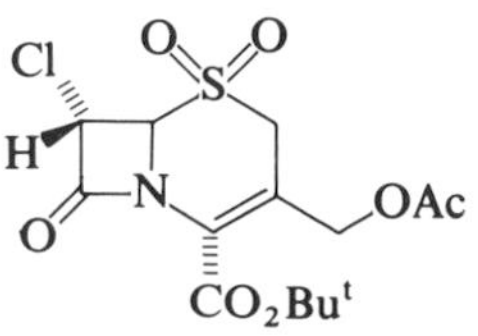

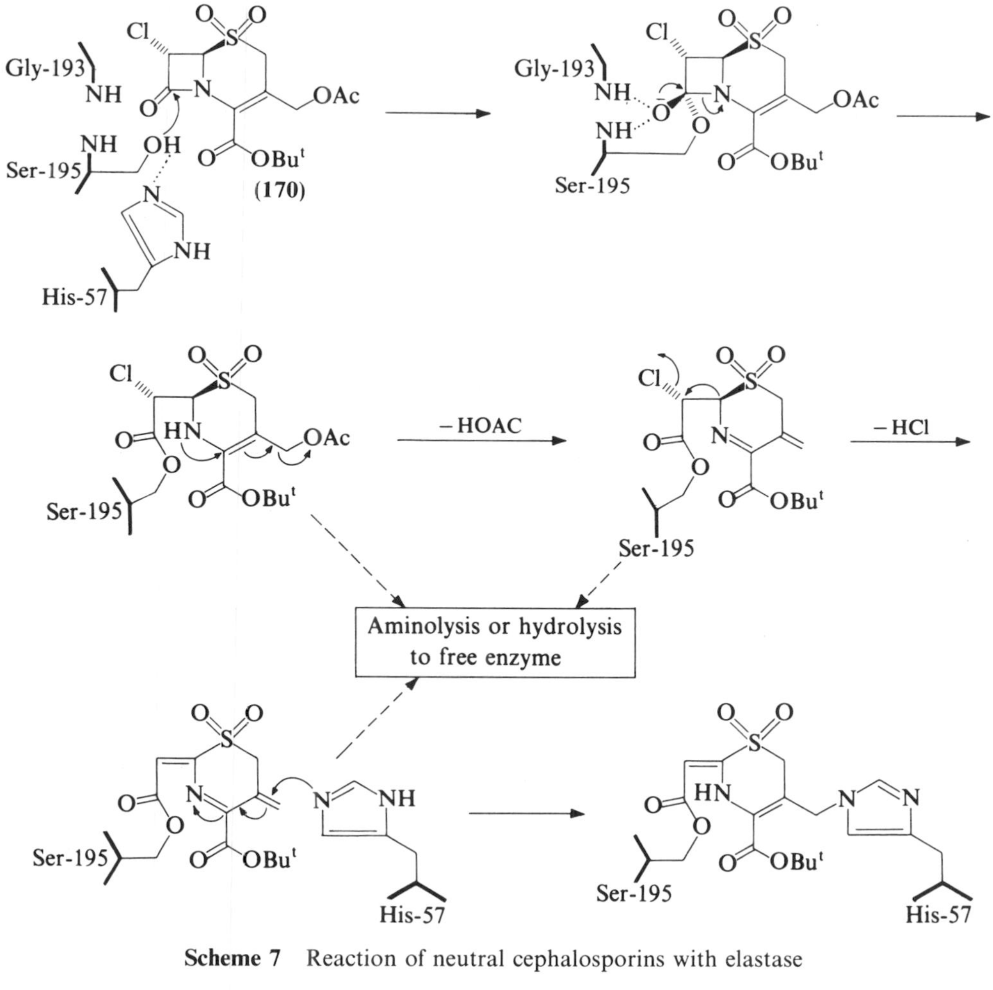

Scheme 7 Reaction of neutral cephalosporins with elastase

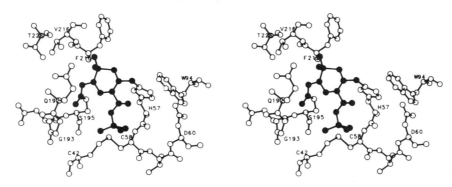

Figure 14 Overall view of the active-site region of the porcine pancreatic elastase complex with neutral cephalosporin inhibitor (**150**). Open circles, atoms of the enzyme; filled circles, inhibitor atoms[178]

8.2.3.6 Aminopeptidase Inhibitors

The aminopeptidases [EC 3.4.11.0], a multivariant group of zinc-containing exopeptidases[179] that specifically cleave polypeptide chains at the amino terminus, probably play a role in the metabolism of many biologically active peptides. This process has been extensively studied with respect to the biological effects produced as a result of inhibiting the degradation of the enkephalins.[13] Aminopeptidases are believed to catalyze hydrolysis of substrates *via* a mechanism in which the substrate carbonyl group is activated by the active-site zinc ion for general base-catalyzed attack of substrate water to form a tetrahedral intermediate, a mechanism which may be related to that depicted in Scheme 5 (Section 8.2.3). Aminopeptidases recognize the N-terminal amine group in substrates, but also can differentiate between the different side chains present in the N-terminal amino acid of substrates. This specificity has been used to classify these enzymes. Aminopeptidase B removes N-terminal basic residues (Arg, Lys), aminopeptidase A removes acidic residues (Glu, Asp), and aminopeptidase M prefers hydrophobic side chain residues. This nomenclature is outmoded and these enzymes are properly referred to respectively as arginyl aminopeptidase, aspartyl aminopeptidase, and membrane-bound leucine aminopeptidase.[180]

A variety of aminopeptidase inhibitors has been synthesized that contain metal chelating groups designed to interact with the active site zinc ion. These include the amino acid hydroxamic acid derivatives, *e.g.* (**171**),[181-183] and the α-mercapto ketone compounds[184,185] derived from α-amino acids, *e.g.* (**172, 173**). Other inhibitors are known which are designed to mimic the tetrahedral geometry of the transition state for enzyme catalyzed hydrolysis. These include the α-aminophosphinic acid derivatives (**174**),[186] the α-aminoaldehydes (**175**), and the α-aminoboronic acid derivative (**176**), the latter two of which hydrate in solution.[187,188]

(171)   (172)   (173)

(174)   (175)   (176) R = Me; Bu^i, Pe^i

The most widely investigated aminopeptidase inhibitors have been discovered as a result of screening microbial fermentations for inhibition of various peptidases. Bestatin (**177**) was the first of the aminopeptidase inhibitors discovered through these methods, followed by amastatin (**178**) and the arphamenines (**179, 180**).[189-191] These compounds have been extensively studied due to their reported immuno-stimulating activities.

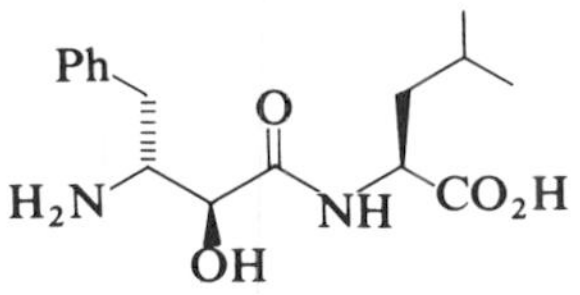

(**177**)

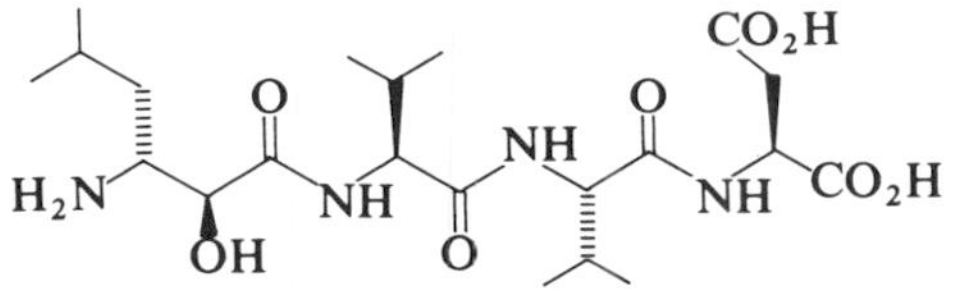

(**178**)

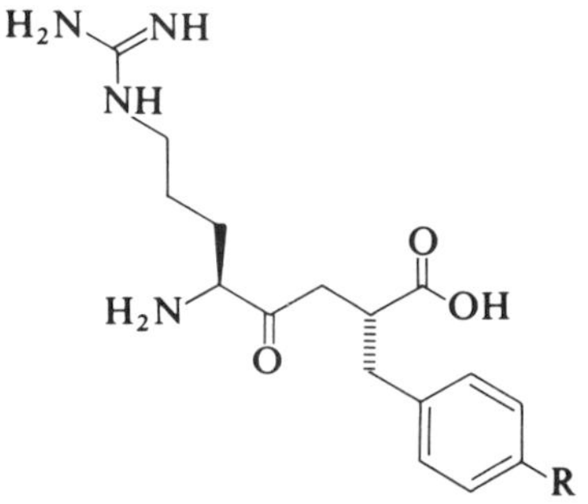

(**179**) R = H
(**180**) R = OH

The mechanisms of action of these compounds as aminopeptidase inhibitors have only recently been elucidated. The arphamenines (**179, 180**) are naturally occurring ketomethylene analogs of dipeptides which contain the side chain elements of the amino acid sequence Arg-Phe or Arg-Tyr, sequences closely related to known substrates for arginyl aminopeptidase.[191] These inhibitors are very effective competitive inhibitors of this enzyme with inhibition constants in the nM range, but they are much poorer inhibitors of other aminopeptidases. Potent inhibitors require the presence of a basic side chain at the N-terminus corresponding to either arginine or lysine. By analogy with the normal arginyl aminopeptidase substrates which contain basic side chains at P_1, the basic side chain is presumed to bind in the arginyl aminopeptidase S_1 subsite; the aromatic ring then binds in the S_1' subsite. This latter interaction is quite important because removal of the phenyl ring from (**179**) weakens binding to this enzyme by a factor of 500.

The relationship between bestatin's unusual amino acid, (2S,3R)-2-hydroxy-3-amino-4-phenyl-butanoic acid (**181**), and its mechanism of inhibition of arginyl aminopeptidase, membrane-bound aminopeptidase and leucine aminopeptidase has been the subject of several inquiries.[192] The apparent relationship between bestatin's C-2 carbon and critical (2S)-hydroxyl group to the probable tetrahedral intermediate for amide bond hydrolysis (Scheme 5) led to the idea that bestatin might be a transition-state analog inhibitor of aminopeptidases is which the sp^3 geometry and alcohol at the C-2 carbon of the inhibitor mimics the tetrahedral intermediate formed during substrate hydrolysis. However, the proposed mechanisms placed the phenyl group in bestatin in the S_1 enzyme subsite, a site presumably designed to attract basic side chains in substrates for arginyl aminopeptidase and, as we have seen, basic side chains in the arphamenines. Furthermore, bestatin analogs which contain a basic side chain in place of the benzyl group are not the improved arginyl aminopeptidase inhibitors relative to bestatin that the structure–activity work with the arphamenines would have suggested. Thus the data indicated that bestatin and the arphamenines bind differently to arginyl aminopeptidase.

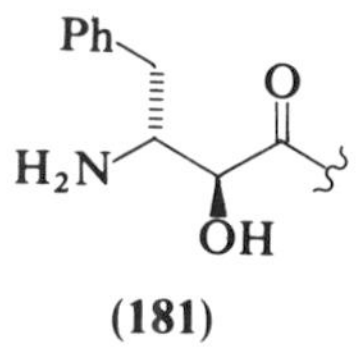

(**181**)

The thiol derivative of bestatin (**182**) was synthesized in an attempt to determine if the (2S)-hydroxyl group in bestatin was binding to the enzyme active-site zinc ion.[193] It was anticipated, based upon the precedents with thiol inhibitors of ACE, thermolysin, collagenase and CPA, and compounds (**172**) and (**173**), that the thiol group in (**182**) would bind more tightly to the active-site zinc ion in aminopeptidases than the (2S)-hydroxyl group in (177), and therefore that (182) would be a more potent inhibitor. In fact, this compound was found to be only equipotent with bestatin for inhibition of several aminopeptidases. Detailed analysis of the kinetics of inhibition of arginyl aminopeptidase by a reduced amide analog of bestatin (183; note that this is a derivative of the amino hydroxy inhibitors 134) suggested that (183) binds to aminopeptidases at the S_1'-S_2' binding sites (Figure 15), not at the assumed S_1-S_1' sites.[194] By analogy, bestatin is thought to bind to the S_1'-S_2' enzyme subsites of arginyl aminopeptidase and possibly other aminopeptidases. This mode of binding provides a rationalization for the presence of a nonbasic, aromatic side chain at the N-terminus of bestatin, rather than the basic side chain expected on the basis of substrate specificity, and is consistent with the binding of the arphamenines in that both inhibitors place an aromatic group in the S_1' binding site. It has been suggested that the essential (2S)-hydroxyl group mimics the second substrate, water, in the enzyme–inhibitor complex.[194]

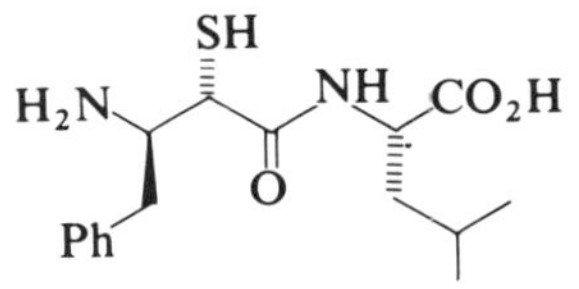

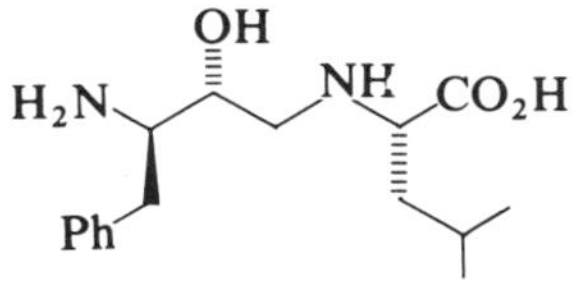

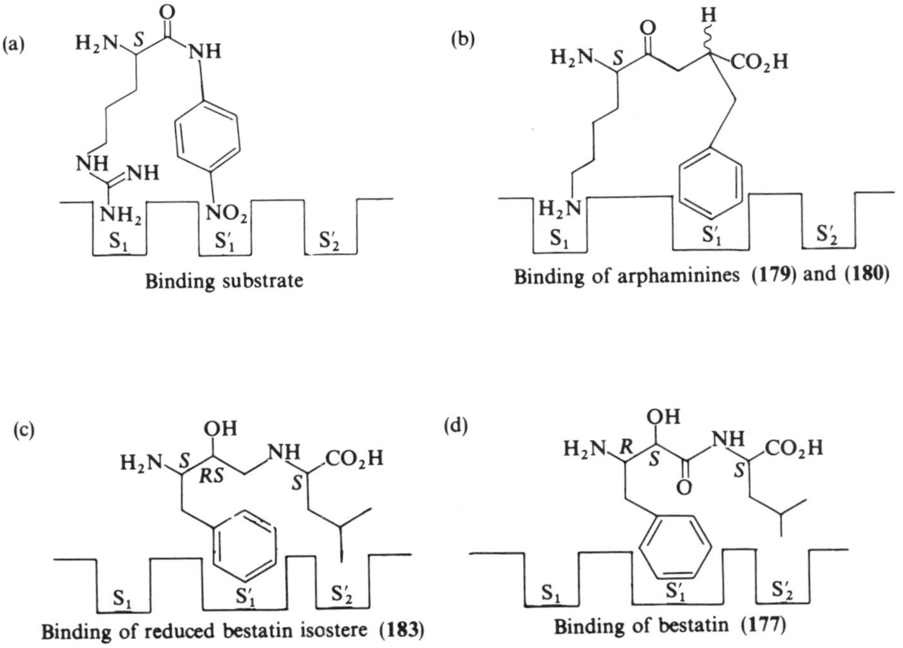

Figure 15 Orientations of substrates and inhibitors in the active site of arginylaminopeptidase. Note the binding of the aromatic side chain in bestatin to the S_1' subsite

The bestatin results illustrate how difficult it can be to rationalize inhibitor structures solely by analogy with the structures of reasonable reaction pathway intermediates. Detailed enzyme kinetic studies of inhibitor binding often are needed to clarify these interactions. Recent work has suggested that multiple forms of the enzyme–inhibitor complexes are possible and that several of these closely related inhibitors may bind to aminopeptidases in slightly different conformations depending on the inhibitor structure,[192] and that these multiple forms are kinetically visible. In conjunction with their work on the inhibition mechanism of bestatin, Ocain and Rich found that substituents α to the thiol group in leucinethiol (184) significantly reduced binding to aminopeptidases.[193] Replacing the missing components of bestatin, *e.g.* carboxamide (185) or carboxamide-Leu (186), restored the inhibitor potency to that of bestatin. Gordon and colleagues observed a similar phenomenon in their studies of diaminothio inhibitors.[195] Compound (187) is an α-substituted derivative of L-leucinethiol and is only a slightly better inhibitor of rat brain aminopeptidase than L-leucinethiol. Stepwise elaboration of this unit into bestatinthiol (182) or dipeptide derivatives, *e.g.* (188), did not produce improved inhibitors. However, very potent inhibitors were obtained with the tripeptide derivative (189). Furthermore, the tripeptide thiol derivative is 350 times more potent than the corresponding hydroxyl derivative (190), which is consistent with the thiol inhibitors interacting with the active site zinc. These results clearly demonstrate the effects that remote binding groups can have on inhibitor potency, and that multiple binding modes are likely. Much work needs to be done before the exact binding modes can be established, but one reasonable rationalization for the data is that the dipeptidyl inhibitors bind to the S_1'-S_2' subsites, as found for the inhibition of arginyl aminopeptidase by the dipeptidyl analog (183), and that the tripeptidyl derivatives bind to the S_1-S_2' binding sites on the enzyme.

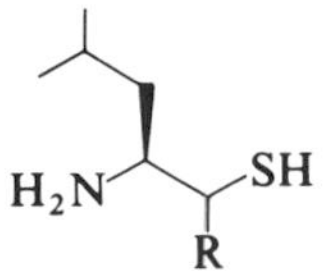

(184) R = H
(185) R = CO$_2$Me
(186) R = CO–Leu–OH

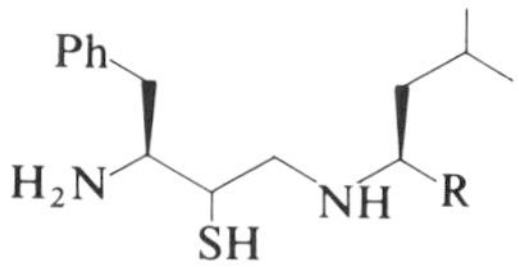

(187) R = H
(188) R = CONH$_2$

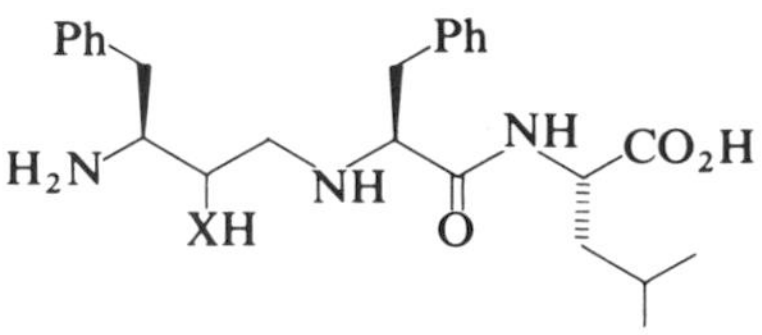

(189) X = S
(190) X = O

8.2.3.7 Inhibitors of Cysteine Proteinases

The cysteine proteinases [EC 3.4.22] comprise a group of peptidases that share the catalytic mechanism shown in Scheme 3 (Section 8.2.2).[20] The most widely studied cysteine proteinase is papain, the enzyme isolated from the latex of the papaya. Related mammalian cysteine proteinases are the cathepsins (B, H and L)[196] and the calcium-activated neutral proteases or calpains.[197] The former are active at acid pH, as required by their location in acidic lysozomes, whereas the latter, which are found in the cytoplasmic fractions of cells, are most active at slightly basic pHs.

Low molecular weight inhibitors of cysteine proteinases have been used to characterize the binding sites, catalytic functional groups and transition-state geometries of cysteine proteinases. The effects produced by these inhibitors on isolated enzymes and in intact systems suggest that cysteine proteinases are involved in a variety of biological processes that may ultimately be linked to specific diseases. At present, most of the known effects of cysteine proteinase inhibitors are linked by a probable role in the processing of biologically active peptides.

Cysteine proteinases are selectively inhibited by a variety of peptide-derived inhibitors. Peptidyl chloromethanes (peptide chloromethyl ketones; 191) were the first such inhibitors designed.[198] These act by alkylating the active site cysteine thiol group of Cys-25 (papain numbering) but also can inhibit serine proteinases (recall the elastase inhibitor 140). Some selectivity for cysteine proteinases can be achieved by modeling the peptide sequence after known substrates for either enzyme. For example, many of the good inhibitors of papain contain phenylalanine at P$_2$.

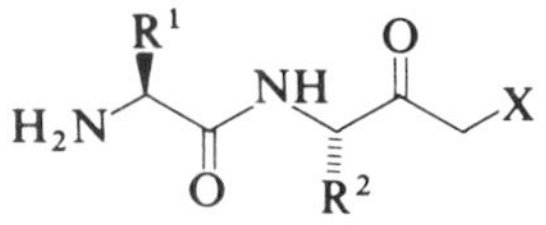

(191) X = Cl
(192) X = F
(193) X = N$_2$

Peptidyl fluoromethanes (**192**) show improved selectivity for cysteine proteinases over serine proteinases.[199] This selectivity apparently arises because the fluoride compounds react slower with serine proteinases but at about the same rate with cysteine proteinases as the peptidyl chloromethanes.

The peptidyl diazomethanes (**193**) are thought to inhibit irreversibly only cysteine proteinases.[198,200] The selectivity probably results from the necessity to protonate the diazomethyl ketone group prior to reaction with the active-site thiol group, a process favored in the active sites of cysteine proteases. These reagents have been utilized extensively for characterizing the active sites of cysteine proteinases but the inhibitors are unstable and decompose readily; this high intrinsic chemical reactivity is likely to render these irreversible inhibitors too toxic for *in vivo* use as drugs.

Peptidyl aldehydes are known to inhibit cysteine proteinases. Antipain (**194**) is a naturally occurring inhibitor,[120] and several peptidyl-aldehydes (**195**) and (**196**) derived from substrate sequences also are efficient inhibitors of cysteine proteinases.[201] ^{13}C NMR studies have shown that the thiol group in papain adds to the aldehyde group to generate a hemithiol acetal (**197**) which is considered to be structurally similar to the transition state for amide bond hydrolysis (Scheme 3).[202] The corresponding semicarbazones (**198**) also can inhibit cysteine proteinases, although in general not as efficiently as the aldehydes. However, peptidyl semicarbazones are useful ligands for affinity chromatography.[203] ^{13}C NMR studies have shown that the active site thiol in papain adds directly to the semicarbazone and that these inhibitors are not converted to the corresponding aldehydes at neutral pH.[204] Nitriles such as (**199**) also are good reversible covalent inhibitors of papain. Here again, ^{13}C NMR methods were used to show that the enzyme active-site thiol adds reversibly to the nitrile.[205,206]

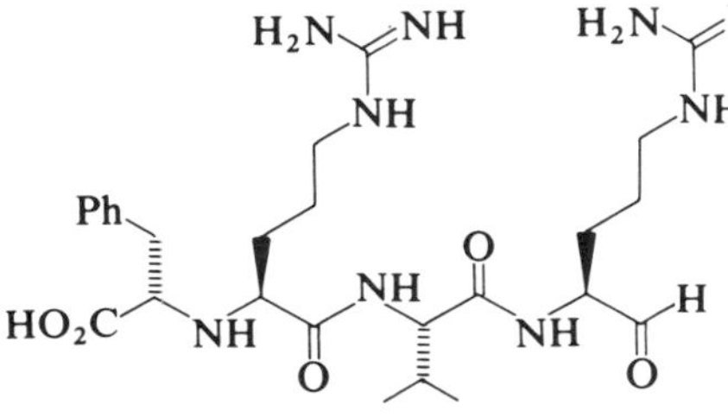

(**194**)

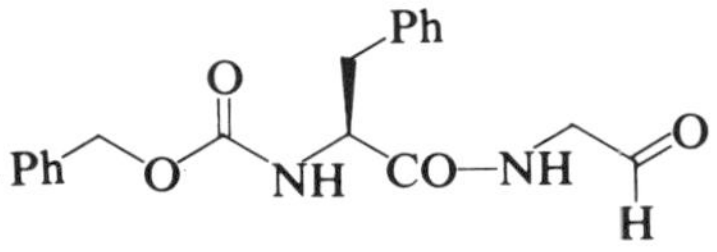

(**195**) 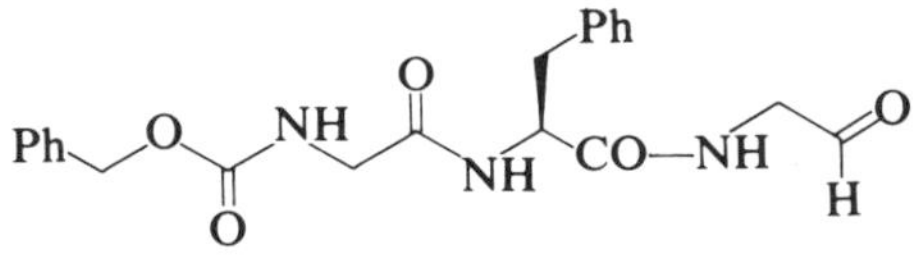   (**196**)

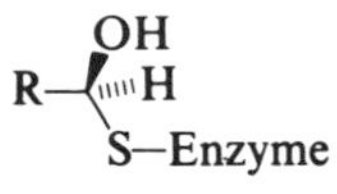

(**197**) 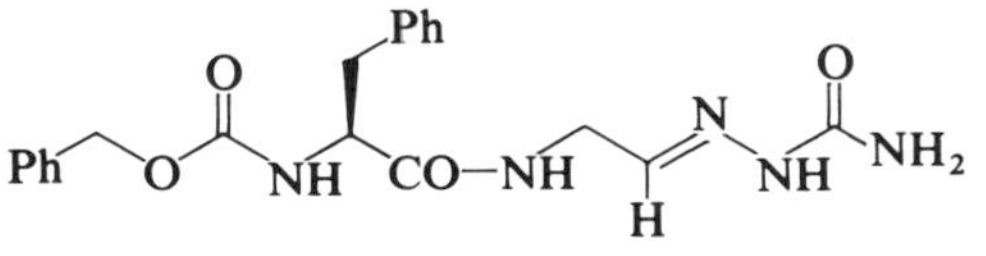   (**198**) 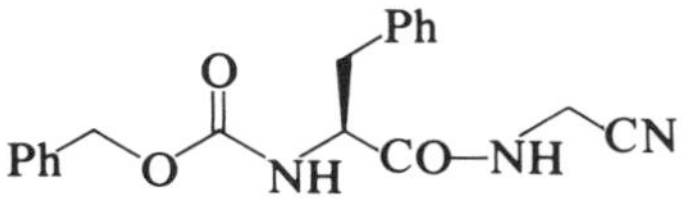   (**199**)

At the present time, analogs of E-64, a natural product isolated and characterized by Hanada *et al.*, are the cysteine proteinase inhibitors most likely to be developed into drugs.[207] E-64 (**200**) is an irreversible inhibitor of cysteine proteinases that alkylates the active site Cys[25]—SH group in papain and (presumably other cysteine proteinases) by reaction on C-3 in the epoxysuccinyl group of the inhibitor (Scheme 8).[223] The reaction of E-64 derivatives with the active-site thiol in cysteine proteinases is remarkably fast (second-order rate constant $= 10^5 \text{ M}^{-1}\text{s}^{-1}$), yet these compounds do not alkylate simple thiols at neutral pH.[26] These results suggest that the reaction of the epoxide group with the active-site thiol group may be an enzyme-catalyzed process.

Efficient inhibition of cysteine proteinases by E-64 analogs requires the presence of the *trans*-epoxide succinic acid group (*S,S* configurations) but the rest of the inhibitor structure can be modified considerably. The ethyl ester derivative (**201**; E-64d) is a prodrug form of E-64 that is hydrolyzed *in vivo* to the free acid. Shoji-Kasai *et al.* have used this membrane-permeable form of E-64 in a study which demonstrated that E-64 arrests human epidermal carcinoma A431 cells at

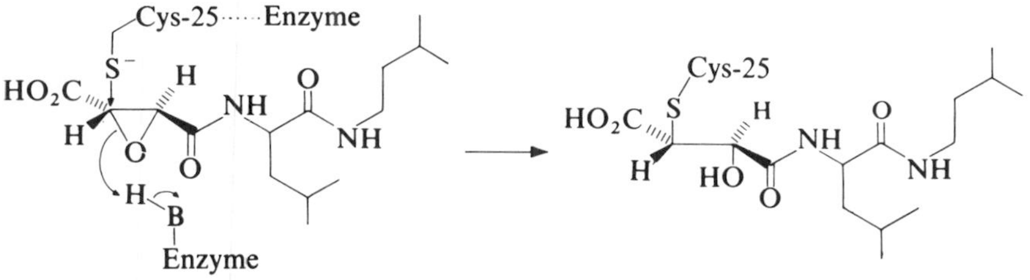

Scheme 8 Schematic representation of the reaction of papain with Ep-475[223]

mitotic metaphase.[208] They suggest that E-64 prevents proteolysis of cyclin A or a related factor that must be eliminated before the cell can proceed past metaphase.

Major changes in the leucyl-agmatine portion of E-64 have been reported. NCO-700 (**202**; note the *R,R* configurations) is reported to inhibit cathepsin B and/or calpain activities in cultured myocardial cells and to inhibit the release of labeled amino acids from isolated rabbit heart muscle.[209] Thus NCO-700 may prove useful for reducing irreversible proteolysis in the ischemic myocardium. Taisho Pharmaceuticals reports that the ethyl ester derivative of Ep-475 (**203**), designated loxistatin (**204**), is orally active in hamsters and may be developed as a drug for muscular dystrophy.[224]

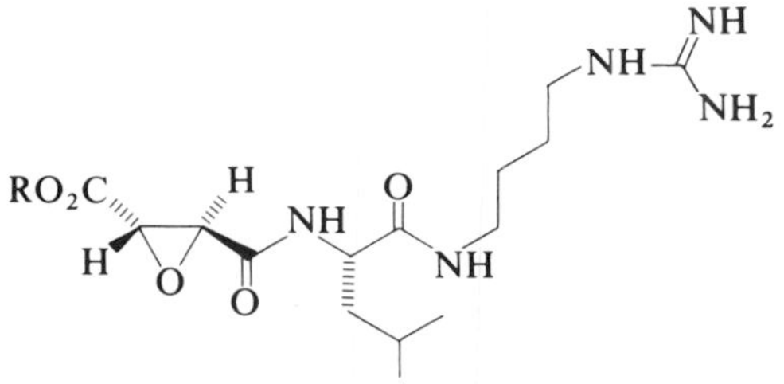

(**200**) R = H
(**201**) R = Et

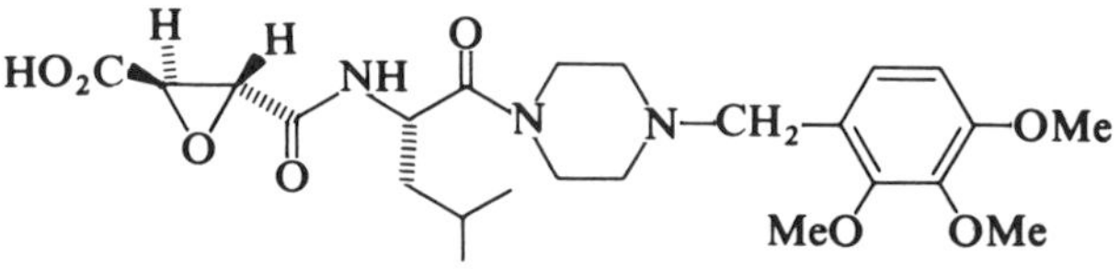

(**202**)

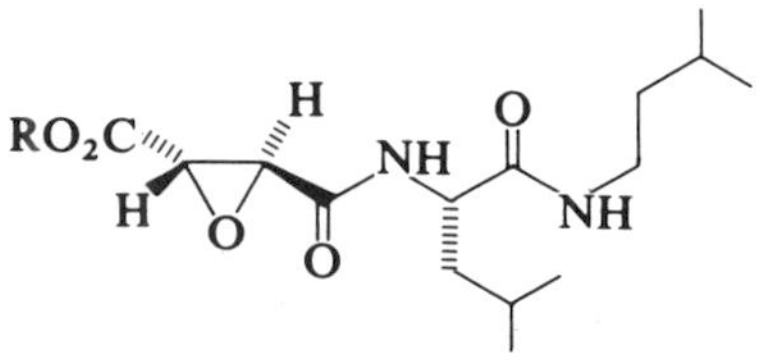

(**203**) R = H; Ep-475
(**204**) R = Et; Loxistatin

8.2.3.8 Inhibitors of Prolyl Endopeptidases

Prolyl endopeptidase ([EC 3.4.21.26]; postproline cleaving enzyme) is a serine proteinase that catalyzes the hydrolysis of peptides at the carboxyl side of L-proline residues. It has been suggested that this enzyme may be important for metabolism of peptides in the CNS. Prolyl endopeptidases are inhibited by aldehyde dervatives of proline. (Z)-L-Pro-L-Prolinal (**205**) is a specific enzyme inhibitor from rabbit brain. (Z)-L-Val-L-Prolinal (**206**; K_i = 2.4 nM) and related analogs are very potent inhibitors of other prolyl endopeptidases.[210,211]

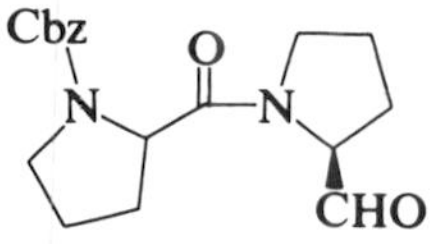

(**205**)

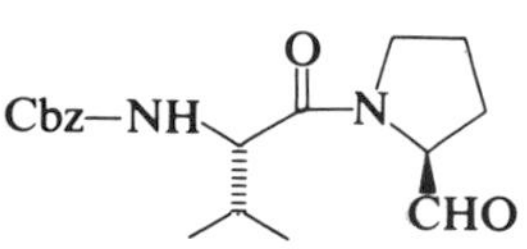

(**206**)

8.2.3.9 Thermolysin Inhibitors

Thermolysin, a metallo endopeptidase [EC 3.4.24.4] isolated from *Bacillus thermoproteolyticus*, is probably the most extensively characterized metallopeptidase with respect to catalytic mechanism and substrate specificity and has served as an important model for the active site of angiotensin converting enzyme (*vide supra*) and other potentially therapeutically important metallopeptidases. Many inhibitors of thermolysin have been prepared to test transition-state analog or collected-product inhibition design concepts. Crystallographic studies of the native enzyme and of the enzyme complexed with a variety of natural and synthetic inhibitors have provided some of the best information we have about the effects that metal chelation, hydrogen bonding and desolvation have upon the stability of enzyme–inhibitor complexes.

The structures of some of the more important inhibitors of thermolysin are shown below. As expected for metallopeptidases, all these inhibitors contain groups that can chelate to zinc as well as bind to S_1'-S_2' enzyme subsites. The crystal structure of phosphoramidon (54) bound to thermolysin established that the *N*-phosphoryl group interacted with the active-site zinc and led to the development of numerous phosphorus containing inhibitors.[27] Bartlett and coworkers have prepared a series of simplified phosphorus containing inhibitors which are very effective inhibitors of thermolysin.[212] These include the phosphonamidate inhibitor (207) and the phosphonate inhibitor (208). A series of five pairs of inhibitors (N *vs.* O substitution) were compared and the oxygen substituted inhibitors were found to be weaker thermolysin inhibitors by a factor of 840-fold relative to the nitrogen inhibitors. The loss of binding to the oxygen analog has been attributed to the loss of the hydrogen bond between the inhibitor and enzyme. The crystal structures of both (207) and (208) bound to thermolysin have been determined by Matthews *et al.* and the enzyme-bound conformations of the inhibitors have been found to be identical within experimental error.[213] The calculated differences in free energy between the NH and O analogs (207) and (208), which were determined by Bash *et al.* means of equilibrium pertubation methods, agreed remarkably closely with the experimental values.[214] However, the difference between the N and O series results from more than the removal of a hydrogen bond since the oxygen in the O series is also forced into an electrostatic environment wherein the partial negative charge on oxygen results in a repulsive interaction. This remarkable series of papers provides unusual insight into the importance and strength of hydrogen bonding in enzyme–inhibitor interactions.

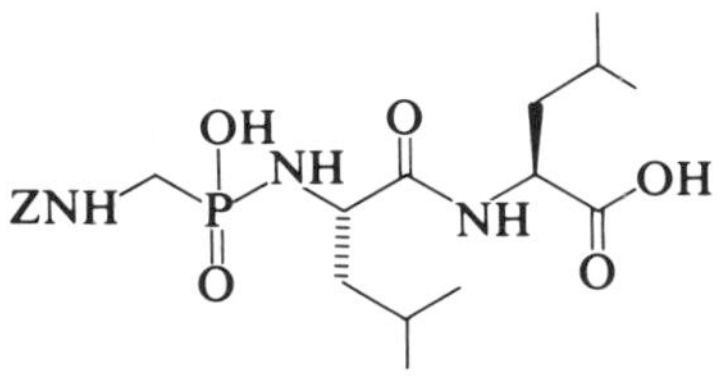

(207) 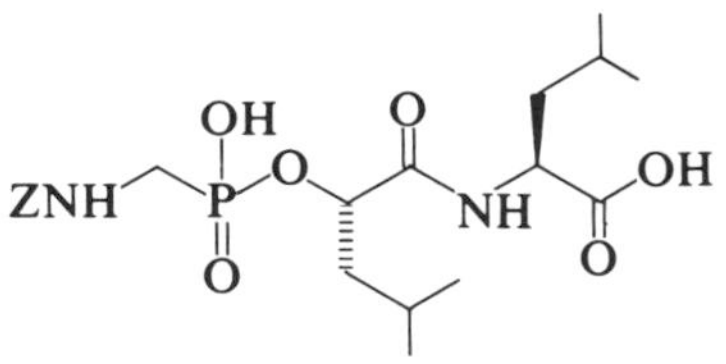   (208)

The phosphorus based inhibitors of thermolysin have provided insight into another type of desolvation that affects the strength of enzyme–inhibitor interactions. The phosphonamidate analogs (209) and (210) are both effective inhibitors of thermolysin but (210) binds to the enzyme much more slowly than the Gly analog (209).[215] Slow binding inhibition is often observed with tight binding enzyme inhibitors but the causes of this phenomenon are not well understood.[216] Slow desolvation of serendipitously bound water molecules or other cosubstrates, or slow interconversions between different enzyme–inhibitor forms, have been suggested.[192] The slow binding of pepstatin to pepsin has been rationalized in terms of the slow escape of an enzyme bound water molecule that must be displaced in order for the critical (3S) statine hydroxyl group to hydrogen bond to the catalytically active aspartic acid carboxyl groups.[121,126] Matthews and coworkers found that the slow and fast binding inhibitors (209) and (210) bound differently in the active site of thermolysin.[217] The fast binding inhibitor formed a complex in which a water molecule bridged the inhibitor and enzyme, whereas the tight, slow binding inhibitor formed a complex in which the water was not present and the hydrogen bond was between enzyme and inhibitor. Matthews and Bartlett and their respective coworkers suggested that the tight-binding complex was formed only when the water was eliminated, either as water slowly escaped from the enzyme–inhibitor complex as postulated for the pepstatin–pepsin model or by a process in which the inhibitor binds to a rare, desolvated form of the enzyme.

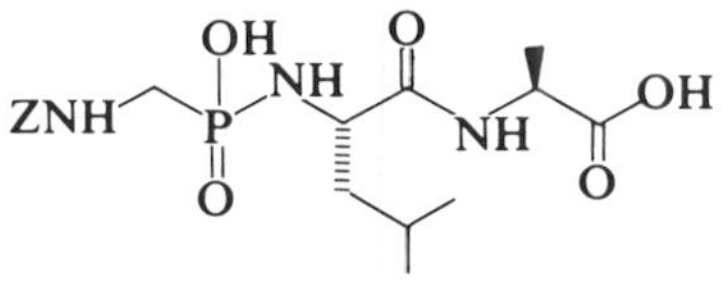

(209) 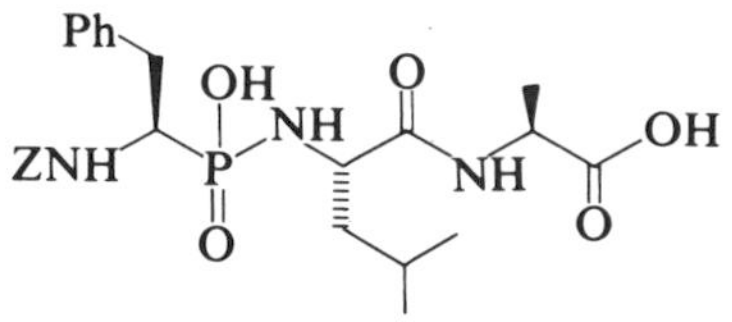 **(210)**

N-Hydroxamic acid derivatives of amino acids and peptides are good inhibitors of thermolysin. The crystal structure of **(211)** bound to thermolysin[218] shows that the benzyl group binds to the S_1' enzyme subsite and that the hydroxamic acid moiety binds as a bidentate complex. This was the first time a pentacoordinate inhibitor–zinc complex was detected in an enzyme.

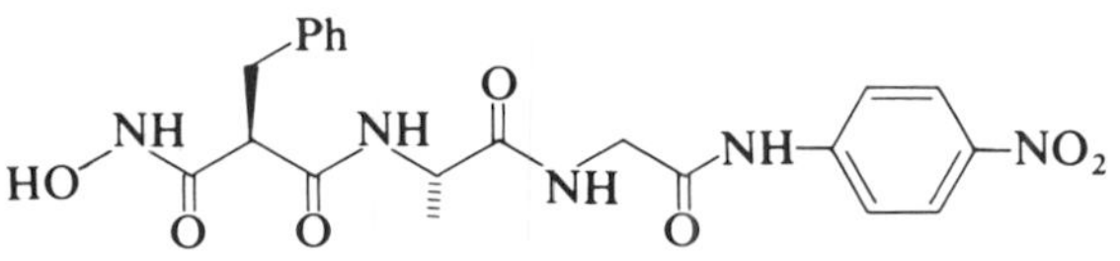

(211)

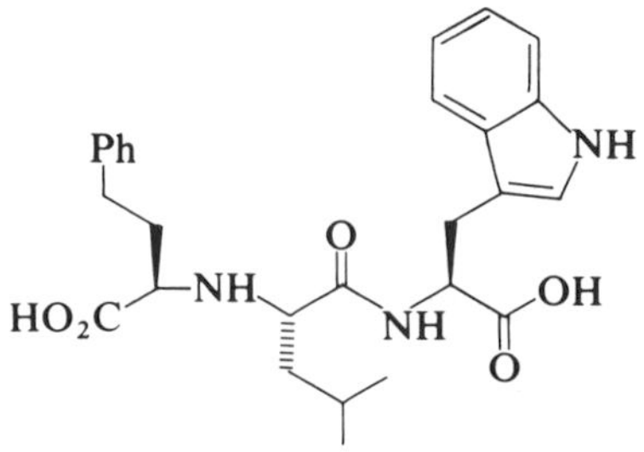

(212)

The crystal structure of the *N*-carboxyalkyl inhibitor **(212)** (which is structurally related to phosphoramidon) has also been solved at 1.9 Å resolution.[219] The *N*-carboxy group is bound to the zinc in an overall pentacoordinate geometry, and the Leu and Trp side chains are in the S_1' and S_2' pockets, respectively. This enzyme–inhibitor complex has served as the template for an interactive molecular modeling study of the catalytic mechanism of thermolysin.[24]

Thiol inhibitors related to the structure of captopril also inhibit thermolysin. The crystal structure of the thiol inhibitor **(213)** bound in the active site of the enzyme has been solved at 1.9 Å resolution.[220] The benzyl group is in the S_1' subsite of the enzyme and the sulfur atom is 1.9 Å away from the active-site zinc atom. Interestingly, the closely related thiol compound **(214)** is a substrate for thermolysin.[221] The product Val-Trp is formed in the active site of the crystalline enzyme by catalytic removal of the α-thiol acetic acid unit. This fortuitous result has provided the first crystal structure for the thermolysin–*product* complex. The structure shows that the Val nitrogen is 2.8 Å from Glu-143 and participates in hydrogen bonds to enzyme and solvent. The proximity of the product amine to the carboxylate of Glu-143 is consistent with the catalytic mechanism proposed on the basis of the inhibitor **(212)**–thermolysin molecular modeling study.[24]

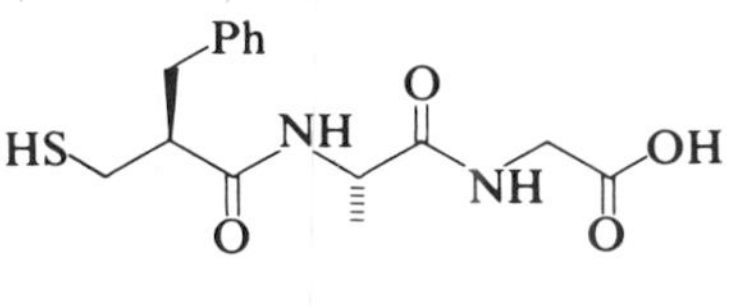

(213) 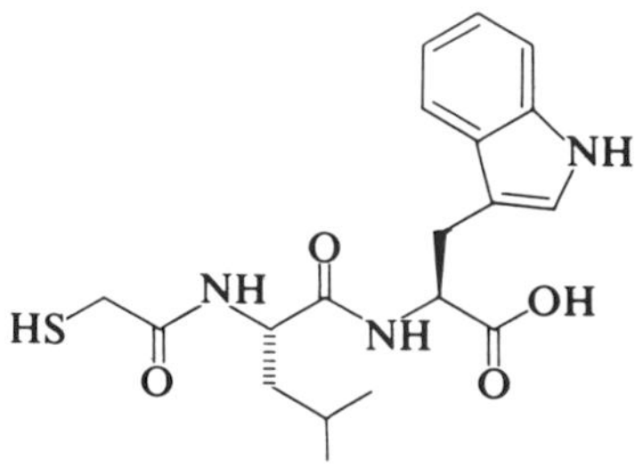 **(214)**

8.2.4 PROSPECTIVES

Modern design of peptidase inhibitors depends critically upon a variety of clever and sophisticated synthetic, enzyme kinetic, conformational and computer graphic techniques that are needed to optimize an inhibitor rapidly for a target peptidase. Two recurring problems limit the successful development of drug candidates: (1) the difficulty in identifying novel 'lead' structures, and (2) the inability to incorporate oral activity or cell permeability in a rational way into inhibitors derived from peptide-like structures.

The discovery of lead compounds remains a constant challenge. As is noted throughout this article, all the lead structures described were either natural products (teprotide, pepstatin, phosphoramidon, bestatin, neutral cephalosporins, *etc.*) discovered by means of screening methods, or else were derived from the substrate for the target enzyme (itself a natural product), usually by applying some version of the transition state analog inhibitor hypothesis. These peptide related inhibitors have proven to be difficult to modify in order to gain suitable pharmacophoric properties without compromising inhibitor potency. Even when peptidase resistant derivatives of the inhibitors are produced by total synthesis, oral activity can be difficult to obtain because of the propensity for these peptide derived inhibitors to be excreted in the bile. Removal of peptide bonds in principle could circumvent this mode of inactivation. However, removal of amides is particularly difficult to do with substrate derived inhibitors because of the tendency of these structures to utilize β-pleated sheets (see Figure 2) to stabilize the enzyme–inhibitor interaction. Because β-structures are most important for larger substrates, the problem becomes most acute with precisely the peptides which need to have amide bonds replaced. Smaller substrates, such as those important in exopeptidases (*e.g.* ACE), are likely to be easier to convert into inhibitors rather than the larger substrates, such as those used by endopeptidases (*e.g.* renin).

It would appear that significant advances in this field could come with innovations in three areas. Elucidation of the fundamental peptide transport mechanisms utilized *in vivo* (either naturally or with successful drugs, *e.g.* lisinopril) could lead to rational methods for delivering peptide derivatives *in vivo*. Utilization of more sophisticated screening methods to identify novel secondary metabolites that inhibit a target peptidase will provide new lead compounds that, in turn, can be rationally altered into clinical candidates by application of the many methods described (*vide supra*). As noted already, such an approach has provided many of the initial starting points for developing peptidase inhibitors. More recently, this approach has proven extremely productive for developing novel inhibitors of peptide receptors for CCK[225] and the extension to many other peptide receptors and enzymes appears likely. Finally, use of modern computer graphic methods, especially novel methods for designing inhibitors *de novo* on the basis of X-ray crystal structures of enzyme active sites or of enzyme active sites characterized by NMR or other methods, would enable chemists to begin the structure optimizations with templates likely to possess better pharmacokinetic potential.

It is clear that the methods developed to date are very powerful and have proven extremely useful for elucidating enzyme catalytic mechanisms, and for providing novel lead structures and even clinical candidates. Additional therapeutic agents are almost certain to be found in the future that work by inhibition of peptidases.

ACKNOWLEDGEMENTS

Special thanks to D. A. Trainor, E. M. Gordon and D. E. Hangauer for critically reading the manuscript, and to J. L. Stanton, Y. Yabe and A. A. Patchett for their help compiling the literature of ACE inhibitors.

8.2.5 REFERENCES

1. Peptidase is the accepted term for a peptide bond hydrolase (EC 3.4). The enzyme class is broken into *exo* and *endo* peptidases. The former group acts on the chain ends, *e.g.* aminopeptidase or carboxypeptidase. The latter group of enzymes hydrolyze internal bonds. Proteinases are peptidases that cleave protein substrates. Proteases is a generic term for peptidases that cleave protein substrates.
2. D. R. Lynch and S. H. Snyder, *Annu. Rev. Biochem.*, 1986, **55**, 773; J. F. McKelvy, *Annu. Rev. Neurosci.*, 1986, **9**, 415.
3. H. Neurath, *Science*, 1984, **224**, 350.
4. E. D. Thorsett and M. J. Wyvratt, in 'Neuropeptidase Inhibitors', ed. A. J. Turner, Ellis Horwood, Chichester, 1987, 229.
5. See chapters 7–20 in 'Proteinase Inhibitors', ed. A. J. Barrett and G. Salvesen, (Research Monographs in Cell and Tissue Physiology, vol.12), Elsevier, Amsterdam, 1986, p. 301.
6. R. J. Baugh and H. P. Schnebli, in 'Proteinases in Tumor Invasion', ed. P. Sträubli, A. J. Barrett and A. Baici, Raven Press, New York, 1986, p. 157.
7. J. M. Frasca, O. Auerbach, H. W. Carter, and V. R. Parks, *Am. J. Pathol.*, 1983, **111**, 11.
8. G. L. Snider, *Drug Dev. Res.*, 1987, **10**, 235.
9. W. N. Schwartz and J. W. C. Bird, *Biochem. J.*, 1977, **167**, 811.
10. M. A. Ondetti and D. W. Cushman, *CRC Crit. Rev. Biochem.*, 1984, **16**, 381.
11. (a) M. J. Wyvratt and A. A. Patchett, *Med. Res. Rev.*, 1985, **5**, 483. (b) A. A. Patchett and E. H. Cordes, *Adv. Enzymol.*, 1985, **57**, 1.
12. D. A. Trainor, *Trends Pharmacol. Sci.*, 1987, **8**, 303.
13. B. P. Roques and M.-C. Fournié-Zaluski, National Institutes of Drug Administration Research Monograph Series, 1986, vol. 70, 128.

14. W. J. Greenlee, *Pharm. Res.*, 1987, **4**, 364.
15. W. H. Johnson, N. A. Roberts and N. Borkakoti, *J. Enzyme Inhibition*, 1987, **2**, 1.
16. A. J. Barrett and G. Salvesen (eds.), 'Proteinase Inhibitors', (Research Monographs in Cell and Tissue Physiology, vol. 12), Elsevier, Amsterdam, 1986.
17. J. Kraut, *Annu. Rev. Biochem.*, 1977, **46**, 331; see also J. C. Powers and J. W. Harper, 'Inhibitors of Serine Proteinases' in ref. 16.
18. S. J. Gardell, C. S. Craik, D. Hilvert, M. S. Urdea and W. J. Rutter, *Nature (London)*, 1985, **317**, 551.
19. W. Kullmann, 'Enzymatic Peptide Synthesis' CRC Press, Boca Raton, FL, 1987.
20. D. H. Rich in ref 16, p. 153.
21. D. H. Rich in ref 16, p. 179.
22. K. Suguna, E. A. Padlan, C. W. Smith, W. D. Carlson and D. R. Davies, *Proc. Natl. Acad. Sci. USA*, 1987, **84**, 7009.
23. T. Hofmann, B. M. Dunn and A. L. Fink, *Biochemistry*, 1984, **23**, 5253.
24. D. G. Hangauer, A. F. Monzingo and B. W. Matthews, *Biochemistry*, 1984, **23**, 5730.
25. A. J. Barrett, in ref. 16, p. 3.
26. (a) A. J. Barrett, A. A. Kembhavi, M. A. Brown, H. Krischke, C. G. Knight, M. Tamai and K. Hanada, *Biochem. J.*, 1982, **201**, 189; (b) C. Parkes, A. Kembhavi and A. J. Barrett, *Biochem. J.*, 1985, **230**, 509.
27. H. Suda, T. Aoyagi, T. Takeuchi and H. Umezawa, *J. Antibiot.*, 1973, **26**, 621.
28. (a) I. Schechter and A. Berger, *Biochem. Biophys. Res. Commun.*, 1967, **27**, 157; (b) 1968, **32**, 898.
29. T. A. Steitz, R. Henderson and D. M. Blow, *J. Mol. Biol.*, 1969, **46**, 337.
30. J. K. Fawcett, N. Camerman and A. Camerman, *Acta Crystallogr., Sect. B*, 1975, **31**, 658.
31. G. D. Rose, L. M. Gierasch and J. A. Smith, *Adv. Protein Chem.*, 1985, **37**, 1.
32. T. Blundell, B. L. Sibanda and L. Pearl, *Nature (London)*, 1983, **304**, 273.
33. R. L. Soffer (Ed.), 'Biochemical Regulation of Blood Pressure', Wiley, New York, 1981.
34. Z. P. Horovitz (ed.), 'Angiotensin Converting Enzyme Inhibitors: Mechanism of Action and Clinical Implications', Urban & Schwarzenberg, Baltimore, 1981.
35. S. H. Ferreira, D. C. Bartelt and L. J. Greene, *Biochemistry*, 1970, **9**, 2583.
36. M. A. Ondetti, N. J. Williams, E. F. Sabo, J. Pluscec, E. R. Weaver and O. Kocy, *Biochemistry*, 1971, **10**, 4033.
37. M. A. Ondetti, B. Rubin and D. W. Cushman, *Science*, 1977, **196**, 441.
38. (a) L. D. Byers and R. Wolfenden, *J. Biol. Chem.*, 1972, **247**, 606; (b) *Biochemistry*, 1973, **12**, 2070.
39. Y. Funae, T. Komori, D. Sasaki and K. Yamamoto, *Biochem. Pharmacol.*, 1980, **29**, 1543.
40. M. Oya, T. Baba, E. Kato, Y. Kawashima and T. Watanabe, *Chem. Pharm. Bull.*, 1982, **30**, 440.
41. J. R. Powell, B. Rubin, D. W. Cushman, J. Krapcho and M. J. Antonaccio, presented at the International Society of Hypertension, 1st European Meeting on Hypertension, Milan, Italy, May 1983, Abstr. 355.
42. (a) D. H. Kim, C. J. Guinosso, G. C. Buzby, D. R. Herbst, R. J. McCaully, T. C. Wicks and R. L. Wendt, *J. Med. Chem.*, 1983, **26**, 394; (b) J. L. Stanton, N. Gruenfeld, J. E. Babiarz, M. H. Ackerman, R. C. Friedmann, A. M. Yuan and W. Macchia, *J. Med. Chem.*, 1983, **26**, 1267.
43. A. Schwab, I. Weinryb, R. Macerata, W. Rogers, J. Suh and A. Khandwala, *Biochem. Pharmacol.*, 1983, **32**, 1957.
44. D. W. Cushman and M. A. Ondetti, in *Prog. Med. Chem.*, ed. G. P. Ellis and G. B. West, Elsevier, Amsterdam, 1979, **17**, 41.
45. S. Kuromaru, S. Tanaka, T. Fujimura, K. Matsunaga, Y. Hinohara, N. Kawamura, H. Nabata and R. Aono, presented at the 104th Annual Meeting, Pharmaceutical Society of Japan, Sendai, Japan, March 1984, Abstr. 29A9-5.
46. K. Hosoki, K. Takeyama, H. Okamoto, H. Minato and T. Kadokawa, *Jpn. J. Pharmacol.*, 1981, **31** (Suppl.), Abstr 0–107.
47. F. J. McEvoy, F. M. Lai and J. D. Albright, *J. Med. Chem.*, 1983, **26**, 381.
48. *US Pat.* 4 442 038; see also C. T. Huang, R. R. Brooks, S. F. Pong, J. R. Skuster, R. L. White, Jr. and A. F. Moore, *Drug Dev. Res.*, 1986, **7**, 141.
49. A. A. Patchett, E. Harris, E. W. Tristram, M. J. Wyvratt, M. T. Wu, D. Taub, E. R. Peterson, T. J. Ikeler, J. ten Broek, L. G. Payne, D. L. Ondeyka, E. D. Thorsett, W. J. Greenlee, N. S. Lohr, R. D. Hoffsommer, H. Joshua, W. V. Ruyle, J. W. Rothrock, S. D. Aster, A. L. Maycock, F. M. Robinson, R. Hirschmann, C. S. Sweet, E. H. Ulm, D. M . Gross, T. C. Vassil and C. A. Stone, *Nature (London)*, 1980, **288**, 280.
50. H. H. Rotmensch, P. H. Vlasses, B. N. Swanson, J. D. Irvin, K. E. Harris, D. G. Merrill and R. K. Ferguson, *Am. J. Cardiol.*, 1984, **53**, 116.
51. H. Metzger, R. Maier, C. Sitter and H.-O. Stern, *Arzneim.-Forsch.*, 1984, **34** (II), 1402.
52. E. J. Sybertz, T. Baum, H. S. Ahn, S. Nelson, E. Eynon, D. M. Desiderio, K. Pula, F. Becker, C. Sabin, R. Moran, G. Vander Vliet, B. Kastner and E. Smith, *J. Cardiovasc. Pharmacol.*, 1983, **5**, 643.
53. K. Nishikawa, Y. Inada, Z. Terashita, M. Tanabe, K. Kawazoe, Y. Oka and S. Kikuchi, *Jpn. J. Pharmacol.*, 1983, **33** (Suppl.), 273P.
54. H. R. Kaplan, D. M. Cohen, A. D. Essenburg, T. C. Major, T. E. Mertz and M. J. Ryan, *Fed. Proc., Fed. Am. Soc. Exp. Biol.*, 1984, **43**, 1326.
55. N. Gruenfeld, J. L. Stanton, A. M. Yuan, F. H. Ebetino, L. J. Browne, C. Gude and C. F. Huebner, *J. Med. Chem.*, 1983, **26**, 1277.
56. M. Vincent, G. Remond, B. Portevin, B. Serkiz and M. Laubie, *Tetrahedron Lett.*, 1982, **23**, 1677.
57. L. H. Weaver, W. R. Kester and B. W. Matthews, *J. Mol. Biol.*, 1977, **114**, 119.
58. (a) R. E. Galardy, *Biochem. Biophys. Res. Commun.*, 1980, **97**, 94; (b) *Biochemistry*, 1982, **21**, 5777.
59. E. W. Petrillo, Jr. and M. A. Ondetti, *Med. Res. Rev.*, 1982, **2**, 1.
60. (a) J. R. Powell, J. M. DeForrest, D. W. Cushman, B. Rubin and E. W. Petrillo, *Fed. Proc., Fed. Am. Soc. Exp. Biol.*, 1984, **43**, P733; (b) J. Krapcho, C. Turk, D. W. Cushman, J. R. Powell, J. M. DeForrest, E. R. Spitzmiller, D. S. Karanewsky, M. Duggan, G. Rovnyak, J. Schwartz, S. Natarajan, J. D. Godfrey, D. E. Ryono, R. Neubeck, K. S. Atwal, and E. W. Petrillo, Jr., *J. Med. Chem.*, 1988, **31**, 1148.
61. D. S. Karanewsky, M. C. Badia, D. W. Cushman, J. M. DeForrest, T. Dejneka, M. J. Loots, M. G. Perri, E. W. Petrillo, Jr. and J. R. Powell, *J. Med. Chem.*, 1988, **31**, 204.
62. (a) For a review of the use of molecular modeling techniques to design conformationally restricted ACE inhibitors, see D. Hangauer in 'Computer Aided Drug Design', ed. T. Perun and C. Probst, Dekker, New York, 1989, p. 253; (b) E. D. Thorsett, *Actual. Chim. Ther.*, 1986, **13**, 257.

63. (a) D. F. Veber, R. M. Freidinger, D. Schwenk Perlow, W. J. Paleveda, F. W. Holly, R. G. Strachan, R. F. Nutt, B. H. Arison, C. Homnick, W. C. Randall, M. S. Glitzer, R. Saperstein and R. Hirschmann, *Nature (London)*, 1981, **292**, 55; (b) H. Kessler, *Angew. Chem., Int. Ed. Engl.*, 1982, **21**, 512.

64. T. K. Sawyer, W. L. Cody, J. J. Knittel, V. J. Hruby, M. E. Hadley, M. D. Hirsch and T. L. O'Donohue, in 'Peptides: Structure and Function. Proceedings of the 8th American Peptide Symposium', ed. V. J. Hruby and D. H. Rich, Pierce Chemical Co., Rockford, IL, 1984, p. 323.

65. (a) P. W. Schiller and J. DiMaio, in ref. 64, p. 269; (b) V. J. Hruby, *Trends Pharm. Sci.*, 1985, 259; (c) H. I. Mosberg, R. Hurst, V. J. Hruby, K. Gee, H. I. Yamamura, J. J. Galligan and T. F. Burks, *Proc. Natl. Acad. Sci. USA*, 1983, **80**, 5871.

66. E. D. Thorsett, E. E. Harris, S. Aster, E. R. Peterson, D. Taub, A. A. Patchett, E. H. Ulm and T. C. Vassil, *Biochem. Biophys. Res. Commun.*, 1983, **111**, 166.

67. W. C. Still and I. Galynker, *Tetrahedron*, 1981, **37**, 3981.

68. E. D. Thorsett, E. E. Harris, S. D. Aster, E. R. Peterson, E. W. Tristram, J. P. Snyder, J. P. Springer and A. A. Patchett, in ref. 64, p. 555.

69. M. J. Wyvratt, M. H. Tischler, T. J. Ikeler, J. P. Springer, E. W. Tristram and A. A. Patchett, in ref. 64, p. 551.

70. W. H. Parsons, J. L. Davidson, D. Taub, S. D. Aster, E. D. Thorsett, A. A. Patchett, E. H. Ulm and B. I. Lamont, *Biochem. Biophys. Res. Commun.*, 1983, **117**, 108.

71. J. L. Stanton, J. W. H. Watthey, M. N. DeSai, B. M. Finn, J. E. Babiarz and H. C. Tomaselli, *J. Med. Chem.*, 1985, **28**, 1603.

72. K. Itoh, M. Kori, Y. Inada, K. Nishikawa, Y. Kawamatsu and H. Sugihara, *Chem. Pharm. Bull.*, 1986, **34**, 3747.

73. H. Yanagisawa, S. Ishihara, A. Ando, T. Kanazaki, S. Miyamoto, H. Koike, Y. Iijima, K. Oizumi, Y. Matsushita and T. Hata, *J. Med. Chem.*, 1987, **30**, 1984.

74. C. H. Hassall, A. Krohn, C. J. Moody and W. A. Thomas, *FEBS Lett.*, 1982, **147**, 175.

75. C. H. Hassall, A. Krohn, C. J. Moody and W. A. Thomas *J. Chem. Soc., Perkin Trans. 1*, 1984, 155.

76. M. R. Attwood, C. H. Hassall, A. Krohn, G. Lawton and S. Redshaw, *J. Chem. Soc., Perkin Trans. 1*, 1986, 1011.

77. A. F. Monzingo and B. W. Matthews, *Biochemistry*, 1984, **23**, 5724.

78. P. R. Andrews, J. M. Carson, A. Caselli, M. J. Spark and R. Woods, *J. Med. Chem.*, 1985, **28**, 393.

79. R. Ciabatti, G. Padova and E. Bellasio *et al.*, *J. Med. Chem.*, 1986, **29**, 411.

80. D. Mayer, C. B. Naylor, I. Motac and G. R. Marshall, *J. Computer Aided Mol. Design*, 1987, **1**, 3.

81. (a) R. G. Almquist, W.-R. Chao, M. E. Ellis and H. L. Johnson, *J. Med. Chem.*, 1980, **23**, 1392; (b) R. G. Almquist, J. Crases, D. Jennings-White, R. F. Meyer, M. L. Hoefle, R.D. Smith, A. D. Essenburg and H. R. Kaplan, *J. Med. Chem.*, 1982, **25**, 1292.

82. (a) S. Natarajan, E. M. Gordon, E. F. Sabo, J. D. Godfrey, H. N. Weller, J. Pluscec, M. B. Rom and D. W. Cushman, *Biochem. Biophys. Res. Commun.*, 1984, **124**, 141; (b) E. M. Gordon, S. Natarajan, J. Pluscec, H. N. Weller, J. D. Godfrey, M. B. Rom, E. F. Sabo, J. Engebrecht and D. W. Cushman, *Biochem. Biophys. Res. Commun.*, 1984, **124**, 148.

83. E. M. Gordon, J. D. Godfrey, J. Pluscec, D. Von Langen and S. Natarajan, *Biochem. Biophys. Res. Commun.*, 1985, **126**, 419.

84. (a) K. Hayashi, K.-I. Nunami, K. Sakai, Y. Ozaki, J. Kato, K. Kinashi and N. Yoneda, *Chem. Pharm. Bull.*, 1983, **31**, 3553; (b) K. Hayashi, K. Nunami, K. Sakai, Y. Ozaki, J. Kato, K. Kinashi and N. Yoneda, *Chem. Pharm. Bull.*, 1985, **33**, 2011.

85. *Drugs Future*, 1987, **512**, 860.

86. D. M. Cohen, A. D. Essenburg, B. J. Olszewski, R. M. Singer, M. J. Ryan, D. B. Evans and H. R. Kaplan, *Pharmacologist*, 1984, **26**, Abstr. 266.

87. BW-A575C (Wellcome) presented as abstract at International Society of Hypertension, September 1986.

88. G. A. Flynn, E. L. Giroux and R. C. Dage, *J. Am. Chem. Soc.*, 1987, **109**, 7914.

89. M. A. Kerr and A. J. Kenny, *Biochem. J.*, 1974, **137**, 477.

90. J. Kenny, *Trends Biochem. Sci.*, 1986, **11**, 40.

91. B. P. Roques, M. C. Fournié-Zaluski, E. Soroca, J. M. Lecomte, B. Malfroy, C. Llorens and J. C. Schwartz, *Nature (London)*, 1980, **288**, 286.

92. B. P. Roques, E. Lucas-Soroca, P. Chaillet, J. Costentin and M. C. Fournié-Zaluski, *Proc. Natl. Acad. Sci. USA*, 1983, **80**, 3178.

93. R. A. Mumford, M. Zimmerman, J. ten Broeke, D. Taub, H. Joshua, J. W. Rothrock, J. M. Hirshfield, J. P. Springer and A. A. Patchett, *Biochem. Biophys. Res. Commun.*, 1982, **109**, 1303.

94. (a) J. G. Berger, *US Pat.* 4 610 816 (1986); (b) J. G. Berger, *US Pat.* 4 721 726 (1988).

95. E. M. Gordon, D. W. Cushman, R. Tung, H. S. Cheung, F. L. Wang and N. G. Delaney, *Life Sci.*, 1983, **33**, Suppl. I, 113.

96. (a) L. B. Hersh, *Biochemistry*, 1981, **20**, 2345; (b) J. M. Hambrook, B. A. Morgan, M. J. Rance and C. F. C. Smith, *Nature (London)*, 1976, **262**, 782; (c) C. Gorenstein and S. H. Snyder, *Life Sci.*, 1979, **25**, 2065.

97. N. Nishino and J. C. Powers, *Biochemistry*, 1978, **17**, 2846.

98. (a) G. Waksman, R. Bouboutou, J. Devin, S. Bourgoin, F. Cesselin, M. Hamon, M.-C. Fournié-Zaluski and B. P. Roques, *Eur. J. Pharmacol.*, 1985, **117**, 233; (b) R. Bouboutou, G. Waksman, J. Derin, M. C. Fournié-Zaluski and B. P. Roques, *Life Sci.*, 1984, **35**, 1023.

99. G. Waksman, E. Hamel, M.-C. Fournié-Zaluski and B. P. Roques, *Proc. Natl. Acad. Sci. USA*, 1986, **83**,1523.

100. T. Turpeenniemi-Hujanen, U. P. Thorgeirsson and L. A. Wotta, *Annu. Rep. Med. Chem.*, 1984, **19**, 231.

101. J. F. Woessner, Jr., in 'Collagenases in Normal and Pathological Tissue', ed. D. E. Woolley and J. J. Evanson, Wiley, Chichester, 1980, p. 223.

102. L. Seltzer, J. J. Jeffrey and A. Z. Eisen, *Biochim. Biophys. Acta*, 1977, **485**, 179.

103. R. D. Gray, R. B. Miller and A. F. Spatola, *J. Cell. Biochem.*, 1986, **32**, 71–77.

104. K. Darlak, R. B. Miller, M. S. Stack, A. F. Spatola and R. D. Gray, unpublished results.

105. K. McCullagh, H. Wadsworth and M. Hann, *Eur. Pat.*, 1984, 159 396; J. P. Dickens, D. K. Donald, G. Kneen and W. R. McKay, *US Pat.*, 1986, 4 599 361.

106. K. A. Mookhtiar, C. K. Marlowe, P. A. Bartlett and H. E. Van Wart, *Biochemistry*, 1987, **26**, 1962.

107. K. McCullagh, H. Wadsworth and M. Hann, *Eur. Pat.*, 1984, 126 974.

108. J. Boger, *Annu. Rep. Med. Chem.*, 1985, **20**, 257.

109. L. Skeggs, K. Lentz, J. Kahn and H. Hochstrasser, *J. Exp. Med.*, 1968, **128**, 13.

110. D. A. Tewksbury, R. A. Dart and J. Travis, *Biochem. Biophys. Res. Commun.*, 1981, **99**, 1311.
111. T. Kokubu, E. Ueda, S. Fujimoto, K. Hiwada, A. Kato, H. Akutsu, Y. Yamamura, S. Saito and T. Mizoguchi, *Nature(London)*, 1968, **217**, 456.
112. (a) J. Burton, R. J. Cody, Jr., J. A. Herd and E. Haber, *Proc. Natl. Acad. Sci. USA*, 1980, **77**, 5476; (b) R. J. Cody, J. Burton, G. Evin, K. Poulsen, J. A. Herd and E. Haber, *Biochem. Biophys. Res. Commun.*, 1980, **97**, 230.
113. L. Pauling, *Chem. Eng. News*, 1946, **24**, 1375.
114. (a) R. Wolfenden, *Annu. Rev. Biophys. Bioeng.*, 1976, **5**, 271; (b) R. Wolfenden, *Acc. Chem. Res.*, 1972, **5**, 10.
115. (a) G. E. Leinhard, *Science*, 1973, **180**, 149; (b) G. E. Leinhard, *Annu. Rep. Med. Chem.*, 1972, **7**, 249.
116. M. J. Parry, A. B. Russell and M. Szelke, 'Chemistry and Biology of Peptides', Proceedings 3rd American Peptide Symp., Ann Arbor Sci. Publ., Ann Arbor, 1972, p. 541.
117. M. Szelke, B. Leckie, A. Hallett, D. M. Jones, J. Sueioras, B. Atrash and A. F. Lever, *Nature(London)*, 1982, **299**, 555.
118. M. Szelke, D. M. Jones, B. Atrash, A. Hallett and B. J. Leckie, in ref. 64. p. 579.
119. J. Boger, in 'Aspartic Proteinases and Their Inhibitors', ed. V. Kostka, de Gruyter, Berlin, 1985, p. 401.
120. (a) H. Umezawa, *Annu. Rev. Microbiol.*, 1982, **36**, 75; (b) H. Umezawa, T. Aoyagi, H. Morishima, M. Matsuzaki, M. Hamada and T. Takeuchi, *J. Antibiot.*, 1970, **23**, 259.
121. For a review of pepstatin structure–activity relationships, see D. H. Rich, *J. Med. Chem.*, 1985, **28**, 263.
122. G. R. Marshall, *Fed. Proc., Fed. Am. Soc. Exp. Biol.*, 1976, **35**, 2494.
123. J. Marciniszyn, Jr., J. A. Hartsuck and J. Tang, *J. Biol. Chem.*, 1976, **251**, 7088.
124. R. Bott, E. Subramanian and D. R. Davies, *Biochemistry*, 1982, **21**, 6956.
125. (a) M. N. G. James, A. Sielecki, F. Salituro, D. H. Rich and T. Hofmann *Proc. Natl. Acad. Sci. USA*, 1982, **79**, 6137; (b) S. I. Foundling, J. Cooper, F. E. Watson, A. Cleasby, L. H. Pearl, B. L. Sibanda, A. Hemmings, S. P. Wood, T. L. Blundell, J. M. Vallner, C. G. Norey, J. Kay, J. Boger, B. M. Dunn, B. J. Leckie, D. M. Jones, B. Atrash, A. Hallett and M. Szelke, *Nature(London)*, 1987, **327**, 349.
126. D. H. Rich, M. S. Bernatowicz, N. S. Agarwal, M. Kawai, F. G. Salituro and P. G. Schmidt, *Biochemistry*, 1985, **24**, 3165.
127. J. Boger, N. S. Lohr, E. H. Ulm, M. Poe, E. H. Blaine, G. M. Fanelli, T. -Y. Lin, L. S. Payne, T.W. Schorn, B. I. LaMont, T. C. Vassil, I.I. Stabilito, D. F. Veber, D. H. Rich and A. S. Boparai, *Nature (London)*, 1983, **303**, 81.
128. J. Maibaum and D. H. Rich, *J. Med. Chem.*, 1988, **31**, 625.
129. D. F. Veber, M. G. Bock, S. F. Brady, E. H. Ulm, D. W. Cochran, G. M. Smith, B. I. Lamont, R. M. DiPardo, M. Poe, R. M. Freidinger, B. E. Evans and J. Boger, *Biochem. Soc. Trans.*, 1984, **12**, 956.
130. J. M. Wood, N. Gulati, P. Forgiarini, W. Fuhrer and K. G. Hofbauer, *Hypertension*, 1985, **7**, 797.
131. J. Boger, L. S. Payne, D. S. Perlow, N. S. Lohr, M. Poe, E. H. Blaine, E. H. Ulm, T. W. Schorn, B. I. LaMont, T.-Y. Lin, M. Kawai, D. H. Rich and D. F. Veber, *J. Med. Chem.*, 1985, **28**, 1779.
132. J. J. Plattner, J. Greer, A. K. L. Fung, H. Stein, H. D. Kleinert, H. L. Sham, J. R. Smital and T. J. Perun, *Biochem. Biophys. Res. Commun.*, 1986, **139**, 982.
133. V. Rasetti, J. L. Stanton, P. Buhlmayer, W. Fuhrer, R. Goshke and J. Wood, Abstracts, Ninth International Symposium on Medicinal Chemistry (EFMC), West Berlin, Sept. 14–18, 1986, p. 130.
134. P. Bühlmayer, A. Caselli, W. Fuhrer, R. Göschke, V. Rasetti, R. Rüeger, J. L. Stanton, L. Sriscione and J. M. Wood, *J. Med. Chem.*, 1988, **31**, 1839.
135. (a) J.-A. Fehrentz, A. Heitz, B. Castro, C. Cazaubon and D. Nisato, *FEBS Lett.*, 1984, **167**, 273; (b) T. Kokubu, K. Hiwada, Y. Sato, T. Iwata, Y. Imamura, R. Matsueda, Y. Yabe, H. Kogen, M. Yamazaki, Y. Iijima and Y. Baba, *Biochem. Biophys. Res. Commun.*, 1984, **118**, 929.
136. D. H. Rich, A. S. Boparai and M. S. Bernatowicz, *Biochem. Biophys. Res. Commun.*, 1982, **104**, 1127.
137. M. Holladay, F. G. Salituro and D. H. Rich, *J. Med. Chem.*, 1987, **30**, 374.
138. D. H. Rich, M. S. Bernatowicz and P. G. Schmidt, *J. Am. Chem. Soc.*, 1982, **104**, 3535.
139. M. H. Gelb, J. P. Svaren and R. H. Abeles, *Biochemistry*, 1985, **24**, 1813.
140. S. Thaisrivongs, D. T. Pals, W. M. Kati, S. R. Turner, L. M. Thomasco and W. Watt, *J. Med. Chem.*, 1986, **29**, 2080.
141. H. N. Weller, E. M. Gordon, M. R. Rom and J. Plušcec, *Biochem. Biophys. Res. Commun.*, 1984, **125**, 82.
142. G. Bolis and J. Greer, in 'Computer Aided Drug Design', ed. T. Perun and C. Probst, Dekker, New York, 1989, p. 297.
143. M. De Claviere, C. Cazaubon, C. Lacour, D. Nisato, J. P. Gagnol, G. Evin and P. Corvol, *J. Cardiovasc. Pharmacol.*, 1985, **7** (Suppl. 4), S58.
144. M. Tree, B. Atrash, B. Donovan, J. Gamble, A. Hallet, M. Hughes, D. M. Jones, B. Leckie. A. F. Lever, J. J. Morton and M. J. Szelke, *Hypertension*, 1983, **1**, 399.
145. P. A. Bartlett and W. B. Kezer, *J. Am. Chem. Soc.*, 1984, **106**, 4282.
146. (a) J. G. Dann, D. K. Stammers, C. J. Harris, R. J. Arrowsmith, D. E. Davies, G. W. Hardy and J. A. Morton, *Biochem. Biophys. Res. Commun.*, 1986, **134**, 71; (b) D. E. Ryono, C. A. Free, R. Neubeck, S. G. Sammaniego, J. D. Godfrey and E. W. Petrillo, Jr., in 'Peptides, Structure and Function. Proceedings of the Ninth Amer. Peptide Symposium', ed. C. M. Deber, V. J. Hruby and K. D. Kopple, Pierce Chem. Co., Rockford, IL, 1985, p. 739.
147. S. Thaisrivongs, D. T. Pals, D. W. Harris, W. M. Kati and S. R. Turner, *J. Med. Chem.*, 1986, **29**, 2088.
148. S. H. Rosenberg, J. J. Plattner, K. W. Woods, H. H. Stein. P. A. Marcotte, J. Cohen and T. J. Perun, *J. Med. Chem.*, 1989, **30**, 1224.
149. C. R. Nakaie, M. C. F. Oliveira, L. Juliano and A. C. M. Piava, *Biochem. J.*, 1982, **205**, 43.
150. H. L. Sham, G. Bolis, H. H. Stein, S. W. Fesik, P. A. Marcotte, J. J. Plattner, C. A. Rempel and J. Greer, *J. Med. Chem.*, 1988, **31**, 284.
151. G. L. Snider, *Drug Dev. Res.*, 1987, **10**, 235.
152. J. C. Powers, *Am. Rev. Respir. Dis.*, 1983, **127**, 554.
153. A. Yasutake and J. C. Powers, *Biochemistry*, 1981, **20**, 3675.
154. J. C. Powers and D. L. Carroll, *Biochem. Biophys. Res. Commun.*, 1975, **67**, 639.
155. A. Renaud, P. Lestienne, D. L. Hughes, J. G. Bieth and J.-L. Dimicoli, *J. Biol. Chem.*, 1983, **258**, 8312.
156. N. E. MacKenzie, J. P. G. Malthouse and A. I. Scott, *Science*, 1984, **225**, 883.
157. M. N. G. James, A. R. Sielecki, G. D. Brayer, L. T. J. Delbaere and C. A. Bauer, *J. Mol. Biol.*, 1980, **144**, 43.
158. H. Hassall, W. H. Johnson, A. J. Kennedy and N. A. Roberts, *FEBS Lett.*, 1985, **183**, 201.
159. E. Meyer and D. A. Trainer, unpublished data.
160. M. H. Gelb, J. P. Svaren and R. H. Abeles, *Biochemistry*, 1985, **24**, 1813.

161. B. Imperiali and R. H. Abeles, *Biochemistry*, 1986, **25**, 3760.
162. R. H. Abeles, *Drug Dev. Res.*, 1987, **10**, 221.
163. U. Brodbeck, K. Schweikert, R. Gentinetta and M. Rottenberg, *Biochim. Biophys. Acta*, 1979, **567**, 357.
164. K. A. Koehler and G. E. Lienhard, *Biochemistry*, 1971, **10**, 2477.
165. D. A. Matthews, R. A. Alden, J. J. Birktoft, S. T. Freer and J. Kraut, *J. Biol. Chem.*, 1975, **250**, 7120.
166. D. S. Matteson, K. M. Sadhu and G. E. Lienhard, *J. Am. Chem. Soc.*, 1981, **103**, 5241.
167. C. A. Kettner and A. B. Shenvi, *J. Biol. Chem.*, 1984, **259**, 15106.
168. M. Laskowski, Jr., M. Tashiro, M. W. Empie, S. J. Park, I. Kato, W. Ardelt and M. Wieczorek, in 'Proteinase Inhibitors', ed. N. Katunuma, H. Umezawa and H. Holzer, Springer-Verlag, Berlin, 1983, p. 55.
169. D. A. Trainor, M. M. Stein, J. J. Lewis, A. Strimpler and R. Stein, Abstr. 29 Med. Chem. presented at New Orleans meeting Am. Chem. Soc., Sept. 1987.
170. R. W. Spencer, L. J. Copp, B. Bonaventura, T. F. Tam, T. J. Liak, R. J. Billedeau and A. Krantz, *Biochem. Biophys. Res. Commun.*, 1986, **140**, 928.
171. R. L. Stein, A. M. Strimpler, B. R. Viscarello, R. A. Wildonger, R. C. Mauger and D. A. Trainor, *Biochemistry*, 1987, **26**, 4126.
172. J. W. Harper, K. Hemmi and J. C. Powers, *Biochemistry*, 1985, **24**, 1831.
173. J. W. Harper and J. C. Powers, *Biochemistry*, 1985, **24**, 7200.
174. L. J. Copp, A. Krantz and R. W. Spencer, *Biochemistry*, 1987, **26**, 169.
175. R. R. Rando, *Acc. Chem. Res.*, 1975, **8**, 281.
176. J. B. Doherty, B. M. Ashe, L. W. Argenbright, P. L. Barker, R. J. Bonney, G. O. Chandler, M. E. Dahlgren, C. P. Dorn, Jr., P. E. Finke, R. A. Firestone, D. Fletcher, W. K. Hagmann, R. Mumford, L. O'Grady, A. L. Maycock, J. M. Pisano, S. K. Shah, K. R. Thompson and M. Zimmerman, *Nature(London)*, 1986, **322**, 192.
177. S. J. Cartwright and S. G. Waley, *Med. Res. Rev.*, 1983, **3**, 341.
178. M. A. Navia, J. P. Springer, T.-Y. Lin, H. R. Williams, R. A. Firestone, J. M. Pisano, J. B. Doherty, P. E. Finke and K. Hoogsteen, *Nature(London)*, 1987, **327**, 79.
179. R. J. Delange and E. L. Smith, in 'The Enzymes', 3rd edn., Academic Press, New York, 1971, vol. 3, p. 81.
180. J. R. McDermott, D. Mantle, B. Lauffant, A. M.Gibson and J. A. Biggins, *J. Neurochem.*, 1988, **50**, 176.
181. (a) J. O. Baker, S. H. Wilkes, M. E. Bayliss and J. M. Prescott, *Biochemistry*, 1983, **22**, 2098; (b) M. E. Bayliss and J. M. Prescott, *Biochemistry*, 1986, **25**, 8113.
182. W. W.-C. Chan, P. Dennis, W. Demmer and K. Brand, *J. Biol. Chem.*, 1982, **257**, 7955.
183. S. H. Wilkes and J. M. Prescott, *J. Biol. Chem.*, 1983, **258**, 13517.
184. W. W.-C. Chan, *Biochem. Biophys. Res. Commun.*, 1983, **116**, 297.
185. T. D. Ocain and D. H. Rich, *Biochem. Biophys. Res. Commun.*, 1987, **145**, 1038.
186. P. P. Giannousis and P. A. Bartlett, *J. Med. Chem.*, 1987, **30**, 1603.
187. A. B. Shenvi, *Biochemistry*, 1986, **25**, 1286.
188. L. Andersson, T. C. Isley and R. Wolfenden, *Biochemistry*, 1982, **21**, 4177.
189. H. Umezawa, T. Aoyagi, H. Sada, M. Hamada and T. Takeuchi, *J. Antibiot.*, 1976, **29**, 97.
190. T. Aoyagi, H. Tobe, F. Kojima, M. Hamada, T. Takeuchi and H. Umezawa, *J. Antibiot.*, 1978, **31**, 636.
191. H. Umezawa, T. Aoyagi, S. Ohuchi, A. Okuyama, H. Suda, T. Takita, M. Hamada and T. Takeuchi, *J. Antibiot.*, 1984, **36**, 1572.
192. For a review of the mechanism of inhibition of aminopeptides and a discussion of slow-binding inhibition, see D. H. Rich and D. B. Northrop, in 'Computer Aided Drug Design', ed. T. Perun and C. Probst, Dekker, New York, 1988, p. 185.
193. T. D. Ocain and D. H. Rich, *J. Med. Chem.*, 1988, **31**, 2193.
194. S. L. Harbeson and D. H. Rich, *Biochemistry*, 1988, **27**, 7301.
195. E. M. Gordon, J. D. Godfrey, N. G. Delaney, M. Asaad, D. Von Langen, S. Natarajan and D. W. Cushman, Abstract presented at 10th American Peptide Symposium, St. Louis, MO, May 1987.
196. A. J. Barrett and J. K. McDonald, 'Mammalian Proteinases: A Glossary and Bibliography', vol. 1, 'Endopeptidases', Academic Press, New York, 1980.
197. T. Murachi, *Trends Biochem. Sci.*, 1983, **8**, 295.
198. (a) E. Shaw, in 'The Enzymes', 3rd edn., ed. P. D. Boyer, Academic Press, New York, 1970, vol. 1, p. 91; (b) E. Shaw, *J. Protein Chem.*, 1984, **3**, 109.
199. P. Rauber, H. Angliker, B. Walker and E. Shaw, *Biochem. J.*, 1986, **239**, 633.
200. G. D. J. Green and E. Shaw, *J. Biol. Chem.*, 1981, **256**, 1923.
201. (a) J. A. Mattis, J. B. Henes and J. S. Fruton, *Biochem.*, 1977, **252**, 6776; (b) J. B. Henes, J. A. Mattis and J. S. Fruton, *Proc. Natl. Acad. Sci. USA*, 1979, **76**, 1131.
202. M. P. Gamcsik, J. P. G. Malthouse, W. Primrose, N. E. Mackenzie, A. S. F. Boyd, R. A. Russell and A. I. Scott, *J. Am. Chem. Soc.*, 1983, **105**, 6324.
203. D. H. Rich, M. A. Brown and A. J. Barrett, *Biochem. J.*, 1986, **235**, 731.
204. D. H. Rich, R. Shute, M. Christopherson and D. Guillame, unpublished data.
205. J. B. Moon, R. S. Coleman and R. P. Hanzlik, *J. Am. Chem. Soc.*, 1986, **108**, 1350.
206. T.-C. Liang and R. H. Abeles, *Arch. Biochem. Biophys.*, 1987, **252**, 626.
207. (a) K. Hanada, M. Tamai, M. Yamagishi, S. Ohmura, J. Sawada and I. Tanaka, *Agric. Biol. Chem.*, 1978, **42**, 523; (b) K. Hanada, M. Tamai, S. Ohmura, J. Sawada, T. Seki and I. Tanaka, *Agric. Biol. Chem.*, 1978, **42**, 529; (c) K. Hanada, M. Tamae, S. Morimoto, T. Adachi, S. Ohmura, J. Sawada and I. Tanaka, *Agric. Biol. Chem.*, 1978, **42**, 537.
208. Y. Shoji-Kasai, M. Senshu, S. Iwashita and K. Imahori, *Proc. Natl. Acad. Sci. USA*, 1988, **85**, 146.
209. H. Sashida and Y. Abiko, *Biochem. Pharmacol.*, 1985, **34**, 3875.
210. M. Nishikata, H. Yokosawa and S.-I. Ishii, *Chem. Pharm. Bull.*, 1986, **34**, 2931.
211. S. Wirk and M. Orlowski, *J. Neurochem.*, 1983, **41**, 69.
212. P. A. Bartlett and C. K. Marlowe, *Science*, 1987, **235**, 569.
213. D. E. Tronrud, H. M. Holden and B. W. Matthews, *Science*, 1987, **235**, 571.
214. P. A. Bash, U. C. Singh, F. K. Brown, R. Langridge and P. A. Kollman, *Science*, 1987, **235**, 574.
215. P. A. Bartlett and C. K. Marlowe, *Biochemistry*, 1987, **26**, 8553.
216. J. F. Morrison and C. T. Walsh, *Adv. Enzymol. Relat. Areas Mol. Biol.*, 1988, **61**, 201; see also ref. 192 for a discussion of

the molecular basis for slow-binding inhibition.

217. H. M. Holden, D. E. Tronrud, A. F. Monzingo, L. H. Weaver and B. W. Matthews, *Biochemistry*, 1987, **26**, 8542.
218. M. A. Holmes and B. W. Matthews, *Biochemistry*, 1981, **20**, 6912.
219. A. F. Monzingo and B. W. Matthews, *Biochemistry*, 1984, **23**, 5724.
220. A. F. Monzingo and B. W. Matthews, *Biochemistry*, 1982, **21**, 3390.
221. H. M. Holden and B. W. Matthews, *J. Biol. Chem.*, 1989, **263**, 3256.
222. N. Borkakoti, M. J. Broadhurst, P. A. Brown, B. K. Handa, W. H. Johnson, G. Lawton, P. J. Machin and N. A. Roberts, Poster presented at 'Peptides as Targets for Drug Research', Meeting of the Society for Drug Research, University of York, July, 1988.
223. Y. Yabe, D. Guillaume and D. H. Rich, *J. Am. Chem. Soc.*, 1988, **110**, 4043.
224. M. Tamai, S. Omura, M. Kimura, K. Hanada and H. Sugita, *J. Pharmacobio-Dyn.*, 1987, **10**, 678.
225. B. E. Evans, K. E. Rittle, M. G. Bock, R. M. DiPardo, R. M. Freidinger, W. L. Whitter, G. F. Lundell, D. F. Veber, P. S. Anderson, R. S. L. Chang, V. J. Lotti, D. J. Cerino, T. B. Chen, P. J. Kling, K. A. Kunkel, J. P. Springer and J. Hirshfield, *J. Med. Chem.*, 1988, **31**, 2235.

8.3

Enzyme Cascades: Purine Metabolism and Immunosuppression

RICHARD B. GILBERTSEN AND JAGADISH C. SIRCAR
Warner-Lambert Company, Ann Arbor, MI, USA

8.3.1	INTRODUCTION	444
	8.3.1.1 Scope of Chapter	444
	8.3.1.2 Cascades of Purine Metabolism	445
8.3.2	PURINE METABOLIC ENZYMES AND IMMUNODEFICIENCIES	446
	8.3.2.1 Adenosine Deaminase (ADA)	447
	8.3.2.1.1 ADA deficiency	447
	8.3.2.1.2 Biochemistry of ADA and ADA deficiency	448
	8.3.2.1.3 ADA inhibitors	449
	8.3.2.1.4 In vitro immunological and biochemical studies employing ADA inhibitors	450
	8.3.2.1.5 Effects of ADA inhibitors in animals	452
	8.3.2.1.6 Clinical studies on ADA inhibitors	452
	8.3.2.2 S-Adenosyl-L-homocysteine Hydrolase (SAHH)	453
	8.3.2.2.1 Biochemistry	453
	8.3.2.2.2 SAHH inhibitors and findings	454
	8.3.2.3 Ecto-5′-nucleotidase (ENT)	456
	8.3.2.3.1 Biochemistry and immunology	456
	8.3.2.3.2 ENT inhibitors	457
	8.3.2.4 Purine Nucleoside Phosphorylase (PNP)	457
	8.3.2.4.1 PNP deficiency	458
	8.3.2.4.2 Biochemistry of PNP and PNP deficiency	458
	8.3.2.4.3 PNP inhibitors	459
	8.3.2.4.4 In vitro immunological and biochemical studies employing PNP inhibitors	463
	8.3.2.4.5 In vivo studies on PNP inhibitors	465
	8.3.2.5 Hypoxanthine–guanine Phosphoribosyltransferase (HGPRT)	466
8.3.3	AUTOIMMUNE DISEASES	466
	8.3.3.1 Rheumatoid Arthritis	467
	8.3.3.1.1 Clinical description and pathogenesis	467
	8.3.3.1.2 Immunosuppressive therapy of rheumatoid arthritis	468
	8.3.3.1.3 Experimental therapy	468
	8.3.3.2 Other Autoimmune Diseases	468
8.3.4	HYPERURICEMIA	468
	8.3.4.1 Gout and Xanthine Oxidase Inhibitors	469
	8.3.4.2 Hyperuricemia and PNP Inhibition	469
8.3.5	MISCELLANEOUS CONDITIONS	469
	8.3.5.1 Organ Transplantation	469
	8.3.5.2 Psoriasis	474
	8.3.5.3 Parasitic Diseases	474
8.3.6	SUMMARY	475
8.3.7	REFERENCES	475

8.3.1 INTRODUCTION

8.3.1.1 Scope of Chapter

In the chapter on coagulation, fibrinolysis and hemostasis (Chapter 8.4), a detailed review and discussion of a 'classical' enzyme cascade are presented. Along with the complement cascade, those systems are perhaps the best-characterized and most representative of enzyme cascades. In coagulation and complement activation each substrate in turn becomes an enzyme capable of acting on multiple substrates in the next step, with the process continuing, resulting in tremendous amplification of an initial event. The cascades to be discussed in this chapter are functionally and organizationally distinct. Substrates of one step do not become enzymes of a subsequent step; rather, the substrates flow down a series of steps, each being an enzymatic reaction that catalyzes a specific step in a sequence of highly regulated, feedback inhibition-controlled reactions. As an example, adenosine triphosphate is catabolized to the diphosphate, to the monophosphate, then to adenosine, which is then broken down by deamination to inosine, and from there to hypoxanthine, xanthine and uric acid, thus creating a cascading of substrates (Figure 1). These systems can be compared to those of the arachidonic acid cascade (Chapter 6.2) in that the substrates of that cascade are metabolized by a sequence of enzymes, but have no enzymatic activity of their own. The enzymes of the purine cascade could be, and historically have been, described in the context of a metabolic sequence. However, either convention can be adopted, as long as the reader is clear that there are conceptual differences between the cascades of fibrinolysis, complement, renin–angiotensin, *etc.*, and those of arachidonic acid and purine metabolism.

In Chapter 7.3 of this work Dr. Hobbs presented details of purines and pyrimidines as drug targets, concentrating on applications as antineoplastic and antiviral agents. Although there is some overlap with the contents of this chapter, the aims of our work are different. In this chapter, the concept will be presented for a novel approach to the rational design of immunosuppressive agents through analysis of immune deficiency syndromes which result from deficiencies in enzymes in the purine metabolic cascade (Section 8.3.2). The specific impairments in immune function associated with each enzyme deficiency will be described. The association between the immune impairment and the metabolic alterations which most likely produce these deficiencies will then be discussed. We will then present, at some length, a number of inhibitors of various enzymes in the purine metabolic cascade, along with pertinent *in vitro* immunological and biochemical data obtained using these inhibitors. *In vivo* and clinical data will also be presented where available.

With this background, a very brief description of disorders of immune aggressiveness, the so-called autoimmune diseases, will be presented (Section 8.3.3). These represent diseases which could benefit from therapy which is directed at the blockade of some step in the purine metabolic cascade. Many agents, such as azathioprine, cyclophosphamide and methotrexate, which are used in the treatment of autoimmune diseases, evolved from cancer chemotherapy research, and were developed through the rationale that they would affect various aspects of purine or pyrimidine metabolism.[1-4] Their development for autoimmune diseases was strictly fortuitous. Other agents, including the 'standard' disease-modifying antirheumatic agents (DMARDs), the gold compounds and D-penicillamine, do not have a primary effect on purine metabolism, but nevertheless achieve some of the same therapeutic objectives through different, albeit largely unproven, mechanisms. Our primary objectives, therefore, are to describe potential targets (enzymes within the purine cascade) which may serve as the basis for development of novel immunosuppressive agents, and to describe some of the more interesting inhibitors of these enzymes, along with relevant data.

While these are the main objectives of this chapter, we will devote some time to discussion of other conditions and diseases which could, or already do, benefit from therapy with some types of purine enzyme inhibitors. Not all of these conditions require immunosuppression. However, each one is included in this chapter because the pathogenesis is intimately linked to purine metabolism, the immune system, or both. These conditions include disorders known broadly as 'hyperuricemic states', which are covered in Section 8.3.4, and miscellaneous conditions, which are covered in Section 8.3.5. Hyperuricemic states are those conditions in which there is either excessive production of purines, excessive nucleotide breakdown, or inadequate salvage of hypoxanthine or guanine, leading to elevated levels of uric acid in the circulation and urine. Some of this material will overlap with Chapters 7.2 and 7.3.

Section 8.3.5 on miscellaneous conditions will deal with one situation clearly requiring immuno-suppression, that being organ and bone marrow transplantation, and with two diseases (psoriasis and parasitic diseases) which might benefit from purine inhibitor therapy. The rationale for the treatment of psoriasis is somewhat convoluted, and is linked to the marked efficacy of cyclosporin, a

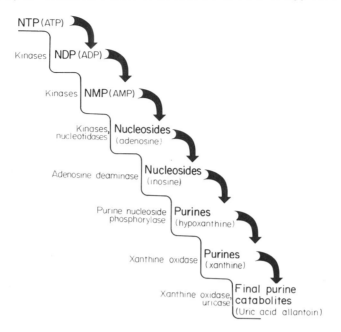

Figure 1 Cascades of purine metabolism. A diagrammatic representation depicting the flow of purine nucleotides and nucleosides down a purine metabolic cascade. These pathways are shown in greater detail in Figure 2

T cell modulator, in that disease, and also to recent findings strengthening the role of the immune system in psoriasis. The rationale of this approach for the treatment of parasitic diseases is clear and based solely on biochemical considerations.

8.3.1.2 Cascades of Purine Metabolism

The overall system of purine (and pyrimidine) metabolism involves an enormously complex and intricate series of metabolic events that has been pieced together over several decades. Coverage of this system is clearly beyond the scope of any single chapter, and hence we will not attempt it here. Instead, reference will be made to current reviews in the sections discussed, as well as to related topics within this work (Chapters 7.2–7.4). The reader should have an awareness that the enzymes of purine metabolism are involved in the synthesis of both RNA and DNA, in the production of regulatory nucleotides (*e.g.* cAMP, cGMP), in energy transfer reactions involving energy rich nucleotides (*e.g.* ATP), in kinase reactions, nucleotide reductions, purine salvages, purine elimination, regulation of methylation reactions, polyamine synthesis and so on. Hence, what we provide here is only a segment of a large and complex metabolic scheme, and the reader is accordingly urged to consider this material in its broader context.

The metabolism of purines can be said to start at inosine monophosphate (IMP), the first fully formed or *de novo* purine (Figure 2). IMP can be metabolized by IMP dehydrogenase and guanylate synthetase to GMP, by adenylosuccinate synthetase and adenylosuccinate lyase to AMP, and to inosine *via* 5′-nucleotidase. Even at this juncture it is immediately apparent that not all of the enzymes of purine metabolism fall under the umbrella of the overall heading of this part (Part 8) because they are neither hydrolases nor peptidases. Therefore, while we will try to give appropriate attention to those enzymes which are peptidases and hydrolases [*i.e.* adenosine deaminase (ADA; EC 3.5.4.4), 5′-nucleotidase (ENT; EC 3.1.3.5), and *S*-adenosyl-L-homocysteine hydrolase (SAHH; EC 3.3.1.1)], we will also discuss other enzymes which do not belong to those categories in order to unify the concept of a purine cascade. Some enzymes will be highlighted simply because they represent more interesting targets from a medicinal chemistry perspective. In particular, considerable time is devoted to purine nucleoside phosphorylase (PNP; EC 2.4.2.1) because of the key role of this enzyme in immune function, because of recent insights into the structure of PNP, and because of recent advances in the development of inhibitors of PNP.

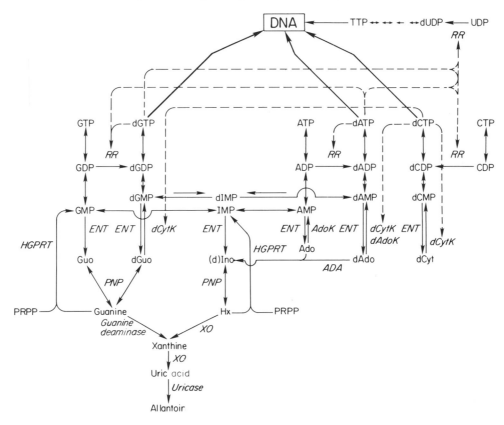

Figure 2 Pathways of purine nucleotide and nucleoside metabolism. Although complicated, this figure is not intended to be complete. Metabolic pathways are shown in solid lines, pathways mediating inhibition are depicted by dotted lines and enzymes by italic script. *De novo* purine biosynthesis starts with inosine monophosphate (IMP), the first fully formed purine nucleotide. IMP is readily converted to adenosine monophosphate (AMP) by the sequential action of adenylosuccinate synthetase and adenylosuccinate lyase, and in the reverse direction to guanosine monophosphate (GMP) by IMP dehydrogenase and guanylate synthetase. IMP can also be reformed from AMP by adenylate deaminase, and from GMP by GMP reductase. The dIMP is readily converted to dGMP and dAMP, but the reverse reactions proceed much more slowly. Breakdown of ribo- and deoxyribo-nucleoside (d) monophosphates (*e.g.* dGMP, dAMP, *etc.*) is catalyzed by nucleotidases, including ecto-5′-nucleotidase (ENT), giving rise to guanosine (Guo), adenosine (Ado), inosine (Ino), and their deoxy forms (*e.g.* dGuo). Guo, dGuo, Ino and dIno are phosphorolyzed by purine nucleoside phosphorylase (PNP) to guanine, hypoxanthine (Hx), and the corresponding deoxyribose 1-phosphates. Guanine deaminase and xanthine oxidase (XO) complete purine nucleotide degradation in higher primates and humans, culminating in formation of uric acid. In all other mammalian species tested, uricase oxidizes uric acid one step further to allantoin. Purine salvage is mediated by hypoxanthine–guanine phosphoribosyltransferase (HGPRT), which utilizes phosphoribosyl pyrophosphate (PRPP) and generates GMP and IMP. Adenine can also be salvaged by adenine phosphoribosyltransferase (not shown). Ado and dAdo are irreversibly deaminated to Ino and dIno, respectively, by adenosine deaminase (ADA). Phosphorylation of deoxyribonucleosides, *e.g.* by adenosine kinase (Adok) and deoxycytidine kinase (dCytk), generates dAMP, dGMP and dCMP. Further phosphorylation eventually generates the corresponding triphosphates, *e.g.* dGTP and dATP, which are potent inhibitors of ribonucleotide reductase (RR). RR catalyzes the reduction of ribonucleoside diphosphates to their corresponding deoxy forms needed for DNA synthesis. In particular, inhibition of RR depletes thymidine triphosphate (TTP) and deoxycytidine triphosphate (dCTP). Addition of deoxycytidine (dCyt), an excellent substrate for dCytK, generates dCTP, a feedback inhibitor of dCytK. Transmethylation reactions, which could be integrated with this scheme at Ado and dAdo, are shown in Figure 4

8.3.2 PURINE METABOLIC ENZYMES AND IMMUNODEFICIENCIES

Two immunodeficiency diseases which clearly result from the absence of a key enzyme in purine metabolism, namely ADA deficiency and PNP deficiency, are emphasized. The objective of doing this is to examine in some detail the effects produced on the immune system as a consequence of these deficiencies, and to develop an appreciation of the effects an inhibitor of each enzyme might therefore be expected to have on immune function. PNP, as noted, is included because of its novelty as a target for selective immunosuppressive therapy.

8.3.2.1 Adenosine Deaminase (ADA)

Adenosine deaminase (EC 3.5.4.4) is a hydrolase acting on nonpeptide C—N bonds. It irreversibly catalyzes the hydrolytic deamination of adenosine (Ado) and deoxyadenosine (dAdo) to inosine (Ino) and deoxyinosine (dIno), respectively.[5] Purified human erythrocyte ADA is a soluble monomeric protein with a molecular weight of about 38 000 Da.[6,7] The amino acid sequence of purified human erythrocyte ADA has been reported,[8] as has been the amino acid sequence of human T cell ADA.[9] The biochemistry associated with ADA is discussed under Section 8.3.2.1.2.

8.3.2.1.1 ADA deficiency

In 1972, Giblett *et al.* described two unrelated patients with severely impaired cellular immune function and abnormal immunoglobulin synthesis, who possessed no detectable ADA activity in their erythrocytes.[10] The occurrence of patients with the near absence of cell-mediated immunity was extremely rare, as was the absence of ADA in red blood cells, and the coexistence of the two findings suggested cause and effect. The result was a landmark observation, establishing the requirement for one purine-metabolizing enzyme for normal immune function.

The diseases resulting from ADA deficiency belong to a class known as combined immune deficiency diseases.[11,12] They are heterogeneous in terms of the extent of immune impairment seen both clinically and in *in vitro* studies. They affect both thymus-derived lymphocyte ('T' cell) function, known as cellular immunity or delayed-type hypersensitivity, and humoral immunity, which is responsible for production of antibodies and is directly a function of 'B' (for bone-marrow-derived) lymphocytes. ADA deficiency has a frequency below 1 in 100 000 live births, and ADA-negative combined immune deficiency accounts for one-quarter to one-third of all autosomal forms of the combined disease.[11-13] One form of ADA-negative deficiency, called severe combined immuno-deficiency (SCID), results not only in strikingly reduced T cell numbers and T cell function, but generally in similar defects in humoral immunity. ADA-positive combined immunodeficiency diseases, like the ADA-negative forms, are rare, but their cause(s) and etiology(ies) are completely unknown.

It is very important to note that combined immunodeficiency in ADA-deficient patients appears to result only when there is a near total lack of this enzyme. Although it has not been shown exactly to what extent there must be deficiency in ADA in order for the severe immune deficiency to be manifested, most patients have less than 1.5% of normal enzyme activity.[14] The impact of this finding is profound, and suggests that one may have to achieve a similar degree of enzyme inhibition in order to produce a pharmacological effect. Conversely, this finding suggests that one may also be able to inhibit ADA substantially before producing deleterious side effects. However, the extent of immune impairment resulting from the congenital absence of ADA may depend on the extreme importance of ADA during neonatal life, when the immune system is developing. The immuno-suppression achieved by administering an ADA inhibitor to a subject with a fully developed immune system may be much less pronounced, and furthermore should be reversible, or at least controllable.

Patients afflicted with combined immune deficiency disease appear normal at birth, but soon present with vomiting, diarrhea, and severe recurrent respiratory infections. Several viral and fungal infections have been described in these patients; candidiasis is particularly common. Trans-placentally derived immunoglobulins and mother's milk may help protect them in the earliest months. Thus, plasma immunoglobulin levels can be normal early in the disease course, but they eventually decline. The ability to respond to antigen challenge, however, is greatly impaired, regardless of plasma immunoglobulin levels, as shown following immunization with standard antigens like tetanus, pertussis and diphtheria. The degree of impairment in humoral immunity is quite variable. Lymphopenia is almost always observed and is generally pronounced, especially later in the disease course. *In vitro* lymphocyte transformation to the T cell mitogens phytohemagglutinin (PHA) and concanavalin A (Con A) is generally greatly reduced. The overall picture is clearly one of severe immune deficiency, and most patients die from infection early in life.

In the following section, we will discuss biochemical mechanisms which might account for the combined immune deficiency diseases. Although numerous hypotheses have been advanced, no satisfactory explanation can be given to account for the diversity in the syndromes, nor to explain why B cell function may range from a state of being totally absent to one of being near normal. As noted, similar dysfunction has been observed in ADA-positive individuals. However, in ADA deficiency it has been shown in several patients that red cell transfusions (red cells contain large quantities of ADA), or even partial bone marrow engraftment, can restore, at least temporarily, some aspects of immune function in these patients.[14,15] Improvement in immune function furthermore correlated with changes in biochemical features predicted from ADA replacement therapy.

The diversity of effects produced on the immune system, and the numerous associated biochemical abnormalities, must be understood fully before endeavoring to design therapeutic agents which would work by inhibition of ADA. Such inhibitors, and their *in vitro*, *in vivo* and clinical effects, are discussed in Sections 8.3.2.1.3–8.3.2.1.6. Before proceeding to these sections, however, the biochemical alterations associated with ADA deficiency will be described, as these can explain the immune impairment and may also be used to monitor the effects of ADA inhibitors in humans.

8.3.2.1.2 *Biochemistry of ADA and ADA deficiency*

The biochemistry associated with ADA and its substrates has been described extensively.[5,10,12-21] The pathogenesis of ADA deficiency in theory could result from the absence of a product of an ADA dependent reaction or from the accumulation of a substrate (or metabolite of a substrate) of this reaction. Of the products, Ino is readily formed from IMP, and dIno is apparently unimportant. Reduced formation of either *via* the ADA pathway is unlikely to be of any consequence.

Ado is a superior substrate for adenosine kinase, compared to ADA, with a K_m of about 2.0 µM in the direction of adenosine monophosphate (AMP) formation (Figure 2).[13-16,18] Hence, metabolism of Ado by ADA is not the preferred pathway. However, the actual metabolism of Ado in lymphocytes depends on the Ado concentration and the state of activation of the lymphocyte.[22] Resting lymphocytes express little adenosine kinase and utilize ADA over a broad range of Ado concentrations, while mitogen-stimulated lymphocytes metabolize Ado *via* adenosine kinase at low Ado concentration, and by ADA at Ado concentrations above 5.0 µM.[22] When formed, AMP can be deaminated by adenylate deaminase, giving rise to IMP and bypassing any block produced by the absence of ADA (Figure 2). Excretion of Ado in urine is not elevated in ADA deficiency, presumably owing to this mechanism. Ado is also a good substrate for SAHH, with a K_m of 1.0 µM in the direction of *S*-adenosylhomocysteine (AdoHcy) synthesis.[14,15,17] The effects of Ado and dAdo on methylation reactions potentially represent a major consequence of ADA deficiency, and are noted in this and in a separate section (Section 8.3.2.2) in which SAHH inhibitors are described. Despite the effects of Ado, most of the consequences of ADA deficiency are felt to depend on dAdo accumulation.

dAdo is a poor substrate for deoxycytidine kinase, with the K_m in the direction of dAMP formation being about 400 µM (Figure 2).[14,15] Rather, the direction of metabolism of dAdo proceeds by deamination *via* ADA. Absence of ADA should lead to accumulation of dAdo, and, in fact, urinary and occasionally plasma dAdo levels have been reported to be elevated in these patients. Ado is not excreted for the reasons noted above. As other routes of metabolism of dAdo proceed primarily by phosphorylation, the dAMP, when formed, is phosphorylated to dADP and dATP. Also, dAMP is a poor substrate for adenylate deaminase. Thus, patients lacking ADA have been found to have elevated levels of dATP in their erythrocytes and lymphocytes.[15,23] Therefore, blockade of dAdo metabolism is felt to be the major biochemical abnormality in *ADA* deficiency.

The mechanism whereby dAdo is toxic to lymphocytes is felt in large part to be dependent on its phosphorylation to dATP. Importantly, dAdo-phosphorylating activity is found primarily in lymphocytes, as opposed to Ado-phosphorylating activity, which is associated with many tissues and types of cells.[24] That dATP is most likely the toxic metabolite in ADA deficiency is supported by clinical studies. As noted, red cell transfusions were found to reduce dATP levels, and this was associated with concomitant improvement in immune function.[14,15] It is well-established that dATP is a potent allosteric inhibitor of the enzyme ribonucleoside diphosphate reductase (ribonucleotide reductase; EC 1.17.4.1).[25,26] Ribonucleotide reductase reduces all of the purine and pyrimidine ribonucleoside diphosphates to their corresponding deoxyribonucleotides, which are needed for DNA synthesis. dATP can block the reduction of all the nucleotides needed for DNA synthesis, including the conversion of ADP to dADP. As is noted in the section on PNP (Section 8.3.2.4), the triphosphate (dGTP) of deoxyguanosine (dGuo) also inhibits ribonucleotide reductase but does not block the reduction of ADP.

Accumulation of AdoHcy is favored in the absence of ADA, a process which has important consequences because of the ability of AdoHcy to inhibit transmethylation reactions (see Section 8.3.2.2).[14,15,17] Some ADA deficient patients also have reduced SAHH activity, possibly owing to elevated dAdo, which is a potent suicide inhibitor of this enzyme.[14,15,17] These mechanisms may contribute to the B cell defects in ADA deficiency. Other mechanisms have been offered to account for the immune impairment, and include blockade of pyrimidine biosynthesis,[27] elevation of cAMP,[28] ATP depletion,[29,30] depletion of nicotinamide adenine dinucleotide (NAD) *via* activation of poly(ADP-ribose) synthetase,[31] inhibition of methionine synthesis,[32] and inhibition of RNA synthesis.[33,34] More evidence exists to support some of these mechanisms than others. For instance, Ado was shown several years ago to inhibit growth of murine fibroblasts and other types of cells at low concentration. The toxicity produced by Ado, even at high concentration, could be prevented by coaddition of millimolar uridine, an observation which suggested that toxicity was mediated by pyrimidine starvation. Mutants deficient in HGPRT (see Section 8.3.2.5) remained sensitive to Ado, suggesting that toxicity proceeded through a pathway probably involving phosphorylation of Ado to AMP. However, subsequent work showed that adenosine kinase deficient mutants were equally susceptible to Ado toxicity, indicating that other metabolic pathways (*e.g.* transmethylation, PRPP synthesis) were affected by Ado accumulation.[14,27]

8.3.2.1.3 ADA inhibitors

ADA inhibitors are of interest for cancer chemotherapy, in addition to their potential as immunosuppressive agents for autoimmune diseases. They should also enhance the activity of adenine nucleoside analogs, which are rendered ineffective by deamination, for the treatment of viral diseases and cancer.[35] These possibilities have triggered the search for potent ADA inhibitors.[36,37]

A number of ADA inhibitors with K_i values ranging from 10^{-6} M to 10^{-12} M have been either synthesized or isolated from microbial broths.[36] The following inhibitors represent diverse chemical structures and have been studied extensively (Figure 3). These inhibitors provide tools to study pharmacological models of ADA deficiency in animals, and have helped in understanding the biochemical mechanisms underlying the associated immunodeficiency. In Sections 8.3.2.1.4–8.3.2.1.6, the *in vitro*, *in vivo* and clinical effects of a few of these compounds are described.

(i) erythro-9-(2-Hydroxy-3-nonyl)adenine (EHNA)

EHNA is a rationally designed, reversible inhibitor of ADA with a K_i value of 1.6×10^{-9} M.[38] EHNA is a racemic mixture of D-EHNA [*erythro*-(−)-9-(2R-hydroxy-3S-nonyl)adenine and *erythro*-(+)-9-(2S-hydroxy-3R-nonyl)adenine]. The two *erythro* (EHNA) and two *threo* isomers (THNA) have now been synthesized.[39,40] The (+)-(2S,3R)-EHNA isomer was found to be 250-fold more potent as an inhibitor of human erythrocytic ADA ($K_i = 2.0$ nM) than was the (−)-(2R,3S)-EHNA isomer ($K_i = 500$ nM). Both of the *threo* isomers [(+)-(2R,3R)-THNA; $K_i = 122$ nM and (−)-(2S,3S)-THNA; $K_i = 80$ nM] were less potent as ADA inhibitors than (+)-(2S,3R)-EHNA, but were more potent than (−)-(2R,3S)-EHNA.[41] EHNA significantly inhibits replication of herpes simplex virus (HSV), whereas the more potent 2′-deoxycoformycin (pentostatin) does not.[42] The plasma half-life of 9-β-D-arabinofuranosyladenine (Ara-A) in mice was increased from 15–20 min to 39 min when given along with EHNA (2.0 mg kg^{-1}, i.p.), demonstrating *in vivo* inhibition of ADA and its potential use in combination chemotherapy.[42]

(ii) Coformycin

Coformycin is an antibiotic isolated from the culture filtrates of *Norcardia interforma*, *Streptomyces kaniharaensis SF-557*, and other microorganisms.[43] It is a unique and potent ADA inhibitor containing a homopurine (8-(R)-3,6,7,8-tetrahydroimidazo[4,5-d][1,3]diazepin-8-ol) aglycon pharmacophore which can be viewed as a transition state analog between Ado and Ino. Thus, it is a very tight-binding inhibitor and has a K_i value of 1.0×10^{-11} M.[44,45] Although coformycin was discovered in 1967, this compound and other ADA inhibitors were not made available for evaluation until 1974 when EHNA and 2′-deoxycoformycin were first tested. Since then, several studies have been published (summarized in Sections 8.3.2.1.4–8.3.2.1.6).

(iii) 2′-Deoxycoformycin (pentostatin)

2′-Deoxycoformycin contains the same homopurine aglycone part of coformycin along with a deoxyribose sugar. Thus, it may be considered to be a transition state analog between dAdo and dIno. It is essentially an irreversible inhibitor and is the most potent ADA inhibitor known, with a

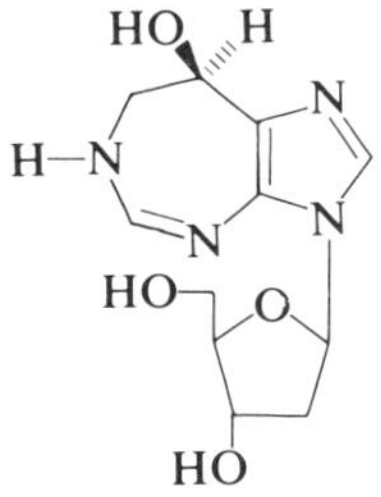

Pentostatin (2'-deoxycoformycin; DCF)

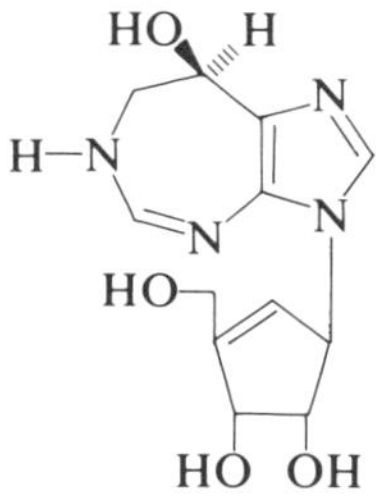

Adecypenol

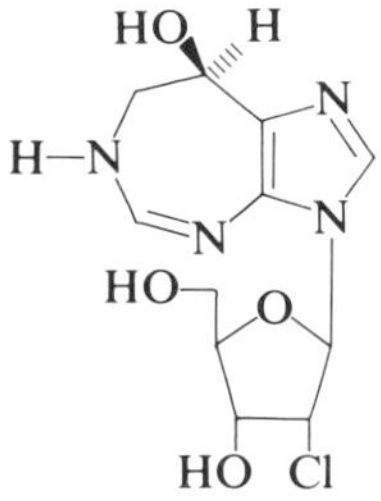

(R)-2'-Chloropentostatin (adechlorin)

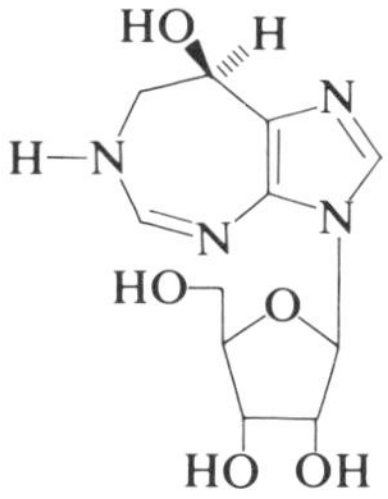

Coformycin (CF)

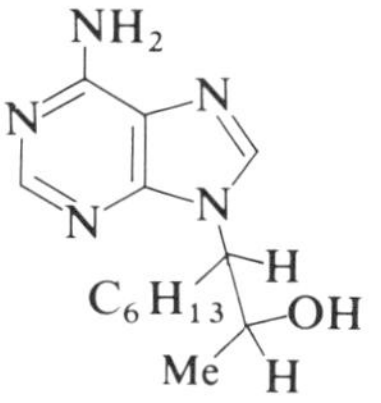

erythro-9-(2-Hydroxy-3-nonyl)adenine (EHNA) 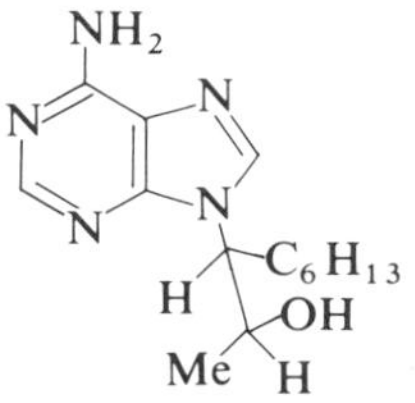   *threo*-9-(2-Hydroxy-3-nonyl)adenine (THNA)

Figure 3 Inhibitors of adenosine deaminase (ADA)

K_i of 2.5×10^{-12} M.[46,47] 2'-Deoxycoformycin has no significant activity against the mouse leukemia L-1210 *in vivo*.[48] It is the most extensively studied ADA inhibitor, both in preclinical and clinical studies, and its activity is discussed in subsequent sections.

(iv) 2'-Chloropentostatin (adechlorin)

This compound is a nucleoside antibiotic produced by *Actinomadura sp. ATCC 39365* and by *Actinomadura sp. OMR-37*.[49] 2'-Chloropentostatin belongs to the same class as coformycin and 2'-deoxycoformycin and is also a tight-binding ADA inhibitor with a K_i of 1.1×10^{-10}–5.3×10^{-10} M.[48,50,51] It is a slightly weaker inhibitor of rat and human ADA than is coformycin, and is considerably weaker than 2'-deoxycoformycin. 2'-Chloropentostatin has no significant antimicrobial activity or antitumor activity against mouse leukemia L1210 *in vivo*, but enhanced the antiviral activity of Ara-A *in vitro* against HSV-1. It showed no significant acute toxicity in mice when given at a dose of 100 mg kg^{-1}, i.v.

(v) Adecypenol

Adecypenol is an antibiotic ADA inhibitor isolated from *Streptomyces sp. OM-3223* broth.[52] It contains the homopurine aglycone moiety of coformycin, 2'-deoxycoformycin and 2'-chloro-pentostatin, and 5-(3-hydroxymethyl-3-cyclopentene-1,2-diol) as the ribose replacement found in Neplanocin-A, a SAHH inhibitor. It is a potent, tight-binding ADA inhibitor with a K_i value of 4.7×10^{-9} M.[53] It has no antimicrobial or antifungal activity, and is nontoxic to mice at a dose of 100 mg kg^{-1} i.v.

8.3.2.1.4 In vitro *immunological and biochemical studies employing ADA inhibitors*

The objective of this section is to review briefly experimentation in which ADA inhibitors have been used to elaborate biochemical mechanisms, and to evaluate the effects of ADA substrates on aspects of lymphocyte or lymphoblast function. The reader is also referred to several reviews on the subject.[5,13-16]

(i) Adenosine

Ado is known to inhibit DNA synthesis, in the absence of ADA inhibition, in PHA-stimulated human lymphocytes with an IC_{50} of about 100 μM.[54] The finding that Ado is a weak inhibitor of [³H]thymidine uptake has been confirmed in similar systems.[55] Concentrations of EHNA (0.3 μM) that produce 90% inhibition of ADA are ineffective at blocking thymidine uptake, but substantially higher EHNA concentrations (100 μM) do have this effect, suggesting that endogenous Ado reserves are low in cultured lymphocytes.[55] Furthermore, noninhibitory EHNA concentrations potentiated the Ado-mediated inhibition of [³H]thymidine uptake.[55,56] Potentiation of Ado-mediated inhibition of T cell blast transformation was also demonstrated using coformycin.[22] These studies dissected out the importance of both Ado concentration and the state of lymphocyte activation in determining route of metabolism. Ado is preferentially metabolized by ADA in resting cells and, at high concentration, by ADA in PHA-stimulated cells. However, at low concentrations Ado is phosphorylated rather than deaminated in PHA-stimulated lymphocytes. The understanding of the effects of Ado on protein synthesis has also been enhanced by the use of ADA inhibitors.[22,55] Ado inhibits [³H]leucine uptake by mitogen-stimulated human lymphocytes when added early in culture, suggesting a mechanism independent of ribonucleotide reductase inhibition. This inhibitory effect of Ado is not enhanced by 2'-deoxycoformycin nor inhibited by dCyt.[57] Mechanisms of inhibition involving cAMP have also been investigated using ADA inhibitors.[55] In the presence of 2'-deoxycoformycin, high concentrations of Ado and homocysteine thiolactone were cytotoxic to dividing human lymphocytes.[58] Considerable work has also been done using murine systems.[13-16] Murine T lymphocyte cytotoxicity is blocked by the addition of Ado to the culture medium, and the inhibition is enhanced by coaddition of ADA inhibitors.[28] Hence, ADA inhibitors have proved highly useful in defining the effects of Ado on lymphocyte functions.

(ii) Deoxyadenosine

The ADA substrate dAdo is significantly more inhibitory to human lymphocytes than is Ado. Simmonds *et al.* reported that PHA, Con A and pokeweed mitogen responses of human lymphocytes were abrogated with much lower dAdo than Ado concentrations, and that inhibition was potentiated by EHNA treatment.[56] Potentiation of dAdo toxicity was also shown by Gelfand *et al.*, who furthermore demonstrated that inhibition could be prevented by coaddition of deoxycytidine (dCyt).[59] dAdo was most inhibitory of protein synthesis ([³H]leucine uptake) in PHA-stimulated human lymphocytes when added early to culture, and this effect was enhanced by 2'-deoxycoformycin and prevented by dCyt and thymidine.[57] Inhibition of the PHA response of dAdo-treated, ADA-inhibited human lymphocytes correlated well with dATP accumulation.[60] In these studies, inhibition was reversed in a concentration dependent manner by coaddition of dGuo, dCyt and thymidine. The importance of dATP accumulation is suggested from other studies which explored the effects on human lymphoblastoid cell lines of the inactivation of SAHH by dAdo. In these studies, T lymphoblasts deficient in deoxyadenosine kinase activity were spared the toxic effects of dAdo despite inhibition of SAHH by dAdo.[13] Effects of dAdo and 2'-deoxycoformycin on suppressor T cell function and B cell differentiation have also been demonstrated *in vitro*.[61]

Similar observations on the effects of dAdo have been made using human lymphoblastoid cell lines.[62-64] MOLT-4 and CEM T lymphoblasts are highly susceptible to the inhibitory effects of dAdo, and inhibition is potentiated by ADA inhibitors. Also, cytotoxicity is associated with accumulation of dATP and the concomitant reduction in dCTP. Each of the activities of cytotoxicity, dATP accumulation and reduction in dCTP could be reversed by adding dCyt to the culture medium. In addition to the work already cited, reference is made to a review on the toxicity of natural deoxyribonucleosides.[65]

(iii) Resting lymphocytes

The toxic effects of Ado and dAdo on resting lymphocytes might contribute to the clinical manifestations of severe combined immunodeficiency diseases due to ADA deficiency. Carson *et al.*, using mature, resting human lymphocytes treated with 2'-deoxycoformycin, showed that dAdo was toxic to human helper and suppressor–cytotoxic T lymphocytes, but not to B lymphocytes.[58] Toxicity was associated with phosphorylation of dAdo and formation of dATP. Cell death was preceded by depletion of ATP, but not by inhibition of SAHH.[58] Kefford *et al.* also studied nonstimulated human lymphocytes and observed accumulation of dATP both in T lymphocytes and non-T lymphocytes.[66,67] The mechanism for the G_0 phase arrest appears to be distinct from ribonucleotide reductase inhibition.

(iv) B cell dATP

While most of the above work suggested that dAdo was more toxic to T cells than was Ado, and that toxicity was associated in many studies with formation of dATP, recent work provides evidence showing a direct effect on B lymphocytes. Goday *et al.* reported that malignant B cell lines accumulated dATP even in the absence of ADA inhibition.[68] However, dAdo was cytotoxic to B cells only upon coaddition of 2'-deoxycoformycin, suggesting that accumulation of dATP alone was inadequate to cause cell death.[68] These observations have been extended to normal human B cells and other cell types.[69] It has also been reported that normal human peripheral blood T cells accumulated little dATP when treated with dAdo (10 or 60 μM) plus 2'-deoxycoformycin, whereas B cells accumulated large amounts under the same conditions.[70] Thus, while considerable evidence has accumulated implicating dATP inhibition of ribonucleotide reductase, dAdo blockade of methylation, and ATP depletion as mechanisms of immunotoxicity, it is possible that other mechanisms may be operative. In the following section, data showing the immunosuppressive potential of ADA inhibitors *in vivo* are presented.

8.3.2.1.5 Effects of ADA inhibitors in animals

Most *in vivo* studies have employed either 2'-deoxycoformycin or EHNA. Neither compound is absorbed orally, and both have been given intraperitoneally (i.p.), intravenously (i.v.), or intramuscularly (i.m.). A single i.m. injection of 2'-deoxycoformycin produced marked inhibition of ADA in all tissues examined (spleen, kidney, heart and liver) when given to rats at doses as low as 0.1 mg kg^{-1}.[71] Moreover, the inhibition lasted in some cases up to 15 d.

Many studies have been done using mice. Mice given a single i.p. dose of 2'-deoxycoformycin at 1.0 mg kg^{-1} showed similar reduction in ADA activity in their tissues 24 h after dosing.[72] Interestingly, lymphocytes (splenocytes and thymocytes) from mice dosed in this manner for 4–5 d showed no impairment in their ability to respond to Con A or allogeneic cells. The lack of effect of 2'-deoxycoformycin *ex vivo* may relate to a species effect, to incomplete inhibition of the enzyme, and/or to the duration of dosing and hence the turnover and maturation of lymphocytes. However, it has been shown that continuous i.p. infusion of 2'-deoxycoformycin for 5 d produced: (i) marked or total inhibition of ADA in all tissues examined; (ii) equally profound reduction in the response of lymphocytes from dosed mice to Con A, PHA and *E. coli* lipopolysaccharide; (iii) reduced delayed-type hypersensitivity responses; and (iv) prolonged survival of skin allografts.[73,74] Similar effects were also reported using a daily single i.p. bolus-dosing protocol.[75] Carson *et al.* demonstrated that dAdo-phosphorylating activity was widespread in the mouse, as opposed to humans, in which this activity is localized primarily in the thymus.[76] This observation is particularly interesting in view of the finding of Tedde *et al.* that 2'-deoxycoformycin-dosed mice appeared normal histologically except for the lymphoid tissues.[74] Analogous to the findings on ADA-deficient patients, administration of 2'-deoxycoformycin to mice resulted in elevation of dATP in mouse erythrocytes.[77] The accumulation of dATP was dependent both on dose and time following dosing.

2'-Deoxycoformycin has been reported to be immunosuppressive in a rat model of arthritis produced by immunization with *Mycobacterium*,[78] but not following immunization with type II collagen.[79] Continuous i.p. infusion of 2'-deoxycoformycin was also reported to produce complete suppression of skin allograft rejection in rats.[80] Administration of 2'-deoxycoformycin to rats blocked both the proliferation and the differentiation of thymocytes in the subcapsular region of the thymus.[81] As these thymocytes are precursors of cortical and medullary thymocytes, it was suggested that this may be the anatomical site of action resulting in immune deficiency in ADA deficient patients. 2'-Deoxycoformycin has been reported to prolong skin graft survival in dogs.[82]

8.3.2.1.6 Clinical studies on ADA inhibitors

The only ADA inhibitor to be evaluated in the clinic is 2'-deoxycoformycin. Because of the marked antiproliferative activity of 2'-deoxycoformycin for lymphoblasts, and its ability to produce profound lymphopenia in animals, the drug was regarded as having potential for treatment of certain types of cancer. Phase I and II clinical trials of 2'-deoxycoformycin in T cell malignancies and solid tumors were reported in 1980.[83,84] 2'-Deoxycoformycin was shown to induce complete remission in two patients with relapsed acute T-lymphoblastic leukemia who were unresponsive to other chemotherapeutic agents.[85] This compound was later tested on 27 children with some positive results observed, but not without significant toxicity.[86] 2'-Deoxycoformycin was used to treat 17 patients with a variety of lymphoid neoplasms, both T and B cell in nature, that were refractory to other forms of therapy. Several partial responses and two complete remissions were reported. [87] Mitchell *et al.* also treated a patient with refractory T cell leukemia in blast crisis.[88] Although the patient died three days after discontinuation of therapy, he was found on postmortem examination to be totally free of leukemic cells in all organs. Biochemical effects produced by 2'-deoxycoformycin therapy were followed in that patient, and were found to parallel closely those observed in ADA-deficient patients: (i) dAdo was elevated in plasma, while Ado was not; (ii) dATP accumulated in the leukemic cells; and (iii) lymphoblast SAHH was markedly inhibited. Plasma dAdo and lymphoblast dATP increased in parallel, and the increase was inversely correlated with lymphoblast count and SAHH activity. [88] These observations were confirmed on a larger number of patients.[89] Major *et al.* studied the effects of 2'-deoxycoformycin in 13 patients with a variety of disorders.[90] As in the studies of Mitchell *et al.*,[88,89] dAdo was elevated in plasma and urine. Major *et al.* also found Ado to be elevated. Partial remissions were observed in one patient with T cell acute lymphocytic leukemia and in one patient with mycosis fungoides. In a larger group of acute leukemics, it was reported that seven with T cell leukemia went into complete remission.[91,92] Elevated dATP was observed in the lymphoblasts of some patients.

Efficacy of 2'-deoxycoformycin in other types of cancer has been reported, but coverage of this material is beyond the scope of this review. However, most recently 2'-deoxycoformycin has been found to be highly effective in the treatment of hairy-cell leukemia. In one study of 27 patients with hairy-cell leukemia, who met all the study's entry criteria, the response rate to 2'-deoxycoformycin was 96% and the remission rate was 59%.[93,94] 2'-Deoxycoformycin (pentostatin) is presently under review by the FDA for its potential use in the treatment of this disease.

2'-Deoxycoformycin produces a broad spectrum of toxicities, which probably account for its not having been evaluated for efficacy in other types of diseases (*e.g.* rheumatoid arthritis). These adverse reactions include renal and liver dysfunction, conjunctivitis, hemolysis, toxicity to the central nervous system, bronchitis, arthralgias and myalgias.[88-90,92] Expectedly, another side effect of 2'-deoxycoformycin therapy is severe infection. An analysis of the first 300 patients treated with 2'-deoxycoformycin for lymphoid malignancies showed a high incidence of severe infections (8%) during a relatively brief period of treatment. Two-thirds of the infections were fatal.[95] Some effects (hypotension, nausea, sedation) possibly result from Ado accumulation and Ado–adenosine receptor interaction.[96,97] Studies designed to demonstrate the immunosuppressive potential of ADA inhibitors in nonmalignant/nonleukemic states have not been done. However, based on the vast knowledge of the biochemistry associated with ADA, and because of the primary selectivity of the effect of ADA deficiency for the immune system, ADA inhibitors have the potential to produce broad-based immunosuppression in humans. Whether or not this immunosuppression can be regulated, pulsed as in cancer chemotherapy, or targeted to achieve a more selective effect remains to be determined.

8.3.2.2 *S*-Adenosyl-L-homocysteine Hydrolase (SAHH)

8.3.2.2.1 *Biochemistry*

S-Adenosyl-L-homocysteine hydrolase (EC 3.3.1.1) is a hydrolase in the purine cascade that catalyzes the reversible cleavage of AdoHcy to Ado and L-homocysteine (Hcy; Figure 4). SAHH has been purified to homogeneity from various sources and characterized. It has a molecular weight of 180 000–190 000 Da and consists of four identical subunits of 45 000–48 000 Da. The metabolism of AdoHcy is of importance because AdoHcy is a potent feedback inhibitor of *S*-adenosylmethionine (SAM) dependent transmethylation reactions.[17,98] When SAHH is inhibited, AdoHcy accumulates and consequently methylation reactions utilizing SAM as a methyl donor are inhibited. At high substrate concentrations (> 1.0 μM), the SAHH reaction favors synthesis of AdoHcy,[99] but physiologically the reaction proceeds in the hydrolytic direction because the Ado and Hcy are effectively removed by further metabolism. Ado is either metabolized to Ino by ADA or phosphorylated to AMP by adenosine kinase (Figure 4). The Hcy produced can be condensed with serine to form β-cystathionine, or methylated to methionine and recycled to SAM. In cellular systems SAM serves as a methyl donor for the enzymatic methylation of various small molecules, proteins, phospholipids and nucleic acids.[100] SAM-mediated transmethylation reactions are involved in the differentiation, secretion and chemotaxis of neutrophils and in lymphocyte killing. Transmethylation has also been recognized as being essential to many general cell activities, including the processing and functional aspects of DNA, rRNA and tRNA. Inhibition of methylation would lead to inhibition of cell functions. An inhibitor of SAHH could be a useful immunosuppressive agent, provided acceptable selectivity for target cells and tissues can be achieved. However, since these agents will most likely have broader effects than other types of agents (*e.g.* PNP inhibitors), SAHH inhibitors might be more useful as anticancer and antiviral agents where proliferation of cells can be checked despite major mechanistic side effects, rather than for autoimmune diseases.

8.3.2.2.2 SAHH inhibitors and findings

Since the discovery of SAHH and the realization of its critical role in transmethylation, a large number of SAHH inhibitors have been rationally designed and synthesized. Many of these have been evaluated in enzyme and other *in vitro* and *in vivo* systems. A few of the most potent SAHH inhibitors (Figure 5) are discussed here; reviews with extensive references have been published.[101,102]

(i) 3-Deazaadenosine (3-DAA)[103]

This is one of the most extensively studied SAHH inhibitors. It has K_i values of 4.0 and 1.0 μM in beef liver SAHH and lupin seed SAHH, respectively.[104] In addition to its ability to inhibit SAM dependent methylation reactions, 3-DAA markedly enhances the cAMP response of lymphocytes by inhibiting cAMP phosphodiesterase (as 3-DAA-homocysteine; 3-DAAHcy) and *via* activation of adenylate cyclase.[105] 3-DAA was reported to inhibit a variety of cell functions, including chemotaxis of neutrophils, monocytes and macrophages,[106] and lysis of tumor cells by sensitized lymphocytes, activated macrophages, and by human NK cells.[107] 3-DAA was shown to be immunosuppressive in mice[108] and to modulate antibody dependent cellular cytotoxicity and antibody dependent phagocytosis.[109] 3-DAA was also shown to be a potent, long-acting and orally active antiinflammatory agent. It was found to be effective in adjuvant-induced polyarthritis in rats ($ID_{50} = 8$ mg kg^{-1} d^{-1}, prophylactic dosing, and $ID_{50} = 18$ mg kg^{-1} d^{-1}, therapeutic dosing). It was also active in the acute carrageenan pleurisy assay ($ED_{50} = 2.8$ mg kg^{-1}).[110] 3-DAA has no antipyretic or analgesic activity, unlike known nonsteroidal antiinflammatory agents. 3-DAA has been shown to affect the growth of a number of viruses including Rous sarcoma, vesicular stomatitis, Sindbis and Newcastle disease virus.[111,112] However, 3-DAA is neither very potent nor very specific in its antiviral activity. 3-DAA

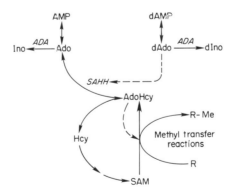

Figure 4 Partial scheme depicting metabolism of purines, related to transmethylation. In transmethylation reactions *S*-adenosylmethionine (SAM) donates a methyl group to proteins and nucleic acids, becoming *S*-adenosyl-L-homocysteine (AdoHcy) in the process. AdoHcy is a potent feedback inhibitor of SAM-mediated transmethylation reactions. AdoHcy is broken down to adenosine (Ado) and L-homocysteine (Hcy) in a reversible reaction catalyzed by *S*-adenosyl-L-homocysteine hydrolase (SAHH). Ado is phosphorylated by adenosine kinase or deaminated by adenosine deaminase (ADA). Hcy can be metabolized further by condensation with serine, giving rise to β-cystathionine (not shown), or methylated by methionine synthetase, forming methionine (not shown). SAM is reformed from Hcy *via* methionine by methionine adenosyltransferase. Deoxyadenosine (dAdo) is a potent suicide inhibitor of SAHH. Solid lines depict metabolic pathways; dotted lines depict inhibition and italic script, enzymes.

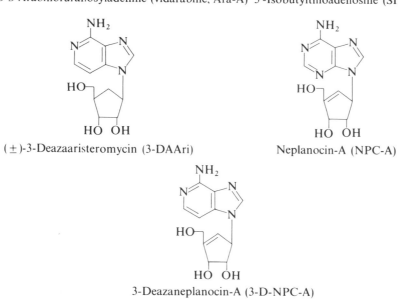

3-Deazaadenosine (3-DAA) Carbocyclic adenosine (aristeromycin; C-ADO)

9-β-D-Arabinofuranosyladenine (vidarabine; Ara-A) 5'-Isobutylthioadenosine (SIBA)

(±)-3-Deazaaristeromycin (3-DAAri) Neplanocin-A (NPC-A)

3-Deazaneplanocin-A (3-D-NPC-A)

Figure 5 Inhibitors of S-adenosyl-L-homocysteine hydrolase (SAHH)

was shown to be toxic for host cells at $40 \, \mu g \, mL^{-1}$, which is 2–20 times the minimum antiviral concentration.[113] More recently, evidence has been obtained showing that elevation of AdoHcy and 3-DAAHcy probably does not contribute to the biological activity of 3-DAA.[106]

(ii) Carbocyclic adenosine (aristeromycin)[114]

This compound is a competitive inhibitor and one of the most potent SAHH inhibitors known, having a K_i value of 5.0 nM both in beef liver SAHH and lupin seed SAHH.[104] Aristeromycin induces a marked elevation of AdoHcy in intact cells, and is devoid of antitumor activity when tested *in vivo* against the L-1210 leukemia (see also Section 7.3.4.1.4).

(iii) 9-β-D-arabinofuranosyladenine (vidarabine; Ara-A)

Ara-A was synthesized[115] before it was discovered as an antibiotic in the cultures of *Streptomyces antibioticus*.[116] The fermentation process is the adopted commercial procedure for preparation of this compound. Ara-A irreversibly inactivates SAHH, both in isolated enzyme preparations and in intact cells.[117] *In vivo*, a single injection of Ara-A ($50 \, mg \, kg^{-1}$) in mice rapidly inactivated the SAHH in liver, kidney, spleen, lung, heart, skeletal muscle and brain, demonstrating a striking effect in these tissues.[118] It potently inhibits bacterial growth.[119] The effective metabolite seems to be the 5'-phosphate (ara-ATP), which may be an inhibitor of mammalian DNA polymerase, although the mechanism of its action is not fully understood. Ara-A is also a potent antiviral agent, especially against DNA viruses.[119] It is presently indicated for the treatment of herpes simplex virus encephalitis, where it has proven to be a life-saving drug.

(iv) 5'-Isobutylthioadenosine

This compound was shown to be an irreversible inhibitor of rat liver and rabbit erythrocyte SAHH. It is also a potent methylation inhibitor and inhibits polyamine synthesis.[120]

(v) (±)-3-Deazaaristeromycin (3-DAAri)[121]

This compound is a carbocyclic analog of 3-deazaadenosine (3-DAA).[122] 3-DAAri is a potent, reversible, competitive inhibitor of SAHH with K_i values of 3.0 μM and 1.0 nM for beef and hamster liver SAHH, respectively, but is not a SAHH substrate. 3-DAAri was shown to be a very potent and effective antiviral agent, as well as being nontoxic to host cells.[113] It inhibits the replication of a variety of viruses, such as vaccinia, reo, measles, parainfluenza and vesicular stomatitis at concentrations of 0.2 to 1.0 μg mL^{-1}, while being nontoxic to host cells at concentrations up to 400 μg mL^{-1}. *In vivo*, it significantly decreased the mortality rate of newborn mice infected with vesicular stomatitis virus when given at a single 20, 100 or 500 μg per mouse dose given 1 h after virus infection.[113] 3-DAAri is active against herpes simplex virus type 1 and HL-23, a C-type virus isolated from human acute myelogenous leukemia cells. 3-DAAri is also a potent anticancer agent and is active against P388D$_1$ cells, NIH3T3 cells, and RAW264 cells. The inhibitory effects are felt to be exerted partly through the inhibition of SAHH.[123,124] It has been shown to induce differentiation and to inhibit proliferation of HL-60 cells.[124]

(vi) Neplanocin-A (NPC-A)

NPC-A, a cyclopentenyl analog of Ado, is one of the antibiotics isolated from a soil-derived strain of *Ampullariella regularis A11079*.[125] It is a potent inhibitor of SAHH having K_i values of 8.4 nM and 0.2 nM for purified bovine liver enzyme and mouse L929 cells, respectively.[126,127] NPC-A inactivates purified bacterial (*A. faecalis*) SAHH in a time and concentration dependent manner. Moreover, SAHH activity could not be recovered after dialysis.[128] It is the most potent *in vivo* antitumor agent known against the L1210 murine leukemia.[129] It also proved to be particularly effective *in vitro* in inhibiting the multiplication of DNA viruses (*e.g.* vaccinia), (−) RNA viruses (*e.g.* parainfluenza, measles and vesicular stomatitis), and double-stranded RNA viruses [*e.g.* reo (in the range of 0.01–4.0 μg mL^{-1})]. These effects seemed to be related to inhibition of SAHH and consequent inhibition of transmethylation reactions.[130]

(vii) 3-Deazaneplanocin-A

This compound is the 3-deaza analog of the naturally occurring antibiotic Neplanocin-A.[131] It is the most potent SAHH inhibitor synthesized to date and has a K_i value of 5.0×10^{-11} M for purified hamster liver SAHH. 3-Deazaneplanocin-A is 250-fold more potent than 3-deazaaristeromycin (3-DAAri), the previously known most potent inhibitor of this enzyme. It was found to be very toxic to the human colon carcinoma cell line HT-29, and also inhibited rRNA and tRNA methylation.[132]

8.3.2.3 Ecto-5'-nucleotidase (ENT)

8.3.2.3.1 Biochemistry and immunology

Ecto-5'-nucleotidase (EC 3.1.3.5) is a hydrolase acting on ester bonds. ENT cleaves the 5'-phosphate from the ribose and deoxyribose moiety of nucleotides, producing the (deoxy)nucleoside and inorganic phosphate (Figure 2). Because cleavage occurs at the ester bond, ENT lacks specificity for the purine portion of the molecule and is able to cleave IMP, GMP and AMP. This enzyme is widely distributed in nature and has been reviewed extensively, including its purification and molecular weight determination.[20,133] ENT is associated with the cell membrane, where the concentrations of nucleotides are normally low. It is therefore felt to act primarily on external 5′-nucleotides, the major function of ENT being to transport AMP into cells.[19,20,134] A cytoplasmic 5′-nucleotidase has been described and regulates the levels of internal 5′-nucleoside monophosphates.[134]

There are no laboratory or clinical features of ENT deficiency which are unique to it.[19,20] That is, reduced ENT levels are noted in a number of immune deficiency syndromes, among which are acquired and congenital hypogammaglobulinemia, IgA deficiency, combined immunodeficiency, sex-linked agammaglobulinemia and Wiskott–Aldrich syndrome. The levels of this enzyme have generally been reported to be low in immature or poorly differentiated lymphocytes (thymocytes, B cells of newborns, chronic lymphocytic leukemia cells) as well as in the lymphocytes of ostensibly normal, but aged, humans. ENT is found in highest concentration in B cells and B lymphoblasts, in quite low levels in monocytes, and is nonexistent in erythrocytes. Because of the number of conditions which are associated with low levels of ENT, it is felt that its levels are a manifestation of immune deficiency, rather than a cause, and may reflect the particular leukocyte type or lymphocyte subpopulation which predominates in the circulation.

8.3.2.3.2 ENT inhibitors

ENT inhibitors would appear to have little value in terms of drug development, but they are excellent pharmacological probes. Several inhibitors, which are adenosine derivatives, have been described and used as tools to study metabolic processes.[135,136] The structures of these compounds are shown in Figure 6.

8.3.2.4 Purine Nucleoside Phosphorylase (PNP)

Purine nucleoside phosphorylase (EC 2.4.2.1) is an integral component of the purine enzyme cascade. It is neither a hydrolase nor a peptidase, but is included in this chapter because of its importance in purine metabolism. PNP inhibitors also offer a novel and testable approach to the development of immunosuppressive agents, which should show more selectivity for T cells than do ADA inhibitors. In addition, development of potent inhibitors of PNP has lagged behind similar efforts in the ADA area.

PNP catalyzes the reversible phosphorolysis of Ino, dIno, Guo and dGuo to their respective bases, hypoxanthine (Hx) and guanine, and the corresponding (deoxy)ribose 1-phosphate (Figure 2). PNP has been isolated from several mammalian cell types, including erythrocytes[137,138] and neutrophils,[139] as well as from placenta,[140] thyroid,[141] spleen,[142] kidney,[142] brain[142] and liver.[143] PNP has also been characterized in bacteria[144] and the malaria parasite.[145] PNP is present in high

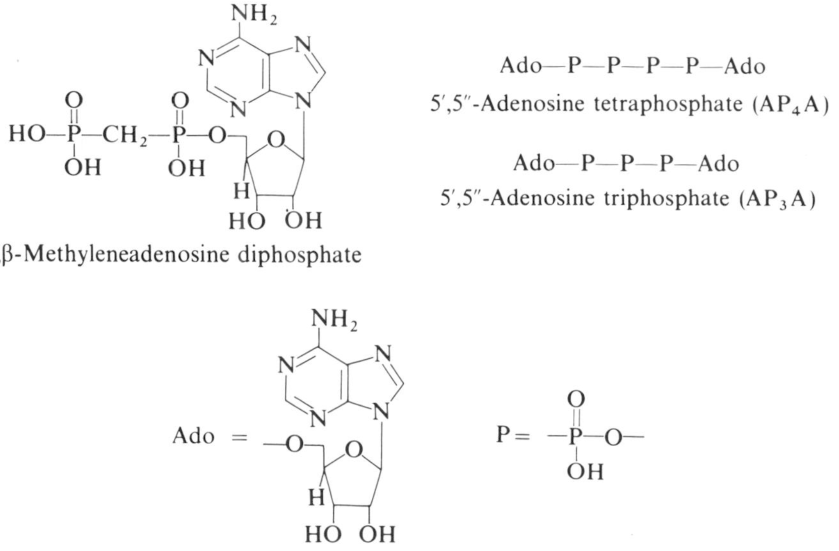

Figure 6   Inhibitors of ecto-5′-nucleotidase (ENT)

concentration in erythrocytes, and may serve to salvage Hx and guanine in muscle and other tissues and convert them back to soluble nucleosides. It is not known, however, why the concentration of PNP is so high in erythrocytes. In humans, PNP is a trimer with a molecular weight of 94 000 Da.[146,147] Considerable study has been made of mutant forms of the enzyme in relatives of PNP deficient patients.[142,148] The amino acid sequence of human PNP[149] and the quaternary structure, based on X-ray diffraction crystallography, have been reported.[146,147] It is highly probable that 'designed' inhibitors will soon be forthcoming based on these latter findings.

8.3.2.4.1 *PNP deficiency*

Congenital absence of PNP has been reported in a small number of patients since the initial description by Giblett *et al.* in 1975.[150] Compared with ADA deficiency, PNP deficiency is even rarer and, as was noted in ADA deficient patients, the PNP immunodeficiency syndrome is manifested only in those patients who show total deficiency in the PNP enzyme. Relatives with as little as 5% of normal adult PNP activity are immunologically normal. While these findings suggest that essentially total inhibition of PNP will have to be effected pharmacologically in order to achieve immunosuppression, the extent of immune impairment in congenital deficiency may be due to the fact that the biochemical effects are manifested before the immune system has developed fully. Hence, varying degrees of immune suppression might be achieved with a PNP inhibitor, depending on its potency and the dosing regimen.

PNP deficient patients present with features of immune deficiency early in life, but generally later than ADA deficient patients do.[148,150-162] The case report described by Giblett *et al.* comprises features of most PNP deficient patients and will be briefly described.[150] This patient was first seen at the age of 4.5 years with a history of lymphopenia, anemia, infections (pneumonia, candida, otitis media) and diarrhea. The child had normal plasma immunoglobulin levels and low titers of red cell agglutinins. Following immunization with pneumococcal polysaccharide and keyhole limpet hemocyanin, antibody responses were normal. However, delayed-type hypersensitivity skin test responses (T cell responses) to mumps, candida, purified protein derivative of tuberculin (PPD), and streptokinase–streptodornase, were negative, despite earlier immunization. *In vitro* blast transformation in response to PHA and allogeneic cell stimulation was depressed. The patient had no detectable erythrocyte PNP. In general, these patients have recurrent upper respiratory infections and other infections of bacterial, viral and fungal origin. Immunization against the typical battery of childhood infectious diseases is reported to be tolerated by these children. Additional features include autoimmune hemolytic anemia, which may be the consequence of deficient T cell suppression and unregulated B cell function, as evidenced by the presence of a number of autoantibodies, and central nervous system abnormalities, which may be due to a CNS infection.[162]

The immune impairment in PNP deficiency is restricted to T lymphocyte function. As noted above, some patients have essentially normal T cell immunity early in life, but this declines with time and eventually little or no T cell function remains. The ability to respond to immunization through formation of specific antibodies is not restricted to thymus independent antigens, as responses to thymus dependent antigens have been documented. In addition, classic delayed-type hypersensitivity skin test responses have been demonstrated, at least early in life. Hence, some components of T cell function are intact early in life. It could be speculated that elevated antibody/immunoglobulin production is the result of the absence of NK cells or of too little interleukin 2 (IL 2).[163-165] The clinical picture, however, is one of reduced T cell number and impaired T cell function.

8.3.2.4.2 *Biochemistry of PNP and PNP deficiency*

The biochemistry associated with PNP, and the abnormalities found in PNP deficient patients, have been thoroughly described elsewhere.[13-16,18,19,21] Patients with PNP deficiency exhibit a number of biochemical changes, most of which are readily explained by the absence of PNP. Of the PNP substrates, Ino and Guo are elevated in plasma and in urine, while the respective deoxynucleosides are rarely elevated, and only in urine. Blockade at PNP is further demonstrated by the reduced levels of oxypurines in plasma and urine, along with hypouricosuria and hypouricemia. No biochemical mechanism has been offered to account for immune impairment due to a reduction in oxypurines and uric acid. From an immunological perspective, elevation of dGTP in erythrocytes,[166] which is a result of dGuo elevation, is probably most telling as it is felt to mirror the same biochemistry occurring in lymphocytes. Although dGTP has been found to be elevated in the lymphocytes of only one PNP deficient patient,[167] this may be due to lymphopenia and the actual death of cells in which dGTP has accumulated. Of the other PNP substrates, Ino, dIno and Guo are much less likely to contribute to the pathogenesis of PNP deficiency because their nucleotides are either less important or are formed (*e.g.* IMP) even in the presence of PNP. GMP can be formed from IMP by the action of IMP dehydrogenase and guanylate synthetase, thereby bypassing the salvage circuit involving PNP (Figure 2). Of interest, Simmonds *et al.* reported that the level of GTP in the erythrocytes of one PNP deficient patient was markedly depressed, most likely as a result of abnormal IMP dehydrogenase.[167] NAD$^+$ levels were similarly raised in the erythrocytes of this patient. Alterations in methylation reactions have also been documented. While these other mechanisms may contribute to the features of PNP deficiency, most support is given to the concept that PNP deficiency results from the accumulation of the triphosphate of dGuo.

As with ADA deficiency, the mechanism for T cell elimination in PNP deficiency probably relates to inhibition of ribonucleotide reductase by, in PNP deficiency, dGTP.[168-170] The dGuo is phosphorylated to dGTP predominantly in cells that have a high capability for doing this, *i.e.* replicating T lymphocytes. The enzyme mediating this activity, deoxycytidine kinase,[171] is found predominantly in lymphoid tissues, mainly in T lymphocytes and thymocytes.[24] Like dATP, dGTP is a potent allosteric inhibitor of ribonucleotide reductase, blocking the formation of three of the ribonucleoside diphosphate precursors of DNA (see also Section 8.3.2.1.2).[25,26] The depletion of dCTP is most critical, and in subsequent portions of this chapter we will describe studies in which dCyt has been added to cell cultures to abrogate the PNP inhibition effect by: (i) replenishing reduced levels of dCTP,[171] and (ii) inhibiting further dGuo phosphorylation by dCTP-mediated inhibition of deoxycytidine kinase.[172]

Implicit in the mode of action of a PNP inhibitor is the expectation that only replicating T cells should be killed. Nondividing T cells and most other cell types which are synthesizing DNA should not be affected. Hence, unstimulated and memory lymphocytes should not be killed. This situation is clearly distinct from that seen with dAdo, which is inhibitory even to resting lymphocytes.[58,66,67] As a result, PNP inhibitors should have some 'antigen specificity' because uninvolved lymphocytes will not be affected through this mechanism. It should be noted that the mechanism(s) by which B lymphocytes resist killing due to dGTP is not fully understood. It has been hypothesized that B cells do not phosphorylate dGuo as extensively, or that they have additional regulatory mechanisms which reduce dGTP levels after they are formed.

8.3.2.4.3 PNP inhibitors

Since the description of PNP deficiency in children and its association with selective T cell deficiency, it was reasoned that a potent PNP inhibitor could be of value in the treatment of certain autoimmune diseases and in organ transplantation because of its lymphocyte class selectivity. This possibility is indirectly supported by the high degree of effectiveness of cyclosporin, a T cell modulator, particularly in transplantation.[173] PNP inhibitors might also be useful in the treatment of lymphoproliferative diseases such as T cell leukemia, in gout and in parasitic diseases. PNP inhibitors have been designed, synthesized and developed using basically two approaches: (i) modifying known inhibitors, and (ii) modifying the structure of substrates and designing compounds which mimic them.[174-177] Although a variety of inhibitors have been synthesized and evaluated, only those inhibitors which are either very potent in the enzyme assay, or which have been tested further in *in vitro* and *in vivo* assays will be described (Figures 7 and 8, Tables 1-4).

(i) Modification of the purine

8-Aminoguanine and 8-aminoguanosine are modified purines that are derivatives of the substrates guanine and Guo. The PNP inhibitory activities of these two compounds, which have been known for a long time, have only recently been described.[177] 8-Aminoguanine and 8-aminoguanosine are potent inhibitors with IC_{50} values of 1.55 and 1.33 μM, respectively.[178] Both compounds are also substrates of PNP. 8-Aminoguanosine has become a standard for PNP inhibitor research and has been tested extensively (see following sections). 8-Aminoguanosine is a prodrug of 8-aminoguanine and is cleaved quickly *in vivo*. Thus, the *in vivo* effects of 8-aminoguanosine may be due to 8-aminoguanine. 8-Amino-3-deazaguanine is a modified 8-aminoguanine analog that was found to be less potent than 8-aminoguanine (IC_{50} = 9.9 μM). It is cytotoxic to both T and B lymphoblastoid cells *in vitro*.[179] Two other noteworthy modified purines are the antibiotic formycin B and allopurinol riboside. These compounds are weak inhibitors with K_i values of 100

Substrates of PNP

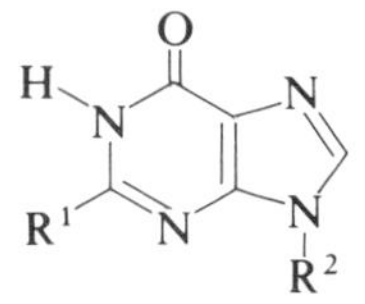

Inosine; $R^1 = H$, $R^2 = $ ribose
Guanosine; $R^1 = NH_2$, $R^2 = $ ribose
Deoxyinosine; $R^1 = H$, $R^2 = $ 2'-deoxyribose
Deoxyguanosine; $R^1 = NH_2$, $R^2 = $ 2'-deoxyribose
Hypoxanthine; $R^1 = R^2 = H$
Guanine; $R^1 = NH_2$, $R^2 = H$

Modified purines

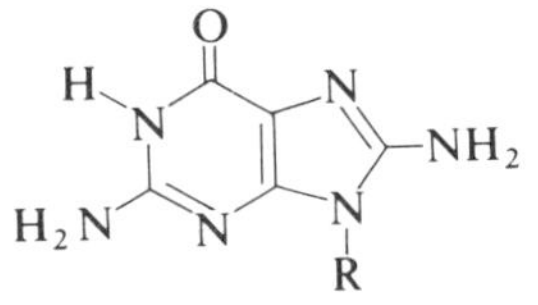

8-Aminoguanine; R = H
8-Aminoguanosine; R = ribose
8-Amino-2'-deoxyguanosine; R = 2'-deoxyribose

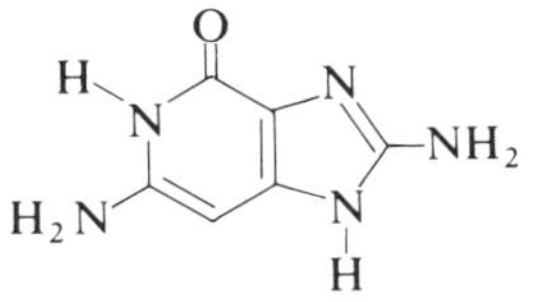

2,6-Diamino-1,5-dihydro-4H-imidazo[4,5-c]pyridin-4-one (8-amino-3-deazaguanine)

Modified purine nucleosides

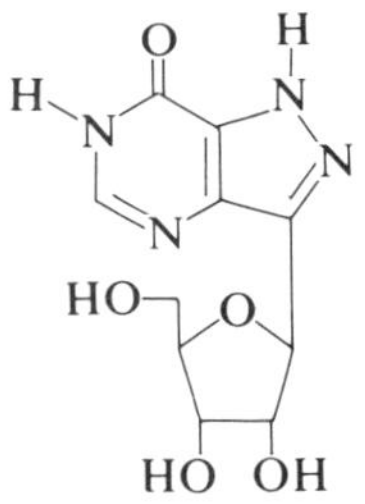

Formycin B

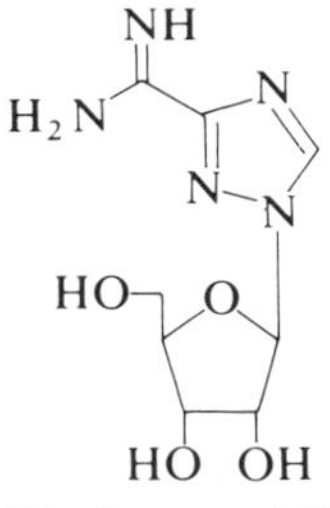

1-β-D-Ribofuranosyl-1H-1,2,4-triazole-3-carboxamidine (TCNR)

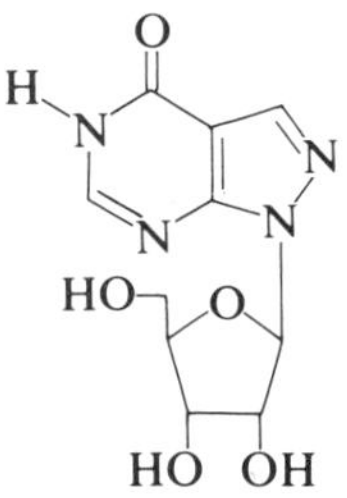

Allopurinol riboside

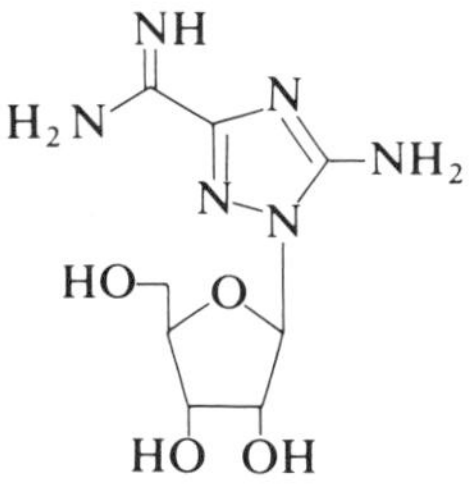

5-Amino-TCNR

Figure 7 Inhibitors of purine nucleoside phosphorylase (PNP)

and 277 μM, respectively.[177,180] Another important modified purine is TCNR (1-β-D-ribofuranosyl-1H-1,2,4-triazole-3-carboximidamidine) with a K_i of 5.0 μM.[181] This compound has been tested extensively *in vitro* and will be discussed later. Recently, several TCNR derivatives were reported, but only 5-amino-TCNR was found to be potent (K_i of 10 μM, *versus* a K_i of 30 μM for TCNR in the same assay).[182] These compounds are shown in Figure 7.

(ii) Modification of the sugar moiety

Since the sugar moiety is necessary for binding and is involved in the PNP reaction, a series of sugar-modified nucleosides were synthesized and tested for PNP inhibitory activity (Figure 8). Among the furanosyl nucleosides, 5'-deoxy-5'-iodo- or 5'-deoxy-5'-chloro-9-deazainosines were the most potent inhibitors, having K_i values of about 0.2 μM.[183] 5'-Deoxy-5'-iodoformycin B is also a potent inhibitor with a K_i value of 7.0 μM.[177] 9-Deazainosine and 9-deazaguanosine were also found to be potent inhibitors with K_i values of 2.0 and 2.9 μM, respectively.[183] 8-Amino-2'-deoxyguanosine was less potent than 8-aminoguanosine and had an IC_{50} value of 4.3 μM.[176]

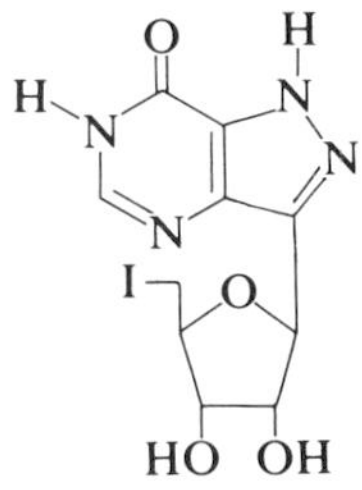

5′-Deoxy-5′-iodoformycin B

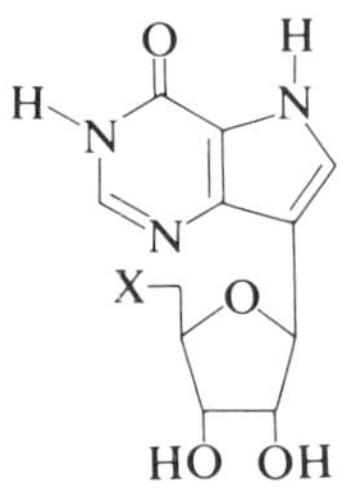

(a) X = I; 7-(5-Deoxy-5-iodo-β-D-ribofuranosyl)-3,5-dihydro-4*H*-pyrrolo[3,2-*d*]pyrimidin-4-one (5′-deoxy-5′-iodo-9-deazainosine)

(b) X = Cl; 7-(5-Deoxy-5-chloro-β-D-ribofuranosyl)-3,5-dihydro-4*H*-pyrrolo[3,2-*d*]pyrimidin-4-one (5′-deoxy-5′-chloro-9-deazainosine)

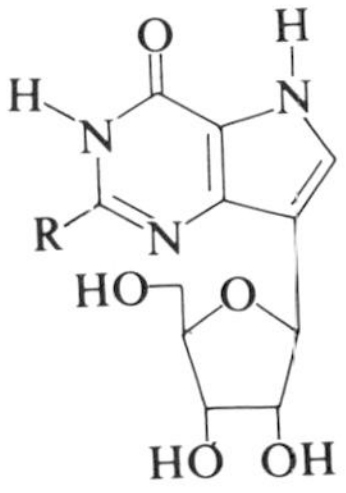

R = H; 3,5-Dihydro-7-β-D-ribofuranosyl-4*H*-pyrrolo[3,2-*d*]pyrimidin-4-one (9-deazainosine)

R = NH$_2$; 2-Amino-3,5-dihydro-7-β-D-ribofuranosyl-4*H*-pyrrolo[3,2-*d*]pyrimidin-4-one (9-deazaguanosine)

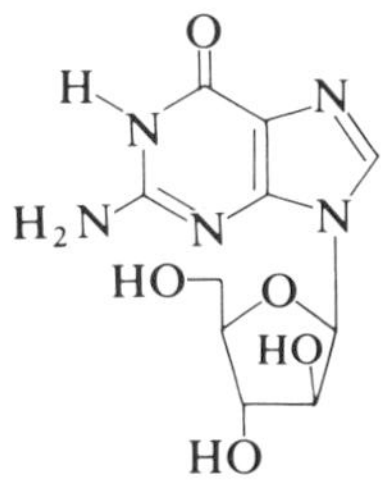

2-Amino-9-β-D-arabinofuranosyl-1,9-dihydro-6*H*-purin-6-one (arabinosylguanine)

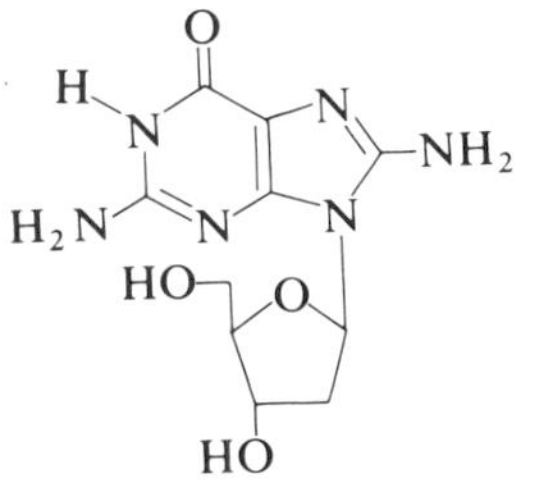

8-Amino-2′-deoxyguanosine

Figure 8 Modified sugar nucleosides as inhibitors of purine nucleoside phosphorylase (PNP)

Although the above compounds are very potent PNP inhibitors, very little other data have been reported on them. Another modified sugar nucleoside is 2-amino-9-β-D-arabinofuranosyl-1,9-dihydro-6*H*-purin-6-one (arabinosylguanine), which was tested in a human lymphoblastoid cell line assay and found to be toxic to T cells only.[184,185] Only in T lymphoblasts was arabinosylguanine phosphorylated to the triphosphate, and activity was blocked by coaddition of dCyt.[184,185]

Acyclosugars are known to mimic the ribose portion of nucleosides. Thus, a variety of acyclic nucleosides were synthesized and evaluated for PNP inhibitory activity.[176] These modified sugar nucleosides were found to be potent PNP inhibitors (Table 1). Substitution at the 8 position of the purine gave the most potent compounds. 8-Amino- and 8-hydroxy-acyclovir have K_i values of 3.8 µM and 4.6 µM, respectively.[186] 8-Amino-2′-nordeoxyguanosine, (8-amino-2′-NDG; PD 116,124) was the most potent compound in this series ($K_i = 0.42$ µM).[176] Replacement of the oxygen atom with carbon in the acyclosugar did not enhance the potency of the compound (*e.g.* carbaacyclovir and carba-2′-NDG).[186] Among all these compounds, only 8-amino-2′-NDG was evaluated in other *in vitro* and *in vivo* systems.

Table 1 Acyclic Nucleosides as PNP Inhibitors

Name	R^1	R^2	X	K_i (μM)	IC_{50}	Ref.
Acyclovir	H	H	O	91[a]	151	176, 186
8-Hydroxyacyclovir	H	OH	O	4.6[a]		186
8-Aminoacyclovir	H	NH$_2$	O	3.8	15.6	176, 186
2'-Nordeoxyguanosine (2'-NDG)[b]	CH$_2$OH	H	O	30	113	176, 186
8-Bromo-2'-NDG	CH$_2$OH	Br	O	34	91	176, 186
8-Thio-2'-NDG	CH$_2$OH	SH	O	13		186
8-Amino-2'-NDG	CH$_2$OH	NH$_2$	O	0.26,0.42		176, 186
Carbaacyclovir[c]	H	H	CH$_2$	42		186
Carba-2'-NDG[d]	CH$_2$OH	H	CH$_2$	38		186

[a] Values obtained at 1.0 mM phosphate concentration. [b] 2-Amino-1,9-dihydro-9-([2-hydroxy-1-(hydroxymethyl)-ethoxy]methyl)-6*H*-purin-6-one. [c] 2-Amino-1,9-dihydro-9-(4-hydroxybutyl)-6*H*-purin-6-one. [d] 2-Amino-1,9-dihydro-9-[4-hydroxy-3-(hydroxymethyl)butyl]-6*H*-purin-6-one.

(iii) Multisubstrate analog inhibitors

Acyclovir disphosphate (acyclo-GDP), a metabolite of acyclovir and the most potent inhibitor of PNP known, has a K_i value of 0.0087 μM when assayed at a low inorganic phosphate concentration (1.0 mM).[187] The K_i value increased to 0.51 μM when the compound was assayed under a phosphate concentration of 50.0 mM, indicating that acyclo-GDP is competing with both the purine binding site and the binding site of the second substrate (inorganic phosphate). Thus, acyclo-GDP is a multisubstrate analog inhibitor (Table 2). Although 2'-NDG is much more potent than acyclovir, the corresponding (*S*)- or (*R*)-diphosphates were found to be equipotent.[186] Phosphate esters of acyclovir and 2'-NDG have improved potency over the parent compounds but their activity was dependent on the inorganic phosphate concentration. As the phosphate esters are multisubstrate analog inhibitors, it was hypothesized that they also bound to the phosphate binding site along with

Table 2 Acyclic Nucleotides as Multisubstrate Analog Inhibitors of PNP

Name	R^1	R^2	K_i^a (μM)	Ref.
Acyclo-GMP	H	P$_i$[b]	6.6	187
Acyclo-GDP	H	PP$_i$[c]	0.0087	187
Acyclo-GTP	H	PPP$_i$[d]	0.31	187
(*S*)-2'-NDG monophosphate	CH$_2$OH	P$_i$[b]	2.2	187
(*S*)-2'-NDG diphosphate	CH$_2$OH	PP$_i$[c]	0.009	186
(*R*)-2'-NDG diphosphate	CH$_2$OH	PP$_i$[c]	0.018	186
(*S*)-2'-NDG triphosphate	CH$_2$OH	PPP$_i$[d]	0.16	186

[a] Values were obtained at 1.0 mM phosphate concentration. [b] P$_i$ is —P(O)(OH)$_2$. [c] PP$_i$ is —P(O)(OH)—O—P(O)(OH)$_2$. [d] PPP$_i$ is —P(O)(OH)—O—P(O)(OH)—O—P(O)(OH)$_2$.

the purine binding site. This hypothesis has been confirmed from the preliminary X-ray analysis of the PNP–acyclovir diphosphate complex.[186] Although the acyclonucleotides are extremely potent PNP inhibitors, they lack the physicochemical properties permitting penetration through the cell membrane, and hence are of little value as drugs. These compounds are, however, useful pharmacological probes and aid in the design of drugs.

A series of 9-(phosphonoalkyl)hypoxanthines were synthesized as multisubstrate analog inhibitors (Table 3). These compounds were also found to be potent PNP inhibitors, but again only when tested at a low phosphate concentration. The most potent compounds in this series were alkyl phosphonates of five-, six- and seven-methylene chains. No inhibition was observed for any of these compounds when tested at a 50 mM phosphate concentration.[188]

(iv) Modification of known inhibitors

Since 8-aminoguanine and 8-aminoguanosine are both substrates and inhibitors of PNP, a large number of 8-amino-9-substituted guanines have been synthesized in order to develop a nonsubstrate inhibitor which is metabolically stable.[189,190] 8-Amino-9-benzylguanine and 8-amino-9-thienylmethylguanines were found to be the most potent compounds in the series (Table 4). 8-Amino-9-benzylguanine has an IC_{50} value of 0.47 μM and has been evaluated further.[189,190] 8-Amino-9-(2-thienylmethyl)guanine (PD 119,229) and 8-amino-9-(3-thienylmethyl)guanine have IC_{50} values of 0.17 μM and 0.085 μM, respectively.[190,191] PD 119,229 has also been evaluated further and is discussed in the following sections.

8.3.2.4.4 In vitro *immunological and biochemical studies employing PNP inhibitors*

Only a few PNP inhibitors have been studied extensively *in vitro*, and therefore each will be discussed separately. As with the discussion on ADA inhibitors, the objective here will be to document major changes in biochemistry or the function of lymphocytes produced *in vitro* by PNP inhibitors.

(i) 8-Aminoguanosine

Many of the early observations on the effects of PNP inhibitors were obtained using 8-aminoguanosine. Most of this material has been extensively reviewed recently and will not be covered in detail here.[176] An early and important paper on the effects of 8-aminoguanosine on human T and B lymphoblast growth was produced by Kazmers *et al.* in 1981.[192] This group observed that the growth of human T lymphoblasts (MOLT-4) and B lymphoblasts (MGL-8) was unaffected by 8-aminoguanosine at concentrations up to 100 μM. However, 8-aminoguanosine potentiated the cytotoxicity of dGuo for human MOLT-4 but not MGL-8 lymphoblasts.[192] dGuo alone caused a concentration dependent increase in intracellular dGTP in MOLT-4 lymphoblasts, and addition of 8-aminoguanosine (100 μM) augmented dGTP elevation significantly. It was shown

Table 3 9-(Phosphonoalkyl)hypoxanthines as Multisubstrate Analog
Inhibitors of PNP

Name	R	n	K_i^a (μM)
9-(3-Phosphonopropyl)hypoxanthine	H	3	2700
9-(5-Phosphonopentyl)hypoxanthine	H	5	1.1
9-(6-Phosphonohexyl)hypoxanthine	H	6	2.2
9-(7-Phosphonoheptyl)hypoxanthine	H	7	0.9

ª Values obtained at 1.0 mM phosphate concentration.

Table 4   8-Amino-9-substituted Purines as PNP Inhibitors

Ar	IC_{50} (µM)	Ref.	Ar	IC_{50} (µM)	Ref.
2-Th[a]	0.17	190	3-Py[c]	21.9	190
2-Th-5-Et	0.93	190	3-Th	0.085	191
2-Th-3-Me	4.05	190	Ph	0.47, 0.2	189, 190
2-Fu[b]	0.25	190	Ribose[d]	1.40	190

[a] Thiophene. [b] Furan. [c] Pyridine. [d] 8-Aminoguanosine; C-1 of ribose attached directly to N-9 of purine.

that the greater sensitivity of MOLT-4, *versus* MGL-8, to inhibition by dGuo and 8-aminoguanosine was not due to a difference in the specific activities of PNP in these cells, nor to other mechanisms, including the ability of these cells to take up the inhibitor. The biochemical effects observed in MOLT-4 and MGL-8 following treatment with 8-aminoguanosine and dGuo were fully consistent with the observations made on PNP deficient patients. Subsequent work showed that human T lymphoblasts have higher dGuo kinase activity than B lymphoblasts.[169] In addition, it was shown that dCyt prevented accumulation of dGTP.[169,172]

However, recently Goday *et al.*[68] and Simmonds *et al.*[193] have reported that some human B lymphoid cell lines, even when cultured without 8-aminoguanosine, accumulate dGTP to levels reported earlier in T lymphoblasts. Furthermore, dGuo inhibited the growth of one cell line which accumulated no dGTP, implicating GTP as the toxic mediator. Sidi and Mitchell observed that the growth of mature human T_4^+ cell lines, as well as that of some B lymphoblast cell lines, was inhibited by dGuo plus 8-aminoguanosine.[194] Inhibition of growth occurred in the absence of any increase in dGTP, and cytotoxicity was observed when there was a three- to five-fold increase in GTP.[194] B lymphoblasts deficient in HGPRT were resistant to dGuo toxicity even at dGuo concentrations approaching 1 mM. Inhibition of B lymphoblast growth could be prevented by Hx or adenine, while dCyt was ineffective, confirming that toxicity was mediated through the salvage pathway.[194] Sidi and Mitchell concluded that 8-aminoguanosine was too weak a PNP inhibitor to block dGuo and Guo phosphorolysis. Thus, toxicity of dGuo and Guo to B cells is mediated by a mechanism which shows no selectivity for cell type.

The metabolism of Guo and dGuo has been studied extensively in normal lymphocytes. As noted above, in the absence of PNP inhibition, additional avenues of guanine nucleoside metabolism are available, and therefore studies prior to the development of potent PNP inhibitors must be interpreted while keeping this caveat in mind. Early work demonstrated that Guo was almost as toxic to PHA-stimulated human[195] and Con A-stimulated mouse[196] lymphocytes as was dGuo, suggesting the importance of the salvage pathway. In support of this, dCyt had no effect on dGuo toxicity in mice in the absence of PNP inhibition.[196] More extensive work analyzing intracellular nucleotides has recently been reported.[197,198] In the absence of PNP inhibition, addition of dGuo to normal human lymphocytes stimulated with mitogens or antigens produced elevation of intracellular dGTP and GTP, and blocked [³H]thymidine uptake.[197] Coaddition of dCyt to these cultures reduced dGTP levels and partially prevented toxicity. Addition of Hx to cultures treated with dGuo reduced GTP, but not dGTP, and essentially restored [³H]thymidine uptake to normal. Of interest, even when PNP was inhibited by 8-aminoguanosine, addition of dGuo still resulted in some elevation of GTP because of incomplete PNP inhibition by 8-aminoguanosine, as reported elsewhere.[194] High concentrations of 8-aminoguanosine protected T cells from dGuo toxicity by more effectively blocking dGuo phosphorolysis and limiting GTP accumulation. This work was extended using PNP deficient and HGPRT deficient cell lines, which did not accumulate GTP and were resistant to Guo toxicity.[197] These observations demonstrate the role of the salvage pathway of purine metabolism in mediating the toxic effects of Guo and, with inadequate PNP inhibition, of dGuo. Mitogenic responses of B cells were also depressed *via* GTP formation from Guo and dGuo, and 8-aminoguanosine protected these cells from toxicity by blocking guanine formation and reducing GTP accumulation.[198]

(ii) 8-Amino-9-heteroarylguanine analogs, including PD 119,229

The most potent inhibitors of PNP have only recently been described. Among these is 8-amino-9-benzylguanine,[189] which has been found to potentiate the toxicity of dGuo for human T lymphoblasts similarly to 8-aminoguanosine. Among the most potent and best-studied members of this class of compounds is PD 119,229 [8-amino-9-(2-thienylmethyl)guanine], which has an apparent K_i of 0.067 μM for human erythrocyte PNP.[199,200] In a test system comparable to the one described by Kazmers *et al.*,[192] PD 119,229 was found to be nontoxic for either of the T or B lymphoblast lines studied, at concentrations up to 100 μM. However, in the presence of a nontoxic concentration of dGuo (10 μM), PD 119,229 had an IC_{50} of 0.9 μM for MOLT-4 lymphoblasts and > 100 μM for MGL-8 lymphoblasts. In MOLT-4 cells, coaddition of dGuo and PD 119,229 was associated with a marked increase in intracellular dGTP (40-fold control) and a minor decrease in intracellular GTP (twofold). These alterations in nucleotides were not observed in MGL-8 cells even at the highest concentrations tested. Addition of dCyt restored MOLT-4 growth and totally reduced dGTP levels under these conditions.[199]

(iii) Acycloguanosine PNP inhibitors including PD 116,124

PD 116,124 (8-amino-2'-NDG; 2-[(2,8-diamino-6-hydroxy-9H-purin-9-yl)methoxy]-1,3-propane-diol) has recently been reported to be a competitive, reversible inhibitor of human PNP with a K_i of 0.42 μM.[176,186] PD 116,124 is noninhibitory to human MOLT-4 and CEM T lymphoblasts, and to MGL-8 or NC-37 B lymphoblasts, at concentrations up to 500 μM.[201] In the presence of a noninhibitory concentration of dGuo (10 μM), however, PD 116,124 inhibited the growth of human MOLT-4 and CEM T lymphoblasts ($IC_{50} < 2$ μM), but not MGL-8 or NC-37 B lymphoblasts ($IC_{50} > 500$ μM). Cytotoxicity correlated with dGTP accumulation. PD 116,124 produced greater elevation of dGTP (49-fold control) than was observed when only dGuo was added to T lymphoblasts, and elevation of dGTP was sustained when PNP was inhibited. As was noted for 8-aminoguanosine, both the inhibition of growth of T cells and the accumulation of dGTP were prevented by the addition of 10 μM dCyt to culture medium.[201]

8.3.2.4.5 In vivo *studies on PNP inhibitors*

(i) 8-Aminoguanosine

The first reports on the *in vivo* effects of PNP inhibition were performed using 8-aminoguanosine. These studies were designed to determine only if some of the biochemical features of PNP deficiency could be produced pharmacologically, and did not attempt to demonstrate immunosuppression.[202] 8-Aminoguanosine was found to produce dose-related elevation of both Ino and Guo in plasma, when administered either orally or intravenously to normal rats. Statistically significant effects were observed following administration of 8-aminoguanosine over the dose range of 0.5 to 5 mg kg^{-1} intravenously, and orally in the range of 0.5 to 150 mg kg^{-1}. Ino elevation was most striking, reaching 5.3 μM (177-fold the vehicle control) at 150 mg kg^{-1}, by mouth. Similar observations have been reported by others.[203] 8-Aminoguanosine has also been reported to elevate plasma nucleosides in dogs following daily intravenous injections at 10 mg kg^{-1}.[204] 8-Aminoguanosine is rapidly converted to 8-aminoguanine in dogs[205] and rats.[203] When 8-aminoguanosine and dGuo were each injected subcutaneously at 36 mg kg^{-1} for 5 d, there was a 65% reduction in the number of circulating lymphocytes in dogs, but no effect on other blood components.[205] As a result of this treatment, there was elevation of Ino and exogenously administered dGuo in plasma. Erythrocyte GTP levels were monitored and were found to be normal.[205]

Few studies of immune function following administration of 8-aminoguanosine have been reported. As noted above, peripheral blood lymphocyte counts declined in dogs given 8-aminoguanosine and dGuo.[205] In mice treated with high intravenous concentrations of 8-aminoguanosine and dGuo (100 μM each), no lymphoid histopathology was observed, nor were blastogenic responses to mitogenic or allogeneic cell stimulation suppressed.[206] 8-Aminoguanosine has been reported to prolong skin graft survival in dogs (see Section 8.3.5.1).[207]

(ii) PD 119,229

PD 119,229 has been tested *in vivo* and produced a statistically significant elevation of plasma Ino in normal male rats 1 h after a 3 mg kg^{-1} oral dose.[200] Elevation of plasma Ino concentration peaked 10 h after oral dosing, reaching 2.13 μM (14.2-fold vehicle). Plasma Guo was maximal 3 h post dosing, reaching 0.77 μM. Moreover, the concentration of Ino in plasma remained statistically significantly elevated for up to 6 h after a single oral dose, which indicates that PNP was effectively inhibited for this duration.[200] Of interest, while the Ino concentration reached a mean elevation of 1.71 μM in the dose–response studies, concentrations as high as 15 μM were reported in some rats.[200] PD 119,229 has also been evaluated in an animal model of gout, which is based on nucleoside elevation and hyperallantoinemia following infusion of fructose.[208] These studies are described in detail in Section 8.3.4.2.

(iii) Acycloguanosine PNP inhibitors including PD 116,124

PD 116,124 caused significant elevation of plasma Ino in normal rats 1 h after dosing (5 mg kg^{-1}, by mouth).[201] Nucleoside elevation was dose-related, being maximal at 500 mg kg^{-1}, by mouth in rats (1.39 μM, 17.4-fold vehicle). When PD 116,124 was administered intravenously at 50 mg kg^{-1}, both Ino and Guo were statistically significantly elevated 15 min after dosing (1.69 μM, 56.3-fold vehicle and 1.43 μM, 71-fold, respectively). The Ino concentration increased through 1 h post dosing (the last time point evaluated), reaching 2.87 μM.[201]

8.3.2.5 Hypoxanthine–guanine Phosphoribosyltransferase (HGPRT)

Hypoxanthine–guanine phosphoribosyltransferase (EC 2.4.2.8) is a transferase which is noted briefly in this chapter to complete the concept of a purine metabolic cascade. HGPRT catalyzes the irreversible salvage, or recovery, of Hx and guanine, utilizing phosphoribosyl pyrophosphate (PRPP) and magnesium in the process and generating IMP and GMP, respectively. This process requires only one ATP molecule, in contrast to *de novo* purine biosynthesis, which requires at least six ATP molecules.[209] Hence, it is a highly energy efficient means of generating nucleotides.

Congenital absence of HGPRT causes Lesch–Nyhan syndrome and hyperuricemia (see Section 8.3.4).[209] Two mechanisms contribute to the hyperuricemia. The first simply results from the inability to salvage guanine and Hx due to the absence of HGPRT. The second is due to the concomitant accumulation of PRPP, which is normally utilized on an equimolar basis with each purine salvaged, leading to IMP or GMP (Figure 2). PRPP stimulates *de novo* purine synthesis, leading to further production of IMP, thus further amplifying the genetic defect. In addition to the abnormalities in purine metabolism, patients with Lesch–Nyhan disease are characterized by striking neurological features including spasticity, choreoathetosis, and varying degrees of mental retardation. Children possessing as little as 1% of normal HGPRT activity do not show neurological symptoms. HGPRT deficient children do not manifest any immune deficiency, and hence from the rationale that agents modifying purine metabolism might be immunosuppressive, HGPRT seems to be a poor approach. No inhibitors of HGPRT are known.

8.3.3 AUTOIMMUNE DISEASES

A number of diseases are classified as 'autoimmune' because of their pathology and the finding of antibodies or lymphoid cells which are reactive against 'self' components. Drugs (steroids, immuno-suppressants, cytotoxic agents) and procedures (total lymph node irradiation, leukapheresis) which suppress immune function are therapeutic in these diseases. Most current forms of therapy suffer from a broad range of serious side effects and lack of efficacy. Hence, agents such as those described in the preceding sections (or analogs of them) have the potential to offer therapy superior to that currently available. Recently, cyclosporin, a T cell modulator, has been found to be efficacious in a number of autoimmune diseases including rheumatoid arthritis[210-212] and juvenile diabetes.[213-215] However, cyclosporin suffers from a unique spectrum of side effects including nephrotoxicity, hypertension and malignancies, and therefore there is still need for safer and more effective therapy. One of the more prevalent autoimmune diseases in need of better therapy is rheumatoid arthritis. In order to exemplify the types of abnormalities in immune function which might be affected by an inhibitor of purine metabolism, a brief description of the pathogenesis and immunological features of rheumatoid arthritis follows. It can be emphasized that most of the abnormalities of immune function seen in rheumatoid arthritis are the very features of the immune system which are lacking in immunodeficient patients lacking, *e.g.* ADA or PNP. The notable exception to this is immunoglobulin production in PNP deficient patients, which can be normal to elevated, including the presence of autoantibodies.

8.3.3.1 Rheumatoid Arthritis

Although clearly far from being fully comprehended, rheumatoid arthritis (RA) is nevertheless one of the most-studied and best-understood autoimmune diseases in terms of the immunological features.[216-218] Discussion of other autoimmune diseases (Section 8.3.3.2) will be greatly reduced in scope for brevity, and because much of what is presented here can be extrapolated to other autoimmune diseases.

8.3.3.1.1 Clinical description and pathogenesis

RA is a disease primarily of joints and connective tissues, known from ancient times.[216-218] Its name was appropriately taken from the Greek 'rheum', which means 'tears'. Roughly 1% of the world's population (i.e. about 50 million people) suffers from definite RA.[219]

There are several interesting immunological features documented for RA, including the rheumatoid pannus, deficient IL 2 production, and the presence of autoantibodies, including rheumatoid factors. Rheumatoid factors are antibodies of any immunoglobulin (Ig) class which can react with the Fc region of IgG, and are frequently observed in RA.[220-222] Most rheumatoid factors are IgM or IgG, although IgA rheumatoid factors have also been described. It is well documented that rheumatoid factors can activate complement, thus generating chemotactic factors which draw in inflammatory cells, primarily polymorphonuclear leukocytes. However, rheumatoid factors are not necessarily arthritogenic in and of themselves, because their existence is well documented in a number of nonarthritic disease states (syphilis, leprosy, viral infections) as well as in 'normal' individuals. Nevertheless, there is considerable support for the hypothesis that the continued production of rheumatoid factors in RA most likely contributes to the severity and chronicity of this disease. It should be noted that the reason(s) for the production of rheumatoid factors in this disease is unknown. One theory maintains that there is a deficiency in IL 2 production in RA, resulting in inadequate natural killer (NK) cell activity.[163-165] NK cells are responsible for down-regulating antibody production, and therefore this scenario could help account for the presence of autoantibodies, including rheumatoid factors, in this disease. Contrary to this, however, is the documented efficacy of cyclosporin (which potently suppresses IL 2 release) in RA, although it is unknown if the mode of action of cyclosporin in RA is through this action on IL 2.

More impressive than the existence of rheumatoid factors, however, is the joint pathology of RA. The synovial tissues lining joints become enlarged and invaded with mononuclear cells, resulting in hypertrophic villous synovitis and pannus formation.[216-218] The infiltrating lymphoid cells form structures within the pannus with the architecture of lymph nodes, complete with lymphoid follicles, even at times containing germinal centers with active immunoglobulin-secreting plasma cells. Elaboration of cytokines by these tissues is well documented. The pannus is juxtaposed to regions of the cartilage that literally dissolve away from underneath it. Synovial lining cells also contain a morphologically heterogeneous population of cells, some highly capable of producing proteolytic enzymes (collagenase, proteoglycanase, *etc.*), which can erode joint tissues. It has been hypothesized that these cells may be under continuous stimulation by IL 1 and other cytokines produced in adjacent tissues, as well as by other factors (complement fragments), resulting in a continual release of proteolytic enzymes.

In addition to the classical immunological and histopathological studies on human joint tissue, considerable indirect evidence exists to implicate lymphocytes in the pathogenesis of RA and other diseases. It is well documented that a number of 'heroic' measures aimed at reducing the lymphocyte load in severely afflicted patients can produce dramatic relief. These procedures include total lymph node irradiation,[223] thoracic duct cannulation,[224] and leukapheresis.[225,226] Hence, while it is still unknown what triggers RA and what the events immediately ensuing are, there is little doubt that lymphocytes play a major role in the pathogenesis of this disease.

Classical immunological research has further demonstrated the requirement for T lymphocytes for production of antibodies to thymus dependent antigens in a number of species and systems. T cells are felt to be essential for the switch from IgM to IgG in the normal antibody response. Overwhelming evidence exists to show that T lymphocytes are a major component in regulation of the immune system, as exemplified in numerous collaborative interactions and the elaboration of cytokines such as IL 2.[227-229] Hence, medicinal agents which can selectively suppress T lymphocytes, or certain aspects of T cell function, could provide therapy which is superior to the forms presently available, which are not specific. This concept is supported by the efficacy of cyclosporin in RA and juvenile diabetes.

The area of greatest speculation for the use of inhibitors of the purine metabolic cascade (*e.g.* ADA, SAHH, or PNP inhibitors) to treat autoimmune diseases lies in the actual site (particular T cell subset or immunologic process) which would best be inhibited to achieve immunosuppression. Based on the published findings of Talal and others,[163-165] it could be argued that T cells blocking release of IL 2 should be suppressed or eliminated so that IL 2 production and NK cell activity can normalize. T cells inhibiting IL 2 release belong to the class of suppressor T cells, and it is generally felt that the functions of this particular subset are more dependent on proliferation than are the functions of helper T cells. However, lack of suppressor T cells has also been postulated to account for the abnormal B cell function in PNP deficient patients. Thus, while autoantibodies and abnormal immunoglobulin production are features both of some immunodeficiency diseases and autoimmune diseases, it is not known if these abnormalities result from impairment at the same biochemical step or in the same cell type.

8.3.3.1.2 *Immunosuppressive therapy of rheumatoid arthritis*

Present therapy for RA, other than the numerous agents which provide symptomatic relief [nonsteroidal antiinflammatory drugs (NSAIDS), including aspirin, piroxicam, sulindac, naproxen, *etc.*] is directed towards reducing lymphocyte activity.[1-4] NonNSAID therapy has been termed 'DMARD', for disease-modifying antirheumatic drugs, and 'SAARD', for slow-acting antirheumatic drugs, to differentiate both the clinical effects and the onset of action of these drugs from NSAIDs. DMARDs generally did not come from ràtional drug discovery programs, but rather evolved through serendipity (*e.g.* gold compounds,[230] chloroquine,[230] D-penicillamine[231]) or were initially developed for the treatment of cancer (cyclophosphamide, methotrexate, azathioprine[232]) and were subsequently tested in RA. These agents have been extensively reviewed elsewhere and will only be summarized here (Table 5). Most have a number of effects, both biochemical and biological, and it is uncertain what the precise mode of action of many of them is. However, multiple effects on various components of the immune system have been documented.

8.3.3.1.3 *Experimental therapy*

NonNSAID therapies are under continuous development for the treatment of RA. A few of these are summarized in Table 6. For many, the mode of action is only hypothesized at present.

8.3.3.2 Other Autoimmune Diseases

A number of other diseases with unknown etiology and uncertain pathogenesis have autoimmune features. Altered immune responsiveness, abnormal lymphocyte subset ratios, autoantibodies, or pathology suggesting lymphoid involvement have been noted in these diseases, and as a result they are often considered autoimmune. Included in this category are systemic lupus erythematosus (SLE),[249,250] juvenile diabetes,[213-215] myasthenia gravis,[251] multiple sclerosis,[252] aplastic anemia,[253] Sjogren's syndrome,[254] and others. In some of these, such as SLE, the weight of evidence supporting the concept that they are autoimmune is substantial. As in SLE, autoantibodies to a number of tissue and cell components have been observed, along with abnormal lymphocyte function (impaired production of IL 2).[163-165]

8.3.4 HYPERURICEMIA

Hyperuricemia is the clinical finding of elevated urinary uric acid levels[255-258] (see also Section 7.3.3). Elevation of uric acid is common, usually asymptomatic, and is associated with a number of procedures and activities (exercise, ethanol ingestion, other dietary means, fructose infusion).[258] Hyperuricemia can also result from congenital disorders like HGPRT deficiency and glycogen storage disease and is considered the hallmark of gout.[256] It is also seen in certain acute situations such as adult respiratory distress syndrome and myocardial infarction. In the latter disorders, a common biochemical feature is a reduction in net intracellular ATP, and hence hyperuricemia has recently been proposed as a marker for cell energy crisis.[258] In some forms of glycogen storage disease, even moderate exercise produces hyperuricemia and elevation of urate precursors in plasma, due to the inability of muscle to utilize glucose to resynthesize ATP. ADP accumulates and initiates a cascading of ADP to AMP, Ado, and so on. Similarly, following ethanol ingestion, ATP is consumed in metabolizing acetate to acetyl CoA. Hypoxia also prevents ATP resynthesis from ADP, accounting for hyperuricemia in adult respiratory distress syndrome. In conditions in which hyperuricemia is a marker for cell energy crisis, therapy is symptomatic and/or is directed towards the insult or defect which results in ATP catabolism or the inability to resynthesize ATP. In conditions like gout, PNP inhibitors may provide a novel means of preventing urate accumulation and the resulting arthritis and nephropathy. The following subsections will outline this rationale and describe the effects of a PNP inhibitor in an animal model of gout.

8.3.4.1 Gout and Xanthine Oxidase Inhibitors

It is of considerable significance that in all mammals, except some higher primates and humans, the final purine degradative product is allantoin.[256] Uric acid is considerably less soluble than allantoin, but the latter cannot be produced in humans because of the absence of the enzyme uricase. Hence, purine overproduction results in the formation and deposition of insoluble uric acid in joint tissues, producing gout. Gout is generally treated prophylactically with uricosuric agents or allopurinol.[255,256,259] Allopurinol (1*H*-pyrazolo[3,4-*d*]pyrimidin-4-ol) is a xanthine oxidase inhibitor which prevents urate elevation by blocking oxidation of Hx and xanthine to uric acid. Acute attacks of gout are highly responsive to colchicine therapy through a mechanism totally unrelated to purine metabolism. Allopurinol is oxidized extensively to oxypurinol [1*H*-pyrazolo-[3,4-*d*]pyrimidine-4,6-diol] by xanthine oxidase. Oxypurinol is a noncompetitive inhibitor of xanthine oxidase and is excreted largely unmetabolized.[259] Allopurinol therapy is effective for gout and is well tolerated by most patients, with hypersensitivity reactions being the main adverse reaction.

8.3.4.2 Hyperuricemia and PNP Inhibition

PNP inhibitors should be effective in preventing hyperuricemia because of their ability to block formation of both purines leading to uric acid formation (guanine and, primarily, Hx). As noted earlier, PNP deficient patients have reduced levels of uric acid and oxypurines in plasma and urine. While xanthine oxidase inhibitors achieve the same effect on uric acid output, the negative aspect of xanthine oxidase inhibition is the accumulation of xanthine, which: (i) is very poorly soluble in plasma, (ii) can aggravate gout, and (iii) lead to xanthine gout. The substrates of PNP by contrast are considerably more soluble owing to the ribose or deoxyribose moiety. Moderate accumulation of either Ino or Guo in plasma should be inconsequential because these nucleosides are readily excreted in the urine.

Fox and Kelly described a model for study of hyperuricemia which is induced by infusion of fructose (Figure 9; reviewed in ref. 260). The phosphorylation of fructose consumes ATP, which initiates the cascade that ultimately results in hyperuricemia (in humans) and hyperallantoinemia (in rats). Studies were recently conducted to determine if PNP inhibitors would be efficacious in this animal model of gout. The recently described PNP inhibitor PD 119,229 was used for these studies.[208] Excretion of allantoin in control rats averaged 27.2 μg h^{-1} over the duration of study (6 h; Figure 10). Allantoin excretion reached 64.9 μg h^{-1} 30 min after fructose infusion in nondrug-treated rats, and remained elevated 4 h post fructose infusion. When PD 119,229 was administered at 50 mg kg^{-1} by mouth 20 min before fructose infusion, elevation of allantoin was totally prevented over the duration of study. Allantoin levels in rats treated with PD 119,229 or allopurinol (100 mg kg^{-1}, by mouth) were comparable.[208] These observations suggest that PNP inhibitors may have utility in some hyperuricemic states.

8.3.5 MISCELLANEOUS CONDITIONS

8.3.5.1 Organ Transplantation

The ultimate success of organ transplantation is clearly as dependent on immunosuppression as on the surgical procedure. The drugs used to prolong organ transplant survival are immunosuppressive and consist mostly of cyclosporin, steroids and azathioprine. The advent of cyclosporin in the

Table 5 Approved DMARD: Immunosuppressive or Immunomodulatory Drugs for Rheumatoid Arthritis (RA)

Drugs	Structure	Route of administration and dose	Mechanism of action in RA	Side effects	Ref.
Corticosteroids, *e.g.* prednisone (X = O) prednisolone (X = OH)		Injectable or PO[a], 5–10 mg d^{-1}	Modulation of lymphocyte and macrophage function, inhibition of IL 1 production, inhibition of PLA$_2$ indirectly	Osteoporosis, cataracts, poor wound healing, GI bleeding, immuno-suppression	233
Gold salts gold sodium thiomalate	CH$_2$CO$_2$Na AuSCHCO$_2$Na	Injectable 25–50 mg week^{-1}	Inhibition of mononuclear phagocyte function	Bone marrow depression, proteinuria, MI, leukopenia, aplastic anemia, thrombo-cytopenia, hepatitis	234
gold thioglucose					
Auranofin		PO, 6 mg d^{-1}	Possible mechanism as deactivator of singlet oxygen formation Inhibits cell proliferation and has multiple effects on tissues, neutrophils, macrophages *etc.*	Multiple: fatal thrombocytopenia, anorexia, anemia, proteinuria	235
Azathioprine		PO, 50–100 mg d^{-1}	T and B lymphocyte suppression. Suppression of natural killer cell activity	Nausea, vomiting, abdominal pain, hepatitis, leukopenia, severe bone marrow depression, carcinogenesis	236
Cyclophosphamide		PO, 100 mg d^{-1}	B and T lymphocyte suppression	Alopecia, hemorrhagic cystitis, severe bone marrow depression, sterility, bladder cancer, malignant diseases	237

Drug	Structure	Dose[a]	Mechanism	Side effects	Ref.
D-Penicillamine	Me H HS—C—C—CO₂H Me NH₂	PO, 10 mg kg⁻¹ d⁻¹	Sulfide–disulfide exchange, altered redox equilibrium, inhibits T cell function	Skin rashes, dysgensia, proteinuria, leukopenia, aplastic anemia, thrombocytopenia, teratogenic, lupus-like illness, fatal bronchiolitis, bone marrow depression	237
Hydroxychloroquine (R = OH), chloroquine (R = H)	(quinoline structure with Cl, N, NHCH(CH₃)₃N(Et)CH₂CH₂R)	PO, 6 mg kg⁻¹ d⁻¹	Stabilizes lysosomal membranes, free radical scavenger, inhibits PLA_2 indirectly, inhibition of Il 1 production, inhibits PMN, B lymphocytes and mononuclear phagocyte functions	Retinal toxicity, GI disturbances, fatal bone marrow depression nephrotoxicity, CNS side effects	238, 239
Methotrexate	(pteridine structure with NH_2, H_2N, N, Me, CH₂, CONHCH, CO_2H, $HO_2C(CH_2)_2$)	PO, 7.5 mg week⁻¹	Folate antagonist, immunosuppressive, inhibits DNA, RNA and protein synthesis	Teratogenic, severe hepatic toxicity, intestinal pneumonitis, bone marrow suppression, severe ulceration and bleeding	236, 240
Sulfasalazine	(structure with pyridine-NHSO₂—C₆H₄—N=N—C₆H₃(CO₂H)(OH))	PO, 1.5 g d⁻¹	Depresses T lymphocytes function and number, reduces B lymphocytes. Metabolite, 5-ASA, is a weak inhibitor of PGE_2 synthesis; metabolite, SP, reduces antigen absorption from gut	CNS and GI disturbances, vomiting, nausea, agranulocytosis, leukopenia, thrombocytopenia, anorexia, cyanosis, anemia, skin rashes	241
Levamisole	(bicyclic thiazolo-imidazole structure with S, N, N, phenyl, H)	PO, 150 mg d⁻¹	Unknown: restored IL 2 activity, lowered antibody production, induces expression of T cells	High incidence of GI, neurological and hematological side effects, agranulocytosis	242
Lobenzarit (CCA)	(structure with CO_2H, NH, phenyl, Cl, CO_2H)	PO, 240 mg d⁻¹	Immunomodulator, affects suppressor T and helper T cell induction	GI disorder, skin disorder, renal dysfunction, CNS effects	243
Orgotein	Cu–Zn-containing SOD enzyme (M = 32 000)	Injectable 4 mg week⁻¹	Superoxide dismutase (SOD), inhibits PG synthesis (?)	Not known	244

[a] PO, by mouth.

Table 6 Experimental Immunosuppressive or Immunomodulatory Drugs for RA

Drugs	Structure	Route of administration and dose	Mechanism of action in RA	Side effects	Ref.
Cyclosporin	11 amino acid cyclic peptide, $M = 1202.63$	PO, 10–6 mg kg^{-1} d^{-1}	T cell modulator, IL 2 inhibitor	Renal function impairment and toxicity, hypertension, lymphoma, convulsion	210, 245
Therafectin		PO, 100 mg kg^{-1} d^{-1}	Immunomodulator, selective suppression of T_8 cells	Not known	246
NPT 15392			Enhances T effector cells, immunomodulator	Not known	247
TCSP	T cell suppressive peptide		Controls T cells involved in autoimmune diseases	Not known	248

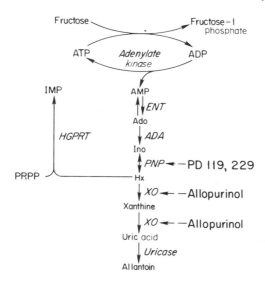

Figure 9 Schematic diagram for nucleotide degradation following fructose infusion. Infusion of fructose consumes adenosine triphosphate (ATP) in the formation of fructose 1-phosphate and gives rise to ADP. ATP can be reformed by at least three mechanisms, one of which involves adenylate kinase. With extensive ATP utilization, ADP accumulates and initiates a cascading of purines down through AMP, adenosine (Ado), inosine (Ino), hypoxanthine (Hx), eventually leading to allantoin excretion in rats. In large part, Hx is salvaged by the action of hypoxanthine–guanine phosphoribosyltransferase (HGPRT), which utilizes phosphoribosyl pyrophosphate (PRPP) and forms IMP. Enzymes involved include ecto-5'-nucleotidase (ENT), adenosine deaminase (ADA), purine nucleoside phosphorylase (PNP), and xanthine oxidase (XO).

PD 119,229, a PNP inhibitor, is described in the text. Allopurinol is a XO inhibitor used in the treatment of gout

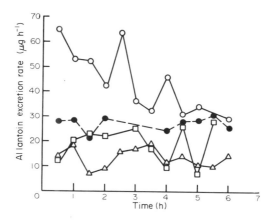

Figure 10 Effects of purine metabolic enzyme inhibitors on allantoin excretion in rats following fructose infusion. Rats were pretreated with 5 mL water i.p. 20 min prior to drug administration to enhance urine specimen collection. Compounds (PD 119,229, 50 mg kg^{-1}, and allopurinol, 100 mg kg^{-1}) were then administered by mouth. 15 min after oral dosing, fructose (800 mg kg^{-1}) was administered i.v. Urine specimens were collected at 30 min intervals for 6 h post fructose infusion, starting 30 min post infusion. Allantoin excretion in untreated normal rat urine averaged 27.2 µg h^{-1} over the collection period (●). Excretion of allantoin in fructose-infused, nondosed rats (O) peaked at 64.9 µg h^{-1} 30 min after fructose infusion. Excretion of allantoin following treatment with allopurinol (△) or PD 119,229 (□) was reduced to levels below those present in normal rat urine

early 1980s[173] marked the beginning of an era in which organ transplantation was no longer viewed as a short-term cure.

Several recent reviews on cyclosporin are available.[173,261] While the precise mechanism of action of cyclosporin is unproven, cyclosporin clearly prevents synthesis of IL 2,[261] which is required for the generation of cytotoxic T lymphocytes that directly mediate graft rejection. Cyclosporin prevents synthesis of mRNA coding for IL 2, while sparing the synthesis of other mRNAs.[262,263] Cyclosporin has also been reported to block IL 2 receptor gene activation.[264] Receptors for cyclosporin, called cyclophilin, have been demonstrated on lymphocytes.[265] A major biochemical effect of cyclosporin appears to be mediated through its binding to calmodulin,[266,267] although this activity alone is not sufficient to explain its immunosuppressive effect because nonimmunosuppressive cyclosporin analogs also bind calmodulin.[268] Cyclosporin binds to the prolactin receptor on lymphocytes and displaces prolactin, and this property has been hypothesized to contribute to its immunosuppressive activity.[269] Cyclosporin also affects phospholipid metabolism,[270] ornithine decarboxylase,[271] and probably exerts its effects *in vivo* through a variety of mechanisms.[268,272]

PNP inhibitor therapy theoretically offers a novel approach towards this end, as noted earlier in this chapter, by inhibiting replication of stimulated (antigen-activated) T lymphocytes. Terasaki hypothesized that the reason blood transfusions were able to prolong graft survival in humans was because they sensitized host lymphocytes. Primed lymphocytes, upon restimulation by the foreign graft, begin to divide, at which point they are more susceptible to the cytotoxic effects of immunosuppressive drugs.[273] If timed appropriately, this procedure can result in elimination of cells which are able to reject the graft. It is well established that antilymphocyte serum (ALS) is beneficial adjunct therapy in organ transplantation, and the effects of ALS are felt to derive from its ability to eliminate graft reactive lymphocytes. Cobbold *et al.* reported that various anti-T cell monoclonal antibodies were highly effective in depleting T cell subpopulations *in vivo*, and could significantly prolong skin allograft survival in mice.[274] Use of PNP inhibitors, possibly in conjunction with dGuo, could mimic the effects of anti-T lymphoblast antibodies because only replicating T cells would be killed. This form of therapy could have significant additional utility during rejection crisis episodes, which result from vigorous lymphocyte assault. Further, it has been demonstrated that dGuo treatment of mouse thymus grafts *in vitro* eliminates resident thymocytes (and perhaps dendritic cells) without affecting the viability of the thymic epithelium.[275] Transplantation of these grafts into histoincompatible mice was followed by full repopulation of the thymus graft with host lymphocytes. As PNP inhibition was not attempted in those experiments, it is uncertain if the effect of dGuo was due to accumulation of dGTP or GTP. If the biochemical mechanism for this effect was due to the accumulation of dGTP, PNP inhibitors should enhance this activity. Similarly, in allogeneic bone marrow transplantation, donor bone marrow could be stimulated with recipient leukocytes in the presence of a PNP inhibitor and dGuo to eliminate cells capable of producing graft *versus* host disease. The PNP inhibitor 8-aminoguanosine,[207] as well as the ADA inhibitors 2'-deoxycoformycin[82] and EHNA,[276] have been found to be active in transplant models.

8.3.5.2 Psoriasis

Psoriasis is a disease characterized by hyperproliferation and lack of terminal differentiation of skin cells.[277] The characteristic pathology consists of neutrophilic infiltration and microabscesses. Abnormalities of arachidonic acid metabolism have been documented in psoriasis, and include markedly elevated PGE_2, 12-HETE and 5-HETE.[278] Preparations of psoriatic lesions have also been shown to contain platelet-activating factor and other neutrophilic chemoattractants, including C5a, as well as interleukin 1 (IL 1).[279] The pathological changes are generally felt to be due to a lack of regulation of skin proliferation, with the neutrophilic involvement being secondary, although there is controversy in this regard.

Although historically little support has been given to the role of lymphocytes in the pathogenesis of psoriasis, new data are accumulating which are changing this concept. Psoriatic lesions have been shown to contain a higher percentage of activated T cells than normal skin or nonlesional skin from psoriatic patients.[280-283] The presence in skin of IL 1, which is a known activator of T lymphocytes, is consistent with this documentation of activated T cells in these lesions. The skin immune system is highly complex, containing a variety of accessory cells and lymphocytes.[280-283] In addition, the proposal has been advanced that abnormal keratinocyte proliferation is induced by T lymphocytes. Consistent with the concept that lymphocytes are important in psoriasis is the recent documentation of the high degree of efficacy of cyclosporin in psoriasis.[284] As noted earlier, cyclosporin produces several biochemical effects, and it is uncertain at this point if the efficacy of cyclosporin in psoriasis is due to blockade of IL 2 release. If further evidence is generated implicating T cells in the pathogenesis of psoriasis, the rationale for testing agents like PNP inhibitors will be given greater impetus.

8.3.5.3 Parasitic Diseases

The role of purines in certain parasitic diseases has been discussed in Section 7.3.6, and will only be briefly mentioned in the context of this chapter. The biochemistry of the malaria parasite has been described in detail.[285] It is known that malaria parasites lack the machinery needed for *de novo* synthesis of purines, and therefore utilize host-derived guanine and Hx to synthesize guanine and adenine nucleotides in the pathway involving ADA, PNP and HGPRT.[145,286] This dependency of the malaria parasite on the salvage pathway offers a rational approach to the development of specific chemotherapeutic agents. The malaria parasite also expresses a distinct species of ADA and PNP.[145] *In vitro* cultures of human erythrocytes infected with *Plasmodium falciparum* showed a three-fold increase in ADA activity over uninfected human erythrocytes, and this was shown to be due to expression of the malarial ADA in these cells.[287] Malaria ADA: (i) is not polymorphic, as is human ADA; (ii) is not modified posttranslationally; and (iii) is distinct from human ADA in its affinity for the ADA inhibitor, EHNA, while having affinity comparable to human ADA for 2'-deoxycoformycin. The reduced affinity for EHNA suggests structural differences which might be exploited to develop an inhibitor specific for malarial ADA.

The ADA inhibitor 2'-deoxycoformycin has been shown to have a marked effect on the growth of *P. knowlesi* in rhesus monkeys.[287] Injection of a single i.v. bolus dose of 2'-deoxycoformycin at 250 µg kg^{-1} after the level of parasite-infected erythrocytes (PRBC) had reached 5.9% was followed by a rapid and profound decrease in the percentage of PRBCs. Infected erythrocytes increased after about 3 d, but it is possible that chronic administration would have a more sustained effect.

The native PNP of the malarial parasite, *Plasmodium lophurae*, is a pentamer of molecular weight 125 300 Da and subunit molecular weight of 23 900 (contrast with human PNP which is a trimer with molecular weight of 94 000 Da). The K_m of the malarial PNP is 33 µM for Ino and 82 µM for Hx. Guo and dGuo are poor substrates for the parasite PNP, again in contrast to the human enzyme. These functional differences suggest that there are structural differences in the active sites of the PNP from these two species which could be exploited in the design of selective inhibitors. Consistent with this possibility, formycin B is a poor inhibitor of human PNP (see Section 8.3.2.4.3) but is a potent inhibitor of malarial PNP with a K_i of 0.39 µM.[145] Conversely, 8-aminoguanosine is not an effective inhibitor of the malaria parasite PNP (8.5% inhibition at 100 µM), but is a potent inhibitor of human PNP ($K_i = 1.33$ µM). The PNP inhibitors 9-deazainosine and formycin B were found to suppress an acute experimental *Trypanosoma brucei* infection in mice. 9-Deazainosine, a nontoxic purine analog, when tested in combination with the ornithine decarboxylase inhibitor, eflornithine, was capable of late stage cures of murine African trypanosomiasis.[288]

8.3.6 SUMMARY

In this chapter, we have described the rationale for the design of novel agents, inhibitors of enzymes in the purine metabolic cascade, which have potential utility in a wide variety of disorders. Among these are autoimmune diseases, such as rheumatoid arthritis, systemic lupus erythematosus and others; organ transplantation; certain hyperuricemic states; parasitic diseases; and psoriasis; in addition to the widely held interest in testing them in leukemias and lymphomas. The rationale for most aspects of this concept is based on study of immunodeficient children who are deficient in certain enzymes within the purine metabolic cascade. These patients lack some aspects of immune function which appear to be the very elements of the immune system that are abnormal and unregulated in patients with autoimmune diseases. Hence, novel immunosuppressive agents developed as inhibitors of purine metabolism, offer the potential for superior therapy. Immunosuppressive agents like cyclosporin, steroids, azathioprine, methotrexate and others are known to be therapeutic in autoimmune diseases, transplantation, psoriasis and some forms of cancer. Therefore, included in this chapter is a summary of novel agents, structural analogs of which may offer more efficacious, or better-tolerated, forms of therapy for these diseases. Whether inhibitors of enzymes within this cascade, such as ADA, SAHH, or PNP inhibitors, will have utility in these diseases, remains to be determined.

8.3.7 REFERENCES

1. G. W. Cannon, in 'Biologically Based Immunomodulators in the Therapy of Rheumatic Diseases', ed. S. H. Pincus, D. S. Pisetsky and L. J. Rosenwasser, Elsevier, 1986, p. 79.
2. F. Spreafico, A. Tagliabue and A. Vecchi, in 'Immunopharmacology', ed. P. Sirois and M. Rola-Pleszczynski, Elsevier Biomedical Press, 1982, p. 315.
3. G. C. Tsokos, *Semin. Arthritis Rheum.*, 1987, **17**, 24.
4. J. J. Solovera, *Med. Actual.*, 1987, **23**, 575.
5. I. H. Fox and W. N. Kelley, *Annu. Rev. Biochem.*, 1978, **47**, 655.
6. W. P. Schrader, A. R. Stacy and B. Pollara, *J. Biol. Chem.*, 1976, **251**, 4026.
7. P. E. Daddona and W. N. Kelley, *J. Biol. Chem.*, 1977, **252**, 110.

8. P. E. Daddona, S. H. Orkin, D. S. Shewach and W. N. Kelley, *Ann. N.Y. Acad. Sci.*, 1985, **451**, 238.
9. D. A. Wiginton, G. S. Adrian and J. J. Hutton, *Nucleic Acids Res.*, 1984, **12**, 2439.
10. E. R. Giblett, J. E. Anderson, F. Cohen, B. Pollara and H. J. Meuwissen, *N. Engl. J. Med.*, 1972, **2**, 1067.
11. R. H. Buckley, *J. Allergy Clin. Immunol.*, 1983, **72**, 627.
12. B. S. Mitchell, *Medical Grand Rounds*, 1982, **1**, 85.
13. D. A. Carson, E. Lakow, D. B. Wasson and N. Kamatani, *Immunol. Today*, 1981, **2**, 234.
14. L. F. Thompson and J. E. Seegmiller, *Adv. Enzymol. Relat. Areas Mol. Biol.*, 1980, **51**, 167.
15. D. W. Martin, Jr. and E. W. Gelfand, *Annu. Rev. Biochem.*, 1981, **50**, 845.
16. B. S. Mitchell and W. N. Kelley, *Ann. Intern. Med.*, 1980, **92**, 826.
17. N. M. Kredich and M. S. Hershfield, in 'The Metabolic Basis of Inherited Disease', ed. J. B. Stanbury, J. B. Wyngaarden, D. S. Fredrickson, J. L. Goldstein and M. S. Brown, McGraw-Hill,' New York, 1983, 5th edn., p. 1157.
18. I. H. Fox, *Metabolism*, 1981, **30**, 616.
19. M. J. Cowan and A. J. Ammann, *Clin. Haematol.*, 1981, **10**, 139.
20. G. R. Boss and J. E. Seegmiller, *Annu. Rev. Genet.*, 1982, **16**, 297.
21. E. W. Gelfand and A. Cohen, *Adv. Host Defense Mechanisms*, 1983, **2**, 43.
22. F. F. Snyder, J. Mendelsohn and J. E. Seegmiller, *J. Clin. Invest.*, 1976, **58**, 654.
23. A. Cohen, R. Hirschhorn, S. D. Horowitz, A. Rubinstein, S. H. Polmar, R. Hong and D. W. Martin, Jr., *Proc. Natl. Acad. Sci. USA*, 1978, **75**, 472.
24. D. A. Carson, J. Kaye and J. E. Seegmiller, *Proc. Natl. Acad. Sci. USA*, 1977, **74**, 5677.
25. E. C. Moore and R. B. Hurlbert, *J. Biol. Chem.*, 1966, **241**, 4802.
26. L. Thelander and P. Reichard, *Annu. Rev. Biochem.*, 1979, **48**, 133.
27. H. Green and T.-S. Chan, *Science (Washington, D.C.)*, 1973, **182**, 836.
28. G. Wolberg, T. P. Zimmerman, K. Hiemstra, M. Winston and L. C. Chu, *Science (Washington, D.C.)*, 1975, **187**, 957.
29. M. F. E. Siaw, B. S. Mitchell, C. A. Koller, M. S. Coleman and J. J. Hutton, *Proc. Natl. Acad. Sci. USA*, 1980, **77**, 6157.
30. H. A. Simmonds, R. J. Levinsky, D. Perrett and D. R. Webster, *Biochem. Pharmacol.*, 1982, **31**, 947.
31. S. Seto, C. J. Carrera, M. Kubota, D. B. Wasson and D. A. Carson, *J. Clin. Invest.*, 1985, **75**, 377.
32. G. R. Boss and R. B. Pilz, *J. Clin. Invest.*, 1984, **74**, 1262.
33. S. S. Matsumoto, J. Yu and A. L. Yu, *J. Immunol.*, 1983, **131**, 2762.
34. J. Yu, S. S. Matsumoto and A. L. Yu, *Cancer Treat. Symp.*, 1984, **2**, 75.
35. S. S. Cohen, *Ann. N.Y. Acad. Sci.*, 1985, **451**, 204.
36. R. P. Agarwal, *Pharmacol. Ther.*, 1982, **17**, 399.
37. R. I. Glazer, *Cancer Chemother. Pharmacol.*, 1980, **4**, 227.
38. H. J. Schaeffer and C. F. Schwender, *J. Med. Chem.*, 1974, **17**, 6.
39. D. C. Baker, J. C. Hanvey, L. D. Hawkins and J. Murphy, *Biochem. Pharmacol.*, 1981, **30**, 1159.
40. G. Bastian, M. Bessodes, R. P. Panzica, E. Abushanab, S. F. Chen, J. D. Stoeckler and R. E. Parks, Jr., *J. Med. Chem.*, 1981, **24**, 1383.
41. M. D. Bessodes, G. Bastian, E. Abushanab, R. P. Panzica, S. F. Berman, E. J. Marcaccio, Jr., S. F. Chen, J. D. Stoeckler and R. E. Parks, Jr., *Biochem. Pharmacol.*, 1982, **31**, 879.
42. T. W. North and S. S. Cohen, *Proc. Natl. Acad. Sci. USA*, 1978, **75**, 4684.
43. T. Niida, T. Niwa, T. Tsuruoka, N. Ezaki, T. Shomura and H. Umezawa, *153rd Meeting of Japan Antibiotics Research Association*, January 27, 1967.
44. R. P. Agarwal, S. M. Sagar and R. E. Parks, Jr., *Biochem. Pharmacol.*, 1975, **24**, 693.
45. H. Nakamura, G. Koyama, Y. Iitaka, M. Ohno, N. Yagisawa, S. Kondo, K. Maeda and H. Umezawa, *J. Am. Chem. Soc.*, 1974, **96**, 4327.
46. D. G. Johns and R. H. Adamson, *Biochem. Pharmacol.*, 1976, **25**, 1441.
47. P. W. K. Woo, H. W. Dion, S. M. Lange, L. F. Dahl and L. J. Durham, *J. Heterocycl. Chem.*, 1974, **11**, 641.
48. R. C. Jackson, W. R. Leopold and D. A. Ross, *Adv. Enzyme Regul.*, 1986, **25**, 125.
49. S. Omura, N. Imamura, H. Kuga, H. Ishikawa, Y. Yamazaki, K. Okano and K. Kimura, *J. Antibiot. (Tokyo)*, 1985, **38**, 1008.
50. J. B. Tunac and M. Underhill, *J. Antibiot. (Tokyo)*, 1985, **38**, 1344.
51. J. P. Schaumberg, G. C. Hokanson, J. C. French, E. Smal and D. C. Baker, *J. Org. Chem.*, 1985, **50**, 1651.
52. S. Omura, H. Tanaka, H. Kuga and N. Imamura, *J. Antibiot. (Tokyo)*, 1986, **39**, 309.
53. S. Omura, H. Ishikawa, H. Kuga, N. Imamura, S. Taga, Y. Takahashi and H. Takana, *J. Antibiot. (Tokyo)*, 1986, **39**, 1219.
54. R. Hirschhorn, J. Grossman and G. Weissmann, *Proc. Soc. Exp. Biol. Med.*, 1970, **133**, 1361.
55. D. A. Carson and J. E. Seegmiller, *J. Clin. Invest.*, 1976, **57**, 274.
56. H. A. Simmonds, G. S. Panayi and V. Corrigall, *Lancet*, 1978, **1**, 60.
57. J. Uberti, J. J. Lightbody and R. M. Johnson, *J. Immunol.*, 1979, **123**, 189.
58. D. A. Carson, D. B. Wasson, E. Lakow and N. Kamatani, *Proc. Natl. Acad. Sci. USA*, 1982, **79**, 3848.
59. E. W. Gelfand, J. J. Lee and H. M. Dosch, *Proc. Natl. Acad. Sci. USA*, 1979, **76**, 1998.
60. H. G. Bluestein, L. F. Thompson, D. A. Albert and J. E. Seegmiller, *Adv. Exp. Med. Biol.*, 1980, **122A**, 427.
61. A. H. Cohen, H. G. Bluestein and D. Redelman, *J. Immunol.*, 1984, **132**, 1761.
62. B. S. Mitchell, E. Mejias, P. E. Daddona and W. N. Kelley, *Proc. Natl. Acad. Sci. USA*, 1978, **75**, 5011.
63. D. A. Carson, J. Kaye, S. Matsumoto, J. E. Seegmiller and L. Thompson, *Proc. Natl. Acad. Sci. USA*, 1979, **76**, 2430.
64. J. M. Wilson, B. S. Mitchell, P. E. Daddona and W. N. Kelley, *J. Clin. Invest.*, 1979, **64**, 1475.
65. J. F. Henderson, F. W. Scott and J. K. Lowe, *Pharmacol. Ther.*, 1980, **8**, 573.
66. R. F. Kefford and R. M. Fox, *Cancer Res.*, 1982, **42**, 324.
67. R. F. Kefford and R. M. Fox, *Br. J. Haematol.*, 1982, **50**, 627.
68. A. Goday, H. A. Simmonds, G. S. Morris and L. D. Fairbanks, *Clin. Exp. Immunol.*, 1984, **56**, 39.
69. A. Goday, H. A. Simmonds, L. D. Fairbanks and G. S. Morris, in 'Purine and Pyrimidine Metabolism In Man', ed. W. L. Nyhan, L. F. Thompson and R. W. E. Watts, Plenum Press, New York, 1986, vol. 5, part A, p. 515.
70. H. E. Gruber, A. H. Cohen, G. S. Firestein, D. Redelman and H. G. Bluestein, in 'Purine and Pyrimidine Metabolism In Man', ed. W. L. Nyhan, L. F. Thompson and R. W. E. Watts, Plenum Press, New York, 1986, vol. 5, part A, p. 503.

71. P. E. Borondy, T. Chang, E. Maschewske and A. J. Glazko, *Ann. N.Y. Acad. Sci.*, 1977, **284**, 9.
72. P. W. Burridge, V. Paetkau and J. F. Henderson, *J. Immunol.*, 1977, **119**, 675.
73. P. P. Trotta, A. Tedde, S. Ikehara, R. Pahwa, R. A. Good and M. E. Balis, *Cancer Res.*, 1981, **41**, 2189.
74. A. Tedde, M. E. Balis, S. Ikehara, R. Pahwa, R. A. Good and P. P. Trotta, *Proc. Natl. Acad. Sci. USA*, 1980, **77**, 4899.
75. E. M. Sordillo, S. Ikehara, R. A. Good and P. P. Trotta, *Cell. Immunol.*, 1981, **63**, 259.
76. D. A. Carson, J. Kaye and D. B. Wasson, *J. Immunol.*, 1980, **124**, 8.
77. C. M. Smith and J. F. Henderson, *Biochem. Pharmacol.*, 1982, **31**, 1545.
78. F. W. Burgess, M. Mell and R. E. Parks, Jr., *The Pharmacologist*, 1984, **26**, 189.
79. R. B. Gilbertsen, *J. Immunopharmacol.*, 1985, **7**, 325.
80. T. J. M. Ruers, W. A. Buurman, C. J. van der Linden and G. Kootstra, *Transplantation*, 1985, **40**, 137.
81. R. W. Barton, *Cell. Immunol.*, 1985, **95**, 297.
82. R. B. Epstein, T. Fey and S. Sarpel, *Transplantation*, 1982, **33**, 208.
83. D. Kufe, P. Major, R. Agarwal, E. Reinherz and E. Frei, *Proc. Am. Soc. Clin. Oncol.*, 1980, **21**, 328.
84. R. S. Benjamin, W. Plunkett, M. J. Keating, L. G. Feun, V. Hug, J. A. Nelson, G. P. Bodey and E. J. Freireich, *Proc. Am. Soc. Clin. Oncol.*, 1980, **21**, 337.
85. H. G. Prentice, K. Ganeshaguru, K. F. Bradstock, A. H. Goldstone, J. F. Smyth, B. Wonke, G. Janossy and A. V. Hoffbrand, *Lancet*, 1980, **2**, 170.
86. N. Winick, G. Buchanan, S. B. Murphy and A. Yu, *Proc. Am. Soc. Clin. Oncol.*, 1987, **6**, A599.
87. A. S. Spiers, J. C. Ruckdeschel and J. Horton, *Scand. J. Haematol.*, 1984, **32**, 130.
88. B. S. Mitchell, C. A. Koller and R. Heyn, *Blood*, 1980, **56**, 556.
89. B. S. Mitchell, N. L. Edwards and C. A. Koller, *Blood*, 1983, **62**, 419.
90. P. P. Major, R. P. Agarwal and D. W. Kufe, *Blood*, 1981, **58**, 91.
91. N. H. Russell, H. G. Prentice, N. Lee, A. Piga, K. Ganeshaguru, J. F. Smyth and A. V. Hoffbrand, *Br. J. Haematol.*, 1981, **49**, 1.
92. H. G. Prentice, N. Lee, H. Blacklock, J. F. Smyth, N. H. Russell, K. Ganeshaguru, A. Piga and A. V. Hoffbrand, *Lancet*, 1981, **2**, 1250.
93. A. S. D. Spiers, D. Moore, P. A. Cassileth, D. P. Harrington, F. J. Cummings, R. S. Neiman, J. M. Bennett and M. J. O'Connell, *N. Engl. J. Med.*, 1987, **316**, 825.
94. B. D. Cheson and A. Martin, *Ann. Intern. Med.*, 1987, **106**, 871.
95. P. J. O'Dwyer, A. S. D. Spiers and S. Marsoni, *Cancer Treat. Rep.*, 1986, **70**, 1117.
96. J. R. S. Arch and E. A. Newsholme, *Essays Biochem.*, 1978, **14**, 82.
97. J. W. Daly, R. F. Bruns and S. H. Snyder, *Life Sci.*, 1981, **28**, 2083.
98. G. L. Cantoni, in 'Biochemistry and Biology of DNA Methylation', ed. G. L. Cantoni and A. Razin, Liss, New York, 1985, p. 47.
99. G. De La Haba and G. L. Cantoni, *J. Biol. Chem.*, 1959, **234**, 603.
100. R. T. Borchardt, *J. Med. Chem.*, 1980, **23**, 347.
101. P. M. Ueland, *Pharmacol. Rev.*, 1982, **34**, 223.
102. R. T. Borchardt, *Biochem. Adenosylmethionine, Proc. Int. Symp.*, 1977, 152.
103. J. A. Montgomery, A. T. Shortnacy and S. D. Clayton, *J. Heterocycl. Chem.*, 1977, **14**, 195.
104. A. Guranowski, J. A. Montgomery, G. L. Cantoni and P. K. Chiang, *Biochemistry*, 1981, **20**, 110.
105. T. P. Zimmerman, C. J. Schmitges, G. Wolberg, R. D. Deeprose, G. S. Duncan, P. Cuatrecasas and G. B. Elion, *Proc. Natl. Acad. Sci. USA*, 1980, **77**, 5639.
106. T. P. Zimmerman, M. Iannone and G. Wolberg, *J. Biol. Chem.*, 1984, **259**, 1122.
107. T. Hoffman, F. Hirata, P. Bougnoux, B. A. Fraser, R. H. Goldfarb, R. B. Herberman and J. Axelrod, *Proc. Natl. Acad. Sci. USA*, 1981, **78**, 3839.
108. J. L. Medzihradsky, T. P. Zimmerman, G. Wolberg and G. B. Elion, *J. Immunopharmacol.*, 1982, **4**, 29.
109. J. L. Medzihradsky, *J. Immunol.*, 1984, **133**, 946.
110. R. Vinegar and G. Wolberg, *US Pat.* 4 322 411 (1982) (*Chem. Abstr.*, 1982, **96**, 205 444b).
111. A. J. Bonder, G. L. Cantoni and P. K. Chiang, *Biochem. Biophys. Res. Commun.*, 1981, **98**, 476.
112. P. K. Chiang, G. L. Cantoni, J. P. Bader, W. M. Shannon, H. J. Thomas and J. A. Montgomery, *Biochem. Biophys. Res. Commun.*, 1978, **82**, 417.
113. E. De Clercq and J. A. Montgomery, *Antiviral Res.*, 1983, **3**, 17.
114. Y. F. Shealy and J. D. Clayton, *J. Am. Chem. Soc.*, 1969, **91**, 3075.
115. W. W. Lee, A. Benitez, L. Goodman and R. R. Baker, *J. Am. Chem. Soc.*, 1960, **82**, 2648.
116. G. A. Lepage, in 'Antineoplastic and Immunosuppressive Agents', ed. A. C. Sartorelli and D. G. Jones, Springer, Berlin, 1975, vol. 2, p. 426.
117. S. Helland and P. M. Ueland, *Cancer Res.*, 1982, **42**, 2861.
118. S. Helland and P. M. Ueland, *Cancer Res.*, 1983, **43**, 1847.
119. G. A. Lepage, in 'Antineoplastic and Immunosuppressive Agents', ed. A. C. Sartorelli and D. G. Jones, Springer, Berlin, 1975, vol. 2, p. 430.
120. F. D. Ragione and A. E. Pegg, *Biochem. J.*, 1983, **210**, 429.
121. J. A. Montgomery, S. J. Clayton, H. J. Thomas, W. M. Shannon, G. Arnett, A. J. Bodner, I. Kion, G. L. Cantoni and P. K. Chiang, *J. Med. Chem.*, 1982, **25**, 626.
122. I. Kim, B. Shim, S. Yang and O. Kwon, *Hanguk Saenghwa Hakhoe Chi*, 1987, **19**, 1.
123. R. K. Gordon, N. D. Brown and P. K. Chiang, *Biochem. Biophys. Res. Commun.*, 1983, **114**, 505.
124. J. Aarbakke, G. A. Miura, P. S. Prytz, A. Bessesen, L. Slordal, R. K. Gordon and P. K. Chiang, *Cancer Res.*, 1986, **46**, 5469.
125. S. Yaginuma, M. Tsujino, N. Muto, M. Hayashi, F. Ishimura, T. Fugi, S. Watanabe, T. Matsuda, T. Watanabe and J. Abe, *Curr. Chemother. Infect. Dis. Proc. Int. Congr. Chemother.*, 11th, 1979, **2**, 1558.
126. R. T. Borchardt, B. T. Kellar and U. Patel-Thombre, *J. Biol. Chem.*, 1984, **259**, 4353.
127. R. L. Bartel and R. T. Borchardt, *Anal. Biochem.*, 1985, **149**, 191.
128. B. Matuszewska and R. T. Borchardt, *Arch. Biochem. Biophys.*, 1987, **256**, 50.
129. S. Yaginuma, N. Muto, M. Tsujino, Y. Sudate, M. Hayashi and M. Otani, *J. Antibiot.*, *(Tokyo)*, 1981, **34**, 359.

130. E. De Clercq, *Antimicrob. Agents Chemother.*, 1985, **28**, 84.
131. R. I. Glazer, K. D. Hartman, M. C. Knode, M. M. Richard, P. K. Chiang, C. K. Tseng, V. E. Marquez, *Biochem. Biophys. Res. Commun.*, 1986, **135**, 688.
132. R. I. Glazer, M. C. Knode, C. K. Tseng, D. R. Haines and V. E. Marquez, *Biochem. Pharmacol.*, 1986, **35**, 4523.
133. G. I. Drummond and M. Yamamoto, in 'The Enzymes', ed. P. D. Boyer, Academic Press, New York, 1971, vol. 4, p. 337.
134. N. L. Edwards, D. Recker, J. Manfredi, R. Rembecki and I. H. Fox, *Am. J. Physiol.*, 1982, **243**, C270.
135. M. J. Conklyn and R. Silber, *Leuk. Res.*, 1982, **6**, 203.
136. A. Ogilvie and J. Luthje, *Hoppe-Seyler's Z. Physiol. Chem.*, 1987, **368**, 1089.
137. J. D. Stoeckler, R. P. Agarwal, K. C. Agarwal and R. E. Parks, Jr., *Methods Enzymol.*, 1978, **51**, 530.
138. W. R. A. Osborne, *J. Biol. Chem.*, 1980, **255**, 7089.
139. D. A. Wiginton, M. S. Coleman and J. J. Hutton, *J. Biol. Chem.*, 1980, **255**, 6663.
140. G. Ghangas and G. H. Reem, *J. Biol. Chem.*, 1979, **254**, 4233.
141. J. D. Carlson and A. G. Fischer, *Biochim. Biophys. Acta*, 1979, **571**, 21.
142. I. H. Fox, C. M. Andres, J. Kaminska and R. L. Wortmann, in 'Enzyme Defects and Immune Dysfunction', Ciba Foundation Series 68, Excerpta Medica, 1979, p. 193.
143. A. S. Lewis and M. D. Glantz, *J. Biol. Chem.*, 1976, **251**, 407.
144. K. F. Jensen and P. Nygaard, *Eur. J. Biochem.*, 1975, **51**, 253.
145. C. M. Schimandle, L. Tanigoshi, L. A. Mole and I. W. Sherman, *J. Biol. Chem.*, 1985, **260**, 4455.
146. S. E. Ealick, T. J. Greenhough, Y. S. Babu, D. C. Carter, W. J. Cook, C. E. Bugg, S. A. Rule, J. Habash, J. R. Helliwell, J. D. Stoeckler, S. F. Chen and R. E. Parks, Jr., *Ann. N.Y. Acad. Sci.*, 1985, **451**, 311.
147. W. J. Cook, S. E. Ealick, C. E. Bugg, J. D. Stoeckler and R. E. Parks, Jr., *J. Biol. Chem.*, 1981, **256**, 4079.
148. R. L. Wortmann, C. Andres, J. Kaminska, E. Mejias, E. Gelfand, W. Arnold, K. Rich and I. H. Fox, *Arthritis Rheum.*, 1979, **22**, 524.
149. S. R. Williams, J. M. Goddard and D. W. Martin, Jr., *Nucleic Acids Res.*, 1984, **12**, 5779.
150. E. R. Giblett, A. J. Ammann, D. W. Wara, R. Sandman and L. K. Diamond, *Lancet*, 1975, **1**, 1010.
151. A. Cohen, D. Doyle, D. W. Martin, Jr. and A. J. Ammann, *N. Engl. J. Med.*, 1976, **295**, 1449.
152. L. H. Siegenbeek van Heukelom, G. E. J. Staal, J. W. Stoop and B. J. M. Zegers, *Clin. Chim. Acta*, 1976, **72**, 117.
153. J. W. Stoop, B. J. M. Zegers, G. F. M. Hendrickx, L. H. Siegenbeek van Heukelom, G. E. J. Staal, P. K. de Bree, S. K. Wadman and R. E. Ballieux, *N. Engl. J. Med.*, 1977, **296**, 651.
154. W. R. A. Osborne, S. H. Chen, E. R. Giblett, W. D. Biggar, A. A. Ammann and C. R. Scott, *J. Clin. Invest.*, 1977, **60**, 741.
155. N. L. Edwards, E. W. Gelfand, D. Biggar and I. H. Fox, *J. Lab. Clin. Med.*, 1978, **91**, 736.
156. J. L. Virelizier, M. Hamet, J. J. Ballet, P. Reinert and C. Griscelli, *J. Pediatr. (St. Louis)*, 1978, **92**, 358.
157. W. D. Biggar, E. R. Giblett, R. L. Ozere and B. D. Grover, *J. Pediatr. (St. Louis)*, 1978, **92**, 354.
158. E. W. Gelfand, H. M. Dosch, W. D. Biggar and I. H. Fox, *J. Clin. Invest.*, 1978, **61**, 1071.
159. E. Carapella-de Luca, F. Aiuti, P. Lucarelli, L. Bruni, C. D. Baroni, C. Imperato, D. Roos and A. Astaldi, *J. Pediatr. (St. Louis)*, 1978, **93**, 1000.
160. K. C. Rich, W. J. Arnold, T. Palella and I. H. Fox, *Am. J. Med.*, 1979, **67**, 172.
161. P. A. Ostergaard, A. Deding, J. Eriksen and J. Mejer, *Acta Pathol. Microbiol. Scand., Sect. C*, 1980, **88**, 299.
162. G. Rijksen, W. Kuis, S. K. Wadman, L. J. Spaapen, M. Duran, B. S. Voorbrood, G. E. Staal, J. W. Stoop and B. J. Zegers, *Pediatr. Res.*, 1987, **21**, 137.
163. N. Miyasaka, T. Nakamura, I. J. Russell and N. Talal, *Clin. Immunol. Immunopathol.*, 1984, **31**, 109.
164. R. M. Pope, in 'Biologically Based Immunomodulators in the Therapy of Rheumatic Diseases', ed. S. H. Pincus, D. S. Pisetsky and L. J. Rosenwasser, Elsevier, 1986, p. 307.
165. N. Talal, in 'Biologically Based Immunomodulators in the Therapy of Rheumatic Diseases', ed. S. H. Pincus, D. S. Pisetsky and L. J. Rosenwasser, Elsevier, 1986, p. 291.
166. A. Cohen, L. J. Gudas, A. J. Ammann, G. E. J. Staal and D. W. Martin, Jr., *J. Clin. Invest.*, 1978, **61**, 1405.
167. H. A. Simmonds, A. R. Watson, D. R. Webster, A. Sahota and D. Perrett, *Biochem. Pharmacol.*, 1982, **31**, 941.
168. A. Cohen, J. W. W. Lee, H. M. Dosch and E. W. Gelfand, *J. Immunol.*, 1980, **125**, 1578.
169. W. R. A. Osborne and C. R. Scott, *Biochem. J.*, 1983, **214**, 711.
170. B. Ullman, L. J. Gudas, S. M. Clift and D. W. Martin, Jr., *Proc. Natl. Acad. Sci. USA*, 1979, **76**, 1074.
171. T. A. Krenitsky, J. V. Tuttle, G. W. Koszalka, I. S. Chen, L. M. Beacham, III, J. L. Rideout and G. B. Elion, *J. Biol. Chem.*, 1976, **251**, 4055.
172. R. L. Momparler and G. A. Fischer, *J. Biol. Chem.*, 1968, **243**, 4298.
173. J. F. Borel, 'Progress in Allergy, Ciclosporin', Karger, New York, 1986, vol. 38.
174. R. E. Parks, Jr., J. D. Stoeckler, C. Cambor, T. M. Savarese, G. W. Crabtree and S. H. Chu, in 'Molecular Actions and Targets for Cancer Chemotherapeutic Agents', ed. A. Sartorelli, J. S. Lazo, J. R. Bertino, Academic Press, 1981, p. 229.
175. J. D. Stoeckler, in 'Developments in Cancer Chemotherapy', ed. R. I. Glazer, CRC Press, Boca Raton, FL, 1984, p. 35.
176. J. C. Sircar and R. B. Gilbertsen, *Drugs of the Future*, 1988, **13**, 653.
177. J. D. Stoeckler, C. Cambor, V. Kuhns, S. H. Chu and R. E. Parks, Jr., *Biochem. Pharmacol.*, 1982, **31**, 163.
178. J. C. Sircar, M. J. Suto, M. E. Scott, M. K. Dong and R. B. Gilbertsen, *J. Med. Chem.*, 1986, **29**, 1804.
179. D. A. Berry, R. B. Gilbertsen and P. D. Cook, *J. Med. Chem.*, 1986, **29**, 2034.
180. Y. Nishida, N. Kamatani, K. Tanimoto and I. Akaoka, *Agents Actions*, 1979, **9**, 549.
181. R. C. Willis, R. K. Robins and J. E. Seegmiller, *Mol. Pharmacol.*, 1980, **18**, 287.
182. Y. S. Sanghvi, N. B. Hanna, S. B. Larson, J. M. Fujitaki, R. C. Willis, R. A. Smith, R. K. Robins and G. R. Revankar, *J. Med. Chem.*, 1988, **31**, 330.
183. J. D. Stoeckler, J. B. Ryden, R. E. Parks, Jr., M. Y. Chu, M. I. Lim, W. Y. Ren and R. S. Klein, *Cancer Res.*, 1986, **46**, 1774.
184. A. Cohen, J. W. W. Lee and E. W. Gelfand, *Blood*, 1983, **61**, 660.
185. B. Ullman and D. W. Martin, Jr., *J. Clin. Invest.*, 1984, **74**, 951.
186. J. M. Stein, J. D. Stoeckler, S. Y. Li, R. L. Tolman, M. MacCoss, A. Chen, J. D. Karkas, W. T. Ashton and R. E. Parks, Jr., *Biochem. Pharmacol.*, 1987, **36**, 1237.
187. J. V. Tuttle and T. A. Krenitsky, *J. Biol. Chem.*, 1984, **259**, 4065.
188. C. E. Nakamura, S. H. Chu, J. D. Stoeckler and R. E. Parks, Jr., *Biochem. Pharmacol.*, 1986, **35**, 133.
189. D. S. Shewach, J. W. Chern, K. E. Pillote, L. B. Townsend and P. E. Daddona, *Cancer Res.*, 1986, **46**, 519.

190. J. C. Sircar, C. R. Kostlan, G. W. Pinter, M. J. Suto, T. P. Bobovski, T. Capiris, C. F. Schwender, M. K. Dong, M. E. Scott, M. K. Bennett, L. M. Kossarek and R. B. Gilbertsen. *Agents Actions*, 1987, **21**, 253.
191. J. C. Sircar and G. W. Pinter (Warner-Lambert), *Eur. Pat.* 178 178 (1986) (*Chem. Abstr.*, 1987, **106**, 49 885p).
192. I. S. Kazmers, B. S. Mitchell, P. E. Daddona, L. L. Wotring, L. B. Townsend and W. N. Kelley, *Science (Washington, D.C.)*, 1981, **214**, 1137.
193. H. A. Simmonds, A. Goday, G. S. Morris and M. F. J. Brolsma, *Biochem. Pharmacol.*, 1984, **33**, 763.
194. Y. Sidi and B. S. Mitchell, *J. Clin. Invest.*, 1984, **74**, 1640.
195. U. H. Ochs, S. H. Chen, H. D. Ochs, W. R. A. Osborne and C. R. Scott, *J. Immunol.*, 1979, **122**, 2424.
196. R. Martineau and J. Willemot, *Immunopharmacology*, 1983, **6**, 289.
197. L. J. M. Spaapen, G. T. Rijkers, G. E. J. Staal, G. Rijksen, S. K. Wadman, J. W. Stoop and B. J. M. Zegers, *J. Immunol.*, 1984, **132**, 2311.
198. L. J. M. Spaapen, G. T. Rijkers, G. E. J. Staal, G. Rijksen, M. Duran, J. W. Stoop and B. J. M. Zegers, *J. Immunol.*, 1984, **132**, 2318.
199. R. B. Gilbertsen, M. E. Scott, M. K. Dong, L. M. Kossarek, M. K. Bennett, D. J. Schrier and J. C. Sircar, *Agents Actions*, 1987, **21**, 272.
200. R. B. Gilbertsen, M. K. Dong and L. M. Kossarek, *Agents Actions*, 1987, **22**, 379.
201. M. K. Dong, M. J. Suto, J. C. Sircar and R. B. Gilbertsen, *The Pharmacologist*, 1988, **30**. A122.
202. R. B. Gilbertsen and M. K. Dong, *Ann. N.Y. Acad. Sci.*, 1985, **451**, 313.
203. W. R. A. Osborne and R. W. Barton, *Immunology*, 1986, **59**, 63.
204. D. L. Frederick, R. B. Epstein and J. Benear, *Clin. Chem.*, 1985, **31**, 939.
205. W. R. A. Osborne, H. J. Deeg and S. J. Slichter, *Clin. Exp. Immunol.*, 1986, **66**, 166.
206. M. J. Herrmann, P. M. Bealmear and J. Epstein, in 'Germfree Research, Microflora Control and Its Application to the Biomedical Sciences', Liss, 1985, p. 455.
207. J. B. Benear, D. Frederick, L. Townsend and R. B. Epstein, *Transplantation*, 1986, **41**, 274.
208. R. B. Gilbertsen, M. K. Dong and J. C. Sircar, *Arthritis Rheum.*, 1988, **31**, S109.
209. J. E. Seegmiller, *Ann. Rheum. Dis.*, 1980, **39**, 103.
210. M. E. Weinblatt, J. S. Coblyn, P. A. Fraser, R. J. Anderson, J. Spragg, D. E. Trentham and K. F. Austen, *Arthritis Rheum.*, 1987, **30**, 11.
211. M. Dougados and B. Amor, *Arthritis Rheum.*, 1987, **30**, 83.
212. O. Forre, F. Bjerkhoel, C. F. Salvesen, K. J. Berg, H. E. Rugstad, G. Saelid, O. J. Mellbye and E. Kass, *Arthritis Rheum.*, 1987, **30**, 88.
213. C. R. Stiller, J. Dupre, M. Gent, M. R. Jenner, P. A. Keown, A. Laupacis, R. Martell, N. W. Rodger, B. V. Graffenried and B. M. J. Wolfe, *Science (Washington, D.C.)*, 1984, **223**, 1362.
214. R. Assan, G. Feutren, M. Debray-Sachs, M. C. Quiniou-Debrie, C. Laborie, G. Thomas, L. Chatenoud and J. F. Bach, *Lancet*, 1985, **1**, 67.
215. P. F. Bougneres, J. C. Carel, L. Castano, C. Boitard, J. P. Gardin, P. Landais, J. Hors, M. J. Mihatsch, M. Paillard, J. L. Chaussain and J. F. Bach, *N. Engl. J. Med.*, 1988, **318**, 663.
216. N. J. Zvaifler, in 'Arthritis and Allied Conditions. A Textbook of Rheumatology', ed. D. J. McCarty, 9th edn., Lea & Febiger, 1979, p. 417.
217. L. Sokoloff, in 'Arthritis and Allied Conditions. A Textbook of Rheumatology', ed. D. J. McCarty, 9th edn., Lea & Febiger, 1979, p. 429.
218. S. M. Krane, in 'Arthritis and Allied Conditions. A Textbook of Rheumatology', ed. D. J. McCarty, 9th edn., Lea & Febiger, 1979, p. 449.
219. M. C. Hochberg, *Epidemiol. Rev.*, 1981, **3**, 27.
220. R. M. Pope and S. J. McDuffy, *Arthritis Rheum.*, 1979, **22**, 988.
221. N. J. Zvaifler, *Adv. Immunol.*, 1973, **16**, 265.
222. R. W. Lightfood, Jr. and C. L. Christian, in 'Textbook of Immunopathology', ed. P. A. Miescher and H. J. Muller-Eberhard, Grune and Stratton, 1969, vol. 2, p. 733.
223. E. H. Field, S. Strober, R. T. Hoppe, A. Calin, E. G. Engleman, B. L. Kotzin, A. S. Tanay, H. J. Calin, C. P. Terrell and H. S. Kaplan, *Arthritis Rheum.*, 1983, **26**, 937.
224. H. E. Paulus, H. I. Machleder, S. Levine, D. T. Y. Yu and N. S. MacDonald, *Arthritis Rheum.*, 1977, **20**, 1249.
225. J. Tennenbaum, M. B. Urowitz, E. C. Keystone, I. L. Dwosh and J. E. Curtis, *Ann. Rheum. Dis.*, 1979, **38**, 40.
226. R. L. Wilder, C. H. Yarboro and J. L. Decker, *Prog. Clin. Biol. Res.*, 1982, **106**, 48.
227. E. R. Unanue, *N. Engl. J. Med.*, 1980, **303**, 977.
228. A. S. Fauci, S. A. Rosenberg, S. A. Sherwin, C. A. Dinarello, D. L. Longo and H. C. Lane, *Ann. Intern. Med.*, 1987, **106**, 421.
229. C. A. Dinarello and J. W. Mier, *N. Engl. J. Med.*, 1987, **317**, 940.
230. N. J. Zvaifler, in 'Arthritis and Allied Conditions. A Textbook of Rheumatology', ed. D. J. McCarty, 9th edn., Lea & Febiger, 1979, p. 315.
231. I. A. Jaffe, in 'Arthritis and Allied Conditions. A Textbook of Rheumatology', ed. D. J. McCarty, 9th edn., Lea & Febiger, 1979, p. 368.
232. A. D. Steinberg and J. L. Decker, in 'Arthritis and Allied Conditions. A Textbook of Rheumatology', ed. D. J. McCarty, 9th edn., Lea & Febiger, 1979, p. 375.
233. O. Foerre, in 'Ciclosporin in Autoimmune Diseases', 1st International Symposium, Basel, 1985, p. 279.
234. V. J. Stecher, J. A. Carlson, K. M. Connolly and D. M. Bailey, *Med. Res. Rev.*, 1985, **5**, 371.
235. E. J. Corey, M. M. Mehrotra and A. U. Khan, *Science (Washington, D.C.)*, 1987, **236**, 68.
236. P. N. Shambrook, G. D. Champion, C. D. Browne, M. L. Cohen, P. Compton, R. O. Day and J. DeJager, *Br. J. Rheumatol.*, 1986, **25**, 372.
237. J. D. O'Duffy and H. S. Luthra, *Drugs*, 1984, **27**, 373.
238. E. J. Brewer, E. H. Giannini, N. Kuzmina and L. Alekseev, *N. Engl. J. Med.*, 1986, **314**, 1269.
239. N. P. Hurst, J. K. French, L. Gorjatschko and W. H. Betts, *J. Rheumatol.*, 1988, **15**, 23.
240. D. E. Furst and J. M. Kremer, *Arthritis Rheum.*, 1988, **31**, 305.
241. T. Pullar and H. A. Capell, *Br. J. Rheumatol.*, 1984, **23**, 26.

242. E. C. Huskisson and J. G. Adams, *Drugs*, 1980, **19**, 100.
243. I. Yamamoto, H. Ohmori and M. Sasano, *Jpn. J. Pharmacol.*, 1983, **33**, 859.
244. K. Goebel, U. Storck and F. Neurath, *Lancet*, 1981, **1**, 1015.
245. M. Dougados, H. Awada and B. Amor, *Ann. Rheum. Dis.*, 1988, **47**, 127.
246. SCRIP, ed. P. Brown, P. J. B. Publications, London, Nov. 13, 1985, No. 1051, p. 24.
247. J. W. Hadden, E. M. Hadden, T. Spira, R. Settineri, L. Simon and A. Giner-Sorolla, *Int. J. Immunopharmacol.*, 1982, **4**, 235.
248. FDC Reports ("Pink Sheet"), ed. C. P. Werble, FDC Reports, Maryland, April 6, 1987, p. 12.
249. E. M. Tan, in 'Arthritis and Allied Conditions. A Textbook of Rheumatology', ed. D. J. McCarty, 9th edn., Lea & Febiger, 1979, p. 715.
250. H. G. Kunkel, *Arthritis Rheum.*, 1977, **20**, 5139.
251. C. G. Reiness, C. B. Weinberg and Z. W. Hall, *Nature (London)*, 1978, **274**, 68.
252. P. W. Lampert, *Am. J. Pathol.*, 1978, **91**, 176.
253. N. C. Zoumbos, P. Gascon, J. Y. Djeu, S. T. Trost and N. S. Young, *N. Engl. J. Med.*, 1985, **312**, 257.
254. N. Talal, in 'Arthritis and Allied Conditions. A Textbook of Rheumatology', ed. D. J. McCarty, 9th edn., Lea & Febiger, 1979, p. 810.
255. J. B. Wyngaarden and E. W. Holmes, in 'Arthritis and Allied Conditions. A Textbook of Rheumatology', ed. D. J. McCarty, 9th edn., Lea & Febiger, 1979, p. 1193.
256. J. Kovarsky and E. W. Holmes, *Handb. Exp. Pharmacol.*, 1979, **50**, 579.
257. I. H. Fox, *J. Lab. Clin. Med.*, 1985, **106**, 101.
258. I. H. Fox, T. D. Palella and W. N. Kelley, *N. Engl. J. Med.*, 1987, **317**, 111.
259. R. J. Flower, S. Moncada and J. R. Vane, in 'The Pharmacological Basis of Therapeutics', ed. A. G. Gilman, L. S. Goodman and A. Gilman, 6th edn., Macmillan, 1980, p. 682.
260. I. H. Fox, *Handb. Exp. Pharmacol.*, 1978, **51**, 93.
261. P. M. Colombani and A. D. Hess, *Biochem. Pharmacol.*, 1987, **36**, 3789.
262. J. F. Elliott, Y. Lin, S. B. Mizel, R. C. Bleackley, D. G. Harnish and V. Paetkau, *Science (Washington, D.C.)*, 1984, **226**, 1439.
263. M. Kronke, W. J. Leonard, J. M. Depper, S. K. Arya, F. Wong-Staal, R. C. Gallo, T. A. Waldmann and W. C. Greene, *Proc. Natl. Acad. Sci. USA*, 1984, **81**, 5214.
264. J.-F. Gauchat, E. W. Khandjian and R. Weil, *Proc. Natl. Acad. Sci. USA*, 1986, **83**, 6430.
265. A. J. Koletsky, M. W. Harding and R. E. Handschumacher, *J. Immunol.*, 1986, **137**, 1054.
266. P. M. Colombani, A. Robb and A. D. Hess, *Science (Washington, D.C.)*, 1985, **228**, 337.
267. S. J. LeGrue and C. G. Munn, *Transplantation*, 1986, **42**, 679.
268. S. J. LeGrue, R. Turner, N. Weisbrodt and J. R. Dedman, *Science (Washington, D.C.)*, 1986, **234**, 68.
269. P. C. Hiestand, P. Mekler, R. Nordmann, A. Grieder and C. Permmongkol, *Proc. Natl. Acad. Sci. USA*, 1986, **83**, 2599.
270. M. Szamel, P. Berger and K. Resch, *J. Immunol.*, 1986, **136**, 264.
271. R. K. Fidelus and A. H. Laughter, *Transplantation*, 1986, **41**, 187.
272. E. W. Gelfand, R. K. Cheung and G. B. Mills, *J. Immunol.*, 1987, **138**, 1115.
273. P. I. Terasaki, *Transplantation*, 1984, **37**, 119.
274. S. P. Cobbold, A. Jayasuriya, A. Nash, T. D. Prospero and H. Waldmann, *Nature (London)*, 1984, **312**, 548.
275. A. R. Ready, E. J. Jenkinson, R. Kingston and J. J. T. Owen, *Nature (London)*, 1984, **310**, 231.
276. C. T. Lum, D. E. R. Sutherland, J. Eckhardt, A. J. Matas and J. S. Najarian, *Transplantation*, 1979, **27**, 355.
277. A. N. Domonkos, 'Andrews' Diseases of the Skin', 6th edn., Saunders, Philadelphia, 1971.
278. S. Hammarstrom, M. Hamberg, B. Samuelsson, E. Z. Duell, M. Stawiski and J. J. Voorhees, *Proc. Natl. Acad. Sci. USA*, 1975, **72**, 5130.
279. R. D. R. Camp, N. J. Fincham, F. M. Cunningham, M. W. Greaves, J. Morris and A. Chu, *J. Immunol.*, 1986, **137**, 3469.
280. H. Valdimarsson, B. S. Baker, I. Jonsdottir and L. Fry, *Immunol. Today*, 1986, **7**, 256.
281. R. H. Cormane, *Arch. Dermatol. Res.*, 1981, **270**, 201.
282. D. N. Saunder, P. L. Bailin, J. Sundeen and R. S. Krakauer, *Arch. Dermatol.*, 1980, **116**, 51.
283. J. D. Bos and M. K. Kapsenberg, *Immunol. Today*, 1986, **7**, 235.
284. C. N. Ellis, D. C. Gorsulowsky, T. A. Hamilton, J. K. Billings, M. D. Brown, J. T. Headington, K. D. Cooper, O. Baadsgaard, E. A. Duell, T. M. Annesley, J. G. Turcotte and J. J. Voorhees, *JAMA, J. Am. Med. Assoc.*, 1986, **256**, 3110.
285. I. E. Sherman, *Microbiol. Rev.*, 1979, **43**, 453.
286. P. E. Daddona, W. P. Wiesmann, W. Milhouse, J. W. Chern, L. B. Townsend, M. S. Hershfield and H. K. Webster, *J. Biol. Chem.*, 1986, **261**, 11667.
287. W. P. Wiesmann, H. K. Webster, C. Lambros, W. N. Kelley and P. E. Daddona, in 'The Red Cell: Sixth Ann Arbor Conference', ed. G. J. Brewer, Liss, New York, 1984, p. 325.
288. C. J. Bacchi, R. L. Berens, H. C. Nathan, R. S. Klein, I. A. Elegbe, K. V. Rao, P. P. McCann and J. J. Marr, *Antimicrob. Agents Chemother.*, 1987, **31**, 1406.

8.4

Enzyme Cascades: Coagulation, Fibrinolysis and Hemostasis

MICHAEL D. TAYLOR
Warner-Lambert Company, Ann Arbor, MI, USA

8.4.1	INTRODUCTION: ENZYME CASCADES	482
8.4.1.1	*Definition and Properties of Enzyme Cascades*	482
8.4.1.2	*Enzyme Cascades as Targets for Drug Design*	482
8.4.2	HEMOSTASIS	483
8.4.2.1	*Hemostatic Mechanisms*	483
8.4.2.2	*Hemostatic Defects and Diseases*	483
8.4.3	COAGULATION (FIBRIN CASCADE)	484
8.4.3.1	*Overview*	484
8.4.3.2	*Detailed Description of Fibrin Cascade*	486
8.4.3.2.1	*Extrinsic pathway*	486
8.4.3.2.2	*Intrinsic pathway*	486
8.4.3.2.3	*Common pathway*	486
8.4.3.2.4	*Vitamin K dependent factors*	487
8.4.3.2.5	*Thrombin*	487
8.4.3.3	*Regulation and Control of Coagulation*	487
8.4.3.4	*Coagulation Defects and Resulting Diseases*	488
8.4.3.5	*Drugs Affecting Coagulation*	488
8.4.3.5.1	*Coagulation factors*	488
8.4.3.5.2	*Oral anticoagulants*	489
8.4.3.5.3	*Heparins*	490
8.4.3.5.4	*Thrombin inhibitors*	490
8.4.3.5.5	*Other mechanisms*	491
8.4.4	FIBRINOLYSIS CASCADE	492
8.4.4.1	*Overview*	492
8.4.4.2	*Plasmin and Plasminogen Activators*	492
8.4.4.2.1	*Plasmin*	492
8.4.4.2.2	*Plasminogen activators*	493
8.4.4.3	*Endogenous Inhibitors of Fibrinolysis*	494
8.4.4.3.1	*Inhibitors of plasmin*	494
8.4.4.3.2	*Inhibitors of plasminogen activators*	494
8.4.4.4	*Defective Fibrinolysis and Thrombosis*	494
8.4.4.5	*Drugs Affecting Fibrinolysis*	495
8.4.4.5.1	*Plasminogen activators*	495
8.4.4.5.2	*Agents increasing plasminogen activator biosynthesis*	496
8.4.4.5.3	*Agents increasing plasminogen activator release*	496
8.4.4.5.4	*Agents affecting plasminogen activator or plasmin inhibition*	496
8.4.4.5.5	*Agents acting by unknown mechanisms*	497
8.4.4.5.6	*Synthetic inhibitors of fibrinolysis*	497
8.4.5	REFERENCES	497

8.4.1 INTRODUCTION: ENZYME CASCADES

8.4.1.1 Definition and Properties of Enzyme Cascades

McFarlane first used the term 'enzyme cascade' in 1964 in describing a proposal for the mechanism of blood clotting. A cascade is a sequence of enzyme reactions, each reaction activated by the previous one, which, once initiated, proceeds to the final one. The essential advantage inherent in this process is rapid biochemical amplification of a response.[1] The cascade also allows exquisite control of the response, first localizing and then ultimately terminating the process. McFarlane saw a resemblance between electronic devices — amplifiers, circuit breakers and feedback loops — and the regulatory mechanisms controlling the initiation, localization and termination of the fibrin system. Use of the analogy of a rapid flow of water to describe the fibrin system was also employed by Davie and Ratnoff,[2] who described a similar mechanism for blood clotting and termed it a 'waterfall sequence'.

Protease cascades share several properties. They are complex multistep processes by definition and are highly specific since each protease usually activates only the next zymogen in the sequence. This activation occurs by a process of limited proteolysis.[3] In such systems, proteins operate in pairs, one acting as enzyme, the other as substrate in turn.[1,3] Another common feature of cascades is a high degree of regulation by specific inhibitors and both positive and negative feedback loops to control and focus the process.[3,4] In the absence of such controlling mechanisms, the explosive power of the response would damage both uninvolved tissues and plasma proteins.

The term cascade has been applied to many sequential enzyme processes, for example the arachidonic acid cascade, the renin–angiotensin system, glycolysis, as well as metabolic sequences. The modulation of metabolic sequences, focusing on the metabolism of purines, is the subject of Chapter 7.3. While these other enzyme sequences share some of the properties of protease cascades, the power of the cascade system results from cumulative amplification. This aspect is most evident in the coagulation and complement systems, which are the most highly developed examples of enzyme cascades. These two cascades share many features and a number of specific enzymes are involved in both cascades.[3,5] Both the fibrin and complement cascades involve large numbers of enzymes (more than 10 for fibrin, more than 20 for complement), proteolytic activation, regulation by protease inhibitors, and result in large rapid responses to physiological stimuli. The human complement system has been reviewed elsewhere.[6]

8.4.1.2 Enzyme Cascades as Targets for Drug Design

The multitude of interdependent enzymes, cofactors, substrates and second messengers involved in enzyme cascades provides a degree of complexity found in few other physiological processes. In contrast to metabolic sequences, most of the reactions of peptidase cascades serve to amplify or regulate the process. Often, only the final step of the sequence is physiologically relevant. This complexity permits increased specificity, decreased reaction time and amplification of response. Complexity, however, also results in greater vulnerability in that failure of any one component may result in failure of the system as a whole. This is the case in the fibrin cascade where absence of, or defects in, the activity of any one of several coagulation enzymes, cofactors, or inhibitors results in hemorrhagic or thrombotic disorders.

The complexity of enzyme cascades similarly presents opportunities for pharmacodynamic intervention by the multitude of potential targets. Rather than being limited to a single critical enzyme, the medicinal chemist may be able to choose the most attractive target, based on a number of variables, maximizing the chances for success. For inhibition of a metabolic sequence it is generally desirable to target the rate-limiting step, since that is the reaction most often controlled by feedback mechanisms.[7] In a system such as the fibrin cascade, where each step is amplified by the previous step, no one reaction (except perhaps the initiating step) is rate limiting and inhibition of any individual step may lead to inhibition of the entire sequence. Thus, other factors such as ease of developing *in vitro* test systems, availability of chemical leads, or specificity (some enzymes are involved in more than one process) may lead the medicinal chemist to target one particular enzyme for inhibition.

The objective of pharmacologic intervention in an enzyme cascade is to enhance or normalize a system which is out of balance. Achieving the objective may involve either inhibiting or potentiating the process, depending on the desired therapeutic effect. For the treatment or prevention of thrombosis, inhibition of coagulation or potentiation of fibrinolysis will have comparable results.

Historically, inhibition of enzymes is a more achievable objective, involving a direct interaction of the enzyme and inhibitor. Potentiation of an enzyme process is more difficult. Potentiation of an enzyme cascade may be possible by eliminating or preventing a negative feedback or inhibitory process. Since endogenous inhibitors represent an important negative regulatory mechanism of enzyme cascades, modulating an inhibitor–enzyme relationship may provide such a potentiating effect. Regardless of whether inhibition or potentiation is the desired result, a fundamental and complete understanding of the relevant biochemistry is essential for design of effective modulators.

8.4.2 HEMOSTASIS

8.4.2.1 Hemostatic Mechanisms

Several distinct but interrelated thrombotic and antithrombotic systems exist to prevent formation of intravascular clots except in response to vascular trauma. Hemostasis is the sum of these mechanisms and serves to limit blood loss following injury. Once coagulation is initiated, these same mechanisms combine first to localize the clot at the site of injury, then terminate coagulation and finally remove the clot once it has served its purpose. These hemostatic mechanisms include platelet activation, coagulation, fibrinolysis and local vascular effects.[8,9] While platelets play an essential role in primary hemostasis,[10] only coagulation, fibrinolysis and agents which modulate these systems are covered in this chapter. The reader may also refer to recent reviews for a complete treatment of platelet function and the role of platelets in hemostasis.[11-13] For an overview of hemostasis and thrombosis, the comprehensive work of Colman *et al.*[14] is strongly recommended and is extensively cited in this chapter. In addition, other chapters discuss inhibition of platelet function, including Chapters 6.2 and 8.5.

8.4.2.2 Hemostatic Defects and Diseases

In the past, patients with thrombotic disorders were thought to suffer from a 'hypercoagulable' state.[15,16] This vague term obscures any number of specific problems relating to either coagulation or fibrinolysis. The processes of coagulation and fibrinolysis represent a dynamic balance involving the formation and removal of fibrin. If the balance of these processes is upset, either excessive bleeding or thrombosis may result.[17] Although bleeding disorders have historically been given much attention, particularly hemophilias, the problem of thrombosis is a much more common problem.[10,17]

Thrombosis causes or contributes to a variety of diseases and conditions. Thrombus formation is believed to be a key process in the initiation, development and terminal occlusion of the atherosclerotic lesion,[18] in unstable angina[19] and myocardial infarction.[19,20] Nearly all patients exhibit thrombotic occlusion of the infarct-related artery in the first hours of myocardial infarction.[21] Venous thromboembolism remains the most preventable cause of hospital death.[22] Each year in the US between 650 000 and 700 000 patients suffer from acute pulmonary thromboembolism and 10% of these patients die as a direct result. In addition, pulmonary thromboembolism may be a contributing cause in as many as 150 000 deaths. Deep-vein thrombosis may occur in up to 27% (2.5 million cases annually) of general surgical cases.[23] Although recent efforts at primary and secondary prevention have decreased the incidence of pulmonary embolism and thrombophlebitis, morbidity and mortality remain high.[24] Thromboembolism is a major complication where the blood is in contact with foreign surfaces such as extracorporeal circulation systems, prosthetic valves, intravascular catheters, and vascular suture.[25] Valvular heart disease, cardiac arrhythmias, congestive heart failure and intimal damage to arteries also are frequently associated with thromboembolism.[26,27]

The coagulation and fibrinolytic systems are extremely complex. However, one can simplify consideration of the process by focusing on the key events which are (i) the transformation of soluble fibrinogen to insoluble fibrin (the clot) by thrombin and (ii) the subsequent degradation of fibrin into soluble fragments by plasmin. All other reactions relate to controlling and focusing the location and timing of these central processes. This simple picture is illustrated in Figure 1.

The proteins involved in coagulation and fibrinolysis are highly related, possessing many common structural features that are shared with other plasma proteins and proteins of other tissues. Some of these common domains are important for binding interactions with substrates, cofactors or surfaces (Figure 2). The structures known as kringles[28] are particularly noteworthy features present

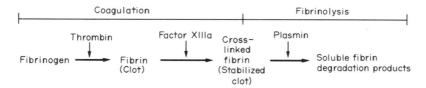

Figure 1 A simplified view of coagulation and fibrinolysis

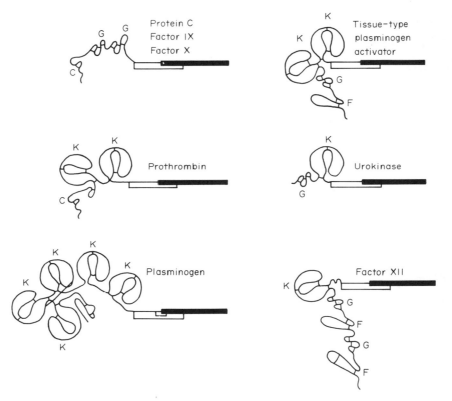

Figure 2 Two-dimensional models of proproteases of the coagulation and fibrinolysis enzyme systems. The black bars represent the catalytic region (the C-terminal catalytic chain of the active proteases). Disulfide bonds are indicated by straight lines. C = calcium binding, Gla-containing, vitamin K dependent module; K = kringle module; G = growth factor module homologous with epidermal growth factor; F = fibronectin homologous modules (S. Müllerts, *Fibrinolysis*, 1987, **1**, 3. Reproduced by permission).

in prothrombin, factor XII, plasminogen, tissue plasminogen activator and urokinase.[29] Most of the cascade enzymes are serine proteases which activate their respective substrates *via* specific limited proteolysis. Three types of activation occur in coagulation and fibrinolysis. The first results in exposure of the catalytic site producing an active enzyme, the second in exposure of binding sites which permit catalysis *via* complexation, and the third in exposure of binding sites leading to self-association or polymerization of fibrin monomers.[5]

8.4.3 COAGULATION (FIBRIN CASCADE)

8.4.3.1 Overview

The fibrin cascade is outlined in detail in Figure 3. The coagulation system has been the subject of several recent reviews.[5,17,30-32] The cascade involves a series of stepwise interactions of more than 20 proteases, cofactors and inhibitors which terminates in the production of insoluble fibrin from soluble fibrinogen. The principal proteins of the cascade are termed factors, each having a Roman numeral designation. The factors and their synonyms are listed in Table 1. The role of the fibrin cascade is to maintain hemostasis by forming fibrin clots in response to vascular injury. The proteases of the coagulation system circulate in the blood in their inactive, zymogen forms. Two separate initiating systems exist which are referred to as the intrinsic pathway and the extrinsic pathway.[30,31] They are so called because all the necessary components for activation of the intrinsic system are contained within the blood, while the extrinsic system requires a factor, thromboplastin, present in exposed tissues or macrophages. As shown in Figure 3, these separate systems converge at factor X. Platelets play an important role in initiating and accelerating coagulation by secreting protein cofactors and providing surfaces which promote formation of enzyme–substrate and enzyme–cofactor complexes. Under normal conditions of course, clotting must not occur and to maintain blood in a fluid state, coagulation is checked at several levels by antithrombotic lipids (prostacyclin), proteins (thrombomodulin), polysaccharides (heparins) and inhibitors of various activated factors.[33]

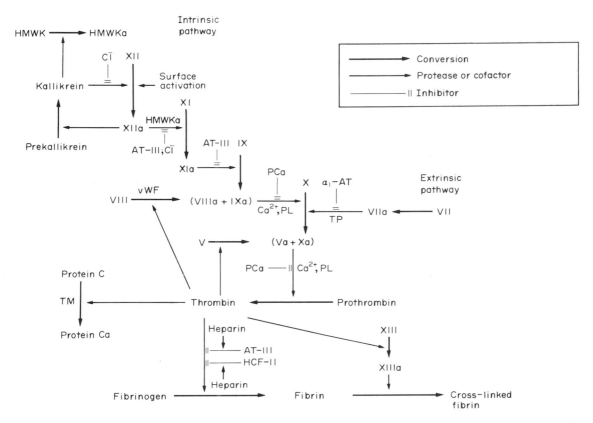

Figure 3 Schematic overview of the fibrin cascade. Abbreviations: AT-III, antithrombin-III; α_1-AT, α_1-antitrypsin; C$\bar{\text{I}}$, C$\bar{\text{I}}$ inhibitor; HCF-II, heparin cofactor-II; HMWK, high molecular weight kininogen; PCa, activated protein C; PL, phospholipid; TM, thrombomodulin; TP, thromboplastin; vWF, von Willebrand factor

Table 1 Coagulation Factors and Their Synonyms

Factor	Pathway	Synonym(s)
Factor I	Common	Fibrinogen
Factor II	Common	Prothrombin
Factor III	Extrinsic	Thromboplastin, tissue factor
Factor IV	Common	Calcium ions
Factor V	Common	Proaccelerin, proaccelerator globulin
Factor VII	Extrinsic	Proconvertin
Factor VIII	Extrinsic	Antihemophilic factor, platelet cofactor I
Factor IX	Intrinsic	Christmas factor, plasma thromboplastin component
Factor X	Common	Stuart–Prower factor
Factor XI	Intrinsic	Plasma thromboplastin antecedent
Factor XII	Intrinsic	Hageman factor, contact factor
Factor XIII	Common	Fibrin stabilizing factor, fibrinoligase, fibrinase
Prekallikrein	Intrinsic	Fletcher factor
High molecular weight kininogen	Intrinsic	HMWK; Williams–Fitzgerald–Flaujeac factor

8.4.3.2   Detailed Description of Fibrin Cascade

8.4.3.2.1   Extrinsic pathway

Coagulation *via* the extrinsic system is significantly more rapid than *via* the intrinsic system due to the fewer number of steps leading to prothrombin activation.[5] Upon vascular injury, thromboplastin (tissue factor) is exposed to the blood and acts as a regulating factor in the activation of factor X by factor VIIa. (By convention, activated forms of coagulation factors are denoted by 'a' appended to the number or acronym.) Factor VII participates only in the extrinsic system. Endothelial cells, smooth muscle cells and fibroblasts all contain thromboplastin, which is a cofactor only and has no intrinsic enzymatic activity. Factor VII, following complexation with thromboplastin, proteolytically activates factor X. Positive feedback occurs as factor Xa activates factor VII to VIIa. Factor VIIa is only 50 times more active than its precursor. The uncommon zymogen reactivity of factor VII permits the coagulation process to begin without initial proteolysis. However, the level of activation is not significant in the absence of thromboplastin and the positive feedback provided by factor Xa. This amplification suggests that thromboplastin is essential to the hemostatic process following injury, and serves as a key regulator of normal hemostasis. Factor Xa, as well as factor XIIa and factor IXa, will increase the activity of factor VII, providing a regulatory link between the intrinsic and extrinsic pathways.[33]

8.4.3.2.2   Intrinsic pathway

In the intrinsic pathway, initiation of coagulation begins with autoactivation of factor XII following binding to a negatively charged surface.[34] This pathway is responsible for the blood clotting observed following contact of blood with nonphysiologic surfaces such as glass. This pathway is therefore referred to as contact or surface activation and involves factor XII, factor XI, prekallikrein and high molecular weight kininogen (HMWK).[5] As shown in Figure 3, factor XIIa can activate prekallikrein to kallikrein. Reciprocal activation of factor XII by kallikrein occurs which is several orders of magnitude faster than autoactivation, markedly accelerating the process. The formation of kallikrein also leads to activation of its substrate, HMWK, producing HMWKa and releasing the potent vasodilator bradykinin. The kininogens are zymogens of vasoactive peptides, including bradykinin. They exist in two forms, a high molecular weight kininogen (HMWK) and a low molecular weight kininogen (LMWK). Plasma prekallikrein circulates predominantly bound to HMWK. Factor XIIa will also cleave HMWK, but at a substantially slower rate.[35] HMWKa attaches to surfaces and forms a noncovalent bimolecular complex with factor XI and prekallikrein, facilitating their activation. Factor XIIa activates the HMWKa–factor XI complex to produce HMWKa–factor XIa which remains tightly bound to the activating surface, while kallikrein, which is also activated by XIIa, circulates. Factor XIa, together with Ca^{2+}, activates factor IX. Factor IXa combines with factor VIIIa, calcium and phospholipid to form the factor IXa complex, which activates factor X to factor Xa. In the absence of factor VIIIa, factor IXa will slowly activate factor X, but the conversion is dramatically accelerated by factor VIIIa. Associated with factor VIII is a large glycoprotein, von Willebrand factor (vWF), which forms a noncovalent complex with factor VIII in plasma. vWF acts as a carrier and prevents clearance of factor VIII. The association of vWF and factor VIII in plasma may increase the concentration of factor VIII at the site of vascular injury.[36]

8.4.3.2.3   Common pathway

The intrinsic and extrinsic pathways converge at factor X to a common pathway leading to thrombin. Activation of prothrombin by factor Xa also involves a complex, the prothrombinase complex, consisting of factor Xa, factor Va, calcium and phospholipids. The phospholipids, phosphatidyl-L-serine and phosphatidylinositol, are derived principally from platelets. Factor V is a procofactor synthesized in the liver and activated by thrombin. Factor Va accelerates the conversion of prothrombin to thrombin by more than 10^5 and has been termed proaccelerator globulin.[37]

Thrombin hydrolyzes fibrinogen at specific arginyl–glycyl bonds to produce fibrin monomers which spontaneously form insoluble fibrin polymers composed of monomers assembled noncovalently end-to-end and side-to-side. The final step in coagulation is stabilization of fibrin by the formation of covalent cross-bridges catalyzed by factor XIIIa. Factor XIII, which is activated by thrombin, is the only enzyme in the cascade which is not a serine protease.

8.4.3.2.4 Vitamin K dependent factors

As noted above several of the coagulation factors, prothrombin, factor VII, factor IX and factor X, are involved in complexes with phospholipid and calcium ions. Modified glutamate residues on these proteins are essential for binding calcium ions. The modified residues are carboxylated at the γ position of glutamate in the liver *via* a vitamin K dependent process, thus these factors are referred to as vitamin K dependent factors. Inhibition of this process is the basis of the activity of the oral anticoagulants (see Section 8.4.3.5.2). Without such post-translational biosynthesis the activity of the factors is markedly reduced.[38]

8.4.3.2.5 Thrombin

As shown in Figure 4, thrombin is a key enzyme in coagulation, not only in its principal role in fibrin formation but also as a regulator of coagulation.[39] Thrombin is involved in several positive feedback relationships. It activates factors V and VIII, both of which in turn will accelerate the production of thrombin. In activating factor XIII, thrombin plays a role in stabilizing fibrin. In addition, thrombin is a potent initiator of platelet activation. Since activated platelets provide a surface for generation of thrombin which is markedly superior to plasma, this represents an additional self-amplification mechanism.[32] Thrombin has important negative influences on coagulation. One mechanism is activation of the anticoagulant, protein C. Thrombin also stimulates release of plasminogen activator.[33] Because of its central importance, inhibition of thrombin has attracted much attention as an approach to preventing thrombosis (see Section 8.4.3.5.4).

8.4.3.3 Regulation and Control of Coagulation

Coagulation is regulated at several levels. Already mentioned are several positive and negative feedback mechanisms. The transience of activated clotting factors is itself an important control mechanism. Following activation, coagulation factors are subject to a gauntlet of metabolic degradation (principally hepatic), plasma proteolysis and inactivation by specific inhibitors (Table 2). The intrinsic phase of the coagulation system is subject to inhibition by the plasma protease inhibitors[40] antithrombin III, C$\bar{\text{I}}$ inhibitor, α_2-macroglobulin and α_1-antitrypsin. In the contact phase most of the relevant inhibition is due to the effect of C$\bar{\text{I}}$ inhibitor on factors XIIa and kallikrein,[41] although α_1-antitrypsin inhibits factor XIa.[42] Overall, the most important inhibitor of the coagulation system is antithrombin III.[43] The inhibitory activity of antithrombin III is dramatically increased by complexation with heparin. The resulting complex inhibits thrombin and factors IXa, Xa, XIa and XIIa. A related inhibitor, heparin cofactor II, is also dependent on heparin and inhibits thrombin.[44]

Protein C and protein S are vitamin K dependent plasma proteins with anticoagulant activity.[45] Protein C was first described by Seegers, who called it autoprothrombin II-A.[46] The active form of protein C (protein Ca) proteolytically inactivates factors Va and VIIIa. Protein S and lipoprotein serve as cofactors and markedly increase the rates of the reactions.[47,48] Protein Ca regulation of

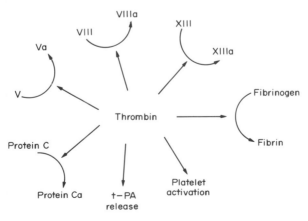

Figure 4 Actions of thrombin in coagulation and fibrinolysis

Table 2   Inhibitors of Coagulation and Fibrinolysis[40]

Inhibitor	System	Enzymes inhibited
Antithrombin III	Coagulation	Factor IXa, factor Xa, factor XIa, factor XIIa, thrombin
α_1-Antitrypsin (α_1-proteinase inhibitor)	Coagulation	Factor XIa
Heparin cofactor II	Coagulation	Thrombin
Protein Ca	Coagulation	Factor V, factor VIII
Protein Ca inhibitor	Coagulation	Protein Ca, thrombin, factor Xa
α_2-Macroglobulin	Coagulation/fibrinolysis	Kallikrein, thrombin, plasmin
CĪ inhibitor	Coagulation/fibrinolysis	Factor XIa, factor XIIa, kallikrein, plasmin
α_2-Antiplasmin	Fibrinolysis	Plasmin
Plasminogen activator inhibitor-1	Fibrinolysis	Tissue plasminogen activator, urokinase
Plasminogen activator inhibitor-2	Fibrinolysis	Tissue plasminogen activator, urokinase

coagulation represents a negative feedback loop, since thrombin is the only known protein C activator,[49] although *in vitro* other proteases will also activate it.[50] Activation of protein C occurs at endothelial surfaces by a complex of thrombin and the thrombin-binding cofactor thrombomodulin, which greatly accelerates the activation of protein C.[51] Protein C may in this way serve as a scavenger of intravascular thrombin.[52] Protein C also enhances fibrinolytic activity by decreasing the activity of an inhibitor of plasminogen activators (see Section 8.4.4.3).[53,54] The importance of proteins C and S is suggested by the fact that deficiencies of either protein may lead to thrombotic disease.[55] An inhibitor of protein Ca has been identified.[56] A link to the complement system of unknown significance is the ability of human C4b-binding protein to complex with and inhibit the anticoagulant activity of protein S.[57] Another regulatory link between complement and coagulation is suggested by the *in vitro* inactivation of factor VII (Hageman factor) by Clq.[58]

8.4.3.4   Coagulation Defects and Resulting Diseases

Abnormalities resulting from the absence or defective activity of specific factors have been found involving virtually every protein in the fibrin cascade.[59] These defects in coagulation factors lead to bleeding disorders. Among the more common disorders include deficiencies of factor VIII (hemophilia A), factor IX (hemophilia B), and vWF (von Willebrand's disease). vWF plays a dual role in mediating platelet adhesion and complexing with factor VIII in coagulation. Patients with von Willebrand disease either have decreased levels of vWF or produce a dysfunctional protein.[60] In addition, since several factors require post-ribosomal modification *via* vitamin K dependent carboxylation, defects in the carboxylating system give rise to a syndrome in which these factors (II, VII, IX and X) are present but inactive.

Deficiencies of coagulation inhibitors such as antithrombin III or protein C result in thrombosis.[61] In the extreme case of homozygous protein C deficiency, the outcome is fatal.[62]

8.4.3.5   Drugs Affecting Coagulation

8.4.3.5.1   Coagulation factors

Many diseases resulting from defective coagulation result from deficiencies of coagulation factors, as described above. Treatment for such conditions usually involves replacement therapy of the missing factor. Replacement may take the form of plasma, plasma concentrates, individual factors or factor complexes.[63] Most of the factors involved in commonly found disorders are available for therapeutic use. Although it is now more common to use concentrates of individual factors, plasma is still preferred for hemophilia B (Christmas disease, factor IX deficiency), and for deficiencies of factors, V, XI or XIII.[64] Use of factor concentrates and cryoprecipitates achieves greater plasma concentrations, but carries the risk of transmitting diseases such as hepatitis and AIDS.

Preclinical studies suggest that protein C is effective in the treatment of intravascular thrombosis.[65] An enhancement of fibrinolytic activity may contribute to its antithrombotic effect.[66]

Some of the more important factors have been cloned and expressed, including factor VIII[67,68] and protein C.[69] These ultimately may be available in pure form *via* this technology.

8.4.3.5.2 *Oral anticoagulants*

Warfarin (1), a coumarin derivative, is the most commonly used orally active anticoagulant.[70] The coumarin anticoagulants interrupt the fibrin cascade by inhibiting post-ribosomal, vitamin K dependent γ-carboxylation of glutamic acid residues of coagulation factors prothrombin, factor VII, factor IX and factor X, and the anticoagulant protein C.[71,72] The coumarins and related indandiones, such as phenindione (2), are mechanism-based, irreversible inhibitors of the hepatic microsomal $NAD(P)^+$ reductase which converts vitamin K epoxide to vitamin K. Inhibition of this enzyme prevents normal turnover of the vitamin and causes the epoxide to accumulate (Figure 5). Silverman[71] proposed that coumarins inhibit the reductase enzyme by acid-catalyzed acylation of the active site.

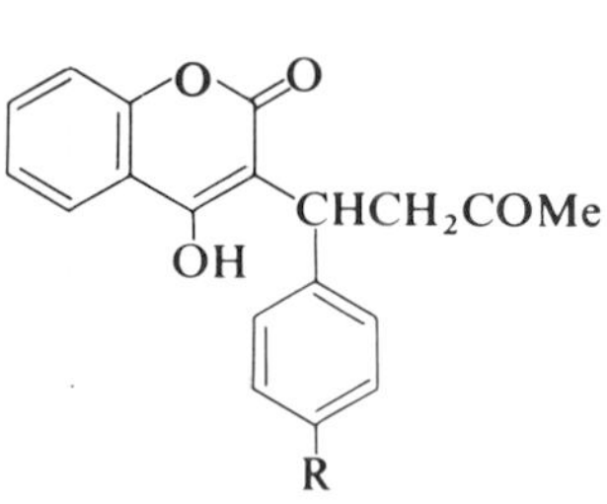

(1) R = H
(3) R = SOMe

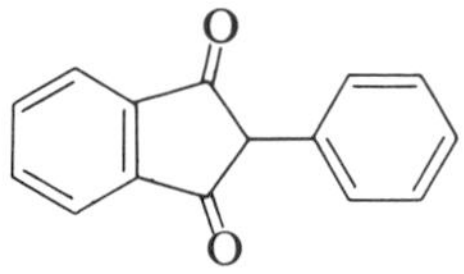

(2)

Since the coumarins have no effect on existing coagulation factors, a minimum of 12–24 h is required before the anticoagulant effect appears to allow for depletion of existing active factors. Although the individual vitamin K dependent factors have different half-lives, coumarin therapy depresses all to a similar degree at steady state.[73]

Treatment with oral anticoagulants requires regular monitoring to assure a safe level of anticoagulation. Considerable interpatient variability and a variety of adverse drug interactions result from extensive protein binding of the compounds in plasma,[74] alterations in the bio-availability of vitamin K and/or warfarin by other compounds, or alterations in the metabolism of the anticoagulant.[72]

The structure–activity relationships (SAR) for anticoagulant activity (prothrombin time) of both coumarin and indandione derivatives have been summarized. Several factors are important, including an acidic proton and the presence of bulky substituents at position 3 of the coumarins and position 2 of the indandiones.[75,76] A polar analog of warfarin, methylsulfinylwarfarin (3) was designed to minimize protein binding. The compound was found to exhibit equivalent anticoagulant activity, a shorter duration of action and a substantially greater free plasma concentration compared to warfarin.[77]

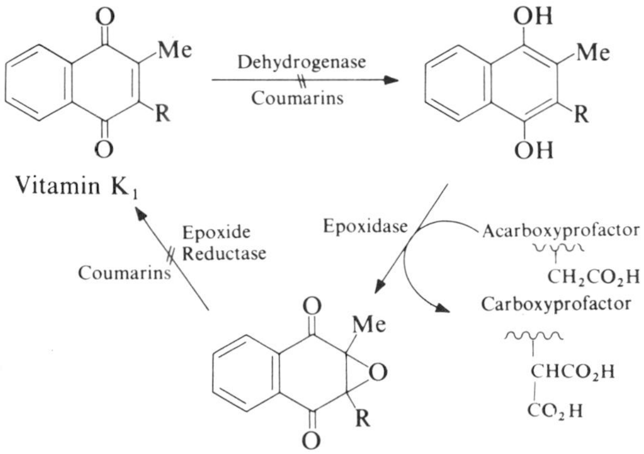

$$R = CH_2CH=CCH_2(CH_2CH_2CHCH_2)_3H$$

Figure 5 The vitamin K cycle and coumarin inhibition

8.4.3.5.3 Heparins

Heparin is the most widely used injectable anticoagulant for the treatment and prophylaxis of thrombosis. Heparin is a heterogeneous mixture of highly sulfated glycosaminoglycan polymers with molecular weights ranging from 3000 to 100 000 Da. The characteristic subunits of the heparin polymers are alternating derivatives of D-glucosamine (*N*-sulfated or *N*-acetylated) and uronic acid (L-iduronic acid with varying sulfates of D-glucuronic acid).[78] Individual heparin molecules have unique primary sequences, which differ in degree and placement of sulfation and acetylation, as well as chain length. Naturally occurring, heparins are found in virtually all animal species and in a variety of tissues, including lungs, heart, kidneys, intestine and liver.[79]

Heparin exerts its anticoagulant effects by binding to antithrombin III, producing a conformational change in the inhibitor that markedly enhances its activity. The major targets of the heparin–antithrombin III complex are factors Xa and thrombin; however, factors XIIa, XIa, IXa and kallikrein may also be inhibited.[80,81] In addition to its association with antithrombin III, heparin will complex with a second cofactor, heparin cofactor II, which also inhibits thrombin.[44,82] Heparin and heparin fractions also promote fibrinolysis by stimulating the release of tissue plasminogen activator. This effect may contribute to its antithrombotic activity.[83]

The anticoagulant effect of heparin can be neutralized by administration of protamine sulfate, a highly basic low molecular weight protein isolated from salmon sperm. Protamine forms a salt with the acidic heparin molecule.[64]

In any given heparin preparation, the individual components may exhibit different biological activities. Even though heparin preparations have been in use for decades, recent understanding of its mechanism of action and structure–activity relationships has led to greater interest in heparin fractions and related oligosaccharides which differ in their specific pharmacological actions. These agents are defined by molecular weight and biologic activity. Because they are more homogeneous, their pharmacologic action is more predictable.[80] Studies using highly purified heparin fractions suggest that the anticoagulant activity can be separated from hemorrhagic side effects, thus improving safety.[81-86] Semisynthetic heparin fraction derivatives, in which the degree of sulfation is increased, maintain good antithrombotic activity, while anticoagulant activity is decreased.[87]

Most 'low molecular weight' heparins are still large molecules, but derivatives as small as pentasaccharides have biological activity.[88] One such derivative (**4**) mimics the heparin binding site for antithrombin III. The compound binds to antithrombin III, inhibits factor Xa and shows some limited antithrombotic activity *in vivo*.[89] Totally synthetic oligosaccharides with anticoagulant activity have been reported.[89-92] Because heparin is suitable for parenteral use only, considerable effort is directed to alternative delivery forms for heparin and heparin fractions focusing on oral formulations.[93]

(**4**)

8.4.3.5.4 Thrombin inhibitors

A variety of compounds will reversibly inhibit the fibrinogenolytic activity of thrombin. Hirudin, a naturally occurring peptide ($M = 10\,800$) isolated from medicinal leeches, is a potent and selective inhibitor of thrombin.[94] Because of the importance of thrombin in producing fibrin and in regulating coagulation, synthetic thrombin inhibitors have attracted much attention. DAPA (**5**) is a potent inhibitor ($K_D = 43$ nM) that exhibits a fluorescence enhancement and blue shift upon binding to thrombin. Its fluorescence properties make it a useful tool in studying the biochemistry of thrombin inhibition.[95] MD-805 (MQPA; MCI-9038; argatroban; **6**) is among the most studied of the synthetic thrombin inhibitors. This compound was selected for clinical evaluation from a series of L-arginine derivatives which possess modifications of the α-amino and carboxyl groups.[96] It is highly specific for thrombin *versus* other proteases.[97] MD-805 also inhibits thrombin-mediated platelet aggregation, without affecting collagen, ADP, epinephrine or arachidonate-induced aggregation.[98] Recently, MD-805 has been shown to accelerate thrombolysis of clots by t-PA or urokinase.[99]

Binding to the active site of thrombin is dependent on the conformation of the piperidine carboxylic acid.[100,101] Inserting a glycine spacer between the arylsulfonyl and arginine α-amine produced a series of markedly more potent inhibitors. One of these is (7), which is the most potent synthetic competitive inhibitor of thrombin reported to date ($K_i = 6$ nM). The compound is very selective for thrombin, its K_i values for inhibition of trypsin, plasmin and factor Xa are 115, 5000 and 1320 times larger, respectively, than for inhibition of thrombin.[102] Another arginine derivative, RGH-2958 (8), also is a highly specific, potent inhibitor of thrombin ($K_i = 10$ nM). This compound binds to thrombin, whereupon the active site serine hydroxyl reacts with the aldehyde to form a stable, but reversible, hemiacetal, which is an analog of the transition state.[103]

While no thrombin inhibitors have yet reached the market, they have several potential advantages over the traditional anticoagulants, *i.e.* coumarins and heparins. Unlike coumarins, thrombin inhibitors exert an anticoagulant effect immediately. They do not need the presence of a cofactor as heparin requires antithrombin III. In addition, they affect bleeding time less than heparin does and therefore may elicit fewer hemorrhagic complications.[104]

8.4.3.5.5 *Other mechanisms*

(i) *Effects on biosynthesis*

The anabolic steroids danazol and stanozolol will increase levels of various coagulation factors and antithrombin III,[64] presumably due to their effect on protein biosynthesis. Use of oral contraceptives and pregnancy are associated with increased incidence of thrombosis. Estrogen and progestogen contraception will increase circulating levels of fibrinogen, factor VII and factor X, and will decrease antithrombin III. Fibrinolytic activity is increased.[105]

(ii) *Inhibition of coagulation factors*

Small molecular weight inhibitors of coagulation factors, such as the thrombin inhibitors described above, have been proposed as potentially useful anticoagulants.[106] A series of iso-coumarin derivatives have been shown to be potent inhibitors of coagulation enzymes in addition to thrombin. Basic groups at the 7-position improved selectivity, for example 4-chloro-7-guanidino-3-methoxyisocoumarin (**9**) is reasonably selective for trypsin, thrombin and human plasma kallikrein. These compounds act as suicide substrates of the serine proteases involved in coagulation and are active as anticoagulants in human plasma *in vitro*.[107]

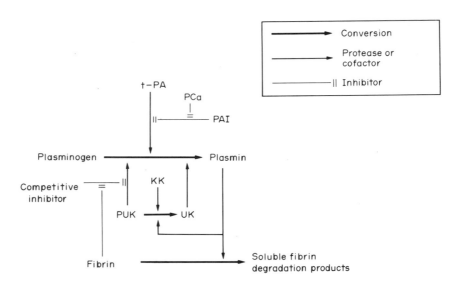

(9)

8.4.4 FIBRINOLYSIS CASCADE

8.4.4.1 Overview

Fibrinolysis has been the subject of a number of recent reviews.[4,86,108,109] The principal function of the fibrinolytic system is to dissolve blood clots by degrading fibrin to soluble fragments. Fibrinolysis under normal conditions prevents uncontrolled thrombosis, while ensuring that hemostatic plugs are maintained. Fibrinolysis also contributes to other physiologic processes, including tissue repair, ovulation and embryo implantation, metastasis and macrophage function.[108] The fibrinolytic system involves a protease cascade (Figure 6), which is considerably shorter than the coagulation cascade, but shares the features of proteolytic activation, feedback activation and inhibition, and regulation by specific protease inhibitors. As in the fibrin cascade, regulatory mechanisms focus fibrinolysis at the site of the thrombus, while minimizing destruction of uninvolved blood and structural proteins.[110]

8.4.4.2 Plasmin and Plasminogen Activators

8.4.4.2.1 Plasmin

Plasmin, a serine protease, is the principal enzyme system involved in the degradation of fibrin, although other proteolytic enzymes may also be involved under physiologic conditions.[4,111] Plasmin is derived from its zymogen form, plasminogen, by the action of various proteases called plasminogen activators. Native plasminogen has glutamine as the N-terminus and is referred to as Glu-plasminogen. Limited proteolysis will convert Glu-plasminogen to Lys-plasminogens, which have

Figure 6 Schematic overview of fibrinolysis. Abbreviations: KK, Kallikrein; PAI, plasminogen activator inhibitor; PCa, activated protein C; PUK, prourokinase; t-PA, tissue plasminogen activator; UK, urokinase

lysine, methionine or valine as the N-terminus. Either Glu- or Lys-plasminogen can be converted to plasmin, although conversion of the Lys form is faster. Activation of plasminogen requires hydrolysis of a single arginyl–valyl bond of the single chain zymogen to form the two-chain plasmin.[112] Plasminogen contains structures which have high affinity for lysine. These lysine binding sites give plasminogen affinity for fibrin, which is rich in lysine. During the formation of fibrin clots, plasminogen binds to fibrin, where it is later activated. This affinity for fibrin effectively increases the concentration of plasminogen at its site of action. Lys-plasminogen has a greater affinity for fibrin than does Glu-plasminogen. Lysine binding sites are also important for the interaction of plasmin with inhibitors.[109] Where plasmin is found in the circulation, either by activation of plasminogen in plasma or following release into the circulation from degraded fibrin, it is rapidly inactivated by α_2-antiplasmin, which is present in a concentration approximately half that of plasminogen. In highly activated states, such as during thrombolytic therapy, the production of circulating plasmin will overwhelm the inhibitory capacity of α_2-antiplasmin and fibrinogen levels are substantially depleted.

8.4.4.2.2 *Plasminogen activators*

A number of plasma proteins can convert plasminogen to plasmin. Based on immunochemical classification there are two main types of plasminogen activators, those related to tissue-type plasminogen activator (t-PA) and those related to urokinase. These are discussed in detail below. An important difference between these classes of activators is their relative fibrin specificity.[113]

(i) t-PA

This protein was first identified in the 1940s, but not isolated until the late 1970s.[114] t-PA is a single-chain serine protease with a molecular weight of 68 000.[115] It is synthesized in vascular tissue and released into the circulation. Although the single-chain t-PA can be converted to a two-chain form by plasmin, this does not increase its activity. There is no known zymogen form of t-PA. The mechanism of release of t-PA is not known, however, it is increased by exercise, venous occlusion and by the action of vasoactive drugs.[116] The activity of t-PA is controlled primarily by its affinity for fibrin and through inhibition by specific inhibitors. Clearance by the liver is also important. Like plasminogen, t-PA contains regions which bind to the lysine rich regions of fibrin.[117] These fibrin binding sites are located on the kringle domains of the molecules (Figure 2). This mutual affinity serves to localize both t-PA and plasminogen, enzyme and substrate, on the target molecule, resulting in high efficiency and minimizing proteolysis of plasma proteins, particularly fibrinogen. Further limiting nonfibrin proteolysis, t-PA is relatively inactive in the absence of fibrin ($K_m = 65 \, \mu M$), but once binding occurs, the activity is substantially increased ($K_m = 0.14 \, \mu M$).[118] Lysine binding sites are important in the reaction of plasmin and plasminogen activators with endogenous inhibitors. The fibrin affinity of t-PA coupled with fibrin activation produce its 'clot selectivity'.[119]

The fibrinolytic system is linked to other cascades. For example, t-PA will hydrolyze prorenin to renin and angiotensin I to angiotensin II. While the significance of this finding is unclear, it suggests that t-PA may exert a local effect on circulation in addition to its fibrinolytic actions.[120]

(ii) (Pro)urokinase

Urokinase is a plasminogen activator consisting of two peptide chains joined by a disulfide bridge. It is produced by the kidneys and found primarily in urine. A marked difference between urokinase and t-PA is that urokinase is considerably less specific for fibrin. The two enzymes are highly homologous, except that t-PA contains a different N-terminal sequence. This sequence contains the lysine binding sites and is essential for the fibrin specificity of t-PA.[109]

In plasma, a single chain form of urokinase has been identified, prourokinase, which exhibits greater specificity for fibrin than the two-chain form of urokinase.[121] Although prourokinase was initially thought to be a true proenzyme, it is now believed that prourokinase is not a true zymogen since proteolytic activity has been demonstrated *in vitro*. Lijnen and coworkers[122] have proposed a mechanism for the activity and fibrin specificity of prourokinase in which the inactivity of prourokinase is explained by the presence of an unidentified plasma inhibitor. This inhibition is competitive and reversed by fibrin. In this mechanism, prourokinase circulates bound to the inhibitor until at the site of a clot, fibrin disassociates the complex and activity is restored. Plasmin will produce urokinase by hydrolysis of the Lys 158-Ile 159 peptide bond of prourokinase.[122]

Because prourokinase is not a true proenzyme, the new terms single-chain urokinase-type plasminogen activator (scu-PA) and two-chain urokinase-type plasminogen activator (tcu-PA) have been suggested for prourokinase and urokinase, respectively.[123] An alternative mechanism explaining the fibrin specificity of prourokinase has been proposed by Gurewich and Pannell.[124] Hepatic clearance, and not plasma inhibition, is the principal route of inactivation of plasma urokinase.[109]

(iii) Factor XII dependent pathway

The contact activation system will initiate fibrinolytic activity, although the physiologic relevance of this pathway is unclear. Much of the activation may be due to the conversion of prourokinase to urokinase by kallikrein.[116] This pathway is referred to as intrinsic activation of fibrinolysis, while activation *via* t-PA or urokinase is extrinsic activation.[109]

8.4.4.3 Endogenous Inhibitors of Fibrinolysis

Several independent mechanisms exist to control fibrinolysis. Plasma proteinase inhibitors (Table 2) play the most important role, although regulation of biosynthesis, release, clearance (hepatic) and fibrin specificity also contribute.

8.4.4.3.1 *Inhibitors of plasmin*

Plasmin generated in the plasma is rapidly inactivated ($t_{1/2} = 0.1$ s) by the inhibitor α_2-antiplasmin. α_2-Macroglobulin also inactivates plasmin but at a slower rate. α_2-Antiplasmin is a protease inhibitor of the serine protease inhibitor family and is thus closely related to α_2-antiplasmin, C$\bar{1}$ inhibitor, antithrombin III, ovalbumin and α_1-antichymotrypsin.[113] Inhibition of plasmin by α_2-antiplasmin occurs stepwise. A fast, reversible second-order reaction is followed by a second slower, irreversible first-order reaction. In addition to enhancing activation of plasminogen, fibrin also protects plasmin from inhibition. Binding of plasmin to either α_2-antiplasmin or fibrin requires a free active site and unoccupied lysine binding sites. Once bound to fibrin, neither the active site nor the lysine binding sites are easily accessible to α_2-antiplasmin and fibrin-bound plasmin is only slowly neutralized.[125] Fibrin may be made more resistant to fibrinolysis by the factor XIIIa-catalyzed crosslinking of α_2-antiplasmin to fibrin.[126]

8.4.4.3.2 *Inhibitors of plasminogen activators*

Like plasmin, the activity of t-PA is highly regulated in order to localize its activity. t-PA is released as free t-PA, which, in the presence of a fibrin clot, is absorbed and becomes relatively inaccessible to inhibitors. In the absence of fibrin, t-PA is subject to complexation by inhibitors and metabolic degradation resulting in a half-life of a few minutes.[127] Although the elimination of t-PA from plasma is not well understood, it is known that a number of protease inhibitors play an important role (Table 2). The common protease inhibitors α_2-antiplasmin, C$\bar{1}$ inhibitor, α_1-antitrypsin, and α_2-macroglobulin react with t-PA, although the importance of these reactions is unclear since they are slow. Recently, highly specific, rapid inhibitors of t-PA have been isolated from a variety of tissues and plasma.[113] These inhibitors fall into two related but immunologically distinct types, those from endothelial cells (PAI-1) and from placenta (PAI-2).[128] PAI-1 may exist in a latent form which is activated by negatively charged phospholipids, phosphatidylinositol and phosphatidyl-L-serine. These phospholipids, produced by platelets, may inhibit fibrinolysis during coagulation by activating latent PAI-1.[129] PAI-2 concentration in plasma increases during pregnancy and probably causes the observed depression of fibrinolysis.[130] PAI-2 is believed to assist maintenance of hemostasis in the child during birth and in the mother during separation of the placenta.[131]

8.4.4.4 Defective Fibrinolysis and Thrombosis

Since therapy of thrombosis has for decades focused on inhibiting coagulation, thrombosis is most often thought to be due to hypercoagulation. With a more precise understanding of the fibrinolytic system it is now clear that defective fibrinolysis also leads to thrombotic disease.[132] Defective fibrinolysis has been associated with a variety of thrombotic conditions, including post-surgical thrombosis,[133] coronary artery disease,[134] myocardial infarction,[135,136] and venous thrombosis.[137,138] In many cases, depressed fibrinolysis associated with thrombotic states is accompanied by increased levels of PAI, for example in idiopathic thromboembolic disease,[139] deep-vein thrombosis,[140,141] myocardial infarction[142] and pregnancy.[143,144] The basal level of PAI increases with age as does the incidence of thrombosis.[145] Despite these highly suggestive observations, the precise role of PAIs in the regulation of fibrinolysis and the pathogenesis of thrombosis still is not clearly defined.[128]

Conditions involving excessive fibrinolytic activity resulting in bleeding disorders are rare but known.[132] These can result from a deficiency of fibrinolytic inhibitors, such as α_2-antiplasmin,[146] or very high levels of plasminogen activators.[147]

8.4.4.5 Drugs Affecting Fibrinolysis

8.4.4.5.1 *Plasminogen activators*

Thrombolytic therapy has become a common component of aggressive treatment of thrombotic disease, particularly acute myocardial infarction.[148,149,150] Enzyme thrombolytic agents used as drugs comprise several classes, including the naturally occurring plasminogen activators of human origin such as t-PA, urokinase and prourokinase. Another type of agent, streptokinase, is isolated from bacteria.[151] The first generation agents streptokinase and urokinase have been in limited use since the 1960s. The development of fibrin selective second-generation agents, t-PA and prourokinase, expanded the use of thrombolytic therapy.[152]

(i) *t-PA*

Recombinant DNA technology has permitted the large scale preparation and use of t-PA. t-PA, when used in the treatment of acute myocardial infarction, will recanalyze occluded coronary arteries in nearly 70% of cases.[153-156] t-PA also has been used successfully in the treatment of other thrombotic conditions.[156] Compared to other thrombolytic agents it is nonantigenic, relatively fibrin (clot) specific and more effective.[157] Human recombinant t-PA (rt-PA) is identical to the enzyme isolated from cell culture.[158]

(ii) *(Pro)urokinase*

Unlike t-PA, urokinase is not specific for fibrin. Since it is a human enzyme it is not antigenic as is streptokinase. Use of urokinase has been limited by its high cost relative to streptokinase and it is used most often in patients with a sensitivity to streptokinase. Prourokinase, in humans, has been shown to have fibrinolytic activity comparable to urokinase, but with less systemic fibrinolytic activation, reflecting the greater fibrin specificity of prourokinase.[159] Genes for both prourokinase[160] and urokinase[109,150] have been cloned and expressed. An oral formulation of urokinase has been reported.[161]

Combinations of plasminogen activators used simultaneously have been reported to produce a synergistic action, particularly t-PA plus prourokinase.[162] The synergism may be enhanced by the order in which the agents are administered and is related to their different and complementary mechanisms of fibrin specificity.[163]

(iii) *Streptokinase*

Streptokinase, which has no intrinsic proteolytic activity, is a protein produced by strains of streptococci. Plasminogen forms a 1:1 complex with streptokinase which becomes proteolytically effective through a conformational effect on the active site of the plasminogen. The active complex then activates other plasminogen molecules producing plasmin. Streptokinase is the most commonly used thrombolytic agent, having the advantages of ready availability and relatively low cost and reasonable efficacy.[164] Because streptokinase has little specificity for fibrin, it is associated with fibrinogen depletion and antigenicity due to its bacterial origin.[113,165] Despite these disadvantages, use of intravenous streptokinase within 6 h of myocardial infarction results in lasting benefit up to 12 months following thrombolysis.[166]

Eminase (anisoylated plasminogen–streptokinase activator complex, APSAC) is a 1:1 complex of streptokinase and plasminogen in which the active site has been blocked by acylation.[167] The inactive complex circulates in plasma and binds to fibrin *via* the lysine binding sites of the plasminogen. Once bound, the complex becomes activated by slow deacylation of the active site.[168] Changing the acylating group will modulate the activity. An analog which deacylates more slowly (BRL-33575) has been shown to have improved activity against older established clots.[169]

(iv) Modified plasminogen activators

Current research is directed to improving the selectivity, potency or duration of action of plasminogen activators. In addition to expression of plasminogen activator genes, recombinant gene technology has resulted in improved forms of plasminogen activators which may have advantages over the native molecules, such as increased activity and fibrin specificity, and decreased susceptibility to inhibitor inactivation.[170] Approaches to improved activators include deleting portions of the activator (deletion mutants) or combining regions from different activators (chimeric molecules).[171,172]

Conjugation of t-PA to a fibrin-specific antibody improved the potency 2.8–10 fold *in vitro* and *in vivo*, respectively. Selectivity as measured by the relative consumption of fibrinogen and α_2-antiplasmin was also enhanced.[173] A similar approach has been successful with antibody-linked urokinase. Coupling of urokinase with the fibrin antibody not only increased specificity, but activity was increased 100 times *in vitro* compared to urokinase alone.[174] Administration of t-PA is currently limited to the intravascular route. Alternative formulations, which allow administration outside a hospital setting, would permit treatment closer in time to the thrombotic event. Time to treatment is critical to avoid irreversible damage. Intramuscular injection of t-PA, which would allow self-medication, produces thrombolytic plasma levels of drug in dogs when facilitated with hydroxylamine.[177]

8.4.4.5.2   Agents increasing plasminogen activator biosynthesis

Several anabolic steroids have been shown to increase the biosynthesis of t-PA, resulting in an increase in fibrinolytic activity.[40] Stanozolol has been studied as a treatment for thrombosis with mixed results.[176,177] Its nonspecific effect on protein synthesis produces both procoagulant and profibrinolytic effects. Dexamethasone increases the biosynthesis of PAI, resulting in a depression of fibrinolytic activity.[178]

8.4.4.5.3   Agents increasing plasminogen activator release

A number of vasoactive compounds increase fibrinolytic activity by enhancing the release of t-PA into the plasma.[116,179] Most of these, including prostacyclin,[180] prostacyclin analogs (CG 4203),[181] pentoxifylline[182] and piretanide[183] are known to increase intracellular cyclic adenosine monophosphate (cAMP), which may represent a mechanism for enhancing the release of t-PA.

The vasoactivity can be separated from the profibrinolytic activity. Desmopressin (1-deamino-8-D-arginine vasopressin, DDAVP) is a nonvasoactive analog of vasopressin used principally for the treatment of central diabetes insipidus. It also has profound effects on coagulation and fibrinolysis, producing increases in factor VIII and plasminogen activator.[184,185] These paradoxical effects have led to the study of desmopressin in the treatment of hemophilia[184] and in the prevention of post-operative depression of fibrinolysis.[185]

8.4.4.5.4   Agents affecting plasminogen activator or plasmin inhibition

Von Kaulla has described the use of synthetic fibrinolytic agents which stimulate fibrinolytic activity through a variety of different mechanisms, including some of those described above.[186] A number of arylaliphatic acids will stimulate fibrinolysis as measured by the euglobulin lysis time. This effect is believed to be related to the agents interfering with the binding of plasmin and perhaps plasminogen activators to inhibitors.

8.4.4.5.5 *Agents acting by unknown mechanisms*

Tizabrin (CP-2129; **10**) stimulates fibrinolysis after intravenous administration in human volunteers due to an increase in t-PA activity. Tizabrin has no intrinsic fibrinolytic activity *in vitro*.[187] S-1623 (**11**), in contrast, will generate a plasmin-like activity *in vitro*. However, the concentrations necessary are very high (4–5 mM for maximum activity).[188]

8.4.4.5.6 *Synthetic inhibitors of fibrinolysis*

Antifibrinolytic therapy is useful in the treatment of hyperfibrinolysis following thrombolytic therapy, in potentiating hemostatic plug formation in hemophiliacs, and in certain patients with unusual localized bleeding due to local hyperfibrinolysis. ε-Aminocaproic acid (EACA; **12**) and transexamic acid (*trans*-4-aminomethylcyclohexane-1-carboxylic acid; AMCA; AMCHA; **13**) are compounds which inhibit the fibrinolytic process. These agents are analogs of lysine (**14**) and occupy the lysine binding sites of plasminogen, thus interfering with the binding of plasminogen to fibrin.[189] SAR suggest that close resemblance to lysine is important. AMCA is six to ten times more potent than EACA, but the *cis* form is inactive.[190,191] Other analogs more potent than EACA include *p*-aminomethylbenzoic acid (PAMBA; **15**) and 4-(2-aminoethyl)bicyclo[2.2.2]octane-1-carboxylic acid (AMBOCA; **16**).[191]

8.4.5 REFERENCES

1. R. G. MacFarlane, *Nature (London)*, 1964, **202**, 498.
2. E. W. Davie and O. D. Ratnoff, *Science (Washington, D.C.)*, 1964, **145**, 1310.
3. H. Neurath, in 'Proteases and Biological Control,' ed. E. Reich, D. B. Rifkin and E. Shaw, Cold Spring Harbor Laboratory, Cold Spring Harbor, NY, 1975, p. 51.
4. S. Müllertz, *Fibrinolysis*, 1987, **1**, 3.
5. C. M. Jackson and Y. Nemerson, *Annu. Rev. Biochem.*, 1980, **49**, 765.
6. L. F. Fries, III and M. M. Frank, in 'The Molecular Basis of Blood Diseases', ed. G. Stamatoyannopoulos, A. W. Nienhuis, P. Leder and P. W. Mejerus, Saunders, Philadelphia, 1987, p. 450.
7. E. J. Ariëns and A.-M. Simonis, *Top. Curr. Chem.*, 1974, **52**, 1.
8. V. J. Marder and C. W. Francis, *Drugs*, 1987, **33** (Suppl. 3), 13.
9. W. H. Seegers, *Semin. Thromb. Hemostasis*, 1981, **7**, 180.
10. P. W. Majerus, in 'Molecular Basis of Blood Diseases', ed. G. Stamatoyannopoulos, A. W. Nienhuis, P. Leder and P. W. Mejerus, Saunders, Philadelphia, 1987, p. 689.
11. V. Fuster, P. C. Adams, J. J. Badimon and J. H. Chesebro, *Prog. Cardiovasc. Dis.*, 1987, **29**, 325.
12. R. W. Colman and P. N. Walsh, in ref. 14, p. 594.
13. J. F. Mustard, R. L. Kinlough-Rathbone and M. A. Packham, *Agents Actions (Suppl.)*, 1987, **21**, 23.
14. R. W. Colman, J. Hirsh, V. J. Marder and E. W. Salzman (eds.), 'Hemostasis and Thrombosis', 2nd edition, Lippencott, Philadelphia, 1987.
15. J. E. Ansell, *Am. Heart J.*, 1987, **114**, 910.
16. C. S. Kitchens, *Semin. Thromb. Hemostasis*, 1985, **11**, 293.
17. E. Briet, L. Engesser and E. J. P. Brommer, *Haemostasis*, 1985, **15**, 228.

18. A. B. Chandler, K. Eurenius, G. C. McMillan, C. B. Nelson, C. J. Schwartz and S. Wessler (eds.), 'The Thrombotic Process in Atherogenesis', Plenum, New York, 1978.
19. J. B. Kruskal, P. J. Commerford, J. J. Franks and R. E. Kirsch, *N. Engl. J. Med.*, 1987, **317**, 1361.
20. J. B. Mandelkorn, N. M. Wolf, S. Singh, J. A. Shechter, R. I. Kersh, D. M. Rodgers, M. B. Workman, L. G. Bentivoglio, S. M. LaPorte and S. G. Meister, *Am. J. Cardiol.*, 1983, **52**, 1.
21. M. A. DeWood, J. Spores, R. Notske, L. T. Mouser, R. Burroughs, M. S. Golden and H. T. Lang, *N. Engl. J. Med.*, 1980, **303**, 897.
22. J. E. Dalen, J. A. Paraskos, I. S. Ockene, J. S. Alpert and J. Hirsh, *Chest*, 1986, **89**, 370S.
23. A. Ansari, *Clin. Cardiol.*, 1986, **9**, 398.
24. R. F. Gillum, *Am. Heart J.*, 1987, **114**, 1262.
25. N. Ellison, D. R. Jobes and A. J. Schwartz, *Int. Anesthesiol. Clin.*, 1982, **20** (4), 121.
26. S. Wessler and S. N. Gitel, *J. Neurosurg.*, 1981, **54**, 1.
27. J. D. Easton and D. G. Sherman, *Stroke*, 1980, **11**, 433.
28. N. Sugiyama, M. Iwamoto and Y. Abiko, *Thromb. Res.*, 1987, **47**, 459.
29. E. W. Davie, A. Ichinose and S. P. Leytus, *Cold Spring Harbor Symp. Quant. Biol.*, 1986, **51**, 509.
30. R. D. Rosenberg, in 'The Molecular Basis of Blood Diseases', ed. G. Stamatoyannopoulos, A. W. Nienhuis, P. Leder and P. W. Mejerus, Saunders, Philadelphia, 1987, p. 534.
31. B. Osterud, *Scand. J. Haematol.*, 1984, **32**, 337.
32. A. I. Schafer, *Annu. Rev. Med.*, 1987, **38**, 211.
33. R. W. Colman, V. J. Marder, E. W. Salzman and J. Hirsh, in ref. 14, p. 3.
34. M. Silverberg and S. V. Diehl, *Biochem. J.*, 1987, **248**, 715.
35. A. H. Schmaier, M. Silverberg, A. P. Kaplan and R. W. Colman, in ref. 14, p. 18.
36. L. H. Hoyer, in ref. 14, p. 48.
37. R. W. Colman, in ref. 14, p. 120.
38. R. E. Olson, in ref. 14, p. 846.
39. K. G. Mann and R. L. Lundblad, in ref. 14, p. 148.
40. G. Murano (ed.), 'Protease Inhibitors of Human Plasma. Biochemistry and Pathophysiology', PJD Publications, Westbury, NY. 1986.
41. R. A. Pixley, A. Schmaier and R. W. Colman, *Arch. Biochem. Biophys.*, 1987, **256**, 490.
42. H. Nishikado, Y. Komiyama, M. Masuda, H. Egawa and K. Marata, *Thromb. Res.*, 1987, **48**, 145.
43. P. C. Harpel, in ref. 14, p. 219.
44. D. M. Tollefsen and M. K. Blank, *J. Clin. Invest.*, 1981, **68**, 589.
45. C. T. Esmon, S. Vigano-D'Angelo, A. D'Angelo and P. C. Comp, *Adv. Exp. Biol. Med.*, 1987, **214**, 47.
46. E. F. Mammen, W. R. Thomas and W. H. Seegers, *Thromb. Diath. Haemorrh.*, 1960, **5**, 218.
47. C. T. Esmon, *Blood*, 1983, **62**, 1155.
48. J. Stenflo, *Semin. Thromb. Hemostasis*, 1984, **10**, 109.
49. C. T. Esmon and N. L. Esmon, *Semin. Thromb. Hemostasis*, 1984, **10**, 122.
50. R. A. Marlar, A. J. Kleiss and J. H. Griffin, *Blood*, 1982, **59**, 1067.
51. E. A. Thompson and H. H. Salem, *Prog. Hematol.*, 1987, **15**, 51.
52. W. C. Owen, in ref. 14, p. 235.
53. V. W. M. van Hinsbergh, R. M. Bertina, A. van Wijngaarden, N. H. van Tilburg, J. J. Emeis and F. Haverkate, *Blood*, 1985, **65**, 444.
54. F. B. Taylor, Jr. and M. S. Lockhart, *Thromb. Res.*, 1985, **37**, 155.
55. B. S. Coller, J. Owen, J. Jesty, D. Horowitz, M. J. Reitman, J. Spear, T. Yeh and P. C. Comp, *Arteriosclerosis*, 1987, **7**, 456.
56. K. Suzuki, J. Nishioka and S. Hashimoto, *J. Biol. Chem.*, 1983, **258**, 163.
57. B. Dahlback, *J. Biol. Chem.*, 1986, **261**, 12 022.
58. E. H. Rehmus, B. M. Greene, B. A. Everson and O. D. Ratnoff, *J. Clin. Invest.*, 1987, **80**, 516.
59. S. Shapiro and J. Martinez, in 'Hemostasis and Thrombosis'. ed. E. J. W. Bowie and A. A. Sharp, Butterworths, London, 1985, p. 197.
60. Z. M. Ruggeri and T. S. Zimmerman, *Blood*, 1987, **70**, 895.
61. E. Marciniak, *Adv. Exp. Biol. Med.*, 1987, **214**, 175.
62. U. Seligsohn, A. Berger, M. Abend, L. Rubin, D. Attias, A. Zivelin and S. J. Rapaport, *N. Engl. J. Med.*, 1984, **310**, 559.
63. E. F. Mammen, M. I. Barnhart, J. M. Lusher and R. T. Walsh (eds.), 'Treatment of Bleeding Disorders with Blood Components', PJD Publications, Westbury, NY, 1980.
64. O. D. Ratnoff, in ref. 14, p. 1026.
65. C. T. Esmon, F. B. Taylor, Jr., L. B. Hinshaw, A. Chang, P. C. Comp, G. Ferrell and N. L. Esmon, *Dev. Biol. Stand.*, 1987, **67**, 51.
66. P. C. Comp and C. T. Esmon, *J. Clin. Invest.*, 1981, **68**, 1221.
67. J. J. Toole, J. L. Knopf, J. M. Wozney, L. A. Sultzman, J. L. Buecker, D. D. Pittman, R. J. Kaufman, E. Brown, C. Shoemaker, E. C. Orr, G. W. Amphlett, W. B. Foster, M. L. Coe, G. J. Knutson, D. N. Fass and R. M. Hewick, *Nature (London)*, 1984, **312**, 342.
68. W. I. Wood, D. J. Capon, C. C. Simonsen, D. L. Eaton, J. Gitschier, B. Keyt, P. H. Seeburg, D. H. Smith, P. Hollingshead, K. L. Wion, E. Delwart, E. G. D. Tuddenham, G. A. Vehar and R. M. Lawn, *Nature (London)*, 1984, **312**, 330.
69. B. W. Grinnell, D. T. Berg, J. Walls and S. B. Yan, *Bio/Technology*, 1987, **5**, 1189.
70. J. H. Winter and A. S. Douglas, *Clin. Haematol.*, 1981, **10**, 459.
71. R. B. Silverman, *J. Am. Chem. Soc.*, 1980, **102**, 5421.
72. D. S. Whitlon, J. A. Sadowski and J. W. Suttie, *Biochemistry*, 1978, **17**, 1371.
73. E. Chan, L. Aarons, M. Serlin and M. Rowland, *Br. J. Clin. Pharmacol.*, 1987, **24**, 621.
74. R. A. O'Reilly, *J. Clin. Invest.*, 1969, **48**, 193.
75. I. Chmielewska and J. Cieslak, *Tetrahedron*, 1958, **4**, 135.
76. R. B. Arora and C. N. Mathur, *Br. J. Pharmacol.*, 1963, **20**, 29.

77. K. Rehse, *Drugs Future*, 1985, **10**, 205.
78. E. Coyne, *Semin. Thromb. Hemostasis*, 1985, **11**, 10.
79. R. D. Rosenberg, in ref. 14, p. 1373.
80. J. Fareed, *Semin. Thromb. Hemostasis*, 1987, **11**, 1.
81. J. Hirsh, F. Ofosu and M. Buchanan, *Semin. Thromb. Hemostasis*, 1985, **11**, 13.
82. M. Andrew, F.Ofosu, F. Fernandez, A. Jefferies, J. Hirsh, L. Mitchell and M. R. Buchanan, *Thromb. Haemostasis*, 1986, **55**, 342.
83. J. Fareed, J. M. Walenga, D. A. Hoppensteadt and H. L. Messmore, *Semin. Thromb. Hemostasis*, 1985, **11**, 199.
84. H. K. Takahashi, H. B. Nader and C. P. Dietrich, *Arzneim.-Forsch.*, 1985, **35**, 1620.
85. P. Bianchini, B. Osima, B. Parma, H. B. Nader, C. P. Dietrich, B. Casu and G. Torri, *Arzneim.-Forsch.*, 1985, **35**, 1215.
86. I. D. Bradbrook, H. N. Magnani, H. C. T. Moelker, P. J. Morrison, J. Robinson, H. J. Rogers, R. G. Spector, T. van Dinther and H. Wijnand, *Br. J. Clin. Pharmacol.*, 1987, **23**, 667.
87. A. Naggi, G. Torri, B. Casu, J. Pangrazzi, M. Abbadini, M. Zametta, M. B. Donati, J. Lansen and J. P. Maffrand, *Biochem. Pharmacol.*, 1987, **36**, 1895.
88. J. M. Walenga, J. Fareed, M. Petitou, M. Samama, J. C. Lormeau and J. Choay, *Thromb. Res.*, 1986, **43**, 243.
89. J. Choay, *Semin. Thromb. Hemostasis*, 1985, **11**, 81.
90. J. Choay, J.-C. Lormeau, M. Petitou, P. Sinay and J. Fareed, *Ann. N.Y. Acad. Sci.*, 1981, **370**, 644.
91. J. Choay, J. C. Lormeau, M. Petitou, P. Sinay, B. Casu, P. Oreste, G. Torri and G. Gatti, *Thromb. Res.*, 1980, **18**, 573.
92. J. M. Walenga and J. Fareed, *Semin. Thromb. Hemostasis*, 1985, **11**, 89.
93. S. E. Lasker, *Semin. Thromb. Hemostasis*, 1985, **11**, 37.
94. P. Walsmann and F. Markwardt, *Thromb. Res.*, 1985, **40**, 563.
95. M. E. Nesheim, F. G. Prendergast and K. G. Mann, *Biochemistry*, 1979, **18,**, 996.
96. A. Hijikata, S. Okamoto, E. Mori, K. Kinjo, R. Kikumoto, S. Tonomura, Y. Tamao and H. Hara, *Thromb. Res.*, 1976, **8** (Suppl. II), 1976.
97. A. Hijikata, S. Okamoto, R. Kikumoto and Y. Tamao, *Thromb. Haemostasis*, 1979, **42**, 1039.
98. D. Green, C. Ts'ao, N. Reynolds, D. Kahn, H. Kohl and I. Cohen, *Thromb. Res.*, 1985, **37**, 145.
99. Y. Tamao, T. Yamamoto, R. Kikumoto, H. Hara, J. Itoh, T. Hirata, K. Mineo and S. Okamoto, *Thromb. Haemostasis*, 1986, **56**, 28.
100. R. Kikumoto, Y. Tamao, T. Tezuka, S. Tonomura, H. Hara, K. Ninomiya, A. Hijikata and S. Okamoto, *Biochemistry*, 1984, **23**, 85.
101. S. Okamoto, A. Hijikata, R. Kikumoto, S. Tonomura, H. Hara, K. Ninomiya, A. Maruyama, M. Sugano and Y. Jamao, *Biochem. Biophys. Res. Commun.*, 1981, **101**, 440.
102. J. Stürzebecher, F. Markwardt, B. Voigt, G. Wagner and P. Walsmann, *Thromb. Res.*, 1983, **29**, 635.
103. D. Bagdy, S. Bajusz and G. Rabloczky, *Drugs Future*, 1985, **10**, 829.
104. B. Kaiser and F. Markwardt *Thromb. Res.*, 1986, **43**, 613.
105. J. Bonnar, *Am. J. Obstet. Gynecol.*, 1987, **157**, 1042.
106. F. Markwardt, *Ann. N.Y. Acad. Sci.*, 1981, **370**, 757.
107. C.-M. Kam, J. C. Copher and J. C. Powers, *J. Am. Chem. Soc.*, 1987, **109**, 5044.
108. D. Collen and H. R. Lijnen, in 'The Molecular Basis of Blood Diseases', ed. G. Stamatoyannopoulos, A. W. Nienhuis, P. Leder and P. W. Mejerus, Saunders, Philadelphia, 1987, p. 662.
109. D. Collen and H. R. Lijnen, *CRC Crit. Rev. Oncol. Hematol.*, 1986, **4**, 249.
110. C. W. Francis and V. J. Marder, *Hum. Pathol.*, 1987, **18**, 263.
111. L. A. Moroz, *Semin. Thromb. Hemostasis*, 1984, **10**, 80.
112. K. C. Robbins, in ref. 14, p. 340.
113. D. Collen, in 'Tissue Plasminogen Activator in Thrombolytic Therapy', ed. B. Sobel, D. Collen and E. B. Grossbard, Dekker, New York, 1987, p. 3.
114. D. C. Rijken, G. Wijngaards, M. Zaal-de Jong and J. Welbergen, *Biochim. Biophys. Acta*, 1979, **580**, 140.
115. K. C. Robbins, G. H. Barlow, G. Nguyen and M. M. Samama, *Semin. Thromb. Hemostasis*, 1987, **13**, 131.
116. F. Bachmann, in ref. 14, p. 318.
117. B. Wiman, H. R. Lijnen and D. Collen, *Biochim. Biophys. Acta*, 1979, **579**, 142.
118. A. J. Tiefenbrunn and B. E. Sobel, in 'Tissue Plasminogen Activator in Thrombolytic Therapy', ed. B. E. Sobel, D. Collen and E. B. Grossbard, Dekker, New York, 1987, p. 25.
119. M. Hoylaerts, D. C. Rijken, H. R. Lijnen and D. Collen, *J. Biol. Chem.*, 1982, **257**, 2912.
120. S. S. Tang, J. Loscalzo and V. J. Dzau, *Council for High Blood Pressure Res., 41st Annu. Fall Conf.*, 1987, abstr. 2.
121. D. Collen, *Prog. Hemostasis Thromb.*, 1986, **8**, 1.
122. H. R. Lijnen, C. Zamarron, M. Blaber, M. E. Winkler and D. Collen, *J. Biol. Chem.*, 1986, **261**, 1253.
123. Subcommittee on Fibrinolysis of the ICTH, *Thromb. Haemostasis*, 1985, **54**, 893.
124. V. Gurewich and R. Pannell, *Semin. Thromb. Hemostasis*, 1987, **13**, 893.
125. B. Wiman and D. Collen, *Eur. J. Biochem.*, 1978, **84**, 573.
126. T. Tamaki and N. Aoki, *J. Biol. Chem.*, 1982, **257**, 14767.
127. T.-C. Wun and A. Capuano, *Blood*, 1987, **69**, 1354.
128. E. D. Sprengers and C. Kluft, *Blood*, 1987, **69**, 381.
129. J. W. J. Y. Lambers, M. Cammenga, B. W. König, K. Mertens, H. Pannekoek and J. A. van Mourik, *J. Biol. Chem.*, 1987, **262**, 17492.
130. I. Lecander and B. Astedt, *Br. J. Haematol.*, 1986, **62**, 221.
131. I. Lecander and B. Astedt, *J. Lab. Clin. Med.*, 1987, **110**, 602.
132. H. R. Lijnen and D. Collen, *Fibrinolysis*, 1989, **3**, 67.
133. I. Juhan-Vague, B. Moerman, F. De Cock, M. F. Aillaud and D. Collen, *Thromb. Res.*, 1984, **33**, 523.
134. J. A. Paramo, M. Colucci, D. Collen and F. van de Werf, *Br. Med. J.*, 1985, **291**, 573.
135. O. Johnson, G. Mellbring and T. Nilsson, *Int. J. Cardiol.*, 1984, **6**, 380.
136. A. Hamsten, B. Wiman, U. de Faire and M. Bolmbäck, *N. Engl. J. Med.*, 1985, **313**, 1557.
137. C. Korninger, K. Lechner, H. Niessner, H. Gössinger and M. Kundi, *Thromb. Haemostasis*, 1984, **52**, 127.
138. L. Häggroth, C. Mattsson, P. Felding and I. M. Nilsson, *Thromb. Res.*, 1986, **42**, 585.

139. N. Bergsdorf, T. Nilsson and P. Wallen, *Thromb. Haemostasis*, 1983, **50**, 740.
140. J. Chmielewska, M. Ranby and B. Wiman, *Thromb. Res.*, 1983, **31**, 427.
141. I. Juhan-Vague, J. Valadier, M. C. Alessi, M. F. Aillaud, J. Ansaldi, C. Philip-Joet, P. Holvoet, A. Serradimigni and D. Collen, *Thromb. Haemostasis*, 1987, **57**, 67.
142. J. Gram, C. Kluft and J. Jespersen, *Thromb. Haemostasis*, 1987, **58**, 817.
143. M. Jorgensen, M. Philips, S. Thorsen, J. Selmer and J. Zeuthen, *Thromb. Haemostasis*, 1987, **58**, 872.
144. B. Wiman, G. Csemiczky, L. Marsk and H. Robbe, *Thromb. Haemostasis*, 1984, **52**, 124.
145. M. F. Aillaud, F. Pignol, M. C. Alessi, J. R. Harle, M. Escande, M. Mongin and I. Juhan-Vague, *Thromb. Haemostasis*, 1986, **55**, 330.
146. L. A. Miles, E. F. Plow, K. J. Donnelly, C. Hougie and J. H. Griffen, *Blood*, 1982, **59**, 1246.
147. J. Aznar, A. Estelles, V. Vila, E. Reganon, F. Espana and P. Villa, *Thromb. Haemostasis*, 1984, **52**, 196.
148. S. D. Rogers, L. B. Riemersma and S. D. Clements, *Pharmacotherapy*, 1987, **7**, 111.
149. V. J. Marder and S. Sherry, *N. Engl. J. Med.*, 1988, **318**, 1512, 1585.
150. A. J. Tiefenbrunn and B. E. Sobel, *Fibrinolysis*, 1989, **3**, 1.
151. U. Schmitz-Huebner and J. van de Loo, *Clin. Haematol.*, 1981, **10**, 481.
152. E. Braunwald, *J. Am. Coll. Cardiol.*, 1987, **10**, 1B.
153. B. E. Sobel, *J. Am. Coll. Cardiol.*, 1987, **10**, 40B.
154. D. O. Williams *et al.*, *Circulation*, 1986, **73**, 338.
155. A. D. Guerci *et al.*, *N. Engl. J. Med.*, 1987, **317**, 1613.
156. S. J. Crabbe and C. C. Cloniger, *Clin. Pharmacol.*, 1987, **6**, 373.
157. M. Verstraete *et al.*, *Lancet*, 1985, **1**, 842.
158. D. Collen, J. M. Stassen, B. J. Marafino, Jr., S. Builder, F. De Cock, J. Ogez, D. Tajiri, D. Pennica, W. F. Bennett, J. Salwa and C. F. Hoyng, *J. Pharmacol. Exp. Ther.*, 1984, **231**, 146.
159. G. Trübestein, S. Popov, H. Wolf and D. Welzel, *Haemostasis*, 1987, **17**, 238.
160. C. Zamarron, H. R. Lijnen, B. Van Hoef and D. Collen, *Thromb. Haemostasis*, 1984, **52**, 19.
161. H. Sumi, K. Sasaki, N. Toki and K. C. Robbins, *Thromb. Res.*, 1980, **20**, 711.
162. D. Collen, D. C. Stump and F. Van de Werf, *Am. Heart J.*, 1986, **112**, 1083.
163. D. Collen, J. M. Stassen and F. De Cock, *Thromb. Haemostasis*, 1987, **58**, 943.
164. GISSI, *Lancet*, 1986, **1**, 397.
165. M. Gonzalez-Gronow, G. E. Siefring, Jr. and F. J. Castellino, *J. Biol. Chem.*, 1978, **253**, 1090.
166. GISSI, *Lancet*, 1987, **2**, 871.
167. J. P. Monk and R. C. Heel, *Drugs*, 1987, **34**, 25.
168. V. J. Marder, R. L. Rothbard, P. G. Fitzpatrick and C. W. Francis, *Ann. Intern. Med.*, 1986, **104**, 304.
169. R. Fears, *J. R. Soc. Med.*, 1985, **78**, 687.
170. N. L. Haigwood, G. Mullenbach, E. G. Afting and E. P. Paques, *Eur. Pat. Appl.*, 227 462 (1985); *Chem. Abstr.*, 1987, **107**, 230 570t.
171. M. Verstraete, *J. Am. Coll. Cardiol.*, 1987, **10**, 4B.
172. D. Collen, *J. Am. Coll. Cardiol.*, 1987, **10**, 11B.
173. M. S. Runge, C. Bode, G. R. Matsueda and E. Haber, *Proc. Natl. Acad. Sci. USA*, 1987, **84**, 7659.
174. C. Bode, G. R. Matsueda, K. Y. Hui and E. Haber, *Science (Washington, D.C.)*, 1985, **229**, 765.
175. B. E. Sobel, L. E. Fields, A. K. Robison, K. A. A. Fox and S. J. Sarnoff, *Proc. Natl. Acad. Sci. USA*, 1985, **82**, 4258.
176. G. Noll, B. Lämmle and F. Duckert, *Thromb. Res.*, 1985, **37**, 529.
177. H. M. Sue-Ling, J. A. Davies, C. R. M. Prentice, J. H. Verheijen and C. Kluft, *Thromb. Haemostasis*, 1985, **55**, 141.
178. P. A. Andreason, C. Pyke, A. Riccio, P. Kristensen, L. S. Nielsen, L. R. Lund, F. Blasi and K. Dano, *Mol. Cell. Biol.*, 1987, **7**, 3021.
179. F. Markwardt and H.-P. Klöcking, *Thromb. Res.*, 1976, **8**, 217.
180. A. Szczeklik, M. Kopec, K. Sladek, J. Musial, J. Chmielewska, E. Teisseyre, G. Dudek-Wojciechowska and M. Palester-Chlebowczyk, *Thromb. Res.*, 1983, **29**, 655.
181. J. Schneider, *Thromb. Res.*, 1987, **48**, 233.
182. H.-P. Klöcking, A. Hoffmann and F. Markwardt, *Thromb. Res.*, 1987, **46**, 747.
183. I. S. Chohan, *Thromb. Haemostasis*, 1986, **56**, 360.
184. D. W. Richardson and A. G. Robinson, *Ann. Intern. Med.*, 1985, **103**, 228.
185. E. Melissari, M. F. Scully, T. Paes, V. Ellis and V. V. Kakkar, *Thromb. Haemostasis*, 1986, **55**, 54.
186. K. N. von Kaulla, in 'Burger's Medicinal Chemistry', 4th edn., ed. M. E. Wolff, Wiley-Interscience, New York, 1979, Part II, p. 1081.
187. J. Roba, F. De Paermentier, G. Defreyn, F. Dessy and P. Niebes, *Abstr. 9th Int. Cong. Thromb. Haemostasis*, 1981, abstr. 245; see also abstracts: 161, 162, 163 and 244.
188. M. H. Chevallet-Prandini, A. Chapel, G. Hudry-Clergeon and M. Suscillon, *Thromb. Res.*, 1981, **24**, 197.
189. V. J. Marder, F. O. Butler and G. H. Barlow, in ref. 14, p. 380.
190. L. Andersson, I. M. Nilsson, J.-E. Nilhen, U. Hedner, B. Granstrand and B. Melander, *Scand. J. Haematol.*, 1965, **2**, 230.
191. S. Okamoto, S. Oshiba, H. Mihara and U. Okamoto, *Ann. N.Y. Acad. Sci.*, 1968, **146**, 414.

8.5

Selective Inhibitors of Phosphodiesterases

RONALD E. WEISHAAR and JAMES A. BRISTOL
Warner-Lambert Company, Ann Arbor, MI, USA

8.5.1	INTRODUCTION	501
8.5.2	MULTIPLE MOLECULAR FORMS OF CYCLIC NUCLEOTIDE PHOSPHODIESTERASE	502
8.5.3	SELECTIVE INHIBITORS OF PHOSPHODIESTERASES	505
	8.5.3.1 Type I Phosphodiesterase Inhibitors	505
	8.5.3.2 Type II Phosphodiesterase Inhibitors	506
	8.5.3.3 Type III Phosphodiesterase Inhibitors	506
8.5.4	RELATIONSHIP BETWEEN SELECTIVE INHIBITION OF PHOSPHODIESTERASES AND PHARMACOLOGICAL RESPONSES	509
	8.5.4.1 Myocardial Contractility	510
	8.5.4.2 Pre- and Post-synaptic Neurotransmission	511
8.5.5	CONCLUSIONS	512
8.5.6	REFERENCES	513

8.5.1 INTRODUCTION

Until recently the involvement of cyclic nucleotide phosphodiesterases in modulating the different second messenger responses to cyclic AMP and cyclic GMP was poorly understood. This lack of appreciation was due in part to the limits of early analytical procedures, which suggested that the K_m for hydrolysis of cyclic nucleotides by phosphodiesterase was high, implying that the enzyme did not play a pivotal role in regulating the intracellular response to cyclic AMP and cyclic GMP. In addition, the available phosphodiesterase inhibitors, *e.g.* methylxanthines such as theophylline, were relatively impotent, and also exerted significant effects on other enzymes and receptors.[1-3]

An awareness of the pivotal role which phosphodiesterase plays in regulating the second messenger response to cyclic nucleotides came in the early 1970s when Thompson, Strada, Wells and others identified multiple molecular forms of phosphodiesterase.[4-6] These different phosphodiesterases were found to vary with regard to substrate specificity (cyclic AMP *vs.* cyclic GMP), kinetic characteristics (K_m and V_{max}), intracellular location (soluble *vs.* membrane-bound) and response to calcium and calmodulin.[3,6] The subsequent discovery of selective inhibitors of several of these different phosphodiesterases has made it possible to characterize their involvement in regulating a number of physiological processes, including cardiac muscle contraction, lipolysis, platelet aggregation and postsynaptic nerve transmission.[7-10] This new information suggests that hydrolysis of cyclic nucleotides is a highly regulated event and that the second messenger responses to cyclic AMP and cyclic GMP can be modulated at the level of degradation as well as synthesis.

The paragraphs to follow will describe recent developments in the field of multiple forms of cyclic nucleotide phosphodiesterase, summarizing what is known regarding the enzymatic and pharmacological properties of each of these different enzymes, and proposing an integrated role for the various phosphodiesterases in regulating the second messenger responses to cyclic AMP and cyclic GMP. In addition, a survey of selective inhibitors of the different subclasses of phosphodiesterase will be provided, with emphasis on the chemistry of representative inhibitors of each class. The final section will focus on two of the molecular forms of phosphodiesterase—the cyclic AMP specific, cyclic GMP-inhibited phosphodiesterase and the cyclic AMP specific, cyclic GMP insensitive phosphodiesterase—and the role which these two enzymes play in regulating myocardial contractility and synaptic nerve transmission, respectively.

8.5.2 MULTIPLE MOLECULAR FORMS OF CYCLIC NUCLEOTIDE PHOSPHODIESTERASE

Enzymatic degradation of cyclic AMP was first described by Sutherland and Rall three decades ago.[11] The enzyme which catalyzed this degradation, later termed phosphodiesterase, was shown to be magnesium dependent, inhibited by methylxanthines and stimulated by imidazole.[11,12] This enzymatic activity was presumably due to a single protein, which was widely distributed.[13] Figure 1 represents an early model for the role which phosphodiesterase was thought to play in regulating cyclic AMP degradation, although presumably the single phosphodiesterase hydrolyzed both cyclic AMP and cyclic GMP.

In the early 1970s Thompson, Appleman and others demonstrated the existence of multiple molecular forms of phosphodiesterase.[15-17] The lack of uniform isolation procedures employed in these early studies often made it difficult to compare the results obtained in different laboratories with regard to the exact number of phosphodiesterases present in various tissues. Thus, whereas in 1973 Amer and Mayol described two forms of phosphodiesterases in human platelets,[18] a later study by Hidaka and Asano reported that human platelets contain three distinct forms of phosphodiesterase.[19] Similar inconsistencies have been reported for cardiac muscle,[8,20,21] and for tracheal smooth muscle.[22,23] These difficulties have been overcome by the widespread adoption of diethylaminoethyl(DEAE)-cellulose anion exchange chromatography.[3,4] Subsequent studies have also employed affinity chromatography[24] and monoclonal antibodies[25] to isolate highly purified phosphodiesterases.

Figure 2 shows a typical isolation chromatograph, in which application of a linear salt gradient following adsorption of a supernatant from guinea pig left ventricular muscle onto DEAE-cellulose results in the elution of three distinct peaks of phosphodiesterase activity. Most investigators in the area have chosen to label these different phosphodiesterases based upon their order of elution, a

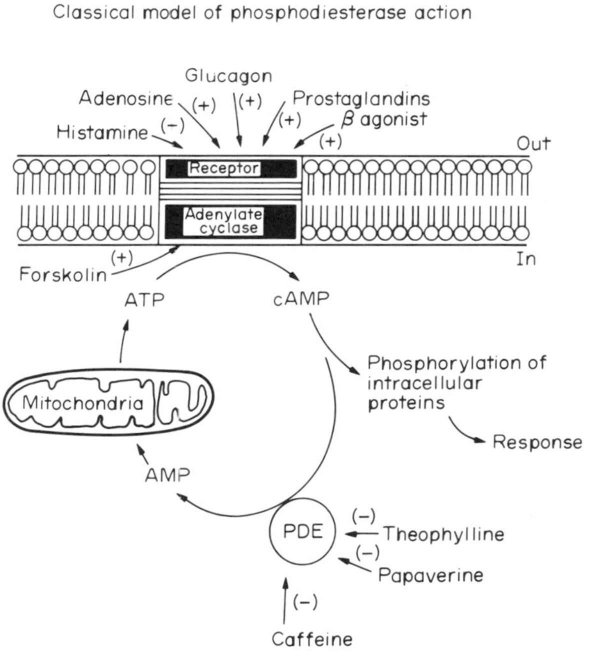

Classical model of phosphodiesterase action

Figure 1 Early model for the involvement of phosphodiesterase in the regulation of intracellular cyclic AMP. Also shown are several of the best-known nonselective phosphodiesterase inhibitors (reproduced from ref. 14 by permission of the editor)

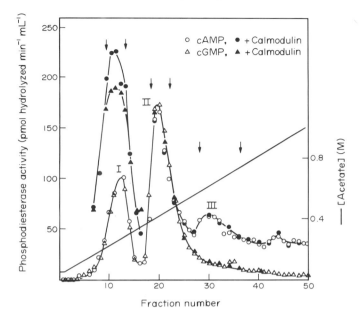

Figure 2 Separation of the three molecular forms of cardiac phosphodiesterase using DEAE-cellulose anion exchange chromatography. Phosphodiesterases were absorbed onto the DEAE-cellulose and then eluted with a 70–1000 mM sodium acetate gradient. Column fractions were collected and cyclic AMP and cyclic GMP phosphodiesterase activity was measured in the presence (filled symbols) and absence (open symbols) of 0.1 U of calmodulin and 10 μM CaCl$_2$ (reproduced from ref. 14 by permission of the editor)

nomenclature system which has on occasion led to difficulty in comparing results obtained from different laboratories. As Figure 2 shows, the three different phosphodiesterases present in guinea pig left ventricular muscle vary in several important ways. First, whereas the type I and type II phosphodiesterases hydrolyze cyclic AMP and cyclic GMP to a comparable extent, the type III phosphodiesterase selectively hydrolyzes cyclic AMP. Figure 2 also illustrates that although the activity of the type I phosphodiesterase can be stimulated two- to three-fold by calmodulin (in the presence of calcium), the type II and type III phosphodiesterases are insensitive to calcium and calmodulin. Based upon this latter difference, the type I phosphodiesterase is typically referred to as the 'calmodulin-stimulated phosphodiesterase'. The different phosphodiesterases also vary considerably in their response to cyclic GMP. Beavo and others have shown that low concentrations of cyclic GMP stimulate the cyclic AMP hydrolytic activity of the type II phosphodiesterase, while potently inhibiting the cyclic AMP hydrolytic activity of the type III phosphodiesterase.[26,27] These differences have led many investigators to refer to the type II phosphodiesterase as the 'cyclic GMP-stimulated phosphodiesterase', and the type III phosphodiesterase as the 'cyclic GMP-inhibited phosphodiesterase'.

The discovery of multiple molecular forms of phosphodiesterase and subsequent reports demonstrating clear differences in substrate specificity, regulation by allosteric effectors, response to selective phosphodiesterase inhibitors, *etc.* has invalidated the model for the involvement of phosphodiesterase in regulating cyclic nucleotide degradation depicted in Figure 1. Figure 3 illustrates an evolving model depicting the manner by which the various molecular forms of phosphodiesterase could regulate intracellular levels of cyclic nucleotides, demonstrating alternative pathways for degrading cyclic AMP and cyclic GMP, and also the potential for one pathway to influence the activity of another pathway, *e.g.* for cyclic GMP to influence the degradation of cyclic AMP.

Several recent discoveries, however, have suggested that this latter scheme only represents an approximation of the actual number of pathways for controlling intracellular cyclic nucleotide degradation. The need for further modification is necessitated by the identification of functional subclasses of the type I or calmodulin-stimulated phosphodiesterase, and the type III or cyclic GMP-inhibited form of phosphodiesterase. Although the calmodulin-stimulated phosphodiesterase present in ventricular muscle from most species hydrolyzes both cyclic AMP and cyclic GMP (Figure 2), in rat ventricular muscle and in smooth muscle from several vascular beds, this form of phosphodiesterase selectively hydrolyzes cyclic GMP.[4,8,17,28] In addition, Weishaar *et al.*[28] have recently shown that human platelets contain a cyclic GMP specific form of PDE I which is

Evolving model of phosphodiesterase

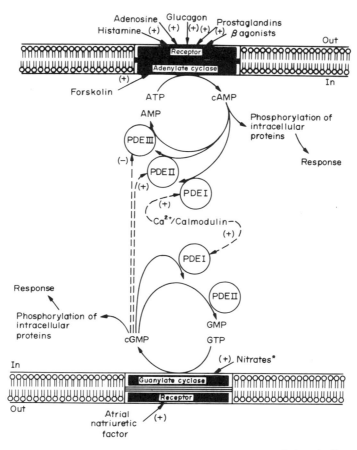

Figure 3 Evolving model for the involvement of the multiple molecular forms of phosphodiesterase in the regulation of intracellular cyclic AMP and cyclic GMP, demonstrating the alternative pathways for degradation of cyclic AMP and cyclic GMP, and the potential for one pathway to influence the activity of another pathway; nitrates stimulate the soluble form of guanylate cyclase (reproduced from ref. 14 by permission of the editor)

insensitive to calmodulin. Hidaka and coworkers have also shown that bovine aortic smooth muscle contains both a calmodulin sensitive and a calmodulin insensitive form of PDE I, and that both enzymes selectively hydrolyze cyclic GMP.[2] These latter two subclasses of PDE I have also been identified in aortic smooth muscle from several species.[29]

Sharma and Wang[30] have recently shown that bovine brain contains two distinct subclasses of calmodulin dependent phosphodiesterase, with subunit molecular masses of 60 and 63 kDa, respectively. The 60 kDa subunit isozyme can be phosphorylated by a cyclic AMP dependent protein kinase, while the 63 kDa subunit isozyme is phosphorylated by a calmodulin dependent protein kinase. Taken together, the results described above suggest the existence of several subclasses of PDE I, including (i) a calmodulin-stimulated form which hydrolyzes both cyclic AMP and cyclic GMP; (ii) a calmodulin-stimulated form which selectively hydrolyzes cyclic GMP; and (iii) a calmodulin insensitive form which selectively hydrolyzes cyclic GMP. Whether any of these subclasses is related to the light-activated cyclic GMP phosphodiesterase present in the rod outer segment is not known.[31]

In addition to the type I phosphodiesterase, functional subclasses of the type III or cyclic GMP-inhibited phosphodiesterase have also been identified. In 1984, Yamamoto *et al.*[32] identified two subclasses of the type III phosphodiesterase in calf liver, one which was potently inhibited by cyclic GMP, and one which was insensitive to cyclic GMP. The authors labeled the cyclic GMP insensitive and cyclic GMP sensitive subclasses of the cyclic AMP specific phosphodiesterase type IIIB and type IIIC, respectively, based upon their order of elution during gel chromatography. These two subclasses of the type III phosphodiesterase have subsequently been characterized in adipocytes and

Table 1 Biological Characteristics of the Multiple Molecular Forms of Phosphodiesterase, Including Subclasses

Type of phosphodiesterase	Functional subclasses	Location (partial list)	Refs.
Light-activated cyclic GMP specific phosphodiesterase	None identified	Rod outer segment	31
Calmodulin-stimulated phosphodiesterase (type I)	Calmodulin-stimulated, cyclic GMP/cyclic AMP nonspecific phosphodiesterase	Left ventricle (guinea pig, hamster, dog, rhesus monkey); brain (cow); aorta (rat, rabbit, human)	8, 21, 28, 29, 30
	Calmodulin-stimulated, cyclic GMP specific phosphodiesterase	Left ventricle (rat); coronary artery (cow, pig); aorta (rat, rabbit, human); brain (cow)	2, 4, 16, 28, 29, 30
	Calmodulin-insensitive, cyclic GMP specific phosphodiesterase	Platelets (human); aorta (cow, human, rat)	2, 19, 28, 29
Cyclic GMP-stimulated phosphodiesterase (type II)	None identified	Left ventricle (guinea pig, rat, hamster, dog, rhesus monkey); platelet (human); liver (rat)	3, 4, 8, 14, 25, 26, 27, 28
Cyclic AMP specific phosphodiesterase (type III)	Cyclic AMP specific, cyclic GMP-inhibited phosphodiesterase (type IIIC)	Left ventricle (guinea pig, dog, cow, rhesus monkey, human); platelets (human); liver (cow)	8, 9, 25, 28, 32, 34, 35, 36
	Cyclic AMP specific, cyclic GMP insensitive phosphodiesterase (type IIIB)	Left ventricle (dog, rat); liver (cow); adipocytes	8, 24, 32, 33

in ventricular muscle, and their differential involvement in regulating lipolysis and myocardial contractility has recently been described.[8,33]

The biological characteristics of these different molecular forms of cyclic nucleotide phosphodiesterase are summarized in Table 1.

8.5.3 SELECTIVE INHIBITORS OF PHOSPHODIESTERASES

In addition to variations in their response to calmodulin, calcium and cyclic GMP, the different molecular forms of phosphodiesterase are also differentially influenced by a variety of pharmacologic agents. These agents have proven useful in characterizing the intracellular responsibilities of the different molecular forms of phosphodiesterase. For simplicity, the selective inhibitors to be described will be divided into three broad classes: the type I phosphodiesterase inhibitors, the type II phosphodiesterase inhibitors and the type III phosphodiesterase inhibitors. Where appropriate, these agents will also be broken into subclass specific phosphodiesterase inhibitors.

8.5.3.1 Type I Phosphodiesterase Inhibitors

Type I phosphodiesterase inhibitors have been identified from several different structural classes. In most cases, however, structure–activity relationships have not been studied and the selective inhibitors represent isolated examples. The vincamine alkaloid vinpocetine (TCV-3B; **1**) and the quinazoline HA-558 (**2**) are relatively selective inhibitors of the calmodulin-stimulated, cyclic GMP specific phosphodiesterase (Figure 4).[2,29,37] Zaprinast (M & B 22,948; **3**)[29,38] and MY-5445 (**4**)[2,37] (Figure 4) are selective inhibitors of the calmodulin insensitive, cyclic GMP specific phosphodiesterase. The dibenzoquinazolinediones constitute a new class of PDE I inhibitor which has not been further subclassified. From an analysis of trequinsin-like agents, Booth *et al.*[38] reported that the dione (**5**) is > 400-fold selective for PDE I *vs.* PDE III, relative to trequinsin which is 100-fold selective for type III PDE. Interestingly, Lugnier *et al.*[29] showed that (**5**) produced an antihypertensive response *in vivo*, which they suggested was due to vascular smooth muscle relaxation *via* elevation of intracellular cyclic GMP.

Certain xanthines have also been shown to be relatively selective inhibitors of PDE I.[39] Included in this group are 7-benzyl-IBMX (**6**), which is 25-fold selective for PDE I *vs.* PDE III. The 'benzo-separated' analog (**7**), however, was only 10-fold selective.[39] These results were rationalized on the basis that binding of the five-membered ring in these xanthines was critical for enzyme selectivity.

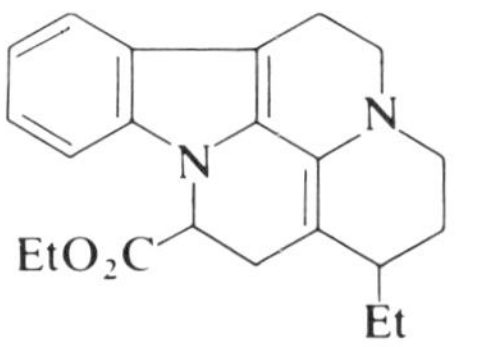

(1) Vinpocetine, TCV-3B

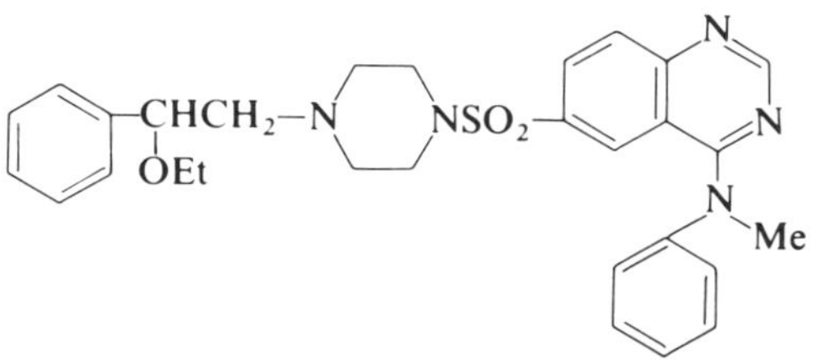

(2) HA-558

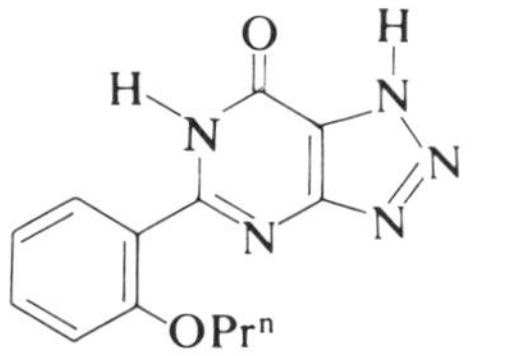

(3) Zaprinast; M&B 22,948

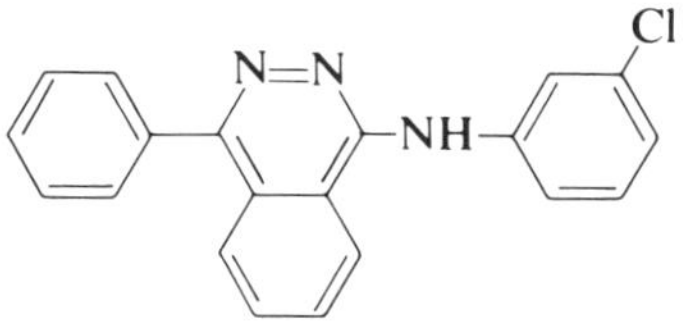

(4) MY-5445

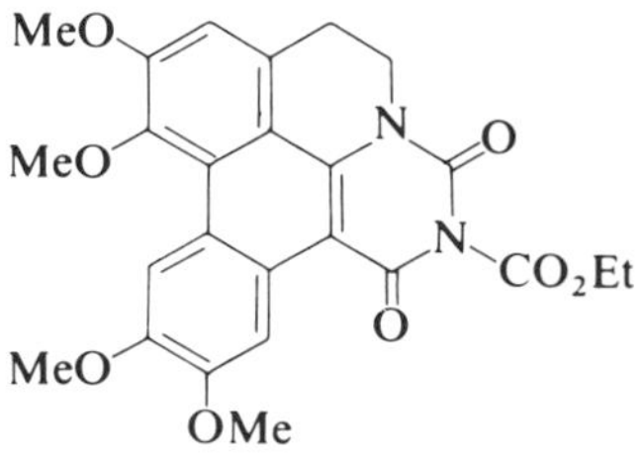

(5)

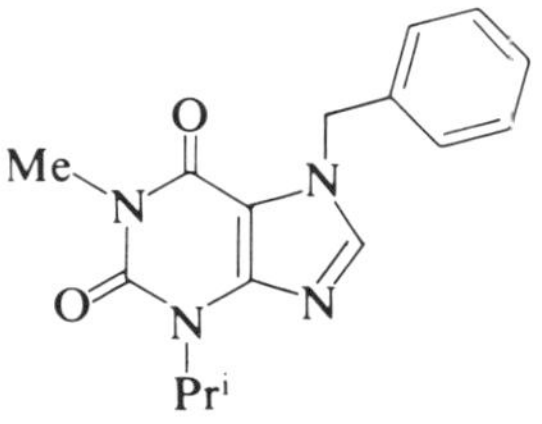

(6)

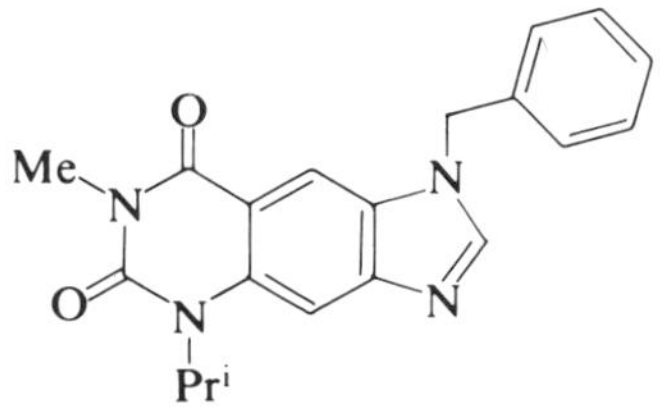

(7)

Figure 4 Type I phosphodiesterase inhibitors

8.5.3.2 Type II Phosphodiesterase Inhibitors

Research on this isozyme of phosphodiesterase continues to be hampered by the lack of selective type II inhibitors. Very modest selectivity is achieved with dipyridamole (8), and the benzimidazoles AR-L57 (9) and sulmazole (10; Figure 5).[28] These agents, however, are not sufficiently selective to represent useful pharmaceutical tools.

8.5.3.3 Type III Phosphodiesterase Inhibitors

By far the most extensively studied group of selective phosphodiesterase inhibitors are the agents which inhibit the cyclic AMP specific, cyclic GMP-inhibited phosphodiesterase (PDE IIIC),

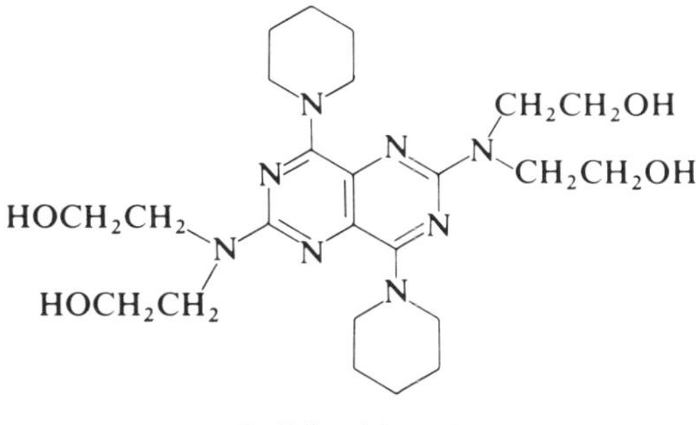

(8) Dipyridamole

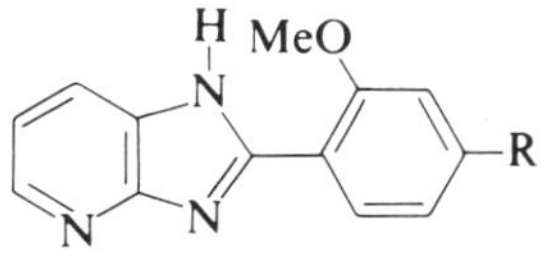

(9) AR-L 57; R = OMe
(10) Sulmazole; R = SOMe

Figure 5 Type II phosphodiesterase inhibitors

primarily due to the utility of these agents for the treatment of congestive heart failure.[3] A number of these selective inhibitors are shown in Figure 6. Within this group the structure–activity relationships for the 4,5-dihydro-3(2H)-pyridazinones have been the most thoroughly studied. The prototypes of this class are imazodan (CI-914; **11**) and CI-930 (**12**).[40,41] In the 4,5-dihydro-6-[4-(1H-imidazole-1-yl)phenyl]-3(2H)-pyridazinone (CI-914) series, it was established that the 5-methyl analog CI-930 (**12**) was the most selective agent, while the tetrahydrobenzimidazole (**13**) was the most potent PDE IIIC inhibitor ($IC_{50} = 0.51$ µM). Within that series, the imidazole ring was found to be primarily responsible for selectivity among the several PDE isozymes and the 4,5-dihydropyridazinone ring was responsible for inhibitory potency. In addition, the 5-methyl substituent was critical for potency.

A related series of 4,5-dihydropyridazinones containing an indolone replacement for the imidazolylphenyl moiety was recently reported to be selective for PDE IIIC.[42] The lead structure of this series is LY195115 (**14**), which is three times as potent as imazodan. Formal conversion of the geminal dimethyl moiety to spirocyclopropyl and higher spirocycloalkyl analogs produced a series of highly potent compounds.[43] The spirocyclopropyl analog (**15**) had the highest potency against cardiac membrane-bound PDE IIIC ($IC_{50} = 13$ µM). Higher spirocycloalkyl derivatives were only slightly less potent. Y-590 (**16**), another dihydropyridazinone containing a quinolone moiety, is a potent inhibitor of rabbit platelet PDE IIIC ($IC_{50} = 90$ nM), and was 1000-fold selective for this isozyme.[44]

In an investigation of heterocyclic analogs of the 4,5-dihydropyridazinone nucleus, Sircar *et al.*[45] reported that among geminal disubstituted pyrazol-3-ones, (**17**) is equipotent to imazodan. Of interest is the observation that the monosubstituted analogs were completely devoid of activity, presumably due to the propensity of these agents to enolize.

Of the many compounds developed for congestive heart failure whose principal mechanism of action is inhibition of PDE IIIC, the two furthest advanced clinically are amrinone (**18**), which is marketed in parenteral form in many countries, and milrinone (**19**), which has been approved for parenteral use in the United States. Curiously, structure–activity relationships for these pyridinones have not been reported, despite their advanced stage of development. Moreover it was some time after the initial disclosure of these agents that their activity as PDE IIIC inhibitors was appreciated.[41] Nevertheless, some structure–activity relationship data are available comparing milrinone and amrinone with other selective PDE IIIC inhibitors. For example, replacement of the pyridyl ring in milrinone with an imidazolyl moiety leads to compound (**20**), with comparable PDE IIIC inhibitory potency ($IC_{50} = 1.85$ and 2.5 µM, respectively).[46] Relative to the 4,5-dihydropyridazinones, the pyridones are less selective for PDE IIIC than to the PDE I and PDE II isozymes. Whether this reduced selectivity is manifested *in vivo* has not been established.

Cilostamide (**21**) is a potent carbostyryl-derived inhibitor of the type IIIC phosphodiesterase isolated from platelets and from cardiac tissue.[47] Comparison of a limited series of carboxylic acid and amide side chain analogs of cilostamide for inhibition of calmodulin dependent phosphodiesterase (PDE I) relative to inhibition of PDE IIIC demonstrated that cilostamide and its higher homolog were approximately fivefold selective for PDE IIIC.[47]

A new series of PDE IIIC inhibitors has been recently disclosed based upon a hybridized combination of cilostamide and the tetrahydroimidazo[2,1-*b*]quinazolin-2-one anagrelide (**22**).[48,49] Thus, (**21**) and (**22**) are potent and selective inhibitors of the platelet PDE IIIC with IC_{50} values of 0.17 and 0.08 µM, respectively. Formal combination of the bulky lipophilic side chain of cilostamide with the nucleus of anagrelide provided RS 82856 (**23**), a highly potent and selective PDE IIIC inhibitor ($IC_{50} = 10$ nM).[49] RS 82856 also significantly inhibited cardiac PDE IIIC isolated from several species. In evaluating the SAR of this series, Venuti *et al.*[49] reported that phosphodiesterase inhibitory potency was correlated with increasing lipophilicity of the side chain. Other modifications generally resulted in reduced potency.

Two extensive and complementary studies have been published which define pharmacophoric models derived from computational chemistry and molecular graphics.[50,51] It is generally agreed that PDE IIIC inhibitors of the extended type, *e.g.* the dihydropyridazinones, cilostamide, anagrelide and RS 82856, adopt a favored planar conformation that mimics the anticonformation of cyclic AMP. This was first reported by Bristol *et al.*,[40] with the proposal of a five-point pharmacophoric model to describe the activity of CI-930. This model was further refined by Moos *et al.*,[50] who showed that for most compounds the overall planar conformation was favored, although torsional distortion of up to 20° out of plane was possible with negligible energy cost.

Two classes of compounds, notably represented by milrinone and enoximone, possess favored conformations 40 to 50° out of plane, and the energy barrier to bring these molecules to a planar conformation is high (*e.g.* 10 kcal mol^{-1} is required to bring milrinone to 20° out of plane;

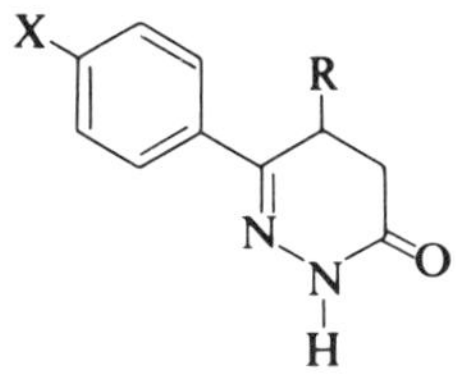

	X	R
(11) Imazodan, CI-914	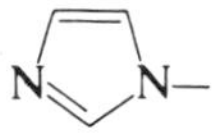	H
(12) Cl-930	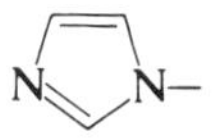	Me
(13)	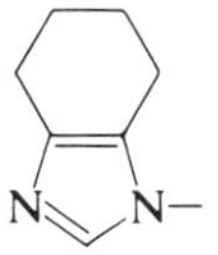	Me

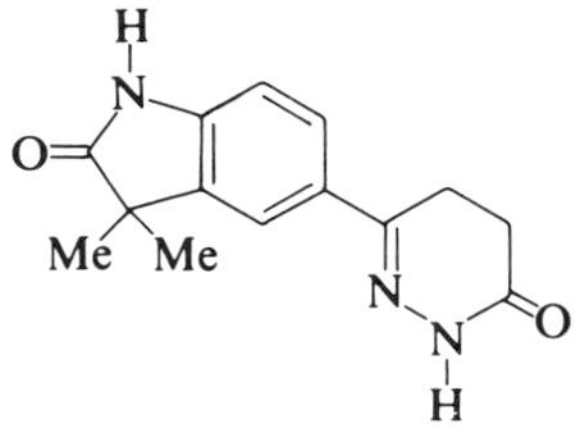

(14) LY 195115

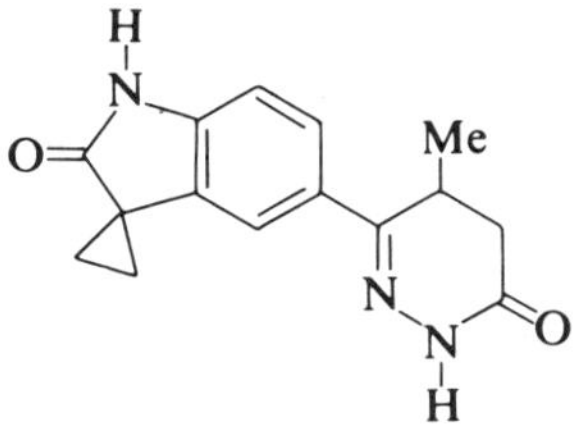

(15)

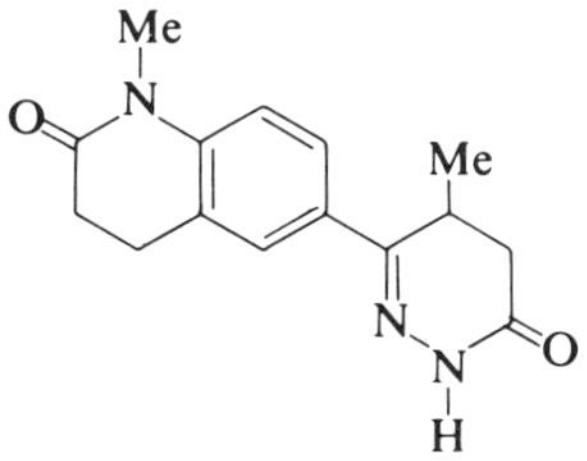

(16)

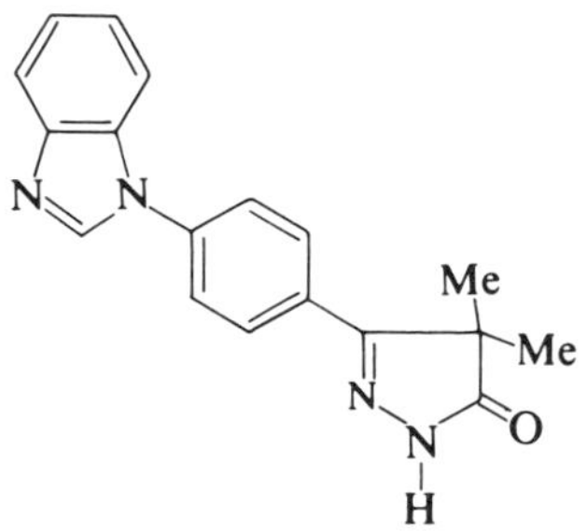

(17)

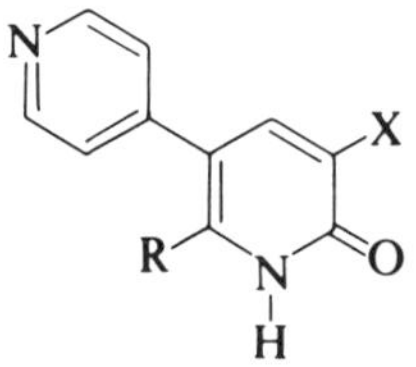

(18) Milrinone; X = CN, R = Me
(19) Amrinone; X = NH_2, R = H

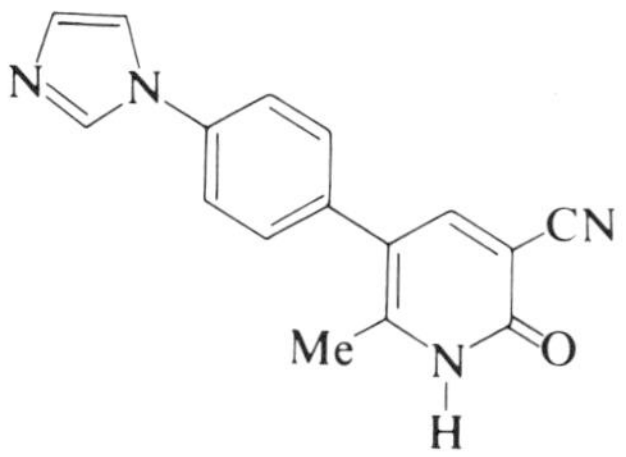

(20)

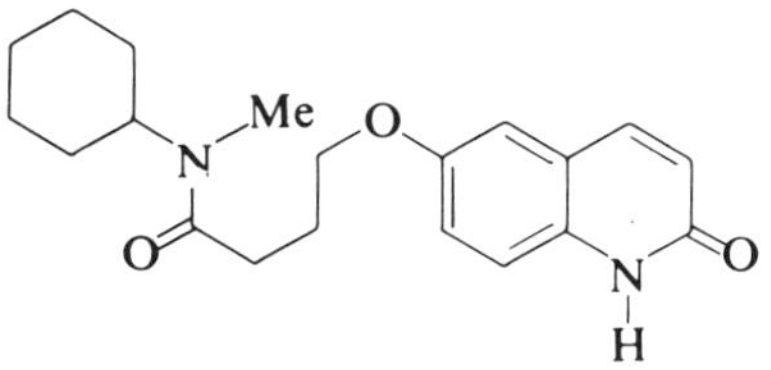

(21)

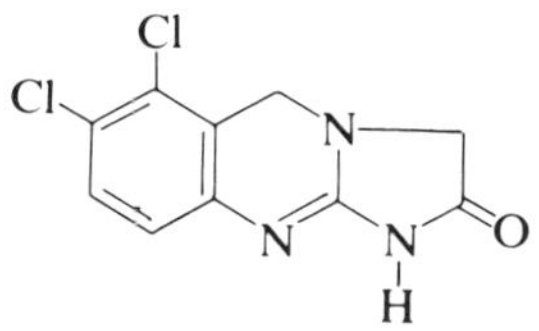

(22)

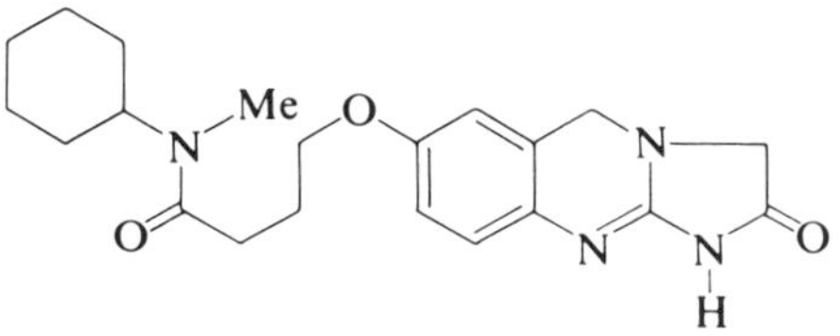

(23)

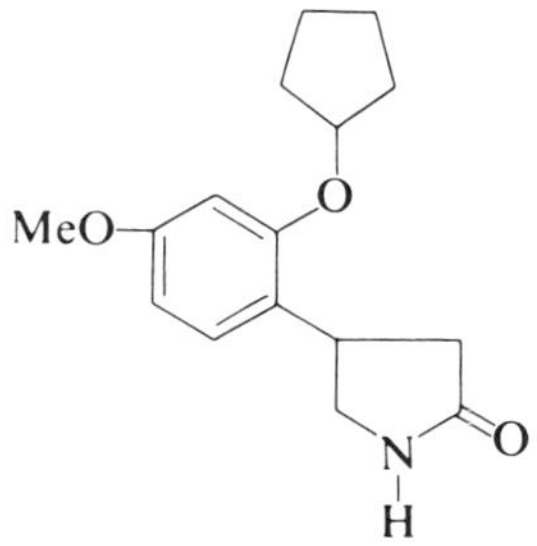

(24)

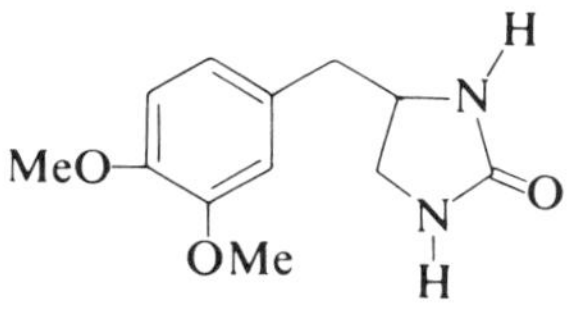

(25)

Figure 6 Type III phosphodiesterase inhibitors

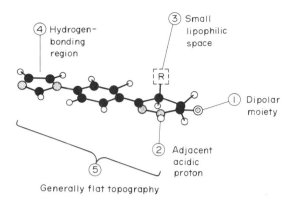

④ Hydrogen-
bonding
region

③ Small
lipophilic
space

R

① Dipolar
moiety

② Adjacent
acidic
proton

⑤

Generally flat topography

Figure 7 Refined five-point model illustrated with cyclic AMP (reproduced from ref. 50 by permission of the American Chemical Society)

1 kcal = 4.18 kJ). On the basis of these studies, a refined pharmacophore model based upon cyclic AMP was proposed (Figure 7). In this model, the planar molecules fit well, with attractive potential between the phosphate and the amidic regions and between N-6/N-1 and the terminal dipolar region. It was postulated that the dipole created by the nitrile of milrinone in the vicinity of the cyclic AMP phosphate group partially accounts for the nonplanarity of this compound. Similarly, it was suggested that enoximone introduces a favorable interaction in the region of the 2'-hydroxyl group of cyclic AMP which could compensate for the binding energy lost due to nonplanarity.

Davis *et al.*[51] described a similar pharmacophore for the planar inhibitors. For the nonplanar inhibitors, they suggested that compounds like milrinone and amrinone bond along the axis of the cyclic AMP pharmacophore and at N-3 rather than at N-1 where planarity was of less importance. It should also be noted that Venuti *et al.*[49] proposed a similar pharmacophore for the RS 82856 series.

A further subdivision of the type III PDE inhibitors is embodied in the cyclic AMP specific, cyclic GMP insensitive phosphodiesterase inhibitors (type IIIB PDE). Examples of this subclass include rolipram (**24**) and Ro 20-1724 (**25**). As with the selective PDE I inhibitors, there are no SAR data for analogs of either of these compounds which describe structural features necessary to impart selective PDE IIIB inhibitory activity. Nevertheless, rolipram and Ro 20-1724 do represent valuable pharmacological tools to study the effects of inhibitors of this subclass of the type III phosphodiesterase.

The evidence that Ro 20-1724 is selective for the type IIIB phosphodiesterase is derived from initial studies reported by Yamamoto *et al.*,[32] who showed that this agent was 120 times more selective for PDE IIIB isolated from calf liver than PDE IIIC derived from the same source. Weishaar *et al.*[8] provided further definition by characterizing soluble and membrane-bound forms of PDE III from canine left ventricular muscle. In this study both rolipram and Ro 20-1724 were found to be selective inhibitors of the soluble form of PDE III (PDE IIIB subclass), whereas imazodam (CI-914; **11**) and cilostamide (**21**) were selective for the membrane-bound form of PDE III (PDE IIIC subclass). Schultz *et al.*[52] subsequently reported that the (−) enantiomer of rolipram inhibited partially purified calmodulin independent phosphodiesterase from rat brain, while the (+) isomer was markedly less active. Interestingly, the neurotropic effects of rolipram are also stereospecific.[53]

8.5.4 RELATIONSHIP BETWEEN SELECTIVE INHIBITION OF PHOSPHODIESTERASES AND PHARMACOLOGICAL RESPONSES

Although many questions remain to be answered regarding the role of the different molecular forms of phosphodiesterase in regulating intracellular cyclic nucleotide metabolism and the second messenger responses to cyclic AMP and cyclic GMP, recent studies have shown that these enzymes play important roles in modulating such diverse metabolic processes as platelet aggregation inhibition[2,9] lipolysis[7,33] and mediator release.[54,55] In the following sections, the involvement of the two subclasses of the type III phosphodiesterase—the cyclic AMP specific, cyclic GMP-inhibited subclass (PDE IIIC), and the cyclic AMP specific, cyclic GMP insensitive subclass (PDE IIIB)—in regulating myocardial contractility and synaptic neurotransmission, respectively, will be described.

8.5.4.1 Myocardial Contractility

The involvement of cyclic nucleotides in regulating myocardial contractility has been well documented, and cyclic AMP is widely believed to be the second messenger involved in modulating the inotropic response to β-adrenergic receptor agonists, histamine and glucagon.[56-58] Although nonselective inhibitors of cyclic nucleotide phosphodiesterase, such as theophylline and papaverine, have been employed as cardiac stimulants in a variety of animal models,[34,59,60] profound species differences have recently been noted regarding the positive inotropic response to selective inhibitors of the cyclic AMP specific, cyclic GMP-inhibited form of phosphodiesterase, such as imazodan and amrinone.[8,61,62] Such differences are surprising since PDE IIIC has been identified in ventricular muscle from a number of species.[6,8,35,36]

In a recent study, Weishaar *et al.*[8] isolated the cyclic AMP specific phosphodiesterase from left ventricular muscle of various species which differ in their inotropic response to selective PDE IIIC inhibitors. Although these enzymes were similar in many respects, major differences were observed with regard to intracellular localization (soluble *vs.* membrane-bound), and also the response of these enzymes to cyclic GMP and to various selective type III phosphodiesterase inhibitors. The authors found that canine ventricle contains both subclasses of PDE III, with the PDE IIIC subclass being membrane bound and the PDE IIIB subclass being soluble. Rat left ventricular muscle also contains both subclasses of PDE III. However, unlike the dog, both subclasses of PDE III in the rat ventricle are soluble. Guinea pig left ventricle apparently contains only the PDE IIIC subclass, which is predominantly a soluble enzyme.

The identification of subclasses of the cyclic AMP specific phosphodiesterase in ventricular muscle, as well as the availability of selective inhibitors of each subclass, makes possible experiments aimed at clarifying the role which each subclass plays in regulating myocardial contractility. Of the various selective type III phosphodiesterase inhibitors administered to the anesthetized dog, only the selective PDE IIIC inhibitors imazodan, CI-930 and amrinone exerted a direct positive inotropic effect.[8] The selective PDE IIIB inhibitors rolipram and Ro 20-1724 did not produce a direct positive inotropic response when administered to the anesthetized dog.[8] In addition, a significant correlation was observed between the *in vitro* inhibitory effects of imazodan, amrinone and several other cardiotonic agents on ventricular PDE IIIC and their positive inotropic effect in the anesthetized dog (Figure 8).

The finding that the imazodan-sensitive subclass of the type III phosphodiesterase (PDE IIIC) can exist in either a membrane-bound form (in the dog ventricle) or in a soluble form (in the guinea pig

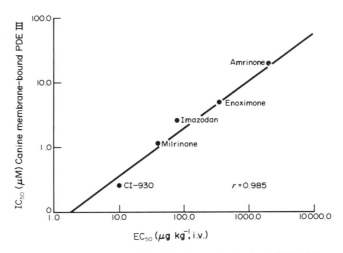

Figure 8 Correlation between the *in vitro* inhibitory effect of various selective PDE III inhibitors on the membrane-bound cyclic AMP specific phosphodiesterase present in canine ventricular muscle (type IIIC phosphodiesterase), and the *in vivo* positive inotropic response to these agents in the anesthetized dog. IC_{50} values refer to the concentration that inhibits cyclic AMP hydrolysis by 50%. ED_{50} refers to the dose which increases basal contractility by 50% (reproduced from ref. 8 by permission of the American Heart Association)

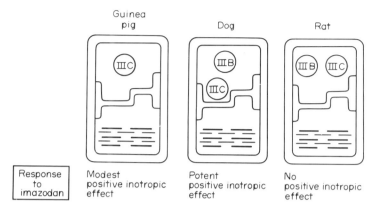

Figure 9 Relationship between the intracellular location of the imazodan sensitive cAMP specific phosphodiesterase and the positive inotropic response to imazodan (reproduced from ref. 8, by permission of the American Heart Association)

and rat ventricle) also allowed for an examination of the role which intracellular localization of this subclass of the cyclic AMP-specific phosphodiesterase plays in the regulation of myocardial contractility. Thus, the finding that imazodan exerts a potent positive inotropic effect in the anesthetized dog, while exerting only a minimal-to-modest inotropic effect in the guinea pig, and no positive inotropic effect in the rat, suggests that the membrane-bound form of PDE IIIC exerts a much greater influence on ventricular contractility than does the soluble form.[8] This hypothesis is supported by the finding that imazodan also exerts a potent positive intropic effect in the Rhesus monkey, in which a membrane-bound PDE IIIC is also present in ventricular muscle, but exerts little or no inotropic effect in the hamster, in which no membrane-bound PDE IIIC could be identified in ventricular muscle.[8] The relationship between intracellular location of the type IIIC subclass of phosphodiesterase and inotropic activity is illustrated in Figure 9.

8.5.4.2 Pre- and Post-synaptic Neurotransmission

In addition to the heart, cyclic nucleotides also influence the response to several neurotransmitters, including histamine,[63] norepinephrine[64] and dopamine.[65] Alterations in cyclic nucleotide metabolism have also been implicated in several neuropathological conditions, including psychosis, depression, mania and Parkinsonism.[66]

The link between depression and alterations in central monoaminergic metabolism has been the subject of considerable investigation for more than 30 years,[67-69] and many of the drugs currently available for treating depression act to increase monoamine levels in the brain.[66,70] These agents include the monoamine oxidase inhibitors, which prevent the metabolism of norepinephrine and other monoamines, the tricyclic antidepressants, which block monoamine uptake, and the amphetamines, which block monoamine uptake and also exert 'false' neurotransmitter actions. All of these agents apparently act presynaptically to increase the amount of monoamines available to post-synaptic receptors.[66]

Recently, Wachtel suggested that the selective cyclic AMP specific, cyclic GMP insensitive phosphodiesterase (PDE IIIB) inhibitors rolipram and Ro 20-1724 possess antidepressant activity.[10] These agents were effective in two animal models predictive of antidepressant activity; antagonism of reserpine-induced hypothermia or hypokinesia, and potentiation of yohimbine lethality.[10] Additional evidence to support the central effects of these inhibitors include the observation that all three agents increase cyclic AMP levels in the brain following systemic administration,[71] and the finding that both the cyclic AMP phosphodiesterase inhibitory effects of rolipram and its effects on behavior in the rat are stereospecific.[53] In addition, the potency of the three agents for inhibiting rat brain cyclic AMP phosphodiesterase correlates with their ability to induce a discrete behavioral syndrome in rats, characterized by hypoactivity, grooming and hypothermia.[53] Preliminary clinical results with rolipram in patients indicate that it does possess antidepressant activity.[10]

The antidepressant mechanism of action of rolipram and other selective PDE IIIB inhibitors is believed to be unique relative to other antidepressant drugs. According to Wachtel, selective

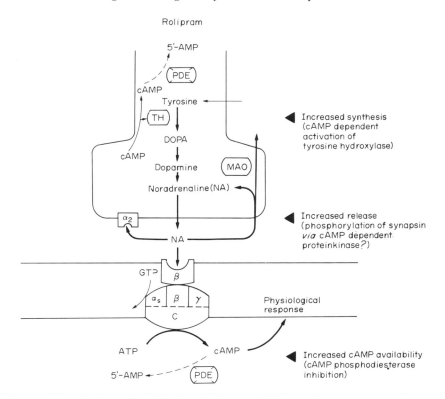

Figure 10   Influence of the selective cyclic AMP phosphodiesterase inhibitor rolipram on monoaminergic neurotransmission. In contrast to classic antidepressants, rolipram influences neutrotransmission both presynaptically and postsynaptically by a direct, receptor independent intraneuronal mechanism. Presynaptically, rolipram enhances noradrenaline synthesis by cyclic AMP dependent activation of tyrosine hydroxylase and also increases noradrenaline, perhaps *via* cyclic AMP-dependent phosphorylation of synapsin I. Postsynaptically, inhibition of cyclic AMP phosphodiesterase by rolipram potentiates the cellular response to noradrenaline and other neurotransmitters which utilize cyclic AMP as a second messenger (reproduced from ref. 72 by permission of the author)

phosphodiesterase inhibitors such as rolipram produce their antidepressant effect by increasing both pre- and post-synaptic levels of cyclic AMP, leading to an increase in the presynaptic turnover of norepinephrine, and also potentiation of the postsynaptic response to norepinephrine.[10] Coccaro and Siever[70] have previously suggested that depression may arise from either an actual decrease in the level of synaptic transmitter, *e.g.* 'low output' depression, or a decrease in the sensitivity or density of the postsynaptic receptor for the transmitter, *e.g.* 'high output/low sensitivity' depression. Agents such as rolipram which act both presynaptically and postsynaptically might restore the balance between these two pathways.[72]

The proposed pre- and post-synaptic effects of rolipram are shown in Figure 10.

8.5.5   CONCLUSIONS

The identification by many investigators of multiple molecular forms of phosphodiesterase suggests that modulation of cyclic nucleotide hydrolysis is a complex process, and that the second messenger responses to cyclic AMP and cyclic GMP can be regulated at the level of degradation as well as at the level of synthesis. This realization implies that selective inhibitors of the different forms of phosphodiesterase can mimic the response to a variety of endogenous substances which act *via* stimulation of cyclic AMP or cyclic GMP synthesis, including epinephrine, norepinephrine, prostaglandins, adenosine and atrial natriuretic factor. Furthermore, differences in the distribution of phosphodiesterases in various tissues, as well as differences in the intracellular localization of the various phosphodiesterases, implies that selective phosphodiesterase inhibitors may produce true tissue specific responses, a characteristic not generally associated with most agents which act to increase cyclic nucleotide synthesis (*e.g.* β-receptor stimulants). Although much insight has been gained in the 15 years since Thompson and Appleman first demonstrated the existence of multiple forms of phosphodiesterase, revelations regarding new subclasses of phosphodiesterase and the identification of new classes of selective phosphodiesterase inhibitors suggest that an understanding of the overall role which the different forms of phosphodiesterase play in regulating the second messenger responses to cyclic AMP and cyclic GMP has yet to be obtained. Developments are likely to be rapid in the next decade.

8.5.6 REFERENCES

1. G. A. Robinson, R. W. Butcher and E. W. Sutherland, *Annu. Rev. Biochem.*, 1968, **37**, 149.
2. H. Hidaka, T. Tanaka and H. Itoh, *Trends Pharmacol. Sci.*, 1984, **5**, 237.
3. R. E. Weishaar, M. H. Cain and J. A. Bristol, *J. Med. Chem.*, 1985, **28**, 537.
4. W. J. Thompson, W. L. Terasaki, P. N. Epstein and S. J. Strada, *Adv. Cyclic Nucleotide Res.*, 1979, **10**, 69.
5. J. N. Wells and J. G. Hardman, *Adv. Cyclic Nucleotide Res.*, 1977, **8**, 119.
6. V. C. Manganiello, *J. Mol. Cell. Cardiol.*, 1987, **19**, 1037.
7. M. L. Elks and V. C. Manganiello, *Endocrinology*, 1985, **116**, 2119.
8. R. W. Weishaar, D. C. Kobylarz-Singer, R. P. Steffen and H. R. Kaplan, *Circ. Res.*, 1987, **61**, 539.
9. H. Hidaka, H. Hayaski, H. Kohri, Y. Kimura, T. Hosokawa, T. Igawa and Y. Saitoh, *J. Pharmacol. Exp. Therap.*, 1979, **211**, 26.
10. H. Wachtel, *Neuropharmacology*, 1983, **22**, 267.
11. E. W. Sutherland and T. W. Rall, *J. Biol. Chem.*, 1958, **232**, 1077.
12. R. W. Butcher and E. W. Sutherland, *Pharmacologist*, 1959, **1**, 63.
13. R. W. Butcher and E. W. Sutherland, *J. Biol. Chem.*, 1962, **237**, 1244.
14. R. E. Weishaar, *J. Cyclic Nucleotide Protein Phosphorylation Res.*, 1987, **11**, 463.
15. W. J. Thompson and M. M. Appleman, *Biochemistry*, 1971, **10**, 311.
16. T. R. Russell, W. L. Terasaki and M. M. Appleman, *J. Biol. Chem.*, 1973, **248**, 1334.
17. J. N. Wells, C. E. Bairo, Y. J. Wu and J. G. Hardman, *Biochim. Biophys. Acta*, 1975, **384**, 430.
18. M. S. Amer and R. F. Mayol, *Biochim. Biophys. Acta*, 1973, **309**, 149.
19. H. Hidaka and T. Asano, *Biochim. Biophys. Acta*, 1976, **429**, 485.
20. V. A. Tkachuk, V. G. Cazarevich and S. E. Severin, *Adv. Myocardiol.*, 1982, **3**, 541.
21. R. E. Weishaar, M. M. Quade, J. A. Schenden, D. B. Boyd and D. B. Evans, *Eur. J. Pharmacol.*, 1985, **119**, 205.
22. B. B. Fredholm, K. Brodin and K. Strandberg, *Acta Pharmacol. Toxicol.*, 1979, **45**, 336.
23. J. B. Polson, J. J. Krzanowski and A. Szentivanyi, *Biochem. Pharmacol.*, 1982, **31**, 3403.
24. S. Fougier, G. Nemoz, A. F. Prigent, M. Marivet, J. J. Bourguignon, C. Wermuth and H. Pacheco, *Biochem. Biophys. Res. Commun.*, 1986, **138**, 205.
25. S. A. Harrison, D. H. Reifsnyder, B. Gallis, G. G. Cadd and J. A. Beavo, *Mol. Pharmacol.*, 1986, **29**, 506.
26. J. A. Beavo, J. G. Hardman and E. W. Sutherland, *J. Biol. Chem.*, 1971, **246**, 3840.
27. W. L. Terasaki and M. M. Appleman, *Metabolism*, 1975, **324**, 311.
28. R. E. Weishaar, S. D. Burrows, D. C. Kobylarz, M. M. Quade and D. B. Evans, *Biochem. Pharmacol.*, 1986, **35**, 787.
29. C. Lugnier, P. Schoeffter, A. LeBec, E. Strouthou and J. C. Stoclet, *Biochem. Pharmacol.*, 1986, **35**, 1743.
30. R. K. Sharma and J. H. Wang, *Biochem. Cell. Biol.*, 1986, **64**, 1074.
31. R. Yee and P. A. Liebman, *J. Biol. Chem.*, 1978, **253**, 8902.
32. T. Yamamoto, F. Lieberman, J. C. Osborne, J. C. Manganiello, M. Vaughan and H. Hidaka, *Biochemistry*, 1984, **23**, 670.
33. M. L. Elks and V. C. Manganiello, *Endocrinology*, 1984, **115**, 1262.
34. R. E. Weishaar, M. M. Quade, J. A. Schenden and D. B. Evans, *J. Cyclic Nucleotide Protein Phosphorylation Res.*, 1985, **10**, 551.
35. T. Kariya, L. T. Willie and R. D. Dage, *J. Cardiovasc. Pharmacol.*, 1982, **4**, 509.
36. P. Siegl, G. Morgan, D. Bohn, N. Tonkonoh and H. Bull, *Circulation*, 1985, **72** (suppl. III), III-1244.
37. H. Hidaka and T. Endoh, in 'Advances in Nucleotide and Protein Phosphorylation Research', ed. S. J. Strada and W. J. Thompson, Raven Press, New York, 1984, p. 245.
38. R. F. G. Booth, S. P. Buckham, D. O. Lunt, P. W. Manley and R. A. Porter, *Biochem. Pharmacol.*, 1987, **36**, 3517.
39. S. W. Schneller, A. C. Ibay, E. A. Martinson and J. N. Wells, *J. Med. Chem.*, 1986, **29**, 972.
40. J. A. Bristol, I. Sircar, W. H. Moos, D. B. Evans and R. E. Weishaar, *J. Med. Chem.*, 1984, **27**, 1099.
41. I. Sircar, B. L. Durell, J. Bibowski, J. A. Bristol and D. B. Evans, *J. Med. Chem.*, 1985, **28**, 1405.
42. R. F. Kauffman, J. G. Crowe, B. G. Utterback and D. W. Robertson, *Mol. Pharmacol.*, 1987, **30**, 609.
43. D. W. Robertson, J. H. Krushinski, G. D. Pollock, H. Wilson, R. F. Kauffman and J. S. Hayes, *J. Med. Chem.*, 1987, **30**, 824.
44. H. Mikashima, T. Nakao, K. Gogo, H. Oehi, H. Yasuda and T. Tsumagari, *Thromb. Res.*, 1984, **35**, 589.
45. I. Sircar, G. C. Morrison, S. E. Burke and R. E. Weishaar, *J. Med. Chem.*, 1987, **30**, 1724.
46. I. Sircar, B. L. Durell, J. A. Bristol, R. E. Weishaar and D. B. Evans, *J. Med. Chem.*, 1987, **30**, 1023.
47. C. Lugnier, M. Bruch, J.-C. Stoclet, M.-P. Strub, M. Marivet and C. G. Wermuth, *Eur. J. Med. Chem.*, 1985, **20**, 121.
48. J. H. Jones, M. C. Venuti, R. Alvarez, J. J. Bruno, A. H. Buks and A. Prince, *J. Med. Chem.*, 1987, **30**, 295.
49. M. C. Venuti, J. H. Jones, R. Alvarez and J. J. Bruno, *J. Med. Chem.*, 1987, **30**, 303.
50. W. H. Moos, C. C. Humblet, I. Sircar, C. Rithner, R. E. Weishaar, J. A. Bristol and A. T. McPhail, *J. Med. Chem.*, 1987, **30**, 1963.
51. A. Davis, B. H. Warrington and J. G. Vinter, *J. Computer-Aided Mol. Des.*, 1987, **1**, 97.
52. J. E. Schultz and B. H. Schmidt, *Naunyn-Schmiedebergs Arch. Pharmacol.*, 1986, **333**, 23.
53. H. Wachtel, *J. Pharmacol.*, 1983, **35**, 440.
54. C. J. Coulson, R. E. Ford, S. Marshall, J. L. Walker, K. R. H. Wooldridge, K. Bowden and T. J. Coombs, *Nature (London)*, 1977, **265**, 545.
55. N. Frossard, Y. Landry, G. Pauli and M. Ruckstuhl, *Br. J. Pharmacol.*, 1981, **73**, 933.
56. R. W. Tsien, *Adv. Cyclic Nucleotide Res.*, 1977, **8**, 363.

57. A. M. Katz, *Adv. Cyclic Nucleotide Res.*, 1979, **11**, 303.
58. R. E. Weishaar and D. B. Evans, in 'Cardiovascular Drugs', ed. J. A. Bristol, Wiley, New York, 1986, p. 3.
59. S. R. Elek and L. N. Katz, *J. Pharmacol. Exp. Ther.*, 1942, **74**, 335.
60. S. Krop, *J. Pharmacol. Exp. Ther.*, 1942, **82**, 48.
61. A. Schwartz, I. Grupp, G. Grupp, C. Johnson, P. Berner, E. Wallick, K. Imai and A. Alousi, *Circulation*, 1979, **60**, 16.
62. A. A. Alousi and A. E. Farah, *Circ. Res.*, 1980, **46**, 887.
63. G. C. Palmer, M. J. Schmidt and G. A. Robinson, *J. Neurochem.*, 1972, **19**, 2251.
64. A. Dolphin, M. Hamont and J. Bockaert, *Brain Res.*, 1979, **179**, 305.
65. M. E. Gnegy, A. Lucchelli and E. Costa, *Naunyn-Schmiedebergs Arch. Pharmacol.*, 1977, **301**, 121.
66. G. E. Palmer, *Prog. Neurobiol.*, 1983, **21**, 1.
67. H. P. Loomer, J. C. Saunders and N. S. Kline, *Psychiatr. Res., Rep. Am. Psychiatr. Assoc.*, 1957, **8**, 129.
68. J. J. Schildkraut and S. S. Kety, *Science (Washington, D.C.)*, 1967, **156**, 21.
69. A. Coppen, *Br. J. Psychiatr.*, 1967, **113**, 1237.
70. E. F. Coccaro and L. J. Siever, *J. Clin. Pharmacol.*, 1985, **25**, 241.
71. G. J. Kant, J. L. Meyerhorf and R. H. Lenox, *Biochem. Pharmacol.*, 1980, **29**, 369.
72. H. Wachtel, in 'New Concepts in Depression', ed. M. Briley and G. Fillian, Macmillan, London, 1988, p. 373.

8.6

Agents Acting Against Phospholipases A$_2$

HENK VAN DEN BOSCH

State University of Utrecht, The Netherlands

8.6.1 ROLE OF PHOSPHOLIPASES IN FORMATION OF BIOACTIVE LIPIDS   515

8.6.2 PROPERTIES OF INTRACELLULAR PHOSPHOLIPASES A$_2$ 517

  8.6.2.1 Calcium Dependent Phospholipase A$_2$ 517
  8.6.2.2 Calcium Independent Phospholipase A$_2$ 519
  8.6.2.3 Do Arachidonate and/or Ether Phospholipid Specific Phospholipases A$_2$ Exist?   519

8.6.3 REGULATION OF PHOSPHOLIPASES A$_2$ 520

  8.6.3.1 Zymogens   521
  8.6.3.2 Activator, Inhibitor and G-proteins   521
  8.6.3.3 Calcium Ions and Calmodulin   522
  8.6.3.4 Membrane Phospholipid Structure   522

8.6.4 PHOSPHOLIPASE A$_2$ INHIBITORS   523

  8.6.4.1 Lipocortins and Related Proteins   523
  8.6.4.2 Other Proteins   524
  8.6.4.3 Antibodies   524
  8.6.4.4 Plant Extracts   525
  8.6.4.5 Lipidic Inhibitors   526
  8.6.4.6 Substrate Mimics   526
  8.6.4.7 Chemicals   527

8.6.5 CONCLUSIONS   527

8.6.6 REFERENCES   528

8.6.1 ROLE OF PHOSPHOLIPASES IN FORMATION OF BIOACTIVE LIPIDS

Membrane phosphoglycerides contain three types of ester bonds that are susceptible to hydrolysis by phospholipases.[1] Acyl ester bonds at the *sn*-1 and *sn*-2 position of the glycerol backbone can be hydrolyzed by phospholipases A$_1$ and A$_2$, respectively. Phospholipase C attacks the internal glycerophosphate and phospholipase D the phosphate–external grouping (choline, ethanolamine, serine, glycerol, inositol) phospho ester bonds. The formation of bioactive lipids upon stimulation of a large variety of cell types is thought to be initiated by phospholipase attack of parent membrane phosphoglycerides. Thus, 1-*O*-alkyl-2-acyl-*sn*-glycero-3-phosphocholine can be converted by a phospholipase A$_2$-catalyzed reaction into 1-*O*-alkyl-2-lyso-*sn*-glycero-3-phosphocholine, the immediate precursor from which the bioactive platelet activating factor (PAF) is formed by acetylation at the *sn*-2-position. This PAF can also be formed by introduction of phosphocholine into 1-alkyl-2-acetyl-*sn*-glycerol, but the prevailing opinion about PAF formation in stimulated cells is that it follows the outlined (Figure 1) sequence of reactions involving phospholipase A$_2$ and 1-*O*-alkyl-2-lyso-*sn*-glycero-3-phosphocholine:acetyl-CoA acetyltransferase. Upon appropriate stimulation PAF can be synthesized by basophils, neutrophils, eosinophils, monocytes, natural killer cells, platelets, peritoneal and alveolar macrophages, mast cells, kidney, lung, liver and endothelial cells. PAF exerts

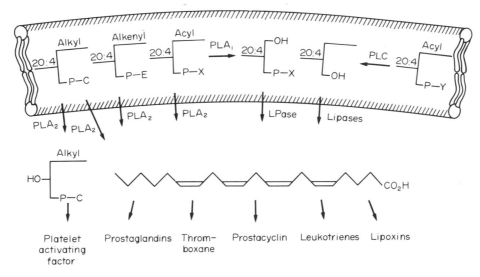

Figure 1 Schematic representation highlighting the initiating role of phospholipases in the formation of bioactive lipids from structural membrane phosphoglycerides. In a single hydrolytic step phospholipase A_2 (PLA_2) can form two precursors for bioactive lipids from alkyl-20:4-P-C, *i.e.* 1-*O*-alkyl-2-lyso-*sn*-glycero-3-P-C and arachidonate. Arachidonate can also be released from alkenyl-20:4-P-E (the main representative of mammalian plasmalogens, E = ethanolamine) and from diacyl-P-X (X = choline, ethanolamine, serine, inositol) by phospholipase A_2. Alternative pathways for arachidonate release include the sequential action of phospholipase A_1 (PLA_1) and lysophospholipase (LPase) on diacyl-P-X or of phospholipase C (PLC) and lipases on diacyl-P-Y (Y = inositol, inositol 4-phosphate, inositol 4,5-diphosphate)

antihypertensive effects and has been implicated in many (patho)physiological processes, including inflammation, allergy, anaphylaxis, bronchoconstriction, platelet and leukocyte aggregation and degranulation. Recent developments suggest that PAF may also function in ovoimplantation as well as during labor. For the chemistry,[2] biochemistry[3-5] and (patho)physiology[3,5,6] of PAF the reader is referred to recent comprehensive reviews.

Arachidonate is localized almost exclusively at the *sn*-2-position of phosphoglycerides and an enrichment of this fatty acid in alkylacyl- and alkenylacyl- in comparison to diacyl-phospholipid species in many cell types has been well documented in recent years. These findings have led to the proposal that 1-*O*-alkyl-2-arachidonoyl-*sn*-glycero-3-phosphocholine may serve as a common source of PAF and arachidonate.[7,8] Other pathways for arachidonate release include the concerted action of phospholipase A_1 and lysophospholipase and of phospholipase C and diglyceride and monoglyceride lipases (Figure 1). Although these mechanisms are less direct than phospholipase A_2-catalyzed release of arachidonate, the phospholipase C–lipase pathway in particular has received considerable attention. Soon after the discovery of this pathway it was claimed[9,10] to have sufficient activity to account for arachidonate release in human platelets after thrombin stimulation, a condition which was not met by measurements of phospholipase A_2 activity in cell free systems. However, the activities of both phospholipase C and diglyceride lipase in these studies were measured *in vitro* with saturating substrate levels, which may not be valid for the *in vivo* situation. This objection holds especially for the diglyceride substrate, which only transiently increases upon stimulation and is rapidly phosphorylated to phosphatidic acid. The relative contributions of the phospholipase A_2 and phospholipase C–lipase pathways to the release of arachidonate in stimulated platelets has been reviewed in detail.[11-13] It has been pointed out[11,13] that results from studies using cells prelabeled with arachidonate should be interpreted with caution in view of the possibility of heterogeneous labeling of different pools. This problem was circumvented in a well-conducted study by Broekman[14] who quantitated thrombin-induced changes in endogeneous arachidonate content of platelet phospholipids. It was concluded that the rise in free and oxygenated arachidonate exceeded the release of free arachidonate from inositol phospholipids and derived phosphatidic acid by a factor of three and that phospholipase A_2 is quantitatively more significant than the phospholipase C–lipases pathway as a mechanism for release of arachidonate. The importance of the phospholipase C pathway resides primarily in signal transduction (see Chapter 11.4) through formation of the second messengers diglyceride and inositol 1,4,5-triphosphate rather than in the direct release of arachidonate. In all likelihood these events occur prior to activation of the phospholipase A_2, which is responsible for arachidonate release.

Free, but not esterified, arachidonate can be converted to a number of oxygenated bioactive products (Figure 1) in reactions initiated by cyclooxygenase and lipoxygenases. Apart from erythrocytes, cyclooxygenase is present in all cells, yielding endoperoxide intermediates. The latter are converted in a cell or tissue specific manner to prostaglandin species in numerous cells and are believed to contribute to inflammation, fever and pain.[15–17] In platelets most of the endoperoxide is used for thromboxane A$_2$ biosynthesis, while vascular endothelial cells produce mainly prostacyclin. Thromboxane synthesis has also been described in lung and derived fibroblasts and in brain,[18] whereas prostacyclin synthesis also occurs in macrophages, vascular and nonvascular smooth muscle cells and cultured endothelial cells.[16] Thromboxane and prostacyclin have opposing effects with thromboxane causing contraction of aortic tissue and triggering platelet aggregation, while prostacyclin relaxes aortic tissue and prevents platelet aggregation.[15–17] Other physiological and pathophysiological effects of these compounds have recently been reviewed.[15–20]

Leukotrienes and lipoxins can be formed from arachidonate in reaction sequences that are initiated by 5-lipoxygenase and 15-lipoxygenase, respectively. 5-Lipoxygenases have been found in rat, guinea pig, porcine and human polymorphonuclear leukocytes, rat basophilic leukemia cells and murine mast cells. The enzymes from the latter four sources and from potatoes not only catalyze the formation of 5-hydroperoxytetraenoic acid but also the conversion of this intermediate to leukotriene A$_4$. Rabbit, murine and human leukocytes and human epithelial and endothelial cells have been reported to contain 15-lipoxygenase activity. In human leukocytes both 5- and 15-lipoxygenase are involved in trihydroxytetraene lipoxin formation. Major processes in which leukotrienes have been implicated include chemotactic and degranulation processes of polymorphonuclear leukocytes, suggesting an important role in inflammation. The sulfidopeptide leukotrienes have pronounced contractive effects, causing bronchoconstriction, and are thought to be major mediators in asthmatic processes. The structure, biosynthesis and major biological effects of leukotrienes[21,22] and lipoxins[22] have recently been reviewed. The arachidonic acid cascade is discussed in detail in Chapter 6.2, while Chapters 12.11 to 12.13 deal with prostanoid, PAF and leukotriene receptors, respectively.

The above introduction emphasizes the central role of phospholipase A$_2$ as an initiator of pathways leading to a variety of bioactive lipids. The relevance of phospholipase A$_2$ in disease states has recently also been discussed by Chang *et al.*[23] Detailed information concerning the properties of phospholipases A$_2$, including X-ray structures, has been obtained for pancreatic and venom enzymes as reviewed by Verheij *et al.*[24] and Dennis.[25] These enzymes are commercially available and hence are used frequently in mechanistic studies and in screening procedures to test possible inhibitors. These studies have been extremely helpful in developing models and strategies for intervening with phospholipase A$_2$ activity, although it remains to be established whether intracellular phospholipases A$_2$ behave similarly in all aspects. A detailed knowledge of the structural and enzymological properties of the intracellular phospholipases A$_2$ is required for this evaluation and still has to be developed. Earlier information on this point has been reviewed by van den Bosch[1,11] and Waite.[26,27]

8.6.2 PROPERTIES OF INTRACELLULAR PHOSPHOLIPASES A$_2$

8.6.2.1 Calcium Dependent Phospholipase A$_2$

Structural investigations on phospholipases A$_2$ from sources other than venoms and pancreas have been hampered by the low abundance of these proteins and a frequently encountered lability during their purification. It is only in the last decade that some progress in this area has been made. Table 1 summarizes current knowledge on mammalian Ca^{2+} dependent phospholipases A$_2$ that have been purified to near homogeneity, *i.e.* to the stage that structural investigations can be started. In all cases the isolated proteins gave one band upon SDS-PAGE, although it has not always been shown unequivocally that this band contained the phospholipase A$_2$ activity. This may provide one explanation for the deviating molecular weights of some enzymes. However, most enzymes exhibit molecular weights that are close to that of the venom and pancreas phospholipases, *i.e.* 13.5 to 18.5 kDa. In those cases where it has been investigated the enzymes were inhibited by *p*-bromophenacyl bromide, an active-site-directed inhibitor alkylating the invariant His-48 residue in pancreatic and venom phospholipases A$_2$.[24] Whereas the pancreatic enzyme showed less than 1% activity when Ca^{2+} was replaced by Sr^{2+}, the purified enzymes from rat liver mitochondria,[43] rabbit parotid gland,[37] partially purified enzyme from human platelets[44] and a synaptic membrane[45] phospholipase A$_2$ showed from 10 to 40% remaining activity in the presence of Sr^{2+}. This may be indicative

Table 1 Current Knowledge on Structural Properties of Ca^{2+} Dependent Mammalian Phospholipases A_2

Source	Localization	M (kDa)	Amino acid composition[a]	Sequence[b]	Ref.
Rat spleen	Membranous	14	+	1–37	134
Rat spleen	Soluble	14	+	1–32	28, 133
Human lung exudate	Secreted	75	+	—	29
Rabbit neutrophils	Membranous	14	+	—	30
Sheep erythrocytes	Membranous	12–18.5	–	—	31
Human platelets	Unknown	44	–	—	32
Sheep platelets	Soluble	30	–	—	33
Bovine brain	Microsomes	18.5	+	—	34
Rat liver	Mitochondria	15	+	1–24	35
Pig intestine	Secreted	16	+	1–48	36
Rabbit parotid gland	Secreted	16	+	—	37
Human synovial fluid	Secreted	15, 17	–	—	38
Rabbit exudate	Secreted	14	+	1–39	39
Rat exudate	Secreted	13.5	+	1–10	40
Rat platelets	Secreted	13.5	+	1–24	41
Human lung	Unknown	14	+	1–126	42
Rat gastric mucosa	Membranous	14	–	1–15	163

[a] Amino acid composition (+) or unknown (−). [b] N-terminal sequence for indicated residues known. For rat liver mitochondrial enzyme: Aarsman and van den Bosch, unpublished.

of small variations in the architecture of the active sites that should be kept in mind when using the pancreatic phospholipase A_2 as a reference enzyme in drug-screening procedures.

Surface differences between pancreatic and other mammalian phospholipases A_2, as revealed in immunochemical approaches, have also been noticed. Antibodies against porcine pancreatic phospholipase A_2 did not cross-react with human seminal plasma[46] and human synovial fluid[38] phospholipase A_2. However, intraspecies cross-reactivity with rat spleen and pancreas enzyme has been reported.[47] Antisera against *Naja naja* phospholipase A_2 recognized the enzyme from guinea pig alveolar macrophages and rat lymphocytes, albeit mainly 30 and 45 kDa bands in immunoblots,[48] as well as phospholipase A_2 from bull spermatozoa.[49] All antibodies used in these experiments were elicited with pancreatic or venom phospholipases. A different approach was taken by de Jong *et al.*,[50] who prepared monoclonal antibodies against phospholipase A_2 isolated from rat liver mitochondria. These cross-reacted with phospholipase A_2 from other rat liver fractions and with rat platelet phospholipase A_2, but did not recognize the enzyme from porcine or rat pancreas or from *Crotalus atrox* venom. These findings clearly demonstrate that while some intracellular phospholipases A_2 share common epitopes with the extracellular enzymes from pancreas and venoms they also possess unique structural surface features.

As indicated in Table 1 a large number of the purified mammalian phospholipases A_2 are recovered in the soluble fraction or in secretions after appropriate challenges. It has become clear in recent years that even the membranous enzymes cannot be considered as integral membrane proteins. Most of them can be extracted in buffers containing 1 M salt and can be purified without the necessity of having detergents present. Sheep erythrocyte phospholipase A_2 is the only notable exception in this respect.[31] Hence, membrane-associated phospholipases A_2 have to be considered mainly as peripheral membrane proteins.

Investigations on mammalian phospholipases have now developed to the stage that amino acid compositions and partial N-terminal sequences are available (Table 1). These data allow the synthesis of oligonucleotide probes and with the availability of antibodies it will soon be possible to apply the tools of molecular biology to get further insights into the structure of these enzymes. Using probes synthesized on the basis of the sequence of porcine pancreatic phospholipase A_2, Seilhamer[42] deduced the complete amino acid sequence for a human lung enzyme. The human lung enzyme sequence was 80% conserved with respect to porcine pancreatic enzyme, but enzymological properties of the gene product were not studied. Human cDNA libraries from liver, testis, lymphocytes, intestine and placenta gave no signals with the probe used, possibly because of lower sequence homology. In this respect it may be worthwhile to mention that sequence homology for the first 24 amino acids between the lung and porcine pancreas enzyme was 67%. In contrast, rat liver mitochondrial phospholipase A_2 for this part of its sequence showed only 25% homology with both rat and porcine pancreas enzyme (Aarsman and van den Bosch, unpublished). Similar data were reported[41] for rat platelet phospholipase A_2.

8.6.2.2 Calcium Independent Phospholipase A$_2$

A development that must be mentioned is the description of an increasing number of Ca^{2+} independent phospholipase A$_2$ activities in recent reports. Ca^{2+} independent acidic enzymes, presumably of lysosomal origin, are not included in the compilation (Table 2). Neutral or alkaline active forms of Ca^{2+} independent phospholipase A$_2$ were described in soluble fractions isolated from rat lung,[51] chicken erythrocytes[52] and canine myocardium.[54] An enzyme with selectivity for phosphatidylethanolamine was recently reported for human platelets and could be separated from a Ca^{2+} dependent enzyme acting on phosphatidylcholine.[57] A phospholipase A$_2$-type enzyme with specificity for the acetyl–ester bond in PAF, *i.e.* PAF-acetyl hydrolase, has been detected in plasma and the cytosolic fraction of a variety of rat tissues,[62] including lung, and also appeared to be independent of Ca^{2+} ions. The rat lung enzyme could be distinguished from the Ca^{2+} independent phospholipase A$_2$ acting on long chain phosphoglycerides by separation on DEAE-cellulose and differential effects of *p*-bromophenacyl bromide and diisopropyl fluorophosphate.[56] Hence, only enzymes acting on long chain phosphoglycerides are included in Table 2. Using 1-*O*-alkyl-2-[3H-oleoyl]-*sn*-glycero-3-phosphocholine as substrate, this type of lipolytic activity was found to be present in the cytosol of all rat tissues tested, *i.e.* kidney, heart, pancreas, spleen, liver, lung and brain.[164] A Ca^{2+} independent phospholipase A was purified to homogeneity from hamster heart cytosol.[61] This enzyme showed both phospholipase A$_1$ and phospholipase A$_2$ activity. The molecular weight was estimated to be 140 kDa from gel filtration experiments, but SDS-PAGE showed the enzyme to be composed of identical subunits of 14 kDa. Antibodies against this enzyme cross-reacted with the supernatant obtained from Triton X-100-treated mitochondrial and microsomal fractions, suggesting that the enzyme occurred in membrane-associated form as well. Membrane-bound forms of Ca^{2+} independent phospholipase A$_2$ have also been reported for rabbit neutrophil plasma membranes,[53] amnion tissue microsomes,[55] bovine endothelial membranes[58] and both guinea pig[59] and rat[60] brush-border membranes. It has been shown in my laboratory that the enzyme from rat lung cytosol is capable of releasing the *sn*-2-acyl group from membrane-associated phosphatidylcholines.[164] In theory, these enzymes could play a role in the basal turnover of *sn*-2-acyl groups of membrane phosphoglycerides and the release of arachidonate under the conditions of submicromolar free Ca^{2+} concentrations in the cytoplasma of unstimulated cells. Much research is, however, still required to further characterize these enzymes and to establish their regulation and role in cellular lipid metabolism.

8.6.2.3 Do Arachidonate and/or Ether Phospholipid Specific Phospholipases A$_2$ Exist?

This section briefly discusses the evidence for the existence of specific phospholipase A$_2$ that could be involved in the release of arachidonate and/or the formation of lyso-PAF.

Selective release of arachidonic acid from prelabeled human platelets in response to thrombin stimulation was noted some time ago.[63,64] However, in cells prelabeled with glycerol this specificity was not observed and all phosphatidylcholine species were hydrolyzed to similar extents.[13,64] These results suggested that preincubation of cells with fatty acids or glycerol may label specific phospholipid pools that are not representative for total mass distributions. Studies with isolated platelet membranes have also provided indications for the selective release of arachidonate from endogenous phospholipids.[65] This selectivity was lost in platelet homogenates[65] or sonicates.[66] Arachidonate selectivity in isolated membranes was observed more clearly with minimal disruption of membrane structure.[67] These results, in combination with the observations made in prelabeled cells, suggest that the phospholipase A$_2$, rather than being selective for arachidonate, is present in a domain of

Table 2 Occurrence of Ca^{2+} Independent Phospholipases A$_2$

Year	Source	Ref.	Year	Source	Ref.
1972	Rat lung cytosol	51	1986	Human platelet	57
1983	Chicken erythrocytes	52	1987	Bovine endothelium	58
1984	Rabbit neutrophil	53	1987	Guinea pig brush border	59
1985	Canine myocardium	54	1987	Rat brush border	60
1986	Human amniotic fluid	55	1987	Hamster heart cytosol	61
1986	Rat lung cytosol	56	1987	Rat tissue cytosols	164

arachidonate-enriched phospholipid classes or species. Loss of this structural arrangement in homogenates or disrupted membranes could then result in an apparent loss of selectivity for arachidonate. To circumvent the problems introduced by prelabeling cells, Purdon et al.[68] determined the release of fatty acids from thrombin-stimulated platelets as mass changes in individual molecular species. These authors arrived at the conclusion that the deacylation was selective for arachidonate species of diacylphosphatidyl-choline and -ethanolamine and explained this finding in intact cells by assuming that a certain proportion of arachidonoyl species of phospholipid are compartmentalized with the platelet membrane proximal to the site of action of the phospholipase A_2. The alternative explanation of specific acyl chain recognition by the enzyme's active site was considered unlikely in view of the fact that acyl chain specificity was not observed with partially purified platelet phospholipase A_2 and micellar or aggregated phospholipids.[69]

Recent experiments with rat polymorphonuclear leukocytes[70] have shown a highly reduced rate of PAF synthesis in cells depleted of arachidonic acid by dietary means. A preferential association of arachidonate in alkylacylphosphatidylcholine, the precursor from which lyso-PAF is formed by phospholipase A_2 action, is well documented.[7,71] The leukocyte findings, therefore, also suggest the existence of a phospholipase A_2 specific for arachidonate release. Presumably, in arachidonate deficient cells lyso-PAF cannot be formed in sufficient amounts. In this case the putative arachidonate rich species surrounding the phospholipase A_2 have to be replaced by other species. It remains a puzzling observation that these nonarachidonate species are not susceptible to phospholipase A_2 for lyso-PAF production. This still leaves the possibility open that phospholipases A_2 exist that specifically recognize the arachidonate structure.

The Ca^{2+} independent phospholipases A_2 in canine myocardium cytosol[54] and amnion tissue microsomes[55] were claimed to have two- to five-fold higher activities with 1-O-alkenyl and 1-O-alkyl analogs, respectively, when compared with diacylphosphatidylcholine. A Ca^{2+} dependent phospholipase A_2 from sheep platelets[33] demonstrated a 100-fold selectivity in catalyzing the release of oleate from 1-O-hexadecenyl-2-oleoyl in comparison to 1-palmitoyl-2-oleoyl species of glycerophosphocholine. Since all these selectivities were determined with phospholipid vesicles consisting in each case of single molecular species, it remains to be seen to what extent the specificities were influenced by possible differences in the quality of the interface caused by different packing of the phospholipid species.

8.6.3 REGULATION OF PHOSPHOLIPASES A_2

Phospholipases A_2, especially the membrane-associated forms, occur in the presence of excess substrate and their regulation is of the utmost importance in view of the membrane-perturbing properties of lysophospholipids and free fatty acids. Models for the regulation of phospholipase A_2

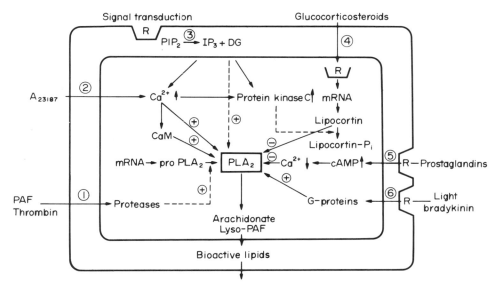

Figure 2 Summary of proposed mechanisms involved in phospholipase A_2 regulation. For details see text. Abbreviations: R, receptor; PIP_2, phosphatidylinositol 4,5-diphosphate; IP_3, inositol triphosphate; DG, diacylglycerol; LC, lipocortin; CaM, calmodulin

have been put forward for the Ca^{2+} dependent enzymes but no clues are currently available as to how the recently discovered group of Ca^{2+} independent enzymes is controlled. Possible models for phospholipase A$_2$ regulation have been reviewed before[1,11,23,26] and will only briefly be updated in this section. Undoubtedly, the regulation is complex, may vary with the cell type and is likely to consist of superimposed mechanisms. A summary of the mechanisms is given in Figure 2.

8.6.3.1 Zymogens

Although the proteolytic activation of an inactive zymogen, as first described for pancreatic phospholipase A$_2$,[72] represents a seemingly irreversible modulation of enzyme activity, enhanced phospholipase A$_2$ activity upon proteolytic treatment of isolated membranes or intact cells has been reported for a variety of rat tissues, plasma, leukocytes, platelets, macrophages and fibroblasts.[1,11] Activation of phospholipase A$_2$ in platelets, as induced by thrombin or collagen, was blocked when the cells had first been treated with the protease inhibitor phenylmethanesulfonyl fluoride.[73] PAF-induced platelet aggregation and secretion, including arachidonate release, was blocked in the presence of chymotryptic-type serine protease inhibitors, but not when the cells were first incubated with the inhibitors and then washed prior to PAF treatment.[74] These results suggested that an early event in PAF-receptor-mediated responses in platelets involves the generation of a chymotryptic-type serine protease prior to receptor-mediated activation of phospholipases C and A$_2$. Such an endogenous protease could then act through the release of a phospholipase A$_2$ activator, through the inactivation of a phospholipase A$_2$ inhibitor or through a transformation of inactive proenzyme into an active form of the enzyme. Until now, proenzymes from sources other than pancreas have not been isolated so their proteolytic activation has not been demonstrated unequivocally.

Screening[42] of a human lung cDNA library with a clone encoding the entire phospholipase A$_2$ from porcine pancreas yielded a clone encoding a 126-residue peptide with 80% sequence homology to porcine pancreatic phospholipase A$_2$ and up to residue 122 identical to human pancreatic phospholipase A$_2$. Interestingly, the sequence of the human lung enzyme was preceded by a seven-residue propeptide and a 15-residue prepeptide. It remains to be established where in the lung and at what stage of the posttranslational processing these parts are cleaved off to yield the mature enzyme.

8.6.3.2 Activator, Inhibitor and G-proteins

Reversible modulations of phospholipase A$_2$ activity can be envisaged through interactions with nonenzymatic inhibitor or activator proteins. Of the inhibitory proteins the glucocorticosteroid inducible lipocortins have received most attention. These will be discussed in Section 8.6.4.1.

Evidence for the occurrence of activator proteins was obtained in studies on leukotriene D$_4$ (LTD$_4$) stimulation of prostanoid synthesis in smooth muscle and endothelial cells.[75] LTD$_4$ caused an increased phospholipase A$_2$ activity and this induction occurred just prior to the appearance of prostanoids. Protein and mRNA synthesis were required for the increase in phospholipase A$_2$ activity. While this could still be explained in various ways, such as synthesis of new enzyme molecules, zymogen conversion into active enzyme or synthesis of activator proteins, the latter possibility was favored in subsequent studies. It was found[76] that antibodies against mellitin, a 2800 kDa phospholipase A$_2$ stimulatory peptide from bee venom, cross-reacted with a 28 000 kDa protein from several cultured cells. The protein from smooth muscle cells was purified by immunoaffinity chromatography and was found to selectively stimulate phospholipase A$_2$ from these cells when phosphatidylcholine was used as substrate. Neither phospholipase C nor pancreatic or venom phospholipase A$_2$ was stimulated. As pointed out by the authors,[76] these data could be interpreted to mean that the activator protein, which increases the apparent V_{max} of the cellular phospholipase A$_2$, is involved in converting inactive enzyme to active enzyme by proteolysis or by competition with an inhibitor protein such as lipocortin. Alternatively, the stimulatory protein could exert its effect by converting substrate in a more susceptible form. The observation that stimulation is only observed with phosphatidylcholine and not with phosphatidylethanolamine as substrate hints at this possibility. LTD$_4$ stimulation of rabbit leukocyte phospholipase A$_2$ acting on *E. coli* membrane phospholipids has also been reported.[77]

Receptor-mediated signal transduction *via* phosphatidylinositol 4,5-diphosphate hydrolysis by phospholipase C may affect phospholipase A$_2$ activity in several ways (Figure 2). Firstly, by perturbation of membrane structure through diacylglycerol formation (Section 8.6.3.4). Secondly, by activation of protein kinase C through formation of the second messenger diacylglycerol and subsequent phosphorylation of lipocortin, thus relieving the inhibitory action of lipocortin (Section 8.6.4.1). Thirdly, by increasing the intracellular free Ca^{2+} concentration through the action of the second messenger inositol 1,4,5-triphosphate.

A rapidly increasing body of evidence suggests that G-proteins are involved in receptor-mediated breakdown of inositol phospholipids[78] and recent experiments in this area suggest that phospholipase A_2 may also be activated directly through G-proteins (see also Chapter 11.4).[79] Based on differential effects of inhibitors, it has been concluded that distinct G-proteins activate phospholipases C and A_2 in cultured thyroid cells[80] and human platelets.[81,82] It has also been suggested that more than one G-protein may couple receptors to phospholipase A_2 activation, since the G-proteins mediating phospholipase A_2 activation in thyroid cells,[80] platelets[81,82] and bovine rod outer segments[83] were pertussis toxin sensitive, whereas that in Swiss 3T3 fibroblasts[84] was not. Whether the phospholipase A_2 activation is elicited by direct interaction of G-proteins with the enzyme or if still other mediators are involved is unknown at present and the exact membrane localization of the activatable phospholipase A_2 in relation to the G-protein involved has not yet been determined. Only in the case of rod outer segments is there compelling evidence that enzyme and G-protein are present in the same membrane. In this case it could be shown directly that phospholipase A_2 activation was brought about by the $\beta\gamma$-subunit of transducin.[83]

8.6.3.3 Calcium Ions and Calmodulin

Many intracellular phospholipases A_2 require Ca^{2+} for activity (Section 8.6.2.1). Thrombin-induced arachidonate release in platelets has been explained in terms of activation of phospholipase A_2 by the enhanced cytosolic free Ca^{2+} concentration as reviewed previously.[11] Along these lines the inhibition of platelet phospholipase A_2 by agents which increase the cellular cAMP level has been ascribed to sequestering of Ca^{2+} induced by cAMP.[11,85] However, a straightforward explanation for the activation of phospholipase A_2 by increased free Ca^{2+} is hampered by the facts that the free Ca^{2+} concentrations in activated cells at best reaches micromolar levels, whereas the enzyme in isolated membranes or after partial purification requires at least 100 μM Ca^{2+} for half-optimal activity.

Small, *i.e.* less than 30%, increases in phospholipase A_2 activity of human platelet membranes upon addition of calmodulin led to the conclusion that phospholipase A_2 was regulated by calmodulin.[86] This conclusion was corroborated by the finding that trifluoperazine, a calmodulin antagonist, inhibited arachidonate release from phosphatidylcholine in prelabeled platelets.[87] Phospholipase A_2 inhibition by calmodulin antagonists has also been reported in Kupffer,[88] renal medullary[89] and aortic endothelial[90] cells. However, the usefulness of trifluoperazine in unraveling calmodulin-mediated processes has been questioned, in view of the findings that this drug exerts additional membrane effects next to calmodulin inactivation.[91,92] Studies with partially purified enzymes from platelets,[92,93] rat spleen[94] and rat liver mitochondria[43] clearly indicated that the Ca^{2+} requirement of these phospholipases A_2 was not mediated by calmodulin. These results contrast with the stimulation of *Naja naja* phospholipase A_2 by calmodulin, which led Moskowitz *et al.*[95] to conclude that phospholipase A_2 is one of the calmodulin-regulated enzymes. The reason for the deviant behavior of this venom phospholipase A_2 is currently difficult to reconcile, especially since the Ca^{2+} requirement of pancreatic phospholipase A_2 did not appear to be mediated by calmodulin.[96]

8.6.3.4 Membrane Phospholipid Structure

Numerous papers have dealt with effects of phospholipid composition, cholesterol content, surface charge and fluidity of biological or model membranes on either endogenous or exogenous phospholipase A_2 activity. The present discussion will be limited to a few attempts that have been made to correlate phospholipase A activity with the organizational structure of phospholipids, *i.e.* bilayer, isotropic or hexagonal H_{II} phases. Initial studies along this line established that in phosphatidylcholine bilayers the organizational defects as introduced at the transition temperature or by membrane perturbants are preferred sites of enzymatic attack.[97] Using phospholipase A_1 from rat liver cytosol Dawson *et al.*[98] found that hydrolysis of the bilayer-forming phosphatidylcholine was greatly enhanced when this substrate was admixed with sufficient hexagonal H_{II} phase-forming phosphatidylethanolamine. While this result may suggest that a hexagonal phase is the preferred organization for phospholipase A_1, recent experiments with mitochondrial phospholipase A_2

showed that phosphatidylcholine hydrolysis in mixtures with phosphatidylethanolamine proceeded optimally in bilayers and rapidly declined upon appearance of isotropic and hexagonal phases as deduced from ^{31}P NMR spectra of the lipid mixtures.[99] Phosphatidylethanolamine hydrolysis increased in going from bilayer *via* isotropic to hexagonal structures. These results could, therefore, be interpreted as suggesting that a disturbance of bilayer phospholipid organization leads to a decreased hydrolysis of the bilayer-preserving phospholipid and to increased hydrolysis of the bilayer-disturbing phospholipid in an attempt to maintain the bilayer organization. Much work will be required to validate this assumption and currently it is completely unknown how the enzyme might sense these structural changes. Interestingly, the induction of a hexagonal phase of phosphatidylcholine by addition of over 40 mol% diacylglycerol also led to a total loss of susceptibility to an intestinal mucosa phospholipase A₂.[100] These experiments aimed at determining whether the transient formation of diacylglycerol by inositol phospholipid hydrolysis after receptor occupation (Figure 2, item 3) could promote hydrolysis of phospholipids that were initially present in bilayer configuration and thought to be largely resistant to phospholipase A₂ attack. Stimulation of phospholipase activity *in vitro* was indeed observed, reaching a maximum at 30 mol% diacylglycerol, both with intestinal mucosa[100] and platelet phospholipase A₂.[101] In intact platelets the addition of diacylglycerols at concentrations sufficient to produce maximal phosphorylation of a 40 kDa protein did not potentiate phospholipases A₂ and C.[102] However, the diacylglycerols used in these experiments were also those that promoted phospholipase A₂ in the vesicle assay the least.[100] Minor modulation of phospholipase A₂ activity by transient formation of diacylglycerols can, therefore, not be completely excluded. Perhaps the process involves translocation of the enzyme from cytosol to membranes as recently reported to be induced in macrophages by 1-oleoyl-2-acetylglycerol.[103]

8.6.4 PHOSPHOLIPASE A₂ INHIBITORS

8.6.4.1 Lipocortins and Related Proteins

The inhibition of prostaglandin production by glucocorticosteroids in susceptible cells was initially pinpointed at the level of arachidonate release from cellular phospholipids. The effect depended on mRNA and protein synthesis and was thought to be due to the steroid-induced synthesis of inhibitory proteins, now termed lipocortins, for phospholipase A₂.[104,105] Subsequent experiments showed that lipocortin was subject to phosphorylation/dephosphorylation and that only the dephosphorylated form acted as an inhibitor to phospholipase A₂.[106,107] Inhibition of pancreatic phospholipase A₂ by lipocortin in a dose dependent manner led to the suggestion that inhibition took place by formation of stoichiometric complexes between enzyme and inhibitor.[106] At the same time the observation that venom phospholipase A₂, bacterial phospholipases C and plant phospholipase D also were inhibited by lipocortin[106] cast doubt as to the specific role of lipocortins as modulators of intracellular phospholipases A₂. Due to insufficient amounts of purified lipocortin and intracellular phospholipases A₂ available, a direct stoichiometric protein–protein interaction could never be demonstrated.

Partial amino acid sequence information obtained for lipocortin purified from rat peritoneal exudates[108] was used for oligonucleotide probe synthesis and subsequent cloning and expression of human lipocortin.[109] This breakthrough had important consequences. The amino acid sequence data of lipocortins I and II from human placenta indicated lipocortin I to be the same protein as the 35 kDa substrate for the EGF-receptor/kinase in a human carcinoma cell line and indicated a 50% sequence homology for lipocortin I and II.[110] Lipocortin II appeared to be the human analog of p36, a major target protein for retrovirus protein-tyrosine kinases characterized in chicken embryo fibroblasts and bovine brush border.[110,111] This protein was also known as calpactin I, because of its calcium dependent phospholipid and actin-binding properties. A number of other Ca^{2+}/phospholipid binding proteins such as calelectrin and endonexin appeared to share sequence homology with lipocortins and contained multiple repeats of a 17 amino acid consensus sequence believed to be involved in Ca^{2+}- and phospholipid-binding.[112–115] Human 67 kDa calelectrin consists of eight repeats of about 70 amino acids that are highly similar to the four of these repeats in lipocortins I and II.[116] Subsequent experiments showed that calpactins,[117] calelectrin[118] and endonexin[119] all inhibited pancreatic phospholipase A₂.

The unexpected structural homology between the Ca^{2+}/phospholipid-binding proteins, such as calpactins, and lipocortins suggested that the inhibition of phospholipase A₂ by lipocortins, hitherto considered specific, might be due to the artefactual sequestering of substrate phospholipid.[117] This was confirmed experimentally in that inhibition of pancreatic phospholipase A_2 with *E. coli* membranes, or derived phospholipid vesicles, as substrate was only observed at low substrate concentrations and could be completely overcome by increasing the amount of substrate.[120] Similar findings were made using intracellular phospholipases A_2 in their action towards phosphatidyl-ethanolamine.[121] In line with the interpretation that inhibition was caused by sequestering of the substrate it was found that lipocortins bound in a micromolar Ca^{2+} dependent manner to negatively charged phospholipid vesicles and *E. coli* membranes, whereas a direct interaction between lipocortins and phospholipase A_2 under the conditions of the assay could not be demonstrated.[120,122] It is noteworthy that phosphorylation of Tyr-21 in lipocortin I resulted in a fivefold reduction of the Ca^{2+} concentration required for half-maximal binding of the protein to phosphatidylserine vesicles.[123] Hence, phosphorylated lipocortin is equally capable of sequestering phospholipid substrate and this finding does not provide an easy explanation for the earlier observation[106,107] that the phosphorylated form of lipocortin is no longer active as an inhibitor of phospholipase A_2. These developments challenged the initial model that glucocorticosteroid inhibition of eicosanoid production was mediated by lipocortin inhibition of arachidonate release. The finding[124] that hydrocortisone inhibited prostaglandin formation, but not arachidonate release, in cultured mouse macrophages likewise questioned the involvement of lipocortin at the level of phospholipases in the antiinflammatory effects of glucocorticosteroids. However, recent experiments have indicated that human recombinant lipocortin I is capable of inhibiting thromboxane release from guinea pig isolated perfused lung[125] and prostacyclin production by human umbilical artery.[126] These and other experiments suggested that it may be the outer cell surface associated lipocortin, rather than the intracellular pool, which controls eicosanoid production. Whether this lipocortin exerts its effect by blocking Ca^{2+} entry, by causing structural changes in the cell, by interfering with signal transduction or by a completely different mechanism remains to be elucidated.

8.6.4.2 Other Proteins

To avoid the isolation of proteins that inhibit phospholipase A_2 action at the substrate level Nevalainen and Evilampi[127] used immobilized pancreatic phospholipase A_2 to purify the inhibitory principle from porcine serum. Substances that bound to the enzyme in the presence of 2 mM $CaCl_2$ were eluted with EDTA and further purified by gel filtration to give a protein fraction with a molecular weight of 60 kDa. This fraction inhibited phosphatidylcholine hydrolysis by pancreatic phospholipase A_2. However, complete inhibition required a 500-fold molar excess of inhibitor over enzyme and the mode of inhibition was not further investigated. Phospholipase A_2 inhibitor proteins were also described in bovine, human and porcine plasma.[128] The inhibitor from bovine plasma was purified 300-fold and appeared to be a lipoprotein with a molecular weight nearly the same as that of bovine plasma high density lipoprotein when determined by gel filtration. SDS-PAGE indicated the presence of two protein subunits of 10 and 42 kDa, respectively. Phospholipase A_2 activity was only suppressed when the inhibitor was preincubated with the enzyme, which was interpreted as being due to formation of an enzyme–inhibitor complex. This lipoprotein inhibited phospholipase A_2 from various sources but not phospholipase C or lipase.[129] Similarly, partially purified proteins from amniotic fluid inhibited phospholipase A_2 but not phospholipase C.[130] Gel filtration indicated inhibitory activity at 70 to 80 kDa and 150 to 165 kDa that were otherwise indistinguishable with respect to heat stability, trypsin sensitivity, HPLC retention time and dose dependency, suggesting a monomer to dimer relationship. The potency of these inhibitors, when expressed in molar concentrations, was superior to that of lipocortin by two orders of magnitude. Protein inhibitors of phospholipase A_2 were also detected in the cytosol of myometrium from pregnant ewes.[131] It was suggested that such proteins may be involved in suppression of eicosanoid biosynthesis by uterine tissue throughout gestation, thus inhibiting contractile activity. Since the proteins eluted with an apparent molecular weight of 35 to 45 kDa and their inhibitory action was assessed in assays using low substrate concentration, they could well belong to the lipocortin family and exhibit a similar mode of action at the substrate level.

8.6.4.3 Antibodies

Specific inhibition at the protein level can be obtained by using specific inhibiting antibodies. In studies with antipancreatic phospholipase A_2 antibodies Meijer *et al.*[132] showed that Fab fragments directed against the N-terminal region of the enzyme were more inhibitory than Fab fragments that recognized two other epitopes. The former Fab fragments completely abolished hydrolysis of dioctanoylphosphatidylcholine at a molar Fab/enzyme ratio of four. Antirat pancreatic phospholipase A$_2$ antibodies were found to show cross-reactivity with rat spleen enzymes.[47] Quite recent results have indicated that rat spleen contains two genetically distinct 14 kDa Ca^{2+} dependent phospholipases A$_2$. The sequence of the N-terminal 32 residues of the cytosolic enzyme was identical to that in the rat pancreatic phospholipase A$_2$ and also its catalytic and immunochemical properties were similar, if not identical. Both the pancreatic and splenic soluble phospholipase A$_2$ were completely inhibited by the antibodies.[133] A second and membrane-associated splenic phospholipase A$_2$ showed a different N-terminal sequence lacking Cys-11 and did not cross-react with the antipancreatic phospholipase A$_2$ antibodies.[134]

Antibodies raised against *Naja naja* phospholipase A$_2$ proved not only inhibitory to this enzyme but also to the phospholipase A$_2$ activity in 1 M NaCl extracts from guinea pig alveolar macrophages and rat lymphocytes.[48] Monoclonal antibodies against rat liver mitochondrial phospholipase A$_2$ also caused complete inhibition of this intracellular phospholipase A$_2$ (Aarsman and van den Bosch, unpublished). Interestingly, and unexpectedly, monoclonal antibodies generated against human lecithin-cholesterol acyltransferase not only caused the inhibition of the esterolytic and cholesterol-esterifying activities of the enzyme but also inhibited a number of pancreatic and snake venom phospholipases A$_2$.[135] The use of highly specific inhibiting antibodies is likely to provide new insights into the role of their target enzymes in phospholipid metabolism and eicosanoid production.

8.6.4.4 Plant Extracts

Manoalide, a sesterpenoid product of the sponge *Luffariella variabilis*, has been found to antagonize inflammation induced by phorbol ester but not that induced by arachidonate.[136] This suggested that manoalide acted prior to oxygenation reactions, possibly by inhibiting arachidonate release. Indeed, manoalide prevented hydrolysis of phosphatidylcholine by bee venom phospholipase A$_2$ due to an irreversible binding to the enzyme.[136] The mechanism of action of manoalide was studied in more detail by Dennis and coworkers,[137-139] using cobra venom phospholipase A$_2$. The rate of inhibition, but not the total extent, which amounted to maximally 85% despite a 500-fold molar excess of manoalide, was increased in the presence of millimolar Ca^{2+}. The irreversible inactivation could be prevented by substrate micelles and was ascribed to modification of lysine residues. The inhibition followed the appearance of carbonyl groups on manoalide resulting from opening of a six-membered hemiacetal ring and/or a γ-lactone ring to form α,β-unsaturated aldehyde functions on carbon atoms 24 or 25 in the open form of manoalide.[137] Studies with a synthetic manoalide analog containing only the α,β-unsaturated γ-lactone ring revealed that while the γ-lactone ring may play an important role in the covalent modification of phospholipase A$_2$ by manoalide, it alone does not produce irreversible inhibition.[138] Unexpectedly, while hydrolysis of phosphatidylcholine by manoalide-treated enzyme was suppressed, that of phosphatidylethanolamine was enhanced. Apparently, when the bulky manoalide residue is attached to lysines on the surface of the enzyme, the specificity of the enzyme is changed in opposite directions for the two substrates. Manoalide inhibition of purified cellular phospholipases A$_2$ has not yet been reported.

Tanacetum parthenium (feverfew) leaf extracts were reported to inhibit arachidonate release and phospholipase A$_2$ activity in human platelets.[140] The inhibitor was heat stable, was recovered in the water phase after lipid extraction and did not act as a Ca^{2+} chelator. Feverfew extract was also inhibitory to phospholipase A$_2$ activity from rat aorta smooth muscle cells[141] and rat liver mitochondria (unpublished experiments). The mode of action has not been further investigated in these cases. Although no experimental details were provided, it has been claimed that feverfew extract acted also at the substrate level and that it had no effect on the hydrolysis of monomeric substrate.[142]

Garlic extracts were found to inhibit platelet aggregation[143] and early attempts to clarify the mechanism suggested that garlic not only reduced the formation of thromboxane from exogenous arachidonate but also inhibited phospholipase activity.[144] Further experiments, summarized by Jain and Apitz-Castro,[145] led to the isolation of ajoene as a nonvolatile antiplatelet compound. Ajoene was found to inhibit platelet aggregation induced by a wide variety of agents, including ADP, collagen, thrombin, Ca^{2+} ionophore, PAF and arachidonate. Ajoene does not influence metabolism of arachidonate and other eicosanoids and its inhibitory effect on platelet aggregation has been ascribed to one of the last steps that lead to the exposure of fibrinogen receptors.[145] Whether garlic contains additional compounds that exert their action at the phospholipase A_2 level remains to be seen.

8.6.4.5 Lipidic Inhibitors

Marked increases in the phospholipase A_2 activity from human platelets during purification yielded the first indications for the presence of an endogenous inhibitor.[93] The inhibitor activity was extractable with chloroform/methanol and has later been identified[146] as a mixture of mono-, di- and poly-unsaturated fatty acids. An IC_{50} value of 0.5 µM was reported, although this value almost certainly is a function of the phospholipid substrate concentration. Evidence was provided to show that the endogenous level of unsaturated fatty acids in platelets is sufficient to suppress a large part of the phospholipase A_2. A small phospholipase inhibitory factor, with properties that made it plausible that it was of the unsaturated fatty acid type, appeared to be released by and to act upon various cell types.[147] It has been questioned by Ballou and Cheung[146] whether phospholipase A_2 inhibition by unsaturated fatty acids is an *in vitro* phenomenon or does have physiological relevance by constituting a mechanism for feedback control. In this respect it is worth noting that the analogous inhibitor from cultured macrophages[147] suppressed the action of distinct inflammatory stimuli. Acid extraction of phospholipase A_2 from human polymorphonuclear leukocytes led to a 10-fold increase in activity and the detection of endogenous inhibitors in the pellet.[148] The bulk of suppressor activity was lipid extractable and also appeared to be due to unsaturated fatty acids and some of their metabolites, suggesting that these could be involved in the careful modulation of the expression of phospholipase A_2 activity in intact cells.[148] Another chloroform soluble phospholipase A_2 inhibitor was detected following the observation that enzyme activity extracted from acetone powders of previously frozen rat liver mitochondria was strongly reduced compared to the activity manifest before acetone powder preparation.[149] This lipidic inhibitor was identified as monolysocardiolipin and effectively suppressed mitochondrial, pancreatic and venom phospholipase A_2 at levels less than 0.1% of the substrate. Although the mode of inhibition was not further investigated, this high potency seems to rule out an effect on the physical state of the substrate. The fact that mono- and di-lysocardiolipin prepared from bovine heart cardiolipin was found to be 10- and 30-fold, respectively, less potent than the material arising from rat liver cardiolipin suggests that acyl group composition or structural isomerism may be an important determinant in the inhibitory action.

8.6.4.6 Substrate Mimics

Early studies on the minimal substrate requirements for phospholipase A_2 by Bonsen *et al.*[150] showed that replacement of the *sn*-2-ester function by an amide linkage rendered the glycerophospholipid resistant to enzyme attack, while showing 5- to 10-fold higher affinity than the parent substrates. Further modification of the substrate by introducing an ether bond at the *sn*-1-position yielded a 1-alkyl-2-(acylamino)-2-deoxyglycerophosphocholine that bound 30- to 100-fold more tightly to the enzyme than phosphatidylcholine substrates in assays using mixed micelles of substrate, inhibitor and excess of Triton X-100.[151] This system was chosen to minimize the effect of addition of the inhibitor on the surface properties of the aggregated substrate. The fact that different K_m/K_i ratios were found depending on the acyl chain composition of the phosphatidylcholine substrates suggests that the surface properties might still have been influenced differently by the inhibitor that had yet another acyl group composition than the substrates. These problems, inherent to kinetic analysis of lipolytic enzymes, were circumvented by de Haas and colleagues[152] by using an amido analog and a substrate bearing identical fatty acids and changing the mole ratio of inhibitor to substrate while keeping the total phospholipid/detergent ratio constant. Such studies indicated K_m/K_i ratios of 11 and 27 for a pancreatic and a venom phospholipase A_2, respectively.

To mimic the putative tetrahedral intermediate that forms during phospholipase A_2-catalyzed hydrolysis, a phospholipid analog containing a difluoromethylene ketone grouping at the 2-position was synthesized.[153] In this compound the oxygen atom at the 2-position was replaced by a methylene group and the methylene group on the other side of the ester carbonyl was replaced by a difluoromethylene group. A short chain phosphatidylethanolamine, containing the difluoromethylene ketone unit as an isosteric replacement of the ester linkage, competitively inhibited phospholipase A_2-catalyzed hydrolysis of monomerically dispersed dibutyrylphosphatidylcholine with a K_m/K_i ratio of 280. The much tighter binding of the inhibitor was ascribed to its ability to form a tetrahedral species upon hydration.[153] Further inhibition and NMR studies with a series of compounds, including those with a reversed positioning of the carbonyl ester and difluoromethylene group, *i.e.* compounds in which the oxygen at the 2-position was replaced by a difluoromethylene group, provided evidence for a correlation between the inhibitory potency of a compound and its ability to become hydrated in the substrate/Triton X-100 mixed micelle.[154] Single chain fluoro ketones more readily adapted a tetrahedral hydrated structure than diacyl analogs and proved more potent inhibitors. A phosphoethanolamine derivative of glycol bearing a long alkyl chain with a difluoromethylene ketone grouping appeared most potent and bound some 1000-fold better to the enzyme than dipalmitoylphosphatidylcholine. Phosphatidylethanolamine hydrolysis was hardly affected by this inhibitor but this may be related to the use of cobra venom phospholipase A_2 in these studies. It remains to be seen what effect the substrate mimics described in this section will have on cellular phospholipases A_2. None of the compounds discussed has so far been tested in cellular systems to study their effects on eicosanoid production.

8.6.4.7 Chemicals

A large number of compounds, detected in a screening program, inhibiting pancreatic and venom phospholipase A_2 were described by Wallach and Brown.[155] IC_{50} values, determined in assays containing 200 μM substrate, generally ranged from 500 to 5 μM. It follows from these data that an effect at the substrate rather than the enzyme level cannot be ruled out. Importantly, however, some exemplars of the compounds inhibited collagen-induced platelet aggregation and eicosanoid production in isolated perfused guinea pig lung. Neither effect was observed when arachidonate was added, suggesting that release of esterified arachidonate had been prevented by the compounds.

Inhibition of phospholipase A_2 by frequently used compounds such as chlorpromazine, mepacrine and local anesthetics has been shown to be primarily due to perturbation of lipid–protein interaction rather than to direct interaction with the catalytic site of the free or lipid-bound enzyme.[142,156] In agreement with this mode of action mepacrine was also reported to inhibit phosphatidylinositol specific phospholipase C.[157]

Indomethacin inhibited rabbit polymorphonuclear leukocytes phospholipase A_2 with an IC_{50} of 30 μM in assays where three venom enzymes and pancreatic phospholipase A_2 remained unaffected by 50 μM concentrations of the drug.[158] Similar IC_{50} values were found for the enzyme from human platelets and rabbit alveolar macrophages.[159] Indomethacin is not a specific inhibitor of phospholipase A_2 in that several other enzymes, listed by Kuehl and Egan,[15] are inhibited. The concentrations required to inhibit these enzymes and phospholipase A_2 are 100- to 1000-fold those necessary to affect prostaglandin formation by cyclooxygenase inhibition. However, the action of other nonsteroidal antiinflammatory agents was found to depend dramatically on the Ca^{2+} concentration.[159] The IC_{50} for sodium meclofenamate dropped from 400 μM at 2.5 mM Ca^{2+} to 50 nM in the presence of 0.5 mM Ca^{2+}. Thus, inhibition of phospholipase A_2 at physiological Ca^{2+} concentrations may still contribute to the antiinflammatory action of these drugs.

An active site directed inhibitor for phospholipase A_2 became available when Volwerk *et al.*[160] discovered in 1974 that *p*-bromophenacyl bromide alkylated an essential histidine of the enzyme. Although this inhibitor has been very valuable in studies on purified phospholipases A_2, it should be remembered that this modifying agent also lacks specificity. It has since been found that *p*-bromophenacyl bromide inhibits other enzymes, including alcohol dehydrogenase,[161] α-chymotrypsin,[161] phosphatidylinositol specific phospholipase C[157,161] and enzymes responsible for arachidonate incorporation into cellular phospholipids.[162] This calls for a cautious interpretation of studies in which the role of phospholipase A_2 in cellular processes is evaluated based on the effect of this inhibitor.

8.6.5 CONCLUSIONS

During the last two decades much has been learned on the initiating role of phospholipase A_2 in the production of eicosanoids, leukotrienes and platelet-activating factor through formation of precursors for these bioactive lipids. These studies have established an enormous heterogeneity among phospholipases with respect to structural properties, Ca^{2+} dependency, substrate specificity, subcellular localization and mode of membrane association. Despite the efforts of many scientists working on these fundamental aspects of phospholipase A_2 action, many open questions remain. The regulation of these potentially membrane-perturbing enzymes under resting conditions as well as the processes that lead to phospholipase A_2 activation upon cell stimulation are not well understood and are likely to vary with the cell type. Complex as this matter is, it is not known which phospholipase A_2, *i.e.* whether soluble, membrane-associated, intracellular or secreted, is actually involved in inflammatory processes or other disease states. An answer to these questions would be highly beneficial to those colleagues aiming to develop therapeutic drugs having phospholipase A_2 as their target enzyme. Kinetic and structural investigations on these enzymes should provide insights into how the active sites of cellular phospholipases A_2 relate to those of pancreatic and venom phospholipases A_2, that are now frequently used as model enzymes. The recent findings, indicating that pancreatic-type phospholipase A_2 occurs in non-pancreatic tissues suggests that much of what has been learned on pancreatic phospholipase A_2 may at least partially be applicable to cellular phospholipases A_2.[42,133,163]

8.6.6 REFERENCES

1. H. van den Bosch, in 'New Comprehensive Biochemistry', ed. J. N. Hawthorne and G. B. Ansell, Elsevier, Amsterdam, 1982, vol. 4, p. 313.
2. D. J. Hanahan and R. Kumar, *Prog. Lipid Res.*, 1987, **26**, 1.
3. D. J. Hanahan, *Annu. Rev. Biochem.*, 1986, **55**, 483.
4. F. Snyder, T.-C. Lee and R. L. Wykle, in 'The Enzymes of Biological Membranes', ed. A. Martonosi, Plenum Press, New York, 1985, vol. 2, p. 1.
5. F. Snyder, *Med. Res. Rev.*, 1985, **5**, 107.
6. P. Braguet, T. Y. Shen, L. Touqui and B. Vargaftig, *Pharmacol. Rev.*, 1987, **39**, 97.
7. F. H. Chilton, J. M. Ellis, S. C. Olsen and R. L. Wykle, *J. Biol. Chem.*, 1984, **259**, 12 014.
8. M. Robinson, M. L. Blank and F. Snyder, *J. Biol. Chem.*, 1985, **260**, 7889.
9. G. Mauco, H. Chap and L. Douste-Blazy, *FEBS Lett.*, 1979, **100**, 367.
10. R. L. Bell, D. A. Kennerly, N. Standford and P. W. Majerus, *Proc. Natl. Acad. Sci. USA*, 1979, **76**, 3238.
11. H. van den Bosch, *Biochim. Biophys. Acta*, 1980, **604**, 191.
12. R. F. Irvine, *Biochem. J.*, 1982, **204**, 3.
13. B. J. Holub, *Can. J. Biochem. Cell Biol.*, 1984, **62**, 341.
14. M. J. Broekman, *J. Lipid Res.*, 1986, **27**, 884.
15. F. A. Kuehl, Jr. and R. W. Egan, *Science (Washington, D.C.)*, 1980, **210**, 978.
16. P. Needleman, J. Turk, B. A. Jakschik, A. R. Morrison and J. B. Lefkowith, *Annu. Rev. Biochem.*, 1986, **55**, 69.
17. J. Vane and R. Botting, *FASEB J.*, 1987, **1**, 89.
18. J. B. Smith, *Fed. Proc., Fed. Am. Soc. Exp. Biol.*, 1987, **46**, 139.
19. M. L. Ogletree, *Fed. Proc., Fed. Am. Soc. Exp. Biol.*, 1987, **46**, 133.
20. S. Chierchia and C. Patrono, *Fed. Proc., Fed. Am. Soc. Exp. Biol.*, 1987, **46**, 81.
21. A. W. Ford-Hutchinson, *Fed. Proc., Fed. Am. Soc. Exp. Biol.*, 1985, **44**, 25.
22. B. Samuelsson, S.-E. Dahlén, J. A. Lindgren, C. A. Rouzer and C. N. Serhan, *Science (Washington, D.C.)*, 1987, **237**, 1171.
23. J. Chang, J. H. Musser and H. McGregor, *Biochem. Pharmacol.*, 1987, **36**, 2429.
24. H. M. Verheij, A. J. Slotboom and G. H. de Haas, *Rev. Physiol. Biochem. Pharmacol.*, 1981, **91**, 91.
25. E. A. Dennis, in 'The Enzymes', ed. P. Boyer, Academic Press, New York, 1983, vol. 16, p. 307.
26. M. Waite, *J. Lipid Res.*, 1985, 1379.
27. M. Waite, 'Handbook of Lipid Research, The Phospholipases', Plenum Press, New York, 1987, vol. 5, p. 111.
28. H. Tojo, T. Teramoto, T. Yamano and M. Okamoto, *Anal. Biochem.*, 1984, **137**, 533.
29. S. Sahu and W. S. Lynn, *Biochim. Biophys. Acta*, 1977, **489**, 307.
30. P. Elsbach, J. Weiss, R. S. Franson, S. Beckerdite-Quagliata, A. Schneider and L. Harris, *J. Biol. Chem.*, 1979, **254**, 11 000.
31. R. M. Kramer, C. Wüthrich, C. Bollier, P. R. Allegrini and P. Zahler, *Biochim. Biophys. Acta*, 1978, **507**, 381.
32. R. J. Apitz-Castro, M. A. Mas, M. R. Cruz and M. K. Jain, *Biochim. Biophys. Res. Commun.*, 1979, **91**, 63.
33. L. A. Loeb and R. W. Gross, *J. Biol. Chem.*, 1986, **261**, 10 467.
34. N. C. C. Gray and K. P. Strickland, *Can. J. Biochem.*, 1982, **60**, 108.
35. J. M. de Winter, G. M. Vianen and H. van den Bosch, *Biochim. Biophys. Acta*, 1982, **712**, 332.
36. R. Verger, F. Ferrato, C. M. Mansbach and G. Pieroni, *Biochemistry*, 1982, **81**, 6883.
37. A. M. Castle and J. D. Castle, *Biochim. Biophys. Acta*, 1981, **666**, 259.
38. E. Stefanski, W. Pruzanski, B. Sternby and P. Vadas, *J. Biochem.*, 1986, **100**, 1297.
39. S. Frost, J. Weiss and P. Elsbach, *Biochemistry*, 1986, **25**, 8381.
40. H. W. Chang, I. Kudo, M. Tomita and K. Inoue, *J. Biochem.*, 1987, **102**, 147.
41. M. Hayakawa, K. Horigome, I. Kudo, M. Tomita, S. Nojima and K. Inoue, *J. Biochem.*, 1987, **101**, 1311.
42. J. J. Seilhamer, T. L. Randall, M. Yamanaka and L. K. Johnson, *DNA*, 1986, **5**, 519.
43. J. M. de Winter, J. Korpancova and H. van den Bosch, *Arch. Biochem. Biophys.*, 1984, **234**, 243.
44. R. M. Kramer, G. C. Checani, A. Deykin, C. R. Pritzker and D. Deykin, *Biochim. Biophys. Acta*, 1986, **878**, 394.
45. A. Baba, H. Onoe, A. Ohta and H. Iwata, *Biochim. Biophys. Acta*, 1986, **878**, 25.
46. M. Wurl and H. Kunze, *Biochim. Biophys. Acta*, 1985, **834**, 411.
47. M. Okamoto, T. Ono, H. Tojo and T. Yamano, *Biochim. Biophys. Res. Commun.*, 1985, **128**, 788.
48. J. Masliah, C. Kadiri, D. Pepin, T. Rybkine, J. Etienne, J. Chambaz and G. Bereziat, *FEBS Lett.*, 1987, **222**, 11.
49. S. Weinmann, C. Ores-Carton, D. Rainteau and S. Puszkin, *J. Histochem. Cytochem.*, 1986, **34**, 1171.
50. J. G. N. de Jong, H. Amesz, A. J. Aarsman, H. B. M. Lenting and H. van den Bosch, *Eur. J. Biochem.*, 1987, **164**, 129.
51. M. Ohta, H. Hasegawa and K. Ohno, *Biochim. Biophys. Acta*, 1972, **280**, 552.

52. I. Adachi, S. Toyoshima and T. Osawa, *Arch. Biochem. Biophys.*, 1983, **226**, 118.
53. B. J. Borman, C. K. Huang, W. M. Mackin and E. L. Becker, *Proc. Natl. Acad. Sci. USA*, 1984, **81**, 767.
54. R. A. Wolf and R. W. Gross, *J. Biol. Chem.*, 1985, **260**, 7295.
55. C. Ban, M. Billah, C. T. Truong and J. M. Johnston, *Arch. Biochem. Biophys.*, 1986, **246**, 9.
56. J. G. Nijssen, C. F. P. Roosenboom and H. van den Bosch, *Biochim. Biophys. Acta*, 1986, **876**, 611.
57. L. R. Ballou, L.M. DeWitt and W. Y. Cheung, *J. Biol. Chem.*, 1986, **261**, 3107.
58. T. W. Martin, R. B. Wysolmerski and D. Luganoff, *Biochim. Biophys. Acta*, 1987, **917**, 296.
59. A. Diagna, S. Mitjavila, J. Fauvel, H. Chap and L. Douste-Blazy, *Lipids*, 1987, **22**, 33.
60. S. Pind and A. Kuksis, *Biochim. Biophys. Acta*, 1987, **901**, 78.
61. Y. Cao, S. W. Tam, G. Arthur, H. Chen and P. C. Choy, *J. Biol. Chem.*, 1987, **262**, 16 927.
62. M. L. Blank, T. Lee, V. Fitzgerald and F. Snyder, *J. Biol. Chem.*, 1981, **256**, 175.
63. T. K. Bills, J. B. Smith and M. J. Silver, *J. Clin. Invest.*, 1977, **60**, 1.
64. V. G. Mahadevappa and B. J. Holub, *J. Biol. Chem.*, 1984, **259**, 9369.
65. A. Derksen and P. Cohen, *J. Biol. Chem.*, 1975, **250**, 9342.
66. O. Colard, M. Breton and G. Bereziat, *Biochim. Biophys. Acta*, 1987, **921**, 333.
67. R. Kannagi, K. Koizumi and T. Masuda, *J. Biol. Chem.*, 1981, **256**, 1177.
68. A. D. Purdon, D. Patelunas and J. B. Smith, *Biochim. Biophys. Acta*, 1987, **920**, 205.
69. R. Kannagi and K. Koizumi, *Arch. Biochem. Biophys.*, 1979, **196**, 534.
70. C. S. Ramesha and W. C. Pickett, *J. Biol. Chem.*, 1986, **261**, 7592.
71. R. M. Kramer, G. M. Patton, C. R. Pritzker and D. Deykin, *J. Biol. Chem.*, 1984, **259**, 13 316.
72. G. H. de Haas, N. M. Postema, W. Nieuwenhuizen and L. L. M. van Deenen, *Biochim. Biophys. Acta*, 1968, **159**, 118.
73. M. B. Feinstein, E. L. Beeker and C. Fraser, *Prostaglandins*, 1977, **14**, 1075.
74. J. Sugatani, M. Miwa and D. J. Hanahan, *J. Biol. Chem.*, 1987, **262**, 5740.
75. M. A. Clark, D. Littlejohn, T. M. Conway, S. Mong, S. Steiner and S. T. Crooke, *J. Biol. Chem.*, 1986, **261**, 10 713.
76. M. A. Clark, T. M. Conway, R. G. L. Shorr and S. T. Crooke, *J. Biol. Chem.*, 1987, **262**, 4402.
77. B. J. Borrmann, C.-K. Huang, W. M. Mackin and E. L. Becker, *Proc. Natl. Acad. Sci. USA*, 1984, **81**, 767.
78. S. Cockcroft, *Trends Biochem. Sci.*, 1987, **12**, 75.
79. R. D. Burgoyne, T. R. Cheek and A. J. O'Sullivan, *Trends Biochem. Sci.*, 1987, **12**, 332.
80. R. M. Burch, A. Luini and J. Axelrod, *Proc. Natl. Acad. Sci. USA*, 1986, **83**, 7201.
81. S. Nakashima, H. Hattori, L. Shirato, A. Takenaka and Y. Nozawa, *Biochem. Biophys. Res. Commun.*, 1987, **148**, 971.
82. I. Fuse and H.-H. Tai, *Biochem. Biophys. Res. Commun.*, 1987, **146**, 659.
83. C. J. Jelsema and J. Axelrod, *Proc. Natl. Acad. Sci. USA*, 1987, **84**, 3623.
84. R. M. Burch and J. Axelrod, *Proc. Natl. Acad. Sci. USA*, 1987, **84**, 6374.
85. A. Imai, H. Hattori, M. Takahashi and Y. Nozawa, *Biochem. Biophys. Res. Commun.*, 1983, **112**, 693.
86. P. Y. K. Wong and W. Y. Cheung, *Biochem. Biophys. Res. Commun.*, 1979, **90**, 473.
87. R. W. Walenga, E. E. Opas and M. B. Feinstein, *J. Biol. Chem.*, 1981, **256**, 12 523.
88. M. Birmelin, D. Marme, E. Ferber and K. Decker, *Eur. J. Biochem.*, 1984, **140**, 55.
89. P. A. Craven and F. R. De Rubertis, *J. Biol. Chem.*, 1983, **258**, 4814.
90. A. R. Whorton, S. L. Young, J. L. Data, A. Barchovsky and R. S. Kent, *Biochim. Biophys. Acta*, 1982, **712**, 79.
91. T. Watanabe, Y. Hashimoto, T. Teramoto, S. Kume, C. Naito and H. Oka, *Arch. Biochem. Biophys.*, 1986, **246**, 699.
92. M. T. Withnall, T. J. Brown and B. K. Diocee, *Biochem. Biophys. Res. Commun.*, 1984, **121**, 507.
93. L. R. Ballou and W. Y. Cheung, *Proc. Natl. Acad. Sci. USA*, 1983, **80**, 5303.
94. T. Teramoto, H. Tojo, T. Yamano and M. Okamoto, *J. Biochem.*, 1983, **93**, 1353.
95. N. Moskowitz, L. Shapiro, W. Schook and S. Puszkin, *Biochem. Biophys. Res. Commun.*, 1983, **115**, 94.
96. M. T. Whitnall and T. J. Brown, *Biochem. Biophys. Res. Commun.*, 1982, **106**, 1049.
97. G. C. Upreti and M. K. Jain, *J. Membrane Biol.*, 1980, **55**, 113.
98. R. M. C. Dawson, R. F. Irvine, N. L. Hemington and K. Hirasawa, *Biochem. J.*, 1983, **209**, 865.
99. H. B. M. Lenting, K. Nicolay and H. van den Bosch, *Biochim. Biophys. Acta*, 1988, **958**, 405.
100. R. M. C. Dawson, R. F. Irvine, J. Bray and P. J. Quinn, *Biochem. Biophys. Res. Commun.*, 1984, **125**, 836.
101. R. M. Kramer, G. C. Checani and D. Deykin, *Biochem. J.*, 1987, **248**, 779.
102. S. P. Watson, B. R. Ganong, R. M. Bell and E. G. Lapetina, *Biochem. Biophys. Res. Commun.*, 1984, **121**, 386.
103. T. Schonhardt and E. Ferber, *Biochem. Biophys. Res. Commun.*, 1987, **149**, 769.
104. R. J. Flower, J. N. Wood and L. Parente, *Adv. Inflammation Res.*, 1984, **7**, 61.
105. F. Hirata, *Adv. Inflammation Res.*, 1984, **7**, 71.
106. F. Hirata, *J. Biol. Chem.*, 1981, **256**, 7730.
107. F. Hirata, K. Matsuda, Y. Notsu, T. Hattori and R. del Carmine, *Proc. Natl. Acad. Sci. USA*, 1984, **81**, 4717.
108. R. B. Pepinsky, L. K. Sinclair, J. L. Browning, R. J. Mattaliano, J. E. Smart, E. P. Chow, T. Falbel, A. Ribolini, J. L. Garwin and B. P. Wallner, *J. Biol. Chem.*, 1986, **261**, 4239.
109. B. P. Wallner, R. J. Mattaliano, C. Hession, R. L. Cate, R. Tizard, L. K. Sinclair, C. Foeller, E. P. Chow, J. L. Browning, K. L. Ramachandran and R. B. Pepinsky, *Nature (London)*, 1986, **320**, 77.
110. K. S. Huang, B. P. Wallner, R. J. Mattaliano, R. Tizard, C. Burne, A. Frey, C. Hession, P. McGray, L. K. Sinclair, E. P. Chow, J. L. Browning, K. L. Ramachandran, J. Tang, J. E. Smart and R. B. Pepinsky, *Cell*, 1986, **46**, 191.
111. C. J. M. Saris, B. F. Tack, T. Kristensen, J. R. Glenney, Jr. and T. Hunter, *Cell*, 1986, **46**, 201.
112. R. H. Kretsinger and C. E. Kreuz, *Nature (London)*, 1986, **320**, 573.
113. M. J. Geisow, K. Fritsche, J. M. Hexham, B. Dash and T. Johnson, *Nature (London)*, 1986, **320**, 636.
114. K. Weber and N. Johnsson, *FEBS Lett.*, 1986, **203**, 95.
115. M. J. Geisow, *FEBS Lett.*, 1986, **203**, 99.
116. T. Südhof, C. A. Slaughter, I. Leznicki, P. Barjon and G. A. Reynolds, *Proc. Natl. Acad. Sci. USA*, 1988, **85**, 664.
117. J. R. Glenney, Jr., *Bio Essays*, 1987, **7**, 173.
118. J. Fauvel, P. Vicendo, V. Roques, J. Ragab-Thomas, C. Granier, I. Vilgrain, E. Chambaz, H. Rochat, H. Chap and L. Douste-Blazy, *FEBS Lett.*, 1987, **221**, 397.
119. J. Fauvel, J. P. Salles, V. Roques, H. Chap, H. Rochat and L. Douste-Blazy, *FEBS Lett.*, 1987, **216**, 45.
120. F. F. Davidson, E. A. Dennis, M. Powell and J. R. Glenney, Jr., *J. Biol. Chem.*, 1987, **262**, 1698.

121. A. J. Aarsman, G. Mijnbeek, H. van den Bosch, B. Rothhut, B. Prieur, C. Comera, L. Jordan and F. Russo-Marie, *FEBS Lett.*, 1987, **219**, 176.
122. H. T. Haigler, D. D. Schlaepfer and W. H. Burgess, *J. Biol. Chem.*, 1987, **262**, 6921.
123. D. D. Schlaepfer and H. T. Haigler, *J. Biol. Chem.*, 1987, **262**, 6931.
124. J. N. Wood, P. R. Coote and J. Rhodes, *FEBS Lett.*, 1984, **174**, 143.
125. G. Cirino, R. J. Flower, J. L. Browning, L. K. Sinclair and R. B. Pepinsky, *Nature (London)*, 1987, **328**, 270.
126. G. Cirino and R. J. Flower, *Prostaglandins*, 1987, **34**, 59.
127. T. J. Nevalainen and O. S. Evilampi, *J. Biochem.*, 1984, **96**, 1303.
128. M. Miwa, I. Kubota, T. Ichihashi, H. Motojima and M. Matsumoto, *J. Biochem.*, 1984, **96**, 761.
129. M. Miwa, T. Ichihashi, H. Motojima, I. Onodera-Kubota and M. Matsumoto, *J. Biochem.*, 1985, **98**, 157.
130. T. Wilson, G. C. Liggins, G. P. Aimer and S. J. M. Skinner, *Biochem. Biophys. Res. Commun.*, 1985, **131**, 22.
131. G. E. Rice, M. H. Wong and G. D. Thorburn, *Prostaglandins*, 1987, **34**, 563.
132. H. Meijer, M. J. M. Meddens, R. Dijkman, A. J. Slotboom and G. H. de Haas, *J. Biol. Chem.*, 1978, **253**, 8564.
133. H. Tojo, T. Ono, S. Kuramitsu, H. Kagamiyama and M. Okamoto, *J. Biol. Chem.*, 1988, **263**, 5724.
134. T. Ono, H. Tojo, S. Kuramitsu, H. Kagamiyama and M. Okamoto, *J. Biol. Chem.*, 1988, **263**, 5732.
135. A. Khalil, J. Faroogui and A. M. Scanu, *Biochim. Biophys. Acta*, 1986, **878**, 127.
136. R. S. Jacobs, P. Culver, R. Langdon and T. O'Brien, *Tetrahedron*, 1985, **41**, 981.
137. D. Lombardo and E. A. Dennis, *J. Biol. Chem.*, 1985, **260**, 7234.
138. R. A. Deems, D. Lombardo, B. P. Morgan, E. D. Mihelich and E. A. Dennis, *Biochim. Biophys. Acta*, 1987, **917**, 258.
139. E. A. Dennis, *Drug Dev. Res.*, 1987, **10**, 205.
140. A. N. Makheja and J. M. Bailey, *Prostaglandins Leukotrienes Med.*, 1982, **8**, 653.
141. J. K. Thakkar, N. Sperelakis, D. Pang and R. C. Franson, *Biochim. Biophys. Acta*, 1983, **750**, 134.
142. M. K. Jain and D. V. Jahagerdar, *Biochim. Biophys. Acta*, 1985, **814**, 319.
143. A. Bordia, *Artherosclerosis*, 1978, **30**, 355.
144. K. C. Srivastava, *Prostaglandins Leukotrienes Med.*, 1986, **22**, 313.
145. M. K. Jain and R. Apitz-Castro, *Trends Biochem. Sci.*, 1987, **12**, 252.
146. L. R. Ballou and W. Y. Cheung, *Proc. Natl. Acad. Sci. USA*, 1985, **92**, 371.
147. J. N. Wood, P. R. Coote, J. Salmon and J. Rhodes, *FEBS Lett.*, 1985, **189**, 202.
148. F. Märki and R. Franson, *Biochim. Biophys. Acta*, 1986, **879**, 149.
149. M. Reers and D. R. Pfeiffer, *Biochemistry*, 1987, **26**, 8038.
150. P. P. M. Bonsen, G. H. de Haas, W. A. Pieterson and L. L. M. van Deenen, *Biochim. Biophys. Acta*, 1972, **270**, 364.
151. F. F. Davidson, J. Hajdu and E. A. Dennis, *Biochem. Biophys. Res. Commun.*, 1986, **137**, 587.
152. M. G. van Oort, Thesis, University of Utrecht, 1988.
153. M. H. Gelb, *J. Am. Chem. Soc.*, 1986, **108**, 3146.
154. W. Yuan, R. J. Berman and M. H. Gelb, *J. Am. Chem. Soc.*, 1987, **109**, 8071.
155. D. P. Wallach and V. J. R. Brown, *Biochem. Pharmacol.*, 1981, **30**, 1315.
156. C. A. Dise, J. W. Burch and D. B. P. Goodman, *J. Biol. Chem.*, 1982, **257**, 4701.
157. S. L. Hofmann, S. M. Prescott and P. W. Majerus, *Arch. Biochem. Biophys.*, 1982, **215**, 237.
158. L. Kaplan, J. Weiss and P. Elsbach, *Proc. Natl. Acad. Sci. USA*, 1978, **75**, 2955.
159. R. C. Franson, D. Eisen, R. Jesse and C. Lanni, *Biochem. J.*, 1980, **186**, 633.
160. J. J. Volwerk, W. A. Pieterson and G. H. de Haas, *Biochemistry*, 1974, **13**, 1446.
161. E. K. Kyger and R. C. Franson, *Biochim. Biophys. Acta*, 1984, **794**, 96.
162. A. A. Abdel-Latif and J. P. Smith, *Biochim. Biophys. Acta*, 1982, **711**, 478.
163. H. Tojo, T. Ono and M. Okamoto, *Biochem. Biophys. Res. Commun.*, 1988, **151**, 1188.
164. A. J. Pierik, J. G. Nijssen, A. J. Aarsman and H. van den Bosch, *Biochim. Biophys. Acta*, 1988, **962**, 345.

8.7

Protein Kinases

KENNETH J. MURRAY and BRIAN H. WARRINGTON

Smith Kline & French Research Ltd, Welwyn, UK

8.7.1	INTRODUCTION	531
8.7.1.1	*General Properties of Protein Kinases*	531
8.7.1.2	*Protein Kinases as Drug Targets*	532
8.7.1.3	*Modulation of Kinase Activity*	532
8.7.2	CYCLIC AMP DEPENDENT PROTEIN KINASE	534
8.7.2.1	*Enzyme Characteristics*	534
8.7.2.2	*Agonists and Antagonists*	535
8.7.2.3	*Inhibitors*	538
8.7.2.4	*Therapeutic Utility*	539
8.7.3	CYCLIC GMP DEPENDENT PROTEIN KINASE	541
8.7.3.1	*Enzyme Characteristics*	541
8.7.3.2	*Agonists and Antagonists*	541
8.7.3.3	*Inhibitors*	542
8.7.3.4	*Therapeutic Utility*	542
8.7.4	PROTEIN KINASE C	543
8.7.4.1	*Enzyme Characteristics*	543
8.7.4.2	*Agonists and Antagonists*	543
8.7.4.3	*Inhibitors*	545
8.7.4.4	*Therapeutic Utility*	545
8.7.5	MYOSIN LIGHT CHAIN KINASE	546
8.7.5.1	*Enzyme Characteristics and Therapeutic Utility*	546
8.7.5.2	*Antagonists*	546
8.7.5.3	*Inhibitors*	547
8.7.6	TYROSINE KINASES	548
8.7.6.1	*Enzyme Characteristics and Therapeutic Utility*	548
8.7.6.2	*Inhibitors*	548
8.7.7	REFERENCES	549

8.7.1 INTRODUCTION

8.7.1.1 General Properties of Protein Kinases

The protein kinases are a group of related enzymes which, in response to a signal, catalyze the transfer of the terminal phosphoryl group of adenosine triphosphate (ATP) to the hydroxyl group of specific hydroxy amino acids in proteins. Their biochemical properties have been the subject of recent reviews.[1,2]

Protein kinases can be classified into two groups: those which phosphorylate serine or threonine residues, and those for which tyrosine is the specific acceptor. The kinases in each group show individual activation requirements and substrate protein specificities. Phosphorylation generally modulates the function of the substrate protein by altering, for example, the kinetic properties of enzymes or the metal-ion-binding properties of nonenzymes. Protein kinases themselves can also be phosphorylated, either by an intramolecular reaction (autophosphorylation) or by another protein kinase. These processes may play a part in their physiological regulation. The overall change in the function of a cell depends on the particular kinases activated and the proteins phosphorylated.

8.7.1.2 Protein Kinases as Drug Targets

Although receptor agonists and antagonists are known to act indirectly on protein kinases by means of intracellular second messengers (see Volume 3), very few drugs with a direct effect on a protein kinase as their established mode of action have been clinically evaluated. More than 80 protein kinases are known.[3] Not all are suitable targets for novel therapeutic agents, but those playing an identifiable role in the regulation of cell function have the greatest therapeutic potential. The protein kinases chosen for discussion in later sections of this chapter, *i.e.* cAMP dependent protein kinase (cA-PrK), cGMP dependent protein kinase (cG-PrK), protein kinase C (PrK-C), myosin light chain kinase (MLCK), and the tyrosine kinases (TyrK) have been selected because their known biology permits reasonable prediction of the therapeutic benefits obtainable from their modulation and there is enough chemical information to consider rational drug design. Other factors, such as the fact that some tyrosine kinases are oncogene products, have also been considered.

Because the major, and sometimes only, role of many second messengers is to modulate protein kinase activity, a drug acting directly on a kinase will often have a pharmacological effect similar to agents which modulate the intracellular levels of the regulatory second messenger of the kinase (*e.g.* receptor agonists, antagonists or metabolic inhibitors).[4,5] However, direct activation of the kinase could have advantages where a protein kinase is activated by several classes of receptor, or the receptor is subject to severe down-regulation, or where responses in more than one tissue are required.

8.7.1.3 Modulation of Kinase Activity

Protein kinases generally contain separate regulatory and catalytic domains. These can even be distinct proteins as in cA-PrK. The regulatory domain is unique to a kinase as it is here that the second messenger or signal molecule is recognized, whereas the catalytic domains are well conserved and every kinase has a recognizable ATP binding site.[3] Either domain can be a target for drug action and, for clarity, the terms agonist and antagonist are used for agents which act at the regulatory domain to respectively increase or decrease phosphotransferase activity, whilst agents which act at the catalytic site to reduce activity are termed inhibitors.

The choice of the regulatory site of a kinase as a target for drug design could show several advantages. Potentially, either agonist or antagonist actions are available and a drug designed on the basis of its natural ligand could show high selectivity for this specialized binding site. However, where the signal molecule has many other actions, as in the case of the Ca^{2+}, calmodulin dependent enzymes (*e.g.* MLCK), the selectivity achievable could be limited.

Therapeutically useful inhibition of catalytic sites has proved to be difficult to achieve. Close analogues of ATP show similar structure–activity relationships in a number of kinases,[6] although minor differences in the substituent[7] and spatial[8] requirements of substrate nucleotides have been demonstrated. Inhibitors which compete with ATP alone are therefore unlikely to show significant ability to distinguish the ATP-handling sites of kinases and the possibility of interaction with other ATP-utilizing proteins[9] must also be considered. An inhibitor required to compete with physiological (*ca.* 5 mM) levels of ATP will also need to show either very high affinity or a substantial noncompetitive element. Compounds designed to inhibit the binding of the substrate protein at the catalytic site are potentially more selective. Studies[10] with synthetic peptides suggest that recognition of target proteins by individual protein kinases depends largely on the sequence of amino acid residues around the phosphorylation site and that each protein kinase has its own recognition sequence. However, substrates that do not conform to these sequences are also known, indicating that the three-dimensional structure can also be important. Replacement of the target hydroxy amino acid in peptide substrates by a nonphosphorylating residue (usually alanine) can yield inhibitors, but these lack the cell penetrant properties necessary for use in intact tissue.

A large number of compounds of diverse structure have been reported to inhibit the activity of the kinases considered in this chapter and it is impossible to describe each in detail in a work of this length. Emphasis will be given to compounds showing high potency and evidence of some ability to distinguish individual kinases. Groups of compounds which have been systematically studied, such as the naphthalene- and isoquinoline-sulfonamides (Tables 1 and 2) described by Hidaka and coworkers[11] will also receive attention. An attempt will also be made to classify compounds according to their site of action and the natural ligand with which they compete, but this is not always clear-cut. For example, the naphthalenesulfonamide W-7 (Table 1), an ATP competitive inhibitor at the active site of several protein kinases,[12] also potently antagonizes the actions of CaM (the protein, calmodulin) at the regulatory domain of CaM dependent protein kinases.[13] An additional problem, seen particularly with ATP competitive inhibitors, is that an agent often shows little or no structural relationship to the natural ligand it apparently competes with. True competition at the exact site used by the natural ligand may therefore not always be occurring and binding at other positions in the reaction pathway could cause the inhibitor to appear noncompetitive with respect to the phosphate acceptor.[14]

Table 1 Inhibition of Protein Kinases by Naphthalenesulfonamides (1) and (2)

| | Name | \multicolumn{5}{c|}{K_i (µM)} | |
		cA-PrK	cG-PrK	PrK-C	MLCK[a]	Ref.
(1a)	A-3	4.3	3.8	47	7.0	15
(1b)	—	23	15	92	18	15
(1c)	—	24	44	120	25	15
(1d)	—	110	55	230	60	15
(1e)	W-7	170	130	340	110	15
(2a)	HH-3209	—	—	—	48	16
(2b)	HH-3709				13	16
(2c)	ML-8	—	—	—	7.0	16
(2d)	HH-3909	—	—	—	3.0	16
(2e)	ML-9	32	—	54	3.8	16
(2f)	HH-3509	—	—	—	1.7	16
(2g)	ML-7	—	—	—	0.3	16

[a] Trypsin treated.

Table 2 Inhibition of Protein Kinases by Isoquinolinesulfonamides (3), (4) and (5)

| | Name | \multicolumn{4}{c|}{K_i (IC_{50}) (µM)} | Ref. | *Mesenteric artery relaxation*[19,20] ED_{50} (µM) |
		cA-PrK	cG-PrK	PrK-C	MLCK		
(3a)	H-9	1.9	0.87	18	70	14	5.0
(3b)	—	4.3	1.7	—	—	14	7.0
(3c)	—	5.4	2.1	—	—	14	11
(3d)	—	12	16	—	—	14	10
(3e)	H-8	1.2	0.48	15	68	14	12
(3f)	HA-1004	2.3	1.3	40	150	14	1.0
		(4.0)	(3.0)	—	(>100)	17	—
(3g)	—	(22)	—	—	—	17	2.0
(3h)	—	(10)	—	—	—	17	2.0
(3i)	—	(40)	—	—	—	17	4.0
(3j)	—	(>3000)	—	—	—	17	20
(4a)	HA-100	—	—	6.5	61	18	0.6
		(8.0)	(4.0)	(12)	(240)	18	—
(4b)	HA-156	—	—	7.2	7.3	18	—
		(8.0)	(6.0)	(11)	(22)	18	—
(4c)	H-7	3.0	5.8	6.0	97	14	5.0
(4d)	—	1.8	7.7	75	170	14	1.5
(4e)	—	2.9	8.4	45	100	14	9.5
(4f)	—	4.7	1.7	29	120	14	13
(4g)	—	(1000)	—	—	—	17	—
(5a)	HA-142	—	—	96	70	18	—
		(>500)	(>500)	(>500)	(270)	18	—
(5b)	HA-140	—	—	68.0	8.7	18	—
		(450)	(300)	(180)	(19)	18	—

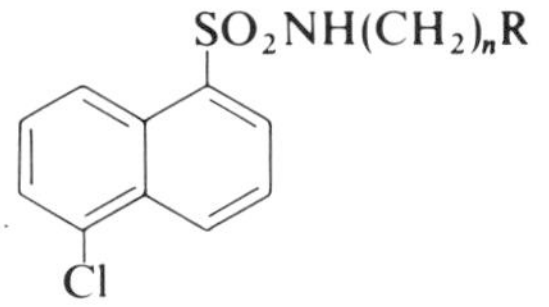

		n	R	
(1)	a:	2	NH$_2$	(A-3)
	b:	3	NH$_2$	
	c:	4	NH$_2$	
	d:	5	NH$_2$	
	e:	6	NH$_2$	(W-7)
	f:	6	Ph	(SC-9)
	g:	6	Me	(SC-10)

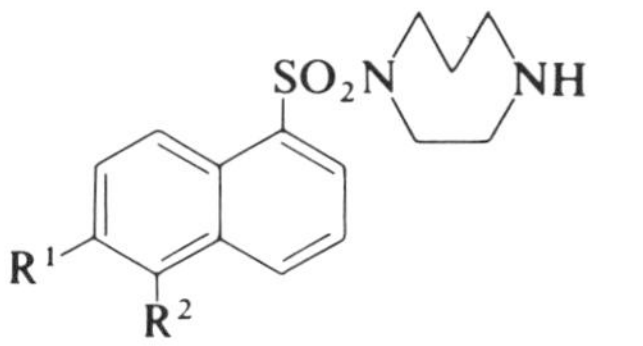

		R^1	R^2	
(2)	a:	H	H	(HH-3209)
	b:	Cl	H	(HH-3709)
	c:	Br	H	(ML-8)
	d:	I	H	(HH-3909)
	e:	H	Cl	(ML-9)
	f:	H	Br	(HH-3509)
	g:	H	I	(ML-7)

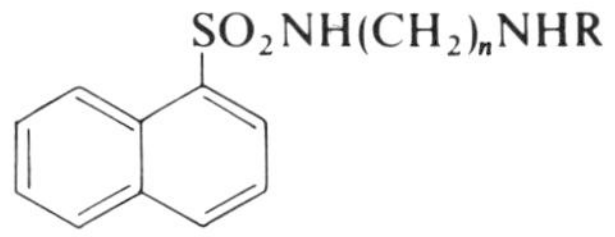

		n	R	
(3)	a:	2	H	(H-9)
	b:	3	H	
	c:	4	H	
	d:	6	H	
	e:	2	Me	(H-8)
	f:	2	C(=NH)NH$_2$	(HA-1004)
	g:	3	C(=NH)NH$_2$	
	h:	4	C(=NH)NH$_2$	
	i:	6	C(=NH)NH$_2$	
	j:	2	2-(4,5-dihydroimidazolyl)	

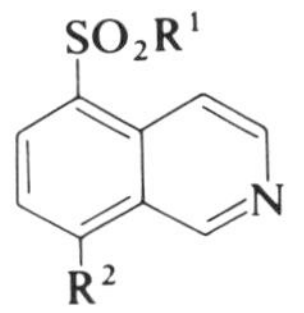

		R^1	R^2	
(4)	a:	1-piperazinyl	H	(HA-100)
	b:	1-piperazinyl	Cl	(HA-156)
	c:	1-(2-Me-1-piperazinyl)	H	(H7)
	d:	1-(2,3-Me$_2$-1-piperazinyl)	H	
	e:	1-(3,5-Me$_2$-1-piperazinyl)	H'	
	f:	NMe$_2$	H	
	g:	C(=NH)NH$_2$	H	

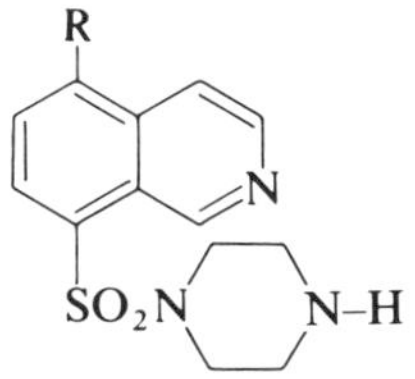

(5a) R = H (HA-142)

(5b) R = Cl (HA-140)

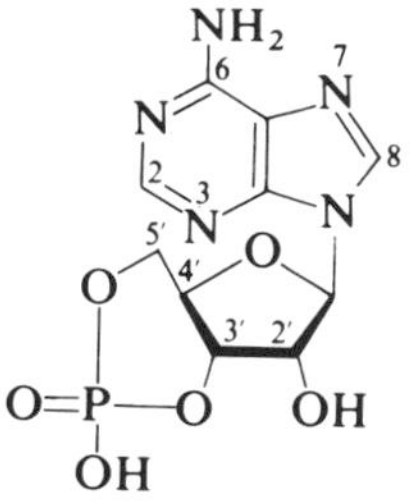

(6)

8.7.2 CYCLIC AMP DEPENDENT PROTEIN KINASE

8.7.2.1 Enzyme Characteristics

The biochemistry and nucleotide-mediated modulation of cAMP dependent protein kinase (cA-PrK) have been reviewed in several places.[1,21,22,23,24] cA-PrK is distributed widely at different concentrations in mammalian tissues and has a large number of known substrates, most of which contain the phosphorylation site sequences Arg-Arg-X-Ser/Thr-X or Lys-Arg-X-X-Ser/Thr-X, where X can be a variety of different residues.[21,25] The enzyme occurs as two microheterogeneous isoenzymes, designated type I and type II. Both isoenzymes contain a dimeric regulatory unit (R$_2$) and two catalytic subunits (C), but are distinguished by the different properties of their R subunits. These include molecular weight (R^I = 49 kDa; R^{II} = 55 kDa), order of elution from DEAE-cellulose and capacity for autophosphorylation. In addition, the isoenzymes interact differently with cyclic nucleotides and ATP. Tissues usually contain both types, although the exact proportions are species dependent. Two distinct genes for each R subunit have been reported.[26] Recently, two forms of the C subunit have been discovered,[27] but their physiological significance is not known. Receptor agonists and PDE inhibitors that increase intracellular levels of cAMP (6) indirectly cause activation of cA-PrK in accordance with equation (1).

$$R_2C_2 \text{ (inactive)} + 4cAMP \rightleftharpoons R_2cAMP_4 + 2C \text{ (active)} \tag{1}$$

8.7.2.2 Agonists and Antagonists

Photoaffinity labelling,[28] using 8-N$_3$-[^{32}P]cAMP, has shown that R^I and R^{II} each contain two contiguous cyclic nucleotide binding domains close to the carboxy terminus. Although many features are conserved at these sites, each has a unique amino acid sequence.[29] cAMP is bound with similar affinity at all four types of site ($K_d \approx 10^{-8}$ M at physiological kinase concentrations of 0.2–0.7 μM), but the sites show differences in kinetic behaviour.[22] Biphasic time courses for the displacement of bound [^{3}H]cAMP by unlabelled cAMP, show that each isoenzyme contains a 'labile'[30] site, with an off rate of about 0.25 min^{-1}, and a 'stable'[30] site, with a slower off rate of about 0.025–0.05 min^{-1}. If a cAMP derivative is present during the binding of [^{3}H]cAMP, the isoenzymes can show significantly different time courses during subsequent displacement of [^{3}H]cAMP by unlabelled cAMP. These arise from the different affinities, relative to cAMP, shown by the derivative at the four structurally different types of cyclic nucleotide binding sites.[23]

At suitably low concentrations, some derivatives can show selectivity for only one type of binding site. Studies with these site selective cAMP derivatives used singly, and in pairs, have shown that occupation of both labile and stable regulatory binding sites of an isoenzyme is required for activation.[31]

In the presence of the catalytic subunit, the binding of a cAMP derivative at one type of site enhances affinity at the other,[32] the positive cooperativity being most pronounced when a pair of derivatives with a high degree of complementary selectivity for the two types of site is used.[33] Tryptic digestion of bovine heart holoenzyme yields a dimeric enzyme (R'C) which shows identical properties to the parent tetramer (R$_2$C$_2$), suggesting that occupation of labile and stable sites in the same chain is necessary for activation.[34,35] These observations have suggested a model for binding in which interaction of a ligand at an exposed stable site induces a short-lived change in cA-PrK conformation and exposes the intrachain labile site. If further binding occurs at the labile site, then the ligand interaction at the stable site is stabilized by loss of the catalytic unit, otherwise the ligand at the stable site dissociates. Additional support for this model is the finding[32,36] that binding at the labile site of R^I is inhibited by the presence of the catalytic subunit and further inhibited by the additional presence of ATP. Only the type I holoenzyme binds ATP with high affinity. The ATP adenine moiety binds at a site in the catalytic subunit, whilst the ribose triphosphate group is bound to residues in the hinge region of the regulatory subunit,[21] which has the effect of stabilizing the holoenzyme. cAMP and other full agonists have lower affinity for the ATP-bound form than for the unbound form, whereas the affinity of antagonists is unaffected,[37] suggesting that agonists and antagonists prefer to bind to the active and inactive forms respectively of the regulatory protein.

The relative activator potencies of many cAMP analogues are known,[38] but there is a paucity of binding site affinity data, particularly for the type II isoenzyme. Consequently, detailed mechanistic studies which consider the ligand properties of cA-PrK-stimulating analogues at the different regulatory sites of isoenzymes and the causal relationship between site occupation and activation are comparatively rare.

The relative activator potencies (K'_a) and affinities (K'_i) of a range of singly modified cAMP analogues are given in Table 3. In most cases, K'_a reflects similar, and approximately equal, orders of relative affinity (K'_i) at both types of binding site, and at high concentration these analogues are full agonists. The principal exceptions are (R$_P$)-cAMPS (7a), which displays the properties of an antagonist in that it evenly displaces [^{3}H]cAMP binding at both types of binding site but, even at high concentration, fails to elicit activation, and (R$_P$)-cAMPNMe$_2$ (8a), which is probably a partial agonist.[37]

Other single modifications of the ribose cyclophosphate region of cAMP lead to large reductions in K'_a and K'_i, often of several orders of magnitude, whilst the changes accompanying single isosteric changes in cAMP adenine structure do not exceed an order of magnitude, indicating that the ribose cyclophosphate moiety rather than adenine makes the major contribution to cAMP binding. The similarity of the dissociation constants (about 10^{-8} μM) found for cAMP at all types of binding

(7a) B = adenosine [(R$_P$)-cAMPS]
(7b) B = guanine [(R$_P$)-cGMPS]

(8a) B = adenosine [(R$_P$)-cAMPNMe$_2$]
(8b) B = guanine [(R$_P$)-cGMPNMe$_2$]

(9a) B = adenosine [(S$_P$)-cAMPS]
(9b) B = guanine [(S$_P$)-cGMPS]

(10a) B = adenosine [(S$_P$)-cAMPNMe$_2$]
(10b) B = guanine [(S$_P$)-cGMPNMe$_2$]

Table 3 Relative Activation and Affinity of Singly Modified Analogues of cAMP (6)

Analogue	K_i' (Type I)[a]		K_a' [b]		Maximal activation type	Ref.
	Labile	Stable	(Type I)	(Type II)		
cAMP (6)	1.0	1.0	1.0	1.0	1.0	—
2'-Deoxy-cAMP	0.000313	0.000313	0.000784	0.0048	1.15	37, 39
2'-O-Methyl-cAMP	—	—	0.0023	0.0017	—	37
Ara-cAMP	—	—	0.0032	0.0029	—	37
3'-Thio-cAMP	—	—	0.01	0.0048	—	37
3'-Amido-cAMP	0.000724	0.000108	0.00096	0.0043	0.9	37, 39
3'-Methylene-cAMP	—	—	< 10^{-4}	< 10^{-4}	—	39
Xylo-cAMP	—	—	< 10^{-4}	< 10^{-4}	—	39
4'-Thio-cAMP	—	—	0.65	0.43	—	39
4'-Methylene-cAMP	—	—	0.032	0.017	—	39
5'-Thio-cAMP	—	—	0.10	0.66	—	39
5'-Amido-cAMP	0.00122	0.00104	0.023	0.0016	0.95	37, 39
5'-Methylene-cAMP	—	—	0.00087	0.0057	—	39
Adenosine 3',5'- cyclic sulfate	—	—	< 10^{-4}	< 10^{-4}	—	39
cAMP ethyl ester	< 10^{-5}	(Total)	< 10^{-4}	< 10^{-4}	—	40, 39
(R$_P$)-cAMPS (7a)	0.00174	0.00231	< 10^{-4}	—	< 0.03	37
(S$_P$)-cAMPS (9a)	0.00234	0.0253	0.0297	—	1.0	37
(R$_P$)-cAMPNMe$_2$ (8a)	< 10^{-4}	0.00201	10^{-4}	—	> 0.4	37
(S$_P$)-cAMPNMe$_2$ (10a)	0.00651	0.0123	0.0134	—	0.75	37
1-Deaza-cAMP	0.23	1.8	1.4	0.98	—	41, 39
3-Deaza-cAMP	0.31	0.191	0.13	—	1.0	33
7-Deaza-cAMP	1.3	0.88	0.83	0.90	—	42, 38
cNMP	0.73	0.34	0.18	0.25	—	42, 38
2-Aza-cAMP	0.708	1.0	0.85	0.16	1.0	33, 38
8-Aza-cAMP	0.818	0.269	0.95	0.63	1.0	33, 38

[a] K_i(cAMP): labile = 25 nM, stable = 53 nM.[42] [b] K_a(cAMP) = 27–49 nM.[39]

site suggest that it is probably bound by the ribose cyclophosphate moiety in a uniform manner at these sites.[21]

The reduced affinity shown by the 2'-altered cAMP analogues in Table 3 provides evidence for H-donating, directional binding by the acidic proton of the 2'-hydroxy group of cAMP, but the nature of binding in the cyclophosphate ring locale is less obvious. Various tentative models for binding have been suggested[37] to explain (i) the poor affinity of the 3'- and 5'-altered species; (ii) the similar levels of affinity and activation shown by formally charged and uncharged species, such as (S$_P$)-cAMPS (9a) and (S$_P$)-cAMPNMe$_2$ (10a); and (iii) the apparent specific requirement for phosphorus in the ring as suggested by the negligible activation and affinity shown by adenosine 3',5'-cyclic sulfate. These have included H-acceptor interactions by the 3'- and 5'-oxygens of cAMP (6), and the covalent binding of phosphorus to the receptor in a pentavalent, trigonal bipyramidal configuration unavailable to the cyclic sulfate. However, much evidence points towards a charged interaction made by an exocyclic phosphate oxygen of cAMP, particularly as the phosphoramidates (**8a**, **10a**), although lacking a formal charge, will show a relatively high charge density on exocyclic oxygen. In addition, arginine and glutamic acid residues which can interact with the phosphate charge and the 2'-hydroxy group of cAMP respectively are conserved in all cyclic nucleotide binding sites.[29]

The cyclophosphate ring probably also contains structural features which prompt or trap the conformational changes of the regulatory subunit associated with activation. 1',2'-Dideoxyribose 3',5'-cyclic monophosphate shows weak agonist properties[33] and the agonist or antagonist properties shown by the isomer pairs (**7a**, **9a** and **8a**, **10a**) show consistency with the relative positions of their centres of charge as charge in the phosphorothioates (**8a**, **10a**) probably resides preferentially on sulfur.[43] However, the recent finding that cAMP phosphorodithioate (cAMPS$_2$), which has a K_i value of 4 µM, similar to that of (R$_P$)-cAMPS ($K_i = 8$ µM), also shows antagonism had led to the suggestion that steric factors may also play a role in antagonist properties.[44]

The high level of regiospecific binding in the ribose cyclophosphate region of cAMP probably ensures its recognition by R subunit receptors in the presence of high physiological concentrations of ATP and, notably, disruption of this critical structure by PDE is the primary physiological mechanism of cAMP deactivation. Probably for these reasons, most of the synthetic effort towards cAMP-based drugs has centred on the substitutions at the cAMP adenine moiety.

The relative activation (K_a') and affinity (K_i') of examples of the wide range of cAMP derivatives known are given in Table 4. Although the binding interactions of the cAMP adenine moiety itself are weak and nonspecific, substitutions at the purine ring can cause wide variations in affinity and site selectivity. As far as is known, all adenine-modified cAMP derivatives are full agonists. In the type I isoenzyme it has been shown that a logarithmic expression of the relative activation shown by 80 derivatives correlates more closely with a logarithmic expression of their mean subsite affinities, $(K_{i(labile)}' K_{i(stable)}')^{1/2}$, than with expressions describing affinity at only one of the sites.[33] This agrees with the qualitative observation that the occupation of both subsites is necessary for activation.

The relationship between the nature of purine ring substitution and site selectivity has been described in detail in the literature,[36,42] but the exact manner in which substituents interact directly, or modify the conformation and interactions of the cAMP moiety, to enhance or reduce affinity at individual binding sites is not well understood. Broadly, the presence of a 6-substituent in cAMP

Table 4 Relative Activation and Affinity of Adenine-modified Derivatives of cAMP (**6**)

| Derivative | K_i' | | | | K_a' | | Ratio I/II[a] | Ref. |
| | Type I | | Type II | | | | | |
	Labile	Stable	Labile	Stable	Type I	Type II		
cAMP (**6**)	1.0	1.0	1.0	1.0	1.0	1.0	1.0	—
cGMP	0.005	0.012	0.0039	0.00092	0.012	0.007	2.6	41, 38
cIMP	0.13	0.027	0.099	0.0056	0.084	0.028	1.4	42, 41, 38
2-*n*-Butyl-cAMP	0.263	0.73	0.084	0.93	0.5	0.63	1.9	42, 41, 38
2-Chloro-cAMP	0.279	2.92	0.43	3.1	1.9	2.5	1.3	42, 41, 38
2-Dimethylamino-cAMP	0.007	0.029	—	—	0.029	0.12	0.64	42, 38
2-Phenyl-cAMP	0.056	0.065	—	—	0.074	0.12	1.0	42, 38
2-Phenyl-ε-cAMP (**11**)	—	—	0.035	3.1	0.046	0.62	0.13	42, 38
N^6-Benzoyl-cAMP	3.8	0.22	4.1	0.034	0.93	0.45	2.3	38
N^6-Benzyl-cAMP	0.79	0.21	2.1	0.72	0.36	0.72	0.67	38
N^6-(*t*-Butylcarbamoyl)-cAMP	0.48	0.068	16	0.13	0.79	1.4	0.81	38
N^6-(1-Methyl-2-phenylethyl)-cAMP	2.3	0.056	0.079	0.12	0.36	0.95	1.0	38
N^6-(1-Methyl-3-phenylpropyl)-cAMP	2.9	0.037	6.8	0.086	0.50	1.6	0.68	38
N^6-Phenyl-cAMP	18	0.48	40	0.44	1.8	2.4	0.93	38
8-Amino-cAMP	0.15	3.9	0.08	3.0	2.5	0.62	3.7	38
8-(6-Aminohexylamino)-cAMP	0.11	1.6	0.021	0.29	1.1	0.12	9.2	38
8-(5-Aminopentylamino)-cAMP	0.16	2.3	0.027	0.47	1.6	0.17	15.0	41, 38
8-Benzylthio-N^6-*n*-butyl-cAMP	1.78	0.722	—	—	1.5	2.9	0.78	42, 38
8-Bromo-cAMP	1.29	1.05	0.11	6.8	1.8	1.7	1.2	42, 41, 38
8-Bromo-2-*n*-butyl-cAMP	2.2	0.23	0.0084	5.0	0.63	0.59	1.9	38
2-Chloro-8-methylamino-cAMP	0.34	5.1	0.0095	0.72	0.63	1.2	3.9	38
8-Methylamino-cAMP	0.21	3.5	0.030	2.0	1.9	0.4	5.1	38
8-Piperidino-cAMP	2.3	0.065	0.046	3.2	0.76	0.80	1.4	38
8-(2-Propyl)thio-cAMP	2.1	0.9	0.016	3.9	1.8	0.32	6.6	38

[a] Average for a range of tissues.[38]

directs binding to the labile sites, whilst electron-donating substituents at the 2-position or electron-withdrawing substituents at the 8-position direct binding to the stable sites, suggesting that the interactions responsible for recognition of the base moieties are similar. However, these effects can differ in degree between the two isoenzymes. For example, cAMP derivatives with an 8-substituent connected through nitrogen show a greater preference for the stable site of R^I than that of R^{II}, whereas those connected through sulfur show the converse preference.[36] These and other differences allow some cAMP derivatives to distinguish isoenzyme types. Derivatives with electron-withdrawing 2-substituents, for example, are better than cAMP as activators of type I, but are less active as activators of type II.[45] The degree of isoenzyme discrimination shown by present analogues is modest, the greatest levels of selectivity for types I and II being shown by 2-phenyl-ε-cAMP (11) and 8-(5-aminopentylamino)-cAMP respectively (Table 4).[38] Many workers have therefore preferred the higher levels of isoenzyme discrimination achieved by the synergistic interaction of a pair of derivatives with complementary selectivity for the binding sites of a target isoenzyme. Combinations of 8-piperidino-cAMP with 2-chloro-8-methylamino-cAMP and with N^6-phenyl-cAMP have been used to stimulate types I and II respectively,[38] although other isoenzyme selective combinations of the compounds shown in Table 4 are possible. Obviously, concentrations must be gauged to ensure that neither compound saturates both sites. As the degree of stimulation of binding and activation shown by pairs of cAMP derivatives correlates well with the difference in their mean site selectivities, the structural factors influencing the stimulatory effect are largely the same as those which influence site selectivity.[31,33]

(11)

(12a) R^1 = H, R^2 = NH$_2$ (amrinone)
(12b) R^1 = Me, R^2 = CN (milrinone)

8.7.2.3 Inhibitors

The determinants of cA-PrK substrate binding have been studied and several mechanisms for the phosphotransferase process have been proposed.[1]

The preferred metal–nucleotide substrate is Mg–ATP, which binds to the enzyme in an *anti* conformation.[1,7] ATP competitive inhibition of cA-PrK is shown by the naphthalene- and iso-quinoline-sulfonamides shown in Tables 1 and 2. The presence of a basic moiety, protonated at physiological pH, is important for potent inhibition of cA-PrK and activity is maximal in both series of compounds when this is at a distance of two carbon atoms from the sulfonamide function (*cf.* 1a–e, 3a–d and 3f–i). A relationship between the pK_a of amines at this position and inhibitory potency has been shown.[14] As the naphthalenesulfonamides show the additional property of CaM antagonism, the isoquinoline series can be considered to be selective for non-CaM dependent kinases, but the degree of selectivity obtained is not great. HA-1004 (3f; Table 2) shows the greatest selectivity towards the cyclic nucleotide dependent kinases, but is unable to distinguish cA-PrK and cG-PrK.

The metabolites of *Nocardiopsis* spp.[46,47] and the related compound staurosporine[48] shown in Table 5 are also potent inhibitors of cA-PrK and other kinases. Only KT-5270, however, shows a high degree of specificity for cA-PrK. Lower levels of inhibition have been reported[49] for amrinone (12a) and milrinone (12b; K_i = *ca.* 200 and 842 μM, respectively) and amrinone has been shown not to inhibit PrK-C.

The structural determinants for substrate protein selectivity have been studied using short synthetic peptide substrates. Primary structure is of great importance and many peptides containing the sequence Arg-Arg-X-Ser-X have kinetic properties comparable to intact substrate proteins. However, some are poor substrates, showing that conformation is also important and a study of a series of six- and seven-residue peptides has shown that it is likely that an extended coil conformation is effective.[50] Whilst in many cases inhibitors can be derived by replacing serine in the substrate

Table 5 Inhibition of Protein Kinases by K-252a and Related Compounds [46,47,48]

| Name | Structure | | | | | K_i (nM) | | | | |
	n	R^1	R^2	R^3	cA-PrK	cG-PrK	PrK-C	MLCK	Ref.
K-252a	0	CO_2Me	H	H	18.0	20.0	25.0	20.0	46, 47
K-252b	0	CO_2H	H	H	90.0	100	20.0	—	46
KT-5720	0	$CO_2(CH_2)_5Me$	H	H	60.0	> 2000	> 2000	—	46
KT-5822	0	CO_2Me	Me	H	37.4	2.4	79.0	—	46
Staurosporine	1	H	Me	NHMe	*ca.* 1	—	2.7	—	48

recognition sequence of these peptides by a nonphosphorylating residue, these usually also inhibit cG-PrK.[51,52,53] However, another approach is to base inhibitory peptide design on portions of the sequence of the naturally found specific inhibitor protein, usually termed PKI.[54] A 20-residue peptide, IP_{20}-amide (Thr-Thr-Tyr-Ala-Asp-Phe-Ile-Ala-Ser-Gly-Arg-Thr-Gly-Arg-Arg-Asn-Ala-Ile-His-His-Asp-NH$_2$) shows an affinity (K_i = 2.3 nM) for cA-PrK approaching that of intact PKI (K_i = *ca.* 0.1 nM) and does not inhibit other protein kinases. Comparison with other inhibitory peptides suggests that critical structural features for high affinity are an Arg-Arg basic subsite, typical of the -Arg-Arg-X-Ser-Y- sequence frequently found in cA-PrK substrates, and an additional near proximal arginine, similar to the arginine cluster found in the autophosphorylation site of R^{II}. A further critical contribution from a residue towards the N-terminal region also appears to be necessary.[54]

8.7.2.4 Therapeutic Utility

A wide variety of hormones increase cAMP in a range of tissues[55] and cA-PrK is the only known receptor for cAMP in mammalian cells. The enzyme is therefore a crucial link in the mechanism whereby agents that increase intracellular concentrations of cAMP alter cell function by protein phosphorylation. Agonists of cA-PrK will therefore produce effects similar to those caused by increases in cAMP.

The actions of cA-PrK agonists have been studied mainly in the heart, where they may show advantages over therapies in current use. In heart, increases in cAMP cause a positive inotropic response by increasing both the rates of contraction and relaxation. At present, cardiac cAMP levels are increased by using β agonists to increase the rate of synthesis of cAMP or by PDE III inhibitors to decrease its rate of degradation.[56] However, both classes of agent are reported to be less effective in the failing myocardium than in the healthy heart.[57] This may be due to a decrease in the number or types of the β receptor in the diseased state,[58] diminishing the effectiveness of β agonists. The decreased rate of cAMP synthesis, in turn, makes PDE inhibition less effective.[57] In contrast, cA-PrK agonists are independent of receptors and PDE.

cAMP penetrates cells poorly and is hydrolyzed rapidly in the cell by cAMP phosphodiesterases but derivatives of cAMP with lipophilic adenine substituents, often presented in the form of prodrug esters, show improved cell penetration. By judicious selection of substituents the rate of hydrolysis and site or isoenzyme selectivity (which may be important—see below) can also be manipulated. Derivatives containing N^1,N^6-etheno- and large 8-substituents are particularly poor substrates for PDE, especially when a 2-substituent is also present.[59]

Table 6 Effect of 8-Substituted cAMP Derivatives in Isolated Guinea Pig Papillary Muscle

| | $Log\, EC_{50}$ | | $Log\, P$ | |
Derivative	Na salt	Benzyl ester	Na salt	Benzyl ester
8-(4-Chlorophenyl)thio-cAMP	− 4.40	− 5.66	− 0.98	+ 2.24
8-*t*-Butylthio-cAMP	− 4.00	− 5.38	− 1.05	+ 2.05
8-Benzylseleno-cAMP	− 3.18	− 4.92	− 1.18	+ 1.67
8-Benzylthio-cAMP	− 3.14	− 4.80	− 1.15	+ 1.92
8-Methylthio-cAMP	− 2.96	− 4.95	− 2.03	+ 1.62
8-Bromo-cAMP	− 2.82	− 4.89	− 2.27	+ 1.58

Several cAMP derivatives have been shown to be positive inotropes *in vitro* and *in vivo*, including N^6,2′-*O*-dibutyryl-cAMP (DB-cAMP), 8-benzylthio-N^6-butyl-cAMP (BTB-cAMP) and the series of 8-substituted-cAMP derivatives[60] shown in Table 6. DB-cAMP, infused into man at 0.2 mg kg^{-1} min^{-1} reduced total system resistance and mean arterial blood pressure, and slightly increased heart rate and contractility.[61] At 3.7 mg kg^{-1} min^{-1} in anaesthetized dogs, BTB-cAMP increased dP/dt_{max} by 84% and improved mesenteric, renal and carotid arterial flow with no significant changes in mean arterial blood pressure or heart rate.[62,63] Both compounds caused release of insulin, which was considered to make them particularly suitable for the treatment of hemorrhagic shock. DB-cAMP was launched by Daiidii in 1984 as 'Actosin' for use during heart transplant surgery.

With compounds such as DB-cAMP and BTB-cAMP determination of the mechanism of these actions can be difficult. Although cAMP derivatives are generally more potent activators of cA-PrK than inhibitors of PDE, at the concentrations used in the above experiments, significant inhibition of PDE, which leads to the same physiological end-point as direct action on the kinase, cannot be ruled out. High concentrations of drug outside the cell can also interact with adenosine receptors,[64] which additionally obscures mechanism. However, it can be noted that incubation of the related cyclic nucleotide 8-(4-chlorophenyl)thio-cAMP results in a *lowering* of endogenous cAMP, presumably due to an activation of PDE.[65]

Direct activation of cA-PrK as a mode of action can be more clearly identified in the case of the 8-substituted cAMP benzyl esters in Table 6, which in guinea pig papillary muscle show substantial inotropic effects at concentrations where significant inhibition of PDE III is unlikely to be achieved.[60] Furthermore, these effects are not antagonized by carbachol, indicating that PDE inhibition is not involved.[66] The results in Table 6 also highlight the importance of further enhancing the lipophilicity of cAMP derivatives, as the inotropic activities of all the compounds shown correlate to a high degree with their log P values.

The actions of cAMP derivatives have been studied in many other cell types and results indicate that inhibition of cholesterol esterification[67] and inhibition of growth in human cancer cells[68] (particularly by cAMP derivatives showing selectivity for the stable site of R^{II}) are possibly further areas of therapeutic usefulness, although the effects of cAMP on cell growth and differentiation are complex.[69] There is evidence that some hormones are selective in their activation of the cA-PrK isoenzymes[70] and an isoenzyme selective agonist could be therapeutically advantageous. However, in studies involving the use of a single cAMP derivative it has not been possible to positively identify the specific functions of isoenzymes present in intact cells. The synergistic action of the analogue pairs may provide a more reliable means of determining isoenzyme function. For example, the synergism shown by N^6-benzoyl-cAMP and 8-isopropylthio-cAMP (Table 4) in specifically activating cA-PrK^{II} reflects the synergism shown by the two analogues in activating lipolysis in isolated adipocytes.[71]

Inhibitors of the catalytic unit and antagonists acting at the regulatory sites should show similar actions and, in theory, should be beneficial in situations where a reduction in the β agonist drive or the effects of elevated cAMP levels is required. The synthesis and pharmacology of antagonists, although potentially more selective than inhibitors in their actions, have received little study. However, (R$_P$)-cAMPS (**7a**) has been shown to antagonize the effects of increased cAMP in a number of cells.[72]

The effects of ATP competitive inhibitors of the catalytic site have received greater attention, although the mechanistic aspects of their effects are not yet well understood. Paradoxically, rather than oppose the effects of elevated cAMP, HA-1004 (Table 2) was found to relax smooth muscle and studies with a related series of related sulfonamides showed that this vasodilating ability could be correlated with their inhibition of cA-PrK.[17] However, it can be noted that for all compounds considered, smooth muscle relaxation was observed at concentrations at which significant inhibition of cA-PrK was unlikely and that smooth muscle relaxation is a general property of all isoquinoline sulfonamides.[19,20] In anaesthetized and pithed rats, HA-1004 has long-lasting vasodilating properties with little effect on heart rate. The compound preferentially blocks α_2-receptor-mediated vasoconstrictor responses, with less effect on α_1-receptor-mediated responses, and appears to have considerable vascular selectivity. The less potent inhibitors, amrinone (**12a**) and milrinone (**12b**), could not be demonstrated to inhibit cA-PrK in well-oxygenated cells, presumably because of high ATP levels.[49]

8.7.3 CYCLIC GMP DEPENDENT PROTEIN KINASE

8.7.3.1 Enzyme Characteristics

Cyclic GMP dependent protein kinase (cG-PrK) has been the subject of several reviews.[1,73] The enzyme shows homology with cA-PrK and is a dimer of identical 78 kDa subunits, each containing two cyclic nucleotide binding sites and a catalytic domain. Full activation occurs when cGMP (**13**) is bound at each of the four sites. The binding sites are kinetically distinguishable using the protocol described for cA-PrK, but the catalytic domain does not dissociate on activation. Most tissues contain considerably more cA-PrK than cG-PrK, the latter being found in the highest concentration in smooth muscle, lung, platelets and certain brain cells. Isoforms of cG-PrK have been described.[74]

(13)

8.7.3.2 Agonists and Antagonists

cG-PrK can bind ATP at the catalytic domain or undergo autophosphorylation. These events affect the kinetic and thermodynamic properties of the cyclic nucleotide binding sites. Each subunit contains a site which exchanges [³H]cGMP rapidly, usually termed site A, and a slowly exchanging site, termed site B. In the presence of Mg–ATP at 0 °C, sites A and B respectively show rates of association of 2.2×10^{-6} and 2.5×10^{-6} $M^{-1} s^{-1}$, rates of dissociation of 0.23 s^{-1} and 0.01 s^{-1}, and apparent dissociation constants of 70 and 4.4 nM for cGMP binding.[75] The rates of association are doubled in the absence of Mg–ATP and, in the autophosphorylated enzyme, the affinity of site B is enhanced 30-fold. However, unlike cA-PrK, the catalytic and regulatory units of cG-PrK do not dissociate on activation and the binding of cyclic nucleotides to cG-PrK shows positive cooperativity in one direction only. Thus, whilst cGMP analogues can show, relative to cGMP, selective affinity for either of the two types of binding site, only those analogues showing a high degree of selectivity for site A stimulate the binding of cGMP. Pairs of cGMP analogues binding at intrachain sites show synergism in their activation of the kinase, particularly if a high degree of complementary site selectivity is shown, suggesting that occupation of both types of site is required for activation.

All reported direct activators of cG-PrK retain a cyclic nucleotide structure. Alterations of the 2'-, 3'- and 5'-heteroatoms in the ribose cyclophosphate region produce large decreases in affinity at both types of binding site. The analogues 2'-*O*-butyryl-cGMP, N^2,2'-dibutyryl-cGMP, 5'-NH-cGMP, 3'-NH-cGMP, (R_P)-cGMPS (**7b**), (S_P)-cGMPS (**9b**), (R_P)-cGMPNMe$_2$ (**8b**) and (S_P)-cGMPNMe$_2$ (**10b**) have a K_i value of < 0.0042 for either site, suggesting that these moieties, and possibly also the phosphate charge, are important contributors to binding. These ribose cyclophosphate altered analogues are ineffective as cG-PrK activators ($K_a < 0.02$).

Table 7 Relative Activation and Affinity of Analogues of cGMP (13)

| Analogue | $K_i'(\mu M)$ | | $K_a'^a(\mu M)$ | Ref. |
	Site A	Site B		
cGMP (13)	1.0	1.0	1.0	76
1-Methyl-cGMP	0.44	0.049	0.37	76
N^2-Butyryl-cGMP	0.12	0.014	0.053	76
N^2-Hexyl-cGMP	0.11	0.039	0.22	76
6-Thio-cGMP	0.45	1.2	1.3	76
7-Deaza-cGMP	0.061	1.4	0.041	76
8-Acetyl-cGMP	0.027	0.035	0.024	76
8-Benzoyl-cGMP	0.0191	0.11	0.22	76
8-Bromo-cGMP	0.43	3.4	4.3	76
8-Hydroxy-cGMP	0.0043	0.066	0.12	76
8-(1-Hydroxy-2-propyl)-cGMP	0.0055	0.019	0.037	76
cAMP	0.0065	0.0031	0.0037	76
2-Amino-cAMP	0.0017	0.0021	0.030	76
cIMP	0.021	0.0045	0.011	76
2-Butyl-cIMP	0.0088	0.0013	0.016	76
2-Fluoro-cIMP	0.0058	0.0014	0.016	76
2-Methylthio-cIMP	0.0023	0.0009	0.0021	76

^{a} K_a(cGMP) = 110 nM.[75]

Few purine-modified cGMP derivatives exceed cGMP in affinity at either type of binding site or in their ability to stimulate cG-PrK. The relative affinity (K_i') and activation (K_a') of a selection of derivatives is given in Table 7. A study[76] of 46 analogues has shown that modifications in the pyrimidine ring, particularly at C-1, cause selectivity for site A, whilst modifications at C-7 and C-8 give rise to selectivity for site B, the selectivity arising mainly from reductions in affinity at the complementary site. In addition, a logarithmic expression of their relative activation correlated with a logarithmic expression of their mean site selectivity in a manner similar to that described for cA-PrK.

Comparison of the data for cGMP, cAMP, cIMP and 2-amino-cAMP in Table 7 shows that the 2-amino group and 6-carbonyl function are both required for high affinity in cG-PrK, but comparison of cAMP, cIMP and cGMP in Table 4 shows that the 2-amino group is the primary feature by which cGMP and cAMP are distinguished by their respective kinases.

The binding of cGMP to cG-PrK is fully antagonized by (R_P)-cAMPS (**7a**), which is also an antagonist of cA-PrK, but the cA-PrK full agonist (S_P)-cAMPS (**9a**) is only a partial agonist for cG-PrK.[77]

8.7.3.3 Inhibitors

The range of ATP competitive compounds showing inhibition of cG-PrK largely parallels that which inhibits cA-PrK. Only KT-5822 (Table 5) shows significant potent specificity for cG-PrK.[46] The substrate protein specificities of cG-PrK and cA-PrK *in vitro* are similar and no specific substrate-based inhibitor is known for either kinase. However, the substrate specificities are not identical and the design of a selective inhibitor of cG-PrK might be possible. Thus, whilst both enzymes require multiple basic residues on the amino terminal side of the serine phosphorylation site, C-terminal basic and N-terminal prolyl residues immediately adjacent to the phosphate acceptor are positive determinants for cG-PrK and negative determinants for cA-PrK.[51,52,53] In addition, cG-PrK shows a greater than 700-fold selectivity over cA-PrK in the phosphorylation of Leu-Arg-Arg-Ala-(N-Me)Ser-Leu-Gly.[8]

8.7.3.4 Therapeutic Utility

The role of cG-PrK in mediating physiological responses in nonretinal tissues has not been defined precisely. Firstly, the substrate specificities of cA-PrK and cG-PrK are similar, with most cA-PrK substrates being kinetically poorer substrates for cG-PrK. Proteins that are specifically phosphorylated by cG-PrK have been demonstrated, although their function is usually unknown. Secondly, in contrast to cAMP, all the actions of cGMP may not be mediated by kinase activation. For example, a cGMP-stimulated PDE has been described,[78] and, acting through this, elevations in cGMP will decrease cAMP levels. 8-Bromo-cGMP has been used to discriminate between these two mechanisms[79] as this derivative retains full kinase activation but is inactive with respect to the PDE.[80] Thirdly, autophosphorylation of cG-PrK increases its affinity for cAMP[81] and there is a possibility that the cG-PrK may be physiologically activated by both cAMP and cGMP. In the retina, cGMP acts by direct effects on ion channels[82] and there is no involvement of cG-PrK.

The therapeutic benefit of a cG-PrK modulator is therefore impossible to determine exactly, although an agonist, rather than an antagonist, will probably be more useful. There is growing evidence that cGMP and cG-PrK play a role in smooth muscle relaxation;[83] nitrovasodilators,[84,85] ANF[86] and EDRF[87] cause relaxation that is associated with increases in cGMP. Importantly, similar effects are obtained with 8-bromo-cGMP[88,89] and increases in protein phosphorylation are observed,[85] suggesting that cG-PrK activation (as opposed to PDE inhibition) is the necessary event.

8.7.4 PROTEIN KINASE C

8.7.4.1 Enzyme Characteristics

Protein kinase C is the commonly used name for a family of Ca^{2+} and phospholipid dependent, diacylglycerol-activated protein kinases. The enzyme phosphorylates serine or threonine and usually requires these to be flanked by basic residues. A large number of protein substrates are known, but it is not clear how many of these are phosphorylated in the cell. Many proteins are substrates for both PrK-C and cA-PrK, and in some instances the same site is phosphorylated,[90] but it is clear that the two kinases have differing specificities.[91]

There are at least four isoforms of PrK-C.[92,93] Each consists of a single polypeptide chain of *ca.* 80 kDa, containing a hydrophobic regulatory domain,[94] which can bind to membranes,[95] and a hydrophilic catalytic domain.[94] It is likely that the enzyme is active only when bound to membranes.[96] PrK-C has a ubiquitous tissue distribution, with brain, platelets and spleen showing the highest levels.[96] Recent reports indicate that the isoforms are not uniformly expressed in every tissue[98] and may show different Ca^{2+} requirements for activation. This has important implications for the physiological activation of the enzyme. A Ca^{2+} dependent PrK-C will be activated by the hydrolysis of phosphatidylinositol diphosphate, which generates both diacylglycerol (DAG) and Ca^{2+} (*via* inositol trisphosphate—see Chapter 11.4) for full activity, whereas it is possible that a Ca^{2+} independent isoform could be activated by the hydrolysis of any membrane lipid that results in the production of diacylglycerol. Several reviews of PrK-C exist[99,100,101] but, as here, mainly contain information that was gathered when PrK-C was believed to be a single form.

8.7.4.2 Agonists and Antagonists

Triacylglycerols and monoacylglycerols are ineffective in stimulating PrK-C. Only the *sn*-1,2-DAGs (**14**), as obtained from physiological turnover of phosphatidylinositols, and not the stereo-isomeric *sn*-2,3-DAGs, activate PrK-C. The necessary three-point interaction with the enzyme is provided by donation of a hydrogen bond from the *sn*-3-hydroxyl and acceptance of hydrogen bonds by the *sn*-1 and *sn*-2 CO groups, all of which have been shown to be essential for activation.[102,103,104] The stoichiometry required for activation, as determined using a Triton X-100 mixed micellar assay technique,[105] is one molecule of DAG and four molecules of phosphatidyl-serine per PrK-C molecule. It has been postulated[104,106] that a surface bound, but inactive, 1:4:1 complex of PrK-C, phosphatidylserine (interacting by means of its polar head groups), and Ca^{2+} is formed and that DAG (or its mimics) causes an activating conformational change by bonding with the Ca^{2+} and the kinase.

In vitro at fixed Ca^{2+} and phosphatidylserine levels, the concentrations of 1,2-dioleoyl-, -didecanoyl-, -oleoylacetyl- and -dioctanoyl-glycerols necessary to half-maximally stimulate PrK-C are close to 10^{-6} M, whereas 1,2-dihexanoylglycerol is about 10-fold less potent and greater than 100 μM 1,2-dibutyrylglycerol is required.[105,107] However, *in vivo* DAGs with long or short fatty acid chains, such as dioleoyl and dibutyryl, are not effective in stimulating PrK-C or displacing phorbol

12,13-dibutyrate (see below) from the enzyme in cells.[105] An optimal fatty acid chain length, as found in 1,2-dioctanoylglycerol, is necessary to permit sufficient solubility for delivery, yet be long enough to partition into the bilayer. An early conclusion[100] that an unsaturated acyl group is required may have reflected the influence of unsaturation on the physical properties of the lipid.

The *cis* unsaturated fatty acids, oleic acid and arachidonic acid have also been reported to show activation of a PrK-C with K_a values of 50 and 53 μM respectively in the absence of Ca^{2+} and phospholipid.[108,109] The corresponding *trans* and saturated acids lacked activity.

The tumour-promoting esters of phorbol (15a) compete with the presumed natural ligand (DAG) for binding and are 20–30 000-fold more potent activators of PrK-C, depending on their specific side chain,[110,111] although, to some extent, this greater potency could derive from their longer persistence in the membrane.

Other diterpenoids, such as mezerein (16), ingenol (17) and gnidimacrin (18), and less-related structures, such as the indole alkaloid teleocidin (19) and the polyacetate derivative aplysiatoxin (20) similarly act as tumour promoters, compete with [³H]PDB (15b), and enhance the enzymic activity of PrK-C. Computer-assisted comparison of the spatial arrangement of functional groups of the phorbol ester TPA (15c), mezerein (16), ingenol 3-tetradecanoate (17b), teleocidin (19) and aplysiatoxin (20) shows that in the active isomers of these compounds the oxygens at C-3, C-4, C-9 and C-20 in the diterpenoids (15b; 16; 17), the O-11, N-13, N-1 and O-24 atoms of teleocidin (19) and the O-27, O-3, O-11 and O-30 oxygens in aplysiatoxin (20) respectively probably interact in a similar manner in binding to PrK-C.[112] A similar study,[113] which confirmed the pharmacophoric importance of the C-4, C-9 and C-20 hydroxy groups of phorbol (15a) and the N-13, N-1 and O-24 atoms of teleocidin (19), considered in more detail the pharmacophoric contribution of the presence and orientation of hydrophobic regions relative to the main structure in these molecules.

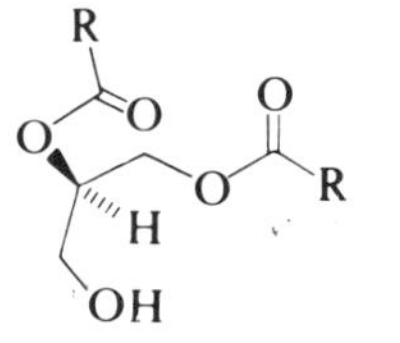

(14) R = acyl

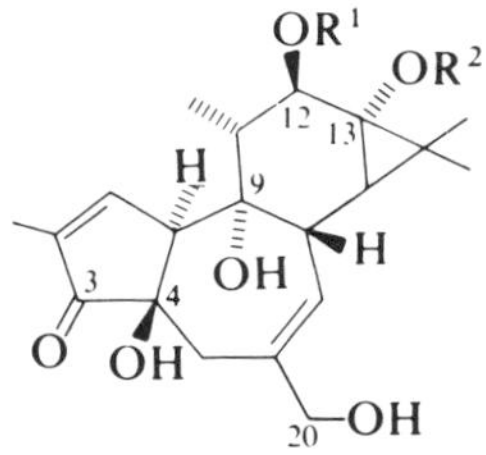

(16)

	R^1	R^2	
(15) a:	H	H	(phorbol)
b:	CO(CH₂)₂Me	CO(CH₂)₂Me	(PDB)
c:	CO(CH₂)₁₂Me	COMe	(TPA)

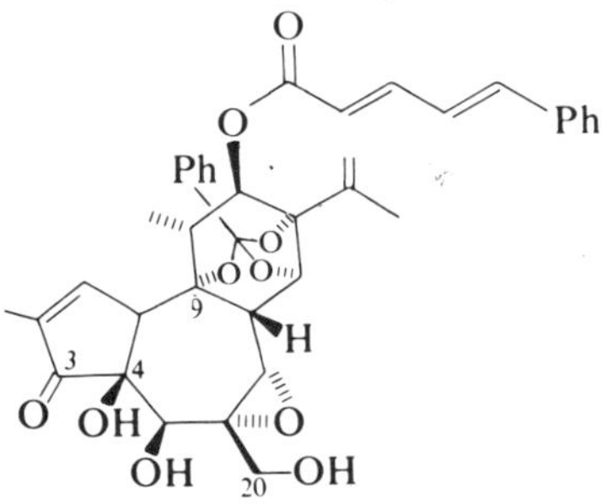

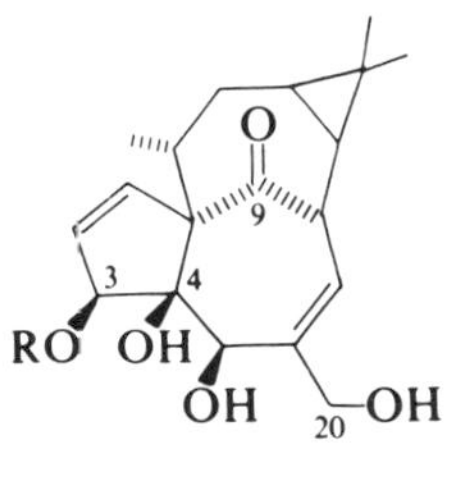

(17a) R = H (ingenol)
(17b) R = CO(CH₂)₁₂Me

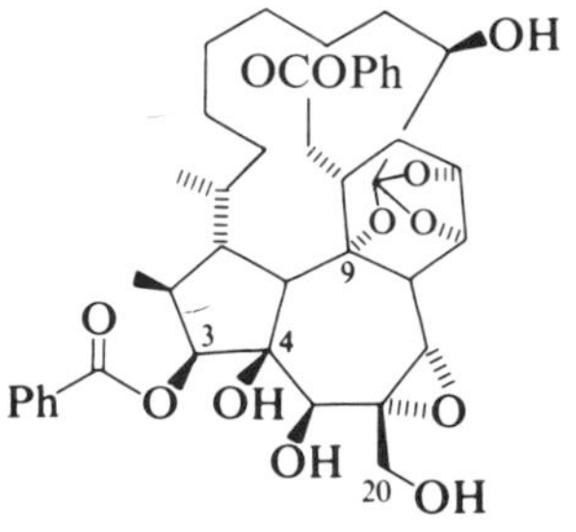

(18)

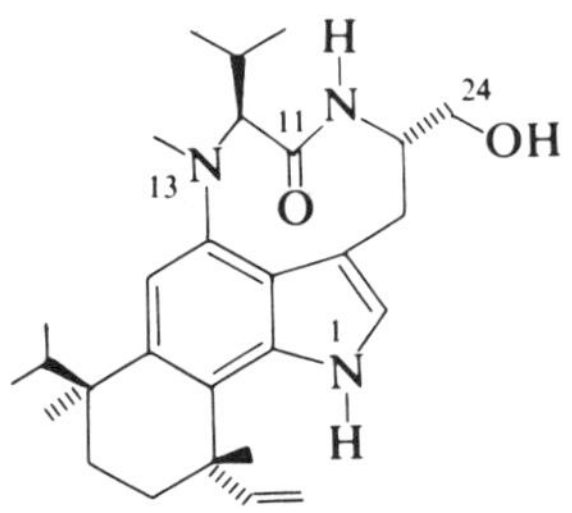

(19)

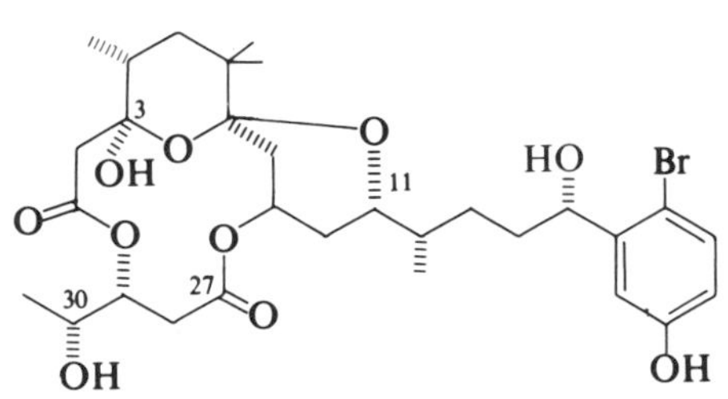

(20)

Minimized structures of the DAGs do not significantly overlay with any of the pharmacophore models,[112,113] particularly with respect to the hydrophobic areas. The best fits are for conformers of the DAG enantiomers about 4 kcal (1 kcal = 4.18 kJ) above the global minimum, in which overlays of the 1- and 2-ester carbonyl oxygens and the 3-hydroxy group of DAG and the C-9, C-4 and C-20 oxygens of phorbol are equally good, but the alkyl groups of the biologically active (S)-enantiomer (sn-1,2-DAG) are more favourably oriented.[113]

The development of agonists, and, more importantly, antagonists of PrK-C has been impeded by the difficulty of synthesis of phorbol congeners. However, the pharmacophoric model[113] has been used to predict simple structures, exemplified by DHI (**21a**), DMHI (**21b**), 3-amino-DMB (**22a**) and 3-acetamido-DMB (**22b**), which show IC_{50} values for the inhibition of [^{3}H]PBu$_2$ binding of 12, 2, 21 and 22 µM respectively, and acetamido-DMB has been shown to cause the phosphorylation of 40 kDa protein in platelets.

(21a) R = H (DHI)
(21b) R = Me (DMHI)

(22a) R = H
(22b) R = COMe (ADMB)

(23)

Other compounds reported to cause activation of PrK-C are sulfatide (cerebroside sulfate),[114] lipopolysaccharide,[115] Lipid A,[115] retenoic acid[116] and the naphthalenesulfonamides[117] SC-9 (**1f**) and SC-10 (**1g**). In all cases, these compounds activate in a manner similar to phosphatidylserine and to approximately the same extent. SC-9 and SC-10 are distinguished from the structurally related naphthalenesulfonamides (**1a–c**), which inhibit protein kinases (Table 1) and antagonize CaM, by the lack of a basic function in the sulfonamide side chain. A positive hydrophobic fragmental constant for this substituent is considered important for PrK-C activation.[117]

PrK-C is antagonized by a wide range of drugs which, in general, have nonspecific actions on membranes or may act by occluding the cavity containing the Ca^{2+}-phosphatidylserine complex. These include the aminoacridines,[106] phenothiazines,[118] W-7[119] (**1e**), calmidazolium,[120] cyto-toxins,[121] polymixin B,[122] neurotoxins,[122] spermine,[123] 1,2-diaminododecane,[123] dibucaine,[124] adriamycin,[118] palmitoylcarnitine,[118] alkyllyophospholipid,[125] gangliosides,[126] sphingosine,[127] staurosporine (Table 5),[48] quercetin,[128] triphenylethylenes such as tamoxifen (**23**), and CP-46,665-1.[129] Of these, the most potent is staurosporine, which has an IC_{50} of 2.7 nM, but its site of interaction with PrK-C is undefined and it is also known to inhibit cA-PrK.[48] In view of the multiple and nonspecific actions shown by the above compounds, the prospect for their use as selective antagonists of PrK-C appears poor and an approach to a specific antagonist would be better based on the structures of the well-defined agonists described above.

8.7.4.3 Inhibitors

The most clearly defined inhibitors of PrK-C are the isoquinolinesulfonamides shown in Table 2. These are competitive with ATP, the most potent being the piperazinylisoquinolinesulfonamides H-7 (**4c**), HA-100 (**4a**) and HA-156 (**4b**) with K_i values in the region of 6.0–7.5 µM. However, the structure–activity relationships of isoquinolinesulfonamides are generally similar for the cyclic nucleotide dependent kinases and PrK-C and little selectivity is shown.

Other potent ATP competitive inhibitors are K-252b (Table 5) and sangivamycin ($K_i = 11$ µM), but again little specificity is shown (sangivamycin inhibits cA-PrK with $K_i = 50$ µM). Other, less potent, compounds reported to interact at the catalytic site include TLCK ($IC_{50} = 1$ mM),[131] amiloride ($K_i = 1$ mM)[132] and bilirubin,[133] which interacts irreversibly.

8.7.4.4 Therapeutic Utility

PrK-C has been suggested, but not usually conclusively proved, to be a regulator of a number of physiological processes.[101] Prediction of therapeutic usefulness of modulating this enzyme is therefore difficult. As the phorbol esters and other PrK-C activators are tumour promoters, it is likely that only inhibitors or antagonists will be of therapeutic benefit, but even these may be limited in their utility. Physiological agents that stimulate PIP_2 hydrolysis (Chapter 11.4) cause elevation of intracellular Ca^{2+} and activation of PrK-C. However, experiments[134] using phorbol esters to activate PrK-C in cells do not completely mimic natural effectors as there is no increase in intracellular Ca^{2+}. A drug that antagonizes the effects of phorbol esters will therefore not necessarily antagonize fully the effects of the natural hormone. A PrK-C inhibitor or antagonist will be effective only in situations where the physiological response is due to PrK-C activation alone and not to Ca^{2+} mobilization. However, this could provide an opportunity for selective actions by an inhibitor or antagonist.

The actions of the isoquinolinesulfonamides have been studied in a number of tissues[11] and the effects noted for H-7 (**4c**) include inhibition of superoxide generation in neutrophils[135] and inhibition of the phorbol ester induced differentiation of the promyelocytic leukemia cell HL60.[136,137] Inhibition of PrK-C inhibition was suggested to be the primary mechanism for these actions as protein phosphorylation was shown to be decreased in a number of experiments and less potent PrK-C inhibitors were ineffective.[11,136,137] However, it can be noted that the isoquinoline-sulfonamides are nonselective for a number of kinases (Table 2) and in therapeutic use these compounds would be expected to show additional effects such as smooth muscle relaxation (see Section 8.7.2.4).[19,20] In addition, an intracellular concentration of 6 mM H-7 would be theoretically required to cause a 50% inhibition of activity, assuming a competitive interaction with physiological ATP (5 mM). It is possible, therefore, that the isoquinolinesulfonamides have an alternative mode of action not presently understood.

The nonsteroidal antiestrogen drug tamoxifen (**23**) has been shown to inhibit neutrophil super-oxide formation.[138] Whether PrK-C inhibition plays a role in its antiproliferative action is unproven, but tamoxifen derivatives show the same rank order of potency for PrK-C inhibition and cytotoxic effect on MCF7 cells, and a different rank order for antiestrogen activity.[139] Sphingosine has also been reported to inhibit superoxide generation,[140] differentiation of HL-60 cells[141] and protein phosphorylation.[142] The natural occurrence of sphingosine in cells also suggests a possible regulatory role.[127] Staurosporine (Table 5) has been shown to inhibit the growth of a number of cancer cell lines,[48] whereas K-252a (Table 5) has effects on platelet function and protein phosphorylation.[143]

In general the above results indicate that a PrK-C inhibitor or antagonist will probably show antiinflammatory and antiproliferative properties and the discovery of isoforms allows selectivity to be approached. As with other kinases, maximum benefit will be obtained when inhibition in response to several hormones or in different cell types is required.

8.7.5 MYOSIN LIGHT CHAIN KINASE

8.7.5.1 Enzyme Characteristics and Therapeutic Utility

The myosin light chain kinases (MLCKs) are dependent on the Ca^{2+}–calmodulin complex for activity and show specificity for a single type of substrate, the myosin P-light chain (P-LC). The kinases exist in three structurally distinct forms in skeletal, cardiac, and smooth and nonmuscle tissues.[2,144,145] The effect of P-LC phosphorylation is tissue dependent. In striated muscle phosphorylation modulates the degree of developed tension, whereas in smooth muscle P-LC phosphorylation is required to actually initiate contraction.[146] In nonmuscle cells, P-LC phosphorylation is thought to be involved in processes such as cell motility and secretion.[145]

The structural differences and high substrate specificity of the various forms of MLCK may provide an opportunity for the design of potent and specific smooth muscle relaxants, acting as either antagonists of CaM or inhibitors of the active site of the smooth muscle enzyme.

8.7.5.2 Antagonists

MLCK shows phosphotransferase activity only when the Ca^{2+}–calmodulin (CaM) complex is bound to a specific binding site on the enzyme,[147] therefore activity will be reduced by agents which antagonize this interaction by binding to either the MLCK binding site of CaM or to the CaM binding site of MLCK, although this latter approach is currently unexplored.

CaM antagonists, such as TFP (**24**) and W-7 (**1e**) are known to antagonize CaM binding to MLCK and show relaxant effects. However, these agents can also inhibit Ca^{2+} influx,[148] increase cyclic nucleotide concentrations by inhibition of CaM sensitive PDEs[149] and have effects on other smooth muscle contractile proteins.[150] All of these effects will also cause relaxation so that the contribution of direct action on MLCK to their overall pharmacological profile is not obvious. Selective antagonism of CaM binding to MLCK may, however, be possible. Studies with calmodulin fragments, genetically engineered calmodulin and the norchlorpromazine isothiocyanate adduct of CaM have shown that MLCK, cAMP PDE, calcineurin and CaM kinase either interact with different domains on CaM or respond differently to the interaction with the same domain.[151] In addition, a structurally diverse group of 26 CaM antagonists could be divided, on the basis of their K_i values for the inhibition of MLCK and the CaM-activated PDE and their K_d values for binding to CaM, into at least three distinguishable groups, which probably selectively bind to different sites on CaM and inhibit specifically the activation of distinct enzymes.[152] However, despite the differences in the relative effectiveness of these antagonists in several CaM-utilizing systems, no compound in this study showed overall selective affinity for MLCK and a large proportion of the antagonists were more effective as relaxants of intact smooth muscle, where all relaxation mechanisms are effective, than in skinned smooth muscle, where only direct interaction with MLCK is possible. Clearly, a rational design of selective MLCK antagonists is better based on compounds that relax skinned strips or inhibit actomyosin interactions at concentrations similar to their effects on intact tissue.

At present the only compound known to affect MLCK specifically is Ro 22-4839 (**25**), which shows slight selectivity for MLCK against CaM-activated PDE, adenylate cyclase and Ca^{2+}, Mg^{2+}-ATPase ($IC_{50} = 3.1, 20, > 20$ and > 30 μM respectively) and has no significant effect on the non-CaM dependent kinases cA-PrK and PrK-C.[153] In addition, its effects are highly specific to vascular smooth muscle contractile systems.

8.7.5.3 Inhibitors

Trypsin treatment of MLCK results in a fragment of the enzyme that is catalytically active but no longer binds calmodulin and this has been used to investigate active site inhibitors.[154] ATP competitive and substrate competitive inhibitors have been reported, but as seen for other kinases, ATP competitors are relatively unselective and must compete with high ATP levels, and the substrate competitors are peptides and therefore provide poor leads to cell penetrant drugs.

The chloro-, bromo- and iodo-substituted naphthalenesulfonamides ML-9 (**2e**), ML-8 (**2c**), ML-7 (**2g**), HH-3209 (**2a**), HH-3709 (**2b**), HH-3909 (**2f**) and W-7 (**1e**; Table 1) inhibit both Ca^{2+}, CaM dependent and independent smooth muscle MLCK in an ATP competitive manner and their inhibitory activity correlates well with their hydrophobicity. Greatest potency against MLCK is observed with ML-9 (**2e**) and ML-8 (**2c**), where the sulfonamide side chain is diazepin.[16] The microbial metabolite K252a (Table 5) also shows ATP competitive inhibition.[47]

ML-9 relaxes both intact and skinned mesenteric artery and the intact preparation shows a parallel dose dependence between relaxation and inhibition of myosin light chain (MLC) phosphorylation.[16] ML-9 also inhibits dopamine release and MLC phosphorylation in pheochromocytoma cells.[155] However, at the concentrations used, ML-9 would not be expected to cause inhibition of MLCK in intact tissues and it has been suggested that ML-9 may interact with other enzymes.[16]

There is now considerable evidence that contraction in smooth muscle is subject to dual regulation[146] and it is possible that ML-9 and related compounds may interact with a second (nonMLCK) regulatory system.

The substrate protein determinants for skeletal and smooth muscle MLCKs have been examined using synthetic peptide analogues of the phosphorylation sites[156] and show significant differences. In an analogue of the phosphorylation site in smooth muscle P-LC, Lys^{11}-Lys-Arg-Ala-Ala-Arg-Ala-Thr-Ser^{19}(P)-Asn^{20}-Val^{21}-Phe^{22}-Ala^{23}, the hydrophobic residues Phe^{22} and Val^{21} were found to be particularly important for a high V_{max} of serine phosphorylation.[157] The shortened peptide 11–19 was an effective inhibitor of smooth muscle MLCK with a K_i of 10 μM and a V_{max} less than 1% of the values of P-LCs.

8.7.6 TYROSINE KINASES

8.7.6.1 Enzyme Characteristics and Therapeutic Utility

The normal and aberrant roles of tyrosine specific phosphorylation have been reviewed.[158,159] In normal cells, tyrosine specific protein kinase activity is associated with receptors for insulin and certain growth factors, such as the epidermal growth factor (EGF) and platelet-derived growth factors (PDGF). These bind to extracellular receptors and cause the activation of a tyrosine kinase present in the intracellular portion of the same receptor protein molecule. The substrates for these protein kinases are not well defined, although in most cases the kinases themselves are autophosphorylated. In the absence of intracellular messengers, therapeutic modulation of TyrKs must be by drug-induced changes of events at the extracellular receptors (not discussed here) or at the phosphotransferase catalytic site.

Tyrosine specific protein kinase activity is also shown by a number of oncogene products and the possibility that tyrosine phosphorylation plays an important role in cell proliferation and transformation suggests that a specific tyrosine kinase inhibitor would have significant therapeutic potential as an antitumour agent. However, as the tyrosine kinase activity of many oncogene products, such as the *src* and *abl* gene families, appears to be regulated only by the amount of protein expressed, inhibition at the catalytic site is the only mechanism possible for decreasing kinase activity.

8.7.6.2 Inhibitors

The Na^+/H^+ antiporter amiloride,[160] the Ca^{2+} antagonists chlorpromazine, imipramine and dibucain,[161] quercetin[162] and other flavanoids,[163] genistein,[164] and a series of 4-hydroxycinnamides[165] and α-cyanocinnamides[166] have been reported to show TyrK inhibitory activity. However, only genistein (26) and the 4-hydroxycinnamides (27–30) have shown evidence of selectivity for the TyrKs.

Genistein inhibited *in vitro* the tyrosine specific protein kinase activity of the epidermal growth factor (EGF) receptor, pp60[v-src] and pp110[gag-fes], with IC_{50} in the range 6.0–8.0 μM. The inhibition was competitive with respect to ATP and noncompetitive with a phosphate acceptor, histone H2B, and the enzyme activities of serine/threonine specific protein kinases such as cA-PrK, phosphorylase kinase and PrK-C were unaffected. When the effect of genistein was examined on the phosphorylation of the EGF receptor in cultured A431 cells, EGF-stimulated phosphorylation of serine, threonine and tyrosine residues was decreased and phosphoamino acid analysis of total cell proteins

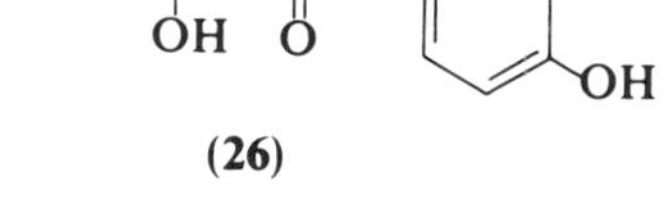

(26)

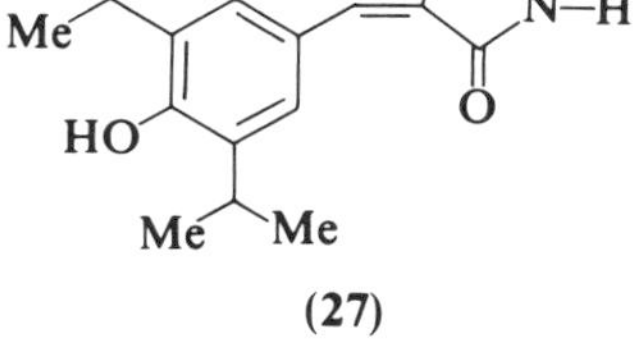

(27)

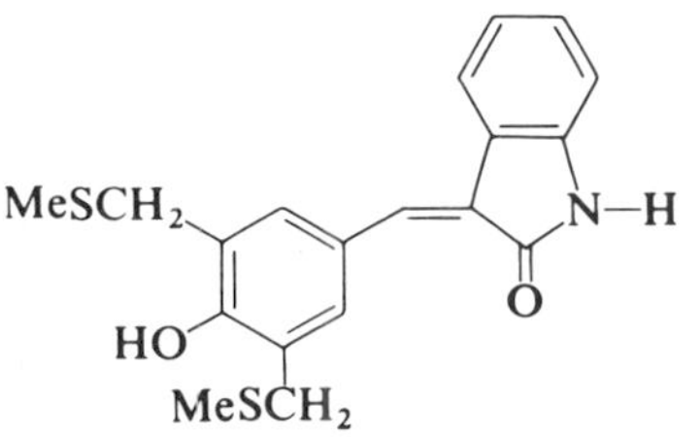

(28)

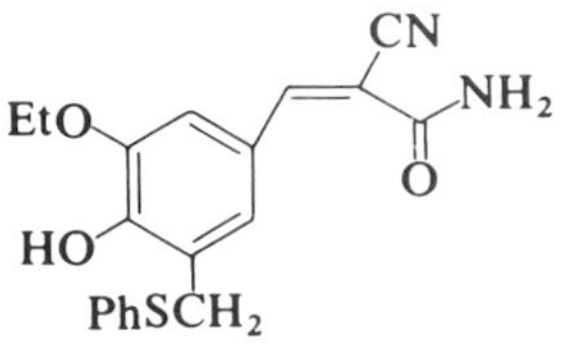

(29)

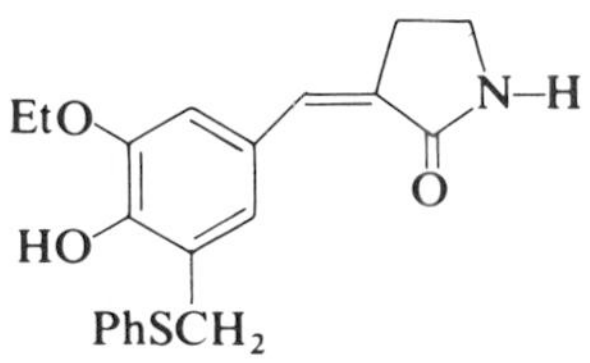

(30)

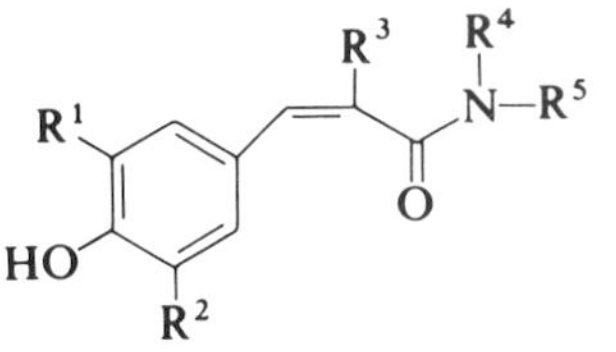

(31)

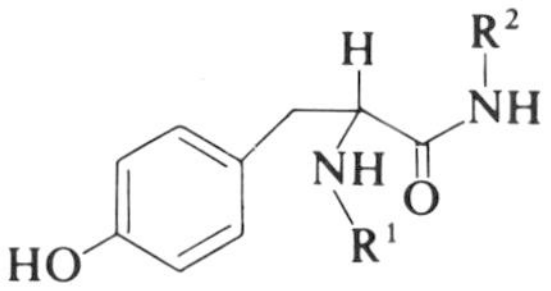

(32)

revealed that genistein inhibited the EGF-stimulated increase in the phosphorylation level in A431 cells.

The 4-hydroxycinnamides ST-280 (27), ST-458 (28), ST-638 (29) and ST-642 (30) inhibited *in vitro* tyrosine specific protein kinase activity in the EGF receptor with IC_{50} values of 0.44, 0.44, 0.37 and 0.85 μM respectively.[165] The inhibitory actions of these and other related compounds was believed to arise from the ability of the 4-hydroxycinnamide skeleton (31) to mimic the structure of the tyrosine residue (32).

8.7.7 REFERENCES

1. A. M. Edelman, D. K. Blumenthal and E. G. Krebs, *Annu. Rev. Biochem.*, 1987, **56**, 567.
2. E. G. Krebs and P. D. Boyer (eds.), 'The Enzymes', Academic Press, New York, 1986, vol. 17.
3. T. Hunter, *Cell*, 1987, **50**, 823.
4. J. D. Corbin and R. A. Johnson, *Methods Enzymol.*, 1988, **159**.
5. R. H. Mitchell, A. H. Drummond and C. P. Downes (eds.), 'Inositol Lipids in Cell Signaling', Academic Press, London, 1988.
6. A. P. Kwaitkowski and M. M. King, *Biochemistry*, 1987, **26**, 7636.
7. D. A. Flockhart, W. Freist, J. Hoppe, T. M. Lincoln and J. D. Corbin, *Eur. J. Biochem.*, 1984, **140**, 289.
8. H. Konodo, J. Kinoshita, T. Matsuba and J. Sunamoto, *FEBS Lett.*, 1986, **201**, 94.
9. G. L. Kenyon and G. A. Garcia, *Med. Res. Rev.*, 1987, **7**, 389.
10. N. E. Thomas, H. N. Branson, A. C. Nairn, P. Greengard, D. C. Fry, A. S. Mildvan and E. T. Kaiser, *Biochemistry*, 1987, **26**, 4471.
11. H. Hidaka and M. Hagiwara, *Trends Pharmacol. Sci.*, 1987, **8**, 162.
12. T. Tanaka, T. Ohmura, T. Yamakado and H. Hidaka, *Mol. Pharmacol.*, 1982, **22**, 408.
13. H. Hidaka, T. Yamiki, M. Naka, T. Tanaka, H. Hayashi and R. Kobayashi, *Mol. Pharmacol.*, 1980, **17**, 66.
14. H. Hidaka, M. Inagaki S. Kawamoto and Y. Sasaki, *Biochemistry*, 1984, **23**, 5036.
15. M. Inagaki, S. Kawamoto, H. Ito, M. Saitoh, M. Hagiwara, J. Takahashi and H. Hidaka, *Mol. Pharmacol.*, 1986, **29**, 577.
16. M. Saitoh, T. Ishikawa, S. Matsushima, M. Naka and H. Hidaka, *J. Biol. Chem.*, 1987, **262**, 7796.
17. T. Ishikawa, M. Inagi, M. Watanabe and H. Hidaka, *J. Pharmacol. Exp. Ther.*, 1985, **235**, 495.
18. M. Hagiwara, M. Inagaki, M. Watanabe, M. Ito, K. Onoda, T. Tanaka and H. Hidaka, *Mol. Pharmacol.*, 1987, **32**, 7.
19. H. Hidaka, T. Sone, Y. Sasaki, T. Sugihara (Asahi Chemical Industry Co. Ltd.), *Eur. Pat.* 61 673 (1982) (*Chem. Abstr.*, 1983, **98**, 71 954y).
20. H. Hidaka and T. Sone (Asahi Chemical Industry Co. Ltd.), *Eur. Pat.* 109 023 (1984) (*Chem. Abstr.*, 1984, **101**, 191 715y).
21. J. Hoppe, *Trends Biochem. Sci.*, 1985, **10**, 29.
22. D. A. Flockhart and J. D. Corbin, *CRC Crit. Rev. Biochem.*, 1982, **12**, 133.
23. S. R. Rannels and J. D. Corbin, *Methods Enzymol.*, 1983, **99**, 183.

24. S. S. Taylor, J. Bubis, J. Toner-Webb, L. D. Saraswat, E. A. First, J. A. Buechler, D. R. Knighton and J. Sowadski, *FASEB. J.*, 1988, **2**, 2677.
25. M.-Y. Yoon and P. F. Cook, *Biochemistry*, 1987, **26**, 4118.
26. C. H. Clegg, G. C. Cadd and G. S. McKnight, *Proc. Natl. Acad. Sci. USA*, 1988, **85**, 3703.
27. M. D. Uhler, J. C. Chirva and G. S. McKnight, *J. Biol. Chem.*, 1986, **261**, 15360.
28. J. Bubis and S. S. Taylor, *Biochemistry*, 1987, **26**, 3478.
29. I. T. Weber, T. A. Steitz, J. Bubis and S. S. Taylor, *Biochemistry*, 1987, **26**, 343.
30. These sites have also been termed sites 1 (or B) and 2 (or A) respectively in the literature.
31. A. M. Robinson-Steiner and J. D. Corbin, *J. Biol. Chem.*, 1983, **258**, 1032.
32. L. E. Kochevar, L. C. Huang and C.-H. Huang, *Int. J. Biochem.*, 1986, **18**, 519.
33. D. Ogreid, S. O. Doskeland and J. P. Miller, *J. Biol. Chem.*, 1983, **258**, 1041.
34. P. A. Connely, T. G. Hastings and E. M. Reimann, *J. Biol. Chem.*, 1986, **261**, 2325.
35. S. R. Rannels, C. E. Cobb, L. R. Landis and J. D. Corbin, *J. Biol. Chem.*, 1985, **260**, 3423.
36. J. D. Corbin, S. R. Rannels, D. A. Flockhart, A. M. Robinson-Steiner, M. C. Tigane, S. O. Doskeland, R. H. Suva and J. P. Miller, *Eur. J. Biochem.*, 1982, **125**, 259.
37. R. J. W. deWit, D. Hekstra, B. Jastorff, W. J. Stec, J. Baraniak, R. van Driel and P. J. M. van Haastert, *Eur. J. Biochem.*, 1984, **142**, 255.
38. D. Ogreid, R. Ekanger, R. H. Suva, J. P. Miller, P. Sturm, J. D. Corbin and S. O. Doskeland, *Eur. J. Biochem.*, 1985, **150**, 219.
39. T. S. Yagura and J. P. Miller, *Biochemistry*, 1983, **20**, 879.
40. B. Jastorff, J. Hoppe, J. Mato and T. M. Konijn, *Nucleic Acids Res.*, 1979, **4**, 237.
41. D. Ogreid, R. E. Kanger, R. H. Suva, J. P. Miller and S. O. Ooskeland, *Eur. J. Biochem.*, 1989, **181**, 19.
42. S. O. Doskeland, D. Ogreid, R. Ekanger, P. A. Sturm, J. P. Miller and R. H. Suva, *Biochemistry*, 1983, **22**, 1094.
43. P. A. Frey and R. D. Sammons, *Science (Washington, D.C.)*, 1985, **228**, 541.
44. L. H. Parker Bothello, L. C. Webster, J. D. Rothermel, J. Barniak and W. J. Stec, *J. Biol. Chem.*, 1988, **263**, 5301.
45. T. S. Yagura and J. P. Miller, *Biochim. Biophys. Acta*, 1980, **630**, 463.
46. H. Kase, K. Iwahashi, S. Nakanishi, Y. Matsuda, Y. Yamada, M. Takahashi, C. Murakata, A. Sato and M. Kaneko, *Biochem. Biophys. Res. Commun.*, 1987, **142**, 436.
47. S. Nakanishi, K. Yamada, H. Kase, S. Nakamura and Y. Nonamura, *J. Biol. Chem.*, 1988, **263**, 6215.
48. T. Tamaoki, H. Nomoto, I. Takahashi, Y. Kato, M. Morimoto and F. Tomoita, *Biochem. Biophys. Res. Commun.*, 1986, **135**, 397.
49. C. Q. Earl, J. Linden and B. Weglicki, *Life Sci.*, 1986, **39**, 1901.
50. H. N. Branson, N. E. Thomas, W. T. Miller, D. C. Fry, A. S. Mildvan and E. T. Kaiser, *Biochemistry*, 1987, **26**, 4466
51. D. B. Glass and S. B. Smith, *J. Biol. Chem.*, 1983, **258**, 14797.
52. D. B. Glass, M. R. El-Maghrabi and S. J. Pilkis, *J. Biol. Chem.*, 1986, **261**, 12166.
53. D. B. Glass, *Biochem. J.*, 1983, **213**, 159.
54. H.-C. Cheng, B. E. Kemp, R. B. Pearson, A. J. Smith, L. Misconi, S. M. Van Patten and D. A. Walsh, *J. Biol. Chem.*, 1986 **261**, 989.
55. D. K. Granner, in 'Harper's Review of Biochemistry', 20th edn., ed. D. W. Martin, P. A. Mayes, V. W. Rodwell and D. K. Granner, Lange Medical Publications, Los Altos, CA, 1985, p. 505.
56. D. B. Evans, *J. Cardiovasc. Pharmacol.*, 1986, **8** (suppl. 9), S22.
57. M. D. Feldman, L. Copelas, J. K. Gwathmey, P. Phillips, S. E. Warren, F. J. Schoen, W. Grossman and J. P. Morgan *Circulation*, 1987, **75**, 331.
58. M. R. Bristow, R. Ginsburg, V. Umans, M. Fowler, W. Minobe, R. Rasmussen, P. Zera, R. Menlove, P. Shah S. Jamieson and E. B. Stinson, *Circ. Res.*, 1986, **59**, 297.
59. J. P. Miller, T. S. Yagura, R. B. Meyer, R. K. Robins and H. Uno, *J. Carbohydr. Nucleosides Nucleotides*, 1980, **7**, 167
60. M. Korth and J. Engels, *Naunyn-Schmiedeberg's Arch. Pharmacol.*, 1987, **310**, 103.
61. S. Matsui, E. Murakimi, N. Takekosh, J. Emoto and M. Matoba, *Am. J. Cardiol.*, 1983, **51**, 1364.
62. J. P. Miller, K. H. Boswell, R. B. Meyer, L. F. Christensen and R. K. Robins, *J. Med. Chem.*, 1980, **23**, 242.
63. M. Nakazawa, K. Takeda, Y. Nakagawa, H. Tamatsu, K. Matsui, H. Nakahara, T. Kimura, T. Kawanda and S. Imai *J. Cardiovasc. Pharmacol.*, 1985, **7**, 862.
64. T. Kimura, S. Imai and T. Nagatomo, *J. Mol. Cell. Cardiol.*, 1985, **17** (suppl. 1), 51.
65. T. W. Gettys, A. J. Vine, M. F. Simmonds and J. D. Corbin, *J. Biol. Chem.*, 1988, **263**, 10359.
66. M. Korth, J. Engels and M. Schafer-Korting, *Naunyn-Schmiedeberg's Arch. Pharmacol.*, 1987, **335**, 166.
67. D. P. Hajjar, *Arch. Biochem. Biophys.*, 1986, **247**, 49.
68. D. Katsaros, G. Tortora, P. Tagliaferri, T. Clair, S. Ally, L. Neckers, R. K. Robins and Y. S. Cho-Chung, *FEBS Lett.* 1987, **223**, 97.
69. S. M. Lohmann and U. Walter, *Adv. Cyclic Nucleotide Protein Phosphorylation Res.*, 1984, **18**, 63.
70. C. S. Chew, *J. Biol. Chem.*, 1985, **260**, 7540.
71. J. D. Corbin and R. A. Johnson, *Methods Enzymol.*, 1988, **159**, 118.
72. J. D. Rothermel, B. Jastorff and L. H. P. Botelho, *J. Biol. Chem.*, 1984, **259**, 8151.
73. T. M. Lincoln and J. D. Corbin, *Adv. Cyclic Nucleotide Res.*, 1983, **15**, 139.
74. H. R. de Jonge, *Adv. Cyclic Nucleotide Res.*, 1984, **14**, 315.
75. S. O. Doskeland, O. K. Vintermyr, J. D. Corbin and D. Ogreid, *J. Biol. Chem.*, 1987, **262**, 3534.
76. J. D. Corbin, D. Ogreid, J. P. Miller, R. H. Suva, B. Jastorff and S. O. Doskeland, *J. Biol. Chem.*, 1986, **261**, 1208.
77. F. Hofmann, H.-P. Gensheimer, W. Landgraf, R. Hullin and B. Jastorff, *Eur. J. Biochem.*, 1985, **150**, 85.
78. T. J. Martins, M. C. Mumby and J. A. Beavo, *J. Biol. Chem.*, 1982, **257**, 1973.
79. R. Fischmeister and H. C. Hartzell, *J. Physiol. (London)*, 1987, **387**, 453.
80. C. Erneux, D. Couchie, J. E. Dumant and B. Jastorff, *Adv. Cyclic Nucleotide Protein Phosphorylation Res.*, 1984, **16**, 107
81. W. Landgraf, R. Hullin, C. Gobel and F. Hofmann, *Eur. J. Biochem.*, 1986, **154**, 113.
82. T. D. Lamb, *Trends Neurosci.*, 1986, **9**, 224.
83. L. J. Ignarro and P. J. Kadowitz, *Annu. Rev. Pharmacol. Toxicol.*, 1985, **25**, 171.
84. K.-D. Schultz, K. Schultz and G. Schultz, *Nature (London)*, 1977, **256**, 750.

85. R. M. Rapoport, M. B. Draznin and F. Murad, *Proc. Natl. Acad. Sci. USA*, 1982, **79**, 6470.
86. R. J. Winquist, E. P. Faison, S. A. Waldman, K. Schwartz, F. Murad and R. M. Rapoport, *Proc. Natl. Acad. Sci. USA*, 1984, **81**, 7661.
87. R. F. Furchgott, P. D. Cherry, J. V. Zawadzki and D. Jothianadan, *J. Cardiovasc. Pharmacol.*, 1984, **6**, S336.
88. H. Kai, H. Kanaide, T. Matsumoto and M. Nakamura, *FEBS Lett.*, 1987, **221**, 284.
89. R. J. Bing and M. Saeed, *J. Am. Coll. Cardiol.*, 1986, **8**, 342.
90. A. Kishimoto, K. Nishiyama, H. Nakanishi, Y. Uratsuji, H. Nomura, Y. Takeyama and Y. Nishizuka, *J. Biol. Chem.*, 1985, **260**, 12492.
91. C. House, R. E. H. Wettenhall and B. E. Kemp, *J. Biol. Chem.*, 1987, **262**, 772.
92. Y. Ono, T. Fujii, K. Ogita, U. Kikkawa, K. Igarashi and Y. Nishizuka, *J. Biol. Chem.*, 1988, **263**, 6927.
93. L. Coussens, P. J. Parker, L. Rhee, T. L. Yang-Feng, E. Chen, M. D. Waterfield, U. Francke and A. Ullrich, *Science (Washington, D.C.)*, 1986, **233**, 859.
94. P. J. Parker, L. Coussens, N. Totty, L. Rhee, S. Young, E. Chen, S. Stabel, M. D. Waterfield and A. Ullrich, *Science (Washington, D.C.)*, 1986, **233**, 853.
95. H. Brockerhoff, *FEBS Lett.*, 1986, **201**, 1.
96. M.-H. Lee and R. M. Bell, *J. Biol. Chem.*, 1986, **261**, 14867.
97. K.-P. Huang and F. L. Huang, *J. Biol. Chem.*, 1986, **261**, 14781.
98. F. L. Huang, Y. Yoshida, N. Nakabayashi and K.-P. Huang, *J. Biol. Chem.*, 1987, **262**, 15714.
99. G. L. Ashedel, *Biochim. Biophys. Acta*, 1985, **822**, 219.
100. Y. Nishizuka, *Nature (London)*, 1984, **308**, 693.
101. U. Kikkawa and Y. Nishizuka, *Annu. Rev. Cell Biol.*, 1986, **2**, 149.
102. L. T. Boni and R. R. Rando, *J. Biol. Chem.*, 1985, **260**, 10819.
103. R. R. Rando and N. Young, *Biochem. Biophys. Res. Commun.*, 1984, **122**, 818.
104. B. Ganong, C. R. Loomis, Y. A. Hannun and R. M. Bell, *Proc. Natl. Acad. Sci. USA*, 1986, **83**, 1184.
105. P. M. Conn, B. R. Ganong, J. G. Ebeling, D. Staley, J. E. Niedel and R. M. Bell, *Methods Enzymol.*, 1986, **124**, 57.
106. Y. A. Hannun and R. M. Bell, *J. Biol. Chem.*, 1988, **263**, 5124.
107. J. G. Ebeling, G. R. Vandenbark, L. J. Katin, B. R. Ganong, R. M. Bell and J. E. Niedel, *Proc. Natl. Acad. Sci. USA*, 1985, **82**, 815.
108. K. Murakami and A. Routtenberg, *FEBS Lett.*, 1985, **192**, 189.
109. L. C. McPhail, C. C. Clayton and R. Snyderman, *Science (Washington, D.C.)*, 1984, **224**, 622.
110. M. Castagna, Y. Takai, K. Kaibuchi, H. Sano, U. Kikkawa and Y. Nishizuka, *J. Biol. Chem.*, 1982, **257**, 7847.
111. N. A. Sharkey and P. M. Blumberg, *Carcinogenesis*, 1986, **7**, 677.
112. A. M. Jeffrey and R. M. J. Liskamp, *Proc. Natl. Acad. Sci. USA*, 1986, **83**, 241.
113. P. A. Wender, K. F. Koehler, N. A. Sharkey, M. L. Dell'Aquila and P. M. Blumberg, *Proc. Natl. Acad. Sci. USA*, 1986, **83**, 4214.
114. H. Fujiki, K. Yamashita, M. Suganuma, T. Horiuchi, N. Tanigichi and M. Makita, *Biochem. Biophys. Res. Commun.*, 1986, **138**, 153.
115. P. D. Wightman and C. R. H. Raetz, *J. Biol. Chem.*, 1984, **259**, 10048.
116. S. Ohkubo, E. Yamada, T. Endo, H. Itoh and H. Hidaka, *Biochem. Biophys. Res. Commun.*, 1984, **118**, 460.
117. M. Ito, T. Tanaka, M. Inagaki, K. Nakanishi and H. Hidaka, *Biochemistry*, 1986, **25**, 4179.
118. M. B. C. Wise and J. F. Kuo, *Biochem. Pharmacol.*, 1983, **32**, 1259.
119. T. Tanaka, T. Ohmura, T. Yamakado and H. Hidaka, *Mol. Pharmacol.*, 1982, **22**, 408.
120. G. J. Mazzei, R. C. Schatzman, R. S. Turner, W. R. Vogler and J. F. Kuo, *Biochem. Pharmacol.*, 1984, **33**, 125.
121. J. F. Kuo, R. L. Raynor, G. J. Mazzei, R. C. Schatzman, R. S. Turner and W. R. Kem, *FEBS Lett.*, 1983, **153**, 183.
122. G. J. Mazzei, N. Katoh and J. F. Kuo, *Biochem. Biophys. Res. Commun.*, 1982, **109**, 1129.
123. D. F. Qi, R. C. Schatzman, G. J. Mazzei, R. S. Turner, R. L. Raynor, S. Liano and J. F. Kuo, *Biochem. J.*, 1983, **213**, 281.
124. T. Mori, Y. Takai, T. Minakuchi, B. Yu and Y. Nishizuka, *J. Biol. Chem.*, 1980, **255**, 8378.
125. D. M. Helfman, K. C. Darnes, J. M. Kinkade, Jr., W. R. Volger, M. Shoji and J. F. Kuo, *Cancer Res.*, 1983, **43**, 2955.
126. J. Y. H. Kim, J. R. Goldenring, R. J. DeLorenzo and R. K. Yu, *J. Neurosci. Res.*, 1986, **15**, 159.
127. Y. A. Hannun, C. R. Loomis, A. H. Merrill, Jr. and R. M. Bell, *J. Biol. Chem.*, 1986, **261**, 12604.
128. A. K. Srivastava, *Biochem. Biophys. Res. Commun.*, 1985, **131**, 1.
129. M. Shoji, W. R. Vogler and J. F. Kuo, *Biochem. Biophys. Res. Commun.*, 1985, **127**, 590.
130. G. R. Loomis and R. M. Bell, *J. Biol. Chem.*, 1988, **263**, 1682.
131. D. H. Solomon, C. A. O'Brien and I. B. Weinstein, *FEBS Lett.*, 1985, **190**, 342.
132. J. M. Besterman, W. S. May, Jr., H. LeVine, III, E. J. Cragoe, Jr. and P. Cautrecasas, *J. Biol. Chem.*, 1985, **260**, 1155.
133. K. Sano, H. Nakamura and T. Matsuo, *Pediatr. Research*, 1985, **19**, 587.
134. T. J. Rink, A. Sanchez and T. J. Hallam, *Nature (London)*, 1983, **305**, 317.
135. I. Fujita, K. Takeshige and S. Minakami, *Biochem. Pharmacol.*, 1986, **35**, 4555.
136. T. Matsui, Y. Nakao, T. Koizumi, Y. Katakami and T. Fujita, *Cancer Res.*, 1986, **46**, 583.
137. M. Nishikawa, Y. Uemara, H. Hidaka and S. Shirakawa, *Life Sci.*, 1986, **39**, 1101.
138. K. Horgan, E. Cooke, M. B. Hallet and R. E. Mansel, *Biochem. Pharmacol.*, 1986, **35**, 4463.
139. C. A. O'Brien, R. M. Liskamp, D. H. Solomon and I. B. Weinstein, *J. Natl. Cancer Inst.*, 1986, **76**, 1243.
140. E. Wilson, M. C. Olcott, R. M. Bell, A. H. Merrill, Jr. and J. D. Lambeth, *J. Biol. Chem.*, 1986, **261**, 12616.
141. A. H. Merrill, A. M. Serini, V. L. Stevens, Y. A. Hannun, R. M. Bell and J. M. Kincade, Jr., *J. Biol. Chem.*, 1986, **261**, 12610.
142. Y. A. Hannun, C. R. Loomis, A. H. Merrill, Jr. and R. M. Bell, *J. Biol. Chem.*, 1986, **261**, 12604.
143. K. Yamada, K. Iwahashi and H. Kase, *Biochem. Biophys. Res. Commun.*, 1987, **144**, 35.
144. J. T. Stull, M. H. Nunnally and C. H. Michinoff, in 'The Enzymes', ed. E. G. Krebs and P. D. Boyer, Academic Press, New York, 1986, vol. 17, p. 114.
145. J. R. Sellers and R. S. Adelstein, in 'The Enzymes', ed. E. G. Krebs and P. D. Boyer, Academic Press, New York, 1986, vol. 17, p. 382.
146. K. E. Kamm and J. T. Stull, *Annu. Rev. Pharmacol. Toxicol.*, 1985, **25**, 593.
147. K. Takio, D. K. Blumenthal, A. M. Edelman, K. A. Walsh, E. G. Krebs and K. Titani, *Biochemistry*, 1985, **24**, 6028.

148. E. Winslow, S. Farmer, M. Martora and R. J. Marshall, *Eur. J. Pharmacol.*, 1986, **131**, 219.
149. T. Tanaka, M. Ito, T. Ohmura and H. Hidaka, *Biochemistry*, 1985, **24**, 5281.
150. P. J. Silver, *J. Cardiovasc. Pharmacol.*, 1988, **8** (suppl. 9), S34.
151. J. A. Putkey, G. F. Draetta, G. R. Slaughter, C. B. Klee, P. Cohen, J. T. Stull and A. R. Means, *J. Biol. Chem.*, 1986, **261**, 9896.
152. M. Zimmer and F. Hofmann, *Eur. J. Biochem.*, 1987, **164**, 411.
153. T. Nakajima and A. Katoh, *Mol. Pharmacol.*, 1987, **32**, 140.
154. T. Tanaka, M. Naka and H. Hidaka, *Biochem. Biophys. Res. Commun.*, 1980, **92**, 313.
155. T. Nagatsu, H. Suzuki, K. Kiutchi, M. Saitoh and H. Hidaka, *Biochem. Biophys. Res. Commun.*, 1987, **143**, 1045.
156. C. H. Michnoff, B. E. Kemp and J. T. Stull, *J. Biol. Chem.*, 1986, **261**, 8320.
157. R. B. Pearson, L. Y. Misconi and B. E. Kemp, *J. Biol. Chem.*, 1986, **261**, 25.
158. J. S. Brugge and M. Chinkers, *Annu. Rep. Med. Chem.*, 1983, **18**, 213.
159. T. Hunter and J. A. Cooper, *Annu. Rev. Biochem.*, 1985, **54**, 897.
160. P. Presek and C. Reuter, *Biochem. Pharmacol.*, 1987, **36**, 2821.
161. S. Ito, N. Richert and I. Pastan, *Biochem. Biophys. Res. Commun.*, 1982, **107**, 670.
162. D. W. End, R. A. Look, N. L. Shaffer, E. A. Balles and F. J. Perisco, *Res. Commun. Chem. Pathol. Pharmacol.*, 1987, **56**, 75.
163. M. Hagiwara, S. Inoue, T. Tanaka, K. Nunoki, M. Ito and H. Hidaka, *Biochem. Pharmacol.*, 1988, **37**, 2987.
164. T. Akiyama, J. Ishida, S. Nakagawa, H. Ogawara, S. Watanabe, N. Itoh, M. Shibuya and Y. Fukami, *J. Biol. Chem.*, 1987, **262**, 5592.
165. T. Shiraishi, T. Domoto, N. Imai, Y. Shimada and K. Watanabe, *Biochem. Biophys. Res. Commun.*, 1987, **147**, 322.
166. T. Shiraishi, K. Kameyama, N. Imai, T. Domoto, I. Katsumi and K. Watanabe, *Chem. Pharm. Bull.*, 1988, **36**, 974.

9.1
Cell Wall Structure and Function

J. BARRIE WARD

Glaxo Group Research Ltd, Greenford, UK

9.1.1	INTRODUCTION	553
9.1.2	STRUCTURE OF BACTERIAL WALLS	554
	9.1.2.1 *Structure of Wall Polymers*	557
	9.1.2.1.1 *Peptidoglycan*	557
	9.1.2.1.2 *Secondary wall polymers in Gram-positive bacteria*	559
	9.1.2.1.3 *Outer membrane of Gram-negative bacteria*	561
	9.1.2.1.4 *Major proteins of the outer membrane*	563
9.1.3	BIOSYNTHESIS OF PEPTIDOGLYCAN	564
9.1.4	ANTIBIOTICS INHIBITING PEPTIDOGLYCAN BIOSYNTHESIS	568
	9.1.4.1 *Inhibitors of UDP-N-Acetylmuramoylpentapeptide Biosynthesis*	568
	9.1.4.1.1 *Phosphonomycin (fosfomycin)*	568
	9.1.4.1.2 *Inhibitors of D-alanine metabolism in peptidoglycan synthesis*	569
	9.1.4.2 *Inhibitors of the Lipid Cycle in Peptidoglycan Biosynthesis*	571
	9.1.4.2.1 *Bacitracin*	571
	9.1.4.2.2 *Tunicamycin*	572
	9.1.4.3 *Inhibitors of Polymerization of Peptidoglycan*	573
	9.1.4.3.1 *Glycopeptide antibiotics (vancomycin-like)*	573
	9.1.4.3.2 *Phosphoglycolipid antibiotics (moenomycin-like)*	578
9.1.5	ANTIBIOTICS INHIBITING BIOSYNTHESIS OF OUTER MEMBRANE COMPONENTS	580
	9.1.5.1 *Inhibitors of Lipopolysaccharide (Lipid A) Biosynthesis*	580
	9.1.5.1.1 *Bicyclomycin (bicozamycin)*	581
	9.1.5.1.2 *Globomycin*	583
9.1.6	STRUCTURE OF FUNGAL WALLS	583
9.1.7	INHIBITORS OF FUNGAL WALL POLYMER BIOSYNTHESIS	585
	9.1.7.1 *Polyoxins and Nikkomycins: Inhibitors of Chitin Synthase*	585
	9.1.7.2 *Echinocandins and Papulacandins: Inhibitors of Glucan Synthase*	587
9.1.8	THE CYTOPLASMIC MEMBRANE—STRUCTURE	589
	9.1.8.1 *Antibiotics Causing Disorganization of Membrane Structure*	590
	9.1.8.1.1 *Polymyxins and octapeptins*	590
	9.1.8.1.2 *Tyrocidins and gramicidin S*	592
	9.1.8.1.3 *Gramicidins*	592
	9.1.8.1.4 *Polyenes*	594
	9.1.8.1.5 *Alamethicin, suzukacillin*	596
	9.1.8.2 *Antibiotics Producing Specific Changes in Membrane Permeability*	598
	9.1.8.2.1 *Ionophores including valinomycin, enniatins and macrotetralides*	598
9.1.9	REFERENCES	603

9.1.1 INTRODUCTION

The cell walls of bacteria and fungi are complex structures containing a range of macromolecules, many of which are unique to the microorganisms. Although walls may differ widely in their gross structure (the differences between Gram-positive and Gram-negative bacteria are discussed in more detail below), they all serve to protect the underlying protoplast from mechanical and physico-chemical damage (osmotic damage). The rigidity of the wall confers the characteristic shape of the organism—isolated walls possess the same shape as the intact bacterium or fungal cell. Removal of the structural component of the wall by enzymic digestion results in the organisms forming spherical 'protoplasts' in osmotically protected environments or in lysis and death when no such protection is present. The walls of all bacteria, with the exception of the Archaebacteria, contain peptidoglycan as their main structural component. It is the unique nature of this polymer, in terms of its structure, biosynthesis and function, that focuses attention on the bacterial wall as a target for selective attack by chemotherapeutic agents. Among the most successful of these agents are the β-lactam antibiotics, the penicillins and cephalosporins, described in subsequent chapters.

Fungal walls, although chemically dissimilar to bacterial walls, also contain unique structural macromolecules providing shape and mechanical strength to the organism. These structural polymers, chitin and β-glucan, provide a crystalline framework on which a matrix of amorphous polysaccharides and protein–polysaccharide complexes is arranged. Again, the structure and function of chitin and glucan provide selective targets suitable for attack.

This chapter will describe the structure and function of bacterial and fungal walls. It will also discuss agents other than β-lactam antibiotics that inhibit the synthesis of bacterial wall polymers. In addition, compounds that are known to cause lysis and death of bacteria and fungi by disruption of cytoplasmic membrane function will also be described.

9.1.2 STRUCTURE OF BACTERIAL WALLS

The Eubacteria can be divided into two groups according to the way they stain in the Gram reaction. Although this staining procedure was formulated over a century ago,[1] it has only recently been established that the differential staining reaction is based on differences in the chemical structure of the bacterial walls.[2] Viewed in the electron microscope, walls of Gram-positive bacteria such as *Staphylococcus aureus* are simple in structure, consisting of a fairly thick (20–60 nm) continuous layer, whereas walls of Gram-negative organisms such as *Escherichia coli* and *Pseudomonas aeruginosa* are more complex, consisting of multilayered structures. These differences are equally obvious when isolated walls are chemically analyzed, both in terms of the variety and quantity of the various macromolecules present (Table 1). For a detailed review see Rogers *et al.*[3]

As mentioned briefly in the introduction, both types of wall contain peptidoglycan as the polymer responsible for the shape, structural integrity and insolubility of bacterial walls. The relative amounts of peptidoglycan present vary widely. In Gram-positive bacteria it may account for up to 80% of the total weight of the wall, although values of 40–50% are more commonly found. The remainder of the wall is usually made up of an anionic polymer, a teichoic or teichuronic acid containing either acidic phosphodiester groups (teichoic acids) or acidic carboxy groups on uronic acids (teichuronic acids). These long flexible macromolecules are covalently linked to the peptido-glycan either directly or through a short linkage group. Protein may also be present either as a wall constituent (*e.g.* protein A in *Staph. aureus* and a variety of proteins in streptococci) or at the outer surface of the wall as regular surface arrays such as those occasionally found in *Bacillus sphaericus* and certain clostridia.

The wall of a Gram-negative bacterium is, because of its structural complexity, more commonly described as the envelope. In contrast to the situation in Gram-positive organisms, peptidoglycan represents only a minor fraction of the wall, accounting for perhaps 1–5% of the total weight. In thin sections it can be seen as an electron dense layer lying between the inner and outer membranes,

Table 1 Major Components of Bacterial Walls

Component	Gram-positive	Gram-negative
Peptidoglycan	+	+
Teichoic acid and/or teichuronic acid	+	−
Polysaccharides	+/−	−
Phospholipid	−	+
Lipopolysaccharide	−	+
Lipoprotein	−	+
Proteins	+/−	+

which are of similar appearance. The inner (cytoplasmic) membrane is analogous to the cytoplasmic membrane of Gram-positive bacteria (Figure 1). Physical techniques have now been developed that allow efficient isolation of Gram-negative cell envelopes and their separation into the individual components of peptidoglycan, inner and outer membranes. Analysis of these fractions has allowed the composition of the outer membrane to be determined. It contains approximately equivalent amounts of protein, phospholipid and lipopolysaccharide. However, one of the interesting features of the outer membrane is the asymmetric distribution of these components. In smooth strains of Gram-negative bacteria (particularly members of the Enterobacteriacae, where much of this work has been done) the lipopolysaccharide is confined to the outer leaflet of the membrane with phospholipid being present on the inner face. This distribution results in completely different permeability and fluidity properties of the outer membrane when compared with the cytoplasmic (inner) membrane.[4,5] This property of the Gram-negative envelope with regard to the permeability of hydrophilic substances, including antibiotics, is discussed in detail in a later section. In other Gram-negative bacteria, the so-called deep rough strains, where the lipopolysaccharide lacks the O-antigenic side chains, some phospholipid is also present in the outer leaflet of the membrane. As a consequence the permeability characteristics of these mutants show major differences from the smooth strains. The outer membrane is anchored to the peptidoglycan by a low molecular weight lipoprotein which is covalently linked to the peptidoglycan.[6] This lipoprotein is also present in the outer membrane of *E. coli* in a nonlinked form. Analogous lipoproteins are probably present in the envelopes of other Gram-negative bacteria. They are, however, not the only points of attachment between peptidoglycan and the outer membrane. A number of specific proteins, including porin proteins, have been shown to be noncovalently associated with peptidoglycan.[7,8] Further points of attachment are the adhesion zones described originally by Bayer,[9] which appear to link the cytoplasmic and outer membranes. In *E. coli* these zones of adhesion have been associated with the transfer of newly synthesized lipopolysaccharide from the cytoplasmic to the outer membrane and as attachment sites for various bacteriophages. They do not appear to be involved in the uptake of antibiotics or nutrients. Rather the porin proteins, referred to above, provide selective channels for the uptake of small hydrophilic molecules. These channels are not selective in terms of the molecules that enter by this route other than their limitation on relative molecular mass and the effects of charge characteristics of the molecule crossing the membrane. However, there are, in addition, specific proteins in the outer membrane required for the transport and uptake of substances such as iron–siderophore complexes and maltodextrins. Thus, the outer membrane provides an effective permeability barrier affecting the entry of antibiotics into the Gram-negative bacterium. The importance of this barrier coupled with the high concentrations of β-lactamase that can occur in the periplasm has been of particular interest in determining the resistance of certain Gram-negative pathogens to otherwise extremely effective β-lactam antibiotics. This function of the outer membrane is discussed in greater detail below. Thus the outer face of the outer membrane is composed largely of porin proteins present as triplets to form the channels allowing nonspecific ingress of low

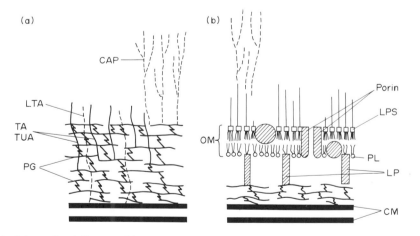

Figure 1 Model of the walls of Gram-positive (a) and Gram-negative (b) bacteria. PG, peptidoglycan; CM, cytoplasmic membrane; OM, outer membrane; LTA, lipoteichoic acid; TA/TUA, teichoic or teichuronic acid; LPS, lipopolysaccharide; LP, bound and free lipoprotein; PL, phospholipid; porin, porin proteins forming aqueous channel across outer membrane; CAP, capsular polysaccharide

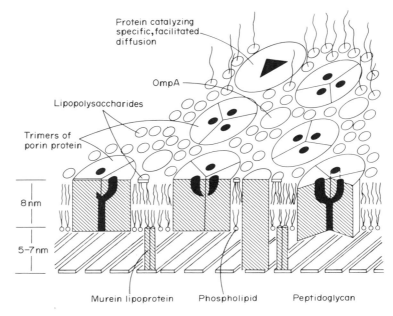

Figure 2 Model of the outer membrane of *E. coli* and *S. typhimurium*. The peptidoglycan sacculus underlies the outer membrane and is attached to it *via* the bound form of lipoprotein. Free lipoprotein may also be present (not shown). The OmpA protein appears to span the thickness of the membrane, it serves as a phage receptor and is labelled in intact cells by nonpenetrating agents while at the same time it can be cross-linked to the peptidoglycan. The porin proteins forming the channels involved in the nonspecific influx of low molecular weight nutrients and antibiotics are shown as trimers and can be extracted in this form by SDS. The specific uptake protein is drawn as a monomer but the LamB protein involved in the uptake of maltodextrins also forms a stable trimer. A variety of other specific transport proteins have been described, including ones required for the uptake of nucleosides and of iron siderophores, but their structure is not yet clear. Lipopolysaccharide is located exclusively in the outer leaflet of the membrane (the saccharide side chain is omitted in many cases and where present is not drawn to scale), while phospholipid is present in the inner leaflet. Further details of outer membrane composition and structure will be found in the reviews of Nikaido *et al.*[4,7,8,33,46] (redrawn from *Microbiol. Rev.*, 1985, **49**, 1 with permission)

molecular weight molecules, together with specific transport proteins, the whole being surrounded by lipopolysaccharide chains (O-antigens) of varying lengths (Figure 2).[7]

Capsules and slime layers are present in certain Gram-positive and Gram-negative bacteria in loose association with the walls.[10,11] In general much, if not all, of this material is lost during the isolation of 'clean' walls. However, the importance of capsules and exopolysaccharides in organisms isolated from natural habitats or grown *in vivo* has been emphasized by Costerton and Marrie.[12] These authors described a glycocalyx surrounding individual bacteria or microcolonies that is thought to help the organisms adhere to, colonize and survive in hostile environments. Such a situation occurs in vegetation seen in staphylococcal endocarditis where the organisms survive antibiotic treatments that would otherwise prove lethal. Clearly such observations present something of a paradox since *in vitro* the capsules do not appear to act as a barrier to the penetration of nutrients. Capsulated and noncapsulated strains are capable of equivalent growth rates and in general do not show significant differences in their susceptibility to antibiotics. In structural terms both homo- and hetero-polysaccharides are found: they are often branched and may contain acidic carboxy residues. Among the former type are the levans and glucans produced in copious amounts by certain streptococci, particularly cariogenic strains. Heteropolysaccharides often contain uronic acids and/or pyruvyl ketal groups, giving the polysaccharide a net negative charge. The acidic nature of these capsules allows them to be visualized in the electron microscope by treatment with cationic dyes (ruthenium red, alcian blue) or cationic ferritin. The initial examples of heteropolysaccharides studied were capsular materials derived from medically important organisms such as *Streptococcus pneumonia* and *Klebsiella aerogenes*. Subsequent work has investigated the alginates produced by *Pseudomonas aeruginosa* (mucoid strains are a major pathogen associated with cystic fibrosis) and the exopolysaccharides of *Xanthomonas campestris*. This polysaccharide has in recent years received considerable attention from commercial laboratories because of its physical properties, which make it suitable as an emulsifying or thickening agent in the food and oil industries.[10,11]

9.1.2.1 Structure of Wall Polymers

9.1.2.1.1 *Peptidoglycan*

The basic structure of peptidoglycan has now been determined for many species and although there is considerable variation in detailed composition, the general structure remains the same. As the name implies, peptidoglycan consists of glycan chains with peptide substituents that are cross-linked. Thus, bacteria are enclosed in a network, the bag-shaped murein of Weidel and Pelzer,[13] which is responsible for the shape of the organism.

The glycan chains contain alternating residues of glucosamine and muramic acid, the 3-O-lactyl ether derivative of glucosamine, in β-1→4 linkage. In most bacteria the amino groups of these hexosamines are N-acetylated, although minor modifications, *e.g.* unsubstituted glucosamine, N-glycoyl- rather than N-acetyl-muramic acid and O-acetylation of N-acetylmuramic acid, are known (for a review see Rogers *et al.*[3]). The glycan chains vary considerably in length from tens to several hundred disaccharide repeating units. Even the shorter chains would be too long to be arranged radially in peptidoglycan and there is now microscopic and physical evidence that the glycan chains of *E. coli* lie more or less perpendicular to the length of the cell.[14,15]

The carboxyl groups of muramyl residues are substituted with short peptides (specific to the bacterium), a proportion of which are cross-linked either directly or through a second short peptide. It is this cross-linkage which joins the glycan chains into a macromolecular network of high tensile strength and rigidity. The primary peptide contains the following amino acid sequence: L-Ala(Gly, L-Ser)-D-Glu-L-R^3-D-Ala.

The R^3 residues are species specific and can be *meso-* or LL-diaminopimelic acid, L-lysine, L-ornithine, L-diaminobutyric acid and, more rarely, L-homoserine or L-alanine. Linkages between the amino acids are α-peptide bonds except for the one between D-glutamic acid and R^3, which is formed by the γ-carboxyl group of the glutamyl residue. In Gram-negative bacteria and *Bacillus* spp. the majority of cross-links are formed directly between the carboxyl group of the C-terminal D-alanine of one peptide side chain and the free amino group on the D centre of *meso*-diaminopimelic acid in a second peptide. However, recent studies using high performance liquid chromatography (HPLC) of muramidase-solubilized peptidoglycan of *E. coli*[16] and *Neisseria gonorrhoeae* revealed a more complex picture.[17] In *E. coli* classical monomers and dimers accounted for only about 70% of the total digest. The remaining 30% was made up of other disaccharide peptides, some of which contained an additional diaminopimelic acid residue. About 20% of the dimers present had this unusual structure, which is associated with the linkage of lipoprotein to the peptidoglycan.

When cross-linkage is indirect, as occurs in the remaining Gram-positive bacteria, one or a number of amino acids are present as the cross-bridge. Cross-bridging usually occurs between the carboxyl group of the C-terminal D-alanine residue of one peptide to the free amino group of a dibasic amino acid. One such example is the pentaglycine cross-bridge in *Staph. aureus* (Figure 3). Where R^3 is not a dibasic amino acid, the bridging peptide contains a dibasic amino acid which is linked to the free α-carboxyl group of glutamic acid. Two major attempts to classify peptidoglycans on the basis of their chemical structure have given major importance to the type of cross-linking present.[18,19]

The extent to which cross-links are formed varies considerably from one bacterium to another. In Gram-negative bacteria and bacilli about 30–50% of the peptides are cross-linked as dimers, whereas in *Staph. aureus* cross-linkage is almost complete and a high proportion of peptides may be isolated as oligomers containing an average of five to eight repeating units.[20]

However, as mentioned briefly above, the overall structure of peptidoglycan is essentially constant, providing a cross-linked lattice structure responsible for the integrity and shape of the bacterial wall. It remains to be determined whether there is any relationship between glycan chain length and extent of cross-linkage. In *Bacillus* spp. and *E. coli*, relatively long glycan chains (45–75 disaccharide units) are associated with lower cross-linking values (30–50%), whereas in *Staph. aureus* short glycan chains (< 10 disaccharides) are highly cross-linked. Such a relationship would allow the hydrolysis of more peptide and glycan bonds without affecting the rigidity of the macromolecule. In this context, recent observations showed that *Staph. aureus* can grow apparently normally with much-reduced cross-linkage.[21] Undoubtedly, a complex situation exists *in vivo*. In Gram-positive bacteria the peptidoglycan is made up of many layers in which the glycan chains are not ordered and cross-linkage occurs within as well as between layers. Even in Gram-negative bacteria there may be two or three layers of peptidoglycan, although the actual value has been disputed.[14]

Uncross-linked peptides are probably interspersed throughout the peptidoglycan; there is no evidence that suggests any specific location. Peptides that do not become involved in cross-links

(a) *E. coli*

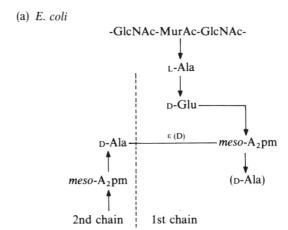

```
            -GlcNAc-MurAc-GlcNAc-
                      │
                      ▼
                    L-Ala
                      │
                      ▼
                    D-Glu ──────────┐
                                    │
                                    ▼
  D-Ala ──────ε (D)────── meso-A₂pm
    ▲
    │                              │
    │                              ▼
  meso-A₂pm                     (D-Ala)

  2nd chain    │    1st chain
```

(b) *M. luteus*

```
            -GlcNAc-MurAc-GlcNAc-
                      │
                      ▼
                    L-Ala
                      │
                      ▼
                    D-Glu ───γ───┐
                      │          │
                      ▼          │
                     Gly         │
                                 │
         L-Lys ──→ D-Ala ──ε──→ L-Lys
           ▲                      │
           │                      ▼
  D-Ala ─→ L-Ala ─→ D-Glu ─→ Gly  D-Ala

  L-Lys
    ▲
    │
  2nd chain        Bridge       1st chain
```

(c) *S. aureus* Copenhagen

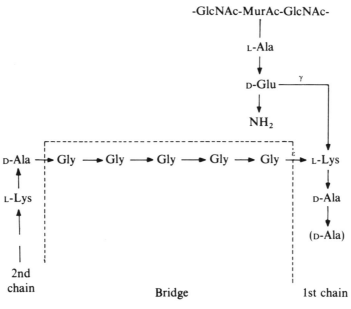

```
            -GlcNAc-MurAc-GlcNAc-
                      │
                      ▼
                    L-Ala
                      │
                      ▼
                    D-Glu ───γ───┐
                      │          │
                      ▼          │
                     NH₂         │
                                 ▼
  D-Ala ─→ Gly → Gly → Gly → Gly → Gly ─ε─→ L-Lys
    ▲                                        │
    │                                        ▼
  L-Lys                                     D-Ala
    ▲                                        │
    │                                        ▼
  2nd                                      (D-Ala)
  chain              Bridge              1st chain
```

Figure 3 Examples of peptidoglycan structure

often lose either one or both terminal D-alanine residues through the action of carboxypeptidases. Hence, peptidoglycan isolated from walls of many bacteria contains on average one or less D-alanine residues per glutamic acid residue.

9.1.2.1.2 *Secondary wall polymers in Gram-positive bacteria*

The walls of Gram-positive bacteria contain, in addition to peptidoglycan, a variety of other polymers. Some of these, such as the teichoic and teichuronic acids together with certain polysaccharides found particularly in streptococci and lactobacilli, are covalently linked to peptidoglycan (for reviews see refs. 3 and 22). Others, including various protein antigens of streptococci and staphylococci are found in close association with the wall but are not linked and can be removed by detergent extraction of wall preparations.

The most abundant of secondary wall polymers isolated from Gram-positive bacteria are the teichoic acids. Originally this term was used to describe polymers of glycerol phosphate or ribitol phosphate substituted to various extents with ester-linked D-alanine and often glycosyl residues. It rapidly became clear, however, that many variations existed in bacteria and the definition has been broadened in recent years to describe phosphate-containing wall polymers in general. A detailed description of the many teichoic acids characterized would be out of place in a chapter such as this: some examples and generalized structures are shown in Figure 4.

The second group of anionic polymers present in walls of Gram-positive bacteria are the teichuronic acids. The teichuronic acids of *B. subtilis* and *B. licheniformis* 6346 contain equimolar amounts of D-glucuronic acid and *N*-acetylgalactosamine, the repeating unit being 4-D-glucuronyl-(β-1 → 3)-*N*-acetylgalactosamine (Figure 5).[23] Recently, as a result of detailed chemical and NMR studies, the teichuronic acid of *B. licheniformis* 9945 was characterized as having the following tetrasaccharide repeating unit: → 4)GlcU(β-1 → 4)GlcU(β-1 → 3)GalNAc(α-1 → 6)GalNAc-(β-1 →).[24] The primary structure of the teichoic acid of *B. megaterium* M46 is unusual in that the uronic acid residues do not belong to the backbone of the polymer.[25]

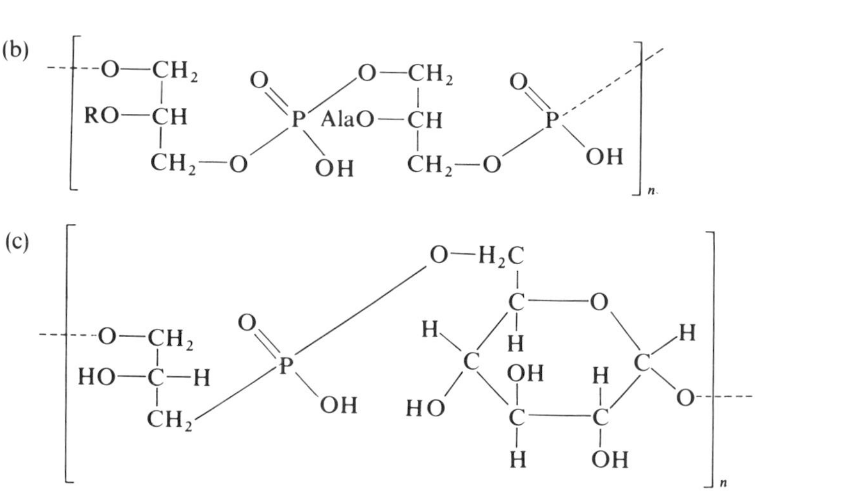

Figure 4 Generalized structures of some teichoic acids: (a) ribitol teichoic acid where R = β-glucosyl residues in *Bacillus subtilis* W23 and α- or β-linked *N*-acetylglucosamine in various strains of *Staphylococcus aureus*; (b) glycerol teichoic acid where R = α-glucosyl residues in *B. subtilis* 168; and (c) glucosylglycerol phosphate teichoic acid found in the wall of *B. licheniformis* 9945A. In each case ester-linked D-alanyl substituents are also present

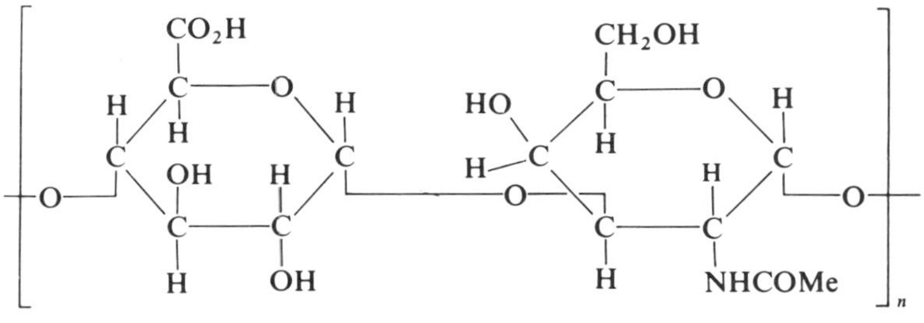

Figure 5   Generalized structure of the teichuronic acid present in the walls of *B. licheniformis* 6346

Teichoic and teichuronic acid are both linked by phosphodiester bonds to between 5 and 10% of the muramyl residues in peptidoglycan.[22] Attachment of teichoic acids involves a short linkage unit consisting of glycerol phosphate residues, *N*-acetylglucosamine and *N*-acetylaminomannuronic acid, the phosphate residue which forms the phosphodiester bond to the 6-hydroxyl of muramic acid being derived biosynthetically from UDP-*N*-acetylglucosamine. In contrast, those teichuronic acids studied, with the exception of the polymer in walls of *Micrococcus luteus*, are attached directly to peptidoglycan: the phosphodiester bond is formed directly between the reducing terminus of the polymer and the glycan. A number of lines of evidence based on growth of bacteria under various nutrient limitations and the effect of β-lactam antibiotics on wall synthesis suggest that teichoic and teichuronic acids are only linked to concomitantly synthesized and cross-linked peptidoglycan. In a cell free system from *B. licheniformis*, teichoic and teichuronic acids were linked to different glycan chains; whether this represents the situation *in vivo* remains unknown.[26] When compared to the glycan chains, the lengths of these anionic polymers are considerably shorter, possessing average chain lengths of 20–40 residues for teichoic acids in staphylococci and *Bacillus* spp. and the teichuronic acid of *B. licheniformis* and *B. subtilis*. Little is known concerning the organization of the various polymers within the wall. Teichoic acids act as receptors for certain bacteriophages and can also interact with lectins such as concanavalin A and with antibodies. In fact antisera raised against *Staph. aureus* and *B. licheniformis* contain antibodies directed chiefly against their respective teichoic acids. Clearly, some polymer chains are exposed and extend from the cell surface, whereas others are undoubtedly buried within the depth of the wall. Evidence for this latter statement comes from the finding that binding of concanavalin A increases on partial autolysis of the wall (for a review see ref. 22).

The incidence of teichoic and teichuronic acids in the wall seems in many bacteria to be regulated by the availability of phosphate in the medium. Thus a deficiency of phosphate leads to the loss of teichoic acids from the wall and their replacement by teichuronic acid. On the other hand, growth under conditions of magnesium limitation results in the maximal presence of teichoic acids.

The majority of Gram-positive bacteria also contain a second form of teichoic acid, which, although not covalently linked to peptidoglycan, is intimately associated with the wall. These are the lipoteichoic acids (LTA), which, as their name implies, consist of a teichoic acid chain, commonly D-alanyl-substituted 1,3-poly(glycerol phosphate) attached to a glycolipid. Glycosyl substituents are frequently present in addition. The glycolipid serves to anchor the polymer in the cytoplasmic membrane and the hydrophilic teichoic acid chain extends away from the membrane into and in some cases through the wall. Thus, LTA is exposed on the surface of certain group D streptococci and lactobacilli. Lipoteichoic acid continues to be synthesized even under conditions of phosphate limitation. In *M. luteus* and certain other micrococci lipoteichoic acid is replaced by an alternative negatively charged polymer, succinylated lipomannan.[27]

By virtue of their negatively charged groups, the anionic polymers are able to bind bivalent cations, particularly Mg^{2+}. This is thought to be required for the correct functioning of many wall synthetic enzymes located in the cytoplasmic membrane. Various studies have sought to confirm this hypothesis but conclusive evidence is still lacking. Other functions ascribed to these polymers relate to the maintenance of cell shape. Mutants lacking (partially) anionic polymers in their walls grew in aberrant shapes, although the degree of cross-linkage of their peptidoglycan was unaffected. Whether these morphological responses result from some control by the anionic polymers of wall-bound autolytic activity remains unclear. In several bacteria lipoteichoic acid has been shown to regulate autolytic enzyme activity *in vitro*.[3,28]

Walls of other organisms, particularly lactobacilli and streptococci, contain polysaccharides that are covalently linked to peptidoglycan. These represent some of the major antigenic determinants of these bacteria and have been used in their classification as group specific antigens.[29] Some chemical analysis has been carried out and, for example, the streptococcal group A polysaccharide was shown to be composed of *N*-acetylglucosamine and rhamnose with the antigenic determinant being the terminal *N*-acetylglucosamine β-linked to rhamnose.[30]

9.1.2.1.3 *Outer membrane of Gram-negative bacteria*

The earliest studies of the chemistry of walls of Gram-negative bacteria showed them to be more complex than those of Gram-positive organisms. In addition to peptidoglycan (which, as described above, is a minor component) they contained protein, phospholipid and complex polysaccharide material, the lipopolysaccharide. Electron microscopic investigation revealed the presence of an outer membrane, which in section had the typical appearance of a unit membrane, that is outer and inner electron dense leaflets separated by an intermediate electron transparent zone. Subsequent detailed investigations have estimated the number of molecules of the components present in 1 μm^2 of outer membrane to be 10^5 molecules of protein, 10^6 molecules of phospholipid and 10^5 molecules of lipopolysaccharide.[4,5,31]

The presence of this membrane presents a barrier to the penetration of a wide variety of molecules, making Gram-negative bacteria less permeable (and in the case of noxious agents often less susceptible) than Gram-positive organisms. Thus, the permeability properties of the outer membrane play a major role in the resistance of many Gram-negative bacteria to antibiotics and antimetabolites by preventing access of these molecules to their sites of action in the cytoplasmic membrane. The outer membrane is also vital in protecting enteric bacteria from enzymic attack and the detergent action of bile salts in the gut.

Detailed studies have been made of the outer membranes of certain enteric bacteria, particularly *E. coli* and various *Salmonella* spp. This work was stimulated by the finding that LPS determined the O-antigen specificity of the organism. Lipopolysaccharide is also responsible for the endotoxic activity of these bacteria and acts as the receptor for many bacteriophages. Thus, LPS is of considerable importance, not only for its involvement in the pathogenicity of many Gram-negative bacteria, but also from both the immunological and taxonomic points of view.[3,32]

Evidence for the barrier function of the outer membrane came first from studies of the action of ethylenediaminetetraacetic acid (EDTA) on Gram-negative bacteria. Treatment of *E. coli* with low concentrations of EDTA results in a marked increase in the permeability of the outer membrane and the bacteria become sensitive to a wide variety of antibiotics to which they were previously resistant. The organisms also become sensitive to detergents and to peptidoglycan-degrading enzymes such as lysozyme. On the other hand, the bacteria remain viable and appear to grow normally. The EDTA treatment extracts about 30–50% of the total lipopolysaccharide and 5–20% of the phosphatidyl-ethanolamine present in the outer membrane but relatively little protein is removed. The EDTA is believed to act by chelating Mg^{2+} present in the outer membrane that is responsible for maintaining phospholipid–LPS and protein–LPS interactions.[33] Mutants of *E. coli* lacking the outer polysaccharide and much of the core regions of LPS (see below) show similar permeability characteristics to the EDTA-treated organisms.

In structure, LPS is a complex amphipathic molecule that can be subdivided into three distinct regions (Figure 6). These are the O side chain, a long polysaccharide composed of specific repeating units showing great variability among different organisms, linked through a short oligosaccharide, the core region, to a glucosamine-containing lipid, lipid A.[32,34] The sugars making up the core region, which can be further subdivided into the outer core oligosaccharide and the backbone region (inner core), include L-glycero-D-mannoheptose and 3-deoxy-D-manno-2-octulosonate, components apparently unique to LPS. Attached to these are phosphorylethanolamine and pyrophosphoryl-ethanolamine substituents. Smooth strains of enteric bacteria, which tend to be the more virulent forms of the organisms, possess complete LPS. However, large numbers of rough mutants have been isolated which contain varying amounts of the LPS molecule. These range from a reduction in the amount of the O side chain to those having only one or two sugar residues of the inner core region. Lipid A, the component of LPS responsible for endotoxic activity, is always present. The only mutants described in lipid A biosynthesis are temperature sensitive and fail to grow at the restrictive temperature. The ketoside linkage between the terminal KDO residue of the core and lipid A is acid labile, hence the two parts of the molecule can be readily separated by mild acid hydrolysis and solvent extraction. The structure of lipid A from a variety of enteric bacteria and *Pseudomonas* spp. is conserved. It is a glucosaminyl-β-1 → 6-glucosamine disaccharide substituted with amide or ester-linked hydroxy fatty acids, particularly β-hydroxymyristic acid, straight chain fatty acids, phosphate, ethanolamine and 4-aminoarabinose phosphate (Figure 7).[34-37]

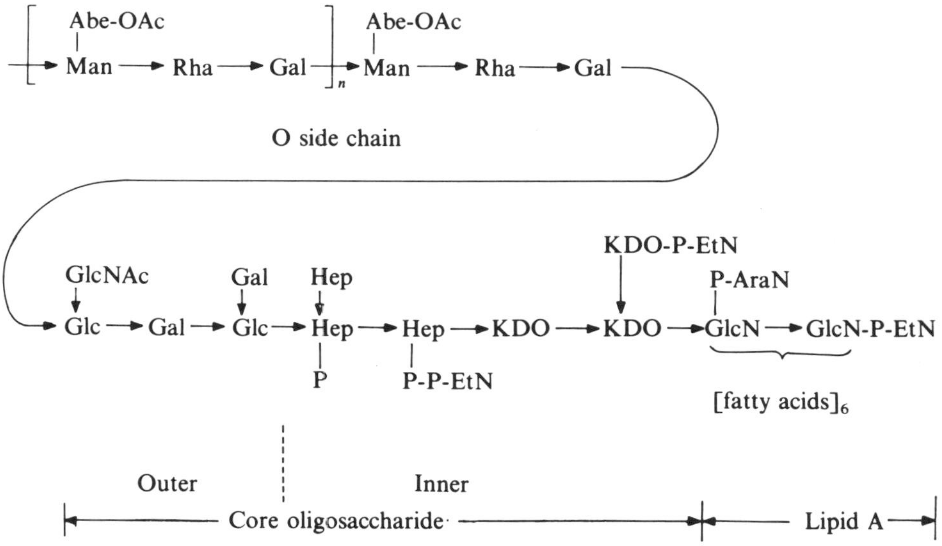

Figure 6 Structure of the lipopolysaccharide of *Salmonella typhimurium*. The molecule is made up of the O side chain (Abe, abequose; Man, D-mannose; Rha, L-rhamnose; OAc, O-acetyl) linked *via* the core oligosaccharide (Gal, D-galactose; GlcNAc, N-acetylglucosamine; Glc, D-glucose; Hep, L-glycero-D-mannoheptose; KDO, 3-deoxy-D-mannooctulosonic acid; EtN, ethanolamine) to lipid A (GlcN, D-glucosamine; AraN, 4-aminoarabinose). The fatty acid substituents are present in both amide and ester linkage

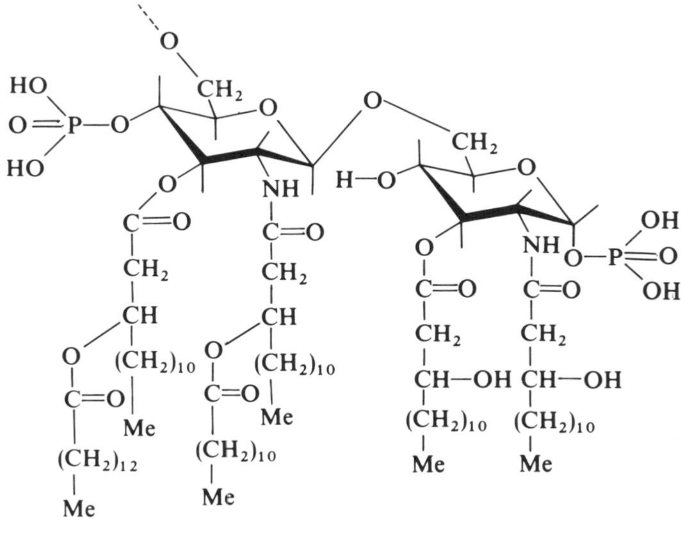

Figure 7 Structure of lipid A from the lipopolysaccharide of *S. typhimurium*. The inner core oligosaccharide is linked to C-6 of the nonreducing glucosamine residue. The glucosaminyl–glucosamine disaccharide is substituted with six (shown here) or possibly seven fatty acid residues in amide and ester linkage. Some fatty acid residues are linked to the 3-hydroxy group of another fatty acid, giving the characteristic 3-acyloxyacyl structure. The phosphate residues are further substituted with ethanolamine and 4-aminoarabinose as shown in Figure 5

One of the characteristics of LPS is its microheterogeneity. Polysaccharide separated from lipid A by mild acid hydrolysis was shown to be heterogeneous with respect to size, with molecules differing in the number of saccharide repeating units from zero to approximately 40.[38,39] Changes in the amount of LPS present have been shown to alter antibiotic sensitivity, presumably by affecting the permeability characteristics of the outer membrane. As the amount of LPS is decreased in rough mutants, increasing amounts of phospholipid are found in the outer leaflet of the outer membrane. In these organisms the permeability of the outer membrane to hydrophobic molecules begins to approach that of the cytoplasmic membrane.[33] The effects of EDTA on outer membrane integrity were described above. Clearly, the acidic lipopolysaccharide molecules may bind magnesium ions and therefore fulfil a similar function in Gram-negative organisms to the teichoic and teichuronic acids in Gram-positive bacteria.

Gram-negative bacteria lacking almost all their lipopolysaccharide (*i.e.* deep rough mutants) remain viable and grow relatively normally in the laboratory. These observations suggest that much of the LPS molecule is not essential for survival and therefore not a target for selective antibiotic attack. However, the inner core and lipid A appear to be essential and it is particularly interesting that selective inhibitors of KDO and lipid A biosynthesis have antibacterial activity. The first of these, a diazoborane 844674, inhibited the incorporation of lipids into lipid A, but proved to be too toxic for chemotherapeutic use.[40] More recently, Hammond *et al.*[41] reported that 2,6-anhydro-3-deoxy-D-glycero-D-talooctonic acid and the 8-amino-8-deoxy derivative were potent inhibitors of CMP-KDO synthetase ($K_i = 4\ \mu M$). The compounds alone were only weakly active against whole organisms due to their inability to cross the cytoplasmic membrane. However, antibacterial activity was dramatically increased by linking a short peptide to the 8-amino moiety, thereby facilitating transport across the cytoplasmic membrane by the bacterial peptide permeases. A similar strategy was adopted earlier by investigators from Roche for alafosfalin, where 1-aminoethylphosphonic acid, an inhibitor of alanine racemase, was linked to L-alanine to facilitate active transport into Gram-negative bacteria.[42]

9.1.2.1.4 *Major proteins of the outer membrane*

The proteins of the outer membrane fall essentially into three categories: a low molecular weight lipoprotein first described in *E. coli* but present in a wide range of Gram-negative bacteria, a family of transmembrane proteins providing channels in the membrane through which low molecular weight hydrophilic substances can diffuse and proteins concerned with the uptake of specific nutrients. Many of this latter group are inducible or derepressible proteins that would normally not be described as 'major' proteins. However, the amounts present in the membrane can dramatically increase under certain conditions of nutrient limitation.

(i) Lipoprotein

This is the most abundant protein in the cell envelope, there being about 7.5×10^5 copies per cell. One-third of the molecules ($M_r \approx 7000$) are covalently linked to peptidoglycan and the remaining two-thirds are free in the outer membrane. Attachment of the lipoprotein to peptidoglycan was reported to occur through linkage of the ε-amino group of the terminal lysine residue of the polypeptide to the α-carboxyl group of the L centre of a *meso*-diaminopimelic acid residue that forms the C terminus of a tripeptide subunit linked in the usual way to muramic acid.[6,43] More recently, lipoprotein was found attached to dimeric and trimeric subunits of peptidoglycan. These studies, which included pulse-labelling experiments and HPLC analysis of peptidoglycan, also suggested that attachment of lipoprotein was preceded by the formation of novel diaminopimelyl–diaminopimelic acid cross-links found in the peptidoglycan of *E. coli*.[16] The lipophilic properties of the protein result from the presence of a fatty acid and a diglyceride substituent on the terminal cysteine residue.

Treatment with trypsin cleaves the bound lipoprotein from peptidoglycan. Similar treatment of cell envelopes results in detachment of the outer membrane from the peptidoglycan. Thus, *in vivo* bound lipoprotein may serve to anchor the outer membrane to the peptidoglycan and hence stabilize it as a barrier to the environment. This conclusion is supported by the observation that mutants lacking lipoprotein grow and divide normally but are very sensitive to detergents and EDTA and tend to leak periplasmic proteins. Moreover, many surface blebs are seen in electron micrographs, presumably as a consequence of the decreased stability of the outer membrane.[44]

(ii) Porin proteins

Porins in most Gram-negative bacteria are major outer membrane proteins of molecular weight 35 000–40 000. Rosenbusch[45] estimated that *E. coli* contained about 10^5 porin protein molecules per cell. Subsequently, Nikaido and his colleagues, using closed lipid vesicles, demonstrated the role of these proteins in the uptake of hydrophilic low molecular weight nutrients across the outer membrane.[4] The porin channels of *E. coli* and *Salmonella typhimurium* were shown to be nonspecific but with a size exclusion limit of about 700 Da. Multiple species of porins are produced by enteric bacteria; *E. coli* K-12 contains Omp F (outer membrane protein F) and Omp C porins, while proteins of similar size and function are known to be present in a wide range of other Gram-negative bacteria. However, the size exclusion limit may be quite different; that of the *P. aeruginosa* porin was reported to be 4000 Da. In the outer membrane the porin proteins are thought to be present as trimers, the individual transmembrane porin molecules surrounding a central aqueous channel with estimated diameters of 1.16 (Omp F) and 1.04 nm (Omp C) in *E. coli* (Figure 2). Those outer membranes with small exclusion limits (*E. coli*, *Salmonella* spp.) will exclude large antibacterial agents like vancomycin, bacitracin, polymyxins and ionophores on the basis of size. In *P. aeruginosa* the channel diameter is about 1.75 nm. However, only about 100–300 porins per cell form functional channels, resulting in a very low uptake by this route. Moreover, charged amino acids are present at the protein/channel interface of all porins, which further restrict the movement of even small hydrophilic compounds. Thus, by different methods the two forms of bacteria have established the outer membrane as a barrier, while allowing the passage of essential nutrients.

The role of porins in the uptake of β-lactam antibiotics has been investigated in detail by Nikaido and his colleagues. Using a series of mutants of *E. coli* constructed to contain only one type of porin, the effect of charge and hydrophobicity on the permeability of a series of penicillins and cephalosporins was determined. The presence of a net positive charge markedly enhanced the rate of diffusion across outer membranes, whereas an increase in hydrophobicity resulted in a decreased rate of diffusion. Thus, the outer membrane acts as a barrier to the uptake of β-lactam antibiotics, while preventing the release of β-lactamases into the surrounding environment. As a consequence the β-lactamase becomes concentrated in the periplasmic space, an area operationally defined as being between the outer membrane and the peptidoglycan. The role of the two factors—rate of uptake and rate of hydrolysis by periplasmic β-lactamase as a mechanism of resistance in Gram-negative bacteria—has received considerable attention in the last few years. For further information the reader is directed to the excellent reviews by Nikaido.[7,8,46]

9.1.3 BIOSYNTHESIS OF PEPTIDOGLYCAN

Despite wide differences in bacterial wall structure from both a chemical and morphological point of view, peptidoglycan synthesis shows an overall similarity which confirms the fundamental nature of the process. Three stages of the biosynthetic pathway can be identified, each occurring at a different cellular site. For detailed information the reader is directed towards a number of recent reviews;[3,47,48] the following account provides a brief outline.

The first stage of peptidoglycan biosynthesis, catalyzed by soluble enzymes in the cytoplasm, is the synthesis of the nucleotide-linked precursors, UDP-*N*-acetylglucosamine and UDP-*N*-acetylmuramoyl-L-Ala-D-Glu-R^3-D-Ala-D-Ala (UDP-*N*-acetylmuramoyl pentapeptide) where R^3 is usually *meso*-A$_2$ pm or L-Lys (Figure 8). The initial reaction is the formation of UDP-*N*-acetylglucosamine from UTP and *N*-acetylglucosamine 1-phosphate. Since *N*-acetylglucosamine is a common constituent of other bacterial wall polymers and the glycosyl substituents of mammalian proteins, this reaction is not specific to peptidoglycan. However, muramic acid is a unique constituent of peptidoglycan and the reactions involving the transfer of enolpyruvate to UDP-*N*-acetylglucosamine and the subsequent reduction of the enolpyruvyl ether to give UDP-*N*-acetylmuramic acid are the first specific steps of peptidoglycan biosynthesis. The inhibition of the transferase by phosphonomycin is described in detail below.

Conversion of UDP-*N*-acetylmuramic acid to UDP-*N*-acetylmuramoylpentapeptide occurs by the sequential addition of three amino acids and the preformed dipeptide D-alanyl-D-alanine. In each case the addition is catalyzed by a specific ligase which requires a divalent cation (Mn^{2+} or Mg^{2+}) and the hydrolysis of ATP for activity. The ordered formation of the peptide is dependent upon the substrate specificities of the ligases and thus is in marked contrast to protein biosynthesis, where the addition of amino acids is directed by a nucleic acid template. D-Glutamic acid, one of the amino acids characteristic of peptidoglycan, is produced by the transamination of D-alanine and α-ketoglutamate in a reaction inhibited by β-halo- (*i.e.* chloro- and fluoro-) D-alanine, D-cycloserine and DL-gabaculine (5-amino-1,3-cyclohexyldienylcarboxylic acid).[49,50]

The synthesis of the dipeptide from L-alanine and its addition to UDP-*N*-acetylmuramoyltripeptide has been studied in detail by Neuhaus and his colleagues. The formation of the dipeptide is unique to bacteria and is of particular interest since a number of inhibitors have been described.[51] The conversion of L- to D-alanine is catalyzed by alanine racemase in a reaction which is irreversibly inhibited by cycloserine, the haloalanines and *O*-carbamoyl-D-serine.[52] L-1-Aminoethylphosphonic acid also inhibits alanine racemase[53] and in recent studies was incorporated into a peptide as L-alanyl-L-1-aminoethylphosphonic acid (alafosfalin) to facilitate transport across the bacterial mem-

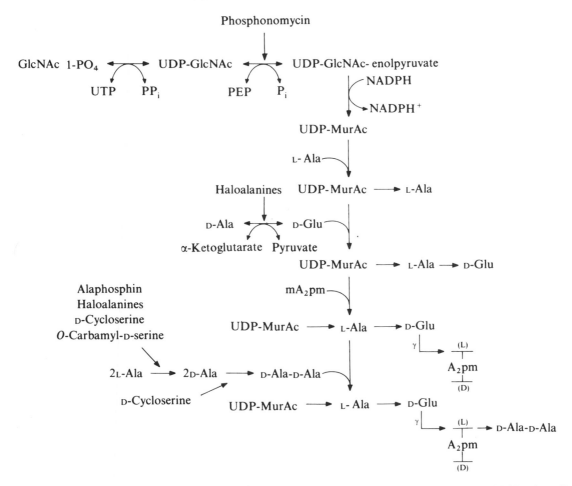

Figure 8 The reactions involved in the biosynthesis of UDP-*N*-acetylmuramoyl pentapeptide and their inhibition by wall antibiotics. The reactions shown are those for the biosynthesis of *meso*-diaminopimelic acid containing UDP-MurAc-pentapeptide which is present in most Gram-negative bacteria, bacilli and certain clostridia. In *Staphylococcus aureus* L-lysine replaces *meso*-diaminopimelic acid. Each of the ligases requires either Mn^{2+} or Mg^{2+} and the hydrolysis of ATP for activity.

brane. The peptide, after being transported into the bacteria, was hydrolyzed intracellularly to yield the antibacterial agent.[42] D-Alanyl-D-alanine synthetase catalyzes the formation of the dipeptide and shows absolute specificity for D-amino acids or glycine. In fact, D-aminobutyric acid was the only analogue found to bind at the donor (N-terminal) binding site. Incorporation of other D-amino acids occurred only in the presence of D-alanine to produce mixed dipeptides with N-terminal D-alanine. The enzyme is inhibited by D-cycloserine. Addition of the dipeptide to UDP-*N*-acetylmuramoyltri-peptide shows the opposite specificity, *i.e.* D-alanine is required as the C-terminal residue with a low specificity for the N terminus. Acting in concert, the synthetase and the adding enzyme ensure that the completed UDP-*N*-acetylmuramoylpentapeptide almost always ends in D-alanyl-D-alanine.

The second stage of peptidoglycan biosynthesis involves the transfer of *N*-acetylglucosamine and *N*-acetylmuramoylpentapeptide from the intracellular sites of synthesis to the membrane (presumably the exterior surface), where they can be incorporated into the growing peptidoglycan (Figure 9). Undecaprenyl phosphate, acting as a lipophilic carrier, plays a central role in this second stage, the reactions of which are catalyzed exclusively by membrane-bound enzymes. The first reaction involves the transfer of phospho-*N*-acetylmuramoylpentapeptide from the nucleotide to undecaprenyl phosphate. The products of the reaction are undecaprenylpyrophosphoryl-*N*-acetylmuramoylpentapeptide and UMP. The reaction, called a translocase because it involves an exchange of carriers by the cleavage and synthesis of pyrophosphoryl bonds of equivalent reactivity,[54] is inhibited by tunicamycin and amphomycin.[55-57] *N*-Acetylglucosamine is then added in a transglycosylation reaction to form undecaprenylpyrophosphoryl-*N*-acetylmuramoylpentapeptide-*N*-acetylglucosamine. Nisin, a peptide antibiotic produced by *Streptococcus lactis*, was reported to

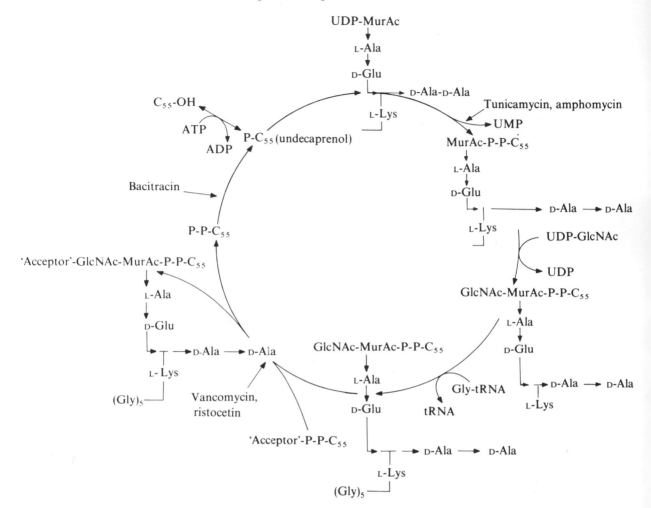

Figure 9 The biosynthesis of uncross-linked peptidoglycan in *S. aureus* and its inhibition by wall antibiotics. Phospho-*N*-acetylmuramoyl pentapeptide is translocated from the nucleotide-linked precursor to the membrane-bound lipid carrier, undecaprenyl phosphate (C_{55}-P). The transfer of *N*-acetylglucosamine follows to complete formation of the disaccharide pentapeptide repeating unit. Addition of the pentaglycine cross-bridge and amidation of the α-CO_2H of D-glutamate (not shown) then occurs before polymerization of the newly synthesized disaccharide–peptide takes place. Covalent linkage to the wall requires the formation of cross-links by transpeptidation

inhibit the addition of *N*-acetylglucosamine by complexing with undecaprenylpyrophosphoryl-*N*-acetylmuramoylpentapeptide.[58] Modifications of the pentapeptide side chain, such as the amidation or glycylation of the α-carboxyl group of D-glutamic acid or the addition of cross-bridge amino acids, generally occur at the lipid disaccharide stage of the biosynthetic cycle.

   The third stage of the biosynthetic process, also catalyzed by membrane-bound enzymes, involves the polymerization of the newly synthesized subunits and their incorporation into the growing peptidoglycan. In bacilli and *M. luteus* polymerization occurs by the transfer of the *N*-acetylmuramoyl terminus of the growing glycan chain to the nonreducing *N*-acetylglucosamine residue of the newly synthesized disaccharide peptide subunit, which remains attached to the membrane by an acid labile bond (presumably to undecaprenyl pyrophosphate, the lipid carrier on which synthesis had occurred). In this way the glycan chain of peptidoglycan is increased by one disaccharide unit and undecaprenyl pyrophosphate is released. In order to reenter the biosynthetic cycle undecaprenyl pyrophosphate must first be dephosphorylated to the monophosphate. It is this dephosphorylation reaction that is inhibited by bacitracin.[59] The formation of nascent (uncross-linked) peptidoglycan as a result of transglycosylation has been reported to occur in several organisms (for review see ref. 48), but the involvement of polymeric intermediates in peptidoglycan biosynthesis remains unclear. It is clear, however, that a number of bacteria continue to synthesize uncross-linked peptidoglycan when cross-linkage (transpeptidation) is inhibited by β-lactam antibiotics.[60-63] Moenomycin and related antibiotics inhibit the synthesis of uncross-linked peptidoglycan in

E. coli[64,65] by direct inhibition of the transglycosylase (peptidoglycan polymerase). Vancomycin, ristocetin and related glycopeptide antibiotics also inhibit this second stage of peptidoglycan synthesis by formation of complexes between the antibiotic and the acyl-D-alanyl-D-alanine terminus of the nascent peptidoglycan.[66]

The nascent peptidoglycan is attached to the preexisting peptidoglycan by the formation of cross-links between peptide side chains of both polymers. Even when uncross-linked peptidoglycan has glycan chains several hundred disaccharide units long, it remains water soluble. Only when cross-links are formed does peptidoglycan become an insoluble matrix capable of maintaining the structural integrity of the wall. Cross-linkage occurs by a transpeptidase reaction in which a peptide bond is formed between the penultimate D-alanine residue of one peptide side chain and the free amino group of the R³ residue or its substituent amino acid (or amino acid chain) of the other peptide. Detailed studies on transpeptidases from a range of bacteria have established that the first stage of the reaction involves the cleavage of the terminal D-alanyl-D-alanine bond of the donor peptide to form an acyl-D-alanyl–enzyme intermediate *via* the carbonyl group of the penultimate D-alanine residue and the immediate release of the terminal D-alanine.[67,68] In the second stage of the reaction the acyl-D-alanyl substituent reacts either with the amino group of a second acceptor peptide to form a cross-link (transpeptidation) or with water to release acyl-D-alanine carboxypeptidase (Figure 10). Both transpeptidation and carboxypeptidases are inhibited by β-lactam antibiotics acting as analogues of acyl-D-alanyl-D-alanine, as originally proposed by Tipper and Strominger.[67]

The donor peptide can be located in either the newly synthesized material or the preexisting peptidoglycan. In bacilli the newly synthesized pentapeptide side chains act as the donor peptides and cross-linkage occurs to free amino groups in the peptide side chains of the preexisting wall.[69,70] In *Gaffkia homari* the opposite situation exists and cross-linked peptidoglycan can be synthesized

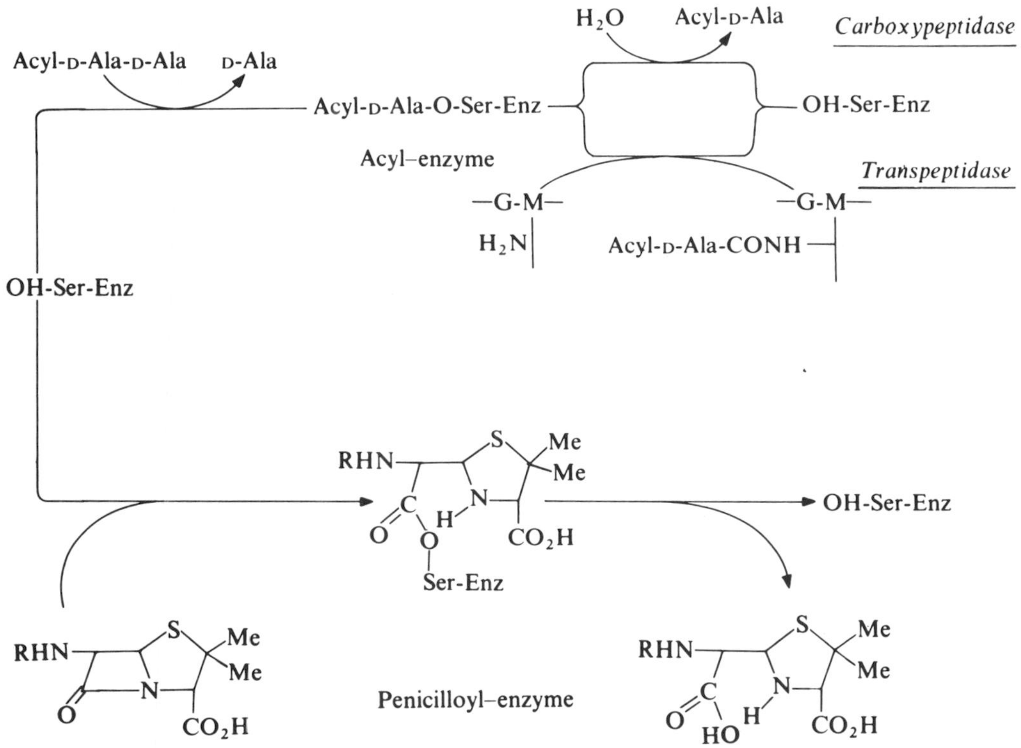

Figure 10 Formation of acyl and penicilloyl intermediates in transpeptidase and D, D-carboxypeptidase reactions. A seryl residue in the donor site of the enzyme reacts either with acyl-D-alanyl-D-alanine or penicillin to yield either acyl-D-alanyl–enzyme and D-alanine or penicilloyl–enzyme intermediates. If water binds to the acceptor site the enzyme functions as a DD-carboxypeptidase; acyl-D-alanine is released and the active enzyme regenerated. Alternatively if a suitable amino acceptor binds to the acceptor site then transpeptidation ensues. Although both functions are shown for the same enzyme individual enzymes tend to favour one reaction, *i.e.* they are transpeptidases with inefficient D,D-carboxypeptidase activity or the reverse may be the case. In the case of penicilloyl–enzyme intermediates these may also react with water to regenerate active enzyme and yield penicilloic acid or alternatively the penicillin nucleus may undergo further degradation to *N*-formylpenicillamine and in the case of benzylpenicillin, phenylacetylglycine

by cell free preparations from UDP-*N*-acetylmuramoyl-tetra- or -tri-peptide and UDP-*N*-acetyl-glucosamine as the only precursors. Thus, the acyl-D-alanyl-D-alanine terminus required for cross-linkage must be provided by the preexisting peptidoglycan.[71-74]

All bacteria studied contain multiple penicillin-binding proteins and these proteins have been shown to catalyze penicillin sensitive transpeptidase and DD-carboxypeptidase reactions (for recent reviews see refs. 48, 75-79). Evidence from biochemical, genetic and morphological studies has established that in *E. coli* individual penicillin-binding proteins are associated with cell elongation (PBP-1), maintenance of cell shape (PBP-2) and separation (PBP-3).[75,80] Moreover, these proteins catalyze both a penicillin insensitive transglycosylation resulting in the synthesis of linear uncross-linked peptidoglycan and a penicillin sensitive transpeptidation reaction.[81,83,84] Thus the three high molecular weight PBPs of *E. coli* are bifunctional enzymes with separate functional domains. The low molecular weight penicillin-binding proteins (PBPs 4, 5 and 6) all catalyze DD-carboxypeptidase activity and, in addition, PBP-4 is a secondary transpeptidase catalyzing the formation of additional cross-links following the initial attachment of nascent peptidoglycan to the preexisting wall.[82] Much less is known about the enzymic activities of penicillin-binding proteins of other Gram-negative and Gram-positive bacteria. However, by analogy with *E. coli*, other organisms contain multiple transpeptidases, although in almost all cases only DD-carboxypeptidase and, occasionally, model transpeptidase activities have been demonstrated.[79]

Thus peptidoglycan synthesis is a vectorial process leading from intracellular soluble precursors to an extramembranal insoluble cross-linked polymer capable of withstanding a wide range of unfavourable environments. The complex series of reactions involved in this process affords a large number of potential sites of attack for specific inhibitors, since peptidoglycan is unique to bacteria. In succeeding sections various numbers of these inhibitors are discussed in detail, while consideration of others, in particular cycloserine and the β-lactam antibiotics, will be found in other chapters of this volume.

9.1.4 ANTIBIOTICS INHIBITING PEPTIDOGLYCAN BIOSYNTHESIS

9.1.4.1 Inhibitors of UDP-*N*-Acetylmuramoylpentapeptide Biosynthesis

9.1.4.1.1 *Phosphonomycin (fosfomycin)*

This antibiotic (Figure 11), isolated as an inhibitor of peptidoglycan synthesis in 1969,[85,86] inhibits what might be regarded as the first step in the biosynthetic pathway, that is the transfer of the enolpyruvate residue from phosphoenolpyruvate to UDP-*N*-acetylglucosamine. Inhibition of the enzyme phosphoenolpyruvate:UDP-*N*-acetylglucosamine-3-*O*-enolpyruvyl transferase is competitive, the antibiotic acting as an analogue of phosphoenolpyruvate. Preincubation of the enzyme with radiolabelled fosfomycin in the presence of UDP-*N*-acetylglucosamine results in complete irreversible inactivation of the transferase with covalent binding of the antibiotic. The transferase was also shown to be rapidly inactivated by thiols leading to the suggestion that inactivation by fosfomycin was through covalent binding of the antibiotic to a cysteine residue in the active site of the enzyme. This was confirmed by the isolation of 2-(*S*)-6-cysteinyl-1-hydroxypropyl-phosphonate.[87]

The proposed reaction of the antibiotic and substrate with the active site of the enzyme is shown in Figure 12. Other enzymes utilizing phosphoenolpyruvate, including pyruvate kinase, enolase, phosphoenolpyruvate carboxykinase and phosphoenolpyruvate:shikimate-5-phosphate-enolpyruvyl transferase are not (or only weakly) inhibited by fosfomycin. The selectivity of fosfomycin appears to be based on the differing enzyme mechanisms. In the case of the UDP-*N*-acetylglucosamine-3-enolpyruvyl transferase this can be regarded as the addition of a thiol group across the C(2)—O bond of fosfomycin (analogous to addition across the C=C bond of phosphoenolpyruvate), whereas the carboxykinase and pyruvate kinase catalyze a nucleophilic attack at the phosphorus atom. Should enolase or shikimate-5-phosphate-enolpyruvyl transferase react with fosfomycin, then

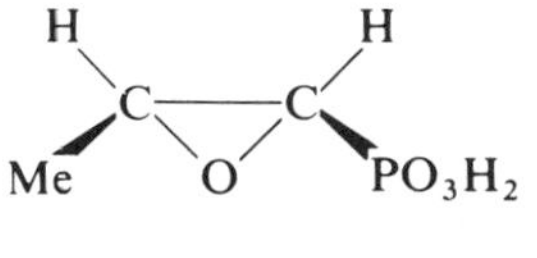

Figure 11 Structure of phosphonomycin

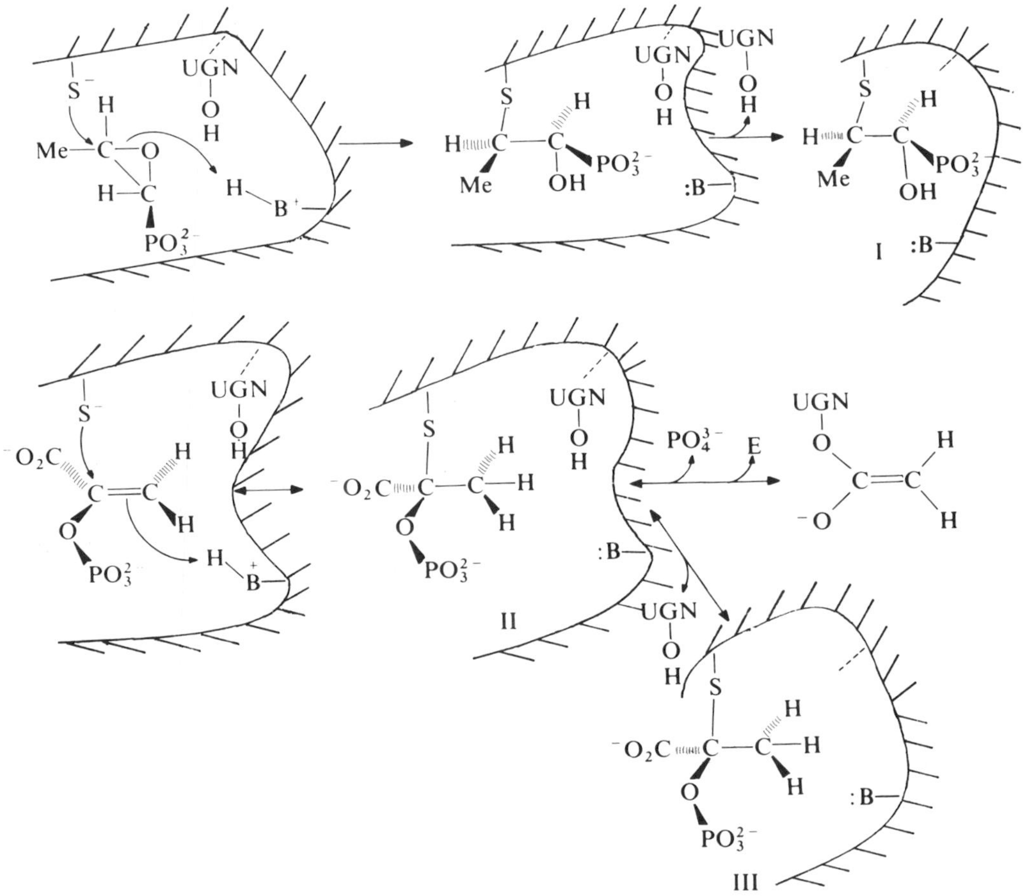

Figure 12 A schematic representation of a possible relationship between the reactions of phosphonomycin and phospho-enolpyruvate at the active site of UDP-*N*-acetylglucosamine-3-*O*-enolpyruvyl transferase. In the initial reaction of both substrate and inhibitor with the enzyme, UDP-*N*-acetylglucosamine (UGN-OH; the 3-hydroxy is shown) is an obligatory cofactor which does not react with either the enzyme or the other substrate. In the case of phosphonomycin the proton donor site H^+—B is thought to activate the epoxide, which is then subject to attack by a cysteinyl sulfur at C-2. This leads to formation of an inactive enzyme–phosphonomycin adduct (I). Similar steps occur with phosphoenolpyruvate and a substrate–enzyme complex (II) is formed, which can become stabilized by removal of UDP-*N*-acetylglucosamine (III). In the final stages the enolpyruvyl residue is transferred with release of phosphate and enzyme. All these reactions are reversible in contrast to the reactions with phosphonomycin[87]

the mechanisms proposed would result in hydrolysis of the antibiotic by enolase and the formation of a shikimate 5-phosphate–fosfomycin adduct, neither situation leading to inactivation of the enzyme.[87]

Early studies on the inhibition of bacterial growth encountered wide variations in susceptibility of the bacteria and these were shown to be related to transport and accumulation of fosfomycin. Sensitive bacteria accumulate the antibiotic either by the L-α-glycerophosphate or by the inducible hexose phosphate uptake systems. The former system is inhibited or repressed by phosphate and glucose respectively, which explains why these substances antagonize fosfomycin activity *in vitro*. On the other hand, the induction of the hexose phosphate transport system explains the increased sensitivity of organisms when blood is present in the growth medium.[88]

The majority of strains isolated as resistant to fosfomycin have been defective in one of the transport systems described above. However, the existence of more than one system transporting fosfomycin has minimized the problems of resistance *in vivo* and the antibiotic is effective as an antibacterial.

9.1.4.1.2 Inhibitors of D-alanine metabolism in peptidoglycan synthesis

The inhibition of bacterial growth by D-cycloserine (which could be reversed by the addition of D-alanine) and the conversion of bacteria to spheroplasts when incubated with the antibiotic in osmotically stabilized media was known before the structure of D-cycloserine was elucidated. These observations led Park[89] to suggest that cycloserine might act by preventing the 'normal incorporation of D-alanine into the wall'. Confirmation of this hypothesis was soon provided by Park and others who described the accumulation of UDP-*N*-acetylmuramoyltripeptide lacking the terminal D-alanyl-D-alanine dipeptide of the peptidoglycan precursor UDP-*N*-acetylmuramoylpentapeptide. These findings were then extended to investigate the effects of cycloserine on the enzymes involved in the incorporation of D-alanine into the pentapeptide. This pathway is shown in Figure 13. At this time only D-cycloserine and *O*-carbamoyl-D-serine were known to inhibit these stages of peptidoglycan biosynthesis. Further consideration of the mechanism of action of these antibiotics is given in Chapter 9.3. However, in recent years alaphosphin and the halogenated derivatives of D-alanine have been added to this group (Figure 14). All these antibiotics inhibit alanine racemase, while D-cycloserine also inhibits D-alanine:D-alanine ligase (D-alanyl-D-alanine synthetase; for reviews see refs. 3, 51).

Alaphosphin (alafosfalin), L-alanyl-L-1-aminoethylphosphonic acid, is the first inhibitor of peptidoglycan synthesis designed as a result of a rational campaign to detect inhibitors of the process. L-1-Aminoethylphosphonic acid (Ala-P) was known to be an inhibitor of alanine racemase of *E. coli* and *S. faecalis* but addition of the compound to growth media did not inhibit growth. The key to its use as an antibacterial agent was the finding that incorporation of the analogue into a suitable dipeptide facilitates active transport and hence accumulation in the target organism. Thus, sensitive bacteria are those having an active dipeptide permease system. Once inside the bacteria, alaphosphin is rapidly hydrolyzed by aminopeptidases to release Ala-P. The lack of transport of

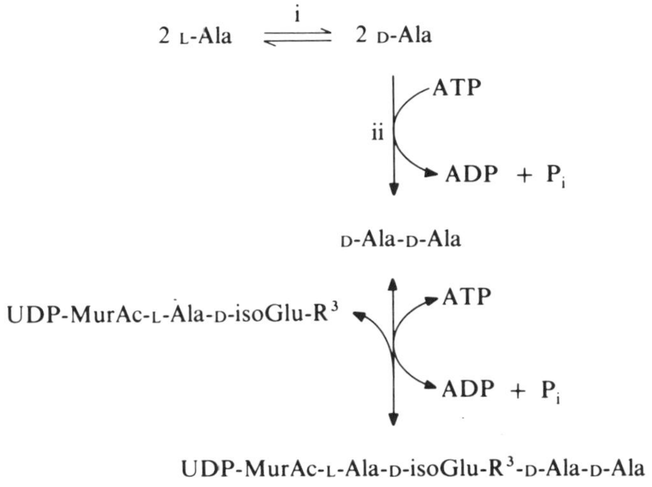

Figure 13 Pathway of alanine in the biosynthesis of UDP-*N*-acetylmuramoylpentapeptide. Alanine racemase (reaction i) converts L-alanine to D-alanine in a reaction inhibited by D-cycloserine, *O*-carbamoyl-D-serine, the haloalanines and alaphosphin, while D-cycloserine alone appears to inhibit D-alanine:D-alanine ligase (ii)

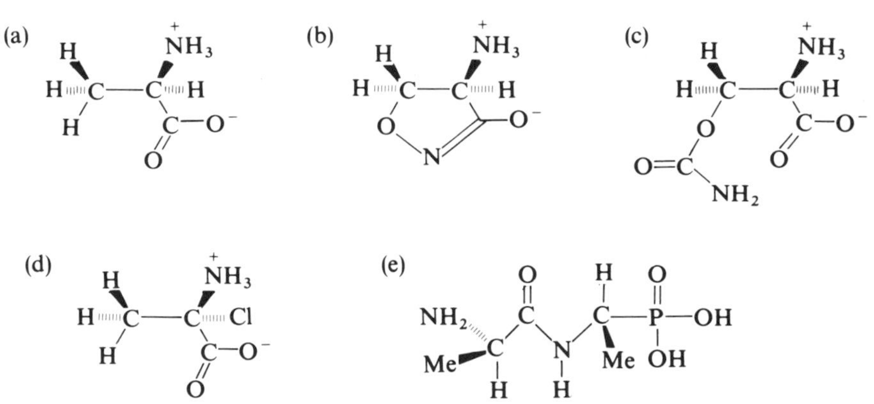

Figure 14 Inhibitors of alanine metabolism: (a) D-alanine; (b) D-cycloserine; (c) *O*-carbamoyl-D-serine; (d) β-chloro-D-alanine; (e) alafosphin (L-alanyl-L-1-aminoethylphosphonic acid)

Ala-P probably prevents the inhibitor leaving the cells and as a consequence very high intracellular concentrations (100–1000-fold in excess of alaphosphin concentration in the medium) are achieved. A variety of other L-amino acids could replace L-alanine to form active phosphonodi- and phosphonooligo-peptides. These peptides and their uptake by peptide permeases have been discussed in detail by Ringrose.[42]

L-1-Aminoethylphosphonic acid inhibits alanine racemase reversibly in Gram-negative bacteria, whereas the inhibition of the racemase from *S. aureus* and *S. faecalis* became irreversible with time.[90] This may result from covalent binding of Ala-P to pyridoxal phosphate, the prosthetic group of alanine racemase. Moreover, these results support the earlier findings of Lambert and Neuhaus[53] of differences in the binding sites and mechanism of inhibition by D-cycloserine in alanine racemases of Gram-negative and Gram-positive bacteria. The major peptidoglycan precursor accumulated by alaphosphin-treated *S. aureus* was UDP-*N*-acetylmuramoyltripeptide. However, a secondary target is the UDP-*N*-acetylmuramoyl-L-alanine ligase and UDP-*N*-acetylmuramoyl-L-Ala-P was shown to accumulate in *E. coli*. *In vitro* Ala-P acts as a weak competitive reversible inhibitor of the ligase and also as a false substrate, to give the resulting phosphonic acid substituted precursor, which is not a substrate for the UDP-*N*-acetylmuramoyl-L-alanine:D-glutamic acid ligase. The weak activity against the L-alanine ligase may, however, be important in whole organisms where the intracellular concentrations of Ala-P can reach 30–40 mM.[91]

Alaphosphin has a broad spectrum of activity and was effective in clinical trials against Gram-negative bacteria causing urinary tract infection. However, the need to optimize peptide length and amino acid sequence for transport and rapid intracellular cleavage, particularly in refractory organisms such as *Pseudomonas* spp. and streptococci, together with the need to achieve stability in human plasma, militated against the clinical development of alaphosphin or related compounds. Resistance, where it was seen in *E. coli*, appeared to involve changes in the peptide permeases, whereas in certain streptococci one of the peptidases may be modified.[92,93]

9.1.4.2 Inhibitors of the Lipid Cycle in Peptidoglycan Biosynthesis

9.1.4.2.1 *Bacitracin*

The bacitracins are a mixture of cyclic peptide antibiotics isolated from *B. licheniformis*, the major component being bacitracin A. This compound contains a thiazoline ring formed between the L-cysteine and L-isoleucine residues and an amide bond between the ε-amino group of lysine and asparagine (Figure 15). The free amino group of L-leucine adjacent to the thiazoline ring is essential. Deamination of this residue, as in bacitracin F, leads to loss of antimicrobial activity (Table 2). In addition, the presence of a divalent metal ion is essential. The activity of the antibiotic is enhanced by Zn^{2+}, Cd^{2+} and Mg^{2+} in decreasing order of effectiveness and activity can be inhibited by metal-chelating agents such as EDTA.[94–96]

In common with several other antibiotics discussed in this chapter, bacitracin was first shown to inhibit peptidoglycan synthesis in *S. aureus* and cause the accumulation of UDP-*N*-acetylmuramoyl-peptides.[97] However, in cell free systems variable inhibition was observed even at high concentrations relative to those giving growth inhibition and when individual peptidoglycan synthetic

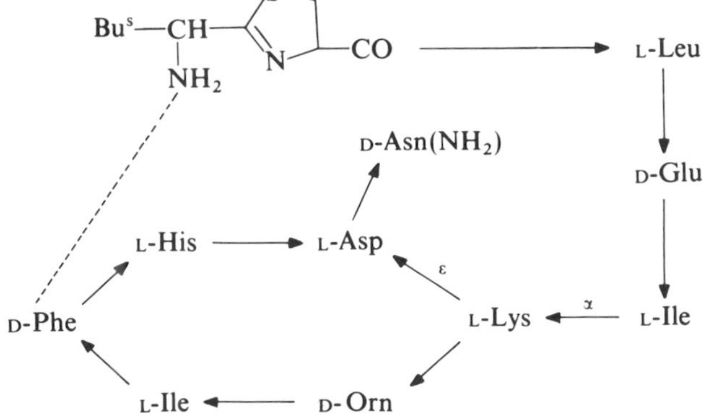

Figure 15 Structure of bacitracin A

Table 2 Activity of Bacitracins Against *Micrococcus luteus* and Binding Constants with C_{55}-isoprenyl Pyrophosphate[96]

Bacitracin	MIC^a (M)	Binding constant $(K)^b$ (M^{-1})
Bacitracin A	1×10^{-7}	1.1×10^6
Bacitracin A, B	2×10^{-7}	1.0×10^6
Bacitracin F	5×10^{-5}	9.8×10^3
Mono-DNP-bacitracin	3×10^{-7}	2.8×10^4
Deamidobacitracin	7×10^{-4}	1.4×10^3

^{a} MIC = minimum inhibitory growth concentration. ^{b} Association constants (K) were determined in 0.1 M Tris–HCl buffer, pH 7.5, containing 1 mM $MgCl_2$ and 0.05 mM mercaptoethanol; ionic strength 0.08 M.

Table 3 Association Constants for the Interaction of Lipid Analogues with Bacitracin A

Substrate	$K_a{}^a(M^{-1})$	Substrate	$K_a{}^a(M^{-1})$
Inorganic phosphate	480	Farnesyl phosphate (C_{15})	5590
Inorganic pyrophosphate	10 400	Farnesyl pyrophosphate (C_{15})	829 000
Isopentenyl pyrophosphate (C_5)	8090	Undecaprenyl pyrophosphate (C_{55})	1 050 000

^{a} Association constants were determined in 0.1 M Tris (pH 7.5) in the presence of 1 mM $MgCl_2$, 0.05 mM mercaptoethanol; ionic strength 0.08 M.[100]

reactions were examined separately, none were inhibited by bacitracin.[59] These findings led Siewert and Strominger[59] to investigate the lipid cycle of peptidoglycan biosynthesis and to the discovery that bacitracin specifically inhibited the dephosphorylation of undecaprenyl pyrophosphate. In systems synthesizing peptidoglycan from undecaprenyl-P-^{32}P-N-acetylmuramoyl-(^{14}C-pentapeptide)-N-acetylglucosamine the formation of ^{14}C-peptidoglycan was unaffected by increasing concentrations of bacitracin, whereas the release of ^{32}P inorganic phosphate was progressively inhibited and a simultaneous increase in undecaprenyl-P-^{32}P was observed. Moreover, the dephosphorylation of undecaprenyl pyrophosphate by membrane preparations of *M. luteus* (*lysodeikticus*) and *S. aureus*, *E. coli* alkaline phosphomonoesterase and alkaline phosphatase from calf intestinal mucosa were all inhibited by low concentrations of bacitracin. On the other hand, dephosphorylation of glucose 1-phosphate or *p*-nitrophenyl phosphate was unaffected, suggesting that the antibiotic might bind to the substrate (*i.e.* lipid pyrophosphate) rather than to the enzyme.

These findings were extended by Storm and Strominger,[100] who demonstrated complex formation between bacitracin, a divalent cation and undecaprenyl pyrophosphate. A study of molecular models suggested that the lipid pyrophosphate fitted into a pocket in bacitracin, with the metal ion forming a bridge between the antibiotic and the polar end of the lipid pyrophosphate. The effect of pH on complex formation indicated that interaction of a number of ionic groups in bacitracin and the lipid pyrophosphate were of importance. These included the α-amino group of bacitracin (but not the δ-NH_2 group of ornithine, which could be dinitrophenylated without loss of activity), the imidazole group of the histidine residue and the third ionization of the pyrophosphate moiety. With these specific requirements, maximum stability of the complex occurred at pH 7–7.5. NMR spectroscopy of the complex indicated the direct involvement of the N-1 of the histidine residue in formation of the metal ion–lipid pyrophosphate complex.[94] In addition, the length of the alkyl side chain is also of some importance and at least three isoprenyl units are necessary for maximum binding to occur (Table 3).

As a consequence of the mechanism of action outlined above, bacitracin should inhibit the biosynthesis of other polymers where recycling of the lipid pyrophosphate is involved. This is the case in the biosynthesis of the O side chains of lipopolysaccharide, which is inhibited by bacitracin.[101] Bacitracin also inhibits certain reactions in squalene biosynthesis where farnesyl pyrophosphate is an intermediate and the antibiotic also inhibits the dolichol pyrophosphate dependent glycosylation reactions involved in glycoprotein biosynthesis.[95]

9.1.4.2.2 *Tunicamycin*

Tunicamycin, a family of nucleoside antibiotics, was isolated from *Streptomyces lysosuperificus* as an antiviral agent, although later it was shown to be active against many Gram-positive bacteria and

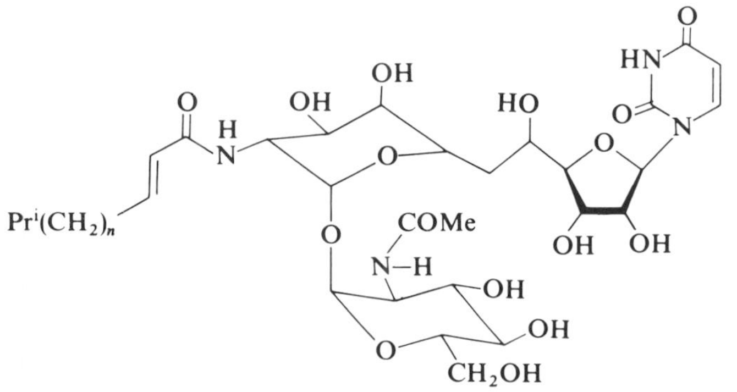

Figure 16 Structure of tunicamycin. Tunicamycin is a family of homologues differing in the chain length of the fatty acid; $n = 8$–11. Whether these compounds differ in their activity is not known

fungi[102] and to inhibit the synthesis of glycoproteins in yeast and mammalian systems (for review see ref. 103).

Analysis showed that tunicamycin contained uracil, N-acetylglucosamine, a complex novel amino sugar (tunicamine—a C_{11} aminodideoxydialdose) and unsaturated fatty acids. The structure given in Figure 16 has now been assigned and revealed that the various members of the tunicamycin group differ only in the chain length of the fatty acids esterified to the amino group of tunicamine. Tunicamycin appears to be identical to mycospocidin described earlier as an antiviral agent[104] and a closely related complex (MM19290) is produced by a strain of *Streptomyces clavuligerus*.[105]

Tunicamycin was first shown to selectively inhibit glycoprotein synthesis in Newcastle disease virus, where it was without effect on protein or nucleic acid synthesis.[106] Subsequently the antibiotic was found to inhibit the synthesis of fungal wall mannan and invertase by preventing the incorporation of glucosamine. The reaction inhibited appeared to be the transfer of N-acetyl glucosamine 1-phosphate from UDP-N-acetylglucosamine to dolichol phosphate. The transfer of subsequent N-acetylglucosamine and mannose residues was unaffected.[107] Moreover, the synthesis of chitin was unaffected. Studies on the formation of dolichol-P-P-N-acetylglucosamine in porcine aorta suggested that tunicamycin may act as a substrate–product transition state analogue, binding irreversibly to the enzyme and thus inactivating it.[108]

In bacteria, tunicamycin inhibits the formation of undecaprenyl-P-P-N-acetylmuramoyl-pentapeptide, the first lipid intermediate involved in peptidoglycan synthesis.[57,109] There are conflicting reports as to whether the transferase catalyzing the formation of the second lipid intermediate undecaprenyl-P-P-N-acetylmuramoyl(pentapeptide)-N-acetylglucosamine is also inhibited.[57,109,110] Inhibition of the translocase alone would be consistent with the findings concerning inhibition of mannan biosynthesis in *Saccharomyces cerevisiae*.[107]

N-Acetylglucosamine-containing lipid intermediates have been described in *Staph. aureus* and several bacilli, where they participate in the linkage of teichoic acids to peptidoglycan. A similar situation exists in *M. luteus* where an N-acetylglucosamine-containing lipid intermediate is involved in the linkage of teichuronic acid to peptidoglycan (for a review see ref. 22). In each case formation of the first lipid intermediate, polyprenyl-P-P-N-acetylglucosamine, by translocation of N-acetylglucosamine 1-phosphate to polyprenyl phosphate is inhibited by tunicamycin. The antibiotic is without effect, however, on the translocation of N-acetylgalactosamine 1-phosphate to form a lipid intermediate involved in the biosynthesis of the teichuronic acid in *B. licheniformis*.[26] Thus, some of the specificity of tunicamycin to inhibit lipid intermediate synthesis involves the phospho-N-acetylglucosamine residue or a close analogue. In this context, N-acetylmuramoylpentapeptide (the peptide-substituted 3-O-lactyl ether of N-acetylglucosamine) but not N-acetylgalactosamine is accepted.

Mutants of *B. subtilis* resistant to tunicamycin have been described but it is not known whether these involve changes in the phospho-N-acetylmuramoylpentapeptide translocase to make it resistant to the antibiotic.[111]

9.1.4.3 Inhibitors of Polymerization of Peptidoglycan

9.1.4.3.1 Glycopeptide antibiotics (vancomycin-like)

Vancomycin and ristocetin, isolated in 1956 and 1957 respectively, were the first members of a structurally related group of glycopeptide antibiotics to be described.[112,113] They were found to be extremely active against Gram-positive bacteria but with the exception of some strains of *Neiserria* were inactive against Gram-negative organisms. Vancomycin was used clinically for the treatment of staphylococcal and other Gram-positive infections prior to the introduction of methicillin and other β-lactamase resistant penicillins and cephalosporins. However, there were considerable problems with associated toxicity and the use of vancomycin was limited. It now appears that much of the toxicity was related to impurities present in the antibiotic preparation and the elimination of these by a process involving preparative HPLC has allowed increased usage. In particular, the continuing and increasing problem of methicillin resistant staphylococci has provided an important role for vancomycin in hospital infection. In addition, oral vancomycin has become the treatment of choice for pseudomembranous colitis caused by *Clostridium difficile*.

Of the other members of the glycopeptide group of antibiotics teichoplanin (teichomycin A_2)[114] is in clinical trial, while avoparcin, actaplanin and A35512 are, or may be used as, animal food additives. Ristocetin was withdrawn from clinical use when it was found to cause platelet aggregation.[113] The aridicins,[115] kibdelins[116] and antibiotics OA-7053[117] and A47934[118] have also been described but their future use is as yet unclear.

Early chemical analysis of vancomycin showed the presence of glucose, aspartic acid, *N*-methylleucine, phenolic and chlorophenolic residues, whereas ristocetin contained, in addition, mannose, arabinose and rhamnose and a novel amino sugar, subsequently identified as ristosamine. Some of the antibiotics of the group contain an unusual amino sugar substituted at varying sites on the molecule, vancosamine, 2,3,6-trideoxy-3-*C*-methyl-L-lyxohexapyranose being present in vancomycin. The removal of sugars to form the aglycone does not totally abolish the antibiotic activity of either vancomycin or the ristocetins, suggesting that these portions of the molecule are not directly involved in binding to the specific peptide structures involved in peptidoglycan biosynthesis (see below).

The early studies on the structure of vancomycin leading to the three-dimensional structure of the crystalline vancomycin degradation product CDP-1 were described in detail by Williams *et al.*[119] It was then established that the conditions used to prepare CDP-1 (long incubation at pH 4.2) resulted in the loss of an amide group with the conversion of asparagine to isoaspartic acid. At the same time the chlorophenyl ring (ring 2) had rotated through approximately 180°. The revised structure for vancomycin is shown in Figure 17 (for a review see ref. 113). Further studies by Williams and others have shown the three-dimensional structural similarities of other members of this group of antibiotics, including ristocetin A (Figure 17), A35512B, aridicin A and the teichomycins.[113,120–124]

Antibiotics of this group are linear heptapeptides with the unusual amino acids held by several phenolic oxidative linkages (Figure 18). Vancomycin has five aromatic rings, whereas all other members of the group have seven. The structural details where known are given in Table 4 based on that of Barna and Williams (ref. 113; Table 2). In many cases the antibiotics occur as families of closely related compounds. Thus ristocetin exists as A and B forms differing in the sugars substituted on ring 4, while actinoidins A and B and avoparcin α and β differ in their chlorine substituents. More complex mixtures occur in the teichoplanins, aridicins and kibdelins, where varying fatty acids are present on the *N*-acylglucosamine and *N*-acylglucosaminuronic acid substituents of ring 4. However, the basic structural similarity of this group of antibiotics is clear and the relationship of basic structure to the stereochemical requirements of their mode of action is well established.

The first observations that vancomycin acted by inhibition of peptidoglycan biosynthesis came from the studies of Reynolds[125] and Jordan,[126] both of whom showed that vancomycin treatment of *Staphylococcus aureus* resulted in the accumulation of UDP-*N*-acetylmuramoylpentapeptide. Similar results were soon obtained with ristocetin. It was also found that bacteria and isolated bacterial walls rapidly bound vancomycin, which could not be released by simple washing procedures.

The affinity of walls for vancomycin was further demonstrated by Sinha and Neuhaus,[127] who showed that addition of walls abolished the inhibition by vancomycin of peptidoglycan synthesis by a particulate (membrane) enzyme preparation from *Micrococcus luteus*. Using membrane preparations capable of synthesizing peptidoglycan *in vitro*, it was soon established that vancomycin inhibited the transglycosylation reaction required for glycan chain polymerization. Inhibition of these systems led to the accumulation of the lipid intermediate, undecaprenyl disaccharide pentapeptide (for reviews see refs. 3, 112).

At approximately the same time Chatterjee and Perkins[128] showed that vancomycin and ristocetin bound to UDP-*N*-acetylmuramoylpentapeptide. The formation of this complex required the presence of the acyl-D-alanyl-D-alanine terminus—both alanine residues had to have the D-configuration and the terminal carboxyl group had to be free.[129] These observations led to a systematic examination of the binding of a series of peptides to vancomycin and ristocetin. These studies, reviewed by Perkins,[112] established that the binding sites of the antibiotics were similar but

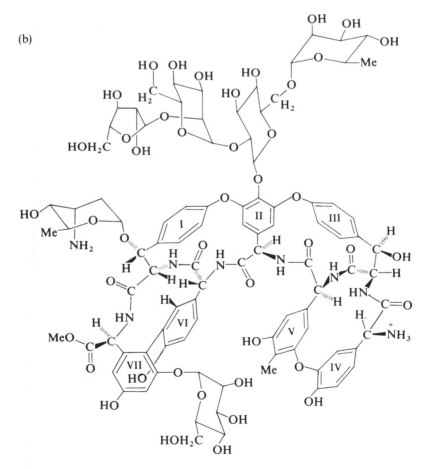

Figure 17 Structures of (a) vancomycin and (b) ristocetin showing stereochemical detail

Table 4 Substituents of the Vancomycin Group Antibiotics[113]

Substituent	Vancomycin	Ristocetin	Avoparcin	Actinoidin	Actaplanin	A35512B	Aridicins	Kibdelins	Teichoplanins	A47934
X¹	Me–CH(–CH₂–)–Me (with CONH₂ side, isobutyl)	[aromatic structure: –O– linked ring, OH, Me]	[aromatic structure with O-α-L-rhamnosyl; α, β rings; Cl, OH]	[two aromatic rings A (OH) and B (Cl, OH), –CH₂–phenyl]	[aromatic structure; OH/X^c; Me]	[aromatic structure; Cl; X^c/HO]	[aromatic structure; Cl, HO, OH]	[aromatic structure; Cl, HO, OH]	[aromatic structure; HO, OH]	[aromatic structure; HO, OSO₃H]
X³	–CH₂, CONH₂									
R¹	H	Me	H	H	Me	Me	H	H	H	H
R²	Cl	H	Cl	Cl	Cl	H	Cl	Cl	Cl	Cl
R³	Cl	H	H	H	H	H	Cl	Cl	Cl	Cl
R⁴	Me	H	Me	H	H	H	Me	Me	H	H
R⁵	H	H	H	H	H	H	H	H	H	H
R⁶	H	O-α-D-mannosyl	H	O-D-Mannosyl	Glucosyl, mannosyl, rhamnosyl, H	Fucosyl/mannosyl	O-D-Mannosyl	O-D-Mannosyl	O-D-Mannosyl	H
R⁷	H	H	H	H		H	H	H	O-D-Mannosyl	H
R⁸	O-α-L-Vancosaminyl-(1 → 2)-O-β-D-Glucosyl	A ristotetrose^a B ristobiose^b	O-α-L-Ristosaminyl-(1 → 2)-O-β-D-Glucosyl	O-L-Acosaminyl-(1 → 2)-O-D-Glucosyl		Rhamnosyl-glucosyl?	N-Acylglucosaminuronyl	N-Acylglucosaminyl	N-Acylglucosaminyl	OH
R⁹	H	O-α-L-Ristosaminyl	O-α-L-Ristosaminyl	O-L-Actinosaminyl	O-L-Ristosaminyl	3-epi-L-Vancosaminyl	H	H	N-Acetylglucosaminyl	H
R¹⁰	OH	OH	O-α-D-Mannosyl	OH	H	OH	OH	OH	H	H
R¹¹	H	H	H	H	H	H	Cl	Cl	H	Cl
R¹²	H	H	H	H	H	H	H	H	H	Cl

^a O-α-D-Arabinosyl-(1 → 2)-O-α-D-mannosyl-(1 → 2)-O-[α-L-rhamnosyl-(1 → 2)-O-[α-L-rhamnosyl-(1 → 6)]-O-β-D-glucosyl. ^b O-α-L-Rhamnosyl-(1 → 6)-O-β-D-glucosyl. ^c X represents possibility that sugar is attached at this phenolic position.

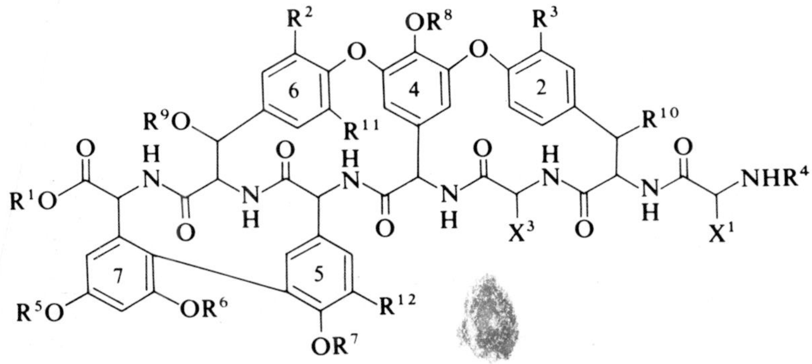

Figure 18 General structure of the vancomycin group antibiotics

not identical. In general terms, vancomycin appeared to have more restrictive binding requirements than ristocetin. However, both antibiotics required D-amino acids (or glycine) to be present in the two C-terminal residues and a long chain L-amino acid in position three from the terminus. Thus D-alanine was preferred as the C-terminal residue for vancomycin; the presence of D-leucine decreased binding for vancomycin but not ristocetin. At position two, D substituents were beneficial for binding to both antibiotics, but here vancomycin accepted larger side chains more readily than did ristocetin. For example Ac-L-Ala-D-Glu-Gly, which represents part of the peptide side chain and cross-bridge of the peptidoglycan of *M. luteus*, bound well to vancomycin but only poorly to ristocetin. This observation explains the fact that walls of *M. luteus*, which contain many such peptides, bind vancomycin avidly but ristocetin less well.

These binding studies were carried out well before the detailed stereostructures of the antibiotics became known. Now it is possible to see how the acyl-D-alanyl-D-alanine terminus fits into a pocket on the antibiotic molecule and how the slight modifications in antibiotic structure explain the observations of Nieto and Perkins (Figures 19 and 20).[130,131] The ristocetin structure can accommodate larger side chains in the D configuration since this side chain projects above ring 4, whereas in vancomycin the chlorine substituents interfere with binding of peptides terminating in D-amino acids larger than D-alanine. The structures also show that sugar substituents do not appear to be important for complex formation, in agreement with the biological studies showing that aglycones of several members of the group retain antibacterial activity.

Clearly, the vancomycin group of antibiotics inhibit peptidoglycan biosynthesis as a consequence of their ability to form complexes with the acyl-D-alanyl-D-alanine terminus. However, the precise site of action remains unclear. Vancomycin does not penetrate the bacterial cytoplasmic membrane[132] and, therefore, complex formation with the nucleotide precursor UDP-*N*-acetyl-muramoylpentapeptide is not involved. This leaves binding to the lipid intermediate or to the nascent uncross-linked peptidoglycan as possible sites of action. Polymerization of disaccharide peptapeptide subunits from the disaccharide pentapeptide lipid intermediate to make linear uncross-linked peptidoglycan was studied by Van Heijenoort *et al.*,[65] using membrane preparations of *E. coli*. The molar proportions of vancomycin and ristocetin that inhibited polymerization (I_{50}, 0.05 and 0.1 mol prop. respectively) imply that the lipid intermediate is not the primary site of action and point to binding to the growing nascent peptidoglycan as being of primary importance.[133]

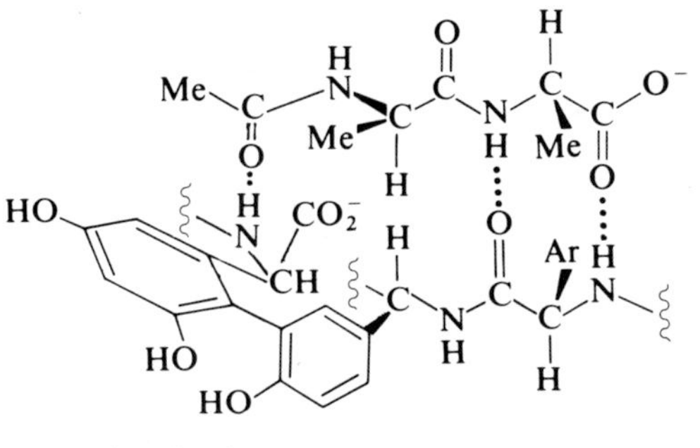

Figure 19 Model of some of the hydrogen-bonding interactions formed between vancomycin and acetyl-D-alanyl-D-alanine

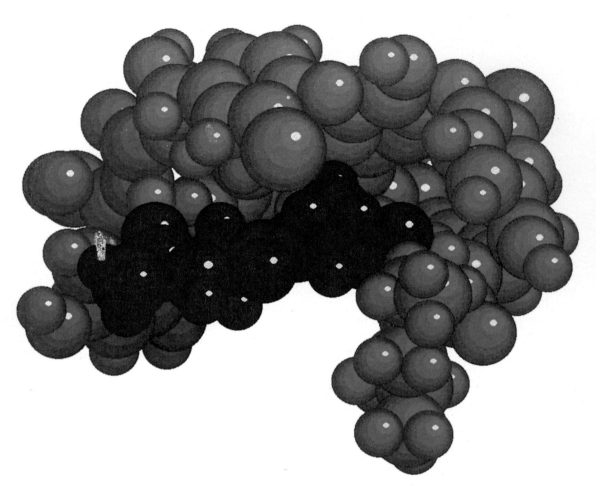

Figure 20 A model of the interaction between vancomycin aglycone and acetyl-D-alanyl-D-alanine. The model shows the acetylated dipeptide held in a pocket of the vancomycin molecule by the hydrogen bonding shown in Figure 19. The figure was prepared by Dr. M. Hahn, Glaxo Group Research Ltd., using a Chem-X, developed and designed by Chemical Design Ltd, Oxford, England

Gram-negative bacteria, with the exception of *N. gonorrhoea*, are resistant to the vancomycin group of antibiotics (including the aglycones) due to the inability of these large polar molecules to cross the outer membrane. However, a series of lipophilic esters of the aglycone of teichoplanin have recently been synthesized and shown to have modest activity against *E. coli*.[134] Since these are of almost identical molecular weight to teichoplanin they presumably cross the outer membrane by diffusion or some other mechanism than the porin pathway. The outer membrane barrier is not present in Gram-positive bacteria and here the development of resistance has been reported only in one or two instances.[135–137] Presumably, these reflect changes in the ability of vancomycin to reach the target since the acyl-D-alanyl-D-alanine terminus is essential for transpeptidation to occur. Resistance by a change in the target would prevent the synthesis of cross-linked peptidoglycan and would, therefore, be lethal to the bacterium. There is no evidence for the enzymic inactivation of vancomycin.

9.1.4.3.2 *Phosphoglycolipid antibiotics (moenomycin-like)*

Moenomycin, marcarbomycin, the diumycins, prasinomycins and antibiotic 11837RP are members of a related group of phosphorus-containing glycolipid antibiotics known to inhibit the biosynthesis of peptidoglycan.[48,138] The antibiotics have similar antimicrobial spectra, being active only against Gram-positive bacteria. Again the resistance of Gram-negative organisms is related to the inability of these compounds to cross the outer membrane and reach the target site.

The structures of moenomycin and the related antibiotic pholipomycin have now been established and are shown in Figure 21.[139-141] More recently, Welzel *et al.*[142] have attempted to define the structural basis of antimicrobial activity by the systematic chemical degradation of moenomycin A. From these studies it was shown that *in vitro* activity was retained in a derivative comprising a disaccharide of *N*-acetylglucosaminyl and moenuronic acid linked through a phosphodiester group to moenocinol-C_{25} lipid. In these experiments it was also established that the lipid could be saturated, the carboxyl function attached to the lipid had to be free and the carbamoyl group in the moenuronic acid moeity must be present. The antibiotic activity of the various derivatives both against whole organisms and in cell free systems is shown in Table 5.

Early investigation of moenomycin and other antibiotics of the group showed inhibition of peptidoglycan synthesis by membrane preparations but not the incorporation of radioactivity from the nucelotide precursors into one or both of the lipid intermediates, undecaprenyl-P-P-*N*-acetyl-muramoyl (or disaccharide) pentapeptide. Subsequently, van Heijenoort *et al.*[64,65] showed that in

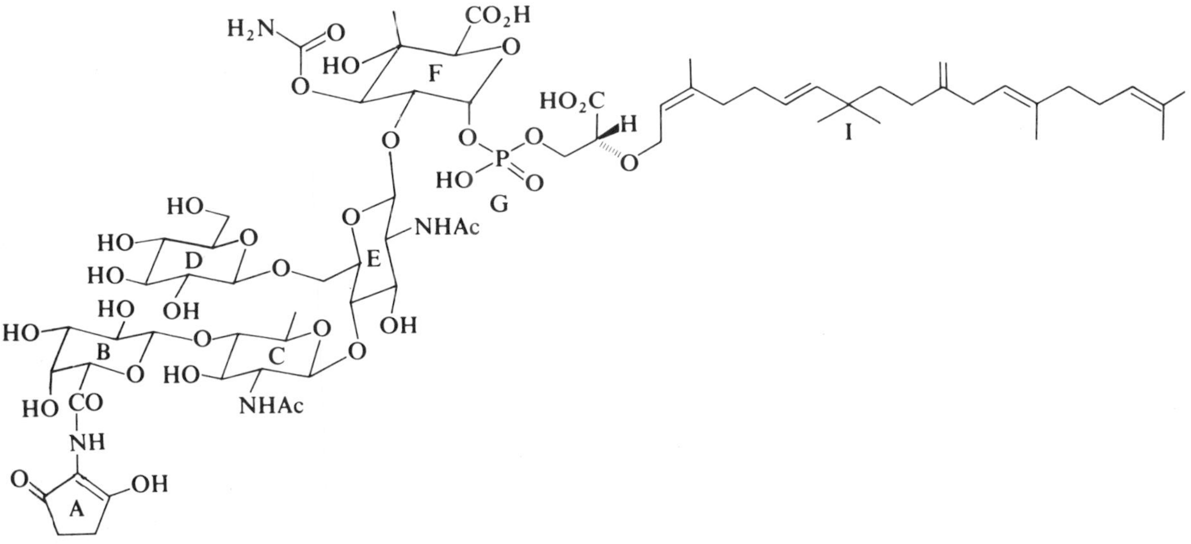

Figure 21 Structure of moenomycin A. Pholipomycin lacks the glucosyl substituent D and substituent C is *N*-acetyl-glucosamine rather than *N*-acetylquinovosamine (*N*-acetyl-6-deoxyglucosamine)

Table 5 Inhibition of Peptidoglycan Transglycosylase (Polymerase) and Growth of *Staph. aureus* by Moenomycin and Certain Degradation Products[a]

Compound	Concentration ($\mu g\,mL^{-1}$)	Inhibition of transglycosylation (%)	Staph. aureus MIC ($\mu g\,mL^{-1}$)
Moenomycin A	1	100	0.19
	0.1	54	
Decahydromoenomycin A	1	100	<0.19
	0.1	0	
−Unit A	1	100	1.56
	0.1	8	
−Unit A, methyl ester of CO_2H group of unit H	100	100	—
	10	64	
−Units A, D	1	100	3.13
	0.1	75	
−Units A, B, D	1	100	3.13
	0.1	30	
−Units A, B, C, D	1	100	12.5
	0.1	45	
−Units A, B, C, D, carbamoyl on F	100	100	>50.0
	10	56	
−Units A, B, C, D, E	10	100	>50.0
	1	53	

[a] Data from ref. 142.

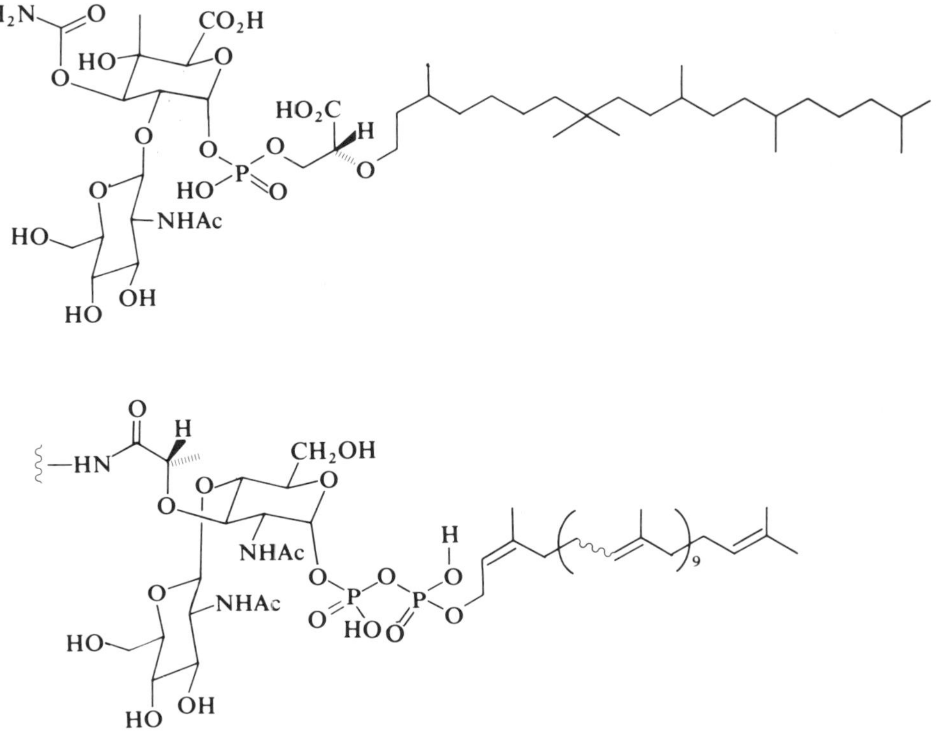

Figure 22 Comparison of the structures of the peptidoglycan precursor, undecaprenylpyrophosphoryl-*N*-acetyl-muramoyl(peptide)-*N*-acetylglucosamine (C$_{55}$-disaccharide peptide) and a derivative of moenomycin A that retains anti-bacterial activity (see Table 5)[142]

E. coli the transglycosylation reaction synthesizing linear uncross-linked peptidoglycan from the lipid intermediate undecaprenyl-P-P-*N*-acetylmuramoyl(pentapeptide)-*N*-acetylglucosamine was inhibited by low concentrations of moenomycin and related antibiotics. These observations were extended to show inhibition of the transglycosylase (polymerase) activity of penicillin-binding proteins 1b, 1a and 3 of *E. coli*.[81,143,144] Earlier, on the basis of structural similarities between the lipid components, Linnett and Strominger[145] suggested that these antibiotics may act as analogues of the C$_{55}$-isoprenyl phosphates involved as lipid carriers in peptidoglycan synthesis. The structural studies of Welzel *et al.*[142] and recent results of van Heijenoort *et al.* (reported by Welzel *et al.*, ref. 142) showing that moenomycin A is a competitive inhibitor of peptidoglycan polymerase, support this analogy (Figure 22).

9.1.5 ANTIBIOTICS INHIBITING BIOSYNTHESIS OF OUTER MEMBRANE COMPONENTS

9.1.5.1 Inhibitors of Lipopolysaccharide (Lipid A) Biosynthesis

As described earlier, lipopolysaccharide is a major component of the outer membrane of Gram-negative bacteria. It thus is directly involved in conferring on the organism properties crucial to its pathogenicity, drug resistance and survival in natural (as opposed to laboratory) habitats. Lipopolysaccharide is unique to Gram-negative bacteria and consists of three regions: lipid A, the core oligosaccharide and, in 'smooth' strains, a long carbohydrate O side chain. Lipid A, an acylated β-(1 → 6)-linked glucosamine disaccharide, anchors the lipopolysaccharide in the outer leaflet of the outer membrane. In most Gram-negative organisms the core oligosaccharide is attached to lipid A through an α-(2 → 6)-linked 3-deoxy-D-*manno*-2-octulosonate (KDO) dimer to which a third KDO residue may be linked. Mutants of *Salmonella typhimurium* with temperature sensitive defects in KDO-8-phosphate synthase and CTP:CMP-KDO cytidyl transferase (CMP-KDO synthetase) have been described. In each case transfer of the organisms to the restrictive temperature resulted in accumulation of lipid A precursors lacking KDO and ester-linked lauric and myristic acid residues, and ultimately in cell death.

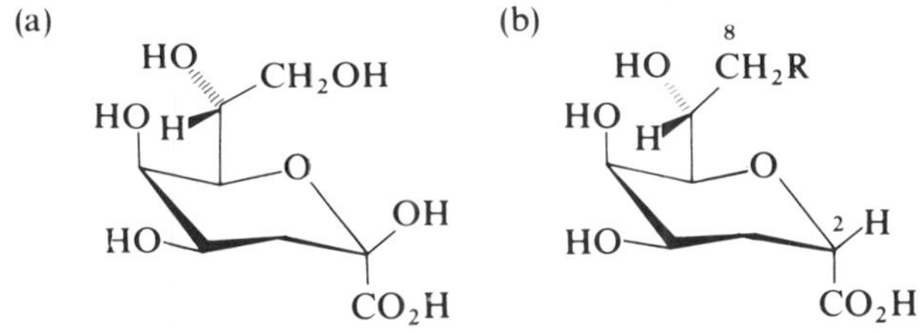

R = OH; 2,6-anhydro-3-deoxy-D-*glycero*-D-*talo*-octonic acid

R = NH$_2$; 8-amino-2,6-anhydro-3,8-dideoxy-D-*glycero*-D-*talo*-octonic acid

Figure 23 Structure of (a) 3-deoxy-β-D-*manno*-2-octulosonic acid (β-KDO) and (b) two analogues that inhibit CMP-KDO synthase

Recently, inhibitors of CMP-KDO synthetase were described independently by two groups of workers.[41,146] These analogues of KDO (Figure 23) were effective competitive inhibitors of the isolated enzyme with K_i values of about 4 μM for 2-deoxy- and 8-amino-2-deoxy-KDO and 2.5 μM for 8-aminomethyl-2-deoxy-KDO. However, they showed little antibacterial activity, due to their inability to be transported across the cytoplasmic membrane. Coupling of the 8-amino analogue to an LL-dipeptide facilitated transport *via* the bacterial peptide permeases and inhibition of bacterial growth was achieved (MIC values of 1.5–2 μM).[41] Accumulation of 8-amino-2-deoxy-KDO occurred within the bacteria following rapid hydrolysis of the peptide bonds. Thus, the system is analogous to that described earlier for alaphosphin, where hydrolysis of the transported phosphonopeptide releases the enzyme inhibitor.[42] In that case the inhibitor L-1-aminoethylphosphonic acid is active against both Gram-positive and Gram-negative bacteria, whereas the KDO analogues can only be active against Gram-negative organisms. This, together with the requirement for a means of transporting them across the cytoplasmic membrane, raises the question as to whether they will be developed further for chemotherapeutic usage.

9.1.5.1.1 Bicyclomycin (bicozamycin)

Bicyclomycin is a novel antibacterial agent produced by various *Streptomyces* spp., including *S. sapporonensis*, *S. aizunensis* and *S. griseoflavus*.[147–149] It is identical to bicozamycin and antibiotic S879.[148,149] The structure has been determined and found to be of the bridged diketopiperazine type; it represents a highly functionalized cyclo-L-leucinyl-L-isoleucine dipeptide (Figure 24).[150] Biosynthetic studies confirmed this structure, showing the incorporation of equivalent amounts of the two radiolabelled amino acids into bicyclomycin.[151]

The compound is active both *in vitro* and *in vivo* against Gram-negative bacteria but is inactive against Gram-positive organisms. It does not show cross-resistance to other known antibacterial agents. Morphological studies showed that bicyclomycin induces the formation of surface blebs and a highly undulated outer membrane of *E. coli*.[152] When tested against anaerobic bacteria, the majority of Gram-negative anaerobes were sensitive to bicyclomycin concentrations of 256 mg L^{-1} or less, whereas all the anaerobic cocci and the majority of the clostridia (40/43) were resistant.[153] Bicyclomycin is only poorly absorbed from the gastrointestinal tract and because of its spectrum of activity it has been evaluated in the clinic as an oral antidiarrhoeal agent. The early results of these trials are promising.[154,155]

Early studies showed that bicyclomycin treatment resulted in the morphological changes described above and that the synthesis of envelope proteins was inhibited to a greater extent than that

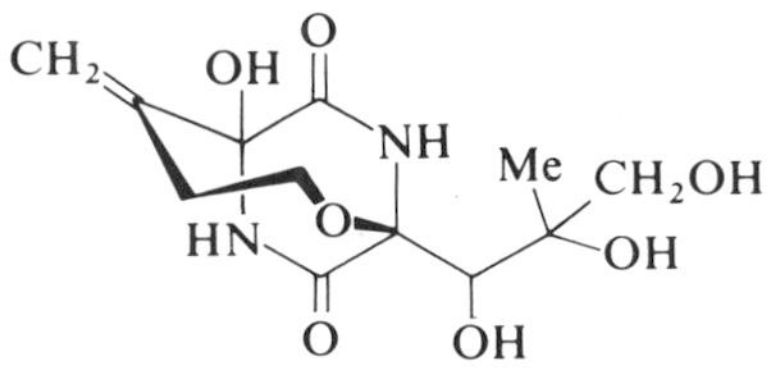

Figure 24 Structure of bicyclomycin

of cytoplasmic proteins. In cell free systems inhibition of lipoprotein synthesis was observed. Finally, it was concluded that the antibiotic acted by inhibiting lipoprotein biosynthesis and its linkage to peptidoglycan.[156] More recently Vasquez and his colleagues[157] questioned this conclusion since *E. coli* mutants lacking lipoprotein can grow normally under a variety of conditions.[158] Detailed analysis of *E. coli* peptidoglycan revealed that the main effect of bicyclomycin treatment was an increase in the number of diaminopimelyl–diaminopimelyl bridged subunits present. Lipoprotein is linked only to subunits containing this cross-bridge, which later becomes hydrolyzed to release disaccharide–tripeptide subunits containing lipoprotein.[16] In bicyclomycin-treated bacteria the number of DAP–DAP containing subunits with attached lipoprotein also increased with a concomitant decrease in the number of disaccharide–tripeptide–lipoprotein subunits. Thus bicyclomycin appeared to inhibit the hydrolysis of DAP–DAP bonds.

The formation of the diaminopimelyl–diaminopimelyl bonds involves the two D centres of the *meso*-diaminopimelic acid residues. The lipoprotein is then linked to the L centre of one of the residues. Pisabarro *et al.*[157] argued that the formation of the first amide bond involving one —CO_2H and one —NH_2 group results in a favourable energetic state for the formation of a second amide bond. The final result could be a six-membered lactam, *i.e.* equivalent to a diketo piperazine ring (Figure 25).

Earlier Someya *et al.*[152] showed that treatment of *E. coli* membranes with radioactive bicyclomycin resulted in binding of the antibiotic. These bicyclomycin-binding proteins were distinct from penicillin-binding proteins but were not otherwise characterized. Binding was inhibited by thiols and in the presence of excess reagent an adduct containing equal molar ratios of bicyclomycin and thiol was formed. The authors concluded that a thiol enzyme might be the target and react similarly with bicyclomycin. However, this seemed less likely when active derivatives were made in which the terminal alkenic group was replaced by an oxime.[159] Such compounds do not react readily with a thiol. More recently, Williams *et al.*[160] proposed that the target, a protease or transpeptidase-like enzyme, catalyzed a nucleophilic attack on the ketone 9-carbon of the diketopiperazine ring of the antibiotic. As a consequence, the 9,10 amide bond would be cleaved and an acyl–enzyme intermediate would be formed leading to inactivation of the enzyme.

Alternatively, Pisabarro *et al.*[157] proposed that the constrained diketopiperazine ring of bicyclomycin acted as a structural analogue of the six-membered lactam involved in the diaminopimelyl–diaminopimelyl bonds (see above; Figure 25). As such, the antibiotic would inactivate the amidase responsible for the hydrolysis of these bonds, leading to the increase in diaminopimelyl–diaminopimelyl bonds found in peptidoglycan of treated organisms. How such an increase, or rather the inability to hydrolyze these subunits to the tripeptide, impedes the normal synthesis of peptidoglycan and growth of the organism remains unclear.

Chemical studies on bicyclomycin have shown that the whole molecule is required for antibacterial activity.[152] The majority of the analogues synthesized in which activity has been retained show little or no improvement over the parent compound.[159,160] Some minor changes in the spectrum of activity were observed, the oxime analogues showed additional activity against *Proteus* spp., while an *N*-benzyl derivative showed minimal activity against some Gram-positive bacteria. Otherwise the antibacterial efficacy of bicyclomycin was clearly very sensitive to change.

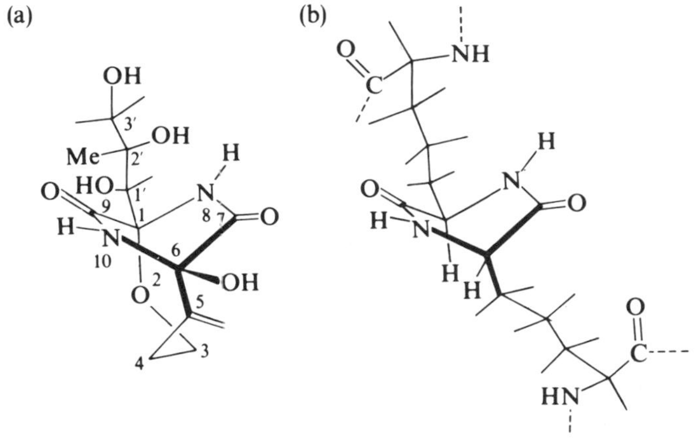

Figure 25 (a) Structure of bicyclomycin based on X-ray crystallography and (b) the proposed structure of the diaminopimelyl–diaminopimelic acid linkage in the peptidoglycan of some Gram-negative bacteria[157]

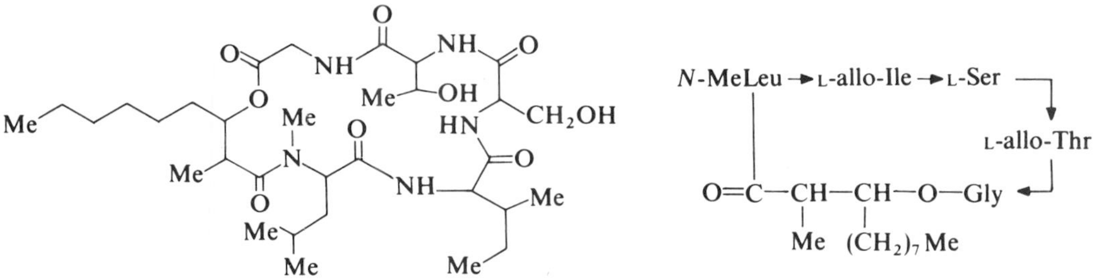

Figure 26 Structure of globomycin

9.1.5.1.2 *Globomycin*

Globomycin is a cyclic pentapeptide antibiotic produced by various *Streptomyces* spp. that is also active only against Gram-negative bacteria.[161] The structure has been determined and shown to contain *N*-methylleucine, L-alloisoleucine, L-allothreonine, serine and glycine (Figure 26).[162] Other members of the globomycin group have been isolated as minor components from cultures of *Streptomyces hygroscopicus*. These compounds shared four amino acids; the fifth was either valine or alloisoleucine and varied in the fatty acid substituent from 3-hydroxy-2-methylheptanoic acid to 3-hydroxy-2-methylundecanoic acid. Maximum activity was shown by the homologue with the longest alkyl chain.[163]

The early studies showed globomycin inhibited the synthesis of bound lipoprotein to a much greater extent than the synthesis of cytoplasmic proteins. Moreover, globomycin treatment led to the accumulation in the membrane of prolipoprotein, a precursor of both the bound and membrane forms of lipoprotein.[164] Further studies on the biosynthesis of lipoprotein, facilitated by the isolation of globomycin resistant mutants, established that the antibiotic inhibited the processing peptidase signal peptidase II. This enzyme has been isolated and purified from *E. coli* B and shown to be globomycin sensitive *in vitro*.[165]

9.1.6 STRUCTURE OF FUNGAL WALLS

As mentioned briefly in the introduction to this chapter, fungal walls, like those of bacteria, consist of a rigid layer outside the protoplast, which they protect from osmotic and other changes in the environment. They are also responsible for the shape of the organism and thus are subject to modification when morphological changes such as the initiation of hyphal branches or the change from yeast to mycelial growth in *Candida albicans* occurs. In recent years, fungal walls have received increasing attention and several detailed accounts of wall structure and biosynthesis of fungal wall polymers have appeared.[166-172] Fungal wall polymers, particularly β-glucan and chitin, remain attractive targets for antifungal drugs, although the most successful agents currently in use, the antifungal azoles and polyenes, disrupt membrane integrity.

Fungal walls are composed largely of polysaccharides, including soluble mannopeptides, embedded in a matrix of alkali soluble and insoluble glucan. Rigidity is conferred by a fibrillar network of chitin and β-glucan, although in yeast forms the vast majority of chitin appears to be confined as rings or discs to the bud scars. In mycelial fungi the chitin–glucan fibrillar network forms a microcrystalline sleeve that covers the lateral walls. However, the biosynthesis of chitin is located predominately in the apical region of the hyphae.

The detailed structure of the mannopeptides has been elucidated largely through the studies of Ballou and his colleagues.[166,167] Briefly, the polysaccharide is made up of a backbone of α-(1 → 6) mannosyl units to which are attached shorter side chains containing both α-(1 → 2)- and α-(1 → 3)-linked units. Linkage to the peptide which makes up 5–10% of the complex occurs either *via* di-*N*-acetylchitibiose to an asparagine residue or in the case of shorter oligosaccharides by an *O*-glycosidic bond to a serine or threonine residue (Figure 27). These latter linkages are subject to β-elimination in weak alkali. As described earlier, tunicamycin inhibits the synthesis of *N*-glycosidically linked mannan because of its action on the formation of the *N*-acetylglucosamine-containing lipid intermediate rather than by inhibiting the transfer of mannosyl residues.

Glucans are a major constituent of the majority of fungal walls. The main component isolated from walls of *Saccharomyces cerevisiae* by chemical extraction and enzymic digestion is β-(1 → 3)-glucan containing some 3–4% β-(1 → 6)-D-glucosidic interchain linkages. Using a three-stage

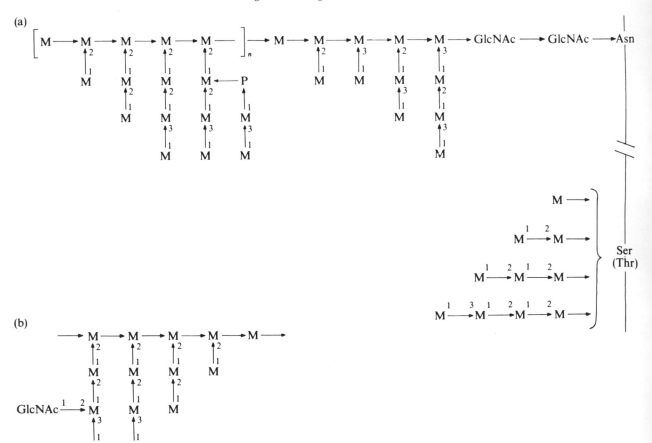

Figure 27 The structure of wall mannan from (a) *Saccharomyces cerevisiae* and (b) *Kluyveromyces lactis*. All anomeric linkages have the α-configuration with the exception of those between the reducing terminal mannose and the two *N*-acetylglucosamine residues which are β-(1 → 4) linkages. The configuration of the mannose attached directly to serine or threonine has not been established. For details see refs. 166 and 167

extraction procedure involving hot dilute alkali, hot 0.5 M acetic acid and further alkali treatment, three fractions of glucan can be distinguished.[173] Initially an alkali soluble, water insoluble β-(1 → 3)-glucan is released, followed in stage two by an alkali insoluble, acetic acid soluble, highly branched β-(1 → 6)-glucan, leaving an alkali insoluble β-(1 → 3)-glucan as the residual wall material. The difference in alkali solubility of the two β-(1 → 3)-glucans is related to the presence in the soluble glucan of 8–12% β-(1 → 6) linkages occurring in repeat sequences of at least three such bonds and as branch points. Linkages between glucan and protein or between glucan and mannan in the walls of certain fungi are suspected but not proven.

Chitin, a homopolymer of β-(1 → 4)-*N*-acetylglucosamine, is a major constituent of the walls of filamentous fungi but in yeasts is only a minor component (less than 1% of the total wall), where it appears to be located predominantly in the bud scar.[174] However, a minor fraction (about 8% of the total) was found uniformly distributed over the lateral walls of *S. cerevisiae* and *C. albicans*.[175] Despite being such a minor component it plays a critical role in yeast morphogenesis, as illustrated by the effects of inhibition of chitin synthesis by the polyoxins or nikkomycins (see below). The nonacetylated analogue, chitosan, is found in the walls of the Mucorales, where it is accompanied by glucuronic acid polymers not generally found in other fungi and also by phosphate that may serve to neutralize the free amino groups of the chitosan.[176]

A general model for the organization of the fungal wall based on the evidence described briefly above is shown in Figure 28. An inner structural layer contains the microcrystalline sleeve of chitin fibrils interwoven with peptides rich in lysine and citrulline. Linked to the exterior of these are shorter thicker fibrils of β-(1 → 3)-glucan, the whole forming a fibrillar glucan–chitin matrix in which are embedded the soluble peptidomannans. The wall is also the location of certain soluble enzymes and glycoenzymes such as invertase and acid phosphatase, which may under certain circumstances be secreted into the medium.[172,177]

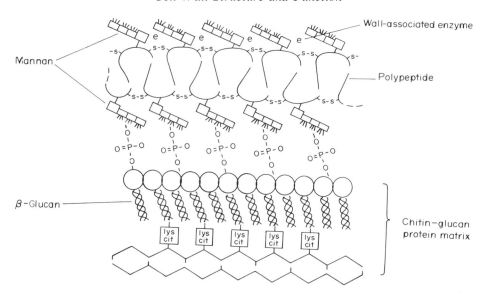

Figure 28 Hypothetical model for the organization of polymers in the wall of *Candida albicans*. The basal structural component of the wall is a fibrillar β-(1 → 3) glucan–chitin matrix. In comparison with many other fungi the chitin content of *C. albicans* is low and only recently has its presence in lateral walls been established. Previously it was thought to be localized exclusively in bud scars. Linkages between the polymers have not been established. However, in *Saccharomyces cerevisiae* and other fungi, enzymic digestion of fibrillar cell wall residues containing glucan and chitin has yielded *N*-acetylglucosamine linked to citrulline and/or lysine. The nature and origin of the protein from which these amino acids are derived remains unclear. The mannans are embedded in the glucan–chitin matrix together with polypeptides. Whether they are linked *via* phosphodiester bonds to glucan has not been established. The occurrence of such mannan–phosphodiester glucan linkages has been reported but not confirmed. Mannan is, however, linked *via* chitobiosyl units to asparagine and directly to serine/threonine residues in the polypeptides. Chelating and sulfhydryl reagents potentiate the activity of wall-degrading glucanases, suggesting that the polypeptides probably contain disulfide bridges and that divalent cations are probably involved as counter ions (? to negative charges of phosphodiesters) in stabilizing the wall structure. Further details of structure and organization can be found in the reviews of Ballou[166] and Reiss[172] (reproduced from ref. 172 by permission of Elsevier)

The biosyntheses of chitin, mannan and glucan have been investigated in a number of fungi, but particularly *S. cerevisiae* and to a lesser extent *C. albicans*. Details of the biosynthesis of fungal wall polymers are to be found in several recent reviews.[3,166,167,178-180] However, recent genetic and biochemical evidence has established the presence of two chitin synthases in the membrane of *S. cerevisiae*, only one of which appears to be essential for chitin synthesis under normal growth conditions.[181-183] Much of the earlier work was carried out using the other apparently nonessential enzyme and although the evidence accumulated undoubtedly describes many of the important features of chitin synthesis a reevaluation of the conclusions drawn should be made in the light of these new findings. Whether there are also multiple enzymes involved in the synthesis of β-(1 → 3)-glucan remains to be clarified.

9.1.7 INHIBITORS OF FUNGAL WALL POLYMER BIOSYNTHESIS

9.1.7.1 Polyoxins and Nikkomycins: Inhibitors of Chitin Synthase

The polyoxins and nikkomycins (neopolyoxins) are two closely related series of peptidyl nucleosides produced by *Streptomyces cacaoi* var. *asoensis* and *Streptomyces tendae*, respectively.[184-188] The structures of individual members of the series are known and examples of polyoxin D, uracil polyoxin C and nikkomycin Z are given in Figure 29. Early investigations established the antifungal activity of the polyoxins, whereas they were apparently without effect on bacteria or yeast.[189] However, both polyoxins and nikkomycins were shown to be competitive inhibitors of chitin synthase from yeasts, including *Saccharomyces cerevisiae* and *C. albicans* and a range of other fungi.[180,190] In each case the K_i values for the enzyme show close similarity (*e.g. S. cerevisiae* chitin synthase polyoxin D, $K_i = 0.5$ μM; UDP-*N*-acetylglucosamine, $K_m = 0.6$–0.9 μM).[180] In the case of *S. cerevisiae* these studies utilized the chitin synthase recently shown to be 'nonessential' for growth. However, both this enzyme and the 'essential' chitin synthase II are known to be inhibited by polyoxin D.[181-183]

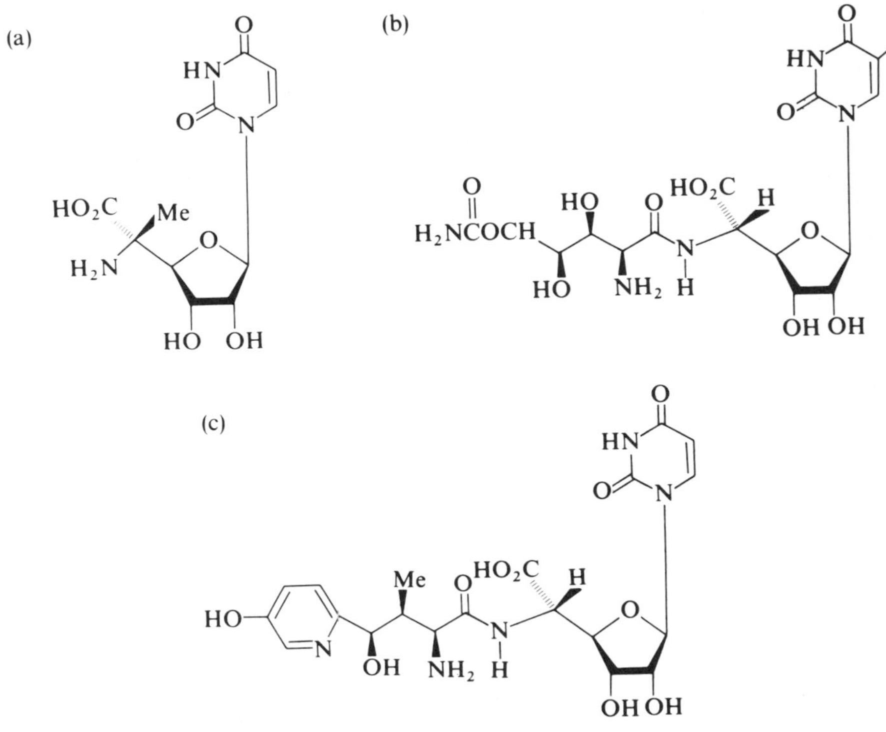

Figure 29   Structure of (a) uracil polyoxin C, (b) polyoxin D and (c) nikkomycin Z

The apparent anomaly that the polyoxins were without effect on the growth of yeast has now been explained. Peptides present in the growth medium antagonize the uptake of polyoxin D.[191,192] The incorporation of polyoxin D at relatively high concentrations in a synthetic growth medium (*i.e.* not containing peptides) resulted in the inhibition of primary septum synthesis and other morphological effects and at higher concentrations the inhibition of growth. In contrast nikkomycin Z is active in rich media containing peptides. The difference between the two compounds relates to their utilization of different peptide permeases to cross the cytoplasmic membrane of *C. albicans*. Thus nikkomycin Z (and presumably the other nikkomycins) enters *via* a peptide permease system that normally mediates the transport of natural di- and tri-peptides; polyoxin D only slightly inhibits nikkomycin Z uptake, indicating a low affinity for this transport system. The nature of the system utilized by polyoxin D remains to be established.

A number of analogues of polyoxin D containing simple amino acid substitution in the side chain were synthesized but all were inactive as antifungal agents. Emmer *et al.*[193] prepared a polyoxin derivative with (2S,3R)-3-hydroxyhomotyrosine, an amino acid component of echinocandin (see below), as the side chain. This was an effective inhibitor of chitin synthase ($I_{50} < 5 \mu M$) but was inactive against whole organisms. Other dipeptidyl and tripeptidyl derivatives and analogues of uracil polyoxin C and polyoxin D have been described.

The earlier dipeptidyl compounds were potent inhibitors of chitin synthase with activity equivalent to that of polyoxin D but showed poor activity against *C. albicans*. Antifungal activity was only obtained when the organisms were incubated with millimolar concentrations of the compounds (Table 6). Whether this lack of *in vitro* activity was due to inefficient transport into the fungi or as a consequence of rapid degradation by intracellular peptidases was not established.[194,195] Although polyoxin D was resistant to the intracellular peptidases, the synthetic compounds were rapidly hydrolyzed. The tripeptidyl analogues were inactive against the membrane-bound chitin synthase but showed similar antifungal activity to the dipeptidyl compounds *in vivo*, suggesting that they were acting as prodrugs, releasing the active moiety on hydrolysis by the intracellular peptidases.

Attempts to increase the stability of polyoxin analogues by modifying the peptide bonds has met with variable success. Poor activity against chitin synthase was obtained by Emmer *et al.*[193] and Boehm and Kingsbury,[196] although intracellular stability was not investigated. On the other hand Naider and his colleagues[197,198] have described a series of dipeptidyl and tripeptidyl analogues resistant to hydrolysis by peptidases. Although good inhibitory activity was obtained against chitin synthase (octanoylphenylalanylpolyoxin D, $K_i \simeq 7 \times 10^{-6}$ M), millimolar concentrations were still

Table 6 Polyoxin Analogues: Activity Against Chitin Synthase and Growth of
C. albicans[a]

Compound	Chitin synthase (I_{50}) (μM)	Growth inhibitory concentration (mM)
Polyoxin D	1.8	0.06
Uracil polyoxin C	100	>1.0
N-Methylnorleucine D	120	>1.0
N-Leucyl D	200	>1.0
Phenylalanyldehydro phenylalanyl D	40	>2.0
N-Octanoylphenylalanyl D	47.5	>1.0
N-Octanoyllysyl-UPOC	0.75	>2.0
N-Octanoylglutaminyl-UPOC	1.25	>2.0

[a] Data from refs. 197 and 198. Chitin synthase—a membrane preparation from *C. albicans* synthesized chitin from [^{14}C]-*N*-acetylglucosamine. The growth inhibitory concentration was determined in yeast nitrogen base medium.[195]

required to inhibit growth of *C. albicans*. Since the compounds appear stable to intracellular peptidases, their relatively poor activity against whole organisms presumably reflects inefficient transport into the cell. As described above, polyoxin D is a poor substrate for the peptide permease(s) of *C. albicans*.[194,195,199] It remains to be seen whether other analogues can be synthesized to utilize alternate transport systems to penetrate *C. albicans*.

9.1.7.2 Echinocandins and Papulacandins: Inhibitors of Glucan Synthase

The echinocandins, aculeacins and mulandocandin and the papulacandins and chaetiacandin are two groups of recently described compounds that inhibit glucan synthesis in fungi. The echinocandin group of compounds are related cyclic lipopeptides produced by various *Aspergillus* spp.[200-205] Their structures have been determined and echinocandin B was found to contain two rare β-hydroxyamino acids, (2*S*,2*R*)-3-hydroxyhomotyrosine (see previous section) and (2*S*,3*S*,4*S*)-3-hydroxy-4-methylproline (Figure 30). The cyclohexapeptide contains, in addition, threonine (2 mol), 4,5-dihydroxyornithine and 4-hydroxyproline. Other members of the echinocandin group differ in their degree of hydroxylation and in the fatty acid substituents; linolenic, myristic and palmitic acids are present.

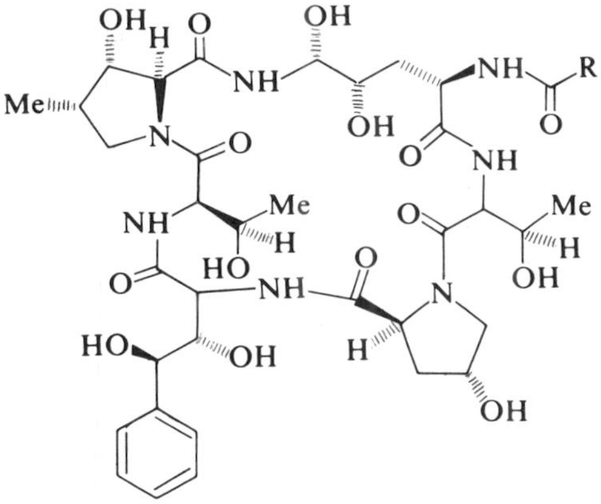

(a) Echinocandin B; R = (CH$_2$)$_4$CH=CHCH$_2$CH=CH(CH$_2$)$_7$Me

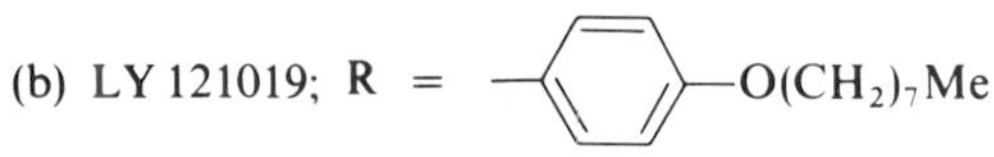

(b) LY 121019; R = —⟨benzene ring⟩—O(CH$_2$)$_7$Me

Figure 30 Structure of (a) echinocandin B and (b) the *p*-(octyloxy)benzoyl analogue (cilufungin, LY121019)

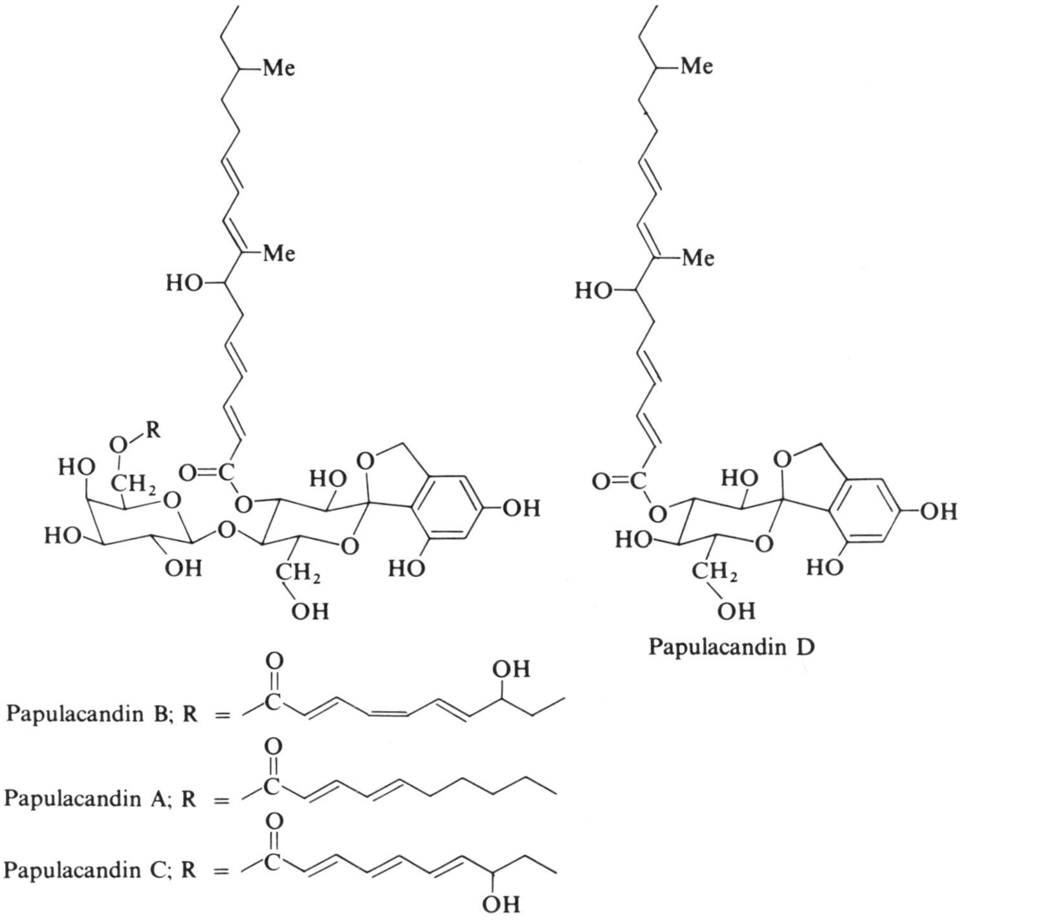

Figure 31   Structure of papulacandins A, B, C and D

The papulacandins consist of a galactosyl-β-(1 → 4)-glucose moiety linked *via* C—O and C—C bonds to an aromatic ring to form a spirocyclic system. Two unsaturated fatty acids are attached to the diglucoside nucleus by ester linkages: a C_{10} acid to position 6 of the galactose residue and a C_{16} acid to position 3 of glucose (Figure 31).[206-208] Other members of the papulacandin group show minor differences in the nature of the shorter fatty acid substituent. Chaetiacandin is a closely related compound differing only in that the spirocycle is opened.[209]

The echinocandins and papulacandins show good activity against *C. albicans* and related yeasts but have little or no activity against other fungi and bacteria. They are potent inhibitors of 1,3-β-glucan synthases of *C. albicans* and a range of other fungi and yeasts including *S. cerevisiae*, *Schizosaccharomyces pombe*, *Geotrichium lactis* and *Neurospora crassa*.[210-215]

A large number of analogues of echinocandin and papulacandin have been synthesized. In the former case the N-terminal linolenic substituent of echinocandin B was removed enzymatically using *Actinoplanes utahensis* to give the cyclic hexapeptide which lacked anti-*Candida* activity. The core peptide was reacylated with a wide variety of acylating agents to give active analogues of echinocandin B. The analogues were tested for anti-*Candida* activity and haemolytic potential (the chief toxic property of echinocandin B). On the basis of these tests LY121019, *p*-(octyloxy)-benzoylechinocandin (cilofungin), was chosen for further evaluation as being the least toxic.[216] Recently, a number of reports of *in vitro* studies with this compound have been published.[217, 217a]

Similarly, a large number of analogues of papulacandin have been synthesized without, in general, significant improvement in either their ability to inhibit glucan synthesis in spheroplasts of *C. albicans* or the growth of the organism.[218]

Briefly, the galactosyl residue and both fatty acids in the unsaturated form are essential for activity. However, the spirocycle can be opened; chaetiacandin has a similar spectrum of activity to papulacandin B. Other derivatives synthesized involved *O*-alkylation of the two phenolic hydroxyl groups but these showed no significant improvement over the parent compound. On the other hand,

Table 7 Activity of Papulacandin Derivatives[a]

	Compound			C. albicans	C. albicans
R^1	R^2	R^3	R^4	MIC ($\mu g\,mL^{-1}$)	ED_{50} ($mg\,kg^{-1}$)
—	—	H	—	0.05–0.1	80
—	—	H	—CH$_2$COMe	0.1–0.2	30
—NHSO$_2$Me	—	H	H	0.5–1	30–50
—NHOC—NN—H (imidazolidinone, C=O)	—	H	H	0.5–1	14

[a] Data from ref. 218.

some derivatives with a substituted amino group located directly on the aromatic ring and the phenolic hydroxyl groups free, showed good *in vitro* activity and some improvement *in vivo* (Table 7). Other findings where derivatives retained inhibitory activity against glucan synthesis but were inactive against the whole organism showed that the two properties can be separated. Whether compounds with improved inhibitory activity and an improved ability to penetrate the organism can be synthesized remains to be established.

9.1.8 THE CYTOPLASMIC MEMBRANE—STRUCTURE

Much of the early work on membrane structure was dominated by the bilayer 'unit membrane' hypothesis of Davson and Danielli.[219] The original proposal suggested a fairly rigid three-layered structure in which there was a middle lipid bilayer with hydrocarbon chains orientated side by side and in end to end contact. The hydrophilic end groups were directed to the outside and were in contact with extended protein molecules on the surface. However, this model took no account of the thermodynamics of the proteins and lipids involved. This situation was corrected in the fluid mosaic model of Singer and Nicholson,[220] proposed primarily on thermodynamic grounds as the arrangement of proteins and lipids that possessed minimum free energy. This model suggests that the membrane consists of a lipid matrix with randomly distributed globular proteins. Some of the proteins, which may form specific aggregates or be present as single polypeptides, are partially buried in one half of the membrane but remain able to interact with the aqueous phase, while others are present as transmembrane proteins. The majority of the lipids are arranged in a bilayer, which may represent up to 70–80% of the membrane area, although a value of 40% is probably more usual. There is good evidence that the bilayer is asymmetric. In *Bacillus megaterium* approximately twice as much phosphatidylethanolamine is present in the inner leaflet of the membrane as in the outer leaflet.[221] Asymmetric distribution of lipids was also reported in *M. lysodeikticus* (*M. luteus*) with the bulk of phosphatidylglycerol and phosphatidylinositol found in the outer and inner leaflets respectively. Phosphatidylglycerol appeared to be more evenly distributed. The most important property of the lipid bilayer is its fluidity with individual lipids free to move rapidly within the leaflet unless they are constrained by interaction with the membrane proteins.

In general two groups of proteins, the 'integral' and 'peripheral' proteins, are distinguished by their degree of association with the lipid bilayer. The 'integral' or 'intrinsic' proteins cannot easily be removed from the membrane, whereas 'peripheral' or 'extrinsic' proteins are readily dissociated by, for example, repeated washing with low ionic strength buffers or by treatment with chelating agents such as EDTA. The peripheral proteins are usually relatively soluble in aqueous media and do not require lipids for activity. In contrast, drastic treatment with reagents breaking hydrophobic bonds (*e.g.* detergents, organic solvents) is required to solubilize integral proteins which frequently remain associated with lipids after extraction. Subsequent removal of the lipid often results in denaturation of the proteins as they become highly insoluble.

Among the major functions of the cytoplasmic membrane proteins are the active and facilitated transport of nutrients and the export of potential toxic metabolites, energy generation through electron-transport chain proteins and a variety of ATPase proton pumps, and the synthesis of cell wall and envelope components. The membrane bilayer provides the necessary environment for the proteins involved in these activities and acts as a permeability barrier to separate the aqueous environments of medium and cytoplasm to allow vectorial processes to occur. Antibiotics affecting the cytoplasmic membrane commonly disrupt the permeability barrier either by major disorganization of membrane structure, by the formation of aqueous pores or by causing specific changes in cation permeability. Other antibiotics inhibit enzymes involved in energy generation or lipid synthesis. The following sections of this chapter will discuss examples of the former types of antibiotic.

9.1.8.1   Antibiotics Causing Disorganization of Membrane Structure

9.1.8.1.1   *Polymyxins and octapeptins*

The polymyxins are a family of polycationic decapeptide antibiotics with a fatty acid 'tail' produced by various species of *Bacillus*. They contain a heptapeptide ring, multiple residues of α,γ-diaminobutyric acid and a peptide side chain substituted with a fatty acid residue which varies from methyl C_8 to methyl C_{10} (Table 8). The octapeptins are similar in structure except that, as the name implies, they are octapeptides (the side chain contains one diaminobutyric residue rather than the two or three residues in the polymyxins) and the fatty acid substituents are usually β-hydroxymethyl fatty acids. The circulins and colistins can also be classified as polymyxins and the structures and activities of the group as a whole have been reviewed by Storm *et al.*[222] The structural features essential for antibacterial activity are the cyclic heptapeptide and a fatty acid of length C_8–C_{14}.

Table 8 Structure of Some Polymyxins and Octapeptins

Polymyxin	X	Y	Z	X'	FA[a]
A	Leu	Thr	Thr	D-DAB	aiC_9
B	Phe	Leu	Thr	L-DAB	aiC_9, iC_8
D	Leu	Thr	Thr	D-Ser	aiC_9, iC_8
E (Colistin)	Leu	Leu	Thr	L-DAB	aiC_9, iC_8
S	Phe	Thr	Thr	D-Ser	aiC_9
T	Phe	Thr	Leu	L-DAB	aiC_9
Octapeptins (lack the dipeptide shown in brackets)					
A	Leu	Leu	Leu	D-DAB	$\beta\text{-OH-}aiC_{11}$
B	Leu	Phe	Leu	D-DAB	$\beta\text{-OH-}iC_{10}$
C	Phe	Leu	Leu	D-DAB	$\beta\text{-OH-}nC_{10}$

[a] Fatty acids: aiC_9, (+)-6-methyloctanoic acid; iC_8, 6-methylheptanoic acid; $\beta\text{-OH-}aiC_{11}$, β-hydroxy-8-methyldecanoic acid; $\beta\text{-OH-}iC_{10}$, β-hydroxy-8-methylnonanoic acid; $\beta\text{-OH-}nC_{10}$, β-hydroxy-6-methyloctanoic acid.

The polymyxins are generally more effective against Gram-negative bacteria and until the discovery of gentamycin and carbenicillin were the only antibiotics with useful activity against *Pseudomonas aeruginosa*. However, despite their structural similarity, differences have been found when their antibacterial activity was compared with that of the octapeptins. Octapeptins A and B were three to 10 times more active than polymyxin B against Gram-positive bacteria.[223]

The ultimate target of these antibiotics is thought to be the cytoplasmic membrane to which they could become bound through interaction with the negatively charged phospholipids. Such an interaction represents no problem in Gram-positive bacteria, where access to the membrane is believed to be 'unrestricted'. However, polymyxins as a group are more active against Gram-negative bacteria and for this binding to happen the antibiotic must first cross or disrupt the outer membrane barrier. With a molecular weight of 1000–1200 polymyxins are too large to go through the porin channels of enteric bacteria.[33] Studies with polymyxin B and octapeptin immobilized on agarose beads showed that contact of the beads with the outer surface of protoplasts of *B. subtilis* and cells and spheroplasts of *E. coli* was sufficient to inhibit growth and respiration.[224] Moreover, electron microscopic studies revealed that polymyxin treatment of *E. coli* caused extensive alteration to the bacterial surface with the extrusion of surface blebs and the release of outer membrane material into the medium.[225] Disruption of the outer membrane permeability barrier was shown by the increased sensitivity of treated organisms to detergents, hydrophobic antibiotics (novobiocin, macrolides and penicillins) and lysozyme. Differential damage to the outer membrane was also shown by the rapid release of periplasmic proteins, whereas longer treatment was required to release cytoplasmic material.[226]

Mutants resistant to polymyxin have been isolated in *S. typhimurium* and the resistance shown to be accompanied by changes in lipopolysaccharide (LPS) composition. The LPS of the mutants contained four- to six-fold increased amounts of 4-aminoarabinose and more ethanolamine than wild-type LPS.[227] Thus the mutant polymer is less acidic and presumably has a lower binding affinity for the cationic polymyxin. However, the mutants were not resistant to octapeptin, suggesting that hydrophobic interactions may be more important in the disorganization of the outer membrane by these agents.[228]

More recently, the papain-cleaved derivative of polymyxin B that has lost the terminal di-aminobutyric acid residue with its fatty acid substituent (polymyxin nonapeptide, PMBN) was found to be very effective at sensitizing *E. coli* to a variety of hydrophobic antibiotics (novobiocin, fusidic acid, macrolides, substituted penicillins and ionophores, including valinomycin and A23187).[229–232] Similar results have been obtained with colistin (polymyxin E) nonapeptide (CNP) and heptapeptide (CHP).[233]

However, PMBN showed only poor antibacterial activity against *E. coli* and *S. typhimurium*. The sensitization to other antibiotics occurred very rapidly but was not accompanied by loss of LPS into the medium. Electron microscopic studies revealed that long finger-like projections were produced from the outer leaflet of the outer membrane.[230] Thus PMBN appears to interact with the LPS to expand the area of the outer leaflet with a consequent increase in permeability. A range of other Gram-negative bacteria, including *Pseudomonas* and *Enterobacter* spp., were also sensitized, although polymyxin resistant strains were also resistant to PMBN.[234] Using ^{3}H-labelled PMBN it was found that PMBN and polymyxin compete with each other for their binding to the outer membrane and approximately $1–2 \times 10^6$ molecules of PMBN are bound with high affinity ($K_d = 1.3 \, \mu M$) to a single *S. typhimurium* cell.[235]

Earlier studies showed that the effects of polymyxin on Gram-negative bacteria can be antagonized by divalent cations, particularly Ca^{2+} and Mg^{2+}. These cations are known to play an important role in maintaining the structural integrity of the outer membrane and treatment with Tris–EDTA disrupts the permeability barrier in a similar manner to the antibiotics. Storm *et al.*[222] proposed that polymyxins and octapeptins acted by competitively displacing Mg^{2+} and Ca^{2+} from negatively charged groups on membrane phospholipids. The advances in our understanding of the asymmetry of outer membrane structure make it more likely that the cations are involved in the stabilization of LPS–LPS interactions. In this situation the antibiotics may disrupt these interactions to increase permeability through the outer membrane by increasing the area of phospholipid bilayer. Clearly, the results obtained with PMBN demonstrate that incorporation of the fatty acid side chain of polymyxin into the lipid bilayer is not required for disorganization of the outer membrane to occur, whereas the fatty acid is required for antibacterial activity. The mechanism whereby the antibiotics interact with the cytoplasmic membrane remains unknown, but incorporation of the fatty acid into the lipid bilayer, together with the possible displacement of divalent cations, may be responsible for the loss of the permeability barrier of this inner membrane.

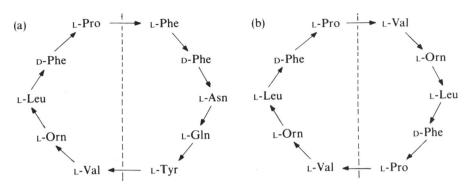

Figure 32   Structure of (a) tyrocidin A and (b) gramicidin S

9.1.8.1.2   Tyrocidins and gramicidin S

These are a related group of cyclic peptide antibiotics produced by *Bacillus* spp. The structure of tyrocidin A is shown in Figure 32; gramicidin S, which is a tyrocidin and should be distinguished from the other gramicidins A, B and C, is a dimer of the left-half of tyrocidin A. The other members of the tyrocidin group have variable structures in the right-half of the molecule. These compounds are not active against Gram-negative bacteria, being unable to cross the outer membrane, but they disrupt protoplast membranes, causing lysis. Structural studies established that the cyclic structure is important for activity as are the free amino groups; acetylation resulted in loss of activity.[236,237] Pache *et al.*[238] concluded from physicochemical studies that gramicidin S interacted with the negatively charged groups of phospholipid, presumably to perturb the phospholipid bilayer sufficiently to induce marked changes in the permeability barrier function of the cytoplasmic membrane.

9.1.8.1.3   Gramicidins

Gramicidins A, B and C (**1**; not to be confused with gramicidin S—see above) form a group of related linear peptides, 15 amino acids in length with the C terminus substituted with ethanolamine and the N terminus formylated.[239] The amino acids are alternately D and L in configuration.

HCO
|
L-Val-Gly-L-Ala-D-Leu-L-Ala-D-Val-L-Val-D-Val-L-Trp-D-Leu-X-D-Leu-L-Trp-D-Leu-L-Trp
|
NHCH₂CH₂OH

$$\text{HCO}-\text{L-Val-Gly-L-Ala-D-Leu-L-Ala-D-Val-L-Val-D-Val-L-Trp-D-Leu-X-D-Leu-L-Trp-D-Leu-L-Trp}-\text{NHCH}_2\text{CH}_2\text{OH}$$

(**1**)   X = L-Trp; gramicidin A
           X = L-Phe; gramicidin B
           X = L-Tyr; gramicidin C

Early work showed that treatment of a variety of natural and artificial membrane systems promoted exchange of K⁺ and H⁺ across these membranes. At first it was thought that gramicidin was acting as an ionophore but more detailed studies argued against this conclusion. Rather, Hladky *et al.*[240] proposed that gramicidin formed open pores in the membranes through which protons and univalent cations can diffuse. In support of this they noted that (i) the diffusion coefficient for an ion across a gramicidin-treated membrane is greater than that expected for an ion–gramicidin complex; (ii) the conductivity of protons and univalent cation permeabilities are in the same order as their mobilities in aqueous solution; (iii) conductance is independent of membrane thickness; (iv) freezing of the membrane does not prevent gramicidin-induced conductance but does prevent ionophore uptake; and (v) no gramicidin–ion complexes have been detected.

The linear peptide of gramicidin with alternating D and L residues can adopt a helical conformation held together by hydrogen bonds lying almost parallel to the axis of the cylinder (Figure 33). The inside of the cylinder is lined with polar groups around a central pore of about 0.4 nm diameter, while the fatty acid side chains of the amino acids form a lipophilic shell to the outside. Conductivity measurements suggest that in membranes the pores have a transient existence with only a fraction of them open at any one time. The gramicidin channel has proved to be of particular interest because unlike most other channels its molecular structure is known. There is detailed experimental evidence suggesting that in membrane bilayers the channel exists as an N-terminus–N-terminus right-handed dimer of single helices, whereas in organic solvents gramicidin forms double helical dimers.[241] Chemical modification of gramicidin at the N terminus prevents dimer formation and the ability to form transmembrane channels (Table 9). Gramicidin which has been modified by replacing the formyl group with an acetyl group (*N*-acetyl) has reduced activity, while the deformyl- and deformylvalyl-gramicidin analogues are essentially inactive. Synthetic dimers produced by chemical linkage of the N termini retain and show improved activity over the native antibiotic. Malonyl–bis(deformyl)gramicidin A dimer showed similar ion selectivity and conductance to gramicidin A but the lifetime of the channel was considerably longer.[242] Gramicidin which has been *O*-acetylated at the C terminus retains full activity (Table 9).

Ion translocation through these channels was thought to involve a 'mechanism' whereby the ion overcame a series of electrostatic barriers arising from ion–channel interactions (Jordan, ref. 243 and

(a)

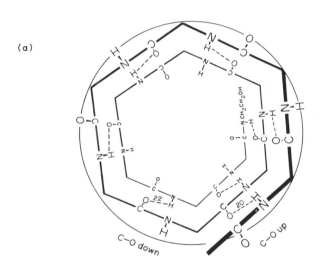

(b)

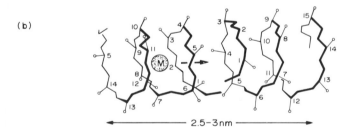

Figure 33 Schematic model of the helical structure of the gramicidin A pore (a) in cross section[302] and (b) functioning as a pore in the membrane.[303] The channel (approx. 0.4 nm wide) along the helical axis is lined with the oxygen atoms of the peptide carbonyl groups. The amino acid side chains located to the outside of the helix are not shown. The transmembrane pore consists of two helices linked head to head by their N termini

Table 9 Conductance Properties of Gramicidin Analogues[241]

Analogue	Relative conductance (steady-state)	Analogue	Relative conductance (steady-state)
Gramicidin	1.0	Deformylgramicidin	0.002
O-Acetylgramicidin	1.2	Deformylvalylgramicidin	0.0001
N-Acetyldeformylgramicidin	0.11		

refs. cited therein). More detailed models can now be calculated[244] which may help to delineate the roles of the various parts of the gramicidin structure and the effects of changes in the structure of both the backbone amino acids and side chains on conductance.

9.1.8.1.4 *Polyenes*

The polyenes are a group of macrolide antibiotics characterized by a large lactone ring containing both a system of conjugated double bonds all in the *trans* configuration and a hydrophilic region with a number of hydroxyl groups. In individual polyenes (over 200 have been described) this basic structure is substituted at specific positions with aliphatic or aromatic side chains, carboxyl groups or the amino sugar mycosamine (3-amino-3,6-dideoxymannose). The structural formulae of certain of the clinically important polyenes are shown in Figure 34. The number of carbon atoms in the lactone ring and the number of hydroxyl groups vary among individual compounds and range from twelve to 37 and six to 14, respectively. Detailed accounts of the chemistry and properties of many polyenes have been reviewed elsewhere.[245,246]

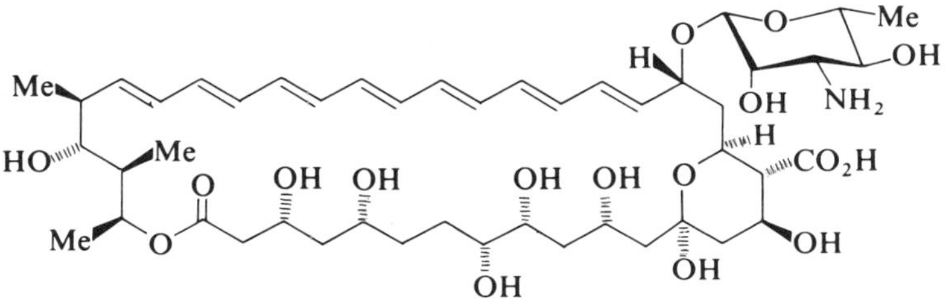

Amphotericin B

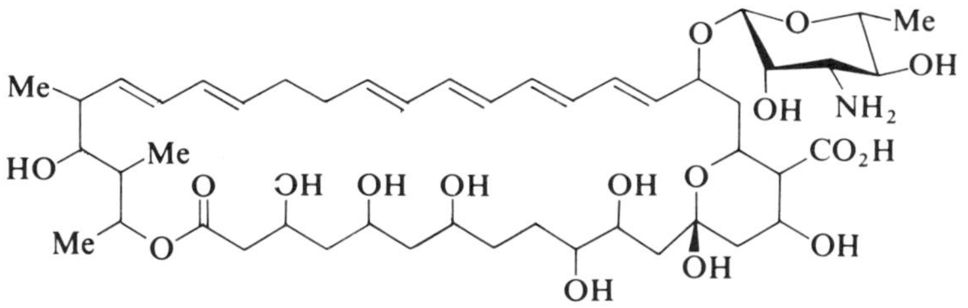

Nystatin

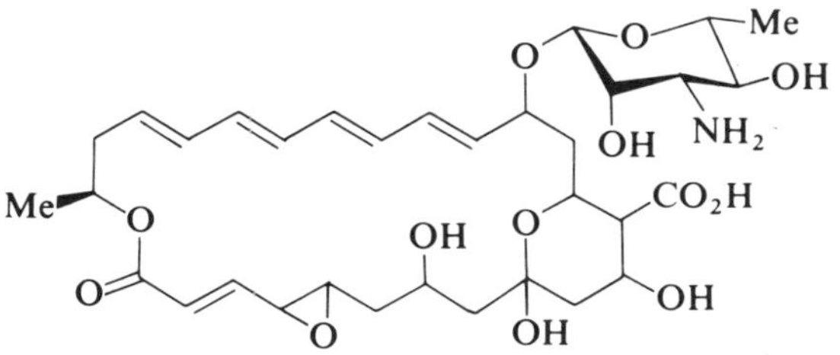

Pimaricin

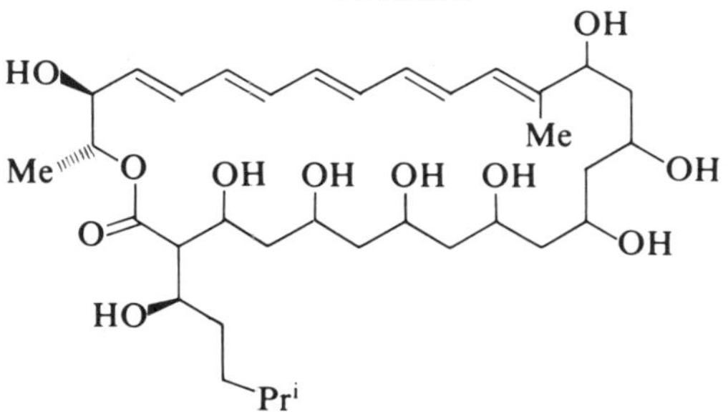

Filipin

Figure 34 Structures of certain clinically important polyene macrolides and filipin

The three-dimensional structure of amphotericin B has been determined by X-ray crystallography.[247] It is a rigid rod-shaped structure held by the polyene structure with the hydroxylated portion lying alongside, hence the molecule has opposing hydrophobic and hydrophilic faces. The amino sugar is present at one end of the molecule and there is a single hydroxyl group at the other. The overall length of amphotericin B is 2.1 nm, roughly equivalent to the length of a phospholipid molecule.

The polyenes act to alter the permeability of the membranes of sensitive cells. These are restricted to those organisms whose membranes contain sterols, *i.e.* a wide range of fungi including *Candida albicans* and other eukaryotic cells, whereas bacteria and blue-green algae are unaffected. Exceptions among prokaryotes are those organisms (mycoplasmas and acholeplasmas) that incorporate sterols into their membranes. The importance of membrane sterols was supported by the observation of Gottleib *et al.*[248] that addition of exogenous sterol to the growth medium can protect fungi from the inhibitory action of certain polyenes. Underlying this protective effect was a hydrophobic interaction between antibiotic and sterol to lower the effective drug concentration.[249] The structural requirements of the sterol necessary to make acholeplasmas sensitive to polyenes has been studied.[250] Using a wide variety of exogenously supplied sterols, these authors concluded that the incorporated sterol must have a 3-β-hydroxyl group, a planar ring structure and a hydrophobic side chain attached to C-17 to render the cells sensitive to the permeabilizing effects of polyenes. In artificial lipid bilayers, the incorporation of ergosterol makes the membranes more sensitive to the action of amphotericin B than does the incorporation of cholesterol; the reverse is true for filipin.[251,252] Similar results were obtained when the sensitivity of *C. albicans* (ergosterol-containing) and mouse LS cells (cholesterol-containing) to amphotericin B was compared.[253,254] The greater affinity of amphotericin B for the ergosterol-containing fungal membranes appears to underlie its clinical effectiveness as an antifungal agent. However, the interactions with cholesterol undoubtedly give the polyene a small therapeutic index and the use of amphotericin B as a systemic antifungal agent has to be carefully monitored. In fact, amphotericin B is the only polyene with sufficient selectivity to be used systematically for the treatment of serious fungal infections. Since it is not absorbed from the gut, it is given intravenously. The selectivity, solubility and clinical utility of amphotericin B is improved in the methyl ester[255] but the drug was abandoned because it seemed to be neurotoxic.[256] More recently *N*-D-ornithylamphotericin B methyl ester, a water soluble congener and liposomal amphotericin B have been evaluated as a further means of reducing the toxicity.[257-259] Other polyenes, in particular nystatin, pimaricin and candicidin are used topically or given orally where, because they are not absorbed, they act exclusively within the digestive tract.

Physical studies with lipid bilayers containing ergosterol or cholesterol suggested that one polyene molecule associated with one sterol. Moreover, the binding constant of amphotericin B to ergosterol was significantly greater than that to cholesterol, a finding again in agreement with the selectivity of amphotericin *in vivo*.[260]

As with the gramicidins, one of the first detectable effects of polyene treatment was the loss of K^+ from sensitive cells.[261] The consequent uptake of protons was thought to result in acidification of the cell, leading to cell death. Subsequently, views changed to conclude that the primary effect is to render the plasma membrane of sensitive cells more permeable to protons; the loss of potassium is therefore a secondary effect.[253,262] Later, electrical conductance experiments provided evidence for the presence of polyene-induced aqueous pores in artificial lipid bilayers. The presence of sterol (cholesterol in most cases examined) was necessary for pore formation to occur.[263,264]

Several molecular models have been proposed to explain the structure of the postulated pores produced in membranes.[265-267] In general, these are similar, suggesting that the hydrophobic face of the polyene molecule reacts either with sterol molecules or phospholipid acyl side chains in one leaflet of the lipid bilayer. Complexes of this type then interact to form extended polymers in which sterol can be interposed between two polyene molecules. Aggregates containing eight polyene molecules could form a cylinder in which the hydrophilic face of the polyenes was directed inwards to surround an aqueous pore of approximately 0.8 nm diameter. The mycosamine group was located at the membrane/medium interface and the length of the amphotericin molecule would produce a pore with a length approximately equal to half that of the membrane.[265] Translocation of the polyene across the membrane was proposed to allow the formation of equivalent structures within the inner leaflet of the membrane. Together an active pore through the complete membrane would be formed by the hydrogen bonding of the terminal hydroxyl groups in a tail to tail configuration. However, such a translocation appears unlikely since neither movement of amphotericin B across multilamellar systems[268] nor the effect of nystatin on intracellular membranes[269] have been detected.

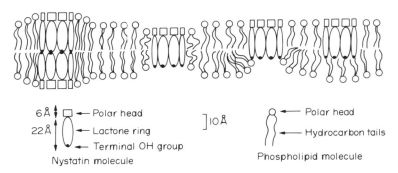

Figure 35 Schematic model of the interactions of nystatin with a lipid bilayer. The polyene forms pores or half-pores; the insertion of the polyene into the bilayer can result in extension of hydrocarbon tails of the phospholipids as shown[266]

A modified model was proposed by Marty and Finkelstein[266] in which either 'double pores' or 'single pores' could span the lipid bilayer. Their studies demonstrated ion selectivity of the pore depending on whether the polyene was added to one or both sides of the membrane. Addition of amphotericin to both sides of the membrane resulted in the formation of anion selective pores, whereas addition to one side, albeit at higher concentrations, led to cation selective pores. Thus half-pores formed in this way appeared to have the selectivity found in fungal membranes treated with the antibiotic. From these observations Marty and Finkelstein concluded that formation of a 'half-pore' within the membrane either altered the thickness of the membrane within its vicinity or so disorganized the inner leaflet as to destroy its permeability barrier properties (Figure 35). Support for this view came from studies of the effect of bilayer thickness on pore formation.[267] In artificial membranes formed from phosphatidylcholine containing mainly oleic acid side chains, addition of amphotericin B to one side resulted in potassium leakage. However, when the membrane was prepared from synthetic lipid containing didocosanoyl rather than oleic acid, polyene had to be added to both sides of the membrane to effect ion leakage. The length of the polyene (2.1 nm) is less than that of the hydrophobic cores of either artificial membrane (oleic acid containing, 3.5 nm; didocosanoyl, 4.5 nm) but the authors concluded that formation of a half-pore was sufficient to perturb the structure of the oleoyl membrane, allowing leakage to occur. The difference in ion selectivity observed was assumed to result from the presence of a ring of terminal hydroxyl groups at the hydrophobic end of the pore. These would form a cation selective gate at the entrance to the pore, whereas in a double pore the hydroxyl groups within would impart an anion selective positive charge relative to the surrounding medium.

However, the formation of aqueous pores comprising polyene–sterol octamers is not proved with certainty. At subgrowth inhibitory concentrations, polyenes have been shown to enhance the uptake of diverse substances (DNA, rifampicin, 5-fluorocytosine) by yeast and animal cells.[270,271] It is difficult to see how penetration of these compounds could occur through aqueous pores without leakage of cellular contents occurring. It remains possible that polyenes become incorporated into membranes and in so doing disrupt the membrane integrity, influencing fluidity and thus altering the permeability properties of the membrane. Clearly, at higher antibiotic concentrations gross disorganization of the membrane occurs, often associated with cell lysis. In the presence of small amounts of antibiotic, reduction in membrane fluidity may be sufficient to enhance permeability of the membrane to protons and hence trigger a cascade of secondary effects.

The development of resistance to polyenes during chemotherapy appears only rarely and in general these resistant strains have reduced amounts of ergosterol in their membranes (for review see Scholer and Polak,[272]). The sterol was often replaced by precursors from earlier stages in the biosynthetic pathway, *e.g.* Δ^8-sterols and some with 4- and 14-methyl groups more closely related to lanosterol. Cross-resistance to other polyenes may be complete or only partial, depending upon the affinity of the antibiotic for the precursors present.

9.1.8.1.5 *Alamethicin, suzukacillin*

Alamethicin is a family of water soluble peptides of 20 amino acids isolated from *Trichoderma viride.*[273] These compounds fall into a special class of peptide antibiotics together with the antiamoebins, emerimicins and suzukacillin called 'peptaibophols'. Peptides of this class contain multiple aminoisobutyric acid residues, an acetylated amino terminus and a phenylalaninol (Phol) substituted carboxyl terminus. After some early confusion, the amino acid sequence of the major component of alamethicin has been determined as shown in (2).[274]

Ac-Aib-Pro-Aib-Ala-Aib-(-Ala)-Gln-Aib-Val-Aib-Gly-Leu-(-Aib)-Aib-Pro-Val- Aib-Aib-Glu-Gln-Phol

(2)

Alamethicin is active against Gram-positive but not Gram-negative bacteria.[273] Interest in alamethicin stems largely from its voltage-gated ionic conductance in artificial membrane systems but it also causes haemolysis of erythrocytes and uncouples oxidative phosphorylation in mitochondria.[275,276] Lau and Chan[277] demonstrated that ionic pores could be formed with alamethicin in vesicles when a transmembrane potential was generated. Thus the pores formed by molecules of this type provide a model for the voltage dependent channels of the electroexcitable membranes of nerve and muscle.

While a large amount of electrical conductivity data have been generated (for a review see Gale *et al.*[278]) only a few structural studies have attempted to relate the structure of alamethicin to its biological activity. Martin and Williams[279] suggested that the molecule could form head to tail oligomers and so cross the lipid bilayer, forming pores if the molecules aggregated side by side. More recently, the NMR studies of Bauerjee *et al.*[280] have suggested the occurrence of dimer formation by alamethicin.

The dimer contained a rigid extended β-pleated structure involving amino acid residues 15 through 20, while residues 3 to 9 were folded into an α-helix. Proline at position 14 disrupts the continuity of the β-pleated sheet and forces amino acids 10 through 14 into an open structure. A dimer of this conformation is highly amphiphilic with one face completely hydrophobic and the other hydrophilic, particularly where it is lined with polar groups towards the C-terminal half of the complex. In aqueous environments the dimers readily aggregate and the formation of such aggregates may underlie the formation of pores in membranes.

Suzukacillin is a related peptide of 23 amino acids also produced by certain strains of *Trichoderma viride* that shows membrane effects similar to those obtained with alamethicin. The structure (3) has been proposed for suzukacillin[281] but in view of the confusion that occurred in determining the structure of alamethicin it may require some revision.

Ac-Aib-Pro-Val-Aib-Val-Ala-Aib-Ala-Aib-Aib-Gln-Aib-Leu-Aib-Gly-Leu-Aib-
Pro-Val-Aib-Aib-Glu(Phol)-Glu

(3)

The molecule has a higher α-helical content than alamethicin and this together with the increased number of aliphatic residues means that suzukacillin aggregates more strongly than alamethicin in aqueous environments. In addition, the greater proportion of aliphatic groups will result in the formation of more stable dimers should this feature be important in the mode of action of compounds of this type. The increased lipophilicity may also explain the observation that suzuka-cillin treatment of artificial lipid bilayers induced both voltage dependent (gated) and voltage independent conductances.[282] The latter finding may result from the increased α-helical content of the molecule facilitating increased lipid–antibiotic interactions. These could in turn lead to a greater chance of insertion of the peptide into the membrane occurring in the absence of an applied potential. A model for the conformation of suzukacillin within a lipid bilayer has been proposed[282] and shows two parts: a curved portion in which β-turns are stabilized by hydrogen bonds and a region of α-helix which may be apart from the conducting channel. Experiments in which artificial membranes were treated with mixtures of alamethicin and suzukacillin suggested that mixed pore formation might occur, *i.e.* pores containing both antibiotics. Certainly, pore formation was not hindered by interaction of the antibiotics.[282]

However, as stated briefly above, interest in compounds of this type stems from their ability to form voltage-gated channels across lipid bilayers rather than from any intrinsic antimicrobial activity. In gated channels there is a requirement for an applied potential to form the pores either by forcing the antibiotic into the membrane, or by forming dimers prior to their insertion into the lipid bilayer. Thus, they can be distinguished from peptides of the gramicidin family or from polyenes which form pores in the absence of an applied potential.

9.1.8.2 Antibiotics Producing Specific Changes in Membrane Permeability

9.1.8.2.1 *Ionophores including valinomycin, enniatins and macrotetralides*

Valinomycin is a cyclic depsipeptide originally isolated from a *Streptomyces* fermentation that contains alternating amino acids and α-hydroxy acids; six residues of valine (three of the L and three of the D configuration), three residues of L-lactic acid and three of D-α-hydroxyvaleric acid.[283] The structure shown in Figure 36 was determined by Shemyakin *et al.*[284]

Early studies established that valinomycin is a powerful uncoupler of oxidative phosphorylation in mitochondria by a mechanism that differed from that of the classical uncouplers.[285] Subsequently, Pressman and his colleagues (for a review see ref. 286) showed that valinomycin induced a selective energy-linked K^+ for H^+ exchange. Moreover, the selectivity was marked in that Na^+ was an ineffective substitute for K^+. Although these early studies were interpreted to show the activation of a specific mitochondrial K^+ transport receptor, further work with artificial membrane systems clearly devoid of any such receptors pointed to molecules of this type being mobile cation carriers. At this time the term ionophore was suggested as a generic name for compounds showing this property and to emphasize the dynamic nature of the transport mechanism.[287]

A large number of analogues and derivatives of valinomycin have been synthesized and the K^+ affinity constants and antibiotic spectra reported.[288,289] Linear analogues, decreasing or increasing the ring size or inverting a single asymmetric centre gave inactive compounds or greatly decreased their transport activity.[290] The enniatins are a family of cyclic hexadepsipeptides, *i.e.* half the ring size of valinomycin, that contain alternating residues of D-α-hydroxyvaleric acid and either *N*-methylisoleucine (enniatin A), *N*-methylvaline (enniatin B), *N*-methylleucine (enniatin C) or *N*-methylphenylalanine (beauvericin; Figure 36). Other ionophores include the macrotetralide nactins. These are a series of cyclic esters produced by various strains of actinomycetes. They differ in the number of ethyl substituents that replace methyl groups in the parent molecule (nonactin), which is made up of alternating residues of D- and L-nonactinic acid. Hence, the series contains nonactin, monactin, di-, tri- and tetra-actin (Figure 36).[291] In general the higher homologues show increased antibiotic activity with increasing affinity constants and ratios for $Na^+:K^+$ complexes.[286,292]

A further group of ionophores that may be considered collectively are the carboxylic ionophores. All are variations of a common theme of polyether, polyalcohol monocarboxylic acids and while they are not covalently cyclized the common features of a carboxyl group at one end of the molecule and one or two hydroxyl groups (or NH in the case of calcimycin) allow cyclization to occur through head to tail hydrogen bonding. This conformation is further stabilized by the twists in the rings of the backbone and the asymmetric centres. The group is exemplified by nigericin. Monensin resembles nigericin except for the added chain length at the tail of the molecule (Figure 37). In a three-dimensional structure this increased length has the effect of pushing the carboxyl moiety out of the complexation plane with the result that although monensin has six oxygen atoms capable of acting as ligands, none form ionic bonds and the molecule forms weaker complexes than nigericin.[293] Dianemycin is a larger carboxylic ionophore with a total of seven liganding oxygens (Figure 37).[294]

Harold and Baarda[295,296] studied the effects of a variety of ionophores on the growth of *Streptococcus faecalis*. Valinomycin (10^{-6} M) inhibited growth of the bacteria but this effect could be reversed by excess K^+. Exchange experiments with bacteria and artificial membranes[297] revealed both types of membrane to become selectively permeable to $H^+ > Rb^+ > K^+ > Cs^+ \gg Na^+ > Li^+$. Similar results were obtained with nigericin and monactin and bacteria preloaded with $^{86}Rb^+$ showed rapid exchange for H^+, K^+ or Rb^+ in the presence of either antibiotic. Harold and Baarda[295,296] proposed that the inhibitory action of these antibiotics results from increased permeability to cations with a consequent depletion of K^+, resulting in inactivation of cellular macromolecular synthesis. Moreover, this selective change in permeability to cations occurs without any direct action on the generation or utilization of ATP, *i.e.* the action of these antibiotics is readily distinguished from the effect on energy-linked transport or accumulation of cations induced by uncouplers of oxidative phosphorylation.

As mentioned briefly above the ionophores act as mobile carriers of cations forming complexes of differing affinity with the various ions. Mueller and Rudin[298] first suggested that the macrocyclic structures could form complexes by interaction of the ester carbonyl oxygen atoms acting as proton acceptor functions with hydrated or naked cations held by multiple dipole interactions or hydrogen bonds. A complex of this type would not only sequester the cation but also provide a strongly hydrophobic exterior formed by the side chains of the ionophore (methyl and isopropyl groups in the case of valinomycin). Examples of all the groups of ionophores mentioned have now been shown to form complexes of this type with K^+ and in each case the structure and conformation of the

(a)

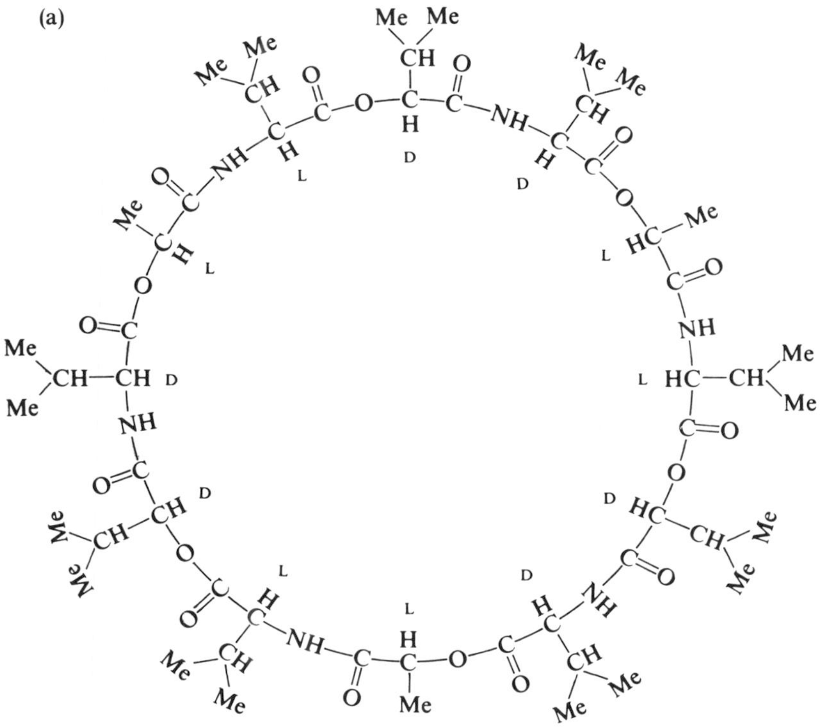

(b)

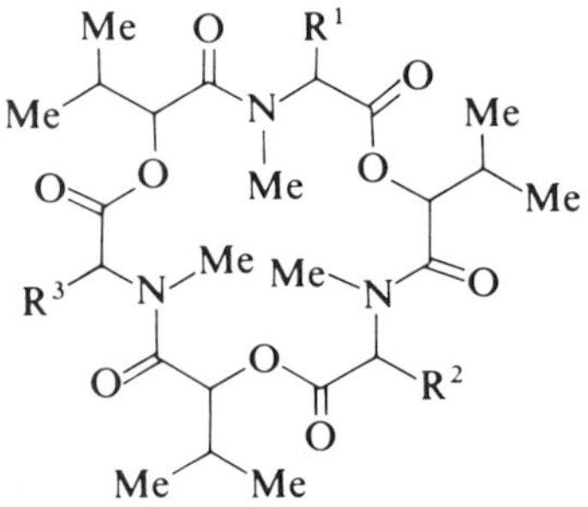

Antibiotic	R^1	R^2	R^3
Enniatin A	Bu^s	Bu^s	Bu^s
Enniatin A$_1$	Pr^i	Bu^s	Bu^s
Enniatin B	Pr^i	Pr^i	Pr^i
Enniatin B$_1$	Bu^s	Pr^i	Pr^i
Enniatin C	Bu^i	Bu^i	Bu^i
Beauvericin	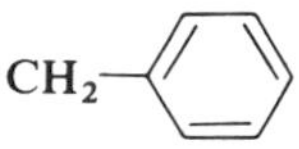 CH$_2$	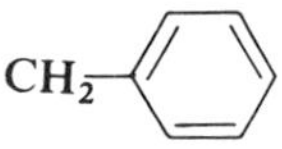 CH$_2$	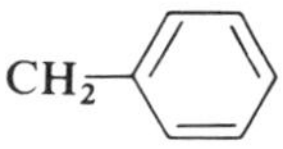 CH$_2$

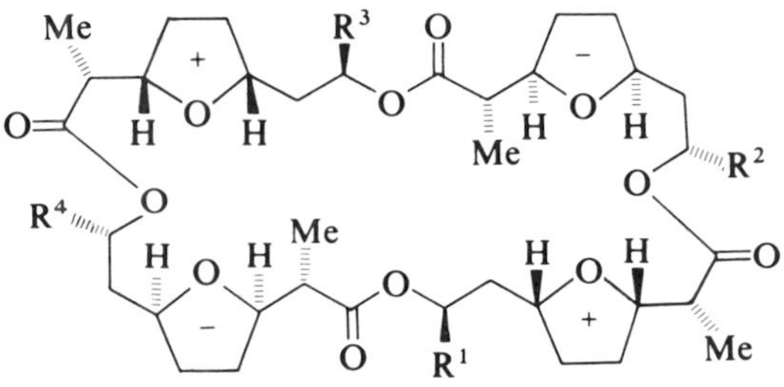

(c)

Antibiotic	R^1	R^2	R^3	R^4
Nonactin	Me	Me	Me	Me
Monactin	Me	Me	Me	Et
Dinactin	Me	Et	Me	Et
Trinactin	Me	Et	Et	Et
Tetranactin	Et	Et	Et	Et

Figure 36 Structures of (a) valinomycin, (b) the enniatins and (c) the nactins

(a) Monensins

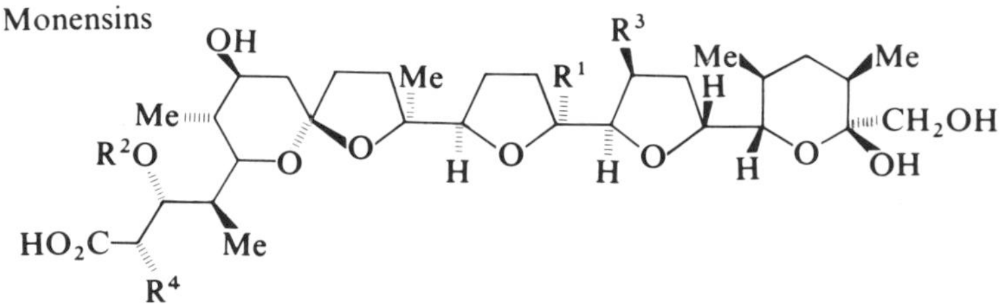

Antibiotic	R^1	R^2	R^3	R^4	←
Monensin A	Et	Me	Me	Me	Glucosyl (A-27106)
Monensin B	Me	Me	Me	Me	
Monensin C	Et	Me	Me	Et	

(b) Nigericin

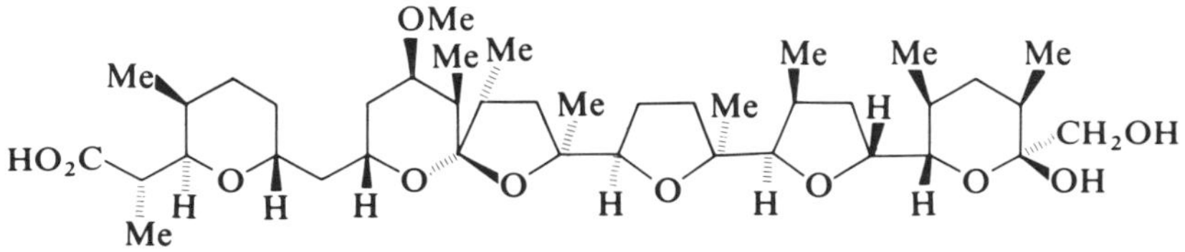

(c) Dianemycin

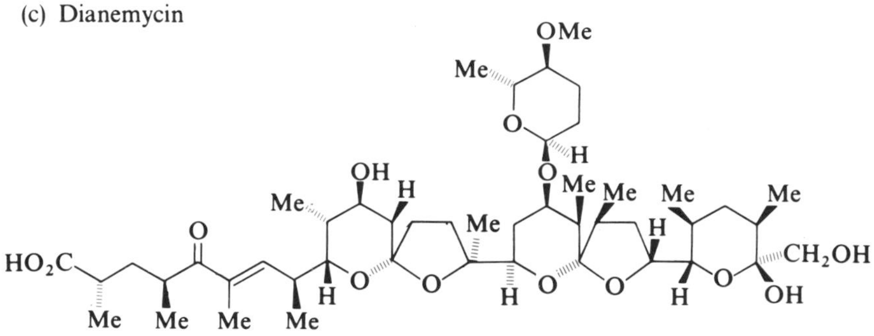

Figure 37 Structures of macrotetralide ionophores

cation–ionophore complex has been established (for reviews see refs. 278, 286, 289, 299). In all the complexes the potassium ion is held by multiple dipole interactions with the oxygen atoms (carbonyl, carboxylate, ether or alcohol) positioned to the interior of the antibiotic molecule. Thus, a lipophilic complex soluble in the lipid bilayer of the membrane is formed.

The structure of the valinomycin–K^+ complex was determined from spectroscopic data and X-ray analysis.[286,299,300] The backbone of the molecule is folded into six loops held by hydrogen bonds formed between NH and C=O groups of the amino acids. Thus, valinomycin is held in a bracelet conformation 4 Å high and 8 Å high in diameter with three ester carboxyl groups held centrally above and three below the bracelet (Figure 38). These define a sphere that can form strong interactions with ions of radius 1.33 Å (K^+), 1.45 Å (Rb^+) and with some stretching 1.69 Å (Cs^+) but not with ions of smaller radius (Na^+, $r = 0.95$ Å; Li^+, $r = 0.60$ Å). The free energy of complex formation by the smaller ions is limited by the number of carbonyl oxygen atoms they can contact at any one time. Hence, the high selectivity (several thousand-fold) of valinomycin for K^+ rather than Na^+ in both biological and model systems can be explained at the molecular level (Table 10).

In the absence of the cation, valinomycin has a hydrophobic and a hydrophilic face so that it tends to accumulate at the lipid membrane surface. When the K^+ complex is formed, the carbonyl groups move towards the centre, increasing the lipophilicity of the complex which then becomes more soluble in the membrane. However, the ability to form the K^+ complex is not the only property essential for ionophore activity. Ovchinnikov *et al.*[300] synthesized analogues of valinomycin and enniatin that formed K^+ complexes as efficiently as the parent compounds but were devoid of antibiotic activity. Clearly, the conformational state must play a major role in determining ionophore and antibiotic activity.

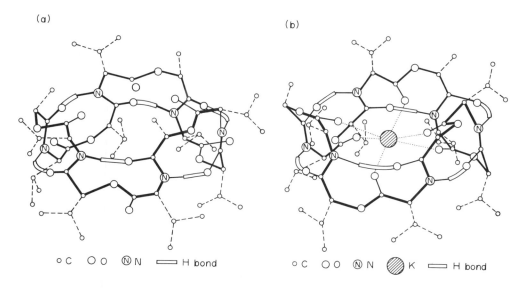

Figure 38 Conformations of (a) valinomycin, (b) the valinomycin–potassium complex and (c) the nigericin–potassium complex. In valinomycin the central K^+ ion is held by the six oxygen atoms of the ester carbonyls. The complex is further stabilized by hydrogen binding between the NH and CO groups of the six amino acids. The nigericin complex shows the K^+ ion held by the oxygen atoms of the ionophores and the requirement for deprotonation of the carboxyl group to enable complex formation to occur

Table 10 Ion-selectivity of Ionophores[286,289]

Compound	Monovalent cations	K^+:Na^+	Divalent cations
Neutral ionophores			
Valinomycin	Rb > K $\geqslant$ NH$_4$ > Na > Li	10 000	Weak
Nactins	NH$_4$ > K > Rb > Na > Li	210	—
Enniantin B	NH$_4$ > K > Rb > Na	37	—
Boromycin	K > Na > Li	1.7	—
Carboxylic ionophores			
Nigericin	H; K > Rb > Na > Li	45	—
Monensin	H; Na > K > Rb > Li	0.1	—
Dianemycin	H; Na > K > Rb > Li	0.5	—
Lasalocid A (A537A)	H; K > Rb > Na > Li	—	Ca > Mn > Sr > Mg
Calcimycin (A23187)	H; Li $\geqslant$ Na > K	0.1	Mn > Ca > Mg > Sr > Ba

The structure of the enniatin–K^+ complex shows that because of their small ring size these compounds form more planar complexes rather than the cage-like complex of valinomycin. This conformation is reflected in the reduced selectivity, the K^+:Na^+ ratio is approximately 10 (*cf.* valinomycin), and their ability to form complexes with a wide range of ions, including transition cations such as Ag^+, Ti^+, Cd^{2+}, Mn^{2+} and Zn^{2+}.[286]

The backbone of the nonactin–K$^+$ complex was described as forming a pattern 'resembling the seam of a tennis ball'. The K$^+$ is held at the centre of the ball with the heterocyclic ether oxygens and ester carbonyl oxygens forming the apices of a cube.[301]

Nigericin and monensin form complexes in which the carboxyl group participates directly along with dipole interactions with oxygen atoms (Figure 38). However, the longer chain length of monensin pushes the carboxyl group out of its plane of complex formation and with only the six liganding oxygen atoms involved it forms somewhat weaker complexes than nigericin.[293,294] In either case the carboxyl group must be deprotonated before complex formation can take place.

The ionophores described above show selectivity for monovalent cations and induce changes in the membrane permeability of these ions. A second group of carboxylic ionophores will bind both mono- and di-valent cations with, in the case of calcimycin, A23187, selectivity for the divalent ions. On the other hand, lasalocid, A537A, shows little selectivity and will form complexes with almost all known cations (Figure 39; Table 10).

Analogues of calcimycin have been synthesized, particularly modifications of the benzoxazole group, and tested for biological activity.[303] None proved as effective as calcimycin at transporting calcium and magnesium ions across rat liver mitochondrial membranes. However, substantial activity was retained when the methylamino moiety on the benzoxazole ring was replaced by hydrogen and a methyl group was introduced at the *para* position. Overall these studies established that the two major liganding sites, the pyridine-like nitrogen and the carboxylic group, must be located in the *ortho* position for complex stability and biological activity. Transfer of the carboxylic group to the *meta* position relative to the oxazolic nitrogen resulted in an almost complete loss of activity. In addition, the carboxylic group must be coplanar with the aromatic ring, otherwise the coordination sphere around the divalent cation is destabilized again with the loss of activity. Whether a chemical approach of this type will result in improved ionophores or in compounds suitable for further investigation of their mechanism of action remains to be established.

As might be expected from the above brief account, the ionophores in general have little or no medical application since their activity is not limited to microbial membranes. A number have, however, found considerable use in animal health. Monensin is used widely as a growth promotant in ruminants, where its use results in a change in rumen microflora to improve feed efficiency, and as a coccidostat in chickens. Coccidiosis is a parasitic disease caused by protozoa of the genus *Eimeria*, which infect the intestinal epithelium. All the polyether ionophores exhibit coccidostatic activity to some degree but because of the poor therapeutic index of the majority only a few such as monensin, lasalocid and maduramicin have proved useful. Tetranactin is used as an agricultural miticide.

In addition to the above specific applications a number of ionophores, including valinomycin and calcimycin, have proved to be very useful in the investigation of ion transport systems, particularly

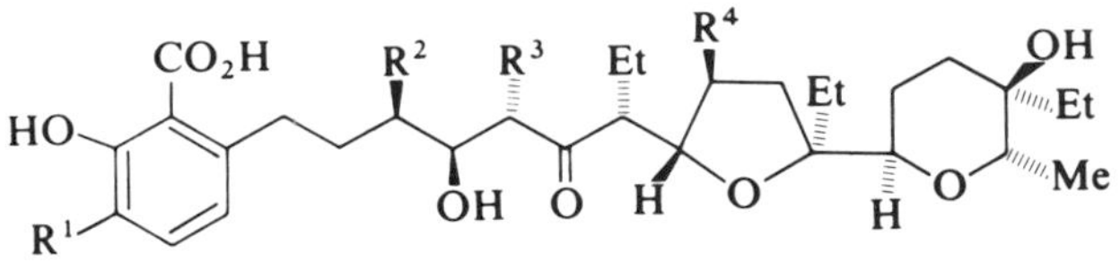

Antibiotic	R^1	R^2	R^3	R^4
Lasalocid A	Me	Me	Me	Me
Lasalocid B	Et	Me	Me	Me
Lasalocid C	Me	Et	Me	Me
Lasalocid D	Me	Me	Et	Me
Lasalocid E	Me	Me	Me	Et

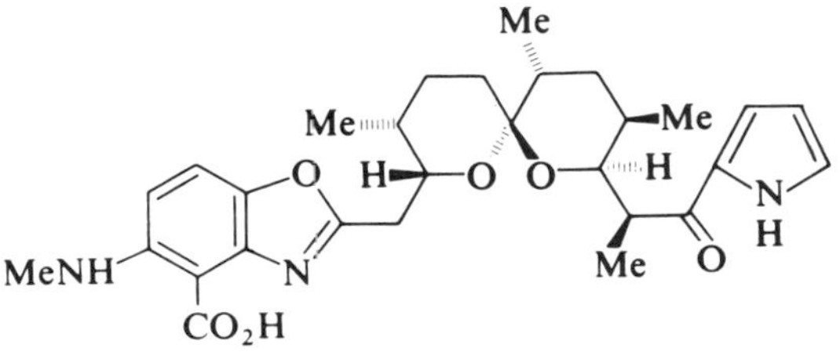

Calcimycin (A23187)

Figure 39 Structures of ionophores binding mono- and di-valent cations. Lasalocid A was referred to earlier as A537A and calcimycin as A23187

calcium fluxes in the case of calcimycin, and of wider use in the study of the biochemistry and pharmacology of membranes and organ systems. These aspects have been reviewed in detail by Pressman.[286]

9.1.9 REFERENCES

1. C. Gram, *Fortschr. Med.*, 1884, **ii**, 85.
2. T. J. Beveridge and J. A. Davies, *J. Bacteriol.*, 1983, **156**, 846.
3. H. J. Rogers, H. R. Perkins and J. B. Ward, 'Microbial Cell Walls and Membranes', Chapman and Hall, London, 1980, p. 564.
4. H. Nikaido and T. Nakae, *Adv. Microb. Physiol.*, 1979, **20**, 163.
5. B. Lugtenberg and L. Van Alphen, *Biochim. Biophys. Acta*, 1983, **737**, 51.
6. V. Braun, *Biochim. Biophys. Acta*, 1975, **415**, 335.
7. H. Nikaido, in 'Bacterial Outer Membranes', ed. M. Inouye, Wiley, New York, 1979, p. 361.
8. H. Nikaido, in 'β-Lactam Antibiotics. Mode of Action, New Developments and Future Prospects', ed. M. R. J. Salton and G. D. Shockman, Academic Press, New York, 1981, p. 249.
9. M. E. Bayer, in 'Membrane Biogenesis, Mitochondria, Chloroplasts and Bacteria', ed. A. Tzagoloff, Plenum Press, New York, 1975, p. 393.
10. I. W. Sutherland, in 'Surface Carbohydrates of the Prokaryote Cell', Academic Press, London, 1977, p. 472.
11. I. W. Sutherland, *Adv. Microb. Physiol.*, 1982, **23**, 79.
12. J. W. Costerton and T. J. Marrie, in 'Medical Microbiology', ed. C. S. F. Easmon, J. Jeljaszewicz, M. R. W. Brown and P. A. Lambert, Academic Press, New York, 1983, p. 63.
13. W. Weidel and H. Pelzer, *Adv. Enzymol.*, 1964, **26**, 193.
14. R. H. W. Verwer, N. Nanninga, W. Keck and U. Schwarz, *J. Bacteriol.*, 1978, **136**, 723.
15. R. H. W. Verwer, E. J. Beachey, W. Keck, A. M. Stohb and J. E. Polermans, *J. Bacteriol.*, 1980, **141**, 327.
16. B. Glauner and U. Schwarz, in 'The Target of Penicillin', ed. R. Hakenbeck, J. V. Holtje and H. Labischinski, de Gruyter, Berlin, 1983, p. 29.
17. T. J. Dougherty, *J. Bacteriol.*, 1985, **163**, 69.
18. J. M. Ghuysen, *Bacteriol. Rev.*, 1968, **32**, 425.
19. K. H. Schleifer and O. Kandler, *Bacteriol. Rev.*, 1972, **36**, 407.
20. D. J. Tipper and M. F. Berman, *Biochemistry*, 1969, **8**, 2183.
21. A. W. Wyke, J. B. Ward, M. V. Hayes and N. A. C. Curtis, *Eur. J. Biochem.*, 1981, **119**, 389.
22. J. B. Ward, *Bacteriol. Rev.*, 1981, **45**, 211.
23. R. C. Hughes and P. F. Thurman, *Biochem. J.*, 1970, **117**, 441.
24. M. R. Lifely, E. Tarelli and J. Baddiley, *Biochem. J.*, 1980, **191**, 305.
25. P. J. White and C. Gilvarg, *Biochemistry*, 1977, **16**, 2428.
26. J. B. Ward and C. A. M. Curtis, *Eur. J. Biochem.*, 1982, **122**, 125.
27. A. J. Wicken and K. W. Knox, *Biochim. Biophys. Acta*, 1980, **604**, 1.
28. J. B. Ward and R. Williamson, in 'Microbial Cell Wall Synthesis and Autolysis', ed. C. Nombela, Elsevier, London, 1984, p. 159.
29. K. W. Knox and A. J. Wicken, *Bacteriol. Rev.*, 1973, **37**, 215.
30. R. M. Krause, *Bacteriol. Rev.*, 1963, **27**, 369.
31. M. Inouye, in 'Bacterial Outer Membranes', Wiley, New York, 1979, p. 534.
32. O. Lüderitz, M. A. Freudenberg, C. Galanos, V. Lehman, E. T. Rietschel and D. H. Shaw, *Curr. Top. Membr. Transp.*, 1983, **17**, 79.
33. H. Nikaido and M. Vaara, *Microbiol. Rev.*, 1985, **49**, 1.
34. E. T. Reitschel (ed.), 'The Chemistry of Endotoxins', Elsevier-North Holland, Amsterdam, 1984.
35. K. Takayama, N. Qureshi and P. Mascagni, *J. Biol. Chem.*, 1983, **258**, 12801.
36. M. Imoto, S. Kusumoto, T. Shiba, H. Naoki, T. Iwashita, E. T. Reitschel, H. W. Wollenweber, C. Galanos and O. Lüderitz, *Tetrahedron Lett.*, 1983, **24**, 4017.
37. H. W. Wollenweber, K. W. Broady, O. Lüderitz and E. T. Reitschel, *Eur. J. Biochem.*, 1982, **124**, 191.
38. P. C. Goldman and L. Leive, *Eur. J. Biochem.*, 1980, **107**, 145.
39. A. M. Kropinski, J. Kuzio, B. L. Angus and R. E. W. Hancock, *Antimicrob. Agents Chemother.*, 1982, **21**, 310.
40. G. Högenauer and M. Woisetschalager, *Nature (London)*, 1981, **293**, 662.
41. S. M. Hammond, A. Claesson, A. M. Jansson, L.-G. Larsson, B. G. Pring, C. M. Town and B. Ekström, *Nature (London)*, 1987, **327**, 730.
42. P. Ringrose, in 'The Scientific Basis of Antimicrobial Chemotherapy', ed. D. Greenwood and F. O'Grady, Cambridge University Press, Cambridge, 1985, p. 219.
43. V. Braun, in 'Relations between Structure and Function in the Prokaryotic Cell', ed. R. Y. Stanier, H. J. Rogers and J. B. Ward, Cambridge University Press, Cambridge, 1978, p. 111.
44. H. Suzuki, Y. Nishimura, S. Yasuda, A. Nishimura, M. Yamada and Y. Hirota, *Mol. Gen. Genet.*, 1978, **167**, 1.
45. J. P. Rosenbusch, *J. Biol. Chem.*, 1974, **249**, 8019.
46. H. Nikaido, *Pharmacol. Ther.*, 1985, **27**, 197.
47. D. J. Tipper and A. Wright, in 'The Bacteria', ed. J. R. Sokatch and L. N. Ornston, Academic Press, New York, 1980, vol. 7, p. 291.
48. J. B. Ward, *Pharmacol. Ther.*, 1984, **25**, 327.
49. T. S. Soper and J. M. Manning, *J. Biol. Chem.*, 1981, **256**, 4263.
50. T. S. Soper and J. M. Manning, *J. Biol. Chem.*, 1982, **257**, 13930.
51. F. C. Neuhaus and W. P. Hammes, *Pharmacol. Ther.*, 1981, **14**, 265.
52. E. Wang and C. Walsh, *Biochemistry*, 1978, **17**, 1313.

53. M. P. Lambert and F. C. Neuhaus, *J. Bacteriol.*, 1972, **110**, 978.
54. F. C. Neuhaus, *Acc. Chem. Res.*, 1972, **4**, 297.
55. G. Tamura, T. Sasaki, M. Matsuhashi, T. Takatsuki and M. Yamasaki, *Agric. Biol. Chem.*, 1976, **40**, 447.
56. H. Tanaka, R. Oiwa, S. Matsukura and S. Omura, *Biochim. Biophys. Res. Commun.*, 1978, **86**, 902.
57. J. B. Ward, *FEBS Lett.*, 1977, **78**, 151.
58. P. Reisinger, H. Seidl, H. Tschesche and W. P. Hammes, *Arch. Microbiol.*, 1980, **127**, 187.
59. G. Siewert and J. L. Strominger, *Proc. Natl. Acad. Sci. USA*, 1967, **57**, 767.
60. D. Keglevic, B. Ladesie, O. Hadzija, J. Tomasic, Z. Valinger, M. Pokotny and R. Naumski, *Eur. J. Biochem.*, 1974, **42**, 389.
61. D. Mirelman, R. Bracha and N. Sharon, *Biochemistry*, 1974, **13**, 5045.
62. Z. Tynecka and J. B. Ward, *Biochem. J.*, 1975, **146**, 253.
63. D. J. Waxman, W. Yu and J. L. Strominger, *J. Biol. Chem.*, 1980, **255**, 11 577.
64. Y. Van Heijenoort and J. Van Heijenoort, *FEBS Lett.*, 1980, **110**, 241.
65. Y. Van Heijenoort, M. Derrien and J. Van Heijenoort, *FEBS Lett.*, 1978, **89**, 141.
66. H. R. Perkins, *Pharmacol. Ther.*, 1982, **16**, 181.
67. D. J. Tipper and J. L. Strominger, *Proc. Natl. Acad. Sci. USA*, 1965, **54**, 1133.
68. D. J. Waxman and J. L. Strominger, in 'Chemistry and Biology of β-Lactam Antibiotics', ed. R. B. Morin and M. Gorman, Academic Press, New York, 1982, vol. 3, p. 210.
69. J. B. Ward and H. R. Perkins, *Biochem. J.*, 1974, **139**, 781.
70. A. F. Giles and P. E. Reynolds, *FEBS Lett.*, 1979, **101**, 244.
71. W. P. Hammes, *Eur. J. Biochem.*, 1976, **79**, 107.
72. W. P. Hammes, *Eur. J. Biochem.*, 1978, **91**, 501.
73. W. P. Hammes and H. Seidl, *Eur. J. Biochem.*, 1978, **84**, 141.
74. W. P. Hammes and H. Seidl, *Eur. J. Biochem.*, 1978, **91**, 509.
75. B. G. Spratt, *J. Gen. Microbiol.*, 1983, **127**, 1247.
76. D. J. Waxman and J. L. Strominger, *Annu. Rev. Biochem.*, 1983, **52**, 825.
77. P. E. Reynolds, in 'The Scientific Basis of Antimicrobial Chemotherapy', ed. D. Greenwood and F. O'Grady, Cambridge University Press, Cambridge, 1985, p. 13.
78. J. M. Frère and B. Joris, *CRC Crit. Rev. Microbiol.*, 1985, **11**, 299.
79. M. V. Hayes and J. B. Ward, in 'Antibiotics in Laboratory Medicine', 2nd edn., ed. V. Lorian, Williams and Wilkins, Baltimore, 1986, p. 722.
80. B. G. Spratt, *Eur. J. Biochem.*, 1977, **72**, 341.
81. M. Matsuhashi, F. Ishino, J. Nakagawa, K. Mitsui, S. Nakajima-Iijima, S. Tamki and T. Hashizume, in 'β-Lactam Antibiotics. Mode of Action, New Developments and Future Prospects', ed. M. R. J. Salton and G. D. Shockman, Academic Press, New York, 1981, p. 169.
82. M. A. De Pedro, U. Schwarz, Y. Nishimura and T. Hirota, *FEMS Microbiol. Lett.*, 1980, **9**, 219.
83. M. Matsuhashi, J. Nakagawa, S. Tamaki, F. Ishino, S. Tomioka and W. Park, in 'The Target of Penicillin', ed. R. Hakenbeck, J. V. Holtje and H. Labishinski, de Gruyter, Berlin, 1983, p. 499.
84. F. Ishino, W. Park, S. Tomioka, S. Tamaki, I. Takase, K. Kunugita, H. Matsuzawa, S. Asoh, T. Ohta, B. G. Spratt and M. Matsuhashi, *J. Biol. Chem.*, 1986, **261**, 7024.
85. B. G. Christensen, W. J. Leanza, T. R. Beattie, A. A. Patchett, B. H. Arison, R. E. Ormond, F. A. Kuehl, G. Albers-Schonberg and O. Jardetzky, *Science (Washington, D.C.)*, 1969, **166**, 123.
86. D. Hendlin, E. O. Stapeley, M. Jackson, H. Wallick, A. K. Miller, F. J. Wolf, T. W. Miller, L. Chaiet, F. M. Kahan, E. L. Foltz, H. B. Woodruff, J. M. Mata, S. Hernandez and S. Mochales, *Science (Washington, D.C.)*, 1969, **166**, 122.
87. F. M. Kahan, J. S. Kahan, P. J. Cassidy and H. Kropp, *Ann. N.Y. Acad. Sci.*, 1974, **235**, 364.
88. R. J. Kadner and H. H. Winkler, *J. Bacteriol.*, 1973, **113**, 895.
89. J. T. Park, *Symp. Soc. Gen. Microbiol.*, 1958, **8**, 49.
90. F. R. Atherton, M. J. Hall, C. H. Hassall, R. W. Lambert, W. J. Lloyd and P. S. Ringrose, *Antimicrob. Agents Chemother.*, 1979, **15**, 696.
91. F. R. Atherton, M. J. Hall, C. H. Hassall, R. W. Lambert, W. J. Lloyd, A. V. Lord, P. S. Ringrose and D. Westmacott, *Antimicrob. Agents Chemother.*, 1983, **24**, 522.
92. F. R. Atherton, M. J. Hall, C. H. Hassall, S. W. Holmes, R. W. Lambert, W. J. Lloyd and P. S. Ringrose, *Antimicrob. Agents Chemother.*, 1980, **18**, 897.
93. T. M. Nisbett and J. M. Payne, *J. Gen. Microbiol.*, 1982, **128**, 1357.
94. D. R. Storm, *Ann. N.Y. Acad. Sci.*, 1974, **235**, 387.
95. D. R. Storm and W. A. Toscano, in 'Antibiotics V-1', ed. F. E. Hahn, Springer, New York, 1979, p. 1.
96. W. A. Toscano and D. R. Storm, *Pharmacol. Ther.*, 1982, **16**, 199.
97. J. T. Park, *Biochem. J.*, 1958, **70**, 2P.
98. J. S. Anderson, M. Matsuhashi, M. A. Haskin and J. L. Strominger, *J. Biol. Chem.*, 1967, **242**, 3180.
99. W. Katz, M. Matsuhashi, C. P. Dietrich and J. L. Strominger, *J. Biol. Chem.*, 1967, **242**, 3207.
100. D. R. Storm and J. L. Strominger, *J. Biol. Chem.*, 1973, **248**, 5208.
101. M. J. Osborne, *Annu. Rev. Biochem.*, 1969, **38**, 501.
102. A. Takatsuki, K. Arima and G. Tamura, *J. Antibiot.*, 1971, **24**, 215.
103. J. S. Tkacz, in 'Antibiotics. Volume 6, Modes and Mechanisms of Bacterial Growth Inhibitors', ed. F. E. Hahn, Springer, New York, 1981, p. 255.
104. J. S. Tkacz and A. Wong, *Fed. Proc., Fed. Am. Soc. Exp. Biol.*, 1978, **37**, 1766.
105. M. Kenig and C. Reading, *J. Antibiot.*, 1979, **32**, 549.
106. A. Takatsuki and G. Tamura, *J. Antibiot.*, 1971, **24**, 785.
107. L. Lehle and W. J. Tanner, *FEBS Lett.*, 1976, **71**, 167.
108. A. Heifetz, R. W. Keenan and A. D. Elbein, *Biochemistry*, 1979, **18**, 2186.
109. G. Tamura, T. Sasaki, M. Matsuhashi, A. Takatsuki and M. Yamasaki, *Agric. Biol. Chem.*, 1976, **40**, 447.
110. G. E. Bettinger and F. E. Young, *Biochem. Biophys. Res. Commun.*, 1975, **67**, 16.
111. S. Nomura, K. Yamane, T. Sasaki, M. Yamasaki, G. Tamura and B. Marino, *J. Bacteriol.*, 1978, **136**, 818.
112. H. R. Perkins, *Pharmacol. Ther.*, 1982, **16**, 181.

113. J. C. J. Barna and D. H. Williams, *Annu. Rev. Microbiol.*, 1984, **38**, 339.
114. F. Parenti, G. Beretta, M. Berti and V. Arioli, *J. Antibiot.*, 1978, **31**, 276.
115. M. C. Shearer, P. Actor, B. A. Bowie, S. F. Grapper, C. H. Dash, D. J. Newman, Y. K. Oh, C. H. Pan and L. J. Nisbet, *J. Antibiot.*, 1985, **38**, 555.
116. M. C. Shearer, A. J. Giovenella, S. F. Grapper, R. D. Hedde, R. J. Mehta, Y. K. Oh, C. H. Pan, D. H. Pitkin and L. J. Nisbet, *J. Antibiot.*, 1986, **39**, 1386.
117. T. Kamogashira, T. Nishida and M. Sugawara, *Agric. Biol. Chem.*, 1983, **47**, 499.
118. L. D. Boeck and F. P. Mertz, *J. Antibiot.*, 1986, **39**, 1533.
119. D. H. Williams, M. P. Williamson and G. Bojesen, in 'Topics in Antibiotic Chemistry', ed. P. G. Sammes, Ellis Horwood, Chichester, 1980, p. 119.
120. A. H. Hunt, *J. Am. Chem. Soc.*, 1983, **105**, 4463.
121. J. C. J. Barna, D. H. Williams, D. J. M. Stone, T. W. C. Leung and D. M. Doddrell, *J. Am. Chem. Soc.*, 1984, **106**, 4895.
122. P. W. Jeffs, L. Mueller, C. De Brosse, S. L. Heald and R. Fisher, *J. Am. Chem. Soc.*, 1986, **108**, 3063.
123. G. Folena-Wasserman, B. L. Poehland, E. W. K. Yeung, D. Staiger, L. B. Killmer, K. Snader, J. J. Dingerdissen and P. W. Jeffs, *J. Antibiot.*, 1986, **39**, 1395.
124. A. Malabarba, P. Ferrari, G. G. Gallo, J. Kettenring and B. Cavalleri, *J. Antibiot.*, 1986, **39**, 1430.
125. P. E. Reynolds, *Biochim. Biophys. Acta*, 1961, **52**, 403.
126. D. C. Jordan, *Biochem. Biophys. Res. Commun.*, 1961, **6**, 167.
127. R. K. Sinha and F. C. Neuhaus, *J. Bacteriol.*, 1968, **96**, 374.
128. A. N. Chatterjee and H. R. Perkins, *Biochem. Biophys. Res. Commun.*, 1966, **24**, 489.
129. H. R. Perkins, *Biochem. J.*, 1969, **111**, 195.
130. M. Nieto and H. R. Perkins, *Biochem. J.*, 1971, **123**, 789.
131. M. Nieto and H. R. Perkins, *Biochem. J.*, 1971, **124**, 845.
132. H. R. Perkins and M. Nieto, *Biochem. J.*, 1970, **116**, 83.
133. J. B. Ward, *Biochem. J.*, 1974, **141**, 227.
134. A. Malabarba, A. Trani, P. Ferrari, R. Pallanza and B. Cavalleri, *J. Antibiot.*, 1987, **40**, 1572.
135. R. S. Schwalbe, J. S. Stapelton and P. H. Gilligan, *New Engl. J. Med.*, 1987, **316**, 927.
136. V. Arioli and R. Pallanza, *Lancet*, 1987, **1**, 39.
137. G. Goleman and A. Efstration, *J. Hosp. Infect.*, 1987, **10**, 1.
138. G. Huber, in 'Antibiotics. Volume 6, Mechanism of Action of Antibacterial Agents', ed. F. E. Hahn, Springer, New York, 1979, p. 135.
139. P. Welzel, F. J. Witteler, D. Müller and W. Reimer, *Angew. Chem.*, 1981, **93**, 130.
140. P. Welzel, B. Weitfeld, F. Kunisch, T. Schubert, K. Hobert, H. Duddeck, D. Müller, G. Huber, J. E. Maggio and D. H. Williams, *Tetrahedron*, 1983, **39**, 1583.
141. S. Takahashi, K. Serita and M. Arai, *Tetrahedron Lett.*, 1983, **24**, 499.
142. P. Welzel, F. Kunisch, F. Kruggel, H. Stein, J. Scherkenbeck, A. Hiltman, H. Duddeck, D. Müller, J. E. Maggio, H. W. Fehlhaber, G. Seibert, Y. Van Heijenoort and J. Van Heijenoort, *Tetrahedron*, 1987, **43**, 585.
143. H. Suzuki, Y. Van Heijenoort, T. Tamura, J. Mizoguchi, Y. Hirota and J. Van Heijenoort, *FEBS Lett.*, 1980, **110**, 245.
144. J. Nakagawa, S. Tamaki, S. Tomoika and M. Matsuhashi, *J. Biol. Chem.*, 1984, **259**, 13937.
145. P. E. Linnett and J. L. Strominger, *Antimicrob. Agents Chemother.*, 1973, **4**, 231.
146. J. O. Capobianco, R. P. Darveau, R. C. Goldman, P. A. Lartey and A. G. Pernet, *J. Bacteriol.*, 1987, **169**, 4030.
147. T. Miyoshi, N. Miyairi, H. Aoki, M. Kosaka and H. Sakai, *J. Antibiot.*, 1972, **25**, 569.
148. S. Miyamura, N. Ogasawara, H. Otsuka, S. Niwayama and H. Tanaka, *J. Antibiot.*, 1973, **26**, 479.
149. K. Ochi, Y. Saito, K. Umehara, I. Ueda and M. Kohsaka, *J. Gen. Micrbiol.*, 1984, **130**, 2007.
150. T. Kamiya, S. Maeno, M. Hashimoto and Y. Mine, *J. Antibiot.*, 1972, **25**, 576.
151. M. Iseki, T. Miyoshi, T. Konomi and H. Imanaka, *J. Antibiot.*, 1980, **33**, 488.
152. A. Someya, M. Iseki and N. Tanaka, *J. Antibiot.*, 1979, **32**, 402.
153. B. Watt and F. V. Brown, *J. Antimicrob. Chemother.*, 1983, **12**, 549.
154. C. D. Ericsson, H. L. Dupont, E. Galindo, J. J. Mathewson, D. R. Morgan, L. V. Wood and J. Mendiola, *Gastroenterology*, 1985, **88**, 473.
155. H. L. Dupont, C. D. Ericsson, P. C. Johnson and F. J. Canada, *Rev. Infect. Dis.*, 1986, **8** (suppl. 2), 167.
156. N. Tanaka, M. Iseki, T. Miyoshi, H. Aoki and H. Imanaka, *J. Antibiot.*, 1976, **29**, 155.
157. A. G. Pisabarro, F. J. Canada, D. Vazquez, P. Arriaga and A. Rodriguez-Tebar, *J. Antibiot.*, 1986, **39**, 914.
158. Y. Hirota, H. Suzuki, Y. Nishimura and S. Yasuda, *Proc. Natl. Acad. Sci. USA*, 1977, **74**, 1417.
159. B. W. Müller, O. Zak, W. Kump, W. Tosch and O. Wacker, *J. Antibiot.*, 1979, **32**, 689.
160. R. M. Williams, R. W. Armstrong and J. S. Dung, *J. Med. Chem.*, 1985, **28**, 733.
161. M. Inukai, R. Enokita, A. Torikata, M. Nakahara, S. Iwado and M. Arai, *J. Antibiot.*, 1978, **31**, 410.
162. M. Nakajima, M. Inakai, T. Haneishi, A. Terahara, M. Arai, T. Kinoshita and C. Tamura, *J. Antibiot.*, 1978, **31**, 426.
163. S. Omoto, H. Ogino and S. Inouye, *J. Antibiot.*, 1981, **34**, 1416.
164. M. Inukai, M. Takeuchi, K. Shimizu and M. Arai, *J. Antibiot.*, 1978, **31**, 1203.
165. I. K. Dev and P. H. Ray, *J. Biol. Chem.*, 1984, **259**, 11 114.
166. C. E. Ballou, *Adv. Microb. Physiol.*, 1976, **14**, 93.
167. C. E. Ballou, in 'Fungal Polysaccharides', ed. P. A. Sandford and K. Matsuda, American Chemical Society, Washington, DC, 1980, p. 1.
168. J. S. D. Bacon, in 'Yeast Cell Envelopes, Biochemistry, Biophysics and Ultrastructure', ed. W. N. Arnold, CRC Press, Boca Raton, 1981, p. 85.
169. E. Cabib and E. M. Shematek, in 'Biology of Carbohydrates', ed. V. Ginsburg and P. W. Robbins, Wiley, New York, 1981, vol. 1, p. 52.
170. R. H. Marchessault and Y. Deslandes, in 'Fungal Polysaccharides', ed. P. A. Sandford and K. Matsuda, American Chemical Society, Washington, DC, 1980, p. 221.
171. E. Reiss, in 'Fungi Pathogenic for Humans and Animals', ed. D. H. Howard and L. F. Howard, Dekker, New York, 1985, part B 11, p. 5.
172. E. Reiss, in 'Molecular Immunology of Mycotic and Actinomycotic Infections', Elsevier, New York, 1986, p. 5.
173. G. H. Fleet and D. J. Manners, *J. Gen. Microbiol.*, 1976, **94**, 180.

174. E. Cabib and B. Bowers, *J. Biol. Chem.*, 1971, **246**, 152.
175. G. Tronchin, D. Poulain, J. Herbaut and J. Biquet, *Eur. J. Cell Biol.*, 1981, **26**, 121.
176. S. Bartnicki-Garcia, *Annu. Rev. Microbiol.*, 1968, **22**, 87.
177. D. K. Kidby and R. Davies, *J. Gen. Microbiol.*, 1970, **61**, 327.
178. E. Cabib, *Annu. Rev. Microbiol.*, 1975, **29**, 191.
179. V. Farkas, *Microbiol. Rev.*, 1979, **43**, 117.
180. G. W. Gooday, *J. Gen. Microbiol.*, 1977, **99**, 1.
181. C. E. Bulawa, M. Slater, E. Cabib, J. Au-Young, A. Sburlati, W. L. Adair and P. W. Robbins, *Cell*, 1986, **46**, 213.
182. A. Sburlati and E. Cabib, *J. Biol. Chem.*, 1986, **261**, 15147.
183. P. Orlean, *J. Biol. Chem.*, 1987, **262**, 5732.
184. U. Dähn, H. Hagenmaier, H. Höhne, W. A. König, G. Wolf and H. Zähner, *Arch. Microbiol.*, 1976, **107**, 143.
185. H. Hagenmaier, A. Keckeisen, H. Zähner and W. A. König, *Liebigs Ann. Chem.*, 1979, 1494.
186. H. Hagenmaier, A. Keckeisen, W. Dehler, H. P. Feidler, H. Zähner and W. A. Konig, *Liebigs Ann. Chem.*, 1981, 1018.
187. K. Isono and S. Suzuki, *Heterocycles*, 1979, **13**, 333.
188. K. Kobinata, M. Uramoto, M. Nishii, H. Kusakabe, G. Nakamura and K. Ishino, *Agric. Biol. Chem.*, 1980, **44**, 1709.
189. A. Endo, K. Kakiki and T. Misato, *J. Bacteriol.*, 1970, **104**, 189.
190. H. Müller, R. Furter, H. Zähner and D. M. Rast, *Arch. Microbiol.*, 1981, **130**, 195.
191. M. Mitani and Y. Inoue, *J. Antibiot.*, 1968, **21**, 492.
192. J. M. Becker, N. L. Covert, P. Shenbagamurthi, A. S. Steinfeld and F. Naider, *Antimicrob. Agents Chemother.*, 1983, **23**, 926.
193. G. Emmer, N. S. Ryder and M. A. Grassberger, *J. Med. Chem.*, 1985, **28**, 278.
194. P. Shenbagamurthi, H. A. Smith, J. M. Becker, A. Steinfeld and F. Naider, *J. Med. Chem.*, 1983, **26**, 1518.
195. F. Naider, P. Shenbagamurthi, A. S. Steinfeld, H. A. Smith, C. Boney and J. M. Becker, *Antimicrob. Agents Chemother.*, 1983, **24**, 787.
196. J. C. Boehm and W. D. Kingsbury, *J. Org. Chem.*, 1986, **51**, 2307.
197. H. A. Smith, P. Shenbagamurthi, F. Naider, B. Kundu and J. M. Becker, *Antimicrob. Agents Chemother.*, 1986, **29**, 33.
198. P. Shenbagamurthi, H. A. Smith, J. M. Becker and F. Naider, *J. Med. Chem.*, 1986, **29**, 802.
199. J. C. Yadan, M. Gonneau, P. Sarthou and F. Le Goffe, *J. Bacteriol.*, 1984, **160**, 884.
200. F. Benz, F. Knüsel, J. Nuesch, H. Treichler, W. Voser, R. Nyfeler and W. Keller-Schierlein, *Helv. Chim. Acta*, 1974, **57**, 2459.
201. C. Keller-Juslén, M. Kuhn, H. R. Loosli, T. J. Petcher, H. P. Weber and A. von Wartburg, *Tetrahedron Lett.*, 1976, 4147.
202. K. Mizuno, A. Yagi, S. Satoi, M. Takada, M. Hayashi, K. Asano and T. Matsuda, *J. Antibiot.*, 1977, **30**, 297.
203. S. Saito, A. Yagi, K. Asano, K. Mizuno and T. Watanabe, *J. Antibiot.*, 1977, **30**, 303.
204. K. Roy, T. Mukhopadhyay, G. C. S. Reddy, K. R. Desikan and B. N. Ganguli, *J. Antibiot.*, 1987, **40**, 275.
205. T. Mukhopadhyay, B. N. Ganguli, H. W. Feldhaber, H. Kogler and L. Vertesy, *J. Antibiot.*, 1987, **40**, 281.
206. P. Traxler, J. Gruner and J. A. L. Anden, *J. Antibiot.*, 1977, **30**, 289.
207. P. Traxler, H. Fritz, H. Fuhrer and W. J. Richter, *Helv. Chim. Acta*, 1977, **60**, 578.
208. P. Traxler, H. Fritz, H. Fuhrer and W. J. Richter, *J. Antibiot.*, 1980, **33**, 967.
209. T. Komori and Y. Itoh, *J. Antibiot.*, 1985, **38**, 544.
210. B. C. Baguley, G. Römmele, J. Grunner and W. Wehrli, *Eur. J. Biochem.*, 1979, **97**, 345.
211. P. Pérez, R. Varona, I. Garcia-Acha and A. Durán, *FEBS Lett.*, 1981, **129**, 249.
212. G. Römmele, P. Traxler and W. Wehrli, *J. Antibiot.*, 1983, **36**, 1539.
213. R. Varona, P. Pérez and A. Durán, *FEMS Microbiol. Lett.*, 1983, **20**, 243.
214. H. Yamaguchi, T. Hiratani, M. Baba and M. Osumi, *Microbiol. Immunol.*, 1985, **29**, 609.
215. D. R. Quigley and C. P. Selitrennikoff, *Exp. Mycol.*, 1984, **8**, 320.
216. R. S. Gordee, D. J. Zeckner, L. F. Ellis, A. L. Thakker and L. C. Howard, *J. Antibiot.*, 1984, **37**, 1054.
217. G. S. Hall, C. Myles, K. J. Pratt and J. A. Washington, *Antimicrob. Agents Chemother.*, 1988, **32**, 1331.
217a. L. H. Hanson and D. A. Stevens, *Antimicrob. Agents Chemother.*, 1989, **33**, 1391.
218. P. Traxler, W. Tosch and O. Zak, *J. Antibiot.*, 1987, **40**, 1146.
219. H. A. Davson and F. C. Danielli, in 'The Permeability of Natural Membranes', Cambridge University Press, Cambridge, 1943.
220. S. J. Singer and G. L. Nicholson, *Science (Washington, D.C.)*, 1972, **175**, 720.
221. J. E. Rothman and E. P. Kennedy, *Proc. Natl. Acad. Sci. USA*, 1977, **74**, 1821.
222. D. R. Storm, K. S. Rosenthal and P. E. Swanson, *Annu. Rev. Biochem.*, 1977, **46**, 723.
223. E. Meyers, F. E. Pansy, H. I. Basch, R. J. McRipley, D. S. Slusarchyk, S. Graham and W. H. Trejo, *J. Antibiot.*, 1973, **26**, 457.
224. D. La Porte, K. S. Rosenthal and D. R. Storm, *Biochemistry*, 1977, **16**, 1642.
225. K. S. Rosenthal, P. E. Swanson and D. R. Storm, *Biochemistry*, 1976, **15**, 5783.
226. G. Cerny and M. Teuber, *Arch. Microbiol.*, 1971, **78**, 166.
227. M. Vaara, T. Vaara, M. Jensen, I. Helander, M. Nurminen, E. T. Reitschel and P. H. Mäkelä, *FEBS Lett.*, 1981, **129**, 145.
228. M. Vaara, *J. Bacteriol.*, 1981, **148**, 426.
229. M. Vaara and T. Vaara, *Nature (London)*, 1983, **303**, 526.
230. M. Vaara and T. Vaara, *Antimicrob. Agents Chemother.*, 1983, **24**, 107.
231. M. Vaara and T. Vaara, *Antimicrob. Agents Chemother.*, 1983, **24**, 114.
232. T. Alatossava, M. Vaara and M. Baschong, *FEMS Microbiol. Lett.*, 1984, **19**, 253.
233. M. Ito-Kagawa and Y. Koyama, *J. Antibiot.*, 1984, **37**, 926.
234. P. Viljanen and M. Vaara, *Antimicrob. Agents Chemother.*, 1984, **25**, 701.
235. M. Vaara and P. Viljanen, *Antimicrob. Agents Chemother.*, 1985, **27**, 548.
236. R. D. Hotchkiss, *Adv. Enzymol.*, 1944, **4**, 143.
237. M. A. Ruttenberg, T. P. King and L. P. Craig, *Biochemistry*, 1966, **5**, 2857.
238. W. Pache, D. Chapman and R. Hillaby, *Biochim. Biophys. Acta*, 1972, **255**, 358.
239. P. E. Hunter and L. S. Schwartz, in 'Antibiotics', ed. D. Gottlieb and P. D. Shaw, Springer, New York, 1967, vol. 1, p. 631.

240. S. B. Hladky, L. G. M. Gordon and D. A. Haydon, *Rev. Phys. Chem.*, 1974, **25**, 11.
241. A. S. Arseniev, I. L. Barsukov, V. F. Bystrov, A. L. Lomize and Y. A. Ovchinnikov, *FEBS Lett.*, 1985, **186**, 168.
242. E. Bamberg and K. Janko, *Biochim. Biophys. Acta*, 1977, **465**, 486.
243. P. C. Jordan, *J. Membr. Biol.*, 1984, **78**, 91.
244. A. Pullman, *Methods Enzymol.*, 1986, **127**, 250.
245. J. M. T. Hamilton-Miller, *Bacteriol. Rev.*, 1973, **37**, 166.
246. J. F. Ryley, R. G. Wilson, M. B. Gravestock and J. P. Poyser, *Adv. Pharmacol. Chemother.*, 1981, **18**, 49.
247. W. Mechlinski and C. P. Schaffner, *Tetrahedron Lett.*, 1970, **44**, 3873.
248. D. Gottleib, H. E. Carter, J. H. Soneker and A. Ammann, *Science (Washington, D.C.)*, 1958, **125**, 361.
249. J. O. Lampen, P. M. Arnow and R. S. Safferman, *J. Bacteriol.*, 1960, **80**, 200.
250. B. De Kruijff, W. J. Gerritsen, A. Oerlemans, R. A. Demel and L. L. M. Van Deenen, *Biochim. Biophys. Acta*, 1974, **339**, 30.
251. D. B. Archer and E. F. Gale, *J. Gen. Microbiol.*, 1975, **90**, 187.
252. D. B. Archer, *Biochim. Biophys. Acta*, 1976, **436**, 68.
253. E. F. Gale, *J. Gen. Microbiol.*, 1974, **80**, 451.
254. W. C. Chen, D. L. Chou and D. S. Feingold, *Antimicrob. Agents Chemother.*, 1978, **13**, 914.
255. W. C. Chen, I. J. Sud, D. L. Chou and D. S. Feingold, *Biochem. Biophys. Res. Commun.*, 1977, **74**, 480.
256. W. G. Ellis, R. A. Sobel and S. L. Nielsen, *J. Infect. Dis.*, 1982, **146**, 125.
257. H. Kim, D. Loebenberg, A. Marco, S. Symchowicz and C. Lin, *Antimicrob. Agents Chemother.*, 1984, **26**, 446.
258. J. D. Perfect and D. T. Durack, *Antimicrob. Agents Chemother.*, 1985, **28**, 751.
259. G. Lopez-Berestein, *Antimicrob. Agents Chemother.*, 1987, **31**, 675.
260. J. B. Readio and R. Bittman, *Biochim. Biophys. Acta*, 1982, **685**, 219.
261. J. O. Lampen, *Symp. Soc. Gen. Microbiol.*, 1966, **16**, 11.
262. S. M. Hammond, R. A. Lambert and B. N. Kliger, *J. Gen. Microbiol.*, 1974, **81**, 325.
263. L. N. Ermishkin, K. M. Kasumov and Y. M. Potseluyev, *Biochim. Biophys. Acta*, 1977, **470**, 357.
264. K. M. Kasumov, M. P. Borisova, L. N. Ermishkin, V. M. Potzeluyev, A. Y. Silberstein and V. A. Vainshtein, *Biochim. Biophys. Acta*, 1979, **551**, 229.
265. B. De Kruijff and R. A. Demel, *Biochim. Biophys. Acta*, 1974, **339**, 57.
266. A. Marty and A. Finkelstein, *J. Gen. Physiol.*, 1975, **65**, 515.
267. P. Van Hoogevest and B. De Kruijff, *Biochim. Biophys. Acta*, 1978, **511**, 397.
268. Y. Aracava, S. Schreirer, R. Phadke, R. Deslauriers and I. P. C. Smith, *Biophys. Chem.*, 1981, **14**, 325.
269. M. Pesti, E. K. Novak, L. Ferenczy and A. Svoboda, *Sabouraudia*, 1981, **19**, 17.
270. G. Medoff, G. S. Kobayashi, C. N. Kwan, D. Schlessinger and P. Venkov, *Proc. Natl. Acad. Sci. USA*, 1972, **69**, 196.
271. B. V. Kumar, G. Medoff, G. S. Kobayashi and D. Schlessinger, *Nature (London)*, 1974, **256**, 323.
272. H. J. Scholer and A. M. Polak, in 'Antimicrobial Drug Resistance', ed. L. E. Bryan, Academic Press, New York, 1984, p. 393.
273. P. Meyer and F. Reusser, *Experientia*, 1967, **23**, 85.
274. T. M. Balasubramanian, N. C. E. Kendrick, M. Taylor, G. R. Marshall, J. E. Hall, I. Vodyanoy and F. Reusser, *J. Am. Chem. Soc.*, 1981, **103**, 6127.
275. R. Latorre and D. Alvarez, *Physiol. Rev.*, 1981, **61**, 71.
276. M. K. Matthew, R. Nagaraj and P. Balaram, *Biochem. Biophys. Res. Commun.*, 1981, **98**, 548.
277. A. L. Y. Lau and S. I. Chan, *Biochemistry*, 1976, **15**, 2551.
278. E. F. Gale, E. Cundliffe, P. E. Reynolds, M. H. Richmond and M. J. Waring, in 'The Molecular Basis of Antibiotic Action', 2nd edn., Wiley, London, 1981, p. 220.
279. D. R. Martin and R. J. P. Williams, *Biochem. Soc. Trans.*, 1975, **3**, 166.
280. U. Bauerjee, F. P. Tsui, T. N. Balasubramanian, G. R. Marshall and S. I. Chan, *J. Mol. Biol.*, 1983, **165**, 757.
281. G. Jung, W. A. Konig, D. Leibfritz, T. Ooka, K. Janko and G. Boheim, *Biochim. Biophys. Acta*, 1976, **433**, 764.
282. G. Boheim, K. Janko, D. Leibfritz, T. Ooka, W. A. Konig and G. Jung, *Biochim. Biophys. Acta*, 1976, **433**, 182.
283. H. Brockmann and G. Schmidt-Kastner, *Chem. Ber.*, 1955, **88**, 57.
284. M. M. Shemyakin, N. A. Aldanova, E. I. Vinogradova and M. Y. Feigina, *Tetrahedron Lett.*, 1963, **28**, 1921.
285. B. C. Pressman, in 'Energy-Linked Functions of Mitochondria,' ed. B. Chance, Academic Press, New York, 1963, p. 181.
286. B. C. Pressman, *Annu. Rev. Biochem.*, 1976, **45**, 501.
287. B. C. Pressman, E. J. Harris, W. S. Jagger and J. H. Johnson, *Proc. Natl. Acad. Sci. USA*, 1967, **58**, 1949.
288. Y. A. Ovchinnikov, V. T. Ivanor and A. M. Shkrob, in 'Membrane Active Complexones', Elsevier, New York, BBA Library, vol. 12, 1974.
289. E. P. Bakker, in 'Antibiotics. Volume 6, Mechanism of Action of Antibacterial Agents', ed. F. E. Hahn, Springer, Berlin, 1979, p. 67.
290. B. C. Pressman, *Proc. Natl. Acad. Sci. USA*, 1965, **53**, 1076.
291. J. Beck, H. Gerlach, V. Prelog and W. Voser, *Helv. Chim. Acta*, 1962, **45**, 620.
292. E. Meyers, F. E. Pansy, D. Perlman, D. A. Smith and F. L. Weisenborn, *J. Antibiot.*, 1965, **18**, 128.
293. L. K. Steinrauf, M. Pinkerton and J. W. Chamberlin, *Biochem. Biophys. Res. Commun.*, 1968, **33**, 29.
294. L. K. Steinrauf, E. W. Czerwinski and M. Pinkerton, *Biochem. Biophys. Res. Commun.*, 1971, **45**, 1279.
295. F. M. Harold and J. R. Baarda, *J. Bacteriol.*, 1967, **94**, 53.
296. F. M. Harold and J. R. Baarda, *J. Bacteriol.*, 1968, **95**, 816.
297. T. E. Andreoli, M. Tieffenberg and D. C. Tosteson, *J. Gen. Physiol.*, 1967, **50**, 2527.
298. P. Müller and D. O. Rudin, *Biochem. Biophys. Res. Commun.*, 1967, **26**, 398.
299. P. Läuger, *Angew. Chem., Int. Ed. Engl.*, 1985, **24**, 905.
300. M. M. Sheniyakin, Y. A. Ovchinnikov, V. T. Ivanov, V. K. Antonov, E. I. Vinnogradova, A. M. Shkrob, G. C. Malenkov, A. V. Estratov, I. A. Laine, E. I. Melnik and I. D. Ryabova, *J. Membr. Biol.*, 1969, **1**, 402.
301. B. T. Kilbourn, J. D. Dunitz, L. A. Pioda and W. Simon, *J. Mol. Biol.*, 1967, **30**, 559.
302. D. W. Urry, M. C. Goodall, J. D. Glickson and D. F. Mayers, *Proc. Natl. Acad. Sci. USA*, 1972, **68**, 1907.
303. Y. A. Ovchinnikov, *Eur. J. Biochem.*, 1979, **94**, 321.
304. M. Prudhomme, J. Guyot and G. Jeminet, *J. Antibiot.*, 1986, **39**, 934.

9.2

β-Lactam Antibiotics: Penicillins and Cephalosporins

CHRISTOPHER E. NEWALL AND PAUL D. HALLAM

Glaxo Group Research, Greenford, Middlesex, UK

9.2.1	INTRODUCTION	609
9.2.2	BIOSYNTHESIS AND TOTAL SYNTHESIS	610
9.2.3	THE TARGET: THE BACTERIAL CELL WALL	612
9.2.4	PHARMACOLOGICAL CONSIDERATIONS	617
9.2.5	EARLY PENICILLINS	618
9.2.6	EARLY CEPHALOSPORINS	623
9.2.7	AMIDINOPENICILLINS	627
9.2.8	β-LACTAMASE STABLE CEPHALOSPORINS	629
9.2.9	ANTIPSEUDOMONAL PENICILLINS	633
9.2.10	THIRD GENERATION CEPHALOSPORINS	635
9.2.11	ANTIPSEUDOMONAL CEPHALOSPORINS	642
9.2.12	NEW ORAL COMPOUNDS	644
9.2.13	FUTURE PROSPECTS	648
9.2.14	REFERENCES	650

9.2.1 INTRODUCTION

At the time of Fleming's original observation in 1929[1] the rational chemotherapy of bacterial infections was limited to the use of salvarsan for the treatment of syphilis. During the years that elapsed before the first clinical use of penicillin in 1941, sulfanilamide was developed, but many serious bacterial infections were still untreatable.

The early chapters in the history of penicillin[2] and cephalosporin C[3] need not be repeated here. This review is more concerned with the development of the natural products, which had limited clinical utility, into the β-lactamase-stable, broad-spectrum agents of today. With this in mind only a brief summary of the biosynthesis and total synthesis of the natural products is provided, the main thrust of the review relating to the chemistry and biology of compounds in clinical use or undergoing clinical trials. These may all be regarded as derivatives of 6-aminopenicillanic acid (6-APA; **1**) and 7-aminocephalosporanic acid (7-ACA; **2**) which contain the penam[4] and cephem ring systems, respectively.

The literature relating to these compounds is vast, with well over 17 000 entries in *Chemical Abstracts* in the past 25 years. The state of the art has been reviewed in several notable compilations and the works edited by Flynn,[5] Morin and Gorman,[6] Sammes,[7] Mitsuhashi[8] and Salton and Shockman[9] provide excellent background reading. The published proceedings of the successive

(1) 6-APA   (2) 7-ACA

Royal Society of Chemistry Symposia[10-12] summarize the progress in the chemistry of penicillins and cephalosporins since 1976. These and other major reviews are referred to frequently in the following pages.

The length of this review does not permit a full discussion of the relationships between structure, chemical reactivity and biological activity. Many studies of structure–activity relationships have been published and we refer briefly to some of these with general comments as appropriate.

9.2.2 BIOSYNTHESIS AND TOTAL SYNTHESIS

Since the early years of β-lactam research a great deal of effort has been expended on the elucidation of the biochemical origins of these compounds. Whilst the exact details of some of the individual steps are still unknown, the overall biosynthetic scheme is now well established. The subject has been extensively reviewed.[13-17]

It is accepted that the biosynthesis of both penicillins and cephalosporins begins with the construction of the tripeptide δ-(L-α-aminoadipoyl)-L-cysteinyl-D-valine (the 'Arnstein tripeptide',[18] (3) in Scheme 1) from the L-isomers of α-aminoadipic acid, cysteine and valine, the latter being incorporated with inversion. This precursor is converted into the primary penicillin, isopenicillin N (4) by the action of the enzyme isopenicillin N synthetase (IPNS). The exact details of this step are unclear. However, all the evidence appears to suggest that it is the β-lactam ring which is formed first in an oxygen dependent reaction.[19]

The pathway to isopenicillin N is common to all organisms producing penicillins and cephalosporins; the subsequent steps are characteristic for each organism. The natural penicillins are synthesized by *Penicillium chrysogenum* by exchanging the aminoadipoyl side chain of isopenicillin N for one of the several monosubstituted acetic acids present in the cell by the action of an acyltransferase. Thus incorporation of phenylacetic acid, from phenylacetyl-CoA, produces penicillin G (5). The presence of exogenous acetic acids can increase the production of particular penicillins by different strains of *P. chrysogenum*; thus adding phenylacetic acid to one strain stimulates the production of penicillin G and the introduction of phenoxyacetic acid to another results in the production of penicillin V (6). In the absence of side-chain precursors, isopenicillin N (4) is hydrolyzed to the penicillin nucleus 6-APA (1).

The cephalosporins are synthesized from isopenicillin N in *Cephalosporium acremonium* and in the streptomycetes. The aminoadipoyl moiety is epimerized to give penicillin N (7). This is followed by an oxidative ring expansion which produces the primary cephalosporin, deacetoxycephalosporin C (8). Subsequent hydroxylation at the C-3 methyl group gives deacetylcephalosporin C (9), which is acetylated by acetyl-CoA to give cephalosporin C (10). The enzymes responsible for the ring expansion (expandase) and hydroxylation (hydroxylase) both appear to be α-ketoglutarate-linked dioxygenases with similar cofactors. A recent study[20] using a highly purified enzyme preparation showed that the expandase activity remained associated with the hydroxylase activity throughout the purification, suggesting that either the two enzymes are very similar or that the activities are associated with a single, bifunctional enzyme.

The biosynthetic pathway in *C. acremonium* terminates with the conversion of deacetylcephalosporin C into cephalosporin C. In the streptomycetes, however, further reactions occur including carbamoylation of the 3-hydroxymethyl group and, most importantly, the methoxylation at the C-7 position which leads to the class of compounds known as the cephamycins. Thus *Streptomyces clavuligerus* converts deacetylcephalosporin C (9) into cephamycin (11).

Carbamoylation occurs by transfer of a carbamoyl group from carbamoyl phosphate under the influence of an *O*-carbamoyltransferase. The methoxy group is derived from molecular oxygen and methionine[21] and is probably introduced by oxygenation at C-7, producing a 7α-hydroxy intermediate which is methylated by a methyltransferase. The order in which these events occurs is unclear.

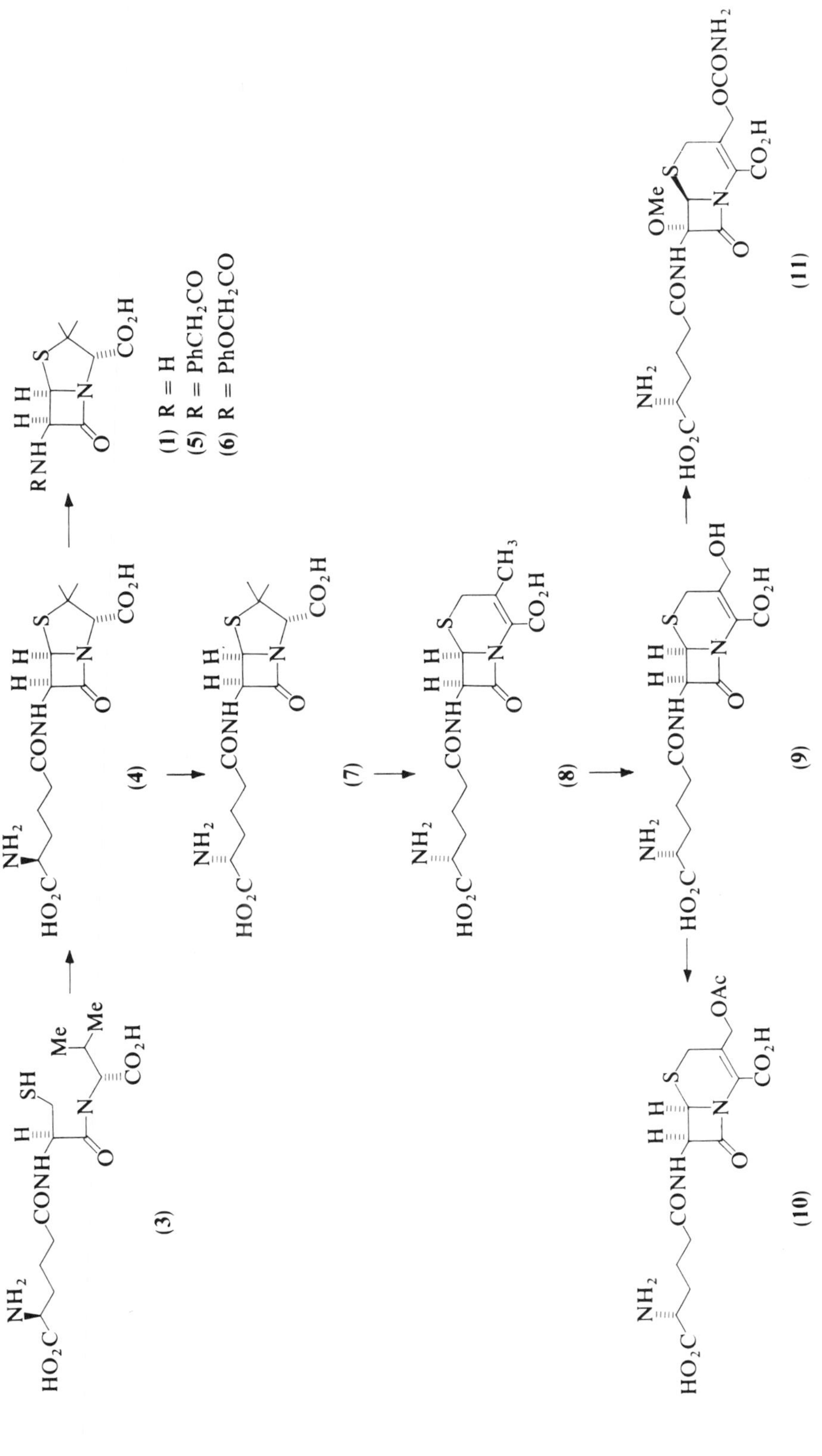
(3)
(4)
(7)
(8)
(9)
(10)
(11)
(1) R = H
(5) R = PhCH2CO
(6) R = PhOCH2CO

Scheme 1 Biosynthesis of penicillins and cephalosporins

The publication of the first general total synthesis of penicillins[22,23] was the culmination of over 15 years work by Sheehan and co-workers at MIT.[24-26] This synthesis, in its final form, is shown in Scheme 2. D-Penicillamine (12) was condensed with the phthalimido aldehyde (13) to give the thiazolidine (14). The reaction was not stereospecific but only two of the four possible isomers were formed and the desired α-isomer (14) could be obtained by fractional crystallization; further amounts were isolated from equilibrium mixtures produced by isomerization of the epimeric δ-isomer with pyridine.[27] The thiazolidine was transformed in three steps to the tritylamine (15), which was cyclized using diisopropylcarbodiimide to give (16). Careful saponification of the methyl ester followed by detritylation afforded the penicillin nucleus, 6-APA (1), which was acylated with phenylacetyl chloride to give penicillin G (5). Likewise, phenoxyacetylation of (1) yielded penicillin V (6).

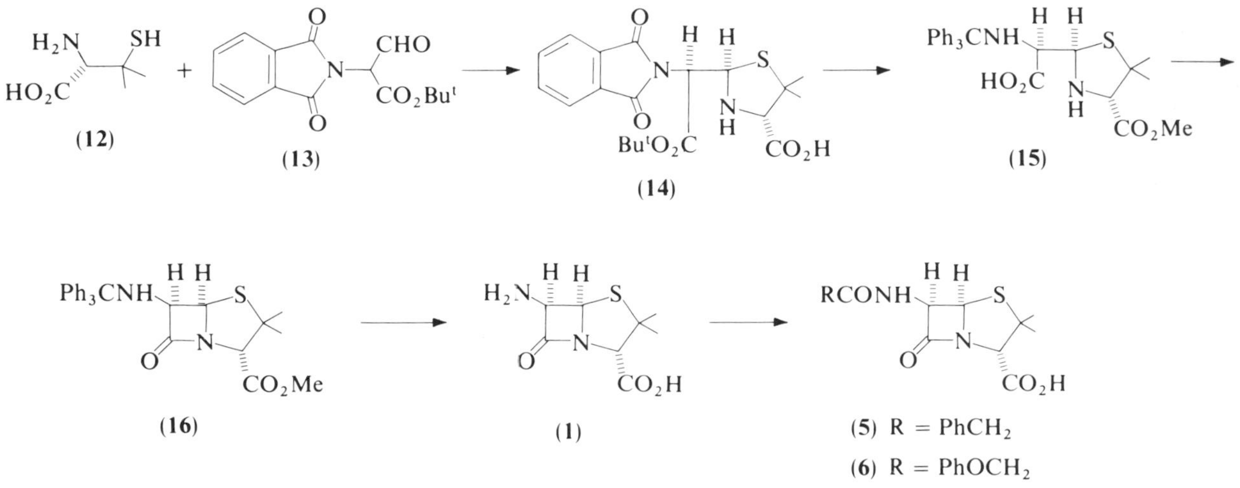

Scheme 2 Sheehan's total synthesis of penicillins

The first total synthesis of cephalosporin C was announced by Woodward in his Nobel lecture in 1965.[28,29] The strategy used in this classic synthesis differed markedly from Sheehan's approach to penicillins in that the β-lactam ring was constructed first, starting from L-cysteine (Scheme 3). The choice of starting material was significant: not only did it provide a large portion of the β-lactam ring but served as the source of one of the two asymmetric centres, obviating the need for a resolution step. The second centre was introduced stereoselectively starting with the conversion of (17) into the cyclic intermediate (18). Treatment of (18) with dimethyl diazodicarboxylate followed by oxidation with lead tetraacetate and hydrolysis of the intermediate acetate gave the *trans*-hydroxy ester (19). After mesylation, reaction with azide ion gave the *cis*-azide (20) which was reduced to the *cis*-amino ester (21) with aluminum amalgam. Triisobutylaluminum effected cyclization of (21) to the β-lactam (22), which was condensed with the highly reactive dialdehyde (23) to give (24). Subsequent treatment with trifluoroacetic acid realised the amino aldehyde (25). Reacylation of (25) with protected D-α-aminoadipic acid, using dicyclohexylcarbodiimide, followed by esterification gave (26) which was transformed into (27) by reduction with diborane, acetylation and isomerization with pyridine. Finally, deprotection with zinc in acetic acid afforded totally synthetic cephalosporin C (10). Also, in an analogous series of reactions, (25) was converted into cephalothin (28), previously prepared semisynthetically from cephalosporin C (*vide infra*).

Subsequently, several other total syntheses of penicillins and cephalosporins have been reported.[25,30] However, despite the increased efficiency that has been achieved, none of these offers a viable alternative to the semisynthesis of commercial penicillins and cephalosporins from starting materials obtained from fermentation.[30]

9.2.3   THE TARGET: THE BACTERIAL CELL WALL

Bacteria have to survive in a changing and potentially hostile environment and most species have a structurally strong cell wall capable of withstanding osmotic shock. Peptidoglycan, the main

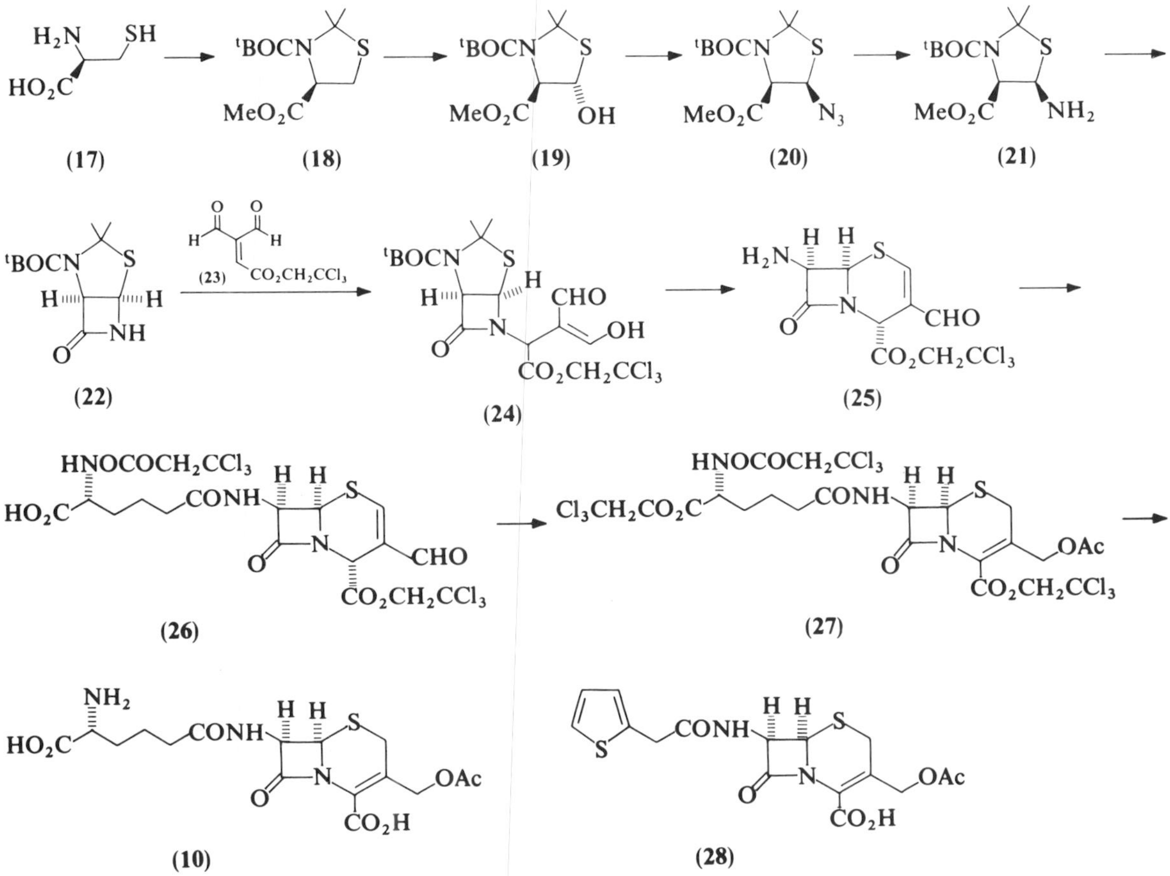

Scheme 3 Woodward's total synthesis of cephalosporins

structural component of the cell wall, is a macromolecule comprising strands of alternating *N*-acetylglucosamine and *N*-acetylmuramyl pentapeptide units, crosslinked *via* the peptide side-chains.[31] This structure is unique to prokaryotic cells and represents an excellent selective target for antibacterial agents. In Gram-positive bacteria the wall consists of a cytoplasmic membrane surrounded by a thick highly crosslinked layer of peptidoglycan (Figure 1a); in Gram-negative cells the peptidoglycan layer is thinner and is surrounded by a complex outer membrane which has protein molecules and lipopolysaccharides embedded in its phospholipid bilayer (Figure 1b).

The inner, cytoplasmic membrane carries the enzyme systems responsible for the synthesis and maintenance of peptidoglycan. The latter is readily penetrated by small molecules and in Gram-positive bacteria these membrane-bound proteins are accessible to antibacterial agents. In Gram-negative organisms the outer membrane forms an effective penetration barrier, access to the cytoplasmic membrane by passive diffusion being largely controlled by outer-membrane proteins (porins) embedded in the phospholipid bilayer.[32] The porins appear to create diffusion channels for small molecules; in *E. coli* the porin channels are believed to exclude compounds with molecular weights in excess of about 600 Da.

The structure of peptidoglycan (Figure 2) and the effect of penicillins and cephalosporins on the crosslinking of the peptidoglycan strands were first investigated by Tipper and Strominger[33] and subsequently by Ghuysen,[34] whose elegant studies involved investigation of the effects of the antibiotics on soluble D-alanyl-D-alanine carboxypeptidases isolated from strains of *Streptomyces* and *Actinomadura*. These studies enabled the mode of action of the β-lactam antibiotics to be worked out in the considerable detail in model systems. Thus, the antibiotic acts as a false substrate for D-alanyl-D-alanine transpeptidases, one normal function of which is to mediate a transamination reaction in which the terminal D-alanine of one pentapeptide side-chain is replaced by an amino group from an adjacent chain, forming a cross link.

A number of groups have studied the conformation of the penam and cephem nuclei with respect to the conformations of D-Ala-D-Ala peptides, a relationship first pointed out by Tipper and Strominger. These studies have been reviewed by Boyd.[35] The acylamino side chain also makes a

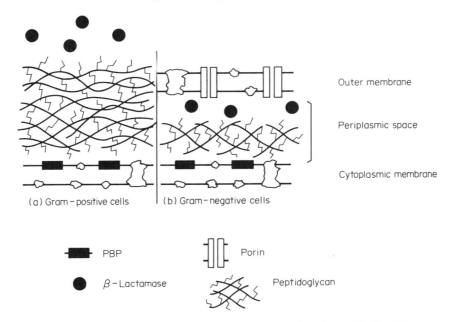

Outer membrane

Periplasmic space

Cytoplasmic membrane

(a) Gram-positive cells   (b) Gram-negative cells

PBP   Porin

β-Lactamase   Peptidoglycan

Figure 1 Schematic representation of the structure of the bacterial cell wall

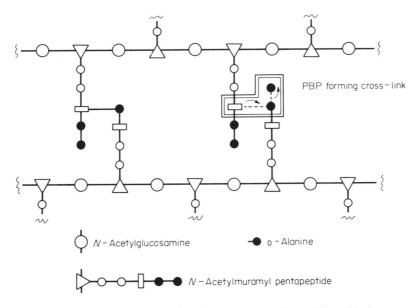

PBP forming cross-link

○ N-Acetylglucosamine   ● D-Alanine

◁○○□●● N-Acetylmuramyl pentapeptide

Figure 2 Schematic representation of the structure of bacterial peptidoglycan

major contribution to the antibacterial activity of the penicillins and cephalosporins, probably by providing additional binding sites at the target enzyme. The nature of this side chain is less important in the nonclassical bicyclic β-lactams (Chapter 9.3).

Further progress was made possible by the development of a competitive binding assay by Spratt,[36] who characterized a family of membrane-bound enzymes of the cytoplasmic membrane as penicillin binding proteins (PBPs). The cellular function of the PBPs in some organisms has now been determined in considerable detail from morphological studies of temperature sensitive PBP mutants (Table 1). Representative examples of the affinity of penicillins and cephalosporins for the PBPs of *E. coli* are shown in Table 2.[37]

Many bacteria produce enzymes capable of inactivating these antibiotics by cleaving the β-lactam ring. These β-lactamases may be chromosomally coded or carried on plasmids or transposons—small pieces of circular DNA which may be transferred from cell to cell. β-Lactamases were classified functionally according to their substrate profiles by Richmond and Sykes.[38] More recently,

Table 1 Properties of the β-Lactam Binding Proteins of *E. coli*

PBP	Mol. wt. (Da)	Cellular function
1a	91 000	PBPs 1a and 1b are involved in cell elongation. Inhibition of both proteins results
1b		in spheroplasting and lysis but inhibition of either on its own does not
2	66 000	PBP2 is an essential protein involved in 'shape' determination. Inhibition of PBP2 leads to 'round-form' production and eventually death
3	60 000	PBP3 is essential and involved in cell division. Inhibition leads to filamentation and eventually cell death.
4	49 000	PBP4 is 'non-essential' as mutants lacking this protein (dacB) are viable
5/6	42 000	PBPs 5/6 are 'non-essential'. Mutants in PBP 5/6 are viable
	40 000	

Table 2 Competition of Penicillins and Cephalosporins for the PBPs of *E. coli* DC0[37]

Compound	I_{50} binding concentration (µg mL^{-1}) for protein							MIC (µg mL^{-1})	
	1a	1b	2	3	4	5	6	DC0	DC2[a]
Penicillin G	0.5	3.0	0.8	0.9	1.0	24	19	16	1.6
Penicillin V	1.4	9.5	7.7	10.5	14	50	17	50	12.5
Cloxacillin	2	23	15	3	125	8	30	>250	4
Methicillin	9.0	250	27	5.5	60	>250	>250	>250	4
Carbenicillin	2.1	5.0	4.0	2.1	3.5	130	120	2	1
Ampicillin	1.4	3.9	0.7	0.9	2.0	140	9	3.2	0.5
Amoxycillin	0.7	2.2	0.9	4.1	2.5	110	9	3.2	1.6
Mecillinam	>250	>250	<0.25	>250	>250	>250	>250	0.05	0.01
Mezlocillin	1.5	8	0.9	0.025	>25	>25	>25	6.4	0.025
Cephalothin	<0.25	16	37	1	60	125	130	1.6	0.8
Cephaloridine	0.25	2.5	50	8	17	>250	>250	2	2
Cefazolin	<0.25	4.7	4.6	5.8	38	>250	>250	4	4
Cefamandole	<0.25	2.9	61	<0.25	38	>250	>250	1.6	0.1
Cephalexin	4	240	>250	8	30	>250	>250	8	4
Cefaclor	1.0	4.4	130	2	22	>250	>250	3.2	1.6
Cefoxitin	0.1	3.9	>250	5.8	7	0.6	0.9	4	2
Cefuroxime	0.12	1.6	13.7	0.09	200	>250	180	2	0.1
Cefotaxime	0.05	0.7	5	<0.05	30	<50	>50	0.08	0.01
Cefoperazone	0.5	1.5	0.9	0.05	>50	>50	>50	0.4	0.05
Cefsulodin	0.47	3.7	>250	>250	>250	>250	>250	25	6.4
Ceftazidime[b]	0.9	3.4	240	0.06	500	500	500	<0.13	<0.13

[a] *E. coli* DC2 is a permeability mutant of the wild type strain DC0. [b] Unpublished data.

Ambler has classified β-lactamases on the basis of their primary structure and mode of action.[39] A recent review by Medeiros summarizes the classification of β-lactamases and their contribution to antibiotic resistance.[40]

Several important β-lactamases are now known to be serine proteases which open the four-membered ring of a β-lactam antibiotic to form a readily hydrolyzed acyl-enzyme. It has been demonstrated that a similar product is formed when β-lactams react with the soluble carboxypeptidase from *Actinomadura* R61;[31,34] in this case the acyl-enzyme is stable and the enzyme is inactivated. The binding of several β-lactams to the R61 enzyme has been studied by X-ray crystallography.[41] Sequence analysis of some of the PBPs from *E. coli* indicates that these may also function as serine proteases and sequence similarities with class A β-lactamases (*e.g.* TEM-1) have been noted.[36,42]

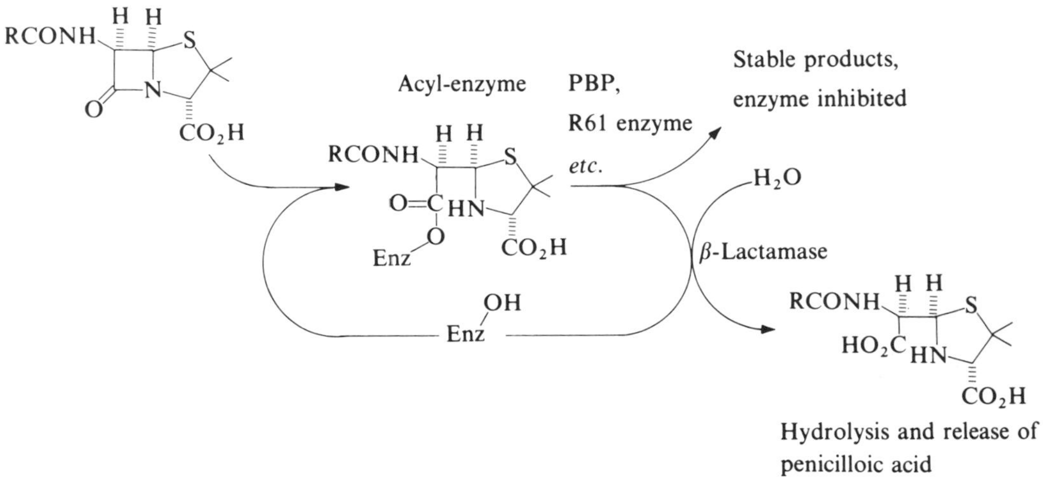

Scheme 4 Interaction of penicillins with D-Ala-D-Ala serine proteases and class A β-lactamases

Scheme 4 illustrates the reaction of these serine enzymes with a penicillin; the hydrolytic step characteristic of β-lactamase activity releases the corresponding penicilloic acid. With cephalosporins a more complex sequence of reactions usually occurs, as nucleophilic attack on the β-lactam ring is followed by rearrangement or fragmentation of the dihydrothiazine ring, often with expulsion of the 3′-substituent (Scheme 5).

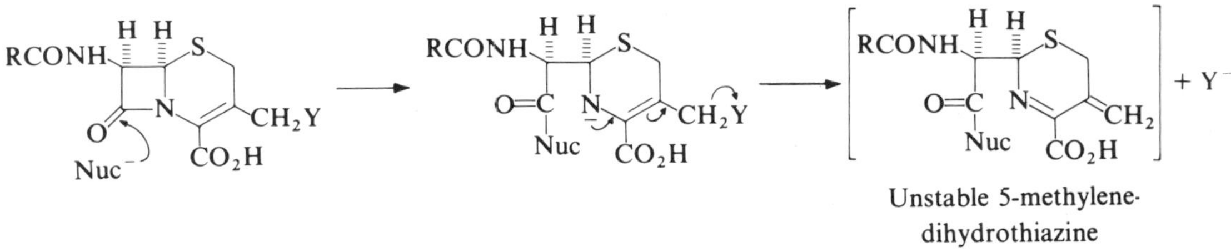

Scheme 5 Nucleophilic attack leading to the fragmentation of cephalosporins

Many species of Gram-positive bacteria produce large amounts of extracellular β-lactamase, which will hydrolyze any β-lactamase-sensitive antibiotics before they reach the cell wall. In Gram-negative cells the β-lactamase molecules are located in the periplasmic space between the inner and outer membranes. They are thus strategically placed to intercept incoming molecules before they reach the target PBPs bound to the cytoplasmic membrane, and comparatively small amounts of β-lactamase provide an effective protection for these organisms.

Thus, Gram-negative cells often have a twofold defence against β-lactam antibiotics, as the porin channels control the diffusion of many molecules into the periplasmic space and β-lactamases may lie in wait within. Resistance to β-lactams in Gram-negative bacteria has been reviewed by Nayler.[43] In Gram-positive cells there is no outer membrane and access to the target enzymes is comparatively unrestricted. Nevertheless, resistance persists and in certain species, most notably *Staphylococci*, has become a major clinical problem; the incidence of infections caused by methicillin-resistant *Staphylococci* having reached epidemic proportions in some hospitals. The nature of the resistance mechanisms in these organisms has been studied in depth and there is now good evidence that some methicillin-resistant strains of *S. aureus* produce an additional protein, PBP 2′, which has a very low affinity for β-lactam antibiotics.[44,45] Low affinity PBPs have also been detected in penicillin-resistant strains of other bacteria.[46]

Although there are abundant biological data in the general literature, the test systems described are usually not strictly comparable and it is difficult to assess the relative merits of compounds from different sources. To provide consistency, all the minimum inhibitory concentration (MIC) data presented are from strains in the Glaxo Group Research culture collection; the organisms used are summarized in Table 3.

Table 3 Representative Bacterial Strains

Organism	Gram type	
S. aureus Pen-S	+	Penicillin susceptible strain
S. aureus Pen-R	+	Penicillinase producing strain
St. faecalis	+	Group D streptococcus
E. coli (DC0)	−	Representative wild type
E. coli TEM-1	−	Resistant strain carrying the TEM-1 plasmid
E. cloacae	−	Representative wild type
E. cloacae P99	−	Resistant strain producing constitutive P99 β-lactamase
K. aerogenes	−	Nonlactamase producing mutant strain
K. aerogenes K1	−	Strain producing K1 β-lactamase
Pr. mirabilis	−	Representative proteus strain
Pr. vulgaris	−	β-Lactamase producing indole-positive proteus
Ps. aeruginosa	−	Representative wild type
Ps. aeruginosa Gen-R	−	Gentamicin resistant β-lactamase producing strain
H. influenzae	−	Ampicillin susceptible strain
H. influenzae Amp-R	−	Ampicillin resistant strain carrying the TEM-1 plasmid

Another self-consistent set of data has been published by Rolinson, who collated the antibacterial activity and pharmacokinetic properties of a large number of β-lactam antibiotics in a recent review.[47]

9.2.4 PHARMACOLOGICAL CONSIDERATIONS

Although the activity of a chemotherapeutic agent at its target is of fundamental importance, many other factors must be taken into account. Thus, the transport of the compound to the target and its distribution in other tissues, metabolism, elimination, side effects and toxicity must all be understood.

The bacterial cell wall and its associated membranes described in Section 9.2.3 differ significantly from those found in the host. These differences provide ways of achieving selective toxicity but may also complicate drug delivery, as the requirements for penetration through the membranes of the pathogen and the host are not always the same. As an example, orally active compounds must survive the hostile environment of the stomach and gut, be absorbed across the brush border membrane, be freely available in the systemic circulation and other body compartments and, finally, be capable of penetrating the bacterial cell wall. Only a relatively few penicillins and cephalosporins meet all these criteria.

The effect of antibacterial agents on the natural flora of the body should be taken into account when new antibiotics are being designed. The flora normally comprise many types of bacteria and fungi and disturbance of the natural balance may lead to side effects. Thus, in extreme cases, treatment with an antibiotic may result in a superinfection caused by overgrowth of a resistant organism; this is most likely to occur in patients who are immunosuppressed or who have serious underlying disease.

The flora of the gastrointestinal (GI) tract, and in particular that of the colon, are most likely to be affected by antibiotic treatment. Anaerobic bacteria, such as *Bacteroides fragilis*, predominate in the gut flora and these often produce powerful β-lactamases capable of hydrolyzing most penicillins and many cephalosporins, which are rapidly destroyed on reaching the colon. Even so, gastrointestinal disturbances caused by changes in the gut flora are a common side effect of antibiotic therapy; abdominal pain and mild diarrhea being common during treatment with oral antibiotics, particularly those which are incompletely absorbed. A more serious complication is pseudomembranous colitis, a potentially fatal condition, which may be associated with superinfections caused by *Clostridium difficile*, an anaerobic toxin producing organism sometimes found in the gut flora.[48]

Most penicillins and cephalosporins are eliminated from the body in the urine and hence many injectable compounds do not reach the lumen of the gut. However, some compounds, particularly lipophilic molecules with a fairly high molecular weight, tend to be excreted in the bile. Wright and Line[49] have demonstrated that, in rats, biliary excretion tends to increase with molecular weight above a threshold value of about 450 Da. In humans the threshold molecular weight for biliary excretion is thought to be closer to 600 Da and some of the higher molecular weight β-lactam antibiotics have been associated with GI tract side effects. On the other hand, therapeutically effective concentrations of antibiotic in the gall bladder and bile ducts are obviously desirable for the treatment of biliary tract infections. Fortunately, effective concentrations may often be achieved when a relatively small proportion of the dose is excreted in the bile and in such cases the risk of GI side effects is slight.

Once a compound has reached the systemic circulation its efficacy may be governed by a number of factors. For example, binding to serum proteins will reduce the amount of free drug available to exert an antibacterial effect and if the level of unbound drug falls below the MIC it is likely to be ineffective. In a recent review, Wise has discussed the relevance of protein binding, molecular size and lipophilicity to the clinical efficacy of antibiotics.[50]

The effects of serum protein binding are not necessarily always negative, as protein-bound antibiotics are less readily eliminated by the kidneys and thus highly bound compounds tend to have a long serum elimination half-life (SE $t_{1/2}$). Providing that the binding is readily reversible this can have the beneficial effect of maintaining the concentration of unbound drug above the MIC for a prolonged time, reducing the need for frequent administration.

Metabolism by enzymes present in the blood or body tissues may also play an important role; again this may be turned to good account. Early studies with cephalosporins (see Section 9.2.6) demonstrated the instability of some ester linkages to serum esterases. This phenomenon has been used to provide prodrug forms of penicillins and cephalosporins which enhance the absorption following oral administration. These prodrugs are usually 'double esters', in which the carboxyl group of the parent compound has been converted into an acyloxyalkyl ester. Several such derivatives exhibit increased absorption with respect to that of the parent compounds and after absorption the active drug is rapidly regenerated by the action of esterases in the blood or liver.[51]

Penicillins and cephalosporins are generally well tolerated but they are not entirely free from toxicity and side effects; their safety has been reviewed by Brown and Martin[52] and by Norrby.[53] As indicated above, mild GI disturbances are not uncommon and superinfections can occur in some groups of patients. The allergic response to penicillins can be dramatic and has occasionally been fatal. Mild allergic reactions were common in the early days and in some cases may have been caused by high molecular weight impurities arising from the isolation process or by partial polymerization on storage; modern, highly purified preparations are much more satisfactory in this respect. Cephalosporins, which are generally more stable than penicillins, are less likely to lead to an allergic reaction and may be used in some patients who are sensitive to penicillins.

Nephrotoxicity is very rare and appears to be restricted to a few early instances in the use of cephaloridine, which is actively secreted by the renal tubules. It has been demonstrated that the nephrotoxicity of cephaloridine can be prevented by the coadministration of probenecid.[54] Neurotoxic effects have occasionally been observed following high intravenous doses of penicillins.

Coagulation defects have been observed with two structural sub-types of β-lactam antibiotic. Thus, high concentrations of compounds with a phenylmalonyl or similar side chain, for example carbenicillin (Section 9.2.5), interfere with platelet aggregation. Another effect is hypoprothrombinemia, apparently caused by a deficiency of vitamin K. It was originally thought that this resulted from the elimination of vitamin K synthesizing bacteria from the GI tract; more recently, cephalosporins carrying a 1-methyltetrazolylthiomethyl group at the 3-position, such as cefoperazone (Section 9.2.10), have been directly implicated. In several cases it has been demonstrated that the prolonged prothrombin times caused by these compounds can be reversed with vitamin K therapy. The presence of this 3-substituent also disposes towards a disulfuram effect, probably *via* release of 5-mercapto-1-methyltetrazole.

Most of the undesirable effects detailed above may be avoided or minimized by the using the most suitable antibiotic at an appropriate dose. Thus, the basic principle of selective toxicity is fulfilled better by the penicillins and cephalosporins than by any other group of antibiotics.

9.2.5   EARLY PENICILLINS

At the time of their introduction in the early 1940s the first fermentation derived penicillins were rightly regarded as 'wonder drugs'. However, resistance developed rapidly and by 1946 hospitals were reporting resistant penicillinase producing strains, particularly of *S. aureus*.

Although the inclusion of suitable side-chain precursors permits the 6-substituent to be modified to some extent, only a limited range of compounds may be produced by this means and the phenylacetyl (**5**) and phenoxyacetyl (**6**) derivatives, penicillins G and V respectively, remained the products of choice for several years.

In 1959, in the same year that Sheehan and Henery-Logan announced their total synthesis of 6-APA,[22] Batchelor and coworkers at Beecham reported the isolation of the penicillin nucleus from fermentation broths devoid of side-chain precursors.[55] Within a year, several groups, working independently, reported that good yields of 6-APA could be obtained by enzymatic deacylation of naturally occurring penicillins such as penicillin G, using a variety of amidases.[56] More recently, a chemical method for the deacylation of penicillins has been developed which utilizes phosphorus pentachloride. This reaction is discussed in more detail in Section 9.2.6 in relation to the deacylation of cephalosporins.

The ready availability of 6-APA provided the medicinal chemist with the opportunity to synthesize a virtually unlimited number of new penicillins with novel side chains and the early 1960s saw rapid advances in the development of more effective antibiotics.

A variety of methods have been developed for the acylation of 6-APA utilizing acid chlorides, mixed anhydrides and, less commonly, acyl azides.[57] In general, reactions are carried out in aqueous acetone in the presence of sodium bicarbonate or in nonaqueous solvents such as chloroform using triethylamine as the base. Side chains have also been coupled directly to the sodium salt of 6-APA using the condensing agent dicyclohexylcarbodiimide in tetrahydrofuran.[58]

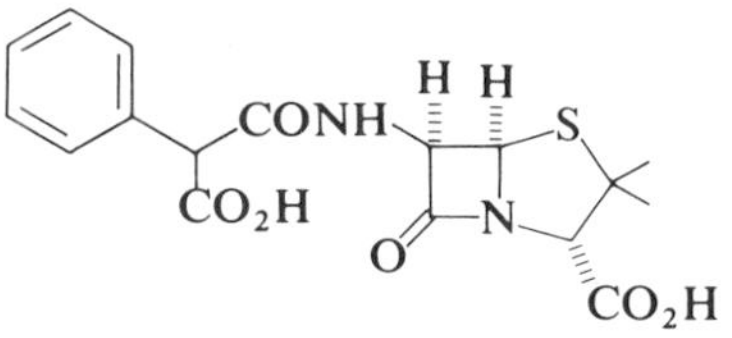

(29) Carbenicillin

Probably the most noteworthy advances were made in the Beecham laboratories, whence came a series of penicillinase stable, narrow-spectrum compounds, the antipseudomonal compound carbenicillin (29) and the orally absorbed, broad-spectrum antibiotic, ampicillin (30). In the latter case the side chain contains an amino group which requires protection prior to the acylation step. This has most commonly been achieved by use of the *N*-benzyloxycarbonyl group which may be removed subsequently by catalytic hydrogenation. This method suffers, however, in that large amounts of catalyst are often required as a result of the poisoning of the catalyst by the sulfur in the penicillin,[59] and in some cases hydrogenation can effect undesired side reactions such as dehalogenation.[60] Alternative protection is afforded by β-dicarbonyl compounds such as ethyl acetoacetate, which convert the amino group into an enamine under mildly basic conditions. Deprotection is achieved by treatment with mild acid.[61]

Some of the above points are illustrated by the syntheses of ampicillin (30) and methicillin (31) (Scheme 6). Condensation of 6-APA with the mixed anhydride derived from *N*-benzyloxycarbonyl-D-phenylglycine (32) in aqueous bicarbonate solution gave the protected penicillin (33), which was

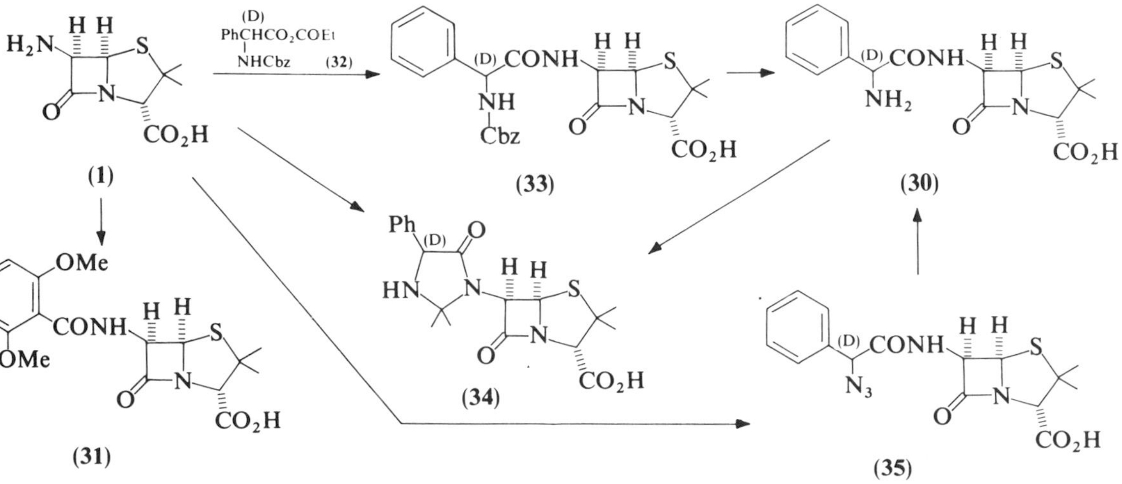

Scheme 6 The conversion of 6-APA into ampicillin (30), hetacillin (34) and methicillin (31)

Table 4 Antibacterial Activity of Early Penicillins[a]

Organism	Pen G	Pen V	Clox	Meth	Carb	Amp	Amox	Mecil
S. aureus Pen-S	<0.06	<0.06	0.13	1	0.5	<0.06	0.25	31
S. aureus Pen-R	125	62	0.5	2	4	8	31	>125
St. faecalis	8	16	125	>125	62	2	4	>125
E. coli (DC0)	31	>125	>125	>125	8	2	8	0.25
E. coli TEM-1	>125	>125	>125	>125	>125	>125	>125	2
E. cloacae	31	>125	>125	>125	4	4	8	0.25
E. cloacae P99	>125	>125	>125	>125	31	>125	>125	0.25
K. aerogenes	4	125	125	>125	4	0.5	2	0.5
K. aerogenes K1	>125	>125	>125	>125	>125	>125	>125	>125
Pr. mirabilis	8	>125	>125	>125	0.5	2	4	31
Pr. vulgaris	>125	>125	>125	>125	2	>125	>125	>125
Ps. aeruginosa	>125	>125	>125	>125	62	>125	>125	>125
Ps. aerug. Gen-R	>125	>125	>125	>125	>125	>125	>125	>125
H. influenzae	0.5	31	2	1	2	0.5	4	125
H. influ. Amp-R	31	125	8	1	2	16	31	>125

[a] MICs in $\mu g\,mL^{-1}$. Penicillin G (Pen G), penicillin V (Pen V), cloxacillin (Clox), methicillin (Meth), carbenicillin (Carb), ampicillin (Amp), amoxycillin (Amox), mecillinam (Mecil).

converted into ampicillin by catalytic hydrogenation.[59] When 6-APA was reacted with the acid chloride of unprotected D-phenylglycine in aqueous acetone the product was the imidazolidinone penicillin, hetacillin (**34**); the same product was formed by the direct reaction of ampicillin with acetone.[62] Condensation of 6-APA with 2,6-dimethoxybenzoyl chloride in chloroform, in the presence of triethylamine, gave methicillin (**31**).[63]

An alternative strategy for the preparation of α-aminobenzylpenicillins was developed by Ekström,[60] who acylated 6-APA with α-azido intermediates. The resulting azidopenicillins (*e.g.* **35**) were readily reduced by hydrogenation using palladium on calcium carbonate or Raney nickel.

Quite commonly, the carboxyl group of the penicillin nucleus may be protected with a readily removable group prior to acylation, thus enabling the reaction to be carried out in organic solvents. The trimethylsilyl ester, formed for example by reaction of 6-APA with hexamethyldisilazane in chloroform, has proved very useful in this respect as the silyl group is readily removed by treatment with water or methanol on work-up after the acylation.

Ampicillin has little stability to β-lactamases but has an extended spectrum, with activity which includes *E. coli*, some *Proteus* and *H. influenzae*, the latter being a common cause of otitis media and upper respiratory tract infections in young children (Table 4). It is very active against group D streptococci such as *St. faecalis*, which are often responsible for urinary tract infections (UTI).

(**36**) Cyclacillin

Dixon and Mizen[64] established that cyclacillin (**36**) is actively transported across the rat intestine, but firm evidence for the active transport of other β-lactams was not forthcoming until Tsuji's[65] and Kimura's[66] studies, which demonstrated that ampicillin and closely related compounds may cross the brush-border membrane of rats by carrier-mediated transport systems.

Despite active transport from the GI tract the oral bioavailability of ampicillin is only about 30% and the unabsorbed portion of the dose sometimes disturbs the gut flora, leading to mild GI side effects. Much effort has been devoted to improving the bioavailability of ampicillin and two successful strategies have emerged.

One of these was the double-ester prodrug approach, first described by Jansen and Russell.[67] They realised that lipophilic derivatives of penicillins such as esters would be more effectively absorbed from the GI tract than the parent acids. However, because simple ester derivatives of penicillins are relatively stable and have no useful antibacterial activity in their own right, a novel type of derivative which would be rapidly cleaved after absorption was required.

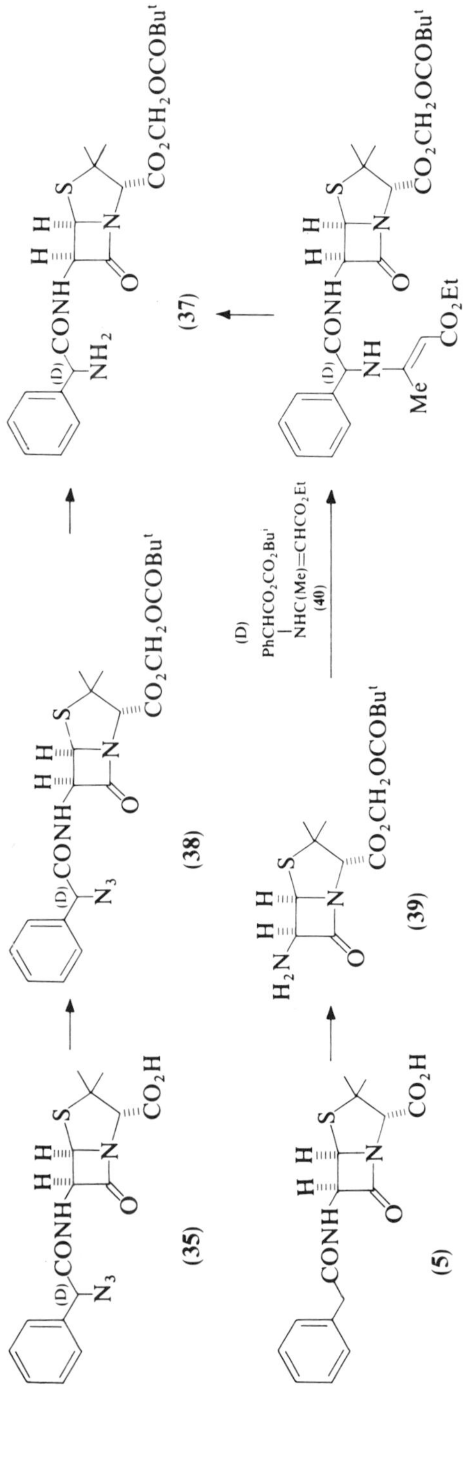

Scheme 7 The synthesis of pivampicillin (37)

It was known that esterases present in the serum, liver and other tissues cleave certain simple esters, particularly acetates, but do not cleave esters of more hindered carboxy compounds, such as penicillins. These observations led to the use of double esters of aldehyde hydrates, in which one ester function would be rapidly cleaved by serum esterases, releasing an unstable hemiacetal which would afford the free form of the penicillanic acid.

Scheme 7 depicts two approaches to the synthesis of the ampicillin prodrug pivampicillin (**37**).[68] In the first, the α-azidopenicillin (**35**) was esterified with chloromethyl pivalate (prepared from pivalyl chloride and paraformaldehyde) in the presence of a catalytic amount of sodium iodide, and the resulting ester (**38**) was converted into pivampicillin (**37**) by catalytic hydrogenation. The second route started from penicillin G (**5**), which was esterified and then deacylated using the phosphorus pentachloride method to give the pivalyloxymethyl ester of 6-APA (**39**). Acylation with the mixed anhydride of the D-phenylglycine Dane salt derivative (**40**) followed by mild acid hydrolysis of the enamine protecting group then gave (**37**). Alternatively, acylating (**39**) with D-phenylglycinyl chloride hydrochloride gave pivampicillin in a single step.

The other prodrugs of ampicillin, talampicillin (**41**) and bacampicillin (**42**), may be prepared using similar strategies.[69, 70] In both cases the esterification reaction results in the formation of 1:1 mixtures of diastereoisomers epimeric at the chiral centre of the ester moiety. The individual isomers of both drugs have been separated and found to exhibit virtually identical bioavailability.

This concept provided a series of prodrugs of ampicillin, some of which had improved bioavailability (*cf.* pivampicillin 70%; **37**). Similar prodrugs of mecillinam (Section 9.2.7) and cephalosporins (Sections 9.2.8 and 9.2.10) have been developed.

Another interesting compound is sarpicillin (**43**), the methoxymethyl ester of the ampicillin acetone aminal hetacillin, which provides an unusual example of a double prodrug.

(**41**) R = O
　　　, Talampicillin

(**42**) R = –CHOCO$_2$Et, Bacampicillin
　　　　　Me

(**43**) Sarpicillin

(**44**) Amoxycillin

(**45**) Cloxacillin

Analogues of ampicillin with modified 6-substituents were also studied in the search for improved activity and bioavailability. The most successful of these derivatives was the 4-hydroxy-D-phenylglycyl analogue amoxycillin (**44**), which has approximately twice the oral bioavailability of ampicillin and is more rapidly bactericidal.

Although methicillin (**31**) and the isoxazolylpenicillins, *e.g.* cloxacillin (**45**), are very active against penicillinase producing bacteria, their activity is restricted to Gram-positive organisms (Table 4). Cloxacillin and its congeners are highly serum bound and thus their *in vivo* activity is somewhat lower than might be expected from their MICs. Nevertheless, they are very effective in the treatment of infections caused by penicillinase producing Gram-positive bacteria.

The stability of these compounds to penicillinases has been attributed to the considerable steric crowding in the vicinity of the β-lactam ring created by the presence of the bulky 6-substituent. The poor activity of methicillin, cloxacillin and other sterically hindered penicillins against Gram-negative bacteria is probably a reflection of poor diffusion across the outer membrane, as both methicillin and cloxacillin bind to PBP 3 of *E. coli* DC0 (Table 2) and are active against the permeability mutant *E. coli* DC2.[37]

9.2.6 EARLY CEPHALOSPORINS

Cephalosporin C (**10**) was stable to staphylococcal penicillinase and was rapidly effective in the treatment of urinary tract infections; clinical use was limited by its generally weak antibacterial activity and the fact that intramuscular injections were very painful. It was inactive when given orally.[3]

With the penicillins as a precedent the 7-acylamino group was the first target for variation in the search for improved activity. It was anticipated that 7-aminocephalosporanic acid (7-ACA; **2**), the required intermediate, might be readily available by enzymatic cleavage of the aminoadipoyl side chain but cephalosporin C proved impervious to such treatment.[71] However, tiny amounts of (**2**) could be prepared by careful acid hydrolysis[72] and subsequently Morin and coworkers developed a commercially viable process which employed nitrosyl chloride.[73] The most commonly used method uses phosphorus pentachloride which converts fully protected cephalosporin C (**46**) into an imino chloride (**47**). Treatment with an alcohol gives the imino ether (**48**) which on hydrolysis and deprotection yields 7-ACA (Scheme 8).[74]

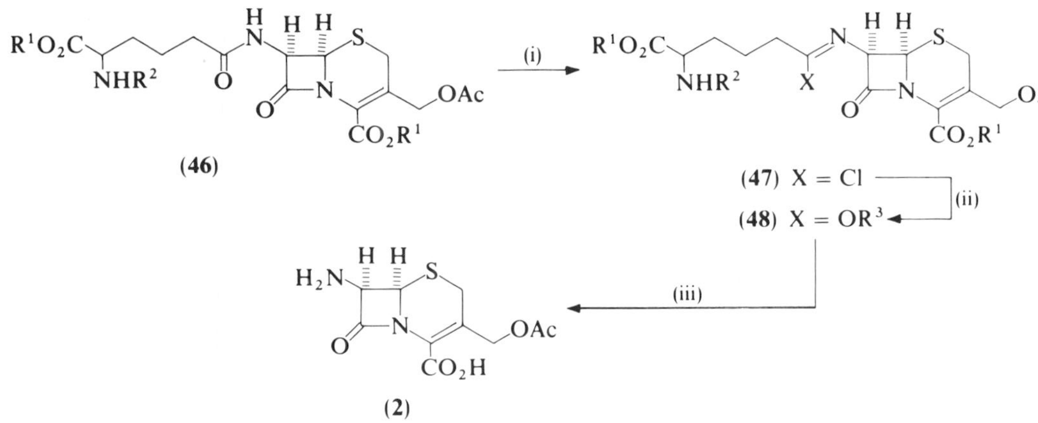

(**46**)

(**47**) X = Cl
(**48**) X = OR³

(**2**)

(i) PCl₅/pyridine; (ii) R³OH; (iii) hydrolysis followed by deprotection

Scheme 8   Conversion of cephalosporin C into 7-ACA (**2**)

As soon as 7-ACA became available, numerous novel cephalosporins were prepared by reacylation[75,76] using the methodology developed in the penicillin field. This work provided cephalothin (**28**),[75] the first semisynthetic cephalosporin to reach the market.

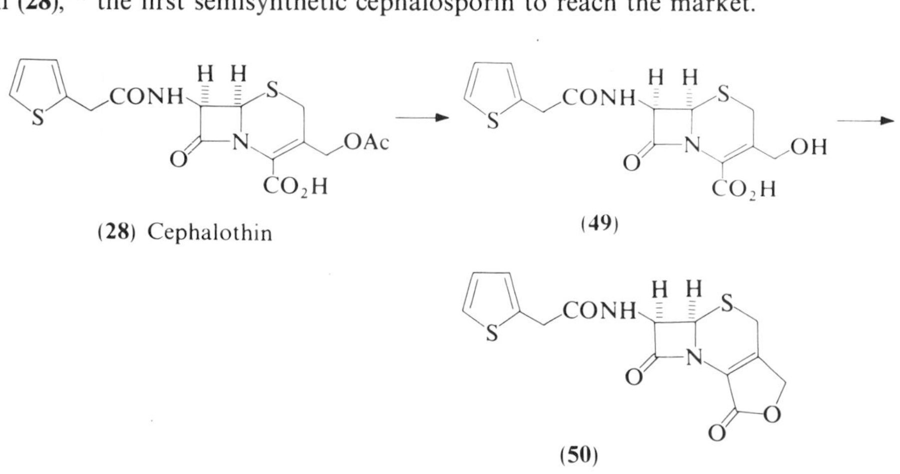

(**28**) Cephalothin

(**49**)

(**50**)

Cephalothin has a broader range of activity than the early penicillins and has been used extensively. It is readily deacetylated by serum esterases to give the less active deacetylcephalothin (**49**), which may cyclize to the inactive lactone (**50**). To overcome these problems, attention turned to chemical modification of the 3-substituent. Scheme 9 summarizes some of the more important

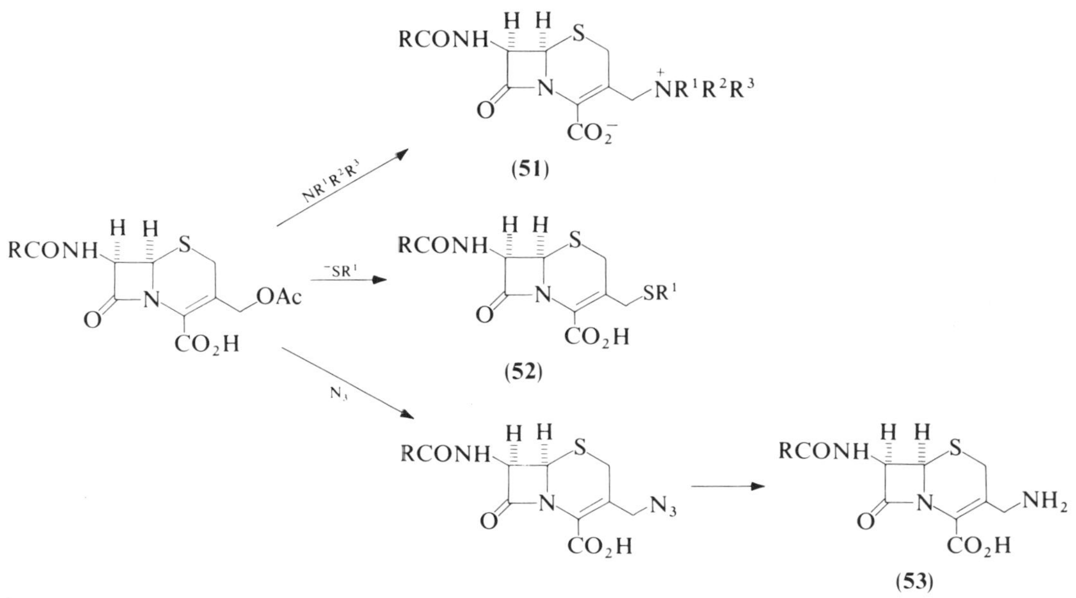

Scheme 9   Variation of the cephalosporin 3-substituent

reactions. Direct displacement of the acetoxy group with nitrogen[77] and sulfur[76] nucleophiles yielded (51) and (52), respectively. Displacement with azide ion followed by reduction afforded the amine (53)[76] which could be modified further, for example by acylation.

Protection of the carboxyl group by esterification confers improved solubility in organic solvents. The *t*-butyl, benzyl, *p*-nitrobenzyl, diphenylmethyl and trichloroethyl esters are the most widely used and methods compatible with the sensitive cephalosporin nucleus have been developed for their introduction and cleavage.[78]

The dihydrothiazine double bond in cephalosporanic acids is prone to isomerization under basic conditions and some of the above reactions are accompanied by the formation of the biologically inactive Δ^2-isomers (54). Cephalosporanic esters are particularly prone to this isomerization but on oxidation with *m*-chloroperbenzoic acid or peracetic acid yield exclusively the Δ^3-sulfoxides (55); these can be converted into the desired sulfides by reduction with either phosphorus tribromide or potassium iodide/acetyl chloride (Scheme 10). This oxidation/reduction sequence provides a convenient method for the conversion of mixtures of double-bond isomers into pure Δ^3-isomers.[79]

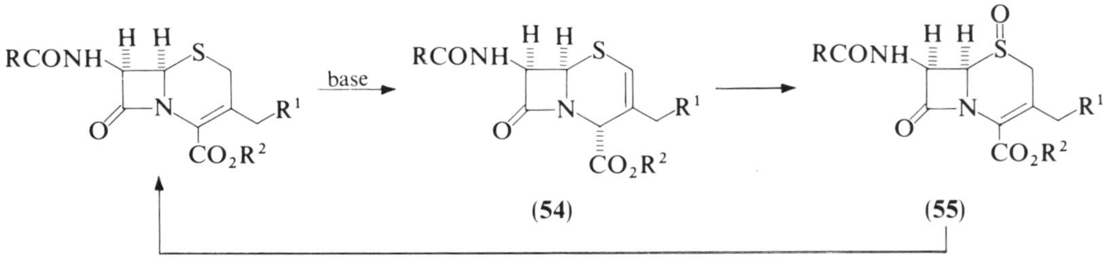

Scheme 10   The interconversion of Δ^2- and Δ^3-cephalosporins

Replacement of the labile acetoxy group by pyridine as described above provided cephaloridine (56),[77] a metabolically stable compound with an antibacterial spectrum similar to that of cephalothin. Cephaloridine penetrates rapidly into bacterial cells and is rapidly bactericidal, binding preferentially to PBP 1a and 1b (Table 2); its activity against *S. aureus* remains unsurpassed by any cephalosporin in clinical use (Table 5). It is well tolerated in normal patients but must be used with caution in cases of renal impairment when very high serum levels are likely to be achieved for long periods. Under these circumstances some nephrotoxicity has occasionally been seen; this may be avoided by carefully monitoring the dose.

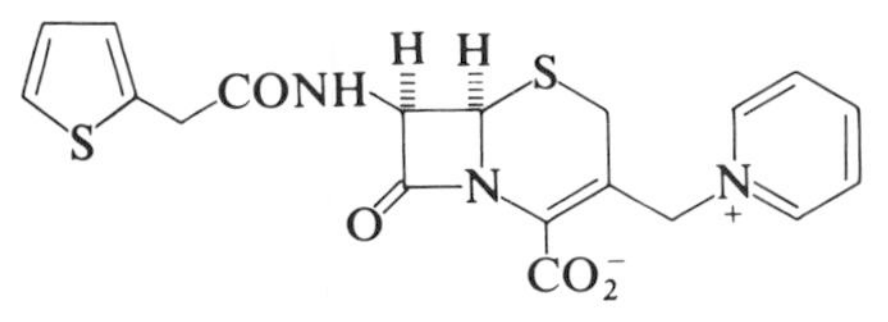

(56) Cephaloridine

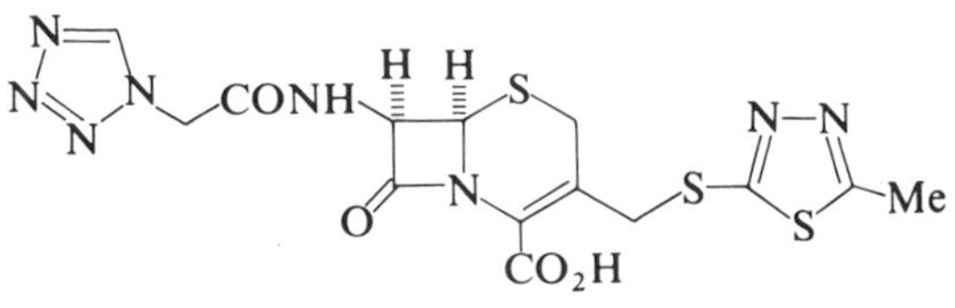

(57) Cefazolin

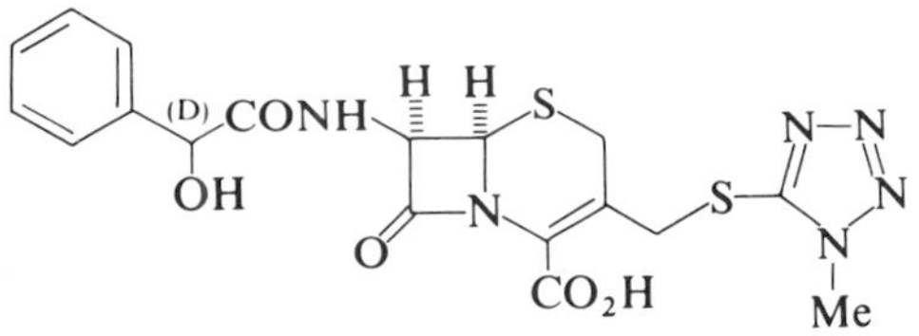

(58) Cefamandole

Table 5 Antibacterial Activity of Early Cephalosporins[a]

Organism	CET	CER	CEZ	CMD	CEX	CED	CDX
S. aureus Pen-S	<0.06	<0.06	0.13	<0.06	4	4	2
S. aureus Pen-R	0.25	0.25	0.5	0.5	8	4	4
St. faecalis	62	8	62	125	125	125	62
E. coli (DC0)	4	4	1	0.25	16	16	8
E. coli TEM-1	31	31	16	16	16	16	31
E. cloacae	8	4	2	2	8	31	16
E. cloacae P99	>125	>125	>125	>125	>125	>125	>125
K. aerogenes	2	4	1	0.5	8	16	8
K. aerogenes K1	>125	>125	>125	>125	>125	>125	125
Pr. mirabilis	8	4	2	1	8	16	16
Pr. vulgaris	>125	>125	>125	>125	>125	>125	>125
Ps. aeruginosa	>125	>125	>125	>125	>125	>125	>125
Ps. aerug. Gen-R	>125	>125	>125	>125	>125	>125	>125
H. influenzae	<0.25	1	0.5	<0.25	4	16	16
H. influ. Amp-R	0.5	2	1	0.5	4	16	16

[a] MICs in $\mu g\,mL^{-1}$. Cephalothin (CET), cephaloridine (CER), cefazolin (CEZ), cefamandole (CMD), cephalexin (CEX), cephradine (CED), cefadroxil (CDX).

Following the success of these early compounds, many variations on the 3- and 7-substituents were prepared and tested, but few compounds reached the marketplace. The most successful were cefazolin (57) and cefamandole (58),[80] which are 2–4 times more active than cephalothin against susceptible Gram-negative bacteria (Table 5). They were still the leading injectable cephalosporins in the mid-1980s.

Although the early cephalosporins were generally stable to staphylococcal penicillinase, most of them were rapidly destroyed by the β-lactamases from Gram-negative bacteria. This instability has been turned to good use in the case of the chromogenic compound nitrocefin (59),[81] which undergoes a colour change when the β-lactam ring is hydrolyzed.

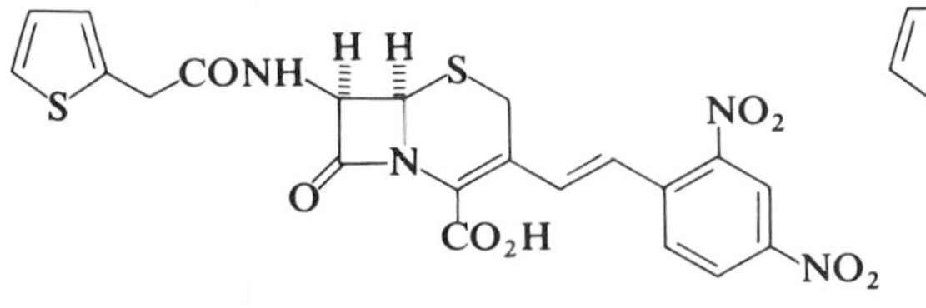

(59) Nitrocefin

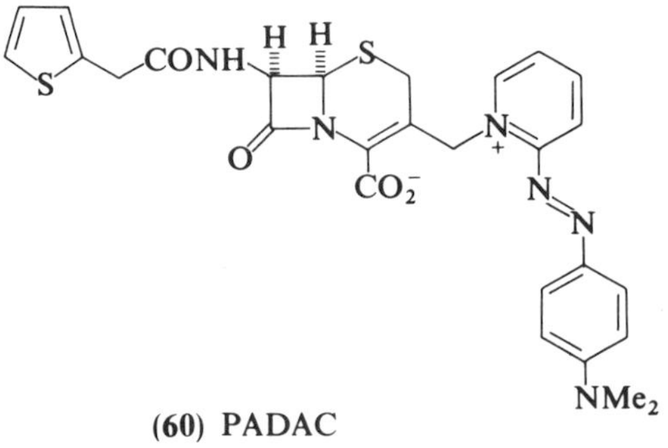

(60) PADAC

Nitrocefin, prepared by conversion of the appropriate 3-halomethylcephalosporin into a triphenylphosphonium ylide followed by a Wittig reaction (*q.v.* Section 9.2.12) with 2,4-dinitrobenzaldehyde,[82] and similar compounds, for example PADAC (60),[83] are now used routinely as diagnostic agents for the detection of β-lactamase producing strains of bacteria.

The success of ampicillin as an oral antibiotic stimulated a search for orally absorbed cephalosporins and it was discovered that compounds with small lipophilic 3-substituents and the D-phenylglycyl or closely related side chains, such as cephalexin (61) and cephradine (62), were well absorbed from the gastrointestinal tract.

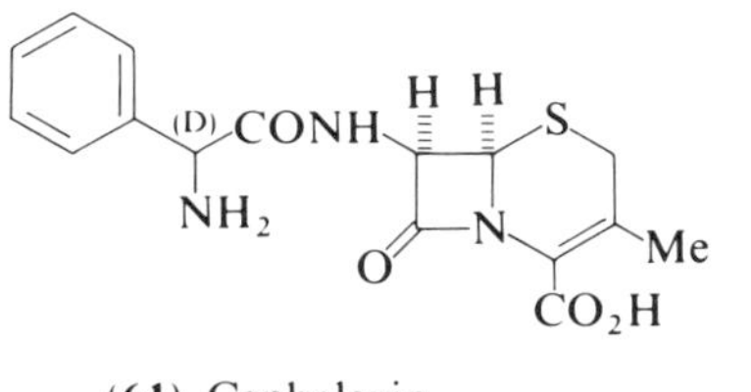

(61) Cephalexin 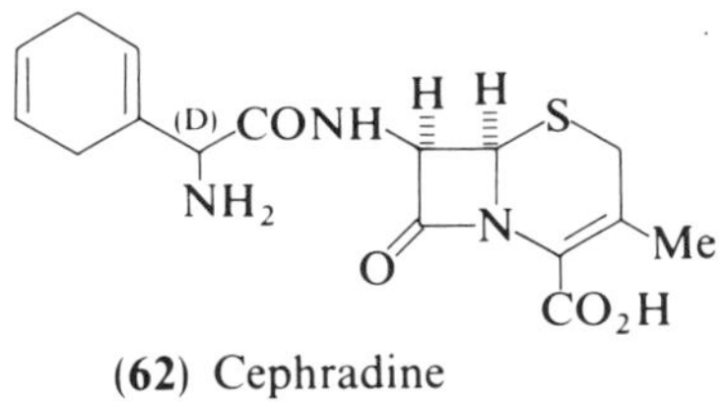   (62) Cephradine 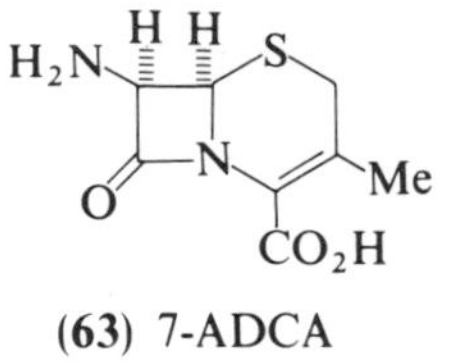   (63) 7-ADCA

Both cephalexin and cephradine were first prepared by acylation of 7-aminodeacetoxycephalosporanic acid (7-ADCA; 63) obtained *via* high-pressure catalytic hydrogenation of cephalosporin C.[84] However, this route suffered from a low yield and the large amounts of catalyst required for the reduction limited its commercial utility.

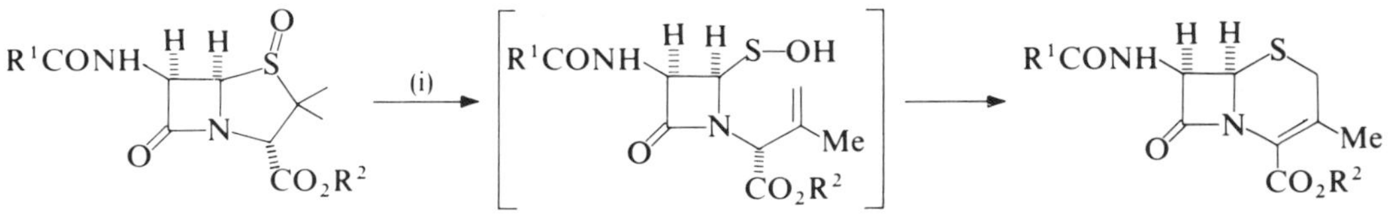

(i) Toluene-*p*-sulfonic acid, xylene, reflux

Scheme 11 The Morin rearrangement

A major development in the synthesis of these compounds was the discovery by Morin and coworkers[85] that penicillin sulfoxides (64) undergo a thermal rearrangement to 3-methylcephalosporins (65; Scheme 11). This reaction, which proceeds *via* the ring-opened sulfenic acid (66), forms the basis for a commercial preparation of cephalexin from penicillin (Scheme 12).[86]

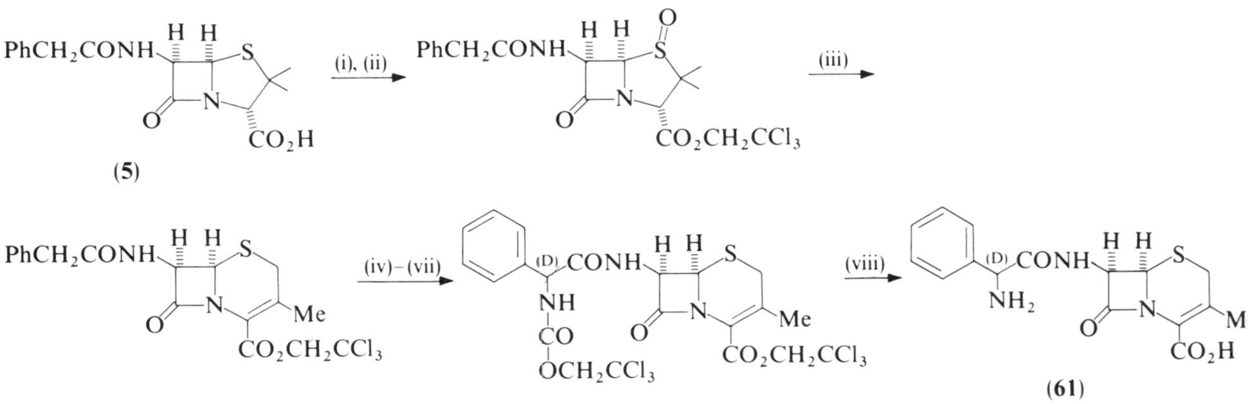

(i) ClCO$_2$CH$_2$CCl$_3$/pyridine, acetone (on potassium salt); (ii) *m*CPBA; (iii) Ac$_2$O, DMF, 130 °C; (iv) PCl$_5$/pyridine; (v) MeOH; (vi) H$_2$O; (vii) reacylation with protected phenylglycine using mixed anhydride method; (viii) Zn/90% aq. HCO$_2$H

Scheme 12 Conversion of penicillin G into cephalexin (61)

Cephaloglycine (**67**),[57,87] another orally absorbed compound, retains the 3-acetoxymethyl group of 7-ACA. Cephaloglycine can be unstable in aqueous solution, having a tendency to degrade *via* nucleophilic attack by the side-chain amino group on the β-lactam ring. It is relatively poorly absorbed, leading to low blood levels.

	R^1	R^2	
(**67**)	H	CH$_2$OAc	Cephaloglycine
(**68**)	HO	CH$_3$	Cefadroxil
(**69**)	H	Cl	Cefaclor

Cephalexin and the closely related cephradine lack the readily eliminated acetoxy group at the 3'-position and are much more stable to nucleophilic attack. Thus, cephalexin has a half-life of several hours in a pH 7 phosphate buffer at 37 °C whereas that of cephaloglycine is only 40 minutes.

The oral absorption of cephalexin and cephradine is virtually 100%, resulting in high blood levels and both compounds are excreted unchanged in the urine. It has been shown that the absorption of these compounds from the GI tract is assisted by an energy dependent amino acid transport system.[88] Cefadroxil (**68**), the 4'-hydroxy analogue of cephalexin, is also well absorbed and has an extended SE $t_{1/2}$ (1.2 h) relative to cephalexin (SE $t_{1/2}$ 0.6 h).[80] The improved pharmacokinetics of cefradroxil appear to be a consequence of lower secretion by the renal tubules.

The stability of the β-lactam ring of the 3-methylcephalosporins is reflected in a somewhat reduced bactericidal activity relative to that of ampicillin; this is offset by their relatively greater stability in the presence of β-lactamases, resulting in a generally broader spectrum of activity. Cephalexin and cephradine are very well tolerated, with minimal side effects and dominated the oral cephalosporin market from 1969 until the introduction of cefaclor (**69**) in 1979.

The structure–activity relationships of penicillins and cephalosporins have been studied extensively and the literature up to 1981 has been reviewed comprehensively by Boyd.[35] Not surprisingly, the 7-acylamino groups of the early cephalosporins closely resembled the side chains of the semisynthetic penicillins. However, the activity and pharmacokinetics of cephalosporins is also influenced by the nature of the substituent at C-3 and the two series soon diverged.

The contribution of the cephalosporin 3-substituent to antibacterial activity has been illustrated by Boyd and his coworkers.[89] They calculated transition state energies for opening the β-lactam ring after attack by hydroxide ion and correlated these with antibacterial activity for a series of 7-thienylacetylcephalosporins. This study and others[35] provided confirmation for the already widely held view that the activity of cephalosporins is related to the ability of the 3-methylene substituent to expel a leaving group (*q.v.* Scheme 5).

However, the increase in activity caused by the presence of a good leaving group at C-3 must be balanced against decreasing chemical stability, especially where the 7-substituent contains a nucleophilic group capable of attacking the β-lactam ring (*cf.* cephaloglycine).

9.2.7 AMIDINOPENICILLINS

The 6β-acylamino side chain is a characteristic of almost all penicillin antibiotics and attempts to replace this feature usually result in a substantial loss of activity. However, in 1972 Lund and Tybring[90] reported that 6-amidinopenicillanic acids had good activity against Gram-negative bacteria and in 1974 Petersen[91] described a series of 6-cyanoamidinopenicillins, designed as bioisosteres of more conventional compounds.

Extensive structure–activity studies[92] revealed that *N*,*N*-dialkylformamidines are very active against many Gram-negative bacteria but relatively inactive against Gram-positive species, whilst compounds with side chains derived from other carboxylic acids tend to resemble the analogous conventional penicillins, regardless of the nature of the *N*,*N*-dialkyl group. For example, the homopiperidine derived formamidine, mecillinam (**70**), is very active against a range of Gram-negative bacteria (Table 4), whereas the spectrum of the phenoxyacetamidine (**71**) is similar to that of penicillin V.

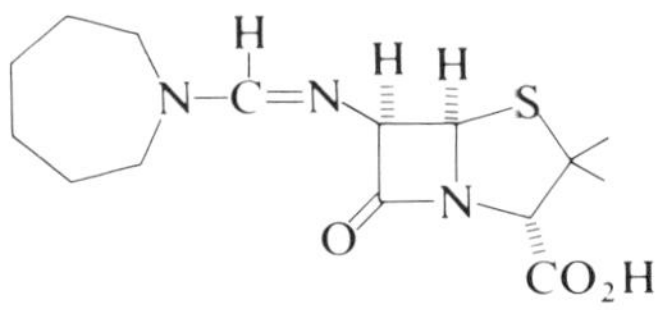

(**70**) Mecillinam 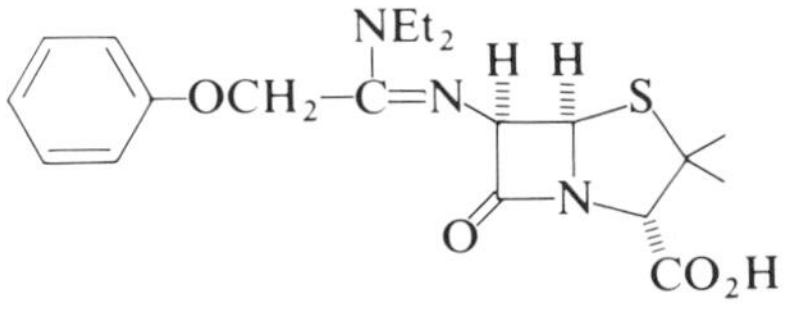   (**71**)

The striking difference between the spectrum of mecillinam and that of conventional 6-acylamino-penicillins is also seen at the target enzymes, mecillinam binding exclusively to PBP 2 of *E. coli* DC0, in contrast to penicillin G and ampicillin which bind to PBPs 1a, 2 and 3 (Table 2). The clinical use and pharmacology of mecillinam has been reviewed by Neu.[93]

Various methods for the preparation of 6β-amidinopenicillins have been described by Lund.[92] Two strategies were investigated: (i) reaction of a 6-APA derivative with a reactive derivative of a tertiary amide or (ii) treatment of a suitably activated penicillin derivative with a secondary amine (Scheme 13).

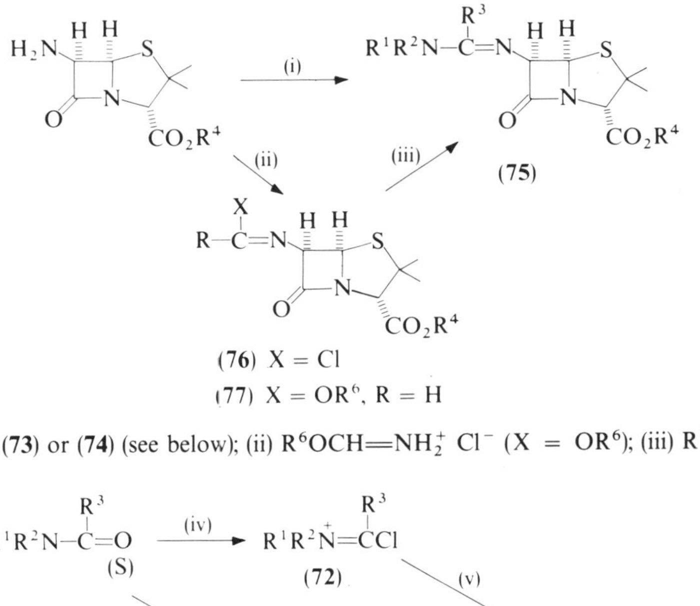

(**76**) X = Cl
(**77**) X = OR^6, R = H

(i) (**72**), (**73**) or (**74**) (see below); (ii) R^{6}OCH=NH$_2^+$ Cl$^-$ (X = OR^6); (iii) R^{1}R^{2}NH

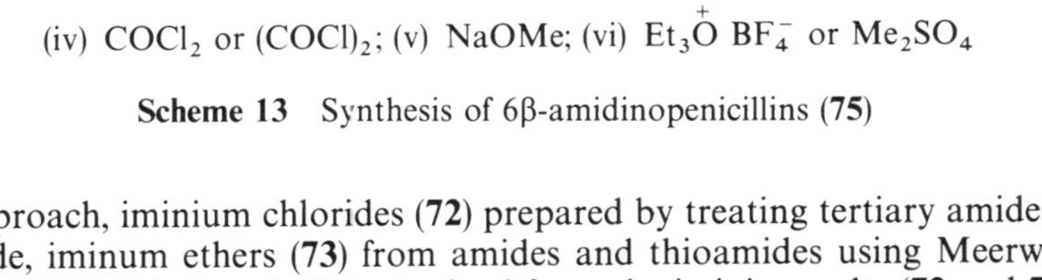

(iv) COCl$_2$ or (COCl)$_2$; (v) NaOMe; (vi) Et$_3$O$^+$ BF$_4^-$ or Me$_2$SO$_4$

Scheme 13   Synthesis of 6β-amidinopenicillins (**75**)

In the first approach, iminium chlorides (**72**) prepared by treating tertiary amides with phosgene or oxalyl chloride, iminum ethers (**73**) from amides and thioamides using Meerwein's reagent or dimethyl sulfate, and amide acetals (**74**) obtained from the iminium salts (**72** and **73**) on treatment with sodium methoxide, have all been found to react with esters of 6-APA to give the amidinopeni-cillins (**75**). The amide acetals had an advantage over the iminium salts in that they were readily purified by distillation and were the reagents of choice for the reaction with the trimethylsilyl ester of 6-APA.

Penicillin imidoyl chlorides (**76**), formed in the first step of the phosphorus pentachloride deacylation of 6β-amidopenicillins, reacted with secondary amines according to the second approach, although the reactions were accompanied by epimerization at C-6. This problem was successfully overcome by using penicillin imidic esters (**77**), prepared by treatment of a 6-APA derivative with 1,1-dichloromethyl ether or, more efficiently, with a formamidic ester hydrochloride.

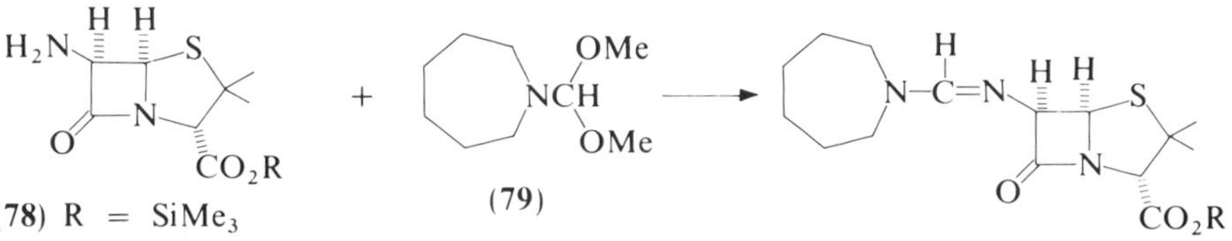

(**78**) R = SiMe$_3$

(**39**) R = CH$_2$OCOBu^t

(**79**)

(**70**) R = H

(**80**) R = CH$_2$OCOBu^t

Scheme 14 Synthesis of mecillinam (**70**) and pivmecillinam (**80**)

Thus treatment of the trimethylsilyl ester of 6-APA (**78**) with 1-hexamethyleniminecarbaldehyde dimethyl acetal (**79**) in ether afforded mecillinam (**70**; Scheme 14). Similar treatment of the pivaloyl-oxymethyl ester of 6-APA (**39**) gave the orally absorbed prodrug pivmecillinam (**80**).[90]

The high activity of amidinopenicillins such as mecillinam stimulated studies of the analogous cephalosporins, but these were found to be much less active.

9.2.8 β-LACTAMASE STABLE CEPHALOSPORINS

The early cephalosporins were successful because of their stability to staphylococcal penicillinases. However, resistant organisms soon appeared and in many cases they were found to be Gram-negative bacteria producing β-lactamases. These were often carried on plasmids, which sometimes conferred resistance to other classes of antibiotics as well.

With the exception of the 3-methyl compounds, like cephalexin, which have a relatively unreactive β-lactam ring, the first generation cephalosporins are unstable to β-lactamases produced by Gram-negative bacteria. Two strategies were used to overcome this resistance: these were to develop compounds with an intrinsic resistance to β-lactamases or to coadminister a β-lactamase inhibitor. The latter approach is dealt with in Chapter 9.3.

Cephamycin C (**11**) is a natural product produced by strains of *Streptomyces clavuligerus*; it has the same ring system and side chain as cephalosporin C but a new 3-substituent. It also carries a 7α-methoxy group and this feature confers excellent stability to β-lactamases. As before, modification of the 7-amino substituent provided compounds with improved activity and one of these, cefoxitin (**81**), has been in clinical use for a number of years.

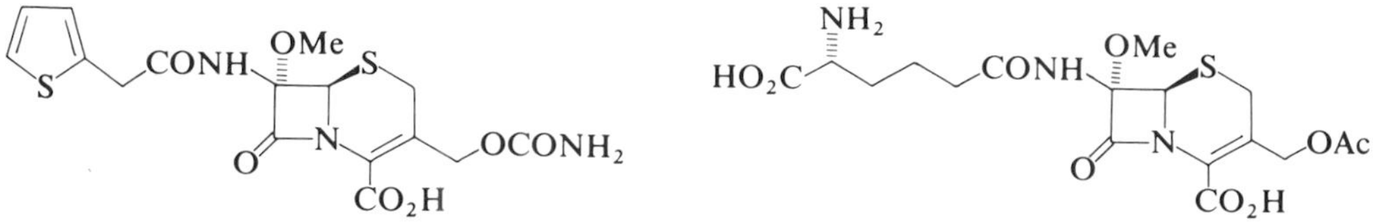

(**81**) Cefoxitin

(**82**)

The apparently straightforward task of replacing the aminoadipoyl side chain of the naturally occurring cephamycins proved unexpectedly difficult. The presence of the carbamoyl group in cephamycin C (**11**) precluded the use of the standard phosphorus pentachloride deacylation procedure and its application to the deacylation of other cephamycins, such as 7α-methoxycephalo-sporin C (**82**), resulted in extensive epimerization at C-7.[94] In addition, the 7α-methoxy-7β-aminocephalosporins proved to be less stable than the unmethoxylated congeners and their acylation was considerably more difficult owing to the reduced nucleophilicity of the amino group and increased steric hindrance.

A novel solution to these problems was devised by the Merck group,[95] who effected the replacement of the cephamycin C side chain by means of a transacylation process (Scheme 15). Cephamycin C (**11**) was protected by acylation with trichloroethyl chloroformate followed by esterification with diphenyldiazomethane. The product (**83**) was then acylated with thienylacetyl

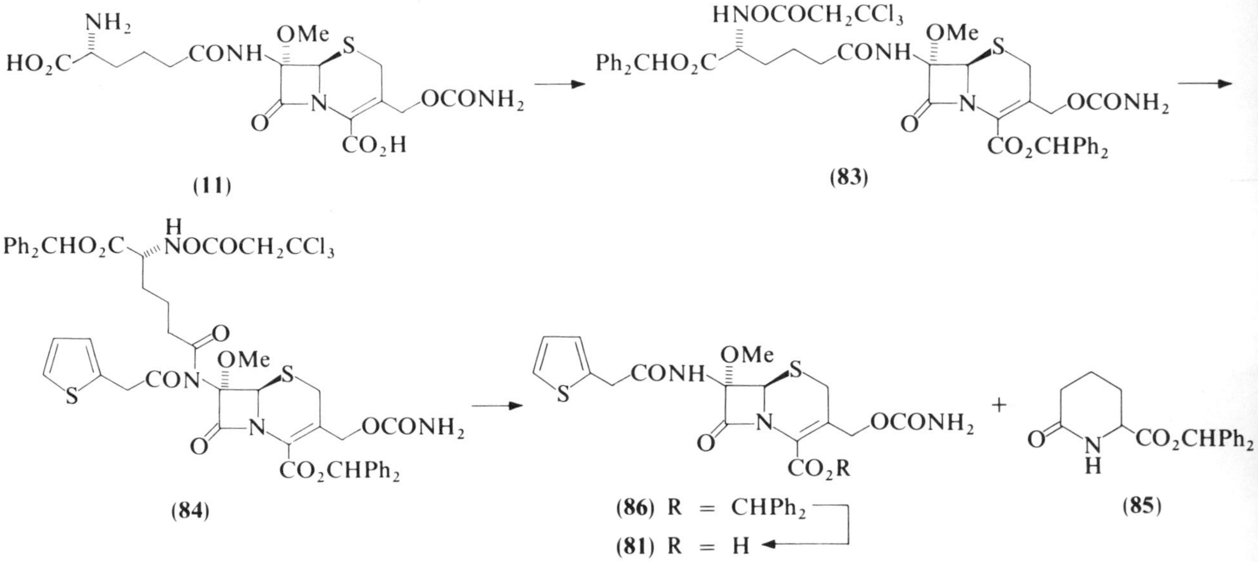

Scheme 15 Synthesis of cefoxitin (**81**) *via* transacylation of cephamycin C (**11**)

Table 6 Antibacterial Activity of β-Lactamase Stable Cephalosporins[a]

Organism	CFX	CMZ	CXM
S. aureus Pen-S	1	0.25	0.25
S. aureus Pen-R	1	0.25	0.25
St. faecalis	125	>125	>125
E. coli (DC0)	2	0.5	4
E. coli TEM-1	8	0.5	4
E. cloacae	2	2	4
E. cloacae P99	>125	62	>125
K. aerogenes	0.5	0.5	2
K. aerogenes K1	>125	0.5	>125
Pr. mirabilis	1	0.5	2
Pr. vulgaris	125	2	>125
Ps. aeruginosa	>125	>125	>125
Ps. aerug. Gen-R	>125	>125	>125
H. influenzae	<0.25	2	<0.25
H. influ. Amp-R	0.5	1	<0.25

[a] MICs in μg mL^{-1}. Cefoxitin (CFX), cefmetazole (CMZ), cefuroxime (CXM).

chloride and *N*-trimethylsilyltrifluoroacetamide in methylene chloride to give the imide (**84**). On deblocking the amino group (zinc/acetic acid) the aminoadipoyl substituent underwent an intramolecular cyclization, forming the piperidone (**85**) and thereby realising (**86**) which was converted into cefoxitin (**81**) under standard deprotection conditions.

Cefoxitin has broad-spectrum activity against many bacteria (Table 6). It is very stable to β-lactamases and the 3-carbamoyloxymethyl group is stable to esterases. In common with many cephalosporins it is active against many aneorbic bacteria, for example *Clostridia*, and its activity against *Bacteroides fragilis*, a common pathogen of abdominal wounds, is claimed to be similar to that of metronidazole.

A considerable effort has been directed towards the preparation of cephamycins by chemical modification of cephalosporins. Numerous methods have been developed for the introduction of the methoxy and other groups and these have been thoroughly reviewed.[96] Many are based on the same principle: the cephalosporin derivative is converted into a planar imine to which is added methanol, the addition occurring from the least hindered α-face thereby ensuring the correct stereochemistry.

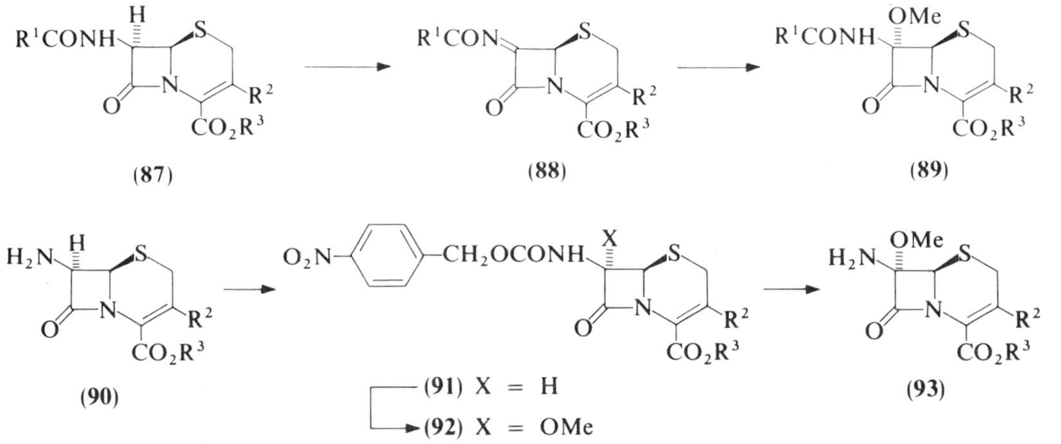

Scheme 16 Preparation of 7α-methoxycephalosporins *via* acylimines

The use of 7-acyliminocephalosporins in this respect has been successfully exploited by a number of workers, notably Koppel and Koehler[97] whose one-pot methoxylation of cephalosporins is depicted in Scheme 16. Treatment with lithium methoxide in tetrahydrofuran, followed by *t*-butyl hypochlorite, converts the cephalosporin (87) into an *N*-chloro derivative. Subsequent elimination gives the imine (88) which adds methanol, giving the 7α-methoxycephalosporin (89). The methodology has been extended to provide a facile general synthesis of 7-methoxy-7-aminocephalosporins.[98] A 7-aminocephalosporin ester (90) is converted into the *p*-nitrobenzyl carbamate (91) and thence the methoxy derivative (92). Catalytic hydrogenation followed by hydrolysis of the unstable intermediate *p*-aminobenzyl carbamate affords the amine (93).

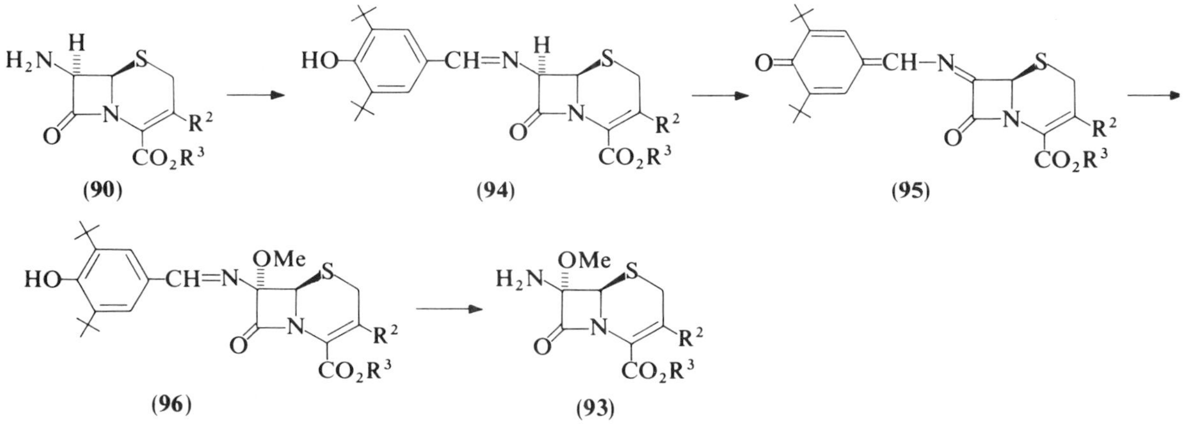

Scheme 17 Preparation of 7β-amino-7α methoxycephalosporins *via* quinoid imines

An alternative approach to 7α-methoxy-7β-aminocephalosporins, which utilizes a quinoid imine intermediate, has been developed by Sankyo (Scheme 17).[99] The 7-aminocephalosporin (90) is converted into the Schiff's base (94) which on oxidation with lead dioxide in benzene is transformed into the quinoidal derivative (95). Stereospecific addition of methanol affords the 7α-methoxy Schiff's base (96) which is converted into the amine (93) by treatment with Girard's reagent T. The application of this simple, yet versatile, procedure to the synthesis of the semisynthetic cephamycin, cefmetazole (97),[100] is shown in Scheme 18.

Studies directed towards more conventional cephalosporins provided another series of compounds with excellent stability to β-lactamases.[101] These carry a 2-aryl-2-oximinoacetamido side chain. The geometry of the oximino group is important, the *syn* isomers (98) generally being considerably more active than the *anti* isomers (99). The *syn*-oximes and their acylated derivatives generally have good activity, but the latter rapidly hydrolyze *in vivo* and thus offer no real advantage over the free oximes. In contrast, the alkoximes are usually metabolically stable and the first such compound to reach the clinic was cefuroxime (100).

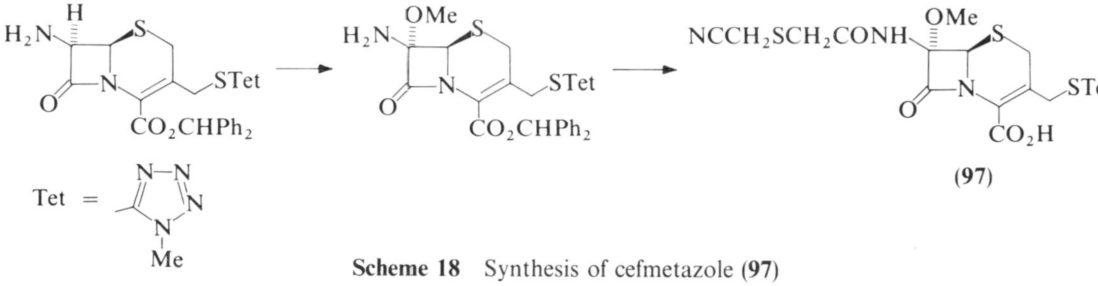

(97)

Tet =

Scheme 18 Synthesis of cefmetazole (**97**)

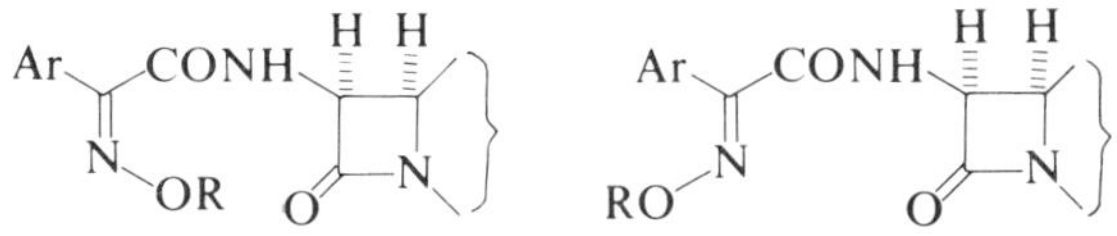

(**98**) *syn*-Oxime (**99**) *anti*-Oxime

Routes to the α-methoxyiminofurylacetamido side chain[101] and the synthesis of cefuroxime[102] developed by Glaxo workers are shown in Scheme 19. Oximation of 2-furylglyoxylic acid (**101**) with hydroxylamine at pH 4–5 and room temperature gave mainly the *syn*-oxime (**102**). The rate of reaction was accelerated considerably in the presence of magnesium hydroxide and the amount of *syn* isomer was increased through formation of the chelated tetrahedral intermediate (**103**). The hydroxyimino acid (**102**) was treated with potassium *t*-butoxide in dimethyl sulfoxide and the resulting dianion reacted with methyl iodide to give the required methoxime (**104**). Alternatively, the glyoxylic acid (**101**) was converted directly into the methoxime (**104**) by reaction with methoxylamine. Again, the *syn* isomer was the major product and any *anti* isomer was removed by recrystallization.

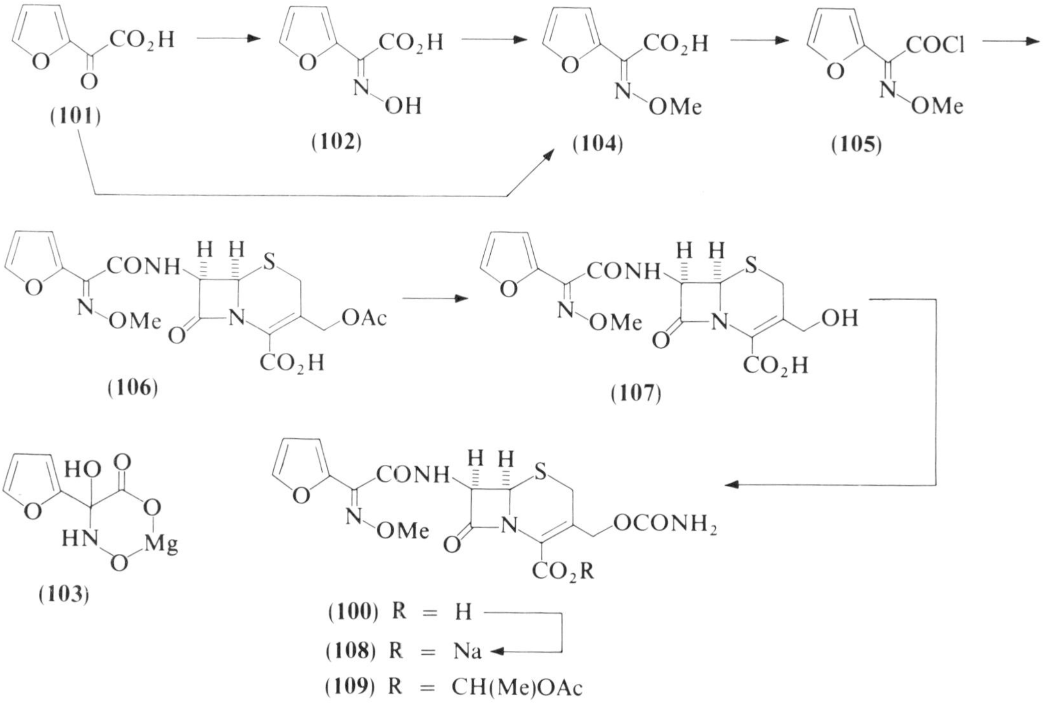

(**101**) (**102**) (**104**) (**105**) (**106**) (**107**) (**103**)

(**100**) R = H

(**108**) R = Na

(**109**) R = CH(Me)OAc

Scheme 19 Synthesis of cefuroxime (**100**)

Treatment of the sodium or triethylammonium salt of the acid (**104**) with oxalyl chloride gave the acid chloride (**105**),[101] which was reacted with 7-ACA in methylene chloride in the presence of triethylamine to give the cephalosporin (**106**).[102] The acetoxymethyl group was hydrolyzed with yeast and the resulting hydroxymethylcephalosporin (**107**) was reacted with either trichloroacetyl isocyanate, chlorosulfonyl isocyanate or dichlorophosphinyl isocyanate, followed by an aqueous work-up,[103] to give cefuroxime (**100**) which was obtained as its sodium salt (**108**) upon treatment with sodium 2-ethylhexanoate.

Cefuroxime, which is very stable to β-lactamases, metabolically stable and remarkably well tolerated, has been in widespread clinical use since 1978.[104] Its efficacy in relation to other second-generation cephalosporins has been reviewed by Tartaglione and Polk.[105] It is not absorbed from the GI tract and must be given by injection. An orally absorbed ester, cefuroxime axetil (**109**),[106] was approved for clinical use in 1987.

9.2.9 ANTIPSEUDOMONAL PENICILLINS

Ps. aeruginosa and other opportunistic pathogens tend to infect patients with poor immune function. Aminoglycosides such as gentamicin are very active against *Pseudomonas* but are relatively toxic, limiting their use against resistant bacteria.

Carbenicillin (**29**) and the closely related ticarcillin (**110**) are much less active against *Pseudomonas* than gentamicin but have low toxicity; they have been widely used at doses of up to 40 g per day given by intravenous infusion. Nevertheless, the modest activity of these compounds against *Pseudomonas* and other intractable Gram-negative bacteria stimulated further research in the penicillin field.

Alkylation or acylation of the amino group of ampicillin generally reduces the antibacterial activity against Gram-negative bacteria. However, several groups discovered that reaction of ampicillin with acyl isocyanates afforded ureido derivatives with improved activity against some Gram-negative species. Further investigations revealed that acylureido derivatives were very active against a wide range of Gram-negative bacteria, including *Ps. aeruginosa*.[107] Pyridone analogues of the acylureidopenicillins have also been described.[108,109] A large number of compounds were prepared and several, including azlocillin (**111**), the closely related mezlocillin (**112**), piperacillin (**113**) and the hydroxynaphthyridine apalcillin (**114**), have reached the market. The antibacterial (Table 7) and pharmacokinetic properties of these and some related compounds have been summarized by Webber and Wheeler[80] and the pharmacokinetics and therapeutic use of piperacillin have been reviewed in detail by Holmes *et al.*[110]

(**110**) X = H   Ticarcillin
(**115**) X = OMe Temocillin

(**111**) R = HN⟨ ⟩N—   Azlocillin

(**112**) R = MeSO₂N⟨ ⟩N—   Mezlocillin

(**113**) R = Et—N⟨ ⟩N—   Piperacillin

(**114**) R =   Apalcillin

The 2,3-diketopiperazine function of piperacillin is now found in a number of very active compounds, including some cephalosporins.[111] Piperacillin analogues with different 2,3-diketo-piperazine N-4 substituents have been studied.[112] Their activity was found to increase with the length of the N-4 alkyl chain; this effect was a result of increased binding to the target enzymes, especially PBP 3.

The acylureidopenicillins are generally active against Gram-positive bacteria, excluding penicillinase producing strains. They are very active against Gram-negative bacteria, including

Table 7   Antibacterial Activity of Antipseudomonal Penicillins[a]

Organism	Carb	Ticar	Azlo	Mezlo	Piper	Temo	Form
S. aureus Pen-S	0.5	0.5	<0.13	>125	<0.13	>125	>125
S. aureus Pen-R	16	16	250	>125	62	>125	>125
St. faecalis	62	8	8	8	4	>125	>125
E. coli (DC0)	8	4	8	2	1	8	<0.06
E. coli TEM-1	125	>125	>125	>125	>125	8	0.13
E. cloacae	4	4	8	2	2	4	1
E. cloacae P99	31	>125	>125	62	125	8	8
K. aerogenes	4	8	8	4	2	4	0.13
K. aerogenes K1	>125	>125	>125	>125	>125	8	0.13
Pr. mirabilis	0.5	0.5	2	0.5	0.5	1	0.5
Pr. vulgaris	2	8	16	8	4	1	0.5
Ps. aeruginosa	62	31	1	16	1	>125	1
Ps. aerug. Gen-R	>125	>125	31	>125	4	>125	4
H. influenzae	2	<0.25	<0.5	<0.25	<0.5	<0.25	0.5
H. influ. Amp-R	2	0.5	<0.5	<0.25	0.5	<0.25	0.5

[a] MICs in $\mu g\ mL^{-1}$. Carbenicillin (Carb), ticarcillin (Ticar), azlocillin (Azlo), mezlocillin (Mezlo), piperacillin (Piper), temocillin (Temo), formidacillin (Form).

Ps. aeruginosa and the indole-positive *Proteus vulgaris*, but like other conventional penicillins are rapidly hydrolyzed by Gram-negative β-lactamases.

In general, the ureidopenicillins may be synthesized by acylating the α-amino group of penicillins, such as ampicillin (**30**) or amoxycillin (**44**), with carbonyl chlorides derived from amides and phosgene.[107] The synthesis of azlocillin (**111**) is a typical example (Scheme 20).

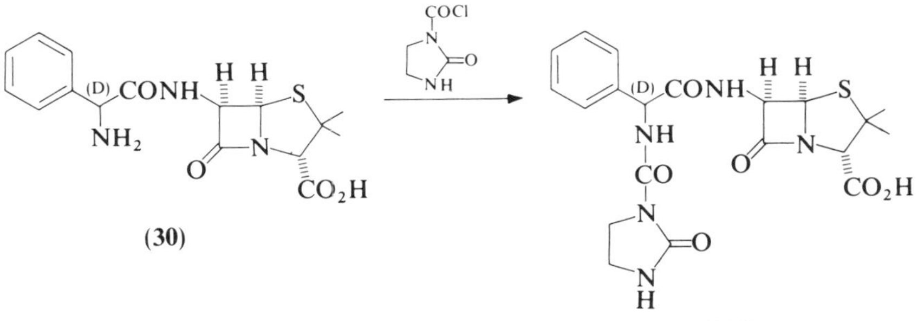

Scheme 20   Synthesis of azlocillin (**111**)

Simultaneously with the investigation of the cephamycin analogues, several groups sought to improve the β-lactamase stability of penicillins by introduction of the methoxy group at C-6, utilizing reactions similar to those described in Section 9.2.8.[96] In general the modified penicillins showed poorer levels of antibacterial activity than the parent compounds but workers at Beecham found that temocillin (**115**), the 6α-methoxy analogue of ticarcillin, had good activity against the *Enterobacteriacae* and was very stable to β-lactamases.[113] It has no useful activity against *Pseudomonas*.

The poor level of activity exhibited by most 6α-methoxyureidopenicillins prompted the Beecham group to investigate other 6α-substituents which might provide increased β-lactamase stability whilst retaining the original levels of antibacterial activity.[114] Amongst the large number of groups examined the 6α-formamido group was found to confer the right combination of stability and activity. The key intermediates for the synthesis of these modified penicillins were the 6α-methylthio derivatives,[96] which readily underwent nucleophilic displacement in dimethylformamide in the presence of mercury(II) acetate (Scheme 21). Under these conditions, (**116**) reacted with dry ammonia and the resulting 6α-amino derivative (**117**) was treated with formic acetic anhydride to give the 6α-formamidopenicillin (**118**) which on deprotection gave formidacillin (BRL 36650; **119**).[114]

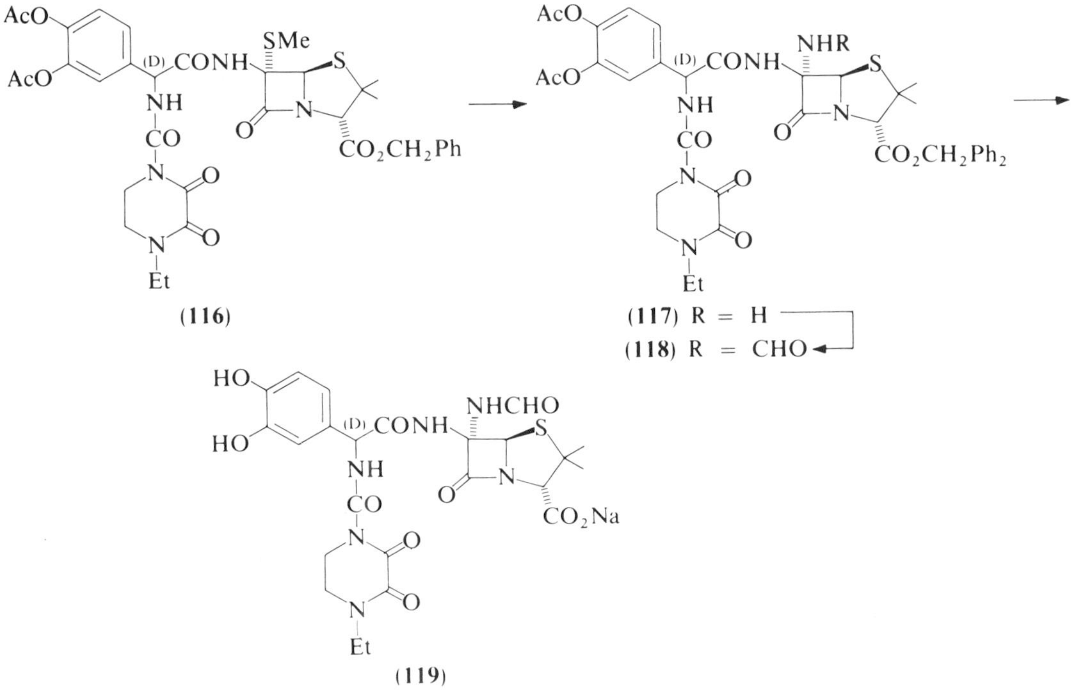

(116)

(117) R = H
(118) R = CHO

(119)

Scheme 21 Synthesis of formidacillin **(119)**

Formidacillin has poor activity against Gram-positive organisms but its activity against Gram-negative bacteria is comparable with that of ceftazidime (*cf.* Section 9.2.11). A series of cephalosporin analogues of formidacillin has been described.[115,116]

9.2.10 THIRD GENERATION CEPHALOSPORINS

Although the compounds described in the preceding sections provided cover for the majority of infections not caused by β-lactamase producing strains of the more intractable bacteria, no single agent possessed the breadth of spectrum of the aminoglycosides.

However, in 1977 Roussel-UCLAF announced the first of a new series of cephalosporins which exhibited most of the properties desirable in a compound to be used for empiric therapy. This was cefotaxime **(120)**, a methoximinocephalosporin bearing a 2-aminothiazol-4-yl group. Cefotaxime combines excellent activity against the *Enterobacteriacae* (*E. coli*, *Enterobacter*, *Klebsiella* and *Proteus* spp. *etc.*) with acceptable activity against Gram-positive organisms.[117] It has moderate activity against *Ps. aeruginosa* (Table 8). The binding of cefotaxime to the PBPs of *E. coli* (Table 2) and *Ps. aeruginosa* has been studied by several groups.[37,118,119] In common with other penicillins and cephalosporins, its activity against methicillin-resistant *Staphylococci* is relatively poor.

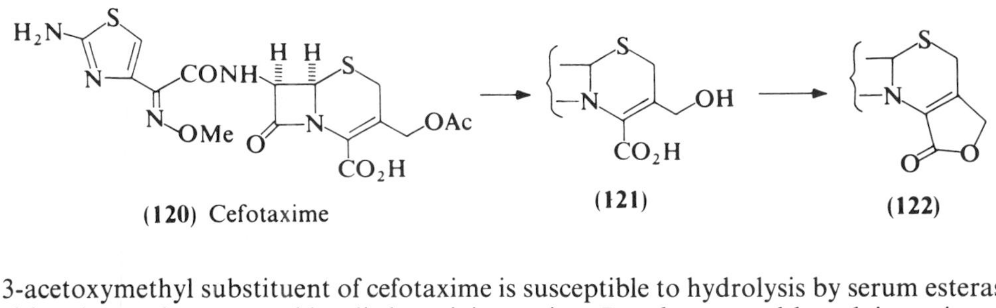

(120) Cefotaxime

(121)

(122)

The 3-acetoxymethyl substituent of cefotaxime is susceptible to hydrolysis by serum esterases and the resulting deacetyl compound has little activity against *Pseudomonas*, although it retains much of the activity of the parent compound against other bacteria. Hydrolysis of the acetoxy group *in vivo* is confirmed by urinary recovery experiments. Thus after a 1 g intramuscular dose the urine contained

Table 8 Antibacterial Activity of Third Generation Cephalosporins[a]

Organism	CTM	CTX	CMN	CTZ	CTR	CEF	MOX	CTT	CPZ
S. aureus Pen-S	1	1	0.5	1	1	1	4	4	8
S. aureus Pen-R	1	1	0.5	1	1	1	4	4	8
St. faecalis	>62	125	250	>62	>31	>125	125	>125	16
E. coli (DC0)	0.13	<0.06	<0.13	<0.03	0.06	0.13	<0.13	<0.06	<0.03
E. coli TEM-1	0.5	<0.06	<0.13	<0.03	0.06	0.25	<0.13	<0.06	16
E. cloacae	0.5	<0.06	<0.13	<0.03	0.13	0.13	<0.13	0.25	0.13
E. cloacae P99	>62	125	62	>62	>31	>125	16	62	>62
K. aerogenes	0.25	<0.06	<0.13	<0.03	0.06	0.13	<0.13	<0.06	0.25
K. aerogenes K1	>62	4	16	0.25	16	62	<0.13	<0.06	>62
Pr. mirabilis	1	<0.06	<0.13	<0.03	0.01	<0.06	<0.13	<0.06	1
Pr. vulgaris	>62	0.13	<0.13	<0.03	0.01	<0.06	<0.13	<0.06	16
Ps. aeruginosa	>62	16	16	16	8	16	8	125	4
Ps. aerug. Gen-R	>62	125	125	>62	>31	62	125	>125	31
H. influenzae	0.25	<0.06	<0.13	<0.13	0.01	<0.25	<0.13	0.5	<0.13
H. influ. Amp-R	0.25	<0.06	<0.13	<0.13	0.01	<0.25	<0.13	0.5	<0.13

[a] MICs in $\mu g\,mL^{-1}$. Cefotiam (CTM), cefotaxime (CTX), cefmenoxime (CMN), ceftizoxime (CTZ), ceftriaxone (CTR), cefodizime (CEF), moxalactam (MOX), cefotetan (CTT), cefoperazone (CPZ).

49% of unchanged cefotaxime and 37% of a mixture of the deacetyl metabolite (**121**) and the poorly active lactone (**122**).[120]

This metabolism limits the use of cefotaxime for the treatment of *Pseudomonas* infections but has been claimed to exert a beneficial effect against some bacteria because of synergy between the parent compound and the deacetyl metabolite.[121] As with cefuroxime, the *anti* isomer of cefotaxime exhibits reduced activity and a lower affinity for PBPs 1a, 1b and 3 of *E. coli* K12 relative to that of the *syn* isomer.[118]

Cefotaxime was swiftly followed by a number of very similar methoximes with different 3-substituents. The antibacterial activities of cefotaxime (**120**), ceftizoxime (**123**), cefmenoxime (**124**) and ceftriaxone (**125**) have been compared with those of other β-lactam antibiotics.[122] The methoximes had very similar antibacterial activity (Table 8) but exhibited different pharmacokinetic profiles.[123,124]

Ceftriaxone is unusual in that it is 95% serum bound and exhibits an extended serum profile (SE $t_{1/2}$ 7.3 h), making it suitable for once or twice daily administration.[125] The advantage of a long serum elimination time may be offset to some extent by the deleterious effect of the high protein binding on its antibacterial activity.[50,125] In severe renal failure the binding of ceftriaxone to serum proteins is reduced by competitive binding of 2-hydroxybenzoylglycine, a compound which accumulates in the serum of uremic patients.[127] A substantial proportion of the dose is eliminated in the bile and GI tract side effects have been reported.

The chemistry and biology of the aminothiazole cephalosporins and related compounds have been extensively reviewed by Dürckheimer[128,129] and others.[130-134]

The preparation of methoxyiminoaminothiazolylcephalosporins is best illustrated by the work of Bucourt and coworkers (Scheme 22).[135] Oximation of ethyl acetoacetate with sodium nitrite in acetic acid followed by chlorination with sulfuryl chloride gave ethyl 4-chloro-2-hydroxyimino-3-oxobutyrate (**126**) which, on treatment with thiourea for one hour in aqueous ethanol, gave the *syn*-hydroxyiminoaminothiazole (**127**). Tritylation followed by methylation with dimethyl sulfate gave (**128**) which was saponified to give the *N*-protected methoxyiminoaminothiazole side chain (**129**).

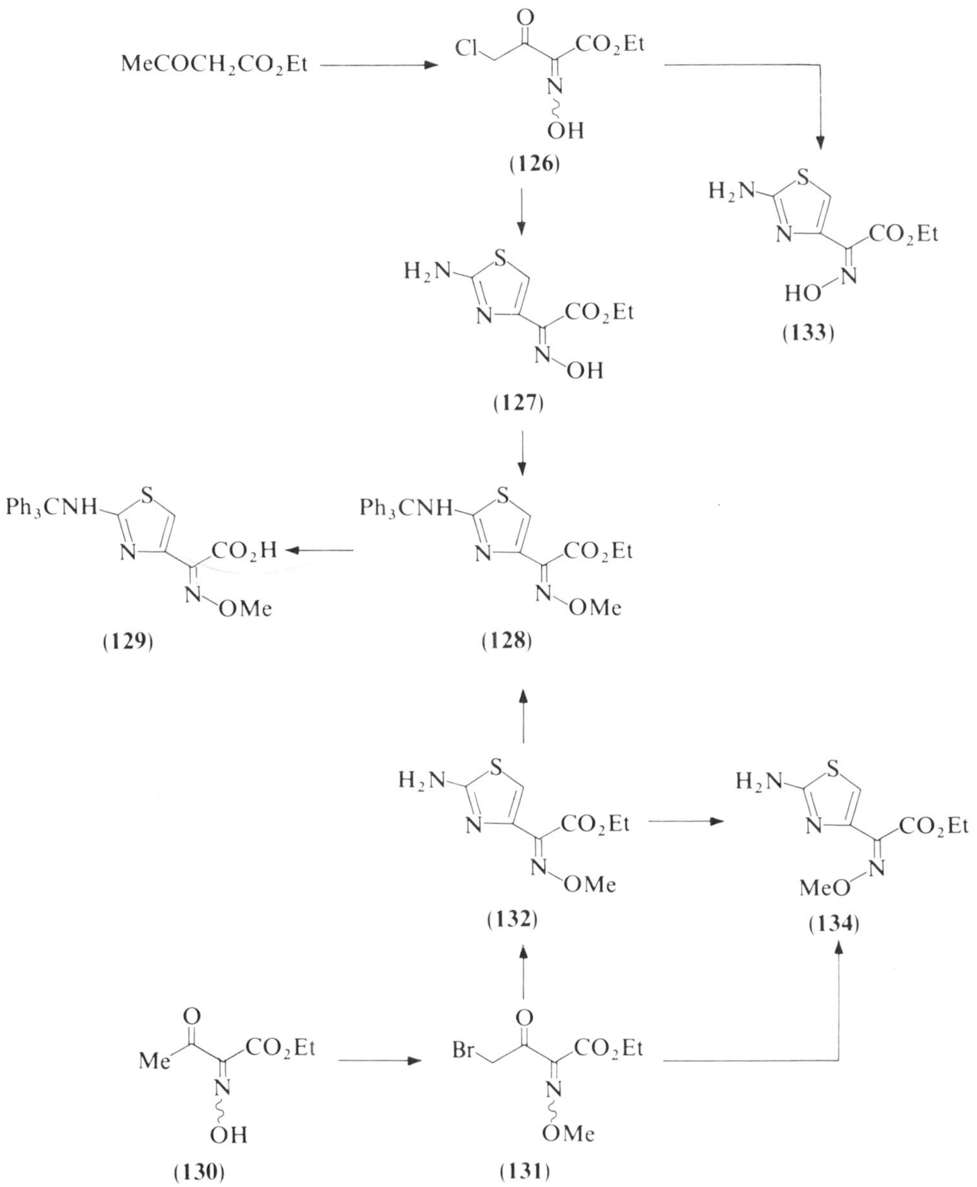

Scheme 22 Preparation of the methoxyiminoaminothiazolyl side chain (129)

Alternatively, *O*-methylation of the oxime (130), obtained from ethyl acetoacetate, followed by bromination gave ethyl 4-bromo-2-methoxyimino-3-oxobutyrate (131). Subsequent treatment with thiourea in dry ethanol at room temperature gave the *syn*-methoxime (132) which was tritylated to give (128).

Under the conditions described above the major product from the cyclization reactions with thiourea are those in which the oxime group has the *syn* configuration. However, on extending the reaction time or on increasing the reaction temperature the geometry of the oxime is reversed and the *anti* isomers (133) and (134) are formed. Thus the cyclization appears to be equilibrium controlled and the *anti* isomer the thermodynamically favoured product. In addition, isomerization of the *syn* to the *anti* isomer is catalyzed by the acid produced in the reaction and is suppressed by the addition of base; the *syn* isomer (132) may be converted into the *anti* isomer (134) on treatment with ethanolic hydrogen bromide.

It is possible to distinguish between the *syn* and *anti* isomers by NMR spectroscopy as the proton on the 5-position of the aminothiazole ring usually resonates at higher field for the *syn* isomer. However, as a means of identification this method is only reliable when both isomers are available for comparison.

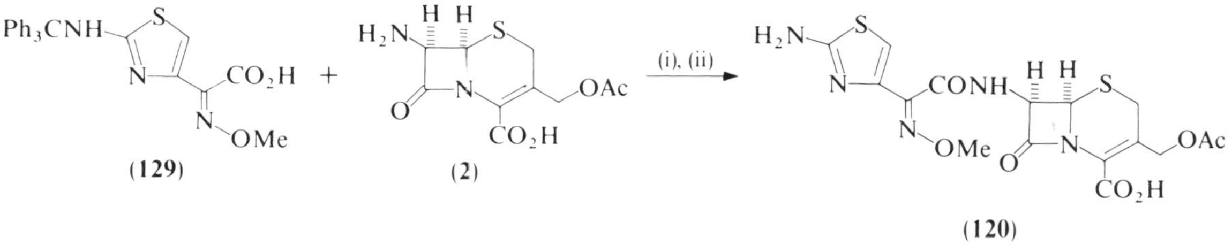

(i) Mixed anhydride acylation; (ii) 50% aq. HCO₂H

Scheme 23   Conversion of 7-ACA into cefotaxime (**120**)

In general, the synthesis of methoxyiminoaminothiazolylcephalosporins is completed by acylating the appropriate 7-aminocephalosporin with the *N*-protected side-chain, *e.g.* (**129**), followed by deprotection. In the case of cefotaxime (**120**) this was effected by acylation of 7-ACA followed by deprotection with mild acid (Scheme 23). However, workers at Takeda have developed a methodology for forming the aminothiazole ring in the final stage of the synthesis—an approach analogous to their synthesis of cefotiam (*vide infra*)—and have applied it to the preparation of radio-labelled cefmenoxime (**124**; Scheme 24).[136] The 4-chloro-3-oxobutyrylcephalosporin (**135**) was oximated with sodium nitrite and the resulting oxime (**136**) was methylated with dimethyl sulfate and then esterified with diphenyldiazomethane to give (**137**). Treatment of (**137**) with radio-labelled thiourea followed by deprotection with trifluoroacetic acid gave cefmenoxime (**124**) labelled at the aminothiazole C-2 position.

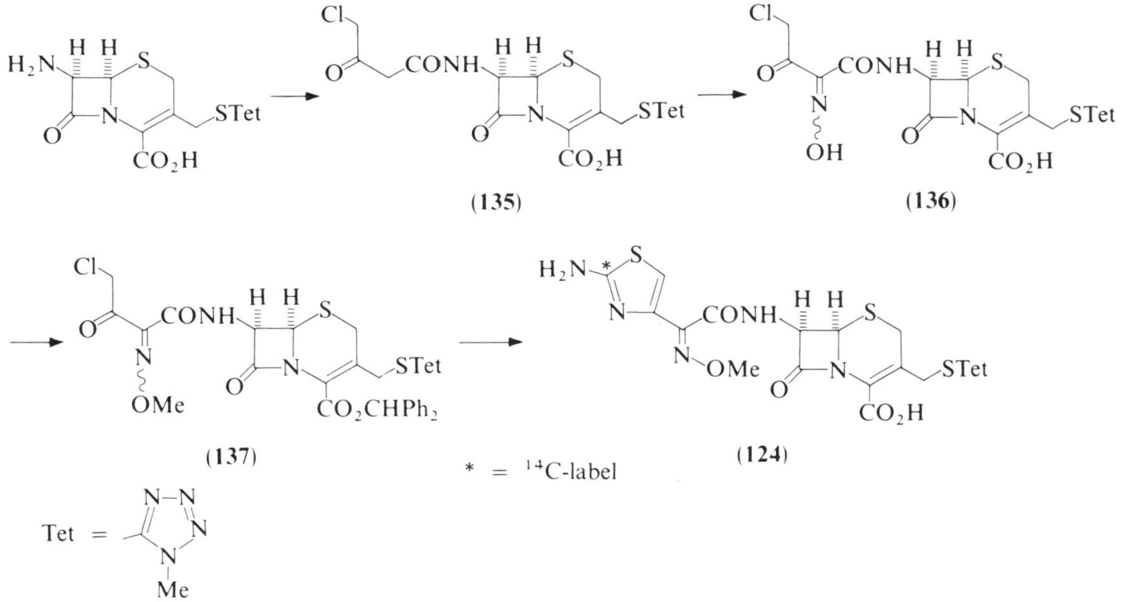

Scheme 24   Synthesis of radio-labelled cefmenoxime (**124**)

In principle, many of the methoxyiminoaminothiazolylcephalosporins might be prepared from cefotaxime by nucleophilic displacement (*cf.* cephalothin, Section 9.2.6). However, in practice it is usually preferred to synthesize the appropriate 7-amino compound from 7-ACA prior to the acylation step in order to provide a more convergent synthesis. The preparation of ceftriaxone (**125**; Scheme 25)[137] illustrates this approach. 3-Mercapto-2-methyl-1,2,4-triazinedione (**138**), prepared in two steps from methylhydrazine as shown, was condensed with 7-ACA to give the 3-substituted 7-aminocephalosporanic acid (**139**). Acylation with the *N*-protected side-chain acyl chloride followed by deprotection with thiourea afforded ceftriaxone (**125**).

The synthesis of the 7-aminocephalosporin (**140**)[138] required for the synthesis of ceftizoxime (**123**)[139] was a less straightforward task (Scheme 26). The key intermediate was the 3-hydroxy-cephalosporin (**141**) prepared by ozonolysis of the 3-exomethylene compound (**142**; see Section

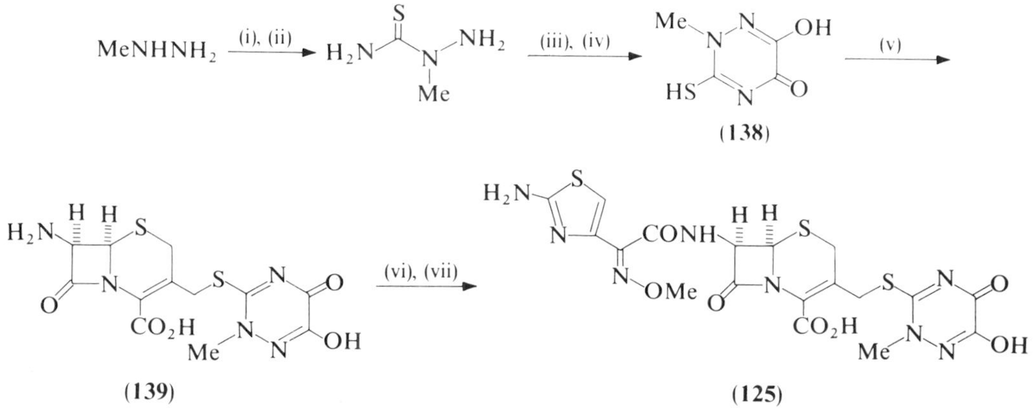

(i) KSCN, H$_2$O; (ii) toluene, heat; (iii) (CO$_2$Me)$_2$/NaOMe, MeOH; (iv) HCl; (v) 7-ACA, H$_2$O; (vi) acylation with acid chloride of *N*-chloroacetyl protected side-chain acid; (vii) H$_2$NCSNH$_2$

Scheme 25 Synthesis of ceftriaxone (**125**)

9.2.12). Reduction of the enol (**141**) with sodium borohydride in aqueous methanol gave the 3-hydroxycepham (**143**). Acetylation with acetic anhydride and pyridine followed by elimination of acetic acid with triethylamine then gave the 3-unsubstituted cephem (**144**), from which the required 7-amino compound (**140**) was obtained by the standard phosphorus pentachloride deacylation procedure.[138]

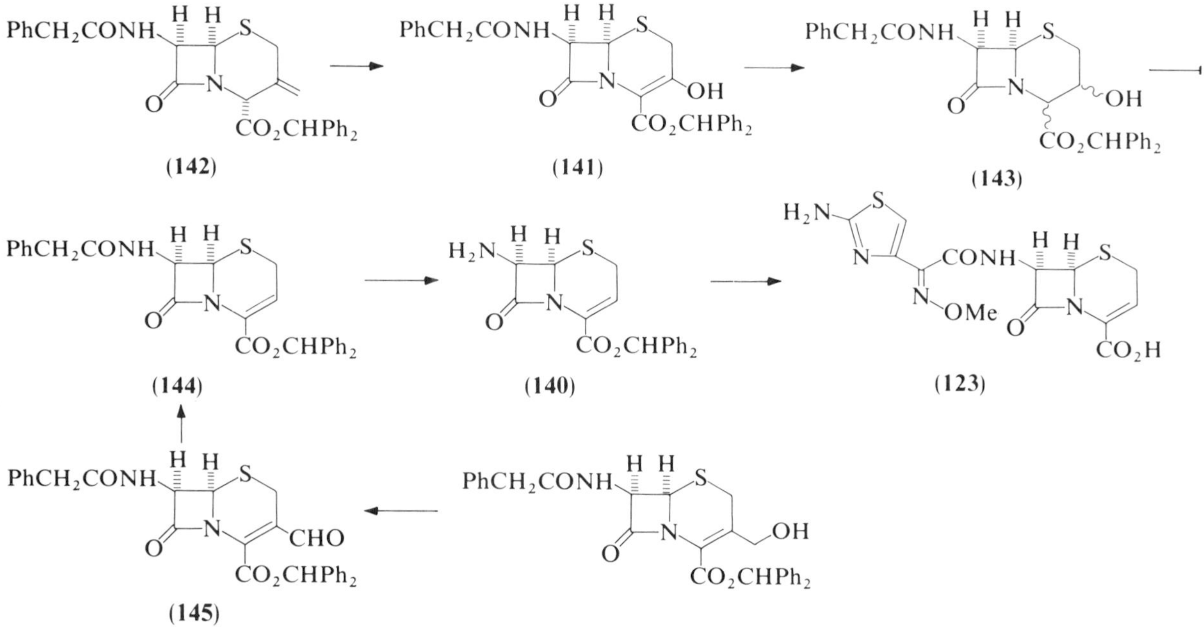

Scheme 26 Synthesis of ceftizoxime (**123**)

An alternative procedure for 3-unsubstituted cephalosporins involves the decarbonylation of 3-formylcephalosporins, *e.g.* (**145**) prepared by oxidation of the corresponding 3-hydroxymethyl compounds, with tris(triphenylphosphine)rhodium chloride.[140]

Not all the broad-spectrum cephalosporins investigated in the late 1970s carry the oxime substituent. Indeed, cefotiam (**146**), a direct precursor of cefmenoxime, bears an unsubstituted aminothiazolylacetamido group. Although it exhibits good antibacterial activity, its spectrum is limited by instability to β-lactamases. Scheme 27 shows the synthesis of cefotiam developed by the Takeda group.[141] 7-ACA was acylated with 4-chloro-3-oxobutyryl chloride to give (**147**). Treatment of (**147**) with thiourea, in the presence of base, gave the aminothiazole (**148**) which was converted into

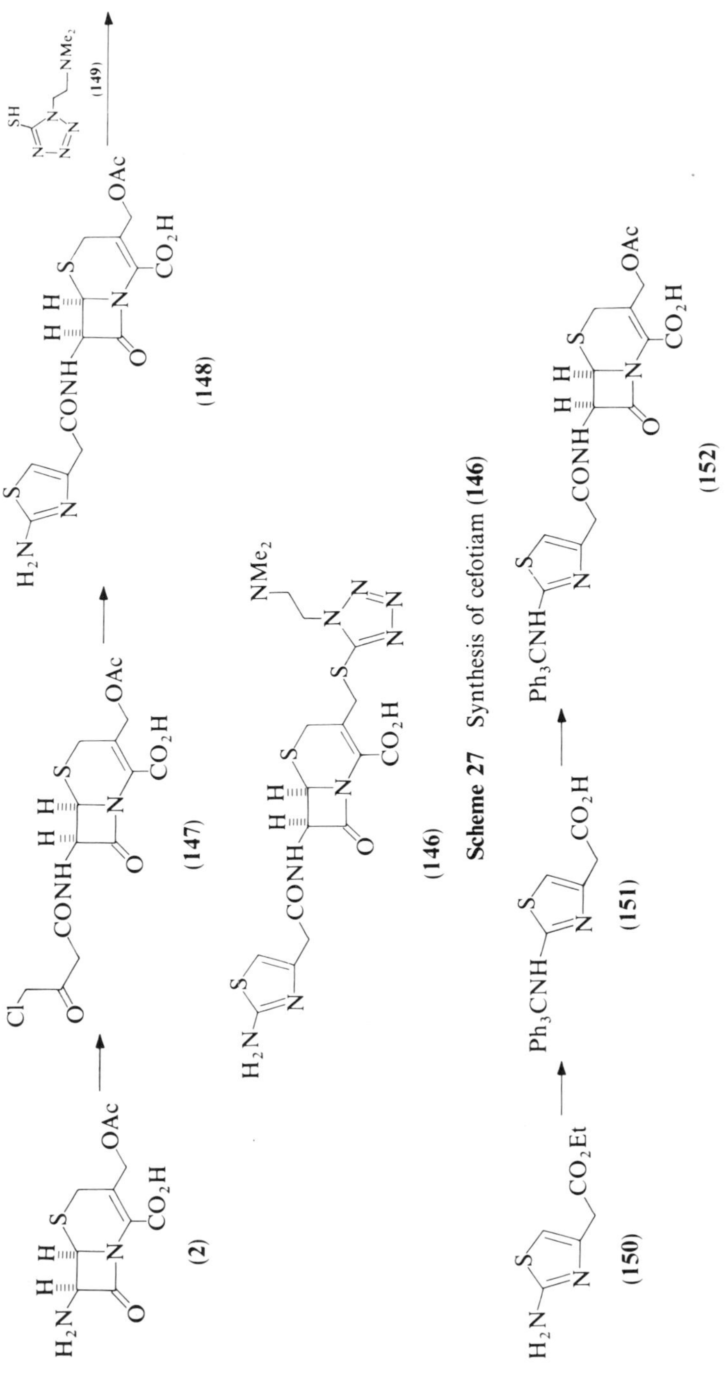

Scheme 27 Synthesis of cefotiam (146)

cefotiam by displacement of the acetoxy group with *N*-(2-dimethylaminoethyl)-1*H*-tetrazole-5-thiol (**149**).

A more orthodox approach to the 7-substituent has been reported by Bucourt and coworkers.[135] Thus, ethyl 2-amino-4-thiazolylacetate (**150**)[142] was tritylated and then saponified to give the *N*-protected side-chain acid (**151**). Acylation of 7-ACA with (**151**), using dicyclohexylcarbodiimide or a mixed anhydride formed by reacting (**151**) with ethyl chloroformate, gave (**152**).

Cefmenoxime (**124**) was developed from cefotiam in a programme in which α-amino substituted derivatives of the latter were being investigated.[143]

The excellent β-lactamase stability of the cephamycins led to the investigation of the analogous 1-oxa compounds by the Shionogi group. Racemic 1-oxacephalothin (**153**) and 1-oxacefamandole (**154**) had first been prepared at Merck[144,145] and their activities were found to be slightly better than the corresponding cephalosporins. The Beecham group developed an optically active synthesis of mandelamidooxacephalosporins such as (**154**), starting from penicillin V, and adjudged them to be similar in activity to cephalexin.[146,147] Optically active 1-oxacephalothin was synthesized by workers at Shionogi and, in contrast to the findings of the Merck group, was found to be 4–8 times more active than cephalothin.[148]

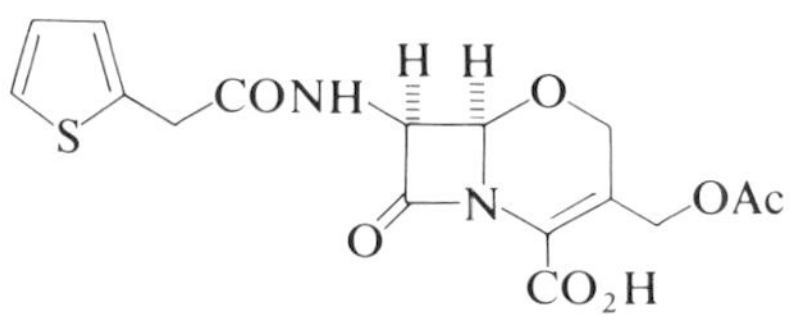

(**153**) Oxacephalothin 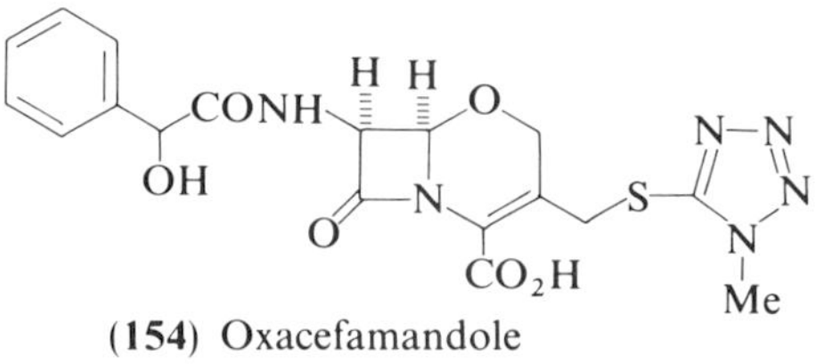 (**154**) Oxacefamandole

Extensive structure–activity studies of oxacephems resulted in the development of the commercial compound, moxalactam (**155**).[148-150]

A number of stereoselective syntheses of (**155**) have been developed: Scheme 28 shows a particularly efficient synthesis which utilizes all of the carbon atoms of the starting material, 6-APA, and

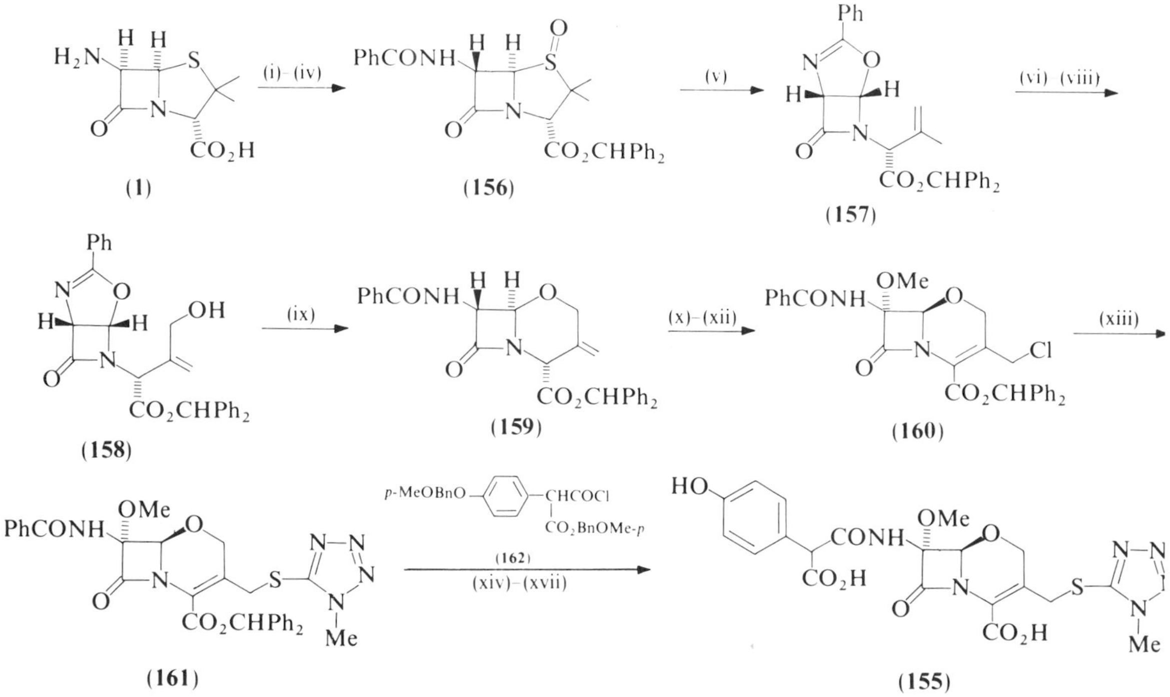

(i) PhCOCl/NaHCO$_3$; (ii) peroxide; (iii) Ph$_2$CN$_2$; (iv) bis(trimethylsilyl)acetamide, 1,5-diazobicyclo[4.3.0]-non-5-ene (DBN); (v) PPh$_3$; (vi) Cl$_2$; (vii) NaI; (viii) Cu$_2$O, H$_2$O, DMSO; (ix) BF$_3$·Et$_2$O; (x) Cl$_2$/*hv*; (xi) DBN; (xii) Bu^{t}OCl, LiOMe; (xiii) *N*-methyltetrazolylthiolate; (xiv) PCl$_5$/pyridine; (xv) MeOH then Et$_2$NH; (xvi) (**162**); (xvii) AlCl$_3$/PhOMe

Scheme 28 Conversion of 6-APA into moxalactam (**155**)

which allows a wide range of variation at both the 3- and the 7-positions.[151] 6-APA was converted into (156) in four steps, including epimerization at C-6. Treatment of (156) with triphenylphosphine opened the thiazolidine ring, forming the oxazoline (157). Allylic chlorination followed by halogen exchange and hydrolysis with copper(II) oxide in aqueous dimethyl sulfoxide gave the allylic alcohol (158) which was stereoselectively cyclized with boron trifluoride etherate. Photochemical chlorination of the product (159) followed by elimination and methoxylation at the C-7α position afforded (160) and subsequent reaction with sodium 1-methyltetrazole-5-thiolate gave (161). Finally, conversion of (161) into moxalactam was effected by removal of the acyl group with phosphorus pentachloride and reacylation with an appropriately protected side-chain acyl chloride (162), followed by a novel deprotection using aluminum trichloride in anisole.

Moxalactam is highly stable to β-lactamases and exhibits excellent broad-spectrum activity. Unfortunately, it possesses both structural features likely to give rise to hematological side effects and episodes of bleeding have occasionally accompanied its use.[152]

More conventional studies of 7α-methoxycephalosporins afforded cefotetan (163), the unusual dithietane tautomer of an isothiazolyl (164) ring system.[153] It has good activity against Gram-negative bacteria apart from *Pseudomonas* and in common with other 7α-methoxycephalosporins is very stable to β-lactamases.[154]

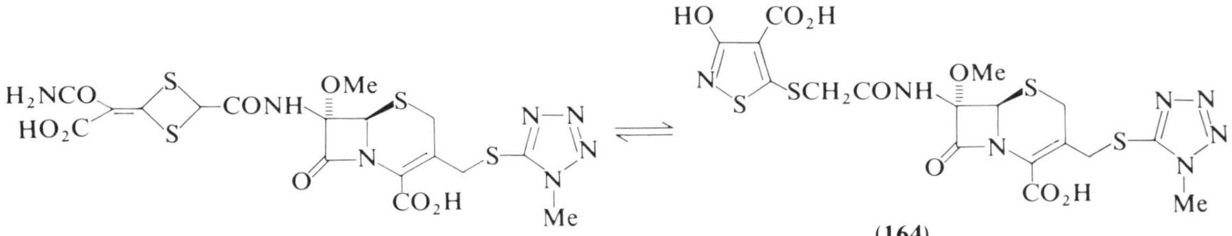

(163) Cefotetan

(164)

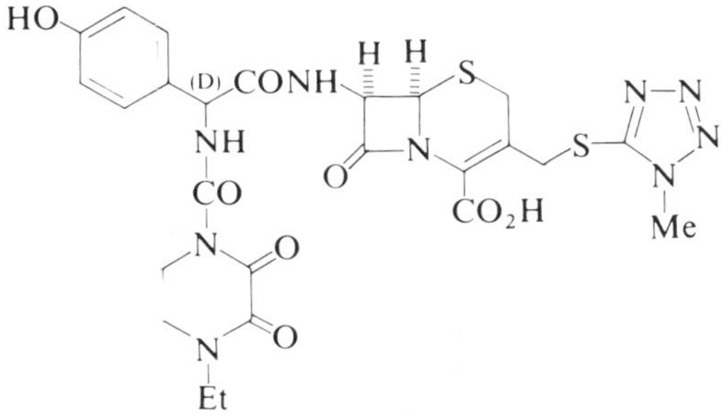

(165) Cefoperazone

Exploitation of ureido side chains similar to those developed for the broad-spectrum penicillins afforded a series of highly active cephalosporins, of which cefoperazone (165) affords the best known example.[111,155] In common with other cephalosporins, cefoperazone is stable to staphylococcal penicillinase but it is generally susceptible to the β-lactamases produced by Gram-negative organisms and this limits its usefulness. Cefoperazone carries the 1-methyltetrazolylthio group and thus may give rise to hypoprothrombinemia. It is 90% serum bound, which may account for some treatment failures.[50] A substantial proportion of the dose (19–36%) is excreted in the bile and GI side effects have been reported.

Corresponding 7α-methoxy compounds are claimed to possess good activity and β-lactamase stability but do not appear to have been developed for the market.[96] The Beecham group has described a series of 7α-formamidocephalosporins with very high activity against Gram-negative bacteria.[116] These compounds are related to the chitinovorins[156] and cephabacins,[157] naturally occurring 7α-formamidocephalosporins with complex 3-substituents, disclosed soon after the discovery and exploitation of the 7α-formamido substituent by Beecham. This is a remarkable example of simultaneous discovery by independent groups of medicinal and natural product chemists.

9.2.11 ANTIPSEUDOMONAL CEPHALOSPORINS

With the exception of the narrow-spectrum antipseudomonal cefsulodin (166),[158] the early cephalosporins had no useful activity against *Pseudomonas aeruginosa*. The aminothiazolyl oximes

Table 9 Antibacterial Activity of Antipseudomonal Cephalosporins[a,b]

Organism	CFS	CAZ	CPIR	CPIM	GR	Genta
S. aureus Pen-S	4	4	0.5	1	0.13	0.13
S. aureus Pen-R	4	8	1	1	0.25	0.13
St. faecalis	>125	>125	>125	>125	62	31
E. coli (DC0)	31	<0.06	<0.06	<0.06	0.13	0.25
E. coli TEM-1	125	0.13	<0.06	<0.06	0.25	0.25
E. cloacae	62	<0.06	<0.06	<0.06	0.5	0.25
E. cloacae P99	>125	31	2	2	16	0.25
K. aerogenes	62	<0.06	<0.06	<0.06	0.13	0.25
K. aerogenes K1	>125	0.13	2	1	4	0.25
Pr. mirabilis	62	<0.06	0.13	<0.06	<0.06	0.13
Pr. vulgaris	125	<0.06	0.13	<0.06	0.5	0.06
Ps. aeruginosa	0.5	1	2	1	2	1
Ps. aerug. Gen-R	62	2	62	31	16	16
H. influenzae	16	<0.06	<0.06	<0.06	<0.06	0.25
H. influ. Amp-R	16	<0.06	<0.06	<0.06	<0.06	0.5

[a] MICs in $\mu g\,mL^{-1}$. Cefsulodin (CFS), ceftazidime (CAZ), cefpirome (CPIR), cefepime (CPIM), GR 32620 (GR), gentamicin (Genta). [b] An organism is considered to be resistant to gentamicin if its MIC exceeds $4\,\mu g\,mL^{-1}$.

described above were moderately active but none approached the activity of the aminoglycoside antibiotics against this organism.

However, the Glaxo group discovered that the introduction of a hindered carboxyalkyl oxime substituent in combination with certain types of 3-substituent provided compounds with greatly improved activity, albeit at the cost of some loss of *in vitro* activity against *S. aureus*. These studies led to the development of ceftazidime (**167**), which combines potent broad-spectrum activity (Table 9) and high stability to β-lactamases with an excellent pharmacokinetic profile. Ceftazidime is well distributed in the body and is very effective against *S. aureus in vivo*, despite only moderate ($4-8\,\mu g\,L^{-1}$) MIC values. It is as active as gentamicin against aminoglycoside-sensitive *Ps. aeruginosa* and is highly active against many gentamicin-resistant strains.[134,159,160]

(**166**) Cefsulodin

(**167**) Ceftazidime

The aminothiazole side chain (**168**) required for the synthesis of ceftazidime was prepared in similar fashion to that of the corresponding methoximes described in Section 9.2.10. Thus alkylation of the hydroxime (**169**; Scheme 29) with *t*-butyl 2-bromoisopropionate using potassium carbonate in dimethyl sulfoxide followed by selective hydrolysis of the ethyl ester afforded the protected side-chain acid (**168**).[134]

The selectivity observed in the hydrolysis is a consequence of the steric hindrance provided by the *gem*-dimethyl groups in the oxime substituent. The analogous side-chain acid (**170**), which lacks these groups, cannot be prepared in this way as the corresponding hydrolysis yields a mixture of monoesters and diacid. Instead, this compound may be prepared directly from the hydroxime acid (**171**) and *t*-butyl bromoacetate using the dianion approach described earlier (Section 9.2.8).[134]

As in the case of many of the aminothiazolylcephalosporins, the synthesis of ceftazidime could proceed *via* acylation of 7-ACA followed by manipulation of the 3-substituent and, indeed, this was the approach first used in the laboratory.[102] The development of an efficient synthesis of 7-amino-3-pyridiniummethylcephalosporanate (**172**) from cephaloridine (**56**) allowed the more practical, convergent synthesis (Scheme 30). Thus acylation of (**172**) with the acid chloride obtained on treating the side chain (**168**) with phosphorus pentachloride, followed by deprotection of the intermediate cephalosporin with a mixture of formic and hydrochloric acids, gave ceftazidime which was obtained, ultimately, as the free betaine (**167**).[102]

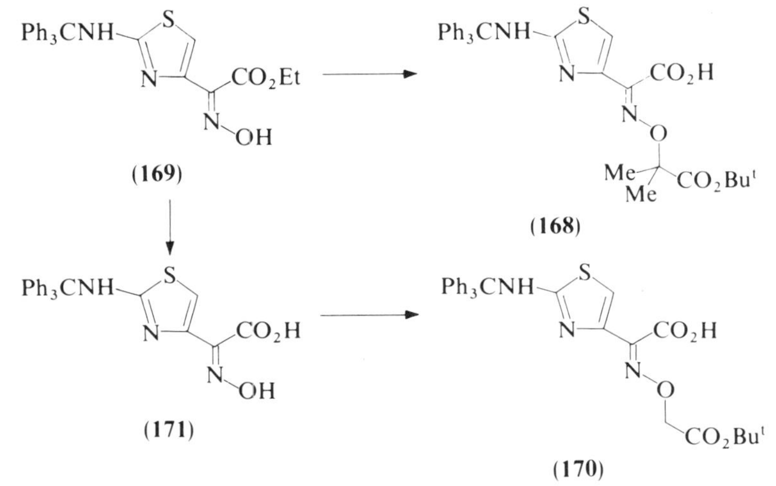

Scheme 29 Preparation of carboxyalkoximinoaminothiazolyl side chains

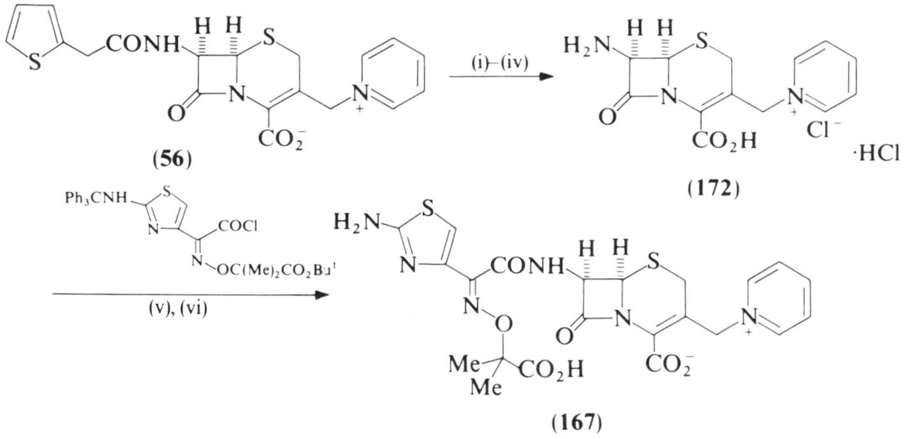

(i) TMSCl/PhNMe$_2$, CH$_2$Cl$_2$; (ii) PCl$_5$; (iii) HO(CH$_2$)$_4$OH; (iv) HCl, Pr^{i}OH; (v) acylation with the acid chloride of (**168**); (vi) HCl/HCO$_2$H then conversion to zwitterion

Scheme 30 Synthesis of ceztazidime (**167**)

Following the success of ceftazidime, a number of methoximes with novel quaternary ammonium 3-substituents have been pursued. Prominent amongst these are cefpirome (HR 810; **173**)[161,162] and cefepime (BMY 28142; **174**).[163–165] A number of higher alkoximes have also been described;[163] these include the cyclopropyl oxime GR 32620 (**175**),[134] DN-9550 (**176**),[166] BO-1236 (**177**)[167] and YM-13115 (**178**).[168]

9.2.12 NEW ORAL COMPOUNDS

Cephalexin and its close analogues, cephradine and cefadroxil (Section 9.2.6), have only moderate antibacterial activity but are well absorbed from the GI tract. Early studies with other 7-D-phenylglycylcephalosporins revealed that many compounds with small lipophilic 3-substituents were well absorbed after oral administration but that some were unstable in solution (*cf.* cephaloglycine). However, exploitation of the 3-exomethylene compounds described below afforded 3-chloro and 3-alkoxy analogues which were stable in solution and in some cases somewhat more active than the corresponding 3-methylcephalosporins (Table 10). This work, conducted primarily in the Lilly and Ciba-Geigy laboratories, culminated in the development of cefaclor (**179**) and cefroxadine (**180**).[169]

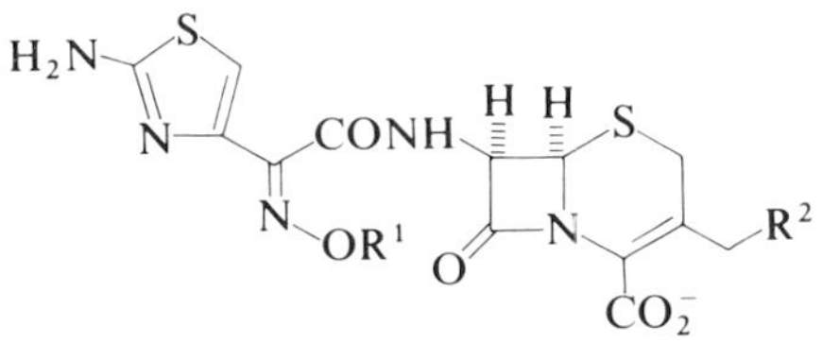
H2N
S
N
CONH
H H
S
N
OR1
O
N
R2
CO2-

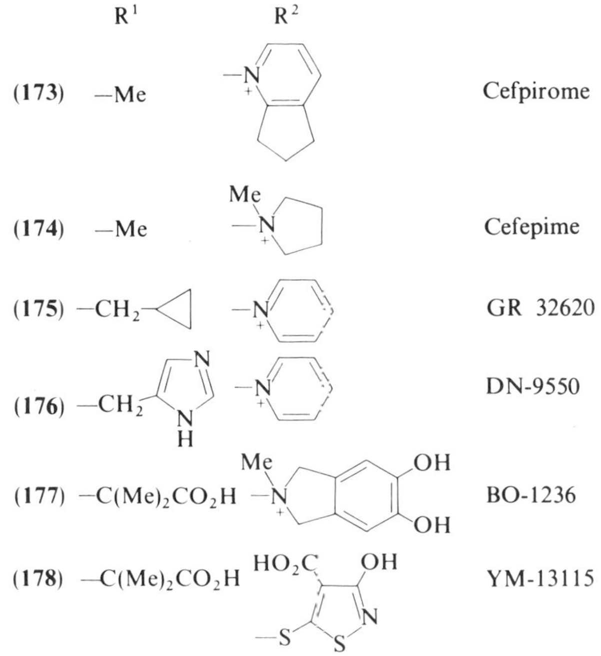
R1
R2
(173) —Me
Cefpirome
(174) —Me
Me
Cefepime
(175) —CH2
GR 32620
(176) —CH2
N
H
DN-9550
(177) —C(Me)2CO2H
Me
OH
OH
BO-1236
(178) —C(Me)2CO2H
HO2C
OH
N
S
YM-13115

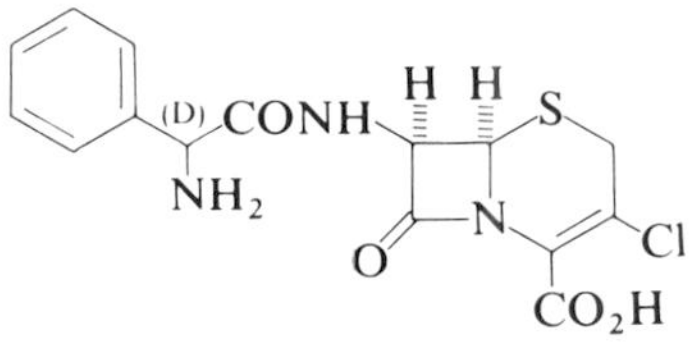
(D) CONH
H H
S
NH2
N
Cl
O
CO2H

(179) Cefaclor

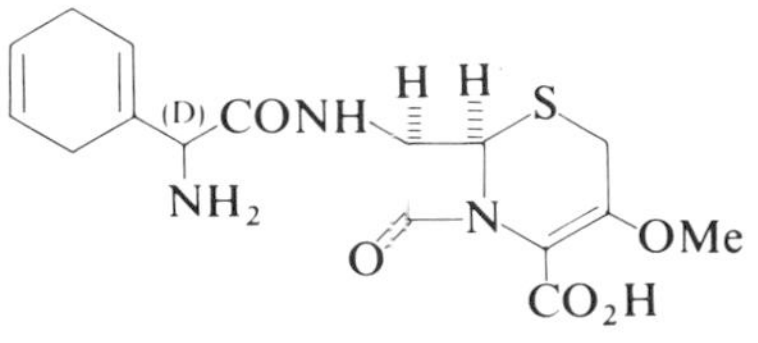
(D) CONH
H H
S
NH2
N
OMe
O
CO2H

(180) Cefroxadine

Table 10 Antibacterial Activity of Recent Oral Cephalosporins[a]

Organism	*CEX*	*CCL*	*CXD*	*CFIX*	*GR-T*	*GR-C*
S. aureus Pen-S	4	2	16	4	4	4
S. aureus Pen-R	8	8	16	4	8	8
St. faecalis	125	31	125	>125	>125	>125
E. coli (DC0)	16	1	4	0.25	<0.06	0.25
E. coli TEM-1	16	4	16	0.25	0.13	0.25
E. cloacae	8	2	4	<0.06	0.25	0.25
E. cloacae P99	>125	>125	>125	>125	31	>125
K. aerogenes	8	2	4	<0.06	<0.06	0.03
K. aerogenes K1	>125	>125	>125	<0.06	2	0.5
Pr. mirabilis	8	1	8	<0.06	<0.06	0.03
Pr. vulgaris	>125	>125	>125	<0.06	<0.06	0.03
Ps. aeruginosa	>125	>125	>125	31	16	62
Ps. aerug. Gen-R	>125	>125	>125	>125	>125	>125
H. influenzae	4	2	4	<0.06	<0.06	0.03
H. influ. Amp-R	4	2	8	<0.06	<0.06	0.03

[a] MICs in $\mu g\,mL^{-1}$. Cephalexin (CEX), cefaclor (CCL), cefroxadine (CXD), cefixime (CFIX), GR 32622 *trans*-isomer (GR-T), GR 32621 *cis*-isomer (GR-C).

As in the case of the 3-unsubstituted cephalosporins (Section 9.2.10), the 3-chloro and 3-methoxy analogues are prepared from the corresponding 3-hydroxy compounds obtained by ozonolysis of the 3-exomethylenecephalosporins. These have been prepared from 3-acetoxymethylcephalosporins by reductive elimination either with chromium acetate[170] or electrochemically,[171] and from 3-thiomethylcephalosporins by reductive desulfurization using Raney nickel.[169] Alternatively, these compounds may be prepared from penicillins using the rearrangement procedure developed by Kukolja and coworkers[172,173] shown in Scheme 31.

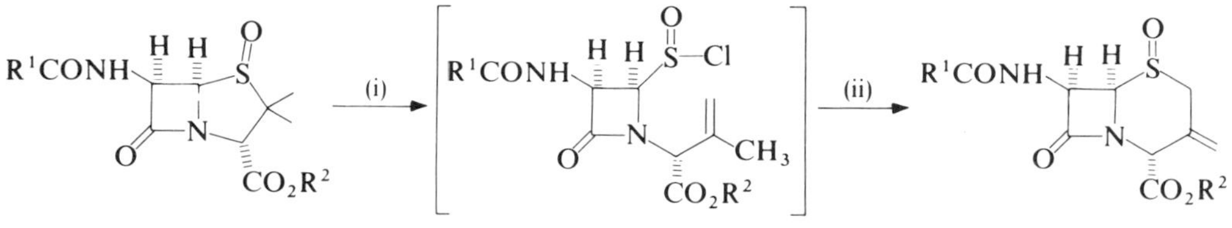

(i) *N*-Halogenating agent (*e.g.* *N*-chlorosuccinimide), acid scavenger; (ii) Lewis acids (*e.g.* SnCl$_4$)

Scheme 31   The Kukolja rearrangement

The conversion of the exomethylene cephalosporin (**181**), obtained from penicillin V (**6**), *via* the rearrangement reaction, into the 3-chlorocephalosporin, cefaclor (**179**), is shown in Scheme 32.[169] Ozonolysis of (**181**) followed by reduction of the sulfoxide with phosphorus tribromide gave the enol (**182**). Treatment of (**182**) with two equivalents of triphenylphosphite dichloride, followed by pyridine, concomitantly effected chlorination of the hydroxyl group and converted the amido group into an imidoyl chloride. The latter was hydrolyzed on work-up with isobutyl alcohol to give, after treatment with dry hydrogen chloride, the 7-amino-3-chlorocephalosporin (**183**) as its hydrochloride.[74] Standard reacylation and deprotection procedures then gave cefaclor (**179**). The reaction of 3-hydroxycephalosporins with diazomethane affords the corresponding 3-methoxy derivatives, which have been used to prepare cefroxadine (**180**).[138]

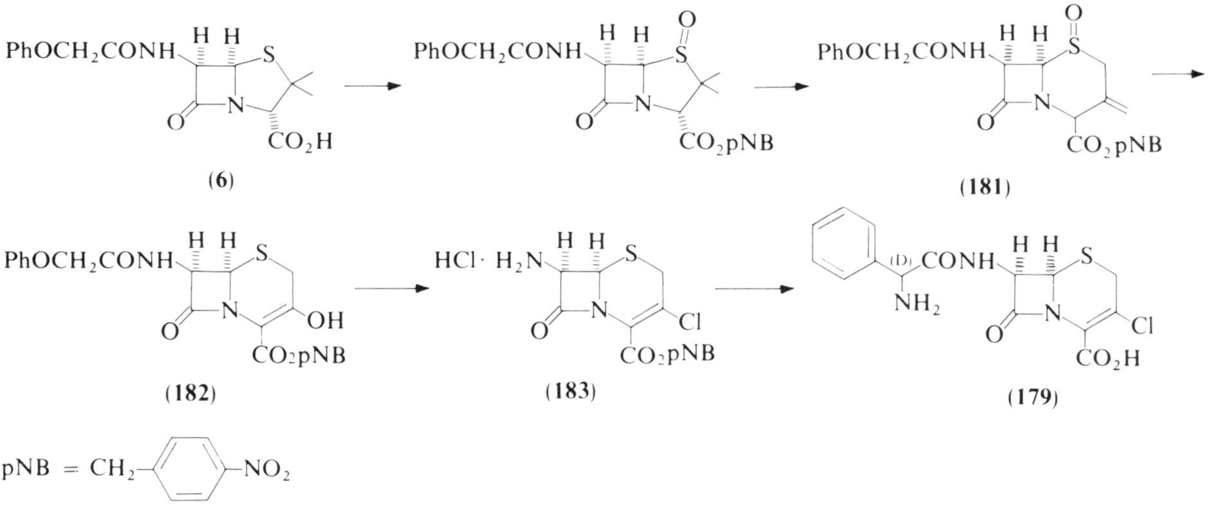

pNB = CH$_2$—⟨benzene ring⟩—NO$_2$

Scheme 32   Conversion of penicillin V (**6**) into cefaclor (**179**)

The 7-D-phenylglycyl derivative of the 3-ethylcephalosporin and the 3-vinyl compound from which it is derived (*vide infra*) were well absorbed and had antibacterial activity similar to that of cephalexin.[82] Bristol Meyers have reported that the 3-*cis*-propenyl analogue (**184**; BMY 28100) of cefadroxil is more active than cephalexin against some important bacteria, including *S. aureus*.[174-176]

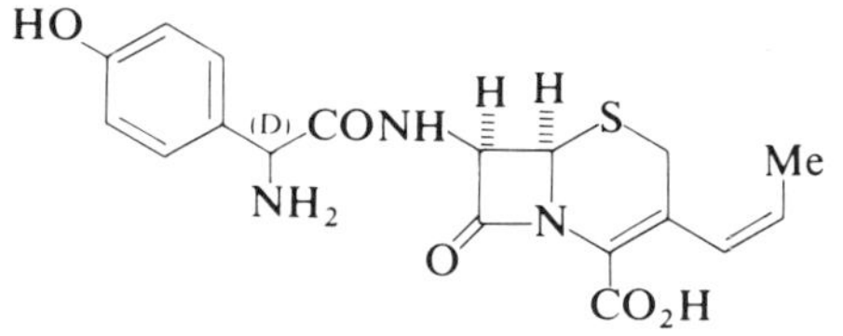

(**184**)   BMY 28100

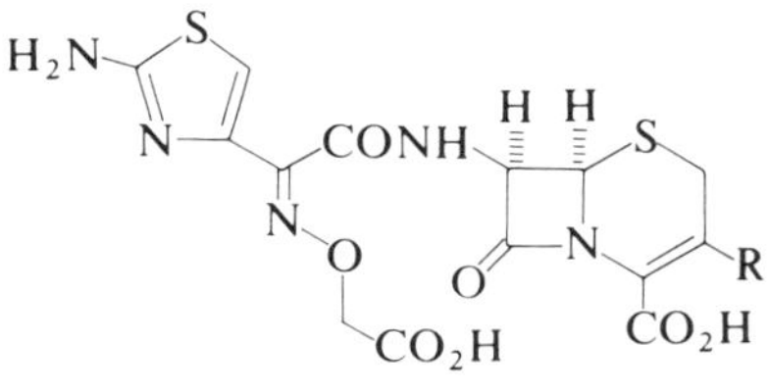

(**185**) —CH=CH$_2$ Cefixime
(**186**) —H   FK 089
(**187**) ⟨C=C—Me⟩   GR 32622
(**188**) ⟨C≡C—Me⟩   GR 32621

Most of the aminothiazolylcephalosporins are not absorbed from the GI tract. The exceptions are certain carboxymethoximes with small lipophilic 3-substituents. Two such compounds, (**185**; cefixime)[177–181] and (**186**; FK089),[182,183] have been studied in detail and are believed to cross the brush-border membrane in rats *via* a carrier-mediated transport mechanism. The antibacterial activities of the corresponding 3-(pent-1-en-3-ynyl)cephalosporins[184] were very similar to that of cefixime (Table 10) but their oral absorption depended on the geometry of the 3-substituent. Thus the *trans* isomer (**187**; GR 32622) showed good absorption in rats and mice whereas the *cis* isomer (**188**; GR 32621) did not.

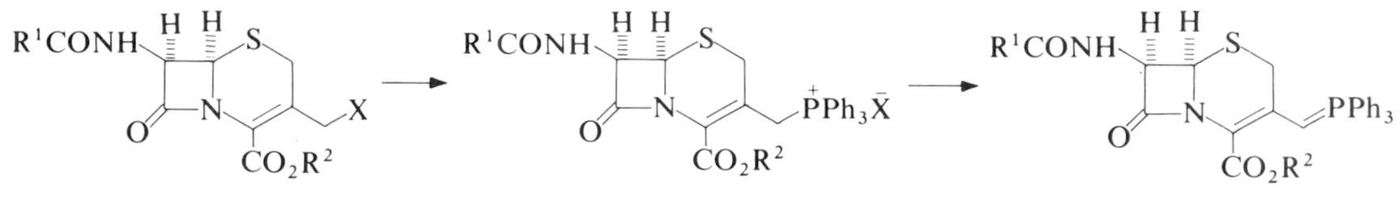

Scheme 33 Preparation of cephalosporin ylides

The 3-substituents of cefixime (**185**),[177] GR 32622 (**187**)[184] and BMY 28100 (**184**)[174] are introduced by means of a Wittig reaction of a cephalosporin ylide (*cf.* nitrocefin, Section 9.2.6) and a number of methods for the generation of these ylides have been developed.[82] Most conveniently they may be prepared by treating a solution of the corresponding phosphonium salt, obtained from 3-halomethylcephalosporins by reaction with triphenylphosphine, with aqueous base (Scheme 33). Some ylides are isolable but they are more usually generated and reacted *in situ* with the appropriate aldehyde. Thus, for example, treatment of the salt (**189**) with aqueous sodium carbonate in the presence of aqueous formaldehyde effected the introduction of the 3-vinyl group into the cefixime intermediate (**190**) (Scheme 34).[177]

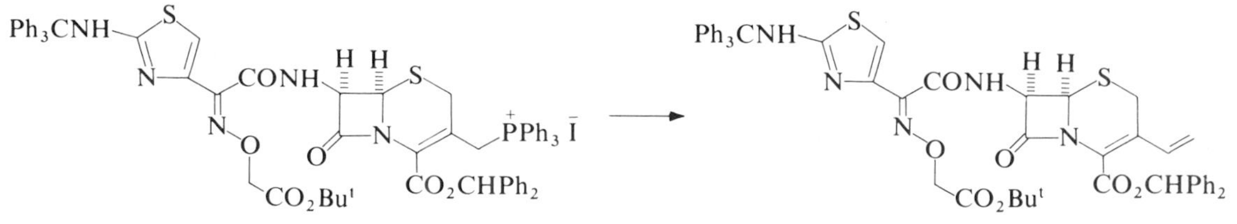

Scheme 34 Synthesis of cefixime intermediate (**190**)

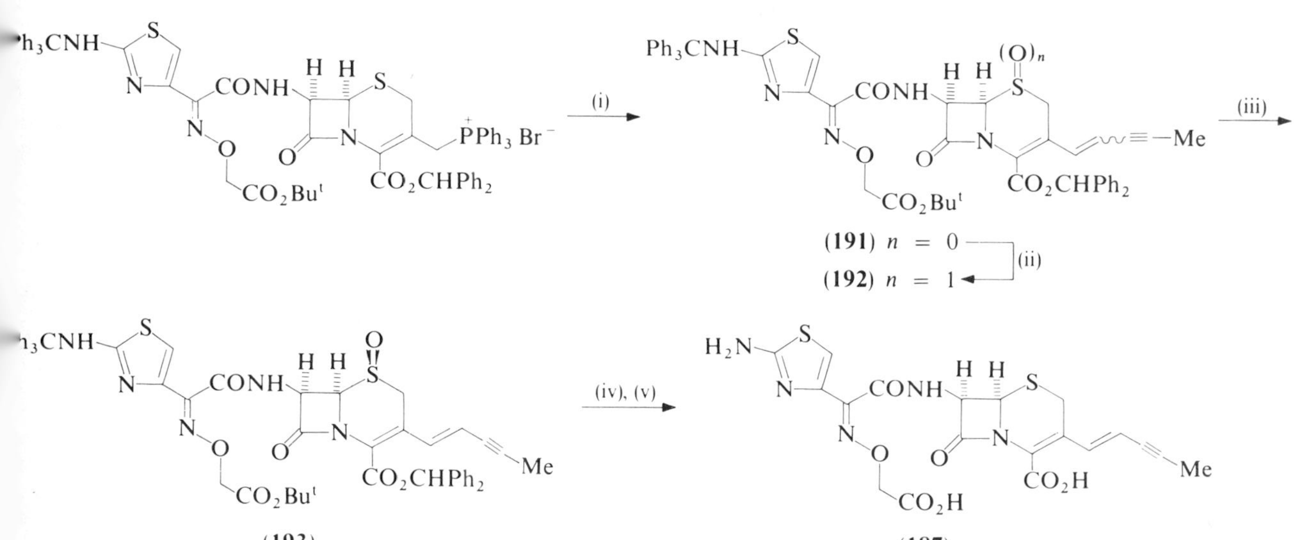

(i) MeC≡CCHO/aq. NaHCO₃, CH₂Cl₂; (ii) mCPBA; (iii) chromatography (separate isomers); (iv) KI/AcCl, DMF; (v) TFA/PhOMe

Scheme 35 Synthesis of GR 32622 (**187**)

With higher aldehydes the Wittig reaction of cephalosporin ylides generally yields a mixture of *cis* and *trans* isomers in which the former is predominant.[82] As a consequence of this, the syntheses of the substituted vinylcephalosporins (184) and (187) both require a separation step in order to remove the product with the wrong stereochemistry. In the case of GR 32622 (187) this was achieved by oxidizing the isomeric sulfides (191), obtained from the Wittig reaction, to the corresponding sulfoxides (192) which were separated by chromatography. The required *trans* isomer (193) was converted into GR 32622 (187) by reduction of the sulfoxide followed by a standard deprotection (Scheme 35).[184]

Prodrug forms of several aminothiazolyl methoximes are under development; these include the pivaloyloxymethyl esters cefteram (194; T-2588)[185,186] and cefetamet pivoxyl (195; Ro 15-8075),[187,188] and the 1-isopropoxycarbonyloxyethyl ester CS-807 (196).[189]

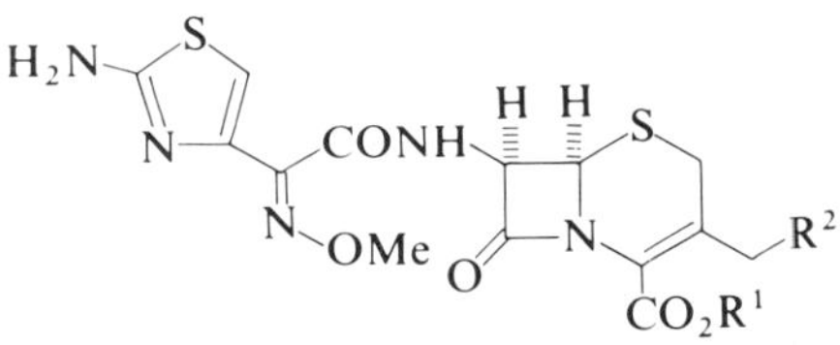

	R^1	R^2	
(194)	—CH_2OCOBu^t	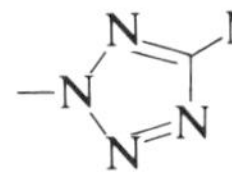 —N⟨ N-N / Me (tetrazole)	Cefteram
(195)	—CH_2OCOBu^t	—H	Cefetamet pivoxyl
(196)	—CHOCO₂CHMe₂ / Me	—OMe	CS-807

9.2.13 FUTURE PROSPECTS

The clinical pharmacology of some of the compounds described above and processes used for their manufacture have been highlighted in a recent book.[190] The world market for antibiotics is enormous (more than £8000 m in 1986) and penicillins and cephalosporins take the lion's share. The 6(7)-D-phenylglycyl penicillins and cephalosporins, their acylureido derivatives and the aminothiazolylalkoximinocephalosporins account for an increasing proportion of this and the success of these compounds continues to stimulate research. Desirable features for future antibiotics include better activity, especially against *Pseudomonas* and methicillin-resistant *Staphylococci*, improved pharmacokinetics permitting once or twice daily dosage, and even higher stability to β-lactamases.

Some methicillin-resistant strains of *S. aureus* (MRSA) produce PBPs with a low affinity for β-lactams (Section 9.2.3) and it is likely that the presence of these modified target enzymes is responsible for their resistance to penicillins and cephalosporins. Some recent compounds, in particular penems and carbapenems (Chapter 9.3), and the oxacephamycin flomoxef (197)[191-193] are claimed to have useful activity against MRSA and may foreshadow compounds with high affinity for their modified PBPs.

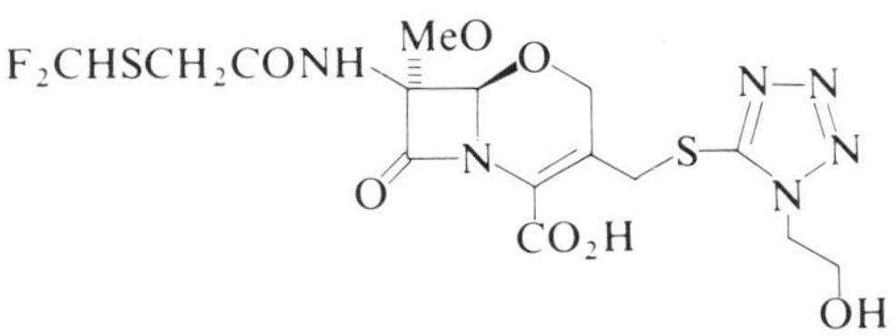

(197) Flomoxef

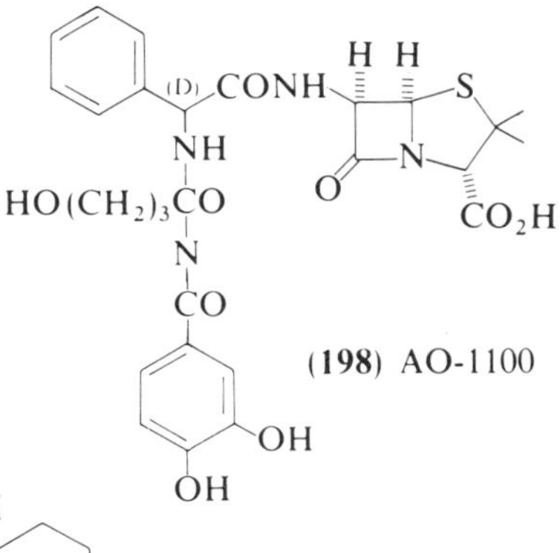

(198) AO-1100

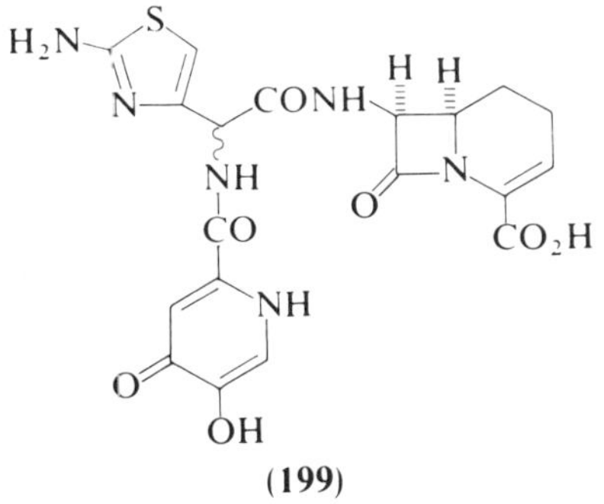

(199)

The impermeability of the outer membrane of *Ps. aeruginosa* to most penicillins and cephalosporins seems to be an important factor in the resistance of this organism. These and other Gram-negative bacteria have a requirement for iron and produce specific scavenging agents (siderophores)[194] capable of removing iron from mammalian transferrins. The siderophore/iron complexes are transported across the bacterial outer membrane by specific receptor proteins, which in *E. coli* appear to be under the control of the *tonB* inner membrane protein.[195] The iron(III) ion chelating groups of several bacterial siderophores are provided by 2,3-dihydroxybenzoyl or similar catechol containing groups. Ohi and his collaborators have postulated that penicillins carrying catechol groups might be taken into the cell *via* the siderophore receptors and have described a series of ureidopenicillins, for example (**198**), with enhanced activity against *Ps. aeruginosa*.[196,197] Unfortunately their early compounds are substrates for catechol-*O*-methyltransferase (COMT) and some activity is lost *in vivo*. This may be overcome by the introduction of chlorine atoms into the catechol function.[198]

Catechols or similar functional groups occur in several classes of recent compounds, including formidacillin (**119**),[114] a series of carbacephems (*e.g.*, **199**),[199,200] and the cephalosporins BO-1236 (**177**),[167] M-14659 (**200**)[201] and E-0702 (**201**).[202] Watanabe and his coworkers have studied the activity of the last compound in a series of mutants of *E. coli* and have concluded that E-0702 is incorporated into the bacterial cells *via* the *tonB*-dependent iron transport system.[203]

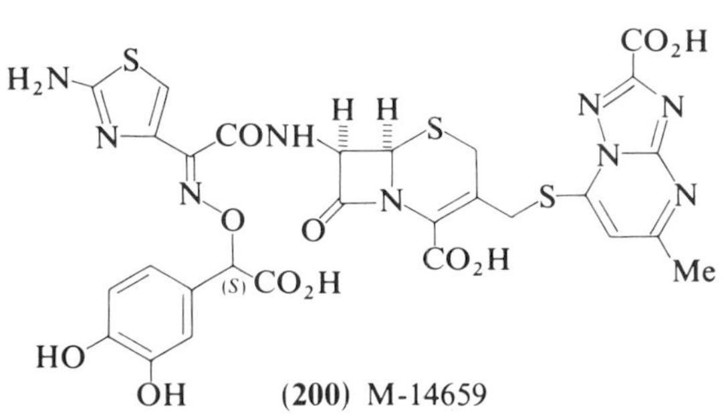

(**200**) M-14659

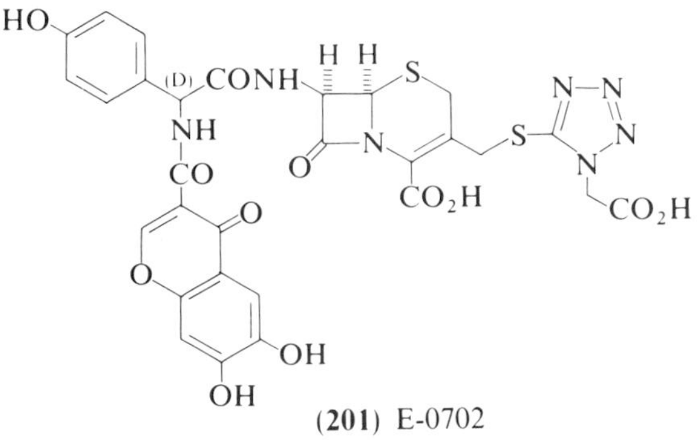

(**201**) E-0702

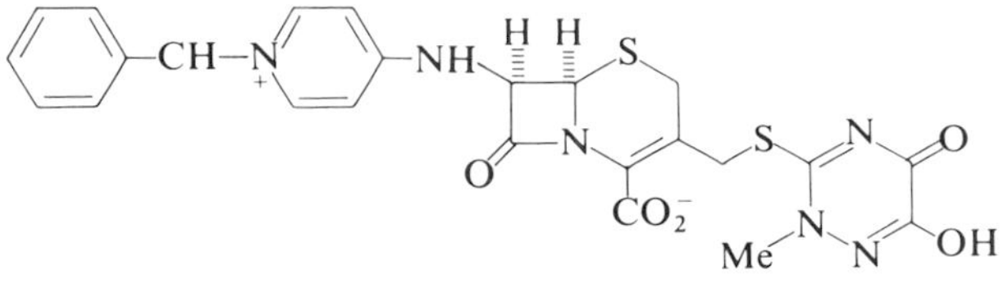

(**202**) L-652,813

Following the success of the aminothiazolylcephalosporins, several groups have investigated compounds containing aminothiazole isosteres. These have included aminopyridines,[204,205] aminopyrimidines,[206,207] aminothiadiazoles[207-210] and aminooxadiazoles;[210-212] generally these derivatives were similar to or less active than their aminothiazolyl counterparts. Carbon analogues of the oxime function have also been investigated and some have been reported to have potent antibacterial activity.[213,214]

Studies of less conventional compounds are also in progress and the Merck group have reported a series of 7-(4-pyridinium)aminocephalosporins,[215] for example L-652,813 (**202**); this has been compared with cefuroxime and ceftriaxone.[216]

The biosynthetic pathways to penicillins and cephalosporins are being studied with a view to producing new 'natural products' and several groups have demonstrated that modified penicillins may be produced from analogues of the Arnstein tripeptide in cell-free systems.[217-219] Baldwin's group have demonstrated that the penicillin to cephalosporin ring expansion reaction may be conducted with a variety of side-chain acids, in particular *m*-carboxyphenylacetic acid.[220] Although these studies are a long way from commercial application, it is possible that they could provide alternatives to the conventional fermentation processes in the long term.

The 1988 Royal Society of Chemistry β-lactam symposium highlighted the most recent trends in the chemistry and biochemistry of the penicillins and cephalosporins and provided pointers to the future development of these and related compounds.[221]

9.2.14 REFERENCES

1. A. Fleming, *Br. J. Exp. Pathol.*, 1929, **10**, 226.
2. 'Penicillin: its practical application', ed. A. Fleming, Butterworth, London, 1946; D. Wilson, 'Penicillin in Perspective', Faber & Faber, London, 1976; J. C. Sheehan, 'The Enchanted Ring: The Untold Story of Penicillin', MIT Press, Cambridge, MA, 1982.
3. G. W. Ross, *Interdiscip. Sci. Rev.*, 1986, **11**, 19.
4. J. C. Sheehan, K. R. Henery-Logan and D. A. Johnson, *J. Am. Chem. Soc.*, 1953, **75**, 3292.
5. 'Cephalosporins and Penicillins', ed. E. H. Flynn, Academic Press, New York, 1972.
6. 'Chemistry and Biology of β-Lactam Antibiotics', ed. R. B. Morin and M. Gorman, Academic Press, New York, 1982.
7. 'Topics in Antibiotic Chemistry', ed. P. G. Sammes, Ellis Horwood, Chichester, 1980.
8. 'β-Lactam Antibiotics', ed. S. Mitsuhashi, Springer-Verlag, Heidelberg, 1981.
9. 'β-Lactam Antibiotics: Mode of Action, New Developments and Future Prospects', ed. M. R. J. Salton and G. D. Shockman, Academic Press, New York, 1981.
10. 'Recent Advances in the Chemistry of β-Lactam Antibiotics', ed. J. Elks, Special Publication No. 28, The Chemical Society, London, 1977.
11. 'Recent Advances in the Chemistry of β-Lactam Antibiotics', ed. G. I. Gregory, Special Publication No. 38, The Royal Society of Chemistry, London, 1981.
12. 'Recent Advances in the Chemistry of β-Lactam Antibiotics', ed. A. G. Brown and S. M. Roberts, Special Publication No. 52, The Royal Society of Chemistry, London, 1985.
13. S. W. Queener and N. Neuss, in ref. 6, vol. 3, p. 1.
14. A. L. Demain, J. Kupka, Y.-Q. Shen and S. Wolfe, in 'Trends in Antibiotic Research. Genetics, Biosynthesis, Actions and New Substances', ed. H. Umezawa, A. L. Demain, T. Hata and C. R. Hutchinson, Japan Antibiotics Research Association, Tokyo, 1982.
15. S. Wolfe, A. L. Demain, S. E. Jensen, and D. W. S. Westlake, *Nature (London)*, 1984, **226**, 1386.
16. J. E. Baldwin, in ref. 12, p. 62.
17. J. F. Martin and P. Liras, *Trends Biotechnol.*, 1985, **3**, 39.
18. H. R. V. Arnstein and D. Morris, *Biochem. J.*, 1960, **76**, 357.
19. J. E. Baldwin, E. P. Abraham, C. G. Lovel and H.-H. Ting, *J. Chem. Soc., Chem. Commun.*, 1984, 902.
20. J. E. Dotzlaf and W.-K. Yeh, *J. Bacteriol.*, 1987, **169**, 1611.
21. J. O'Sullivan and E. P. Abraham, *Biochem. J.*, 1980, **186**, 613.
22. J. C. Sheehan and K. R. Henery-Logan, *J. Am. Chem. Soc.*, 1959, **81**, 5838.
23. J. C. Sheehan and K. R. Henery-Logan, *J. Am. Chem. Soc.*, 1962, **84**, 2983.
24. K. Heusler, in ref. 5, p. 255.
25. K. G. Holden, in ref. 6, vol. 2, p. 101.
26. M. S. Manhas and A. K. Bose, in 'Synthesis of Penicillin, Cephalosporin C and Analogues', Marcel Dekker, New York, 1969, p. 101.
27. J. C. Sheehan and K. R. Henery-Logan, *J. Am. Chem. Soc.*, 1957, **79**, 1262.
28. R. B. Woodward, *Science (Washington, D.C.)*, 1966, **153**, 487.
29. R. B. Woodward, K. Heusler, J. Gosteli, P. Naegeli, W. Oppolzer, R. Ramage, S. Ranganathan and H. Vorbruggen, *J. Am. Chem. Soc.*, 1966, **88**, 852.
30. R. Bucourt, in ref. 11, p. 1.
31. J.-M. Ghuysen, in ref. 7, vol. 5, part A, p. 9.
32. H. Nikaido, in ref. 9, p. 249.
33. D. J. Tipper and J. L. Strominger, *Proc. Natl. Acad. Sci. USA*, 1965, **54**, 1133.
34. J.-M. Ghuysen, J.-M. Frere, M. Leyh-Bouille and O. Dideberg, in ref. 9, p. 127.
35. D. B. Boyd, in ref. 6, vol. 1, p. 437.
36. B. G. Spratt, *J. Gen. Microbiol.*, 1983, **129**, 1247.
37. N. A. Curtis, D. Orr, G. W. Ross and M. G. Boulton, *Antimicrob. Agents Chemother.*, 1979, **16**, 533.
38. M. H. Richmond and R. B. Sykes, in 'Advances in Microbial Physiology', ed. A. H. Rose and D. W. Tempest, Academic Press, New York, 1973, vol. 9, p. 31.
39. R. P. Ambler, *Philos. Trans. R. Soc. London, Ser. B*, 1980, **289**, 321.
40. A. A. Medeiros, *Br. Med. Bull.*, 1984, **40**, 18.
41. J. A. Kelly, J. R. Know, P. C. Moews, G. J. Hite, J. B. Bartolone, H. Zhao, B. Joris, J.-M. Frere and J.-M. Ghuysen, *J. Biol. Chem.*, 1985, **260**, 6449.
42. D. J. Waxman and J. L. Strominger, *J. Biol. Chem.*, 1980, **255**, 3964.
43. J. H. C. Nayler, *J. Antimicrob. Chemother.*, 1987, **19**, 713.
44. D. F. J. Brown and P. E. Reynolds, *FEBS Lett.*, 1980, **122**, 275.
45. K. Ubukata, N. Yamashita and M. Konno, in 'Recent Advances in Chemotherapy', ed. J. Ishigami, University of Tokyo Press, 1985, p. 369; Y. Utsui, M. Tajima and T. Yokota, *ibid.*, p. 371.
46. A. Tomasz, *Rev. Infect. Dis.*, 1986, **8**, supp. 3, S260.
47. G. N. Rolinson, *J. Antimicrob. Chemother.*, 1986, **17**, 5.
48. B. Aronsson, R. Möllby and C.-E. Nord, *J. Infect. Dis.*, 1985, **151**, 476.
49. W. E. Wright and V. D. Line, *Antimicrob. Agents Chemother.*, 1980, **17**, 842.
50. R. Wise, *Clin. Pharmacokinet.*, 1986, **11**, 470.
51. H. Ferres, *Chem. Ind. (London)*, 1980, 435.
52. K. R. Brown and C. M. Martin, in ref. 9, p. 445.
53. S. R. Norrby, *Rev. Infect. Dis.*, 1986, **8**, supp. 3, S358.
54. K. J. Child and M. G. Dodds, *Br. J. Pharmacol. Chemother.*, 1967, **30**, 354.
55. F. R. Batchelor, F. P. Doyle, J. H. C. Nayler and G. N. Rolinson, *Nature (London)*, 1959, **183**, 257.
56. F. M. Huber, R. R. Chauvette and B. G. Jackson, in ref. 5, p. 27.
57. G. V. Kaiser and S. Kukolja, in ref. 5, p. 74.
58. D. C. Hobbs and A. R. English, *J. Med. Pharm. Chem.*, 1961, **4**, 207.

59. F. P. Doyle, G. R. Fosker, J. H. C. Nayler and H. Smith, *J. Chem. Soc.*, 1962, 1440.
60. B. Ekström, A. Gomez-Revilla, R. Mollberg, H. Thelin and B. Sjoberg, *Acta Chem. Scand.*, 1965, **19**, 281.
61. E. Dane and T. Dockner, *Chem. Ber.*, 1965, **98**, 789.
62. G. A. Hardcastle, Jr., D. A. Johnson, C. A. Panetta, A. L Scott and S. A. Sutherland, *J. Org. Chem.*, 1966, **31**, 897.
63. F. P. Doyle, K. Hardy, J. H. C. Nayler, M. J. Soulal, E. R. Stove and H. R. J. Waddington, *J. Chem. Soc.*, 1962, 1453.
64. C. Dixon and L. W. Mizen, *J. Physiol. (London)*, 1977, **269**, 549.
65. A. Tsuji, E. Nakashima, I. Kagami and T. Yamana, *J. Pharm. Sci.*, 1981, **70**, 768 and 772.
66. T. Kimura, H. Endo, M. Yoshikawa, S. Muranishi and H. Sezaki, *J. Pharmacobio-Dyn.*, 1978, **1**, 262.
67. A. B. A. Jansen and T. J. Russell, *J. Chem. Soc.*, 1965, 2127.
68. W. von Daehne, E. Frederiksen, E. Gundersen, F. Lund, P. Mørch, H. J. Peterson, K. Roholt, L. Tybring and W. O. Godtfredsen, *J. Med. Chem.*, 1970, **13**, 607.
69. J. P. Clayton, M. Cole, S. W. Elson, H. Ferres, J. C. Hanson, L. W. Mizen and R. Sutherland, *J. Med. Chem.*, 1976, **19**, 1385.
70. N.-O. Bodin, B. Ekström, U. Forsgren, L.-P. Jalar, L. Magni, C.-H. Ramsay and B. Sjoberg, *Antimicrob. Agents Chemother.*, 1975, **8**, 518.
71. F. M. Huber, R. R. Chauvette and B. G. Jackson, in ref. 5, p. 27.
72. B. Loder, G. G. F. Newton and E. P. Abraham, *Biochem. J.*, 1961, **79**, 408.
73. R. B. Morin, B. G. Jackson, E. H. Flynn and R. W. Roeske, *J. Am. Chem. Soc.*, 1962, **84**, 3400.
74. L. D. Hatfield, W. H. W. Lunn, B. G. Jackson, L. R. Peters, L. C. Blaszczak, J. W. Fisher, J. P. Gardner and J. M. Dunigan, in ref. 11, p. 109.
75. R. R. Chauvette, E. H. Flynn, B. G. Jackson, E. R. Lavagnino, R. B. Morin, R. A. Mueller, R. P. Pioch, R. W. Roeske, C. W. Ryan, J. L. Spencer and E. M. Van Heyningen, *J. Am. Chem. Soc.*, 1962, **84**, 3401.
76. J. D. Cocker, B. R. Cowley, J. S. G. Cox, S. Eardley, G. I. Gregory, A. G. Long, J. C. P. Sly and G. A. Somerfield, *J. Chem. Soc.*, 1965, 5015.
77. J. L. Spencer, F. Y. Siu, B. G. Jackson, H. M. Higgins and E. H. Flynn, *J. Org. Chem.*, 1967, **32**, 500.
78. T. W. Greene, 'Protective Groups in Organic Synthesis', Wiley-Interscience, New York, 1981.
79. G. V. Kaiser, R. D. G. Cooper, R. E. Koehler, C. F. Murphy, J. A. Webber, I. G. Wright and E. M. Van Heyningen, *J. Org. Chem.*, 1970, **35**, 2430.
80. J. A. Webber and W. J. Wheeler, in ref. 6, vol. 1, p. 371.
81. C. H. O'Callaghan, A. Morris, S. M. Kirby and A. H. Shingler, *Antimicrob. Agents Chemother.*, 1972, **1**, 283.
82. A. H. Shingler and N. G. Weir, in ref. 10, p. 153.
83. P. Schindler and G. Huber, in 'Enzyme Inhibitors', ed. U. Brodbeck, Verlag Chemie, Weinheim, 1980, p. 169.
84. C. W. Ryan, R. L. Simon and E. M. Van Heyningen, *J. Med. Chem.*, 1969, **12**, 310.
85. R. B. Morin, B. G. Jackson, R. A. Mueller, E. R. Lavagnino, W. B. Scanlon and S. L. Andrews, *J. Am. Chem. Soc.*, 1963, **85**, 1896.
86. R. R. Chauvette, P. A. Pennington, C. W. Ryan, R. D. G. Cooper, F. L. Jóse, I. G. Wright, E. M. Van Heyningen and G. W. Huffman, *J. Org. Chem.*, 1971, **36**, 1259.
87. W. E. Wick, in ref. 5, p. 496.
88. A. Tsuji and T. Yamana, in ref. 8, p. 235.
89. D. B. Boyd, *J. Med. Chem.*, 1983, **26**, 1010.
90. F. Lund and L. Tybring, *Nature (London), New Biol.*, 1972, **236**, 135.
91. H. J. Petersen, *J. Med. Chem.*, 1974, **17**, 101.
92. F. Lund, in ref. 10, p. 25.
93. H. C. Neu, *Pharmacotherapy*, 1985, **5**, 1.
94. W. H. W. Lunn, R. W. Burchfield, T. K. Elzey and E. V. Mason, *Tetrahedron Lett.*, 1974, 1307.
95. S. Karady, S. H. Pines, L. M. Weinstock, F. E. Roberts, G. S. Brenner, A. M. Hoinowski, T. Y. Cheng and M. Sletzinger, *J. Am. Chem. Soc.*, 1972, **94**, 1410.
96. E. M. Gordon and R. B. Sykes, in ref. 6, vol. 1, p. 199.
97. G. A. Koppel and R. E. Koehler, *J. Am. Chem. Soc.*, 1973, **95**, 2403.
98. W. H. W. Lunn and E. V. Mason, *Tetrahedron Lett.*, 1974, 1311.
99. H. Yanagisawa, M. Fukushima, A. Ando and H. Nakao, *Tetrahedron Lett.*, 1975, 2705.
100. H. Nakao, H. Yanagisawa, B. Shimizu, M. Kaneko, M. Nagano and S. Sugawara, *J. Antibiot.*, 1976, **29**, 554.
101. P. C. Cherry, M. C. Cook, M. W. Foxton, G. I. Gregory and G. B. Webb, in ref. 10, p. 145.
102. E. M. Wilson, *Chem. Ind. (London)*, 1984, 217.
103. D. C. Humber, S. B. Laing and G. G. Weingarten, in ref. 11, p. 38.
104. R. N. Brogden, R. C. Heel, T. M. Speight and G. S. Avery, *Drugs*, 1979, **17**, 233.
105. T. A. Tartaglione and R. E. Polk, *Drug Intell. Clin. Pharm.*, 1985, **19**, 188.
106. S. M. Harding, P. E. O. Williams and J. Ayrton, *Antimicrob. Agents Chemother.*, 1984, **25**, 78; see also *Anon.*, *Drugs Future*, 1985, **10**, 113.
107. H. König, K. G. Metzget, H. A. Offe and W. Schröck, in ref. 10, p. 78.
108. B. Wetzel, E. Woitun, R. Reuter, R. Maier, U. Lechner, H. Goeth and R. Werner, in ref. 11, p. 26.
109. H. Noguchi, Y. Eda, H. Tobiki, T. Nakagome and T. Komatsu, *Antimicrob. Agents Chemother.*, 1976, **9**, 262.
110. R. Holmes, D. M. Richards, R. N. Brogden and R. C. Heel, *Drugs*, 1984, **28**, 375.
111. I. Saikawa, in ref. 8, p. 99.
112. J. Mitsuyama, M. Takahata, T. Yasuda and I. Saikawa, *J. Antibiot.*, 1987, **40**, 868.
113. B. Slocombe, M. J. Basker, P. H. Bentley, P. Clayton, M. Cole, K. R. Comber, R. A. Dixon, R. A. Edmondson, D. Jackson, D. J. Merrikin and R. Sutherland, *Antimicrob. Agents Chemother.*, 1980, **20**, 38.
114. R. J. Ponsford, M. J. Basker, G. Burton, A. W. Guest, F. P. Harrington, P. H. Milner, M. J. Pearson, T. C. Smale and A. V. Stachulski, in ref. 12, p. 32.
115. A. W. Guest, C. L. Branch, S. C. Finch, A. C. Kaura, P. H. Milner, M. J. Pearson, R. J. Ponsford and T. C. Smale, *J. Chem. Soc., Perkin Trans. 1*, 1987, 45.
116. M. J. Basker, C. L. Branch, S. C. Finch, A. W. Guest, P. H. Milner, M. J. Pearson, R. J. Ponsford and T. C. Smale, *J. Antibiot.*, 1986, **39**, 1788.

117. A. A. Carmine, R. N. Brogden, R. C. Heel, T. M. Speight and G. S. Avery, *Drugs*, 1983, **25**, 223.
118. K. Seeger, W. Dürckheimer, F. Wengenmayer and H. Strecker, *Drugs Exp. Clin. Res.*, 1983, **9**, 427.
119. S. Masuyoshi, M. Inoue, M. Takaoka and S. Mitsuhashi, *Arzneim.-Forsch.*, 1981, **31**, 1070.
120. S. M. Harding, A. J. Munro, J. E. Thornton, J. Ayrton and M. I. J. Hogg, *J. Antimicrob. Chemother.*, 1981, **8**, supp. B, 263.
121. S. Selwyn and M. Bakhtiar, *Drugs Clin. Exp. Res.*, 1986, **12**, 953.
122. C. Thornsberry, *Am. J. Med.*, 1985, **79**, suppl. 2A, 14.
123. S. M. Harding, *Am. J. Med.*, 1985, **79**, supp. 2A, 21.
124. G. E. Schumacher, *Clin. Pharm.*, 1987, **6**, 59.
125. D. M. Richards, R. C. Heel, R. N. Brogden, T. M. Speight and G. S. Avery, *Drugs*, 1984, **27**, 469.
126. R. N. Jones and A. L. Barry, *Antimicrob. Agents Chemother.*, 1987, **31**, 818.
127. C. Fiset, F. Vallée, M. LeBel and M. G. Bergeron, *Ther. Drug Monit.*, 1986, **8**, 483.
128. W. Dürckheimer, J. Blumbach, R. Lattrell and K. H. Schheunemann, *Angew. Chem., Int. Ed. Engl.*, 1985, **24**, 180.
129. W. Dürckheimer, J. Blumbach, R. Heymes, E. Schrinner and K. Seeger, *Symp. Front. Pharmacol.*, 1982, **1**, 3 (*Chem. Abstr.*, 1982, **97**, 119 899).
130. R. Bucourt, R. Heymes, J. Perronnet, A. Lutz and L. Pénasse, *Eur. J. Med. Chem. Chim. Ther.*, 1981, **16**, 307.
131. R. Reiner, *Drugs Exp. Clin. Res.*, 1986, **12**, 299.
132. J. Blumbach, W. Dürckheimer, E. Ehlers, K. Fleischmann, N. Klesel, M. Limbert, B. Mencke, J. Reden, K. H. Scheunemann, E. Schrinner, G. Seibert, M. Wieduwilt and M. Worm, *J. Antibiot.*, 1987, **40**, 29.
133. C. H. O'Callaghan and J. Elks, in 'Encyclopedia of Chemical Technology', ed. R. E. Kirk and D. F. Othmer, Wiley, New York, 1984, suppl. vol., p. 64.
134. C. E. Newall, in ref. 12, p. 1.
135. R. Bucourt, R. Heymes, A. Lutz, L. Pénasse and J. Perronnet, *Tetrahedron*, 1978, **34**, 2233.
136. M. Ochiai, A. Morimoto and T. Miyawaki, *J. Antibiot.*, 1981, **34**, 186.
137. R. Reiner, U. Weiss, U. Brombacher, P. Lanz, M. Montavon, A. Furlenmeier, A. Angehrn and P. J. Probst., *J. Antibiot.*, 1980, **33**, 783.
138. R. Scartazzini and H. Bickel, *Helv. Chim. Acta*, 1974, **57**, 1919.
139. T. Takaya, H. Taksugi, T. Masugi, T. Chiba, H. Kochi, T. Takano and H. Nakano, *Nippon Kagaku Kaishi*, 1981, 785 (*Chem. Abstr.*, 1981, **95**, 150 563).
140. H. Peter and H. Bickel, *Helv. Chim. Acta*, 1974, **57**, 2044.
141. M. Numata, I. Minamida, M. Yamaoka, M. Shiraishi, T. Miyawaki, H. Akimodo, K. Naito and M. Kida, *J. Antibiot.*, 1978, **31**, 1262.
142. M. Steude, *Liebigs Ann. Chem.*, 1891, **261**, 30.
143. S. Mitsuhashi and M. Ochiai, *Chemotherapy*, 1985, **33**, 519.
144. L. D. Cama and B. G. Christensen, *J. Am. Chem. Soc.*, 1974, **96**, 7582.
145. R. A. Firestone, J. L. Fahey, N. S. Maciejewicz, G. Patel and B. G. Christensen, *J. Med. Chem.*, 1977, **20**, 551.
146. E. G. Brain, C. L. Branch, A. J. Eglington, J. H. C. Nayler, N. F. Osbourne, M. J. Pearson, T. C. Smale, R. Southgate and P. Tolliday, in ref. 10, p. 204.
147. C. L. Branch and M. J. Pearson, *J. Chem. Soc., Perkin Trans. 1*, 1979, 2268.
148. M. Narisada, T. Yoshida, H. Onoue, M. Ohtani, T. Okada, T. Tsuji, I. Kikkawa, N. Haga, H. Satoh, H. Itani and W. Nagata, *J. Med. Chem.*, 1979, **22**, 757.
149. H. Otsuka, W. Nagata, M. Yoshioka, M. Narisada, T. Yoshida, Y. Harada and H. Yamada, *Med. Res. Rev.*, 1981, **1**, 217.
150. W. Nagata, M. Narisada and T. Yoshida, in ref. 6, vol. 2, p. 1.
151. M. Yoshioka, T. Tsuji, S. Uyeo, S. Yamamoto, T. Aoki, Y. Nishitani, S. Mori, H. Satoh, Y. Hamada, H. Ishitobi and W. Nagata, *Tetrahedron Lett.*, 1980, **21**, 351.
152. A. A. Carmine, R. N. Brogden, R. C. Heel, J. A. Romankiewicz, T. M. Speight and G. S. Avery, *Drugs*, 1983, **26**, 279.
153. M. Iwanami, T. Maeda, M. Fujimoto, Y. Nagano, N. Nagano, A. Yamazaki, T. Shibanuma, K. Tamazawa and K. Yano, *Chem. Pharm. Bull.*, 1980, **28**, 2629.
154. A. Ward and D. M. Richards, *Drugs*, 1985, **30**, 382.
155. R. N. Brogden, A. A. Carmine, R. C. Heel, P. A. Morley, T. M. Speight and G. S. Avery, *Drugs*, 1981, **22**, 423.
156. J. Shoji, T. Kato, R. Sakazaki, W. Nagata, Y. Terui, Y. Nakagawa, M. Shiro, K. Matsumoto, T. Hattori, T. Yoshida and E. Kondo, *J. Antibiot.*, 1984, **37**, 1486.
157. H. Ono, Y. Nozaki, N. Katamaya and H. Okazaki, *J. Antibiot.*, 1984, **37**, 1528; S. Harada, H. Tsubotani, H. Ono and H. Okazaki, *ibid.*, 1536; S. Tsubotani, T. Hida, F. Kasahara, Y. Wada and S. Harada, *ibid.*, 1546; Y. Nozaki, K. Okonogi, N. Katayama, H. Ono, S. Harada, M. Kondo and H. Okazaki, *ibid.*, 1555.
158. K. Tsuchiya, in ref. 8, p. 107.
159. H. C. Neu, *Am. J. Med.*, 1985, **79**, supp. 2A, 2.
160. D. M. Richards and R. N. Brogden, *Drugs*, 1985, **29**, 105.
161. R. N. Jones, C. Thornsberry and A. L. Barry, *Antimicrob. Agents Chemother.*, 1984, **25**, 710.
162. W. Dürckheimer and E. Schrinner, in 'Recent Advances in Chemotherapy: Antimicrobial Section 1, ed. J. Ishigami, University of Tokyo Press, 1985, p. 51.
163. T. Naito, S. Aburaki, H. Kamaci, Y. Narita, J. Okumura and H. Kawaguchi, *J. Antibiot.*, 1986, **39**, 1092.
164. S. T. Forgue, W. C. Shyu, C. R. Gleason, K. A. Pittman and R. H. Barbhaiya, *Antimicrob. Agents Chemother.*, 1987, **31**, 799.
165. R. E. Kessler, M. Bies, R. E. Buck, D. R. Chisholm, T. A. Pursiano, Y. H. Tsai, M. Misiek, K. E. Price and F. Leitner, *Antimicrob. Agents Chemother.*, 1985, **27**, 207.
166. T. Une, T. Otani, M. Sato, T. Ikeuchi, Y. Osada, H. Ogawa, K. Sato and S. Mitsuhashi, *Antimicrob. Agents Chemother.*, 1985, **27**, 473.
167. S. Nakagawa, M. Sanada, K. Matsuda, N. Hazumi and N. Tanaka, *Antimicrob. Agents Chemother.*, 1987, **31**, 1100.
168. H. Matsui, M. Komiya, C. Ikeda and I. Tachibana, *Antimicrob. Agents Chemother.*, 1984, **26**, 204.
169. S. Kukolja and R. R. Chauvette, in ref. 6, vol. 1, p. 93.
170. M. Ochiai, O. Aki, A. Morimoto, T. Okada and H. Shimazu, *J. Chem. Soc., Chem. Commun.*, 1972, 800.
171. M. Ochiai, O. Aki, A. Morimoto, T. Okada, K. Shinozaki and Y. Asahi, *Tetrahedron Lett.*, 1972, 2345.
172. S. Kukolja, in ref. 10, p. 181.
173. S. Kukolja, S. R. Lammert, M. R. Gliessner and A. I. Ellis, *J. Am. Chem. Soc.*, 1976, **98**, 5040.

174. T. Naito, H. Hoshi, S. Aburaki, Y. Abe, J. Okamura, K. Tomatsu and H. Kawaguchi, *J. Antibiot.*, 1987, **40**, 991.
175. F. Leitner, T. A. Pursiano, R. E. Buck, Y. H. Tsai, D. R. Chisholm, M. Misiek, J. V. Desiderio and R. E. Kessler, *Antimicrob. Agents Chemother.*, 1987, **31**, 238.
176. N.-X. Chin and H. C. Neu, *Antimicrob. Agents Chemother.*, 1987, **31**, 480.
177. H. Yamanaka, T. Chiba, K. Kawabata, H. Takasugi, T. Masugi and T. Takaya, *J. Antibiot.*, 1985, **38**, 1738.
178. H. Yamanaka, K. Kawabata, K. Miyai, H. Takasugi, T. Kamimura, Y. Mine and T. Takaya, *J. Antibiot.*, 1986, **39**, 101.
179. H. Sakamoto, T. Hirose and Y. Mine, *J. Antibiot.*, 1985, **38**, 496.
180. A. Tsuji, T. Terasaki, I. Tamai and H. Hirooka, *J. Pharmacol. Exp. Ther.*, 1987, **241**, 594.
181. A. Tsuji, H. Hirooka, T. Terasaki, I. Tamai and E. Nakashima, *J. Pharm. Pharmacol.*, 1987, **39**, 272.
182. A. Tsuji, H. Hirooka, I. Tamai and T. Terasaki, *J. Antibiot.*, 1986, **39**, 1592.
183. A. Tsuji, I. Tamai, H. Hirooka and T. Terasaki, *Biochem. Pharmacol.*, 1987, **36**, 565.
184. R. Bell, P. D. Hallam and M. W. Foxton, *Ger. Offen.*, DE 3 516 777 (1985) (*Chem. Abstr.*, 1986, **104**, 224 783).
185. *Anon., Drugs Future*, 1986, **11**, 732.
186. S. Okamoto, Y. Hamana, M. Inoue and S. Mitsuhashi, *Antimicrob. Agents Chemother.*, 1987, **31**, 1111.
187. R. N. Jones, P. C. Fuchs, A. L. Barry, L. W. Ayers, E. H. Gerlach and T. L. Gavan, *Antimicrob. Agents Chemother.*, 1986, **30**, 961.
188. R. Wise, J. M. Andrews and L. J. V. Piddock, *Antimicrob. Agents Chemother.*, 1986, **29**, 1067.
189. K. Fujimoto, S. Ishihara, H. Yanagisawa, J. Ide, E. Nakayama, H. Nakao, S.-I. Sugawara and M. Iwata, *J. Antibiot.*, 1987, **40**, 370.
190. 'β-Lactam Antibiotics for Clinical Use', ed. S. F. Queener, J. A. Webber and S. W. Queener, Marcel Dekker, New York, 1986.
191. T. Tsuji, H. Satoh, M. Narisada, Y. Hamashima and T. Yoshida, *J. Antibiot.*, 1985, **38**, 466.
192. T. Yoshida, in 'Recent Advances in Chemotherapy: Antibiotics Section 1', ed. J. Ishigami, University of Tokyo Press, 1985, p. 48.
193. *Anon., Drugs Future*, 1986, **11**, 452.
194. R. C. Hider, in 'Siderophores from Microorganisms and Plants', ed. M. J. Clarke *et al.*, Springer Verlag, Berlin, 1984, p. 25.
195. K. Hantke and V. Braun, *J. Bacteriol.*, 1978, **135**, 190.
196. N. Ohi, B. Aoki, T. Shinozaki, K. Moro, T. Noto, T. Nehashi, H. Okazaki and I. Matsunaga, *J. Antibiot.*, 1986, **39**, 230.
197. N. Ohi, B. Aoki, K. Moro, T. Kuroki, N. Sugimura, T. Noto, T. Nehashi, M. Matsumoto, H. Okazaki and I. Matsunaga, *J. Antibiot.*, 1986, **39**, 242.
198. N. Ohi, B. Aoki, T. Kuroki, M. Matsumoto, K. Kojima and T. Nehashi, *J. Antibiot.*, 1987, **40**, 22.
199. K. Mochida, C. Shiraki, M. Yamasaki, T. Hirata, K. Sato and R. Okachi, *J. Antibiot.*, 1987, **40**, 14.
200. K. Mochida, Y. Ono, M. Yamasaki, C. Shiraki, T. Hirata, K. Sato and R. Okachi, *J. Antibiot.*, 1987, **40**, 182.
201. H. Mochizuki, Y. Oikawa, H. Yamada, S. Kasukabe, T. Shiihara, K. Murakami, K. Kato, J. Ishiguro and H. Kosuzume, *J. Antibiot.*, 1988, **41**, 377; see also H. Mochizuki, H. Yamada, Y. Oikawa, J. Ishiguro, H. Kosuzume, N. Aizawa and E. Mochida, *Antimicrob. Agents Chemother.*, 1988, **32**, 1648.
202. K. Katsu, K. Kitoh, M. Inoue and S. Mitsuhashi, *Antimicrob. Agents Chemother.*, 1982, **22**, 181.
203. N.-A. Watanabe, T. Nagasu, K. Katsu and K. Kitoh, *Antimicrob. Agents Chemother.*, 1987, **31**, 497.
204. K. Kawabata, H. Yamanaka, H. Takasugi and T. Takaya, *J. Antibiot.*, 1986, **39**, 404.
205. J. Goto, K. Sakane, T. Teraji and T. Kamiya, *J. Antibiot.*, 1984, **37**, 532.
206. J. Goto, K. Sakane, Y. Nakai, T. Teraji and T. Kamiya, *J. Antibiot.*, 1984, **37**, 546.
207. H. Yamanaka, H. Takasugi, T. Masugi, H. Kochi, K. Miyai and T. Takaya, *J. Antibiot.*, 1985, **38**, 1068.
208. J. Goto, K. Sakane, and T. Teraji, *J. Antibiot.*, 1984, **37**, 557.
209. I. Csendes, B. W. Muller and W. Tosch, *J. Antibiot.*, 1983, **36**, 1020.
210. K. Sakagami, T. Mishina, T. Kuroda, M. Hatanaka and T. Ishimaru, *J. Antibiot.*, 1983, **36**, 1205.
211. W. J. Wheeler, J. B. Deeter, D. R. Finley, M. D. Kinnick, R. Koehler, H. E. Osborn, J. T. Ott, J. K. Swartzendruber and D. G. Wishka, *J. Antibiot.*, 1986, **39**, 111.
212. W. J. Wheeler, D. R. Finley, R. J. Messenger, R. Koehler and J. T. Ott, *J. Antibiot.*, 1986, **39**, 121.
213. M. Boberg, K. G. Metzger and H. J. Zeiler (Bayer A.-G.), *Ger. Offen.* DE 3 224 866 (*Chem. Abstr.*, 1984, **100**, 174 531).
214. K. Nishide, T. Kobori, D. Tunemoto and K. Kondo, *Heterocycles*, 1987, **26**, 633.
215. J. Hannah, C. R. Johnson, A. F. Wagner and E. Walton, *J. Med. Chem.*, 1982, **25**, 457.
216. B. A. Pelak, E. C. Gilfillan, B. Weissberger and H. H. Gadebusch, *J. Antibiot.*, 1987, **40**, 354.
217. S. Wolfe, A. L. Demain, S. E. Jensen, and D. W. Westlake, *Science (Washington, D.C.)*, 1984, **226**, 1386.
218. J. E. Baldwin, R. M. Adlington, A. E. Derome, H.-H. Ting and N. J. Turner, *J. Chem. Soc., Chem. Commun.*, 1984, 1211.
219. J. E. Baldwin, E. P. Abraham, G. L. Burge and H.-H. Ting, *J. Chem. Soc., Chem. Commun.*, 1985, 1808.
220. J. E. Baldwin, R. M. Adlington, J. B. Coates, M. J. C. Crabbe, J. W. Keeping, G. C. Knight, T. Nomoto, C. J. Schofield and H.-H. Ting, *J. Chem. Soc., Chem. Commun.*, 1987, 374.
221. 'Recent Advances in the Chemistry of β-Lactam Antibiotics', ed. P. H. Bentley and R. Southgate, Special Publication No. 70, The Royal Society of Chemistry, London, 1989.

9.3

Other β-Lactam Agents

ALLAN G. BROWN, MICHAEL J. PEARSON and ROBERT SOUTHGATE
Beecham Pharmaceuticals, Betchworth, Surrey, UK

9.3.1	INTRODUCTION	655
9.3.2	THE NOCARDICINS AND MONOBACTAMS	656
	9.3.2.1 Nocardicins	656
	9.3.2.2 Monobactams	658
9.3.3	CLAVULANIC ACID AND ANALOGUES	662
9.3.4	CARBAPENEMS	671
9.3.5	PENEMS	674
9.3.6	OTHER FUSED β-LACTAM SYSTEMS	677
	9.3.6.1 Penams and Cephems	677
	9.3.6.2 1-Oxadethiapenams	679
	9.3.6.3 1-Carbapenams	679
	9.3.6.4 1-Azadethiapenams and 1-Azadethiapenems	680
	9.3.6.5 2-Hetero-1-dethiapenams	681
	9.3.6.6 1-Oxadethiacephems	684
	9.3.6.7 1-Carbacephems	685
	9.3.6.8 1-Azadethiacephems	687
	9.3.6.9 1-Phosphadethiacephem 1-Oxides	687
	9.3.6.10 2-Hetero-1-dethiacephems	688
	9.3.6.11 3-Hetero-1-dethiacephams	690
	9.3.6.12 3-Hetero-cephems and -cephams	691
	9.3.6.13 2,3-Diaza-1-dethiacephems	691
	9.3.6.14 Diverse Bicyclic Systems	692
	9.3.6.15 Fused 1,2-Diazetidinones	693
9.3.7	NON-β-LACTAM β-LACTAM ANALOGUES	693
9.3.8	β-LACTAM ACTIVITY, REACTIVITY AND STRUCTURAL FEATURES	694
9.3.9	REFERENCES	696

9.3.1 INTRODUCTION

During the 1950s and 1960s, effort on the chemistry and biology of β-lactams was generally devoted to developing penicillins and cephalosporins with improved antibacterial spectra, increased potency, better stability to β-lactamases and more useful pharmacokinetic properties. In the late 1960s and early 1970s, further advances evolved around the discovery of the cephamycins. The chemotherapeutic properties of these penicillins/cephalosporins and C-6/C-7 substituted derivatives is reviewed in Chapter 9.2. Reports of other naturally occurring β-lactams in this period were restricted to the tabtoxins[1] from the bacterium *Pseudomonas tabaci* and the pachystermines[2] from the plant *Pachysandra terminalis*. Up to the mid-1960s the general chemistry of β-lactams, apart from the acylation of 6-aminopenicillanic acid (6-APA) and 7-aminocephalosporanic acid (7-ACA), had been explored in a very limited manner, despite the early work of Staudinger.[3] Extensive investigations during the 1970s and 1980s have resulted in the discovery of a variety of new β-lactam antibiotics and β-lactamase inhibitors as natural products. These developments have inspired a vast amount of synthetic organic chemistry aimed at the modification of the naturally occurring compounds and at the total synthesis of these products and a wide range of analogues. Some of the history of these discoveries and progress in their development is to be found in a number of major review volumes and specialist publications.[4-10]

The year 1976, in particular, heralded a new era in the history of β-lactams with the description of the nocardicins (**1**),[11] clavulanic acid (**2**),[12,77] thienamycin (**3**)[13] and the olivanic acids (**4**).[13] These compounds represented naturally occurring examples of the azetidinone (**5**), the oxapenam/clavam (**6**) and the carbapenem (**7**) β-lactam ring systems and, together with subsequently reported new systems, *e.g.* the monobactams (**8**),[14] resulted from screening programmes designed to detect β-lactam antibiotics, β-lactamase inhibitors, and inhibitors of bacterial cell wall synthesis, produced by fungi, actinomycetes and bacteria.

This chapter will briefly outline the properties of these new β-lactams and, in addition, those of the variety of synthetic β-lactam entities prepared as alternative products for evaluation as analogues of the older penicillins/cephalosporins or 'classical' types and the newer 'nonclassical' β-lactams. It is not possible to give extensive details on the chemical, biochemical and biological properties of all structures but the principal features of the various 'nonclassical' β-lactam containing systems will be discussed.

9.3.2 THE NOCARDICINS AND MONOBACTAMS

9.3.2.1 Nocardicins

Fujisawa research workers derived a mutant strain of *Escherichia coli* which was highly sensitive to a variety of β-lactams. Application of this strain in a screening programme, designed to highlight β-lactams in fermentation broths obtained from cultures isolated from a range of soil samples, led to the identification of the nocardicin family of β-lactam antibiotics.[15] The *E. coli* assay was used in conjunction with tests for β-lactamase susceptibility and effect on an *in vitro* peptidoglycan synthesizing system. The nocardicins A–G (**9**)–(**15**) were isolated from *Nocardia uniformis tsuyamanensis* and were shown to be substituted azetidinones with the same stereochemical features as penicillins at the site of the substitution of the β-lactam ring and at the carbon atom containing the carboxylate function; the presence of an oximino function on some of the nocardicins was a further unusual feature. Other actinomycetes, *e.g. Streptomyces alcalophilus*, *Actinosynnema mirum*, *Nocardiopsis atra* and *Microtetraspora caesia*, have been claimed to produce nocardicins. The biosynthetic precursors of nocardicin A have been identified as L-methionine, L-*p*-hydroxyphenylglycine and L-serine, suggesting it is derived from a peptide, with β-lactam

Nocardicin		R^1	R^2, R^3
A	(9)	HO$_2$CĊH(CH$_2$)$_2$ (D), NH$_2$	=N–OH
B	(10)	HO$_2$CĊH(CH$_2$)$_2$ (D), NH$_2$	=N–OH
C	(11)	HO$_2$CĊH(CH$_2$)$_2$ (D), NH$_2$	H, NH$_2$
D	(12)	HO$_2$CĊH(CH$_2$)$_2$ (D), NH$_2$	=O
E	(13)	H	=N–OH
F	(14)	H	=N–OH
G	(15)	H	H, NH$_2$

(16) 3-ANA; R = H
(17) R = Bu^{t}OCO

(18)

formation occurring *via* nucleophilic displacement of an activated serine hydroxyl group by an amide nitrogen atom.[16]

Nocardicin A possesses weak *in vitro* antibacterial activity with greatest effect against certain strains of *Pseudomonas aeruginosa* and indole-positive and -negative strains of *Proteus*. On *in vivo* evaluation against *Proteus* and *Pseudomonas* infections in mice the therapeutic effects seen are not only related to the cellular damage caused by nocardicin A but also to the effect of serum components on already damaged bacteria. Phagocytosis by polymorphonuclear leukocytes also plays a role in bactericidal effects observed *in vivo*.

Removal of the *N*-acyl function of nocardicins was examined in order to yield 3-aminonocardinic acid (3-ANA; **16**). The most efficient chemical method involves preparation of the *t*-butyloxy-carbonyl derivative (**17**) of 3-ANA from nocardicin A. Hydrolysis of (**17**) gives 3-ANA in high yield. An earlier method depended upon the hydrolysis of the bisthiourea derivative of nocardicin C. Enzymatic deacylation of nocardicin C has also been successful. 3-ANA has been prepared from penicillin G, as have a range of derivatives *via* the cycloaddition of ketenes to imines. A variation on this last procedure using the hexahydro-*sym*-triazine (**18**), derived from *p*-hydroxyphenylglycine, and phthalimidoacetyl chloride led to a variety of nocardicin analogues. Despite the preparation of several hundred compounds in the nocardicin series, none was found to be more effective than nocardicin A.[11]

Recently reported naturally occurring nocardicins include chlorocardicin (**19**) from a *Streptomyces* spp.[17] and the formadicins (**20**)–(**23**) derived from the bacterium *Flexibacter alginoliquefaciens*.[18]

(**19**)

Formadicin		R^1	R^2	R^3
A	(**20**)	NHCHO	OH	
B	(**21**)	H	NHCHO	
C	(**22**)	NHCHO	OH	H
D	(**23**)	H	NHCHO	H

9.3.2.2 Monobactams

Use of a β-lactam hypersensitive strain of *Pseudomonas aeruginosa* by Takeda workers[19] and application of a β-lactamase induction assay based on a strain of *Bacillus licheniformis* by the Squibb group[20] resulted in the discovery of examples of the monobactam family of β-lactams.[14,21] The compounds (**24**)–(**27**) were produced by certain Gram-negative bacteria (*e.g. Pseudomonas, Gluconobacter, Acteobacter, Chromobacterium* and *Agrobacterium* spp.) and are acyl derivatives of 3-amino-2-oxoazetidine-1-sulfonic acid (**28**). A characteristic feature of this class of β-lactams, relative to the nocardicins, is the presence of a sulfonic acid residue on the β-lactam nitrogen atom instead of the phenolic acid unit of the nocardicins. These naturally occurring compounds have only weak antibacterial activity but prompted a great amount of interest in that, until their discovery, all attempts to produce antibacterially active azetidinones by total synthesis had failed to produce any compounds with significant activity. The simplest monobactam (**26**) has been synthesized from 7-aminodeacetoxycephalosporanic acid.[22] Biosynthetic studies[23] on SQ 26,180 (**26**) and sulfazecin (**24**) indicate that the β-lactam carbon atoms come from serine, while the source of the sulfamate sulfur atom can be either inorganic or organic depending upon the monobactam producing organism used. The origin of the β-lactam nitrogen atom is unknown but the methoxy methyl group is methionine derived.

None of the naturally occurring monobactams are useful precursors for semisynthetic modification,[24] but their novelty and structural simplicity provided stimulus for the development of a variety of innovative, totally synthetic approaches. Although the 3α-methoxy group imparts greater β-lactamase stability to the natural monobactams, it has a detrimental effect on their chemical

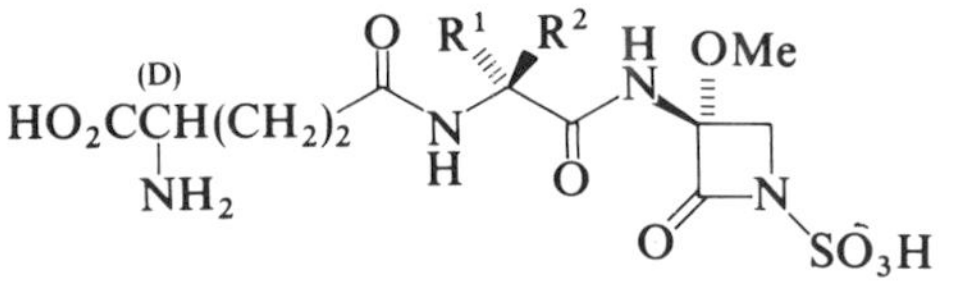

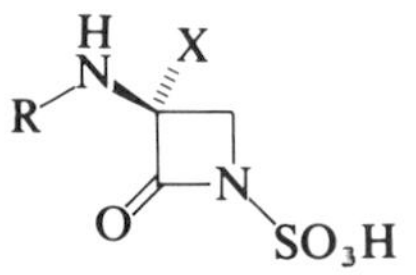

(24) Sulfazecin; R^1 = Me, R^2 = H
(25) Isosulfazecin; R^1 = H, R^2 = Me

(26) SQ 26,180; R = COMe; X = OMe
(28) X = R = H

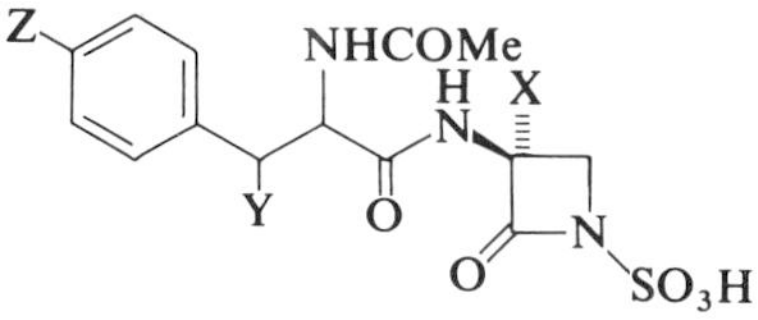

X = H or OMe
Y, Z = H, OH or OSO$_3$H

(27)

stability.[25] Replacement of methoxy by methyl[26] or formamido[27] has also been reported, although the main thrust of structure–activity studies has been directed towards 3α-unsubstituted derivatives.

In view of the common enzyme targets it is not surprising that the side-chain structure–activity relationship in monobactams[14,28–30] essentially parallels those of the penicillins and cephalosporins. The 2-oxoazetidine-1-sulfonate skeleton was retained whilst side-chain alteration at C-3 and substitution at C-4 were examined. Intrinsic activity was improved by modification of the acyl side chain, although the derivatives suffered from enzyme inactivation. This deficiency was best overcome by the addition of an alkyl group at the C-4 position (29), 4β-alkylmonobactams showing better β-lactamase stability than their 4α counterparts. 4-Alkyl substitution also increased Gram-negative potency but compromised the already poor Gram-positive activity. Increasing the size of the alkyl group beyond methyl diminished intrinsic activity.[31]

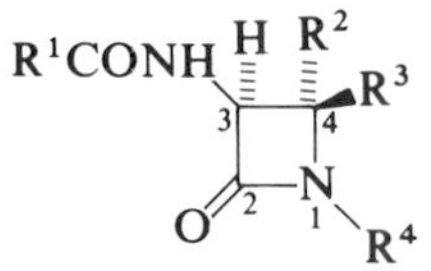

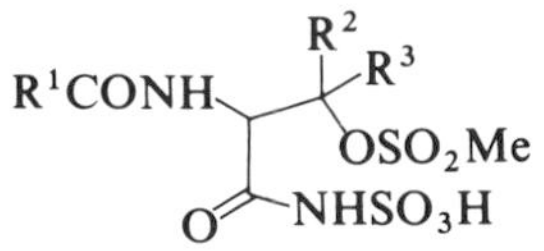

(29) R^4 = SO$_3$H (32) R^4 = CO$_2$H
(30) R^4 = H (33) R^4 = CONHSO$_2$NR^{5}R^6
(31) R^4 = OH (34) R^4 =

(35)

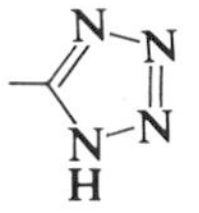

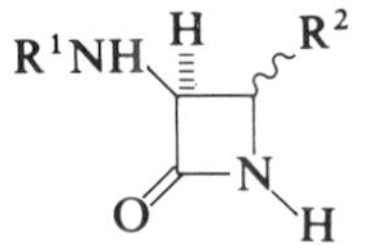

(36) R^2 = SO$_2$Ph, OCOMe

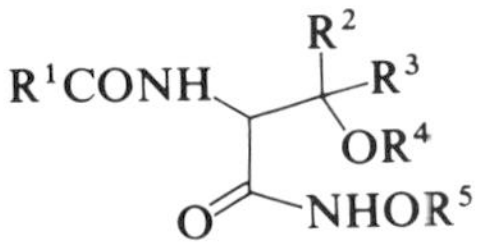

(37) R^4 = H
(38) R^4 = SO$_2$Me

Two general strategies were adopted for the synthesis of 4-alkylated monobactams.[31–34] These involve sulfonation of 4-alkyl-3-(protected)-2-azetidinones (30) with various complexes of sulfur trioxide, or cyclization of β-methanesulfonyloxyacyl sulfamates (35). With respect to the first approach, 4-alkylazetidinones have been prepared *via* displacements on penicillin-derived β-lactams (36), or Mitsunobu-type ring closure of β-hydroxyamino acid hydroxamates (37). Hydroxamic acid activation is crucial to the success of this cyclization, which proceeds with clean inversion at the carbinol centre and retention of configuration at C-3.[35] The versatility of the method has been exemplified by its extension to the base-induced cyclization of methanesulfonates (38).[36,37] N—O reduction either directly in the case of *N*-methoxy derivatives,[36] or *via* the *N*-hydroxy derivatives (31) with *N*-benzyloxy compounds,[35] then provided access to the required *N*-unsubstituted β-lactams. The second, more efficient protocol, is illustrated in Scheme 1.[38] The use of

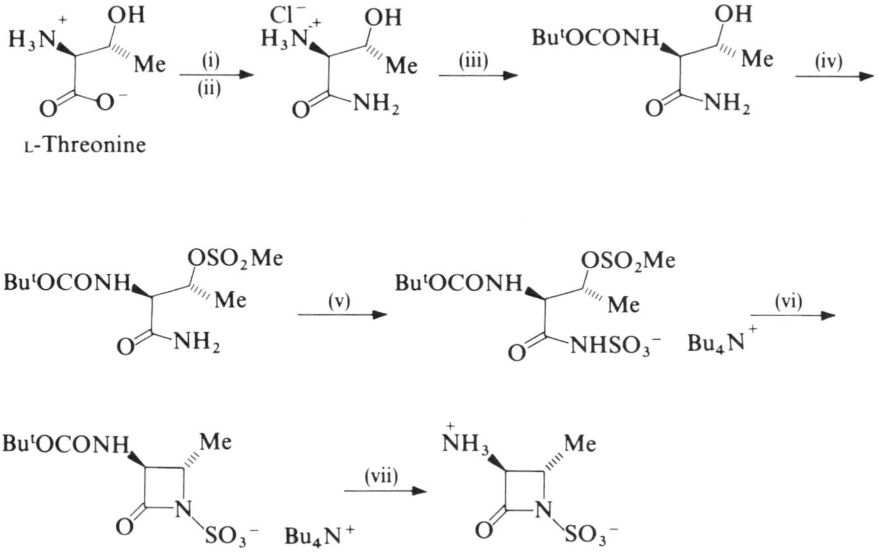

(i) SOCl$_2$/MeOH; (ii) NH$_3$, H$_2$O; (iii) (Bu^{t}OCO)$_2$O; (iv) MeSO$_2$Cl;

(v) 2-Pic/SO$_3$, Bu$_4$N$^+$ HSO$_4^-$; (vi) KHCO$_3$; (vii) 97% HCO$_2$H

Scheme 1

2-picoline–sulfur trioxide was the result of an involved study employing several amine–sulfur trioxide complexes, and the mildness of the cyclization procedure means that a wide variety of amino protecting groups can be tolerated. The conversion of L-threonine to zwitterion (**39**) proceeds in approximately 50% overall yield in seven steps with *t*-butoxycarbonyl protection of the amino function, without the necessity for chromatography.

Synthetic considerations and the structure–activity studies led to the development of aztreonam (**40**) as a clinically useful monobactam.[39,40] Aztreonam exhibits a high degree of stability to β-lactamases and is specifically active against aerobic Gram-negative bacteria including *Pseudomonas aeruginosa*. Inactivity against Gram-positive organisms is due to lack of affinity for the essential target proteins and not enzyme inactivation.[41]

Other groups have described related monobactams with similar activity profiles to aztreonam.[42–47] In particular, carumonam (**41**) has been selected for clinical evaluation.[48,49] The trifluoromethyl analogue (**42**) was devoid of significant antibacterial activity, a factor attributed to chemical instability.[50] Within a series of monobactams with heteroatom bound substituents at the 4-position, antibacterial activity decreases in the order SR > N$_3$ > OR > SO$_2$R.[51,52] In the most active series of sulfur-linked derivatives, the 4β-carboxamidomethylthio analogue (**43**) is the most potent. Compared to its 4β-methyl counterpart, it displays similar β-lactamase stability but less intrinsic activity. The 3-(1-hydroxyethyl) derivative (**44**) was devoid of antibacterial activity, indic-

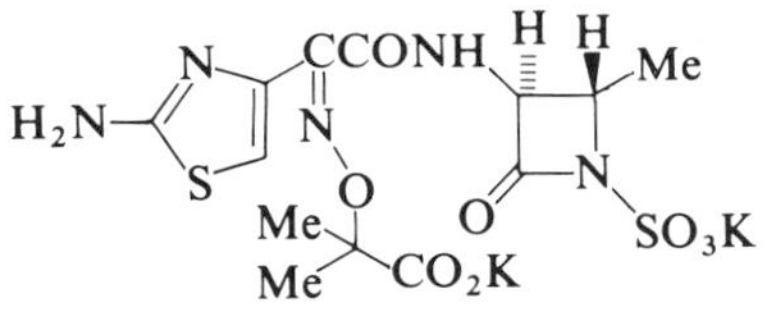

(**40**) Aztreonam

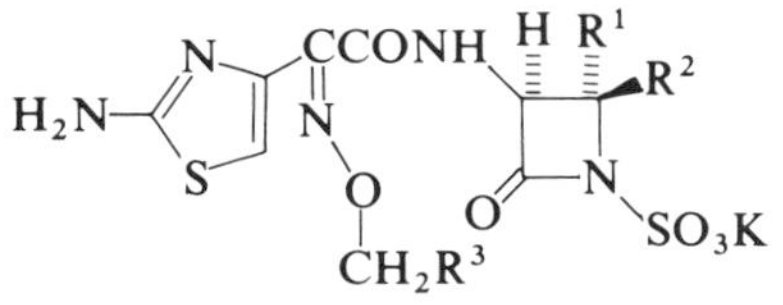

(**41**) R^1 = H, R^2 = CH$_2$OCONH$_2$, R^3 = CO$_2$K

(**42**) R^1 = CF$_3$, R^2 = R^3 = H

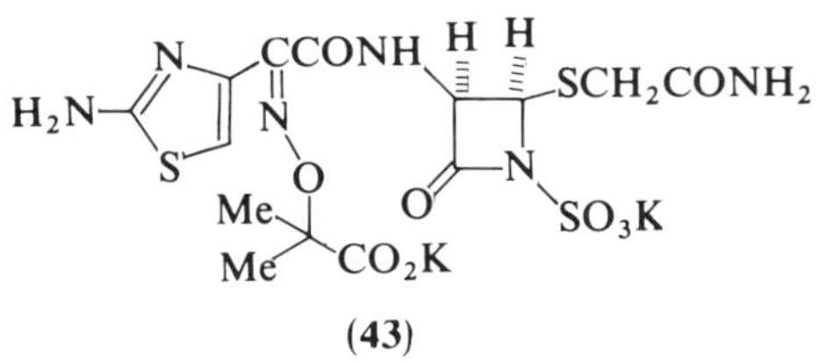

(**43**)

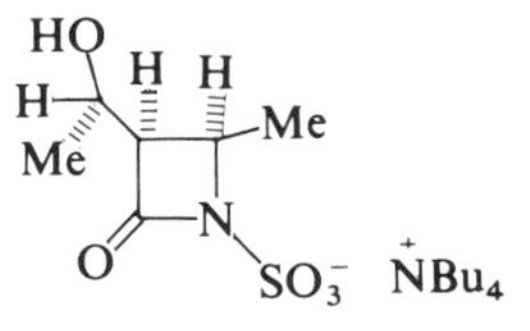

(**44**)

ative that, in common with penicillins and cephalosporins, the acylamino side chain is essential for useful antibacterial activity.[53]

Monocyclic β-lactams can be activated by replacing the sulfonate moiety of monobactams by other peripheral ionizable groups.[54] The ubiquitous aminothiazolyloxime side chains were preferred in all series. Phosphorylation of azetidinones provides monophosphams (**45**).[54,55] In comparison with the monobactams, the 1-phosphonic acid residue confers less intrinsic activity, particularly against *Pseudomonas*, but greater β-lactamase stability. The residue attached to the phosphorus profoundly influences antibacterial activity (OMe > OPh > OH > Me ≈ Ph).

Attempts to prepare the carboxylates (**32**) were frustrated by their decarboxylation to give N-1 unsubstituted azetidinones and carbon dioxide.[56] Examination of other carbonyl activating groups led to the monocarbams (**33**). Reaction of a suitably protected azetidinone (**46**) with chlorosulfonyl isocyanate affords the key intermediate (**47**), from which all the other derivatives are prepared.[57,58] The presence of a sufficiently acidic —NH— is mandatory for antibacterial activity. 4-Unsubstituted compounds are favoured and exhibit potent activity against aerobic Gram-negative, β-lactamase producing bacteria. The hydroxypyridone (**48**) was particularly active against *Pseudomonas*.[59]

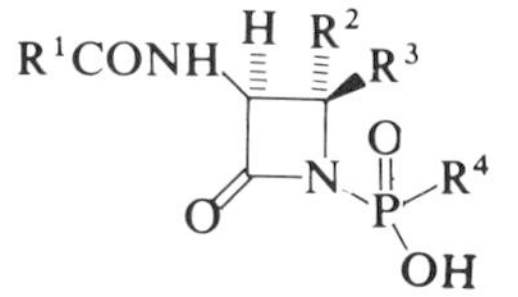

(**45**)

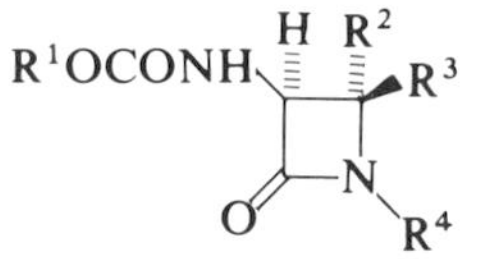

(**46**) R^1 = Bu^t, CH$_2$Ph, R^4 = H

(**47**) R^1 = Bu^t, CH$_2$Ph, R^4 = CONHSO$_2$Cl

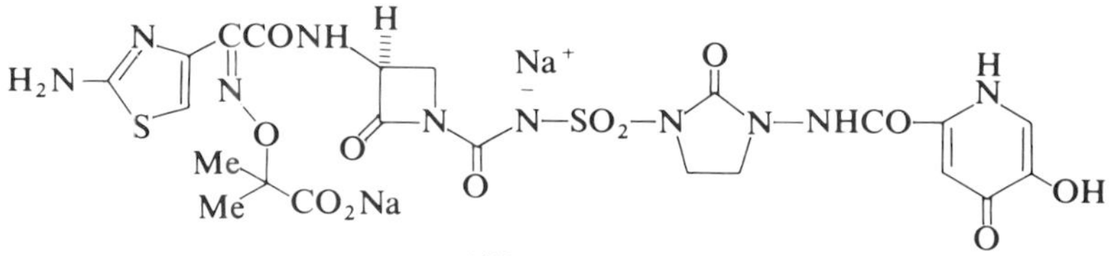

(**48**)

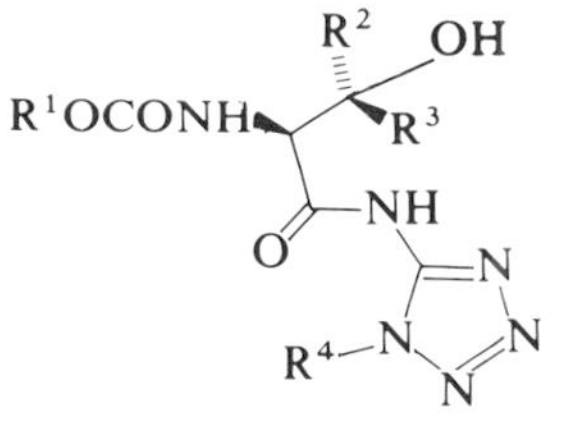

(**49**) R^4 = Protecting group

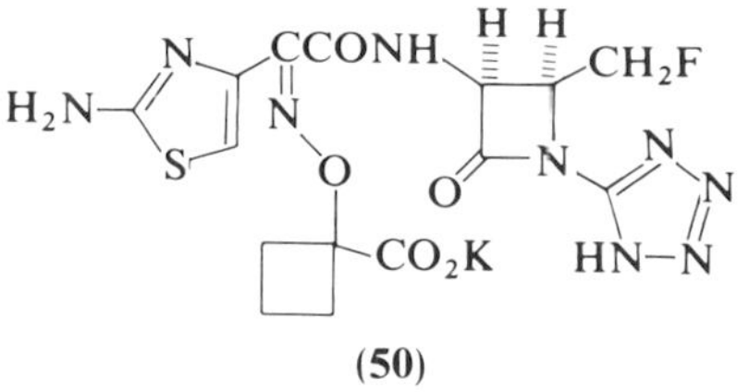

(**50**)

The tetrazole moiety possesses comparable electron-drawing power and acidity to a carboxylic acid, and N-(tetrazol-5-yl)azetidinones (**34**) have been extensively investigated.[60-62] Intramolecular ring closure of appropriately protected amides (**49**) affords the desired derivatives without difficulty. Structure–activity relationships broadly parallel those of the monobactams. The aminothiazole (**50**) is reported to be more active than aztreonam, although sufficient potency against Gram-positive organisms remained elusive.[60]

The availability of 1-hydroxyazetidinones (**31**)[35] facilitated the synthesis of the monosulfactams (**51**).[54,63-65] The early derivatives showed good intrinsic activity against Gram-negative bacteria, but even the C-4 monomethylated compounds were chemically unstable and susceptible to enzymic inactivation. 4,4-Dimethyl substitution (**54**) overcomes these drawbacks and, in contrast to the monobactams,[31] does not compromise the antibacterial activity. Removal of the *gem*-dimethyl groups from the C-3 side chain of (**54**) minimally affects β-lactamase stability, but affords an orally absorbed monosulfactam (**55**).[66] Related oxygen-activated azetidinones (*vide infra*) are substantially less absorbed. It appears that the aza analogues (**52**) are only moderately active,[67] and the sulfur ones (**53**) inactive.[68]

N-1 *O*-Phosphorus activated azetidinones (**56**) are more active than the corresponding phosphonates, the most dramatic improvement being against *Pseudomonas*. However, they are intrinsically less stable to β-lactamases, a feature that is little modified by 4-substitution.[25,54]

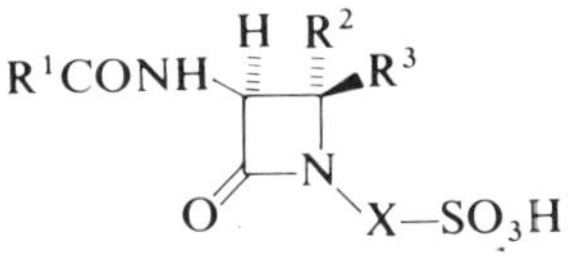

(51) X = O
(52) X = NR⁴
(53) X = S

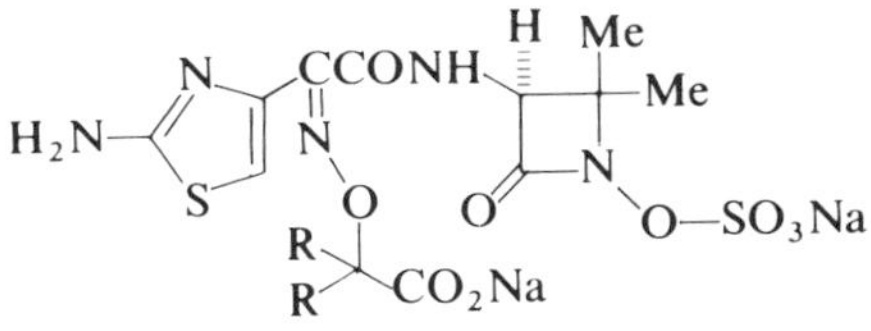

(54) R = Me

(55) R = H

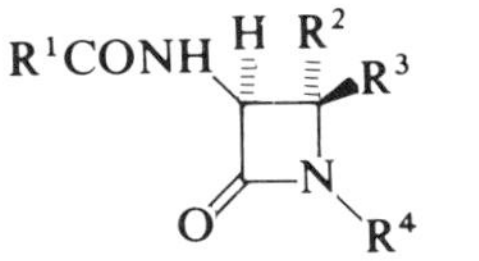

(56) R⁴ = O—P(=O)(R⁵)(OH)

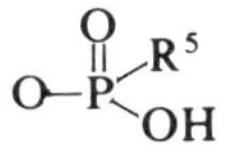

(57) R⁴ = OCH₂CO₂H

(58) R⁴ = SCH₂CO₂H

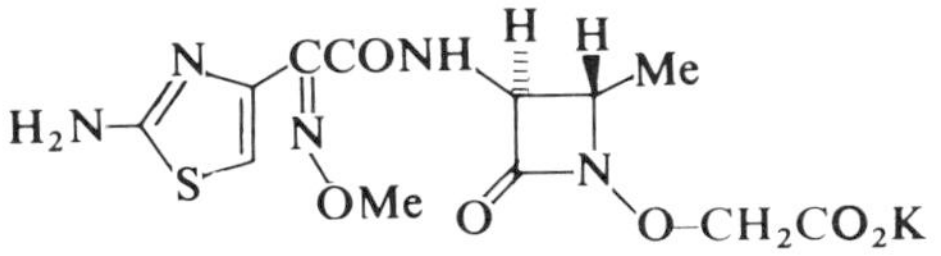

(59)

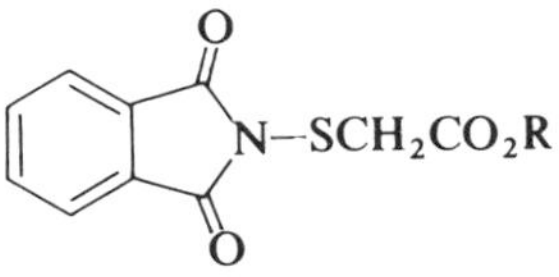

(60)

To test the compatibility of remote positioning of the ionizable group in heteroatom-activated β-lactams, the oxamazins (57) were synthesized.[69,70] Direct alkylation of *N*-hydroxyazetidinones (31) with bromoacetates proved the most convenient method of synthesis. Structure–activity relationships in this series are closely related to those in the monobactams, and the amino-thiazolyloxime (59) represents the optimal combination of antibacterial activity and β-lactamase stability.[69,71] Replacement of the oxygen by a sulfur atom led to the thiamazins (58).[72] Direct sulfenylation of *N*-unsubstituted β-lactams with substituted phthalimides (60) provided the basic nucleus most efficiently. The various thiamazins tested were devoid of antibacterial activity, a feature that has been largely attributed to a poor fit in the active site of the bacterial enzymes due to the molecule's inherent bond lengths and bond angles.[73]

The principle of heteroatom activation of monocyclic β-lactams is now well established. However, despite intensive investigation the N-1 sulfonates still provide the most favourable match of antibacterial potency and resistance to enzyme inactivation. Although oral absorption has been achieved, the concomitant incorporation of good Gram-positive activity remains an attractive goal.

9.3.3 CLAVULANIC ACID AND ANALOGUES

Application of a screen[74] to detect inhibitors of the β-lactamase produced by the bacterium *Klebsiella aerogenes* to culture filtrates, derived from various *Streptomyces* spp. isolated from soil samples collected from different locations, resulted in the detection of the group of antibiotic β-lactamase inhibitors called the olivanic acids (see below). Before these β-lactam metabolites had been fully identified and in order to show that they were not the cephamycins (see Chapter 9.2), the known cephamycin producing *Streptomyces* spp. were screened for β-lactamase inhibitor production. One such culture, *Streptomyces clavuligerus*, was found to produce a metabolite (initially called BRL 14151) with different chromatographic properties to the known olivanic acids and cephamycins. Subsequent isolation and characterization identified BRL 14151 as the bicyclic β-lactam (2) which was named clavulanic acid.[75–77] The novel structural features of clavulanic acid (relative to the penicillins and cephalosporins) include the presence of oxygen instead of sulfur, no acylamino side chain at C-6 and the presence of an exocyclic β-hydroxyethylidene substituent at C-2. The stereochemistry of C-3 and C-5 was the same as that of the penicillins. The basic ring system of clavulanic acid, by analogy with the use of penam (61) and cephem (62), has been given the trivial name clavam (6). Systematic nomenclature defines clavulanic acid as (Z)-(2R,5R)-3-(β-hydroxyethylidene)-7-oxo-4-oxa-1-azabicyclo[3.2.0]heptane-2-carboxylic acid. Clavulanic acid has also been isolated from other strains and genera of *Streptomyces*, e.g. *S. jimonjinensis* and *S. katsurahamanus*.[78]

Clavulanic acid possesses weak, though broad-spectrum, activity as an antibiotic, but it is a potent progressive inhibitor of a wide range of β-lactamase types.[79] The production of β-lactamase is the

(61)   **(62)**

major mechanism whereby Gram-positive and Gram-negative bacteria develop resistance to β-lactam antibiotics. β-Lactamase was recognized, initially by Abraham and Chain,[80] as an enzyme capable of degrading penicillin into penicilloic acid. Cephalosporins are similarly broken down but a more complex reaction sequence can ensue. The various types of β-lactamase (Figure 1) have been classified by a number of systems (see Chapter 9.2).[81-83] The classification developed by Richmond and Sykes[81] uses Classes or Types depending upon the substrate and inhibitor profiles, isoelectric points, molecular weights and genetic origin of the enzyme (Table 1). That of Ambler[82] is based upon amino acid sequence and defines three classes A, B and C (Table 2) (see later). The β-lactamase inhibitory properties of clavulanic acid are given in Table 3.[79] The data show that clavulanic acid inhibits various β-lactamases produced by the Gram-negative bacteria, including the plasmid-mediated β-lactamases from isolates of *Escherichia coli, Klebsiella pneumoniae (aerogenes), Proteus mirabilis, Pseudomonas aeruginosa, Haemophilus influenzae, Neisseria gonorrhoeae* and *Branhamella catarrhalis*. While over 20 different plasmid-mediated β-lactamases have been characterized, it is the Class III or TEM Type in particular that have been recognized as the principal causative factor in outbreaks of resistant bacteria. Clavulanic acid also inhibits the chromosomally mediated β-lactamases from *Klebsiella pneumoniae, Proteus mirabilis Proteus vulgaris* and *Bacteroides fragilis*; it does not, however, inhibit the chromosomal (Class I type) β-lactamases (cephalosporinases) produced by certain strains of *Enterobacter cloacae, Citrobacter freundii, Morganella morganii, Proteus retgerii* and *Pseudomonas aeruginosa*. The β-lactamase from the Gram-positive organism *Staphylococcus aureus*, and some produced by certain *Bacillus* spp. (*e.g. B. cereus* I β-lactamase), are readily inhibited by clavulanic acid.

In conjunction with certain β-lactam antibiotics, such as penicillin G (63), ampicillin (64), amoxycillin (65), ticarcillin (66) and cephaloridine (67), which show varying degrees of susceptibility to β-lactamases, clavulanic acid shows a pronounced synergistic effect and protects these β-lactams from inactivation.[85] Tables 4 and 5 show the *in vitro* activity of amoxycillin and ticarcillin in combination with clavulanic acid against a variety of β-lactamase producing bacteria. The synergistic effect seen was in general bactericidal with, in the case of amoxycillin, spheroplast formation and lysis; it was not adversely affected by media components, high inocula, or serum; there

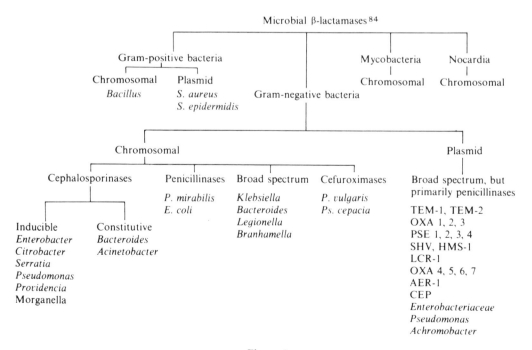

Figure 1

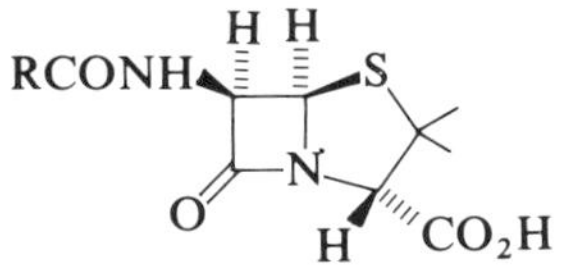

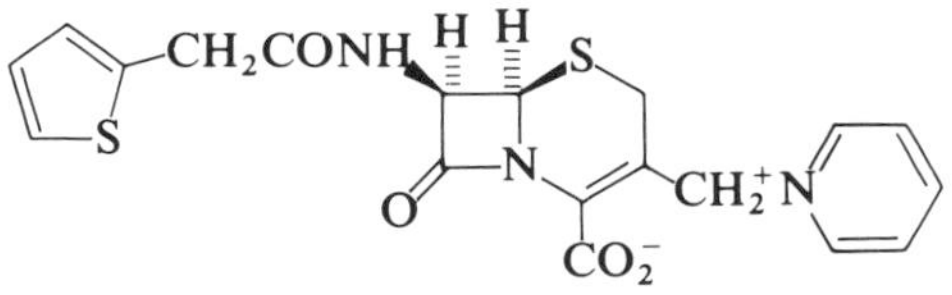

(63) R = phenyl-CH₂

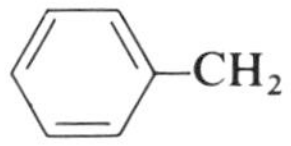

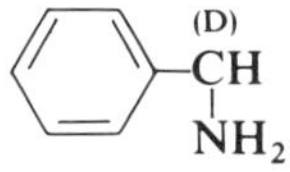

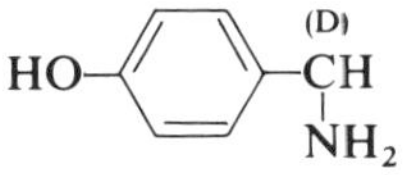

(63) R = ⟨phenyl⟩—CH₂

(64) R = ⟨phenyl⟩—CH(D)—NH₂

(65) R = HO—⟨phenyl⟩—CH(D)—NH₂

(66) R = ⟨thiophene⟩—CH(D)—CO₂H

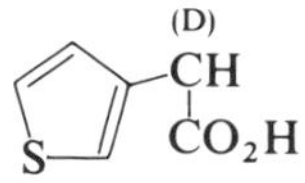

Table 1 Richmond–Sykes Classification of β-Lactamase with a Sixth Class Added for Bacteroides[84]

TYPE I	Enzymes active principally against cephalosporins, inhibited by isoxazoyl penicillins and partially by carbenicillin, but not by clavulanic acid or sulbactam Molecular weight: 24 000–46 000 Genetic origin: Chromosomal Profile:[a] Pen 100, Amp 100, Carb 5, Clox 0, CER 150–8000 Trivial names: P99 of *Enterobacter*. Sabath–Abraham enzyme *P. aeruginosa* Found in: *Enterobacter, Morganella, Proteus vulgaris, Providencia, Pseudomonas, Klebsiella, Serratia, Citrobacter*
TYPE II	Enzymes active against penicillins inhibited by isoxazoylpenicillins but not by carbenicillin; inhibited by clavulanic acid Molecular weight: 25 000–30 000 Profile:[a] Pen 100, Amp 150, Carb 40, Clox 0, CER 10 Found in: *Proteus mirabilis, E. coli*
TYPE III	Enzymes that equally hydrolyze penicillins and cephalosporins, inhibited by pCMB, isoxazoylpenicillins; low activity against carbenicillin; inhibited by clavulanic acid, sulbactam Molecular weight: 17 000–29 000 Genetic origin: Plasmid Profile:[a] Pen 100, Amp 110, Carb 10, Clox 0, CER 75 Trivial names: TEM-1, TEM-2, SHV-1, HMS Found in: *E. coli, Haemophilus, Neisseria, Salmonella, Shigella, Pseudomonas*. Most common enzyme worldwide.
TYPE IV	Enzymes equally hydrolyze penicillins and cephalosporins; resistant to inhibition by isoxazoylpenicillins, carbenicillin and pCMB but inhibited by clavulanic acid and sulbactam Molecular weight: 18 000–25 000 Genetic origin: Chromosomal Profile:[a] Pen 100, Amp 150, Carb 50, Clox 20, CER 70 Trivial names: K-1 Found in: *Klebsiella*
TYPE V	Equally hydrolyze penicillins and cephalosporins; hydrolyze isoxazoylpenicillins better than cephalosporins; inhibited by clavulanic acid Molecular weight: 12 000–32 000 Genetic origin: Plasmid Profile:[a] Pen 100, Amp 200, Carb 90–250, Clox 200, CER 10–50 Trivial names: OXA 1, 2, 3; PSE 1, 2, 3, 4 Found in: *E. coli, Pseudomonas, Serratia*
TYPE VI	Hydrolyze cephalosporins better than penicillins; inhibited by cloxacillin or carbenicillin; hydrolyze cefamandole, cefuroxime, cefotaxime, but not cefoxitin. Profile:[a] Pen 3, Amp 1.5, Carb 0.5, CER 100 Found in: *Bacteroides*

[a]Pen = penicillin, Amp = ampicillin, Carb = carbenicillin, Clox = cloxacillin, CER = cephaloridine. Rates are relative.

Table 2 Classification of β-Lactamases Based on Amino Acid Sequences[84]

Class A	*S. aureus, B. licheniformis, B. cereus, E. coli* TEM Key amino acid: serine
Class B	Metalloenzymes; zinc; inactivate moxalactam, imipenem and cefoxitin Occurs in *B. cereus*
Class C	Chromosomally encoded Amp C gene of *E. coli* K12; *P. aeruginosa* Key amino acid: serine

Table 3 β-Lactamases; Their Classification and Inhibition by Clavulanic Acid[a]

β-Lactamase	Class type	Example	Amoxycillin hydrolysis rate[b]	Effect of clavulanic acid	IC_{50}[c] (μg mL^{-1})
Chromosomal cephalosporinase	I	*Enterobacter cloacae* P99	5	−	110
		Pseudomonas aeruginosa A	110	−	
		Escherichia coli JT410	2	−	
		Bacteroides fragilis	> 100	+	
Chromosomal penicillinase	II	*Proteus mirabilis* C889	142	+	0.015
Chromosomal broad spectrum	IV	*Klebsiella pneumoniae* E70	127	+	0.007
		Branhamella catarrhalis 1908	120	+	
Plasmid mediated	III	TEM-I and 2; *E. coli* JT4	86	+	0.06
		SHV	126		
	V	OXA type, *e.g.* OXA-2	162	+	
		PSE type, *e.g.* PSE-4	92	+	
		Pseudomonas aeruginosa Dalgleish			
Gram-positive penicillanase		*Staphylococcus aureus* Russell	168	+	0.03

[a] Adapted from ref. 79. [b] Pen G = 100. [c] Amount required to give 50% protection of nitrocefin (at 250 μg mL^{-1}) substrate after a 5 min preincubation of clavulanic acid with β-lactamase.

Table 4 Effect of Clavulanic Acid (CA) on the Activity of Amoxycillin Against β-Lactamase-producing Organisms[a,102]

Organism[b]	Amoxycillin + CA (μg mL^{-1})			
	0	1.0	5.0	CA alone
Bacteroides fragilis (28)	33	0.48	0.14	13.1
Escherichia coli[c] (100)	> 5000	94.5	13.2	24.8
Haemophilus influenzae[c] (15)	150	0.72	0.44	36.8
Klebsiella aerogenes (45)	315	1.75	0.89	33.2
Klebsiella aerogenes[c] (32)	> 5000	126	20	33.6
Neisseria gonorrhoeae[c] (6)	> 40	0.18	—	5.6
Proteus species (23)	433	11.6	4.2	62.9
Staphylococcus aureus (35)	106	0.72	0.17	17.1

[a] MIC values. [b] Numbers in parentheses indicate numbers of strains tested. [c] Strains producing a plasmid-mediated β-lactamase.

was no indication of a rapid emergence of resistance. Amoxycillin- and ticarcillin-sensitive strains did not show increased sensitivity in the presence of clavulanic acid. Methicillin-resistant strains of *Staphylococcus aureus* did show some increased sensitivity but this was not significant as the resistance of such strains is due to a non-β-lactamase mediated mechanism, *i.e.* an altered penicillin binding protein (PBP). Clavulanic acid itself reacts most readily with PBP-2.[86]

Good results are obtained with both amoxycillin and ticarcillin plus clavulanic acid against β-lactamase producing isolates in a variety of animal models. Clavulanic acid is well absorbed and distributed in a number of animal species and in man.[87] Successful toxicological and metabolism studies coupled with detailed evaluation against clinical isolates and in clinical trials have resulted in potassium clavulanate, in combination with amoxycillin, being marketed worldwide as oral and injectable presentations. The formulation is of use against urinary tract, skin and soft tissue, and respiratory tract infections in adults and children. Potassium clavulanate plus ticarcillin is available

Table 5 Antibacterial Activity of Ticarcillin Alone and in the Presence of Clavulanic Acid Against a Variety of β-Lactamase-producing Organisms[85]

Organism	MIC (mg L⁻¹) MIC ticarcillin + CA (mg L⁻¹)					CA alone
	0	1	5	10	20	
Bacteroides fragilis B72	> 80	2.5	1.25	—	—	31
Citrobacter freundii E8	16	16	16	16	—	31
Citrobacter diversus	500	8	4	2	—	31
Enterobacter cloacae N1	16	16	16	16	—	62
Escherichia coli JT410[a]	31	31	31	31	—	31
Escherichia coli JT39 R^b_{TEM}	> 2000	125	31	8	—	31
Klebsiella aerogenes E70[a]	500	10	5	2	—	31
Klebsiella aerogenes Ba95[a,b]	2000	500	62	16	—	31
Proteus mirabilis C889	500	62	16	8	—	125
Providentia alcalifaciens Hd[a,b]	> 1000	16	4	2	—	62
Pseudomonas aeruginosa A[c]	16	—	16	16	16	250
Pseudomonas aeruginosa Dalgleish	> 4000	—	62	—	31	125
Pseudomonas aeruginosa 1822	> 4000	—	125	—	31	250
Pseudomonas cepacia	1000	125	31	—	1.0	31
Pseudomonas pseudomallei	500	31	8	—	1.0	250
Serratia marcescens US20[a]	16	16	16	16	—	62
Serratia marcescens US39[a,b]	2000	125	62	31	—	31
Staphylococcus aureus Russell	62	4	2	1	—	15

[a] Producing a chromosomally mediated β-lactamase. [b] Plasmid carrying (R+) strain. [c] Carbenicillin-sensitive strain producing Sabath enzyme only.

as an injectable preparation which is of use against similar infections, plus bone and joint disorders caused by β-lactamase producing bacteria.[88]

The chemistry of clavulanic acid has been explored in depth. Clavulanic acid contains a number of structural features which can be readily modified and a large number of derivatives have been prepared and evaluated as potential β-lactamase inhibitors with improved properties over clavulanic acid. Detailed references can be found in a number of papers and extensive reviews.[89-95]

In summary it can be manipulated as follows. The double bond can be reduced to yield dihydroclavulanic acid as a mixture of epimers, oxidized to form epoxides, cleaved to give a lactone, and isomerized *via* photolysis to derivatives in the (*E*) or isoclavulanic acid (**68**) series. The carboxylate function can be reduced, converted into an amide and removed, while position 6 can be alkylated and aminated. The β-lactam carbonyl group can undergo the Wittig reaction, forming an azetidine.[96] All derivatives of the above types were poor inhibitors of β-lactamase. In general terms, only modification of the allylic alcohol function resulted in compounds with significant β-lactamase inhibitor activity, although oxidation to the aldehyde gave a poor inhibitor.

The allylic alcohol function can be transformed to give deoxyclavulanic acid (**69**), acylated with ester (**70**) and carbamate formation, converted into primary and secondary amines (**71**) and an azido compound (**72**), obtained *via* the chloro derivative (**73**); (**72**) can be further reacted to yield a range of heterocyclic derivatives, *e.g.* (**74**).[97] Ethers (**75**) are readily prepared, as are thioethers (**76**) and the corresponding sulfoxides and sulfones; tetrazoles (**77**) can be produced, while the formation of vinylclavams (**78**) led to the preparation of higher homologues of clavulanic acid. The diene (**79**), a clavem derivative, has been prepared from acyl compounds. Betaine intermediates (**80**), obtained from C-9 substituted clavulanate derivatives, have been transformed into racemic penem compounds (**81**).[98] The ring expanded products (**82**) and (**83**) were also prepared from (**2**).[99] (±)-Methyl clavulanate has been prepared by total synthesis, as have a number of clavams; none, however, was of interest as a useful β-lactamase inhibitor.[100]

The β-lactamase inhibitory profiles[101-103] of a range of clavulanic acid derivatives are given in Table 6. To determine if any had better chemotherapeutic properties than clavulanic acid in conjunction with other β-lactams, such as amoxicillin, required considerable detailed *in vitro* and *in vivo* studies. In addition to being a potent inhibitor a compound has also to penetrate the outer membrane of a Gram-negative cell, be relatively unaffected by serum, give rise to suitable blood, tissue and urine levels, and not be readily metabolized, especially to more toxic products. Deoxyclavulanic acid (**69**), while showing β-lactamase inhibitory characteristics similar to clavulanic acid,

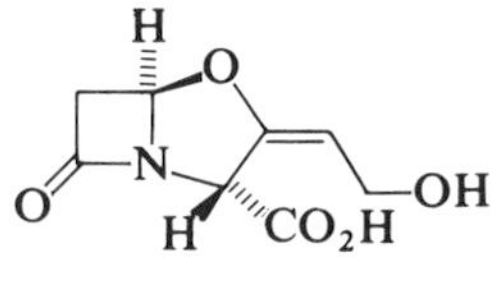

(68)

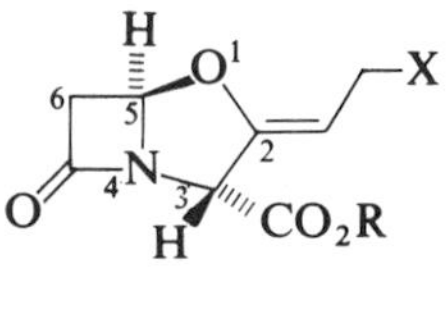

(69) X = H
(70) X = OCOR, OCONR¹R²
(71) X = NR¹R²
(72) X = N₃
(73) X = Cl
(74) X = (imidazole)
(77) X = (tetrazole isomers)
(75) X = OR
(76) X = SR; SOR; SO₂R

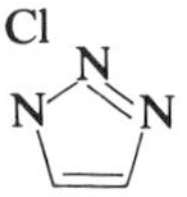

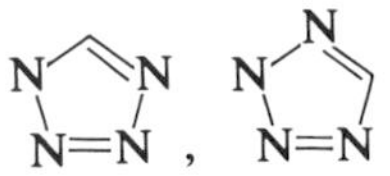

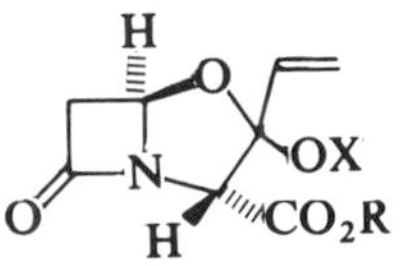

(78) X = CH₂CH₂R¹, COR¹

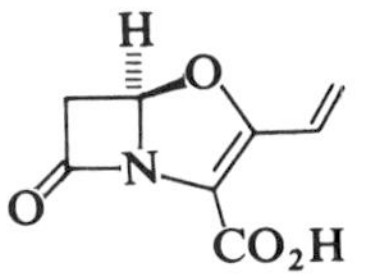

(79)

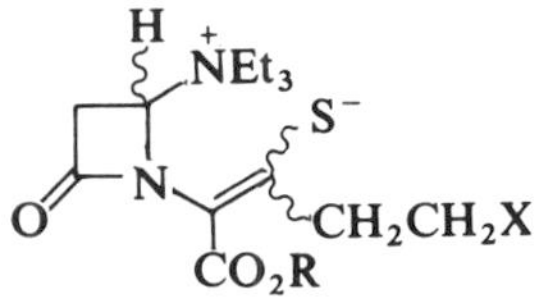

(80)

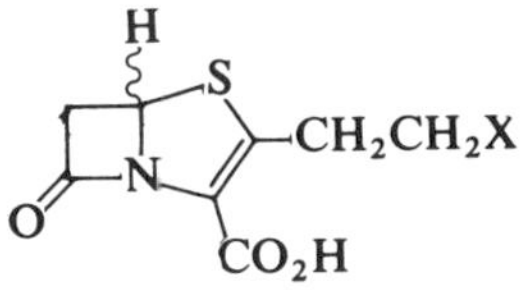

(81)

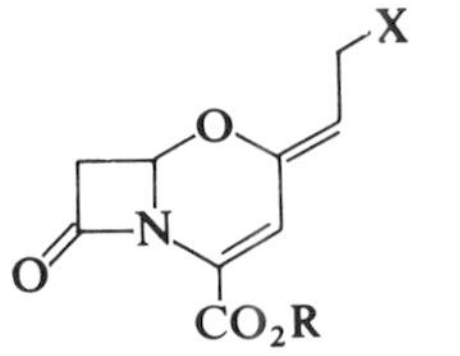

(82)

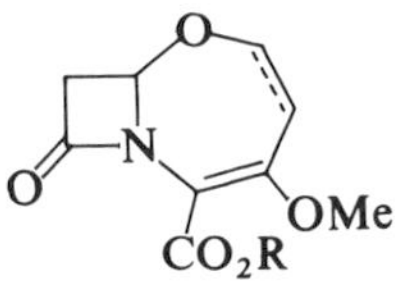

(83)

with amoxycillin (65) showed poorer synergistic activity and bioavailability and hence was less effective in infection models.[102] 9-Aminodeoxyclavulanic acid (71; R¹ = R² = H), on the other hand, performed better than clavulanic acid as an inhibitor and synergist with amoxycillin, gave better blood levels in mice by the oral and subcutaneous routes, and was therefore more effective in experimental infections. Detailed toxicological studies on the amine (71; R¹ = R² = H) highlighted excessive changes in the serum chemistry characteristic of kidney and liver damage, which precluded the progression of this compound as an alternative to clavulanic acid.[102] On metabolism in man and a number of animal species, clavulanic acid has been shown to be converted into the amine (84) and the reduced pyrrole (85).[104] In addition to carbon dioxide and the amine (84), pyrazines (86) have been detailed as degradation products from clavulanic acid.[105]

Detailed investigation[106] on the biosynthesis of clavulanic acid by *Streptomyces clavuligerus* show that the three carbon units of the β-lactam fragment is derived from glycerol or glycerate while the other carbon atoms come from glutamate *via* ornithine. The azetidinone, proclavaminic acid (87), and the amine, clavaminic acid (88), have been recently implicated as intermediates between glycerate, glycerol and the final product.[107] Proclavaminic acid (87) is converted into clavaminic acid (88) *via* an iron dependent dioxygenase. The stereochemistry at the two chiral centres in clavaminic acid (88) is opposite to that found in clavulanic acid (2), and the detail of this conversion of (88) into (2) is under study by Elson *et al.* within Beecham. By means of various vectors, chromosomal DNA fragments have been cloned into nonproducing mutants of *S. clavuligerus* with

Table 6 β-Lactamase Inhibitory Activity of Selected Clavulanic Acid Derivatives[101]

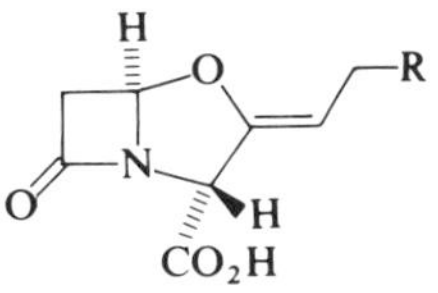

	Inhibitory activity, ID_{50} (μg mL^{-1}) of β-lactamase from				
	Staphylococcus aureus Russel!	*Escherichia coli JT4 III (TEM)*	*Klebsiella aerogenes E70 II*	*Proteus mirabilis C889 IV*	*Citrobacter freundii Mantio I*
Clavulanic acid; R = OH	0.06	0.07	0.03	0.03	10
Deoxyclavulanic acid; R = H	0.12	0.09	0.05	—	5
Isoclavulanic acid	0.6	1.0	0.45	—	5
Acetate; R = OCOMe	0.04	—	> 0.4	—	0.4
Carbamate; R = OCONHMe	1.5	2.5	2.5	—	0.45
Methyl ether; R = OMe	0.05	0.18	0.07	0.01	8.5
Benzyl ether; R = OCH$_2$Ph	0.005	0.1	0.04	0.02	4.4
Thioether; R = SMe	0.11	0.04	0.13	0.01	≫ 10^a
Amine; R = N(CH$_2$Ph)$_2$	0.002	0.04	0.08	0.01	0.62

^{a} β-Lactamase from *Enterobacter cloacae* P99.

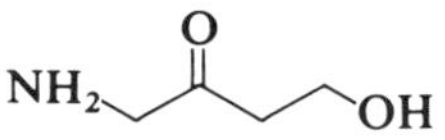

(84)

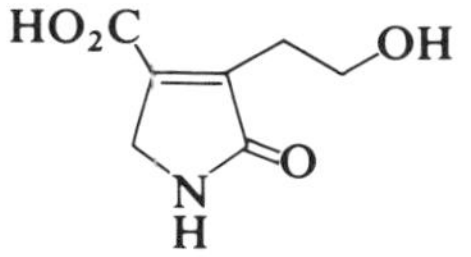

(85)

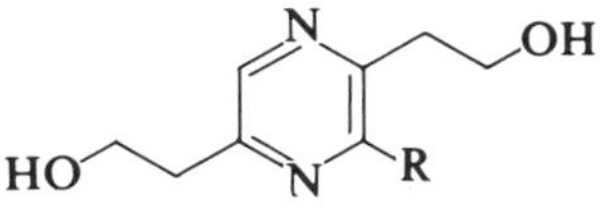

(86) R = H, Et

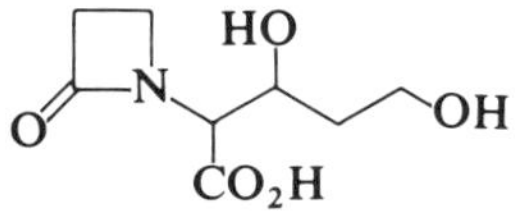

(87)

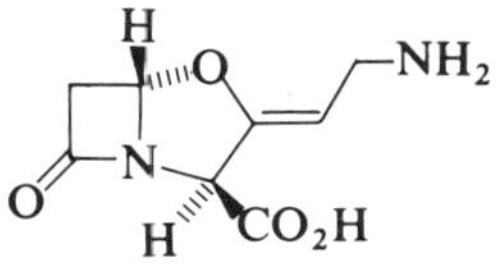

(88)

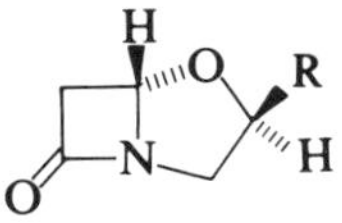

(89) R = CO$_2$H, CH$_2$OH, CH$_2$OCHO, CH$_2$CHCO$_2$H
                                       NH$_2$

ex. *S. clavuligerus*

(90) R = CH$_2$CH$_2$OH, CH$_2$CHCHCO$_2$H
                            OH
                                 NHCOCHCHMe$_2$
                                 NH$_2$

ex. *S. antibioticus*

(91) R = CHCHCO$_2$H
          OH NHR^1 Clavamycins A–F

ex. *S. hygroscopicus*

the restoration of clavulanic acid production.[108] Knowledge of the chromosomal regions involved in clavulanic acid synthesis is being used to generate improved production strains. It is worth noting that other clavams (89)–(91) have been reported from *Streptomyces* spp. and all of these possess the opposite configuration to clavulanic acid at C-5.[109] They lack β-lactamase inhibitory activity, though some have antifungal properties.

    The discovery of clavulanic acid and the olivanic acids (see below) prompted considerable effort on the preparation of analogues as β-lactamase inhibitors by total synthesis, in particular from penicillin G or 6-aminopenicillanic acid. The penicillanic acid derivatives (92)–(105) have been

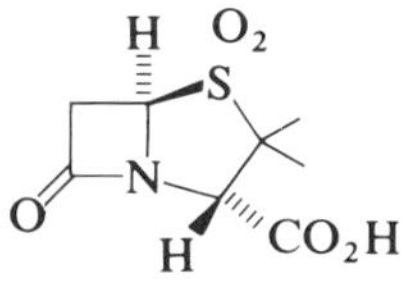

(92)

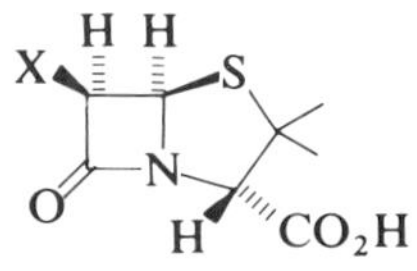

(93) X = Br
(94) X = I

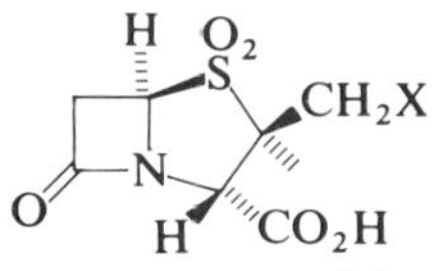

(95) X = N\N=N (triazole)

(96) X = Cl

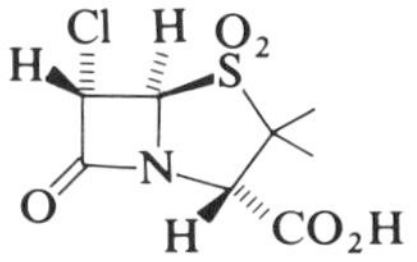

(97)

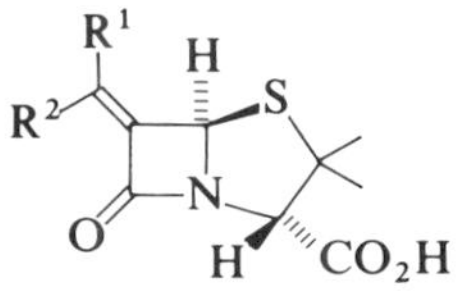

(98) R¹ = OMe, R² = H
(99) R¹ = COCH₃, R² = H
(100) R¹ = H, R² = COMe

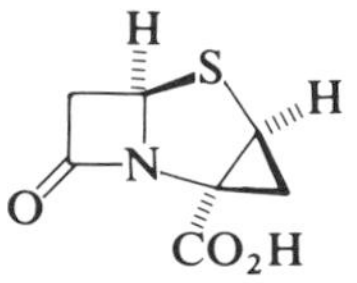

(101)

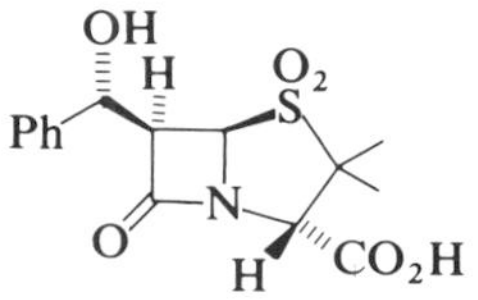

(102)

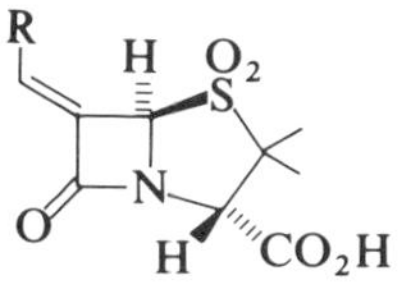

(103) R = heterocycle

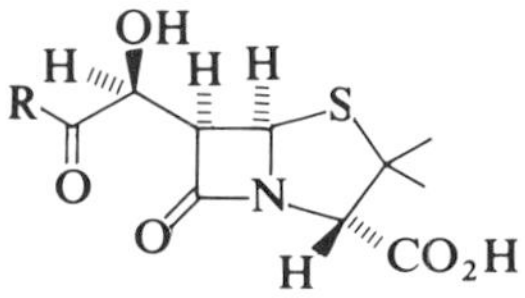

(104)

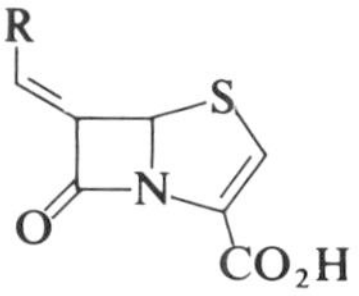

(105) R = Me or heterocycle

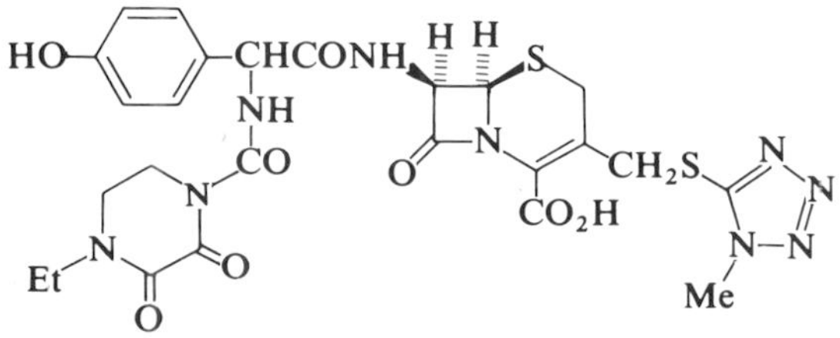

(106)

reported[110] as β-lactamase inhibitors, with penicillanic acid sulfone, sulbactam (92),[111] β-bromo-penicillanic acid (93)[102,103,112] and YTR-830 (95)[113] being of greatest potential. Compounds in the substituted 6-methylenepenem (105) series inhibit a wider range of β-lactamases than clavulanic acid.[114] A combination of sulbactam plus cefoperazone (106) is under development by Pfizer.[115] A mutual prodrug of sulbactam and ampicillin is also of some interest.[116]

As indicated earlier, the various β-lactamases can be grouped by an alternative system to the Richmond and Sykes classification.[81] The different classification, proposed by Ambler,[82] is based upon grouping the β-lactamases into three classes A, B and C depending upon amino acid sequence. Class A contains the *E. coli* RTEM and pBR 322, *Staphylococcus aureus*, *Bacillus licheniformis* and *B. cereus* I (569/H) enzymes. These enzymes have a high degree of homology around the active site which interacts with a β-lactam antibiotic or a β-lactamase inhibitor, a common serine residue (serine 70 or 44) being involved in formation of an initial acyl–enzyme complex.[117] Detailed X-ray crystallographic studies on these enzymes and their interaction with β-lactamase inhibitors are under active investigation in a number of laboratories, as is their comparison with the various penicillin binding proteins (PBPs) derived from the related parent organisms.[118] Class C enzymes are chromosomally mediated and are obtained from *E. coli via* the Amp C gene and *Pseudomonas aeruginosa*. They are also serine proteases but lack extensive homology with Class A enzymes. The β-lactamases of Class A and C do show a certain degree of homology around the active site, in amino acid sequence Phe-X-X-X-Ser-X-X-Lys, where the phenylalanine, serine and lysine residues are conserved. Class B enzymes are metalloenzymes and are represented by the zinc-dependent β-lactamase II from *B. cereus* and *Pseudomonas maltophilia* β-lactamase. Enzymes in this class readily destroy cephalosporins and the carbapenem nucleus. The mode of action of *B. cereus* II

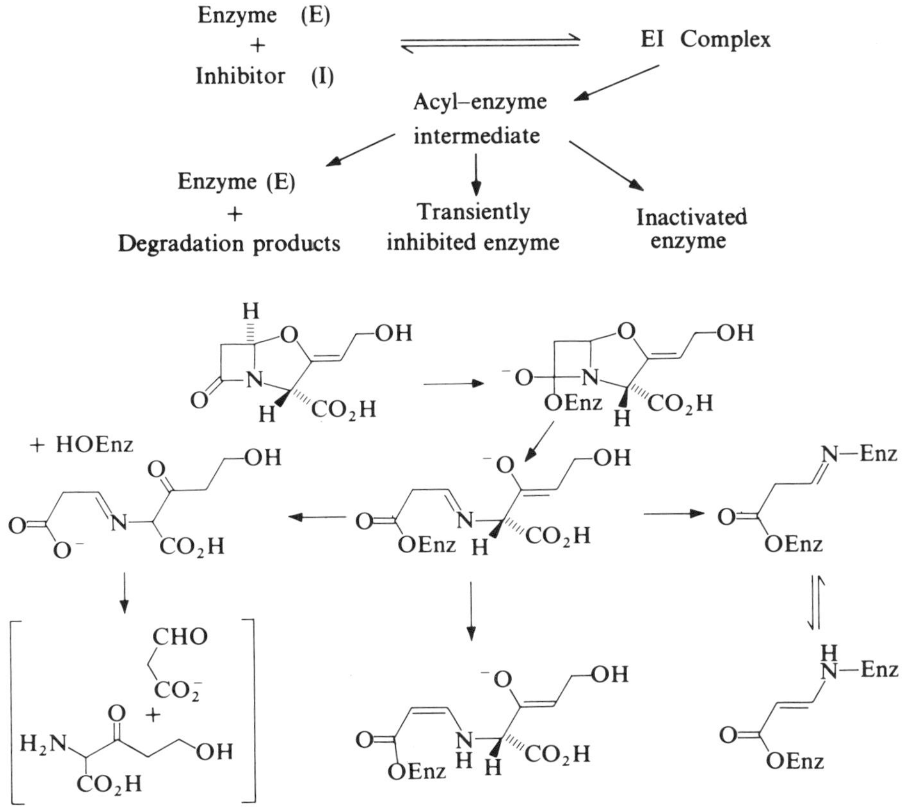

Scheme 2   Inhibition of β-lactamase by clavulanic acid

enzyme has been investigated by Waley and coworkers,[119] and crystallographic studies by Phillips *et al.*[120] show that it is not closely related to the zinc-containing D-alanyl-D-alanine cleaving carboxypeptidase of *Streptomyces albus* or other zinc metalloenzymes.[121]

The detailed mechanisms of the interaction of clavulanic acid, sulbactam and certain carbapenems have been elucidated by various groups, but in particular by Knowles and coworkers,[117] using the TEM β-lactamase. The kinetics of the reaction with *Staphylococcus aureus*, *Bacillus cereus* I, *Klebsiella pneumoniae*, *K. aerogenes* and *Branhamella catarrhalis* enzymes have also been examined.[122] The inhibitory activity of clavulanic acid and sulbactam can be illustrated by the sequences shown in Scheme 2. Following the formation of an acyl–enzyme intermediate between the inhibitor and the hydroxyl group of the active site serine residue, three separate sequences can occur. Inhibitor acts as a substrate for the enzyme and the β-lactam ring is hydrolytically cleaved. In addition, enzyme is trapped as a transiently inhibited form, leading to enzyme plus degradation products, and enzyme is also irreversibly inactivated. With clavulanic acid some 115 molecules are hydrolyzed for each molecule of inactivated enzyme formed; with sulbactam, 7000 turnovers occur.

With β-bromopenicillanic acid (**93**) the mechanism is slightly different.[123,124] For *Bacillus cereus* I β-lactamase with (**93**) a stable dihydrothiazine residue (**108**), derived from a thiolate intermediate (**107**), is formed on the active site serine residue, whereby every turnover leads to destroyed enzyme.

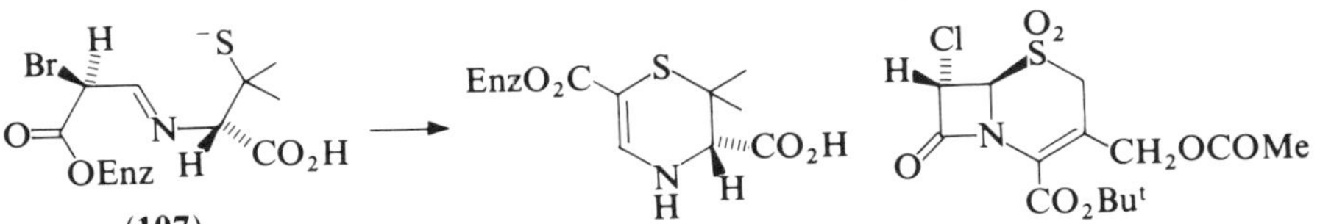

Recently, in contrast to the above inhibitors, the cephalosporin ester (**109**) has been shown to be a progressive, irreversible inhibitor of the enzyme elastase.[125] Mechanistic studies, including X-ray examination of an inhibitor–enzyme complex, indicate that the detailed chemistry of this inhibitor is considerably different to that outlined for a β-lactamase inhibitor.

Non-β-lactam β-lactamase inhibitors which have been studied include the macrolide-like alkyl sulfates izumenolide and dotriacolide,[126] and certain phenylboronic acid derivatives.[127] None, however, showed a significant effect as a synergist in combination with a β-lactamase sensitive β-lactam on *in vitro* evaluation against β-lactamase producing bacteria. Lack of activity was probably due to these inhibitors not penetrating the bacterial cell wall.

9.3.4 CARBAPENEMS

Another family of novel β-lactams to emerge from natural sources is that based on the 1-carbapenem ring system (110). The first compounds to be reported were thienamycin (3)[128] from *Streptomyces cattleya*, and a group of interrelated metabolites from *S. olivaceus* known as the olivanic acids,[129] *e.g.* MM 13902 (111; R = SO$_3$H). Thienamycin has the (8R) configuration of the hydroxy group with a *trans* arrangement of β-lactam protons, whereas in the olivanic acids the stereochemistry is (8S) with a *cis* substituted β-lactam in the sulfated series and both *cis*- and *trans*-β-lactams in the hydroxy cases. While thienamycin was detected in a screen for inhibitors of peptidoglycan synthesis, the olivanic acids were discovered using the β-lactamase inhibition screen described for clavulanic acid.

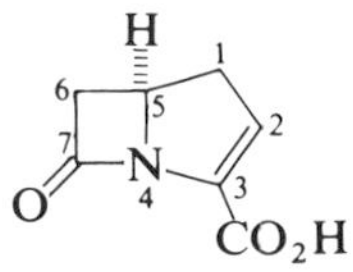

(110) 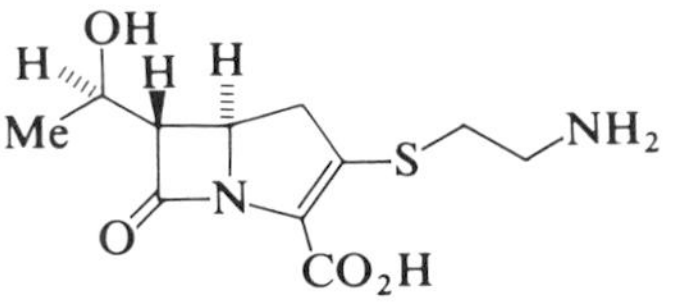   (3)

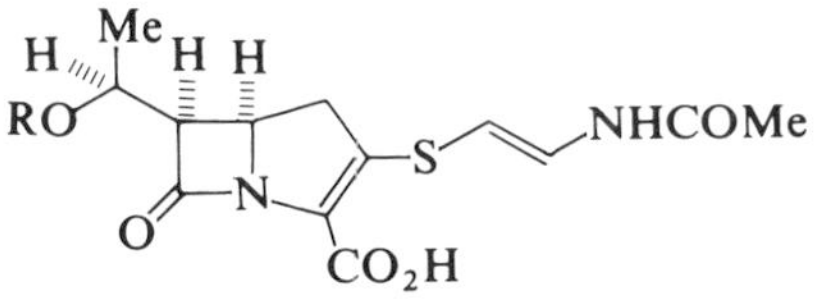

(111) R = SO$_3$H, H 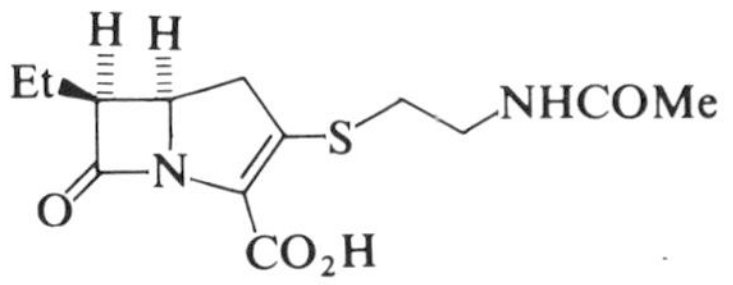   (112)

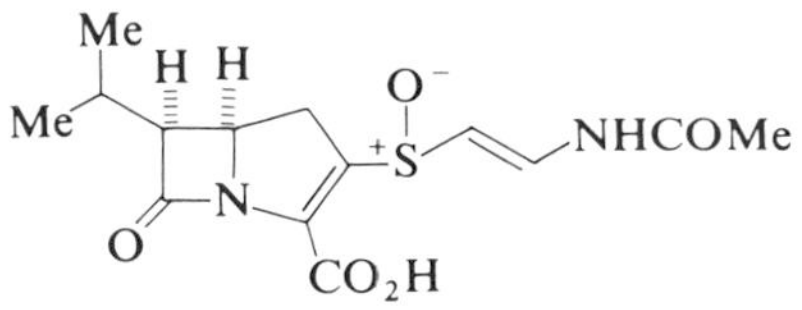

(113) 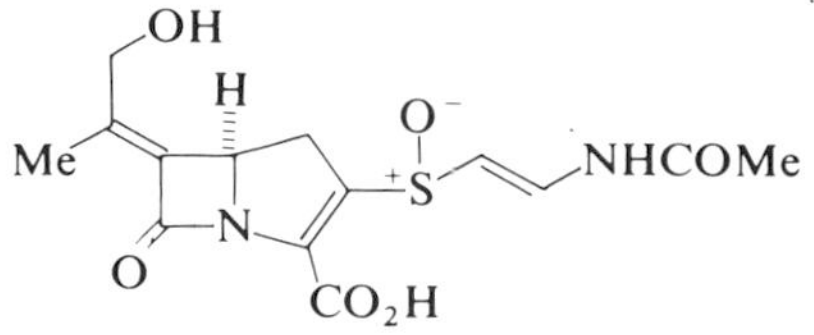   (114)

Subsequently a number of other structural types represented by PS-5 (112),[130] carpetimycin A (113)[131] and asparenomycin A (114)[132] were isolated from other streptomycete strains. To date over 40 variations with differing C-6 and C-2 substituents have been described.[133] Production of the rather unstable parent nucleus (110) has also been shown to occur in certain bacterial strains of *Erwinia* and *Serratia*.[134]

Early studies on the biosynthesis of thienamycin established that the pyrroline ring is derived from glutamic acid.[135] Subsequently, radioactive and stable isotope substrates were used to demonstrate that the C-6 and C-7 atoms of the β-lactam ring originate from acetate units, the cysteaminyl side chain from cysteine and the two carbon atoms of the hydroxyethyl group from methionine.[136]

A more recent disclosure has demonstrated the incorporation of glutamate into the parent ring system (110); in this study the corresponding saturated carbapenam nucleus (115) was also isolated.[137] Some streptomycete carbapenem producers have also been found to give the reduced form (116) of the corresponding unsaturated structure.[138] The possible role of the carbapenam ring system, together with a unified approach to explain the diversity of carbapenem structures, has been the subject of some speculation.[137-139] The biosynthetic relationship of several metabolites of *S. olivaceus*[140] and the role of the OA-6129[141] group of carbapenems on the biosynthetic pathway have also been described.

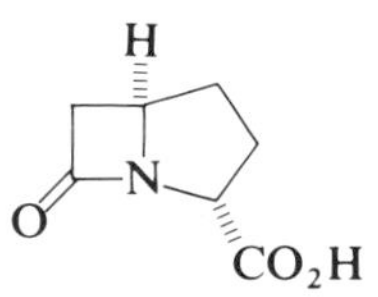

(115)

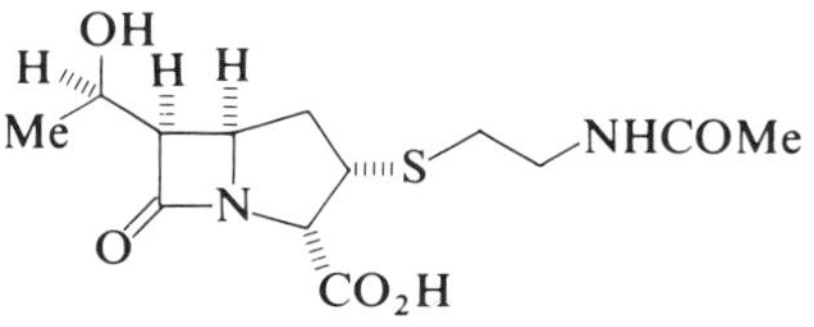

(116)

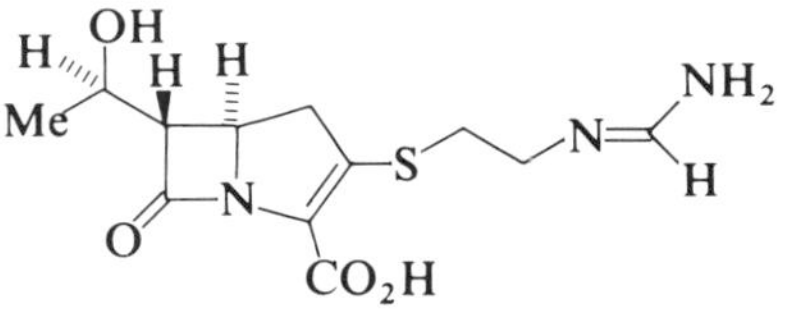

(117) MK 0787

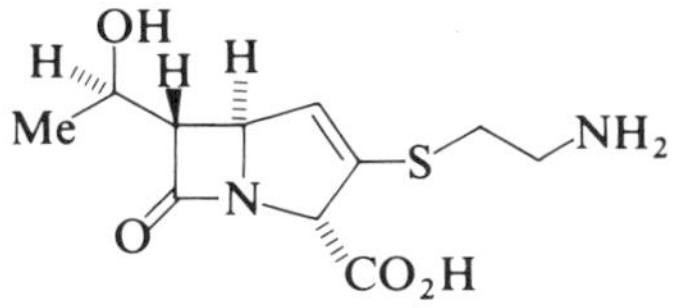

(118)

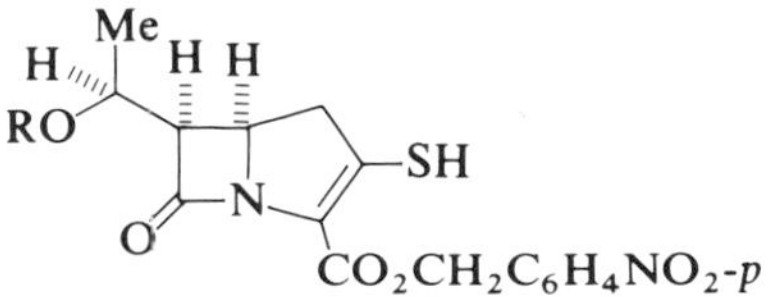

(119)

Fermentation yields of the natural products are low, and the metabolites are extremely sensitive to both acid and alkaline reaction conditions. Nevertheless many useful and interesting chemical modifications have been carried out. Early studies on thienamycin were designed to improve or maintain the antibiotic properties while imparting chemical stability to the nucleus. Derivatization of the hydroxy, amino and carboxy groups together with other reactions have been extensively studied by Merck.[13] Of the many amino derivatives the *N*-formimidoyl compound (117) (MK 0787) fulfilled the objectives of the initial programme and was selected for clinical evaluation. Isomerization of the Δ^2 double bond can also occur in some instances to give the more stable, but biologically inactive, Δ^1 isomers such as (118).[142] Reductive removal of the cysteamine residue has also been reported.[143]

In the hydroxyolivanic acid series, inversion of the (8*S*) stereochemistry provides an entry to the (8*R*) isomers of the thienamycin type.[144] Other reactions of the olivanic acids include isomerization[145] of the (*E*) double bond of the acetamidovinyl group to give the (*Z*) geometry and also complete removal of the grouping to give the thiol (119),[146] an intermediate eminently suitable for the synthesis of other alkyl and vinyl derivatives. Carbapenem sulfoxides are also susceptible to displacement reactions with a variety of thiols,[147] while microbial deacylation[148] provides a source of amino compounds.

Total synthesis methodology has been extensively developed to provide analogues and also the natural products. This can be seen particularly with the Merck approach to MK 0787, where an efficient synthesis was a prerequisite for obtaining the large supplies required for production and clinical evaluation. Since the synthetic approaches have been extensively reviewed,[13,133,149,150] only a selected number of examples relating to the natural products are illustrated here. The methods, however, are widely applicable to analogue synthesis.

Formation of the [2,3] double bond by way of an intramolecular Wittig reaction was an approach used by both the Merck and Beecham groups to obtain the racemic unsubstituted nucleus (110). This was prior to the discovery of (110) as a natural product. Thus oxidation of the alcohol (120)[151] or allyl derivative (121)[150,152] provided the aldehyde (122) which spontaneously cyclized to give the protected nucleus. Deprotection afforded the unstable sodium salt. Intramolecular cyclizations with thiol esters (*e.g.* 123) are also possible and provide a route to racemic *N*-acetyldehydrothienamycin and the olivanic acid metabolite MM 22383.[153] Thiols also add readily to the double bond of phosphorane derived unsubstituted derivatives such as (124) and this approach has been used for the synthesis of (±)-PS-5.[154]

Of the many Merck approaches,[13] formation of the C-3—N-4 bond by a diazo insertion process affords a highly efficient route to thienamycin, MK 0787 and other analogues.[155] Thus elaboration of the aspartate derived β-lactam (125) to (126) followed by rhodium acetate catalyzed cyclization affords the bicyclic β-keto ester (127). Activation as the enol phosphate and reaction with the protected cysteamine side chain and hydrogenation produced thienamycin (3). Other Merck

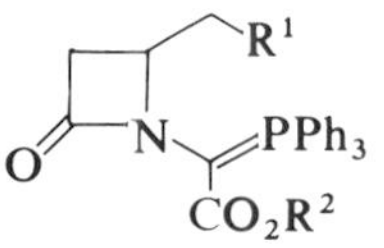

(120) R^1 = CH$_2$OH
(121) R^1 = =CH$_2$
(122) R^1 = CHO

R^2 = CH$_2$C$_6$H$_4$NO$_2$-*p*, CH$_2$COMe

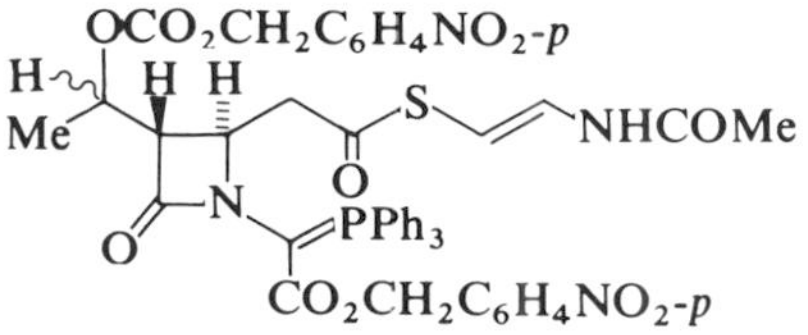

(123)

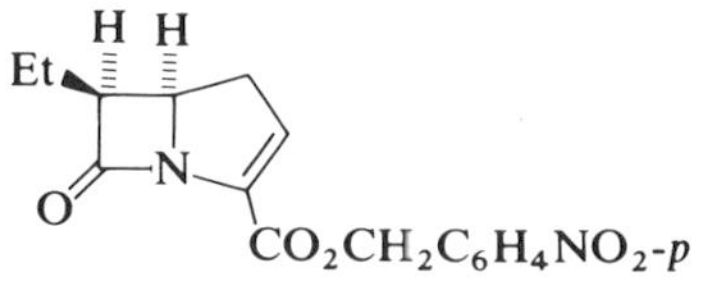

(124)

approaches to (127) started from the racemic or chiral lactone (128)[156,157] or incorporation of (129)[158] at the C-4 position of a functionalized β-lactam derived from the penicillin nucleus. More recent effort has been directed towards the synthesis of the C-1 β-methylcarbapenem (131)[159] as a successor to MK 0787. This was achieved using ketone (130).

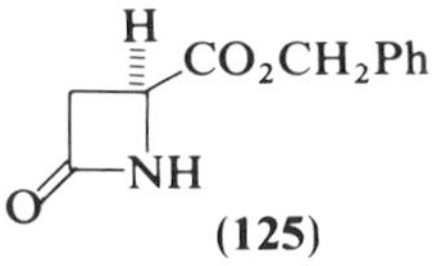

(125)

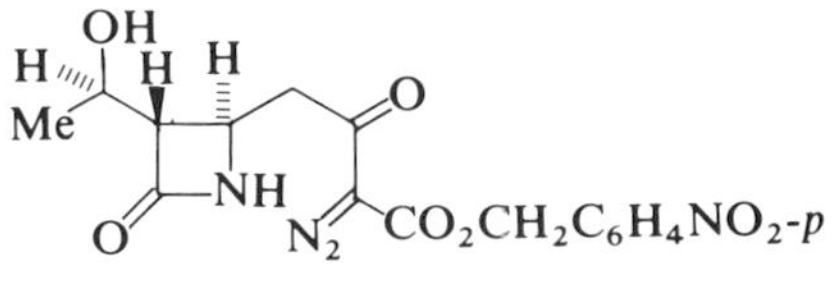

(126)

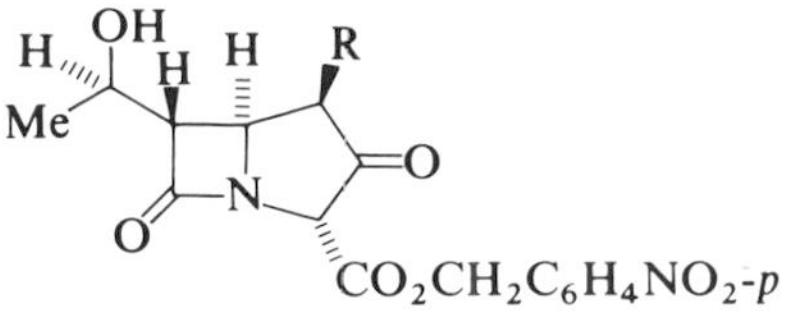

(127) R = H
(130) R = Me

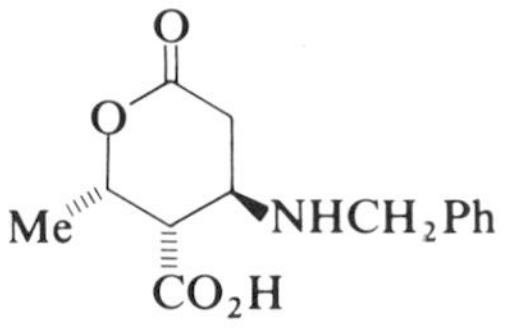

(128)

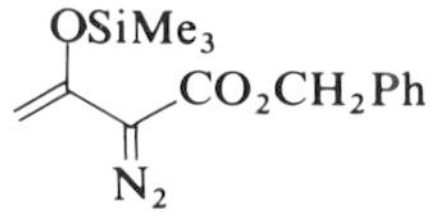

(129)

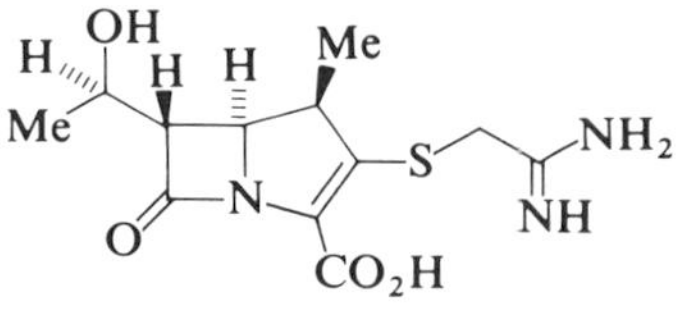

(131)

Numerous asymmetric approaches for synthesizing β-lactam precursors of thienamycin and other carbapenems have been described.[160] Most strategies have focused on the elaboration of the correct stereochemistry at the three chiral centres of a monocyclic precursor. Subsequent progression has usually been by way of the diazo-ketone approach. Often the hydroxyethyl side chain has been introduced stereoselectively by aldol condensation.[161] In other cases the substituent has been incorporated before construction of the β-lactam ring using chiral building blocks derived from D-glucose,[162] L-threonine[163] and D-allothreonine.[164]

Other chiral syntheses leading to key intermediates for (+)-thienamycin make use of the chiral isoxazolidine (132),[165] derived from an enantioselective [3+2] cycloaddition of a chiral nitrone with benzyl crotonate, while an approach based on oxidation of the tricarbonyliron lactam complex

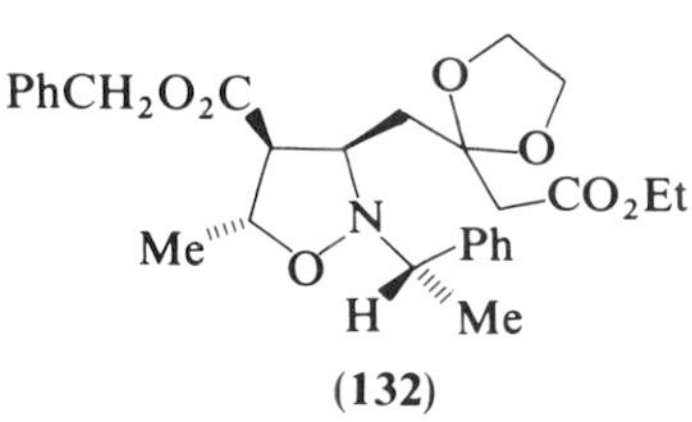

(132)

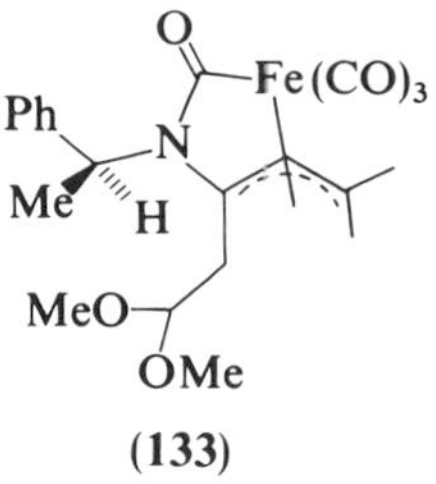

(133)

(133)[166] also leads to a suitable β-lactam precursor. One of the most recent methods relies on an addition–cyclization reaction of the dianion of (S)-(+)-ethyl 3-hydroxybutanoate with N-aryl-aldimines.[167]

Thienamycin and the majority of carbapenems are highly active broad spectrum antibacterial agents; in many cases they are also potent inhibitors of bacterial β-lactamases.[92,103,117] Unlike the classical penicillins and cephalosporins, a rigid adherence to the *cis* arrangement of β-lactam protons is not necessary for activity. Thus both (3) and (111) are extremely active. Structure–activity studies have provided the following conclusions for the natural products.[168] Thienamycin is the most potent antibiotic, including the best activity against *Pseudomonas* strains. All of the N-acylated derivatives have reduced antipseudomonal activity. Thienamycin is also the most stable to β-lactamases. The order of antibiotic potency in the nonsulfated series appears to be *trans* substituted β-lactam with (8R) stereochemistry > *cis* (8S) > *trans* (8S).

Overall, thienamycin has an extremely broad spectrum of activity, surpassing all of the 'third generation' cephalosporins. The spectrum covers both Gram-positive and Gram-negative cocci, and a wide range of other Gram-negative organisms including *Pseudomonas aeruginosa*. Anaerobes such as bacteroides are also susceptible. The solution stability of thienamycin however made it unsuitable for formulation as a clinical product. Fortunately the previously described N-formimidoyl derivative (117) emerging from the derivatization programme not only met the requirements of being chemically stable, but in many cases was even more potent than thienamycin. It is this derivative, known as imipenem, that was chosen for clinical development. Like other β-lactam antibiotics, thienamycin and imipenem interfere with the synthesis of the bacterial cell wall. They have the greatest affinity for penicillin binding protein 2 in *E. coli* and *P. aeruginosa*, although also binding well to proteins 1, 4, 5 and 6.[86,169] The possible mode of action of carbapenems as β-lactamase inhibitors has also been described.[117]

Of the naturally occurring carbapenems, thienamycin is the most effective on evaluation after parenteral administration in animal models. None are absorbed by the oral route. However, with thienamycin, imipenem and other derivatives only a low recovery of antibiotic was evident in the urine. Similarly, in man with imipenem this ranged from 6–40% although the plasma half-life (1 h) was considered satisfactory.[170] Subsequently, detailed studies established that this class of antibiotics is extensively degraded by a renal dipeptidase enzyme (DHP-I) located in the brush borders of the kidney.[171] It is ironic, therefore, that although stable to bacterial β-lactamases, the β-lactam ring of the carbapenems is ruptured by an enzyme which has no effect on penicillins and cephalosporins.

In order to overcome the effect of DHP-I, Merck workers looked for inhibitors of the enzyme. They found that dehydroamino acid derivatives such as MK 0791 (cilastatin) (134) had a dramatic effect on the urinary recovery of imipenem. It also reduced the nephrotoxic potential of the antibiotic. These results led to the evaluation of an imipenem–cilastatin combination[172] as a commercial broad-spectrum injectable product. At present, imipenem is the only carbapenem in clinical use. Current research is concentrated on the synthesis of 1-substituted analogues such as (131) which have a greatly reduced susceptibility to DHP-I and could thus be administered as a single entity.[159]

(134) MK 0791 (Cilastatin)

9.3.5 PENEMS

Penicillins and cephalosporin antibiotics possess a reactive β-lactam ring which is generally considered to be essential for biological activity. In the penicillins this reactivity is attributed to the pyramidal geometry of the β-lactam nitrogen, resulting in a destabilizing effect on the amide bond. In cephalosporins a more planar geometry is apparent but the presence of the double bond in the six-membered ring can also influence the delocalization of the nitrogen unshared pair of electrons and increase the reactivity of the β-lactam carbonyl group. To combine both structural elements which appear to be necessary for the biological activity of these compounds, workers at the Woodward Research Institute synthesized the new bicyclic penem ring system shown by structure (135).[173] This can be considered as a hybrid of the penicillin and cephalosporin ring systems. No natural products of this type have been discovered.

These initial penem compounds were obtained from penicillin derived β-lactam intermediates. Again a Wittig reaction was used to form the sensitive bicyclic structure under mild and neutral conditions. The key intermediate phosphorane (136) was available in a number of steps from the appropriate penicillin *S*-oxide. Deprotection to provide (135) gave very unstable products, but some antibacterial activity was observed.

(135) R = H, Me

(136)

(137) R = H, Me, CH₂Ph,
CH₂CH₂NH₂, *S*-alkyl

(138) X = O, R = alkyl
(139) X = S, R = *S*-alkyl

(140)

Further work was then undertaken to explore the penem ring system in the 6-unsubstituted series (137) using a totally synthetic approach from phosphorane precursors such as (138) and (139).[174] Surprisingly, and unlike 6-unsubstituted penicillanic acid derivatives, the penems of type (137) revealed good *in vitro* antibacterial activity. Of additional significance was the greatly enhanced chemical stability of these compounds compared to the acylamino derivatives (135). A chiral synthesis also demonstrated that it was the (5*R*) component of these racemic compounds which exhibited the activity.[175] This is the stereochemistry common to all naturally occurring fused bicyclic β-lactams. Subsequently many other approaches[174] to 6-unsubstituted penems were reported, including the use of clavulanic acid as a starting material.[176] Overall the activity was good but some deficiencies were apparent, including a lack of stability to many β-lactamase producing strains. In order to improve activity and β-lactamase stability, effort was directed by many research groups to introducing new substituents into the 2- and 6-positions of the penem nucleus. In particular the (6*R*)-hydroxyethyl side chain of thienamycin has proved to be widely used, leading to a variety of compounds of structure (140).

As with the carbapenems, an extensive variety of synthetic methods for penem synthesis has been disclosed.[174,177] The 6-hydroxyethyl compounds are broad-spectrum β-lactamase stable antibiotics including activity against anaerobes, but few possess good antipseudomonas activity. Overall the structure–activity relationships are broadly similar to those found with the carbapenems. They are also susceptible to degradation by the DHP-I enzyme. As described earlier (Section 9.3.3), some alkylidene penems have been reported as β-lactamase inhibitors.

No penem is yet in clinical use but several have been progressed to a considerable extent along the development pathway. The orally administered Schering compound Sch 29482 (141) was the first to undergo extensive *in vitro* and *in vivo* evaluation.[178] This derivative has good β-lactamase stability and a high level of activity against staphylococci, most streptococci and the majority of common Gram-negative pathogens, excluding *P. aeruginosa*. It is also active against anaerobic bacteria.

A much used and readily available chiral starting material for the synthesis of penems such as (141) is the 6,6-dibromopenicillanic acid ester (144) obtainable from 6-aminopenicillanic acid. Reaction with methylmagnesium bromide gives an α-bromoenolate which can be reacted with acetaldehyde to form predominantly isomer (145); zinc reduction removes the bromine with inversion to provide the *trans*-hydroxyethylpenam (146).[179] The elaboration of (146) to a variety of penems has been achieved by several methods.[177] With (141) the initial approach was to transform (146) to the phosphorane (147) followed by cyclization in the normal manner; several other

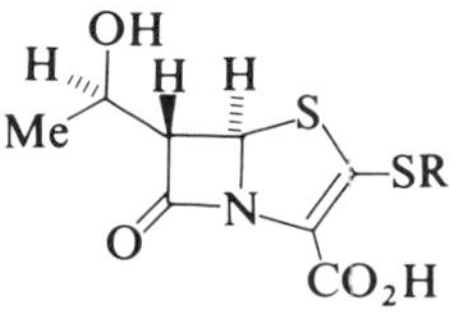

(141) Sch 29482, R = Et

(142) Sch 34343, R = CH₂CH₂OCONH₂

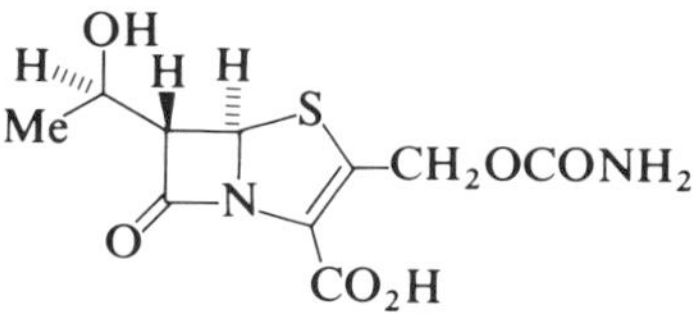

(143) FCE 22101

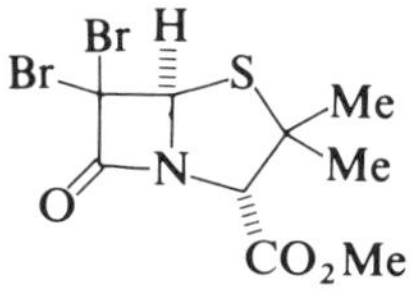

(144) 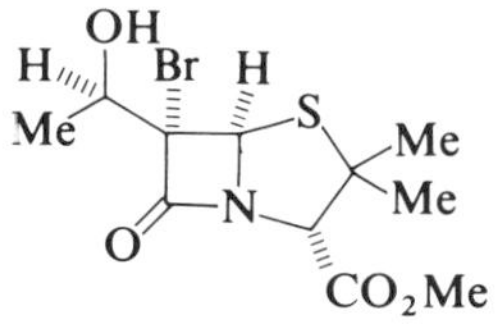   **(145)** 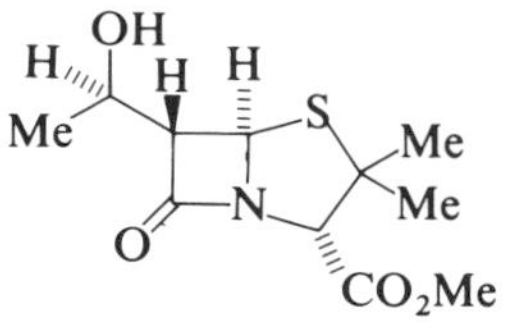   **(146)**

approaches have been subsequently described.[180] These authors also described a nonphosphorane synthesis of the key intermediate **(148)** which can be converted to a variety of 2-thiapenems. The allyl group was found to be extremely convenient for protecting the acidic function of the penem ring system, being readily removed under extremely mild conditions with palladium reagents.

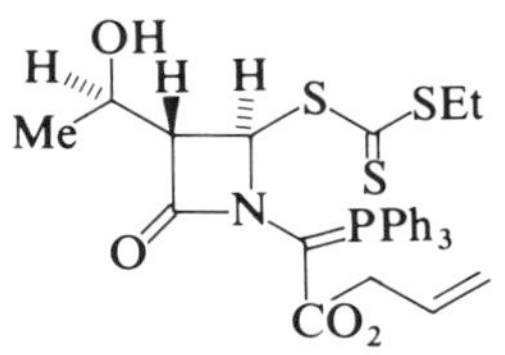

(147) 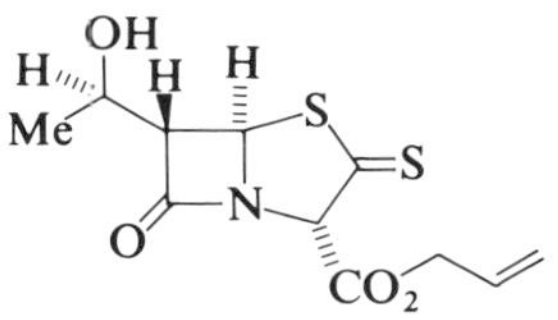   **(148)** 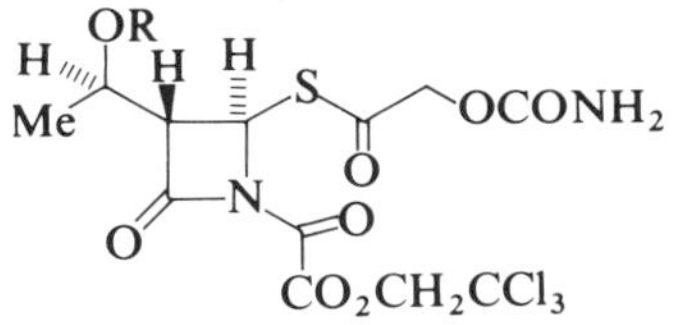   **(149)**

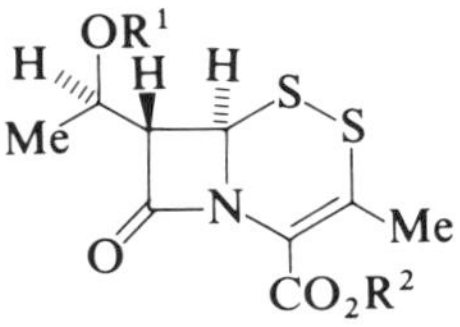

(150) 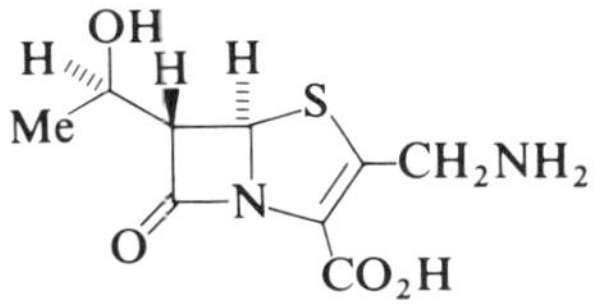   **(151)** CGP 31608

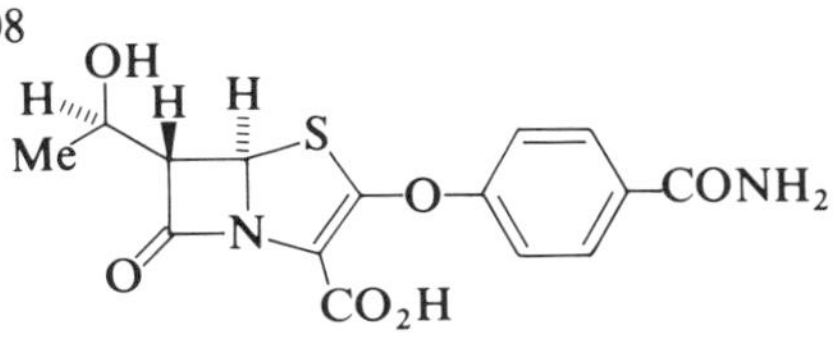

(152) HRE 664

The development of Sch 29482 was halted due to unacceptable metabolic by-products and attention was focused on the parenteral product Sch 34343 **(142)**.[181] The synthetic strategy was again based on the use of **(146)**. Sch 34343 is resistant to staphylococcal β-lactamases and the common β-lactamases of Gram-negative bacteria. Again it is a broad-spectrum agent, lacking only in activity against *Pseudomonas* strains. It interacts primarily with penicillin binding proteins 2, 1A and 1B of *E. coli* and *B. fragilis*.[181] One interesting facet of the mode of action of many penems and carbapenems is seen in their ability to kill nongrowing cells, which is not apparent with penicillins and cephalosporins.[182]

The Farmitalia group has also devoted a great deal of attention to the penem area and chose FCE 22101 **(143)** and its orally absorbed prodrug esters as compounds for progression.[183] Many approaches to the synthesis of **(143)** and related penems have been described.[184] One method, which incorporates virtually all the atoms of the penicillin skeleton, relies on conversion of a 6-hydroxy-ethylpenam to the oxalimide **(149)**. Reductive cyclization with triethyl phosphite and deprotection afforded FCE 22101 **(143)**. Another interesting variation involves the ring contraction of 2-thiacephems **(150)**. This method can also be used for analogues in which sulfur is replaced by selenium, although the compounds were less active. Two of the more recent penems to arouse interest are CGP 31608 (151)[185] and HRE 664 (152).[186] The former is one of the few compounds to have acceptable activity against *Pseudomonas aeruginosa*, and also against methicillin resistant staphylococci. Overall the penems represent an area of synthetic β-lactams which are extremely active, and which could be extremely useful in providing broad spectrum antibiotic cover against a variety of infections.

9.3.6 OTHER FUSED β-LACTAM SYSTEMS

The synthesis and structure–activity relationships of nuclear analogues of penicillins and cephalosporins have been comprehensively covered in reviews which cover the field up to about 1983.[5,187,188] The Royal Society of Chemistry Special Publications, which have appeared every four years since 1977, have also provided a useful topical insight into how the subject has rapidly unfolded over the past decade.[8-10] This section considers both the newer structural types and the major developments of previous review periods. The emphasis is perhaps more on the chemistry than the biology because, although considerable chemical ingenuity has been evident in the synthesis of many novel systems, all too frequently the final product was antibacterially inactive, or was never obtained in a form suitable for biological evaluation. Space prevents a detailed discussion of every new analogue, and many, only remotely related to the natural products, have been intentionally excluded.

In general the most successful strategies generate monocyclic β-lactams (azetidinones) early in the synthetic sequence. Initially, partial synthesis *via* selective cleavage of the penicillin thiazolidine ring was the only source of optically active derivatives. Chiral control[189,190] in the Staudinger reaction (Scheme 3),[3] and the use of chiral synthons,[160,191] in particular β-hydroxy acids,[32] have resulted in the situation where almost any chiral β-lactam can be constructed. More specific references are given in the relevant subsections. Most monocyclic β-lactams have sufficient chemical stability to withstand a wide range of functional group manipulation, and cyclization to a more reactive fused β-lactam ring system can then be accomplished as almost the final synthetic step. In many syntheses where the final product is racemic, for convenience only one enantiomer has been depicted.

Chapter 9.2 discussed the evolution of structure–activity relationships of penicillins and cephalosporins and thus concentrated on the acyl derivatives of 6-APA and 7-ACA, with an indication of relevant 6α- and 7α-functionalization. Substitution of C-3 of cephalosporins, which has led to better biological properties, was also covered, but other transformations of both nuclei which retain the natural skeleton were not reviewed. In this connection, past reviews and a few more topical papers are now briefly summarized, before the true nuclear analogues are discussed.

9.3.6.1 Penams and Cephems

The wealth of heterocyclic chemistry and biological evaluation that has flowed from the modification of the penam (61) and cephem (62)[192] nuclei has been reported in some depth.[4,8-10,193-195] The topic is particularly well covered in the review edited by Sammes.[5] Some more recent lead references are also worthy of consultation.[196-198]

Most penicillins and cephalosporins in use as antibiotics bear a 6β(7β)-acylamino substituent, since this classical prerequisite for sufficient antibacterial activity has allowed few exceptions. Mecillinam (153), in which an amidine group replaces the normal substituent, retains activity against only Gram-negative bacteria.[199] More recently the quaternary heterocyclylamino β-lactam (154) has been shown to be superior to mecillinam in potency and breadth of spectrum.[200,201]

Appropriate modification at the C-2 position of penicillins and C-3 of cephalosporins has retained or improved antibacterial activity, but all other transformations, with the exception of certain specific 6α- and 7α-substitution of the respective nuclei, compromise biological properties. The most favourable modification at C-2 of penicillin is the removal of the *gem*-dimethyl group. Bisnorpiperacillin (155) was found to possess superior activity against certain Gram-negative

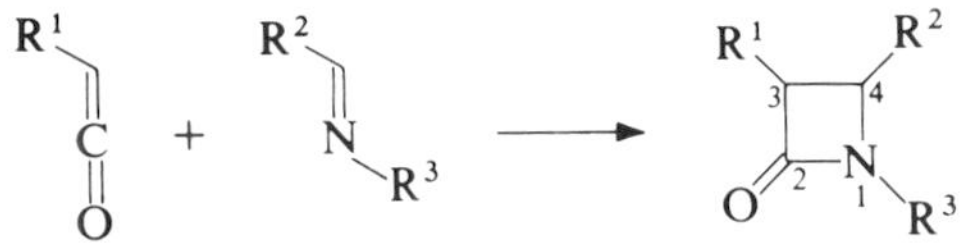

Scheme 3

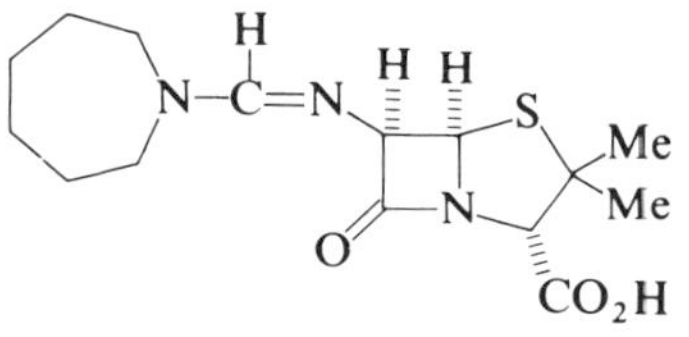

(153)

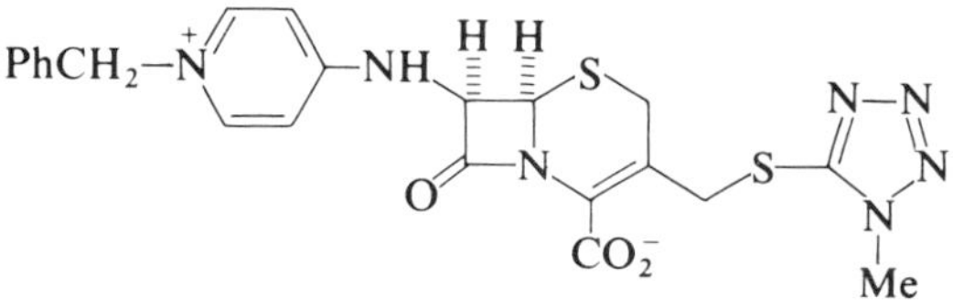

(154)

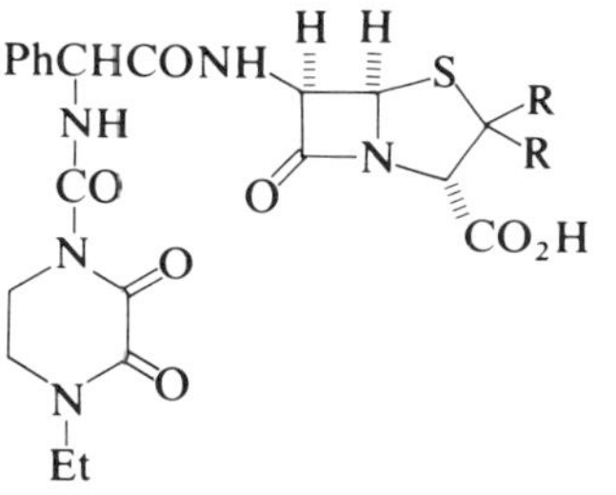

(155) R = H
(156) R = Me

bacteria compared with its conventional counterpart piperacillin (**156**), notably *Escherichia coli* and *Klebsiella pneumoniae*,[202] the improved activity being due to an increase in stability to the β-lactamases produced by these organisms.

Cyclopropyl analogues of the acylaminopenams have recently been of some interest. The antibacterial potency of the (2,3)-β-methylenepenicillin (**157**) was considerably less than penicillin G (**63**) itself, and led to the conclusion that the 'open' conformation of penicillins is the biologically active form.[203] This in turn suggested that the (2,3)-α-methylenepenams might be chemically reactive and biologically active. This was confirmed by the synthesis and evaluation of various acylamino derivatives. Generally, (2,3)-α-methylenepenams are more active against Gram-negative bacteria than the corresponding penams, but slightly weaker in potency toward Gram-positive bacteria.[204] In the penicillin G analogue (**158**) the absolute configuration at C-3 is opposite to the naturally occurring penicillins, placing the carboxyl group in a position similar to cephalosporins. Consistent with this, the compound was more susceptible to cephalosporinase than penicillinase.[205] The instability of (2,3)-α-methylenepenams to nonenzymatic hydrolysis reduces their effectiveness as antibacterials. Replacement of the cephem double bond by a cyclopropane ring gave (**159**), which was devoid of antibacterial activity.[206]

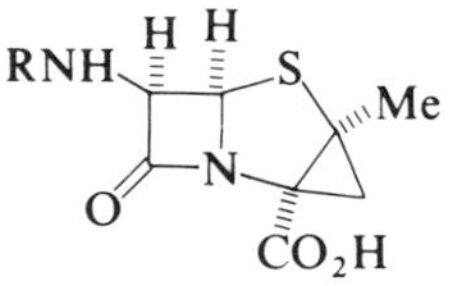

(157)

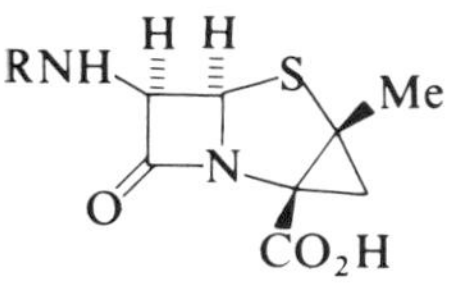

(158) R = PhCH₂CO

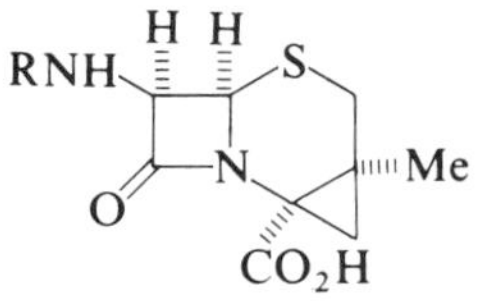

(159)

Compared with efforts made at varying the acyl group, relatively few studies have been made at replacing the 6β- or 7β-amido group by other substituents. The discovery of new naturally occurring β-lactams with widely differing side chains rekindled interest in structural variation at this site. The results have been largely disappointing, since the less reactive penam and cephem nuclei show a lower tolerance to chemical change than the more reactive carbapenem and penem nuclei.[207] Thus homothienamycin (**160**) was devoid of activity.[208] Derivatives in which the amide group was replaced by substituents of similar shape have been investigated. The alkene (**161**) has 1/1600 of the activity of penicillin G (**63**) against penicillin sensitive *Staphylococcus aureus*, and other double bond derivatives were less potent.[207] Analogues in which the NH of the side chain has been replaced by

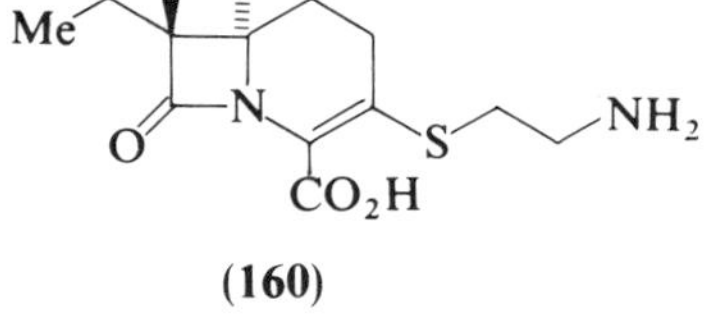

(160)

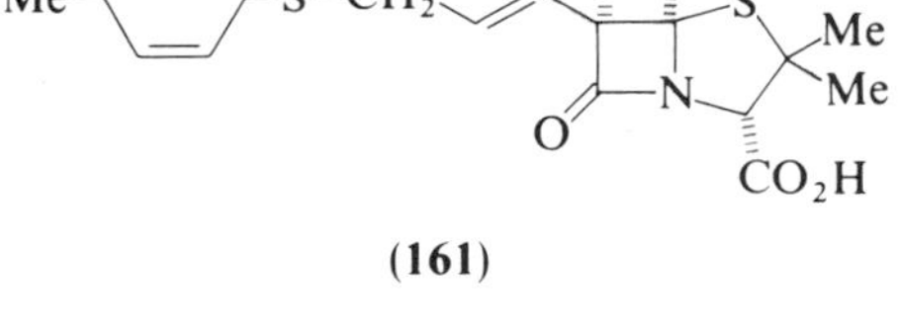

(161)

CH_2, O or S have been evaluated. In almost all cases the compounds were considerably less active than the corresponding penicillins or cephalosporins.[209] Some of the penicillin analogues were penicillinase inhibitors, as are certain penicillanic acid derivatives (see Section 9.3.3).

9.3.6.2   1-Oxadethiapenams

Replacement of the penicillin thiazolidine sulfur by oxygen gives derivatives which are less active,[210] and in the case of the bisnor derivative (163), less stable in aqueous solution.[211] Both partially and totally synthetic procedures have proved to be successful. The synthesis of the optically active analogue (164) illustrates how the penicillin molecule can be used as a source of versatile intermediates which retain the β-lactam ring.[212] Thiazolidine ring cleavage of the penicillanate (165) with mercury(II) acetate gave the *trans*-acetate (168), from which the dehydrovalinyl moiety was removed by oxidation with potassium permanganate to provide the azetidinone (169). Zinc acetate mediated displacement of the acetoxy group with benzyl 2-bromo-3-methylbutyrate gave a mixture of *cis* and *trans* ethers. The *cis* isomer (170) was cyclized to afford both C-3 epimers (166) and (167) of the oxapenam. The salt (164) was prepared conventionally, and the methodology extended to give the bisnor analogue (163).

Synthetic and naturally occurring derivatives lacking an acylamino side chain are covered in Section 9.3.3.

(162) R = H
(163) R = CH_2Ph

(164) R^1 = CO_2Na, R^2 = H, X = O
(165) R^1 = CO_2Me, R^2 = H, X = S
(166) R^1 = H, R^2 = CO_2CH_2Ph, X = O
(167) R^1 = CO_2CH_2Ph, R^2 = H, X = O

(168)

(169) R^1 = OAc, R^2 = H
(170) R^1 = H, R^2 = $OCMe_2CHBrCO_2CH_2Ph$

9.3.6.3   1-Carbapenams

Substituted carbapenams with a protected carboxylic acid functionality at C-3 have been widely reported as intermediates in the synthesis of carbapenems (see Section 9.3.4). Inevitably such carbapenams are generally progressed without conversion to the corresponding acids, but in limited cases where the latter were prepared and evaluated, no antibacterial activity of consequence was found.[213,214] The strain of *Serratia* yielding carbapenem (110) has now been shown to produce carbapenams (115) and (171), which were isolated as *p*-nitrobenzyl esters.[137]

A variety of innovative syntheses has appeared specifically targeted at carbapenam carboxylates (172). Pyrrolidines have proved to be versatile intermediates,[215] and C-2 to C-3 bond formation has been achieved *via* intramolecular alkylation of monocyclic β-lactams (173).[216] The nucleophilicity of the β-lactam nitrogen atom has been exploited in intramolecular N-heterocyclization procedures that provide derivatives (174), (176) and (177).[217–219] Homolytic cyclization of the alkenyl β-lactam (178) affords carbapenams (180), annelation occurring exclusively *via* the *exo* mode.[220] No carbapenams were detected in the case of the terminal alkene (179), *endo* cyclization giving the carbacepham (181) as the major product.

(171)   (172)   (173)   (174) R = CH$_2$Ph
                                             (175) R = K

(176)   (177)

(178) R = Ph   (180)   (181)
(179) R = H

The potassium salt (175)[217] was devoid of antibacterial activity and all attempts to isolate the acid (115) by deblocking the esters (172) were unsuccessful, probably due to the instability of the final product.[216] The incorporation of a 6β-acylamino group has not led to derivatives suitable for antibacterial testing.[221,222] In common with earlier observation in the carbapenem series,[142] it therefore appears that the substituent further enhances the reactivity of the already sensitive β-lactam unit and causes self-destruction of the ring system.

9.3.6.4 1-Azadethiapenams and 1-Azadethiapenems

Monocyclic β-lactams (182) derived from the ubiquitous 4-acetoxyazetidinone (183)[223,224] have been particularly useful for the synthesis of these analogues. Replacement of the alkylthio group by chlorine, followed by reaction with the appropriate nucleophile, provides key intermediates such as the isothiocyanate (184) and the azides (185) and (186).

(182) R^1 = Me, Bu^t (183)
      R^2 = CO$_2$Me, C≡CR^3,
             CH=CH$_2$

(184)   (185) R = CH=CH$_2$
               (186) R = C≡CR^1

Intramolecular base cyclization of the isothiocyanate (184) gave the thioamide (187), from which the amide (188) and the thioimidate (190) were prepared.[225] The potassium salt (189) was devoid of antibacterial activity. In the *trans*-6-hydroxyethyl series the acid (191) was reported to possess weak antibacterial activity, restricted to *Staphylococcus aureus*.[224]

The azides (185) and (186) gave the imine (192)[227] and the tricyclic triazoles (193)[228] respectively, on intramolecular thermal cyclization. The triazoles can be considered as aza analogues of clavulanic acid (2), and although antibacterially inactive were weak β-lactamase inhibitors.

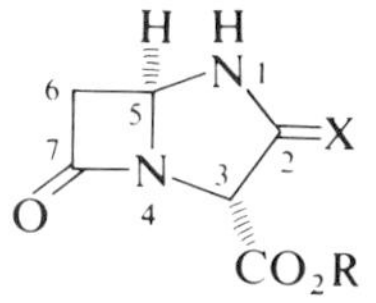

(187) X = S, R = Me
(188) X = O, R = Me
(189) X = O, R = K

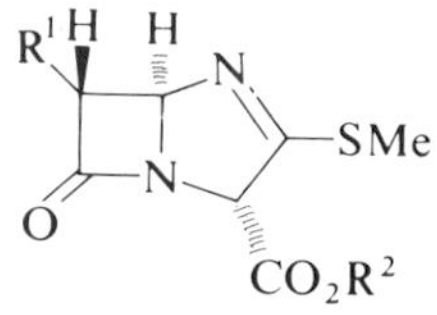

(190) R^1 = H, R^2 = Me
(191) R^1 = CH(OH)Me,
         R^2 = H

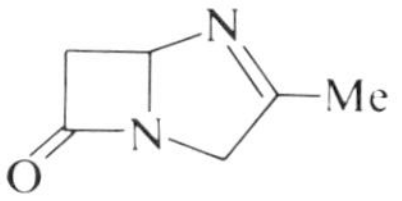

(192)

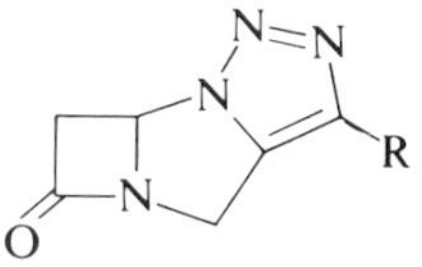

(193) R = H, CHO, CH$_2$OH,
         CH=CHCO$_2$Me

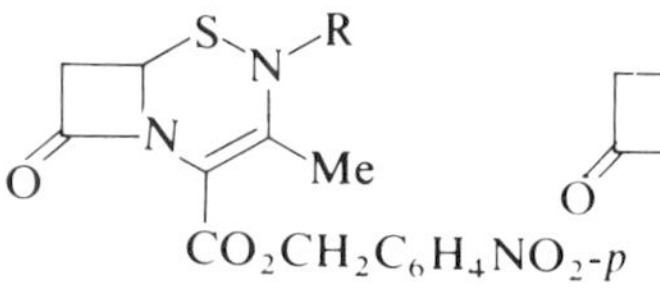

(194) R = Ph, Et

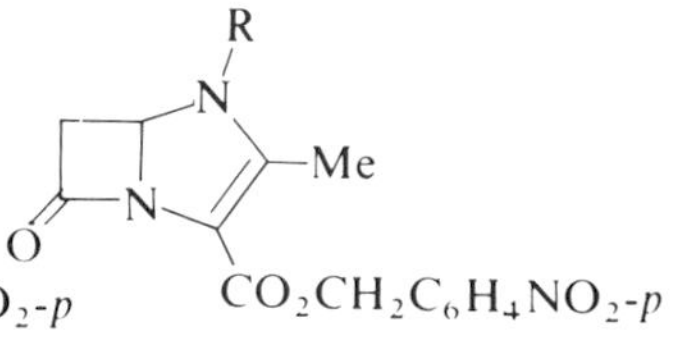

(195) R = Ph, Et

Δ^2-Azapenems are less stable than penems (see Section 9.3.5). Thus triphenylphosphine mediated desulfurization of the 2-azacephems (194) gave (195), which decomposed on attempted removal of the acid protecting group.[229]

9.3.6.5 2-Hetero-1-dethiapenams

Analogues in which the heteroatom is sulfur, oxygen or nitrogen have all been prepared. Derivatives containing sulfur, more commonly called isopenams, are the most chemically stable and biologically active of the three types.

The racemic isopenam (196) was one of the earliest nuclear analogues of the penam ring system, and displays broad-spectrum antibacterial activity comparable with the corresponding penicillin, particularly against Gram-negative organisms.[230] The monocyclic derivative (197) is also active, and is thought to be converted *in situ* to the bicyclic system. Introduction of a 3α-methoxy group decreased activity.[231]

Several key factors in the synthesis of (196) are also relevant to the preparation of many other novel systems: the development of an appropriate protecting group for the β-lactam nitrogen, the utility of an azido group as a progenitor of the acylamino side chain, and the application of methodology for rebuilding functionality on the β-lactam nitrogen after removal of the blocking group. Thus ketene–imine cycloaddition gave exclusively the *cis*-azide (198) which was deblocked by oxidation with potassium persulfate to give the azetidinone (199).[232] Other typical aromatic protecting groups include 4-(methoxymethoxy)phenyl[233] and 4-methoxyphenyl,[234] both of which are removed with cerium(IV) ammonium nitrate. The ester (199) was converted into the iodide (200) in several steps during which the azido group was reduced using zinc–acetic acid.[235] Several alternative methods for the reductive step are also compatible with the presence of a β-lactam ring.[236,237] Refunctionalization of the β-lactam nitrogen was modelled on a procedure originally described by Woodward.[238] An α-hydroxy ester residue was attached using benzyl glyoxylate, and the hydroxy group was in turn replaced by chlorine, and then by thioacetate (Scheme 4).[230] Chloro esters of type (201) are key intermediates in the synthesis of many nuclear analogues *via* displacement of chlorine with an appropriate nucleophile. Attempts to cleave the thioacetate group selectively

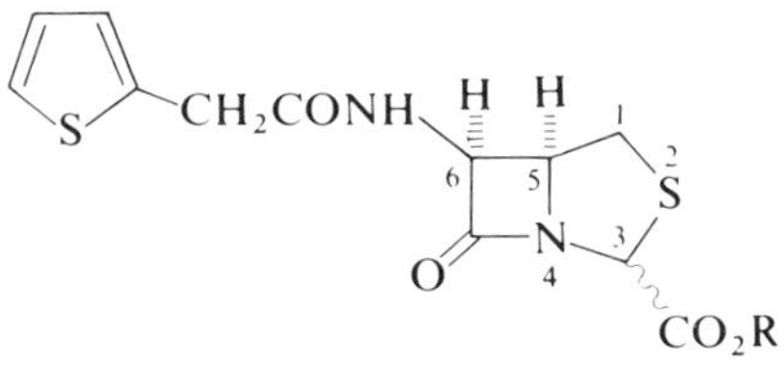

(196) R = $H_3NC_6H_{11}$

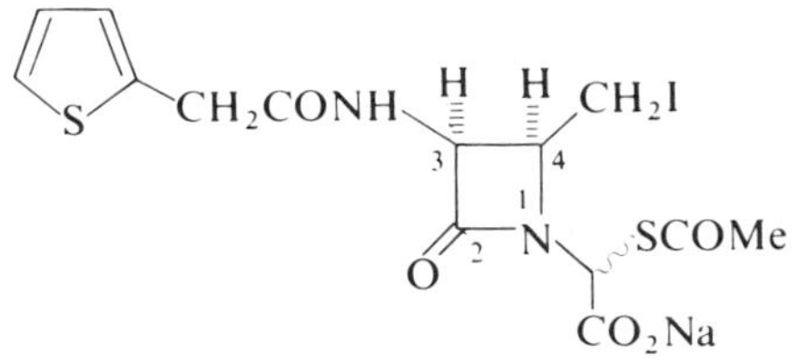

(197)

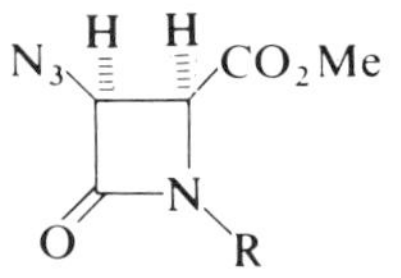

(198) R = CH_2—⟨aryl⟩—OMe, OMe

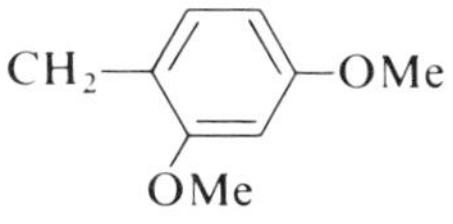

(199) R = H

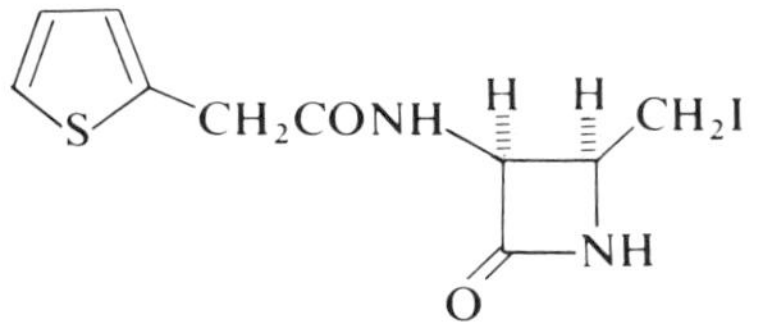

(200)

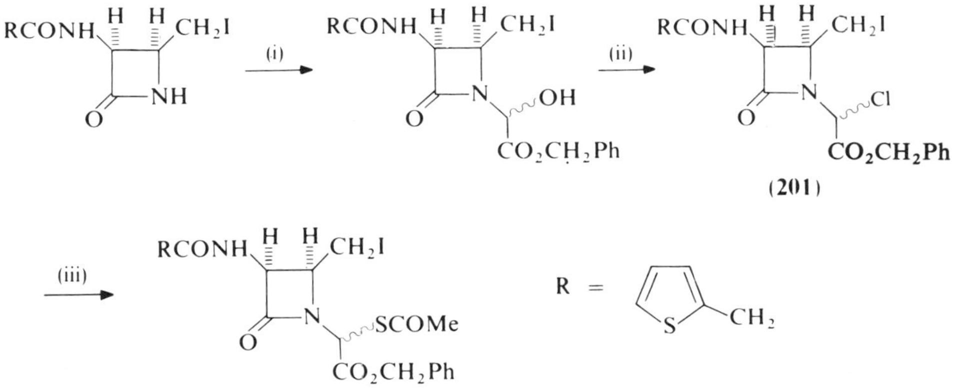

(201)

(202)

(i) $OHCCO_2CH_2Ph$, $BF_3 \cdot Et_2O$; (ii) $SOCl_2/py$; (iii) $KSCOMe$, THF/DMF

Scheme 4

from (202) unexpectedly gave the sodium salt (197), which was cyclized to the racemic 2-isopenam (196) with cyclohexylamine.

Neither the unsubstituted analogue (203)[239] nor the *trans*-hydroxyethyl derivative (204)[240] exhibit antibacterial activity. The structural similarity of the sulfone (205) to sulbactam (92) did not manifest in its biological properties and the compound (205) was a poor inhibitor of β-lactamases.[241]

The 7-oxo-3-oxa-1-azabicyclo[3.2.0]heptane skeleton (206) is trivially referred to as 'isoclavam'. Because the C—O bond is significantly shorter than the C—S bond (1.43 *vs.* 1.81 Å), isoclavams are more strained entities than isopenams.[239] The β-lactam linkage of the former nuclear analogues would therefore be expected to be more reactive. This has been confirmed by synthesis.

Strategies[242,243] involving formation of the 2–3 bond provide access to therapeutically less interesting 2,2-disubstituted derivatives such as (207), which could not be deprotected without destruction of the β-lactam ring. Alternatively, construction of the bicyclic system *via* 3–4 bond closure, as with isopenams, led to the synthesis of the salt (208).[244,245] The instability of (208) in aqueous solution (80% decomposition within 12 h) precludes meaningful biological evaluation, and claims of β-lactamase inhibition have not been substantiated.

Azetidinone intermediates of type (198) proved sufficiently versatile to allow the synthesis of the 2-aza analogue (209) *via* acylative cyclization of the imine (210) with acetyl chloride, followed by hydrogenolysis.[232] The acid (209) was a mixture of two epimeric racemates and the mediocre

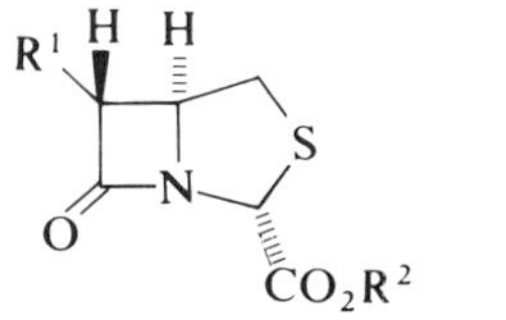

(203) R^1 = H, R^2 = Na
(204) R^1 = CH(OH)Me, R^2 = Na

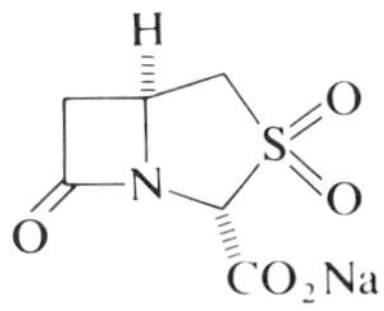

(205)

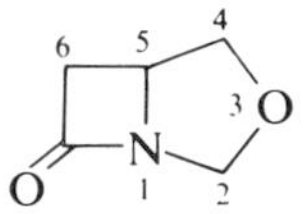

(206)

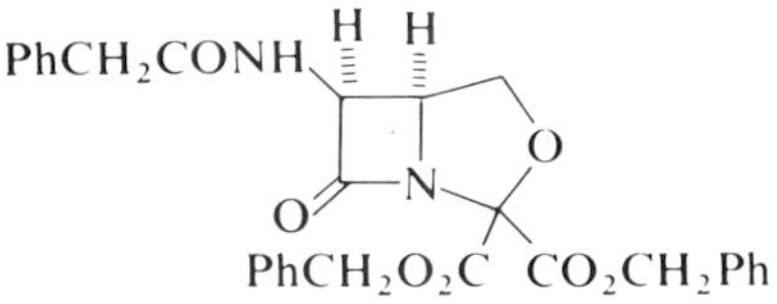

(207)

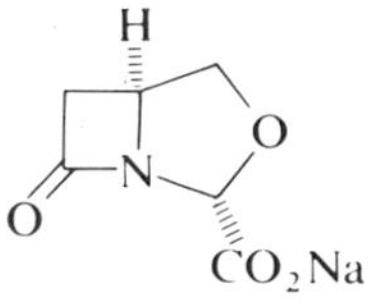

(208)

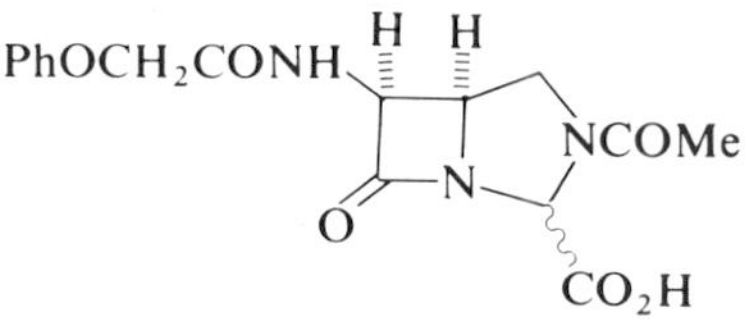

(209)

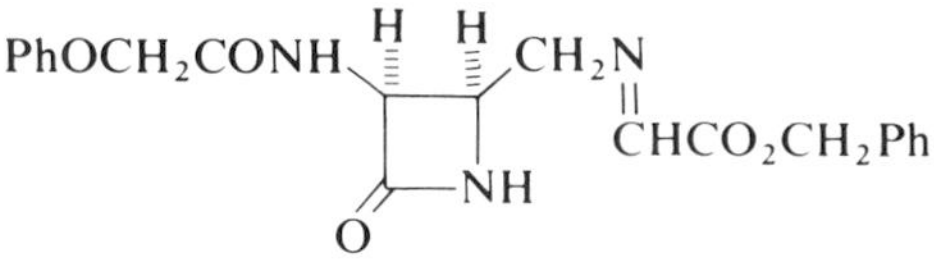

(210)

antibacterial activity, restricted to Gram-positive organisms, was consistent with the short half-life of two hours in aqueous solution at pH 7.

The synthesis of optically active analogues is now achievable by utilization of recent protocols essentially targeted at the chiral synthesis of monobactams (see Section 9.3.2.2).[48,246-248] It is doubtful whether the activity and stability shown by any of the nuclei justifies such an undertaking.[48]

An entirely different approach to further aza analogues was adopted by Beecham.[249] The intramolecular 1,3-dipolar cycloaddition of an azido group to a multiple bond proved a versatile method of synthesizing a range of derivatives. Thus when the azido ester (211) was heated in refluxing toluene for 23 h, the vinylogous urethane (213) was isolated as separable carboxylate epimers. The enamine double bond was cleaved with ozone to provide the amide (215). Cycloaddition of the unactivated alkene (212) gave the imine (217).[227,249] The derived acids (214), (216) and (218) with the natural configuration at the carboxyl function were all antibacterially inactive. Extension of the methodology to derivatives containing a *cis*-acylamino side chain (for example, 219) offered no improvement in biological properties. The double bond of the imine proved resistant to isomerization, and 2-aza-1-dethiapenems (220) still remain unrepresented in the literature. They are likely to be very reactive entities and isolation may prove impractical.

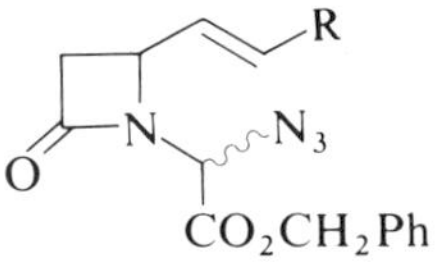

(211) R = CO_2Me
(212) R = H

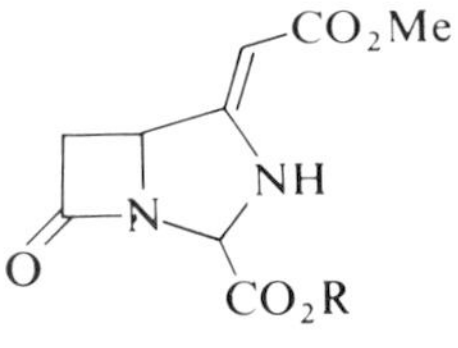

(213) R = CH_2Ph
(214) R = H

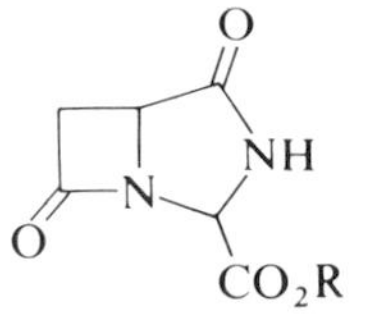

(215) R = CH_2Ph
(216) R = H

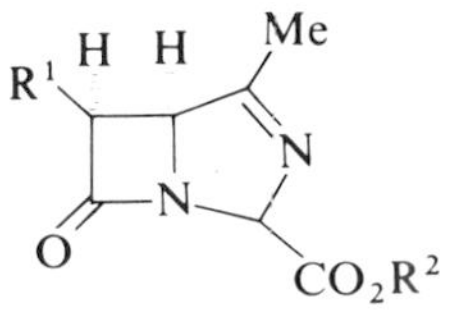

(217) R^1 = H, R^2 = CH_2Ph
(218) R^1 = H, R^2 = H
(219) R^1 = NHCOCH$_2$OPh, R^2 = H

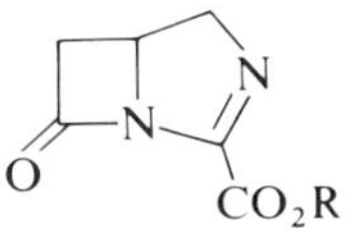

(220)

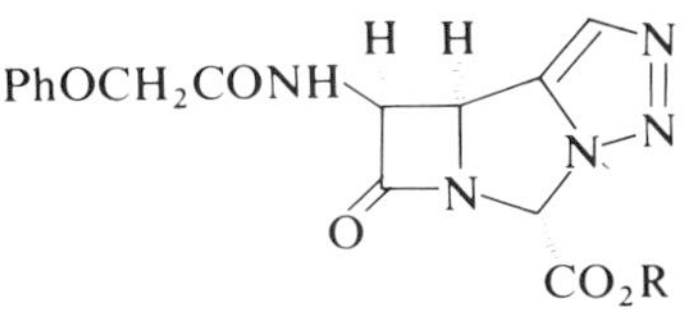

(221) R = CH_2Ph
(222) R = H

The highly strained tricyclic triazole (221) was also obtained by intramolecular cycloaddition methodology, but the corresponding acid (222) was not active.[250]

9.3.6.6 1-Oxadethiacephems

The discovery that racemic 1-oxacephalothin (223) possessed antibacterial activity of the same magnitude as its thia congener[251,252] led to an intense investigation of the 1-oxadethiacephem class of β-lactam antibiotics. They have never been isolated as natural products and as a consequence several totally synthetic approaches have been devised, which have been reviewed in some detail.[93,252-255] The construction of the required skeleton involves oxazine ring cyclization either between C-6 and O-1 or C-3 and C-4 as the key steps. The former ring closure has been exemplified in Section 9.2.9 of Chapter 9.2.

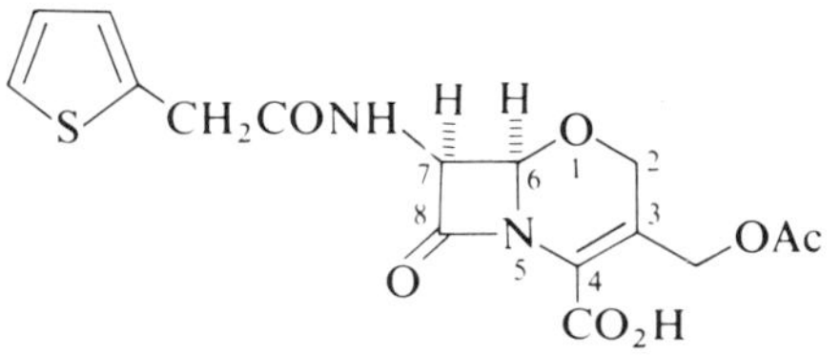

(223) 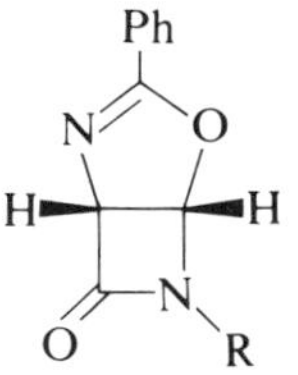 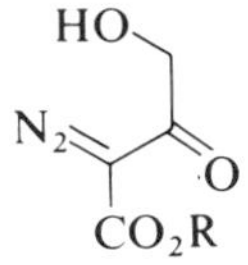   (224) R = H, SiMe₂Bu^t   (225) R = Bu^t, CHPh₂

$$(224)\ R = H,\ SiMe_2Bu^t \qquad (225)\ R = Bu^t,\ CHPh_2$$

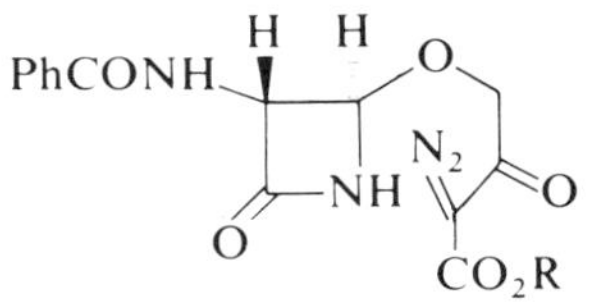

(226) R = Bu^t, CHPh₂   (227) R = Bu^t, CHPh₂   (228)

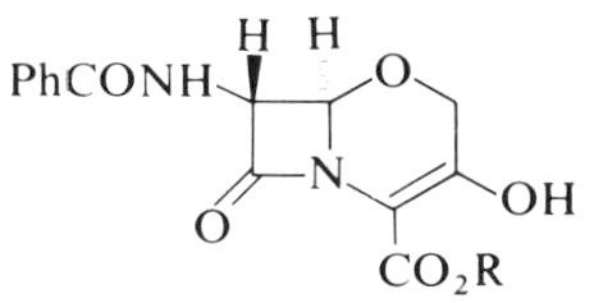

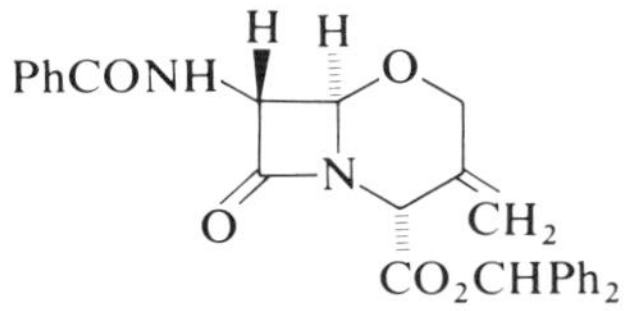

The earlier linear multistep procedures have now been supplemented with convergent syntheses, in which the operations performed in the presence of a β-lactam ring are reduced to a minimum.[256,257] The crucial transformation in each synthesis was the reaction of two building blocks (224) and (225) to form intermediates (226), followed by intramolecular carbene insertion to provide (227). The appropriate choice of alcohol building block also allowed the preparation of the exomethylene derivative (228), the primary product of many previous syntheses, and the precursor of a range of C-3 substituted analogues. The lack of stereoselectivity and practicality of previous methods led to the development of a one-pot procedure, which is outlined in Scheme 5.[258] The

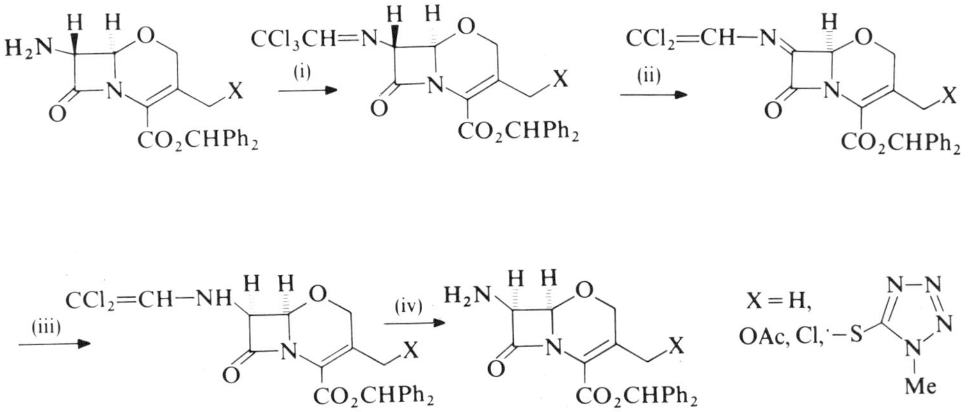

(i) CCl₃CHO, PhH, mol. sieves, reflux; (ii) EtNPr^i₂, THF, −40 °C;
(iii) KBH₄, aqueous THF, 0 °C; (iv) 2NHCl/MeCN

Scheme 5

method is compatible with a wide range of C-3 functionality, including the reactive 3-chloro-methyl and base-sensitive 3-exomethylene groups, and can be applied generally to other β-lactam derivatives.

1-Oxadethiacephems are more susceptible to hydrolysis by β-lactamases than the corresponding cephalosporins. This deficiency can be overcome by the incorporation of a 7α-methoxy substituent. Modification of the exomethylene derivative (**228**) at C-3 followed by stereospecific 7-methoxylation from the less-hindered α-face provides the key intermediate (**229**) (see Section 9.2.9 of Chapter 9.2). A derivatization programme[254,259] then identified moxalactam (**230**) as the preferred member of the series and resulted in its commercial exploitation.[260] Unfortunately the compound has some adverse side effects. Further chemical modification of the 7β-acyl moiety as well as the 3-substituent resulted in the selection of 6315-S (**231**) for further evaluation.[261] Although inactive against *Pseudomonas*, relative to moxalactam, 6315-S exhibits a more potent and broader antibacterial spectrum, particularly against *Staphylococci*, and is devoid of any disulfiram-like action.

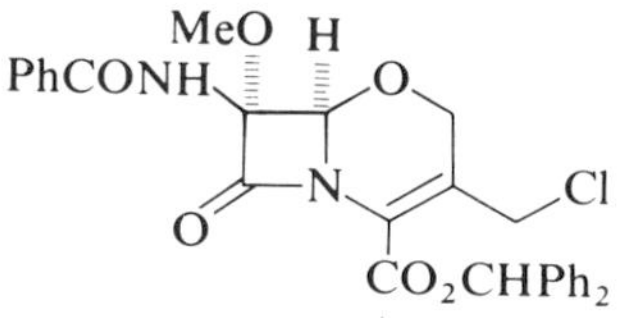

(**229**)

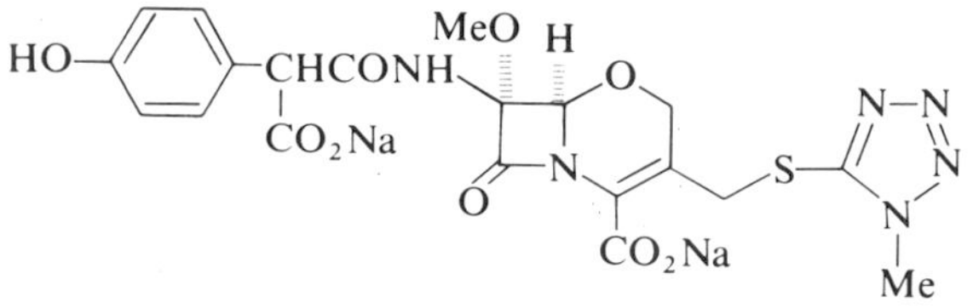

(**230**) Moxalactam

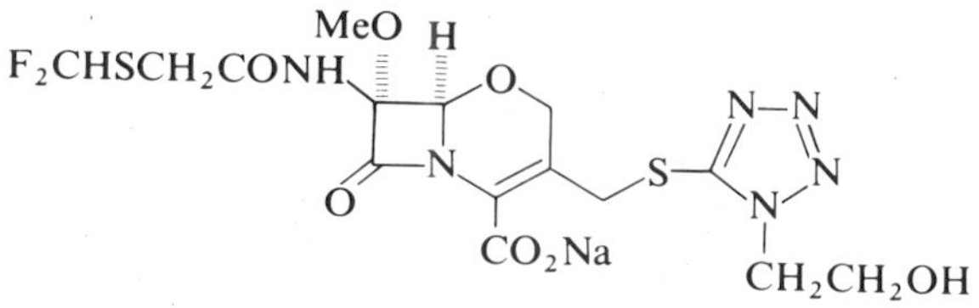

(**231**) 6315-S

Attempts have been made to relate intrinsic antibacterial activity and β-lactam chemical reactivity, to gain a greater understanding of the effect of the 3-substituent on the antibacterial activity.[262,263] Degradation mechanisms of 1-oxacephems in alkaline solution have also been studied.[264]

Recent reports[265,266] indicate that 7α-formamido substitution provides highly active, β-lactamase stable derivatives, and also confers the ability to inactivate cephalosporinases, an effect not observed in the 7α-methoxy series.

9.3.6.7   1-Carbacephems

The early approaches to the carbacephem ring system have been reviewed,[267] and interest in the series has continued up to the present day. Derivatives with a wide range of functionality have been reported,[268,269] but in all cases the presence of a 7β-acylamino side chain is necessary for antibacterial activity.[208,270] Most routes are modelled on the original Merck procedure,[271] which involves imine annulation to form a monocyclic β-lactam of type (**232**), followed by intramolecular Wittig–Horner cyclization.

Simple Dieckmann condensation of the key intermediate (**233**) provides a convenient synthesis of racemic carbacephems (**234**) with a hetero substituent at C-3.[272] Enantioselective synthesis of analogous derivatives has been achieved by the development of asymmetric ketene–imine cycloaddition employing (4*S*)-phenyloxazolidylacetyl chloride (**235**).[273] Thus hydrogenation of the initial adduct (**236**) gave the dihydro derivative (**237**). The crucial step in the further elaboration of (**237**) was a dissolving metal reduction, which removed the oxazolidone auxiliary, with concomitant reduction of the anisole ring, and debenzylation of the azetidinone nitrogen. *In situ* acylation provided the dihydroanisole (**238**), which proved to be an effective equivalent synthon for β-keto ester (**239**). Sequential diazo transfer, transesterification, and carbene insertion was followed by *in situ* trapping of the 3-hydroxycarbacephem (**240**) to give the trifluoromethanesulfonate (**241**), the precursor of a range of C-3 and C-7 substituted analogues.

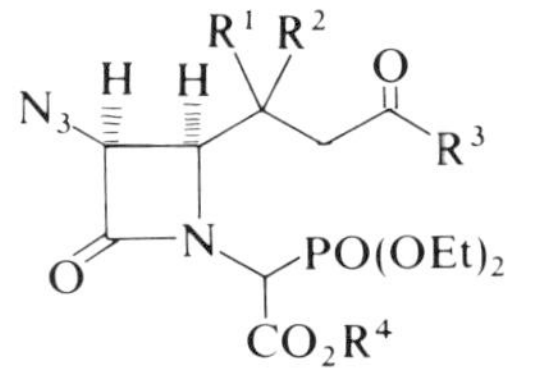

(232)

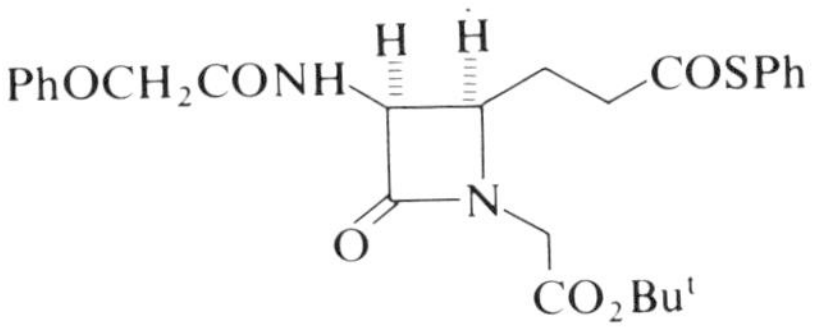

(233)

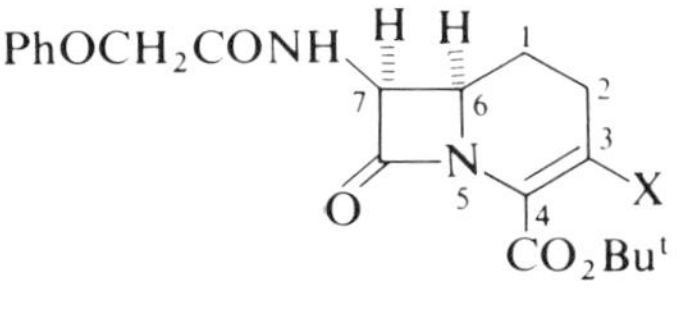

(234)

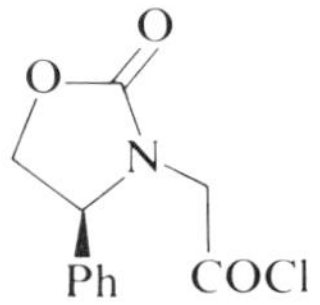

(235)

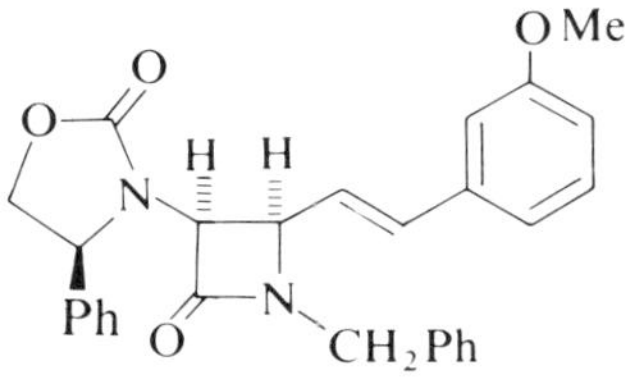

(236)

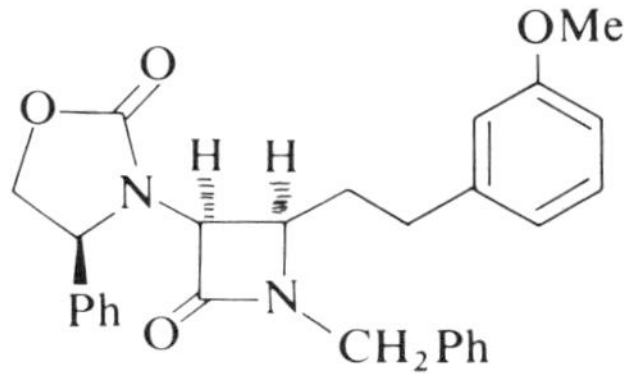

(237)

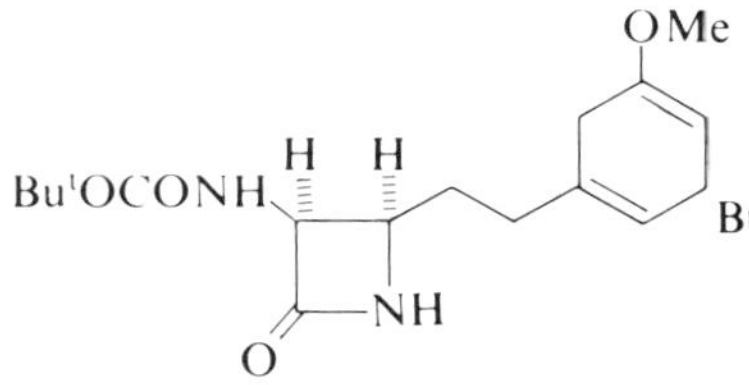

(238)

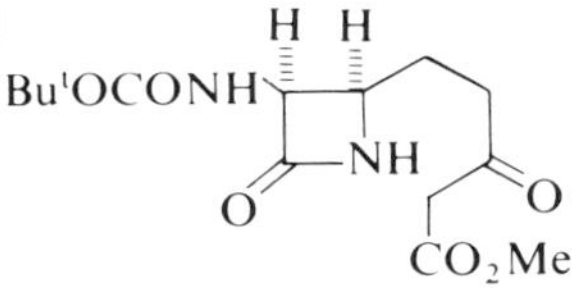

(239)

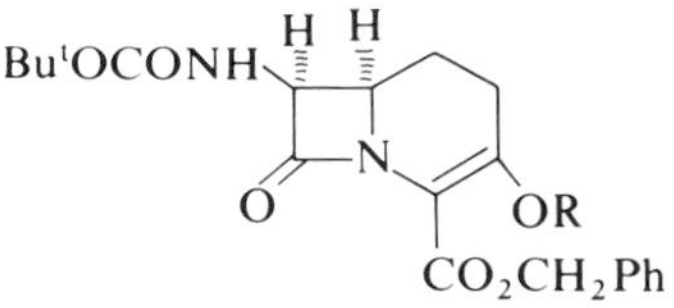

(240) R = H
(241) R = SO$_2$CF$_3$

1-Carbacephem derivatives are less active than the corresponding cephems.[267,269] Although substitution at C-1 or C-2 is generally detrimental to antibacterial activity, a 2β-acetoxy group slightly increases potency, and a 2-oxo functionality considerably improves intrinsic activity, but at the expense of β-lactamase stability.[274] Structure–activity studies have identified the analogue (242) as possessing the highest overall activity.[269] The (6R,7S) enantiomer of (242) was synthesized enzymatically from the racemate and has an antibacterial spectrum comparable with the third-generation cephalosporins.[275] The catechol (243) is a broad-spectrum agent with high potency against *Pseudomonas*.[276] Evaluation of a range of catechol isosteres indicated that only the 5-hydroxy-4-pyridone derivative (244) possessed equivalent activity.[277]

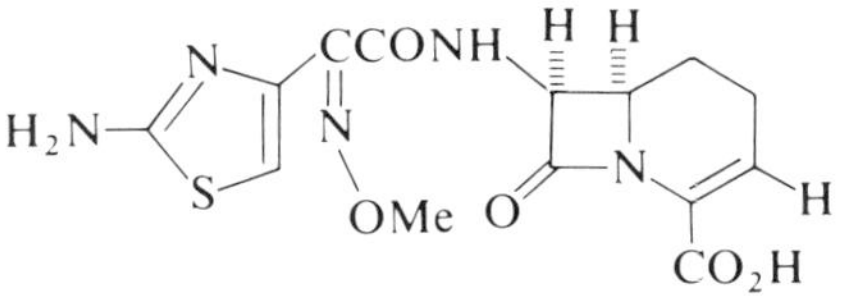

(242)

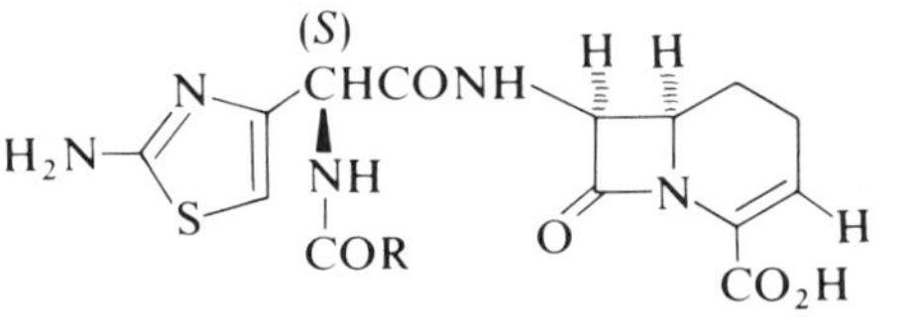

(243) R =

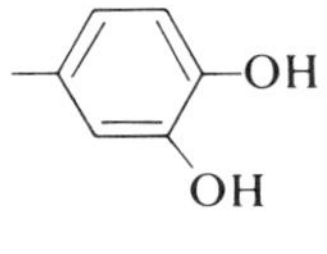

(244) R =

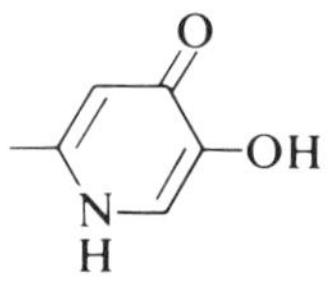

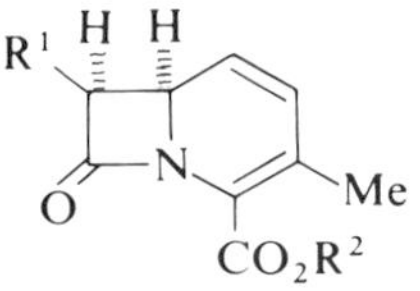

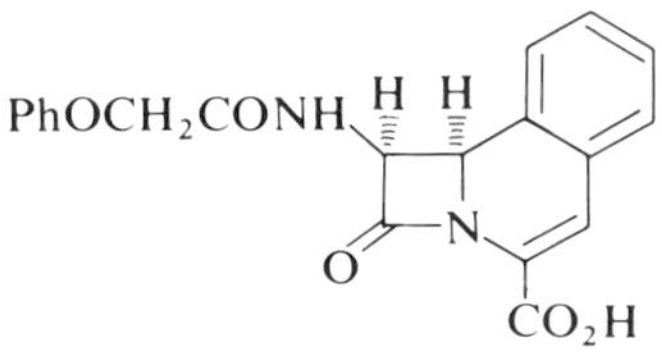

(245) R^1 = NHCOCH$_2$Ph, R^2 = H (247)
(246) R^1 = OMe, R^2 = K

In order to introduce increased strain into the β-lactam ring, various groups have investigated the synthesis of the 1-carbacephadiene system.[278,279] The analogues (245) and (246) were prepared, but both were devoid of antibacterial activity. The benzo-fused β-lactam (247) was also inactive.[280]

9.3.6.8 1-Azadethiacephems

Although a patent[281] has described the synthesis of several racemic 1-azadethiacephems (248), a definitive paper has never appeared in the literature. Their synthesis has been reviewed[282] but little is known about their biological properties. Derivatives (249) with a fused triazole ring system are antibacterially active. The tricyclic ring system (251) was constructed *via* the azide (252) using an intramolecular 1,3-dipolar cycloaddition reaction.[283] Conventional methodology then provided a range of acylamino carboxylic acids, of which the 2-thienyl derivative (250) showed the best broad-spectrum activity with good stability to staphylococcal β-lactamase.[284] Substitution of the triazole ring was detrimental to antibacterial activity.

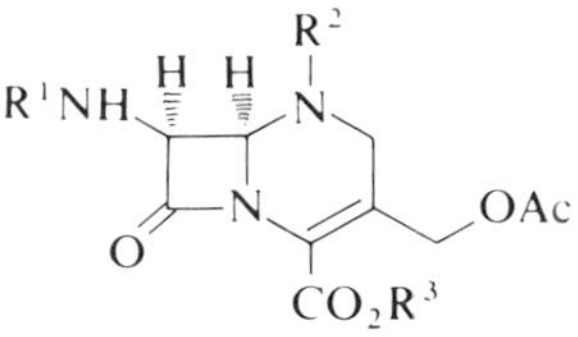

(248)

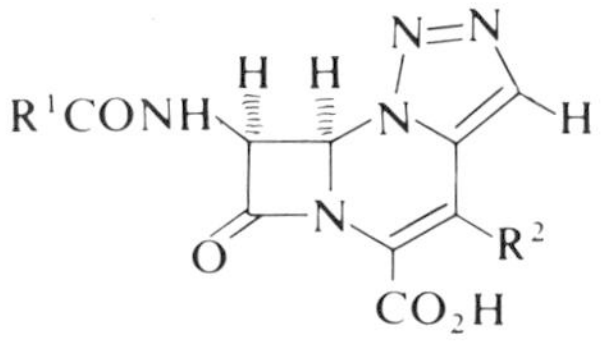

(249)

(250) R^1 = CH$_2$—〈thienyl〉, R^2 = Me

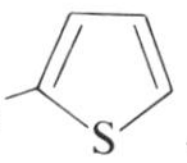

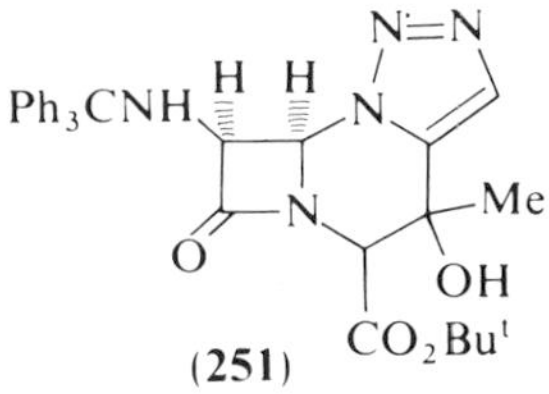

(251)

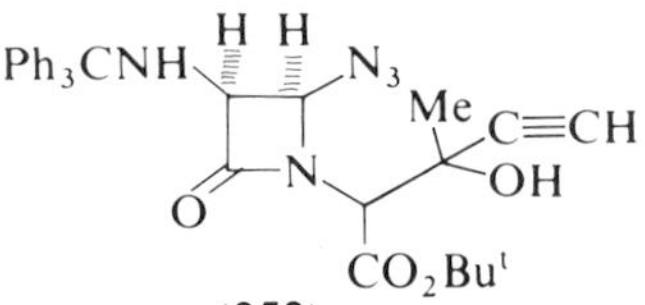

(252)

9.3.6.9 1-Phosphadethiacephem 1-Oxides

Replacement of the sulfur atom in the cephem nucleus with the electron-withdrawing P(O)OEt group was designed to enhance the reactivity of the β-lactam carbonyl and improve the biological activity. Diethyl allylphosphonite displaced the acetoxy group from the azetidinone (183) to provide the phosphinate (253). The bicyclic system (254) was then prepared conventionally *via* an intramolecular Wittig cyclization.[285] Neither of the two epimeric isomers at phosphorus showed antibacterial activity. Introduction of a *trans*-hydroxyethyl side chain at C-7 offered no improvement. Attempts to obtain 7β-acylamino derivatives were thwarted when the amine (255) could not be isomerized to the 7β-isomer. 7α-Methoxylation of the amide (256) also failed.[286]

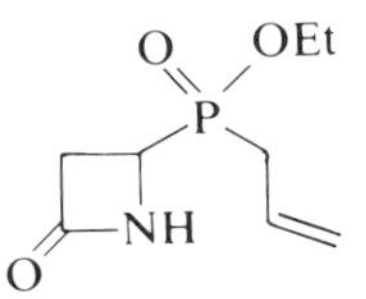

(253)

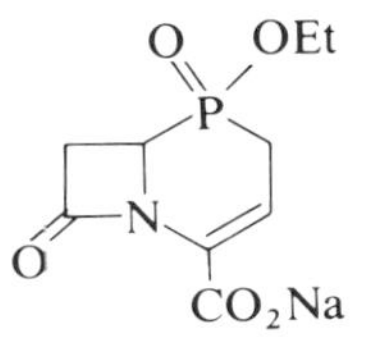

(254)

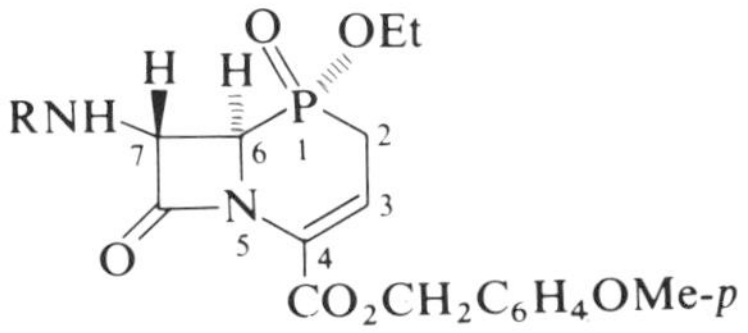

(255) R = H
(256) R = PhOCH$_2$CO

9.3.6.10 2-Hetero-1-dethiacephems

The Bristol group have extensively investigated 2-oxa, 2-thia and 2-aza analogues of 1-dethia-cephems, and a substantial part of their work has been reviewed.[287,288] The 2-oxa derivatives have received the most attention because they exhibit the best biological activity.[274] In the first of a classic series of papers, Doyle[289] described the preparation of certain key monocyclic intermediates, such as the *cis*-azide (257). The exclusive formation of (257) *via* a cycloaddition reaction between azidoacetyl chloride and the imine (258) was unexpected, and resulted in considerable speculation as to the reaction mechanism.

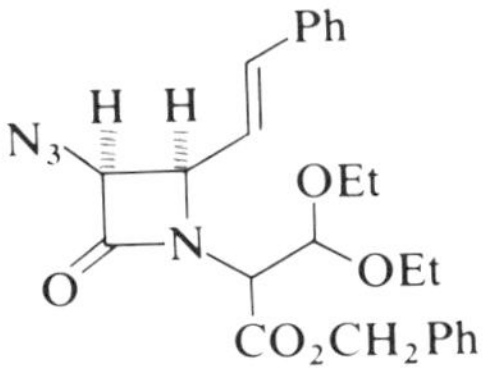

(257) 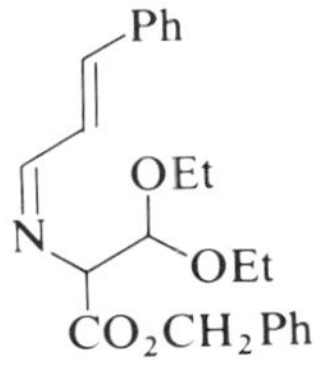   (258) 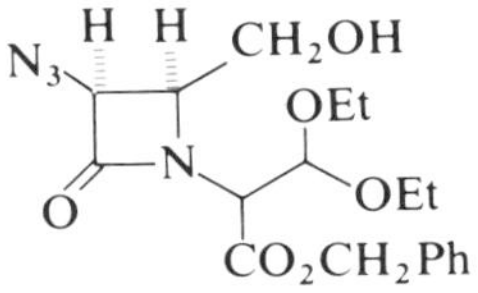   (259) 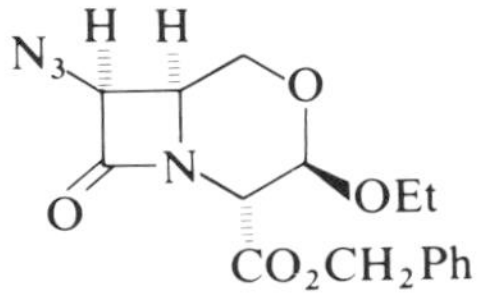   (260)

Further derivatization of the azide (257) gave the alcohol (259), which was cyclized to the single isomer (260) with boron trifluoride etherate. All efforts to convert (260) into the 2-oxa-1-dethiacephem ring system failed. An alternative mode of ring closure was therefore developed which involved initial conversion of the alcohol (259) to the enol (261).[290] Internal nucleophilic displacement of methanesulfonate by the enolate anion then resulted in the direct formation of the desired ring system (263). The synthesis of the enol (261) allowed the preparation of both 2-thia and 2-aza analogues by minor adjustments to the subsequent synthetic steps (*vide infra*). The aforementioned methodology only allowed the synthesis of derivatives in which the substituent at the 3-position was hydrogen, methyl or aralkyl. In view of the pronounced improvement in activity observed when the 3-substituent is acetoxymethyl in the natural series, entry into these systems was explored. This necessitated some modification in synthetic procedure.[291] Thus the allene (264) was generated *in situ* from the bis-methanesulfonate (262) by treatment with triethylamine, and reacted with iodine to give the diiodide (265). Cyclization in the presence of excess potassium acetate in dimethylformamide gave the desired intermediate (266), which was conventionally progressed to provide a range of 7β-acylamino derivatives (267). 2-Oxa-1-dethiacephem and cephem nuclei have about the same

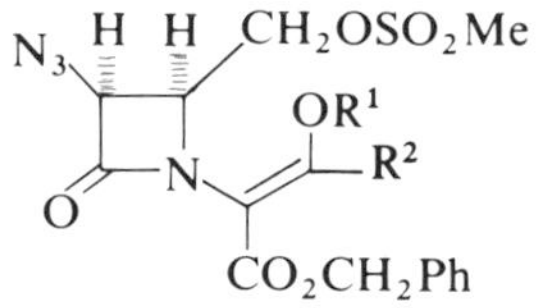

(261) R^1 = R^2 = H   (263)   (264)
(262) R^1 = SO$_2$Me, R^2 = Me

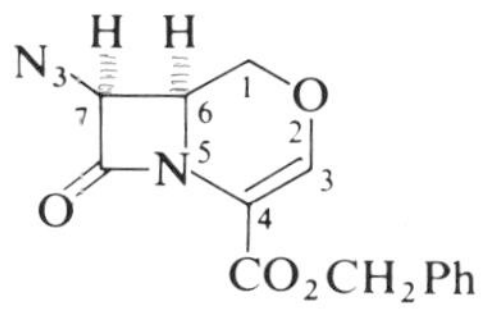

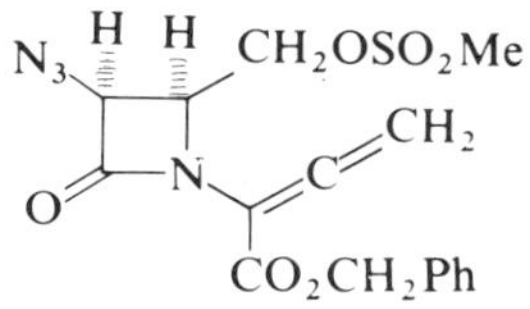

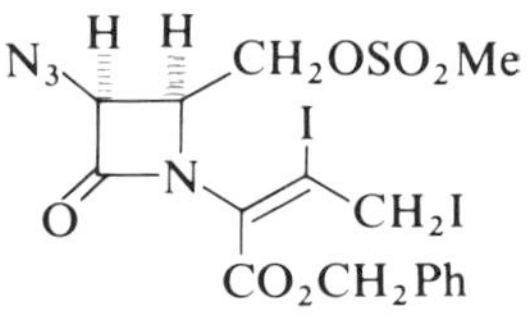

(265)

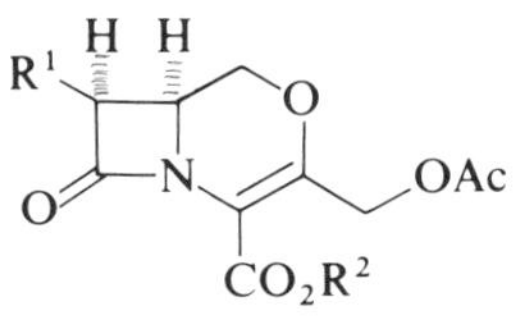

(266) R^1 = N$_3$, R^2 = CH$_2$Ph
(267) R^1 = R^{3}CONH, R^2 = H

inherent activity, but the former is more susceptible to β-lactamase, in common with 1-oxadethia-cephems. A 7α-methoxy substituent appears to give some protection against β-lactamases, but overall activity is severely compromised.[292] Substitution at C-1 also decreases potency.[274]

The most versatile synthetic approach to 2-thia analogues involved hydrogen sulfide mediated cyclization of bis-methanesulfonates (268).[293] Concomitant reduction of the azido group gave the amine (269) which was treated *in situ* with the appropriate acylating agent. Deprotection gave analogues suitable for testing. The mediocre level of antibacterial activity shown in this series discouraged detailed investigation and the range of 3-substituents was restricted to hydrogen and methyl.[274] Oxidation of sulfur marginally improved potency, but the 7α-methoxy derivative (270) was devoid of activity,[292] in common with the corresponding cephalosporin.[294]

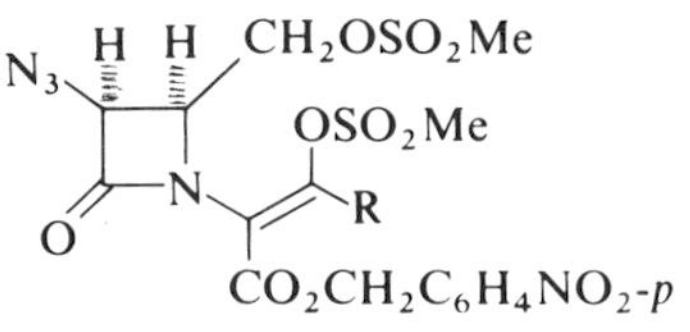

(268)

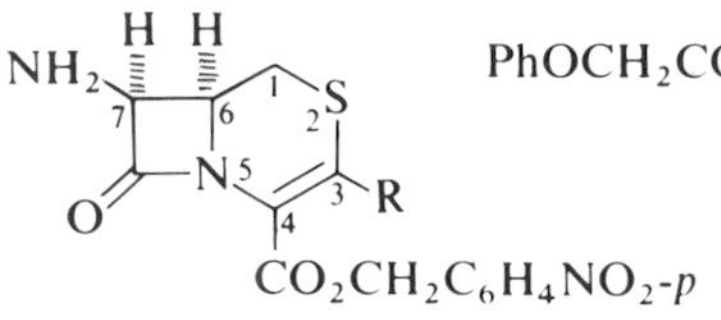

(269) R = Me, H

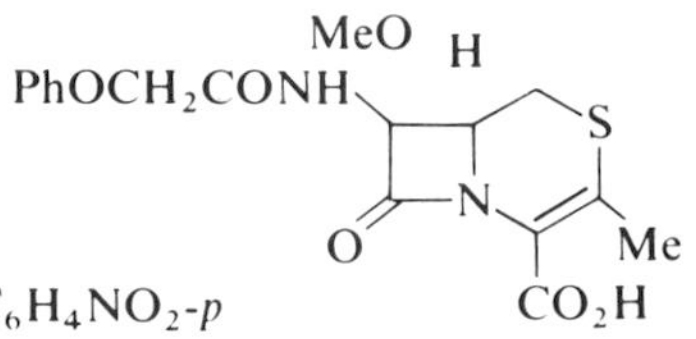

(270)

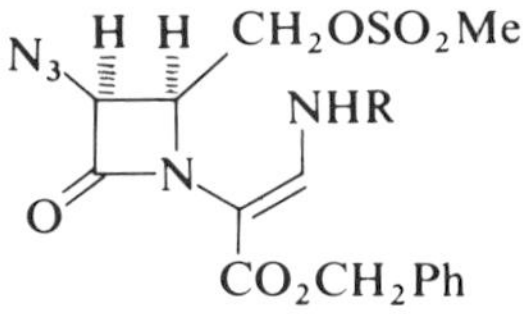

(271) R = H
(272) R = Me
(273) R = CO$_2$Et

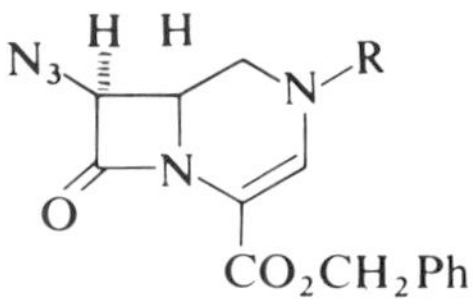

(274) R = Me
(275) R = CO$_2$Et

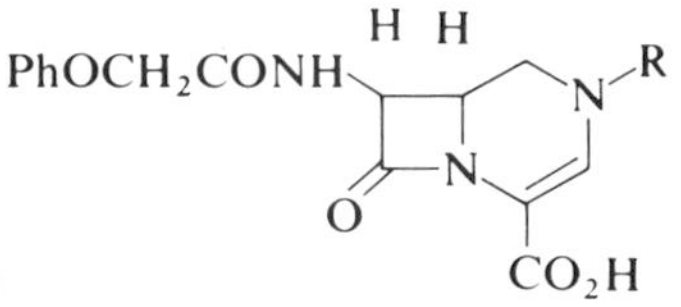

(276) R = Me
(277) R = CO$_2$Et

The preferred route to 2-aza-1-dethiacephems is *via* the intermediate (268) used to synthesize the 2-thia analogues.[295] The amine (271) and its monomethyl derivative (272) were prepared, but only the latter afforded a cyclized product (274), albeit in low yield, on treatment with sodium hydride in dimethyl sulfoxide. Reaction of the amine (271) with ethyl chloroformate and triethylamine gave the carbamate (275) *via* spontaneous cyclization of the putative intermediate (273). The phenoxyacet-amido compounds (276) and (277) were prepared using standard methodology, but were the least active and stable of such hetero analogues. It appears that both electron donation to the double bond and steric bulk at the 2-position decrease antibacterial activity.[274]

More recently, alternative methods for the synthesis of key precursors of these 2-hetero analogues have been reported,[296] as well as stereocontrolled synthesis of optically active nuclei.[297-300] The unstable 2-aza-1-dethiacephem ring system can be effectively stabilized by the addition of a further ring.[250] Thus the tricyclic triazole (278) was prepared by intramolecular cycloaddition of the vinyl azide (280). The derived acid (279) possessed modest Gram-positive activity, but was ten-fold less active than the isomeric triazole (281). This can be rationalized in part by a substantial lowering in the reactivity of the β-lactam carbonyl system, manifested by a 35 cm^{-1} decrease in the β-lactam carbonyl stretching frequency in the IR spectrum of (279) compared with (281).

The aziridine (282) was obtained when the vinyl azide (284) was heated in benzene for 20 min.[301] The use of 2,2,2-trichloroethyl as an acid protecting group proved particularly amenable to the synthetic methodology, and deblocking was achieved with activated zinc dust in tetrahydrofuran containing potassium dihydrogen phosphate. The acid (283) was antibacterially inactive.

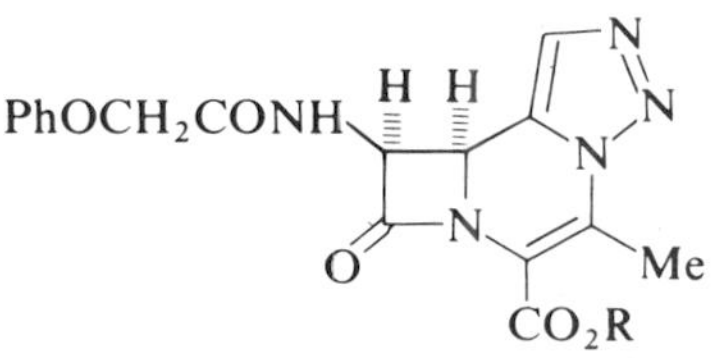

(278) R = CH$_2$Ph
(279) R = H

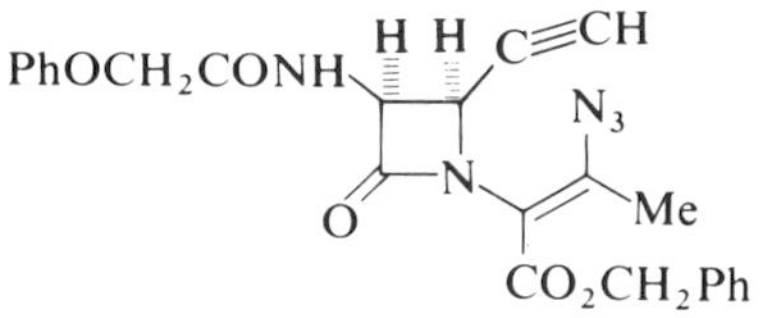

(280)

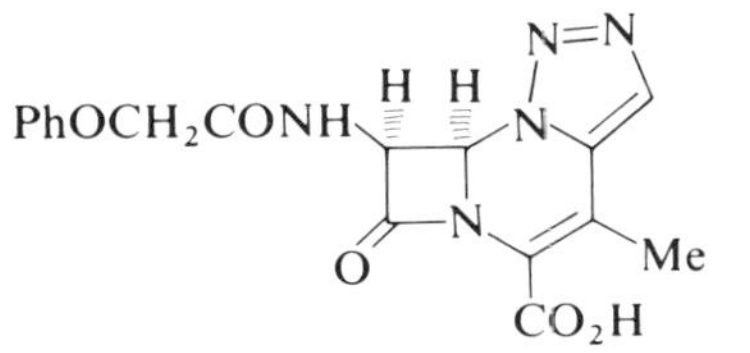

(281)

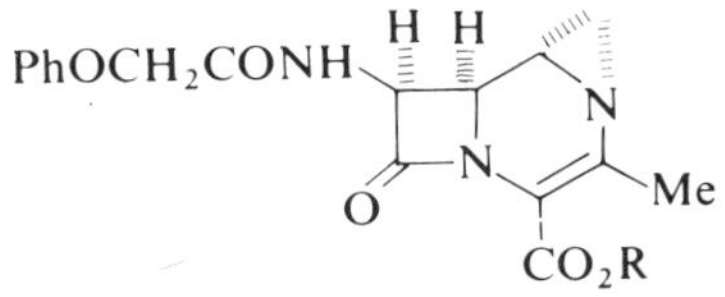

(282) R = CH$_2$CCl$_3$
(283) R = H

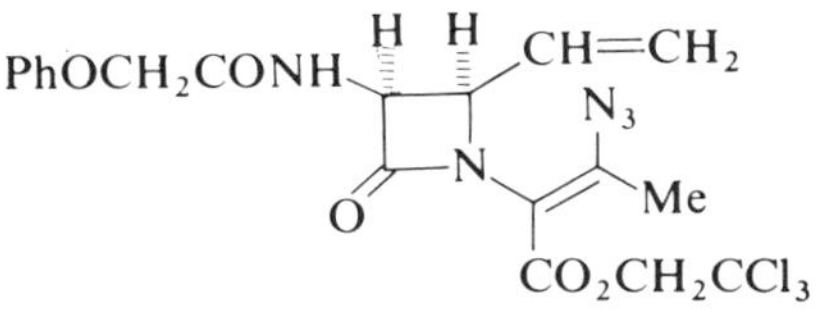

(284)

9.3.6.11 3-Hetero-1-dethiacephams

The versatile monocyclic β-lactam (198) (*cf.* isopenams, Section 9.3.6.5) has been used to prepare 3-oxa-, 3-thia- and 3-aza-1-dethiacephalosporins. The methoxycarbonyl group was converted into the desired ketal functionality in several steps and, after deblocking and refunctionalization of the β-lactam nitrogen, intramolecular acid catalyzed cyclization of the appropriate intermediates (285) provided the required bicyclic systems.

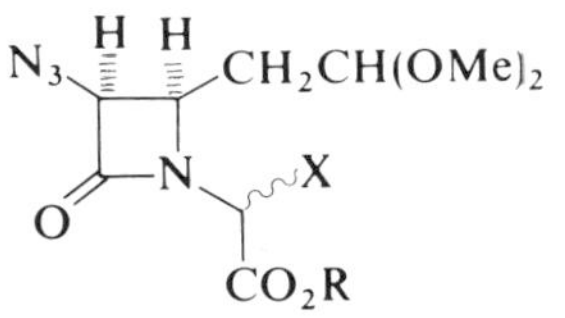

(285) X = OH, SCOMe, NHC$_6$H$_4$Me-*p*
       R = CH$_2$Ph, Me

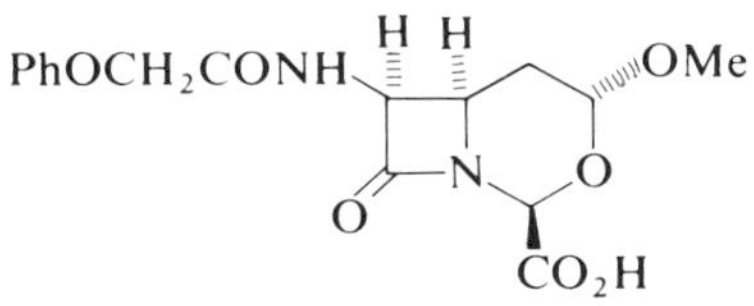

(286)

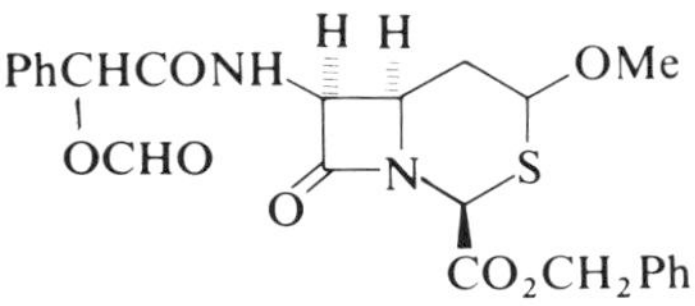

(287)

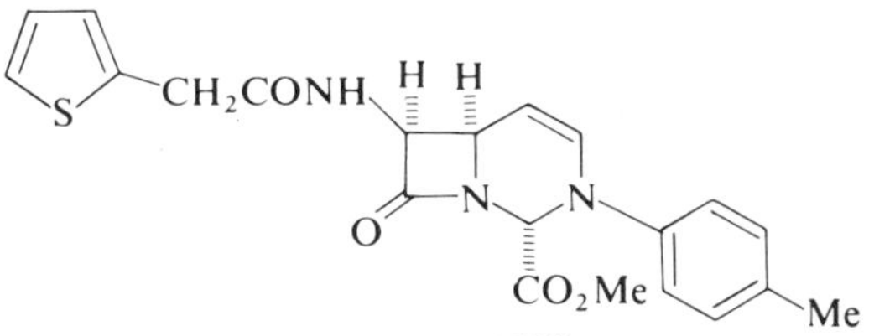

(288)

In the 3-oxa series, acid (286) was the first example of a saturated 4,6-bicyclic system with good antibacterial activity.[302] Such activity was unexpectedly restricted to the isomer with the unnatural β-configuration of the carboxylic acid. The presence of a 2-methoxy group was also obligatory. It has been speculated that fragmentation of the oxazine ring plays a role in activating the bicyclic β-lactam for antibacterial activity. In the sulfur and nitrogen series only the respective β- and α-carboxylates (287) and (288) were obtained.[303] The corresponding acids were not reported but, as might be expected, the esters were devoid of biological activity.

3-Aza-1-dethiacephems have also been synthesized using intramolecular dipolar cyclo-additions.[304] A range of structural types (289), (290) and (292) were evaluated but none showed antibacterial activity. Of the corresponding derivatives with an acylamino side chain, only the amides (293) and (294) exhibited activity. Although potency was mediocre, in common with the 3-oxa series the more active epimer possessed the unnatural carboxylic acid stereochemistry (293). The introduction of further unsaturation into the enamine (291) provided the azadethiacephem (295), but attempted removal of the acid protecting group caused rapid cleavage of the β-lactam ring. A similar observation has been reported in the 3-azacephem series (*vide infra*).

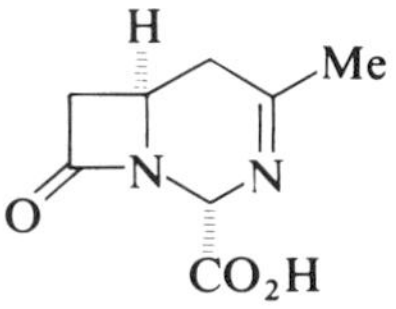

(289)

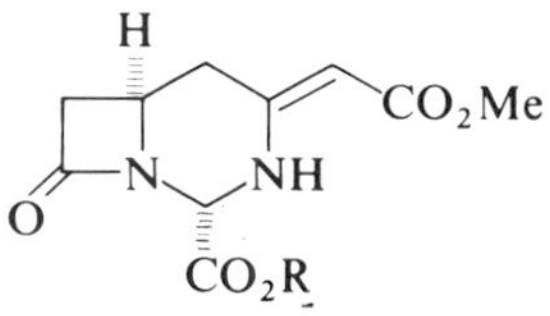

(290) R = H
(291) R = CH₂Ph

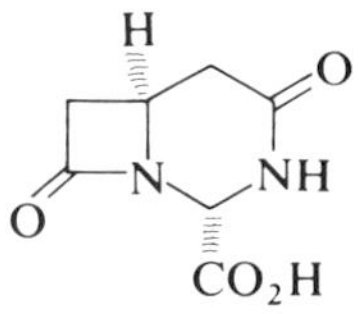

(292)

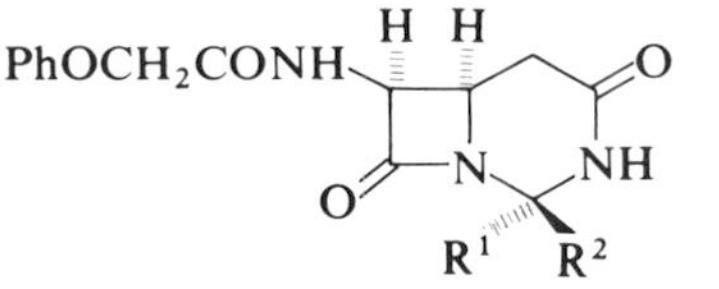

(293) R¹ = H, R² = CO₂H
(294) R¹ = CO₂H, R² = H

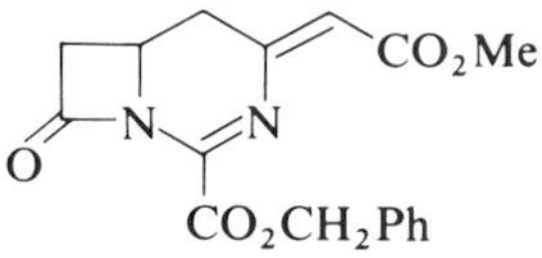

(295)

9.3.6.12 3-Hetero-cephems and -cephams

In principle, replacement of the carbon atom at C-3 of the cephem nucleus with the electron-withdrawing nitrogen atom should enhance the reactivity of its β-lactam carbonyl. This was observed in practice, and the analogue (296) underwent rapid β-lactam cleavage in methanol.[305] The pivaloyloxymethyl ester (297), which was expected to be versatile for antibacterial testing as in the case of penicillins and cephalosporins, showed no significant antibacterial activity. The acid (298) and the N-acetyl derivative (300) exhibited some activity against Gram-positive bacteria, but the oxa analogue (299) was not active.[306]

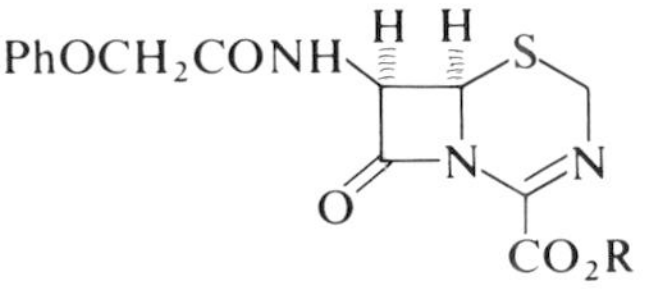

(296) R = CH₂Ph
(297) R = CH₂OCOBuᵗ

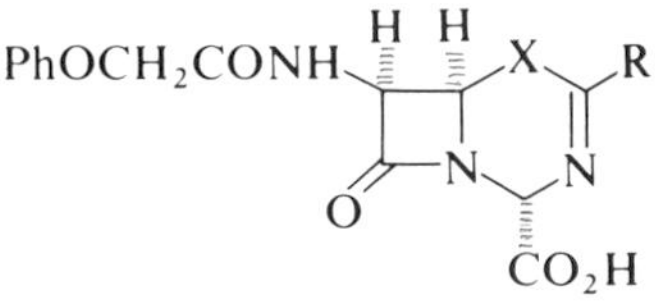

(298) X = S, R = H
(299) X = O, R = Et

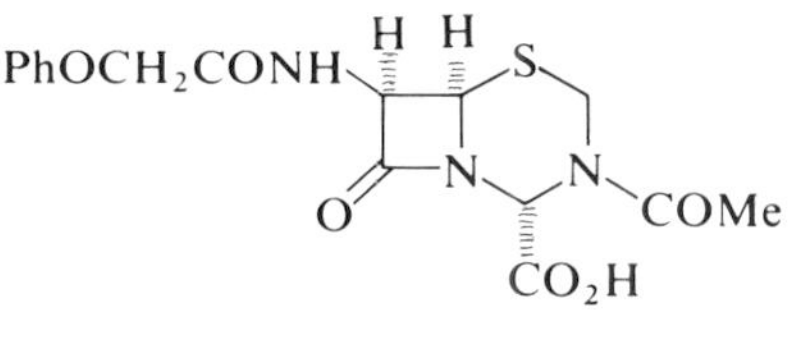

(300)

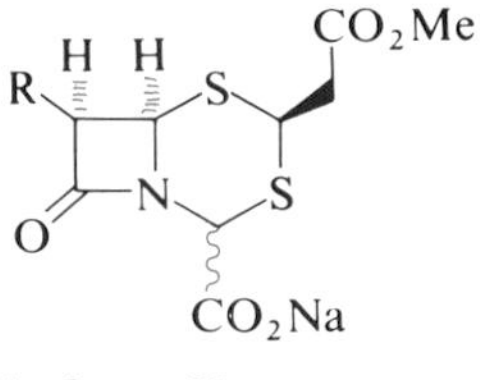

(301) R = H
(302) R = PhCH₂CONH

The first representatives of the 3-thiacepham family have been reported. The racemic sodium salt (301) was inactive, but although of limited stability in aqueous solution, the 7β-acylamino analogue (302) was endowed with weak antibacterial properties.[307]

9.3.6.13 2,3-Diaza-1-dethiacephems

In view of the lack of a suitably positioned double bond, the inactivity of the bicyclic derivative (303) was not unexpected.[308] More appropriate substitution has been incorporated by generation of the analogue (304) from the azetidinone (306) by silver oxide mediated oxidative elimination. The salt (305) was stable in deuterium oxide over a 24 h period, but showed no significant antibacterial activity.[309]

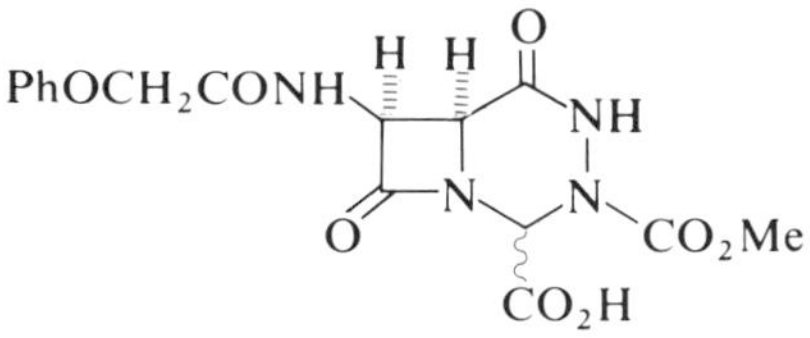

(303)

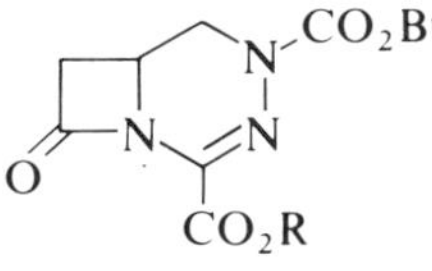

(304) R = CH₂Ph
(305) R = Na

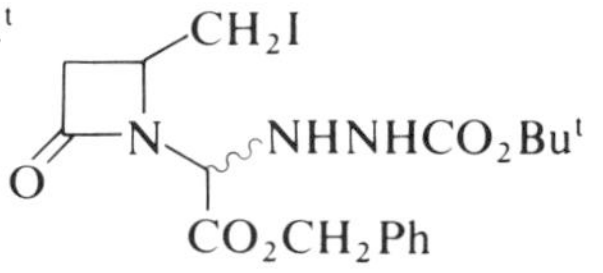

(306)

9.3.6.14 Diverse Bicyclic Systems

The 4,7-bicyclic system (307) was synthesized as early as 1972, but was devoid of antibacterial activity.[238] Further analogues have been reported, but both the oxathiazepine (308) and the thiadiazepine (309) were inactive.[310] The imidate (310) and the amidine (311) were similarly lacking in good antibacterial activity.[311]

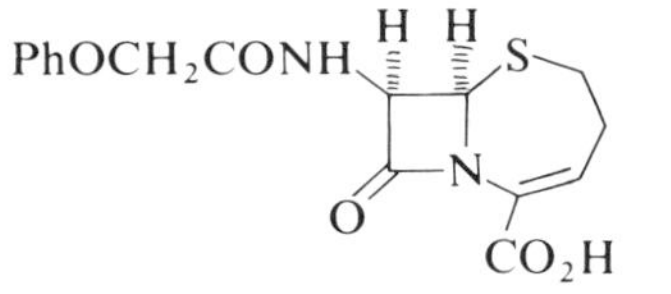

(307)

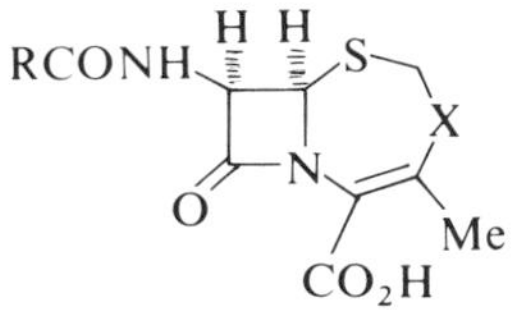

(308) R = H₂N—⟨thiazole⟩, X = O

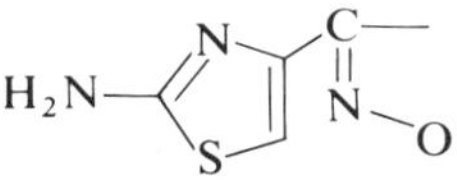

(309) R = PhCH₂, X = S

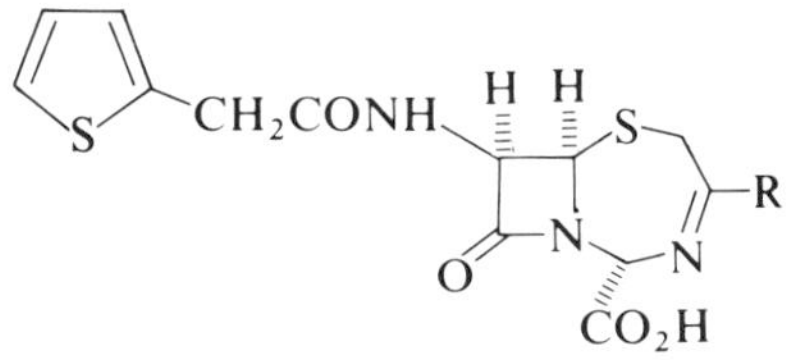

(310) R = OMe
(311) R = N(Me)Ph

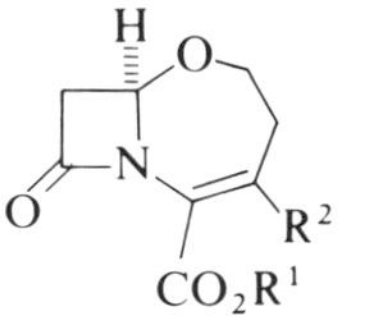

(312) R¹ = Buᵗ, R² = H
(313) R¹ = CH₂Ph, R² = H
(314) R¹ = Li, R² = OMe

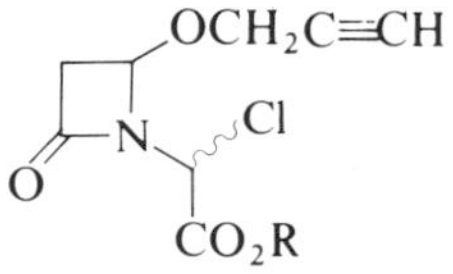

(315) R = Buᵗ
(316) R = CH₂Ph

The racemic 1-oxadethiahomocephem (312) was obtained *via* regiochemically controlled intramolecular addition of the free radical derived from the alkyne (315).[312] Alternatively, palladium catalyzed ene–halogenocyclization of the corresponding benzyl ester (316) gave the analogue (313).[313] The closely related lithium salt (314),[99] derived from clavulanic acid (Section 9.3.2.2) was of no biological interest.[314]

The synthesis of the novel 1,3-bridged β-lactam (317) has been disclosed.[315] The relatively labile diazo-β-keto ester (318) was prepared conventionally from ethyl acetoacetate, and cyclized with rhodium(II) acetate to provide the solid bicyclic compound (317). No biological evaluation was reported. The relatively surprising stability of (317) suggests that many stable 'anti-Bredt' β-lactams can be synthesized and investigated for novel chemical and possibly biological properties.

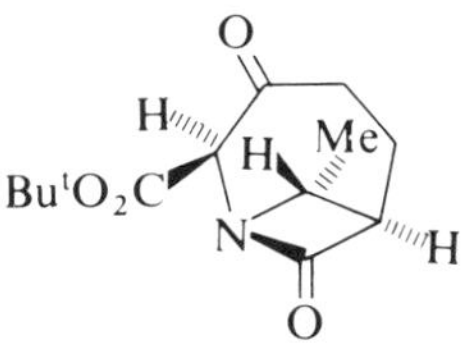

(317)

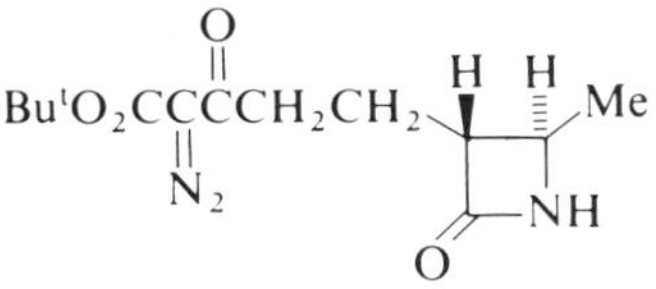

(318)

9.3.6.15 Fused 1,2-Diazetidinones

Aza β-lactams based on 1,2-diazetidinones, in which the additional nitrogen is incorporated into the four-membered ring itself, have been investigated by Taylor and Moody. Ylides of type (319) are of particular synthetic utility.[316] The structural similarity of 1,2-diazetidinones with β-lactams is more apparent than real, and even simple extensions of standard β-lactam chemistry to its aza analogues can be frustrated through the interception of intermediates by N-1.[317] Despite these problems the bicyclic aza β-lactam (320) has been prepared, but biological evaluation was not reported.[318]

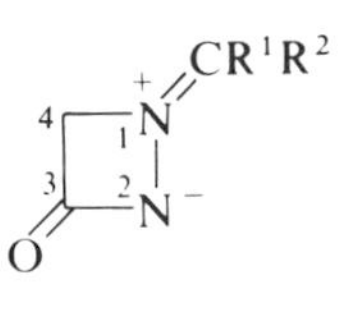

(319) 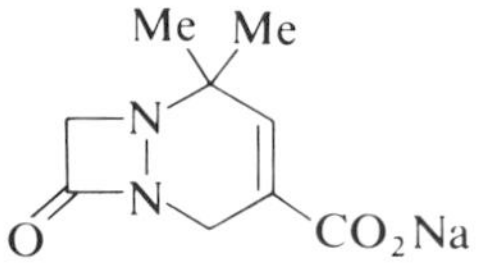   (320) 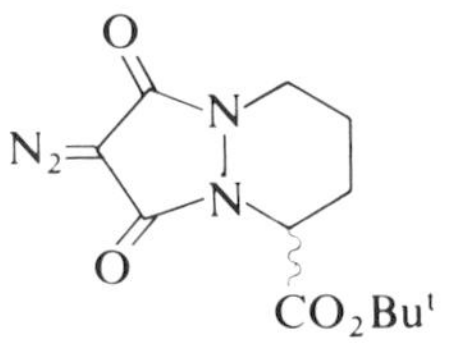   (321) 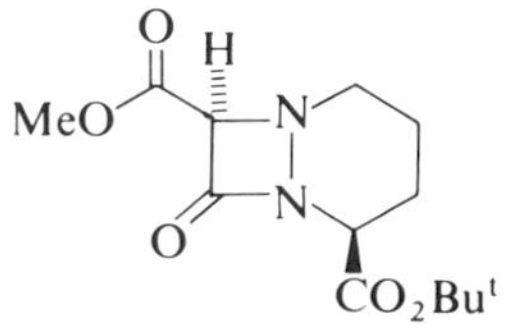   (322)

In the ring contraction of the bicyclic diazo compound (321), the Wolff rearrangement was highly regio- and stereo-selective, and only one (322) of the possible regioisomeric aza β-lactams was isolated, albeit in low yield.[319] These studies are at an interesting stage, and no doubt the evaluation of more aptly substituted derivatives will appear some time in the future.

9.3.7 NON-β-LACTAM β-LACTAM ANALOGUES

Compounds containing some of the structural features which are present in penicillins and other β-lactams have been considered as possible antibacterials. Targets, generally defined as acyl dipeptides lacking the β-lactam ring or having an alternative to the β-lactam ring, have been prepared,[320] such as the aziridine (323), acylsuccinimido (324) and heterocyclic derivatives (325). The thio (326) and β-sultam (327) analogues have been reported[321] and considerable effort has been directed towards the preparation of analogues whereby the β-lactam ring has been replaced by cyclobutanone, resulting in bicyclo[3.2.0]heptan-2-one and 2-oxabicyclo[3.2.0]heptan-6-one derivatives, *e.g.* (328), (329) and (330).[322] Recently, γ-lactams have been considered as alternatives to β-lactams, although (331) was prepared over 40 years ago.[323] Compounds with well defined stereochemistry which have been reported include the bicyclic pyrazolidinones (332),[324] the penam, penem and carbapenem analogues (333) and (334), and the azetidine (335).[325] Of these compounds the bicyclic pyrazolidinone of the type (332) possessed significant antibacterial activity and was found to bind to PBPs; none was reported to have any β-lactamase inhibitory activity of interest.[324]

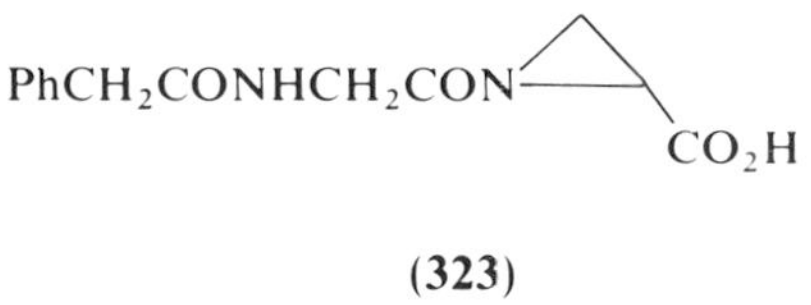

(323) 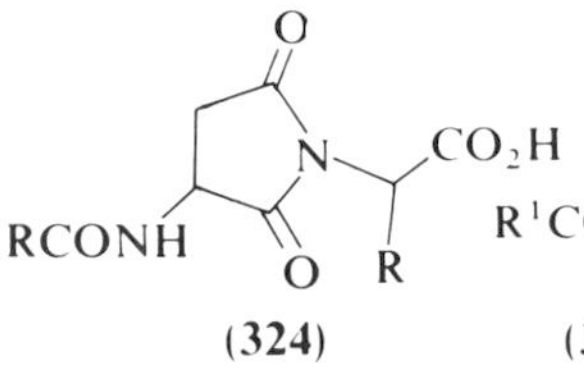   (324) 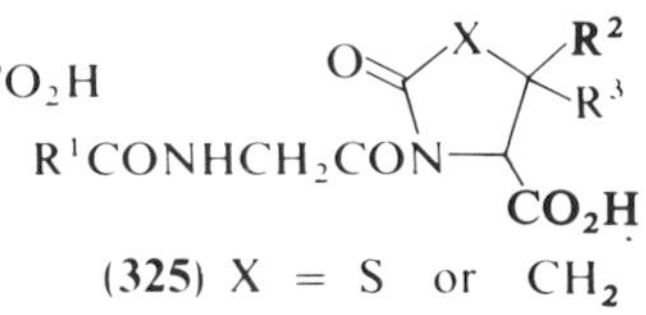   (325) X = S or CH$_2$

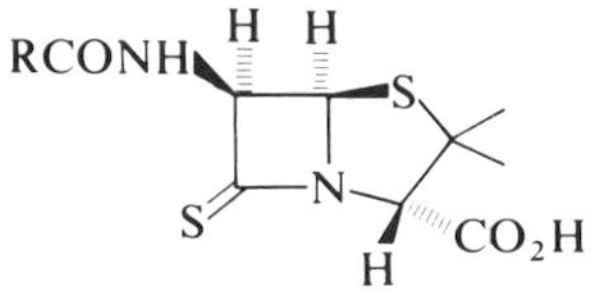

(326) 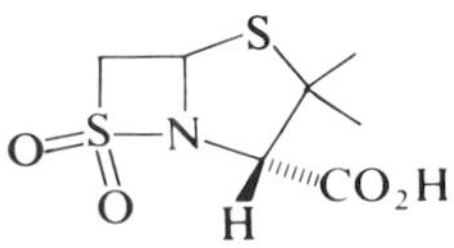   (327) 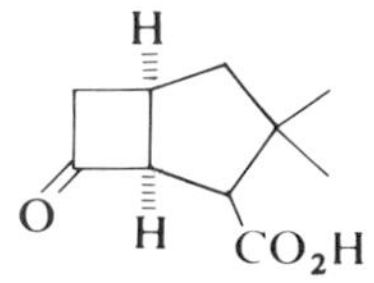   (328)

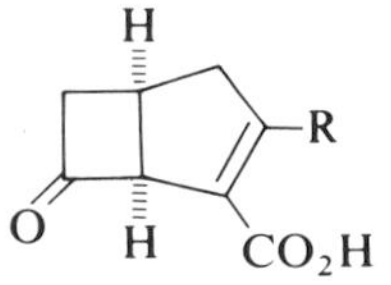

(329) 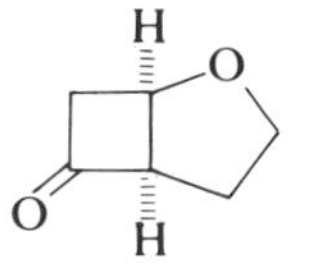   (330) 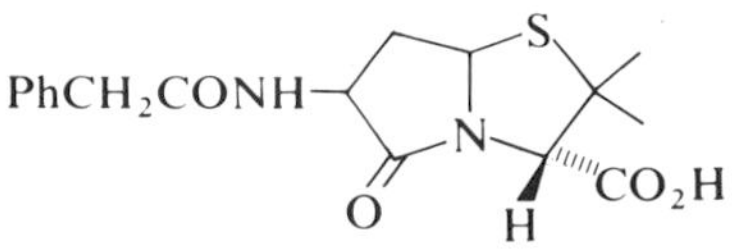   (331)

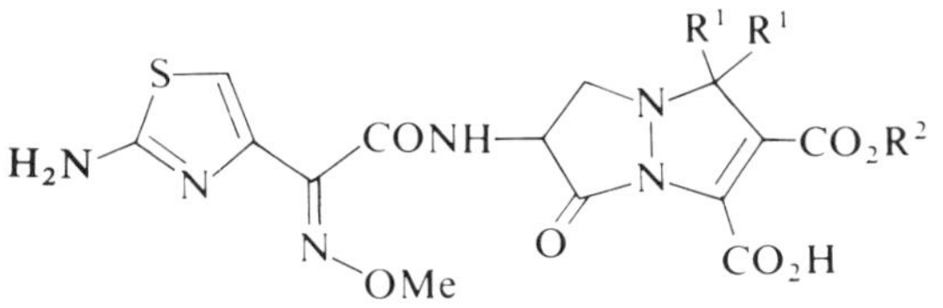

(332) R^1 = H,   R^2 = Me
       R^1 = Me,   R^2 = H

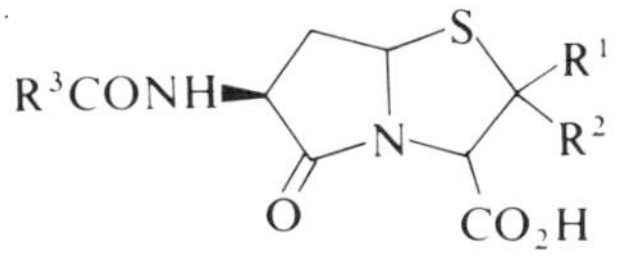

(333)

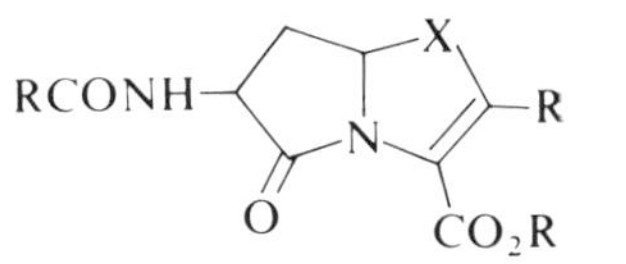

(334) X = S or CH$_2$

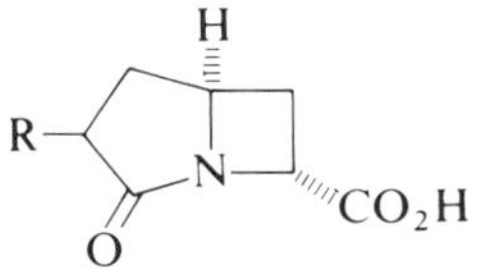

(335)

Takeda workers have recently described the identification of an L-cycloserine derivative, lactivicin (336), from the bacteria *Empedobacter lactamgenus* and *Lysobacter albus*.[326] It was moderately active against Gram-positive organisms; it bound to the PBPs of *E. coli* and *B. subtilis*, and was sensitive to certain types of β-lactamases. Lactivicin and derivatives have been synthesized.[327] The Squibb group have isolated[328] β-lactones, *e.g.* obofluorin (337), as a spin-off from their search for β-lactams, but it is not evident if these compounds interact with PBPs.

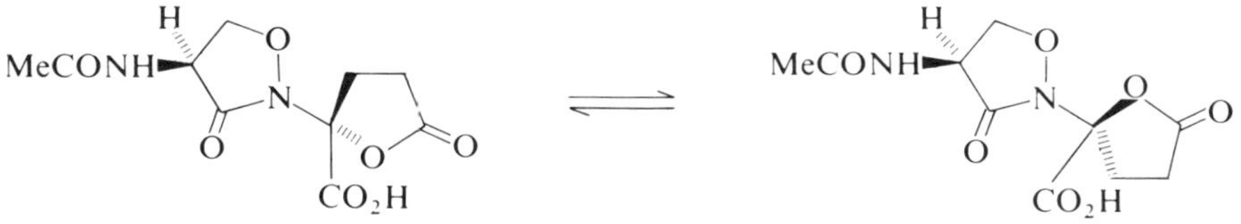

(336) Lactivicin

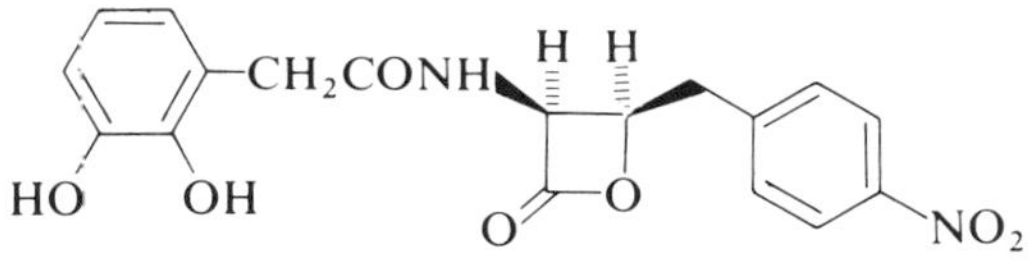

(337) Obofluorin

9.3.8 β-LACTAM ACTIVITY, REACTIVITY AND STRUCTURAL FEATURES

Much has been written on the mode of action of β-lactam antibiotics and β-lactamase inhibitors. Comprehensive accounts of all the different aspects involved are available in reviews by Ghuysen,[329,330] Matsuhashi,[331] Tipper,[332] Tomasz,[333] Waxman and Strominger,[334] Frère,[335] Coulson,[336] Hamilton-Miller and Smith,[337] and by Knowles.[117] The application of modern computational techniques (quantum and molecular mechanics, *etc.*) to the study of structure–activity relationships between the physicochemical properties of β-lactams and antibacterial activity has been examined and discussed in detail by Boyd.[338]

Woodward proposed[339] that the stability or reactivity of a fused β-lactam ring resulted from the increased strain in the four-membered ring. In the β-lactam ring of a penicillin, delocalization of the lone pair of electrons on the bridgehead nitrogen atom is greatly reduced compared with that in a simple amide. This results in an increase in the ketonic character of the β-lactam carbonyl group and an increase in the pyramidyl nature of the nitrogen atom. With Δ^3-cephem derivatives (and penem and carbapenems) the lone pair of nitrogen interacts with the double bond, decreasing the stability of the amide function. This enamine resonance also explains the greater reactivity of Δ^3-cephem derivatives over the analogous Δ^2 compounds. The increased ketonic nature of the β-lactam

Table 7 X-Ray and IR Absorption Data for Representative β-Lactams

Structural type[a]	Example of derivative of	ν_{max} (cm^{-1})[b]	C=O[c] (Å)	OC—N[d] (Å)	h(Å)[e]	Σ(°)[f]	CO···C[g] (Å)	Ref.
Penam	Penicillin G (63)	1770–1780	1.20	1.39	0.40	337	4.26[h]	346
	Ampicillin (65)		1.20	1.36	0.38	339	3.90	340
	Sulbactam (92)		1.20	1.39	0.38	339	3.90	348
Penem	(137) R = Me	1785–1805	1.20	1.42	0.44	332	3.61	174, 347
Δ²-Cephem	(338)	1765–1775	1.22	1.39	0.06	359	4.11	340, 341
Δ³-Cephem	Cephaloridine (67)	1765–1775	1.21	1.38	0.24	351	3.20	340, 341
Clavam	Clavulanic acid (2)	1790–1815	1.19	1.39	0.47	328	4.40	89, 350
1-Oxadethiacephem	(339)	1770–1780	1.20	1.39	0.30	346	3.53	254, 338
Δ¹-Carbapenem	(118)	1750–1770		1.40	0.54	320	4.28	142
Δ²-Carbapenem	Thienamycin (3)	1750–1770	1.22	1.42	0.49	326	3.57	128
Monobactam	Sulfazecin (24)	1775	1.21	1.37	0.13	358	3.56[i]	324
Free amide	—	1600–1680	1.24	1.32		360		341

Structure (338): PhOCH₂CONH–, H H, S, Me, H, CO₂H, N, O

Structure (339): PhCH₂CONH–, H H, O, N, O, S, N–N, N, Me, CO₂CHPh₂

[a] Ref. 192. [b] β-Lactam carbonyl IR absorption frequency. [c] Length of the β-lactam carbonyl bond. [d] Length of the bond between the β-lactam nitrogen atom and the carbon atom of the carbonyl group. [e] Distance of β-lactam N atom from the plane of the three attached atoms. [f] Sum of the angles around the β-lactam N atom. [g] Distance between the oxygen atom of the β-lactam carbonyl group and the carbon atom of the carboxylate function. [h] Distance with the thiazolidine ring in the CO₂H axial conformation; this distance is ~3.9 Å in the CO₂H equatorial position. [i] Distance in the CO₂H equatorial position, which is considered to be the one associated with activity.[338] Distance between the β-lactam carbonyl group and the sulfur atom of the sulfonic acid residue.

carbonyl group is reflected in the characteristic carbonyl stretching frequency found in the IR spectrum of β-lactams, bicyclic β-lactams showing a higher C=O absorption than that of a monocyclic β-lactam (Table 7).[340] Detailed X-ray crystallographic studies on a variety of penicillin and cephalosporin derivatives have confirmed that the more potent and reactive compounds have a shorter C=O bond, a longer C—N bond and a more pyramidyl bridgehead nitrogen compared with monocyclic derivatives.[340,341] These characteristics are also evident from the IR and structural features of the newer 'non-classical' clavam, penem and carbapenem compounds. With the monobactam ring system, X-ray data show there is a lack of steric strain in the ring, with the electron-withdrawing effect of the sulfonate group on nitrogen being responsible for the activation of the β-lactam ring.[342] Here the sulfonate oxygen atoms are equivalent to the carboxylate anion of the other β-lactams. The presence of other substituents (*e.g.* at the C-3 position on a cephem) can also effect activity/reactivity, electron-withdrawing or electron-donating properties being transmitted to the β-lactam ring. A summary of various geometric parameters for a range of β-lactam structures, determined *via* X-ray crystallographic studies, is given in Table 7. In addition to these parameters, base hydrolysis of β-lactams has been used to measure their susceptibility to nucleophilic attack to relate such data to antibacterial or β-lactamase inhibitory activity.[349] A detailed discussion of the differences between all these β-lactams is beyond the scope of this chapter, and the reader is referred in particular to the comprehensive review by Boyd.[338]

Earlier in this chapter (see also Section 9.2.3) it was indicated that an active β-lactam compound interacted with a serine hydroxy residue of a cell wall transpeptidase (PBP) for antibacterial activity, and a similar residue within a β-lactamase for inhibitory activity. Cohen has examined a variety of β-lactam structures and attempted to relate features of their 3-D geometry to their biological properties.[343] A comparison of the distance separating the β-lactam carbonyl oxygen atom and the carbon atom of the carboxylate function is given in Table 7. In general terms it was noted that for active β-lactam compounds this distance (3.0–3.9 Å) is shorter than for inactive derivatives (>4.1 Å) Such data help, for example, to explain the activity of the monobactam family of β-lactams, the related oxamazin series, and the poor activity of the thiamazin (58) analogues.[73]

The reactivity of the β-lactam, as estimated by data from X-ray studies, from rates of hydrolysis or the application of quantum mechanical calculations, has to relate to a biological parameter. While the minimum inhibitory concentration (MIC) is the parameter most readily measured, other points, such as passage through the outer membrane of bacterial cells and the effect of β-lactamase, have to be considered.[344,345] This may result in an explanation of the *in vitro* behaviour of a particular β-lactam, but may contribute little to the understanding of its activity (or lack of activity) *in vivo*, where further aspects such as the effect of serum, metabolism, solubility and toxicity come into play.

9.3.9   REFERENCES

1. W. W. Stewart, *Nature (London)*, 1971, **229**, 174.
2. T. Kikuchi and S. Uyeo, *Chem. Pharm. Bull.*, 1967, **15**, 549.
3. G. A. Koppel, in 'The Chemistry of Heterocyclic Compounds', ed. A. Hassner, Wiley, New York, 1983, vol. 42, pt. 2, p. 219.
4. 'Cephalosporins and Penicillins', ed. E. H. Flynn, Academic Press, New York, 1972.
5. F. A. Jung, W. R. Pilgrim, J. P. Poyser and P. J. Siret, in 'Topics in Antibiotic Chemistry', ed. P. G. Sammes, Ellis Horwood, Chichester, 1980, vol. 4.
6. 'Chemistry and Biology of β-Lactam Antibiotics', ed. R. B. Morin and M. Gorman, Academic Press, New York, 1982.
7. 'β-Lactam Antibiotics: Mode of Action, New Developments and Future Prospects', ed. M. R. J. Salton and G. D. Shockman, Academic Press, New York, 1981.
8. 'Recent Advances in the Chemistry of β-Lactam Antibiotics', ed. J. Elks, Special Publication No. 28, The Chemical Society, London, 1977.
9. 'Recent Advances in the Chemistry of β-Lactam Antibiotics', ed. G. I. Gregory, Special Publication No. 38, The Royal Society of Chemistry, London, 1981.
10. 'Recent Advances in the Chemistry of β-Lactam Antibiotics', ed. A. G. Brown and S. M. Roberts, Special Publication No. 52, The Royal Society of Chemistry, London, 1985.
11. T. Kamiya, H. Aoki and Y. Mine, in ref. 6, vol. 2, p. 165.
12. P. C. Cherry and C. E. Newall, in ref. 6, vol. 2, p. 361.
13. R. W. Ratcliffe and G. Albers-Schönberg, in ref. 6, vol. 2, p. 227.
14. W. H. Koster, C. M. Cimarusti and R. B. Sykes, in ref. 6, vol. 3, p. 339.
15. H. Aoki, H. Sakai, M. Kohsaka, T. Konomi, J. Hosoda, Y. Kubochi, E. Iguchi and H. Imanaka, *J. Antibiot.*, 1976, **29**, 492.
16. J. Hosoda, N. Tani, T. Konomi, S. Ohsawa, H. Oaki and H. Imanaka, *Agric. Biol. Chem.*, 1977, **41**, 2008; C. A. Townsend and A. M. Brown, *J. Am. Chem. Soc.*, 1983, **105**, 913; C. A. Townsend, A. M. Brown and L. T. Nguyen, *ibid.*, p. 919.
17. L. J. Nisbet, R. J. Mehta, Y. Oh, C. H. Pan, C. G. Phelen, M. J. Polansky, M. C. Shearer, A. J. Giovenella and S. F. Grappel, *J. Antibiot.*, 1985, **38**, 133.

18. N. Katayama, Y. Nozaki, K. Okonogi, H. Ono, S. Harada and H. Okazaki, *J. Antibiot.*, 1985, **38**, 1117.
19. A. Imada, K. Kitano, K. Kintaka, M. Muroi and M. Asai, *Nature (London)*, 1981, **289**, 590.
20. R. B. Sykes and J. S. Wells, *J. Antibiot.*, 1985, **38**, 119.
21. W. L. Parker, J. O'Sullivan and R. B. Sykes, *Adv. Appl. Microbiol.*, 1986, **31**, 181.
22. W. L. Parker, W. H. Koster, C. M. Cimarusti, D. M. Floyd, W.-C. Liu and M. L. Rathnum, *J. Antibiot.*, 1982, **35**, 189.
23. J. O'Sullivan, A. M. Gillum, C. A. Aklonis, M. L. Sonser and R. B. Sykes, *Antimicrob. Agents Chemother.*, 1982, **21**, 558; J. O'Sullivan, M. L. Sonser, C. C. Kao and C. A. Aklonis, *ibid.*, 1983, **23**, 598.
24. C. M. Cimarusti, H. E. Applegate, H. W. Chang, D. M. Floyd, W. H. Koster, W. A. Slusarchyk and M. G. Young, *J. Org. Chem.*, 1982, **47**, 179.
25. D. P. Bonner and R. B. Sykes, *J. Antimicrob. Chemother.*, 1984, **14**, 317.
26. T. Hirose, J. Nakano and H. Uno, *J. Pharm. Soc. Jpn.*, 1983, **103**, 1210.
27. R. J. Ponsford, M. J. Pearson and S. C. Finch (Beecham), *Jpn. Pat.* 58 134 074 (1983) (*Chem. Abstr.*, 1984, **100**, 51 359).
28. C. M. Cimarusti and R. B. Sykes, *Med. Res. Rev.*, 1984, **4**, 1.
29. T. Matsuo, T. Sugawara, H. Masuya, Y. Kawano, N. Noguchi and M. Ochiai, *Chem. Pharm. Bull.*, 1983, **31**, 1874.
30. X.-H. Tung and Z.-H. Li, *Chinese J. Antibiot.*, 1987, **12**, 223.
31. C. M. Cimarusti, D. P. Bonner, H. Breuer, H. W. Chang, A. W. Fritz, D. M. Floyd, T. P. Kissick, W. H. Koster, D. Kronenthal, F. Massa, R. H. Mueller, J. Pluscec, W. A. Slusarchyk, R. B. Sykes, M. Taylor and E. R. Weaver, *Tetrahedron*, 1983, **39**, 2577.
32. M. J. Miller, *Acc. Chem. Res.*, 1986, **19**, 49.
33. S. Hanessian, S. P. Sahoo, C. Couture and H. Wyss, *Bull. Soc. Chim. Belg.*, 1984, **93**, 571.
34. P. Berilacqua and J. L. Roberts, *Synth. Commun.*, 1983, **13**, 797.
35. M. J. Miller, P. G. Mattingly, M. A. Morrison and J. F. Kerwin, Jr, *J. Am. Chem. Soc.*, 1980, **102**, 7026.
36. D. M. Floyd, A. W. Fritz, J. Pluscec, E. R. Weaver and C. M. Cimarusti, *J. Org. Chem.*, 1982, **47**, 5160.
37. M. A. Krook and M. J. Miller, *J. Org. Chem.*, 1985, **50**, 1126.
38. D. M. Floyd, A. W. Fritz and C. M. Cimarusti, *J. Org. Chem.*, 1982, **47**, 176.
39. *J. Antimicrob. Chemother.*, 1981, **8**, suppl E.
40. R. N. Brogden and R. C. Heel, *Drugs*, 1986, **31**, 96.
41. C. M. Cimarusti and R. B. Sykes, *Chem. Br.*, 1983, 302.
42. G. Teutsch and A. Bonnet, *Tetrahedron Lett.*, 1984, **25**, 1561.
43. J. S. Skotnicki, T. J. Commons, R. W. Rees and J. L. Speth, *J. Antibiot.*, 1983, **36**, 1201.
44. Y. Yoshioka, T. Miyawaki, S. Kishimoto, T. Matsuo and M. Ochiae, *J. Org. Chem.*, 1984, **49**, 1427.
45. K. Matsuda, S. Nakagawa, F. Nakano, M. Inoue and S. Mitsuhashi, *J. Antimicrob. Chemother.*, 1987, **19**, 753.
46. S. Kishimoto, M. Sendai, M. Tomimoto, S. Hashiguchi, T. Matsuo and M. Ochiai, *Chem. Pharm. Bull.*, 1984, **32**, 2646.
47. K. S. Kim and J. H. Chambers, *J. Antibiot.*, 1987, **40**, 124.
48. S. Hashiguchi, Y. Maeda, S. Kishimoto and M. Ochiai, *Heterocycles*, 1986, **24**, 2273.
49. M. Sendai, S. Hashiguchi, M. Tomimoto, S. Kishimoto, T. Matsuo, M. Kondo and M. Ochiai, *J. Antibiot.*, 1985, **38**, 346.
50. P. F. Bevilacqua, D. D. Keith and J. L. Roberts, *J. Org. Chem.*, 1984, **49**, 1430.
51. V. D. Treuner, T. Denzel, H. Breuer and D. P. Bonner, '22nd Interscience Conference on Antimicrobial Agents and Chemotherapy, Miami', 1982, Abstr. 676.
52. N. Noguchi, H. Masuya, T. Sugawara, Y. Kawano, T. Matsuo and M. Ochiai, *J. Antibiot.*, 1985, **38**, 1387.
53. C. J. Ashcroft, J. Brennan, C. E. Newall and S. M. Roberts, *Tetrahedron Lett.*, 1984, **25**, 877.
54. W. A. Slusarchyk, T. Dejneka, E. M. Gordon, E. R. Weaver and W. H. Koster, *Heterocycles*, 1984, **21**, 191.
55. W. H. Koster, R. Zahler, D. P. Bonner, H. W. Chang, C. M. Cimarusti, G. A. Jacobs and M. Perri, ref. 51, abstr. 674.
56. W. A. Slusarchyk, H. E. Applegate, D. P. Bonner, H. Breuer, T. Dejneka and W. H. Koster, ref. 51, abstr. 670.
57. H. Breuer, T. Denzel, H. Höhn, V. D. Treuner, K. R. Lindner, D. P. Bonner and W. A. Slusarchyk, ref. 51, abstr. 671.
58. H. Breuer, T. Denzel, H. Straub, K. R. Lindner, D. P. Bonner and W. A. Slusarchyk, ref. 51, abstr. 672.
59. S. K. Tanaka, K. Bush, D. P. Bonner, R. Schwind, F. Liu, B. Minassian, S. A. Smith, D. Visnic and R. B. Sykes, '25th Interscience Conference of Antimicrobial Agents and Chemotherapy, Minneapolis', 1985, abstr. 372.
60. C. Yoshida, K. Tanaka, R. Hattori, Y. Fukuoka, M. Komatsu, S. Kishimoto and I. Saikawa, *J. Antibiot.*, 1986, **39**, 215.
61. M. Klich and G. Teutsch, *Tetrahedron*, 1986, **42**, 2677.
62. A. Andrus, B. Partridge, J. V. Heck and B. G. Christensen, *Tetrahedron Lett.*, 1984, **25**, 911.
63. E. M. Gordon, M. A. Ondetti, J. Pluscec, C. M. Cimarusti, D. P. Bonner and R. B. Sykes, *J. Am. Chem. Soc.*, 1982, **104**, 6053.
64. M. J. Miller, A. Biswas and M. A. Krook, *Tetrahedron*, 1983, **39**, 2571.
65. C. Yoshida, T. Hori, K. Momonoi, K. Nagumo, J. Nakano, T. Kitani, Y. Fukuoka and I. Saikawa, *J. Antibiot.*, 1985, **38**, 1536.
66. W. A. Slusarchyk, T. Dejneka, J. Gougoutas, W. H. Koster, D. R. Kronenthal, M. Malley, M. G. Perri, F. L. Routh, J. E. Sundeen, E. R. Weaver and R. Zahler, *Tetrahedron Lett.*, 1986, **27**, 2789.
67. H. Breuer, H. Straub and V. D. Treuner (Von Heyden GmbH), *Ger. Pat.* 3 328 047 (1984) (*Chem. Abstr.*, 1984, **101**, 38 265); W. V. Curran, A. A. Ross and V. J. Lee, *J. Antibiot.*, 1988, **41**, 1418.
68. J. Iwagami, S. R. Woulfe and M. J. Miller, *Tetrahedron Lett.*, 1986, **27**, 3095.
69. S. R. Woulfe and M. J. Miller, *J. Med. Chem.*, 1985, **28**, 1447.
70. F. R. Atherton and R. W. Lambert, *Tetrahedron*, 1984, **40**, 1039.
71. H. Breuer, H. Straub, V. D. Treuner, J.-M. Drossard, H. Höhn and K. R. Lindner, *J. Antibiot.*, 1985, **38**, 813.
72. S. R. Woulfe and M. J. Miller, *J. Org. Chem.*, 1986, **51**, 3133.
73. D. B. Boyd, C. Eigenbrot, J. M. Indelicato, M. J. Miller, C. E. Pasini and S. R. Woulfe, *J. Med. Chem.*, 1987, **30**, 528.
74. A. G. Brown, D. Butterworth, M. Cole, G. Hanscomb, J. D. Hood, C. Reading and G. N. Rolinson, *J. Antibiot.*, 1976, **29**, 668.
75. T. T. Howarth, A. G. Brown and T. J. King, *J. Chem. Soc., Chem. Commun.*, 1976, 266.
76. C. Reading and M. Cole, *Antimicrob. Agents Chemother.*, 1977, **11**, 852.
77. A. G. Brown, in 'Medicinal Chemistry', ed. S. M. Roberts and B. J. Price, Academic Press, London, 1985, p. 227; D. Butterworth, in 'Biotechnology of Industrial Antibiotics', ed. E. J. Vandamme, Dekker, New York, 1984, p. 225.
78. S. J. Box (Beecham Group Ltd.), *Ger. Pat.* 2 646 001 (1977) (*Chem. Abstr.*, 1977, **87**, 403); M. Arai, A. Terahara and T. Haneishi (Sankyo Co. Ltd.), *Jpn. Pat.*, 76 118 890 (*Chem. Abstr.*, 1977, **86**, 119 252); K. Kitano, K. Kintaka and

K. Katamoto (Takeda Chemical Industries Ltd.), *Jpn. Pat.* 78 10 476 (1978) (*Chem. Abstr.*, 1979, **90**, 119 758); Sanraku-Ocean Co. Ltd., *Jpn. Pat.* 80 162 993 (1980) (*Chem. Abstr.*, 1981, **94**, 137 803).
79. C. Reading and T. Farmer, in 'Antibiotics', ed. A. D. Russell and L. B. Quensel, Academic Press, London, 1983, p. 141; M. Cole, *Drugs Future*, 1981, **5**, 697.
80. E. P. Abraham and E. Chain, *Nature (London)*, 1940, **146**, 837.
81. R. B. Sykes and M. Matthew, *J. Antimicrob. Chemother.*, 1976, **2**, 115; M. H. Richmond and R. B. Sykes, in 'Advances in Microbial Physiology', ed. A. H. Rose and D. W. Tempest, Academic Press, London, 1973, vol. 9, p. 31.
82. R. P. Ambler, *Philos. Trans. R. Soc. London, Ser. B*, 1980, **289**, 321.
83. S. Mitsuhashi and M. Inque, in 'Microbial Drug Resistance', ed. S. Mitsuhashi, Tokyo University Press, 1982, p. 41.
84. H. C. Neu, *Am. J. Med.*, 1985, **79**, (513), 2.
85. P. A. Hunter, K. Coleman, J. Fisher and D. Taylor, *J. Antimicrob. Chemother.*, 1980, **6**, 455; P. A. Hunter, C. Reading and D. A. Witting, in 'Current Chemotherapy', ed. W. Siegenthaler and R. Lüthy, American Society of Microbiology, Washington, DC, 1978, p. 978.
86. B. G. Spratt, V. Jobanputra and W. Zimmerman, *Antimicrob. Agents Chemother.*, 1977, **12**, 406.
87. 'Augmentin', ed. G. N. Rolinson and A. Watson, Excerpta Medica, Amsterdam, 1980; 'Augmentin', ed. D. A. Leigh and O. P. W. Robinson, Excerpta Medica, Amsterdam, 1982; 'Augmentin', ed. E. A. P. Croydon and M. F. Michel, Excerpta Medica, Amsterdam, 1983; *Chemotherapy*, 1983, **31** (S-2); *Postgrad. Med.*, 1984, Sept./Oct.
88. R. Sutherland, A. S. Beale, R. J. Boon, K. E. Griffin, B. Slocombe, D. H. Stokes and A. R. White, *Am. J. Med.*, 1985, **79** (5B), 13; *J. Antimicrob. Chemother.*, 1986, **17**, suppl. C.
89. A. G. Brown, D. F. Corbett, J. Goodacre, J. B. Harbridge, T. T. Howarth, R. J. Ponsford, I. Stirling and T. J. King, *J. Chem. Soc., Perkin Trans. 1*, 1984, 635.
90. A. G. Brown, J. Goodacre, J. B. Harbridge, T. T. Howarth, R. J. Ponsford and I. Stirling, in ref. 8, p. 295.
91. G. Brooks, G. Bruton, M. J. Finn, J. B. Harbridge, M. A. Harris, T. T. Howarth, E. Hunt, I. Stirling and I. I. Zomaya, in ref. 10, p. 221.
92. A. G. Brown, in 'Encyclopedia of Chemical Technology', Wiley, New York, 1984, suppl. vol., p. 83.
93. R. D. G. Cooper, in 'Topics in Antibiotic Chemistry', ed. P. G. Sammes, Ellis Horwood, Chichester, 1980, vol. 3, p. 39.
94. C. E. Newall, in ref. 7, p. 287.
95. C. Newall, in 'Proceedings of VIIth International Symposium on Medicinal Chemistry', ed. R. Dahlbon and J. L. G. Nilsson, Swedish Pharmaceutical Press, Stockholm, 1985, vol. 2, p. 69.
96. M. L. Gilpin, J. B. Harbridge, T. T. Howarth and T. J. King, *J. Chem. Soc., Chem. Commun.*, 1981, 929; M. L. Gilpin, J. B. Harbridge and T. T. Howarth, *J. Chem. Soc., Perkin Trans. 1*, 1987, 1369.
97. R. C. Mearman, C. E. Newall and A. P. Tonge, *J. Antibiot.*, 1984, **37**, 885.
98. C. E. Newall, in ref. 9, p. 151.
99. G. Brooks, B. C. Gasson, T. T. Howarth, E. Hunt and K. Luk, *J. Chem. Soc., Perkin Trans. 1*, 1984, 1599; G. Brooks and E. Hunt, *J. Chem. Soc., Perkin. Trans. 1*, 1983, 115.
100. P. H. Bentley, P. D. Berry, G. Brooks, M. L. Gilpin, E. Hunt and I. I. Zomaya, *J. Chem. Soc., Chem. Commun.*, 1977, 748; P. H. Bentley, G. Brooks, M. L. Gilpin and E. Hunt, *Tetrahedron Lett.*, 1979, 1889; P. H. Bentley, P. D. Berry, G. Brooks, M. L. Gilpin, E. Hunt and I. I. Zomaya, in ref. 9, p. 175.
101. A. G. Brown, *J. Antimicrob. Chemother.*, 1981, **7**, 15.
102. P. A. Hunter, *Pharm. Weekbl.*, 1984, **119**, 650.
103. C. Reading and M. Cole, *J. Enzyme Inhibition*, 1986, **1**, 83.
104. G. C. Bolton, G. D. Allen, B. E. Davies, C. W. Filer and D. J. Jeffery, *Xenobiotica*, 1986, **16**, 853.
105. M. J. Finn, M. A. Harris, E. Hunt and I. I. Zomaya, *J. Chem. Soc., Perkin Trans. 1*, 1984, 1345.
106. S. W. Elson and R. S. Oliver, *J. Antibiot.*, 1978, **31**, 586; I. Stirling and S. W. Elson, *ibid.*, 1979, **32**, 1125; S. W. Elson, R. S. Oliver, B. W. Bycroft and E. A. Faruk. *ibid.*, 1982, **35**, 81; S. W. Elson, in ref. 9, p. 142; C. A. Townsend and M.-f. Ho, *J. Am. Chem. Soc.*, 1985, **107**, 1065 and 1066; C. A. Townsend, M.-f. Ho and S.-s. Mao, *J. Chem. Soc., Chem. Commun.*, 1986, 638.
107. S. W. Elson, K. H. Baggaley, J. Gillett, S. Holland, N. H. Nicholson, J. T. Sime and S. R. Woroniecki, *J. Chem. Soc., Chem. Commun.*, 1987, 1739; S. W. Elson. S. R. Woroniecki and K. H. Baggaley (Beecham Group), *Eur. Pat. Appl.* 213 914 (1986).
108. C. R. Bailey, M. J. Butler, I. D. Normansell, R. T. Rowlands and D. J. Winstanley, *Biotechnology*, 1984, **2**, 808.
109. D. Brown, J. R. Evans and R. A. Fletton, *J. Chem. Soc., Chem. Commun.*, 1979, 282; S. W. Elson and T. J. King, personal communication; D. L. Preuss and M. Kellit, *J. Antibiot.*, 1983, **36**, 208; R. H. Evans, Jr, H. Ax, A. Jacoby, T. H. Williams, E. Jenkins and J. P. Scannell, *ibid.*, 1983, **36**, 213; J.-L. Muller, V. Toome, D. L. Preuss, J. F. Blount and M. Weigle, *ibid.*, 1983, **36**, 217; M. Wanning, H. Zähner, B. Krone and A. Zeeck, *Tetrahedron Lett.*, 1981, **22**, 2539; H. Zähner and H. Drantz, in 'Proceedings of VIIth International Symposium on Medicinal Chemistry', ed. R. Dahlbon and J. L. G. Nilsson, Swedish Pharmaceutical Press, Stockholm, 1985, vol. 2, p. 49; H. D. King, J. Langhärig and J. J. Sanglier, *J. Antibiot.*, 1986, **39**, 510; H. V. Naegeli, H.-R. Loosli and A. Nussbaumer, *ibid.*, 1986, **39**, 516; F. Röhl, J. Rabenhorst and H. Zähner, *Arch. Microbiol.*, 1987, **147**, 315.
110. A. R. English, J. A. Retsema, A. E. Girard, J. E. Lynch and W. E. Barth, *Antimicrob. Agents Chemother.*, 1978, **14**, 414; R. F. Pratt and M. J. Loosemore, *Proc. Natl. Acad. Sci. USA*, 1978, **75**, 4145; W. von Daehne, *J. Antibiot.*, 1980, **33**, 451; D. I. John (Beecham Group), *Eur. Pat. Appl.* 00 13 617 (1980); W. von Daehne, E. T. Hansen and N. Rastrup-Anderson, in ref. 10, p. 375; B. A. Moore and K. W. Brammer, *Antimicrob. Agents Chemother.*, 1981, **20**, 327; S. J. Cartwright and A. F. W. Coulson, *Nature (London)*, 1979, **278**, 360; W. J. Gottstein, L. B. Crast, Jr, R. G. Graham, V. J. Haynes and D. N. McGregor, *J. Med. Chem.*, 1981, **28**, 518; T. W. Hall, S. N. Maiti, R. G. Micetich, P. Spevak, N. Ishida, M. Kajitani, M. Tanaka and T. Yamasaki, in ref. 10, p. 242; M. Arisawa and S. Adam, *Biochem. J.*, 1983, **211**, 447; M. Arisawa and R. Then, *Biochem. J.*, 1983, **209**, 609; S. Adam, R. L. Then and P. Angehrn, *J. Antibiot.*, 1987, **40**, 108; D. G. Brenner, *J. Org. Chem.*, 1985, **50**, 18; D. G. Brenner and J. R. Knowles, *Biochemistry*, 1984, **23**, 5839; Y. L. Chen, C.-W. Chang, K. Hedberg, K. Guarino, W. Welch, L. Kiessling, J. A. Retsema, S. L. Haskill, M. Anderson, M. Manousos and J. F. Barrett, *J. Antibiot.*, 1987, **40**, 803; Y. L. Chen, C.-W. Chang and K. Hedberg, *Tetrahedron Lett.*, 1987, **27**, 3449; S. Adam, *Heterocycles*, 1984, **22**, 1509; D. D. Keith, J. Tengi, P. Rossman, L. Todaro and M. Weigele, *Tetrahedron*, 1983, **39**, 2445; L. A. Reed, D. A. Charleson and R. A. Volkmann, *Tetrahedron Lett.*, 1987, **28**, 3431.
111. *Rev. Infect. Dis.*, 1986, **6**, suppl. 5; D. M. Campoli-Richards and R. N. Brogden, *Drugs*, 1987, **33**, 577.
112. R Wise, J. M. Andrews and N. Patel, *J. Antimicrob. Chemother.*, 1981, **7**, 531.

113. S. Aronoff, M. R. Jacobs, S. Johenning and S. Yamabe, *Antimicrob. Agents Chemother.*, 1984, **26**, 580.
114. N. F. Osborne (Beecham Group), *Eur. Pat. Appl.* 41 768 (1981) (*Chem. Abstr.*, 1982, **96**, 181 068); J. H. C. Nayler, in 'Proceedings of VIIth International Symposium on Medicinal Chemistry', ed. R. Dahlbon and J. L. G. Nilsson, Swedish Pharmaceutical Press, Stockholm, 1985, vol. 2, p. 33.
115. R. N. Jones, A. L. Barry, C. Thornsberry and H. W. Wilson, *Am. J. Clin. Pathol.*, 1985, **84**, 496.
116. B. Baltzer, E. Binderup, W. von Daehne, W. O. Godtfredsen, K. Hansen, B. Nielsen, H. Sorensen and S. Vangedal, *J. Antibiot.*, 1980, **33**, 1183; H. J. Rogers, I. D. Bradbrook, P. J. Morrison, R. G. Spector, D. A. Cox and L. J. Lees, *J. Antimicrob. Chemother.*, 1983, **11**, 435; A. M. Emmerson, D. A. Cox and L. J. Lees, *Eur. J. Clin. Microbiol.*, 1983, **2**, 340.
117. J. R. Knowles, *Acc. Chem. Res.*, 1985, **18**, 97.
118. J. B. Bartolone, G. J. Hite, J. A. Kelly and J. R. Knox, in ref. 10, p. 318; J. A. Kelly, P. C. Moews, J. R. Knox, J. M. Frère and J. M. Ghuysen, *Science (Washington, D.C.)*, 1982, **218**, 475; J. A. Kelly, J. R. Knox, P. C. Moews, G. J. Hite, J. B. Bartolone, H. Zhao, B. Joris, J. M. Frère and J. M. Ghuysen, *J. Biol. Chem.*, 1985, **260**, 6449; J. A. Kelly, O. Dideberg, P. Charlier, J. P. Wery, M. Libert, P. C. Moews, J. R. Knox, C. Duez, C. Fraipont, B. Joris, J. Dusart and J. M. Ghuysen, *Science (Washington, D.C.)*, 1986, **231**, 1429; B. Samraoni, B. J. Sutton, R. J. Todd, P. J. Artymuik, S. G. Waley and D. C. Phillips, *Nature (London)*, 1986, **320**, 378; O. Herzberg and J. Moult, *Science (Washington, D.C.)*, 1987, **236**, 694.
119. C. Little, E. L. Emanuel, J. Gagnon and S. G. Waley, *Biochem. J.*, 1986, **223**, 465.
120. D. C. Phillips, A. Cordero-Borboa, B. J. Sutton and R. J. Todd, *Pure Appl. Chem.*, 1987, **57**, 279.
121. O. Dideberg, P. Charlier, G. Dive, B. Joris, J. M. Frère and J. M. Ghuysen, *Nature (London)*, 1982, **249**, 569.
122. C. Reading and P. Hepworth, *Biochem. J.*, 1979, **179**, 67; J. P. Durkin and T. Viswanatha, *J. Antibiot.*, 1978, **31**, 1162; C. Reading and T. Farmer, *Biochem. J.*, 1981, **199**, 779; T. Farmer and C. Reading, *Antimicrob. Agents Chemother.*, 1982, **21**, 506.
123. V. Knott-Hunziker, B. S. Orlek, P. G. Sammes and S. G. Waley, *Biochem. J.*, 1979, **177**, 365; S. A. Cohen and R. F. Pratt, *Biochemistry*, 1980, **19**, 396; B. S. Orlek, P. G. Sammes, V. Knott-Hunziker and S. G. Waley, *J. Chem. Soc., Perkin Trans. 1*, 1980, 2322.
124. S. Waley, *Chem. Ind. (London)*, 1981, 131.
125. M. A. Navia, J. P. Springer, T.-Y. Lin, H. R. Williams, R. A. Firestone, J. M. Pisano, J. B. Doherty, P. E. Finke and K. Hoogsteen, *Nature (London)*, 1987, **327**, 79.
126. W. C. Liu, G. Astle, J. S. Wells, W. H. Trejo, P. A. Principe, M. L. Rathnum, W. L. Parker, O. R. Koey and R. B. Sykes, *J. Antibiot.*, 1980, **33**, 1256; Y. Ikeda, S. Kondo, T. Sawa, M. Tsuchiya, D. Ikeda, M. Hamada, T. Takeuchi and H. Umezewa, *J. Antibiot.*, 1981, **34**, 1628.
127. T. Beesley, N. Gascogne, V. Knott-Hunziker, S. Petursson, S. G. Waley and B. Jaurin, *Biochem J. Mol. Aspects*, 1983, **209**, 229.
128. G. Albers-Schönberg, B. H. Arison, O. T. Hensens, J. Hirshfield, K. Hoogsteen, E. A. Kaczka, R. E. Rhodes, J. S. Kahan, R. W. Ratcliffe, E. Walton, L. J. Rushwinkle, R. B. Morin and B. G. Christensen, *J. Am. Chem. Soc.*, 1978, **100**, 6491.
129. A. G. Brown, D. F. Corbett, A. J. Eglington and T. T. Howarth, *J. Chem. Soc., Chem. Commun.*, 1977, 523; *J. Antibiot.*, 1979, **32**, 961.
130. K. Yamamoto, T. Yoshioda, Y. Kato, N. Shibamoto, K. Okamura, Y. Shimauchi and T. Ishikura, *J. Antibiot.*, 1980, **32**, 796.
131. M. Nakayama, S. Kimura, S. Tanabe, T. Mizoguchi, I. Watanabe, T. Mori, K. Miyahara and T. Kawasaki, *J. Antibiot.*, 1981, **34**, 818.
132. N. Tsuji, K. Nagashima, M. Kobayashi, J. Shoji, T. Kato, Y. Terui, H. Nakai and M. Shiro, *J. Antibiot.*, 1982, **35**, 24.
133. R. Southgate and S. Elson, in 'Progress in the Chemistry of Organic Natural Products, ed. W. Herz, H. Grisebach, G. W. Kirby and Ch. Tamm, Springer-Verlag, Heidelberg, 1985, vol. 47, p. 1.
134. W. L. Parker, M. L. Rathnum, J. S. Wells, W. H. Trejo, P. A. Principe and R. B. Sykes, *J. Antibiot.*, 1982, **35**, 653.
135. G. Albers-Schönberg, B. H. Arison, E. Kaczka, F. M. Kahan, J. S. Kahan, B. Lago, W. M. Maiese, R. E. Rhodes and J. L. Smith, in 'Abstracts of the 16th Interscience Conference on Antimicrobial Agents and Chemotherapy, Chicago', American Society for Microbiology, Washington, DC, 1976, abstr. 228.
136. J. M. Williamson, E. Inamine, K. E. Wilson, A. W. Douglas, J. M. Liesch and G. Albers-Schönberg, *J. Biol. Chem.*, 1985, **260**, 4637.
137. B. W. Bycroft, C. Maslen, S. J. Box, A. G. Brown and J. W. Tyler, *J. Chem. Soc., Chem. Commun.*, 1987, 1623.
138. T. Haneishi, M. Nakajima, N. Serzawa, M. Inukai, Y. Takiguchi, M. Arai, S. Satoh, H. Kowano and C. Tamura, *J. Antibiot.*, 1983, **36**, 1581.
139. J. M. Williamson, *CRC Crit. Rev. Biotechnol.*, 1986, **4**, 111.
140. S. J. Box, J. D. Hood and S. R. Spear, *J. Antibiot.*, 1979, **32**, 1239.
141. Y. Fukagawa, M. Okabe, T. Yoshioka and T. Ishikura, in ref. 10, p. 165.
142. D. H. Shih and R. W. Ratcliffe, *J. Med. Chem.*, 1981, **24**, 639.
143. D. H. Shih, J. Hannah, and B. G. Christensen, *J. Am. Chem. Soc.*, 1978, **100**, 8004.
144. D. F. Corbett, S. Coulton and R. Southgate, *J. Chem. Soc., Perkin Trans. 1*, 1982, 3011.
145. A. G. Brown, D. F. Corbett, A. J. Eglington and T. T. Howarth, in ref. 9, p. 255.
146. D. F. Corbett, *J. Chem. Soc., Chem. Commun.*, 1981, 803.
147. K. Yamamoto, T. Yoshioka, Y. Kato, K. Isshiki, M. Nishino, F. Nakamura, Y. Shimauchi and T. Ishikura, *Tetrahedron Lett.*, 1982, **23**, 897.
148. Y. Fukagawa, K. Kubo, T. Ishikura and K. Kouno, *J. Antibiot.*, 1980, **33**, 543.
149. T. Kametani, *Heterocycles*, 1982, **17**, 463.
150. J. H. Bateson, A. J. G. Baxter, K. M. Dickinson, R. I. Hickling, R. J. Ponsford, P. M. Roberts, T. C. Smale and R. Southgate, in ref. 9, p. 291.
151. L. D. Cama and B. G. Christensen, *J. Am. Chem. Soc.*, 1978, **100**, 8006.
152. J. H. Bateson, A. J. G. Baxter, P. M. Roberts, T. C. Smale and R. Southgate, *J. Chem. Soc., Perkin Trans. 1*, 1981, 3242.
153. R. J. Ponsford and R. Southgate, *J. Chem. Soc., Chem. Commun.*, 1980, 1085.
154. J. H. Bateson, R. I. Hickling, P. M. Roberts, T. C. Smale and R. Southgate, *J. Chem. Soc., Chem. Commun.*, 1980, 1084.
155. T. N. Salzmann, R. W. Ratcliffe, B. G. Christensen and F. A. Bouffard, *J. Am. Chem. Soc.*, 1980, **102**, 6161.
156. D. G. Melillo, I. Shinkai, T. Liu, K. Ryan and M. Sletzinger, *Tetrahedron Lett.*, 1980, **21**, 2783.
157. D. G. Melillo, R. J. Cuetovich, K. M. Ryan and M. Sletzinger, *J. Org. Chem.*, 1986, **51**, 1498.

158. S. Karady, J. S. Amato, R. A. Reamer and L. M. Weinstock, *J. Am. Chem. Soc.*, 1981, **103**, 6765.
159. D. H. Shih, F. Baker, L. Cama and B. G. Christensen, *Heterocycles*, 1984, **21**, 29.
160. R. Labia and C. Morin, *J. Antibiot.*, 1984, **37**, 1103.
161. F. A. Bouffard and T. N. Salzmann, *Tetrahedron Lett.*, 1984, **25**, 6285.
162. A. Knierzinger and A. Vasella, *J. Chem. Soc., Chem. Commun.*, 1985, 9.
163. H. Maruyama, M. Shiozaki and T. Hiraoka, *Bull. Chem. Soc. Jpn.*, 1985, **58**, 3264.
164. M. Shiozaki, N. Ishida, H. Maruyama and T. Hiraoka, *Tetrahedron*, 1983, **39**, 2399.
165. T. Kametami, T. Nagahara and T. Honda, *J. Org. Chem.*, 1985, **50**, 2327.
166. S. T. Hodgson, D. M. Hollinshead and S. V. Ley, *J. Chem. Soc., Chem. Commun.*, 1984, 494.
167. G. I. Georg, J. Kant and H. S. Gill, *J. Am. Chem. Soc.*, 1987, **109**, 1129.
168. W. J. Leanza, K. J. Wildonger, J. Hannah, D. H. Shih, R. W. Ratcliffe, L. Borash, E. Walton, R. A. Firestone, G. F. Patel, F. M. Kahan, J. S. Kahan and B. G. Christensen, in ref. 85, p. 240.
169. T. Hashizume, F. Ishino, J. Nakagawa, S. Tamaki and M. Matsuhashi, *J. Antibiot.*, 1984, **37**, 394.
170. S. R. Norrby, B. Björnegard, F. Ferber and K. H. Jones, *J. Antimicrob. Chemother.*, 1983, **12**, suppl. D, 109.
171. H. Kropp, J. G. Sundelof, R. Hajdu and F. M. Kahan, *Antimicrob. Agents Chemother.*, 1982, **22**, 62.
172. F. M. Kahan, H. Kropp, J. G. Sundelof and J. Birnbaum, *J. Antimicrob. Chem.*, 1983, **12**, suppl. D, 1.
173. R. B. Woodward, in ref. 8, p. 167.
174. I. Ernest, in ref. 6, vol. 2, p. 315.
175. I. Ernest, J. Gosteli and R. B. Woodward, *J. Am. Chem. Soc.*, 1979, **101**, 6301.
176. P. C. Cherry, C. E. Newall and N. S. Watson, *J. Chem. Soc., Chem. Commun.*, 1979, 663.
177. W. Dürckheimer, J. Blumbach, R. Lattrell and K. H. Scheunemann, *Angew. Chem., Int. Ed. Engl.*, 1985, **24**, 180.
178. I. Phillips, R. Wise and H. C. Neu (ed.) *J. Antimicrob. Chemother.*, 1982, **9**, suppl. C.
179. W. J. Leanza, F. DiNinno, D. A. Muthard, R. R. Wilkening, K. J. Wildonger, R. W. Ratcliffe and B. G. Christensen, *Tetrahedron*, 1983, **39**, 2505.
180. A. S. Afonso, A. K. Ganguly, G. Girijavallabhan and S. McCombie, in ref. 10, p. 266.
181. R. Wise and I. Phillips, *J. Antimicrob. Chemother.*, 1985, **15**, suppl. C.
182. E. Tuomanen, *Rev. Infec. Dis.*, 1986, **8**, suppl. 3, S279.
183. G. Franceschi, M. Foglio, M. Alpegiani, C. Battistini, A. Bedeschi, E. Perrone, F. Zarini, F. Arcamone, C. Della Bruna and A. Sanfilippo, *J. Antibiot.*, 1983, **36**, 938.
184. G. Franceschi, M. Alpegiani, C. Battistini, A. Bedeschi, E. Perrone and F. Zarini, *Pure Appl. Chem.*, 1987, **59**, 467.
185. H. C. Neu, N. Chin and N. M. Neu, *Antimicrob. Agents Chemother.*, 1987, **31**, 558.
186. G. Sibert, D. Isert, N. Klesel, M. Limbert, A. Pries, E. Schrinner, M. Cooke, J. Walmsley and P. H. Bentley, *J. Antibiot.*, 1987, **40**, 660.
187. G. L. Dunn, in 'Comprehensive Heterocyclic Chemistry', ed. A. R. Katritzky, C. W. Rees and W. Lwowski, Pergamon Press, Oxford, 1984, vol. 7, p. 341.
188. Ref. 6, vol. 2.
189. D. A. Evans and E. B. Sjorgen, *Tetrahedron Lett.*, 1985, **26**, 3783.
190. R. D. G. Cooper, B. W. Dougherty and D. B. Boyd, *Pure Appl. Chem.*, 1987, **59**, 485.
191. C. C. Wei, S. de Bernardo, T. P. Tengi, J. Borgese and M. Weigele, *J. Org. Chem.*, 1985, **50**, 3462.
192. A. G. Brown, *J. Antimicrob. Chemother.*, 1982, **10**, 365.
193. Ref. 6, vol. 1.
194. K. G. Holden, in ref. 187, p. 285.
195. D. M. McGregor, in ref. 187, p. 299.
196. J. E. Baldwin, J. E. Cobb and L. N. Sheppard, *Tetrahedron*, 1987, **43**, 1003.
197. R. G. Micetich, R. Singh, W. O. Merlo, D. M. Tetteh, C.-C. Shaw and R. B. Morin, *Heterocycles*, 1984, **22**, 2757.
198. R. G. Micetich, S. N. Maiti, M. Tanaka, T. Yamazaki and K. Ogawa, *Heterocycles*, 1985, **23**, 325.
199. F. J. Lund, in ref. 8, p. 25.
200. J. Hannah, C. R. Johnson, A. F. Wagner and E. Walton, *J. Med. Chem.*, 1982, **25**, 457.
201. L. R. Koupal, B. Weissberger, D. L. Shungu, E. Weinberg and H. H. Gadeshbusch, *J. Antibiot.*, 1983, **36**, 47.
202. M. J. Driver, P. H. Bentley, R. A. Dixon, R. A. Edmondson, A. C. Kaura and A. W. Taylor, *J. Antibiot.*, 1984, **37**, 297.
203. D. D. Keith, J. Tengi, P. Rossman, L. Todaro and M. Weigele, *Tetrahedron*, 1983, **39**, 2445.
204. C.-C. Wei, K. C. Luk, K. F. West, J. L. Roberts, D. Pruess, P. Rossman, M. Weigele and D. D. Keith, in 'Abstracts of 26th Interscience Conference on Antimicrobial Agents and Chemotherapy', American Society for Microbiology, Washington, DC, 1986, abstr. 1292.
205. D. D. Keith, J. Christensen, N. Georgopapdakou, V. Madison, D. Pruess, P. Rossman and L. Todaro, in ref. 204, Abstr. 1293.
206. A. G. Long, in ref. 8, p. 214.
207. C. D. Foulds, A. A. Jaxa-Chamiec, A. C. O'Sullivan and P. G. Sammes, *J. Chem. Soc., Perkin Trans. 1*, 1984, 21.
208. T. N. Salzmann, R. W. Ratcliffe and B. G. Christensen, *Tetrahedron Lett.*, 1980, **21**, 1193.
209. J. C. Sheehan, E. Chalko, T. J. Commons, Y. S. Lo, D. R. Ponzi and A. Schwabacker, *J. Antibiot.*, 1984, **37**, 1441.
210. Ref. 5, p. 51.
211. L. D. Cama and B. G. Christensen, *Tetrahedron Lett.*, 1978, 4233.
212. R. G. Alexander and R. Southgate, *J. Chem. Soc., Chem. Commun.*, 1977, 405.
213. S. Oida, A. Yoshida and E. Ohki, *Chem. Pharm. Bull.*, 1980, **24**, 3494.
214. J. G. de Vries, G. Hauser and G. Sigmund, *Tetrahedron Lett.*, 1984, **25**, 5989.
215. T. Nagasaka, A. Tsukada and F. Hamaguchi, *Heterocycles*, 1986, **24**, 2015.
216. S. M. Schmitt, D. B. R. Johnston and B. G. Christensen, *J. Org. Chem.*, 1980, **45**, 1135.
217. M. Shibuya, M. Kuretani and S. Kubota, *Tetrahedron Lett.*, 1981, **22**, 4453.
218. A. G. M. Barrett, G. G. Graboski and M. A. Russell, *J. Org. Chem.*, 1985, **50**, 2603.
219. F. Dumas and J. D'Angelo, *Tetrahedron Lett.*, 1986, **27**, 3725.
220. M. D. Bachi, A. De Mesmaeker and N. Stevenart-De Mesmaeker, *Tetrahedron Lett.*, 1987, **28**, 2637.
221. G. H. Hakimelahi, A. Ugolini and G. Just, *Helv. Chim. Acta*, 1982, **65**, 1374.
222. P. Herdewijn, P. J. Claes and H. Vanderhaeghe, *Can. J. Chem.*, 1982, **60**, 2903.
223. K. Clauss, D. Grimm and G. Prossel, *Liebigs Ann. Chem.*, 1974, 539.

224. S. Mickel, *Aldrichim. Acta*, 1985, **18**, 95.
225. G. Johnson, P. M. Rees and B. C. Ross, *J. Chem. Soc., Chem. Commun.*, 1984, 970.
226. T. Shibata, Y. Sugimura, S. Sato and K. Kawazoe, *Heterocycles*, 1985, **23**, 3069.
227. I. Nagakura, *Heterocycles*, 1981, **9**, 1495.
228. D. Davies and M. J. Pearson, *J. Chem. Soc., Perkin Trans. 1*, 1981, 2539.
229. G. Johnson and B. C. Ross, *J. Chem. Soc., Chem. Commun.*, 1981, 1269.
230. W. H. Huffman, R. F. Hall, J. A. Grant and K. G. Holden, *J. Med. Chem.*, 1978, **21**, 413.
231. R. M. DeMarinis and W. M. Bryan, *Tetrahedron Lett.*, 1982, **23**, 731.
232. W. F. Huffman, K. G. Holden, T. F. Buckley, III, J. G. Gleason and L. Wu, *J. Am. Chem. Soc.*, 1977, **99**, 2352.
233. T. Fukuyama, R. K. Frank and C. F. Jewell, Jr., *J. Am. Chem. Soc.*, 1980, **102**, 2122.
234. D. R. Kronenthal, C. Y. Han and M. K. Taylor, *J. Org. Chem.*, 1982, **47**, 2765.
235. D. B. Bryan, R. F. Hall, K. G. Holden, W. F. Huffman and J. G. Gleason, *J. Am. Chem. Soc.*, 1977, **99**, 2353.
236. M. D. Bachi and J. Vaya, *J. Org. Chem.*, 1979, **44**, 4393.
237. K. Hirai, K. Fujimoto and Y. Iwano, *Ann. Rep. Sankyo Res. Lab.*, 1985, **37**, 133.
238. R. Scartazzini, H. Peter, H. Bickel, K. Heusler and R. B. Woodward, *Helv. Chim. Acta*, 1972, **55**, 408.
239. P. H. Crackett, C. M. Pant and R. J. Stoodley, *J. Chem. Soc., Perkins Trans. 1*, 1984, 2785.
240. K. Hirai, Y. Iwano and K. Fujimoto, *Tetrahedron Lett.*, 1982, **23**, 4025.
241. J. Brennan, in ref. 10, p. 151.
242. G. H. Hakimelahi and G. Just, *Can. J. Chem.*, 1981, **59**, 941.
243. B. Y. Lee, A. Biswas and M. J. Miller, *J. Org. Chem.*, 1986, **51**, 106.
244. J. Brennan, G. Richardson and R. J. Stoodley, *J. Chem. Soc., Perkin Trans. 1*, 1983, 649.
245. D. Häbich, P. Naab and K. Metzger, *Tetrahedron Lett.*, 1983, **24**, 2559.
246. Y. Takanashi, H. Yamashita, S. Kobayashi and M. Ohno, *Chem. Pharm. Bull.*, 1986, **34**, 2732.
247. P. G. Mattingley and M. J. Miller, *J. Org. Chem.*, 1983, **48**, 3556.
248. M. Sendai, S. Hashiguchi, M.Tomimoto, S. Kishimoto, T. Matsuo, M. Kondo and M. Ochiai, *J. Antibiot.*, 1985, **38**, 346.
249. C. L. Branch and M. J. Pearson, *J. Chem. Soc., Perkin Trans. 1*, 1986, 1077.
250. C. L. Branch, S. C. Finch and M. J. Pearson, *J. Chem. Soc., Perkin Trans. 1*, 1985, 1491.
251. L. D. Cama and B. G. Christensen, *J. Am. Chem. Soc.*, 1974, **96**, 7582.
252. M. Narisada, H. Onoue and W. Nagata, *Heterocycles*, 1977, **7**, 839.
253. H. Otsuka, W. Nagata, M. Yoshioka, M. Narisada, T. Yoshida, Y. Harada and H. Yamada, *Med. Res. Rev.*, 1981, **1**, 217.
254. W. Nagata, M. Narisada and T. Yoshida, in ref. 6, vol. 2, p. 3.
255. Y. Hamashima, H. Matsumura, S. Matsumura, W. Nagata, M. Narisada and T. Hoshida, in ref. 9, p. 57.
256. D. Häbich and W. Hartung, *Tetrahedron*, 1984, **40**, 3667.
257. S. Yamamoto, H. Itani, H. Takashashi, T. Tsuji and W. Nagata, *Tetrahedron Lett.*, 1984, **25**, 4545.
258. T. Aoki, N. Haga, Y. Sendo, T. Konoike, M. Yoshioka and W. Nagata, *Tetrahedron Lett.*, 1985, **26**, 339.
259. M. Narisada, T. Yoshida, H. Onoue, M. Ohtani, T. Okada and W. Nagata, *J. Antibiot.*, 1982, **35**, 463.
260. A. A. Carmine, R. N. Brogden, R. C. Heel, J. A. Romankiewicz, T. M. Speight and G. S. Avery, *Drugs*, 1983, **26**, 279.
261. T. Tsuji, H. Satoh, M. Narisada, Y. Hamashima and T. Yoshida, *J. Antibiot.*, 1985, **38**, 466.
262. M. Narisada, J. Nishikawa, F. Watanabe and Y. Terui, *J. Med. Chem.*, 1987, **30**, 514.
263. M. Narisada, *Pure Appl. Chem.*, 1987, **59**, 459.
264. J. Nishikawa, F. Watanabe, M. Shudon, Y. Terui and M. Narisada, *J. Med. Chem.*, 1987, **30**, 523.
265. C. L. Branch, M. J. Basker and M. J. Pearson, *J. Antibiot.*, 1986, **39**, 1792.
266. Z. Murakami, M. Doi and T. Yoshida, *Antimicrob. Agents Chemother.*, 1986, **30**, 447.
267. Ref. 5, p. 82.
268. P. Herdewijn, P. J. Claes and H. Verderhaeghe, *J. Med. Chem.*, 1986, **29**, 661.
269. S. Uyeo and H. Ona, *Chem. Pharm. Bull.*, 1980, **28**, 1563.
270. C. W. Greengrass, D. W. T. Hoople and M. S. Nobbs, *Tetrahedron Lett.*, 1982, **23**, 2419.
271. R. N. Guthikonda, L. D. Cama and B. G. Christensen, *J. Am. Chem. Soc.*, 1974, **96**, 7584.
272. M. Hatanaka and T. Ishimaru, *Tetrahedron Lett.*, 1983, **24**, 4837.
273. D. A. Evans and E. B. Sjogren, *Tetrahedron Lett.*, 1985, **26**, 3787.
274. T. W. Doyle, J. L. Douglas, B. Belleau, T. T. Conway, C. F. Ferrari, D. E. Horning, G. Lim, B.-Y. Luh, A. Martel, M. Menard and L. R. Morris, *Can. J. Chem.*, 1980, **58**, 2508.
275. Y. Hashimoto, S. Takasawa, T. Ogasa, H. Saito, T. Hirata and K. Kimura, *Ann. N. Y. Acad. Sci.*, 1984, **434**, 206.
276. K. Mochida, C. Shiraki, M. Yamasaki and T. Hirata, *J. Antibiot.*, 1987, **40**, 14.
277. K. Mochida, Y. Ono, M. Yamasaki, C. Shiraki and T. Hirata, *J. Antibiot.*, 1987, **40**, 182.
278. S. Uyeo and H. Ona, *Chem. Pharm. Bull.*, 1980, **28**, 1578.
279. M. W. Foxton, R. C. Mearman, C. E. Newall and P. Ward, *Tetrahedron Lett.*, 1981, **22**, 2497.
280. J. Finkelstein, K. G. Holden and C. D. Perchonock, *Tetrahedron Lett.*, 1978, 1629.
281. Merck and Co., *Br. Pat.* 1 455 016 (1973) (*Chem. Abstr.*, 1974, **81**, 37 560).
282. Ref. 5, p. 79.
283. M. J. Pearson, *J. Chem. Soc., Perkin Trans. 1*, 1981, 2544.
284. D. Davies and M. J. Pearson, in ref. 9, p. 88.
285. H. Satoh and T. Tsuji, *Tetrahedron Lett.*, 1984, **25**, 1733.
286. H. Satoh and T. Tsuji, *Tetrahedron Lett.*, 1984, **25**, 1737.
287. Ref. 5, p. 89.
288. K. G. Holden, in ref. 6, p. 125.
289. T. W. Doyle, B. Belleau, B.-Y. Luh, C. F. Ferrari and M. P. Cunningham, *Can. J. Chem.*, 1977, **55**, 468.
290. T. W. Doyle, B Belleau, B.-Y. Luh, T. T. Conway, M. Menard, J. L. Douglas, D. T.-W. Chu, G. Lim, L. R. Morris, P. Rivest and M. Casey, *Can. J. Chem.*, 1977, **55**, 484.
291. T. T. Conway, G. Lim, J. L. Douglas, M. Menard, T. W. Doyle, P. Rivest, D. Horning, L. R. Morris and D. Cimon, *Can. J. Chem.*, 1978, **56**, 1335.
292. J. L. Douglas, D. E. Horning and T. T. Conway, *Can. J. Chem.*, 1978, **56**, 2879.
293. T. W. Doyle, J. L. Douglas, B. Belleau, J. Meunier and B.-Y. Lu, *Can. J. Chem.*, 1977, **55**, 2873.
294. T. Jen, J. Frazee and J. R. E. Hoover, *J. Org. Chem.*, 1973, **38**, 2857.

295. T. W. Doyle, B.-Y. Luh, D. T.-W. Chu and B. Belleau, *Can. J. Chem.*, 1977, **55**, 2719.
296. A. K. Bose, M. S. Manhas, J. E. Vincent, K. Gala and I. F. Fernandez, *J. Org. Chem.*, 1982, **47**, 4075.
297. A. K. Bose, M. S. Manhas, J. M. van der Veen, S. S. Bari, D. R. Wagle, V. R. Hegde and L. Krishnan, *Tetrahedron Lett.*, 1985, **26**, 33.
298. A. K. Bose, J. E. Vincent, I. F. Fernandez, K. Gala and M. S. Manhas, in ref. 9, p. 80.
299. H. Nitta, M. Hatanaka and T. Ishimaru, *J. Chem. Soc., Chem. Commun.*, 1987, 51.
300. S. M. Tenneson and B. Belleau, *Can. J. Chem.*, 1980, **58**, 1605.
301. M. J. Pearson and J. W. Tyler, *J. Chem. Soc., Perkin Trans. 1*, 1985, 1927.
302. J. G. Gleason, T. F. Buckley, K. G. Holden, D. B. Bryan and P. Siler, *J. Am. Chem. Soc.*, 1979, **101**, 4730.
303. J. G. Gleason, D. B. Bryan and K. G. Holden, *Tetrahedron Lett.*, 1980, 3974.
304. C. L. Branch and M. J. Pearson, *J. Chem. Soc., Perkin Trans. 1*, 1986, 1077.
305. M. Aratani and M. Hashimoto, in ref. 9, p. 97.
306. C. L. Branch and M. J. Pearson, *J. Chem. Soc., Perkin Trans. 1*, 1986, 1097.
307. P. H. Crackett, C. W. Greengrass and R. J. Stoodley, *Tetrahedron Lett.*, 1986, **27**, 1301.
308. J. Finkelstein, K. G. Holden, R. Sneed and C. D. Perchonock, *Tetrahedron Lett.*, 1977, 1855.
309. R. J. Stoodley, in ref. 10, p. 183.
310. D. O. Spry, *Tetrahedron Lett.*, 1981, **22**, 3695.
311. D. O. Spry, A. R. Bhala, W. A. Spitzer, N. D. Jones and J. K. Swartzendruber, *Tetrahedron Lett.*, 1984, **25**, 2531.
312. M. D. Bachi, F. Frolow and C. Hoornaert, *J. Org. Chem.*, 1983, **48**, 1841.
313. M. Mori, N. Kanda and Y. Ban, *J. Chem. Soc., Chem. Commun.*, 1986, 1375.
314. Beecham Pharmaceuticals, unpublished results.
315. R. M. Williams and B. H. Lee, *J. Am. Chem. Soc.*, 1986, **108**, 6431.
316. E. C. Taylor and H. M. L. Davies, *J. Org. Chem.*, 1984, **49**, 113.
317. E. C. Taylor, H. M. L. Davies and J. S. Hinkle, *J. Org. Chem.*, 1986, **51**, 1530.
318. E. C. Taylor and H. M. L. Davies, *J. Org. Chem.*, 1986, **51**, 1537.
319. G. Lawton, C. J. Moody, C. J. Pearson and D. J. Williams, *J. Chem. Soc., Perkin Trans. 1*, 1987, 885.
320. K. R. Henery-Logan and A. M. Lunburg, *Tetrahedron Lett.*, 1966, 4915; M. J. Mardle, J. H. Nayler, D. W. Rustidge and H. R. J. Waddington, *J. Chem. Soc. (C)*, 1968, 237; D. B. Miller, J. H. C. Nayler and H. R. J. Waddington, *ibid.*, 1968, 242.
321. P. W. Wojtkowski, J. E. Dolfini, O. Kocy and C. M. Cimarusti, *J. Am. Chem. Soc.*, 1975, **97**, 5628; F. Cavagna, W. Koller, A. Linkies, H. Rehling and D. Reuschling, *Angew. Chem., Int. Ed. Engl.*, 1982, **21**, 548.
322. G. Lowe and S. Swain, in ref. 10, p. 209; E. M. Gordon, J. Pluscec and M. A. Ondetti, *Tetrahedron Lett.*, 1981, **22**, 1871; O. Meth-Cohn, A. J. Reason and S. M. Roberts, *J. Chem. Soc., Chem. Commun.*, 1982, 90.
323. V. Du Vigneaud and F. H. Carpenter, in 'The Chemistry of Penicillin', ed. H. T. Clarke, J. R. Johnson and R. Robinson, Princeton University Press, Ithaca, New York, 1949, p. 1004.
324. L. N. Jungheim, S. K. Sigmund, N. D. Jones and J. K. Swartzendruber, *Tetrahedron Lett.*, 1987, **28**, 289; L. N. Jungheim, S. K. Sigmund and J. W. Fisher, *ibid.*, 1987, **28**, 285.
325. D. B. Boyd, T. K. Elzey, L. D. Hatfield, M. D. Kinnick and J. M. Morin, Jr., *Tetrahedron Lett.*, 1986, **27**, 3453; D. B. Boyd, B. J. Foster, L. D. Hatfield, W. J. Hornback, W. D. Jones, J. E. Munroe and J. K. Swartzendruber, *ibid.*, 1986, **27**, 3457; J. E. Baldwin, C. Lowe and C. J. Schofield, *ibid.*, 1986, **27**, 3461.
326. Y. Nozaki, N. Katayama, H. Ono, S. Tsubotani, S. Harada, H. Okazaki and Y. Nakao, *Nature (London)*, 1987, **325**, 179; S. Harada, S. Tsubotani, T. Hida, H. Ono and H. Okazaki, *Tetrahedron Lett.*, 1986, **27**, 6229.
327. H. Natsugari, Y. Kawano, A. Morimoto, K. Yoshioka and M. Ochiai, *J. Chem. Soc., Chem. Commun.*, 1987, 62.
328. J. S. Wells, W. H. Trejo, P. A. Principe and R. B. Sykes, *J. Antibiot.*, 1984, **37**, 802; A. A. Tymiak, C. A. Culver, M. F. Malley and J. Z. Gougoutas, *J. Org. Chem.*, 1985, **50**, 5491.
329. J.-M. Ghuysen, J.-M. Frère, M. Leyh-Bouille, H. R. Perkins and M. Nieto, *Philos. Trans. R. Soc. London, Ser. B*, 1980, **289**, 167.
330. J.-M. Ghuysen, in 'Topics in Antibiotic Chemistry', ed. P. G. Sammes, Ellis Horwood, Chichester, 1980, vol. 5, p. 9.
331. M. Matsuhashi, F. Ishino, S. Tamaki, S. Nakajima-Jijima, S. Tomioka, J. Nakagawa, A. Hirata, B. G. Spratt, T. Tsuruoka, S. Inouye and Y. Yamada, in 'Trends in Antibiotic Research; Genetics, Biosynthesis, Actions and New Substances', ed. H. Umezawa, A. L. Demain, T. Hata and C. R. Hutchinson, Japanese Antibiotics Research Association, Tokyo, 1982, p. 99.
332. D. J. Tipper, *Pharmacol. Ther.*, 1985, **27**, 1.
333. A. Tomasz, *Res. Infect. Dis.*, 1986, **8**, suppl. 3, S260.
334. D. J. Waxman, R. R. Yocum and J. L. Strominger, *Philos. Trans. R. Soc. London, Ser. B*, 1980, **289**, 257; D. J. Waxman and J. L. Strominger, in ref. 6, vol. 3, p. 209; D. J. Waxman and J. L. Strominger, *Annu. Rev. Biochem.*, 1983, **52**, 825.
335. J.-M. Frère and B. Joris, *CRC Crit. Rev. Microbiol.*, 1985, **11**, 299.
336. A. F. W. Coulson, *Biotechnol. Genet. Eng. Rev.*, 1985, **3**, 219.
337. 'β-Lactamases', ed. J. M. T. Hamilton-Miller and J. T. Smith, Academic Press, New York, 1979.
338. D. B. Boyd, in ref. 6, vol. 1, p. 437.
339. R. B. Woodward, A. Newberger and N. R. Trenner, in 'Chemistry of Penicillins', ed. H. T. Clarke, J. R. Johnson and R. Robinson, Princeton University Press, Ithaca, New York, 1949, p. 415.
340. R. M. Sweet, in ref. 4, p. 280.
341. R. M. Sweet and L. F. Dahl, *J. Am. Chem. Soc.*, 1970, **92**, 5489.
342. K. Kamiya, M. Takamoto, Y. Wada and M. Asai, *Acta Crystallogr., Sect. B*, 1981, **37**, 1626.
343. N. C. Cohen, *J. Med. Chem.*, 1983, **26**, 259.
344. J. H. C. Nayler, *J. Antimicrob. Chemother.*, 1987, **19**, 713.
345. W. W. Nichols, *Biochem. J.*, 1987, **244**, 509.
346. D. D. Dexter and J. M. van der Veen, *J. Chem. Soc., Perkin Trans. 1*, 1978, 185.
347. H. R. Pfaendler, J. Gosteli, R. B. Woodward and G. Rihs, *J. Am. Chem. Soc.*, 1981, **103**, 4526.
348. D. G. Brenner and J. R. Knowles, *Biochemistry*, 1981, **20**, 3680.
349. J.-M. Frère, J. A. Kelly, D. Klein and J.-M. Ghuysen, *Biochem. J.*, 1982, **203**, 223.
350. V. S. R. Rao and T. K. Vasudevan, *CRC Crit. Rev. Biochem.*, 1985, **14**, 173.

10.1

DNA Intercalating Agents

LAURENCE P. G. WAKELIN

Peter MacCallum Cancer Institute, Melbourne, Victoria, Australia

and

MICHAEL J. WARING

University of Cambridge, UK

10.1.1	INTRODUCTION	704
	10.1.1.1 Overview of the Development of Intercalating Agents	704
	10.1.1.2 Structure of DNA	705
	10.1.1.3 Types of Reversible Drug Binding	708
10.1.2	GENERAL FEATURES OF INTERCALATION COMPLEXES	709
	10.1.2.1 Evidence for Intercalation	709
	10.1.2.2 Equilibrium Measurements	711
	10.1.2.3 Rates and Mechanisms of Binding	712
	10.1.2.4 Sequence Specificity of Binding	712
	10.1.2.5 Crystallographic Studies	712
	10.1.2.6 Theoretical Studies	713
	10.1.2.7 NMR Studies	714
10.1.3	SURVEY OF MONOFUNCTIONAL INTERCALATORS	714
	10.1.3.1 Antitumour Antibiotics	714
	10.1.3.1.1 Actinomycins	714
	10.1.3.1.2 Anthracyclines	714
	10.1.3.2 Synthetic Antitumour Agents	716
	10.1.3.2.1 Acridines	716
	10.1.3.2.2 Anthraquinone derivatives	716
	10.1.3.2.3 Ellipticine and bisantrene	717
	10.1.3.3 Other Intercalating Ligands	717
10.1.4	POLYFUNCTIONAL INTERCALATING AGENTS	717
	10.1.4.1 Concepts of Polyfunctional Intercalation	717
	10.1.4.2 Bifunctional Antibiotics	718
	10.1.4.2.1 Quinoxalines	718
	10.1.4.2.2 Luzopeptin	718
	10.1.4.2.3 Bisanthracyclines	719
	10.1.4.3 Diacridines	719
	10.1.4.4 Other Bisintercalators	720
	10.1.4.5 Triacridines	721
10.1.5	CELLULAR CONSEQUENCES OF INTERCALATION	721
	10.1.5.1 Biochemical Pharmacology	721
	10.1.5.2 Effects on Chromatin	722
	10.1.5.3 Unwanted Toxicities	722
10.1.6	CONCLUSIONS AND PERSPECTIVES	722
10.1.7	REFERENCES	723

10.1.1 INTRODUCTION

10.1.1.1 Overview of the Development of Intercalating Agents

Nowadays we are accustomed to considering DNA as the central controlling element of cellular physiology. This perception stems from our appreciation that DNA embodies the genetic specification of an organism, and from our rapidly expanding knowledge of the biochemical mechanisms whereby it directs cellular homeostasis, proliferation and differentiation. At the most basic level its two major functions are self-replication and the provision of a template for the synthesis of RNA, the latter process being termed transcription. Transcription involves the synthesis of various RNA species, including messenger RNA, which carries to ribosomes the instructions for the correct assembly of amino acids into proteins. Many proteins are involved in the replication and transcription processes themselves, including those that manipulate DNA topology. Inhibition of any of these activities involving DNA could have lethal consequences for the cell, and accordingly DNA is a natural drug target for the development of agents cytotoxic to either prokaryotic or eukaryotic organisms.

Thus it is not surprising to find that many of the active compounds discovered during the early years of the search for chemotherapeutic agents,[1] when these perceptions were not realized, were later shown to be DNA-binding ligands. We now recognize that several important examples belong to classes of compounds, notably acridines and phenanthridines, identifiable as intercalating agents (Figure 1). Predating antibiotics and the sulfonamides, the aminoacridines proflavine and 9-aminoacridine played a decisive role in wound asepsis and they still have topical antibacterial uses today. Other aminoacridines, such as quinacrine and its quinoline analogue, chloroquine, were the mainstay in antimalarial therapy. Quinacrine was also used in the treatment of giardiasis and helminthic infections. Quaternary phenanthridines, such as ethidium, found use as trypanocidal agents in cattle, and the 1940s saw the discovery of the intercalating actinomycin antibiotics, which have since proven useful in the treatment of some childhood cancers.

There was little progress in studies of drug–DNA interactions until Peacocke and Skerrett investigated the DNA-binding properties of proflavine.[2] However, interest lapsed until it was discovered that aminoacridines are powerful frameshift mutagens,[2] a finding that was to prove important in revealing the triplet nature of the genetic code. In 1961 Lerman made a seminal contribution by proposing that proflavine intercalates into DNA, a process whereby the planar acridine ring comes to be inserted between the base pairs of the Watson–Crick helix.[2] The 1960s saw a rapid expansion both in physicochemical studies of DNA–intercalator complexes and in studies of the capacity of intercalating agents to inhibit the template functions of DNA *in vivo* and *in vitro*.[2] The introduction of hydrodynamic measurements, using covalently closed circular DNA and sonicated rodlike DNA fragments, made for more facile and reliable assessment of intercalative potential.[2] The compounds particularly well studied at this time were the aminoacridines, ethidium, actinomycin and the anthracycline, daunomycin.[2]

In the 1970s much attention was devoted to efforts to develop intercalating antitumour agents. Major achievements here were the anilinoacridine amsacrine,[3] the 9-aminoacridine derivative nitracrine,[3] derivatives of the pyridocarbazole ellipticine,[4] and most importantly the broad spectrum anthracycline antibiotic adriamycin and its analogues (Figure 1).[5] This period also saw the discovery that the antitumour quinoxaline antibiotics, known since the 1950s, were bifunctional intercalating agents.[6] Concomitantly, many groups began investigating the antitumour potential and DNA-binding properties of synthetic polyfunctional intercalating compounds.[6] New dimensions were added to our understanding of DNA–intercalator interactions as a result of kinetic studies,[2] the solution of the crystal structures of dinucleotide–drug complexes, and the advent of molecular mechanics simulations of the structure of intercalated complexes.[2,7] Spectroscopic measurements and molecular modelling with oligonucleotide complexes drew attention to the sequence dependent nature of the binding of simple intercalators,[2,7] and the development of filter–elution assays for evaluating drug-induced strand breaks in cellular DNA led to the important discovery that intercalating drugs are poisons of the enzyme topoisomerase II.[8]

The 1980s have witnessed the greatest efforts so far to find clinically useful antitumour intercalating drugs. The drugs with an established place in the clinic are actinomycin, daunomycin, adriamycin, amsacrine and mitoxantrone. The list of agents recently in clinical trial is extensive and includes echinomycin, ditercalinium, bisantrene, ametantrone, amsalog, menogaril, epidaunorubicin, chryseneaminodiol, oxanthrazole and nafidimide (Figure 2). At present the newest insights into the nature of DNA–intercalation complexes derive from NMR and crystallographic studies using defined sequence oligonucleotides, computer graphics simulations and molecular mechanics

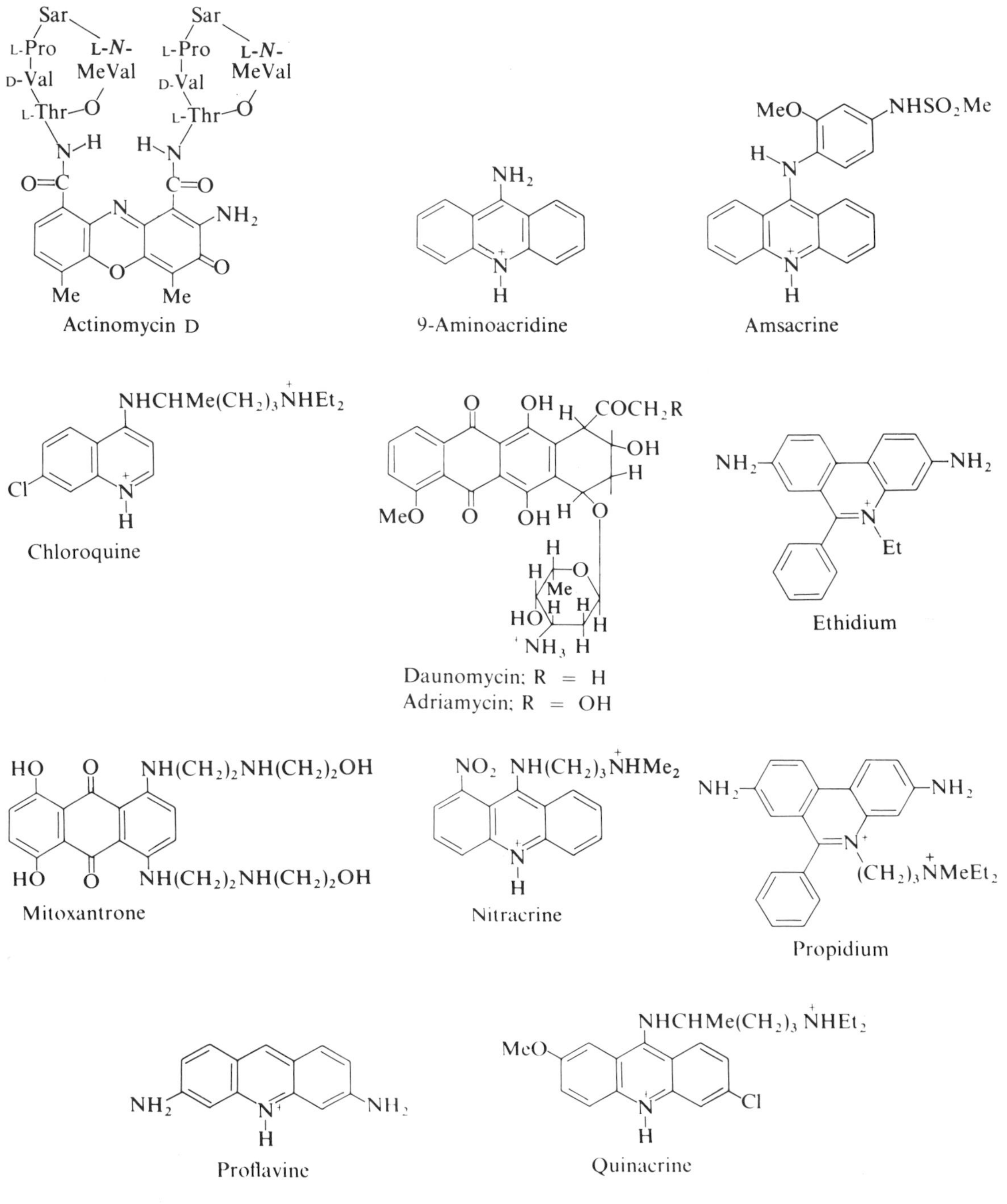

Figure 1 Intercalating agents with therapeutic uses

and dynamics modelling, stopped flow and NMR kinetic studies and the use of Maxam–Gilbert sequencing methods[9] to locate drug binding sites on mixed sequence DNA molecules of natural origin.

10.1.1.2 Structure of DNA

The three-dimensional structure and dynamic properties of DNA are known in greater detail than is common for a drug receptor.[10] This has enabled a rational approach to the design of intercalating agents and has permitted meaningful interpretation of experimental data in structural and mechanistic terms. Before the advent of synthetic polynucleotides and single-crystal diffraction studies, the

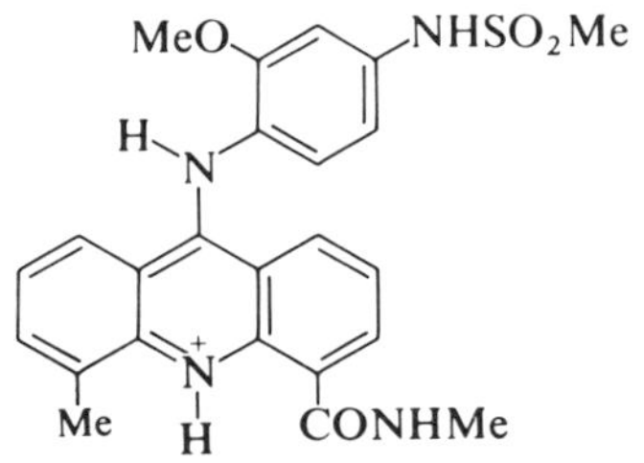

Amsalog (CI-921)

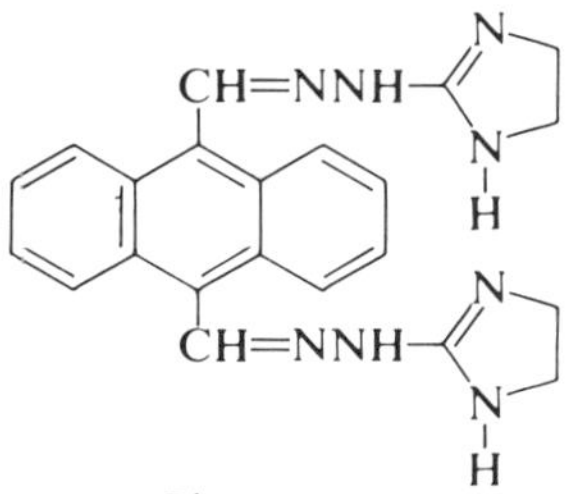

Bisantrene

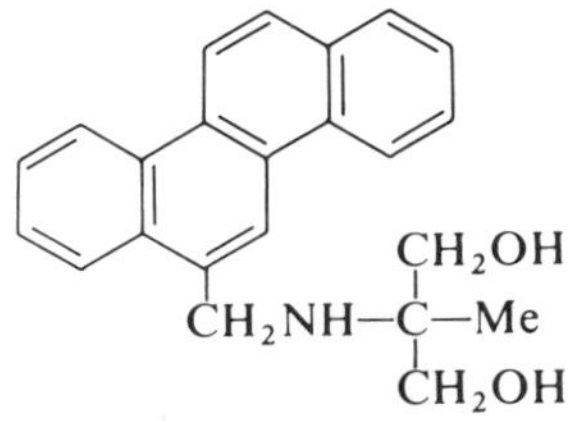

Chryseneaminodiol

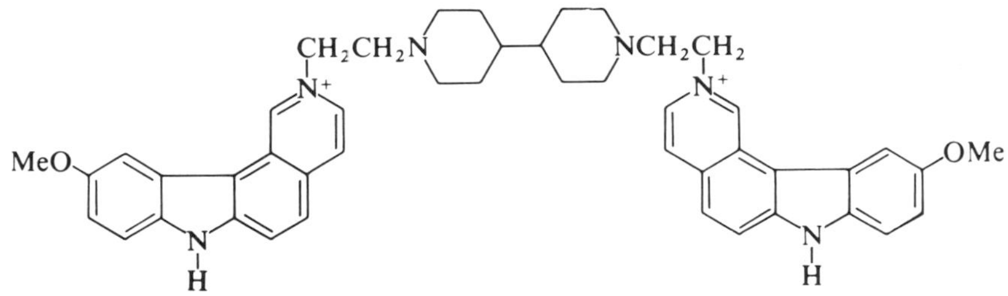

Ditercalinium

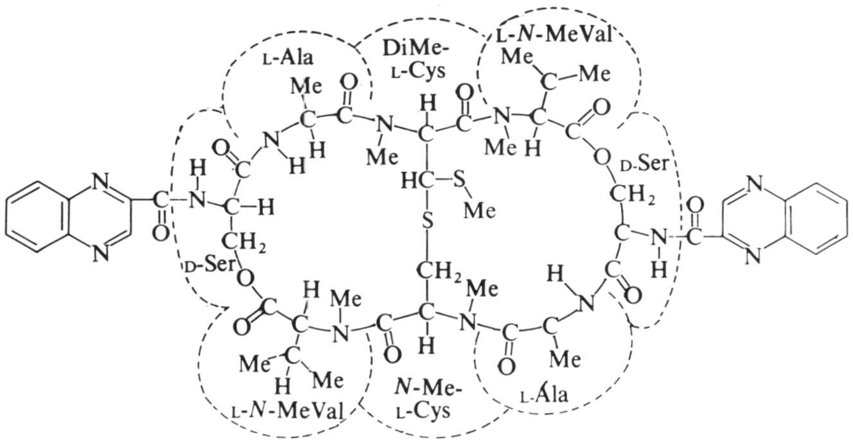

Echinomycin

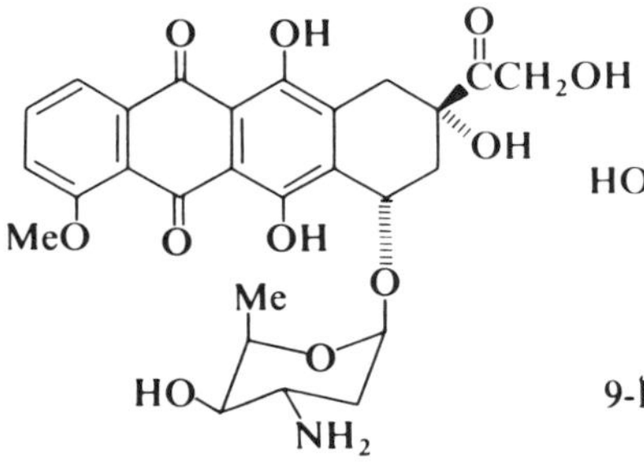

Epirubicin

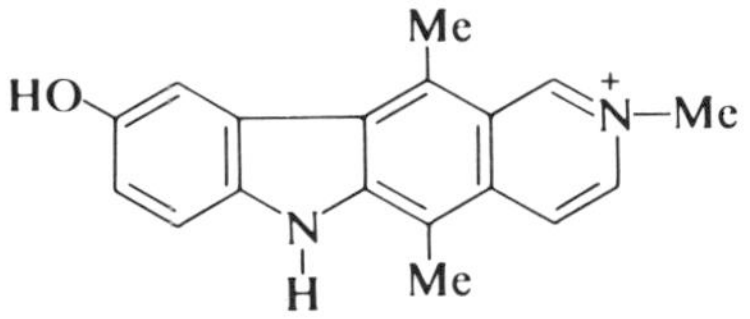

9-Hydroxy-*N*-methylellipticinium

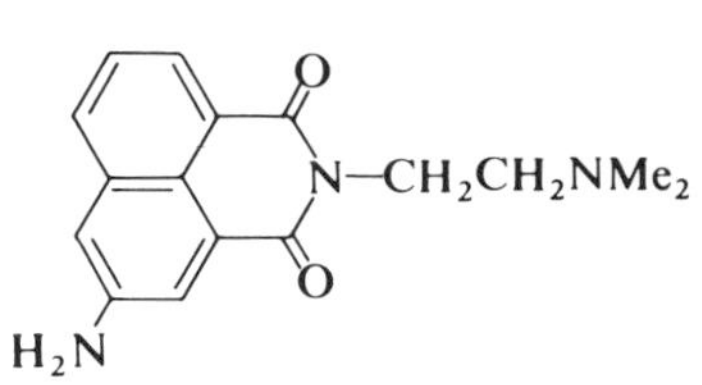

Nafidimide

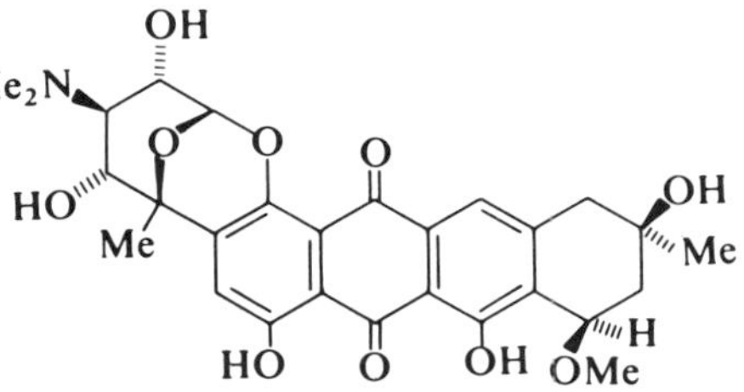

Menogaril

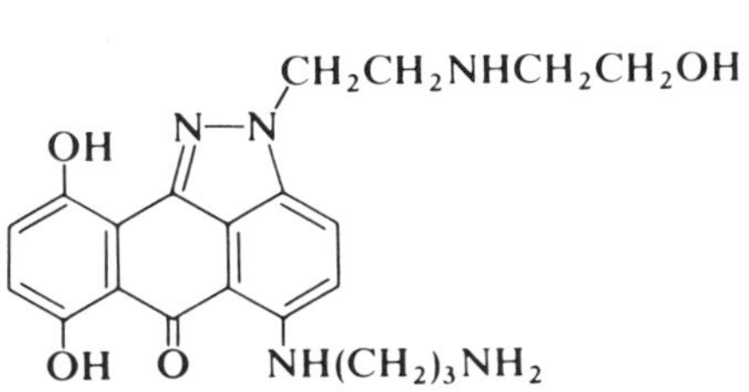

Oxanthrazole

Figure 2 Intercalating antitumour agents on clinical trial

general view of DNA was that presented by model-building refinements of the Watson–Crick double helix. Thus, natural DNA was seen to exist, both in solution and in fibres, as a right-handed antiparallel double helix, fixed in the A or B motif largely as a result of the degree of hydration.[10] The helical parameters were considered to be independent of nucleotide sequence, and the structural basis for the A to B transition was attributed to a change in sugar ring pucker attendant upon changes in water activity.[10] From the point of view of drug binding, perhaps the most significant differences between the A and B forms are the dimensions of the helical grooves and the disposition of the base pairs with respect to the helix axis. The major and minor grooves of B-DNA are approximately equally deep but have very different widths. By contrast A-DNA has a deep yet narrow major groove and a shallow wide minor groove. The base pairs are practically orthogonal to the helix axis and lie in the centre of the molecule in the B form, whereas in the A form they are tilted and pulled away from the helix axis so that the polymer has a cylindrical hole running down its middle. Later, fibre diffraction studies with synthetic DNAs of simple repetitive sequence revealed a variety of additional helical forms, making it clear that DNA structure can be sequence dependent.[10]

Details of just how the helical parameters depend on sequence have been provided by single-crystal X-ray diffraction studies of DNA oligodeoxynucleotides. A variety of duplexes have been crystallized in the A form series, a dodecamer and close homologues have been crystallized in the B motif, and several others of alternating pyrimidine–purine sequence have been revealed in the left-handed Z form.[10,11] For the right-handed helices Calladine and Drew have formulated rules based on cross-strand purine–purine steric clashes for relating structure to sequence in terms of base pair twist, roll and slide parameters.[10,11] The most notable findings of the B form studies are the high propeller twist, helical twist and tilt angles of AT compared to GC sequences. These features result in runs of As or Ts having a relatively narrow minor groove and a wide major groove compared to the fibre models. Conversely, runs of Gs or Cs have a relatively wide minor groove and a narrow major groove. Sugar ring puckers and glycosidic bond angles are also found to be sequence dependent: only rarely are they actually identical to the classical B parameters.

The left-handed forms of DNA are largely confined to regular alternating pyrimidine–purine sequences, CG being the best known, and their biological importance is a subject of controversy.[10,11] Their stability with respect to right-handed DNA, in solution at least, is critically dependent upon solvent composition, base modification and DNA torsional stress. The bases pair according to the Watson–Crick hydrogen-bonding scheme but, unlike right-handed DNA, the purine nucleosides adopt the *syn* rather than the *anti* conformation and the polymer repeat unit is a dinucleotide. The Z duplex has a very deep and narrow minor groove, the major groove being almost nonexistent, since the base pairs present a convex face on this surface of the helix. Intercalating agents are generally unable to bind to Z-DNA, but can be powerful modifiers of its morphology since they induce it to undergo the phase transition to the B form.

These new insights into the polymorphic character and flexibility of DNA will undoubtedly have great influence on our perceptions about the structure and dynamics of DNA–ligand complexes. In the past, molecular models were mostly constructed within a conceptual framework built on minimally distorted B type helices but, clearly, future models will have to take into account a wider range of structural possibilities.

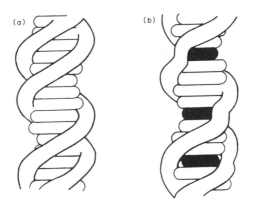

Figure 3 The intercalation model: (a) is a representation of B type DNA with the base pairs shown in edgewise projection and the sugar–phosphate backbone represented as a smooth coil; (b) shows the same DNA containing intercalated drug molecules (reproduced from ref. 2 by permission of Wiley)

10.1.1.3   Types of Reversible Drug Binding

Intercalating agents come in many guises, and may interact with DNA with or without sequence selectivity. They may be simple or complex chemical structures, and may bind with their chromophore side chains lodged in the minor or major groove, or even both.[2] Complex formation is usually freely reversible, its driving force being a combination of electrostatic, entropic, hydrogen-bonding,

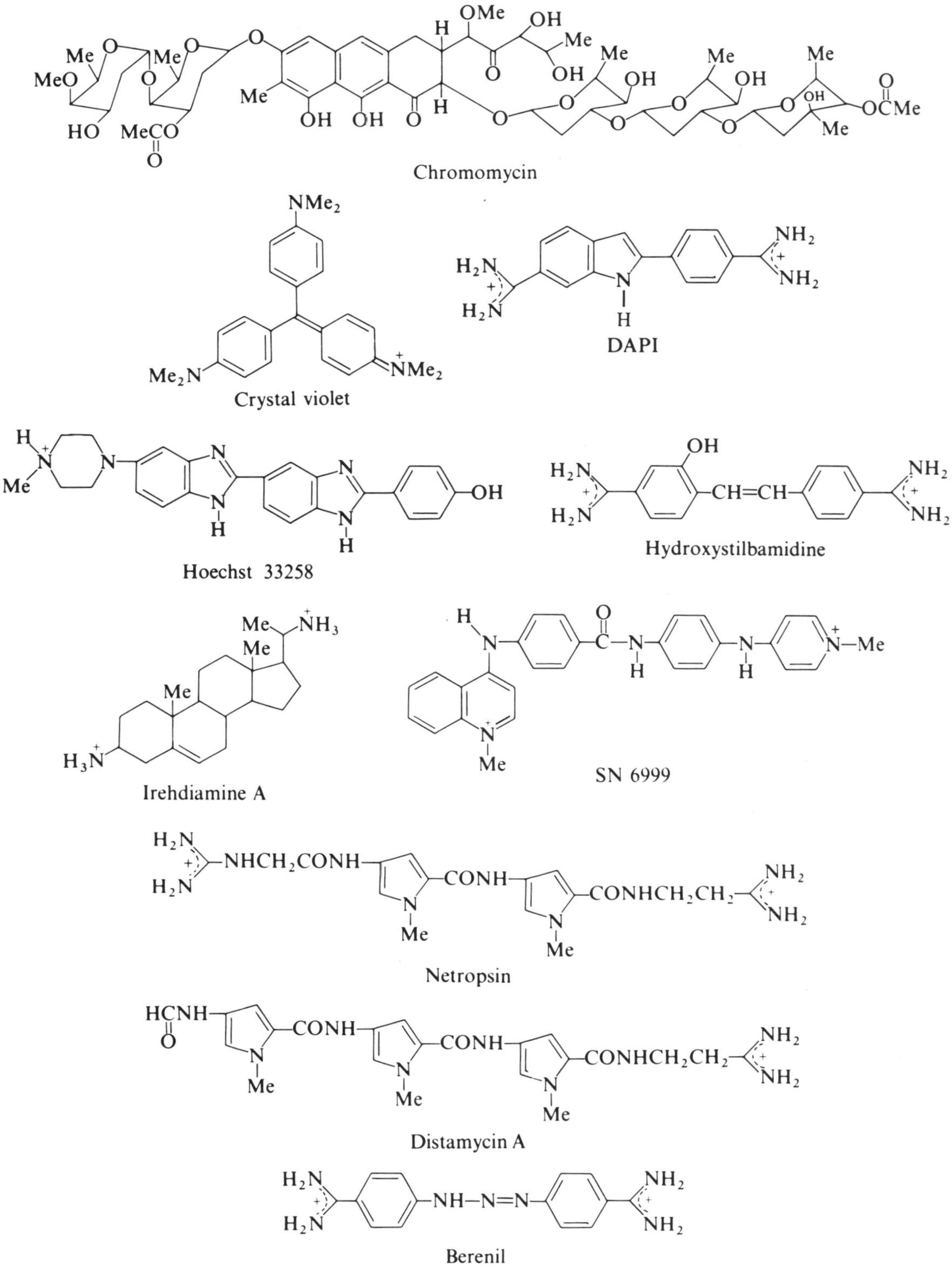

Figure 4   Nonintercalating DNA-binding ligands

van der Waals and hydrophobic interactions. The foundations of the intercalation concept were laid by Lerman as a result of his X-ray fibre diffraction and hydrodynamic studies of the proflavine–DNA complex.[2] The fibre diffraction pattern revealed the loss of the long range regularity of the helix, but showed retention of the 3.4 Å meridional reflections, due to the regular stacking of base pairs perpendicular to the helix axis. Lerman found that proflavine binding enhanced the viscosity of DNA but lowered its sedimentation coefficient. He rationalized these observations by proposing that the planar proflavine molecule becomes inserted between the DNA base pairs as indicated in Figure 3. The major structural distortions to the DNA are an increase in length by the thickness of an aromatic ring, 3.4 Å, and a local unwinding of the helix at the intercalation site. In Lerman's model, proflavine was oriented so that its major axis was parallel to the major axes of the base pairs, the DNA unwinding angle being 36°. Subsequently, the model was developed further and extended to describe the intercalative binding of ethidium by Fuller and Waring.[2] The accumulated evidence of the past 25 years has proved the essential correctness of the basic tenets of the intercalation hypothesis, and many dozens of planar aromatic ligands have since been shown to bind in this manner.[2] However, intercalation is not the only means by which drugs may interact reversibly with DNA, and below we briefly survey compounds that bind by nonintercalative mechanisms in order to provide the reader with insights into other potential modes of binding.

The best characterized of the nonintercalating ligands is a group of compounds known as the 'minor groove binding agents'. This group includes, for example, the polypyrrole peptide antibiotics netropsin and distamycin and their synthetic derivatives, the diamidine trypanocides berenil and hydroxystilbamidine, the benzimidazole and indole chromosome stains Hoechst 33258 and DAPI (4′,6-diamidino-2-phenylindole), and the diquaternary antitumour agent SN 6999 (Figure 4).[2,11,12] These compounds share the structural similarity of being flexibly linked aromatic systems whose preferred configuration is a flat, banana-shaped crescent that is complementary in shape to the minor groove of the B type helix. There is a wealth of information concerning their interaction with DNA, the most conspicuous feature of which is the almost absolute specificity for binding in the minor groove at runs of three or four consecutive AT base pairs. This is confirmed by equilibrium, footprinting and NMR measurements with natural and synthetic DNAs, and by molecular mechanical calculations and crystallographic studies of complexes with the dodecanucleotide d(CGCGAATTCGCG)$_2$.[2,11,12] Netropsin, for example, nestles in the minor groove of the AATT segment of the dodecamer, replacing the spine of hydration found in this region of the drug free DNA. It completely fills the groove, its amide groups making bifurcated hydrogen bonds to the adenine N-3 and thymidine O-2 atoms. The sequence specificity of minor groove binding agents has its origins in the complementarity of fit between ligand and DNA, the complex being stabilized by electrostatic, hydrogen-bonding and van der Waals interactions. Chromomycin A3, mithramycin and olivomycin are closely related antitumour antibiotics (Figure 4) that bind to GC rich regions of DNA.[2,12] They exhibit pronounced sequence selectivity, high affinity binding requiring two contiguous GC base pairs with many sites comprising CCC or GGG triplets.[12,13] Methylation protection experiments with mithramycin[13] and NMR measurements with chromomycin[14] suggest that they bind in the major groove, few details being known other than that the chromomycin chromophore lies along the floor of the groove. Steroidal diamines like irehdiamine A and dipyrandium, and the triphenylmethane dye crystal violet (Figure 4), have the unusual property among nonintercalators of causing DNA to unwind. The steroidal diamines are too bulky to insert between the base pairs, but they nevertheless attempt to intercalate and become lodged across the minor groove of a partially unwound kinked DNA structure.[2] Crystal violet binds selectively to sequences containing two adjacent AT base pairs by a mechanism that remains obscure.[15]

10.1.2 GENERAL FEATURES OF INTERCALATION COMPLEXES

10.1.2.1 Evidence for Intercalation

Acridines and phenanthridines are the archetypal DNA intercalating agents, and we shall use the best known among them, 9-aminoacridine, proflavine, ethidium and propidium (Figure 1), to illustrate the general features of intercalation complexes. The most direct tests for intercalative binding derive from the changes that intercalation forces upon DNA structure, *i.e.* the modifications it causes to helix winding and contour length. The extent of these modifications depends on ligand type, unwinding angles falling in the range 10 to 26°, and increases in contour length varying between 1.7 and 3.4 Å for the binding of monofunctional compounds.[2]

Helix unwinding angles are measured by observing the effects of ligand binding on the super-coiling of covalently closed circular DNA.[2] This is usually achieved by monitoring hydrodynamic properties such as sedimentation rate and viscosity, or electrophoretic mobility, which sensitively reflect superhelical density. The topological character of circular DNA causes a direct coupling between the winding of the Watson–Crick duplex and the degree of supercoiling.[2] Consequently, in a drug titration, intercalation-induced unwinding is first seen as a reduction in the number of naturally occurring right-handed supercoils.[2] As the binding level increases, the point is reached where the supercoiling is completely removed, termed the equivalence point: here the sedimentation coefficient and electrophoretic mobility fall to minimum values and viscosity rises to a maximum.[2] Binding additional drug causes the introduction of left-handed supercoils. Since the total unwinding of the Watson–Crick duplex required to reach the equivalence point is a constant for a given circular DNA, the helix unwinding angle of a ligand may be evaluated by comparing its equivalence binding ratio with that of a reference compound, *e.g.* ethidium, whose unwinding angle is independently known. Originally the unwinding angle of ethidium was taken to be 12°, as a result of the theoretical study by Fuller and Waring,[2] but subsequently it was shown experimentally to be 26°.[2] By titration with circular DNA the unwinding angle of propidium is found to be the same as that of ethidium, whereas that of 9-aminoacridine or proflavine is in each case 17°.[2]

Drug effects on contour length are most conveniently monitored by measuring the viscosity of sonicated rodlike fragments of DNA ($M_r = 250\,000$) as a function of binding ratio. Cohen and Eisenberg have shown that the viscosity of short fragments of this type depends primarily on contour length, and they have deduced that the relative increase in contour length in the presence of bound drug is approximated by the cube root of the ratio of the intrinsic viscosity of the DNA–drug complex to that of the free DNA.[16] This treatment, plus the assumption that an intercalation event is equivalent to adding another monomer unit to DNA, led to the very useful relation

$$L/L_0 \ = \ 1 \ + \ mr \tag{1}$$

where L/L_0 is the relative increase in contour length in the presence of the drug, usually further approximated to the cube root of measured reduced viscosities, r is the ligand binding ratio and m is the slope of the plot of L/L_0 *versus* r, sometimes known as the helix extension parameter. The theoretical value of m, characterizing ideal monofunctional intercalation, depends on the choice of units of r, being 2 if the binding ratio is expressed with respect to nucleotides or 1 if it is defined in base pairs: both conventions are found in the literature. Extension parameters are often interpreted in terms of a distance, by identifying a value of $m = 1$ (or 2) with an increase in contour length of 3.4 Å and, in this way, ethidium and proflavine are each found to cause an extension of 2.7 Å on binding, whereas 9-aminoacridine increases contour length by 3.0 Å.[17,18]

That measured helix extension parameters frequently fall short of the theoretical value is usually ascribed: (i) to inadequacies attendant upon the use of many hydrodynamic approximations, both theoretical and experimental; or (ii) to the possibility that the ligand induces kinking or bending of the DNA helix at the intercalation site; or (iii) to the adoption by bound ligand of a tilted geometry with respect to the helix axis, so that the increase in contour length is genuinely less than the thickness of the drug chromophore. Thus, in respect of the latter, for example, whilst flow dichroism and fluorescence polarization studies on complexes oriented by flow show that the plane of the intercalated chromophore is approximately parallel to the plane of the base pairs,[2] transient electric dichroism measurements reveal that the bound ligand may be tilted and twisted with respect to the helix axis.[19]

In addition to inducing conformational changes, intercalative binding affects other physical properties of DNA: thus aminoacridines and phenanthridines stabilize DNA towards thermal denaturation and reduce its buoyant density in CsCl gradients.[2] Intercalation also modifies spectroscopic and chemical properties of the bound ligand. Electronic absorption spectra universally show a bathochromic and hypochromic shift on binding, whereas fluorescence characteristics vary in a more complex manner.[2] For example, in the presence of DNA, fluorescence emission intensity is greatly enhanced for ethidium and propidium, whereas for 9-aminoacridine and proflavine it is diminished; conversely, for some acridines, like quinacrine, it may be enhanced or quenched, depending upon nucleotide sequence.[2] Although synthetic intercalators do not usually have chiral centres and so are optically inactive, when intercalated they form part of an asymmetric complex by virtue of the stereochemistry of DNA and acquire an induced circular dichroism spectrum.[2] The reduced accessibility of the intercalated ligand to solvent is revealed by the strong shielding of the amino groups of DNA-bound proflavine to reaction with nitrous acid.[2] Similarly, the hydrogen atoms of the amino groups of ethidium are protected from H–D exchange in D_2O/H_2O mixtures.[20]

10.1.2.2 Equilibrium Measurements

The affinity and extent of DNA binding of intercalating agents are usually measured by partition methods, such as equilibrium dialysis and solvent partition analysis, and by spectroscopic methods that take advantage of the metachromatic changes in absorption and fluorescence spectra that accompany binding;[2] a fluorescent ethidium displacement assay also finds substantial use.[21] Having obtained the free ligand concentration c that maintains the binding ratio r (expressed in the early literature in nucleotide units but now frequently expressed in terms of base pairs), the data are invariably plotted as r/c *versus* r, in the manner of Scatchard,[2] and fitted to the equation of McGhee and von Hippel[22]

$$r/c = K(0) (1 - nr) \left(\frac{1 - nr}{1 - (n - 1)r} \right)^{n-1} \tag{2}$$

to yield values of the intrinsic association constant, $K(0)$, and the number of bases or base pairs occluded per bound ligand molecule, n. The McGhee and von Hippel treatment is but one of a variety of statistical mechanical methods that give the same analytical expression, which reduces to the familiar Scatchard equation when n has the value 1.[2,23,24] Clearly, when the bound ligand occludes more than one lattice unit, the binding isotherm is not linear in r, but is bent, concave upwards. The origins of this nonlinearity can be traced to entropy effects, associated with the rearrangement of bound ligand molecules on the DNA necessary to remove gaps of less than a site size that accumulate as the lattice saturates. Isotherms for the binding of ethidium and 9-amino-acridine to natural DNA are well-fitted by equation (2), giving values of n equal to two base pairs, and association constants in the range 10^5 to 10^6 M^{-1} in buffers of ionic strength 0.01 to 0.1 M.[23-25] Two base pairs per binding site is a common finding for simple intercalators and has been interpreted in terms of a neighbour exclusion model, in which the space between each base pair is viewed as a potential intercalation site, but where binding of a ligand molecule precludes simultaneous occupation of adjacent potential sites.[2,23] Despite some plausible and elegant hypotheses, it is fair to say that the physical basis of the neighbour exclusion phenomenon remains unclear.

Propidium-binding curves with natural DNA are more concave than expected for a simple excluded site model, and have been analyzed using a modification of equation (2) that includes a third parameter, ω, that allows for cooperative ($\omega > 1$) or anticooperative ($\omega < 1$) effects between bound ligands.[22] Thus propidium binding to calf thymus DNA is characterized by values of $n = 2$ and $\omega = 0.2$, showing that the affinity for binding a propidium molecule in the next allowable site adjacent to an already occupied one is reduced by a factor of 5 compared to binding to an isolated site.[26] Cooperative effects which enhance affinity between bound ligands give rise to 'humped' Scatchard plots.[22] Proflavine, whose affinity for isolated DNA sites is similar to that of ethidium, exhibits yet another binding behaviour, its isotherms tending to become parallel to the r axis at binding ratios above 0.4 ligand molecules per base pair.[2] Together with other information, this is interpreted as evidence for an additional binding mode, probably involving stacking of proflavine aggregates on the surface of DNA, which becomes populated as intercalation sites saturate.[2]

The affinity of intercalating agents diminishes at elevated ionic strengths,[2,25,26] an effect having its provenance in the polyelectrolyte character of DNA.[27] Since DNA is a polyanion, cations such as Na^+ condense on to its surface in a delocalized manner in order to reduce the polymer charge density to a level that is stable in aqueous solution.[27] Condensation results in a local Na^+ concentration of about 1 M in the vicinity of the surface of DNA, a value that is approximately constant and independent of bulk salt concentration.[27] Site binding of a charged intercalator causes the expulsion of a Na^+ ion from the condensation sheath into the solvent; the resultant entropy of mixing contributes to the free energy of ligand binding.[27] The entropy of dilution of the Na^+ ion is large if the bulk Na^+ concentration is low (a transfer from 1 to 0.01 M, for example) but is reduced to zero when the bulk concentration rises to 1 M. Thus measurements at 1 M Na^+ reveal intrinsic affinities without polyelectrolyte contributions and are typically in the region of 10^3 to 10^4 M^{-1} for the binding of aminoacridines and phenanthridines to natural DNA.[2,24-26] Theory predicts,[27] and experiments confirm,[2,25,26] a linear relationship between $\log K(0)$ and $\log [Na^+]$, whose slope is directly proportional to the charge of the intercalator. It has been mooted that anticooperativity resulting from polyelectrolyte effects may lie at the heart of the neighbour exclusion phenomenon.[28]

Lastly in this section, it is worth noting that the binding of intercalators to supercoiled DNA has a feature that is widely exploited in the isolation of closed circular DNAs, and may be important in the selective toxicity of these compounds towards mitochondria and similar organelles containing circular DNA.[2] At low binding densities, where intercalation is accompanied by progressive removal of right-handed supercoils, affinity is enhanced compared to binding to an equivalent linear or nicked circular DNA because the free energy of ligand binding contains a positive contribution resulting from the release of superhelical strain.[2] Conversely, at binding ratios which introduce left-handed supercoils, affinity is diminished compared to relaxed circular DNA, since now energy is required to wind up the (reversed) supercoiling.[2]

10.1.2.3   Rates and Mechanisms of Binding

Intercalation of simple acridines and phenanthridines is a rapid process. Temperature jump relaxation measurements show that proflavine intercalates *via* a mechanism involving initial fast bimolecular attachment to the surface of the helix, followed by one or more slower insertion steps.[2,29] It appears that the intercalation event does not require preopening of the duplex, and at equilibrium a proportion of the ligand remains externally bound. Similar studies with ethidium revealed quite a different binding mechanism, in which the ligand intercalated directly into two distinguishable sites, there being no externally bound intermediates within either pathway.[30] Of special importance is the finding that ethidium bound to a site on one DNA molecule can transfer directly to an empty site of the second variety on another DNA molecule, without passing through a free solution state. This 'direct ligand transfer' step proceeds very rapidly, enabling ethidium to distribute among its binding sites much more quickly than if it had first to dissociate fully. The nature of the intercalated forms remains unclear, although it has been suggested that the complexes represent ethidium bound in each of the two grooves of DNA.[31] Rate constants vary with ionic strength in a manner which suggests that ethidium and propidium first exchange with condensed cations on the surface of DNA as a prelude to intercalation.[26]

10.1.2.4   Sequence Specificity of Binding

The four nucleotides and the polarity of the sugar–phosphate backbone combine to create 10 distinguishable intercalation sites in DNA, differentiated by their arrangement of base pairs.[2] Thus, given the sequence dependent nature of DNA structure and the different capacity of AT and GC pairs to engage in intermolecular stacking interactions there is the potential for a fair degree of sequence specificity in ligand binding. Nevertheless, the parameters of binding to natural DNA for ethidium, proflavine and 9-aminoacridine show little, if any, dependence on base composition.[2] However, footprinting studies, where the binding site of a ligand is revealed by its ability to inhibit the activity of a DNA-cleaving agent made visible on a Maxam–Gilbert sequencing gel,[9] indicate that ethidium and proflavine bind best to mixed sequences of alternating pyrimidines and purines, and are relatively excluded from runs of adenines and thymines.[32] Similarly, the cutting patterns of the intercalating cleavage reagent MPE-Fe[II], which is a phenanthridine structurally related to ethidium, are not sequence neutral and also show a distaste for runs of As and Ts.[33] Studies of ethidium[24] and propidium[34] binding to synthetic DNAs again reveal a strong preference for interaction with alternating purine–pyrimidine sequences compared to homopolymeric duplexes, polydA · polydT being an especially poor substrate. Optical and NMR measurements using di- and oligo-nucleotides confirm these findings and further indicate a relative preference for binding to pyrimidine–3′,5′-purine over purine–3′,5′-pyrimidine sequences for ethidium, propidium and pro-flavine.[2,7] 9-Aminoacridine behaves similarly, but has only a weak preference for pyrimidine–purine over purine–pyrimidine sequences.[7]

In addition to having high affinity for B-DNA, aminoacridines and ethidium bind well to local double helical regions in heat-denatured DNA and to the A type duplex of RNA and DNA/RNA hybrids, but fail to intercalate into left-handed Z-DNA.[2,24,25] However, unlike its binding to natural DNA, ethidium frequently binds cooperatively to synthetic polynucleotides, and neighbouring site exclusion extends to three base pairs for those polymers adopting the A conformation.[24] Thus, given the available data, it seems that the origins of the evident site selectivity among simple intercalators do not lie in the ability to recognize any specific arrangement of base pairs, but in the capacity to discriminate between local polynucleotide structures which are themselves determined by base sequence. For example, the feeble binding of intercalators to polydA · polydT no doubt stems from the peculiarly rigid structure of this sequence recently revealed by X-ray diffraction analysis.[35]

10.1.2.5   Crystallographic Studies

A wealth of detailed information concerning the stereochemistry of intercalation complexes has been obtained from X-ray diffraction studies of drugs bound to ribo- and deoxyribopyrimidine–3',5'-purine dinucleoside monophosphates.[2,7] The central feature of each complex is an intercalated ligand sandwiched between two antiparallel Watson–Crick paired dinucleotides. Ethidium complexes have the side chains of the ligand lying in the minor groove, its amino groups making hydrogen bonds to purine furanose oxygens.[2,7] The amino groups of proflavine are found in the major groove, hydrogen bonding to phosphate oxygens in some complexes, but not in others.[2,7] 9-Aminoacridine may be intercalated symmetrically with its amino group pointing into the minor groove, or it may be rotated through 180° and stacked asymmetrically between the bases of one strand only.[2,7] The dinucleotide complexes have a common backbone geometry belonging to the A-RNA or A'-DNA motifs, the conformational parameters being essentially independent of ligand type, although base pair turn angles are variable.[2,7] The main chain torsion angles and the glycosidic angle at the purine 3' end are somewhat greater than is found in the unperturbed dinucleotide structures, changes which serve to separate the bases on each strand to a distance of 6.8 Å and to maintain the base pairs parallel to one another. The structures display $C(3')$-*endo*-3',5'-$C(2')$-*endo* patterns of sugar puckering, except in the case of proflavine complexes where the pucker at the 3' end appears a 'soft' parameter being either $C(2')$-*endo* or $C(3')$-*endo*.

The dinucleotide structures have, naturally enough, been used as a basis to formulate models of intercalation complexes in polymeric DNA. The major difficulties with this approach are that the crystal structures belong to the A class of helices, and they lack true helical symmetry. Sobell and coworkers described geometrical models in which they relaxed the crystal parameters so that the dinucleotide complex could be inserted into a regular B type helix, with some distortion extending up to two base pairs from the site.[2,7] An important feature of their models is alternating $C(3')$-*endo*-3',5'-$C(2')$-*endo* sugar puckering, which they suggested was the cause of the neighbouring site exclusion effect. Neidle and Abraham took a different tack by restraining the geometry of the intercalation site to that found in the proflavine–CpG complex and, by using molecular modelling techniques, built the unstacked dinucleotide into an A type RNA polymer.[7] They found it necessary to modify severely the geometry of one or two base pairs on either side of the dinucleotide before a regular helical structure could be attached. These structural distortions, which do not necessarily involve mixed sugar puckering, have also been offered as an explanation of neighbouring site exclusion.[7]

10.1.2.6 Theoretical Studies

Computational studies have provided insights into the extent and reasons for preferential binding at pyrimidine–3',5'-purine sequences by simple intercalators, and led to predictions of the structure of intercalation sites in oligo- and poly-nucleotides.[2,7] Molecular mechanics calculations suggested that the binding preference of ethidium and proflavine is a consequence of the ease of unstacking pyrimidine–3',5'-purine sequences compared to others, although molecular orbital calculations imply that specificity results from the particularly favourable energetics of unwinding these steps.[2,7] Extensive molecular mechanics calculations using hexanucleotide models confirmed the preference for ethidium binding to 5'-CpG over 5'-GpC, but indicated an equality of affinity among the corresponding AT isomers.[11] The latter studies also predicted the experimentally observed preference for binding to alternating over homopolymeric AT sequences.[24]

Miller and coworkers used molecular mechanical methods to construct intercalation sites at the centre of DNA tetranucleotides.[7] Their models have the general property that they can be inserted into regular polynucleotide helices, a feat achieved by holding the base pairs parallel and creating the intercalation cavity by modifying backbone conformations. The authors described three classes of stable structures, characterized by distinct unwinding and glycosidic angles at the drug binding site, which they suggest may account for the experimentally observed range of unwinding angles.[2,7] Rao and Kollman[36] used molecular mechanics and dynamics simulations of the binding of 9-aminoacridine to a heptanucleotide in an attempt to elucidate the origins of neighbouring site exclusion. In contrast to others (see below), they found no stereochemical reasons for the phenomenon and speculated that the restriction may be a consequence of polyelectrolyte or vibrational entropy effects. At the polymer level, the proflavine–DNA complex has been modelled in both an A and B type duplex, true symmetry and helical continuity being maintained throughout the structure.[2,7] In the B helix no intermolecular hydrogen bonding between drug and backbone is proposed and ligand-induced unwinding is distributed over several base pairs. The A type structure is remarkably similar to that seen in the proflavine–5'-CpG crystalline complex, at least so far as intermolecular hydrogen bonding, sugar puckering and the absence of helix unwinding go.[2,7]

10.1.2.7 NMR Studies

Difficulties arising from multiple site occupancy and rapid kinetics preclude the use of the powerful 2-D NOESY method for investigating atomic details of simple intercalator–DNA complexes in solution. Nevertheless, perturbations to the chemical shift values of ligand and base pair aromatic protons on binding, as well as to those of nucleotide sugar and imino resonances, have enabled identification of drug binding sites and provided some details of intercalation geometries and binding kinetics. Direct information about main chain torsional angles and sugar puckers has been obtained from ^{31}P NMR spectra and from vicinal coupling constants of sugar ring protons.[2,7,37,38] Ethidium, propidium, proflavine and 9-aminoacridine have been extensively studied using di-, oligo- and synthetic poly-nucleotides as well as natural DNA.[2,7,37,38] These investigations made a special contribution to the determination of sequence selectivity (see above) and have shown that intercalation complexes in solution generally have chromophore orientations as expected from crystallographic and model-building studies. They have also provided direct evidence that sugar puckers may change on drug binding,[37] that at lattice saturation neighbour exclusion is a regular phenomenon,[38] and that helix unwinding accompanying ethidium and propidium binding is not localized at the intercalation site but is spread over adjoining base pairs.[38]

10.1.3 SURVEY OF MONOFUNCTIONAL INTERCALATORS

10.1.3.1 Antitumour Antibiotics

10.1.3.1.1 Actinomycins

Actinomycin D (Figure 1) is the best-studied example of the group of antibiotics comprising two cyclic pentapeptide lactones attached to a phenoxazinone chromophore.[2] It is used in the treatment of Wilms' tumour in children but has its widest application as a specific inhibitor of DNA-directed RNA synthesis.[2] Actinomycin intercalates selectively into B form DNA, its affinity, unwinding and extension parameters being similar to those of ethidium.[2,39] In natural DNA it binds at some, but not all, GpC sequences, with the peptide rings covering three base pairs on either side of the bound chromophore.[39,40] Using information obtained from a crystal structure of actinomycin bound to deoxyguanosine, Sobell and coworkers proposed a detailed molecular model for its DNA complex.[2] The major feature of the model is an actinomycin molecule intercalated in the minor groove at the sequence GpC, with specific hydrogen-bonding interactions between the N-3 atoms and 2-amino groups of the guanines and the amide groups of the threonine residues of the symmetrically disposed peptides. Many spectroscopic, kinetic and NMR studies attest to the essential correctness of the model,[2,7,37,39] with particularly compelling support coming from NMR experiments which show the peptide rings bound in the DNA minor groove appropriately positioned for participating in the guanine to threonine hydrogen-bonding scheme.[41] The kinetics of actinomycin binding to natural DNA are complex,[39] best interpreted as though the ligand binds, without undergoing conformational change, to different GpC dinucleotides with variable affinity depending on the nature of the surrounding sequences.[41,42] There is good evidence that binding occurs initially to relatively abundant weak sites and that the ligand subsequently redistributes by a process of dissociation and reassociation ('shuffling') to sites of higher affinity.[43] Binding kinetics are an important determinant of the biological activity of actinomycins; correlations have been noted between dissociation rates and the capacity to inhibit the chain elongation reaction of RNA polymerase activity.[2,39]

10.1.3.1.2 Anthracyclines

Adriamycin and daunomycin (Figure 1) are the best known members of the anthracycline antibiotics and are the most commonly used intercalating agents in the treatment of cancer.[2,5] Adriamycin has a broad spectrum of activity, being particularly efficacious against solid tumours, whereas daunomycin is the agent of choice for some leukaemias. Both antibiotics inhibit DNA and RNA synthesis and cause protein-associated strand breaks *in vivo*.[2,5] Whilst there is substantial evidence that the biological activity of the anthracyclines results from their capacity to perturb the biochemical functions of DNA, there is also a body of data suggesting other sites of action.[2,5] We shall return to this point later (Section 10.1.5.1), as we shall to a discussion of the dose-limiting cardiomyopathy associated with the use of the anthracycline drugs (Section 10.1.5.3).

Adriamycin and daunomycin bind tightly to B form DNA, but have little or no affinity for A-DNA, left-handed Z-DNA or duplex RNA. They extend and unwind DNA in the manner expected of intercalating agents.[2,44,45] Daunomycin intercalates *via* a mechanism reminiscent of that of proflavine, there being rapid external attachment to the helix followed by an insertion step and a monomolecular rearrangement.[46] Sequence selectivity is complex and the data are probably best summarized by concluding that alternating pyrimidine–purine tracts of mixed sequence provide the highest affinity sites.[44,46] Early model-building studies have the ligand intercalated from the major groove with maximal chromophore–base pair overlap and the amino sugar making contact with phosphate oxygens.[44] Later, molecular mechanics simulations suggested that adriamycin intercalates from the minor groove with the major axis of the chromophore lying at right angles to the major axes of the base pairs.[44] The crystal structure of a 2:1 daunomycin–d(CGTACG)$_2$ complex provides striking confirmation for the latter mode of binding, the ligand intercalating at the CpG steps with its amino sugar lying in the minor groove, and substituents on the anthracycline A ring hydrogen bonding to the bases lining the binding site.[47] The DNA is asymmetrically distorted, the sugar puckers and glycosidic torsion angles being variable along the sequence and generally not equivalent to those typical of the B motif. The base pairs sandwiching daunomycin have the normal B-DNA twist angle, and drug-induced unwinding is confined to the adjacent base pair step. NMR data are consistent with this structure in solution,[7,44] and molecular modelling studies based on the crystallographic coordinates have provided insights into the likely origins of the sequence preferences of the anthracyclines.[48]

Nogalamycin is distinguished from other anthracyclines by possessing, in addition to a nogalose sugar in position 7, a bicyclic amino sugar attached *via* a glycosidic bond at position 1 and a C–C linkage at C-2 (Figure 5).[2,5,11] Thus it has bulky sugars at either end of its chromophore and its charged amino group is in an unusual location. It is endowed with antitumour activity in animal models and is a selective inhibitor of RNA synthesis.[2,5] A derivative lacking the nogalose sugar, menogaril (Figure 2), is currently on clinical trial. The structure of nogalamycin has aroused special interest as regards its mechanism of DNA binding since, for it to intercalate, as has been established

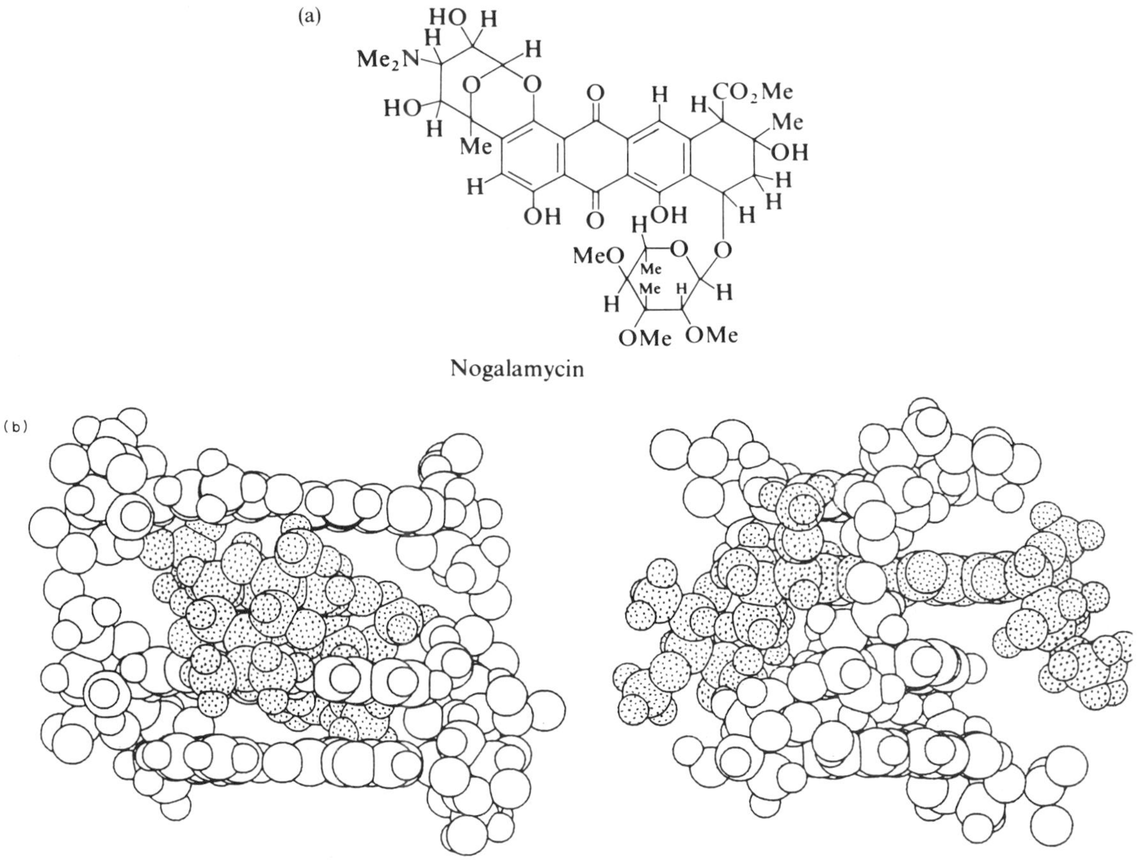

(a)

Nogalamycin

(b)

Figure 5 Illustration of the structure of (a) nogalamycin and (b) the nogalamycin-d(GCA)·d(TGC) complex (nogalamycin shaded)

by the usual criteria,[2,11] each of its sugars must necessarily lie in a different DNA groove, and to thread a bulky sugar through the helix requires that base pairing be transiently disrupted. Kinetic studies support this view,[49] and, as found for actinomycin, indicate a 'shuffling' process whereby nogalamycin binds initially to readily meltable AT rich sequences and later redistributes to higher affinity sites from which it dissociates slowly.[43] These favoured sites occur at TpG (CpA) and GpT (ApC) dinucleotides in regions of alternating purine–pyrimidine sequence, the most common site comprising GpCpA (TpGpC).[50] The orientation of nogalamycin at this preferred site is revealed by NMR studies of its complex with d(GCATGC)$_2$ wherein two symmetry-related antibiotic molecules are bound at the CpA and TpG steps.[51] The major axis of the anthracycline chromophore is aligned at right angles to the major axes of the base pairs (Figure 5) and the nogalose sugar occupies the DNA minor groove. The bicyclic sugar lies in the major groove: its charged dimethylamino group and hydroxyl functions hydrogen bond to the O-6 and one of the N-7 atoms of the guanines, and its bridge atoms are in van der Waals contact with the thymine methyl group.

10.1.3.2 Synthetic Antitumour Agents

10.1.3.2.1 Acridines

Many derivatives of 9-aminoacridine have experimental antitumour activity, and several find use in the treatment of human disease.[3] Here we briefly describe some important examples. Nitracrine (Figure 1) was developed in Poland, where clinical trials showed it to be active against advanced mammary and ovarian carcinomas.[3] In the absence of reducing agents nitracrine intercalates reversibly into DNA but in the presence of thiols it also binds irreversibly, forming interstrand crosslinks.[3] It appears that the biological activity of nitracrine can be directly ascribed to the consequences of crosslinking DNA *in vivo*, the facile reduction of nitracrine being attributed to activation resulting from steric interaction between the 1-nitro group and the 9-amino side chain.[3] Nitracrine is selectively toxic to hypoxic cells in culture, as expected for a nitroaromatic compound activated by reduction, which may account for its efficacy against solid tumours.

Amsacrine (Figure 1) is the best-known example of the many hundreds of anilinoacridines that have been investigated for antitumour activity, and is an established agent in the treatment of acute leukaemias and malignant lymphomas.[3] Other derivatives and analogues have shown promising activity against solid tumours in animal models and one, amsalog (Figure 2),[52] is currently on clinical trial. The anilinoacridines intercalate selectively into DNA compared to RNA,[25] and extensive QSAR studies can be rationalized in molecular terms by assuming that intercalation is a necessary but not sufficient condition for biological activity.[3] Amsacrine inhibits topoisomerase II activity, causing the appearance of protein-associated DNA strand breaks *in vivo*, and it seems likely that the critical event in its mode of action may prove to be the formation of a ternary DNA–drug–topoisomerase complex.[3,8]

In contrast to the wide variation in structure that can be tolerated while retaining biological activity in the 9-anilinoacridines, structural variations compatible with antileukaemic activity in the series of intercalating 9-aminoacridinecarboxamides are limited. Activity is confined to *N*-[2-(dimethylamino)ethyl]-9-aminoacridine-4-carboxamide and its 2 and 5 derivatives.[53] Correlations have been noted between DNA-binding kinetics, ligand structure and biological activity.[54] Active compounds, but not inactive ones, cause protein-associated DNA strand breaks *in vivo*, suggesting that the mechanism of action of these agents also involves inhibition of topoisomerase II activity.[55] Removing the 9-amino group broadens the spectrum of activity to include solid tumours, and the structure–activity relations of the resulting acridinecarboxamides are quite different from those of the 9-aminoacridine derivatives.[56]

10.1.3.2.2 Anthraquinone derivatives

Development of the aminoalkylanthraquinones and related compounds resulted from efforts to produce anthracycline analogues that lack cardiac toxicity. Mitoxantrone (Figure 1) and its dedihydroxy analogue ametantrone were evaluated in the clinic, and the former has found a regular place in the treatment of breast cancer. Numerous studies have identified mitoxantrone as an intercalating agent, all indications being that its biological activity is attributable to its DNA-binding properties.[57] It binds at mixed composition alternating pyrimidine–3′,5′-purine sequences,[58] with its chromophore 'skewering' DNA in the same manner as daunomycin, and both side chains lying in the major groove.[57] A wide spectrum of experimental antitumour activity is reported for both the anthrapyrazoles,[59] oxanthrazole (Figure 2) being on clinical trial, and the pyrazoloacridines,[60] the latter of which in a way combine structural features of the anthrapyrazoles with those of nitracrine.[3]

10.1.3.2.3 *Ellipticine and bisantrene*

Ellipticine is a 6*H*-pyridocarbazole alkaloid with experimental antitumour activity. A derivative, *N*-methyl-9-hydroxyellipticinium (Figure 2), is effective against metastatic breast cancer in the clinic.[2,4] The ellipticines are good intercalating agents, their biological properties generally being ascribed to their capacity to inhibit topoisomerase II activity.[4] The anthracenedicarbaldehyde derivative bisantrene (Figure 2) is active against a range of tumours, but clinical trials have revealed that it can cause serious vascular damage. Despite its unusual topology it binds tightly to DNA by intercalation, inhibits DNA template functions and causes protein-associated DNA strand breaks *in vivo*.[61]

Lastly, it is appropriate to recall that, in addition to having antitumour activity, intercalating agents may also be selectively toxic to single and multicellular parasitic organisms, and that chloroquine, quinacrine and lucanthone are used in the chemotherapy of infections caused by such organisms.[2,62]

10.1.3.3 Other Intercalating Ligands

Here we briefly draw attention to a few other classes of compound which illustrate further the structural diversity among intercalating agents, and make special reference to their sequence selectivity. A large range of substituted phenazines related to phenyl neutral red bind selectively to GC base pairs, the most discriminating among them being as specific as actinomycin.[63] GC selective binding is also a characteristic of platinum-containing intercalators.[64] It has been suggested that both classes bind to DNA from the major groove, selectivity being ascribed to coupled steric and electronic interactions between ligand chromophore and GC base pairs. In contrast AT selectivity has been noted for the antiviral fluorenone derivative tilorone, which has been mooted to intercalate 'sideways' with one side chain lying in each groove.[18] Anthracene and other aromatic hydrocarbon derivatives bearing amide and ester side chains containing cationic groups also intercalate preferentially at AT base pairs.[65] In each of these cases AT specificity has been attributed to water-mediated hydrogen bonding between ligand side chains and thymine O-2 oxygen atoms.[65] Finally, we make mention of the intercalating porphyrin derivatives whose pattern of sequence selectivity is complex, with many sites centred around the dinucleotide steps CpG and TpA.[66]

10.1.4 POLYFUNCTIONAL INTERCALATING AGENTS

10.1.4.1 Concepts of Polyfunctional Intercalation

The early desire to construct ligands which competitively inhibit DNA binding of regulatory proteins and RNA polymerases, and would thereby provide a measure of selective control over gene expression, led in part to the development of polyfunctional intercalating agents. The DNA affinity, sequence selectivity and dissociation rates of simple intercalators are inadequate for the task because they are confined within limits imposed by the size of the ligand and the nature of the forces underlying the intercalating phenomenon. However, in principle, substantial sequence selectivity can be achieved by incorporating two or more intercalators in a polyfunctional ligand since, now the DNA binding site is much enlarged, the selectivity that attends intercalation of a single chromophore is amplified in the polymer, and new opportunities for specificity arise by virtue of potential interactions between the base pairs and the interchromophore linker(s). Moreover, thermodynamic and kinetic considerations imply that polyfunctional ligands should have enhanced affinities and slower dissociation rates than their monomeric counterparts. In parallel with the development of synthetic polyfunctional compounds came the discovery that two classes of DNA-binding antibiotics, the quinoxalines and the luzopeptins, are bifunctional intercalators.[2,6] Polyfunctional ligands have been comprehensively reviewed[6] and here we give an overview of the main classes of these compounds.

10.1.4.2 Bifunctional Antibiotics

10.1.4.2.1 Quinoxalines

Echinomycin (Figure 2) and triostin A are the best-studied examples of the quinoxaline anti-tumour antibiotics; they differ only in the nature of their sulfur-containing crossbridge.[2,6,11,67] Echinomycin is active against experimental tumours and is currently on phase II clinical trial. All available data are consistent with the notion that the quinoxaline antibiotics can adopt a conformation shaped like a staple, their cyclic depsipeptides forming a more or less flat rectangular disc, with the chromophores disposed on one side separated by a distance of 10–11 Å. The sulfur-containing crossbridge projects away from the molecule on the other face of the disc. They intercalate into DNA from the minor groove, with the quinoxaline rings lying parallel to one another, spanning two base pairs in the complex.[2,6,11,67] The quinoxaline antibiotics are highly sequence selective, the actual sequence preference of individual examples being determined both by the nature of the peptide ring and the chromophore substitution pattern.[2,6,11,67] Echinomycin and triostin binding sites in natural DNA centre largely on the sequence CpG, the immediate flanking sequences being quite variable.[6,11,67] However, not all CpG dinucleotides provide acceptable sites and there is evidence of modifications to the DNA conformation in the vicinity of the bound ligand.[6,11,67] Kinetic studies show that the chromophores of echinomycin intercalate in a concerted manner and that the multiple binding sites in natural DNA are characterized by widely different affinities.[6,11,67] As found for actinomycin and nogalamycin, binding proceeds by a 'shuffling' mechanism in which echinomycin molecules initially bind nonselectively, but subsequently rearrange to locate the highest affinity sites.[6,11,67]

The origins of the sequence selectivity of echinomycin and triostin A are revealed in the crystal structures of their complexes with oligonucleotides.[6,11,67] In the cases studied, the quinoxaline chromophores intercalate from the minor groove, spanning the dinucleotide CpG. The alanine residues of the peptide ring play a commanding role, their CO and NH groups forming hydrogen bonds to the 2-amino groups and N-3 atoms respectively of the sandwiched guanines, and many stabilizing van der Waals contacts can be perceived elsewhere in the complexes. The DNAs display an unusual structural feature in that the bases flanking the sandwiched dinucleotide are paired according to the Hoogsteen scheme. In consequence, the helix diameter is reduced by 2 Å at these points and the number of intermolecular van der Waals contacts is much increased. Drug-induced unwinding in the crystal compares favourably with that determined in solution, and is distributed between all four base pairs forming the binding site. Sugar puckers are variable, few being found in the C(2')-*endo* conformation characteristic of classical B-DNA. Whether ligand-induced Hoogsteen pairing occurs in solution is a moot point. The demonstration that echinomycin promotes reaction with diethyl pyrocarbonate at adenines adjacent to its binding sites has been cited as evidence in its favour,[68] but the evidence is not unequivocal.[89,90] Furthermore, molecular mechanics calculations did not find a triostin A–hexanucleotide complex to be more stable when Hoogsteen pairing was incorporated uinto the model.[69]

10.1.4.2.2 Luzopeptin

Luzopeptin is the principal component of the complex of antibiotics isolated from the organism *Actinomadura luzonensis*. These antibiotics resemble the quinoxaline family, being cyclic depsipeptides comprising 10 amino acid residues bearing two 3-hydroxy-6-methoxyquinaldinic acid chromophores.[6] They contain the hitherto biologically unknown amino acid *trans*-(3S,4S)-4-acetoxy-2,3,4,5-tetrahydropyridazine-3-carboxylic acid. X-Ray studies show that luzopeptin adopts a right-handed twisted rectangle shape with a twofold symmetry axis, very reminiscent of the shape of the quinoxalines.[6] The depsipeptide is stabilized by a pair of transannular hydrogen bonds involving sarcosine and glycine residues. The tetrahydropyridazine rings and their acetyl groups stretch across the peptide ring, lying parallel to its minor axis, and occlude one face of the molecule. The quinoline chromophores lie on the obverse and may be rotated to give the staple-like structure suitable for bisintercalation, the interchromophore separation being about 14 Å. Luzopeptin is a bifunctional intercalator with high affinity for DNA, heat-denatured DNA, rRNA and tRNA and has the capacity to form reversible DNA–DNA crosslinks.[6] The solution data, such as there are, together with the crystal structure seem to imply that the chromophores of luzopeptin may span three base pairs in the bisintercalated complex. Footprinting experiments reveal that this antibiotic binds best to AT-rich regions of DNA, though no consensus di- or tri-nucleotide sequence can be discerned, and its pattern of sequence selectivity is quite different from that of echinomycin.[91]

Luzopeptin has a wide spectrum of antitumour activity and is toxic towards Gram-positive bacteria.[6] Interestingly, the antitumour and bactericidal activity, as well as acute toxicity, of the luzopeptins correlate positively with the number of tetrahydropyridazine acetyl groups on the molecule.[6]

10.1.4.2.3 *Bisanthracyclines*

Reported dimers of the anthracycline antibiotics are semisynthetic and comprise two molecules of either daunomycin, 4-demethoxydaunomycin, or adriamycin, linked through their C-13 ketones with dicarboxylic acid hydrazide bridges.[6] It is a common finding with these compounds that the hydrazone linkage is unstable at 37 °C. Measurements with rodlike and circular DNA confirm the bifunctional mode of binding of various series of polyamine- and carboxamidoalkyl-linked bis-daunomycins, kinetic studies showing that some examples dissociate from DNA many thousands of times more slowly than the parent molecule.[6,70] The dissociation rates depend on chain length in some cases but not in others: cytotoxicity was not found to correlate with kinetic properties.[6,70] Daunomycin dimers bridged with alkyl chains have similar or improved antitumour activity compared to daunomycin, and preferentially inhibit RNA synthesis over DNA synthesis.[6] However, there is no obvious relationship between biological activity and linker chain length. Many similar compounds have been synthesized in the 4-demethoxydaunomycin series, one example being more efficacious than adriamycin against a variety of tumour types and substantially less cardiotoxic.[6]

10.1.4.3 Diacridines

Much information has been gathered concerning the geometrical requirements for bifunctional intercalation of dimers composed of two 9-aminoacridines linked *via* their amino groups.[2,3,6] Given the widespread occurrence of neighbour exclusion binding it might be expected that bisintercalation would be confined to those compounds in which the interchromophore separation permits the ligand to span two, or more, base pairs (Figure 6). However, the evidence suggests that bifunctional reaction may be found even among compounds which can form only a single base pair sandwich, in violation of neighbouring site exclusion (Figure 6).[2,3,6] In consequence, special attention has been given to the effects of linker chain composition and acridine substitution pattern on the mode of binding of diacridines whose maximum interchromophore separation lies in the range 7.5–11.3 Å (*i.e.* five to eight methylene groups).[18] Classical measurements show unsubstituted 9-aminoacridine dimers to be bifunctional when bridged by flexible alkane or carboxamidoalkanes containing six linking atoms.[2,3,6] In the absence of severe DNA bending or kinking, these compounds can span only a single base pair given their fully extended interchromophore separation of about 8.7 Å. However, it is clear that the rigidity and conformation of linkers of this size have an important influence on performance, since whereas a bis-9-anilinoacridine whose chromophores are constrained to lie within about 7 Å of each other is bifunctional, a six-atom glycine hyrazide-linked diacridine is not.[6] Chromophore substituents in the 4 position do not hinder bisintercalation of the simple hexane-linked diacridine but those in the 2, 3 and 6 positions do, the latter preventing bifunctional reaction until the linker chain is long enough to span two base pairs.[2,3,6] The importance of chromophore shape and size as a determinant of bifunctional binding is also made evident by the failure of flexibly linked bisquinolines to bisintercalate.[71] Thus, it appears that the capacity to form single base pair bisintercalation complexes depends sensitively on the linker conformation and the size of the major axis of the acridine chromophore, whereas the structural requirements for forming a two base pair sandwich complex are somewhat less demanding.

Many dimers of 9-aminoacridine and 2-methoxy-6-chloro-9-aminoacridine have been studied in which the linker comprises simple alkanes, polyamines, piperazine, amides, ethers, pyrazole and oligopeptides.[3,6,18] It is a general finding that bisintercalation of these compounds is associated with enhanced affinity for DNA, although the degree of enhancement is usually much less than expected theoretically. This appears to be due to unfavourable entropy effects. The substantial binding constants found for polyamine-linked diacridines are largely a consequence of their polycationic character. Where studied, the binding of diacridines is associated with little or no sequence selectivity, although there is often a slight preference for interaction with poly(dA–dT) over poly(dG–dC). NMR measurements provide direct evidence for the formation of two base pair bisintercalation complexes of several ligands,[6] one study showing that the linker lies in the DNA major groove.[72] These studies also indicated that the six-atom alkyl- and carboxamide-linked

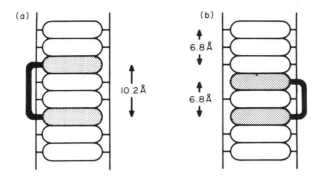

Figure 6 Schematic illustration of bifunctional intercalation by diacridines (a) subject to the constraint of neighbouring site exclusion or (b) in violation of that principle. The intercalated chromophores (shaded) and the base pairs are shown as discs 3.4 Å thick, the DNA being represented as a ladder. The chain linking the chromophores is shown as a heavy line, its length determining whether the separation between chromophores is sufficient to allow sandwiching of two base pairs (a) or one (b) (reproduced from ref. 3 by permission of Macmillan)

9-aminoacridine dimers were monofunctional,[6] a finding probably related to peculiarities in the structure of the $d(AT)_5$ DNA substrate used. Bifunctional diacridines dissociate very much more slowly from DNA than their corresponding monomeric compounds *via* complex pathways.[6] Comparison of dissociation rates and chromophore residence times determined by stopped-flow and NMR methods shows that flexibly linked dimers are able to 'creep' along the DNA helix, whereas this tendency is minimized in rigidly linked diacridines.[6] The latter bind more in the manner of a staple, a binding mechanism that promotes cytotoxicity and antitumour activity.[6] It appears that although bifunctional diacridines are usually cytotoxic, their antitumour efficacy is generally rather poor.

10.1.4.4 Other Bisintercalators

Binuclear phenanthridines related to ethidium have attracted attention, the chromophores being tethered *via* different ring positions.[6] Dimers bridged at the quaternary nitrogen with alkyl, diamine and rigid diphenyl ether linkers all show bifunctional behaviour, the data indicating the formation of two base pair sandwich complexes.[6] The diamine-linked dimer also interacts with tRNA, and has been reported to intercalate bifunctionally into left-handed Z-DNA. It binds 100-fold more tightly than ethidium to native DNA in $1 M Na^+$, without sequence selectivity.[6] The DNA-binding properties of a *p*-carboxyphenylmethidium dimer linked *via* the carboxyl groups with spermine are similar,[6] although this ligand is sensitive to both DNA sequence and polynucleotide conformation. Bifunctional reaction has also been found for carboxyphenylmethidium dimers bridged with alkyl and ether linkers, the mode of binding being in part determined by the position of attachment to the phenyl ring.[6] These latter compounds selectively inhibit RNA synthesis compared to DNA synthesis and some are active antitumour agents.[6]

The antitumour activity of ellipticine inspired the synthesis of several series of 7*H*-pyrido-carbazole dimers in which the chromophores are linked *via* the pyridine or carbazole nitrogens with alkanes, polyamines, polyalcohols, *N*-alkylpiperazines and bis(*N*-alkylpiperazines).[6,72] Bisintercalation is confined to polyamine and rigidly linked compounds at pH values where their backbone linkages are charged.[6,72] Those dimers whose linkers are both rigid and attached to position 2 have good antitumour activity, as distinct from their monomeric analogues and isomeric biscarbazoles which have little or no activity.[6,72] A bisethylbispiperidine derivative has entered clinical trial under the name of ditercalinium (Figure 2). NMR studies reveal that it bisintercalates with its linker lying in the major groove, the chromophores sandwiching two base pairs.[72] Ditercalinium shows a delayed toxicity towards cells in culture, with some indication of a novel mechanism of action, since it neither inhibits DNA or RNA synthesis at toxic concentrations, nor does it arrest cells in the G2 + M phase of the cell cycle or poison topoisomerase II activity in the manner typical of intercalating agents.[6,72] There is the suggestion that it may interfere with mitochondrial activity.[72]

Attempting to produce GC selective ligands by exploiting intercalators of proven specificity, McFadyen *et al.*[73] synthesized a series of dimers comprising two platinum terpyridine chromophores linked *via* α,ω-bisthioalkanes. Bifunctional binding was noted at an interchromophore separation sufficient to span only a single base pair. Although the dimers are GC selective, they are much less so than expected, the specificity of the component chromophores being severely compromised, presumably for geometrical or electronic reasons.[73]

10.1.4.5 Triacridines

With the twin objectives of enhancing affinity yet further and improving sequence selectivity by enlarging the binding site size, several groups have investigated trimeric ligands comprising three flexibly linked 9-aminoacridines.[6,72,92] Trimers bridged by carboxamidoalkyl chains do not show evidence of trisintercalation until the distance between the central and terminal chromophores reaches about 17 Å, a separation that provides the opportunity for straddling up to three base pairs in each arm of the complex. Notwithstanding the potential larger binding site, and the possible involvement of the linker amides in hydrogen-bonding interactions with the DNA, the triacridines are nonsequence selective with affinities only slightly greater than those of comparable diacridines.[6,72] Where studied, the capacity to inhibit RNA polymerase *in vitro* and the cytotoxicity of these compounds do not correlate with structure.[6] Some homologues have marginal antitumour activity.[6] Increasing the ligand charge by replacing amide linkages with amines enhances affinity and promotes trisintercalation at smaller interchromophore distances, some examples being trifunctional when juxtaposed chromophores can seemingly span only a single base pair.[6,72] The latter finding suggests double violation of neighbour exclusion binding in these instances. A trisintercalating polyamine-bridged trimer of 2-methoxy-6-chloro-9-aminoacridine (interchromophore distance 17 Å, charge $+7$) has a DNA-binding constant of about 10^{14} M^{-1} at physiological salt concentrations, a value equivalent to that of DNA regulatory proteins.[72] Tri- and tetra-acridines linked by oligopeptides and polyamines have also been described.[6,74]

10.1.5 CELLULAR CONSEQUENCES OF INTERCALATION

10.1.5.1 Biochemical Pharmacology

Intercalating agents are cytotoxic to proliferating, but not quiescent, mammalian cells and cause a range of biochemical effects. For example, they generally prevent progression past the G2 + M phase of the cell cycle, induce chromosome abnormalities, concentrate in the nucleus or nucleolus, condense chromatin, and inhibit DNA and RNA synthesis.[2-6] RNA synthesis is generally the more sensitive of the template functions, and some agents have been reported to inhibit preferentially transcription from nucleolar ribosomal RNA genes.[2,5,6] Studies with purified bacterial enzymes *in vitro* show that intercalating drugs variously inhibit different stages of transcription; the diacridines and phenanthridines, for example, are more effective antagonists of initiation, whereas actinomycin, echinomycin and nogalamycin selectively block chain elongation.[2,5,6,39,75] Some agents that also bind to RNA interfere with tRNA function and the processing of primary RNA transcripts.[6,76] Much evidence has accumulated to suggest that the critical cytotoxic event is related to the capacity of intercalating ligands to inhibit the functions of topoisomerase II.[3-6,8] This enzyme is involved in modifying the topological properties of DNA, and its action appears important in both DNA replication and transcription.[8] Attention has focused on the ability of drugs to trap the enzyme–DNA complex in a state known as the 'cleavable complex', the effect being revealed as protein-associated DNA strand breaks.[3-6,8] These strand breaks are rapidly repaired on removal of drug, and it is not clear why they should be lethal to the cell even when they may be only transient.[8] Topoisomerase II activity diminishes when cells enter the stationary phase, as does the ability of drugs to cause protein-associated DNA breaks, a correlation that could provide a molecular explanation for the selective toxicity of intercalators towards proliferating cells.[8]

Although the anthracycline antibiotics inhibit DNA template functions and topoisomerase II activity,[2,5,8] additional mechanisms have been sought to account for their cytotoxicity.[77] The anthracyclines interact with cellular membranes, which aroused interest in what role, if any, this might play in their mode of action.[77] Intriguingly, adriamycin retains its activity against cells in culture when bound covalently to agarose beads,[77] indicating that the drug can exert a toxic effect even when held extracellularly on the plasma membrane. That these drugs may also act by yet other means not involving DNA binding can be seen with *N*-acetyl derivatives of adriamycin, *e.g.* AD 32, which are unable to bind DNA but are nevertheless cytotoxic, inhibit RNA synthesis and cause protein-associated DNA strand breaks.[5,77] It seems that the toxicity of AD 32 may result from direct inhibition of the topoisomerase II enzyme, in a manner reminiscent of the epipodophyllotoxins.[5,7,8]

Aminoacridines and ethidium display extraordinary selectivity for organelles containing extra-chromosomal circular DNA in a variety of cell types. For example, they cause loss of plasmids from bacteria, kinetoplasts from trypanosomes, mitochondria from yeast and mammalian cells, and chloroplasts from plant tissues, at concentrations which otherwise show no interference with cell metabolism.[2] The origins of these phenomena are unknown, although there is evidence that the DNA and RNA polymerases of mitochondria are more susceptible to inhibition by intercalating agents than are the nuclear enzymes.[2] It has been suggested that the selectivity may be partly a result of preferential binding to circular DNA (see Section 10.1.2.2).[2] It is appropriate to recall here that aminoacridines and phenanthridines may also be potent frameshift mutagens, though the mechanism of mutagenicity is still a matter of debate.[2,78]

10.1.5.2 Effects on Chromatin

Studies on the binding of intercalating drugs to chromatin have essentially addressed two questions: how does the wrapping of DNA into nucleosomes and higher order structures affect its capacity to bind ligands? And what is the effect of ligand binding on the structure and dynamics of chromatin? With respect to the first question it seems, by and large, that intercalators prefer to bind to DNA at the ends of nucleosomes or in the linker region.[79-81] Their binding to nucleosome core particles may be diminished or, in some cases, completely abolished compared to the same DNA free in solution.[80-82] More explicit information is available concerning drug-induced modifications to chromatin structure. For example, ethidium causes the release of histone H1 from the 30 nm chromatin fibre,[83] unfolding of nucleosome core particles[84] and dissociation of the latter with selective loss of histone H2A and H2B proteins.[85] Disruption of nucleosome core particles by high concentrations of intercalating ligand is in fact a common finding. Echinomycin, but not actinomycin or nogalamycin, has the curious property, along with DNA minor groove binders, of causing a rotation of DNA on the surface of isolated nucleosomes.[83,84,86] Thus the data to hand imply that *in vivo* intercalating drugs would bind preferentially to the internucleosome linker and at high doses may cause widespread disruption to chromatin structure.

10.1.5.3 Unwanted Toxicities

The clinical effectiveness of adriamycin and daunomycin is restricted by a peculiar dose-limiting cardiomyopathy which leads to irreversible heart failure.[5] Biochemical evidence suggests that this phenomenon may have its origins in the quinone redox chemistry of the anthracycline ring. Cardiac sarcosomes and mitochondria reduce adriamycin and daunomycin to their semiquinones and in the presence of oxygen generate reactive free radicals which cause peroxidative damage to cardiac membrane lipids.[57,77] More universal toxicities of anticancer drugs include bone marrow suppression, nausea and vomiting, diarrhoea, alopecia, hypotension, hearing loss, kidney and neurological damage. The hypotensive and neurological effects of intercalating drugs may arise from direct interaction with neurotransmitter receptors,[87] or by nonspecific membrane-binding phenomena.[77] This list of side effects is a serious concern since one or more of the entries frequently becomes dose-limiting in the clinic and as such represents a critical life-threatening toxicity.

10.1.6 CONCLUSIONS AND PERSPECTIVES

It is evident from the foregoing that the greatest interest in intercalating agents lies in their potential as anticancer drugs. In recent times medicinal chemists have developed numerous new chromophore types endowed with impressive experimental antitumour activity. Indeed, so successful have they been that with some of the new agents it is possible to cure mice completely of many of the tumours imposed upon them. Similarly impressive developments in the fields of NMR spectroscopy, crystallography, computational chemistry, rapid kinetics and DNA chemistry are bringing us to the point where we can expect to obtain full and precise descriptions of the structure, thermodynamics and kinetic properties of DNA intercalation complexes. The examples of actinomycin, echinomycin, nogalamycin, adriamycin/daunomycin, mitoxantrone and ditercalinium perhaps illustrate best the progress made to date in these areas with clinically important compounds. However, this intimate knowledge of drug–DNA complexes at the molecular level has not, in general, translated into an understanding of pharmacological efficacy, and what distinguishes an active from an inactive member in a series of related antitumour agents remains largely a mystery. Clearly much needs to be done to establish a better link between DNA-binding characteristics, ligand structure and antitumour activity. Given the number of compounds that need to be examined, perhaps the most useful binding data will emerge from footprinting, stopped-flow, NMR and computational studies, since these are likely to provide the quickest answers to the questions: where, for how long, in which groove and to what does it bind? The answer to the question 'how tightly?' does not seem to have been very illuminating to date.

In a parallel vein the nature of the biochemical responses to intercalation that lead to tumour selective toxicity are also largely unknown. The relationships between DNA-binding properties, ligand structure and inhibition of DNA synthesis, transcription and topoisomerase activity need to be better defined. Doubtless there are also other processes modified by intercalating agents about which we have yet to learn. Fortunately, the accelerating advances in molecular and cellular biology are beginning to provide the necessary tools for these tasks as well as a biochemical framework in which to work. There is the additional difficult, but soluble, problem of discovering how DNA-binding properties measured *in vitro* are modified *in vivo*, where the physical setting of DNA is very different from that studied in test tubes. Notwithstanding these formidable problems, there are encouraging signs of progress in all areas of the biochemical pharmacology of intercalating drugs.

With regard to the immediate further development of antitumour drugs it seems that interest is growing in agents having mixed functionality, so as to combine intercalating properties with alkylating functions on the one hand and with DNA-cleaving groups on the other. In the arena of sequence specific polyfunctional molecules, efforts are beginning to concentrate on compounds incorporating groove selective linkers and intercalating chromophores known to have defined sequence preferences, as well as intercalators linked to oligonucleotides.[88] The advent of agents specific for a turn of the helix, say, would open the door to their use in the treatment of viral and parasitic infections and may also lead to more selectively toxic anticancer drugs.

10.1.7 REFERENCES

1. A. Albert, 'Selective Toxicity', Wiley, New York, 1979.
2. E. F. Gale, E. Cundliffe, P. E. Reynolds, M. H. Richmond and M. J. Waring, 'The Molecular Basis of Antibiotic Action', 2nd edn., Wiley, London, 1981, p. 258.
3. W. A. Denny, B. C. Baguley, B. F. Cain and M. J. Waring, in 'Molecular Aspects of Anti-Cancer Drug Action', ed. S. Neidle and M. J. Waring, Macmillan, London, 1983, p. 1.
4. C. Esnault, B. Lambert, J. Markovits, E. Segal-Bendirdjian, G. Muzard, C. Garbay-Jaureguiberry, B. P. Roques and J. B. Le Pecq, in 'Molecular Mechanisms of Carcinogenic and Antitumour Activity', ed. C. Chagas and B. Pullman, Adenine Press, Schenectady, 1987, p. 295.
5. J. R. Brown, in 'Molecular Aspects of Anti-Cancer Drug Action', ed. S. Neidle and M. J. Waring, Macmillan, London, 1983, p. 57.
6. M. J. Waring and L. P. G. Wakelin, *Nature (London)*, 1974, **252**, 653; L. P. G. Wakelin, *Med. Res. Rev.*, 1986, **6**, 275.
7. S. Neidle and Z. Abraham, *CRC Crit. Rev. Biochem.*, 1984, **17**, 73.
8. B. S. Glisson and W. E. Ross, *Pharmacol. Ther.*, 1987, **32**, 89.
9. R. J. Wilkins, *Anal. Biochem.*, 1985, **147**, 267.
10. W. Saenger, 'Principles of Nucleic Acid Structure', Springer-Verlag, New York, 1984.
11. S. Neidle, L. H. Pearl and J. V. Skelly, *Biochem. J.*, 1987, **243**, 1.
12. C. Zimmer, G. Luck, G. Burckhardt, K. Krowicki and J. W. Lown, in 'Molecular Mechanisms of Carcinogenic and Antitumour Activity', ed. C. Chagas and B. Pullman, Adenine Press, Schenectady, 1987, p. 339.
13. K. R. Fox and N. R. Howarth, *Nucleic Acids Res.*, 1985, **13**, 8695.
14. M. A. Keniry, S. C. Brown, E. Berman and R. H. Shafer, *Biochemistry*, 1987, **26**, 1058.
15. L. P. G. Wakelin, A. Adams, C. Hunter and M. J. Waring, *Biochemistry*, 1981, **20**, 5779.
16. G. Cohen and H. Eisenberg, *Biopolymers*, 1969, **8**, 45.
17. K. E. Reinert, *Biochim. Biophys. Acta*, 1973, **329**, 135.
18. R. G. McR. Wright, L. P. G. Wakelin, A. Fieldes, R. M. Acheson and M. J. Waring, *Biochemistry*, 1980, **19**, 5825.
19. M. Hogan, N. Dattagupta and D. M. Crothers, *Biochemistry*, 1979, **18**, 280.
20. C. Mandal, S. W. Englander and W. R. Kallenbach, *Biochemistry*, 1980, **19**, 5819.
21. B. C. Baguley, W. A. Denny, G. J. Atwell and B. F. Cain, *J. Med. Chem.*, 1981, **24**, 170.
22. J. D. McGhee and P. H. von Hippel, *J. Mol. Biol.*, 1974, **86**, 469.
23. D. M. Crothers, *Biopolymers*, 1968, **6**, 575.
24. J. L. Bresloff and D. M. Crothers, *Biochemistry*, 1981, **20**, 3547.
25. W. R. Wilson, B. C. Baguley, L. P. G. Wakelin and M. J. Waring, *Mol. Pharmacol.*, 1981, **20**, 404.
26. W. D. Wilson, C. R. Krishnamoorthy, Y.-H. Wang and J. C. Smith, *Biopolymers*, 1985, **24**, 1941.
27. G. S. Manning, *Q. Rev. Biophys.*, 1978, **11**, 179.
28. R. A. G. Friedman and G. S. Manning, *Biopolymers*, 1984, **23**, 2671.
29. H. J. Li and D. M. Crothers, *J. Mol. Biol.*, 1969, **39**, 461.
30. J. L. Bresloff and D. M. Crothers, *J. Mol. Biol.*, 1975, **95**, 103.
31. L. P. G. Wakelin and M. J. Waring, *J. Mol. Biol.*, 1980, **144**, 185.

32. K. R. Fox and M. J. Waring, *Nucleic Acids Res.*, 1987, **15**, 491.
33. M. W. Van Dyke and P. B. Dervan, *Science (Washington, D.C.)*, 1984, **225**, 1122.
34. W. D. Wilson, Y.-H. Wang, C. R. Krishnamoorthy and J. C. Smith, *Chem.-Bio. Interact.*, 1986, **58**, 41.
35. H. C. M. Nelson, J. T. Finch, B. F. Luisi and A. Klug, *Nature (London)*, 1987, **330**, 221.
36. S. N. Rao and P. A. Kollman, *Proc. Natl. Acad. Sci. USA*, 1987, **84**, 5753.
37. T. R. Krugh and M. E. Nuss, in 'Biological Applications of Magnetic Resonance', ed. R. G. Shulman, Academic Press, New York, 1979, p. 113.
38. S. Chandrasekaran, R. L. Jones and W. D. Wilson, *Biopolymers*, 1985, **24**, 1963.
39. W. Muller and D. M. Crothers, *J. Mol. Biol.*, 1968, **35**, 251.
40. M. W. Van Dyke, R. P. Hertzberg and P. B. Dervan, *Proc. Natl. Acad. Sci. USA*, 1982, **79**, 5470.
41. S. C. Brown, K. Mullis, C. Levenson and R. Shafer, *Biochemistry*, 1984, **23**, 403.
42. T. R. Krugh, J. W. Hook, Jr., M. S. Balakrishnan and F. M. Chen, in 'Nucleic Acid Geometry and Dynamics', ed. R. H. Sarma, Pergamon Press, Oxford, 1980, p. 351.
43. K. R. Fox and M. J. Waring, *Nucleic Acids Res.*, 1986, **14**, 2001.
44. S. Neidle and M. R. Sanderson, in 'Molecular Aspects of Anti-Cancer Drug Action', ed. S. Neidle and M. J. Waring, Macmillan, London, 1983, p. 35.
45. J. B. Chaires, *Mol. Pharmacol.*, 1986, **29**, 74.
46. J. B. Chaires, N. Dattagupta and D. M. Crothers, *Biochemistry*, 1985, **24**, 260.
47. A. H.-J. Wang, G. Ughetto, G. J. Quigley and A. Rich, *Biochemistry*, 1987, **26**, 1152.
48. B. Pullman, in 'Molecular Mechanisms of Carcinogenic and Antitumour Activity', ed. C. Chagas and B. Pullman, Adenine Press, Schenectady, 1987, p. 3.
49. K. R. Fox, C. Brassett and M. J. Waring, *Biochim. Biophys. Acta*, 1985, **840**, 383.
50. K. R. Fox and M. J. Waring, *Biochemistry*, 1986, **25**, 4349.
51. M. S. Searle, J. G. Hall, W. A. Denny and L. P. G. Wakelin, *Biochemistry*, 1988, **27**, 4340.
52. B. C. Baguley, W. A. Denny, G. J. Atwell, G. J. Finlay, G. W. Rewcastle, S. J. Twigden and W. R. Wilson, *Cancer Res.*, 1984, **44**, 3245.
53. G. W. Rewcastle, G. J. Atwell, D. Chambers, B. C. Baguley and W. A. Denny, *J. Med. Chem.*, 1986, **29**, 472.
54. L. P. G. Wakelin, G. J. Atwell, G. W. Rewcastle and W. A. Denny, *J. Med. Chem.*, 1987, **30**, 855.
55. W. A. Denny, I. A. G. Roos and L. P. G. Wakelin, *Anticancer Drug Design*, 1986, **1**, 141.
56. G. J. Atwell, G. W. Rewcastle, B. C. Baguley and W. A. Denny, *J. Med. Chem.*, 1987, **30**, 664.
57. J. W. Lown, K. Reszka, P. Kolodziejczyk and W. D. Wilson, in 'Molecular Mechanisms of Carcinogenic and Antitumour Activity', ed. C. Chagas and B. Pullman, Adenine Press, Schenectady, 1987, p. 243.
58. K. R. Fox, M. J. Waring, J. R. Brown and S. Neidle, *FEBS Lett.*, 1986, **202**, 289.
59. W. R. Leopold, J. M. Nelson, J. Plowman and R. C. Jackson, *Cancer Res.*, 1985, **45**, 5532.
60. J. C. Sebolt, S. V. Scavone, C. D. Pinter, K. L. Hamelehle, D. D. Von Hoff and R. C. Jackson, *Cancer Res.*, 1987, **47**, 4299.
61. W. A. Denny and L. P. G. Wakelin, *Anticancer Drug Design*, 1987, **2**, 71.
62. W. C. Bowman and M. J. Rand, 'Textbook of Pharmacology', 2nd edn., Blackwell, Oxford, 1980.
63. W. Muller, H. Bunemann and N. Dattagupta, *Eur. J. Biochem.*, 1975, **54**, 279.
64. L. P. G. Wakelin, W. D. McFadyen, A. Walpole and I. A. G. Roos, *Biochem. J.*, 1984, **222**, 203.
65. W. D. Wilson, Y.-H. Wang, S. Kusuma, S. Chandrasekaran and D. W. Boykin, *Biophys. Chem.*, 1986, **24**, 101.
66. K. Ford, K. R. Fox, S. Neidle and M. J. Waring, *Nucleic Acids Res.*, 1987, **15**, 2221.
67. M. J. Waring, in 'Molecular Mechanisms of Carcinogenic and Antitumour Activity', ed. C. Chagas and B. Pullman, Adenine Press, Schenectady, 1987, p. 317.
68. D. Mendel and P. B. Dervan, *Proc. Natl. Acad. Sci. USA*, 1987, **84**, 910.
69. U. C. Singh, N. Pattabiraman, R. Langridge and P. A. Kollman, *Proc. Natl. Acad. Sci. USA*, 1986, **83**, 6402.
70. R. T. C. Brownlee, P. Cacioli, C. J. Chandler, D. R. Phillips, P. A. Scourides and J. A. Reiss, *J. Chem. Soc., Chem. Commun.*, 1986, 659.
71. W. D. McFadyen, W. A. Denny and L. P. G. Wakelin, *FEBS Lett.*, 1988, **228**, 235.
72. P. Laugaa, M. Delepierre, P. Leon, C. Garbay-Jaureguiberry, J. Markovits, J. B. Le Pecq and B. P. Roques, in 'Molecular Mechanisms of Carcinogenic and Antitumour Activity', ed. C. Chagas and B. Pullman, Adenine Press, Schenectady, 1987, p. 275.
73. W. D. McFadyen, L. P. G. Wakelin, I. A. G. Roos and B. L. Hillcoat, *Biochem. J.*, 1987, **242**, 177.
74. P. O.-L. Mack, D. P. Kelly, R. F. Martin and L. P. G. Wakelin, *Aust. J. Chem.*, 1987, **40**, 97.
75. D. C. Straney and D. M. Crothers, *Biochemistry*, 1987, **26**, 1987.
76. J. Kapuscinski and Z. Darzynkiewicz, *Proc. Natl. Acad. Sci. USA*, 1986, **83**, 6302.
77. H. S. Schwartz, in 'Molecular Aspects of Anti-Cancer Drug Action', ed. S. Neidle and M. J. Waring, Macmillan, London, 1983, p. 93.
78. L. R. Ferguson, W. A. Denny and D. G. McPhee, *Mutat. Res.*, 1985, **157**, 29.
79. I. L. Cartwright, R. P. Hertzberg, P. B. Dervan and S. C. R. Elgin, *Proc. Natl. Acad. Sci. USA*, 1983, **80**, 3213.
80. J. B. Chaires, N. Dattagupta and D. M. Crothers, *Biochemistry*, 1983, **22**, 284.
81. J. Portugal and M. J. Waring, *Nucleic Acids Res.*, 1987, **15**, 885.
82. J. Portugal and M. J. Waring, *Nucleic Acids Res.*, 1986, **14**, 8735.
83. H. Fenske, I. Eichorn, M. Bottger and R. Lindigkeit, *Nucleic Acids Res.*, 1975, **2**, 1975.
84. H.-M. Wu, N. Dattagupta, M. Hogan and D. M. Crothers, *Biochemistry*, 1980, **19**, 626.
85. C. T. McMurray and K. E. van Holde, *Proc. Natl. Acad. Sci. USA*, 1986, **83**, 8472.
86. C. M. L. Low, H. R. Drew and M. J. Waring, *Nucleic Acids Res.*, 1986, **14**, 6785.
87. A. Adams, B. Jarrott, B. C. Elmes, W. A. Denny and L. P. G. Wakelin, *Mol. Pharmacol.*, 1985, **27**, 480.
88. C. Chagas and B. Pullman (eds.), 'Molecular Mechanisms of Carcinogenic and Antitumour Activity', Adenine Press, Schenectady, 1987.
89. J. Portugal, K. R. Fox, M. J. McLean, J. L. Richenberg and M. J. Waring, *Nucleic Acids Res.*, 1988, **16**, 3655.
90. C. Jeppesen and P. E. Nielsen, *FEBS Lett.*, 1988, **231**, 172.
91. K. R. Fox, H. Davies, G. R. Adams, J. Portugal and M. J. Waring, *Nucleic Acids Res.*, 1988, **16**, 2489.
92. J. B. Hansen, T. Koch, O. Buchardt, P. E. Nielsen, B. Norden and M. Wirth, *J. Chem. Soc., Chem. Commun.*, 1984, 509.

10.2
DNA Binding and Nicking Agents

DAVID I. EDWARDS

Polytechnic of North East London, UK

10.2.1	INTRODUCTION	725
10.2.1.1	*Antibacterial and Antiprotozoal Agents*	726
10.2.1.2	*Anticancer and Antitumor Agents*	727
10.2.1.3	*Medical Applications of DNA Strand-breaking Drugs*	728
10.2.1.3.1	*Antimicrobial agents*	728
10.2.1.3.2	*Anticancer and antitumor agents*	729
10.2.2	METHODS OF MEASUREMENT OF DNA STRAND BREAKAGE	731
10.2.2.1	*Viscometry*	731
10.2.2.2	*Spectrophotometric Thermal Denaturation and Renaturation*	732
10.2.2.3	*Gradient Sedimentation*	732
10.2.2.4	*Agarose Gel Electrophoresis*	733
10.2.2.5	*Hydroxyapatite Chromatography*	733
10.2.2.6	*Filter Elution*	733
10.2.3	FREE RADICAL REACTIONS IN DNA STRAND BREAKAGE	734
10.2.3.1	*Formation of Oxygen Radicals*	734
10.2.3.2	*Radical-induced DNA Damage*	735
10.2.3.3	*Redox Cycling*	736
10.2.3.4	*Action of Nitro Homocyclic and Heterocyclic Drugs*	737
10.2.4	OTHER DRUGS WHICH BIND AND NICK DNA	742
10.2.4.1	*m-AMSA*	742
10.2.4.2	*Platinum-based Drugs*	742
10.2.4.3	*Other Metal Complexes*	743
10.2.5	REPAIR OF DNA STRAND BREAKS	744
10.2.5.1	*Biological Effects of Strand Breaks*	744
10.2.5.2	*Base Damage and Apurinic Sites*	745
10.2.5.3	*Mechanisms of Repair of DNA Strand Breaks*	745
10.2.5.4	*Inhibition of DNA Repair*	746
10.2.6	REFERENCES	748

10.2.1 INTRODUCTION

There are at least 35 anticancer drugs currently available to treat various types of neoplastic disease and of these about 28 act directly or indirectly on DNA. Drugs which act at the level of DNA are also used as antibacterial agents, antiprotozoals and antivirals. Whereas it may be commonly thought that DNA is not a target that can be exploited in the search and design of selectively toxic drugs, nevertheless, much of our knowledge of how such drugs exert their effect has come from the study of compounds which, initially, have shown promise *in vitro* and which require further study as to their mechanism of action in order to develop more effective drugs. Indeed, current anticancer drugs are not particularly selectively toxic and so a study of the molecular basis of their action may provide clues to identify exploitable differences between normal and malignant cells or between human cells and a microbial pathogen. In the following discussion, consideration of antiviral drugs will be omitted as these are dealt with in Chapter 7.3. Consideration of the mutagenic and possible carcinogenic effects of the drugs to be discussed will also be omitted, although attention should be drawn to the cost effectiveness of short-term tests for carcinogenicity which has recently been reviewed and is relevant to all who are concerned in the synthesis, development, research and production of chemotherapeutic agents.[1]

Drugs which bind to and nick DNA, *i.e.* cause strand breaks, may be classified into two large groups which reflect their selective toxicity or spectrum of activity, *viz.* those which act predominantly against bacterial or protozoal infections and those which act as antitumor or anticancer agents.

10.2.1.1 Antibacterial and Antiprotozoal Agents

The majority of these comprise nitro heterocyclic or homocyclic compounds and an ever-increasing range of quinolone drugs which will be discussed in Chapter 10.3.

(1)   (2)

The only homocyclic compounds of note are chloramphenicol (1) and thiamphenicol (2). Although these drugs are usually regarded as being selective inhibitors of protein synthesis in prokaryotic cells, the rare but significant occurrence of aplastic anemia during or after prolonged or high-dose chloramphenicol therapy may be attributed to damage to DNA characterized by strand breakage. Chloramphenicol causes a dose-related reversible bone marrow suppression by a mechanism involving inhibition of mitochondrial protein synthesis.[2] A more serious side effect, however, is aplastic anemia which is not dose related and is irreversible. Thiamphenicol (in which the *p*-nitro group of chloramphenicol is replaced by a methylsulfonyl group) also produces a dose-related hematopoietic depression, but no cases of aplastic anemia have been reported even though over 50 million patients have received the drug since 1972. The inference therefore is that the nitro group is responsible for aplastic anemia. In this respect it has been shown that chloramphenicol induces DNA damage in bacteria,[3] which occurs only as a result of nitro group reduction and that this leads to extensive strand breaks in DNA. The possible mechanism and the nature of the damaging species will be further discussed in Section 10.2.3.4.

Nitro heterocycles which act as antibacterial or antiprotozoal drugs are numerous and comprise nitrofurans, nitrothiazoles and, most importantly, the nitroimidazoles. Nitrofurans have been used clinically for over 45 years for a range of Gram-negative infections, whereas nitrothiazoles have very limited clinical applicability other than the drug entramin, 2-amino-5-nitrothiazole (enheptin).

The nitroimidazole group of drugs acting against bacteria and protozoa are almost invariably based on the 5-nitroimidazole nucleus, substituted in the N-1 position of the nitroimidazole ring. The exception, in this respect, is satranidazole (3), which is a 5-nitroimidazole but which behaves as a 2-nitroimidazole. All of these drugs appear to act against bacteria or protozoa which inhabit an anaerobic or microaerophilic environment and are therefore selectively toxic to organisms which inhabit a low-redox environment and, in general, have no activity against aerobes.[4] These drugs are amongst the most important antimicrobial agents ever produced and the most widely used, metronidazole, accounts in the UK for about 30% of the cost of the hospital antibiotics drug bill, and a significant proportion of the cost of drugs prescribed by the general practitioner and, albeit more rarely, by dentists. These drugs include metronidazole (4), tinidazole (5), ornidazole (6), nimorazole (7), carnidazole (8) and secnidazole (9). In veterinary medicine, dimetridazole (10), ipronidazole (11) and ronidazole (12) are frequently used.

(3)   (4)   (5)

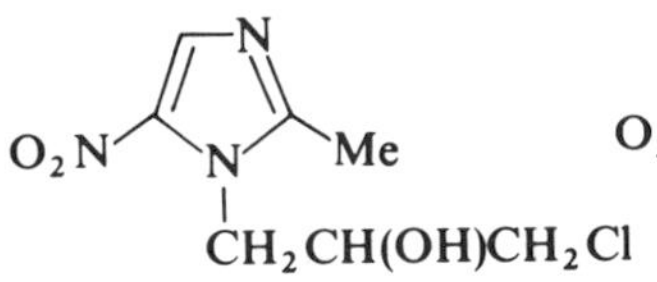

Structures (6)–(12):

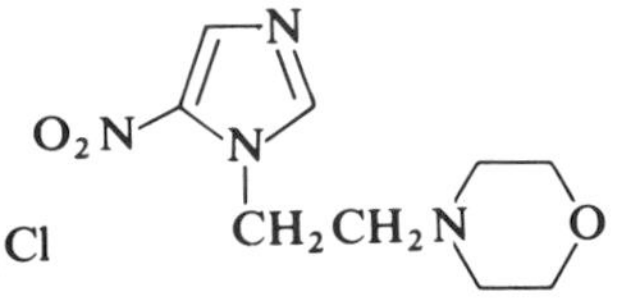

O_2N — imidazole — Me
$CH_2CH(OH)CH_2Cl$
(6)

O_2N — imidazole
CH_2CH_2N — morpholine (O)
(7)

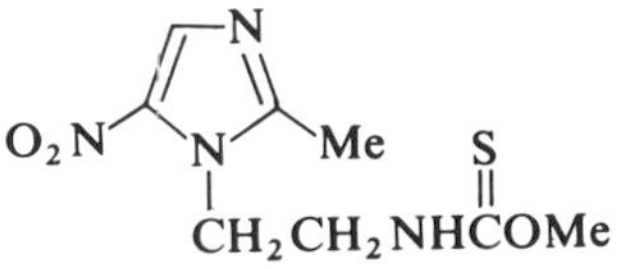

O_2N — imidazole — Me   S ‖ C
$CH_2CH_2NHCOMe$
(8)

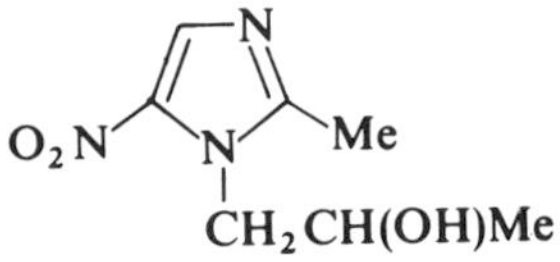

O_2N — imidazole — Me
$CH_2CH(OH)Me$
(9)

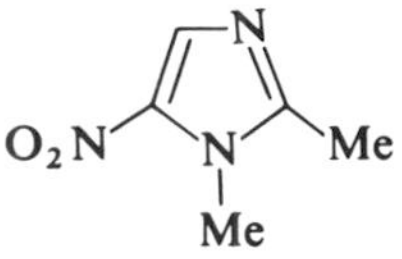

O_2N — imidazole — Me
Me
(10)

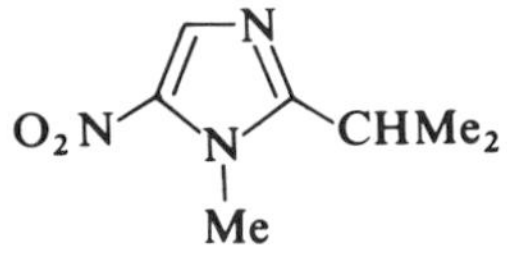

O_2N — imidazole — $CHMe_2$
Me
(11)

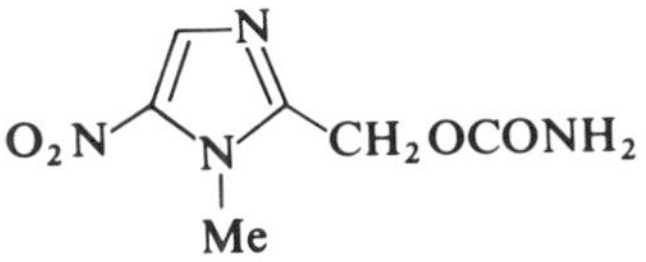

O_2N — imidazole — CH_2OCONH_2
Me
(12)

10.2.1.2 Anticancer and Antitumor Agents

Whereas those drugs which cause DNA strand breaks in bacteria and protozoa fall into a well-defined class, those which produce similar effects on DNA but which act primarily as antineoplastic agents exhibit a wide range of chemical diversity. Again, the nitroimidazoles are represented, but as the 2-nitro derivatives, rather than the 5-nitro compounds. In contrast to these, well-established drugs include aromatic quinones, alkylating aziridines, those based on the rare metals platinum, rhodium or ruthenium, aromatic *N*-oxides and many others. At this point it may be worth stating that the state of anticancer chemotherapy at present most probably equates to the state of antimicrobial chemotherapy at the turn of this century where mercury, antimonial and arsenic compounds were used. These, of course, were better than nothing, but hardly very effective and certainly accompanied by very serious side effects.

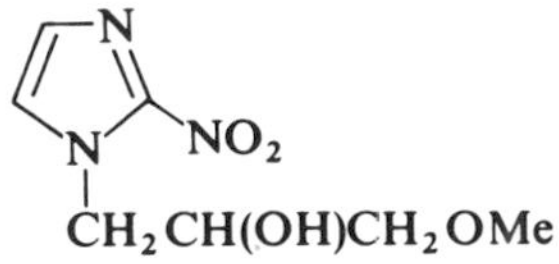

Structures (13)–(17):

imidazole — NO_2
$CH_2CH(OH)CH_2OMe$
(13)

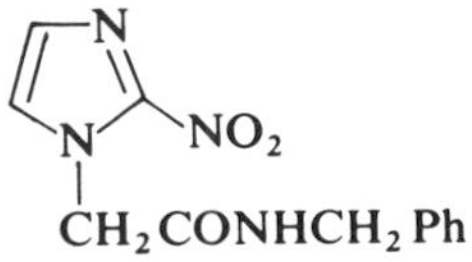

imidazole — NO_2
$CH_2CONHCH_2Ph$
(14)

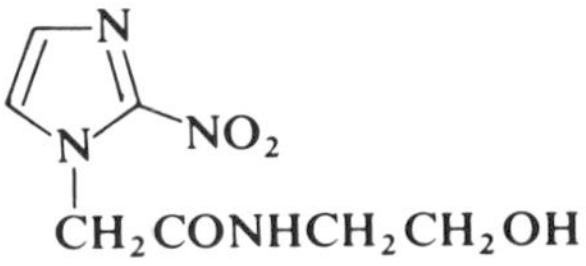

imidazole — NO_2
$CH_2CONHCH_2CH_2OH$
(15)

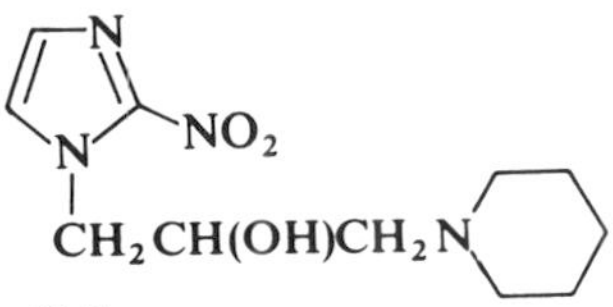

imidazole — NO_2
$CH_2CH(OH)CH_2N$ — piperidine
(16)

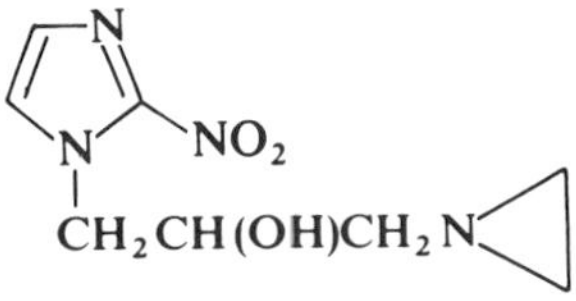

imidazole — NO_2
$CH_2CH(OH)CH_2N$ — aziridine
(17)

The 2-nitroimidazoles of current interest include misonidazole (**13**) and its demethyl derivative benznidazole (**14**), SR-2508 (**15**) and Ro-03-8799 (**16**). A new group of drugs in which the 2-nitroimidazole nucleus is combined with an aziridine-containing side-chain, *e.g.* RSU-1069 (**17**) and derivatives thereof, may prove to be more effective drugs for the treatment of hypoxic tumors but research in this area is in a very early stage. The aromatic quinones, including the anthracycline adriamycin (**18**) (daunomycin, doxorubicin), the anthracenediones mitoxantrone (**19**) and ametantrone (**20**), and recent derivatives based on the anthrapyrazole structure, the benzoquinones mitomycin C (**21**), porfiromycin and naphthoquinones, all exhibit antineoplastic activity and, together with bleomycin, are used in the treatment of various cancers. Very recently, a group of benzotriazine *N*-oxides has been shown to posses antitumor activity *in vitro* and this effect appears to be a consequence of DNA strand-break induction upon reduction of the *N*-oxide group.

The study of anticancer platinum drugs is also of continuing interest and much effort has been put into synthesizing derivatives of cisplatin, *cis*-diamminedichloroplatinum(II), which shows useful anticancer activity. Using this as a lead compound a range of compounds has been tested which, whilst maintaining the general structure, has sought to incorporate other metals, including rhodium and ruthenium. These compounds show useful properties and may be expected to reach the clinical stage in the future. In many cases, drugs with functional groups known to be essential for anticancer or antitumor activity have been combined in a single molecule. Thus, nitroimidazoles containing an aziridine grouping in the N-1 side chain, and platinum, ruthenium and rhodium drugs containing a 2-nitroimidazole ligand have been tested for activity and valuable information obtained as to the essential features of such molecules to maintain anticancer or antitumor activity whilst minimizing toxic side effects.

10.2.1.3   Medical Applications of DNA Strand-breaking Drugs

10.2.1.3.1   *Antimicrobial agents*

Nitrobenzenoid drugs, including chloramphenicol and the methylsulfonyl congener thiamphenicol, are wide-spectrum antibiotics and have long been used for those life-threatening diseases such as typhoid, bacterial endocarditis and acute bronchial disease. Although both are still regarded as the drug of choice for typhoid, chloramphenicol-resistant infections may be treated with vancomycin; however, because of the rare blood dyscrasias associated with chloramphenicol therapy, thiamphenicol should be the drug of choice.

Nitro heterocyclic drugs are widespread and are used for the treatment of a wide range of bacterial and protozoal disease. The 5-nitroimidazole metronidazole is the drug of choice for the treatment of vaginitis caused by the protozoan *Trichomonas vaginalis*, or the bacterium *Gardnerella vaginalis* (*Haemophilus vaginalis*). Many other drugs of the same type are also effective against these diseases, for example tinidazole (**5**), secnidazole (**9**) and ornidazole (**6**), all of which have a longer half-life than metronidazole (**4**). These drugs are also useful against other anaerobic microbial diseases including amoebic infections of the gut and Vincent's disease, an acute ulcerative gingivitis, and infections caused by *Bacteroides*; hence their use in post-operative sepsis. Other drugs of this type, for example nimorazole (**7**), carnidazole (**8**) and secnidazole (**9**), although clinically available, show no improved activity over the longer-established drugs. There is, however, one newcomer to this list which is of interest: the drug satranidazole (3) differs from the other 5-nitroimidazole drugs in that C-2 of the imidazole ring is linked *via* a nitrogen to a substituted imidazolidinone moiety. The drug appears to have superior activity to metronidazole (the standard drug for any comparisons), both for amoebiasis and trichomoniasis *in vitro* and *in vivo* in animal tests, and clinical data are awaited with interest because the drug, on theoretical grounds, should display less resistance to the more conventional nitroimidazoles.

In the veterinary field, dimetridazole (10) is used extensively against histomoniasis in turkeys and other poultry and, together with ipronidazole (11) and ronidazole (12), is also used in the treatment of swine dysentery and bovine trichomoniasis. Benznidazole (14) is the most valuable drug for *Trypanosoma cruzi* infections (Chagas' disease), but also shows antitumor activity for the treatment of hypoxic tumors, in common with a number of other 2-nitroimidazoles (*vide infra*). The only nitrothiazole of note is 2-amino-5-nitrothiazole (entramin, enheptin), which displays antiamoebic activity.

The nitrofuran group of drugs is very large and displays, equally, a very large range of clinical and nonclinical use. Since the 1940s they have been used for the treatment of urinary tract disease, nitrofurazone and nitrofurantoin being used primarily, if not exclusively, against Gram-negative infections typified by *Escherichia coli*, *Pseudomonas* and *Proteus*. Resistant organisms usually succumb to these drugs under anaerobic conditions. Nitrofurans are, unfortunately, ubiquitously used as antimicrobial agents as stabilizers to wine, in the preservation of food, as animal feedstuffs additives and even to prevent microbial degradation of clothing materials. Such use is a legitimate cause for concern because many nitrofurans are known to be mutagenic and putatively carcinogenic.

10.2.1.3.2 Anticancer and antitumor agents

Of the antitumor and anticancer drugs the anthracycline antibiotics, *e.g.* adriamycin (= daunorubicin and doxorubicin) are considered to have the widest range of antitumor activity of any anticancer drug in clinical use.[5] Adriamycin (18) and daunomycin are in widespread and routine use for treatment of small-cell carcinoma of the lung, breast cancer and acute lymphocytic and myelocytic leukemia. The major drawback which limits their clinical applicability is primarily cardiotoxicity, which can exhibit as an insidious cardiomyopathy (often delayed), through to irreversible congestive heart failure. While second-generation anthracyclines, such as aclacinomycin A, have significantly less cardiotoxicity, they also possess a decreased range of antitumor activity. The modern derivatives of this group, the anthracenediones, exemplified by mitoxantrone (19) and ametantrone (20), show a lowered incidence of cardiotoxicity accompanied by a narrowed spectrum of activity such that they are of use only in the treatment of the acute leukemias and carcinoma of the breast. Further modifications of this type of approach have resulted in the recent synthesis of anthrapyrazoles, which are chromophore-modified anthracenediones. These drugs show a broad range of anticancer and antitumor activity and may be of future interest.[6]

In 1971 the first clinical trial of cisplatin (22) showed it had promising activity against several human tumors, but kidney toxicity was a major dose limitation. These disadvantages were partially solved by using pre- and post-hydration schedules in conjunction with slow intravenous infusion over 24 h, an approach which made it a useful drug for ovarian and testicular cancers. Cisplatin is now also used for bladder tumors and those in the head and neck; a recent technique involves the instillation of the drug directly into the brain tumor, with promising results.[7] Nevertheless, to decrease the nephrotoxicity, derivatives of cisplatin were synthesized and amongst them carboplatin (JM-8) (23) was evaluated. This drug has dose-limiting bone-marrow toxicity, but overall is equal in activity to cisplatin in treating ovarian cancer, is minimally nephrotoxic, ototoxic and neurotoxic and causes less gastrointestinal disturbance. It is currently licensed in the UK for the treatment of ovarian cancer and small-cell carcinoma of the lung. Other derivatives, including iproplatin (JM-9) (24) and spiroplatin (TNO-6) (25), show no improved activity over the others and are not clinically available.[8] No rhodium, ruthenium or gold complexes are currently available clinically although such compounds are actively being researched.

Mitomycin C (21) and its methyl congener porfiromycin have been used for the treatment of a variety of cancers, with no outstanding success. However, as a result of its mechanism of action involving reduction of the quinone group to activate the aziridine side chain, it is currently being re-evaluated for possible use in the treatment of solid hypoxic tumors, particularly those in inoperable sites. In this respect porfiromycin shows better activity than mitomycin C under conditions of hypoxia.

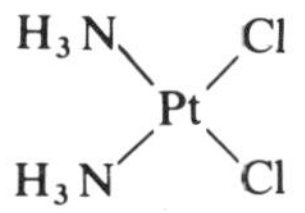

(22)

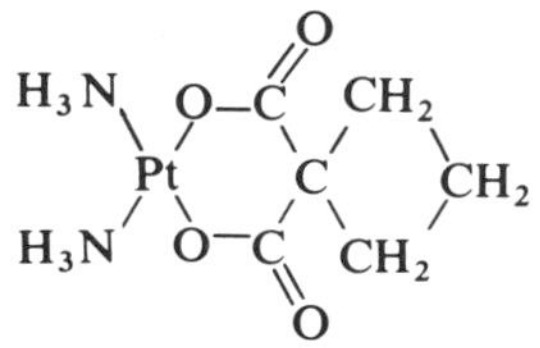

(23)

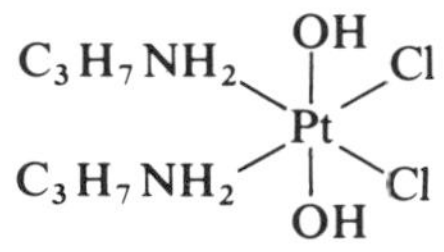

(24)

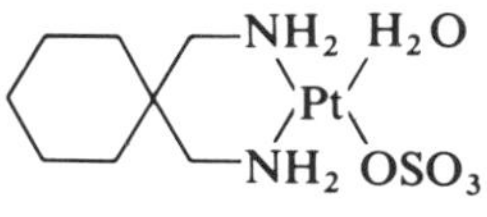

(25)

At this point it is relevant to review the present state of cancer therapy. In general, surgery may be effective in 20–27% of cases, radiotherapy in 12–20% of cases (with or without surgery), chemotherapy (alone or in conjunction with radiotherapy or surgery) about 4%, and chemotherapy alone, for the treatment of metastases, has only a 2% success rate. These figures are quoted for cures, defined as no treatment required and no recurrence of the original disease five years after treatment, which yields an average 45% five-year survival rate for all cancers. As radiotherapy is the most effective treatment regimen in nonoperable situations, and one which has the most flexibility, it is significant that this form of treatment displays about a 30% failure to control the primary tumor. This is due, it is thought, to the hypoxic cell component of most human tumors which is resistant to the effects of radiation (*i.e.* radioresistance).

Hypoxia develops especially in rapidly growing tumors, where the rate of growth of tumor cells outstrips the growth of new capillaries delivering oxygen to the tumor. The 2-nitroimidazoles, their aziridine derivatives and the recently developed benzotriazine di-*N*-oxides (26) have been designed in order to overcome the hypoxic cell problem by specifically sensitizing hypoxic cells to the effects of radiation while leaving normal, oxic cells relatively unscathed. The challenge is one which models itself upon those drugs such as the 5-nitroimidazoles, *e.g.* metronidazole, which displays a specific cytotoxic effect to anaerobic pathogens alone: it should therefore be possible to produce cytotoxic drugs which are specific for hypoxic cells and which would be effective either as an adjunct to radiotherapy or as cytotoxic agents *per se*. This problem is by no means an easy one since it can be shown that in order to demonstrate a 20% improvement in (five-year) cure rates within 95% confidence limits for any particular, defined, cancer, one would need a population of *ca.* 500 patients matched for age, sex, *etc.* Little wonder that trials of any new anticancer drug are carried out as multicentre trials often involving international cooperation.

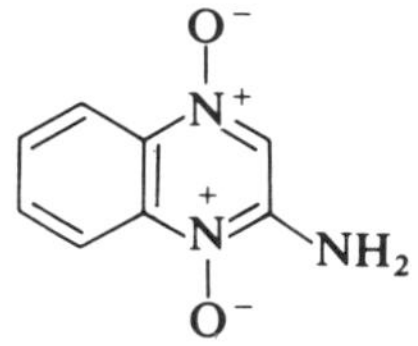

(26)

In the early 1970s it was discovered that several nitro compounds were able to sensitize hypoxic cells to radiation. The ability of metronidazole to sensitize hypoxic cells selectively both *in vitro* and *in vivo*, coupled with its known clinical safety, rapidly led to radiotherapy trials. Although these generally were disappointing, it was soon realized that better sensitization could be achieved by using the more electron-affinic 2-nitroimidazoles of which misonidazole, its demethyl derivative, SR-2508 and Ro-03-8799 were the main contenders. Although misonidazole has shown encouraging results in head and neck cancers, it also displays unacceptable neurotoxicity at the multiple high doses needed to obtain sufficient tumor penetration. The demethyl derivative also shows some peripheral neuropathy at the maximum tolerated dose of misonidazole, but no central effects, and despite its decreased lipophilicity is unlikely to show a significant advantage over misonidazole. SR-2508 shows better tumor penetration and can be administered at higher doses without neurological side effects. Ro-03-8799 is, again, an analogue of misonidazole but shows better tumor penetration and a higher radiosensitization efficiency. Benznidazole, originally used for and still the drug of choice for Chagas' disease, is also at present undergoing trials for treatment of metastatic malignant melanoma in combination with CCNU (lomustine), and for the treatment of recurrent cerebral gliomas, but no results are yet available. Clinical trials of all these drugs are in progress in the UK and the USA. A more recent compound, RSU-1069, is also a 2-nitroimidazole, but carries a side chain which contains an aziridine ring with potential alkylating function. Unlike the other 2-nitroimidazoles, this drug shows a 10-fold increase in sensitizing efficiency compared with misonidazole and compounds of this type are being actively studied. Several reviews of these drugs have appeared.[9-11]

Recently a new class of non-nitro radiosensitizing cytotoxic drug has been developed. These are the benzotriazine di-*N*-oxides, which show an enormously selective cytotoxicity towards hypoxic cells.[12]

The bleomycins (**27**) and the almost identical phleomycins are antibiotics isolated from *Streptomyces verticillus* in the form of copper chelates. Bleomycin is an effective antitumor agent but its action is often accompanied by pulmonary toxicity. Second-generation bleomycin analogues include tallisomycin S10b and peplomycin, a semisynthetic bleomycin analogue with lower pulmonary toxicity. Bleomycin is used, often in combination with other cytotoxic drugs, for the treatment of disseminated testicular cancers, Hodgkins disease, non-Hodgkins lymphoma, squamous cell carcinomas and a variety of teratomas in combination with vinblastine and cisplatin. The activity of all of these drugs can be enhanced by the addition of local anesthetics, including lidocaine, or with several calmodulin antagonists including pimozide which produce true pharmacological synergy *in vitro*.[13] All of the drugs mentioned so far have one thing in common even though there is no structural commonality: they all induce strand breaks in DNA.

(**27**)

10.2.2 METHODS OF MEASUREMENT OF DNA STRAND BREAKAGE

There is a range of methods that can be used to detect strand breaks in DNA. These include general methods that are not specific for breaks but may also indicate other types of DNA damage, *e.g.* the use of viscometric methods or the application of spectrophotometric thermal melting and renaturation of DNA. Those methods which show specificity for the assay of the number of strand breaks include gradient sedimentation, hydroxyapatite chromatography and filter elution.

10.2.2.1 Viscometry

In very general terms, if a macromolecule is added to a solvent the solution will have a higher viscosity than the solvent. The change in viscosity is expressed as the ratio of the viscosity of the solution to that of the solvent, *i.e.* h/h_0, which is known as the relative viscosity, h_{rel}. Frequently the specific viscosity, h_{sp}, is used which is $h_{rel} - 1$ and is a concentration-dependent term. If h_{sp} is measured at several concentrations (c), and h_{sp}/c plotted *versus* c, an extrapolation to $c = 0$ gives the intrinsic viscosity $[h]$, which is related to the molecular weight of the macromolecule. DNA shows considerable viscosity at low concentrations and thus the intrinsic viscosity is easily measured. For double-stranded DNA molecules, $[h]$ is related to the molecular weight (M) as shown by equation (1).

$$0.665 \log M = 2.863 + \log([h] + 5) \tag{1}$$

A comparison of the molecular weight of DNA before (M_0) and after treatment (M_t) with a strand-breaking drug enables the number of strand breaks (n) to be calculated from the relationship $n = M_0/M_t - 1$. All viscosity measurements need to be carried out at closely controlled temperatures, usually $\pm 0.001\,°C$, as viscosity varies by about 2% per degree at $20\,°C$. The viscosity of solutions of DNA at low concentrations may be easily measured using glass suspended-level capillary viscometers of the Ubbelohde type which are designed to be independent of the volume of solution used. For more viscous solutions the Couette or Zimm–Crothers viscometer may be used. These are particularly good for proteins of the viscosities found in blood or serum. In practice, for capillary viscometers one needs only to measure the flow times of the solutions, a decrease in which is indicative of strand breakage. However, decreased viscosity of DNA solutions after drug treatment is not necessarily indicative of strand breakage alone as such results may be brought about by strand separation. Increased viscosity as a result of drug interaction may result from intercalation or helix 'bending' effects. Other techniques should be used to demonstrate unambiguously that strand breaks are the sole or predominant effect.

10.2.2.2 Spectrophotometric Thermal Denaturation and Renaturation

If DNA is heated in a spectrophotometric cuvette, nothing much happens until sufficient thermal energy is reached to begin to disrupt the hydrogen bonding between the bases of the DNA, causing the helix to unwind. This transition from a helix to a random coil is accompanied by an increase of the absorbance of DNA at 260 nm (a hyperchromic effect). When absorbance is plotted *versus* temperature a characteristic sigmoid curve results, the mid-point of which is known as the T_m value (temperature of melting). The T_m value depends upon the ionic strength of the medium, the %GC content of the DNA, and presence or absence of drugs which may alter its value. Many strand-breaking drugs decrease the T_m value and frequently increase the melting range, *i.e.* the temperature over which, say, 90% of the helix–coil transition occurs. Like viscometry the technique may be used to determine a variety of effects on DNA. Intercalators, for example, tend to increase the T_m value. If the DNA is cooled very slowly from a high temperature at which the DNA exists solely in a random coil state (single stranded), it will anneal or renature and this process is accompanied by a decrease in absorbance (a hypochromic effect). Fast cooling prevents renaturation and the absorbance remains high. However, if drugs cause cross linking of the DNA helix the absorbance may decrease. The technique, therefore, is a useful one to determine if drugs acting on DNA do so by strand breakage, intercalation or by helix cross linking.

A more sensitive assessment of strand breakage may be made if the DNA is heated to the T_m value and then rapidly cooled. If the DNA is intact, renaturation will occur; if strand breaks are present in the sample, only partial renaturation will occur. The Rowley equation which describes this behaviour enables the degree of strand breakage to be expressed as a decrease in the amount of the original intact helix present.[14]

10.2.2.3 Gradient Sedimentation

In this technique, DNA which has been radioactively labelled is placed on the top of a sucrose gradient and centrifuged for a few hours until the DNA reaches a point in the gradient at which its density equals that of the surrounding sucrose. Strand breaks will result in a decrease in the molecular weight of the DNA and sedimentation at a less dense region of the gradient. Normally, 5–20% linear sucrose gradients are constructed either by using an automatic gradient maker or by hand layering the gradient. If single-strand breaks are to be measured the DNA must also be in the single-strand condition and this is achieved by preparing the gradient at pH 12.5. For the measurements of double-strand breaks, gradients at or near pH 7 are used. Samples of 100 μl are usually hand layered on to a 5 ml gradient and although the centrifugation speed will vary depending upon the nature of the gradient, the rotor and the DNA, speeds which develop in the region of $250\,000g$ will give reasonable separation after *ca.* 2.5 h. Gradients are subsequently removed from the tubes as fractions of 0.25 ml by upward displacement from a very dense, inert liquid (*e.g.* Maxidens) and the fractions counted for their radioactive content in a liquid scintillation counter. If the gradient is truly linear, and this can be checked by a refractometer, the banding of the DNA can be related to the distance travelled, *i.e.* sedimented, from the point of origin. The simple relationship $n = (D_0/D_t) - 1$, where n is the number of strand breaks and D is the relative distance travelled by the untreated (D_0) and treated (D_t) DNA, gives an approximation to the number of strand breaks produced.[15] Computer programs are available to convert positions of the sample of DNA in the gradient to molecular weights, sedimentation values and strand breaks.[16]

10.2.2.4 Agarose Gel Electrophoresis

This technique can only be used to measure double strand breaks in DNA. In general terms, the concentration of agarose gel used will depend upon the molecular weight of the DNA, as the higher the molecular weight the lower the gel concentration needed for separation. If gels are prepared in sodium dodecyl sulfate, charge effects are nullified and separation occurs solely on the basis of molecular weight differences. A singular advantage of the technique is that very small amounts of DNA are needed and 0.001–0.01 µg of sample is usually sufficient. In order to visualize the separated bands an intercalating dye is either added to the gel during its preparation, or added after separation is complete. Both ethidium bromide and acridine orange are commonly used. The gel is then examined under ultraviolet light, which causes the dye to fluoresce, and the gel may then be photographed using special film.

The application of agarose gel electrophoresis to measure strand breaks in DNA needs care. Strand-broken DNA has a lowered molecular weight and thus migrates further than high molecular weight material. However, strand breaks also cause local helix unwinding and, consequently, less of the visualizing dye intercalates. Thus, low molecular weight material may be difficult to detect. If this is not a problem the number of strand breaks may be calculated from the migration distances, which are related to the molecular weights of the DNA and which are also influenced by the time of electrophoresis, the gel concentration and the nature and ionic strength of the buffer. The relationship is described by the equation $D = a - b\log M$, where D is the migration distance, M the molecular weight and a and b are constants influencing the separation as described above. Nevertheless, if molecular weight markers are included in the gel, a plot of the migration distance *versus* the log of the DNA molecular weight may be used to calculate the molecular weights of unknown DNA samples. Again, as detailed above, the number of strand breaks (n) produced is calculated from the difference in molecular weight before (D_0) and after treatment (D_t), and is given by the expression $n = (D_0/D_t) - 1$.[17]

10.2.2.5 Hydroxyapatite Chromatography

Hydroxyapatite is a form of calcium phosphate which, under the right conditions, enables the very efficient separation of single- from double-stranded DNA. Both forms of DNA will bind to a hydroxyapatite column at low phosphate concentrations, but if the phosphate concentration is increased slightly, single-stranded DNA elutes leaving the double-stranded material still bound to the column. The double-stranded DNA may be eluted at still higher phosphate concentrations. Separation is made more efficient if the process is carried out at about 60 °C and eluted with an increasing phosphate gradient. Under these conditions, RNA and proteins fail to bind so the technique can be used on cell extracts without the need for purification. Strand breaks are therefore detected as an increase in the single-strand component eluted and may be calculated from the Rowley equation.[18]

10.2.2.6 Filter Elution

This technique, developed by Kohn and co-workers,[19] is based on the principle that DNA will elute from a filter at a rate which is related to its molecular weight. Thus, low molecular weight DNA elutes faster than high molecular weight material. Again the technique has the advantage that cell extracts can be used if the DNA is suitably labelled and purification of the DNA is not required. However, the system requires calibration and the average rate elution constant K should be calculated to indicate that the kinetics of the elution are unchanged.[20] During the elution, fractions of the eluate (*ca.* 0.5 ml) are taken and the DNA content assayed. Each fraction should represent a fixed elution time, say 15 min, and not a fixed volume which may vary with elution time. Results are generally expressed as a plot of the fraction of DNA retained on the filter (F_r) *versus* the eluate volume or fraction number (F). The initial slopes of the curves may then be compared to assess whether strand breakage or cross linking has occurred. To assess the number of strand breaks induced in DNA the system is usually calibrated by running a range of DNA samples which have been exposed to radiation. Here the dose can be measured with a high degree of accuracy and the energy of the radiation known. Thus the number of strand breaks can be accurately calculated. The technique is very popular and has been used to study both DNA damage and repair in a variety of cells, both prokaryotic and eukaryotic.

10.2.3   FREE RADICAL REACTIONS IN DNA STRAND BREAKAGE

10.2.3.1   Formation of Oxygen Radicals

The role that free radicals play in biological processes cannot be underestimated. There is little doubt that they are frequently involved in causing DNA damage. Indeed, a recent suggestion that the majority of all amoebicidal drugs exert their action *via* electron transfer and formation of superoxide causing oxidative stress,[21] although too general a hypothesis to be considered as a mode of action panacea, emphasizes the increasing importance that free radicals have in biology and medicine. However, some basic free radical reactions of biological relevance should be introduced at this point. Free radicals, defined as molecules with an unpaired number of electrons, are formed as intermediate species in many biochemical processes, *e.g.* in respiration, synthesis, metabolism, and oxidative and reductive reactions. Furthermore, free radicals (more particularly oxygen radicals) have been implicated in a wide range of pathological events including exposure to hyperbaric oxygen or hypo-oxygenation, many chemical toxicities including that from paraquat, carbon tetrachloride, anticancer agents like adriamycin, the pulmonary toxicity attributed to bleomycin and nitrofurans, drug-induced hemolytic anemia, some vitamin deficiencies, certainly aging and most certainly any inflammatory condition, including burns, various infections and edemas, arthritis, emphysema, many cancers, radiation-induced effects, such as sunburn, radiation sickness and side effects caused by radiation during radiotherapy, and, lastly, but by no means a complete list, the effects caused by environmental pollution, for example smoking (people as well as factory chimneys and vehicle engine exhaust emissions), ozone, and various nitrogen oxides.[22]

Oxygen radicals are usually formed *via* redox reactions because oxygen is the most efficient biological electron acceptor known and, by doing so, forms the superoxide radical $O_2^{\cdot-}$, as shown in equation (2).

$$R{-}X^{\cdot-} + O_2 \rightarrow R{-}X + O_2^{\cdot-} \qquad (2)$$

Thus, for example, a compound will be reduced by the addition of one electron to a reactive intermediate which is able to transfer its electron to oxygen, because of its high electron affinity, and forms superoxide in the process. The superoxide radical is remarkably unreactive despite assumptions to the contrary that it must be toxic because there exists an enzyme, superoxide dismutase, which 'inactivates' it. Certainly the superoxide radical can undergo a variety of reactions, the most common of which is the dismutation to hydrogen peroxide and oxygen as shown in equation (3). This reaction is commonly catalyzed by the Cu-containing enzyme superoxide dismutase. The hydrogen peroxide formed can be removed by catalase (equation 4) or by glutathione peroxidase (equation 5). However, if a transition metal is present, commonly Fe or Cu, the metal can be reduced and a hydroxy radical formed, $OH^{\cdot}$ (equation 6). This is the most reactive oxygen radical and is responsible for a wide variety of cellular damage.

$$2O_2^{\cdot-} + 2H^+ \rightarrow H_2O_2 + O_2 \qquad (3)$$

$$2H_2O_2 \rightarrow 2H_2O + O_2 \qquad (4)$$

$$H_2O_2 + 2GSH \rightarrow GSSG + H_2O \qquad (5)$$

$$H_2O_2 + Fe^{2+} \rightarrow Fe^{3+} + OH^{\cdot} + OH^- \qquad (6)$$

$$O_2^- + Fe^{3+} \rightarrow Fe^{2+} + O_2 \tag{7}$$

$$O_2^- + H_2O_2 \rightarrow OH^\cdot + OH^- + O_2 \tag{8}$$

Equation (6) is the Fenton reaction and equations (6) and (7) describe the Haber–Weiss reaction, summarized in equation (8). Thus, any redox reaction involving oxygen is potentially damaging to the cell if superoxide formed in the process takes part in a metal-catalyzed reaction which generates the hydroxy radical.[23]

Despite the efficiency of utilization of molecular oxygen by aerobic cells, up to 95% of which is used in the mitochondrial electron transport system, the terminal cytochrome oxidase appears to be totally efficient in reducing oxygen to water because oxygen radicals have not been detected in this process. Various other pathways accounting for the production of superoxide have been detected and these include phagocytosis, microsomal electron transport involving cytochrome P-450, various enzyme-catalyzed oxidations, and various substrate autoxidations.[24] It is this oxidative stress which is capable of damaging cells and causes a variety of effects ranging from drug-induced toxicity to cancer and death. Nevertheless, the cell also has several mechanisms to protect against free radical-induced oxidative stress. These include ferrooxidase enzymes, such as caeruloplasmin, radical scavengers, such as glutathione and other aminothiols, and enzymes such as catalase and GSH peroxidase which will remove hydrogen peroxide (see equations 4 and 5). Redox cycling is thus not only an important part of normal cellular biochemistry but these mechanisms are also the means whereby many drugs able to accept electrons at biologically relevant redox potentials may cause cellular damage and death.

10.2.3.2 Radical-induced DNA Damage

As the OH radical is of such importance *vis-à-vis* DNA damage, it is relevant to discuss the general mechanism by which DNA damage is caused. This is shown in Figure 1 where the hydroxy radical, irrespective of whether it is formed by radiolysis (as depicted) or by drug-induced mechanisms, interacts with DNA bases to form a hydroxy radical adduct. It appears that thymidine and, to a lesser extent, adenine may be rather more sensitive bases for hydroxy radical attack than the others,

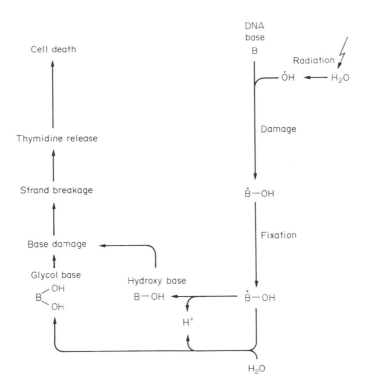

Figure 1 DNA damage induced by hydroxy radicals (B refers to a base in DNA). The transition (fixation) step from $\dot{B}$—OH to $\overset{+}{B}$—OH requires the abstraction of an electron normally carried out by oxygen or, in hypoxic cells, by an electron-affinic, oxygen-mimetic drug

but in any event a neutral hydroxy radical adduct is formed. Although this represents a damaging process, the adduct can be removed, *i.e.* repaired by several pathways, the most common of which involves the reversal by glutathione (GSH), a nonspecific radical scavenger with which all cells appear to be well supplied. If an electron is now removed from the neutral hydroxy–base radical adduct to form a positively charged base adduct, the damage is fixed, *i.e.* can no longer be repaired. In well-oxygenated cells this process is easily accomplished by oxygen, which accepts an electron to form superoxide. At this stage the hydroxy–base cation adduct can undergo two chemical transformations: loss of a proton to form a hydroxy base, or a condensation reaction in which a second hydroxy group is added to the base (usually across the 4,5 double bond) to form a glycol derivative of the base which is rapidly recognized as aberrant by the cell's repair enzymes and the base is efficiently removed. The result of removal of the damaged base is strand breakage, the extent of which, if unable to be repaired, will result in cytotoxicity, possible mutagenicity and/or carcinogenicity and/or cell death.[25]

10.2.3.3 Redox Cycling

It is apparent from the previous examples given that many drugs capable of being reduced in biological systems may also have a real possibility of interacting with oxygen, certainly in oxic or aerobic systems, to produce superoxide, the further reactions of which may be damaging to the DNA of the cell. If a drug requires reductive activation to produce its damaging effect upon the cell, this reaction may be nullified in the presence of oxygen which is the most electron affinic of all biological molecules. The overall effect is one of redox cycling which, if it prevents the drug from being reductively activated, is also called, 'futile cycling',[26] an example of which is shown in equation (9). Here the example is that of a nitro compound which, upon being reduced by a single electron, forms the nitro radical anion. Oxygen, being more electron affinic than the nitro radical, abstracts the electron, regenerating the parent drug molecule and forms superoxide. The reaction is very similar to that shown in equation (2). Because the formation of superoxide is readily detectable, drugs capable of forming it are assumed to undergo redox cycling and often are erroneously further assumed to produce their toxic effects *via* consequent formation of the toxic hydroxy radical as depicted in equations (3) and (6)–(8). That the hydroxy radical is toxic is in no doubt—not only in causing DNA damage, but also in causing the peroxidation of lipids in membranes,[27] or the oxidation of amino acids in proteins leading to conformational changes and inactivation of enzymes.[28]

$$R-NO_2 + 1e^- \rightarrow R-NO_2^{\overline{\cdot}} + O_2 \rightarrow R-NO_2 + O_2^{\overline{\cdot}} \qquad (9)$$

Since the hydroxy radical is very reactive and has a short half-life, it must be generated very close to the target with which it interacts. On the basis of experiments with OH radical scavengers, Chapman and coworkers have shown that the reaction rate of OH$^{\cdot}$ with the target is 7.2×10^8 s^{-1}. The reaction rate constant for DNA bases in double-stranded DNA in solution is about 2×10^8 M^{-1} s^{-1} and the concentration of DNA bases within the target volume is 3.6 M, indicating that the target volume cannot extend more than a few Ångstrom away from the DNA;[29] a radius of about 34 Å or one turn of the helix has been calculated. However, although these observations are of interest and relevance to the mechanism of OH$^{\cdot}$ killing, other interpretations are possible.[30]

The anthracyclines (adriamycin and daunomycin) show at least two different mechanisms of causing strand breaks in DNA. The first is the ability to intercalate between DNA bases and, by inhibiting DNA topoisomerases (gyrases), may induce strand breaks. The second is a consequence of bioreductions which lead to free radical formation and which can lead to strand breaks and other types of cell damage. Free radical formation by anthracyclines occurs after the parent drug has been reduced by a flavin-dependent oxidoreductase which reduces the quinone to the semiquinone.[31] Under aerobic conditions the semiquinone can undergo futile redox cycling with oxygen to re-form the parent drug and generate superoxide.[32] Under hypoxic or anaerobic conditions, however, further reactions can occur including the formation of an alkylating species (the C-7 radical) capable of causing DNA strand breaks.[32] It seems that the free radicals produced by redox cycling with oxygen may be the predominant cause of the cardiotoxic effects of these drugs, because cardiac tissue has very low contents of superoxide dismutase and catalase. However, the major antitumor effects of the drugs are not dependent on free radical formation, since the presence of free-radical scavengers do not seem to affect tumor cell killing.[33]

Many classic intercalaters are now known to produce protein-associated cross-links which cause strand breaks. The most effective of these antitumor drugs are those which exhibit relatively long residence times, *i.e.* show slow binding and concomitantly slow dissociation kinetics.[34]

The mechanism of cytotoxicity of the bleomycins is regarded as a consequence of its ability to bind to DNA and cause both single and double strand breaks. There is a good correlation between the degree of cell kill produced by these drugs, the number of strand breaks induced in DNA, and the degree of chromosome damage observed.[35-37] Double-strand breaks occur as a result of staggered single-strand breaks about two base-pairs apart. Single-strand breaks are also accompanied by an equal number of alkali-labile sites which correspond to the release of a base. In this respect thymine seems to be more susceptible to removal than other bases, although at higher drug concentrations all four bases in DNA may be removed. The strand scission and base removal is specific for DNA; RNA is not degraded even when present in a DNA-RNA hybrid, and GpT and GpC sites appear particularly susceptible to attack.[38]

It is known that, for bleomycin to cause its effect, oxygen is required as no DNA damage occurs under anaerobic conditions, superoxide and OH radicals are produced, free-radical scavengers and superoxide dismutase inhibit strand breakage, and peroxidation products of DNA deoxyribose have been identified. It is further known that metal ions are required for activity of the drug, notably Fe^{II},[39] and the drug species responsible for DNA damage may be an oxygen-liganded iron(III)–drug complex, because it has been shown that neither free OH radicals[40,41] nor superoxide is involved in bleomycin-induced degradation of DNA, although the possibility that they are formed in close proximity to DNA or are bound in some way to the drug complex cannot be discounted. If, as seems probable, the strand breaks in DNA are caused by OH radicals produced at nonrandom sites on DNA directed by the specificity of binding of bleomycin, then the mechanism of production of the damaging radicals would seem to be as follows: the drug (BLM) forms an initial complex with Fe^{II} which, in the presence of oxygen, forms a ternary complex (equation 10). The ternary complex undergoes reduction yielding a bound superoxide ternary complex (equation 11) which is capable of generating OH radicals (equation 12), presumably in a reaction similar to a Fenton one (equation 6). The oxidized drug complex undergoes redox cycling with a nuclear flavoprotein-linked NADPH cytochrome P-450 reductase (equation 13),[42] although xanthine oxidase is also effective in this respect. Presumably the radicals are produced close to the C-3′—C-4′ deoxyribose bond of the sugar, preferentially at GpT or GpC sequences to produce a base-propenal, a glycolic acid ester, a free base and the degradation product of the base-propenal malondialdehyde. This scheme is consistent with free-radical abstraction of the C-4′ proton of the deoxyribose sugar which results in the oxidative cleavage of the C-3′—C-4′ bond, a reaction which requires the intermediacy of a radical ion.[43].

$$BLM + Fe^{II} \rightarrow BLM—Fe^{II} + O_2 \rightarrow BLM—Fe^{II}—O_2 \tag{10}$$

$$BLM—Fe^{II}—O_2 \rightarrow BLM—Fe^{III}—O_2^- \tag{11}$$

$$BLM—Fe^{III}—O_2^- \rightarrow BLM—Fe^{III} + OH^{\cdot} \tag{12}$$

$$BLM—Fe^{III} + FAD_{red} \rightarrow BLM—Fe^{II} + FAD_{ox} \tag{13}$$

Many quinones are also known to undergo redox cycling. Mitomycin C, AZQ (aziridinylbenzoquinone) and other similar drugs are reduced by mitochondrial microsomal and nuclear NADPH-cytochrome P-450 reductases.

The anthracenediones, for example mitoxantrone, have also been demonstrated to partake in redox cycling involving microsomal NADPH-cytochrome P-450 and mitochondrial NADH dehydrogenase. All of these drugs are known to cause DNA strand breaks.[44]

10.2.3.4 Action of Nitro Homocyclic and Heterocyclic Drugs

These drugs comprise a very large group with useful clinical activity as antibacterial, antiprotozoal and anticancer agents. With all nitro compounds their activity is solely dependent upon reduction of the nitro group, the products of which are responsible for DNA damage. The key to understanding the basis of selective toxicity of these drugs lies in the knowledge of the range of reduction potentials exhibited by such compounds. Figure 2 shows the redox spectrum or scale of electron affinity exhibited by nitroaromatic compounds.[45] It shows the redox potential of oxygen (the best biological electron acceptor known) as the most electron affinic agent. What the diagram does not show is that there is a boundary, or hiatus, between those redox reactions capable of being carried out under aerobic (or oxic) and anaerobic (or hypoxic) conditions. The boundary falls at about − 0.35 V. To the right of this division, redox reactions are easily performed under aerobic conditions; in fact, the most negative of redox reactions carried out by the aerobic cell is that

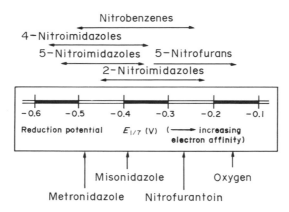

Figure 2 The redox 'spectrum' showing the electron affinity ($E_{1/7}$) of nitroaromatic drugs.[45] Oxygen is the most electron affinic (most readily reduced) and drugs of lower (more negative) potential are less readily reduced. Nitro compounds of redox potentials below *ca.* -500 mV are not reduced by biological systems

involving NAD(NADP)–NADH(NADPH) couples at *ca.* -0.35 V. Thus, redox reactions more negative (lower) than *ca.* -0.35 V will not be possible under aerobic conditions, but are possible under anaerobic conditions. Consequently anaerobes have evolved special electron-transfer proteins to deal with low-potential redox reactions characterized by the possession of iron–sulfur proteins such as the ferredoxins and rubredoxins. Since the voltage span for the acceptance of a single electron is 60 mV, one may expect some weak electron transfer reactions down to a potential of *ca.* -0.4 V in aerobes, but below that we enter the domain of the anaerobic organism. Even between the values of *ca.* -0.35 and -0.4 V the aerobic cell would be considerably acid and hypoxic, so it is not surprising that the 2-nitroimidazoles come into their own in this region, being selectively reduced under hypoxia but having redox potentials too positive to be usefully considered as possible antianaerobe drugs.

The redox potential of bioreducible drugs is therefore a crucial beginning in understanding how these compounds may be applied to combat diseases which flourish in certain redox environments. Indeed, this relationship has been used to explain the anomalous behaviour of satranidazole, a 5-nitroimidazole drug which behaves as a 2-nitroimidazole because its reduction potential is relatively high compared with others of the same type.[46] Wardman and colleagues have performed the same type of quantitative structure–activity relationship analysis on the 2-nitroimidazole radiosensitizers which show a good correlation between the electron affinity (measured as the one-electron redox potential, $E_{1/7}$, or the polarographic half-wave potential, $E_{1/2}$) and the concentration required to produce a fixed, relevant biological response. The relationship is given by the general equation (14). As we shall see later the redox or, more correctly, the reduction potential of these drugs not only determines their selective toxicity, but also the stability and/or reactivity of the reduction products which have an important bearing on their biological effects in the presence or absence of oxygen.[47]

$$\log(1/C) = \log C = b_0 + b_1 E \tag{14}$$

As mentioned in Section 10.2.1.1, the only nitro homocyclic drug of note is chloramphenicol (CAP), the reduction products of which are responsible for aplastic anaemia. In this respect it has been demonstrated that nitrosochloramphenicol (NOCAP; **28**) is far more toxic than CAP,[48,49] but, because NOCAP is unstable in blood or serum with a half-life of seconds only, it is therefore highly unlikely to be the toxic agent. Very recently, however, one of the reduction metabolites produced by gut bacteria, the reduced form of dehydrochloramphenicol (DHCAP; **29**) has been shown to be a more likely candidate as this has a half-life sufficient to account for the observed damage to blood cells and, moreover, has a toxicity equal to that of NOCAP. Whatever the agent responsible for aplastic anaemia it is undoubtedly a reduction product which is capable of causing extensive DNA damage by strand scission, although in this case the mechanism is quite unknown.[50,51]

Nitrofurans are known to produce DNA strand breaks, again as a consequence of reduction of the nitro group, and this area of research has been expertly reviewed in recent years.[52-54] Nitrofurans are known to undergo futile cycling and there appear to be two effects of these drugs relevant to DNA damage, both of which depend upon reduction of the nitro group. The first is that related to resistance patterns, in which it can be shown that in *Escherichia coli* two nitroreductases operate. One is insensitive to oxygen (type I), is known to have three components, is possibly responsible for nitro reduction under aerobic conditions, and which produces the six-electron reduction product, the amine, without passing through a detectable one-electron nitro radical anion.[55,56] This enzyme system is known also to have a NADPH requirement. In contrast, the type II reductases are highly oxygen-sensitive and undergo futile cycling with the concomitant production of superoxide. In the absence of oxygen, however, further reduction occurs, possibly by a nonenzymatic decay involving disproportionation to the nitroso and nitro derivatives,[57] as shown in equation (15). It would appear that reduction of nitrofurans by a type I mechanism leads to a DNA–nitrofuran adduct which may lead to the production of a mutagenic product, whereas the type II reduction leads to classic OH radical damage caused by futile cycling. As we shall see later, it is the ratio between the degree of adduct *versus* radical damage that not only characterizes the type of damage, but also the relative mutagenicity that the compounds may be expected to exhibit. Because nitrofurans have a relatively high (*i.e.* positive) redox potential and are effective in aerobic infections, the major damage induced to DNA is that of adduct formation rather than free-radical effects. With nitroimidazoles the opposite is the case.

$$2RNO_2^{\cdot-} + 2H^+ \rightarrow R-NO_2 + R-NO + H_2O \tag{15}$$

Although it is now recognized that 5-nitroimidazoles behave rather differently from 2-nitroimidazoles, the following discussion will centre upon their reactions thought to be important in causing DNA damage. Nitroimidazoles generally exhibit a preferential toxicity to anaerobic, hypoxic or anoxic cells; toxicity to aerobic or oxic cells also occurs, but only after longer exposure times or with high drug concentrations. In all susceptible cells, be they bacteria, protozoa or mammalian, cytotoxicity and death is only produced after reduction of the nitro group. The rate of reduction can vary over three orders of magnitude, depending on the reduction potential of the drug, and in mammalian cells reduction is mediated by xanthine oxidase, NADPH cytochrome P-450 reductase, or NADPH cytochrome *c* reductase, all of which are capable of reducing the more electron-affinic 2-nitroimidazoles, nitrofurans and nitro homocycles.[58]

In anaerobic microorganisms which are capable of reducing the less electron-affinic 5-nitroimidazoles, the electron source is the pyruvate dehydrogenase–ferredoxin oxidoreductase complex and ferredoxin itself.[59-61] All of these drugs cause a decrease in the viscosity of DNA, a decrease in the helix content as shown by thermal denaturation and renaturation experiments, and can be shown to cause strand breaks by a wide range of methods listed in Section 10.2.2, and in intact microorganisms and mammalian cells.[47] Considerable effort has been directed towards establishing the nature of the reduction product(s) responsible for the DNA damage, which appears to be nonrandom. The first indication of this came from studies which showed that the degree of damage induced by nitroimidazoles related to the base composition of the DNA in that DNAs of high %A + T content were more susceptible to damage than those of low %A + T, and that strand breakage was associated with the specific release of thymidine phosphates. Damage was maximal with poly(d[AT]) and absent with poly(d[GC]) polymers.[62] Indeed, although it is known that binding of reduced nitroimidazoles occurs to G-residues in DNA, the binding is several orders of magnitude lower than the extent of observed damage,[63] which correlates with the %A + T content of the DNA. A clear demonstration that binding and damage are unrelated was shown in an experiment in which misonidazole was reduced chemically. The amount of drug which bound to DNA was the same irrespective of whether the DNA was present during reduction or added later.[64,65] Since it can be shown that strand breaks occur during reduction and not when DNA is added a few seconds later, the reduction product responsible for damage must be a short-lived species and those which bind are other reduction products which are not involved in the damage process.

The reduction of nitroimidazoles follows the reaction sequences shown in equations (16)–(24), which is summarized in equation (25). As can be seen, the reduction process is highly complex and proceeds, generally, by a single-electron addition to form a radical anion, followed by a protonation which yields a neutral radical, followed by a disproportionation. The overall reaction sequence shown in equation (25) illustrates that a one-electron addition gives the nitro radical anion, two electrons would give a nitroso species, four electrons the hydroxylamine derivative, and six electrons the corresponding amine. Despite considerable efforts, however, no stable reduction products of either 5-nitro- or 2-nitro-imidazoles have ever been shown to give the toxicity and target interaction typical of the drugs. Certainly, since the final reduction products are inactive, and 2-nitroimidazoles do not proceed beyond a four-electron reduction, the species responsible for DNA damage must be one of less than four electrons.

$$R\text{—}NO_2 + 1e^- \rightarrow R\text{—}NO_2^{\bar{}} \tag{16}$$

$$R\text{—}NO_2^{\bar{}} + H^+ \rightarrow R\text{—}NO_2H^{\cdot} \tag{17}$$

$$2R\text{—}NO_2H^{\cdot} \rightarrow R\text{—}NO_2 + R\text{—}N(OH)_2 \tag{18}$$

$$R\text{—}N(OH)_2 \rightarrow R\text{—}NO + H_2O \tag{19}$$

$$R\text{—}NO + 1e^- \rightarrow R\text{—}NO^{\bar{}} \tag{20}$$

$$R\text{—}NO^{\bar{}} + H^+ \rightarrow R\text{—}NOH^{\cdot} \tag{21}$$

$$R\text{—}NOH^{\cdot} + R\text{—}NO_2H^{\cdot} \rightarrow R\text{—}NHOH + R\text{—}NO_2 \tag{22}$$

$$R\text{—}NHOH + 1e^- \rightarrow R\text{—}NHOH^{\bar{}} \tag{23}$$

$$R\text{—}NHOH^{\bar{}} + 2H^+ \rightarrow R\text{—}NH_2 + H_2O \tag{24}$$

$$R\text{—}NO_2 \rightarrow R\text{—}NO_2^{\bar{}} \rightarrow R\text{—}NO \rightarrow R\text{—}NHOH \rightarrow R\text{—}NH_2 \tag{25}$$

Recently, McClelland's group has shown that reduction of misonidazole to the hydroxylamine is followed by a Bamberger-type rearrangement to yield a glyoxal fragment.[66,67] The fragment has been shown to interact with guanine and guanosine in DNA but no binding has been associated with DNA damage.[68,69] The chemistry of glyoxal formation from 2-nitroimidazoles (it is not possible from 5-nitroimidazoles) has recently been reviewed.[70] The hydroxylamine derivative of the 2-nitroimidazole SR-2508 has been prepared and found to cause strand breaks in DNA, but the reaction only occurs in air and damage most probably occurs because of oxidation of the hydroxylamine to the nitroso species with the concomitant production of hydrogen peroxide, superoxide and hydroxy radicals.[71] Also a major reduction product of misonidazole produced by anaerobic radiolysis has been identified. It is a cyclic guanidinium ion, the 2-amino-4,5-dihydro-4,5-dihydroxy-1-(2'-hydroxy-3'-methoxy-1'-propyl)imidazolium ion, which is produced in over 80% yield and can readily fragment to form the glyoxal derivatives mentioned earlier. However, it is less toxic than the parent drug and free glyoxal and cannot be the agent responsible for DNA damage and associated cytotoxicity.[72]

In the search for more effective radiosensitizing drugs with hypoxic-specific cytotoxicity, 2-nitroimidazoles carrying an aziridine side chain have been developed. These drugs, exemplified by RSU-1069, when reduced bind to DNA about a 1000-fold more than misonidazole, and produce more strand breaks upon reduction than in the unreduced form. Derivatives of RSU-1069 in which the aziridine ring has undergone alkyl substitution to decrease the associated gut toxicity are less effective in that less DNA damage is seen with compounds in which the aziridine ring is progressively substituted.[73]

One of the approaches to study this challenging problem of identifying the short-lived agent responsible for causing DNA damage is the use of coulometry. Whereas nitro homocyclic, nitrofurans and 2-nitroimidazoles show a four-electron stoichiometry, the 5-nitroimidazoles give nonintegral electron values typically between 3 and 4. The exception to this is satranidazole, a 5-nitroimidazole which behaves as a 2-nitroimidazole in this and other respects.[46] The reason for this anomalous behaviour of the 5-nitroimidazoles is that they also produce significant amounts of nitrite upon reduction, which arises as a result of the decomposition of the nitro radical anion, as indicated in equation (26). Importantly, however, the electron stoichiometry decreases in the presence of DNA and the decrease is related to the %A + T content of the DNA.[74] This effect is seen most clearly with the 2-nitroimidazoles and nitrophenyl compounds, but is complicated in the case of 5-nitroimidazoles because of the production of nitrite which can reach over 30% on a molar basis.[74-76]

$$R\text{—}NO_2^{\bar{}} \rightarrow R^{\cdot} + NO_2^- \tag{26}$$

The inference is that the short-lived reduction intermediate is capable of abstracting electrons from DNA (accounting for the decreased electron stoichiometry), thereby oxidizing it and causing strand breakage and the release of thymidine. If it is assumed that the agent responsible for DNA damage is a free radical, it becomes necessary to consider the possible reactions that it could undergo and, more importantly, their relative stability and reactivity. The nitro radical anion can decompose to form nitrite and the imidazole radical (equation 26), or in the presence of oxygen it can undergo futile redox cycling (equations 2 and 9), or it can be further reduced, or it can react with itself to produce the parent drug and a further reduced product, *i.e.* disproportionation as shown in equation (15). The decay of the nitro radical anion of 5-nitroimidazoles is a second-order one; that of 2-nitroimidazoles is, however, a first-order one.[77] The possibility exists, nevertheless, that 2-nitroimidazoles may undergo a second-order disproportionation if preceded by a first-order protonation as shown in equation (17). In this respect it is likely that the damaging agent is a protonated one because more DNA damage is produced by these drugs at low pH than at high pH.[78] If disproportionation were a major pathway and the damaging species produced at the level of the nitroso or its further reduction products, one would expect that DNA damage would increase with an increased reduction rate. This does not occur; indeed, the opposite is the case.[79]

Recently it has been demonstrated that the metronidazole radical anion is sufficiently stable to diffuse across membranes since it can be detected outside preparations of intact *T. vaginalis* hydrogenosomes inside which it is generated.[80] A model proposed by Edwards' group has attempted to take all these observations into account and envisages the protonated nitro radical anion as the damaging agent,[47,62] which is assumed to react with DNA in a manner analogous to a nitrobenzene radical with an alkyl halide[81] as shown in equation (27). The protonated radical accepts electrons from DNA, becoming further reduced to the nitroso. This interaction not only explains the decreased electron stoichiometry observed when reduction is carried out in the presence of DNA, but also explains the lack of damage which occurs at increased reduction rates because under these conditions the high concentration of radicals produced would enable the disproportionation reaction to predominate. The model (Figure 3) further predicts that, because the mechanism involves abstraction of electrons from DNA, binding of drug to the target would not occur and that any binding observed would be due to further reduction products not associated with the damage process.

$$Ar\!-\!NO_2^- + R\!-\!X \rightarrow X^- + Ar\!-\!\underset{\underset{RO}{|}}{N}\!-\!O^{\cdot} \rightarrow ArNO + RO^{\cdot} \tag{27}$$

While it may be easily inferred that much further work needs to be done as regards the mechanism of action of these drugs, not only from the point of view of the intellectually stimulating challenge they offer but also because the knowledge obtained could result in a real advance in the applications of these drugs in cancer chemotherapy, probably the biggest challenge is that presented by the nitroaromatic aziridines. These drugs have multiple mechanisms of action on DNA and the interaction of two different functional groups requires considerably more research. One of the most interesting examples of this type is CB-1954 (2,4-dinitro-5-aziridinylbenzamide), which appears to be highly active against the Walker carcinoma but no other cell types. The aziridine group may be expected to cause cross linking of DNA but, curiously, has no effect upon several cell lines, including the rat Yoshida sarcoma, which are known for their sensitivity to bifunctional alkylating agents of the nitrogen mustard type.[82] Until recently it has been regarded, variously, as an inhibitor of ribonucleotide reductase because of its benzamide structure[83] or as an antimetabolite.[84] Recent work, however, shows that at least one of the nitro groups can be reductively activated such that,

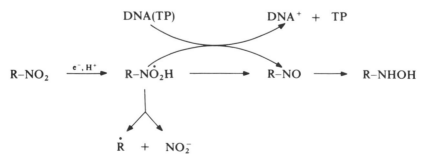

Figure 3 Mechanism of DNA damage induced by nitro heterocyclic drugs. TP refers to thymidine phosphates in DNA and DNA$^+$ is an electron deficient (oxidized) form of DNA, leading to strand breaks and release of TP. The decomposition of R—$N\dot{O}_2$H forms a non-(DNA) damaging imidazole radical ($\dot{R}$) and nitrite (NO$_2^-$). The model envisages the protonated one-electron radical anion (R—NO$_2$H) abstracting electrons from DNA and being reduced in the process (see text for details)

together with the aziridinyl alkylating function, it behaves as a difunctional agent, but only in the sensitive Walker cell line.[85] It would therefore appear that some cancer cells have the ability to activate drugs of this type selectively by reduction of the nitro group and the cellular basis of this action deserves further investigation.

10.2.4   OTHER DRUGS WHICH BIND AND NICK DNA

In this section we shall consider the action of platinum and other metal-based drugs, and *m*-AMSA (4′-(9-acridinylamino)methanesulfon-*m*-anisidide). There are other compounds which are known to bind and cause strand breaks in DNA, but these will not be considered as they have no clinical application.

10.2.4.1   *m*-AMSA

This drug, used as an antileukemic agent, was initially thought to act as a classic intercalator, as indeed, it does show such properties, based on a comparison of the unwinding angle of the DNA helix of 21° compared with that of ethidium which exhibits an unwinding angle of 26°.[86] However, many drugs of this type (reviewed in Chapter 10.1) exhibit additional effects which are not explained on the basis of intercalation alone. Whereas a number of intercalators cause strand breakage, thought to be the result of oxygen radical production when bound to DNA (for example, adriamycin, daunomycin), there are some drugs which are known not to do so; ellipticine is an example.[87,88] *m*-AMSA was established as producing a covalent interaction between DNA and nuclear proteins using the then recently established alkali elution assay of Kohn *et al.*,[89] and *m*-AMSA was reported as having the highest frequency of DNA-protein-associated cross links.[90] The reaction was sensitive to temperature and saturable, but not with regard to drug uptake, suggesting that the effect was enzyme mediated.[91,92]

The drug not only induces strand breaks, it also alters the linking number of the DNA helix. The DNA molecule as a plectonemically coiled double helix can be described by the linking number *Lk*, which is the number of times one strand winds around the other when both strands lie in a single plane, and is related to the number of base pairs in the molecule, *N*, and the helix pitch, *p*, which is the number of base pairs per turn of the helix. These parameters are related to the number of twists, *Tw*, of the helix as shown in equation (28).

$$Tw = Lk - N/p \tag{28}$$

Lk is an integer and cannot be altered other than by strand breakage. Whereas many intercalators alter *Tw* they have no effect upon *Lk*, but *m*-AMSA changes *Lk*.[93] The cell possesses two types of topoisomerase enzyme, also known as DNA gyrase. The first, topoisomerase I, produces single strand breaks and the second, topoisomerase II, produces double strand breaks.[94] *m*-AMSA appears to interact both with DNA and topoisomerase II by reacting either with the DNA–topoisomerase complex or with the enzyme first and subsequently with the DNA,[95] producing stabilized strand breaks in a manner not dissimilar to the 4-quinolone antibacterial drugs described in Chapter 10.3. That the topoisomerase II enzyme is the major target of *m*-AMSA has recently been demonstrated by the isolation of human leukemia cell lines which are highly resistant to *m*-AMSA. The topoisomerase enzyme in these cells is less amenable to the drug insofar as a greatly decreased amount of strand breaks occurs in the presence of the drug.[96] The effects of *m*-AMSA as a strand breaking agent is also enhanced by ara-C, hydroxyurea and 5-azacytidine,[97,98] and α-difluoromethylornithine (DFMO), an irreversible inhibitor of ornithine decarboxylase, which results in the depletion of intracellular polyamines.[99]

10.2.4.2   Platinum-based Drugs

The mechanism of action of these compounds has been expertly reviewed by Roberts and Thomson.[100] In a previous section (10.2.1.3.2) in this chapter the medical applications of the platinum group of drugs were briefly outlined. In addition to the cytotoxic effects of these drugs *per se*, recent research has focused upon their activity as radiosensitizers and in combination chemotherapy. Cisplatin has been studied in combination with alkylating agents in the treatment of solid tumours[101] because their toxicities do not overlap and their action on DNA appears to be due to two separate and independent effects. This is important because it is known that tumor cell kill is correlated with cross links induced by these drugs.[102] Moulder and coworkers have recently demonstrated that renal irradiation in rats decreases the clearance of cisplatin and this correlates with an increase in the toxicity of the drug. This is not surprising considering that the object of mannitol diuresis is to reduce the toxicity of the drug,[103] but the observation is significant in view of the fact that kidney irradiation is used in a range of treatment protocols including combined therapy for Wilm's tumor, and both lower hemi-body and total body irradiation for bone marrow transplantation. In this respect the use of a radioprotector, WR-2721, an organic thiophosphate, is of interest because it protects against cisplatin's nephrotoxicity but has no effect upon its antitumor activity in humans.[104] As a result of the considerable data amassed on cisplatin, not only as an antitumor agent *per se* but also as a radiosensitizer,[105,106] the other clinically promising drugs carboplatin and iproplatin have been studied for their radiosensitization potential and *in vitro* in mammalian cells were found to be better than cisplatin.[107] In the following the chemical characteristics relevant to their mode of action will be described.

Platinum compounds have only two oxidation states in aqueous solution, $2+$ and $4+$, both of which dictate the stereochemical arrangement of the ligands in either a square planar (Pt^{II}) or octahedral (Pt^{IV}) configuration. The most effective compounds *vis-à-vis* their anticancer activity are those having a *cis* configuration about a Pt^{II} nucleus. CisPt, or cisplatin, is *cis*-diammine-dichloroplatinum(II), a bifunctional compound with both chlorides available for substitution provided the incoming groups are in excess or are thermodynamically more favourable. In contrast, the ammine groups are thermodynamically stable and are inert to substitution. The *trans* isomers have no significant biological activity even though they bind equally well to cellular macromolecules;[108] they will not be discussed further.

DNA is the primary target for platinum drugs, selectively inhibiting DNA synthesis at concentrations which leave RNA or protein synthesis unaffected. Although the drugs interact with all bases of DNA, the reaction with guanine is preferential and occurs at the fastest rate.[109] The interaction with guanine occurs preferentially with the N-7 position and can form a bifunctional (bidentate) chelate whereas the *trans* isomers can form only a monofunctional adduct. The *cis* isomers are also known to cause cross linking of DNA,[110] the linking occurring presumably between two N-7 positions on guanine residues after the initial binding to form a monofunctional adduct causing distortion of the helix. This favours a second bifunctional adduct which results in cross linking of the helix.[100] Such cross links are not limited to DNA but also include DNA–protein cross links which lead to strand breaks.[111–113]

Despite the inhibition of DNA synthesis and the inactivation of the template for replication, some cells are proficient in the repair of platinum-induced cross linking. This is primarily by excision repair processes, as has been demonstrated both in prokaryotic[113] and mammalian cell systems.[114] It would appear that the active *cis* platinum derivatives exert their lethal action on DNA by interstrand and intrastrand cross links which are repaired with difficulty and therefore persist, in contrast to those formed by the *trans* derivatives which are easily repaired. The persistence of the *cis*-induced cross links are believed to be due to conformational changes induced in DNA which make it difficult for repair enzymes to recognize the lesion.[115]

10.2.4.3 Other Metal Complexes

As mentioned above, the *cis* platinum group of compounds which have chloride ligands are most amenable to substitution and advantage has been taken of this property to prepare a number of *cis* platinum drugs in which 2- or 5-nitroimidazole ligands have been introduced. Of particular interest is the *cis*-$[Pt^{II}Cl_2(\text{metronidazole})_2]$ compound known as FLAP, which was reported as having remarkable radiosensitization characteristics.[116] However, further research did not confirm the earlier results insofar as it was demonstrated that the sensitization achieved was that indicated by a combination of free cisplatin and metronidazole.[117] Nevertheless, if the ammine groups are replaced by 2-nitroimidazoles the radiosensitization efficiency is greater than the corresponding 5-nitro-imidazoles,[117] and ruthenium(II)–2-nitroimidazole complexes show even better sensitization.[117,118] Such drugs may show promise for the future if their antitumor activity can be separated from their unwanted organ toxicity.

10.2.5 REPAIR OF DNA STRAND BREAKS

There is considerable evidence that cell death resulting from the action of drugs which are capable of redox cycling with oxygen involve DNA and that the major lesion involves strand breaks, particularly double strand breaks.[119]

10.2.5.1 Biological Effects of Strand Breaks

Although the number of single or double strand breaks required to inactivate a molecule of DNA is not known with any certainty for the drugs described in the previous sections, considerable knowledge has been amassed regarding the radiation dose required. Such data give clues as regards the likely amount of damage required to inactivate a DNA molecule, but it must be borne in mind that the efficiency of repair depends on the nature of the lesion even though the result may be the same, *i.e.* the production of a strand break. What is important therefore is to construct experiments which yield data concerning biologically active DNA and not merely that which gives relative numbers of strand breaks using solely physical data.

Schulte-Frohlinde's group has shown that a double strand break may not be lethal *per se* but that it depends on the nature of the break and the bases involved. In other words the quality of the break is important in survival of the cell. For example, in the circular double-stranded plasmid pBR 322, a single double-strand break may be introduced by the restriction endonucleases Bam Hl or Pvu II. The survival of the plasmid is, however, dependent upon the nature of the break: that produced by Bam Hl leads to a survival level more than 20 times higher than breaks produced by Pvu II even though survival is low (2.5% and 0.11% respectively).[120] Similar results have been obtained with other plasmids and endonucleases. The reason for the differential cell survival lies in the nature of the strand break: Bam Hl produces a staggered strand break with an overlap of four base pairs, whereas Pvu II produces a blunt-ended break. It is also known that blunt-end cuts can result in chromosome aberrations, whereas staggered or cohesive end cuts do not.[121] In the case of double strand breaks produced by restriction endonucleases there is no loss of genetic information, but mutation and loss can result from the further action of exonucleases. It has been calculated that, for linearized DNA carrying overlapping or staggered ends, the percentage carrying deleted genetic material is about 1%, but blunt-ended DNA results in a population of molecules containing over 40% deletions.

Survival of such DNA therefore depends upon the relative competing reactions in which the DNA is degraded by exonucleases and the repair of the strand breaks by ligases. Several groups have used the biologically active single-stranded DNA from the bacteriophage φX174 and shown that single strand breaks are 100% lethal (*i.e.* a single break is sufficient to inactivate the molecule completely).[122] Apurinic sites are also 100% lethal,[123] but thymine glycols are only 30% lethal.[124] The situation in double stranded DNA is different because there is a greater possibility of repair. Nevertheless, it can be shown that single strand breaks plus base damage are less than 5% lethal, as are those induced by pancreatic nuclease; apurinic sites induced by heat are about 6% lethal, but those induced by restriction endonucleases are more than 96% lethal. Thus, lethality appears to occur only if both strands of the DNA are involved. However, any type of damage is potentially lethal; it depends on how the lesion may be further processed, as we shall see.

There seems to be no direct correlation between lethality and strand breaks even though it has frequently been assumed that single strand breaks are easily repaired but double strand breaks are not. Schulte-Frohlinde has shown that survival of an irradiated plasmid does not correlate with double strand breaks, nor single strand breaks, nor base damage.[120] However, strand breaks may arise not only as the direct result of the scission of a phosphodiester bond but may result from initial damage to a base or sugar which is translated into a strand break. The amount of damage necessary

Table 1 DNA Damage Required to Kill 63% of Cells

Agent	DNA lesion	No. of lesions/cell
Ionizing radiation	SSB	1 000
	DSB	40
Bleomycin	SSB	150
	DSB	30
UV light	SSB	100
Hydrogen peroxide	SSB	400 000

to kill cells varies with the nature of the damage. For example, the number of double (DSB) or single strand breaks (SSB) per cell required to kill 63% of the cell population (a figure derived from the exponentiality of the kill curve) with different agents is shown in Table 1 (taken from Ward *et al.*[125]).

According to Wallace,[126] the relative contribution of various types of DNA damage produced by X-ray inactivation of the PM2 bacteriophage is: single-strand breaks 4.2%, thymine ring saturation products (glycols and urea) 5.2%, double strand breaks 11%, alkali-labile lesions 13.8%, and other stable base damage accounts for 65.8%. There is thus a large amount of base damage, the nature of which needs to be accounted for.

10.2.5.2 Base Damage and Apurinic Sites

Many drugs can induce strand breaks in DNA by undergoing redox cycling with oxygen, producing oxygen radicals which form thymine glycols and urea residues. Agents known to cause cell death *via* this mechanism include most of the bioreducible quinones, nitro heterocyclic drugs, several carcinogens, ionizing radiation and UV light. Apurinic sites are usually produced by oxidative stress and detected because of their alkaline lability in which the lesion is transformed into a strand break.

The formation of thymine glycols can be measured with relative ease, either by the acetol fragment assay,[127] chromatographically,[128,129] by enzyme assays,[130] or by antibody assays,[131] including a monoclonal assay.[132,133] Urea residues can be assayed using enzyme susceptibility.[134] Both thymine glycols and urea residues and apurinic sites in DNA constitute replicative blocks to DNA polymerases, and glycols are not mutagenic lesions.[126] Thus, all of these lesion types are potentially lethal if not repaired.

It is fairly common that DNA damage by whatever agent rarely results in a single lesion, but multiple locally damaged sites may occur consisting of damaged bases and strand breaks. The relative lethality of these depends upon the efficiency and fidelity of repair.

10.2.5.3 Mechanisms of Repair of DNA Strand Breaks

There is now good evidence that double strand breaks can be repaired. Part of the evidence comes from studies involving radiosensitive cell lines which have defects in the mechanism for the repair of double strand breaks,[135] although these cell lines are all proficient in the repair of single strand breaks. Several excellent reviews covering all aspects of DNA repair have appeared in recent years.[136,137]

If the damage to DNA consists of an altered base this will be recognized by a specific repair enzyme, either a glycosylase that removes the base resulting in an apurinic or apyrimidinic site (AP) which, *via* the action of an AP endonuclease, cleaves the phosphodiester bonds of the DNA, or an endonuclease that cleaves the DNA backbone close to the original damaged site. If the damage is to a sugar a deletion occurs which results in strand breakage. This repair requires the cleaning up of the termini prior to the restoration of the damaged site. Repair is then completed by the replacement of the missing nucleotides and its insertion and ligation into the DNA. All of these mechanisms are part of the excision repair pathway that occurs both in prokaryotic and eukaryotic cells and are highly efficient for individual single lesions because of the presence of the intact template on the opposite strand. However, the efficiency of repair is markedly decreased if these lesions occur close to each other on both strands of the helix.

An example of such damage would be if there were two damaged sites opposite each other; the removal of both bases results in an irreparable lesion. When damage occurs on both strands but offset by a few bases, the damage may also not be able to be repaired. If the lesions consist of a strand break and a damaged base, for example, there will be loss of genetic material if both lesions occur within the size (*i.e.* length) of the repair patch.[138] Again the efficiency of repair will be determined by the rate and fidelity of the repair enzymes involved in competition with the fixation of the damage involving hydrolysis of labile sugar bonds, endonuclease nicking of base-damaged sites, separation of the ends of the double strand break, or the rupture of the hydrogen bonds between breaks on opposite strands.

Oxidative base damage can be repaired by at least four enzymes in *E. coli*, all of which operate as an excision repair system. Uracil-DNA glycosylase recognizes uracil formed by the oxidative deamination of cytosine either in single or double stranded DNA and releases the free base.[139] The abasic site is recognized by an apurinic endonuclease which cuts adjacent to the damaged site. An exonuclease removes the abasic residue, a polymerase inserts the missing bases, and a ligase seals the new patch into the DNA, completing the repair. Formamidopyrimidine–DNA glycosylase recognizes the formamidopyrimidine and its methyl derivative formed from partial fragmentation of a purine.[140] This type of damage is produced by alkylating agents followed by alkaline hydrolysis.

Endonuclease III is specific for double stranded DNA containing thymine glycols or thymine-derived urea residues. It releases thymine glycol, urea residues and other pyrimidine breakdown products by producing a strand break on the 3′ side of the damage.[126] Recently, *E. coli* mutants (*nth*) have been isolated which lack endonuclease III activity. These are proving to be valuable tools in the detection of agents which induce thymine glycol formation in DNA.

Exonuclease III also is specific for double stranded DNA and recognizes urea residues. It produces a single strand break 5′ to the damage to yield a urea residue. Mutants lacking exonuclease III activity (*xth*) have also been isolated.[126] The action of these enzymes are shown in Figures 4 and 5.

In Figure 4 the action of endonuclease III is depicted and illustrates the glycolytic action of the enzyme. In Figure 5 the action of exonuclease III is shown in which the action is on urea residues; the action of endonuclease III on removal of urea residues is also depicted. Recent work has shown that the action of nitroimidazole–aziridines involves the action of exonuclease III in repair of these lesions insofar as DNA damage is increased in the absence of this enzyme in mutants of *E. coli* using the transfection technique.[158]

10.2.5.4 Inhibition of DNA Repair

One of the ways in which the chemotherapy of tumors may be improved is by the specific inhibition of DNA repair processes. The concept is not confined to cancer, but applies also to any pathogenic disease including parasitic disease and those caused by bacteria. Indeed, a model for this

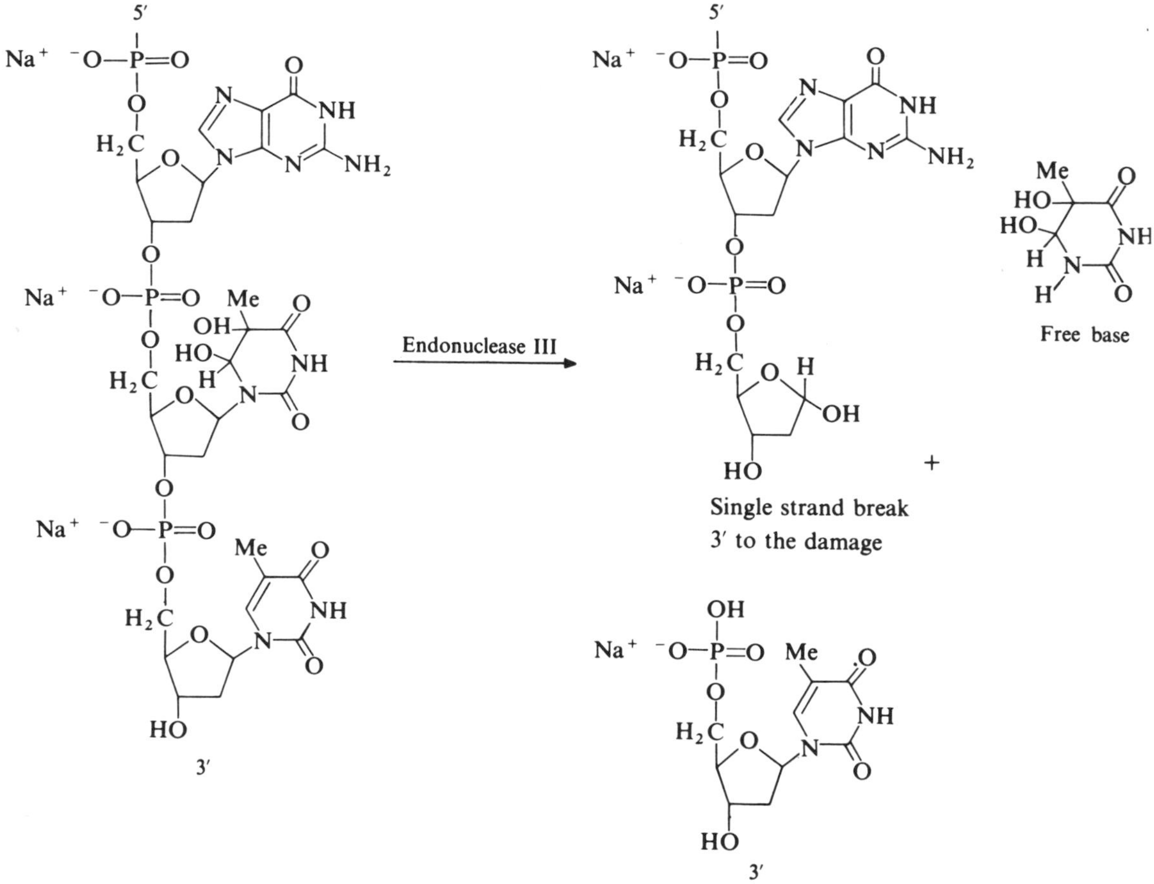

Figure 4 The action of endonuclease III on DNA containing thymine glycols.[126] The base sequence as depicted, 5′-pCpTgpT-3′ (where Tg is a thymidine glycol), is hydrolyzed by endonuclease III to yield pCp-deoxyribose plus a free thymine glycol and pT (see text for further details)

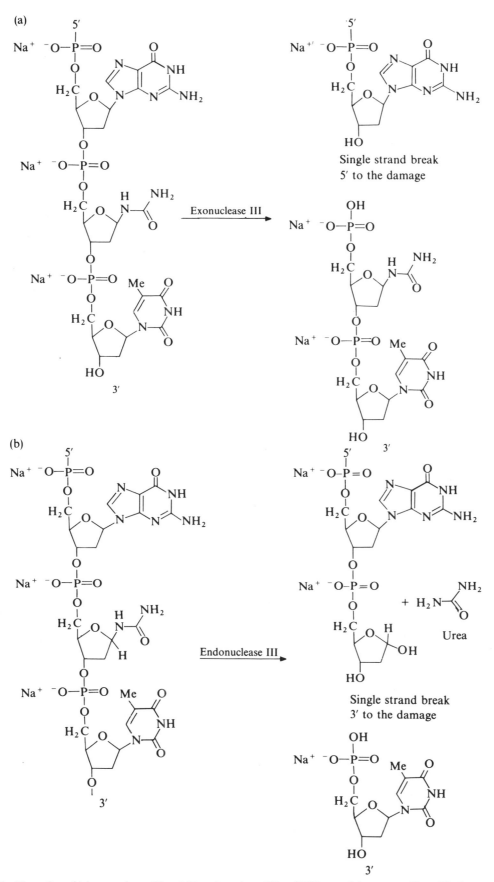

Figure 5 The action of (a) exonuclease III and (b) endonuclease III on DNA containing urea residues. The base sequence 5'-pCpUreapT-3' is hydrolyzed by exonuclease III to yield C-5'-P and pUreapT. However, if the sequence is hydrolyzed by endonuclease III the products are pCp deoxyribose, urea and T-t-P (see text for further details)

chemotherapeutic approach may reside in the use of the 4-quinolones, which appear to have a selective action against the topoisomerase II (DNA gyrase) enzyme in bacteria. Most DNA inhibitors may be expected to be analogues designed to compete with normal repair substrates. The problems of this type of approach are large, as one must take into account the relative pool size of substrates with which the inhibitor must compete and the problem of delivery of the drug to the cancer or pathogen site.

One of the enzymes involved in the repair of DNA strand breaks is poly(ADP-ribose) polymerase, which transfers ADP-ribose from NAD to an acceptor protein in the cell nucleus. Inhibitors of the enzyme retard the repair of strand breaks caused by X-rays,[141] bleomycin[142] and alkylating agents.[143] Amongst the known inhibitors of the enzyme, 3-aminobenzamide appears to be the most potent of a series of benzamides; others include nicotinamides and methylxanthines. The inhibition of the enzyme by benzamides has been used to enhance the cytotoxicity of other drugs, including L-phenylalanine mustard both *in vitro* and *in vivo*.[144] More recent work with human tumor cell lines has shown that 3-aminobenzamide sensitizes normal human fibroblasts and Ewing's sarcoma, but has no effect upon lung adenoma or melanoma or osteosarcoma. These results correlate with the known sensitivities of these cell types to radiation. Thus the Ewing sarcoma is considered to be radiocurable clinically, whereas the others are considered to be radioresistant and not radiocurable.[145] However, for such an approach to be effective clinically the level of the poly(ADP-ribose) polymerase enzyme in tumor or cancer cells must be significantly higher than in normal cells for any therapeutic gain to be realized.

Hydroxyurea, cytosine arabinoside (ara-C) and adenine arabinoside (ara-A) have been used to inhibit DNA repair. Unlike inhibitors of poly(ADP-ribose) polymerase, these act more as inhibitors of DNA synthesis. These compounds inhibit the repair of DNA single strand breaks.[146] Even though the formation of single strand breaks are not lethal lesions, the process of single strand repair is also concerned with that for the repair of double strands. As mentioned earlier, while this may be true the converse is not; cells deficient in double strand repair are proficient in single strand repair.[147,148] Hydroxyurea has been shown to inhibit the repair of single strand breaks in HeLa cells by 63%.[146] The precise mechanism by which it does this has not been established, but because the compound is known to decrease the pool size of DNA synthesis precursors, *i.e.* the pool of triphosphonucleotides,[149] a decrease in the size of the pool would consequently reduce the available material for DNA repair. A second property of the agent could be equally relevant: that of increasing the amount of DNA synthesis precursors necessary to repair the damage,[150,151] a concept that is known as the repair patch size. What this means in practice is that hydroxyurea may have the property of increasing the size of the repair patch such that lesions on opposite strands, normally repairable by short patches, now become irreparable because the increased length of the patch may encounter a second lesion. The overall effect, therefore, may result in an increased number of double strand breaks or increased degradation lengths upon a single strand, which increase the chances of encountering a second lesion or break.[152]

Ara-C only exerts its effect after its incorporation into DNA. Its structure influences the final ligation step in repair such that this is inhibited because of steric effects.[153] It causes 53% inhibition of the repair of single strand breaks in HeLa cells,[138] presumably by prolonging the life of the strand break. Ara-A is even more effective insofar as it causes 88% inhibition of the repair of single strand and 16% inhibition of double strand breaks in Ehrlich ascites tumor cells.[154,155]

Dideoxythymidine also inhibits the repair of DNA single strand breaks by inhibiting chain elongation and ligation because the molecule lacks a 2'-hydroxy group on the sugar.[156] Irrespective of the specificity of such agents, if they are to be used as effective drugs they will need to be present during the strand breaking process, whether induced by other drugs or radiation. The reason for this is an obvious one: repair processes by their very nature are rapid. Ward's group has calculated the rate of repair of single strand and double strand breaks as having a half-life of about 4 min in mammalian cells. This is in contrast to the rate of removal of base damage, which has a half-life of about 7 min.[157] Nevertheless, those double strand breaks left unrepaired by the fast (4 min) process are subsequently repaired by a slower process.[138] Thus, any drug with useful properties as a DNA repair inhibitor would not only have to be fast acting, but also be present during the strand breaking process, which if radiation induced would be in a period of less than about 10 min—chemotherapy at its most challenging.

10.2.6 REFERENCES

1. L. B. Lave and G. S. Omenn, *Nature (London)*, 1986, **324**, 29.
2. F. C. Firkin, *J. Clin. Invest.*, 1972, **51**, 2085.

3. S. F. Jackson, B. R. Wentzell, D. R. McCalla and K. B. Freeman, *Biochem. Biophys. Res. Commun.*, 1977, **78**, 151.
4. D. I. Edwards, *Biochem. Pharmacol.*, 1986, **35**, 53.
5. P. H. Wiernik, in 'Anthracyclines: Current Status and Development', ed. S. T. Crook and S. D. Reich, Academic Press, New York, 1980, p. 273.
6. H. D. H. Showalter, D. W. Fry, W. R. Leopold, J. W. Lown, J. A. Plambeck and K. Reska, *Anti-Cancer Drug Design*, 1986, **1**, 73.
7. G. Bouvier, *Neurosurgery*, 1987, **20**, 286.
8. C. F. J. Barnard, M. J. Cleare and P. C. Hydes, *Chem. Br.*, 1986, **22**, 1001.
9. P. Alexander, J. Gielen and A. C. Sartorelli (ed.), 'Bioreduction in the Activation of Drugs', Pergamon Press, Oxford, 1986.
10. E. M. Zeman, J. M. Brown, M. J. Lemmon, V. K. Hirst and W. W. Lee, *Int. J. Radiat. Oncol. Biol. Phys.*, 1986, **12**, 1239.
11. J. M. Brown (ed.), *Int. J. Radiat. Oncol. Biol. Phys.*, 1986, **12**.
12. E. M. Fielden, J. F. Fowler, J. H. Hendry and D. Scott (ed.), 'Radiation Research: Proceedings of the 8th International Congress of Radiation Research,' Taylor and Francis, London, 1987, vol. 1.
13. K. A. Kennedy, W. N. Hait and J. S. Lazo, *Int. J. Radiat. Oncol. Biol. Phys.*, 1986, **12**, 1367.
14. D. A. Rowley, R. C. Knight, I. M. Skolimowski and D. I. Edwards, *Biochem. Pharmacol.*, 1979, **28**, 3009.
15. D. Rickwood (ed.), 'Centrifugation: A Practical Approach', IRL Press, London, 1978.
16. B. D. Young, in ref. 15, p. 93.
17. D. Freifelder, 'Physical Biochemistry: Applications to Biochemistry and Molecular Biology', 2nd edn., Freeman, San Francisco, 1982.
18. D. A. Rowley, R. C. Knight, I. M. Skolimowski and D. I. Edwards, *Biochem. Pharmacol.*, 1980, **29**, 2095.
19. K. W. Kohn, R. A. G. Ewig, L. C. Erickson and L. A. Zwelling, in 'DNA Repair—A Laboratory Manual of Research Procedures', ed. E. C. Friedberg and P. C. Hanawalt, Dekker, New York, 1981, vol. 1, part B, p. 379.
20. S. Parodi, M. Pala, P. Russo, A. Zunino, C. Balbi, A. Albini, F. Valerio, M. Cimberle and L. Santi, *Cancer Res.*, 1982, **42**, 2277.
21. J. R. Ames, U. Hollstein, A. R. Gagneux, M. D. Ryan and P. Kovacic, *Free Rad. Biol. Med.*, 1987, **3**, 85.
22. M. Tamba, in 'Oxygen and Sulfur Radicals in Chemistry and Medicine', ed. A. Breccia, M. A. J. Rodgers and G. Semerano, Lo Scarabeo, Bologna, 1986, p. 21.
23. S. D. Aust, L. A. Morehouse and C. E. Thomas, *Free Rad. Biol. Med.*, 1985, **1**, 3.
24. J. M. C. Gutteridge and B. Halliwell, in ref. 22, p. 47.
25. H. M. Pinedo and B. A. Chabner (ed.), 'Cancer Chemotherapy Annual', Elsevier, Amsterdam, 1984, vol. 6.
26. E. Perez-Reyes, B. Kalyanaraman and R. P. Mason, *Mol. Pharmacol.*, 1980, **17**, 239.
27. H. Kappus, in 'Oxidative Stress', ed. H. Sies, Academic Press, New York, 1985, p. 273.
28. J. S. Bus and J. E. Gibson, in 'Drug Metabolism and Drug Toxicity', ed. J. R. Mitchel and M. G. Horning, Raven Press, New York, 1984, p. 21.
29. J. D. Chapman, A. P. Reuvers, J. Borsa and C. L. Greenstock, *Radiat. Res.*, 1973, **56**, 291.
30. G. Czapski and J. Israel, *Chemistry*, 1984, **24**, 29.
31. J. Doroshow, *Proc. Am. Assoc. Cancer Res.*, 1981, **22**, 203.
32. L. Gianni, B. J. Corden and C. E. Myers, *Rev. Biochem. Toxicol.*, 1983, **5**, 1.
33. C. E. Myers, W. P. McGuire, R. H. Liss, J. Ifrim, K. Grotzinger and R. C. Young, *Science (Washington, DC)*, 1977, **197**, 165.
34. W. A. Denny, I. A. Roos and L. A. G. Wakelin, *Anti-Cancer Drug Design*, 1986, **1**, 141.
35. J. M. Clarkson and R. M. Humphrey, *Cancer Res.*, 1976, **36**, 2345.
36. C. H. Huang, C. K. Mirabelli, Y. Yan and S. T. Crooke, *Biochemistry*, 1981, **20**, 233.
37. M. A. Sognier, W. N. Hittelman and M. Pollard, *Mutat. Res.*, 1982, **93**, 149.
38. E. F. Gale, E. Cundliffe, P. E. Reynolds, M. H. Richmond and M. J. Waring, 'The Molecular Basis of Antibiotic Action', 2nd edn., Wiley, Chichester, 1981, p. 366.
39. E. A. Sausville, J. Peisach and S. B. Horwitz, *Biochem. Biophys. Res. Commun.*, 1976, **73**, 814.
40. R. M. Burger, J. Peisach and S. B. Horwitz, *J. Biol. Chem.*, 1981, **256**, 11 636.
41. L. O. Rodriguez and S. M. Hecht, *Biochem. Biophys. Res. Commun.*, 1982, **104**, 1470.
42. H. Kappus and I. Mahmutoglu, in 'Proceedings of the Third International Symposium on Biological Reactive Intermediates', Plenum Press, New York, 1986.
43. L. Giloni, M. Takeshita, F. Johnson, C. Iden and A. P. Grollman, *J. Biol. Chem.*, 1981, **256**, 8608.
44. G. M. Cohen and M. d'Arcy Doherty, *Br. J. Cancer*, 1987, **55**, suppl. 8, 46.
45. P. Wardman, in 'Chemotherapeutic Strategy', ed. D. I. Edwards and D. R. Hiscock, Macmillan Press, London, 1983, p. 173.
46. A. Zahoor, R. C. Knight, P. W. Whitty and D. I. Edwards, *J. Antimicrob. Chemother.*, 1986, **18**, 17.
47. D. I. Edwards, *Biochem. Pharmacol.*, 1986, **35**, 53.
48. A. A. Yunis, A. M. Miller, Z. Salem and M. D. Corbett, *J. Lab. Clin. Med.*, 1980, **96**, 36.
49. A. A. Yunis, *Sex. Trans. Dis., Suppl.*, 1984, **11**, 340.
50. S. F. Jackson, B. R. Wenzell, D. R. McCalla and K. B. Freeman, *Biochem. Biophys. Res. Commun.*, 1977, **78**, 151.
51. I. M. Skolimowski, R. C. Knight and D. I. Edwards, *J. Antimicrob. Chemother.*, 1983, **12**, 535.
52. D. R. McCalla, in 'Antibiotics', ed. F. E. Hahn, Springer, Heidelberg, 1979, vol. 5, p. 176.
53. D. R. McCalla, in 'Short-term Tests for Chemical Carcinogens', ed. H. F. Stich and R. H. C. San, Springer, Heidelberg, 1981, p. 36.
54. R. McCalla, *Environ. Mutagens*, 1983, **5**, 745.
55. D. W. Bryant, D. R. McCalla, M. Leeksma and P. Laneuville, *Can. J. Microbiol.*, 1981, **27**, 81.
56. F. J. Peterson, R. P. Mason, J. Housepian and J. L. Holtzman, *J. Biol. Chem.*, 1979, **254**, 4009.
57. R. P. Mason and J. L. Holtzman, *Biochem. Biophys. Res. Commun.*, 1975, **67**, 1267.
58. J. M. Brown, *Int. J. Radiat. Oncol. Biol. Phys.*, 1982, **8**, 675.
59. D. I. Edwards, M. Dye and H. Carne, *J. Gen. Microbiol.*, 1973, **76**, 135.
60. R. Marczak, T. E. Gorrell and M. Muller, *J. Biol. Chem.*, 1983, **258**, 12 427.
61. N. Yarlett, T. E. Gorrell, R. Marczak and M. Muller, *Mol. Biochem. Parasitol.*, 1985, **14**, 29.
62. D. I. Edwards, J. H. Tocher, L. D. Dale, D. A. Widdick and N. S. Virk, in 'Selective Activation of Drugs by Redox Processes', ed. G. E. Adams, A. Breccia, E. M. Fielden and P. Wardman, Plenum, New York, 1990, in press.

63. R. J. Knox, R. C. Knight and D. I. Edwards, *Biochem. Pharmacol.*, 1981, **30**, 1925.
64. A. J. Varghese and G. F. Whitmore, *Cancer Clin. Trials*, 1980, **3**, 43.
65. A. J. Varghese and G. F. Whitmore, *Chem. Biol. Interact.*, 1981, **36**, 141.
66. R. A. McClelland, J. R. Fuller, N. E. Seaman, A. M. Rauth and R. Battistella, *Biochem. Pharmacol.*, 1984, **33**, 303.
67. R. A. McClelland, R. Panicucci and A. M. Rauth, *J. Am. Chem. Soc.*, 1985, **107**, 1762.
68. A. J. Varghese and G. F. Whitmore, *Cancer Res.*, 1983, **43**, 78.
69. A. J. Varghese and G. F. Whitmore, *Int. J. Radiat. Oncol. Biol. Phys.*, 1984, **10**, 1361.
70. G. F. Whitmore and A. J. Varghese, *Biochem. Pharmacol.*, 1986, **35**, 97.
71. K. R. Laderoute, E. Eryavec, R. A. McClelland and M. Rauth, *Int. J. Radiat. Oncol. Biol. Phys.*, 1986, **12**, 1215.
72. R. Panicucci, R. A. McClelland and A. M. Rauth, *Int. J. Radiat. Oncol. Biol. Phys.*, 1986, **12**, 1227.
73. A. R. J. Silver, P. O'Neill, T. C. Jenkins and S. S. McNeil, *Int. J. Radiat. Oncol. Biol. Phys.*, 1986, **12**, 1203.
74. R. J. Knox, D. I. Edwards and R. C. Knight, *Int. J. Radiat. Oncol. Biol. Phys.*, 1984, **10**, 1315.
75. E. Gattavecchia, D. Tonelli, A. Breccia and S. Roffia, *Int. J. Radiat. Biol.*, 1982, **42**, 1009.
76. E. Gattavecchia, D. Tonelli and P. G. Fuochi, *Int. J. Radiat. Biol.*, 1984, **45**, 469.
77. P. Wardman, *Life Chem. Rep.*, 1985, **3**, 22.
78. D. I. Edwards, R. C. Knight and A. Zahoor, *Int. J. Radiat. Oncol. Biol. Phys.*, 1986, **12**, 1207.
79. A. Zahoor, M. V. M. Lafleur, R. C. Knight, H. Loman and D. I. Edwards, *Biochem. Pharmacol.*, 1987, **36**, 3299.
80. A. Chapman, R. Cammack, D. Linstead and D. Lloyd, *J. Gen. Microbiol.*, 1985, **131**, 2141.
81. J. H. Wagenknecht, *J. Org. Chem.*, 1977, **42**, 1836.
82. L. M. Cobb, T. A. Connors, L. A. Elson, A. H. Khan, B. C. Mitchley, W. C. J. Ross and M. E. Whison, *Biochem. Pharmacol.*, 1969, **18**, 1519.
83. M. J. Tisdale and M. J. Habberfield, *Biochem. Pharmacol.*, 1980, **29**, 1947.
84. T. A. Connors, H. G. Mandel and D. H. Melzack, *Int. J. Cancer*, 1972, **9**, 126.
85. J. J. Roberts, F. Friedlos and R. J. Knox, *Biochem. Biophys. Res. Commun.*, 1986, **140**, 1073.
86. M. J. Waring, *Eur. J. Cancer*, 1976, **12**, 995.
87. W. E. Ross, D. L. Glaubiger and K. W. Kohn, *Biochem. Biophys. Acta*, 1978, **519**, 23.
88. W. E. Ross, D. L. Glaubiger and K. W. Kohn, *Biochem. Biophys. Acta*, 1979, **562**, 41.
89. K. W. Kohn, R. A. G. Ewig, L. C. Erickson and L. A. Zwelling, in 'DNA Repair: A Laboratory Manual of Research Procedures', ed. E. C. Friedberg and P. C. Hanawalt, Dekker, New York, 1981, p. 379.
90. L. A. Zwelling, S. Michaels, L. C. Erickson, R. S. Ungerleider, M. Nichols and K. W. Kohn, *Biochemistry*, 1981, **20**, 6553.
91. Y. Pommier, R. E. Schwartz, K. W. Kohn and L. A. Zwelling, *Biochemistry*, 1984, **23**, 3194.
92. L. A. Zwelling, D. Kerrigan, S. Michaels and K. W. Kohn, *Biochem. Pharmacol.*, 1982, **31**, 3269.
93. Y. Pommier, M. R. Mattern, R. E. Schwartz, L. A. Zwelling and K. W. Kohn, *Biochemistry*, 1984, **23**, 3194.
94. J. C. Wang, *Annu. Rev. Biochem.*, 1985, **54**, 665.
95. E. M. Nelson, K. M. Tewey and K. F. Liu, *Proc. Natl. Acad. Sci. USA*, 1984, **81**, 1361.
96. L. A. Zwelling, L. Silberman and E. Estey, *Int. J. Radiat. Oncol. Biol. Phys.*, 1986, **12**, 1041.
97. J. Minford, D. Kerrigan, M. Nichols, S. Shackney and L. A. Zwelling, *Cancer Res.*, 1984, **44**, 5583.
98. L. A. Zwelling, J. Minford, M. Nichols, R. I. Glazer and S. Shackney, *Biochem. Pharmacol.*, 1984, **33**, 3903.
99. L. A. Zwelling, D. Kerrigan and L. J. Marton, *Cancer Res.*, 1985, **45**, 1122.
100. J. J. Roberts and A. J. Thomson, *Prog. Nucl. Acid Res. Mol. Biol.*, 1979, **22**, 71.
101. L. H. Einhorn and S. D. Williams, *New Engl. J. Med.*, 1979, **300**, 289.
102. J. M. Ducore, L. C. Erickson, L. A. Zwelling, G. Laurent and K. W. Kohn, *Cancer Res.*, 1982, **42**, 897.
103. J. E. Moulder, J. S. Holcenberg, B. A. Kamen, M. Cheng and B. L. Fish, *Int. J. Radiat. Oncol. Biol. Phys.*, 1986, **12**, 1415.
104. D. Glover, J. H. Glick, C. Weiler, J. Yuhas and M. M. Kligerman, *Int. J. Radiat. Oncol. Biol. Phys.*, 1984, **10**, 1781.
105. C. T. Couglin and R. C. Richmond, *Int. J. Radiat. Oncol. Biol. Phys.*, 1985, **11**, 915.
106. E. A. Douple, *Platinum Met. Rev.*, 1985, **29**, 118.
107. J. O'Hara, E. A. Douple and R. C. Richmond, *Int. J. Radiat. Oncol. Biol. Phys.*, 1986, **12**, 1419.
108. J. M. Pascoe and J. J. Roberts, *Biochem. Pharmacol.*, 1974, **23**, 1345.
109. L. Munchausen and R. O. Rahn, *Biochem. Biophys. Acta*, 1975, **414**, 242.
110. J. J. Roberts and J. M. Pascoe, *Nature (London)*, 1972, **235**, 282.
111. L. A. Zwelling, K. W. Kohn and T. A. Anderson, *Proc. Am. Assoc. Cancer Res.*, 1978, **19**, 233.
112. L. A. Zwelling, K. W. Kohn, R. A. G. Ewig and T. Anderson, *Cancer Res.*, 1978, **38**, 1762.
113. A. Zahoor, M. V. M. Lafleur, E. J. Pluijmackers-Westmijze and D. I. Edwards, *Br. J. Cancer*, 1986, **53**, 829.
114. N. A. A. Fraval, C. R. Rawlings and J. J. Roberts, *Mutat. Res.*, 1978, **51**, 121.
115. O. Vrana, V. Brabec and V. Kleinwachter, *Anti-Cancer Drug Design*, 1986, **1**, 95.
116. J. R. Bales, P. J. Sadler, C. J. Coulson, M. Laverick and A. H. W. Nias, *Br. J. Cancer*, 1982, **46**, 701.
117. R. Chibber, I. J. Stratford, I. Ahmed, A. B. Robbins, D. Goodgame and B. Lee, *Int. J. Radiat. Oncol. Biol. Phys.*, 1984, **10**, 1213.
118. P. K. I. Chan, K. A. Skov, B. R. James and N. P. Farrell, *Int. J. Radiat. Oncol. Biol. Phys.*, 1986, **12**, 1059.
119. R. B. Painter, in 'Radiation Biology in Cancer Research', ed. R. E. Meyn and H. R. Withers, Raven Press, New York, 1979, p. 59.
120. D. Schulte-Frohlinde, *Br. J. Cancer*, 1987, **55**, suppl. 8, 129.
121. P. E. Bryant, *Int. J. Radiat. Biol.*, 1984, **46**, 57.
122. H. Dertinger and H. Jung, 'Molekulare Strahlenbiologie', Springer, Berlin, 1969.
123. M. V. M. Lafleur, J. Woldhuis and H. Loman, *Nucleic Acid Res.*, 1981, **9**, 6591.
124. P. A. Cerutti, in 'Photochemistry and Photobiology of Nucleic Acids', ed. S. Y. Wang, Academic Press, New York, 1976, vol. 2, p. 375.
125. J. F. Ward, J. W. Evans, C. L. Limoli and P. M. Calabro-Jones, *Br. J. Cancer*, 1987, **55**, suppl. 8, 105.
126. S. S. Wallace, *Br. J. Cancer*, 1987, **55**, suppl. 8, 118.
127. P. V. Hariharan, *Radiat. Res.*, 1980, **81**, 496.
128. L. H. Breimer and T. Lindahl, *Biochemistry*, 1985, **24**, 4018.
129. M. Dizdaroglu, *Anal. Biochem.*, 1985, **144**, 593.
130. S. S. Wallace, *Environ. Mutagen.*, 1983, **5**, 769.

131. R. Rajagopalan, R. J. Melamede, M. F. Laspia, B. F. Erlanger and S. S. Wallace, *Radiat. Res.*, 1984, **97**, 499.
132. S. A. Leadon and P. C. Hanawalt, *Mutat. Res. DNA Repair Rep.*, 1983, **112**, 191.
133. S. A. Leadon, *Br. J. Cancer*, 1987, **55**, suppl. 8, 113.
134. Y. W. Kow and S. S. Wallace, *Proc. Natl. Acad. Sci. USA*, 1985, **82**, 8354.
135. L. M. Kemp, S. G. Sedgwick and P. A. Jeggo, *Mutat. Res.*, 1984, **132**, 189.
136. G. C. Walker, *Mutat. Res.*, 1984, **48**, 60.
137. E. C. Friedberg, 'DNA Repair', Freeman, New York, 1985.
138. J. F. Ward, *Int. J. Radiat. Oncol. Biol. Phys.*, 1986, **12**, 1027.
139. B. K. Duncan and B. Weiss, *J. Bacteriol.*, 1982, **151**, 750.
140. L. H. Breimer, *Nucleic Acids Res.*, 1984, **12**, 6359.
141. L. A. Zwelling, D. Kerrigan and Y. Pommier, *Biochem. Biophys. Res. Commun.*, 1982, **104**, 897.
142. E. G. Miller, *Fed. Proc., Fed. Am. Soc. Exp. Biol.*, 1976, **36**, 906.
143. B. W. Durkacz, O. Omidiji, D. A. Gray and S. Shall, *Nature (London)*, 1980, **283**, 593.
144. D. M. Brown, M. R. Horsman, D. G. Hirst and J. M. Brown, *Int. J. Radiat. Oncol. Biol. Phys.*, 1984, **10**, 1665.
145. P. J. Thraves, K. L. Mossman, D. M. T. Frazier and A. Dritschilo, *Int. J. Radiat. Oncol. Biol. Phys.*, 1986, **12**, 1541.
146. J. F. Ward, E. I. Joner and W. F. Blakely, *Cancer Res.*, 1984, **44**, 59.
147. A. Giaccia, R. Weinstein, J. Hu and T. D. Stamato, *Somatic Cell Mol. Genet.*, 1985, **11**, 485.
148. P. A. Jeggo and L. M. Kemp, *Mutat. Res.*, 1984, **112**, 313.
149. L. Skoog and G. Bjursell, *J. Biol. Chem.*, 1974, **249**, 6434.
150. A. A. Francis, R. D. Blevins, W. L. Carrier, D. P. Smith and J. D. Regan, *Biochem. Biophys. Acta*, 1979, **563**, 385.
151. A. A. Francis, R. D. Snyder, W. C. Dunn and J. D. Regan, *Mutat. Res.*, 1981, **83**, 159.
152. J. F. Ward, in 'Mechanisms of DNA Damage and Repair', ed. M. G. Simic, L. Grossman and A. D. Upton, Academic Press, New York, 1985, p. 135.
153. N. R. Cozzarelli, *Annu. Rev. Biochem.*, 1977, **46**, 641.
154. G. Iliakis, *Radiat. Res.*, 1980, **83**, 537.
155. P. E. Bryant and D. Blocher, *Int. J. Radiat. Biol.*, 1982, **42**, 385.
156. E. Ben-Hur, *Radiat. Res.*, 1981, **88**, 155.
157. M. R. Mattern, P. V. Hariharan and P. A. Cerutti, *Biochem. Biophys. Acta*, 1975, **395**, 48.
158. L. D. Dale, D. A. Widdick and D. I. Edwards, *Int. J. Radiat. Oncol. Biol. Phys.*, 1989, **16**, 995.

10.3
Agents Interfering with DNA Enzymes

ROBERT P. HERTZBERG

Smith Kline & French Laboratories, King of Prussia, PA, USA

10.3.1	THERAPEUTIC POTENTIAL OF DNA ENZYME INHIBITORS	753
10.3.2	DNA POLYMERASES	755
10.3.2.1	*Enzymology and Functions of DNA Polymerases*	755
10.3.2.1.1	*Prokaryotic DNA polymerases*	755
10.3.2.1.2	*Eukaryotic DNA polymerases*	756
10.3.2.1.3	*Viral DNA polymerases and reverse transcriptase*	758
10.3.2.2	*Inhibitors of DNA Polymerases*	759
10.3.2.2.1	*Azidothymidine and dideoxynucleosides*	760
10.3.2.2.2	*Acyclovir*	761
10.3.2.2.3	*Arabinosylcytosine and arabinosyladenine*	762
10.3.2.2.4	*Aryl derivatives of purines and pyrimidines*	763
10.3.2.2.5	*Aphidicolin*	764
10.3.2.2.6	*Phosphonoformate and phosphonoacetic acid*	765
10.3.2.2.7	*Anionic polymerase inhibitors*	766
10.3.2.2.8	*DNA binding agents*	766
10.3.3	DNA TOPOISOMERASES	769
10.3.3.1	*Enzymology and Functions of Topoisomerases*	769
10.3.3.1.1	*Prokaryotic DNA topoisomerases*	770
10.3.3.1.2	*Eukaryotic DNA topoisomerases*	771
10.3.3.2	*Inhibitors of DNA Topoisomerases*	773
10.3.3.2.1	*Quinolone antibacterial agents*	773
10.3.3.2.2	*Novobiocin*	775
10.3.3.2.3	*Camptothecin*	775
10.3.3.2.4	*Amsacrine*	776
10.3.3.2.5	*Epipodophyllotoxins*	778
10.3.3.2.6	*Anthracyclines*	779
10.3.3.2.7	*Ellipticine*	781
10.3.3.2.8	*Actinomycin*	781
10.3.4	DNA METHYLTRANSFERASES	782
10.3.4.1	*Enzymology and Functions of DNA Methyltransferases*	782
10.3.4.2	*5-Azacytidine*	783
10.3.5	MICROTUBULE ASSEMBLY	783
10.3.5.1	*Properties and Functions of Microtubules*	783
10.3.5.2	*Mitotic Inhibitors*	784
10.3.5.2.1	*Colchicine*	784
10.3.5.2.2	*Podophyllotoxin*	785
10.3.5.2.3	Vinca *alkaloids*	785
10.3.5.2.4	*Taxol*	786
10.3.6	REFERENCES	787

10.3.1 THERAPEUTIC POTENTIAL OF DNA ENZYME INHIBITORS

Enzymes that catalyze the synthesis and modification of DNA are critical for the maintenance and propagation of all life forms. The fact that genes coding for nucleic acid enzymes are found within the genomes of almost every single living organism testifies to their importance. Even primitive viruses that contain very few gene products often code for at least one enzyme that can synthesize or modify nucleic acids. Because DNA enzymes are so vital to living organisms, agents that inhibit such enzymes interfere with critical cellular processes and often have potent therapeutic effects.

DNA enzymes have two primary functions: to perpetuate the genetic information by DNA replication, and to express the genetic information by transcription into mRNA. There are several classes of DNA enzymes responsible for carrying out these cellular processes. DNA and RNA polymerases are the primary enzymes that catalyze the assembly of monomer units into new nucleic acid sequences complementary to the DNA template. In addition, there are several classes of secondary enzymes that are important during replication and transcription. DNA topoisomerases are enzymes that alter DNA topology and facilitate the helix unwinding required for DNA processing. DNA helicases unwind the double helix and prepare the template for copying. DNA methyltransferases modify the nucleotide bases and are thought to play a role in gene regulation. DNA repair enzymes remove nucleotides that have been damaged by mutagens and carcinogens and restore the natural sequence. DNA transposases and recombinases move genetic material from one place to another within the genome; this may be important for the control of gene expression. Endonucleases and exonucleases cleave DNA and are important to such processes as DNA repair, proofreading, and recombination. The restriction enzymes, which cleave specific sequences of DNA, have proven to be important tools for the manipulation of genes in molecular biology research.

Since the functions of these DNA enzymes are central to the survival and propagation of life forms, agents that interfere with these enzymes generally lead to cell death. Therefore the primary therapeutic use of such agents occurs in those instances where unwanted cells impact humans in a negative fashion. Infectious diseases (bacterial, viral, fungal, *etc.*) and cancer are examples where abnormal cells need to be destroyed to obtain therapeutic gain. To use a DNA enzyme for a target, a chemotherapeutic agent must be selective for the enzyme of the invading cells since inhibition of the corresponding human enzyme will produce host toxicity. This task is much easier in the case of infectious diseases, since the infectious agent is very different from a human cell. In most cases the target DNA enzyme of the infecting organism differs structurally and pharmacologically from the complementary human enzyme, and this difference can be exploited to produce a drug with selective toxicity.

The quinolone antibacterial agents represent a recent successful example of this strategy.[1, 2] These agents are potent inhibitors of bacterial topoisomerase II (DNA gyrase), but are very weak inhibitors of human topoisomerase II and bind poorly to DNA.[3] As a result of this selectivity the quinolone antibacterials have a large therapeutic index and are currently being used clinically to treat a variety of bacterial infections. The high selectivity of the quinolones is unusual; most agents that inhibit DNA enzymes are less selective and demonstrate toxic side effects in addition to killing the infectious agent and producing the desired therapeutic effect.

In the case of human cancer, the distinction between tumor and normal DNA enzymes is small and there are no unambiguous qualitative differences that can be exploited. Consequently, a quantitative difference in enzyme content or activity is often used to differentiate between tumor and normal cells. Nevertheless, many (in fact most) antitumor agents are thought to exert their therapeutic effects by interfering with DNA processing. These cytotoxic antitumor drugs are probably selective because the proliferation rate of certain tumors is greater than that of normal tissue. Cell proliferation requires high DNA enzyme activity, and agents targeted at these enzymes can have some degree of selective cytotoxicity. In the case of certain tumors, this selectivity translates into clinically useful drugs that successfully treat human disease. Unfortunately, many cancers are resistant to drugs that target DNA enzymes. These tumors, which are generally slower growing, may contain levels of DNA enzyme activity that more closely resemble those in adjacent normal cells. The quantity and quality of DNA enzymes in these cells must be characterized if antitumor agents efficacious against these resistant tumors are to be developed. Examples of studies directed toward this goal are discussed throughout this chapter within each enzyme class.

The chemical aspects of DNA enzyme inhibition begin with the identification of various classes of inhibitory mechanisms. All such enzymes bind to DNA to form a binary complex, and some DNA enzymes require a second substrate such as dNTP (dioxynucleotide triphosphate) or a cofactor such as ATP. Agents can inhibit DNA enzymes by several mechanisms: (1) binding to the substrate DNA; (2) binding to the free enzyme; (3) binding to the binary enzyme–DNA complex; (4) binding to the nucleotide triphosphate binding site; or (5) binding to the product binding site. While examples of all of these inhibitory mechanisms are known, some offer more potential to obtain the selectivity required to limit host toxicity. Since all DNA enzymes are likely to be inhibited by DNA binding agents, such inhibitors are usually not very selective. Similarly, since all polymerases use dNTPs as substrates, nucleotide analogs usually inhibit host enzymes to some degree and are partially selective at best. The same is true for product analogs that mimic the pyrophosphate removed from dNTPs during the DNA polymerization process. Among the mechanisms noted above, those that offer the greatest therapeutic potential involve specific binding to enzyme or enzyme–DNA complexes.

This chapter discusses the biology, biochemistry and medicinal chemistry of DNA enzyme inhibition. The cellular functions and catalytic mechanisms of the most important DNA enzymes are briefly reviewed. The therapeutically relevant inhibitors for each enzyme class are discussed, with emphasis on mechanisms of inhibition and structure–activity relationships. The relationship between biochemical mechanism, target selectivity and therapeutic efficacy for each group of inhibitors is also presented.

10.3.2 DNA POLYMERASES

10.3.2.1 Enzymology and Functions of DNA Polymerases

DNA replication, the process of copying the DNA template prior to cellular division, requires multienzyme complexes in both prokaryotes and eukaryotes. These complexes consist of a large DNA polymerase subunit and several smaller proteins and enzymes. The chemistry of DNA polymerization, as well as the chemical and structural properties of the DNA helix, are virtually identical in all living organisms. The synthesis of new DNA strands occurs through the formation of new phosphodiester bonds. The DNA polymerase mediates the addition of monomer nucleotides to the end of a growing chain by catalyzing a chemical reaction between a deoxyribose 3′-OH group and a nucleotide triphosphate. The polymerization reaction almost always requires a short nucleic acid primer that is base paired to a DNA template. DNA synthesis always proceeds in the 5′→3′ direction, and base pairing with the template strand dictates the nature of the nucleotide that is added to the chain.

Although the chemistry of DNA synthesis is similar among all organisms, the mechanistic details and the auxiliary proteins in the multienzyme complexes often differ. In bacteria the molecular structure and enzymology of DNA polymerases have been well studied. In eukaryotic systems, however, the complex picture of DNA replication is just beginning to become clear.

10.3.2.1.1 Prokaryotic DNA polymerases

The most well-studied prokaryotic replication system is that of *E. coli*, which has three different DNA polymerases capable of carrying out DNA synthesis. DNA polymerase I is the simplest and best understood DNA polymerase. Although it is not the major replication polymerase in *E. coli*, it plays an important role in DNA replication and DNA repair. DNA polymerase III, a multisubunit assembly, synthesizes most of the new DNA in *E. coli*. DNA polymerase II is a third DNA polymerase in *E. coli*; its function is not yet known.

E. coli DNA polymerase I (Pol I) provides an excellent model system for understanding all DNA polymerases. The recent crystal structure analysis of the large fragment of Pol I has led to an appreciation of the molecular details of DNA binding and synthesis.[4] Pol I is a single polypeptide of 103 kDa that catalyzes the addition of dNTPs to the 3′-end of a growing chain.[5, 6] The enzyme is dependent on a single-stranded DNA template and requires a divalent metal such as Mg^{2+} to bind nucleotides and polymerize DNA. It is now accepted that Pol I does not contain a tightly bound Zn^{2+} ion.[7] Pol I is a moderately processive enzyme, catalyzing the addition of about 20 dNTPs during each template binding event. Pol I has two enzymatic activities in addition to DNA polymerization: a 3′→5′ exonuclease that removes terminal nucleotides not complementary to the template, and a 5′→3′ exonuclease that removes DNA ahead of the growing point of a DNA chain. The 3′→5′ exonuclease dramatically increases the fidelity of DNA replication since it removes mismatched residues before the polymerase proceeds, thus acting as a proofreading function.

The active site for the polymerase and editing activities are known to be clearly separate from the 5′→3′ exonuclease. Proteolysis of Pol I affords two separate domains: a 35 kDa N-terminal fragment containing the 5′→3′ exonuclease activity, and a 68 kDa fragment (Klenow fragment) containing the polymerization and proofreading functions.[8] High-resolution structural information on the DNA binding and catalytic mechanism of the Klenow fragment has been deduced from crystal structure analysis.[4, 9] The polymerase active site has been shown to be separated by about 30 Å from the 3′→5′ exonuclease domain. The small N-terminal domain binds a dNMP (the product of DNA hydrolysis) and two divalent metal ions, consistent with its assignment as the exonuclease site. The large domain contains a fascinating deep cleft, 25 Å wide and 30 Å deep, which appears to be the binding site for duplex DNA (the product of polymerization). This cleft has been said to resemble a right hand holding a rod; a flap over the cleft has been proposed to facilitate processivity of polymerization.[9]

The mechanism of DNA polymerization by Pol I has been elucidated by a variety of kinetic and structural analyses.[4,10,11] First, the enzyme binds to the primer template, and then dNTP binds to the active site. A conformational change, which is rate limiting and possibly important for ensuring fidelity, occurs prior to the DNA polymerization step.[10] The required Mg^{2+} ion is thought to bind to the enzyme and to the β and γ phosphates of the dNTP, which activates the departure of the pyrophosphate group.[11] Polymerization occurs by removal of the 3'-OH proton at the end of the growing chain and attack of the 3'-oxygen on to the α-phosphate of the dNTP, thereby forming the new phosphodiester bond and liberating pyrophosphate. Many of the features of this mechanism may be shared by other DNA polymerases.

DNA polymerase III, the primary replicative polymerase of *E. coli*, is significantly more complex than Pol I.[5,6,12] The holoenzyme of DNA polymerase III contains at least eight polypeptides and has a mass of about 800 kDa. It is extremely processive, more so than most DNA polymerases, catalyzing the addition of thousands of nucleotides in a single binding event. The high processivity facilitates a very fast polymerization rate, about 500 nucleotides per second, compared with only 10 per second for Pol I. Even at this high rate the enzyme is very accurate since it contains a 3'→5' exonuclease activity for proofreading. In addition to the DNA polymerase III holoenzyme, efficient DNA synthesis in *E. coli* requires several auxiliary enzymes such as DNA helicase (for DNA unwinding), primase (to synthesize the RNA primer), DNA ligase (to seal DNA nicks) and DNA gyrase (to introduce negative supercoils). DNA gyrase is an important target for antibacterial agents and is discussed in Section 10.3.3.

10.3.2.1.2 Eukaryotic DNA polymerases

Eukaryotic cells contain several different DNA polymerases. The major polymerase activity in eukaryotic cells is DNA polymerase α, which plays a major role in the replication of genomic DNA and in repair synthesis. DNA polymerase α contains several subunits, including a DNA primase that is capable of synthesizing short oligoribonucleotides complementary to the DNA template, and a major polymerase subunit responsible for chain elongation. Other eukaryotic DNA polymerases include DNA polymerase β, which functions in DNA repair; DNA polymerase γ, which replicates mitochondrial DNA; and the newly discovered DNA polymerase δ.

DNA polymerase α levels have been shown to increase dramatically upon the induction of cell division, indicating that it is associated with the replication of genomic DNA.[13] Of DNA polymerases α, β and γ, polymerase α is the only animal polymerase demonstrated to have this property. DNA polymerase α is located in the nucleus and is associated with newly replicating DNA. In addition, selective inhibitors of this enzyme (for example, aphidicolin) reversibly block genomic DNA replication without affecting mitochondrial DNA replication. Furthermore, monoclonal antibodies specific for DNA polymerase α were shown to arrest DNA replication specifically in a dose-dependent manner and were concentrated in the cell nucleus.[14] These findings provide a very convincing argument that DNA polymerase α is the primary enzyme in animal cells responsible for the replication of genomic DNA.

In addition to its role in DNA replication, several studies have indicated that DNA polymerase α also functions in DNA repair synthesis. Specific inhibitors of this enzyme have been shown to block the repair of DNA lesions induced by UV light and other DNA-damaging agents.[13,15] In addition, mutant hamster cells that contain an altered polymerase α have been shown to contain diminished DNA repair capability. Other DNA polymerases, notably polymerase β, are also important for DNA repair synthesis; these two enzymes may cooperate to repair certain kinds of DNA lesions.[16]

The structure and biochemical properties of DNA polymerase α are only partially characterized. This is because, until recently, there has been no general method for purification of the intact multisubunit enzyme complex. The complex has now been purified from several different eukaryotic organisms; in each case it contains four subunits of approximately 180, 70, 60 and 50 kDa.[12,17] The 180 kDa subunit has DNA polymerase activity and is thought to be the catalytic subunit. Until recently it was thought that this subunit lacked 3'→5' exonuclease activity, but recent studies with Drosophila embryo polymerase α demonstrated exonuclease activity in the catalytic polypeptide when separated from the other subunits. Similar to its function in prokaryotic polymerases, this exonuclease activity may provide a proofreading function by removing nucleotides that are not complementary to the template.

The mechanism of DNA polymerase α was recently reviewed by Loeb and coworkers.[13,16] It should be noted that many studies of the catalytic properties of DNA polymerase α were obtained with enzyme preparations that contained only part of the multienzyme complex. There is a possibility that mechanistic conclusions reached on the basis of these results do not reflect precisely the situation *in vivo*, where the catalytic polypeptide interacts with several auxiliary subunits and cofactors. This caveat also pertains to experiments involving the inhibition of DNA polymerase α (see Section 10.3.2.2).

DNA polymerase α uses DNA templates most efficiently; unlike *E. coli* DNA polymerase I, it has not been shown to copy a natural RNA template. According to a model for binding and catalysis proposed by Korn,[18,19] the enzyme binds to a single-stranded region on the template and undergoes an allosteric transition that activates a second DNA binding site. This second site binds to the primer stem and allows binding of a dNTP molecule. Although these initial kinetic studies were performed with the catalytic subunit alone, recent experiments with an immunopurified DNA polymerase α preparation composed of a more complete multisubunit complex confirmed this ordered reaction mechanism.[20]

DNA polymerase α binds to and extends primers composed of either deoxyribonucleotides or ribonucleotides. Unlike DNA polymerases β or γ, polymerase α can extend natural RNA primers, consistent with its function *in vivo* as the polymerase that elongates RNA-primed Okazaki fragments at the replication fork.[13] A single mismatch at the end of the primer stem significantly lowers the efficiency of binding and extension; this helps to ensure fidelity of DNA replication.[21] Studies of enzyme processivity have shown that DNA polymerase α polymerizes between 5 and 20 nucleotides per primer-template binding event.[13] This low degree of processivity may reflect the lack of appropriate auxiliary factors in these assays.

Kinetic studies of the catalytic subunit have led to an estimation of the K_m for dNTPs of between 1 and 20 μM.[13] This relatively large value suggests that, *in vivo*, the multienzyme complex contains proteins that bind more tightly to dNTPs and deliver them to the catalytic core subunit. Like prokaryotic polymerases, catalysis by polymerase α requires a divalent cation, with magnesium being the most efficient activating metal. The most likely role of the magnesium ion involves coordination with both the incoming dNTP and the active site of the enzyme. In addition, it has been shown that magnesium ions affect enzyme–DNA binding independent of catalysis, suggesting that more than one type of metal binding site may be present in the enzyme active site.

The other subunits of the multienzyme replication complex have been recently isolated but are not well understood. The 50 and/or 60 kDa subunit provides DNA primase activity, which is defined as the synthesis of short oligoribonucleotides that can be used by the catalytic subunit for extension.[12] DNA primase is a critical enzyme that is responsible for initiating the synthesis of Okazaki fragments on the discontinuous side of the replication fork. Aphidicolin is a strong inhibitor of DNA polymerase α but has no effect on primase. Selective inhibitors of DNA primase would be invaluable tools for mechanistic and functional studies of this important enzyme.

DNA polymerase β was the second enzyme discovered in eukaryotic cells that is capable of synthesizing DNA.[22] It has a much more restricted function that DNA polymerase α, playing a major role in the repair of nuclear DNA.[16] DNA polymerases of the β class consist of a single subunit of size ranging from 32 to 50 kDa. DNA polymerases β is relatively resistant to thiol reagents and is insensitive to aphidicolin. Even though DNA polymerases α and β differ in structure and function, their enzymatic mechanisms for DNA synthesis are similar.[16] Both require divalent metal ions and follow the same ordered mechanism for substrate binding.[23] Consistent with its role in DNA repair, polymerase β prefers gapped DNA as a template; short gaps of approximately ten nucleotides are optimal and are filled to completion.

DNA polymerase γ is a eukaryotic polymerase whose major function appears to be the replication of mitochondrial DNA.[16] While it is located both in the nucleus and the mitochondria, its role in nuclear DNA synthesis (if any) is not known. The primary catalytic property of DNA polymerase γ that distinguishes it from other DNA polymerases is its preference for an oligo(dT)·poly(rA) primer template. DNA polymerase γ is highly processive, more so than polymerase α or β, synthesizing long stretches of DNA during a single binding event.

DNA polymerase δ is the most recently discovered DNA polymerase and its function is unknown. DNA polymerase δ most closely resembles DNA polymerase α in its properties; for example, these two are the only eukaryotic polymerases sensitive to aphidicolin.[24] Mammalian DNA polymerase δ contains a $3' \rightarrow 5'$ exonuclease activity and may function in DNA repair and/or proofreading. Consistent with this notion, a recent report indicated that purified DNA polymerase δ, when added to permeabilized repair-deficient human fibroblast cells, restored DNA repair synthesis.[25] Studies of the catalytic and structural properties of the rabbit enzyme showed that both the exonuclease and polymerase functions appear to reside on a single 122 kDa polypeptide.[16]

Recently, the gene for the catalytic polypeptide of human DNA polymerase α has been cloned.[26] These investigators demonstrated that the expression of polymerase α mRNA, like its enzymatic activity, increased with cell proliferation and transformation. Comparison of the amino acid sequences among several DNA polymerases, including those from viruses, yeast, and human, revealed several conserved domains.[26] Although homology was found between human DNA polymerase α and polymerases from several different species, there was no homology with non-replicative eukaryotic DNA polymerases such as DNA polymerase β.

10.3.2.1.3 *Viral DNA polymerases and reverse transcriptase*

While the strategy of DNA replication among higher organisms is relatively constant, viruses use a wide range of mechanisms for genomic replication. This is due to the diversity of genetic material contained in viruses. Viral nucleic acid can be single stranded or double stranded; it can be RNA or DNA, or both; it can be linear, circular or of complex structure. Furthermore, the nature of the viral genetic material often changes upon infection of the host. Therefore, different viruses encode several different kinds of nucleic acid polymerases important for viral replication, including DNA-directed DNA polymerases, RNA-directed DNA polymerases, and RNA-directed RNA polymerases. Many viruses use host enzymes to begin transcription of their genes and code for their own polymerases for replication. Other viruses do not code for any polymerases and depend totally on the host for nucleic acid enzymes. A description of the diversity of viral replication mechanisms is beyond the scope of this discussion; the reader is directed to recent reviews.[27-29] Instead, a brief description of two therapeutically relevant DNA polymerases, herpes simplex virus DNA polymerase and HIV reverse transcriptase, is presented.

The DNA polymerase encoded by herpes simplex virus 1 (HSV-1) is a single polypeptide of molecular mass 136 kDa containing both polymerase and $3' \rightarrow 5'$ exonuclease activities, but lacking a $5' \rightarrow 3'$ exonuclease. The gene for HSV-1 polymerase has been cloned and sequenced.[30] The enzyme is extremely processive (>5000 nucleotides per binding event) due to its tight binding affinity for primer terminus ($K_d < 1$ nM).[31] Although HSV-1 DNA polymerase is extraordinarily slow at polymerization *in vitro* (0.25 nucleotide per second), it is likely that auxiliary proteins (either host or viral) stimulate the DNA synthesis rate *in vivo*. Consistent with this model, it has been shown that *E. coli* single-stranded binding protein increases the rate of polymerization by HSV-1 DNA polymerase 20-fold.[31] Unlike *E. coli* DNA polymerase I, the exonuclease active site appears to be close to the polymerization site since phosphonoacetic acid inhibits both activities with equal potency (see Section 10.3.2.2.6).

There are several important differences between herpes DNA polymerase and DNA polymerase α from mammalian cells.[31] The different subunit structure, molecular mass and DNA affinity suggest that these two enzymes contain active sites with different conformations. It is therefore possible to find inhibitors with a preference for the viral polymerase over the host enzyme. This selectivity has been realized with both nucleoside analogs and non-nucleoside drugs and is discussed in Section 10.3.2.2.

Reverse transcriptase is a DNA polymerase contained in RNA viruses called retroviruses. Its primary distinguishing property is its ability to use RNA as a template. The function of reverse transcriptase is to produce double-stranded DNA from the retroviral single-stranded RNA, so that the genetic information can be integrated into a host cell chromosome. Recently, interest in reverse transcriptases has increased dramatically due to the AIDS epidemic. Human immunodeficiency virus (HIV), which causes AIDS, is a retrovirus and depends on this enzyme to replicate its genetic material. HIV reverse transcriptase, being virally encoded, is likely to differ in structure and mechanism from human DNA polymerases and has become a prime target for antiretroviral therapy. Because there is no evidence for a human-derived reverse transcriptase, it should be possible to find inhibitors that will be selective for the retroviral enzyme. The most important example of this strategy is azidothymidine, a reverse transcriptase inhibitor with partial selectivity that is the only drug currently used for AIDS therapy (see Section 10.3.2.2.1).

Reverse transcriptase is packaged into the core of retroviral virions along with other enzymatic activities such as ribonuclease H, endonuclease and protease.[27] All of these enzymatic activities are encoded by the *pol* gene and translated as a precursor protein, which is subsequently processed by the viral protease into three polypeptides. One polypeptide contains both the ribonuclease H and reverse transcriptase activities, another product is the endonuclease (also called integrase), and the third is the free protease enzyme. The integrase is responsible for cleaving the retroviral DNA and integrating it into the genome of the host cell.

HIV reverse transcriptase is most likely composed of two closely related polypeptides of molecular mass 66 and 51 kDa that share a common amino terminus.[32] It has several associated activities including an RNA-directed DNA polymerase, a DNA-directed DNA polymerase, an RNAse H activity, an endonuclease activity, and a DNA unwinding activity. All of these different activities are used during the viral replication process.[27] First, the enzyme copies the single-stranded RNA template to produce an RNA–DNA hybrid. The RNA strand of the hybrid is digested by the RNAse H activity. The resulting single-stranded DNA is copied by the polymerase to produce a double-stranded DNA molecule with the viral nucleotide sequence. That all of these functions reside on a single enzyme reflects the striking efficiency of the retroviral genome.

Recently, the HIV reverse transcriptase has been cloned in a number of laboratories.[33–35] In one case, a single 66 kDa protein has been demonstrated to have authentic reverse transcriptase activity, casting some doubt on the relevance of the 51 kDa polypeptide also observed in infected cells.[33] Sequence analysis comparisons with known RNAses have shown that the C-terminus of this polypeptide may contain the RNAse H function. Initial site-specific mutagenesis studies of HIV reverse transcriptase have been initiated and several important functional regions of the enzyme have been tentatively identified.[36] The recent progress towards crystallization of HIV reverse transcriptase may provide further information on the relationship of structure to function.[37]

Much of what is known about HIV reverse transcriptase has been deduced from earlier studies with the avian myeloblastosis virus (AMV) enzyme.[28] AMV reverse transcriptase is a heterodimer, containing equal amounts of two polypeptides designated α and β. The mature form of the enzyme is thought to result from the cleavage of a $\beta\beta$ dimer to yield $\alpha\beta$. The α subunit of AMV reverse transcriptase has full polymerase and ribonuclease activity but does not specifically bind the tRNA primer or have endonuclease activity. The enzyme polymerizes about 26 nucleotides on average per template binding event and thus is a moderately processive enzyme. It requires a divalent cation such as Mg^{2+} or Mn^{2+} and may contain a tightly bound Zn^{2+} ion in or near the active site. AMV reverse transcriptase binds to intact duplex DNA, to nicked duplex DNA, to circular DNA, to single-stranded DNA and RNA, and to tRNA. The enzyme binds selectively to tRNA[trp], the primer that initiates AMV proviral DNA synthesis. Although AMV reverse transcriptase does not contain a $5'\rightarrow3'$ or a $3'\rightarrow5'$ DNA exonuclease activity, it can carry out pyrophosphorolysis and pyrophosphate exchange reactions.

Reverse transcriptase, like all template-directed DNA polymerases, requires a polynucleotide template that is base paired to a presynthesized primer (a nucleic acid fragment containing a free 3'-OH group). All reverse transcriptases obviate the need for primase activity by using a host tRNA molecule as a primer for DNA synthesis. The initiation complex for HIV DNA synthesis consists of three macromolecules and one small molecule: a) the HIV RNA template; b) a tRNA[lys] molecule from the host cell; c) the reverse transcriptase enzyme; and d) the first dNTP for DNA synthesis. Although the mechanism of DNA polymerization is not yet established, it is likely that a Mg^{2+} ion binds to the β and γ phosphates of the dNTP and activates the departure of the pyrophosphate group. Attack of the terminal 3'-OH group of the primer on to the α-phosphate of the dNTP produces the phosphodiester bond. After dissociation of pyrophosphate, the enzyme catalyzes another round of synthesis. Recent studies of HIV reverse transcriptase have suggested that the enzyme has an extremely high error rate and often misincorporates nucleotides.[38,39] This finding not only suggests a mechanism for genetic variation by the AIDS virus, but also may provide the basis for selective inhibition by nucleoside analogs that are not tolerated by host polymerases.

10.3.2.2 Inhibitors of DNA Polymerases

When designing an enzyme inhibitor, the first considerations for a medicinal chemist are the structure of the substrates and the nature of the chemical reaction catalyzed by the enzyme. For DNA polymerases, the substrates are DNA and nucleoside triphosphates. DNA analogs have proven to be of limited utility for DNA polymerase inhibition, but recent progress in this field has resulted in interesting activity for antisense DNA as inhibitors of translation *via* binding to complementary mRNA. In contrast, nucleoside analogs are the most successful DNA polymerase inhibitors used clinically.

In addition to substrate analogs, product analogs such as phosphonoformate, a pyrophosphate mimic, have also proven useful for inhibition of DNA polymerases. Compounds that bind to the enzyme–DNA complex, such as aphidicolin, and compounds that bind to DNA, such as distamycin and anthramycin, also have activity as DNA polymerase inhibitors. All of these strategies of inhibition are discussed in the following sections, using some of the most therapeutically useful compounds as examples.

10.3.2.2.1 _Azidothymidine and dideoxynucleosides_

The strategy of designing nucleoside analogs that are selective for viral DNA polymerases is the most well-studied and successful approach to viral chemotherapy, and has led to the discovery of several clinically useful antiviral drugs. This strategy, however, has inherent limitations. Human DNA polymerases also require dNTPs, and the chemical mechanisms of polymerization by the viral and human enzymes are similar. Nucleoside analogs often have significant host toxicity that is probably related to inhibition of host cell DNA synthesis. Nevertheless, these compounds constitute the major class of antiviral drugs, and this approach is likely to yield additional active compounds in the near future. For the long term, however, other strategies may ultimately lead to more selective agents with lower toxicity.

Obviously, the key to selectivity is to design analogs with a lower affinity for the host enzyme than the viral enzyme, which requires that there be structural differences between the enzyme active sites. For reverse transcriptase, the most well-studied inhibitor is 3'-azido-3'-deoxythymidine (AZT; 1), which is currently used clinically to treat AIDS.[40,41] Kinases from the host cell phosphorylate AZT to its 5'-mono-, 5'-di- and 5'-tri-phosphate forms. The rate-limiting step in the metabolism of AZT is the conversion of the monophosphate to the diphosphate form by cellular thymidylate kinase.[42] The efficiency with which AZT is converted to its triphosphate form is dependent on the cell species, suggesting that kinases from different species vary with respect to affinity for AZT. Phosphorylation is required for enzyme inhibition, since only the triphosphate derivative of AZT inhibits reverse transcriptase.

(1)

AZT-triphosphate inhibits reverse transcriptase by competition with the normal substrates (dNTPs). In addition, it can incorporate into the growing DNA chain and result in the termination of DNA synthesis, since it lacks a 3'-OH group necessary for attachment of the next dNTP.[40,42,43] AZT inhibits HIV reverse transcriptase with an IC_{50} of 40 nM,[43] but is 100–300 times less active against mammalian DNA polymerase α[40] and DNA polymerase γ.[43] The reason for this selectivity is not clear, since AZT is a chain terminator for mammalian DNA polymerases and inhibits normal cellular DNA synthesis.[44] It is possible that both enzymes bind and incorporate AZT into growing DNA chains but that reverse transcriptase is less capable of removing the false nucleotide and continuing DNA synthesis. Reverse transcriptase has no 3'→5' exonuclease activity; such activity in prokaryotic polymerases is postulated to serve as a proofreading function during DNA synthesis. This is consistent with the recent observation that HIV reverse transcriptase is one of the least accurate DNA polymerases known.[38,39]

Analyses of the three-dimensional structure of AZT by X-ray diffraction have recently been reported.[45,46] It was suggested that the azido group plays a role in enzyme binding, since two of the azido nitrogens can be superimposed upon two oxygen atoms of a thymidine 3'-phosphate moiety.[45] This raises the possibility that functional groups on reverse transcriptase hydrogen bond to these azido nitrogens in a manner similar to interactions with the natural substrates. That the K_i value for AZT-triphosphate is 2–3 orders of magnitude lower than the K_m value for dTTP[43] is consistent with the hypothesis that the azido group contributes binding energy. In addition, it is possible that AZT binds the enzyme–DNA complex in an unusual conformation, since one of the crystals was found to contain a C-3'-_exo_/C-4'-_endo_ pucker in the furanose ring.[46] There is no evidence, however, that this high-energy conformation is the biologically active form.

Several other dideoxynucleoside analogs have been shown to be potent inhibitors of HIV replication _in vitro_.[47,48] In general, these compounds have the same mechanism of action as AZT, that is, intracellular conversion to the triphosphate derivative and subsequent inhibition of HIV reverse transcriptase. Some of these compounds are simply analogs of the natural 2'-deoxynucleo-side in which the 3'-OH group has been replaced with a hydrogen, such as 2',3'-dideoxycytidine (2), 2',3'-dideoxyadenosine (3) and 2',3'-dideoxythymidine (4). Other analogs contain a 2'–3' double bond, such as 2',3'-didehydro-2',3'-dideoxythymidine (**5**). Several related analogs with other modifications to the ribose ring or the heterocyclic base moiety have also been reported to have activity against HIV or HIV reverse transcriptase.[49,50]

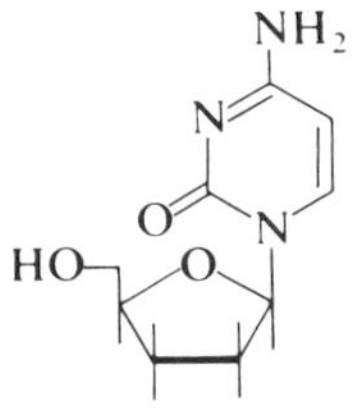

(2) 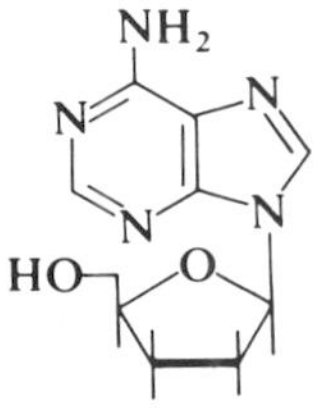   (3) 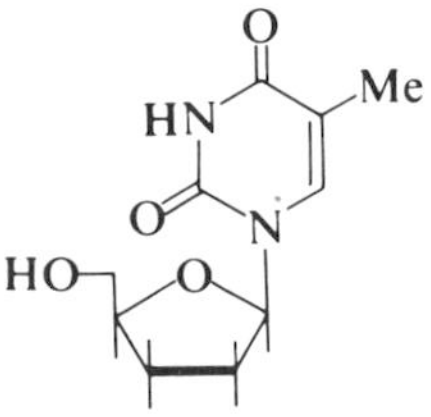   (4) 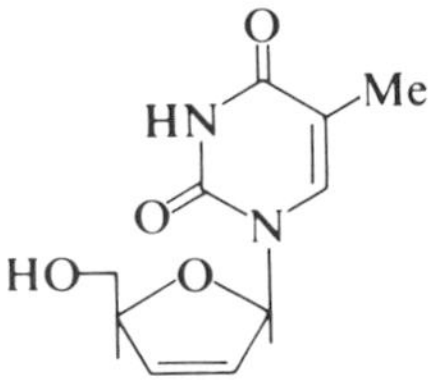   (5)

Although the general mechanism of action among the dideoxynucleoside analogs is the same, the potency and selectivity in the series can differ with structure. Importantly, some derivatives are not efficiently converted to the active triphosphate form. For example, compound (**4**) is a poor substrate for human thymidine kinase and has little activity against HIV.[51] AZT does not have this problem (see above), suggesting that the azido group at the 3'-position facilitates phosphorylation by human thymidine kinase.

Among dideoxynucleoside analogs that have been evaluated, dideoxycytidine (**2**) is efficiently converted to its triphosphate form in human cells and is one of the most potent compounds against HIV.[47] Interestingly, this compound has little activity against murine retroviruses,[52] most likely because it is poorly phosphorylated in murine cells.[53] Thus, there is apparent species-specificity among cellular kinases, and possibly cell-type-specificity within one species. This may turn out to be important, considering recent evidence that human cells other than T lymphocytes are infected with HIV. Compound (**2**) is currently in clinical trials.

10.3.2.2.2 *Acyclovir*

While the nucleoside analogs discussed above achieve selectivity at the level of the viral DNA polymerase or reverse transcriptase, another major class of antiviral drugs exploit differences between viral and cellular thymidine kinase to obtain a selective advantage. The most successful and well-understood compound in this class is acyclovir (**6**), which is highly active against herpes simplex virus (HSV) and varicella zoster virus infections, and is used clinically.[54,55] Acyclovir is a 2'-deoxyguanosine analog that contains an acyclic side chain in the place of the deoxyribose ring. It contains all of the atoms of 2'-deoxyguanosine, with the exception of the 2'- and 3'-carbons and 3'-hydroxy group of the deoxyribose ring.

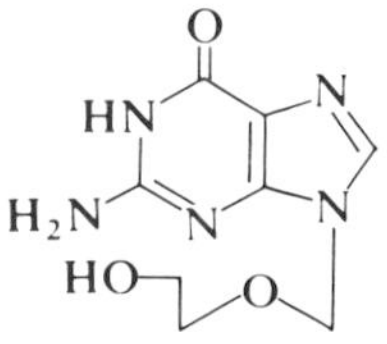

(6)

Like most nucleoside analogs, acyclovir is a prodrug. Unlike other nucleoside analogs such as AZT, however, prodrug activation is one of the mechanisms by which acyclovir obtains selectivity. The compound has a selectivity index of approximately 3000; the IC_{50} is 0.1 μM for inhibition of HSV-1, while the IC_{50} is 300 μM for cytotoxicity to host Vero cells *in vitro*.[54] This selectivity is derived from two different mechanisms: differential activation of acyclovir in infected cells and increased potency against HSV DNA polymerase. Acyclovir is phosphorylated 30–120 times faster with extracts from HSV-1-infected cells compared with uninfected cell extracts.[55] Moreover, the K_i values for HSV-encoded DNA polymerases are 10–40 times lower than for cellular DNA polymerases.[54] Thus the active form of acyclovir is present in greater quantities within virally infected cells and is also more potent against viral DNA polymerase.

Activation of acyclovir to its triphosphate form is mediated by both viral and cellular enzymes. Virally encoded thymidine kinase produces acyclovir monophosphate, which is further phosphorylated by cellular GMP kinase to the diphosphate derivative. The diphosphate is converted to acyclovir triphosphate, the active derivative, by cellular kinases such as phosphoglycerate kinase. There are several lines of evidence that support HSV thymidine kinase as a key enzyme for the antiviral effect of acyclovir.[54] Perhaps the most convincing evidence is derived from studies with thymidine kinase-deficient mutants of HSV, which are resistant to acyclovir because no phosphorylation occurs within infected cells.

The triphosphate form of acyclovir inhibits DNA polymerase. The mechanism of inhibition is complex and may involve enzyme inactivation.[56] Initially, the compound competes with dGTP for binding to the polymerase–template complex. Following binding, it has been suggested that enzyme inactivation occurs by the formation of a tight, irreversible enzyme–DNA–acyclovir triphosphate ternary complex. Addition of dGTP to this complex does not release active viral DNA polymerase. Interestingly, DNA polymerase α from the host cell is not inactivated in this manner. In addition to dGTP competition and enzyme inactivation, acyclovir triphosphate inactivates the viral DNA template by chain termination, since it does not contain the 3'-OH group necessary for the addition of subsequent dNTPs.[57] In virally infected cells, incomplete DNA replication occurs upon treatment with acyclovir.[58]

Several analogs of acyclovir have been investigated, both at the level of the target enzymes and for antiviral effect. Various alterations to the aliphatic side chain of acyclovir are tolerated by the HSV thymidine kinase, as long as the terminal hydroxy group is retained for phosphorylation.[59] However, the enzyme is far less tolerant of modifications to the heterocyclic base moiety. Since pyrimidines are the natural substrates for the viral thymidine kinase, it seemed logical to predict that pyrimidine analogs of acyclovir might be good kinase substrates and have antiviral activity. However, this is not the case. Similarly, among purine analogs, only those containing guanine are good substrates for the viral enzyme.[59]

Ganciclovir (7) is an acyclovir analog that contains an additional hydroxymethyl group in the side chain.[60] Because this compound contains groups that mimic both the 5'-OH and 3'-OH groups of deoxyguanosine, it does not act as a chain terminator for DNA synthesis. Nevertheless, compound (7) is selectively phosphorylated by the viral thymidine kinase and, as the triphosphate form, inhibits HSV-1 and HSV-2 polymerase.[61] Ganciclovir has been reported to be superior to acyclovir against human cytomegalovirus (HCMV). Although the compound is converted to the triphosphate form in HCMV-infected cells, the enzymes responsible for this transformation are not yet known. There is no evidence that HCMV encodes a thymidine kinase.

(7)   (8)   (9)   (10)

Among acyclovir analogs, the cyclic monophosphate derivative (8) has been reported to have the broadest spectrum of activity against DNA viruses.[55,62] Within cells, compound (8) is converted to the (S) enantiomer of ganciclovir monophosphate (9), the same compound that is obtained from phosphorylation of (7) by HSV-1 thymidine kinase. Other acyclovir analogs that have been prepared include compound (10) and a series of related derivatives.[55,63] These compounds have a similar mechanism as acyclovir, and several have comparable antiviral activity.

10.3.2.2.3 *Arabinosylcytosine and arabinosyladenine*

Nucleoside analogs have held a place in the field of cancer chemotherapy for many years. Among nucleoside analogs with antitumor activity that inhibit DNA synthesis as their primary mechanism of action, 1-β-D-arabinofuranosylcytosine (Ara-C; 11) is one of the most useful. Ara-C has been used for the treatment of acute non-lymphocytic leukemia for several years.[64] The usefulness of Ara-C for the treatment of cancer derives from its S-phase specificity. Like many antineoplastic agents, it is selectively cytotoxic to cells that are actively proliferating.[65] Ara-C is converted to its monophosphate derivative by cellular deoxycytidine kinase and is subsequently converted to the triphosphate form (Ara-CTP) by other cellular kinases. The variation of deoxycytidine kinase activity throughout the cell cycle may contribute to the S-phase specificity of Ara-C. In addition, the compound is rapidly deaminated by cellular deaminases to arabinosyluracil, a less active derivative.

(11)   (12)

Ara-C is a potent inhibitor of DNA synthesis in several different types of cells.[65] This inhibition results from two different mechanisms: competition with dCTP for binding to DNA polymerase α, and incorporation into the growing chain. *In vitro*, DNA polymerase α from a number of cell types is inhibited by Ara-CTP with K_i values in the low micromolar range.[16] In mammalian cells, however, the IC_{50} for DNA synthesis inhibition is significantly lower than the concentration required for competitive inhibition of purified DNA polymerase.[66] This suggests that competition with dCTP is not the only mechanism by which Ara-C affects DNA synthesis.

Ara-C has been shown to become incorporated into DNA of mammalian cells. Although the compound does possess a 3′-OH group, it acts as a chain terminator *in vitro*, possibly because the arabinose sugar is a poor substrate for chain elongation.[67] In mammalian cells, however, most of the Ara-C that is incorporated into DNA is found in internucleotide linkages rather than at 3′-termini.[65] The relationship between these observations and the potent inhibition of DNA synthesis is not clear. Incorporation of Ara-C within DNA may alter the overall conformation of the DNA helix and affect interaction with other DNA processing enzymes.[68] Alternatively, Ara-C may inhibit DNA synthesis by slowing down the progression of the multienzyme replication complex.[69]

A related nucleoside analog, 9-β-D-arabinofuranosyladenine (Ara-A; **12**), has significant antiviral activity against herpes virus and vaccinia virus.[70] In cells, Ara-A is phosphorylated to its triphosphate derivative, which preferentially inhibits HSV-1 and HSV-2 DNA polymerase relative to mammalian polymerases α and β.[71] However, Ara-A is toxic to host cells because of mammalian DNA polymerase inhibition. In addition, Ara-A is metabolized to arabinofuranosylhypoxanthine, a significantly less active derivative, by adenosine deaminase.[70] Halogenation of Ara-A at position 2 on the adenine ring reduces deamination.[72]

10.3.2.2.4 *Aryl derivatives of purines and pyrimidines*

Derivatives of nucleic acid bases have been developed as inhibitors of DNA polymerases by Brown and coworkers.[73] Compounds (**13**)–(**15**) contain two domains: a heterocyclic base that hydrogen bonds to the DNA template and an aryl group that interacts with the enzyme. One of the first compounds synthesized in this series (**13**) was found to be a selective inhibitor of *B. subtilis* DNA polymerase III. This compound binds to the DNA template using the N-1 hydrogen, the C-2 carbonyl oxygen and the C-6 NH groups to form hydrogen bonds with cytosine. Substitution in this domain or removal of cytosine from the template destroys inhibition.[73] In addition, structure–activity relationships demonstrate that the aryl group is required, suggesting that this group interacts with the DNA polymerase.

(13)   (14)

(15)

Compounds (14) and (15) were designed as inhibitors of DNA polymerase α.[73] For compound (14) the side chain on the aryl ring was found to impact on the selectivity for bacterial DNA polymerase III *vs.* mammalian polymerase α. As the size of the alkyl chain was increased from one carbon to four carbons, inhibition of polymerase III decreased while inhibition of polymerase α increased. Compound (14) inhibits polymerase α with a K_i value of 40 μM whereas polymerases β and γ are not inhibited.[16] Compound (15), a guanine base with an aryl substituent on the 2-amino group, inhibits polymerase α with a K_i value of about 10 μM. Interestingly, addition of a 2′-deoxyribose moiety to compound (15) at position 9 had little effect on inhibitory potency, but subsequent phosphorylation to the di- and tri-phosphate derivatives produced inhibitors with K_i values in the low nanomolar range.[74] The deoxyribose triphosphate derivative of (15) was shown to be competitive with dGTP, but does not incorporate into DNA. This DNA polymerase inhibitor is very selective; no inhibition to polymerases β and γ was observed at concentrations as high as 400 μM. In addition, a recent report indicated that the deoxyribose triphosphate derivative of (15) does not inhibit DNA polymerase δ, even though this polymerase closely resembles polymerase α in its properties and sensitivity to other inhibitors.[25]

10.3.2.2.5 Aphidicolin

Aphidicolin (16), a natural product isolated from fungi, is a potent and highly specific inhibitor of DNA polymerase α.[16,75] Consistent with inhibition of the primary replicative polymerase in eukaryotic cells, aphidicolin at 0.04–2.0 μg mL^{-1} arrests nuclear DNA replication in several different cell types. In purified systems the compound inhibits DNA polymerase α from a variety of cell types with K_i values in the range of 2 μM and below.[76-78] Although aphidicolin does not inhibit purified DNA polymerases β or γ, recent evidence has shown that the compound is a potent inhibitor of DNA polymerase δ.[24] In eukaryotic cells, aphidicolin has no effect on mitochondrial DNA replication or RNA synthesis, consistent with its selectivity for DNA polymerases α and δ. The compound has also been shown to be a potent inhibitor of viral polymerases, including enzymes from herpes simplex virus and vaccinia, but does not inhibit reverse transcriptase, terminal deoxynucleotidyl transferase, or bacterial DNA polymerases.[16]

(16)

The mechanism by which aphidicolin inhibits DNA polymerases is not fully understood. Several initial mechanistic studies, reviewed by Fry and Loeb,[16] indicated that inhibition of DNA polymerase by aphidicolin is competitive with respect to dCTP, noncompetitive with respect to the other three dNTPs, and uncompetitive with the DNA template and magnesium. These observations suggest that compound (16) binds to the enzyme–DNA complex at or near the dCTP binding site, and that the binding site for the other dNTPs is different from the dCTP site. Recent evidence, however, suggests that the mechanism of inhibition is likely to be more complex. Using certain types of DNA homopolymers as templates, inhibition by aphidicolin was shown to be competitive with dTTP. In addition, kinetic studies suggested that the drug acts as an allosteric inhibitor of DNA polymerase α.[79] The conformation of the enzyme–DNA complex, and consequently the shape of the binding site(s) for dNTPs and aphidicolin, probably varies with the nature of the primer template. One hypothesis is that the dNTP and aphidicolin binding sites overlap with each other, with the greatest overlap being between the aphidicolin and dCTP site. The nature of the binding site overlap may vary with conformational changes dictated by DNA sequences flanking the active site and by auxiliary proteins.

Structure–activity relationships of aphidicolin have recently been carried out with semisynthetic derivatives.[78,80,81] The 17-OH group has been acetylated to produce 17-acetylaphidicolin, which was orginally reported to inhibit DNA polymerase α with similar potency as (16).[81] Although one group reported 17-acetylaphidicolin to be inactive,[80] careful kinetic studies by Wright and coworkers demonstrated that the compound inhibited Chinese hamster ovary polymerase α with a K_i value of 5 μM, which is about 10-fold higher than the K_i value for aphidicolin. Furthermore, 17-acetylaphidicolin, like (16), was found to be competitive with dCTP and noncompetitive with the other dNTPs.[78] Removal of both the 16-hydroxy and hydroxymethyl groups, leaving 16-oxo or 16-(H$_2$) substituents, reduced activity 100-fold relative to (16).[78] Some modifications to the other end of the molecule are tolerated; the 3-deoxy and 3'-oxo derivatives of (16) were reported to be one-third as active as (16). In contrast, however, the 3-β-OH isomer and the 2,3-didehydro-3-deoxy derivative were inactive, possibly because of conformational changes induced in the A-ring that lead to movement of the 18-OH group.[80] Bulky substituents on the 18-OH group were found to eliminate activity.

Strong evidence indicating that DNA polymerase α is the relevant target for aphidicolin is derived from studies with drug-resistant cells. In some cases, aphidicolin-resistant cells have been found to contain elevated quantities of polymerase α.[16] In other systems, DNA polymerase α purified from aphidicolin-resistant cells displayed resistance to inhibition by the compound *in vitro*.[82,83] That mammalian cells can alter their DNA polymerase α in order to resist the cytotoxic effects of aphidicolin is consistent with this enzyme being the relevant target *in vivo*.

10.3.2.2.6 *Phosphonoformate and phosphonoacetic acid*

Phosphonoformate (PFA; 17) and phosphonoacetic acid (PAA; 18) are pyrophosphate analogs that inhibit a wide variety of DNA polymerases. The structures shown represent the trisodium salts; the actual charge of the acid groups depends on the pH of the solution. PFA has activity against several DNA and RNA viruses and has been tested clinically for the treatment of HSV, HIV and other viral infections. The antiviral properties and mechanism of action of PFA have recently been reviewed.[84]

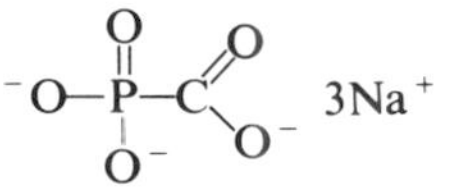

(17) 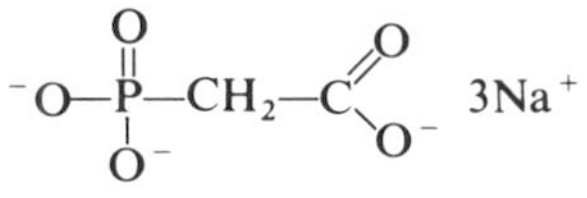   (18)

PFA is more selective than PAA with respect to inhibition of viral polymerases and has been used more frequently as an antiviral agent. PFA is especially potent against reverse transcriptases; for example, it inhibits HIV reverse transcriptase by 50% at 0.1–1 μM.[85-87] It does not inhibit the RNAse H activity of reverse transcriptase. Interestingly, PAA does not inhibit HIV reverse transcriptase up to 500 μM, suggesting that the longer distance between the phosphono and carboxyl groups is not tolerated by the enzyme.[87]

Both PFA and PAA inhibit DNA polymerases from several DNA viruses, including HSV-1, HSV-2, Epstein–Barr virus and human cytomegalovirus, at concentrations of about 1 μM or below.[84] PAA has been reported to inhibit the polymerase and exonuclease activities of the HSV-1 enzyme with similar potencies, suggesting close communication between these two catalytic sites.[31] Inhibition of mammalian DNA polymerases by PFA and PAA occurs at much higher concentrations than inhibition of most viral polymerases. For example, concentrations of PFA in the 30–50 μM range are required for 50% inhibition of polymerase α. PFA does not inhibit DNA polymerase β, possibly because this enzyme does not carry out pyrophosphate exchange or pyrophosphorolysis.[16]

The mechanism by which PFA and PAA inhibit DNA polymerization is generally thought to involve binding at the pyrophosphate binding site.[88,89] Inhibition of DNA polymerase by PFA[89] and PAA[90] has been shown to be noncompetitive with respect to dNTPs and uncompetitive with respect to the template nucleic acid. The best evidence that these compounds bind at the pyrophosphate site, however, is derived from studies showing competitive inhibition of the dNTP–pyrophosphate exchange reaction.[88,91] Furthermore, PFA-resistant DNA polymerases from HSV-1 mutants were shown to be cross-resistant to inhibition by both PAA and pyrophosphate, suggesting that these compounds bind to the same or overlapping sites.[84,91]

Dual inhibitor studies using HSV-1 DNA polymerase have shown that inhibition by PFA prevents inhibition by either acyclovir triphosphate, ganciclovir triphosphate or aphidicolin.[92] This pattern of mutually exclusive inhibition may mean that all of these compounds bind at the same or overlapping sites on the enzyme–DNA complex. This is not surprising for nucleoside triphosphate analogs and PFA, considering that PFA binds at the pyrophosphate site. For aphidicolin, however, allosteric effects may be responsible for mutually exclusive inhibition.

Structure–activity relationships have indicated that only a narrow range of structural modifications to PFA or PAA are tolerated.[84] For example, insertion of another methylene group between the phosphorus and the carboxylate of PAA eliminates activity against DNA polymerases. Similarly, addition of substituents to the methylene of PAA significantly reduces inhibitory potency. Esterification of either the carboxylate or phosphorus oxygen of PFA eliminates activity. Several other analogs of pyrophosphate have been examined, but no compounds have been found that inhibit DNA polymerases with greater potency than PFA or PAA.[84] Nucleoside-PFA derivatives, in which the phosphonate group of PFA is attached to the 5′-hydroxy group of a nucleoside, have shown antiviral activity.[93] Whether these compounds act as prodrugs or intact derivatives is not currently known.

10.3.2.2.7 Anionic polymerase inhibitors

Several anionic compounds have recently been investigated as inhibitors of DNA polymerases. The most well-studied compound in this class is suramin (19), which has been used for the treatment of trypanosomiasis and was found to block the cytopathic effect of HIV *in vitro*.[94] As DNA polymerase inhibitors, however, anionic compounds are generally not very selective. For example, suramin has been shown to inhibit DNA primase, DNA polymerase α, reverse transcriptase, and *E. coli* RNA polymerase I with K_i values in the range 0.3–2.6 μM.[95] DNA polymerases β and γ from eukaryotic cells and *E. coli* DNA polymerase I are relatively resistant to inhibition. In addition, suramin has been reported to inhibit a variety of enzymes and to bind nonspecifically to proteins such as albumin and globulins. This lack of specificity was probably responsible for the severe toxicity observed when suramin was tested in clinical trials for HIV infections.

(19)

The mechanism by which suramin inhibits reverse transcriptase and DNA polymerase α involves competition with the primer template, suggesting that the compound binds to free enzyme at or near the DNA binding site.[94,95] Electrostatic interactions between the anionic sulfonic acid groups of suramin and basic side chains on the DNA polymerase are likely, analogous to electrostatic attraction between the negatively charged phosphate groups of DNA and positively charged residues on many DNA enzymes. Thus the anionic character of suramin may explain its lack of selectivity. However, it may be possible to modify the structure of (19) and include moieties that confer selectivity. Toward this end, several analogs related to suramin have been investigated for inhibition of reverse transcriptase.[96] A recent study in which 90 suramin analogs were tested for inhibition of HIV reverse transcriptase suggested that molecular size and number of acidic groups were important determinants of inhibitory potency. Several compounds more potent than suramin were found, but inhibition of mammalian DNA polymerases was not tested.

10.3.2.2.8 DNA binding agents

Several DNA binding compounds that are useful as antitumor agents act by inhibiting DNA enzymes. These agents inhibit DNA and RNA polymerases by preventing enzyme–DNA binding or blocking chain elongation. Some of the most well-characterized DNA binding agents are intercalators, which insert between the base pairs of DNA, unwind the helix, and generally inhibit DNA and RNA polymerases nonselectively. Examples of such compounds include ethidium bromide, acridines, anthracyclines and actinomycin. Interruption by intercalating agents of DNA polymerase-mediated nick-translation[97] and RNA polymerase-mediated transcription[98] has been used to identify the sequence-specificity and kinetics of intercalator–DNA binding. A subset of the intercalating agents also inhibit DNA topoisomerases and are discussed in Section 10.3.3.2. For more information regarding compounds that bind to DNA by intercalation, the reader is also directed to Chapter 10.1.

Agents that bind to DNA in a non-intercalative manner can also have potent effects on DNA processing. Compounds in this class include agents that bind in the minor groove of DNA, such as netropsin (**20**), distamycin (**21**) and Hoechst 33258.[99] Netropsin (**20**), the most well-characterized non-intercalative DNA binding agent, interferes with both replication and transcription. The molecular details of netropsin–DNA interaction have been elucidated by a number of studies, including an X-ray analysis of a complex between the compound and a B-DNA dodecamer.[100] Netropsin binds within the minor groove of A–T regions of DNA, and each of its three amide NH groups forms a bifurcated hydrogen bond with adjacent adenine N-3 or thymine O-2 atoms. Close van der Waals contacts also provide binding energy and specificity. The related antibiotic, distamycin (**21**), is a longer groove binding molecule that binds to DNA in a similar manner. A crystal structure of a distamycin–DNA complex shows that the compound covers 5–6 base pairs and is bound by hydrogen bonds, van der Waals interactions and electrostatic bonds.[101] The sequence specificity of distamycin and netropsin is derived from two major factors: A–T regions of DNA have a narrow groove that provides closer van der Waals surfaces for interaction with the pyrrole rings, and the amino group at the 2-position of guanine sterically blocks the minor groove and prevents the drugs from contacting the floor of the groove at G–C sequences.

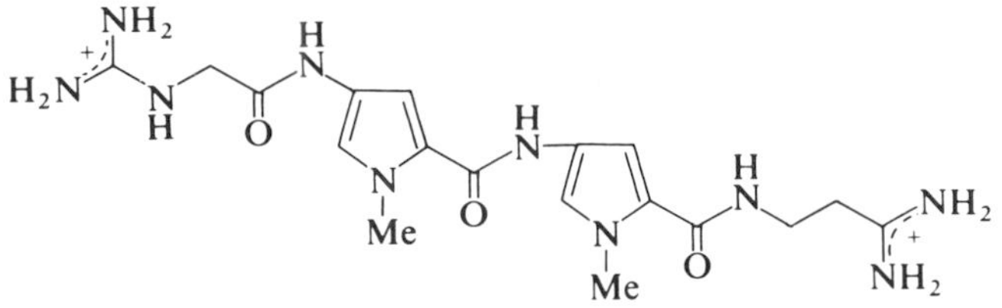

(**20**)

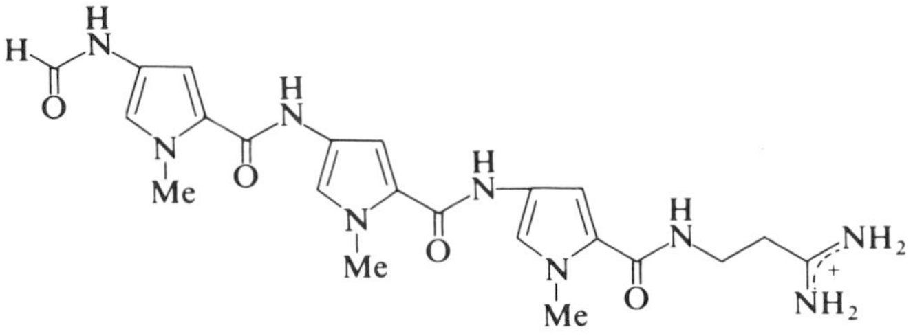

(**21**)

Another class of DNA binding agents form covalent bonds with nucleic acids. These include mitomycin, anthramycin, *cis*-diamminedichloroplatinum(II) and CC-1065.[102] All of these agents nonselectively inhibit most DNA and RNA polymerases, as well as DNA topoisomerases and other DNA-dependent enzymes. Mitomycin is a potent antitumor antibiotic that binds covalently to DNA and forms cross links. Bonding of mitomycin to DNA occurs primarily at the N-2 position of guanine and requires preactivation by reduction or protonation.[103]

Anthramycin (**22**) is a carbinolamine-containing antitumor agent that binds in the minor groove of DNA and forms reversible covalent bonds with the N-2 of guanine.[102] The ultimate DNA-reactive species is thought to be the N-10—C-11 imine form, which is in equilibrium with the carbinolamine form and undergoes nucleophilic attack from the N-2 of guanine to form an aminal linkage (**23**). Although this aminal linkage is reversible, binding interactions between anthramycin and the minor groove of DNA contribute to the stability of the covalent drug–DNA complex. No dissociation occurs while the helix is intact, but upon DNA denaturation the aminal bond breaks and the anthramycin dissociates. Regarding the sequence specificity of DNA bonding, anthramycin requires a guanine base and prefers to bond to 5′-PuGPu sequences.

(22)

(23)

Structure–activity studies have shown that chirality at position 11 of anthramycin is not important for DNA bonding or for biological activity, because both enantiomeric carbinolamines react with DNA via the N-10—C-11 imine and are biologically equivalent.[104] At the C-11a position, however, only those derivatives with the natural configuration (11aS) bond to DNA and have biological activity. This configuration provides the molecule with a right-handed twist that facilitates helix interaction. Molecular twist is also affected by the degree of unsaturation in the five-membered ring, and consequently the nature of this ring affects DNA bonding, cytotoxicity and antitumor activity. In contrast, modifications to the phenyl ring are generally well tolerated, except at the 9-position; this position appears to be close to the floor of the minor groove and cannot accommodate groups larger than hydroxy.

Another compound that forms covalent adducts with DNA is *cis*-diamminedichloroplatinum(II) (*cis*-DDP; cisplatin; **24**), which is one of the most important and widely used antitumor agents.[105,106] *cis*-DDP, which contains two moderately labile chloride leaving groups, bonds to the N-7 atoms of purines within DNA *via* its platinum atom. The compound forms several different kinds of DNA adducts including intrastrand crosslinks, interstrand crosslinks and monoadducts. Its principal adduct is thought to be an intrastrand crosslink in which the N-7 atoms of two successive guanines are linked by the platinum metal. It also bonds to d(ApG) and d(GpNpG) sequences and, less frequently, links guanines on opposite strands. *trans*-DDP, which contains chloride atoms in the *trans* configuration, is clinically inactive. Although *trans*-DDP forms DNA crosslinks, stereochemical constraints preclude the formation of the same kind of adducts formed by *cis*-DDP. The inactivity of *trans*-DDP has been attributed to two possible reasons: its DNA adducts may be less effective at inhibition of DNA replication than *cis*-DDP adducts, or the *trans*-DDP adducts may be more easily repaired by cellular enzymes than the *cis*-DDP adducts. Decreased repair of *cis*-DDP adducts within tumor cells has been suggested as a rationale for the tumor cell selectivity of this compound.

(24)

Several lines of evidence suggest that DNA binding is responsible for the biological activity of *cis*-DDP.[105] The compound is a potent inhibitor of DNA replication in mammalian cells. In purified systems, *cis*-DDP causes DNA polymerases to stop at sequences opposite G_n sites. The compound does not inactivate purified DNA polymerase, but is thought to inactivate the template *via* covalent attachment. Besides producing a steric blockade for the DNA enzyme, alterations in DNA conformation induced by drug bonding may contribute to enzyme inhibition. This mechanism is rather nonselective; several different DNA enzymes are inhibited by *cis*-DDP.

Most biologically active platinum analogs contain two moderately labile leaving groups in a *cis* orientation and two *cis* nitrogen donor ligands. All *trans* derivatives are inactive, as are monofunctional complexes that contain only one leaving group. Platinum complexes with more labile leaving groups are generally more toxic and have less antitumor activity. In contrast, platinum adducts with less labile leaving groups are often less toxic; for example, carboplatin (**25**), which contains two *cis* carboxylate ligands, retains much of the antitumor activity of *cis*-DDP but has fewer side effects. Clinical trials of carboplatin have shown it to be a very useful antitumor agent.

CC-1065 (**26**) is an extremely potent cytotoxic antitumor agent that binds to A–T rich regions of DNA in the minor groove and alkylates the N-3 of adenine.[102] The adduct (**27**) between CC-1065 and DNA results from attack of an adenine N-3 on to the cyclopropyl ring of (**26**) while the drug is

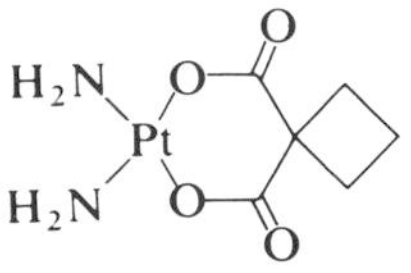

(25)

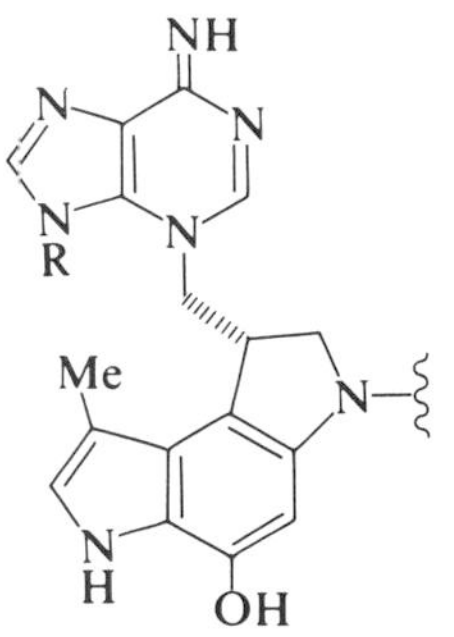

(27)

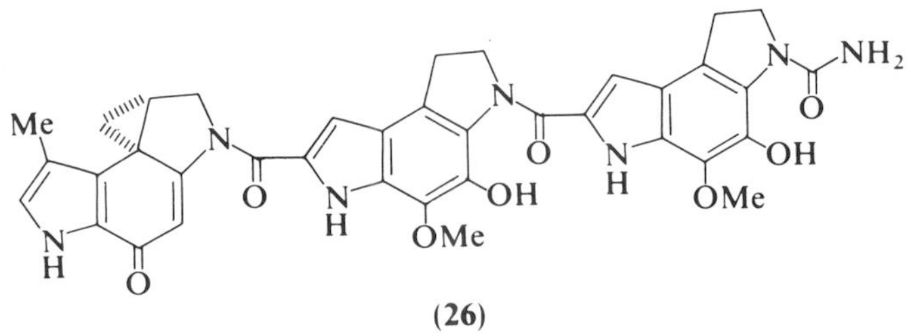

(26)

bound in the minor groove. Thermal treatment of the covalent adduct results in cleavage of the N–glycosidic bond and subsequent breakage of the DNA backbone. This thermal strand-breakage reaction was used to determine the sequence specificity of CC-1065–DNA bonding. The compound bonds covalently to 5'-AAAAA and 5'-PuNTTA sequences, with the A at the 3'-end of the site being covalently modified. Studies with fragments of CC-1065 have suggested that the sequence specificity is determined by the covalent bonding reaction, since fragments consisting of only the cyclopropyl-pyrroloindole moiety alkylate the same DNA sequences. Higher concentrations of these CC-1065 fragments are required for the same degree of DNA alkylation, compared with CC-1065, implying that the alkylation rate is slower. Consistent with their slower DNA alkylation rates, the fragments are less potent cytotoxic agents. Derivatives that lack the cyclopropyl group do not alkylate DNA and are generally less cytotoxic.

Streptonigrin (28) is an antitumor antibiotic that degrades DNA and inhibits DNA synthesis in mammalian cells. Interest in this agent has recently increased due to the findings that it is a potent inhibitor of reverse transcriptase.[107,108] The compound has been shown to bind to and cleave DNA in a manner that is dependent on metal ions and molecular oxygen.[109] Although the mechanism by which (28) inhibits reverse transcriptase is not fully understood, studies have suggested that oxidation–reduction reactions are involved.[107, 108] Preincubation of streptonigrin with the enzyme or template primer enhanced inhibition, suggesting the possibility that radical-mediated inactivation of enzyme and/or template is responsible for the observed inhibition. Consistent with this hypothesis, the quinoline quinone moiety was shown to be the minimum structure required for activity. Whether there is a specific interaction between streptonigrin and the reverse transcriptase–DNA complex, or whether these observations simply represent nonselective inactivation, is not yet clear.

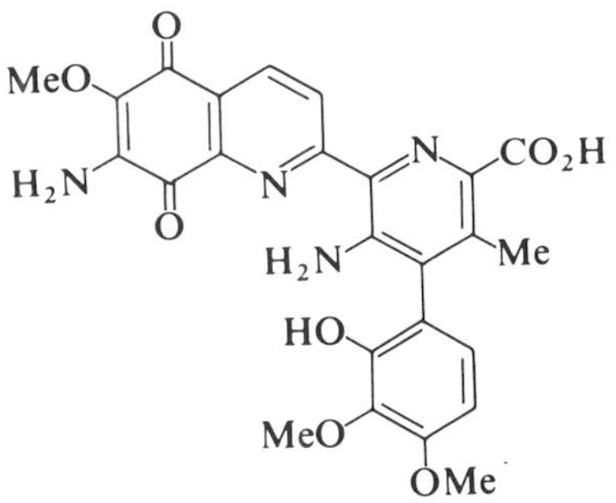

(28)

10.3.3 DNA TOPOISOMERASES

10.3.3.1 Enzymology and Functions of Topoisomerases

DNA topoisomerases are enzymes that are capable of altering the topology of circular DNA without permanently changing its chemical structure. The topological state of a circular DNA molecule cannot be changed unless a DNA strand is broken. The hallmark of all DNA topoisomerases is their ability to break transiently one or both strands of DNA concomitant with the formation of a covalent enzyme–DNA intermediate. The chemistry that leads to DNA strand breakage involves the attack of an enzyme tyrosine residue on a phosphodiester bond of the DNA backbone.[110] The enzyme–DNA intermediate undergoes conformational changes that result in the alteration of DNA topology, the strand breakage reaction is reversed, and the chemical structure of the DNA backbone is restored. While all DNA topoisomerases share these fundamental properties, the mechanistic and structural details of catalysis, as well as the sensitivity to inhibitors, varies among enzymes from different sources.

Cells require topoisomerases because of DNA packaging. Within the nucleus, DNA is bound to histones and other chromosomal proteins that enforce topological constraints. The ends of the DNA helix are held by these proteins, and so the nucleic acid behaves, in a topological sense, as though it were circular. The separation of the individual DNA strands, which is required for DNA replication and transcription, cannot occur without the aid of enzymes that alter topology. It has long been recognized that these processes would be facilitated by a swivel point that could relieve the superhelicity generated by the unwinding of the DNA strands.[111] The ability of topoisomerases to alter DNA topology by creating transient breaks in the DNA backbone imparts these enzymes with the necessary activity to fulfill this role. Consequently, DNA topoisomerases are important for DNA replication and gene expression, and are critical during cell proliferation.

The cells of all prokaryotic and eukaryotic organisms examined to date contain DNA topoisomerases.[112] In addition, the genomes of some viruses and bacteriophages have been shown to code for these enzymes. In prokaryotic and eukaryotic organisms there are two classes of topoisomerases, type I and type II. The type I enzymes are characterized by the production of transient single-strand breaks in DNA and the lack of an ATP requirement. Enzymes in this class are *E. coli* topoisomerase I (formerly called ω-protein) and mammalian topoisomerase I. The type II enzymes produce transient double-strand breaks in DNA and catalyze the ATP-dependent passage of another DNA segment through the break. Type II topoisomerases include DNA gyrase (prokaryotic topoisomerase II) and mammalian topoisomerase II. While all topoisomerases are capable of removing supercoils from a circular DNA molecule *in vitro*, only DNA gyrase has the ability to add supercoils.

10.3.3.1.1 *Prokaryotic DNA topoisomerases*

Interest in the enzymology of DNA gyrase has been stimulated by the finding that the quinolone antibacterial agents are potent inhibitors of this enzyme (see Section 10.3.3.2.1). *E. coli* DNA gyrase is composed of two A subunits encoded by the gyrA gene and two B subunits encoded by the gyrB gene.[113] The A subunit is responsible for the DNA cleavage activity and interacts with the quinolone antibacterial agents, while the B subunit contains the ATPase activity and is sensitive to novobiocin. That DNA gyrase is the target of the quinolones is supported by studies of drug-resistant bacteria that contain mutations in the gyrA gene; enzyme purified from these bacteria is resistant to quinolone inhibition *in vitro*.[114,115]

DNA gyrase can introduce negative supercoils into DNA and can join DNA circles into interlocking circles (catenation).[110] These reactions require the hydrolysis of ATP and occur by the passage of DNA strands through transient double-strand breaks. DNA gyrase can also remove DNA supercoils from circular DNA in a reaction that does not require ATP hydrolysis. Most of the intermediate steps of these reactions are well understood. DNA gyrase: (1) binds to DNA, (2) cleaves both strands of the helix concomitant with the covalent attachment of a tyrosine residue of the A subunit to the 5′-phosphoryl group of each DNA strand, (3) binds and hydrolyzes ATP, (4) undergoes a conformational change that results in the passage of a double-stranded DNA segment through the break and a change in the DNA linking number by an increment of two, (5) catalyzes the rejoining of the DNA backbone, and (6) dissociates from DNA. The exact order of these steps, particularly with respect to the timing of ATP binding and hydrolysis, is not definitively established.

The binding of DNA gyrase to double-stranded DNA is thought to involve the wrapping of DNA around the protein in a structure similar to a nucleosome.[110] DNase I protection experiments have shown that gyrase protects a region of 120–155 base pairs (bp) of DNA; 40–50 bp in the center of the binding site are most strongly protected. The site of cleavage produced by DNA gyrase in the presence of oxolinic acid is located within the binding region most strongly protected from nuclease digestion. Regions of DNA flanking the 120–155 bp binding site contain DNase I hypersensitive sites spaced 10–11 bp apart. The size of the nuclease-resistant segment, as well as the 10 bp periodicity of the cleavage pattern, are indicative of a model in which the DNA is wrapped around the enzyme. Electron microscopy studies are consistent with this model and reveal that the gyrase–DNA complex is heart shaped, with the A subunits forming the upper and larger lobes of the structure.[116]

DNA cleavage mediated by gyrase, an intermediate step in the overall reaction mechanism, can be observed in the presence of a quinolone antibacterial agent. The cleavage occurs at selective DNA sequences,[117] but it is not as sequence specific as cleavage by restriction enzymes. The double-strand cleavage occurs with a four-base stagger between the cuts. The strand scission does not require ATP binding or hydrolysis, but ATP or nonhydrolyzable ATP analogs such as 5′-adenylylimidodiphosphate (AMP-PNP) increase the efficiency of cleavage and alter the pattern of sequence selectivity.[110] Because DNA supercoiling is an energy-requiring process, ATP hydrolysis is necessary for the catalytic activity of DNA gyrase. Although the ATPase activity of gyrase is significantly enhanced in the presence of a DNA substrate, DNA-independent ATPase activity can be observed for the B subunit, suggesting that it contains the ATP binding site.[118]

Prokaryotic topoisomerase I is a simpler enzyme than DNA gyrase; it is a monomer of about 100 kDa that breaks only one strand of DNA and does not require ATP.[110,112] Bacterial topoisomerase I cannot introduce DNA supercoils, but instead removes negative (but not positive) supercoils by transiently breaking a single DNA strand, catalyzing DNA unwinding, and resealing the break. Like all topoisomerases, the strand break involves covalent attachment of a tyrosine residue on the enzyme to the newly formed phosphoryl group on the DNA. Bacterial topoisomerase I prefers to bind to single-stranded DNA, and preferentially cleaves single-stranded gaps or the DNA strand across from a nick. The DNA cleavage is somewhat sequence selective, but several synthetic DNA homopolymers and short oligomers consisting of dG, dA, dT or dC have been shown to be substrates for cleavage.[119]

Bacterial topoisomerases provide several functions during the processes of DNA replication and transcription.[112] In *E. coli* it is thought that there exists a dynamic interplay between type I and type II topoisomerases that maintains the correct degree of supercoiling. Genetic studies using various mutants in the topA, gyrA or gyrB genes have suggested that gyrase produces negative superhelicity inside bacteria while bacterial topoisomerase I removes superhelicity. The presence of both enzymes maintains the proper topological state necessary for DNA transcription. Furthermore, the synthesis of both subunits of DNA gyrase are regulated by DNA supercoiling, providing a feedback mechanism.

In addition to the transcriptional regulation, DNA gyrase also performs a critical role during DNA replication. It is required both for the initiation of replication as well as during the terminal stages. After the chromosome divides, the progeny DNA molecules are multiply intertwined. The activities of gyrase, transient DNA breakage and strand passage, resolve the daughter strands into their individual components. Bacterial topoisomerases may also be active during earlier stages of bacterial DNA replication, providing a swivel function during helix unwinding that permits the single-stranded template to be copied.

10.3.3.1.2 *Eukaryotic DNA topoisomerases*

There is considerable evidence suggesting that human topoisomerases are the cellular targets for certain antineoplastic agents. Several structurally diverse, clinically useful antitumor agents have been shown to interrupt the normal function of mammalian DNA topoisomerases.[120] Alkaline elution studies of the intercalating agents adriamycin and ellipticine, for example, have shown that leukemia cells exposed to these drugs display a specific type of protein-associated DNA break. The breaks were shown to be double stranded and covalently associated with a nuclear protein at their 5′-termini. Subsequent experiments with purified mammalian topoisomerase II showed that the formation of a covalent enzyme–DNA intermediate was enhanced in the presence of adriamycin, daunomycin, mitoxantrone, *m*-AMSA, etoposide and actinomycin.[121,122] In addition, protein–DNA complexes produced by these drugs in intact cells were immunoprecipitated with antitopoisomerase II polyclonal antibodies, demonstrating that the protein associated with drug-induced breaks is in fact topoisomerase II.[123] Subsequently, mammalian topoisomerase I was shown to be the target for the antitumor agent camptothecin.[124]

Although the eukaryotic topoisomerases share some of the fundamental enzymatic properties of the prokaryotic enzymes, there are significant differences. For example, the protein homology between prokaryotic and eukaryotic topoisomerase I is very low,[125] and the prokaryotic enzyme is not inhibited by camptothecin. Mammalian topoisomerase I is a monomeric enzyme of approximately 100 kDa that can catalyze the interconversion of certain DNA topoisomers; for example, it can relax supercoiled DNA to circular DNA and can join single-stranded rings. The enzyme binds to DNA and introduces a transient single-strand break, unwinds the double helix (or allows it to unwind), and then reseals the break before dissociating.[110] The enzyme–DNA intermediate is covalently linked, with the 3'-phosphate end of the broken DNA bound to a tyrosine residue on the enzyme. This link conserves the energy of the phosphodiester bond and allows the resealing step to proceed without the need for external energy in the form of ATP.

The intermediate covalent enzyme–DNA complex can be trapped by the addition of alkali or protein denaturants, and usually represents only a small percentage of the total DNA. Antitumor drugs such as camptothecin stabilize this covalent enzyme–DNA complex and consequently produce enzyme-mediated DNA cleavage (see Section 10.3.3.2.3). DNA cleavage by topoisomerase I shows moderate sequence selectivity, generating DNA fragments of discrete sizes. Sequence analysis of the cutting sites by two different groups revealed a loose consensus sequence for eukaryotic enzymes from several different species.[126,127] Recent studies have revealed a 16 base-pair sequence element that is a particularly strong cleavage site for topoisomerase I.[128]

Topoisomerase I also binds to and cleaves single-stranded DNA. In this case the covalent enzyme–DNA complex is formed spontaneously without the addition of protein denaturants. Enzyme-mediated cleavage of single-stranded DNA occurs in regions with the potential for intramolecular base pairing.[129] These results indicate that eukaryotic topoisomerase I requires at least a short region of DNA duplex to effect cleavage.

Topoisomerase II differs in several ways from its type I counterpart. The type II enzyme of eukaryotic cells consists of two identical subunits of molecular mass 170 kDa.[110,122] Recent evidence has suggested the existence of a second, distinct form of mammalian topoisomerase II with an apparent molecular mass of 180 kDa.[130] Topoisomerase II transiently breaks both strands of the helix and passes another double-stranded segment through the break. The strand passing step requires ATP hydrolysis. The enzyme–DNA intermediate contains two covalent links, since both 5'-phosphate ends at the double-strand DNA break are bound to tyrosine residues on individual topoisomerase II subunits. The topoisomerization reactions catalyzed by purified topoisomerase II include the relaxation of supercoiled DNA (in linking number steps of two) and the knotting/ unknotting or catenating/decatenating of double-stranded rings. Unlike DNA gyrase, there is no evidence that eukaryotic topoisomerase II introduces supercoils into circular DNA *in vitro*.

The covalent DNA–topoisomerase II complex, which is an intermediate in the catalytic reaction mechanism, is stabilized by several different antitumor agents. This complex differs from the DNA–topoisomerase I complex in that all four ends of the DNA are held tightly by the enzyme. While the 5'-ends are covalently linked to topoisomerase II, the 3'-ends have limited mobility and cannot rotate freely. Topoisomerization takes place by the passage of a second DNA strand through the DNA break, concomitant with enzyme conformational changes driven by ATP binding.[131,132] ATP hydrolysis may be necessary for the enzyme–DNA dissociation and turnover. This strand-passage mechanism changes the DNA linking number by steps of two; in contrast, topoisomerase I changes the DNA linking number by steps of one (for a discussion of DNA linking number, see a recent review on topoisomerase mechanism[110]).

The functions of topoisomerases in mammalian cells are not completely established. Evidence has been presented suggesting critical roles during replication, transcription, recombination and repair.[112,122] For topoisomerase I, several lines of evidence demonstrate that the enzyme acts as a swivel during transcription. Immunofluorescence studies have shown that topoisomerase I is restricted to actively transcribing loci of the polytene chromosomes,[133] while topoisomerase II is broadly distributed within the nucleus.[134] Furthermore, DNA strand breaks produced by topoisomerase I in intact cells occur within transcriptionally active genes, while little cleavage occurs in nontranscribed flanking regions.[135,136]

Topoisomerases are also critical for DNA replication, as shown by studies in SV40 replication systems.[137,138] DNA–topoisomerase I complexes were found at or very near the replication fork, implying that topoisomerase I acts as a swivel for SV40 replication. Topoisomerase II is responsible for the decatenation of newly replicated SV40 daughter chromosomes, since inhibitors of this enzyme produce catenated dimers rapidly and reversibly. In the absence of topoisomerase I it is likely that topoisomerase II can provide the necessary swivel functions for replication.

These results are consistent with recent genetic studies of topoisomerase function in yeast. Yeast mutants that lack topoisomerase I are viable, implying that the topoisomerase II can substitute for the activities of topoisomerase I.[139,140] Topoisomerase II mutants are lethal; they are defective in the termination of DNA replication and the segregation of daughter chromosomes, but otherwise appear to replicate and transcribe DNA normally. The simplest explanation consistent with these and other results is that either topoisomerase I or II can function as a swivel to allow replication fork movement, but only topoisomerase II can separate daughter DNA strands.

Recently, the genes for human topoisomerase I[125] and topoisomerase II[141] have been cloned. The protein sequence of mammalian topoisomerase I is very different from that of *E. coli* topoisomerase I, but shares 42% identity with the yeast sequence. Human topoisomerase I has a predicted molecular mass of 90 649 Da, 43% charged residues, and a calculated isoelectric point of 10.05. The gene for human topoisomerase II encodes a protein with a calculated molecular mass of 174 000 Da. Human topoisomerase II shares significant homology with the yeast and Drosophila enzymes. In addition, the amino-terminal sequence of Drosophila topoisomerase II is homologous to the B subunit of *B. subtilis* DNA gyrase, suggesting that the ATPase activity resides in this region of the enzyme.[142] The central portion of the Drosophila enzyme is similar to the A subunit of DNA gyrase, implying that it contains the DNA cleavage activity.

10.3.3.2 Inhibitors of DNA Topoisomerases

There are two primary mechanisms by which compounds can inhibit topoisomerases: inhibition of catalytic DNA topoisomerization and stabilization of enzyme-mediated DNA cleavage. These two mechanisms are not mutually exclusive, since agents that induce cleavage often inhibit catalytic activity. Although it is not known which mechanism leads to cell death, evidence suggests that induction of DNA damage by topoisomerase inhibitors is responsible. Topoisomerases share a characteristic that distinguishes them from most chemotherapeutic targets. The sensitivity of a cell to topoisomerase inhibitors increases as the level of enzyme in the cell increases. In the case of most drug targets, cells with low enzyme levels are hypersensitive to the drug, and cells that overproduce the target enzyme (for example, by gene amplification or overexpression) are resistant. Topoisomerase activity appears to be required for drug cytotoxicity, implying that the enzyme is subverted during its catalytic cycle to produce a DNA lesion that ultimately leads to cell death.

10.3.2.2.1 *Quinolone antibacterial agents*

The quinolone and 1,8-naphthyridine antibacterial agents are clinically important drugs that possess broad-spectrum activity and oral efficacy.[1, 2] The first clinically important molecule in this class was nalidixic acid (**29**), which is limited to urinary tract infections and has a limited spectrum of antibacterial activity. Several new compounds in this class, notably the fluoroquinolones such as compounds (**30**)–(**33**), are significantly more potent than nalidixic acid and have much broader spectra of activity. Several fluoroquinolones have recently been approved for clinical use and represent the most important new class of antibacterial agents.

The target of the quinolone antibacterial agents has been identified as DNA gyrase. Early studies demonstrated that nalidixic acid and oxolinic acid (**34**) inhibited DNA gyrase purified from wild-type bacteria. DNA gyrase that was reconstituted with an A subunit purified from nalidixic acid-resistant bacteria was found to be resistant to inhibition by nalidixic acid and oxolinic acid.[114,115] It was subsequently shown that mutations leading to significant quinolone resistance in bacteria occur exclusively in the gyrA gene (formerly called the nalA gene), demonstrating that the A subunit of DNA gyrase is a target for this class of antibacterial agent.[2]

The mechanism by which quinolones inhibit DNA gyrase involves stabilization of an intermediate in the normal DNA gyrase reaction scheme such that rejoining of the broken DNA duplex is prevented. Nalidixic acid and oxolinic acid induce DNA cleavage when added to a mixture of purified gyrase and double-stranded DNA.[113-115] The 5'-ends of the broken duplex DNA are each covalently attached to a gyrase A subunit. The site of DNA cleavage induced by the quinolones is in the center of the gyrase binding site, which covers approximately 120 base pairs.[110] Induction of DNA cleavage is not restricted to reactions with purified enzyme, since treatment of intact bacteria with nalidixic acid results in the production of gyrase-linked DNA cleavage.[117,143]

The molecular details of quinolone–gyrase–DNA binding are not fully understood. Norfloxacin (**30**) has been shown to bind to purified DNA in a nonintercalative fashion, but not to bind to purified DNA gyrase.[3,144] However, DNA binding by norfloxacin occurs with extremely low frequency, approximately one drug per 2500 nucleotides. It was suggested that norfloxacin binds to single-stranded regions of DNA, and then gyrase binds to the drug–DNA complex to form a ternary complex. Alternatively, binding to free DNA may be nonspecific; quinolones may inhibit gyrase by specifically binding to a gyrase–DNA complex.

Since the quinolones bind to DNA, albeit with low frequency, their specificity for other DNA enzymes should be considered. Inhibition of mammalian topoisomerase II by quinolones has been reported, but only at concentrations approximately 100 times the K_i values for DNA gyrase inhibition.[145] Certain quinolones, however, have been reported to be more potent against mammalian topoisomerase II.[146] Prokaryotic and eukaryotic DNA topoisomerase I and DNA polymerase α are not inhibited by the quinolones, or are inhibited only at very high concentrations.

The relationship between gyrase inhibition and bacterial killing is not fully understood. The quinolones inhibit the DNA supercoiling activity of DNA gyrase in addition to stabilizing enzyme-mediated DNA cleavage. Bacterial killing may be mediated by DNA damage, by a disruption in DNA topology, or both. DNA damage is likely to play a role, since nalidixic acid is a potent inducer of the SOS DNA repair system.[143] It has been suggested the SOS system leads to a series of events that ultimately ends in cell death.[147]

Numerous quinolones have been synthesized and evaluated for antibacterial activity and DNA gyrase inhibition, and so the structure–activity relationships are reasonably well understood. Two factors contribute to antibacterial activity: inhibition of DNA gyrase and bacterial cell penetration. A comparison between antibacterial potency and DNA gyrase inhibition by several quinolones has been reported.[148] A C-6 fluorine atom was found to increase antibacterial potency dramatically. This substituent was reported to enhance inhibition of DNA gyrase by 2- to 17-fold, and to increase cell penetration by 1- to 70-fold. Almost all of the clinically useful quinolones contain a fluorine at the C-6 position.

Several modifications at the C-7 position of the quinolones have been reported, with a piperazinyl group being the most common substituent.[148,149] Norfloxacin (**30**) and ciprofloxacin (**31**), which contain a C-6 fluorine and C-7 piperazinyl group, are among the most potent quinolones and possess broad spectrum activity. Ciprofloxacin is one of the most potent inhibitors of DNA gyrase; it induces DNA cleavage at 0.5 μg mL^{-1}. However, a reasonable variety of structural modifications at the 7-position are tolerated by the enzyme. Five- or six-membered rings are among the best substituents, resulting in more potent enzyme inhibition than linear or small substituents. Basic groups, such as piperazine or 3-(aminomethyl)pyrrolidine, provide increased activity against Gram-positive bacteria but are not required for DNA gyrase inhibition.[148,149]

Among substituents at position N-1, the cyclopropyl group affords maximum activity in all regards. Quinolones containing N-1 cyclopropyl groups are 2–5 times more potent at inducing gyrase-mediated DNA cleavage than quinolones containing N-1 ethyl groups.[150] Bulky groups that occupy space above and below the plane of the quinolone nucleus, such as isopropyl or cyclohexyl groups, are not well tolerated by the enzyme. In addition, the (S) enantiomer of ofloxacin (**32**) is significantly more potent than the (R) enantiomer, consistent with bulk tolerance on only one side of the quinolone ring system.[151] In addition to steric bulk at position N-1, electronic effects have been implicated to influence biological activity, as shown by the potency of certain N-1, phenyl-substituted quinolones.[150,152]

Functional groups on the quinolone structure that are thought to be required for activity are the 4-keto group and the 3-carboxylic acid. Several modifications to the carboxylic acid group have all resulted in inactive compounds. However, a recently reported quinolone (**33**), in which the carboxylic acid was replaced with a isothiazolo ring, was found to be significantly more potent than ciprofloxacin.[153] The nitrogen proton of (33) is acidic, and the isothiazolo ring was suggested as a carboxylic acid mimic.

10.3.3.2.2 Novobiocin

Novobiocin (35) and the related derivative coumermycin are substituted coumarin antibiotics isolated from *Streptomyces* that inhibit type II topoisomerases. The most potent and well-characterized activity of novobiocin is inhibition of DNA gyrase.[154,155] Novobiocin competitively inhibits the ATPase activity of DNA gyrase and is thought to bind to the ATP site on the B subunit. The apparent K_i value for inhibition of the ATPase activity is 10 nM, about four orders of magnitude lower than the K_m for ATP.[155] The compound does not prevent DNA cleavage by gyrase, but blocks the shift in the oxolinic-acid-induced cleavage pattern that occurs upon ATP addition. Furthermore, novobiocin has been shown to prevent covalent attachment of the 2',3'-dialdehyde derivative of ATP to the B subunit of DNA gyrase.

(35)

The structural features of novobiocin required for inhibition of DNA gyrase have been studied with fragments of the compound.[156,157] A fragment consisting of the coumarin residue plus the noviose sugar was shown to inhibit DNA gyrase. Similarly, a fragment containing a chlorinated coumarin residue plus the 3-isopentyl-4-hydroxybenzoic acid moiety inhibits the enzyme. Although the coumarin residue by itself was not active, these studies suggest that it is an important structural feature of novobiocin.

Novobiocin also inhibits mammalian topoisomerase II, but at much higher concentrations than are necessary for gyrase inhibition.[131,145,158] Inhibition of the mammalian enzyme is competitive with ATP.[131] Novobiocin has been reported to have a number of effects on eukaryotic cells, including inhibition of DNA synthesis, inhibition of transcription, and cleavage of active gene sequences.[159] Many of these effects occur at relatively high concentrations of novobiocin and may be the result of nonspecific binding, since the compound appears to interact with many basic proteins.[160] In addition, novobiocin inhibits the activity of DNA polymerase α *in vitro*.[145]

10.3.3.2.3 Camptothecin

Camptothecin (36) is a cytotoxic alkaloid with significant antitumor activity in several animal tumor models.[161] Camptothecin inhibits purified mammalian topoisomerase I by stabilizing a covalent enzyme–DNA complex, thereby producing single-strand DNA breaks.[124] Within mammalian cells, camptothecin produces DNA damage mediated by the action of topoisomerase I.[162,163] Proof that this enzyme is the target for camptothecin comes from studies with yeast cells in which the topoisomerase I gene has been deleted; these cells have no detectable enzyme and are completely resistant to the drug.[164,165] Furthermore, camptothecin-resistant mammalian cells have been found to contain topoisomerase I that was altered in such a way as to become resistant to the effects of the drug in purified systems.[166,167] That camptothecin-resistance in mammalian cells is manifested by the appearance of a qualitatively different topoisomerase I strongly implicates this enzyme as the target for camptothecin.

The details of camptothecin–topoisomerase I–DNA interaction that lead to enzyme inhibition are partially understood. Most of the available evidence suggests that camptothecin binds to the DNA–topoisomerase I complex subsequent to the DNA cleavage step in a reversible manner.[124,168–170] The production of enzyme-mediated DNA cleavage is reversible, both in purified systems and in intact cells. Furthermore, [³H]camptothecin was shown to bind reversibly

(36)   (37)

to an enzyme–DNA complex, but not to isolated enzyme or isolated DNA, suggesting a drug binding site is created as the enzyme–DNA covalent complex is formed.[168] As a consequence of binding to the DNA–topoisomerase I complex, camptothecin also inhibits the catalytic activity of topoisomerase I.

Topoisomerase I cleaves DNA at selective nucleotide sequences. Camptothecin has been shown to stimulate cleavage at topoisomerase I recognition sequences and not to significantly alter the sequence specificity of enzyme-mediated cleavage. However, camptothecin enhances DNA cleavage at some sites much more than at other sites.[169, 170] The sequences of camptothecin-induced cleavages have been mapped in intact cells and found to occur primarily within transcriptionally active genes.[135, 136] In SV40 replication systems, camptothecin-induced DNA breaks have been located at or near the replication fork, and it has been suggested that collision of the replication fork with a topoisomerase I–DNA complex converts it into a permanent DNA lesion.[137]

The structure–activity relationships of camptothecin show that the hydroxylactone ring is the most critical structural feature. Several camptothecin derivatives containing a modified hydroxylactone ring have been synthesized and evaluated for inhibition of mammalian topoisomerase I and cytotoxicity to mammalian cells.[171] Each of the groups of the hydroxylactone moiety (the carbonyl oxygen, the ring lactone oxygen, and the 20-hydroxy group) are critical for enzyme inhibition. For example, the lactol, lactam, thiolactone and 20-deoxy derivatives do not stabilize the covalent DNA–topoisomerase I complex and are not cytotoxic to mammalian cells. Similarly, Wall and coworkers have established that the hydroxylactone ring is critical for antitumor activity *in vivo*.[172] The correlation between topoisomerase I activity and antitumor activity is consistent with topoisomerase I being the relevant target for camptothecin.

While lactone rings are usually quite stable at neutral pH, the presence of the α-hydroxy group imparts an unusual electrophilicity to the lactone carbonyl of camptothecin, perhaps because of an intramolecular hydrogen bond.[173] The lactone ring hydrolyzes at pH 7.5 and an equilibrium is established between the closed-ring form (36) and the open-ring form (37). The sodium salt of the open form (37) has similar antitumor activity as compound (36) but is about 10-fold less potent.[172] Compound (37) is most likely a prodrug for compound (36), since only the lactone form inhibits topoisomerase I.

In addition to the structure–activity relationships in the E-ring, several camptothecin derivatives with modifications in the A-ring have been synthesized.[172–174] Camptothecin has a reasonably large degree of bulk tolerance at position 9. Most analogs in this class inhibit topoisomerase I and have potent antitumor activity. Bulk tolerance at the 10-position is somewhat limited. Derivatives with flexible side chains that can occupy space in the plane of the A-ring are able to produce enzyme-mediated DNA cleavage, but derivatives with side chains that protrude above or below the plane of the A-ring are much less active. The 11-position does not tolerate large substituents, and the 12-position does not tolerate any substituents that have been investigated. In general, A-ring modified camptothecins that do not inhibit topoisomerase I are not cytotoxic and lack antitumor activity.

10.3.3.2.4 *Amsacrine*

The clinically active antitumor agent 4'-(9-acridinylamino)methanesulfon-*m*-anisidine (*m*-AMSA; **38**) is a DNA intercalator that inhibits purified mammalian topoisomerase II and produces DNA strand breaks in mammalian cells.[121] Topoisomerase II is suggested to be the primary target for *m*-AMSA, since *o*-AMSA (**39**), a closely related compound that lacks antitumor activity, does not inhibit the purified enzyme or produce DNA breaks in mammalian cells. Interestingly, both *m*-AMSA and *o*-AMSA intercalate into DNA with similar affinities,[175,176] implying that intercalation is not sufficient for topoisomerase II inhibition among acridine derivatives.

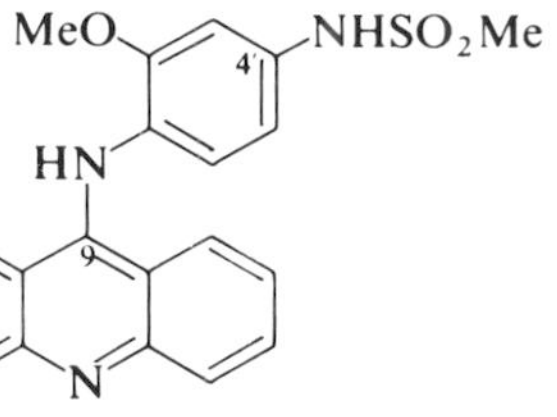

(38)

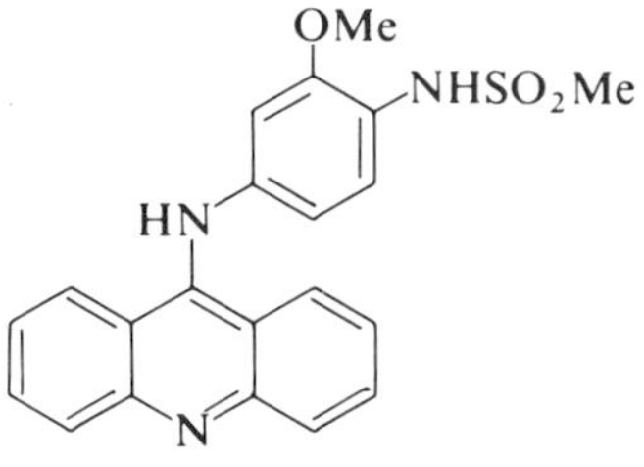

(39)

m-AMSA stimulates the formation of a DNA–topoisomerase II complex and consequently inhibits DNA rejoining; the broken DNA contains topoisomerase II subunits covalently linked to each 5'-phosphoryl end.[121] The cleavage is reversed by the addition of high salt or by dilution, suggesting that the drug binds reversibly to the DNA–topoisomerase II complex. Supercoiled DNA, when used as a substrate for DNA cleavage, remains supercoiled after salt reversal, indicating that all four ends of the DNA are bound tightly to the enzyme within the putative drug–enzyme–DNA complex. The breaks occur at specific DNA sequences, which are a subset of those DNA sequences normally broken by topoisomerase II.[177] DNA cleavage by antitumor drugs of different structural classes produce different patterns of sequence-specific DNA cleavage. Studies of DNA cleavage induced by *m*-AMSA in intact cells have shown that both single- and double-strand breaks occur.[175] The breaks are protein-linked, sequence-specific and immunoprecipable with antitopoisomerase II antibodies.[123] DNA breakage induced by *m*-AMSA has been reported to be enhanced by ATP,[178] but at high ATP concentrations DNA cleavage is inhibited.

The chemical interactions that lead to topoisomerase II inhibition by *m*-AMSA are not fully understood. As mentioned above, *m*-AMSA binds to DNA by intercalation with an association constant of about 10^5 and a binding site size of about two base pairs.[175,176] However, intercalation is not sufficient for topoisomerase II inhibition, as shown by the inactivity of *o*-AMSA, and so drug–enzyme interactions may play a role. One possibility is that the pendant aniline moiety interacts with topoisomerase II within the complex, and that the presence of an *ortho* methoxy group disrupts binding due to steric repulsion. An analog that lacks a methoxy group on the aniline ring is equal in potency to *m*-AMSA, suggesting that the *meta* methoxy group does not contribute binding energy.[179]

A large number of derivatives of 9-anilinoacridines have been synthesized and tested for antitumor activity by Denny and coworkers, and quantitative analyses of the structure–activity relationships have been published.[180–182] It had been suggested that, within a series of *m*-AMSA analogs, there is a correlation between DNA binding affinity and antitumor dose potency.[181] The relationship is incomplete, however, possibly because inhibition of topoisomerase II depends on other structural features in addition to those that contribute to DNA binding.

Only a handful of *m*-AMSA derivatives have been evaluated for inhibition of purified topoisomerase II.[179,183] These studies have shown that there is a general correlation between the ability to produce enzyme-mediated DNA breaks and cytotoxic potency. Removal of the sulfonamide group from *m*-AMSA reduces activity. Similarly, 9-aminoacridine, which does not contain the pendant aniline moiety, is significantly less potent at inducing enzyme-mediated DNA cleavage and is not cytotoxic. As mentioned above, removal of the *meta* methoxy group has no effect, but an *ortho* methoxy group as in (39) reduces activity. A derivative that contains an additional anilino group on the sulfonamide was the most potent acridine derivative tested for topoisomerase II inhibition. Consistent with its potent enzyme inhibition, this compound is 100 times more cytotoxic than *m*-AMSA. Finally, compound (40), which is currently in clinical trials, was similar in potency to *m*-AMSA.

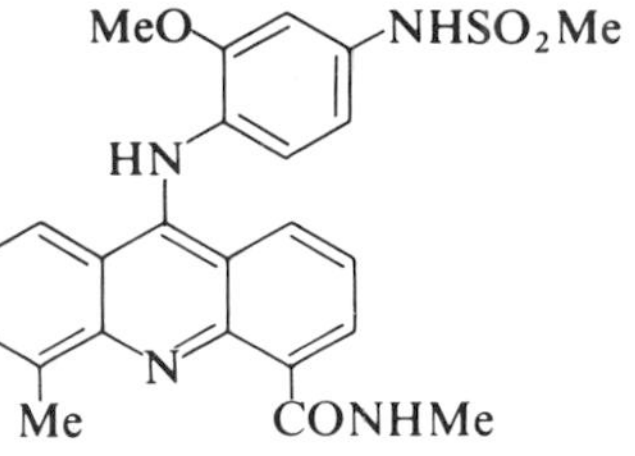

(40)

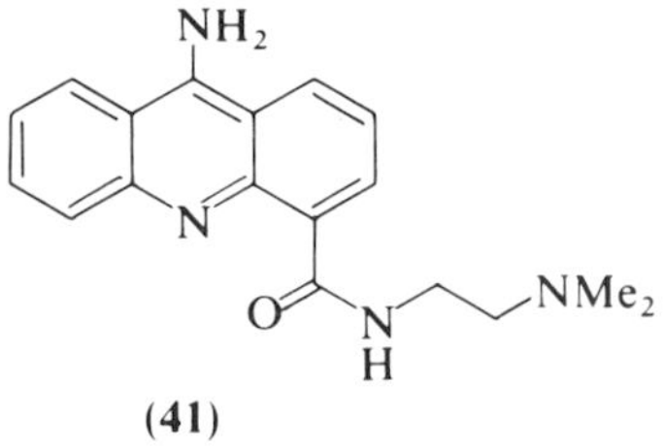

(41)

Denny and coworkers have synthesized a series of new acridine derivatives with antitumor activity that contain carboxamide substituents at the 4-position, such as compound (41). The position of the cationic amino group was found to be critical for antitumor activity and for the induction of DNA cleavage in cultured cells.[184] It is interesting that these compounds retain biological activity in spite of the fact that they do not contain a pendant phenyl ring, since 9-aminoacridine is inactive. The carboxamide side chain may facilitate topoisomerase II inhibition; alternatively, this side chain may bind to DNA and reduce the kinetics of drug–DNA dissociation.[185]

10.3.3.2.5 *Epipodophyllotoxins*

The epipodophyllotoxins represent a class of topoisomerase II inhibitors that do not intercalate into DNA.[186] Within this class, teniposide (42) and etoposide (43) are the most well studied, and etoposide is used clinically against several different types of tumors.[187] These compounds are semisynthetic derivatives of podophyllotoxin, a natural product that inhibits microtubule assembly. In an effort to produce more efficacious antitumor agents, the podophyllotoxins were modified by epimerization at C-4 and glycosylation. The resultant compounds had little antitumor activity, but the cyclic acetals made from their condensation with aldehydes (etoposide and teniposide) were found to have activity in several animal tumor models. In contrast to the podophyllotoxins, teniposide and etoposide do not act by a mitotic arrest mechanism and do not bind to tubulin, but instead have been shown to stabilize the covalent DNA–topoisomerase II intermediate and inhibit the strand-passing activity of the enzyme.

(42) R = [thiophene structure]

(43) R = Me

Several pieces of evidence indicate that topoisomerase II is the relevant target for etoposide and related derivatives.[122] Initial studies demonstrated that these compounds produce protein-associated DNA strand breaks in intact cells. Subsequent studies have shown that these strand breaks are sequence specific, and are bound covalently at their 5′-termini to a protein that can be immunoprecipitated with antitopoisomerase II antibodies. Furthermore, a Chinese hamster ovary cell line that was selected for resistance to epipodophyllotoxins, which exhibits cross resistance to several intercalative topoisomerase II inhibitors, was shown to contain an altered form of the enzyme that no longer produces enhanced DNA cleavage in response to the drug.[188] The finding that mammalian cells modify their topoisomerase II in order to become resistant to the epipodophyllotoxins strongly implicates this enzyme as the primary cellular target.

The mechanism of topoisomerase II inhibition by etoposide is not completely clear. Although initial studies indicated that the drug does not bind to DNA, recent experiments have demonstrated nonintercalative binding by etoposide in the presence of high concentrations of plasmid or kinetoplast DNA.[189] The binding of etoposide to DNA is similar to that observed for norfloxacin; the frequency of binding sites is very low and binding is greater to single-stranded DNA than to duplex DNA. The DNA binding of teniposide (42) is similar to that of etoposide (43), even though teniposide has 10-fold greater potency as a topoisomerase II inhibitor. It was suggested that the epipodophyllotoxins have two domains: the polycyclic array that interacts with DNA, and the glycoside moiety that interacts with the enzyme.[189]

Structure–activity relationships of the epipodophyllotoxins for topoisomerase II inhibition have been studied by measuring the induction of DNA cleavage in intact cells.[190] Another study made use of mutant Chinese hamster ovary cells that contain a drug-resistant topoisomerase II (see above); cytotoxicity to these cells and to wild-type cells was compared.[191] Both studies found that teniposide (42) is about 10 times more potent than etoposide (43). Furthermore, replacement of the thienyl group in teniposide with a phenyl group resulted in no loss in potency, suggesting that a planar hydrophobic group located off the glycoside provides a favorable interaction with the enzyme–DNA complex. Cleavage of the acetal and removal of this terminal group, leaving a glucoside moiety, significantly decreased activity.

At the other end of the epipodophyllotoxin structure, the presence of a free 4′-hydroxy group has been shown to enhance significantly both DNA breakage activity and cytotoxicity.[190,191] Opening of the lactone moiety reduces cytotoxic potency, and the picro isomer, in which the *trans* lactone is isomerized to a *cis* lactone, is not cytotoxic. Studies of epipodophyllotoxins that measure cytotoxicity may not reflect inhibitory potency towards topoisomerase II, since inhibition of microtubule formation can also lead to cytotoxicity. However, derivatives that contain a glycoside moiety at the C-4 position do not inhibit microtubules.[187]

Besides inhibiting topoisomerase II, etoposide has been implicated to undergo metabolic activation in the dimethoxyphenol ring to afford a catechol derivative.[187] The catechol of etoposide can be oxidized to an *ortho* quinone *via* a semiquinone free radical. These derivatives have been suggested to react covalently with DNA, topoisomerase II, or other cellular macromolecules in a manner that contributes to cytotoxicity.

10.3.3.2.6 Anthracyclines

Daunomycin (44) and adriamycin are anthracycline derivatives that are among the most efficacious and widely used anticancer agents. Although the anthracyclines have been used clinically for some time and have been extensively studied, there is still no consensus as to the mechanism by which they exhibit antitumor activity. Several possible mechanisms have been suggested, including inhibition of topoisomerase II,[177] cleavage of DNA mediated by iron and radicals,[192] inhibition of DNA and RNA polymerases,[193] and interaction with membranes.[194] This section will concentrate on anthracycline–DNA binding and the resultant inhibition of DNA enzymes. For additional information on the role of iron- and radical-mediated processes, the reader is directed toward a recent review.[192]

(44) $R^1 = R^2 = H$
(45) $R^1 = OH$, $R^2 = COCF_3$

The most well-known property of the anthracyclines is their ability to intercalate between DNA base pairs.[193,195] DNA binding appears to be related to the biological activity of the anthracyclines, since adriamycin intercalates into DNA within the nucleus of drug-sensitive cells but not within adriamycin-resistant cells. Regarding the molecular details of DNA binding, the crystal structure of a daunomycin–DNA complex has been solved recently by X-ray diffraction analysis.[196] The most interesting feature of this complex is that the daunomycin chromophore is oriented perpendicular to the long axis of the base pairs. The amino sugar almost completely occupies the minor groove, with the hydroxy and amino groups of the sugar pointing away from the DNA molecule. The C-9 hydroxy group, which has been shown to be essential for biological activity, forms two hydrogen bonds with guanine nitrogen atoms. The conformation of this drug–DNA complex has implications for the inhibition of DNA enzymes by anthracyclines (see below).

Like many DNA intercalators, the anthracyclines produce protein-linked DNA breaks in cultured mammalian cells that are mediated by the action of topoisomerase II. *In vitro*, anthracyclines have been shown to inhibit purified topoisomerase II by stabilizing a covalent enzyme–DNA intermediate.[177] However, the relationship between topoisomerase II inhibition, cytotoxicity and antitumor activity is not as well established for the anthracyclines as it is for the acridines or epipodophyllotoxins. Recent evidence supporting this mechanism of action was obtained using lymphocytes derived from patients with B-cell chronic lymphocytic leukemia (CLL). These cells were found to contain no detectable topoisomerase II, while measurable levels of the enzyme were found in other lymphomas and leukemias.[197] Adriamycin has found very little use for the treatment of CLL, possibly because its putative target, topoisomerase II, is required for cytotoxicity but not found in this tumor.

The interaction of anthracyclines with the DNA–topoisomerase II complex is not clear. At low drug concentrations (10 ng mL^{-1}), anthracyclines bind to DNA and produce enzyme-mediated cleavage, but at higher concentrations they inhibit DNA cleavage.[177] High drug:DNA ratios may prevent enzyme–DNA binding by steric occlusion and/or DNA conformational changes. The adriamycin analog AD41 (**45**), which contains an acetylated amino group on the sugar and therefore has no positive charge, has been shown to bind to DNA with much lower affinity than adriamycin and yet still induces DNA cleavage mediated by topoisomerase II.[189] This analog does not exhibit the autoinhibition phenomenon, consistent with potent DNA binding being responsible for the prevention of cleavage.

The activity of AD41 suggests that, among anthracyclines, potent DNA binding is not necessary for inhibition of topoisomerase II. These compounds have been suggested to contain one domain that binds to DNA and another that interacts with the enzyme.[189] Along these lines, Wang and coworkers have suggested that functional groups on the daunomycin sugar, which were shown by X-ray analysis to protrude out of the minor groove of DNA and into the solvent, are in a position to interact with DNA enzymes.[196]

In a similar manner, the amino sugar of anthracyclines may interact with DNA and RNA polymerases and inhibit polynucleotide synthesis. Adriamycin and daunomycin have been shown to inhibit purified DNA and RNA polymerases from several eukaryotic and prokaryotic sources.[193] The mechanism by which daunomycin inhibits *E. coli* RNA polymerase is specifically targeted at the initiation complex. The drug prevents the addition of nucleotides to the initial dinucleotide, but has no effect on the binding of enzyme to DNA.[198] Regarding mammalian cytotoxicity and antitumor activity for the anthracyclines, the relative importance of RNA/DNA polymerase inhibition *vs.* topoisomerase II inhibition is not established.

Numerous analogs of the anthracyclines have been prepared and evaluated for antitumor activity, and the subject has been reviewed.[195,199] Regarding antitumor activity, changes to the structure of the aglycone moiety frequently reduces activity, and the 9-hydroxy group is essential. However, the C-4 methoxy group on daunomycin and adriamycin is not required for antitumor activity, as several active derivatives contain a C-4 hydroxy group. The carbohydrate moiety is much more tolerant of alterations; numerous derivatives with modifications of the daunosamine sugar have antitumor activity. Several active anthracycline derivatives, such as marcellomycin and aclacinomycin, have di- or tri-saccharide carbohydrate chains.[195,199]

(**46**)

A serious side effect of the anthracyclines is cardiotoxicity, which may be mediated by oxidation–reduction reactions leading to the generation of oxygen radicals.[192] In an attempt to prepare analogs with a lower probability for cardiotoxicity, intercalative anthracenediones such as mitoxantrone (**46**) have been synthesized and evaluated for antitumor activity.[200] Mitoxantrone and its derivatives intercalate into DNA and are potent inhibitors of topoisomerase II.[177] NMR studies have suggested that, like daunomycin, mitoxantrone binds to DNA at right angles to the axis of the base pairs. The sequence specificity of DNA cleavage induced by mitoxantrone is very similar to that induced by daunomycin and adriamycin, suggesting a similar mechanism of enzyme inhibition. Regarding structure–activity relationships, the side chains of mitoxantrone have been shown to be critical for biological activity. There is a requirement for two basic groups separated by two carbon atoms,[201] which is similar to structural requirements observed for acridines with carboxamide side chains (see Section 10.3.3.2.4).

10.3.3.2.7   Ellipticine

Ellipticine (**47**) is a plant alkaloid that has antitumor activity in animal tumor models.[202] It is a potent DNA intercalating agent with a structure that resembles the shape of a purine–pyrimidine base pair. Ellipticine is thought to elicit its antitumor activity through an active metabolite, 9-hydroxyellipticine, which can undergo oxidation to a quinone imine. Since 9-hydroxyellipticine was found to be inactivated *in vivo* by glucuronide formation at N-2, a derivative with an N-2 methyl group was prepared. This compound, 9-hydroxy-2-methylellipticinium acetate (**48**), has been shown to have antitumor activity in humans.

(47)   (48)

The mechanism of action of ellipticine derivatives is thought to be related to DNA intercalation and topoisomerase II inhibition.[203] The compounds produce reversible protein-linked breaks in mammalian cells and stabilize the formation of a DNA–topoisomerase II complex in purified systems. With the purified enzyme,[203] as well as in isolated nuclei,[204] DNA cleavage induced by ellipticine derivatives is strongly inhibited at high drug concentrations. Similarly, compound (**48**) above 5 μM was shown to inhibit enzyme-mediated DNA cleavage induced by *m*-AMSA. This phenomenon is observed with other intercalative topoisomerase II drugs, but not with noninter-calative inhibitors. Compound (**48**) was shown to be about 25 times more potent than compound (**47**) at producing DNA cleavage in the presence of purified topoisomerase II, consistent with its higher potency for cytotoxicity to mammalian cells.[203] The presence of a positive charge on the quaternary nitrogen of (**48**) may contribute to this increase in potency.

The 9-hydroxy group of ellipticine derivatives and metabolites is thought to be required for biological activity. 9-Hydroxyellipticine derivatives may behave as potential alkylating agents following biooxidation to a quinone imine.[205] Compound (**48**) has been shown to be oxidized by peroxidases to an electrophilic quinone imine that reacts with cellular macromolecules. If topo-isomerase II is a cellular target for alkylation by compound (**48**), irreversible DNA strand breaks may result.[205] This hypothesis has not yet been tested, and although the majority of DNA strand breaks induced by ellipticine derivatives in cells are reversible,[120] a small proportion may be permanent and associated with covalently bound drug.

10.3.3.2.8   Actinomycin

Actinomycin D (**49**) is an antitumor antibiotic composed of a phenoxazone chromophore with two pentapeptide lactone rings. Actinomycin has very poor selective toxicity because it intercalates into DNA and strongly inhibits DNA-directed RNA synthesis in many different cell types.[206] Structure–activity relationships show significant correlations between DNA binding affinity, inhibitory potency towards RNA polymerase, and antibiotic activity. These studies have shown that the two pentapeptide lactones are required for activity, although variations to the amino acid sequence are tolerated. As expected for a DNA intercalating agent, the planarity of the phenoxazone chromophore is essential for DNA binding and biological activity.

(49)

The molecular details of DNA intercalation by actinomycin have been studied extensively.[206] The compound is thought to intercalate with the phenoxazone ring parallel to the axis of the base pairs and the pentapeptide lactone rings occupying the minor groove of DNA. An X-ray diffraction study of a complex between actinomycin and two deoxyguanosine molecules suggests the existence of hydrogen bonds between the 2-amino group of guanine and the threonine residue of each peptide.[207] In addition to providing DNA binding energy, it is possible that the pentapeptide moieties interact directly with DNA enzymes.

Actinomycin has been implicated in the inhibition of several different DNA enzymes. The most well-studied enzyme target for actinomycin is RNA polymerase. The compound is thought to provide a long-lived block to RNA synthesis by interfering with the chain-elongation reaction; it has little effect on initiation or termination processes.[208] In addition, actinomycin has been shown to inhibit mammalian topoisomerase II by stabilizing the covalent enzyme–DNA intermediate.[177] DNA cleavage induced by the drug occurs at sites that are different from all of the other known topoisomerase II inhibitors. Finally, inhibition of mammalian topoisomerase I by actinomycin has been suggested as a mechanism for transcription arrest.[209] Inhibition of any or all of these suggested enzyme targets by actinomycin could lead to the observed effects in intact cells: potent inhibition of RNA synthesis and the production of protein-linked DNA strand breaks. The relative importance of these putative DNA–enzyme targets to the biological activity of actinomycin is not established.

10.3.4 DNA METHYLTRANSFERASES

10.3.4.1 Enzymology and Functions of DNA Methyltransferases

The DNA of mammalian cells contains cytosines that are methylated at the 5-position.[210-212] 5-Methylcytosine is the only modified base found in eukaryotic DNA and represents about 4% of the total cytosines in a mammalian cell. The presence of methylated cytosines at specific DNA sites has been shown to regulate gene expression in several eukaryotic systems, possibly by altering DNA conformation and modifying protein–DNA interactions. Methylation of critical sites in genetic sequences, usually located at or near the 5′-ends, often leads to gene inactivation.

Enzymes that add methyl groups to the 5-position of cytosines within DNA are called DNA methyltransferases. These enzymes are responsible for maintaining specific patterns of DNA methylation within cells.[210-212] Although cytosine methylation occurs primarily at CpG sequences, not all CpG sequences are methylated. Consequently, tissue-specific patterns of methylation exist within mammalian cells. These patterns are maintained throughout cycles of DNA replication by a process called maintenance methylation. DNA methyltransferases preferentially act upon hemi-methylated DNA, the product of semiconservative replication, and thus conserve the original pattern in the daughter chromosome. This is possible because the primary site of methylation, CpG, is palindromic and methylation is symmetrical on both strands. For example, ^{me}CpG:^{me}CpG is produced from ^{me}CpG:CpG by the action of the DNA methyltransferase.

Cytosine methylation occurs while the base is already incorporated into the DNA, not at the level of free nucleoside or nucleotide.[211] Hemimethylated DNA is a 10–100 times better substrate than unmethylated DNA. The source for the methyl group is the universal methyl donor, *S*-adenosyl-methionine. The chemistry of the methylation reaction appears to be similar to that of several other enzymes that catalyze electrophilic substitution at the 5-position of pyrimidines. The paradigm among such enzymes is thymidylate synthetase. This enzyme contains a nucleophile in its active site that adds to the pyrimidine C-6 to form an enzyme–pyrimidine covalent intermediate.[213] The 5-position of the pyrimidine is thereby activated for reaction with an electrophile, affording a 5,6-dihydropyrimidine that is covalently bound to enzyme (**50**). Proton abstraction and β-elimination gives the methylated pyrimidine and releases active enzyme. The mechanism of mammalian DNA methyltransferase has been proposed to proceed by this mechanism, with the activated methyl group of *S*-adenosylmethionine being the electrophile.[214]

(50) (51) (52)

10.3.4.2 5-Azacytidine

5-Azacytidine (aza-C; **51**) is a cytotoxic agent that inhibits DNA methyltransferases; it has been widely used to study the role of DNA methylation in mammalian cells. The compound was found to prevent the methylation of newly replicated DNA in cultured mammalian cells and to affect gene expression.[211] Aza-C, when incorporated into the DNA of mammalian cells, results in a significant loss of DNA methyltransferase activity.[215] This loss of enzyme activity is concentration- and time-dependent, suggesting that enzyme inhibition is irreversible. Experiments have shown that hemi-methylated DNA containing aza-C inhibits purified DNA methyltransferase in a manner that cannot be reversed by high salt concentrations. Free aza-C or aza-CPT does not inhibit DNA methyltransferase, showing that the compound must be incorporated into DNA to be acted upon by the enzyme.

Santi and coworkers have proposed a mechanism for inhibition of DNA methyltransferases by aza-C based on previous work with other pyrimidine inhibitors such as 5-fluoro-2'-deoxy-uridylate.[214] It is thought that after the enzyme binds to DNA containing aza-C (aza-C–DNA), a nucleophilic group attacks the 6-position of the pyrimidine. Usually, this step would be followed by proton abstraction and β-elimination to release active enzyme (see above). With aza-C–DNA, however, the putative enzyme–DNA intermediate (**52**) is thermodynamically favorable, even though it could be reversed. This inhibitory mechanism has been confirmed with bacterial *Hpa* II methylase, which becomes covalently and irreversibly attached to aza-C–DNA.[216] The covalent adduct (**52**) might be expected to fragment and give an inactive formylated enzyme. In the case of *Hpa* II methylase, however, aza-C fragmentation does not occur since covalent complexes containing aza-C–DNA were isolated. Thus, aza-C is a mechanism-based inhibitor of *Hpa* II methylase and probably inhibits mammalian DNA methyltransferase in a similar manner.

10.3.5 MICROTUBULE ASSEMBLY

10.3.5.1 Properties and Functions of Microtubules

Microtubules are long, hollow cylinders composed primarily of the protein tubulin. They are important components of eukaryotic cells and are used for several different functions, including chromosome movement and cell motility associated with flagella and cilia.[217] Since microtubules are a component of the mitotic spindle used for chromosome separation, they are required for cell division. Consequently, compounds that perturb microtubule assembly inhibit mitosis and have a profound effect on cell growth. Mitotic inhibitors are selectively toxic to proliferating cells, because nondividing cells do not require mitotic spindles and are resistant. Several mitotic inhibitors have found clinical utility as antitumor agents.

Microtubules are composed primarily of tubulin, a dimeric protein containing two different acidic polypeptides (α and β) each with a molecular weight of 50 000 Da.[218] Microtubule assembly occurs *via* tubulin polymerization, a dynamic process that requires GTP. There are three phases: nucleation, the formation of short microtubule seeds; elongation, the addition of either tubulin $\alpha\beta$ dimers or tubulin oligomers to the growing microtubule chain; and steady state, a dynamic equilibrium involving no net change in microtubule length.[218] At steady state it is thought that there is constant addition of tubulin at one end of the microtubule and a balanced loss of tubulin at the other end. GTP binding is required for tubulin polymerization *in vitro*, and GTP hydrolysis occurs during the assembly process. However, nonhydrolyzable analogs of GTP can lead to microtubule assembly *in vitro*, and so GTP hydrolysis is not an absolute requirement. There are two GTP binding sites on tubulin: a nonexchangeable site that binds GTP tightly but noncovalently, and an exchangeable site that binds a molecule of GTP that is subsequently hydrolyzed to GDP.[219] GTP hydrolysis is not coupled to tubulin polymerization, but appears to occur following incorporation of tubulin into the microtubule. In cells, GTP hydrolysis may regulate the dynamic behavior of microtubule assembly.[220]

There are a few different mechanisms by which mitotic inhibitors can disrupt microtubules. Some compounds bind to tubulin and prevent its incorporation into polymers, targeting the dynamic process of steady-state microtubule assembly. These compounds effect microtubule depolymerization and thereby cause mitotic arrest. Other compounds bind to assembled microtubules and interfere with the formation of the mitotic spindle. The mechanisms, structure–activity relationships and antitumor activity of the most important mitotic inhibitors are discussed in the following sections.

10.3.5.2 Mitotic Inhibitors

10.3.5.2.1 Colchicine

Colchicine (**53**), a naturally occurring complex tropolone derivative, is a potent inhibitor of cell mitosis.[221, 222] It has limited antitumor activity and is mainly used for the treatment of gout. Colchicine is the classical example of a mitotic inhibitor that disrupts microtubule assembly by binding to a high-affinity site on the tubulin $\alpha\beta$ dimer. Colchicine has no effect on intact microtubules, but rather perturbs the dynamic equilibrium of steady-state microtubule assembly that exists within cells. After colchicine binds to tubulin (at a site distinct from the GTP site), it is thought that a conformational change occurs to form a stable tubulin–colchicine complex that may bind to the growing end of the microtubule and 'cap' it for further assembly. By preventing the attachment of further tubulin dimers to the growing end, colchicine perturbs the steady state in cells and effects microtubule depolymerization.

(53)

The most important structural features of colchicine that contribute to tubulin binding are the A and C rings. A particular spatial relationship between these rings is required: the so-called 'biaryl' moiety must be in the (S) configuration.[221] Unnatural (7R)-colchicine, which exists predominantly in the (R) configuration with respect to its biaryl moiety, does not bind to tubulin. The B ring is not required for tubulin binding, since derivatives that contain only the A and C rings retain activity as long as the biaryl moiety can achieve the required conformation. Furthermore, the methoxy groups at C-1 and C-10 are required for tubulin binding, although a thiomethyl group at C-10 is tolerated. The methoxy groups at positions 2 and 3 are not required, but bulk tolerance is limited at these positions.

10.3.5.2.2 Podophyllotoxin

Podophyllotoxin (**54**) is a naturally occurring plant alkaloid that binds to tubulin at a site overlapping the colchicine binding site.[223, 224] Like colchicine, podophyllotoxin is a mitotic spindle poison that prevents microtubule assembly in cells. That podophyllotoxin and colchicine bind to overlapping sites on tubulin is supported by two findings: podophyllotoxin prevents the binding of colchicine to tubulin, and tropolone (the c ring of colchicine) inhibits the binding of colchicine but not podophyllotoxin.[224] Podophyllotoxin is thought to bind to tubulin *via* its trimethoxyphenyl ring, a structural feature shared by colchicine.

(54)

The stereochemical configuration of podophyllotoxin results in a strained *trans*-γ-lactone. This feature contributes to mitotic inhibition and tubulin binding, since the thermodynamically more stable picropodophyllotoxin (with a *cis*-lactone) is inactive.[223] Although the podophyllotoxins have not found utility as antitumor agents, epipodophyllotoxins (epimerized at C-4 and glycosylated, compounds **42** and **43**) are very useful against cancer. The epipodophyllotoxins, however, have a very different mechanism of action; they do not bind to tubulin or have any effects on microtubules but rather inhibit mammalian topoisomerase II (see Section 10.3.3.2.5).

10.3.5.2.3 Vinca alkaloids

The *Vinca* alkaloids, vinblastine (**55**) and vincristine (**56**), are naturally occurring anticancer agents that bind to tubulin and prevent microtubule assembly.[225] Both compounds are important anticancer drugs; vinblastine is used clinically to treat Hodgkin's disease, and vincristine is used to treat acute lymphocytic leukemia and other neoplasms. In mammalian cells, vincristine and vinblastine cause mitotic arrest by disrupting microtubules and are selectively toxic to proliferating tumor cells.

(**55**) R = Me
(**56**) R = CHO

The *Vinca* alkaloids have a mechanism of action that is different from that of colchicine and podophyllotoxin. First, they bind to a different site on tubulin than either colchicine or podophyllotoxin. Second, they have the capability of inducing the formation of highly ordered tubulin crystals that have a different conformation than microtubules.[226] The *Vinca* alkaloids have been observed to have two effects in mammalian cells: at low concentrations, they disrupt cellular microtubules in a fashion similar to colchicine, but at high concentrations the compounds induce the formation of intracellular tubulin crystals. Studies of vinblastine–tubulin interactions in purified systems have suggested a mechanism by which the highly ordered crystals may form.[225-227] At concentrations of vinblastine slightly higher than those required for depolymerization of steady-state microtubules, tubulin aggregation is observed. The formation of tubulin aggregates and crystals is thought to occur because the drug binds to the protein and may alter the conformation of one or more of its surfaces. Thus, tubulin–tubulin interactions are perturbed, but not abolished, and a polymer forms that is different from a normal microtubule.[225] If vinblastine-induced tubulin aggregation occurs at the ends of microtubules, normal tubulin polymerization will be blocked.

Most studies of the binding of vinblastine and vincristine to tubulin have shown that there are two binding sites. Accurate measurements of the binding stoichiometry is very difficult, however, because of the drug-induced aggregation.[225,227] Maytansine, a macrocyclic compound isolated from plants, binds to tubulin in a manner that is competitive with the binding of the *Vinca* alkaloids.[225] Maytansine inhibits microtubule assembly but does not induce the formation of ordered tubulin polymers. Although maytansine has antitumor activity, it is too toxic to be therapeutically useful.

The selective action of vincristine has been suggested to be the result of differential uptake and efflux in tumor cells relative to normal cells.[228] In spite of this, vincristine has dose-limiting neurotoxicity, and so analogs of both *Vinca* alkaloids have been prepared semisynthetically in an effort to reduce toxicity. Most of these analogs are deacetylated and have an amide group in place of the ester group on the vindoline moiety.[229] Although some of the analogs retain antitumor activity and have lower toxicity, none has replaced vincristine or vinblastine in the clinic.

10.3.5.2.4 *Taxol*

Taxol (**57**) is a novel diterpenoid isolated from plants in the *Taxus* species that affects microtubule formation in cells.[230] Taxol has demonstrated antitumor activity in animal tumor models and is currently in clinical trials. The compound is unique in several ways: taxol has a complex structure that contains an unusual oxetane ring; it affects microtubules differently than all other mitotic inhibitors; and in cells it induces the formation of stable microtubule bundles that lead to mitotic arrest.

(**57**)

In the presence of purified tubulin, taxol enhances the rate of microtubule assembly.[230] This is opposite to the effects caused by other mitotic inhibitors (colchicine, podophyllotoxin, and the *Vinca* alkaloids), which inhibit microtubule assembly. Taxol reduces the critical concentration of tubulin required for assembly, and microtubules formed in its presence are more stable to calcium-induced or podophyllotoxin-induced depolymerization. [³H]Taxol has been shown to bind reversibly to assembled microtubules, with about 1 mol of taxol bound per mole of tubulin αβ dimer in the polymer.[231] There appears to be no taxol binding site on the free tubulin dimer, suggesting that a binding site is created as one molecule of tubulin binds to another.[230] The taxol binding site on the microtubule is different from the binding site for GTP, colchicine, podophyllotoxin or vinblastine. While GTP is usually required for microtubule assembly, taxol can induce assembly in the absence of GTP but inhibits neither the binding of GTP nor its hydrolysis.

The relationship between the effects of taxol on microtubules *in vitro* and in cells, and the consequences of these effects that lead to antitumor activity, are only partially understood. Treatment of mammalian cells with taxol results in the formation of aberrant microtubule bundles that are not associated with the microtubule organizing center.[230] These bundles are formed by taxol in a time- and concentration-dependent manner. While taxol binding to cells reaches saturation after 60 min, the formation of the microtubule bundles requires up to 24 h. The mitotic arrest caused by taxol may be a consequence of these secondary sites of microtubule bundle formation, which may lead to a depletion or rearrangement of the microtubules in the mitotic spindle required for cell division.

Taxol is very insoluble in water and has been difficult to formulate for clinical trials. Semisynthetic derivatives have been prepared in an effort to improve the compound's water solubility and antitumor activity.[232, 233] These derivatives have been evaluated for their effects on microtubules, as well as for their cytotoxicity to mammalian cells. Although there is a reasonable correlation between the two measurements, certain derivatives are cytotoxic but have no effect on microtubule assembly. The intact taxane ring and the ester side chain at C-13 are both essential for the induction of microtubule assembly and for cytotoxicity. Small changes to taxol, such as removal of the acetyl group at position 10 or alteration of the *N*-acyl substituent, have little effect on the drug's activity. The hydroxy group at C-7 is not required and can be acetylated, but the hydroxy group at C-2' appears to be very important since its acetylation eliminates activity *vs.* microtubules. However, 2'-acetyltaxol is cytotoxic, and it has been suggested that 2'-acyltaxol derivatives could serve as prodrugs for taxol.[232]

10.3.6 REFERENCES

1. P. B. Fernandes and D. T. W. Chu, *Annu. Rep. Med. Chem.*, 1988, **23**, 133.
2. D. C. Hooper and J. S. Wolfson, *Rev. Infect. Dis.*, 1988, **10**, S14.
3. L. L. Shen and A. G. Pernet, *Proc. Natl. Acad. Sci. USA*, 1985, **82**, 307.
4. D. L. Ollis, P. Brick, R. Hamlin, N. G. Xuong and T. A. Steitz, *Nature (London)*, 1985, **315**, 762.
5. A. Kornberg, 'DNA Replication', Freeman, San Francisco, 1980.
6. A. Kornberg, *Trends Biochem. Sci.*, 1984, **9**, 122.
7. K. E. Walton, P. C. Fitzgerald, M. S. Herrmann and W. D. Behnke, *Biochem. Biophys. Res. Commun.*, 1982, **108**, 1353.
8. H. Klenow and I. Henningsen, *Proc. Natl. Acad. Sci. USA*, 1970, **65**, 168.
9. C. M. Joyce and T. A. Steitz, *Trends Biochem. Sci.*, 1987, **12**, 288.
10. R. D. Kuchta, V. Mizrahi, P. A. Benkovic, K. A. Johnson and S. J. Benkovic, *Biochemistry*, 1987, **26**, 8410.
11. L. J. Ferrin and A. S. Mildvan, in 'DNA Replication and Recombination', UCLA Symposia in Molecular and Cellular Biology, ed. R. McMacken and T. J. Kelly, Liss, New York, 1987, vol. 47, p. 75.
12. A. Kornberg, *J. Biol. Chem.*, 1988, **263**, 1.
13. L. A. Loeb, P. K. Liu and M. Fry, *Prog. Nucleic Acid Res. Mol. Biol.*, 1986, **33**, 57.
14. M. R. Miller, R. G. Ulrich, T. S.-F. Wang and D. Korn, *J. Biol. Chem.*, 1985, **260**, 134.
15. G. Ciarrocchi, J. G. Jose and S. Linn, *Nucleic Acids Res.*, 1979, **7**, 1205.
16. M. Fry and L. A. Loeb, 'Animal Cell DNA Polymerases', CRC Press, Boca Raton, FL, 1986.
17. J. A. Huberman, *Cell*, 1987, **48**, 7.
18. P. A. Fisher, T. S.-F. Wang and D. Korn, *J. Biol. Chem.*, 1979, **254**, 6128.
19. P. A. Fisher, J. T. Chen and D. Korn, *J. Biol. Chem.*, 1981, **256**, 133.
20. S. W. Wong, L. R. Paborsky, P. A. Fisher, T. S.-F. Wang and D. Korn, *J. Biol. Chem.*, 1986, **261**, 7958.
21. B. Reckman, F. Grosse and G. Krauss, *Nucleic Acids Res.*, 1983, **11**, 7251.
22. A. Weissbach, A. Schlabach, B. Fridlender and A. Bolden, *Nature (London)*, 1977, **231**, 167.
23. T. S.-F. Wang and D. Korn, *Biochemistry*, 1982, **21**, 1597.
24. M. Y. W. T. Lee, C.-K. Tan, K. M. Downey and A. G. So, *Biochemistry*, 1984, **23**, 1906.
25. C. Nishida, P. Reinhard and S. Linn, *J. Biol. Chem.*, 1988, **263**, 501.
26. S. W. Wong, A. F. Wahl, P.-M. Yuan, N. Arai, B. E. Pearson, K. Arai, D. Korn, M. W. Hunkapiller and T. S.-F. Wang, *EMBO J.*, 1988, **7**, 37.
27. J. M. Coffin, in 'AIDS and Other Manifestations of HIV Infection', ed. G. P. Wormser, R. E. Stahl and E. J. Bottone, Noyes, Park Ridge, NJ, 1987, p. 130.
28. G. F. Gerard, in 'Enzymes of Nucleic Acid Synthesis and Modification', ed. S. T. Jacob, CRC Press, Boca Raton, FL, 1983, p. 1.
29. R. P. Bercoff, in 'The Molecular Basis of Viral Replication', ed. R. P. Bercoff, NATO ASI Series, Series A, Plenum Press, New York, 1987, vol. 136, p. 197.
30. J. S. Gibbs, H. C. Chiou, J. D. Hall, D. W. Mount, M. J. Retondo, S. K. Weller and D. M. Coen, *Proc. Natl. Acad. Sci. USA*, 1985, **82**, 7969.
31. M. E. O'Donnell, P. Elias and I. R. Lehman, *J. Biol. Chem.*, 1987, **262**, 4352.
32. F. DiMarzo Veronese, T. D. Copeland, A. L. De Vico, R. Rahman, S. Oroszlan, R. C. Gallo and M. G. Sarngadharan, *Science*, 1986, **229**, 1402.
33. B. Larder, D. Purifoy, K. Powell and G. Darby, *EMBO J.*, 1987, **6**, 3133.
34. W. G. Farmerie, D. D. Loeb, N. C. Casavant, C. A. Hutchinson, M. H. Edgell and R. Swanstrom, *Science*, 1987, **236**, 305.
35. J. Hansen, T. Schulze and K. Moelling, *J. Biol. Chem.*, 1987, **262**, 12393.
36. B. Larder, D. Purifoy, K. Powell and G. Darby, *Nature (London)*, 1987, **327**, 716.

37. D. M. Lowe, A. Aitken, C. Bradley, G. K. Darby, B. A. Larder, K. L. Powell, D. J. M. Purifoy, M. Tisdale and D. K. Stammers, *Biochemistry*, 1988, **27**, 8884.
38. B. D. Preston, B. J. Poiesz and L. A. Loeb, *Science*, 1988, **242**, 1168.
39. J. D. Roberts, K. Bebenek and T. A. Kunkel, *Science*, 1988, **242**, 1171.
40. E. DeClercq, *Trends Pharmacol. Sci.*, 1987, **8**, 339.
41. P. A. Furman and D. W. Barry, *Am. J. Med.*, 1988, **85** (2A), 176.
42. P. A. Furman, J. A. Fyfe, M. H. St. Clair, K. Weinhold, J. L. Rideout, G. A. Freeman, S. N. Lehrman, D. P. Bolognesi, S. Broder, H. Mitsuya and D. W. Barry, *Proc. Natl. Acad. Sci. USA*, 1986, **83**, 8333.
43. Y. Cheng, G. E. Dutschman, K. F. Bastow, M. G. Sarngadharan and R. Y. C. Ting, *J. Biol. Chem.*, 1987, **262**, 2187.
44. E. DeClercq, E. Balzarini, J. Descamps and F. Eckstein, *Biochem. Pharmacol.*, 1980, **29**, 1849.
45. A. Camerman, D. Mastropaolo and N. Camerman, *Proc. Natl. Acad. Sci. USA*, 1987, **84**, 8239.
46. G. I. Birnbaum, J. Giziewicz, E. J. Gabe, T.-S. Lin and W. H. Prusoff, *Can. J. Chem.*, 1987, **65**, 2135.
47. H. Mitsuya and S. Broder, *Nature (London)*, 1987, **325**, 773.
48. E. DeClercq, in 'Antiviral Drug Development, A Multidisciplinary Approach', ed. E. DeClercq and R. T. Walker, NATO ASI Series, Series A, Plenum Press. New York, 1988, vol. 143, p. 97.
49. C.-H. Kim, V. E. Marquez, S. Broder, H. Mitsuya and J. S. Driscoll, *J. Med. Chem.*, 1987, **30**, 862.
50. Y.-C. Cheng, G. E. Dutschman, K. F. Bastow, M. G. Sarngadharan and R. Y. C. Ting, *J. Biol. Chem.*, 1987, **262**, 2187.
51. H. Mitsuya and S. Broder, *Proc. Natl. Acad. Sci. USA*, 1986, **83**, 1911.
52. J. E. Dahlberg, H. Mitsuya, S. B. Blam, S. Broder and S. A. Aaronson, *Proc. Natl. Acad. Sci. USA*, 1987, **84**, 2469.
53. D. A. Cooney, M. Dalal, H. Mitsuya, J. B. McMahon, M. Nadkarni, J. Balzarini, S. Broder and D. G. Johns, *Biochem. Pharmacol.*, 1986, **35**, 2065.
54. G. B. Elion, *Top. Med. Chem.*, 1988, **65**, 163.
55. R. K. Robins and G. R. Revankar, in 'Antiviral Drug Development, A Multidisciplinary Approach', ed. E. DeClercq and R. T. Walker, NATO ASI Series, Series A, Plenum Press, New York, 1988, vol. 143, p. 11.
56. P. A. Furman, M. H. St. Clair and T. Spector, *J. Biol. Chem.*, 1984, **259**, 9575.
57. G. B. Elion, *Am. J. Med.*, 1982, **73** (1A), 7.
58. P. V. McGuirt, J. E. Shaw, G. B. Elion and P. A. Furman, *Antimicrob. Agents Chemother.*, 1984, **25**, 507.
59. P. M. Keller, L. Beauchamp, C. M. Lubbers, P. A. Furman, H. J. Schaeffer and G. B. Elion, *Biochem. Pharmacol.*, 1981, **30**, 3071.
60. T. Matthews and R. Boehme, *Rev. Infect. Dis.*, 1988, **10**, S490.
61. K. B. Frank, J. G. Chiou and Y. C. Cheng, *J. Biol. Chem.*, 1984, **259**, 1556.
62. M. MacCoss, R. L. Tolman, W. T. Ashton, A. F. Wagner, J. Hannah, A. K. Field, J. D. Karkas and J. T. Germershausen, *Chem. Scr.*, 1986, **26**, 113.
63. A. Larsson, K. Stenberg, A.-C. Ericson, U. Haglund, W.-A. Yisak, N.G. Johansson, B. Oberg and R. Datema, *Antimicrob. Agents Chemother.*, 1986, **30**, 598.
64. J. H. Burchenal, in 'Cancer: Achievements, Challenges, and Prospects for the 1980's', ed. J. H. Burchanal and H. F. Oettgen, Grune and Stratton, New York, 1981, p. 249.
65. M. G. Pallavicini, *Pharmacol. Ther.*, 1984, **25**, 207.
66. F. L. Graham and G. F. Whitmore, *Cancer Res.*, 1971, **30**, 2636.
67. R. Momparler, *Mol. Pharmacol.*, 1972, **8**, 362.
68. D. W. Kufe and P. P. Major, *Med. Pediat. Oncol.*, 1982, **1**, 49.
69. N. H. Heintz and J. L. Hamlin, *Biochemistry*, 1983, **22**, 3557.
70. T. W. North and S. S. Cohen, *Pharmacol. Ther.*, 1979, **4**, 81.
71. W. E. G. Muller, R. K. Zahn, K. Bittlingmaier and D. Falke, *Ann. N. Y. Acad. Sci.*, 1977, **284**, 34.
72. J. A. Montgomery, in 'Nucleosides, Nucleotides, and Their Biological Applications', ed. J. L. Rideout, D. W. Henry and L. M. Beacham, III, Academic Press, New York, 1983, p. 19.
73. N. C. Brown, L. W. Dudycz and G. E. Wright, *Drugs Exp. Clin. Res.*, 1986, **12**, 555.
74. N. N. Kahn, G. E. Wright, L. W. Dudycz and N. C. Brown, *Nucleic Acids Res.*, 1984, **12**, 3695.
75. S. Spadari, F. Sala and C. Pedrali-Noy, *Trends Biochem. Sci.*, 1982, **7**, 29.
76. M. Longiaru, J. E. Ikeda, Z. Jarkovsky, S. B. Horwitz and M. S. Horwitz, *Nucleic Acids Res.*, 1979, **6**, 3369.
77. G. Pedrali-Noy and S. Spadari, *Biochem. Biophys. Res. Commun.*, 1979, **88**, 1194.
78. L. Arabshahi, N. Brown, N. Khan and G. Wright, *Nucleic Acids Res.*, 1988, **16**, 5107.
79. K. Ono, Y. Iwata and H. Nakane, *Biomed. Pharmacother.*, 1983, **37**, 27.
80. S. Hiranuma, T. Shimizu, H. Yoshioka, K. Ono, H. Nakane and T. Takahashi, *Chem. Pharm. Bull.*, 1987, **35**, 1641.
81. T. Haraguchi, M. Oguro, H. Nagano, A. Ichihara and S. Sakamura, *Nucleic Acids Res.*, 1983, **11**, 1197.
82. A. Sugino and K. Nakayamara, *Proc. Natl. Acad. Sci. USA*, 1980, **77**, 7049.
83. P. K. Liu, C. C. Chang, J. E. Trosko, D. K. Dube, G. M. Martin and L. A. Loeb, *Proc. Natl. Acad. Sci. USA*, 1983, **80**, 797.
84. B. Oberg, *Pharmacol. Ther.*, 1989, **40**, 213.
85. P. S. Sarin, Y. Taguchi, D. Sun, A. Thornton, R. C. Gallo and B. Oberg, *Biochem. Pharmacol.*, 1985, **34**, 4075.
86. E. G. Sandstrom, R. E. Byington, J. C. Kaplan and M. S. Hirsch, *Lancet*, 1985, **1**, 1480.
87. L. Vrang and B. Oberg, *Antimicrob. Agents Chemother.*, 1986, **29**, 867.
88. S. S. Leinbach, R. M. Reno, L. F. Lee, A. F. Isbell and J. A. Boezi, *Biochemistry*, 1976, **15**, 426.
89. M. Ostrander and Y.-C. Cheng, *Biochim. Biophys. Acta*, 1980, **609**, 232.
90. J. C.-H. Mao and E. E. Robishaw, *Biochemistry*, 1975, **14**, 5475.
91. D. Derse, K. F. Bastow and Y.-C. Cheng, *J. Biol. Chem.*, 1982, **257**, 10 251.
92. K. B. Frank and Y.-C. Cheng, *Antimicrob. Agents Chemother.*, 1985, **27**, 445.
93. M. M. Vaghefi, P. A. McKernan and R. K. Robins, *J. Med. Chem.*, 1986, **29**, 1389.
94. E. DeClercq, *Antiviral Res.*, 1987, **7**, 1.
95. K. Ono, H. Nakane and M. Fukushima, *Eur. J. Biochem.*, 1988, **172**, 349.
96. K. D. Jentsch, G. Hunsmann, H. Hartmann and P. Nickel, *J. Gen. Virol.*, 1987, **68**, 2183.
97. M. Robbie and R. J. Wilkins, *Chem. Biol. Interactions*, 1984, **49**, 189.
98. D. R. Phillips and D. M. Crothers, *Biochemistry*, 1986, **25**, 7355.

99. C. Zimmer and U. Wahnert, *Prog. Biophys. Mol. Biol.*, 1986, **47**, 31.
100. M. L. Kopka, C. Yoon, D. Goodsell, P. Pjura and R. E. Dickerson, *Proc. Natl. Acad. Sci. USA*, 1985, **82**, 1376.
101. M. Coll, C. A. Frederick, A. H.-J. Wang and A. Rich, *Proc. Natl. Acad. Sci. USA*, 1987, **84**, 8385.
102. M. A. Warpehoski and L. Hurley, *Chem. Res. Toxicol.*, 1988, **1**, 315.
103. M. Tomasz, R. Lipman, B. F. McGuinness and K. Nakanishi, *J. Am. Chem. Soc.*, 1988, **110**, 5892.
104. L. H. Hurley, T. Reck, D. E. Thurston, D. Langley, K. G. Holden, R. P. Hertzberg, J. R. E. Hoover, G. Gallagher, L. F. Faucette, S.-M. Mong and R. K. Johnson, *Chem. Res. Toxicol.*, 1988, **1**, 258.
105. S. E. Sherman and S. J. Lippard, *Chem. Rev.*, 1987, **87**, 1153.
106. A. Eastman, *Pharmacol. Ther.*, 1987, **34**, 155.
107. Y. Inouye, Y. Take, K. Oogose, A. Kubo and S. Nakamura, *J. Antibiot.*, 1987, **40**, 105.
108. H. Okada, Y. Inouye and S. Nakamura, *J. Antibiot.*, 1987, **40**, 230.
109. R. Cone, S. K. Hasan, J. W. Lown and A. R. Morgan, *Can. J. Biochem.*, 1976, **54**, 219.
110. A. Maxwell and M. Gellert, *Adv. Protein Chem.*, 1986, **38**, 69.
111. J. C. Wang, in 'DNA Synthesis In Vitro', ed. R. D. Wells and R. B. Inman, University Park Press, Baltimore, 1973, p. 163.
112. J. C. Wang, *Annu. Rev. Biochem.*, 1985, **54**, 665.
113. M. Gellert, *Annu. Rev. Biochem.*, 1981, **50**, 879.
114. A. Sugino, C. L. Peebles, K. N. Kreuzer and N. R. Cozzarelli, *Proc. Natl. Acad. Sci. USA*, 1977, **74**, 4767.
115. M. Gellert, K. Mizuuchi, M. H. O'Dea, T. Itoh and J.-I. Tomizawa, *Proc. Natl. Acad. Sci. USA*, 1977, **74**, 4772.
116. T. Kirchhausen, J. C. Wang and S. C. Harrison, *Cell*, 1985, **41**, 933.
117. D. Lockshon and D. R. Morris, *J. Mol. Biol.*, 1985, **181**, 63.
118. A. Maxwell and M. Gellert, *J. Biol. Chem.*, 1984, **259**, 14472.
119. Y.-C. Tse-Dinh, *J. Biol. Chem.*, 1986, **261**, 10931.
120. W. E. Ross, *Biochem. Pharmacol.*, 1985, **34**, 4191.
121. E. M. Nelson, K. M. Tewey and L. F. Liu, *Proc. Natl. Acad. Sci. USA*, 1984, **81**, 1361.
122. B. S. Glisson and W. E. Ross, *Pharmacol. Ther.*, 1987, **32**, 89.
123. L. Yang, T. C. Rowe, E. M. Nelson and L. F. Liu, *Cell*, 1985, **41**, 127.
124. H.-S. Hsiang, R. P. Hertzberg, S. M. Hecht and L. F. Liu, *J. Biol. Chem.*, 1985, **260**, 14873.
125. P. D'Arpa, P. S. Machlin, H. Ratrie, III, N. F. Rothfield, D. W. Cleveland and W. C. Earnshaw, *Proc. Natl. Acad. Sci. USA*, 1988, **85**, 2543.
126. K. A. Edwards, B. D. Halligan, J. L. Davis, N. L. Nivera and L. F. Liu, *Nucleic Acids Res.*, 1982, **10**, 2565.
127. M. D. Been, R. R. Burgess and J. J. Champoux, *Nucleic Acids Res.*, 1984, **12**, 3097.
128. H. Busk, B. Thomsen, B. J. Bonven, E. Kjeldsen, O. F. Nielsen and O. Westergaard, *Nature (London)*, 1987, **327**, 638.
129. M. D. Been and J. J. Champoux, *J. Mol. Biol.*, 1985, **180**, 515.
130. F. H. Drake, J. P. Zimmerman, F. L. McCabe, H. F. Bartus, S. R. Per, D. M. Sullivan, W. E. Ross, M. R. Mattern, R. K. Johnson, S. T. Crooke and C. K. Mirabelli, *J. Biol. Chem.*, 1987, **262**, 16739.
131. N. Osheroff, E. R. Shelton and D. L. Brutlag, *J. Biol. Chem.*, 1983, **258**, 9536.
132. N. Osheroff, *J. Biol. Chem.*, 1986, **261**, 9944.
133. G. Fleishmann, G. Pflugfelder, E. K. Steiner, K. Javaherian, G. C. Howard, J. C. Wang and S. C. R. Elgin, *Proc. Natl. Acad. Sci. USA*, 1984, **81**, 6958.
134. M. Berrios, N. Osheroff and P. A. Fisher, *Proc. Natl. Acad. Sci. USA*, 1985, **82**, 4142.
135. D. S. Gilmour and S. C. R. Elgin, *Mol. Cell. Biol.*, 1987, **7**, 141.
136. A. F. Stewart and G. Schutz, *Cell*, 1987, **50**, 1109.
137. R. M. Snapka, *Mol. Cell. Biol.*, 1986, **6**, 4221.
138. L. Yang, M. S. Wold, J. J. Li, T. J. Kelly and L. F. Liu, *Proc. Natl. Acad. Sci. USA*, 1987, **84**, 950.
139. S. J. Brill, S. DiNardo, K. Voelkel-Meiman and R. Sternglanz, *Nature (London)*, 1987, **326**, 414.
140. T. Goto and J. C. Wang, *Proc. Natl. Acad. Sci. USA*, 1985, **82**, 7178.
141. M. Tsai-Pflugfelder, L. F. Liu, A. A. Liu, K. M. Tewey, J. Whang-Peng, T. Knutsen, K. Huebner, C. M. Croce and J. C. Wang, *Proc. Natl. Acad. Sci. USA*, 1988, **85**, 7177.
142. E. Wyckoff, D. Natalie, J. M. Nolan, M. Lee and T.-S. Hsieh, *J. Mol. Biol.*, 1989, **205**, 1.
143. K. Drlica, *Microbiol. Rev.*, 1984, **48**, 273.
144. L. L. Shen, W. E. Kohlbrenner, D. Weigl and J. Baranowski, *J. Biol. Chem.*, 1989, **264**, 2973.
145. P. Hussy, G. Maass, B. Tummler, F. Grosse and U. Schomburg, *Antimicrob. Agents Chemother.*, 1986, **29**, 1073.
146. P. B. Fernandes and D. T. W. Chu, *Annu. Rep. Med. Chem.*, 1988, **23**, 133.
147. L. J. V. Piddock and R. Wise, *FEMS Microbiol. Lett.*, 1987, **41**, 289.
148. J. M. Domagala, L. D. Hanna, C. L. Heifetz, M. P. Hutt, T. F. Mich, J. P. Sanchez and M. Solomon, *J. Med. Chem.*, 1986, **29**, 394.
149. J. M. Domagala, C. L. Heifetz, T. F. Mich and J. B. Nichols, *J. Med. Chem.*, 1986, **29**, 445.
150. J. M. Domagala, C. L. Heifetz, M. P. Hutt, T. F. Mich, J. B. Nichols, M. Solomon and D. F. Worth, *J. Med. Chem.*, 1988, **31**, 991.
151. L. A. Mitscher, P. N. Sharma, D. T. W. Chu, L. L. Shen and A. G. Pernet, *J. Med. Chem.*, 1987, **30**, 2283.
152. D. Chu, P. Fernandes, A. Claiborne, E. Pihuleac, C. Nordeen, R. Maleczka and A. Pernet, *J. Med. Chem.*, 1985, **28**, 1558.
153. D. T. W. Chu, P. B. Fernandes, A. K. Claiborne, L. Shen and A. G. Pernet, *Drugs Exp. Clin. Res.*, 1988, **14**, 379.
154. K. Mizuuchi, M. H. O'Dea and M. Gellert, *Proc. Natl. Acad. Sci. USA*, 1978, **75**, 5960.
155. A. Sugino, N. P. Higgins, P. O. Brown, C. L. Peebles and N. R. Cozzarelli, *Proc. Natl. Acad. Sci. USA*, 1978, **75**, 4838.
156. F. Reusser and L. A. Dolak, *J. Antibiot.*, 1986, **39**, 272.
157. I. W. Althaus, L. Dolak and F. Reusser, *J. Antibiot.*, 1988, **41**, 373.
158. K. G. Miller, L. F. Liu and P. T. Englund, *J. Biol. Chem.*, 1981, **256**, 9334.
159. B. Villeponteau, T. M. Pribyl, M. H. Grant and H. G. Martinson, *J. Biol. Chem.*, 1986, **261**, 10359.
160. M. W. Van Dyke and R. G. Roeder, *Nucleic Acids Res.*, 1987, **15**, 4365.
161. M. Suffness and G. A. Cordell, in 'The Alkaloids: Chemistry and Pharmacology', ed. A. Brossi, Academic Press, Orlando, 1985, vol. 25, p. 73.
162. M. R. Mattern, S.-M. Mong, H. F. Bartus, C. K. Mirabelli, S. T. Crooke and R. K. Johnson, *Cancer Res.*, 1987, **47**, 1793.

163. Y.-H. Hsiang and L. F. Liu, *Cancer Res.*, 1988, **48**, 1722.
164. W.-K. Eng, L. Faucette, R. K. Johnson and R. Sternglanz, *Mol. Pharmacol.*, 1988, **34**, 755.
165. J. Nitiss and J. C. Wang, *Proc. Natl. Acad. Sci. USA*, 1988, **85**, 7501.
166. R. S. Gupta, R. Gupta, B. Eng, R. B. Lock, W. E. Ross, R. P. Hertzberg, M. J. Caranfa and R. K. Johnson, *Cancer Res.*, 1988, **48**, 6404.
167. T. Andoh, K. Ishii, Y. Suzuki, Y. Ikegami, Y. Kusunoki, Y. Takemoto and K. Okada, *Proc. Natl. Acad. Sci. USA*, 1987, **84**, 5565.
168. R. P. Hertzberg, M. J. Caranfa and S. M. Hecht, *Biochemistry*, 1989, **28**, 4629.
169. E. Kjeldsen, S. Mollerup, B. Thomsen, B. Bonven, L. Bolund and O. Westergaard, *J. Mol. Biol.*, 1988, **202**, 333.
170. J. J. Champoux and R. Aronoff, *J. Biol. Chem.*, 1989, **264**, 1010.
171. R. P. Hertzberg, M. J. Caranfa, K. G. Holden, D. R. Jakas, G. Gallagher, M. R. Mattern, S.-M. Mong, J. O'Leary Bartus, R. K. Johnson and W. D. Kingsbury, *J. Med. Chem.*, 1989, **32**, 715.
172. M. C. Wani, P. E. Ronman, J. T. Lindley and M. E. Wall, *J. Med. Chem.*, 1980, **23**, 554.
173. C. R. Hutchinson, *Tetrahedron*, 1981, **37**, 1047.
174. M. C. Wani, A. W. Nicholas, G. Manikumar and M. E. Wall, *J. Med. Chem.*, 1987, **30**, 1774.
175. B. Marshall and R. K. Ralph, *Adv. Cancer Res.*, 1985, **44**, 267.
176. W. R. Wilson, B. C. Baguley, L. P. G. Wakelin and M. J. Waring, *Mol. Pharmacol.*, 1981, **20**, 404.
177. K. M. Tewey, T. C. Rowe, L. Yang, B. D. Halligan and L. F. Liu, *Science (Washington, DC)*, 1984, **226**, 466.
178. J. Minford, Y. Pommier, J. Filipski, K. W. Kohn, D. Kerrigan, M. Mattern, S. Michaels, R. Schwartz and L. A. Zwelling, *Biochemistry*, 1986, **25**, 9.
179. T. C. Rowe, G. L. Chen, Y.-H. Hsiang and L. F. Liu, *Cancer Res.*, 1986, **46**, 2021.
180. W. A. Denny, B. F. Cain, G. J. Atwell, C. Hansch, A. Panthananickal and A. Leo, *J. Med. Chem.*, 1982, **25**, 276.
181. B. C. Baguley, W. A. Denny, G. J. Atwell and B. F. Cain, *J. Med. Chem.*, 1981, **24**, 520.
182. G. J. Atwell, G. W. Rewcastle, B. C. Baguley and W. A. Denny, *J. Med. Chem.*, 1987, **30**, 652.
183. J. M. Covey, K. W. Kohn, D. Kerrigan, E. J. Tilchen and Y. Pommier, *Cancer Res.*, 1988, **48**, 860.
184. W. A. Denny, I. A. G. Roos and L. P. G. Wakelin, *Anti-Cancer Drug Design*, 1986, **1**, 141.
185. L. P. G. Wakelin, G. J. Atwell, G. W. Rewcastle and W. A. Denny, *J. Med. Chem.*, 1987, **30**, 855.
186. G. L. Chen, L. Yang, T. C. Rowe, B. D. Halligan, K. M. Tewey and L. F. Liu, *J. Biol. Chem.*, 1984, **259**, 13560.
187. J. M. S. van Maanen, J. Retel, J. de Vries and H. M. Pinedo, *J. Natl. Cancer Inst.*, 1988, **80**, 1526.
188. B. Glisson, R. Gupta, S. Smallwood-Kentro and W. Ross, *Cancer Res.*, 1986, **46**, 1934.
189. K.-C. Chow, T. L. MacDonald and W. E. Ross, *Mol. Pharmacol.*, 1988, **34**, 467.
190. B. H. Long, S. T. Musial and M. G. Brattain, *Biochemistry*, 1984, **23**, 1183.
191. R. S. Gupta, P. C. Chenchaiah and R. Gupta, *Anti-Cancer Drug Design*, 1987, **2**, 1.
192. C. Myers, L. Gianni, J. Zweier, J. Muindi, B. K. Sinha and H. Eliot, *Fed. Proc., Fed. Am. Soc. Exp. Biol.*, 1986, **45**, 2792.
193. S. Neidle, *Prog. Med. Chem.*, 1979, **16**, 151.
194. T. R. Tritton and G. Yee, *Science (Washington, DC)*, 1982, **217**, 248.
195. S. T. Crooke and S. D. Reich, 'Anthracyclines: Current Status and New Developments', Academic Press, New York, 1980.
196. A. H.-J. Wang, G. Ughetto, G. J. Quigley and A. Rich, *Biochemistry*, 1987, **26**, 1152.
197. M. Potmesil, Y.-H. Hsiang, L. F. Liu, B. Bank, H. Grossberg, S. Kirschenbaum, T. J. Forlenzar, A. Penziner, D. Kanganis, D. Knowles, F. Traganos and R. Silber, *Cancer Res.*, 1988, **48**, 3537.
198. T. Kriebardis, D. Meng and S. Aktipis, *J. Biol. Chem.*, 1987, **262**, 12632.
199. F. Arcamone, *Med. Res. Rev.*, 1984, **4**, 153.
200. J. W. Lown, K. Reszka, P. Kolodziejczyk and W. D. Wilson, *Pontif. Acad. Sci. Scr. Varia*, 1986, **70**, 243.
201. K. C. Murdock, R. G. Child, P. F. Fabio, R. B. Angier, R. E. Wallace, F. E. Durr and R. V. Citarella, *J. Med. Chem.*, 1979, **22**, 1024.
202. K. W. Kohn, W. E. Ross and D. Glaubiger, in 'Antibiotics', ed. F. E. Hahn, Springer-Verlag, New York, 1979, vol. 5, part 2, p. 195.
203. K. M. Tewey, G. L. Chen, E. M. Nelson and L. F. Liu, *J. Biol. Chem.*, 1984, **259**, 9182.
204. Y. Pommier, J. K. Minford, R. E. Schwartz, L. A. Zwelling and K. W. Kohn, *Biochemistry*, 1985, **24**, 6410.
205. G. Meunier, D. De Montauzon, J. Bernadou, G. Grassy, M. Bonnafous, S. Cros and B. Meunier, *Mol. Pharmacol.*, 1988, **33**, 93.
206. E. F. Gale, E. Cundliffe, P. E. Reynolds, M. H. Richmond and M. J. Waring, in 'The Molecular Basis of Antibiotic Action', Wiley, Chichester, 1981, p. 314.
207. S. C. Jain and H. M. Sobell, *J. Mol. Biol.*, 1972, **68**, 1.
208. H. M. Sobell, *Proc. Natl. Acad. Sci. USA*, 1985, **82**, 5328.
209. D. K. Trask and M. T. Muller, *Proc. Natl. Acad. Sci. USA*, 1988, **85**, 1417.
210. A. Razin and A. D. Riggs, *Science*, 1980, **210**, 604.
211. A. Razin, H. Cedar and A. D. Riggs, 'DNA Methylation, Biochemistry and Biological Significance', Springer-Verlag, New York, 1984.
212. W. Doerfler, *Annu. Rev. Biochem.*, 1983, **52**, 93.
213. D. V. Santi and P. V. Danenberg, in 'Folates and Pterins', ed. R. L. Blakely and S. J. Benkovic, Wiley, New York, 1984, p. 345.
214. D. V. Santi, C. E. Garrett and P. J. Barr, *Cell*, 1983, **33**, 9.
215. S. M. Taylor and P. A. Jones, *J. Mol. Biol.*, 1982, **162**, 679.
216. D. V. Santi, A. Norment and C. E. Garrett, *Proc. Natl. Acad. Sci. USA*, 1984, **81**, 6993.
217. D. Soifer, 'Dynamic Aspects of Microtubule Biology', *Ann. N.Y. Acad. Sci.*, 1986, **466**.
218. S. N. Timasheff and L. M. Grisham, *Annu. Rev. Biochem.*, 1980, **49**, 565.
219. E. Hamel, in 'Developments in Cancer Chemotherapy', ed. R. I. Glazer, CRC Press, Boca Raton, FL, 1984, p. 131.
220. D. Pantaloni and M.-F. Carlier, in ref. 217, p. 496.
221. A. Brossi, H. J. C. Yeh, M. Chrzanowska, J. Wolff, E. Hame, C. M. Lin, F. Quin, M. Suffness and J. Silverton, *Med. Res. Rev.*, 1988, **8**, 77.
222. J. M. Andreu and S. N. Timasheff, in ref. 217, p. 676.

223. J. M. Hartwell and A. W. Schrecker, *Fortschr. Chem. Org. Naturst.*, 1958, **15**, 83.
224. F. Cortese, B. Bhattacharyya and J. Wolff, *J. Biol. Chem.*, 1977, **252**, 1134.
225. R. F. Luduena, W. H. Anderson, V. Prasad, M. A. Jordan, K. C. Ferrigni, M. C. Roach, P. M. Horowitz, D. B. Murphy and A. Fellous, in ref. 217, p. 718.
226. G. C. Na and S. N. Timasheff, *J. Biol. Chem.*, 1982, **257**, 10 387.
227. G. C. Na and S. N. Timasheff, *Biochemistry*, 1986, **25**, 6214.
228. J. A. Houghton, L. G. Williams, P. M. Torrance and P. J. Houghton, *Cancer Res.*, 1984, **44**, 582.
229. W. A. Remers, in 'Antineoplastic Agents', ed. W. A. Remers, Wiley, New York, 1984, p. 213.
230. J. J. Manfredi and S. B. Horwitz, *Pharmacol. Ther.*, 1984, **25**, 83.
231. J. Parness and S. B. Horwitz, *J. Cell Biol.*, 1981, **91**, 479.
232. W. Mellado, N. F. Magri, D. G. I. Kingston, R. Garcia-Arenas, G. A. Orr and S. B. Horwitz, *Biochem. Biophys. Res. Commun.*, 1984, **124**, 329.
233. S. B. Horwitz, L. Lothstein, J. J. Manfredi, W. Mellado, J. Parness, S. N. Roy, P. B. Schiff, L. Sorbara and R. Zeheb, in ref. 217, p. 733.

10.4

Inhibitors of the Transcribing Enzymes: Rifamycins and Related Agents

PIERO SENSI

Università di Milano, Italy

and

GIANCARLO LANCINI

Lepetit Research Center, Gerenzano, Italy

10.4.1	THE PROCESS OF TRANSCRIPTION	793
	10.4.1.1 Transcription in Prokaryotes	794
	10.4.1.2 Transcription in Archaebacteria	794
	10.4.1.3 Transcription in Eukaryotes	795
	10.4.1.3.1 Nuclear enzymes	795
	10.4.1.3.2 Organelle enzymes	795
	10.4.1.4 Transcription and RNA Replication in Viruses	796
10.4.2	INHIBITORS OF TRANSCRIPTION	796
	10.4.2.1 Inhibitors of Prokaryotes RNA Polymerase	797
	10.4.2.1.1 Ansamycins	797
	10.4.2.1.2 Streptolydigin and tyrandamycin	797
	10.4.2.1.3 Lipiarmycin	798
	10.4.2.1.4 Myxopyronins and corallopyronins	799
	10.4.2.1.5 Sorangicins	799
	10.4.2.2 Inhibitors of Eukaryotic RNA Polymerases	800
10.4.3	RIFAMYCINS	800
	10.4.3.1 Origin and Nature of Rifamycins	800
	10.4.3.2 Mechanism of Action	801
	10.4.3.3 Structural Requirement for Antibacterial Activity	802
	10.4.3.4 Structural Modifications	803
	10.4.3.5 Modulation of Chemotherapeutic Properties	806
	10.4.3.6 Rifamycins in Clinical Use	807
	10.4.3.6.1 Rifamycin SV	807
	10.4.3.6.2 Rifamide	807
	10.4.3.6.3 Rifampicin	807
	10.4.3.6.4 Rifaximin	808
	10.4.3.6.5 Rifamycins under clinical evaluation	808
	10.4.3.7 Effect of Rifamycins on Nonbacterial Polymerases	808
10.4.4	REFERENCES	809

10.4.1 THE PROCESS OF TRANSCRIPTION

Transcription is the central process in cell metabolism by which genes, that is specific sequences of DNA, are selectively transcribed by DNA-dependent RNA polymerases (EC 2.7.7.6, in short RNA polymerase) to produce messenger, ribosomal or transfer RNAs. The basic reaction catalyzed by

RNA polymerases is the transfer of a ribonucleoside monophosphate to the 3'-hydroxy group of the terminal nucleotide of a growing RNA chain, using ribonucleoside triphosphates as substrate. In order for this polymerization process to occur in a meaningful way the enzyme must: (a) recognize specific sequences of a DNA strand (called promoters) from where to start the transcription; (b) insert the correct nucleotide on the basis of the complementary rule; (c) repeatedly perform the synthetic reaction; and (d) recognize termination signals in order to end the RNA molecule at appropriate points. Whereas this process is basically the same in all cells, the structure of RNA polymerases as well as the recognition and termination sequences are quite different in the various classes of organisms and are thus briefly separately described.

10.4.1.1   Transcription in Prokaryotes

RNA polymerases from all the eubacteria studied, as well as blue-green algae and actinomycetes, are similar in structure and size.[1-3] They consist of a 'core' enzyme, constituted by four subunits denoted α, α, β and β'. In *Escherichia coli*, whose enzyme is the most deeply studied, M of the subunits is respectively 36 000 (329 amino acids), 150 000 (1342 amino acids) and 155 000. Two atoms of Zn^{2+} are bound to the enzyme. These are believed to participate in the catalytic action and are generally found in other RNA polymerases. The core enzyme can carry on the polymerization reactions, transcribing DNA sequences, but is unable to correctly recognize the starting points. Most of the functions appear to be performed by the β subunit but only in the presence of all the other components.

The 'holoenzyme' consists of the core enzyme plus a protein, the σ factor (M 70 000, 613 amino acids in the case of *E. coli*), whose presence is essential for the recognition of promoters. In several bacteria (*B. subtilis* for example) there are more than one σ factors recognizing different promoters, so that the transcription of sets of genes is switched on or off according to the cell development phase. The promoters are segments of DNA where RNA polymerase binds with high affinity and is activated to a form capable of initiating transcription at the correct DNA nucleotide.

Analysis of DNA sequences has identified two regions, around 10 and 35 nucleotides respectively upstream to the starting point of transcription, which appear to be involved in RNA polymerase recognition and binding.[4] A spacing of 17 bases between specific sites in the -10 and -35 regions appears a characteristic of good promoters. In the adjacent regions, other sequences are specifically recognized by repressor or activator proteins, whose presence or absence determine whether a given segment of DNA will be transcribed. These regulatory genes are denoted 'operators'. An operon is a unit of transcription, that is the segment of DNA including the promoter, the operator and the genes to be transcribed in one single RNA molecule.

The sequence of the events constituting transcription in *E. coli* can be summarized as follows:[4] RNA polymerase, which may be loosely bound to DNA, tightly binds to the promoter region, forming a 'closed complex' with double strand DNA. Partial separation of the strands occurs (melting-in) with formation of an 'open complex'. The first nucleotide triphosphate (usually a purine triphosphate in bacteria) is inserted and a stable ternary complex is thus formed. The second nucleotide triphosphate binds to the complex and the first phosphodiester bond is formed. At this point, or after the addition of a few more nucleotides, the σ factor is released and begins the elongation phase, in which the polymerization is completed. Termination occurs at specific sequences, generally recognized with the help of a protein, the ρ factor.

10.4.1.2   Transcription in Archaebacteria

The molecular biology of the archaebacteria has been studied only recently and our knowledge of transcription in these organisms is incomplete. The DNA-dependant RNA polymerase of archaebacteria is composed of 8–11 peptides, generally denoted with capital letters A, B, *etc.* Only one of the smaller peptides occurs in more than one copy per enzyme. Two major types can be recognized, corresponding to the major phyla of the archaebacteria, one comprising the anaerobic methanogenes and the extreme aerobic halophiles and the second including the extremely thermoacidophilic genera.[5]

RNA polymerases of the first group[6] are generally composed of four large components (M 40 000–140 000) and five small components (M 10 000–23 000). Those of thermoacidophiles[7] have only three large peptides (M 40 000–130 000) and again five small peptides (M 10 000–25 000). The component pattern of archaebacteria polymerases thus appears quite different from that of eubacteria and strikingly similar to that of some eukaryotic polymerases (see below).

Other evidence of similarity with eukaryotic polymerases are: (a) the rate of transcription is increased by silybin (a flavolignan from *Silybium marianum*) which stimulates transcription by eukaryote polymerase I (see below) but not II or III or *E. coli* polymerase; (b) antibodies against the heaviest component of yeast polymerases I and II react with component A of the RNA polymerase of *Halobacterium halobium*, and other similar immunochemical cross reactions have been observed. However, as discussed later, the polymerases of archaebacteria differ from both eubacteria and eukaryote polymerases in respect to inhibitors, being insensitive to rifampicin, streptolidigin or α-amanitin.

10.4.1.3 Transcription in Eukaryotes

10.4.1.3.1 Nuclear enzymes

The eukaryotic transcribing system is considerably more complex than that of prokaryotes and much more difficult to analyze *in vitro*.[8] In all eukaryotic cells there are three RNA polymerases associated with the nucleus.[9] In addition, specific RNA polymerases are present in the cytoplasmic organelles, mitochondria and chloroplasts. The RNA polymerases of the nucleus (designed I, II, III or A, B, C) perform different functions and differ also in their localization and in the DNA sequences they recognize as initiation and consensus signals. Their molecular weights range from 500 000 to 600 000 and each enzyme is composed of 9–11 subunits.[10] Two nonidentical large subunits (M from 130 000 to over 200 000) are always present and are similar in the various organisms. Several small subunits (M below 50 000) are present and some of them are in common in the polymerases of a single organism. Only in polymerase III is a subunit of intermediate size (M about 90 000) consistently found.

RNA polymerase I (or A) is found in association with the nucleolus and catalyzes the synthesis of a large ribosomal RNA precursor which is then processed to give the 18S and the 28S rRNAs. In mammalian cells there are about 200 copies of the genes for ribosomal RNA, arranged in tandemly repeating units consisting of a transcribed segment (about 13 000 bases) and a nontranscribed segment (about 30 000 bases).

RNA polymerase II (or B) synthesizes messenger RNAs. The promoters of the mRNA genes include a sequence TAT (often TATAAT) 25–30 bases upstream from the start site of transcription.[11] Other sequences having a control function are found further upstream, even at a distance of 250–300 nucleotides from the initiation site.

The mechanism of initiation and polymerization is believed to be similar to that described for *E. coli*. Termination appears a complex process, in which several proteins, or termination factors, intervene. Transcription is only a part of the process of formation of eukaryotic mRNAs. The following additional steps are also generally performed:[8] (a) capping, which consists of the addition of 7-methylguanosine to the 5'-triphosphate of the first nucleotide; (b) poly A addition, where a specific enzyme (poly A polymerase) adds from 40 to 200 adenosines to the 3' terminus; (c) splicing, where specific endonucleases cut the primary transcripts; sequences coded by introns are eliminated and the others are ligated to give a shorter RNA molecule.

RNA polymerase III makes a variety of small stable RNA molecules, including the transfer RNAs and the 5S ribosomal RNA. In contrast to the RNA polymerases I and II, it initiates transcription at the correct point by recognizing a gene regulatory protein that binds to DNA about 45 nucleotides downstream of the start of transcription. The genes coding for tRNAs and 5S rRNAs are tandemly arranged and include nontranscribed sequences, similarly to those transcribed by RNA polymerase I.

10.4.1.3.2 Organelle enzymes[9,12]

RNA polymerases from mitochondria have been purified from a variety of organisms but have not yet been well characterized. They appear to differ from both the bacterial and the nuclear enzyme in size and in sensitivity to inhibitors.

RNA polymerase from *Zea mays* chloroplasts has been purified to apparent homogeneity. This enzyme possesses two large subunits (M 200 000 and 180 000) plus several smaller polypeptides. From *Euglena gracilis* chloroplasts a transcriptionally active chromosome has been isolated which *in vitro* produces ribosomal RNA. The enzyme has a relatively simple structure with two major chains (M 125 000 and 50 000) and a minor one (M 47 000).

10.4.1.4 Transcription and RNA Replication in Viruses

Transcription and RNA replication in viruses are processes whose complexity depends on the viral genome composition and may be carried out either by cellular or by virus specific enzymes. In respect to transcription, viruses can be divided in six major groups.[13,14]

(1) *Double strand DNA viruses.* Messenger RNAs specific to viruses of this group, as in the case of papova, adeno and herpes viruses, are synthesized by cell nuclear RNA polymerases. Similarly, the T4 bacteriophage messengers are synthesized by the host enzyme, eventually being modified.[15] The pox viruses, on the contrary, possess a specific RNA polymerase which is carried in the virion itself. It is constituted by seven subunits (M from 13 500 to 135 000) and preferably transcribes single strand DNA.[16] In T7 bacteriophages, early transcription is made by the host enzyme.[15] Late transcription is made by a virus coded RNA polymerase containing a single polypeptide (M about 100 000) which very specifically recognizes viral promoters.

(2) *Single strand DNA viruses.* In this group, which includes parvoviruses, cell polymerases are required for transcription.

(3) *Retroviruses.* A virion associated enzyme, the reverse transcriptase, synthesizes a DNA strand using RNA as the template. After replication, the double stranded DNA is integrated in the cell genome and is transcribed by the nuclear enzymes.

(4) *Double strand RNA viruses.* Members of this group, including reoviruses and rotaviruses, produce mRNA by transcribing viral RNA with virus specified enzymes which form part of the virion particle. Host enzymes are not involved.

(5) *Negative strand RNA viruses.* With rhabdoviruses the replication is cytoplasmic, performed by virus coded enzymes. With influenza virus, replication occurs in the nucleus, and requires both cellular and viral enzymes.

(6) *Positive single strand RNA viruses.* With picorna and coronaviruses the viral genome is directly translated to produce proteins which include a transcriptase concerned with viral RNA replication.

10.4.2 INHIBITORS OF TRANSCRIPTION

A vast number of transcription inhibitors are reported in the literature. However, most of these products do not interact with the transcribing enzymes but act by binding to DNA. As a consequence, their effect is not selective on RNA production since at the same time both transcription and DNA replication are inhibited. Moreover, they do not present organism specificity due to the similar DNA structure in all living cells. Thus these cannot be used as antiinfective agents but in some cases, as for actinomycin D or doxorubicin, have been used for their antitumor activity.

We describe in this chapter products which inhibit RNA polymerases activity by directly interfering with the enzymes. These products are both selective in their action and, due to the large differences of RNA polymerases composition, are specific for each class of organisms. Inhibition of RNA polymerase in bacterial growing cells gives origin to a sequence of events which can be clearly visualized by examining the kinetics of macromolecular synthesis.[17,18] RNA synthesis is blocked almost immediately. Protein synthesis continues at a rapidly decreasing rate for about ten minutes, corresponding to the degradation time of existing mRNAs. DNA synthesis is not directly inhibited but, since its reinitiation requires RNA synthesis, DNA replication ceases in about one generation time.

In eukaryotes the sequence of events should be basically similar, but it is difficult to examine because of the long life of messenger RNAs and the presence of different RNA polymerases. Since in bacteria a temporary inhibition of RNA synthesis is not *per se* lethal, the effect of inhibitors can be either bacteriostatic or bactericidal according to the nature of the complex formed and the reversibility of the binding.

10.4.2.1 Inhibitors of Prokaryotes RNA Polymerase

10.4.2.1.1 Ansamycins

Ansamycins are a class of microbial and plant metabolites characterized by an aromatic ring spanned by an aliphatic chain (*ansa* from the latin word for handle). One of the junctions of the ring with the chain is a carbon–carbon bond; the other is an amide. Ansamycins can be divided in four groups, characterized by similar structural and biological properties:[19] (a) naphthalenic with a 23 atom ansa; (b) naphthalenic with a 17 atom ansa; (c) benzenic with a 17 atom ansa; (d) benzenic with a 15 atom ansa.

The members of group (b) are specific inhibitors of bacterial RNA polymerase. The other groups possess different activities and mechanism of action. Two major families constitute the (b) group: rifamycins, which are discussed in detail in the Section 10.4.3, and streptovaricins. The latter are characterized by an uninterrupted carbon ansa chain, whilst that in the rifamycins includes an oxygen atom.

Streptovaricins were isolated from *Streptomyces spectabilis*[20] as a complex of structurally related antibiotics (1)[21,22] active against Gram-positive bacteria and *M. tuberculosis*. Antibacterial activity is observed at concentrations around 1 $\mu g\,mL^{-1}$ with Gram-positive organisms and mycobacteria. Higher concentrations (about 10 $\mu g\,mL^{-1}$) are required for inhibiting *E. coli* and some other Gram-negative species, whereas some fungi are inhibited at concentrations over 100 $\mu g\,mL^{-1}$.[20,23,24]

	R^1	R^2	R^3	R^4
Streptovaricin A	OH	OH	COMe	OH
B	H	OH	COMe	OH
C	H	OH	H	OH
D	H	OH	H	H
E	H	=O	H	OH
G	OH	OH	H	OH
J	H	OCOMe	H	OH

(1)

Efficacy in curing *in vivo* tuberculous infections was demonstrated in mice and guinea pigs and also confirmed in clinical trials, but the high doses required and some toxicity problems excluded practical applications.[25] The mechanism of action was studied in some detail[23,26] and, as for the rifamycins, was found to consist of inhibition of bacterial RNA polymerase activity.

A marked protection from inhibition is observed when the enzyme and DNA are preincubated with a nucleotide; this was interpreted as evidence of a primary effect on initiation. No inhibition was observed with mammalian or yeast polymerases. RNA polymerase extracted from a resistant *E. coli* strain was insensitive to these antibiotics, and presented a modified subunit, indicating that this is the site of action.

Streptovaricins were found to have some activity on reverse transcriptase and to prevent the transformation of cells infected with RNA oncogenic viruses. However, they do not affect the proliferation of experimental tumors. The numerous and sometimes conflicting results of the studies[24,27] in this complex field make it difficult to draw clear cut conclusions, but it appears that there is little hope of practical applications in this area.

10.4.2.1.2 Streptolydigin and tyrandamycin

Streptolydigin (2) is an antibiotic isolated from cultures of *Streptomyces lydicus*.[28] It is very active on clostridia (MICs from 0.1 to 1 $\mu g\,mL^{-1}$) and on streptococci (MICs from 0.3 to 3 $\mu g\,mL^{-1}$), less active on mycobacteria and staphylococci, requiring 10–25 $\mu g\,mL^{-1}$ for inhibition. It is practically ineffective in curing experimental infections in mice.[29] Chemically it belongs to the alkyltetramic group of antibiotics.[30] The chemically related antibiotic tyrandamycin (3),[30] produced by *Streptomyces tirandis*,[31] presents a similar spectrum of activity but higher concentrations are required for inhibition.[32,33]

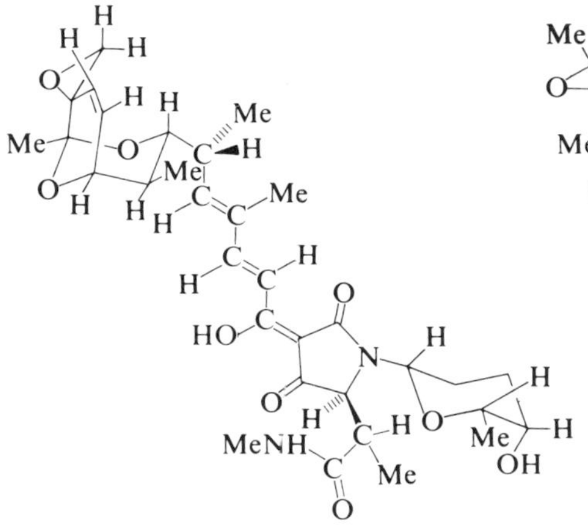

(2)

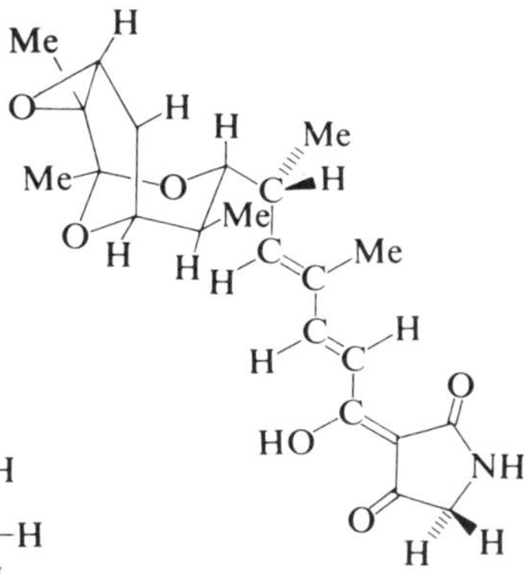

(3)

Streptolydigin in cell free systems inhibits both initiation and elongation steps of transcription by interfering with RNA polymerase, as shown by the finding that the enzyme from streptolydigin-resistant mutants is insensitive to the antibiotic and presents an altered subunit.[34,35] Apparently streptolydigin affects the conformation of the RNA polymerase–template complex, preventing the formation of the phosphodiester bonds.[36] Tyrandamycin acts in a similar manner and *in vitro* inhibits transcription catalyzed by *E. coli* RNA polymerase.[32] Both streptolydigin and tyranda-mycin have no effect on RNA synthesis in a system catalyzed by RNA polymerases extracted from rat liver.[33]

10.4.2.1.3 *Lipiarmycin*

Lipiarmycin is an antibiotic isolated from *Actinoplanes deccanensis*,[37] active on Gram-positive bacteria at concentration of about 1 μg mL^{-1}. Further studies revealed that it can be resolved into four closely related components,[38] the main component being lipiarmycin A3 (4). Similar complexes, the clostomycins[39] and tiacumycins,[40] were subsequently described. Low concentrations of lipiarmycin selectively inhibit RNA synthesis in growing bacteria.[37] In cell-free systems, lipiarmycin inhibits transcription by holoenzyme RNA polymerase much more efficiently than it inhibits transcription by core enzyme.[41]

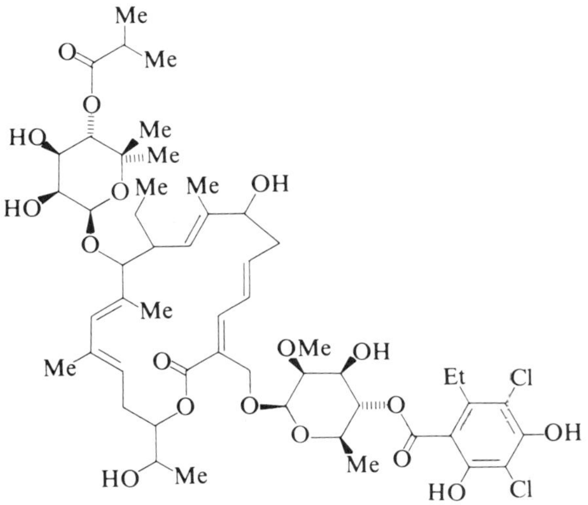

(4)

Addition of lipiarmycin to an actively transcribing cell-free system leads to residual RNA synthesis for several minutes, whereas when the antibiotic is present at the beginning, no RNA chain is formed.[41] These data are consistent with an effect on initiation rather than on elongation and in fact lipiarmycin prevents the synthesis of the first dinucleotide, that is the formation of the first phosphodiester bond. This property has been exploited to study the RNA polymerase specificity in *B. subtilis* bacteriophage infections.[42]

10.4.2.1.4 *Mixopyronins and corallopyronins*

The mixopyronins (**5**), a family of antibiotics moderately active on Gram-positive bacteria, were isolated from cultures of *Myxococcus fulvus*.[43] A chemically related family, the corallopyronins (**6**), are produced by another mixobacterium, *Corallococcus coralloides*. Among these, corallopyronin A appears the most active compound, inhibiting many Gram-positive organisms at 0.1–1 $\mu g\,mL^{-1}$.[44]

(**5**) Mixopyronin A: R $= n$-C_3H_7
Mixopyronin B: R $= n$-C_4H_9

(**6**) Corallopyronin A: R $=$ H
Corallopyronin B: R $=$ Me

Both mixopyronins and corallopyronins rapidly block RNA synthesis in growing bacteria, with a delayed effect on protein synthesis and little effect on DNA replication.[43,44] At low concentrations they inhibit *E. coli* RNA polymerase activity in cell-free systems whereas eukaryotic polymerase II appears insensitive. The kinetics of inhibition suggests that the elongation steps of transcription are affected.[43,44]

10.4.2.1.5 *Sorangicins*

Sorangicins (**7**) are also antibiotics produced by a mixobacterium, *Sorangium cellulosum*.[45] Their antibacterial activity is rather interesting since Gram-positive microorganisms and mycobacteria

(**7**) Sorangicin A: R $=$ OH
Sorangicin B: R $=$ H

are inhibited at 0.01–0.3 $\mu g\,mL^{-1}$ and several Gram-negative bacteria at less than 20 $\mu g\,mL^{-1}$. In experiments with growing cultures, sorangicins present a pattern of macromolecular inhibition very similar to that of the corallopyronins.[45] In cell-free systems, *E. coli* RNA polymerase activity is specifically inhibited but only when the antibiotic is added before initiation of transcription, a kind of interference similar to that of the rifamycins or lipiarmycin.[45]

10.4.2.2 Inhibitors of Eukaryotic RNA Polymerases

α-Amanitin, a bicyclic octapeptide (**8**) isolated from the fungus *Amanita phalloides*, is the best known member of a family of fungal toxins (α-, β- and γ-amanitins, amanin and amanullin) which are potent inhibitors of mammalian RNA polymerase II.[46] α-Amanitin inhibits this enzyme *in vitro* at 0.05 $\mu g\,mL^{-1}$. Concentrations of 5–10 $\mu g\,mL^{-1}$ are required for inhibition of RNA polymerase III, but RNA polymerase I is insensitive. In contrast, yeast RNA polymerase I and II are weakly inhibited and polymerase III is resistant.[9] It was found that α-amanitin, suitably derivatized, binds covalently to the subunit of polymerase II of M 140 000 which is then believed to be the site of action.[47].

(**8**)

In cell cultures, α-amanitin is scarcely active. Since activity is increased by the addition of products enhancing cell membrane permeability, the inactivity is attributed to lack of penetration.[48] By comparison of the activity of the naturally occurring amatoxins and of several semisynthetic derivatives, structural requirements for activity were established.[49] The second amino acid (proline) must have an hydroxy group on carbon 3. The third amino acid must have a methyl side chain. Absence of a side chain in the fifth amino acid and isoleucine as the sixth amino acid are important for binding to the enzyme. Moreover, derivatives with high activity can be obtained by modifications at the tryptophan ring.[50]

10.4.3 RIFAMYCINS

10.4.3.1 Origin and Nature of Rifamycins

Among the inhibitors of the transcribing enzymes, some members of the rifamycin family of antibiotics are of remarkable interest for their chemotherapeutic efficacy against bacterial infections.[51-53] The rifamycins are secondary metabolites synthesized by a microorganism originally considered to belong to the genus *Streptomyces* and subsequently reclassified as a *Nocardia* (*N. Mediterranea*, Margalith and Beretta). In normal fermentation conditions the strain produces a complex of several related substances.

Rifamycin B was isolated as a crystalline pure substance and is practically the only component of the rifamycin complex produced when sodium diethylbarbiturate is added to the fermentation media.[54] Rifamycin B has the unusual property that, in aqueous solution, it tends to change spontaneously into other products with higher antibacterial activity.[55] The structures of rifamycin B (**9**) and of the related compounds involved in the activation process (rifamycin O, **10**; rifamycin S, **11**) have been elucidated by chemical and X-ray crystallographic methods.[56-58] Rifamycin SV (**12**), obtained by mild reduction from rifamycin S, was the first member of the rifamycin family to find clinical application.[59,60]

Besides the few rifamycins originally isolated from *N. mediterranea*, several other rifamycins, active and inactive, have been found in cultures of the same strain, of its mutants or of other actinomycetales strains. Closely related substances, the tolypomycins and the halomycins, have been isolated from *Streptomyces tolypophorus* and *Micromonospora halophytica* respectively.[19,61] Rifamycin B is the only rifamycin produced industrially by fermentation.[19,62,63] Several hundred semisynthetic rifamycins have been obtained through chemical modifications starting from rifamycin B. A few of them (rifamycin SV, rifamide, rifampicin, rifaximin) have found clinical application and others (rifapentine, rifabutin) are under clinical investigation.

The rifamycins derive biogenetically from an aromatic moiety (3-amino-5-hydroxybenzoic acid) which initiates a single polyketide chain composed of eight propionate and two acetate units, as indicated in Figure 1. The aromatic moiety derives from a modification of the aromatic pathway and the C-37 methyl group from methionine. The biosynthetic pathway of the other ansamycins is similar to that of rifamycins.[19,61,63] The total synthesis of rifamycin S has been performed after several studies on stereoselective synthesis of the ansa chain and on its closure on the aromatic unit.[64,65]

10.4.3.2 Mechanism of Action[66,67]

The mechanism of action of the clinically used rifampicin is described here as a prototype of the antibacterial rifamycins. Rifampicin specifically inhibits the bacterial DNA-dependent RNA polymerase at low concentrations. As a consequence of the block of RNA synthesis, the protein synthesis in the bacterial cell ceases, leading eventually to the death of the cell.[18] The corresponding mammalian enzyme is not affected by rifampicin.[68]

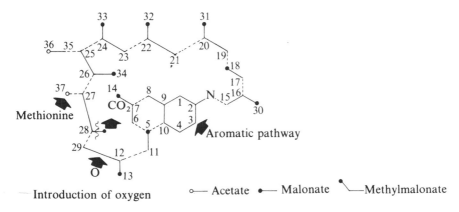

Figure 1 Summary of the biogenesis of rifamycin S

The inhibition of the action of RNA polymerase by rifampicin is due to the formation of a rather stable, noncovalent complex between the antibiotic and the enzyme with a binding constant of 10^{-9} M at 37 °C.[69,70] One molecule of rifampicin (M 823) is bound with one molecule of the enzyme (M 455 000).[71] The drug is not bound covalently to the protein because the complex dissociates in the presence of guanidinium chloride.[72]

The binding site of rifampicin to the RNA polymerase has been particularly well studied in *E. coli*. The enzyme comprises five subunits (α, α, β, β' and σ) and rifampicin binds to the β subunit. In fact, rifampicin-resistant strains of *E. coli* contain a modified subunit, as shown by reconstitution experiments.[73] The enzyme reconstituted from the α and β' subunits, isolated from a resistant strain, and the β subunit from a sensitive strain is still sensitive to rifampicin. On the other hand, the reconstituted enzyme is not sensitive to the drug if only the β subunit comes from a resistant strain. It is not known which region of the β subunit is the binding site for rifampicin, because the complex has been not analyzed by X-ray crystallography. Change of one amino acid in the sequence of the β subunit leads to bacterial resistance to the drug.

Genetic studies on the resistant strains indicate that most mutations affect a sequence of about 50 amino acids in the center of the 1342-long amino acid chain of the β subunit.[74] In the sensitive strains this region is protected against proteolytic cleavage by bound rifampicin.[75] When rifampicin is bound to the enzyme, the complex can still attach to the DNA template and catalyzes the initiation of the RNA synthesis with formation of the first phosphodiester bond, *e.g.* of the dinucleotide pppApU. However, the formation of a second phosphodiester bond and therefore the synthesis of long-chain RNAs is inhibited. The action of rifampicin is to lead to an abortive initiation of RNA synthesis.[76-78] This fact, together with the observation that RNA strongly inhibits the catalytic activity of the enzyme when added prior to the template, suggests an interference of the bound drug with the binding of the newly formed RNA chains.[67]

It has been shown that the penetration of rifampicin into the cells does not involve an active mechanism but a passive diffusion process.[71] Therefore the effect of rifampicin on the bacteria, measured as minimal inhibitory concentration, depends on two factors: the degree of sensitivity of the RNA polymerase to the antibiotic, which is of the same order in Gram-positive and Gram-negative bacteria, and cell permeability, which is lower in Gram-negative than in Gram-positive.

Mutations leading to resistance occur in one step and the order of the rate of mutation is 10^{-8} or less for various pathogens. They involve modifications of the target enzyme.[69,79] The variety of locations and nature of the amino acid substitutions in the enzyme explains the different degree of sensitivity of the enzyme to rifampicin in various mutants. The increase of the minimal inhibitory concentrations on the cell generally corresponds to a decrease of the enzyme sensitivity although the two parameters do not run strictly parallel.[66] Although the main mechanism of resistance is the modification of the target enzyme, some mutagenic treatments yield resistant mutants in which the RNA polymerase is still highly sensitive to the drug but the rate of rifampicin uptake is reduced. The mechanism of this permeability mutation is not yet clear.[67]

10.4.3.3 Structural Requirements for Antibacterial Activity[80-83]

Most natural and semisynthetic rifamycins are very active against Gram-positive bacteria and mycobacteria and show also a moderate activity against Gram-negative bacteria. Microorganisms resistant to one rifamycin are generally resistant to all the naphthalenic ansamycins. However, there are a few semisynthetic rifamycins which show some activity against rifampicin resistant strains, but this activity is not related to the RNA polymerase inhibition.

With few exceptions, the rifamycins are active against bacteria when they are active as RNA polymerase inhibitors and *vice versa*. The exceptions are constituted by rifamycins bearing a free carboxyl group. The latter are active against the target enzyme but have little or no activity against intact bacterial cells because a permeability barrier exists which the polar derivatives cannot pass. Rifamycin B belongs to this group but, although inactive *per se* on the intact bacterial cells, it is easily transformed into rifamycins which are very active.

A few rifamycins show activity in the intact cells and no activity on the RNA polymerase, and this is due to the fact that they undergo some chemical modification during the antibacterial test (*e.g.* 8-acylrifamycins, inactive against the enzyme, but active against bacteria due to deacylation).[84]

The evaluation of the biological properties of a large number of natural and semisynthetic rifamycins made it possible to draw the following clear information on the essential structural requirements of the rifamycin molecule for inhibition of the bacterial RNA polymerase.

(a) The presence of two free hydroxy groups in positions C-21 and C-23 is essential for activity. In fact, substitution or elimination of these two groups yields inactive products. Inversion of the configuration at C-23 leads to an inactive compound, and the inversion at C-21 strongly reduces the activity.[85,86] Functional modifications which, although leaving the hydroxy groups at C-21 and C-23 unaltered, produce important changes in the conformation of the ansa chain, also give inactive or only moderately active products. This suggests that the ansa conformation of the natural rifamycins is needed for keeping hydroxy groups at C-21 and C-23 in a position suitable for biological activity.

(b) The presence of a free hydroxy or keto group at position C-1 and a free hydroxy group at C-8 is also essential for the activity. Modifications at positions C-3 and C-4 of the aromatic nucleus do not affect dramatically the antibacterial activity, indicating that this side of the molecule does not play an important role in the binding.

(c) The four oxygenated functions at C-1, C-8, C-21 and C-23 not only must be unhindered and underivatized but they must display well defined spatial relationships with one another. The absolute requirement for these four functions to be in a correct geometrical relationship suggests that they are involved in the noncovalent attachment of the antibiotic to the bacterial RNA polymerase. This contention is supported by the observation that these four functional groups lie on the same side of the molecule and almost on the same plane, as can be seen by the spatial model derived from X-ray studies (Figure 2).[87] Spatial models derived from X-ray studies performed on some active and inactive rifamycins unequivocally show that all the active and only the active products have the same conformation in the middle part of the ansa chain, with identical interatomic distances between the oxygens located on C-1, C-8, C-21 and C-23.[88] However, conformational differences have been observed at the joining points of the ansa chain to the naphthoquinone chromophore according to the nature of the substituents in position 3.[89] It is important to point out that ^{1}H NMR studies of rifamycin S have confirmed that the conformation of the molecule in solution corresponds well to that obtained in the solid state.[90]

10.4.3.4 Structural Modifications[80-83]

Rifamycin SV (**12**) is very active *in vitro* on a variety of pathogenic bacteria, but its clinical use is limited by the fact that it is poorly absorbed in the gastrointestinal tract and when administered parenterally it is quickly excreted through the liver in the bile.[60] A systematic study of chemical modifications of rifamycin has been made with the aim of obtaining a derivative presenting a greater absorption in the gastrointestinal tract and a slower rate of biliary excretion than rifamycin SV.[80] In fact, improvements in these two pharmacokinetic parameters were considered more important than the intrinsic antibacterial activity (very high in rifamycin SV) for the potential achievement of therapeutic efficacy in the oral treatment of various infections, including tuberculosis.

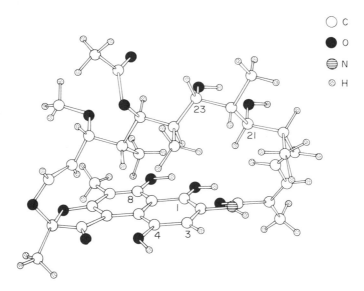

Figure 2 Stereo model of rifamycin SV according to X-rays

Chemical modifications have been made on most of the parts of the molecule. Changes on the aliphatic ansa have generally led to rifamycins with lower activity than rifamycin SV, with the exception of the 27-*O*-demethyl and 25-*O*-deacetyl derivatives, which have antibacterial activity only slightly inferior than that of the parent compounds.[91,92] The activity is strongly depressed in the derivatives where the conformation of the ansa around the C-23 and C-21 is modified or where the hydroxyls on these two carbons are substituted.

Modifications on the glycolic moiety of rifamycin B and on the aromatic nucleus of rifamycin S or SV on position 3 and/or 4, leaving the structure and conformation of the aliphatic chain unaltered, gave several series of active derivatives, some of which with improved activity *in vitro* and *in vivo* in comparison with rifamycin SV. The general structures of most derivatives are indicated in Figure 3. Modifications of the glycolic chain were made following the hypothesis that the low activity of rifamycin B could be due to the presence of a free carboxyl group in the molecule, which in fact was shown later to affect the penetration through the bacterial cell wall.[93] Most of the amides, hydrazides and esters prepared showed antibacterial activity of the same order as rifamycin SV.[94] The diethylamide (rifamide, **13**; R = NEt$_2$) was found to possess the highest *in vivo* activity and the lowest toxicity, but its absorption and elimination is similar to rifamycin SV.

Chemical modifications of the chromophoric nucleus of rifamycin on position 3 and/or 4, gave a large number of derivatives. Rifamycins modified at position 4 include a series of 4-amino derivatives obtained by condensation of rifamycin O with amines, hydrazides, amidrazones and aminoguanidines (**14**) and (**15**).[95,96] The *in vitro* activity of many of these derivatives is extremely high, *e.g.* the aminoguanidine compound (rifamycin AG, **14**; R = —NH—C(=NH)—NH$_2$) has a MIC of 0.001 µg mL^{-1} against *S. aureus*, 0.005 µg mL^{-1} against *M. tuberculosis* and 0.75 µg mL^{-1} against *E. coli*. However, their *in vivo* activity is negligible due to their insolubility in water!

Various rifamycin derivatives bear substitutions on both positions 3 and 4. Products obtained by condensation of rifamycin S with *o*-phenylenediamines (**16**) and *o*-aminophenols (**17**) have good *in vitro* and *in vivo* activity.[91,97,98] Rifazine (**16**; R = H) showed activity also when administered orally, but was discarded for toxicological reasons. By condensation of rifamycin S with aminoacyl esters and amides or with monoimino derivatives of β-diketones, a series of semisynthetic rifamycins with a pyrrolic nucleus fused on position 3–4 (**18**) were obtained.[99] Most of the pyrrole derivatives have excellent activity against Gram-positive bacteria except those substituted on the pyrrole nitrogen with large or hydroxylated groups. One pyrrole derivative (**18**; R^1, R^2 = Me; R^3 = CO$_2$Bu^t) has a surprisingly higher activity when administered orally than subcutaneously.

Thiazorifamycins (**19**) were isolated from fermentation broths but can also be obtained chemically.[100,101] They are active *in vitro* and one of them, rifamycin P (**19**; R = H), also shows good *in vivo* activity when administered orally and is the most active natural rifamycin so far isolated.

The condensation of 3-amino-4-iminorifamycin S with aldehydes yields the imidazorifamycins (**20**), and with 4-piperidones yields the spiroimidazorifamycins,[102] among which rifabutin (**21**; R = Bu^i) has a moderate activity on some mycobacterial species or mutants resistant to rifampicin.[103,104]

A derivative with two heterocyclic nuclei fused at position 3–4 of rifamycin S, rifaximin (**22**; R = Me),[105] has a broad antibacterial spectrum and negligible gastrointestinal absorption and has been studied for its effect on the fecal flora.

A great number of semisynthetic rifamycins have substitutions only on position 3. A few thioethers of rifamycin SV (**23**) were obtained by reaction of rifamycin S with selected thio compounds and showed good *in vitro* activity but limited *in vivo* activity.[106]

The aminomethyl derivatives of rifamycin SV (**24**) were obtained by a Mannich reaction, treating rifamycin S with formaldehyde and an amine.[107] In comparison with rifamycin SV, some of them showed an improvement of activity against Gram-negative microorganisms and better protection in the experimental *S. aureus* infections on oral administration.

Most of the 3-aminomethylrifamycins are unstable in the presence of oxidizing agents. The oxidation product is 3-formylrifamycin SV, which opened the way for the preparation of a large series of derivatives (**25**), including imines, oximes, hydrazones, hydrazide-hydrazones and semicarbazones.[108-111] All these derivatives showed high antibacterial activity and the majority of the *N,N*-disubstituted hydrazones and some oximes presented exceptionally high *in vivo* activity on staphylococcal infections in mice, with the oral and subcutaneous ED$_{50}$ values almost equal. Among the most active derivatives the hydrazone of 3-formylrifamycin SV with *N*-amino-*N'*-methyl-piperazine (**25**; X = N—NC$_4$H$_8$NMe) was developed for its favorable chemotherapeutic, pharmacokinetic and toxicological properties. It was introduced in therapy in 1968 with the name of rifampicin (rifampin, USAN) for the treatment of tuberculosis and other severe bacterial infections by oral administration.[112,113]

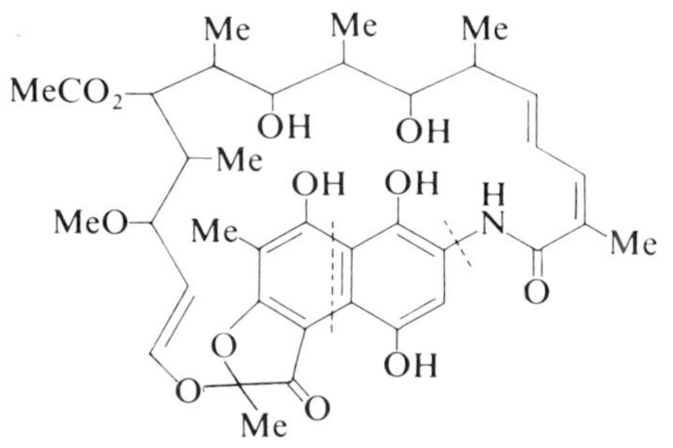

(12)

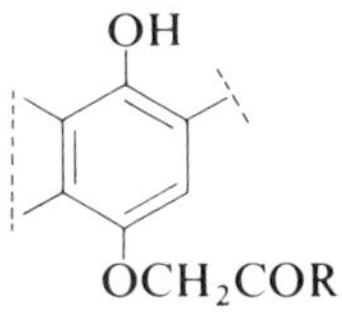

(13)

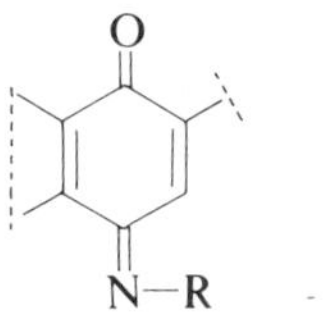

(14)

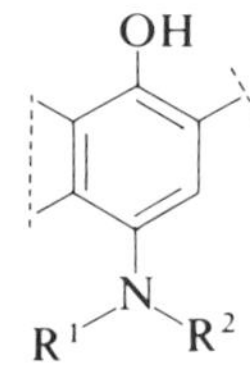

(15)

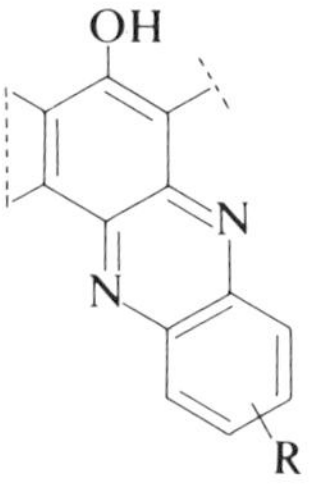

(16)

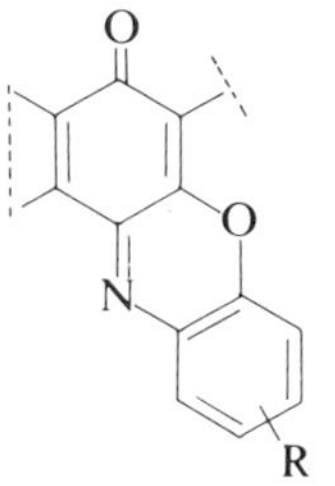

(17)

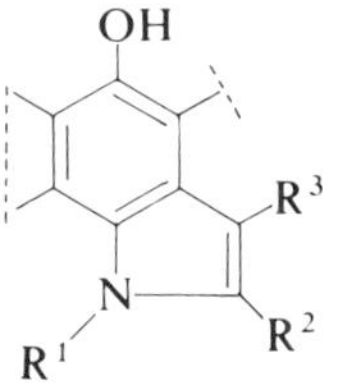

(18)

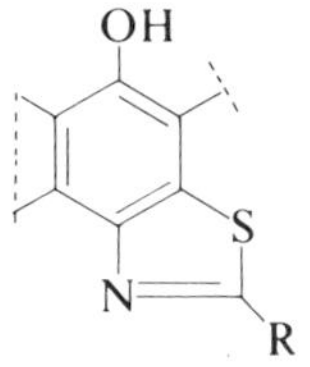

(19)

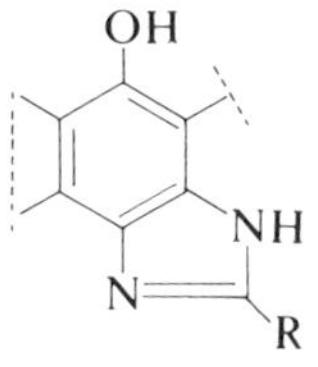

(20)

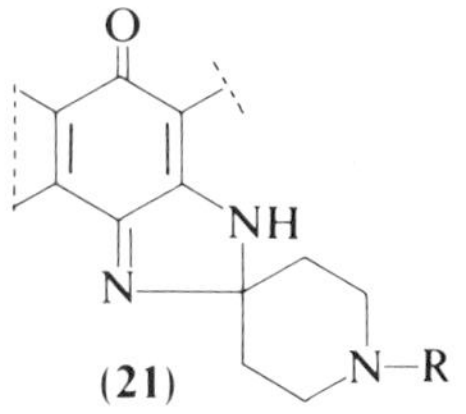

(21)

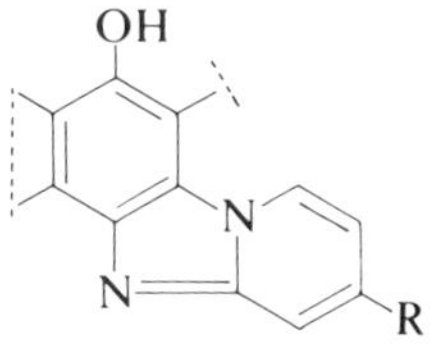

(22)

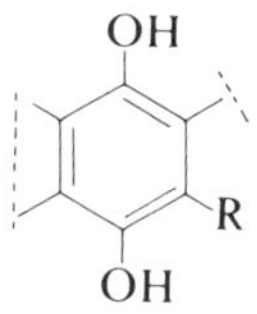

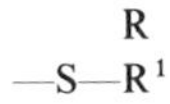

(23)

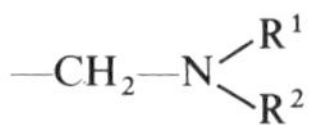

(24)

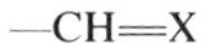

(25)

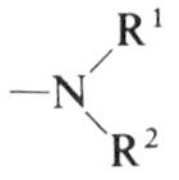

(26)

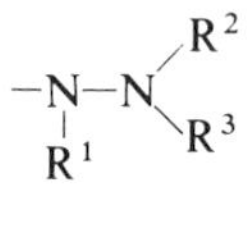

(27)

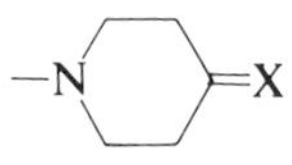

(28)

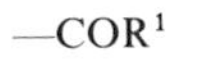

(29)

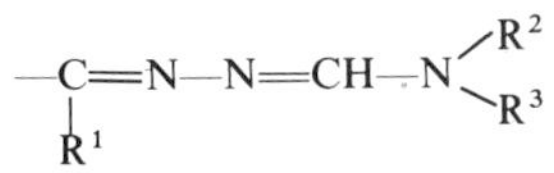

(30)

Figure 3 General structures of rifamycin derivatives

Rifampicin possesses most of the properties sought in the program of chemical modification of rifamycin. However, synthesis and evaluation of other rifamycin derivatives continued, with the aim of finding a product with improved chemotherapeutic properties. In the series of derivatives of 3-formylrifamycin SV, the hydrazone with *N*-amino-*N'*-cyclopentylpiperazine (rifapentine; **25**; $X = N—NC_4H_8N—C_5H_{10}$) has an antibacterial spectrum similar to that of rifampicin but longer serum half-life.[114] The 3-amino derivatives (**26**) have a remarkable activity against Grampositive bacteria.[91,98,115] Of particular interest is the activity against a rifampicin resistant *S. aureus* mutant strain shown by 3-(4'-isopropylpiperidinyl)rifamycin SV (**26**; $R^1, R^2 = C_5H_{10}Pr^i$). Some of the 3-amino derivatives administered orally have a good protective action against murine tuberculosis.

A series of 3-hydrazinorifamycin derivatives (**27**) was obtained from 3-bromorifamycin S.[116] From the same compound the 3-*N*-(piperidin-4'-one)rifamycin SV was prepared from which a series of oximes, hydrazones and semicarbazones (**28**) were obtained.[117] All the compounds (**27**) and (**28**) showed good activity *in vitro* but not *in vivo* by oral administration. Esters, amides and hydrazides of 3-carboxyrifamycin S (**29**) were synthesized from the 3-(cyanocarbonyl)rifamycin S.[118] The *in vitro* microbiological activity of these derivatives is quite low due to the poor penetration of the compounds through the bacterial cell wall. Another series of derivatives are the azinomethyl compounds (**30**) and among them 3-(piperidinomethylazinomethyl)rifamycin SV has a particular interest for its long serum half-life.[119]

10.4.3.5 Modulation of Chemotherapeutic Properties

All the chemical modifications of the rifamycin structure in position 3 and/or 4 practically always give derivatives with very active *in vitro* activity against bacteria but with some variations in the degree of activity, because the nature of the substituents modifies the physicochemical properties and consequently their ability to cross biological membranes. Various studies have evaluated the quantitative structure–activity correlations in the rifamycin derivatives. *In vitro* antibacterial activity of some groups of semisynthetic rifamycins, including 40 oximes of 3-formylrifamycin SV, 16 rifamycin B amides and 13 rifamycin SV iminomethylpiperazines, was correlated with their lipophilicity expressed by the chromatographically determined R_m constants, which correlate well with log P values and Hansch π constants.[120] Using only the R_m terms in the regression analysis, the antibacterial activity against *Staphylococcus aureus* of all the examined rifamycin derivatives was parabolically related to the lipophilicity of the compounds. However, when each series of rifamycin derivatives was examined individually, three slightly different parabolic relationships were obtained, probably because lipophilicity is not the only factor determining the degree of antibacterial activity of the various groups of rifamycins. These results are in agreement with other correlation equations between antibacterial activities and structures, as expressed by their calculated physicochemical parameters, for a large series of amides and hydrazides and for some hydrazones of 3-formylrifamycin SV.[121,122]

The nature of the substituents in position 3 to some extent also affects the activity on the target enzyme. It has been shown that the inhibition of RNA polymerase is markedly increased by electron attracting substituents in position 3 of rifamycin S. In the case of the corresponding derivatives of rifamycin SV the effect is negligible. It has been suggested that the naphthoquinone ring is involved in a donor–acceptor π-complex interaction with the enzyme, which is affected by the introduction of electronegative substituents on the aromatic nucleus.[123] However, subsequent NMR studies indicate that these substitutions affect the conformation of the ansa and thus probably the fitting with the enzyme.[124]

Besides the *in vitro* activity on pathogens, pharmacokinetics in the host is a factor of primary relevance for the chemotherapeutic usefulness of the rifamycin derivatives. Rifamycin SV is partially absorbed from the gastrointestinal tract, but oral treatment does not give rise to detectable blood levels because the drug is concentrated in the liver and rapidly eliminated in the bile. Absorption from the gastrointestinal tract, binding to the plasma proteins and rate of biliary excretion are the pharmacokinetic parameters which are affected by the nature of the substituents. Correlations of these parameters with therapeutic efficacy have been demonstrated in some selected groups of derivatives. In a series of dialkylamides of rifamycin B, the rate of biliary excretion decreases and the oral efficacy in staphylococcal infections increases with the length of alkyl chains on the nitrogen.[125] In a small series of hydrazones of 3-formylrifamycin SV with *N*-amino-*N'*-alkyl- or -aralkyl-piperazines, the rate of biliary excretion decreases with the increase of alkyl/aralkyl chains of the substituents but the contrary happens for the binding to serum albumin.[126]

These few examples give an indication of the possibility of obtaining semisynthetic rifamycins with a large range of pharmacokinetic properties. Rifampicin was selected from among several hundred rifamycin derivatives because the (4-methyl-1-piperazinyl)iminomethyl group added in position 3 to rifamycin SV confers to the resulting molecule very good properties concerning gastrointestinal absorption, distribution in the tissues and rate of elimination, optimal for the daily oral treatment of various infectious diseases. Other derivatives with a longer serum half-life (*e.g.* rifapentine) are in clinical evaluation for weekly treatment.

10.4.3.6 Rifamycins in Clinical Use

10.4.3.6.1 *Rifamycin SV*

This product (**12**; INN, rifamycin) is in clinical use in the form of its water soluble sodium salt.[60] It is very active *in vitro*, with bactericidal effect against Gram-positive bacteria, both sensitive and resistant to other antibiotics (MIC from 0.005 to 0.1 µg mL^{-1}), Gram-negative cocci and *M. tuberculosis* (MIC of the order of 0.1 µg mL^{-1}). It has a limited activity against Gram-negative bacteria (MIC from 10 to 200 µg mL^{-1}) and is inactive against fungi, protozoa and spirochetes. It is effective in Gram-positive experimental infections and has a low toxicity. Parenteral administration of 250 mg of rifamycin in man gives therapeutic levels for about 6 hours with peaks of about 1 µg mL^{-1}, but oral absorption is poor. The product is excreted mainly through the liver in the bile. It is effective in a variety of infections due to Gram-positive organisms, especially staphylococci, but because of its pharmacokinetic properties is poorly effective in pulmonary tuberculosis. An important therapeutic indication is the treatment of infections of the biliary duct, in which it reaches very high concentrations, where it is effective also against Gram-negative pathogens. Another widespread use is the topical application on wounds, especially in surgery.

10.4.3.6.2 *Rifamide*

The diethylamide of rifamycin B (**13**; R = NEt$_2$) has been introduced into clinical use in some countries in the form of its injectable sodium salt, because laboratory studies indicated a therapeutic index somewhat better than that of rifamycin.[127,128] Clinically it is similar to rifamycin in antibacterial spectrum, absorption and elimination. The therapeutic indications and effectiveness of rifamide are practically the same as those of rifamycin.

10.4.3.6.3 *Rifampicin*

Rifampicin is 3-((4-methyl-1-piperazinyl)iminomethyl)rifamycin SV (**25**; X = N—NC$_4$H$_8$NMe). It has an amphoteric nature (pK_a 1.7, 7.9) and is soluble in most organic solvents, slightly soluble in water at neutral pH, but more soluble in acidic and alkaline solutions.[112] Rifampicin has a broad spectrum of antibacterial activity.[129-132] It is active at very low concentrations (0.01–0.1 µg mL^{-1}) on Gram-positive bacteria, some Gram-negative (Neisseria, Haemophilus, Brucella), and on anaerobes (Clostridia, *Bacteroides fragilis*, Chlamydia and Legionella). Against other Gram-negative bacteria, as Enterobacter and Pseudomonas, it is active at higher concentrations (1–50 µg mL^{-1}). Rifampicin is also very active on *M. tuberculosis* (MIC 0.1–1.0 µg mL^{-1}) and *M. leprae* (MIC 0.1–0.5 µg mL^{-1}) as well as on a high percentage of atypical mycobacteria.

Rifampicin exerts a bactericidal activity at concentrations near to the MIC. The frequency of resistant mutants in sensitive bacterial populations is of the order of 10^{-7} for *E. coli* and of 10^{-8} for *M. tuberculosis*. No transferable resistance has been observed so far. Rifampicin is rapidly absorbed from the upper part of the GI tract and diffuses well into all organs, tissues and fluids of the body.[133,134] In man, peak serum levels of the order of 10 µg mL^{-1} are achieved between 1.5 to 3.0 h after oral administration of 600 mg of the drug on an empty stomach. The serum half-life is of the order of 2–3 hours. It is excreted mainly in the bile and at a lesser extent in the urine, in a proportion increasing with the dose. The principal metabolite is 25-*O*-deacetylrifampicin, which is nearly as active as the parent compound.[135]

Rifampicin is a first-line drug for the treatment of tuberculosis.[136-139] The so called SCT (short course chemotherapy) of tuberculosis is a very effective treatment based on the combined administration of rifampicin with isoniazid and other antituberculous drugs for 6–9 months.[136–139] Rifampicin is also very useful in the treatment of leprosy.[140,141]

Besides the mycobacterial infections, rifampicin has been found effective in the prophylaxis of meningococcal infections and in the treatment of a variety of diseases caused by sensitive bacteria (*e.g.* staphylococcal infections, respiratory tract infections including Legionnaire's disease, gonorrhea, brucellosis, chlamydial infections), and infections due to intracellular sensitive pathogens.[129–132] In case of severe infections sustained by large populations of pathogens, rifampicin should be used in combination with other chemotherapeutic agents to avoid the emergence of resistant mutants.[142]

Adverse effects associated with daily rifampicin administration are infrequent and trivial and the risk of hepatoxicity is small in patients with no previous history of liver disease.[143] Influence syndrome (fever, headache, malaise and bone pain) is an adverse reaction of immunological origin peculiar of intermittent therapy, appearing with a certain frequency when high doses of the drug are used at weekly intervals. Other untoward effects of an immunological nature (thrombocytopenic purpura, hemolytic anemia, renal failure, shock) are very rare, almost exclusively associated with intermittent or irregular administration.

Rifampicin produces a reddish discoloration of the urine and in some cases of sputum and tears. Rifampicin competes with the excretion of bilirubin. It also stimulates the overproduction of metabolizing enzymes and this enzyme induction is likely responsible for the decrease of serum concentrations of cardiac glycosides, oral contraceptives, anticoagulants, narcotics and analgesics, oral antidiabetics, and corticosteroids.[144]

10.4.3.6.4 *Rifaximin*

This derivative containing an imidazopyridine nucleus fused with rifamycin in positions 3,4 (**22**; R = Me) has an *in vitro* activity against Gram-positive and Gram-negative bacteria very similar to that of rifampicin. The absorption of rifaximin by the gastrointestinal tract is negligible.[105,145] The product has been introduced in clinical use for the oral treatment of intestinal infections.

10.4.3.6.5 *Rifamycins under clinical evaluation*

Rifapentine (**25**; R = N—NC$_4$H$_8$N— C$_5$H$_{10}$) is more lipophilic and has a serum half-life about 5 times longer than rifampicin.[114,126,146] This is due to slower hepatic clearance and possibly to its stronger binding to serum proteins than rifampicin.[147] In comparison with the latter, rifapentine is slightly less active *in vitro* and *in vivo* against Gram-positive and Gram-negative bacteria but more active against *M. tuberculosis*, particularly *in vivo*. In experimental tuberculous infections, rifapentine administered once a week has practically the same therapeutic efficacy as rifampicin administrated daily.[114] Rifapentine is under clinical experimentation as a potential drug for the therapy of tuberculosis and leprosy at lower dosage and frequency of administration than rifampicin.

Rifabutin (**21**; R = Bu^i) has an antibacterial spectrum similar to rifampicin but appears to possess incomplete cross-resistance with rifampicin *in vitro*.[103,104,148] In fact some strains resistant to rifampicin are still moderately sensitive to rifabutin. Another characteristic of rifabutin is its activity against *Mycobacterium avium intracellulare* is higher than that of rifampicin[149] and therefore is in clinical evaluation for the potential therapeutic use in infections due to this pathogen which, although generally rare, are particularly frequent in immunodepressed patients. *In vitro*, rifabutin has been reported to have an effect against the human immunodeficiency virus type I at low concentrations[150] but no data on clinical efficacy are so far available.

10.4.3.7 Effect of Rifamycins on Nonbacterial Polymerases

The discovery of the inhibitory effect of rifamycins on the bacterial RNA polymerase suggested the hypothesis that structural modifications of the rifamycin molecule could yield products with an inhibitory effect on other RNA or DNA polymerizing enzymes such as the DNA-directed RNA polymerase of cytoplasmic DNA viruses and the RNA-directed DNA polymerase of oncogenic RNA viruses. The RNA polymerase contained in the virions of some large mammalian cytoplasmic DNA viruses, such as pox virus, has been thought to be a possible target for rifamycins. In fact, rifampicin itself was found to inhibit the growth of pox viruses but its viral inhibitory dose is of some orders of magnitude higher than the antibacterial one.[151,152] Attempts to increase the antiviral activity through chemical modifications had practically no success. It was later discovered that, in contrast to the original assumption, the antivaccinia activity of rifampicin is not related to inhibition of viral polymerase but is due to a block in the virus maturation process.[153]

The other area of possible usefulness for rifamycin derivatives is their inhibitory effect on the polymerase which catalyzes the complementary transcription from a polyribonucleotide strand into a polydeoxyribonucleotide (reverse transcriptase). This enzymatic activity has been identified in the RNA oncogenic viruses and also in leukocytes of patients with acute leukemia. Inhibitors of the reverse transcriptase could constitute a powerful tool for understanding the role of this enzyme in viral carcinogenesis and perhaps could have an inhibitory effect on tumor induction or on tumor growth. The first experiments performed in different laboratories showed that some rifamycin derivatives were active both on the reverse transcriptase of oncogenic viruses and on transformed cells.[154-156] The two effects are not necessarily correlated, and in fact rifampicin itself, which is not active against reverse transcriptase, inhibits focus formation by Rous sarcoma virus, probably because it is preferentially toxic to transformed cells.

A large number of rifamycin derivatives were prepared and tested for their inhibitory effect on reverse transcriptase.[157,158] Among them, various derivatives with a rather high activity were found. They have generally a common physicochemical character: a bulky, lipophilic side chain. They also inhibit, at comparable concentrations, other nucleotide polymerases, such as the calf thymus RNA and DNA polymerases.[159] Therefore the higher activity against reverse transcriptase is generally associated with a broader spectrum of activity against other polymerases and with poor selectivity and thus increased toxicity.[160,161] This could not apply for all the semisynthetic rifamycins, as proven by the finding of specific activity of rifabutin against the human immunodeficiency virus type I.[150]

10.4.4 REFERENCES

1. M. Chamberlin, *Enzymes*, 1982, **15**, 61.
2. R. Doi, *Bacteriol. Rev.*, 1977, **41**, 568.
3. T. Yura and A. Ishihama, *Annu. Rev. Genet.*, 1979, **13**, 59.
4. P. H. von Hippel, D. G. Bear, W. D. Morgan and J. A. McSwiggen, *Annu. Rev. Biochem.*, 1984, **53**, 389.
5. W. Zillig, R. Schnabel and J. Tu, *Naturwissenschaften*, 1982, **69**, 197.
6. J. Madon, U. Leser and W. Zillig, *Eur. J. Biochem.*, 1983, **135**, 279.
7. D. Prangishvilli, W. Zillig, A. Gierl, L. Biesert and I. Holz, *Eur. J. Biochem.*, 1982, **122**, 471.
8. B. Alberts, D. Bray, J. Lewis, M. Raff, K. Roberts and J. D. Watson, 'Molecular Biology of the Cell', Garland, New York, 1983, p. 406.
9. M. K. Lewis and R. R. Burgess, *Enzymes*, 1982, **15**, 109.
10. M. R. Paule, *Trends Biochem. Sci.*, 1981, **6**, 128.
11. J. R. Nevins, *Annu. Rev. Biochem.*, 1983, **52**, 441.
12. E. Wintersberger, in 'Regulation of Transcription and Translation in Eukaryotes', ed. E. K. F. Bautz, Springer-Verlag, Berlin, 1973, p. 179.
13. J. J. Skehel, in 'Topley and Wilson's Principles of Bacteriology, Virology and Immunity', 7th edn., ed. F. Brown and G. Wilson, Arnold, London, 1984, vol. 4, p. 49.
14. E. L. Smith, R. L. Hill, J. R. Lehman, R. J. Lefkowitz, P. Handler and A. White, in 'Principles of Biochemistry: General Aspects', 7th edn., McGraw-Hill, New York, 1983, p. 791.
15. M. Chamberlin and T. Ryan, *Enzymes*, 1982, **15**, 87.
16. E. Spencer, S. Shuman and J. Hurwitz, *J. Biol. Chem.*, 1980, **255**, 5388.
17. G. C. Lancini and F. Parenti, 'Antibiotics — An Integrated View', Springer-Verlag, New York, 1982, p. 54.
18. G. C. Lancini and G. Sartori, *Experientia*, 1968, **24**, 1105.
19. G. C. Lancini, in 'Biotechnology', ed. H. J. Rehm and G. Reed, VCH Verlagsgesellschaft, Weinheim, 1986, vol. 4, p. 431.
20. P. Siminoff, R. Smith, W. T. Sokolski and G. M. Savage, *Am. Rev. Tuberc. Pulm. Dis.*, 1957, **75**, 576.
21. K. L. Rinehart, M. L. Maheshwari, F. J. Antosz, H. Matur, K. Sasaki and R. J. Schacht, *J. Am. Chem. Soc.*, 1971, **93**, 6273.
22. K. L. Rinehart and F. J. Antosz, *J. Antibiot.*, 1972, **25**, 71.
23. H. Yamazaki, *J. Antibiot.*, 1968, **21**, 209.
24. K. L. Rinehart, F. J. Antosz, K. Sasaki, P. K. Martin, M. L. Maheshwari, F. Reusser, L. H. Li, D. Moran and P. F. Wiley, *Biochemistry*, 1974, **13**, 861.
25. B. Milavetz and W. A. Carter, in 'Inhibitors of DNA and RNA Polymerases', ed. P. S. Sarin and R. C. Gallo, Pergamon Press, New York, 1980, p. 191.
26. S. Mizuno, H. Yamazaki, K. Nitta and H. Umezawa, *Biochim. Biophys. Acta*, 1968, **157**, 322.
27. W. A. Carter, V. W. Brockman, L. Li and F. Reusser, *Clin. Res.*, 1971, **19**, 490.
28. C. De Boer, A. Dietz, W. S. Silver and G. M. Savage, *Antibiot. Annu.*, 1955–1956, 868.
29. L. Lewis, J. R. Wilkins, D. F. Schwarz, L. T. Nikitas, *Antibiot. Annu.*, 1955–1956, 897.
30. D. J. Duchamp, A. R. Bramfam, A. C. Button and K. L. Rinehart, *J. Am. Chem. Soc.*, 1973, **95**, 4077.
31. C. E. Mayer, *J. Antibiot.*, 1971, **24**, 558.
32. F. Reusser, in 'Antibiotics', ed. F. E. Hahn, Springer-Verlag, New York, 1979, vol. V, part 1, p. 361.

33. F. Reusser, *Antimicrob. Agents Chemother.*, 1976, **10**, 618.
34. G. Cassani, R. Burger, H. M. Goodman and L. Gold, *Nature (London)*, 1971, **230**, 197.
35. Y. Iwakura, A. Ishihama and T. Yura, *Mol. Gen. Genet.*, 1973, **121**, 181.
36. W. R. McClure, *J. Biol. Chem.*, 1980, **255**, 1610.
37. C. Coronelli, R. J. White, G. C. Lancini and F. Parenti, *J. Antibiot.*, 1975, **28**, 253.
38. B. Cavalleri, A. Arnone, E. Di Modugno, G. Nasini and B. P. Goldstein, *J. Antibiot.*, 1988, **41**, 308.
39. S. Omura, N. Imamura, R. Oiwa, H. Kuga, R. Iwata, R. Masuma and Y. Iwai, *J. Antibiot.*, 1986, **39**, 1407.
40. J. E. Hochlowski, S. J. Swanson, L. M. Ranfranz, D. N. Whittern, A. M. Buko and J. B. McAlpine, *J. Antibiot.*, 1987, **40**, 575.
41. A. L. Shonenshein and H. B. Alexander, *J. Mol. Biol.*, 1979, **127**, 55.
42. M. S. Osburne and A. L. Shonenshein, *J. Virol.*, 1980, **33**, 945.
43. H. Irschik, K. Gerth, G. Höfle, W. Kohl and H. Reichenbach, *J. Antibiot.*, 1983, **36**, 1651.
44. H. Irschik, R. Jansen, G. Höfle, K. Gerth and H. Reichenbach, *J. Antibiot.*, 1985, **38**, 145.
45. H. Irschik, R. Jansen, K. Gerth, G. Höfle and H. Reichenbach, *J. Antibiot.*, 1987, **40**, 7.
46. T. Wieland and M. Faulstich, *Crit. Rev. Biochem.*, 1978, **5**, 185.
47. O. G. Brodner and T. Wieland, *Biochemistry*, 1976, **15**, 3480.
48. T. J. Lindell, in 'Inhibitors of DNA and RNA Polymerases', ed. P. S. Sarin and R. C. Gallo, Pergamon Press, New York, 1980, p. 111.
49. T. Wieland, C. Götzendörfer, G. Zanotti and A. C. Vaisins, *Eur. J. Biochem.*, 1981, **117**, 161.
50. E. Falck Pederson, W. Neuman and P. W. Morris, *Biochemistry*, 1982, **21**, 5164.
51. P. Sensi, P. Margalith and M. T. Timbal, *Farmaco, Ed. Sci.*, 1959, **14**, 146.
52. P. Margalith and G. Beretta, *Mycopathol. Mycol. Appl.*, 1960, **8**, 321.
53. J. E. Thiemann, G. Zucco and G. Pelizza, *Arch. Mikrobiol.*, 1969, **67**, 147.
54. P. Margalith and H. Pagani, *Appl. Microbiol.*, 1961, **9**, 325.
55. P. Sensi, R. Ballotta, A. M. Greco and G. G. Gallo, *Farmaco, Ed. Sci.*, 1961, **16**, 165.
56. W. Oppolzer, V. Prelog and P. Sensi, *Experientia*, 1964, **20**, 336.
57. M. Brufani, W. Fedeli, G. Giacomello and A. Vaciago, *Experientia*, 1964, **20**, 339.
58. J. Leitich, W. Oppolzer and V. Prelog, *Experientia*, 1964, **20**, 343.
59. P. Sensi, M. T. Timbal and G. Maffii, *Experientia*, 1960, **16**, 412.
60. M. Bergamini and G. Fowst, *Arzneim.-Forsch.*, 1965, **15**, 951.
61. G. C. Lancini and M. Grandi, in 'Antibiotics', ed. J. W. Corcoran, Springer-Verlag, New York, 1981, vol. 4, p. 12.
62. P. Sensi and J. R. Thiemann, *Prog. Ind. Microbiol.*, 1967, **6**, 21.
63. O. Ghisalba, J. A. L. Auden, T. Schupp and J. Nüesch, in 'Biotechnology of Industrial Antibiotics', ed. E. J. Vandamme, Dekker, Basel, 1984, p. 281.
64. Y. Kishi, *Pure Appl. Chem.*, 1981, **53**, 1163.
65. S. Hanessian, J. R. Pougny, I. K. Boessenkool, *J. Am. Chem. Soc.*, 1982, **104**, 6164.
66. W. Wehrli, *Rev. Infect. Dis.*, 1983, **5** (suppl. 3), S407.
67. G. R. Hartmann, P. Heinrich, M. C. Kollenda, B. Skrobranek, M. Tropschug and W. Weiss, *Angew. Chem., Int. Ed. Engl.*, 1985, **24**, 1009.
68. G. Hartmann, K. O. Honikel, F. Knüsel and J. Nüesch, *Biochim. Biophys. Acta*, 1967, **145**, 843.
69. W. Wehrli, F. Knüsel, K. Schmidt and M. Staehelin, *Proc. Natl. Acad. Sci. USA*, 1968, **61**, 667.
70. W. Wehrli, *Eur. J. Biochem.*, 1977, **80**, 325.
71. R. J. White and G. C. Lancini, *Biochim. Biophys. Acta*, 1971, **240**, 429.
72. U. I. Lill and G. R. Hartmann, *Eur. J. Biochem.*, 1973, **38**, 336.
73. A. Heil and W. Zillig, *FEBS Lett.*, 1970, **11**, 165.
74. Y. A. Ovchinnikov, G. S. Monastyrskaya, S. O. Guriev, N. F. Kalinina, E. D. Sverdlov, A. I. Gragerov, I. A. Bass, I. F. Kiver, E. P. Moiseyeva, V. N. Igumnov, S. Z. Mindlin, V. G. Nikiforov and R. B. Khesin, *Mol. Gen. Genet.*, 1983, **190**, 344.
75. H. Mi and G. R. Hartmann, *Eur. J. Biochem.*, 1983, **131**, 113.
76. W. R. McClure and C. L. Cech, *J. Biol. Chem.*, 1978, **253**, 8949.
77. C. Kessler, M. Huaifeng and G. R. Hartmann, *Eur. J. Biochem.*, 1982, **122**, 515.
78. W. Schulz and W. Zillig, *Nucleic Acids Res.*, 1981, **9**, 6889.
79. W. Zimmermann and W. Wehrli, *Experientia*, 1977, **33**, 132.
80. P. Sensi, N. Maggi, S. Furesz and G. Maffii, *Antimicrob. Agents Chemother.*, 1967, 699.
81. P. Sensi, *Pure Appl. Chem.*, 1975, **41**, 15.
82. G. C. Lancini and W. Zanichelli, in 'Structure–Activity Relationships among the Semisynthetic Antibiotics', ed. D. Perlman, Academic Press, New York, 1977, p. 531.
83. M. Brufani, in 'Topics in Antibiotic Chemistry', ed. P. G. Sammes, Ellis Horwood, Chichester, 1977, vol. 1, p. 91.
84. N. Maggi, S. Furesz and P. Sensi, *J. Med. Chem.*, 1968, **11**, 368.
85. M. Brufani, G. Cecchini, L. Cellai, M. Federici, M. Guiso and A. Segre, *J. Antibiot.*, 1985, **38**, 259.
86. M. Brufani, L. Cellai, L. Cozzella, M. Federici, M. Guiso and A. Segre, *J. Antibiot.*, 1985, **38**, 1359.
87. M. Brufani, S. Cerrini, W. Fedeli and A. Vaciago, *J. Mol. Biol.*, 1974, **87**, 409.
88. S. K. Arora, *Mol. Pharmacol.*, 1983, **23**, 133.
89. M. Brufani, L. Cellai, S. Cerrini, W. Fedeli, A. Segre and A. Vaciago, *Mol. Pharmacol.*, 1981, **21**, 394.
90. G. G. Gallo, E. Martinelli, V. Pagani and P. Sensi, *Tetrahedron*, 1974, **30**, 3093.
91. H. Bickel, F. Knüsel, W. Kump and L. Neipp, *Antimicrob. Agents Chemother.*, 1967, 352.
92. N. Maggi, A. Vigevani and R. Pallanza, *Experientia*, 1968, **24**, 209.
93. N. Maggi, S. Furesz and P. Sensi, *J. Med. Chem.*, 1968, **11**, 368.
94. P. Sensi, N. Maggi, R. Ballotta, S. Furesz, R. Pallanza and V. Arioli, *J. Med. Chem.*, 1964, **7**, 596.
95. P. Sensi, M. T. Timbal and A. M. Greco, *Antibiot. Chemother.*, 1962, **12**, 488.
96. R. Cricchio and G. Tamborini, *J. Med. Chem.*, 1971, **14**, 721.
97. G. G. Gallo, C. R. Pasqualucci, N. Maggi, R. Ballotta and P. Sensi, *Farmaco, Ed. Sci.*, 1966, **21**, 68.
98. F. Kradolfer, L. Neipp and W. Sackmann, *Antimicrob. Agents Chemother.*, 1967, 359.

99. N. Maggi, V. Arioli and G. Tamborini, *Farmaco, Ed. Sci.*, 1969, **24**, 263.
100. R. Cricchio, P. Antonini, G. C. Lancini, G. Tamborini, R. J. White and E. Martinelli, *Tetrahedron*, 1980, **36**, 1415.
101. R. Cricchio, P. Antonini and G. Sartori, *J. Antibiot.*, 1980, **33**, 342.
102. L. Marsili, C. R. Pasqualucci, A. Vigevani, B. Gioia, G. Schioppacassi and G. Oronzo, *J. Antibiot.*, 1981, **34**, 1033.
103. C. Della Bruna, G. Schioppacassi, D. Ungheri, D. Jabes, E. Morvillo and A. Sanfilippo, *J. Antibiot.*, 1983, **36**, 1502.
104. D. Ungheri, C. Della Bruna and A. Sanfilippo, *G. Ital. Chemioter.*, 1984, **31**, 211.
105. E. Marchi, G. Mascellani, L. Montecchi, M. Brufani and L. Cellai, *Chemioterapia*, 1983, **2** (suppl. 5), 48.
106. N. Maggi and R. Pallanza, *Farmaco, Ed. Sci.*, 1967, **22**, 307.
107. N. Maggi, V. Arioli and P. Sensi, *J. Med. Chem.*, 1965, **8**, 790.
108. N. Maggi, R. Pallanza and P. Sensi, *Antimicrob. Agents Chemother.*, 1966, 765.
109. S. Furesz, V. Arioli and R. Pallanza, *Antimicrob. Agents Chemother.*, 1966, 770.
110. R. Cricchio, G. Cietto, E. Rossi and V. Arioli, *Farmaco, Ed. Sci.*, 1975, **30**, 695.
111. R. Cricchio, V. Arioli and G. C. Lancini, *Farmaco, Ed. Sci.*, 1975, **30**, 605.
112. N. Maggi, C. R. Pasqualucci, R. Ballotta and P. Sensi, *Chemotherapy*, 1966, **11**, 285.
113. P. Sensi, in 'Chronicles of Drug Discovery', ed. J. S. Bindra and D. Lednicer, Wiley, New York, 1982, vol. 1, p. 201.
114. V. Arioli, M. Berti, G. Carniti, E. Randisi, E. Rossi and R. Scotti, *J. Antibiot.*, 1981, **34**, 1026.
115. F. Knüsel, H. Bickel and W. Kump, *Experientia*, 1969, **25**, 1207.
116. L. Marsili, G. Franceschi, B. Gioia, G. Oronzo, G. Schioppacassi and A. Vigevani, *Farmaco, Ed. Sci.*, 1982, **37**, 641.
117. L. Marsili, G. Franceschi, C. Galliani, A. Sanfilippo and A. Vigevani, *Farmaco, Ed. Sci.*, 1987, **37**, 781.
118. P. Bellomo, E. Marchi, G. Mascellani and M. Brufani, *J. Med. Chem.*, 1981, **24**, 1310.
119. C. Della Bruna, D. Ungheri, G. Sebben and A. Sanfilippo, *J. Antibiot.*, 1985, **38**, 779.
120. G. Pelizza, G. C. Lancini, G. C. Allievi and G. G. Gallo, *Farmaco, Ed. Sci.*, 1973, **28**, 298.
121. F. R. Quinn, J. S. Driscoll and C. Hansch, *J. Med. Chem.*, 1975, **18**, 332.
122. J. A. Kiritsy, D. K. Yung and D. E. Mahony, *J. Med. Chem.*, 1978, **21**, 1301.
123. M. F. Dampier and H. W. Witlock, Jr., *J. Am. Chem. Soc.*, 1975, **97**, 6254.
124. M. F. Dampier, C. W. Chen and H. W. Witlock, Jr., *J. Am. Chem. Soc.*, 1976, **98**, 7064.
125. P. Schiatti, N. Maggi, P. Sensi and G. Maffii, *Chemotherapia*, 1967, **12**, 155.
126. A. Assandri, T. Cristina and L. Moro, *J. Antibiot.*, 1978, **31**, 894.
127. R. Pallanza, S. Furesz, M. T. Timbal and G. Carniti, *Arzneim.-Forsch.*, 1965, **15**, 800.
128. S. Furesz, V. Arioli and R. Scotti, *Arzneim.-Forsch.*, 1965, **15**, 802.
129. G. Binda, E. Domenichini, A. Gottardi, B. Orlandi, E. Ortelli, B. Pacini and G. Fowst, *Arzneim.-Forsch.*, 1971, **21**, 1907.
130. A. Blasi, L. Donatelli and C. Zanussi, 'Rifampicina', Minerva Medica, Torino, 1978.
131. B. Farr and G. L. Mandell, *Med. Clin. North Am.*, 1982, **66**, 157.
132. A. Kucers and N. McK. Bennett, 'The Use of Antibiotics', Heinemann, London, 1987, p. 914.
133. G. Acocella, *Clin. Pharmacokinet.*, 1978, **3**, 108.
134. M. T. Kenny and B. Strates, *Drug Metab. Rev.*, 1981, **12**, 159.
135. N. Maggi, S. Furesz, R. Pallanza and G. Pelizza, *Arzneim.-Forsch.*, 1969, **19**, 651.
136. W. Fox and D. A. Mitchison, *Am. Rev. Respir. Dis.*, 1975, **111**, 325.
137. W. Fox, *Proc. R. Soc. Med.*, 1977, **70**, 4.
138. A. K. Dutt, L. Jones and W. W. Stead, *Chest*, 1979, **75**, 441.
139. British Thoracic and Tuberculosis Association, *Lancet*, 1976, **2**, 1102.
140. Report of a WHO Study Group, *Tech. Rep. Ser.*, 1982, no. 675.
141. R. J. W. Rees, *Lepr. Rev.*, 1983, **54**, 81.
142. G. L. Archer, J. L. Johnston, G. L. Vazquez and H. B. Haywood, *Rev. Infect. Dis.*, 1983, **55** (suppl. 3), 538.
143. D. J. Girling, *J. Antimicrob. Chemother.*, 1977, **3**, 115.
144. G. Acocella and R. Conti, *Tubercle*, 1980, **61**, 171.
145. A. P. Venturini, E. Marchi, G. Blandino and R. Mattina, *Chemioterapia*, 1983, **2** (suppl. 5), 50.
146. A. T. Birmingham, A. J. Coleman, M. L'E Orme, B. K. Park, N. J. Pearson, A. H. Short and P. J. Southgate, *Br. J. Clin. Pharmacol.*, 1978, **6**, 455.
147. A. Assandri, A. Perazzi and M. Berti, *J. Antibiot.*, 1977, **30**, 409.
148. A. Sanfilippo, C. Della Bruna, L. Marsili, E. Morvillo, C. R. Pasqualucci, G. Schioppacassi and D. Ungheri, *J. Antibiot.*, 1980, **33**, 1193.
149. C. L. Woodley and J. O. Kilbum, *Am. Rev. Respir. Dis.*, 1982, **126**, 586.
150. R. Anand, J. L. Moore, J. W. Curran and A. Srinivasan, *Antimicrob. Agents Chemother.*, 1988, **32**, 684.
151. E. Heller, M. Argman, H. Levy and M. Goldblum, *Nature (London)*, 1969, **222**, 273.
152. J. H. Subak-Sharpe, M. C. Timbury and J. F. Williams, *Nature (London)*, 1969, **222**, 341.
153. B. Moss, E. N. Rosenblum and P. M. Grimley, *Virology*, 1971, **45**, 123.
154. C. Gurgo, R. K. Ray, L. Thiry and M. Green, *Nature (London)*, 1971, **229**, 111.
155. R. C. Gallo, *Nature (London)*, 1971, **234**, 194.
156. H. Diggelman and C. Weissman, *Nature (London)*, 1969, **224**, 1277.
157. S. S. Yang, M. F. Herrera, R. G. Smith, M. S. Reitz, G. C. Lancini, R. C. Ting and R. C. Gallo, *J. Natl. Cancer Inst.*, 1972, **49**, 7.
158. C. Gurgo, R. K. Ray and M. Green, *J. Natl. Cancer Inst.*, 1972, **49**, 61.
159. M. Meilhac, Z. Tysper and P. Chambon, *Eur. J. Biochem.*, 1972, **28**, 291.
160. W. Wehrli and M. Staehelin, in 'Antibiotics', ed. J. W. Corcoran and F. E. Hahn, Springer-Verlag, Berlin, 1975, vol. 3, p. 252.
161. C. Gurgo, *Pharmacol. Ther.*, 1977, **2**, 139.

10.5

Agents which Interact with Ribosomal RNA and Interfere with its Functions*

MICHAEL CANNON

King's College London, UK

10.5.1 INTRODUCTION   814

10.5.2 STRUCTURE OF RIBOSOMES   814

10.5.3 THE MECHANISM OF PROTEIN SYNTHESIS   815

    10.5.3.1 Initiation of Protein Synthesis   815
    10.5.3.2 Elongation Reactions in Protein Synthesis   815
    10.5.3.3 Termination of Protein Synthesis   816
    10.5.3.4 Summary   816

10.5.4 CATALYTIC INHIBITORS OF PROTEIN SYNTHESIS   816

    10.5.4.1 Bacteriocins   816
    10.5.4.2 α-Sarcin   818
    10.5.4.3 Ricin   819
    10.5.4.4 Summary   820

10.5.5 METHYLATED RESIDUES IN RIBOSOMAL RNA AND THE USE OF KASUGAMYCIN
        TO PROBE A PHYSIOLOGICAL ROLE   820

    10.5.5.1 Methylated Residues in 16S Ribosomal RNA   821
    10.5.5.2 The Mode of Action of Kasugamycin   822
    10.5.5.3 Resistance to Kasugamycin   822

10.5.6 THE USE OF ANTIBIOTICS FOR THE MORE PRECISE LOCALIZATION OF RIBOSOMAL
        FUNCTIONS   823

    10.5.6.1 The Mode of Action of Thiostrepton — an Antibiotic of Central Importance in Studies of
            Ribosomal Structure–Function Relationships   823
    10.5.6.2 Resistance to Thiostrepton in Bacteria   824
    10.5.6.3 Resistance to Thiostrepton in Drug-producing Organisms   825
    10.5.6.4 Some Conclusions   826

10.5.7 FUNCTIONALLY IMPORTANT INTERACTIONS BETWEEN DISTINCT RIBOSOMAL
        DOMAINS   827

    10.5.7.1 The Antibiotics Erythromycin and Chloramphenicol   827
    10.5.7.2 The Ribosomal Receptor Sites for Erythromycin and Chloramphenicol and the Exploitation of
            Point Mutations within Ribosomal RNA   827
    10.5.7.3 Structurally Complex Functional Domains that Interact within Ribosomes   829
    10.5.7.4 The Use of Chemical Footprinting in Locating Binding Sites for Antibiotics on Ribosomal RNA 830

10.5.8 THE AMINOGLYCOSIDE ANTIBIOTICS AND THE RIBOSOMAL DECODING SITE   831

    10.5.8.1 Mode of Action of Streptomycin and Mechanisms that Underlie Resistance to Aminoglycoside
            Antibiotics   831

* I should like to dedicate this chapter to the memory of two close friends and colleagues—Gary Craven and David Vazquez—whose interest in ribosomes never waned and whose work inspired others in the task of unravelling the workings of this fascinating organelle.

10.5.8.2   *Resistance to Aminoglycoside Antibiotics in Drug-producing Organisms*   832
10.5.8.3   *The Use of Chemical Footprinting for the Identification of Ribosomal RNA Receptor Sites for*
               Aminoglycoside Antibiotics   832
10.5.8.4   *The Ribosomal Decoding Site*   833

10.5.9   RIBOSOMAL RNA—THE FUNDAMENTAL DETERMINANT OF RIBOSOMAL FUNCTIONS?   834

10.5.10   REFERENCES   837

10.5.1 INTRODUCTION

The nucleic acids represent, within cells, target sites that can be selectively inhibited by a variety of agents. There are several levels at which inhibition can be expressed, including interference with the synthesis of nucleic acid precursors, prevention of the assembly of precursors into mature macromolecules and impairment of a specific structural or functional role(s) associated with replication, transcription or protein synthesis. This chapter will focus on agents acting on RNA. Specifically a limited number of agents will be considered, including selected toxins and antibiotics, the coverage being restricted to compounds that act, at least in part, upon ribosomal RNA. Although at first sight this might be considered a somewhat blinkered approach, rapid advances have been made in recent years in understanding the structures and functions of the various ribosomal RNA species. It is, accordingly, a suitable time to consider the impact of these studies on modern day thinking in molecular biology. Central to recent advances has been the use of highly specific chemical probes which have aided the elucidation of crucial functional roles in protein synthesis that are thought to be attributable to ribosomal RNA. Indeed, there is now a growing conviction that ribosomal RNA is the fundamental determinant of ribosomal functions and it might be expected, therefore, that agents interfering with such functions may well do so by directly affecting ribosomal RNA. Before the evidence for this viewpoint is assembled, it is pertinent to review very briefly our current knowledge of the structure of ribosomes and outline the mechanisms that are central to protein synthesis. For simplicity, the discussion will concentrate mainly on prokaryotic systems, although eukaryotic systems will be considered where important differences are apparent.

10.5.2 STRUCTURE OF RIBOSOMES

The ribosome is the site of protein synthesis within cells and the particle comprises several ribosomal RNAs and a large number of ribosomal proteins. In prokaryotes, the ribosome sediments at 70S and has two subunits sedimenting at 50S and 30S. There are three ribosomal RNA molecules in each ribosome (23S, 16S, 5S) and in *Escherichia coli*, for example, there are 52 unique ribosomal proteins per particle. Each ribosomal protein is identified by a lettering and numbering system that describes its subunit localization—small (S) or large (L)—and its electrophoretic mobility. Thus, ribosomal proteins from *E. coli* are identified as L1–L34 and S1–S21. Note that L26 and S20 are identical, L7 and L12 differ only by the latter being acetylated at its amino terminus and L8 is a complex of L10 with a L7/L12 tetramer. Cytoplasmic eukaryotic ribosomes are larger than their prokaryotic counterparts and their components are more numerous. The intact eukaryotic particle and its isolated subunits sediment, respectively, at 80S, 60S and 40S. In *Saccharomyces cerevisiae*, for example, the 80S particle contains four ribosomal RNA species (25S, 18S, 5.8S, 5S) and approximately 80 ribosomal proteins. Ribosomes from other sources vary in their size and in the number of their constituent components. Ribosomal proteins from *E. coli* have been extensively characterized using chemical, physical and immunological techniques and all the proteins have been sequenced. Primary structures have also been determined for many ribosomal RNAs isolated from a variety of organisms and some of these sequences will be considered in detail at a later stage. For a more extensive coverage of ribosomal structure and the citation of relevant references the reader should consult articles by Gale *et al.*,[1] Wool[2] and Wittmann.[3]

The ultimate goal in research on ribosomes is to define with precision the exact location and function of every component contained within the particle. Such studies were facilitated by the development of ribosomal reconstitution systems. Partial reconstitution uses 'core particles' and the 'split proteins' released from ribosomes exposed to high ionic strength under which conditions the 'core particles' are formed. In total reconstitution, ribosomes are re-assembled *in vitro* from isolated ribosomal RNAs and ribosomal proteins. Relevant references to this important technique are cited in Gale *et al.*[1] Many other approaches have also been used but a particularly fruitful method is the application of neutron-scattering analysis. This enables measurements to be made of the distances between proteins within the ribosome and allows the construction of a refined model for the structure of the particle. This important technique and other approaches are outlined in several articles in a recent publication to which the reader is referred.[4]

10.5.3 THE MECHANISM OF PROTEIN SYNTHESIS

Protein synthesis is a subject that is covered in detail in a large number of both textbooks and reviews (see, for examples, refs. 4, 5 and 6) and only the major points will be outlined here. The process involves the three phases of initiation, elongation and termination and these can be considered separately.

10.5.3.1 Initiation of Protein Synthesis

In prokaryotes, ribosomal subunits are either combined within the polyribosome complex or exist free within the cells. The free 30S subparticle is associated with initiation factor 3 (IF-3) and this prevents its spontaneous association with a 50S partner. IF-3 is also credited with promoting the binding of natural messenger RNA to the 30S subparticle. The subunit–factor complex selects the initiation signal on messenger RNA (see later) and an AUG (or rarely GUG) codon, located 3′ to the initiation signal, is correctly positioned on the ribosome. The AUG codon is recognized by an initiator transfer RNA (tRNA) molecule (tRNA$_F^{met}$) which has previously been charged with methionine by the appropriate aminoacyl-tRNA synthetase. After its attachment to tRNA the methionine is formylated at its amino terminus by a transformylase enzyme. The selection by the ribosome–messenger RNA complex of the fmet-tRNA$_F^{met}$ requires that the latter first forms a ternary complex with initiation factor 2 (IF-2) and GTP. Functional binding of this complex to the 30S ribosomal subunit involves hydrolysis of GTP to GDP and the subsequent release of IF-2, now linked to GDP, from the ribosome. A third initiation factor (IF-1) stimulates the activities of IF-2 and IF-3, although its precise function remains elusive.

Initiation in eukaryotes also involves initiation factors, of which there appear to be nine, and a specific initiator tRNA which carries methionine and is again selected by an AUG codon. In eukaryotes, however, the methionine remains unformylated. Furthermore, in eukaryotes, there appears to be no strong initiation signal between messenger RNA and the ribosome. Rather, the 40S ribosomal subunit binds to the messenger RNA near to the latter's 5′ terminus and scans the messenger RNA sequence until it reaches the first internal AUG codon that signifies initiation of protein synthesis. This difference between prokaryotes and eukaryotes is particularly striking and the mechanisms involved were elucidated by exploiting the specific action of certain inhibitors, as will become apparent later in the text.

10.5.3.2 Elongation Reactions in Protein Synthesis

Following the formation of the initiation complex it is joined to a 50S subparticle in a reaction again involving GTP hydrolysis. It seems likely that release of IF-3 occurs during this association of the two ribosomal subunits. The second messenger RNA codon is now correctly positioned and it attracts the relevant aminoacyl-tRNA to its ribosomal binding site. Functional binding again requires the formation of a ternary complex, this time between the aminoacyl-tRNA, GTP and an elongation factor (EF-Tu). Hydrolysis of GTP once again occurs and GDP is released in a complex with EF-Tu. The active factor is regenerated in a series of reactions involving GTP and a further elongation factor (EF-Ts). The two aminoacyl-tRNAs are now located adjacent to each other in two distinct functional ribosomal-binding sites — the P site, holding fmet-tRNA$_F^{met}$ and the A site, holding the aminoacyl-tRNA specified by the second messenger RNA codon. A peptide bond is formed by interaction of the carboxyl group of methionine with the amino group of the adjacent amino acid. This reaction is catalyzed by the peptidyl transferase centre of the ribosome leaving deacylated tRNA$_F^{met}$ in the ribosomal P site and the dipeptidyl-tRNA in the ribosomal A site.

The dipeptidyl-tRNA is then translocated from the A site to the P site in a reaction involving elongation factor G (EF-G) and the deacylated tRNA is ejected. Hydrolysis of GTP is again required and EF-G is subsequently released in a complex with GDP. The translocation reaction is poorly understood at the molecular level and a third tRNA binding site (the E site) is thought to be involved. Nevertheless, translocation clears the A site so that the third messenger RNA codon can move into place and an appropriate aminoacyl-tRNA can be selected in a reaction involving EF-Tu as described above. The processes of peptide bond formation and translocation continue alternately throughout protein synthesis and the peptide chain is thus progressively extended. Internal methionine residues are also coded for by AUG but $tRNA_F^{met}$ is not selected internally. Instead, a second tRNA molecule ($tRNA_M^{met}$) is involved and the carried methionine is not formylated.

The elongation processes in eukaryotes are essentially identical to those described above, although there is a different nomenclature for the various elongation factors. Thus, the eukaryotic elongation factor EF-1 (eEF-Tu or EF-1α) corresponds to EF-Tu and the eukaryotic translocation factor EF-2 corresponds to EF-G.

From the foregoing discussion, the importance of a GTPase reaction associated with the ribosome during protein synthesis will be apparent. This reaction is of particular relevance to the present article and it will be considered in more detail at a later stage in the text. It represents an exquisitely sensitive ribosomal target site for the interaction of certain antibiotics.

10.5.3.3 Termination of Protein Synthesis

Elongation of the peptide chain from its amino-terminal methionine residue continues on the ribosome until one of three termination codons (UAA, UAG, UGA) becomes located in the ribosomal A site. These codons attract release factors of which there are three in prokaryotes (RF-1, RF-2, RF-3), although eukaryotes appear to possess a single complex only. The release factors modify the action of peptidyl transferase in such a way that the bond linking the polypeptide on to the tRNA molecule that binds it to the ribosomal P site is hydrolyzed. The completed polypeptide diffuses from the ribosome, the deacylated tRNA is discharged (presumably under the influence of EF-G), the ribosome dissociates to allow IF-3 access to its binding site on the 30S subparticle and further rounds of protein synthesis can thus be initiated as described in the preceding paragraphs.

10.5.3.4 Summary

It will be evident from the above brief survey that protein synthesis on ribosomes involves the sequential and highly specific interaction of a large number of components with the particles. The ribosome associates with many macromolecules, including both RNA and proteins, and such macromolecules must interact with various functional domains created within the ribosome's structure. These domains include the binding sites for tRNA and messenger RNA, the binding sites for protein factors, the peptidyl transferase centre and the GTPase centre. A detailed knowledge of ribosomal architecture is essential for a complete understanding of how these various domains operate, but despite a vast reservoir of experimental data, enormous gaps still remain in our knowledge of ribosomal structure–function relationships. Nevertheless, progress has been made and is being made in this intriguing research area and several compounds, which will now be considered in detail, have been of particular use in this respect.

10.5.4 CATALYTIC INHIBITORS OF PROTEIN SYNTHESIS

A number of proteins isolated from bacteria, fungi or plants are highly effective inhibitors of protein synthesis and, in many cases, they act catalytically on ribosomes. Three of these proteins — colicin E3 (a bacteriocin), α-sarcin and ricin — have been selected from this wide range of natural products. A knowledge of their precise inhibitory effects has aided very considerably our understanding of how ribosomal RNA functions within cells.

10.5.4.1 Bacteriocins

The bacteriocins (colicin E3 and cloacin DF13) are proteins that act catalytically against prokaryotic ribosomes. They are particularly effective *in vitro* but are more selective in their action *in vivo*. Colicin E3 is produced by various strains of *E. coli* and can be fractionated into two components designated protein A and protein B. The former is the active enzyme, whereas the latter (the immunity protein) combines with protein A and inhibits its action against the producing strain. It seems likely that protein B promotes the binding of protein A to cell membrane receptors in colicin sensitive bacteria. Protein A then enters the cell leaving its inactive partner at an extracellular location. Protein B affords protection only to those strains that elaborate the toxin. The reader is directed to Gale *et al.*[1] for references.

Colicin E3 was employed in early studies on bacterial protein synthesis and was found to inhibit the process in sensitive cells. Analysis of ribosomes from such inhibited strains revealed no gross structural changes. Nevertheless, ribosome reconstitution studies carried out at a later stage showed conclusively that colicin E3 inactivates 16S ribosomal RNA within the 30S ribosomal subunit — a result[7] that provided the foundation for the subsequent elucidation of the precise action of the toxin. Colicin E3 cleaves 16S ribosomal RNA at a position approximately 50 nucleotides from the 3′ terminus. This cleavage does not cause the release of any ribosomal proteins from the 30S subparticle. Nevertheless, if reconstitution experiments are carried out using the cleaved 16S ribosomal RNA, ribosomal protein S21 fails to incorporate into the structure. It is now known[8] that this protein has one of its two crosslinking sites on 16S ribosomal RNA located at the latter's 3′ terminus. It is significant, however, that if S21 is omitted under normal reconstitution conditions using intact 16S ribosomal RNA a subparticle is formed that shows only a partial loss of activity in functional tests. This protein is not, therefore, indispensable *per se* for ribosome function. Ribosomal protein S1 also crosslinks to the 3′ end of 16S ribosomal RNA and binds to the colicin E3 RNA fragment. However, this does not appear to be the primary binding site for the protein and the area is thought to provide only a weak site that is available in inactivated 30S ribosomal subunits.[9] For additional references see refs. 4 and 10.

The (complete) inhibitory action of colicin E3 on ribosomes is indicative of a crucial physiological function for ribosomal RNA and the toxin has been used to great effect in showing that this is indeed

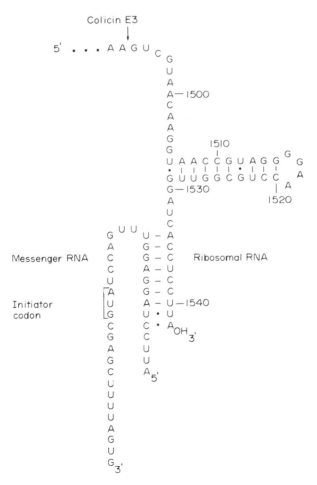

Figure 1 Initiation complex between part of the 3′ terminus of *E. coli* 16S ribosomal RNA and the bacteriophage R17 A protein initiator region. The colicin E3 cleavage site is indicated and the figure also shows the position of the AUG initiator codon in the messenger RNA. Residues are numbered according to ref. 36

the case. As described earlier, initiation of protein synthesis in bacteria requires a specific interaction between the 30S ribosomal subunit and the messenger RNA. It is crucial that the AUG codon that will select fmet-tRNA$_F^{met}$ is correctly positioned on the ribosome. In polycistronic messenger RNAs there are several such codons. They are all located internally within the messenger RNA sequence and are independently selected by individual ribosomes. It was proposed by Shine and Dalgarno[11] that correct positioning might involve certain sequences located 5' to the AUG start codons in messenger RNA that could base pair with complementary sequences in 16S ribosomal RNA. This hypothesis was experimentally tested by Steitz and Jakes[12] in an elegant series of experiments that involved the use of colicin E3. Initiation complexes, formed between 30S ribosomal subunits and a messenger RNA fragment of known sequence (containing the putative initiator region of the A protein cistron of R17 bacteriophage RNA), were treated with colicin E3 and then disassembled by exposure to sodium dodecyl sulfate. Subsequent gel electrophoresis allowed the resolution of a stable complex comprising the initiator region of the messenger RNA and the 3' terminal fragment of ribosomal RNA released by colicin E3 cleavage. This base-paired complex was dissociated by heating in the presence or absence of urea and the two RNA components can then be sequenced. These experiments clearly established a functional role for ribosomal RNA in protein synthesis. The structure of part of the 3' terminus of *E. coli* 16S ribosomal RNA is illustrated in Figure 1 with the colicin E3 cleavage site indicated and a Shine–Dalgarno[11] interaction shown involving a messenger RNA initiator region identified by Steitz and Jakes.[12]

Although it is now widely accepted that the above mechanism is important in prokaryotes, the interactions described are not sufficient or necessary in themselves for the correct selection and positioning of messenger RNA initiator sites (see Stormo *et al.*[13]). Indeed, there is often no simple relationship between the translational efficiency of a particular messenger RNA and the complementarity of its initiator signal for the corresponding area of 16S ribosomal RNA. Other (nonrandom) sequences, located close to AUG start codons in messenger RNA, may well be involved and the important role of initiation factors must not be overlooked. Shine–Dalgarno interactions may occur at a ribosomal site that is distinct from and independent of the site with which initiation factors associate and may be of particular importance in producing high local concentrations of an AUG start codon. Initiation factors themselves may influence the correct positioning and subsequent use of the AUG codon in protein synthesis, thus being largely concerned with kinetic selection of messenger RNA.[14]

10.5.4.2 α-Sarcin

α-Sarcin is another interesting compound that has a particularly exquisite action—this time on eukaryotic ribosomes. It is a small, basic protein produced by *Aspergillus giganteus* and its primary structure has been determined.[15] The protein is exceedingly toxic and the basis of this action is a potent inhibition of protein synthesis. *In vitro*, α-sarcin inhibits the enzymic binding of aminoacyl-tRNA into the ribosomal A site and interferes with the uncoupled hydrolysis of GTP that can be catalyzed by ribosomes in the presence of the elongation factors EF-1 and EF-2 (for references see Gale *et al.*[1] and Wool[2]). α-Sarcin acts catalytically on ribosomes as a highly specific nuclease[16,17] and in this respect it resembles colicin E3. Unlike the latter, however, α-sarcin can also cleave free RNA, for which it displays an entirely different specificity in comparison with its action on ribosomes. The toxin causes extensive degradation of both 28S and 5S ribosomal RNAs, for example, and appears to hydrolyze at the 3' position of almost every adenosine and guanosine residue within the sequences.[18] The nuclease acts equally well on either single-stranded or double-stranded regions within the molecules. In a remarkable contrast to such relatively low specificity, α-sarcin has a highly selective action on 28S ribosomal RNA contained within either 60S subparticles or 80S ribosomes and hydrolyzes a single phosphodiester bond that is located 459 nucleotides from the RNA's 3' terminus (see later).[18] The ribosomal domain that is cleaved is thought to be a single-stranded loop and since it is accessible to the nuclease it is presumably located on the surface of the 60S subparticle. The proposed loop structure is purine rich and appears to represent a highly conserved region in the large ribosomal RNAs from a variety of sources, both eukaryotic and prokaryotic. These facts make the specificity of α-sarcin perhaps even more remarkable. Because of the action of the toxin in inhibiting the EF-1-catalyzed binding of aminoacyl-tRNA,[19,20] it may well be that the ribosomal RNA structural area that embraces the α-sarcin cleavage site is crucial for this particular function in protein synthesis. Such an interpretation should, however, be viewed with some caution. In eukaryotic ribosomes (rat liver, for example) 28S ribosomal RNA associates noncovalently with 5.8S ribosomal RNA and there is evidence that the former has two contact sites for the latter—both contact sites being located close to the 5' terminus of the 28S ribosomal RNA species.[21] Cleavage of 60S subparticles with·α-sarcin markedly destabilizes the above association and it has been suggested recently by Wool[22] that the effect observed represents a dramatic collapse in the 28S ribosomal RNA structure. It remains possible, therefore, that the α-sarcin cleavage site is well removed spatially from the ribosomal site that controls EF-1-mediated functions. In his very interesting article Wool discusses the possible implications of α-sarcin action for switch structures which other authors[23,24] have suggested might occur in the large ribosomal RNAs. Such switches could, in principle, control ribosomal functions such as translocation or messenger RNA movement, both of which occur during protein synthesis. Switches could be represented by alternate helical structures that form within ribosomal RNA by breakage and rejoining (elsewhere) of hydrogen-bonded regions.

10.5.4.3 Ricin

A large number of extremely toxic proteins and glycoproteins have been isolated from various plant sources and their actions combine, in many cases, certain of those that are observed for the bacteriocins and α-sarcin. These plant-derived inhibitors are the subject of excellent reviews elsewhere (for examples see refs. 25 and 26) and only one example will be considered in detail here. This is the compound ricin — a toxic lectin derived from the beans of *Ricinus communis*. This toxin, in addition to its specific inhibitory action on ribosomal RNA (as will be considered shortly), has been central to other fascinating scientific developments. Thus in the early 1900s, Paul Ehrlich fed seeds of *Ricinus communis* to rabbits and established the basic immunological principle of specific antibody precipitation by observing the production of serum proteins that inactivated and precipitated ricin (see Olsnes and Pihl[27] for references). Events followed a more sinister turn when, in 1979, a Bulgarian broadcaster standing in a London street was reputed to have been murdered by receiving an injection of a ricin-containing pellet (see ref. 28) supposedly delivered from the tip of a passing umbrella. Encouragingly, however, the highly toxic properties of ricin are being employed more humanely and with some success in cancer chemotherapy (see refs. 29 and 30 for examples).

Ricin is composed of two protein subunits (A and B) that are linked by a single disulfide bond, and the latter has a crucial role to play in the toxin's action *in vivo*. The intact protein binds to a suitable cell surface receptor and intracellular entry is somehow facilitated in an energy-requiring process. The inhibitory A chain is subsequently released by reduction of the disulfide bond, whereupon it associates with its target site — the ribosome. Ricin is a catalytic inhibitor of either 60S ribosomal subunits or 80S ribosomes and, in common with α-sarcin, the toxin primarily affects reactions in eukaryotic protein synthesis that require the participation of factors EF-1 and EF-2. Ricin has no effect on prokaryotic ribosomes and ribosomes from plants and protozoa are less sensitive to the compound's action than are those from animal cells. The precise target site for ricin was not identified until very recently when some elegant experiments carried out by Endo *et al.*[31] indicated that the ricin A chain can modify either or both of two nucleoside residues located within the sequence of 28S ribosomal RNA present in rat liver 80S ribosomes. It seemed possible that the action of the toxin resembles that of a *N*-glycosylase and involves the removal of a purine base(s), thereby leaving the relevant position(s) resistant to ribonuclease action but causing the phosphodiester bond(s) adjacent to the modified base(s) to become very susceptible to hydrolysis. Endo and Tsurugi[32] have now provided direct evidence that the ricin A chain inactivates ribosomes by cleaving the *N*-glycosidic bond of residue A_{4326} within 28S ribosomal RNA in a hydrolytic fashion. Ricin also appears to act in the same way, albeit slowly, on isolated 28S ribosomal RNA and its specificity in this respect is clearly more complete than is that of α-sarcin. What is particularly exciting is that this residue is adjacent to the α-sarcin cleavage site — itself located between G and A residues at positions 4327 and 4328. As discussed above, this sequence is highly conserved in the large ribosomal RNAs from a variety of both prokaryotes and eukaryotes, although eubacteria, for example, are not inactivated by either α-sarcin or ricin. The sequence is also highly conserved in Archaebacteria,[33] a kingdom comprising the methanogens, sulfur-metabolizing thermophiles and the halophiles. Ribosomes from Archaebacteria are, in general, sensitive to inhibition by α-sarcin although, in comparison with eukaryotic systems, relatively high concentrations of toxin are required and there is also differential sensitivity between the various Archaebacteria tested. The thermophile *Sulfolobus solfataricus*, for example, is described as moderately sensitive to α-sarcin and the RNA fragment cleaved by the toxin appears to have the same 5' sequence as that found in the α-sarcin fragment generated from eukaryotic ribosomes.[33] Figure 2 illustrates the sequence determined for rat liver 28S ribosomal RNA in the immediate vicinity of the α-sarcin and ricin cleavage sites which are themselves indicated in the figure. Residues are numbered according to Wool (personal communication).

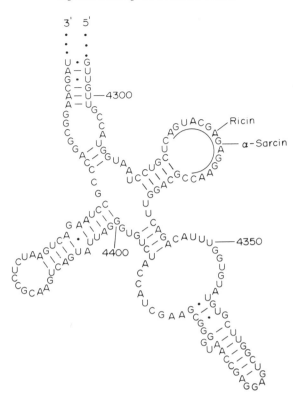

Figure 2 The sequence of rat liver 28S ribosomal RNA that is in the immediate vicinity of the α-sarcin and ricin cleavage sites. These sites are indicated in the figure. The complete secondary structure postulated for the RNA is illustrated in Wool.[22] Residue numbers according to Wool (personal communication)

It is interesting to speculate as to why the ribosomal domain under discussion is insensitive in eubacteria, for example, to the action of α-sarcin. It may be that the toxin fails to cleave ribosomes from *E. coli* because, as is known (see ref. 4), the potential target site on 70S ribosomes has a single base change in the relevant ribosomal RNA sequence. This may not, of course, be the complete answer. As this sequence is highly conserved in all known ribosomes and as cleavage by α-sarcin or ricin can cause complete inactivation of certain classes of ribosomes, a crucial functional role for the sequence is indicated. As discussed above, such a role in eukaryotes might involve, either directly or indirectly, functions mediated by the elongation factors EF-1 or EF-2 and an equivalent role might well be associated with the corresponding domain in prokaryotic ribosomes. The precise structures of these two (equivalent) domains are, however, unlikely to be identical and will almost certainly be controlled, perhaps subtly, by the various ribosomal proteins that are located at or close to this ribosomal area. Since prokaryotic and eukaryotic ribosomal proteins are different, the latter presumably stabilize a particular conformation that allows the catalytic actions of α-sarcin and ricin. The selectivity shown by these compounds is in line with the above interpretation.

10.5.4.4 Summary

As will be evident from the preceding discussion, catalytic inhibitors have been employed successfully to probe the mechanisms of protein synthesis and their use has convincingly demonstrated that ribosomal RNA plays a number of fundamental roles in the process. Although primary sequences in ribonucleic acid molecules ultimately determine their functions, other features of these macromolecules are crucial to their roles. Such a feature is chemical modification and one such modification, that of methylation, will now be considered.

10.5.5 METHYLATED RESIDUES IN RIBOSOMAL RNA AND THE USE OF KASUGAMYCIN TO PROBE A PHYSIOLOGICAL ROLE

Methylation of macromolecules is an important covalent modification that can be involved in the regulation of cellular processes. Although precise physiological roles associated with methylation have rarely been easy to assign, the modification can affect enzymic activities, and methylation of proteins controls the chemotactic response in bacteria. Furthermore, methylation of DNA in eukaryotes has been linked to the control of gene expression and differentiation. Ribosomal RNAs are also subject to methylation and the modification is confined to specific sites within the molecules. The presence of these methyl groups may be particularly important for controlling the biogenesis of ribosomes. Other roles are not, of course, excluded and the relevance of methylation to the functioning of mature ribosomal RNA and its importance in determining whether or not certain drugs can react with the macromolecule will become increasingly apparent throughout the remainder of this chapter. Initially, however, the discussion will concentrate on the methylated residues that are located close to the 3' end of prokaryotic 16S ribosomal RNA.

10.5.5.1 Methylated Residues in 16S Ribosomal RNA

Methylated residues in 16S ribosomal RNA are present in sequences that are highly conserved and in this macromolecule from *E. coli*, for example, 13 such methyl groups can be detected. Of these,

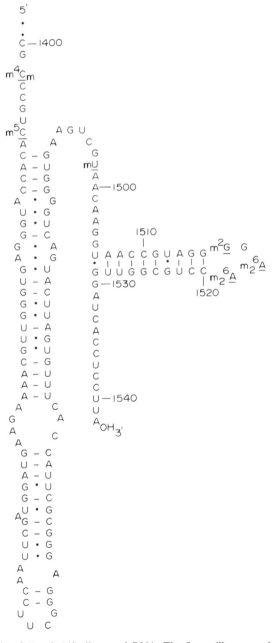

Figure 3 The 3' terminal domain of *E. coli* 16S ribosomal RNA. The figure illustrates the positions of the methylated residues that are located in this domain. The complete secondary structure postulated for the RNA is illustrated in Figure 7

nine are located[34] within the 140 nucleotide 3′ terminal sequence (see Figure 3). Residues C_{1400} and A_{1500} are found in this region and the sequences centred at these locations appear to be conserved universally[35] and are thought to be associated with the ribosomal A site where coded interactions between the transfer RNA and messenger RNA triplets occur during protein synthesis. There is convincing experimental evidence for the above interpretation and of particular relevance is the action of colicin E3, which cleaves between residues A_{1493} and G_{1494} in 16S ribosomal RNA (see Section 10.5.4) and causes ribosomes to lose their ability to bind tRNA. (for references see Noller *et al.*[36]). Precisely how ribosomal RNA stabilizes the codon–anticodon interaction is unknown but Noller *et al.*[36] have proposed an interesting model for the process which will be discussed in Section 10.5.8.4. The RNA fragment derived from ribosomes by colicin E3 treatment itself contains six methyl groups of which five are present in the hairpin loop that is thought to form closely adjacent to the 3′ end of 16S ribosomal RNA (see Figure 3). Two residues located here — both N^6,N^6-dimethyladenines (positions 1518 and 1519) — are of particular interest and their possible function(s) have been probed using the aminoglycoside antibiotic kasugamycin. The reader is referred to Gale *et al.*[1] and Van Knippenberg[37] for relevant articles concerning this drug.

10.5.5.2 The Mode of Action of Kasugamycin

Many aminoglycosides, for example streptomycin, cause the selection, by messenger RNA codons, of incorrect aminoacyl-tRNA molecules at the ribosomal A site. Such miscoding is not, however, caused by kasugamycin and neither does cross resistance between this antibiotic and other aminoglycosides occur. Kasugamycin is an inhibitor of protein synthesis in both prokaryotic and eukaryotic systems and *in vitro* studies using the former have pinpointed the action of the antibiotic on initiation. Kasugamycin inhibits the binding of fmet-tRNA$_F^{met}$ to ribosome–messenger RNA complexes, being effective when either 70S ribosomes or 30S ribosomal subunits are used. Furthermore, the drug causes the release of fmet-tRNA$_F^{met}$ from preformed initiation complexes involving 30S subparticles, although this action is not observed if such complexes are preformed using 70S ribosomes. Very recently, kasugamycin has been shown to decrease the translational errors that can occur when protein synthesis is programmed *in vitro* by MS2 bacteriophage RNA.[37] The ribosomal target site for kasugamycin is likely to be involved, either directly or indirectly, with those functions, described above, that are inhibited by the drug and the nature of this target site has been probed using bacterial strains resistant to the antibiotic.

10.5.5.3 Resistance to Kasugamycin

Three loci — *ksg*A, *ksg*B and *ksg*C — have been described that confer resistance to kasugamycin, and *ksg*A strains are of particular relevance to the present discussion. Such strains have a slightly slower growth rate than have their drug sensitive parents, and certain frameshift mutations or nonsense mutations in the *lac* operon become increasingly leaky *in vivo* in the additional presence of the *ksg*A mutation. This effect is related, presumably, to the drug's ability to suppress translational errors *in vitro* in drug sensitive strains, as discussed above. Other differences are shown *in vitro*. For example, binding of fmet-tRNA$_F^{met}$ to initiation complexes containing ribosomes from *ksg*A strains requires relatively high levels of IF-3, although the effect is only observed in the absence of IF-1. In addition, 30S ribosomal subunits from *ksg*A strains show a decreased affinity *in vitro* for their 50S partners.

The determinant of kasugamycin resistance in *ksg*A strains is the 16S ribosomal RNA and the resistance phenomenon is associated with a methyltransferase that is the product of the *ksg*A locus. The gene that encodes this enzyme has recently been cloned and its nucleotide sequence determined.[38] The methyltransferase derived from drug sensitive strains of *E. coli* can methylate core particles prepared from 30S ribosomal subunits previously isolated from *ksg*A mutants. Following reconstitution, the 30S ribosomal subunit is sensitive, *in vitro*, to the inhibitory action of kasugamycin, as is the 70S ribosome formed from reconstituted (methylated) 30S subparticles and 50S subparticles isolated from *ksg*A strains. An identical result is achieved if the intact 30S ribosomal subunits from *ksg*A strains are methylated by purified methyltransferase and used for fmet-tRNA$_F^{met}$ binding studies. The methyltransferase introduces four methyl groups into 16S ribosomal RNA by dimethylation of the two adenine residues located at positions A_{1518} aand A_{1519} in the sequence. These N^6,N^6-dimethyladenines are, of course, closely adjacent to the 3′ terminus of 16S ribosomal

RNA and are located 16 and 17 residues 5' to the Shine–Dalgarno sequence in the colicin E3 fragment. As considered earlier, this ribosomal RNA domain is of crucial importance to the functioning of ribosomes, being concerned with several events that are necessary for initiation of protein synthesis, including binding of IF-3, messenger RNA selection and ribosomal subunit interaction. Furthermore, the involvement of this domain with ribosomal A site functions that are important for correct selection of aminoacyl-tRNA residues during protein synthesis has also been alluded to. It seems reasonable to conclude, therefore, that the dimethyladenine residues located within this well-defined ribosomal domain are important for some, if not all, of the above functions. They form part of the ribosomal target site for kasugamycin and this drug can, of course, perturb many of the events just considered when it acts as an inhibitor of protein synthesis in drug sensitive cells. The importance of the dimethyladenines is also indicated by the pleiotropic effects on the initiation and elongation reactions of protein synthesis in *E. coli* strains carrying the *ksg*A mutation.

In his very recent article, Van Knippenberg[37] has not only considered the above points but has also speculated on other possible roles for these methylated residues. 30S ribosomal subunits isolated from *ksg*A strains have a high affinity for methyltransferase, and when this enzyme is bound to them they fail to associate with 50S subparticles. The methyltransferase cannot chemically modify 70S ribosomes from *ksg*A strains. Furthermore, the enzyme fails to methylate 30S subparticles isolated from *ksg*A strains if these subparticles are first crosslinked to S21, a ribosomal protein whose presence is particularly important for the initiation of protein synthesis (see ref. 4). In the process of ribosome biogenesis, certain ribosomal precursor particles have been shown to contain tightly bound methylases. These enzymes prevent binding to the particles of both ribosomal protein S21 and IF-3, although maturation proceeds normally. Van Knippenberg[37] suggests that the methyltransferase might act as a gatekeeper in that its binding to immature particles prevents their premature participation in protein synthesis. In *ksg*A strains, this control does not operate, immature particles can participate in protein synthesis and a slightly handicapped A site function is displayed which diminishes the accuracy of decoding. In Van Knippenberg's article, to which the reader is directed for further details and references, the importance of other methylated residues in 16S ribosomal RNA is also reviewed briefly. A ribosomal domain associated functionally with messenger RNA and IF-3 binding is centred at nucleotide 980 in 16S ribosomal RNA—a site at which there are closely adjacent methylated residues—m^2Gm^5C (966,967). In addition, the residues m^4Cm_{1402} and m^5C_{1407} are in the region where the anticodon of tRNA molecules bound at the ribosomal P site can be crosslinked to C_{1400} in 16S ribosomal RNA (for references see Ofengand *et al.*[39]). Finally, of course, the functional importance of methylated residues within tRNA itself must not be overlooked.

10.5.6 THE USE OF ANTIBIOTICS FOR THE MORE PRECISE LOCALIZATION OF RIBOSOMAL FUNCTIONS

From the work already described in this chapter, it will be apparent that the use of certain inhibitors has provided crucial information on the mechanisms of protein synthesis and has aided very considerably our understanding of structure–function relationships within ribosomes. The ribosomal target sites for the compounds considered so far have been relatively easy to identify although, unfortunately, this has not always been the case for other inhibitors. The ribosome is an extremely complex macromolecular structure and the majority of antibiotic binding sites within the particles are likely to be created only after highly specific and cooperative interactions between the constituent components. It is pertinent to ask, therefore, how such interactions can create a functional domain and in so doing create, fortuitously, a specific drug target site whose character-ization facilitates the functional study. A particularly striking example of the success of this approach has involved the use of thiostrepton—a modified peptide antibiotic—and it is this compound that will now be considered in detail. Relevant references relating to the antibiotic are cited in Gale *et al.*[1]

10.5.6.1 The Mode of Action of Thiostrepton—an Antibiotic of Central Importance in Studies of Ribosomal Structure–Function Relationships

The thiostrepton group of antibiotics (including thiostrepton, siomycin, sporangiomycin and thiopeptin) are selective inhibitors of bacterial protein synthesis. The target site for the inhibitors is the 50S ribosomal subunit and binding between drug and particle is extremely strong

($K_A > 10^9 \, M^{-1}$) and very difficult to reverse. Such a strong binding is unusual for antibiotics which, in general, associate rather weakly with their ribosomal receptor sites.[40] Thiostrepton inhibits those reactions in protein synthesis that involve the binding to ribosomes of IF-2, EF-Tu and EF-G, and are accompanied by the hydrolysis of GTP. The drug can also inhibit termination of peptide chain synthesis as mediated by release factors, and also inhibits the 'uncoupled' hydrolysis of GTP that can be catalyzed by ribosomes and GTP in isolation together. It seems likely that thiostrepton interferes with a ribosomal domain that includes the A site and that can functionally interact with all the protein factors mentioned above. The site is assumed to be associated with most, if not all, ribosome-mediated GTPase reactions.

Early studies, using a gel filtration assay, indicated that ribosomal core particles lacking ribosomal protein L11 failed to bind thiostrepton, although binding was restored if this protein was included in the reaction mixture. An important role for L11 was further indicated by the work of Thompson et al.,[41] who showed that thiostrepton bound with high affinity to a complex containing L11 in combination solely with 23S ribosomal RNA. Furthermore, digestion of the RNA–L11 complex with ribonuclease produced a protein-protected fragment of RNA comprising approximately 60 nucleotides that itself retained the ability to bind thiostrepton with high affinity.[41] Nevertheless, L11 in isolation does not appear to bind the antibiotic, although a relatively weak interaction ($K_A = 2 \times 10^6 \, M^{-1}$) between the drug and 23S ribosomal RNA can, in fact, be demonstrated.[40]

At the time that the above experiments were being carried out, it was generally assumed that the majority of antibiotics that inhibited ribosomal functions would have ribosomal proteins as their target sites. Nucleic acids were not readily associated with an ability to interact specifically with small molecules. Rather, thinking was dominated by a vast literature describing interactions between enzymes and their substrates with its emphasis on the central roles played, in such cases, by protein molecules. If ribosomal proteins are indeed essential determinants of ribosomal function, it would be expected that certain mutants with changes in one or more of these components would show functional defects both *in vivo* and *in vitro*. A large number of temperature sensitive or cold sensitive mutants with altered ribosomal proteins have indeed been isolated. However, rather than being conditionally lethal (as might be expected if the relevant mutated protein controlled a crucial function like, for example, peptidyl transferase activity) many of these mutants simply show defects in the mechanisms of ribosome assembly (see ref. 10). Such observations might indicate, therefore, that a large number of ribosomal proteins, whilst being important for the correct construction of the functional ribosome, play only minor supporting roles in the catalytic reactions of protein synthesis.[10]

10.5.6.2 Resistance to Thiostrepton in Bacteria

As considered already (Section 10.5.5.3), antibiotic resistant mutants can be extremely valuable in probing ribosomal structure–function relationships and it is particularly relevant now to consider bacterial mutants resistant to thiostrepton. As outlined above, thiostrepton inhibits the ribosomal GTPase centre and a role for ribosomal protein L11 in this centre is suggested. This protein cannot be the sole component of the GTPase functional domain. Indeed, the L8 ribosomal protein complex (comprising one molecule of L10 and two dimers of L7/L12) is also involved.[40] In 1979 Cundliffe et al.[42] isolated a mutant of *Bacillus megaterium* that was resistant to thiostrepton. Ribosomes from the strain were found to lack a protein (BM-L11) that was shown to be related immunologically to *E. coli* ribosomal protein L11. The absence of this protein from the mutant's ribosomes was unambiguously proved by the use of several electrophoretic and immunochemical techniques. Independent, but similar, results were obtained from other laboratories[43] using thiostrepton resistant strains of *Bacillus subtilis*. As expected, ribosomes from the drug resistant *B. megaterium* strains bound thiostrepton only weakly ($K_A = 5 \times 10^6 \, M^{-1}$) and this accounts for their relative insensitivity to drug inhibition.[40] Drug sensitivity can, however, be restored to wild-type levels using ribosomes from thiostrepton resistant strains of *B. megaterium* that have been reconstituted in the presence of added BM-L11.[1,40] Studies on these mutants provided information that was to prove crucial for the development of current interpretations as to how thiostrepton interacts with its ribosomal receptor site. The thiostrepton resistant strains of *B. megaterium* grow very poorly and this sickness is reflected by a low protein-synthetic activity of their ribosomes *in vitro*. Significantly, however, these strains are not totally insensitive to the antibiotic — a situation that applies also to other antibiotic resistant mutants isolated under laboratory conditions. Two important conclusions can be drawn. Firstly, even though ribosomal protein L11 influences very considerably the ability of ribosomes to interact with thiostrepton, the receptor site retains a residual drug-binding capacity in the complete absence of the protein. Secondly, ribosomal protein L11 is not indispensible for the functioning of the ribosome in protein synthesis. The latter conclusion is also applicable to other ribosomal proteins, as implied from the earlier brief discussion of cold sensitive and temperature sensitive mutations. Furthermore, Dabbs[44] has isolated at least 16 individual *E. coli* mutants each lacking a different ribosomal protein. Many of these mutants resemble the thiostrepton resistant mutants of *B. megaterium* in that they grow very poorly; but they do, nevertheless, grow.

10.5.6.3 Resistance to Thiostrepton in Drug-producing Organisms

Although the above results emphasize that cooperative interactions between ribosomal components are required for functional expression, the precise nature of such interactions remains, mostly, to be defined. Once more, however, work with thiostrepton has provided important information in this area and has confirmed the suspicion that ribosomal RNA is a key component in directing ribosomal functions. Thiostrepton is produced by the Gram-positive bacterium *Streptomyces azureus*. Unlike the *B. megaterium* strains already considered, this organism is totally resistant to the inhibitory action of thiostrepton both *in vivo* and *in vitro*. Ribosomes isolated from *S. azureus* show no detectable binding of the drug, although they do contain a protein that is related immunologically to the ribosomal proteins BM-L11 and L11 isolated, respectively, from *B. megaterium* and *E. coli*. Since streptomycetes are, in general, potently inhibited by thiostrepton, the total resistance shown by *S. azureus* towards the antibiotic provides this organism with a logical self-defence mechanism and also provides the scientist, fortuitously, with an ideal system for the study of a particular drug resistance phenomenon. It was established that methylation of 23S ribosomal RNA controls resistance to the drug in *S. azureus*. The organism possesses an RNA-pentose methylase and this enzyme introduces a 2'-*O*-methyladenosine residue into the macromolecule. The enzyme, which has now been purified 11 000-fold,[40] also modifies 23S ribosomal RNA from other organisms, including *E. coli* and certain thiostrepton sensitive streptomycete strains, but fails to act on isolated 70S ribosomes or 50S ribosomal subunits. It is assumed that in *S. azureus* the 23S ribosomal RNA is modified at an early stage during or after transcription. The organisms *Streptomyces laurentii* and *Streptomyces sioyaensis* elaborate, respectively, the compounds thiostrepton and siomycin and possess a similar thiostrepton resistance methylase to that characterized in *S. azureus*. Furthermore, an enzyme with comparable specificity is also detected in the organism that produces sporangiomycin, *Planomonospora parontospora*.[1]

The site of action of the *S. azureus* thiostrepton resistance methylase is nucleotide A_{1067} in the 23S ribosomal RNA sequence and according to current models this residue is located within a looped region of the macromolecule (see Figure 4). Since methylation of the nucleotide causes ribosomes to become completely resistant to thiostrepton, A_{1067} clearly is an important component of the antibiotic-binding ribosomal domain. For thiostrepton binding to occur, position 1067 within 23S ribosomal RNA needs to be a purine residue. Thus, site-directed mutagenesis indicates that whereas the replacement of A_{1067} by a G residue has little or no effect on thiostrepton binding, A to U or A to C substitutions cause ribosomes to become highly resistant to the drug when they are assayed *in vitro*.[45] Recent work with halophilic Archaebacteria confirms the importance of A_{1067}. These organisms have a single set of ribosomal RNA genes in their chromosomes and mutations within the RNA are easily achieved. Results again indicate[46] that not only methylation of A_{1067} but also base changes at this position can cause high level resistance to thiostrepton. It may be that A_{1067} is recognized directly by thiostrepton. Alternatively, methylation of this residue or a purine to a pyrimidine conversion at position 1067 may affect adversely a closely adjacent ribosomal domain that itself forms the drug receptor site. Thiostrepton is thought to assume a rigid conformation and its size ($M_r = 1660$) suggests that it recognizes and spans more than one nucleotide within the ribosomal RNA structure. Residue A_{1067} could, of course, be a central component in a fairly complex ribosomal-recognition site.[45]

As discussed earlier, ribosomal proteins L8 and L11 are considered to be central components of the ribosomal GTPase centre and this knowledge has allowed the importance, for thiostrepton action, of residue 1067 within 23S ribosomal RNA to be confirmed. Within the L8 complex it is ribosomal protein L10 that is thought to bind directly to the RNA and experiments have been carried out to see which sequences within this latter macromolecule can be protected by L8(L10) and/or L11 against nuclease attack.[40] Under conditions of limited enzymic digestion, L8(L10) protects residues 1028–1124 and L11 protects residues 1052–1110 within the 23S ribosomal RNA sequence. It is suggested, therefore, that these residues, along with their associated ribosomal proteins, form the ribosomal GTPase centre.[40] It is also clear from these nuclease protection studies

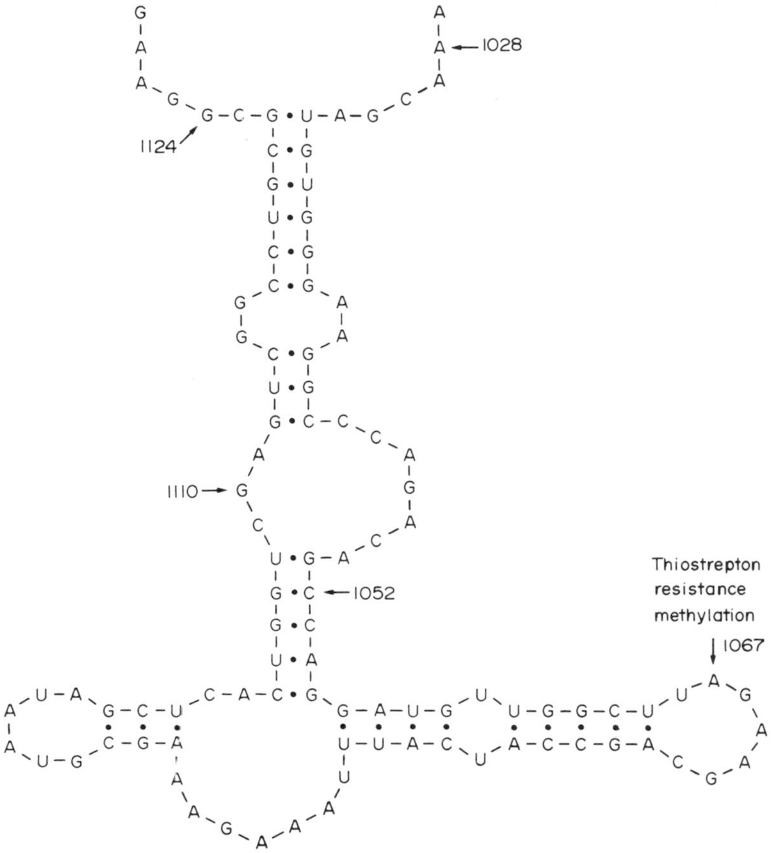

Figure 4 A portion of *E. coli* 23S ribosomal RNA that is postulated to be present in the ribosomal GTPase centre. The site of action of the thiostrepton resistance methylase is indicated in the figure. The complete secondary structure postulated for the RNA is illustrated in Noller *et al.*[36] (reproduced from ref. 40 by permission of Springer)

that L10 and L11 bind to ribosomal RNA at closely adjacent or overlapping sites. The precise interactions, between proteins and RNA, that are required to form, on 50S ribosomal subunits, the GTPase centre and the thiostrepton receptor site remain, of course, to be elucidated. Nevertheless, it can now be envisaged why ribosomes lacking ribosomal protein L11 can function to a limited extent in protein synthesis; the ribosomal RNA itself has a crucial role in defining a ribosomal function.

10.5.6.4 Some Conclusions

The use of thiostrepton in elucidating the topography of an exquisite ribosomal domain has been a triumph for those who, over the years, have kept faith in the potential power of antibiotics as specific probes to study ribosomal structure–function relationships. Nevertheless, although the experimental work inspired by the availability of thiostrepton strongly suggests that ribosomal RNA forms the drug target site, the importance of some ribosomal proteins for the partial reactions of protein synthesis must not be undervalued. Indeed, binding of ribosomal protein L11 to a specific region of ribosomal RNA is an event that confers upon the latter the correct conformational state that is so important for functional expression. Neither of the macromolecules can operate adequately without the other and cooperativity must, ultimately, be the order of the day. However, the importance of methylation reactions is clearly apparent. It is a curious coincidence that L11 is itself the most highly methylated of the *E. coli* ribosomal proteins (see Cannon *et al.*,[47] for references) and the possible roles of methylated residues at the 3′ terminus of 16S ribosomal RNA have, of course, been considered earlier. Studies of the ribosomal GTPase centre may provide the lead that ultimately will allow methylation of RNA components to be linked with a wider range of functional activities. Some of these other possibilities can now be expanded upon by considering the inhibitory actions of other classes of antibiotics that have provided invaluable information on the relationships that exist, within ribosomes, between proteins and RNA.

10.5.7 FUNCTIONALLY IMPORTANT INTERACTIONS BETWEEN DISTINCT RIBOSOMAL DOMAINS

The antibiotics chloramphenicol and erythromycin have been of particular use in demonstrating that distinct structural domains within ribosomes need to interact in order to allow certain functions to be expressed. Accordingly, the following discussion will concentrate on these two drugs although other compounds will be alluded to where appropriate. The reader is again directed to Gale *et al.*[1] for references.

10.5.7.1 The Antibiotics Erythromycin and Chloramphenicol

Erythromycin is a macrolide antibiotic whose mode of inhibitory action has, over the years, been the subject of some controversy. An idea initially favoured was that the drug inhibited translocation, but erythromycin and certain other macrolides were later claimed to prevent the dissociation, from ribosomes, of peptidyl-tRNA, possibly during translocation of the latter from the ribosomal A to the ribosomal P site. Very recently, Vester and Garrett[48] have concluded that erythromycin acts immediately after initiation of protein synthesis. The drug destabilizes 70S ribosomes that have bound to messenger RNA and prevents them from recycling by causing degradation of 50S ribosomal subunits. The implication is that these subunits undergo an erythromycin-induced conformational change. Andersson and Kurland[49] have presented evidence that erythromycin inhibits at, or immediately after, initiation of protein synthesis and claim that the step affected is the transition from the initiation mode to the elongation mode.

A particularly interesting feature of erythromycin is that it competes efficiently with the antibiotic chloramphenicol for the latter's ribosomal receptor site. Chloramphenicol is an inhibitor of the peptidyl transferase centre on ribosomes but erythromycin has no direct effect on the reaction catalyzed by this centre. Despite this striking difference in action, the fact that the two drugs bind to ribosomes in mutually exclusive fashion indicates that the ribosomal receptor sites for the two drugs are closely overlapping and the possibility is even raised that the drugs may share the same receptor site. In either case, binding of one drug sterically excludes the other and in the latter model interaction between drug and ribosome might bestow upon the (single) receptor site a conformation unique for the interaction of each inhibitor that subsequently controls the expression of a function associated with one of two separate, but possibly closely adjacent, ribosomal domains.

10.5.7.2 The Ribosomal Receptor Sites for Erythromycin and Chloramphenicol and the Exploitation of Point Mutations within Ribosomal RNA

The precise characterization of the erythromycin receptor site on ribosomes has once again benefitted from the analysis of drug resistant bacterial strains. Certain *E. coli* mutants resistant to erythromycin have alterations in ribosomal protein L4 and this alteration cotransduces in genetic crosses with erythromycin resistance. Other erythromycin resistant *E. coli* strains have an altered ribosomal protein L22 but this can be segregated in genetic crosses from the resistance phenotype and the protein may not, therefore, be directly involved in the resistance phenomenon. A third, but less well-characterized erythromycin resistance mutation has also been discovered that may affect the 30S ribosomal subunit. It may be, therefore, that the action of erythromycin is indeed more complex than considered earlier and that its action somehow involves the participation of both ribosomal particles.

Further information on the nature of the erythromycin receptor site has come from a study of MLS resistance. This phenomenon was first described in *Staphylococcus aureus* but it is, in fact, found in Gram-positive strains generally. Treatment of such cells with subinhibitory concentrations of erythromycin rapidly induces high level resistance not only to erythromycin and other macrolides but also to lincomycin and streptogramin B. The induced resistance involved dimethylation of an adenine residue located within 23S ribosomal RNA. The study of MLS resistance thus reveals interesting parallels to those mechanisms discussed above that are associated with resistance to kasugamycin and thiostrepton and these parallels have been extended by studies on the erythromycin-producing organism *Streptomyces erythreus*.

In 1979 Graham and Weisblum[50] had detected N^6,N^6-dimethyladenine in 23S ribosomal RNA from *S. erythreus* and more recently the methylase responsible for this modification has been purified 16 000-fold.[40] Like the thiostrepton resistance methylase of *S. azureus*, the *S. erythreus* methylase can

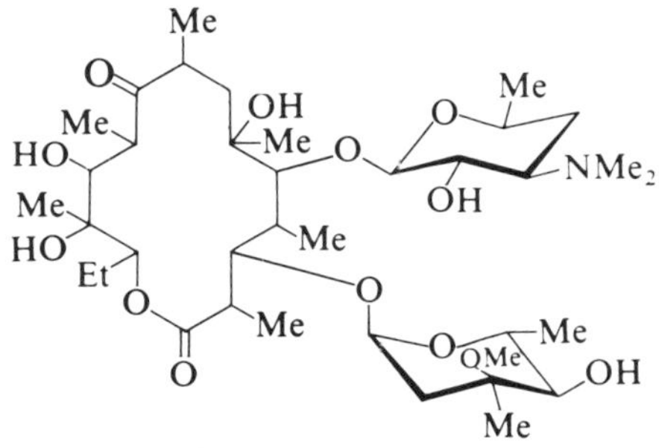

Erythromycin A

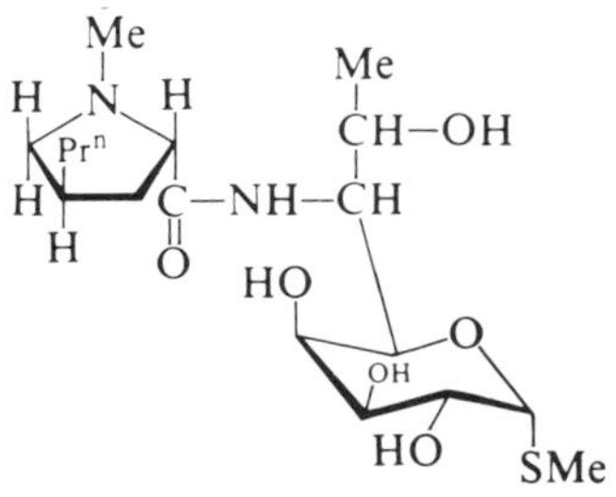

Lincomycin

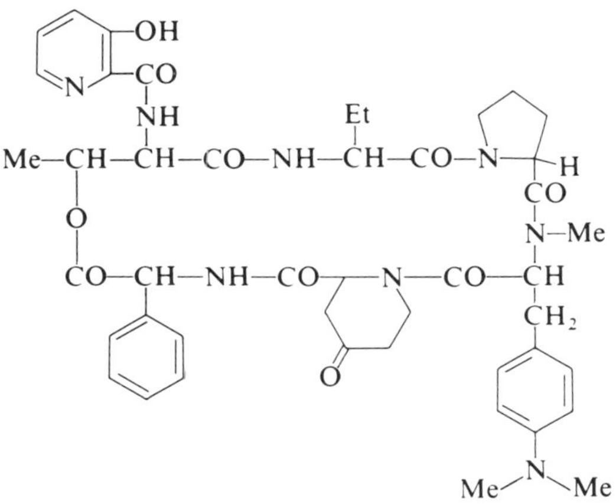

Streptogramin B

Figure 5   The structural formulae of the antibiotics erythromycin A, lincomycin and streptogramin B

modify 23S ribosomal RNA from a variety of organisms but fails to act on either intact ribosomes or ribosomal subunits. Ribosome reconstitution experiments carried out using components from *Bacillus stearothermophilus* have confirmed that methylation of 23S ribosomal RNA confers resistance to MLS antibiotics and pinpoints the dimethylation site as A_{2058} (see ref. 40). The importance of this residue is also indicated by studies involving mutagenesis of a ribosomal RNA operon (rrnH) which, after modification, can be incorporated into a multicopy plasmid and subsequently reintroduced into *E. coli*. Selection and analysis of erythromycin resistant transformants reveals the presence of a point mutation located at A_{2058} within the 23S ribosomal RNA sequence.[51] These studies are particularly interesting when one considers the chemical structures of the antibiotics erythromycin, lincomycin and streptogramin B (see Figure 5). Residue A_{2058} clearly has a central role to play in controlling the functional binding of not only these three drugs but also of other macrolides. It will be a difficult but fascinating challenge to elucidate how such structurally diverse compounds can be sterically accommodated within a single ribosomal domain. Indeed, the nature of this domain has already been probed further by exploiting the fact that ribosomal RNA genes within mitochondria are present in this organelle as single copies and can be easily subjected to point mutations. Such an approach has not been successful using cytoplasmic ribosomes. As the ribosomal RNA genes are present in multiple copies (seven in *E. coli*), isolation of discrete RNA mutations can not be achieved although, as considered earlier, mutations causing base modifications or their absence have been used to great effect in elucidating ribosomal structure–function relationships. Point mutations of mitochondrial ribosomal RNA genes have now produced resistance to several antibiotics including erythromycin, spiramycin and chloramphenicol (see ref. 40) and the nature of the mutation has, in all cases, been identified by sequence analysis. Because of the structural similarities between ribosomal RNA genes from different sources, it is possible to transpose, albeit tentatively, the changes identified in mitochondria on to the corresponding sequence within *E. coli*. Using this approach, erythromycin resistance is associated with changes at either A_{2058} or C_{2611} (yeast mitochondria) and chloramphenicol resistance is associated with changes at five locations — G_{2447} and A_{2503} (yeast mitochondria), A_{2451} and C_{2452} (mouse mitochondria) and U_{2504} (human and mouse mitochondria). Residue C_{2611} is particularly interesting since base changes at this site cause resistance to either erythromycin plus spiramycin (a macrolide antibiotic) or to spiramycin alone.

10.5.7.3 Structurally Complex Functional Domains that Interact within Ribosomes

Not only do the above data again indicate that erythromycin (and chloramphenicol) might bind directly to ribosomal RNA but they also go some way towards explaining why the two drugs compete for their respective receptors. All seven positions associated within 23S ribosomal RNA with erythromycin and chloramphenicol resistance are located within a discrete folded region that exists in a secondary structure model proposed for this *E. coli* macromolecule (see later and ref. 36). Furthermore, since chloramphenicol is such a highly selective and well characterized inhibitor of the ribosomal peptidyl transferase centre, it might be expected that the folded region embracing the five sites within ribosomal RNA that can each control resistance to the drug is itself an integral part of the enzymic centre.[40] For functional expression, the peptidyl transferase centre requires cooperative interactions between RNA and ribosomal proteins and evidence, from affinity labelling from the 3′ end of aminoacyl-tRNA or from studies using antibiotics known to be peptidyl transferase inhibitors, has identified at least eight ribosomal proteins in the centre's vicinity: L2, L11, L14, L15, L16, L18, L23 and L27.[39] Three of these — L2, L16 and L23 — have most consistently been claimed, using several experimental approaches, to be necessary for peptidyl transferase actvity.[52] In addition, certain of the proteins cited above have been variously linked with the action of chloramphenicol and hence its ribosomal receptor site.[1] Thus, protein L16 restores the ability of core particles to bind not only this drug but also erythromycin , and direct-affinity immune electron microscopy of ribosome–chloramphenicol crosslinked complexes suggests that L15, L18 and L27 are in the vicinity of the peptidyl transferase centre.

Further consideration of the ribosomal target for erythromycin makes the situation even more intriguing. The drug's receptor is in a position that, at the least, must be very close to the chloramphenicol binding site. Indeed, the drugs can reasonably be considered as interacting with the looped ribosomal domain described above, although, as discussed earlier, such a binding domain could include components that are associated with either of at least two functional domains within the particle. Ribosomal protein L16 is clearly involved in binding both erythromycin and chloramphenicol (see ref. 1) to ribosomes, and it is striking that ribosomal protein L15 has been claimed to bind erythromycin in free soluton,[53] since this protein is also strongly implicated in both the peptidyl transferase centre and the chloramphenicol binding site on ribosomes (see above). By carrying out experiments with radioactively labelled drug derivatives which can be covalently linked to ribosomal proteins, Tejedor and Ballesta[54] have concluded that a major target site for the macrolides carbomycin A, niddamycin and tylosin is ribosomal protein L27. In contrast to this finding, however, Tangy *et al.*[55] have shown that the macrolide rosaramycin can be crosslinked on ribosomes to proteins L1, L5, L6 and S1. The fact that certain erythromycin resistant bacterial strains have an altered ribosomal protein L4 indicates strongly that this protein too must be associated, either directly or indirectly, with the drug's receptor site. Protein L4 is a ribosomal core protein and, along with other such proteins, is thought to be important in allowing 23S ribosomal RNA to acquire, during ribosome assembly, a conformation that is closely similar to that found in the mature ribosome. Ribosomal protein L4 is not significantly related to any of the others, although it does crosslink, albeit relatively weakly, to several of them, including L11, L14, L15, L16 and L18 (see ref. 4). As considered above, these proteins are regarded as being in the vicinity of the peptidyl transferase centre and it is likely, therefore, that L4 is on its periphery. In fact, Tejedor and Ballesta[54] consider that chloramphenicol and certain macrolides (including erythromycin) select closely adjacent ribosomal receptor sites with the former drug and the latter drugs binding, respectively, within the A and P site areas of the peptidyl transferase centre. If indeed L27 forms an important part of the centre's P site structure and is the principal ribosomal protein that crosslinks with macrolides, one would expect L4 to be in its vicinity. Evidence for this is not completely convincing, but in the macrolide binding site envisaged by Tejedor and Ballesta,[54] L27 is adjacent to L16 and the latter can crosslink weakly with L4. Other possible indirect interactions may well be the order of the day. Although the position is hardly crystal clear it is, nevertheless, apparent that certain ribosomal proteins implicated in peptidyl transferase activity can associate with both erythromycin and chloramphenicol, although it must always be remembered that erythromycin is not an inhibitor of the peptide bond-forming step in protein synthesis.

As considered earlier, the ribosomal domain that binds erythromycin might include components of at least two separate functional domains and evidence for this interpretation is, in fact, available.

Ribosomal RNA can itself be divided into a number of structural domains.[56] The mutations in 23S ribosomal RNA that confer resistance to chloramphenicol are located in the central loop of domain V. Barta *et al.*[57] have proposed that peptidyl transferase activity requires interaction of this domain (V) with domain II. This model is supported by the fact that UV-induced crosslinks can be formed between, respectively, the 575 and 745 regions of domain II and the 2030 and 2615 regions of domain V[58] and even more direct evidence has recently been forthcoming.[59] Using *in vitro* mutagenesis, resistance to erythromycin was found to be conferred by a change in a 500 base pair fragment of the gene corresponding to domain II in 23S ribosomal RNA. The mutation results in the deletion of 12 nucleotides (positions 1219–1230) and this results in the shortening of a base-paired stem in the secondary structure model. This work indicates clearly that erythromycin resistance can be associated with two distal regions of 23S ribosomal RNA and other, perhaps more subtle, interactions cannot be excluded at this stage. The general situation is consistent with the idea that two drugs with different inhibitory actions, in this case chloramphenicol and erythromycin, can bind within a specific structural domain in ribosomes that can itself interact with one or more additional structural domains, thereby creating more than one functional domain that can be differentially perturbed by one or other antibiotic. Further elucidation of such exquisite interactions within the ribosome has recently been forthcoming and has involved the technique of chemical footprinting.

10.5.7.4 The Use of Chemical Footprinting in Locating Binding Sites for Antibiotics on Ribosomal RNA

The technique of chemical footprinting has very recently been used with considerable success to confirm and extend some of the data so far considered. By using suitable chemical probes, such as dimethyl sulfate and kethoxal, it is possible to monitor interactions of ligands with RNA and study conformational changes. If antibiotics do indeed bind to specific bases within the RNA sequence, such interactions should interfere with the binding of the relevant chemical probe. Thus, if erythromycin and chloramphenicol associate with the peptidyl transferase centre, these drugs should cause the appearance of 'footprints' in the highly conserved regions of ribosomal RNA when the latter, in combination with the drugs, is exposed to the chemical probe. By applying this basic approach Moazed and Noller[60] have now studied the sites on 23S ribosomal RNA that are protected, against chemical modification, by the antibiotics chloramphenicol, erythromycin, carbomycin (a macrolide antibiotic) and the streptogramin antibiotic vernamycin B. The drugs all protect overlapping nonequivalent sites in the central loop of domain V within the RNA. The protected sites, along with those sites implicated in erythromycin resistance and chloramphenicol resistance using mutational analyses, are illustrated in Figure 6. This figure is a representation of the central

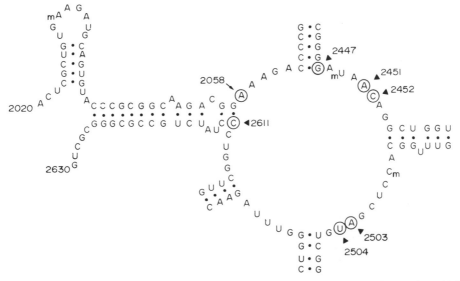

Figure 6 A portion of *E. coli* 23S ribosomal RNA that is postulated to be present in the ribosomal peptidyl transferase centre. The arrow indicates the (circled) site (residue 2058) where dimethylation confers resistance to MLS antibiotics. Other circled residues represent the sites that after mutation cause resistance to certain antibiotics and/or are protected by certain antibiotics against chemical modification. Details are provided in the text (reproduced from ref. 40 by permission of Springer)

loop of the ribosomal RNA domain V. Carbomycin, like chloramphenicol, is an inhibitor of peptidyl transferase activity and Moazed and Noller[60] have concluded from their data that these drugs inhibit the centre by binding strongly to residue A_{2451}. This site is regarded, therefore, as a central component of the ribosomal RNA region controlling the centre's activity (see also ref. 40). Erythromycin and vernamycin B are considered to interact within the ribosomal RNA region that has A_{2058} as its central component. This region is possibly associated with the translocation step of protein synthesis. It is of interest too that vernamycin B also gives strong protection of residue A_{752}. This residue is located within the ribosomal RNA domain II and this result with vernamycin B again indicates that domains II and V are closely adjacent within the ribosome and are functionally interdependent. The data considered show conclusively how a combination of mutational studies and chemical probing can provide vital information concerning ribosomal structure–function relationships and again pinpoint elegantly the fundamental importance of ribosomal RNA in controlling ribosomal functions. One of these functions is the correct binding of aminoacyl-tRNA to the ribosomal A site during protein synthesis and certain antibiotics will now be considered that have elucidated how this selection is controlled.

10.5.8 THE AMINOGLYCOSIDE ANTIBIOTICS AND THE RIBOSOMAL DECODING SITE

Many of the aminoglycoside antibiotics have been particularly well characterized as specific inhibitors of protein synthesis in bacteria, although others inhibit eukaryotes, and it is this group of antibiotics that will be the ones mainly under discussion in this final section of the chapter. Work with aminoglycosides has once again implicated ribosomal RNA as a major functional structure within ribosomes. In addition, the compounds have aided the further elucidation of the mechanisms that underlie translational accuracy and have cast more light on the ribosomal decoding site. Certain aminoglycosides have also been used recently in chemical footprinting experiments and such studies have confirmed the power of this approach for assigning functional roles to specific ribosomal RNA domains. The text will concentrate initially, albeit superficially, on the mode of action of the aminoglycoside streptomycin and the role of ribosomal proteins in determining the bacterial response to the antibiotic.

10.5.8.1 Mode of Action of Streptomycin and Mechanisms that Underlie Resistance to Aminoglycoside Antibiotics

Streptomycin exerts a range of effects in bacteria both *in vivo* and *in vitro* and two actions of the drug have been particularly well characterized: an induction of miscoding during protein synthesis and an inhibition of the initiation step. Both effects have been well documented elsewhere.[1] Misreading results from the drug's interference with the proof-reading step in translation and studies on the ribosomal receptor site for streptomycin were thought likely to provide important information on this central event of protein synthesis. Bacteria can be sensitive to, resistant to or dependent upon streptomycin. High level resistance to the drug is acquired by a single-step mutation that affects ribosomal protein S12 but this mutation does not induce resistance to aminoglycosides of the neomycin, gentamicin and kasugamycin families, although these compounds all cause miscoding during protein synthesis. The fact that streptomycin resistance is associated with a mutation in S12 suggests that this ribosomal protein is linked with the control of translational fidelity. Indeed, certain amino acid substitutions in S12 can reduce ribosomal error levels *per se*. In contrast, certain mutants (Ram mutants) with alterations in ribosomal protein S4 show elevated levels of miscoding. Such Ram mutations can suppress streptomycin dependence, which itself is associated with a mutated form of ribosomal protein S12.

All the above data demonstrate convincingly that certain ribosomal proteins can modulate translational accuracy. This relatively early work with streptomycin indicated strongly that ribosomal protein S12 was an extremely important target site for the antibiotic. However, as has been shown for certain other antibiotics considered earlier (Section 10.5.6.1), this interpretation is not strictly correct. Although 30S ribosomal subunits resistant to streptomycin failed to bind the antibiotic, neither did ribosomal protein S12 isolated from streptomycin sensitive strains. Furthermore, although S12 deficient particles were poorly active in protein synthesis *in vitro*, some activity was, nevertheless, detectable. Indeed, these protein deficient particles were extremely active in synthesizing polyphenylalanine under the direction of polyuridylic acid. Although later work implicated other ribosomal proteins, including S3, S4, S5, S7, S9, S10, S14, S16 and S17, as forming part of a complex streptomycin-binding site, it was work with drug-producing organisms that once again was to prove crucial in clarifying the situation. In discussing the current position, the different aminoglycoside antibiotic families will be considered, since a more comprehensive coverage of these drugs is necessary to allow a full appreciation of their contribution towards furthering our knowledge of ribosomal structure–function relationships. The reader is again directed to Gale *et al.*[1] for references relating to the above sections and to the following section.

10.5.8.2   Resistance to Aminoglycoside Antibiotics in Drug-producing Organisms

Organisms that elaborate aminoglycoside antibiotics produce enzymes that have closely similar actions to those present in clinical isolates of those bacteria that display high level resistance to the drugs. In addition, however, certain of the drug-producing strains also have drug resistant ribosomes and this latter resistance phenomenon is mediated by the state of methylation of 16S ribosomal RNA. Examples of such strains are *Streptomyces tenjimariensis*, *Micromonospora purpurea*, *Streptomyces tenebrarius* and *Streptomyces kanamyceticus* which produce, respectively, istamycin, gentamicin, the nebramycin complex and kanamycin. These strains are resistant to the drugs they elaborate and each strain is tolerant of other aminoglycoside antibiotics. Nevertheless, their resistance patterns are not identical and the differences shown are maintained *in vitro*. Furthermore, as indicated earlier, although streptomycin resistant bacterial strains are usually coresistant to other members of the streptomycin group of aminoglycoside antibiotics, they are not coresistant to neomycin, gentamicin or kanamycin. Different ribosomal target sites are thus indicated and, indeed, aminoglycosides of the last three groups bind to ribosomes at multiple sites, whereas streptomycin has a single receptor site only (see ref. 40).

The molecular basis underlying drug resistance in some of the strains that produce aminoglycosides have now been clarified. Methylation of specific residues in 16S ribosomal RNA induces resistance to certain combinations of aminoglycosides. Thus, a methylase in *Micromonospora purpurea* modifies residue G_{1405} (to 7-methylguanosine) to give ribosomes that are resistant to both gentamicin and kanamycin and a further methylase activity in *Streptomyces tenjimariensis* modifies residue A_{1408} (to 1-methyladenosine) to give ribosomes resistant to both kanamycin and apramycin.[61] Streptomycin resistance can also be associated with other covalent changes in RNA. Thus, streptomycin resistant ribosomes have been characterized in mutated chloroplasts from *Euglena* where a C to U transition was detected in a position corresponding to residue 912 of *E. coli* 16S ribosomal RNA.[62] Further mutations in ribosomal RNA have been described that are associated with resistance to other aminoglycoside antibiotics that can inhibit eukaryotic protein synthesis. Thus, paromomycin resistance results from a single base change in either 15S ribosomal RNA from yeast mitochondria[63] or 18S ribosomal RNA from *Tetrahymena*.[64] The corresponding positions in *E. coli* 16S ribosomal RNA are, respectively, at residues 1409 and 1491 with a C to G change at the former and a G to A change at the latter. In addition, hygromycin B resistance can be generated in *Tetrahymena*[64] by a change of U to C at residue 1495, again using the *E. coli* numbering system.

10.5.8.3   The Use of Chemical Footprinting for the Identification of Ribosomal RNA Receptor Sites for Aminoglycoside Antibiotics

The results described above have, in the main, been obtained by analysis of drug resistant mutants and indicate yet again that ribosomal RNA can provide target sites for the interaction of a wide range of antibiotics. Very recently, the technique of chemical footprinting (see above) has been employed to confirm and extend some of the data so far considered. Using this experimental approach, streptomycin can be shown to bind to 70S ribosomes where it strongly protects the adenosine residues 913- 915 of 16S ribosomal RNA from chemical attack.[65] The drug also provides weak, but significant, protection of U_{911} and C_{912}. These results are consistent with the results from mutational analyses which, as indicated earlier, pinpoint C_{912} as a streptomycin target site in 16S ribosomal RNA. Similar analyses show that hygromycin B strongly protects G_{1494} (target site claimed at U_{1495})[64] but actually increases the chemical reactivity of A_{1408}. The compounds neomycin, paromomycin, gentamicin and kanamycin cause miscoding during protein synthesis and, along with many other aminoglycoside antibiotics, they also inhibit translocation (see ref. 65). The four antibiotics protect *E. coli* 16S ribosomal RNA from chemical attack at sites A_{1408} and G_{1494} and increase the chemical reactivity of C_{525}. Additional weak effects are observed elsewhere in the macromolecule. As stressed by Noller *et al.*,[36] C_{525} is in a looped region of the ribosomal RNA that is protected by the messenger RNA dependent binding of tRNA, although this region, located at a considerable distance from the decoding site, is likely to be brought into the picture only as a result of a conformational change(s) elsewhere. In contrast, A_{1408} and G_{1494} are closely adjacent to the C_{1409}–G_{1491} base pair proposed to be present in the decoding domain of 16S ribosomal RNA. As stated earlier, the latter two positions are associated with paromomycin resistance where disruption of the base pair must occur. Predictably, a tRNA protection site including A_{1492} and A_{1493} is also located in this area. A_{1493} and G_{1494} are of particular interest since this dinucleotide is the one specifically cleaved, within ribosomes, by colicin E3 (see Section 10.5.4.1). Cleavage is prevented by gentamicin and aminoacyl-tRNA and also, more surprisingly, by streptomycin, since resistance to the latter is associated with residues 913–915 in chemical protection studies and with C_{912} in mutational studies. However, as pointed out by Moazed and Noller[65] (and see refs. contained therein) these observations might indicate that the 900 and 1490 regions within 16S ribosomal RNA are proximal in the ribosome, thus allowing streptomycin, which also weakly protects G_{1494} from chemical attack, to influence the decoding site. Such an interpretation is supported by studies on the antibiotic tetracycline which, in addition to giving strong protection of A_{892}, prevents, like streptomycin, colicin E3 cleavage reactions on ribosomes. Tetracycline is a potent inhibitor of the binding of aminoacyl-tRNA to the ribosomal A site (see Gale *et al.*,[1] for refs.). Certainly, the 900 region seems likely to occupy a key position within ribosomes. Thus, C_{912} lies approximately midway between the binding site for ribosomal protein S4 (the 500 region)—a domain likely to be involved in proof-reading—and the decoding site itself that is located in the 1400 region.[36]

Consideration of the aminocyclitol antibiotic spectinomycin makes the situation even more intriguing, although hardly clearer. This compound does not cause misreading or phenotypic suppression but seems likely to inhibit translocation on 70S ribosomes.[1] Although resistance to the drug can be associated with mutations in ribosomal protein S5,[66] resistance is also conferred by a C to U transition at position 1192 in 16S ribosomal RNA.[51] More recently,[67] C_{1192} has been mutated to A, G or U and although the mutants are resistant to different levels of spectinomycin, their growth rates are unaffected in the absence of the drug. Remarkably, however, a change of C_{1192} to G_{1192} makes the isolated strain resistant to 40 mg mL^{-1} of spectinomycin. The drug protects the 16S ribosomal RNA residues C_{1063} and G_{1064} from chemical modification by dimethyl sulfate and G_{1064} pairs with C_{1192} in a proposed secondary structure model for the RNA.[65] However, despite this interesting relationship there are no indications to date as to how this ribosomal domain functions in translocation.

10.5.8.4 The Ribosomal Decoding Site

Although these recent data, discussed above, on mutations within RNA excitingly demonstrate a key role for this macromolecule in forming, at least in part, antibiotic receptor sites on ribosomes a mass of structure–function relationships remain to be elucidated. Although speculation is at the forefront, there are, nevertheless, some fascinating theories and a recent one put forward by Noller and his colleagues[36] deserves special mention. These authors consider the nature of the ribosomal decoding site by piecing together experimental evidence, much of which has been outlined above, and combining these hard facts with impressive logic. Their model is outlined very superficially here but the reader is referred to their original article[36] to allow a more detailed and critical consideration of the ideas expressed. In keeping with present day trends, the primitive ribosome is regarded as a RNA machine. This not unreasonable idea is given credence by the recent discovery of enzymic reactions that can be catalyzed by RNA alone[68,69] and by the highly conserved nature of ribosomal RNA sequences. Indeed, the crucial importance not only of such sequences but also of specific bases within the RNA in forming exquisite antibiotic receptor sites should already be apparent to the reader of this chapter. The selection of the correct aminoacyl-tRNA is, in many ways, the most critical event in protein synthesis. The evidence is now very strong that the universally conserved sequences contained within 16S ribosomal RNA and located around positions 1400 and 1500 are intimately concerned with codon–anticodon interactions, with position 1400 seeming likely to be of particular importance. A (central) unanswered question asks how triplet–triplet interactions, which are relatively unstable *per se*, are, nevertheless, stabilized at the ribosomal decoding site. An important role for ribosomal proteins could well be indicated but the primitive ribosome is assumed to contain no protein components. Noller *et al.*[36] propose that induced coaxial base stacking could provide the answer. Stacking of a two or three base-pair auxiliary helix coaxially on to a cognate codon–anticodon complex would increase the stability of the latter interaction. It is suggested that the auxiliary helix is provided by the 1400/1500 regions of 16S ribosomal RNA. This intriguing model places residue C_{1400} directly adjacent to the wobble base within the associated transfer RNA. It will indeed be interesting to see how and where ribosomal proteins fit in when considering the modern-day sophisticated ribosome. Furthermore, it is relevant to speculate as to how error free the decoding reaction was on the very first ribosome that came into existence.

10.5.9   RIBOSOMAL RNA — THE FUNDAMENTAL DETERMINANT OF RIBOSOMAL FUNCTIONS?

Certain technological advances that have been perfected in recent years now make it easier, in many ways, to work with ribosomal RNAs rather than with ribosomal proteins. Of particular importance has been the refinement of sequencing techniques, since their application to the various ribosomal RNAs has allowed the construction of secondary structure models for these macromolecules. The fact that base sequences have been highly conserved within the ribosomal RNAs suggests, in itself, that the RNA components of the ribosome play a crucial role in at least some of the reactions of protein synthesis. More recent data outlined in the present chapter confirm this interpretation by indicating that many of the antibiotics that inhibit protein synthesis and have the ribosome as their target site appear to bind directly to a specific location on one or other of the ribosomal RNAs that are contained within the particle. Recent and significant progress has also been made in understanding how ribosomal proteins fit spatially into the ribosome's structure. As considered earlier, neutron scattering has been a particularly fruitful technique in this respect and results obtained using this experimental approach have allowed a model to be constructed that shows the positions, within the 30S ribosomal subunit of *E. coli*, occupied by all the proteins contained therein.[70,71] A proposed secondary structure for 16S ribosomal RNA is illustrated in Figure 7. The corresponding structures for 23S ribosomal RNA and the eukaryotic ribosomal RNAs are not reproduced here but can be viewed in Noller *et al.*,[36] and in Wool[22] (and see refs. contained therein).

If indeed ribosomal RNA has a truly functional role(s) to play within ribosomes, it must have catalytic activity. Such a concept has only become conceivable in recent years. Previously, it was difficult for many to consider the ribosome as a RNA enzyme stabilized by proteins (see Moore[72]). Two experimental breakthroughs were crucial in changing the intellectual climate. Thus, in 1982 Kruger *et al.*[68] showed that an intervening sequence contained within the initial ribosomal RNA transcript of *Tetrahymena* was excised in a self-catalyzed transesterification reaction. A short time later, Guerrier-Tedaki *et al.*[69] identified the RNA portion of the enzyme ribonuclease P as having nucleolytic activity. There seems a strong possibility too that the small nuclear ribonucleoprotein particles (snRNPs) that control messenger RNA splicing in eukaryotes may prove to have this important activity catalyzed by the RNA component. Therefore, although more evidence is required to prove the point conclusively, the potential is clearly there for ribosomal RNA to catalyze specific ribosomal reactions such as peptide bond formation, translocation, GTPase activity and release of completed polypeptide chains. Translocation is, perhaps, a prime candidate for occurring as a result of a conformational switch within the ribosome, a concept discussed briefly earlier in the chapter and considered recently by Moore.[72]

It is important to consider how the various ribosomal RNA domains acquire the conformational states that are needed to allow these macromolecules to carry out their proposed catalytic roles and to speculate concerning the extent to which such states can exist in the absence of ribosomal proteins. It is apparent that a large number of very structurally diverse compounds, only a relatively small selection of which have been considered in the present text (see Figure 8 for examples), can interact with ribosomes in an exquisitely specific manner. Such considerations imply that the number of conformational variations that result in the generation of antibiotic receptor sites on ribosomes and whose formation is influenced directly by ribosomal proteins must be greatly in excess of the number of conformational states that can be taken up by RNA alone. The converse situation can, nevertheless, apply also. Thus, the number of sites that can potentially be recognized on ribosomal RNA by an antibiotic can be reduced as a result of RNA–protein interactions. This point is well illustrated by the action of α-sarcin which, as discussed earlier, has a truly remarkable specificity for intact ribosomes within which it cleaves a single bond but loses this absolute specificity when it reacts with isolated ribosomal RNA. The proteins are, therefore, crucial in allowing the RNA to be packaged within the ribosome in such a way as to present only one site for the action of α-sarcin that is accessible to the nuclease in the single-stranded RNA loop that is positioned on the surface of the particle. Ribosomal proteins must also be of importance in bringing together, within

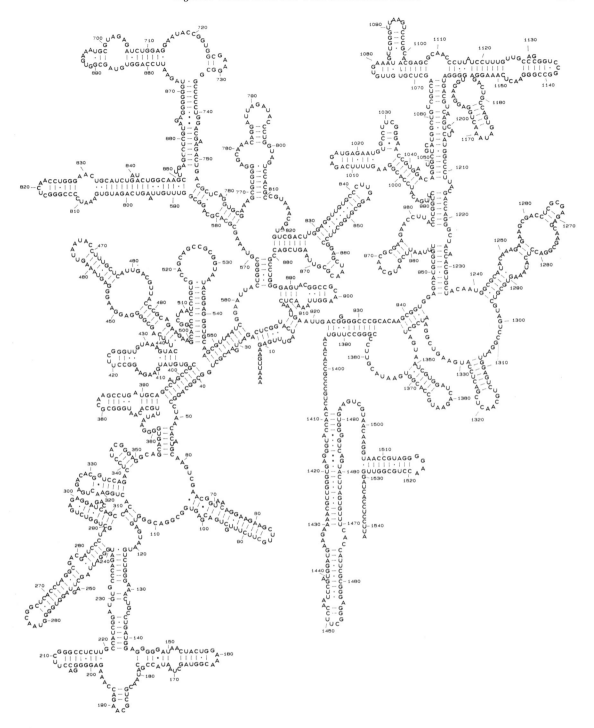

Figure 7 The proposed secondary structure of *E. coli* 16S ribosomal RNA (reproduced from ref. 36 by permission of Springer)

the ribosome, otherwise distal areas of ribosomal RNA. This particular phenomenon is typified by the interaction, discussed earlier, that occurs between domains II and V of 23S ribosomal RNA. These two domains must initially assume the correct conformational states and then be brought closely adjacent so that functional organization of the ribosomal peptidyl transferase centre can be completed. Indeed, although certain ribosomal core particles that lack several ribosomal proteins can still retain peptidyl transferase activity, there comes a point when removal of a further ribosomal protein abolishes the activity completely. Cooperative interactions between protein and RNA are clearly, therefore, of central importance to ribosomal functions and also to the creation of antibiotic

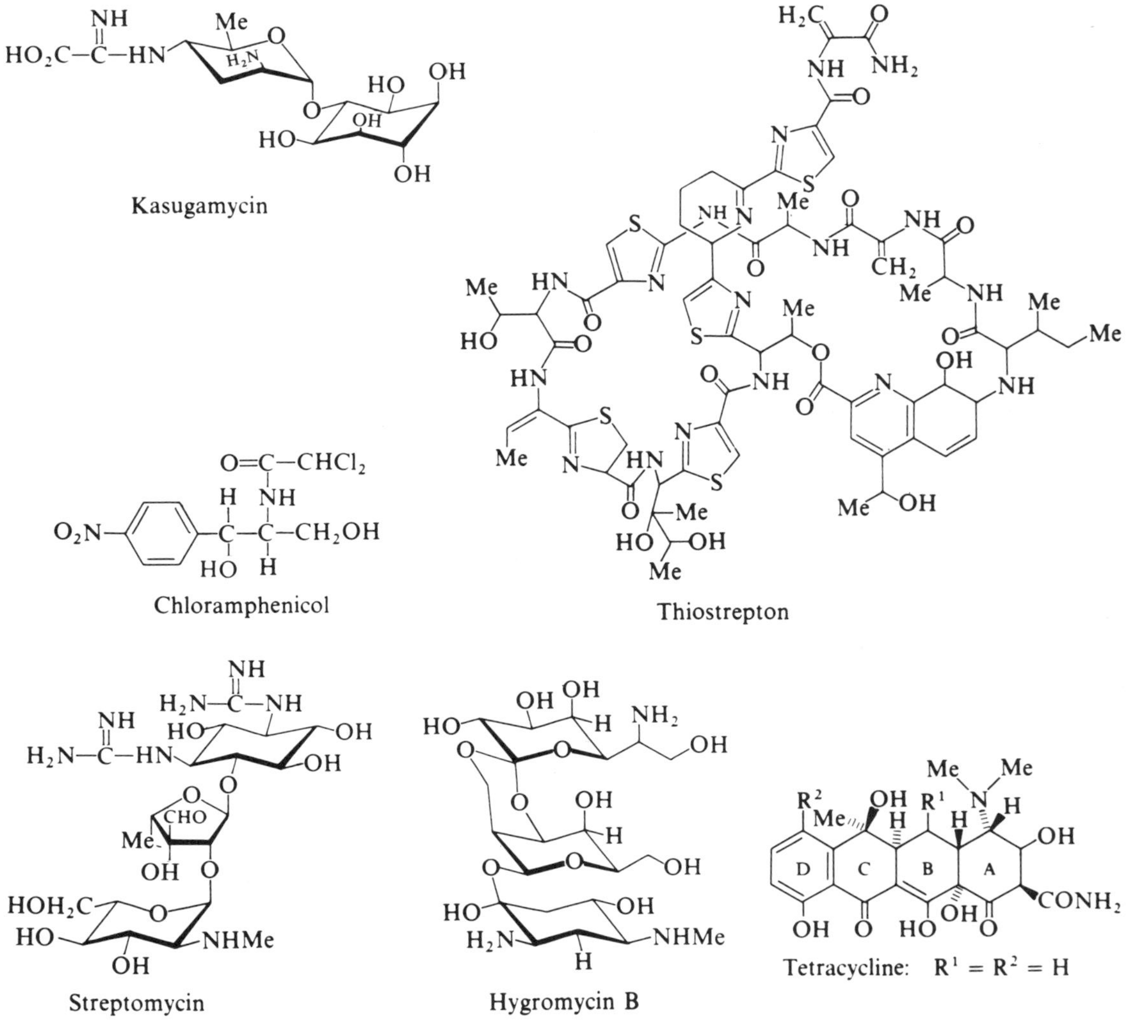

Figure 8 The structural formulae of selected antibiotics whose modes of inhibitory action and ribosomal RNA receptor sites have been considered in detail in the text

receptor sites within the particle. An excellent example of such cooperativity with respect to the latter phenomenon is exemplified by the studies on the ribosomal receptor site for thiostrepton considered earlier in the text. Although the antibiotic can bind weakly to 23S ribosomal RNA, high affinity binding of the drug to this macromolecule within the ribosome requires the presence of ribosomal protein L11 to ensure the formation and/or stabilization of the RNA conformational state that maximizes drug interaction.

The above considerations stress the roles of ribosomal proteins and support the premise that within a macromolecular assembly as complex as the ribosome, single components are unlikely to act in isolation and independently of each of the others. Clearly, however, there is now ample proof to show that sequences within the ribosomal RNAs themselves have crucial functional roles in protein synthesis. Indeed, single bases have now been implicated directly in some of these functions and in forming antibiotic receptor sites. Furthermore, not only point mutations within the ribosomal RNAs alter the above properties but also chemical modification at key points within the structure can control, often more subtly, the functional roles of the macromolecules. Particularly striking in this respect is the interaction that takes place between 16S ribosomal RNA and kasugamycin. Although ribosomal proteins will influence the position that is taken up within the ribosome by the 3' terminal domain of 16S ribosomal RNA, fine control of the latter's functions and its ability to contribute to the formation of the kasugamycin receptor site is exquisitely controlled by methylation of two bases central to the domain. It is becoming easier to accept the view that ribosomal RNA is indeed the fundamental determinant of both the structure and function of the ribosome.

10.5.10 REFERENCES

1. E. F. Gale, E. Cundliffe, P. E. Reynolds, M. H. Richmond and M. J. Waring, in 'The Molecular Basis of Antibiotic Action', 2nd edn., Wiley, London, 1981.
2. I. G. Wool, *Annu. Rev. Biochem.*, 1979, **48**, 719.
3. H. G. Wittmann, in 'Structure, Function, and Genetics of Ribosomes', ed. B. Hardesty and G. Kramer, Springer, New York, 1986, p. 1.
4. M. Oakes, E. Henderson, A. Scheinman, M. Clark and J. A. Lake, in 'Structure, Function, and Genetics of Ribosomes', ed. B. Hardesty and G. Kramer, Springer, New York, 1986, p. 47.
5. K. Moldave, *Annu. Rev. Biochem.*, 1985, **54**, 1109.
6. L. Stryer, 'Biochemistry', 2nd edn., Freeman, San Francisco, 1981.
7. C. M. Bowman, J. E. Dahlberg, T. Ikemura, J. Konisky and M. Nomura, *Proc. Natl. Acad. Sci. USA*, 1971, **68**, 964.
8. R. Brimacombe, J. Atmadja, A. Kyriatsoulis and W. Stiege, in 'Structure, Function, and Genetics of Ribosomes', ed. B. Hardesty and G. Kramer, Springer, New York, 1986, p. 184.
9. M. Laughrea and P. B. Moore, *J. Mol. Biol.*, 1978, **121**, 411.
10. M. Nomura, *Cold Spring Harbor Symp. Quant. Biol.*, 1987, **52**, 653.
11. J. Shine and L. Dalgarno, *Proc. Natl. Acad. Sci. USA*, 1974, **71**, 1342.
12. J. A. Steitz and K. Jakes, *Proc. Natl. Acad. Sci. USA*, 1975, **72**, 4734.
13. G. D. Stormo, T. D. Schneider and L. M. Gold, *Nucleic Acids Res.*, 1982, **10**, 2971.
14. C. O. Gualerzi, C. L. Pon, R. T. Pawlik, M. A. Canonaco, M. Paci and W. Wintermeyer, in 'Structure, Function, and Genetics of Ribosomes', ed. B. Hardesty and G. Kramer, Springer, New York, 1986, p. 621.
15. G. Sacco, K. Drinkamer and I. G. Wool, *J. Biol. Chem.*, 1983, **258**, 5811.
16. D. G. Schindler and J. E. Davies, *Nucleic Acids Res.*, 1977, **4**, 1097.
17. Y. Endo and I. G. Wool, *J. Biol. Chem.*, 1982, **257**, 9054.
18. Y. Endo, P. W. Huber and I. G. Wool, *J. Biol. Chem.*, 1983, **258**, 2662.
19. C. Fernandez-Puentes and D. Vazquez, *FEBS Lett.*, 1977, **78**, 143.
20. A. N. Hobden and E. Cundliffe, *Biochem. J.*, 1978, **170**, 57.
21. Y. L. Chan, J. Olvera and I. G. Wool, *Nucleic Acids Res.*, 1983, **11**, 7819.
22. I. G. Wool, in 'Structure, Function, and Genetics of Ribosomes', ed. B. Hardesty and G. Kramer, Springer, New York, 1986, p. 391.
23. C. Glotz and R. Brimacombe, *Nucleic Acids Res.*, 1980, **8**, 2377.
24. H. F. Noller and C. R. Woese, *Science (Washington D.C.)*, 1981, **212**, 403.
25. L. Barbieri and F. Stirpe, *Cancer Surveys*, 1982, **1**, 489.
26. A. Jimenez and D. Vazquez, *Annu. Rev. Microbiol.*, 1985, **39**, 649.
27. S. Olsnes and A. Pihl, in 'The Specificity and Action of Animal, Bacterial and Plant Toxins (Receptors and Recognition Series B)', ed. P. Cuatrecasas, Chapman and Hall, London, 1976, vol. 1, p. 129.
28. R. C. Hughes, *Nature (London)*, 1979, **281**, 526.
29. G. Griffiths, A. Leith and M. Green, *New Sci.*, 1987, **115**, 59.
30. E. S. Vitetta and J. W. Uhr, *Cell*, 1985, **41**, 653.
31. Y. Endo, K. Mitsui, M. Motizuki and K. Tsurugi, *J. Biol. Chem.*, 1987, **262**, 5908.
32. Y. Endo and K. Tsurugi, *J. Biol. Chem.*, 1987, **262**, 8128.
33. R. Amils and J. L. Sanz, in ' Structure, Function, and Genetics of Ribosomes', ed. B. Hardesty and G. Kramer, Springer, New York, 1986, p. 605.
34. H. F. Noller and P. H. Van Knippenberg, in 'Horizons in Biochemistry and Biophysics', ed. A. M. Kroon, Wiley, New York, 1983, vol. 7, p. 71.
35. P. H. Van Knippenberg, J. M. A. Van Kimmenade and H. A. Heus, *Nucleic Acids Res.*, 1984, **12**, 2595.
36. H. F. Noller, M. Asire, A. Barta, S. Douthwaite, T. Goldstein, R. R. Gutell, D. Moazed, J. Normanly, J. B. Prince, S. Stern, K. Triman, S. Turner, B. Van Stolk, V. Wheaton, B. Weiser and C. R. Woese, in 'Structure, Function, and Genetics of Ribosomes', ed. B. Hardesty and G. Kramer, Springer, New York, 1986, p. 143.
37. P. H. Van Knippenberg, in 'Structure, Function, and Genetics of Ribosomes', ed. B. Hardesty and G. Kramer, Springer, New York, 1986, p. 412.
38. C. P. J. J. Van Buul and P. H. Van Knippenberg, *Gene*, 1985, **38**, 65.
39. J. Ofengand, J. Ciesiolka, R. Denman and K. Nurse, 'Structure, Function, and Genetics of Ribosomes', ed. B. Hardesty and G. Kramer, Springer, New York, 1986, p. 473.
40. E. Cundliffe, in 'Structure, Function, and Genetics of Ribosomes', ed. B. Hardesty and G. Kramer, Springer, New York, 1986, p. 586.
41. J. Thompson, E. Cundliffe and M. Stark, *Eur. J. Biochem.*, 1979, **98**, 261.
42. E. Cundliffe, P. Dixon, M. Stark, G. Stöffler, R. Ehrlich, M. Stöffler-Meilicke and M. Cannon, *J. Mol. Biol.*, 1979, **132**, 235.
43. B. Wienen, R. Ehrlich, M. Stöffler-Meilicke, G. Stöffler, I. Smith, D. Weiss, R. Vince and S. Pestka, *J. Biol. Chem.*, 1979, **254**, 8031.
44. E. R. Dabbs, in 'Structure, Function, and Genetics of Ribosomes', ed. B. Hardesty and G. Kramer, Springer, New York, 1986, p. 733.
45. E. Cundliffe, *Biochimie*, 1987, **69**, 863.
46. H. Hummel and A. Böck, *Biochimie*, 1987, **69**, 857.
47. M. Cannon, D. Schindler and J. Davies, *FEBS Lett.*, 1977, **75**, 187.
48. B. Vester and R. A. Garrett, *Biochimie*, 1987, **69**, 891.
49. S. Andersson and C. G. Kurland, *Biochimie*, 1987, **69**, 901.
50. M. Y. Graham and B. Weisblum, *J. Bacteriol.*, 1979, **137**, 1464.
51. C. D. Sigmund, M. Ettayebi and E. A. Morgan, *Nucleic Acids Res.*, 1984, **12**, 4653.
52. R. R. Traut, D. S. Tewari, A. Sommer, G. R. Gavino, H. M. Olson and D. G. Glitz, in 'Structure, Function, and Genetics of Ribosomes', ed. B. Hardesty and G. Kramer, Springer, New York, 1986, p. 286.
53. H. Teraoka and K. H. Nierhaus, *J. Mol. Biol.*, 1978, **126**, 185.
54. F. Tejedor and J. P. G. Ballesta, *J. Antimicrob. Chemother.*, 1985, **16**, Suppl. A, 53.

55. F. Tangy, M. L. Capmau and F. Le Goffic, *Eur. J. Biochem.*, 1983, **131**, 581.
56. H. F. Noller, *Annu. Rev. Biochem.*, 1984, **53**, 119.
57. A. Barta, G. Steiner, J. Brosius, H. F. Noller and E. Kuechler, *Proc. Natl. Acad. Sci. USA*, 1984, **81**, 3607.
58. W. Stiege, C. Glotz and R. Brimacombe, *Nucleic Acids Res.*, 1983, **11**, 1687.
59. S. Douthwaite, J. B. Prince and H. F. Noller, *Proc. Natl. Acad. Sci. USA*, 1985, **82**, 8330.
60. D. Moazed and H. F. Noller, *Biochimie*, 1987, **69**, 879.
61. A. A. D. Beauclerk and E. Cundliffe, *J. Mol. Biol.*, 1987, **193**, 661.
62. P. E. Montandon, P. Nicolas, P. Schürmann and E. Stutz, *Nucleic Acids Res.*, 1985, **13**, 4299.
63. M. Li, A. Tzagloloff, K. Underbrink-Lyon and N. C. Martin, *J. Biol. Chem.*, 1982, **257**, 5921.
64. E. A. Spangler and E. H. Blackburn, *J. Biol. Chem.*, 1985, **260**, 6334.
65. D. Moazed and H. F. Noller, *Nature (London)*, 1987, **327**, 389.
66. A. Bollen, J. Davies, M. Ozaki and S. Mizushima, *Science (Washington, D.C.)*, 1969, **165**, 85.
67. P. C. Makosky and A. E. Dahlberg, *Biochimie*, 1987, **69**, 885.
68. K. Kruger, P. J. Grabowski, A. Zaug, J. Sands, D. E. Gottschling and T. R. Cech, *Cell*, 1982, **31**, 147.
69. C. Guerrier-Takada, K. Gardiner, T. Marsh, N. Pace and S. Altman, *Cell*, 1983, **35**, 849.
70. P. B. Moore, M. S. Capel, M. Kjeldgaard and D. M. Engelman, *Cold Spring Harbor Symp. Quant. Biol.*, 1987, **52**, 23.
71. M. S. Capel, D. M. Engelman, B. R. Freeborn, M. Kjeldgaard, J. A. Langer, V. Ramakrishnan, D. G. Schindler, D. K. Schneider, B. P. Schoenborn, I. Y. Sillers, S. Yabuki and P. B. Moore, *Science (Washington. D.C.)*, 1987, **238**, 1403.
72. P. B. Moore, *Nature (London)*, 1988, **331**, 223.

Subject Index

AA-861
  5-lipoxygenase
    inhibitor, 164
7-ACA — *see* Cephalosporanic acid, 7-amino-
ACE — *see* Angiotensin converting enzyme
Acetamide, allylisopropyl-
  cytochrome *P*-450
    inhibition, 136
Acetaminophen (paracetamol)
  cyclooxygenase
    inhibition, 161
Acetic acid, aminoxy-
  aminobutyrate transaminase
    inhibitor, 242
Acetic acid, diethylenetriaminepenta- (DTPA)
  iron decorporation, 183
  plutonium decorporation, 185
Acetic acid, ethylenediaminetetra- (EDTA)
  Gram-negative bacteria, 561
  lead decorporation, 185
Acetic acid, nitrilotri- (NTA)
  iron decorporation, 186
Acetidinone, 4-acetoxy-
  synthesis, 680
Acetoacetate decarboxylase
  inhibition
    β-diketones, 79
Acetohydroxamic acids
  asthma, 166
  lipoxygenase inhibitors, 166
    inflammation, 167
Acetohydroxamic acids, phenoxycinnamyl-
  lipoxygenase
    selective inhibitors, 166
Acetopyruvate
  inhibitor
    acetoacetate decarboxylase, 79
Acetylcholine
  hydrolysis
    acetylcholinesterase, 383
  receptors
    gastric acid secretion, 194
Acetylcholinesterase
  electric organs
    fish, 383
  inhibition, 381
  structure, 383
Acetylene
  cytochrome *P*-450
    inhibition, 350, 352
Acid thiol ligases
  nomenclature, 36
Aclacinomycin
  antitumor activity, 729, 780
Acridines
  antitumor agents, 716
  DNA binding agents, 767
  intercalating agents
    DNA, 704, 712
Acridines, 9-amino-
  dimers

DNA intercalating agents, 719
DNA topoisomerase II
  inhibition, 778
intercalating agent
  DNA, 709, 712
Acridines, 9-anilino-
  antitumor activity, 777
Actaplanin
  peptidoglycan polymerization
    inhibition, 574
Actinomadura R61
  carboxypeptidase, 615
Actinomycin
  antitumor agents, 781
  DNA binding agents, 767
  intercalating agent
    DNA, 704
Actinomycin D
  antitumor agents, 781
    DNA intercalating agents, 714
  molecular model
    DNA complex, 714
Actinomycins
  antitumor antibiotics
    DNA intercalating agents, 714
Aculeacins
  glucan synthase inhibitors, 587
Acycloguanosine
  antiviral agent, 320, 321
  monophosphate
    antiviral agent, 321
Acyclovir
  antiviral agent, 320
  DNA polymerases
    inhibitor, 761
Acyclovir, 8-amino-
  purine nucleoside phosphorylase
    inhibitor, 461
Acyclovir, carba-
  purine nucleoside phosphorylase
    inhibitor, 461
Acyclovir, 8-hydroxy-
  purine nucleoside phosphorylase
    inhibitor, 461
Acyclovir diphosphate
  purine nucleoside phosphorylase
    inhibitor, 462
Acylases
  inhibition, 102
Acylcholesterol acyltransferase
  cholesterol
    biosynthesis, 334
Acyl hydrolases
  arachidonic acid
    release, 149
AD41
  DNA topoisomerase II
    inhibition, 780
Adechlorin — *see* Pentostatin, 2′-chloro-
Adecypenol
  adenosine deaminase inhibitor, 450

Adenine, 9-β-D-arabinofuranosyl- (vidarabine)
  adenosine deaminase inhibitor, 449
  adenosyl-L-homocysteine hydrolase
    inhibitor, 455
  antiviral activity, 319, 763
  purine antagonist, 309
Adenine, 9-(2-chloro-6-fluorobenzyl)-
  antiparasitic agent, 325
Adenine, 9-(2,3-dihydroxypropyl)-
  antiviral agent, 321
Adenine, *erythro*-9-(2-hydroxy-3-nonyl)- (EHNA)
  adenosine deaminase inhibitor, 449
  organ transplantation, 474
Adenine, N^6-methyl-
  antiparasitic agent, 325
Adenine arabinoside
  DNA strand-break repair
    inhibition, 748
Adenine deaminase
  purine metabolism, 303
Adenine phosphoribosyltransferase
  nitrogen metabolism, 301
Adenosine
  adenosine deaminase inhibitor
    biochemistry, 451
Ara-Adenosine
  antiviral agent, 319, 320
  purine antagonist, 309, 310
Ara-Adenosine, 2-amino-
  antiviral agent, 320
Ara-Adenosine, 2′-amino-2′-deoxy-
  purine antagonist, 310
Ara-Adenosine, 2′-azido-
  antiviral agent, 320
Ara-Adenosine, 2′-azido-2′-deoxy-
  purine antagonist, 310
Adenosine, 2-bromo-2′-deoxy-
  purine antagonist, 310
Adenosine, 2-chloro-2′-deoxy-
  purine antagonist, 310
Adenosine, 3-deaza-
  antiviral agent, 321
  purine antagonist, 310
  S-adenosyl-L-homocysteine hydrolase
    inhibitor, 454
Adenosine, 9-deaza-
  purine antagonist, 310
Adenosine, deoxy-
  adenosine deaminase inhibitor
    biochemistry, 451
Adenosine, 3′-deoxy-
  antiparasitic agent, 324
Adenosine, 2′-deoxy-2-fluoro-
  purine antagonist, 310
Adenosine, 2′,3′-dideoxy-
  HIV replication
    inhibition, 760
Ara-Adenosine, 2-fluoro-
  antiviral agent, 320
Adenosine, 5′-isobutylthio-
  S-adenosyl-L-homocysteine hydrolase
    inhibitor, 456
Adenosine, 2′-C-methyl-
  antiviral agent, 320
Adenosine, 3′-C-methyl-
  antiviral agent, 320
Adenosine deaminase
  biochemistry, 448
  deficiency
    biochemistry, 448
    purine metabolism, 446
    symptoms, 447
  inhibitors, 449
    animal studies, 452

clinical studies, 452
    in vitro biochemical studies, 450
    in vitro immunological studies, 450
  malaria parasite, 475
  purine metabolism, 302, 447–453
Adenosine kinase
  purine metabolism, 301
Adenosine monophosphate
  cyclic
    adenine-modified derivatives, 537
    analogues, 536
    protein kinases, 535, 539
    protein kinases isoenzymes, 538
    second messenger response, 501
    t-PA release, 496
  cyclic 8-benzylthio-N^6-butyl-
    protein kinases, 540
  cyclic 8-(4-chlorophenyl)thio-
    protein kinases, 540
  cyclic N^6,2′-O-dibutyryl-
    protein kinases, 540
Adenosine 2′-monophospho-5′-diphosphoribose
  HMGCoA reductase
    inhibition, 337
Adenosine triphosphate
  binding
    protein kinases, 535
S-Adenosyl-L-cysteine hydrolase
  biochemistry, 453
  inhibitors, 454
  purine metabolism, 453–456
Adenylate cyclase
  activation
    3-deazaadenosine, 454
Adenylate deaminase
  purine metabolism, 303
Adenylate kinase
  competitive inhibitor, 63
  inhibition
    salicylate, 63
  multisubstrate analog inhibitors, 65
Adhesin
  drug resistance, 119
Adrenocorticotropic hormone
  inhibition, 347
Adriamycin (doxorubicin)
  antitumor drug, 187, 727, 729
  DNA polymerases
    inhibition, 780
  DNA strand-breakage, 736
  DNA synthesis
    inhibition, 714
  DNA topoisomerase
    inhibition, 779
  intercalating agent
    DNA, 704
  protein kinase C
    antagonist, 545
  toxicity, 722
  trimetrexate
    synergism, 286
Adult respiratory distress syndrome
  hyperuricemia, 468
Affinity chromatography
  enzymes, 7
Agarose gel electrophoresis
  DNA strand-breakage
    measurement, 733
Agonists
  definition, 532
  protein kinases
    regulatory site, 535
AIDS

treatment
  AZT, 760
trimetrexate, 287
virus
  castanospermine, 378
Alafosfalin
  inhibitor
    lipid A biosynthesis, 563
Alamethicin
  structure, 596
D-Alanine
  peptidoglycan synthesis
    antibiotic inhibition, 569
Alanine, 3-chloro-
  alanine racemase
    inhibitor, 246
  alanine transaminase, 241
    inhibitor, 241
  amino acid transaminase
    inhibitors, 244
  threonine deaminase
    inhibitor, 247
Alanine, dichloro-
  tryptophan synthase
    inhibition, 247
Alanine, difluoro-
  tryptophan synthetase
    inhibitor, 247
Alanine, 3,3-difluoro-
  alanine racemase
    inhibitor, 246
Alanine, 3-fluoro-
  alanine racemase
    inhibitor, 246
  amino acid transaminase
    inhibitors, 244
  serine hydroxylmethyltransferase
    inhibitor, 247
D-Alanine, β-fluoro-
  inhibitor
    alanine racemases, 82
D-Alanine, β-halo-
  inhibitor
    bacterial cell wall biosynthesis, 564
Alanine, trifluoro-
  alanine racemase
    inhibitor, 246
  γ-cystathionase
    inhibition, 248
  tryptophan synthase
    inhibition, 247
  tryptophanase
    inhibition, 247
Alanine racemase
  inhibition, 240
  inhibitors, 246
    β-fluoro-D-alanine, 82
  pyridoxal-dependent systems, 228, 229
  swinging door mechanism, 228
Alanine transaminase
  pyridoxal-dependent systems, 216
L-Alanine transaminase
  inhibition, 241
D-Alanyl-D-alanine carboxypeptidases
  β-lactam antibiotics, 613
D-Alanyl-D-alanine synthetase
  drug resistance, 114
D-Alanyl-D-alanine transpeptidases
  β-lactam antibiotics, 613
Alanyl-tRNA ligase
  nomenclature, 36
L-Alanyl-tRNA synthetase
  nomenclature, 36
Alaphosphin

peptidoglycan synthesis
  antibiotic inhibition, 570
Albumin
  metal ion transport, 180
Aldolase
  binding
    suicide inhibition, 82
Aldose reductase
  inhibition, 271
  inhibitors, 292
  structure, 292
Aldosterone
  biosynthesis, 347
  cytochrome *P*-450, 352
Alginates
  cell walls, 556
Alkali metals
  biochemical competition, 179
Alkenes
  dopamine β-hydroxylase
    inhibitors, 129
Alkylating agents
  enzyme inhibition, 78
Alkynes
  cytochrome *P*-450
    inhibition, 136
  dopamine β-hydroxylase
    inhibitors, 129
Allantoinase
  purine metabolism, 303
Allenyldiamine
  polyamine oxidase
    inhibitors, 128
Allopurinol
  antiparasitic agent, 323, 325
  azathioprine, 307
  gout, 143, 469
  reperfusion injury, 143
  uric acid
    inhibition, 305
Allopurinol oxidase
  xanthine oxidase, 305
Allopurinol ribonucleoside
  antiparasitic agent, 323
Allopurinol ribonucleoside, 3-bromo-
  antiparasitic agent, 324
Allopurinol riboside
  purine nucleoside phosphorylase
    inhibitor, 459
Alloxanthine
  antiparasitic agent, 325
  uric acid
    inhibition, 305
Allylamine
  monoamine oxidases
    inhibitor, 127
Almond emulsin β-glucosidase
  castanospermine, 376
Aluminum
  poisoning, 185
α-Amanitin
  RNA polymerases
    inhibition, 800
Amastatin
  aminopeptidase inhibitor, 429
Ametantrone
  antitumor drug, 727
  cardiotoxicity, 729
  intercalating agent
    DNA, 704
Amide synthetases
  nomenclature, 36
Amikacin
  resistance, 95, 107

Amiloride
  diuretics, 267
  inhibitor
    protein kinase C, 545
Amine oxidases
  activation, 125
Amino acid aminotransferase
  structure, 219
ω-Amino acid pyruvic transaminase
  stereochemistry, 220
Amino acid racemase
  pyridoxal-dependent systems, 228
  specificity, 229
Amino acids
  proteins
    sequence analysis, 16
  reactions at C-3, 233
Amino acids, α-aza-
  peptides
    acyl enzymes, 79
D-Amino acid transaminase
  inhibitors, 243
  pyridoxal-dependent systems, 222
L-α-Amino acid transaminases
  inhibitors, 240
  pyridoxal-dependent systems, 215
ω-Amino acid transaminases
  pyridoxal-dependent systems, 220
Aminoacyl-tRNA synthetases
  kinetics, 50
γ-Aminobutyrate transaminase
  inhibition
    gabaculine, 82
  inhibitors, 242
  pyridoxal-dependent systems, 221
α-Amino-ε-caprolactam
  pyridoxal-dependent systems, 229
1-Amino-1-carboxycyclopropane synthetase
  inhibition, 248
Aminocyclitols
  resistance, 107
1-Aminocyclopropane-1-carboxylate deaminase
  reactions at C-4, 239
Aminoglycoside antibiotics
  resistance, 831, 832
  ribosomal decoding site, 831
  ribosomal RNA receptor sites
    chemical footprinting, 832
Aminoglycoside 6-phosphotransferase
  streptomycin resistance, 91
Aminoglycosides
  resistance, 107, 110, 111
  susceptibility, 92
    cell envelope, 95
Aminoidoxuridine
  antiviral agent, 315
Aminopeptidases
  classification, 428
  inhibitors, 428
  metalloproteases, 71
  substrates and inhibitors
    active site binding, 430
Aminopterin
  dihydrofolate reductase
    inhibitor, 275, 278
  dihydropterin reductase
    inhibitors, 291
Aminopterin, 5-deaza-
  dihydrofolate reductase
    inhibitor, 279
Aminopterin, 5,8-deaza-
  dihydrofolate reductase
    inhibitor, 279
Aminopterin, 10-ethyl-10-deaza-

binding
  dihydrofolate reductase, 278
Amoxycillin
  synthesis, 622
Amphetamines
  cytochrome *P*-450
    inhibition, 135
  neurotransmission
    inhibition, 511
Amphomycin
  inhibitor
    bacterial cell wall biosynthesis, 565
Amphotericin B
  antifungal agent, 326
  fungal resistance, 114
  sterols, 361
  structure, 595
  toxicity, 595
Ampicillin
  activity, 620
  enzyme inhibition, 104
  history, 619
  oral bioavailability, 620
  resistance, 67, 92, 93, 95
Amrinone
  inhibitor
    phosphodiesterases, 507, 510
    protein kinases, 538, 541
m-AMSA
  antileukemic agent, 742
Amsacrine
  DNA cleavage, 781
  intercalating agent
    DNA, 704, 716, 776
Amsalog
  intercalating agent
    DNA, 704, 716
Amyloglucosidases
  fungal
    castanospermine, 376
Anaerobic bacteria
  cefoxitin, 630
Anagrelide
  inhibitor
    phosphodiesterase, 507
Anaphylaxis
  slow reacting substance
    lipoxygenase, 158
Androgens
  biosynthesis, 353
Androsta-1,4-diene-3,17-dione
  steroid hormones
    inhibition, 351
5β,14β-Androstane-3β,14-diol
  structure–activity relationship, 207
Androst-4-ene-3,17-dione, 7α-(4′-aminophenylthio)-
  steroid hormones
    inhibition, 352
Androstenedione, 4-hydroxy-
  steroid hormones
    inhibition, 351
Androstenedione, 19-mercapto-
  steroid hormones
    inhibition, 352
Androstenedione, 19-methylthio-
  steroid hormones
    inhibition, 352
Angiotensin converting enzyme (ACE)
  inhibition, 394
    kinetics, 71
  inhibitors, 400
    conformation, 408
    miscellaneous, 409
    metalloproteases, 71

Anhydroelastase
  binding, 70
Ansamycins
  RNA polymerase
    inhibition, 797
Antagonists
  definition, 532
  protein kinases
    regulatory site, 535
Anthracenediones
  antitumor drugs, 727
  cardiotoxicity, 729, 780
Anthracycline antibiotics
  cardiotoxicity, 729
Anthracyclines
  antitumor activity, 780
  DNA binding agents, 767
  DNA synthesis
    inhibition, 714
  DNA template functions
    inhibition, 721
  DNA topoisomerase
    inhibition, 779
  free radical formation
    DNA strand-breakage, 736
  razoxane, 187
Anthramycin
  DNA binding agents, 767
  structure–activity relationships, 768
Anthrapyrazole
  anticancer activity, 727, 729
Anthraquinones
  intercalating agents
    DNA, 716
Antiamoebins
  structure, 596
Antibacterial agents
  DNA
    binding agents, 726
Antibiotics
  bifunctional
    DNA intercalating agents, 718
  history, 90
  ribosomal functions, 823
Antibodies
  phosphilipase A$_2$, 524
Anticancer agents
  phospholipase A$_2$, 727, 729
  metal complexes, 186
Anticoagulants
  oral, 489
Antidepressants
  monoamine oxidases
    inhibitors, 125
Antifungal agents
  14α-demethylase, 139
  design, 357
  purines/pyrimidines, 326
  sterol metabolism inhibition, 354
Antihypertensive agents
  peptidase inhibitors, 394
Antiinflammatory agents
  oral
    3-deazaadenosine, 454
Antilymphocyte serum
  organ transplantation, 474
Antimicrobial agents
  DNA strand-breaking, 728
  resistance, 89–119
Antineoplastic agents
  DNA topoisomerases, 771
  purine targets, 306
Antipain
  cysteine proteinases

  inhibition, 432
Antiparallel β-pleated sheet
  peptidases
    inhibition, 399
Antiparasitic agents
  purines/pyrimidines, 323
α2-Antiplasmin
  plasmin
    inhibitor, 494
Antiprotozoal agents
  DNA
    nicking agents, 726
Antithrombin III
  coagulation
    inhibitor, 487
  deficiency
    thrombosis, 488
Antithrombotic properties
  arachidonic acid
    metabolism, 162
α1-Antitrypsin
  oligosaccharides, 374
Antitumor agents
  antibiotics
    DNA intercalating agents, 714
  oligosaccharides, 727, 729
  synthetic, 716
Antiviral agents
  drug targets, 314
  protein synthesis suppression, 326
  purine nucleosides, 319
  pyrimidine nucleosides, 315
6-APA — *see* Penicillanic acid, 6-amino-
Apalcillin
  activity, 633
Aphidicolin
  DNA polymerase α
    inhibitor, 757
  DNA polymerases
    inhibition, 757, 764
    inhibition, mechanism, 764
  structure–activity relationships, 764
Aplastic anemia
  autoimmune disease, 468
  chloramphenicol
    DNA damage, 726
Aplysiatoxin
  protein kinase C, 544
Apramycin
  resistance, 107
D-Arabinitol, 1,4-dideoxy-1,4-imino-
    glycosidases
    inhibition, 371
Arabinosyladenine
  DNA polymerases
    inhibition, 762
Arabinosylcytosine
  DNA polymerases
    inhibition, 762
Arachidonate
  phosphilipase A$_2$, 516
  specific, 519
Arachidonic acid
  cascade, 147–170
  eicosanoids from, 149
  enzyme cascades, 482
  metabolism, 160
    cyclooxygenase, 149
    diet, 168
    inhibitors, 160
    lipoxygenase, 156
    psoriasis, 474
    protein kinase C, 544
Arachidonic acid, dehydro-

lipoxygenase
  inhibition, 164
Archaebacteria
  transcription, 794
Argatroban — *see* MD-805
Arginine racemase
  pyridoxal-dependent systems, 229
Aristeromycin
  adenosyl-L-homocysteine hydrolase
    inhibitor, 455
  purine antagonist, 310
Aristeromycin, 3-deaza-
  adenosyl-L-homocysteine hydrolase
    inhibitor, 456
  antiviral agent, 321
AR-L57
  inhibitor
    phosphodiesterase, 505
Arnstein tripeptide
  modified penicillins, 649
  synthesis, 610
Aromatase
  synthesis, 137
  inhibition, 348, 350, 358
Aromatic amino acid decarboxylase
  inhibitors, 246
  pyridoxal-dependent systems, 227
Aromatic amino acid transaminase
  structure, 219
Arphamenines
  aminopeptidase inhibitors
    binding, 430
    discovery, 429
Arprinocid
  antiparasitic agent, 325
Arridicins
  peptidoglycan polymerization
    inhibition, 574
Arsenic
  essential element, 176
Asialoglycoproteins
  receptor
    swainsonine, 374
Asparenomycin A
  isolation, 671
Aspartate aminotransferase
  inhibitors, 240
  nomenclature, 35
  pyridoxal-dependent systems, 215
  site specific mutagenesis, 220
  structure, 219
Aspartate aminotransferase isoenzymes
  structure, 219
Aspartate β-decarboxylase
  pyridoxal-dependent systems, 223
Aspartate kinase
  nomenclature, 38
L-Aspartate: 2-oxoglutarate aminotransferase
  nomenclature, 35
Aspartate transaminase
  pyridoxal-dependent systems, 216
Aspartate transcarbamoylase
  pyrimidine metabolism, 303
Aspartate transcarbamylase
  multisubstrate analog inhibitors, 65
  structure, 64
Aspartic acid, 3-chloro-
  amino acid transaminase
    inhibitor, 240
  glutamate decarboxylase
    inhibitor, 240
Aspartic acid, 3-methenyl-
  alanine transaminase
    inhibitor, 241

amino acid transaminase
  inhibitor, 240
Aspartic acid, 2-methyl-
  amino acid transaminase
    inhibitor, 240
Aspartic peptidases
  classification, 396
Aspartic protease
  inhibitors, 72
Aspartic proteinases
  inhibition, 397
Aspartokinase
  nomenclature, 38
Aspirin
  anti-inflammatory properties, 160
  arachidonic acid
    inhibitors, 160
  cyclooxygenase
    inhibition, 160, 161
  thrombosis, 162
Aspiritrexem
  dihydrofolate reductase
    inhibitor, 288
Asthma
  lipoxygenase, 158
Atherosclerosis
  cholesterol, 334
  prostacyclin, 154
Atherosclerotic peripheral vascular disease
  prostacyclin, 155
ATPase, Ca^{2+}-
  sarcoplasmic reticulum, 194
ATPase, H$^+$,K$^+$-
  competitive inhibitors, 196
  gastric acid secretion, 194
  inhibitors
    biological screening, 196
  noncompetitive inhibitors, 198
  reaction mechanism, 194
  structure, 195
ATPase, Na$^+$,K$^+$-
  function, 205
  reaction mechanism, 194
  structure, 205
ATPases
  inhibition, 68
Atrial natriuretic factor
  phosphodiesterases, 512
Atropine
  nerve gases
    antidotes, 385
Autoimmune diseases, 466
  adenosine deaminase inhibitors, 449
  purine metabolism, 444
Automatic sequence analysis
  proteins, 18
Avian myeloblastosis virus
  reverse transcriptase, 759
Avoparcin
  peptidoglycan polymerization
    inhibition, 574
AXQ
  DNA strand-breakage
    redox cycling, 737
AY-9944
  hypocholesterolemic, 343
2-Azaadenosine
  purine antagonist, 309
8-Azaadenosine
  purine antagonist, 308
Azacholesterols
  fungicide, 360
5-Azacytidine
  DNA methyltransferases, 783

pyrimidine antagonist, 311
3-Aza-1-dethiacephalosporins
  properties, 690
1-Azadethiacephems
  properties, 687
2-Aza-1-dethiacephems
  synthesis, 688, 689
3-Aza-1-dethiacephems
  synthesis, 690
1-Azadethiapenams
  synthesis, 680
8-Azaguanine
  purine antagonist, 308
8-Azaguanosine, O^6-methyl-
  purine antagonist, 309
8-Azaguanosine 5'-monophosphate
  purine antagonist, 308
8-Azaguanylic acid
  purine antagonist, 308
2-Azahypoxanthine
  purine antagonist, 309
8-Azahypoxanthine
  purine antagonist, 308
2-Azainosine
  purine antagonist, 309
8-Azainosine
  purine antagonist, 308
8-Azainosine, O^6-ethyl-
  purine antagonist, 309
8-Azainosine, O^6-methyl-
  purine antagonist, 309
8-Azainosinic acid
  purine antagonist, 308
5-Azaorotate
  antiparasitic agent, 325
5-Azaorotic acid
  pyrimidine antagonist, 311
5-Azaorotidylic acid
  pyrimidine antagonist, 311
9,11-Azaprosta-5,13-dienoic acid
  thromboxane receptor antagonist, 163
Azapurines
  purine antagonist, 308
Azapyrimidines
  pyrimidine antagonists, 311
Azaribine
  pyrimidine antagonist, 311
Azaserine
  glutamine amino transfer
    inhibition, 301
8-Aza-6-thioguanine
  purine antagonist, 309
8-Aza-6-thioinosine
  purine antagonist, 309
8-Aza-6-thioinosine, S-methyl-
  purine antagonist, 309
Azathioprine
  autoimmune diseases, 444
  organ transplantation, 469
  prodrug
    6-mercaptopurine, 306
  purine antagonist, 307
  rheumatoid arthritis, 468
5-Azathymidine, 5,6-dihydro-
  antiviral agent, 318
5-Azauracil
  antiparasitic agent, 325
  pyrimidine antagonist, 311
6-Azauracil
  antiparasitic agent, 325
  antiviral agent, 318
6-Azauracil, N^1-aryl-
  antiparasitic agents, 325
6-Azauracil, 2-thio-

antiviral agent, 318
6-Azauracil, 4-thio-
  antiviral agent, 318
5-Azauridine
  pyrimidine antagonist, 311
6-Azauridine
  antiviral agent, 318
  pyrimidine antagonist, 311
6-Azauridine, 2',3',4'-tri-O-acetyl-
  pyrimidine antagonist, 311
5-Azauridylic acid
  pyrimidine antagonist, 311
Azetidinones
  naturally occurring, 656
  synthesis, 677
  N^1,O-phosphorus activated
    activity, 661
Azetidinones, N-(tetrazol-5-yl)-
  synthesis, 661
Azidothymidine
  DNA polymerases
    inhibitor, 760
Aziridines
  nitro aromatic
    DNA strand-breakage, 741
Azlocillin
  activity, 633
  resistance, 95, 114
Azole nucleosides
  antiviral agents, 322
Azoles
  antifungal agents
    mode of action, 356
    molecular action, 357
AZT
  AIDS
    treatment, 760, 761
  antiviral agent, 319
  human immunodeficiency virus, 319
  structure, 760
Aztreonam
  activity, 660
  resistance, 95, 102

B-cell chronic lymphocytic leukemia
  anthracyclines, 780
B cells
  adenosine deaminase inhibitors, 452
Bacampicillin
  synthesis, 622
Bacitracin A
  association constants
    lipid analogs, 572
Bacitracins
  activity
    Micrococcus luteus, 572
  inhibitor
    bacterial cell wall biosynthesis, 566
  peptidoglycan biosynthesis
    lipid cycle inhibition, 571
Bacteria
  drug resistance
    mechanism, 98–114
  drug susceptibility, 95
  strains, 617
Bacteriocins
  protein synthesis
    inhibition, 816
Bacteroides
  DNA strand-breaking drugs, 728
Bacteroides fragilis
  antibiotics, 617
  cefoxitin, 630
Baicalein

lipoxygenase
  inhibition, 164
BAL — *see* Propanol, 2,3-dimercapto-
Benoxaprofen
  LTB$_4$
    synthesis, 164
Benzamide, 3-amino-
  DNA strand-break repair
    inhibition, 748
Benzene, methylenedioxy-
  cytochrome *P*-450
    inhibition, 135
Benzimidazole sulfoxides, pyridylmethyl-
  chemical rearrangement, 198
  H$^+$,K$^+$-ATPase
    inhibition, 198
  mode of action, 198
  structure–activity relationships, 201
Benznidazole
  antitumor drug, 727
  Chagas' disease, 729
  hypoxic cells, 731
Benzoic acid, *p*-amino- (PABA)
  biological role, 258
  competitive antagonists
    sulfonamides, 256
Benzoic acid, *p*-aminomethyl-
  fibrinolysis
    inhibitor, 497
Benzoic acid, *p*-chloromercuri-
  enzyme inhibition, 104
  penicillinases, 104
Benzoisothiazolone, *N*-acyl-
  enzyme acylation, 79
Benzoquinones
  antitumor drugs, 727
Benzotriazine
  N-oxides
    antitumor drugs, 728
Benzotript
  gastrin antagonist, 196
Benzo-1,2,3-triazole, 1-amino-
  cytochrome *P*-450
    inhibition, 136
Benzoxazinones
  elastase inhibition, 79
Benzylamines
  deamination
    monoamine oxidase, 125
Berenil
  nonintercalating agent
    DNA, 709
Bestatin
  structure–activity relationship
    aminopeptidase inhibitor, 429
Bestatin, 2-thio-
  aminopeptidase inhibitor, 430
Bicozamycin
  antibiotic, 581
Bicyclomycin
  antibiotic, 581
Bicyclo[2.2.2]octane-1-carboxylic acid,
    4-(2-aminoethyl)-
  fibrinolysis
    inhibitor, 497
Biguanides
  resistance, 110
Bile acids
  cholesterol
    biosynthesis, 335
Biliary excretion
  β-lactam antibiotics, 617
Bilirubin
  inhibitor

protein kinase C, 545
Binding
  metal ions, 180
    pharmaceuticals, 181
Binding agents
  DNA, 766
Binding energy
  enzymes, 51
Binding sites
  enzymes, 47
Biotin
  cofactor
    enzymes, 5
Bis(allenyl)diamines
  polyamine oxidase
    inhibitors, 128
Bisanthracyclines
  intercalating agents
    DNA, 719
Bisantrene
  antitumor agent, 717
  intercalating agent
    DNA, 704
Bisintercalators
  DNA, 720
Bismuth subcitrate
  peptic ulcers, 195
Bisnorpiperacillin
  activity, 677
Bleomycin
  antitumor agents, 187, 327, 728
  cytotoxicity, 737
  medical applications, 731
Blood clotting
  enzyme cascades, 482
Blood plasma
  fractionation
    proteins, 7
Bluensomycin
  resistance, 107
BM 13.177
  thromboxane receptor antagonist, 163
Bond angle
  deformation
    enzymes, 57
Bond lengths
  deformation
    enzymes, 57
Bone marrow transplantation
  immunosuppression, 444
Borane, difluoro-
  enzyme inhibitors, 71
Boronic acid
  elastase inhibitor, 425
  enzyme inhibitors, 70
Bovine mastitis
  streptococci
    antibiotic susceptibility, 91
Bradykinin potentiating peptides
  ACE inhibitors, 400
Breast cancer
  aromatase, 138
Briggs–Haldane kinetics
  enzymes, 49
BRL 14151
  discovery, 662
p-Bromophenacyl bromide
  inhibitors
    phosphilipase A$_2$, 527
Bronchodilators
  prostaglandins, 151
Butaconazole
  fungal resistance, 114
Butakacin

resistance, 109
1-Butanamine, 4-nitro-
GABA-T
inhibitors, 243
Butenoic acid, 4-amino-3-aryl-
GABA-T
inhibitors, 243
2-Butenoic acid, 4-amino-2-methyl-
GABA-T
inhibitors, 243
But-3-ynoate, 2-hydroxy-
inhibitor
L-lactate oxidase, 83
Butyric acid, γ-amino- (GABA)
ω-amino acid transaminases, 220
biosynthesis, 222
Butyric acid, γ-amino-γ-ethynyl-
GABA-T
inhibitors, 242
Butyric acid, γ-amino-5-fluoromethyl-
GABA-T
inhibitors, 242
Butyric acid, γ-amino-5-trifluoromethyl-
GABA-T
inhibitors, 242
Butyric acid, γ-amino-γ-vinyl-
GABA-T
inhibitors, 242
BW A4C
bronchospasm, 167
BW755C
inhibitor
cyclooxygenase/lipoxygenase, 165

Cadmium
bioaccumulation, 176
pollution
bone-brittleness, 184
Caffeic acid
lipoxygenase
inhibition, 164
Calcimycin
alkali metal complexes, 602
cell membrane
permeability, 598
Calcium
biochemical competition, 178
phosphilipase A_2, 522
phosphodiesterases, 503
Calelectrin
inhibitor
phospholipase A_2, 523
Calmidazolium
protein kinase C
antagonist, 545
Calmodulin
cyclosporin binding, 473
phosphilipase A_2
regulation, 522
phosphodiesterases, 503
Calpactin I
inhibitor
phospholipase A_2, 523
Calpains
inhibition, 431
Camphor hydroxylase
oxygen insertion, 133
Camptothecin
DNA polymerase inhibition, 775
structure–activity relationships, 776
Campylabacter pylori
antibacterial agents, 195
Cancer
2′-deoxycoformycin, 453

chemotherapy
adenosine deaminase inhibitors, 449
Cancer cells
growth
inhibition, 540
Candicidin
absorption, 595
fungal resistance, 114
Canrenone
Na^+,K^+-ATPase
inhibition, 207
Cap structure
antiviral chemotherapy, 179
Caproic acid, ε-amino-
fibrinolysis
inhibitor, 497
Capsules
cell walls, 556
Captopril
analogs
ACE inhibitors, 404
binding, crystal structure, 435
angiotensin converting enzyme
inhibition, 72, 400
discovery
peptidase inhibition, 392
P_2' variations
ACE inhibitors, 401
1-Carbacephadiene
activity, 687
Carbacepham
synthesis, 679
1-Carbacephems
activity, 686
properties, 685
synthesis, 685
1-Carbapenams
synthesis, 679
Carbapenem, β-methyl-
synthesis, 673
Carbapenems
activity, 674
mechanism of action, 670
naturally occurring, 656
properties, 671
total synthesis, 672
Carbazides
aminobutyrate transaminase
inhibitor, 242
Carbenicillin
activity, 633
cell membrane disorganization, 591
history, 619
resistance, 114
Carbocyclic adenosine — *see* Aristeromycin
Carbomycin
chemical footprinting
ribosomal binding sites, 830
Carbonic anhydrase
inhibition, 63
inhibitors, 267
parietal cell, 195
Carbonium ions
alkylating agents
enzyme inhibition, 78
Carboplatin
DNA binding agent, 768
radiosensitization, 743
side-effects, 729
Carboxylases
nomenclature, 36
Carboxypeptidase A
energetics, 57
inhibition

mechanism, 71
Carboxypeptidases
  metalloproteases, 71
  penicillin binding proteins, 99
Carcinoma
  colon
    3-deazaneplanocin-A, 456
Cardenolides
  structure–activity relationship, 207
Cardiac glycosides
  clinical applications, 206
  computer graphics
    activity, 209
  inotropic action, 206
  structure–activity relationship, 207
Cardiolipin, monolyso-
  inhibitor
    phosphilipase A_2, 526
Cardiomyopathy
  adriamycin, 722
Carnidazole
  antimicrobial agent, 726
  DNA strand-breaking, 728
Carnitine, palmitoyl-
  protein kinase C
    antagonist, 545
Carnitine acetyl transferase
  multisubstrate analog inhibitors, 65
Carpetimycin A
  isolation, 671
Carrier-mediated transport
  penicillins, 620
Carumonam
  activity, 660
Cassaine
  Na^+,K^+-ATPase
    inhibition, 207
Castanospermine
  glucosidases
    inhibition, 376
  glycoproteins, 376
  occurrence, 370
Catalase
  superoxide dismutases, 306
Catalytic domains
  protein kinases, 532
Catechol-*O*-methyl transferase
  penicillins, 649
Cathepsin D
  inhibitors, 72
  renin inhibitor, 417
Cathepsins
  inhibition, 431
CB-1954
  Walker carcinoma, 741
CBS-1108
  leukotriene production
    inhibition, 164
CC-1065
  DNA binding agent, 767, 768
Cefaclor
  activity, 644
  synthesis, 645
Cefadroxil
  activity, 644
  oral absorption, 627
Cefamandole
  activity, 625
  resistance, 94
Cefazolin
  activity, 625
  hydrophobicity
    crypticity, 97
Cefepime

  activity, 644
Cefetamet pivoxyl
  activity, 648
Cefixime
  activity, 647
Cefmenoxime
  activity, 636
  synthesis, 638
Cefoperazone
  hypoprothrombinemia, 618
  side-effects, 642
Cefotaxime
  activity, 635
  resistance, 114
Cefotetan
  activity, 642
Cefotiam
  activity, 639
Cefoxitin
  clinical use, 629
  resistance, 94, 110
Cefpirome
  activity, 644
Cefroxadine
  activity, 644
  synthesis, 645
Cefsulodin
  resistance, 95
Ceftazidime
  activity, 643
  resistance, 114
Cefteram
  activity, 648
Ceftizoxime
  activity, 636
  synthesis, 638
Ceftriaxone
  activity, 636
Ceftriazone
  resistance, 114
  synthesis, 638
Cefuroxime
  clinical use, 631
  resistance, 94, 114
Cefuroxime axetil
  oral absorption, 633
Cell envelope
  drug susceptibility, 95
  mutations
    drug resistance, 98
Cell membranes
  disorganization
    antibiotics, 590
  permeability
    antibiotics, 598
Cellular immunity
  adenosine deaminase deficiency, 447
Cell walls
  bacteria
    O-antigens, 556
    components, 554
    β-lactam antibiotics, 612
    proteins, 563
    structure, 554
  fungal, 554
  microbial
    antibiotics inhibiting biosynthesis, 553
  polymers
    structure, 557
Cephabacins
  occurrence, 642
Cephacetrile
  hydrophobicity
    crypticity, 97

Cephalexin
  activity, 626, 644
  oral absorption, 627
Cephaloglycine
  stability, 627
Cephaloram
  hydrophobicity
    crypticity, 97
Cephaloridine
  ceftazidime synthesis from, 643
  enzyme inhibition, 104
  hydrophobicity
    crypticity, 98
  nephrotoxicity, 618
  synthesis, 624
Cephalosporanate, 7-amino-3-pyridiniummethyl-
  synthesis, 643
Cephalosporanic acid, 7-amino-
  cephem ring system, 609
  synthesis, 623
Cephalosporinases
  classification, 104
Cephalosporin C
  activity, 623
  biosynthesis, 610
  synthesis, 623
  total synthesis, 612
Cephalosporin C, deacetoxy-
  biosynthesis, 610
Cephalosporin ester
  elastase inhibitor, 670
Cephalosporins
  antibacterial activity, 625, 630
  antipseudomonal, 642
    antibacterial activity, 643
  binding proteins
    Escherichia coli, 615
  biosynthesis, 610
  cell wall agents, 609–650
  elastase inhibitors, 426
  enzyme inhibition, 104
  history, 623
  inhibitors
    transpeptidases, 81
  β-lactamase stable, 629
  nuclear analogs, 677
  oral
    antibacterial activity, 645
  structure–activity relationships, 627
  susceptibility, 92
  third generation, 635
    antibacterial activity, 636
Cephalosporins, 7-amino-
  synthesis, 638
Cephalosporins, 7α-chlorosulfoxide
  inhibition
    porcine pancreatic elastase, 81
Cephalosporins, 3-exomethylene-
  ozonolysis, 645
Cephalosporins, 7α-formamido-
  activity, 642
Cephalosporins, 7α-methoxy-
  stability, 629
  synthesis, 631
Cephalosporins, 7α-methoxy-7β-amino-
  synthesis, 631
Cephalosporins, methoxyiminoaminothiazolyl-
  preparation, 636
Cephalosporins, oximino-
  properties, 631
Cephalosporins, ureido-
  activity, 642
Cephalosporium acremonium
  cephalosporin biosynthesis, 610

Cephalothin
  activity, 641
  hydrophobicity
    crypticity, 97
  synthesis, 623
  total synthesis, 612
Cephalothin, deacetyl-
  synthesis, 623
Cephams, 3-hetero-
  properties, 691
Cephamycin C
  structure, 629
Cephamycins
  biosynthesis, 610
  detection, 662
  discovery, 655
Cephems
  4,7-bicyclic system
    activity, 692
  conformation, 613
  nuclear modification, 677
  ring system
    β-lactam antibiotics, 609
Cephems, 3-hetero-
  properties, 691
Cephradine
  activity, 626, 644
  oral absorption, 627
Ceruloplasmin
  copper poisoning, 184
CGP 31608
  activity, 677
Chaetiacandin
  glucan synthase inhibitors, 587
Chagas' disease
  benznidazole, 729
Chang cells
  metal ion binding, 180
Chelation
  use in therapy, 181
Chelation therapy
  synergistic, 185
Chemical footprinting
  antibiotics
    ribosomal binding sites, 830
  ribosomal RNA receptor sites
    aminoglycoside antibiotics, 832
Chicken heart mitochondrial enzyme
  structure, 219
Chimeric proteins
  bifunctional enzymes, 38
Chitin
  cell walls
    fungal, 554, 583
  structure, 584
Chitin synthase
  inhibition, 585
    polyoxin, 579
Chitin synthetase
  antifungal agent, 357
Chitinovorins
  occurrence, 642
Chitosan
  fungi
    walls, 584
Chloramphenicol
  DNA damage, 726
  DNA strand-breakage, 738
  mode of action, 827
  resistance, 92,–95, 101, 828
    Shigella dysenteriae, 93
  ribosomal receptor site, 827
  specificity, 106
  susceptibility, 92

Chloramphenicol, dehydro-
　DNA strand-breakage, 738
Chloramphenicol, nitroso-
　DNA strand-breakage, 738
Chlorguanide
　dihydrofolate reductase
　　inhibitor, 285
Chlormadinol acetate
　Na^+,K^+-ATPase
　　inhibition, 207
Chlorocardicin
　occurrence, 658
Chloroquine
　intercalating agent
　　DNA, 704
　rheumatoid arthritis, 468
Chlorpromazine
　inhibitor
　　phosphilipase A_2, 527
Chlortetracycline
　resistance, 111
5α-Cholest-8(14)-en-3β-ol-15-one
　HMGCoA reductase
　　inhibition, 339
Cholesterol
　atherosclerosis, 334
　biosynthesis
　　HMG-CoA reductase, 335
　　inhibition, 334, 343
　14α-demethylase
　　inhibition, 141
　esterification
　　inhibition, 540
Cholesterol, (22R)-22-amino-
　cholesterol
　　inhibition, 349
Cholesterol, 25-aza-
　hypocholesterolemic, 345
Cholesterol, 20,25-diaza-
　hypocholesterolemic, 343
Cholesterol, (24S),25-epoxy-
　cholesterol
　　biosynthesis, 342
Cholesterol, 25-hydroxy-
　HMGCoA reductase
　　inhibition, 339
Cholesterol, 7-keto-
　HMGCoA reductase
　　inhibition, 339
Cholesterol metabolism
　prostacyclin, 155
Cholesterol side-chain cleavage enzyme
　inhibition, 358
Cholestyramine
　low density lipoproteins, 335
Cholinergic synapse
　description, 381
Cholinesterases
　inhibitors, 384
Chou–Fasman method
　secondary structure
　　prediction, 22
Chromatin
　intercalating drugs
　　binding, 722
Chromomycin A3
　nonintercalating agent
　　DNA, 709
Chryseneaminodiol
　intercalating agent
　　DNA, 704
Chymosin
　inhibitors, 72
Chymostatin

binding, 70
　elastase inhibitor, 423
Chymotrypsin
　binding, 70
　folding, 47
　inhibition
　　boranes, 71
α-Chymotrypsin
　amide hydrolysis
　　intermediates, 67
　binding, 70
　inhibition
　　acylation, 81
　inhibitors
　　6-chloro-2-pyrones, 80
　kinetics, 54
α-Chymotrypsin, 4-aminobenzoyl-
　enzyme inactivation, 80
Cigarette smoking
　atherosclerosis
　　prostacyclin, 154
　emphysema, 422
Cilastatin
　imipenem combination
　　activity, 674
　metabolism, 674
Cilazapril
　angiotensin converting enzyme
　　inhibition, 406
Cilostamide
　inhibitor
　　phosphodiesterase, 507
Cimetidine
　peptic ulcers, 204
Cinnamic acid
　4-amidinophenyl esters
　　enzyme inhibition, 79
Cinnamide, 4-hydroxy-
　inhibitor
　　tyrosine kinase, 548
Ciprofloxacin
　DNA gyrase
　　inhibitor, 774
　resistance, 101
Circular dichroism
　proteins
　　secondary structure, 20
Circulins
　cell membrane disorganization, 590
Cisplatin
　anticancer activity, 186, 728, 743
　clinical trials, 729
　dichloromethotrexate
　　dihydrofolate reductase inhibition, 280
　DNA binding, 767, 768
　　biological activity, 768
CI-914
　inhibitor
　　phosphodiesterases, 507
CI-930
　inhibitor
　　phosphodiesterases, 507, 510
Classification
　enzymes, 33
　　requirements, 34
Clavam
　configuration, 668
　naturally occurring, 656
　ring system, 662
Clavaminic acid
　biosynthesis, 667
Clavulanic acid
　activity, 662, 665
　amoxycillin activity, 665

benzyl ester
    elastase inhibitor, 426
biosynthesis, 667
chemistry, 666
derivatives, 666
detection, 662
enzyme inhibition, 104
history, 656
inhibitors
    β-lactams, 81
β-lactamase
    classification, 665
    inhibition, 663, 668
    mechanism of action, 670
metabolism, 667
structure, 662
synergy with β-lactams, 663
total synthesis, 668
Clavulanic acid, 9-aminodeoxy-
    activity, 667
Clavulanic acid, deoxy-
    activity, 666
Clorgyline
    monoamine oxidases
        inhibition, 125, 126
Clostridia
    cefoxitin, 630
Clostridium difficile
    antibiotics, 617
Clotrimazole
    antifungal agent, 140, 354
    fungal resistance, 114
Cloxacillin
    activity, 622
    enzyme inhibition, 104
    penicillinases, 104
    resistance, 99, 103
Clumping inducing agent
    drug resistance, 119
Coagulation
    common pathway, 486
    control, 487
    defects, 488
    diseases, 488
    drug therapy, 488
    extrinsic pathway, 486
    factors, 484
        inhibition, 491
        therapeutic use, 488
    fibrin cascade, 484
    fibrinolysis, 483
    inhibitors, 487
    intrinsic pathway, 486
    regulation, 487
Coaxial base-stacking
    ribosomal RNA
        triplet–triplet interactions, 833
Codecarboxylase — *see* Coenzyme B₆
Coenzyme B₆
    history, 214
    mechanism, 214
    stereochemistry, 214
    structure, 214
Coenzymes
    specificity, 83
Coformycin
    adenosine deaminase inhibitor, 449
    antiviral agent, 320
Coformycin, 2′-deoxy- (pentostatin)
    adenosine deaminase inhibitor, 73, 449
        animals, 452
    antiparasitic agents, 302
    antiviral agent, 320
    clinical studies, 452

immunosuppressive agent
    rats, 452
malaria, 475
organ transplantation, 474
purine antagonist, 309
T-lymphoblastic leukemia
    remission, 453
toxicities, 453
Coformycin, 2′-chloro-2′-deoxy-
    adenosine deaminase inhibition, 450
Colchicine
    gout, 469
    mitotic inhibitor, 784
Colestipol
    HMGCoA reductase
        inhibition, 338
Colicin E3
    ribosomes
        inactivation, 817
Colistins
    cell membrane disorganization, 590
Collagenase
    inhibitors, 394, 412
    metalloproteases, 71
    substrates
        molecular modeling, 412
Colley's anemia
    iron chelator, 190
Combination therapy
    drug resistance, 93
Combined immune deficiency diseases
    adenosine deaminase deficiency, 447
    symptoms, 447
Compactin
    fungicide, 360
    HMGCoA reductase
        inhibition, 336, 337, 339
Computer graphics
    cardiac glycosides
        activity, 209
    chemotherapy, 326
Concanavalin A
    binding
        cell walls, 560
Conduritol, bromo-
    glycosidase
        inhibition, 380
Conduritol epoxide
    glycosidase
        inhibition, 380
Contractile response
    prostaglandins, 151
Copper
    chelation therapy, 183
    free ion concentration
        plasma, 178
    oxygen
        activation, 125
    transport proteins, 180
Copper penicillinamine
    superoxide dismutases
        inhibitor, 306
Corallopyronins
    antibacterial activity, 799
Cordycepin
    antiparasitic agent, 324
Core particles
    ribosomal reconstitution, 814
Corneal ulceration
    collagenase
        inhibitors, 412
Coronavirus mouse hepatitis virus A59
    E-2 glycoprotein
        inhibition, 377

Cortisol
  biosynthesis, 347
COSY
  NMR
    protein structure, 26
Coulometry
  DNA damage
    measurement, 740
Coumarins
  anticoagulant, 489
Coumermycin
  DNA polymerase II
    inhibition, 775
  resistance, 102
CP-46,665-1
  protein kinase C
    antagonist, 545
p-Cresol
  dopamine β-hydroxylase
    inhibitors, 129
Crosslinking
  proteins, 27
Crypticity
  bacterial cell walls
    definition, 97
  drugs, 97
Crystal violet
  nonintercalating agent
    DNA, 709
Crystallization
  enzymes, 8
Crystallography
  intercalation
    DNA, 712
CS-514
  HMGCoA reductase
    inhibition, 338
Cushing's syndrome
  metyrapone, 347
Cyclacillin
  transport, 620
Cyclaridine
  antiviral agent, 320
Cycloglutamic acid
  alanine transaminase
    inhibitor, 241
Cyclohexanone oxygenases
  suicide inhibitors, 83
Cyclohexene polyol epoxides
  inhibition
    glucosidases, 76
Cyclooxygenase
  arachidonic acid
    metabolism, 149
  lipoxygenase
    dual inhibitors, 164
  products
    biological properties, 150
    metabolism, 149
Cyclophilin
  cyclosporin receptors, 473
Cyclophosphamide
  autoimmune diseases, 444
  rheumatoid arthritis, 468
  trimetrexate
    synergism, 286
Cyclopropanes
  dopamine β-hydroxylase
    inhibitors, 129
Cyclopropylamine
  cytochrome *P*-450
    inhibition, 136
Cyclopropylamine, *trans*-2-phenyl-
  monoamine oxidases

    inhibitor, 126
Cycloserine
  alanine racemase
    inhibitor, 246
  alanine transaminase
    inhibitor, 241
  amino acid transaminase
    inhibitors, 244
D-Cycloserine
  drug resistance, 111
  inhibitor
    bacterial cell wall biosynthesis, 564
  peptidoglycan synthesis
    antibiotic inhibition, 569
Cyclosporin
  autoimmune diseases, 466
  organ transplantation, 469
  psoriasis, 474
  purine nucleoside phosphorylase
    inhibitor, 459
  T cell modulator, 444
γ-Cystathionase
  inhibition, 247, 248
Cystathione γ-synthetase
  inhibition, 248
Cystathionine
  inhibition, 248
γ-Cystathionine synthase
  pyridoxal-dependent systems, 238
Cysteine proteases
  alkylation
    peptidyl halomethyl ketones, 76
Cysteine proteinases
  classification, 396
  inhibition, 397
  inhibitors, 431
Cytarabine
  pyrimidine antagonist, 313
Cytaramin
  pyrimidine antagonist, 313
Cytarazid
  pyrimidine antagonist, 313
Ara-Cytidine
  antiviral agent, 318
  pyrimidine antagonist, 313
  vincristine combinations
    pyrimidine antagonists, 314
Cytidine, 3-amino-2′,3′-dideoxy-
  pyrimidine antagonist, 314
Cytidine, 5-bromo-2′-deoxy-
  antiviral agent, 315
Ara-Cytidine, 2′-chloro-2′-deoxy-5-iodo-
  antiviral agent, 315
Ara-Cytidine, 2′-chloro-2′-deoxy-5-methyl-
  antiviral agent, 318
Cytidine, 3-deaza-
  antiviral agent, 318
Cytidine, 2′,3′-dideoxy-
  HIV replication, 319
  inhibition, 760
Cytidine, 2′,3′-dideoxy-2′,3′-didehydro-
  antiviral agent, 319
Cytidine, 5-fluoro-
  pyrimidine antagonist, 312, 314
Cytidine, 5-fluoro-2′-deoxy-
  pyrimidine antagonist, 312
Ara-Cytidine, 2′-fluoro-2′-deoxy-5-iodo-
  antiviral agent, 315
Cytidine, 5-iodo-2′-deoxy-
  antiviral agent, 315
Cytidine deaminase
  inhibitors, 313
  pyrimidine metabolism, 304
Cytochrome *P*-450, 133

catalysis
  reactions, 133
hydroxylation
  mechanism, 348
  inactivators, 349
inhibition
  mechanism, 134
isozymes
  inhibition, 347
  substrate specific, 137
  xenobiotic metabolism, 134
substrates, 134
Cytoplasmic membrane
  structure, 589
Cytoprotection
  prostacyclin, 155
Cytoprotective agents
  peptic ulcers, 195
Cytosine, 2,2′-anhydro-1-β-D-arabinofuranosyl-
  pyrimidine antagonist, 313
Cytosine, 1-β-D-arabinofuranosyl-
  DNA polymerases
    inhibition, 762
  pyrimidine antagonist, 313
Cytosine, 5-fluoro-
  antifungal agent, 326
  antiparasitic agents, 326
  fungal resistance, 115
Cytosine, 5-methyl-
  DNA methyltransferases
    inhibition, 782
Cytosine arabinoside
  DNA strand-break repair
    inhibition, 748
Cytosine deaminase
  pyrimidine metabolism, 304
Cytotoxic agents
  chelates, 186
Cytotoxins
  protein kinase C
    antagonist, 545

Danazol
  coagulation factors, 491
DAPA
  thrombin inhibitor, 490
Dapsone
  antibacterial action, 260
  resistance, 265
  structure, 258
Daunomycin
  antitumor drug, 187, 727
  DNA polymerases
    inhibition, 780
  DNA strand-breakage, 736
  DNA synthesis
    inhibition, 714
  DNA topoisomerase
    inhibition, 779
  intercalating agent
    DNA, 704
  toxicity, 722
Dazmegrel
  microalbuminuria, 163
Dazoxiben
  thromboxane synthase
    inhibition, 163
Deaminases
  pyrimidine metabolism, 303
trans-Decal-3β-ol, 4,4,10β-trimethyl-
  cholesterol
    biosynthesis, 342
Decarboxylases
  inhibitors, 244

pyridoxal-dependent systems, 222
Decoding site
  ribosomes, 833
Defamandole
  resistance, 114
Demeclocycline
  resistance, 111
14α-Demethylase
  resistance, 138
Demethylases
  purine metabolism, 302
Denaturation
  proteins, 9
  spectrophotometric thermal
    DNA strand-breakage, 732
Deoxycytidine kinase
  T cells
    replication, 459
Deoxyribonucleases
  purine metabolism, 302
Deprenyl
  monoamine oxidases
    inhibitor, 125, 126
Depression
  phosphodiesterases, 511
Desferrioxamine
  DNA degradation
    free radical scavengers, 187
  iron decorporation, 183, 186
  plutonium decorporation, 185
Desmopressin
  coagulation
    and fibrinolysis, 496
Desmosterol
  hypocholesterolemic, 343
1-Dethiacephams, 3-hetero-
  properties, 690
1-Dethiacephems, 2-hetero-
  properties, 688
1-Dethiapenams, 2-hetero-
  synthesis, 681
Dexamethasone
  fibrinolysis, 496
DHI
  protein kinase C, 545
Diabetes
  prostacyclin, 154
  sulfonamides, 268
  zinc deficiency, 181
Diacridines
  intercalating agents
    DNA, 719
Dialkylamino acid transaminase
  pyridoxal-dependent systems, 215
α,ω-Diaminopimelate racemase
  pyridoxal-dependent systems, 228
α,ω-*meso*-Diaminopimelic acid decarboxylase, 222
Dianemycin
  cell membrane
    permeability, 598
2,3-Diaza-1-dethiacephems
  synthesis, 691
1,2-Diazetidinones
  fused
    synthesis, 693
Diazonium salts
  alkylating agents
    enzyme inhibition, 78
Dibekacin
  resistance, 107
Dibenzoquinazolinediones
  inhibitor
    phosphodiesterase, 505
Dibucaine

protein kinase C
   antagonist, 545
Dicloxacillin
   resistance, 103
Digitoxigenin
   ATPase
      dephosphorylation inhibition, 381
   structure–activity relationship, 207
Digitoxin
   clinical use, 206
Digoxigenin
   glycosylation
      activity, 209
   structure–activity relationship, 207
Digoxin
   clinical use, 206
Dihomo-γ-linolenic acid
   arachidonic acid
      metabolism, 168
Dihydrofolate reductase
   binding
      methotrexate, 74
   biochemistry, 272
   drug interactions, 277
   inhibition, 37, 258, 271
   inhibitors
      classical, 275
      clinical uses, 290
      nonclassical, 282
      synthesis, 288
      toxicities, 290
   structure, 274
Dihydrolipoamide
   nomenclature, 38
Dihydropteridine reductase
   inhibition, 271
Dihydropterin reductase
   inhibitors, 290
7,8-Dihydropteroate synthetase
   competitive inhibitor, 63
β-Diketones
   inhibitors
      acetoacetate decarboxylase, 79
Dimethyl sulfate
   chemical footprinting
      ribosomal binding sites, 830
Dimetridazole
   antitumor drug, 726, 727
   histomoniasis, 729
Diniconazole
   antifungal agent, 357
Dipeptidyl carboxypeptidases
   metalloproteases, 71
Diphtheria toxin
   group transfer reactions
      enzymes, 3
Dipole moment
   cardiac glycosides, 209
Dipyrandium
   nonintercalating agent
      DNA, 709
Dipyridamole
   inhibitor
      phosphodiesterase, 505
Disease treatment
   drug resistance, 92, 95
Disodium cromoglycate
   aldose reductase
      inhibitor, 293
Distamycin
   DNA binding agents, 767
   nonintercalating agent
      DNA, 709
Disulfuram effect

β-lactam antibiotics, 618
Ditercalinium
   antitumor activity, 720
   intercalating agent
      DNA, 704
Diumycins
   phosphoglycolipid antibiotic, 578
DMB, 3-acetamido-
   protein kinase C, 545
DMB, 3-amino-
   protein kinase C, 545
DMHI
   protein kinase C, 545
DNA
   apurinic sites
      repair, 745
   base damage
      repair, 745
   binding
      intercalating agents, 711
   binding agents, 725–748, 766
   cleavage
      quinolones, 771
   complexes
      fluorescence polarization, 710
   double strand
      viruses, 796
   enzymes
      agents against, 753–787
      inhibitors, 753
   helix
      uncoiling, 710
   induced circular dichroism spectra, 710
   intercalating agents, 703–723
   intercalation
      actinomycin, 782
      anthracyclines, 779
   nicking agents, 725–748
   packaging
      DNA topoisomerases, 770
   radical-induced damage, 735
   repair
      inhibition, 746
   repair synthesis
      DNA polymerases, 756
   sequence analysis, 16, 18
   single strand
      viruses, 796
   sites
      intercalation, 712
   strand breakage
      biological effects, 744
      free radical reactions, 734
      measurement, 731
      medical application, 728
      repair mechanisms, 744, 745
   structure, 705
   topology
      DNA topoisomerases, 770
   types, 707
DNA gyrase
   DNA binding, 770
   inhibitors, 754
   quinolones
      binding, 774
   type II topoisomerase, 770
DNA helicases
   functions, 754
DNA methyltransferases
   anthracyclines, 782
   5-azacytidine, 783
   enzymology, 782
   functions, 754, 782
DNA polymerase III

properties, 756
DNA polymerase α
  acyclovir, 762
  aphidicolin
    allosteric inhibitor, 764
    inhibition, 764, 765
  arabinosylcytosine
    inhibition, 763
  functions, 756
  gene, 758
  suramin
    inhibition, 766
DNA polymerase β
  DNA repair synthesis, 756
  DNA synthesis, 757
DNA polymerase γ
  mitichondrial DNA replication, 757
DNA polymerase δ
  aphidicolin
    inhibition, 764
  function, 757
DNA polymerases
  anionic compounds
    inhibition, 766
  anthracyclines
    inhibition, 779, 780
  enzymology, 755
  Escherichia coli, 755
  eukaryotic, 756
  functions, 755
  herpes simplex virus
    acyclovir, 761
  human immunodeficiency virus
    DNA-directed, 759
    RNA-directed, 759
  inhibitors, 755–769
  phosphonoformate
    inhibition, 765
  prokaryotic, 755
  pyrimidine metabolism, 304
  viral, 758
    DNA-directed, 758
    RNA-directed, 758
DNA primase
  Okazaki fragments, 757
  suramin
    inhibition, 766
DNA replication
  DNA polymerases, 756
  enzymes, 755
  fidelity
    DNA polymerases, 757
DNA synthesis
  inhibition
    adenosine deaminase deficiency, 448
DNA synthetases
  nomenclature, 36
DNA topoisomerase I
  camptothecin
    inhibition, 775
  Escherichia coli, 770
  human
    genes, 773
  mammalian, 770, 772
    actinomycin, 782
  prokaryotic, 771
DNA topoisomerase II
  acridines
    inhibition, 777
  actinomycin
    inhibition, 782
  amsacrine, 777
  anthracyclines
    inhibition, 779

ellipticine, 781
etoposide
  inhibition, 778
human
  genes, 773
mammalian, 770, 772
prokaryotic, 770
DNA topoisomerases
  bacterial
    functions, 771
  enzymology, 769
  eukaryotic, 771
  functions, 754, 769
  inhibition, 769
  inhibitors, 773
  mammalian
    functions, 772
  prokaryotic, 770
  yeast
    function, 772
DNA virus
  antiviral chemotherapy, 314
Docosahexaenoic acid
  arachidonic acid
    metabolism, 169
Dodecane, 1,2-diamino-
  protein kinase C
    antagonist, 545
Dolichol cycle
  glycosidases, 367
Domains
  proteins, 13
DOPA, α-allyl-
  aromatic amino acid decarboxylase
    inhibitor, 246
DOPA, α-chloromethyl-
  aromatic amino acid decarboxylase
    inhibitor, 246
DOPA, α-difluoromethyl-
  aromatic amino acid decarboxylase
    inhibitor, 246
DOPA, α-ethynyl-
  aromatic amino acid decarboxylase
    inhibitor, 246
DOPA, α-fluoromethyl-
  aromatic amino acid decarboxylase
    inhibitor, 246
DOPA, α-vinyl-
  aromatic amino acid decarboxylase
    inhibitor, 246
DOPA decarboxylase
  pyridoxal-dependent systems, 227
Dopamine
  biosynthesis, 222
  hydroxylation
    dopamine β-hydroxylase, 128
Dopamine β-hydroxylase
  hydroxylation, 128
  inhibitors, 129
    multisubstrate, 131
Dotriacolide
  activity, 671
Doxorubicin — *see* Adriamycin
Doxycycline
  resistance, 111
DPI 201-106
  Na^+,K^+-ATPase
    inhibition, 210
Drug binding
  reversible
    intercalating agents, 708
Drug dependence
  reversible, 115
Drug design

enzyme cascades, 482
Drug detoxification
  enzyme cascades, 102
Drug interactions
  P-450 inhibition, 134
Drug resistance
  antibiotics, 90
  bacteria
    mechanism, 98
  cell envelope, 95
  disease treatment, 92
  gene mobility, 116
  genes, 91
  metabolic by-pass mechanisms, 113
  prevalence, 91
Drug susceptibility
  cell envelope, 95
  microorganisms, 95
Drug targets
  metal ions, 177
Drug tolerance
  metal ions, 97
Drug transport
  drug resistance, 110, 111
DTPA — *see* Acetic acid, diethylenetriaminepenta-
Dwarfism
  zinc deficiency, 180

Echinocandin B
  sterols, 361
Echinocandins
  antifungal activity, 588
  glucan synthase inhibitors, 587
Echinomycin
  intercalating agent
    DNA, 704, 718
Econazole
  fungal resistance, 114
Ecto-5'-nucleotidase
  biochemistry and immunology, 456
  inhibitors, 457
Edema
  prostaglandins, 151
Edman degradation
  proteins, 17
EDTA — *see* Acetic acid, ethylenediaminetetra-
Eglin
  inhibitor
    emphysema, 71
EHNA — *see* Adenine, *erythro*-9-(2-hydroxy-3-
    nonyl)-
Eicosanoids
  from arachidonic acid, 149
Eicosapentaenoic acid
  arachidonic acid
    metabolism, 168
  hyperlipidemia, 169
  inflammatory disease, 169
Eicosatrienoic acid, epoxy-
  arachidonic acid
    metabolism, 160
Elases
  inhibition
    boranes, 71
Elastase
  binding, 70
  inhibition, 71
    benzoxazinones, 79
  inhibitors, 79, 422
  peptidyl inhibitors, 422
Elastatinal
  binding, 70
Elastin
  elastase substrate

structure, 422
Elastinal
  elastase inhibitor, 423
Electron density map
  proteins
    structure, 24
Electron microscopy
  proteins
    quaternary structure, 27
Electrophoresis
  enzymes, 6
Electrophorus electricus
  acetylcholinesterase, 383
Ellipticine
  antitumor activity, 717, 720, 781
  DNA polymerases
    inhibition, 781
  intercalating agent
    DNA, 704
Ellipticine, 9-hydroxy-
  antitumor activity, 781
Ellipticinium acetate, 9-hydroxy-2-methyl-
  antitumor activity, 781
Ellman's reagent
  thiol reactivity
    protein modification, 30
Elongation
  protein synthesis, 815
Elongation factor
  eukaryotic, 816
  protein synthesis, 815
Emerimicins
  structure, 596
Emetine
  history, 90
Eminase
  thrombolytic agent, 496
Emphysema
  treatment, 422
Enalapril
  ACE inhibition, 394, 401
  analogs
    ACE inhibitors, 402, 404
Enalaprilat
  inhibitors
    angiotensin converting enzymes, 72
Endoglycosidases
  classification, 355
Endometrial
  aromatase, 138
Endometriosis
  aromatase, 138
Endonexin
  inhibitor
    phospholipase A_2, 523
Endonuclease III
  strand-break repair, 746
Endopeptidase
  neutral
    inhibition, 410
Endothelium-derived relaxing factor
  prostacyclin, 153
Enheptin
  antiamoebic activity, 729
  DNA damage, 726
Enkephalin
  conformation
    ACE inhibitors, 404
Enkephalinase
  inhibition, 71
  inhibitors, 394, 410
Enniatins
  cell membrane
    permeability, 598

potassium complex
  structure, 601
Enol esters
  mechanism based inhibitors
    enzymes, 80
Enoximone
  conformation, 507
Enterochromaffin-like cells
  gastric acid secretion, 195
Entramin
  antiamoebic activity, 729
  DNA damage, 726
Enzyme cascades
  definition, 482
  DNA damage, 481–497
  drug design, 482
  pharmacological regulation, 482
  properties, 482
  purine metabolism, 445
  purines, 443–475
Enzymes
  acidic groups
    modification, 30
  active site directed irreversible inhibitors, 76
  amino acids
    polymers, 2
  aromatic residues
    modification, 28
  assay, 6
  basic groups
    modification, 31
  binding energy, 55
  catalysts, 45
  catalytic efficiency, 51
  classification, 5, 33
    requirements, 34
  competitive inhibition
    kinetics, 62
  energetics, 51
  flexibility, 59
  folding, 46
  free energy of activation, 66
  hydroxyl groups
    modification, 29
  inhibition, 61–83
    transition state analogs, 66
  irreversible inhibition, 74
  isofunctional
    nomenclature, 37
  isolation, 5
  kinetics, 48
  lifetime, 46
  mechanism-based irreversible inhibitors, 79
  molecular fit, 56
  multisubstrate analogs, 63
  nomenclature, 33
    formalization, 34
  physical properties, 9
  purification, 5
  reversible inhibition
    kinetics, 62
  reversible inhibitors, 62
  salting out, 6
  solubility, 6
  specific activity, 6
  specificity, 2, 48, 66
    induced-fit mechanism, 51
  structure, 1–31, 45–59
  thiol groups
    modification, 29
  transition state
    stability, 51
Enzymic cleavage
  proteins

sequence analysis, 16
6-Epicastanospermine
  glycosidases
    inhibition, 378
Epidaunorubicin
  intercalating agent
    DNA, 704
Epidermal growth factor receptor
  swainsonine, 374
Epidermolysis byllosa
  collagenase
    inhibitors, 412
16-Epigitoxigenin
  glycosylation
    activity, 209
Epimerases
  classification, 36
Epipodophyllotoxins
  DNA topoisomerase II inhibition, 721, 778
    structure–activity relationship, 779
  mitotic inhibitor, 785
  resistance, 778
Epoxides
  glycosidase
    inhibition, 380
Ergosta-5,7,24(28)-trien-3β-ol
  biosynthesis, 354
Ergosterol
  binding
    amphotericin B, 595
  biosynthesis
    inhibition, 356
  biosynthesis inhibitors
    fungicides, 360
Erythema
  prostaglandins, 151
Erythromycin
  mode of action, 827
  resistance, 100, 101, 828
  ribosomal receptor site, 827
Erythromycin A
  resistance, 109
Erythromycin esterase
  production
    drug resistance, 109
Erythropoietin
  prostaglandins, 150
Escherichia coli
  cell envelope
    drug susceptibility, 96
  DC2
    penicillins, 622
  β-lactam binding proteins, 615
  penicillin binding proteins, 622
    cefotaxime, 636
  porins, 613
  siderophores
    tonB inner membrane protein, 649
Esculetin
  lipoxygenase
    inhibition, 164
Eserine
  cholinesterase
    inhibitors, 385
Esterases
  inhibition, 102
  mechanism based inhibitors, 80
Estrogen
  prostacyclin, 155
Estrone
  aromatase, 137
Ethanol
  drug dependence, 115
Ethidium bromide

DNA binding agent, 767
  intercalating agent
    DNA, 709, 712
Ethylenes, triphenyl-
  protein kinase C
    antagonist, 545
Ethylphosphonic acid, L-1-amino-
  alanine racemase
    inhibition, 571
Etoposide
  DNA topoisomerase II
    inhibition, 778
  metabolic activation, 779
Exoglycosidases
  classification, 365
Exons
  relation to domains, 15
Exonuclease III
  strand-break repair, 746
Exopolysaccharides
  cell walls, 556

Familial hypercholesterolemia
  LDL receptors, 334
Fatty acids
  unsaturated
    phosphilipase A$_2$, inhibition, 526
FCE 22101
  synthesis, 676
Fenoctimine
  proton pump blockers, 205
Fenton reaction
  DNA strand-breakage, 735
  free radicals, 187
Ferredoxins
  DNA strand-breakage, 738
Ferritin
  iron binding, 180
Fibrin
  enzyme cascade
    coagulation, 484
    description, 486
Fibrinolysis
  coagulation, 483
  defective, 494
  drugs, 495
  enzyme cascade, 492
  fibrinolysis
    defective, 494
  function, 492
  inhibitors
    endogenous, 494
    synthetic, 497
  prostacyclin, 155
Fibronectin
  swainsonine, 374
Filipin
  sterols, 361
Filter elution
  DNA strand-breakage
    measurement, 733
Firamycin SV
  clinical application, 801
Fish
  ischaemic heart disease, 169
FLAP
  radiosensitization, 743
Flavin
  oxygen
    activation, 123, 125
Flavoenzymes
  inhibition, 83
Flomoxef
  activity, 648

Floxuridine
  pyrimidine antagonist, 312
Flucloxacillin
  resistance, 103
Fluconazole
  antifungal activity, 140, 358
Flucytosine
  antifungal agent, 326, 327
  fungal resistance, 115
Fluorescent ethidium displacement
  DNA
    intercalating agent binding, 711
Fluorophosphonate, isopropoxymethyl
  nerve gases
    antidotes, 385
Flurbiprofen
  cyclooxygenase
    inhibition, 160
Folding
  proteins, 15
Folic acid
  biochemistry, 272
  biological role, 258
  biosynthesis, 258
  structure, 258
Folic acid, dihydro-
  reduction
    dihydrofolate reductase, 258
Folic acid, tetrahydro-
  biosynthesis, 258
  drug resistance, 113
Formadicine
  occurrence, 658
Formamidopyrimidine-DNA glycosylase
  strand-break repair, 746
Formidacillin
  synthesis, 634
Formycin A
  antiparasitic agent, 324
Formycin B
  antiparasitic agent, 296, 324
  malaria, 475
  purine nucleoside phosphorylase
    inhibitor, 459
Formycin B, 5'-deoxy-5'-iodo-
  purine nucleoside phosphorylase
    inhibitor, 460
Fortimicin A
  resistance, 107
Fosfomycin (phosphonomycin)
  resistance, 111
  transferase
    inhibition, 564
  UDP-*N*-acetylmuramoyl pentapeptide biosynthesis
    inhibition, 568
Fosfopril
  angiotensin converting enzyme
    inhibition, 403
FPL 52694
  histamine release
    inhibition, 196
Free radical reactions
  DNA strand-breakage, 734
  metal ions, 181
Fructose-bisphosphate aldolase
  nomenclature, 36
D-Fructose-1,6-bisphosphate D-glyceraldehyde-3-
      phosphate lyase
  nomenclature, 36
Ftorafur
  pyrimidine antagonist, 312
L-Fucitol, 1,5-dideoxy-1,5-imino-
  glycosidases
    inhibition, 371

Fumarate hydratase
  nomenclature, 36
Fumarate reductase
  inhibition, 271
  inhibitors, 294
Fungi
  cell envelope
    drug susceptibility, 97
  drug resistance, 114
  drug susceptibility, 95
  wall polymer biosynthesis
    inhibitors, 585
  walls
    structure, 583
Furans, nitro-
  coulometry
    DNA damage, 740
  DNA damage, 726
  DNA strand-breakage, 739
  mutagenicity, 729
Furantoin, nitro-
  Gram-negative infections, 729
Furazone, nitro-
  Gram-negative infections, 729
Furosemide
  diuretics, 267
Fusidic acid
  resistance, 93, 101, 110
Futile cycling
  DNA strand-breakage, 736

GABA — *see* Butyric acid, γ-amino-
Gabaculine
  amino acid transaminase
    inhibitors, 244
  GABA-T
    inhibitors, 82, 242
DL-Gabaculine
  inhibitor
    bacterial cell wall biosynthesis, 564
D-Galactal
  galactosidase inhibition, 74
D-Galactitol, 1,5-dideoxy-1,5-imino-
  glycosidases
    inhibition, 371
β-Galactosidases
  inhibition
    alkylation, 78
    cyclohexene polyol epoxides, 76
    D-galactal, 74
Ganciclovir
  DNA polymerases
    inhibition, 762
Gangliosides
  protein kinase C
    antagonist, 545
Gardnerella vaginalis
  metronidazole, 728
Garlic
  platelet aggregation inhibition
    phosphilipase A2, 525
Gastric acid
  H+,K+-ATPase
    role in secretion, 194
  secretion
    inhibitors, 195
    prostaglandins, 151
Gastrin
  gastric acid secretion, 194, 195
Gastrointestinal tract
  metal ions
    uptake, 179
  sulfonamides, 266
Gel filtration

enzymes, 7
Gene amplification
  p-aminobenzoic acid enzyme, 265
Genes
  mobility
    drug resistance, 116
  transfer
    drug resistance, 118
Genetic engineering
  chemotherapy, 326
Genistein
  inhibitor
    tyrosine kinase, 548
Gentamycin
  cell membrane disorganization, 591
  Pseudomonas, 633
  resistance, 91–93, 95, 100, 107
Gitaloxigenin
  structure–activity relationship, 207
Gitaloxin
  clinical use, 206
Gitoxigenin
  glycosylation
    activity, 209
  structure–activity relationship, 207
Glaucoma
  sulfonamides, 267
Glc-swainsonine
  synthesis, 375
Glipizide
  diabetes, 268
Globomycin
  antibiotic, 583
Glomerular disease
  prostacyclin, 154
β-Glucan
  cell walls
    fungal, 554, 583
Glucans
  structure, 583
Glucan synthase
  inhibitors, 587
D-Glucitol, 2-acetamido-1,5-imino-1,2,5-trideoxy-
  glycosidases
    inhibition, 371
Glucocorticoids
  diguanyl hydrazone derivatives
    Na+,K+-ATPase inhibition, 207
Glucosephosphate isomerase
  nomenclature, 36
D-Glucose-6-phosphate ketol isomerase
  nomeclature, 36
α-D-Glucose-1,6-phosphomutase
  nomenclature, 35
α-Glucosidase
  yeast
    inhibition, 379
β-Glucosidase
  inhibition
    almond emulsin, 379
    cyclohexene polyol epoxides, 76
Glucosidases
  inhibition
    castanospermine, 376
    membrane-bound, 360
Glutamate decarboxylase
  inhibition, 240
  inhibitors, 244
  pyridoxal-dependent systems, 225
Glutamate racemase
  pyridoxal-dependent systems, 222, 229
L-Glutamic acid
  biosynthesis, 215
Glutamic acid, α-chloromethyl-

glutamate decarboxylase
inhibitor, 245
Glutamic acid, α-difluoromethyl-
glutamate decarboxylase
inhibitor, 245
Glutamic acid, α-fluoromethyl-
glutamate decarboxylase
inhibitor, 245
Glutamic-pyruvic transaminase
pyridoxal-dependent systems, 215
Glutaric acid, tetramethylene-
aldose reductase
inhibitor, 293
Glutathione reductase
inhibition, 271
inhibitors, 294
Glutethimide, amino-
aromatase
inhibition, 138
steroid hormones
inhibition, 347
L-Glycerol-3-phosphate oxidase
nomenclature, 37
Glycine, aminoethoxyvinyl-
1-amino-1-carboxycyclopropane synthetase
inhibition, 248
Glycine, ethynyl-
alanine racemase
inhibitor, 246
Glycine, 3-methoxyvinyl-
alanine transaminase
inhibitor, 241
glutamate decarboxylase
inhibitor, 240
Glycine, propargyl-
alanine transaminase
inhibitor, 241
amino acid transaminase
inhibitor, 244
cystathione γ-synthetase
inhibitor, 248
pyridoxal phosphate enzymes
inhibitor, 82
Glycine, vinyl-
amino acid transaminase
inhibitors, 244
reactions at C-4, 238
Glycocalyx
cell walls, 556
Glycogen
metabolism
castanospermine, 378
Glycogen phosphorylase
pyridoxal-dependent systems, 239
Glycogen storage disease
hyperuricemia, 468
Glycollate oxidase
inhibition
2-hydrobut-3-ynoate, 83
Glycoproteins
castanospermine, 376
deoxynojirimycin, 376
hybrids
swainsonine, 373
processing
neutral glycosidases, 366
processing inhibition
swainsonine, 372
viral
swainsonine, 373
Glycosidases
hydrolysis, 366
inhibition, 68
inhibitors, 365

biochemistry, 371
occurrence, 369
in metabolism, 366
neutral
glycoprotein processing, 366
occurrence, 366
Gnidimacrin
protein kinase C, 544
Gold compounds
rheumatoid arthritis, 181, 468
Gout
hyperuricemia, 468
uric acid, 305
xanthine oxidase, 143
inhibitors, 469
Gradient sedimentation
DNA strand-breakage
measurement, 732
Gramicidin S
antibiotic, 592
Gramicidins
conductance, 593
structure, 592
Gram-negative bacteria
cell wall structure
antibiotic resistance, 561
EDTA, 561
outer membrane, 561
peptidoglycan
β-lactam antibiotics, 613
Gram-positive bacteria
cell walls
secondary wall polymers, 559
peptidoglycan
β-lactam antibiotics, 613
Griseofulvin
fungal resistance, 115
Growth factors
prostacyclin, 154
GTPase reaction
ribosomes, 816
Guaiaretic acid, nordihydro-
lipoxygenase
inhibition, 164
Guaneran
prodrug
6-thioguanine, 308
Guanine, 8-amino-
purine nucleoside phosphorylase
inhibitor, 459
Guanine, 8-amino-3-deaza-
purine nucleoside phosphorylase
inhibitor, 459
Guanine, 8-amino-9-benzyl-
in vitro studies, 465
purine nucleoside phosphorylase
inhibitor, 463
Guanine, 8-amino-9-heteroaryl-
in vitro studies, 465
Guanine, 8-amino-9-(2-thienylmethyl)- (PD
119,229)
hyperuricemia, 469
in vitro studies, 465
in vivo studies, 465
purine nucleoside phosphorylase
inhibitor, 463
Guanine, 8-amino-9-(3-thienylmethyl)-
purine nucleoside phosphorylase
inhibitor, 463
Guanine, 3-deaza-
antiviral agent, 321
purine antagonist, 310
Guanine, 9-(3,4-dihydroxybutyl)-
antiviral agent, 321

Guanine, 9-(1,3-dihydroxy-2-propoxymethyl)-
antiviral agent, 321
Guanine, 9-(2-hydroxyethoxymethyl)-
antiviral agent, 320
Guanine, 6-thio-
trimetrexate
synergism, 286
Guanine, 9-β-D-xylofuranosyl-
antiviral agent, 320
Guanine deaminase
purine metabolism, 303
Guanosine, acyclo-
in vitro studies, 465
in vivo studies, 466
Guanosine, 8-amino-
in vitro studies, 463
in vivo studies, 465
organ transplantation, 474
purine nucleoside phosphorylase
inhibitor, 459
Guanosine, 8-amino-2′-deoxy-
purine nucleoside phosphorylase
inhibitor, 460
Guanosine, 8-amino-2′-nordeoxy-
purine nucleoside phosphorylase
inhibitor, 461
Guanosine, carba-2′-nordeoxy-
purine nucleoside phosphorylase
inhibitor, 461
Guanosine, 3-deaza-
antiviral agent, 321
Guanosine, 9-deaza-
purine nucleoside phosphorylase
inhibitor, 460
Guanosine, 2′-nor-2′-deoxy-
antiviral agent, 321
Guanosine monophospate
cyclic
protein kinases, 541–543
second messenger response, 501
Guanylic acid, 3-deaza-
antiviral agent, 321
Gynecomastia
aromatase, 138

H-142
renin inhibitor, 416
H-261
renin inhibitor, 416
HA-558
inhibitor
phosphodiesterase, 505
HA-1004
inhibitor
protein kinases, 540
Haber–Weiss reaction
DNA strand-breakage, 735
Haemophilus vaginalis
metronidazole, 728
Hairy-cell leukemia
2′-deoxycoformycin, 453
Halocarbons
cytochrome *P*-450
inhibition, 135
Halomycins
isolation, 801
Harmaline
monoamine oxidases
inhibitor, 127
Harmine
monoamine oxidases
inhibitor, 127
α-Helix
proteins, 9

Helix–coil transition
DNA strand-breakage
measurement, 732
Hemochromatosis
metal ions, 180
Hemophilia A
coagulation defect, 488
Hemoprotein cofactor
oxygen
activation, 124
Hemostasis
definition, 483
enzyme cascades, 481–497
mechanism, 483
Hemostatis
defects, 483
diseases, 483
Heparin
anticoagulant
mechanism, 490
use in thrombosis, 490
antithrombin III, 487
fractions
anticoagulant, 490
low molecular weight derivatives
anticoagulant, 490
oral formulation
development, 490
semisynthetic derivatives
anticoagulant, 490
structure, 490
Herpes simplex virus
acyclovir, 761
phosphonoformate, 765
thymidine kinase, 762
Herpes simplex virus 1
Ara-A
inhibition, 763
DNA polymerase, 758
gene, 758
Hetacillin
synthesis, 620
12-HETE
psoriasis, 158
Hexitols, 1,5-dideoxy-1,5-imino-
glycosidases
inhibition, 371
Hexokinase
isoenzymes
nomenclature, 37
protein
dynamics, 26
5-Hexyne-1,4-diamine
ornithine decarboxylase
inhibitor, 245
5-Hexyne-1,4-diamine, 1-methyl-
ornithine decarboxylase
inhibitor, 245
Hex-5-ynoic acid, 4-amino-
glutamate decarboxylase
inhibitor, 244, 245
HH-3209
inhibitor
myosin light chain kinase, 547
HH-3709
inhibitor
myosin light chain kinase, 547
HH-3909
inhibitor
myosin light chain kinase, 547
High performance liquid chromatography
enzymes, 7
Hinge-bending enzymes
protein

dynamics, 26
Hirudin
  thrombin inhibitor, 490
Histamine
  biosynthesis, 222
  gastric acid secretion, 194
Histamine-secreting cells
  gastric acid secretion, 195
Histidine, α-chloromethyl-
  histidine decarboxylase
    inhibitor, 245
Histidine, α-fluoromethyl-
  histidine decarboxylase
    inhibitor, 227, 245
Histidine decarboxylase
  inhibitors, 245
  pyridoxal-dependent systems, 227
Histone H1
  release
    ethidium, 722
HMGCoA reductase — *see* Hydroxymethylglutaryl-
    CoA reductase
Hoechst 33258
  DNA binding agents, 767
  nonintercalating agent
    DNA, 709
Homoserine dehydrase
  pyridoxal-dependent systems, 238
Homoserine dehydrogenase
  nomenclature, 38
Homothienamycin
  activity, 678
Hormone sensitive lipase
  classification, 36
HRE 664
  activity, 677
Human cytomegalovirus
  ganciclovir
    treatment, 762
Human immunodeficiency virus
  antiviral agents, 318
  endonuclease activity, 759
  function, 758
  phosphonoformate, 765
  reverse transcriptase, 759
    error rate, 759
    gene, 759
    inhibitors, 761
  RNAse H activity, 759
    gene, 759
Human neutrophil elastase
  elastase inhibitor, 422
  inhibitors
    heterocyclic covalent, 426
Humoral immunity
  adenosine deaminase deficiency, 447
Hydralazine
  prostacyclin, 155
Hydrazides
  monoamine oxidases
    inhibitor, 127
Hydrazines
  aminobutyrate transaminase, 242
  cytochrome *P*-450
    inhibition, 135
  dopamine β-hydroxylase
    inhibitors, 129
  monoamine oxidases
    inhibitor, 127
  pyridoxal-dependent systems
    inhibitors, 239
Hydrazines, phenyl-
  monoamine oxidases
    inhibitor, 127

Hydrochlorothiazide
  structure, 257
Hydrogen bonding
  enzymes
    specificity, 59
Hydrolases
  inhibitors, 365–386
  nomenclature, 34
  purine metabolism, 302
Hydrophobic core
  proteins, 9
Hydrophobicity
  enzymes, 9
Hydrophobicity plot
  proteins
    structure, 11
Hydroxamic acid
  collagenase
    inhibitors, 412
  lipoxygenase
    selective inhibitors, 166
Hydroxyapatite chromatography
  DNA strand-breakage
    measurement, 733
Hydroxylamine
  aminobutyrate transaminase
    inhibitor, 242
  nerve gases
    antidotes, 385
  pyridoxal-dependent systems
    inhibitors, 239
11β-Hydroxylase
  inhibition, 347, 358
Hydroxyl radicals
  DNA strand-breakage, 734, 736
Hydroxymethylglutaryl-CoA reductase
  cholesterol
    biosynthesis, 334, 335
  enzymology, 336
  genomic regulation, 339
  inhibition, 336
  inhibitors, 337
Hyperallantoinemia
  fructose infusion
    model for hyperuricemia, 469
Hypergonadism
  zinc deficiency, 180
Hypertension
  control
    angiotensin II, 400
  monoamine oxidases
    inhibitors, 125
  prostaglandins, 151
Hypertrophic villous synovitis
  rheumatoid arthritis, 467
Hyperuricemia
  hypoxanthine-guanine phosphoribosyltransferase,
      466
  purine metabolism, 468
  purine nucleoside phosphorylase
    inhibition, 469
  uric acid, 305
Hyperuricemic states
  purine metabolism, 444
Hypocholesterolemics
  history, 343
Hypoprothrombinemia
  β-lactam antibiotics, 618
Hypouricemia
  purine nucleoside phosphorylase deficiency, 458
Hypouricosuria
  purine nucleoside phosphorylase deficiency, 458
Hypoxanthine
  antiparasitic agents, 323

catabolism
  xanthine oxidase, 143
Hypoxanthine, 9-β-D-arabinofuranosyl-
  antiviral agent, 319
  DNA polymerases
    inhibition, 763
Hypoxanthine, 9-(phosphonoalkyl)-
  purine nucleoside phosphorylase
    inhibitor, 463
Hypoxanthine-guanine phosphoribosyltransferase
  nitrogen metabolism, 301
  purine metabolism, 466
Hypoxic cells
  intercalation
    DNA, 716

Ibuprofen
  arachidonic acid
    metabolism, 160
Ido-swainsonine
  synthesis, 375
Idoxuridine
  antiviral agent, 315
Imazodan
  inhibitor
    phosphodiesterases, 507, 510, 511
Imidazole-5-carboxamide, 4-amino-
  antiparasitic agent, 325
Imidazole-4-carboxamide, 5-fluoro-1-β-D-
    ribofuranosyl-
  antiviral agent, 322
Imidazole radicals
  DNA strand-breakage, 741
Imidazoles
  fungal resistance, 114
  susceptibility, 92
Imidazoles, nitro-
  DNA damage, 726
  DNA strand-breakage, 739
Imidazoles, 2-nitro-
  antitumor drugs, 727
  coulometry
    DNA damage, 740
  nitro radical anion, 741
  radiosensitizing drugs, 740
  ruthenium complexes
    radiosensitization, 743
  selective reduction under hypoxia, 738
Imidazoles, 5-nitro-
  coulometry
    DNA damage, 740
  nitro radical anion, 741
Imidazoles, N-trityl-
  antifungal activity, 140
Imidazo[1,2-a]pyrazines
  structure–activity relationships, 197
Imidazopyridines
  H+,K+-ATPase
    inhibition, 197
Imidazo[1,2-a]pyridines
  structure–activity relationships, 197
Imipenem
  cilastatin combination
    activity, 674
  mode of action, 674
  resistance, 95, 102, 110
Immobilization
  enzymes, 8
Immune deficiency
  purine nucleoside phosphorylase
    deficiency, 458
Immune impairment
  mechanisms
    adenosine deaminase deficiency, 449

Immunodeficiency
  purine metabolism, 446–466
Immunosuppression
  adenosine deaminase inhibitors, 453
  enzyme cascades, 443–475
Immunosuppressive agents
  adenosine deaminase inhibitors, 449
  3-deazaadenosine, 454
Immunosuppressive diseases
  swainsonine, 374
Immunotoxicity
  mechanisms
    ATP depletion, 452
Imuran
  purine antagonist, 307
Indole, 4',6-diamidino-2-phenyl-
  nonintercalating agent
    DNA, 709
Indomethacin
  cyclooxygenase
    inhibition, 160
  inhibitors
    cyclooxygenase/lipoxygenase, 165
    phosphilipase A2, 527
Inflammation
  prostaglandins, 151
Inflammatory disease
  collagenase inhibitors, 394
Influenza virus
  glycoproteins
    swainsonine, 373
  hemagglutinin
    castanospermine, 376
Infrared spectroscopy
  proteins
    secondary structure, 22
Ingenol
  protein kinase C, 544
Inhibitors
  definition, 532
Initiation complexes
  protein synthesis, 818
Initiation factor
  protein synthesis, 815
Inosine, 7-deaza-
  antiparasitic agent, 324
Inosine, 9-deaza-
  antiparasitic agent, 324
  malaria, 475
  purine nucleoside phosphorylase
    inhibitor, 460
Inosine, 3'-deoxy-
  antiparasitic agent, 324
Inosine, 5'-deoxy-5'-halo-9-deaza-
  purine nucleoside phosphorylase
    inhibitor, 460
Inosinic acid
  biosynthesis, 301
Insulin receptors
  swainsonine, 374, 377
Integral proteins
  cell walls, 589
Interactive molecular graphics
  proteins
    tertiary structure, 25
Intercalating agents
  binding
    DNA, 711
  cytotoxicity, 721
  DNA, 703–723
  overview, 704
  polyfunctional, 717
Intercalation
  binding rate, 712

cellular consequences, 721
complexes
  general features, 709
  DNA
    crystallography, 712
    molecular mechanical methods, 713
    NMR studies, 714
    sites, 712
    theoretical studies, 713
  evidence, 709
  mechanism of binding, 712
  shuffling mechanism, 718
Intercalators
  DNA
    monofunctional, 714
Interleukin 1
  psoriasis, 474
  pyrexia, 151
Interleukin 2
  cyclosporin, 473
  rheumatoid arthritis, 467
International Union of Biochemistry
  enzymes
    nomenclature, 34
Intrinsic resistance
  drugs, 97
Ion exchange chromatography
  enzymes, 7
Ionophores
  cell membrane
    permeability, 598
  ion selectivity, 601
  metal binding, 188
Ion transport
  active, 193–210
  passive systems, 175–190
Iproniazide
  monoamine oxidases
    inhibitor, 127
Ipronidazole
  antitumor drug, 727
  veterinary use, 729
Iproplatin
  anticancer activity, 729
  radiosensitization, 743
Irehdiamine A
  nonintercalating agent
    DNA, 709
Iron
  chelation therapy, 182
    synergism, 186
  deficiency, 181
  free ion concentration
    plasma, 178
  Gram-negative bacteria
    scavenging agents, 649
  metal ion transport, 180
  transport protein, 180
Isoclavam
  ring system, 682
Isoconazole
  fungal resistance, 114
Isocoumarin, 3-alkoxy-7-amino-4-chloro-
  alkylation
    enzymes, 80
Isocoumarin lactone, dichloro-
  enzyme inhibition, 81
Isoenzymes
  nomenclature, 37
Isogabaculine
  GABA-T
    inhibitors, 243
Isomerases
  nomenclature, 34

Isomorphic replacement
  X-ray crystallography
    proteins, 24
Isopenam
  synthesis, 681
Isopenicillin N
  penicillin biosynthesis, 610
Isopenicillin N synthetase
  penicillin biosynthesis, 610
Isoquinolinesulfonamides
  inhibitors
    protein kinases, 533, 538
Isoquinolinesulfonamides, piperazinyl-
  inhibitors
    protein kinase C, 545
Isothiocyanate, allyl
  proton pump blockers, 205
Itraconazole
  antifungal activity, 140, 358
Ivermectin
  fumarate reductase
    inhibitor, 294
Izumenolide
  activity, 671

Juvenile diabetes
  autoimmune disease, 468
  cyclosporin, 466

K-252a
  inhibitor
    protein kinases, 539
K-252b
  inhibitor
    protein kinase C, 545
Kanamycin
  resistance, 92, 107
Kasugamycin
  drug dependence, 116
  mode of action, 822
  resistance, 100, 101, 107, 822
  ribosomal RNA
    methylated residues, 820
Kelatorphan
  enkephalinase inhibitor, 411
Kethoxal
  chemical footprinting
    ribosomal binding sites, 830
Ketoconazole
  antifungal agent, 140, 354, 356–358
  cholesterol
    inhibition, 340
  fungal resistance, 114
  steroid hormones
    inhibition, 348
Ketones, chloromethyl
  binding, 70
Ketones, α-diazomethyl
  pyroglutamyl peptide hydrolase
    inhibition, 76
Ketones, α-halomethyl
  enzyme inhibition
    alkylation, 78
Ketones, peptidyl halomethyl
  cysteine proteases
    inhibition, 76
Ketones, peptidyl methyl
  inhibitors, 70
Ketones, trifluoromethyl
  enzyme inhibitors, 70
Kibdelins
  peptidoglycan polymerization
    inhibition, 574
Klebsiella spp.

drug resistance, 95
Klenow fragment
  DNA polymerases, 755
KT-5270
  inhibitor
    protein kinases, 538
KT-5822
  inhibitor
    protein kinases, 542
Kukolja rearrangement
  penicillins, 645
Kynureninase
  pyridoxal-dependent systems, 225

β-Lactam antibiotics
  biosynthesis, 610, 655–696
  β-bridged
    synthesis, 692
  cell wall agents, 609–650
  chiral synthesis, 677
  double esters
    prodrugs, 618
  history, 655
  mode of action, 694
  monocyclic
    activation, 661
  non-β-lactam analogs, 693
  pharmacology, 617
  porins, 564
  side effects, 618
  total synthesis, 610
  X-ray and IR absorption data, 695
Lactamases
  cephalosporins, 625
  chromosomally-mediated, 663
  classification, 104, 663–665, 669
    clavulanic acid inhibition, 665
  clavulanic acid, 662
  Gram-negative bacteria, 104
  Gram-positive bacteria, 104, 616
  inhibition, 102
  inhibitors
    mechanism based, 81
    mode of action, 694
  plasmid mediated, 663
  production, 614
    drug resistance, 114
β-Lactams
  isoxazoyl-
    inhibition, 104
  resistance, 102
Lactate dehydrogenase
  isoenzymes
    nomenclature, 37
  multisubstrate analog inhibitors, 65
  nomenclature, 35
D-Lactate dehydrogenase
  inhibition
    2-hydrobut-3-ynoate, 83
L-Lactate:NAD$^+$ oxidoreductase
  nomenclature, 35
L-Lactate oxidase
  inhibition
    2-hydrobut-3-ynoate, 83
Lactivicin
  activity, 694
Lactones
  haloenol
    alkylation, 81
  mechanism based inhibitors
    enzymes, 80
Lanostene, 32-oxy-
  HMGCoA reductase
    inhibition, 340

Lanosterol
  demethylation
    14α-demethylase, 138
  HMGCoA reductase
    inhibition, 340
Lanosterol 14α-demethylase
  HMGCoA reductase
    inhibition, 340
  inhibition, 354, 357
Lasalocid
  alkali metal complexes, 602
LDL receptors
  cholesterol homeostasis, 334
Lead
  bioaccumulation, 176
  poisoning, 184
    chelation therapy, 185
Lesch–Nyhan syndrome
  hypoxanthine-guanine phosphoribosyltransferase, 466
Leucine aminopeptidase
  binding, 70
Leucomycin A$_3$
  resistance, 109
Leucovorin
  from folic acid
    inhibition, 275
Leukemia
  2′-deoxycoformycin, 453
  L1210 murine leukemia
    neplanocin A, 456
Leukotriene D$_4$
  phospholipase A$_2$, 521
Leukotrienes
  biosynthesis, 517
  chemotactic properties, 159
  inflammation, 158
  inflammatory pain, 158
  5-lipoxygenases, 149
  peptido
    bronchoconstrictors, 158
  structure, 157
Leupeptin
  elastase inhibitor, 423
Ligases
  nomenclature, 34
Lincomycin
  structure, 828
Lincosamides
  resistance, 101
Lipiarmycin
  antibacterial activity, 798
Lipid A
  biosynthesis
    inhibition, 580
  lipopolysaccharides
    Gram-negative bacteria, 561
  protein kinase C, 545
Lipids
  phospholipases, 515
Lipocortins
  inhibitors
    phosphilipase A$_2$, 523
Lipolysis
  phosphodiesterases, 509
Lipomannan
  succinylated
    cell walls, 560
Lipopolysaccharides
  biosynthesis
    inhibition, 580
  cell walls, 555
  endotoxic activity
    bacteria, 561

protein kinase C, 545
Lipoprotein lipase
  nomenclature, 35
Lipoproteins
  cell walls
    bacteria, 563
    diaminopimelyl–diaminopimelyl bridges, 582
Lipoteichoic acids
  Gram-positive bacteria
    cell walls, 560
Lipoxin A
  arachidonic acid
    metabolism, 160
  protein kinase C, 160
Lipoxins
  arachidonic acid
    metabolism, 160
  biosynthesis, 517
Lipoxygenase
  arachidonic acid
    metabolism, 156
  cyclooxygenase
    dual inhibitors, 164
  inhibitors, 163
    bronchial anaphylaxis, 166
    inflammation, 167
    mechanism, 163
  products
    airway disease, 158
    biological properties, 158
    inflammatory disease, 158
    vasculature, 159
  selective inhibitors, 166
5-Lipoxygenase
  calcium dependence, 157
  substrate specificity, 157
Lisinopril
  angiotensin converting enzyme
    inhibition, 401
  structure, 404
Liver alcohol dehydrogenase
  inhibition
    salicylate, 63
Liver pyruvate kinase
  nomenclature, 37
Local anesthetics
  inhibitors
    phosphilipase A2, 527
Lomustine
  hypoxic cells, 731
Lumphopenia
  combined immune deficiency diseases, 447
Lupus erythematosus
  prostacyclin, 154
Luzopeptin
  intercalating agent
    DNA, 717, 718
LY195115
  inhibitor
    phosphodiesterase, 507
Lyases
  inhibition, 248, 358
  nomenclature, 34
Lymphocytes
  3-deazaadenosine, 454
  psoriasis, 474
  resting
    adenosine deaminase inhibitors, 451
Lyophospholipid, alkyl-
  protein kinase C
    antagonist, 545
Lysine
  ω-amino acid transaminases, 220
Lysozyme

folding, 47
Gram-negative bacteria
  cell walls, 561
  structure, 2, 366
D-Lyxitol, 1,4-dideoxy-1,4-imino-
  glycosidases
    inhibition, 371

Macrolide antibiotics
  cytochrome *P*-450
    inhibition, 135
Macrolides
  susceptibility, 92
Macrotetralides
  cell membrane
    permeability, 598
Maduramicin
  coccidostatic activity, 602
Magnesium
  biochemical competition, 178
Malaria
  purine metabolism, 474
L-Malate hydrolase
  nomenclature, 36
Maleic acid
  amino acid transaminase
    inhibitor, 240
Manganese
  biochemical competition, 179
Mania
  phosphodiesterases, 511
D-Mannitol, 1,4-diamino-1,4-dideoxy-
  glycosidase
    inhibition, 379
D-Mannitol, 1,4-dideoxy-1,4-imino-
  glycosidases
    inhibition, 371
D-Mannoheptose, L-glycero-
  lipopolysaccharides
    Gram-negative bacteria, 561
Mannojirimycin, deoxy-
  coronavirus mouse hepatitis virus
    E-2 glycoprotein, 377
  glycosidases
    inhibition, 370, 378
D-*Manno*-2-octulosonates, deoxy-
  inhibitors
    lipopolysaccharide biosynthesis, 580, 581
  lipopolysaccharides
    Gram-negative bacteria, 561
Mannopeptides
  fungi
    walls, 583
Mannosidases
  inhibition
    swainsonine, 371
  isolation, 369
β-Mannosidases
  inhibitors, 68
α-Mannosidosis
  swainsonine, 371, 375
Manoalide
  inhibitor
    phosphilipase A2, 525
Marcarbomycin
  phosphoglycolipid antibiotic, 578
Marcellomycin
  antitumor activity, 780
Mass spectrometry
  proteins, 18
Maxam–Gilbert method
  DNA
    sequence analysis, 18
Maytansine

microtubule assembly, 786
M&B 693 — *see* Sulfapyridine
MD-805
  thrombin inhibitor, 490
MDL 72145
  monoamine oxidases
    inhibition, 125
Mebendazole
  fumarate reductase
    inhibitor, 294
Mecillinam
  activity, 628, 677
  resistance, 99
Mediator release
  phosphodiesterases, 509
Melanoma cells
  lung colonization
    swainsonine, 374
α-Melanotropin
  conformation
    ACE inhibitors, 404
Melarsen oxide
  glutathione reductase
    inhibitor, 295
Menogaril
  intercalating agent
    DNA, 704, 715
Mepacrine
  inhibitor
    phosphilipase A$_2$, 527
Mercury
  bioaccumulation, 176
  poisoning, 184
Messenger RNAs
  polycistronic
    protein synthesis, 818
Metal complexes
  antitumor drugs, 743
Metal ions
  binding, 180
    pharmaceuticals, 181
  biochemical competition, 178
  biological systems, 175
  drug targets, 177
  physiological role, 176
  physiological transport, 179
  transport proteins, 180
Metalloenzymes
  metal ions, 176
Metallopeptidases
  classification, 396
Metalloproteases
  inhibitors, 71
Metalloproteinases
  inhibition, 397
Metallothionein
  metal ion binding, 180
Metal poisoning
  non-essential metals, 184
Methacycline
  resistance, 111
Methanesulfonyl fluoride
  cholinesterase
    inhibitors, 385
Methazolamide
  glaucoma, 268
Methicillin
  enzyme inhibition, 104, 106
  β-lactamases, 104
  resistance, 94, 99, 103
  staphylococci resistance, 616
  synthesis, 619
Methionine, *S*-adenosyl-
  decarboxylation, 222

transmethylation
  inhibition, 453
Methionine γ-synthetase
  inhibition, 248
Methotrexate
  autoimmune diseases, 444
  binding
    dihydrofolate reductase, 74
  t-butyl ester
    dihydrofolate reductase inhibitors, 281
  clinical uses
    toxicity, 290
  dihydrofolate reductase
    inhibition, 258
    inhibitor, 271, 275
  esters
    dihydrofolate reductase inhibitors, 281
  interactions
    dihydrofolate reductase, 277
  membrane transport, 278
  rheumatoid arthritis, 468
  stereospecificity
    dihydrofolate reductase, inhibition, 279
  synergistic combinations
    sulfonamides, 265
Methotrexate, 5-deaza-
  dihydrofolate reductase
    inhibitor, 279
Methotrexate, dichloro-
  dihydrofolate reductase
    inhibitor, 280
Methylases
  purine metabolism, 302
Methylation
  adenosine deaminase deficiency, 448
  S-adenosylmethionine, 454
    3-deazaadenosine, 454
  aminoglycoside antibiotic resistance, 832
  inhibition
    immunotoxicity, 452
5,10-Methylenetetrahydrofolate dehydrogenase
  pyridoxal-dependent systems, 230
Methylmalonyl-CoA CoA-carbonylmutase
  nomenclature, 36
Methylmalonyl-CoA mutase
  nomenclature, 36
Methyltransferase
  kasugamycin
    resistance, 822
Metronidazole
  activity, 630
  antimicrobial agent, 726
  DNA damage, 726
  DNA strand-breaking, 728
  platinum complexes
    radiosensitization, 743
  radical anion, 741
  susceptibility, 92
Metyrapone
  steroid hormone
    inhibition, 347
Mevacor
  HMGCoA reductase
    inhibition, 339
Mevinolin
  HMGCoA reductase
    inhibition, 337, 338
Mezerein
  protein kinase C, 544
Mezlocillin
  activity, 633
  resistance, 114
Michaelis–Menten equation
  enzymes

kinetics, 48
Miconazole
    antifungal agent, 140, 354, 356
    fungal resistance, 114
Micrococcus luteus
    bacitracins activity, 572
Microtubules
    asssembly
        bacitracins activity, 783
    functions, 783
Milrinone
    conformation, 507
    inhibitor
        phosphodiesterase, 507
        protein kinases, 538, 541
Minocycline
    resistance, 111
Misonidazole
    antitumor drug, 727
    hypoxic cells, 730
    reduction, 740
Mithramycin
    nonintercalating agent
        DNA, 709
Mitomycin
    DNA binding agents, 767
Mitomycin C
    anticancer activity, 729
    antitumor drug, 728
    DNA strand-breakage
        redox cycling, 737
Mitotic inhibitors
    DNA strand-breakage, 784
Mitoxantrone
    antitumor activity
        cardiotoxicity, 780
    antitumor agent, 727
        clinical use, 716
    cardiotoxicity, 729
    DNA strand-breakage
        redox cycling, 737
    intercalating agent
        DNA, 704
Mixopyronins
    antibacterial activity, 799
MK 0787
    synthesis, 672
MK 0791
    metabolism, 674
MK 906
    steroid 5α-reductase
        inhibition, 353
ML-7
    inhibitor
        myosin light chain kinase, 547
ML-8
    inhibitor
        myosin light chain kinase, 547
ML-9
    inhibitor
        myosin light chain kinase, 547
MLS antibiotics
    dimethylation, 828
MM 13902
    activity, 671
Modules
    enzymes
        binding sites, 47
Moenomycin
    peptidoglycan biosynthesis
        inhibition, 566
    peptidoglycan transglycosylase inhibition, 579
    phosphoglycolipid antibiotic, 578
    structure, 579

Molecular dynamic simulation
    protein dynamics, 27
Molecular modeling
    collagenase
        substrates, 412
    proteinases, 396
Molecular replacement
    proteins
        tertiary structure, 25
Monensin
    alkali metal complexes, 602
    growth promoter, 602
Monoamine oxidase
    activation, 125
    inhibitors, 126, 511
    type A, 125
    type B, 125
Monobactams
    inhibitors, 656
    detection, 658
    discovery, 656
    hetero-substituted, 660
    resistance, 114
    structure–activity relationships, 659
Monobactams, 4-alkyl-
    synthesis, 659
Monocarbams
    activity, 661
Monoclonal antibodies
    chemotherapy, 326
Monophosphams
    activity, 661
Monosulfactams
    synthesis, 661
Morin rearrangement
    penicillins, 626
Moxalactam
    activity, 641
    resistance, 102, 114
    side effects, 642, 685
Mulandocandin
    glucan synthase inhibitors, 587
Multienzyme systems
    nomenclature, 38
Multiple sclerosis
    autoimmune disease, 468
Murein
    cell walls, 557
Muscarinic receptors
    gastric acid secretion, 196
Muscle pyruvate kinase
    nomenclature, 37
Mutases
    classification, 36
Mutations
    chromosomal
        drug resistance, 101
    ribosomes
        drug resistance, 100
Myasthenia gravis
    autoimmune disease, 468
Mycobacterium lufu
    sulfonamide resistance, 265
Mycoplasmas
    sterols
        biosynthesis, 334
Mycosis fungoides
    2′-deoxycoformycin, 453
Myocardial contractility
    phosphodiesterases, 510
Myocardial infarction
    hyperuricemia, 468
    prostacyclin, 155
Myocardial injury

prostacyclin, 155
Myoinositol, 1,2-anhydro-
    glycosidase
        inhibition, 380
Myokinase
    purine metabolism, 301
Myosin light chain kinase
    antagonists, 546
    characteristics, 546
    inhibitors, 547
    purine metabolism, 546
    therapeutic utility, 546

NADPH
    binding
        dihydrofolate reductase, 284
Nafcillin
    resistance, 103
Nafidimide
    intercalating agent
        DNA, 704
Naftifine
    squalene oxidase
        inhibition, 359
Nalidixic acid
    antibacterial agents, 773
    resistance, 101
Naphthalenesulfonamides
    inhibitors
        protein kinases, 533, 538
Naphthoquinones
    antitumor drugs, 728
1,8-Naphthyridine
    antibacterial agents, 773
Natamycin
    fungal resistance, 114
Natural killer cells
    rheumatoid arthritis, 467
NC 1300
    H⁺,K⁺-ATPases
        inhibition, 204
Neamin
    resistance, 100
Neighbor exclusion model
    intercalation, 711
Neisseria gonorrhoeae
    drug resistance, 93
Neisseria meningitidis
    drug resistance, 95
Neocarzinostatin
    antitumor agents, 327
Neomycin
    resistance, 100, 107
Neosalvarsan
    history, 90
Neostigmine
    cholinesterase
        inhibitors, 385
Nephrotoxicity
    cephaloridine, 618
Neplanocin
    purine antagonist, 311
Neplanocin-A
    adenosyl-L-homocysteine hydrolase
        inhibitor, 456
Neplanocin-A, 3-deaza-
    adenosyl-L-homocysteine hydrolase
        inhibitor, 456
Nerve gases
    antidotes, 385
Netilmicin
    resistance, 107, 109
Netropsin
    DNA binding agents, 767

nonintercalating agent
    DNA, 709
Neurospora crassa
    cell envelope
        drug susceptibility, 97
Neurotoxins
    protein kinase C
        antagonist, 545
Neurotransmission
    postsynaptic
        phosphodiesterases, 511
    presynaptic
        phosphodiesterases, 511
Neutron diffraction
    protein dynamics, 27
Nicotinamide
    Na⁺,K⁺-ATPase
        inhibition, 210
Nifedipine
    prostacyclin, 155
Nigericin
    alkali metal complexes, 602
Nikkomycin Z
    structure, 585
    transport, 586
Nikkomycins
    chitin synthesis inhibition, 584, 585
Nimorazole
    antimicrobial agent, 726
    DNA strand-breaking, 728
Nisin
    inhibitor
        bacterial cell wall biosynthesis, 565
Nitracrine
    antitumor agents, 716
    intercalating agent
        DNA, 704
Nitric oxide
    endothelium-derived relaxing factor, 153
Nitriles
    dopamine β-hydroxylase
        inhibitors, 129
Nitrites
    5-nitroimidazoles, 740
Nitro aromatic compounds
    DNA strand-breakage, 737
Nitrocefin
    β-lactam hydrolysis
        assay, 625
Nitro heterocyclic drugs
    DNA strand-breakage, 728, 737
Nitro homocyclic drugs
    coulometry
        DNA damage, 740
    DNA strand-breakage, 737
Nitro radical anion
    decomposition, 741
Nocardicin A
    antibacterial activity, 657
    biosynthesis, 656
Nocardicin C
    hydrolysis, 657
Nocardicins
    activity, 658
    detection, 656
    history, 656
    hydrolysis, 656
Nocardinic acid, 3-amino-
    synthesis, 657
NOESY
    NMR
        protein structure, 26
Nogalamycin
    intercalating agent

DNA, 715
Nojirimycin
glycosidases
inhibition, 370
Nojirimycin, deoxy-
glycoproteins, 376
glycosidases
inhibition, 370
Nolinium bromide
proton pump blockers, 205
Nomenclature
enzymes, 33
formalization, 34
Nonactin
potassium complex
structure, 602
structure, 598
Nonsteroid anti-inflammatory drugs
prostaglandin synthesis
inhibition, 160
Norepinephrine
deamination
monoamine oxidase, 125
hydroxylation
dopamine β-hydroxylase, 128
neurotransmission
inhibition, 512
Norfloxacin
DNA binding, 774
DNA gyrase
inhibitor, 774
resistance, 101
Norleucine, 6-diazo-5-oxo-
glutamine amino transfer
inhibition, 301
L-Norvaline, 5-nitro-
GABA-T
inhibitors, 243
Novobiocin
DNA polymerase II
inhibition, 775
resistance, 93, 102
NTA — *see* Acetic acid, nitrilotri-
Nuclear enzymes
eukaryotes, 795
Nuclear magnetic resonance spectroscopy
chemotherapy, 326
proteins
tertiary structure, 25
Nuclear Overhauser effect
proteins
tertiary structure, 26
Nucleic acid
sequencing
chemotherapy, 326
synthesis
mutations, 101
Nucleoside diphosphokinases
purine metabolism, 301
pyrimidine metabolism, 303
Nucleoside kinases
salvage pathway, 301
Nucleoside monophosphokinases
purine metabolism, 301
pyrimidine metabolism, 303
Nucleoside phosphorylases
purine metabolism, 302
Nucleosides, dideoxy-
antiviral agents, 319
DNA polymerases
inhibitor, 760
Nucleotidases
purine metabolism, 302
Nucleotides

prodrugs, 326
Nystatin
absorption, 595
fungal resistance, 114

Obofluorin
activity, 694
Octapeptins
cell membrane disorganization, 590
resistance, 110
structure, 590
Ofloxacin
DNA gyrase
inhibitor, 774
resistance, 101
Okazaki fragments
DNA polymerases, 757
synthesis
DNA primase, 757
OKY-1581
myocardial infarction, 163
Old yellow enzyme
nomenclature, 34
Oleandomycin
resistance, 109
Oleic acid
protein kinase C, 544
Oligonucleoside methylphosphonates
antiviral agents, 327
Oligonucleotides
synthesis
chemotherapy, 326
Oligosaccharides
anticoagulant, 490
Olivanic acid
activity, 671
biosynthesis, 671
detection, 662, 671
history, 656
Olivomycin
nonintercalating agent
DNA, 709
Omeprazole
H^+,K^+-ATPase
inhibition, 198
mode of action, 198
structure–activity relationships, 202
therapeutic activity, 204
Operators
regulatory genes, 794
Optical rotatory dispersion
proteins
secondary structure, 21
Oral contraceptives
platelet aggregation, 155
Organophosphorus compounds
serine protease inhibition, 76
Organosulfonates
cholinesterase
inhibitors, 385
Organ transplantation
immunosuppression, 444, 469
Ornidazole
antimicrobial agent, 726
DNA strand-breaking, 728
Ornithine
ω-amino acid transaminases, 220
Ornithine, α-chloromethylornithine decarboxylase
inhibitor, 245
Ornithine, α-vinyl-
ornithine decarboxylase
inhibitor, 245
Ornithine aminotransferase

inhibitors, 243
pyridoxal-dependent systems, 221
structure, 219
Ornithine decarboxylase
inhibitors, 245
pyridoxal-dependent systems, 226
Orotate phosphoribosyltransferase
pyrimidine metabolism, 303
Orotic acid, 5-fluoro-
pyrimidine antagonist, 312
Osyterol
binding protein, 339
endogenous regulators, 339
Oubain
ATPase
dephosphorylation inhibition, 381
structure–activity relationship, 207
1-Oxacefamandole
activity, 641
1-Oxacephalothin
activity, 641
Oxacillin
resistance, 103
Oxacillinase
occurrence, 106
1-Oxadethiacephems
activity, 684
β-lactamase stability, 685
synthesis, 684
3-Oxa-1-dethiacephams
properties, 690
2-Oxa-1-dethiacephems
synthesis, 688
1-Oxadethiahomocephem
synthesis, 692
1-Oxadethiapenams
synthesis, 679
Oxamazin
activity, 696
structure–activity relationships, 662
Oxanthrazole
antitumor agent
clinical use, 717
intercalating agent
DNA, 704
Oxapenam
naturally occurring, 656
Oxidoreductases
nomenclature, 34
Oxidosqualene cyclase
cholesterol
biosynthesis, 342
Oxolinic acid
DNA gyrase
inhibition, 773
Oxyanion hole
proteinases, 395
Oxygen
activation
enzyme prosthetic groups, 123
Oxygenases
DNA strand-breakage, 123–143
Oxygen radicals
DNA strand-breakage, 734
Oxyntic cells
gastric acid secretion, 195
Oxysterol
HMGCoA reductase, 339
Oxytetracycline
resistance, 111

PABA — *see* Benzoic acid, *p*-amino-
Pachystermines
discovery, 655

PADAC
chromogenic cephalosporin, 626
Papain
binding, 70
inhibition, 431
nomenclature, 34
Papaverine
inhibitor
phosphodiesterases, 510
Papulacandins
activity
Candida albicans, 589
antifungal activity, 588
glucan synthase inhibitors, 587
Paracetamol — *see* Acetaminophen
Parasitic diseases
purine inhibitor therapy, 444
purine metabolism, 474
Pargyline
monoamine oxidases
inhibitor, 126
squalene oxidase
inhibition, 360
Parietal cells
gastric acid secretion, 195
Parkinsonism
phosphodiesterases, 511
Paromomycin
drug dependence, 115
resistance, 107
PD 116,124 — *see* 1,3-Propanediol, 2-[(2,8-
diamino-6-hydroxy-9*H*-purin-9-yl)-
methoxy]-,
PD 119,229 — *see* Guanine, 8-amino-9-(2-
thienylmethyl)-
Penam
conformation, 613
ring system
β-lactam antibiotics, 609
Penam sulfones
inhibitors
β-lactams, 81
Penems
activity, 675
β-lactamase stability, 675
nuclear modification, 677
properties, 674
ring system, 675
synthesis, 676
Penems, (2,3)-α-methylene-
activity, 678
Penicillamine
amino acid transaminase
inhibitors, 244
copper chelation, 184
rheumatoid arthritis, 468
Penicillanic acid, 6-amino-
penam ring system, 609
Penicillanic acid, β-bromo-
activity, 669
mechanism of action, 670
Penicillanic acid sulfone
activity, 669
Penicillinase
cephalosporins, 625
classification, 104
Penicillin binding protein 2
mecillinam binding, 628
Penicillin binding protein 2'
Staphylococcus aureus, 616
Penicillin binding proteins
bacterial cell walls, 568
cellular function, 614
mechanism, 103

structure, 669
Penicillin G
  biosynthesis, 610
  history, 618
  total synthesis, 612
Penicillin N
  biosynthesis, 610
Penicillins
  allergic response, 618
  antibacterial activity, 620
  antipseudomonal, 633
    antibacterial activity, 634
  binding proteins
    Escherichia coli, 615
  biosynthesis, 610
  cell wall agents, 609–650
  history, 90, 618
  β-lactamases, 104
  modified
    Arnstein tripeptide, 649
  nuclear analogs, 677
  protein binding, 98
  resistance, 91, 93, 102, 618
  structure–activity relationships, 627
  susceptibility, 92
  toleration, 98
Penicillins, amidino-
  activity, 627
Penicillins, azido-
  reduction, 620
Penicillins, benzyl-
  enzyme inhibition, 106
  resistance, 93, 94
Penicillins, 6α-formamido-
  activity, 634
Penicillins, 6β-halo-
  inhibitors
    β-lactamase, 81
Penicillins, isoxazolyl-
  activity, 622
  enzyme inhibition, 106
  β-lactamases, 104
Penicillins, ureido-
  activity, 633
Penicillin V
  biosynthesis, 610
  history, 618
  total synthesis, 612
Penicillium chrysogenum
  penicillin biosynthesis, 610
Penicillopepsin
  renin inhibitor, 417
Pentostatin — *see* Coformycin, 2′-deoxy-
Peplomycin
  medical applications, 731
Pepsin
  inhibitors, 72, 74
  renin inhibitor, 417
Pepstatin
  aspartic protease inhibitor, 73
  binding
    pepsin, 434
  renin inhibitor, 416
Peptaibophols
  structure, 596
Peptic ulcers
  proton pump blockers, 195
Peptidase
  classification, 397
  inhibitors, 391–436
  types, 394
Peptides
  bonds, 9
  sequential degradation, 16

Peptide synthetases
  nomenclature, 36
Peptidoglycan
  bacterial cell walls
    β-lactam antibiotics, 612
  biosynthesis, 228, 229
    antibiotic inhibition, 568
    bacterial cell walls, 564
    lipid cycle inhibition, 571
  cell walls, 554
  constitution, 557
  polymerization
    antibiotic inhibition, 573
  structure, 613
Peptidoglycan transglycosylase
  inhibition
    moenomycin, 579
Peptidyl aldehydes
  cysteine proteinases
    inhibition, 432
Peptidylchloromethanes
  cysteine proteinases
    inhibition, 431
Peptidyldiazomethanes
  cysteine proteinases
    inhibition, 432
Peptidylfluoromethanes
  cysteine proteinases
    inhibition, 432
Peptidyl semicarbazones
  cysteine proteinases
    inhibition, 432
Peptidyl transferase
  chloramphenicol
    inhibition, 827
Peptidyl transferase center
  components, 829
  protein synthesis
    ribosomes, 815
Peptiglycan
  cell walls, 554
  structure, 557
Peripheral proteins
  cell walls, 589
Permeability
  drug resistance, 110
Permeation
  drug resistance, 110
Phage-mediated conjugation
  drug resistance, 118
Phase problem
  X-ray crystallography
    proteins, 23
Phenanthridines
  binuclear
    DNA intercalating agents, 720
  intercalating agents
    DNA, 704, 712
1,10-Phenanthroline
  bacteriocidal action, 186
Phenethylamines
  deamination
    monoamine oxidase, 125
Phenindione
  anticoagulant, 489
Phenothiazines
  Na⁺,K⁺-ATPase
    inhibition, 210
  protein kinase C
    antagonist, 545
Phenylalanine, α-allyl-
  aromatic amino acid decarboxylase
    inhibitor, 246
Phleomycins

medical applications, 731
Pholipomycin
    structure, 579
Phorbol
    protein kinase C, 544
1-Phosphadethiacephem 1-oxides
    synthesis, 687
Phospharamidon
    enzyme inhibitor, 71
Phosphatases
    inhibition, 68
    purine metabolism, 302
Phosphatidylglycerol
    cytoplasmic membrane, 589
Phosphatidylinositol
    cytoplasmic membrane, 589
Phosphinic acids
    pepsin inhibition, 74
Phosphodiesterase
    calmodulin dependent, 504
    cAMP, inhibition
        3-deazaadenosine, 454
    forms, 501
    isolation, 502
    role, 501
    selective inhibitors, 501
        pharmacology, 509
    type I inhibitors, 505
    type II inhibitors, 505
    type III inhibitors, 505
        phosphodiesterase, 506
Phosphofluoridate, diisopropyl
    cholinesterases
        inhibitors, 384
Phosphoglucomutase
    nomenclature, 35
Phosphoglycolipids
    antibiotics, 578
Phospholipases
    amino acid composition, 518
    lipid formation
        role, 515
Phospholipases A$_2$
    activation, 521
    activator protein, 521
    agents acting against, 515
    antibodies, 518
    arachidonate or lyso-PAF specific, 519
    calcium-dependent
        properties, 517
        structure, 518
    calcium independent
        occurrence, 519
    inhibition, 521
    inhibitor protein, 521
    inhibitors, 523
    intracellular
        properties, 517
    membrane-associated, 518
    regulation, 520
    substrate mimics, 526
Phospholipids
    membrane
        structure, 522
N-Phosphonacetyl-L-aspartate
    structure, 64
Phosphonamidates
    enzyme inhibitors, 71
Phosphonate, 1-aminoethyl-
    alanine racemase
        inhibitor, 246
Phosphonoacetic acid
    DNA polymerases
        inhibition, 765

structure–activity relationships, 766
Phosphonoformate
    DNA polymerases
        inhibition, 765
    structure–activity relationships, 766
Phosphonomycin — *see* Fosfomycin
Phosphoramidates
    enzyme inhibitors, 71
Phosphoramidic acid
    collagenase
        inhibitors, 412
Phosphoramidon
    angiotensin converting enzyme
        inhibition, 403
    crystal structure, 434
Phthalazineacetic acids
    aldose reductase
        inhibitor, 292
Physostigmine
    cholinesterase
        inhibitors, 385
Picoprazole
    H$^+$,K$^+$-ATPase
        inhibition, 198
Pimaricin
    absorption, 595
Piperacillin
    activity, 633, 678
    resistance, 95, 114
Piperazine, bisdioxo-
    anticancer agents, 187
Piperidines, *n*-alkyl-
    Na$^+$,K$^+$-ATPase
        inhibition, 210
Piperidines, 4-phenyl-
    dihydropterin reductase
        inhibitors, 291
Piperidines, tetrahydro-
    Parkinson-like symptoms, 128
Pirenzepine
    M$_1$-blockers
        gastric acid secretion, 196
Piritrexem
    dihydrofolate reductase
        inhibitor, 283, 286, 288
Pivampicillin
    synthesis, 622
Pivmecillinam
    oral absorption, 629
Plasmids
    drug resistance, 116
    β-lactamases
        coding, 614
    penicillin resistance, 629
Plasmin
    derivation from plasminogen, 492
    fibrinolysis, 492
    inhibition
        drugs, 496
    inhibitors, 494
Plasminogen activators, 492
    action on plasminogen, 492
    biosynthesis, 496
    drugs, 496
    immunochemical classification, 493
    inhibitors, 494
    modified activators, 496
    prourokinase
        mechanism of fibrin specificity, 493
        therapeutic uses, 495
    release, 496
    thrombosis, 495
    tissue type
        fibrinolysis, 493

fibrin-specific antibodies, 496
  inhibitors, 494
  mechanism of fibrin specificity, 493
  release, 496
  therapeutic uses, 495
  urokinase
    fibrin-specific antibodies, 496
    therapeutic uses, 495
Plasmodia
  sulfonamides
    resistance, 265
Platelet activating factor
  arachidonic acid, 520
  phospholipase A_2, 515
Platelet aggregation
  inhibition, 509
  β-lactam antibiotics, 618
Platelets
  hemostasis, 483
Platinum(II), *cis*-diamminedichloro- — *see* Cisplatin
Platinum(II), *trans*-diamminedichloro-
  DNA binding agent, 768
Platinum, *cis*-dichlorobis(metronidazole)-
  radiosensitization, 743
Platinum-based drugs
  anticancer drugs, 742
  antitumor drugs, 728
Platinum compounds
  intercalating agents
    DNA, 717
Plutonium
  biological concentration, 176
  poisoning, 185
  transport
    transferrin, 180
Podophyllotoxin
  DNA topoisomerase II, 778
  mitotic inhibitor, 785
Poly(ADP-ribose)polymerase
  DNA strand-break repair
    inhibition, 748
Polyamine oxidase
  mitotic inhibitor, 128
Polyamines
  biosynthesis, 226
Polyene antibiotics
  resistance, 596
    fungal, 114
  sterols, 361
  structure, 594
  susceptibility, 92
Polylysine
  secondary structure, 21
Polymerase α
  nucleic acid base derivatives
    inhibition, 764
Polymers
  amino acids
    enzymes, 2
Polymorphonuclear leukocytes
  5-lipoxygenase, 157
Polymyxin B
  papain-cleaved derivative
    antibiotic synergism, 591
  protein kinase C
    antagonist, 545
Polymyxins
  antagonism
    metal ions, 591
  cell membrane disorganization, 590
  resistance, 110
    lipopolysaccharides, 591
  structure, 590
  susceptibility, 92

Polyoxin D
  structure, 585
Polyoxins
  analogs
    stability, 586
  chitin synthase
    inhibition, 579
  chitin synthase inhibitors, 585
Polypeptides
  enzymes, 2
Porcine pancreatic elastase
  inhibition
    7α-chloro sulfoxide cephalosporin, 81
Porfiromycin
  anticancer activity, 729
  antitumor drug, 728
Porins
  bacterial cell walls, 555, 563, 613
Post-translational modification
  polypeptides, 3
Potassium
  biochemical competition, 178
Potassium clavulanate
  clinical trials, 665
Prasinomycins
  phosphoglycolipid antibiotic, 578
Primaxin
  clinical use, 674
Prinzmetal's angina
  thromboxane B_2, 156
Proclavaminic acid
  biosynthesis, 667
Prodrugs
  β-lactam antibiotics, 618
  nucleotides, 326
  penicillins, 620
Proflavine
  DNA binding
    affinity for isolated sites, 711
  intercalating agent
    DNA, 704, 709, 712
Proglumide
  gastrin antagonist, 196
Prolinal, (*Z*)-L-Pro-L-
  prolyl endopeptidase
    inhibition, 433
Prolinal, (*Z*)-L-Val-L-
  prolyl endopeptidase
    inhibition, 433
Proline, 3-hydroxy-
  collagen
    post-translational modification, 3
Prolyl endopeptidases
  inhibitors, 433
Promoters
  DNA segments
    RNA polymerase binding, 794
Prontosil
  discovery, 90, 255
  metabolism, 256
1,3-Propanediol
  in vivo studies, 466
1,3-Propanediol, 2-[(2,8-diamino-6-hydroxy-9*H*-
    purin-9-yl)methoxy]- (PD 116,124)
  in vitro studies, 465
Propanol, 2,3-dimercapto-
  chelation therapy, 184
  lead decorporation, 185
Propidium
  intercalating agent
    DNA, 709, 712
Propionic acid, hydrazino-
  aminobutyrate transaminase
    inhibitor, 242

Prostacyclin
  biological properties, 152
  biosynthesis, 517
  cAMP, 152
  cardiothrombotic conditions, 170
  discovery, 149
  dosage form, 155
  endothelium-derived relaxing factor, 153
  fibrinolytic activity, 154
  hypotensive agent, 152
  pathological implications, 154
  plasminogen activator, 154
  platelet aggregation, 152
  prostaglandin H_2
    metabolism, 150
Prostacyclin synthase
  prostacyclin biosynthesis, 150
Prostaglandin G_2
  isolation, 149
Prostaglandin H_2
  isolation, 149
Prostaglandins
  anti-inflammatory properties
    aspirin, 160
  biosynthesis, 517
  cardiac output, 150
  cytoprotective agents, 195
  from arachidonic acid, 150
  hypertension, 151
  isolation, 147
  pathological implications, 151
  platelets, 150
  stable
    biological properties, 150
Prostaglandin synthetase
  unsaturated fatty acid affinity, 149
Protamine sulfate
  heparin antidote, 490
Proteases
  enzyme cascades, 482
  mechanism based inhibitors, 80
  regulation, 74
Protein binding
  β-lactam antibiotics, 618
Protein C
  coagulation
    inhibitor, 487
  deficiency
    thrombosis, 488
Protein G
  phospholipase A_2, 521
  swainsonine, 373
Protein kinase C
  agonists, 543
  antagonists, 543
  characteristics, 543
  deficiency, 543
  inhibitors, 545
    therapeutic utility, 545
  isoforms, 543
  substrates, 543
Protein kinases, 531–549
  activity modulation, 532
  agonists
    therapeutic utility, 539
  cAMP dependent, 534
    characteristics, 534
    inhibitors, 538
    structure, 534
    substrates, 534
  cGMP dependent, 541
    agonists, 541
    antagonists, 541
    characteristics, 541

    inhibitors, 542
  drug targets, 532
  inhibitors
    therapeutic utility, 542
  properties, 531
Protein L4
  ribosomal core protein, 829
Protein L11
  thiostrepton
    resistance, 824
Protein L15
  erythromycin, 829
Protein L16
  erythromycin
    binding, 829
Protein L27
  macrolides
    cross-linking, 829
Protein S
  coagulation, 487
Proteins
  architecture
    hierarchy, 12
  dynamics, 26
  β-pleated sheet, 11
  stability, 46
  structure
    classification, 12
    primary, determination, 16
    secondary, 12, 20, 22, 26
    super-secondary, 13
    tertiary, 15, 23
  β-turn
    structure, 11
  weak interactions, 9
α-Proteins
  structure, 12
Protein synthesis
  catalytic inhibitors, 816
  elongation, 815
  initiation, 815
  mechanism, 815
  miscoding
    aminoglycoside antibiotics, 831
  termination, 816
Protoporphyrin IX, *N*-(2-oxoethyl)-
  steroid hormones
    inhibition, 350
Proton pump blockers
  peptic ulcers, 195
Prourokinase
  plasminogen activator, 493
PS-5
  isolation, 671
Pseudomembranous colitis
  antibiotics, 617
Pseudomonas aeruginosa
  cephalosporins, 642
  drug resistance, 95
  drug susceptibility
    cell envelope, 95
  penicillin binding proteins
    cefotaxime, 635
  penicillins, 633
Psoriasis
  penicillins, 474
  purine inhibitor therapy, 444
Psychosis
  phosphodiesterases, 511
Pteridine, 2,4-diamino-6,7-dimethyl-
  binding
    dihydrofolate reductase, 284
  dihydropterin reductase
    inhibitors, 291

Pteridine, 2,4-diamino-6,7-diphenyl-
 dihydrofolate reductase
 inhibitor, 283
Pteridine, 2,4-diamino-6-methyl-
 dihydrofolate reductase
 inhibitor, 280
Pulmonary hypertension
 prostacyclin, 155
Purine, 2-amino-6-chloro-1-deaza-
 purine antagonist, 310
Purine, aryl-
 DNA polymerases
 inhibition, 763
Purine, 6-(cyclopentylthio)-9-ethyl-
 purine antagonist, 308
Purine, 9-β-D-(2-deoxyribofuranosyl)-6-methyl-
 antiparasitic agent, 325
Purine, 2,6-diamino-9-(2-hydroxyethoxymethyl)-
 antiviral agent, 321
Purine, 9-ethyl-6-mercapto-
 purine antagonist, 308
Purine, 6-mercapto-
 antineoplastic purine drugs, 306
 purine antagonist, 306
Purine, 6-methyl-
 antiparasitic agent, 325
Purine nucleoside kinases
 purine metabolism, 301
Purine nucleoside phosphorylase
 biochemistry, 458
 deficiency
 biochemistry, 458
 purine metabolism, 446
 symptoms, 458
 inhibition
 hyperuricemia, 469
 inhibitors, 459
 biochemical studies, 463
 immunological studies, 463
 in vivo studies, 465
 malaria parasite, 475
 purine metabolism, 457
 salvage pathway, 301
Purine nucleosides
 antiviral agents, 319
Purine nucleotides
 anabolism, 301
 interconversions, 301
Purine ribonucleotide, 6-(methylthio)-
 purine antagonist, 307
Purines
 antagonists, 306
 antiparasitic agents, 323
 drug targets, 299–327
 enzymes
 inhibitors, 444
 metabolism, 300
 enzyme cascades, 443–475
 immunodeficiencies, 446
6H-Purin-6-one, 2-amino-9-β-D-arabinofuranosyl-
 1,9-dihydro-
 purine nucleoside phosphorylase
 inhibitor, 461
Purity
 enzymes, 6
Putrescine
 biosynthesis, 222, 226
 ornithine decarboxylase
 inhibitor, 245
 oxidation
 polyamine oxidase, 128
Putrescine, α-fluoromethyl-
 ornithine decarboxylase
 inhibitor, 245

Putrescine, α-fluoromethyldehydroornithine decarboxylase
 inhibitor, 245
Pyrazofurin
 antiviral agent, 322
Pyrazole-5-carboxamide, 4-hydroxy-3-β-D-
 ribofuranosyl-
 antiviral agent, 322
Pyrazolidinones
 bicyclic
 activity, 693
Pyrazoline, phenyl-
 leukotriene production
 inhibition, 164
Pyrexia
 prostaglandins, 151
Pyridazinones, dihydro-
 inhibitors
 phosphodiesterase, 507
Pyridine, 4-phenyl-
 hydroxylated
 dihydropterin reductase inhibitors, 291
Pyridine, 4-phenyl-1,2,3,6-tetrahydro-
 dihydropterin reductase
 inhibitors, 291
Pyridine 2-aldoxime dodeciodide
 nerve gases
 antidotes, 385
2-Pyridine aldoxime methiodide
 nerve gases
 antidotes, 385
Pyridoxal
 dependent systems, 213–250
 inhibitors, 239
Pyridoxal 5′-phosphate — *see* Coenzyme B$_6$
Pyridoxamine-pyruvate transaminase
 multisubstrate analog inhibitors, 65
 pyridoxal-dependent systems, 215
Pyrimethamine
 binding
 dihydrofolate reductase, 277
 clinical uses
 toxicity, 290
 dihydrofolate reductase
 inhibitor, 271, 283
 enzyme inhibition, 37
 synergistic combinations
 sulfonamides, 265
 synthesis, 288
Pyrimidine, aryl-
 DNA polymerases
 inhibition, 763
Pyrimidine, 2,4-diamino-5-(4-hydroxy-3,5-
 diisopropylbenzyl)-
 dihydrofolate reductase
 inhibitor, 285
Pyrimidine, 5,6-dihydro-
 thymidylate synthetase, 783
Pyrimidine, fluoro-
 pyrimidine antagonists, 312
Pyrimidine arabinonucleosides
 antiviral agents, 317
Pyrimidine 2′-deoxynucleosides
 antiviral agent, 316
Pyrimidine nucleosides
 antiviral agents, 315
Pyrimidine nucleotides
 biosynthesis, 303
Pyrimidines
 antagonists, 311
 antiparasitic agents, 323
 drug targets, 299–327
 metabolism, 303
Pyroglutamyl peptide hydrolase

inhibition
α-diazomethyl ketones, 76
2-Pyrone, 5-benzyl-6-chloro-
inhibitors
α-chymotrypsin, 80
2-Pyrone, 6-chloro-
inhibitors
α-chymotrypsin, 80
Pyrophosphate, tetraethyl alkyl
nerve gases
antidotes, 385
Pyrrolidine, 2,5-dihydroxymethyl-3,4-dihydroxy-
glycosidases
inhibition, 370, 379
Pyruvate:carbon dioxide ligase
nomenclature, 36
Pyruvate carboxylases
nomenclature, 36
Pyruvate dehydrogenase
nomenclature, 38
quaternary structure, 15

Quantitative structure–activity relationships
chemotherapy, 326
sulfonamides, 261
Quaternary structure
proteins, 15, 27
Quercetin
protein kinase C
antagonist, 545
Quinacrine
intercalating agent
DNA, 704
Quinazoline, 2-amino-4-thio-
dihydrofolate reductase
inhibitor, 279
Quinazoline, 2,4-diamino-
dihydrofolate reductase
inhibitor, 279, 286
Quinine
history, 90
Quinoline, 8-hydroxy-
bacteriocidal action, 186
Quinolones
antibacterial agents, 773
DNA gyrase inhibitors, 754
DNA gyrase, 770
DNA gyrase inhibition
structure–activity relationships, 774
susceptibility, 92
Quinolones, fluoro-
antibacterial agents, 773
Quinones
dihydropterin reductase
inhibitors, 291
redox cycling
DNA strand-breakage, 737
Quinoxalines
intercalating agents
DNA, 704, 717, 718

Racemases
classification, 36
inhibitors, 246
Raman spectroscopy
proteins
secondary structure, 22
Ranitidine
peptic ulcers, 204
Rat liver ornithine decarboxylase
half-life, 46
Raynaud's syndrome
prostacyclin, 155
tromboxane synthase

inhibition, 163
Razoxane
anthracyclines, 187
Receptor-mediated endocytosis
LDL receptors, 334
Receptor-mediated signal transduction
phospholipase C, 521
Redox cycling
DNA strand-breakage, 736
Reductases
inhibition, 271–295
Regulatory domain
protein kinases, 532
Relaxation times
NMR
proteins, 25
Release factors
protein synthesis
termination, 816
Renaturation
spectrophotometric thermal
DNA strand-breakage, 732
Renin
inhibitors, 72, 394, 414
conformationally restricted, 421
statine-containing inhibitors, 416
Renin inhibitory peptide
development, 414
Repairase
nomenclature, 34
Replicase
nomenclature, 34
Resistance
antimicrobial drugs, 89–119
Resorptive cells
glucocorticoid
swainsonine, 374
Retenoic acid
protein kinase C, 545
Retrovir — *see* AZT
Retroviral proteases
inhibitors, 72
Retroviruses
transcription, 796
Reverse phase chromatography
enzymes, 6
Reverse transcriptase
AZT
inhibition, 760
function, 758, 759
genes, 758
HIV, 319
rifamycins, 809
RNA viruses
antiviral chemotherapy, 315
suramin
inhibition, 766
viral DNA polymerases, 758
Reverse turn
proteins
structure, 11
RGH-2958
thrombin inhibitor, 491
Rheumatoid arthritis
clinical description, 467
collagenase
inhibitors, 412
cyclosporin, 466
gold compounds, 181
immunosuppressive/immunomodulatory drugs
approved, 470
experimental, 472
immunosuppressive therapy, 468
joint pathology, 467

pathogenesis, 467
prostaglandins, 151
Rheumatoid factors
  role in rheumatoid arthritis, 467
Rhizobiotoxin
  cystathionase
    inhibition, 247
Rhodanese
  nomenclature, 34
Rhodanine-3-acetic acid
  aldose reductase
    inhibitor, 292
Ribavirin
  antiviral agent, 322
  phosphorylated, 322
Ribavirin, 2′,3′,5′-tri-*O*-acetate
  antiviral agent, 322
Ribonucleases
  purine metabolism, 302
Ribonucleoside diphosphate reductase
  inhibition, 448
Ribonucleotide reductases
  inhibition, 459
    immunotoxicity, 452
  purine metabolism, 301
  pyrimidine metabolism, 303
Ribonucleotides
  biosynthesis, 301
Ribose, 1′,2′-dideoxy-
    3′,5′-cyclic monophosphate
  protein kinases, 537
Ribosomal proteins
  eukaryotes, 814
  prokaryotes, 814
  S1
    crosslinking, colicin E3, 817
  S12
    streptomycin resistance, 831
  S21
    cleavage, colicin E3, 817
Ribosomal RNA
  agents interacting with, 813–837
  drug target
    bases, 831
  functions, 834
  methylated residues, 820
  receptor sites
    aminoglycoside antibiotics, 832
  16S
    methylated residues, 821
  secondary structure models, 834
  sedimentation coefficients, 814
Ribosomes
  bacterial, 100
  decoding site, 833
    aminoglycoside antibiotics, 831
  domain II, 830
  domain V, 830
  domains
    interactions, 827
  functional domains, 829
    interaction, 829
  functions
    antibiotics, 823
  mutations
    drug resistance, 100
  reconstitution systems, 814
  structure, 814
Ricin
  ribosomal RNA, 819
Rifabutin
  activity, 804
  clinical application, 801
  clinical evaluation, 808

Rifamide
  activity, 804
  clinical application, 801, 807
Rifampicin
  activity, 806, 807
  adverse effects, 808
  clinical application, 801, 807
  mechanism of action, 801
  resistance, 91, 93, 102, 802
Rifamycin AG
  activity, 804
Rifamycin B
  antibacterial activity
    structure, 802
  modifications, 804
  production, 801
  properties, 800
Rifamycin O
  structure, 800
Rifamycin S
  modifications, 804
  structure, 800
Rifamycin S, 3-carboxy-
  synthesis, 806
Rifamycin SV
  activity, 803
  aminomethyl derivatives
    activity, 804
  clinical application, 800, 807
  modifications, 804
  thioethers
    activity, 804
Rifamycin SV, 3-formyl-
  activity, 806
Rifamycins
  activity, 808
  antibacterial activity
    structure, 802
  chemotherapeutic properties
    modulation, 806
  clinical use, 807
  origin, 800
  pharmacokinetic relationships, 806
  properties, 801
  resistance, 110
  RNA polymerase inhibition, 102
  structural modifications, 803
  structure–activity relationships, 806
  synthesis, 800
  transcribing enzymes
    inhibition, 793–809
  X-ray structure, 803
Rifamycins, 3-amino-
  activity, 806
Rifamycins, azinomethyl-
  activity, 806
Rifamycins, 25-*O*-deacetyl-
  activity, 804
Rifamycins, 27-*O*-demethyl-
  activity, 804
Rifamycins, imidazo-
  activity, 804
Rifamycins, 3-hydrazino-
  activity, 806
Rifapentine
  activity, 806
  clinical application, 801
  clinical evaluation, 807, 808
Rifaximin
  activity, 804
  clinical application, 801, 808
Rifazine
  activity, 804
Ristocetin

peptidoglycan biosynthesis
  inhibition, 567
peptidoglycan polymerization
  inhibition, 573
RNA
  base mutation
    aminoglycoside resistance, 832
  cooperative interactions
    protein, 835
  double strand
    viruses, 796
  negative strand
    viruses, 796
  positive single strand
    viruses, 796
RNA polymerase I
  suramin
    inhibition, 766
RNA polymerases
  anthracyclines
    inhibition, 779, 780
  antibiotic action, 102
  Archaebacteria, 794
  DNA-directed
    pyrimidine metabolism, 303
  Escherichia coli
    daunomycin inhibition, 780
  eukaryotes
    inhibitors, 800
    nucleus, 795
    organelles, 795
  functions, 754
  prokaryotes, 794
    inhibitors, 797
  viral
    RNA-directed, 758
RNA replication
  viruses, 796
RNase L
  protein synthesis suppression, 326
RNA synthetases
  nomenclature, 36
RNA viruses
  antiviral chemotherapy, 315
Ro-03-8799
  antitumor drug, 727
  hypoxic cells, 730
Ro 20-1724
  inhibitor
    phosphodiesterases, 509–511
Ro 22-4839
  myosin light chain kinase, 547
Rolipram
  inhibitor
    phosphodiesterases, 509–511
Ronidazole
  antimicrobial agent, 726
  veterinary use, 729
Rowley equation
  DNA strand-breakage, 732
RP 40749
  proton pump blockers, 205
RRM-188
  renin inhibitor, 418
RS-2039
  gastrin antagonist, 196
RS 82856
  inhibitor
    phosphodiesterase, 507
RSU-1069
  antitumor drug, 727
  hypoxic cells, 731
  radiosensitizing drugs, 740
Ruthenium complexes
  radiosensitization, 743

S-1623
  fibrinolysis, 497
Saccharin, *N*-acyl-
  enzyme acylation, 79
Saccharomyces cerevisiae
  cell envelope
    drug susceptibility, 97
Salicylates
  inhibitor
    dehydrogenase NADH, 63
Salmonella typhi
  drug resistance, 93
Salmonella typhimurium
  antibiotic resistance, 92
  cell envelope
    drug susceptibility, 96
Salvage pathways
  ribonucleotides, 301
Salvarsan
  history, 90
Sanger method
  DNA
    sequence analysis, 19
Sangivamycin
  purine antagonist, 310
α-Sarcin
  ribosomes
    eukaryotic, 818
Sarpicillin
  synthesis, 622
Satranidazole
  DNA damage, 726
  DNA strand-breakage, 729, 738
SC-9
  protein kinase C, 545
SC-10
  protein kinase C, 545
SCH 28080
  H^+,K^+-ATPase
    inhibitor, 196
SCH 29382
  clinical trials, 675
SCH 32651
  H^+,K^+-ATPase
    inhibition, 197
SCH 34343
  activity, 676
Schering 32615
  enkephalinase inhibitor, 411
SCRIP — *see* Statine-containing renin inhibitors
Secnidazole
  antimicrobial agent, 726
  DNA strand-breaking, 728
Secondary structure
  proteins, 12
    determination, 20
    NMR, 26
    prediction, 22
Second messenger responses
  cyclic AMP
    phosphodiesterases, 501
  phosphodiesterases, 512
Selenazole
  antiviral agent, 322
Selenocysteine β-lyase
  pyridoxal-dependent systems, 236
Sequence analysis
  DNA, 16
Serine
  sulfate
    amino acid transaminase inhibitor, 240
    glutamate decarboxylase inhibitor, 244

threonine deaminase
  inhibitor, 247
Serine, *O*-acetyl-
  alanine racemase
    inhibitor, 246
D-Serine, *O*-carbamoyl-
  peptidoglycan synthesis
    antibiotic inhibition, 570
Serine, *O*-carbamyl-
  alanine racemase
    inhibitor, 246
D-Serine dehydratase
  structure, 219
Serine hydroxymethyltransferase
  inhibition, 247
  pyridoxal-dependent systems, 222, 230
Serine proteases
  acylation, 79
  chemical modification, 28
  inhibition
    organophosphorus compounds, 76
  inhibitors, 69
  β-lactamases, 615
Serine proteinases
  classification, 394
  inhibition, 397
Severe combined immunodeficiency
  adenosine deaminase deficiency, 447
Sex-pheromone-induced conjugation
  drug resistance, 118
Shigella boydii
  drug resistance, 93
Shigella dysenteriae
  drug resistance, 93
Shigella flexneri
  drug resistance, 93
Shigella sonnei
  drug resistance, 93
Siderophores
  Gram-negative bacteria, 649
Sindbis virus
  formation
    inhibition, 378
Siomycin
  protein synthesis
    inhibition, 823
Sisomicin
  resistance, 95, 107
Site-directed mutagenesis
  enzymes, 52
Sjogren's syndrome
  autoimmune disease, 468
Skin allograft
  rejection suppression
    2'-deoxycoformycin, 452
Smiles rearrangement
  pyridylmethylbenzimidazole sulfoxides, 198
SN 6999
  nonintercalating agent
    DNA, 709
Sodium
  biochemical competition, 178
Sodium-potassium ATPase
  properties, 381
Somatostatin
  conformation
    ACE inhibitors, 404
Sorangicins
  antibacterial activity, 799
Sorbinil
  aldose reductase
    inhibitor, 292
SOS DNA repair system
  nalidixic acid, 774

Soybean
  glycoproteins
    castanospermine, 377
Specific inhibitor protein
  protein kinases, 539
Spectinomycin
  drug dependence, 116
  resistance, 94, 100, 107
  translocation
    inhibition, 833
Spermidine
  ornithine decarboxylase
    inhibitor, 245
  oxidation
    polyamine oxidase, 128
Spermidine acetyl transferase
  multisubstrate analog inhibitors, 65
Spermine
  ornithine decarboxylase
    inhibitor, 245
  oxidation
    polyamine oxidase, 128
  protein kinase C
    antagonist, 545
Sphingosine
  inhibitor
    protein kinase C, 546
  protein kinase C
    antagonist, 545
Spiramycin
  resistance, 109, 828
Spiroimidazorifamycins
  activity, 804
Spironolactone
  cytochrome *P*-450, 352
Spiroplatin
  anticancer activity, 729
Split proteins
  ribosomal reconstitution, 814
Sporangiomycin
  protein synthesis
    inhibition, 823
SQ 26180
  biosynthesis, 658
SQ 29852
  angiotensin converting enzyme
    inhibition, 404
Squalene
  biochemistry, 359
Squalene dioxide
  cholesterol
    biosynthesis, 342
Squalene oxidase
  inhibition, 359
Squalene oxide
  cholesterol
    biosynthesis, 342
Squalene synthetase
  cholesterol
    biosynthesis, 336
SR-2508
  antitumor drug, 727
  DNA strand-breakage, 740
  hypoxic cells, 730
ST-280
  inhibitor
    tyrosine kinase, 549
ST-458
  inhibitor
    tyrosine kinase, 549
ST-638
  inhibitor
    tyrosine kinase, 549
ST-642

inhibitor
  tyrosine kinase, 549
Stanozolol
  coagulation factors, 491
  fibrinolysis, 496
Staphylococcal nuclease
  binding, 52
Staphylococcus aureus
  methicillin resistance, 665
  penicillin binding protein 2′, 616
Statil
  aldose reductase
    inhibitor, 294
Statine
  renin inhibitor, 417
  structure, 73
Statine-containing renin inhibitors (SCRIP)
  discovery, 417
Statone
  renin inhibitor, 418
Statone, 2,2-difluoro-
  porcine pepsin inhibition, 73
Staurosporine
  inhibitor
    protein kinase C, 546
    protein kinases, 538
  protein kinase C
    antagonist, 545
Steroid hormones
  biosynthesis
    inhibition, 347
Steroid 18-hydroxylase
  inhibition, 352
Steroid 5α-reductase
  inhibition, 353
Steroids
  organ transplantation, 469
  sidechain cleavage
    cytochrome *P*-450, 141
$\Delta^{5,7}$-Sterol 7-reductase
  inhibition, 343
Sterol 24-reductase
  inhibition, 343
Sterols
  biosynthesis, 333
  cell membranes, 595
  metabolism
    inhibition, 333–362
Stilbamidine, hydroxy-
  nonintercalating agent
    DNA, 709
Stilbestrol, diethyl-
  inhibition, 353
Strain
  enzyme binding, 53
Streomycin
  ribosome binding site, 832
Streptogramin B
  resistance, 101
  structure, 828
Streptokinase
  complex with plasminogen, 496
  therapeutic uses, 495
Streptolydigin
  antibacterial activity, 797
  resistance, 102
Streptomyces cattleya
  thienamycin, 671
Streptomyces clavuligerus
  BRL 14151, 662
  cephalosporin biosynthesis, 610
  cephamycin C, 629
Streptomyces olivaceus
  olivanic acids, 671

Streptomycin
  drug dependence, 115
  mode of action, 831
  resistance, 91–93, 100, 107, 111
    Shigella dysenteriae, 93
Streptonigrin
  antitumor antibiotic, 769
Streptothricin acetyltransferase
  antitumor antibiotic, 110
Streptovaricins
  antibacterial activity, 797
  RNA polymerase inhibition, 102
Streptozotocin
  diabetes
    thromboxane B_2, 156
Stroke
  prostacyclin, 155
Strophantidin
  structure–activity relationship, 207
Structure–activity relationships
  sulfonamides, 261
Styrene
  Na^+,K^+-ATPase
    inhibition, 210
Succinate dehydrogenase
  inhibition, 271
Succinic acid, hydrazino-
  amino acid transaminase
    inhibitor, 240
Succinyl-CoA-acetoacetate transferase
  kinetics, 49
Sucralfate
  peptic ulcers, 195
Suicide inhibition
  cholesterol
    inhibition, 349
Suicide inhibitors
  enzymes, 79
  pyridoxal-dependent systems, 240
Sulbactam
  activity, 669
  mechanism of action, 670
Sulfadimethoxine
  resistance
    plasmodia, 265
Sulfaguanidine
  structure, 257
Sulfamethoxazole
  trimethoprim combination
    antibactericide, 290
Sulfanilamide
  competitive inhibitor, 63
  discovery, 90
  prontosil metabolism, 256
Sulfanilamides, N^1-phenyl-
  whole cell inhibitory activity, 262
Sulfanilamides, N^1-pyridyl-
  whole cell inhibitory activity, 262
Sulfapyridine
  history, 256
  structure, 257
Sulfathiazole
  folic acid analog, 258
Sulfatide
  protein kinase C, 545
Sulfazecin
  biosynthesis, 658
Sulfonamides
  active forms, 262
  antibacterial action, 258, 260
  assay, 256
  biosynthesis, 255
  cell free studies, 264
  chemotherapeutic applications, 266

competitive binding agent
    p-aminobenzoic acid, 258
competitive inhibitor, 63
crystalluria, 266
discovery, 255
diuretics, 266
folate inhibition
    rate-limiting step, 264
metabolism, 266
permeability, 263
pharmacokinetics, 266
pK_a, 257
protein binding, 266
resistance, 92, 113, 264
    Shigella dysenteriae, 93
selectivity, 260
side effects, 266
structure, 257
structure–activity studies, 260
susceptibility, 92
whole cell inhibitory activity, 262
Sulfones
    folate inhibition
        rate-limiting step, 264
    whole cell inhibitory activity, 255–268
Sulfones, diaryl
    cell free studies, 264
Sulfoxides, allyl
    cystathionine γ-synthetase
        inhibitor, 248
    methionine γ-synthetase
        inhibitor, 248
Sulmazole
    inhibitor
        phosphodiesterase, 505
Sulperazone
    activity, 669
Sultamicillin
    activity, 669
Sulthiame
    structure, 257
Superoxide dismutases
    inhibition, 306
Superoxide ions
    xanthine oxidase, 306
Superoxides
    DNA strand-breakage, 734
Suramin
    DNA polymerases
        inhibitor, 766
    enzyme inhibition, 37
Susceptibility
    antibiotics, 90
Suzukacillin
    structure, 596
SV40
    replication
        DNA topoisomerases, 772
Swainsonine
    animals, 374
    coronavirus mouse hepatitis virus
        E-2 glycoprotein, 377
    glycoprotein processing
        inhibition, 372
    glycosidase
        inhibition, 369
    insulin receptors, 377
    mannosidases
        inhibition, 371
    occurrence, 369
    structure, 370
    synthesis, 375
Synapse
    description, 381

Synchrotron radiation
    proteins
        tertiary structure, 24
Syncytium
    formation
        castanospermine, 378
Synthetases
    inhibition, 248
    nomenclature, 34
Synvinolin
    HMGCoA reductase
        inhibition, 338
Systemic lupus erythematosus
    autoimmune disease, 468

T-cell acute lymphocytic leukemia
    2′-deoxycoformycin, 453
T cells
    adenosine deaminase deficiency, 447
    purine nucleoside phosphorylase deficiency
        immune impairment, 458
        mechanism of elimination, 459
    rheumatoid arthritis, 467
Tabtoxins
    discovery, 655
Talampicillin
    synthesis, 622
Tallisomycin S10b
    medical applications, 731
Talooctonic acid, 8-amino-8-deoxy-D-glycero-
    inhibitor
        lipid A biosynthesis, 563
D-Talooctonic acid, 2,6-anhydro-3-deoxy-D-glycero-
    inhibitor
        lipid A biosynthesis, 563
Tamoxifen
    inhibitor
        protein kinase C, 546
    protein kinase C
        antagonist, 545
Tanacetum parthenium
    inhibitor
        phosphilipase A$_2$, 525
Taxol
    antitumor activity, 786
Tegafur
    pyrimidine antagonist, 312
Teichoic acids
    Gram-positive bacteria
        cell walls, 559
    roles, 560
Teichoplanin
    peptidoglycan polymerization
        inhibition, 574
Teichuronic acids
    Gram-positive bacteria
        cell walls, 559
Teleocidin
    protein kinase C, 544
TEM-1
    penicillin binding proteins, 615
Temocillin
    activity, 634
Temperature factors
    protein dynamics, 26
Temperature of melting
    DNA strand-breakage, 732
Teniposide
    DNA binding, 778
    DNA topoisomerase II
        inhibition, 778
Teprotide
    angiotensin converting enzyme
        inhibition, 400

Terbinafine
squalene oxidase
inhibition, 359, 360
Testosterone
inhibition, 353
Testosterone, 5α-dihydro-
inhibition, 353
Tetracyclines
metal ion binding, 189
resistance, 92, 110, 111
efflux enhancement, 111
Shigella dysenteriae, 93
ribosomal A site
binding, 833
susceptibility, 92
Tetraenoic acid, 5-hydroperoxy-
biosynthesis, 517
5,6,7,8-Tetrahydrofolate:NADP$^+$ oxidoreductase
inhibition, 37
Tetranactin
agricultural miticide, 602
Tetroxoprim
resistance, 113
TFP
calcium calmodulin antagonist, 546
β-Thalassemia
metal chelation therapy, 183
Theophylline
inhibitor
phosphodiesterases, 510
Thermitase
inhibitors, 70
Thermolysin
inhibition, 71
inhibitors, 72, 434
metallopeptidase, 397
metalloproteases, 71
Thiabendazole
fumarate reductase
inhibitor, 294
2-Thiacephems
ring contraction, 676
3-Thia-1-dethiacephams
properties, 690
2-Thia-1-dethiacephems
synthesis, 688, 689
[1,2,3]Thiadiazolo[5,4-*d*]pyrimid-7-amine
purine antagonist, 309
Thiamazins
activity, 662, 696
Thiamphenicol
DNA damage, 726
DNA strand-breaking, 728
2-Thiapenems
synthesis, 676
Thiazole, 2-amino-5-nitro-
anti-amoebic activity, 729
DNA damage, 726
Thiazole, nitro-
antiamoebic activity, 729
DNA damage, 726
Thiazole-4-carboxamide, 2-β-D-ribofuranosyl-
pyrimidine antagonist, 314
Thiazorifamycins
activity, 804
Thienamycin
activity, 671, 674
biosynthesis, 671
chiral synthesis, 673
derivatives, 672
detection, 671
history, 656
metabolism, 674
mode of action, 674

precursors
synthesis, 673
Thiiranes
aromatase
inhibition, 138
6-Thioguanine
purine antagonist, 306, 307
6-Thioguanylic acid
purine antagonist, 307
Thiolactones
suicide inhibitors
cyclohexanone oxygenases, 83
Thiopeptin
protein synthesis
inhibition, 823
Thiopurines
purine antagonists, 306
Thiopurinol
antiparasitic agent, 324
Thiorphan
enkephalinase inhibitor, 410
Thiostrepton
mode of action, 823
resistance, 91, 100, 101, 824
drug-producing organisms, 825
ribosomal functions, 823
Threonine deaminase
inhibition, 247
Threonine synthase
reactions at C-4, 238
Thrombin
coagulation, 487
inhibitor
natural, 490
synthetic, 490
Thromboplastin
coagulation, 486
Thrombosis
diseases, 483
therapy, 494
Thrombotic thrombocytopenic purpura
prostacyclin, 154
Thromboxane
antagonists, 163
Thromboxane A$_2$
biological properties, 155
biosynthesis, 517
from arachidonic acid, 150
isolation, 149
myocardial infarction, 156
pathological implications, 156
production, 156
ulcerogen, 156
vasoconstrictor, 155
Thromboxane B$_2$
diabetes, 156
isolation, 149
Thromboxane synthase
blood, 150
inhibition, 163
Ara-Thymidine
antiviral agent, 317
pyrimidine antagonist, 314
Thymidine, 3′-amino-3′-deoxy-
pyrimidine antagonist, 314
Thymidine, 3′-azido-2′,3′-dideoxy- — *see* AZT
Thymidine, 2′,3′-dideoxy-
DNA strand-break repair
inhibition, 748
HIV replication
inhibition, 760
Thymidine, 2′,3′-dideoxy-2′,3′-didehydro-
HIV replication
inhibition, 761

Thymidine kinase
  herpes simplex virus
    acyclovir, 761
  pyrimidine metabolism, 304
Thymidine phosphorylase
  pyrimidine metabolism, 304
Thymidylate synthetase
  functions, 783
  inhibition
    5-fluoro-2′-deoxyuridylic acid, 312
  pyrimidine metabolism, 303
Thymine glycols
  DNA strand-breaks
    repair, 745
Tiazofurin
  antiviral agent, 322
  pyrimidine antagonist, 314
Ticarcillin
  activity, 633, 634
  resistance, 114
Tilorone
  intercalating agent
    DNA, 717
Timegadine
  leukotriene production
    inhibition, 164
Timoprazole
  chemical rearrangement, 198
  H⁺,K⁺-ATPase
    inhibition, 198
  structure–activity relationships, 201
Tinidazole
  antimicrobial agent, 726
  DNA strand-breaking, 728
Tizabrin
  fibrinolysis, 497
TLCK
  inhibitor
    protein kinase C, 545
Tobramycin
  resistance, 107
α-Tocopherol
  DNA degradation
    free radical scavengers, 187
Tolbutamide
  diabetes, 268
Tolerance
  antimicrobial drugs, 89–119
Tolnaftate
  squalene oxidase
    inhibition, 359
Tolrestat
  aldose reductase
    inhibitor, 293
  synthesis, 293
Tolypomycins
  isolation, 801
Topoisomerase II
  inhibition
    amsacrine, 716
  intercalating agent
    DNA, 704, 721
  mutations
    drug resistance, 101
Torasemide
  structure, 257
Torpedo marmorate
  acetylcholinesterase, 383
Toyocamycin
  purine antagonist, 310
Tranexamic acid
  fibrinolysis
    inhibitor, 497
Transaminases

coenzyme, 214
  inhibitors, 240
  pyridoxal-dependent systems, 215
Transcription
  Archaebacteria, 794
  eukaryotes, 795
  inhibitors, 796
  process, 793
  prokaryotes, 794
  viruses, 796
Transduction
  drug resistance, 118
Transferases
  nomenclature, 34
Transferrin
  iron transport, 180
Transformation
  drug resistance, 118
Transfusional siderosis
  metal ions, 180
Transglycosylases
  penicillin binding proteins, 99
Transition metals
  biochemical competition, 178
  excretion, 180
Translational fidelity
  ribosomal protein S12, 831
Translocation
  protein synthesis
    ribosomes, 815
  spectinomycin, 833
Transmethylation
  S-adenosylmethionine, 453
Transpeptidases
  inhibition
    cephalosporins, 81
    penicillin binding proteins, 98
Transposons
  drug resistance, 117
Trehalase
  insect
    inhibition, 379
Trequinsin
  inhibitor
    phosphodiesterase, 505
Triacridines
  DNA intercalating agents, 721
Triacylglycerol acylhydrolase
  nomenclature, 35
Triacylglycerol lipase
  nomenclature, 35
Triacylglyceroprotein acylhydrolase
  nomenclature, 35
Triamterene
  diuretics, 267
Triazene, glycosylmethyl-*p*-nitrophenyl-
  glycosidase
    inhibition, 381
s-Triazines, 4,6-diamino-1,2-dihydro-2,2-dialkyl-1-
      phenyl-
  dihydrofolate reductase
    inhibitor, 285
1,2,4-Triazole-3-carboxamide
  antiviral agent, 322
1,2,4-Triazole-3-carboxamide, 1-β-D-ribofuranosyl-
  antiviral agent, 322
1*H*-1,2,4-Triazole-3-carboximidamidine, 5-amino-1-
      β-D-ribofuranosyl-
  purine nucleoside phosphorylase
    inhibitor, 460
1*H*-1,2,4-Triazole-3-carboximidamidine, 1-β-D-ribo-
      furanosyl-
  purine nucleoside phosphorylase
    inhibitor, 460

Trichomonas vaginalis
  metronidazole, 728
Tricyclic antidepressants
  neurotransmission
    inhibition, 511
Tridemorph
  fungicide, 360
Triethylenetetramine
  copper chelation, 184
Trifluoperazine
  phospholipase A_2
    calmodulin antagonist, 522
  proton pump blockers, 205
Trifluridine
  antiviral agent, 316
  thymidylate synthetase
    inhibitor, 313
Trimethoprim
  clinical uses
    toxicity, 290
  dihydrofolate reductase
    inhibition, 258
    inhibitor, 271, 283
  enzyme inhibition, 37
  nonclassical folate antagonist, 272
  resistance, 92, 113
  susceptibility, 92
  synergistic combinations
    sulfonamides, 265
  synthesis, 288
Trimetrexate
  clinical pharmacology
    pediatric patients, 287
  dihydrofolate reductase
    inhibitor, 271, 283, 286
  elimination, 287
  intravenous infusion
    dihydrofolate reductase, 286
  nonclassical folate antagonist, 272
Triostin A
  intercalating agent
    DNA, 718
Triparanol
  fungicide, 360
  hypocholesterolemic, 343
Triplet–triplet interactions
  ribosomal RNA
    stabilization, 833
Trypanothione reductase
  inhibitors, 294
Tryptamine, 5-hydroxy-
  deamination
    monoamine oxidase, 125
L-Tryptophan, 2,3-dihydro-5-fluoro-
  reactions at C-3, 234
L-Tryptophan, 5-fluoro-
  reactions at C-3, 234
Tryptophanase
  inhibition, 247
  stereochemistry, 234
Tryptophan synthase
  inhibition, 247
  reactions at C-3, 234
Tubercidin
  antiparasitic agent, 324
  purine antagonist, 310
Tubulin
  microtubules, 783
Tunicamycin
  coronavirus mouse hepatitis virus
    E-2 glycoprotein, 377
  inhibitor
    bacterial cell wall biosynthesis, 565
  peptidoglycan biosynthesis

  lipid cycle inhibition, 572
Tylosin
  resistance, 109
Tyrandamycin
  antibacterial activity, 797
Tyrocidin A
  structure, 592
Tyrocidins
  antibiotics, 592
Tyrosine, α-allyl-
  aromatic amino acid decarboxylase
    inhibitor, 246
Tyrosine, diiodo-
  thyroglobulin
    post-translational modification, 3
Tyrosine aminotransferase
  structure, 219
Tyrosine kinases
  structure, 548
  characteristics, 548
  inhibitors, 548
  therapeutic utility, 548

U14804
  elastase inhibitor, 424
U18666A
  cholesterol
    biosynthesis, 342
UDP-*N*-acetylmuramoyl pentapeptide
  biosynthesis
    inhibitors, 568
Undecaprenyl phosphate
  peptidoglycan biosynthesis
    bacterial cell walls, 565
Uracil, arabinosyl-
  DNA polymerases
    inhibition, 763
Uracil, 2′-deoxy-2′-fluoro-5-methyl-1- β-D-arabino-
  furanosyl-
    pyrimidine antagonist, 314
Uracil, 5-fluoro-
  antifungal agent, 326
  pyrimidine antagonist, 312
Uracil, 5-fluoro-1-(tetrahydro-2-furyl)-
  pyrimidine antagonist, 312
Uracil, 1-phenyl-
  antiparasitic agents, 325
Uracil-DNA glycosylase
  strand-break repair, 745
Uracil *N*-glycosylase
  pyrimidine metabolism, 304
Uracil polyoxin C
  structure, 585
Urea
  DNA strand-breaks
    assay, 745
Urea, hydroxy-
  DNA strand-break repair
    inhibition, 748
Uric acid
  biochemistry, 305
Uricase
  purine metabolism, 303
Ara-Uridine
  antiviral agent, 317
  pyrimidine antagonist, 314
Uridine, 5′-amino-5-bromo-2′,5′-dideoxy-
  antiviral agent, 315
Uridine, 5′-amino-5-chloro-2′,5′-dideoxy-
  antiviral agent, 315
Uridine, 5′-amino-5-iodo-2′,5′-dideoxy-
  antiviral agent, 315
Uridine, 5-aminomethyl-2′-deoxy-
  pyrimidine antagonist, 314

Uridine, 5-azidomethyl-2'-deoxy-
pyrimidine antagonist, 314
Ara-Uridine, 5-bromo-
pyrimidine antagonist, 314
Uridine, 5-bromodeoxy-
antiviral agent, 315
Uridine, 5-bromo-2'-deoxy-
5'-monophosphates
pyrimidine antagonists, 312
Uridine, 5-bromo-6-methoxydihydro-5-fluoro-2'-
deoxy-
pyrimidine antagonist, 312
Ara-Uridine, 5-(2-bromovinyl)-
antiviral agent, 317
Uridine, 5-(2-bromovinyl)-2'-deoxy-
antiviral agent, 316
Uridine, 5-chloro-2'-deoxy-
5'-monophosphates
pyrimidine antagonists, 312
Uridine, 5-chloro-5'-trifluoromethyl-2',5'-dideoxy-
antiviral agent, 315
Uridine, 5-cyanovinyl-2'-deoxy-
antiviral agent, 316
Uridine, 3-deaza-
antiviral agent, 318
Ara-Uridine, 5-ethyl-
antiviral agent, 317
Uridine, 5-ethyl-2'-deoxy-
antiviral agent, 316
Uridine, 5-ethynyl-2'-deoxy-
antiviral agent, 316
Uridine, 5-fluoro-
pyrimidine antagonist, 312
Lyxo-Uridine, 5-fluoro-2'-deoxy-
pyrimidine antagonist, 312
Ara-Uridine, 2'-fluoro-2'-deoxy-5-methyl-
antiviral agent, 318
Uridine, 5-formyl-2'-deoxy-
antiviral agent, 317
Uridine, 5-hydroxymethyl-2'-deoxy-
pyrimidine antagonist, 314
Ara-Uridine, 5-iodo-
pyrimidine antagonist, 314
Uridine, 5-iodo-2'-deoxy-
antiviral agent, 315
5'-monophosphates
pyrimidine antagonists, 312
Uridine, 5-(2-iodovinyl)-2'-deoxy-
antiviral agent, 316
Uridine, 5-isopropyl-2'-deoxy-
antiviral agent, 316
Ara-Uridine, 5-nitro-
antiviral agent, 317
Uridine, 5-(2-propenyl)-2'-deoxy-
antiviral agent, 316
Uridine, 5-propyl-2'-deoxy-
antiviral agent, 316
Uridine, 5-(1-propynyl)-2'-deoxy-
antiviral agent, 316
Uridine, tetrahydro-
cytidine deaminase
inhibitor, 313
Uridine, 5-trifluoromethyl-2'-deoxy-
antiviral agent, 316
thymidylate synthetase
inhibitor, 313
Uridine, 5-(3,3,3-trifluoropropenyl)-2'-deoxy-
antiviral agent, 316
Uridine, 5-vinyl-2'-deoxy-
antiviral agent, 316
Uridine phosphorylase
pyrimidine metabolism, 304
Uridylic acid, 5-fluoro-2'-deoxy-
antifungal agent, 326

thymidylate synthetase
inhibitor, 312
Urokinase
plasminogen activator, 493

Valinomycin
cell membrane
permeability, 598
potassium binding, 188
structure, 600
Vancomycin
peptidoglycan biosynthesis
inhibition, 567
peptidoglycan polymerization
inhibition, 573
resistance, 110
structure, 576
Varicella zoster virus infections
acyclovir, 761
Vasodilators
prostaglandins, 150
Verdamicin
resistance, 109
Vernamycin B
chemical footprinting
ribosomal binding sites, 830
Vesicular stomatitis virus
glycoproteins
swainsonine, 373
Vidarabine — *see* Adenine, 9-β-D-arabinofuranosyl-
Vinblastine
anticancer agents, 785
ara-cytidine combinations
pyrimidine antagonists, 314
Vinca alkaloids
anticancer agents, 785
Vincent's disease
DNA strand-breaking drugs, 728
Vincristine
anticancer agents, 785
ara-cytidine combinations
pyrimidine antagonists, 314
Vinpocetine
inhibitor
phosphodiesterase, 505
Viomycin
resistance, 100
Viral diseases
adenosine deaminase inhibitors, 449
Virazole
antiviral agent, 322
Viruses
cell envelope
drug susceptibility, 97
drug susceptibility, 95
transcription, 796
Viscometry
DNA strand-breakage
measurement, 731
Vitamin K
coagulation, 487
hypoprothrombinemia
β-lactam antibiotics, 618
von Willebrand's Diseases
coagulation defect, 488

W-7
calcium calmodulin antagonist, 546
inhibitor
myosin light chain kinase, 547
protein kinase C
antagonist, 545
Warfarin
anticoagulant, 489

Wilson's disease
  copper poisoning, 183
  metal ions, 180
WR-2721
  antitumor activity, 743

Xanthine oxidase
  inhibitors
    gout, 469
  protein kinase C, 143
  purine metabolism, 303
  pyrimidine metabolism, 305
  tissue injury, 143
Xanthines
  inhibitor
    phosphodiesterase, 505
X-ray crystallography
  R61 enzyme
    β-lactam binding, 615
  tertiary structure, 23
XU 62-320
  HMGCoA reductase
    inhibition, 339

Y-590

inhibitor
  phosphodiesterase, 507
Yeast
  DNA topoisomerase
    function, 772

Zaprinase
  inhibitor
    phosphodiesterase, 505
Zinc
  deficiency, 180
  excretion, 180
  free ion concentration
    plasma, 178
  transport proteins, 180
Zollinger–Ellison syndrome
  omeprazole, 204
Zovirax
  antiviral agent, 320
Zyloric
  uric acid
    inhibition, 305
Zymogens
  phospholipase A_2, 521

DATE DUE
